2008 IEEE Custom Integrated Circuits Conference

(CICC 2008)

San Jose, California
21-24, September 2008

Pages 1-406

IEEE Catalog Number: CFP08CIC-PRT
ISBN: 978-1-4244-2018-6

Copyright © 2008 by the Institute of Electrical and Electronic Engineers, Inc
All Rights Reserved

Copyright and Reprint Permissions: Abstracting is permitted with credit to the source. Libraries are permitted to photocopy beyond the limit of U.S. copyright law for private use of patrons those articles in this volume that carry a code at the bottom of the first page, provided the per-copy fee indicated in the code is paid through Copyright Clearance Center, 222 Rosewood Drive, Danvers, MA 01923.

For other copying, reprint or republication permission, write to IEEE Copyrights Manager, IEEE Service Center, 445 Hoes Lane, Piscataway, NJ 08854. All rights reserved.

***This publication is a representation of what appears in the IEEE Digital Libraries. Some format issues inherent in the e-media version may also appear in this print version.**

IEEE Catalog Number:	CFP08CIC-PRT
ISBN 13:	978-1-4244-2018-6
Library of Congress No:	85-653738

Additional Copies of This Publication Are Available From:

Curran Associates, Inc
57 Morehouse Lane
Red Hook, NY 12571 USA
Phone: (845) 758-0400
Fax: (845) 758-2633
E-mail: curran@proceedings.com

TABLE OF CONTENTS

FRONT MATTER

SESSION 1- KEYNOTE PRESENTATION

KEYNOTE PRESENTATION
More than Moore
Dave Bergeron

SESSION 2- STATISTICAL MODELING

Process Variability at the 65nm Node and Beyond (INVITED)..1
    Sani Nassif

Mismatch Analysis and Statistical Design at 65 nm and Below (INVITED)............................9
    Larry Pileggi, Gökçe Keskin, Xin Li, Ken Mai, Jon Proesel, Carnegie Mellon

Statistical Prediction of Circuit Aging under Process Variations..13
    Wenping Wang, Vijay Reddy, Bo Yang, Varsha Balakrishnan, Srikanth Krishnan, Yu Cao

SESSION 3 – POWER MANAGEMENT

**A Fully-Integrated 0.18μm CMOS DC-DC Step- Down Converter, Using a Bondwire
Spiral Inductor** 19
    Mike Wens and Michiel Steyaert

**A 90-240MHz Hysteretic Controlled DC-DC Buck Converter with Digital PLL Frequency
Locking**..23
    Pengfei Li, Deepak Bhatia, Xue Lin and Rizwan Bashirullah

**Fast Transient Technique (FTT) in Buck Current- Mode DC-DC Converters for Low-
Voltage SoCv Systems**..27
    Chia-Hsiang Lin, Hong-Wei Huang, Ke- Horng Chen

**A Process Variation Compensation Scheme Using Cell-Based Forward Body-biasing
Circuits Usable for 1.2V Design**...31
    *Fumihiko Tachibana, Hironori Sato, Takahiro Yamashita, Hiroyuki Hara, Takeshi Kitahara, Shuou
Nomura, Fumiyuki Yamane, Yoshiro Tsuboi, Keiko Seki, Shuuji Matsumoto, Yoshinori Watanabe,
Mototsugu Hamada*

**Built-in Resistance Compensation (BRC) Technique for Fast Charging Li-Ion Battery
Charger** 35
    Chia-Hsiang Lin, Hong-Wei Huang, Ke-Horng Chen

SESSION 4- HIGH-SPEED TEST, CHARACTERIZATION, AND DEBUG

A Voltage Drop Aware Crosstalk Measurement with Multi-Aggressors in 65 nm Process 41
    Genichi Tanaka, Kan Takeuchi, Minoru Ito, Hiroaki Matsushita

Measurements of the Silicon Die Characteristics of Packaged Drivers for High-Speed I/O (INVITED) 45
  Gerry Talbot and Edoardo Prete

Inductor-Based ESD Protection under CDM-like ESD Stress Conditions for RF Applications 53
  Steven Thijs, Mototsugu Okushima, Jonathan Borremans, Philippe Jansen, Dimitri Linten, Mirko Scholz, Piet Wambacq and Guido Groeseneken

Non-Destructive IC Defect Localization Using Optical Beam-Based Imaging (INVITED) 57
  Edward Cole Jr.

SESSION 5- BROADBAND CIRCUIT TECHNIQUES FOR EMERGING WIRELESS COMMUNICATIONS

Emerging Application Opportunities for SiGe Technology (INVITED) 63
  John Cressler

A 0.8GHz-10.6GHz SDR Low-Noise Amplifier in 0.13-μm CMOS 71
  Shuzuo Lou and Howard Luong

A 1.3-6 GHz Triple-Mode CMOS VCO Using Coupled Inductors 75
  Zahra Safarian and Hossein Hashemi

SESSION 6- ADVANCED SoC/SiP INTEGRATION AND CO-DESIGN

A SOC/SOP Co-design Approach for mmW CMOS in QFN Technology (INVITED) 81
  Joy Laskar, Stephane Pinel, Padmanava Sen, Bevin Perumana, Debasis Dawn, David Yeh and Francesco Barale

Chip to Carrier C4 Technology Challenges with Pb-free Solders (INVITED) 89
  Eric Perfecto, Brian Sundlof, Kamalesh Srivastava and Minhua Lu

A Study on Process-compatibility in CMOS-first MEMS-last Integration 93
  Kazuhiro Takahashi, Makoto Mita, Hiroyuki Fujita, Kazuhiro Suzuki, Hideyuki Funaki, Kazuhiko Itaya, Hiroshi Toshiyoshi

SESSION 7- HIGH RESOLUTION CONVERTERS

A 101-dB SNR Hybrid Delta-Sigma Audio ADC using Post Integration Time Control 99
  Moo-Yeol Choi, Sung-No Lee, Seung-Bin You, Wang-Seup Yeum, Ho-Jin Park, Jae-Whui Kim and Hae-Seung Lee

An 8.1 mW, 82 dB Delta-Sigma ADC with 1.9 MHz BW and -98 dB THD 103
  Kyehyung Lee, Matthew R. Miller and Gabor C. Temes

A 2.5mhz Bw and 78db Sndr Delta-sigma Modulator Using Dynamically Biased Amplifiers 107
  Yan Wang, KyeHyung Lee and Gabor Temes

74dB SNDR Multi-Loop Sturdy-MASH Delta-Sigma Modulator Using 35dB Opamp Gain 111
  Nima Maghari, Sunwoo Kwon and Un-Ku Moon

A/D Converter Trends: Power Dissipation, Scaling and Digitally Assisted Architectures (INVITED) 115
  Boris Murmann

A 16b 10MS/s Digitally Self-Calibrated ADC with Time Constant Control..................123
Tae-Hwan Oh, Ho-Young Lee, Ju-Hwa Kim, Ho-Jin Park, Kyoung-Ho Moon, Jae-Whui Kim and Hae-Seung Lee

A 15b Power-Efficient Pipeline A/D Converter Using Non-Slewing Closed-Loop Amplifiers..................127
Shoji Kawahito, Kazutaka Honda, Zheng Liu, Keita Yasutomi and Shinya Itoh

SESSION 8- CHARACTERIZATION AND TEST METHODS FOR DEVICE VARIABILITY IN NANOSCALE TECHNOLOGIES

An Array-Based Test Circuit for Fully Automated Gate Dielectric Breakdown Characterization..................133
John Keane, Shrinivas Venkatraman, Paulo Butzen, Chris Kim

A High Sensitivity Process Variation Sensor Utilizing Sub-threshold Operation..................137
Mesut Meterelliyoz, Peilin Song, Franco Stellari, Jaydeep P. Kulkarni and Kaushik Roy

Measurement and Analysis of Variability in 45nm Strained-Si CMOS Technology..................141
Liang-Teck Pang, Borivoje Nikolic

Within-Die Gate Delay Variability Measurement using Re-configurable Ring Oscillator..................145
Bishnu Prasad Das, Bharadwaj Amrutur, H.S. Jamadagni, N.V. Arvind and V. Visvanathan

Expected Vectorless Teacher-Student Swap (TSS) Test Method with Dual Power Supply Voltages for 0.3V Homogeneous Multi-core LSI's..................149
Taro Niiyama, Koichi Ishida, Makoto Takamiya and Takayasu Sakurai

SESSION 9- BROADBAND CIRCUIT TECHNIQUES FOR EMERGING WIRELESS COMMUNICATIONS

MIMO Techniques for High Data Rate Radio Communications (INVITED)..................155
Yorgos Palaskas, Ashoke Ravi and Stefano Pellerano

Wireless Interconnection within a Hybrid Engine Controller Board..................163
Swaminathan Sankaran, Kyujin Oh, Hsinta Wu and Kenneth O

SESSION 10- PANEL DISCUSSION
Sure, Moore's Law Can Continue, But Should It:

A 0.6-to-1V Inverter-Based 5-bit Flash ADC in 90nm Digital CMOS..................171
Jonathan Proesel and Lawrence Pileggi

A 10~15b 60MS/s Floating-Point ADC with Digital Gain and Offset Calibration..................175
Yun-Shiang Shu, Moon- Jung Kyung, Wei-Ming Lee, Bang-Sup Song and Bedabrata Pain

Digital Correction of Dynamic Track-and-Hold Errors Providing SFDR > 83 dB up to fin = 470 MHz..................179
Parastoo Nikaeen and Boris Murmann

A 65nm CMOS 1.2V 12b 30MS/s ADC with Capacitive Reference Scaling..................183
Kang-Jin Lee, Kyoung-Jun Moon, Kwang-Sung Ma, Kyoung-Ho Moon and Jae-Whui Kim

A Continuous-time Input Pipeline ADC..................187
David Gubbins, Bumha Lee, Pavan Kumar Hanumolu and Un-Ku Moon

A 0.8 V Asynchronous ADC for Energy Constrained Sensing Applications..................191
Michael Trakimas and Sameer Sonkusale

Modeling, Design and Optimization of Hybrid Electromagnetic and Piezoelectric MEMS Energy Scavengers 195
Xiaochun Wu, Alireza Khaligh and Yang Xu

Amorphous Silicon Logic Circuits on Flexible Substrates 199
Rahul Shringarpure, Lawrence Clark, Sameer Venugopal, David Allee and Shrinivas Uppili

Frequency Tunable Silicon Carbide Resonators for MEMS Above IC 203
Frederic Nabki, Tomas A. Dusatko, Mourad N. El-Gamal

MEMS Wafer-Level Vacuum Packaging with Transverse Interconnects for CMOS Integration 207
Dominique Lemoine, Paul-Vahé Cicek, Frederic Nabki, Mourad El-Gamal

Variation-Tolerant Spin-Torque Transfer (STT) MRAM Array for Yield Enhancement 211
Jing Li, Haixin Liu, Sayeef Salahuddin and Kaushik Roy

Pure Logic CMOS Based Embedded Non-Volatile Random Access Memory for Low Power RFID Application 215
Liyang Pan, Xian Luo, Yaru Yan, Jirong Ma, Dong Wu and Jun Xu

A High-speed, Low-power 3D-SRAM Architecture 219
H. Henry Nho, Mark Horowitz and S. Simon Wong

Early Prediction of Product Performance and Yield Via Technology Benchmark 223
Choongyeun Cho, Daeik Kim, Jonghae Kim, Daihyun Lim and Sangyeun Cho

A FPGA Vernier Digital-to-Time Converter with 3.56ps Resolution and -0.23~+0.2LSB Inaccuracy 227
Poki Chen, Juan-Shan Lai and Po-Yu Chen

RASP 2.8: A New Generation of Floating-gate based Field Programmable Analog Array 231
Arindam Basu, Christopher M. Twigg, Stephen Brink, Paul Hasler, Csaba Petre, Shubha Ramakrishnan, Scott Koziol and Craig Schlottman

Modeling of Triple-Well Isolation and the Loading Effects On Circuits up to 50 GHz 235
Piljae Park and C. Patrick Yue

A General Weak Nonlinearity Model for LNAs 239
Wei Cheng, Anne Johan Annema, Jeroen Croon, Dick Klaassen and Bram Nauta

Modeling and Synthesis of Wide-B, Switched- Resonators for VCOs 243
Bodhisatwa Sadhu, Umaikhe Omole and Ramesh Harjani

Faster Statistical Cell Characterization using Adjoint Sensitivity Analysis 247
Ben Gu, Kiran Gullapalli, Yun Zhang and Savithri Sundareswaran

An ESD-Protected 5-GHz Differential Low-Noise Amplifier in a 130-nm CMOS Process 251
Yuan-Wen Hsiao and Ming-Dou Ker

A 65nm 3.4Gbps HDMI TX PHY with Supplyregulated Dual-tuning PLL and Blending Multiplexer 255
Jongshin Shin, Jaehyun Park, Bongjin Kim, Jongjae Ryu, Chiwon Kim, JiYoung Kim, Seung- Hee Yang, Hyungoo Kim and Jaewhui Kim,

A Scalable Digitalized Buffer for Gigabit I/O, ...259
Hung Wen Lu, Chau Chin Su, Chien-Nan Liu

Broadband, Transimpedance Amplifier in 0.35-um SiGe BiCMOS Technology for 10-Gb/s Optical Receiver Analog Front-End Application 263
Ji-Chen Huang, Yu-Sheng Lai and Klaus Yune-Jane Hsu

A Multifunction Transceiver RFIC for 802.11abg WLAN and DVB-H Applications 267
Yin Shi, Fa Foster Dai, Jun Yan, Xueqing Hu, Hua Xu, Ming Gu, Xuelian Zhang, Qiming Xu, Bei Chen, Fangxiong Chen, Peng Yu, Heping Ma, Fang Yuan and Richard Jaeger

A Fully Integrated Zero-IF Mobile TV Tuner RFIC for S-b, CMMB Application 271

Yin Shi, Fa Foster Dai, Jun Yan, Xueqing Hu, Hua Xu, Ming Gu, Xuelian Zhang, Qiming Xu, Bei Chen, Fangxiong Chen, Peng Yu, Heping Ma, Fang Yuan and Richard Jaeger

SESSION 11- COMPACT MODELING

Compact Modeling of Multiple-Gate MOSFETs (INVITED) 277

Yuan Taur, Jooyoung Song and Bo Yu

Compact Modeling and Simulation of PD-SOI MOSFETs: Current Status and Challenges (INVITED) 285

Jung-Suk Goo, Richard Williams, Glenn Workman, Qiang Chen, Sungjae Lee, Edward Nowak

Modeling Ionizing Radiation Effects in Solid State Materials and CMOS Devices (INVITED) 293

Hugh Barnaby, Michael Mclain, Ivan Esqueda and X. J. Chen

Characterization, Simulation,Modeling of FET Source/Drain Diffusion Resistance 301

Ning Lu and Bill Dewey

Analysis of the Impact of Interfacial Oxide Thickness Variation on Metal-Gate High-K Circuits 305

Minki Cho, Kingsuk Maitra and Saibal Mukhopadhyay

SESSION 12- HIGH SPEED A/D CONVERTERS

Time-Interleaved Analog-to-Digital Converters (INVITED) 311

David Nairn

A 12b 50MSPS 34mW Pipelined ADC 319

Hao Yu, Sing Chin and Bill Wong

Background ADC Calibration in Digital Domain, ...323

Cheongyuen Tsang, Yun Chiu, Johan Vanderhaegen Sebastian Hoyos, Charles Chen, Robert Brodersen and Borivoje Nikolic

A 1.2v 11b 100Msps 15mW ADC Realized using 2.5b Pipelined Stage Followed by Time Interleaved SAR in 65nm Digital CMOS Process 327

Pratap Narayan Singh, Ashish Kumar, Chandrajit Debnath and Rakesh Malik

A 52mW 10b 210MS/s Two-Step ADC for Digital-IF Receivers in 0.13µm CMOS 331

Zhiheng Cao and Shouli Yan

A 24GS/s 5-b ADC with Closed-Loop THA in 0.18µm SiGe BiCMOS 335

Jaesik Lee, Joseph Weiner, Pascal Roux, Andreas Leven and Young-Kai Chen, Alcatel-Lucent

SESSION 13: BIOMEDICAL, SENSORS AND MEMS

A Biomedical Implantable FES Battery-Powered Micro-Stimulator (INVITED) 341

Eusebiu Matei, Edward Lee, John Gord, Patrick Nercessian, Phil Hess, Howard Stover, Taihu Li and James Wolfe

CMOS LSI-based Multi-chip Flexible Retinal Prosthesis Device for Subretinal Implantation 349

Takashi Tokuda, Shigeki Sawamura, Yasuo Terasawa, Yasuo Tano and Jun Ohta,

A CMOS TDC-Based Digital Magnetic Hall Sensor Using the Self Temperature Compensation 353
Young- Jae Min and Soo-Won Kim

A Micro-Power Neural Spike Detector and Feature Extractor in .13µm CMOS................357
Jeremy Holleman, Apurva Mishra, Chris Diorio and Brian Otis

A Compact and Programmable High-Frequency Oscillator Based on a MEMS Resonator 361
Frederic Nabki, Tomas A. Dusatko, Mourad N. El-Gamal

SESSION 14- ADVANCED SoCs- TECHNIQUES AND APPLICATIONS

Characterization and Design for Variability and Reliability (INVITED).................................367
Kevin Nowka, Sani Nassif and Kanak Agarwal

A 9Gbit/s Serial Transceiver for On-chip Global Signaling over Lossy Transmission Lines...................373
JunYoung Park, Joshua Kang, Sunghyun Park, Michael Flynn

A 28mW OFDM Baseb, Receiver Chip for DVBT/ H with All Digital Synchronization 377
Ting-Chen Wei, Wei-Chang Liu, Chi-Yao Tseng, Syu-Siang Long, Shyh-Jye Jou and Muh-Tian Shiue

Nonvolatile Magnetic Flip-Flop for Stanby-powerfree SoCs 381
Noboru Sakimura, Tadahiko sugibayashi, Ryusuke Nebshi, Naoki Kasai

A Fully Integrated Pulsed-LASER Time-of-Flight Measurement System with 12ps Single-Shot Precision 385
Tino Copani, Bert Vermeire, Anuj Jain, Habib Karaki, Kailash Chandrashekar, Sushmit Goswami, Jennifer Kitchen, Hoon Hee Chung, Ilker Deligoz, Bertan Bakkaloglu, Hugh Barnaby and Sayfe Kiaei

A Low-Power IC Design for the Wireless Monitoring System of the Orthopedic Implants....................389
Hong Chen, Jia Chen, Yi Chen, Ming Liu, Chun Zhang and Zihua Wang

Tessellation-Enabled Shader for a Bandwidth- Limited 3D Graphics Engine 393
Kyusik Chung, Chang- Hyo Yu, Donghyun Kim and Lee-Sup Kim

SESSION 15- IC TECHNOLOGY – MORE MOORE AND MORE THAN MOORE

Lithography Options for the 32nm Half Pitch Node and Beyond (INVITED) 399
Kurt Ronse, Philippe Jansen, Roel Gronheid, Eric Hendrickx, Mireille Maenhoudt, Mieke Goethals and Geert Vandenberghe

45nm High-k + Metal Gate Strain-Enhanced CMOS Transistors (INVITED) 407
Chris Auth

Will BiCMOS Stay Competitive for mmW Applications (INVITED).......................................415
Patrice Garcia, Alain Chantre, Sebastien Pruvost, Pascal Chevalier, Sean Nicolson, David Roy, Sorin Voinigescu, Christophe Garnier

Microelectronics for the Real World: "Moore" versus "More than Moore" (INVITED) 423
John Kent and Jagdish Prasad

SESSION 16- EMBEDDED MEMORY

A 512-KB Level-2 Cache Design in 45 nm for sub- 2W Low Power IA Processor Silverthorne 433
Mohammed Taufique, Alex Okpisz, Haseeb Ahmed, John Riley, Mohammad Hasan, Gian Gerosa

A Voltage Scalable 0.26V, 64kb 8T SRAM with Vmin Lowering Techniques and Deep Sleep Mode 437
Tae-Hyoung Kim, Jason Liu and Chris H. Kim

Compensation of Systematic Variations Through Optimal Biasing of SRAM Wordlines 441
Andrew Carlson, Zheng Guo, Liang-Teck Pang, Tsu-Jae King Liu and Borivoje Nikolic

Variation-Tolerant SRAM Sense-Amplifier Timing Using Configurable Replica Bitlines 445
Umut Arslan, Mark P. McCartney, Mudit Bhargava, Xin Li, Ken Mai,Lawrence T. Pileggi

A 135mV 0.13µW Process Tolerant 6T Subthreshold DTMOS SRAM in 90nm Technology 449
Myeong-Eun Hwang and Kaushik Roy

Robust Ultra-Low Voltage ROM Design, ...453
Mingoo Seok, Scott Hanson, Jae-Sun Seo, Dennis Sylvester,David Blaauw

A Million Cycle 0.13µm 1Mb Embedded SONOS Flash Memory Using Successive Approximated Read Calibration 457
Nan Wang, Xiang Yao, Yu Lei, Guoyou Feng, Qiaohua Dong, Liang Xu, Lu Guo and Zi Wang

SESSION 17- CLOCKING CIRCUITS

A Wide Tuning Range (1 GHz-to-15 GHz) Fractional-N All-Digital PLL in 45nm SOI 463
Alexander Rylyakov, Jose Tierno, George English, Michael Sperling and Daniel Friedman

Clocking Circuits for a 16Gb/s Memory Interface, ...467
Ting Wu, Xudong Shi, Kambiz Kaviani, Haechang Lee, Jung-Hoon Chun, TJ Chin, Jie Shen, Rich Perego and Ken Chang

A 1V 15.6mW 1-2GHz -119dBc/Hz @ 200kHz clock multiplying DLL 471
Sander Gierkink

A 0.5-to-2.5GHz Supply-Regulated PLL with Noise Sensitivity of -28dB 475
Abhijith Arakali, Srikanth Gondi,Pavan Kumar Hanumolu

20 GHz Low Power QVCO and De-skew Techniques in 0.13µm Digital CMOS 479
Masum Hossain and Anthony Chan Carusone

A 3 GHz Spread Spectrum Clock Generator for SATA Applications Using Chaotic PAM Modulation 483
Fabio Pareschi, Gianluca Setti and Riccardo Rovatti

A 1.5 GHz Spread Spectrum Clock Generator with 5000ppm Piecewise Linear Modulation 487
Minyoung Song, Sunghoon Ahn, Inhwa Jung, Yontae Kim and Chulwoo Kim

A 8x5 Gb/s Source-Synchronous Receiver with Clock Generator Phase Error Correction 491
Ankur Agrawal, Pavan Kumar Hanumolu and Gu-Yeon Wei

SESSION 18- MILLIMETER-WAVE CIRCUIT TECHNIQUES

Millimeter-wave CMOS Integrated Circuits for Gigabit WPAN Applications (INVITED) 497
Tian-Wei, Huang and Huei Wang

A Zero-IF 60GHz Transceiver in 65nm CMOS with > 3.5Gb/s Links 505
Alexander Tomkins, Ricardo A. Aroca, Takuji Yamamoto, Sean T. Nicolson, Yoshiyasu Doi and Sorin P. Voinigescu

Low-Cost Fully Integrated BiCMOS Transceiver for Pulsed 24-GHz Automotive Radar Sensors 509
Laurence Moquillon, Patrice Garcia, Sebastien Pruvost, Stephane Le Tual, Maxime Marchetti, Laurent Chabert, Nicole Bertholet, Angelo Scuderi, Salvatore Scaccianoce, Alberto Serratore, Nicolo Ivan Piazzese, Cedric Dehos and Domique Morche

A 0.13um CMOS Fully Differential Receiver with On-Chip Baluns for 60GHz Broadb, Wireless Communications 513
Chao-Shiun Wang, Juin-Wei Huang, Kun-Da Chu and Chorng-Kuang Wang

A Low power 20 GHz 1.5 Gb/s CMOS Injection- Pulling FSK Modulator and Frequency Discriminator for 60GHz Links 517
Shon-Hang Wen, Chao-Shiun Wang and Chorng-Kuang Wang

A 24/77GHz Dual-B, BiCMOS Frequency Synthesizer 521
Vipul Jain, Babak Javid and Payam Heydari

An X/Ku-B, Frequency Synthesizer Using A 9- Bit Quadrature DDS 525
Xuefeng Yu, Fa Foster Dai, Dayu Yang, J. David Irwin, Richard C. Jaeger,

A Dynamic Offset Control Technique for Comparator Design in Scaled CMOS Technology................531
Xiaolei Zhu, Yanfei Chen, Masaya Kibune, Yasumoto Tomita, Takayuki Hamada, Hirotaka Tamura, Sanroku Tsukamoto, Tadahiro Kuroda,

A Low-Voltage OP Amp with Digitally Controlled Algorithmic Approximation 535
Dong-Woo Jee, Seung- Jin Park, Hong-June Park and Jae-Yoon Sim,

A 105.5 dB, 0.49 mm2 Audio Sigma Delta Modulator using Chopper Stabilization and Fully Randomized DWA..539
Yi-Gyeong Kim, Min-Hyung Cho, Kwi-Dong Kim, Jong-Kee Kwon and Jongdae Kim

An Ultra Low Power 1V, 220nW Temperature Sensor for Passive Wireless Applications 543
Yu- Shiang Lin, Dennis Sylvester and David Blaauw

A 9-Bit Configurable Current Source with Enhanced Output Resistance for Cochlear Stimulators...547
Song Guo, Hoi Lee and Philipos Loizou

Super-Resolution: Imaging beyond the Pixel Size Limit 551
Tamer Elkhatib and Khaled Salama

Polysilicon Vertical Actuator Powered with Waste Heat 555
Jorge Varona, Margarita Tecpoyotl-Torres and Anas Hamoui

Low Noise uWatt Interface Circuits for Wireless Implantable Real-Time Digital Blood Pressure Monitoring 559
Peng Cong, Wen Ko and Darrin Young

A Flexible Decoder IC for WiMAX QC-LDPC Codes..563
Tzu-Chieh Kuo and Alan Willson

Minimizing the Supply Sensitivity of CMOS Ring Oscillator by Jointly Biasing the Supply Control Voltage 567
Ping-Hsuan Hsieh, Jay Maxey, Chih-Kong Ken Yang

The Superchip: Innovative Teaching of IC Design and Manufacture..................................571
Peter Wilson, Reuben Wilcock, Matthew Swabey, Iain McNally and Bashir Al-Hashimi

A 3D Graphics Processor with Fast 4D Vector Inner Product Units and Power Aware Texture Cache 575
Jae-Sung Yoon, Donghyun Kim, Chang-Hyo Yu and Lee-Sup Kim

Timing Yield Enhancement Through Soft Edge Flip-Flop Based Design 579
Michael Wieckowski Young, Min Park, Carlos Tokunaga, Dong Woon Kim, Zhiyoong Foo, Dennis Sylvester and David Blaauw

1/5 Power Reduction by Global Optimization based on Fine-GrainedBody Biasing 583
Yasumi Nakamura, David Levacq, Limin Xiao, Takuya Minakawa, Taro Niiyama, Makoto Takamiya and Takayasu Sakurai

A DC-DC Converter with a Dual VCDL-based ADC and a Self-Calibrated DLL-based Clock Generator for an Energy-Aware EISC Processor 587
Sunghwa Ok, Jungmoon Kim, Gilwon Yoon, Hyunho Chu, Jaegeun Oh, Seon Wook Kim and Chulwoo Kim

Active Autonomous AC-DC Converter for Piezoelectric Energy Scavenging Systems 591
Enrico Dallago, Daniele Miatton, Giuseppe Venchi, Valeria Bottarel, Giovanni Frattini, Giulio Ricotti, Monica Schipani

A 10Gb/s Receiver with Linear Backplane Equalization and Mixer-Based Self-Aligned CDR 595
Simone Erba, Massimo Pozzoni, Matteo Pisati, Riccardo Brama, Davide Sanzogni, Emanuele Depaoli, Paolo Viola and Francesco Svelto

A 12.5-Gbps, 7-bit Transmit DAC with 4-Tap LUTbased Equalization in 0.13μm CMOS 599
Hayun Chung, Andrew Liu and Gu-Yeon Wei

Active Deskew in Injection-Locked Clocking 603
Lin Zhang, Berkehan Ciftcioglu and Hui Wu

A 53 GHz DCO for mm-Wave WPAN, ... 607
Raffaella Genesi, Francesco M. De Paola and Danilo Manstretta

Twisted Inductors for Low Coupling Mixed-Signal and RF Applications 611
Nathan Neihart, David Allstot, Matt Miller, Pat Rakers

A Common-Base Linear RF Power Amplifier for 3G Cellular Applications 615
Flavio Avanzo, Francesco M De Paola and Danilo Manstretta

Design of Low Power CMOS Ultra-Wideb, 3.1- 10.6 GHz Pulse-Based Transmitters 619
Kuan-Yu Lin and Mourad El-Gamal

SESSION 19- LOW POWER AND NON-TRADITIONAL RF TRANCEIVERS

Energy-efficient Wireless Front-end Concepts for Ultra Low Power Radio (INVITED) 625
John R. Long, Wanghua Wu, Yunzhi Dong, Yi Zhao, Mihai A.T. Sanduleanu, John F.M. Gerrits and Gerrit van Veenendaal

A 0.4 nJ/b 900MHz CMOS BFSK Super- Regenerative Receiver 629
James Ayers, Kartikeya Mayaram and Terri Fiez

A 900-MHz Low-Power Transmitter With Fast Frequency Calibration For Wireless Sensor Networks 633
Napong Panitantum, Kartikeya Mayaram, Terri Fiez

A 3.5-mW 15-Mbps O-QPSK Transmitter for Realtime Wireless Medical Imaging Applications 637
Yao- Hong Liu and Tsung-Hsien Lin

A CMOS Direct Conversion Transmitter With IEEE 802.22 Cognitive Radio Applications 641
Jongsik Kim, Seungsoo Kim, Jaewook Shin, Youngcho Kim, Junki Min, Kihong Kim and Hyunchol Shin,

A 5-GHz Wireless LAN Transmitter with Integrated Tunable High-Q RF Filter 645
Robert Wiser, Masoud Zargari, David Su and Bruce Wooley

SESSION 20- ADVANCED WIRELINE TECHNIQUES

A 6.5 Gb/s Backplane Transmitter with 6-tap FIR Equalizer and Variable Tap Spacing 651
Mike Bichan and Anthony Chan Carusone

Phase-Locking in Wireline Systems: Present and Future (INVITED) 655
Behzad Razavi

A 20Gb/s SerDes Transmitter with Adjustable Source Impedance and 4-tap Feed-Forward Equalization in 65nm Bulk CMOS 663
Rick Philpott, James Humble, Robert Kertis, Karl Fritz, Barry Gilbert,Erik Daniel

Wideband, mmWave CML Static Divider in 65nm SOI CMOS Technology (INVITED) 667
Daeik Kim, Choongyeun Cho, Jonghae Kim,Jean-Olivier Plouchart

A 32/16 Gb/s 4/2-PAM Transmitter with PWM Pre- Emphasis and 1.2 Vpp per side Output Swing in 0.13-μm CMOS 675
Horace Cheng and Anthony Chan Carusone

A 5-Gb/s/pin Transceiver for DDR Memory Interface with a Crosstalk Suppression Scheme 679
Kwang-Il Oh, Lee-Sup Kim, Kwang-Il Park, Young- Hyun Jun and Kinam Kim

EMI Resisting Smart-power Integrated LIN Driver with Reduced Slope Pumping 683
Jean-Michel Redouté and Michiel Steyaert

SESSION 21- LEVERAGING THE THIRD DIMENSION

Stacking Technology Based on 8-inch Wafers using Direct Connection between TSV and Microbump 689
Nobuaki Miyakawa, Eiri Hashimoto, Takanore Maebashi, Natsuo Nakamura, Yutaka Sacho, Shigeto Nakayama,Shinjiro Toyoda

Clock Distribution Networks for 3-D Integrated Circuits 693
Vasilis Pavlidis, Ioannis Savidis and Eby Friedman

Inter-Die Signaling in Three Dimensional Integrated Circuits 697
Christopher Mineo, Ravi Jenkal, Samson Melamed and W. Rhett Davis

Variability in 3-D Integrated Circuits, ...701
Filipp Akopyan, Carlos Tadeo Ortega Otero, David Fang, Sandra Jackson and Rajit Manohar

3D Heterogeneous Integrated Systems: Liquid Cooling, Power Delivery,Implementation (INVITED) 705
Muhannad Bakir, Calvin King, Deepak Sekar, Hiren Thacker, Bing Dang, Gang Huang, Azad Naeemi and James Meindl

SiP for GSM/EDGE in CMOS Technology, ...713
Giuseppe Li Puma, Ernst Kristan, Paolo De Nicola, Cyril Vannier, Braam Greyling and Salvatore Piccolella

Heterogeneous Multicore SoC for Secure Multimedia Applications 717
Hiroyuki Kondo, Masami Nakajima, Sugako Otani, Osamu Yamamoto, Norio Masui, Naoto Okumura, Mamoru Sakugawa, Masaya Kitao, Koichi Ishimi, Masayuki Sato, Fumitaka Fukuzawa, Kazuhiro Inaoka, Yoshihiro Saito, Kazutami Arimoto and Toru Shimizu

A 159.2mW SoC Implementation of T-DMB Receiver including Stacked Memories...............................721
Joohyun Lee, Sungdo Kim, Jinkyu Kim, Duckhwan Kim, Youngsu Kwon, Minseok Choi, Kihyuk Park,
Bontae Koo, Nakwoong Eum and Hyuckjae Lee

SESSION 22- NOISE AND OSCILLATOR SIMULATION

Modeling, Measurement and Mitigation of Crosstalk Noise Coupling in 3D-Ics.....................................727
Liuchun Cai, Ramesh Harjani

A Method Using Circuit/Substrate Macro Modeling to Analyze Substrate Noise in a 3.2-
GHz 350Mtransistor Microprocessor...731
Mikiki Sode, Mikihiro Kajita, Naoya Nakayama and Satoshi Nakamoto,

Characterization of Random Decision Errors in Clocked Comparators 735
Brian Leibowitz, Jaeha Kim, Jihong Ren and Christopher Madden

Noise Tolerant Oscillator Design Using Perturbation Projection Vector Analysis 739
Igor Vytyaz, Josh Carnes, Ting Wu, Pavan Hanumolu, Un-Ku Moon and Kartikeya Mayaram

Strong Injection Locking of Low-Q LC Oscillators, ...743
Mozhgan Mansuri, Frank O'Mahony, Ganesh Balamurugan, James Jaussi, Joseph Kennedy, Sudip
Shekhar, Randy Mooney and Bryan Casper

SESSION 23- ANALOG TECHNIQUES

A Low Power 1.3GHz Dual-Path Current Mode Gm-C Filter 749
Manisha Gambhir, Vijay Dhanasekaran, Jose Silva-Martinez and Edgar Sanchez-Sinencio

A 1V Downconversion Filter Using Duty-cycle Controlled Bandwidth Tuning 753
Peter Kurahashi, Pavan Kumar Hanumolu and Un-Ku Moon

A Reconfigurable FIR Filter Embedded in a 9b Successive Approximation ADC 757
Joshua Kang, David Lin, Li Li and Michael Flynn

Voltage References for Ultra-Low Supply Voltages (INVITED) 761
Peter Kinget, Christos Vezyrtzis, Ed Chiang, B. Hung and T.L. Li

Design of Bandgap Voltage Reference Circuit with all TFT Devices on Glass Substrate in a
3-μm LTPS Process 767
Ting-Chou Lu, Ming-Dou Ker, Hsiao-Wen Zan, Chung-Hung Kuo, Chun-Huai Li, Yao-Jen Hsieh and
Chun-Ting Liu

Deep Submicron Effects on Data Converter Building Blocks (INVITED) 771
William Evans and David Burnell

SESSION 24- ADVANCED SUBSYSTEMS FOR CONNECTIVITY AND CELLULAR RADIO

A Highly Linear SAW-less CMOS Receiver Using a Mixer with Embedded Tx Filtering for
CDMA 777
Namsoo Kim, Lawrence E. Larson and Vladimir Aparin

High-power Digital Envelope Modulator for a Polar Transmitter in 65nm CMOS 781
Manel Collados, Paul T.M. van Zeijl and Nenad Pavlovic

A 2.4GHz, 20dBm Class-D PA with Single-Bit Digital Polar Modulation in 90nm CMOS......................785
Jason T. Stauth and Seth R. Sanders

Linearity and Efficiency Enhancement Strategies for 4G Wireless Power Amplifier Designs (INVITED) 789
Larry Larson, Donald Kimball and Peter Asbeck

An Analog Enhanced All Digtial RF Fractional-N PLL With Self-Calibrated Capability 797
Ping-Ying Wang, Jing-Hong Zhan, Hsiang-Hui Chang and Bing- Yu Hsieh

A Delta-Sigma Fractional-N Synthesizer with Customized Noise Shaping for WCDMA/HSDPA Applications 801
Xueyi Yu, Yuanfeng Sun, Woogeun Rhee, Zhihua Wang, Hyung Ki Ahn, Byeong-Ha Park

Author Index

Welcome from the CICC Committee

Welcome to CICC 2008, the 30th annual IEEE Custom Integrated Circuits Conference and leading international conference for integrated circuit development at the DoubleTree Hotel in San Jose, California. In addition to the core technical lecture and poster presentations, CICC offers attendees a total educational experience with educational sessions, a keynote address, a special luncheon lecture, exhibits, panels, tutorials, and stimulating networking events. The conference begins with a keynote address entitled "More Than Moore" by Dave Bergeron, CEO of SVTC. He will describe directions in semiconductor innovation that leverage older technology nodes to develop novel functions in passive devices, high- and low-voltage transistors and a host of other products. Our conference luncheon guest speaker, Dr. Alberto Sangiovanni-Vincentelli will speak on the use of IC's in automotive design.

The conference begins on Sunday, September 21 with educational sessions taught by practicing experts working at the leading edge of their fields. The technical session themes are; The Fundamentals of Analog Design, High-Speed Serial IO Design, and Coping with Technology Scaling. This year we also added a special professional development session on Effective Presentations and Technical Writing.

The Custom Integrated Circuits Conference has grown and changed with the industry and continues to showcase technical papers describing the most advanced analog and digital circuits and their applications. This year's program will start with the keynote address on "More Than Moore" by Dave Bergeron, CEO of SVTC. For over thirty years productivity has been driven primarily by Moore's Law, providing twice the number of circuits on a piece of silicon every 18-24 months. Progress has been measured by increased density and performance. While it's true that scaling continues and 22nm IC's are not far away, progress today is often measured by low power consumption and 3-D integration for portable applications, green technologies that preserve the environment, the ability to operate in hostile high temperature/high vibration environments, and high reliability solutions for life-critical applications. The evolution of technology to solve these issues has been dubbed "More than Moore." A panel discussion on Monday afternoon titled "Sure, Moore's Law Can Continue, But Should It?" will further debate issues around this growth area.

The paper submissions in 2008 allow us to bring you a conference of the highest technical caliber. This year 120 papers were selected from 364 submissions and organized into 24 sessions. The topics addressed by these high quality papers include the latest innovations in 3-D circuits, ADC's, sensors and displays, SOC's, wireless circuits, power management, high performance wired interfaces, PLLs, embedded memories, simulation and modeling, design test and debug issues, and manufacturing developments. Highlights include invited and tutorial papers from leading experts in industry and academia.

Our Monday evening Welcome Reception and Tuesday Conference Reception are professional networking at its very best! Our exhibits area will include booths from prominent industry suppliers and Poster Sessions. The poster session is a unique forum for in-depth discussions with authors.

CICC is co-located with the IEEE Behavior, Modeling, and Simulation Conference 2008. BMAS will take place September 25– 26 at the DoubleTree Hotel, San Jose, California. Visit the BMAS website at www.bmas-conf.org for complete conference information.

Ann Marie Rincon
General Chair

David Nairn
Conference Chair

Jacqueline Snyder
Technical Program
Chair

Steering Committee

Henry Chang
Designer's Guide Consulting

David Nairn
University of Waterloo

Ann Rincon
ON Semiconductor

Jacqueline Snyder
Marvell Semiconductor

Trudy Stetzler
Texas Instruments

Larry Wissel
IBM

Organizing Committee

General Chair
Ann Rincon, ON Semiconductor

Conference Chair
David Nairn, University of Waterloo

Technical Program Chair
Jacqueline Snyder, Marvell Semiconductor

Educational Sessions Chair
Shahriar Mirabbasi, University of British
Columbia

Exhibits Chair
Tom Andre, EverSpin Technologies

Panel Chair
Aurangzeb Khan, Consultant

Publicity Chair
Arif Rahman, Xilinx Research Labs

Sponsorship Chair
Eric Naviasky, Cadence Design

Best Paper Awards
Jennifer Lloyd, Analog Devices

Treasurer
Trudy Stetzler, Texas Instruments

Technical Program Committee

Analog Circuit Design
Seated left to right: George LaRue, Washington State University, David Nairn, University of Waterloo, Donald Thelen, ON Semiconductor, Jennifer Lloyd, Analog Devices, *Standing left to right:* Takahiro Miki, Renesas, Ken Suyama, Epoch Microelectronics, University, Yusuf Haque, Consultant, *Not Pictured :* Yun Chiu, University of Illinois, Eric Naviasky, Cadence

Biomedical, Sensors, Displays, and MEMS
Standing left to right: Ken Szajda, LSI, , Dawn Fitzgerald, Aurora Enterprises, Steve Garverick, Case Western University, *Standing left to right:*, Makoto Nagata, Kobe University, Mourad El Gamal, McGill University, Edward Lee, Alfred Mann Foundation, *Not Pictured:* Sang-Soo Lee, Pixelplus Semiconductor

Characterization, Debug and Test
Seated left to right: Gordon Roberts, McGill University, R Hamid Mahmoodi, San Francisco State University, Jeanne Trinko Mechler, IBM, Mike Li, Altera,

Digital Circuits and SoC-SIP Designs and Methodology
Seated left to right: Charles Thomas, NICTA, Arif Rahman, Xilinx, Rakesh Patel, Altera Corp., Raj Amirtharajah, University of California, Davis, Henry Chang, Designer's Guide Consulting, *Standing left to right:* Ann Marie Rincon, ON Semiconductor, Aurangzeb Khan, Consultant, Paul Billig, Cavendish Kinetics, Osamu Takahashi, IBM, Steve Wilton, University of British Columbia, Ric Williams, Sun Microsystems, Mike Seningen, Intrinsity, *Not Pictured:* Ram Krishnamurthy, Intel

Embedded Memory
Seated left to right: Takashi Akioka, Renesas Technology, Tom Andre, EverSpin Technologies, Kenji Noda, NScore, Jean-Christophe Vial, Infineon, *Not Pictured:* Subramani Kengeri, TSMC, Larry Wissel, IBM

Manufacturing
Seated left to right: Philipe Jansen, Infineaon, Jordan Lai, TSMC, David Sunderland, Boeing Satellite Systems, *Not Pictured:* Rich Liu, Macronix, Alvin Loke, AMD

Power Management
Seated left to right: Jerry Zheng, Iwatt, Makoto Takamiya, University of Tokyo, Gordon Lee, Qualcomm, Vikas Chandra, ARM, Lawrence Clark, Arizona State University

Wireless Designs
Seated left to right: Nobuyuki Itoh, Toshiba, Ramesh Harjani, University of Minnesota, Trudy Stetzler, Texas Instruments, John Rogers, Carleton University, Andrea Mazzanti, Universita di Modena, *Standing left to right:* Earl McCune, Panasonic, Ranjit Gharpurey, University of Texas, Howard Luong, Hong Kong University of Science and Technology, Fa Foster Dai, Auburn University, Stefan Drude, NXP Semiconductors, *Not Pictured:* Payam Heydari, University of California, Irvine, Cicero Vaucher, NXP Semiconductors

Simulation and Modeling
Seated left to right: Gennady Gildenblat, Arizona State University, Hidetoshi Onodera, Kyoto University, Hong-Ha Vuong, LSI, *Standing left to right:* Colin McAndrew, Freescale Semiconductor, Inc., Rob Jones, IBM, Brian Qiang Chen, AMD, *Not Pictured:* Larry Nagel, Omega Enterprises

Wired Communications
Seated left to right: Ed van Tuijl, Philips Research, Tony Chan Carusone, University of Toronto, Kimo Tam, Analog Devices, *Standing left to right:* Ken Chang, Rambus, Shahriar Mirabbasi, University of British Columbia, Dennis Fischette, AMD, *Not Pictured:* Jin Liu, University of Texas, Dallas, Cormac O'Connell, TSMC

CONFERENCE OVERVIEW

SUNDAY, SEPTEMBER 21

Time	OAK BALLROOM	FIR BALLROOM	PINE BALLROOM	CEDAR BALLROOM	OTHER ROOMS	BAYSHORE FOYER	DONNER BALLROOM
8:00 am – 5:00 pm	Ed Session 1 - High Speed I/O	Ed Session 2 - Coping with Technology Scaling	Ed Session 3 – Fundamentals of Analog Design		Silicon Valley Room — 2:00 PM – 5:00 PM Effective Technical Writing and Presentations Session	Ed Session Registration 7:30 am – 2:00 pm; Technical Session Registration 2:00 pm – 5:00 pm	

MONDAY, SEPTEMBER 22

Time	OAK BALLROOM	FIR BALLROOM	PINE BALLROOM	CEDAR BALLROOM	OTHER ROOMS	BAYSHORE FOYER	DONNER BALLROOM
7:30 am – 5:00 pm						Technical Session Registration 7:30 am – 5:00 pm	Exhibits Open 4:00 – 8:00 pm
8:15 am – 9:30 am	1. Keynote Presentation						
10:00 am – 12:00 pm	2. Statistical Modeling	3. Power Management	4. High-Speed Test, Characterization, & Debug.	5. Broadband Circuit Techniques for Emerging Wireless Communications I			
1:30 pm – 5:30 pm	6. Advanced SoC/SiP Integration & Co-Design	7. High Resolution Converters	8. Characterization & Test Methods for Device Variability in Nanoscale Technologies / 10. Panel Discussion 4:00 pm – 5:30 pm	9. Broadband Circuit Techniques for Emerging Wireless Communications II			
5:00 pm – 7:00 pm							Poster Session
5:30 pm – 8:00 pm							Welcome Reception

TUESDAY, SEPTEMBER 23

Time	OAK BALLROOM	FIR BALLROOM	PINE BALLROOM	CEDAR BALLROOM	OTHER ROOMS	BAYSHORE FOYER	DONNER BALLROOM
8:00 am – 5:00 pm						Technical Session Registration 8:00 am – 5:00 pm	Exhibits Open 4:00 – 8:00 pm
8:25 am – 12:00 pm	11. Compact Modeling	12. High Speed A/D Converters	13. Biomedical, Sensors and MEMS	14. Advanced SoCs - Techniques and Applications			
12:00 pm – 1:50 pm					Cedar Ballroom - CICC Luncheon		
2:00 pm – 5:00 pm	15. IC technology – More Moore and More Than Moore	16. Embedded Memory	17. Clocking Circuits	18. Millimeter-Wave Circuit Techniques			
5:00 pm – 7:00 pm							Poster Session
5:30 pm – 8:00 pm							Conference Reception

WEDNESDAY, SEPTEMBER 24

Time	OAK BALLROOM	FIR BALLROOM	PINE BALLROOM	CEDAR BALLROOM	OTHER ROOMS	BAYSHORE FOYER	DONNER BALLROOM
8:00 am – 3:00 pm						Technical Session Registration 8:00 am – 3:00 pm	
8:25 am – 12:00 pm	19. Low Power and Non-traditional RF Tranceivers	20. Advanced Wireline Techniques	21. Leveraging the Third Dimension				
1:30 pm – 5:00 pm	22. Noise and Oscillator Simulation.	23. Analog Techniques	24. Advanced Subsystems for connectivity and Cellular Radio				

EDUCATIONAL SESSION 1   Oak Ballroom

Chairperson: Shahriar Mirabbasi, University of British Columbia

High-Speed I/O

Organizer: Tony Chan Carusone, University of Toronto
Co-Organizer: George LaRue, Washington State University

9:00 am - 10:50 am
E1-1 Multi-Gigabit I/O Design for Microprocessor Platforms
Randy Mooney (Intel)
A discussion of the constraints of the design space for microprocessor platforms, and a look at the state of the technologies required to deliver bandwidth in these platforms. These technologies include analysis tools, interconnect components, modulation and equalization choices that fit the constraints, and the various circuits required in silicon. This is followed by a look at future platform requirements, and the potential solution space to meet those needs.

11:10 am - 1:00 pm
E1-2 Jitter and Signal Integrity at 10 Gbps
Mike Li (Altera Corp.)
In this tutorial presentation, we will first review where the technology is heading to for the multiple Gbps high-speed links and I/O buses for devices and systems in networks and computers. Second, we will discuss why jitter and signal integrity have become the major challenges, as well as limiting factors for developing those high-speed, high performance, high volume, and low-cost I/O devices and systems as the data rate approaches 10 Gbps and beyond. Third, we will discuss the jitter and signal integrity modeling, simulation, verification and characterization methodologies within the context of a serial link. We will cover these ever evolving cutting edge topics from generic perspective, as well as practical application perspective, with real-world examples from multiple Gbps link technologies such as Giga Bit Ethernet (GBE), PCI Express (PCIe), Fibre Channel (FC), with emphasis on their latest generations operating at single lane data rates in the vicinity of 10 Gbps. Emerging challenges such as jitter amplification and mitigation, equalization optimization and verification, on-chip jitter de-embedding will also be covered.

2:00 pm - 3:50 pm
E1-3 Equalization & High-Speed Transceiver Design
Jared Zerbe (Rambus)
Equalization is an ever-critical aspect of serial data systems and is even beginning to expand into high-speed parallel systems. This tutorial provides a basic overview of the serial data transmission problem and the goals of equalization, along with some of the practical challenges at high speeds and some vision of its future. The architecture of equalized systems is explained, with detail on key types of equalizers such as linear receive equalizers, transmitter pre-emphasis, and DFE, along with the pros and cons of each type. The tutorial will also teach how various equalizer components can be used together to mitigate each other's weaknesses. Equalizers will be presented from various viewpoints, including effectiveness and practical circuit design as well as future trends. Effectiveness of different approaches on different practical environments will be compared. Simulation approaches for equalization will be discussed. Finally, as an alternate to equalization, certain modulation approaches such as 4-PAM & duo-binary will be covered and pros and cons reviewed.

4:10 pm - 6:00 pm
E1-4 Clocking and CDRs
Jafar Savoj (Qualcomm)
High-purity clock generation enables longer reach in wired communication systems. Optical and copper standards set an upper bound on the maximum noise and distortion added to the signal at the source. This tutorial describes means of efficient clock generation and distribution in high-performance chips to satisfy requirements imposed by the standards. Clock and data recovery (CDR) circuits are an integral part of wired communication systems. With the accelerated rate of device scaling in recent decades, CDR architectures have transitioned from fully analog into mixed-mode and digital implementations. The tutorial later addresses the evolution of CDR architectures, as well as the design of their building blocks.

EDUCATIONAL SESSION 2   Fir Ballroom

Coping with Technology Scaling

Organizer: Colin McAndrew, Freescale Semiconductor
Co-Organizer: Foster Dai, Auburn University

9:00 am – 10:50 am
E2-1 Technology and Reliability
Paul Packan (Intel)
Continued technology scaling drives not only developments and changes in device designs, characteristics, and performance, but also involves an ever expanding complexity of constraints that must be applied and phenomena, like variability and stress, that must be taken into account to enable the design of billion+ transistor ICs. This tutorial will review nMOS and pMOS scaling trends and their impact on circuit performance benchmarks, including power. Sources of variability and their impact on circuit performance at scaled supply voltages will be reviewed, as will circuit level techniques to mitigate problems caused by variability. The impact of both device and circuit architectures will be discussed, as related to memories, RF circuits, power consumption, and performance. Scaling has also lead to issues with reliability, and these will be reviewed. Finally, because of the growing complexity of restrictions that must be applied to make functions ICs, lithography, layout, and design rule issues that affect the manufacturability of designs will be discussed.

11:10 am – 1:00 pm
E2-2 Logic and Memory Scaling Challenges
Bora Nikolic (University of California, Berkeley)
Digital logic and memory are expected to scale down in area by 50% with each new technology node. This is the only key benefit of technology scaling as the active and leakage power limit the rate of further logic speed increase. This tutorial will address the main challenges and known solutions for keeping the expected scaling rate: increased cost of design and manufacturing, design under power limitation, impact of technology variability, and design with added technology features.

2:00 pm – 3:50 pm
E2-3 CAD and Modeling Issues
Sani Nassif (IBM)
Technology scaling is not just a problem for the manufacturing engineers; it presents unique challenges for those who must use this same technology to produce working high performance chips in volumes that can lead to profit. Activities like OPC and DFM have become common place terms for designers and EDA engineers, and are all part of the response to the increasing complexity of the design/manufacturing interface. This interface has historically been defined by layout design rules and so-called corner models. Both of these representations are unraveling as we enter the 45nm node with thousands of design rules, and with overall manufacturing variability becoming the most significant challenge faced by design. In this tutorial, we will review the design/manufacturing interface and show current trends, explain how technology characterization and modeling leads to specific challenges for the representation of technology in simulation tools, and finally review some of the design responses to technology scaling that leverage adaptivity and regularity.

4:10 pm – 6:00 pm
E2-4 Analog and RF Design Issues in Deep Submicron CMOS Technology
Behzad Razavi (University of California, Los Angeles)
This tutorial presents the challenging issues in analog and RF design as technology nodes go beyond 65 nm and 45 nm. Noise-power-speed and mismatch-power-speed trade-offs resulting from supply scaling are quantified and the effect of switch nonlinearities in sampling circuits is formulated. Phenomena such as output resistance nonlinearity and the gate leakage current are studied and their impact on circuits such as PLLs and op amps is summarized. Noise-linearity trade-offs in passive and active RF mixers and various deep-submicron effects in LC oscillators are also presented and low-voltage circuit techniques are described.

EDUCATIONAL SESSION 3   Pine Ballroom

Fundamentals of Analog Design

Organizer: Sang Soo Lee, Hynix Semiconductor

9:00 am - 10:50 am
E3-1 Amplifiers
Boris Murmann (Stanford University)
This lecture covers a systematic methodology for the design of high performance operational transconductance amplifiers (OTAs) in deep sub-micron CMOS technology. The first part of this presentation reviews the basic design equations and power/speed/noise tradeoffs in OTAs using a two-stage Miller-compensated design as an example. In the second part, Spice-generated look-up tables are introduced as a means to bridge the gap between simulation, hand analysis and Matlab optimization. Using tabulated device data that captures the fundamental tradeoff between speed (gm/Cgg) and transconductance efficiency (gm/ID), the proposed method yields near-optimal designs without the need for iterative Spice simulations or expensive CAD tools.

11:10 am - 1:00 pm
E3-2 References
Wing-Hung Ki (Hong Kong University of Science and Technology)
In this lecture, the treatment of voltage references that is systematic and coherent, rigorous but not excessive is attempted. The talk starts with fundamentals of voltage references. Popular bandgap references (BGRs) are then discussed, with emphasis on CMOS bandgap references using parasitic BJTs in a CMOS process. Performance parameters such as temperature coefficient, power supply rejection, line and load regulation, and loop gain are introduced. For BGR with simple structures, analytic results on loop gain and power supply rejection are presented. The development of op-amp based BGR for reducing effect due to op-amp input offset voltage, folded resistor for lowering power supply voltage (Vdd) requirement, and folded resistor divider for further lowering Vdd requirement, are traced. Non-op-amp based BGRs are discussed, starting with the 4T current-voltage-mirror (CVM) scheme in replacing the op-amp. The principle of symmetrical matching is then introduced to minimize systematic errors due to channel length modulation, and an 8T symmetrically matched CVM is used to realize a BGR with improved power supply rejection. The BGRs discussed are designed using a 0.18 procedure and simulation results are presented. Design issues such as trimming, resistor strings and organization of voltage references in an IC system are also sketched.

2:00 pm - 3:50 pm
E3-3 PLL
Behzad Razavi (UCLA)
This tutorial deals with the analysis and design fundamentals of PLLs. Various voltage-controlled ring oscillator topologies are described that can be used in timing applications up to several gigahertz. Next, type I PLLs and their shortcomings are studied, leading to type II (charge-pump) PLLs as a superior choice. The dynamics of the PLLs are derived, the effect of various charge pump nonidealities is presented, and circuit techniques for alleviating these effects are summarized. Lastly, a design procedure for PLLs is outlined and demonstrated by a transistor-level implementation.

4:10 pm - 6:00 pm
E3-4 DAC
Doug Mercer (Analog Devices)
Modern communication systems have spawned a growing interest in high performance, high speed Digital to Analog Converter designs which can be easily embedded into larger mixed signal systems. Implementing larger systems in addition require peripheral support D/A functions outside the main signal path in applications such as tuning and calibration. The tutorial will concentrate on D/A converter design in MOS process technologies and cover these three topics.

     1) A brief look Digital to Analog conversion first principles including a description of the D/A function and the key specifications that define the performance of a D/A.
     2) Common D/A architectures will be explored with these first principles in mind. The advantages and disadvantages of each will discussed.
     3) Case studies of example CMOS implementations will be included.

SPECIAL AFTERNOON WORKSHOP   Silicon Valley Room

Effective Technical Writing and Presentations

2:00 PM – 5:00 PM
Ann Marie Rincon (ON Semiconductor)

My technical work is outstanding - why didn't my paper get accepted? I thought my description was very clear - why was my thesis misunderstood?

This class will provide answers to these questions and help engineers and programmers write clear, concise technical papers. The writing do's and don'ts covered in this class can be applied to other technical documents such as application notes, product specifications and emails.

The class will provide:
     A standard technical paper outline and a description of each section
     Tips for submitting a paper to an external conference
     General writing tips including do's and don'ts
     Tips for translating your technical paper into an effective presentation
     Several lucky attendees will receive a copy of "The Elements of Style" by William Strunk Jr. and E.B. White.

Keynote Presentation   Session 1

8:15 am   **Welcome and Opening Remarks**
                Awards Presentations
                Keynote Speaker Introduction
                Ann Marie Rincon, General Chairman

8:30 am   **Keynote Presentation**

"More Than Moore"
Dave Bergeron, CEO, SVTC Technologies

There's plenty of life left in Moore's Law, as today's leading-edge semiconductor developments will attest. However, there is a new direction in semiconductor innovation that takes advantage of old technology nodes. It's called "More than Moore."

This approach leverages the CMOS backbone established processes and technologies to develop novel function in passive devices, high- and low-voltage transistors, MEMS and a host of other products.

In this keynote, Dave Bergeron describes the evolving infrastructure and expanding marketplace, and lays out options and strategies for "More than Moore" players in today's semiconductor industry.

Dave Bergeron is the Chief Executive Officer, SVTC Technologies. Dave recently served as Executive-in-Residence at Tallwood Venture Capital, where he evaluated semiconductor chip products and equipment opportunities. Prior to joining Tallwood, Dave held senior management positions at Applied Materials, Candescent Technologies and IBM Microelectronics. These positions included VP and General Management responsibilities for a semiconductor equipment product line; semiconductor fab operation management responsibilities, including CMOS DRAM and CMOS LOGIC product development; and multiple technology product integration responsibilities supporting an advanced display opportunity. Dave has authored 17 U.S. patents grants and has published more than 30 technical bulletins and papers. Dave received a B.S. in Physics and an M.S. in Applied Mathematics from Georgetown University.

Notes

Statistical Modeling   Session 2

Chair: Hidetoshi Onodera, Kyoto University
Co-Chair: Hong-Ha Vuong, LSI

With device dimensions in the nanometer regime, variability is now a serious concern in LSI design. Aggressive scaling, along with ever increasing technology complexity, leads to an explosion in the magnitude of variability while also introducing new sources of variability that need to be characterized and modeled. The topic of this session, statistical modeling of performance variability and circuit reliability, is therefore one of key challenges for achieving robust design of LSIs.

Our first paper reviews in a first part the current status of performance variability. It examines the sources of variability and explains how they can be characterized using test structures. Examples of measured variabilities at 65nm bulk and SOI processes are disclosed, which include the distribution of threshold voltages within a die and a breakdown of spatial variability into each component of "Lot-to-Lot", "Wafer-to-Wafer", "Within Wafer", "Within Die", etc. This paper then examines variability trends for future scaled technologies. It also shows that the impact of variability is changing its character such that the parametric performance variability is moving closer to the region of catastrophic faults.

Our second paper, taking an SRAM sense amplifier at a 65nm bulk CMOS process as a test vehicle, analyzes mismatches in NMOS and PMOS threshold voltages and discusses a statistical design method using transistor sizing. A linear response surface model is derived that relates input offset voltage of the amplifier to threshold voltages variations. A successful application of the model in the statistical sizing is verified by the measured variability of fabricated test circuits.

The last paper discusses statistical modeling of circuit performance degradation over time. Analytical solutions are developed that efficiently predict the statistics of circuit timing and the leakage due to NBTI and process variations. It is shown that the degradation rate and its variance can be predicted from the characteristics of transistor degradation and circuit performance sensitivity to aged parameters, which are independent of the type and the amount of process variations. An aging model is implemented into SPICE and verified by simulation results and 65nm silicon data.

Notes

IEEE 2008 Custom Intergrated Circuits Conference (CICC)

Process Variability at the 65nm node and Beyond

Sani R. Nassif

IBM Austin Research Laboratory, 11501 Burnet Rd., Austin, TX 78758, USA

Abstract

The impact of manufacturing-induced variations has been well established as a first order impediment to modern integrated circuit design [1]. Numerous research efforts are currently underway to (a) understand and characterize variability[2], to (b) predict its impact on circuit behavior[3], and (c) to develop layout and circuit design techniques to reduce the impact of variability[4]. Simultaneously, except for the most advanced high performance designs, the increasing cost of migrating to sub-65nm technology nodes is slowing down adoption of advanced technologies. This slow down is allowing current efforts to catch up and help mitigate variability as an impediment.

The future of CMOS technology scaling, however, is such that the impact of variability on design is going to change in character, not just magnitude. Phenomena that currently affect the *parametric* performance of circuits are expected to reach magnitudes which will impact the *correctness* aspects of circuits.

This paper reviews the sources, characterization, and future trends of manufacturing-induced variability, and shows some examples of the new challenges we are facing when the magnitude of variability is such that its impact is indistinguishable from catastrophic faults.

1 Sources of Variability

The electrical performance of an integrated circuit is impacted by three sources of variation:

- *Environmental factors* which include variations in power supply voltage and temperature. These factors are highly design dependent and exhibit time constants similar on the same scale as that of the clock frequency.

- *Reliability factors* which are usually related to the high electrical fields present in modern devices, and include phenomena like negative bias temperature instability (NBTI)[5] and electromigration. These factors are also design dependent, but typically have a

much longer time constant measured in months or even years of device operation.

- *Physical factors* which result in variations in the electrical parameters which characterize the behavior of active (MOSFET) and passive (interconnect) devices. These variations are caused by the manufacturing process and include a wide variety of mechanisms. Some of these mechanisms are *systematic* in the sense that they repeat over many chips or wafers, an example would be wafer-level variations induced by the rapid thermal annealing processes in common use[6]. Other mechanisms are are *design dependent*, an example is the impact of chemical-mechanical-polishing on interconnect resistance). A few mechanisms are characterized as being *random*, these include phenomena like the impact of random dopant fluctuations on MOSFET threshold voltage.

While the environmental and reliability factors are certainly important, and indeed drive a lot of work in characterization, design as well as in design automation, we choose to focus in this paper on physical variability, since it is the component most severely impacted by technology scaling.

2 Motivation for Technology Characterization

From a design point of view, the most important characteristic of a specific source of variability is whether that variability is *random* vs. *systematic* or design-dependent. This distinction is important because the two types of variability elicit very different design responses.

In the case of systematic variability, where there exists a *model* of variability as a function of design implementation, the designer has the option to *null* out the impact of the variability by suitably modifying his design. For example, if it were known that variability in high-VT devices was lower than that in standard-VT devices, then a designer can decide to switch device types in an application where VT variability would be important. This results

978-1-4244-2018-6/08 $25.00 © 2008 IEEE

in a decrease in variability, and a net improvement in the design.

It is important to remember that some components of systematic variability are not directly addressable during the design phase, and require instead improvements in the manufacturing process. Wafer-level non-uniformity, for example, is a systematic source of variability which cannot be addressed by making changes to the design.

For random variability, on the other hand, the only information available is typically the statistical distribution of that variability. So the designer is required to build enough design margin to accommodate that variability to whatever confidence level is required. So if vias were known to have a certain mean and a certain standard deviation, then the designer would apply *worst case design* principles to insure that this circuit works over a sufficiently broad range of the via resistance distribution. This results generally in *over-design* and means that the design is using more resources: area, power or delay, in order to achieve the desired performance.

So while systematic variability can be accommodated in the design with relatively low cost, random variability typically requires a higher cost to insure against it. Now consider what happens when a new source of variability is identified. If all that is known is that the magnitude of that variability follows some observed distribution, then it must be treated in a random manner. Additional investment in modeling and characterization can, however, provide a model for this variability and its dependence on - say- layout, in which case the designer can then design around it in a systematic manner.

It is this extra investment in technology characterization and variability modeling that can provide understanding that in effect transforms random variability into systematic. That transformation in turn facilitates decreasing design margining and lowering design cost. Of course the cost of such an investment must be considered. The investment can be particularly costly for example in a foundry relationship where expense or the prevailing business models make it difficult for technology users to perform technology characterization. Where the characterization investment is possible, however, development of efficient, effective characterization techniques is a key step toward making the effort pay off in terms of reduced design margins and cost.

The remainder of this paper is organized as follows. First, a review of variability trends establishes the consensus that variability is increasing and provides an estimate of by how much. Next, goals that must be met to achieve efficient, effective technology characterization are

Table 1: The amount of variability in key parameters predicted by the latest edition of ITRS (adapted from [7]).

Year	'08	'09	'10	'11	'12	'13
Pitch (nm)	57	50	45	40	36	32
V_{DD}	10%	10%	10%	10%	10%	10%
V_{TH}	37%	42%	42%	42%	58%	58%
CD	12%	12%	12%	12%	12%	12%
Delay	46%	49%	50%	53%	54%	57%
Power	57%	57%	58%	58%	59%	59%

discussed, followed by examples of characterization that take steps toward meeting those goals, one that focuses on random threshold voltage variation and another on spatial variability. Finally, we look at longer-term variability implications and, in particular, at a potential trend for catastrophic variability-induced events to move beyond the realm of SRAM into logic circuits.

3 Variability Trends

Due to the sensitive nature of manufacturing variability data, there is relatively little published data on measured variability and its trends with scaling. One source of such data is the International Technology Roadmap for Semiconductors, which attempts to define the general future course of technology. The ability to project into the future the amounts of variability to be expected is particularly important for researchers as their research priorities are determined by these numbers. So it is crucial to have an idea of the typical magnitudes of variability we may expect in the future. In this section we address the question of how this projection was made and, given that almost a decade has passed since the first projection, we also provide an updated prediction.

Table 1 below contains a summary of variability projections with respect to some key process (V_{th}, L_{gate}) and circuit (delay, power) level parameters. This table comes from the Design chapter of the 2007 Edition of ITRS. The rather large amount of variability in threshold voltage, driven by the increasing impact of random dopant fluctuations, drives significant variability in core circuit performance such as power and delay.

Note that these predictions are made based on input from manufacturing engineers, who precisely follow the history of various technology capabilities, and are aware of new developments in manufacturing equipment, processes, materials, and methods that can influence said capability. Since these predictions are reviewed by a large

978-1-4244-2018-6/08 $25.00 © 2008 IEEE

number of practitioners, they represent the best information we have about nominal values as well as the expected amount of variability in the future. The expectation is that near term values are quite precise, but that values that are out two or more technology nodes are less so.

One of the most important variability metrics is critical dimension (CD) control, which defines the expected variability in device length. CD control is a complex mixture of systematic, spatial and random components, all of which are highlighted in the ITRS document. Note that the 12% number in table 1 is for within-die identical structures and clearly represents a *goal* which assumes continuing improvement in solving difficult (referred to in the ITRS as *red bricks*) manufacturing problems.

4 Goals for Variability Characterization

Technology development occurs at a frantic pace, with the constant introduction of new technology features to continue the pace of performance and density improvements expected from historical scaling trends. This results in ever increasing technology complexity, with more potential sources of systematic and random variability. Significant effort is required to characterize and model this constant stream of new phenomena. With limited test resources, silicon area, and engineering effort available, variability characterization is emerging as an important challenge to future scaling.

While it is possible to perform first-principles studies of new variability phenomena to understand the sources, magnitude and design dependencies, such activities typically take far too long to perform. Thus most variability characterization is done in a more empirical manner using *test structures*[2]. Test Structures are simply circuits designed to measure or assess a specific source of variability. Such test structures may get fabricated frequently or seldom depending on whether the phenomena in question is one that requires *tracking*[8].

Test structures can be created for a myriad of purposes, and in fact there are conferences devoted to nothing else (e.g. the IEEE International Conference on Microelectronic Test Structures[9]). In the end, however, the efficacy of a test structure needs to be measured in order to justify using one vs. another. Several metrics are important for said efficacy:

1. *Density* measures the amount of silicon area required for the test structure. All other things being equal, a smaller test structure is more useful because it will

have a lower cost. Also, smaller test structures afford the opportunity for replication across the die, allowing for more information about spatial variability and/or statistical distributions.

2. *Manufacturing Complexity* relates to the portion of the overall fabrication *flow* required for the test structures. A so-called *short flow* test structure to measure -say- via resistance would be more efficient than a *full flow* test structure to do the same thing.

3. *Test efficiency* relates to the amount of test time and test complexity required to gather the needed data. Clearly, test structures which can deliver the same amount of information with fewer or less complicated measurements are preferred.

4. *Generality* is the most difficult of the test structure metrics to assess. It measures the applicability of the results found from the test structure to other circuits. A test structure that measures ring oscillator frequency, for example, is not terribly relevant to the performance of a phase-locked-loop. A test structure that measures the threshold voltage of a MOSFET, on the other hand, is very generally applicable.

5 Threshold Voltage Variability Characterization

As an example of the current trend in test structure design, we briefly describe a recently developed an extremely efficient threshold voltage statistical characterization site[10]. The site relies on the observation that the standard deviation of a quantity, calculated using an equation of the form:

$$\sigma \quad \sqrt{\frac{N}{N}\sum_{i=1}^{N} x_i - \mu^{\;2}} \qquad (1)$$

where σ is the standard deviations, μ is the mean, N is the number of samples, and x_i denotes the ith sample; is essentially identical to the root-mean-square of a signal, calculated using an equation of the form:

$$V_{RMS} \quad \sqrt{\frac{M}{M}\sum_{i=1}^{M} V_i^{\;2}} \qquad (2)$$

where V_{RMS} is the RMS voltage, M is the number of measured samples, and V_i is the ith measurement.

The technique uses a self-clocked scan chain to sequentially cycle through a device array, creating a periodic

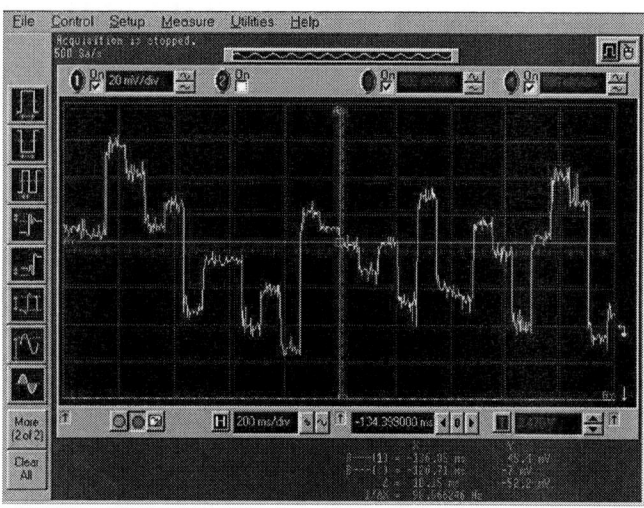

Figure 1: A measured waveform from the threshold voltage characterization test site.

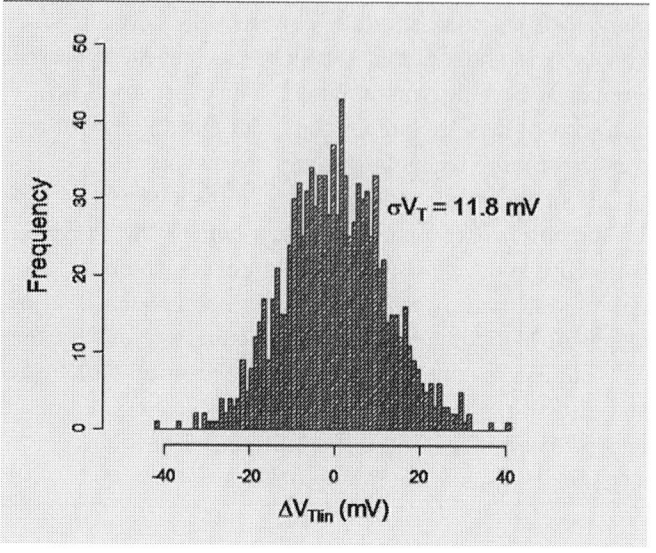

Figure 2: A measured threshold voltage distribution from the characterization test site.

Table 2: Comparison between direct and statistical summary measurements for a threshold voltage characterization test site.

Width μm	Length μm	Mean^i mV	Sigma^i mV	Mean mV	Sigma mV
1.0	0.06	562	12.1	554	11.8
0.50	0.06	559	16.7	561	16.5
0.12	0.06	526	29.0	530	29.2

waveform each value of which corresponds to the threshold voltage of one of the devices. That waveform when filtered using a low-pass filter will produce, as a DC measurement, the *mean* μ of the threshold voltage. Referencing the waveform to the low-pass filtered version of the same waveform (i.e. the mean) accomplishes the subtraction needed, and the result, when measured using an RMS meter, provides a direct measurement of the standard deviation of the threshold voltage.

This characterization site was implemented in a 65nm bulk CMOS process. Figure 1 shows the type of waveforms the site produces, and Figure 2 shows the distribution of P-channel threshold voltage, measured from an array of 1000 devices. To verify the accuracy of the test site, we compared the measurements made from the site directly with the results of measuring each of the 1000 individual devices independently (labeled with the superscript i in the table). The comparison between the two measurements is shown in table 2.

6 Spatial Variability Characterization

Due to the hierarchical nature of the integrated circuit manufacturing process, where fabrication steps happen in units of *lots* which contain multiple *wafers*, which in turn contain multiple *dies*, variability acquires a spatial dependence causing systematic differences across the various spatial domains [11].

In order to measure spatial variability, it is common in IBM to include a number of identical individually measurable ring-oscillators across an integrated circuit [12]. These ring oscillators are measured during final wafer testing, and that data is used for process control as well yield learning. The analysis of such data can provide a breakdown of the various components of spatial variability, and is an area of some considerable interest by current researchers -in spite of the general lack of publicly available data (see for example [13] for a creative solution to that problem).

A simple quantitative model of the spatial variability of identical structures (i.e. the ring oscillators distributed within an integrated circuit) would divide it into the following components:

- Lot to Lot variations, which measure the shift in the mean from one lot to another.

- Wafer to wafer within a lot variations, which measure the shift in the mean from one wafer to another within one lot.

- Within wafer variations, which measure the shift in

Table 3: Approximate spatial variability breakdown.

Component	Percentage of Total
Lot to Lot	35 %
Wafer to Wafer	15 %
Within Wafer	20 %
Within Die	20 %
Random Residual	10 %

the mean from one die to another within one wafer.

- Systematic die variations, which measure the repeatable die-level pattern found across a population of dies.

- Residual random variations, which are the remainder of the overall variability when all the other components are accounted for.

Many of these variations are related to process drift over time, and to the differing properties of equivalent manufacturing *tools* like steppers, furnaces and implanters. The lone standout is the systematic die variations component, which is dominated by lithography and mask-making aspects.

We analyzed a large set of data from IBM's 65nm SOI technology, comprising 23 lots with 24 wafers each, approximately 100 die per wafer, and 14 ring oscillators across each die. The total number of dies (excluding some missing points) was about 36000. While it is quite dangerous to take any one data set as being *typical* or indicative of an overall trend, we believe that the overall breakdown of variability across the various levels is very instructive to practitioners working in the area. Due to space limitations, we do not report here on the details of how the analysis was performed, but report the general overall breakdown of spatial variability in table 3.

The variability breakdown described in table 3 is extremely important in one specific regard. The data shows that 70 % of the total variability is from one die to another, 20 % is due to systematic (i.e. repeating) within-die effects, and only 10 % is due to random or unknown sources of variation. Consider the impact of this breakdown on a yield improvement effort. It shows that the bulk of design variability can be captured using traditional corner-based analysis (also referred to as worst-case design, e.g. [14]). For the remaining within-die variability, the data shows that early in the design cycle, i.e. before the full place and route physical implementation of the design is complete, one must build enough design margin to handle up to 30 % of the variability. However, once the design is complete,

and assuming appropriately accurate models exist for the various mechanisms that contribute to systematic within-die variability, then there remains only 10 % of the total variability to be compensated for or margined against.

7 Variability Implications

In current digital CMOS technologies, the prevailing view of the impact of variability focuses on how it affects important circuit performances like delay, power, and leakage. Much of current published research focuses on either assessing this parametric impact of variability on circuits, or on creating circuits that are somehow less sensitive to variability, either by design, or by the use of adaptation in various forms.

One area in which the impact of variability is qualitatively different is memory circuits. The demand for ever larger on-chip memories pushed SRAM designers to use the smallest possible devices, and to create the densest possible layouts. It is not uncommon for SRAM to be four times more dense in terms of devices per unit area than other parts of a digital CMOS chip. This extreme miniaturization makes SRAM circuits significantly more dense than other parts of the design, causing them also to be much more sensitive to the manufacturing process.

The sensitivity of SRAM to various components of variability is well documented[15, 16], but it is even more interesting to compare the manner with which this variability impacts SRAM with how it impacts other circuits. While a normal CMOS gate such as an inverter would predominantly slow down or speed up as the manufacturing process varies, SRAM cells experience catastrophic faults whereby a given cell can no longer store a "1", i.e. it is somehow *stuck at* "0". The *probability* of such an event occurring, i.e. of a combination of parameters that results in an SRAM cell which is incapable of operating correctly, can be as ten parts per million. Since many SRAM arrays have far more than a million bits in them, *redundancy* is introduced to substitute good cells for bad ones, and thus to insure that the overall array continues to function as required.

Consider the SRAM circuit illustrated in figure 3. An SRAM cell is most prone to failure during read operation. A read operation in an SRAM typically involves precharging the bitlines (labeled BL and BLB in the figure) high, followed by enabling the two access (pass) transistors (labeled PL and PR in the figure) by taking the word line (labeled WL) high. In normal operation, since one of the two storage nodes (A and B) is zero, when the access transistors are turned on, one of the precharged bitlines will

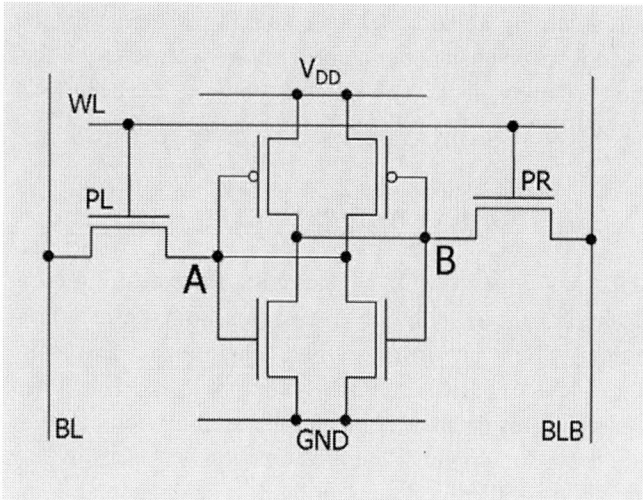

Figure 3: Circuit of an SRAM 6-transistor cell.

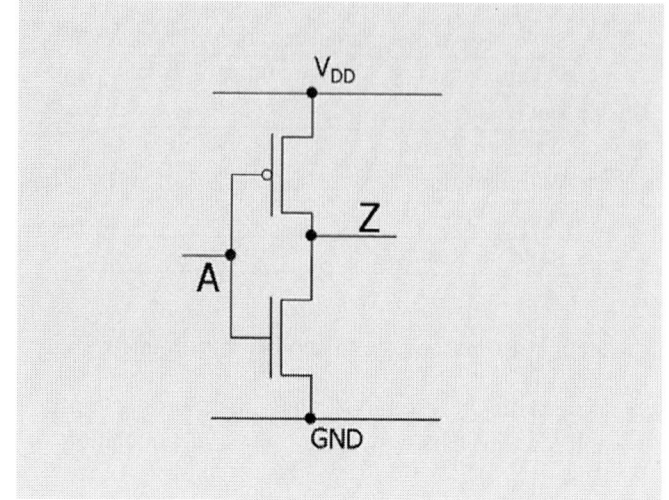

Figure 4: Circuit of a CMOS inverter.

be discharged through the access device and the inverter pull-down transistor. Note that this method of reading cell contents exposes the internal storage nodes A and B to a possible disturbance that would be caused by the resistive voltage division between the access and the pull down devices. Typically, this disturbance is minimized by making the pull-down device larger than the access transistor. However, random variations in the threshold voltages and therefore the strengths of various devices in an SRAM cell can cause the read operation to asymmetrically *flip* the contents of the cell, resulting in a cell that can be only read as a "1" or a "0". These failures are defined as *read stability failures*.

Since SRAM devices are generally very small, and since the threshold voltage variation caused by random dopant fluctuations is inversely proportional to device size, the *probability* of having a cell fail due to a read stability problem is -while small- not insignificant. Coupled with the fact that an array can easily have tens of millions of bits, these types of failures are clearly something the designer must be concerned about.

Consider however a simple CMOS inverter, illustrated in figure 4. When the input is high, the output is supposed to -obviously- be low. However, if the threshold voltage of the N-channel device is *very* large -making the device very weak, and the threshold voltage of the P-channel device is *very* small, making it very strong, then it is possible for the inverter to lost its ability to invert. This would happen if the leakage in the P-channel device is so high (because of the low threshold voltage) that it does not allow the N-channel device to pull the output node Z to ground. Such a situations is equivalent to an inverter *stuck at "1"*.

In any given technology, an inverter uses large devices than an SRAM, and therefore has smaller threshold variations, thus the probability that such a failure can occur is much smaller than that for an SRAM. Since we are dealing with such small values, it is more convenient to represent the probability of such rare occurrences using the distance from the mean, measured in standard deviations. So a distance of zero corresponds to a point at the mean, a distance of one corresponds to a point on the hypersphere of radius one standard deviation, and so on. Larger numbers correspond to smaller probabilities.

Using this distance metrics, table 4 illustrates the failure probability for three types of circuits: an SRAM, a Latch and an inverter. The failure probability is shown for three technologies, 90nm CMOS, 65nm CMOS and 45nm CMOS. We observe the following:

- The failure probabilities for *all* circuits are increasing.

- SRAM is clearly the most failure prone, inverters are the most robust, and latches are somewhere in between.

- At the current scaling trend, techniques applied to SRAM design in 90nm CMOS are likely to be required for latches at the 32nm or 22nm nodes, and for buffers one or more nodes beyond that.

This trend, which is also presented in graphical form in figure 5 implies that future technology nodes will bring a change in the *character* of the impact of variation on circuit performance. Catastrophic phenomena currently visible in SRAM are likely to become more common in other

978-1-4244-2018-6/08 $25.00 © 2008 IEEE

Table 4: Failure probability for various circuits in current and near-future technologies.

Technology	SRAM	LATCH	INVERTER
90nm SOI CMOS	7.37	13.5	22
65nm SOI CMOS	7.69	13	19
45nm SOI CMOS	5.92	10.7	16

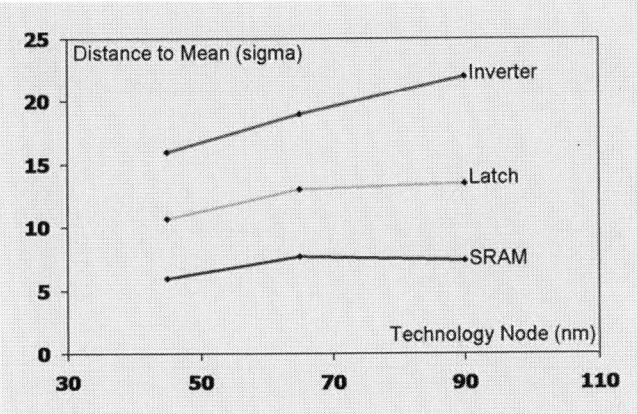

Figure 5: Failure probability trend vs. technology.

types of circuits, necessitating increased attention from designers are the circuit as well as the micro-architecture levels.

8   Conclusions

The impact of variability on circuit design is (a) increasing as we scale technology further, and (b) changing in character its impact shifts from parametric to catastrophic faults. This general increase will require increased attention to (a) accurate technology characterization, to (b) the generation of appropriate models for spatial, systematic and random variability, and to (c) the representation of these models in design flows and computer-aided design tools.

9   Acknowledgments

The author gratefully acknowledges contributions by Adhruva Acharyya, Kanak Agarwal, Jerry Hayes, Anne Gattiker, and Frank Liu.

References

[1] M. Orshansky, S. R. Nassif, and D. Boning. *Design for Manufacturability and Statistical Design: A Constructive Approach*. Springer, 2007.

[2] K. Agarwal and S. R. Nassif. Characterizing process variation in nanometer cmos. In *Proceedings of IEEE DAC*, 2007.

[3] M. Eisele, J. Berthold, D. Schmitt-Landseidel, and R. Mahnkopf. The impact of intra-die device parameter variations on path delays and on the design for yield of low voltage digital circuits. *IEEE Trans. VLSI*, Dec 1997.

[4] A. B. Agarwal and R. O. Tpoaloglu. Performance-aware cmp fill pattern optimization. In *Proceedings of Intl. VLSI/ULSI Multilevel Interconnection*, 2007.

[5] S. V. Kumar, C. H. Kim, and S. S. Sapatnekar. Impact of nbti on sram read stability and design for reliability. In *Proceedings of IEEE ISQED*, 2006.

[6] I. Ahsan et. al. Rta-driven intra-die variations in stage delay, and parametric sensitivities for 65nm technology. In *Proceedings of VLSI Technology Symposium*, 2006.

[7] *The International Technology Roadmap for Semiconductors*, 2007.

[8] K. Agarwal, F. Liu, C. McDowell, S. R. Nassif, K. Nowka, M. Palmer, D. Acharyya, and J. Plusquellic. A test structure for characterizing local device mismatches. In *Proceedings of IEEE VLSI*, 2006.

[9] *International Conference on Microelectronic Test Structures*, http://www.see.ed.ac.uk/ICMTS/.

[10] J. Hayes, K. Agarwal, and S. R. Nassif. Rapid characterization of parametric distributions using a multi-meter. In *Proceedings of IEEE ICMTS*, 2008.

[11] C. Cho, D. Kin, J. Kim, J. Plouchart, D. Lim, S. Cho, and R. Trzcinski. Decomposition and analysis of process variability using constrained principal component analysis. *IEEE Trans. Semiconductor Manufacturing*, Feb 2008.

[12] M. Bhushan, A. Gattiker, M. Ketchen, and K. Das. Ring oscillators for cmos process tuning and variability control. *IEEE Trans. Semiconductor Manufacturing*, July 2006.

[13] B. Hargreaves, H. Hult, and S. Reda. Within-die process variations: How accurately can they be statistically modeled? In *Proceedings of ASPDAC*, 2008.

[14] A. J. Strojwas, S. R. Nassif, and S. W. Director. A methodology for worst case design of integrated circuits. In *Proceedings of IEEE ICCAD*, 1983.

[15] K. Agarwal and S. R. Nassif. Statistical analysis of sram cell stability. In *Proceedings of IEEE DAC*, 2006.

[16] R. Kanj, R. Joshi, and S. R. Nassif. Mixture importance sampling and its application to the analysis of sram designs in the presence of rare failure events. In *Proceedings of IEEE DAC*, 2006.

978-1-4244-2018-6/08 $25.00 © 2008 IEEE

IEEE 2008 Custom Intergrated Circuits Conference (CICC)

Mismatch Analysis and Statistical Design at 65 nm and Below

Larry Pileggi, Gökçe Keskin, Xin Li, Ken Mai and Jon Proesel
Carnegie Mellon University
5000 Forbes Ave. Dept. of ECE, Pittsburgh, PA 15213 USA
{pileggi,gkeskin,xinli,kenmai,jproesel}@andrew.cmu.edu

Abstract- **Transistor sizing to control random mismatch is investigated. Input offset voltage of 65nm bulk CMOS SRAM sense amplifiers are measured to analyze NMOS and PMOS threshold voltage (Vtn, Vtp) variation effects and compare them with statistical models and Pelgrom model predictions. A linear statistical response surface model (RSM) relating input offset to Vtn and Vtp is shown to agree well with measured results. Designs optimized using the RSMs produce circuits with 25% lower input offset voltage spread at a cost of 10% more active device area. Statistical models for post-manufacturing configuration are postulated and shown for sub-65nm technologies.**

Keywords: Mismatch model, sense amplifier, input offset voltage

I. INTRODUCTION

Pelgrom's models [1] are universally applied for determining sufficient sizing of CMOS transistors to control mismatch due to random process variations. It is commonly applied to a wide range of circuit building blocks, such as differential amplifiers, current sources, and sense amplifiers. Importantly, since Pelgrom's analysis describes the mismatch between two elements, its application to circuits with more than one dominant source of mismatch is not straightforward [2] and can result in considerable overdesign.

To investigate Pelgrom's model for capturing mismatch for 65nm bulk CMOS, we consider its application for input offset voltage of the latch type sense amplifier in Fig. 1. This circuit was chosen because: a) the input offset is not easily described by an analytical formula; and b) the input offset is generally dominated by the random mismatch of the input NFETs. Both of which make this a classic circuit for application of the Pelgrom model.

Along with the Pelgrom model, a simple linear response surface model (RSM) for the offset is constructed based on Monte Carlo simulations. One version of the sense amp is designed using this RSM and following a statistical optimization methodology similar to that in [3]. The sense amplifiers are fabricated and tested to compare the statistical measurement results. While the required sizing is reasonable for handling mismatch for 65nm, we will show that below the 65nm node, post manufacturing configuration will eventually be the most effective way to manage the increase random variations that result in mismatch.

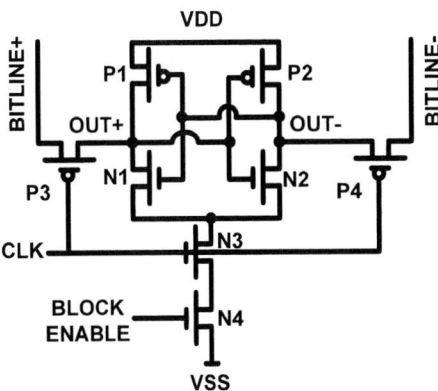

Fig. 1. Latch type sense amplifier (LTSA).

II. PELGROM'S MODEL AND THE LATCH TYPE SENSE AMPLIFIER

When used in SRAM circuits, latch type sense amplifiers (LTSAs) are enabled only after memory bitlines reach a differential voltage swing level that is detectable [4]. For a robust read operation, this swing voltage must be larger than the input offset voltage of the LTSA. Bitlines are discharged by an individual memory cell, therefore, input offset voltage of the LTSA is an important factor in determining the overall parametric yield of the memory at a chosen operating frequency.

LTSAs operate at high speed by using the positive feedback formed by the (N1, P1) and (N2, P2) inverter pairs [5]. Unfortunately, this positive feedback makes it difficult to derive analytical expressions for the offset voltage and yield estimates. Previous work for modeling the offset ranges from simple analyses [4] following Pelgrom's model [1] to complex equations based on imprecise square law models for transistors [2].

Pelgrom's model on the matching of a process parameter P (e.g., threshold voltage) between two circuit elements is given by the equation [1]:

$$\sigma^2(\Delta P) = \frac{A_P^2}{WL} + S_P^2 D_x^2 \qquad (1)$$

where $\sigma^2(\Delta P)$ is the variance of the difference of parameter P between the elements; A_P and S_P are the area and spacing proportionality constants for parameter P; W and L are the dimensions of each element; and D_x is the spacing between them. If the elements are laid out in close proximity with a layout style that eliminates most sources of systematic offset (e.g., common centroid), mismatch is mostly dominated by the $A_P^2/(WL)$ term in (1). Pelgrom's model is particular useful for initial design of circuits where a certain performance specification is dominated by a *single* mismatch source, P, such as the LTSA in Fig. 1. With proper selection of transistor lengths, and careful layout to avoid systematic variations, the input offset voltage is dominated by the threshold voltage mismatch of NMOS transistors N1 and N2 when the input voltages are close to V_{DD}.

In an SRAM, before the read operation, both bitlines are precharged to a voltage V_{pc}. During the first phase of the read operation; CLK is pulled low, one of the bitlines is discharged by the accessed memory cell, and the other is held at V_{pc}. The bitline discharge is relatively slow since SRAM cells are formed by small transistors to maximize memory density. Bitline voltages pass through P3 and P4, then are stored at the output nodes OUT+ and OUT−. Transistor N4 is included to disable a block of LTSAs when used in an SRAM design, hence it will be enabled for our analysis. When CLK is pulled high, the cross coupled inverters turn on and strong positive feedback pulls the output nodes to complementary logic levels quickly. It is desirable to perform CLK low to high transition as early as possible for reduced power consumption and increased memory speed. However, for a correct read operation, voltage differential at OUT+ and OUT− must be larger than the input offset voltage of the LTSA when CLK turns high.

If V_{pc} is kept near V_{DD}, after one of the bitlines is discharged and CLK goes high, (N1, N2) pair is on while (P1, P2) pair is off. A decision is well under way when one of the outputs reaches $V_{DD}-V_{tp}$ and one of the PMOS transistors turns on. Therefore, mismatch of (P1, P2) is less important and mismatch of (N1, N2) is the dominant factor on offset when V_{pc} is near V_{DD}. Such conditions meet requirements for application of Pelgrom's model.

Fig. 2 shows the scatter plot of the standard deviation of input offset voltage (σ_{Offset}) with respect to Diff(Vtn) and Diff(Vtp) from Monte Carlo simulations when $V_{pc}=V_{DD}=1.0V$. The axes are normalized to their respective standard deviations and:

$$Diff(Vtn) = VtN1{-}VtN2 \quad Diff(Vtp){=} VtP2{-}VtP1 \quad (2)$$

$$\sigma_{Diff(Vtn)}= \sigma_{VtN1} \sqrt{2} \qquad \sigma_{Diff(Vtn)}= \sigma_{VtP1} \sqrt{2} \quad (3)$$

For modeling purposes we assume that the input offset is dominated by random mismatch and that the (VtN1, VtN2) and (VtP1, VtP2) pairs are independent, identically distributed random variables representing the threshold voltages.

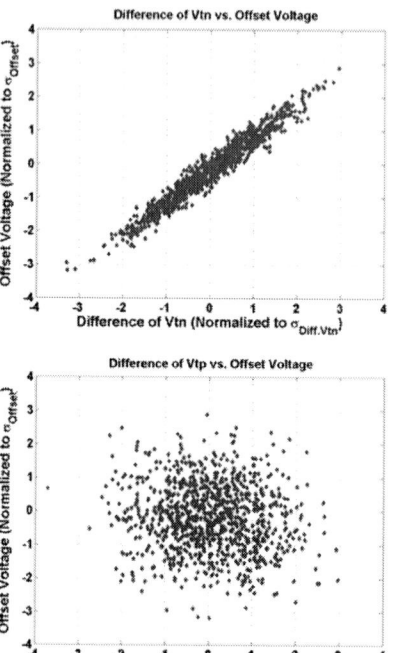

Fig. 2. Offset voltage vs. Diff(Vtn) and Diff(Vtp), $V_{pc}=V_{DD}=1.0V$.

As expected, Fig. 2 shows a linear relationship between offset and Diff(Vtn), while there is little or no correlation with Diff(Vtp). Therefore, we apply the following first order model for offset:

$$Offset = \big(a \times Diff\,(Vtn)\big) + \big(b \times Diff\,(Vtp)\big) + c \quad (4)$$

We characterize this model using data for four different precharge voltages (0.7, 0.8, 0.9 and 1.0V) while keeping V_{DD} constant at 1.0V. Compared to simulations, the modeling error of the linear approximation for each case is shown to be within 3%, thus confirming that input offset is indeed dominated by random threshold voltage mismatch in terms of the simulation models.

Fig. 3 shows the correlation coefficients (CC) of Diff(Vtn) and Diff(Vtp) with offset for different V_{pc}'s. As expected, as V_{pc} is lowered, P1 and P2 turn on sooner during the latching phase and their mismatch has an increasingly stronger impact on the offset. The correlation coefficient between offset and Diff(Vtp) increases correspondingly, and Pelgrom's model is no longer directly applicable when the second source of mismatch becomes significant. Measurement results in the next section will support this expected behavior.

Using the RSM model in (4), the transistors are sized to reduce the mismatch by 25% while minimizing the area and power impact. Simulation results for this statistically optimized design are also shown in Fig. 3. The optimization results in larger NFETs (N1, N2) but smaller PFETs (P1, P2), since the offset impact of the former are greater. We will compare this RSM model and Pelgrom's formula with measurement results.

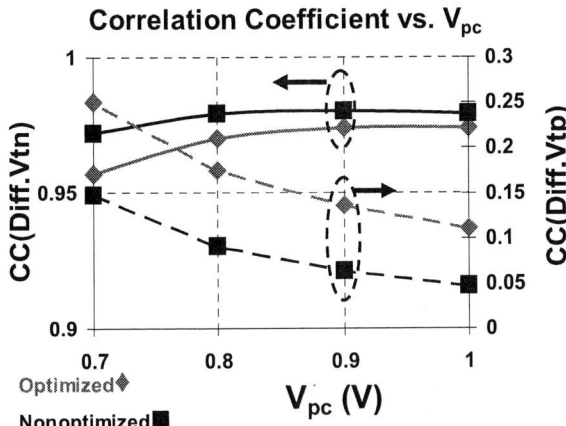

Fig. 3. Correlation Coefficient (CC) of Offset with Diff(Vtn), Diff(Vtp).

III. MEASUREMENT RESULTS

We designed and fabricated arrays of optimized and non-optimized LTSAs in a commercial 65nm bulk CMOS technology (Fig. 4). Each chip includes 2048 LTSAs, 50% based on the original design and 50% based on the statistically optimized design. Each differential output is connected to a D-flipflop, and the flops are connected as a scan chain. The chip is wire-bonded in a PGA package, and mounted on a printed circuit board (PCB) using a socket. Inputs (bitlines) are heavily decoupled with both on and off-chip capacitance, and the input differential voltages are externally applied. After inputs are set, LTSAs and flops are clocked a few times to clear any potential metastability. After clocking, flops are put to scan mode and digital outputs are pushed through the scan chain at a low frequency. The scan chain output is read by a logic analyzer probe on the PCB. An automated test setup sweeps the inputs in steps of a few mV over a wide input range, and the switching point of each LTSA is determined by post processing the outputs [6]. The tests are repeated at four different precharge voltages.

Histograms for the measured input offset distribution for the circuits are shown in Fig. 5. More than 12k samples for each design are collected from 12 different die. Bins are normalized to the standard deviation of the offset voltage of the non-optimized circuit ($\sigma_{Offset,NO}$), and results are shown for $V_{pc}=1.0V$. The improved offset spread for the statistically optimized circuit is apparent in the measurement results. Simulated and measured values of $\sigma_{Offset,NO}$ are within 5%, and this difference is even less for the optimized circuit ($\sigma_{Offset,O}$).

Fig. 6 shows a comparison of the $\sigma_{Offset,NO}/\sigma_{Offset,O}$ ratio for a simple area ratio based on Pelgrom model:

$$SimpleAreaRatio = \sqrt{\frac{Area(N1+N2), OptimizedCircuit}{Area(N1+N2), NonOptimizedCircuit}} \quad (5)$$

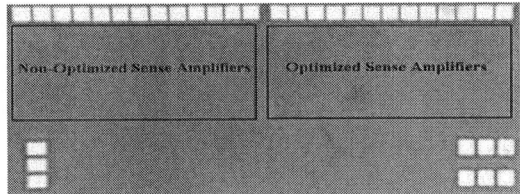

Figure 4. Die photo of the manufactured chip.

The efficacy of Pelgrom's model is apparent in Fig.6; at the region where σ_{Offset} is dominated by the mismatch of N1 and N2 (i.e., $V_{pc}=1.0V$), the Pelgrom model predicts the performance of the circuit very well. As V_{pc} is decreased and mismatch of P1 and P2 further impacts σ_{Offset}, Pelgrom's ratio is no longer applicable. The RSM model in [3] is nearly overlapping with the simulation curve in Figure 6. Such an RSM model can be an effective model of performance (σ_{Offset}) during initial design in general, and can further suggest design trade-offs in the circuit and for statistical optimization.

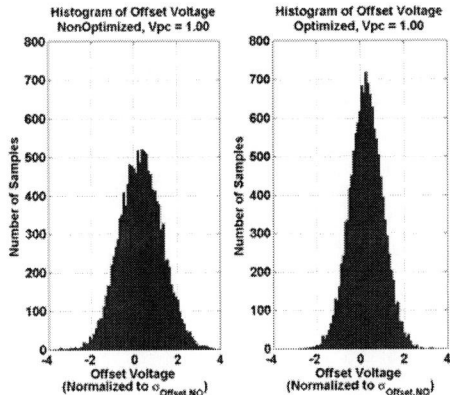

Figure 5. Histogram of input offset voltage (measured).

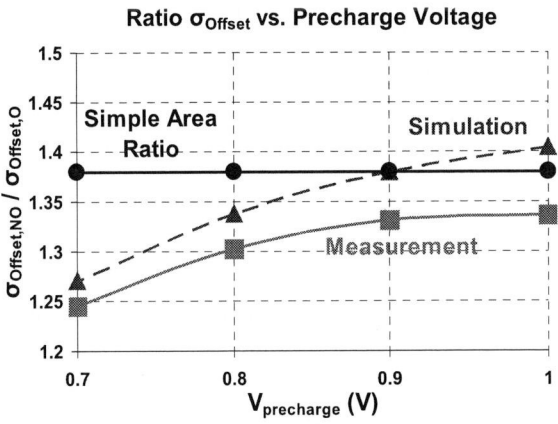

Fig. 6. Comparison of σ_{Offset} ratio with simple area ratio based on Pelgrom model.

IV. STATISTICAL ELEMENT SELECTION

Even with more accurate RSM models, the scaling of analog designs will be limited due to the large device sizes required to accommodate mismatch specifications. For 45nm and below, the random variations will become more dominant for general bulk CMOS. Instead of oversizing devices to *average out* random mismatch, the randomness can be used to improve the ability to provide post-manufacturing configuration to match devices and sub-components with *statistical element selection (SES)*.

Consider the LTSA in Fig. 1 that is designed with multiple sub-components connected in parallel. The number of sub-components would be determined by Pelgrom's model or the RSM in (4). Next consider a post-manufacturing capability of enabling only a subset of the sub-components for matching, but being able to scan through the subset combinations to pick a good, or even the best sets. The random variations help to create a very large population of choices for matching. It can be shown that there is an *exponential* increase in the number of subsets with a linear increase in the number of elements in a set. Correspondingly, there is an exponential decrease in offset voltage spread among a set of selectable elements with a linear increase in area (Fig. 7), as compared to a 1/sqrt(area) relationship following the Pelgrom model [7].

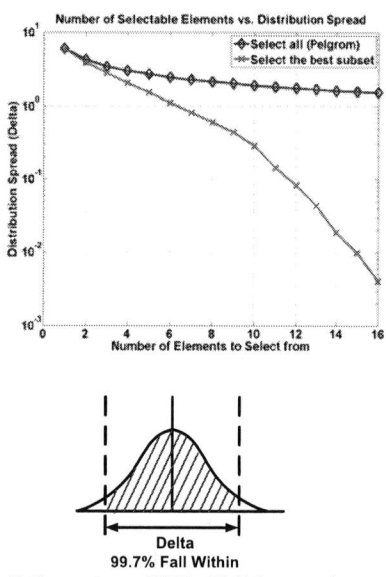

Fig. 7. Comparison of SES with Pelgrom style matching.

At extreme scaling of CMOS, a wide range of circuits can potentially benefit from an SES approach. Sense amplifiers can be made reconfigurable to reduce input offset voltage, resulting in a higher parametric yield for the memory. The selection can be performed with digital logic and stored in the memory. In general, the exponential scaling with area and relatively low analog complexity makes SES extremely competitive with other potential post-manufacturing calibration techniques.

CONCLUSIONS

A linear response surface model for relating input offset voltage of latch type sense amplifiers to threshold voltages variations has been described. The model is compared the design choices specified by the Pelgrom model. The response surface model was used to demonstrate the efficacy of statistical sizing for the LTSA. Measurements from a 65nm bulk CMOS testchip were used to compare the original and optimized designs and their corresponding models. The statistically optimized design resulted in a 25% decrease in the standard deviation of the input offset voltage at a cost of 10% increase in active area for the LTSA. A statistical element selection (SES) algorithm was proposed as the next step to control random variations in scaled CMOS processes.

ACKNOWLEDGEMENTS

The authors would like to thank Umut Arslan and Mark McCartney of Carnegie Mellon for the helpful discussions during the course of this work. We are very grateful for the access to 65nm fabrication from IBM, and the implementation support from John Cohn, Jack Pekarik, Randy Wolf and Ida Pucino. The authors also acknowledge the support of the Focus Center for Circuit & System Solutions (C2S2), one of five research centers funded under the Focus Center Research Program, a Semiconductor Research Corporation Program, and the National Science Foundation under contract CCF-0702278.

REFERENCES

[1] M.J.M. Pelgrom, A.C.J. Duinmaijer, and A.P.G. Welbers, "Matching properties of MOS transistors," *IEEE J. Solid-State Circuits*, vol. 24, pp. 1433-1439, October 1989.

[2] R. Singh and N. Bhat, "An offset compensation technique for latch type sense amplifiers in high-speed low-power SRAMs," *VLSI Systems, IEEE Transactions on*, vol. 12, pp. 652-657, June 2004.

[3] X. Li, P. Gopalakrishnan, Y. Xu, and L.T. Pileggi, "Robust analog/RF circuit design with projection-based performance modeling," *Computer-Aided Design of Integrated Circuits and Systems, IEEE Transactions on*, vol. 26, pp. 2-15, January 2007.

[4] S.J. Lovett, G.A. Gibbs, and A. Pancholy, "Yield and matching implications for static RAM memory array sense-amplifier design," *IEEE J. of Solid-State Circuits*, vol. 35, pp. 1200-1204, August 2000.

[5] B. Wicht, T. Nirschl, and D. Schmitt-Lansiedel, "Yield and optimization of a latch type voltage sense amplifier," *IEEE J. of Solid-State Circuits*, vol. 39, pp. 1148-1158, July 2004.

[6] R. Ho, "On-chip wires: Scaling and efficiency," *Ph.D. Thesis*, Stanford University, 2003.

[7] X. Li, B. Taylor, Y-T. Chien, L. Pileggi, "Adaptive post-silicon tuning for analog circuits: concept, analysis and optimization," *ICCAD*, 2007.

IEEE 2008 Custom Intergrated Circuits Conference (CICC)

Statistical Prediction of Circuit Aging under Process Variations

Wenping Wang[†], Vijay Reddy[‡], Bo Yang[†], Varsha Balakrishnan[†], Srikanth Krishnan[‡], Yu Cao[†]

[†]Department of Electrical Engineering, Arizona State University, Tempe, AZ 85287, USA

[‡]External Development and Manufacturing, Texas Instruments, PO Box 650311, MS 3740, Dallas TX 75243, USA

Abstract—**Accurate prediction of circuit aging and its variability is essential to reliable design and analysis. Such a capability further helps reduce the load in statistical reliability test. Based on compact models of transistor degradation and circuit performance, we develop analytical solutions that efficiently predict the statistics of both circuit timing and the leakage under temporal stress and process variations. These solutions prove that circuit aging and its variance can be fully predicted from the characteristics of transistor degradation and circuit performance sensitivity to aged parameters, independent on the type and the amount of process variations. Specific results include: (1) under variations, the standard deviation of circuit speed declines with the stress time, following a power law of $1/6$; and (2) the logarithmic mean and the standard deviation of leakage current decrease with the stress time as $t^{1/6}$. The results are systematically validated by simulation and measurement data from an industrial 65nm technology, enhancing the predictability and efficiency of statistical reliability analysis.**

I. INTRODUCTION

With the continuous scaling of CMOS technology, device variations and reliability degradation are emerging as fundamental challenges to robust integrated circuit design at 65nm node and below [1]–[3]. As a result, circuit performance metrics, such as the speed, power, and the leakage, exhibit an excessive amount of variability post the fabrication process. Figure 1 illustrates an example that shows the statistical measurement of switching frequency and the leakage ($IDDQ$) of ring oscillators (ROs) before the reliability test. In this 65nm technology, more than 3X and   variability are observed in circuit leakage and the speed, respectively.

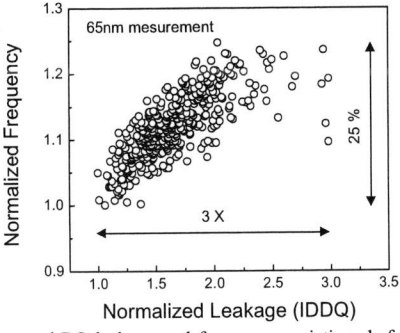

Fig. 1.   Measured RO leakage and frequency variations before the stress.

Circuit performance and its variability not only depend on static process variations, but also change over the period of dynamic operation because of the effect of circuit aging [4], [5]. Figure 2 shows the measured speed degradation from the same set of ROs as those in Figure 1. Since the degradation rate is highly sensitive to process parameters and operation conditions, such as threshold voltage (V_{th}), temperature, and switching activity [4], [5], circuit aging strongly interacts with process variations, dynamically shifting both the mean and the variance of circuit performance. Therefore, accurate prediction of circuit performance distribution during its life time should

consider the impact of static variations, primary reliability mechanisms, and more importantly, their interactions. This prediction is essential for designers to safely guardband the circuit for a sufficient life time. Otherwise, we have to either use an overly pessimistic bound, or resort to expensive stress tests in order to collect enough statistical information.

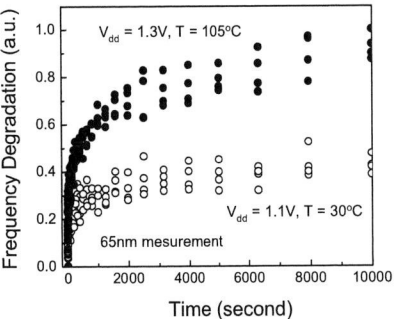

Fig. 2.   Measured frequency degradation of a 11-stage ring oscillator under different stress conditions.

In this work, we leverage compact models of transistor degradation and circuit performance to achieve accurate and efficient reliability prediction. Negative-Bias-Temperature-Instability (NBTI), which is the dominant reliability mechanism in scaled CMOS technology, is incorporated into the analytical framework to account for the aging of circuit speed and the leakage [4], [5]. The specific contribution and conclusions of this work include:

- A statistical predictive methodology of circuit aging is proposed. In these models, only five parameters need to be extracted from fresh data (i.e., before the stress). With the initial information of the transistor and circuit topology, these models provide accurate prediction of circuit performance degradation and the variability.
- The degradation rate of circuit speed and its standard deviation follows a power law of / , and they are independent on the amount and the type of variations in the circuit.
- The mean and the standard deviation of logarithmic $IDDQ$ reduce with the stress time as $t^{1/6}$, with the variance more sensitive to global variations.

The outline of the paper is as follows: The statistical modeling for both transistor and circuit performance degradation is described in Section II. The proposed models are comprehensively verified with silicon data from industrial 65nm technology and SPICE simulation results, as shown in Section II and Section III. Finally Section IV concludes this work.

II. STATISTICAL MODELING OF CIRCUIT AGING

NBTI is the dominant effect of circuit aging in advanced CMOS technology [4]. Based on the reaction-diffusion mechanism, we developed the model of V_{th} shift under NBTI effect. In the presence of process variations, this model is further expanded as a linear function of static V_{th} variation to efficiently predict the performance degradation.

To characterize circuit performance change under the stress, the Alpha-power law based delay model and the leakage model are calibrated for a given gate. Under the condition that the amount of NBTI-induced V_{th} shift is much smaller than the nominal value of V_{th}, both models are simplified to extract the dependence of circuit performance to V_{th} change.

Finally, the gate-level models are integrated into various circuit paths to analytically predict the aging of path timing and the leakage. Both the mean value and the variance of these important metrics are derived as a function of static performance variability, the nominal sensitivity of circuit performance, and other operation conditions, such as voltage and the temperature.

A. Transistor Degradation Model

NBTI occurs when a negative gate bias is applied to PMOS and it has two phases: stress and recovery (Figure 3) [3]. In the stress phase, the holes in the channel weaken the Si-H bonds, which results in the generation of the positive interface charges and hydrogen species. During the recovery, the interface traps can be annealed by the hydrogen species and thus, V_{th} degradation ($\Delta V_{th-nbti}$) is partially recovered.

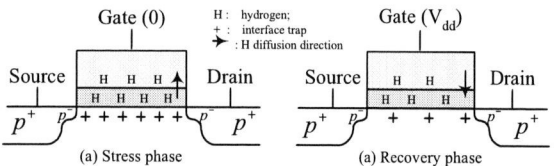

(a) Stress phase   (a) Recovery phase

Fig. 3. The reaction diffusion mechanism of NBTI.

Based on the reaction-diffusion mechanism, we developed compact models to predict long-term $\Delta V_{th-nbti}$ [3],

$$\Delta V_{th-nbti}=\left(\sqrt{K_v^2\cdot T_{clk}\cdot\alpha}\Big/\left(1-\beta_t^{1/2n}\right)\right)^{2n} \quad (1)$$

where

$$\beta_t=1-\frac{2\xi_1\cdot t_e+\sqrt{\xi_2\cdot C\cdot(1-\alpha)\cdot T_{clk}}}{2t_{ox}+\sqrt{C\cdot t}} \quad (2)$$

$$K_v=\left(\frac{qt_{ox}}{\epsilon_{ox}}\right)^3 K_1{}^2 C_{ox}(V_{gs}-V_{th})\sqrt{C}\exp\left(\frac{2E_{ox}}{E_{o1}}\right) \quad (3)$$

Fig. 4. Threshold voltage degradation under different stress conditions.

Figure 4 verifies this model with experimental data under different stress conditions. This model assumes nominal degradation without considering the statistical process variations. If there are global and local process variations, V_{th} in Equation (3) should be expressed as:

$$V_{th}=V_{th0}+\Delta V_{th-g}+\Delta V_{th-l} \quad (4)$$

where V_{th0} is the nominal threshold voltage, ΔV_{th-g} and ΔV_{th-l} represent the change of threshold voltage due to

global and local variations, respectively. This model is further simplified as a variation dependent model, i.e.,

$$\Delta V_{th-nbti}=A\left(1-S_v(\Delta V_{th-g}+\Delta V_{th-l})\right)t^n \quad (5)$$

where the value of A depends on both technology parameters and operating conditions; S_v is the nominal sensitivity of NBTI degradation to V_{th} shift. Figure 5 validates the simplified model with the long term predictive model under different process variations. It provides accurate prediction of threshold voltage degradation.

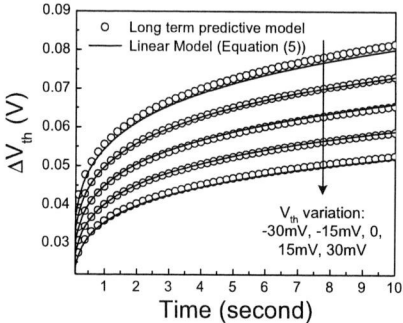

Fig. 5. Verification of process variation dependent NBTI model.

B. Gate Delay Degradation Model

A widely used gate delay (T_{di}) model is based on the Alpha-power law that was proposed in [6],

$$T_{di}=(C_{li}V_{dd})/(\beta_i(V_{dd}-V_{th_i})^\alpha) \quad (6)$$

where C_{li} is the effective load capacitance of the gate; β_i is a parameter depending on gate size. Under both process variations and NBTI effect, V_{thi} of PMOS is given by

$$V_{thi}=V_{th0}+\Delta V_{thi} \quad (7)$$

$$\Delta V_{thi}=\Delta V_{thi-g}+\Delta V_{thi-l}+\Delta V_{thi-nbti} \quad (8)$$

Substituting V_{thi} into Equation (6) and using the approximation of $(1-x)^\alpha\approx(1-\alpha\cdot x)$ (for $x\ll1$), we obtain:

$$T_{di}=\frac{C_{li}V_{dd}}{\beta_i(V_{dd}-V_{th0}-\Delta V_{thi})^\alpha}\approx\frac{C_{li}V_{dd}}{\beta_i(V_{dd}-V_{th0})^\alpha}\left(1+\frac{\alpha\Delta V_{thi}}{(V_{dd}-V_{th0})}\right) \quad (9)$$

We define $T_{d0i}\triangleq C_{li}V_{dd}/(\beta_i(V_{dd}-V_{th0})^\alpha)$, which is the gate delay without process variations and NBTI degradation ($\Delta V_{thi}=0$), and $S_{ti}\triangleq\alpha/(V_{dd}-V_{th0})$, which is the nominal sensitivity of gate delay to PMOS V_{th} shift. These two parameters rely on the process technology and the circuit structure. They can be conveniently extracted from SPICE simulation at the nominal condition. Thus, Equation (9) becomes:

$$T_{di}=T_{d0i}\left(1+S_{ti}\Delta V_{thi}\right) \quad (10)$$

Substitute Equations (5) and (8) into (10),

$$T_{di}=T_{d0}\left(1+S_{ti}\left(At^n+(1-AS_vt^n)\Delta V_{thi-g}+(1-AS_vt^n)\Delta V_{thi-l}\right)\right) \quad (11)$$

We assume ΔV_{thi-g} and ΔV_{thi-l} are modeled as Gaussian random variables. Their mean (μ_g and μ_l) are 0 and their standard deviations (σ_g and σ_l) depends on the manufacturing process [1]. Since gate delay is linearly proportional to the threshold voltage change, the probability distribution function (PDF) of gate delay also follows the normal distribution $N\sim(\mu_{T_{di}},\sigma_{T_{di}}^2)$.

At $t=0$, $\Delta V_{th-nbti}=0$. Assuming global and local variations are uncorrelated random variables, $\mu_{T_{di}}$ and $\sigma_{T_{di}}^2$ are given by:

$$\mu_{T_{di}}(0)=T_{d0i}, \quad \sigma_{T_{di}}(0)=T_{d0i}S_{ti}\sqrt{\sigma_g^2+\sigma_l^2} \quad (12)$$

At $t>0$, from Equation (11), we get

978-1-4244-2018-6/08 $25.00 © 2008 IEEE

$$\mu_{T_{di}}(t)=\mu_{T_{di}}(0)(1+AS_{ti}t^n), \quad \sigma_{T_{di}}(t)=\sigma_{T_{di}}(0)(1-AS_vt^n) \quad (13)$$

Given the initial condition of the process and timing information for the transistor and the gate, Equation (13) predicts the mean and standard deviation with increasing time. From these equations, we have four observations:

(1) The mean of gate delay increases with the stress time, while the variance decreases. Since a lower-V_{th} transistor has a faster degradation rate and thus, larger V_{th} increase, this phenomenon compensates static process variations and reduces the variance during the stress period.

(2) As the stress time increases, the aging of both mean and standard deviation follow the same power law of $t^{1/6}$.

(3) The degradation rate of gate delay and its variance are independent on the amount and the type of variations.

(4) The degradation rate is determined by the sensitivities to V_{th} shift. Process variations only affect the fresh variability, but not the degradation rate.

C. Circuit Performance Degradation

1) Path timing: The PDF of a path comprising n gates corresponds to the linear combination of the n PDFs of gate delays. The mean and the variance of the path delay (T_d) are given by

$$\mu_{T_d}=\sum_i^n \mu_{T_{di}}, \quad \sigma_{T_d}^2=\sum_i^n\sum_j^n \sigma_{T_{di}}\cdot\rho_{ij}\cdot\sigma_{T_{dj}} \quad (14)$$

where ρ_{ij} is the correlation coefficient between two gates. For the simplicity of the demonstration, we assume the inter-gate correlation is the same for all the gates, i.e., ρ_{ij} ρ, while this methodology is general enough for all statistical conditions. Thus, the variance of path delay is derived as

$$\sigma_{T_d}^2=\sum_i^n\sum_j^n \sigma_{T_{di}}\cdot\rho\cdot\sigma_{T_{dj}}+\sum_i^n(1-\rho)\cdot\sigma_{T_{di}}^2 \quad (15)$$

In the case of local variations, ρ , i.e, the variations between two gates are uncorrelated. The case of ρ describes global variations, i.e., the variations between two gates are correlated. With both local and global variations, $\sigma_{T_d}^2$ is given by the linear combination of the variance of the local and global variations. Figure 6 shows the delay distribution of ROs due to circuit aging. The distribution of gate delay becomes increasing narrower under the stress as indicated by Equation (13).

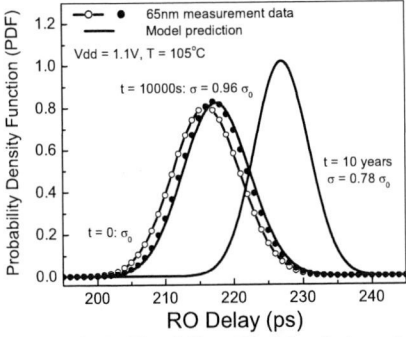

Fig. 6. The PDF's of 65nm RO delay during aging.

The proposed predictive methodology is generated for a path consisting of various types of gates. Figure 7 shows such a circuit example. By stressing the path for different years, Figure 8 compares the model prediction with SPICE simulation results of gate delay. Under different types and amount of variations, the model provides accurate prediction of the mean and standard deviation.

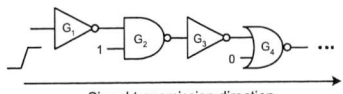

Fig. 7. Circuit example for path timing analysis.

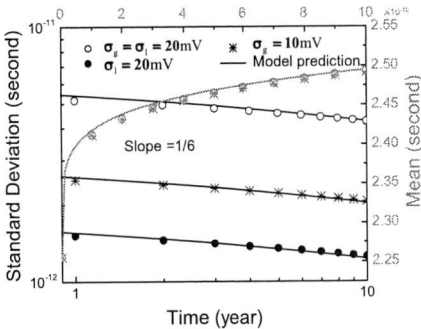

Fig. 8. The temporal increase of the mean and standard deviation of circuit speed (path is shown in Fig. 7).

2) Leakage: $IDDQ$ of a circuit is defined as the total amount of leakage current at the standby. It has an exponential dependence on V_{th}:

$$IDDQ=\sum_i^n I_{0i}\cdot e^{-\frac{V_{thi}}{mv_T}}=\sum_i^n I_{0i}\cdot e^{-\frac{(V_{th0i}+\Delta V_{thi})}{mv_T}} \quad (16)$$

where I_{0i} $\beta_i m -$ $-e^{-V_{ds}/v_T}$, m is the body effect coefficient and v_T is the thermal voltage (kT/q). Substitute Equation (8) into (16), we get

$$IDDQ=\sum_i^n I_{0i}e^{-\frac{V_{th0i}}{mv_T}}e^{-\frac{\Delta V_{thi-g}+\Delta V_{thi-l}+\Delta V_{thi-nbti}}{mv_T}}$$
$$=\sum_i^n IDDQi(0)e^{-\frac{\Delta V_{thi-g}+\Delta V_{thi-l}+\Delta V_{thi-nbti}}{mv_T}} \quad (17)$$

$IDDQi$ is the gate leakage at t , i.e., V_{thi-g} V_{thi-l} $V_{thi-nbti}$. Taking the natural logarithms on both sides of Equation (17), we have

$$Ln(IDDQ)=Ln\left(\sum_i^n IDDQi(0)e^{-\frac{\Delta V_{thi-g}+\Delta V_{thi-l}+\Delta V_{thi-nbti}}{mv_T}}\right) \quad (18)$$

Under global variations, using Equation (5), Equation (18) becomes the following

$$Ln(IDDQ)=-\frac{At^n+(1-AS_vt^n)\Delta V_{thi-g}}{mv_T}+Ln\left(\sum_i^n IDDQi(0)\right) \quad (19)$$

The mean and standard deviation of circuit leakage are

$$\mu_{Ln(IDDQ)}(t)=\mu_{Ln(IDDQ)}(0)-A/(mv_T)t^n \quad (20)$$

$$\sigma_{Ln(IDDQ)}(t)=(1-AS_vt^n)/(mv_T)\cdot\sigma_g \quad (21)$$

where $\mu_{Ln(IDDQ)}$ $Ln\left(\sum_i^n IDDQi \right)$

Under local variations, Equation (18) is approximated as

$$Ln(IDDQ)\approx Ln\left(e^{-\frac{At^n}{mv_T}}e^{-\frac{(1-AS_vt^n)\Delta V_{thi-l}}{(mv_T)\cdot\eta}}\sum_i^n IDDQi(0)\right)$$
$$=-\frac{At^n}{mv_T}+\frac{(1-AS_vt^n)\Delta V_{thi-l}}{(mv_T)\cdot\eta}+Ln\left(\sum_i^n IDDQi(0)\right) \quad (22)$$

where η has the value between 0 and 1, depending on the circuit structure. The mean and standard deviation of circuit leakage are

$$\mu_{Ln(IDDQ)}(t)=\mu_{Ln(IDDQ)}(0)-A/(mv_T)t^n \quad (23)$$

$$\sigma_{Ln(IDDQ)}(t)=(1-AS_vt^n)/(mv_T)\cdot(\sigma_l/\eta) \quad (24)$$

Akin to path timing, logarithmic $IDDQ$ has the same time dependence under either global or local variation. Their impact is only different by a factor of η, which is derived from the circuit structure. Figure 9 shows that the logarithmic mean and

978-1-4244-2018-6/08 $25.00 © 2008 IEEE

the standard deviation degradation of leakage current follow the power law of $t^{1/6}$. The mean is relatively independent of the type of variations, while standard deviation is more sensitive to the global variation.

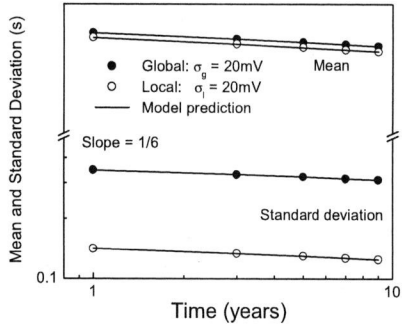

Fig. 9. The mean and standard deviation degradation of leakage.

III. MODEL PREDICTION AND VALIDATION

A. Model Parameter Extraction

In order to accurately predict the circuit performance degradation, there are five parameters need to be characterized at t , including: A, S_v, S_t, T_{d0i} and $IDDQi$.

A: Parameter of long term V_{th} degradation under nominal conditions. Its value is extracted from Equation (1) - (3), with the dependence of temperatures, V_{dd}, and switching activity.

S_v: the sensitivity of NBTI-induced transistor degradation to the nominal value of V_{th}.

S_t: the sensitivity of gate delay to PMOS V_{th} shift.

T_{d0i}: nominal gate delay.

$IDDQi$: nominal gate leakage.

Note these parameters are all extracted from the nominal condition, but not affected by process variations. Variations only change the distribution of T_d and $IDDQ$, which can be obtained from the statistics at t (i.e., before the stress). During the stress, the interaction between variability and reliability follows the prediction of our new models. These statistical reliability models improve the predictability in the design stage, avoiding expensive reliability test at the circuit level.

B. Model Validation with Silicon Data

The proposed statistical model is verified by 65nm technology silicon data under different stress conditions. We use a 11-stage RO as the representative test circuit. Figure 10 shows the delay degradation of ROs. The dots present the mean changes; the error bars are scaled delay distribution of the sample circuits; and the lines are the model predictions. While the mean value increases as $t^{1/6}$, as a signature of the dominance of NBTI effect in circuit aging, the distribution declines with stress time. Figure 11 evaluates the change of both the active current ($IDDA$), which is proportional to the switching frequency, and the leakage current ($IDDQ$). Our predictive models only require the sensitivities of transistor and circuit aging, as well as the statistics before the stress. Then the degradation of circuit performance is fully predicted toward the end of the life time.

IV. CONCLUSIONS

In this work, a statistical methodology is developed to predict circuit performance degradation under both NBTI effect and process variations. These analytical solutions reveal that the degradation rate and its standard deviation are independent on the type and the amount of process variations. Only the characteristics of transistor degradation and performance sensitivity to aged parameters are necessary to predict the mean and the bound of circuit aging. The aging of circuit speed and $IDDQ$ shows a power law dependence on the stress time, as an evidence of the dominance of NBTI effect. The proposed model is implemented into SPICE simulation and comprehensively verified by SPICE simulation results and 65nm silicon data.

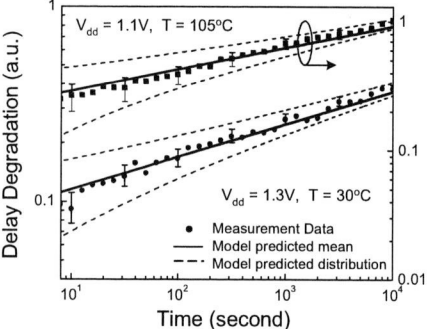

Fig. 10. Delay degradation of ROs under various stress conditions.

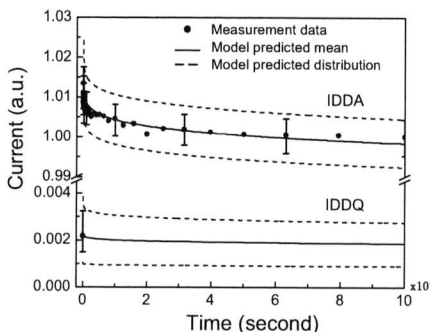

Fig. 11. The prediction of IDDA and $IDDQ$ degradation.

V. ACKNOWLEDGEMENTS

This project is supported by SRC, Task 1609. It is also partially supported by Gigascale System Research Center (GSRC), one of five research centers funded under the Focus Center Research Program, a Semiconductor Research Corporation Program.

REFERENCES

[1] S. Borkar and etc. Parameter variations and impact on circuits and microarchitecture. *DAC*, pages 338–342, 2003.

[2] S. R. Nassif. Modeling and analysis of manufacturing variations. *CICC*, pages 223–228, 2001.

[3] W. Wang and etc. Compact modeling and simulation of circuit reliability for 65nm cmos technology. *IEEE Transactions on Device and Materials Reliability*, 7(4):509–517, 2007.

[4] D. K. Schroder and etc. Negative bias temperature instability: Road to cross in deep submicron silicon semiconductor manufacturing. *Journal of Applied Physics*, 94(1):1–18, Jul. 2003.

[5] W. Wang and etc. The impact of nbti on the performance of combinational and sequential circuits. *DAC*, pages 364–369, Jun. 2007.

[6] T. Sakurai and etc. Alpha-power law mosfet model and its application to cmos logics. *JSSC*, 25(2):584–594, 1990.

Power Management

Session 3

Chair: Gordon Lee, Qualcomm
Co-Chair: Makoto Takamiya, University of Tokyo

With consumer demand for more product sophistication and processing power in smaller form factors, effective power management strategies are becoming increasingly important in mobile and portable devices. Switching regulators are an integral component in most advanced battery powered systems and hence are well represented in this session. The first three papers investigate interesting directions for step down (buck) DC-DC converters relating to contemporary challenges of integration and efficiency, spectral behavior, and fast transient response. In addition to efficient power generation, bias techniques are also an important topic in the area of system power optimization; a paper is included which considers process compensation as a potential path to power reduction. Finally, because much of the mobile / portable electronics industry is powered by Li-Ion batteries, a paper is included which describes a scheme for charging time reduction through series resistance compensation.

The Power Management session opens with a paper describing a step down DC-DC converter which utilizes a bond wire spiral inductor and MOS / MIM capacitors for a fully integrated 0.18µm CMOS implementation that does not require external components. A 65% peak power conversion efficiency is achieved at a conversion ratio of 0.5, significant in that it is the first fully integrated buck switcher known to the authors with a demonstrated efficiency advantage over linear regulators.

The second paper of the session presents a step down DC-DC converter with a DPLL frequency locking technique for high frequency hysteretic control. The programmable DPLL can be locked to a reference clock to keep the switching frequency away from power supply resonance bands. The switching frequency of this 0.13µm CMOS design can be locked in the range of 90MHz to 240MHz, and is constant over a wide output voltage range. A 40ns load response to a 120mA load step is reported.

The third paper presents another step down DC-DC converter innovation with a adaptive switching frequency control technique to achieve fast transient response time with a current-mode topology. During a load transient period, the switching frequency is dynamically adjusted and a two-step positive feedback mechanism for bandwidth control is utilized to ensure stable operation and fast transient response. This 0.18µm CMOS design has a recovery time of less than 10µs for a 100mA to 500mA load step, and undershoot / overshoot below 50mV.

The fourth paper of the session presents a scheme for compensation of slow process corner material using a forward body-biasing scheme extendable above 0.6V with assured latch up free operation. The bias scheme does not need voltages extending beyond the normal supply rail. A 20% reduction in worst case delay on a 65nm CMOS media processor is reported. The speed improvement is noted to enable a 100mV supply reduction for lower overall power consumption. Leakage is constrained to stay below the worst case fast process corner, and latch up is avoided by a bias scheme that breaks the potential positive feedback loop that could trigger a parasitic thyristor.

The Power Management session closes out with a paper presenting a resistance compensation technique proposed to speed the charging time for Li-Ion batteries by compensating for battery pack series resistance. The 0.35µm CMOS design generates a digital code to represent the estimated internal battery resistance for charger compensation. The charger system can then use the information to avoid premature switching from constant current to constant voltage charging. Allowing the fast charge period to be extended before transitioning to constant voltage charging with the ensuring current taper off reduces the overall charging time.

978-1-4244-2018-6/08 $25.00 © 2008 IEEE

Notes

A Fully-Integrated 0.18 μm CMOS DC-DC Step-Down Converter, Using a Bondwire Spiral Inductor

Mike Wens and Michiel Steyaert
ESAT-MICAS, K.U. Leuven
Kasteelpark Arenberg 10
B-3001 Leuven, Belgium
Email: mike.wens@esat.kuleuven.be

Abstract—A fully-integrated DC-DC step-down converter in a 0.18 μm 1.8 V **CMOS technology is realized, with a bondwire spiral inductor and integrated MOS and MIM capacitors. The converter is designed to generate an output voltage of 1.8 V out of a 3.6 V power supply. No external components are required. The maximum power conversion efficiency is 65 %, for a voltage conversion ratio of 0.5. This is a 23 % improvement compared to a linear regulator. Higher values are likely to be reached using a converter with external components, which is not the case here. Stacked transistors are used to cope with the high voltage. A novel control system based on a constant on/off-time keeps the output voltage constant at load conditions from 0 mW to 300 mW. The converter operates in discontinuous, asynchronous switching mode. The switching frequency ranges from 20 Hz to 140 MHz.**

I. Introduction

As technology scales down due to Moore's law, it becomes more cost-efficient to integrate all the building blocks of a system on a single die. These SoC's require multiple supply voltages. For instance the digital signal processing part typically needs a lower supply voltage than the power amplifier, which needs to be able to deliver sufficient output power into the communication medium. Therefore a dedicated power management system is required to accommodate each building block of the SoC with a proper supply voltage and power [1]. Moreover the increasing miniaturization leads to portable, wireless and battery-operated applications. The battery voltage however is usually too high for most parts of the circuitry in deep-submicron technologies. It therefore needs to be decreased. Using a linear regulator for this task will lead to excessive power loss. A switching step-down regulator can minimize this power loss, extending the battery lifetime.

In order to achieve a fully-integrated SoC the voltage converter(s) should be integrated on the same die, leading to distributed power management on-chip. Previous publications report fully-integrated step-down converters [2] [3], only requiring some commonly used technology options to integrate the passives. Other publications require extra (expensive) processing steps such as thick film and stacked-chip's [4] [5]. Clearly the most stringent problem on fully-integrated DC-

DC converters is the integration of the passives. The poor Q-factors of integrated capacitors and inductors make it difficult to achieve power conversion efficiencies which are higher than those achievable with linear voltage regulators. This is the reason why DC-DC converters using external components typically achieve higher efficiencies [6].

Equation 1 gives the power conversion efficiency η_{lin} of a linear voltage regulator, where P_{in} is the input power, P_{out} is the output power, I_{in} is the input current, U_{in} is the input voltage, U_{out} is the output voltage and U_{drop} is the voltage-drop over the linear regulator. In this equation the power consumption of the control system is assumed to be much lower than the output power P_{out} of the converter and is therefore not taken into account.

$$\eta_{lin} = \frac{P_{out}}{P_{in}} = \frac{I_{in}U_{out}}{I_{in}\left(U_{out} + U_{drop}\right)} = \frac{U_{out}}{U_{in}} = k_{lin} \quad (1)$$

As can be observed in equation 1 the power conversion efficiency η_{lin} of linear voltage regulators is only dependant on the voltage conversion ratio k_{lin}, and not of the output power P_{out}. From this point of view it is clear that the power conversion efficiencies for step-down voltage regulators, both linear and switched, should not be compared with each other at different voltage conversion ratios k. This is a common misunderstanding and it should be clear that the power conversion efficiencies of voltage regulators should only be compared at the same voltage conversion ratios k.

Obviously the most important reason to use a switched voltage converter instead of a linear voltage converter is the ability of a switched converter to achieve higher power conversion efficiencies. Therefore the power conversion efficiency of a switched step-down voltage regulator η_{sw} is to be compared with the power conversion efficiency of a linear voltage regulator η_{lin} at the same voltage conversion ratio k.

In order to benchmark a switched step-down regulator, in terms of power conversion efficiency, equation 2 can be used, where k_{sw} is the voltage conversion ratio of a switched voltage regulator. This expression gives the Efficiency Enhancement Factor (*EEF*) for switched step-down voltage regulators ver-

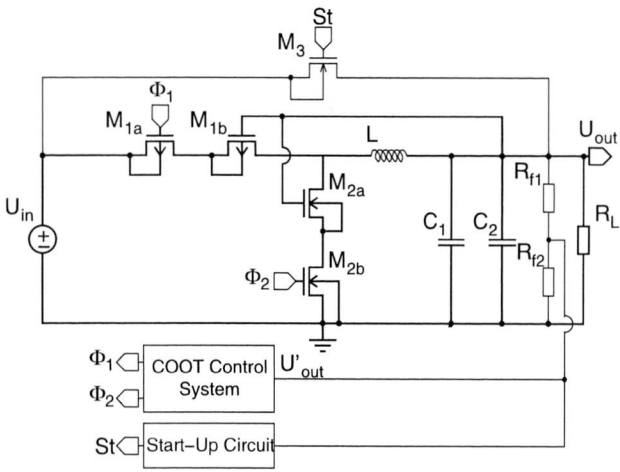

Fig. 1. The proposed architecture of the buck converter with feedback loop, control system and start-up circuit.

sus linear voltage regulators. A negative EEF value indicates that the switched converter has a lower power conversion efficiency than its linear converter equivalent (unwanted) and vice versa (wanted).

$$EEF = 1 - \frac{\eta_{lin}}{\eta_{sw}}|_{k_{lin}=k_{sw}} \quad (2)$$

The presented buck converter achieves an efficiency of 65 %, with a voltage conversion ratio k of 50 %. The efficiency improvement compared to a linear voltage regulator, according to the EEF, is 23 %. The input voltage range is 3 V to 4 V and the output voltage range is 1.5 V to 2.1 V. A novel constant On/Off-time (COOT) control system keeps the output voltage constant and stable at the desired level under varying load conditions, ranging from 0 mW to the maximum output power of 300 mW.

The architecture of the converter and the COOT control system is described section II. In section III the measurement results are discussed and a comparison is made between this work and other work. Finally conclusions are drawn in section IV.

II. ARCHITECTURE

A. Buck converter

Figure 1 shows the proposed architecture of the buck converter with the feedback loop, the COOT-control system and the start-up circuit.

The switch transistors are implemented in a waffle-shaped structure. Two transistors M_{1a}, M_{1b} and M_{2a}, M_{2b} are stacked to implement respectively the high-side and the low-side switch. The gates of M_{1b} and M_{2a} are biased with the output voltage U_{out} in order to divide the input voltage U_{in} over M_{1a}, M_{1b} and M_{2a}, M_{2b}. This approach enables U_{in} to be higher than the nominal technology voltage. The high-side switch transistors M_{1a} and M_{1b} have a width of 3800 μm and the low-side switch transistors M_{2a} and M_{2b} have a width of 2000 μm, all at minimal length.

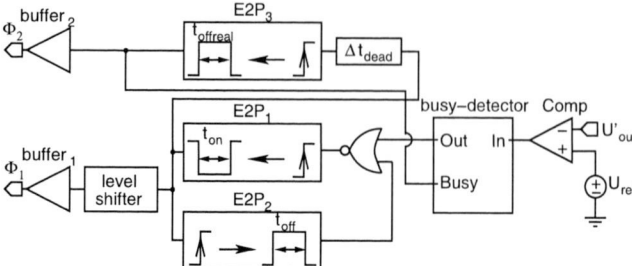

Fig. 2. The architecture of the COOT-control system.

The output inductor L is implemented as a bondwire spiral inductor, according to [7]. The four windings are made from golden bondwires with a diameter of 25 μm, placed on 17 bondpads. This yields an inductance of 18 nH, with a series resistance of 1 Ω.

Output capacitor C_1 is implemented with a MOScap, using the gate capacitance of 1300 parallel nMOS transistors. Each of the individual nMOS transistors has a width of 100 μm and a length of 10 μm. This yields a capacitance of 9 nF, for an area of 1.3 mm². Output capacitor C_2 is realized using a MIMcap, with a capacitance of 1.3 nF. The MIMcap is placed on top of the MOScap, reducing the required area.

As the control system uses the output voltage U_{out} as supply voltage, a start-up circuit is required. The start-up circuit switches transistor M_3 on until the output voltage U_{out} reaches a value of 1.4 V. Hereafter M_3 is switched off again and the COOT-control system takes over.

B. COOT Control system

Figure 2 shows the architecture of the constant on/off-time (COOT) control system, which keeps the output voltage U_{out} constant at the desired level under varying loads P_{out} and varying supply voltages U_{in}.

The output voltage U_{out} is lowered to U'_{out} with the resistive divider formed by R_{f1} and R_{f2}. The comparator then compares U'_{out} with a reference voltage U_{ref}, which determines the desired output voltage U_{out}. The output of the comparator is used as an input for the busy-detector.

If U'_{out} is higher than U_{ref} no action is taken and the switches remain closed. The current through the load is the generated by the output capacitors C_1 and C_2. If U'_{out} is lower than U_{ref} the busy-detector detects wether the input of buffer$_2$ is low, meaning that the converter is in idle mode. If this is true the busy-detector outputs a falling edge, enabling the oscillator formed by E2P$_1$ (edge-to-pulse-generator), E2P$_2$ and the NOR-gate. When enabled the oscillator starts oscillating, with a fixed low-time t_{on} of 4.4 ns and a fixed high-time t_{off} of 2.7 ns, until U'_{out} is higher than U_{ref}.

During an oscillation cycle the high-side switch is turned on, through the level shifter and buffer1, with a fixed low-pulse of duration t_{on}. The inductor L and the output capacitors C_1 and C_2 are than charged.

The output of E2P$_1$ is also fed into a time-delay Δt_{dead} of 0.49 ns, providing a dead-time between the opening of the

TABLE I
SUMMARY OF THE MEASUREMENTS RESULTS.

Input voltage range	3 V-4 V
Output voltage range	1.5 V-2.1 V
Output power range	0 mW-300 mW
Switching frequency range	20 Hz-140 MHz
Power efficiency @ U_{in} =3.6 V and U_{out} =1.8 V	65 %
Efficiency Enhancement Factor	23 %
Maximum output ripple @ P_{out} =0 mW	160 mV
Minimum output ripple @ P_{out} =300 mW	50 mV
Load regulation $\delta u_{out}/\delta i_{out}$	$-0.3\ \Omega$
Line regulation $\delta u_{out}/\delta u_{in}$	0.02

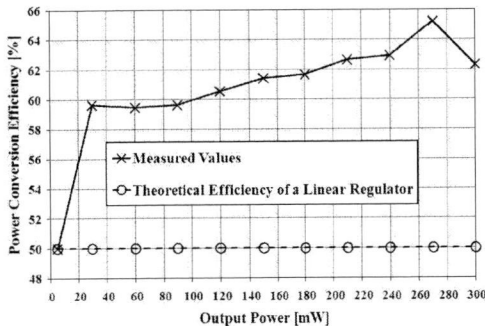

Fig. 3. The measured efficiency of the converter at varying loads compared with the theoretical maximum efficiency of a linear voltage regulator, having the same voltage conversion ratio k. The input voltage and the output voltage are kept constant at 3.6 V and 1.8 V respectively.

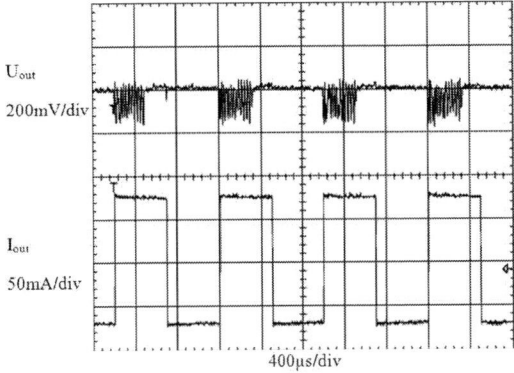

Fig. 4. The measured output voltage (upper curve) at a varying load current (lower curve).

high-side switch and the closing of the low-side switch. Time-delay Δt_{dead} drives E2P3, which triggers on a rising edge and generates a high-pulse with a fixed duration $t_{offreal}$ of 2 ns. The low-side switch is then turned on during a time $t_{offreal}$ through buffer₂, discharging the inductor L into the output capacitors C_1 and C_2 and the load.

The busy-detector avoids a new switching cycle to be started until the previous one is finished.

In essence the COOT-control system makes sure that the converter always transfers a fixed amount of energy to the output during one switching cycle, due to the fact that both the on-time t_{on} and the off-time $t_{offreal}$ are fixed. However the rate at which these fixed amounts of energy are transferred from the input voltage source to the output of the converter is dependant on the desired output power. Clearly when more power needs to be transferred to the output, more energy per unit of time needs to be transferred to the output. Since the transferred energy per cycle is fixed, the switching frequency will be increased. When no output power is needed, ideally the converter will stop switching. In reality it will still switch at a very low frequency, because the control system itself is powered by the output of the converter. The maximum switching frequency is designed to be 140 MHz, which determines the maximum output power of the converter.

Normally current sensing through the low-side switch is required when using asynchronous buck converters in discontinuous mode. However because t_{on} is set to be a constant value, $t_{offreal}$ can be chosen fixed as well. This is due to the fact that $t_{offreal}$ only depends on the value of the inductor L and the output capacitors C_1 and C_2, which in turn are fixed. Therefore no complicated and sensitive current sensing circuitry is required to determine when the low-side switch needs to be opened.

III. MEASUREMENTS AND COMPARISON

A. Measurements

This converter was implemented in a 0.18 μm 1.8 V CMOS technology. It measures 1.5 mm x 1.5 mm. Figure 5 shows a die photograph.

The solid line in figure 3 shows the measured power conversion efficiency of the entire converter at varying loads P_{out}. The input voltage U_{in} and the output voltage U_{out} are kept constant at 3.6 V and 1.8 V respectively. The dashed line in figure 3 denotes the theoretical power conversion efficiency of a linear voltage regulator with the same voltage conversion

ratio $k = 0.5 = 1.8/3.6$. The maximum power conversion efficiency is 65 %. This yields an Efficiency Enhancement Factor (EEF) of +23 % over a linear voltage regulator with the same voltage conversion ratio. It can be observed that the power conversion efficiency of the converter increases with an increasing output power. Clearly this behavior is desired due to the fact that the power losses will become proportionally more important at an increased output power demand.

The worst case output voltage ripple of the converter is 160 mV. This is less than 10 % of the output voltage and therefore acceptable for most applications. The minimum output voltage ripple is 50 mV.

The converter is capable of delivering any output power between 0 mW and 300 mW. Moreover it is intrinsically stable at every load condition in this power range. This is a huge advantage over Pulse Width Modulation control systems, which become intrinsically unstable at a low output power.

Figure 4 shows the measurement of the load regulation. The lower curve represents the modulated current through the load, ranging from 16 mA to 167 mA. This is equivalent to an output power ranging from 30 mW to 300 mW. The upper curve represents the output voltage. Apart from the ripple due to the switching of the converter itself, an extra voltage drop

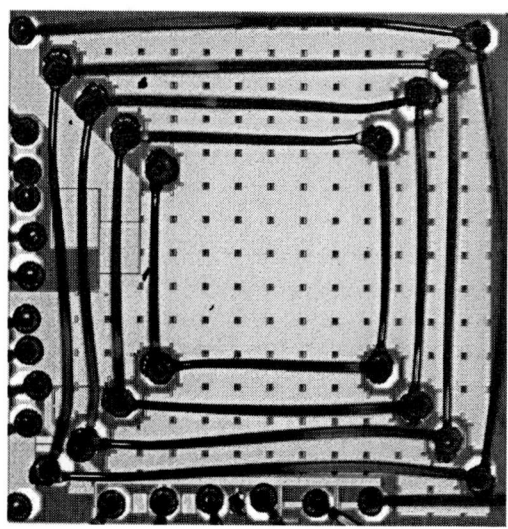

Fig. 5. Micro photograph of the chip.

TABLE II
COMPARISON WITH OTHER WORK.

ref	[4]	[6]	[2]	[3]	[5]	this work
year	2005	2005	2006	2007	2007	2008
process (μm)	1.5 CMOS	0.09 CMOS	0.18 BiCMOS	0.08 CMOS	0.35 CMOS	0.18 CMOS
U_{in} (V)	5	1.2	2.8	1	3.3	3.6
U_{out} (V)	2.5	0.9	1.8	0.52	2.3	1.8
$P_{out,max}$ (mW)	75	270	360	53	161	300
$k=\eta_{lin}$ (%)	50	75	64	52	70	50
η_{sw} (%)	49	83(*)	58	48	62	65
EEF (%)	-2	+9.6(*)	-10	-8.3	-13	+23
f_{sw}	10MHz	233MHz	45MHz	3GHz	200MHz	20Hz-140MHz
L (nH)	80	4x6.8 off-chip(*)	2x11	0.32	22	18
C (nF)	3	2.5	6	0.35	1	10.3
area (mm^2)	16	1.267(*)	6.75	0.81	4	2.25
Power per area (mW/mm^2)	4.7	213(*)	53	65	40	133
level of integration	thick film L	* air core off-chip SMT L	metal track L	metal track L	stacked chip C and L	bondwire L

of 50 mV appears when the output power is shifted to its maximum. This yields a load regulation of $-0.3\ \Omega$. The line regulation is measured using an input voltage varying between 3 V and 4 V, it is measured to be 0.02.

Finally the input voltage range is 3 V to 4 V and the output voltage range is 1.5 V to 2.1 V.

Table I shows a summary of the most important measurements results.

B. Comparison with other work

Table II shows a comparison between the presented work and other publications in the field of integrated DC-DC step-down converters, known to the authors.

As can be observed no other fully-integrated buck converter introduces an efficiency improvement over a linear voltage regulator, having the same voltage conversion ratio k. This can also be seen through the EEF, which is negative for all other fully-integrated buck converters. Moreover the converters using special technology options, such as thick-film inductors or stacked-chip passives, also fail to outperform a linear voltage regulator in terms of power conversion efficiency. The only solutions which present a positive EEF are this work and [6] which uses off-chip inductors. Furthermore the table shows that the presented work has the highest EEF of +23 %, meaning that, at its specific voltage conversion factor, this work proves to deliver the highest power conversion efficiency benefit over a linear voltage regulator.

In absolute figures [2] provides a higher maximum output power, of 360 mW, than the presented work. However the area requirement should also be taken into account, when making this comparison. It is indeed obvious that a larger area will lead to a higher maximum output power. Therefore it is more useful to compare the power per unit of area. From this point of view it is clear that the presented converter outperforms all the other fully-integrated converters. Needless to say however, that [6] does achieve a much higher power per unit of area,

due to the fact that the area of the four off-chip inductors is not taken into account.

IV. CONCLUSIONS

A fully-integrated DC-DC step-down converter with a COOT control system, to keep the output voltage constant at varying loads and supply voltages, is realized. A maximum output power of 300 mW, generated with a power conversion efficiency of 65 %, is achieved without needing any external components. To the authors knowledge this is the first fully-integrated converter which has a power conversion efficiency benefit over a linear voltage regulator. This is translated in a Efficiency Enhancement Factor (EEF) of +23 %.

The novel COOT control system keeps the output voltage constant and stable over the entire load range from 0 mW to 300 mW. Furthermore it enables the converter to work with input voltages from 3 V to 4 V and to generate output voltages between 1.5 V and 2.1 V.

REFERENCES

[1] C. Shi, B. C. Walker, E. Zeisel, B. Hu and G. H. McAllister. "A Highly Integrated Power Management IC for Advanced Mobile Applications", *IEEE Journal of Solid-State Circuits*, vol.42, no.8, pp. 1723-1731, August 2007

[2] S. Abedinpour, B. Bakkaloglu, and S. Kiaei. "A Multistage Interleaved Synchronous Buck Converter With Integrated Output Filter in 0.18 μm SiGe Process", *IEEE ISSCC Dig. Tech. Papers*, pp. 356-357, 2006

[3] M. Alimadadi, S. Sheikhaei, G. Lemieux, S. Mirabbasi and P. Palmer. "A 3 GHZ switching DC-DC converter using clock-tree charge-recycling in 90 nm CMOS with integrated output filter", *IEEE ISSCC Dig. Tech. Papers*, pp. 532-533, 2007

[4] S. Musunuri and P. L. Chapman. "Design of Low Power Monolithic DC-DC Buck Converter With Integrated Inductor", *Power Electronics Specialists*, 36th conference, pp. 1773-1779, September 2005

[5] K. Inizuka, K. Inagki, H. Kawaguchi, M. Takamiya and T. Sakurai. "Stacked-Chip Implementation of On-Chip Buck Converter for Distributed Power Supply System in SiPs", *IEEE Journal of Solid-State Circuits*, vol.42, no.11, pp. 2404-2410, November 2007

[6] P. Hazucha, et al. "A 233-MHz, 80 %-87 % efficient four-phase DC-DC converter utilizing aircore inductors on package", *IEEE Journal of Solid-State Circuits*, vol.40, no.4, pp. 838-845, April 2005

[7] M. Wens, K. Cornelissens and M. Steyaert. "A Fully-Integrated 0.18μm CMOS DC-DC Step-Up Converter, Using a Bondwire Spiral Inductor", *33rd European Solid State Circuits Conference*, pp. 268-271, September 2007

IEEE 2008 Custom Intergrated Circuits Conference (CICC)

A 90-240MHz Hysteretic Controlled DC-DC Buck Converter with Digital PLL Frequency Locking

Pengfei Li, Deepak Bhatia, Lin Xue, and Rizwan Bashirullah

Department of Electrical and Computer Engineering,
University of Florida, Gainesville, FL 32611

Abstract— This paper reports a digital phase locked loop (DPLL) frequency locking technique for high frequency hysteretic controlled dc-dc buck converters. The proposed technique achieves constant operating frequency over a wide output voltage range, eliminating the dependence of switching frequency on duty cycle or output voltage conversion range. The DPLL is programmable over a wide range of parameters and can be locked to a reference clock to ensure the converter switching frequency falls outside power supply resonance bands. We demonstrate a 90-240MHz single phase converter with fast hysteretic control and output conversion range of 33%-80%. The converter achieves an efficiency of 80% at 180MHz, a load response of 40ns for a 120mA current step and a peak-to-peak ripple less than 25mV. The circuit was implemented in 130nm digital CMOS process.

I. INTRODUCTION

Hysteretic controlled multiphase switch-mode converters operated at ultra-high frequencies upwards of 100MHz and with peak load efficiencies greater than 80% have been shown to provide extremely fast load response times of 1-5ns [1]. This enables several orders of magnitude reduction in capacitor (and inductor) size, potentially leading to highly integrated near-load power delivery solutions for high performance microprocessors requiring fast entry and exit strategies from multiple supply domains. Unlike the widely used pulse width modulation (PWM) controller, hysteretic control techniques based on a simple feedback loop achieves a near immediate load response without stability issues. However, if the controller is not properly synchronized, the free-running switching frequency will change with conversion voltage. If kept unchecked, the free-running oscillations may fall in undesired power supply resonance bands created by parasitic package inductance interconnects and on-die decoupling capacitances. This can potentially generate large voltage excursions in the supply network due to high impedance peaks formed by the multi-resonant networks, compromising overall system operation and device reliability [2].

Therefore, ideally it is desirable to synchronize the converter to an on-chip clock generated from within the processor to mitigate noise injection in undesirable frequency bands. Previous implementations of synchronization schemes for hysteretic control techniques have employed injection of synchronization signals in the reference voltage of the hysteretic comparator [1]. However, this approach requires that the amplitude, shape and frequency be carefully controlled to achieve proper frequency lock. In a multiphase voltage-mode hysteretic controller reported in [3] the synchronization scheme limits the output conversion range to 1/N of the input voltage, where N is the number of converter phases, and therefore requires larger minimum input voltage when the number of phases increases. Moreover, this technique does not maintain a constant switching frequency when the conversion ratio changes.

In this work, we present a digital phase lock loop (DPLL) synchronization scheme for hysteretic switch-mode buck dc-dc converters that achieves automatic phase synchronization and constant operating frequency over a wide output conversion range. Owing to the digital nature of the PLL, all parameters including frequency are digitally programmable and locked to an external (or internal) frequency source. Using the proposed techniques, a single phase hysteretic buck converter is designed in 130nm digital CMOS process that achieves 90-240MHz locked frequency operation. In section II we present an overview of the hysteretic buck converter. The proposed DPLL based hysteretic controller is discussed in section III. Measurement results and concluding remarks are presented in section IV and V, respectively.

II. HYSTERETIC BUCK CONVERTER

Fig. 1 shows a current mode hysteretic buck converter consisting of a hysteretic comparator, high-side PMOS and low-side NMOS drivers, a buck bridge with an LC filter and the feedback network made of R_F and C_F. The output voltage V_O in buck converter is given by

$$V_O = \langle V_X \rangle - r_L I_O = D V_{IN} - (R_{BRDG} + r_L) I_O \quad (1)$$

where D is the duty cycle, V_{IN}, V_X are the input voltage and the bridge output respectively, I_o is the average output current, r_L is the equivalent series resistance (ESR) of the inductor and R_{BRDG} is the weighted average of series resistances of the bridge switches. The duty cycle in the buck converter is adjusted by a feedback control loop to satisfy the desired conversion ratio and load current I_O,

$$D = \frac{V_O}{V_{IN}} - \frac{(R_{BRDG} + r_L) I_O}{V_{IN}} \quad (2)$$

The feedback network R_F and C_F works as a high-pass filter from V_O to V_{FB}, and senses the inductor current ramp by integrating the inductor voltage. Fig. 2a shows the idealized voltage waveforms at the feedback node V_{FB} and bridge output V_X. If we neglect the propagation delays of the comparator and high/low side bridge buffers, the switching frequency is given by $f_s = (1-D)D/\tau_{RC}$, where $\tau_{RC} = R_F C_F (V_H / V_{IN})$ is a time constant that depends on the hysteretic window V_H, the input

978-1-4244-2018-6/08 $25.00 © 2008 IEEE

voltage V_{IN} and the high-pass filter time constant. Since the hysteretic window V_H determines the ripple, and V_{IN} is fixed, the operating frequency can be controlled by adjusting the filter time constant. In the presence of finite comparator, buffers and bridge switching propagation delays, the switching frequency decreases and can be expressed as [5],

$$f_S = \left(\frac{\tau_{RC}}{D(1-D)} + \frac{T_{don}}{(1-D)} + \frac{T_{doff}}{D} \right)^{-1} \quad (3a)$$

where T_{don} and T_{doff} correspond to the finite propagation delays of switching the inductor to V_{IN} and ground, respectively (Fig. 2b). Assuming the propagation delays are equal ($T_{don} = T_{doff} = \tau_D$), the switching frequency can be expressed as

$$f_S = \frac{D \cdot (1-D)}{\tau_{RC} + \tau_D} \quad (3b)$$

From Eq. (3b), the maximum switching frequency occurs when D=0.5, or

$$f_{MAX} = \frac{1}{4(\tau_{RC} + \tau_D)} \quad (4)$$

If the conversion range of the converter is D=0.2 to D=0.8, the switching frequency will vary from $0.64 f_{MAX}$ to f_{MAX}. This frequency dependence as a function of conversion range is undesirable. As indicated by measured results for a 4-phase hysteretic controlled buck-converter [6] shown in Fig. 3, the switching frequency exhibits a parabolic dependence on the conversion range and closely follows Eq. (3b).

III. DPLL SYNCHRONIZED DC-DC CONVERTER

A. Architecture

The proposed DPLL synchronized hysteretic controlled dc-dc buck converter is shown in Fig. 4. The control loop defined by the dc-dc converter can be viewed as a free-running oscillator with switching frequency f_s. The oscillation frequency is defined by Eq. (3b), and nominally depends on the duty cycle (D), the time constant (τ_{RC}) and the overall loop delay. Since D sets the conversion range, and τ_{RC} affects both the load response via the high-pass feedback network $R_F C_F$ and the ripple voltage via the hysteretic window V_H, it is desirable to synchronize the oscillating frequency by changing the loop delay. In order to digitally adjust the control loop delay and hence the oscillating frequency, a digitally controlled delay line (DCDL) is inserted in the controller feedback path, as shown in Fig. 4. The output of the hysteretic comparator is divided down and compared against an external reference clock using a bang-bang phase and frequency detector (PFD). The resulting early/late information is filtered using a proportional-integral digital loop filter. The output bits of the loop filter are fed to a sigma delta modulator and a decoder to adjust the DCDL. The first order sigma-delta is used to enhance the delay resolution of the DCDL. A serial interface is used to program DPLL parameters such as the proportional and integral gains, and the divide ratio. The dc-dc converter low/high side buffer sizes, the hysteretic window height V_H and the feedback network $R_F C_F$ are also programmable.

B. Digitally Phased Locked Loop (DPLL)

Fig. 5 shows a block diagram of the DPLL. The bang-bang PFD compares the arrival times of the reference clock

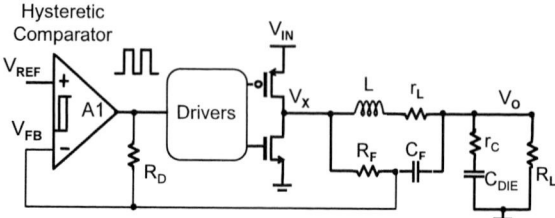

Fig.1 Single phase hysteretic buck converter.

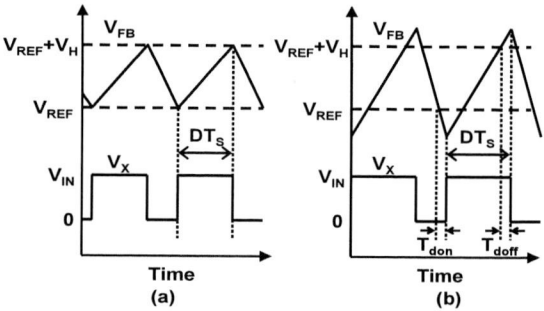

Fig.2 Voltage waveforms of V_{FB} and V_X (a) without converter propagation delay and (b) with finite propagation delay.

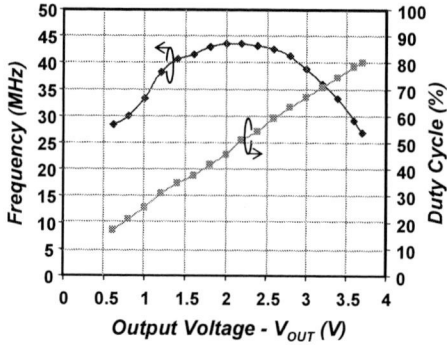

Fig.3 Measured frequency as a function of conversion ratio for a 4-phase hysteretic buck converter with 5V input voltage [6].

and the divided down signal from the output of the hysteretic controller. The PFD generates the early/late signals for the digital loop filter with controllable gains on the integral and proportional paths. The ratio of the gains is set to stabilize the loop. A selector (SEL) at the output of the loop filter selects 11 bits out of the 19 bits available (i.e [10:0], [11:1],...). Out of these 11 bits, 3 bits are used by the sigma-delta to generate a bit-stream which controls one of the fine delay cells in the DCDL. The 8 remaining bits at the output of the selector along with the sigma-delta output form a 9-bit control word (A[0:8]), where A[0] represents the bit-stream of the first order sigma-delta. Two 3-to-8 decoders control the coarse delay line and the three LSB's of the control word control the fine delay cells. Programmable dividers generate the necessary clocks for the PFD, sigma-delta, digital loop filter and the selector (SEL). The sigma delta is clocked at a programmable divide-by-M ratio, whereas the loop filter and PFD operate at a divide-by-NxM ratio locked to the reference clock frequency. The sigma-delta as shown in Fig. 5 is a 3-bit accumulator with the carry out of the final adder producing the bit-stream. The sigma delta can be operated at 1/4, 1/8,

1/16 or 1/32 of the output frequency depending upon the 2-bit control signal (DIV_SEL). With the exception of the PFD and DCDL, automatic synthesis and place and route tools were employed to implement the DPLL.

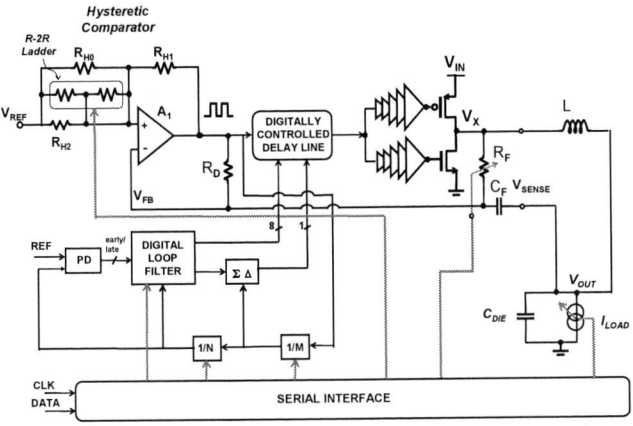

Fig. 4 Proposed DPLL synchronized hysteretic controlled buck converter.

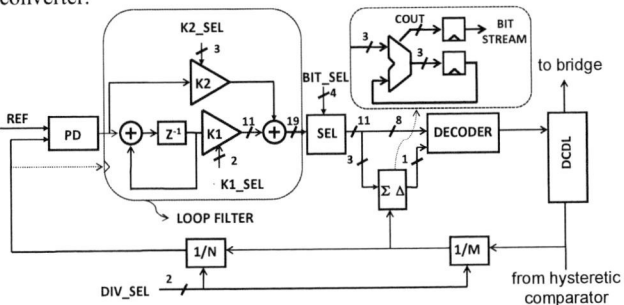

Fig.5 DPLL block diagram.

C. Digitally Controlled Delay Line (DCDL)

The digital-controlled delay line as shown in Fig 6 is employed to supply variable delay to the control loop. The delay line consists of two tuning stages: a coarse tuning stage and a fine tuning stage. The coarse tuning delay chain is made up of 63 delay buffers each of which has a fixed delay of about 40ps. To avoid large load capacitance, two stage tristate inverters are used. The 63 delay cells are divided into 8 groups with the first group having 7 delay cells and the others 8. Two decoders are used to select the required delay from the coarse delay line based on the 6 MSB's (A[3:8]) of the control word. A[3:5] selects the number of delay cells from each group and A[6:8] selects the number of groups. The fine tuning stage consists of 3 tri-states inverters (1X, 1X, 2X) in parallel controlled by the sigma-delta bit-stream (A[0]) and the two LSB's (A[1], A[2]).

IV. EXPERIMENTS

An integrated buck DC-DC converter with hysteretic control loop and DPLL based synchronization was fabricated in standard 130nm digital CMOS process. The die photo is shown in Fig. 7. The converter is operational from 90MHz to 240MHz and occupies approximately 0.34mm^2, which includes the DCDL and a 0.7nF input decoupling capacitor. The DPLL without the DCDL measures 320µm by 200µm.

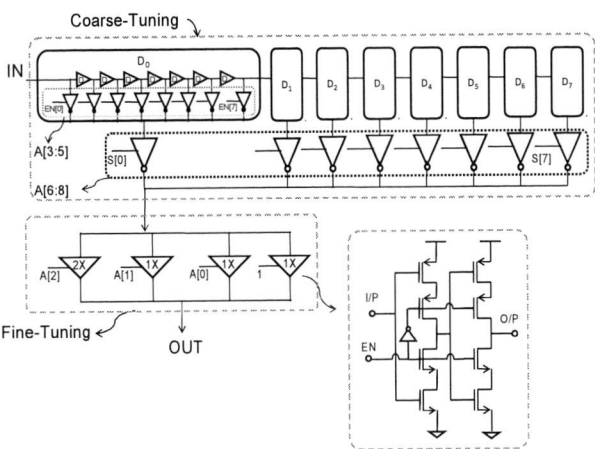

Fig.6 Digital-Controlled Delay Line.

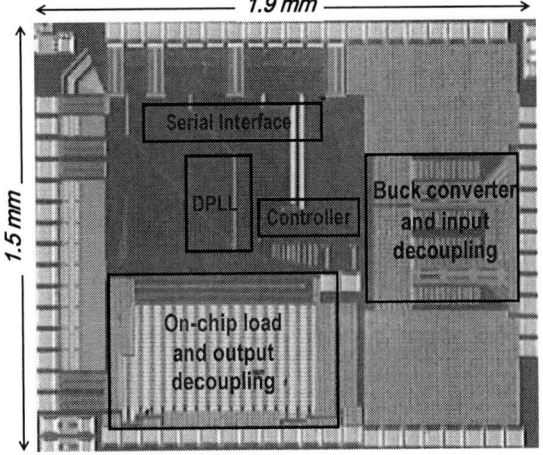

Fig.7 Die Photograph

An SMT size-805 8.2nH air core inductor with quality factor of 25 at 180MHz and a 20nF output decoupling capacitor were mounted on the test board.

In order to evaluate the performance of the DPLL synchronization scheme and frequency locking behavior, the converter output voltage was varied and the output frequency was measured, as shown in Fig. 8. When the DPLL is disabled and the DCDL is programmed for minimum delay using the serial interface, the free-running switching frequency exhibits the parabolic dependence on output voltage for a fixed input voltage (V_{IN}) of 1.2V. To test the DPLL, we set the divide ratio to 128 and varied the reference frequency from approximately 703kHz to 1.875MHz in 30MHz/128 increments. The corresponding measurements show that the output switching frequency is locked to the input reference over a wide switching frequency range of 90MHz to 240MHz. The output voltage range over which the DPLL locks is bounded by the free-running switching frequency of the converter. Thus the output voltage range decreases as the frequency is increased. The jitter histogram of the divided down clock (V_{DIV}) and the multiplied clock or bridge output signal (V_X) are shown in Fig 9a and 9b, respectively. The RMS and peak-to-peak jitter of the bridge output is 42.5ps and 244ps, respectively.

The measured power efficiency at 180MHz versus load current for 1.2V/0.8V conversion is shown in Fig. 10. The peak efficiency of ~80% is achieved for a load current of 200mA at 0.8V. Increasing the load current decreases the efficiency due to higher resistive loss associated with the ripple current in the inductors, whereas at lighter loads the switching frequency loss associated with the bridge is dominant. Fig. 11 shows the converter transient response to a 120mA load step. The current step was generated using a programmable on-chip load with a ramp time of 100ps. The resulting output voltage is about 0.8V for a low current load of 30mA and 0.75V for a high load of 150mA. This low-to-high load transient causes an output voltage droop of 100mV and a response time of approximately 40ns.

V. CONCLUSIONS

We demonstrate a DPLL frequency locking technique for ultra-high frequency hysteretic controlled dc-dc buck converters. The DPLL locks the converter operating frequency to a clock reference to eliminate the dependence of switching frequency on output conversion voltage. Moreover, since the DPLL loop parameters are fully digitally programmable, this technique can also be used to implement a pulse frequency modulation (PFM) controller to improve light load efficiencies [7]. The proposed DPLL converter operates over a wide frequency range of 90-240MHz and achieves a conversion range of 33% to 80%, or 0.4V to 0.96V. Using a single 8.2nH inductor and a 20nF external decoupling capacitance, the converter achieves a load response of 40ns to a 120mA step and ripple voltage less than 25mV. The DC-DC converter was implemented in 130nm 1.2V digital CMOS process and achieves a peak efficiency of 80% at 180MHz for load current of 200mA.

REFERENCES

[1] P. Hazucha et al., "A 233-MHz 80%-87% efficient four-phase DC-DC converter utilizing air-core inductors on package", IEEE J. Solid-State Circuits, vol.40, no.4, pp838–845, 2005

[2] J. Xu, et. al, "On-die Supply Resonance Suppression Using Band-Limited Active Damping," ISSCC, pp. 286-287, Feb 2007.

[3] J. Abu-Qahouq, et al., "Multiphase voltage-mode hysteretic controlled DC-DC converter with novel current sharing", IEEE Trans. Power Electronics, vol.19, no.6, pp.1397 -1407, Nov. 2004

[4] T. Karnik, "High Frequency DC-DC Conversion: Fact or Fiction" ISCAS, pp. 245-24, 2006.

[5] T. Nabeshima et al, "Analysis and design considerations of a buck converter with a hysteretic PWM controller", IEEE PESC, vol.2, pp.1711-1716, 2004.

[6] P. Li, R. Bashirullah, P. Hazucha, T. Karnik "A Delay Locked Loop Synchronization Scheme for High Frequency Multiphase Hysteretic DC-DC Converters," IEEE VLSI Circuits Symp., pp. 26-27, June 2007

[7] T. Man, P. Mok, M. Chan, "An Auto-Selectable-Frequency Pulse-Width Modulator for Buck Converters with Improved Light-Load Efficiency," IEEE ISSCC, pp. 440-441, Feb 2008.

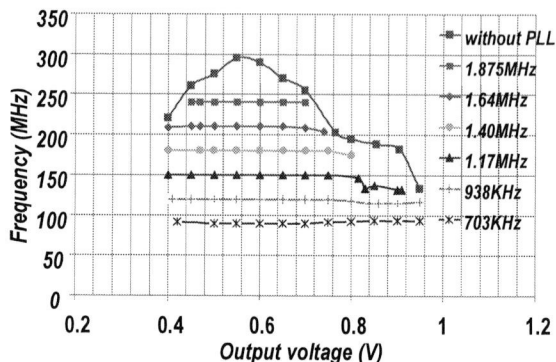

Fig. 8. Converter switching frequency versus output voltage.

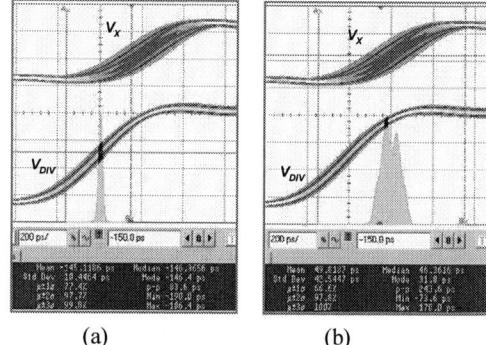

(a)   (b)

Fig. 9 Jitter histogram of (a) divided clock and (b) dc-dc bridge output.

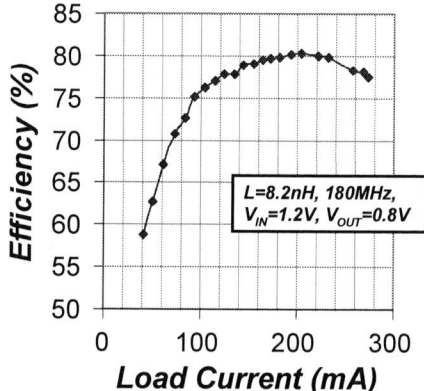

Fig. 10 Measured efficiency.

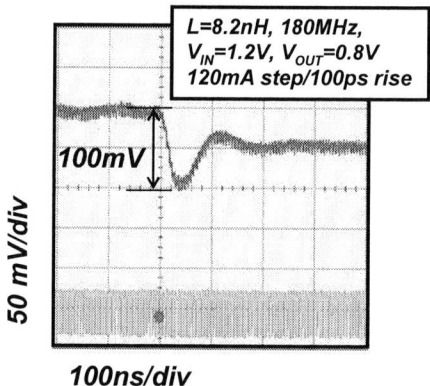

Fig. 11 Measured load response.

IEEE 2008 Custom Intergrated Circuits Conference (CICC)

Fast Transient Technique (FTT) in Buck Current-Mode DC-DC Converters for Low-Voltage SoC Systems

Chia-Hsiang Lin[+], Hong-Wei Huang[*], and Ke-Horng Chen[+]

[+] Department of Electrical and Control Engineering National Chiao Tung University, Hsinchu, Taiwan, R.O.C

[*] RichTek Technology Corporation, Hsinchu, Taiwan, R.O.C

Abstract—**This paper proposes a fast transient technique (FTT) to achieve excellent transient response of current-mode buck DC-DC converters. The proposed technique combines a non-linear control and a linear control mechanism to achieve fast transient response time, low output ripples, and stable transient operation at the same time. The test chip was fabricated in UMC 0.18-µm process. Experimental results show that the transient undershoot/overshoot voltage and the recovery time are not exceed 48mV and 10µs, respectively.**

I. INTRODUCTION

Among the numerous requirements included in the ability to build high-performance system-on-chip (SoC) systems, the imperative demand is to supply a dynamic voltage in terms of the processing throughputs. Dynamic voltage scaling (DVS) technique is the most popular power management technique for reducing power loss of SoC systems. Hence, the design consideration for DC-DC converters with good transient performance is necessary to provide good dynamic performance and simultaneously ensure the regulator's stability. In other words, several techniques are demanded for improving transient response time and voltage ripples in order to make sure low supply voltage ripple and maintain a reliable supply voltage to SoC systems.

Furthermore, for low voltage designs in SoC systems, the signal level is decreased with the same noise level in high voltage systems. The signal-to-noise ratio (SNR) is seriously reduced. It is obvious that the operating range and SNR of 0.35-µm process are larger than those of 0.18-µm. Thus, it is imperative to keep a stable and continuous power to the SoC system in order to react to fast load variations. The most popular literature in today's fast transient technique is to speed up the charging or discharging time of large compensation capacitor [1-3]. Large driving current is sourcing or sinking into the compensation capacitor when load current changes. However, a careful design consideration of the system stability is needed to make sure the stable operation. Thus, a current-mode buck DC-DC converter with fast transient technique is proposed as shown in Fig.1 to solve those problems.

The following section describes the concept of the current-mode buck DC-DC converter with FTT. Section III presents the circuit implementation. Experimental results are shown in section IV. Finally, a conclusion is made in section V.

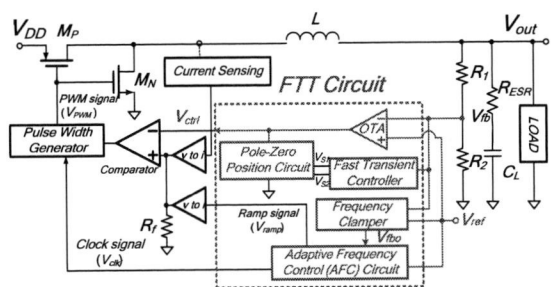

Fig. 1. The proposed buck DC-DC converter with the proposed fast transient technique.

II. CONCEPT OF THE CURRENT-MODE BUCK DC-DC CONVERTER WITH FAST TRANSIENT TECHNIQUE

The proposed FTT is composed of an adaptive frequency control (AFC) and a two-step non-linear control. Adaptive frequency control is the operation control technique that adaptively and smoothly adjusts the switching frequency by dynamically varying the switching frequency. During load transient period, switching frequency will step up (step down) to increase (decrease) duty cycle according to the quantity of the error voltage (ΔV) between the reference voltage V_{ref} and the scaled voltage V_{fb}. As for two-step non-linear control, it moves the compensated pole-zero pair to higher frequency during an optimum transient time in order to get a wide bandwidth and improve transient response.

The concept of FTT is shown in Fig. 2 and it contains three fast transient mechanisms (mode A, B, C) which can be decided by load-dependent crossover frequency (f_c). When load current changes from light load to heavy load, crossover frequency is moved away from f_{c6} to f_{c1} for getting a fast transient response and operation mode is changed from mode A to mode B. Two-step non-linear control and AFC technique are used to immediately recover the output voltage level. After a load-dependent time, operation mode is changed to mode C to prevent the system from ringing ($f_{c2}<0.2\times f_s$) by disabling one part of the non-linear mechanism. Operation mode goes back mode A when output voltage V_{out} stops dropping and trends back to the regulated value, then the control mechanism is fully determined by AFC technique.

978-1-4244-2018-6/08 $25.00 © 2008 IEEE

Similarly, the operation is the same when load current changes from heavy to light. Thus, the advantage of the fast transient technique is that it can smoothly and effectively regulate the output voltage level when load current changes.

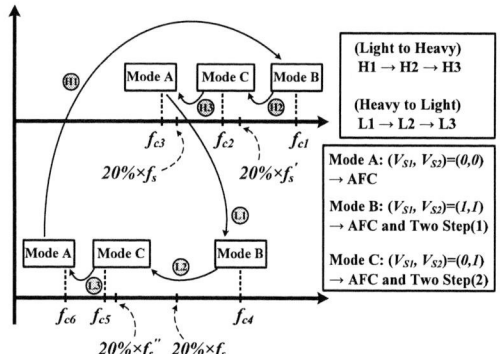

Fig. 2: The concept of the proposed fast transient technique contains three fast transient mechanisms (mode A, B, C).

III. CIRCUIT IMPLEMENTATION

Two-step positive feedback control circuit is composed of OTA, fast transient controller, and pole-zero position circuit as shown in Fig. 3. There are two operation modes in this circuit. At normal operation, the signals V_{S1}, V_{S2}, V_{S1B} and V_{S2B} that are generated by fast transient controller set the switches S_1, S_2 off and S_1', S_2' on. Besides, the aspect ratios of transistors M_{Z2}, M_{Z3}, and M_{Z4} are set mk, nk, and $(1-m-n)k$ times that of transistor M_{Z1}, respectively. Once a large load variation happens, the output capacitor of the error amplifier needs to be charged or discharged as fast as possible. Thus, the fast transient mechanism is started by setting switches S_1, S_2 on and switches S_1', S_2' off.

The redirecting current is partitioned into two current branches. One current branch composed of I_{z2} and I_{z3} is utilized to charge or discharge the small capacitor C_{z1} for speeding up the transient response time. The other one current branch, I_{z4}, is still utilized to compensate the DC-DC converters. Thus, the slew rate of the error amplifier can be increased. As we know, if the compensation circuit is implemented by voltage-mode Miller capacitor, the compensation circuit has no extra current to speed up the transient time for output voltage of error amplifier. It is why we use current-mode Miller capacitor to speed up and compensation current-mode buck DC-DC converters.

The other operation mode is the fast operation which is dependent on the fast transient controller as shown in Fig. 3 (a). Fig. 3 (b) and (c) show the schematics of the error amplifier and the voltage follower. The transistors M_{B1}-M_{B9} constitute a standard bias circuit for a cascode operational transconductance amplifier, which is a high gain single-stage amplifier with one dominant pole [11, 12]. For the voltage follower M_{V1}-M_{V14}, due to the symmetry of voltage follower [13], the input voltage level (the gate of M_{V1}) is equal to the output node (the gate of M_{V2}).

This fast transient controller is composed of an optimum transient time generator with low-pass network

and a one-shot circuit. The optimum transient time generator is used to generate the signal V_{S2} for controlling the equivalent current of transistor M_{Z3}. Similarly, the signal V_{S1} is generated to control the equivalent current of transistor M_{Z2} by one-shot circuit. Node V_{P1} is connected to AFC circuit as shown in Fig. 4 (a) owing to regulate the first step positive feedback time of signal V_{S1} by a correction current I_c. Once a large load variation happens, output voltage V_{out} will be dropped down or pulled high. In order to detect this instant variation, we use two comparators to compare the scaled voltage V_{fb} with the high V_{PH} and low V_{PL} threshold voltages for enabling the FTT circuit into fast operation mode. In design aspect, we should increase the redirecting current as large as possible to charge or discharge output capacitor of error amplifier for enhancing the transient response time and reducing the overshoot/undershoot voltage. However, large current may cause the oscillation scenario. Thus, a two-step positive feedback mechanism is utilized to make sure the stable operation and achieve fast transient response.

Based on node V_{P1} which is connected to AFC circuit as shown in Fig. 4 (a), a correction current i_c can regulate the operation time of mode B according to load variation. i_a and i_b are converted from the error voltage (V_{fbo}-V_{ref}) by the difference voltage to current (ΔV-to-I) converter. These two current sources can automatically and smoothly revise the rising and falling slope of ramp signal for supple more/less energy during load variation. When output voltage is regulated back to expected voltage, i.e. ΔV is small, the revised currents become trickle and are not large enough to affect the switching frequency. Finally, frequency is constant again during steady state. Besides, in order to avoid sub-harmonic oscillation, the maximum and minimum switching frequencies are defined by clamping the feedback voltage V_{fb} between lower voltage V_{bl} and upper voltage V_{bh}, which are shown in Fig. 4(b).

The definition of current i_a, i_b, and i_c is (1).

$$i_a = 2K \frac{V_{ref}-V_{fb}}{R_d}, i_b = 2M \frac{V_{ref}-V_{fb}}{R_d}, \text{ and } i_c = 2T \frac{V_{ref}-V_{fb}}{R_d} \quad (1)$$

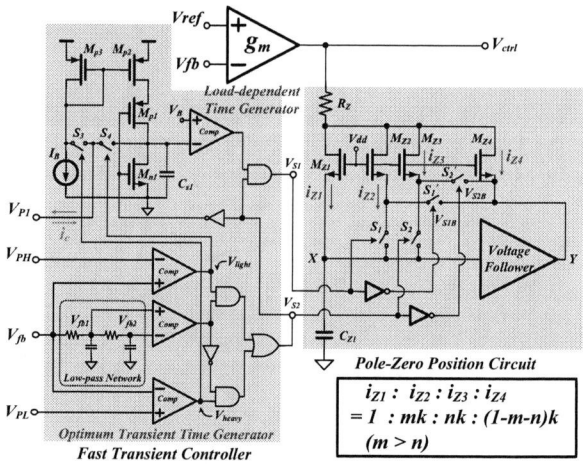

(a)

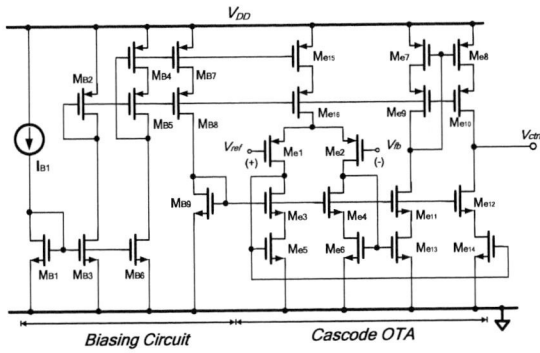

(b)

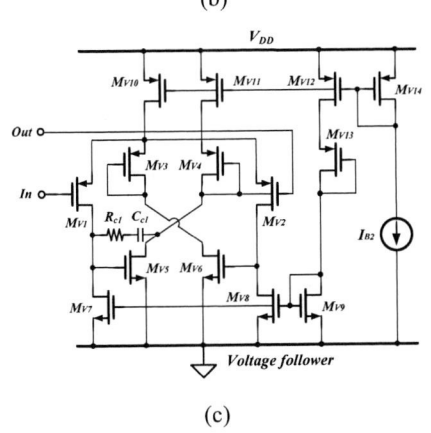

(c)

Fig. 3. Two-step positive feedback control circuit. (a) Architecture of two-step positive feedback control. (b) Schematic of error amplifier. (c) Schematic of the voltage follower.

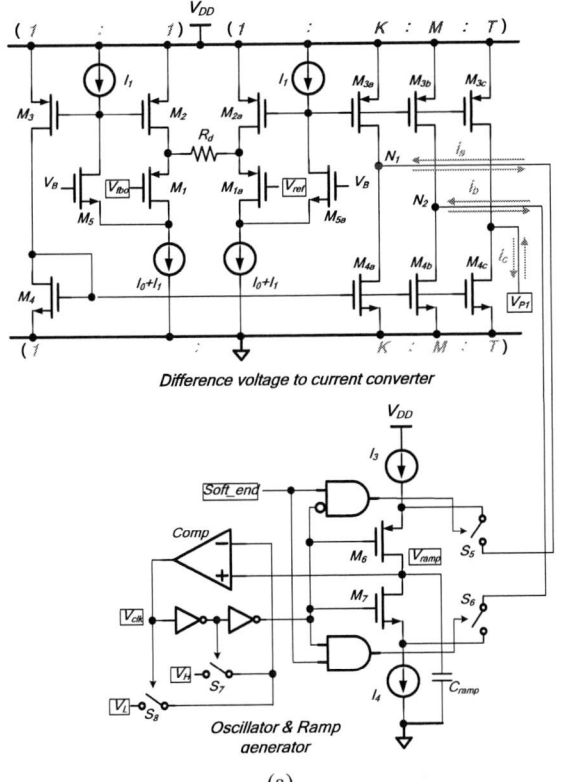

(a)

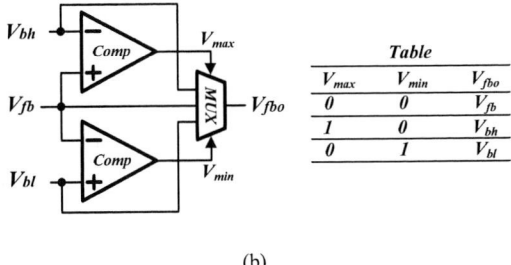

(b)

Fig. 4: Adaptive frequency control circuits. (a) The proposed ΔV-I converter, oscillator, and ramp generator. (b) The operation the proposed frequency clamper. (c) The determination of current I_a, I_b, and I_c.

IV. EXPERIMENTAL RESULTS

The proposed LDO regulator with adaptive reference control technique was implemented in UMC 0.18-μm CMOS technology. The chip micrograph is shown in Fig. 5 and chip area is 400 μm×470 μm. A summary of the LDO performance is shown in Table I.

The proposed current-mode buck DC-DC converter with the proposed fast transient technique is implemented by UMC 0.18-μm process. Specifications of the proposed DC-DC converter are listed in Table I. The efficiency and chip micrograph are shown in Fig. 5 and Fig. 6, respectively. Maximum measured efficiency is 85.6% at 280mA. Transient waveforms are shown in Fig. 7 (a)-(c) when load current changes from 100mA to 500mA within 1μs, or vice versa. The undershoot (overshoot) voltage is about 48mV (44mV) and recovery time is less than 10μs. When load current increases, the switching frequency is increased to 1.2MHz according to the error voltage. On the contrary, switching frequency is decreased to 675kHz when load current decreases rapidly.

Table I: Specification of Proposed DC-DC Converter

Fabrication Process	UMC 0.18-μm process
Supply Voltage	1.8V
Output Voltage	1V
Load Current	< 500mA
Power Efficiency	72.4%~85.6%@V_{out}=1V
Inductor / Capacitor	L=4.7μH; C_L=10μH
Switching Frequency (100mA↔300mA) (100mA↔400mA) (100mA↔500mA)	1.05MHz ~ 800kHz 1.15MHz ~ 730kHz 1.20MHz ~ 657kHz
Max. Quiescent Current	450μA@I_L=500mA
Parameter	M=0.5, n=0.15 and k=100

V. CONCLUSION

A fast transient technique for DC-DC converter to achieve excellent transient performance in low-voltage SoC system is proposed. Fast transient response time, low output ripples, and stable transient operation are demonstrated by experimental results. With the proposed fast transient technique, the performance of DC-DC converters is improved significantly.

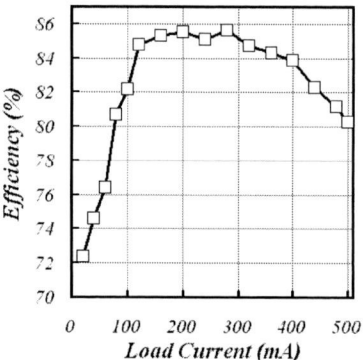

Fig. 5: Power efficiency.

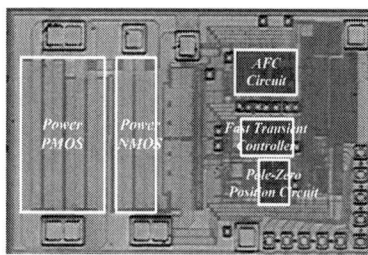

Fig. 6: Chip micrograph.

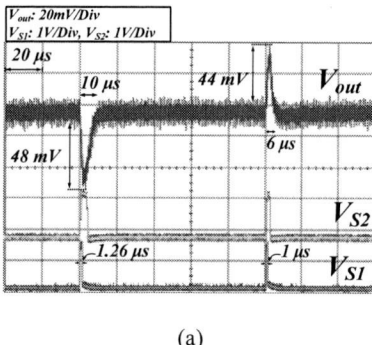

(a)

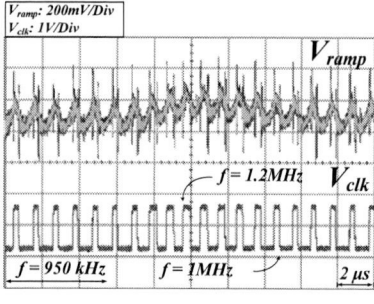

(b)

(c)

Fig. 7. Transient waveforms of proposed DC-DC converter. (a) Load current steps from 100mA to 500mA, or vice versa. (b) The enlarged waveforms of that load current stepping from 100mA to 500mA within 1μs. (c) The enlarged waveforms of that load current stepping from 500mA to 100mA within 1μs. (V_{out} = 1V)

REFERENCES

[1] Edgar Sanchez-Sinencio and Andress G. Andreou, *Low-Voltage/Low-Power Integrated Circuit and System*, IEEE Press, New York 1998.

[2] Hoi Lee, Philip K. T. Mok, and Ka Nang Leung, "Design of low-power analog drivers based on slew-rate enhancement circuits for CMOS low-dropout regulators," *IEEE Trans. Circuits and Systems II*, vol. 52, no. 9, pp. 563-567, Sept. 2005.

[3] Hong-Wei Huang, Hsin-Hsin Ho, Chieh-Ching Chien, Ke-Horng Chen, Gin-Kou Ma, and Sy-Yen Kuo, "Fast Transient DC-DC Converter with On-Chip Compensated Error Amplifier," *32nd European Solid-State Circuits Conference*, pp. 324-327, Sep. 2006.

[4] Jeongjin Roh, "High-performance error amplifier for fast transient DC-DC converters," *IEEE Trans. Circuits and Systems II*, vol. 52, no. 9, pp. 591-595, Sept. 2005.

[5] Kaiwei Yao, Kisun Lee, Ming Xu, and Fred C. Lee, "Optimal design of the active droop control method for the transient response," *Applied Power Electronics Conference and Exposition*, vol. 2, pp. 718-723, Feb. 2003.

[6] Man Siu, Philip K.T. Mok, Ka Nang Leung, Yat-Hei Lam, Wing-Hung Ki, "A voltage-mode PWM buck regulator with end-point prediction," *IEEE TCAS II*, vol. 53, no. 4, pp. 294-298, April 2006.

[7] Lu Chen, Bruno Ferrario, "Adaptive Frequency Compensation for DC-to-DC converter," *US Patent 20060176098*, Aug. 10, 2006.

[8] Andrés Barrado, Antonio Lázaro, Ramón Vázquez, Vicente Salas, and Emilio Olías, "The Fast Response Double Buck DC–DC Converter (FRDB) Operation and Output Filter Influence," *IEEE Trans. Power Electronics*, vol. 20, no. 6, pp. 1261-1270, Nov. 2005.

[9] Robert W. Erickson, Dragan Maksimović, *Fundamentals of Power Electronics*. Norwell, MA: Kluwer, 2001.

[10] Hong-Wei Huang, Chun-Yu Hsieh, Ke-Horng Chen, and Sy-Yen Kuo, "Adaptive Frequency Control Technique for Enhancing Transient Performance of DC-DC Converters," *IEEE 33nd European Solid-State Circuits Conference*, pp. 174-177, Sep. 2007.

[11] Cheung Fai Lee and Philip K. T. Mok, "A Monolithic Current-Mode CMOS DC-DC Converter with On-Chip Current-Sensing Technique," *IEEE J. Solid-State Circuits*, vol. 39, no. 1, pp. 3-14, Jan. 2004.

[12] Behzad Razavi, *Design of Analog CMOS Integrated Circuits*. Boston, MA: McGRAW-HILL, 2001.

[13] G. Palumbo and S. Pennisi, "A high-performance CMOS voltage follower," *IEEE International Conference on Electronics, Circuits and Systems*, vol. 2, pp. 21-24, Sept. 1998.

[14] Werner Hollinger, Manfred Punzenberger, "An Asynchronous 1.8MHz DC/DC Boost Converter Implemented in the Current Domain for Cellular Phone Lighting Management," *IEEE European Solid-State Circuits Conference*, pp. 528-531, Sep. 2006.

IEEE 2008 Custom Intergrated Circuits Conference (CICC)

A process variation compensation scheme using cell-based forward body-biasing circuits usable for 1.2V design

Fumihiko Tachibana[1], Hironori Sato[1], Takahiro Yamashita[1], Hiroyuki Hara[1], Takeshi Kitahara[1], Shuou Nomura[1], Fumiyuki Yamane[1], Yoshiro Tsuboi[1], Keiko Seki[2], Shuuji Matsumoto[2], Yoshinori Watanabe[2], and Mototsugu Hamada[1]

[1]Toshiba Corporation, Kawasaki, Japan
[2]Toshiba Microelectronics Corporation, Kawasaki, Japan

Abstract- **A cell-based forward body-biasing technique to suppress the global process variation and its design flow are proposed. Latch-up free operation is guaranteed by embedded current source cells and limiter cells even when supply voltage is 1.2V with small area overhead. By applying this technique to a media processor, the worst-case delay is reduced by 20% without sacrificing the maximum leakage spec.**

I. INTRODUCTION

With technology scaling, die-to-die variations become larger and larger. Large variations make it difficult to apply the conventional scaling theory to the performance improvement in terms of power dissipation and propagation delay. To compensate for the process variation, circuit techniques adjusting supply voltage [1-2] or adjusting body-bias [3-8] depending on as-fabricated chip performance are reported.

Considering that we may face un-balanced corners, where the threshold voltage of pMOS is high and that of nMOS is low, or vice versa, to utilize the body-biasing to control the threshold voltage of pMOS and nMOS separately, will be a more powerful tool. Among the body-biasing techniques, some papers utilized forward body-biasing and others did reverse body-biasing. The merit in forward body-biasing is that it does not require any voltage levels out of the power rails, so that simple biasing circuits can be implemented in a chip. For example, swapped body biasing technique [5] and self-adjusted forward body bias technique [6,7] are reported. In [5], n-well is connected to ground and p-well is connected to the supply voltage. However, large current flows through pn-junction between p-well and n-well when supply voltage is higher than 0.6V. In [6,7], the current sources connected to n-well and p-well determine the body-biasing voltage by the forward-biased diode. However, as shown in Fig. 1(a), collector current and base current of parasitic bipolar transistors in p-well and n-well form a positive feedback loop when supply voltage is higher than 0.6V. Thus, body-biasing voltage is not controllable at >0.6V. Although some paper [7] reported that latch-up did not occur at low voltage where h_{FE} of a parasitic bipolar transistor is less than unity, it is still an issue to be considered for mass production. Actually, at 1.1-1.3V or higher, for the burn-in mode, measured results show that h_{FE} is more than unity even in 65nm CMOS technology.

In this paper, a forward body-biasing technique with cell-based biasing circuits usable at >0.6V is proposed. The body-

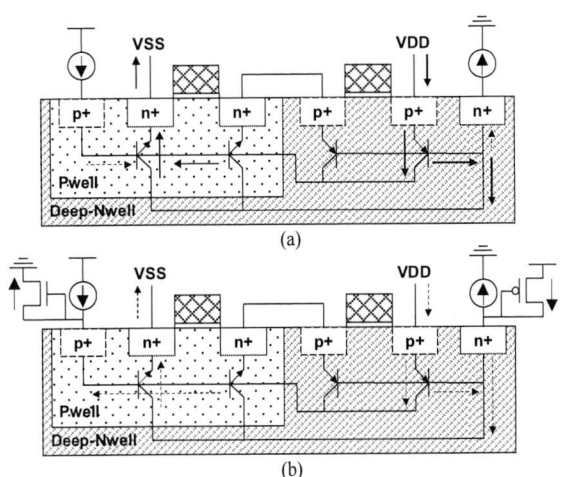

Fig. 1. (a) Latch-up issue using forward body-bias and (b) Proposed technique to realize latch-up free operation

biasing is applied only for the chip at the slow-process corner so that the process variation can be effectively suppressed. A cell-based forward body-biasing circuit can be embedded with small area overhead. This paper is organized as follows. In Section II, we will overview the scheme. In Section III, a design flow to implement the technique will be shown. In Section IV, measurement results are shown. In Section V, conclusions will be drawn.

II. CONCEPT OF THE SCHEME

As-fabricated performance of a chip distributes as the broken line in Fig. 2(a). The solid line shows the performance distribution where forward body-biasing is applied only to the chips at around the slow corner as labeled "conditional forward body-bias", whereas the chips around the fast corner remain the same. With this scheme, corner-to-corner performance variation can be reduced.

While the concept of the scheme is simple, there are some issues, such as (1) where to set the cutoff line between the chips to be biased and the chips to be zero-biased, (2) how to resolve the latch-up issue in a forward biasing scheme above 0.6V operation.

As for the cutoff line, it is defined by the performance of the forward-biased chip originally at the slow corner. As shown in Fig. 2(b), the delay of the chip with forward biasing can be improved from the open circle to the filled circle having the

978-1-4244-2018-6/08 $25.00 © 2008 IEEE

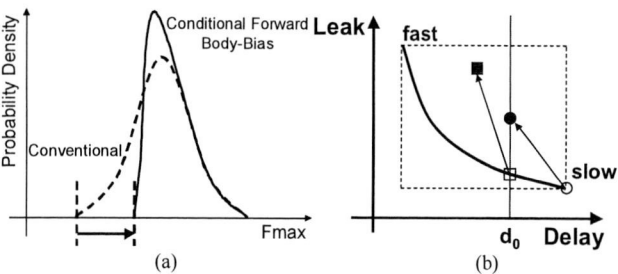

(a)   (b)

Fig. 2. (a) Fmax distribution before and after forward body-bias and (b) Leakage dependence on delay with process variation

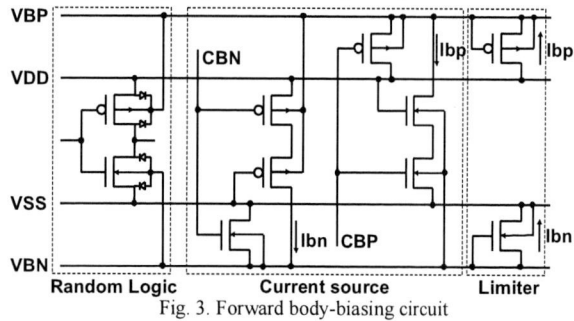

Fig. 3. Forward body-biasing circuit

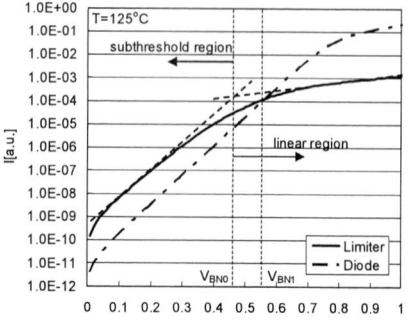

Fig. 4. I-V characteristics of limiter and diode

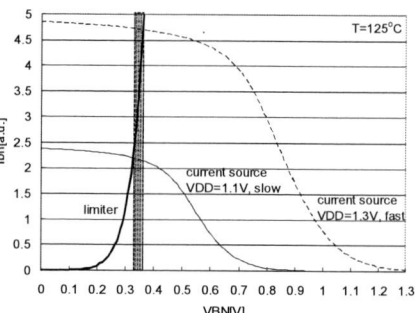

Fig. 5. Load line between current source and limiter

delay of d_0. Therefore, chips having the delay less than d_0 are not required to be forward-biased since the new slow corner is determined by the chip at the filled circle. We have to think about the side-effect, increased leakage by the forward biasing. When we apply forward biasing to the chip having the delay of d_0 without forward biasing, the leakage increases from the open square to the filled square, while the delay decreases as well. The leakage of the filled square includes the substrate biasing current. By choosing the circuit parameters so that the leakage of the filled square does not surpass the leakage of the fast corner, we can reduce the size of the leakage-delay window.

As for the second issue, the mechanism of latch-up is depicted in Fig. 1(a). Forward biasing current source supplies base current of parasitic bipolar transistors. Since h_{FE} of the bipolar transistor is higher than 1 at $V_{DD}>0.6V$, collector current of the bipolar transistor becomes larger than base current. In the circuit shown in Fig. 1(a), the base of NPN is connected to the collector of PNP. Once base current flows, it ends up with latch-up at high V_{DD} through a thyristor configuration.

In order to avoid the situation, we introduce current paths to bypass the base current of PNP/NPN to V_{DD}/V_{SS}, respectively. The current paths are implemented by diode-connected transistors (limiter) as loads of current sources as depicted in Fig. 1(b). If the limiter current is much larger than base current of the parasitic bipolar transistor, collector current of the parasitic bipolar transistor is also limited to much smaller than the limiter current. As a result, the positive feedback loop is not formed and forward body-biasing operation is usable at V_{DD} of higher than 0.6V.

The forward body-biasing circuit consists of a current source and a limiter as shown in Fig. 3. We implement these circuits compatibly with standard logic cells, so that these cells can be placed and routed by commercial CAD tools in

the power grid design stage of the design flow as will be described in Section III. Moreover, these cells are much smaller than the cells used in [8], so that this technique can be applicable to a small design with small area overhead.

Body-biasing voltage is determined by, mostly, the ratio of the number of the current source cells to the number of limiter cells as described in the next subsection. Moreover, body-biasing voltage generated by the limiter cell will be the highest at the slow corner, compensating for the global variation effectively.

A. Design strategy of the number of current source cells and limiter cells

To apply the scheme to a large scale SoC, a top-down strategy to determine the number of current source cells and limiter cells is required.

Once we obtain the gate count to be biased, we can calculate the required number of current source cells and limiter cells as follows. First, we calculate the required number of limiter cells. Fig. 4 shows I-V characteristics of a limiter (solid curve) and a pn-diode (broken curve) at 125°C, which is Tj_{MAX} in this design. As shown in the figure, the operating mode of the limiter is in the subthreshold region below V_{BN0} and in the linear region above V_{BN0}. To minimize the biasing voltage variations, limiters should be in the subthreshold region. Besides, note that the pn-diode current exceeds the limiter current above V_{BN1}, where latch-up can occur. Our design guideline is to make the limiter current one order higher than the pn-diode current at the designated biasing voltage. In this case, our design has the biasing voltage of 0.33V where the limiter current is 10X higher than pn-diode current. Now we can calculate the number of limiter cells since we know the number of gates to be biased, thereby the number of pn-diodes.

978-1-4244-2018-6/08 $25.00 © 2008 IEEE

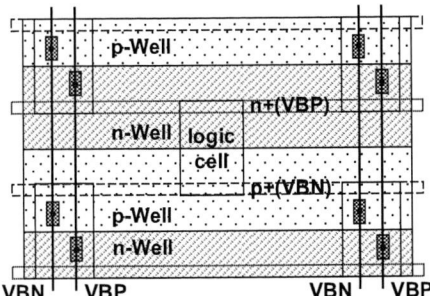

Fig. 6. Connection between body-bias line and wells

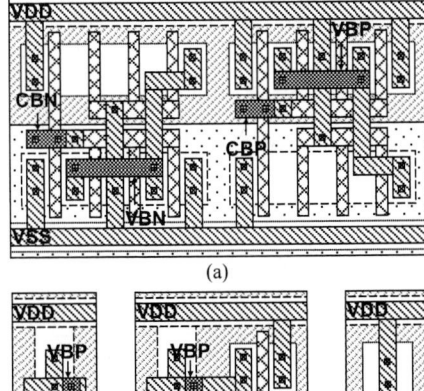

(a)

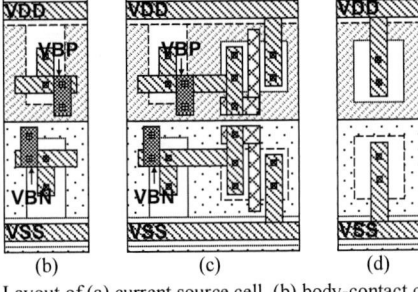

(b)   (c)   (d)

Fig. 7. Layout of (a) current source cell, (b) body-contact cell, (c) body-contact&limiter cell, and (d) filler cell

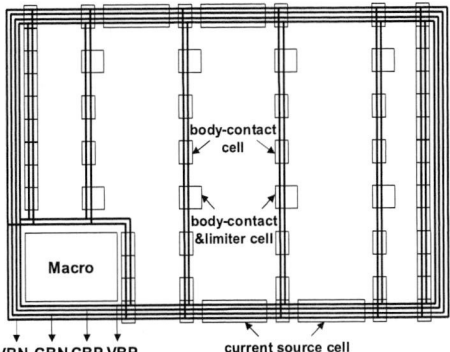

Fig. 8. Body-bias network

Fig. 9. Chip micrograph

Then we calculate the number of current source cells. Fig. 5 shows the I-V characteristics of the limiter cell and that of the current source. As long as the limiter cells are the dominant loads of the current source, the cross-point of two curves is the operating point, that is, the forward biasing voltage. Note that the variation of the operating point in terms of the forward biasing voltage is less than 40mV while the biasing current varies by about 2X as shown in Fig. 5. Since we have already decided the number of the limiter cells, we can calculate the required number of the current source cells by using this graph so that the operating point comes to the designated forward biasing voltage with the designated variation.

Finally, we have to make sure that the sum of substrate current and leakage current per transistor is smaller than the leakage of a transistor at the leakage-worst corner, that is, the fast corner.

B. Distribution of body-biasing current

The body-biasing current is distributed to each well via body-contact cells as shown in Fig. 6. Layouts of the current source cell, the body-contact cell, and the body-contact&limiter cell are shown in Fig. 7(a), (b), and (c), respectively. To distribute the body-biasing current to all the wells in the designated area, body-contact cells are placed

every other row, which makes it easy to place large cells in rows without body-contact cells. The diffusion layer under the power line works as body-biasing network along the row. Since the resistance of the diffusion layer is relatively larger than that of the metal layer, the interval between the body-contact cells along the row should be <50μm to suppress the substrate noise.

Filler cells are placed where logic cells are not placed as conventional cell-based design flow. However, the filler cell is different from the one used in the conventional cell-based flow. The cell has a pn-junction between the power line and the well as shown in Fig. 7(d). Since this pn-junction is also the load of the body-biasing current, the total junction area in a certain chip size can be predictable irrespective of the cell utilization of a design. Moreover, the parasitic capacitance between the power lines and the wells helps reduce the crosstalk noise and the substrate noise where the cell utilization is low.

III. DESIGN FLOW

A design flow for the proposed forward body-biasing scheme is as follows. RTL design goes through a conventional logic synthesis to create the gate-level netlist. In the floor-planning stage, power and ground network is generated and macro cells are placed, and the target size of P&R area is determined. Then, the number of current source cells and limiter cells are determined based on the size of the area. After the floor-planning, the body-contact cells, the

978-1-4244-2018-6/08 $25.00 © 2008 IEEE

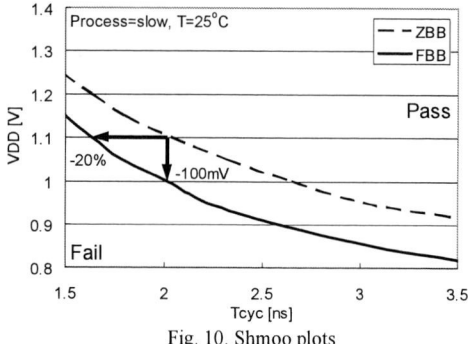

Fig. 10. Shmoo plots

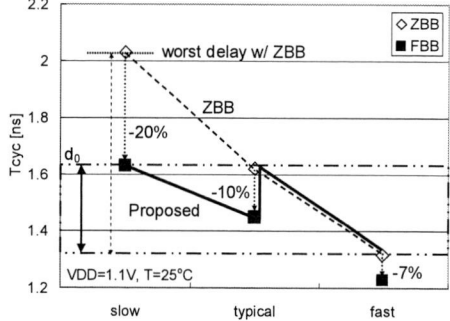

Fig. 11. Tcyc dependence on process variation

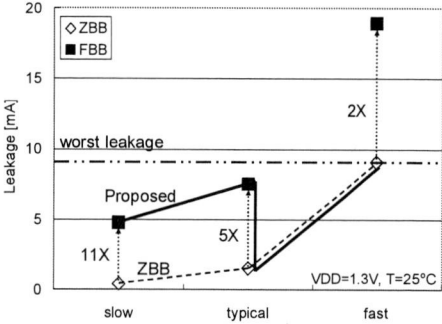

Fig. 12. Leakage dependence on process variation

body-contact&limiter cells, and the current source cells are placed and routed as shown in Fig. 8. The top row and the bottom row are designated to be used for the current source cells. The body-contact cells are placed every other row and at a certain interval to meet the process guideline. After that, standard logic cells and filler cells are placed and routed as conventional designs. The chip micrograph is shown in Fig. 9 [9]. Area penalty is less than 3%.

IV. MEASUREMENT RESULTS

We measured a Media Processing Engine (MPE) in [9] to verify the scheme. The chip was fabricated in 65nm CMOS process. The target biasing voltage at the slow corner is 0.33V at T=125°C, while the biasing voltage at 25°C is 0.38V. Forward body-biasing mode (FBB) and zero body-biasing mode (ZBB) were measured in the same chip.

The achievable cycle time (Tcyc) under different V_{DD}, shmoo plot, is shown in Fig. 10. About 20% cycle time reduction was achieved at V_{DD}=1.1V at the slow corner by this scheme. This effect corresponds to a potential V_{DD} reduction by 100mV, which leads to a lower power operation.

The achievable cycle time and leakage current under different process conditions are shown in Figs. 11 and 12, respectively. These figures show that the cycle time reduction and the leakage current increase by the forward body-biasing become higher as the process condition gets slower. It is because the body-biasing voltage generated by the limiter cell and the body effect factor of the devices are higher at a slower process condition. Note that the leakage at the typical corner with forward biasing is still lower than that at the fast corner without forward biasing. Therefore, by setting the cutoff line of d_0 at the typical corner, the proposed scheme can reduce the worst delay without increasing the worst leakage.

In addition, latch-up did not occur even at V_{DD}=1.3V, T=125°C, which is the most critical condition for latch-up.

V. CONCLUSION

In this paper, a latch-up free forward body-biasing technique with cell-based biasing circuits usable at >0.6V and its design flow are proposed. The technique was employed in a design of Media Processing Engine in 65nm CMOS process. By using forward body-biasing, the delay reduction of 20% is obtained at V_{DD}=1.1V without increasing the worst leakage.

REFERENCES

[1] S. Akui, K. Seno, M. Nakai, T. Meguro, T. Seki, T. Kondo, A. Hashiguchi, H. Kawahara, K. Kumano, and M. Shimura, "Dynamic Voltage and Frequency Management for a Low-Power Embedded Microprocessor," ISSCC Dig. Tech. Papers, pp.64-66, Feb. 2004.

[2] H. Okano, T. Shiota, Y. Kawabe, W. Shibamoto, T. Hashimoto, A. Inoue "Supply Voltage Adjustment Technique for Low Power Consumption and its Application to SOCs with Multiple Threshold Voltage CMOS," Symp. VLSI Circuits, June 2006.

[3] T. Kuroda, T. Fujita, S. Mita, T. Nagamatu, S. Yoshioka, F. Sano, M. Norishima, M. Murota, M. Kako, M. Kinugawa, M. Kakumu, T. Sakurai "A 0.9V 150MHz 10mW 4mm2 2-D Discrete Cosine Transform Core Processor with Variable-Threshold-Voltage Scheme," ISSCC Dig. Tech. Papers, pp.166-167, Feb. 1996.

[4] M. Sumita, S. Sakiyama, M. Kinoshita, Y. Araki, Y. Ikeda, and K. Fukuoka "Mixed Body-Bias Techniques with Fixed Vt and Ids Generation Circuits," ISSCC Dig. Tech. Papers, pp.158-159, Feb. 2004.

[5] S. Narendra, J. Tschanz, J. Hofsheier, B. Bloechel, S. Vangal, Y. Hoskote, S. Tang, D. Somasekhar, A. Keshavarzi, V. Erraguntla, G. Dermer, N. Borkar, S. Borkar, and V. De "Ultra-Low Voltage Circuits and Processor in 180nm to 90nm Technologies with a Swapped-Body Biasing Technique," ISSCC Dig. Tech. Papers, pp.156-157, Feb. 2004.

[6] K. Ishibashi, T. Yamashita, Y. Arima, I. Minematsu, T. Fujimoto "A 9uW 50MHz 32b Adder Using a Self-Adjusted Forward Body Bias in SoCs," ISSCC Dig. Tech. Papers, pp.116-117, Feb. 2003.

[7] Y. Komatsu, K. Ishibashi, M. Yamamoto, T. Tsukada, K. Shimazaki, M. Fukazawa, and M. Nagata "Substrate-Noise and Random-Fluctuations Reduction with Self-Adjusted Forward Body Bias," IEEE Custom Integrated Circuits Conference, pp.35-38, Sept. 2005.

[8] S. Narendra, M. Haycock, V. Govindarajulu, V. Erraguntla, H. Wilson, S. Vangal, A. Pangal, E. Seligman, R. Nair, A. Keshavarzi, B. Bloechel, G. Dermer, R. Mooney, N. Borkar, S. Borkar, and V. De "1.1V 1GHz Communications Router with On-Chip Body Bias in 150nm CMOS," ISSCC Dig. Tech. Papers, pp.270-271, Feb. 2002.

[9] S. Nomura, F. Tachibana, T. Fujita, C. H. Teh, H. Usui, F. Yamane, Y. Miyamoto, C. Kumtornkittikul, H. Hara, T. Yamashita, J. Tanabe, M. Uchiyama, Y. Tsuboi, T. Miyamori, T. Kitahara, H. Sato, Y. Homma, S. Matsumoto, K. Seki, Y. Watanabe, M. Hamada, and M. Takahashi "A 9.7mW AAC-Decoding, 8-Core Media Processor with Embedded Forward-Body-Biasing and Power-Gating Circuit in 65nm CMOS Technology," ISSCC Dig. Tech. Papers, pp.262-263, Feb. 2008.

978-1-4244-2018-6/08 $25.00 © 2008 IEEE

Built-in Resistance Compensation (BRC) Technique for Fast Charging Li-Ion Battery Charger

Chia-Hsiang Lin[+], Hong-Wei Huang[*], and Ke-Horng Chen[+]

[+]Department of Electrical and Control Engineering National Chiao Tung University, Hsinchu, Taiwan
[*]Graduate Institute of Electronics Engineering National Taiwan University, Taipei, Taiwan

Abstract —This paper presents a BRC technique to speed up the charging time of the Li-Ion battery. Based on the physical properties of the battery cell, the charger circuit charges the cell with three stages, which are trickle current (TC), constant current (CC), and constant voltage (CV) stages. Due to the internal parasitic resistance of the Li-Ion battery pack system, the charger circuit switches from the CC stage to the CV stage without fully charging the cell to the rated voltage value. At the CV stage, the cell is needed to be charged by a degrading current. The longer the cell stays at the CV stage, the longer the charging time is due to the degrading current. The BRC technique can dynamically estimate the internal resistance of the battery pack system to extend the period of the CC stage. The test chip was fabricated in TSMC 0.35-μm process. Experimental results show the period of the CC stage can extended to about 40% that of the original design. The charging time can be effectively reduced.

Index Terms— Li-Ion cell, charger, fast-charging.

I. INTRODUCTION

Nowadays, the portable devices have become the main applications of advanced technical products. Due to small-size, light-weight, and rechargeable characteristics, Li-Ion batteries are well suited to portable electronic applications. Besides, Li-Ion batteries can store more energy than that of Ni-Cd batteries with the same weight and volume. However, the life cycles of Li-Ion batteries are easily affected by undercharge or overcharge. The reason is that overcharge may damage the physical component of the battery and undercharge may reduce the energy capacity of the battery. Thus, in order to prevent the battery from being overcharging, the charging process needs to switch from CC stage to CV stage in order to charge the battery by a degrading current. However, the CV stage takes a long time to fully charge the battery. The long charging time is the most severe drawback during charging the Li-Ion battery.

It is not a correct concept that a large charging current can speed up the charging time of the Li-Ion battery. A large charging current may cause a large IR-drop voltage of the external resistance of battery pack. Then, the charger detects the transition point of the CV stage. The charger switches the charging process from CC stage to CV stage more early than that by a small charging current. However, the voltage at the cell is not fully charged to the level that is ready to enter the CV stage. In other words, it takes a long time to charge the battery to the rated voltage at the CV

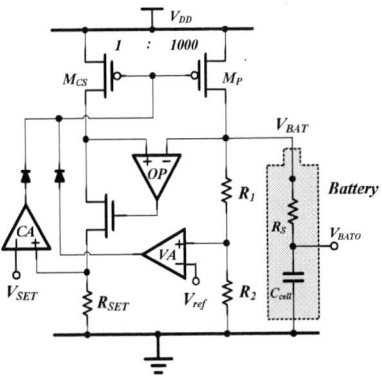

Fig. 1. The structure of the basic charger circuit.

stage when using a large charging current. In previous design of the charger [1], it used the external components to compensate the IR-drop voltage. The disadvantage is that the module of the charger is too large to be compact for portable devices. The built-in resistance compensation technique is proposed in this paper to make the module of the charger be compact and prolong the period of the CC stage. Owing to the longer time at the stage of CC stage, the cell can be charged to more closely approach to the rated voltage. In other words, the period of the CV stage can be reduced. Hence, the proposed BRC technique can shorten the charging time of the battery.

II. THE FUNDAMENTAL CHARGING PROCESS

Generally speaking, the charging process of Li-Ion battery charger is divided into three operation stages, i.e. TC, CC, and CV stage [1, 2]. The simplified diagram of Li-Ion battery charger is shown in Fig. 1 [3]. At the beginning, the charger detects the initial voltage of the battery V_{BAT}. If voltage V_{BAT} is smaller than the specified voltage (normally 2.5V), the charger operates at the TC stage. Once $V_{BAT} >$ 2.5V, the operation stage is switched to the CC stage.

At the TC stage, the battery is charged by a trickle current. The current is decided by an external resistor R_{SET}. It protects the battery from being damaged by using a charging current under low battery voltage. When the battery is charged to an adequate voltage, the charging process is needed to be switched to the CC stage automatically. The charging current at the CC stage is larger than that at TC or CV stage. It means much energy is rapidly stored in battery during CC stage. When the

978-1-4244-2018-6/08 $25.00 © 2008 IEEE

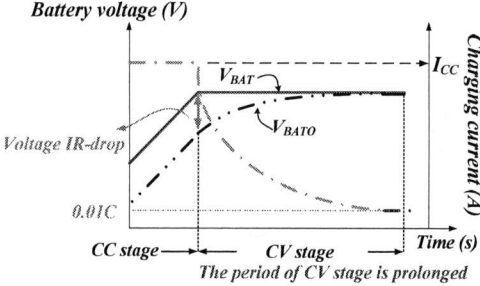

Fig. 2. The waveform of the battery voltage.

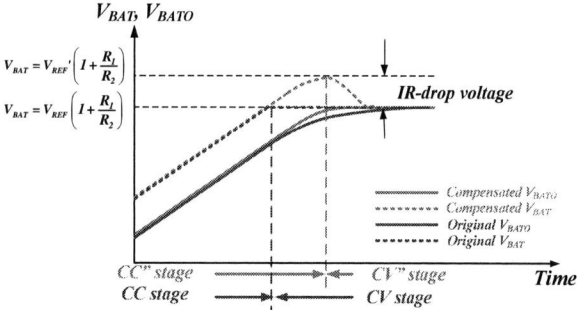

Fig. 3. The improved waveform at the end of charging process.

battery voltage reaches a rated voltage which is close to 4.2V, the charging process is switched to CV mode. The charger keeps the battery at a regulated voltage until the charging time is over. The charging path that is composed of the power MOSFET M_{CS} and R_{SET} is shutdown and the voltage of V_{SET} is gradually decreased to zero.

However, there is a critical problem related to the long charging time. The battery pack system includes the Li-Ion cell, the protection circuits, and the internal resistance (normally about 150mΩ to 300mΩ). The value of the external voltage V_{BAT} is not equal to the value of the voltage V_{BATO} at the Li-Ion cell. There is a IR-drop voltage between the external voltage V_{BAT} and the voltage V_{BATO} at the Li-Ion cell. In order to reach the full charge of battery, the charger needs to extend the charging time at the CV stage. At the CV stage, the charger charges the battery with a degrading current until the voltage V_{BATO} at the Li-Ion cell reach the full-scale voltage. Since the current is degrading, the charging time is very long and thus prolongs the whole charging process. In order to shorten the charging time, the compensation of the IR-drop voltage is a critical problem needs to be solved. The circuit of detecting the internal resistance of battery pack system is proposed to implement a fast charging technique.

III. THE PROPOSED BRC TECHNIQUE

When the voltage V_{BAT} of the battery pack system reaches the rated voltage (normally 4.1 or 4.2V), the voltage V_{BATO} at the Li-Ion cell is smaller than 4.2V. Because the internal resistance results in an IR-drop voltage, the energy of cell is not fully charged. Fig. 2 shows the degraded charging current, the voltage of the battery pack system V_{BAT}, and the voltage V_{BATO} at the Li-Ion cell when the charging process is

switched from CC to CV stage. The IR-drop voltage lengthens the period of the CV stage and thus prolongs the charging time for getting sufficient energy stored at the Li-Ion cell.

To solve the problem, a BRC technique is applied in the charging process. Referring to Fig. 3, since the difference voltage between the voltage V_{BATO} at the Li-Ion cell and the voltage of the battery V_{BAT} is the IR-drop voltage, it implies that the reference voltage can be shifted to a higher level to re-define the final voltage level of V_{BATO}. As a result, the voltage V_{BATO} at the Li-Ion cell can be closely approached to 4.2V at the end of the CC stage. In Fig. 3, the period of the original CC stage is extended to a long period named as CC'' stage after using the BRC technique. The increment voltage V_{INC} at the node of the reference voltage V_{REF} is defined as:

$$V_{INC} = IR_{drop} \times \frac{R_2}{R_1 + R_2} \qquad (1)$$

The increment voltage V_{INC} will be added to the reference voltage V_{REF} to generate a new reference voltage V_{REF}'. Thus, the voltage of the battery pack system is charged to a higher voltage level. The Li-Ion cell can be stored much more energy at the CC'' stage for a longer time than that at the CC stage of the original design. Afterward the reference voltage at the CV stage is decreased back to its original value gradually; the battery can be charged to the designated voltage when the charging process is finished without the overcharging problem.

Before the charger being switched to the CV stage, the voltage V_{BATO} at the Li-Ion cell is still smaller than the specified voltage due to the IR-drop voltage. Based on the fact that charger charges the battery cell tardily when the battery voltage V_{BAT} is close to 4.2V, the voltage V_{BATO} of the Li-ion cell can viewed as a constant during the short test time. The charging current changes from I_{CHRG1} to I_{CHRG2} and causes a voltage difference at the battery voltage V_{BAT} owing to the voltage drop across the internal resistance R_S. Two battery voltages V_1 and V_2 can be written as (2) and (3) for the two different charging current I_{CHRG1} and I_{CHRG2}, respectively.

$$V_1 = I_{CHRG1} \times R_S + V_{BATO1} \qquad (2)$$

$$V_2 = I_{CHRG2} \times R_S + V_{BATO2} \qquad (3)$$

According to the previous assumption that V_{BATO1} and V_{BATO2} are the same within a small charging time, the internal resistance R_S can be estimated by (4).

$$R_S \approx \frac{V_1 - V_2}{I_{CHRG1} - I_{CHRG2}} \quad \text{when } V_{BATO1} \approx V_{BATO2} \qquad (4)$$

Therefore, if the values of constant currents I_{CHRG1}, I_{CHRG2}, and V_1 are pre-defined, the value of R_S can be determined by estimating the value of battery voltage V_2.

The proposed charger with BRC technique is shown in Fig. 4. The compensation circuit is composed of the internal resistance detector, the reference voltage switch, and the reference shift circuit. The voltages V_{r1} and V_{r3} are used to

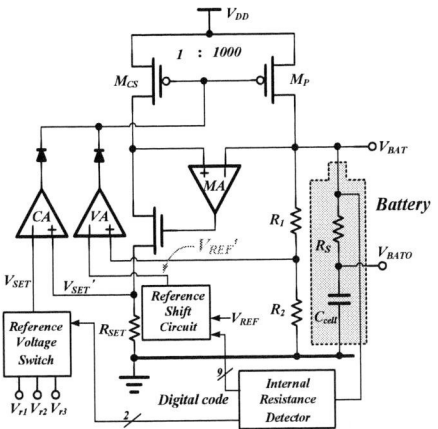

Fig. 4. The proposed fast-charging charger with the internal resistance estimation.

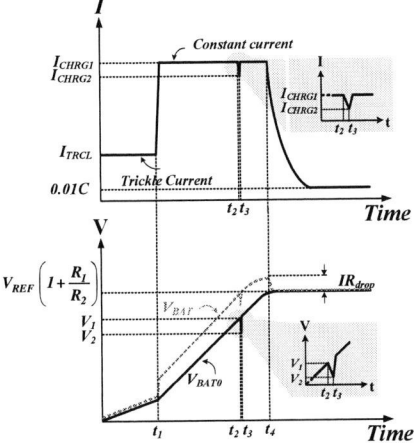

Fig. 5. The waveform of battery voltage during detecting

decide the transitions from TC to CC stage and from CC to CV stage, respectively. A new rated voltage V_{r2} is used to decide another charging current at the CC stage for the internal resistance estimation. The values of the two constant charging currents I_{CHRG1} and I_{CHRG2} are determined by

$$I_{CHRG1} = \frac{V_{r3}}{R_{SET}} \text{ and } I_{CHRG2} = \frac{V_{r2}}{R_{SET}} \quad (5)$$

During the estimation of the internal resistance, the value of V_{SET} is set from V_{r3} to V_{r2} at time (t_2) and back to V_{r3} at time (t_3). The variation of the value of the voltage V_{SET} will cause a voltage drop at the battery voltage V_{BAT}, which is shown in Fig. 5.

By using a sample and hold circuit to reserve the value of V_2 as shown in Fig. 6 (a), the value of R_S can be derived from (4). The internal resistance of the Li-Ion battery may vary from 150 mΩ to 300mΩ, and the IR-drop voltage is defined as:

$$IR_{drop} = R_S \times I_{CHRG} = R_S \times \left(1000 \times \frac{V_{SET}}{R_{SET}} \right) \quad (6)$$

$$= 1000 \times R_S \times I_{DROP}$$

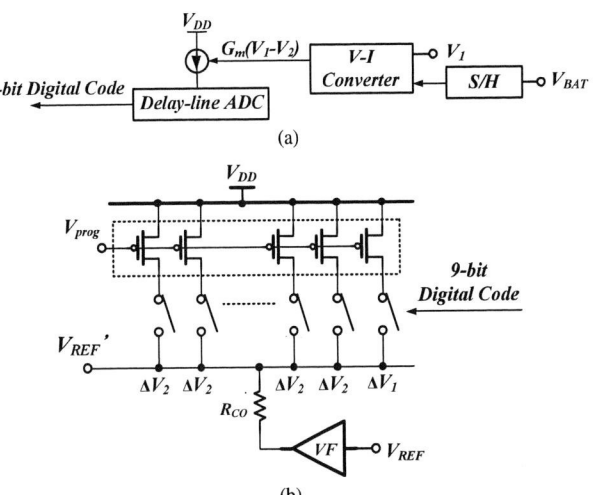

Fig. 6. (a) The internal resistance detector. (b) The reference shift circuit generates the BRC reference voltage V_{REF}' by adding the increment voltage V_{INC} to the original reference voltage V_{REF}.

The V-I converter then transfers the IR-drop voltage to a current I_{DROP}, which is determined by V_{SET}/R_{SET}, to bias the temperature-independence delay-line ADC. Therefore, the delay-line ADC can output a different digital code for representing the internal resistance according to the IR-drop voltage [4]. As we know, the delay time is inversely proportional to the IR-drop voltage. That is a large voltage IR-drop voltage can interpreted as a big digital number. The reference shift circuit will increase the reference V_{REF} to V_{REF}' according to the digital code.

After the estimation the internal resistance of the Li-Ion battery pack system, a nine-bit digital code generated by the internal resistance detector is sent to the reference shift circuit to add the increment voltage V_{INC} to the V_{REF} by the decision of the digital number, which is converted by the delay-line ADC in Fig. 6 (b). The incremental voltages are added to the reference voltage are ΔV_1 and ΔV_2. The voltage drop across the resistor R_{CO} is equal to the voltage of V_{INC}. Thus, the reference shift circuit can generate a reference voltage V_{REF}' to the original charger to compensate the internal resistance of the battery pack system. That is the period of the CC stage can be extended to charge much more energy to the voltage V_{BATO} at the Li-Ion cell.

IV. EXPERIMENTAL RESULTS

The proposed charger with the BRC technique was implemented in TSMC 0.35-µm CMOS technology. The chip micrograph is shown in Fig. 7. The silicon area is 1300×850µm². The set-up of the test chip is shown in Fig. 8. The pin of EN_BRC can be set to low to turn off the BRC technique. The supply voltage of charger is 5.5V and the fully charged voltage is 4.2V. In order to emulate the internal resistance of the battery pack system. One resistor R_s of 300 mΩ is used to stand for the internal resistance. Furthermore, in order to monitor the charging process within a short period. One large capacitor C_{CELL} of 10 mF is

978-1-4244-2018-6/08 $25.00 © 2008 IEEE

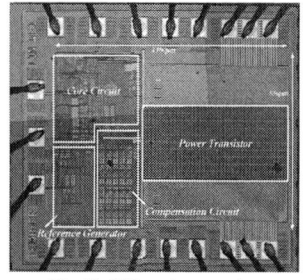

Fig. 7. Chip micrograph.

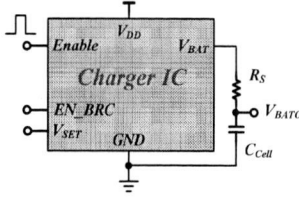

Fig. 8. The set-up of the proposed charger.

used to emulate the large capacity of the Li-Ion battery.

Since the output capacitor is 10mF, it is easy to find out the charging time of the TC stage is 500 ms. In Fig. 9, the estimated result is similar to that got from the calculation. Furthermore, the period of the CC and CC" stages are 22ms and 32ms, respectively. The period of the CC stage is extended to about 40% that of the original charger without the BRC technique.

Due to the internal resistance is 300 mΩ, the digital number of the detector is 111111111. Then, the reference-shift circuit adds the max shift-voltage to the reference voltage V_{REF} for compensating the IR-drop voltage. Thus, the voltage V_{BATO} at the Li-Ion cell can reach the specific voltage (4.2V) more quickly than that of the original design without the BRC technique. From the zoom-in window in Fig. 9, the voltage V_{BATO} can be rapidly raised to 4.2V due to the extension of the CC stage.

In Fig. 10, The IR-drop voltage exists between the voltages V_{BATO} and V_{BAT}. The estimated values of the original V_{BAT} and the compensated V_{BAT} are 4.14V and 4.26V, respectively. The period of the constant current stage can be extended from the CC stage to the CC" stage since the value of the compensated V_{BAT} is larger than that of the original V_{BAT}. Since the voltage variation at the voltage V_{BAT} is very small, the node voltage V_{SET} is connected to output to monitor the detection period in Fig. 8. The estimated waveform of the voltage V_{SET} is shown in Fig. 11. During the detection period, the value of the voltage V_{SET} is pulled low to 0.9V for a short time and back to 1.5V. The detection period is 0.5ms in Fig. 11. During the detection period, the charger acquires the sufficient information of the internal resistance of the battery pack system to compensate the IR-drop voltage.

V. CONCLUSIONS

The BRC technique is proposed to speed up the charging time of the Li-Ion battery. Due to the internal parasitic resistance of the Li-Ion battery pack system, the charger circuit switches from the CC stage to the CV stage without fully charging the cell to the rated voltage value. The longer the charging time is due to the degrading current at the CV stage. The BRC technique can dynamically estimate the internal resistance of the battery pack system to extend the period of the CC stage for achieving a fast charging response. Experimental results show the period of the CC stage can extended to about 40% that of the original design.

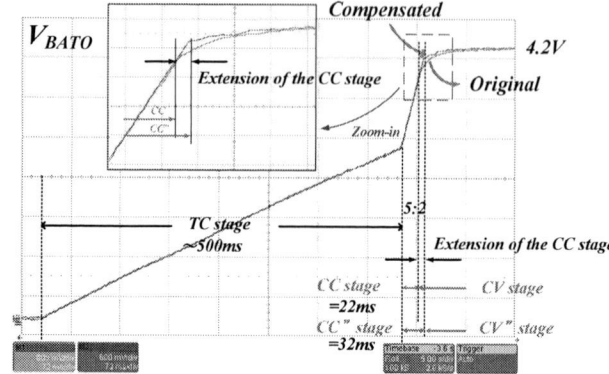

Fig. 9. The waveforms of the voltage V_{BATO} w/i and w/o the BRC technique. The CC stage of the original design is extended to the CC" stage of the BRC design.

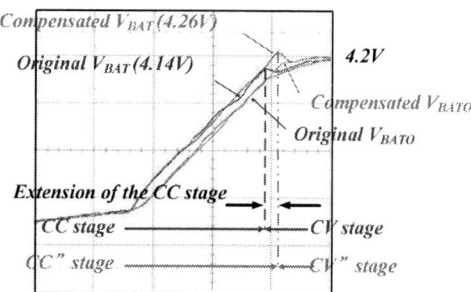

Fig. 10. The waveforms of the voltages V_{BAT} and V_{BATO} w/i and w/o the BRC technique. The voltages of the compensated V_{BAT} and V_{BATO} are got from the BRC technique.

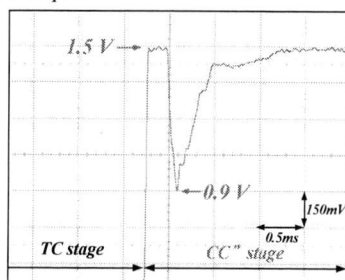

Fig. 11. The waveform of the voltage V_{SET} during the detection period.

The charging time can be effectively reduced.

REFERENCE

[1] Roland Saint-Pierre, "A Dynamic Voltage-Compensation Technique for Reducing Charge Time in Lithium-Ion Batteries," *The Fifteenth Annual Battery Conference on Applications and Advances*, pp. 179-184, Jan. 2000.

[2] Chia-Chun Tsai, Chin-Yen Lin, Yuh-Shyan Hwang, Wen-Ta Lee and Trong- Yen Lee, "A multi-mode LDO-based Li-ion battery charger in 0.35μm CMOS Technology," *IEEE Asia-Pacific Conference on Circuits and Systems*, pp. 49-52, Dec. 2004.

[3] Yuh-Shyan Hwang, Shu-Chen Wang, Fong-Cheng Yang, Jiann-Jong Chen, "New Compact CMOS Li-Ion Battery Charger Using Charge-Pump Techniques for Portable Applications," *IEEE Trans. Circuits and Systems I*, vol. 54, no. 4, pp. 705-712, April 2007.

[4] Min Chen, Gabriel A. Rincón-Mora, "Accurate, Compact, and Power-Efficient Li-Ion Battery Charger Circuit," *IEEE Trans. Circuits and Systems II*, vol. 53, no. 11, pp. 1180-1184, Nov. 2006.

[5] Hong-Wei Huang, Ke-Horng Chen, and Sy-Yen Kuo, "Dithering Skip Modulation, Width and Dead Time Controllers in Highly Efficient DC-DC Converters for System-on-chip Applications," *in IEEE Journal of Solid-State Circuits*, pp. 2451-2465, Nov. 2007.

978-1-4244-2018-6/08 $25.00 © 2008 IEEE

High-Speed Test, Characterization, and Debug

Session 4

Chair: Mike Li, Altera
Co-Chair: Gordon Roberts, McGill University

Challenges associated with high speed I/O including cross-talk, channel characterization, ESD protection, and debug will be addressed in this session. Crosstalk affects circuit performance with three key variables: aggressor noise injection timing, aggressor location and direction, and voltage drop. The first paper provides an in-situ crosstalk delay measurement circuit and results on 65nm CMOS. As I/O data rates move to 10 Gb/s, the challenge of channel characterization is to understand the effect of die, packaging and board parasitics and their impact on signal integrity. This paper provides a test characterization methodology for separating out these effects for modeling. The benefits of inductive CDM ESD protection in 45nm planar CMOS technology is discussed in the third paper for an RF circuit operated at 13GHz. The session concludes with a tutorial overview of non-destructive optical beam failure analysis techniques for defect localization. Increasing IC complexities resulting from dense multi-level metallization and flip-chip packaging have left only the backside of the IC available for interrogation and which further complicates failure and yield analysis, especially in the presence of soft defects.

978-1-4244-2018-6/08 $25.00 © 2008 IEEE

Notes

A Voltage Drop Aware Crosstalk Measurement with Multi-Aggressors in 65nm Process

Genichi Tanaka, Kan Takeuchi, Minoru Ito*), and Hiroaki Matsushita

Renesas Technology Corp., Itami, Hyogo, Japan

*) Hitachi Information & Communication Engineering, Ltd Kodaira, Tokyo, Japan

Abstract-An efficient crosstalk delay degradation measurement method with a 65nm process is proposed. The voltage drop impact on the crosstalk delay is measured. The test module incorporates filters which omit glitches high speed complicated circuits unintentionally create. The module consists of standard cells only, that makes designing very easy. An intensive comparison of measured results with simulations for 64 x 216 patterns of six aggressor activations (timing and combinations) shows precise matching with less than 10% errors.

I. INTRODUCTION

As process technology has advanced and sizes of devices and interconnects have been scaled down, device behavior becomes very sensitive to signal integrity issues like crosstalk and voltage drop. The measurements by using test structures are very effective for understanding and modeling the phenomena quantitatively. The dynamic voltage drop can be measured by direct voltage measurement with probing target points or indirect methods through delay. Recently, many methods by using on-chip monitor of voltage drop have been reported. However, only few have been reported for crosstalk delay measurements [1-4]. This is partly because there are too many variables to be considered in order to make the correlation between measurement results and simulations reliable. The variables include noise injection time, noise direction, and combination of multi-aggressor noise injection. In addition, under the real product circumstances, the victim net may suffer from voltage drop at the same time. In the past, crosstalk delay has been measured under the condition of some limited typical noise injection patterns of aggressors and ideal power supply environment, although the importance of multi-aggressors has been pointed out [5]. At the stand points of reduction of measurement time, optimization methods of crosstalk test pattern are reported [6,7].

In this paper, crosstalk delay under the circumstances probable in the products is investigated. As many as 64 x 216 patterns of crosstalk delay by six aggressors are measured by using a fast method, and the results are modeled. The glitch filter circuit is incorporated in the test module for properly evaluating the delay of rippling signals, which are caused by the synchronized activation of multi-aggressors. The combinational effects of both crosstalk delay and voltage drop are also studied by implementing less rigid power rail structures intentionally. In section 2 we briefly describe our crosstalk measurement test structure. In section 3, measurement results and simulations are compared.

II. MEASUREMENT ARCHITECTURE

A. Circuit Implementation

Delay degradation caused by crosstalk depends on several factors which are 1) multi-aggressors' noise injection time, 2) noise direction (rise/fall), 3) noise injection coupling location, and 4) voltage drop. Fig.1 shows a simplified circuit diagram. To capture those four factors described above, the test circuits mainly consist of following parts.

- Noise injection unit: crosstalk injection to a victim by six aggressors

- Aggressor injection timing control unit: consists of total 6 coarse and fine timing shift units which control the sixth power of two (equal to 64) variations of the noise injection timing of aggressors.

- Aggressor direction select unit: selects transient direction among rise, fall or quiescent for each aggressor individually

- Ring oscillator unit: captures delay degradation by checking frequency shift.

- Counter unit: checks ring oscillator frequency

- Glitch filter: eliminates unintended glitches

For voltage dependency measurement, rigid power rail areas and less rigid power rail areas were implemented. In less rigid power

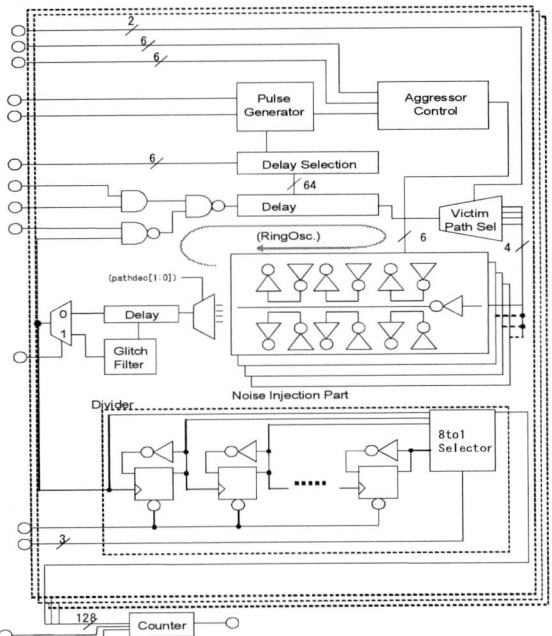

Fig. 1 Circuit Diagram

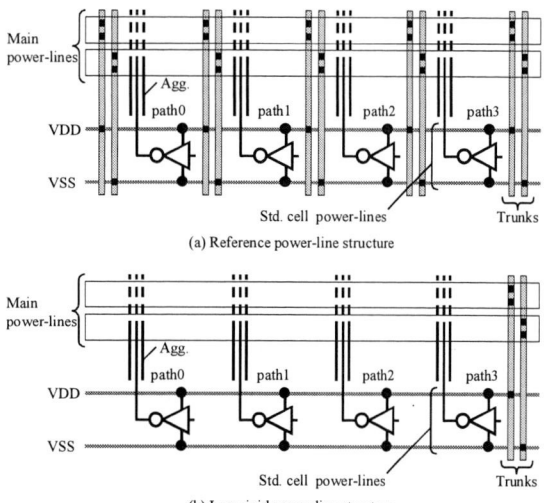

(a) Reference power-line structure

(b) Less rigid power-line structure

Fig. 2 Physical Structure of Power Supply Lines

rail area, one-handed dangling standard cell row rails from a power trunk supply power to circuits as illustrated in Fig. 2.

Four circuit modules were implemented in each row in a less rigid power rail area. Voltage drop is gradually occurred in a row from trunks to the end of the rows. For reference in rigid area rigid power and ground stripes are assigned in both side of each circuit block in Fig. 2 as well. With these structures this circuit can capture voltage drop dependency for crosstalk delay degradation. Drivability variation of aggressor and victim cells are implemented as well. The combination of drivability of 1x, 2x, 4x and 8x are designed.

In addition another advantage of this measurement circuit structure is that the circuit is implemented by standard cells only. Not to need any special analog cells or hard macros is very easy to be designed and implemented.

Figure 3 shows a microphotograph of the fabricated chip and layout of 4-path ring oscillator unit. The crosstalk test module is shared with other two modules in the chip, and has 20 blocks of the 4-path ring oscillator units (Fig.1).

For crosstalk measurement two contradictory factors should be considered. One is a wide range of aggressor noise injection

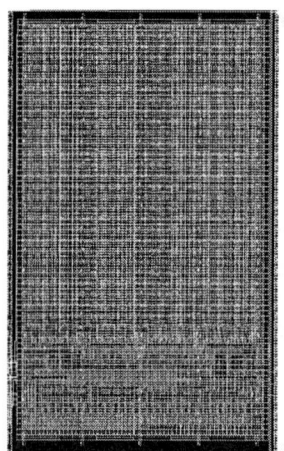

20 blocks of 4-path ring oscillator units are implemented.

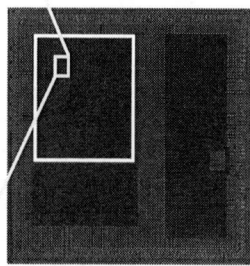

Fig. 3 Chip microphotograph

timing/direction combination and the other is a short measurement time. We have achieved as many as 64 (timing)x 216 (direction combination) measurements for one path of the ring oscillator unit.

Previous work [1] measured crosstalk delay with a frequency monitor but measurement samples were very limited because it used an oscilloscope for the frequency measurement and stabilization time of waveform is needed.

To solve the measurement time issue, we implemented the counter for ring oscillator frequency measurement. This counter enables to count the number of waveform directly and enables to shorten measurement time dramatically by one thousandth or less.

B. Data Reliability and Accuracy

To calculate crosstalk induced delay by a ring oscillator, the frequency under the situation in which all aggressors are quiescent as a reference. During crosstalk measurement the exact same circuit is used, that any environmental disturbance is eliminated as much as possible. In addition to elimination of environmental errors, to collect reliable measurement data for counting the ring oscillator waveform, relatively long measurement time is required. The counter can count, of course, integer only. So one in total count number is resolution. To achieve accuracy less than 0.05% the number of counts should be around 2000 at minimum. To get enough counting information, random fluctuation errors become negligible and precise results can be obtained. This is because the variance of fluctuation is significantly reduced in proportional to the number of measurement points. The measurement time length can be defined by trade off between measurement time and data reliability. Even if a minimum 2000 count is used, much less time is required in comparison with monitoring with oscilloscope measurement. Thus little environmental issues nor rounding issues are existing in this test module.

In another aspect process variations could be an issue for data accuracy. Fig. 4 shows the variations of the R.O. delta-delay by crosstalk among chips on the same wafer. X axis is a relative delay degradation and y axis is sample sorted by normalized standard deviation. In this particular case, -3.5% degradation is average and distribution from -3.2% to -3.7% can be observed in a plus minas 2 sigma range. The median values have been used for the correlation analysis with simulations.

With introducing measurement by counter, a new concern is raised up. Glitches in waveform may cause wrong counting. Typical glitches caused by crosstalk are shown in the top column in Fig.5. The output of the cell victim net drives can be the second column waveform in Fig.5. As easily understood with this

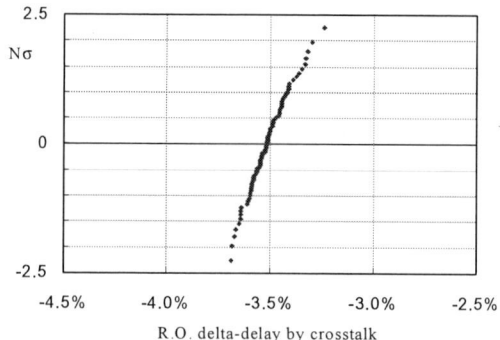

R.O. delta-delay by crosstalk

Fig.4 R.O. Delta Delay Variation among Chip

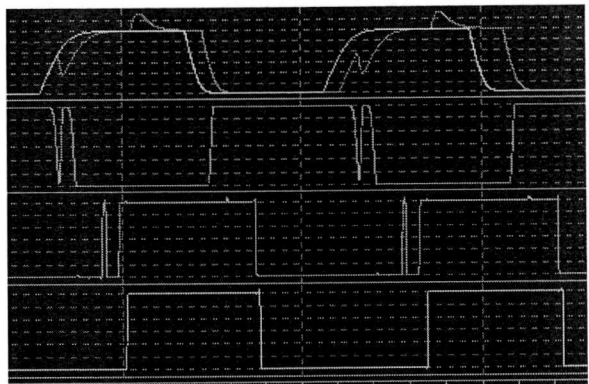

Fig. 5 Waveform of Noise Elimination

waveform diagram, the glitch does not matter to a normal frequency waveform monitor because the frequency itself is still correct. The counter, however, recognizes the glitches as normal transition, that means over-counting. To eliminate these glitches, glitch filter has been introduced and after noise cancellation waveform has no glitches as shown in the bottom in Fig.5.

III. MEASUREMENT RESULTS

Many varieties of comparisons can be captured with the measurements. The dependency of 1) aggressor noise injection timing, 2) aggressor location and direction, and 3) voltage drop are presented in this section.

A. Timing/Location Dependency

What aggressor noise injection timing against victim net affects crosstalk is well known. Figure 6 shows an example of delay degradation with the effective aggressor number of four (one rising and five falling aggressors). Three pairs of aggressors are assigned at the location of front, middle, and rear side of victim net. Aggressors are rise + fall, fall x 2, and fall x 2, respectively. Note that the state of the victim is rise. The crosstalk induced delay dissipation at around unit 52 of aggressor timing is catastrophic. This is because injected noise does not affect to delay any more after injected noise bump does not across the delay threshold. This measurement result confirms that the glitch filter is working properly. Without the filter circuit unintended delay increasing jump can be occurred. For location and direction dependence with 1st order approximation linear dependency can be considered. Physically saying, it should be non-linear function

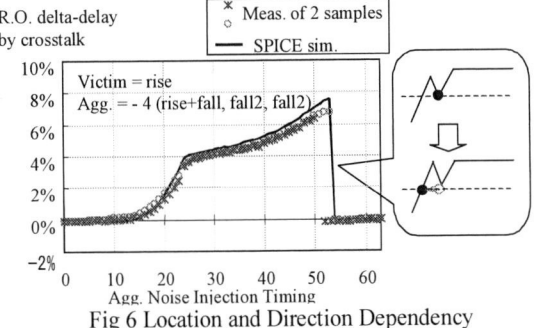

Fig 6 Location and Direction Dependency

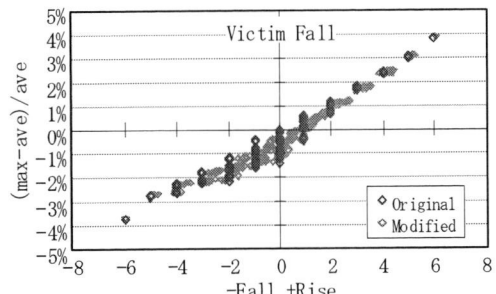

Fig 7 Location and Direction Dependency

Table1 Location Dependency Factor Results

	Front	Middle	Rear
Rise	0.8324	0.8851	0.9431
Fall	1.0205	1.1088	1.2092

because of holding resistance non linearity and many other physical non linear location dependence for crosstalk should be considered [10]. Fig 7 shows how much linearity the crosstalk induced delay shift has against the number of aggressors. X axis is (the number of rising aggressor) minus (the number of falling aggressors). Y axis is delay shift. Blue rhombus is raw data. It is relatively linear but even in the same column some distribution can be found. This is because the location of aggressors is different. To capture location dependency, the least square method with three parameters as multiplier for the number of aggressor is used. The result of the fitting factor for aggressor location is shown in table 1. To apply this location dependency to the data, modified results are shown in figure 7 with red rhombus. Much more linearity can be observed. The error against linear function is 0.01% in average and standard deviation is 0.02%. For location dependency, the aggressors in the middle are 6% to 9% more effective and they in the rear are 13% to 18% more effective than they in the front. The difference of the factor between rise and fall comes from drivability difference between n-mos and p-mos, which is 25% difference in average. We measured crosstalk location dependency individually to check if the table 1 result comes from location dependency only. To measure location dependency crosstalk noise is injected by only an aggressor located in the front, middle or rear location.

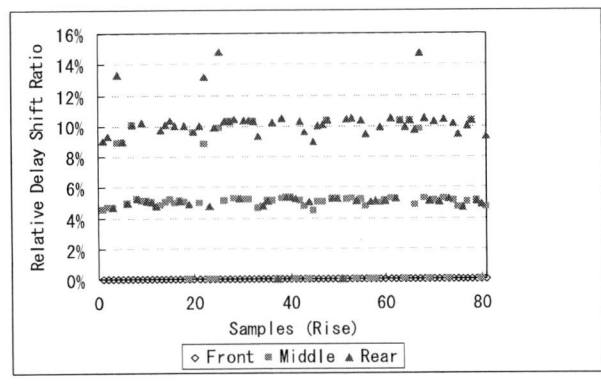

Fig. 8 Location Dependency Results

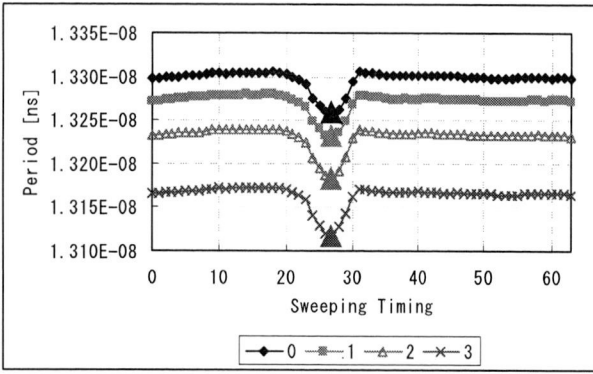

Fig 9 Comparison between measurement and simulation

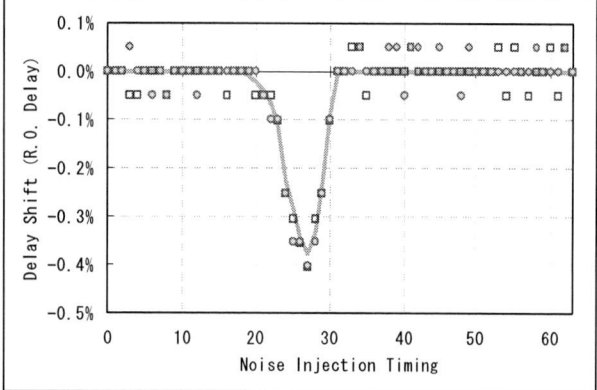

Fig 10 Relative Delay Shift for Less Rigid Nets

The result is shown in Fig.8 and it indicates the direct evidence of location dependency for crosstalk delay. Delay shift relative value of middle and rear locations are plotted with the delay shift at front location as a reference. The absolute number is slightly different from the fitting results but the order is well matched. The reason of the difference comes from multi-aggressor effect which exists in fig.7 only.

B. Voltage Drop Dependency

To capture voltage drop dependency against crosstalk induced delay we implemented four paths which are different from voltage drop as shown in Fig.2 (b). Path0 gets the least rigid power supply. The more the path number increases, the more rigid power is provided from standard cell power rails.

In Fig.9 relative delay shift difference among four paths from path0 to path3 is illustrated. X-axis is noise injection timing and y-axis is oscillator period. Solid lines are simulated results and rectangles are measurement results. The top line of 0 means the result from path0 and so on. Path0 means the cells are affected by voltage drop and the delay becomes slower. Voltage drop makes ring oscillator frequency slower even if crosstalk does not apply to the nets. So the periods even at unit 0 are different among paths.

For rigid power line structure cases, power supplies enough to any paths evenly and little crosstalk delay difference among paths can be found. On the other hand, for less rigid power rail examples, the extra crosstalk delay shift caused by voltage drop can be observed. Regardless of the robustness of power-line structures, the difference between the measured results and the simulation result is far less than 10 %.

Base voltage drop effect for ring oscillator is about 1%. The reason this very small number is occurred is that the occupancy percentage of the cells with less rigid power line is less than 2% in the ring oscillator path because ring oscillator consists of more than fifty inverters and crosstalk affects to the only one cell. What even those relative small effects can be detected means this crosstalk delay shift detection module works very well.

A relative delay degradation difference of least rigid power line path0 against most rigid power line path3 is illustrated in Fig. 10. Y axis is the difference normalized by path3 result. Rectangular points are calculated from measured results and pink solid line is simulated results. Rectangle points of green and blue are different chip results. The figure shows measurement results precisely match simulation results at any sweeping time even for voltage drop is occurred within the resolution level of measurement.

IV.   CONCLUSION

A voltage drop aware multi-aggressors crosstalk in-situ measurement has been proposed. The module using only standard cells is implemented with the 65nm test module. It implemented the on-chip counter for a tremendous number of measurement point and the glitch filter for data reliability. The dependence of crosstalk delay on 1) aggressor noise injection timing, 2) aggressor location and direction, and 3) voltage drop has been measured. Simulation and measurement results are precisely matched with much less than 10% errors even for voltage drop dependency measurement, which is measured for the first time in the area.

ACKNOWLEDGMENT

The authors would like to thank Y. Kanno, T. Watanabe, N.Ikehata, J. Mano, K. Nakashiro, Y. Uchimura, Y. T.Sato, Y. Katsuki, K. Yoshizumi and S. Hamasaki for their valuable discussion.

REFERENCES

[1] T. Sato et al., "Accurate In Situ Measurement of Peak Noise and Delay Change Induced by Interconnect Coupling," IEEE J. of Solid State Circuits, Vol. 36, issue 10, Oct. 2001, pp.1587-1591

[2] Y. Sasaki, et al., "Crosstalk delay analysis of a 0.13-\spi mu/m node test chip and precise gate-level simulation technology", IEEE J. of Solid State Circuits, Vol. 38, issue 5, May 2003 pp.702-708

[3] T. Yamashita et al, "Techniques and Examples for High Speed LSI Design with STA," STARC Advanced Seminar 2004

[4] F. Picot, et al., "Test structure for crosstalk characterization", Electronics Letters, Vol.38, Issue 15, 18 July 2002, p.774-p.776

[5] B. Franzini, et al., "Crosstalk Aware Static Timing Analysis: a Two Step Approach," in proc. of ISQED 2000 pp.499-503

[6] J. Liu, et al., "Crosstalk Test Pattern Generation for Dynamic Programmable Logic Arrays," IEEE J. of Trans. on Instrumentation and Measurement, Vol.55, No.4, August 2006, pp1288-1302

[7] T.Garbolino et al., "Crosstalk-Insensitive Method for Testing of Delay Faults in Interconnects Between Cores in SoCs," in proc. of MIXDES 2007 pp.496-500

IEEE 2008 Custom Intergrated Circuits Conference (CICC)

Measurements of the silicon die characteristics of packaged drivers for high-speed I/O

Gerry Talbot, Edoardo Prete
Advanced Micro Devices

Abstract: **As the data rates of high-speed I/O interfaces with large numbers of I/Os increase beyond 5Gb/s it becomes increasingly difficult to separate out the effects of silicon performance from the interaction of package, socket and test fixture characteristics. This work discusses the need for separating silicon measurements from the channel measurements of a packaged device, the challenges of making these measurements on high-pin-count devices and proposes some methods to achieve this with some experimental results.**

I. INTRODUCTION

Complex high-performance devices such as microprocessors are required to have increasingly high-performance I/O interfaces to provide sufficient I/O throughput to maintain balance with the increasing aggregate performance of high CPU core count implementations. This drives a requirement for other high-performance complex devices that connect directly to the CPU to have comparable high-speed interfaces. This in turn requires in the ecosystem for these semiconductor devices, a need to be able to measure, characterize and test these interfaces in such a way as to ensure interoperability between multiple vendors and in multiple channels.

At microwave frequencies, a typical channel between two silicon devices -- once convolved with the waveform produced at the silicon transmitter die-pad -- introduces so much distortion that it is very difficult to directly measure the fundamental characteristics of a transmitter. These fundamental characteristics need to be determined to establish the performance of the transmitter in a range of different channels. Furthermore, the characteristics of the package and socket of the device need to be determined so they can also be convolved with the channels with which the device needs to inter-operate.

One approach to this problem is to measure a device driving a reference channel at its far end that mimics the characteristics of all channels with which the device needs to inter-operate. For some applications that only have channels that are high-loss or are non-reflective this may be sufficient however, in the commoditized general-purpose computer market, there are a large number of possible channels that need to be supported that typically range from reflection and crosstalk dominated to those dominated by discontinuity and loss. The reflective cases create the greatest difficulty because the reflections from the channel interact with the transmitter so it becomes necessary to be able to model this interaction.

Ideally one would like to measure with high bandwidth the silicon device directly at its die-pad and separately measure the package socket combination of the device at the pads used

for connection to the silicon device. Unfortunately this is impractical with high-pin-count devices because of the difficulties associated with microwave micro-probing of large numbers of I/Os. The state of the art methodology is to measure the packaged device on some form of high-quality electrical test fixture or compliance board that minimizes the loss and discontinuities from a measurement reference plane and the devices pin. Fig. 1 shows an example of such an electrical compliance board for a high-pin-count device, the reference planes use a microwave modular connector that allows simultaneous mating of eight coaxial connections.

Once the silicon device and package are assembled it is no longer practical to measure them separately so there comes a requirement to de-embed the test board and the devices package from measurements made at the reference plane as well as to be able to measure the test board and package separately.

Fig. 1 Compliance test board

This work derives a set of equations and describes a method to use them to extract the full S-parameters of the die pad to reference plane channel. The method uses measurements of the transmitter at two different drive strengths together with an additional vector network analyzer (VNA) measurement. We show how to practically use these equations to make a set of measurements with a suitable test pattern to extract a channel inverse. Then with the transmitter driving a pseudo-random binary sequence (PRBS), we show how to de-embed a transmit waveform measurement at the reference plane back to the device's die pad. We further show by taking

978-1-4244-2018-6/08 $25.00 © 2008 IEEE   45

additional VNA measurements of replica channels how to estimate the S-parameters at different points in the channel.

We will also discuss the fundamental transmitter parameters that must be measured so it is possible to use them in a statistical channel simulator to determine inter-operability between the device and an arbitrary set of channels.

Section II describes the background of the transmitter measurements that are required to simulate its performance in multiple channels; it discusses the use of a statistical tool that will calculate an eye opening at the receiver for the transmitter convolved with the channel at some target bit error ratio (BER). Section III shows the derivation of a set of equations that can be used to extract the channel inverse from a pair of transmitter measurements at different drive strengths. We go on to show in Section IV how to use these equations in a practical measurement and demonstrate experimental results. Section V. applies the calculated channel inverse to a measurement of a transmitter generating a PRBS and extracts the key transmitter parameters. Section VI illustrates how to deconstruct the package and test board sections from the die-pad to reference plane extracted channel. Finally in Section VII we summarize our findings and conclude.

II. BACKGROUND

The current state-of-the-art method for analyzing a high-speed channel with multiple lanes for data rates in excess of 5Gb/s is to use a statistical analysis technique similar to that used in Stateye [1]. The basic approach of this technique is to take a channels impulse or step response and calculate based on statistical convolution a probability density function (PDF) for a two-dimensional representation of an eye opening. This is further enhanced by convolving the noise generated by neighbors' crosstalk onto the victim lane as well as including some representation of the jitter distribution of the transmitter. This leads to two basic requirements to characterize a transmitter: its voltage waveform and the jitter it generates relative to an ideal unit interval (UI) that defines the non-linear stimulus for the channel (the non-linearity comes from the jitter generated in the timing circuitry of the transmitter). Similarly, to generate the response of the channel to be stimulated by the transmitter, we need a model of the device package that can be included in the modeling of different channels.

Fig. 2 shows a typical electrical compliance test board represented as a series of S-parameters blocks (labeled U, V and W) connecting a reference plane (which would be connected to test equipment) to the die pad. The die pad itself is a frequency dependent element in the network; however we can define an abstract frequency-independent driver that sources power into the network with a suitable rise time that is sufficient to support the maximum data rate.

The goal of compliance testing is to extract enough information about the device under test (DUT) that we can determine both its behavior in an operational environment and its ability to generate an open eye at the die pad of a receiver. To achieve this, the frequency-dependent characteristics of the three S-parameter blocks shown in Fig. 2 need to be determined so that a measurement of the DUT driving the reference

plane can be de-embedded back to the ideal driver. The voltage time waveform obtained by this de-embedding allows us to determine the key transmitter characteristics, such as voltage amplitude and timing jitter.

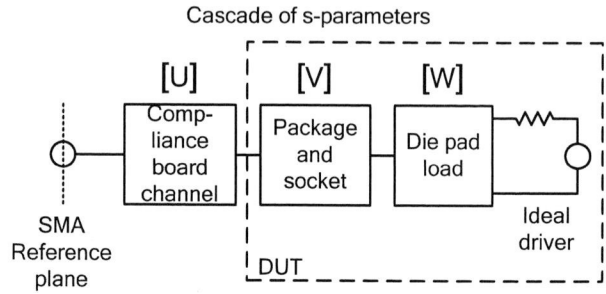

Fig. 2 Model of electrical compliance test board

The current state of the art is to use the DUT pin as the measurement point [7] and the assumption is made that the compliance board channel can be de-embedded moving the reference plane to the DUT pin [2][3][4]. Unfortunately as data rates and I/O counts increase it becomes impractical to de-embed the compliance channel accurately to the device pin and furthermore it's not practical, because of PCB layer transitions, to make the compliance channel's return loss at the device pin ideal over the frequency range of interest. This means the non-ideal return loss of the DUT interacts with the compliance board channel and introduces complex frequency-dependent error terms into the measurements.

The proposed new approach to de-embedding extracts the S-parameters of the complete path from SMA reference plane to the ideal driver. This allows, after de-embedding with this S-parameter information, direct measurement of the ideal driver that is needed to establish DUT compliance. Further measurements using replica compliance board channels and replica compliance board and package channels allow different measurement points to be calculated mathematically. The main advantage of this approach is that sub-dividing the channel from its overall response using replicas avoids the accumulation of errors that is inherent in constructing the overall channel from the individual replicas. This is because the replica channels have significant differences from the sections of the channel that they represent.

III. DERIVATION OF EXTRACTION FORMULA

If we consider the path from die pad of a transmitter through to a reference load as being described by its S-parameters and include within these S-parameters any frequency dependent effects of the transmitter's source, we can describe a transmitter driving a reference load Z_0 as shown in Fig. 3, where the quantities a_n and b_n represent the forward and reverse power at the two ports of the network and v_n represents the voltage at each port of the network. The transmitter is represented by the voltage source V_s and source impedance R_s.

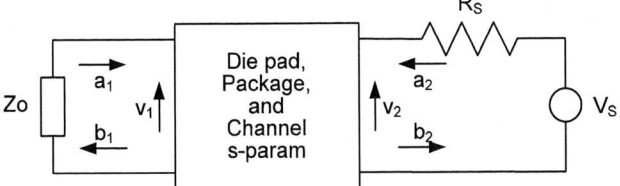

Fig. 3 Representation of an ideal transmitter driving package and test board

We now relate the S-parameters for the path from transmitter to load. Using the notations defined in [5]:

$$\begin{bmatrix} b_1 \\ b_2 \end{bmatrix} = \begin{bmatrix} s_{11} & s_{12} \\ s_{21} & s_{22} \end{bmatrix} \begin{bmatrix} a_1 \\ a_2 \end{bmatrix}$$

Or:

$$b_1 = a_1 s_{11} + a_2 s_{12} \qquad \{1\}$$

$$b_2 = a_1 s_{21} + a_2 s_{22} \qquad \{2\}$$

The impedance looking into port 2 with port 1 terminated by Z_0 is:

$$Z_{IN} = Z_0 \frac{1 + s_{22}}{1 - s_{22}}$$

The voltage on port 2

$$v_2 = V_S \frac{Z_{IN}}{Z_{IN} + R_S}$$

Expressing source impedance in gamma

$$\Gamma_S = \frac{R_S - Z_0}{R_S + Z_0}$$

$$R_S = Z_O \frac{1 + \Gamma_S}{1 - \Gamma_S}$$

$$v_2 = V_S \frac{\dfrac{1 + s_{22}}{1 - s_{22}}}{\dfrac{1 + s_{22}}{1 - s_{22}} + \dfrac{1 + \Gamma_S}{1 - \Gamma_S}}$$

$$v_2 = V_S \frac{(1 + s_{22})(1 - \Gamma_S)}{(1 + s_{22})(1 - \Gamma_S) + (1 - s_{22})(1 + \Gamma_S)}$$

$$v_2 = V_S \frac{(1 + s_{22})(1 - \Gamma_S)}{2 - 2 s_{22} \Gamma_S} \qquad \{3\}$$

From $\{2\}$ with $a_1 = 0$ we can calculate b_2 the reflected wave from the network:

$$b_2 = a_2 s_{22}$$

And for the voltage at port 2 with constant reference impedance for each port the ratios of the power are equivalent to the ratio of the voltages:

$$v_2 = a_2 + b_2$$

$$v_2 = a_2 + a_2 s_{22}$$

$$a_2 = \frac{v_2}{1 + s_{22}}$$

Using $\{3\}$ for v_2

$$a_2 = V_S \frac{(1 - \Gamma_S)}{2 - 2 s_{22} \Gamma_S} \qquad \{4\}$$

And so b_1 from $\{1\}$

$$b_1 = a_2 s_{12}$$

$$b_1 = V_S \frac{(1 - \Gamma_S)}{2 - 2 s_{22} \Gamma_S} s_{12}$$

As $a_1 = 0$, $v_1 = b_1$ and so the transfer function is given by:

$$H = \frac{v_1}{V_S} = \frac{(1 - \Gamma_S)}{2 - 2 s_{22} \Gamma_S} s_{12} \qquad \{5\}$$

For two values of R_S and so Γ_S calculate the ratio A

$$A = \frac{H_1}{H_2} = \frac{\dfrac{(1 - \Gamma_{S1})}{2 - 2 s_{22} \Gamma_{S1}}}{\dfrac{(1 - \Gamma_{S2})}{2 - 2 s_{22} \Gamma_{S2}}}$$

$$A = \frac{(1 - \Gamma_{S1})(1 - s_{22} \Gamma_{S2})}{(1 - \Gamma_{S2})(1 - s_{22} \Gamma_{S1})}$$

Defining K for convenience:

$$A = K \frac{(1 - s_{22} \Gamma_{S2})}{(1 - s_{22} \Gamma_{S1})}$$

$$A - s_{22} \Gamma_{S1} = K - s_{22} K \Gamma_{S2}$$

$$s_{22} = \frac{K - A}{K \Gamma_{S2} - A \Gamma_{S1}}$$

Replacing K:

$$s_{22} = \frac{\dfrac{(1 - \Gamma_{S1})}{(1 - \Gamma_{S2})} - A}{\dfrac{(1 - \Gamma_{S1})}{(1 - \Gamma_{S2})} \Gamma_{S2} - A \Gamma_{S1}}$$

$$s_{22} = \frac{1 - \Gamma_{S1} - A(1 - \Gamma_{S2})}{\Gamma_{S2}(1 - \Gamma_{S1}) - A \Gamma_{S1}(1 - \Gamma_{S2})} \qquad \{6\}$$

Since s_{22} only depends on A and DC measurements of the transmitter, it is independent of the wave shape of V_S as long as V_S is constant for the two measurements.

From $\{5\}$ we can calculate s_{12} with a R_S that does not equal Z_O

$$s_{12} = 2H\frac{1 - s_{22}\Gamma_S}{1 - \Gamma_S} \qquad \{7\}$$

IV. EXTRACTING COMPLIANCE CHANNEL FROM A PAIR OF TRANSMITTER MEASUREMENTS

To be able to extract the s_{12} and s_{22} from a transmitter driven measurement, we need to vary the impedance of the transmitter R_S. Modern transmitter designs typically include calibration circuits for the output impedance of the transmitter so setting it to two different impedances is straightforward. Experimentally we have found using lower impedances to yield better results, with approximately 20% difference in the two impedance values.

Measuring H_1 and H_2 can be accomplished by creating a test data pattern that contains a step from the transmitter that has a high and low period exceeding the settling time of the channel, typically in the range of 20ns. The step waveform can be sampled by a digitizing oscilloscope and averaged to produce a high-fidelity signal with low noise floor. Based on an estimate for the DC resistance of the channel and DC measurements of the transmitter the DC resistance of the driver can be calculated for the two drive strengths. This is then used to calculate Γ_{S1} and Γ_{S2}. The extraction methodology uses these gammas as real and constant numbers with frequency; the frequency-dependent characteristics of the die pad are lumped into the extracted channel.

From the time domain step responses, we can calculate the impulse response as shown in Fig. 4. To calculate H_1 and H_2 we need to approximate the impulse response at the ideal driver. The edge rate used to generate this impulse response is not critical for measuring the transmit amplitude and jitter but it does affect the high-frequency (HF) magnitude of s_{12}. To minimize the impact of this error term on s_{12} the edge rate used for extraction must also be used in subsequent channel modeling to ensure that the correct HF loss is preserved. The remaining residual error from this approximation comes from reflections off of the transmitter which are mitigated by the fact that the amplitude of these reflections at the receive end of the channel are proportional to s_{12}^2 which at HF causes them to be further attenuated.

The edge rate used for the extraction needs to be fast enough to settle within a UI and needs to approximate the actual die pad edge rate to ensure that measurement noise floor does not get unnecessarily amplified. Another consideration is that given a Fourier series is being used to represent this edge it must roll off quickly to avoid a Gibbs effect on the final de-embedded waveform. We chose a Gaussian impulse response to meet these criteria with a 10-90% rise time of approximately 50% of a UI. The corresponding step response height for this Gaussian impulse is set to the calculated DC level of the transmitter for the two transmitter drive strengths. Note the DC measurements of the transmitter need to be at the ideal driver which means the DC resistance of the compliance channel needs to be allowed for. This is typically in the range of 1-2ohms and can be estimated from a suitable replica measurement or can be modeled.

The measured impulse responses and the ideal driver impulse response are then converted into the frequency domain where by division the response for H_1 and H_2 can be calculated. From these two responses and the die pad gammas calculated from DC measurements we can use equations {6} and {7} from Section III to calculate s_{12} and s_{22}. These are shown in Fig. 5.

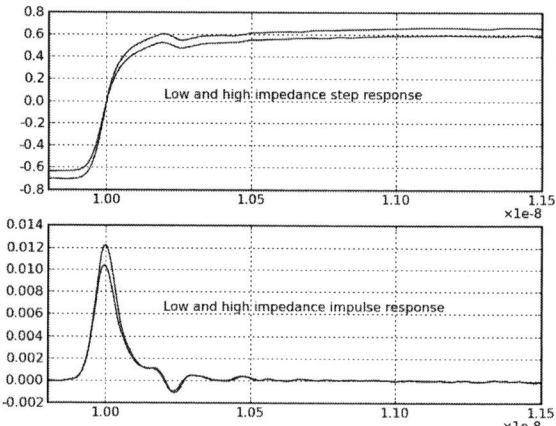

Fig. 4 Step and impulse response of the transmitter driving the compliance channel

We applied a small amount of frequency domain smoothing after calculating A to improve the noise floor in the extraction of s_{22}, this smoothing was implemented to correctly preserve magnitude and continuous phase. In the upper end of the frequency range the accuracy in extracting s_{22} starts to drop off because of the measurement noise floor and the loss between the reference plane and the die pad. However our primary interest is in s_{12} which as can be seen is useable up to 14GHz.

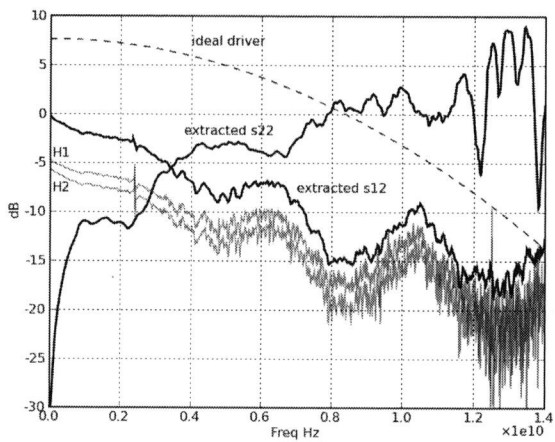

Fig. 5 Extracted S-parameters

The non-physical result for s_{22} above 8GHz is the consequence of using a frequency-independent model for the ideal driver's gamma, moving the real driver's frequency-dependent gamma into the extracted S-parameters. Therefore these S-parameters -- when used with the ideal driver in channel modeling -- will generate the correct response.

Finally, to get s_{11} of the channel, a VNA measurement is made at the reference plane with the transmitter's DC resistance adjusted to 50ohms.

V. DE-EMBEDDING THE TRANSMITTER DIE PAD WAVEFORM

As a check and to validate the correctness of the extracted S-parameters in the time domain the originally measured step response is de-embedded back to the ideal driver by calculating a channel inverse function from the S-parameters and convolving this with the measured step responses.

Fig. 6 shows the results from three different lanes of the DUT. Each lane had its S-parameters extracted and then these were convolved with the input step response to calculate the step at the ideal driver. The ripple seen on the flat portion of the ideal driver waveform comes from the noise floor of the measurement. Note the steady-state voltage levels represent the DC voltage levels for each lane and differ because of variations between lanes.

The extracted S-parameters can be used to de-embed a PRBS transmit waveform that is measured at the reference plane back to the ideal driver. Fig. 7 shows a transmitter running at 8Gb/s measured at the reference plane. Oversampling of the measured waveforms was performed in the frequency domain by zero padding the FFT data from the measurement.

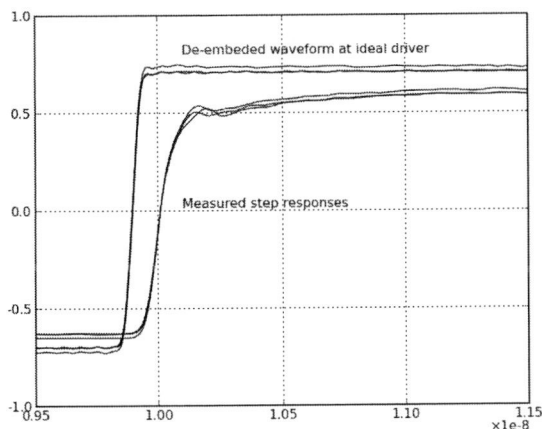

Fig. 6 De-embedded step from reference plane to ideal driver for three lanes

Practical devices operating at these data rates introduce jitter and voltage noise. These effects are further increased by the noise floor of the measurement equipment. It is therefore necessary to quantify these effects using a statistical technique to be able to make consistent and repeatable measurements of a device. It is common to use a statistical representation of the waveform shown as an eye diagram. Constructing the eye diagram requires an assumption about a time reference. Since the final goal is to characterize a transmitter's performance in an operational channel there are two time references that are important, the one that a receiver will use to sample the data and the one that is influenced by the channel.

For this work to model the clock data recovery (CDR)

function of the receiver we use a binary phase detector and integrator in a feedback loop that locks to the zero-volt crossing time of the data transitions. The integration time constant is selected to filter the ISI jitter caused by the data pattern and to track the lower frequency jitter inherent in the transmitter. This provides the equivalent of a reference receiver for the transmitter and as such represents the best-case tracking performance for a real receivers CDR.

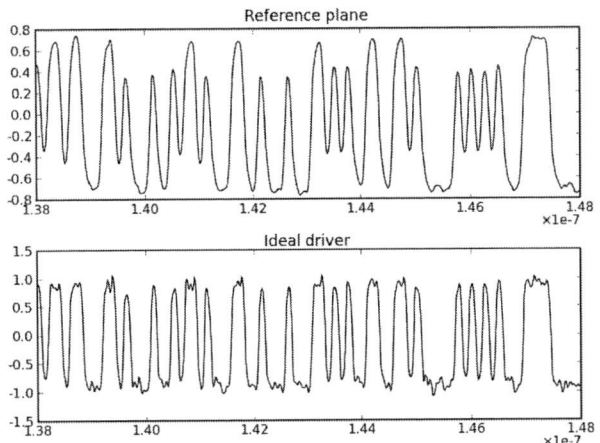

Fig. 7 De-embedded Tx PRBS at ideal driver

Fig. 8 shows a statistical eye diagram measured at the reference plane with FFT interpolation and no other post processing. The eye diagram is plotted as a PDF on a log white-to-black color scale representing probability. To quantify the jitter distribution a dual-Dirac model is fitted to the data using an inverse cumulative normal distribution of the jitters CDF; this is plotted on a Q-scale as described in [7]. This statistical model fit is shown in the eye diagram figures on the lower axes, which yields two variables for the distribution; a random jitter (R_J) and dual-Dirac deterministic jitter (D_{Jdd}) from which the total jitter (T_J) at a particular probability level or BER is given by equation {8} which for a BER of 10^{-12} Q=7.03.

$$T_J(BER) = D_{Jdd} + 2QR_J \qquad \{8\}$$

As can be see in Fig. 8 there is significant ISI visible in the eye diagram which makes it difficult to accurately separate the jitter present at the transmitter from the jitter added by the compliance test channel. Fig. 9 shows the same measurement de-embedded back to the ideal driver using a channel inverse calculated from extracting the compliance channel S-parameters using the method described above. As can be seen the ISI is considerably reduced with the fitted R_J reducing from 34mUI to 20mUI.

The channel HF loss effectively amplifies pulse width shrinkage from an ideal UI caused by transmitter jitter, at higher data rates this becomes a critical modeling parameter. The transmitter pulse width jitter (PWJ) is used to measure this effect and can be characterized by using the zero-volt crossing time of the transmit waveform as the time reference for the eye diagram.

This is shown in Fig. 10; as can be seen this is effectively an n-cycle jitter measurement with the first UI showing a smaller eye height from the HF loss of the compliance channel. Fig. 11 shows the result of de-embedding the measured waveform using the crossing time of the data as the time reference, the fitted R_J decreases from 33mUI to 18mUI for a single UI pulse width.

sary to choose an edge rate for the Gaussian ideal driver model that optimizes both jitter and ISI noise; as described earlier a value of 70ps was found to be optimal although -- as can be seen in Fig. 12 -- any value from 60-80ps yields a significant reduction in distortion compared to the measured waveform at the reference plane.

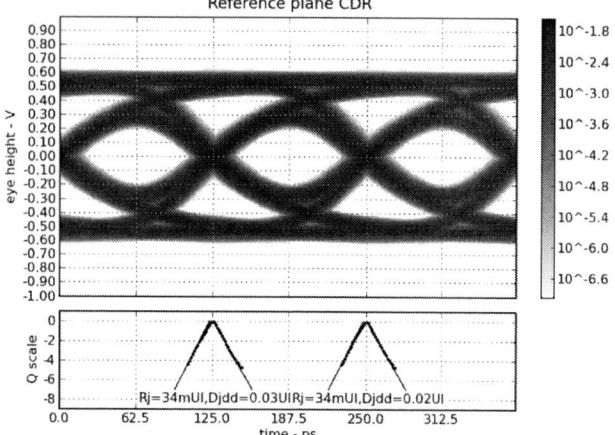

Fig. 8 CDR eye diagram at reference plane

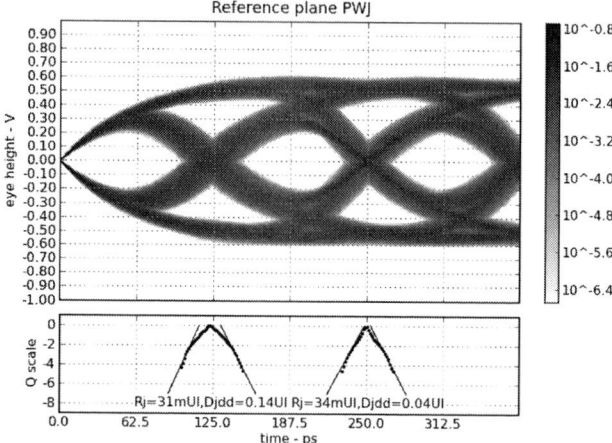

Fig. 10 PWJ eye diagram at the reference plane

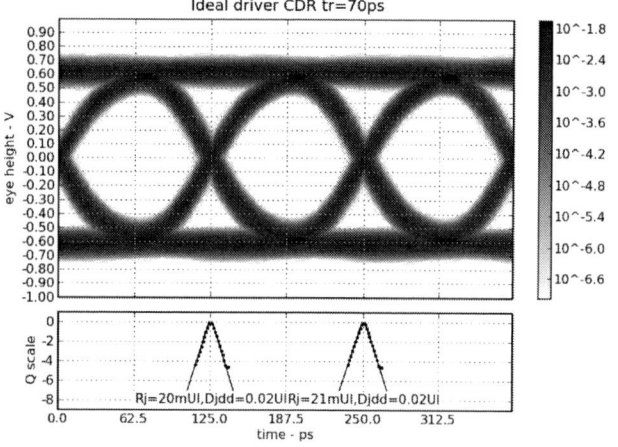

Fig. 9 CDR eye diagram at ideal driver

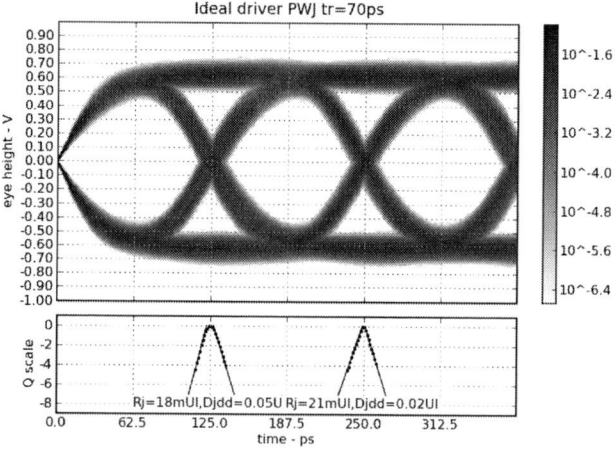

Fig. 11 PWJ eye diagram at ideal driver

As discussed previously there is a tradeoff in the approximation for the ideal driver rise time that is used to extract the compliance channel's S-parameters. Fig. 12 illustrates this tradeoff by plotting the PWJ T_J and the ISI voltage noise against the extraction rise time. Also on this plot shown as horizontal lines are T_J and the ISI voltage noise measured at the reference plane. It can be seen that, for all values of extraction rise time the signal de-embedded back to the ideal driver has lower distortion.

We can also see the tradeoff between using a faster edge rate on T_J, which is caused by two reasons; the real driver actual edge rate and the noise floor of the measurement. Extracting to a very fast rise time will give minimal ISI noise but will amplify measurement noise and potentially invent HF data that is not present in the real driver. Therefore it is neces-

VI. DECONSTRUCTING THE EXTRACTED CHANNEL

Once the full S-parameters from reference plane to die-pad are extracted it is possible to separate the package S-parameters and frequency-dependent die-pad S-parameters from the overall channel response by the use of replica channels. Two replica channels are required to achieve this; one represents the channel from the reference plane to the pin of the device, the second from the reference plane to the C4 pad in the package that the solder bumps of the die attach. Based on the labeling shown in Fig. 2 the two replica channels to be measured are designated U_R and UV_R, with the extracted S-parameters UVW_E.

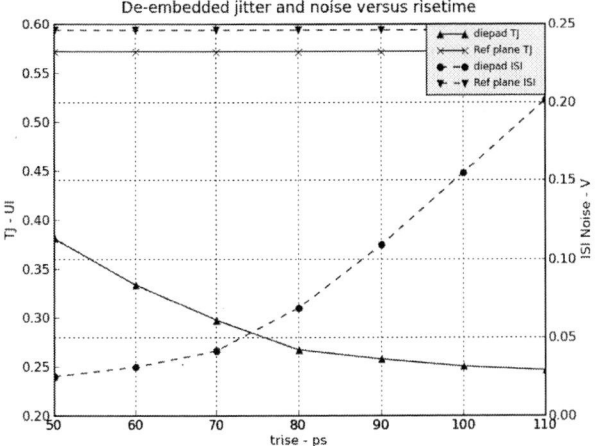

Fig. 12 Variation of jitter and ISI with extraction rise time

Replica test board channels are typically incorporated in compliance test boards and have microwave connectors at each end that mimic the actual test channel that is used to the device pin. Measuring from the C4 pad to the reference plane requires micro-probing of a package with no die attached. In both cases these replicas will differ from the DUT compliance channel that is being measured, as a result of this we can only approximate the frequency-dependent die-pad load W and the package plus die VW.

As described in [6] using, either the transmission coefficient form of the S-parameters or by converting into ABCD parameters we can express the relationship of the S-parameter matrixes and derive an estimate for the S-parameters for the package and die pad VW:

$$[UVW] = [U][VW]$$

$$[VW] \approx [U_R]^{-1}[UVW_E] \qquad \{9\}$$

Similarly, for the die pad S-parameters W:

$$[UVW] = [UV][W]$$

$$[W] \approx [UV_R]^{-1}[UVW_E] \qquad \{10\}$$

Fig. 13 shows the estimate for channel that includes the frequency-dependent die pad characteristics and the package calculated from {9}. It can be seen that the HF loss of UVW is reduced as we effectively de-embed the channel from the device pin to the reference plane.

Fig. 14 shows the estimate for the frequency-dependent die pad characteristics calculated from {10} along with the s_{11} and s_{12} of a fitted parallel RC model shown as bold dashed lines.

The difference between the RC model and the extracted s_{11} at low frequency is caused by the small error in adjusting the transmitter to 50ohms for the VNA measurement of s_{11} for the compliance channel. The other differences come from the real die pad having more complex frequency-dependent behavior caused by the physical structure of the silicon transistors and ESD diodes along with the effect of the C4 bumps that attach

the die to the package that are not included in the measurement of UV.

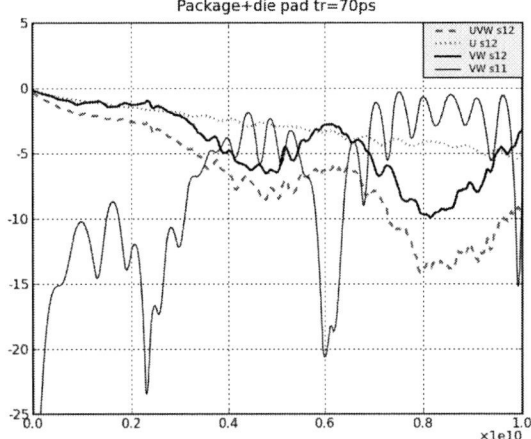

Fig. 13 Extracted S-parameters of package and die pad

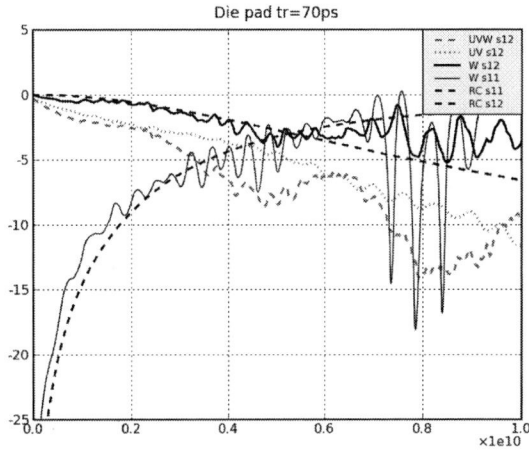

Fig. 14 Extracted S-parameters of die pad

VII. SUMMARY AND CONCLUSIONS

We have discussed the requirement for accurate measurement of packaged transmitters that will allow evaluation of their characteristics in a multitude of operational channels to ensure inter-operability with other devices. To achieve this we need to determine the fundamental characteristics of the transmitter that determine its performance in an operational channel. These characteristics are difficult to separate from the response of the channel that exists between die-pad and reference plane in a typical electrical compliance test board for high-pin-count devices. The performance of a device in an operational channel is also determined by frequency-dependent die pad and package characteristics which also need to be measured so they can be correctly convolved with the characteristics of an operational channel to determine overall system margins.

Using conventional de-embedding techniques for high-pin-count devices is less effective because they necessarily have to be based on replica channels that cannot have the identical characteristics of the individual lanes under test. They are also limited (without micro-probing) in only allowing de-embedding to occur to the pin of a device which makes accurate specifications of parameters like pulse width jitter difficult because of the HF loss that exists in the typical packages of commodity high-pin-count devices.

This work has shown that it is possible to calculate two of the three unique S-parameters of a packaged device from an abstract ideal driver to the reference plane on a compliance test board. This is achieved by making two measurements of the transmitter at two different output impedances using a test waveform that contains a low frequency step that can be time averaged by the measurement equipment. After determining the DC impedance of the driver at the two drive strengths to determine its gamma the s_{22} of the compliance channel can be calculated from the ratio of the step responses. Once this is determined s_{12} from the ideal driver to the reference plane can be calculated. By carrying out a s_{11} VNA measurement of the transmitter calibrated to the reference impedance at the reference plane the full S-parameters can be established.

Once we have the S-parameters from the ideal driver to the reference plane its inverse can be calculated and the waveform that is measured at the reference plane can be convolved with this inverse to calculate the waveform at an ideal driver. This waveform can then be measured to determine the voltage amplitude of the waveform and the transmitters jitter. As the HF loss of the channel is removed from the measurement it is possible to determine the transmitter's cumulative jitter as seen by a receiver's CDR and the transmitter's pulse width jitter at the ideal driver along with the transmitter's voltage amplitude.

Finally with the S-parameters of the compliance channel and measurements of suitable replica channels an estimate can be made of the device's die pad and package frequency-dependent characteristics by deconstruction.

We have shown that, using this methodology, it is possible to calculate a useful channel inverse for a packaged device on a compliance test board with measurements only at the reference plane test connector. It relies on the ability to change the device's output impedance by approximately 20% and a transmitter test pattern that contains a repetitive low frequency step

that can be time averaged with a digital oscilloscope. This channel inverse allows the measurement of any test data pattern to be de-embedded back to a high-bandwidth ideal driver removing ISI distortion created by the HF loss of the compliance channel. From this de-embedded measurement we can determine the transmitter's voltage waveform as well as a broadband measurement of transmitter jitter.

The extracted S-parameters from the transmitter can be used with additional VNA measurements of compliance channel's s_{11} and suitable replica structures to deconstruct the channel from reference plane to the die-pad for modeling of the device in an arbitrary set of operational channels.

ACKNOWLEDGMENTS

The authors owe their gratitude to Alan Lear who has patiently made countless measurements on compliance boards and with micro-probing, package substrates as the authors converged on a viable methodology. We also wish to give credit to Mark Marlett of LSI Logic for first proposing the concept of de-embedding back to an ideal driver, and to Anthony Sanders of Infineon for many helpful discussions on the topic. Finally thanks also go to the PCI Express EWG where we have had numerous discussions on the problem of specifying devices using de-embedding to the die pad for the development of the 8GT/s specification.

REFERENCES

[1] Anthony Sanders, Channel Compliance Testing Utilizing Novel Statistical Eye Methodology, DesignCon 2004
[2] Mihai Rotaru, Electrical Characterization of a High Performance Microprocessor Socket for System Level Simulation, 2003 Electronics Packaging Technology Conference.
[3] Hong Shi, Accurate Calibration and Measurement of Non-Insertable Fixtures in FPGA and ASIC Device Characterization, 2006 DesignCon.
[4] Lih-Tyng Hwang, A Post-Processing De-Embedding Technique and Effect of TRL Measurement Using an Incorrect Fixturing, 2007 Electronic Component and Technology Conference.
[5] David M. Pozar, Microwave Engineering, Second Edition
[6] Simon Ramo, Fields and Waves in Communication Electronics, Third Edition
[7] PCI-SIG, PCI Express Base Specification Revision 2.0, 2007

IEEE 2008 Custom Intergrated Circuits Conference (CICC)

Inductor-Based ESD Protection under CDM-like ESD Stress Conditions for RF Applications

S. Thijs[1], M. Okushima[2], J. Borremans[3], P. Jansen, D. Linten, M. Scholz, P. Wambacq[3] and G. Groeseneken[1]

IMEC vzw, Kapeldreef 75, Leuven, B-3001, Belgium
Email: Steven.Thijs@imec.be, Tel: + 32 16 28 7749, Fax: +32 16 28 1706
(1) also at: Electrical Engineering Dept., Katholieke Universiteit Leuven, Belgium,
(2) on leave from: Core Development Division, NEC Electronics Corporation, Japan,
(3) also at: Vrije Universiteit Brussel, Brussels, Belgium.

Abstract – **Charged Device Model (CDM) ElectroStatic Discharge (ESD) stress is a major concern for inductor-based ESD protection strategies for RF circuits processed in advanced nano-CMOS technologies. The CDM robustness of such protection methodology is investigated in this paper based on Very-Fast Transmission Line Pulse (VFTLP) measurements. Its applicability is discussed for future technologies and RF applications.**

I. INTRODUCTION

In [1], a summary of different ESD protection strategies utilized in RF circuit applications in CMOS and BiCMOS technologies was presented. It was found that traditional dual-diode ESD protection was still suitable up to 5 GHz, but for higher frequencies the parasitic diode capacitance would deteriorate the RF performance of the circuit. Hence, other solutions would be required. One of the most promising solutions for frequencies of 5 GHz and above was found to be the use of "plug-and-play" inductive ESD protection, where an ESD inductor shunts the ESD current from the RF-pin to ground. A schematic showing the ESD inductor L_{ESD} as protection element for the LNA is shown in Figure 1, Box A. "Plug-and-play" means that the ESD inductor is only added after the full RF design is completed, and that it has only a marginal impact on the RF performance [2]-[4]. The applicability of such ESD protection method for higher frequency RF circuits has been demonstrated for a 17 GHz Low Noise Amplifier (LNA) in a BiCMOS technology [5].

Record Human Body Model (HBM), Machine Model (MM) and TLP protection levels were obtained by boosting the ESD performance using additional clamping diodes [2] (Figure 1, Box B) and/or Above-IC Wafer-Level Packaging (WLP) techniques [4] for a 5 GHz LNA processed in 90 nm CMOS technology. Despite these excellent results, the applicability of an inductor as ESD protection for CDM stress remained a big concern due to potential voltage overshoots [6]. This paper addresses this issue.

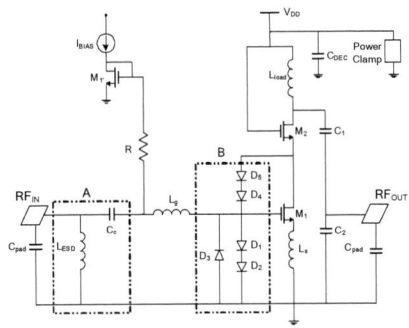

Figure 1. Schematic of inductor based ESD-protected LNA (ESD inductor is located in Box A and additional clamping diodes are located in Box B).

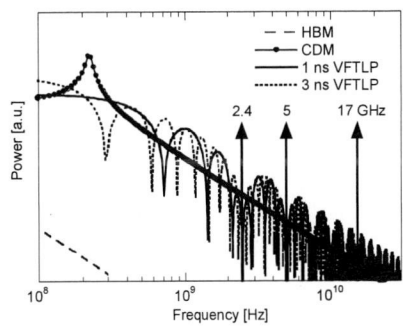

Figure 2. Frequency spectrum of HBM, CDM, 1 ns and 3 ns VFTLP. Different WLAN frequencies are indicated.

CDM is characterized by high (6-10 A) and very short (~1 ns) current pulses. The pulses typically have a rise time of 400 ps or less [7]. As can be seen in Figure 2, the frequency spectrum of CDM and VFTLP extends well into the GHz range and hence they will be coupled much more inside the RF circuit than HBM. At a higher RF operating frequency (e.g. 13 GHz), the coupling is reduced, as will be further discussed in section IV.

978-1-4244-2018-6/08 $25.00 © 2008 IEEE

VFTLP measurements are used to characterize devices in the time and current domain of CDM [8]. VFTLP typically has a rise time of 200 ps and pulse widths of 1 to 10 ns. Even though a one-to-one correlation between VFTLP and CDM does not exist, VFTLP remains a very useful tool to investigate the transient response of a certain ESD protection structure.

In Section II, VFTLP measurements are used to evaluate the ESD performance of an inductor to ground ESD protection strategy (Figure 1), including clamping diodes and Above-IC processing. Section III discusses the behavior of a transformer based ESD protection strategy under CDM-like stress conditions. In this case, L_{ESD} is placed underneath the matching inductor L_G (Figure 1), avoiding the area penalty by L_{ESD} and creating a galvanic separation between the RF input pin and the core LNA circuit [9]. The benefits of inductive ESD protection for CDM in 45 nm planar CMOS technology is discussed in section IV for a RF circuit with increased operating frequency. Finally, the paper is concluded.

II. INDUCTOR TO GROUND

A 5 GHz LNA has been protected with an inductor L_{ESD} between RF input and ground (Figure 1). An ESD robustness of 2.5 kV HBM was measured for positive stress between input and ground. Additional clamping diodes were successfully used to boost the ESD performance up to 5.5 kV by clamping the voltage at the gate oxide of M_1 which forms the weak spot for ESD stress at the RF input. Changing L_{ESD} from Back End Of Line (BEOL) (3 nH inductance with 5.5 Ω series resistance) to Above-IC (3nH, R=0.6 Ω) further boosts the ESD performance above 8 kV due to the reduced series resistance of L_{ESD} [4]. Since the gate oxide is more fragile when stressed in inversion rather than in accumulation, this paper focuses only on positive stress between input and ground. Even though other pin-to-pin ESD stress combinations might yield lower robustness, they can be solved by improving power clamp design and avoiding parasitic current paths in the core circuit. Some of these techniques are discussed in [2]–[4] and [10] and are beyond the scope of this work. Protecting the RF pins remains the most critical task for the ESD design engineer.

The voltage overshoot across L_{ESD} during a fast transient ESD current is defined by:

$$V_{input} = L_{ESD} \frac{dI}{dt} + RI \qquad (1)$$

Very high voltages can be expected due to the fast, high current nature of the CDM pulse because the LdI/dt component cannot be neglected anymore. This voltage overshoot is coupled across the decoupling capacitor C_C (Figure 1) and can cause internal circuit damage at the fragile gate oxide of transistor M_1 if not well understood and controlled.

The rise time of the VFTLP tester (Hanwa HED-T5000) is 200 ps, however this rise time is not seen at the device level. Probe needles were chosen for flexibility and cost reasons

instead of RF-needles and their parasitic inductance lowers the rise time. The rise time reaching the device is around 400 ps, which is according to the CDM specification [7]. Figure 3 shows VFTLP simulations of 2.3 A (around 200 V CDM of Class II [7]) with 3 ns pulse width and 400 ps rise time when stressing input positive to ground. This yields a maximum peak voltage of 30 V for standard BEOL L_{ESD} (3 nH, R=5.5 Ω), according to (1) at 400 ps (point A in Figure 3). The positive dV/dt charges up the floating gate node according to the capacitive division between C_C and the capacitance of M_1:

$$V_{gate} = V_{input} \frac{C_C}{C_C + C_{M1}} \qquad (2)$$

Since C_C is a large decoupling capacitor of about 2.3 pF and M_1 has a gate capacitance of around 150 fF, around 94 % of the voltage overshoot is transmitted across C_C onto M_1 (Vgate – no diodes – BEOL, Figure 3). When the VFTLP pulse reaches its stable region, dI/dt equals zero and the voltage drops down to $R*I$ (point B in Figure 3) at the input, while the gate node follows according to (2). The latter voltage remains at the gate for the full duration of the VFTLP pulse width, after which a negative voltage spike (point C in Figure 3) is measured when the VFTLP pulse switches off. The 90 nm gate oxide cannot withstand such voltage stress and fails already at 0.9 A (Table 1). More information on gate oxide reliability in the ESD time domain can be found in [11] and [12].

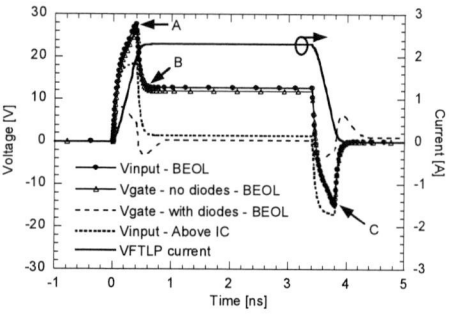

Figure 3. Simulated voltages during a 2.3 A VFTLP pulse of 400 ps rise time and 3 ns pulse width, with and without clamping diodes. The clamping diodes remove the voltage overshoot generated by the ESD inductor L_{ESD} at the gate.

When using clamping diodes at the gate of M_1, the gate voltage is clamped to 7–8 V peak as seen in Figure 3 (Vgate – with diodes – BEOL). Even though this peak voltage still causes gate oxide damage, by using clamping diodes the ESD robustness is improved, Table 1. However, this improvement was smaller than the factor of 3 expected by simulated results. The reason is due to the voltage overshoot of the STI defined diodes used here, which was not considered in the simulation. Further improvement is possible by changing the type of clamping diode from STI to poly defined. These poly defined diodes were found to have less voltage overshoot during ESD pulses than STI defined diodes [13] and thus should be used for

978-1-4244-2018-6/08 $25.00 © 2008 IEEE   54

voltage clamping, especially during fast CDM events. The size of each clamping diode was 5x2 μm^2 yielding a parasitic capacitance of 11 fF. The influence of the clamping diodes on the RF performance was found to be marginal as described in [4]. However, the ESD performance can be further increased by enlarging the clamping diodes at the expense of RF performance. It is preferred to take the diode parasitics into account during circuit design, and as such an optimal ESD-RF performance can be "co-designed".

TABLE 1. SUMMARY OF VFTLP RESULTS (200 PS RISE TIME AND 3 NS PULSE WIDTH).

VFTLP IN+ GND-	Normal BEOL	Above-IC
No clamping diodes	0.9 A	1.7 A
With clamping diodes	1.3 A	2.7 A

Since L_{ESD} is the same, both for standard BEOL and for the Above-IC inductor, only the ohmic part of the voltage overshoot is reduced when using Above-IC. In Figure 3, the decrease in voltage at the input pad can be observed when comparing Above-IC with standard BEOL. This results in a significant improvement up to 2.7 A as seen in Table 1.

These results indicate that an inductor to ground can be used as CDM protection element in combination with clamping diodes when high-quality inductors are available.

III. TRANSFORMER BASED

To avoid the area penalty introduced by L_{ESD}, the ESD inductor can be shifted underneath the gate matching inductor L_G and hence a transformer based ESD protection circuit is created as shown in Figure 4. The circuit was implemented in a 130 nm CMOS technology and details on the RF performance are described in [9]. Besides excellent HBM robustness, we measured also a VFTLP robustness above 3 A (200 ps rise time and 3 ns pulse width), without clamping diodes at the gate of M_{n1}, Table 2. Adding clamping diodes further improved the VFTLP robustness up to 5 A. This improvement cannot be attributed only to the use of a 130 nm CMOS technology and hence a deeper study of the behavior of the transformer under VFTLP stress is needed. L_1 is the inductor of the transformer at the input pad and L_2 is the inductor near the core circuit as seen in Figure 4. R_1 and R_2 are their parasitic resistances (not indicated in the figure) and I_1 and I_2 the currents through them respectively.

The voltage overshoot at the RF input pad V_{input} is equal to:

$$V_{input} = L_1 \frac{dI_1}{dt} + R_1 I_1 + M \frac{dI_2}{dt} \qquad (3)$$

$$M = k\sqrt{L_1 L_2} = nL_1 \qquad (4)$$

where M is the mutual inductance, k the coupling factor and n the turn ratio of the transformer. I_1 equals the ESD current while I_2 is only in the order of mA due to the presence of the MIM decoupling capacitor and gate oxide of M_1. Therefore, the last

term can be neglected in (3) with respect to the first. The voltage right after the transformer V_{trafo} is calculated using:

$$V_{trafo} = nL_1 \frac{dI_1}{dt} + L_2 \frac{dI_2}{dt} + R_2 I_2 \qquad (5)$$

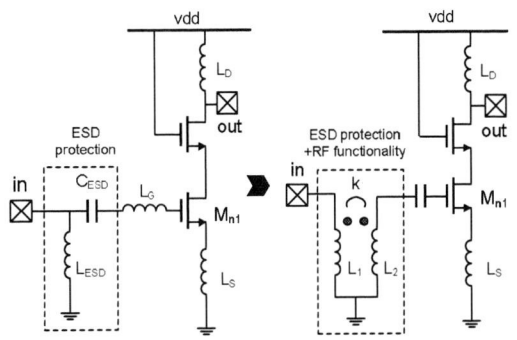

Figure 4. Inductor-based ESD protection (left) and transformer-based ESD protection (right).

TABLE 2. HBM AND VFTLP MEASUREMENT RESULTS OF THE TRANSFORMER BASED ESD PROTECTED LNA, WITH AND WITHOUT ADDITIONAL CLAMPING DIODES.

HBM	Without diodes	With diodes
IN+ GND-	4.5 kV	7.3 kV
VFTLP	Without diodes	With diodes
IN+ GND-	3.2 A	5 A

From (5), it can be seen that V_{trafo} only depends on the transient current through L_1 since the last two terms can be neglected. The inductive overshoot voltage at the input pad is up-converted with a factor n after the transformer. For this application, n and k are respectively 1.6 and 0.82.

For HBM, due to the relative slow rise time (small dI/dt), the voltage overshoot is dominated by $R*I$. Since only the inductive part is coupled across the transformer, the voltage after the transformer is much smaller than at the input pad. However, for VFTLP, dI/dt is dominant for the voltage overshoot, which is then up-converted across the transformer. This results in V_{trafo} being larger than V_{input}, Figure 5. However, the inductor L_1 which has to ground the ESD current, has an inductance value of 0.8 nH [9] and despite the voltage up-conversion, the voltage overshoot is still lower than in the case of the 3 nH inductor to ground. Figure 5 shows the different simulated voltages during 3 A VFTLP stress conditions. It is also important to notice that V_{trafo} is independent of both the parasitic resistances R_1 and R_2 of L_1 and L_2 meaning that for the transformer-based ESD protection Above-IC is not expected to improve CDM robustness.

Interestingly, (5) yields also another important benefit of the use of the transformer for VFTLP. When dI_1/dt equals zero, namely after the rising part of the VFTLP pulse, the full voltage after the transformer V_{trafo} drops to zero as can be seen in Figure 5, and consecutively also the voltage at the gate of transistor M_{n1}

(according to (2) with V_{trafo} instead of V_{input}). Note that in the case of CDM, there will always be a dI/dt component present.

When comparing Figure 3 and Figure 5, in the case of a transformer based ESD protection, the voltage stress at the gate is only present during the transient part of the VFTLP pulse, i.e. during 400 ps, whereas in the case of the inductor to ground, the stress is present during the full 3 ns of the VFTLP pulse. This reduction in voltage stress time is a significant benefit for the transformer based ESD protection solution compared to the inductor to ground as shown by the excellent VFTLP measurement results in Table 2.

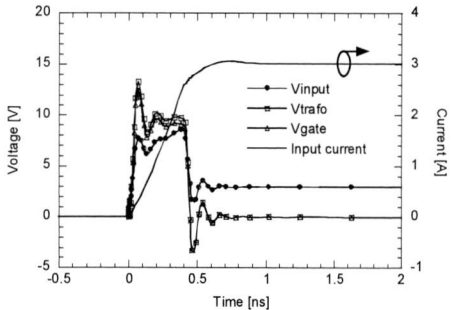

Figure 5. Simulated voltages during a 3 A VFTLP pulse of 400 ps rise time and 3 ns pulse width for the ideal transformer ESD protection. No clamping diodes were present.

IV. FUTURE APPLICATIONS

A LNA with inductor to ground type ESD protection was designed in a 45 nm planar bulk CMOS technology with Above-IC inductor, operating at 13 GHz using a similar topology as described in Figure 1 [14]. Single-contact sized (minimum design rule) clamping diodes were implemented. Due to the increased RF operating frequency, the required inductance of L_{ESD} is only 1 nH which reduces the inductive voltage overshoot (1) with a factor 3 compared to section II. The voltage overshoot (1) caused by the ohmic effect is also greatly reduced due to the use of Above-IC and the fact that inductors with lower inductance require fewer turns and wider metal tracks. Figure 2 also indicates that less ESD signal will be coupled inside the RF circuit at this increased operating frequency.

This reduction in voltage overshoot directly translates into an improved HBM and VFTLP result. When stressing the RF input positively to ground, a robustness of 5 kV HBM and 2.9 A VFTLP was measured, complying with Class II CDM. This result was obtained even with minimum sized clamping diodes. Further ESD improvement can still be obtained by increasing their size (limited by the allowed RF performance degradation) and by changing the clamping diodes from STI to poly defined.

V. CONCLUSIONS

Two different inductor-based ESD protection methodologies, namely inductor to ground and transformer-based, have been studied for their impact on CDM robustness. Analysis of the transient voltage response revealed the benefits of transformer-based protection. Using an Above IC inductor or a transformer as well as additional clamping diodes yields both excellent HBM, MM and TLP results, as well as VFTLP results up to 5 A. This robustness can be further improved by optimizing the diode type and size, making the inductor-based ESD protection methodology efficient against CDM stress.

At a higher RF operating frequency, less ESD signal couples into the RF circuit and due to the decreased inductance of the protection inductor, less voltage overshoot is present. This yields an improved VFTLP robustness and as such increasing the RF operating frequency is beneficial for CDM robustness.

Further, when scaling CMOS technology, the ESD robustness of active ESD protection devices typically decreases. However, inductor-based ESD protection robustness is mainly determined by the back-end quality, which remains rather constant for the RF requirements. Hence the ESD performance using inductor-based ESD protection remains similar for different technology generations as has been shown for a 45 nm CMOS technology.

ACKNOWLEDGEMENTS

This work is part of IMEC's Industrial Affiliation Program. Part of this work is funded by the NANO-RF IST-027150 Project of European Union.

REFERENCES

[1] Ph. Jansen et al., "RF ESD Protection strategies The design and performance trade-off challenges", IEEE Custom Integrated Circuits Conference, pp. 489-496, 2005.
[2] S. Thijs et al., "Implementation of Plug-and-Play ESD Protection in 5.5 GHz 90 nm RF CMOS LNAs - Concepts, Constraints and Solutions", Journal of Microelectronics Reliability, Vol. 46, Issues 5-6, pp. 702-712, May-June 2006.
[3] D. Linten et al., "A 5 GHz Fully Integrated ESD-Protected Low-Noise Amplifier in 90 nm RF CMOS", IEEE Journal of Solid-State Circuits, Vol. 40, Issue 7, pp. 1434-1442, July 2005.
[4] S. Thijs et al., "Class 3 HBM and Class C MM ESD Protected 5.5 GHz LNA in 90 nm RF CMOS using Above-IC Inductors", EOS/ESD Symposium, pp. 25-32, 2005.
[5] D. Linten et al., "Implementation of 6 kV ESD Protection for a 17 GHz LNA in 130 nm SiGeC BiCMOS", IEEE International Conference on Semiconductor Electronics, pp. A7-A12, 2006.
[6] S. Hyvonen et al., "Cancellation Technique to Provide ESD Protection for multi-GHz RF inputs", IEEE Electronics Letters, pp. 285-285, February 2003.
[7] JEDEC standard, "Field-Induced Charged-Device Model Test Method for Electrostatic-Discharge-Withstand Thresholds of Microelectronic Components", available online at http://www.jedec.org/download/search/22c101c.pdf
[8] H. Gieser et al., "Very Fast Transmission Line Pulsing of Integrated Structures and the Charged Device Model", IEEE Trans. Components, Packaging and Manufacturing Technology – Part C, Vol. 21, No. 4, 1998.
[9] J. Borremans et al., "A 5 kV HBM Transformer-Based ESD Protected 5-6 GHz LNA", IEEE Symposium on VLSI Circuits, pp. 100-101, 2007.
[10] D. Linten et al., "T-Diodes - A Novel Plug-and-Play Wideband RF Circuit ESD Protection Methodology", EOS/ESD Symposium, pp. 242-249, 2007.
[11] A. Ille et al., "Ultra-thin Gate Oxide Reliability in the ESD Time Domain", EOS/ESD Symposium, pp. 285-294, 2006.
[12] T.W. Chen et al., "Gate Oxide Reliability Characterization in the 100 ps Regime with Ultra-fast Transmission Line Pulsing System", EOS/ESD Symposium, pp. 102-106, 2007.
[13] M. Scholz et al., "Calibrated Wafer-Level HBM Measurements for Quasi-Static and Transient Device Analysis", EOS/ESD Symposium, pp. 89-94, 2007.
[14] J. Borremans et al., "Perspective of RF design in future planar and FinFET CMOS", accepted at IEEE Radio Frequency Integrated Circuits symposium, 2008

IEEE 2008 Custom Intergrated Circuits Conference (CICC)

Non-Destructive IC Defect Localization Using Optical Beam-Based Imaging

Edward I. Cole Jr.
Sandia National Laboratories
MS 1072, 1515 Eubank SE
Albuquerque, NM 87185-1081, USA

Abstract-**Optical beam failure analysis methods provide unique capabilities to identify and localize defect types that would be difficult or impossible by other methods. By understanding the physics of signal generation, the user gains the insight necessary to optimize technique performance.**

I. INTRODUCTION

Modern IC complexity resulting from reduced feature sizes, circuit density, and sophisticated electrical stimulus has made failure analysis and defect localization extremely difficult. Additionally, dense multi-level metallization and flip-chip packaging may leave only the backside of the IC available for interrogation. Laser-based methods provide some of the powerful tools analysts depend on to overcome these obstacles. This paper describes the signal generation physics and examples of using theses methods.

II. LASER/IC INTERACTION PHYSICS

Different light/semiconductor interactions depending upon the wavelength (λ) used for the light source are exploited for defect localization. Reflected light is used for imaging. Si is relatively transparent to infrared wavelengths >1000 nm so analysis through the backside is possible (Fig. 1). The two main interactions used for analysis are local photocurrent or electron-hole pair generation and heat generation. Wavelengths less (or energies greater) than the indirect Si bandgap (<1100 nm) will produce both photocurrents and heating that can interact with device operation. Normally the photocurrent electrical effects dominate the local heating responses. λs > 1100 nm will not efficiently produce photocurrents in Si, but will generate local heating which becomes the dominate effect. Both effects are used to examine static and dynamic IC operation.

III. DEFECT LOCALIZATION EXAMPLES

Following are defect examples that highlight the differentiating strengths of a given technique. The last example section illustrates the power of combining multiple techniques.

A. Light-Induced Voltage Alteration (LIVA)

LIVA [1] image contrast is based on the changes in power demand from an IC while a photocurrent producing laser is

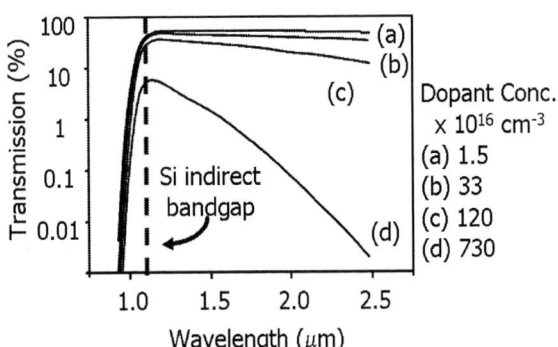

Fig. 1. Percent transmission of light through
625 μm of Si as a function of wavelength and doping.

scanned over the device. The non-random recombination of electron-hole pairs resulting from Si defects can significantly alter the power demands of an IC. In LIVA (as with all the –IVA techniques) the IC is biased with a constant current source with changes in the supply voltage used to make an image. The power demand changes are acquired using a simple ac-coupled amplifier. While sample dependent, the changes in voltage are normally easier to capture and generate a larger signal (100s of mV) than the current changes (100s of nAs) with a constant voltage supply.

The LIVA example in Fig. 2 shows an entire die, backside reflected light image and the LIVA signal in the lower right corner. Fig. 3 is a higher magnification reflected light/LIVA image pair. Subsequent deprocessing showed the defect due to an open metal to silicon contact. Note that the area is completely under the upper metal power bus and invisible to front side optical inspection.

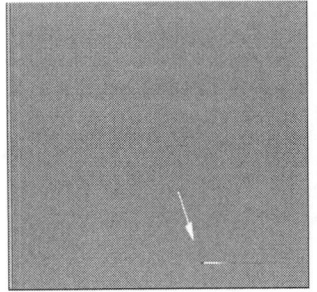 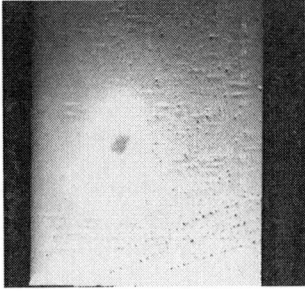

Fig. 2. Whole die, backside LIVA (left) and reflected (right) images of an open metal to silicon contact site.

978-1-4244-2018-6/08 $25.00 © 2008 IEEE   57

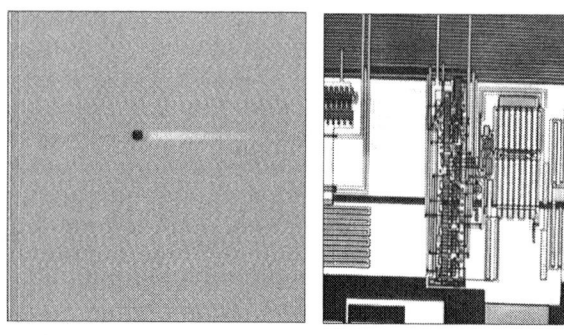

Fig. 3. Backside LIVA (left) and reflected (right) images of the open metal to silicon contact site seen in Fig. 2 but at higher magnification.

B. OBIRCH, TIVA, and SEI

OBIRCH (optical beam induced resistance change), TIVA (thermally-induced voltage alteration), and SEI (Seebeck Effect Imaging) are techniques that use local thermal gradients to localize defects with a 1340 nm laser [2, 3]. OBIRCH and TIVA localize shorts by changing the defect's resistance and hence its power demands when biased. OBIRCH identifies changes in IC current with a constant voltage supply and TIVA identifies changes in IC voltage using a constant current supply. The increased resistivity associated with local heating of a current carrying conductor produces a power demand reduction in an IC with a short defect. The change in power produces the image contrast.

SEI localizes floating conductors using the Seebeck Effect, which results in a long floating conductor possessing a voltage gradient when a thermal gradient is produced. Thermal osmosis of electrons, phonon drag, and differing work functions are the main generators of the Seebeck Effect [2]. The Seebeck Effect voltage gradients (very small, µV/C) will change the power demands of the IC and can be detected with constant current biasing.

Note that work function differences, such as at a metal to polysilicon contact, can also be seen in OBIRCH and TIVA if the change in power demand is not masked by the resistance change.

The TIVA example in Fig. 4 shows a shorted IC conductor with the TIVA signal highlighting the entire short path. A focused ion beam (FIB) cross section of the failure site show the shorting stress extrusion (Fig. 5).

The SEI example in Fig. 6 is a backside image showing a floating conductor segment. Brighter and darker contrast is present at the metal/polysilicon contacts from the thermocouple effect at the different materials junction.

C. Soft Defect Localization (SDL) and Laser-assisted Device Alteration (LADA)

A new development trend is the combination of multiple techniques to produce a more powerful method. In SDL [4], a device is dynamically exercised by continuously looping test vectors while looking for a change in the pass/fail condition as the surface is scanned with a 1340 nm laser. The change in the pass/fail condition with local heating produces

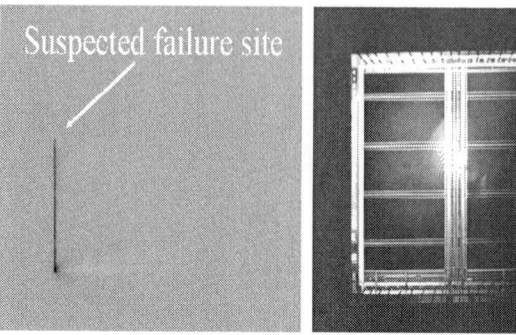

Fig. 4. TIVA (left) and reflected (right) images of a short on an SRAM device.

Fig. 5. FIB cross section of the extrusion causing the short and TIVA signal seen in Fig. 4.

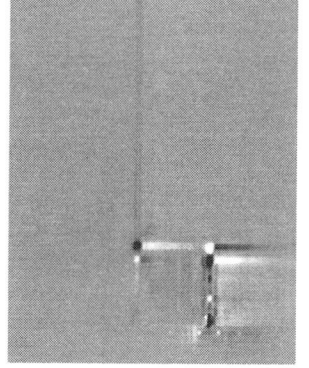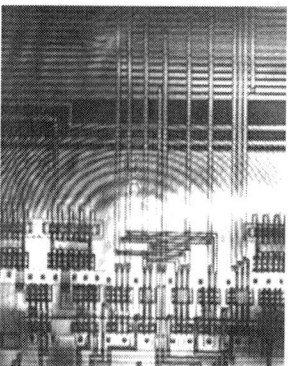

Fig. 6. Backside SEI (left) and reflected (right) images of a floating conductor and its fanout to a transistor. Enhanced contrast at metal to polysilicon junctions occurs due to a thermocouple effect with heating.

the SDL image contrast. The image can localize a host of soft defects such as resistive interconnections, weak parallel and serial timing paths, and any other thermally sensitive defects. These "soft defects" have historically been extremely difficult to find. LADA is a similar method that uses photocurrent generation to activate and localize weak serial timing paths [5].The SDL backside example in Fig. 7 shows a resistive interconnection site. As the laser scans over the site two mechanisms occur; the via's resistivity increases with temperature and the material in the via expands. The net result is that via resistance is lowered by the improved physical contact that decreases the resistance more than the increase in resisitivity with temperature. As a result the cycled vector set passes the test more frequently. The defect site was verified by a FIB cross section (Fig. 8).

Fig. 9 shows a parallel path race condition from the front side. Heating of the "strong" leg slowed it down enough so that the device passed. A higher magnification, combined SDL/reflected light image of the site is seen in Fig. 10. The transistor highlighted slows to yield a passing condition.

Fig. 11 indicates a weak link site in a serial timing path. Almost the entire die is visible in the backside image. Fig. 12 is a higher magnification view of the site. In Fig. 12 the site of interest has been isolated to a single transistor.

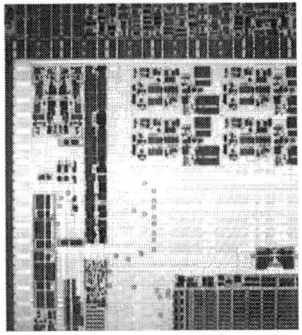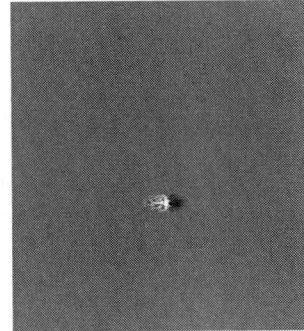

Fig. 7. Backside reflected (left) and SDL (right) images highlighting a resistive interconnection.

IV. SPATIAL RESOLUTION IMPROVEMENTS

There have been significant improvements in the spatial resolution possible with backside infrared inspection of ICs by increasing the effective numerical aperture through SILs (solid immersion lenses). The NAIL (numerical aperture increasing lens) (Fig. 13) is a moveable Si hemisphere that has been demonstrated to generate better than 200 nm spatial resolution using a 1050 nm illumination source [6]. The particulars of the hemisphere depend on the Si doping and substrate thickness.

The quality of the NAIL's contact with the Si backside is critical to good image quality. To address the contact issue two non-movable SILs have been demonstrated that essentially are in perfect "contact" with the Si. The FOSSIL (forming substrate into solid immersion lens) is a hemisphere machined on the backside of the Si with a small lathe [7]. Another approach forms a diffractive or Fresnel lens on the Si backside using an FIB [8]. While not movable, both of these "machined" lenses have short vertical profiles on the order of microns and can be polished smooth and reformed.

V. FUTURE TRENDS AND CONCLUSIONS

Quick defect localization and diagnosis during ramp up of a new process or product is essential for success. At the same

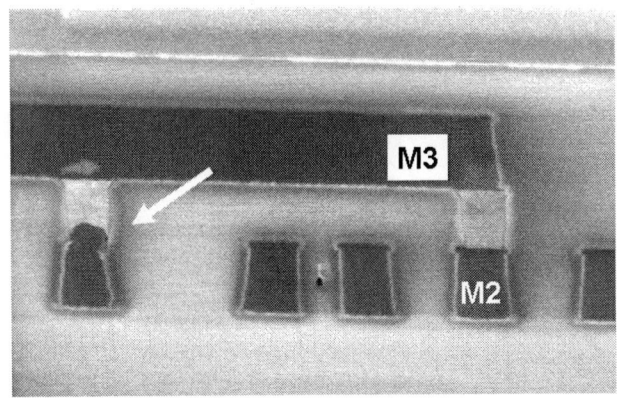

Fig. 8. FIB cross section of the resistive interconnection indicated in Fig. 7. The defect site is indicated by the arrow.

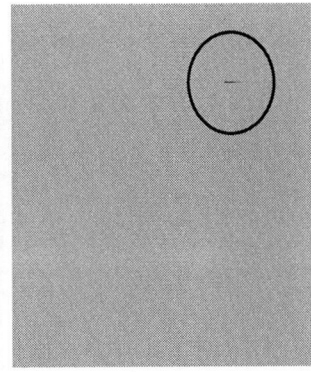

Fig. 9. Whole die front side reflected light (left) and SDL (right) images showing the site associated with a parallel path speed failure. When the highlighted area was illuminated with a 1340 nm laser it slowed enough for the IC to pass the functional test.

Fig. 10. Higher magnification combined reflected/SDL image of the site identified in Fig. 9. The suspect transistor is indicated by an arrow.

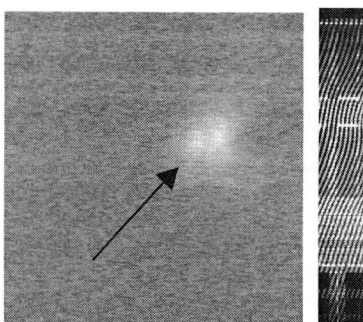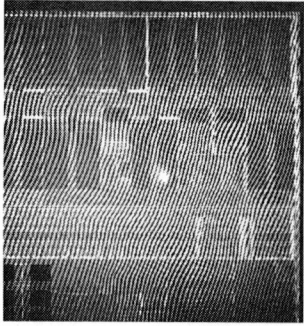

Fig. 11. Backside, SDL (left) and reflected (right) images locating a weak link in a serial timing path.

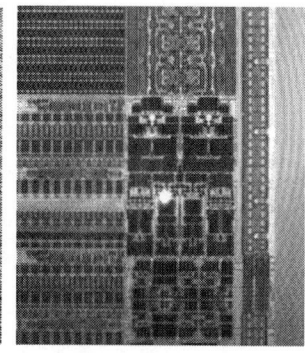

Fig. 12. Higher magnification backside SDL (left) and reflected (right) images of the weak link in a serial timing path seen in Fig. 11.

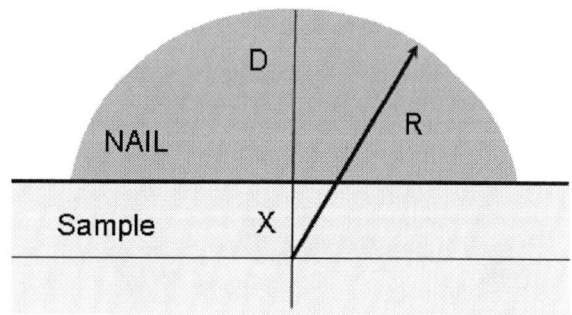

Fig. 13. NAIL solid immersion lens on the backside of a Si sample [6]. The height of the lens, D, depends on the sample thickness and doping.

time, increasing IC complexities have made failure and yield analysis more difficult. In some cases the ability to perform successful failure analysis may define the limits of a robust, reliable technology as well as economic survival. To meet these needs laser-based techniques will continue to develop by evolution and improvement of existing methodologies, by the combination of techniques for defect localization, and by the development of new analysis methods.

ACKNOWLEDGMENTS

Sandia is a multiprogram laboratory operated by Sandia Corporation, a Lockheed Martin Company for the United States Department of Energy's National Nuclear Security Administration under contract DE-AC04-94AL85000.

REFERENCES

[1] E.I. Cole Jr., J.M. Soden, J. L. Rife, D.L. Barton, and C.L. Henderson, "Novel Failure Analysis Techniques Using Photon Probing With a Scanning Optical Microscope," *Int. Rel. Phys. Symp. (IRPS),* pp. 388-399, 1994.

[2] E.I. Cole Jr., P. Tangyunyong, and D.L. Barton, "Backside Localization of Open and Shorted IC Interconnections," Int. Rel. Phys. Symp. (IRPS), pp. 129-137, 1998.

[3] K. Nikawa and S. Inoue, "New Capabilities of OBIRCH Method for Fault Localization and Defect Detection," *Proc. Of Sixth Asian Test Symposium,* pp. 214-219, 1997.

[4] M.R, Bruce, V.J. Bruce, D.H. Eppes, J. Wilcox, E.I. Cole Jr., P. Tangyunyong, C.F. Hawkins, "Soft Defect Localization (SDL) on ICs," Int. Symp. for Testing and Failure Analysis (ISTFA), pp. 21-27, 2002.

[5] J.A. Rowlette and T.M. Eiles, "Critical Timing Analysis in Microprocessors Using Near-IR Laser Assisted Device Alteration (LADA)," *Int. Test Conf.,* pp. 264-273., 2003.

[6] S.B. Ippolito, A.K. Swan, B.B. Goldberg, and M.S. Ünlü, "High-Resolution IC Inspection Technique," *Proceedings of SPIE: Metrology-based Control for Micro-Manufacturing,* Vol. 4275, pp. 126-137, June 2001.

[7] T. Koyama, E. Yoshida, J. Komori, Y. Mashiko, T. Nakasuji and H. Katoh, "High Resolution Backside Fault Isolation Technique Using Directly Forming Si Substrate into Solid Immersion Lens," *Int. Rel. Phys. Symp. (IRPS),* pp. 529-35, 2003.

[8] F. Zachariasse and M.J. Goossens, "Diffractive Lenses for High Resolution Laser Based Failure Analysis," *Int. Symp. for Testing and Failure Analysis (ISTFA),* pp. 1-7, 2005.

Broadband Circuit Techniques for
Emerging Wireless Communications

Session 5

Chair: Fa Foster Dai, Auburn University
Co-Chair: Howard Luong, Hong Kong University of Science and Technology

This session focuses on circuit design techniques for emerging broadband wireless applications, including multiple-input multiple-output (MIMO), software-defined radios (SDR), radar, vehicles, and space communications. The session is slit into two sub-sessions with three papers presented in the morning from 10:00 am to 11:40am and two papers presented in the afternoon from 2:00pm to 3:20pm.

The first paper is an invited paper given by Prof. John D. Cressler from Georgia Institute of Technology. He will talk about new technology trends, ultimate performance limits, and the emerging applications for SiGe technology. He will further demonstrate that SiGe HBTs are capable of extremely high frequency operation, exhibit very low broadband and 1/f noise, and high transconductance per unit area, making them near-ideal devices for high-speed, mixed-signal circuits.

The second paper, from Hong Kong University of Science and Technology, describes a 0.8GHz–10.6GHz ultra-wideband low-noise amplifier for software-defined radios. The paper employs the noise-cancellation common-gate stage and capacitive cross coupling for wideband input impedance matching and small noise figure, T-coils and inductive peaking for extended the output bandwidth and 2nd-order intermodulation injection for improved linearization.

The third paper, entitled "a 1.3-6 GHz triple-mode CMOS VCO using coupled inductors", is from University of Southern California. The paper presents a triple-mode VCO design including a 6th-order resonator based on three coupled inductors, banks of switched varactors, and continuously-tuned varactors, to achieve an ultra-wide tuning range from 1.28 to 6.06GHz.

The first paper after the lunch break is an invited paper from Intel Corporation on MIMO techniques for high data rate radio communications. The paper presents a 2x2 MIMO 5GHz WLAN transceiver RFIC implemented in 90nm CMOS. Architectural and circuit techniques are described for minimizing crosstalk between the multiple radio chains. The paper also discusses the future MIMO evolutions, including collaborative, directional and 60GHz phased-array communications.

The last paper in the session is from University of Florida on wireless interconnection within hybrid engine controller board. Implemented in a 130-nm CMOS process, a receiver consisting of the circuits and an on-chip antenna was demonstrated to be capable of demodulating AM signals around 14-16GHz with a data rate up to 400 Mbps while consuming 60mW.

978-1-4244-2018-6/08 $25.00 © 2008 IEEE

Notes

IEEE 2008 Custom Intergrated Circuits Conference (CICC)

Emerging Application Opportunities for SiGe Technology

John D. Cressler

School of Electrical and Computer Engineering
777 Atlantic Drive, N.W., Georgia Institute of Technology, Atlanta, GA 30332-0250 USA

Invited Paper

Abstract—Bandgap-engineered SiGe HBTs are fully-Si-manufacturing compatible, and can be fabricated in a BiCMOS platform on 200 mm wafers at high yield. SiGe HBTs are capable of extremely high frequency operation, exhibit very low broadband and 1/f noise, and high transconductance per unit area, all at very modest lithographic nodes (e.g., 200 GHz / 285 GHz f_T / f_{max} at 130 nm), making them near-ideal devices for high-speed, mixed-signal circuits. In this work we discuss new technology trends, ultimate performance limits, and emerging application opportunities in SiGe technology.

1. INTRODUCTION

SiGe HBT technology, by any fair accounting, must be judged a remarkable success story. As the first practical bandgap engineered Si-based transistor, the SiGe HBT has gone from the first demonstration of a (barely) functional device in December of 1987, to over 500 GHz peak f_T, in less than 20 years [1-2]. Rapid evolution (Figure 1). Simple device and circuit demonstrations in a select few research laboratories have transitioned to robust commercial production across the globe, and products abound across a wide spectrum of applications and operating frequencies (Figure 2). Time flies. In this paper I will review the state-of-the-art of SiGe HBT technology, with a special emphasis on some of the emerging technology trends and new application opportunities that beckon.

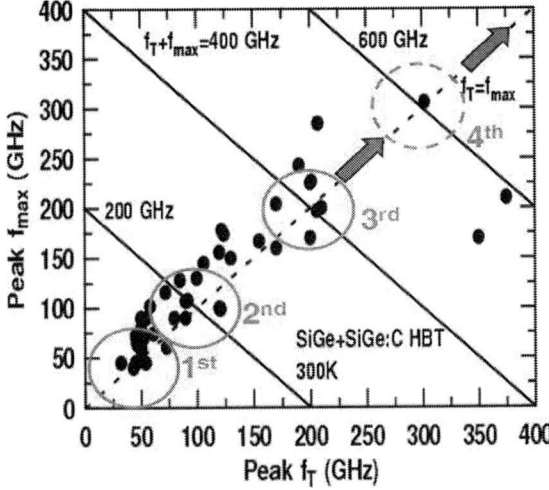

Figure 1. Measured maximum oscillation frequency vs. cutoff frequency for a variety of commercially-available SiGe technology generations.

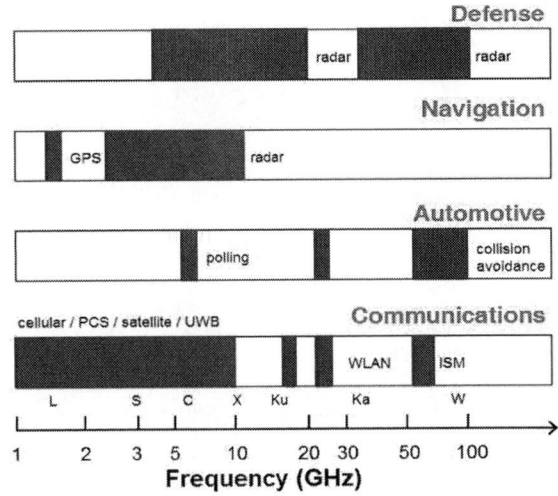

Figure 2. Applications and frequency bands of SiGe HBT technology.

2. SiGe HBT TECHNOLOGY TRENDS

Commercial SiGe HBT BiCMOS technologies presently exist in upwards of two dozen companies worldwide, with HBT speed metrics (f_T / f_{max}) ranging from 50 GHz / 50 GHz at the low end (0.5 µm, first generation), to 200 GHz / 285 GHz at the high end (130 nm, third generation) [2]. Importantly, this extreme level of device performance in a fully Si-compatible platform can be achieved at remarkably conservative lithographic nodes (130 nm in the case of 200 GHz / 285 GHz), giving SiGe a two-node generational scaling advantage over CMOS in terms of raw transistor performance. Prototype fourth generation (300 GHz / 350 GHz) SiGe HBT demonstrations already exist, impressively also achieved at 130 nm. The march for TeraHertz aggregate small-signal bandwidth (f_T + f_{max} = 1,000 GHz = 1 THz) is now aggressively underway. Figures 3, 4 show an SEM cross-section and doping profile for a typical commercial SiGe HBT.

As almost universally practiced today, SiGe technology exists as a BiCMOS implementation (SiGe HBT + Si CMOS), as an "adder" module to a core CMOS technology platform, which conveniently facilitates using the SiGe HBT where it offers the greatest advantage (i.e., in analog, RF, microwave, mm-wave, and very high-speed digital circuits), while using the Si CMOS to its strongest

978-1-4244-2018-6/08 $25.00 © 2008 IEEE

advantage. At present, the commercial applications of SiGe HBT BiCMOS technology rest squarely in the domain of "mixed-signal" ICs, which in general incorporate analog and RF/microwave/mm-wave circuit functions, and the required passive elements and interconnects to implement them, together with highly-integrated digital circuit functionality, to cost-effectively realize either system-on-a-chip (SoC) or system-in-a-package (SiP) solutions. The commercial market for such mixed-signal ICs is exploding, in the ever-expanding quest to build the electronics infrastructure supporting the 21st century communications revolution.

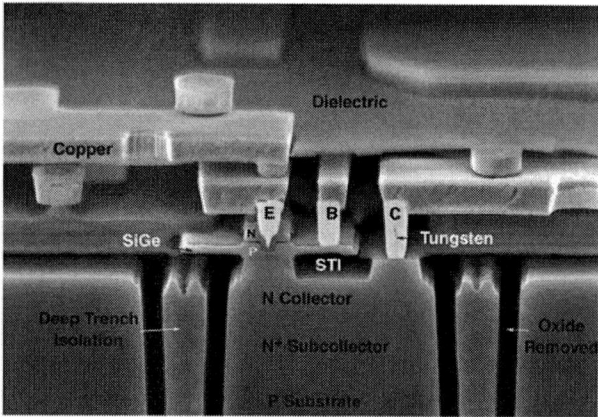

Figure 3. SEM cross-section of a SiGe HBT, showing the epitaxial SiGe base.

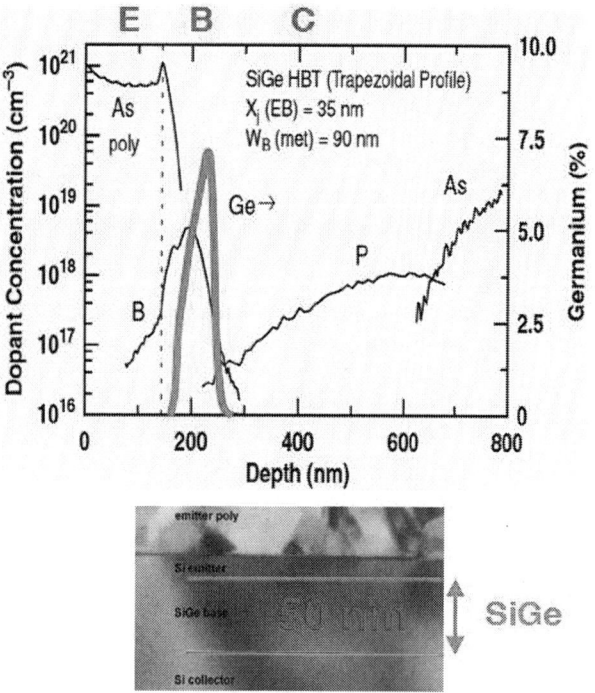

Figure 4. SIMS doping profile and cross-sectional TEM of a first generation SiGe HBT.

While it is no longer a secret how to make manufacturable SiGe HBTs, fundamentally new technology twists have recently emerged, primarily in two domains: the development of complementary SiGe (C-SiGe) HBTs for high-end analog applications, and the marriage of SiGe HBTs with thin-film, CMOS-compatible, SOI wafers.

It has long been recognized that an IC technology platform containing both *npn* and *pnp* device topologies with, ideally, equal performance, can offer compelling advantages for many types of analog circuits [3]. The use of such complementary bipolar (C-bipolar) devices can be used for lowering voltage supply rails, improving current sources, and implementing fast and low distortion driver amplifiers, for instance, and for many years C-bipolar has been the standard paradigm for analog IC design platforms. Not surprisingly, such C-Si BJT analog IC platforms tend to lag the state-of-the-art in terms of raw transistor performance, many times by a substantial margin. Bandgap-engineered SiGe HBTs can easily improve some of the key figure-of-merit bottlenecks faced by analog device designers (e.g., improving the βV_A product at a given breakdown voltage without compromising frequency response, or lowering noise without compromising frequency response), and thus are increasingly being deployed in the core analog market. Such "analog-SiGe" technologies would naturally be amenable to adding a suitable *pnp* SiGe HBT to form a complementary SiGe (C-SiGe) IC design platform for analog applications. While analog-SiGe technologies tend to have transistors with lower f_T and higher breakdown voltage than best-of-breed SiGe HBT technologies (and are often placed on thick film SOI to improve isolation), given the size of the analog/mixed-signal market, where cost margins are typically much more favorably aligned than say for wireless handsets, the extra cost associated with adding a high-speed SiGe *pnp* transistor can be easily justified. In fact, the core high-end analog industry is moving rapidly in this direction, with multiple recent demonstrations of C-SiGe HBT platforms already in production (Figure 5, 6) [4-5]. More will surely soon follow.

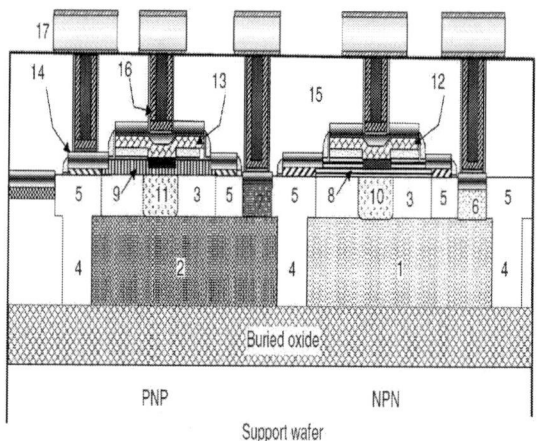

Figure 5. Schematic cross-section of a C-SiGe HBT/SOI BiCMOS technology platform [4].

978-1-4244-2018-6/08 $25.00 © 2008 IEEE

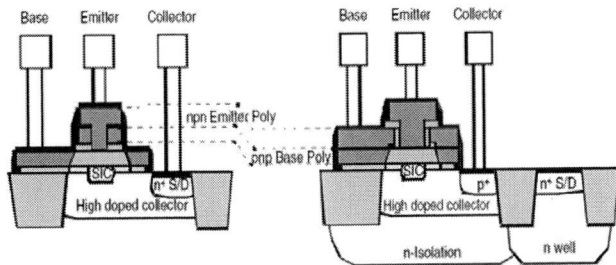

Figure 6. Schematic cross-section of an ultra-high-speed C-SiGe HBT technology platform [5].

SOI CMOS technology has increasingly received commercial attention because it boosts performance, exhibits improved isolation and cross-talk, reduces parasitics and leakage, and importantly within the context of space-and terrestrial radiation, can provide *significant* improvements to single-event upset (SEU) immunity without additional process hardening. Leveraging these virtues of SOI, SiGe technologists have recently successfully married SiGe HBTs with thin-film, CMOS-compatible, SOI substrates, and this appears to be an attractive path for future SiGe BiCMOS scaling [6-8]. In such SiGe HBT on thin-film SOI platforms, the HBT collector structure must be carefully "folded" and presenting optimization challenges for maintaining attractive performance (Figure 7). Novel contacting schemes, and profile optimization have resulted in impressive levels of performance (>100 GHz), and has the added virtue of preserving a pathway for integration of *pnp* SiGe HBTs, potentially yielding C-SiGe HBT on thin-film SOI platforms for all-purpose mixed-signal applications.

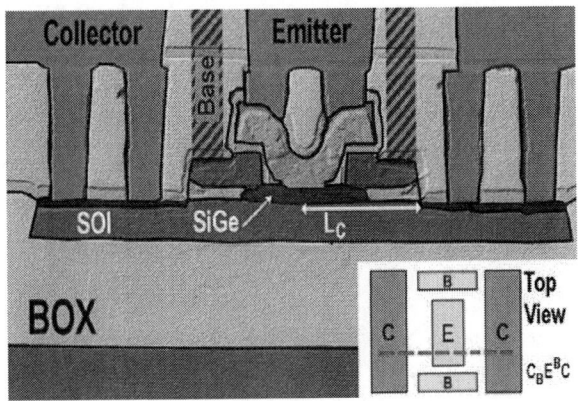

Figure 7. Schematic cross-section of a SiGe HBT on CMOS compatible thin-film SOI technology.

3. SiGe HBT PERFORMANCE LIMITS

Quantifying the ultimate speed limits of SiGe HBTs remains a hot topic. Interest in building mm-wave to TeraHertz (1 THz = 1,000 GHz) remote sensing, radar, and communication systems is growing rapidly, and recent activity has focused on expanding device options from two-terminal (e.g., Schottky diodes) to three-terminal (transistors) to better enable such THz systems. Significant progress has been made in improving the frequency response of III-V HBTs and HEMTs, with the current III-V record speed above 700 GHz. The possibility of using low-cost, highly integrated SiGe IC platforms for THz systems is clearly highly appealing. Cooling is known to improve the speed of SiGe HBTs because it amplifies the favorable effects bandgap engineering, and can thus be used as a convenient scaling lever to better understand the ultimate performance limits of SiGe HBTs. The report of the first half-Terahertz f_T SiGe HBT [9], together with more recent work involving f_{max} optimization, suggest that a 500 GHz / 500 GHz (f_T / f_{max}) – 1 THz aggregate bandwidth – SiGe HBT is likely viable at useful breakdown voltages ($BV_{CEO} > 1.5$ V), as shown in Figures 8, 9 [10-11].

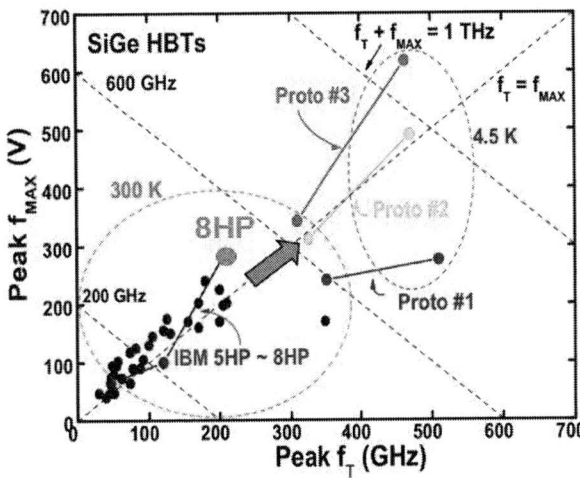

Figure 8. Measured f_{max} vs. f_T for a variety of SiGe technology generations, also showing results from recent 4th generation prototypes that have been cooled to cryogenic temperatures.

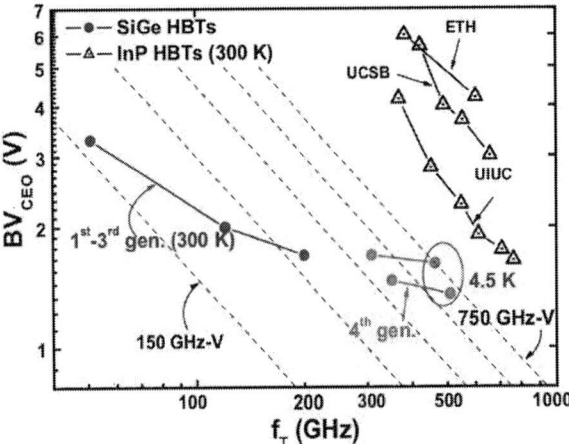

Figure 9. BV_{CEO} voltage vs f_T for various SiGe HBTs compared to best-of-breed III-V HBTs.

978-1-4244-2018-6/08 $25.00 © 2008 IEEE

4. SiGe RADARs, MM-WAVE SYSTEMS, AND ADCs

SiGe HBT BiCMOS technologies can offer competitive monolithic solutions for performance-constrained, high frequency applications such as phased array radar systems, mm-wave communications, and mm-wave imaging systems, long the exclusive domain of III-V technologies. Several recent studies have investigated highly integrated transmit/receive (T/R) modules for DoD phased-array radar systems. Such radar systems can incorporate thousands of T/R modules, making size and cost of the individual modules a crucial consideration. For this reason, an affordable, highly integrated solution is desirable, making a SiGe HBT BiCMOS technology an appealing platform to implement single-chip T/R modules with on-chip digital control functionality. Recently, a single-chip, 4-bit, SiGe T/R module for X-band phased-array radar applications was demonstrated (Figures 10, 11) [12-13]. This SiGe T/R module provides 16 distinct phase states from 0 to 360 degrees, with a *rms* phase error < 4 degrees across band. The receiver achieves a gain of over 15 dB, an operational bandwidth from 8.0 to 10.7 GHz, a noise figure less than 4 dB, and an input-referred, third-order intercept point (IIP3) of -10 dBm, while only dissipating 34 mW of power (Figures 12, 13). The transmitter achieves 18 dB gain, over 20 dBm of output power, 28% PAE, and dissipates 505 mW of power. The entire SiGe T/R module occupies a total of only 13.3 mm^2, orders of magnitude smaller than the original GaAs MMIC implementation it is intended to replace.

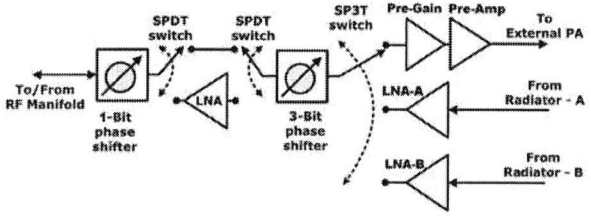

Figure 10. Block diagram for the SiGe X-band T/R module.

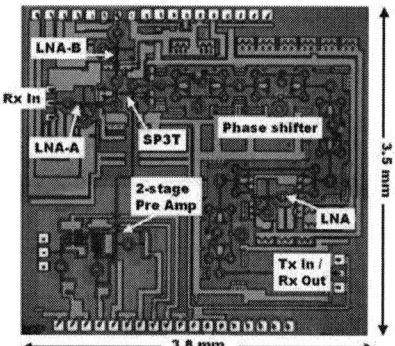

Figure 11. Die photo of the SiGe X-band T/R radar module.

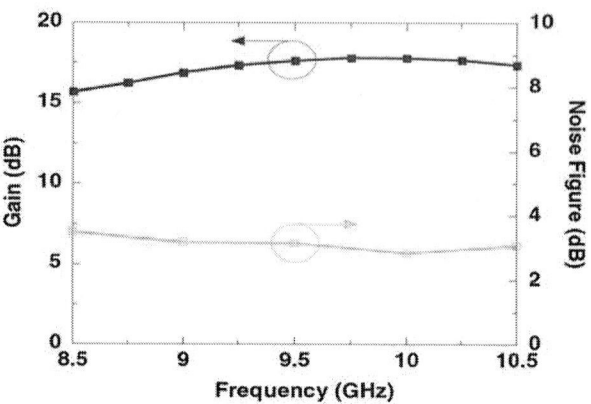

Figure 12. Receiver gain and noise figure for the SiGe X-band T/R radar module.

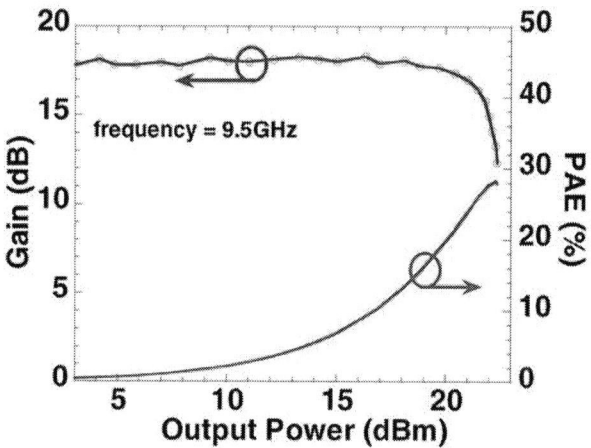

Figure 13. Transmitter power and power added efficiency for the SiGe X-band T/R radar module.

It is of course highly desirable in this context to also implement the output power amplifier on the same SiGe die, and despite limited breakdown voltages compared to III-V devices, design tricks can be played in SiGe HBTs [14], recently yielding an 850 mW, 11 dB gain, 18% PAE, X-band SiGe PA (Figure 14, 15) [15]. These output power levels, while clearly modest compared to what GaAs and GaN can offer, nevertheless are suitable for a wide range of radar applications within the newly emergent low-power-density radar system design paradigm [12], and preserve the attractive features of being realized in a Si-manufacturing compatible platform.

Other recent work in SiGe high-frequency circuits has centered on implementing commercial radar systems in SiGe, typically operating at 24 GHz and 77 GHz and primarily driven by the needs to the automotive industry for collision avoidance systems. Commercial prototypes already exist, and the potential for compact, low-cost SiGe mm-wave radars for the mass production automobile market is clearly foreseen as an attractive volume driver [16-17]. Short range, high data rate mm-wave wireless

978-1-4244-2018-6/08 $25.00 © 2008 IEEE

communications systems, at 60 GHz and/or 94 GHz have also generated significant interest in the SiGe community [18-20]. Single package (with on-board antennae) 60 GHz SiGe WLAN prototypes have already been demonstrated [18-19], as well as circuit block level demonstrations for a potentially viable path forward to 94 GHz SiGe WLAN. Recent work, for instance, using a 130 nm, third-generation SiGe platform demonstrated a SiGe LNA with 13 dB of gain, an IIP3 of -5.4 dBm, and a noise figure of 5.1 dB with DC power consumption of only 8.1 mW at 91 GHz [21]. In such demanding high-frequency applications, particularly when they are cost-constrained, having the ability to implement the mm-wave front-end in SiGe and also use on-die CMOS for DSP, all within a single chip (or at least low chip count) solution is clearly highly appealing.

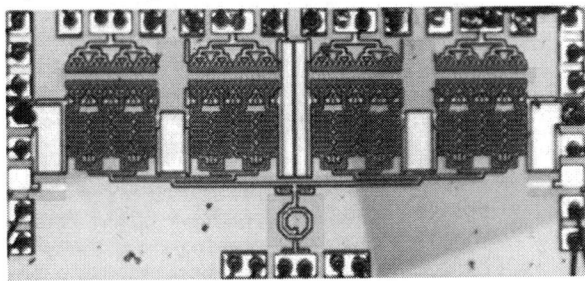

Figure 14. Die photo of the 850 mW X-band SiGe power amplifier.

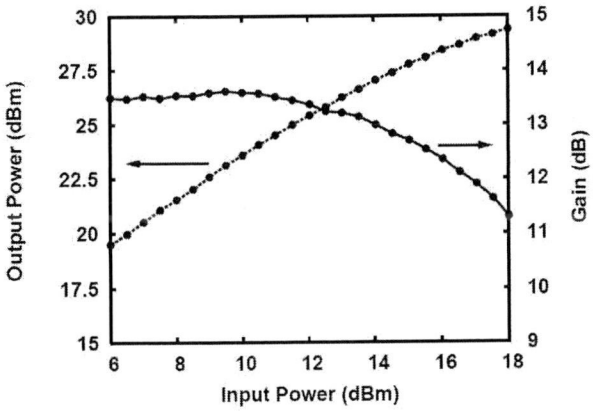

Figure 15. Output power and gain vs. input power for the SiGe X-band power amplifier.

Finally, we point out that SiGe is playing an increasingly important role is ultra-high-speed data conversion. Analog-to-digital converters (ADCs), in particular, represent a key bottleneck for many envisioned future wireless and wireline systems, where direct RF sampling is desirable (e.g., in radars and WLAN), mandating perhaps 10-20 GS/s ADCs at 8-10 bits of resolution, or in the wireline sphere, were 100 G Ethernet is likely to require 50 GS/s ADCs with 4-5 bits of resolution. Being able to accomplish this in a Si-compatible environment for monotlithic implementations is clearly

desirable. Work on SiGe ultra-high speed SiGe ADCs is in progress [22-25], with current records in the range of 40 GS/s (Figures 16, 17), and would seem to be a natural fit for best-of-breed SiGe IC platforms.

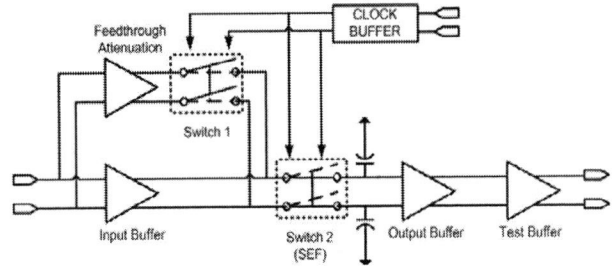

Figure 16. A 40 GS/s SiGe track and hold amplifer (THA) block diagram.

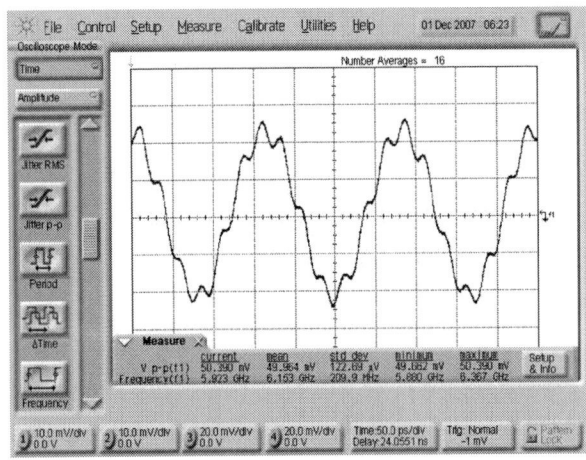

Figure 17. Measured SiGe THA differential output with a 40 GS/s sampling rate and a 6 GHz 1.0 Vpp input.

5. USING SiGe HBTs IN EXTREME ENVIRONMENTS

"Extreme environments" represent an important niche venue for electronic components, and spans the operation of electronics in surroundings lying outside the domain of conventional commercial and military specifications. Such extreme environments would include, for instance, operation down to very low temperatures (e.g., to 77 K or even 4.2 K), 2) operation up to very high temperatures (e.g., to 200ºC or even 300ºC), 3) operation across very wide and cyclic temperature swings, and 4) operation in a radiation-rich environment (e.g., space), or even all four simultaneously. The unique bandgap-engineered features of SiGe HBTs offer great utility to simultaneously satisfy all four of these extreme environment domains, potentially with little or no process modification, ultimately providing compelling advantages at the IC and system level, across a wide class of commercial and defense applications [26-27].

Figure 18 shows transistor-level performance of a third-generation SiGe HBT over temperature. Both gain,

978-1-4244-2018-6/08 $25.00 © 2008 IEEE

frequency response, and noise performance all improve dramatically with cooling. Importantly these improvements can be leveraged at the circuit level, yielding very impressive (sub-0.3 dB/20 K noise figure/temperature) SiGe low-noise amplifiers operating at deep cryogenic temperatures (Figure 19), and which may well be attractive in emerging radio astronomy applications [28-29].

Given that the performance of SiGe HBTs improves with cooling, it seems logical that performance should degrade at high temperatures, and this is indeed the case, although the degradation is modest and very attractive performance can be maintained to up to 200°C (Figure 20). Importantly, preliminary reliability studies of both devices and circuits do not reveal any serious reliability concerns at temperatures at least up to 200°C, perhaps higher.

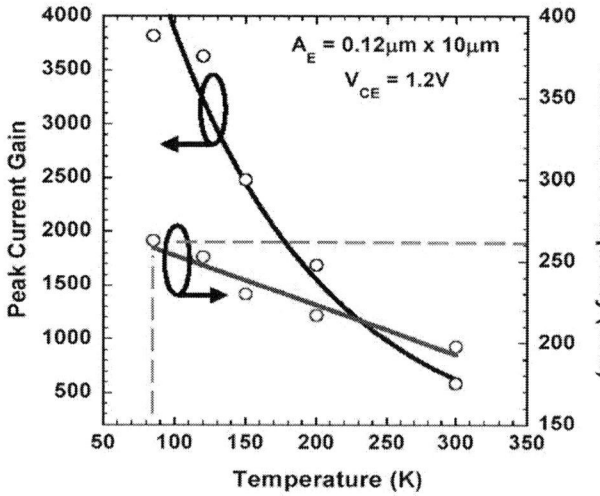

Figure 18. Current gain and peak cutoff frequency as a function of temperature for a thrid-generation SiGe HBT technology.

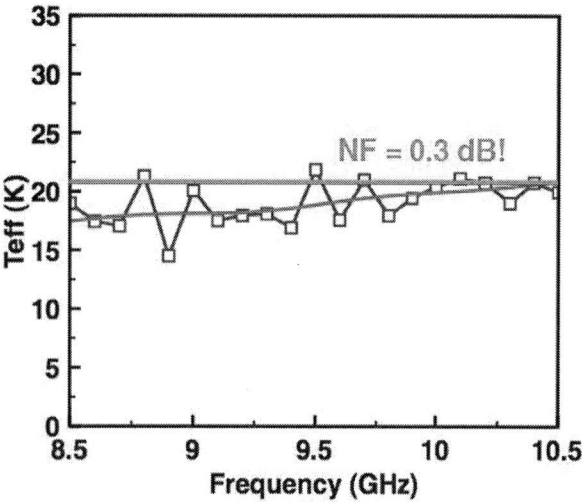

Figure 19. Effective noise tempeature of an X-band SiGe HBT LNA cooled to 15 K.

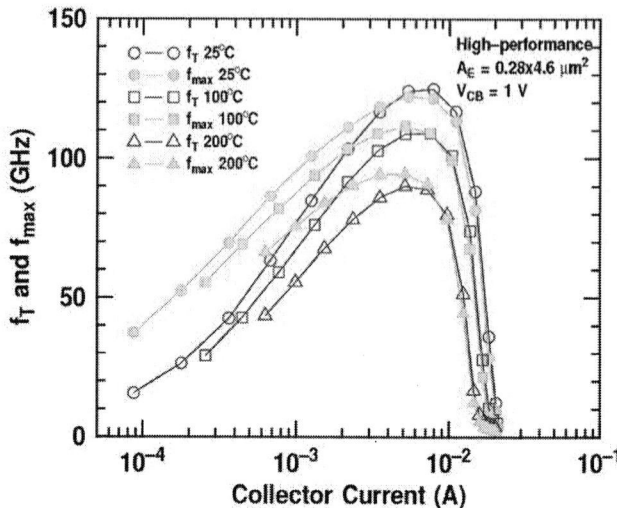

Figure 20. Cutoff frequency and maximum oscillation frequency as a function of bias current up to 200C for a second-generation SiGe HBT.

Radiation tolerance, both for total ionizing dose (TID) and for heavy ion events (i.e., single event effects - SEE) is a key concern for all space-based applications of electronics. Importantly, SiGe HBTs are TID-hard as built to multi-Mrad dose levels (Figure 21) [26]. SEE is a known problem in bulk SiGe technologies, but a combined effort based on improved TCAD for charge collection modeling, ion microbeam data, and improved circuit design using radiation hardening by design (RHBD) techniques (Figure 22), has recently demonstrated that that SEU-hard SiGe digital circuits can in fact be built (Figure 23) [30]. SEE in RF circuits is generally much more benign due to their insensitivity to short transients induced by heavy ion strikes.

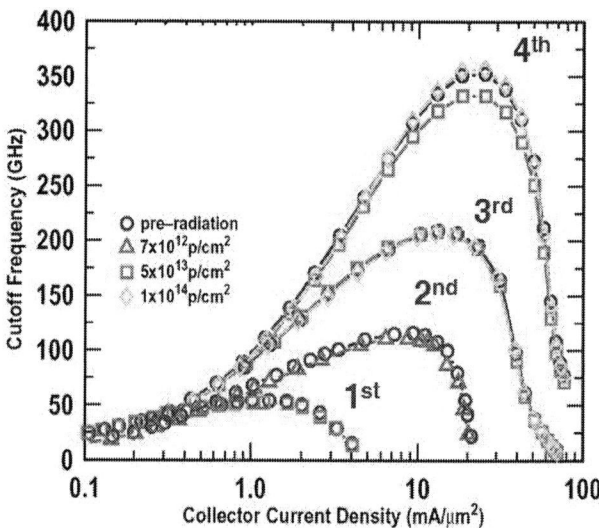

Figure 21. Multi-Mrad total dose radiation response of four generations of SiGe HBTs.

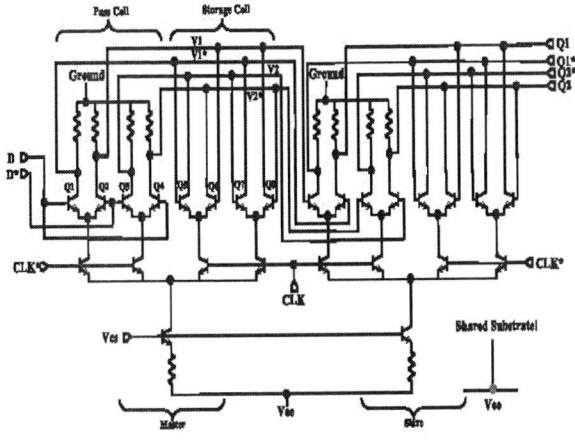

Figure 22. A new SiGe RHBD latch for SEU mitigation.

Figure 23. Measured SEU cross-section as a function of LET for various SiGe shift registers, showing the effectiveness of applying RHBD to SiGe circuits.

6. SiGe FOR LUNAR & PLANETARY EXPLORATION

The extreme environmental conditions on the Moon (at worst case, -230°C in the shadowed polar craters, and ranging from -180°C during lunar night to +120°C during the lunar day) precludes the use of conventional terrestrial electronics for sensing, actuation, and control under ambient lunar conditions. This is problematic for NASA, since the development of modular and reconfigurable human and robotics systems for lunar missions clearly requires electronic components and integrated packaged electronics modules which can operate robustly without external thermal control. Unmanned lunar missions necessarily combine mobility on the lunar surface (e.g., on a rover) with sensing functions, electronics, and motor/actuators for

control on that rover.

The sensor/actuator networks on such a lunar rover provide a distributed system to monitor the health and performance of the rover, in order to sense the environment for scientific exploration or to act on the environment, for example by using a drill to obtain a soil sample for analysis. These rover networks consist of remote "intelligent" nodes. Since these remote electronics nodes are in principle distributed over the entire vehicle, they cannot be efficiently located within protective "warm boxes" to shield them from the ambient (extreme) environment. Currently, this need for protective electronic "warm boxes" critically limits the mission designer's ability to create a truly distributed, modular electronics system for such rovers, resulting in excessive point-to-point wiring, increased system weight and complexity, lack of modularity, and an overall reduction in system reliability.

SiGe technology has been shown to be an ideal IC platform to dramatically change this situation, and the requisite infrastructure for marching this path is presently being developed, including (Figures 24-26): modeling tools (both device and circuit); establishing best-practice circuit design approaches for ultra-wide-temperature range systems; developing mixed-signal circuit libraries for NASA (and DoD) re-use; developing low thermal mismatch lunar packaging technology, complete with physics-of-failure and lifetime prediction models; establishing the robustness of components using wide temperature range cycling, vibration testing, and radiation testing; practicing radiation-hardening-by-design for needed radiation immunity; and demonstrating a general purpose "remote electronics unit" (REU) as a testbed for technology validation [31]. This REU demonstration testbed will be a highly integrated, versatile sensor interface engine, built in both a multi-chip module and as a connector, from validated SiGe components, and ready for immediate use by mission designers in a wide variety of application contexts.

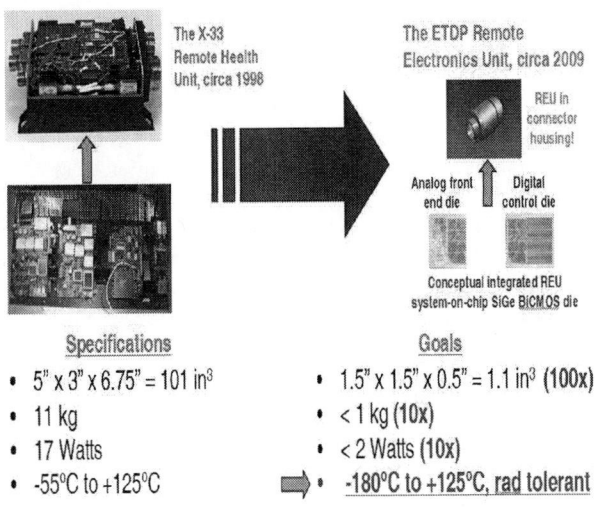

Figure 24. Vision for the SiGe Remote Electronics Unit (REU) for lunar applications.

Figure 25. A 16 channel 10x12 mm^2 SiGe Remote Electronic Unit Sensor Interface (RSI) ASIC.

Figure 26. Specially designed environment packaging platform of SiGe lunar electronic systems.

7. SUMMARY

SiGe HBT BiCMOS technology combines SiGe HBTs possessing extremely high frequency capabilities, very low broadband and 1/f noise, and high transconductance per unit area, all at very modest lithographic nodes, with conventional Si CMOS, to provide a near-ideal IC design platform for implementing many highly-integrated mixed-signal circuits and systems. A host of new application opportunities await, many with a potential to re-define the way electronic systems are implemented and deployed.

ACKNOWLEDGEMENT

This work was supported by NASA ETDP, MDA, GTRI, NASA-GSFC, DTRA, DARPA, IBM, JPL, BAE Systems, TI, ST Microelectronics, and an AFOSR MURI. The author is grateful to his students and colleagues for their many contributions, especially J. Comeau, L. Kuo, C. Grens, J. Yuan, T. Thrivikraman, S. Horst, R. Krithivasan, A. Sutton, J. Andrews, M. Bellini, R. Diestelhorst, L. Najafizadeh, and S. Finn, as well as the NASA SiGe ETDP team.

REFERENCES

[1] J.D. Cressler and G. Niu, *Silicon-Germanium Heterojunction Bipolar Transistors*, Artech, 2003.

[2] J.D. Cressler, Editor, *The Silicon-Heterostructure Handbook: Materials, Fabrication, Devices, Circuits, and Applications of SiGe and Si Strained-Layer Epitaxy*, CRC Press, 2006.

[3] D.L. Harame *et al., Tech. Dig. IEEE VLSI Tech.*, p. 71, 1991.

[4] B. El-Kareh et al., Chapter 3.11 in Silicon Heterostructure Handbook, 2006.

[5] D. Knoll *et al.,* Chapter 3.9 in *Silicon Heterostructure Handbook,* 2006.

[6] J. Cai *et al., Proc. IEEE BCTM*, pp. 215, 2003.

[7] G. Avenier *et al., Proc. IEEE BCTM*, pp. 128-131, 2005.

[8] M. Bellini *et al., Proc. IEEE Trans. Nucl. Sci.*, vol. 53 p. 3182, 2006.

[9] R. Krithivasan *et al., IEEE Elect. Dev. Lett.*, vol. 27, p. 567, 2006.

[10] J. Yuan *et al., Proc. IEEE BCTM*, p. 22, 2007.

[11] P. Chevalier *et al., Proc. IEEE BCTM*, p. 26, 2007.

[12] M. Mitchell *et al., Proc. IEEE Radar Conf.*, p. 664, 2007.

[13] J.P. Comeau *et al., Proc. IEEE BCTM*, p. 172, 2007.

[14] C. Grens *et al., Proc. IEEE BCTM*, p. 135, 2007.

[15] J. Andrews *et al., Proc. IEEE BCTM*, 2008, in press.

[16] L. Wang *et al., Proc. IEEE BCTM*, p. 218, 2007.

[17] G. Fischer *et al., Proc. ECS SiGe Symp.*, 2008.

[18] B. Floyd *et al., IEEE J. Solid-State Circuits*, vol. 40, p. 156, 2005.

[19] S. Reynolds *et al., Proc. IEEE BCTM*, p. 192, 2007.

[20] S. Voinigescu *et al., Proc. IEEE BCTM*, p, 223, 2006.

[21] J. Alvarado *et al., IEEE IMS Tech. Dig.*, 2008, in press.

[22] X. Li *et al., Proc. IEEE BCTM*, 2008, in press.

[23] S. Shahramian *et al., IEEE J. Solid-State Circuits*, vol. 41, p. 2233, 2006.

[24] J. Lee *et al., IEEE ISSCC Tech. Dig.*, p. 466, 2007.

[25] Y. Yao *et al., Proc. IEEE ASSCC*, p. 420, 2007.

[26] J.D. Cressler, *Proc. IEEE*, vol. 93, p. 1559, 2005.

[27] J.D. Cressler, *IEEE IRPS Tech. Dig.*, p. 141, 2007.

[28] S. Weinreb *et al., IEEE Trans. Micro. Theory Tech.*, vol. 55, p. 2306, 2007.

[29] T. Thrivikraman *et al., IEEE Micro. Wireless Comp. Letters*, in press.

[30] R. Krithivasan *et al., IEEE Trans. Nucl. Sci.*, vol. 53, p. 3400, 2006.

[31] R. Berger *et al., Proc. IEEE Aero. Conf.*, 2008.

A 0.8GHz–10.6GHz SDR Low-Noise Amplifier in 0.13-μm CMOS

Shuzuo Lou and Howard C. Luong

Department of Electronic and Computer Engineering,
Hong Kong University of Science and Technology, Kowloon, Hong Kong

Abstract — An ultra-wideband low-noise amplifier (LNA) is designed for software-defined radios (SDR). Noise-cancellation common-gate stage is combined with capacitive cross coupling for wideband input impedance matching and small noise figure (NF). T-coils and inductive peaking are employed to extend the output bandwidth and to reject the noise from the loading resistors. Linearization using second-order intermodulation injection is adopted to improve the IIP₃. Operated at 1.5 V from 0.8GHz to 10.6GHz, the 0.13-μm CMOS LNA measures 16-dB gain, -12dB S_{11}, 3.4-5.6dB NF, and 1.6-dBm IIP₃.

I. INTRODUCTION

Software-defined radios operating over an ultra-wide frequency band have become quite attractive to cover many wireless standards, including GSM, CDMA, WLAN, Bluetooth, UWB, etc. In achieving such a radio, designing an ultra-wideband LNA with low noise, high linearity, and low power remains critical and challenging. Existing LNA solutions cover only few selected wireless standards either from 0.8GHz to 6.0GHz [1], 1GHz to 7GHz [2], or 3GHz to 10GHz [3].

In this paper, an ultra-wideband LNA is proposed to cover the whole frequency band from 0.8GHz to 10.6GHz, which to our best knowledge is the first ever reported. The proposed LNA adopts a two-stage topology to deliver more than 16-dB gain over the whole frequency range. Broadband input matching and low NF are achieved by combining capacitive cross-coupling input stage [4] with noise-cancellation technique [5]. Inductive series peaking and T-coils are used as wideband output loading not only to extend the bandwidth but also to help suppress the noise contribution from the loading resistor. The linearization technique using second-order intermodulation injection [6] is applied to improve IIP₃.

II. CIRCUIT DESIGN

The proposed wideband LNA is shown in Fig. 1. A 2-stage design is employed in order to achieve the required large gain-bandwidth product (GBW).

A. Input Impedance Matching and NF

For the wide input frequency from 0.8GHz to 10.6GHz, low-noise low-power input impedance matching is a big challenge. The conventional LC ladder-filter matching [7] requires many

LC stages, which would be bulky and have large NF due to the low quality factor Q of the on-chip inductors. Resistive feedback technique [2] is popular but would consume large power in particular for high-frequency applications. Capacitive-cross-coupling (CCC) common-gate stage [4] is capable of low power operation. The limitation is the bias of the input devices that could be done by a large inductor, which would need to be large for the minimum frequency but would have a self-resonant frequency too low for the maximum operation frequency, or by an active current source, which would contribute significant noise. A common-gate stage with noise-cancellation technique [5] could be employed to cancel the noise from the common-gate transistor and its current source but would also consume more power.

In this proposed LNA, the CCC common-gate [4] and noise-cancellation techniques [5] are combined and optimized to achieve input impedance matching with low NF and low power. In Fig. 1, M_1 and M_2 together with capacitors C_1 and C_2 form a CCC common-gate stage. Since C_1 and C_2 are effectively shorted at high frequencies, the effective g_m of the M_1 and M_2 is doubled by the cross-coupling connection. Cross-coupled biasing transistor M_3 and M_4 are adopted for noise cancellation. The real part of the input impedance is determined by the tranconductances of M_1 and M_3:

$$Z_{in}\mid_{real} = \frac{1}{2g_{m1} - g_{m3}} \quad (1)$$

Under the impedance-matched condition ($Z_{in} = R_s = 50\Omega$), the NF can be derived as:

$$F = 1 + \gamma \frac{(g_{m1}R_s - 1)^2}{g_{m1}R_s} + \gamma(2g_{m1}R_s - 1) + \frac{1}{g_{m1}^2 R_1 R_s} \quad (2)$$

where γ is the channel thermal noise coefficient and the last three terms represent the noise contribution from M_1, M_3, and R_1, respectively. Considering only the input device (M_1) and the bias current source (M_3), minimum NF is achieved when

$$g_{m1} = 1/(\sqrt{3}R_s) \quad (3)$$

for which $F_{min} = 1 + 0.46\gamma$. As comparison, for a CCC common-gate topology with a transistor as the bias current source, $F \approx 1 + 1.5\gamma$ while for noise-cancellation technique alone [5], $F_{min} = 1 + 0.657\gamma$ and the power is almost doubled. In other words, the proposed combination of the two techniques helps achieve wider frequency band with better noise performance as compared to each individual technique alone.

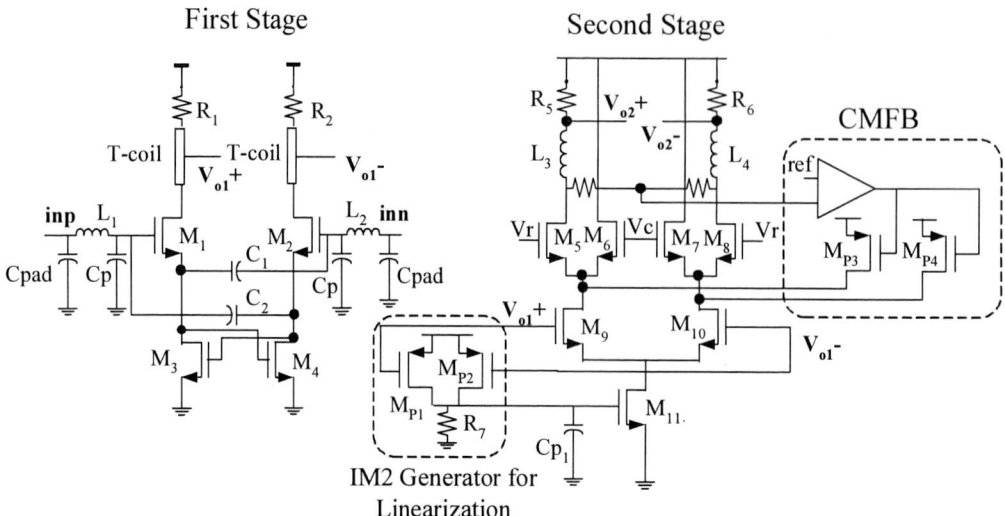

Fig. 1   Schematic of the proposed ultra-wideband LNA

Due to the output bandwidth limitation, the loading resistor R_1 can not be too large and thus contribute significant noise. In this design, a 200-Ω resistor is used for R_1. Taking into account its noise contribution and revisiting Eq. (2), the minimum NF is achieved when

$$g_{m1} = 0.803 / R_s \qquad (4)$$

which is larger than the required value $g_{m1}=0.5/R_s$ in the original CCC common-gate case. Consequently, the proposed noise-cancellation technique has the advantages of not only cancelling noise from the active devices but also increasing the required value of g_{m1} for input impedance matching to further reduce the total NF. With the proposed noise-cancellation technique, the LNA's total NF is improved by more than 0.7 dB across the whole frequency band.

After the noise cancellation in the input devices and under the optimized condition in Eq. (4), the noise contribution from M_1, M_3, and R_1 are 4%, 49%, and 47%, respectively, which implies that R_1's noise contribution becomes significant and comparable to that of M_3.

Inductors L_1 and L_2 of 400 pH each, which are implemented as transmission lines for area efficiency, are inserted at the inputs to compensate for S_{11} degradation at high frequencies due to the input parasitic capacitance C_p.

B. Wideband Loading for Resistive Load Noise Suppression

For wideband output loading, T-coil and inductive series peaking are useful to extend output bandwidth by more than 3 times [8]. T-coil is used for the 1st-stage load for its flatter frequency response while inductive series peaking is used as the 2nd-stage load because of its simpler structure and smaller area.

In addition to bandwidth extension, T-coil and inductive series peaking also help reject the noise contribution of loading resistors at the output, which is particularly useful in this

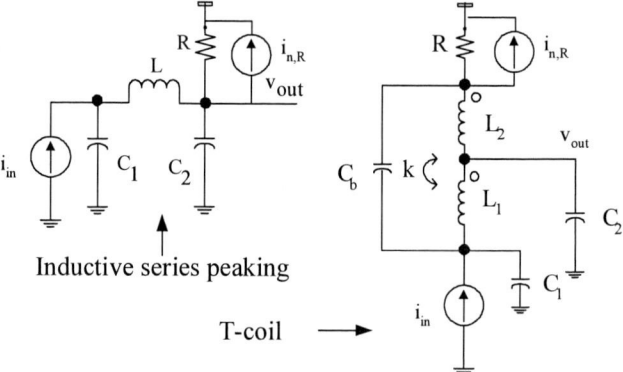

Inductive series peaking

T-coil ⟶

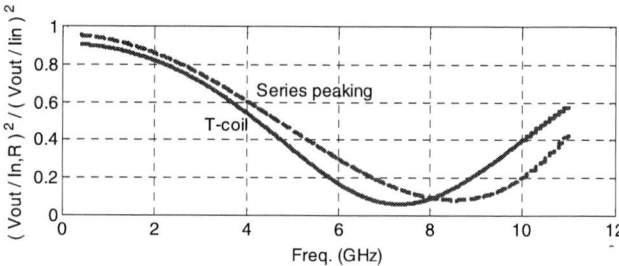

Fig. 2   Suppression of noise contribution from the loading resistor by the inductive series peaking and the T-coil

design with the small loading resistors and their significant noise contribution as described above. As illustrated in Fig. 2, the noise current and the signal current are injected into the peaking networks at different locations and thus experience different transfer functions. As an example, for the inductive series peaking in Fig. 2, the series tank formed by L and C_1 resonates at the mid-frequency, and consequently, the load noise current $i_{n,R}$ sees a low-impedance path, and its noise contribution to the output v_{out} is greatly suppressed. On the

978-1-4244-2018-6/08 $25.00 © 2008 IEEE   72

other hand, most of the current signal i_{in} flows through C_2 due to the low-impedance path formed by the series tank L and C_2. As a result, the signal bandwidth is greatly extended while the noise contribution of the load resistor is suppressed to less than half in the frequency band from 4GHz to 10GHz.

C. Linearity Consideration

IM$_2$-injection technique for transconductance linearization [6] is implemented in the second-stage of the LNA because its linearity is dominant. The basic principle is to generate IM$_3$ tone that is anti-phase and equal-magnitude for cancellation of the intrinsic IM$_3$ from M_9 and M_{10}. A simple squaring circuit, composed of M_{P1}, M_{P2} and R_7, is used to generate a low-frequency IM$_2$ tone and to mix it with the fundamental tone (through 2nd-order nonlinearity of M_9 and M_{10}) to create the required IM$_3$ tone. The generated low-frequency IM$_2$ tone at the gate of M_{11} should be in-phase with the envelope of the two input tones for best IM$_3$ cancellation, which can be satisfied as long as the IM$_2$ frequency is low enough that its phase is not affected by the cut-off frequency of the low-pass filter formed by R_7 and C_{p1}.

Due to IM$_2$-injection linearization, IM$_2$ signal may leak to the differential output in the presence of differential mismatches and degrade IIP$_2$ of the LNA. However, it is worthwhile to sacrifice some IIP$_2$ degradation for IIP$_3$ improvement because for fully-differential circuits as proposed, IIP$_2$ is typically limited by the device mismatches and is therefore much better than IIP$_3$.

A common-mode feedback (CMFB), which is composed of an opamp, M_{P3}, and M_{P4}, serves as current-steering to share part of the current from loading resistor to facilitate low-voltage operation and to set the optimal output DC voltage for maximum linearity.

D. Variable Gain

In the proposed LNA, the T-coils are optimized to achieve maximum output bandwidth with 200-Ω loading resistors for the first stage. In the lower frequency band from 800 MHz to 3 GHz, since the bandwidth is not limited, the loading resistor is designed to be reconfigurable to increase from 200-Ω to 300-Ω by turning off switches S_1 and S_2. This reconfigurability of the loading effectively increases the gain to reduce the noise at low frequencies. In addition, current steering pair $M_5 - M_8$ is employed in the second stage to achieve another 10-dB variable gain to be able to accommodate large input signals and thus achieve better dynamic range.

III. MEASUREMENT RESULTS

Fabricated in a 0.13-μm CMOS process and operated at 1.5-V supply, the differential LNA consumes a total power of 14.4 mW, and occupies chip area of 0.88 x 0.96 mm^2 as shown in Fig. 3.

The LNA drives a total loading capacitance of 80fF from the measurement pad and buffer. With S_1 and S_2 being shorted and M_6 and M_7 being turned off, the LNA measures a nominal gain

larger than 16 dB and S_{11} better than -12 dB in the whole frequency range from 0.8GHz to 10.6GHz, as shown in Fig. 4 and Fig. 5, respectively.

As shown in Fig. 6, NF is measured to be 3.5 dB for frequencies up to 7GHz and less than 5.6 dB over the rest of the frequency band from 7GHz to 10GHz for UWB. The gain drops in the middle of the frequency band while the NF remains low mainly because of the load noise rejection from the peaking networks as discussed in Section II. The NF degradation at high frequencies for UWB is mainly due to the limited output bandwidth and ineffective noise rejection by the peaking networks. Nevertheless, the NF degradation is still acceptable for UWB applications, which require much relaxed NF for full-band operation.

Fig. 7 summarizes IIP$_3$ as a function of RF frequencies. At the gain setting of 12 dB with a 2-tone spacing of 5MHz, IIP$_3$ is measured to be around 4dBm for frequencies below 5GHz and around 2dBm for frequencies up to 10.6GHz. The linearization circuitry provides an IIP$_3$ improvement of 8dB without affecting the gain or the NF while consuming only an extra current of 0.1 mA. The linearization becomes less effective at higher frequencies because the non-linearity of the device capacitors becomes more dominant than the non-linearity of the transconductors and thus cannot be suppressed by the proposed linearization technique.

When two input tones at 4.7 and 5.7 GHz are applied, IIP$_2$ at 1 GHz and 10.4 GHz are measured to be 29dBm and 22 dBm respectively. With the linearization circuitry being turned off, the two IIP$_2$ values become 33dBm and 26dBm, respectively.

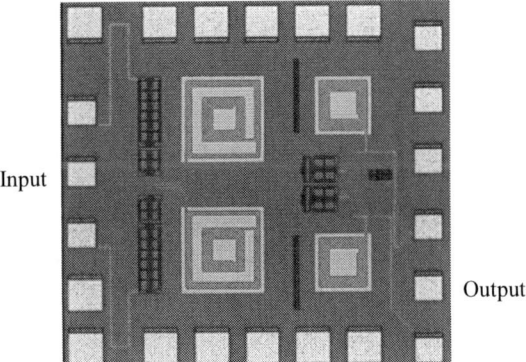

Fig. 3   Chip photo of the proposed ultra-wideband LNA

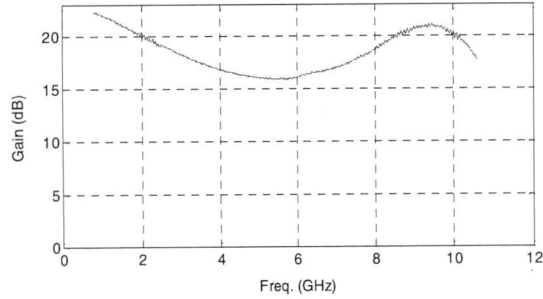

Fig. 4   Measured gain of the proposed LNA

Table I summarizes the performance comparison with other wideband LNAs. The proposed LNA achieves the widest operation frequency range and the best linearity with acceptable NF and comparable gain and power consumption. To the best of our knowledge, this is the first LNA ever reported that can cover the whole frequency band from 0.8GHz to 10.6GHz.

IV. CONCLUSION

An ultra-wideband LNA is demonstrated featuring high performance over an ultra-wide frequency band from 0.8GHz to 10.6GHz for most of the existing wireless standards, including UWB. Noise-cancellation technique is combined with capacitive cross-coupling at the input to achieve good input impedance matching, low NF, and low power. In addition, wideband output peaking networks help extend the output bandwidth and reject the noise from resistive loading to further reduce NF. Finally, linearization technique is embedded to suppress the IM_3 of the LNA without affecting the gain and NF while drawing negligible extra current.

ACKNOWLEDGEMENT

This work was supported by the grant 617407 funded by the Research Grant Council, Hong Kong. The authors wish to acknowledge the assistance from F. Kwok and K. W. Chan with the measurement.

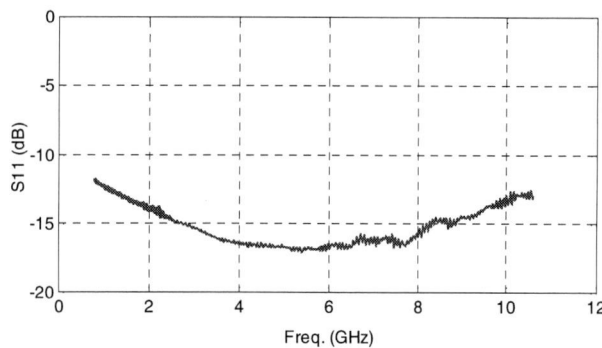

Fig. 5   Measured S_{11} of the proposed LNA

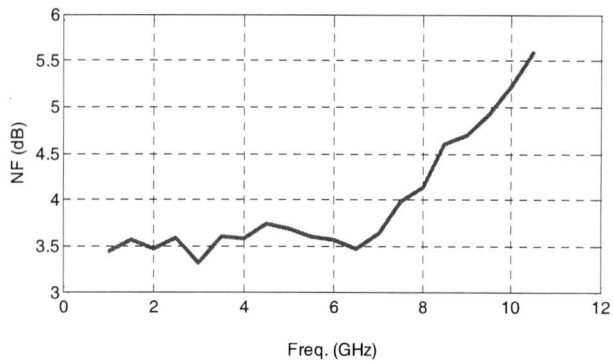

Fig. 6   Measured NF of the proposed LNA

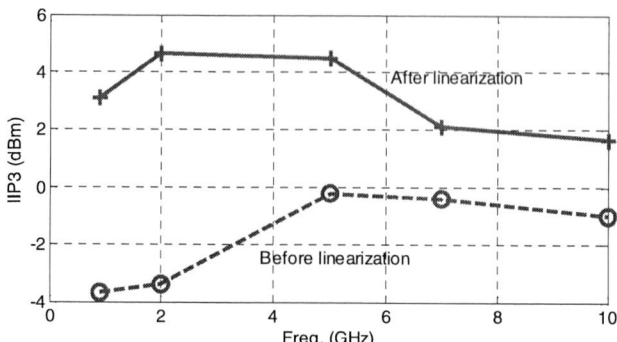

Fig. 7   Measured IIP_3 of the LNA with and without linearization

TABLE I. PERFORMANCE COMPARISON OF WIDEBAND LNAs

	[1]	[2]	[3]	This work
Frequency /GHz	0.8-6	1-7	3.1-10.6	0.8-10.6
S_{11}/dB	<-10	<-10	<-9.9	<-12
Gain/dB	>18	>16	~15.1	>16
NF/dB	2.1-2.5	2.7-3 (1-3 GHz)	2.1-3	3.4-5.6
IIP_3/dBm	--	-4.1	-8.5 - -5.1	>1.6
Supply/V	2.5	1.4	1.2	1.5
Power/mW	12.5	25	9	14.4
Process /CMOS	0.09μm	0.13μm		

REFERENCES

[1] R. Bagheri, A. Mirzaei, S. Chehrazi, M. E. Heidari, M. Lee, M. Mikhemar, W. Tang, and A. A. Abidi, "An 800-MHz-6-GHz software-defined wireless receiver in 90-nm CMOS," *IEEE J. Solid-State Circuits*, Vol. 41, No. 12, pp. 2860-2876, Dec. 2006.

[2] R. Ramzan, S. Andersson, J. Dabrowski, and C. Svensson, "A 1.4V 25mW inductorless wideband LNA in 0.13μm CMOS," *IEEE International Solid-State Circuits Conf.*, Feb. 2007.

[3] M. T. Reiha, and J. R. Long, "A 1.2 V reactive-feedback 3.1-10.6 GHz low-noise amplifier in 0.13 μm CMOS," *IEEE J. Solid-State Circuits*, Vol. 42, No. 5, pp. 1023-1033, May 2007.

[4] W. Zhuo, S. Embabi, J. P. de Gyvez, and E. Sanchez-Sinencio, "Using capacitive cross-coupling technique in RF low noise amplifiers and down-conversion mixer design," *European Solid-State Circuits Conf.*, pp. 77-80, Sep. 2000.

[5] A. Amer, E. Hegazi, and H. F. Ragaie, "A 90-nm wideband merged CMOS LNA and mixer exploiting noise cancellation," *IEEE J. Solid-State Circuits*, Vol. 42, No. 2, pp. 323-328, Feb. 2007.

[6] S. Lou, and H. Luong, "A linearization technique for RF receiver front-end using second-order-intermodulation injection," *IEEE Asian Solid-State Circuits Conf.*, pp. 87-90, Sep. 2007.

[7] A. Bevilacqua, and A. M. Niknejad, "An Ultra-wideband CMOS LNA for 3.1 to 10.6 GHz wireless receivers," *IEEE International Solid-State Circuits Conference*, Feb. 2004.

[8] S. Lou, H. Zheng, and H. Luong, "A 1.5-V CMOS receiver front-end for 9-band MB-OFDM UWB system," *IEEE Custom Integrated Circuits Conf.*, pp. 801-804, Sep. 2006.

IEEE 2008 Custom Intergrated Circuits Conference (CICC)

A 1.3-6 GHz Triple-Mode CMOS VCO Using Coupled Inductors

Zahra Safarian and Hossein Hashemi

Electrical Engineering - Electrophysics, University of Southern California, Los Angeles, CA-90089
Email: zsafaria@usc.edu, hosseinh@usc.edu

Abstract- **An integrated VCO with an ultra-wide tuning range is designed and fabricated in a 0.13μm CMOS technology. The triple-mode VCO uses a 6th-order resonator based on three coupled inductors with a compact common-centric layout, banks of switched varactors, and continuously-tuned varactors to cover the measured 1.28-6.06GHz frequency range. Each mode corresponds to a resonant frequency of the 6th-order resonator. Mode selection is achieved using three independent active cores without using lossy switches in the resonator path. The VCO current consumption is automatically adjusted from 2.9mA to 6.1mA to achieve a low phase noise throughout the frequency range. The measured phase noises at 1MHz offset from carrier frequencies of 1.76, 2.26, 3.3, 4.5, and 5.6 GHz are -120.89, -121.82, -118.46, -117.12, and -114.15 dBc/Hz, respectively. The chip area, including the pads, is 1mm x 1 mm and the supply voltage is 1.5V.**

Index Terms — **CMOS, voltage controlled oscillator (VCO), phase noise, coupled inductor.**

I. INTRODUCTION

Wideband voltage controlled oscillator (VCO) is an essential block in multi-standard, multi-band, and multi-function wireless systems. The VCO must maintain low phase noise and power consumption throughout the entire covered frequency range, and should consume a small chip area.

Tunable active inductors have been proposed to achieve a wide frequency turning range [1]; but, they usually don't result in a low phase-noise VCO. Varactor-tuned LC VCOs are common in CMOS implementations. Unfortunately, a single LC resonator cannot cover a large tuning range due to the limited varactor tuning range. Switched capacitors, switched inductors, and switched resonators have been proposed to enhance the VCO tuning range. In these implementations, the switch loss degrades the resonator quality factor (*Q*) hence and the oscillator phase noise. Moreover, in the latter two cases, the chip area becomes large due to using several on-chip inductors. Coupled inductors (e.g., vertically stacked, common centered, intertwined, etc.) can be used to reduce the chip area [2][3]. But, *Q* degradation due to the switch loss will still remain. A switch-less dual-band transformer-based VCO that covers a wide frequency range has been reported in [4]. A wide range of frequencies can be generated in a compact area by using digital dividers that follow an LC VCO [5], at the expense of divider power consumption. In [6], a dual-mode oscillator based on multiple active cores and coupled inductors was illustrated. In [7], it was showed that an oscillator that uses a higher-order

resonator can have multiple stable oscillation frequencies, and that the desired oscillation frequency can be selected without using *Q*-degrading switches at the resonator. A dual-mode oscillator using this concept was also demonstrated in [7].

In this paper, a 6th-order resonator that is based on three coupled inductors, laid out in an area-efficient common-centric configuration, will be used in a triple-mode wideband VCO. Mode selection is achieved using three independent active cores. The proposed triple-mode operation is covered in section II. Circuit design details and measurement results are presented in sections III and IV, respectively, and are followed by concluding remarks in section V.

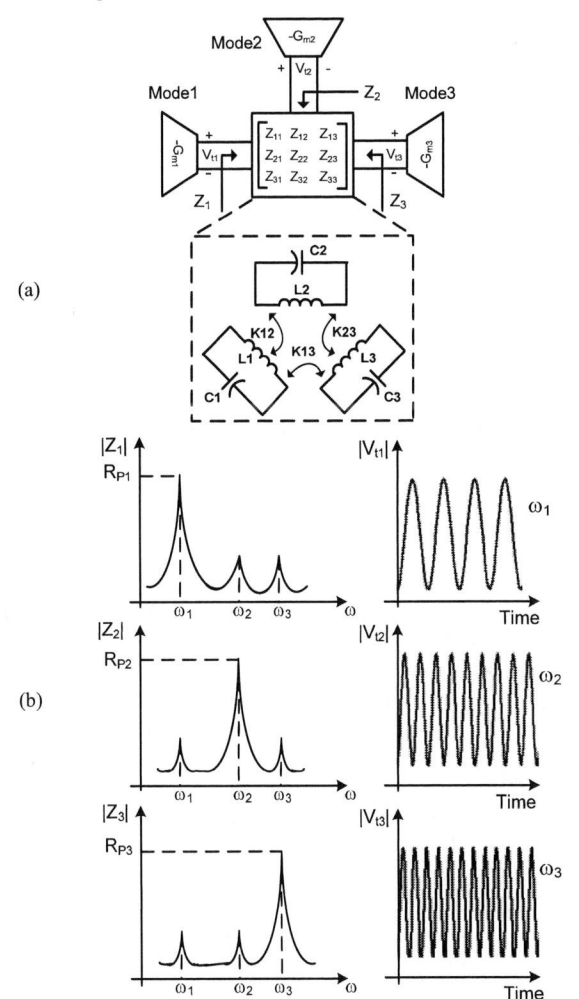

Fig. 1. (a) Triple-mode oscillator, and (b) the impedance magnitude seen at each port and the corresponding steady-state voltage at each port.

978-1-4244-2018-6/08 $25.00 © 2008 IEEE

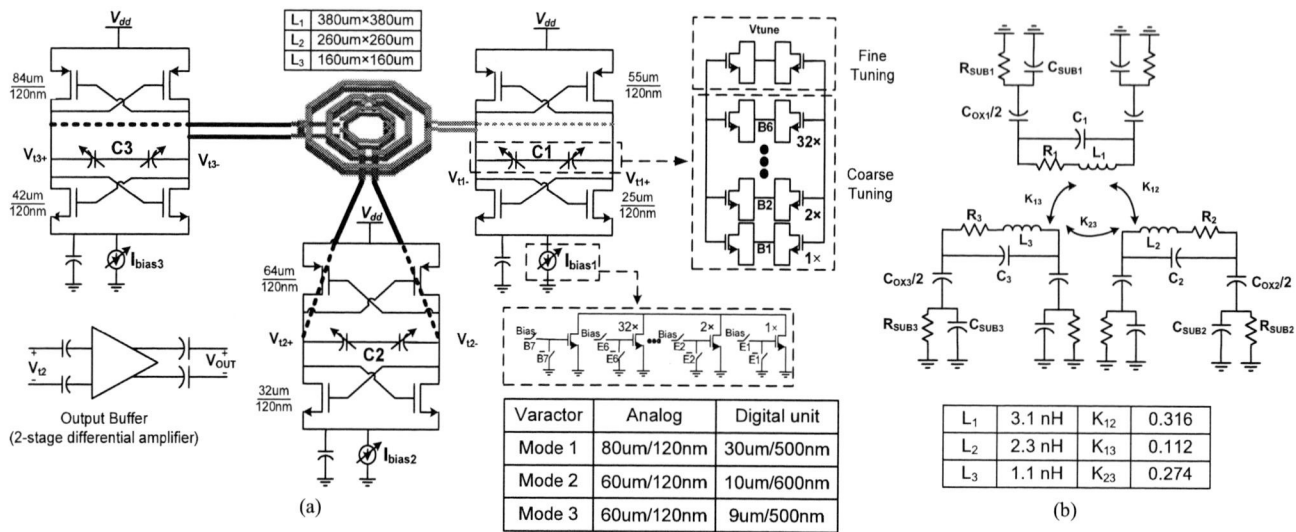

Fig. 2. (a) The detailed schematic of the triple-mode VCO, and (b) the broadband model for the three coupled inductors.

II. TRIPLE-MODE OPERATION USING COUPLED INDUCTORS

The proposed VCO is built around a 6th-order resonator based on three coupled inductors as shown in Fig. 1(a). A 6th-order resonator has three resonant frequencies. In a coupled inductor-based resonator, the relative magnitude peaks depend on the coupling factors, k_{ij}. Conventional wisdom may suggest using higher k between all pairs. However, extensive simulations indicate that having large coupling coefficients make one of the resonant peaks much larger than the others and hence, sustaining steady-state oscillations at all resonant modes becomes difficult or impossible. Therefore, lower coupling factors are preferred in multi-mode oscillator designs. In theory, the maximum achievable Q of a resonator that is based on n coupled identical inductors with identical coupling, k, is $(1+nk)$ times larger than the individual Q of inductors [8]. However, this improvement very much depends on the resonator topology and the value of components [8]. In our implementation, the coupled inductors are designed to have a small mutual coupling, while achieving a large quality factor in all resonant modes. Compared with other 6th-order resonators, the advantage of the proposed triple coupled inductor-based resonator is the presence of three ports at which the impedance magnitude has the highest peak at one of the distinct resonance frequencies (Fig. 1 (b)). At any port, only the oscillation corresponding to the highest peak is stable Therefore, independent active cores can be placed at each port to sustain oscillation at the desired mode. Only one active core is ON at any given time; the corresponding port sustains the highest oscillation swing, while oscillations also appear in the other two ports with smaller amplitude at the same oscillation frequency due to the coupling. In this design, the voltage across the 2nd (middle) coupled inductor is buffered to pick up oscillations for all modes. No switches are used at the resonator, and hence, the phase noise is not degraded.

III. VCO ARCHITECTURE

In order to put the foregoing discussions in use, the principle of triple-mode oscillator is applied in the design of a wideband VCO. The detailed schematic of the VCO is depicted in Fig. 2 (a). The actives at each core are standard cross-coupled CMOS differential pairs with variable current sources to optimize the phase noise performance as the frequency is varied.

A. Triple Coupled Inductor Design

As mentioned in section II, in order to have all the three modes, a relatively low k among the three coupled inductors is desired. In this design, common-centric 3D inductors are used. Each inductor has two turns implemented in the vertical dimension. In order to reduce the series loss, the top turn is realized by stacking the top two 0.55μm-thick metal layers with 0.65μm spacing, and the bottom layer is realized by stacking three 0.32μm-thick metal layers with 0.35μm spacing. Metal layers M2-M4 are chosen for the bottom layer to minimize the capacitance between the vertically coupled turns (8-metal process). The lumped model shown in Fig. 2 (b) with fixed values, except for R_1, fits well with the IE3D electromagnetic simulations over the entire frequency range. The resonator Q looking into each port with EM-simulated inductors and assuming ideal capacitors at all ports is shown in Fig. 3 (a).

B. Varactor Design

In order to tune the oscillator over a wide frequency band for each mode, a large capacitance variation is still required. A low-gain for VCO is often desirable to minimize the AM-PM noise conversion. In this design, a 6-bit binary-weighted switched varactor has been used for the coarse tuning in each mode. Switched varactors have been exploited rather than the more conventional switched finger capacitors due to their higher Q, smaller area, and higher tuning range. For fine tuning, a small NMOS varactor with minimum channel-length has been used for each mode. The same digital bits and analog control voltage are used for all modes, so that all resonant frequencies change together. Same bits control the current sources to optimize the phase noise performance. There is a known trade-off between the tuning range and Q of the varactors; larger channel length results in higher the tuning range (C_{max}/C_{min}) and lower Q. In this design, the channel length has been chosen in such a way that the simulated the varactor Q is higher than 60. As shown in Fig. 3 (a), the resonator Q with ideal capacitors (dashed-line) and the resonator Q with the switched varactors (solid-line) are almost the same.

Fig. 3 (b) shows the peak impedance magnitude looking into each port when all the other ports are terminated with their corresponding switched varactors. The voltage swing is less than the supply voltage (current limited region) to avoid resonator Q degradation when transistors enter the triode region.

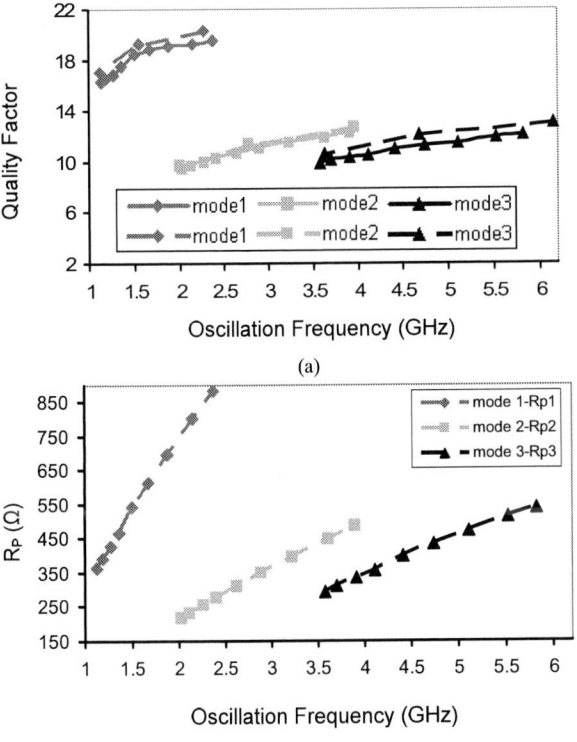

(a)

(b)

Fig. 3. Simulated curves for (a) Q of the 6th-order resonator looking in to each port when other ports are terminated with ideal capacitances (dashed) and with switched-varactors (solid), and (b) the peak magnitude of the impedance at each port.

IV. MEASUREMENT RESULTS

The triple-mode VCO is fabricated in a 0.13μm CMOS technology with 8 metal layers. The chip size, including the pads, is 1mm x 1mm (Fig. 4). Fig. 5 shows the measured VCO frequency range as the control voltage changes from 0 to 1.5 V, for the three modes. Mode 1 covers the 1.28-2.27GHz, mode 2 covers 2.34-4.03GHz, and mode 3 covers 3.65-

6.06GHz. There is about 380MHz overlap between mode 2 and 3. However, there is an unfortunate 70MHz gap between modes 1 and 2, and a few other gaps throughout the bands. These can be easily eliminated by using a larger fine tuning varactor. From the frequency tuning curves, the VCO gain (K_{VCO}) is calculated and plotted in Fig. 6.

The current consumption of the VCO and its bias circuitry for different modes and throughout the frequency range has been shown in Fig. 7. The supply voltage is 1.5V. It should be reminded that the VCO current is automatically adjusted to achieve a low phase noise as the frequency is varied. For instance, a higher Q in mode 1 leads to lower current consumption. Fig. 8 (a) shows a sample measured phase noise versus offset frequency when the VCO is operating at 4.54 GHz. Fig. 8 (b) shows the phase noise at 1 MHz offset frequency throughout the tuning range. Table 1 summarizes the performance of this VCO in comparison with recently reported wideband CMOS VCOs.

V. CONCLUSION

A 6th-order resonator, formed by three coupled inductors, has three resonant frequencies, and as such can be used in a triple-mode oscillator. Switching between modes is achieved by using independent active cores at the three ports of the coupled inductor-based resonator. A compact wideband triple-mode VCO has been designed and fabricated in 0.13μm CMOS technology. The measured 1.28-6.06 GHz frequency tuning range is achieved due to a combination of 6th-order resonator and switched-varactors. Current consumption is adjusted automatically across the frequency tuning range to optimize the phase noise performance. The wideband VCO can be used in multi-standard, multi-mode, and multi-function wireless systems.

ACKNOWLEDGMENTS

This work was partially supported by the National Science Foundation under contract ECS-0621874.

REFERENCES

[1] L. H. Lue, et al., "A wide tuning-range CMOS VCO with a differential tunable active inductor," IEEE Trans. Microwave Theory Tech., vol. 54, no. 9, pp. 3462–3468, Sep. 2006.

[2] M. Demirkan, et al., "11.8 GHz CMOS VCO with 62% tuning range using switched coupled inductors," in Proc. RFIC, 2007, pp. 401–404.

[3] L. Geynet, et al., "Fully-integrated multi-standard VCOs with switched LC tank and power controlled by body voltage in 130nm CMOS/SOI," in Proc. RFIC, 2006.

[4] A. Bevilacqua, et al., "A 3.4-7 GHz transformer-based dual-mode wideband VCO," in Proc. ESSCIRC, 2006, pp. 440–443.

[5] J. Borremans, et al., "A single-inductor dual-band VCO in a 0.06mm² 5.6GHz multi-band front-end in 90 nm digital CMOS," in IEEE ISSCC Dig. Tech. Papers, 2008, pp. 324–325.

[6] R. Gharpurey, et al., "A single-tank dual-band reconfigurable oscillator," in Proc. VLSI Circuits Symp., 2006, pp.176–177.

[7] A. Goel, et al., "Frequency switching in dual-resonance oscillators," IEEE J. Solid-State Circuits, vol. 42, no. 3, pp. 571–582, March 2007.

[8] H. Krishnaswamy, et al., "Inductor- and transformer-based integrated RF oscillators: a comparative study," in Proc. CICC, 2006, pp. 381–384.

[9] D. Hauspie, et al., "Wideband VCO with simultaneous switching of frequency band, active core, and varactor size," IEEE J. Solid-State Circuits, vol. 42, no. 7, pp. 1472–1480, July 2007.

978-1-4244-2018-6/08 $25.00 © 2008 IEEE

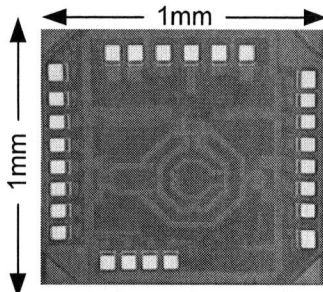

Fig. 4. Chip microphotograph of the triple-mode VCO.

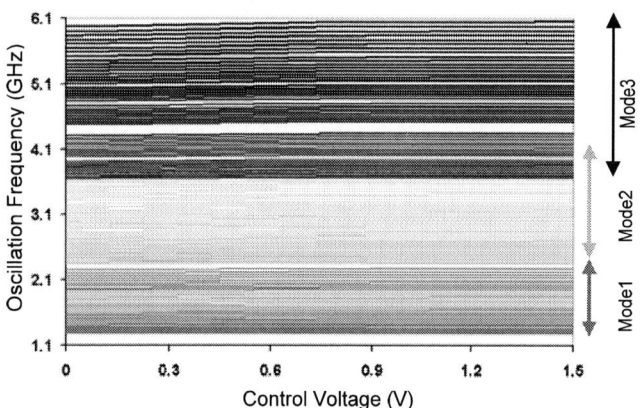

Fig. 5. Measured tuning curves of the triple-mode VCO.

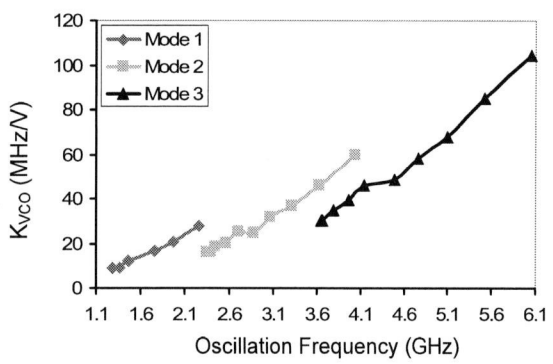

Fig.6. Measured VCO gain versus frequency.

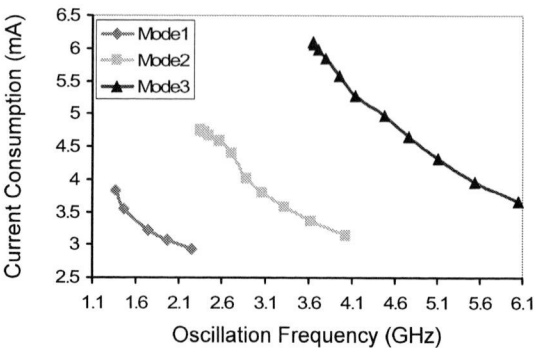

Fig. 7. Current consumption of the VCO and its bias circuitry as a function of oscillation frequency.

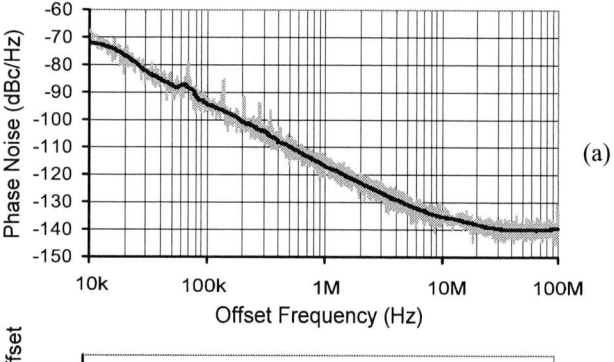

(a)

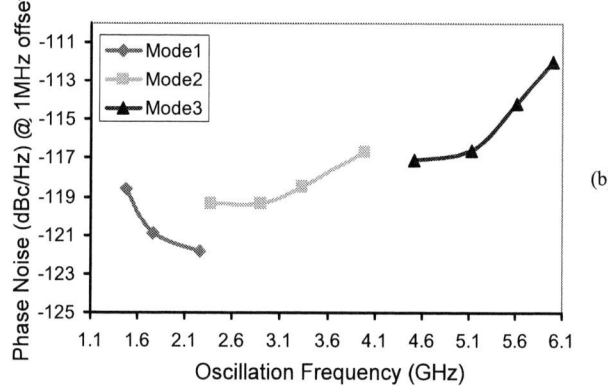

(b)

Fig. 8. (a) Sample phase noise measurement at 4.54 GHz, and (b) phase noise at 1MHz offset from the center frequency throughout all the bands of operation.

TABLE I

PERFORMANCE SUMMARY AND COMPARISON WITH RECENTLY PUBLISHED WIDEBAND VCOS

Ref.	Technology	Tuning Range (GHz)	Power (mW)	PN @ 1 MHz (dBc/Hz)	Area (mm x mm)	Topology
[1]	0.18μm CMOS	0.5 - 3	6 - 28	-102 @ 2.9 GHz	0.4 × 0.63	Active inductor
[4]	0.13μm CMOS	3.4 - 7	1 - 8	-118.6 @ 4.6GHz	0.85 × 0.77	Dual-mode transformer
[5]	90nm CMOS	DC – 5.6 [(1)]	2.16– 9.96 [(2)]	-123 @ 5.4GHz [(3)]	0.034	Single-inductor dual-band + Dividers
[9]	0.13μm CMOS	3.1 – 5.2	2.52 – 8.5	-118 @ 4GHz	NA	LC + switched active core
This Work	**0.13μm CMOS**	**1.3 - 6**	**4.35-9.15** [(4)]	-117.12 @ 4.5GHz	**1 × 1**	**Triple coupled inductors**

[(1)] Several large gaps in the range. [(2)] Does not include divider power consumption. [(3)] At 2.5 MHz offset. [(4)] Includes the bias circuitry.

Advanced SoC/SiP Integration & Co-Design   Session 6

Chair: Rich Liu, Macronix International Co., Ltd.
Co-Chair: Philippe Jansen, IMEC

Higher level system integration to achieve better performance and lower cost has consistently driven new technologies. The solutions are no longer just monolithic; in some cases the best solution involves innovative new packaging approaches and in others on-chip integration of different technologies. This session presents several approaches that may best solve particular application challenges.

The first, invited paper from Georgia Institute of Technology, Atlanta, discusses the recent progress in CMOS-based millimeter-wave technology. Although advances in CMOS technology can achieve 30GHz to 100GHz millimeter-wave performance, however, conventional metal packaging still presents a bottleneck for true low cost single-chip radio. This paper describes how to use SoC/SIP co-design to allow the use of standard QFN package for ultra-high frequency and paves the way for low cost consumer applications of millimeter-wave ICs.

The second, also invited paper from IBM, presents a lead-free solder process that satisfies both regulatory constraints and IC industry's environmental conscience. The higher hardness of Sn-based solder, coupled with softer BEOL low-K dielectric material create unique challenges that demand careful optimization of the process parameters.

The last paper from Toshiba addresses the process compatibility of high voltage devices, MEMS and CMOS technology. MEMS technology often uses different materials and processes than conventional CMOS and thus it is desirable to add MEMS after CMOS. However, MEMS Deep Reactive Ion Etching processes may cause plasma damage and thus affect CMOS devices. This paper demonstrates both the integration of HV DMOS device and MEMS on thick SOI wafers and investigates the effect of long time plasma etching on CMOS circuits.

978-1-4244-2018-6/08 $25.00 © 2008 IEEE

Notes

A SOC/SOP Co-design approach for mmW CMOS in QFN Technology

J. Laskar, S. Pinel, S. Sarkar, P. Sen, B. Perunama, D. Dawn, D. Yeh, and F. Barale.

Georgia Electronic Design Center, School of Electrical and Computer Engineering,
Georgia Institute of Technology, Atlanta, GA 30332-0269 USA

Fax: 404-894-0222 Email: joy.laskar@ece.gatech.edu

ABSTRACT — **The past few years have witnessed the emergence of CMOS based circuits operating at millimeter wave-frequencies. The co-design of fully integrated mmW CMOS single chip digital radios with low cost QFN packages operating from 30 to 100GHz offers the promise for high volume manufacturing which paves the way for a new millimeter-wave industry. As standardization efforts catalyzed the interest and investment of industry and agencies, one can be assured of ubiquitous millimeter-wave technology in the consumer electronic market place in the fairly near future.**

Index Terms —**60GHz, CMOS, single chip radio, 90nm, transceiver, QFN.**

I. INTRODUCTION

In the past few years, the interest in the millimeter wave spectrum at 30–300 GHz has drastically increased. The emergence of low cost high performance CMOS technology and low loss, low cost organic packaging has opened a new perspective for system designers and service providers because it enables the development of millimeter wave radios at the same cost structure of radios operating in the gigahertz range or less. In combination with available ultra wide bandwidths, this makes the millimeter wave spectrum more attractive than ever before for supporting a new class of systems and applications ranging from ultra-high speed data transmission, video distribution, portable radar, sensing, detection and imaging of all kinds [1].

Conventional millimeter-wave technology is based on expensive and bulky on metal housing, high performance ceramic substrates and relies on low volume manual multi-chip hybrid assembly as show in figure 1. Although well proven, this approach leads to prohibitive manufacturing costs of several thousand dollars and does not scale well for high volume applications.

Figure 1: Conventional millimeter wave modules.

Since the early 1990's, many examples of MMIC chip-sets have been reported for millimeter wave radio applications using GaAs FET and InP pHEMT technologies increasing the integration density. Compact 60GHz Wireless 1.25 Gb/s Transceiver Modules utilizing Multilayer Co-fired Ceramic and GaAs pHEMT technologies has been successfully demonstrated as shown in figure 2 [2]. More recently SiGe BiCMOS technology has also been demonstrated to be a viable alternative at these frequencies [3]. However, despite their commercial availability and their performances, these technologies still struggle to enter the market because of their cost and their limited capability to integrate advanced base-band processing.

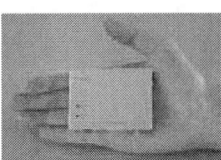

Figure 2: Compact 60GHz Wireless 1.25 Gb/s Transceiver Modules utilizing Multilayer Co-fired Ceramic and GaAs pHEMT technologies [2].

Recently, the availability of standard CMOS technology enabling the design of MMIC circuits operating efficiently up to 100 GHz has revived the interest and investment in the use of the 60GHz spectrum, targeting indoor ultra-high speed short-range wireless communications for multimedia applications [4-5].

The emergence of a multitude of "bandwidth hungry" multimedia applications has exacerbated the need for multi-gigabit wireless solutions, out of reach of conventional WLAN technology (802.11a,b and g), or even more recent emerging UWB and MIMO systems. Uncompressed high-definition video distribution and massive data synchronization are driving data-throughput requirements well beyond gigabit/s, and already demanding up to 10Gbps with the introduction of, for example, the HDMI 1.3 video standard. The availability of the 7 GHz unlicensed bandwidth in the 60GHz spectrum represents a unique opportunity to address such data-throughput requirements. As standardization efforts catalyzed the interest and investment of the industry [6-7], one can

978-1-4244-2018-6/08 $25.00 © 2008 IEEE

anticipate that 60GHz technology will move into the consumer electronic marketplace in the near future.

Similarly, numerous opportunities exist for low cost commercial millimeter-wave systems at higher frequency such as 77GHz for automotive radar, 71-76 and 81-86 GHz outdoor 10Gbps networks, 94GHz for medical and security imaging providing a prelude to the possibility of terabit (in aggregate throughput) systems operating well into the sub-THz regime (several hundred GHz).

In this paper, we discuss the integration of a 60 GHz digitally controlled single-chip 90nm CMOS fully integrated radio [8-9] scalable to phased array systems and co-designed with integrated antenna [10-11] in a standard QFN package. This co-design has been optimized for robustness against process variation and temperature, and verified by statistical measurement results to establish a foundation for high volume manufacturing (HVM).

A wideband super-heterodyne architecture combined with a high-speed digital signal processor has been designed to support the complete 57 to 66 GHz bandwidth and enable data throughput exceeding 7Gbps QPSK and 15Gbps 16QAM for a total DC power budget below 200mW

In addition, an on-chip signal processor provides optional high speed PHY processing (ADC/DAC, pulse shaping, equalization, demodulation, bit synchronization, etc…). Thus, the 60 GHz CMOS single-chip radio is a fully digitally controlled, data "bits-in / bits-out" solution. This is the highest system performance (power consumption, data rate, and bandwidth) and the highest level of integration for a 60GHz (or any radio above 5GHz) single-chip radio reported to date. It provides the lowest energy per bit transmitted wirelessly at multi-gigabit rates, to meet the very stringent low-power specifications for battery operated consumer portable electronic devices, and enable scalability towards digital CMOS millimeter wave radar.

II. CO-DESIGN APPROACH FOR MMW SOC/SOP SOLUTION

The design of millimeter wave circuits and systems presents considerable challenges because of the structural complexity of both active and passives circuits to be optimized. Current empirical models lack range and accuracy, numerical modeling methods tend to be computationally expensive, and analytical methods are difficult to optimize. In addition, the trends toward using low cost manufacturing technologies introduce additional process variability, making design centering difficult, requiring cumbersome iterative hardware prototyping and therefore impacting significantly the time-to-market.

We have developed an efficient design for an effective SOC/SOP co-design of new millimeter wave systems is described in figure 3, establishing a foundation for high volume manufacturing (HVM).

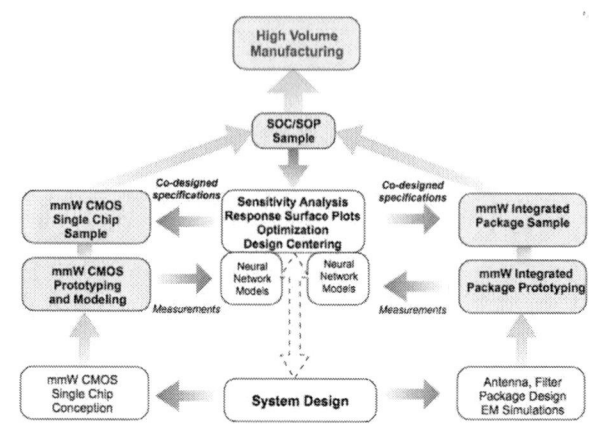

Figure 3: SOC/SOP co-design approach for new millimeter wave systems.

In the first phase, it enables powerful and scalable model development based on a limited number of prototypes. CMOS and passives/package enabling blocks are identified from the system analysis and implemented in a generic prototype form to generate a statistical measurement data base. It is imperative that these prototypes are fabricated using a standard manufacturing process (in this case a standard digital CMOS and a standard PCB process) to guarantee scalability to HVM. Electrical models, complex parasitic effects in 3D structures, significance of components and layout parameters are captured in these systematic measurements. Neural networks have emerged as an attractive technique for modeling complex nonlinear relationships [12], since they can be used to generate accurate models directly from measured data, thus accounting for manufacturing variations and parameter indeterminacy issues. No assumptions about the component behavior are made when developing neural network models.

In the second phase, millimeter wave CMOS single chip radio and millimeter wave integrated package samples are co-designed using neuro-genetic methodology. Genetic algorithms have been extensively developed for a simultaneous multi-parameter search [13]. Therefore, neuro-genetic methodology combines the accuracy of neural networks in nonlinear modeling and efficiency of genetic algorithm in a multi-parameter search. It allows for virtual prototyping, fast optimization of the co-design specifications, design partitioning and centering for the complete SOC and SOP solution. This allows one to develop optimum SOC/SOP samples for first pass design and can be statistically verified against process and temperature variation (PVT). The results of this characterizations can easily be processed by the neuro-genetic tool for further optimization, minimizing drastically the number of development cycles at this stage before fixing the design for high volume manufacturing.

978-1-4244-2018-6/08 $25.00 © 2008 IEEE

This approach maximizes the system integration gain achievable by combining the SOC and SOP approach. The SOC/SOP co-design approach is a key enabler for new mmW markets. It fosters the convergence of wide range of technologies such as a QFN standard package, embedded antennas and filters, CMOS single chip multi-gigabit transceiver with integrated signal processor as described in figure 4. This approach has allowed the GEDC's Millimeter-wave Gigabit Wireless group to establish new benchmarks for the highest data rate transmitted wirelessly ("Bits-in/Bits-out") at 60GHz and at the lowest power budget.

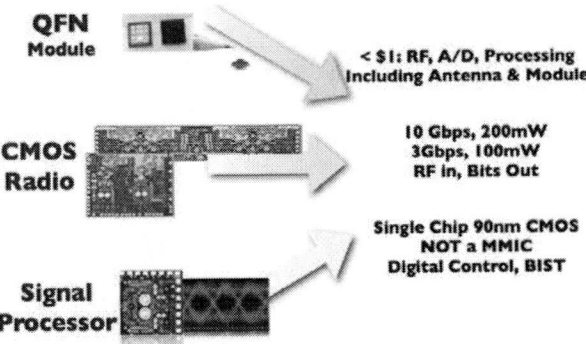

Figure 4: Measured SOC/SOP wideband millimeter wave (60GHz in this example) system benchmarks

III. 60GHZ SINGLE-CHIP CMOS 90NM RADIO ARCHITECTURE AND IMPLEMENTATION

With the proper transistor layout, 90nm digital CMOS technology has been established as a technology of choice to enable robust and low-power implementation of 60GHz CMOS-based multi-gigabit transceivers. The typical 90nm digital CMOS node offers sufficient cut-off frequency performance for the front-end radio and integration levels for the back-end and offers an optimal trade-off in performance and cost. Numerous recent publications [14-16] report examples of efficient implementations of 60GHz CMOS building blocks, thus demonstrating the potential of CMOS technology at 60GHz. When integrated on low-cost organic packaging, this technology promises efficient high-volume fabrication, lowering the cost (same cost structure as radios operating in the gigahertz range or less), and providing the foundation for major commercial impact for consumer electronics.

The block level architecture of the single-chip 60GHz radio is described in Fig. 5 and the fabricated die photo is shown in Fig. 6. The wideband super-heterodyne architecture has been designed to support the complete 57 to 66 GHz band. The band is divided into 4 channels of 2160 MHz. However, there are possibilities for channel bonding to support ultra-high throughput. Thus, in order to support

the ultra-high throughput, the IF frequency needs to be carefully chosen, and needs to be high enough to simultaneously prevent baseband spectrum aliasing. In this particular work, the chosen 7GHz-13GHz IF incorporates ultra-wide-band design techniques.

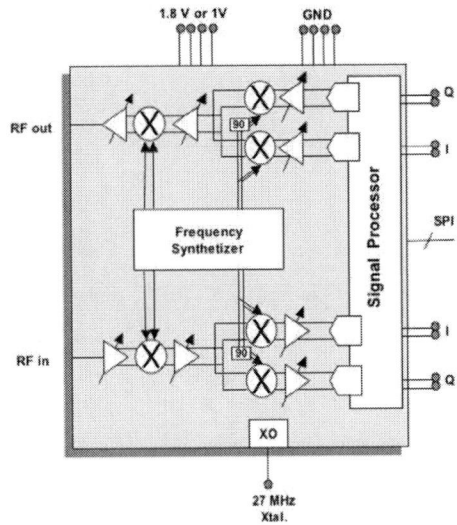

Fig. 5: 60GHz single-chip 90nm CMOS radio architecture.

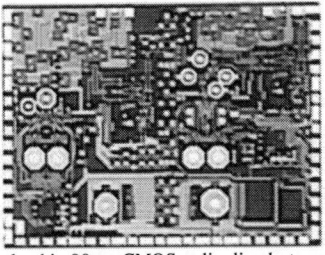

Fig. 6: 60GHz single-chip 90nm CMOS radio die photograph.

In addition, an on-chip signal processor provides optional high speed PHY processing (ADC/DAC, pulse shaping, equalization, demodulation, and bit synchronization). The 60 GHz CMOS single-chip radio is externally controlled by a standard SPI interface and uses a low cost 27MHz crystal. Digital controls include: operation modes (Transmit/Receive/Sleep), channel/frequency selection, Transmitter/Receiver chain gain and phase (providing scalability to phased array), on-chip signal processor controls, etc... The performances of the fully integrated single-chip 60GHz radio are summarized in table I when operated from a 1.8V supply. Further power consumption reduction is achieved when operated from 1V supply and 2dBm output power.

978-1-4244-2018-6/08 $25.00 © 2008 IEEE

TABLE I
MEASURED PERFORMANCE SUMMARY OF 60GHz SINGLE CHIP RADIO

Transceiver Blocks	PA	IQ Modulator QVCO PLL	VCO PLL	LNA Mixer IF amplifier	IQ Demod QVCO PLL
Frequency (GHz)	57-65	7-13	49-55	57-65	7-13
Gain (dB)	17	-	-	32	17
P_{Sat} (dBm)	8.4	-	-	-	-
P1-dB (dBm)	5.1 (output)	3 (output)	-	-30 (input)	-
Matching (dB)	>12 (output)	>25 (input)	-	>15 (input)	>25 (output)
Phase noise @1MHz offset	-	-	-95 dBc /Hz	-	-
DC power consumption (mW)	54	17+22+20	20+40	70	19+20+ 20
Total DC power consumption	Tx Mode: 173 mW Rx Mode: 189mW				

As the die area is largely dominated by the analog design. Up to 200,000 gates can be easily integrated without significant increase in die size. Integrated sensors are also easily integrated and coupled with DSP capabilities for "smart" digital control or digitally assisted analog as shown in figure 7. The digitally assisted analog design approach incorporates built-in self-test (BIST) enabling automatic output power calibration/control, frequency synthesizer calibration, DC offset and IQ imbalance compensation, temperature and process corner compensation.

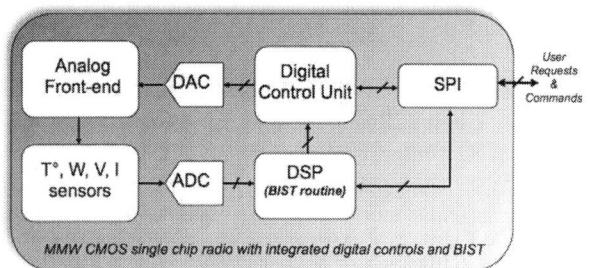

Fig. 7 Digitally assisted analog design approach incorporates built-in self-test (BIST).

The performances and functionality/integration level can be further improved by scaling the digital CMOS radio architecture to a digital single chip CMOS radar. An example of a 1 by 4 single chip 60GHz CMOS radar is shown in figure 8. It should be noted that we expect this type of radio architecture can be used to develop a variety digital architectures from 6 to 100GHz.

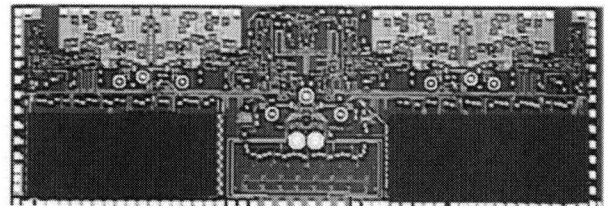

Fig. 8: 60GHz single-chip 90nm CMOS radar die photograph.

IV. FR4 BASED MODULE AND ANTENNA TECHNOLOGY

The combination of FR4 and high frequency laminates appears as a platform of choice for the packaging of the future 60 GHz gigabit radio. Multi-layers substrates (0.61m x 0.46m panels) are fabricated using standard PCB High Volume Manufacturing facilities. An example of a large panel area multi-layer substrate process flow is shown in figure 9. The multi-layer substrates include compact embedded filters, wideband millimeter-wave feed-through transition and antenna phased array, and allows for flexible high performance interconnects between sub-modules integrated on the same panel. The module design enables extended azimuth and elevation coverage, provided by conformal multi-sector configurations which are enabled at no extra mechanical assembly cost unlike conventional approaches [17].

Numerous antenna array solutions have been developed to address various applications scenarios ranging from VSR (very short reach) omni-directional to point-to-point links. Such a generic packaging platform provides a path of choice toward the low cost integration of scalable SISO-MIMO radio systems (SM radio) using a compact multi-sector phased array architecture that overcomes the fundamental limitations of millimeter wave signal propagation. The extended range (including non-LOS scenarios) provided by high gain adaptive phased array and/or multi-sector technologies are breakthrough attributes for future low cost commercial millimeter-wave systems.

978-1-4244-2018-6/08 $25.00 © 2008 IEEE

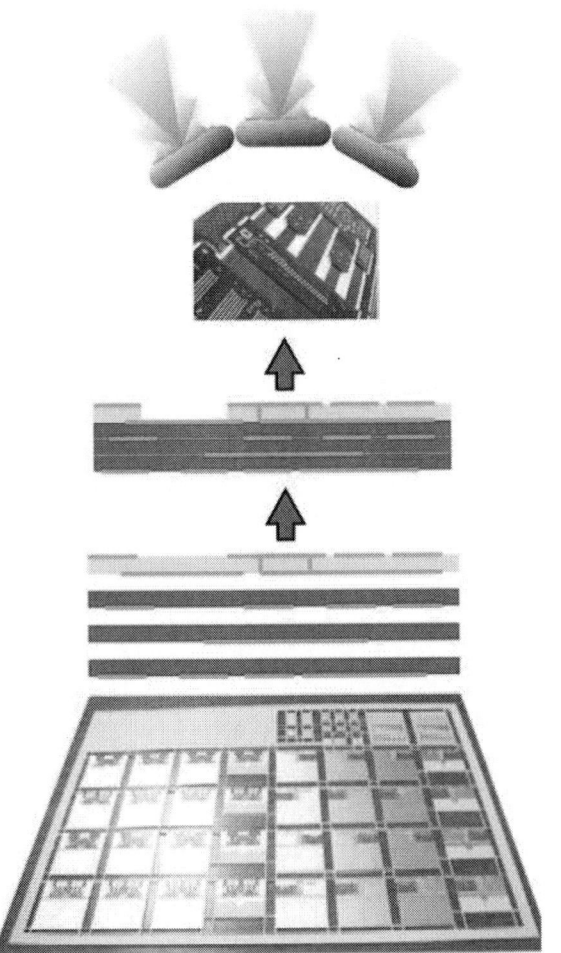

Figure 9: Large panel area multi-layer substrate including compact 60GHz antennas and filters, and conformal multi-sectors phased array architecture for extended azimuth and elevation coverage,

V. EMBEDDED FILTER DESIGN AND CHARACTERIZATION.

The figure 10 shows an example of a backed co-planar wave- guide (BCPW) coupled line topology. The characteristic impedance has been set during the design to higher impedances such as 65 Ohms, leading to a larger line width and spacing and adapted back to 50 Ohms at the input and output. Thus, characteristic impedance, width and spacing has been adjusted and optimized to accommodate HVM requirements.

The simulated and measured performances are reported in figure 11. A very good agreement between simulated and measured results has been obtained. The measured reflection loss have been measured to be better than 15dB. A minimum insertion of 1.85dB has been measured at 60.3 GHz for a relative bandwidth of 11%. Also, a high rejection of 32dB has been measured at 11.3% offset from the center frequency.

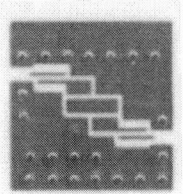

Figure 10: Photograph of the fabricated BCPW LCP filter.

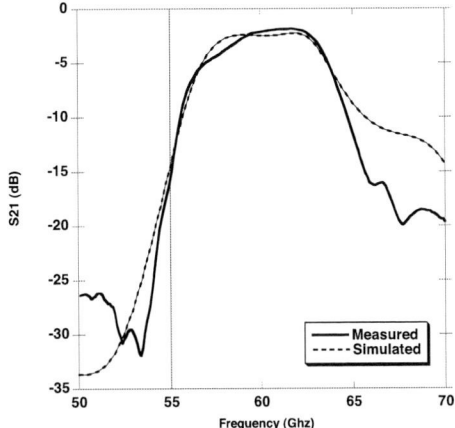

Figure 11: Measured and simulated insertion loss of the BCPW filter.

Figure 12 shows the simulated and measured group delay of the filter. A variation of less than 25 picoseconds has been measured in the pass-band, which makes the filter very attractive for high data rate wireless systems.

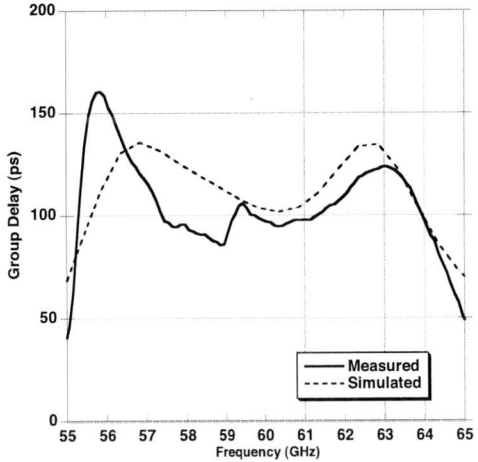

Figure 12: Measured and simulated group delay of the BCPW filter.

In addition, integrated waveguide (IWG) techniques can be used to implement wideband millimeter-wave feed-through transitions. Designs exhibiting less than 0.2 dB insertion loss at 60GHz have been implemented [4]. It replaces expensive and difficult to align co-axial feedthroughs, facilitates the integration with planar interconnect, enables low loss, and enables well controlled millimeter wave vertical transitions between two layers of packaging.

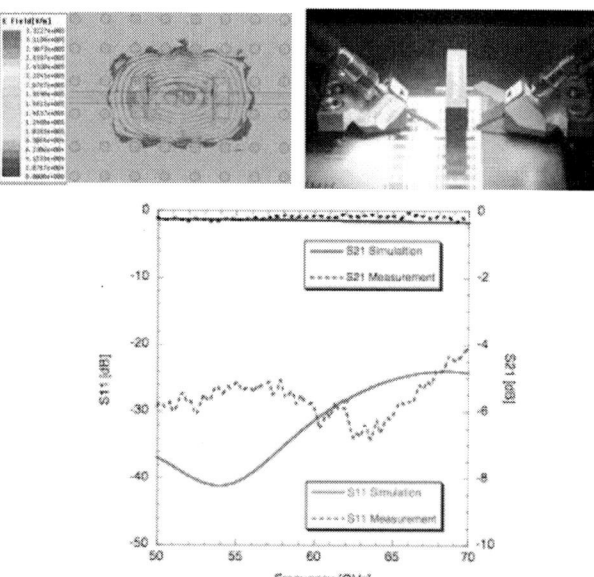

Figure 13: Multi-layer compact embedded wideband millimeter-wave feed-through transition design, test and measurement results.

Planar antennas are very attractive for low cost millimeter wave applications due to their simplicity and good performance. In figure 14 we show an example of a 60GHz 2 by 2 antenna array exhibiting a measured gain of 12dBi. Such an antenna is considered as a sub-array in a larger phased array antenna system. With the proper design, multiple sub-arrays can be driven with different phases to electronically steer the radiated beam as demonstrated in figure 15 for a 16 element phased array. It also increases the overall antenna gain, maintains low side lobe level, and reduces the required number of electronic controls.

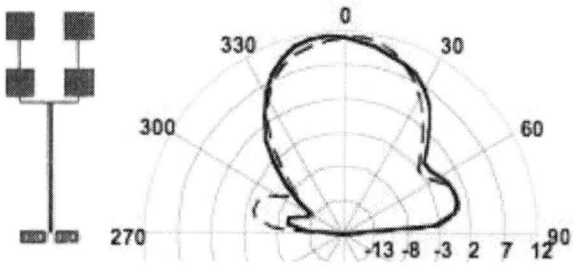

Figure 14: Fabricated compact 60GHz antenna sub-array and measured (vs simulated in dotted line) radiation pattern

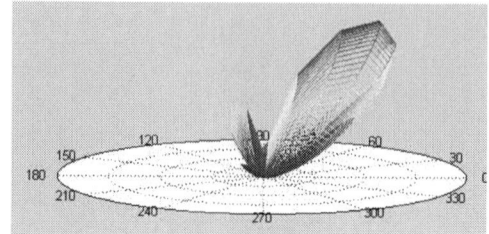

Figure 15: Phased array antenna performance.

VI. MILLIMETER WAVE BGA AND QFN INTEGRATED PACKAGE SOLUTION.

A millimeter wave BGA and air cavity QFN package have been introduced for several years for millimeter wave GaAs MMIC packaging. However the lack of SOC integration in GaAs technology and the challenging interconnection with the mother board at millimeter frequencies have limited this approach for frequencies below 40GHz. In the case of SOC/SOP millimeter wave radio "Bits-in / Bits-out" solution with compact integrated antenna, no such limitation exists and operation at almost any millimeter wave frequency is possible. In figure 16 and 17 we show compact 3D integrated millimeter-wave BGA and QFN modules, respectively, including a CMOS single chip radio, embedded filter and phased array antenna.

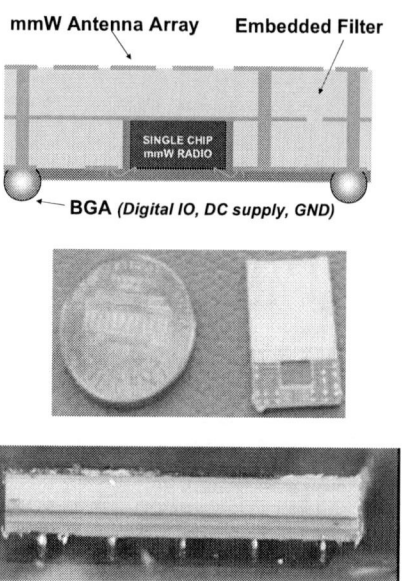

Figure 16: compact 3D integrated millimeter-wave BGA modules, including CMOS single chip radio, embedded filter and phased array antenna.

978-1-4244-2018-6/08 $25.00 © 2008 IEEE

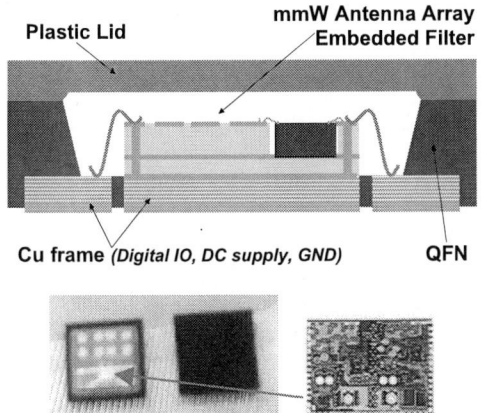

Figure 17: compact 3D integrated millimeter-wave QFN modules, including CMOS single chip radio, embedded filter and phased array antenna.

BGA and QFN low cost, high frequency, high reliability test sockets are widely available, operating at frequencies up to > 40GHz and maintaining performance well beyond 500,000 insertions, as shown in figure 18. For millimeter wave radio "Bits-in / Bits-out" solution integrated into BGA or QFN package, a simple design of the test socket lid allows for sorting tests from antenna radiation pattern to Digital and DC test. The built-in self-test (BIST) in the digitally assisted analog radio design allow on to move beyond Known-Good-Die sorting, to "Known-Good-System" from Antenna to Digital bits.

Figure 18: low cost, high frequency, high reliability test sockets

VII. 15 GBPS AND HD-VIDEO MILLIMETER-WAVE TESTBED.

At the GEDC we have established an experimental millimeter-wave wireless test-bed, using 60GHz as a demonstrator vehicle to study the channel characteristic of a real indoor environment. We have recently established a new world record for the highest data rate transmitted wirelessly at 60GHz, achieving a peak data transfer rate of 15 Gigabit/s at a distance of 1 meter, 10 Gigabit/s at a

distance of 2 meters and 5 Gigabit/s at a distance of 5 meters. In addition, HDMI video streaming has been demonstrated through one inch thick wood table. Special efforts have been dedicated to the complete transceiver module implementation operating at a power budget well below the one hundred pico-joules/bit range. In figure 19, we show the HDMI streaming demonstration in Line-Of-Sight (LOS) and in figure 20, the HDMI streaming demonstration in Non-Line-Of-Sight (NLOS) through a 2.5cm wood obstacle.

Figure 19: HDMI streaming demonstration in LOS.

Figure 20: HDMI streaming demonstration through a 2.5cm thick wood table (NLOS).

VIII. CONCLUSION

The development of millimeter wave digita radios at the same cost structure of consumer electronic radios operating in C-Band opens a new field of innovation for systems designers. The convergence of FR4 based modules, CMOS digital radio, signal processing and high efficiency PHY-MAC technologies will enable a new generation of low cost high performance millimeter-wave systems. The feasibility of ultra high-speed wireless transmission beyond 10Gbps has been demonstrated on a low power platform. Power budgets well below the one hundred pico-joules/bit range have been achieved. One may expect that 100Gbps serial transmission with a femto-joule/bit power budget can be developed in the near future. These advances will enable a variety of volume millimeter-wave CMOS applications including: peer-to-peer ultra fast synchronization and adaptive WPAN, automotive radar, out-door point-to-point/point-to-multi-point links, portable radar, digital radar, security, sensing, imaging and medical applications.

ACKNOWLEDGEMENT

This material is based upon work supported by SPAWAR under Award No. N66001-06-1-2033. Any opinions, findings, and conclusions or recommendations expressed in this publication are those of the author (s) and do not necessarily reflect the views of the SPAWAR. Authors would like to acknowledge DARPA for their support in this work.

REFERENCES

[1]   J. Laskar, S. Pinel, D. Dawn, S. sarkar, B. Perumana and P. Sen, "The Next Wireless Wave is a Millimeter Wave", *Microwave Journal*, pp22-36, August 2007.

[2]   K. Ohata, et al., "Wireless 1.25 Gb/s Transceiver Modules utilizing Multilayer Co-fired Ceramic technology", *ISSCC digest,* February 7-9, 2000.

[3]   S.K. Reynolds, B.A. Floyd, T. Zwick, "60 GHz transceiver circuits in SiGe bipolar technology", *IEEE ISSCC Dig. Tech. Papers,* Feb. 2004, pp 442-443.

[4]   Jean-Pierre Ebert, Eckhard Grass, Ralf Irmer, Rolf Kraemer, and Gerhard Fettweis, Karl Strom, Günther Tränkle, Walter Wirnitzer, Reimund Witmann, Hans-Jürgen Reumerman, Egon Schulz, Martin Weckerle, Peter Egner, and Ulrich Barth, "Paving the Way for Gigabit Networking", Global Communications Newsletter • April 2005 1

[5]   G. Fettweis: "WIGWAM – Wireless Gigabit With Advanced Multimedia Support", In Proc. Wireless World Research Forum (WWRF), New York, 27.-28. October 2003

[6]   IEEE 802.15 Working Group for WPAN; http://www.ieee802.org/15/

[7]   http://www.ecma-international.org/memento/TC32-TG20-M.htm

[8]   S. Pinel, S. sarkar, P. Sen, B. Perumana, D.Yeh, D. Dawn, J. Laskar, "60GHz Single chip 90 CMOS Radio", *ISSCC 2008, Feb 2008.*

[9]   S. Sarkar, P. Sen, B. Perumana, D.Yeh, D. Dawn, S. Pinel, J. Laskar, "60GHz Single chip 90 CMOS Radio with integrated signal processor", to be presented at *IMS 2008.*

[10] KiSeok Yang, Stephane Pinel, Il Kwon Kim, and Joy Laskar. "Low-Loss Integrated Waveguide Passive Circuits Using Liquid Crystal Polymer System-on-Package (SOP) Technology for Millimeter-Wave Applications," IEEE Transactions on MTT, Volume 54, Issue 12, Part 2, Page(s):4572 - 4579 , Dec. 2006.

[11] Il Kwon Kim, Stephane Pinel, Joy Laskar, and Jong-Gwan Yook, "Circularly & Linearly Polarized Fan Beam Patch Antenna Arrays on Liquid Crystal Polymer Substrate for V-band Applications", APMC 2005, Volume 4, 4-7 Dec. 2005.

[12] K. Hornik, M. Stinchcombe, and H. White, "Multilayer feedforward networks are universal approximators," Neural Networks, vol. 2, pp. 359-366, 1989.

[13] J. F. Frenzel, "Genetic algorithms," IEEE Potentials, pp. 21-24, Oct. 1993.

[13] C.Doan et al., "Design Considerations for 60 GHz CMOS Radios", IEEE Communications Magazine, pp132-140, December 2004.

[14] Sohrab Emami et al., "A Highly Integrated 60GHz CMOS Front-End Receiver", *ISSCC Dig. Tech Papers*, pp 190-191, Feb., 2007.

[15] Terry Yao, et al., "Algorithmic Design of CMOS LNAs and PAs for 60-GHz Radio", *IEEE J. Solid-State Circuits*, vol. 42, No. 5, pp. 1044-1057, May 2007.

[16] Babak Heydari, et al., "Low-Power mm-wave Components up to 104GHz in 90nm CMOS", *ISSCC Dig. Tech Papers*, pp 200-201, Feb., 2007.

[17] T. Ihara, et al. 'Switched four-sector beam antenna for indoor wireless LAN systems in the 60 GHz band", Topical Symposium on Millimeter Waves, 7-8 July 1997, pp 115 – 118.

IEEE 2008 Custom Intergrated Circuits Conference (CICC)

Chip to Carrier C4 Technology Challenges with Pb-free Solders

Eric D. Perfecto, Brian Sundlof, Kamalesh Srivastava and Minhua Lu**
IBM Systems & Technology Group (STG)
2070 Route 52 M/S 6C1, Hopewell Junction, N.Y.   12533
**IBM Watson Research
email: perfecto@us.ibm.com, phone: (845-894-4400)

Abstract- **IBM's C4 interconnection technology has continuously evolved over a period of forty years, i.e. from evaporation, to electroplating to C4NP, a C4 New Process. IBM's initial C4NP efforts are focused on Sn-based Pb-free solder technology, in line with client requirements. Currently, all IBM bumped lead-free C4s are produced using the C4NP technology. Sn-based lead-free solders pose unique challenges because of higher microhardness and anisotropy of the tin crystalline structure, as compared to Pb-based solders. The simultaneous design requirements of increased power and current density, increased I/O counts and larger chips, and weak BEOL structure with low-k or ultra-low-k dielectric, demand a careful material interaction optimization between under bump metallurgy (UBM), bump solder, laminate solder, and laminate surface finish. In this paper, we will be discussing the challenges and some solutions of lead-free C4 bumping in terms of mechanical and thermo-electromigration.**

I. INTRODUCTION

Solder connection between the semiconductor chip and the first level package, using controlled chip collapsed connection (C4) concept has been practiced at IBM for more than 40 years, particularly for server applications where high levels of reliability are required. In 1973, IBM started evaporating the UBM and the C4 Solder through a metal mask. The sequence of evaporation was UBM, TiW/Cr-Cu/Cu, followed by Pb and a thin Sn layer to obtain desired solder composition upon a 365 C reflow [1]. Unlike any other interconnection techniques, the C4 allows for high level reliability area array connection which can fully populate the device area providing a high number of I/O connections.

The increased number of C4 connections presented a stability problem for the metal mask and necessitated the introduction of C4 electroplating through a resist pattern. Here, the sputtered TiW/Cr-Cu/ Cu UBM also acted as the seed layer for the subsequent plating of the 97% Pb 3%Sn solder. The seed layer was etched chemically and electrochemically around each solder bump.   Pattern electroplating, introduced at IBM in 1996 on SRAM products, has been adopted for all products; and as the industry migrated to 300mm wafer form factor, pattern electroplating was the C4 deposition method of choice.

In 2007 IBM has implemented C4NP for all its 300mm Pb-free flip chip bumping [2, 3]. The C4NP process offers an environmental friendly solution to solder deposition since the solder is deposited into a mold which then gets transferred to a UBM on the Si wafer. Figure 1 shows a typical pattern of

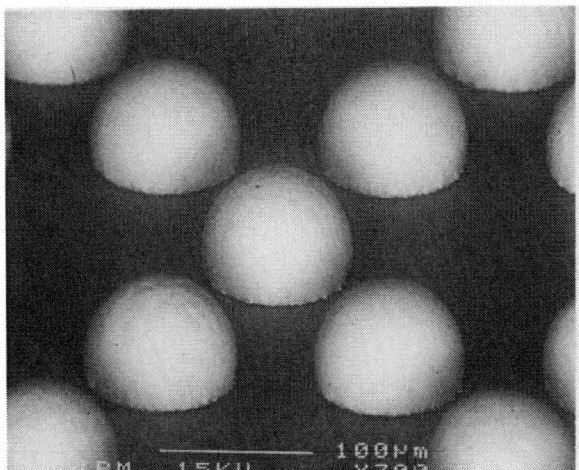

Fig. 1: Pb-free solder bumps on 150 um pitch.

solder bumps built using C4NP, in the post transfer state. Pb-free solder necessitated a change of the UBM barrier from the traditional sputtered Cr-Cu/Cu to electroplated Ni/Cu, where the plated Cu serves as a wetting layer during solder transfer, and the Cu-Sn intermetallic compounds (IMC) slow down the consumption of Ni.

This paper will cover the new mechanical, electrical and thermal challenges on the first level package. It will also detail the new C4NP process.

II. FIRST LEVEL PACKAGING CHALLENGES

A series of technical challenges surfacing simultaneously are influencing the materials and process selection for the first level interconnection, i.e. increased power and thermal requirements, architectural shift to multi-cores with increased I/O count and current density, and weaker BEOL structures. Additionally, to reduce cost, organic laminates are the main substrate base for single chip and dual chip modules. These laminates have a high coefficient of thermal expansion (CTE) compared to the Si chip which further strains the interconnect at chip join.

Thermal demands: The IBM ES9000 was the last server system which used the bipolar technology. In 1990, it required water cooling to support the 13 watts/cm2 module heat flux. With the movement to CMOS the cooling requirement were reduced significantly. However, with increasing level of semiconductor integration, today's modules are requiring heat

978-1-4244-2018-6/08 $25.00 © 2008 IEEE

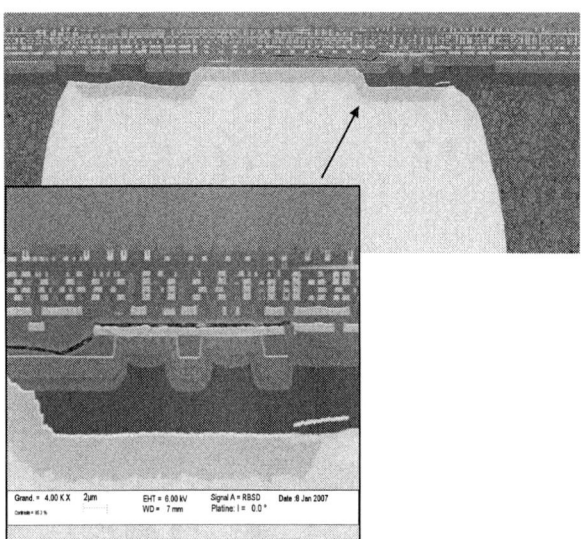

Fig. 2: BEOL cracks caused during reflow joining.

dissipation equivalent to the bipolar case. Cooling hats attached to a thin thermal conduction interface are now commonly used to compensate for high junction temperatures (~ 85 C). This increase in operating temperature is giving rise to electromigration (EM) concern, particularly in Sn-based Pb-free systems.

Increase in I/O Count: To accommodate increased I/O count requirements, the C4 pitch is moving from 200 um to 150 um. Emerging technologies, such as 3D, are further requiring the reduction of the C4 pitch to 50 um [4], and in come cases elimination of the C4 all together and replacing it with metal to metal connections. In addition, the chip size is increasing significantly, i.e., a 20mm x 20mm chip size could be a common occurrence for high performance applications in the near future.

Chip underfill is an essential part of module reliability addressing the high chip/substrate CTE mismatch and/or non-hermetic condition. A post assembly sonoscan of modules is practiced to assure a void-free underfill, or crack-free chip corner. Underfills are available with a variety of Tg and modulus which can be optimized for a particular application. *Weaker BEOL Structures:* Semiconductor industry's continuous drive towards smaller wiring and faster chips comes with a price as it moves to low k and ultra-low k dielectrics. Pb-free C4 interconnect stress builds up during chip join, due to the CTE mismatch between the device and the organic carrier, may exceed device BEOL structure strength for specific design structures, resulting in delamination or cracking. Interfacial adhesion becomes critical to eliminate these stress cracks which occur during the cooling stage of the chip joining process (see Figure 2).

The ability of the solder to permit stress relaxation is measured in terms of its microhardness. Once the chip is joined to the laminate, it is underfilled. Here also underfill selection plays a critical role in distributing the solder stresses between the laminate and the BEOL structures.

III. Pb-FREE IMPLEMENTATION

State of RoHS (Restriction of Use of Certain Hazardous Substances in Electrical and Electronic Equipment) on banning use of lead is different in different countries or regions. For example, EU has legally banned using lead starting from July 2006; China has implemented RoHS in the form of product catalogue using method of exclusions; most of the United States has banned lead in many products like gasoline, plumbing, and paints but not in electronic products as they are exempted. IBM's initial efforts are focused on using lead-free solder with C4NP technology, in line with client requirements. *Choice of Sn:* Choice of Sn-Ag solder composition for device joining purposes is dependent on the design factors including, liquidus temperature being relatively low (about 220 C), elimination of formation of rod-like Ag_3Sn precipitate, reduction in suppression of solidus, relatively low hardness, and longer life expectation from the point of electromigration behavior.

The precipitation of rod-like Ag_3Sn phase during chip-joining process has been observed in alloy containing as low as 2.3%Ag. Therefore Sn-<2%Ag alloys together with Sn-0.7%Cu have been examined for suppression of solidus, micro-hardness of the module level joint using laminates with Sn-0.7%Cu and Sn-3.5%Ag-0/7%Cu solder pastes. Solder composition effects on undercooling have been reported earlier for a BGA solder earlier [5]. Microhardness of the solder joint is sensitive to the percent of Ag content (see Figure. 3). This result is consistent with the improvements on stress relaxation and white bump reduction observed with reduced Ag content [6].

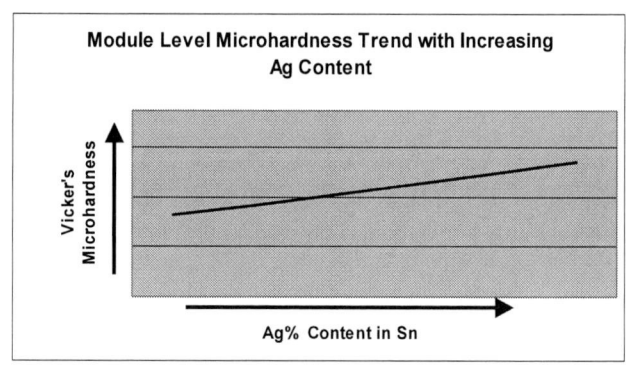

Fig. 3: Module level solder joint microhardness

C4 bumps examined with X-ray radiographic technique for hundreds of wafers were found to be completely void-free. Alpha emission rate of the solder is kept at a low level from the point of minimization of soft error (SER) incidence. In our solder design, alpha-emission–rate is kept below 0.005 counts per cm^2 per hour. This is achieved by a vendor-confidential refining process. Also, we have found that the alpha emission rate of the as received solder sample plates does not increase with time for measurements completed up to 180 days, unlike low-alpha emission type of leaded solder.

978-1-4244-2018-6/08 $25.00 © 2008 IEEE

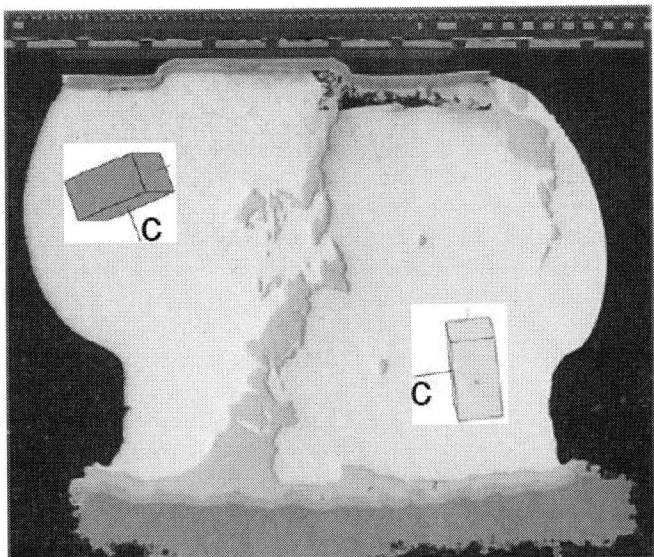

Fig. 4: SEM of an EM stressed C4 with SnCu solder.

IV. C4 ELECTROMIGRATION

Electromigration performance in Pb-free C4 interconnect structure has been the focus of recent studies due to the continuing demand for higher current density and circuit miniaturization. Different from the face center cubic structure of lead, tin (Sn) has a tetragonal crystal structure and tends to form large grains that exhibit highly anisotropic behavior in mechanical, thermal, electrical, and diffusion properties in the high Sn-based Pb-free solder joint. Importantly, the noble and near noble metals are extremely fast diffusers in Sn and the diffusivity is highly anisotropic. For instance, the diffusivity of Ni along the tetragonal (c-) axis is ~7x10E4 times faster than that at right angles (a- or b-axis) at 120°C. Experimentally it has been found that electromigration damage is strongly dependent on the Sn-grain orientation in Pb-free solders. Figure 4 shows a SEM image of an EM stressed C4 with SnCu solder and the insertion of the Sn unit cell indicating the grain orientation. The grain on the right has the c-axis at nearly a right angle with respect to the current direction, where the rates of Cu and Ni diffusion in Sn are slow. Failure is characterized by Sn self-diffusion or lattice diffusion resulting in void formation between the IMC and solder, here referred as mode-I failure. The grain on the left, however, has its c-axis closely aligned with the current direction, which drives very fast Ni and Cu diffusion through the Sn grain. Mode-II failure is due to rapid depletion of intermetallic compounds and UBM. Since Mode-II failure usually occurs early, it is the mode which should be avoided.

In addition to the grain orientation, the solder and UBM interaction and solder alloy composition play an important role as well. Figure 5 is the plot of average time to failure for SnCu and SnAg(Cu) solders and three UBM, Cu, Ni(P)/Au, Ni(P)/Cu, surface finishes [7]. In general SnAg(Cu) solder

performs better than SnCu solders and Ni UBMs are better than Cu UBMs. This study showed that Ag_3Sn IMC network is more stable than that of the Cu_6Sn_5 network under thermal and electrical stress, which is attributed to the much smaller diffusivity of Ag compared to Cu. Grain growth and grain reorientation is common in SnCu solder, while a much more stable cyclic twinning structure is more frequently found in SnAg solders, especially in high Ag solders. Although Cu_6Sn_5 is not stable against electromigration, this study shows that a certain amount of Cu on UBM to form a layer a Cu_6Sn_5 IMC after reflow is important to give an extra protection to Ni barrier layer. Effect of alloy composition such as Ag, Cu and other doping elements and UBM metallurgy in electromigration is complicated and convoluted. Systematic and carefully designed and conducted experiments are essential to obtain correct conclusion.

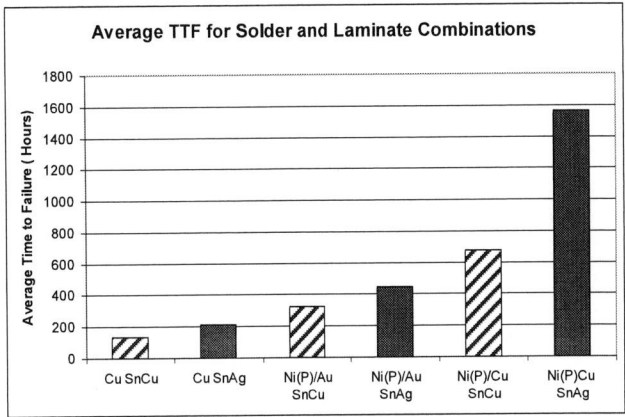

Fig. 5: UBM and solder selection influence on EM failure.

V. C4NP PROCESS

The C4NP process starts with a glass mold in which the bump pattern for an entire wafer is replicated as a mirror image of cavities in the glass mold. These cavities are filled with solder as the mold is scanned below a fill head. The fill head contains a reservoir of molten solder and a slot through which the solder is injected into the mold cavities. The cavity geometry determines the volume of the solder bumps that will be subsequently formed on the wafer. The filled mold is inspected automatically and then aligned below a wafer with exposed UBM pads facing the mold. The mold and wafer are heated above the solder melting point and then brought into soft contact. The solder forms spherical balls which transfer from the mold to the UBM regions on the wafer, where they preferentially wet and solidify. Wafer and mold are separated, and the mold is cleaned for reuse. Figure 6 describes this process flow.

The material of choice for the mold is borosilicate glass with a CTE = $3.25 \pm 0.10 \times 10$-6 /K (3.25 ± 0.010 ppm). This value closely matches the CTE of a silicon wafer. Matched CTE's allow mold design, fabrication and alignment at ambient temperature to use the same dimensions as the wafer,

978-1-4244-2018-6/08 $25.00 © 2008 IEEE

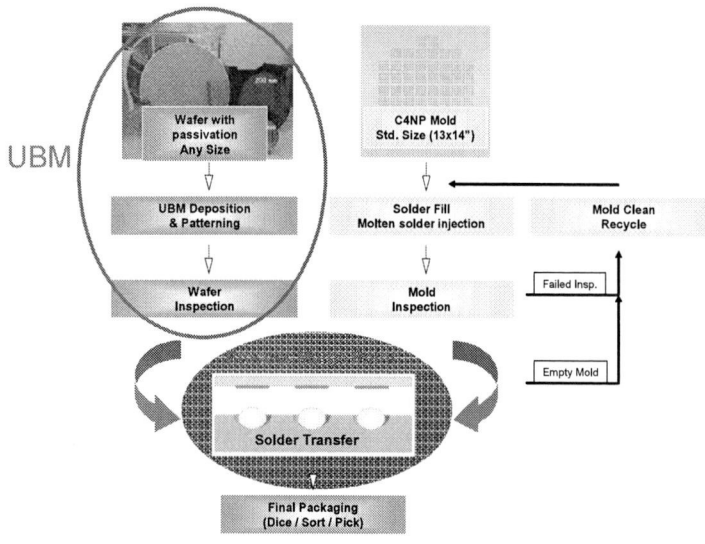

Fig. 4: Module X-section for 150 um pitch

yet features maintain their spatial relationships at solder transfer temperatures. Mold yield and fabrication has been discussed elsewhere [8].

For 130nm and coarser chip wiring applications, C4 bumps are placed on 200um or higher pitch. In advanced chip technology at the 45nm and 32nm nodes, the C4 bump pitches are reduced to 150um. There is no difference in processing or materials between the 200 um and the 150 um C4 pitch. C4NP shows equivalent inspection yields to pattern electroplating and has demonstrated a high level of process maturity.
Reliability Stressing: The test vehicles for C4NP technology developments were specifically designed to evaluate all aspects of C4 reliability: joint integrity, solder fatigue, metal migration, thermal aging stability, tin whiskers, tin pest and electromigration. After chip joining to laminate and post JEDEC reflows, chips were pulled to determine tensile pull strength. The pulled chips along with the substrates were then inspected for voids, non-wets, fracture mode, delamination and any other anomalies. Reliability evaluations encompassed standard JEDEC stress conditions (Thermal Cycle, Temp./Humidity/Bias, High Temp. Storage, Low Temp. Storage) and electromigration. Prior to stressing, all test modules were subjected to JEDEC pre-conditioning (Level-3). All stress conditions have been successfully met for the reliability stressing for 200um pitch and 150 um pitch [9, 10]; and for EM the current structure supports up to 175 watts and 150 watts at 100C, 100KPOH, respectively.

VI. SUMMARY

The first level interconnect for high performance chips is literary a hot area. Increases in thermal requirements generated by semiconductor integration and large count of I/O, coupled with increases in current density demands materials and processes that can support high number of POH. Wire testing simulations have demonstrated the importance of judicious selection of UBM, solder and laminate metallurgy, and supports the decision to Ni UBM and SnAg solder for C4.

The second main area of innovation occurred in improving the chip to package interaction. Here solder composition and underfill selection play a pivotal role in balancing the CTE stresses. Finally, the positive qualification results demonstrated a good UBM and solder selection for the C4NP process.

ACKNOWLEDGMENTS

The authors would like to acknowledge their management team for their support, in particular DY Shih and Glenn Daves. Special thanks are due Hai Longworth and David Hawken for the reliability evaluation and to the FA team: Luc Patry from IBM Bromont, and Charles Goldsmith from East Fishkill.

REFERENCES

[1] P. A. Totta, S. Khadpe, N. Koopman, T. Reiley, and M. Sheaffer, "Chip-to-Package Interconnections," Microelectronics Packaging Handbook, Part II, Second Edition, R. Tummala, E. Rymaszewski, and A. Klopfenstein, Eds., Chapman & Hall, London, 1997, pp. 129–283.

[2] P. Gruber *et al.*, "Low Cost Wafer Bumping", *IBM J. Res. & Dev.* Vol. 49 No. 4/5, July/September 2005.

[3] E. Laine, K. Ruhmer, E. Perfecto, H. Longworth, D. Hawken, "C4NP as a high-volume manufacturing method for fine-pitch and lead-free Flip Chip Solder Bumping" *Proceedings of ESTC, IEEE*, Dresden, Germany, September, 2006

[4] Dang, D-Y Shih, et al., "50μm Pitch Pb-Free Microbumps by C4NP Technology," *Proceedings of 58th ECTC*, Orlando, May, 2008

[5] I. de Sousa, D. Henderson, L. Patry, et al., "The Influence of Low Level Doping in Thermal Evolution of SAC Alloys Solder Joins with Cu pad Structures," *Proceedings of 56th ECTC*, San Diego, May, 2006

[6] J. Sylvestre, et al., "The Impact of Process Parameters on the Fracture of Brittle Structures During Chip Joining on Organic Laminates," *Proceedings of 58th ECTC*, Orlando, May, 2008

[7] M. Lu, D-Y Shih, et al., "Comparison of Electromigration Performance for Pb-free Solders and Surface Finishes," *Proceedings of 58th ECTC*, Orlando, May, 2008

[8] E. Laine, E. Perfecto, B. Campbell, J. Wood, J. Busby, J. Garant, L. Guerin "C4NP Technology for Lead Free Solder Bumping" , *Proceedings of 57th ECTC*, Reno NV, May, 2007.

[9] A. Giri, E. Perfecto, et al., "Development and Implementation of C4NP Technology for 300 mm wafers," *Proceedings of 57th ECTC*, Reno, June, 2007

[10] E. Perfecto, et al., "C4NP Technology: Manufacturability, Yields and Reliability," *Proceedings of 58th ECTC*, Orlando, May, 2008

IEEE 2008 Custom Intergrated Circuits Conference (CICC)

A Study on Process-compatibility in CMOS-first MEMS-last Integration

K. Takahashi[1], M. Mita[1], H. Fujita[1], K. Suzuki[2], H. Funaki[2], K. Itaya[2], and H. Toshiyoshi[1]

[1] Institute of Industrial Science, the University of Tokyo, Tokyo, Japan
[2] Research & Development Center, TOSHIBA Corp., Kanagawa, Japan

Abstract- **We report monolithic integration of MEMS (Micro Electro Mechanical Systems) actuators and high-voltage driver circuits into a silicon chip. Driver circuits of up to 40 V were prepared on an 8-μm-thick SOI (Silicon-on-Insulator) wafer by the DMOS (Double-diffused Metal Oxide Semiconductor) processes, after which MEMS electrostatic actuators were integrated into the identical SOI layer by post-processing using DRIE (Deep Reactive Ion Etching). This technique opens up a new way of high-voltage ASIC (Application Specific Integrated Circuit) for MEMS designer. We investigated the process compatibility between the integrated circuits and MEMS. Studies are made on (1) the design and fabrication of the MEMS-circuit electrical interconnection and (2) the effect of dry-etching plasma on the circuit characteristics in MEMS DRIE process.**

I. INTRODUCTION

MEMS (Micro Electro Mechanical Systems) approach is an enabling technology to develop micromechanical structures by using the semiconductor micro fabrication techniques. Advantages of using the MEMS technology is in the capability of developing intelligent micro machines by the multi-functional integration of micro actuators, micro sensors, and micro processors onto a monolithic chip [1,2]. While the conventional LSI application has spread in the wide range of electronics applications of mass-production, MEMS has not yet found a large quantity market in the consumer electronics but it is rather experiencing difficulty even in small scale market. The difficulty comes from the costly fabrication even in a trial-and-error phase of development; MEMS uses various types of materials other than semiconductors, such as metals plastics, and polymers, and it makes it difficult to standardize the MEMS process. Successful practice of MEMS technology thus depends on the fabrication flexibility for various applications of small quantity.

Based upon such technical and economical background of MEMS, we propose a new designing and fabrication technique of the CMOS-MEMS platform that is adaptive to small-quantity MEMS products. We also point out a possibility of using single crystalline silicon of SOI (silicon-on-insulator) wafer to develop micro actuators of superior performance to those made by the conventional surface micromachined CMOS-MEMS technology. This work also presents a new technology of monolithic integration of high-voltage driver circuits for the sake of high performance and high reliability of MEMS actuators. In particular, we focus on the process compatibility between the MEMS post-processing and IC pre-fabrication.

II. DESIGN of CMOS-MEMS

A. CMOS-first, MEMS-next integration

We have proposed a new MEMS integration process to make the integrated circuits first on the SOI wafer and then post-process the MEMS onto the identical SOI layer, which gives flexibility in designing MEMS variation [3]. Fig. 1(a) shows an example of bulk-micromachined MEMS actuator integrated with a high-voltage circuit chip. While surface micromachining processes are allowed in the conventional post-CMOS MEMS processes, we adopted bulk SOI micromachining in this work to build micromechanical components of higher aspect ratio and more mechanical reliability. Due to the process sequence of the integration, our method uses an array of IC and MEMS devices flattened into a

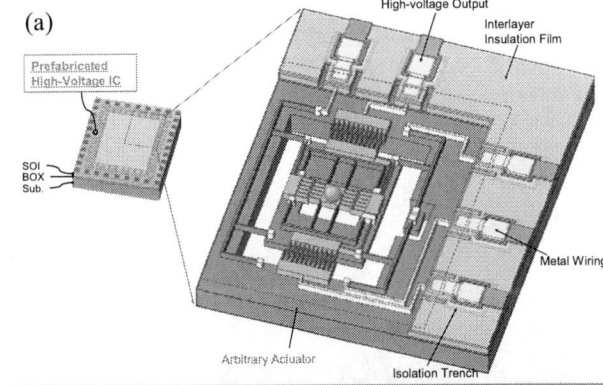

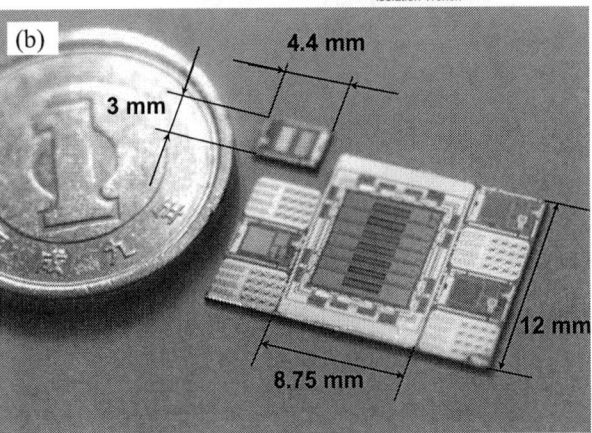

Fig. 1. (a) Conceptual diagram of monolithic integration of CMOS driver circuits and MEMS actuators. (b) Photograph of MEMS actuators post-processed on SOI CMOS chip.

single layer of SOI, and thus it occupies larger device footprint than that of the add-on style of MEMS such as the Texas Instruments DMD [1]. Nevertheless, tiled layout of MEMS and IC is less suffering from electrical interference.

In out recent report [3], we have presented high voltage driver circuits for MEMS integration. The driver circuits were made by using the TOSHIBA R&D Center's CMOS technology upwards of 40V. Fig. 1(b) shows a photograph of the prototype CMOS-MEMS chip (reticle size 12 mm × 8.75 mm). TEG (Test Element Group) circuits were located near the edge of the chip. The high-voltage circuits were placed around a blank field (7 mm × 6 mm) of the SOI layer that was used for making microactuators. Use of high voltage for MEMS drive gives us more margins in designing micro actuators, namely faster mechanical response and larger mechanical displacement.

Fig. 2(a) shows an XY-stage micro actuator has been produced by using a layer-separation designing technique of the SOI wafer [4]. A high fill-factor of device layout became possible by allocating the electrical function such as actuators electrodes and mechanical functions such as suspensions and frames into the different layers of an SOI wafer. The developed micro XY-stage could be used for a movable stage for the optical microscope and for media translating stage of the AFM (atomic force microscope) type data storage [5].

Fig. 2(b) shows another example of the CMOS-MEMS technology, an array of grating light valves integrated with the IC drivers, developed for a digitally controlled image projection display [6]. Each unit cell of the grating could be

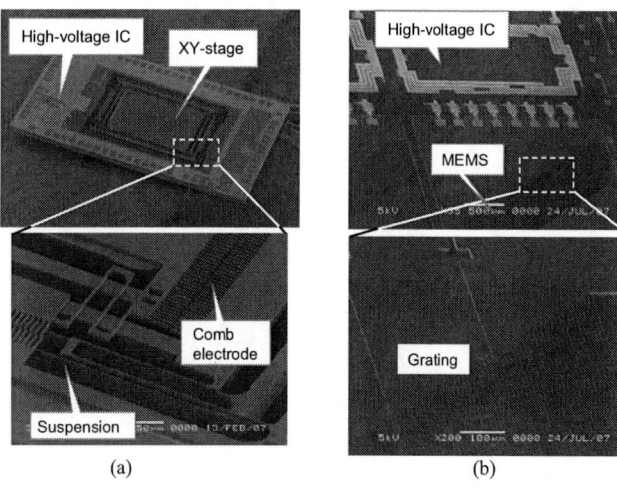

Fig. 2. SEM images of (a) integrated double-deck electrostatic XY-stage and (b) grating light valves integrated with the IC drivers for image projection display.

laterally controlled in an individual manner to tune the diffraction angle of light pixel-wise. The line image produced by the MEMS is then scanned with a galvanometric scanner to create a 2D image. We observed the electrostatic operation and its optical effect of the gratings by applying drive voltage. An improved model of the MEMS gratings has been developed by giving a skew angle to the gratings array and by avoiding the overlap between the 1st and the 0th order

diffraction patterns. A high-speed high-voltage level shifter driver circuit and the MEMS grating have been integrated to show the potential of MEMS approach for constructing the image projection optical system of high definition and high contrast.

B. Circuit design

The driver circuits consisted of low-voltage logic circuits, sample-and-hold circuits, and high-voltage operational amplifiers, as schematically shown in Fig. 3(a). The high-voltage output stage is designed to work on a single voltage supply of 40 V, for the electrostatic force is proportional to the square of applied voltage.

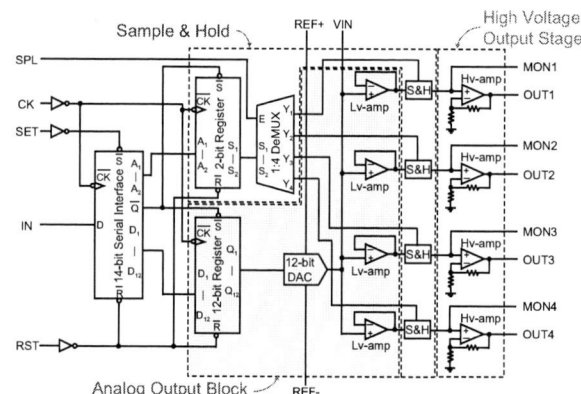

Fig. 3. Block diagram of analog driver circuits

(a)

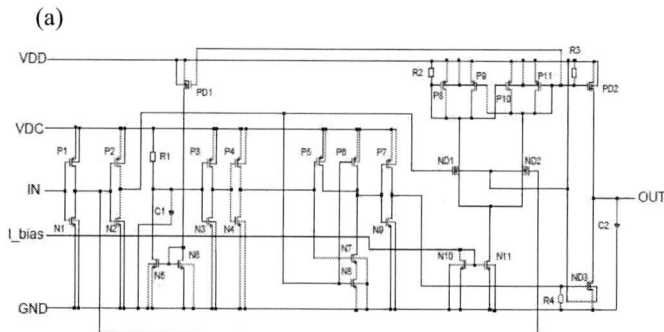

(b)

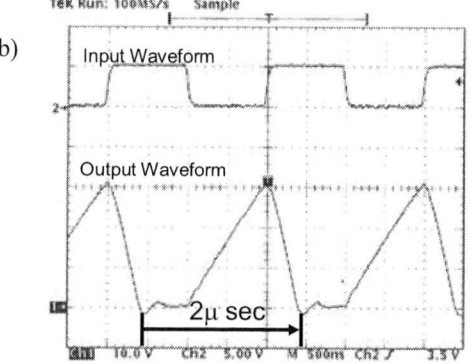

Fig. 4. (a) Circuit diagram of high-voltage level shifter. (b) Response time of high-voltage level shifter.

Fig. 4(a) shows the circuit diagram of the developed high-voltage level-shifter. Only one set of equivalent circuit is shown out of total 8 channels implemented. The voltage level shifter operates with two voltage supplies, namely high voltage (VDD, 40V) and low voltage (VDC, 5V) for a 3V input pulse (IN). A current bias stage (maximum current handling 500 μA) has been newly integrated in the differential amplifier block to reduce the effective resistance of the current path and to extend the cut off frequency. The ON/OFF state of the current bias was designed to be controllable such that the total power consumption could be minimized. Current pass through the output stage (ND3) at high frequency was avoided by carefully pre-setting the ON/OFF timing of the transistors PD2 and ND3. ON/OFF switching of up to 500 kHz was experimentally confirmed, as shown in Fig. 4 (b).

III. DISCUSSION of PROCESS COMPATIBILITY

A. MEMS Post-process

MEMS actuators were post-processed after the TOSHIBA 0.6 μm LDMOS (Laterally Diffused MOS) [7]. Process steps are shown in Fig. 5. The surface of as-delivered IC chip was covered with an interlayer-insulation TEOS film, which was first removed to open a part of SOI field for the MEMS post-process. Aluminum was vacuum deposited and electrical interconnection was made between the output port of the level shifter and MEMS actuation pads. Micro actuators were made by DRIE of the 8-μm-thick SOI layer. Electrical isolation between channels was established at the same time. Protecting the front side, the backside of the substrate was DRIE-processed to the buried oxide (BOX), and then the BOX was selectively removed to release the mechanical structures.

The actuator patterns are formed by the DRIE in the SOI layer. Electrical isolation between the actuator pads is established at the same time; one side of the MEMS actuator's pad is defined by the isolation trench of the high-voltage IC. By having DRIE-trenches to meet the isolation trench, the pattern can be made to be electrically separated from the neighborhood MEMS patterns and isolated from CMOS body at the same time.

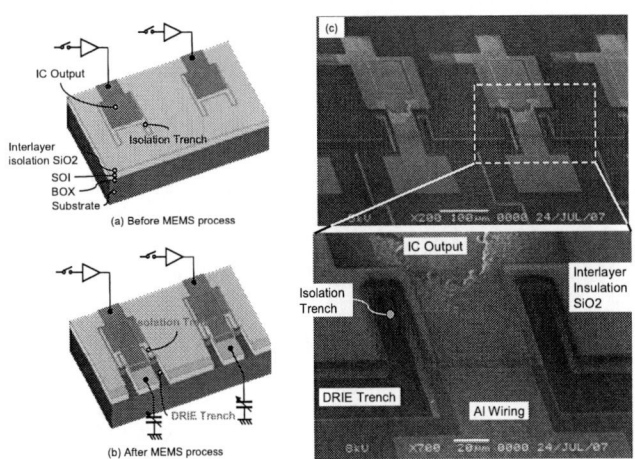

Fig. 6. Electrical isolation between the actuator pads and CMOS body: one side of the MEMS actuator pad is defined by the isolation trench of the high-voltage IC.

B. Reliability evaluation of MEMS post-process

Fig. 7(a) shows a microphotograph of wire-bonded CMOS-MEMS device which was integrated with digital level shifter and grating light valve. The illustration below is a cross sectional view of the chip. A 3 V input pulse signal is applied to input pad at the left hand side of the chip; the output voltage of the level shifter and the MEMS pad are shown in Fig. 9(b). From this plot, we confirmed the electrical interconnection of low resistance between the high-voltage level shifter to the MEMS actuator's pad. We also confirmed the operation of the level shifter as designed.

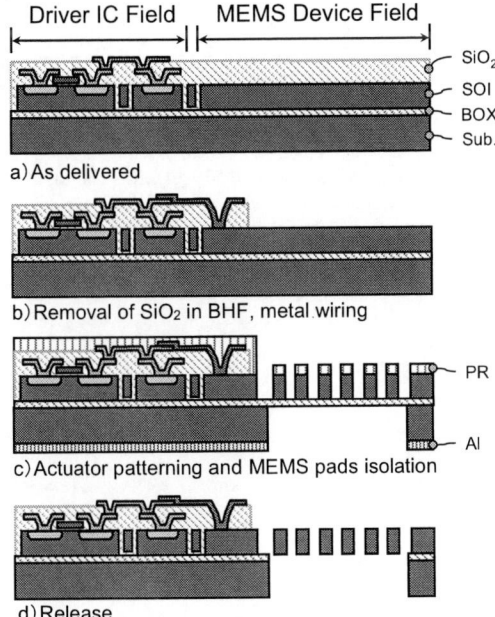

Fig. 5. Simplified MEMS post-process flow

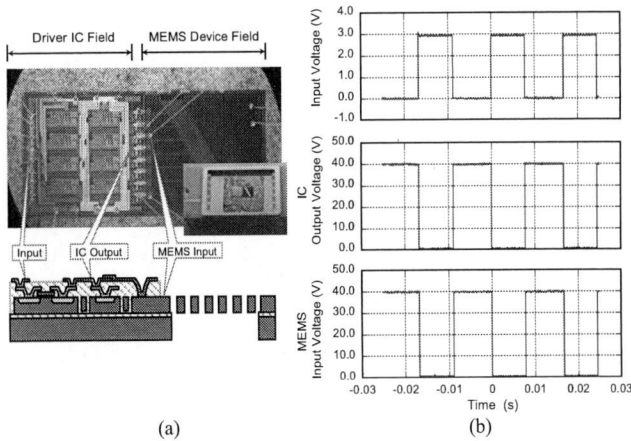

Fig. 7 (a) Microphotograph of wire-bonded CMOS-MEMS chip. (b) Measured output data of high-voltage level shifter.

The MEMS post-processing method used plasma etching processes such as RIE (reactive ion etching) or ICP-RIE (inductive coupled plasma-RIE), and the degradation of transistor associated to the plasma damage was anticipated. We evaluated the characteristic of MOS transistor in the TEG area before and after the plasma processes. Plasma damage was investigated in terms of the drain current at each stage of follows:

a) DRIE process of actuator patterning,
b) Plasma ashing to remove the passivation photoresist, and
c) RIE process for etching of the interlayer insulation SiO2 and BOX.

Fig. 8 shows the Id-Vd characteristics before and after the plasma exposure. The transistors degradation was not visibly observed in any of the plasma conditions. The Id-Vg characteristics are also shown in Fig.9. One can tell the plasma-induced damage by the lateral shift or rotation of the curve. None of the plots in Fig. 9 indicates such shift of S-factors, and thus we concluded the plasma damage from the MEMS post-process was negligible.

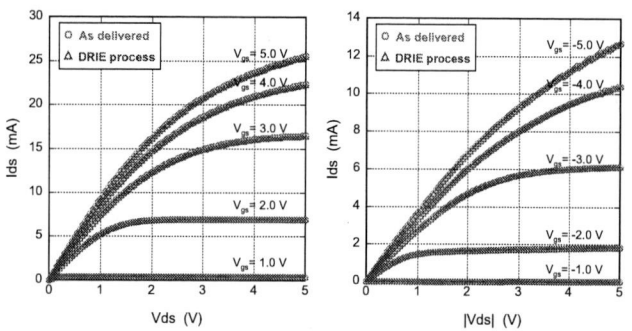

Fig. 8. Id-Vd characteristic (a) NMOS (b) PMOS

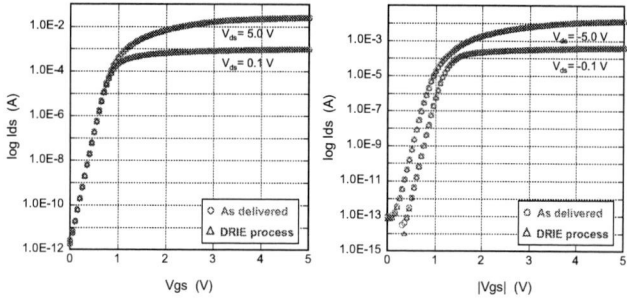

Fig. 9. Id-Vg characteristic (a) NMOS (b) PMOS

IV. CONCLUSIONS

We report fabrication compatibility between the IC pre-fabrication and MEMS actuator post-processing of silicon bulk micromachining. We experimentally confirmed successful electrical intra-chip interconnection between the IC and MEMS actuator as well as the intra-channel isolation between MEMS driver channels. The developed level-shifter of 40 Vdd was successfully demonstrated. Despite the sequential processes of plasma etchings, no electrical degradation was found on the integrated transistors. Thanks to the CMOS-first-MEMS-next fabrication approach, we have obtained flexibility in designing the MEMS actuators on/in the 8-μm-thick SOI layer.

ACKNOWLEDGMENTS

The photomasks used in this work were made using the University of Tokyo VLSI Design & Education Center (VDEC)'s 8-inch EB writer F5112+VD01 donated by ADVANTEST Corporation. This research has been supported in part by the G-COE Program of the Electrical Engineering and Electronics, The University of Tokyo.

REFERENCES

[1] P. F. Van Kessel, L. J. Hornbeck, R. E. Meier, and M. R. Douglass, "A MEMS-Based Projection Display," Proceedings of the IEEE 86(8), 1998, pp.1687-1704.

[2] T. J. Brosnihan, S. A. Brown, A. Brogan, C. S. Gormley, D. J. Collins, S. J. Sherman, M. Lemkin, N. A. Polce, and M. S. Davis, "Optical IMEMS-A Fabrication Process for MEMS Optical Switches with Integrated on-Chip Electronics," The 12th International Conference on Solid-State Sensors, Actuators, and Microsystems (Transducers 03), 8-12 June 2003, Boston, USA, 3E118.P, pp.1638-1642.

[3] K. Takahashi, H. N. Kwon, M. Mita, H. Fujita, H. Toshiyoshi, K. Suzuki, H. Funaki, "Monolithic Integration of High Voltage Driver Circuits and MEMS Actuators by ASIC-like Postprocess," Proc. 13th Int. Conf. on Solid-State Sensors, Actuators and Microsystems (TRANSDUCERS '05), June 5-9, 2005, Seoul, Korea, pp. 417-420.

[4] K. Takahashi, M. Mita, H. Fujita, H. Toshiyoshi, K. Suzuki, H. Funaki, and K. Itaya, "SOI-CMOS PLATFORM FOR MONOLITHICALLY INTEGRATING HIGH-VOLTAGE DRIVER CIRCUITS WITH BULK-MICROMACHINED ACTUATORS," Proc. 14th Int. Conf. on Solid-State Sensors, Actuators and Microsystems (TRANSDUCERS '07), June 10-14, 2007, Lyon, France, pp. 1329-1332

[5] P. Vettiger, G. Cross, M. Despont, U. Drechsler, U. Drurig, B. Gotsmann, W. Haberle, M. A. Lantz, H. E. Rothuizen, R. Stutz, and G. K. Binnig, "The "Millipede"- Nanotechnology Entering Data Storage," IEEE Transactions on nanotechnology, vol. 1, no. 1, pp. 39-55, 2002.

[6] Kazuhiro Takahashi, Hiroyuki Fujita, Hiroshi Toshiyoshi, Kazuhiro Suzuki, Hideyuki Funaki, and Kazuhiko Itaya, "Tunable Light Grating Integrated with High-voltage Driver IC for Image Projection Display," 20th IEEE Int. Conf. on Micro Electro Mechanical Systems (MEMS'07), Jan. 21-25, 2007, Kobe, Japan, pp. 147-150.

[7] H. Funaki, Y. Yamaguchi, Y. Kawaguchi, Y. Terazaki, H. Mochizuki, and A. Nakagawa, "High Voltage BiCDMOS Technology on Bonded 2 μm SOI Integrating Vertical npn pnp, 60V-LDMOS and MPU, Capable of 200 °C Operation," International Electron Devices Meeting (IEDM '95), December 10-13, 1995, Washington, USA.

High Resolution Converters

Session 7

Chair: Yusuf Haque, SliceX Inc.
Co-Chair: George LaRue, Washington State University

The session deals with the techniques that improve performance and lowers power in both over sampled and Nyquist ADC's.

Discrete time delta sigma modulators perform well but require antialiased inputs. Continuous time modulators do not require antialiasing and are low power, but suffer from time constant variation. The first paper presents a hybrid architecture that employs a combination of discrete and continuous time over sampled techniques. RC time constant variation of the continuous time modulator is calibrated and jitter insensitive techniques are demonstrated for an audio application.

The second paper demonstrates a power efficient approach to wideband delta sigma ADC design. Third order noise shaping is achieved using only 2 integrators

Opamps typically are the major consumers of power in discrete time delta sigma ADC's and dynamic biasing of amplifiers can help to reduce this power. The 3rd paper describes a wideband delta sigma ADC implemented with dynamically biased amplifiers that achieves higher bandwidth and lower distortion than previous work.

Multistage Noise Shaping (MASH) has been previously shown to have superior stability for high order modulators but require high accuracy integrators. The fourth paper demonstrates a new MASH architecture that retains its stability properties but can now be built with low gain opamps, a substantial advantage in deep submicron processes.

The fifth paper is a tutorial on low power ADC and technology scaling trends. It also discusses the impact of digital logic to help improve performance while lowering power dissipation.

The next 2 papers present low power techniques for high resolution ADC's.

The sixth paper presents a calibrated 16 bit ADC using time constant control to lower power dissipation.

The last paper demonstrates a power-efficient pipelined high-resolution ADC that employs non-slewing closed-loop amplifiers.

978-1-4244-2018-6/08 $25.00 © 2008 IEEE

Notes

IEEE 2008 Custom Intergrated Circuits Conference (CICC)

A 101-dB SNR Hybrid Delta-Sigma Audio ADC using Post Integration Time Control

Moo-Yeol Choi, Sung-No Lee, Seung-Bin You, Wang-Seup Yeum, Ho-Jin Park, Jae-Whui Kim, Hae-Seung Lee[1]

Samsung Electronics, Yongin, Korea,
[1]Massachusetts Institute of Technology, Cambridge, MA

Abstract-A 3rd-order hybrid (continuous-time and discrete-time) delta-sigma audio ADC, implemented in 65nm CMOS process, dissipates 15mW and occupies an active die area of $0.28mm^2$. A post integration time control (PITC) technique is proposed for calibration of the RC time constant variation of the continuous-time integrator. In addition, a jitter insensitive pulse generator (JIPG) circuit overcomes the degradation of SNR due to the feedback DAC clock jitter. The measured SNR and DR are 101dB and THD is -94dB.

I. INTRODUCTION

Discrete-time (DT) delta-sigma ($\Delta\Sigma$) ADCs using switched-capacitor circuits offer good coefficient matching, scalability with sampling rate and low sensitivity to the feedback DAC clock jitter. Continuous-time (CT) $\Delta\Sigma$ ADCs using CT loop filters such as an active RC filter and a Gm-C filter provide intrinsic anti-alias filtering and lower power consumption.

The DT delta-sigma modulator ($\Delta\Sigma$M) has been widely used because CT $\Delta\Sigma$M needs an additional tuning block of passive element variations and jitter sensitivity. However, the usage of DT $\Delta\Sigma$M is also limited by the requirements of additional anti-aliasing filter (AAF) and the switching noise coupled to the input of the converter. The AAF consumes additional power and it is difficult to integrate the AAF on chip. The switched-capacitor input makes the signal dependent switching noise to the outside of the package and induces the EMI problem in using the off-chip AAF.

The hybrid $\Delta\Sigma$M's comprised of CT and DT integrators have been reported in literature [1], [2], [5] to avoid these drawbacks of the DT $\Delta\Sigma$M. The CT input stage makes no switching noise and functions as an intrinsic AAF, the DT post stages have an advantage of switched-capacitor circuits such as coefficient matching, scalability and low sensitivity to clock jitter. However, the switching circuits were not completely removed from the input and the tuning circuits for RC time constant variations weakened the intrinsic anti-aliasing characteristic of the CT integrator in [1].

In this paper, a hybrid 3rd-order (1st and 2nd CT / 3rd DT integrator) $\Delta\Sigma$M completely removes the switching circuits from the input and the newly proposed post integration time control (PITC) technique calibrates the RC time constant variation efficiently while maintaining the effective anti-alias filtering. The experimental result also shows the effectiveness of the jitter insensitive pulse generator circuit (JIPG) for pulse width jitter of the feedback DAC. The prototype enables

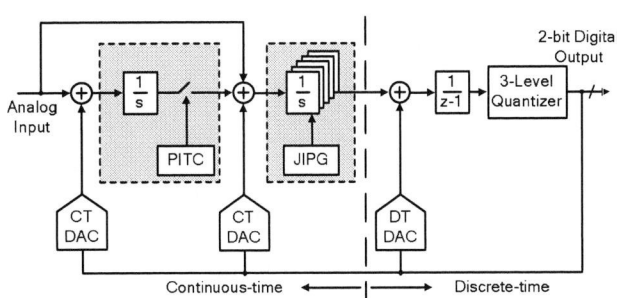

Fig. 1. Architecture of the 3rd-order hybrid $\Delta\Sigma$M.

the small size $\Delta\Sigma$M due to its simple and robust resistor feedback DAC.

II. ARCHITECTURE

Fig. 1 shows the proposed architecture of a 3rd-order, 3-level quantizer hybrid $\Delta\Sigma$M with an oversampling ratio of 256 with the PITC and the JIPG. For more than 100dB SNR audio ADC, the simulated signal-to-quantization-noise ratio (SQNR) is 126dB. This means the quantization noise is under a 10% to total noise power [8] and a negligible amount of the total noise.

The feedforward path is adopted between the input of ADC and the input of the 2nd-stage. This feedforward path reduces the internal output swing of the 1st integrator, and the reduced output swing makes it possible to achieve power saving and better signal linearity [6], [9]. However, since the feedforward path connects the input and the switched-capacitor sampling network of the 2nd integrator [1], this structure couples capacitor switching noise to inputs. In this paper, CT integrators are employed in both the 1st and the 2nd stages to avoid this problem. Only the 3rd stage is a DT integrator which is insensitive to the variation of process and temperature.

The first two CT stages provide effective 2nd-order anti-alias filtering for the main sampling circuits. The additional aliasing due to the integration time control of the 1st integrator is suppressed by the 1st-stage CT loop filter. The simple 3-level internal quantizer is utilized to avoid the nonlinearity of the feedback DAC without an additional dynamic element matching (DEM) block.

The transfer function and the loop filter coefficient of the hybrid $\Delta\Sigma$M are obtained by the impulse invariant transform of the equivalent DT $\Delta\Sigma$M [3]. The loop filter of the DT $\Delta\Sigma$M satisfied the required SQNR is designed and the coefficient of CT $\Delta\Sigma$M is acquired for system equivalence including the

978-1-4244-2018-6/08 $25.00 © 2008 IEEE

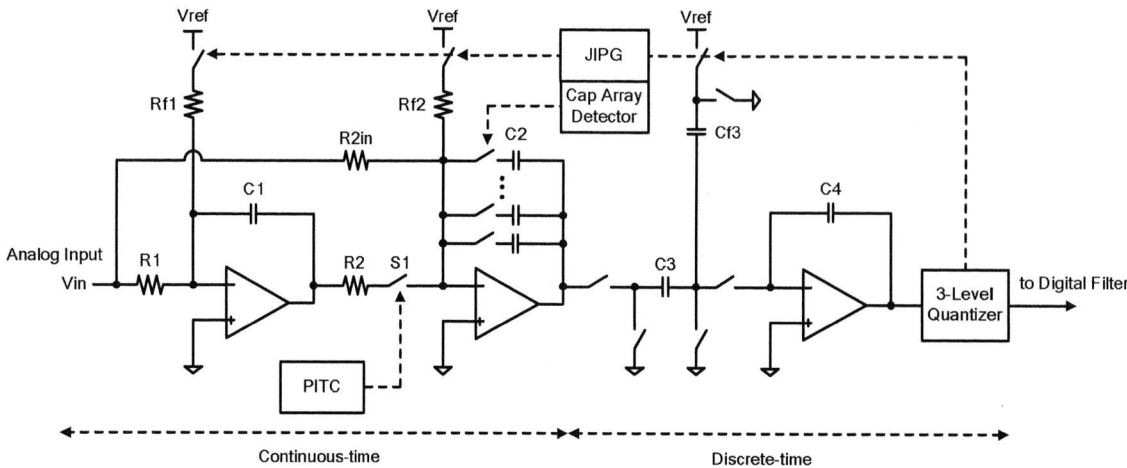

Fig. 2. Circuit diagram of the 3rd-order hybrid (1st and 2nd CT / 3rd DT) $\Delta\Sigma$M.

waveform and the delay of feedback DAC. The equivalence between CT $\Delta\Sigma$M and DT $\Delta\Sigma$M is verified with MATLAB simulation.

III. IMPLEMENTATION

Fig. 2 shows the simplified circuit diagram of the proposed 3rd-order hybrid $\Delta\Sigma$M. In real circuits, the fully-differential scheme is adopted for the better signal linearity and SNR advantage. The active RC loop filter is implemented for the better signal linearity in the 1st and 2nd CT loop filter instead of the Gm-C loop filter.

The PITC block including the switch S1 is added for the calibration of 1st integrator RC time constant variations. The capacitor array detection circuit is added for the calibration of 2nd integrator RC time constant variations. The clock of feedback DAC through JIPG has low pulse width jitter. The 3rd integrator is a general switched-capacitor integrator. The 3-level quantizer is implemented without a DEM block because feedback DAC uses only one resistor and three levels of voltage references. The simple resistor feedback DAC makes for small area and good noise performance.

A. Calibration of RC Time Constant Variations

One drawback of CT $\Delta\Sigma$M is the process and temperature variations of the RC time constants. The absolute value of the passive element varies about ±20% in general CMOS process and the varied absolute value directly changes the RC time constant. The RC time constant variations increase the in-band noise and can cause a stability problem. Fig. 4 (a), (c) show the frequency spectrum of the $\Delta\Sigma$M as RC variations. When the resistor and the capacitor are increased, the RC time constant is decreased. The decreased filter coefficient makes SNR degradation about 5dB in this modulator as shown in Fig. 4 (a). When the resistor and the capacitor are decreased, the RC time constant is increased. This increased filter coefficient makes stability problem as show in Fig. 4 (c) though variations of the passive elements are only -10% individually.

Fig. 3 shows the circuit and timing diagram of the proposed PITC technique which calibrates the RC time

constant variation of the 1st-stage. When the RC time constant varies, the output value of the 1st integrator (Vout) also varies, which is then integrated by the 2nd integrator in conventional CT $\Delta\Sigma$M. In the proposed $\Delta\Sigma$M, the duty rate of the 2nd-stage input switch (S1) is made proportional to the RC time constant

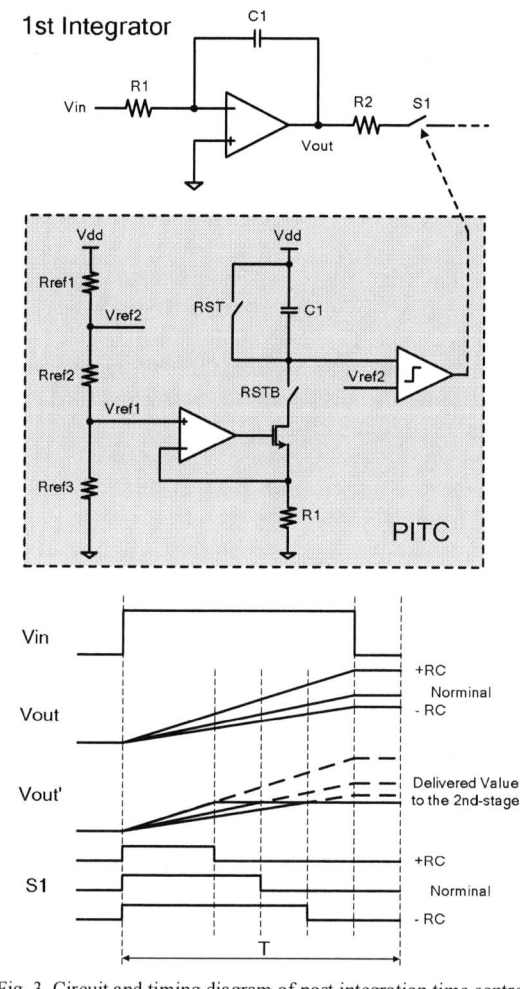

Fig. 3. Circuit and timing diagram of post integration time control.

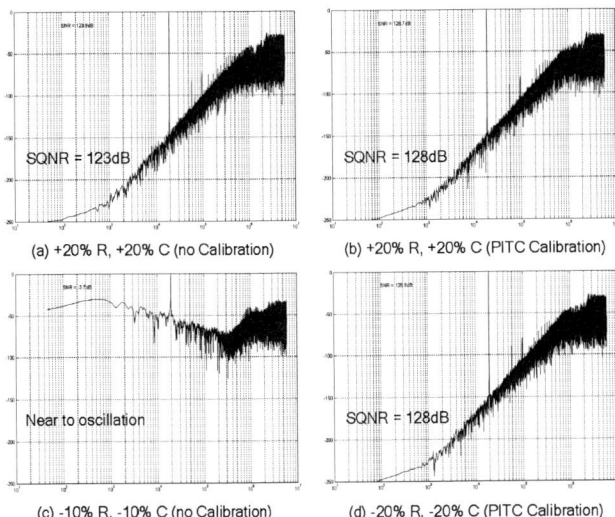

(a) +20% R, +20% C (no Calibration)

(b) +20% R, +20% C (PITC Calibration)

(c) -10% R, -10% C (no Calibration)

(d) -20% R, -20% C (PITC Calibration)

Fig. 4. Frequency spectrum simulation as RC time constants variations and the effectiveness of PITC.

of the 1st-stage to calibrate the 1st-stage RC time constant variation. The effectiveness of the PITC is proved by MATLAB simulation as shown in Fig. 4 (b), (d). The simulation results show that the PITC maintains the SNR performance and the stability in case the passive elements variations are ±20% individually.

This PITC technique not only efficiently calibrate the RC time constant variations but also completely removes switched-capacitor network from the input because of no tuning switch in the 1st-stage. In addition out-of-band noise aliasing is reduced by first-order noise shaping compared with conventional integration time adjustment previously reported [1]. Although the output swing of the 1st-stage can increase due to the RC time constant variation, the input feedforward structure efficiently reduces the output swing to the linear range.

The RC time constant variation of the 2nd-stage can also be removed by making the duty rate proportional to the square of the RC time constant. In this prototype, this was not implemented and programmable integration capacitors are employed for the 2nd-stage instead. The 2nd integrator

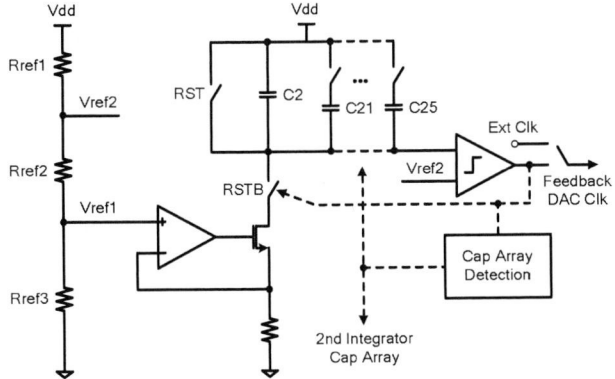

Fig. 5. Calibration block of the 2nd stage RC variation and jitter insensitive pulse generator.

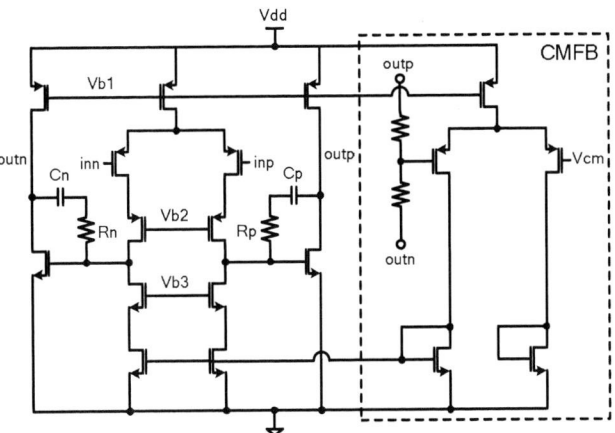

Fig. 6. Schematic of the 1st stage amplifier.

typically consists of a large resistor and a small capacitor for cost saving, because the 2nd-stage thermal noise of the $\Delta\Sigma M$ is negligible due to the noise shaping loop. Therefore it is efficient to calibrate the 2nd-stage RC time constant variations with capacitor arrays, since the size of capacitor array is negligibly small and the non-linear calibration does not affect the total system performance.

Fig. 5 shows the on-chip calibration circuits of the 2nd-stage RC time constant variation. The current through the resistor in the voltage-to-current converter is integrated to the 5-bit capacitor array, and the integrated level and the reference level are compared during the half clock period. The varied RC time constant is detected by the output of the comparator and the capacitor array detection circuits [4].

B. Effectiveness of the Jitter Insensitive Pulse Generator

Another drawback of CT $\Delta\Sigma M$ is the sensitivity to the feedback DAC clock jitter. However, performance of the CT $\Delta\Sigma M$ with the return-to-zero (RTZ) feedback DAC is only affected by the pulse width jitter [1], and the JIPG using the monostable multivibrator can effectively generate the internal clock with the required jitter performance.

In this audio ADC, the feedback DAC waveform is a 50% duty rate RTZ for the avoidance of inter-symbol interference (ISI). The capacitor array detection circuit shown in Fig. 5 for the calibration of the 2nd-stage also functions as a JIPG. Fig. 7 shows the two FFT plots of the SNR comparison of the JIPG on and off, showing the effectiveness of the JIPG as the 16dB SNR improvement.

C. Miscellanies

The small swing of the 1st-stage output by the feedforward structure reduces the slew-rate requirement of the amplifier. This structure reduces the static power consumption of the amplifier and guarantees the linear output swing range. The amplifier of the 1st CT integrator which determines the performance of the modulator is a telescopic amplifier with a common source output stage as shown in Fig. 6.

The complexity of the feedback DAC increases the amount of thermal noise. The transistor current mirror type feedback DAC typically contributes even more thermal noise than

978-1-4244-2018-6/08 $25.00 © 2008 IEEE

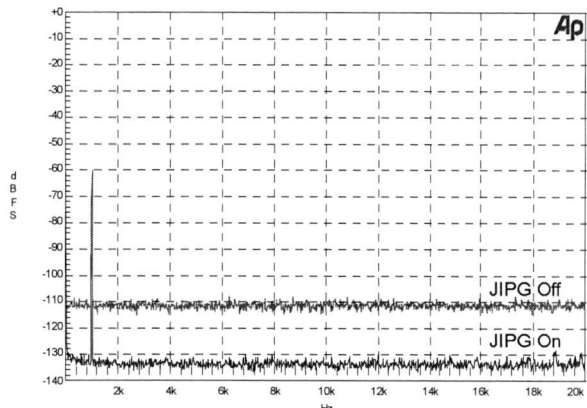

Fig. 7. FFT plots of a -60dBFS 1kHz tone with jitter insensitive pulse generator on and off.

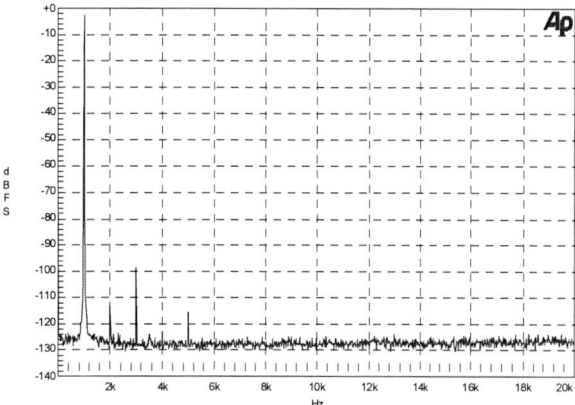

Fig. 8. FFT plot of a -3dBFS 1kHz tone.

resistor type DAC. In the prototype, a simple resistor feedback DAC with three reference voltage level is used for lower thermal noise and smaller overall area.

The excess loop delay, difference between the quantizer clock and feedback DAC pulse, makes SNR degradation in CT $\Delta\Sigma$M. In this audio ADC, the effect of excess loop delay is relieved by the feedback coefficient tuning [7].

The output of $\Delta\Sigma$M is filtered and decimated to produce the final output of ADC by the digital filter consisted of a sinc filter followed by two halfband and compensation filters. The analog sampling clock of $\Delta\Sigma$M is 12.288MHz (48kHz × 256) and provided by the digital filter.

IV. MEASURED RESULTS

This audio ADC is implemented in 65nm 6-metal CMOS process. Fig. 9 shows the die photograph of the 1-channel

TABLE I
Measured Performance

SNR (20-20KHz)	101dB (A-weighted)
DR	101dB (A-weighted)
SNDR	85dB (@ 0dBFS)
THD	-94dB (@ -3dBFS)
Power dissipation	15mW
Active Die Area	0.28mm^2

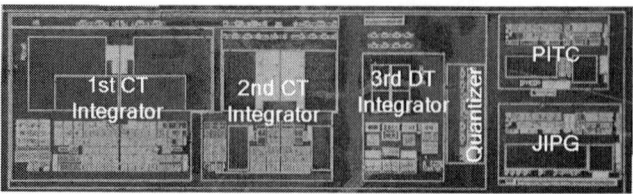

Fig. 9. Die photograph of the 1channel Audio ADC.

audio ADC occupying 0.28mm^2 (=1030µm × 275µm). The area of additional PITC and JIPG blocks is 0.06mm^2 (=220µm × 275µm) and the multi-channel audio ADC shares these calibration blocks.

The ADC dissipates 15mW per analog channel from a 3.3V supply. Fig. 7 shows the DR plot of -60dBFS 1kHz input signal and the 16dB improvement of SNR by using the JIPG block. Fig. 8 shows the FFT plot of a -3dB 1kHz input signal. The Measured SNR and DR are 101dB (A-weighted) and THD is -94dB. The performance of the proposed Audio ADC is summarized in Table I.

V. CONCLUSION

This paper introduces a hybrid 3rd-order $\Delta\Sigma$M with the proposed PITC technique. In addition, the effectiveness of the JIPG is proven by the experimental results. The PITC calibrates the RC time constant variations effectively and JIPG overcomes the degradation of SNR due to the clock jitter of the feedback DAC. The CT integrator using these proposed techniques emits no switching noise to the input while maintaining effective anti-alias filtering.

This audio ADC achieves 101dB SNR and DR, -94dB THD, 15mW power consumption and 0.28mm^2 active die area in 65nm CMOS process.

REFERENCES

[1] K. Nguyen, R. Adams, K. Sweetland and H. Chen, "A 106-dB SNR hybrid oversampling analog-to-digital converter for digital audio," ISSCC J. Solid-State Circuits, vol. 40, no. 12, pp. 2408-2415, Dec. 2005.

[2] S. D. Kulchycki, R. Trofin, K. Vleugels and A. Wooley, "A 1.2-V 77-dB 7.5-MHz continuous-time/discrete-time cascaded $\Sigma\Delta$ modulator," Symposium on VLSI Circuits, pp. 238-239, Jun. 2007.

[3] O. Oliaei, "Design of continuous-time sigma-delta modulators with arbitrary feedback waveform," IEEE Trans. Circuits and Systems, vol. 50, no. 8, pp. 437-444, Aug. 2003.

[4] B. Xia, S. Yan and E. Sanchez-Sinencio, "An RC time constant auto-tuning structure for high linearity continuous-time sigma-delta modulators and active filters," IEEE Trans. Circuits and Systems, vol. 51, no. 11, pp. 2179-2188, Nov. 2004.

[5] B. Del Signore, D. A. Kerth, N. S. Sooch and E. J. Swanson, "A monolithic 20-b delta-sigma A/D converter," ISSCC J. Solid-Stage Circuits, vol. 25, pp. 1311-1317, Dec. 1990.

[6] S. Norsworthy, R. Schreier, and G. Temes, Delta-Sigma Data Converters: Theory, Design, and Simulation. New York: IEEE Press, 1996.

[7] J. A. Cherry and W. M. Snelgrove, Continuous-Time Delta-Sigma Modulators for High-Speed A/D Conversion: Threrory, Practice and Fundamental Performance Limits. Kluwer Academic Publishers, 2000.

[8] R. Schreier, J. Silva, J. Steensgaard and G. C. Temes, "Design-oriented estimation of thermal noise in switched-capacitor circuits," IEEE Trans. Circuits and Systems, vol. 52, no. 11, pp. 2358-2368, Nov. 2005.

[9] R. Schreier and G. C. Temes, Understanding Delta-Sigma Data Converters. IEEE Press, 2005.

IEEE 2008 Custom Intergrated Circuits Conference (CICC)

An 8.1 mW, 82 dB Delta-Sigma ADC with 1.9 MHz BW and -98 dB THD

Kyehyung Lee, Matthew R. Miller[1], and Gabor C. Temes

Oregon State University, Corvallis, OR 97331, USA

[1]Freescale Semiconductor, Lake Zurich, IL 60047, USA

Abstract-A switched-capacitor low-distortion 15-level delta-sigma ADC is described. It achieves third-order noise shaping with only two integrators by using quantization noise coupling. It provides 81 dB SNDR, 82 dB dynamic range, and -98 dB THD in a signal bandwidth of 1.9 MHz. It dissipates 8.1 mW with a 1.5 V power supply (analog power 4.4 mW, digital power 3.7 mW). Its figure-of-merit is among the best reported for wideband ADCs.

I. INTRODUCTION

With increasing demand for wide-band low-power data converters in many wired and wireless applications, both the proper selection of ADC topology and the efficient circuit realization to meet stringent specifications are becoming important. Discrete-time delta-sigma ADCs can provide more than 80 dB SNDR for MHz-range signals with low power consumption. For given bandwidths, dynamic range and resolution, the power dissipation can be minimized through the choice of power-efficient modulator architecture and circuit implementation. A low-distortion delta-sigma ADC topology greatly relaxes the linearity requirements for its loop filter by reducing the internal signal swing [1]. This is achieved by appending to a conventional modulator shown in Fig. 1(a) a direct feedforward path from the input to the quantizer, as illustrated in Fig. 1(b). This makes the signal transfer function (STF) equal to 1, and hence cancels the input signal at the input of loop filter, so that the loop filter processes the quantization noise only. The summation of signals at the quantizer input may be performed by either a passive adder or an active one. Passive summation can be done with a capacitive adder, which does not require an extra opamp. However, the signal attenuation due to parasitics results in increased power dissipation and mismatch effects at the quantizer. On the other hand, active summation requires an extra opamp. Neither of the two approaches is thus power-efficient.

In this paper, an improved low-distortion delta-sigma ADC is described, which uses the injection of shaped quantization noise to boost the noise shaping performance from second to third order. This is achieved without changing the original modulator coefficients and without degrading its loop stability. It demonstrates a power-efficient design approach to wide-band ΔΣ ADC design.

II. PROPOSED MODULATOR ARCHITECTURE

The proposed improved low-distortion ΔΣ modulator architecture is shown in Fig. 1(c). Here, a linearized model is used, where the nonlinear quantizer block is replaced by a

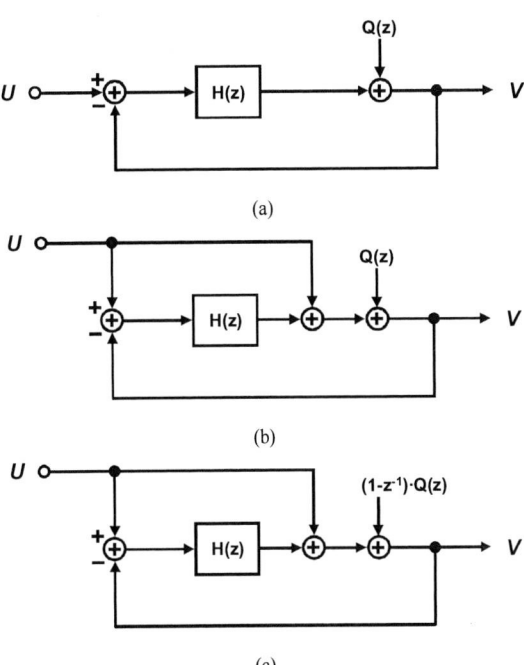

Fig. 1. Steps to get an improved modulator architecture. Injection of shaped quantization noise is shown in Fig.1(c) for a linearized model of ΔΣ modulator.

simple adder, based on an assumption that the quantization noise is random and is decorrelated with the input signal. This assumption is usually valid when a multi-bit quantizer is used, and the input signal condition is sufficiently busy. In the proposed modulator architecture, a shaped quantization noise is injected via noise coupling instead of a white quantization noise. If *STF* is the signal transfer function and *NTF* is the noise transfer function, then the output of the proposed modulator is represented by

$$V(z) = STF(z)\cdot U(z) + NTF(z)\cdot (1-z^{-1})\cdot Q(z). \quad (1)$$

It is clear that the effective noise-shaping order is increased by 1 due to the injection of the shaped quantization noise.

The injection of shaped quantization noise was applied to a second-order low-distortion modulator to get a third-order noise shaping, as illustrated in Fig. 2. The shaped quantization noise is obtained by delaying the quantization noise by one clock cycle, and subtracting it from the summing node before the quantizer. The injected quantization noise effectively raises the modulator order from two to three, without changing its signal transfer function [2]. Thus, the reduced signal swing and low-distortion property are preserved, and the modulator

978-1-4244-2018-6/08 $25.00 © 2008 IEEE

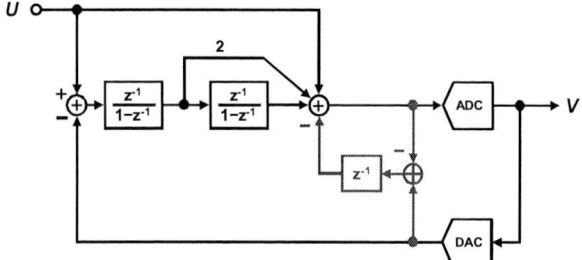

Fig. 2. Block diagram of the proposed modulator with shaped quantization noise injected via noise coupling.

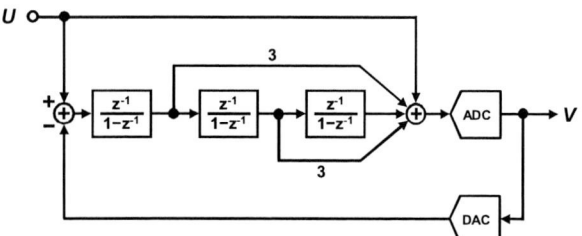

Fig. 3. Conventional third-order low-distortion modulator.

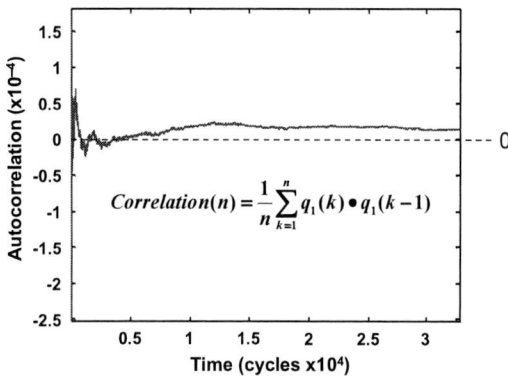

$$Correlation(n) = \frac{1}{n}\sum_{k=1}^{n} q_1(k) \bullet q_1(k-1)$$

Fig. 4. Time evolution of the autocorrelation of the quantization noise with one clock cycle of lag in Fig.2.

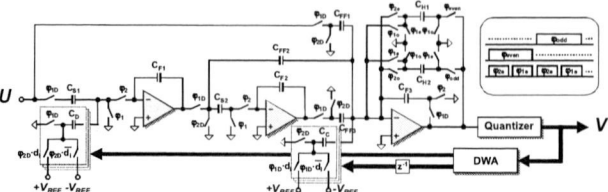

Fig. 5. SC circuit implementation of the noise-coupled modulator.

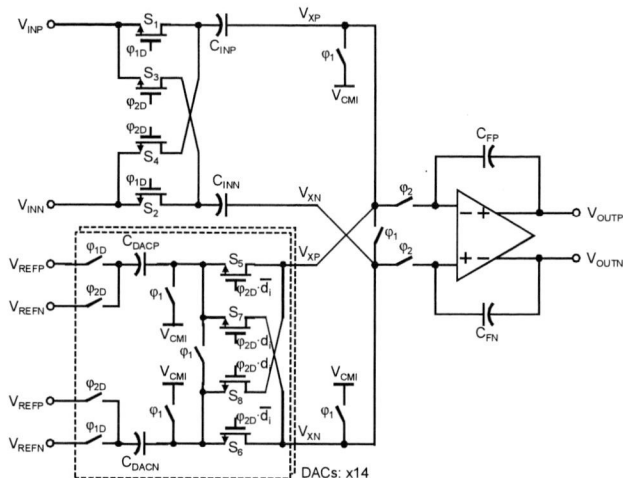

Fig. 6. Detailed diagram of the input branch of the modulator with separate input and DAC capacitors.

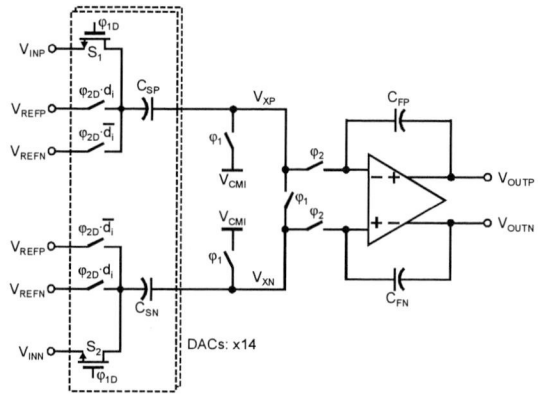

Fig. 7. Input branch of the modulator with shared input and DAC capacitors.

performance and power efficiency are significantly improved. The sampling frequency and oversampling ratio (OSR) can be reduced for the same bandwidth (BW) and SNDR performance, or either the BW or the SNDR performance can be increased for the same sampling frequency, by applying this technique. The use of an extra opamp to realize the active signal summation at the input of quantizer is now justified, since it provides the virtual ground for both active addition and noise coupling. Compared with conventional third-order low-distortion delta-sigma ADC as illustrated in Fig. 3, the proposed topology provides similar noise shaping, but with increased stability as well as reduced tone generation and THD, while using a reduced number of amplifiers. The use of a multi-bit quantizer and DAC reduces the signal swings at the internal nodes of the loop filter further, and improves the modulator stability. The quantization noise of the modulator also becomes more random due to multi-bit quantization and noise coupling. The additional advantage for this modulator architecture is that the injected quantization noise works as a dither signal, and thus idle tones and harmonic spurs are reduced and the modulator linearity is improved. Fig. 4 shows

the time evolution of the correlation between the quantization noise and its replica delayed by one clock cycle in the modulator in Fig. 2. This demonstrates the similarity of the quantization error to white noise, and justifies the linearized model shown in Fig. 1.

III. CIRCUIT IMPLEMENTATION

Fig. 5 illustrates the simplified SC circuit diagram of the noise-coupled modulator. There are two integrators followed by an active adder. The input branch for the first integrator is shown in Fig. 6. The input sampling and DAC capacitors are separated to avoid any reference errors due to signal-dependent loading of the reference, which occurs in an input branch having shared input and DAC capacitors, as demonstrated in Fig. 7. However, the separation of the input

978-1-4244-2018-6/08 $25.00 © 2008 IEEE

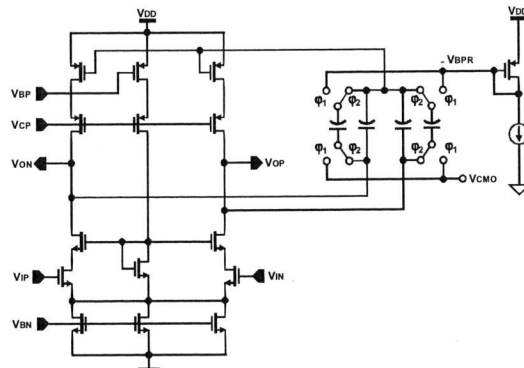

Fig. 8. Telescopic cascode opamp with SC common-mode feedback.

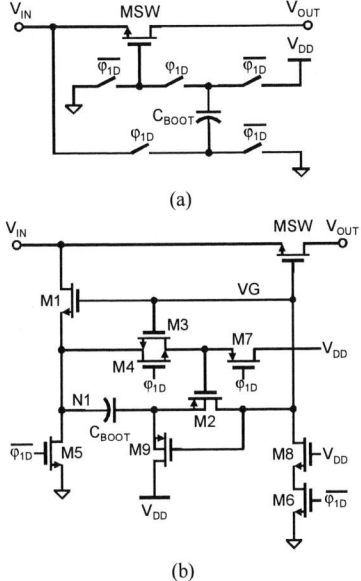

(a)

(b)

Fig. 9. Bootstrapped clock signal generator; (a) simplified diagram, (b) detailed circuit realization.

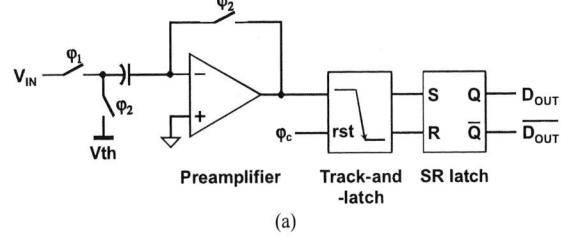

(a)

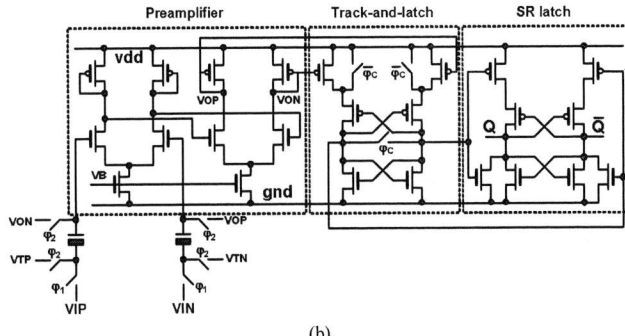

(b)

Fig. 10. Quantizer with input offset sampling; (a) simplified diagram, (b) detailed circuit implementation.

sampling and DAC capacitors slows down the settling of the opamp, by reducing its closed-loop bandwidth that is given by the feedback factor multiplied by the UGBW of the opamp. To get around this adverse effect, the size of the input sampling and DAC capacitors were reduced by half, and the same feedback factor was preserved. The signal swings are also doubled to keep the same SNR through the use of cross-coupled switches between positive and negative rails, as illustrated in Fig. 6. Therefore, even with the separate input sampling and DAC capacitor scheme, the same feedback factor and settling time of the first integrator opamp are achieved with the same power.

The signal swing of the modulator loop filter is significantly reduced, thanks to the low-distortion modulator architecture and multi-bit quantizer. As a result, the limited output signal swing of the telescopic cascode opamp is not an issue for this design. This allows the most power-efficient telescopic cascode opamp with a SC common-mode feedback circuitry, illustrated in Fig. 8, for the integrators and the active adder. The output signal swing of each rail of the designed opamp is ± 0.36 V using a power supply of 1.5 V. The slew

rate requirement for the opamp is also greatly relaxed by using the low-distortion topology and multi-bit quantization. An offset-compensated opamp is used for the active adder.

Bootstrapped clock signals are applied to ensure linear sampling at the critical input switches, as illustrated in Fig. 9 [3]. Figure 10 demonstrates the simplified diagram and the detailed circuit implementation of the quantizer [4]. Each comparator contains a two-stage preamplifier followed by a track-and-latch and an SR latch. Input offset sampling scheme is employed in the two-stage preamplifier to minimize the effect of an offset of the track-and-latch. Thus, the input-referred offset of the latch is reduced by the gain of the two-stage preamplifier. The threshold voltages of the quantizer are generated by a resistor string. Data weighted averaging (DWA) is applied to shape the mismatch errors of the 15-level DAC. The transistors in the quantizer and in the 4-stage logarithmic shifter of the DWA circuitry are optimized to meet the critical timing specifications. The essential timing parameters, such as clock nonoverlap time, are digitally programmable via a 3-wire serial interface to get optimum performance reliably under variations of process, voltage, and temperature.

IV. EXPERIMENTAL RESULTS

A prototype device was fabricated in a 0.18μm 2P4M CMOS process. It uses a 1.5 V power supply for the analog circuitry and a 1.45 V one for the digital stages. The modulator was tested with a single-tone input with frequencies from 10 kHz to 1 MHz. It shows uniform performance in this range with a 60 MHz sampling clock. The measured spectrum at peak SNDR is illustrated in Fig. 11 for a -1.13 dBFS 100 kHz input signal. It shows an SFDR around 100.2 dB. The SNR and SNDR variations with input signal level are shown in Fig. 12. They illustrate linearly increasing, nearly identical SNR and SNDR performances for signal powers from -85

978-1-4244-2018-6/08 $25.00 © 2008 IEEE

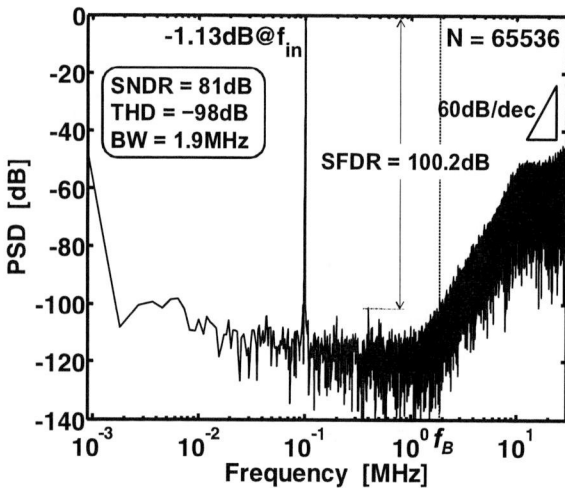

Fig. 11. Measured spectrum at peak SNDR for a -1.13 dBFS 100 kHz input.

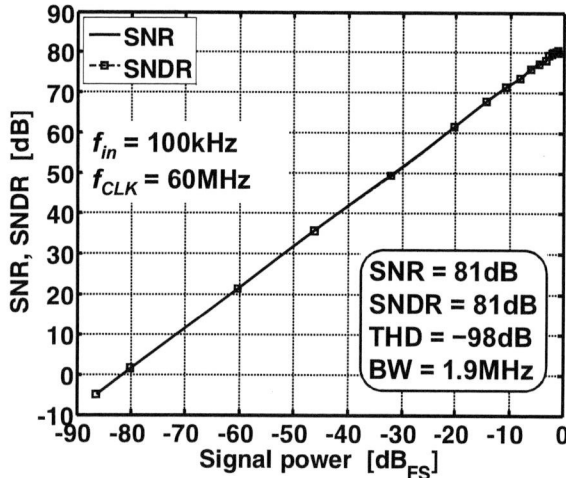

Fig. 12. SNR and SNDR variations with input signal power.

Table 1: Measured performance summary.

Clock frequency	60 MHz
Signal bandwidth	1.9 MHz
OSR	16
Input range (diff.)	1.44 V_{pp}
C_{IN} and C_{DAC}	separate
Dynamic range	82 dB
SNR	81 dB
SNDR	81 dB
THD	−98 dB
FOM	0.25 pJ/conv
Power consumption	4.4 mW (A), 3.7 mW (D)
Power supply	1.5 V (A), 1.45 V (D)
Process	0.18μm 2P4M CMOS
Core area	1.27 mm²

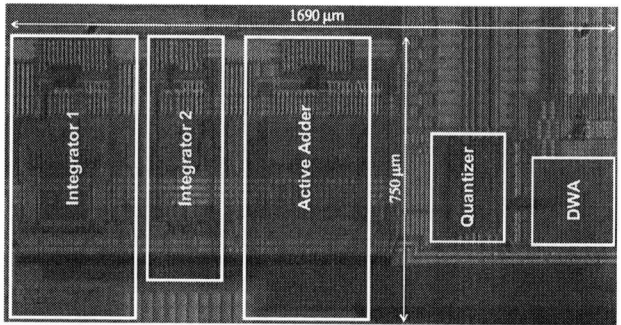

Fig. 13. Chip photograph of the noise-coupled modulator.

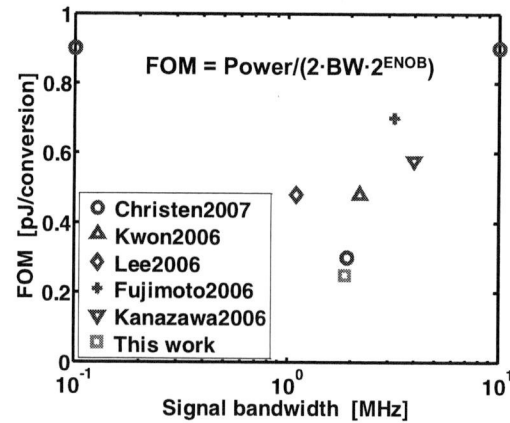

Fig. 14. Comparison with other discrete-time modulators published recently.

dBFS up to -1 dBFS, indicating highly linear modulator operation. The prototype achieves 81 dB peak SNR and SNDR, 82 dB DR, and -98 dB THD. The measured power dissipation is P = 8.1 mW (analog: 4.4 mW, digital: 3.7 mW) and the figure-of-merit (FOM), defined by $P/(2 \cdot BW \cdot 2^{(SNDR-1.76)/6.02})$, is 0.25 pJ/conversion-step, which is among the lowest published for a wideband discrete-time delta-sigma ADC. The measured performance is summarized in Table 1, and the die micrograph is shown in Fig. 13. Comparison with other discrete-time wideband modulators published recently is illustrated in Fig. 14.

V. CONCLUSION

An improved low-distortion ΔΣ ADC architecture with shaped quantization noise injected using a novel noise coupling technique was proposed. The measured results of the prototype demonstrated state-of-the-art performance and verified the effectiveness of the proposed architecture.

ACKNOWLEDGMENTS

The authors would like to thank Asahi Kasei EMD Corp. for providing prototype fabrication and useful discussions.

REFERENCES

[1] J. Silva, U. Moon, J. Steensgaard, G. C. Temes, "Wideband low-distortion ΔΣ ADC topology," *Electronics Letters*, vol. 37, no. 12, pp. 737-738, June 2001.

[2] K. Lee, M. Bonu, and G. C. Temes, "Noise-coupled delta-sigma ADCs," *Electronics Letters*, vol. 42, no. 24, pp. 1381-1382, Nov. 2006.

[3] M. Dessouky and A. Kaiser, "Very low-voltage digital-audio ΔΣ modulator with 88-dB dynamic range using local switch bootstrapping," *IEEE J. Solid-State Circuits*, vol. 36, no 3, pp.349-355, Mar. 2001.

[4] I. Mehr and L. Singer, "A 55-mW, 10-bit, 40-Msample/s Nyquist-rate CMOS ADC," *IEEE J. Solid-State Circuits*, vol. 35, no 3, pp.318-325, Mar. 2000.

IEEE 2008 Custom Intergrated Circuits Conference (CICC)

A 2.5MHz BW and 78dB SNDR Delta-Sigma Modulator Using Dynamically Biased Amplifiers

Yan Wang, Kyehyung Lee, Gabor C. Temes
School of Electrical Engineering and Computer Science
Oregon State University, Corvallis, OR 97331

Abstract- **A new dynamically-biased scheme is proposed to implement a 13-bit delta-sigma modulator with a 2.5 MHz signal bandwidth. It uses the low-distortion architecture, and hence the opamp linearity requirements are greatly relaxed. Its noise-coupled and time-interleaved structure further decreases the power consumption. The prototype chip was fabricated in a 0.18um CMOS technology. Experimental results show that 78dB SNDR is achieved when it is clocked at 60MHz sampling rate. With 1.6V power supply, the power dissipation is 19.2 mW.**

I. INTRODUCTION

Discrete-time delta-sigma A/D converters with 12 to 14 bit resolution and 1 to 5 MHz signal bandwidth find many applications in xDSL,wireless and wire line communication systems. Switched-capacitor circuits are often used to realize the fundamental blocks of discrete-time $\Delta\Sigma$ ADCs. In conventional design, each op-amp in the switched-capacitor circuits is biased with a constant current source, which is required to satisfy the slew-rate and settling accuracy requirement under worst-case conditions. This is wasteful under other operation conditions. To achieve low power consumption, various dynamically-biased schemes [1-3], such as tailed-biased dynamic scheme, clocked current dynamic scheme and switched-capacitor sampling network dynamic biased scheme etc., have been proposed. However, the applications of these schemes are limited by narrow signal bandwidth or large harmonic distortion.

In this paper, we present a 2.5MHz BW 78dB SNDR low power $\Delta\Sigma$ modulator using a new dynamically-biased telescopic amplifier [4] for the input integrator. The dynamic current detection scheme used for general switched-capacitor circuits is shown in Fig. 1.By summing the currents flowing through the differential pair transistors, the bias current can be made proportional to the absolute value of the differential input signal. Since the current detection path is fast, application for high clock frequencies becomes possible. Noise-coupled [5] and time-interleaved techniques, combined with low distortion architecture [6], have been used to achieve good linearity performance with wide signal bandwidth.

II. Dynamically-Biasing Scheme

A. Dynamically Biased Integrator

Fig.1 shows the proposed dynamic biasing scheme for switched-capacitor amplifier and integrator. Fig. 2 is the basic block diagram for a single-ended switched-capacitor integrator. During the input sampling phase (CK1=1), the input capacitor Cs is charged to Vin, while during the integration phase (CK2=1), Cs is discharged, and all its charge

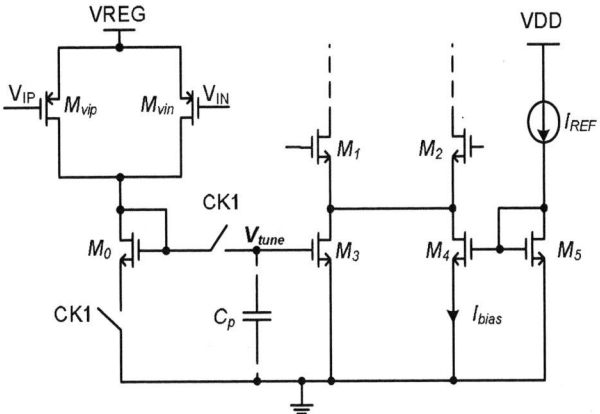

Fig. 1. Dynamic biasing scheme for a differential input signal.

is transferred to Cint. The bias current of the op-amp is the sum of a fixed Ibias and the current flowing through M3, which is controlled by Vin during the sampling phase and held constant during the integration phase. If Vin is large, so the transferred charge is large, M3 will dynamically increase the total bias current. A straightforward extension to differential operation is illustrated in Fig. 1. Let

$$V_{IP} = V_{cmi} + V_{diff} \qquad (1)$$

$$V_{IN} = V_{cmi} - V_{diff} \qquad (2)$$

Assume that VIP and VIN swing significantly from their common mode value $V_{cmi,}$ but the transistors still stay at saturation region. For square-law characteristics, the current flowing through M0 and M3 can be expressed as

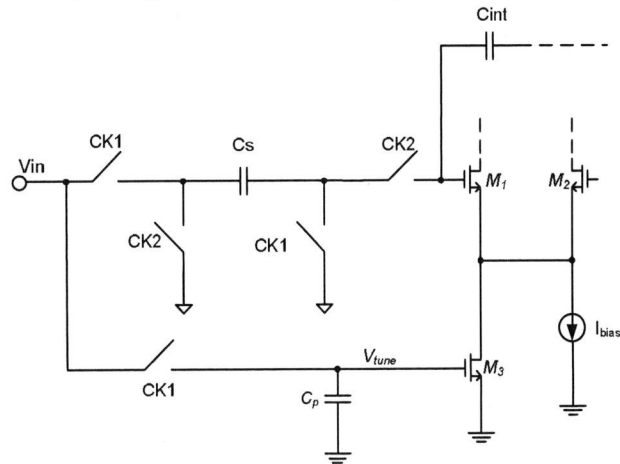

Fig.2. The block diagram of a single-ended SC integrator with dynamic biasing.

978-1-4244-2018-6/08 $25.00 © 2008 IEEE   107

$$I = K \times (V_{REG} - V_{cmi} - V_{th})^2 + K \times V_{diff}^2 \qquad (3)$$

Here, $K = \mu C_{ox} \dfrac{W}{L}$ and V_{diff} is the difference between V_{cmi} and input signals V_{IP} or V_{IN}. The current contains a constant part and a tuned one, and they can be chosen so as to meet the specifications of the stage. If $V_{REG} = V_{cmi} + V_{th}$, the fixed current will be much smaller than the tuning current, and the tuning range is enlarged .For a slew-rate limited opamp,the total bias current at integration phase should be large enough to provide the slew current into the opamp's load capacitor C. This gives

$$I_{bias} + \frac{K}{2} \times V_{diff}^2 > \frac{2C \times V_{diff}}{t_{slew}} \qquad (4)$$

Here, t_{slew} is the allocated slew time for the opamp.Equation (4) can only be satisfied if the condition $KI_{bias} > 2C^2 / t_{slew}^2$ holds. This sets a design equation for I_{bias} and K.

B. Dynamic biasing implementation for delta-sigma modulator

For the input integrator in a delta-sigma modulator, the total charge which needs to be transferred during the integration phase depends not only on the input signal, but also on the DAC feedback voltage. Hence, the current detection circuit is required to extract the difference between them. Fig. 3 shows the circuit realization for this task. When CK1=1, the differential input signals from the ADC input and the feedback DAC are differenced at nodes A and B by capacitors C1-C4. The voltages at node A and B can be expressed as

$$V_A = \frac{V_{IP} - V_{DACP}}{2 + r} + V_{cmi} \qquad (5)$$

$$V_B = \frac{V_{IN} - V_{DACN}}{2 + r} + V_{cmi} \qquad (6)$$

Here, r = Cp/2C, and Cp is the parasitic capacitance at nodes A and B. The current flowing through M0 and M3 will track with the voltage variation at node A and B, and the input signal attenuation can easily be compensated by adjusting the current ratio between M0 and M3. When CK2=1, C1 through C4 are reset to prevent memory effects. The voltage Vtune is held by M3 transistor's parasitic capacitance, and hence the opamp's bias current keeps constant. (Note that the clock timing for the summation capacitors might need to be changed for different modulator architectures.)

III MODULATOR ARCHITECTURE

Fig 4 shows the block diagram of the two-channel noise-coupled time-interleaved modulator. Each channel contains a second-order low-distortion modulator with multi-bit quantizer. The bias current of the telescopic op-amp for the input integrator is dynamically biased with proposed scheme. The time interleaved sampling doubles the effective sampling rate to 120MS/s for the 60MHz clock rate and introduces two notches at Nyquist frequency for the NTF. So quantization noise power at the Nyquist frequency is greatly attenuated and noise folding to signal band due to channel mismatching can be negligible. Meanwhile, the second-order modulator's quantization noise at each channel is extracted and injected into the loop of the other channel's modulator. This cross-coupling between two channels raises the effective noise-shaping order from two to three without degrading the loop stability .And the injected noise can also act as dither signals which can reduce the in band idle tones and harmonic distortion. Finally, with the low distortion architecture, the integrator just needs to deal with the quantization noise, so the harmonic tones due to the dynamically-biased current can be greatly reduced. Furthermore, with 4-bit quantizer, the quantization noise is usually small and the low power consumption with dynamically biasing can be expected

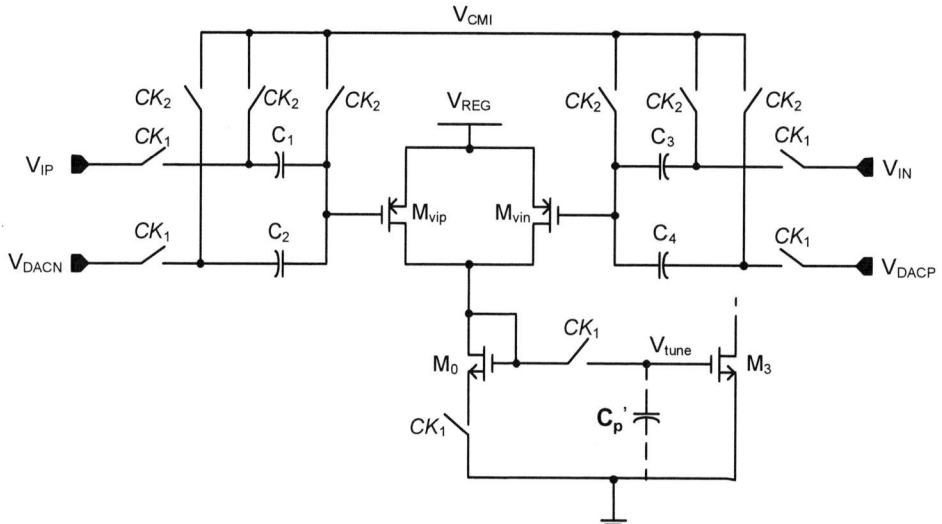

Fig.3. Current detection circuit for a dynamically biased integrator.

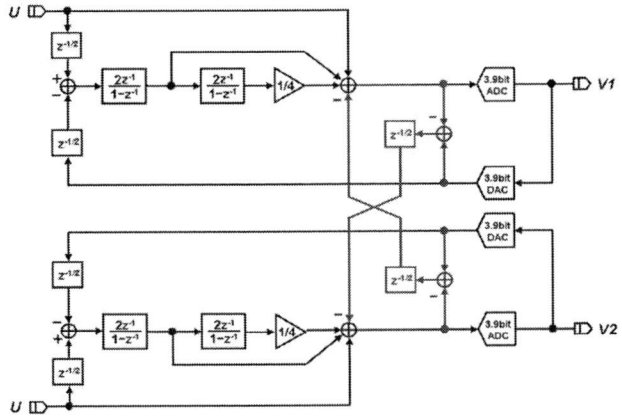

Fig.4 The block diagram of the noise-coupled time-interleaved $\Delta\Sigma$ ADC.

IV. EXPERIMENTAL RESULTS

The prototype chip was fabricated in a 2P4M 0.18um CMOS technology. Fig 5 shows the measured output spectrum with a -1.25dBFS 100 kHz sine-wave input, and a 60MHz clock signal. Fig. 6 illustrates the measured SNR, SNDR and dynamic range, indicating good linearity in spite of the nonlinear biasing algorithm used. The THD was -88.7 dBFS, and the SFDR was 89.2dB. Fig. 7 shows the variations of the peak SNR and SNDR with the input signal frequency. The total power dissipation was 19.2mW, 11.2mW for the analog circuit and 8mW for the digital circuitry. The active die area is 3.66mm^2. Fig 8 shows the chip die photograph. The overall performance is summarized in Table I.

V ACKNOWLEDGMENTS

The authors gratefully acknowledge the financial and fabrication support by AKM.

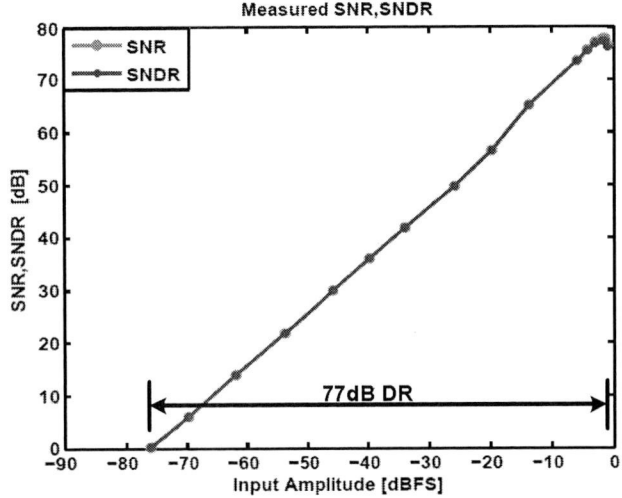

Fig. 6. Measured SNR/SNDR with different input levels.

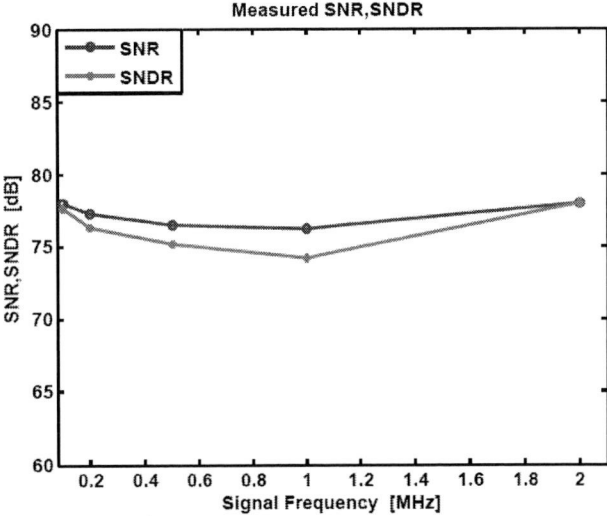

Fig .7. Measured SNR/SNDR with different signal frequencies.

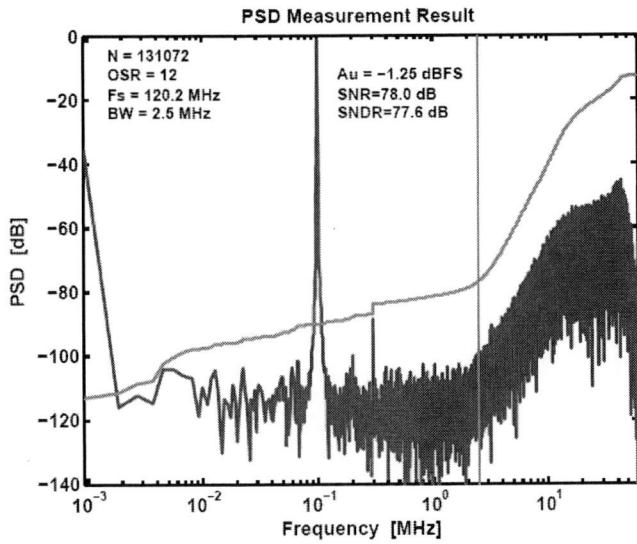

Fig 5. Measured output spectrum at the peak SNDR.

Fig.8. chip die photograph

TABLE I
MEASUREMENT RESULTS SUMMARY

Process	0.18um 2P4M CMOS
Sampling frequency	120MHz
Clock frequency	60MHz
Signal bandwidth	2.5MHz
OSR	12
C_{IN}/C_{DAC}	shared
Power supply	1.6V
Dynamic Range	77dB
SNDR/SFDR	77.6dB/89.2dB
THD	-88.7dB
Power consumption	11.2mW(A) 8mW(D)

REFERENCES

[1] B.J.Hosticka, "Dynamic CMOS Amplifiers," IEEE J.Solid-State Circuit, pp 887-894, Oct. 1980.

[2] D.B.Kasha, W.L.Lee and A.Thomsen,"A 16-mW, 120-dB Linear Switched-Capacitor Delta-Sigma Modulator with Dynamic Biasing," IEEE J.Solid-State Circuit, vol.34,pp 921-925,July. 1999

[3] Z.Cao, T.Song and S Yan ,"A 14mW 2.5MS/s 14bit Sigma-Delta Modulator Using Pseudo-Differential Split-Path Cascode Amplifiers", pp.49-52, IEEE Custom Integrated Circuits Conference 2006

[4] Y.Wang and G.C.Temes ,"Dynamic Biasing Scheme for High-speed/low-power Switched-capacitor Stages," Electronics Letters, vol .43, no.4, pp.214-216,Feb.2007

[5] K.Lee, F.Maloberti and G.C.Temes, " Noise-Coupled Multi-Cell Delta-Sigma ADCs,"Proc. IEEE ISCAS, vol.1,pp.249-252,May 2007

[6] J. Silva,U. Moon, J. Steensgaard and G.C.Temes," Wideband Low-Distortion Delta-Sigma ADC Topology," Electronics Letters, vol.37, no.12, pp.737-738, Jun.2001.

IEEE 2008 Custom Intergrated Circuits Conference (CICC)

74dB SNDR Multi-Loop Sturdy-MASH Delta-Sigma Modulator Using 35dB Opamp Gain

Nima Maghari, Sunwoo Kwon and Un-Ku Moon
School of EECS, Oregon State University, Corvallis, OR 97331

Abstract—**In this paper a new multi-loop delta-sigma modulator is presented. This multi-loop modulator is insensitive to low gain opamps while maintaining the stability advantage. As a proof of concept, a prototype was implemented to show the functionality of this structure. Measurement results shows that with open-loop opamp gain of only 35dB, this prototype achieves over 74dB SNDR at oversampling ratio of 16. The sampling frequency is 20MHz and the total power dissipation is 3.2mW from a 1.2V power supply.**

I. INTRODUCTION

Improvement in digital communication systems as well as popularity of portable systems has increased the demand for low-voltage low-power analog-to-digital converters (ADCs) over the past few years. Among various ADCs, delta-sigma modulators (DSMs) are well suited for low-to-medium bandwidth high-resolution conversion due to their noise shaping plus oversampling property. The oversampling nature of these ADCs will vastly reduce the capacitor size for low-speed high-accuracy application such as audio ADCs, and it will greatly relax the anti-aliasing filter requirements in medium speed applications. The drawback of the oversampled system is that the signal bandwidth is only a small fraction of the overall bandwidth. Consequently, to increase the signal bandwidth, higher sampling speed is required, which increases power dissipation. The alternative approach is to enhance the noise shaping property by increasing the order of the modulator.

The two most commonly used delta-sigma modulator structures are the single loop high order modulator and the multi-stage noise shaping (MASH) structure. The single loop high order DSM can provide high signal-to-noise ratio (SNR) with relaxed circuit elements, but it is prone to instability. In high order modulators, the stable input range, which is defined by the maximum input signal range that leads to proper operation of the modulator, is limited. This range mostly depends on the number of levels used in the quantizer and the order of the modulator. On the other hand, the MASH structure can guarantee the stability by employing extra/cascaded ADCs or modulators to enhance its noise shaping property, but it requires high accuracy integrators to minimize the quantization error leakage resulting from the classic mismatch problem between the analog and the digital transfer functions [1]. This will become more challenging in submicron process due to the intrinsic gain reduction of transistors. Moreover, in low-voltage applications, high opamp gain requirement will often result in multi-stage opamps where both stability and

power dissipation will be a dominant limiting factor of the overall performance.

This work describes the IC implementation of the Sturdy-MASH (SMASH) modulator [2] to overcome the need of high opamp gain requirements and accurate modulator coefficients. It will be shown that similar performance of the MASH structures can be achieved by employing opamps with DC gain around 30dB. Low gain requirements will enable using simple opamps with large available output swing even in very low-voltage applications.

This paper is organized as follows: In Section II, a brief review of SMASH modulator is presented followed by the system level architecture for this implementation. Section III gives details of the circuit implementation of the prototype ADC. Measurements results are provided in Section IV and finally conclusions are drawn in Section V.

II. SMASH DELTA-SIGMA MODULATOR

A general two-loop SMASH DSM is illustrated in Fig. 1, where L_{Si}, L_{Ni} and E_i denote the signal loop filter, noise loop filter and quantization error of the ith stage, respectively. Unlike the traditional MASH structure where the output of the second loop is subtracted from the output of the first loop via digital filters (STF_{2D} and NTF_{1D}) outside the modulator loop, in this architecture the output of the second loop is subtracted from the first stage output inside the loop. In other words, the digital summing block is moved inside the first loop and all digital filters are eliminated. Writing down the signal transfer function (STF) and the noise transfer function (NTF) for the SMASH structure shown in Fig. 1 yields

$$Y_{SMASH} = STF_1X + NTF_1(1-STF_2)E_1 - NTF_1NTF_2E_2. \quad (1)$$

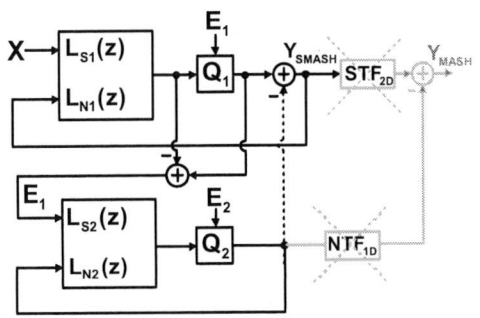

Fig. 1. MASH (shown in gray) and SMASH (shown with the dashed line) structures.

978-1-4244-2018-6/08 $25.00 © 2008 IEEE

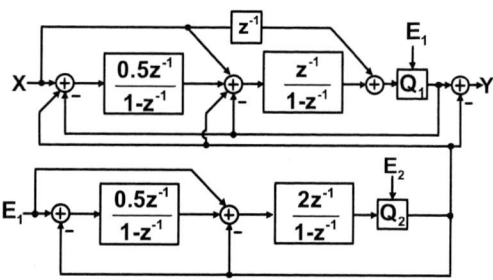

Fig. 2. Implemented 2+2 SMASH modulator.

In order to achieve proper noise shaping for both of the quantization errors, i.e., E_1 and E_2 in a 2+2 SMASH structure, the second stage signal transfer function should satisfy

$$STF_2 = 2z^{-1} - z^{-2}. \qquad (2)$$

It is instructive to note that in the SMASH structure both quantization errors will be shaped by the order of the modulator while in the regular MASH structure, only the last quantization error will be shaped and all other quantization errors will be canceled out via digital filters. Theoretically, it is possible to cancel E_1 in SMASH structure by choosing $STF_2 = 1$ [3]. However, this demands extracting E_1, feeding it to the second loop and finally subtracting it from the output of the first loop, all without any delay which is not feasible.

For this design, the modulator shown in Fig. 2 was chosen. It can be seen that the second loop satisfies the criteria of (2). The first loop uses distributed signal feedforward topology to minimize the swing requirements of the integrators. The delay in the signal feedforward path is due to the non-zero delay of the quantizer and feedback path, and it will result in less than 0.1dB in-band peaking in signal transfer function. The modulator coefficients are chosen to minimize the output swing of integrators while maximizing their feedback factor. Although the second loop should have a specific signal transfer function, there are no specific constrains on the first loop signal and noise transfer functions.

In this architecture, multi-bit quantizers are preferred to alleviate the swing requirements at the output of the integrators. To linearize the input digital-to-analog converter (DAC), dynamic element matching (DEM) should be used since any non-ideality in this DAC will appear at the output of the modulator, unshaped. Depending on the number of quantization levels and the DEM technique employed, this operation might take up to a half of the clock period. To reduce the DEM complexity and eliminate the need of fast digital adder in the modulator loop, the digital adder inside the modulator is replaced with two additional DACs at the input of the first and second integrators and one digital adder outside both loops [4]. Later, it will be shown that the DAC associated with the second loop output at the input of the modulator will only have a negligible effect on total noise and power contributions.

It is worth noting that any error in the second loop will be multiplied by the first loop NTF because the output of the second loop is effectively fed to the output of the first loop,

shown with the dashed line in Fig. 1. This relaxes the matching requirements of the DACs in the second loop.

III. CIRCUIT DESIGN

A. Operational Amplifiers

Due to the fact that no digital filters are used in this topology, the analog transfer function accuracy requirements are relaxed, enabling to use low gain opamps. Matlab simulations shows that 28dB opamp gain is sufficient to provide over 80dB signal-to-quantization noise-ratio (SQNR). To add a safety margin for effective gain degradation due to the feedback factor [5] and also process and temperature variations, opamps with 35dB open loop gain (based on transistor level simulations) are designed. Shown in Fig. 3, simple opamps with NMOS cascode transistors are chosen to satisfy the required gain with large available output swing. In contrast to MASH structure where high gain opamps often result in two-stage or gain boosted topologies with increased power dissipation, in SMASH architecture a single stage opamp with no additional gain boosting techniques allows low-power operation and results in a simpler design. Continuous-time common-mode feedback is used to set the output common-mode of the opamps to the desired value.

B. Quantizers

3-bit quantizers are used for this implementation. This will ensure the modulator stability even for the input signals close to the full-scale range. Multi-bit quantizers will also reduce both the in-band and the out-of-band quantization noise, resulting in improved SQNR. A switched-capacitor (SC) summer is used in front of each pre-amplifier to add the input signal, output of the second integrator and the reference voltages as demonstrated in Fig. 4. It is critically important to reset the pre-amplifier in order to minimize any memory effects from gate-drain capacitance of the input transistors [6]. To eliminate this effect, the input transistor is reset in every period.

The operation of this circuit is as follows: In ϕ_1, the input signal is sampled on C_{Q1} while the comparator offset is sampled on C_{Q2} and the gate-drain capacitance of the input transistor is discharged. In the next phase, the output of the

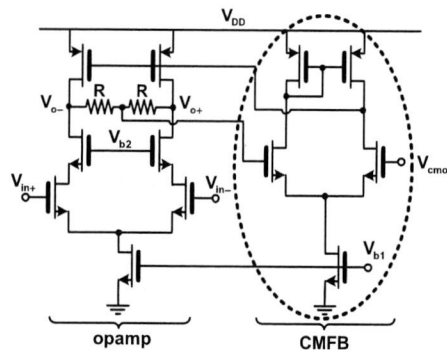

Fig. 3. Opamp used for integrators.

978-1-4244-2018-6/08 $25.00 © 2008 IEEE

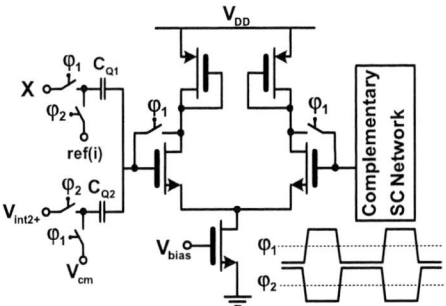

Fig. 4. Passive summer in front of the pre-amplifier.

second integrator is connected to C_{Q2} while C_{Q1} provides the reference voltage generated from the resistor ladder. At the falling edge of ϕ_2, the comparator latches the data of the pre-amplifier (latches are not shown for simplicity).

The same structure is used for the second loop. However, the output of the second loop will not utilize the full dynamic range of its quantizer due to the fact that the input of the second loop is only the quantization error of the first stage. Therefore, extra comparators are removed. This will reduce the complexity of the DAC associated with the second loop output at the input of the modulator by removing the extra capacitors, minimizing the impact of this DAC to the overall noise and power contributions.

C. Error Extraction

It is vital to extract the quantization error of the first loop accurately enough to be contained within the overall noise budget. This is due to the fact that any error in this extraction will be only shaped by the first loop. Since no active adder was used in front of the quantizer, this error cannot be extracted in conventional way (e.g., by subtracting the input of the quantizer from its output). In this architecture, error extraction is done at the input of the second loop using output of the first quantizer (Q_{1_out}), input signal (X) and the second integrator output (V_{int2}). In other words, the first stage quantization error is equal to

$$E_1 = Q_{1\_out} - V_{int2} - z^{-1}X. \tag{3}$$

A conceptual block diagram is depicted in Fig. 5 where only second and third integrators are shown for simplicity. It can

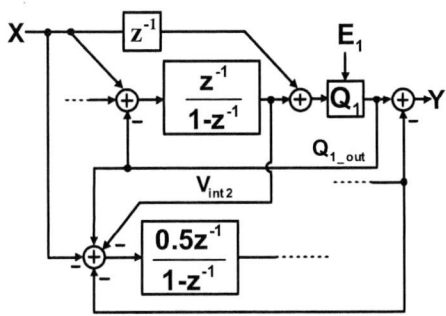

Fig. 5. Extraction of the first loop quantization error.

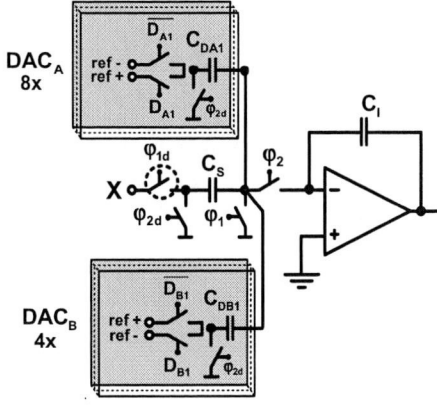

Fig. 6. Input of the modulator. DAC_B uses half as many elements of DAC_A (single-ended shown for simplicity).

be seen that the signal fed to the second loop is the advanced version of the signal fed to the first quantizer. This will only result in a small signal component in the second loop and it will not affect the noise shaping property. In this prototype, all the summations are done at the inputs of the third and fourth integrators to eliminate the need of an extra active element.

D. Input Switches and DACs

To avoid signal dependent loading on the references of the DACs, the sampling capacitors at the input of the modulator and the input DACs are separated. This is shown in Fig. 6 where DAC_A and DAC_B denote the DACs associated with the first loop and second loop outputs at the input of the modulator, respectively. A 1pF sampling capacitor for the input signal, total of 1pF capacitor for DAC_A and a total of 0.5pF capacitor for DAC_B are used. The actual design uses fully-differential implementation. Data weighed averaging (DWA) is used to linearize the input DACs of the modulator. First loop uses a 3-bit DWA and the second loop employs a 2-bit DWA. Both DACs work in ϕ_2 leaving sufficient time for DEM operation to take place. Bootstrap switches presented in [7] are used for sampling the input signal to guarantee the linearity of the sampled signal. The bootstrap switch is circled with the dashed line in Fig. 6.

IV. Measurement Results

The prototype ADC was fabricated in a 0.18μm 2-Poly 5-Metal CMOS process. Fig. 7 shows the die photograph of the prototype IC. The active area is 1.6 x 1.2 mm^2. Fig. 8 shows a 32k-sample measured output spectra with 92kHz -6dB full-scale (FS) sine-wave input signal. The 3rd and 5th harmonics are at -98dB and -94dB, respectively. The noise floor is at -105dB within the signal band. The same opamp gain using MASH topology would result in severe performance degradation.

Fig. 9 illustrates the measured SNR and SNDR of the prototype chip as a function of the input signal power. The maximum measured dynamic range, SNR and SNDR are 76.9dB, 75.6dB and 74.6dB, respectively.

978-1-4244-2018-6/08 $25.00 © 2008 IEEE

Fig. 7.   Die photograph

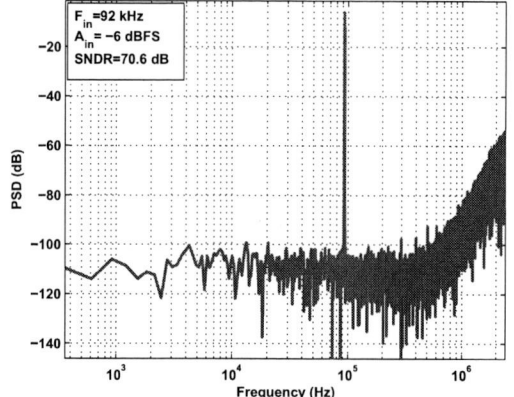

Fig. 8.   Measured output spectrum.

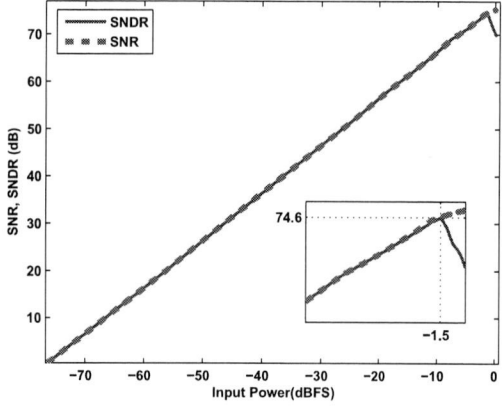

Fig. 9.   SNR and SNDR versus input power.

This prototype was tested at 1.2V analog and digital power supplies. Because a passive adder is used in front of the quantizer, input signals close to the full-scale voltage (FS=1.2V) can be processed by the modulator. In this design, the minimum power supply voltage was limited by the digital power supply. Analog power supply can be further reduced down to 1V with minimal degradation. Table I summarizes the measured performance of the prototype ADC. The total power dissipated in analog part is 2.1mW where 1.2mW of this power is dissipated in the opamps and the rest is dissipated in the

TABLE I
PERFORMANCE SUMMARY

Supply Voltage	1.2V
Sampling Frequency	20MHz
Oversampling Ratio	16
Dynamic Range	76.9dB
SNR	75.6dB
SNDR	74.6dB
Power Dissipation	2.1mW Analog
	1.1mW Digital
Technology	0.18μm 2P5M CMOS
Active Area	1.92 mm^2

quantizers and the resistor ladders. Further power optimization is possible in both digital and analog parts; however, this prototype ADC was primarily intended to demonstrate the feasibility of SMASH modulators.

V. CONCLUSION

A fourth order SMASH modulator prototype IC implementation details and results were presented in this paper. It was shown that even with using very low gain opamps the obtained performance is comparable with that of the MASH structure utilizing costly high gain opamps. This major advantage over traditional multi-loop architectures not only allows low-voltage operation, but it will also result in lower power dissipation. The passive switched-capacitor summer and error extraction, together with the unique modulator topology, enables processing large input signals with robust stability.

ACKNOWLEDGMENT

The authors would like to thank National Semiconductor for providing fabrication of the prototype IC and also R. Gregoire and P. Kurahashi for their useful comments. This work was supported by Semiconductor Research Corporation (SRC) under contract 2005-HJ-1308.

REFERENCES

[1] R. Schreier and G. C. Temes, *Understanding Delta-Sigma Data Converters.* Piscataway, NJ: IEEE Press, 2005, pp. 127-136.
[2] N. Maghari, S. Kwon, G. C. Temes, and U. Moon, "Sturdy-MASH Δ-Σ Modulator" *Electron. Lett.*, vol. 42, pp. 1269-1270, Oct. 2006.
[3] P. Benabes, A. Gauthier, and R. Kielbasa, "New High-Order Universal Σ-Δ Modulators," *Electron. Lett.*, vol. 31, pp. 8-9, Jan. 1995.
[4] S. Kwon and F. Maloberti, "A 14mW Multi-bit Δ-Σ Modulator with 82dB SNR and 86dB DR for ADSL2+," *ISSCC Dig. Tech. Papers*, pp. 68-69, Feb. 2006.
[5] Y. Geerts, M. S. .J. Steyaert, and W. Sansen, "A High Performance multibit $\Delta\Sigma$ CMOS ADC," *IEEE J. Solid-State Circuits*, vol 35, No. 12, pp. 1829-1840, Dec. 2000.
[6] C. Sandner, M. Clara, A. Santner, T. Hartig, and F. Kuttner, "A 6-bit 1.2-GS/s low-power flash-ADC in 0.13-μm Digital CMOS," *IEEE J. Solid-State Circuits*, vol 40, No. 7, pp. 1499-1505, July 2005.
[7] M. Dessouky, and M. Kaiser, "Very low-voltage digital-audio Δ-Σ modulator with 88-dB dynamic range using local switch bootstrapping," *IEEE J. Solid-State Circuits*, vol 36, No. 3, pp. 349-355, Mar. 2001.

A/D Converter Trends: Power Dissipation, Scaling and Digitally Assisted Architectures

(Invited Paper)

B. Murmann
Stanford University, Stanford, CA

Abstract— **This paper summarizes recent trends in the area of low-power A/D conversion. Survey data collected over the past eleven years indicates that the power efficiency of ADCs has improved on average by a factor of two every two years. A closer inspection on the impact of technology scaling is presented to explain the observed trend in the context of shrinking supply voltages and increasing device speed. Finally, a discussion on minimalistic and digitally assisted design approaches is used to sketch a route toward further improvements in ADC power efficiency and performance.**

I. INTRODUCTION

Analog-to-digital converters (ADCs) are important building blocks in modern electronic systems. In many cases, the efficiency and speed at which analog information can be converted to digital signals profoundly affects system architectures and their performance. Even though modern integrated circuit technology can provide very high conversion rates, the associated power dissipation is often incompatible with application constraints. For instance, the high-speed 6-8-bit ADCs of [1, 2] achieve sampling rates in excess 20 GS/s, at power dissipations of 1.2 W and 10 W, respectively. Operating such blocks in a handheld application is impractical, as they would drain the device's battery within a short amount of time. Consequently, it is not uncommon to architect power constrained applications "bottom-up," by determining the analog/RF front-end and ADC specifications based on the available power budget. A discussion detailing such an approach for the specific example of software-defined radio receivers is presented in [3].

With power dissipation being among the most important concerns in mixed-signal/RF applications, it is important to track trends and understand the relevant trajectories. The purpose of this paper is to review the latest developments in low-power A/D conversion and to provide an outlook on future possibilities.

Following this introduction, Section II provides survey data on ADCs published over the past eleven years. These data show that contrary to common perception, extraordinary progress has been made in improving ADC power efficiency. Among the factors that have influenced this trend are technology scaling, and the increasing use of simplified analog sub-circuits with digital correction. Therefore, Section III takes a closer look at the impact of shrinking feature sizes, while Sections IV and V discuss recent ideas in "minimalistic" and "digitally assisted" ADC architectures.

II. ADC PERFORMANCE TRENDS

A. Survey Data and Figure of Merit Considerations

Several surveys on ADC performance are available in literature [3-7]. In this section, we will review recent data from designs presented at the IEEE International Solid-State Circuits Conference (ISSCC) and the VLSI Circuit Symposium. Fig.1. shows a scatter plot of results published at these venues over the past eleven years [8]. Fig. 1(a) plots the energy per Nyquist sample (P/f$_s$, i.e. power divided by Nyquist sampling rate) against the achieved signal-to-noise-and-distortion ratio (SNDR). This plot purposely avoids dividing the conversion energy by the effective number of quantization steps, as done in the commonly used figure of merit [4]

$$FOM = \frac{P}{f_s \cdot 2^{ENOB}} \quad (1)$$

where

$$ENOB = \frac{SNDR(dB) - 1.76}{6.02} \quad (2)$$

Normalizing by the number of quantization steps assumes that doubling precision would double power, which finds only empirical justification [4]. In fact, if a converter were purely limited by thermal noise, its power would quadruple per added bit (see Section III). However, as this assumption is pessimistic for real designs, it is preferable to avoid a fixed relationship between precision and energy when plotting data across a large range of architectures and resolutions.

Fig. 1(a) indicates that some of the lowest energy ADCs were published at this year's ISSCC. Interestingly, most of these designs target only low to moderate resolution; activity in the high-resolution space appears to be lagging. With respect to (1), included as a straight line for the numerical example of FOM = 100 fJ/conversion-step, it is clearly visible that state-of-the-art high resolution designs (SNDR > 85dB) do not obey the implied 2x increase in power per bit. Furthermore, the most recent low-resolution designs also manage to break away from any linear fit to the overall scatter plot that is based on a slope of 2x per bit.

In addition to an ADC's energy efficiency, the available signal bandwidth is an important parameter. Fig 1(b) plots bandwidth against SNDR for the given data set. In this chart, the bandwidth plotted for Nyquist converters is equal to the input frequency used to obtain the stated SNDR; this

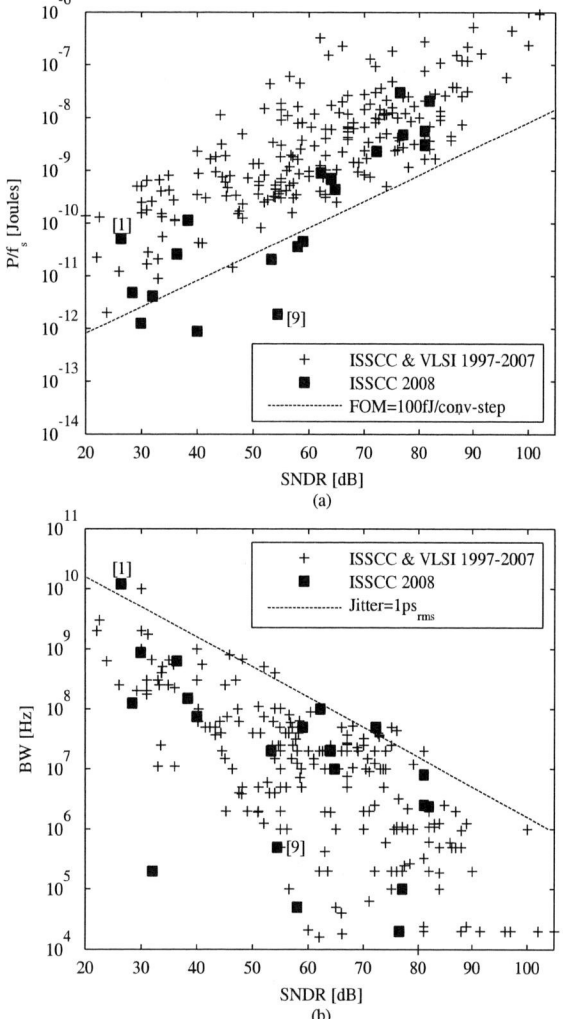

Fig. 1. ADC performance data (ISSCC 1997-2008, VLSI Circuit Symposium 1997-2007). (a) Power efficiency versus SNDR. (b) Conversion bandwidth versus SNDR.

frequency is not necessarily $f_s/2$. The first interesting observation from Fig. 1(b) is that across all resolutions, the parts with the highest bandwidth achieve a performance that is approximately equivalent to an aperture uncertainty of 1 ps$_{rms}$. The dashed line in Fig. 1(b) represents the performance of an ideal sampler with sinusoidal input and 1 ps$_{rms}$ sampling clock jitter. Clearly, any of the ADC designs at this performance front rely on a significantly better clock, to allow for additional nonidealities that reduce SNDR. Such nonidealities include quantization noise, thermal noise, differential nonlinearity and harmonic distortion. From the data in Fig 1(b), it is also clear that any new design aiming to push the speed-resolution envelope will require a sampling clock with jitter on the order of ~100 fs$_{rms}$ or better.

In order to assess the overall merit of an ADC (power efficiency and bandwidth), it is interesting to compare the locations of its particular design points in plots (a) and (b). For example, [1] achieves a bandwidth close to the best designs,

while showing only average power efficiency. The opposite is true for [9]; this part ranks among the lowest energy designs published to date, but achieves only moderate bandwidth. These examples confirm the intuition that pushing a design toward the speed limits of a given technology will sacrifice power efficiency. To date, there exists no single-number figure of merit that captures this tradeoff in a fair and balanced way across all architectures and resolutions. The same holds true for input capacitance. For example, it is possible to improve the SNDR of most ADC architectures by increasing their input capacitance. An ideal figure of merit would take the power needed to drive the converter input into account.

B. Trends in Power Efficiency and Speed

Using the data set discussed above, it is interesting to extract trends over time. Fig. 2(a) is a 3-D representation of the power efficiency data [Fig. 1(a)] with the year of publication included along the y-axis. The resulting slope in time corresponds to an average reduction in power by 2x approximately every 2 years.

A similar 3-D fit could be constructed for bandwidth performance. However, such a fit would not convey interesting information, as the majority of designs published in recent years do not attempt to maximize bandwidth. This contrasts the situation with power efficiency, which is subject to optimization in most modern designs. In order to extract a

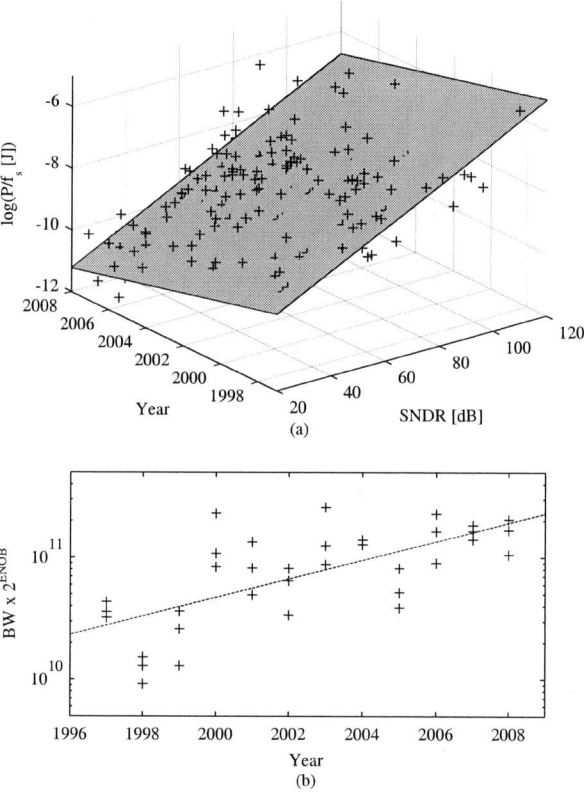

Fig. 2. Trends in ADC performance. (a) 3-D fit to power efficiency. The fit plane has a slope of 0.5x/2 years along the time axis. (b) Fit to speed-resolution product of top 3 designs in each year. The slope of the fit line is 2x/4 years.

978-1-4244-2018-6/08 $25.00 © 2008 IEEE

trend on achievable bandwidth, Fig. 2(b) scatter-plots the speed-resolution products of the top three designs in each year. This metric is justified by the constant speed-resolution boundary observed from Fig. 1(b). A fit to the data in Fig. 2(b) reveals that performance has doubled every 4 years; a rate that is much lower than the improvement in power efficiency. In addition, as evident from the data points, there is no clear trend as far as the top performance point is concerned; designs of the early 2000's are up to par with some of the works published recently. Consequently, the extracted progress rate of speed-resolution performance should be viewed as a relatively weak and error-prone indicator.

III. IMPACT OF TECHNOLOGY SCALING

As shown above, the power dissipation of A/D converters has halved approximately every 2 years over the past decade. Over the same period, CMOS technologies used to implement the surveyed ADCs have scaled from approximately 0.6 μm down to 65 nm. In this section, we will investigate the role of technology scaling in the observed power efficiency trend. Broader discussions on the impact of scaling are presented in [7, 10, 11].

A well-known challenge in designing ADCs using modern processes is the diminishing voltage headroom. Since device scaling requires a reduction in supply voltage (V_{DD}), the noise in the analog signals must be reduced proportionally to maintain the desired signal-to-noise ratio. Since noise trades with power dissipation, this suggests to first order that power efficiency should worsen, and not improve, for ADCs in modern technologies. One way to overcome supply voltage limitations is to utilize thick-oxide I/O devices, which are available in most standard CMOS processes. However, using those devices often reduces speed. Closer inspection of the survey data considered in this paper reveals that most published state-of-the-art designs do not rely on thick oxide devices, and rather cope with supply voltages around 1 V.

To investigate further, it is worthwhile to examine the underlying equations that capture the trade-off between supply voltage and power dissipation via thermal noise constraints. In most analog sub-circuits used to build ADCs, noise is inversely proportional to capacitance

$$N \propto \frac{kT}{C} \qquad (3)$$

where k is Boltzmann's constant and T stands for absolute temperature. For the specific case of a transconductance amplifier that operates linearly, we can write

$$f_s \propto \frac{g_m}{C} \qquad (4)$$

Further assuming that the signal power is proportional to $(\alpha \cdot V_{DD})^2$ and that the circuit's power dissipation is V_{DD} multiplied by the transistor drain current, I_D, we find

$$\frac{P}{f_s} \propto \frac{1}{\alpha^2} \frac{1}{V_{DD}} \cdot \frac{1}{\left(\frac{g_m}{I_D}\right)} kT \cdot SNR \qquad (5)$$

The variable g_m/I_D in (5), is related to the "gate overdrive" of the transistor that implements the transconductance. Assuming MOS square law, $g_m/I_D = 2/(V_{GS}-V_t)$ and in weak inversion $g_m/I_D = 1/(n \cdot kT/q)$, with $n \cong 1.5$. Considering the fractional swing (α) and transistor bias point (g_m/I_D) as constant, it is clear from the above expression that power efficiency in noise-limited transconductors should deteriorate at low V_{DD}. In addition, we see that (5) indicates a very steep tradeoff between SNR and energy; increasing the SNR by 6 dB requires a 4x increase in P/f_s.

Since both of these results do not correlate well with the observations of Section II, it is instructive to examine the assumptions that lead to (5). The first assumption is that the circuit is purely limited by thermal noise. This assumption clearly holds for ADCs with very high resolution, but typically few, if any, low resolution converters are impaired by thermal noise.

To get a feel for a typical SNDR value at which today's converters become "purely" limited by noise, it is helpful to plot the data of Fig. 1(a) normalized to a 4x power increase per bit [12]. Fig. 3 shows such a plot in which the P/f_s values have been divided by

$$\left(\frac{P}{f_s}\right)_{min} = 4 \cdot kT \cdot SNR \qquad (6)$$

while assuming SNR $\cong$ SNDR. The pre-factor of 4 in this expression follows from the power dissipated by an ideal class-B amplifier that drives the capacitance C with a rail-to-rail tone at $f_s/2$ [13]. Therefore, (6) represents a fundamental bound on the energy required to process a charge sample at a given SNR.

The primary observation from Fig. 3 is that the normalized data exhibits a visible "corner" beyond which (P/f_s)/(P/f_s)$_{min}$ approaches a constant value. This corner, approximately located at 75 dB, is an estimate for the SNDR at which a typical state-of-the-art design becomes truly limited by thermal noise. Since ADCs with lower SNDR do not achieve

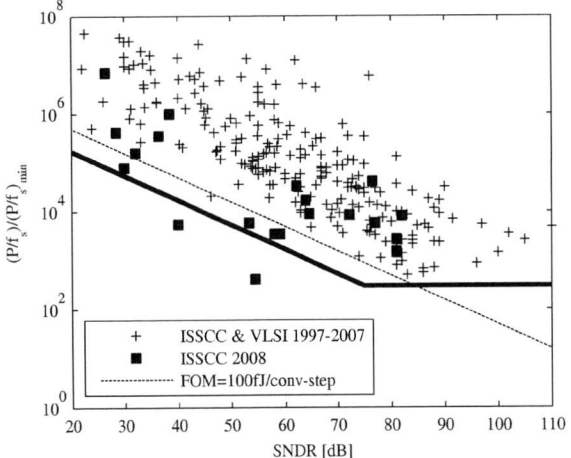

Fig. 3. Data of Fig. 1(a) normalized by (P/f_s)$_{min}$ as given in (6). This illustration suggests the existence of an "SNDR corner." Only ADCs with SNDR > 75dB appear to be primarily limited by thermal noise.

the same noise-limited power efficiency, it can be argued that these designs are at least partially limited by the underlying technology. This implies that over time, technology scaling may have helped improve their power efficiency as opposed to the worsening predicted by (5).

To investigate further, we partitioned the data of Fig. 1(a) into two distinct sets, i.e. high resolution (SNDR > 75 dB) and low-to-moderate resolution (SNDR ≤ 75 dB). We then applied a 3-D fit similar to that shown in Fig. 2(a) to each set and extracted the progress rates over time. For the set with SNDR > 75 dB it was found that P/f_s has halved only every 5.4 years, while for SNDR ≤ 75 dB, P/f_s halves every 1.6 years. The difference in these progress rates confirms the above speculation. For high-resolution designs, (5) applies and scaling technology over time, associated with lower supply voltages, cannot help improve power efficiency. As observed in [7], this has led to a general trend toward lower resolution designs. Since it is very difficult to attain high SNDR at low supply voltages, most applications are steered away from relying on high-resolution ADCs in current fine-line processes. This is qualitatively confirmed in Fig. 4, which highlights the P/f_s data points of ADCs built in CMOS at 90 nm (and $V_{DD} \cong 1$ V) and below.

The above-discussed situation strongly contrasts the impact of scaling on low-to-moderate resolution designs, as evident from the extracted improvement rate. Quantifying the benefits of scaling on low-to-moderate resolution ADCs from first principles is a complex task, primarily because the involved tradeoffs strongly depend on architecture and design specifics. An analysis that highlights the benefits of scaling in flash and folding ADCs is presented in [14]. In the following paragraphs we will discuss qualitatively the scaling behavior of a moderate resolution pipelined ADC.

Consider the 10-bit, 0.6-μm pipelined ADC described in [15, 16]; this design reflects state-of-the-art in 1996. Close inspection of the design details in [16] reveals that about 30% of the total power in this ADC is dissipated by noise-limited amplifiers. The remaining power is consumed by digital gates, comparators and amplifier stages whose component sizes are set by feature size constraints. To first order, the power dissipation in these latter blocks should scale approximately as $C \cdot V_{DD}^2$, i.e. logic gate energy. Since 1997, we have seen a reduction in process $C \cdot V_{DD}^2$ of approximately 300 times [17]. Yet, this change alone cannot explain the vastly larger improvement factor that 10-bit designs have seen over the past decade; the improvement would be limited to no more than 3.3x in terms of total power.

Clearly, the situation is more complex. First, a circuit that is "limited" by noise may still carry overhead that reduces with scaling. Especially in high-speed designs, amplifier self-loading and parasitic loading at intermediate circuit nodes plays an important role. Technology scaling helps mitigate these capacitances and therefore improves overall efficiency. Unfortunately this effect is hard to quantify.

A more transparent factor is the trade-off between g_m/I_D and the transit frequency (f_T) of the active devices. Switched capacitor circuits based on class-A operational transconductance amplifiers typically require transistors with $f_T > 80 f_s$. Even for speeds of several tens of MS/s, it was necessary in older technologies to bias transistors far into strong inversion (V_{GS}-V_t > 200 mV) to satisfy this requirement. In more recent technologies, very large transit frequencies are available in moderate- and even weak-inversion. This is further illustrated in Fig. 5(a) which compares typical minimum-length NMOS devices in 180-nm and 90-nm CMOS.

For a fixed sampling frequency, and hence fixed f_T requirement, newer technologies deliver higher g_m/I_D. This tradeoff is plotted directly, without the intermediate variable V_{GS}-V_t, in Fig. 5(b). In order to achieve f_T = 30 GHz, a 180-nm device must be biased such that $g_m/I_D \cong 9$ S/A. In 90-nm technology, f_T = 30 GHz is achieved in weak inversion, at $\cong$ 18

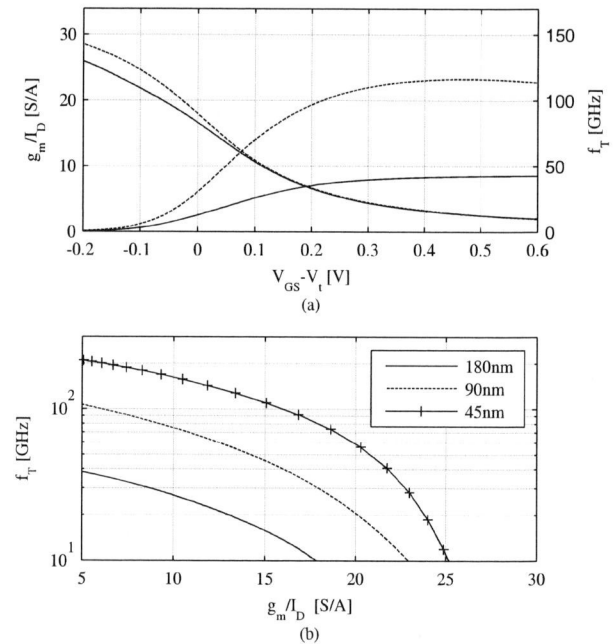

Fig. 5. Tradeoff between g_m/I_D and f_T in modern technologies.

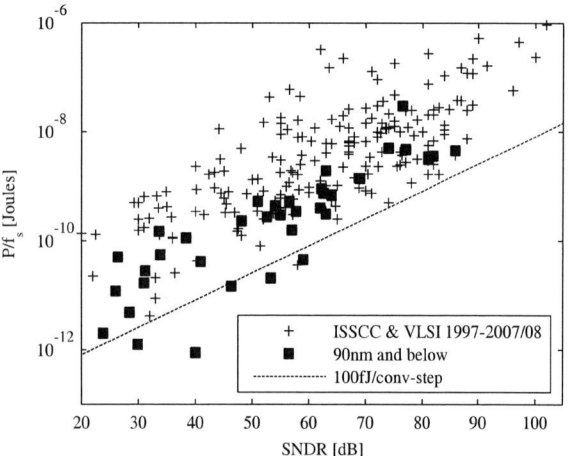

Fig. 4. Power efficiency (P/f_s) for ADCs built in 90-nm CMOS and below.

S/A. From equation (5), it is clear that this advantage can counteract the reduction in V_{DD} when going to a newer process. Note that this advantage can only materialize when the sampling speed is kept constant or at least not scaled proportional to the f_T improvement. This was also one of the observations drawn from Fig. 2(b). A converter that pushes the speed envelope using a new technology typically won't simultaneously benefit from scaling in terms of power efficiency.

A last and perhaps even more significant factor to consider is the accrual of design experience, improved optimization, exploitation of process options and refinement of circuit techniques over many generations of technology. For instance, A/D converters in 5-V technologies used to be relatively wasteful in terms of headroom utilization [α-term in (5)]. Newer designs are typically optimized to accommodate signal swings as large as 1 $V_{pp,diff}$ at $V_{DD} = 1$ V. In addition, we are beginning to see designs that efficiently exploit technology options. For instance, the 10-bit pipelined ADC of [18] uses thin-oxide high-performance analog (HPA) devices to achieve high DC gain using simple, power-efficient telescopic transconductance amplifiers.

Additional directions in the context of design techniques that have recently gained in importance include "minimalistic" and "digitally assisted" approaches. The trend toward such solutions may be explained by the fact that evolutionary grown designs have come very close to practical power limits, imposed by their circuit topologies and associated fundamental constraints. The ideas summarized in the following two sections outline promising directions in this area of research.

IV. Minimalistic Design

Power dissipation in the analog portion of ADCs is strongly coupled to the complexity of the constituent sub-circuits. The goal of minimalistic design is to improve power efficiency and potentially increase speed by utilizing simplified analog sub-circuits.

In architectures that previously relied on op-amp based signal processing, there exists a clear trend toward simplified amplifier structures. Examples include inverter-based sigma-delta modulators [19, 20] and various approaches emphasizing op-amp-less implementation of pipelined ADCs [21-25]. Especially in switched capacitor circuits, eliminating class-A op-amps can dramatically improve power efficiency. This is for two reasons. First, operational amplifiers typically contribute more noise than simple gain stages, as for example resistively loaded open-loop amplifiers. Secondly, the charge transfer in class-A op-amp circuitry is inherently inefficient; the amplifier draws a constant current, while delivering on average only a small fraction of this current to the load. In [26], it was found that the efficiency of a class-A op-amp in a switched capacitor circuit is inversely proportional to the number of settling time constants. For the typical case of settling for approximately 10 or more time constants, the overall efficiency, i.e. charge drawn from the supply versus charge delivered to the load, is only a few percent.

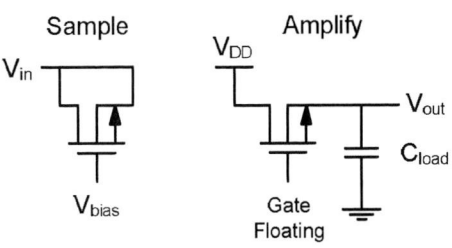

Fig. 6. Dynamic amplifier concept used in [23].

As discussed further in [27], this inherent inefficiency of op-amps contributes to the power overhead relative to fundamental limits. Consider for instance the horizontal asymptote of Fig. 3, located at approximately 300 times the minimum possible P/f_s. The factor of 300 can be explained for op-amp based circuits as follows. First, the noise is typically given by $\beta \cdot kT/C$, where β can range from 5-10, depending on implementation details. Second, charge transfer using class-A circuits, as explained above brings a penalty of approximately 20x. Third, op-amp circuits usually do not swing rail-to-rail as assumed in (6); this can contribute another factor of two. Finally, adding further power contributors beyond one dominant op-amp circuit easily explains a penalty factor greater than 200...400.

A promising remedy to this problem is to utilize circuits that process charge more efficiently and at the same time contribute less thermal noise. A well-know example of an architecture that achieves very high efficiency is the charge-based successive approximation register (SAR) converter [1, 9, 28]. Such converters have seen a renaissance in recent years, primarily because the architecture is well-suited for leveraging the raw transistor speed of new technologies, while being insensitive to certain scaling implications, such as reduced intrinsic gain (g_m/g_{ds}). A problem with SAR architectures is that they cannot deliver the best possible performance when considering absolute speed, resolution and input capacitance simultaneously. This is one reason why relatively inefficient architectures, such as op-amp based pipelined ADCs are still being used and investigated.

In order to make pipelined architectures as power efficient as competing SAR approaches, various ideas are being explored in research. Fig. 6 shows a new single-transistor residue amplification concept that was utilized in the low-power pipelined converter of [23]. In the sampling phase, the transistor is configured as a MOS capacitor in accumulation. During amplification, the drain is switched to V_{DD} and the source drives the discharged capacitive load. At this time, the gate is left floating and the transistor acts as a source follower. V_{out} rises until V_{gs} nears the threshold voltage of the device. Incremental input voltage amplification occurs because of charge conservation. During sampling, signal dependent charge is stored on C_{ox} and gate-source/drain overlap parasitics (C_{ol}). At the end of the amplification phase, the charge on $C_{gs} = C_{ox} + C_{ol}$ is constant (due to $V_{gs} = V_t$) and all signal dependent charge now appears across $C_{gd} = C_{ol}$. To first order, the voltage gain is set by the ratio $(C_{ox} + 2C_{ol})/C_{ol}$. In essence, this scheme mimics charge-redistribution around an operational amplifier, while providing significantly lower noise and highly efficient charge transfer from the supply to the capacitive load.

A general concern with most minimalistic design approaches is that they tend to sacrifice robustness, e.g., in terms of power supply rejection, common mode rejection and temperature stability. It remains to be seen if these issues can be handled efficiently in practice. Improving supply rejection, for instance, could be achieved using voltage regulators. This is custom practice in other areas of mixed-signal design, as for example PLLs [29]. Especially when the power of the ADC's critical core circuitry is lowered by orders of magnitude, implementing supply regulation should be a manageable task.

A second issue with minimalistic designs is the achievable resolution and linearity. Op-amp circuits with large loop gain help linearize transfer functions; this feature is often removed when migrating to simplified circuits. For instance, the amplifier scheme of Fig. 6 is linear only to approximately 9-bit resolution. In cases where simplicity sacrifices precision, it is attractive to consider digital means for recovering conversion accuracy. Digitally assisted architectures are therefore the topic of the next section.

V. DIGITALLY ASSISTED ARCHITECTURES

Technology scaling has significantly reduced the energy per operation in CMOS logic circuits. As explained in [30], the typical 0.7x scaling of features along with aggressive reductions in supply voltage have led to a 65% reduction in energy per logic transition for each technology generation. The survey data presented in [17] suggests that a 2-input NAND gate dissipates roughly 1.3 pJ per logic operation in a 0.5-μm CMOS process. The same gate dissipates only 4.5 fJ in a more recent 90-nm process; this amounts to a ~300x improvement in only 10 years. The corresponding progress in ADC energy (based on Section II) amounts to a 32x reduction over 10 years. This means that the relative "cost" of digital computation has reduced roughly by a factor of ten over the past decade.

To get a feel for how much logic can be used to "assist" a converter for the purpose of calibration and error correction, it is interesting to the divide energy per conversion (P/f_s) figures of ADCs by the energy of a single NAND gate. The numbers compiled in Table 1 use data from the fit plane of Fig. 2(a) for 2008, and assume $E_{NAND} = 4.5$ fJ. At low signal fidelity, e.g. SNDR = 30 dB, a single A/D conversion consumes as much energy as toggling approximately 4,700 logic gates. On the other hand, at 90 dB SNDR, more than two million logic gates would need to switch to consume the energy of an A/D conversion at this level of precision.

The consequence of this observation is that in a low-resolution converter, it is unlikely that tens of thousand of gates can be used for digital error correction without exceeding reasonable energy or power limits. A large number

TABLE I

$E_{ADC} = P/f_s$ in today's ADCs [fit data from Fig. 2(b)], relative to logic gate energy ($E_{NAND} = 4.5$ fJ) in 90-nm CMOS.

SNDR [dB]	E_{ADC}	E_{ADC}/E_{NAND}
30	21 nJ	4,700
50	168 nJ	38,000
70	1.35 μJ	300,000
90	10.8 μJ	2,400,000

of gates may be affordable only if the involved gates operate at a low activity factor or if they can be shared within the system. Conversely, in high resolution ADCs, each analog operation is very energy consuming and even a large amount of digital processing may be accommodated in the overall power budget.

The following sub-sections provide a non-exhaustive list of opportunities for leveraging digital logic gates in A/D converters.

A. Oversampling

The longest standing example of an architecture that efficiently leverages digital signal processing abilities is the oversampling sigma-delta converter. This architecture uses noise shaping to push the analog quantization error, along with other nonidealities, outside the signal band [12]. Subsequent digital filtering creates a high-fidelity output signal, while the constituent analog sub-circuits require only moderate precision. Even in fairly old technologies, it was reasonable to justify high gate counts in the converter's decimation filter, simply because the analog signal processing energy per sample for typical high-SNDR converters is very large.

A new paradigm that might gain significance in the future is the use of oversampling in traditional Nyquist converters. An example of such an ADC is described in [31]. As we have noted from Fig. 5(b), migrating a converter with a fixed sampling rate to technologies with higher f_T can help improve power efficiency. Ultimately, however, there is diminishing return in this trend due to the weak-inversion "knee" of MOS devices [see Fig. 5(a)]. G_m/I_D no longer improves beyond a certain minimum bias; it therefore makes no sense to target a transistor f_T below a certain value. This, in turn, implies that for optimum power efficiency, one should not operate an ADC below a certain clock rate. Consider for example the f_T versus g_m/I_D plot for 45-nm technology in Fig. 5(b). For $g_m/I_D > 20$ S/A, f_T drops sharply without a significant increase in g_m/I_D. At this point, $f_T \cong 50$ GHz, implying that is possible to build a switched capacitor circuit with $f_{clock} = 50$ GHz/80 = 625 MHz.

To date, there exist only a limited number of applications that require such high sampling rates, and there will clearly remain a number of systems in the future that demand primarily good power efficiency at only moderate speeds. A solution to this situation might be to oversample the input signal by a large factor and to remove out-of band noise (thermal noise, quantization noise, and jitter) using a digital filter. Per octave of oversampling, this increases ADC

978-1-4244-2018-6/08 $25.00 © 2008 IEEE   120

resolution by ½ bit. In a situation where a converter is purely limited by noise, this improvement is in line with the fundamental thermal noise tradeoff expressed in (5).

B. Mismatch Correction

Assuming constant gate area (W·L), transistor matching tends to improve in newer technologies. In matching-limited flash ADC architectures, this trend has been exploited in the past to improve the power efficiency by judiciously down-sizing the constituent devices [14]. In order to scale such architectures more aggressively and at the same time address new sources of mismatch in nano-scale technologies, it is desirable to aid the compensation of matching errors through digital means.

In flash-ADCs, there are at least two general trends in this direction. The first is to absorb mismatch errors using a fault-tolerant encoder in conjunction with a comparator bank that contains redundant elements [32]. This approach requires a relatively large number of logic operations per sample. In light of the conclusions from Table 1, such a solution may be most efficient only for technologies below 90 nm.

An alternative approach is to provide redundant comparators that are selectively activated (in a static manner) to minimize the converter's nonlinearity [33]. An extension of this approach is to include digital trimming circuitry in addition to redundant elements [34]. This scheme can yield good power efficiency as it attacks the mismatch problem along two degrees of freedom. An extension of this idea, incorporating redundant channels in a time-interleaved ADC is discussed in [35].

C. Digital Linearization of Amplifiers

As pointed out in Section IV, power-efficient and minimalistic design approaches are typically unsuitable for high-resolution applications, unless appropriate digital error correction schemes are used to enhance conversion linearity. In [20], it was demonstrated that a simple open-loop differential pair used in a pipeline ADC can be digitally linearized to within 12-bit precision. Such a digital correction requires only moderate complexity on the order of a few thousand logic gates. It is foreseeable that the future will bring additional schemes that help improve the nonlinearity of simplified amplifiers; e.g. in the context of sigma-delta modulators.

One key issue in most digital linearization schemes is that the correction coefficients must track changes in operating conditions relatively quickly; preferably with time constants no larger than 1-10 ms. Unfortunately, most of the basic statistics-based algorithms for coefficient adaptation require much longer time constants at high target resolutions [36, 37]. Additional research is needed to extend the recently proposed "split-ADC" [38, 39] and feedforward noise cancellation techniques [40] for use in nonlinear calibration schemes.

D. Digital Correction of Dynamic Errors

Most of the digital correction methods developed in recent years have targeted the compensation of static circuit errors; work on dynamic errors that limit a converter's effective resolution bandwidth has been lagging. In various applications, as for instance sub-sampling base-station receivers, it is desirable to improve the converter's high-frequency linearity beyond raw technology limits [41]. Digital compensation of the relevant frequency dependent nonlinearities in the sampling front-end of ADCs will likely evolve as an attractive area for future research. If digital capabilities in nano-scale technologies continue to improve, dynamic compensation schemes based on relatively complex Volterra series may become feasible [42].

E. System-Synergistic Error Correction Approaches

In the above discussion, ADCs are being viewed as "black boxes" that deliver a set performance without any system level interaction. Given the complexity of today's applications, it is important to realize that there exist opportunities to improve ADC performance by leveraging specific system and signal properties.

For instance, large average power savings are possible in radio receivers when ADC resolution and speed are dynamically adjusted to satisfy the minimum instantaneous performance needs. The design described in [43] demonstrates the efficacy of such an approach.

In the context of digital correction, it is conceivable to leverage known properties of application-specific signals to "equalize" the A/D converter together with the communication channel [44, 45]. For instance, the converter described in [45] uses the system's OFDM pilot tones to extract component mismatch information. In such an approach, existing system hardware, as for instance the FFT block, can be re-used to accommodate ADC calibration.

VI. SUMMARY

This paper has summarized recent trends in the context of low-power A/D conversion. Using survey data from the past eleven years, we have observed that power efficiency in ADCs has improved at an astonishing rate of 2x every 2 years. In part, this progress rate is based on cleverly exploiting the strengths of today's technology. Smaller feature sizes help improve the power dissipation in circuits that are not limited by thermal noise. In circuit elements that are limited by thermal noise, exploiting the high f_T of modern transistors can be of help in mitigating a penalty from low supply voltages.

A promising paradigm is the trend toward minimalistic ADC architectures and digital means of correcting analog circuit errors. Digitally assisted ADCs aim to leverage the low computing energy of modern processes to improve the resolution and robustness of simplified circuits. Future work in this area promises to fuel further progress in optimizing the power efficiency of A/D converters.

Overall, future improvements in ADC power dissipation are likely to come from a combination of aspects that involve improved system embedding and reducing analog sub-circuit complexity and raw precision at the expense of "cheap" digital processing resources.

978-1-4244-2018-6/08 $25.00 © 2008 IEEE

REFERENCES

[1] P. Schvan, et al, "A 24GS/s 6b ADC in 90nm CMOS," *ISSCC Dig. Techn. Papers*, pp. 544-545, Feb. 2008.

[2] K. Poulton, et al., "A 20-GSample/s 8b ADC with a 1-MByte Memory in 0.18-um CMOS," *ISSCC Dig. Techn. Papers*, pp. 318-319, Feb. 2003.

[3] A. A. Abidi, "The Path to the Software-Defined Radio Receiver," *IEEE J. Solid-State Circuits*, vol. 42, no. 5, pp. 954-966, May 2007.

[4] R. H. Walden, "Analog-to-digital converter survey and analysis," *IEEE J. Select. Areas Commun.*, vol. 17, no. 4, pp. 539-550, Apr. 1999.

[5] P. B. Kenington and L. Astier, "Power Consumption of A/D Converters for Software Radio Applications," *IEEE Trans. Vehicular Techn.*, vol. 49, pp. 643-650, Mar. 2000.

[6] K. G. Merkel and A. L. Wilson, "A Survey of High Performance Analog-to-Digital Converters for Defense Space Applications," *Proc. IEEE Aerospace Conf.*, vol. 5, pp. 2415-2427, Mar. 2003.

[7] Y. Chiu, "Scaling of analog-to-digital converters into ultra-deep-submicron CMOS," *Proc. CICC*, pp. 375-382, Sep. 2005.

[8] B. Murmann, "ADC Performance Survey 1997-2008," [Online]. Available: http://www.stanford.edu/~murmann/adcsurvey.html.

[9] M. van Elzakker, et al., "A 1.9μW 4.4fJ/Conversion-step 10b 1MS/s Charge-Redistribution ADC," *ISSCC Dig. Techn. Papers*, pp. 244-245, Feb. 2008.

[10] A.-J. Annema, et al., "Analog Circuits in Ultra-Deep-Submicron CMOS," IEEE J. Solid-State Circuits, vol. 40, no. 1, pp. 132-143, Jan. 2005.

[11] K. Bult, "The Effect of Technology Scaling on Power Dissipation in Analog Circuits," in Analog Circuit Design, M. Steyaert, A. H. M. Roermund, and J. H. v. Huijsing, (eds.), Springer, 2006.

[12] R. Schreier and G. C. Temes, *Understanding Delta-Sigma Data Converters*: IEEE Press, 2005.

[13] E. A. Vittoz, "Future of analog in the VLSI environment," *Proc. IEEE ISCAS*, pp.1372-1375, May 1990.

[14] P. C. S Scholtens, et al., "Systematic power reduction and performance analysis of mismatch limited ADC designs," *Proc. ISLPED*, pp. 78-83, Aug. 2005.

[15] T. B. Cho, et al., "A power-optimized CMOS baseband channel filter and ADC for cordless applications," *Dig. VLSI Circuits Symposium*, pp.64-65, Jun. 1996.

[16] G. Chien, "High-Speed, Low-Power, Low-Voltage Pipelined Analog-to-Digital Converter, MS Thesis, University of California, Berkeley, 1996.

[17] B. Murmann, "Digitally Assisted Analog Circuits – A Motivational Overview," *ISSCC Special-Topic Evening Session* (SE1.1), Feb. 2007.

[18] M. Boulemnakher, et al., "A 1.2V 4.5mW 10b 100MS/s Pipeline ADC in a 65nm CMOS," *ISSCC Dig. Techn. Papers*, pp. 250-251, Feb. 2008.

[19] Y. Chae et al., "A 0.7V 36μW 85dB-DR Audio ΔΣ Modulator Using Class-C Inverter," *ISSCC Dig. Techn. Papers*, pp. 491-492, Feb. 2008.

[20] R. H. M. van Veldhoven et al., "An Inverter-Based Hybrid ΣΔ Modulator," *ISSCC Dig. Techn. Papers*, pp. 493-494, Feb. 2008.

[21] B. Murmann and B. E. Boser, "A 12-bit 75-MS/s Pipelined ADC using Open-Loop Residue Amplification," *IEEE J. Solid-State Circuits*, vol. 38, no. 12, pp. 2040-2050, Dec. 2003.

[22] J. K. Fiorenza *et al.*, "Comparator-Based Switched-Capacitor Circuits for Scaled CMOS Technologies," *IEEE J. Solid-State Circuits*, vol. 41, no. 12, pp. 2658-2668, Dec. 2006.

[23] J. Hu, et al., "A 9.4-bit, 50-MS/s, 1.44-mW Pipelined ADC Using Dynamic Residue Amplification," *Dig. VLSI Circuits Symposium*, Jun. 2008.

[24] M. Anthony, et al., "A Process-Scalable Low-Power Charge-Domain 13-bit Pipeline ADC," *Dig. VLSI Circuits Symposium*, Jun. 2008.

[25] A. Nazemi, et al., "A 10.3GS/s 6bit (5.1 ENOB at Nyquist) Time-Interleaved/Pipelined ADC Using Open-Loop Amplifiers and Digital Calibration in 90nm CMOS, *Dig. VLSI Circuits Symposium*, Jun. 2008.

[26] E. Iroaga and B. Murmann, "A 12b, 75MS/s Pipelined ADC Using Incomplete Settling," *IEEE J. Solid-State Circuits*, vol. 42, no. 4, pp. 748-756, Apr. 2007.

[27] B. Murmann, "Limits on ADC Power Dissipation," in Analog Circuit Design, M. Steyaert, A. H. M. Roermund, and J. H. v. Huijsing, (eds.), Springer, 2006.

[28] D. Draxelmayr, "A 6b 600MHz 10mW ADC array in digital 90nm CMOS," *ISSCC Dig. Techn. Papers*, pp. 264-265, Feb. 2004.

[29] A. Maxim and M. Gheorghe, "A sub-1psrms jitter 1-5GHz 0.13um CMOS PLL using a passive feedforward loop filter with noiseless resistor multiplication," *Dig. RFIC Symposium*, pp. 207–210, Jun. 2005.

[30] S. Borkar, "Design challenges of technology scaling," *IEEE Micro*, vol. 19, pp. 23-29, Apr. 1999.

[31] M. Hesener, et al., "A 14b 40MS/s Redundant SAR ADC with 480MHz Clock in 0.13pm CMOS," *ISSCC Dig. Techn. Papers*, pp. 248-249, Feb. 2007.

[32] C. Paulus, et al. "A 4GS/s 6b flash ADC in 0.13um CMOS," *Dig. VLSI Circuits Symposium*, pp. 420-423, Jun. 2004.

[33] C. Donovan and M. Flynn, "A 'Digital' 6-bit ADC in 0.25-μm CMOS," *IEEE J. Solid State Circuits*, vol. 37, pp. 432-437, Mar. 2002.

[34] S. Park, et al., "A 4-GS/s 5-b Flash ADC in 0.18um CMOS," *IEEE J. Solid-State Circuits*, vol. 42, no. 9, pp. 1865-1872, Sep. 2007.

[35] B. P. Ginsburg and A. P. Chandrakasan, "Highly Interleaved 5b 250MS/s ADC with Redundant Channels in 65nm CMOS," *ISSCC Dig. Techn. Papers*, pp. 240-241, Feb. 2008.

[36] B. Murmann and B. E. Boser, "Digital Domain Measurement and Cancellation of Residue Amplifier Nonlinearity in Pipelined ADCs," *IEEE Trans. on Instrumentation and Measurement*, vol. 56, no. 6, pp. 2504-2514, Dec. 2007.

[37] A. Panigada and I. Galton, "Digital background correction of harmonic distortion in pipelined ADCs," *IEEE Trans. Circuits Syst. I*, vol. 53, no. pp. 1885–1895, Sep. 2006.

[38] J. Li and U.-K. Moon, "Background calibration techniques for multistage pipelined ADCs with digital redundancy," *IEEE Trans. Circuits Syst. II*, vol. 50, no. 9, pp. 531–538, Sep. 2003.

[39] J. McNeill, et al., "'Split ADC' architecture for deterministic digital background calibration of a 16-bit 1-MS/s ADC," *IEEE J. Solid-State Circuits*, vol. 40, no. 12, pp. 2437–2445, Dec. 2005.

[40] K.-W. Hsueh, et al, "A 1V 11b 200MS/s Pipelined ADC with Digital Background Calibration in 65nm CMOS," *ISSCC Dig. Techn. Papers*, pp. 547-548, Feb. 2008.

[41] P. Nikaeen and B. Murmann, "Digital Correction of Dynamic Track-and-Hold Errors Providing SFDR > 83 dB up to fin = 470 MHz," in preparation.

[42] Y. Chiu, et al., "Least-mean-square adaptive digital background calibration of pipelined analog-to-digital converters," *IEEE Trans. Circuits and Systems I*, vol. 51, pp. 38-46, Jan. 2004.

[43] P. Malla et al., "A 28mW Spectrum-Sensing Reconfigurable 20MHz 72dB-SNR 70dB-SNDR DT ΔΣ ADC for 802.11n/WiMAX Receivers," *ISSCC Dig. Techn. Papers*, pp. 496-497, Feb. 2008.

[44] W. Namgoong, "A channelized digital ultrawideband receiver," *IEEE Trans. Wireless Communications*, pp. 502-510, Mar. 2003.

[45] Y. Oh and B. Murmann, "A Low-Power, 6-bit Time-Interleaved SAR ADC Using OFDM Pilot Tone Calibration," *Proc. CICC*, pp. 193-196, Sep. 2007.

IEEE 2008 Custom Intergrated Circuits Conference (CICC)

A 16b 10MS/s Digitally Self-Calibrated ADC with Time Constant Control

Tae-Hwan Oh, Ho-Young Lee, Ju-Hwa Kim, Ho-Jin Park, Kyoung-Ho Moon, Jae-Whui Kim, Hae-Seung Lee*

Samsung Electronics, Co., Ltd, Yongin-City, Korea 446-711
*Massachusetts Institute of Technology, Cambridge, MA 02139, USA

Abstract- **Time constant control (TCC) incorporating on-chip digital self-calibration technique performs two-step calibration of pipeline ADC stage errors and reduces power consumption at the same time. Using the proposed technique, the current of the amplifier in 1st pipeline stage employing TCC is reduced by 93%. The prototype 3.3V 16b 10MS/s ADC based on 65nm CMOS process is implemented in an active die area of 1.32mm² and the on-chip calibration logic occupies only 18% of the die area. The reduction of overall power consumption of the ADC is 36%, from 123.8mW to 79.2mW with TCC. After TCC and digital self-calibration, the measured DNL, INL, and SNDR are ±0.65 LSB, ±5.76 LSB, and 75.4dB, respectively, for a 5MHz and 2.4Vpp differential input signal at 10MS/s.**

I. INTRODUCTION

Conventional ADC's with high resolution and sampling rate (> 14b and > 10MS/s) required in many applications typically consume a large amount of power. Several innovative approaches have been proposed to reduce the power consumption [1]-[2]. However, these approaches have not yet been demonstrated in high resolution ADC's with resolution 14b or higher.

In this paper, a 16b 10MS/s digitally self-calibrated pipe-lined ADC is described. To achieve high resolution of 16b with small active area and low power consumption, the proposed ADC employs an alternative technique, Time Constant Control (TCC) along with simple digital self-calibration. In the proposed technique, a 6b DAC controls the time constant of the amplifier by reducing the bias current and the on-chip calibration logic corrects ADC errors to achieve low power consumption and high resolution at the same time.

II. PROPOSED ARCHITECTURE

A. Basic concept of the proposed Time Constant Control

The example of TCC in unity gain stage is shown in Fig. 1. During the amplification phase Q_2, the stage output V_{out}, settles to the target value V_a, at the desired settling time T_0, assuming an ideal capacitor matching (a=0). In TCC mode, the sampling capacitor Cs is intentionally made larger than the feedback capacitor Cf by a factor of 1+a (a>0), such that the output settles to $V_a \cdot (1+a)$ at T_0 as shown in Fig. 1. In this case, V_{out} reaches the desired output voltage V_a at T_1 before reaching T_0. Now, if the settling time constant τ of the amplifier is increased by reducing the bias current I_{BS} of the amplifier, the accurate output voltage V_a is obtained at the desired settling time, T_0. Therefore, TCC allows the amplifier

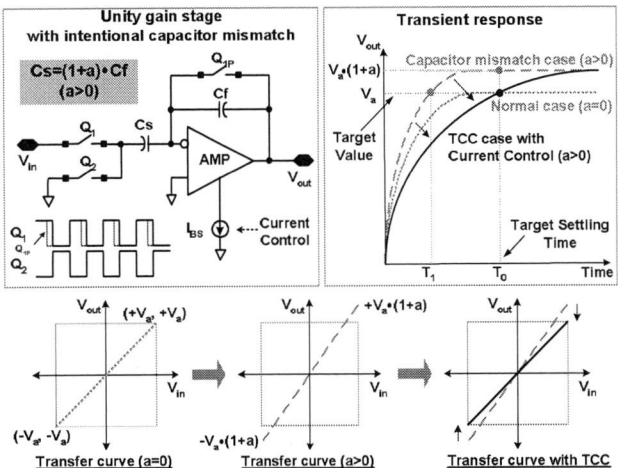

Fig.1. Concept of the proposed TCC technique.

to consume just enough power to reach the desired output with a longer time constant. TCC also eliminates the typical over-design of amplifier time constants to account for process and temperature variations. TCC, applied to individual stages, can calibrate error components such as finite DC gain of the amplifier and capacitor mismatch in the analog domain and reduce the power consumption at the same time.

However, it is difficult to employ TCC alone to ADC's with 14b or higher resolution because the required time constant control accuracy is high. In this work, we propose hybrid calibration that combines TCC with a simple digital self-calibration. TCC performs coarse calibration, and the digital self-calibration calibrates the residual error. TCC permits the use of very simple digital calibration with a nominal radix of 2 [3] despite the large amount of insufficient settling. The advantage of the hybrid calibration is that the digital calibration is simpler leading to smaller and lower power calibration circuitry compared with full digital calibration.

B. Proposed 4b stage with TCC

Fig. 2 shows the proposed 4b pipeline ADC stage with TCC. It consists of 4b Flash ADC, a decoder, 8 sampling capacitors, a feedback capacitor, switches, an amplifier, a comparator and a current control circuit for TCC. Each sampling capacitor is 1.05 times larger than the feedback capacitor in the 4b stage. With this nominal capacitor ratio, the required settling time to reach the ideal value is only 3τ. This compares to 10τ normally required for the full settling

978-1-4244-2018-6/08 $25.00 © 2008 IEEE

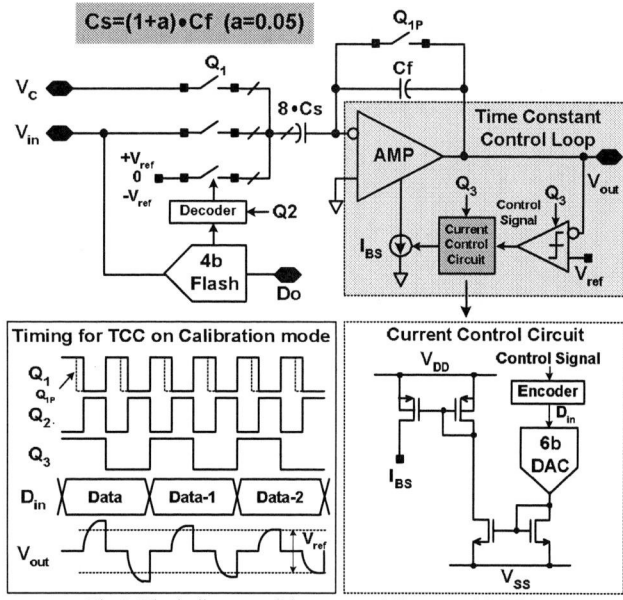

Fig.2. Block diagram of the proposed 4b stage with TCC.

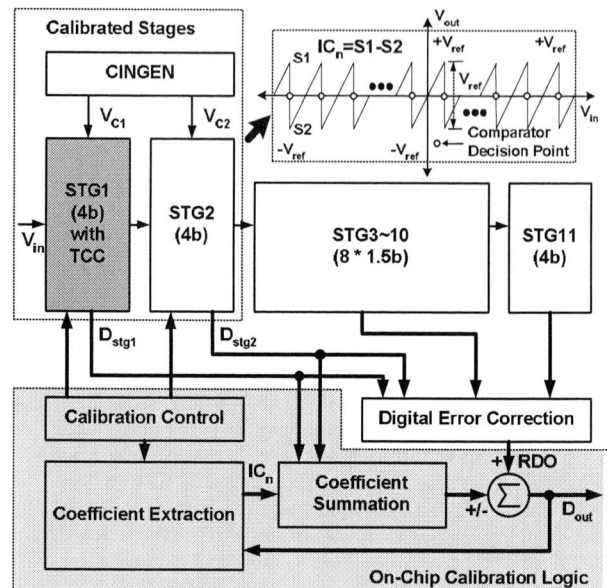

Fig.3. Architecture of the proposed self-calibration ADC.

without TCC. The nonlinear relationship between the amplifier settling time and the bias current yields substantial savings in power consumption when only 3τ settling.

During the TCC calibration, the calibration input signal V_C, is sampled on Q_1 and the residue output of the 1st stage, which slightly exceeds $+V_{ref}/2$ and $-V_{ref}/2$ due to the intentional capacitor mismatch, is generated alternatively on every Q_2. The comparator in TCC loop compares V_{ref} to the difference between two residue values near $+V_{ref}/2$ and $-V_{ref}/2$ on Q_3. Then, the comparator decides whether the difference is larger or smaller than V_{ref}. If the difference is larger than V_{ref}, the current control circuit which consists of an encoder, 6b DAC, and a current mirror decreases the bias current of the amplifier, I_{BS}, by 1 LSB on every Q_3 until the difference is smaller than V_{ref}. When TCC loop determines the value of I_{BS} which changes the polarity of the control signal, TCC mode is terminated. In this design, a 6b DAC is chosen for TCC. The remaining gain error due to the limited DAC resolution is nominally within 0.5% of the full scale range. Although the residual error can be reduced by increasing the resolution of the DAC, it is more efficient to remove such a small error by the simple digital calibration.

C. Proposed ADC architecture

The architecture of the proposed ADC is shown in Fig. 3. The ADC consists of 2 calibrated 4b stages, eight 1.5b stages, a 4b final stage, a calibration input generator, digital error correction logic, on-chip calibration logic and other extra supporting blocks. The ADC is designed to calibrate only the 1st and 2nd stages, because the errors in subsequent stages are relatively small and the required accuracy is easily obtained without any calibration process. The proposed TCC technique is adopted only in the 1st stage to demonstrate the concept in the prototype ADC.

D. Calibration procedure

When calibration is initiated, the ADC executes TCC in the 1st stage for the optimum time constant of the amplifier and for reduced power consumption. After TCC, the remaining errors in 1st and 2nd stages are calibrated first on the 2nd stage, then the 1st stage. The digital calibration removes residual error in the 1st stage and calibrates the 2nd stage errors with a similar algorithm to one used in [3]. Although the previous algorithm was demonstrated on a 1b/stage architecture, the same calibration concept can be easily applied to multi-bit/stage architecture.

Referring to Fig. 3, for each comparator decision point in the 1st and 2nd the stages, the calibration coefficient, IC_n is calculated by subtracting S2 from S1. The inputs for calibration, V_{C1} and V_{C2}, are generated by the calibration input generator, CINGEN. Fig. 4 shows the procedure of the calibration in case of the 1st stage. For simplicity, only gain error component is considered in Fig. 4. To correct the residual error after TCC as shown in Fig. 4(a), the extracted coefficients are added or subtracted to the un-calibrated output, RDO, as the digital output of the 1st stage, D_{stg1}. The extracted calibration coefficients, IC_n's, are summed and used to calibrate the ADC errors during the normal operation as shown in Fig. 4(b). All control and calibration logic circuits are integrated on the chip.

III. CIRCUIT IMPLEMENATION

To further reduce the power consumption and die area, the ADC eliminates a front-end SHA and uses the similar sampling scheme as in [4]. Fig. 5 shows the input sampling block in 1st stage composed of MDAC and FLASH. Since both blocks use the same sampling clock, Q_{SP}, the difference of the sampled values between two blocks is negligible. As a result, the SHA is not required in the proposed ADC.

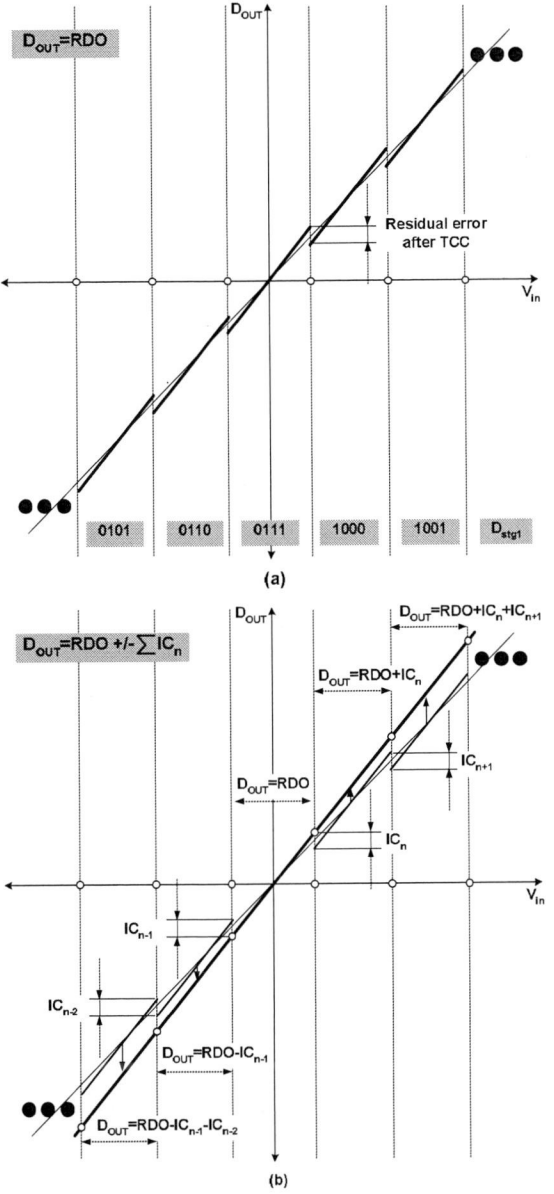

Fig.4. Digital calibration procedure:
(a) before calibration and (b) after calibration.

However, the timing margin for sampling and pre-amplifying is reduced because two operations are executed on the same clock phase. On the other hand, the time required for pre-amplifying can be minimized because the comparator always samples the DC reference voltage rather than AC input signal. To maximize the time for sampling, the pre-amplifying time, PA, is designed to be 2ns.

When the ADC executes TCC, the linear operation of the amplifier under the condition of incomplete settling is crucial for the ADC linearity and TCC. Therefore, a single-stage folded cascode amplifier with gain boosting topology is adopted to ensure sufficient phase margin and large DC gain. The simulated values of phase margin for closed loop and DC gain for open loop are $90°$ and 124dB, respectively.

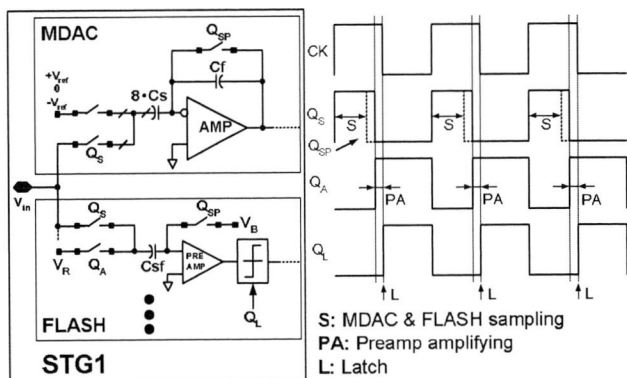

Fig.5. Timing diagram for input sampling.

IV. PROTOTYPE MEASUREMENTS

The prototype ADC is fabricated in a 65nm CMOS process. The micrograph of the prototype ADC is shown in Fig. 6 and the active die area is $1.32mm^2$ (=1.15mm $\times$ 1.15mm), including the on-chip digital self-calibration logic. The calibration logic occupies only 18% of the die area.

After calibration using TCC, the ADC consumes 79.2mW at 10MS/s with 3.3V and 1.2V power supplies. The current of the amplifier in 1st stage is reduced by 93% compared to the current required for settling to full precision without TCC. The reduction of overall power consumption of the ADC is 36%, from 123.8mW to 79.2mW with TCC. The 1.2V power supply is applied only to the calibration logic. The power consumption of calibration logic is negligible because the calibration algorithm is simple, and only minimal digital operations are required during the normal conversion mode.

Fig. 7 shows the measured DNL and INL. After TCC and digital calibration, the measured DNL and INL are improved from ±1.15LSB to ±0.65LSB and from ±66.93LSB to ±5.76LSB, respectively, for a 5MHz and 2.4Vpp differential input signal at 10MS/s. The major spikes in INL plot are originated from the incomplete settling of the calibration input signal generated from calibration input generator due to the under-estimated parasitic resistance and capacitance of the signal path. This incomplete settling of the calibration input limits the conversion rate to 10MS/s. The measured INL at 5MS/s is ±3.32LSB.

Fig. 8 shows the measured output spectra for a 5MHz and 2.4Vpp differential input signal at 10MS/s. With calibration, the measured SNDR and SFDR are improved from 55.1dB to 75.4dB and from 64.5dB to 90.5dB, respectively. The ADC achieves better than 72.3dB SNDR up to a 10MHz input frequency without a front-end SHA at 10MS/s. Table I summarizes the measured ADC performance.

V. CONCLUSION

A 16b 10MS/s digitally self-calibrated pipelined ADC has been presented. In this work, Time Constant Control technique incorporating on-chip digital self-calibration is proposed to achieve low power consumption and high resolution at the same time. With the proposed technique, the power consumption of the amplifier in 1st stage is reduced by

93% and the overall power consumption of the ADC is reduced by 36%. Although TCC was only adopted in 1st stage to demonstrate the concept in the prototype ADC, further reduction of the power consumption may be possible by adopting TCC in more stages. The measured results demonstrate that the proposed technique can be effectively employed in high resolution ADC with low power consumption.

TABLE I
PERFORMANCE SUMMARY

Resolution	16 bits	
Conversion Rate	10MS/s	
Process	65nm 1 poly 6 metal CMOS	
Supply Voltage	3.3V (ADC Core) / 1.2V (Calibration Logic)	
Input Range	2.4Vpp	
Active Die Area	1.32mm^2 (=1.15mm × 1.15mm)	
DNL	Before Calibration	After Calibration
	±1.15LSB	±0.65LSB
INL	±66.93LSB	±5.76LSB
SNDR	55.1dB	75.4dB
SFDR	64.5dB	90.5dB
Power Dissipation	123.8mW	79.2mW

REFERENCES

[1] T. Sepke, et al., "Comparator-Based Switched-Capacitor Circuits For Scaled CMOS Technologies, " ISSCC Dig. Tech. Papers, pp. 220-221, Feb. 2006.

[2] E. Iroaga, B. Murmann, "A 12b, 75MS/s pipelined ADC using incomplete settling, " Dig. Symp.VLSI Circuits, pp. 274-275, June 2006.

[3] H.-Y. Lee, et al., "A 14-b 30MS/s 0.75mm^2 Pipelined ADC with On-Chip Digital Self-Calibration," Proc. CICC, pp. 313-316, Sep. 2007.

[4] Y.-D. Jeon, et al., "A 4.7mW 0.32mm^2 10b 30MS/s Pipelined ADC Without a Front-End S/H in 90nm CMOS," ISSCC Dig. Tech. Papers, pp. 456-457, Feb. 2007

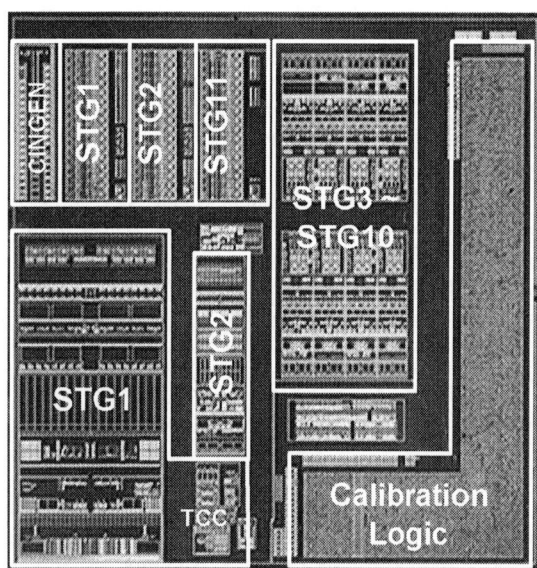

Fig.6. Die micrograph of the ADC.

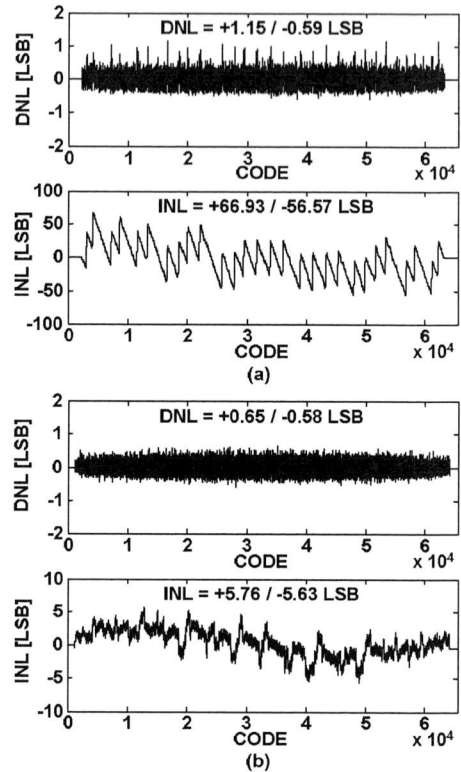

Fig.7. Measured DNL and INL:
(a) without calibration and (b) with calibration.

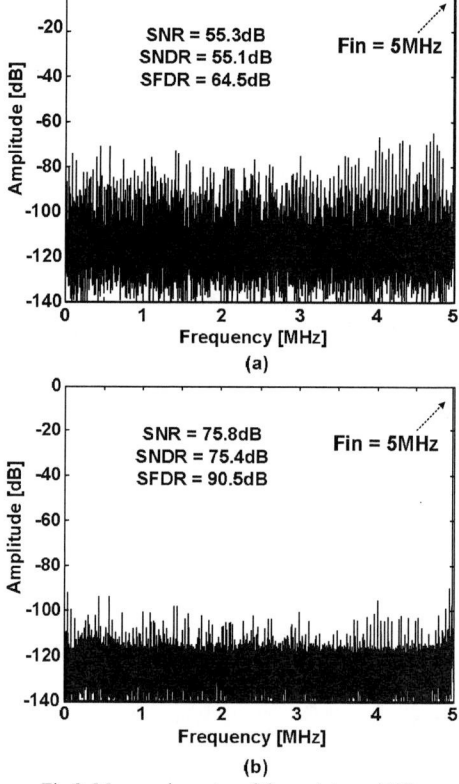

Fig.8. Measured spectra of the prototype ADC:
(a) without calibration and (b) with calibration.

IEEE 2008 Custom Intergrated Circuits Conference (CICC)

A 15b Power-Efficient Pipeline A/D Converter Using Non-Slewing Closed-Loop Amplifiers

Shoji Kawahito, Kazutaka Honda, Zheng Liu, Keita Yasutomi and Sinya Itoh
Shizuoka University

Abstract- **A 15b power-efficient pipeline A/D converter using capacitance-coupling non-slewing amplifiers is presented. A modified 1.5b/stage transfer curve combined with the non-slewing amplifier is useful for the error corrections of incomplete settling error. The relationship between the input signal and the incomplete settling errors can be linearized and the errors can be corrected in digital domain with a simple calculation. A prototype ADC fabricated in 0.25μm process consumes 123mW at 30MSample/s and 2.5V power supply. The SNDR and the SFDR at 30MS/s are 75.0 dB and 86.5 dB, respectively with the incomplete settling error corrections.**

I. INTRODUCTION

Pipeline analog-to-digital converters (ADC's) are widely used for applications which require high speed and high resolution with reasonably low power dissipation. To meet never-ending demands for power saving, various techniques for low power designs have been proposed. Recently-reported low power pipeline ADC's use a power-efficient open-loop amplifier and digital collections of the non-linearity, device mismatch and incomplete settling error [1][2]. The large non-linearity of the open-loop amplifier, however, may not be suitable for pipeline ADC's with high resolution of 14b or more. This paper describes a low power 15b pipeline ADC that exploits power-efficient capacitance-coupling non-slewing amplifiers and a digital correction of incomplete settling error of the amplifier using a modified 1.5b/stage transfer characteristic of pipeline stages.

II. PIPELINE STAGES USING NON-SLEWING AMPLIFIER

The Designed 1.5bit/stage pipeline chain consists of a sample-and-hold amplifier (SHA) saving first stage, the following 15 MDAC stages, and 2b flash at the end. Figs. 1 and 2 show the proposed SHA-saving first stage and non-slewing amplifier using a capacitance-coupling technique, respectively. In the rest of MDAC stages, switches controlled by ϕ_{2p} in Fig. 1 are not used. Removing the dedicated SHA has advantages in power and noise [3]. The SHA-saving first stage has three operating phases: sample, comparison, and hold. After the capacitors C_s sample an input signal, the top plates (bottom plates in physical structure) of C_s are connected to the output for the comparison in a sub-ADC during the comparison phase ϕ_{2p}. This loading-free output responds quickly without an aperture error. Input switches of every stage use a gate-bootstrapping technique and native-nMOS transistors to suppress the input signal dependency. Bootstrapping circuits of MDAC stages are shared between input switches and the previous stage's feedback switches to improve the settling behavior[4].

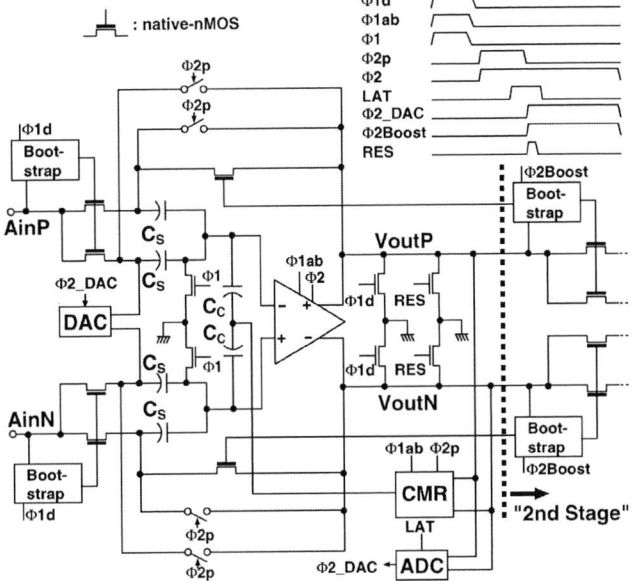

Fig. 1. SHA-saving first stage.

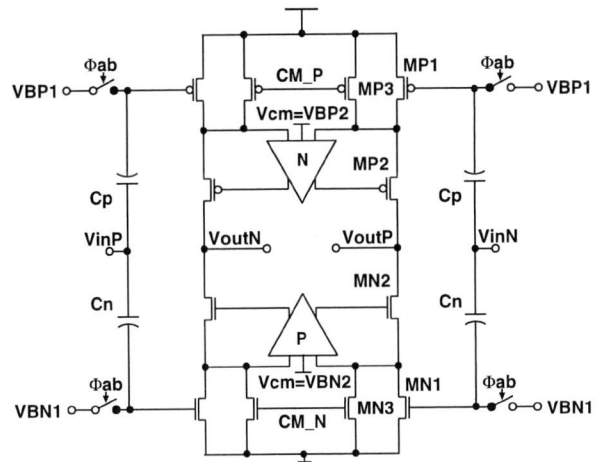

Fig. 2. Non-slewing Amplifier with capacitively-coupled inputs.

Fig. 3 shows the incomplete settling error of an MDAC using a conventional class-A amplifier as a function of the input signal. The non-linear input signal dependency of the settling error is caused by the combination of the slewing behavior and exponential response due to the transconductance of amplifiers. If the incomplete settling error can be expressed as a simple linear function of the input signal, the error can be calibrated in digital domain. In ref. [2], the calibration of the incomplete settling error has been performed in a pipeline ADC using resistor-road open-loop

978-1-4244-2018-6/08 $25.00 © 2008 IEEE   127

amplifiers which do not have slewing behavior. However, because of the non-linear input signal dependency caused by the slewing behavior at the beginning of the response, the conventional closed-loop MDAC with a class-A operational amplifier shows complicated non-linear dependency of the error to the input signal as shown in Fig. 3.

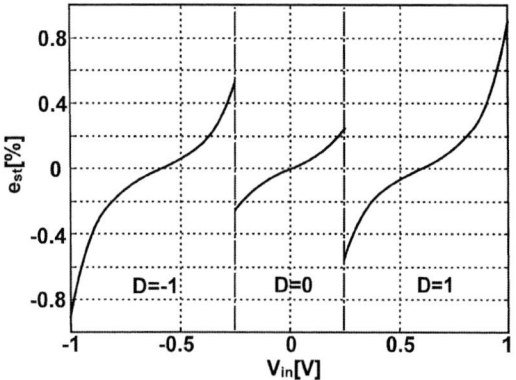

Fig. 3 Non-linear input signal dependency of the incomplete settling error in the MDAC with a class-A amplifier.

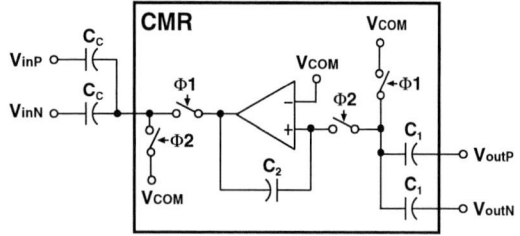

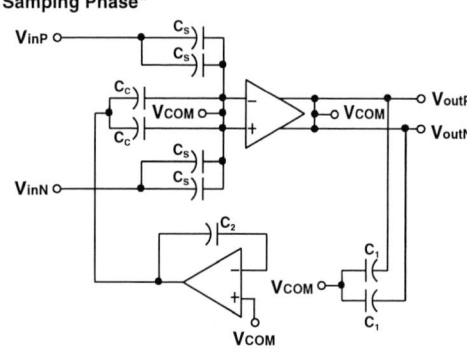

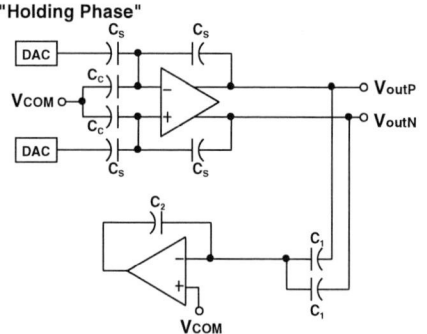

Fig. 4 Common-mode regulator and the operation.

The capacitance-coupling push-pull amplifier used in the MDAC, shown in Fig.2, provides two major benefits for low-power design of high-speed high-resolution pipeline ADC's. The direct push-pull driving of nMOS and pMOS common source transistors provides a large transient output current with little static power dissipation, leading to a power efficient design of the pipelined ADC. The gates of MP1 and MN1 are capacitively coupled to the amplifier input through capacitors C_p and C_n, respectively. These capacitances are pre-charged during the sampling phase ϕ_{ab} and act as level shifters of the input to drive complementarily MP1 and MN1.

Another benefit of this amplifier is the suppression of the slewing in the transient response, which leads to faster response and linearization of the input signal dependency of the incomplete settling error. The technique for incomplete settling error correction using the non-slewing amplifier will be discussed in Section III.

The capacitance-coupling push-pull amplifier is sensitive to the input common-mode (CM) voltage shifting. This problem is overcome by using a common-mode regulator (CMR) as shown in Fig. 4. At the holding phase of the MDAC amplifier, the output CM of the MDAC amplifier is sensed by C_1 and the deviation from the CM is accumulated by a switch-capacitor (SC) integrator. At the sampling phase, C_C is charged by the SC integrator output and simultaneously C_1 is reset. At the next holding phase, Cc regulates the input and output CM voltages of the MDAC amplifier to be unchanged by injecting the stored charge of C_C if the output of the SC integrator deviates from V_{COM}. The loop with the MDAC and the CMR acts as a discrete-time negative feedback system.

III. INCOMPLETE SETTLING ERROR CORRECTION

The non-slewing amplifier linearizes the input signal dependency of the settling error. However, the non-linearity still occurs at large output swing. Fig. 5 shows a modified 1.5bit/stage transfer curve for output swing suppression and the simulated settling error characteristic of the first stage. The output dynamic range (DR) is reduced to [-0.6V_r, 0.6V_r] while the input DR is [-0.8V_r, 0.8V_r] where V_r is reference voltage of the MDAC. This leads to a benefit of increasing input signal full scale range with relatively low power supply. The use of this transfer curve reduces the margin for comparator offsets to +0.05V_r/-0.05V_r. The switches controlled by RES in Fig. 1 also help to settle linearly by forcing the output node of the first stage to be zero at the beginning of the holding phase. As shown in Fig. 5, the settling error is sufficiently linearized in the input range of [-0.8V_r, 0.8V_r]. The error can be corrected in digital domain by subtracting the digital estimate of the error. In the modified 1.5bit/stage transfer curve, available digital codes are limited to 80% of the full scale. However, since 1/0.8 is expressed as $1+2^{-2}$, the code range can be expanded to the full scale by a simple operation of two-bit shifting and addition. This conversion does not introduce any dynamic performance degradation if a few extra bits to 15b resolution are available.

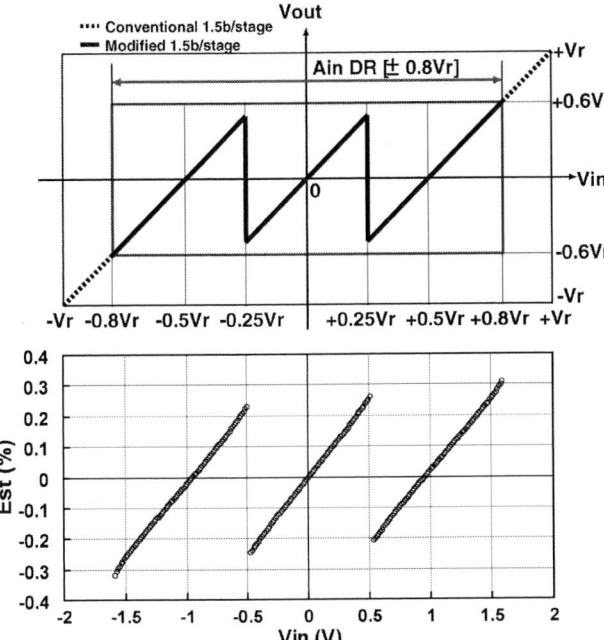

Fig. 5 Modified transfer curve and incomplete settling error as a function of the input signal.

The designed ADC has 16 stages of 1.5b code output and extra two bits. The redundant digital code with the DR of 0.8 x 2^{18} is converted to 15b non-redundant digital code with a 15b full DR of 2^{15} after error correction. The error correction and full-scale conversion are performed by off-chip processing.

IV. MEASUREMENT RESULTS

A prototype ADC (Fig. 6) using the proposed techniques was fabricated in a 2.5V, 0.25μm, 5M1P CMOS technology. Capacitors are implemented using metal-insulator-metal (MIM) structures. It occupies an active area of 6.4mm². The area is not optimized in this experimental chip. All measurements were performed with a chip-on-board (COB) assembly.

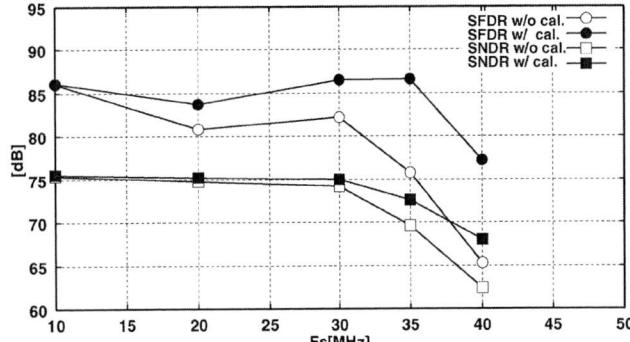

Fig. 7 SNDR and SFDR as a function of the sampling frequency.

Fig. 7 shows measured signal-to-noise-and-distortion ratio (SNDR) and spurious free dynamic range (SFDR) as a function of conversion rate at an input frequency of 10MHz. The plot shows that SNDRs and SFDRs without calibration remain above 74dB and 80dB up to 30MS/s, respectively. At 35MS/s, the SNDR and SFDR are degraded to 69dB and 76dB due to the settling error, respectively. The linearized settling error calibration improves the SNDR and SFDR to 73dB and 87dB at 35MS/s, respectively. Capacitor mismatches and finite gain error are also corrected, but these are not dominant because large capacitances and high-gain amplifiers are utilized.

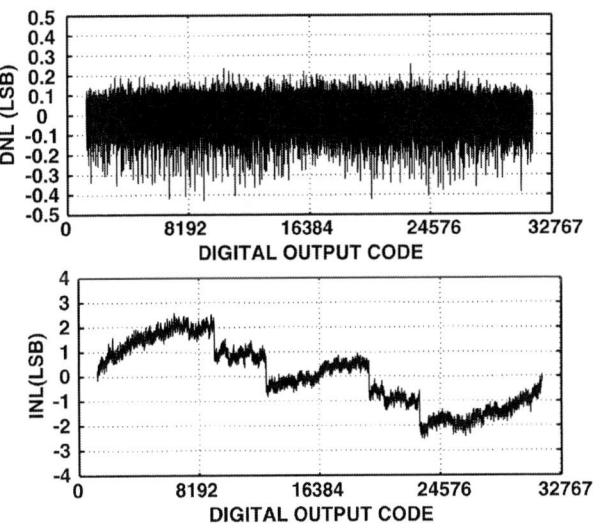

Fig. 8 DNL and INL after full-scale conversion.

Fig.8 shows the differential and integral non-linearity of the ADC after the full scale conversion to 15b full-scale code of 0 to 32767. Though the full scale conversion introduces 25% degradation of the DNL and INL, sufficient linearity is attained.

Fig. 6 Implemented pipeline ADC chip with 0.25um CMOS.

Table I Performance Summary.

Technology	0.25μm CMOS
Resolution	15 bits
Sampling Rate	30 MS/s
Supply Voltage	2.5 V
Analog Input Range	3.2 V$_{p-p}$
INL	-2.5/+2.5 LSB
DNL	-0.42/+0.25 LSB
SNDR @ F$_{in}$=10MHz	75.0 dB (w/ cali.) 74.2dB (w/o cali.)
SFDR @ F$_{in}$=10MHz	86.5 dB (w/ cali.) 82.2 dB (w/o cali.)
Power Consumption	123mW
Packaging	Chip-on-board
Active area	1.6 mm x 4.0 mm

The performance of the ADC is summarized in Table I. The total power dissipation of the chip is 123mW consisting of 117mW for analog and 6mW for digital at 30MS/s excluding the digital output drivers. The DNL and INL are -2.5/+2.5LSB and -0.42/+0.25LSB, respectively. The SNDR and SFDR with error corrections are 75dB and 87dB at 30MS/s, respectively.

Table II Comparison of FOM in recently reported ADC's.

Reso-lution [bit]	V$_{DD}$ [V]	F$_s$ [MHz]	ENOB [bit]	Power [mW]	FOM1 [pJ/ step]	FOM2 [fJ/ step^2]	Reference
15	2.5	50	12.21	780	3.30	0.70	[5]
15	3.3	40	11.95	370	2.33	0.59	[6]
14	1.8	12	12.29	98	1.61	0.32	[7]
14	3.0	75	11.87	350	1.24	0.33	[3]
14	1.5	100	10.97	224	1.12	0.56	[8]
14	2.8	30	11.14	51	0.75	0.35	[9]
15	2.5	30	12.07	123	0.95	0.22	This work
15	2.5	30	12.21	(123)	(0.87)	(0.19)	This work (with Cali.)

Table II shows the comparison of figure of merit (FOM) of recently reported 14b and 15b ADC's. In Table II, FOM1 means the conventional FOM defined by Power/($2^{ENOB} f_s$) where $ENOB$ is the equivalent number of bits and f_s is the sampling frequency. In high-resolution ADC's, the thermal noise based FOM defined by Power/($2^{ENOB})^2 f_s$ is often used[10]. This is based on a fact that increasing SNR by a factor of two requires 4 times larger capacitance to reduce kT/C noise, and 4 times larger power. This FOM is denoted by FOM2 in Table II. The ADC of this work without the error calibration achieves the best FOM2 of 0.22fJ/(conversion-step)^2 in reported 14b and 15b ADC's. Data for the case of calibrating errors including the incomplete settling error are also shown though it does not include the power required for the error correction logic circuits.

V. CONCLUSIONS

This paper presents a 15b pipeline ADC using power-efficient non-slewing amplifiers. The non-slewing amplifier using capacitance-coupling push-pull driving technique reduces the power dissipation of the 15b pipeline ADC, and the best thermal noise based FOM has been achieved. The non-slewing behavior also simplifies the correction of incomplete settling error by linearizing the input signal dependency of the error. In the measurement results, the effect of incomplete settling error correction to the improvement of SNDR is not so drastic. This is because the actual behavior of the incomplete settling is not as simple as that predicted by simulation. More exact but simple modeling of the incomplete settling error for calibration will further improve the dynamic performance of the ADC and the resulting power efficiency.

ACKNOWLEDGMENTS

This work is supported by Semiconductor Technology Academic Research Center (STARC) and VLSI Design and Education Center (VDEC), the University of Tokyo in collaboration with Cadence Design Systems, Inc., Synopsys, Inc., and Mentor Graphics, Inc.

REFERENCES

[1] B. Murman and B. E. Boser, "A 12-b 75MS/s Pipelined ADC Using Open-loop Residue Amplification," Dig. Tech. Papers, ISSCC, pp.328-3290, Feb. 2003.

[2] E. Iroaga, and B. Murmann, "A 12b 75MS/s Pipelined ADC Using Incomplete Settling," Symp. VLSI Circuits, pp.274-275, Jun. 2006.

[3] I. Mehr, and L. Singer, "A 55-mW 10-bit 40-MSample/s Nyquist-rate CMOS ADC," IEEE J. Solid-State Circuits, vol. 35, pp. 318-325, Mar. 2000.

[4] D. Kelly, W. Yang, I. Mehr, M. Sayuk, and L. Singer, "A 3V 340mW 14b 75MSPS CMOS ADC with 85dB SFDR at Nyquist," IEEE Int. Solid-State Circuits Conf., vol. XLIV, pp. 134 - 135, Feb. 2001.

[5] K. Nair and R. Harjani, "A 96dB SFDR 50MS/s digitally enhanced CMOS pipeline A/D converter," IEEE Int. Solid-State Circuits Conf., vol. XLVII, pp. 456 - 457, Feb. 2004.

[6] H. Liu, Z. Lee, and J. Wu, "A 15-b 40-MS/s CMOS pipelined analog-to-digital converter with digital background calibration," IEEE J. Solid-State Circuits, vol. 40, pp. 1047 - 1056, May 2005.

[7] Y. Chiu, P. R. Gray, and B. Nikolic, "A 14-b 12-MS/s CMOS pipeline ADC with over 100-dB SFDR," IEEE J. Solid-State Circuits, vol. 39, pp. 2139 - 2151, Dec. 2004.

[8] P. Bogner, F. Kuttner, C. Kropf, T. Hartig, M. Burian, and H. Eul, "A 14b 100MS/s digitally self-calibrated pipelined ADC in 0.13μm CMOS," IEEE Int. Solid-State Circuits Conf., vol. XLIX, pp. 224 - 225, Feb. 2006.

[9] H. Matsui, M. Ueda, M. Daito, and K. Iizuka, "A 14bit digitally self-calibrated pipelined ADC with adaptive bias optimization for arbitrary speeds up to 40MS/s," Symp. VLSI Circuits Dig. , pp. 330 - 333, June 2005.

[10] A. M. A. Ali, C. Dillon, R. Sneed, A. S. Morgan, S. Bardsley, J. Kornblum, and L. Wu, "A 14-bit 125MS/s IF/RF Sampling Pipelined ADC With 100dB SFDR and 50fs Jitter," IEEE J. Solid-State Circuits, vol.41, pp.1846-1855, Aug.2006.

Characterization and Test Methods for
Device Variability in Nanoscale Technologies

Session 8

Chair: Hamid Mahmoodi, San Francisco State University
Co-Chair: Jeanne Trinko Mechler, IBM

With scaling of CMOS towards nano-scale technologies, device variability and reliability is emerging as a major challenge for circuit design in such technologies. Hence, accurate measurement and characterization of sources of device variations and reliability degradations is very critical for coming up with appropriate circuit design techniques to mitigate such device non-idealities, as well as to insure the device manufacturability and yield. This session presents cutting-edge papers on effective measurement methods for characterization of gate dielectric breakdown, threshold voltage variations, delay variations, and an effective testing of multi-core SOC under process variations.

The first paper proposes an array-based circuit for characterizing gate dielectric breakdown. Such a design is beneficial when studying this statistical process, where up to thousands of samples are needed to create an accurate time to breakdown distribution. Measurement results are presented from a 32x32 test array implemented in a 130nm process.

The second paper, proposes a novel low-power, bias-free, high-sensitivity process variation sensor for monitoring random variations in threshold voltage. This design utilizes the exponential current-voltage relationship of sub-threshold operation for improved sensitivity. Measurement results on 28 dies fabricated in 65nm technology demonstrate the effectiveness of the proposed sensor.

The third paper presents results from a test-chip in a low-power 45nm technology, featuring uniaxial strained-Si, built to study variability in circuits. Systematic layout-induced variation, die-to-die, wafer-to-wafer and within-die variability has been measured and analyzed. Delay is characterized using an array of ring-oscillators and transistor leakage current is measured with an on-chip ADC. Results show that systematic variations are small and layout-induced variation is dominated by strain effects.

The fourth paper reports a circuit technique to measure the on-chip delay of an individual logic gate using digitally reconfigurable ring oscillator. Experimental results from a test chip in 65nm technology show the feasibility of measuring the delay of an individual inverter to within 1 ps accuracy. Delay measurements of different nominally identical inverters in close physical proximity show variations of up to 26% indicating the large impact of local or within-die variations.

The last paper presents a test method with the dual supply voltage (VDD) for the ultra low VDD homogeneous multi-core LSI's in 90 nm technology. In this method, two same cores with different power supply voltages test each other by comparing their outputs, which eliminates the need for the expected vector.

978-1-4244-2018-6/08 $25.00 © 2008 IEEE

Notes

An Array-Based Test Circuit for Fully Automated Gate Dielectric Breakdown Characterization

John Keane   Shrinivas Venkatraman   Paulo Butzen*   Chris H. Kim

*State University of Rio Grande do Sul, Porto Alegre, Brazil

University of Minnesota, Minneapolis

Abstract- **We propose an array-based test circuit for efficiently characterizing gate dielectric breakdown. Such a design is highly beneficial when studying this statistical process, where up to thousands of samples are needed to create an accurate time to breakdown distribution. The proposed circuit also facilitates investigations of any spatial correlation of dielectric failures, and can monitor a progressive decrease in gate resistance. Measurement results are presented from a 32x32 test array implemented in a 130nm process.**

I. INTRODUCTION

While scaling CMOS device dimensions allows designers to pack more, and faster, transistors on a die, it also leads to an increased susceptibility to variations and reliability mechanisms. One such reliability issue is time-dependent dielectric breakdown (TDDB). This mechanism causes a conduction path to form through a gate dielectric layer placed under electrical stress, leading to parametric or functional failure. Breakdown has been a cause for increasing concern as gate dielectric thicknesses are scaled down to the one nanometer range, because a smaller critical density of traps is needed to form a conducting path through these thin layers, and stronger electric fields are formed across gate insulators when voltages are not scaled as aggressively as device dimensions.

Although many of the physical details behind TDDB are still under debate, the percolation model is widely used to describe the gradual accumulation of electrical defects through a stressed oxide, which eventually form a current conduction path resulting in the breakdown [1]. Some studies have used the time to the first breakdown event (defined as an increase in gate current to some pre-determined level) to extrapolate predicted device lifetimes from accelerated stress experiments [2]. A range of currents can be detected after this first event, and the distinction between current paths with low and high conduction levels led to the classification of "soft" and "hard" breakdowns. However, the definitions of those terms are contentious, and some authors claim that all breakdown events are more correctly described as progressive in nature [3]. In addition, it has become apparent that transistors can continue to function in certain cases after that first breakdown event, and the progressive, post-breakdown current evolution must also be taken into consideration to obtain a less pessimistic lifetime projection [3-5]. This is particularly true when operating at lower voltages and with thinner dielectrics, making an observable progressive breakdown current

more likely before final device failure [5].

TDDB is a function of a number of variables, including the gate voltage and oxide thickness as mentioned earlier, as well as temperature, device area, and dielectric materials and purity. Several models have been used to describe the relationship between the time to failure due to breakdown and these variables, but additional work is needed to more fully characterize TDDB in general so that the correct predictive models can be selected. The breakdown behavior of each new CMOS process must also be thoroughly tested during the process characterization phase in order to obtain a detailed understanding of the technology reliability.

In this paper, we present a circuit design that performs automated measurements in a test array to efficiently gather the breakdown characteristics that define this statistical process. The proposed circuit can monitor a progressive decrease in gate resistance, or simply an abrupt failure often referred to as a hard breakdown. This structure greatly reduces the required process characterization testing time, which may involve continuously monitoring the current through a single Device Under Test (DUT) per experiment with a finely tuned parametric test system. Given the need for up to thousands of test samples to correctly define the Weibull slope of the time to breakdown (T_{BD}) distribution [2], that serial testing process quickly becomes cumbersome. In addition, the array format is also a convenient method to study any spatial correlation of gate dielectric breakdown characteristics, without requiring sophisticated testers or elaborate test setups. Test array structures of this type are gaining popularity as a fast and efficient way to gather process technology information that is statistically meaningful, since individual device probing is not practical when large numbers of readings are required [6-7].

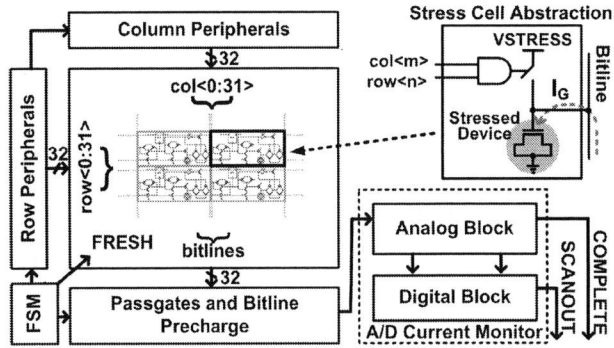

Fig. 1. 32x32 array for fully automated gate dielectric breakdown characterization.

II. Breakdown Characterization Array Design

The proposed test circuit design consists of a 32x32 array of stressed NMOS transistors, whose gate currents (I_G) are periodically measured using an A/D current monitor and on-chip control logic (Fig. 1). (Note that PMOS transistors could be tested within the same framework with the addition of a slightly modified stress cell design.) After an initialization sequence, cells are cycled through automatically without the need to send or decode cell addresses, in order to simplify the logic and attain faster measurement times. A single external clock signal is asserted each time that the controlling software is ready for a new I_G measurement. Although we chose to simplify and speed up the circuit in this manner, we do have the ability to select any one portion of the array for measurements while turning off stress in the rest of the test cells, as will be described later. The finite state machine (FSM) pictured in Fig. 1 controls the initialization sequence timing, as well as that of the subsequent measurements. The row and column peripheral circuits contain D flip-flops and multiplexers used to select a particular cell, as well as level shifters to boost signals from the 1.2V digital supply domain to the stress voltage (VSTRESS) level, which is used as the supply voltage within the array.

The stress cell structure shown in Fig. 2 was implemented to facilitate the accelerated stressing of the DUTs, by using thick oxide I/O transistors in the supporting circuitry to avoid excessive aging or breakdown in these other devices. The row<n> and col<m> signals are used to select one stress cell when both are logic high. When a cell is selected, devices M1 and M2 are turned off, and the transmission gates connecting the gate of the DUT to its bitline are turned on. M2 precharges the node between the two transmission gates to VCC, which matches the bitline precharge level, until this cell selection event. After the transmission gates are turned on, the gate current through the stressed DUT is measured using circuits in the A/D current monitor.

The FRESH signal is used to permanently gate off stress on a broken device when a high gate current is detected, in order to avoid excessive current drawn from the VSTRESS supply. After a sufficiently high breakdown current is measured in a selected cell, FRESH is automatically set to 0V by the controlling software before row<n> goes low, which latches a low value on Q. This isolates the DUT from VSTRESS by turning off device M1. When a cell is not being

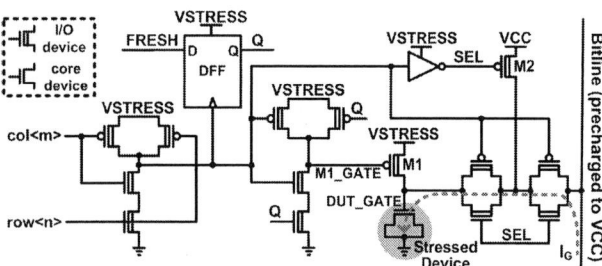

Fig. 2. Stress cell with bitline leakage compensation and stress/no-stress capability.

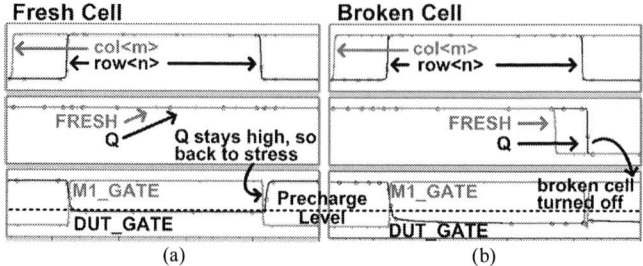

Fig. 3. Signal waveforms for one measurement in (a) a fresh cell and (b) a broken cell. The precharge level on the bitline (VCC) is indicated by the dashed line.

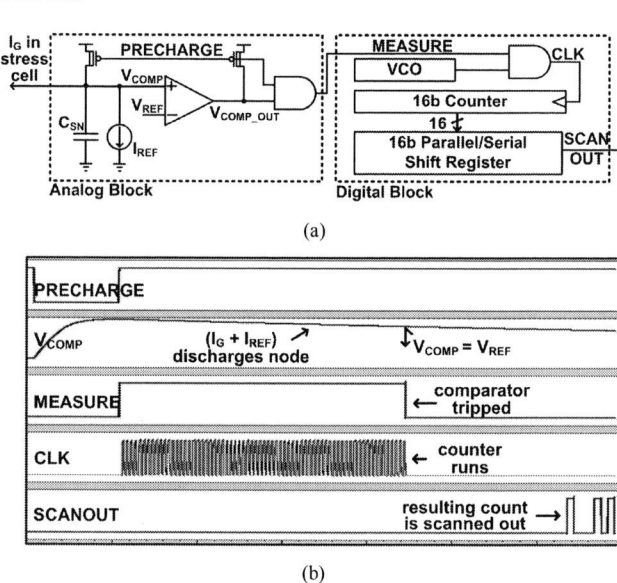

Fig. 4. (a) Analog and digital blocks of the A/D current monitor for measuring breakdown current progression. (b) Simulation of these blocks during a breakdown cell measurement

selected and Q is still high, M1 is turned on and the DUT is placed under constant voltage stress. Simulation waveforms demonstrating the measurement procedure for a fresh cell (low gate current) and a broken cell (high gate current) are presented in Fig. 3. In the unbroken, or "fresh" cell, the DUT_GATE node voltage quickly drops to the 1.2V precharge level before slowly decaying to $\sim V_{REF}$ (discussed later in this section), and then being charged to VSTRESS again when the row<n> signal drops to 0V. When the "broken" cell is accessed for a measurement the DUT_GATE node quickly discharges to 0V through the breakdown path. In this case, the FRESH signal is set to 0V before the cell is deselected, so M1_GATE remains high, and no further stressing occurs in this cell. The FRESH signal can also be used during circuit initialization to gate off stress in a range of unused cells which may be tested at a later time, or cells that are already broken from a previous experiment. This feature allows us to measure any one portion of the array during a single test, rather than the entire 1024 cells, if so desired.

The two transmission gates were placed between the gate of each DUT and its bitline, with the internal node held at VCC when the cell is not selected, to cut off the leakage in all unselected stress cells from the A/D current monitor. Simulations show that the leakage sourced by all 1023 unselected cells in the array during a measurement is limited to roughly 108nA. This worst-case leakage on the discharge path occurs when the selected DUT's gate node has discharged to ~1.1V (the V_{REF} level) at 30°C and VCC = 1.2V.

The analog block shown in Fig. 4(a) contains a comparator whose precharged output drops to 0V when the precharged input voltage (stored on an 80pF metal capacitor, C_{SN}) falls to the reference voltage (V_{REF}) level. That discharge rate is determined by I_G in the selected cell, plus an external reference current (I_{REF}) that is used for calibration purposes. The digital block (Fig. 4(a)) contains a 16 bit counter that runs at a rate set by a voltage controlled oscillator (VCO), starting at the end of the precharge event until the comparator's output falls, indicating that a measurement is complete. Therefore, less I_G (i.e., a larger gate resistance, R_{GATE}) translates to a higher count. The final result is latched into a parallel/serial shift register and subsequently scanned off-chip after the software interface detects a COMPLETE signal, which is asserted by the analog block. The results are stored in a convenient spreadsheet format for post-processing. The simulation waveforms presented in Fig. 4(b) illustrate the basic outline of this measurement procedure.

III. TEST CHIP MEASUREMENTS

A test chip was fabricated in a 1.2V, 130nm process, and automatic measurements were completed via LabVIEW™. The calibration procedure and measurement results are illustrated in Fig. 5. In order to obtain the final count vs. total discharge path resistance characteristic, the A/D current monitor was gated off from the breakdown array, and an adjustable external resistor (R_{EXT}) was attached to the I_{REF} path. Each output count during measurements can be

translated into a gate path resistance by using this calibration curve, and the relationship $R_{TOTAL} = R_{EXT} \| R_{GATE}$, with R_{EXT} being fixed during measurements.

The resistance of the transmission gates located on the path from V_{COMP} to the DUT gates is not accounted for in this simple procedure, and therefore introduces a measurement error that becomes more severe as the DUT gate resistance drops into the hard breakdown region. That is, our measured R_{GATE} result will be larger than the correct value because of the transmission gate resistances. However, due to the relatively high value of R_{GATE} during the progressive degradation stage leading up to the final hard breakdown, this error is small in the region of interest. The measurement error is less than 5% for R_{GATE} values of 64kΩ and greater, corresponding to gate currents up to 18.8µA at a sensing voltage of 1.2V. Several authors have indicated that the soft to progressive breakdown regimes are well within this current limit [4-5]. Also note that a more detailed calibration path could be included in future implementations to take the transmission gate resistances and other circuit parasitics into account during calibration.

As seen in the direct device probing results of Fig. 6, we did not typically observe progressive dielectric breakdown in the CMOS process used here during accelerated measurements with stress voltages of ≥4V, when recording four measurements per second. Therefore, although the proposed design is capable of monitoring progressive breakdowns, we were specifically looking for hard breakdowns in the automated array measurements presented here. These events were defined as a sudden decrease in the scanned out discharge time count of at least an order of magnitude.

Cumulative distribution functions (CDF) of the time to breakdown for a range of stress voltages, both on a standard percentage scale as well as the Weibull scale, are displayed in

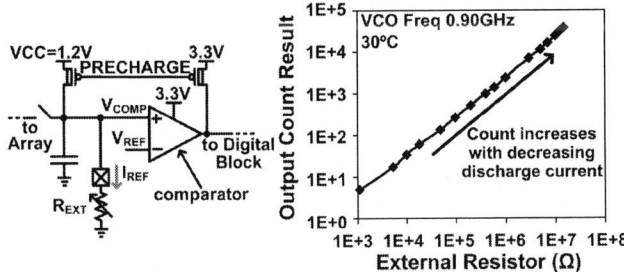

Fig. 5. Calibration setup and measured results.

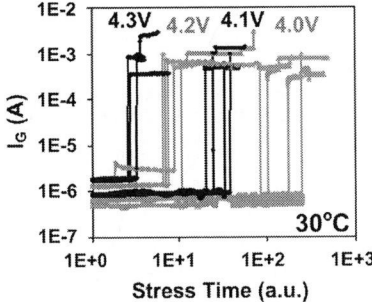

Fig. 6. Discrete device probing results showing hard breakdown behavior.

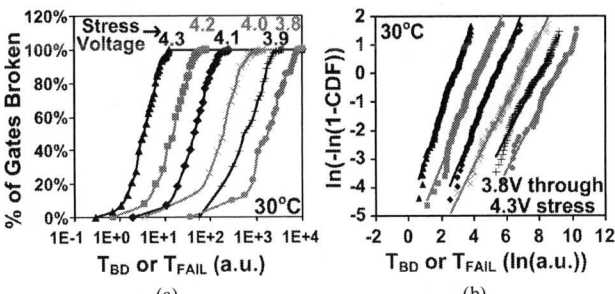

Fig. 7. Measured T_{BD} CDF on (a) a percentage scale and (b) a Weibull scale.

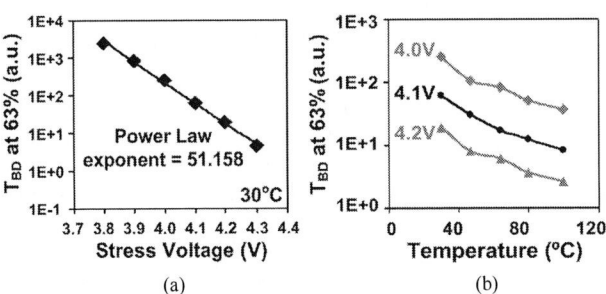

Fig. 8. (a) Voltage acceleration of T_{BD} at the 63% point. (b) TBD at the 63% point vs. temperature.

978-1-4244-2018-6/08 $25.00 © 2008 IEEE

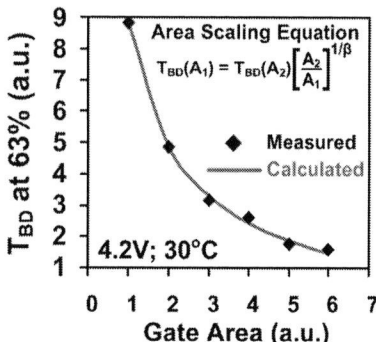

Fig. 9. Area scaling data, computed from measurement results using the weakest link characteristic of TDDB, compared with theoretical results [2].

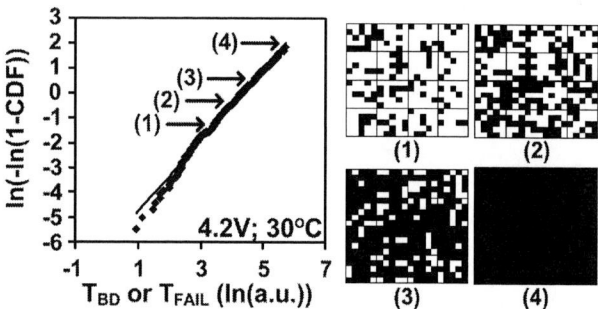

Fig. 10. Spatial T_{BD} distribution in a stress cell array at four time points on the Weibull scale CDF.

Fig. 7. The Weibull slope factor (β) for 4.2V stress was 1.443, with that factor slightly decreasing for lower stress voltages, and increasing at 4.3V. The exponential relationship of the Weibull characteristic life (time at which 63% of the devices have failed) with voltage is illustrated in Fig. 8(a). The power law exponent of this plot is ~51, which is slightly larger than that reported in previous work where the time to the first (soft) breakdown was recorded [8]. The dependency of the time to breakdown on stress temperature is shown in Fig. 8(b) for a range of voltages. Fig. 9 compares measured and calculated area scaling characteristics [2, 8]. The measured numbers were obtained by combining the results for a given number of spatially adjacent DUTs and selecting the smallest T_{BD} from each group, due to the weakest-link character of dielectric breakdown. In Fig. 10, the spatial distribution of T_{BD} in a 20x20 portion of a test array stressed at 4.2V is plotted along with the corresponding Weibull distribution. The four spatial diagrams correspond to the four divisions of the Weibull plot representing 25% of the cells each. As indicated in this figure, no spatial correlation was detected in these experiments. A test chip microphotograph and summary of the chip characteristics is shown in Fig. 11.

V. CONCLUSION

We have presented a circuit design for the efficient characterization of gate dielectric breakdown during process characterization. The proposed design consists of a large array of test cells that facilitate the accelerated stressing of the devices under test and an A/D current monitor that translates

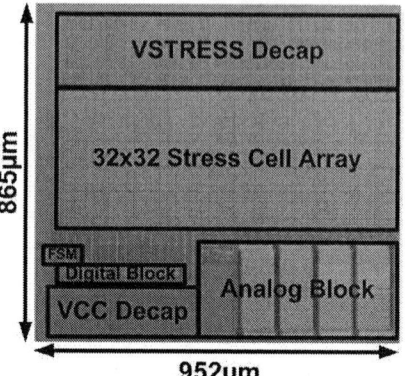

Technology	0.13µm CMOS
Digital Supply	1.2V
Dimensions	952x865µm2
Gate Resisistance Measurement Range	~1kΩ - 10MΩ

Fig. 11. Microphotograph and summary of the test chip characteristics.

the gate current of each measured device into a digital count that is scanned off chip for post processing. A simple automated design such as this could greatly reduce testing times, as up to thousands of samples are needed to correctly define the statistical characteristics of TDDB. A range of test chip measurements from a 32x32 array implemented in a 1.2V, 130nm CMOS process were presented to demonstrate the functionality and flexibility of this design.

ACKNOWLEDGMENTS

The authors would like to thank Intel and IBM for financial support, as well as United Microelectronics Corporation (UMC) for the foundry design kit and chip fabrication.

REFERENCES

[1] R. Degraeve, et al., "New Insights in the Relation Between Electron Trap Generation and the Statistical Properties of Oxide Breakdown," Trans. on Electron Devices, vol. 45, no. 4, pp. 904-911, 1998.

[2] E. Y. Wu, et al., "CMOS scaling beyond the 100-nm node with Silicon-Dioxide-Based Gate Dielectrics," IBM J. of R & D, pp. 287-298, Vol. 46, No. 2/3, 2002.

[3] J. H. Stathis, "Gate Oxide Reliability for Nano-Scale CMOS," Int. Conf. on Microelectonics, pp. 78-83, 2006.

[4] J. Suñé, et al., "Statistics of Competing Post-Breakdown Failure Modes in Ultrathin MOS Devices," Trans. on Electron Devices, vol. 53, no. 2, pp. 224-234, 2006.

[5] A. Kerber, "Lifetime Prediction for CMOS Devices with Ultra Thin Gate Oxides Based on Progressive Breakdown," Int. Reliability Physics Symp., pp. 217-220, 2007.

[6] L. Pang and B. Nikolic, "Impact of Layout on 90nm CMOS Process Parameter Fluctuations," Symp. on VLSI Circuits, pp. 69-70, 2006.

[7] K. Agarwal, et al., "A Test Structure for Characterizing Local Device Mismatches," Symp. on VLSI Circuits, pp. 67-68, 2006.

[8] E. Y. Wu, et al., "Experimental Evidence of T_{BD} Power-Law for Voltage Dependence of Oxide Breakdown in Ultrathin Gate Oxides," Trans. On Electron Devices, vol. 49, no. 12, pp. 2244-2253, 2002.

IEEE 2008 Custom Intergrated Circuits Conference (CICC)

A High Sensitivity Process Variation Sensor Utilizing Sub-threshold Operation

Mesut Meterelliyoz[1*], Peilin Song[2], Franco Stellari[2], Jaydeep P. Kulkarni[1] and Kaushik Roy[1]

[1]School of Electrical and Computer Engineering, Purdue University, West Lafayette, IN
[2]IBM T. J. Watson Research Center, Yorktown Heights, NY

Abstract— In this paper, we propose a novel low-power, bias-free, high-sensitivity process variation sensor for monitoring random variations in the threshold voltage. The proposed sensor design utilizes the exponential current-voltage relationship of sub-threshold operation thereby improving the sensitivity by 2.3X compared to the above-threshold operation. A test-chip containing 128 PMOS and 128 NMOS devices has been fabricated in 65nm bulk CMOS process technology. A total of 28 dies across the wafer have been fully characterized to determine the random threshold voltage variations.

Index Terms— process variations, sensor, sub-threshold operation, random threshold voltage variations

I. INTRODUCTION

The ever increasing need for high performance and increased chip functionality with lower costs has resulted in aggressive scaling of transistor dimensions. As transistor dimensions are scaled down, the effect of parameter variations on circuit robustness has aggravated. Increased process variations result in lower circuit performance and can potentially lead to functional/parametric failures degrading manufacturing yield. Hence it is important to monitor/track the effect of process variations and to tune the process for improving the manufacturing yield [1] [2].

Process variations can be classified into two categories: inter-die and intra-die [2]. Under inter-die variations, we assume that all transistors in a die are affected in a similar way. On the other hand, intra-die variations can be further divided into "random" and "systematic" components based on their spatial correlation. Systematic variations typically have strong spatial correlation which results in similar variations in transistor characteristics for the devices that are close to each other in a die. Due to the spatial correlation, techniques such as adaptive body biasing can be employed in order to compensate systematic intra-die variations and to improve the design yield [3] [4].

In contrast to systematic intra-die variations, random process variations tend to have no spatial correlation which may result in different transistor characteristics even for neighboring devices. Line Edge Roughness (LER) and Random Dopant Fluctuations (RDF) are the main cause of random process variations. The impact of random variations is severe for SRAM bitcells where minimum geometry transistors

are often used. The increased random variations can affect the SRAM bitcell stability leading to increased memory failures.

The random process variations can result in large fluctuations in circuit parameters, mainly in the threshold voltage (V_{th}) of devices. Accurate characterization, measurement, and estimation of random V_{th} fluctuations in closely spaced devices are crucial for yield learning, enhancement and process optimization. Traditionally, this characterization was done by determining the I-V curves of single neighboring devices. However, this is not practical for a large number of devices since separate pad would be required for individual transistor characterization. Recently, researchers have proposed techniques in which multiplexed devices under test (DUT) are characterized [5-8]. However, these techniques require accurate current measurements followed by extensive data analysis. Another technique utilizes the ring oscillators for measuring variability statistics; however, the large area required in this technique reduces its applicability [9]. Other techniques such as [10] require variation independent bias circuits which introduce large area overhead. In [11] a sense-amplifier based test circuit and measurement method is presented to measure the mismatch between pairs of transistors. This technique has limited use since it can only measure the mismatch between two transistors. Finally, in [12], an on-chip digital circuit is presented to measure the local threshold voltage variations in identical devices. In this technique, the devices under test operate in the saturation region. Since the saturation current of a transistor varies linearly with the threshold voltage, the technique proposed in [12] has low sensitivity (0.74V/V).

In this work, we propose a novel low-power, bias-free, high-sensitivity process variation sensor (PVS) for characterizing the random threshold voltage variation by utilizing the sub-threshold mode operation. Exponential current-voltage relationship in sub-threshold region increases the sensitivity of the DUT current on threshold voltage variations, leading to high sensitivity process variation sensor design. Moreover, since the proposed PVS does not require any external or on-chip bias/bias generation circuitry, it consumes small area. Hence multiple such sensors can be efficiently used across the die for accurate process monitoring.

The rest of the paper is organized as follows. Section II describes the proposed process variation sensor utilizing the sub-threshold operation. The analysis of the proposed PVS is presented in Section III. Section IV presents the test methodology. Results are given in Section V. Finally, Section VI draws the conclusions.

---

* This work was done while Mesut Meterelliyoz was working as an intern at IBM T. J. Watson Research Center during summer 2006.

978-1-4244-2018-6/08 $25.00 © 2008 IEEE

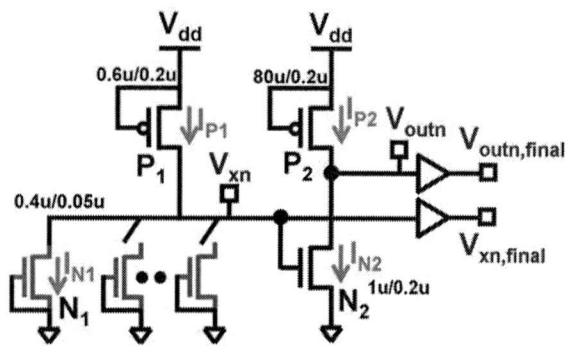

Fig. 1 Process Variation Sensor for NMOS

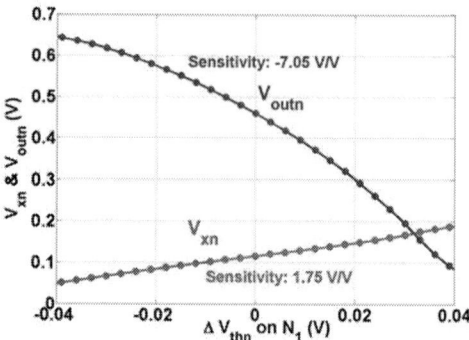

Fig. 2 Sensitivity of NMOS process variation sensor

II. PROPOSED PROCESS VARIATION SENSOR

The proposed process variation sensor designed to characterize the random threshold voltage mismatch between closely space transistors consists of four transistors (selected N_1, P_1, N_2, P_2), all operating in sub-threshold region as shown in Fig. 1. Transistor N_1 is the Device Under Test (DUT). By choosing one of parallel N_1 transistors in Fig. 1, V_{xn} is set depending on the threshold voltage of the active N_1 transistor. When a different transistor is selected from the N_1 set of parallel transistors, V_{xn} varies depending on the threshold voltage of the new DUT. Hence, by measuring V_{xn}, the V_{th} mismatch between all parallel devices can be measured. The second stage in the sensor (N_2 and P_2) is used to improve the sensitivity of the sensor. As the voltage at V_{xn} changes depending on the threshold voltage of the active N1 transistor, the conductivity of N_2 is modified. Since N_2 is in sub-threshold region ($V_{xn} < V_{th}$), a small change in V_{xn} causes a exponential change in the current through N_2. This results in a large swing in V_{out} and hence improves the sensitivity (i.e. $\Delta V_{out}/\Delta V_{th}$).

In the proposed sensor, P_1, N_2 and P_2 are shared for all parallel N_1 transistors, thus, the changes at V_{xn} and V_{out} are solely because of the variations in N_1. Any variations in P_1, N_2 and P_2 would result in offsetting V_{xn} (or V_{out}) node voltage. However, for determining ΔV_{th}, the difference in V_{xn} values is more important rather than the absolute value of V_{xn}. In our design, the size of N_1 is kept small (small width, minimum length) in order to observe process variations. Other transistors (P_1 P_2 and N_2) are sized reasonably large (large width and length) to reduce the variations. More specifically, P_1 is sized such that even when V_{xn} reaches its maximum value for high-V_{th} N_1 transistors, N_2 will still be in the sub-threshold region. Finally, N_2 and P_2 are sized for a large and linear swing at V_{out}. The transistor sizes for NMOS sensor are shown in Fig. 1. Similarly, complementary transistors and circuit topology are used to realize PMOS process variation sensor.

The simulation results for the NMOS sensor at the 65nm bulk CMOS technology are shown in Fig. 2. In our simulations, the threshold voltage of N_1 in Fig. 1 is varied between +/- 40mV and V_{xn} and V_{out} are plotted. It is observed that the sensitivity of NMOS sensor at V_{xn} is around 1.75V/V. Furthermore, the sensitivity can be increased to 7.05V/V with acceptable linearity using the second stage. The corresponding V_{xp} and V_{outp} sensitivities for PMOS sensor are found to be 1.35V/V and 6.12V/V, respectively.

The sensitivity of the proposed process variation sensor utilizing sub-threshold operation is 2.3X better compared to the above-threshold operation as described in [12]. Thus exponential current-voltage relationship in sub-threshold region gives better accuracy for random threshold voltage mismatch measurement.

III. ANALYSIS OF THE PROPOSED SENSOR

The operation of the proposed sensor depends on the change in sub-threshold current as the threshold voltage of DUT changes. However, for the sub-threshold region circuit design, it is very imperative to consider various other leakage components as well. In nano-scaled bulk CMOS technology, the main components of leakage are (i) sub-threshold current, (ii) gate leakage, (iii) junction band-to-band tunneling leakage (BTBT) and (iv) gate induced drain leakage (GIDL) [13]. Sub-threshold leakage and gate leakage are the dominant leakage components for the technology under consideration under the given bias conditions.

Sub-threshold leakage component can be written as [14]:

$$I_{sub} = A \times e^{(V_{gs} - V_{th0} - \gamma V_{sb} + \eta V_{ds})/mv_T} \times (1 - e^{(-V_{ds}/v_T)}) \quad (1)$$

where

$$A = \mu_0 C_{ox} \frac{W}{L_{eff}} v_T^2 e^{1.8} \quad (2)$$

In this equation, μ_0 is the zero bias carrier mobility, C_{ox} is the gate oxide capacitance, L_{eff} is the effective channel length, W is the transistor width, η is the drain induced barrier lowering coefficient, γ is the linearized body effect coefficient, m is the sub-threshold swing coefficient of the transistor and v_T is the thermal voltage given by KT/q.

The gate leakage component can be expressed as [15]:

$$J_{DT} = A(V_{ox}/T_{ox})^2 \exp\left[\frac{-B(1 - (1 - V_{ox}/\phi_{ox})^{3/2})}{V_{ox}/T_{ox}}\right] \quad (3)$$

where J_{DT} is the gate leakage current density, V_{ox} is the potential drop across the oxide, T_{ox} is the oxide thickness and Φ_{ox} is the barrier height for the tunneling particle (electron or hole).

In the proposed design, the sensing mechanism depends on sub-threshold leakage; hence gate leakage of P_1 in Fig. 1 might decrease the sensitivity. As given in Eq. 3, gate leakage in this transistor can be reduced exponentially by employing thick oxide transistors. Similarly, leakage on the non-active parallel DUTs can degrade the sensitivity.

978-1-4244-2018-6/08 $25.00 © 2008 IEEE

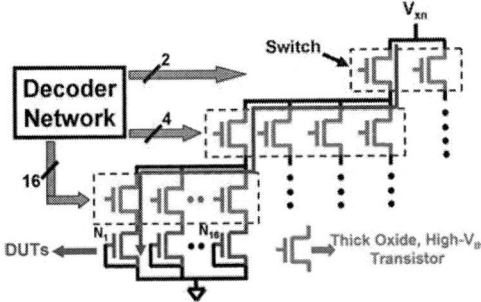

Fig. 3 Hierarchical switch network

To reduce sub-threshold leakage on non-active paths, a hierarchical switch network which utilizes thick oxide, high V_{th} transistors is employed (Fig. 3). In Fig. 3, assuming N_1 is selected, the non-active switch transistors will have negative V_{gs} due to "stacking effect" [13]. Combined with high threshold voltage, negligible sub-threshold leakage is achieved on the non-active paths. Similarly, gate leakage in switch network is minimized using thick oxide devices. As a result, large numbers of DUTs can be tested without degrading the sensitivity.

Based on the discussion above, sensitivity of the sensor can be estimated by solving Eq. 1 for I_{P1} and I_{N1} (sub-threshold leakage of P_1 and N_1, respectively) in Fig. 1. In Eq. 1, as $V_{ds} \gg V_T$ the term $(1- e^{-(V_{ds}/V_T)})$ can be ignored. Similarly body bias voltage $V_{SB} = 0$. Hence γ. We can write:

$$I_{P1} = I_{N1}$$
$$A_{P1} \times e^{(-|V_{th0P1}| + \eta_{P1}(V_{dd} - V_{xn}))/m_{P1}v_T} = A_{N1} \times e^{(-V_{th0N1} + \eta_{N1}V_{xn})/m_{N1}v_T} \quad (4)$$

Now, V_{xn} can be given by:

$$V_{xn} = \frac{(m_{N1}\eta_{P1}V_{DD} + m_{P1}V_{th0N1} - m_{N1}m_{P1}v_T \ln(\frac{A_{N1}}{A_{P1}}) - m_{N1}|V_{th0P1}|)}{m_{N1}\eta_{P1} + m_{P1}\eta_{N1}} \quad (5)$$

Finally, the sensitivity of V_{xn} on V_{th0N1} variations can be written as:

$$\frac{\partial V_{xn}}{\partial V_{th0N1}} = \frac{m_{P1}}{m_{N1}\eta_{P1} + m_{P1}\eta_{N1}} \quad (6)$$

From Eq. 6, the sensitivity of the proposed sensor depends only on the DIBL coefficients and the sub-threshold swing coefficients of P_1 and N_1. As DIBL coefficient can be controlled by modifying the channel length of devices, the sensitivity of the sensor can be adjusted.

As mentioned earlier, the second stage of the sensor is used to amplify the sensitivity. Following a similar analysis for the second stage, and noting that;

$$\frac{V_{outn}}{V_{th0N1}} = \frac{\partial V_{outn}}{\partial V_{xn}} \times \frac{\partial V_{xn}}{\partial V_{th0N1}} \quad (7)$$

The final sensitivity of the sensor can be written as:

$$\frac{\partial V_{outn}}{\partial V_{th0N1}} = \frac{m_{P1}}{m_{N1}\eta_{P1} + m_{P1}\eta_{N1}} \times \frac{m_{P2}}{m_{N2}\eta_{P2} + m_{P2}\eta_{N2}} \quad (8)$$

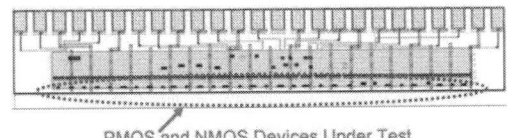

Fig. 4 Layout of 65nm bulk CMOS node test-chip

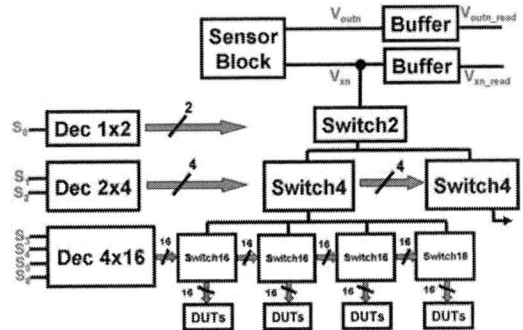

Fig. 5 Block diagram of test chip

In Eq. 8, increasing the channel lengths of P_2 and N_2 to minimize variations on these transistors reduces η_{P2} and η_{N2} which in turn increases the sensitivity as shown in Fig. 2. A similar analysis is performed for the PMOS PVS sensor.

IV. TEST METHODOLOGY

A test-chip containing the proposed sensor design with 128 devices with varying device dimensions has been fabricated in 65nm bulk-CMOS process technology. The layout and the block diagram of the test chip are shown in Fig. 4 and Fig. 5, respectively. 128 NMOS and 128 PMOS devices are closely placed and tested at a supply voltage of 1V. Half of NMOS transistors are sized with 2X width to observe the effects of sizing on process variations. Fig. 5 shows the test methodology for testing of NMOS devices (half of DUTs are shown). To test PMOS devices, high-V_{th}, thick oxide PMOS transistors are used in the switch network. By sharing the decoders, PMOS and NMOS DUTs are tested in parallel. All measurements are done at DC conditions.

As shown in Fig. 5, seven control inputs (S_0-S_6) are used to select one DUT at a time. Nodes V_{xn} and V_{outn} are buffered and final outputs V_{xn_read} and V_{outn_read} are measured for each DUT. The output buffers are characterized and their input-output transfer function is obtained. This transfer function is used to determine the actual V_{xn} and V_{outn} from V_{xn_read} and V_{outn_read} measurements. A similar methodology is followed for the PMOS case.

V. RESULTS

The histogram of the measured V_{xn} and V_{outn} are shown in Fig. 6 for a single die (64 DUTs). In this figure, the spread in V_{xn} and V_{outn} represents the variation in threshold voltage of the DUTs. For this die, the V_{xn} variation is 200mV demonstrating high sensitivity. This sensitivity is further increased at V_{outn} which shows a spread of 560mV.

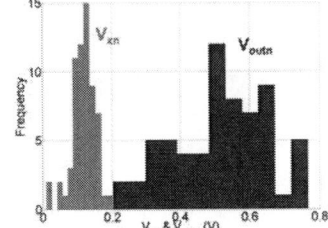

Fig. 6 Histogram of V_{xn} and V_{outn} for 64 devices

978-1-4244-2018-6/08 $25.00 © 2008 IEEE

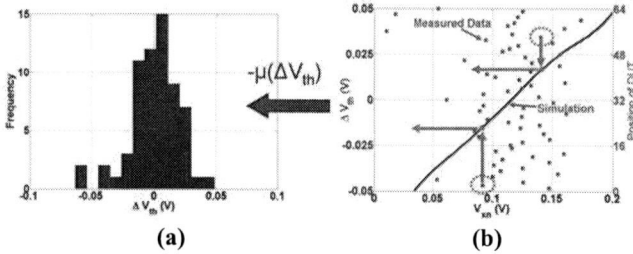

(a)   (b)

Fig. 7a Histogram of ΔV_{th} for one die
Fig. 7b Conversion from V_{xn} (measured) to ΔV_{th}

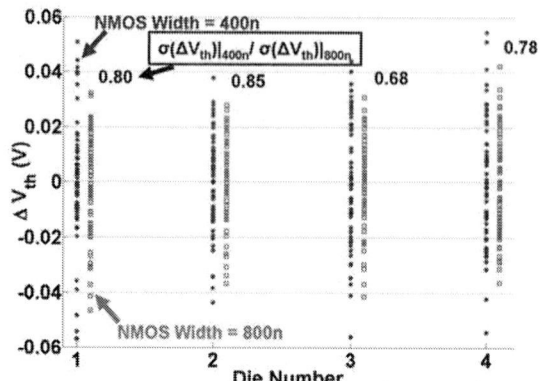

Fig. 9 ΔV_{th} distribution for different transistor widths

Next, the variations at V_{xn} (or V_{outn}) are converted into ΔV_{th} using simulations. In Fig. 7b, both simulation results and measured data are superimposed. The dots show the measured V_{xn} values for each device under test. To identify the corresponding ΔV_{th}, the measured V_{xn} values are mapped on to simulation curve (solid line in Fig. 7b) as illustrated for two DUTs. The histogram of the resulting ΔV_{th} population is shown in Fig. 7a. In order to remove the inter-die variations, the mean of ΔV_{th} is subtracted from each data point. Using the same methodology, ΔV_{th} statistics of all dies are collected. Fig. 8 shows the normalized $\sigma(\Delta V_{th})$ values for NMOS and PMOS devices. Average of all measured $\sigma(\Delta V_{th})$ is used for normalization.

To observe the effect of sizing on process variations, ΔV_{th} population for two different NMOS sizes are compared for four different dies as shown in Fig. 9. Measured data shows that process variations can be reduced with increased device widths. For each die, the ratio $\sigma(\Delta V_{th})|_{400n}/\ \sigma(\Delta V_{th})|_{800n}$ is calculated as shown in Fig. 9. It is observed that, from 400nm to 800nm case, σ value tends to follow the inverse proportionality to the device width ($\sigma(\Delta V_{th}) \propto W^{-1/2}$) [15]. The average reduction in $\sigma(\Delta V_{th})$ with doubling the device width is measured to be 0.77.

VI. CONCLUSIONS

In this work, we propose a novel process variation sensor to monitor random variations in the threshold voltage by effectively utilizing sub-threshold operation. Exponential I-V relationship in sub-threshold region achieves 2.3X better sensitivity compared to the above-threshold operation. Measurement results on 28 test-chips fabricated in 65nm bulk CMOS technology demonstrate the effectiveness of the proposed technique.

VII. ACKNOWLEDGEMENTS

We thank Brian Ji and Rahul Rao from IBM for helpful discussions. Jaydeep P. Kulkarni and Kaushik Roy were partially funded by Boeing.

REFERENCES

[1] S. Borkar et al., "Parameter Variations and Impact on Circuits and Microarchitecture," *Design Automation Conference,* pp.338-342, June, 2003.

[2] K.A. Bowman et al., "Impact of Die-toDie and Within-Die Parameter Fluctuations on the Maximum Clock Frequency Distribution for Gigascale Integration," *IEEE Journal of Solid-State Circuits*, pp. 183-190, 2002.

[3] J. W. Tschanz et al., "Adaptive body bias for reducing impacts of die-to-die and within-die parameter variations on microprocessor frequency and leakage," *IEEE Journal of Solid-State Circuits*, pp. 1396-1402, Nov, 2002.

[4] N. Azizi et al., "Compensation for within-die variations in dynamic logic by using body-bias," *IEEE-NEWCAS*, pp. 167-170, June, 2005.

[5] A. Keshavarzi et al., "Measurement and Modeling of Intrinsic Fluctuations in MOSFET Threshold Voltage," *ISLPED*, pp. 26-29, Aug. 2005

[6] H. Klimach et al., "Characterization of MOS transistor current mismatch," *Symp. Int. Circuits Sys. Design*, pp. 33-38, Sep. 2004.

[7] K. Agarwal et al., "A Test Structure for Characterizing Local Device Mismatches," *Symp. VLSI Circuits*, pp. 67-68, Jun, 2006.

[8] N. Drego et al., "A Test Structure to Efficiently Study Threshold-Voltage Variation in Large MOSFET Arrays," *ISQED*, pp. 281-286, March 2007.

[9] M. Bhushan et al., "Ring Oscillator Based Technique for Measuring Variability Statistics," *ICMTS*, pp. 87-92, March 2006.

[10] C. H. Kim, et al., "Self Calibrating Circuit Design for Variation Tolerant VLSI Systems", *IOLTS*, pp. 100-105, July 2005.

[11] S. Mukhopadhyay et al., "Statistical Characterization and On-Chip Measurement Methods for Local Random Variability of a Process Using Sense-Amplifier-Based Test Structure," *ISSCC*, pp. 400-401, Feb. 2007.

[12] R. Rao et al., "A Completely Digital On-Chip Circuit for Local-Random-Variability Measurement", *ISSCC*, pp. 412-413, Feb. 2008.

[13] K. Roy, et al.,"Leakage Current Mechanisms and Leakage Reduction Techniques in Deep-Submicron CMOS Circuits," *Proceedings of IEEE*, vol.91, No.2, Feb, 2003.

[14] A. Chandrakasan, et al., *Design of High-Performance Microprocessor Circuits*, Wiley-IEEE Press, 2000.

[15] Y. Taur, et al., "Fundamentals of Modern VLSI Devices," *Cambridge University Press*, 1998.

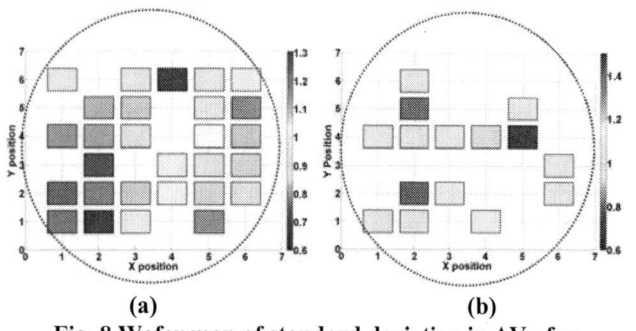

(a)   (b)

Fig. 8 Wafer map of standard deviation in ΔV_{th} for (a) NMOS & (b) PMOS

Measurement and Analysis of Variability in 45nm Strained-Si CMOS Technology

Liang-Teck Pang, Borivoje Nikolić

Electrical Engineering and Computer Sciences, University of California, Berkeley

Abstract-A test-chip in a low-power 45nm technology, featuring uniaxial strained Si, has been built to study variability in CMOS circuits. Systematic layout-induced variation, die-to-die (D2D), wafer-to-wafer (W2W) and within-die (WID) variability has been measured and analyzed. Delay is characterized using an array of ring-oscillators and transistor leakage current is measured with an on-chip ADC. Results show that systematic variations are small and layout-induced variation is dominated by strain effects.

I. INTRODUCTION

Scaling of deep-submicron CMOS devices has increased the impact of process variability on circuits, to the point where it is likely to be the main technological barrier to further scaling. Shrinking of critical dimensions has been possible due to advances in lithography. Resolution enhancement techniques such as optical proximity correction (OPC), phase shift masks, double patterning and immersion lithography have contributed to the ability to scale to the present 45nm node and beyond. In addition, performance enhancing techniques such as the use of strained-Si have been introduced. These techniques have, however contributed to an already complex manufacturing process and compounded the sources of process variability. To mitigate systematic poly-Si density effects on devices, restricted design rules and more complex OPC have been introduced.

In this work, we characterize variability in a modern 45nm process and compare it with a previous 90nm process [1]. We also evaluate the nature of systematic variations in the presence of OPC and restricted layout design rules.

II. LOW-POWER 45NM STRAINED-SI PROCESS

Table 1 summarizes several techniques used in the low-power 45nm process [2, 3, 4]. Transistor channels are oriented in the <100> direction, which increases PMOS transistor mobility and makes it insensitive to stress. The use of sub-atmospheric chemical vapor deposition oxide (SACVD) for trench isolation further reduces stress effects. Instead of a strong compressive strain, these trenches now exert a weak tensile strain on the transistors. Strong uniaxial tensile strain is created by the nitride layer in order to increase NMOS mobility. Resolution is enhanced with immersion lithography and low-k dielectric is used for the copper interconnects. Strain induced by the shallow trench isolation (STI) and the contact etch stop layer (CESL) nitride are layout dependent and are investigated in the test chip.

III. TEST CHIP

To characterize variability and evaluate the impact of layout-induced variations in the process, a 45nm test-chip has been designed that contains an array of structures with different layout styles.

TABLE 1 Summary of the 45nm process. [2,3,4]

PROCESS FEATURE	45NM PROCESS	EFFECT
Si substrate	<100>-oriented channel	Higher PMOS mobility
Shallow trench isolation (STI)	Sub-atmospheric deposited oxide	Reduce STI stress
Contact etch stop layer (CESL)	Nitride layer create high tensile strain	Higher NMOS mobility
Immersion	NA < 1	Improved resolution
Backend dielectric	Low k ~2.5	Low RC delay

Sixteen layout styles were created to study the effects of layout. These are shown in Figure 1 and described in Table 2. Layouts P1, P2, P3, P4 vary the spacing of the poly-Si nearest to the transistor's gate. SP1, SP2, SP3 vary the distance of the poly-Si that is the second nearest neighbor to the gate. S1, S2 are layouts that are symmetrical to each other. D1 has a longer source/drain (S/D) diffusion area than P3, which has been observed to cause different strain in a transistor [5]. M1 has metal-2 coverage over its gate which has been shown to cause different annealing temperatures [6] and T1 has neighboring poly at the ends of its gate. R1, R2 and R3 have regular poly pitches that vary from minimum to maximum. NSTI is the same as R1 except that there is no STI and isolation is achieved by turning off the adjacent transistors [7]. In addition, in layout V1 the inverters of a RO are placed in the vertical direction instead of the horizontal direction as illustrated in Figure 2.

The die photo of the chip is shown in Figure 3. The array contains 18 x 16 tiles, each tile contains 17 ROs and 17 NMOS and PMOS transistors with $V_{GS}=0$ in each of the 17 layouts. The measurement circuits in this chip have been adapted from [1]. In the RO array, row and column bits from a scan-chain enable the RO of interest and select the multiplexer to output its RO frequency which is multiplexed out to a row divider and further divided down before being output to a pad. A local divide-by-2 circuit within each RO allows for the use of small number of stages (13) by reducing the frequency of the signal that is multiplexed out.

PMOS (I_{LEAKP}) and NMOS (I_{LEAKN}) leakage currents are measured using current integrators. I_{LEAKP} measurement is described in Figure 4. Switches P1, P2 and P1b choose the currents of interest to integrate. During integration, the output (*Vout*) of the op-amp will ramp down. As it passes the threshold voltages of on-chip comparators, start and stop signals are generated. By timing the interval between these signals the currents are measured. In this chip, 1.8V thick-oxide transistors are used for all analog components.

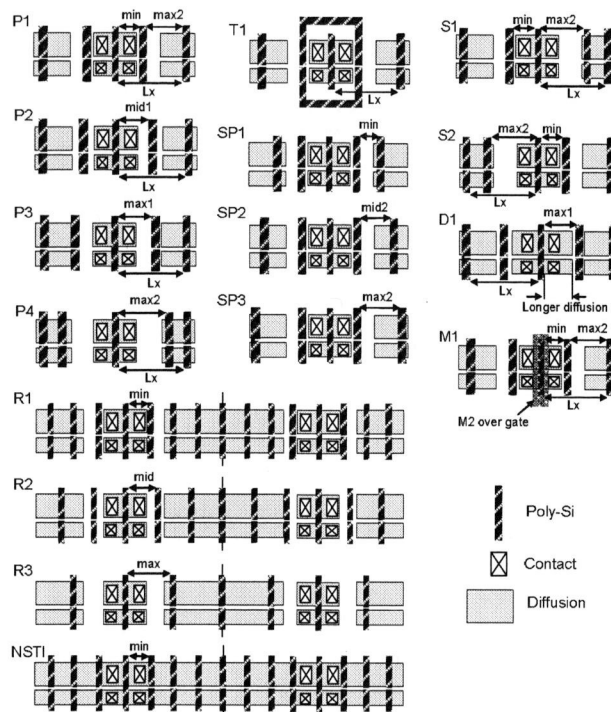

Fig. 1. Layout configurations.

TABLE 2 Characteristics of the layout configurations.

LAYOUTS	TARGETED EFFECT
P1,P2,P3,P4	Primary proximity
SP1,SP2,SP3	Secondary proximity
S1,S2	Symmetry
D1	Larger S/D area
M1	Metal coverage over gate
T1	Poly-Si at extremity of gate
R1,R2,R3	Regular pitch, different density
NSTI	No STI

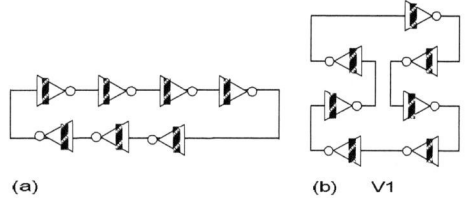

Fig. 2. Illustrates (a)horizontally vs (b)vertically placed inverters.

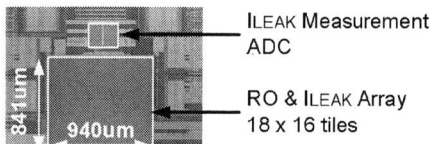

Fig. 3. 45nm test-chip die photo.

IV. MEASUREMENT RESULTS

Measured data shows several important trends. Systematic layout-induced variations, in particular those related to poly-Si density, are significantly reduced in this early 45nm process compared to the early 90nm process. Variations due to strain and other factors are now dominant. Finally, random WID

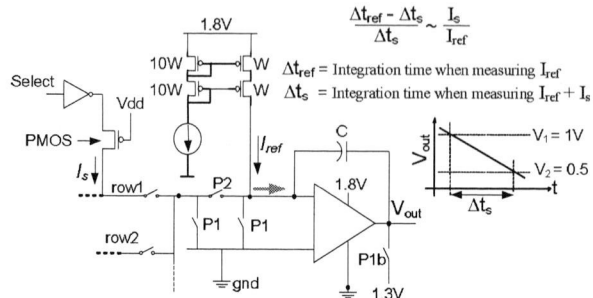

Fig. 4. PMOS leakage current measurement.

TABLE 3 Comparison of 45nm and 90nm technology results.

	90nm	45nm
D2D ($3*\sigma/\mu$)	15%	8.5%
WID ($3*\sigma/\mu$)	3%	6.6%
WID Spatial Correlation	Significant	Insignificant
Proximity Effects (ΔF)	10%	2%

variation has increased proportionally to transistor area reduction, while systematic D2D variation has decreased. Table 3 compares the results of this work with the previous 90nm test-chip [1].

A. W2W, D2D, WID Variations

Figure 5 plots the RO frequency distribution for 4 layouts. Each colored histogram represents the distribution for a die and the overall distribution is plotted as a continuous curve. Measured results of 22 dies from two wafers are plotted. Only the histograms of the fastest and slowest die from each wafer are shown. Vertical lines labeled SS and TT represent simulation results from the extracted layout for SS and TT corners. For each layout, there is a shift of ~2.5% in frequency between the mean of the 2 wafers. D2D 3 × standard-deviation/mean ($3*\sigma/\mu$) is shown in Figure 6. It varies from 6.5% to 7.5% for the 2 wafers and overall D2D variation was 8.5% for all the layouts except for the vertically placed layout V1. In the 90nm test-chip, D2D variation for 36 dies on 1 wafer was 15% in a radial slope.

Figure 7 plots the WID mean and σ/μ of the RO frequency for each of the 22 dies as a function of the layout configuration. The mean frequency has been normalized to the SS corner in order to compare differences in layouts that have not been captured by the layout extraction. WID variation is approximately 2.2% which is more than twice that in 90nm ($\sigma/\mu \sim 1\%$). This is consistent with the transistor area reduction by a factor of 4 between the two processes.

Systematic variations in the mask are investigated by normalizing the data of each chip to zero mean and unity standard deviation and averaging the normalized data from 22 chips to remove random variations. Figures 8a and 8b plot the mean frequency for each column and each row respectively. No significant systematic variation was found, partially due to the fact that random WID variation has increased. Finally, there is no significant spatial correlation in the measurements.

B. Layout Effects

Measurement results show systematic variations for different layouts that are not captured by the layout extraction

978-1-4244-2018-6/08 $25.00 © 2008 IEEE

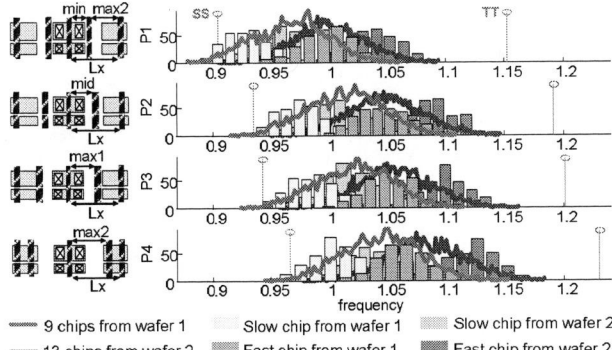

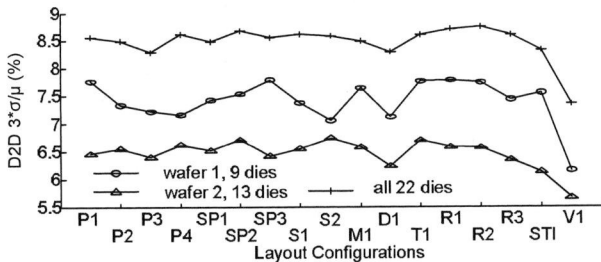

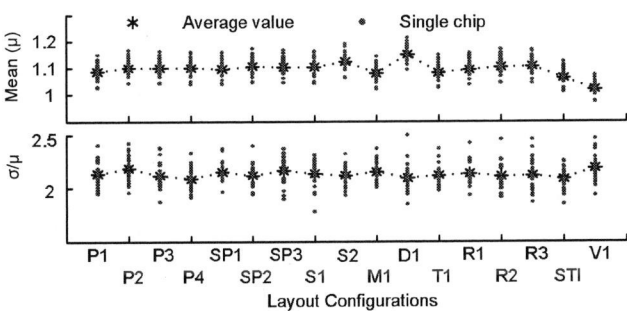

| ‑‑‑‑‑ 9 chips from wafer 1 | Slow chip from wafer 1 | Slow chip from wafer 2 |
| ——— 13 chips from wafer 2 | Fast chip from wafer 1 | Fast chip from wafer 2 |

Fig. 5. RO frequency distribution for 4 layouts showing W2W, D2D and WID variations. Vertical red lines represent as layout extracted simulation results for SS and TT corners.

Fig. 6. Comparison of RO frequency D2D variations for 2 wafers and the overall D2D variations.

Fig. 7. 45nm WID statistics of RO frequency for 22 dies. The frequency is normalized to the SS corner frequency.

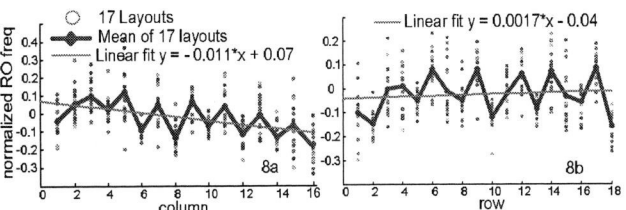

Fig. 8. Plot of RO frequency normalized to zero mean and unit variance showing mask characteristics. Small systematic variation of ~ 0.2 x σ

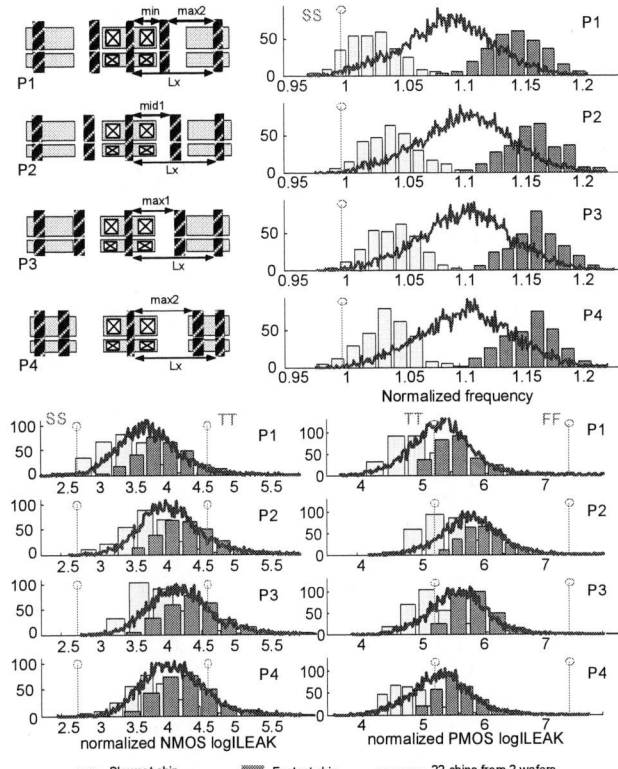

| Slowest chip | Fastest chip | —— 22 chips from 2 wafers |

Fig. 9. Effect of nearest neighbour poly-Si pitch on RO frequency and transistor leakage currents. Plots of RO frequency distribution, NMOS log(ILEAK) distribution and PMOS log(ILEAK) distribution for 22 dies.

tool. The explanations that we propose for the observations are plausible but by no means exact.

In the following analysis, RO frequencies are normalized to the SS corners in order to remove the differences in parasitics that are captured by the layout extraction. Leakage currents are not aligned with the corners as they are independent of parasitics. Distributions of normalized frequency and normalized leakage in the log domain are plotted in figures 9, 10 and 11. Measured results show that the impact of layout on performance is small. 2% shift in frequency ($\Delta F = 2\%$) exists due to proximity of poly-Si. The most significant effects were due to a larger S/D area ($\Delta F = 5\%$) and the removal of STI ($\Delta F = 3\%$).

Figure 9 shows the distributions for 4 layouts with different poly-Si gate pitches. Maximum systematic shift in frequency is ~ 2%. Leakage currents also experience small systematic shifts. This effect could be due to small gate-length

(L) variations and layout dependent variation of the strain caused by the CESL. PMOS transistors have a sharper V_T roll-off and hence PMOS leakage currents are more sensitive to L variation. An isolated gate will generally experience more strain from the CESL. In the plot of PMOS leakage, poly density induced L variation is observed. The P2 layout on the second row likely has a shorter gate length than the others resulting in increased PMOS leakage. The effect on NMOS leakage is weaker and is completely compensated by the CESL stress. As the poly-Si pitch increases, more tensile stress is applied, increasing the mobility of NMOS transistors and raising the amount of NMOS leakage thereby offsetting the effect of increased gate length. PMOS leakage is not affected since it is insensitive to stress in a <100>-oriented channel. At the same time, RO frequency also increases and this offsets the effect of a longer L.

Figure 10 studies the impact of a longer S/D diffusion. Differences in S/D capacitance are captured by the layout

978-1-4244-2018-6/08 $25.00 © 2008 IEEE

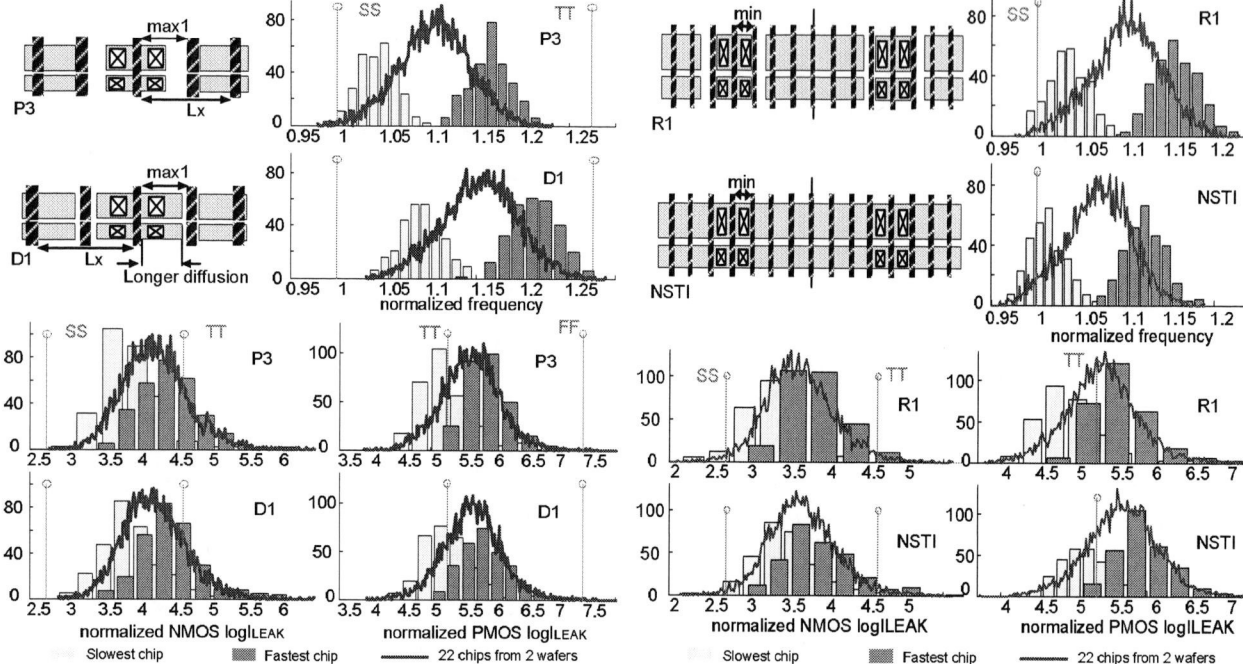

Fig. 10. Effect of a longer S/D diffusion on RO frequency and transistor leakage currents.

Fig. 11. Effect of STI on RO frequency and transistor leakage currents. NSTI uses gate isolation [7] instead of STI.

extraction. After normalization, the layout with larger S/D area is 5% faster, while the leakage currents remain approximately unchanged. This can be explained by the fact that a larger S/D area will allow the CESL to exert more tensile strain on the transistors, thereby increasing the mobility of the NMOS transistors and resulting in faster RO. Since mobility varies linearly with leakage current, its effect on leakage current in the log scale is small.

Figure 11 compares the layouts with and without STI. The layout without STI is slower by around 3% and has higher PMOS leakage current likely due to the effect of L variation on PMOS leakage. As this process uses SACVD trench oxide that generates a low tensile strain, STI stress increases the mobility of NMOS transistors and causes the layout with STI to be faster, thereby compensating the effect of L. Variation in L could be due to the STI step that causes unevenness on the surface of the chip.

V. Conclusion

Restricted design rules reduce the layout induced performance variations by limiting the variations in poly density. We have measured only 2% frequency shifts due to proximity effects. However, other layout-induced variations due to CESL and STI stress have become more significant and are added on to the total variation.

Measurements show a trend towards less systematic but more random variations. From 90nm to 45nm, random WID variation has more than doubled whereas systematic D2D variation has decreased.

Acknowledgments

The authors wish to acknowledge the contributions of the students, faculty and sponsors of the Berkeley Wireless

Research Center, the National Science Foundation Infrastructure Grant No. 0403427, wafer fabrication donation of STMicroelectronics, and the support of the Center for Circuit & System Solutions (C2S2) Focus Center, one of five research centers funded under the Focus Center Research Program, a Semiconductor Research Corporation program. The authors would also like to thank Ernesto Perea, Richard Ferrant and the team of engineers in STMicroelectronics for their support; Professors Andrew Neureuther, Tsu-Jae King, Costas Spanos, and students Andrew Carlson, Zheng Guo for helpful discussions; students Jason Tsai, Kenneth Duong and Emmanuel Adeagbo for their help with layout and data collection.

References

[1] L.T. Pang, B. Nikolić, "Impact of layout on 90nm process parameter fluctuations," VLSI Circ. Dig., 2006, pp 69-70

[2] E. Josse et al, "A Cost-Effective Low Power Platform for the 45-nm Technology Node," International Electron Devices Meeting, IEDM 2006, pp 1-4

[3] C. Le Cam et al., "A Low Cost Drive Current Enhancement Technique Using Shallow Trench Isolation Induced Stress for 45-nm Node," VLSI Tech. Dig., 2006, pp 82-83

[4] B. Le Gratiet et al, "Process Control for 45nm CMOS logic Gate Patterning," Proceedings SPIE, Volume 6922, March 2008

[5] R. A. Bianchi, G. Bouche, O. Roux-dit-Buisson, "Accurate Modeling of Trench Isolation Induced Mechanical Stress Effects on MOSFET Electrical Performance ," International Electron Devices Meeting, IEDM 2002, pp 117-120

[6] H. Tuinhout, M. Pelgrom, R. Penning de Vries and M. Vertregt, "Effects of Metal Coverage on MOSFET Matching," International Electron Devices Meeting, IEDM 1996, pp 735-738

[7] I. Ohkura et al, "Gate isolation – A novel basic cell configuration for CMOS gate arrays," Proc. IEEE 1982 CICC, May 1982, pp 307-310

IEEE 2008 Custom Intergrated Circuits Conference (CICC)

Within-Die Gate Delay Variability Measurement using Re-configurable Ring Oscillator

Bishnu Prasad Das[1], Bharadwaj Amrutur[1], H.S. Jamadagni[1], N.V. Arvind[2], V. Visvanathan[2]

1 Indian Institute of Science, Bangalore,

2 Texas Instruments, Bangalore

bpdas@cedt.iisc.ernet.in, amrutur@ece.iisc.ernet.in, hsjam@cedt.iisc.ernet.in, aravind@ti.com, vish@ti.com.

Abstract—We report a circuit technique to measure the on-chip delay of an individual logic gate (both inverting and non-inverting) in its unmodified form using digitally reconfigurable ring oscillator (RO). Solving a system of linear equations with different configuration setting of the RO gives delay of an individual gate. Experimental results from a test chip in 65nm process node show the feasibility of measuring the delay of an individual inverter to within 1pS accuracy. Delay measurements of different nominally identical inverters in close physical proximity show variations of up to 26% indicating the large impact of local or within-die variations.

I. INTRODUCTION

A typical design today involves a high level of integration along with high data transmission rate in the order of gigabytes per second which necessitates a much more accurate timing analysis of digital system. Timing analysis using corner models lead to over-design. Instead, statistical models are being developed to give realistic picture of timing analysis. To validate statistical models, we need test circuit to measure the process variation effects [1]. In advanced process nodes, other effects like poly-pitch effect, stress effect and neighborhood effect need accurate characterization. These requirements motivated us to develop techniques to measure delay of the individual standard cells inside the chip.

The source of process variations can be divided into the following two categories: 1) inter-die or global or die-to-die variations; and 2) intra-die or local or within-die variations.

In the recent past, attempts were made by several researchers [2] [3] towards the measurement of random local variability. Besides measuring the device parameters like threshold voltage, oxide thickness, length etc, we would like to directly measure the delay of individual standard cells, as that is the parameter of interest in the statistical timing analysis tool.

Picosecond Imaging Circuit Analysis (PICA) [4] is used to obtain quantitative delay information by counting the number of infra-red photons from the back side of a thinned package chip. However, the technique requires infra-red measurement which makes it costly and tedious. Delay measurement based on Delay Lock Loop(DLL) was proposed by [5]. Even this solution looks promising in measuring to sub-nanosecond resolution of the cells under test. However, the complexity of the on-chip measurement circuitry limits its large scale implementation which will be needed for characterizing variations. The authors in [6] proposed random sampling technique to

measure the delay of standard cells. In this technique, two samplers are used to sample the input and output of the cell. Due to local process variation [7], the two samplers will suffer from variation and hence the measurement will be inaccurate.

Normally, the delay of a cell is quite small; delay can be amplified by cascading a number of stages and the delay of the individual cell can be found out by averaging. Traditional way of measurement of propagation delay variation due to process variations is usually performed with ring oscillator (RO) because of its easy on-chip implementation and high sensitivity to process parameter [8] [9] [10] [11]. However, it is difficult to extract gate to gate delay variation from a simple ring oscillator. It might be possible to use very small rings 3-5 stages, but the frequency will be very high leading to complexity in implementation.

Authors in [12] have presented a technique to measure incremental delay i.e. the difference between the two propagation delays of the same cell. In this technique, the authors have modified the cell and formed ring oscillator using these cells. The main drawback of this approach is that the modified cells are not a cell of any standard cell library. Hence, this approach is not suitable for delay mismatch measurement of standard cell library.

A method to measure cell-to-cell delay mismatch due to process variations is presented in [13]. In this approach, the cell is a combination of inverter and transmission gate. The authors in [13], point out the importance of the symmetry in the ring oscillator based measurement technique. However, the layout of the structure is not regular because of multiplexer used outside the ring.

Our contribution in this work is in-situ measurement of standard cell gate delay (both inverting and non-inverting) in its unmodified form. The proposed approach is also useful in measuring delay variation due to local transistor variability, neighborhood effects, voltage and temperature variations.

II. GATE DELAY MEASUREMENT CELL

Fig. 1 shows the basic concept of delay measurement technique. There are two paths between input to output. The difference of delay between two paths conceptually gives the delay of the gate under test. Fig. 2 shows schematic of proposed gate delay measurement cell (GDMC). This cell consists of two different types of inverters and two multiplexers. The inverter I_1 is the gate whose delay is to be measured. The

978-1-4244-2018-6/08 $25.00 © 2008 IEEE   145

inverters I_2, I_3, I_4 and I_5 are used for buffering and load matching. The multiplexer Mux_1 allows bypassing of the logic gate, thus enabling the calculation of the gate's delay by the difference of two period measurements of the ring oscillator. Thus, the calculated delay will be the sum of the delay of the logic gate and the difference in path delays of the multiplexer Mux_1 (i.e between input A to output Y and input B to output Y) due to unequal slew input to two inputs of Mux_1 for the two different mux settings.

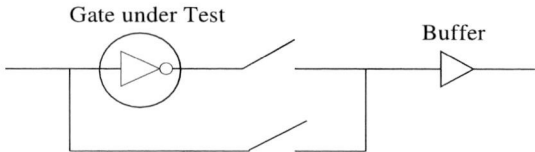

Fig. 1. Concept of gate delay measurement

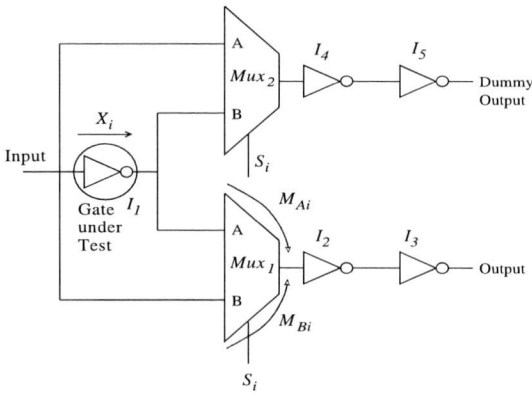

Fig. 2. Schematic of a Gate Delay Measurement Cell

The input signal to the stage is connected both directly and through inverter I_1, to the two inputs of the multiplexer Mux_1. The output Y of the multiplexer is driven out through inverters I_2 and I_3 which buffers the cell from configuration changes in the subsequent stage. Multiplexer Mux_2 and inverters I_4, I_5 are a dummy structure used to provide equal loading to inverter I_3 of the previous stage irrespective of status of the select input of the multiplexers. Symmetric multiplexers of large size which have balanced delays between input A to output Y and input B to output Y are used to reduce any systematic mismatches. The sizes of multiplexer Mux_1, inverters I_2, I_3 are same as of multiplexer Mux_2, inverters I_4, I_5 respectively. The multiplexer Mux_1 and Mux_2 are matched in the layout.

The above structure allows symmetry and load matching which helps in finding the individual gate delay. In the absence of process variation all the T-stage delays (i.e. T=any odd number $\leq N$ where N is number of stages in RO) in ring oscillators should produce equal delays. In the absence of process variation, the delay of inverter I_2 and I_3 should be same irrespective of the status of the control inputs.

III. RECONFIGURABLE RING OSCILLATOR STRUCTURE

For simplicity of description, we have considered a 5-stage ring oscillator below. However, the technique is applicable for any number of stages of ring oscillator. With this circuit setup, it is shown that delay of inverting gate can be measured using this structure.

A. Inverting gate delay measurement

Fig. 3 shows the delay measurement circuit for inverting gate. The technique of delay measurement for inverting gate is presented below. Here X_i is the average of rise and fall delay of the ith inverter I_1 under test, and M_{Ai} and M_{Bi} are the delays between inputs A and B to output Y of the multiplexer Mux_1. K is the delay of the rest of the elements of the ring oscillator like the buffers consisting of inverters I_2, I_3 of all the cells, the wires etc. I_4, I_5 will not contribute to the frequency of the RO, because it is not part of the ring. Note all the delays are average of rise and fall delays. Let S be the configuration vector consisting of 5 bits. If S_i = 0, then input A to output Y of the multiplexer Mux_1 of the ith stage GDMC is selected. If S_i = 1, then input B to output Y of the multiplexer Mux_1 of the ith stage GDMC is selected. T_1, T_2, T_3 and T_4 are the period of RO for four configuration vectors S = 00000, 00011, 00110, 00101 respectively. The delay measurement of inverting gate requires four configuration vectors or control words. Since, each GDMC switches twice during a complete cycle, the sum of all the average cell delays equates to half a clock period of the RO signal.

When select status vector S = 00000, then the input A to output Y of stages 1st, 2nd, 3rd, 4th, 5th are connected. Then,

$$\sum_{i=1}^{5}(X_i + M_{Ai}) + K = T_1/2 \qquad (1)$$

When select status vector S = 00011, then the input A to output Y of stages 1st, 2nd, 3rd are connected and the input B to output Y of stages 4th and 5th are connected.

$$\sum_{i=1}^{3}(X_i + M_{Ai}) + \sum_{i=4}^{5} M_{Bi} + K = T_2/2 \qquad (2)$$

Taking the difference between Eq. (1) and (2), we get

$$X_4 + M_{A4} + X_5 + M_{A5} - M_{B4} - M_{B5} = (T_1 - T_2)/2 \qquad (3)$$

When select status vector S = 00110, then the input A to output Y of stages 1st, 2nd, 5rd are connected and the input B to output Y of stages 3th and 4th are connected.

$$\sum_{i=1,2,5}(X_i + M_{Ai}) + \sum_{i=3}^{4} M_{Bi} + K = T_3/2 \qquad (4)$$

When select status vector S = 00101, then the input A to output Y of stages 1st, 2nd, 4th are connected and the input B to output Y of stages 3rd and 5th are connected.

$$\sum_{i=1,2,4}(X_i + M_{Ai}) + \sum_{i=3,5} M_{Bi} + K = T_4/2 \qquad (5)$$

978-1-4244-2018-6/08 $25.00 © 2008 IEEE

Taking the difference between Eq. (4) and (5), we get

$$-X_4 - M_{A4} + X_5 + M_{A5} + M_{B4} - M_{B5} = (T_3 - T_4)/2 \quad (6)$$

Adding the Equations (3) and (6), we get

$$X_5 + (M_{A5} - M_{B5}) = (T_1 - T_2 + T_3 - T_4)/4 \quad (7)$$

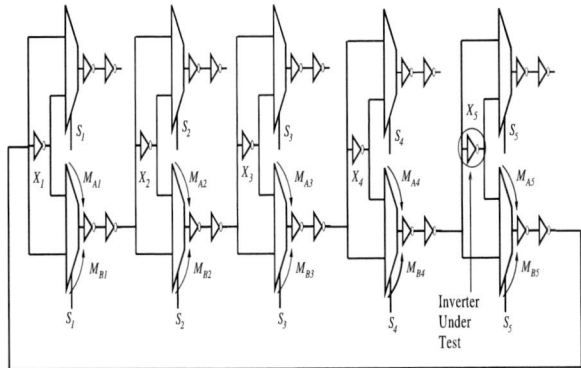

Fig. 3. Inverting Gate Delay Measurement showing Inverter under test

Period measurements of the ring oscillator T_1-T_4, with four different control word settings lead to the Eq. 7 for the delay of inverter I_1 in the 5th stage along with the residual error term, which is the difference between input A to output Y and input B to output Y delays of the multiplexer Mux_1. However, unequal slew rate at the two inputs of Mux_1 for the two different mux settings contributes to a residual delay error between the A, B inputs and mux output, which is observed to be 1.2pS from circuit simulations of extracted layout in the absence of process variation. Since the proposed technique requires the computation of difference of linear equation, the residual error term for the mux delay mismatch cancels out the global transistor variation component term leaving behind the local variation term. However, X_5 includes nominal delay plus global and local variation terms.

The minimum number of stages required for this type of delay measurement is four or five for non-inverting or inverting gate delay measurement respectively. Since odd number of stage ROs can be formed, the number of equations formed using 5-stage RO are $C(5,5) + C(5,3) = 11$ and number of variables (e.g. gate delays to be measured) is 5. Hence the delays can be cross checked across many different measurements similar to [12]. For our test chip, measurements of 22 different gates each with 6 different configurations yields the delay values for the corresponding gates to be within 0.64pS (refer to Fig.7(b)), indicating the robustness of the measurements.

The technique presented here shows the on-chip measurement of inverter. However, if we replace inverter I_1 of Fig. 2 with any inverting gate, then delay of that gate can be measured. Similarly, the delay of non-inverting gate can also be measured with only two configuration vectors or control words.

IV. CHIP RESULTS

The layout of the test chip with the distribution of ring oscillators is shown in Fig. 4. The divider is used to slow down the frequency because the I/O has a frequency limitation of 100MHz. Each gate's delay measurement requires period measurements for four different oscillator configurations for inverting gate as described in Eq. 7.

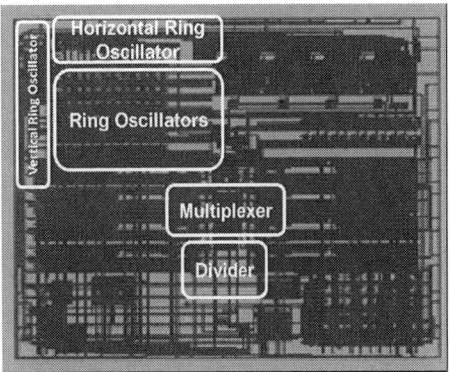

Fig. 4. Layout of test chip showing the distribution of ring oscillators

Fig. 5 shows the measured individual delays of inverter I_1 (refer to Fig. 2) (normalized to a fan-out 1 loaded inverter) in each of the 10 stages of the same ring from two different chips. The inverter to inverter delay in the ring, as a percentage of the mean delay varies by up to 26% for chip 1 and 17.4% for chip 2, indicating the effect of intra-chip local variations. There is no discernible pattern of variation for the delays within the same ring. Between the two chips, the variation pattern has some similarity, for stages 4 to 8, but is different for the rest indicating the randomness of these local or within-die variations.

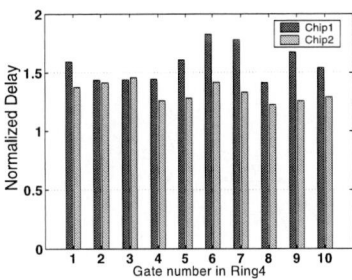

Fig. 5. Measured delays of 10 nominally identical inverters in the same ring from different chips

We also correlated the measured results to results from statistical SPICE simulation and found that out of 20 measured delays from two chips (see Fig. 5), 17 points were well within the distribution whereas the rest of the points lie towards the tail of the distribution.

We apply this gate delay measurement to study the delay variation due to orientation in layout. Fig. 6(a) shows the delays of the 10 inverters in two ring oscillators, one laid out vertically and the other horizontally for chip1, showing a

978-1-4244-2018-6/08 $25.00 © 2008 IEEE

delay spread of 19%. Note that in these two rings, the inverter I_1 has a different size, layout and loading compared to that of Fig. 5. The pattern of local delay variation is different for the vertical and horizontal cells. Fig. 6(b) shows the delays of the 10 identical inverters in horizontal and vertical ring oscillators for chip2. Comparison of Fig. 6(a) and Fig. 6(b) shows the gate number 1,2,3,4,8 and 10 have a pattern of similarity where rest of the gates show opposite pattern signifying the importance of random variation.

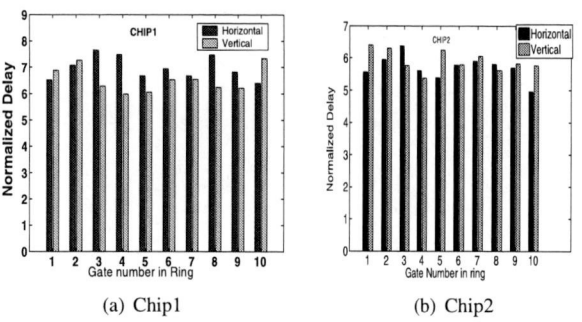

(a) Chip1   (b) Chip2

Fig. 6. Measured delays of 10 nominally identical inverters in horizontal and vertical orientation of RO

Fig. 7(a) shows the measured delay variations of the 10 cells in a ring for two different supply voltages. We observe that the delay and the delay spread at 0.8V are more than for 1V, but the pattern of variation is the same. As we decrease the supply voltage, the spread in delays increases from 17% at 1V to 29% at 0.8V, indicating that the main source of local variations is the process.

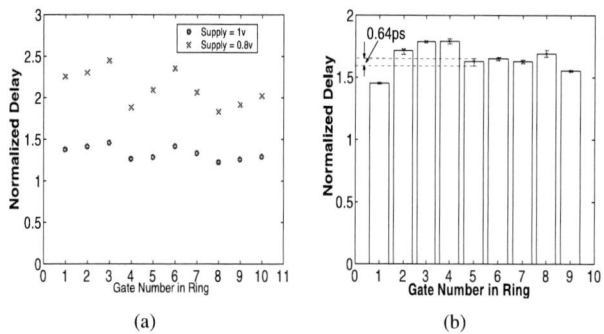

(a)   (b)

Fig. 7. (a) Measured gate delays in a ring for 1V and 0.8V supply (b) Error in Delay measurement of different gates

The error in delay measurement is shown in Fig.7(b). As mentioned earlier, same gate delay can be measured in multiple ways, because the number of equation is more than the number of variables. Here the number of equation means the different RO oscillator configuration setting and the number of variables is the number of gate delay to be measured. Fig.7(b) shows 9 different gates, each gate delay is measured in 6 different ways. The maximum error in delay measurement in Fig.7(b) is found to be 0.64ps. This proves the robustness of delay measurement to be within 1pS accuracy.

These measurements indicate that local gate to gate variations are very significant in advanced process nodes and need to be accounted in the models and design practices. The presented measurement technique is simple yet powerful to study these variations.

V. CONCLUSION

A circuit technique has been developed to measure the delay of individual gates using reconfigurable ring oscillators in standard cell library inside the chip. The technique is all-digital in nature and can be easily embedded within standard digital logic to perform in-situ measurement. The results from 65nm test chip show the efficacy of measurement to within 1pS accuracy. Delay measurements of different nominally identical inverters in close physical proximity shows variations of up to 26% indicating the large impact of local variations. The proposed technique is quite suitable for early process characterization, monitoring mature process in manufacturing, correlating model-to-hardware and study local variation and neighborhood effects.

ACKNOWLEDGMENT

We gratefully acknowledge the help of Pratap Das, V. Janakiraman of IISc and Shankar Karantha, J. Sridhar of TI.

REFERENCES

[1] M. Orshansky, L. Milor, and C. Hu, "Characterization of spatial intrafield gate CD variability, its impact on circuit performance, and spatial mask-level correction," *IEEE Trans. on Semi. Manufacturing*, vol. 17, no. 1, pp. 2–11, Feb 2004.

[2] E. Karl, P. Singh, D. Blaauw, and D. Sylvester, "Compact in-situ sensors for monitoring negative- bias-temperature-instability effect and oxide degradation," in *Proc. of IEEE ISSCC*, 2008, pp. 410–411.

[3] S. Mukhopadhyay, K. Kim, K. A. Jenkins, C. Chuang, and K. Roy, "Statistical characterization and on-chip measurement methods for local random variability of a process using sense-amplifier-based test structure," in *Proc. of IEEE ISSCC*, 2007, pp. 400–401.

[4] P. Sanda, D. Knebel, J. Kash, H. Casal, J. Tsang, E. Seewann, and M. Papermaster, "Picosecond Imaging Circuit Analysis of the power3 clock distribution," in *IEEE ISSCC*, 1999, pp. 372–373.

[5] N. Abaskharoun and et al, "Circuits for on-chip subnanosecond signal capture and characterization," in *Proc. of IEEE Conf on Custom Integrated Circuits*, 2001, pp. 251–254.

[6] S. Maggioni, A. Veggetti, A. Bogliolo, and L.Croce, "Random sampling for on-chip characterization of standard cell propagation delay," in *Proc. of IEEE ISQED*, 2003.

[7] K. Okada, K.Yamaoka, and H.Onodera, "A statistical gate-delay model considering intra-gate variability," in *ICCAD*, 2003, pp. 908–913.

[8] J. Panganiban, "A Ring Oscillator based variation test chip," M.Engg. Thesis, MIT Dept. of Electrical Engineering and Computer Science, May 2002.

[9] M. Bhushan and et al, "Ring oscillators for CMOS process tuning and variability control," *IEEE Trans. on Semiconductor Manufacturing*, vol. 19, no. 1, pp. 10–18, Feb 2006.

[10] H. Masuda, S. Ohkawa, A. Kurokawa, and M. Aoki, "Challenge: Variability characterization and modeling for 65- to 90-nm processes," in *Proc. of IEEE Conf on Custom Integrated Circuits*, 2005, pp. 908–913.

[11] S. Ohkawa, M. Aoki, and H. Masuda, "Analysis and characterization of device variations in an lsi chip using an integrated device matrix array," *IEEE Trans. on Semiconductor Manufacturing*, pp. 155–165, May 2004.

[12] A. Bassi, A. Veggetti, L.Croce, and A. Bogliolo, "Measuring the effects of process variations on circuit performance by means of digitally-controllable ring oscillators," in *Microelectronic Test Structures*, March 2003, pp. 214-217.

[13] B. Zhou and A. Khouas, "Measurement of delay mismatch due to process variations by means of modified ring oscillators," in *ISCAS*, 2005, pp. 5246–5249.

IEEE 2008 Custom Intergrated Circuits Conference (CICC)

Expected Vectorless Teacher-Student Swap (TSS) Test Method with Dual Power Supply Voltages for 0.3V Homogeneous Multi-core LSI's

Taro Niiyama, Koichi Ishida, Makoto Takamiya, and Takayasu Sakurai

University of Tokyo, 4-6-1 Komaba, Meguro-ku, Tokyo 153-8505, Japan

Abstract- **A Teacher-Student Swap (TSS) test method with the dual supply voltage (V_{DD}) for the ultra low V_{DD} homogeneous multi-core LSI's is proposed and the test chips are fabricated in 90 nm CMOS. In this method, two same cores with different power supply voltages test each other by comparing their outputs, which eliminates the need for the expected vector. When V_{DD} is less than 0.3V, the die-to-die power reduction by the dual V_{DD} in the 5 chips was from 18% to 48%. In order to manage the large die-to-die variations at low V_{DD}, the fine grain dual V_{DD} with TSS test method is a promising approach without increasing the test cost.**

I. INTRODUCTION

Both the low power supply voltage (V_{DD}) and multi core are the recent trend for power efficient processors. Many works have been carried out on the subthreshold logic circuits [1-4] and homogeneous multi-core LSI's (e.g. 80 cores [5] and 64 cores [6]). The subthreshold logic circuits are energy efficient [1,4], while they are very slow. Therefore, an ultra parallel processing with homogeneous subthreshold multi cores may be a promising approach for the energy efficient processors with the practical processing speed.

The increasing test cost proportional to the number of cores, however, is among the most serious issues facing the multi-core LSI's. Fig. 1 shows the conventional test method of the homogeneous multi-core LSI's. The outputs from n cores are compared with expected vectors, and pass or fail are determined. The increasing quantity of the outputs with the number of cores raises the test cost.

Meanwhile, the minimum power supply voltage (V_{DDmin}) of logic circuits is determined by the function errors of logic

gates due to device variations, and V_{DDmin} increases with the number of logic gates [7]. Therefore, the fine grain V_{DD} tuning [7] is required to reduce V_{DD} of the subthreshold logic circuits and the tuning also raise the test cost.

This paper presents a new test method to reduce the test cost in ultra low V_{DD} homogeneous multi-core LSI's. The concept of the proposed Teacher-Student Swap (TSS) test method is shown in Section II. Section III presents the chip implementation of the dual V_{DD} homogeneous multi-core LSI's with TTS test method. Measurement results from 90nm CMOS test chips are described in Section IV.

II. PROPOSED TEACHER-STUDENT SWAP TEST METHOD

Fig. 2 shows a proposed TSS test method of the homogeneous multi-core LSI's. Every core is paired with the neighboring cores and the pair is called "teacher and student pair" in this paper. A teacher core and a student core are swappable and they test each other. The teacher core is a function guaranteed core and operates as an expected vector generator for the student core, which eliminates the need for the input of the expected vectors from the tester. The student core is the circuits under test. The outputs from the student core are compared with the expected vectors from the teacher core, and only the pass or fail signals are reported to the tester. Therefore, the quantity of the outputs from the chip is much smaller than that in Fig. 1, thus reducing the test cost for the

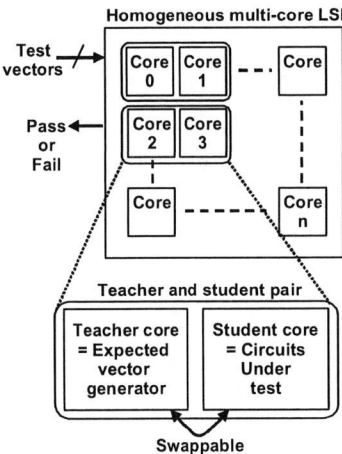

Fig. 2. proposed TSS test method of the homogeneous multi-core LSI's.

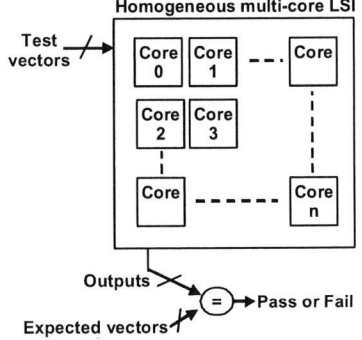

Fig. 1. The conventional test method of the homogeneous multi-core LSI's.

978-1-4244-2018-6/08 $25.00 © 2008 IEEE

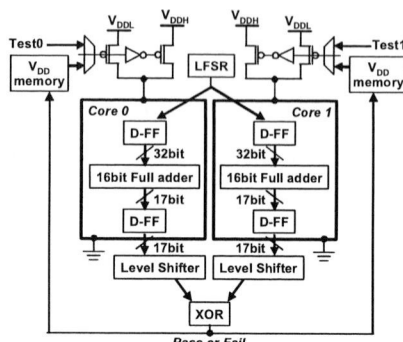

Fig. 5. Schematic of the fabricated teacher and student pair with dual VDD.

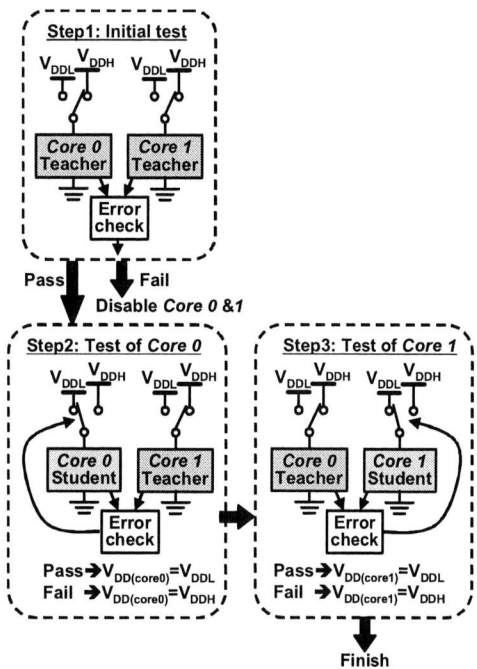

Fig. 3. A flow chart of TSS test.

multi-core LSI's.

Fig. 3 shows a flow chart of TSS test. In order to achieve the fine grain V_{DD} tuning for the ultra low V_{DD} homogeneous multi-core LSI's, dual V_{DD} (a high V_{DD} (V_{DDH}) and a low V_{DD} (V_{DDL})) is adopted. V_{DDH} is the minimum supply voltage where all cores operate correctly and V_{DDL} is the supply voltage where some of the cores fail to operate. Thus the core with V_{DDH} is the teacher core and the core with V_{DDL} is the student core. As shown in Fig. 3, in the initial test (Step1), both core 0 and 1 are teacher-mode with V_{DDH} to find the initial failure of the cores. When the initial failure is found, both core 0 and 1 are disabled and can be replaced with redundant cores. In the test of core 0 (Step2), the core 0 with V_{DDL} is the student mode and the core 1 with V_{DDH} is the teacher mode. When the test is passed, the core 0 operates with V_{DDL}. In contrast, when the test is failed, the core 0

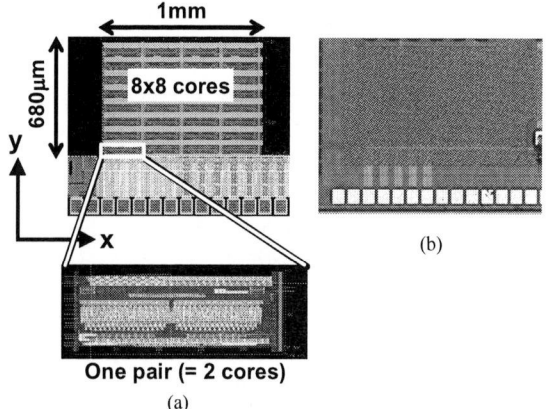

Fig. 4. (a) Layout and (b) micrograph of a homogeneous 64 core LSI with the TSS test method in 90 nm CMOS.

operates with V_{DDH}. Step3 is the swapped version of Step2 and concludes the TSS test.

III. CHIP IMPLEMENTATION OF HOMOGENEOUS MULTI-CORE LSI WITH TEACHER-STUDENT SWAP TEST METHOD

In order to demonstrate the TSS test method with dual V_{DD} for the ultra low V_{DD} homogeneous multi-core LSI's, test chips has been designed and fabricated. Figs. 4 (a) and (b) show the layout and the micrograph of a homogeneous 64 core LSI with the TSS test method in 90 nm CMOS respectively. The core area is 1.0mm×0.68mm. The 64 core LSI has 32 teacher and student pairs. The two cores in a pair have symmetrical layouts.

Fig. 5 shows the schematic of the fabricated teacher and student pair with dual V_{DD}. Each core includes the 16 bit ripple carry adder and D-flip-flops. In the cores, the primitive cells provided by the foundry are used and are not optimized for low V_{DD} operations [1,8]. A 32 bit LFSR generates pseudo random numbers, and the 17 bit adder outputs from core 0 and 1 are compared. The pass or fail signals are stored in a V_{DD} memory. V_{DDH} or V_{DDL} is selected by pMOSFET's. The all circuits except the cores and I/O's use a separate V_{DD}.

IV. MEASUREMENT RESULTS

V_{DDL} distributions within 64 cores are measured by the TSS test method. The power reduction by the dual V_{DD} is also discussed.

A. V_{DDL} Distributions

Fig. 6 shows the measured V_{DDL} dependence of the number of error cores in 64 cores of a chip. The clock frequency is varied from 10kHz to 30MHz. V_{DDH} is defined as V_{DD} where the first error core is observed. The V_{DDL} distributions are divided into 2 modes (a delay error mode ($V_{DD} > 0.3V$) and a function error mode ($V_{DD} < 0.3V$)). At 30MHz and 15MHz, the V_{DDL} distributions are narrow and determined by the classical timing error of the logic gates, which is the delay error mode. In contrast, at 1MHz and 10kHz, the V_{DDL} distributions are broad and do not depend on the clock frequency. The distributions are determined by the function error of the logic gates due to device variations[7],

978-1-4244-2018-6/08 $25.00 © 2008 IEEE

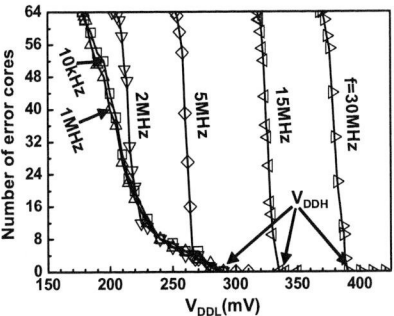

Fig. 6. Measured V_{DDL} dependence of the number of error cores in 64 cores of a chip. The clock frequency is varied from 10kHz to 30MHz.

which is the function error mode. At 5MHz and 2MHz, both the delay error mode and the function error mode are mixed.

Figs. 7 (a) and (b) show the measured V_{DDL} distribution map on 64 cores at 30MHz (delay error mode) and 10kHz (function error mode) respectively. The x- and y-axis in Fig. 7 corresponds to that in Fig. 4 (a). The maximum V_{DDL} is 388mV and the minimum V_{DDL} is 370mV at 30MHz. The maximum V_{DDL} is 178mV and the minimum V_{DDL} is 280mV at 10kHz. The V_{DDL} distribution in Fig. 7 (a) has both the random and the systematic (y-axis direction) component. The systematic component will derive from the on-chip IR-drop. In contrast, the V_{DDL} distribution in Fig. 7 (b) has only the random component. The peak-to-peak variation of V_{DDL} (=102mV) is larger than that (=18mV) in Fig. 7 (a), because the low V_{DD} exposes the effect of the device variations on the circuit variations.

B. Power Reduction by Dual V_{DD}

Fig. 8 (a) shows the V_{DDL} dependence of the power and the number of error cores in 64 cores of a chip at 4 different clock frequencies. The power is simulated by SPICE based on the measured number of error cores. The leakage power accounts

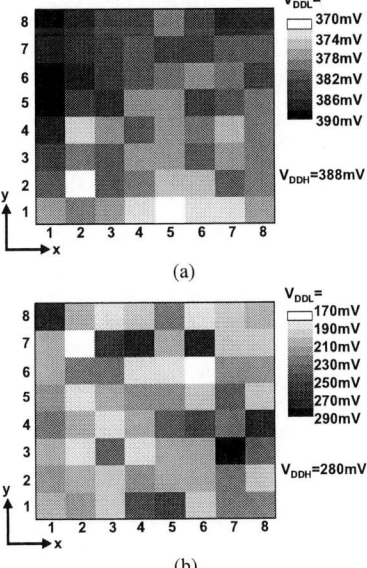

Fig. 7. Measured V_{DDL} distribution map on 64 cores (a) at 30MHz (delay error mode) and (b) 10kHz (function error mode).

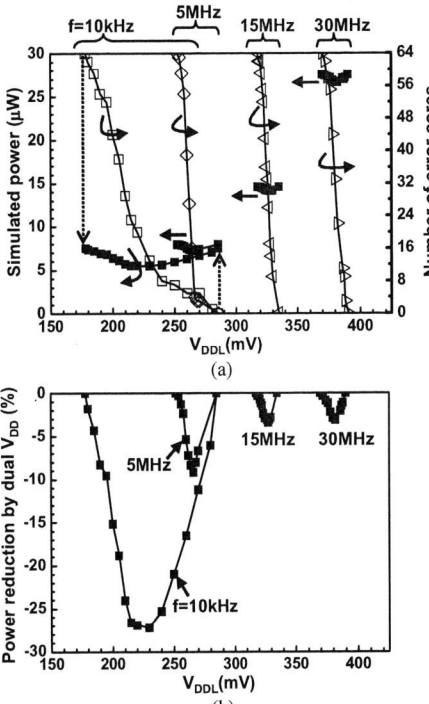

Fig. 8. V_{DDL} dependence of (a) the power and the number of error cores in 64 cores of a chip and (b) the power reduction by the dual V_{DD} at 4 different clock frequencies.

for 48% and 99.9% of the total power shown in Fig. 8 at 30MHz and 10kHz respectively. By using the proposed TSS test method, V_{DDH} is assigned to the error cores and V_{DDL} is assigned to the non-error cores automatically. When all the 64 cores have the errors, all cores operate at V_{DDH}, which is equal to the all V_{DDH} approach. Therefore, the power is minimized by optimizing V_{DDL}. Fig. 8 (b) shows the V_{DDL} dependence of the power reduction by the dual V_{DD} at 4 different clock frequencies. At 30MHz and 15MHz (delay error mode), the power reduction is only 3%, because the V_{DDL} distributions are narrow. In contrast, at 10kHz (function error mode), the power reduction is 27%, because the V_{DDL} distribution is broad.

Finally, the die-to-die variations of the measured VDDL distribution are discussed. Fig. 9 shows the measured V_{DDL} dependence of the number of error cores in 64 cores of 5 chips

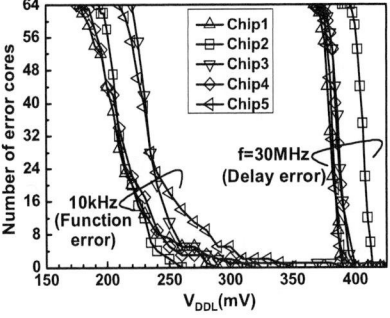

Fig. 9. Measured V_{DDL} dependence of the number of error cores in 64 cores of 5 chips at 10kHz and 30MHz clock.

at 10kHz and 30MHz clock signals. Chip1 is the same as that in Fig. 6. At both 10kHz and 30MHz, the die-to-die variations of the V_{DDL} distribution are observed. An interesting result is that the delay error mode and the function error mode are not correlated, because their mechanisms are different. For example, Chip2 has the highest V_{DDL} at 30MHz, however, Chip3 or Chip5 has the highest V_{DDL} at 10kHz.

Fig. 10 (a) shows the V_{DDL} dependence of the simulated power of 5 chips at 10kHz clock. Fig. 10 (b) shows the V_{DDL} dependence of the power reduction by the dual V_{DD}. The power reduction is from 18% to 48% among 5 chips, and the high V_{DDL} chip (Chip3) achieves the largest power reduction. Fig. 10 (c) shows the V_{DDL} /V_{DDH} dependence of the power reduction by the dual V_{DD}. In the conventional processors, the optimum V_{DDL} /V_{DDH} to achieve the minimum power is 0.7[9]. In contrast, at 0.2 - 0.4V and 10kHz (function error mode), the optimum V_{DDL} /V_{DDH} is from 66% to 82% among 5 chips. In order to manage the large die-to-die variations at low V_{DD}, the fine grain dual V_{DD} with TSS test method is a promising approach without increasing the test cost.

V. CONCLUSION

The Teacher-Student Swap (TSS) test method with the dual V_{DD} for the ultra low V_{DD} homogeneous multi-core LSI's was proposed and the test chips were fabricated in 90 nm CMOS. Depending on V_{DD}, the LSI's have the delay error mode ($V_{DD} > 0.3$V) and the function error mode ($V_{DD} < 0.3$V). At the delay error mode, the power reduction by the dual V_{DD} was only 3%, while the power reduction was 27% at the function error mode. The die-to-die power reduction by the dual V_{DD} in the 5 chips was from 18% to 48%, when the optimum V_{DDL} /V_{DDH} was from 66% to 82%. In order to manage the large die-to-die variations at low V_{DD}, the fine grain dual V_{DD} with TSS test method is a promising approach without increasing the test cost.

ACKNOWLEDGMENTS

This work is partially supported by STARC. The VLSI chips were fabricated through the chip fabrication program of VLSI Design and Education Center (VDEC), the University of Tokyo, with the collaboration by STARC, Fujitsu Limited, Matsushita Electric Industrial Company Limited., NEC Electronics Corporation, Renesas Technology Corporation, and Toshiba Corporation.

REFERENCES

[1] B. Calhoun, and A. Chandrakasan, "Ultra-dynamic voltage scaling (UDVS) using sub-threshold operation and local voltage dithering," IEEE Journal of Solid-State Circuits, Vol. 41, No. 1, pp. 238-245, Jan. 2006.

[2] S. Hanson, B. Zhai, M. Seok, B. Cline, K. Zhou, M. Singhal, M. Minuth, J. Olson, L. Nazhan-dali, T. Austin, D. Sylvester, and D. Blaauw, "Performance and variability optimization strategies in a sub-200mV, 3.5pJ/inst, 11nW subthreshold processor," IEEE Symposium on VLSI Circuits, pp. 152-153, June 2007.

[3] M. Hwang, A. Raychowdhury, K. Kim, and K. Roy, "A 85mV 40nW process-tolerant subthreshold 8x8 FIR filter in 130nm technology," IEEE Symposium on VLSI Circuits, pp. 154-155, June 2007.

[4] H. Kaul, M. Anders, S. Mathew, S. Hsu, A. Agarwal, R. Krishnamurthy, and S. Borkar, "A 320mV 56μW 411GOPS/Watt ultra-low voltage

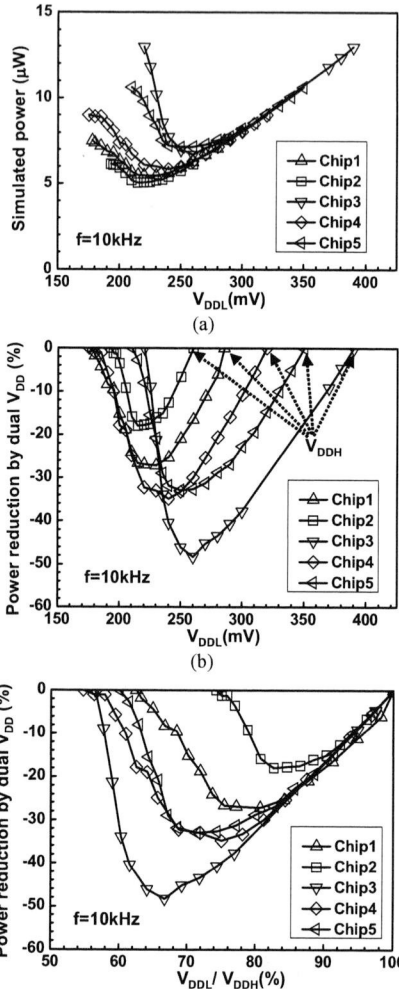

Fig. 10. (a) V_{DDL} dependence of the power of 5 chips at 10kHz clock. (b) V_{DDL} dependence of the power reduction by the dual V_{DD}. (c) V_{DDL} /V_{DDH} dependence of the power reduction by the dual V_{DD}.

motion estimation accelerator in 65nm CMOS," IEEE International Solid-State Circuits Conference, pp. 316-317, Feb. 2008.

[5] S. Vangal, J. Howard, G. Ruhl, S. Dighe, H. Wilson, J. Tschanz, D. Finan, P. Iyer, A. Singh, T. Jacob, S. Jain, S. Venkataraman, Y. Hoskote, and N.Borkar, " An 80-Tile 1.28TFLOPS Network-on-Chip in 65nm CMOS," IEEE International Solid-State Circuits Conference, pp. 98-99, Feb. 2007.

[6] S. Bell, B. Edwards, J. Amann, R. Conlin, K. Joyce, V. Leung, J. MacKay, M. Reif, L. Bao, J. Brown, M. Mattina, C.-C. Miao, C. Ramey, D. Wentzlaff, W. Anderson, E. Berger, N. Fairbanks, D. Khan, F. Montenegro, J. Stickney, and J. Zook, "TILE64 Processor: a 64-Core SoC with mesh interconnect," IEEE International Solid-State Circuits Conference, pp. 88-89, Feb. 2008.

[7] T. Niiyama, P. Zhe, K. Ishida, M. Murakata, M. Takamiya, and T. Sakurai, "Dependence of minimum operating voltage (V_{DDmin}) on block size of 90-nm CMOS ring oscillators and its implications in low power DFM," IEEE International Symposium on Quality Electronic Design, pp. 133-136, March 2008.

[8] A. Wang, and A. Chandrakasan, "A 180-mV subthreshold FFT processor using a minimum energy design methodology," IEEE Journal of Solid-State Circuits, Vol. 40, No. 1, pp. 310-319, Jan. 2005.

[9] T. Kuroda, and M. Hamada, "Low-power CMOS digital design with dual embedded adaptive power supplies," IEEE Journal of Solid-State Circuits, Vol. 35, No. 4, pp. 652-655, April 2000.

Broadband Circuit Techniques for
Emerging Wireless Communications

Session 9

Chair: Fa Foster Dai, Auburn University
Co-Chair: Howard Luong, Hong Kong University of Science and Technology

The first paper is an invited paper from Intel Corporation on MIMO techniques for high data rate radio communications. The paper presents a 2x2 MIMO 5GHz WLAN transceiver RFIC implemented in 90nm CMOS. Architectural and circuit techniques are described for minimizing crosstalk between the multiple radio chains. The paper also discusses the future MIMO evolutions, including collaborative, directional and 60GHz phased-array communications.

The last paper in the session is from University of Florida on wireless interconnection within hybrid engine controller board. Implemented in a 130-nm CMOS process, a receiver consisting of the circuits and an on-chip antenna was demonstrated to be capable of demodulating AM signals around 14-16GHz with a data rate up to 400 Mbps while consuming 60mW.

978-1-4244-2018-6/08 $25.00 © 2008 IEEE

Notes

MIMO Techniques for High Data Rate Radio Communications

Yorgos Palaskas, Ashoke Ravi, Stefano Pellerano

Communications Circuits Lab, Intel Corporation, Hillsboro, OR
palaskas@ieee.org

Abstract: MIMO wireless transceivers employ multiple antennas and advanced digital signal processing to achieve higher data rates and superior link reliability compared to their single-antenna counterparts. After a simplified overview of general MIMO theory, this paper presents a 2x2 MIMO 5GHz WLAN transceiver implemented in 90nm CMOS. Crosstalk between the different MIMO channels is shown to have a detrimental effect on MIMO performance. Such crosstalk can occur at the radio channel, between the antennas, and at the radio IC. Several architectural and circuit techniques are described that minimize crosstalk between the multiple radio chains co-existing on the same silicon die. Measurements results demonstrate the increased data rate of the fabricated system compared with a conventional single-antenna system. The paper concludes with some interesting scenarios for future MIMO evolution, including collaborative, directional and 60GHz phased-array communications.

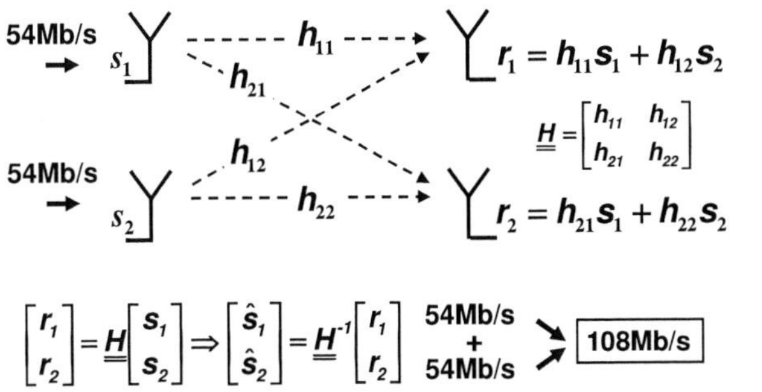

Fig. 1: MIMO Spatial Multiplexing as a means for increasing spectral efficiency: the two transmitters transmit independent bit-streams on spatial channels; the MIMO receiver estimates and equalizes the channel to de-convolve the bit streams.

I. INTRODUCTION

Recent years have witnessed an explosive growth in wireless connectivity. Upcoming wireless applications such as video, interactive multi-media, voice and data and the need for "always-on" connectivity will require higher data rates and improved network coverage and capacity in LAN, WAN and PAN devices. MIMO (Multiple-Input, Multiple-Output) is a promising new technology that uses multiple antennas and advanced DSP to improve the performance of a wireless network [1][2]. This section presents an overview of MIMO theory and some details of common MIMO algorithms.

The transmitted RF signal undergoes multiple reflections as it travels through a medium rich in scatterers, e.g. indoor office environments. For narrowband signals, such as those used in WiFi and WiMax, the receiver cannot distinguish between the different reflections for typical times of arrival and the superposition of these multiple reflections at the receiver results in a signal power following the Rayleigh distribution. This Rayleigh fading occurs in addition to any path losses and shadowing from large obstructions that are factored into the link budget. For such multipath rich environments, the availability of multiple antennas at the receiver, the transmitter, or both, can be used to realize a linear improvement of the channel capacity. The fundamental theoretical limit is governed by the equation: $C \approx B \cdot M \cdot \log(1 + SNR)$, where C is the maximum rate, B is the RF bandwidth and M is the number of antennas at the transmitter and receiver (the smaller of the two if not equal). This increase in capacity is achieved without tightening the

SNR requirements on the radio or degrading the spectral occupancy, but is strongly dependent on the fading properties of the channel.

Let us consider a system with 2 transmit and 2 receive antennas (2x2 MIMO) as shown in Figure 1. The baseband input-output equation for each sub-carrier in an OFDM (orthogonal frequency division multiplexing) system can be written as:

$$\begin{bmatrix} r_1 \\ r_2 \end{bmatrix} = \begin{bmatrix} h_{11} & h_{12} \\ h_{21} & h_{22} \end{bmatrix} \cdot \begin{bmatrix} s_1 \\ s_2 \end{bmatrix} + \begin{bmatrix} n_1 \\ n_2 \end{bmatrix} \Leftrightarrow \underline{r} = \underline{\underline{H}} \cdot \underline{s} + \underline{n} \qquad (1)$$

where $\underline{s}$ is the transmitted signal vector, $\underline{r}$ is the received vector, $\underline{n}$ is the receiver noise, and $\underline{\underline{H}}$ is a 2x2 channel matrix describing the equivalent baseband channel between each pair of transmit and receive antennas for the given sub-carrier. Performance of the MIMO system in equation (1) is maximized when the channel matrix $\underline{\underline{H}}$ is of full rank, i.e. no row or column can be written as a linear combination of other rows or columns. In a multipath environment when antennas are spaced more than half a wavelength apart, received signals tend to exhibit independent statistics resulting in good MIMO performance. In contrast, if the transmitter and receiver are within line of sight (LOS) of each other, i.e. the scattered reflections are insignificant as compared to strong direct paths, the signals on multiple transmit or receive antennas are highly correlated and performance degrades (this will further be discussed in Section II below).

978-1-4244-2018-6/08 $25.00 © 2008 IEEE

MIMO is commonly used in one of two modes: (1) spatial diversity – where the information is spread over multiple space-time resources, thereby improving reliability, and (2) spatial multiplexing – where the information is multiplexed onto the available space-time resources to improve data throughput.

Spatial diversity can be classified into receive or transmit diversity. The optimal receiver diversity algorithm, maximal ratio combining (MRC), aligns the channel phases and weights the channel amplitudes on the receiver antennas in such a way as to maximize the SNR of the superposition. Transmit diversity schemes depend on whether or not channel information is available at the transmitter. In closed-loop transmit diversity, the transmitter uses its knowledge of the channel to adjust the phases and amplitudes such that they add up constructively at the receiver (transmitter beamforming). The channel information can be sent *explicitly* from the receiver to the transmitter on a control channel with appropriate compression techniques to minimize the capacity overhead. An alternative with reduced overhead is to derive the channel state information *implicitly* by using the channel coefficients from packets sent from the receiver to the transmitter and exploiting the reciprocity of the channel over reasonable time intervals. The latter technique (implicit feedback) is best suited for TDD systems where the same frequency is used for transmission and reception. The details and limitations of implicit feedback, as well as techniques to handle non-reciprocal elements (e.g. the transceiver circuits) are described in more detail in [3].

Channel feedback can be problematic in highly mobile environments where, by the time the channel information is fed back to the transmitter, the channel might have already changed. In such cases, or when no channel knowledge is available at the transmitter, open-loop transmitter diversity techniques can be used instead. In open-loop transmitter diversity schemes, specially designed mappings, called space-time codes are used to spread the input data streams onto multiple spatial streams and multiple time slots. Common examples include the Alamouti Code [4] and cyclic delay diversity [5]. These open-loop diversity schemes might offer lower performance but are more amenable to highly mobile environments, are simpler to implement and support backward compatibility, accounting for their popularity.

Spatial multiplexing (Fig. 1) transmits independent bit-streams on the transmit antennas to maximize throughput. The maximum possible number of independent spatial streams is set by the minimum of the number of transmit and receive antennas. The channel matrix can be estimated at the receiver using an appropriate training sequence during the preamble of the packet. The receiver can then deconvolve the transmitted bit-streams by applying the inverse of the channel matrix to the received signals (channel equalization):

$$\hat{\underline{s}} = \underline{\underline{H}}^{-1} \cdot \underline{r} = \underline{s} + \underline{\underline{H}}^{-1} \cdot \underline{n} \qquad (2).$$

The decomposition of Eq. (2) allows transmission of different bit-streams within the same frequency channel. This is similar to establishing multiple parallel "data pipes" between transmitter and receiver, resulting in increased spectral efficiency [1]. For example, if each of the data streams has a data rate of 54Mb/s (64QAM OFDM modulation over a 16MHz channel), the 2x2 MIMO system can achieve double this rate, or 108Mb/s. Higher channel bandwidths and number of antennas can be used to achieve even higher data rates.

The decomposition of Eq. 2 can be explained intuitively as follows. The signal s_1 of transmitter 1 (see Fig. 1) arrives to the two receive antennas with phases $\angle h_{11}$ and $\angle h_{21}$ (the phases of the corresponding channel coefficients). This is equivalent to the signal arriving from a specific direction in space determined by $\angle h_{11}$ and $\angle h_{21}$. Similarly the signal s_2 of transmitter 2 appears to arrive from a direction determined by $\angle h_{22}$ and $\angle h_{12}$. In a fading channel the perceived angles of arrival for the two transmitted signals are generally different because of the random phases involved. The multiple-antenna receiver can be viewed as a phased-array system that can use baseband processing to effectively steer the phased-array in a desired direction. Signal s_1 can then be received by steering the phased-array in the perceived direction of arrival of s_1, which typically results in attenuation of s_2 since it generally appears to be coming from a different direction. Since the phased-array processing is done in the digital baseband, the same operation can be performed for s_2, allowing simultaneous reception of both transmitted signals.

Spatial diversity and spatial multiplexing can be combined to simultaneously improve range and throughput. For example, in a 2x3 transceiver, the 2 spatial channels double the data rate, while the additional receiver can be used to improve the sensitivity.

A number of MIMO radio transceivers have been reported in recent literature. [6] reports a 2x2 SiGe transceiver implementing transmit beamforming and maximum ratio combining. [7] reports a 1x2 CMOS transceiver with MRC implemented in a companion IC. A 2x2 CMOS transceiver is combined in [8] with a 3x3 antenna using an adaptive antenna selection scheme. [9] presents a 1x2 transceiver with integrated power amplifiers. Finally [10] presents a 2x2 MIMO System-on-Chip that incorporates the RFIC and baseband/MAC.

MIMO algorithms have been incorporated into the latest draft of the 802.11n standard [11]. The standard supports different MIMO configurations ranging from single antenna handsets with open-loop transmitter diversity, to full 4x4 transceivers at both ends. 802.11n can support data rates of up to 500Mbps which is achieved by combining the use of 4 spatial channels with reduced redundancy in the forward error correction, shortened guard intervals and bonded channels (using two adjacent channels when available). In legacy WLAN standards, because of variations in channel conditions, the radio often falls back to lower order modulation schemes, resulting in typical throughputs that are significantly lower than the peak supported. 802.11n incorporates support for transmit beamforming and bit-loading in addition to improved

978-1-4244-2018-6/08 $25.00 © 2008 IEEE

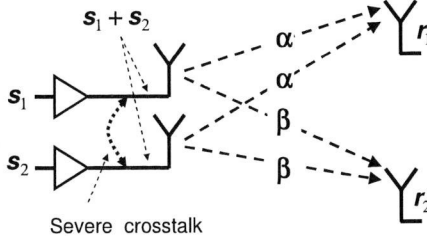

Fig. 2: If there is significant crosstalk between the two transmitters, the channel matrix becomes [α α; β β]. Such a matrix is non-invertible causing MIMO spatial multiplexing to fail. This is an extreme example for illustration purposes. More reasonable crosstalk levels result in suboptimal MIMO performance due to noise amplification.

LDPC error control codes to minimize the drop in throughput. Wide area networks like WiMAX (802.16d/e/m) and 3G-LTE (Long Term Evolution) can also significantly benefit from MIMO ideas, e.g. Alamouti transmit diversity (ideally suited for mobile environments), spatial multiplexing for higher data rates, and MRC/beam-forming to extend range.

We next present a detailed discussion on a spatial multiplexing MIMO IC originally reported in [12]. Many of the ideas discussed here are also applicable to other MIMO modes, e.g, spatial diversity.

II. MIMO CROSSTALK

The key implementation concern in a MIMO spatial multiplexing system is the crosstalk between the different MIMO paths. This crosstalk can be particularly problematic in low-cost single-chip MIMO solutions.

As was explained in the previous section, spatial multiplexing can be explained intuitively by the different perceived angles of arrivals for the two transmitted signals s_1 and s_2 (see Fig. 1). In a LOS environment both transmitted signals arrive at the receiver antennas from the same direction and the MIMO demodulator might not be able to de-convolve the two bit-streams reliably. This might also be the case if there is significant crosstalk between the two TX antennas and/or transmitter chains (similarly for the receivers). To see this consider the case shown in Fig. 2 where there is so much crosstalk between the two TX antennas that they behave as a single antenna transmitting a signal s_1+s_2. In this case the channel matrix becomes $[h_{11}\ h_{12};\ h_{21}\ h_{22}]=[\alpha\ \alpha;\ \beta\ \beta]^1$. This channel matrix is non-invertible (singular) and MIMO equalization according to Eq. (2) fails, thus preventing the system from supporting multiple independent bit-streams and high data rates.

It should be noted that the above is a rather extreme example. More realistic crosstalk levels might result in the channel matrix $\underline{\underline{H}}$ being *close* to singular (rather than exactly singular as in the case of Fig. 2). The determinant of $|\underline{\underline{H}}|$ is

---

[1] More precisely, this is the *effective* channel matrix which includes the effect of the channel, the radio IC and the antennas.

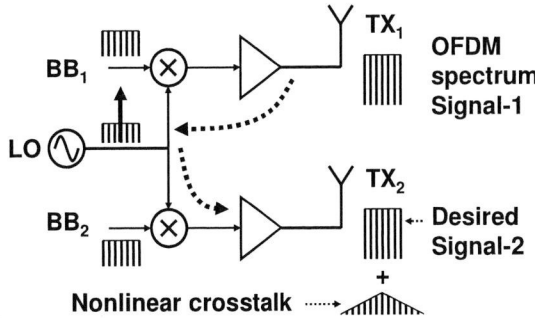

Fig. 3: Nonlinear MIMO crosstalk can occur if the PA output disturbs the LO signal used in the upconversion mixers.

then close to zero, and the noise term $\underline{\underline{H}}^{-1}\underline{n}$ in Eq. (2) becomes significant, resulting in low *SNR* and degraded MIMO performance. This phenomenon is called "noise amplification". Noise amplification can be reduced to some extent by employing more complicated equalization algorithms rather than the simple expression of Eq. (2). For example, in [12] an MMSE (Minimum Mean Square Error) equalizer was used to reduce noise amplification [1].

System-level simulations have shown that 20dB of isolation is adequate for good MIMO performance [12]. This relatively benign isolation requirement is because MIMO is inherently trying to eliminate the crosstalk between the ideally independent parallel channels. This holds regardless of where the crosstalk occurs, be it in the channel, the antennas or the radio IC. The only problem is that in the process of eliminating crosstalk, the MIMO equalizer amplifies the channel noise which eventually limits the performance. Reference [12] shows that crosstalk between the transmitters of the MIMO system typically causes larger degradation in MIMO performance than crosstalk between the receivers. For this reason, transmitter isolation is favored over receiver isolation, e.g. in the floor-plan of the chip.

The required level of isolation is achieved at the radio IC level by employing appropriate architectural, design and layout techniques as explained in Section III below. The crosstalk between the antennas is usually minimized by placing them far apart, (e.g. more than few wavelengths distance). In addition to minimizing crosstalk, this ensures that the multipath signals arrive at the antennas with de-correlated phases, which is required to generate the perception of different angles of arrival as explained in Section I.

The discussion until this point assumed linear crosstalk between the individual MIMO channels. Nonlinear crosstalk is also possible. One possible mechanism of nonlinear crosstalk is depicted in Fig. 3. Here one of the power amplifiers couples a disturbance to the LO signal which is no longer a high purity sinusoid. This disturbed LO signal is then used to upconvert the baseband signal of the second transmitter, resulting in nonlinear crosstalk between the two transmitters. Nonlinear crosstalk is more problematic than the linear crosstalk discussed above, since MIMO equalization is a linear

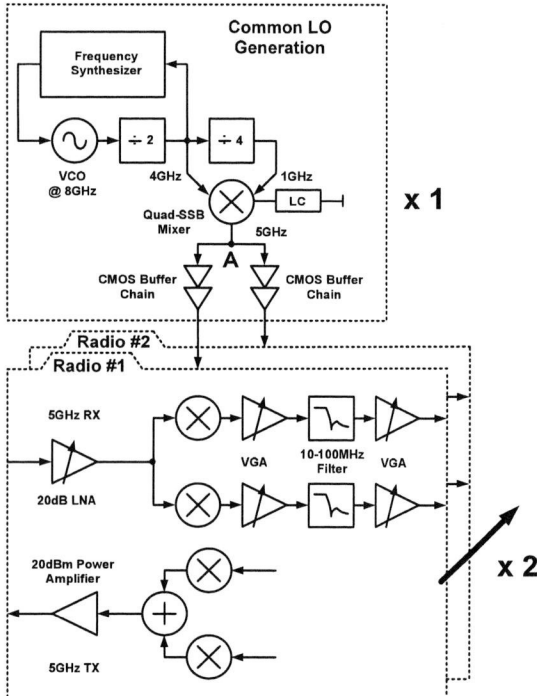

Fig. 4: The architecture of the fabricated 2x2 MIMO transceiver. The chip includes two receive chains, two transmit chains with integrated power amplifiers with P_{1dB}=20.5dBm, and a shared LO generation and distribution scheme specifically optimized for MIMO.

operation and cannot cancel the nonlinear crosstalk term. It is therefore important to keep nonlinear crosstalk small enough so that it does not degrade the system performance to begin with. For the fabricated system the nonlinear crosstalk was specified to be less than -35dB so that it does not raise the EVM floor of -27dB appreciably.

III. CHIP ARCHITECTURE & DESIGN

Figure 4 shows the architecture of the 5GHz direct-conversion 2x2 MIMO transceiver [12]. The system consists of two receivers, two transmitters and a shared LO generation circuit. Each of the two receivers consists of an inductively degenerated differential LNA, a pMOS-based Gilbert mixer, and a 6th order Gm-C elliptic filter interleaved with a 5-stage VGA. Each of the transmitters consists of a Gilbert upconversion mixer and an integrated, class-AB power amplifier with P_{1dB}=20.5dBm. A simple linearization scheme is used to reduce AM-PM distortion [12]. The matching networks of the PAs and the LNAs are realized as microstrips on the top layer of a multi-layer flip-chip package in order to get high quality factors and to minimize external component count (Fig. 5).

In order to minimize detrimental MIMO coupling (see Section II), the two transmitters and associated matching networks are placed at the maximum possible distance on the die and the package, respectively (see Fig. 5). Several other isolation techniques are used to minimize MIMO coupling,

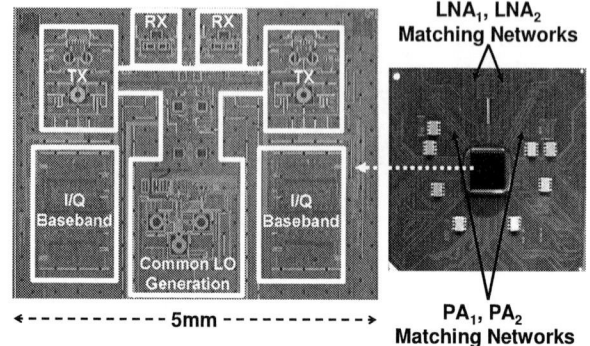

Fig. 5: Die shot (left) and package photo (right).

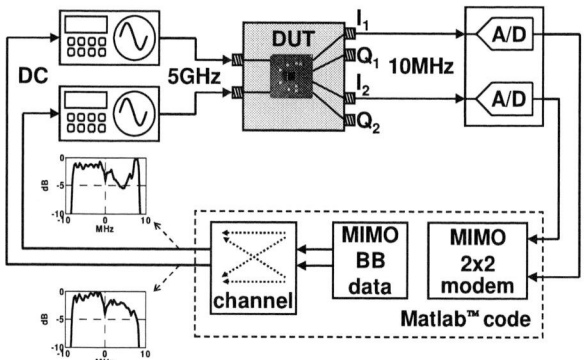

Fig. 6: The test setup used to characterize the MIMO receiver. The channel fading was emulated in software. Multiple different channels were used for statistical analysis.

e.g. separate supplies, extensive decoupling on supplies and bias lines, high-resistivity isolation layers in the substrate; ground shields, differential signaling, etc.

The LO is shared between the two MIMO paths. This reduces chip area overhead and improves pilot-based phase noise tracking since it provides two estimates of the *same* phase noise at the receiver [12]. The 5GHz LO is generated from an 8GHz synthesizer by division and single side-band mixing as shown in Fig. 4. Operating the VCO at a frequency non-harmonically related to the RF frequency minimizes VCO pulling by the integrated PAs, thus reducing nonlinear MIMO crosstalk. We exploit CMOS scaling and use rail-to-rail CMOS buffer chains to drive the mixers. CMOS inverter buffers have larger swings and present a lower impedance than low-swing LC-loaded buffers, thereby minimizing nonlinear coupling through the LO. Two separate chains of inverter buffers are used to drive the two transceivers as shown in Fig. 4. The two buffer chains exhibit significant reverse isolation minimizing undesired crosstalk from the two transmitters to the common LO node *A* in Fig. 4. If a common LO distribution buffer chain had instead been used, there would have been no isolation between the LO ports of the upconversion mixers of the two transmitters.

IV. EXPERIMENTAL RESULTS

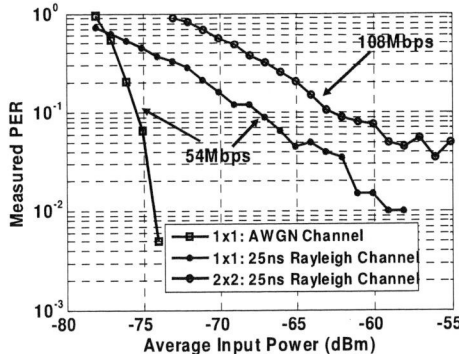

Fig. 7: Measured packet error rate (PER) as a function of average received power. The 2x2 MIMO receiver achieves a sensitivity of -62.5dBm while receiving 108Mb/s in the presence of multipath fading (25ns Rayleigh channel).

The 2x2 MIMO transceiver was fabricated in a 90nm CMOS process and occupies a total die area of 18mm^2 (Fig. 5). The transceiver was tested using a custom MIMO modem implemented in software as shown in Fig. 6. The software generates MIMO OFDM signals and passes them through an appropriate model of a 2x2 radio channel. Random Rayleigh channels with 25ns rms delay spread, typical of office environments [1], were used during the testing. The same effect can be generated by passing the unfaded OFDM signals through a combination of adjustable attenuators and phase shifters to emulate the multipath fading effect. The resulting faded RF signals are then applied to the two receivers of the MIMO chip. The outputs are then digitized, passed through timing and frequency synchronization routines, MMSE channel equalization [1], and Viterbi decoding, all implemented in software. Nowadays commercial solutions exist for characterization of MIMO systems supporting many different available MIMO modes [13].

Fig. 7 shows measured waterfall curves (PER versus input power) obtained from the fabricated prototype. Two hundred random channels per data point were used to obtain statistically significant results. In legacy 54Mb/s mode, each receiver achieves a sensitivity of -75.5dBm (PER=10%) in the presence of additive white Gaussian noise (AWGN). Due to multi-path fading, this sensitivity degrades to -68dBm for a 25ns Rayleigh channel. In the 108Mb/s 2x2 MIMO spatial multiplexing mode the corresponding sensitivity is -62.5dBm. The 2x2 MIMO sensitivity is seen to be somewhat worse than the SISO sensitivity. However, the MIMO system achieves double the data rate of its SISO counterpart. Achieving 108Mb/s in SISO mode would require very high SNR levels. If we account for the different data rates it can be estimated that the 108Mb/s 2x2 MIMO transceiver actually has a 12.5dB *SNR* advantage over a conventional 1x1 transceiver [12].

In legacy 1x1 mode the transmitter delivers an average power of +16dBm (EVM=-25dB) with an efficiency of 7%. In the 2x2 mode each PA delivers an average power of +13dBm while meeting the more stringent EVM of -27dB required for MIMO.

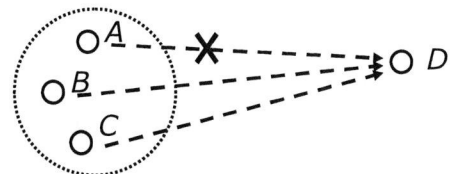

Fig. 8: In collaborative MIMO device *A* collaborates with nearby devices *B* and *C* to transmit a message to *D*.

V. MIMO Evolution

Wireless technologies continue to enjoy an explosive growth. The key trends that are expected to shape the direction of wireless communications in the upcoming years are:

- Mobile devices are becoming smaller for increased mobility.

- Wireless networks are becoming more crowded with more users and more bandwidth intensive applications.

- Mobile devices are becoming smarter and can optimize their performance according to their environment, e.g. presence of interfering signals.

- Higher data rates are required e.g. for synch-&-go applications and wireless HD video.

MIMO techniques have been evolving to adapt to these new requirements. Next we give examples of these new MIMO trends [14].

Collaborative MIMO: As mobile devices become smaller, the number of antennas that can be deployed on a single device is also limited. Packing many antennas in a small volume would result in unacceptable radiation efficiency due to mutual coupling between the antennas, and also detrimental MIMO crosstalk as discussed in Section II. Cost considerations might also limit the number of antennas in a wireless device. Collaborative MIMO is a new concept that can be used to alleviate these issues [15].

In collaborative MIMO independent devices establish virtual MIMO clusters that collaborate to improve the overall network performance (see Fig. 8). As an example consider the case where device *A* is having difficulty communicating with device *D* e.g. due to multipath fading. To overcome this problem, device *A* can share its message with nearby devices *B* and *C* and they can then jointly transmit it to device *D*. In the simplest form of collaborative MIMO, devices *B* and/or *C* can be used to simply relay the message to *D*. In order for this collaboration scheme to be effective, the collaborating nodes need to have a significant communication advantage, e.g. be much closer to each other than the final destination node *D*. Collaborative MIMO represents a paradigm shift in wireless communications in that it has wireless stations collaborate with each other rather than compete/interfere with each other as is the case in conventional wireless networks. As such collaborative MIMO is expected to reshape wireless communications when it actually takes off.

978-1-4244-2018-6/08 $25.00 © 2008 IEEE

Directional Communications: A receiver with multiple antennas can be viewed as a phased-array system. Such a receiver can use baseband or RF processing to effectively steer the phased-array in the direction of a desired incoming signal or away from an interfering signal. The same can be done for the transmitter that can point towards a desired device. Directional communications offer significant advantages: (1) reduce the power dissipation required for achieving certain range since the transmitted power is focused in the direction of the desired station only; (2) increase spatial reuse by allowing multiple independent radio links between independent pairs of devices in the same space and frequency. The concept of directional communications is rather old. Recently it has been receiving increased attention as networks become more crowded and more intelligence (processing power) is becoming available at the mobile station for practical deployment of the scheme. Directional communications are indispensable in 60GHz radio systems due to high path loss, as will explained below.

Pointing the transmitter and receiver phased-arrays towards each other can be a difficult problem. It is shown in reference [16] that this problem can be described as a singular value decomposition problem for the channel matrix. Under this formulation the optimal weight factors for the transmitter and receiver beamformers are the eigenvectors corresponding to the maximum eigenvalue. The problem is solved iteratively in [16] by sending packets back and forth between the transmitter and receiver, progressively steering the beams closer to the optimal directions in every step of the process. Reference [17] presents alternative beam-training algorithms. Directional communication is also particularly challenging at the MAC level. For example, when a device is transmitting or receiving directionally it might be unable to receive network messages (e.g. RTS – Request To Send) sent from different directions, an issue called "device deafness" [18]. Directional communications is an area of active research today.

60GHz Phased-array communications: The large unlicensed band around 60GHz can be used for wireless communications of few Gb/s over relatively short distances. The opportunities for multi-Gb/s data rates are numerous: Wireless Personal Area Networks (WPAN), wireless HDMI, synch-&-go and wireless docking station to name a few. Low-cost technologies like CMOS are already proving to deliver the performance required to build a reliable millimeter-wave wireless link [19][20][21][22][23][24][25]. In order to break through to the consumer market, the mm-wave technology will have to be able to deliver substantially higher data rate compared with competing technologies, e.g. UWB or 802.11n. A 2.5GHz RF channel with a 16QAM OFDM modulation and a standard ¾ coding rate would already be able to deliver a raw data rate of 5Gb/s. However, this opportunity for very high data rates comes with significant challenges. Due to the reduced effective aperture of the antennas, free-space loss as expressed by Friis equation [26] increases considerably at 60GHz. Antenna size, on the other hand, scales with wavelength. This allows the possibility to gain back the additional free-space loss by using multiple antennas

configured as a phased-array [27]. What kind of range can be expected from this technology? Let us assume an array of 36 elements at both the transmitter and receiver. For a P_{1dB} of 5dBm for a CMOS power amplifier [23][24][25] and 10dB back-off due to OFDM peak-to-average ratio (PAPR) as an example, the total average transmitted power is about 11dBm. With an antenna gain of 16dB at both transmitter and receiver corresponding to the 36 element array, and with 68dB of aperture loss at 1m, the total received power is -26dBm. If the receiver has a 10dB NF [28], the link margin at 1m for a 10dB Eb/N0 (energy per bit / noise power density) is about 31dB. The zero-margin distance is about 34m. If we include 3dB implementation loss for the receiver and 10dB margin for shading [29], the achievable distance drops to about 8m. This estimate assumes an AWGN channel. Of course, the range could be increased by increasing the number of antennas, but this might result in impractical power dissipation and array area and might violate maximum transmitted EIRP (Effective Isotropic Radiated Power) regulations. For the previous example, the EIRP is about 27dBm.

Directivity in the antennas introduces a new set of challenges. First of all, the TX and RX beams have to be steered towards each other such that the system can fully exploit the antenna gains. Due to the very large data rates involved, implementing the beamforming in the digital baseband can be very expensive in terms of power consumption. It is therefore preferred to directly implement the beamforming at RF, by phase-shifting each TX signal by the proper amount. This phase shift can be introduced at the LO signal used for the final mm-wave up-conversion [30][31] or after the up-conversion on the RF signal [27][32]. Related issues of phased-array training and directional MAC were touched upon in the earlier discussion on Directional Communications.

Due to its small size, the 60GHz antenna can be integrated on the CMOS IC [34][30][31][33]. The complete antenna array can still be rather large, e.g. approximately 15x15 mm² for a 6x6 element array. This can significantly increase the cost of the system-on-chip, especially if we take into account that the RFIC has to be implemented in a modern deep submicron process due to the high performance required for the transceiver. On-chip antenna performance can also be severely limited by the silicon substrate resistivity, metal thickness and maximum distance from the substrate for the top-most metal layer [33][34]. Instead of integrating the antennas on the CMOS RFIC, it might be more practical to embed the antenna structures on the package that is hosting the RFIC. Losses can be much lower even without using expensive, very low-loss materials. Low-cost package and antenna design at 60GHz is still very challenging, and issues like routing losses, coupling between the antennas, thermal expansion and IC/package interface will have to be carefully addressed. Recently, new organic-based materials like Liquid Crystal Polymer (LCP) are being investigated as a cheaper alternative to more expensive LTCC solutions to build mm-wave system-on-package modules [35].

978-1-4244-2018-6/08 $25.00 © 2008 IEEE

VI. CONCLUSIONS

This paper presented an overview of MIMO techniques for high data rate communications. We have demonstrated a practical implementation of an integrated 2x2 MIMO 5GHz transceiver fabricated in a 90nm CMOS process. We have introduced MIMO spatial multiplexing as a means for increasing spectral utilization and have intuitively established the need for high isolation between the different paths of the MIMO radio, antennas and radio channels. We have presented a radio IC design specifically optimized for high isolation and good MIMO performance, verified with experimental results. Finally, some interesting scenarios for future MIMO evolution have been described.

ACKNOWLEDGMENTS

The authors would like to thank B. R. Carlton, M. Elmala, R. Bishop, G. Banarjee, Rich Nicholls, S. Ling, Nati Dinur, S. S. Taylor, K. Soumyanath, S. Sandhu, Guoqing Li, Rick Boberts, Ed Casas, Menashe Soffer, David Steele, Minnie Ho, Keith Holt, Avi Biran, S. Somayazulu, Hossein Alavi, D. Martin, R. Wang, N. Yaghini, J. Gong, C. Zhan, J. Vu, D. Trammo, A. Crouch, and K. Kahn for their contributions.

REFERENCES

[1] M. Jankiraman, "Space-time codes and MIMO systems", Artech House, 2004.

[2] D. Gesbert, M. Shafi, Da-shan Shiu, P.J. Smith, and A. Naguib, "From theory to practice: An overview of MIMO space-time coded wireless systems," in *IEEE J. Sel. Areas Commun.*, vol. 21, no. 3, pp. 281–302, April 2003.

[3] A. Bourdoux, B. Come, and N. Khaled, "Non-reciprocal transceivers in OFDM/SDMA systems: impact and mitigation," in *Proceedings of IEEE Radio and Wireless Conference (RAWCON '03)*, pp. 183–186, Boston, Mass, USA, August 2003.

[4] S. Alamouti, "A simple transmit diversity technique for wireless communications," in *IEEE Journal on Selected Areas in Communications*, vol. 16, Oct. 1998, pp. 1451 – 1458.

[5] D. Gore, S. Sandhu, A. Paulraj, "Delay diversity codes for frequency selective channels," in *IEEE International Conference on Communications (ICC)*, vol. 3, 28 April-2 May 2002, pp. 1949 – 1953.

[6] J. W. M. Rogers et al, "A Fully Integrated Multi-Band MIMO WLAN Transceiver RFIC", *VLSI Circ. Symposium*, pp. 290 – 293, June 2005, Kyoto.

[7] T. Montalvo et al, "A Wireless Transceiver with Integrated Data Converters for 802.11a/b/g Access Points", in *IEEE ISSCC Dig. Tech. Papers*, 2006, pp. 1440 – 1449.

[8] A. Behzad et al, "A Fully Integrated MIMO Multi-Band Direct-Conversion CMOS Transceiver for WLAN Applications (802.11n)", in *IEEE ISSCC Dig. Tech. Papers*, 2007, pp. 560 – 622.

[9] O. Degani et al, "A 1x2 MIMO Multi-Band CMOS Transceiver with an Integrated Front-End in 90nm CMOS for 802.11a/g/n WLAN Applications", in *IEEE ISSCC Dig. Tech. Papers*, 2008.

[10] L. Nathawad et al, "A Dual-Band CMOS MIMO Radio SoC for 802.11n Wireless LAN", in *IEEE ISSCC Dig. Tech. Papers*, 2008.

[11] P802.11n/D3.00, Draft Standard for Information Technology-Telecommunications and information exchange between systems--Local and metropolitan area networks--Specific requirements-- Part 11: Wireless LAN Medium Access Control (MAC) and Physical Layer (PHY) specifications: Amendment 4: Enhancements for Higher Throughput, Status: Active Unapproved Draft, Publication Data 2007, http://ieeexplore.ieee.org/servlet/opac?punumber=4360106.

[12] Y. Palaskas et al, "A 5GHz 108Mb/s 2x2 MIMO Transceiver with Fully Integrated 20.5dBm P_{1dB} Power Amplifiers in 90nm CMOS", *IEEE J. of Solid State Circuits*, vol. 41, no. 12, Dec. 2006, pp. 2746-2756.

[13] Agilent Technologies, Application Note 1509.

[14] Y. Palaskas, A. Ravi, S. Pellerano, S. Sandhu, "Design considerations for integrated MIMO radio transceivers", in Wireless Technologies: Circuits, Systems, and Devices, CRC Press, Editor Kris Iniewski, 2007.

[15] P. Mitran, H. Ochiai, V. Tarokh, "Space-time diversity enhancements using collaborative communications," in *IEEE Transactions on Information Theory*, vol. 51, June 2005, pp. 2041 – 2057.

[16] Yang Tang, B. Vucetic, Yonghui Li, "An iterative singular vectors estimation scheme for beamforming transmission and detection in MIMO systems", *IEEE Communications Letters,* Vol. 9, No. 6, pp. 505-507, June 2005.

[17] F. Alam, Donghee Shim, B. D. Woerner, "Comparison of low complexity algorithms for MSNR beamforming", *IEEE Vehicular Technology Conference*, Vol. 4, pp. 1776-1780, 2002.

[18] R. R. Choudhury and N. H. Vaidya, "Deafness: a MAC problem in ad hoc networks when using directional antennas," in *IEEE International Conference on Network Protocols*, Berlin, 2004.

[19] B. Floyd, S. Reynolds, U. Pfeiffer, T. Zwick, T. Beukema, B. Gaucher, "SiGe bipolar transceiver circuits operating at 60 GHz", in *IEEE J. of Solid State Circuits*, vol. 40, no. 1, Jan. 2005, pp. 156 – 167.

[20] B. Razavi, "A 60-GHz CMOS receiver front-end", in *IEEE J. of Solid-State Circuits*, pp. 17-22, Jan. 2006.

[21] C.H. Doan, S. Emami, A.M. Niknejad and R.W. Brodersen, "Millimeter-wave CMOS design," in *IEEE J. of Solid-State Circuits*, pp. 144-155, Jan. 2005.

[22] Chieh-Min Lo, Chin-Shen Lin and Huei Wang, "A miniature V-band 3-stage cascode LNA in 0.13um CMOS," in *IEEE ISSCC Dig. Tech. Papers*, 2006, pp. 1254-1263.

[23] T. Yao, M. Q. Gordon, K. K. W. Tang, K. H. K. Yau, M.-T. Yang, P. Schvan and S. P. Voinigescu, "Algorithmic Design of CMOS LNAs and PAs for 60-GHz Radio," in *IEEE J. of Solid-State Circuits*, pp. 1044 - 1057, May. 2007.

[24] Yanyu Jin, Eduardo Alarcon Rivero, Mihai A.T. Sanduleanu and John R. Long, "A Millimeter-Wave Power Amplifier with 25dB Power Gain and +8dBm Saturated Output Power," in *Proc. 2007 European Solid-State Circuit Conf. (ESSCIRC)*, Sept. 2007, pp. 276 – 279.

[25] Mikko Varonen, Mikko Kärkkäinen, and Kari Halonen, "Millimeter-Wave Amplifiers in 65-nm CMOS," in *Proc. 2007 European Solid-State Circuit Conf. (ESSCIRC)*, Sept. 2007, pp. 280 – 283.

[26] D. M. Pozar, Microwave Engineering, 2nd ed., John Wiley & Sons.

[27] Sayf Alalusi and Robert Brodersen, "A 60GHz phased array in CMOS", in *IEEE Proc. of the Custom Integrated Circuits Conference*, 2006.

[28] B. Razavi, "A Millimeter-Wave CMOS Heterodyne Receiver With On-Chip LO and Divider," in *IEEE J. of Solid-State Circuits*, pp. 477-485, Feb. 2008.

[29] Hao Xu, Vikas Kukshya and Theodore S. Rappaport, "Spatial and Temporal Characteristics of 60-GHz Indoor Channels," in *IEEE J. on Selected Areas in Communications*, vol. 20, no. 3, Apr. 2002.

[30] A. Natarajan, A. Komijani, X. Guan, A. Babakhani, A. Hajimiri, "A 77-GHz Phased-Array Transceiver With On-Chip Antennas in Silicon: Transmitter and Local LO-Path Phase Shifting," in *IEEE J. of Solid-State Circuits*, Vol. 41, pp. 2807 – 2819, Dec. 2006.

[31] A. Natarajan, A. Komijani, X. Guan, A. Babakhani, A. Hajimiri, "A 77-GHz Phased-Array Transceiver With On-Chip Antennas in Silicon: Receiver and Antennas," in *IEEE J. of Solid-State Circuits*, Vol. 41, pp. 2795 – 2806, Dec. 2006.

[32] Kwang-Jin Koh, G.M. Rebeiz, "0.13-μm CMOS Phase Shifters for X-, Ku-, and K-Band Phased Arrays," in IEEE J. of Solid-State Circuits, Vol. 42, pp. 2535 - 2546, Nov. 2007.

[33] B. Razavi, "CMOS transceivers for the 60-GHz band," in *Proc. of IEEE Radio Frequency Integrated Circuits (RFIC) Symposium*, 2006.

[34] B. Floyd, Chih-Ming Hung, K. K. O, "Intra-chip wireless interconnect for clock distribution implemented with integrated antennas, receivers, and transmitters", in *IEEE J. of Solid State Circuits*, vol. 37, no. 5, May 2002, pp. 543 – 552.

[35] Manos M. Tentzeris, Joy Laskar, John Papapolymerou, , Stéphane Pinel, V. Palazzari, R. Li, G. DeJean, N. Papageorgiou, D. Thompson, R. Bairavasubramanian, S. Sarkar, and J.-H. Lee "3-D-integrated RF and millimeter-wave functions and modules using liquid crystal polymer (LCP) system-on-package technology," in *IEEE Transactions on Advanced Packaging*, Vol. 27, pp. 332 – 340, May 2004.

978-1-4244-2018-6/08 $25.00 © 2008 IEEE

IEEE 2008 Custom Intergrated Circuits Conference (CICC)

Wireless Interconnection within a Hybrid Engine Controller Board

Swaminathan Sankaran, Kyujin Oh, Hsinta Wu, and Kenneth K. O

Silicon Microwave Integrated Circuits Research Group (SIMICS), Department of Electrical and Computer
Engineering, University of Florida, Florida – 32611.

Abstract- **Wireless interconnection within a hybrid engine controller board has been demonstrated by demodulating an AM signal transmitted with an on-chip antenna using a receiver located in the board. The receiver centered between 14 and 16GHz uses a Schottky diode detector and should be able to support 400-Mbps data rate and ~15cm range. The receiver does not require a crystal frequency reference. The measured peak conversion gain is ~35dB, sensitivity over the band with minimum E_b/N_o of 14dB for ASK is -58dBm. The receiver occupies ~1.5mm^2 and consumes ~60mW.**

I. INTRODUCTION

Hybrid electric vehicles (HEV) with improved fuel efficiency are gaining in popularity. The engines of hybrid electric vehicles are controlled by a PC board which is composed of low voltage control and high voltage motor driver sections (Fig. 1). The floating voltage on the return paths of motor driver section can be greater than 300V [1]. This high voltage is isolated from the return paths of low voltage section using photo-couplers. Wireless interconnects using single chip radios including an on-chip antenna can be a low cost alternative to the photo couplers. In addition, wireless interconnects can support data rate greater than 100's of Mbits/sec and reduce the board size by eliminating wiring traces. Since the board is placed in a metallic enclosure, the frequency bands and channels can be allocated without the concerns for satisfying the emission rules set by government agencies.

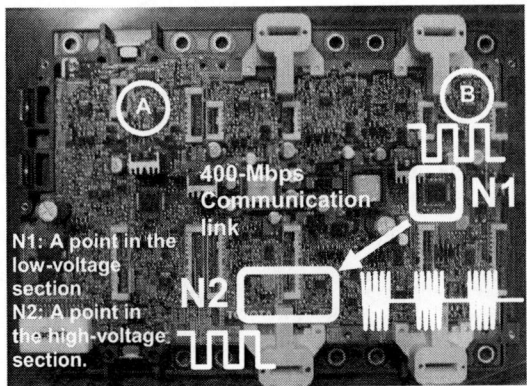

Fig. 1. Hybrid engine driver PC board. Indicated are the high/low voltage sections and typical waveforms along the communication link.

This paper suggests the feasibility of such a system by demonstrating a receiver in 130-nm foundry CMOS, and picking up and demodulating an AM modulated signal in the engine controller board. The receiver operates from 14-16GHz, and should be able to support up to 400-Mbps data

rate. For detection, the receiver uses high cut-off frequency Schottky Barrier Diodes (SBD's) fabricated in CMOS [2] without process modifications. Using diode detection removes a synthesizer and mixer drivers thereby simplifying the radio and reducing the power consumption. More importantly, this also eliminates the need for a crystal frequency reference, which is critical for making the approach a cost effective alternative to the photo-couplers.

II. CIRCUIT DESIGN

Fig. 2 shows the block diagram of receiver, and sample waveforms at various points in the receiver. The receiver employs a 4-mm on-chip dipole antenna constructed using 3 shunted metal layers (Metal 6-8). A duplexer [3] precedes a differential cascode low noise amplifier (LNA) with inductive source degeneration [4]. The duplexer isolates signals above 20GHz from TX which can be incorporated to facilitate full-duplex communication. A low input Q (~1.5) is used for the LNA to accommodate for the broad bandwidth and matching to the duplexer output. A two-section capacitive coupled filter with ~500-Ω termination is used at the LNA output. The 500-Ω termination is chosen to increase voltage gain and make the filter component values suitable for implementation.

The differential LNA output is combined using an active current-mirror balun. The $1/g_m$ low-impedance node of balun is tuned to increase the in-band gain and to make its frequency response flat. Multiple filter stages along the RF signal path are intended to achieve flat in-band gain as well as to reject the signal from TX that will eventually be integrated. The SBD is connected in shunt as a half-wave rectifier for signal detection [5]. The diode is biased using a polysilicon resistor, and its size and bias are set to maximize the output [5]. The down-converted signal is filtered using a 4th order 1.2-GHz wide Chebychev low-pass filter. The filter is followed by a 3-stage amplifier and a 4-stage tapered buffer for driving the 50-Ω input of instruments. A die photo of the receiver is shown in Fig. 3. All the circuits and components are integrated. The receiver is tested on a printed circuit board.

III. MEASUREMENT RESULTS

RF stage by itself has ~14-dB gain, 7-dB noise figure (NF), and is reasonably matched (<-7dB) at the input from 14-16GHz as shown in Fig. 4. The 4th order LPF-baseband stage provides a gain of ~33dB with 3-dB bandwidth of ~0.9GHz (Fig. 5). A receiver without a duplexer is characterized using 12-16GHz input signals amplitude modulated with 200-MHz sine wave and 50% modulation index. The modulation

This work is supported by Toyota Motor Corp. and Toyota InfoTechnology Center Co., Ltd.

978-1-4244-2018-6/08 $25.00 © 2008 IEEE

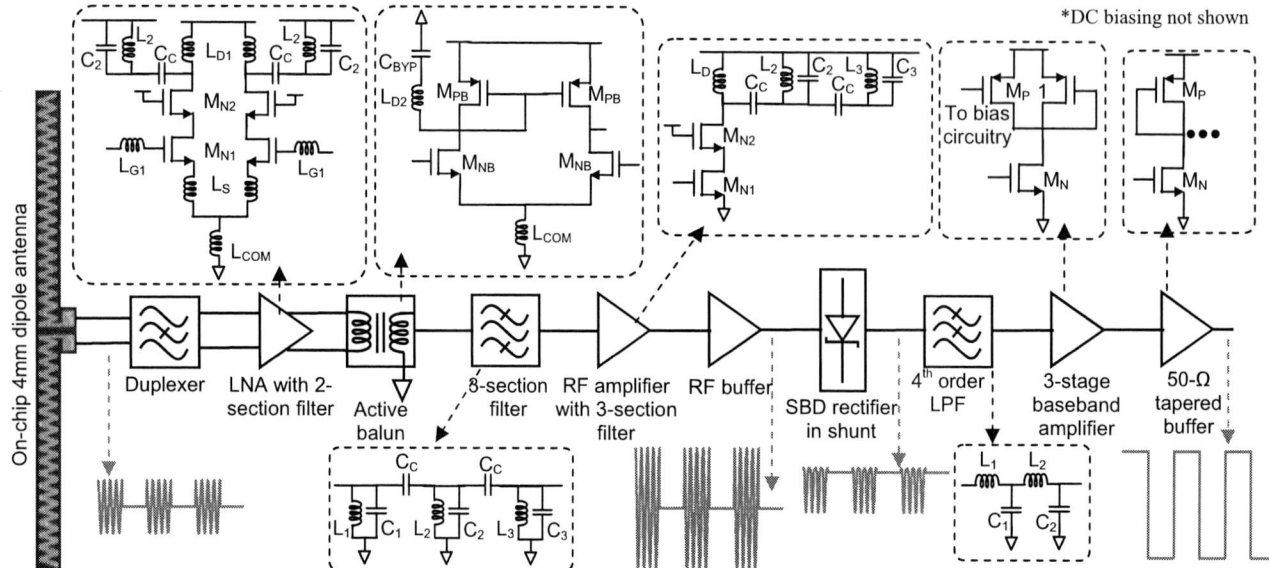

Fig. 2. Block diagram and schematic of the AM receiver.

frequency is selected to emulate the waveforms of 400-Mbps data. The AM signal is generated using a broad-band single-sideband RF mixer.

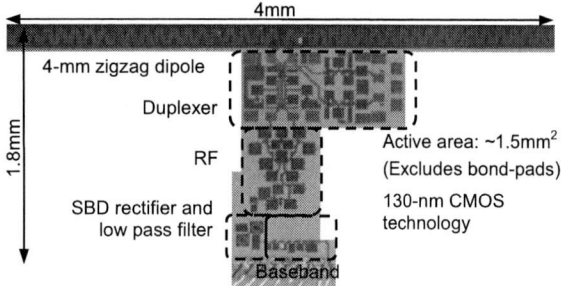

Fig. 3. Die photo of the receiver.

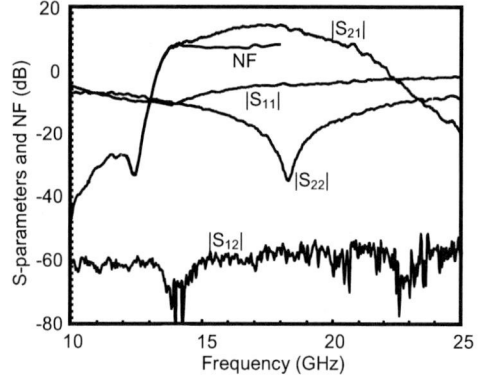

Fig. 4. Measured magnitudes of S-parameters and NF of RF section.

Fig. 6(a) plots the output power (P_{out}) at 200MHz versus diode bias current. The input signal frequency is 14.5GHz. The sideband power (P_{in}) is -54dBm. The output power is maximized at the diode bias current of ~130μA. An optimal SBD bias occurs as a result of the tradeoff between conversion gain and loading of the preceding RF stage with the bias current increase. A plot of P_{out} versus carrier frequency ($f_{carrier}$)

is shown in Fig. 6(b). P_{out} fluctuates by ~3dB over 14-16GHz. The receiver response is centered at a lower frequency and has smaller bandwidth compared to $|S_{21}|$ of the RF section by itself. This is due to the mistuning at the input and output ports as well as the high-frequency limitation of mixer (~18 GHz) used to generate the AM signal.

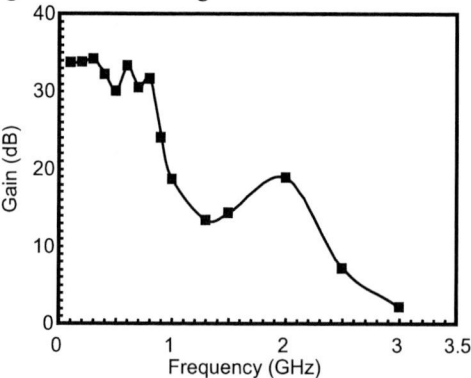

Fig. 5. Frequency response of LPF-baseband stage.

Fig. 7(a) plots P_{out} at 200MHz versus the sideband power, P_{in}. The input is centered at 14.5GHz. The square-law region marked with a slope of 2 is over 37dB at the input [6]. The peak conversion gain occurs when P_{in} is ~-45dBm and is ~35dB. The RX noise is measured by terminating the RF input with 50-Ω and observing the output noise [5] (Fig. 7(b)). The total integrated output noise to 1 GHz is ~-42dBm. From Fig. 7(a), this corresponds to P_{in} of ~-64dBm which is the receiver input noise floor. This translates to a system NF of 20dB and ~-58dBm receiver sensitivity for an E_b/N_o of 14dB (bit error rate of 10^{-13} for ASK) and 400-Mbps data rate.

The input and output signals in time and frequency domain are shown in Figs. 8(a)-(d). P_{in} is -54dBm and is centered at 14.5 GHz. Figs. 8(c) and (d) indicate successful recovery of the 5ns/200MHz baseband signal with adequate rejection of higher order harmonics.

978-1-4244-2018-6/08 $25.00 © 2008 IEEE

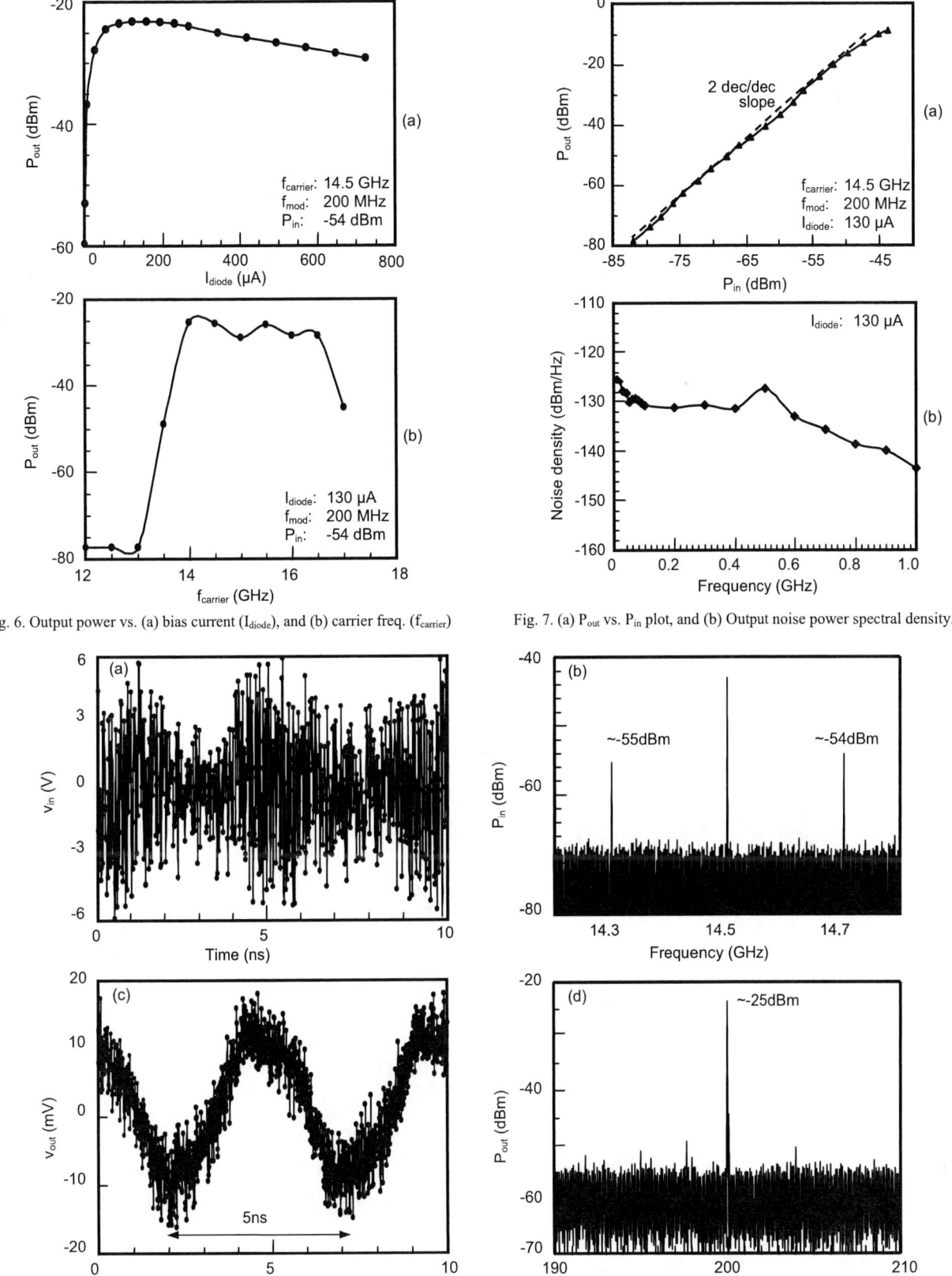

Fig. 6. Output power vs. (a) bias current (I_{diode}), and (b) carrier freq. ($f_{carrier}$)

Fig. 7. (a) P_{out} vs. P_{in} plot, and (b) Output noise power spectral density.

Fig. 8. Input AM signal observed on (a) an oscilloscope, (b) spectrum analyzer and 5ns, 200 MHz output seen on (c) an oscilloscope and (d) spectrum analyzer.

The gain is ~30dB lower than needed to achieve ~100mV peak output at baseband when the input is at ~-58dBm. The RF stage gain can be raised by 10dB using inductors with top two metal layers shunted instead of just the top metal layer. This should increase the detector conversion gain by 20dB. This should also reduce NF by 10dB. Adding another baseband amplification stage should increase the total gain by 10dB.

The presence of metal traces, capacitors, heatsinks, and other components on the controller board is expected to make wireless communication challenging. The measurement setup shown in Fig. 9 is used to evaluate the feasibility of wireless link within the motor control board. A 200-MHz AM signal centered at 14.5GHz is transmitted from point A in Fig. 1 using a 4-mm dipole antenna. A signal with ~2dBm sideband power is converted to differential and fed to the antenna.

A metallic cover is placed ~3.5cm above the controller board to emulate the operation environment. The cover surprisingly reduces the propagation loss and impact of multi-path effects [7]. The signal is picked up at point-B using an RX (with the antenna and duplexer). The separation between points A and B is ~15cm, which is the maximum needed for the system. The transmitted and received signals are also shown in Fig. 9. A 200-MHz received signal with ~-38dBm power indicates a channel loss including those for on-chip antennas of ~60dB (from Fig. 6(a), measured 3-dB duplexer loss). The individual on-chip antenna gain is ~-8dBi with ~10% efficiency. The corresponding channel loss is ~4dB larger than the ideal free-space path loss (~40dB at 15GHz and 15cm range).

IV. CONCLUSION

A receiver capable of demodulating AM signals centered around 14-16GHz and supporting up to 400-Mbps data rate is presented. The receiver implemented in a 130-nm CMOS foundry process consumes ~60mW of power, occupies ~1.5mm^2 of area (circuits + antenna), and does not require a crystal frequency reference. This work suggests that wireless communication within the engine controller board should be feasible. Using this, it should be possible to interface sub-sections with floating return path voltage differences greater than 1000V. The wireless interconnects could potentially be a lower cost and higher data rate alternative to the photo-couplers on a hybrid engine controller board.

REFERENCES

[1] A. K. Kawahashi, "A New-Generation Hybrid Electric Vehicle and Its Supporting Power Semiconductor Devices," *IEEE Power Semiconductor Devices and ICs Symposium Digest of Technical Papers*, pp. 23-29, May 2004.

[2] S S. Sankaran and K. K. O, "Schottky Barrier Diodes for Millimeter Wave Detection in a Foundry CMOS Process," *IEEE Electron Device Letters*, vol. 26, no. 7, pp. 492-494, July 2005.

[3] J. -J. Lin, H. T. Wu and K. K. O, Submitted for publication.

[4] X. Guo and K. K. O, "A Power Efficient Differential 20-GHz Low Noise Amplifier with 5.3 GHz 3-dB Bandwidth," *IEEE Microwave and Wireless Component Letters*, pp. 603-605, Sept. 2005.

[5] S. Sankaran and K. K. O, "A Ultra-Wideband Amplitude Modulation (AM) Receiver Using Schottky Barrier Diodes in Foundry CMOS Technology," *IEEE Journal of Solid State Circuits*, vol. 42, no. 5, pp 1058-1064, 2007.

[6] T. Zhang, W. R. Eisenstadt, R. M. Fox and Q. Yin, "Bipolar Microwave RMS Power Detectors," *IEEE Journal of Solid State Circuits*, pp. 2188-2192, Sept. 2006.

[7] H. T. Wu, J. -J. Lin and K. K. O., Submitted for publication.

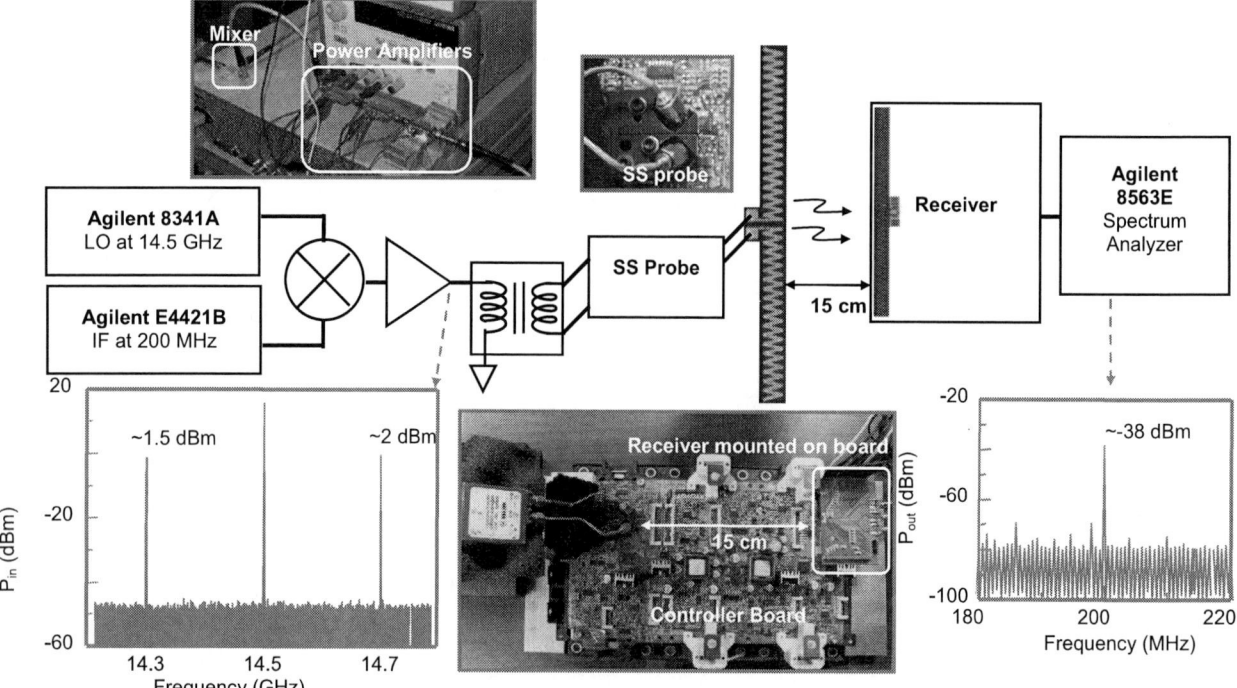

Fig. 9. Wireless link demonstration and setup.

Panel Discussion   **Session 10**

Sure, Moore's Law Can Continue, but Should It:

Panel Moderator: David Sunderland, Boeing Technical Fellow, Boeing Satellite Systems

Panelists:

Kazuyuki Kawauchi
President & CEO
Fujitsu Microelectronics America

John Kent
Vice President
Worldwide Technology R&D
ON Semiconductors

Randy Mooney
Director, Circuits Lab
Intel Corporation

Chuck Moore
Senior Fellow
Advanced Micro Devices

Prof. Clark T.-C. Nguyen
Dept. of EECS
University of California at Berkeley

Moore's Law scaling of semiconductor devices has provided a metronome for semiconductor progress for the past 45 years. While developing semiconductor products on the Moore's Law curve makes economic and technical sense for advanced processors and high-volume SOCs/ICs, increasingly, many products can be created which achieve performance and business goals off the Moore's Law curve. This panel will explore the various dimensions of building products (and the underlying technology) on and off the Moore's Law curve, and the panelists will share their perspective on how the semiconductor business will potentially evolve in taking advantage of design and process/device technologies on these several dimensions, to optimize the business and economic value of new products.

978-1-4244-2018-6/08 $25.00 © 2008 IEEE

Notes

Monday Poster Session

Notes

IEEE 2008 Custom Intergrated Circuits Conference (CICC)

A 0.6-to-1V Inverter-Based 5-bit Flash ADC in 90nm Digital CMOS

Jonathan E. Proesel and Lawrence T. Pileggi
Carnegie Mellon University
5000 Forbes Avenue, Pittsburgh, PA 15213 USA
{jproesel, pileggi}@ece.cmu.edu

Abstract–A 0.6-to-1V inverter-based 5-bit flash ADC in 90nm digital CMOS is presented. Single-ended comparators are formed using digital inverters and resistors. The comparators are designed for compatibility with nanoscale CMOS lithography. A single-ended flash architecture was used without a front-end sample-and-hold. The ADC achieves a low frequency effective number of bits (ENOB) between 4.08 bits and 4.45 bits without calibration. Voltage scaling is demonstrated by 60 MS/s, 300 MS/s, and 600 MS/s operation at 0.6 V, 0.8 V, and 1 V, respectively. Power scales from 1.3 mW to 6.7 mW.

I. INTRODUCTION

Digital CMOS layouts are becoming more regular [1] and are beginning to follow more restricted design rules [2] due to sub-wavelength lithography and manufacturing challenges. Analog circuits have always been laid out in a regular pattern style with controlled pattern neighborhoods to control systematic variations. Now analog circuits must be patterned under additional constraints for compatibility with the lithography and device topologies that will be used for scaled CMOS digital circuits.

Scaled CMOS transistors have problems beyond lithographic restrictions. Systematic variability can be controlled by regular layout, but the impact of random mismatch is then dominant and increasing with scaling. Decreasing supply voltage for digital technologies reduces circuit headroom, hence signal swing range. And the intrinsic gain ($g_m r_o$ product) for 65nm digital CMOS transistors is on the order of 10 V/V, thereby greatly limiting performance margin.

The mask patterns used to create a simple single-ended open-loop CMOS amplifier can be constructed to be very similar to those used to form a digital inverter to ensure lithography compatibility. The short transistor stack for an inverter amplifier topology also makes it compatible with low supply voltages. And the g_m of the inverter is based on the combined gain for both the PMOS and NMOS transistors.

Inverter topologies have been used as an open-loop CMOS amplifiers to provide the aforementioned benefits. In [3], a flash ADC uses an autozeroed inverter as an open-loop comparator [3]. More recently, $\Sigma\Delta$ ADCs have been developed using inverters as amplifiers [4, 5] and comparators

This work was supported by the MARCO/DARPA Center for Circuit and System Solutions (C2S2) and the Industrial Technology Research Institute (ITRI). Chip fabrication was provided by the U.S. Department of Defense Trusted Foundry Access program and IBM.

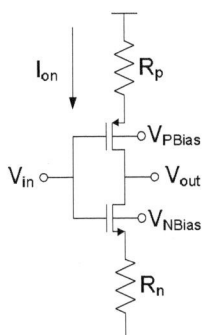

Fig. 1. Comparator circuit schematic.

[5], along with other ADCs in [6, 7].

This paper presents a 5-bit flash ADC built using open-loop inverter-based comparators. The comparator design eliminates the need for the separate resistor ladder used in [3, 5] and allows supply voltage scaling, as in [7].

II. COMPARATOR DESIGN

A. Threshold Point

An open-loop amplifier with an adjustable threshold point is used to form the comparator (Fig. 1). The comparator consists of a CMOS inverter and two resistors, R_p and R_n. The inverter provides open-loop amplification, while the resistors shift the threshold point. The inverter without resistors is ratioed to set the nominal threshold point to $0.5 \cdot V_{DD}$. The current drawn by the comparator when it is near its threshold point, I_{on}, creates voltage drops across the resistors. When the voltage drops are symmetric, the threshold point will be nominally set to $0.5 \cdot V_{DD}$. Asymmetry shifts the threshold point linearly with respect to the difference in voltage drops:

$$V_{threshold} = \frac{V_{DD}}{2} + \frac{R_n I_{on} - R_p I_{on}}{2}. \quad (1)$$

Controlling the difference in voltage drops in a consistent and linear fashion allows comparators with linearly varying threshold points. By setting the total resistance, $R_t = R_n + R_p$, to a fixed value, an array of converters can vary $V_{threshold}$ across a voltage range of $R_t \cdot I_{on}$. In a flash ADC, this voltage range is the full scale range (FSR) of the converter. Linear stepping is achieved by dividing R_t into n unit resistors of R_s, where n is the number of comparators desired. R_p and R_n are realized as integer multiples of R_s. The resulting built-in reference voltages eliminate the need for a reference ladder, similar to [7].

978-1-4244-2018-6/08 $25.00 © 2008 IEEE

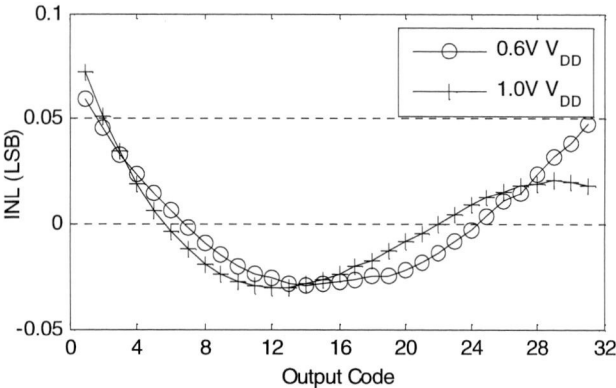

Fig. 2. Simulated INL for 5-bit flash ADC.

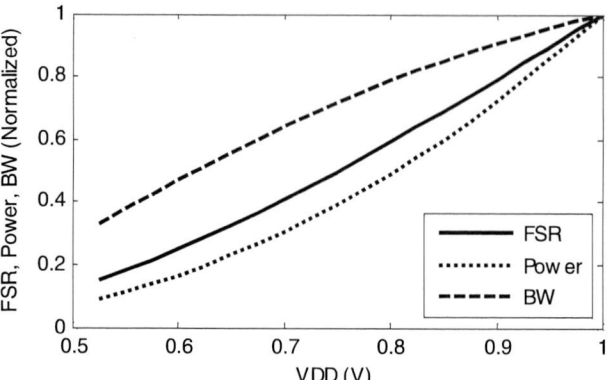

Fig. 4. Simulated scaling of FSR, Power, and BW with V_{DD}.

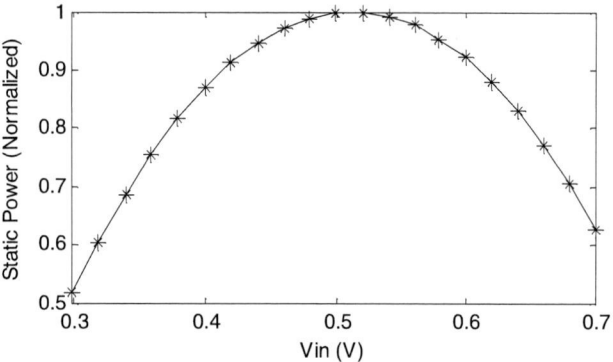

Fig. 3. Simulated array static power v. input voltage @ 1V V_{DD}.

A key assumption of (1) is that the threshold point is the input where the output and input are equal. However, the digital circuitry following the comparator has a threshold point of $0.5 \cdot V_{DD}$. Redefining the threshold point to be the input where the output is $0.5 \cdot V_{DD}$, the finite gain of the comparator shifts the threshold point. Assuming the gain is constant in the region of interest and equal to A,

$$V_{threshold} = \frac{V_{DD}}{2} + \frac{R_n I_{on} - R_p I_{on}}{2} \cdot \frac{A-1}{A} . \quad (2)$$

The FSR is reduced to $R_t \cdot I_{on} \cdot (A - 1)/A$.

The well biases V_{NBias} and V_{PBias} are shared by all comparators to tune for process and environmental variations. Adjusting the well biases changes the transistor threshold voltages, thereby adjusting the comparator threshold points and I_{on}. The nominal voltages for the well biases are GND for V_{NBias} and V_{DD} for V_{PBias}.

Systematic nonlinearity of the threshold points is primarily caused by two mechanisms, variable gain and body effect. The comparator's nonlinear transfer function and the varying source degeneration cause the gain to vary around the threshold point, adding nonlinearity to A in (2). The transistors' bodies are not source-tied, causing the body effect to change the transistors' V_T and adding nonlinearity to both the $0.5 \cdot V_{DD}$ and I_{on} terms in (2). Despite the presence of two systematic nonlinearity sources, the simulated integral nonlinearity (INL) for a 5-bit ADC remains below 0.1 LSB (Fig. 2).

B. Bandwidth

The bandwidth of the comparators is decreased by the resistors, which act as source degeneration. The values for R_p and R_n used in the comparator array result in different bandwidths and delays for each comparator. The differences in bandwidth and delay among the comparators affect the ADC performance for high frequency input signals. The effect of the different bandwidths and delays can be minimized by moving the comparators from continuous to discrete time.

C. Power Consumption

Each comparator draws static power only when the input is near the threshold point, similar to the CMOS inverter. This reduces the power consumption of the comparator array since most comparators will not be near their threshold points. The power reduction of the array is illustrated in Fig. 3, which shows the total static power consumption across the input FSR. This provides a simple means of putting the ADC into standby mode by grounding the input.

D. Voltage Scaling

The comparator exhibits similar behavior to the CMOS inverter with voltage scaling. I_{on} decreases as V_{DD} decreases, causing bandwidth, power, and FSR to drop. Bandwidth drops with V_{DD} due to the rising R_{out}. Power drops due to the drop in I_{on} and V_{DD}. FSR drops due to the drop in $R_t \cdot I_{on}$. Simulated results are presented in Fig. 4. Since the power scales faster than the bandwidth, the ADC is expected to show increased power efficiency as V_{DD} decreases.

E. Layout and Sizing

The CMOS inverter layout is highly regular to minimize systematic variations. A fixed FEOL (front end of line) layout pattern of resistors is added nearby. Metal-mask configuration is used to set R_p and R_n. The inverter and resistors are sized statistically to keep random mismatch below 0.5 LSB. The statistically-sized comparator array accounts for over 90% of the power consumption of the ADC. Comparator calibration techniques to reduce mismatch can considerably reduce the total power.

978-1-4244-2018-6/08 $25.00 © 2008 IEEE

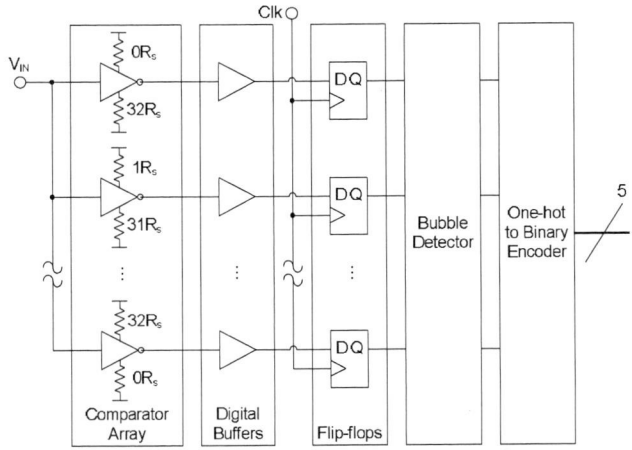

Fig. 5. Architecture of the 5-bit flash ADC.

III. FLASH ADC ARCHITECTURE

Fig. 5 shows the flash ADC architecture. The first stage consists of 33 comparators, stepped from $R_p = 0 \cdot R_s$ to $R_p = 32 \cdot R_s$ and $R_n = 32 \cdot R_s$ to $R_n = 0 \cdot R_s$. The highest and lowest threshold comparators provide over- and under-range detection, while the middle 31 provide the A/D conversion. The outputs of the first stage comparators are fed to digital buffers for amplification to digital levels.

Both the comparators and digital buffers operate in continuous time. As a result, these stages are vulnerable to the bandwidth and delay variations. The addition of a front-end sample-and-hold (S/H) could shift their operation to discrete time. A S/H was not used in this design in order to better characterize the dynamic performance of the inverter-based comparator array.

The digital outputs of the buffers are sampled by low-metastability flip-flops. The bubble detector converts the thermometer code to one-hot while suppressing bubbles in the thermometer code. The one-hot code is encoded as the 5-bit binary output.

IV. EXPERIMENTAL RESULTS

The ADC is fabricated in a 90nm digital CMOS process. Low V_T transistors are used in the comparators for higher bandwidth; all other transistors are standard V_T. Fig. 6 shows the die micrograph and ADC core layout. The ADC occupies 0.11 mm^2 of active area, excluding I/O and decoupling.

Testing was conducted at supply voltages of 0.6 V, 0.8 V, and 1 V V_{DD} for a typical chip. The well biases were at their nominal voltages during testing. The sampling rate is 60 MS/s, 300 MS/s, and 600 MS/s at 0.6 V, 0.8 V, and 1 V V_{DD}, respectively.

Fig. 7 shows the differential nonlinearity (DNL) and INL for 0.6 V, 0.8 V, and 1 V V_{DD}s with a 1 MHz input. DNL and INL are similar across all supply voltages, with DNL increasing as V_{DD} decreases. The increase in DNL occurs because threshold voltage variations are a proportionally larger portion of the FSR. The INL seen is consistent across

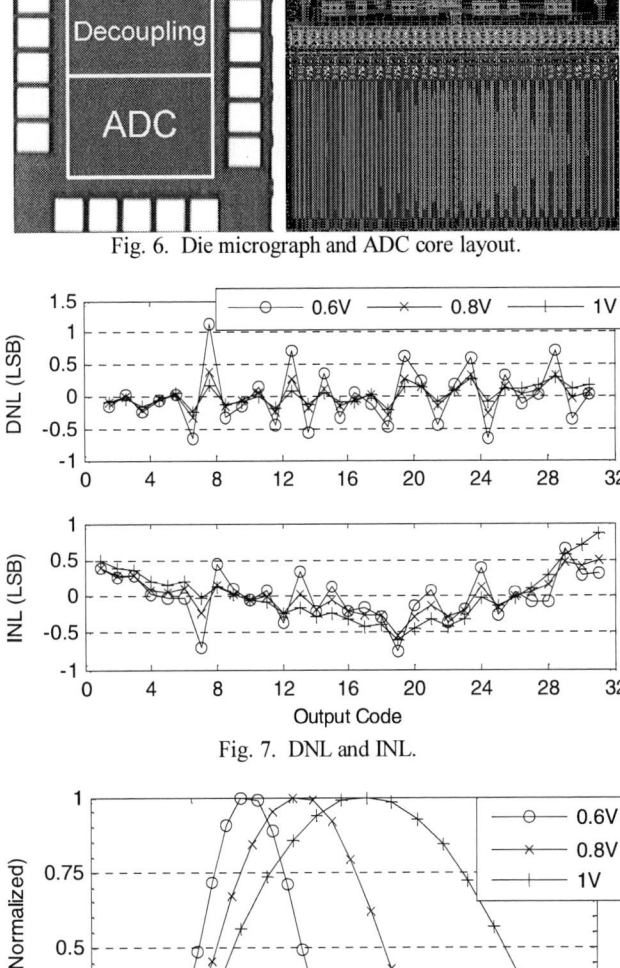

Fig. 6. Die micrograph and ADC core layout.

Fig. 7. DNL and INL.

Fig. 8. Static power of comparator array v. input voltage.

multiple test chips and is an order of magnitude larger than the simulated results. The primary cause is body effect.

Fig. 8 shows the static power consumption of the comparator array across input voltage at 0.6 V, 0.8 V, and 1 V V_{DD}s. The plot shows a minor shift of the peak away from $0.5 \cdot V_{DD}$ due to a slight positive shift of the center comparator's threshold point. The standby power is less than 0.1 mW for 0.6 V and 0.8 V, and less than 0.2 mW for 1 V.

Fig. 9 shows the SNDR v. input freq. at 0.6 V, 0.8 V, and 1 V V_{DD}s. At 0.6 V, the SNDR curve shows the typical flat curve at low frequencies and rolls off after 10 MHz. At 0.8 V, the SNDR shows mild peaking between 5 MHz and 50 MHz, after which the SNDR rolls off. At 1 V, the SNDR shows peaking after 30 MHz. Despite the peaking, the SNDR at 300 MHz is only 1.3 dB worse than the low frequency SNDR. The

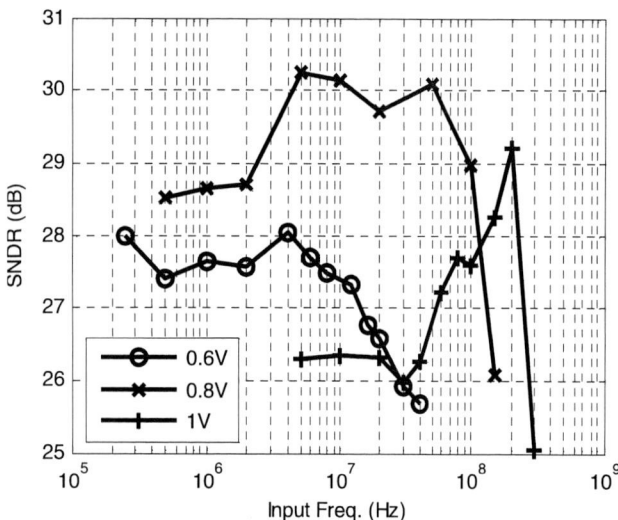

Fig. 9. SNDR v. input frequency.

TABLE I
SUMMARY OF RESULTS

Supply voltage	0.6 V	0.8 V	1 V
Input range	125 mV	263 mV	420 mV
Sampling rate	60 MS/s	300 MS/s	600 MS/s
Power	1.3 mW	3.2 mW	6.7 mW
Standby power	< 0.1 mW	< 0.1 mW	< 0.2 mW
DNL (LSB)	+1.14/−0.67	+0.40/−0.33	+0.31/−0.25
INL (LSB)	+0.65/−0.75	+0.49/−0.54	+0.87/−0.61
ENOB @ DC	4.36 bits	4.45 bits	4.08 bits
SNDR @ Nyquist	25.9 dB	26.1 dB	25.0 dB
FoM	1.06 pJ/conv	0.49 pJ/conv	0.66 pJ/conv

peaking is due to the effects of the varying comparator delays at high frequencies.

The overall performance of the flash ADC is summarized in Table I. The figure of merit (FoM) decreases from 1 V to 0.8 V as previously expected, but increases significantly from 0.8 V to 0.6 V. The increase in FoM from 0.8 V to 0.6 V is caused by the significant reduction in sampling rate. The FoM achieves a low of 0.49 pJ/conv at 0.8 V V_{DD}.

V. FUTURE DIRECTIONS

The flash ADC can be transformed into a pseudodifferential design by using two copies of the flash ADC and digitally combining their outputs, similar to [7]. In this case, the two copies would be half-sized to maintain the same total power as the current design. The INL can significantly improve in the pseudodifferential case. Pseudodifferentially combining the measurement results from 2 different typical chips results in an INL of ±0.32 LSB. The complete INL plot is shown in Fig. 10.

A S/H would alleviate the effect of bandwidth and delay variations in the comparator array by holding the input constant during which time all comparators can settle to their decisions. The comparator can be redesigned to use linear region MOSFETs in place of the resistors. Controlling the gate biases of the linear region MOSFETs allows tuning for both global and local variations and eliminates the need for well biasing. The revised schematic is shown in Fig. 11.

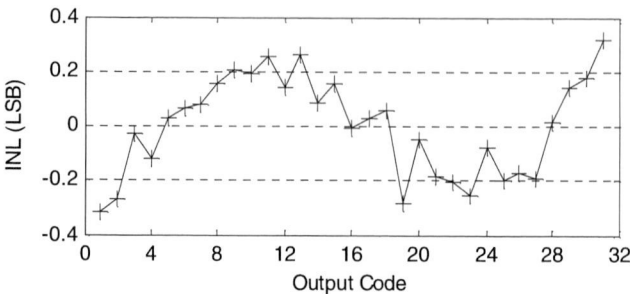

Fig. 10. Predicted pseudodifferential INL from measurement.

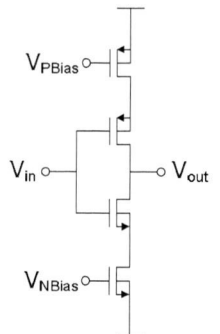

Fig. 11. Redesigned comparator schematic.

VI. CONCLUSION

Inverter-based comparator topologies result in analog designs with simple patterning that is compatible with the sub-wavelength lithography for scaled CMOS. A 5-bit flash ADC demonstrated the potential of inverter-based analog design by achieving a FoM of 0.49 pJ/conv with a simple design and clear directions for future improvement.

ACKNOWLEDGEMENT

The authors thank D. Li, Y.-T. Chien, and G. Keskin for discussion and assistance.

REFERENCES

[1] T. Jhaveri *et al.*, "Maximization of layout printability/manufacturability by extreme layout regularity," *J. of Micro/Nanolithography, MEMS and MOEMS*, vol. 6, no. 3, Jul.–Sept. 2007.

[2] L. W. Liebmann *et al.*, "High-performance circuit design for the RET-enabled 65-nm technology node," in *Proc. of SPIE*, Vol. 5379, May 2004, pp. 20–29.

[3] A. G. F. Dingwall, "Monolithic expandable 6 bit 20 MHz CMOS/SOS A/D converter," *IEEE J. Solid-State Circuits*, vol. 16, no. 6, pp. 926–932, Dec. 1979.

[4] Y. Chae, I. Lee, and G. Han, "A 0.7V 36µW 85dB-DR audio ΔΣ modulator using class-C inverter," in *IEEE Int. Solid-State Circuits Conf. (ISSCC) Dig. Tech. Papers*, 2008, pp. 490–491.

[5] R. van Veldhoven, R. Rutten, and L. Breems, "An inverter-based hybrid ΣΔ modulator," in *IEEE Int. Solid-State Circuits Conf. (ISSCC) Dig. Tech. Papers*, 2008, pp. 492–493.

[6] "An all-digital analog-to-digital converter with 12-µV/LSB using moving-average filtering", *IEEE J. Solid-State Circuits*, vol. 38, no. 1, pp. 120–125.

[7] D. Daly and A. Chandrakasan, "A 6b 0.2-to-0.9V highly digital flash ADC with comparator redundancy," in *IEEE Int. Solid-State Circuits Conf. (ISSCC) Dig. Tech. Papers*, 2008, pp. 554–555.

IEEE 2008 Custom Intergrated Circuits Conference (CICC)

A 10~15b 60MS/s Floating-Point ADC with Digital Gain and Offset Calibration

Yun-Shiang Shu, Moon-Jung Kyung, Wei-Ming Lee, Bang-Sup Song, and Bedabrata Pain[1]

University of California, San Diego, La Jolla, CA 92093
[1]Jet Propulsion Laboratory, Pasadena, CA 91109

Abstract – A variable-gain amplifier (VGA) with pseudo-random noise (PN) signal-dependent dithering and chopping is proposed. It allows the ADC gain and offset errors to be calibrated digitally in background. A 10~15b 60MS/s floating-point ADC (FADC) with variable gains from 1 to 32 enhance the INL from 24 to 0.9LSB at 15b level. Its non-linearity resulting from the VGA gain and offset errors is eliminated after calibration. A chip in 0.18μm CMOS occupies 3.5x2.5mm^2 and consumes 300mW at 1.8V.

I. INTRODUCTION

A floating-point architecture provides an alternative way to achieve wide dynamic range A/D conversion without demanding a high-resolution ADC. It uses variable gain amplifier (VGA) to amplify small input signal before quantization; thereby, the equivalent quantizer resolution is higher when input magnitude is low [1]–[3]. Similar to time-interleaved ADCs, floating-point ADCs (FADCs) suffer from gain and offset mismatch while switching circuit configuration with different input samples [2]. In switched-capacitor circuits, the gain and offset errors are resulting from the device mismatch and finite opamp gain. Digital calibration technique using pseudo-random noise (PN) has been an attractive approach to compensate these non-ideal effects in background. It tracks the long-term process drift, and the extra digital logic cost diminishes as process advances. This work describes a 10~15-b FADC, where the VGA stage is modified to enable background digital calibration with split capacitors and one more comparator. Signal-dependent PN dithering [4], which shortens the calibration time without sacrificing the signal range, and PN-based chopping [5] are used to measure the gain the offset errors, respectively. The proposed VGA can be used as a sample-and-hold (S/H) stage to calibrate the gain and offset mismatch in time-interleaved ADCs.

II. SYSTEM ARCHITECTURE

A. Floating-Point Architecture

A FADC is composed of a VGA, an ADC, and a digital divider as shown in Fig. 1(a). The VGA amplifies the input with gain values of 1, 2,···, 2^{M-1}, 2^M depending on the input levels and provides M-bit exponent as a function of the digital divider. The amplified signal is quantized by the N-bit ADC and converted into N-bit mantissa. Because the VGA gain is set higher when the input is lower, the FADC exhibits a non-uniform quantization nature as shown in Fig. 1(b). The

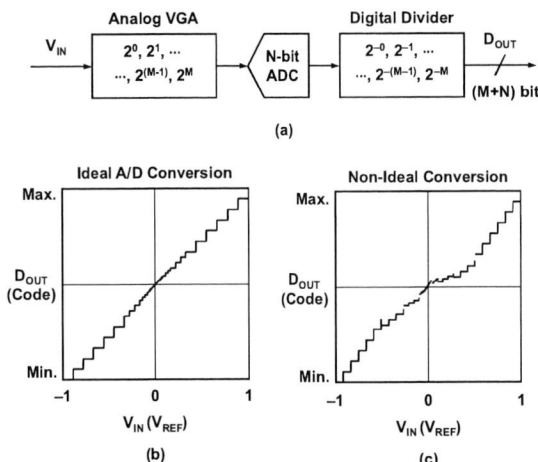

Fig. 1. Floating-point ADC. (a) Architecture. (b) Transfer function. (c) Transfer function with VGA gain and offset errors.

FADC resolution is enhanced to (M+N) bit when the input is close to zero. However, if the gain and offset errors of the FADC vary with different gain settings, large step errors occur in its transfer function as shown in Fig. 1(c). The step errors cause missing code or non-monotonicity and may degrade the linearity of the FADC to even lower than N-bit. In this work, these step errors are digitally calibrated in background. Once calibrated, the FADC linearity is only limited by the linearity of the N-bit ADC.

B. Gain and Offset Calibration

Fig. 2 shows the principle of background gain and offset calibration, where the basic functions of a FADC are depicted with gray blocks. Assume that the VGA and the ADC have gain errors of α and β, respectively, and have a total offset of V_{OS}. The offset is measured by chopping the input with a zero-mean pseudo-random pulse sequence PN_O and averaging the ADC output [5]. The input V_{IN} is translated into a noise, which approaches zero after a large number of samples are averaged. The ADC output is then subtracted by the measured offset and chopped by the same PN_O again to restore the input signal. To exclude the offset resulting from the comparator, the signal goes through the comparator is chopped at the comparator output instead of the input so that the comparator offsets are randomized by PN_O.

The gain error, $(1+\alpha)(1+\beta)$, is measured by injecting a PN_G-modulated calibration signal, V_{CAL}, into the signal path. After correlating the ADC output with the same PN_G and averaging, the value of $(1+\alpha)(1+\beta)V_{CAL}$ is obtained. The

978-1-4244-2018-6/08 $25.00 © 2008 IEEE

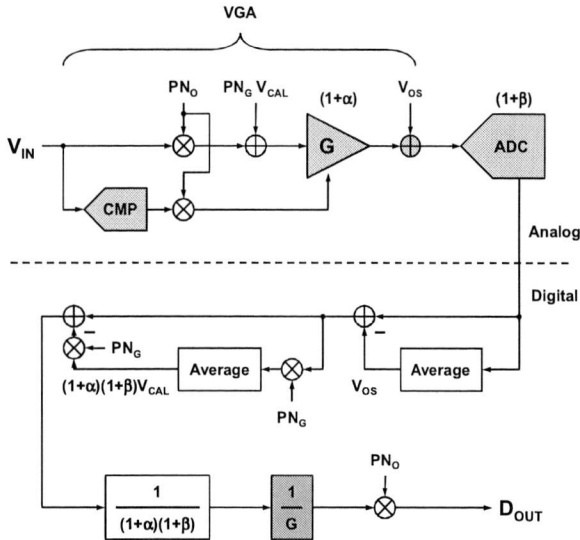

Fig. 2. Principle of FADC background gain and offset calibration.

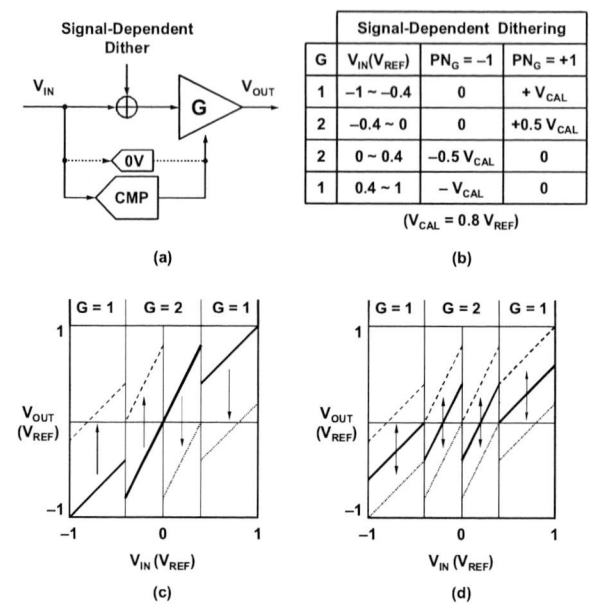

Fig. 3. Signal-dependent dithering. (a) VGA stage. (b) Dither magnitude. (c) V_{OUT} vs. V_{IN} of VGA stage. (d) Equivalent fixed-magnitude dithering.

injected dither is subtracted from the digitized signal, and the gain error is corrected after the signal is multiplied by $1/(1+\alpha)(1+\beta)$. This digital divider is realized with a multiplier by approximating the divider ratio of $1/(1+\varepsilon)$ by $(1-\varepsilon)$ to simplify the digital logic and reduce chip area. Unlike the gain calibration in the pipelined ADC, where a precise V_{CAL} is critical to calibrate the error accurately, inaccurate V_{CAL} in this application doesn't affect the linearity, since the remaining gain errors with different VGA gain settings are well-matched.

C. Signal-Dependent Dithering

The signal-dependent PN dithering scheme is chosen for the background gain calibration. It has an advantage of lowing the signal-to-dither ratio to shorten the calibration time without sacrificing the signal range, which is a common problem in the fixed-magnitude dithering [4]. Fig. 3(a) and (b) shows a VGA example with gain values of 1 and 2 modified for signal-dependent PN dithering. One more comparator is added at 0V to divide the input into more ranges. The dither of $\pm V_{CAL}$, $\pm 0.5V_{CAL}$, or 0 is injected depending on the input level. Fig. 3(c) illustrates the VGA output with the signal-dependent dithering. Negative dithers are used when input is higher than 0V, and positive dithers are injected as input is lower than 0V. The calibration voltage V_{CAL} is set to about $0.8V_{REF}$ in order to allow sufficient room for the comparator offset to vary. Fig. 3(d) shows the equivalent fixed-magnitude dithering of Fig. 3(c). The VGA output is equivalent to $(-0.4\sim0.4) \pm 0.4$ V_{REF} when gain is set to 2, and is equivalent to $(-0.6\sim0.6) \pm 0.4$ V_{REF} when gain is 1. The signal-to-dither ratios are kept at 1 and 1.5, respectively, while a full signal swing is reserved.

Fig. 4 illustrates the proposed switched-capacitor VGA, which enables the gain and offset calibration with arbitrary gain of n. During the sampling phase, the input is chopped by switching the sampling capacitor between $\pm V_{IN}$ according to PN_O, and the feedback capacitor is reset to zero, as shown in

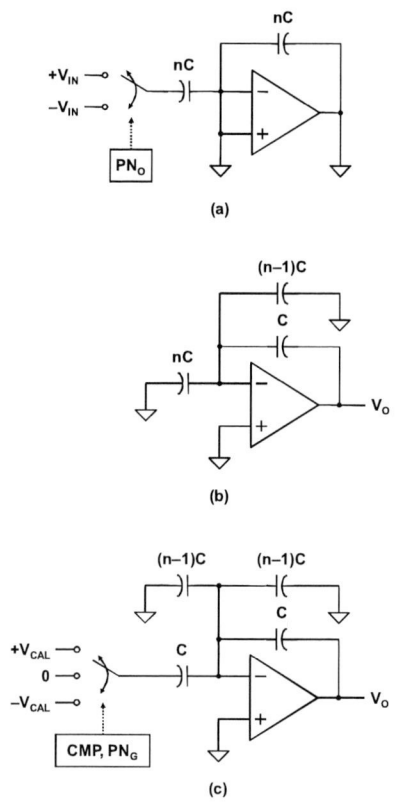

Fig. 4. VGA stage for gain and offset calibration. (a) Sampling phase. (b) Amplification phase. (c) With signal-dependent dithering.

978-1-4244-2018-6/08 $25.00 © 2008 IEEE

Fig. 4(a). In the amplification phase, the sampling capacitor is switched to zero, and only one of the n feedback capacitors is switched to the opamp output when no dithering is injected as shown in Fig. 4(b). While the gain calibration is enabled, the signal-dependent dither is injected by switching one of the n sampling capacitor to $\pm V_{CAL}$ or 0 according to PN_G and comparator outputs, which determine the input range, as shown in Fig. 4(c). Since the circuit configuration in Fig. 4(c) only measures the gain error from one of the split sampling capacitors. The gain errors contributed by all capacitors are measured in sequence and combined to generate the actual gain error.

III. CIRCUIT IMPLEMENTATION

Shown in Fig. 5 is the block diagram of a 10~15b FADC. The VGA is divided into three stages to relax the comparator offset and opamp bandwidth requirements. It has total gain values of $2^{0~5}$, and provides 5b exponents. The 10b ADC has 2 redundant bits to reduce digital truncation error. The digital output is 15b. The comparator outputs in each stage are connected to the digital logic. The digital encoding, digital correction, digital calibration, and also the PN sequence generator functions are all integrated in the digital logic. It provides each VGA stage with the PN values and the calibration enable signals.

One common problem in a switched-capacitor VGA is that its closed-loop bandwidth varies with the gain setting. The closed-loop bandwidth is set by the opamp unity-gain bandwidth and the feedback factor, which is determined by the circuit configuration. Generally, a high gain configuration leads to a small feedback factor, and slows down the settling behavior. When the circuit is configured to have low gain, the loop tends to get unstable. In this work, an opmap with variable open-loop bandwidth is proposed so that the VGA has a constant closed-loop bandwidth with different gain settings. Fig. 6 shows the Miller-compensated two-stage opamp used in the VGA stage, where the unity-gain bandwidth is set by the ratio of the input G_M and the Miller capacitor values. The extra Miller capacitors are connected when the VGA gain is low and disconnected when the gain is high to change the opamp unity-gain bandwidth. As a result, the variation of the feedback factor is compensated. During the sampling phase, the extra Miller capacitors are connected for the charges on

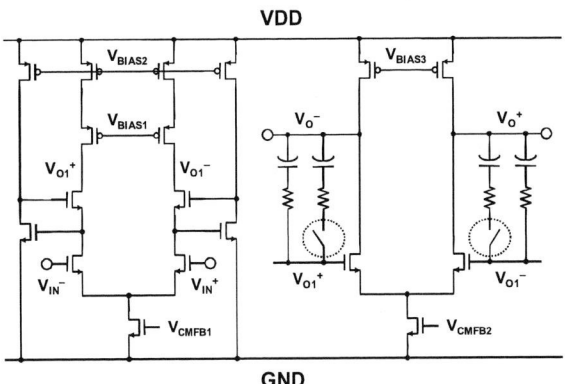

Fig. 6. Operational amplifier for VGA stage.

them to be reset.

A modified clock scheme is used to avoid the sampling mismatch between the switched-capacitor amplifier and the comparators in the first VGA stage [6]. The normal sampling period is divided into two clock phases, ϕ_{S1} and ϕ_{S2}, as shown in Fig. 7. During ϕ_{S1}, the comparator samples the signal with a scaled replica sampling network of the amplifier. Therefore, the signals sampled in the comparator and the amplifier are equal. The sampled signal in the comparator is then amplified by the pre-amp during ϕ_{S2} to suppress the input-referred latch offset. The modified clock scheme imposes a tradeoff between sampling accuracy and comparator offset. Long ϕ_{S1} increases the sampling accuracy but also increases the comparator offset. Short ϕ_{S1} leads to less comparator offset but more sampling error. In the work, the duty cycles of the two extra phases are made adjustable. They are optimized to achieve best performance.

IV. EXPERIMENTAL RESULTS

The prototype fabricated in 0.18-μm CMOS occupies an active area of 3.5x2.5mm^2, and consumes 300mW at 1.8V. The die photo is shown in Fig. 8. The digital calibration algorithm is entirely implemented on-chip. The digital logic occupies 1.5mm^2. Fig. 9 shows the measured DNL and INL at 15b level before calibration when sampling at 60MS/s. The non-uniform quantization nature of the FADC is demonstrated in the DNL plot. The DNL decreases as the input signal becomes smaller. However, the INL plot shows large steps at the VGA gain-range boundaries. The symmetric steps with

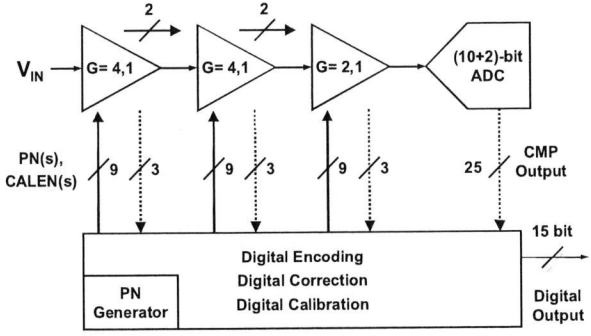

Fig. 5. Floating-point ADC block diagram with digital calibration.

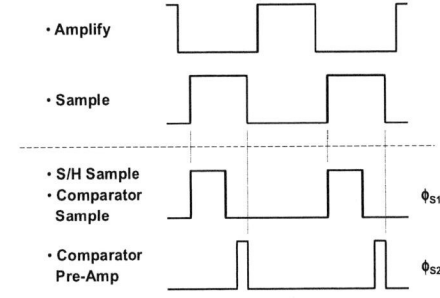

Fig. 7. Additional clock phases for sampling stage.

978-1-4244-2018-6/08 $25.00 © 2008 IEEE   177

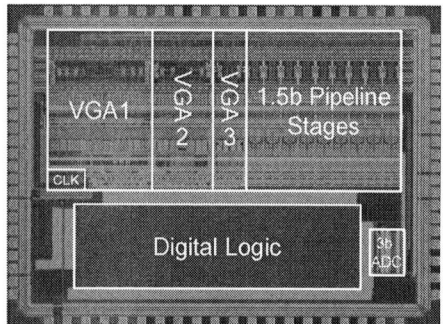

Fig. 8. Chip photograph.

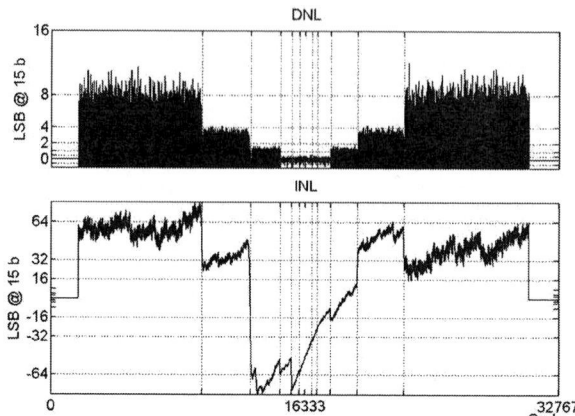

Fig. 9. Measured DNL and INL @15b before calibration.

opposite polarities are resulting from the gain mismatch, and the steps having the same polarity are contributed by the offset mismatch. Both errors are significant in the INL plot, and the largest step is higher than 100LSB. Fig. 10 shows the DNL and INL after calibration. The step errors are greatly reduced. The INL after calibration indicates that the FADC effectively enhances the INL to ±0.9LSB when the input is close to zero. The INL in every gain range is mainly the replica of that of the 10b ADC, which has a measured INL of ±24LSB at a 15b level. That implies the linearity is only limited by the 10b ADC after calibration. It is also proven in the measured SFDR and SNDR by varying the input amplitude as shown in Fig. 11. The FADC exhibits an uniform SFDR over a wide signal ranges after calibration. Without calibration, the SFDR and the SNDR are limited by the harmonics resulting from the large INL step errors. The SNDR after calibration is equivalent to that of an ideal 10-b ADC when the input is large. It remains constant as the SNDR of the ideal 10-b ADC while the input amplitude decreases, and is finally dominated by the −80 dB thermal noise floor of the input VGA stage.

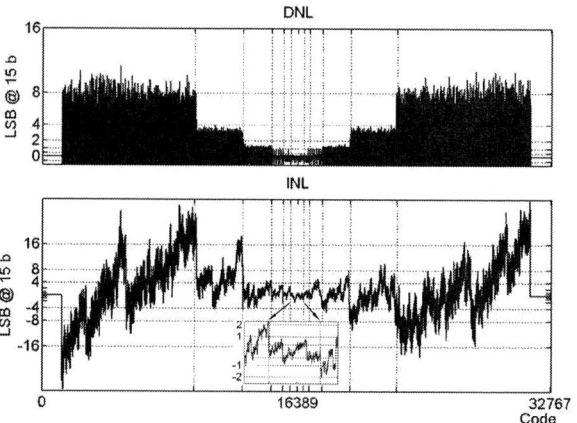

Fig. 10. Measured DNL and INL @15b after calibration.

V. CONCLUSION

A digital background gain and offset calibration with a modified VGA is demonstrated in a 10~15b FADC. The floating-point architecture effectively extends the dynamic range of a 10b ADC by 5b after the gain and offset mismatches are calibrated. Signal-dependent dithering is enabled by adding one additional comparator and split capacitors in the VGA while accurate dither magnitude is not required. The FADC achieves a conversion rate of 60MS/s. A bandwidth-adjustable opamp is used to maintain a large VGA closed-loop bandwidth when the VGA gain is high. The proposed approach relaxes the requirement on analog components, and is suitable for time-interleaved architecture to achieve a higher conversion rate.

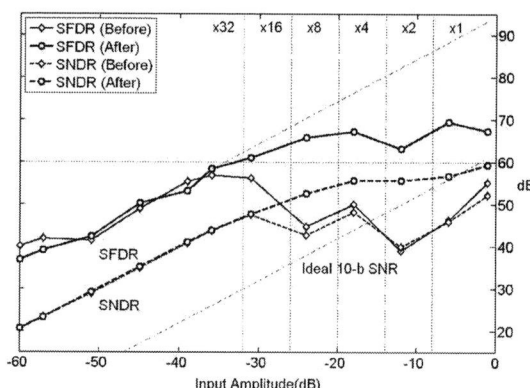

Fig. 11. Measured SFDR and SDNR before and after calibration.

REFERNECE

[1] F. Chen and C. S. Chen, "A 20-b dynamic range floating-point data acquisition system," *IEEE Trans. Ind. Electron.*, vol. 38, pp. 10–14, Feb., 1991.

[2] J. Piper and J. Yuan, "Design considerations of a floating-point ADC with embedded S/H," *Proc. of IEEE ISCAS*, vol. 6, pp. 6166–6169, May 2005.

[3] D. U. Thompson and B. A. Wooley, "A 15-b pipelined CMOS floating-point A/D converter," *IEEE J. Solid-State Circuits*, vol. 36, no. 2, pp. 299–303, Feb. 2001.

[4] Y.-S. Shu and B.-S. Song, "A 15-bit linear 20-MS/s pipelined ADC digitally calibrated with signal-dependent dithering," *IEEE J. Solid-State Circuits*, vol. 43, pp. 342–350, Feb. 2008.

[5] S. M. Jamal, D. Fu, N. C.-J. Chang, P. J. Hurst, and S. H. Lewis, "A 10-b 120-Msample/s time-interleaved analog-to-digital converter with digital background calibration," *IEEE J. Solid-State Circuits*, vol. 37, no. 12, pp. 1618–1627, Dec. 2002.

[6] I. Mehr and L. Singer, "A 55-mW, 10-bit, 40-Msample/s Nyquist-rate CMOS ADC," *IEEE J. Solid-State Circuits*, vol. 35, pp. 318–325, Mar. 2000.

978-1-4244-2018-6/08 $25.00 © 2008 IEEE

Digital Correction of Dynamic Track-and-Hold Errors Providing SFDR > 83 dB up to f_{in} = 470 MHz

Parastoo Nikaeen and Boris Murmann
Stanford University, Stanford, CA

Abstract- **A digital technique for the compensation of dynamic nonlinearities at the front-end of high-speed, high-resolution ADCs is presented. The complexity of the digital post-processing scheme is minimized using judicious modeling of the relevant nonidealities. Applying the method to a 14-bit, 155-MS/s ADC provides > 83 dB SFDR up to f_{in} = 470 MHz. The post-processing block is estimated to consume 52 mW and occupy 0.54 mm^2 in 90-nm CMOS.**

I. INTRODUCTION

In A/D converter applications such as wireless base stations, IF sub-sampling is an attractive method for minimizing component count. By applying this method, one or more steps of down-conversion are removed from the receiver path and some of the analog front-end signal processing functions can be moved to the digital domain. In such a solution, the ADC's linearity at high input frequencies becomes a critical issue. Despite the use of a dedicated track-and-hold amplifier (THA), nonlinearities in the circuit's input network often introduce dynamic errors that limit the performance of the ADC at high input frequencies.

Previous efforts on improving high frequency linearity have mainly focused on analog approaches. In [1] it was shown that integrated BJT input buffers can reduce the distortion across a wide bandwidth. This improvement, however, cannot be realized in standard CMOS. In [2], a modified bootstrapping circuit was used to cancel the backgate effect of the converter's input switch in order to improve linearity. This technique shows good performance up to moderate frequencies, but ultimately suffers from bandwidth limitations in the active bootstrap circuitry.

In recent years, digital correction methods have been introduced to compensate circuit nonlinearity in ADCs [3-5]. However, these techniques mainly address static errors in the converter core and are not effective at removing dynamic nonlinearities in the THA. In this paper, we present a digital enhancement scheme that is specifically tailored to remove high frequency distortion at the sampling front-end of ADCs. The computational complexity of the proposed digital post-processor is minimized using judicious modeling of the dominant distortion mechanisms.

Following this introduction, Section II discusses the relevant circuit nonidealities in a typical THA front-end. Section III develops a suitable digital compensation mechanism for the dominant high-frequency errors. Section IV details measured results that were obtained by post-processing the outputs of a commercially available 14-bit, 155-MS/s ADC [6]. Our measurements show that the proposed scheme improves the converter's high-frequency linearity to more than 83 dB SFDR at frequencies up to 470 MHz. We end this paper by providing additional data on the robustness of the correction scheme as well as an estimate for the digital hardware complexity and power dissipation.

II. CIRCUIT DESCRIPTION

Fig. 1 shows a simplified half-circuit of the commonly used flip-around THA [7] with bottom-plate sampling and bootstrapped series switches [8]. During clock phase φ_1, switches M_1 and M_2 are on. The circuit is in its tracking phase and the voltage across the sampling capacitor follows the input. At the sampling instant, M_2 is turned off first to freeze the total charge at the bottom plate node of the capacitor; this prevents subsequent charge injected by M_1 from disturbing the acquired sample. During φ_2, switch M_3 is turned on and M_1 and M_2 remain off. By charge conservation, V_{out} becomes equal to the input voltage seen at the sampling instant (opening of M_2). V_{out} is now acquired by the ADC core for quantization. In this remaining process, the physical frequency of the input signal plays a minor role, primarily due to aliasing and cycle-to-cycle reset of all states in the ADC core circuitry. The most critical aspect for high-frequency performance is how linearly the continuous time waveform at V_{in} is mapped into discrete samples at the THA output.

In the above-discussed THA circuit, there are two main effects that introduce nonlinear distortion in the acquired samples. The first issue is signal-dependent charge injection and the second problem is tracking nonlinearity. Both of these effects result from input-dependent variations in the on-resistance of M_1 (R_1), which are partly due to backgate effect and finite bandwidth in the bootstrapping circuit. Changes in R_1 alter the amount of charge injected by M_2 at the sampling instant; this effect typically dominates at low frequencies and is usually not the primary concern in a sub-sampling application. At high frequencies, the main issue is the tracking nonlinearity due to the voltage dependence of R_1, which interacts with the sampling capacitor C to introduce frequency

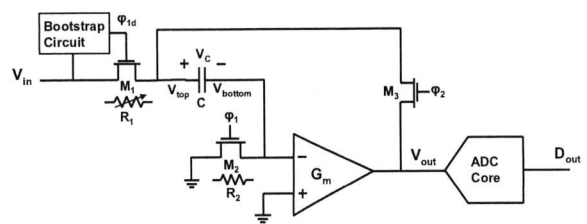

Fig. 1. Schematic of a flip-around THA.

dependent distortion. The resulting nonlinear relationship with memory between the THA input and output at the relevant discrete time instances can be expressed as a Volterra series of the form

$$V_{out}(k) = f_{nonlin}[V_{in}(k), V_{in}(k-1), V_{in}(k-2), ..., V_{in}(k-n)] \quad (1)$$

Unfortunately, a well-known issue with Volterra models is that even for relatively simple circuits, the corresponding inverse functions (for error correction) can be prohibitively complex [9]. In order to address this issue and to minimize hardware requirements, we simplified the distortion model using circuit-specific insight and judicious approximations as discussed next.

In the circuit's tracking phase, the series combination of R_1, R_2 and C form a nonlinear first order differential equation that links the input voltage (V_{in}) to the capacitor voltage (V_C). Modeling the backgate effect of M_1 as the dominant nonlinearity and using first-order device models, we find

$$V_{in} = V_C + (R_1 + R_2)C\frac{dV_C}{dt} \quad (2)$$

$$R_1 = R_1(V_{in}, V_{top}) = \frac{1}{\mu_n C_{ox} \frac{W}{L}[V_{DD} - V_{th}(V_{in}, V_{top})]} \quad (3)$$

where (3) assumes that M_1 is perfectly bootstrapped such that $V_{GS1} = V_{DD} = const$. In a sampling network with sufficiently large bandwidth, V_{top} and V_C will track the input closely and differ only by a weakly nonlinear term determined by the respective RC product and the signal derivative. Therefore, the resistance of M_1 can be approximated by a nonlinear static function of V_{in}, V_C, and consequently V_{out}. Taking these considerations into account, the discrete time distortion model simplifies to a product of a memoryless nonlinear function, expressed below as a power series, and the signal derivative.

$$V_{in}(k) = V_{out}(k) + [a_0 + a_1V_{out}(k) + a_2V_{out}^2(k) + ...] \times \frac{dV_{out}(k)}{dt} \quad (4)$$

As a refinement to this result, we include an additional term $\alpha \cdot V_{out}(k-1)$ that helps absorb errors due to residual charge from the previous sample. Contributions from this term can be significant in a flip-around THA, since the sampling capacitance is not reset between consecutive samples and its charge is dumped into the input circuit. The model becomes

$$V_{in}(k) = V_{out}(k) + \sum_{i=0}^{K} a_i V_{out}^i \times \left[\frac{dV_{out}(k)}{dt} + \alpha V_{out}(k-1)\right] \quad (5)$$

where K is the order of the static nonlinearity.

In contrast to (1), the distortion model of (5) is already in its inverse form, i.e. $V_{in} = f(V_{out})$. Therefore, this expression is directly applicable for correction and no inversion is needed. Assuming that V_{out} is processed by the ADC core with sufficient resolution and linearity, (5) can be applied in the digital domain, i.e. $V_{out} \cong V_{ref} \cdot D_{out}$, where V_{ref} is the reference voltage of the converter.

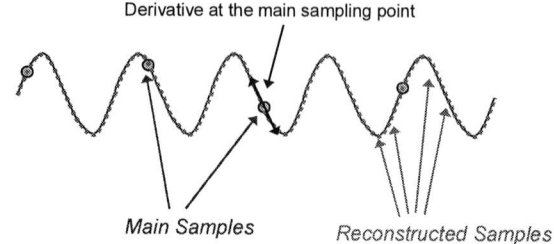

Fig. 2. Subsampled signal with interpolation.

III. DIGITAL CORRECTION

The correction function derived in the previous section requires the signal derivative at the ADC sampling instants. One way to model the derivative of a subsampled signal is to interpolate the samples and use a subset of the reconstructed value to approximate the slope. This is schematically illustrated in Fig. 2. In order to perform the interpolation, the ADC output must be upsampled and processed by a bandpass filter with a bandwidth of $f_s/2$ centered at the Nyquist zone of the input signal. The band-pass filter is implemented in the digital domain and its required number of taps is chosen based on the accuracy we need in the interpolated samples.

In principle, only one reconstructed sample adjacent to the main sample is needed to approximate the derivative. However, due to thermal noise in the ADC and small inaccuracies in the interpolation process, multiple interpolated samples around the sampling point should be used. Furthermore, weighting these interpolated samples independently and multiplying them into the nonlinear power series terms provides additional degrees of freedom to the correction. We found experimentally that this helps absorb second order effects that are not explicitly contained in the basic model of (5). With these considerations, we form the correction filter as

$$D_{out,corrected} = D_{out}(k) + \sum_{i=1}^{K} D_{out}^i(k) \times$$
$$\left[\sum_{j=0}^{2L} h_{ij}D_{out}\left(k - \frac{j-L}{M}\right) + g_i D_{out}(k-1)\right] \quad (6)$$

where L is the number of interpolated samples used on each side of the main sample to model the derivative and M is the upsampling factor. Note that the linear terms are removed from the model (i=0), since they do not contribute to the overall nonlinearity of the system.

Fig. 3 shows a block diagram of the digital post-processor that implements (6). Output samples from the ADC are processed by a digital interpolation filter to reconstruct samples with M=100. From measurements, we found that considering nonlinear terms up to third order (K = 2) and using four interpolated samples (L = 2) was sufficient. The interpolation filter must therefore produce only four additional samples per ADC sample; it is therefore significantly less complex than a conventional interpolator.

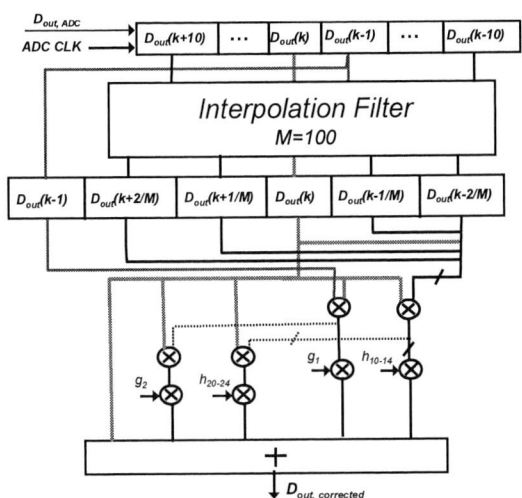

Fif. 3. Block diagram of the digital post processor

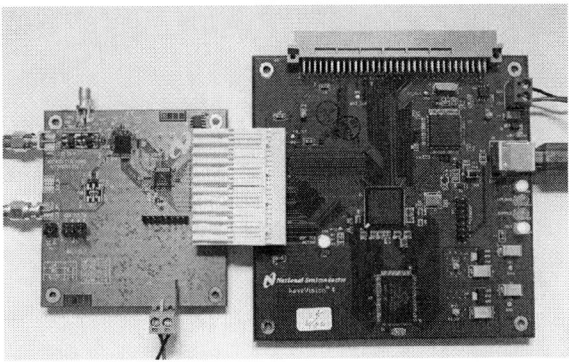

Fig. 4. Photo of the ADC test setup with the data capture board.

The overall post-processor requires K·(2L+2) = 12 coefficients that must be calibrated. This is done through a foreground calibration using training signals. In the experiments described below, we used single tone sine waves to determine the coefficients via a least mean square fit to (6) using the ADC's output samples, $D_{out}(k)$, and an estimate of the corresponding input samples, $V_{in}(k)$. The input sample estimates were obtained by fitting an ideal sine function to the measured output samples.

IV. EXPERIMENTAL RESULTS

For performance evaluation, we applied the post-processing scheme to a commercially available 14-bit, 155-MS/s CMOS ADC that uses a flip-around THA with 1.1 GHz input bandwidth [6]. Fig. 4 shows a photo of the ADC test board together with the data capture board. A block diagram of the overall test setup is shown in Fig. 5. The training sine waves are generated using an HP8644 signal generator and are passed through a bandpass filter to remove unwanted harmonics. The data capture board stores blocks of data that are transferred to a computer that emulates the post-processing block of Fig. 3. Alternatively, the processing could be applied in real time, e.g., using an FPGA. However, since there is no feedback path from the post-processor to the ADC, a real-time implementation would not alter the presented results.

The coefficients of the post-processor were determined using the foreground calibration approach discussed above. Three single tone sine waves (distributed across the respective Nyquist zone) were chosen as the training signals. We found that different sets of coefficients must be extracted for each Nyquist zone; but with each set, any band-limited signal in the corresponding frequency range can be corrected.

Fig. 6 shows measured spectra in the ADC's 5th Nyquist zone (f_s = 155 MHz, f_{in} = 326 MHz). These data indicate that the post-processing reduces the 2nd and 3rd harmonics significantly, yielding an SFDR improvement of 13 dB. Fig. 7 shows SFDR versus f_{in} swept across the 5th and 6th Nyquist

zone of the ADC. The SFDR remains above 83 dB for this frequency range with applied digital post-processing. The maximum input frequency was limited by our test equipment; the achieved SFDR was partly limited by static nonlinearity in the ADC core and interactions therewith.

In order to test the robustness of the calibration, we subjected the ADC to temperature variations of ±30 °C relative to room temperature. For this test, the post-processor coefficients were extracted once (at room temperature) and were kept constant throughout the experiment. Fig. 8 shows the resulting SFDR over temperature. As evident from this plot, the distortion compensation is only weakly affected by temperature and the SFDR of the post-processed converter remains above 80 dB. This result indicates that a simple foreground calibration approach may be feasible in some applications that do not experience large temperature variations without providing a time window for re-calibration. Similar to the scheme implemented in [10], a foreground re-calibration could be triggered by a temperature sensor upon detection of a large change in operating temperature.

As a further test of robustness and general applicability of our algorithm, we applied the post-processing scheme to a second ADC provided by a different vendor. The chosen device is a 14-bit, 80-MS/s CMOS ADC with an input bandwidth of 450 MHz [11]. Measured results on this ADC showed similar improvements and robustness, yielding an SFDR greater than 82 dB over its entire input bandwidth.

In order to gauge the hardware cost of the proposed scheme, we implemented the post-processing block of Fig. 3 in Verilog and synthesized it using a 90-nm CMOS standard cell library. The block was designed to operate at the ADC's clock rate

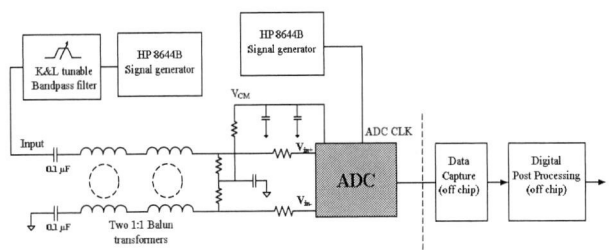

Fig. 5. Block diagram of the test setup.

978-1-4244-2018-6/08 $25.00 © 2008 IEEE   181

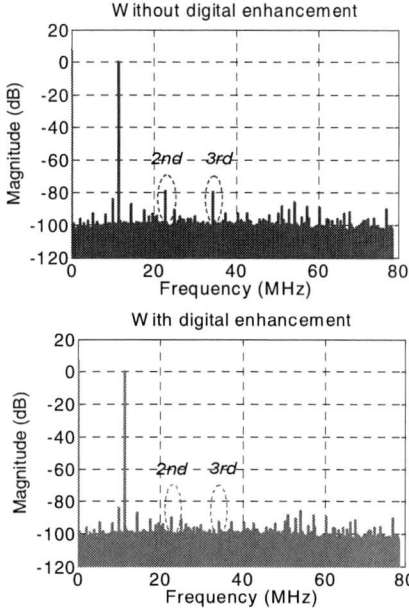

Fig. 6. Frequency spectrum at f_{in}=326 MHz. (f_{clk}=155 MHz).

(f_{clk} = 155 MHz); it generates corrected outputs with 32 cycles latency, which is due to the requirement of post samples for interpolation and several stages of pipelined logic. The obtained area and power estimates for the post-processor are 0.54 mm^2 (61,339 standard cells) and 52 mW (at V_{DD} = 1V), respectively. This estimated power is relatively small compared to the power dissipation of a typical high-speed, high-resolution ADC. The converters used in our experiments dissipate 967 mW [6] and 318 mW [11].

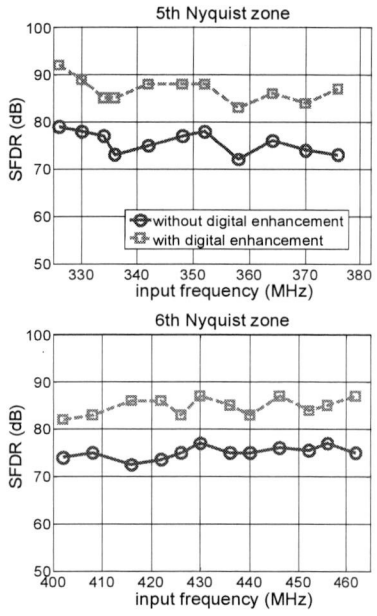

Fig. 6. SFDR vs. input frequency in the 5th and 6th Nyquist zone (f_{clk} = 155 MHz).

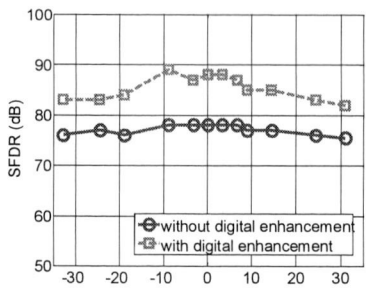

Fig. 8. SFDR vs. temperature at f_{in} = 326 MHz (f_{clk}=155 MHz).

V. CONCLUSIONS

A digital method for improving the linearity of ADCs at high input frequencies was presented. Experimental results show that our technique improves the SFDR of a commercially available ADC to a value greater than 83 dB up to input frequencies of 470 MHz. The complexity and power consumption of the digital algorithm was estimated for a 90-nm CMOS process. The digital post-processor can either be integrated on chip (with a relatively small power overhead) or implemented on an FPGA that may already be part of the overall system, as for instance a typical wireless base station.

ACKNOWLEDGMENTS

This project was funded in part by National Semiconductor and a Texas Instruments Diversity Fellowship. We thank David Boisvert and Bumha Lee of National Semiconductor as well as Marco Corsi and Martin Izzard of Texas Instruments for their help.

REFERENCES

[1] A.M.A. Ali et al., "A 14bit 125Ms/s IF/RF sampling pipelined A/D converter," Proc. *CICC*, pp. 391-94, Sept. 2005.

[2] H. Pan et al., "A 3.3V, 12b, 50MSample/s A/D converter in 0.6um CMOS with over 80dB SFDR," *ISSCC Dig. Tech. Papers*, pp. 40-41, Feb. 2000.

[3] B. Murmann and B. E. Boser, " A 12-bit 75-MS/s pipelined ADC using open-loop residue amplification," *IEEE J. Solid-State Circuits*, pp. 2040-2050, Dec. 2003.

[4] Y. Chiu, et al., "Least Mean Square adaptive digital background calibration of pipelined analog-to-digital converters," *IEEE Trans. Ckts Syst. I*, pp. 38-46, Jan 2004.

[5] C. Grace, P.J. Hurst and S.H. Lewis, "A 12 b 80 MS/s pipelined ADC with bootstrapped digital calibration," *ISSCC Dig. Tech. Papers*, pp. 460-461, Feb. 2004.

[6] National Semiconductor, "ADC14155 Datasheet" [Online]. http://www.national.com/pf/DC/ADC14155.html#Datasheet.

[7] W. Yang, et al., "A 3-V 340-mW 14-b 75-MSample/s CMOS ADC With 85-dB SFDR at Nyquist Input," *IEEE J. Solid-State Circuits,* pp. 1931-1936, Dec. 2001.

[8] A.M. Abo and P.R. Gray, "A 1.5-V, 10-bit, 14.3-MS/s CMOS pipeline analog-to-digital converter," *IEEE J. Solid-State Circuits*, pp.599-606, May 1999.

[9] John Tsimbinos, *Identification and compensation of nonlinear distortion*. Ph.D. dissertation, University of South Australia.

[10] K. Poulton et al., "A 20-GSample/s 8b ADC with a 1-MByte Memory in 0.18-um CMOS," *ISSCC Dig. Tech. Papers*, pp 318-319, Feb. 2003.

[11] Texas Instruments, " ADS6143 Datasheet" [Online]. http://focus.ti.com/lit/ds/symlink/ads6143.pdf.

978-1-4244-2018-6/08 $25.00 © 2008 IEEE

IEEE 2008 Custom Intergrated Circuits Conference (CICC)

A 65nm CMOS 1.2V 12b 30MS/s ADC
with Capacitive Reference Scaling

Kang-Jin Lee, Kyoung-Jun Moon, Kwang-Sung Ma, Kyoung-Ho Moon and Jae-Whui Kim

Samsung Electronics Co., Ltd
San#24 Nongseo-Dong, Giheung-Gu
Yongin-City, Gyeonggi-Do, Korea

Abstract- **A 1.2V 12b 30MS/s pipelined ADC, implemented in a 65nm standard CMOS technology, achieves an SNDR of 65.1dB with a rail-to-rail 4.7MHz input. A capacitive reference scaling technique is proposed to alleviate the high gain requirement of the opamp and a wide input range of 2.4Vp-p differential for low voltage operation in the nanometer domain. The prototype ADC dissipates 18mW and occupies an active die area of 0.34mm². The measured DNL and INL are ±0.44LSB and ±1.33LSB, respectively.**

I. INTRODUCTION

As the processes move into nanometer scale era, the analog performance has been degraded due to lowered output impedance of transistors, reduced input signal power from the scaled supply voltage, and increased on-resistance of switches. Especially, the continuous degradation of opamp gain with the process development is one of the most substantial difficulties in low voltage high resolution ADC designs for the mobile communication systems. Recently, sub-ranging architecture without closed-loop or 3-stage opamp are suggested to avoid or to meet the high gain requirement [1-2]. In this paper, a capacitive reference scaling (CRS) technique is proposed as a viable solution to lower opamp gain requirement in the Multiplying D/A Converter (MDAC), to enable rail-to-rail wide input dynamic range for better SNR, and to improve settling behavior of MDACs from better reference sampling for low voltage low power operation.

II. OP AMP GAIN REQUIREMENT IN MDACS

The minimum opamp gain requirement in the conventional MDACs of pipeline ADCs is derived from [3] as

$$\frac{2^k}{1+(1+2^k+\alpha_p)/A_O} \geq 2^k(1-2^{-(N-m)}) \quad (1)$$

where 2^k is the interstage gain, Ao is a finite gain of non-ideal opamp, α_p is a ratio of summing node parasitic capacitance to feedback capacitance, 2^N is the accuracy of input, and m is the stage resolution. Neglecting α_p, the required gain of opamp is almost same with the accuracy of input even with large m as long as k=m. On the other hand, the gain requirement of opamp can be reduced by 6×(m-k) dB by designing m larger than k. Instead, the output dynamic range is reduced by 2×(m-k) from the input range with the decreased stage resolution of the next. Also, it does not degrade the noise budget for the next stage because the required accuracy of the reduced output range is relaxed as well. The disadvantage is that the next stage demands another reference voltage for the reduced dynamic range.

III. PROPOSED ADC ARCHITECTURE

The proposed 1.2V 12b 30MS/s ADC without a dedicated front-end SHA consists of a 3b first stage followed by three 2.5b stages terminated by a 3.5b flash ADC, along with other supplementary blocks as shown in Fig. 1. The Nyquist sampling is achieved by the use of boot-strapped switch maintaining low on-resistance and by careful layout reducing sampling aperture error of the first stage input path.

The proposed CRS dealing with the different input ranges is depicted in the bottom of Fig. 1. In the first stage, k=2 and m=3 is chosen so that α_p is far less than $1+2^k$ and the gain requirement of opamp for 12 bit resolution takes advantage of reduction of 6dB according to (1). Therefore, the rail-to-rail input is converted to 3-bit digital codes by 7 comparators across the reference voltage. At the same time, the residue is multiplied by 2^2 to meet the gain requirement as well as output dynamic range with headroom margin of the opamp in MDAC1. On the other hand, in the remaining MDACs, the halved input range is converted to 3 bit digital codes including a bit for digital correction by 5 comparators while the residue amplification ratio is same as 2^2. The last stage, FLASH5, consists of 9 comparators with 1 bit interpolation for 12b outputs. All MSBs of output codes corresponding to each comparator from the second stage to the last are used for the over-range detection.

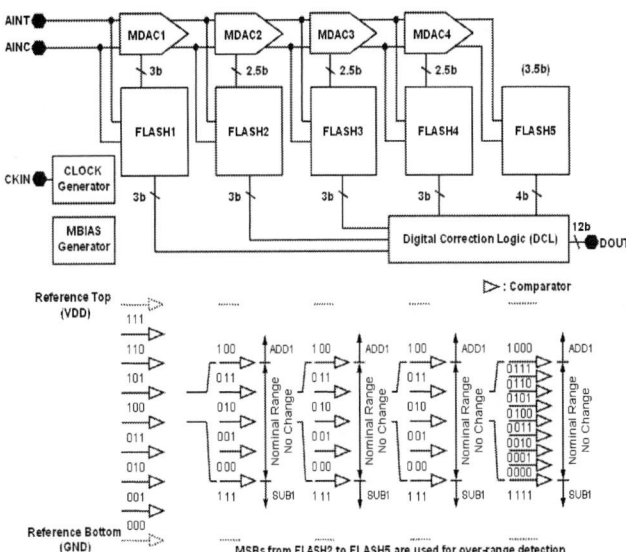

Fig. 1. Block diagram of the proposed ADC with coding technique.

In a subranging flash ADC, comparators are typically implemented along a resistor string in series with same voltage interval suffering from same amount of static and dynamic offsets. Digital correction logic (DCL) can adjust the error from comparator offsets by overlapping one bit of the output digital codes from each stage. In conventional designs, a half interval between decision levels decided by comparators is multiplied and overlapped in the next stage for the correction. In this work, the correction range is designed as 1/12 of the interval, but it has advantage of doubled input dynamic range of conventional one. After all, the effective correction voltage range is one third compared to that of the classical scheme. For the FLASH sub ADCs with low offset margin, the comparator must be designed and laid out carefully. The transitions of all node voltage levels of input pair in comparators, especially, are designed to be minimized so that kick-back noise is reduced thereby lessening the influence on analog signal and reference voltages. From this, the low offset requirement can be satisfied by using power efficient latch-only structure. Therefore, it can save additional stages from more offset margin for the reduction of size and power consumption.

With the CRS, the same reference voltage is applied to the halved input range by capacitive scaling instead of additional appropriate reference selection [4]. In this design, the power supply (VDD) and ground (GND) are used for reference top and bottom, respectively, to increase signal power by rail-to-rail input and to improve reference sampling by setting VDD as gate-source voltage of reference switches. Also, it gets rid of reference output pins and the related elements such as external capacitors as extra benefit.

IV. CIRCUIT DESIGN WITH CAPACITIVE REFERENCE SCALING

The simplified block diagrams of MDAC1 and MDAC2 implementing the proposed CRS are shown in Fig. 2. The MDAC1 is composed of 9 unit capacitors, a switching block, and 2-stage opamp. It operates with two phase non-overlapping clocks denoted as Q1 and Q2; the sampling and the amplification phase in MDAC1, respectively. The Q1H generated by clock booster, in phase with Q1, has magnitude of the sum of input voltage and logic voltage of Q1. The 8 unit capacitors are used for sampling except a dedicated feedback one, CF, in the sampling phase. In amplification phase of Q2, 7 unit capacitors are connected to the reference while the selected one is used as feedback capacitor with the dedicated one to achieve interstage gain 2^2. The selection of one unit capacitor among 8 depends on the input value using a modified commutative feedback capacitor switching technique by control logic to improve linearity [5].

MDAC2 has 8 unit capacitors to scale down the reference value to half even though the number of the related comparators is 5 in the FLASH2. All the capacitors are used for sampling when Q2 is high in MDAC2. In amplification phase, Q1, the selected 5 capacitors are connected to the reference depending on the input and the other two are used as feedback like MDAC1 to maintain the interstage gain. The last unit capacitor, C8, is connected to half VDD to adjust the offset from the over-range detection bit. After all, the quantized DAC analog output corresponding to a single comparator in FLASH2 is not $(2 \times VDD)/4$ but $(2 \times VDD)/8$ like the first stage. The remaining MDACs have the same architecture as MDAC2.

A simple 2-stage folded cascode opamp using a 65nm CMOS digital process, shown in Fig. 3, is able to meet the reduced gain requirement and output dynamic range of opamp in MDAC1. The additional NMOS transistors in the first stage opamp, shown in dotted rectangle, are added and optimized to reduce the gain loss from parallel connections without extra delay. There is no loss in headroom margin from VDD gated NMOS transistors. The second stage opamp adopts a folded cascode structure of 3 CMOS stack to maintain the wide output dynamic range and to get rid of gain loss from input pair of short length. In addition, an improved compensation technique is adopted to reduce the current of opamps [6].

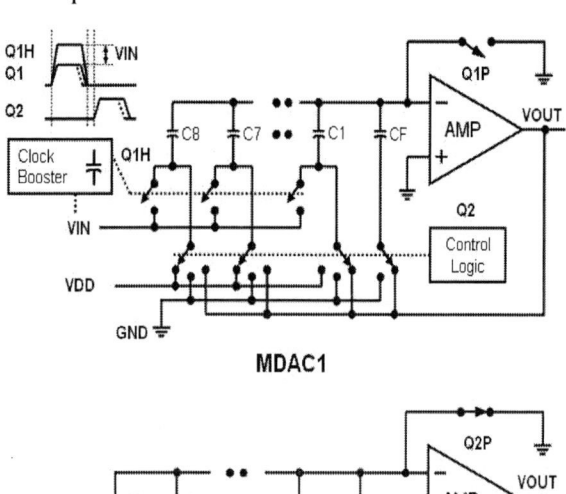

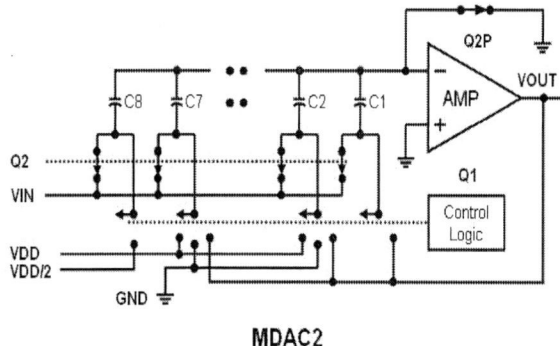

Fig. 2. Block diagrams of MDACs for the proposed CRS.

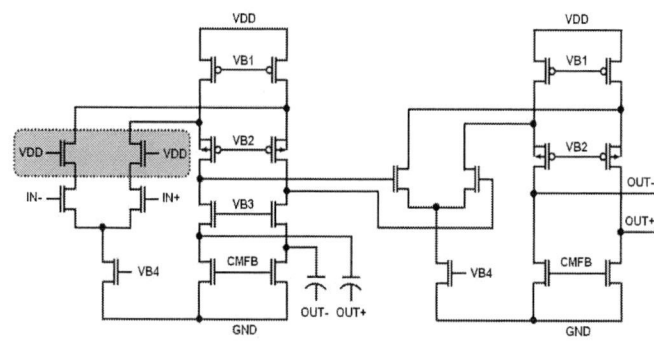

Fig. 3. Schematic of 2-stage folded cascode opamp.

V. LAYOUT CONSIDERATION

As the processes have developed, more layout issues have risen, especially related to poly gate and its channel, and they are needed to be considered for high resolution in small size.

In a subranging flash ADC, two comparators at both ends of the comparator array have relatively large static and dynamic offsets compared with the rest of the comparators due to charge-induced damages during plasma exposure. In this design, narrow poly lines connected to a fixed bias are added between the adjacent comparators and the poly gates are laid out compactly to reduce the undesirable comparator input-referred offsets instead of dummy comparators at both ends of array [7]. As a result, it provides same environment for the critical devices in comparator.

The capacitor arrays used in the entire ADC are based on Vertical Natural Capacitor (VN-Cap) using metal fringing capacitance. In this design, the capacitors in MDACs and flash ADCs consist of MET1 to MET4. They are all laid out in symmetric structure in parallel and isolated by substrate to MET5 barrier to get the same parasitic capacitance illustrated in Fig. 4. Also the employed layout technique provides the same conditions for all the unit capacitor so that it eliminates parasitic capacitance mismatches from the metal routings to get a high-resolution without extra dummy capacitor [8].

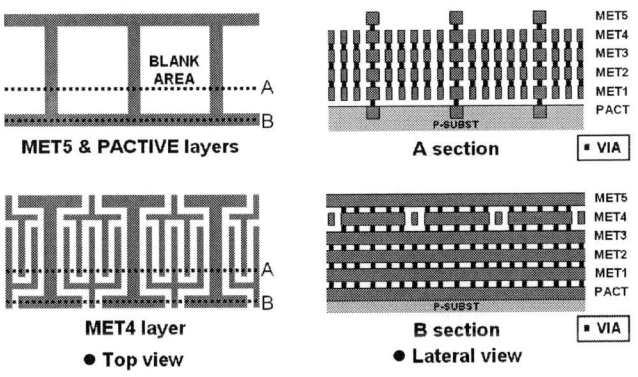

Fig. 4. Layout for high matching VN Capacitor.

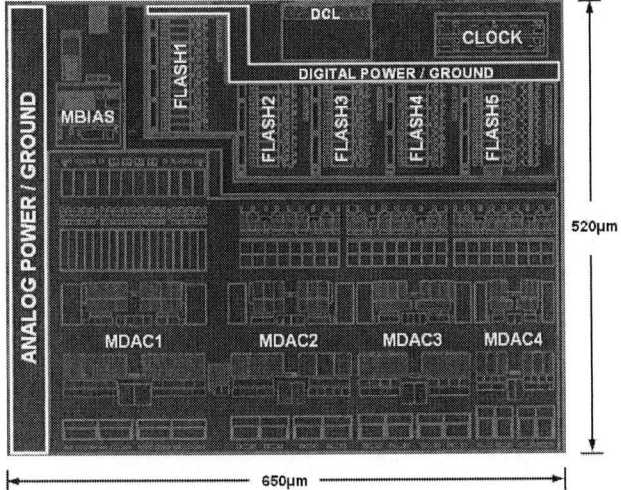

Fig. 5. Die photograph.

VI. EXPERIMENTAL RESULTS

The prototype ADC is fabricated in a 65nm single-poly five-metal standard CMOS process using no analog enhancement, and occupies 0.34mm^2 (650μm × 520μm) active die area as shown in Fig. 5. There is no dummy capacitor and no dummy comparator. It dissipates 18mW at 30MS/s with a 1.2V power supply. The measured DNL and INL are ±0.44LSB and ±1.33LSB, respectively, as plotted in Fig. 6. The dynamic performance is measured from the FFT spectrum as shown in Fig. 7. For a 4.7MHz 2.4Vp-p differential input, the measured SNDR and SFDR are 65.1dB and 74.9dB, respectively, at a clock frequency of 30MHz. The calculated FOM, defined as Power/(2^{ENOB}×Fs), is 0.41pJ/Conversion-step. At the Nyquist sampling rate, the FOM is 0.46pJ/Conversion-step from the measured SNDR of 64.1dB with a 14.2MHz input.

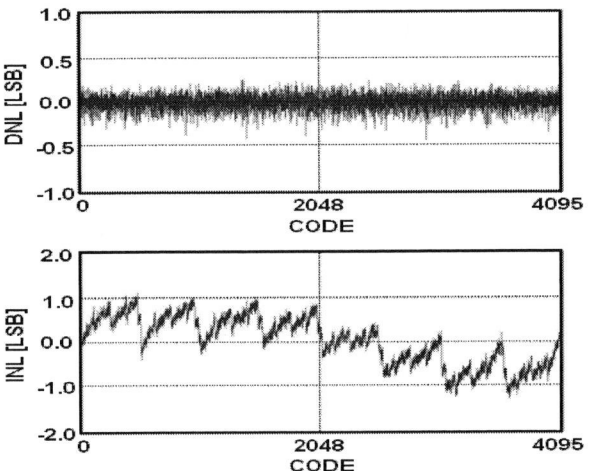

Fig. 6. Measured DNL and INL.

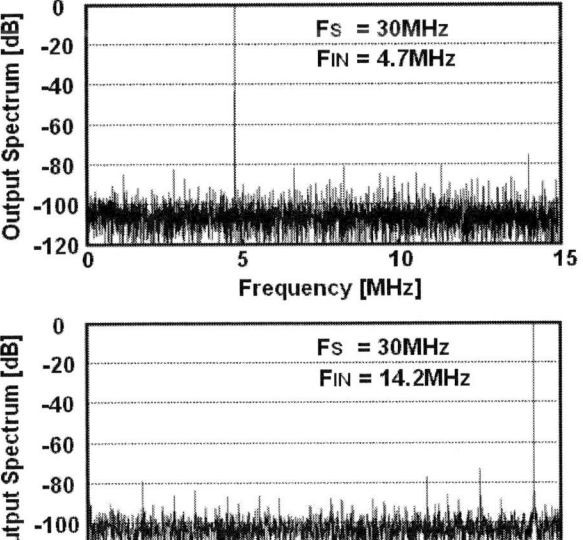

Fig. 7. Measured FFT spectrums.

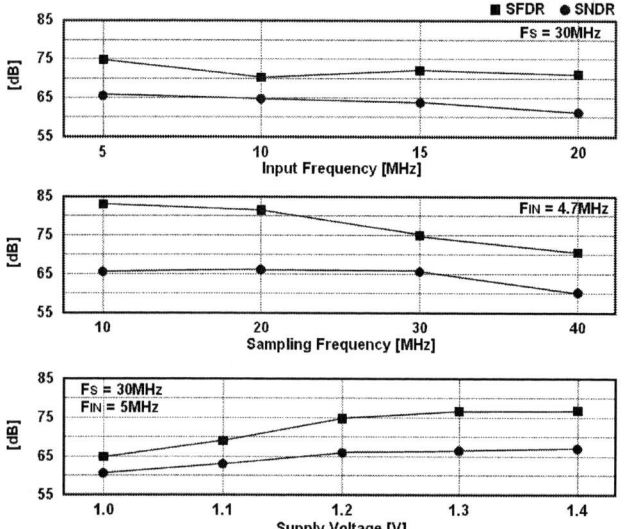

Fig. 8. SNDR and SFDR vs. input frequency, sampling frequency, and supply voltage.

The SNDR and SFDR versus the input frequency, the sampling frequency, or the supply voltage are plotted in Fig. 8. The SNDR is maintained over Nyquist sampling rate and ENOB of more than 10b can be achieved down to 1.0V power supply.

VII. CONCLUSIONS

A 65nm CMOS 1.2V 12b 30MS/s pipelined ADC has been presented. In this work, the capacitive reference scaling (CRS) technique is proposed to achieve high SNR with low gain opamp by utilizing wide input dynamic range at low supply voltage. By adopting the proposed techniques, the prototype ADC achieves over SNDR of 65dB at 1.2V supply showing FOM of less than 0.5pJ/Conversion-Step up to Nyquist sampling rate.

The overall ADC performance is summarized in table I.

Table I.
Performance Summary of the proposed ADC.

Resolution	12 bits
Conversion Rate	30 MS/s
Process	65nm 1 poly 5 metal CMOS
Supply Voltage	1.2 V
Input Range	2.4 Vp-p differential
DNL / INL	± 0.44 LSB / ± 1.33 LSB
SNDR/SFDR (@F_{IN} = 4.7MHz)	65.1 dB / 74.9 dB
SNDR/SFDR (@F_{IN} = 14.2MHz)	64.1 dB / 71.9 dB
Power Dissipation	18 mW
Active Die Area	0.34 mm^2 (=650μum $\times$ 520μum)

REFERENCES

[1] Y. Shimizu, S. Murayama, K. Kudoh, H. Yatsuda, A. Ogawa, "A 30mW 12b 40MS/s Subranging ADC with a High-Gain Offset-Canceling Positive-Feedback Amplifier in 90nm Digital CMOS, " ISSCC Dig. Tech. Papers, pp. 218-219, Feb. 2006.

[2] Y. D. Jeon, S. C. Lee, K. D. Kim, J. K. Kwon, J. Kim, "A 4.7mW 0.32mm2 10b 30MS/s Pipelined ADC Without a Front-End S/H in 90nm CMOS, " ISSCC Dig. Tech. Papers, pp. 456-457, Feb. 2007.

[3] B. S. Song, S. H. Lee, and M. F. Tompsett, "A 10-b 15-MHz CMOS Recycling Two-Step A/D Converter," IEEE J. Solid-State Circuits, vol. 25, no. 6, pp. 1328-1338, Dec. 1990.

[4] O. Stroeble, V. Dias, C. Schwoerer, "An 80MHz 10b Pipeline ADC with Dynamic Range Doubling and Dynamic Reference Selection," ISSCC Dig. Tech. Papers, pp. 462-463, Feb. 2004.

[5] P. C. Yu and H. S. Lee, "A 2.5-V, 12-b, 5-MSample/s Pipelined CMOS ADC," IEEE J. Solid-State Circuits, vol. 31, no. 12, pp. 1854-1861, Dec. 1996.

[6] B. Ahuja, "An improved frequency compensation technique for CMOS operational amplifiers," IEEE J. Solid-State Circuits, vol. 18, no. 6, pp. 629-633, Dec. 1983.

[7] D.-Y. Chang and S.-H. Lee, "Design Techniques for a Low-Power Low-Cost CMOS A/D Converter," IEEE J. Solid-State Circuits, vol. 33, no. 8, pp. 1244-1248, Aug. 1998.

[8] K. J. Lee et al., "A 90nm CMOS 0.28mm^2 1V 12b 40MS/s ADC with 0.39pJ/Conversion-Step," Symp. VLSI Circuits Dig. Of Tech. Papers, pp 198-199, June 2007.

978-1-4244-2018-6/08 $25.00 © 2008 IEEE

IEEE 2008 Custom Intergrated Circuits Conference (CICC)

A Continuous-time Input Pipeline ADC

David Gubbins, Bumha Lee[1], Pavan Kumar Hanumolu, Un-Ku Moon

School of EECS, Oregon State University, Corvallis, OR 97331

[1]National Semiconductor, Santa Clara, CA

Abstract— A new pipeline ADC architecture that employs a continuous-time first stage followed by a conventional switched capacitor pipeline ADC is presented. Such an approach overcomes many of the challenges associated with a pure switched-capacitor architecture and leads to a low area, low power solution with excellent distortion performance. Measured results obtained from a proof of concept test chip fabricated in a $0.18\mu m$ CMOS process validate the effectiveness of proposed techniques.

I. INTRODUCTION

Pipeline ADCs have traditionally been implemented with switched-capacitor methods. However switched-capacitor methods present many challenges to ADC designers when more than 10 bit ENOB, high SFDR and low power are required [1]. These challenges are most acute in the pipeline ADC first stage, mainly due to the fact that the first stage is the main contributor to ADC noise, distortion and consequently power [2]. Sampling the input signal onto large sampling capacitors with low distortion poses a challenge to circuit designers. Such capacitors require large sampling switches with low on resistance, R_{on}, in order that the input sampling network settles in the alloted time period. Large sampling switches demand large switch logic drivers and this increases the digital power consumption. Large sampling switches also bring with them nonlinear junction capacitance which can cause the sampled signal to be distorted. Sampling is further complicated by input signal bondwire inductance which may cause ringing on the input signal sampling network. The large ADC capacitive load also presents a challenge for the reference buffers which need to settle to a required accuracy.

This work addresses these challenges by implementing a continuous-time(CT) first stage followed by a scaled down switched capacitor pipeline back-end ADC. For every extra bit resolved in the CT first stage the back-end switched-capacitors can be scaled down by a factor of four [3]. This new architecture is shown in Fig. 1. Such an approach leads to lower power, reduced sampling distortion, lower capacitance area and allows rail-to-rail input swing. The resistive input load of the continuous-time first stage is also more favorable when compared to a dynamically changing switched-capacitor input load [4]. By resolving two bits in the first stage in a continuous-time fashion and moving the sampling operation to the second stage input the back-end switched-capacitor circuitry can be scaled down appropriately because of the reduced back-end noise requirements. This eases the required switch resistances due to the smaller sampling capacitors and obviates the need for a front end sample-and-hold circuit.

This paper is organized as follows. Section II introduces the

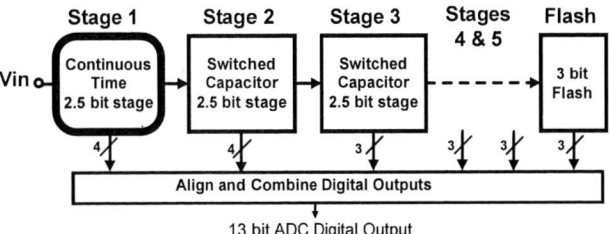

Fig. 1. The continuous-time input pipeline ADC architecture

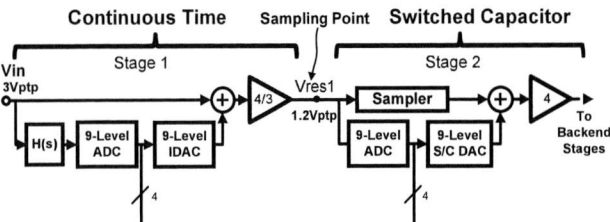

Fig. 2. The first two stages of the continuous-time input pipeline ADC

proposed continuous time front end architecture. Section III describes the prediction filter used in the first stage. Section IV provides the circuits used in the first stage. Section V presents the experimental results followed by conclusions.

II. ARCHITECTURE OF THE CONTINUOUS-TIME INPUT PIPELINE ADC

Before discussing the new ADC architecture it is worth reviewing the basic operation and timing of a simplified 1.5 bit switched capacitor pipeline ADC first stage shown in Fig 3. The continuous-time input signal is denoted by Vin. The input signal is sampled on to the capacitor Cin on the negative edge of $\phi 1$ and the flash ADC quantizes Vin during the ensuing non-overlap period. Thus from Fig. 3 the sampling instant is given by time t2 and the non-overlap time is between t2 and t3. The flash ADC output propagates through to the multiplying DAC (MDAC) switches. At time t3, the positive edge of $\phi 2$, the charges on Cin and Cdac are injected into the virtual ground of the amplifier to create the amplified residual voltage, Vres1 at the stage output. Vres1 is then sampled by the following stage on the negative edge of $\phi 2$ so that further bits can be resolved. The key point from this brief discussion is that there is a half sample delay through the first stage. This half sample delay allows the flash ADC to operate and allows the Vres1 voltage to settle. We will now discuss the proposed continuous-time input stage that circumvents the explicit sampling of the input.

978-1-4244-2018-6/08 $25.00 © 2008 IEEE

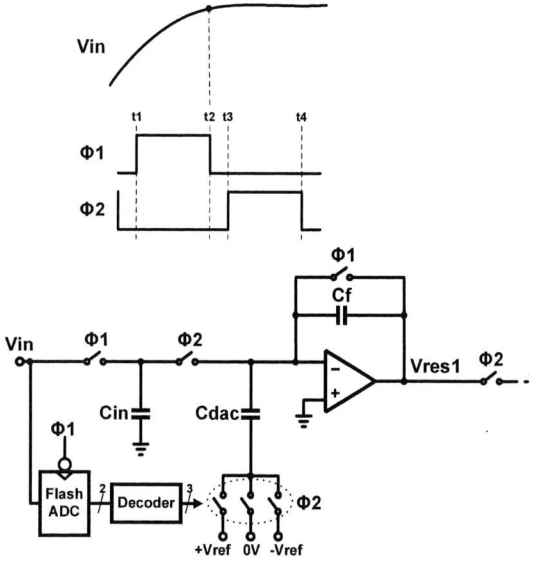

Fig. 3. Simplified diagram of the conventional pipeline ADC first stage with timing diagram

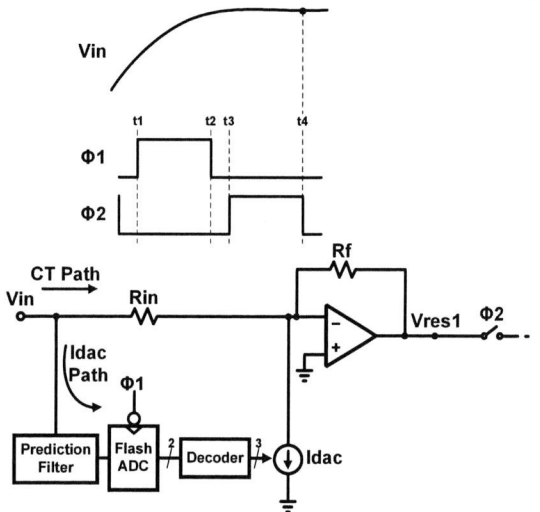

Fig. 4. The continuous-time input pipeline ADC first stage with timing diagram

The continuous-time input pipeline ADC first stage is shown in Fig. 4. The continuous-time signal path from Vin to Vres1 consists of an inverting amplifier whose gain is given by the ratio of resistors Rin and Rf. An NMOS current digital-to-analog converter, labelled as Idac, completes the MDAC. Looking at Fig. 4 it is clear that the sampling instant occurs at the output of the first stage at time t4. However in order to have an accurate residue voltage that is within the second stage input range the Idac component of the continuous-time residue, Vres1 must be settled. The Idac thus needs to be updated at time t3. Therefore at time t2 an estimate of the input voltage at time t4 must be presented to the flash ADC. A prediction filter is used to provide this estimate within a certain error budget that is consistent with this particular

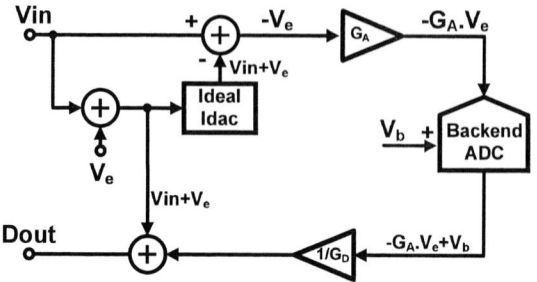

Fig. 5. Mathematical model showing propagation of error Ve

TABLE I
FILTER SPECIFICATIONS BASED ON ADC ARCHITECTURE

Tuning Range	$\pm 40\%$
Frequency Accuracy	$\pm 5\%$
THD	$40dB$
Magnitude Ripple	$\pm 0.3dB$
Phase Ripple	$\pm 2^o$

pipeline implementation. This error budget determines the filter specification which is shown in Table I. The details of such a filter will be dealt with in Section III. The obvious question at this point is why not instead introduce a phase lag in the continuous-time signal path from Vin to Vres1. By placing the prediction filter before the flash ADC the filter accuracy requirements are much reduced due to the presence of redundancy in the ADSC path [2]. Also artifacts of the prediction filter frequency response do not appear in the ADC frequency response. Mathematically this is explained concisely in Fig. 5 where G_A is an analog gain, G_D is a digital gain, V_b is the back-end ADC quantization error and V_e is a term containing both coarse quantization and prediction filter error. It can be seen that if $G_D = G_A$ then

$$Dout = Vin + \frac{V_b}{G_D} \qquad (1)$$

Eq. 1 shows that any frequency shaping introduced by the prediction filter will not be present in the final digital representation of the input signal.

III. PREDICTION FILTER

The motivation of this work is to develop a CT input architecture that efficiently digitizes lowpass wideband signals. It is assumed that the input signal is bandlimited by an anti-aliasing filter, so as to ensure that the signal level at $Fs - Fb$ meets the dynamic range requirements of the ADC [5], where Fb is the signal bandwidth. For this ADC Fb can be as high as $0.8 \times \frac{Fs}{2}$. Given that the input is a lowpass, bandlimited signal an analog prediction filter at the ADC input can use continuous-time information that is usually ignored by conventional discrete time ADCs to estimate the next sample ahead of time. In this work the prediction filter zeros and poles are optimized to provide an approximate half-sample time advancement and constant gain over the bandwidth of interest. This phase advancement is valid for both sinusoidal

978-1-4244-2018-6/08 $25.00 © 2008 IEEE

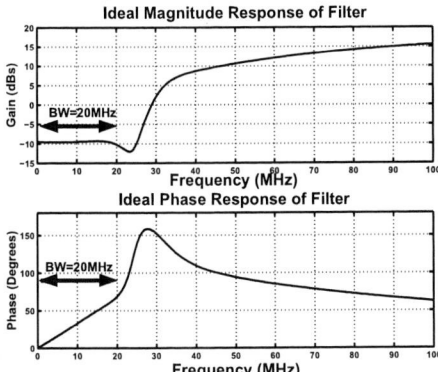

Fig. 6. Prediction filter Frequency Response for sampling rate of 50MHz and a magnitude scaling factor of -9.55dB

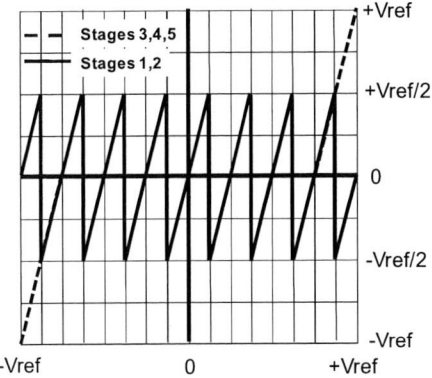

Fig. 7. Transfer Functions of the pipeline stages

tones and general transient signals provided they are lowpass and bandlimited, in this case to $0.8 \times \frac{Fs}{2}$.

The frequency response of such a filter is shown in Fig. 6. The zeros are more dominant at lower frequency. This provides the required phase lead. Poles are necessary not only to make the filter physically realizable but also to limit the out of band gain that can be seen in Fig. 6. This out of band gain is an unavoidable consequence of the in band phase lead [6]. Fortunately, due to practical finite gain bandwidth of the amplifier, the filter gain rolls off at higher frequencies.

IV. CIRCUIT DETAILS

The continuous-time input pipeline ADC architecture is shown in **Figs. 1 and 2.** The first two stages are implemented using two identical 9-level flash ADCs to limit the residue signal swing. This also eases the accuracy requirements of the first stage flash ADC and the prediction filter [7]. Stages 3, 4, and 5 are implemented using 7-level flash ADCs to reduce power in the remaining stages. The resulting stage transfer functions are shown in Fig. 7. The back end ADC is implemented using standard switched-capacitor techniques. Fig. 8 shows the first stage MDAC. The interstage gain is given by the Idac fullscale current, the feedback resistor and the open-loop gain of the amplifier. This stage scales the input voltage range from 3Vptp to 1.2Vptp. This improves the feedback factor of the stage. It is also worth noting that the Idac does not significantly degrade the feedback factor due to its higher impedance with respect to Rin and Rf. Simple digital foreground calibration similar to [8] was used to used to digitally measure the first stage analog jumps in Fig. 7. Once measured these 9 digital values are used during normal operation to create the digitally calibrated output code. This scheme corrects for first stage interstage gain error and mismatch between the Idac current elements.

The prediction filter is implemented with two active-RC low-Q biquads. The biquads are implemented using simple amplifiers with well controlled low gains of 26dB[9]. Such an approach leads to low filter power and ensures that the filter is not a bottleneck in terms of sampling speed. The maximum capacitor ratio is 5 and the maximum resistor ratio is 19.

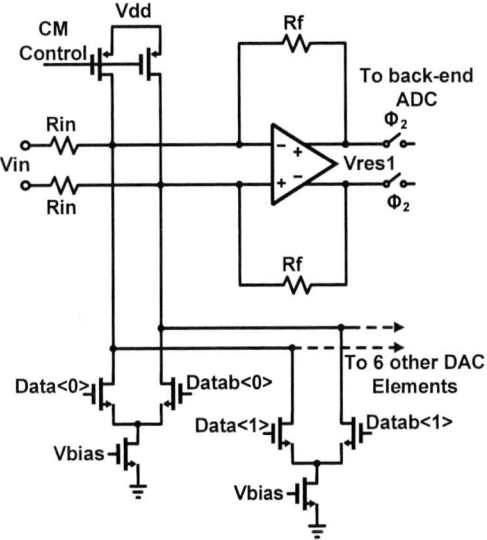

Fig. 8. Simplified first stage circuitry

V. EXPERIMENTAL RESULTS

The converter was fabricated in a $0.18 \mu m$ CMOS process. Fig. 12 shows the die photo with a total active area of $1.6mm^2$. Fig. 9 shows the DNL and INL performance at the 11-bit level. Fig. 10 shows a typical single tone FFT performance plot at a sample rate of 16MSPS. Fig. 11 shows SNR, SNDR and SFDR over input frequency. The complete performance summary is given in Table II. Unfortunately, results from this initial version of silicon were limited by a timing bug in the switched-capacitor back-end circuitry. This limited the sampling rate of the ADC and caused the power of the ADC to be excessive. This also increased the OSR from 1.25 in simulation to 4 on the bench. However the concept of a continuous time input pipeline ADC was verified and an improved performance is anticipated from future versions of this silicon.

VI. CONCLUSIONS

A new ADC architecture employing a continuous-time input stage offers a more favorable trade-off between noise,

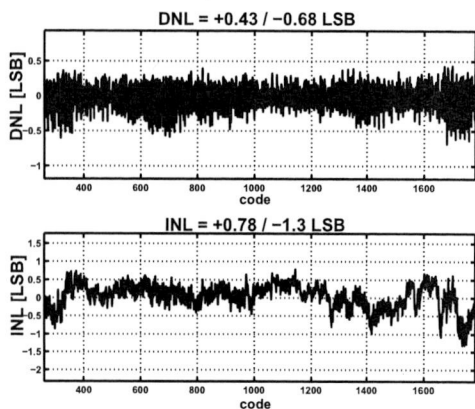

Fig. 9. DNL and INL Performance of the ADC at 11 bit

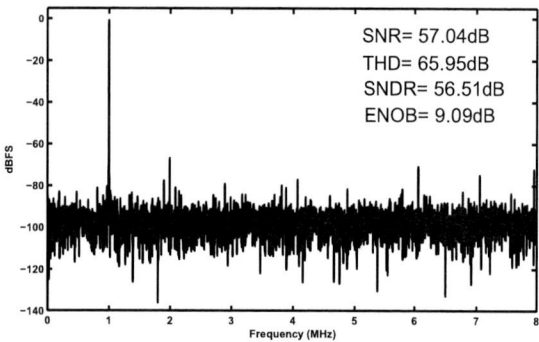

Fig. 10. Single tone FFT plot at 1MHz input and 16MSPS

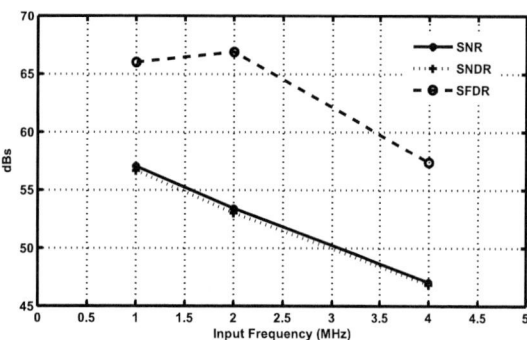

Fig. 11. SNR, SNDR, SFDR over input frequency at 16MSPS

Fig. 12. Photo of the die

TABLE II

PERFORMANCE SUMMARY

Resolution	11 bits
Conversion Rate	16MSPS
Input Range	3 Vptp
SNDR	56.51dB @ Fin=1MHz
SNR	57dB @ Fin=1MHz
THD	65.95dB @ Fin=1MHz
SFDR	66dB @ Fin=1MHz
DNL	0.43/-0.68 LSBs
INL	0.78/-1.3 LSBs
Total Chip Power Consumption	50mW
Supply voltage	1.8V
Die area	1.2×1.36 mm^2
Technology	0.18μm CMOS

distortion and amplifier power when compared to switched-capacitor methods. This architecture also results in a lower area due to back-end capacitor scaling. Consequently switch Ron requirements are very much eased and sampling distortion is reduced. The benign resistive input load easily allows a rail to rail input signal to be coupled into the ADC. This continuous-time input architecture has been enabled by the introduction of signal prediction within the ADC first stage. Future efforts will focus on moving the prediction filter into the digital domain and incorporating anti-alias filtering into the CT first stage.

ACKNOWLEDGMENT

The authors would like to thank National Semiconductor for supporting this work and providing chip fabrication. Much thanks is due to Sunwoo Kwon, Naga Sasidhar Lingam and Nima Maghari for useful discussions and layout help.

REFERENCES

[1] S. Bardsley et al,"A 100dB SFDR 80MSPS 14-bit 0.35um BiCMOS Pipeline ADC", *IEEE J.Solid-State Circuits*, Vol.41,No.9, September 2006.

[2] S. Lewis,"Optimising the stage resolution in Pipelined, multistage, Analog-to-Digital Converters for Video-Rate Applications" *IEEE TCAS 2*, Vol.39, No.8, August 1992.

[3] D. Kelly et al,"A 3-V 340-mW 14-b 75-Msample/s CMOS ADC With 85-dB SFDR at Nyquist Input", *IEEE J.Solid-State Circuits*, Vol.36, No.12, December 2001.

[4] R. Reeder,"AN742 Frequency Domain Response of Switched-Capacitor ADCs", *Analog Devices Application Note*.

[5] D. Robertson,"Selecting Mixed-Signal Components for Digital Communications Systems, Part V, Aliases, images, and spurs", *Analog Dialogue*, 31-3, 1997.

[6] H.W. Bode, "Network analysis and feedback amplifier design", 1945.

[7] D.Y. Chang,"Design Techniques for a Pipelined ADC Without Using a Front-End Sample-and-Hold Amplifier" *IEEE TCAS 1* Vol.51, No.11, November 2004.

[8] T. Kuyel.F. Tsay, "Optimal analog trim techniques for improving the linearity of pipeline ADCs" Proc. International Test Conference, pp.367-375 Oct 2000.

[9] A. Baschirotto et al, "A 150 Msample/s 20 mW BiCMOS switched-capacitor biquad using precise gain op amps" *ISSCC Digest of Technical Papers*,pp. 212-213, Feb 1995.

IEEE 2008 Custom Intergrated Circuits Conference (CICC)

A 0.8 V Asynchronous ADC for Energy Constrained Sensing Applications

M. Trakimas, S. Sonkusale

Tufts University

Abstract-This paper discusses the design of an asynchronous analog-to-digital converter targeted for low-power sensing applications. The asynchronous sampling scheme will save power because it only samples the input signal when it is changing. The idea of using an adaptive resolution to increase the maximum input frequency of the ADC is introduced. A prototype chip has been fabricated in a 0.18 μm CMOS process. Initial measurement results are presented.

I. INTRODUCTION

In wireless sensor networks and portable medical devices, system power efficiency becomes increasingly important to increase the lifetime of the devices. Tsividis [1] suggested the use of continuous time digital signal processing as a viable alternative for aliasing-free high resolution low power signal processing. Such a system will also be energy efficient since their power consumption scales linearly with the input activity. For input signals exhibiting lower activity, synchronous sampling based digital signal processors are not power efficient due to its continuous sampling at a rate twice the highest frequency of the input signal, as stated by Shannon's sampling theory. When the input signal is changing at a rate lower than its maximum value, the synchronous system will continue to process the signal at the same rate burning unnecessary power during such idle periods.

A more efficient architecture uses an asynchronous sampling scheme where samples are taken only when the input signal changes [1]. The highest improvement in power efficiency for such architecture will be seen for signals which are relatively constant with brief periods of activity. This has been shown to be true in applications such as temperature sensors, pressure sensors, electro-cardiograms, and speech signals [2]. Another promising application is for brain implants which use electrodes to probe neurons. When a neuron fires, a voltage spike is seen at the electrode followed by little activity until the next time the neuron fires.

In order to realize an asynchronous system, analog-to-digital converters (ADCs) capable of processing signals asynchronously must be designed. In this paper we present the design of an asynchronous ADC which is targeted for use in energy constrained applications like implantable biomedical sensors. These applications are good candidates for an asynchronous architecture due to the nature of their signals and the critical importance of the system power consumption. In the following section we go into more detail about the asynchronous sampling scheme and its benefits. Section III presents the design of our ADC and introduces the concept of using programmable and adaptive resolution to improve the performance of asynchronous ADCs. We follow this with simulated and initial chip measurement results.

II. Asynchronous Sampling

The asynchronous sampling scheme can be explained using Fig. 1, which shows an asynchronously sampled signal (some examples include ECG or spiking neurons) The dashed lines in the figure represent 2^M-1 quantization levels which are evenly spaced across the full amplitude range, where M is the hardware resolution of the ADC. A sample is taken only when the signal crosses one of the quantization levels. As seen in the figure, the sampling rate adapts to the rate of change of the signal and drops to zero when the signal is not changing. This leads to a very efficient design which only takes samples if they will provide new information about the input signal. In order to preserve the shape of the signal, the time between samples, Δt, must also be kept. This value can be determined by either a local on chip timer or a time-to-digital converter which have been shown in [3] to add little to the overall power consumption of the ADC. The amplitude-time data pairs are output from the ADC and can be processed by an asynchronous DSP as demonstrated in [1]. The ADC can also be integrated with a synchronous DSP by using a synchronizer between the ADC and DSP. This will allow for increasing the power efficiency of ADCs in current systems without requiring a complete system redesign.

In order for asynchronous ADCs to be designed, the effect that the asynchronous sampling has on the system signal-to-noise ratio (SNR) must be analyzed. This is best explained by comparing the frequency responses of a sampled signal using a synchronous and an asynchronous sampling scheme. For a synchronous system, the signal harmonics will be aliased around the constant sampling rate back into the baseband. This effectively raises the noise floor and degrades the SNR which is limited to the theoretical maximum value of 6.02n + 1.76 dB, where n is the resolution of the ADC. The authors' of [1] have shown that in the case of an asynchronously sampled signal, the harmonics will no longer be aliased into the baseband since there is no longer a constant sampling rate. This theoretically eliminates the noise floor and as a result

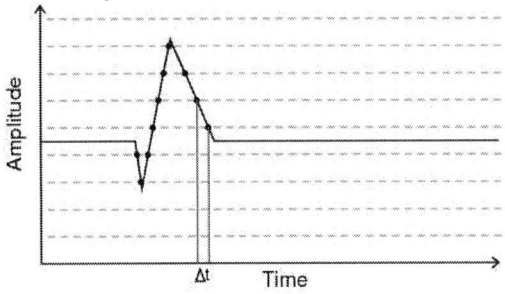

Fig. 1. Example of an asynchronously sampled signal.

only the signal harmonics within the baseband will affect the SNR. By setting the signal close to the system bandwidth, most of the harmonics will be out-of-band and the SNR can be significantly improved compared to the synchronous case.

In practical designs, there will still be quantization noise added to the signal due to the inexact time of the sampling instants resulting in jitter-like noise. This error in time, δt, will cause an error in the output voltage (1), where dV_{in}/dt is the slope of the input signal. Since the SNR of a system is dependant on the ratio of the power in V_{in} to the power in δV, the theoretic value for the asynchronous case can be calculated as (2), where T_C is the timer clock period which determines δt [2]. This shows that the SNR no longer depends on the bit resolution of the ADC but instead depends on the timer period T_C and the statistical properties of the input signal V_{in}. It has been shown in [3] that for speech signals, the ENOB of a 4-bit asynchronous ADC can be increased up to 10-bits by increasing the timer frequency to 1 MHz. In theory, the SNR will continue to increase as T_C increases, but in reality it is limited by the accuracy of the analog components in the ADC which determine the error in the quantization levels.

$$\delta V = \frac{dV_{in}}{dt} * \delta t , \qquad (1)$$

$$SNR_{dB} = 10\log\left(\frac{3P(V_{in})}{P\left(\frac{dV_{in}}{dt}\right)}\right) + 20\log\left(\frac{1}{T_C}\right), \qquad (2)$$

Another important parameter which must be calculated in the case of an asynchronous ADC is the maximum frequency which it can process. This value has been shown to be (3), where δ is the loop delay which is the minimum time the ADC requires between samples [2]. The loop delay is the inverse of the maximum sampling frequency of the ADC. Equation (3) shows that the maximum input frequency will be limited by the number of quantization levels since the ADC must process a sample every time the input signal crosses a level. In most applications, the input signal will never make a full-scale change at its maximum frequency and this requirement can be relaxed.

$$f_{max} = \frac{1}{\delta * 2\pi\left(2^M - 1\right)}, \qquad (3)$$

The asynchronous architecture has several other beneficial characteristics besides eliminating aliasing, improving quantization noise, and increasing power efficiency. The reduction in the sampling speed with low signal activity will reduce the electromagnetic interference caused by switching in the ADC. The ADC will also be immune to metastability since no sampling in the conventional sense takes place. The maximum sampling rate of the ADC can also be increased when compared to some ADCs such as a successive approximation (SAR) converter. The increase in speed is realized because the asynchronous ADC requires only one cycle per sample, while a SAR converter requires M cycles, where M is the resolution of the ADC [2].

III. ASYNCHRONOUS ADC DESIGN

Since the target application for our asynchronous ADC was low-frequency biomedical sensors, the main concern in our design was minimizing the power consumption. In order to do this we used a supply voltage of 0.8 V which biased the analog circuits in the moderate inversion region. This was a good compromise between the strong inversion region which allows for high bandwidths but wastes power, and the weak inversion region which offers the largest g_m/I_D ratio, but has absolute bandwidths too low for most applications.

A. ADC Architecture

The architecture of the designed asynchronous ADC is shown in Fig. 2. A fully-differential design was implemented in order to double the input signal swing which was limited by the low supply voltage. The operation of the ADC can be explained by first assuming that the digital logic has just taken a sample and has stored the digital value D_{OUT} which corresponds to the amplitude of the differential input signal. The logic has an internal timer which keeps track of the time between samples Δt and has been reset to zero at the time of the sample. After storing D_{OUT}, this value is sent to the two digital-to-analog converters (DACs) which output analog voltages equal to (4) and (5), where $\Delta V_{IN,diff}$ is the change in amplitude of the differential input signal since the last sample, and LSB is the least significant bit of the ADC. The input signal has been subtracted within the DAC to keep the output of the DAC $< \pm LSB$. This will reduce the DAC settling time requirements and the loop delay of the ADC, allowing higher bandwidth signals to be processed. Assuming that the input signal amplitude has not changed while the DACs are being updated, $DAC1$ and $DAC2$ will initially output $+LSB$ and $-LSB$ respectively, and the outputs of both comparators will be low. When the differential input signal amplitude changes by $\pm LSB$, one of the comparators will toggle high signaling the digital logic to increment or decrement the value of D_{OUT} by LSB, as well as to store the value of the timer. At this point the digital logic will update the DACs with the new value of D_{OUT} and reset the timer. This cycle will repeat as the ADC tracks the input signal. The values of D_{OUT} and Δt will be output at each sample for subsequent digital processing.

$$V_{DAC1} = +LSB - \Delta V_{IN,diff} , \qquad (4)$$

$$V_{DAC2} = -LSB - \Delta V_{IN,diff} , \qquad (5)$$

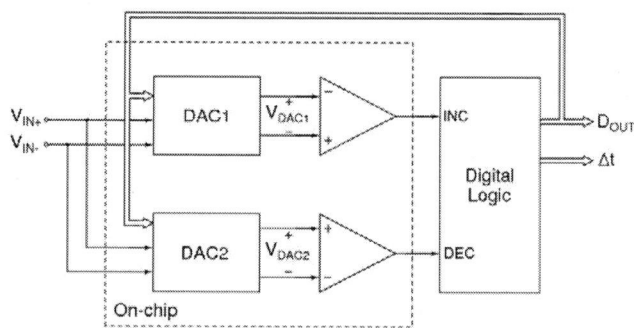

Fig. 2. Asynchronous ADC architecture.

B. DAC

The schematic of the 10-bit DAC designed for our ADC is shown in Fig. 3. A fully-differential operational transconductance amplifier (OTA) based hybrid charge-redistribution resistor string architecture was used for the design. The four least significant bits (LSBs) are resolved by the resistor string which is switched to the unit capacitor C_U closest to the OTA in Fig. 3. Both sides of the DAC share the same resistor string to save power. The next five bits are resolved by the capacitor array which includes 31 thermometer coded unit capacitors. These can be switched to either V_{DD} or GND depending on the most significant bit (MSB) of the DAC which determines the sign. The DAC has been over designed for 10 bits for added flexibility in the initial prototype. The unit capacitors C_U have been set to 150 pF in order to make the kT/C noise irrelevant. This sets the value of C_{FB} to 4.8 pF. The differential input signal is subtracted within the DAC using capacitor C_{IN} which has also been set to 4.8 pF for a gain of 1.

The operation of the DAC is similar to standard charge-redistribution DACs. It is reset by resetting the OTA and switching all the capacitors to V_{REF} which is $V_{DD}/2$. Once the DAC has settled to the required accuracy the OTA is taken out of reset and the capacitors are switched to either V_{DD} or GND and the first unit capacitor is switched to the resistor string. The number of capacitors switched and the location on the resistor string tapped depends on the value of the digital input. The difference between our DAC and a standard one is that the input signal is also switched to capacitor C_{IN} which adds its negative value to the DAC output. As discussed in the previous section, this will keep the DAC outputs within one LSB of V_{REF}, and will allow for a quicker settling time. An added benefit of this is that by keeping the outputs around $V_{DD}/2$, the leakage across the OTA switches is minimized. This reduces the voltage drift seen at the output of the DAC which can degrade the performance of an asynchronous ADC during extended periods of inactivity. The fully-differential architecture reduces the differential leakage seen at the output further and increasing the DAC performance. A second capacitor array and resistor string have been added in parallel with those shown in the schematic for compensating any offsets in the DAC.

C. OTA & Comparator

The schematic of the OTA designed for the DAC is shown in Fig. 4 with its common-mode feedback (CMFB) circuit omitted. The architecture used was a standard 2nd-order Miller compensated OTA. All the transistors were biased in the moderate inversion region to improve the power efficiency while still achieving a useful bandwidth. The CMFB circuit is essentially a replica of the input differential pair of the OTA and sets the bias point of the input stage pMOS loads. The OTA was designed with a bandwidth of 12 MHz for 5 pF loads on each output. The design of the comparator is shown in Fig. 5. It uses a Class-A amplifier structure followed by a pair of inverters to increase the gain and sharpen the transition between a '0' and '1' output.

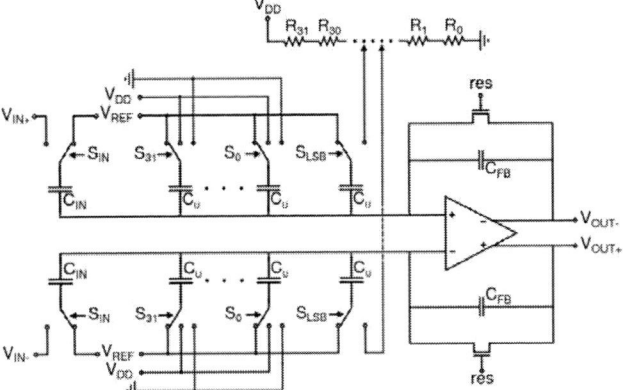

Fig. 3. Schematic of the hybrid charge-redistribution resistor string DAC.

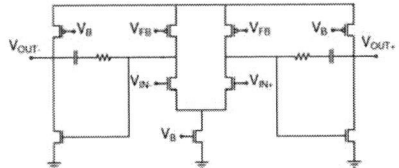

Fig. 4. Schematic of the OTA designed for the DAC (CMFB not shown).

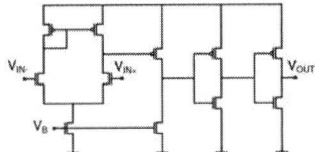

Fig. 5. Schematic of the continuous time comparator.

D. Programmable & Adaptive Resolution

As shown by (3), the maximum input frequency of the ADC is limited by the number of bits of resolution. Previous works have suggested that since the SNR of the ADC is theoretically independent of the bit resolution, the resolution can be set low while still achieving a high ENOB [2]. This is true for certain applications but will cause problems in applications where small amplitude signals must be processed. Take for example the case of an ADC processing the signals from a neural electrode. If the resolution is set too low then the ADC will miss some of the voltage spikes coming from the electrode. A tradeoff must be made between raising the resolution too high so that the ADC can not accurately track the high frequencies in the signal, and lowering it too much so that small pulses will be missed. Our designed ADC allows for programming the resolution anywhere from 2 to 8 bits which will allow the optimal resolution for the application to be used.

Another interesting concept which could be implemented in our ADC in the future is to have the resolution automatically adapt as the characteristics of the input signal change. This is possible since the resolution of the ADC is controlled by the digital logic. The resolution would initially be set high so that no small signals are missed. If the rate of change of the signal exceeds a set amount the resolution will decrease so that the ADC can quickly track the signal. The resolution will increase again as the signal slows down. This would effectively increase the maximum input frequency of the ADC. It also would not decrease the SNR of the ADC since it does not depend on the bit resolution.

IV. RESULTS

A prototype chip of the asynchronous ADC was implemented in the IBM 0.18 μm 7RF CMOS process at the MOSIS fabrication facility. The digital logic was implemented off chip in an FPGA so that different control algorithms could be tested. The supply voltage of the DAC and comparators was set to 0.8 V to reduce the power consumption of the ADC. The digital logic which controlled the switches ran at 1.8 V in order to keep the on resistance of the switches low. The higher digital power supply will not affect the power efficiency of the ADC as the power consumption will be dominated by the analog circuits. A micrograph of the prototype chip is shown in Fig. 6. The die area for the analog section of the ADC was 0.96 mm^2.

Simulation results show that the ADC has a loop delay less than 1 μs, which leads to a maximum sampling rate of 1 MHz. This is achieved due to the low output swing required at the DAC which reduces its settling time requirement. Measurement results on the prototype chip confirm that the DAC reset and settling time will allow for a 1 MHz sampling rate. This will give a maximum input frequency of 10.5 kHz to 600 Hz for bit resolution from 4 to 8-bits.

The limiting factor in testing the chip has been the external FPGA which has limited the loop delay to 100 μs. This has only allowed us to test the prototype chip at low frequencies. A plot of the measured 6-bit ADC output for a 10 Hz 1.2 V peak-to-peak sinusoidal input is shown in Fig. 7. It can be seen that at the top of the sinusoid signal the sampling rate decreases. The ADC has been verified to function up to 8-bits. The maximum input of the ADC is 1.6 V since the input is subtracted within the DAC.

Initial ENOB measurements have been hampered by the large loop delay and the process used to capture the output spectrum. Despite this an ENOB of 7 has been measured for an 8-bit resolution. This number will increase as the FPGA problems are worked out and measurement techniques suited for asynchronous signals are used.

The static power consumption of the ADC has been measured to be 50 μW. The dynamic power could not be measured due to the digital logic being off chip. While the absolute value of the dynamic power could not be measured, the simulation of Fig. 8 shows that it will be much less than in a synchronous design for our targeted applications. Fig. 8 shows a plot of the analog power consumption for a train of pulse inputs which represent neuron voltage spikes. It is seen that dynamic power is only consumed when the signal is changing. This would not be the case for a synchronous design where we would see dynamic power consumed between the pulses.

V. CONCLUSION

A 0.8 V asynchronous ADC has been designed and implemented. Measurements show that the ADC will have a maximum sampling frequency of 1 MHz for a static power

Fig. 6. Micrograph of the prototype chip (shows two copies of the ADC).

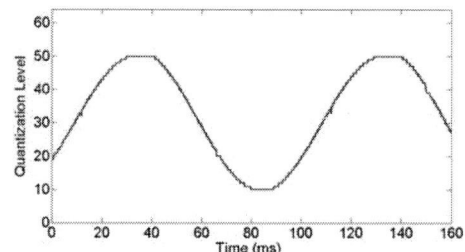

Fig. 7. Measured 6-bit ADC output for 10 Hz sinusoidal input.

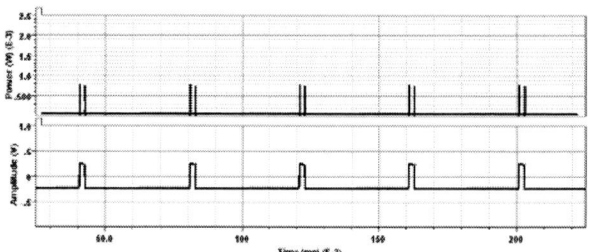

Fig. 8. Plot of ADC power in mW (top trace) for a train of pulses.

consumption of 50 μW. More testing will be required to fully characterize the ADC. Simulation results show that the ADC will save dynamic power in the case of signals with sparse activity. The designed ADC allows for an adaptive resolution control algorithm to be implemented in the future. This will increase the maximum input frequency that the ADC can process.

REFERENCES

[1] Y. W. Li, K. L. Shepard, Y. P Tsividis, "Continuous-Time Digital Signal Processors," IEEE International Symposium on Asynchronous Circuits and Systems, 2005.

[2] E. Allier, G. Sicard, L. Fesquet, M. Renaudin, "A New Class of Asynchronous A/D Converters Based on Time Quantization," IEEE International Symposium on Asynchronous Circuits and Systems, 2003.

[3] E. Allier, J. Foulier, F. Sicard, A. Dessani, E. Ander, m. Renaudin, "A 120 nm Low Power Asynchronous ADC," ISLPED, 2005.

[4] N. Sayiner, H. V. Sorensen, T. R. Viswanathan, "A Level-Crossing Sampling Scheme for A/D Conversion," IEEE TCASII, April 1996.

[5] M. Renaudin, "Asynchronous Circuits and Systems: a Promising Design Alternative," Microelectronic Engineering 54, 2000.

[6] L. Alacoque, M. Renaudin, S. Nicolle, "Irregular Sampling and Local Quantification Scheme A-D Converter," Electronic Letters, Feb. 2003.

[7] F. Akopyan, R. Manohar, A. Apsel, "A Level-Crossing Flash Asynchronous Analog-to-Digital Converter," IEEE International Symposium on Asynchronous Circuits and Systems, 2006.

978-1-4244-2018-6/08 $25.00 © 2008 IEEE

Modeling, Design and Optimization of Hybrid Electromagnetic and Piezoelectric MEMS Energy Scavengers

Xiaochun Wu, Alireza Khaligh, Yang Xu

Electrical and Computer Engineering Department, Illinois Institute of Technology

3301 South Dearborn Street
Chicago, IL 60616

Abstract-**A hybrid energy scavenging technique is introduced to harness ambient energy through electromagnetic and piezoelectric mechanisms to achieve higher power density and higher energy conversion efficiency in a high mechanical damping MEMS structure. By inspecting the second-order mechanical dynamic system for electromagnetic and piezoelectric energy conversions, a unified model is proposed to capture the relation of the recoverable hybrid energy and the input vibration frequency as well as the amplitude. Design trade-offs are considered for the hybrid scavenger size and power maximization. A hybrid MEMS energy scavenger design with optimization results for implementation is shown as an example.**

Keywords: energy scavenging, harvesting, electromagnetic, piezoelectric, MEMS, hybrid energy conversion

I. INTRODUCTION

Energy scavenging from ambient energy sources is emerging as a potential solution to limited battery capacity. More and more electronics using batteries can be powered by energy from solar, electromagnetic fields, human motions, and mechanical vibrations. However, many applications are still limited by the recovered power and conversion efficiency. While low-power circuit design techniques are explored [1] to reduce the power requirements of electronics applications, it is of tremendous interest to develop new energy conversion devices to increase the power density and conversion efficiency [2-9].

Compared to the meso-scale counterparts, MEMS energy scavengers provide low power due to the small size. However, in ultra-low power applications such as wireless sensor nodes and implantable medical devices, MEMS energy scavenging becomes a plausible solution due to its small form factor and low cost in mass production. Since kinetic energy is a common source in the ambient environment, MEMS energy scavengers can be used in applications where other sources such as solar and chemical power are not available. Further, MEMS energy scavengers have the advantage of compatibility with integrated circuits. Fig. 1 shows the system diagram of an integrated self-powered wireless sensor node. The voltage from the MEMS energy scavenger is regulated and then charges a storage battery. The battery supplies the power for the sensor, DSP, and low-power RF transceiver circuits.

This paper presents a hybrid MEMS energy scavenging technique that coverts ambient kinetic energy to electricity both electromagnetically and piezoelectrically. The technique makes use of the commonality of ambient kinetic energy sources and has the advantage of improving the energy conversion efficiency, when the mechanical damping coefficient is high and the amounts of the two kinds of converted energy are comparable.

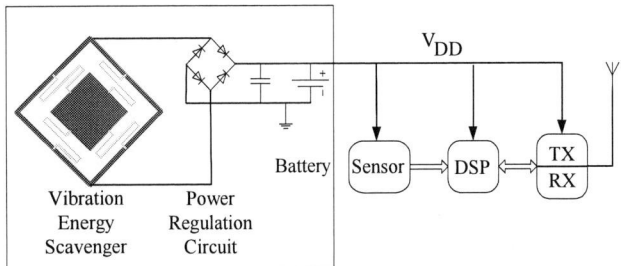

Fig. 1. Block diagram of an integrated self-powered wireless sensor node

II. BACKGROUND

Human motions such as arm swing, horizontal foot motion, and center-of-mass motion are energy sources that have been attracting research on energy extraction for powering wearable electronic devices [11]. Energy extraction from human motion is especially useful for long-term monitoring of human body implantable devices. In [12], researchers have demonstrated a PZT unimorph piezoelectric power generator embedded in shoes with an RMS power of 1.8 mW and used it to power a digital RFID tag successfully. The authors in [13] introduced a meso-scale electro-magnetic energy scavenger for energy extraction from center-of-mass motion of human body. The scavenger has achieved a power density of 0.44 mW/cm^3. But due to the large amplitude (a typical value of 4 cm – 7 cm) and low frequency (typically 2 Hz) for center-of-mass motion of human body, the power density will drop if the scavenger scales down to millimetre-scale or micro-scale, since the scavenger size will limit the amplitude of the proof mass vibration.

Mechanical vibration is a common kinetic energy source in the environment, including vibrations of buildings, machine tools, and car engines. Typical ambient mechanical vibrations have a higher frequency and smaller amplitude than human motions. Researchers in [15] fabricated early prototypes of piezoelectric cantilevers (9 to 25 millimeters in length) with a relatively heavy mass on the free end, capable of generating $375\,\mu W\,/\,cm^3$ from a vibration source of 2.5 m/s^2 at 120 Hz. The smaller vibration amplitude makes MEMS energy scavengers more applicable. Authors in [16] showed a conceptual design of a $5\times5\times1\,mm^3$ electromagnetic micro-generator and predicted power generation of 1 μW at 70 Hz and 0.1 mW at 330 Hz, assuming a deflection of 50 μm.

To improve the power density of MEMS scavengers, hybrid energy scavenging is proposed, which has the advantage of extracting energy from multiple energy sources so that even if one energy source is not available in the environment, alternative energy sources may still supply power and the probability of "power outage" is smaller, increasing the reliability of the systems and enabling the energy diversification. Therefore, hybrid energy scavenging is a promising energy scavenger design trend.

III. ANALYSIS AND MODELING

A. Mass-Damper-Spring Model for Vibration Energy Scavengers

Electromagnetic and piezoelectric energy scavengers based on proof mass vibration can be modelled as a second-order mass-damper-spring dynamic system.

Fig. 2 demonstrates the schematic of a second-order energy scavenger model proposed by Williams and Yates [16]. The system consists a proof mass, m, a spring, k, a mechanical damping coefficient, b_m, and an electrical damping coefficient, b_e. x and y represent the spring deflection and the input displacement, respectively. The differential equation that describes the system is given as Eqn. (1).

$$m\ddot{x} + \left(b_m + b_e\right)\dot{x} + kx = m\ddot{y} \qquad (1)$$

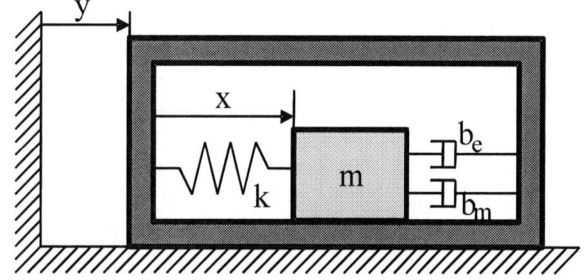

Fig. 2. Schematic of a vibration energy scavenger

B. Electromagnetic Energy Conversion

The extracted output power is maximized, if the resonant frequency of a mass-damper-spring system, described in Eqn. (1), is equal to the input vibration frequency [16]. In this particular case and under the assumption that the power scavenging doesn't hamper the source vibration, the extracted power is given in Eqn. (2), or equivalently, (3) [6].

$$P_{EM} = \frac{m\zeta_e\omega^3 Y^2}{4\zeta_T^2} \qquad (2)$$

$$P_{EM} = \frac{m\zeta_e A^2}{4\omega\zeta_T^2} \qquad (3)$$

where P_{EM} is the magnitude of the electromagnetically generated power, Y is the magnitude of the input displacement, and ζ_e is the electrical damping ratio ($b_e = 2m\zeta_e\omega$), which can be controlled for output power maximization [16]. ζ_T is the total damping ratio ($\zeta_T = \zeta_m + \zeta_e$), ω is the input frequency, ζ_m is the mechanical damping ratio, and A is the input acceleration magnitude. Eqn. (2) and (3) represent the maximum recoverable power for a scavenger. And the maximum extraction power is achieved when the electrical damping ratio matches the mechanical damping ratio.

C. Piezoelectric Energy Conversion

Eqn. (2) and (3) also set an upper bound for piezoelectric energy conversion, where ζ_e represents the effective electrical damping ratio resulting from piezoelectric energy conversion. Referred to Eqn. (15) in [6], the effective damping ratio is limited by the coupling coefficient, k_{31}, of the piezoelectric material. When the effective electrical damping ratio is much lower than the mechanical damping ratio in MEMS structures, the piezoelectric power generation can be calculated as follows.

Fig. 3 shows a piezoelectric energy scavenging structure. A piezoelectric beam with a length L, thickness t, and width of W is clamped on both ends with a seismic mass in the center. When the system resonant frequency is equal to the input vibration frequency, the power generation in the bending mode can be derived from Eqn. (47) in [14] as Eqn. (4) here.

$$P_{PZ} = \frac{C_b \left(\dfrac{t}{2k_2}\right)^2 \dfrac{Y_{PZT}}{\varepsilon} A^2}{2\omega^3 \zeta_m \left[\sqrt{\left(\dfrac{\zeta_m}{k_{31}^2}\right)^2 + 1} + 1\right]} \qquad (4)$$

where P_{pz} is the magnitude of the piezoelectric power generation, $C_b = 2\varepsilon WL/t$ is the equivalent piezoelectric capacitance, $k_2 = L^2/3t$ is the geometric constant relating strain to deflection for a clamped-clamped beam, ε is the dielectric constant of the piezoelectric material, Y_{pz} is the Young's modulus of the material.

D. Hybrid MEMS Energy Conversion

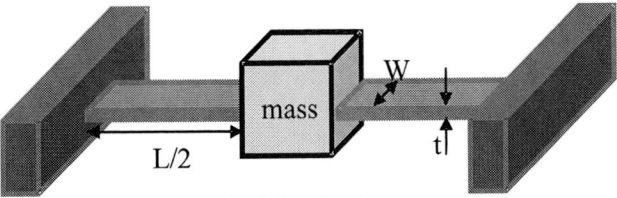

Fig. 3. Schematic of piezoelectric energy scavenger

For a hybrid electromagnetic and piezoelectric MEMS energy scavenger, when the total effective damping coefficient of the system is less than the mechanical damping coefficient, the two separate energy scavengers can be modelled as a single simple system using a unified expression as follows.

Substituting the expressions of the spring constant $k = 2\dfrac{192t^3W}{12L^3}Y_{PZT}$ and the resonant frequency $\omega = \sqrt{k/m}$ into Eqn. (4), we have,

$$P_{PZ} = \frac{9}{128\zeta_m \left[\sqrt{\left(\dfrac{\zeta_m}{k_{31}^2}\right)^2 + 1} + 1 \right]} m\omega^3 Y^2 \qquad (5)$$

As we can see from Eqn. (2) and (5), both the electromagnetic and piezoelectric power generations are proportional to the mass, the cube of the input frequency and the square of the vibration amplitude. Adding the two parts of energy, we have,

$$P_{HB} = \left(\frac{\zeta_{EM}}{4\zeta_T^2} + \frac{9}{128\zeta_m \left[\sqrt{\left(\dfrac{\zeta_m}{k_{31}^2}\right)^2 + 1} + 1 \right]}\right) m\omega^3 Y^2 \qquad (6)$$

where ζ_{EM} is the damping ratio resulting from the electromagnetic conversion, and ζ_T is the effective total damping ratio.

IV. STRUCTURE DESIGN EXAMPLE

Fig. 4 shows a hybrid electromagnetic and piezoelectric energy scavenger designed for vibration energy harvesting. In the center, a movable vibration proof mass is connected to the sides of the stator frame by four piezoelectric serpentine springs. High magnetic flux density permanent magnets are placed on the mass. A copper coil is fixed relative to the stator frame in the middle of the mass. When the mass vibrates up and down, the coil generates an AC voltage, and the piezoelectric springs, electrically connected through the stator frame, convert the strain in the springs into another AC voltage.

For millimetre-scale hybrid structures, silicon, tungsten or PZT can be selected as the material of the vibration mass, stator frame and serpentine springs. A silicon, tungsten or PZT wafer can be patterned to the desired shape using deep reactive ion etching (DRIE). If silicon is used, PZT needs to be deposited on the top and bottom of the serpentine springs.

A multi-turn Cu coil and a pair of Neodymium Iron Boron (NdFeB) magnets can be mounted on the scavenger.

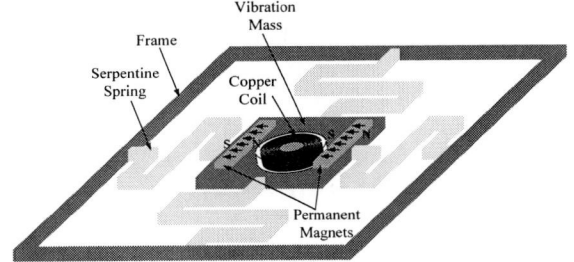

Fig. 4. Schematic of a hybrid energy scavenger

V. DESIGN CONSIDERATIONS

As the energy scavenger size scales down to millimeters and microns that are the typical scales of MEMS structures, the proof mass vibration amplitude is limited if the ambient vibration amplitude is too large. So for constant acceleration vibration, meso-scale energy scavenging structures are suitable for low-frequency large-amplitude vibration applications while MEMS energy scavengers are more desirable for high-frequency small-amplitude applications, as shown in Fig. 5 (the acceleration is 12 m/s² in this example).

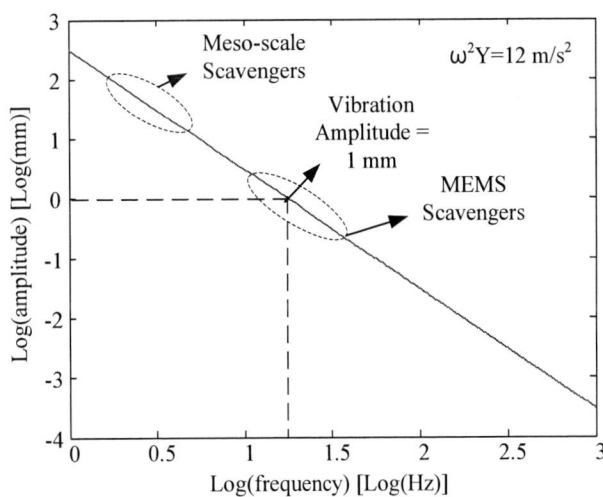

Fig. 5. Design regimes for meso-scale and MEMS energy scavengers

Maximizing the scavenging power also requires the total electrical damping ratio to match the mechanical damping ratio. At meso-scale, the mechanical damping coefficient of energy scavengers is small (e.g., less than 0.1). It is easy to design an electrical damping coefficient as large as the mechanical damping coefficient using either electromagnetic conversion [16] or piezoelectric conversion [6]. In MEMS scavengers, the mechanical damping coefficient is large (e.g., higher than 0.5), and it is difficult to design an electrical damping coefficient for electromagnetic or piezoelectric scavengers as large as the mechanical coefficient, since it is not practical to fabricate too many turns of coils on MEMS structures for electromagnetic conversion, or the electrical damping coefficient is limited by the piezoelectric coupling coefficient. Therefore, hybrid energy scavenging technologies are suitable for MEMS scavengers, where the electromagnetic and piezoelectric power generations are comparable. With a larger total electrical damping coefficient (not exceeding the mechanical damping coefficient), hybrid energy scavenging maximizes the overall output power density and improves the energy conversion efficiency.

VI. OPTIMIZATION RESULTS FOR MEMS IMPLEMENTATION

Most ambient mechanical vibrations have a small acceleration, e.g., 2.25 m/s^2 at a peak frequency of 121 Hz for a small microwave oven [6]. But a car engine compartment has a typical vibration acceleration of 15 m/s^2 at around 50 Hz. Since the corresponding vibration amplitude is about 150 μm, MEMS hybrid energy scavengers can be designed for powering low power electronics, such as wireless sensor nodes on cars.

For a volume constraint of 20 mm^3 for car engine vibration energy harvesting, the dimensions of the frame in Fig. 4 are chosen as $7 \times 7 \times 0.4 \text{ mm}^3$. The proof mass is designed as large as possible [16], as long as the left volume is enough for the serpentine springs to meet the resonant frequency matching requirement.

For compact non-vacuum packaging, a mechanical damping ratio of 0.5 is assumed. But the electrical damping ratio due to the electromagnetic or piezoelectric conversion cannot be higher than 0.5. Otherwise the total damping ratio would be higher than 1. PZT is used as the proof mass and serpentine spring materials, and the thicknesses are both 0.4 mm. The dimensions of the proof mass and each serpentine spring are designed as $5 \times 5 \times 0.4 \text{ mm}^3$ and $40 \times 0.1 \times 0.4 \text{ mm}^3$, respectively. A tungsten mass with dimension of $5 \times 5 \times 0.6 \text{ mm}^3$ is mounted on the PZT mass to increase the total mass. Using Eqn. (6) in [16] to select the load resistor for electromagnetic output maximization, we get an output power of 16.3 μW, or a power density of 0.83 mW/cm^3. Here, a resistive load as low as 10 Ω is selected (lower load resistances will result in higher dissipated power), and a 400 turn coil and a flux density of 0.5 T are used. The estimated electromagnetic damping ratio is about 0.2. Based on the above parameters and assumptions, we get an output power of 6.9 μW, or a power density of 0.35 mW/cm^3 as the piezoelectric contribution. So the total power density for this MEMS scavenger would be 1.18 mW/cm^3.

VII. CONCLUSION

Hybrid energy scavenging is proposed to maximize the MEMS energy density when the vibration amplitude is limited and the MEMS scavenger has a much higher mechanical damping coefficient. A hybrid power expression is derived for modeling hybrid electromagnetic and piezoelectric energy conversion. The hybrid power is proportional to the cube of the input vibration frequency and the square of the amplitude. Practical design considerations are discussed that when the electromagnetic power generation is comparable with the piezoelectric power generation, hybrid MEMS energy scavenging improves the output power. A MEMS energy scavenger example with optimized dimensions for car vibration energy scavenging is presented with an output power density of 1.18 mW/cm^3.

REFERENCES

[1] T. Sterken, K. Baert, C. Van Hoof, R. Puers, G. Borghs, and P. Fiorini, "Comparative modelling for vibration scavengers", *Sensors, Proceedings of IEEE*, Oct. 2004, vol. 3, pp. 1249-1252.

[2] N. S. Shenck and J. A. Paradiso, "Energy scavenging with shoe-mounted piezoelectrics", *IEEE Micro*, vol. 21, pp. 30-42, May-June 2001.

[3] E.M. Yeatman, "Rotating and gyroscopic MEMS energy scavenging", *Proceedings of the International Workshop on Wearable and Implantable Body Sensor Networks (BSN'06)*, Apr. 2006.

[4] J. A. Paradiso, and T. Starner, "Energy scavenging for mobile and wireless electronics," *IEEE Pervasive Compt.*, 4 (2005) (1), pp. 18-27

[5] T. Starner and J. A. Paradiso, "Human generated power for mobile electronics", *Low-Power Electronics Design*, C. Piquet, Ed., CRC Press, 2004, ch. 45, pp. 1-35.

[6] S. Roundy, P.K Wright, and J. Rabaey, "A study of low level vibrations as a power source for wireless sensor nodes", *Comput. Commun.*, 26 (2003), pp. 1131-1144.

[7] E.S. Leland, E. M. Lai, and P.K Wright, "A self-powered wireless sensor for indoor environmental monitoring", *Wireless Networking Symposium*, University of Texas, Austin, Oct. 2004.

[8] D. P. Arnold, S. Das, F. Cros, I. Zana, M. G. Allen and J. H. Lang, "Magnetic induction machines integrated into bulk-micromachined silicon", *Journal of microelectromechanical systems*, vol. 15, pp. 406-414, Apr. 2006.

[9] E. M. Yeatman, "Applications of MEMS in power sources and circuits", *Journal of micromechanics and microengineering*, 17 (2007), pp. S184-S188.

[10] Y. H. Chee, A. M. Niknejad and J. M. Rabaey, "An ultra-low-power injection locked transmitter for wireless sensor networks", *IEEE Journal of Solid-State Circuits*, vol. 41, no. 8, pp. 1740-1748, Aug. 2006.

[11] P. Niu, and P. Chapman, "Design and performance of linear biomechanical energy conversion devices", *Power Electronics Specialists Conference (PESC'06)*, June 2006, pp. 1-6.

[12] J. Kymissis, C. Kendall, J. Paradiso and N. Gershenfeld, "Parasitic power harvesting in shoes", *Second Symposium on Wearable Computers*, Oct. 1998, pp. 132-139.

[13] P. Niu, "Biomechanical energy conversion", Ph.D. Dissertation, University of Illinois at Urbana-Champaign, Urbana, Illinois, Sept. 2007.

[14] S. Roundy and P.K Wright, "A piezoelectric vibration based generator for wireless electronics", *Smart Materials and Structures*, 13 (2004), pp. 1131-1142.

[15] S. Roundy, E.S. Leland, J. Baker, E. Carleton, E. Reilly, E. Lai, B. Otis, J. Rabaey, P.K Wright and V. Sundararajan, "Improving power output for vibration-based energy scavengers", *IEEE Pervasive Compt.*, 4 (2005) (1), pp. 28-36.

[16] C.B. Williams and R.B. Yates, "Analysis of a micro-electric generator for microsystems", *Proceedings of the Transducers 95/Eurosensors IX*, (1995), pp. 369-372.

IEEE 2008 Custom Intergrated Circuits Conference (CICC)

Amorphous Silicon Logic Circuits on Flexible Substrates

Rahul Shringarpure, Lawrence T. Clark, Sameer M. Venugopal, David R. Allee, *and* Shrinivas G. Uppili

Abstract— This paper describes the design of amorphous silicon (a-Si:H) logic circuits using static and dynamic programmable logic arrays (PLA's) as the baseline examples. The PLA's are designed with n-channel a-Si:H thin film transistors (TFT's) manufactured in a low temperature (180°C) process for flexible substrates. Measured and simulated results demonstrate that dynamic circuits are the best design approach, from both a power and delay viewpoint, for a-Si:H flexible logic circuits.

*Index Terms—***a-Si:H TFT, flexible electronics, combinational logic, threshold voltage degradation.**

I. INTRODUCTION

AMORPHOUS silicon (a-Si:H) thin film transistors (TFT's) are widely used in active matrix backplanes for LCD displays on glass. There is current interest in extending the technology to flexible substrates [1] for lightweight, rugged displays conformal to shaped surfaces. Even more applications will be enabled with the development of fully flexible digital electronic systems based on the a-Si:H TFT. These applications beyond displays have already begun with the integration of a-Si:H display row and column drivers [2]. Ultimately, one can envisage a fully flexible system including sensors, microcontroller, display and flexible power supply source that might implement a 'smart' medical bandage. Robust digital a-Si:H TFT circuit building blocks are essential for implementing such flexible systems.

Section II describes the a-Si:H TFT device and highlights its characteristics that pose circuit design challenges, such as single device polarity, low mobility, and threshold voltage (V_T) shifts. Section III focuses on the design of a-Si:H static and dynamic PLA's as fundamental logic circuit building blocks in flexible a-Si:H technologies. Section IV compares the measured results from the fabricated a-Si:H PLA's and determines the best approach to reduce power, increase speed and optimize area while ensuring circuit reliability. Section V concludes.

II. A-SI:H TFT DEVICE

The a-Si:H TFT has a bottom gate inverted staggered structure and is manufactured with a low temperature (180°C) process designed for the development of a-Si:H TFTs on flexible substrates such as stainless steel and heat stabilized polyethylene naphthalate (PEN). The gate metal is molybdenum, the gate dielectric is silicon nitride and the active layer is hydrogenated amorphous silicon (a-Si:H) deposited with plasma enhanced chemical vapor deposition. (see Fig. 1 inset). The source/drain metal is sputtered on as an N+ amorphous silicon/aluminum bilayer. The current voltage characteristics for a typical a-Si:H TFT are shown in Fig. 1.

The device performance is summarized in Table I.

There are several challenges with a-Si:H circuit design [3]: Firstly, the electron mobility is low, ranging from 0.5 to 1 cm²/Vs. This mobility, which is 500x less than that for single crystal silicon, makes a-Si:H circuitry inherently slow. Secondly, only N-type a-Si:H TFT's are available, requiring dynamic NMOS circuit design techniques not widely practiced since the 1970's. Finally, the threshold voltage shift of a-Si:H TFT's poses a significant long term reliability limitation. Electrical stress shifts the a-Si:H TFT V_T significantly. Fig. 1 shows the measured I_{DS} versus V_{GS} curves for a 96μm/9μm a-Si:H TFT stressed for 10000 and 20000 seconds with a stress condition of $V_{GS} = 20$ V and $V_{DS} = 0$ V.

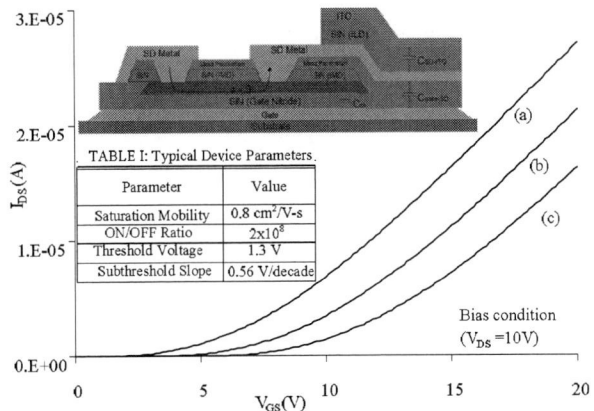

TABLE I: Typical Device Parameters.

Parameter	Value
Saturation Mobility	0.8 cm²/V-s
ON/OFF Ratio	2x10⁸
Threshold Voltage	1.3 V
Subthreshold Slope	0.56 V/decade

Fig. 1. The measured I_{DS} versus V_{GS} curves at $V_{DS} = 10$ V of an 96/9 μm a-Si:H TFT which is a) unstressed and stressed for b) 10000 and c) 20000 secs with $V_{GS} = 20$ V, $V_{DS} = 0$ V. Inset: Bottom-gate inverted a-Si:H TFT.

III. A-SI:H LOGIC CIRCUITS

Since there is no complimentary (PMOS) transistor available, the NMOS load transistors of a-Si:H logic gates provide DC currents when the logic outputs are low. The conventional diode-connected inverter circuit can only drive the output voltage to V_{DD} - V_T. To obtain a rail to rail voltage swing the well known bootstrap inverter [4] is preferred. The bootstrap inverter has the ability to maintain this swing under V_T degradation. Lack of a complementary pull up device, and the DC currents that it entails, mandate that TFT logic circuits minimize the number of pull up transistors. This suggests programmable logic arrays, due to their high fan-in and ability to generate many logic outputs with a minimum number of load devices.

A. Static a-Si:H PLA

The static a-Si:H PLA fabricated in this work is a 12-input PLA supporting 35 minterms and 24 sums. It functions as the instruction decoder in a flexible microcontroller. The PLA

978-1-4244-2018-6/08 $25.00 © 2008 IEEE

NOR gates for both logic planes, as NOR gates with high fan-in, are efficient when implemented in pseudo-NMOS logic [5]. The static NOR gates use diode connected N-channel TFT's (see Fig. 2). The first stage, comprised of bootstrap inverters, buffers the inputs to the PLA. The buffered inputs drive the NMOS NOR gates in the AND plane to generate the minterms. The OR plane is driven by the outputs of the AND plane, generating the sum-of-product terms. The pull-up (load) to pull-down (drive) transistor ratio in the NOR gates is 1-to-8. The circuit nominally operates at $V_{DD} = 20$ V and $V_{SS} = 0$ V. The PLA output buffers, which are bootstrap inverters, can drive a capacitive load of 20 pF. The static PLA is designed with TFT's with L = 20 μm. This channel length guarantees the separation necessary to avoid collapse of the source/drain contacts during de-bonding.

The pseudo-NMOS static a-Si:H PLA is preferred due to its timing simplicity. The static PLA's simple timing comes at the expense of static power consumption and slow low-to-high output transition. It also suffers from poor noise margins, which are further degraded by gate bias induced V_T shift. Increasing the ratio between the driver and diode-connected load detrimentally affects the already large low-to-high propagation delay (t_{pLH}).

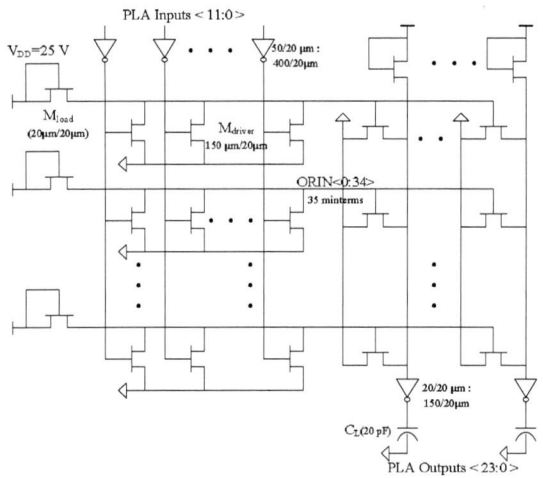

Fig. 2. A schematic of the a-Si:H AND-OR static PLA.

B. Dynamic a-Si:H PLA

Dynamic approaches, which clock the pull up transistors, can limit the static power consumption of a-Si:H circuits. Specific dynamic circuit topologies can also mitigate the V_T shifts [7]. We investigate a dynamic a-Si:H PLA implementing the same functions as the static PLA. The dynamic PLA has footed dynamic NOR gates in the AND and OR planes as shown in Fig. 3. The dynamic NOR gates use two non-overlapping clocks, Φ_1 and Φ_2. (see Fig. 4). When clock Φ_1 is high the AND plane pull-up TFT's are active and the outputs (product terms labeled ORIN) are precharged to $V_{DD}-V_T$. During Φ_2 the product terms evaluate to the state determined by the AND plane pull-down network. The AND plane output drives the OR plane directly, violating monotonicity conditions required by dynamic circuits [5]. Since buffering or replica clocks would create large DC

currents, another set of non-overlapping clocks (Φ_3 and Φ_4) are asserted after the AND plane has completed evaluation. The OR-plane NOR gates pre-charge during Φ_3 and evaluate during Φ_4. The OR-plane outputs must be captured before the AND-plane pre-charges so that the results are not corrupted. This is accomplished by the pass-transistors which isolate the minterms of the AND-plane. The pass-transistors are ON until the next Φ_1 phase pre-charges the ORIN lines and asserted by the Φ_{2X} clock which is essentially the logical OR of clock phases: Φ_2, Φ_3, and Φ_4.

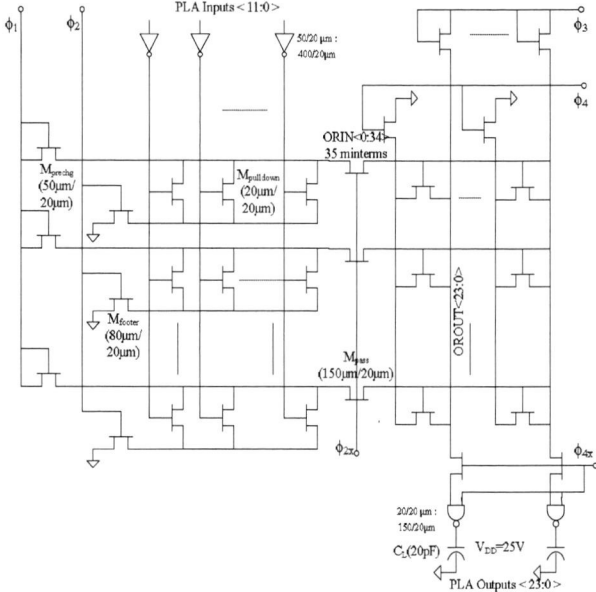

Fig. 3. A schematic of the a-Si:H AND-OR dynamic PLA

For PLA based state machines the OR-plane outputs must be latched to hold the state in the subsequent phase. Delaying the result on the PLA output nodes until Φ_3 is asserted eliminates the need for latches. The additional pass-transistors in the OR-plane are on until the next Φ_3 phase pre-charges the PLA out lines. They are asserted by the Φ_{4X} clock phase which is the logical OR of clock phases: Φ_4, Φ_1 and Φ_2. To reduce charge-sharing noise, the pass transistors are made larger (10x) than the pull-down TFT's. The PLA outputs are logically ANDed with clock phase Φ_{4X} to reduce the time during which the dynamic node might be discharged by leakage current by the Φ_3 clock. During Φ_3 the OROUT nodes are precharged to $V_{DD}-V_T$.

The additional clock phases: Φ_{2x} and Φ_{4x} produce an increased gate-voltage induced V_T shift but are important to the circuit timing. Conventional CMOS dynamic PLA's have a critical timing race which results in incorrect outputs when the delay through their replica clock path is less than that through the signal path [6]. The 6-phase dynamic PLA clocking scheme eliminates this race condition. Fig. 4 shows the operation of the dynamic a-Si:H PLA with $V_{DD} = 25$ V and a clock period of 2 ms. The PLA outputs are buffered with bootstrap inverters. The critical timing edge in dynamic circuits is from high-to-low. Consequently the dynamic NOR gate footer TFT's are 4x larger than the pull-up TFT's.

978-1-4244-2018-6/08 $25.00 © 2008 IEEE

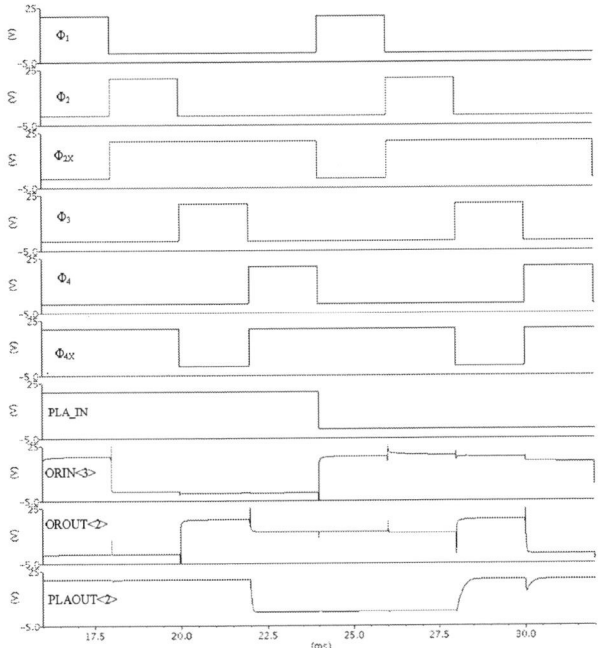

Fig. 4. Simulated operation of the dynamic a-Si:H PLA.

C. Physical Design

The flexible a-Si:H circuits are designed to cover large surfaces as they are intended primarily for portable displays. The a-Si:H PLA layout is partitioned into cells with the pitch matching that of the pull-down transistor. The layout is made compact by sharing ground lines between pairs of minterms and outputs such that each minterm and output occupy the same pitch. The 24-inputs (12 inverted), 35 minterms and 24 output logic function implemented results in 1680 cells for the AND and OR plane pull-down networks and an additional 59 cells for the AND and OR plane pull-up networks.

The static PLA cell size of 150 μm x 20 μm equals that of the AND and OR plane pull-down TFT's. The pull-up load TFT's are 8 times smaller but are aligned to the same grid size. The fabricated static PLA occupies 71.5 mm². The dynamic PLA adds 118 cells for the footer and pass transistors. The dynamic PLA has pull-up and pull-down TFT's of 50μm/20μm and 20 μm/9 μm respectively. The footer transistor is 4x larger than the pull-down requiring multiple grids. The pass-transistors are 200 μm/20 μm and are folded to fit the pitch. The dynamic PLA occupies 110.2 mm². Fig. 5 shows the layout and die picture of the static and dynamic PLA's on flexible stainless steel substrate..

IV. EXPERIMENTAL RESULTS

The static and dynamic PLA's are compared in this section using four performance metrics: area, speed, power and reliability, using both simulation and the measured results from the fabricated circuits.

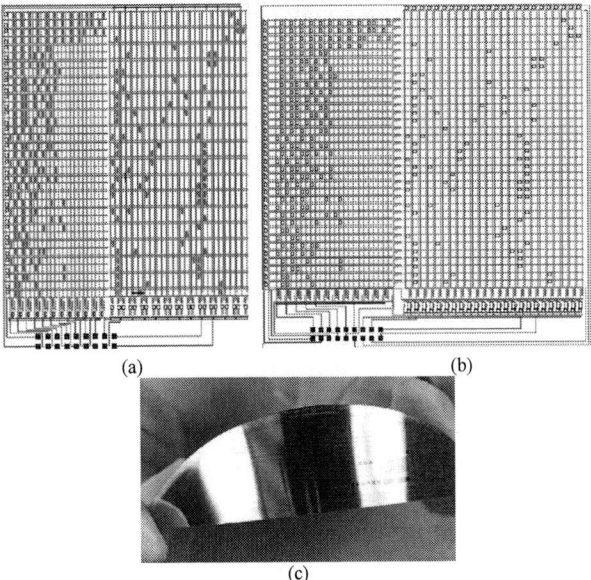

Fig. 5. The layout of a) static PLA, b) dynamic PLA and c) die picture of PLA's on flexible stainless steel.

A. Speed and Power

Fig. 6(a) shows the measured outputs of the static PLA. The measured average t_{pLH} is 125 μs and t_{pHL} is 85 μs. The long propagation delays are primarily due to the large load capacitance of the output pad (20 pF) and the overlap capacitance of the wide inverter buffer TFT's. The measured average DC power of the static PLA is 8.7 mW.

The dynamic PLA critical timing path is dominated by the pull-down response, which is measured from the rising edge of the Φ_4 clock phase. Fig. 6(b) shows the clock phases and the measured dynamic PLA outputs for clock period of 8ms. The dynamic PLA is given an input opcode of all '1's (V_{DD}) which results in the PLA outputs to swing to V_{DD} and V_{SS} respectively. The measured t_{pHL} is 36 μs. The dynamic PLA DC power is measured by measuring power at multiple input clock periods and projecting the measured power versus frequency curve to the y-axis. The average DC power is 3.5 mW, virtually all of which is attributable to the output buffers driving the output pads. With this term only a 2.5 x power reduction vs. the static version is achieved. The active power measured for a clock period of 8 ms is 0.7 mW. Assuming an activity factor of 2, the total switching capacitance in the dynamic PLA is 4.5nF. Table II lists the measured static and dynamic PLA performance values. Fig. 7 shows the simulated power and delay curves versus power supply. The simulated and measured results for both PLA's match within 15%. The static and dynamic PLA's have a minimum power-delay product at V_{DD} = 25 V and 20 V respectively.

B. Reliability

The V_T shift in a-Si:H TFT's is a function of time, the applied gate voltage and hence circuit duty-cycle, operating temperature and transitor mode of operation [7, 8]. Driving the TFT's with a lower duty cycle input improves the circuit lifetime. Moreover, TFT's biased in saturation mode exhibit less degradation. Both of these effects are evident in the pull-down TFT's of the PLA's. Fig. 8 shows the measured output of the both PLA's at time t = 0 and after applying continuous stress for 12 hrs. The dynamic design shows better V_{OH} both before and after stress. Table II also shows the maximum V_T shift experienced by TFT's in the static and dynamic PLA.

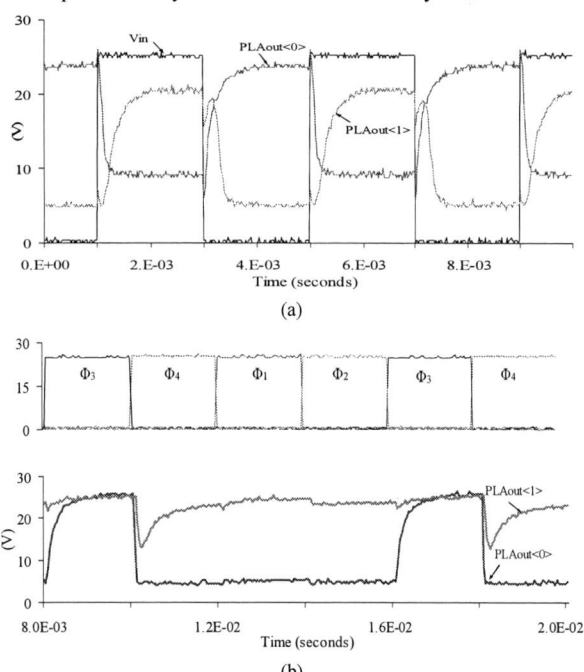

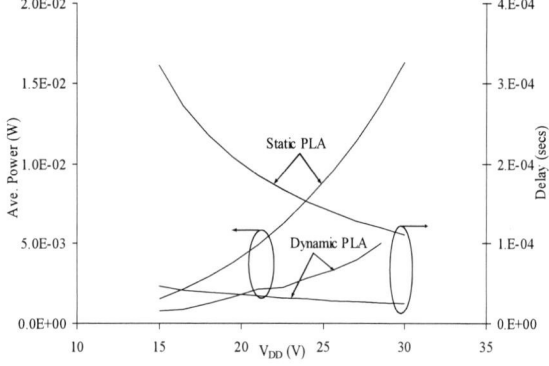

(b)

Fig. 6.. The measured outputs of the a) static PLA and b) dynamic PLA.

TABLE II: FABRICATED STATIC AND DYNAMIC PLA COMPARISON.

Metric (Measured at V_{DD}=25 V)	Static PLA	Dynamic PLA
Circuit Area	71.5 mm²	110 mm²
Propagation Delay	105 µsec	36 µsec
Average. DC. Power	8.7 mW	3.5 mW
Max. V_T shift (after 12 hrs of stress)	5.6 V	4.4 V

Fig. 7. The simulated propagation delay and average power versus supply voltage from the a-Si:H static and dynamic PLA's.

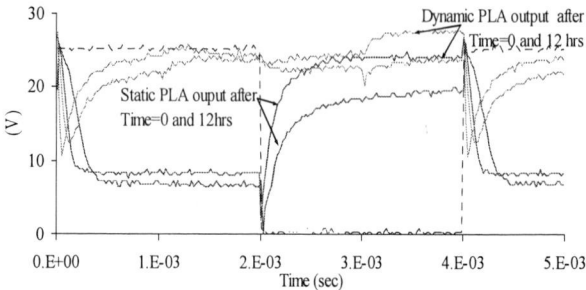

Fig. 8. Measured static and dynamic PLA outputs after 12 hrs of continuous operation.

V. CONCLUSIONS

Since a-Si:H technologies lack a P-type transistor, challenges with circuit power are reintroduced from the 1970's. The low mobility of a-Si:H TFTs requires high supply voltages further increasing power consumption. Because these circuits operate at low frequencies, the dynamic power is small in comparison to static power. Moreover, the dynamic circuits exhibit less V_T shift. These factors favor the dynamic logic style in the design of flexible a-Si:H logic circuits, which are faster and dissipate lower total power than their static counterparts.

VI. ACKNOWLEDGEMENT

The authors gratefully acknowledge the financial support of the Army Research Laboratory (ARL) through Cooperative Agreement W911NF-04-2-0005. The views and conclusions contained in this document are those of the authors and should not be interpreted as representing the official policies, either expressed or implied, of the ARL or the U.S. Government. The U.S. Government is authorized to reproduce and distribute reprints for Government purposes notwithstanding any copyright notation hereon.

REFERENCES

[1] G. Raupp, et. al, "Low-temperature amorphous-silicon backplane technology development for flexible displays in a manufacturing pilot-line environment," J, Society of Information Displays, 15(7), pp. 445-454, July 2007.

[2] S. Venugopal, and D. Allee, "Integrated a-Si:H source drivers for 4" QVGA electrophoretic display on flexible stainless steel substrate," IEEE Journal of Display Technology, 3(1), pp. 57-63, March 2007.

[3] K. Wissmiller, et, al, "Reducing Power in Flexible a-Si Digital Circuits While Preserving State," Proc. CICC, 2005, pp. 219-222, 2005.

[4] J. Uyemura, CMOS Logic Circuit Design, Kluwer Academic Publisher, Norwell, MA, pp. 319-324, 1999.

[5] N. Weste, and D. Harris, CMOS VLSI Design: a circuits and systems perspective, Pearson/Addison-Wesley, Boston, pp. 750-756, 2004.

[6] G. Samson and L. Clark, "A 0.13 µm Race-free Low-power Programmable Logic Array," Proc. CICC, pp. 313-316, 2006.

[7] H. Lebrun, T. Kretz, J. Magarino, and N. Szydlo, "Design of Integrated Drivers with a-Si TFTs for Small Displays: Basic Concepts," J. Society for Information Display,. 36(2), pp. 950-953, 2005.

[8] R. Shringarpure, et al, "Localization of gate biased induced threshold voltage degradation in a-Si:H TFT's", IEEE Elec. Dev. Lett., 29(1), pp. 93-95, Jan. 2008.

IEEE 2008 Custom Intergrated Circuits Conference (CICC)

FREQUENCY TUNABLE SILICON CARBIDE RESONATORS FOR MEMS ABOVE IC

F. Nabki, T. A. Dusatko, and M. N. El-Gamal

McGill University, Montreal, Canada

Abstract- **Micro-electromechanical beam resonators and arrays are fabricated using a custom low-temperature (<300 °C) CMOS-compatible silicon carbide micro-fabrication process. A special feature of this process is that it allows the integration of heaters directly onto the MEMS devices, thus enabling resonant frequency tuning with constant insertion loss and considerable extension of the tuning range. Characteristics for different devices are measured with quality factors of $Q \approx 1000$ and resonant frequencies of up to 21.8 MHz.**

I. INTRODUCTION

Micro electromechanical systems (MEMS) have enabled new possibilities for radio frequency (RF) system integration. More specifically, micro-mechanical resonators have been subject to considerable advancements, exploring their potential in wireless transceivers integration [1]. The concept is illustrated in Fig. 1, where all the typically large off-chip transceiver components in grey could eventually be replaced by their micro-machined equivalent devices [2]. MEMS resonators can replace the quartz reference crystal, and oscillators have been shown to meet the stringent GSM phase noise requirements [3]. In parallel, recent advanced resonator modeling techniques have been proposed for accurate performance estimation in circuit simulators, leading to insightful phase noise optimization [4].

Accordingly, a key element to become a commercially competitive technology, similar to above-IC BAW filters [5], is the monolithic integration of resonators to the back-end of integrated circuit (IC) processes such as CMOS. This would enable the creation of true MEMS/CMOS system-on-chip wireless solutions (e.g. Fig. 1).

This paper presents novel tunable MEMS clamped-clamped beam resonators fabricated in a low-temperature (<300 °C) silicon carbide (SiC) CMOS-compatible process. Details of the process are reported in [6]. SiC has a higher acoustic velocity than that of poly-silicon, making it an excellent candidate for MEMS in general, and in particular resonators. Arrays of resonators are also presented here, and a tuning method based on integrated heaters is introduced and tested [7]. This work represents a critical step towards the integration of MEMS resonators with CMOS electronics.

II. OPERATING PRINCIPLE

Figure 2(a) shows a typical clamped-clamped beam resonator with biasing. A DC bias voltage, V_P, is applied across the electrodes, enabling electrostatic transduction of an input voltage, $v_i(t)$, to excite the beam in its first flexural mode as depicted in Fig. 2(b). This in turn generates a bandpass filtered output current, $i_o(t)$

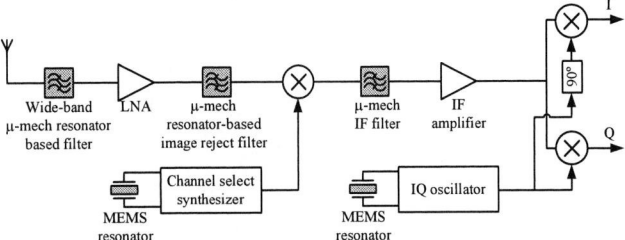

Fig. 1: Super-heterodyne architecture with off-chip components replaced by integrated MEMS resonators (based on [2]).

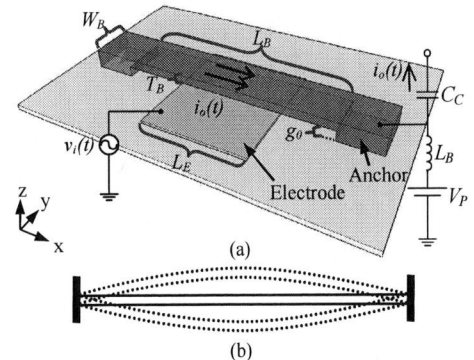

Fig. 2: (a) Clamped-clamped beam resonator, and its (b) first flexural mode.

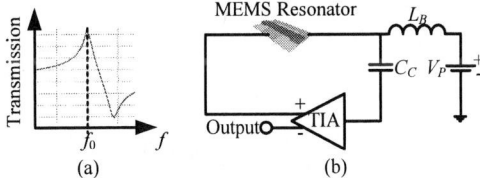

Fig. 3: (a) The typical transfer characteristic of a resonator, and a (b) resonator-based oscillator.

Resonators may be used as simple filters by making use of their high-Q bandpass transfer characteristics (Fig. 3(a)), or they can be used as frequency references when connected to a high-gain trans-impedance amplifier (TIA) in a positive feedback loop, as shown in Fig. 3(b).

III. THERMAL TUNING

During fabrication, MEMS resonators, similarly to ICs, are subject to process variations that alter their resonant frequencies. These frequencies are also, similarly to crystals, prone to drift and temperature variations. Therefore, resonators ideally require a post-fabrication mechanism for resonant frequency tuning in order to improve yield.

978-1-4244-2018-6/08 $25.00 © 2008 IEEE

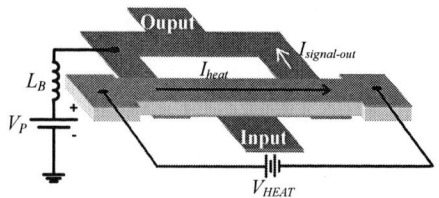

Fig. 4: Clamped-clamped beam resonator design with an integrated heater.

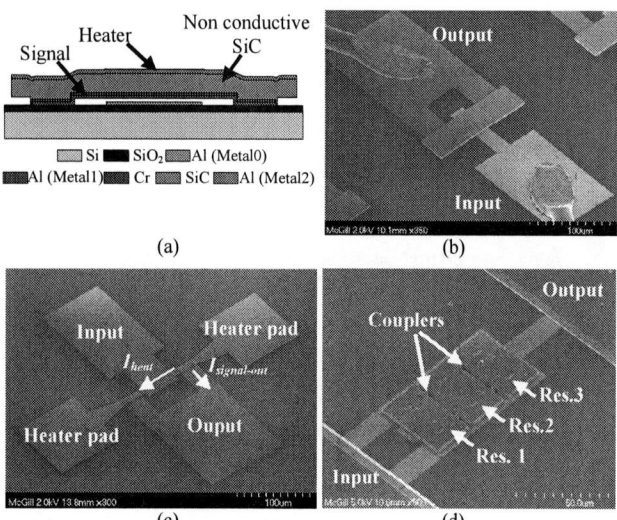

(a)   (b)

(c)   (d)

Fig. 5: (a) A typical clamped-clamped beam resonator cross-section, and SEM micrographs of (b) a clamped-clamped beam resonator, (c) a resonator with an integrated heater, and (d) a three-resonator array.

Spring softening induced by the bias voltage, V_p, can be used to electrostatically tune the resonant frequency; however, the device linearity is deteriorated as the voltage is increased, and large insertion loss variations occur during tuning. Furthermore, prohibitively high voltages may be needed to achieve a wide enough tuning range to account for process corners. The electrostatic tuning range is ultimately limited by device destruction due to electrode pull-in at too elevated voltages.

The resonators proposed here feature a top metal layer that allows thermal heating through a current, I_{heat}, as shown in Fig. 4 and Fig. 5(a, c). The heater enables the tuning of the beam while in use, as it is electrically insulated from the beam's bottom metallization carrying the signal current, $I_{signal-out}$. Direct thermal heating is an alternative frequency tuning method, which only depends on material properties, and is independent of electrostatic transduction. Insertion loss and linearity are thus unchanged during tuning. If electrostatic tuning remains an option, then thermal tuning yields the added benefit of extending the tuning range beyond the pull-in limit. The heating element being in direct contact with the resonator, combined with the use of a good thermal conductor such as SiC, allows for efficient and rapid thermal tuning of the device, compared to techniques where the heaters are in proximity to the device (e.g. [8]). It also allows for a simpler device design and process flow, compared to heated resonators requiring two transducer gaps that electrically isolate the beam to allow for electrical operation (e.g. [9]).

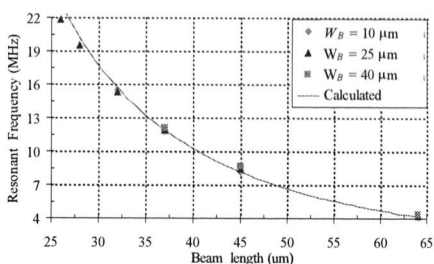

Fig. 6: Resonant frequencies of different clamped-clamped beam resonators.

IV. RESONATOR FABRICATION

Several clamped-clamed beam resonators with different widths and lengths were fabricated, having resonant frequencies ranging from 4.2 to 21.8 MHz. Figure 5(a) shows the structure of a typical resonator with an integrated heater on top. It consists of a 2 μm non-conductive amorphous Hexoloy-SiC film, metallized on its underside by a 140 nm aluminum/chromium bi-layer, and on its topside by a 200 nm aluminum layer. The structure is anchored at both extremities. After the dry release of a 200 nm polyimide sacrificial layer, a gap is created, allowing for electrostatic actuation. Scanning electron microscope (SEM) micrographs of a wire-bonded resonator, and that of a variant with an integrated heater, are shown in Fig. 5(b) and (c), respectively. A three-resonator array with mechanical couplers is shown in Fig. 5(d).

V. EXPERIMENTAL RESULTS

The S-parameters of fifty packaged MEMS devices were measured under vacuum to mitigate the effects of air damping. An Agilent 8753 vector network analyzer was used for parameters extraction, and a Philips SA5211 TIA facilitated measurements.

A. Resonant Frequency

The average resonant frequencies of devices having different widths and lengths are plotted in Fig. 6, along with resonant frequency values estimated using modal analysis, while taking axial deformation of beam supports into account. Wider beams have slightly higher measured resonant frequencies than predicted, which can be attributed to the in-plane dilation suppression that accompanies axial strain. On the other hand, shorter beams exhibit slightly lower frequencies than predicted by theory, due to anticlastic bending.

B. Frequency Tuning

Figure 7 shows a device's measured transfer curve tuning using both the bias voltage, and a heater current. For this design, the heater resistance was measured to be approximately 1.4Ω, and the total tuning range of the resonator was 8.4%. The very desirable near-constant insertion loss achieved by thermal tuning results in a larger usable frequency range, compared to what is attainable with voltage tuning.

C. Motional Resistance

The motional resistances of beams at different bias voltages are plotted in Fig. 8, along with calculated values based on a model developed earlier [4].

978-1-4244-2018-6/08 $25.00 © 2008 IEEE   204

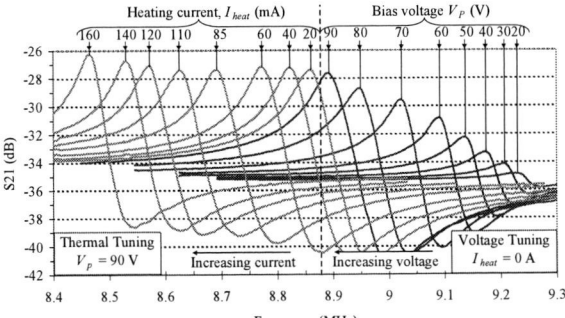

Fig. 7: Frequency tuning of a 10 MHz resonator using the bias voltage and the heater current (W_B=25μm, L_B=45μm).

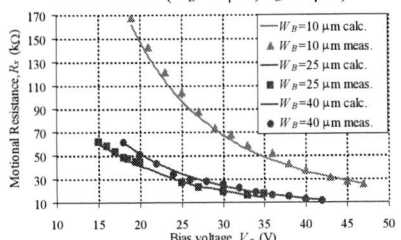

Fig. 8: Motional resistance measurements (L_B=45μm, L_E=35μm).

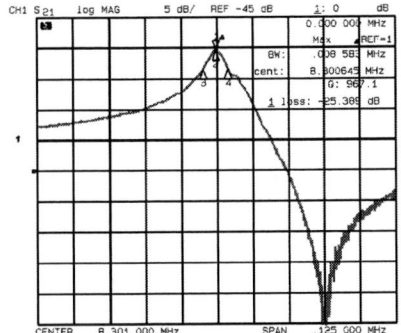

Fig. 9: An 8.3 MHz clamped-clamped resonator measured transfer function. (V_P=25V, W_B=25μm, L_B=45μm, L_E=23μm)

The 10 μm-wide beams have the highest motional resistances, because of their small electrode areas. Wider beams can exhibit lower Q-factors because of increased anchor loss. Consequently, the 40 μm beams have higher motional resistances than the 25 μm-wide beams, irrespective of their larger electrode areas. Beam width can therefore not be increased arbitrarily, as Q-factor deterioration eventually overtakes the benefit of electrode area increase.

D. Q-factor

Q-factors were extracted from the bandwidth of the devices' transmission transfer functions (S_{21}). Figure 9 shows the transfer function of a typical clamped-clamped device with the highest Q-factor measured. The 8.3-MHz device has a Q-factor of 967, which compares favorably with values reported for similar SiC beam-type resonators (e.g. Q=490 in [10]). A parallel resonance due to the feed-through capacitance is present, but is far from the resonant peak, and does not significantly affect measurements or operation.

The average Q-factors for three different beam lengths are plotted against the beam widths, in Fig. 10. The increased anchor losses of the 40 μm-wide beams translate into Q-factor deterioration.

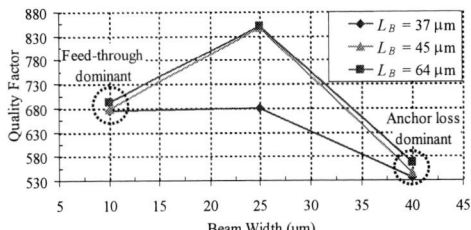

Fig. 10: Q-factor of resonators for different beam widths and lengths.

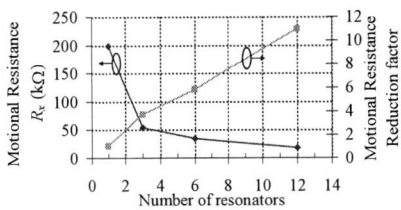

Fig. 11: Motional resistances of different resonator arrays (V_P=32V, W_B=25μm L_B=64μm).

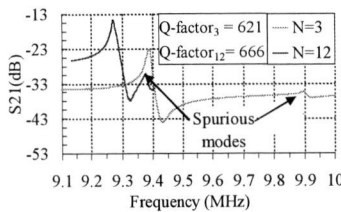

Fig. 12: Transfer functions of similar frequency resonator arrays with different number of beams (V_P=36V, W_B=25μm L_B=64μm).

The lower maxima for the 37 μm-long beams can also be attributed to the larger anchor losses. Interestingly, the reduced current drive of the 10 μm-wide devices causes lower Q-factors as well, because of the feed-through capacitance strongly affecting the series-resonance peak.

E. Resonator Arraying

As the resonators have high motional resistances due to their relatively large transducer gaps, mechanical arraying of multiple beam resonators using beam couplers (Fig. 5(d)) was investigated to reduce the effective motional resistance [3].

The motional resistances of arrays with different numbers of resonators are plotted in Fig. 11. The improvement in motional resistance is nearly linear with the number of resonators. Energy losses caused by the added mechanical couplers lower the Q-factor, and counteract the improvement as the array size is increased beyond a given limit.

The frequency responses of both twelve and three resonator arrays are shown in Fig. 12. The insertion loss of the larger array is reduced by 8.4dB, which is quite desirable; however, its spurious mode is closer to the main resonant mode. This is undesirable when closely spaced signal interferers are present such as in channel select filters, or in oscillators where the phase noise performance is important.

VI. DISCUSSION

Table 1 summarizes the characteristics of other published beam resonators. For meaningful comparison, the theoretical normalized resonant frequencies (f_{0-norm}) of these resonators were numerically estimated, after scaling them to the same dimensions of the resonator listed in the first column.

Table 1: Comparison of clamped-clamped beam resonators.

Resonator	This work	[11]	[12]	[13]	[10]
Material	Amorphous SiC	Doped Poly-silicon		TaN/SiON	Poly-SiC
Young's Mod. (GPa)	258	150		66	436
Thermal budget (°C)	<300	950		290	1280
Bias (V)	25	13	7	5	-
Q-factor	967	1036	3600	2200	490[†]
f_0 (MHz)	8.3	8.6	9.8	11.3	4.0
R_x (kΩ)	54.90	0.34	17.60	10.70	2400[*]
W_B (µm)	25	40	8	12	9
H_B (µm)	2	2	2	2.05	0.64
L_B (µm)	45	40	40	24	51
Gap (nm)	200	100	100	80	100
L_E (µm)	23	32	20	14	-
f_{0-norm} (MHz)	8.3	6.8	7.7	3.1	16.0

[*]Includes interconnect resistance. [†]Indirect measurement in oscillator loop.

Unlike the devices presented here, resonators in [10-12] are fabricated at high temperatures, and are not suitable for CMOS post-processing. The thermal budget of the devices fabricated in [13] is adequate in that respect, but unlike the process used here, which is chemically compatible with CMOS [6], the process in [13] employs Piranha (Sulfuric-Peroxide) as an etchant, imposing serious limitations on CMOS post-processing. Furthermore, the SiON/TaN composite beams have a Young's modulus of 66 GPa, yielding a low normalized frequency.

The poly-SiC resonator presented in [10] has a higher normalized frequency due to the higher Young's modulus of the chemical vapor deposited poly-SiC; however, the beam is not metallized, resulting in a high series electrical resistance which reduces the Q-factor. The Q-factor of the devices reported in this work is somewhat lower than that of the poly-silicon-based beam resonators in [12]. This can be attributed to the fact that no annealing was performed here, as the process temperature budget is limited for compatibility with CMOS post-processing. Annealing of resonators has been shown to improve Q-factors of resonators by reducing defects, such as dislocations [14]. In addition, the thin metal layers that render the SiC conductive can increase the magnitude of thermo-elastic damping and thus reduce the Q-factor [15].

The higher motional resistances of the devices fabricated here, compared to those of other resonators are due to the bigger gaps inherent to the current fabrication process. This process is currently being optimized to reduce the gap to 100 nm, which is expected to bring the insertion loss of the devices on par with other works. Advantageously, this would enable resonators having comparable resonant frequencies to yield lower motional resistances than those of poly-silicon or SiON/TaN resonators, due to the resulting larger beam geometries stemming from the SiC's higher acoustic velocity.

VII. CONCLUSION

Different resonator devices were fabricated using a custom SiC MEMS fabrication process compatible with CMOS monolithic integration. The tested devices exhibited performances similar to those of other conventional processes which are not CMOS-compatible. Q-factors in the order of $Q \approx 1000$ were measured, along with resonant frequencies up to 21.8 MHz. Resonator arrays were fabricated and tested, yielding lower motional resistances, and demonstrating the ability for more complex resonant structures to be implemented. Frequency tuning was investigated, and a novel integrated heater, enabled by the fabrication process, extended the devices' usable tuning range beyond pull-in. Improvements to the process are under investigation to enhance Q-factor performance and insertion loss.

REFERENCES

[1] F. Nabki, T. A. Dusatko, and M. N. El-Gamal, "Microelectromechanical Resonators for RF Applications," in *Emerging Wireless Technologies from Systems & Circuits to Device Technology*, K. Iniewski (Editor), pp. 589-628, CRC Press, Florida, USA, October 2007.

[2] C. T. C. Nguyen, "RF MEMS in wireless architectures," *Proceedings of the IEEE Design Automation Conference*, pp. 416-420, June 2005.

[3] Y.-W. Lin, S.-S. Li, Z. Ren, and C. T. C. Nguyen, "Low phase noise array-composite micromechanical wine-glass disk oscillator," *IEEE International Electron Devices Meeting*, pp. 287-290, December 2005.

[4] F. Nabki and M. N. El-Gamal, "Modeling and simulation of micro electromechanical (MEM) beam resonator-based oscillators," *accepted, the International Symposium on Circuits and Systems*, May 2008.

[5] A. Dubois, et al., "Integration of high-Q BAW resonators and filters above IC," *Digest of Technical Papers of the IEEE International Solid-State Circuits Conference*, pp. 392-393, February 2005.

[6] F. Nabki, T. A. Dusatko, S. Vengallatore, and M. N. El-Gamal, "Low-temperature (<300°C) low-stress silicon carbide surface micromachining fabrication technology," *accepted, the Solid-State Sensors, Actuators and Microsystems Workshop, Hilton Head Island*, 2008.

[7] F. Nabki, T. Dusatko, and M. N. El-Gamal, "Direct contact heat control of micro structures," US Patent Application, December 2007.

[8] C. T.-C. Nguyen and R. T. Howe, "Microresonator frequency control and stabilization using an integrated micro oven," *International Conference on Solid State Sensors and Actuators*, vol. 1, pp. 1040-1043, June 1993.

[9] M. Hopcroft, et al., "Active temperature compensation for micromachined resonators," *Solid-State Sensor, Actuator and Microsystems Workshop*, pp. 364-367, 2004.

[10] R. F. Wiser, M. Tabib-Azar, M. Mehregany, and C. A. Zorman, "Polycrystalline silicon-carbide surface-micromachined vertical resonators - part II: electrical testing and material property extraction," *Journal of Microelectromechanical Systems*, vol. 14, no. 3, pp. 579-589, June 2005.

[11] Y.-W. Lin, S. Lee, Z. Ren, and C. T.-C. Nguyen, "Series-resonant micromechanical resonator oscillator," *Proceeding of the IEEE International Electron Devices Meeting*, pp. 961-964, December 2003.

[12] S. Lee, M. U. Demirci, and C. T.-C. Nguyen, "A 10-MHz micromechanical resonator Pierce reference oscillator for communications," *International Conference on Solid State Sensors, Actuators and Microsystems*, pp. 1094-1097, June 2001.

[13] S. Young, et al., "A novel low-temperature method to fabricate MEMS resonators using PMGI as a sacrificial layer," *Journal of Micromechanics and Microengineering*, vol. 15, pp. 1824-1830, October 2005.

[14] K. Wang, A.-C. Wong, W.-T. Hsu, and C. T.-C.Nguyen, "Frequency trimming and Q-factor enhancement of micromechanical resonators via localised filament annealing," *International Conference on Solid State Sensors, Actuators and Microsystems*, pp. 109-112, June 1997.

[15] S. Vengallatore, "Analysis of thermoelastic damping in laminated composite micromechanical beam resonators," *Journal of Micromechanics and Microengineering*, vol. 15, pp. 2398-2404, September 2005.

MEMS Wafer-Level Vacuum Packaging with Transverse Interconnects for CMOS Integration

D. Lemoine, P.-V. Cicek, F. Nabki, and M. N. El-Gamal

McGill University, Montréal, Québec, Canada

Abstract— A novel vacuum wafer-level packaging technology for micro-electromechanical systems (MEMS) is presented. It supports monolithic integration with electronics, and is suitable for different MEMS processes. Bulk-etched transverse feedthroughs are used to connect with the encapsulated systems. Silicon carbide is successfully used for membrane stress cancellation and improved hermeticity.

I. INTRODUCTION

The deployment of micro-electromechanical systems (MEMS) has experienced a significant growth in recent years, and is expected to maintain this trend for many years to come. Electronics are often combined with MEMS for sensing and control applications. There is a pressing need for full monolithic integration of circuits and MEMS within one process flow, in order to reduce size, cost, and improve performance.

Many MEMS devices have to be encapsulated in a vacuum environment. For example, while the quality (Q)-factor of a MEMS resonator could reach a maximum Q-factor of 50 under atmospheric pressure, the same resonator can reach Q-factors higher than 2000 in high vacuum environments [1]. Packaging is, in general, of critical importance for the successful commercialization of MEMS-based solutions.

Vacuum packaging could be performed at the chip level, where each MEMS chip, or IC+MEMS chip, is individually enclosed in a hermetically sealed package (Fig. 1(a)). Recently, wafer-level packaging (WLP) has emerged as a promising substitute to chip-level packaging (CLP) – it offers many benefits: 1) expensive custom hermetic packages are avoided; 2) encapsulation is incorporated into the device fabrication flow, resulting in fewer processing steps; 3) all chips on a wafer are batch encapsulated, improving throughput; 4) lower capacitive and inductive parasitics are present, which is critical in many applications; and 5) the fragile MEMS devices are protected early on from debris during chip dicing, thereby improving yield.

This paper proposes a novel WLP technology for the encapsulation of MEMS in vacuum [2]. It complements the low-temperature fabrication process detailed in [3], and can also be applied to other surface micromachining MEMS processes, to achieve either vacuum or hermetic sealing.

A key feature of the solution presented here is that it is fully CMOS-compatible, both in terms of the processing temperatures involved and the chemicals used, to allow for monolithic integration of MEMS devices directly on top of CMOS electronics, as illustrated in Fig. 1(b). Transfer of

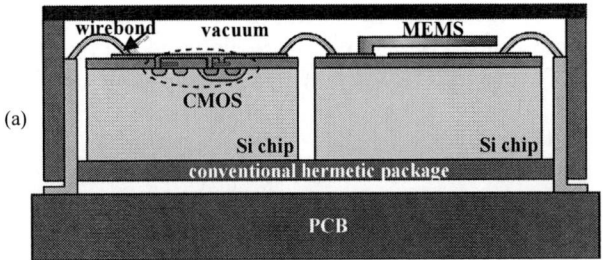

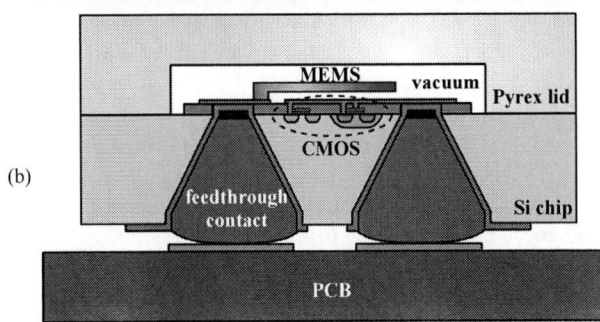

Fig. 1. PCB integration of a CMOS / MEMS system (a) for typical system-in-package CLP, and (b) for the WLP monolithic solution presented here.

MEMS from another substrate (e.g. [4]) is not required, avoiding the costly waste of a temporary carrier wafer.

Furthermore, the use of a transparent Pyrex lid for device encapsulation makes the technology appealing for packaging micro-opto-electromechanical systems (MOEMS), where light signals must reach the MEMS devices (e.g. [5]).

Once packaged, the encapsulated dies can be directly affixed to a printed circuit board (PCB) using standard surface mounting technologies, thus significantly reducing the physical system size and the parasitic effects inherent to conventional chip-level packaging solutions (see Fig. 2).

In the remainder of this paper, the process flow of the proposed technology is presented, followed by a discussion of experimental results, design choices and trade-offs.

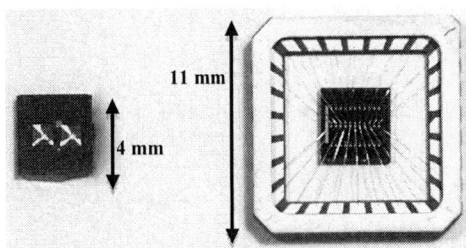

Fig. 2. A diced WLP chip with two contacts (left), next to a state-of-the-art Spectrum Semiconductors package with a wire bonded MEMS chip (right).

II. PROCESS FLOW

A. Process Overview

Fig. 3 illustrates the process flow of the proposed technology, which can be separated into four main stages:

1) A silicon (Si) wafer (which may hold previously fabricated CMOS devices) is pre-processed with transverse feedthrough interconnects to allow electrical interfacing from the back of the substrate (Fig. 3(a)-(d)).

2) MEMS devices are fabricated on the frontside of the substrate, using a surface micromachining process (Fig. 3(e)).

3) Cavities are etched in a Pyrex boro-silicate lid wafer (Fig. 3(f)).

4) The substrate and lid wafers are anodically bonded, hermetically enclosing MEMS devices in vacuum (Fig. 3(g)).

B. Process Description

The flow begins with a <100> Si substrate with 2.5 μm of silicon oxide (SiO_2) on both sides. The wafer may also hold previously fabricated electronic devices (e.g. CMOS). The backside SiO_2 is patterned by CHF_3-CF_4-Ar reactive-ion-etching (RIE), to serve as a hardmask for subsequent bulk Si wet etching for transverse feedthrough formation. These pyramidal vias are anisotropically etched through the complete bulk of the wafer using a CMOS-compatible 25% tetramethylammonium hydroxide (TMAH) etchant [6], releasing 2.5 μm-thick SiO_2 membranes on the frontside (Fig. 3(a)).

Subsequently, 200 nm of aluminum (Al) is DC sputtered and wet patterned on the frontside of the substrate to form membranes and a first interconnect layer. The backside SiO_2 layer is then completely removed by CHF_3-CF_4-Ar RIE, leaving the membranes constituted solely of 200 nm-thick Al at this point (Fig. 3(b)).

Next, 1 μm of Al is DC sputtered and wet patterned on the backside to contact with the frontside interconnects, resulting in thicker 1.2 μm Al membranes. For structural reinforcement of these sealing membranes and improvement of hermeticity, a 2 μm layer of amorphous silicon carbide (a-SiC) is DC sputtered at low temperature ([3]) on the backside of the wafer. The a-SiC is patterned by NF_3 RIE using a chromium (Cr) hardmask to cover the membranes (Fig. 3(c)).

In order to prepare the front surface for anodic wafer bonding, the SiO_2 layer is removed by CHF_3-CF_4-Ar RIE on most of the area of the wafer. To preserve SiO_2 for electrical insulation and as a protection for the Al interconnects, Cr is wet patterned prior to RIE to mask the device regions. This Cr mask is immediately wet removed once the RIE is performed (Fig. 3(d)).

At this point, the MEMS devices are fabricated as detailed in [3], or in any other suitable surface micromachining process. The MEMS devices are subsequently released (Fig. 3(e)).

In parallel, 5 μm-deep cavities are etched into a Pyrex lid wafer by SF_6-Ar RIE [7] using a Cr hardmask, which is then

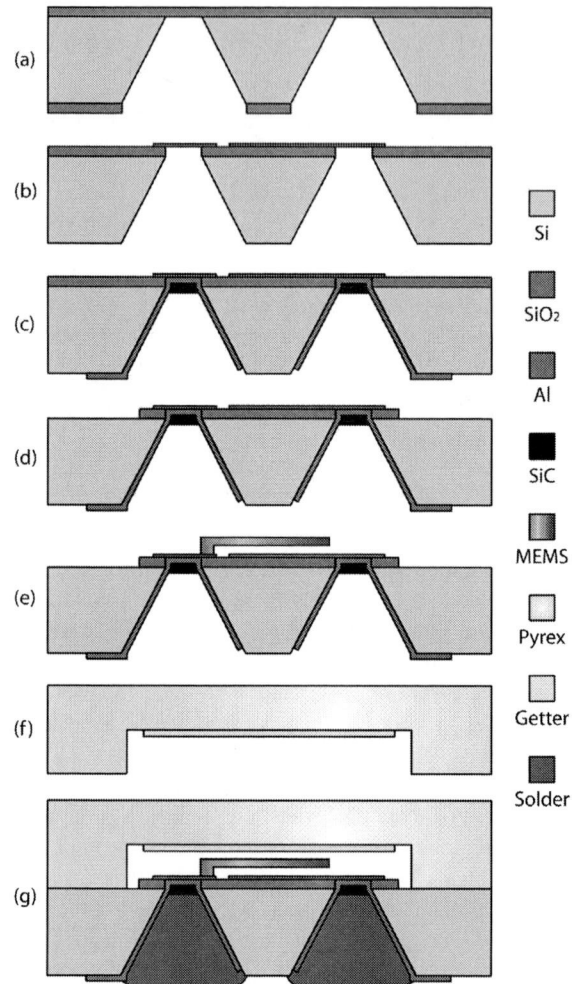

Fig. 3. Process flow to encapsulate a typical MEMS device.

removed by wet etching. Optionally, a getter can be patterned inside the cavities to counter eventual outgassing (Fig. 3(f)).

The Si substrate and Pyrex lid wafers are solvent cleaned, aligned, and anodically bonded in a 0.75 mTorr environment, thus sealing the MEMS in vacuum micro-cavities. Finally, the feedthroughs are filled with solder balls to further toughen the vacuum interface, and the wafers are diced into surface mountable chips (Fig. 3(g)).

Fig. 4 provides microscope pictures of the front and back sides of a dummy device fabricated following the process flow detailed above.

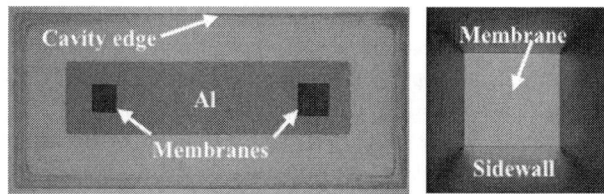

Fig. 4. Microscope pictures of a sealed dummy device on its frontside (left), and of one of its feedthroughs on the backside (right).

III. EXPERIMENTAL RESULTS AND DISCUSSION

A. Transverse Feedthroughs

Using the process described in section II, the MEMS devices are enclosed in hermetic cavities with tranverse feedthrough contacts across the bulk of the wafer. Compared to other processes, such as in [8], where the electrical interconnects are lateral, hermeticity is improved by avoiding topographical irregularities at the bonding interface. This allows the cavities to preserve lower pressure levels, due to the reduction in gas leakages.

The shape of the transverse feedthroughs is an important consideration. Having a working metallic electrical connection all the way to the sealing membranes would present a considerable challenge if feedthroughs were too narrow. Conversely, if they were too wide, the sealing membranes would be excessively large, rendering them mechanically weaker. Therefore, the transverse feedthrough must ideally be wide at the opening, i.e. at the interface to the PCB, but narrow closer to the sealing membrane. This simultaneously allows for an effective interconnection interface, and a sturdy sealing membrane. To achieve this goal, TMAH, modified so that it is CMOS-compatible as detailed in [6], is selected to etch the feedthroughs anisotropically with the desired pyramidal profile. Due to the <100> crystalline orientation of the Si wafer, a square mask opening results in a sidewall angle of 54.7°, since the <111> crystal plane is attacked much slower than the other planes.

SiO$_2$ is used as a mask and as an etch stop for the TMAH anisotropic etch of Si. The measured etch rate of Si in TMAH was of about 25 μm/min, whereas SiO$_2$ was attacked at about 1 nm/min. For a Si wafer of thickness t and an SiO$_2$ hardmask square opening width D on one side of the wafer, the expected width d of the ensuing oxide membrane on the other side can be computed using

$$d = \frac{D - 2t}{\tan(54.7°)} \qquad (1)$$

Accordingly, the membrane for a 675 μm-thick wafer with a square hardmask opening of 1120 μm on the back of the wafer should have a side length of about 166 μm, which agrees with the measured value of 183 μm (Fig. 5). Thus, the anistropic etch profile allows for very precise alignment to a relatively small membrane.

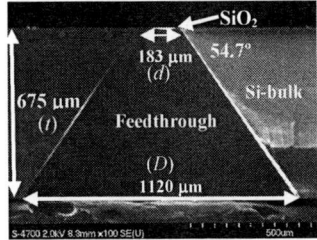 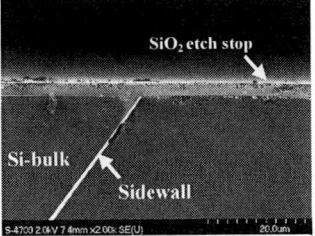

Fig. 5. Feedthrough SEM cross-section (i.e. Fig. 1a). Overall view (left), and close-up view (right), showing the desired anisotropic etch profile.

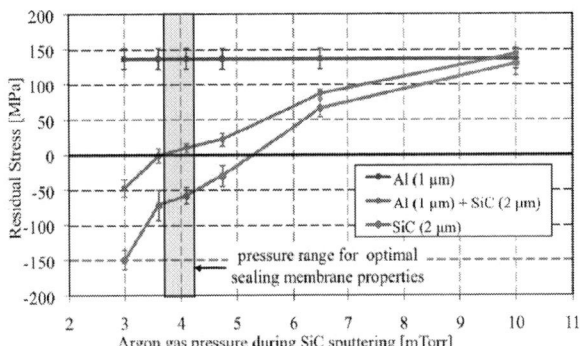

Fig. 6. Residual stress curves of sputtered Al, SiC, and Al-SiC films vs. SiC deposition Ar pressure. The SiC recipe is chosen to compensate for the stress of the Al film (error bars indicate standard deviation).

As discussed earlier, in order to provide an electrical contact with a transverse feedthrough, Al is sputtered on frontside over the SiO$_2$ etch stop membrane, and the SiO$_2$ is then etched away from the backside. Al is then sputtered on the backside, forming a connection with the front Al. The DC parasitic resistance of a single feedthrough is measured to be 0.69 ± 0.23 Ω before solder ball filling. Since the processed chips are directly surface-mountable, reduced parasitics are expected, compared to conventional packaging solutions.

B. Sealing Membranes

The sealing membranes have to withstand a significant pressure difference when vacuum is established inside the cavities. Hence, low-stress amorphous silicon carbide is used to reinforce the membranes. SiC has desirable mechanical and tribological properties in comparison to other materials accessible for microfabrication [9], while its potential for sealing vacuum has been demonstrated in [10]. Fig. 6 shows that the overall residual stress of the membranes can be minimized by tuning the argon (Ar) gas pressure during a-SiC DC sputtering. For an Ar pressure of 4.1 mTorr, the stress of the SiC layer is optimized so as to compensate for the high tensile stress of the Al. This results in an important improvement in the membrane structural resistance to applied pressures.

Fig. 7 displays pictures of the sealing membrane at different stages of the process flow. An SiO$_2$ membrane exhibits a compressive stress, as can be observed from its plaited appearance indicative of buckling (Fig. 7(a)). In contrast, Al has a highly tensile stress since the membrane is taut (Fig. 7(b)). The shape of an optimized membrane composed of Al and SiC shows that its stress is properly balanced, as no significant warping is observed (Fig. 7(c)). Fig. 8 shows the SEM cross-section of a reinforced membrane.

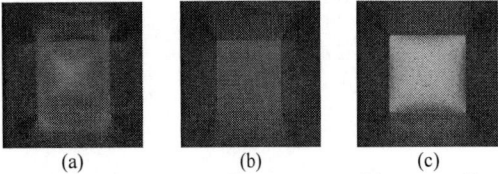

Fig. 7. Sealing membrane at different stages of the process flow:
(a) before SiO$_2$ removal – buckled; (b) after backside Al deposition – taut;
(c) after backside a-SiC deposition – low stress.

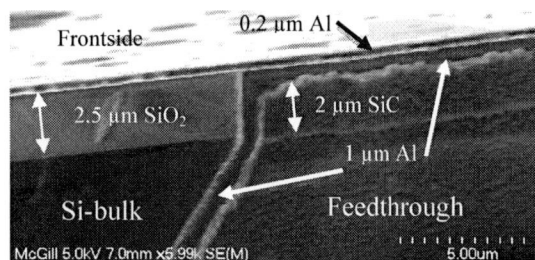

Fig. 8. SEM cross-section of an Al+SiC sealing membrane.

C. Wafer Bonding

Wafer bonding hermetically seals devices in vacuum micro-cavities. To ensure compatibility with most CMOS and MEMS processes, the encapsulation methodology is performed at low temperatures. For a maximum processing temperature of 400 °C, the characteristics of Si-Si direct bonding and Si-Pyrex anodic bonding were compared experimentally. These two bonding methods are the only low-temperature options that neither require adhesive nor intermediary layers.

The quality of a wafer bond can be evaluated i) from the *bond surface energy*, quantifying the force holding the wafers together, and ii) from the *void density*, conveying the uniformity of the bond interface.

The crack test technique introduced in [11] is used here to calculate the bond surface energy γ using

$$\gamma = \frac{2Ed^3y^2}{32L^4}. \tag{2}$$

A thin razor blade of thickness $2y$ is inserted between two bonded wafers, each with thickness d. The Young's modulus is E, and L is the length of the resulting crack measured with an infrared (IR) camera. Fig. 9(a) shows the crack test technique performed on a direct-bonded Si wafer pair. The measured bond surface energy γ is 0.118 J/m^2.

The crack test technique was unable to measure the strength of the anodic Si-Pyrex bond, because the Pyrex wafer broke before the razor blade could even be slightly introduced into the interface. That is however indicative of the quality of the Si-Pyrex anodic bond, which has a typical surface energy of around 1 J/m^2 [12], a full order of magnitude stronger than the Si-Si direct bond.

The void density for each of the two bonding methods was inspected under IR, as illustrated in Fig. 9 (b,c). With anodic bonding, the quality of the bonding interface is superior to that of direct bonding, as implied by the reduced number of voids. Anodic bonding thus exhibits better adhesion and interface uniformity at 400 °C, and also allows the use of a Pyrex transparent lid, making the packages fit for housing MOEMS.

IV. CONCLUSION

A vacuum WLP MEMS technology for full monolithic integration and packaging was presented, featuring transverse feedthroughs for electrical connectivity, and anodic bonding for device encapsulation. The process flow is carried out at low temperatures (< 400 °C), and uses chemicals, which ensure compatibility with a wide range of MEMS and CMOS processes. The packaged chips can be directly surface mounted to a PCB, allowing for low-cost integration.

ACKNOWLEDGMENT

The process reported in this paper was implemented in the NanoTools Microfabrication Facility at McGill University, and was funded by the CFI, NSERC, ReSMiQ, and FQRNT.

REFERENCES

[1] C. T.-C. Nguyen, "Vibrating RF MEMS for Next Generation Wireless Applications", *Proc. Custom Integrated Circuits Conference*, pp. 257-264, Oct. 2004.

[2] D. Lemoine, P. Cicek, F. Nabki, and M. N. El-Gamal, "Low-Temperature Vacuum Wafer-Level Packaging for MEMS Devices," US patent application, Mar. 2008.

[3] F. Nabki, T. Dusatko, S. Vengallatore, and M. N. El-Gamal, "Low Temperature (<300°C) Low-Stress Silicon Carbide Surface Micromachining Fabrication Technology," *in press, Solid-State Sensors, Actuators and Microsystems Workshop, Hilton Head Island*, June 2008.

[4] J. Chae, J. Giachino, and K. Najafi, "Fabrication and Characterization of a Wafer-Level MEMS Vacuum Package with Vertical Feedthroughs," *Journal of Microelectromechanical Systems*, vol. 17, no. 1, pp. 193–200, Feb. 2008.

[5] S. Yuhe, D. Dickensheets, and P. Himmer, "3-D MOEMS Mirror for Laser Beam Pointing and Focus Control", *Selected Topics in Quantum Electronics*, vol. 10, no. 3, pp. 528-535, May 2004.

[6] N. Fujitsuka, K. Hamaguchi, H. Funabashi, E. Kawasaki, and T. Fukada, "Silicon Anisotropic Etching Without Attacking Aluminum with Si and Oxidizing Agent Dissolved in TMAH Solution," *Proceedings of the International Conference on Solid-State Sensors Actuators*, pp. 1667-1670, Boston, MA, 2003.

[7] X. Li, T. Abe and M. Esashi, "Deep Reactive Ion Etching of Pyrex Glass Using SF$_6$ Plasma," *Sensors and Actuators A:Physical*, vol. 87, no. 3, pp. 139-145, Jan. 2001.

[8] J. Mitchell, G. R. Lahiji, and K. Najafi, "Encapsulation of Vacuum Sensors in a Wafer Level Package Using a Gold-Silicon Eutectic," *Proceedings of the International Conference on Solid-State Sensors Actuators*, pp. 928-931, June. 2005.

[9] G. L. Harris, *Properties of Silicon Carbide*, London, Inspec/IEE, 1995.

[10] D. G. Jones, R. G. Azevedo, M. W. Chan, A. P. Pisano, and M. B. J. Wijesundara, "Low Temperature Ion Beam Sputter Deposition of Amorphous Silicon Carbide for Wafer-Level Vacuum Sealing," *Proceedings of the International MEMS Conference*, pp. 275–278, Kobe, Japan, Jan. 2007.

[11] W. P. Maszara, "Silicon-On-Insulator by Wafer Bonding: A Review," *Journal of Electromechanical Society*, vol. 138, no. 1, pp. 341-347, Jan. 1991.

[12] R. Knechtel, M. Knaup, and J. Bagdahn, "A Test Structure for Characterization of the Interface Energy of Anodically Bonded Silicon-Glass Wafers", *Microsystem Technologies*, vol. 12, no. 5, pp. 462-467, Apr. 2006.

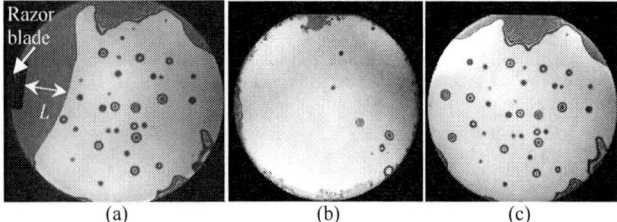

Fig. 9. Infrared picture during crack testing for the Si-Si bond (a) and of the bond interface uniformity for (b) anodic, and (c) direct bonding.

IEEE 2008 Custom Intergrated Circuits Conference (CICC)

Variation-Tolerant Spin-Torque Transfer (STT) MRAM Array for Yield Enhancement

Jing Li, Haixin Liu, Sayeef Salahuddin and Kaushik Roy

\<jingli, haixin, ssalahud, kaushik\>@purdue.edu

School of Electrical and Computer Engineering, Purdue University, West Lafayette, IN 47907

Abstract— Spin-Torque Transfer Magnetic RAM (STT MRAM) has emerged as a promising candidate for future universal memory. It not only combines the desirable attributes of all current memory technologies (SRAM, DRAM and flash memories) but also solves the critical drawbacks of conventional MRAM technology: poor scalability and high write current. However, variations in process parameters can lead to large number of cells to fail, severely affecting the yield of the memory array. In this paper, we provide a thorough understanding of the interrelationship between design parameters and parametric failures of STT MRAM cell in presence of process variations. Based on comprehensive physics-based model, solving the Non-Equilibrium Green's Function (NEGF) formalism in the ballistic regime, we develop an optimization methodology for robust cell design (in 1T1M configuration) to account for both stability and cell area. Further, we propose an efficient circuit design for variation tolerance. The proposed technique can effectively decouple the conflicting design requirements of read/write stability and area in conventional 1T1M cell, leading to considerably improved yield of memory array. Simulation results show that in our proposed cell, the robustness (cell stability) is improved by 36% with only 9% area overhead.

1. Introduction

On chip embedded memory (SRAM, DRAM, and Flash) is increasingly becoming important for future high performance microprocessors. Today, more than 70% die area is occupied by memory and the percentage is increasing with technology scaling. However, all the existing memory technologies are encountering insurmountable obstacle in terms of scalability, manufacturability and yield beyond 45nm. Hence, designers are now looking for alternative technological solutions, including phase-change RAM (PRAM), resistive RAM (ReRAM) and magnetic RAM (MRAM), etc. Among these options, Spin-Torque Transfer Magnetic RAM (STT MRAM), derived from the first-generation MRAM, has become one of the most promising candidates to replace current memory technologies as a universal memory [1]. It combines the fast read and write speed of SRAM (<10ns), the capacity and cost benefits of DRAM, and the non-volatility of Flash (zero standby power), coupled with essentially unlimited endurance (>10^{15} cycle).

Fig. 1 shows the schematic of a typical STT MRAM bit-cell (in 1-Transistor-1-MTJ or 1T1M configuration). It consists of a transistor (NMOS), a magnetic tunneling junction (MTJ), word line (WL), bit line (BL) and source line (SL). The main component of STT MRAM for information storage is the MTJ (Fig.2). An MTJ consists of two ferromagnetic electrodes with a thin insulating layer

in-between. The top magnet is the storage layer (referred as *free layer*) and the bottom magnet is the reference (referred as *fixed layer*). Data writing is performed by injecting spin-polarized current from one of ferromagnetic electrodes to change the magnetic orientation of the free layer. The effective resistance of MTJ is low (R_P or R_L, "0"state) when the two layers are spin aligned (parallel), and high (R_{AP} or R_H, "1"state) when they are not (anti-parallel) (Fig.3). The cell can be read by applying a small bias voltage and sensing the current. Comparison with reference current value determines whether the state is "1"or "0".

By solving the critical drawbacks of conventional MRAM, STT MRAM features excellent writeability and scalability. To increase the integration density and to improve the read/write speed, STT MRAM can be scaled from one processor generation to another. However, a significant challenge facing STT MRAM scaling is due to process variation. The major sources of process variations in MTJ include variations in (a) tunneling oxide thickness (τ) and (b) cross-sectional area (A) [2]. These parameters affect the static as well as the dynamic behavior of MTJ. These variations can result in possible failure of the cell — read failure (flipping of the cell data while read/incorrect read); and inability to write to the cell, severely affecting the yield of memory array. To accelerate successful STT MRAM commercialization, it is imperative that we develop efficient design/manufacturing solutions for yield enhancement. Considerable efforts have been made by device engineers and material scientists to explore various device structures with different material for MTJ. The device performance and uniformity can be improved by proper device design and process control [2]. However, even with the most advanced fabrication technique, process imperfections due to lithography and inherent material variations are truly difficult to overcome from device/technology perspective. To improve memory yield without incurring expensive process modification and to reduce time-to-market, there is a need to develop an efficient circuit level solution to tolerate variation induced parametric failures in STT MRAM.

In this work, we present a variation-tolerant circuit design technique which can effectively improve memory yield at the early design phase. In particular, we employ a comprehensive physics-based device model for circuit simulation. The model employs the Non-Equilibrium Green's Function (NEGF) formalism to solve quantum transport in the ballistic regime [3][4]. Unlike previous static resistance model which incurs significant pessimism (in terms of timing and cell area penalty), the proposed model can accurately capture the coupled electro-magnetic dynamics of spintronics device [3][4]. Specifically, the highlights of this work are as follows:

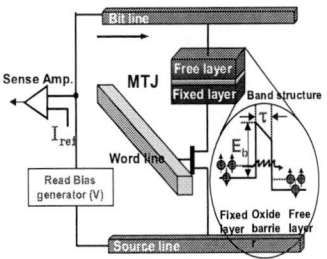

Figure 1. An STT MRAM bit cell includes : 1) MTJ; 2) BL; 3) SL; 4) WL; 5) NMOS and 6)read drive circuitry configuration

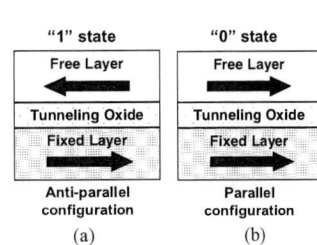

Figure 2. MTJ structure in (a) anti-parallel (high resistance state); and (b) parallel (low resistance state)

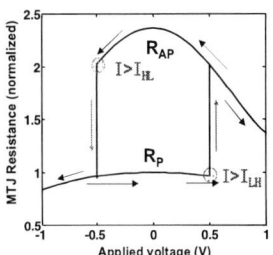

Figure 3. Resistance vs. voltage response of MTJ with τ= 1nm and A= 100×150nm Resistances are normalized to its parallel case with bias voltage (V_{MTJ}) = 0.

978-1-4244-2018-6/08 $25.00 © 2008 IEEE   211

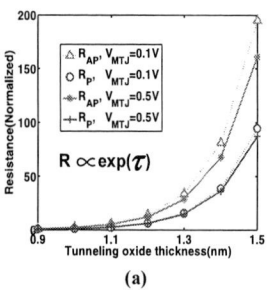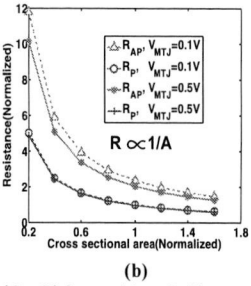

(a)   (b)

Figure 4. Impact of (a) tunneling oxide thickness (τ) and (b) cross sectional area (A) in MTJ resistance (R) in parallel and anti-parallel configuration under different bias voltage (V_{MTJ}). The values are normalized to its parallel case when τ=1.0nm, V_{MTJ}=0.1V

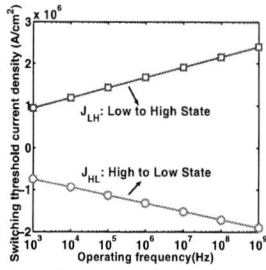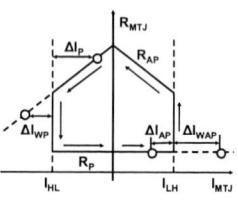

Figure 5. Frequency dependency of switching threshold current density of MTJ [5]. J_{HL} and J_{LH} are switching threshold current density from high to low and from low to high resistive state, respectively.

Figure 6. Read margin for parallel (ΔI_P) and anti-parallel (ΔI_{AP}) direction read. Write margin (ΔI_{WP}) for writing "0" and write margin (ΔI_{WAP}) for writing "1".

- We perform an in-depth discussion and statistical analysis on interrelationship between design parameters and parametric failures of STT MRAM in the presence of process variations. We observed that the size of NMOS transistor not only affects the memory density but also has crucial impact on cell stability.
- Optimal cell design is achieved to account for read stability, write stability and cell area in conventional 1T1M cell. Further, we show that the optimization of 1T1M cell is constrained by the inherent conflicting design requirement between read/write stability, and cell area.
- We propose a variation-tolerant circuit design technique to effectively decouple the conflicting design requirement. It provides flexibility to optimize these metrics (read stability, write stability and cell density) independently. Thus, better robustness is obtained with minimum area penalty.

The organization of the paper is as follows. Section 2 describes sources of process variations in spintronics devices and the associated read/write failure mechanisms in the presence of process variations. Section 3 provides an in-depth discussion on interrelationships between design parameters and cell stability. Cell optimization is performed for 1T1M configuration. In section 4, a variation-tolerant circuit is proposed and compared with optimal cell (in 1T1M configuration) in terms of area and robustness. Finally, we conclude the paper in section 5.

2. STT MRAM Cell Design

2.1. Source of process variations in MTJ

Since MTJ is the key component for information storage, the stability of an STT MRAM bit cell primarily depends on the characteristics of MTJ device. It has been experimentally demonstrated that the sources of variations in process parameters of MTJ mainly arises from variations in (a) tunneling oxide thickness (τ) and (b) cross-sectional area (A) [2]. These variations can lead to random variation in electro-magnetic characteristics of MTJ.

Due to the quantum mechanical tunneling, the thickness of tunneling film (τ) affects the high and low resistance in an exponential manner (Fig. 4(a)). Hence, small variation in τ can lead to large spread in high and low resistances of MTJ. In contrast to τ, cross-sectional area (A) variation affects both electrical static and magnetic dynamic switching characteristics of MTJ. In static states (i.e., high or low resistance states), the resistances are inversely proportional to A (Fig. 4(b)). Due to the time-dependent stochastic magnetic switching process, variations in A introduce a linear shift in switching threshold current (Fig. 5).

2.2. Parametric failures in STT MRAM cell

Variations in process parameters (τ and A) lead to large spread of variations in MTJ resistances and switching threshold current, thereby resulting in failure of a cell during read and write operations. A cell failure can occur due to: (a) destructive read or wrong decision in sense amplifier (read failure); (b) unsuccessful write (write failure).

Read failures — A circuit diagram for reading a cell is shown in Fig. 1. In read operation, a small bias voltage (v_R) is applied to bit line and the source line is grounded (namely, parallel direction reading). After turning on the word line, the cell current is sensed and compared with the reference current in sense amplifier. There is another way in reading (namely anti-parallel direction reading), where the voltage polarities applied to bit line and source line are switched. If the sensed current is lower (or higher) than the reference current when reading a cell storing "0" (or "1"), a wrong decision will be made by sense amplifier, resulting in read failure (referred as *decision failure*). Another read failure mechanism arises from the disturbance of a cell -- a cell can flip during read operation (referred as *disturbance failure*). This can occur only if the reading current is above the threshold current of MTJ. Due to the hysteresis nature of MTJ, there is only one direction disturbance (i.e., "1-to-0 disturbance" in parallel direction reading or "0-to-1 disturbance" in anti-parallel direction reading). The read failure probability (P_{RF}) in parallel direction reading can be estimated using the following equation [4]

$$P_{RF} = \alpha \iint\limits_{i_L(v_R,\tau,A)<i_{ref}} f(A)f(\tau)\,d\tau\,dA$$
$$+ (1-\alpha) \iint\limits_{\substack{i_H(v_R,\tau,A)>i_{ref} \\ i_H(v_R,\tau,A)>I_{HL}(t_R,A)}} f(A)f(\tau)\,d\tau\,dA \qquad (1)$$

where α is probability of a cell storing "0"; $i_H(v_R,\tau,A)$ and $i_L(v_R,\tau,A)$ are measured cell current at bias (v_R) for state "1" and "0", respectively; $I_{HL}(t_R,A)$ is the switching threshold current of MTJ at operating frequency of f_R ($1/t_R$).

The fault model for anti-parallel direction reading is derived in the similar way [4].

Write failures — In contrast to read operation, bi-directional writing is required to write a cell. For example, while writing a "0" to a cell, the bit line is pre-charged to V_{DD} and the source line is grounded (Fig.1). Write process starts immediately after turning on the word line. The voltage polarities applied to bit line and source line are switched when writing a "1" to a cell. Unsuccessful write occurs when the writing current is lower than the switching threshold current. The write failure probability is given by [4]

978-1-4244-2018-6/08 $25.00 © 2008 IEEE

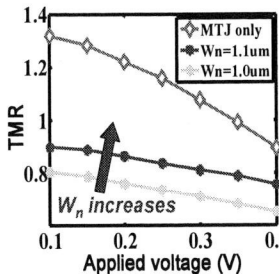
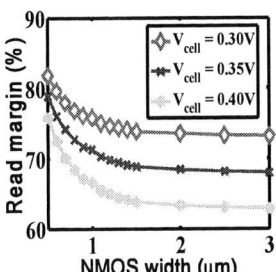
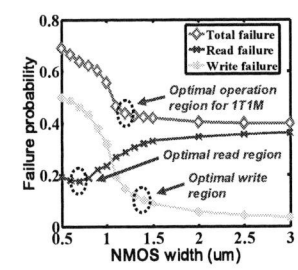

Figure 7. Effective TMR of STT MRAM bit cell vs. Applied voltage with different NMOS width

Figure 8. Read margin vs. NMOS size for different applied voltages

Figure 9. Decision failure and disturbance failure vs. NMOS size

Figure 10. Failure probability vs. NMOS size: 1)read failure; 2)write failure and 3)total failure

$$P_{WF} = \alpha\beta \iint\limits_{i_L(v_{DD},\tau,A)<I_{LH}(t_w,A)} f(A)\,dA$$
$$+ (1-\alpha)(1-\beta) \iint\limits_{i_H(v_{DD},\tau,A)<I_{HL}(t_w,A)} f(A)\,dA \qquad (2)$$

where α is probability of a cell storing "0" and β is probability of writing "1" to a cell.

The above fault model will be used in this paper to evaluate the failure probabilities of different events.

2.3. STT MRAM bit cell design consideration

We studied STT MRAM designed with 130nm BULK CMOS and MTJ device with cross-sectional area (A) of 100×150 (nm^2) and tunneling oxide (τ) of 10 Å. The MTJ geometry is scaled from the parameters given in [5] using 180nm CMOS technology. For simulation of MTJ, we use the Spin-Torque Transfer model proposed in [3]. To enable efficient circuit simulation, we use the CAD tool developed in [4] to accurately capture the static and dynamic behavior of MTJ and the combined magnetic/circuit response in the presence of transistors. In this analysis, we consider parallel direction reading technique. The probability of storing or writing a "1" to a cell is assumed to be 0.5. We use the above device for the *nominal* (i.e. mean $\tau \sim 10$ Å and mean $A \sim 100 \times 150$ nm^2) case. The spread of the variation (σ/μ) is assumed to be 5% in both τ and A (τ and A are assumed to follow Gaussian distribution). Both the read and write operation frequencies are assumed to be 500MHz, which is the maximum speed reported in literature [5].

A) Effect of NMOS sizing on Read Stability

There are two quality metrics to evaluate read stability of a cell: 1) effective Tunneling Magneto-Resistance (TMR) and 2) read margin (RM). In contrast to conventionally TMR defined in [3], effective TMR is proposed to account for the series resistance of NMOS, i.e., $(\tilde{R}_H - \tilde{R}_L)/\tilde{R}_L$, where $\tilde{R}_H$ and $\tilde{R}_L$ are effective cell resistances of MTJ in series with NMOS in "1" and "0" states, respectively. Similar to TMR, effective TMR has strong voltage dependence and decreases with increase in bias voltage. Read margin is defined as the distance between the reading point and the trip point (switching threshold) (Fig. 6). Larger read margin results in less read disturbance, thereby less failure. To reduce read failures, high TMR is desirable for MTJ due to enhanced fractional change in the effective resistance (large noise margin). However, it is shown that NMOS (~size) is equally important as MTJ (~TMR) in determining the cell read failure probability. This is due to the fact that enhanced states change of MTJ (high TMR) can be overshadowed by negative feedback of NMOS (Fig.7). To reduce the strength of negative feedback, NMOS should be upsized. To verify this, we performed Monte-Carlo circuit simulations (~100,000) to estimate the cell read failure probability for different NMOS width. The simulation result in Fig.7 shows that due to

reduced NMOS series resistance by upsizing, the effective TMR can be improved, leading to less decision failure. However, upsizing NMOS can increase reading current. Consequently, read margin is reduced (Fig.8), thereby resulting in more disturbance failure. However, it is seen from our simulation that decision failure is dominant when NMOS size is small while disturbance failure becomes more pronounced with increase in NMOS size (Fig.9). Minimum read failure occurs when NMOS size is 0.7μm (Fig.10).

B) Effect of NMOS sizing on Write Stability

Write stability of the cell is measured using write margin (WM) which is defined as the difference between the writing current and the trip point (switching threshold) (Fig.6). Higher is the write margin, better is the stability. In contrast to read margin, write margin can be improved by upsizing of NMOS transistor. By solving the transport equation of MTJ in conjugate with NMOS self consistently, the voltage division between MTJ and NMOS is achieved. We see that larger NMOS size decreases its Drain-to-Source voltage, resulting in larger voltage drop across MTJ. Consequently, writing current is improved, thereby reducing write failure (Fig.10).

C) Effect of NMOS sizing on Cell Density

We note that the required switching current is of the order of hundreds μA for MTJ, which cannot be provided by a minimum sized transistor. Since MTJ is integrated with NMOS using 3D technology, the memory cell size is dominated by the area of the NMOS. The layout of a bit cell in CMOS layer (1T1M) is illustrated in Fig.12 (a).

D) Cell Optimization for 1T1M Configuration

From the above discussion, we can conclude that NMOS sizing is crucial in determining the stability and the density of STT MRAM. Optimal design to account for both read and write stability is difficult for conventional 1T1M cell structure. The underlying issue with cell failure probability is the conflicting design requirement on NMOS transistor size between read/ write stability and cell area (Fig.10). Enlarging NMOS size can reduce write failure probability at the expense of increased read failures and area penalty. At the circuit level, techniques that optimize read stability can hurt write stability and density, and vice-versa. To achieve optimal trade-off between robustness and density, we plot the total failure probability with NMOS size in Fig. 10. The optimal size to account for both stability and density is chosen to be 1.2μm.

3. Variation-Tolerant Cell Design

As discussed in previous sections, in conventional 1T1M cell structure, NMOS has to be carefully designed to achieve both robustness and high memory density. However, it is challenging to achieve optimal design due to the conflicting design requirement between read stability, write stability and cell density. In this

978-1-4244-2018-6/08 $25.00 © 2008 IEEE

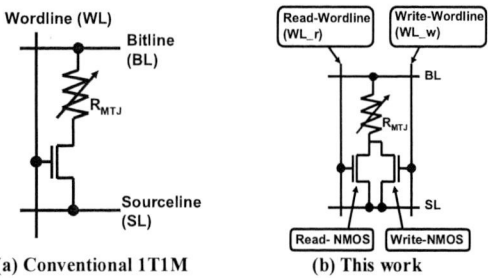

(a) Conventional 1T1M **(b) This work**

Figure 11. Conventional and proposed MRAM schematic comparison

section, an efficient variation-tolerant circuit design technique is proposed to relax the design requirements and compensate process induced variation.

The proposed cell schematic is illustrated in Fig. 11(b). In contrast to conventional 1T1M configuration (Fig. 11(a)), the proposed circuit consists of one MTJ and two NMOS transistors (Read-NMOS and Write-NMOS) in parallel, with independent gate control: Read-Wordline (WL_r) and Write-Wordline (WL_w). The total size of NMOS can be dynamically adjusted by turning on these control signals accordingly during read and write operations. The operating principle is summarized as follows: in read operation, only WL_r is activated, thereby turning on Read-NMOS to achieve optimal read immunity to disturbance; during write operation, both WL_r and WL_w are selected so that Read-NMOS and Write-NMOS are turned on simultaneously for better write stability. This technique effectively decouples the conflicting design requirement between read stability and write stability, resulting in considerably improved robustness.

However, area overhead may increase due to the layout of two transistors in a cell. The layout view of our proposed cell structure is shown in Fig. 12(b) in comparison with conventional 1T1M cell in Fig. 12(a). In this work, we refer to the MOSIS scalable CMOS design rule at 0.13μm technology node [6] which is equally applicable to further scaled technologies (<0.13μm). It is seen that the two transistors (Read-NMOS and Write-NMOS) in the proposed cell are on the same active region, sharing one contact which is connected to the MTJ. Note that the height of the cell (in y direction) is determined by CMOS technology (in this work we choose 0.13μm technology) but the width of the cell (in x direction) can vary with different sizings of these two transistors. Consequently, the total cell area is determined by the larger size of transistor between Read-NMOS and Write-NMOS. As a result, to improve robustness and minimize area overhead, both of the NMOS transistors have to be sized properly.

It is shown in Fig. 10 that the optimal transistor size (Read-

NMOS) for minimal read failure probability is 0.7μm at operating frequency of 500MHz. In contrast, the write failure probability monotonically decreases with increase in NMOS size. To reduce write failure, Write-NMOS has to be upsized. However, increase in Write-NMOS will increase area overhead (Fig. 13). To account for these factors, the size of Write-NMOS is chosen to be 0.7μm to achieve the optimal stability-area trade-off (Fig. 14). With further increase in Write-NMOS size, the total failure probability decreases at the cost of much faster increased area overhead.

From above discussion, size of Read-NMOS to Write-NMOS ratio is chosen as 0.7μm/0.7μm. Compared to the optimized conventional 1T1M cell (section 2.3 D), the read failure probability is reduced by 39% and the write failure probability is reduced by 31% in our proposed cell. Consequently, cell stability (to account for both read stability and write stability) is improved by 36% with minimal area overhead (~9%).

4. Conclusion

In this work, we performed a thorough analysis on the impact of design parameters on stability and density (cell area) of STT MRAM in presence of process variations. Our simulation is based on a comprehensive physics-based MTJ model. The results show that reducing transistor dimension can reduce read disturbance and improve layout efficiency but that would increase write failure. Conflicting design requirement between read/write stability and cell area has been observed. To compensate process variation induced functional failures and to take care of the conflicting read and write requirements, we proposed an efficient circuit technique that improved the cell stability by 36% with 9% area overhead.

Acknowledgments

This work was supported in part by Focus Center Research Program (FCRP) and Nano Research Initiative (NRI).

References

[1] A.D. Smith et al. STT-RAM –A New Spin on Universal Memory, *Future Fab Intl.* Vol. 23, July 2007

[2] E.Y. Chen et al., Comparison of oxidation methods for magnetic tunnel junction material, J. Appl. Phys., vol. 87, pp.6061–6063, 2000.

[3] S. Salahudding et al., Self-Consistent Simulation of Hybrid Spintronic Devices, *IEDM Tech. Dig.*, pp.1-4, Dec., 2006.

[4] J. Li et. al., "Modeling of Failure Probability and Statistical Design of Spin-Torque Transfer Magnetic Random Access Memory (STT MRAM) Array for Yield Enhancement," *DAC*, 2008

[5] M. Hosomi et al., A Novel Nonvolatile Memory with Spin Torque Transfer Magnetization Switching: Spin-RAM, *IEDM Tech. Dig.*, pp. 473-476, Dec., 2006.

[6] Vendor-Independent MOSIS Scalable CMOS Design Rules http://www.mosis.com/Technical/Designrules/scmos/scmos

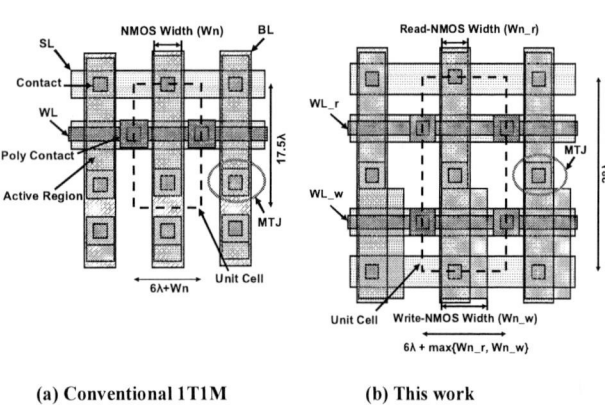

(a) Conventional 1T1M **(b) This work**

Figure 12. Conventional and proposed MRAM cell layout comparison

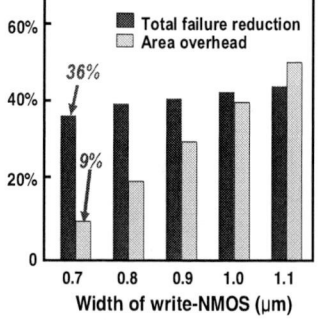

Figure 13. Total failure reduction and area overhead vs. width of Write-NMOS when Read-NMOS width is 0.7μm

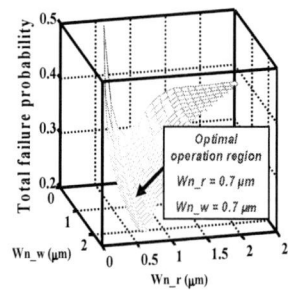

Figure 14. Optimization of Read-NMOS width and Write-NMOS width to minimize the total failure probability and cell area

978-1-4244-2018-6/08 $25.00 © 2008 IEEE

IEEE 2008 Custom Intergrated Circuits Conference (CICC)

Pure Logic CMOS Based Embedded Non-Volatile Random Access Memory for Low Power RFID Application

Liyang PAN [1,2], Xian LUO [1,*], Yaru YAN [1], Jirong MA [1], Dong WU [1,2], Jun XU [1,2]

Institute of Microelectronics of Tsinghua University, Beijing 100084, China [1];

Tsinghua National Laboratory for Information Science and Technology [2];

Email*: x-luo06@mails.tsinghua.edu.cn

Abstract- **Based on a novel two-dimension array architecture, a 1.8V 0.44mm² 1Kbits embedded Non-Volatile Random Access Memory (NVRAM) IP is developed with 0.18μm standard logic CMOS process. Several high voltage solutions and circuits are proposed to improve the reliability and safety of the system. Furthermore, the power consumption for read and write operations are controlled under 312μA and 88μA respectively. The merits make it suitable for low power RFID application.**

I. INTRODUCTION

Embedded non-volatile memory (eNVM) is becoming one of the key components for SOC applications. However, conventional eNVM approaches such as EEPROM or eFlash require additional masks and process steps, which increase the chip cost. In addition, the special processes usually lag behind the state-of-the-art standard logic process by two or three technology generations, therefore compromise the SOC system performance.

A standard CMOS process based NVM cell, which is made up of two PMOS capacitors and a PMOS read transistor with gates coupled together, was firstly described by Katsuhiko Ohsaki [1]. Recently, Virage logic and Impinj also tried to develop CMOS process based eNVM systems for security and SOC applications [2] [3]. Due to the limitation of the CMOS process, these systems adopt low voltage (LV) core CMOS or 3.3V I/O CMOS transistors to design the memory cell, the high voltage (HV) generator and the HV switches. Because the LV transistor cannot be used to select or prohibit the HV signals, a complicated architecture with storage circuit, sensing module, HV switch/controller and LV decoder must be integrated into each memory cell (bit). Moreover, the memory cells can only be laid out in one-dimension, resulting in the inefficiency of the memory array area and inflexibility for applications. Furthermore, the one-dimension array architecture also increases the operation power consumption, making it not suitable for low power SOC applications, such as RFID.

In this paper, a pure logic CMOS process based eNVM cell is firstly proposed to form a novel two-dimension array architecture. Its operation principles, write and read characteristics are then presented. Based on the proposed architecture, a 1Kbits embedded Non-Volatile Random Access Memory (NVRAM) system, along with the HV generator and HV management circuits, is finally discussed and designed for RFID applications.

II. TWO-DIMENSION NVRAM ARRAY ACHITECTURE

A. Cell Structure and Array Architecture

Fig.1(a) illustrates the NVRAM cell structure, which consists of three circuit blocks: A) Basic storage circuit, which is made up of two complementary storage elements (similar to the one describe in [1]) and has two coupled floating gates (FG and FG_n) to store the state "1" or "0" respectively through Fowler-Nordheim (FN) tunneling. The storage element is illustrated in fig.1(b), with the coupling coefficient $\alpha_C = C_C / (C_C + C_T + C_P)$ being about 94%. B) Sensing circuit: senses the stored state "1" or "0" and outputs the data to BL/BL_n under the control of read selection during read operation. C) Write voltage equalization and prohibition circuit (WEPC): equalizes voltages on node NA and NA_n in the unselected cell to prohibit the FN tunneling injection during write operation, wherein four native NNMOS transistors (fabricated directly on P-substrate with V_{TH}~0V) are adopted to isolate the bottom LV control transistors from HV signals. There are also two PMOS-type isolation diodes to isolate V_P/V_N from V_{PP}/V_{PP_n} in the unselected cell.

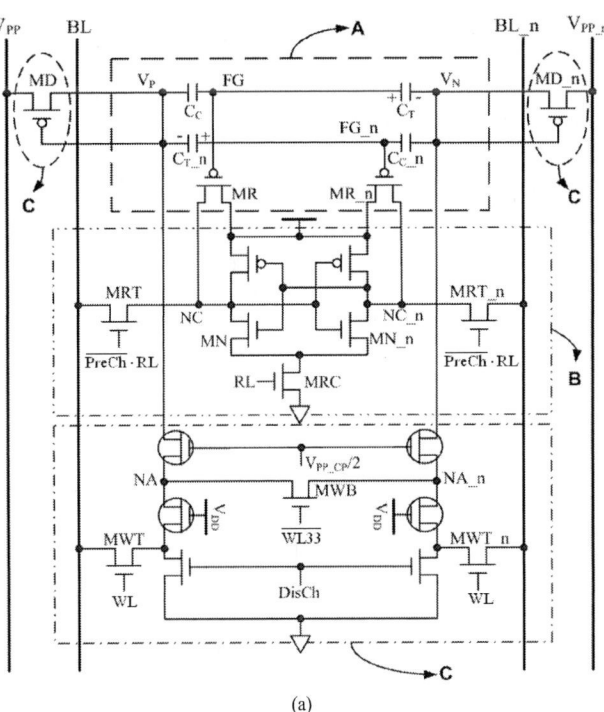

(a)

978-1-4244-2018-6/08 $25.00 © 2008 IEEE   215

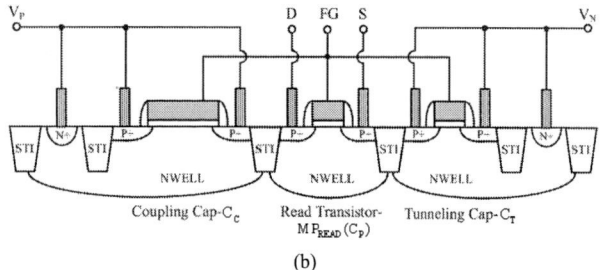

(b)

Fig.1. (a) NVRAM cell schematic, and (b) cross section view of the basic storage element.

Based on the proposed memory cell structure, a 1Kbits two-dimension NVRAM array architecture, which includes memory array, HV switch, HV cascade switch, read pre-charge circuit and data input/output control circuit, is described in fig.2. Wherein, the charge pump generated voltage V_{PP_CP} (7.8V) is sent to the HV switch and HV cascade switch, which then generate and transmit a pair of write voltage V_{PP} and V_{PP_n} to the selected columns during write operation. The 1Kbits array is organized as 16×16 bytes with each byte having four bits and written or accessed in parallel. It needs to be pointed out that a corresponding local byte decoder, which is controlled by both of the row_decoder and column_decoder, is used to generate the required control signals (RL, $\overline{PreCh \cdot RL}$, WL and $\overline{WL33}$ shown in fig.1(a)) for the byte operations. This kind of the two-dimension array architecture and the four-bit byte organization contribute to the reduction of power consumption.

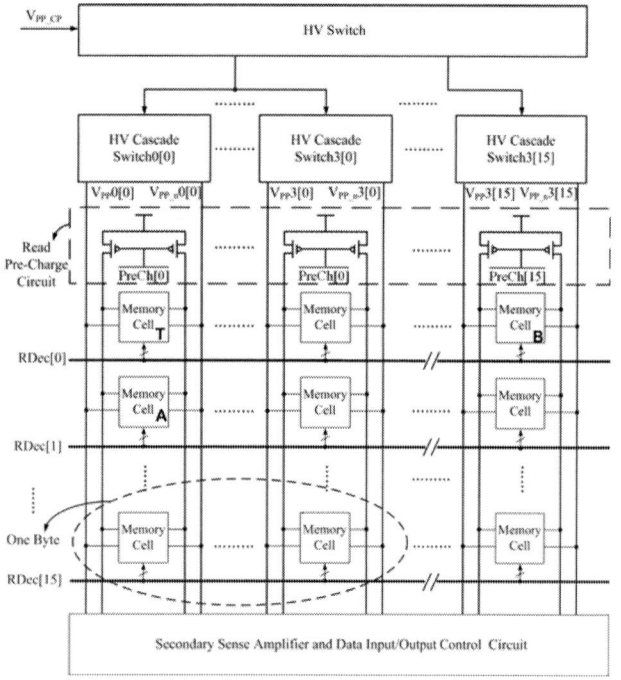

Fig.2. 1Kbits two-dimension NVRAM array architecture.

B. Write and Read Operation Principles

During write operation, the cell control signals (DisCh, RL and $\overline{PreCh \cdot RL}$), which are only involved in the read operation,

are all set to 0V. While writing "1", the column decoder and the data input/output control circuit transfer the write data to the selected BL and BL_n, and set them to 1.8V and 0V respectively. According to the level of the BL and BL_n, the cascade switch then outputs 7.8V and 0V to the corresponding V_{PP} and V_{PP_n} respectively, which are later transferred to V_P and V_N in the selected memory cells through the isolation diodes MD and MD_n. Due to the threshold voltage loss of the MD transistor, the final voltage on node V_P decreases to about 7.5V as shown in fig.3. In the case of writing "0", all the signals of the discussed nodes are reverse to the situations of writing "1".

For the unselected cells (classified as Cell A in fig.2) in the same selected column, the write prohibition is realized by turning on the WEPC shown in fig.1(a). In this case, the two write selection transistors MWT and MWT_n are turned off and the equalization transistor MWB is turned on by biasing $\overline{WL33}$ at $\frac{1}{2}V_{PP\_CP}$. Herein, V_P and V_{NA} are still 7.5V and $\frac{1}{2}V_{PP\_CP}$ respectively, same as the biases in the selected cells, while V_N and V_{NA_n} are equalized to $\frac{1}{2}V_{PP\_CP} - V_{TH\_MWB}$, which is about 3V, and isolated by MD_n from V_{PP_n}. The voltage difference $(V_P - V_N)$ biased on the storage elements is therefore reduced to 4.5V to efficiently prohibit the write operation, as shown in fig.3. The detailed write operation conditions are listed in Table.I.

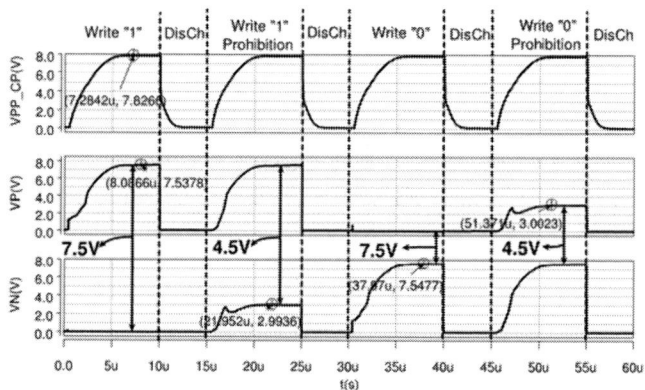

Fig.3. Simulated waves of V_{PP_CP}, V_P and V_N of write "1", write "0" and write prohibition operations.

Table.I. Write operation conditions

Cell	Write Data	V_{PP}/V_{PP_n}	V_P/V_N	$\overline{WL33}$	WL
T	"1"	7.8V/0V	7.5V/0V	0V	1.8V
(Target)	"0"	0V/7.8V	0V/7.5V	0V	1.8V
A	"1"	7.8V/0V	7.5V/3V	3.9V	0V
A	"0"	0V/7.8V	3V/7.5V	3.9V	0V
B	--	0V/0V	0V/0V	3.9V	0V

During read operation, V_{PP}, V_{PP_n}, V_P and V_N are all discharged to 0V to provide the required gate voltages of MR and MR_n for reading. The read operation is divided into three stages. At the first pre-charge and sensing stage, the BL and BL_n of the selected columns are pre-charged to V_{DD} by the read pre-charge circuit shown in fig.2. Meanwhile, NC and NC_n are also pre-charged to V_{DD} by MR and MR_n, and the

978-1-4244-2018-6/08 $25.00 © 2008 IEEE   216

sensing circuits of the selected cells are subsequently turned on to sense the stored data. At the second accessing stage, the MRT and MRT_n of the selected cells are turned on to discharge BL or BL_n and create a voltage difference between the two bit-lines. At the third output stage, the voltage difference between BL and BL_n is amplified by a secondary sense amplifier in the data input/output control circuit and eventually outputs to the data ports.

Compared with the conventional SRAM cell, the proposed architecture has two additional PMOS read transistors (MR and MR_n), which are always in on state during read operation and result in static leakage current. Therefore, the local byte decoder is designed for controlling MRC to realize the byte operation, decrease the read power consumption and improve the stability during the voltage transition. It should be noted that the size of the transistors in the sensing circuit must be elaborately designed to achieve a compromise between the read speed and the static noise margin (SNM).

C. Write and Read Characteristics

Fig.4(a) shows the schematic of the basic storage element. During write operation, the charges $Q(t)$ and $Qn(t)$ on FG and FG_n can be derived from (1a) and (1b), and the threshold voltage $V_{TH}(t)$ and $V_{TH\_n}(t)$ of the two storage elements are guided by (2a) and (2b):

$$-\frac{dQ(t)}{dt} = f\left(\frac{Q(t)}{C_{total}} + \frac{C_C V_P + C_T V_N + C_P V_{DD}}{C_{total}} - V_N\right) \quad (1a);$$

$$-\frac{dQn(t)}{dt} = f\left(\frac{Qn(t)}{C_{total}} + \frac{C_C V_N + C_T V_P + C_P V_{DD}}{C_{total}} - V_P\right) \quad (1b);$$

$$V_{TH}(t) = V_{TH0} - \frac{Q(t)}{C_C + C_T} \quad (2a); \quad V_{TH\_n}(t) = V_{TH0} - \frac{Qn(t)}{C_C + C_T} \quad (2b);$$

Where $C_{total} = C_C + C_T + C_P$, $f(V)$ is the I-V function of C_T, and V_{TH0} is the initial threshold voltage.

Fig.4(b) shows the measured FN tunneling current as a function of the gate voltage on the tunneling capacitor C_T, which is fabricated with $0.3\mu m \times 0.3\mu m$ I/O PMOS transistor. Fig.5 shows the $V_{TH}(t)$ and $V_{TH\_n}(t)$ of the two storage elements which are calculated from the above equations and the measured FN current curves. The V_{TH} and V_{TH_n} are -0.3V and -1.2V, and a 0.9V V_{TH} Window is obtained in 2ms write time. The proposed complementary structure can reduce the read time and improve the retention characteristics [2]. For the unselected cells, it is obvious that the disturbance current is extremely small as the voltage difference (V_P-V_N) is limited at 4.5V, which benefits the system reliability.

Fig.6 illustrates the read static noise margin (SNM) in the supply of 1.8V and 1.2V. After write operation, the complementary structure turns into an asymmetrical circuit due to the V_{TH} difference between the two storage elements. Study shows that the "0" side is easier to be disturbed by the read noise. By elaborately designing the size ratio of MN and MRT, the worst case SNM in the supply of 1.8V and 1.2V can be enlarged to 120mV and 90mV respectively.

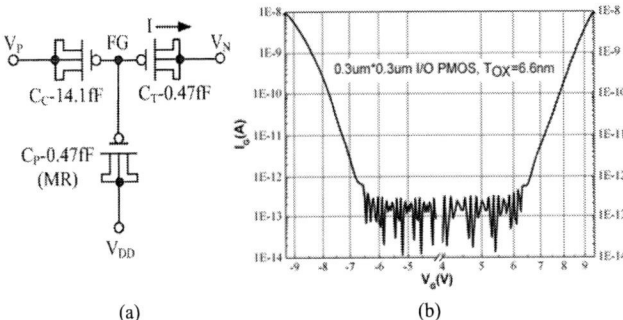

(a)   (b)

Fig.4. (a) Basic storage element schematic, and (b) measured FN tunneling current versus C_T gate voltage characteristics.

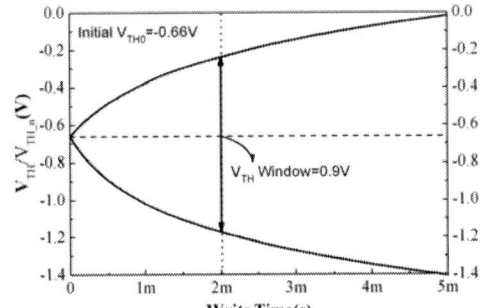

Fig.5. Calculated V_{TH} and V_{TH_n} versus write time characteristics.

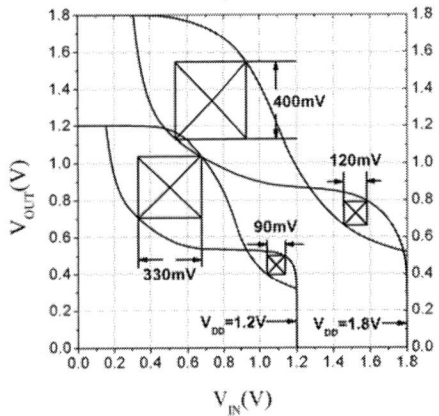

Fig.6. Static noise margin in read operation.

III. NVRAM SYSTEM CIRCUITRIES AND HV SOLUTIONS

The whole NVRAM system is made up of the proposed 1Kbits two-dimension array, a HV generator circuit, hybrid HV switch circuits, row and column decoders, a 10MHz clock circuit and a user interface. The on-chip HV generator circuit, which includes a dual-path Dickson charge pumping circuit [4] (shown in fig.7(a)) and a HV regulator, is designed to generate the write operation needed V_{PP_CP} (7.8V $\pm 7\%$). The generated V_{PP_CP} is then selected and transferred to the memory array through hybrid HV switch circuits, as shown in fig.2. Fig.7(b) illustrates the schematic of a basic unit of the dual-output HV cascade switch.

978-1-4244-2018-6/08 $25.00 © 2008 IEEE

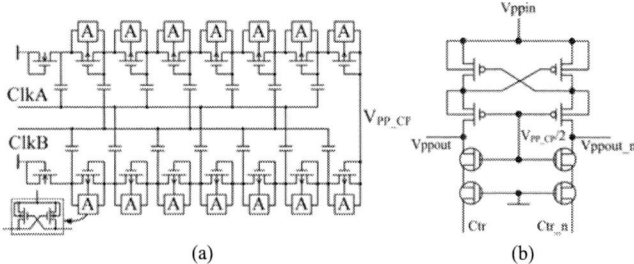

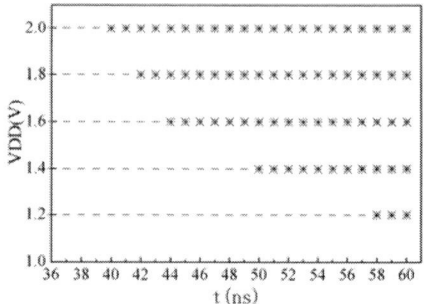

Fig.7. Schematic of (a) the dual-path Dickson charge pumping circuit, and (b) the basic unit of the HV cascade switch circuit.

Fig.9. Read speed schmoo.

Considering the voltage limitations of the CMOS transistors, it should be pointed out that the following techniques are implemented to guarantee the safety of all the HV circuits:

A) 3.3V I/O PMOS transistors are adopted to form the HV circuits, and their voltages are limited to ½V_{PP_CP} to prevent breakdown.

B) The substrate voltage of the PMOS transistors in the charge pumping circuit is biased by a dynamic substrate control circuit [5] to prevent the junction leakage and the latch up effect.

C) Metal-metal parasitic capacitors, similar to [2], are utilized to endure the high voltages in the charge pumping circuit.

D) A special isolation structure with two native NNMOS transistors in series is proposed in the HV circuits and memory array to isolate the LV control transistors from the HV signals. This kind of native NNMOS transistors are fabricated directly on P-substrate with $V_{TH}{\sim}0V$ and completely compatible with the standard CMOS process. Their junction breakdown voltage can be twice as big as the one of the normal NMOS transistors, owing to its light substrate doping.

Fig.8 and fig.9 show the 1Kbits NRAM layout picture and read speed schmoo plot respectively. The size of each memory cell and the NVRAM IP are 153μm^2 and 0.44mm^2 respectively, and the read speed is 42ns and 58ns in the supply of 1.8V and 1.2V. The main features of the proposed NVRAM system are finally listed in Table.II.

Table.II. NVRAM features

Process	0.18μm Standard CMOS Process
Memory Cell Size	153μm^2
IP Size(1Kbits)	0.44mm^2
Write Power Supply	1.6-2.0V
Read Power Supply	1.2-2.0V
Write Speed	2ms /4bits
Write Power Consumption	312μA @ 1.8V /4bits
Read speed	42ns @ 1.8V /4bits
Read Power Consumption	88μA @ 1.8V /4bits

IV. CONCLUSION

In conclusion, a 1Kbits pure logic CMOS process based NVRAM system without additional masks and process steps is developed. By using the equalization method to prohibit the write operation, a novel two-dimension array architecture is realized, which greatly reduces the IP size and improves the flexibility for SOC applications. Owning to the single power supply and the systemic low power design, NVRAM is particularly suitable for low power RFID application.

ACKNOWLEDGMENTS

The research is supported by the National Basic Research Program of China (No. 2006CB302700).

REFERENCES

[1] Ohsaki K., Asamoto N., Takagaki S., "A single poly EEPROM cell structure for use in standard CMOS processes," IEEE Journal of Solid-State Circuits, Vol. 29, Iss. 3, pp.311-316, 1994

[2] Raszka J., Advani M., Tiwari V., Varisco L., Hacobian N.D., Mittal A., et al., "Embedded flash memory for security applications in a 0.13μm CMOS logic process," Digest of Technical Papers of International Solid-State Circuit Conference, Vol. 1, pp.46-512, 2004

[3] http://www.impinj.com/ip/aeon-mtp.aspx?ekmensel=c580fa7b_22_0_10 76_2

[4] Dickson J F., "On-chip high-voltage generation in MNOS integrated circuits using an improved voltage multiplier technique," IEEE Journal of Solid-State Circuits, Vol.11, Iss. 3, pp. 374-378, 1976.

[5] Park Jin-Young, Chung Yeonbae, "A low-voltage charge pump circuit with high pumping efficiency in standard CMOS logic process," Electron Devices and Solid-State Circuits, pp.317-320, 2007.

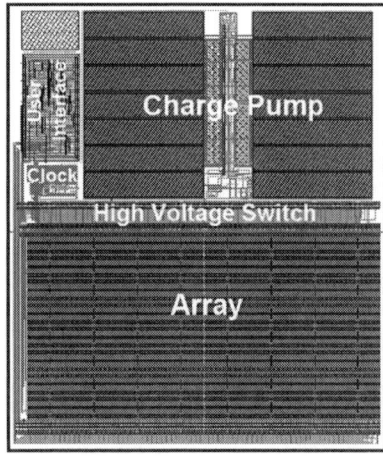

Fig.8. 1Kbits NVRAM Layout.

IEEE 2008 Custom Intergrated Circuits Conference (CICC)

A High-speed, Low-power 3D-SRAM Architecture

H. Henry Nho, Mark Horowitz, S. Simon Wong

Stanford University

Abstract – **This paper presents a novel 3D-SRAM architecture that can be used to extend the scaling of SRAM. This architecture significantly reduces the bit-line capacitance, achieves 3.4 times reduction in active power consumption and 1.8 times reduction in access time. In this architecture, local bit-lines are vertical and connect through select transistors to the global bit-lines routed on the bottom level. A proof-of-concept 32Kb sub-array emulating the critical path of the 3D-SRAM has demonstrated about 5 times improvement in power-delay over conventional 2D-SRAM.**

I. INTRODUCTION

While the demand for integrated SRAM continues to grow, scaling SRAM to sub 45nm technology has become increasingly difficult [1]. Process variations in deep sub-100nm, such as random dopant fluctuation, well proximity effect, and gate line edge roughness, cause significant variations in circuit behavior. As a result, increasingly conservative design is required as the technology scales – for example, trading off delay performance by allocating more timing margin, or trading off silicon area by using larger transistors that suffer less process variations [2].

In this paper, we explore the potential advantages of using 3D integration for future SRAM scaling. We start with monolithic 3D integration, which enables accurate stacking of multiple active layers on a single wafer, each with a thickness of only few hundreds of nanometer [3]. In the resulting design, inter-layer vias are similar in size to typical inter-metal vias, giving this technology a very high density vertical interconnects. This technology is similar to the Stacked Single-Crystal Silicon (S^3) SRAM [4] which enables 3D integrations at the transistor level. Our proposed design has a novel bit-line and decoder architecture that derives the maximum benefits of 3D.

Next, we explore the feasibility of implementing the 3D architecture using wafer-to-wafer bonding technology with various inter-layer alignment precisions. It is shown that the proposed 3D architecture can also be built with the most advanced wafer-to-wafer bonding technology with minimal performance degradation. 65nm technology was assumed for the following analysis.

II. BACKGROUND AND OVERVIEW

A. Bit-line architecture

Bit-line delay usually constitutes the majority of the total access time because a single SRAM cell must discharge a long bit-line. It also occupies major portion of the active power dissipation because a large number of columns discharge every time a WL is asserted. Thus bit-line capacitance is an important performance factor in a SRAM. Although bit-line capacitance can always be reduced by putting fewer cells per bit-line, it is not preferred because the number of sub-banks will increase and the area efficiency will degrade.

A hierarchical bit-line architecture [5] reduces the total bit-line capacitance by isolating the cell junction capacitances from the global bit-lines. However, the overall reduction in bit-line capacitance is limited because cell junction capacitance is only a portion of the total bit-line capacitance. A larger portion of the bit-line capacitance comes from the metal coupling, which is proportional to the length and hence is virtually unchanged for the same number of cells per bit-line.

In our proposed 3D-SRAM architecture shown in Fig. 1, the local bit-lines extend upward, through an inter-layer via that connects SRAM cells vertically. The local bit-line connects through a select transistor to the global bit-line routed on the bottom layer. As a result, the length and capacitance of the global bit-lines depend only on the number of local bit-lines, not the total number of cells. Despite the parasitic capacitances and resistances of the inter-layer vias, overall bit-line capacitance can be reduced significantly because the length of

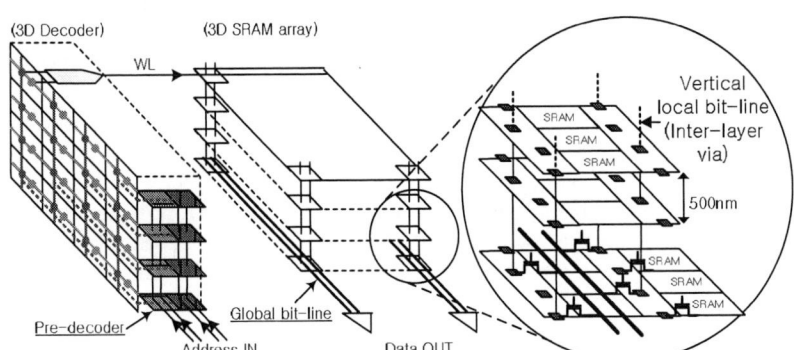

Fig. 1: Array and decoder architecture of 3D-SRAM with 4 active layers.

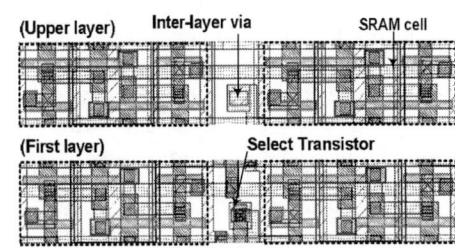

Fig. 2: Layouts of the first and upper layer of 3D-SRAM.

978-1-4244-2018-6/08 $25.00 © 2008 IEEE

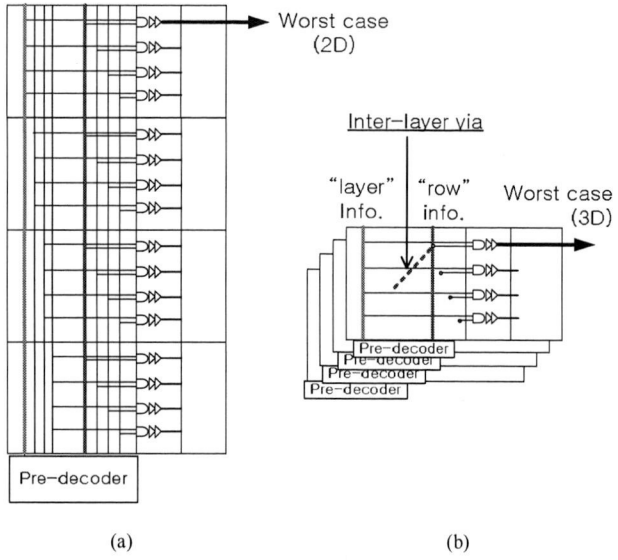

(a)   (b)

Fig. 3: The worst case delay of (a) a typical 2D decoder, and (b) the proposed 3D decoder.

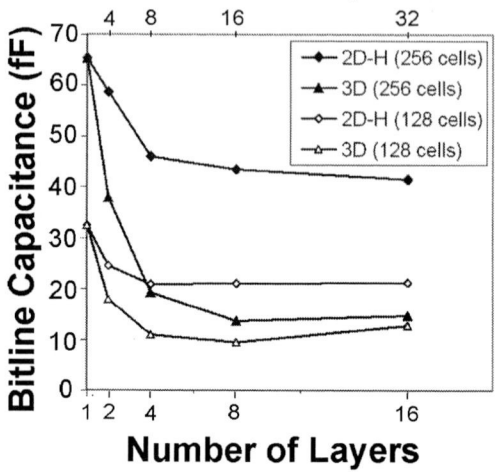

Fig. 4: Comparison of bit-line capacitances of 2D, 2D-hierarchical (2D-H), and 3D-SRAM.

the global bit-line reduces by a factor of the number of layers. Thus this memory will be denser, faster and lower power than a conventional design.

Fig. 2 shows a layout view of the SRAM cells in the 3D-array using 65nm technology. The cell is similar to a conventional 2D-SRAM cell and hence does not require sophisticated 3D technology. Select transistors that connect the vertical local-bitline and global bit-line are located on the first layer, between SRAM cells, adding only 18% area overhead per cell. Area efficiency is maximized by reusing this area for inter-layer vias in the upper layers.

B. Decoder architecture

In a typical 2D decoder (Fig. 3a), it is necessary for the pre-decoder outputs to stretch along the SRAM array to fully decode all the word-lines in the array. Because they swing fully, these long pre-decoder outputs contribute significant power dissipation as well as delay in the decoder operation.

Fig. 3b illustrates the proposed 3D decoder that derives the maximum benefits of 3D. The decoder consists of two sets of pre-decoders, one to decode the "layer" information and the other for the "row" information. The pre-decoder output for the "row" information connects vertically through inter-layer vias to the inputs of main decoders on each layer. These pre-decoders are distributed on every layer, as shown in Fig. 1, and operate concurrently. An "AND" logic then drives the chosen word-line in the selected layer. The word-line signal also controls the access transistor that connects the local bit-line to the global bit-line. Since only the pre-decoders for the selected layer and row need to be activated, the worst case capacitance is reduced to $1/n$ of the 2D design, where n is the number of active layers used.

Note that in the proposed 3D-SRAM design, the layouts for the upper layers are identical and the masks can be reused.

III. 3D-SRAM PERFORMANCE SIMULATION RESULTS

There are various 3D technologies for stacking layers, including chip-level [6], wafer-level [7] and monolithic integration [3]. An important difference between these 3D technologies is the size and pitch of the inter-layer vias, which directly impact the size of the 3D-SRAM cell as shown in Fig. 2. We first start from monolithic 3D integration, assuming perfect alignment between layers such that the inter-layer vias are similar in size to normal inter-metal vias.

Fig. 4 compares the total bit-line capacitance versus the number of active layers for the 3D architecture, and the corresponding number of cells per local bit-line (LBL) in the 2D-hierarchical (2D-H) architecture. Because two adjacent SRAM cells share a bit-line contact and an inter-layer via, the total number of cells per LBL is twice the number of layers. Thus a 3D-SRAM with n layers will have $2n$ cells per LBL. Although the simulation result shows that using 8 layers yields the best result, using 4 layers achieves the best tradeoff between performance and complexity. Also note that there is more reduction in the bit-line capacitance with a total of 256 cells on the bit-line versus with 128 cells. This is because doubling the number of cells per bit-line has only marginal effect on the total bit-line capacitance in 3D-SRAM, whereas in 2D or 2D-hierarchical SRAM, the length of the bit-line and hence the capacitance almost double. Using 4 layers for 3D-SRAM, we achieve 3.4X reduction in capacitance compared to 2D-SRAM, and 2.4X reduction compared to 2D-hierarchical SRAM. Bit-line delay also follows the trend of the bit-line capacitance. In 3D, the delay is reduced by about 1.8X compared to 2D-SRAM, and 1.7X compared to a corresponding 2D-hierarchical SRAM. Overall, 3D-SRAM achieves about 6X improvement in power-delay over 2D-SRAM, and 4X over 2D-hierarchical SRAM.

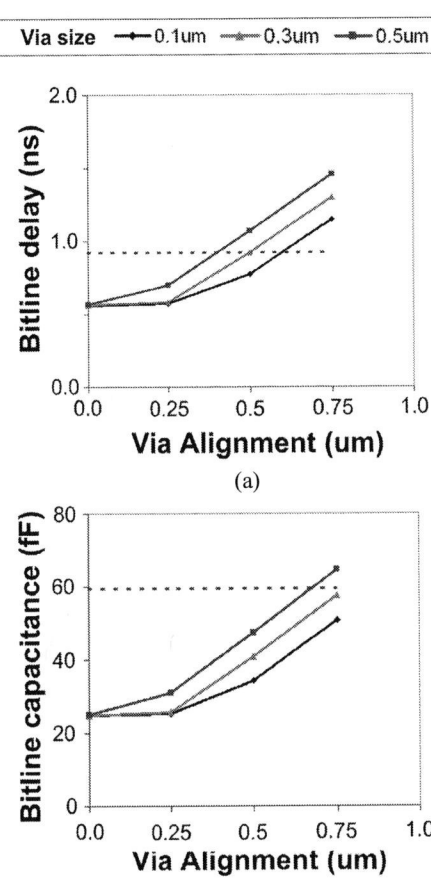

(a)

(b)

Fig. 5: Bit-line (a) capacitance and (b) delay of 3D-SRAM versus alignment accuracy and via size. Horizontal dotted line is the result of 2D-hierarchical SRAM for reference.

Next, we compare the bit-line delay and capacitance when wafer-to-wafer bonding technique is used for 3D integration. Compared to the monolithic integration, wafer-to-wafer bonding requires larger inter-layer vias and a landing pad to accommodate the misalignment between layers. Fig. 5 shows bit-line delay and capacitance versus alignment accuracy and inter-layer via size. As expected, the performance will degrade as the alignment accuracy worsens or inter-layer via size increases. For reference, simulation results of 2D-hierarchical SRAM are shown as horizontal dotted lines. It is important to note that even if the inter-layer via size is relaxed to 0.3 μm and alignment accuracy to 0.25 μm, the performance remain unchanged. This is because the area overhead is originally limited by the select transistors, and consequently, slight increase in the inter-layer size or alignment does not significantly affect the overall capacitance. However, if the alignment degrades to > 0.5um, the benefits start to vanish.

Inter-layer via size of 0.3μm and alignment accuracy of 0.25μm correspond to an inter-layer via pitch of 1μm (=0.3+0.25+0.25+0.2μm), including the pad spacing of 0.2um. Note that inter-layer vias with size of 0.14μm and pitch of

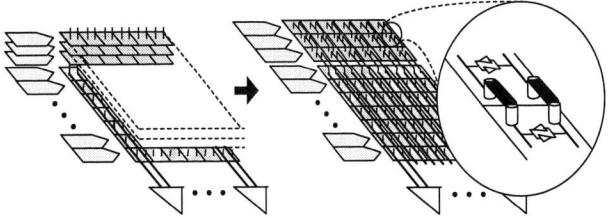

Fig. 6: Emulating 3D-architecture with 2D.

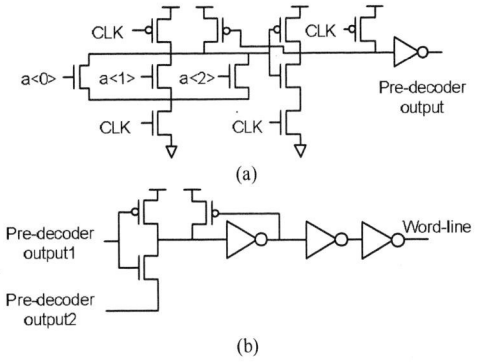

(a)

(b)

Fig. 7: Schematic of (a) Nambu-style dynamic pre-decoder, and (b) source coupled 2-input NAND gate word-line driver. Pre-decoder output1 and 2 are precharged to 0 and 1, respectively.

0.4μm have been demonstrated [8]. Thus our 3D architecture can be fabricated with currently available wafer-to-wafer bonding technology and will still achieve great performance improvement. Of course, the inter-layer via needs to scale correspondingly with the technology node to derive similar benefits in the future.

IV. EXPERIMENTAL RESULTS

In order to verify our simulation, we have created a 32Kb proof-of-concept macro using 0.13μm technology to emulate the 3D architecture. The SRAM array has 128 word-lines, by 256 bit-lines, arranged as a 3D array of 4 layers x 32 word-lines x 256 bit-lines. In the macro, a vertical slice of 4-layer 3D-SRAM cells furthest from the address inputs is transformed into a 2D design and the interlayer via is emulated by two upper level metal vias (Fig. 6). This allows us to evaluate the worst case delay and power of the 3D array. A conventional 2D-SRAM block with the same capacity has also been fabricated for performance comparison. In both designs, current latched sense amplifier (CLSA) is used for data sensing and an 8-to-1 MUX matches the bit-line pitch with the sense amplifier pitch. 3D-decoders that lie on the critical path are also transformed into a 2D design and connected with inter-layer via that is emulated by two upper level metal vias. A Nambu-style dynamic circuit [9] is used to improve performance, power and area. Since the pre-decoder outputs are pulses, a simple source coupled NAND gate is used for the word-line driver (Fig. 7).

978-1-4244-2018-6/08 $25.00 © 2008 IEEE

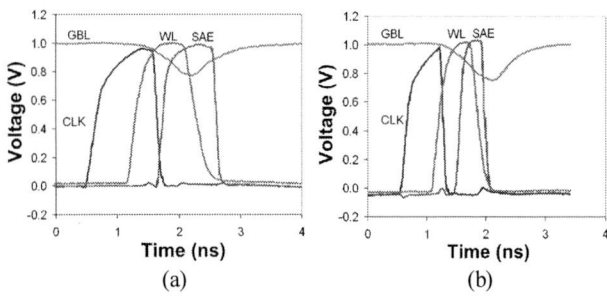

Fig. 8: Reconstructed measured waveforms of (a) 2D and (b) 3D-SRAM

Table 1: Comparison of simulation and measurement results
(Simulation results are calibrated using measured FO4 ring oscillators)

Vdd = 1V		Access time		Energy per cycle (pJ)	
		ns	FO4	Write	Read
Simulation	2D	1.74	14.9	6.56	4.14
	3D	1.27	10.9	2.26	1.86
Measurement	2D	1.86	16.0	6.83	4.23
	3D	1.28	11.0	2.37	2.04
Reduction ratio		1.5	1.5	2.9	2.1

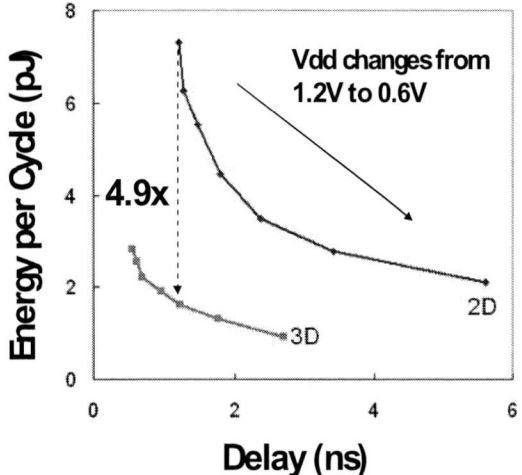

Fig. 9: Energy-delay profile comparison of 2D and 3D-SRAM.

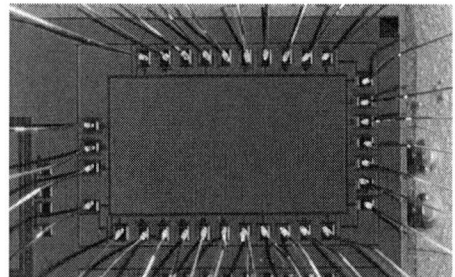

Fig. 10: Die micrograph of the test chip

ACKNOWLEDGEMENT

This work has been supported by DARPA/SPAWAR grant N66001-04-1-8916. Nho has been partially supported by the Samsung Scholarship Foundation.

REFERENCES

[1] Y. Morita et al., "An Area-Conscious Low-Voltage-Oriented 8T-SRAM Design under DVS Environment", VLSI Circ., p.256, 2007

[2] Hiroyuki Yamauchi, "Embedded SRAM design and Nano scale Trend", Advanced Circuit Forum, ISSCC 2008.

[3] S. Wong et al., "Monolithic 3D Integrated Circuits", International Symposium on VLSI Technology, Systems and Applications, p. 66-69, Hsinchu, Taiwan, April 2007.

[4] S. M. Jung et al., "The Revolutionary and Truly 3-D 25F^2 SRAM Technology with the Smallest S3 Cell, 0.16um^2, and SSTFT for Ultra High Density SRAM", Symp. on VLSI Tech., Dig., p.228-229, June, 2004

[5] K. Osada et al., "A 2ns access, 285 MHz, two-port cache macro using double global bit-line pairs", ISSCC Dig., p.402-403, 1997

[6] K. Sakuma et al., "3D Chip Stacking Technology with Low-Volume Lead-Free Interconnections", ECTC, p.627-632, 2007

[7] K. Bernstein et al., "Interconnects in the Third Dimension: Design Challenges for 3D ICs", DAC, p.562-567, 2007

[8] A.W. Topol et al., "Three-dimensional integrated circuits", IEDM, p.352-355, 2005

[9] T. Suzuki et al., "A Stable SRAM Cell Design Against Simultaneously R/W Disturbed Accesses", Symp. On VLSI Circ., Dig., p.11-12. 2006.

[10] Y. Bai, Ph.D. Thesis, 2008

An on-chip waveform sub-sampling circuit [10] has been included to measure the waveforms of clock (CLK), word-line (WL), global bit-line (GBL), and Sense Amplifier Enable (SAE) signals, as shown in Fig 8. At room temperature, with 1V supply, 3D-SRAM achieves 1.5X reduction in access time, while consuming 2.9X less power in write operation and 2.1X less power in read operation. There is more power reduction in write operation because the bit-lines swing fully.

Table 1 compares the simulation and measurement results. When 50% read and write activities are assumed, the 3D SRAM consumes about 5 times less energy per cycle than the 2D SRAM while operating at the same frequency (Fig. 9). These experimental improvements are similar to the 65nm projections discussed earlier. Test chip is shown in Fig. 10.

V. CONCLUSION

A novel 3D-SRAM architecture that derives the maximum benefits of 3D is proposed and demonstrated. This architecture significantly reduces the bit-line capacitance achieving improvement in power and delay over conventional 2D and 2D hierarchical architecture. Experimental emulation of 3D shows that when operating at the same delay performance, 3D-SRAM consumes about 5 times less active power than 2D-SRAM.

IEEE 2008 Custom Intergrated Circuits Conference (CICC)

Early Prediction of Product Performance and Yield Via Technology Benchmark

Choongyeun Cho, Daeik D. Kim
IBM Semiconductor R&D Center
Hopewell Junction, NY
{cycho,dkim}@us.ibm.com

Jonghae Kim
Qualcomm Inc.
San Diego, CA
jonghaek@qualcomm.com

Daihyun Lim
MIT MTL
Cambridge, MA 02139
daihyun@mit.edu

Sangyeun Cho
Univ. of Pittsburgh
Pittsburgh, PA 15260
cho@cs.pitt.edu

Abstract—**This paper presents a practical method to estimate IC product performance and parametric yield solely from a well-chosen set of existing electrical measurements intended for technology monitoring at an early stage of manufacturing. We demonstrate that the components of mmWave PLL and product-like logic performance in a 65nm SOI CMOS technology are predicted within a 5% RMS error relative to mean.**

I. INTRODUCTION

In a deep sub-micron technology, process variability has become a predominant limiter of performance and yield of IC products [1]. At the same time, it becomes increasingly more difficult to model and predict it accurately, partly due to the complicated nature of process variations [2]. While one can use a simplified analytical model or Monte-Carlo circuit simulation to directly estimate a product circuit's performance, this often leads to inaccuracies because the process variation is amplified by specific parameters, rather than the overall parameter distributions. The nonlinearity becomes especially pronounced in modern deep sub-micron devices whose variation is dominated by the device parameters and parasitic components.

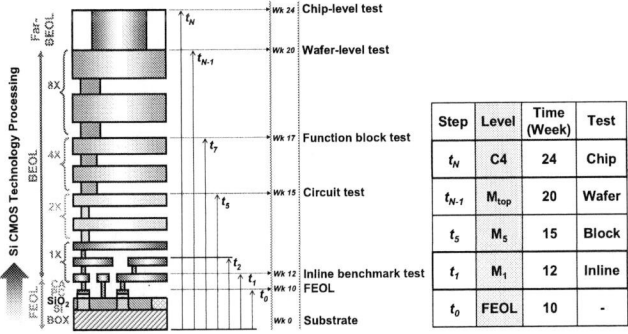

Fig. 1. Schedule for CMOS technology development and testing. The motivation of this work is that the product performance is typically tested only after C4 level is completed, but in-line benchmark measurements at M1, if carefully planned and selected, carry relevant information to reliably predict the product performance.

Fig. 1 shows a typical schedule for CMOS technology development and testing at different levels. Typical elapsed time for key technology process steps (FEOL, M1, M4, top metal and C4) are arranged in the table. The purpose of this

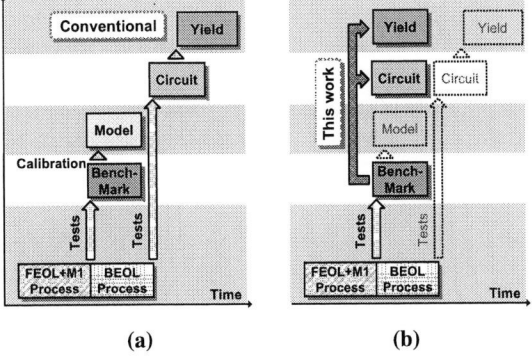

Fig. 2. (a) A typical IC development flow in terms of process, benchmark, model, and circuit/product test. (b) Flow of the proposed early prediction of product performance and yield based on manufacturing in-line benchmark structures. Note that through prediction of product performance/yield from benchmark, the time and cost for testing/qualification reduce significantly.

work is to reliably estimate the performance and yield of a product at an early stage of IC manufacturing, especially at the first metal level (M1).

Often, accurate product performance and yield information becomes available only after dicing and packaging. Fig. 2(a) illustrates a typical IC development time line that involves different development components – process, benchmark, model, and a product circuit. A model and benchmarking are strongly tied together at all times through many rounds of model calibration via model-to-hardware correlation (MHC). Circuit performance and yield are conventionally measured and qualified after the back-end-of-line (BEOL) processes are completed. This long cycle of production to yield hinders fast yield learning and product ramp-up, adding greatly to the product development and manufacturing cost. For early estimation, we utilize a well-chosen set of existing electrical measurements from *manufacturing in-line benchmark structures* (MIBS). The MIBS in this paper collectively refer to assorted test structures that are measured in a manufacturing line using a standard parametric tester for the purpose of defect diagnosis, DC device characterization, and MHC [3]. Typical MIBS include test structures for MOSFET devices having different sizes and layouts, ring oscillators (ROs), resistance/capacitance structures. MIBS data are typically collected

978-1-4244-2018-6/08 $25.00 © 2008 IEEE   223

in an automated fashion for all or most wafers at a lower metalization level, usually at M1. A rationale for utilizing measured MIBS data is that the process-induced variability in key technology parameters, such as threshold voltage (V_{th}), oxide thickness (T_{ox}), and gate length (L_{poly}) dominates the variation of circuit performance. We prudently select a set of MIBS at early stage of manufacturing (e.g. M1) such that the variability of these key physical parameters is embedded therein. The proposed methodology is validated using an ensemble of MIBS data collected from test structures built with a 65nm SOI CMOS technology.

While a prior work used MIBS data to identify the most correlated device characteristics with respect to circuit performance [4], MIBS have seldom been exploited for the purpose beyond what they are intended. FET devices and ROs are popularly used to characterize CMOS circuit performance and variability [5]. However, there has been little practice for a statistical approach to characterize and qualify circuit product's performance using a collection of heterogeneous MIBS measurements, mainly due to the increasing complexity and nonlinearity of the technology-product relationship.

We anticipate that the method proposed in this work will have large, practical impact on product yield-learning acceleration and testing time/cost reduction. An early prediction of a product circuit performance and yield based on MIBS data allows rapid yield learning, thus saving development cost and time (see Fig. 2(b)). Also to design community, the mapping from device parameters to product performance metric can provide pertinent feedback in order to modify circuit more tolerant to particular process variation source.

The rest of this paper is organized as follows. In Section II, we describe the proposed estimation method in detail. In Section III, our estimation method is evaluated using two test vehicles, an RF frequency divider and an RO, which are representative of important building blocks for clocking and logic product, respectively. A summary will follow, in the conclusion.

II. Proposed Performance/Yield Prediction

In this section, we briefly introduce a multivariate statistical technique to estimate circuit performance and yield from MIBS measurements. The method is based on feature extraction and nonlinear estimation, widely used in semiconductor manufacturing industry [6], [7]. Feature vectors are extracted in order to reduce the dimensionality of input data. A nonlinear estimation uses an ensemble-based training of a neural network. Neural nets are used to learn and compute functions for which the relationships between input and output vectors are unknown or computationally complicated which is generally true for product circuit parameters with respect to device-level parameters.

A. Projected Principal Component Analysis

The number of MIBS parameters monitored on a regular basis can be on the order of thousands or tens of thousands, and it is not feasible to stably and reliably train an estimator

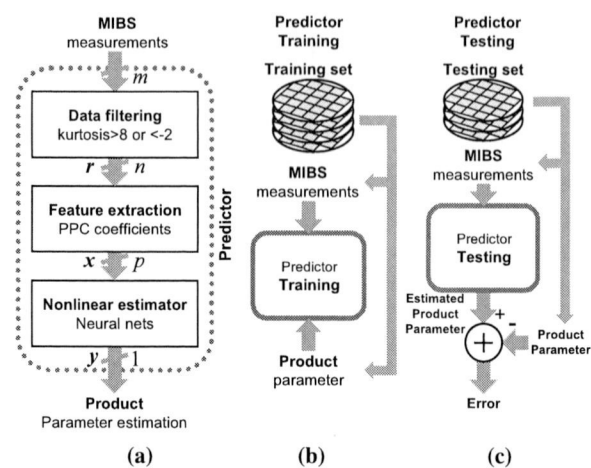

Fig. 3. (a) Flow chart of the proposed nonlinear estimator for circuit parameters. (b) Training of the estimator. (c) Testing scheme.

based on all raw MIBS measurements. A variant of principal component analysis (PCA) is employed to extract features in all MIBS data. The MIBS measurements are projected onto the subspace of circuit performance parameter(s) *while preserving the most information relevant to the circuit parameters.* Hence, this transform is called *projected* PCA.

B. Performance/Yield Prediction Algorithm

Fig. 3(a) illustrates the proposed nonlinear estimation algorithm. First, the input MIBS data of m parameters is screened to include only usable data of size n. In this work, we employed a simple fourth-momentum test to filter out data with abnormal distribution. Second, the resulting data is compressed to a manageable number of parameters, p. Projected PCA is used to extract feature vectors. The selection of how many PC's to retain (p) for the subsequent steps is based on a trade-off between the accuracy of estimation and the stability of the estimation training (fitting). Third, feature vectors are put into a nonlinear estimator to predict circuit performance parameter(s).

Fig. 3(b) shows how the estimation algorithm is trained. Using a pair of *actual* MIBS measurements (input) and product circuit parameter (output), the weights of the neural networks are adjusted. Fig. 3(c) illustrates how the algorithm is tested; the error will be calculated based on predicted and actual circuit parameters. We use percentage RMS error as a metric for estimation error.

Yield estimation (binary detection) is an extension of circuit parameter prediction (continuous estimation). After estimating circuit performance parameter, a yield is calculated as the fraction of samples that meet pre-defined specifications. For example, this method accommodates design specifications such as maximum gate delay and maximum current (power) in the case of an RO.

TABLE I
CATEGORY OF M1 MIBS USED IN THIS EXPERIMENT.

Test structure category	# of parameters (before screening)	# of parameters (after screening)
FET	1,988	759
Ring oscillator	248	83
SRAM	398	159
Capacitance	222	108
Total	2,856	1,109

III. PRODUCT PERFORMANCE/YIELD PREDICTION

For our experiments, 65nm SOI CMOS technology data is used. Table I categorizes the MIBS used in our experiments. All MIBS used in the experiments are at the first metal level (M1).

The number of parameters at each stage in Fig. 3 is: $m=2,856$ (the number of all MIBS parameters), $n=1,109$ (the number of MIBS parameters after screening), and $p = 10$ (the number of PPC's used for the estimation). 390 chips (from 10 wafers and 39 chips per wafer) were used for fitting a linear regression, or training the neural nets, and 2 other wafers (78 chips) were used for the testing purpose. The neural net configuration used in these experiments include one hidden layer with 4 nodes ($k = 4$). The number of training samples are more than 5 times the degrees of freedom in the neural net estimator, exceeding the general rule of thumb for neural net training set size.

We used two common types of product circuits for the validation of the proposed methodology: a CML frequency divider operating at up to 90GHz, and a static RO operating at around 1GHz. They in general represent essential building blocks for clocking and logic products. Fig. 4 illustrates approximate locations of the frequency divider, RO and MIBS used in this experiment. The RF divider and RO's are adjacent within a die, and MIBS are clustered in two large columns.

A. mmWave System Clock Component

A typical phase-locked loop (PLL) block uses a loop structure to lock a free-running voltage-controlled oscillator (VCO) to a desired frequency. A primary frequency divider is one of the key PLL components because it must reliably divide VCO operating frequency into a desired lowered frequency as mmWave circuit [8]. Self-oscillation frequency (f_{SO}) is an important figure-of-merit for a frequency divider because it is closely linked to a maximum dividable bandwidth and a sensitivity curve, and is also practical for measurement.

The percentage error (RMS error normalized to mean value) for f_{SO} is arranged in Table II for linear fit and neural net based nonlinear estimator. Estimation errors are approximately 5% of its mean for all estimation schemes. A yield, here, is defined as percentage of chips that meet the product specification: f_{SO} is higher than 34GHz and active current is lower than 27mA. The neural net estimator is slightly more accurate than linear fit due to its flexibility and robustness to highly nonlinear relationship of input (device-level parameters) and output (circuit-level parameters). Neural net is, however,

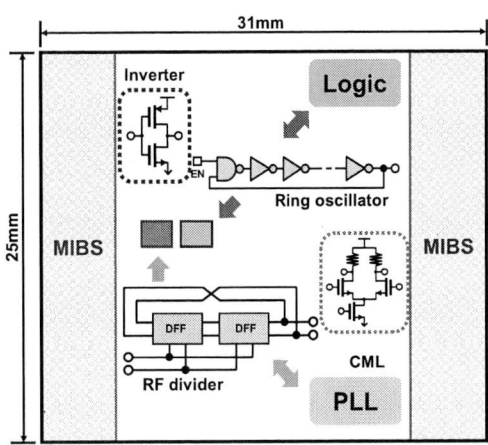

Fig. 4. Die size and approximate locations of product circuits (test vehicles for the validation of the proposed method) and MIBS.

superior for the yield estimation where nonlinearity is more severe. Thus, neural net is preferred overall for estimation of both performance metric and yield.

B. Microprocessor Logic Product

ROs are commonly used as logic-type test vehicles for MHC and variability monitoring. Gate delay (τ) is known to be representative of product environment [9], thus was selected as a subject of the proposed estimation. In this experiment, τ of 101-stage inverter-based ROs was estimated using the same MIBS data as in the previous subsection. (83 RO-related MIBS parameters were excluded in experiment for the fairness of prediction.) Table III presents the normalized RMS estimation errors. A yield for this case is defined as ratio of chips satisfying the specification: gate delay is less than 4.1ps and active current is lower than 2.2mA.

Using mostly FET characteristics, gate delay was estimated within a 4% error for all estimation schemes. The estimation result for yield shows a similar trend as that of the frequency divider: Neural network outperforms linear and second-order regression fits because yield is a highly nonlinear function of design parameters. There are a number of factors contributing to the estimation error: (1) BEOL process variability (above M1) is excluded for estimation since MIBS only up to M1 level were taken into account; (2) Intra-die variation between test vehicles and MIBS was not considered. In spite of these factors, the prediction accuracy obtained with our methodology is reasonably high.

C. Validation of Prediction Robustness

The proposed method uses only the statistics of hardware measurements without considering an underlying physical model. Validity of the estimation errors in Tables II, III can be questioned, especially if training and testing sets are *inbred* in nature, or deliberately selected to minimize estimation error. In this subsection, the robustness of the presented estimation method is evaluated via different combinations of training

TABLE II
ESTIMATION ERROR FOR RF FREQUENCY DIVIDER (SEE SECTION III-A).

Parameter type	Mean	Std dev	Linear fit %err	Neural net %err
Self-oscillation frequency (f_{SO})	34.6GHz	2.9GHz (8.3%)	5.1%	4.6%
I_{dd}	24.0mA	3.6mA (15.1%)	10.1%	10.0%
Yield	27.5%	–	19.3%	17.7%

TABLE III
ESTIMATION ERROR FOR RING OSCILLATOR (SEE SECTION III-B).

Parameter type	Mean	Std dev	Linear fit %err	Neural net %err
RO gate delay (τ)	4.3ps	0.37ps (8.7%)	3.8%	3.5%
$I_{A,RO}$	2.5mA	0.28mA (11.1%)	5.5%	5.1%
Yield	25.6%	–	15.6%	13.1%

TABLE IV
f_{SO} ESTIMATION ERROR FOR FOUR TESTING/TRAINING SCENARIOS.

Training set	Testing set	Percentage error
Slow	Slow	4.5%
Slow	Fast	5.6%
Fast	Slow	5.5%
Fast	Fast	4.9%

set and testing set. Using f_{SO} as a criterion, 12 available wafers are divided into two groups of equal size – faster and slower wafers, relative to the threshold f_{SO}=35GHz. Due to the limited sample size, a linear regression fit was used in this experiment. Better estimation accuracy is expected with a neural net nonlinear estimator if a larger data set is available.

The resulting estimation errors are shown in Table IV. Percentage errors for all training/testing combinations are in a narrow range from 4.5% to 5.6%, showing that the proposed estimation method is robust and can predict circuit performance and yield accurately even when significant wafer variations are present.

D. Feedback to Circuit Design and Technology

Previously, we have validated the proposed methodology via two test circuits, an RF frequency divider and an RO. The mapping from MIBS parameters to product performance metric (learned from training of an estimator) can lead to useful insight as to what device characteristics are most sensitive to product performance, and to what degree. In addition to conventional circuit simulation and analysis, a designer can leverage this information to make circuit more robust to process variation, and to make an educated trade-off between design parameters. For example, the three most dominating device characteristics with relation to the RO delay are found to be FET on / off currents and threshold voltage (obtained by a linear estimation in Section III-B) with the relationship:

$$\tau \propto I_{on} + 0.817 I_{off} + 0.765 V_{th} + \epsilon \qquad (1)$$

where each parameter is normalized to zero-mean and unit-variance.

IV. CONCLUSION

An efficient statistical method was presented to predict circuit performance and yield based on MIBS intended for technology monitoring and device characterization. It predicts the mmWave frequency divider performance within 5% error, and the RO gate delay within 4% error. The significances of this work are: (1) For the first time, we verified the potential of utilizing existing up-to-M1 device-level measurements to directly predict product circuit's performance and yield without assuming any physical model. (2) In-line measurements are available early in the fabrication cycle, e.g., at M1 level. Hence, the performance/yield of a complex, customized RF circuit can be predicted while still in a manufacturing line, allowing significant reduction of the cost and time for product circuit testing. (3) The mapping from device parameters to product performance metric is useful to designers to make a circuit more robust to process variation.

ACKNOWLEDGEMENT

The authors thank IBM SRDC engineers P. O'Neil, C. Schnabel, K. Warren, K. Ginn, and G. Patton for their support, and D. Boning at MIT for valuable comments.

REFERENCES

[1] S. Nassif, "Modeling and analysis of manufacturing variations," in *Proc. IEEE CICC*, 2001, pp. 223–228.

[2] S. Samaan, "The impact of device parameter variations on the frequency and performance of VLSI chips," in *Proc. IEEE/ACM ICCAD*, 2004, pp. 343–346.

[3] M. Ketchen, M. Bhushan, and D. Pearson, "High speed test structures for in-line process monitoring and model calibration," in *Proc. IEEE ICMTS*, 2005, pp. 33–38.

[4] C. Cho, D. Kim, J. Kim, J.-O. Plouchart, and R. Trzcinski, "Statistical framework for technology-model-product co-design and convergence," in *Proc. ACM/IEEE DAC*, San Diego, CA, 2007, pp. 503–508.

[5] M. Ketchen and M. Bhushan, "Product-representative "at speed" test structures for CMOS characterization," *IBM J. Res. & Dev.*, vol. 90, no. 4/5, pp. 451–468, Jul/Sep 2006.

[6] D. White, D. Boning, S. Butler, and G. Barna, "Spatial characterization of wafer state using principal component analysis of optical emission spectra in plasma etch," *IEEE Trans. Semiconduct. Manufacf.*, vol. 10, no. 10, pp. 52–61, Feb 1997.

[7] F. Chen and S. Liu, "A neural-network approach to recognize defect spatial pattern in semiconductor fabrication," *IEEE Trans. Semiconduct. Manufacf.*, vol. 13, no. 3, pp. 366–373, Aug 2000.

[8] D. Lim, J. Kim, J.-O. Plouchart, C. Cho, D. Kim, R. Trzcinski, and D. Boning, "Performance variability of a 90GHz static CML frequency divider in 65nm SOI CMOS technology," in *IEEE ISSCC Dig. Tech. Papers*, 2007, pp. 542–621.

[9] M. Bhushan, M. Ketchen, S. Polonsky, and A. Gattiker, "Ring oscillator based technique for measuring variability statistics," in *Proc. IEEE ICMTS*, 2006, pp. 87–92.

978-1-4244-2018-6/08 $25.00 © 2008 IEEE

A FPGA Vernier Digital-to-Time Converter with 3.56*ps* Resolution and -0.23~+0.2LSB Inaccuracy

Poki Chen, Juan-Shan Lai and Po-Yu Chen

National Taiwan University of Science and Technology

Abstract- A simple but powerful digital-to-time converter, or digital pulse generator, realizable with FPGA chips is proposed. Based on vernier principle, the effective resolution is made equivalent to the period difference of two phase-locked loop (PLL) outputs and is achieved as fine as 3.56*ps* with Altera Stratix II GX FPGA chips. The programmable delay range wider than ever is verified to be 33.4 minutes. The measured integral nonlinearity (INL) is between -0.23LSB to 0.2LSB (-0.8*ps*~0.7*ps*). It ensures every input bit is valid under such fine resolution. Only two embedded PLLs and some standard logic cells are required for circuit realization. Compared with its predecessors', both design effort and chip cost of the proposed converter are lowered substantially.

I. INTRODUCTION

The digital-to-time converter (DTC), one of the most important cores of automatic test equipments (ATE), measurement instruments and digital testing systems, generates a time signal with a width proportional to the programmed input value. It is extensively used by digital IC BIST (built-in self-test) applications for cost reduction. Due to different operational principles adopted, the digital-to-time conversion can be fulfilled by the absolute [1,5,7] or relative [2-4,6] time generation. The absolute time generation can be utilized to provide wide delay range with low offset. However the demerits are comparatively poor resolution and more sensitivity to PVT (process, voltage and temperature) variations. For the relative time generation, the effective resolution equals the delay difference between different transmission paths. The resolution can be made extremely fine, but the performance is easily hampered by the path or device mismatch. With 0 input, the delays of different paths cannot be made exactly the same after fabrication and there always exists a large offset.

For performance enhancement, the conventional DTCs were realized with GaAs or Bipolar processes based on the absolute time generation principle. Limited by the inherent accuracy of the DAC and comparator used, the achieved measurement range and resolution were 15.875ns and 125ps only [1]. Later, the programmable vernier delay line and calibration RAM were utilized to improve the resolution to 40ps [2]. Furthermore, the vernier delay lines were replaced with the delay matrices which composed of multiple delay cells with rather small delay differences and the multiplexers were adopted to vary the effective transmission path according to the programmed delay. The operation ranges were restricted by the maximum achievable delay of the delay matrices. Although an effective resolution of 8*ps* could be realized [3,4], the circuits faced serious device mismatch problems and thus had poor nonlinearity errors.

To boost the circuit integration and cut down the fabrication cost and power consumption, CMOS processes were gradually adopted for DTC design. By using a single cyclic delay line and 8× phase interpolators, a CMOS DTC was proposed to get 37.5*ps* resolution and 5*ms* programmable delay range. However, the INL was as large as 0~7LSB which could only be reduced to ±0.4LSB with chip-by-chip calibration [5]. Similarly, a CMOS DTC was invented to reduce the device mismatch impact by storing calibration data in high speed SRAM. It owned 100~400MHz operation frequency and 19.5*ps* resolution. But the operation range was merely 2.5*ns* and the INL still reached 35*ps* [6]. Due to the use of large quantity of calibration RAM, the chip size was increased tremendously. More address/data bits of calibration RAM are required to achieve finer resolution. A 100-fold increase in the calibration RAM size might be required to achieve a resolution of several picoseconds [7]. Another novel DTC was presented to get rid of the calibration RAM through the utilization of DLL to conquer the problems caused by PVT variations. The resolution and INL reached 1.83 and 3.89*ps* respectively [7]. However, the monotonicity could only be ensured by using 1-stage current-controlled high-linearity fine-delay circuit with delay adjusted by two DACs. Also, the active noise cancellers were required to eliminate I_{DD}/I_{SS} fluctuations and a PLL/DLL multiple feedback system was utilized to eliminate the timing drift and jitter. It made the circuit rather complicated and very hard to design. Moreover, the conventional DTCs depend on full-custom design which is both time and human power consuming. For the reduction of design time and fabrication cost, a vernier DTC realizable with nowadays FPGA chips is proposed to alleviate the need of full-custom design. It will be proven to promise a resolution as fine as 3.56*ps* and an INL less than ±0.23LSB.

The remainder of the paper is organized as follows. Section II describes the operation principle of the proposed circuit. Section III details the circuit structure. Section IV discusses important FPGA implementation issues and presents the measurement results. A summary of the paper is given in Section V.

II. OPERATION PRINCIPLE

The vernier principle is widely applied to time-to-digital converters (TDCs) with a typical timing diagram shown in Fig. 1(a) [8,9]. When Start signal arrives, T_s signal is triggered to oscillate. Similarly, the occurrence of Stop signal activates T_f signal to vibrate with a period slightly shorter than T_s. Since T_f signal oscillates a little bit faster than T_s signal, it will catch up

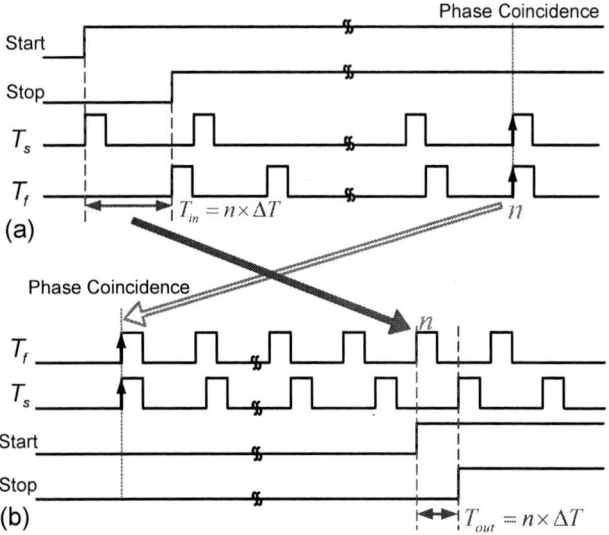

(a)

(b)

Fig. 1 Timing diagram of (a) the vernier TDC, (b) the proposed vernier DTC.

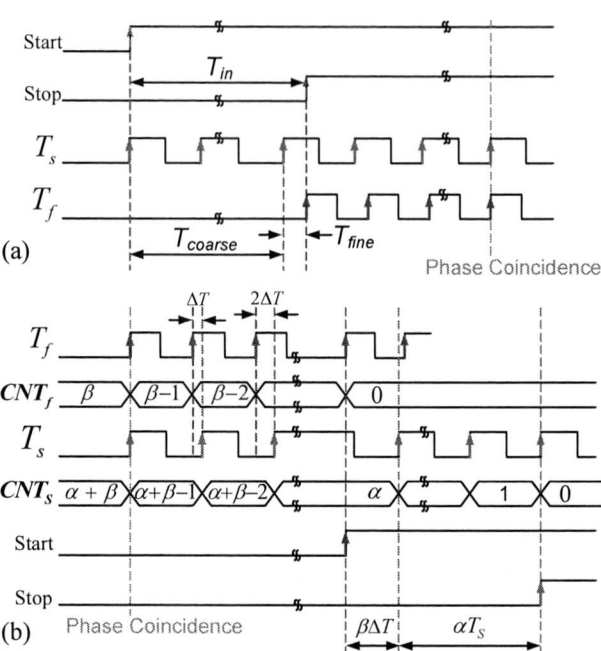

(a)

(b)

Fig. 2 Timing diagram of (a) the single-stage vernier TDC with extended operation range and (b) the modified vernier DTC.

with T_s signal after some oscillations. At the phase coincidence, the input time width can be calculated as

$$T_{in} = n\Delta T , \qquad (1)$$

where n and ΔT are the number of oscillations before phase coincidence and the difference between T_s and T_f respectively. Since the resolution ΔT equals the period difference, it can be made extremely fine even with low operation frequencies. However, the measurement range is limited to one T_s. If the TDC operation is reversed as depicted in Fig. 1(b), the vernier principle can be applied likewise to high accuracy DTC. Both T_s and T_f signals oscillate continuously with small period difference ΔT. After the phase coincidence of T_s and T_f, each oscillation will add one more ΔT delay between T_s and T_f signals. If T_f and T_s signals are programmed to oscillate n cycles to activate Start and Stop signals correspondingly after phase coincidence, the width of output interval can also be derived as

$$T_{out} = n\Delta T , \qquad (2)$$

which is exactly proportional to the input value n and successfully fulfills the function of digital-to-time conversion.

It deserves noticing that the width of output interval is no longer limited to one T_s since no restriction is imposed on the value of n which can be set to any value as required. However, the output latency reaches nT_f which will be turned out to be unbearable for large n. Fortunately, a single-stage vernier TDC with self interpolation was invented to substantially extend the operation range with the timing diagram illustrated in Fig. 2(a) [10]. After the arrival of Start signal, a coarse counter was adopted to count the oscillations of the activated T_s signal until Stop signal arrived. Then the other fine counter was stimulated to count the racing oscillations between T_f and T_s signals before the phase coincidence. The input time width was derived as

$$T_{in} = T_{coarse} + T_{fine} = \alpha T_s + \beta \left(T_s - T_f\right) = \alpha T_s + \beta \Delta T , \quad (3)$$

where α and β represent the counting values of the coarse and fine counters accordingly. The operation principle can also be

reversely applied to DTC for output latency reduction as depicted in Fig. 2(b) where CNT_f and CNT_s represent the down counters clocking by T_f and T_s signals respectively to count the cycles before setting Start and Stop signals. After phase coincidence, both counters start counting and the first β cycles of T_f and T_s signals are used to generate $\beta\Delta T$ delay difference between Start and Stop signals. Then, CNT_s is allowed to count extra α cycles for producing additional αT_s delay difference. The width of output signal becomes

$$T_{out} = \left(\alpha + \beta\right)T_s - \beta T_f = \alpha T_s + \beta \left(T_s - T_f\right) = \alpha T_s + \beta \Delta T . \ (4)$$

For a required output width $n\Delta T$ in Fig. 1(b) or equivalent $\alpha T_s + \beta \Delta T$ in Fig. 2(b), we have

$$\alpha = \left\lfloor \frac{n}{T_s / \Delta T} \right\rfloor, \ \beta = n \ \bmod \ \frac{T_s}{\Delta T}, \qquad (5)$$

where $\lfloor x \rfloor$ denotes the largest integer less than or equal to x and $y \bmod z$ computes the remainder of $y \div z$. Usually, ΔT is designed to be much smaller than T_s. The output latency is reduced to merely βT_f which is tremendously less than nT_f for wide output.

III. CIRCUIT DESCRIPTION

Fig. 3 shows the block diagram of the proposed DTC which realizes the digital-to-time conversion function described in Fig. 2(b). Since the rise edges of T_s and T_f signals can be synchronized to that of T_{ref} signal by phase-locked loops PLL_s and PLL_f, each rise edge of T_{ref} signal indicates one phase coincidence of T_s and T_f signals. A period signal generator is added to set the output repetition rate and two output pulse generators are employed to generate Start and Stop signals with a delay difference set by the DTC input. The period signal generator can be simply realized by a reloadable

978-1-4244-2018-6/08 $25.00 © 2008 IEEE   228

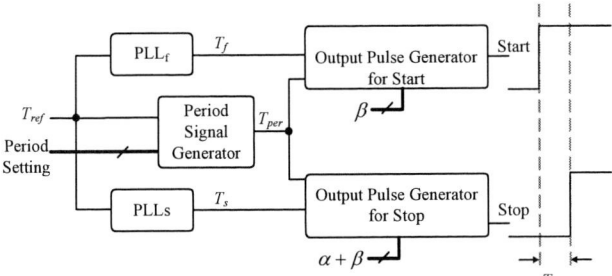

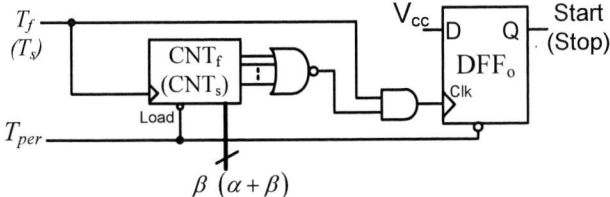

Fig. 3 Block diagram of the proposed DTC.

Fig. 4 Schematic of the output pulse Generator.

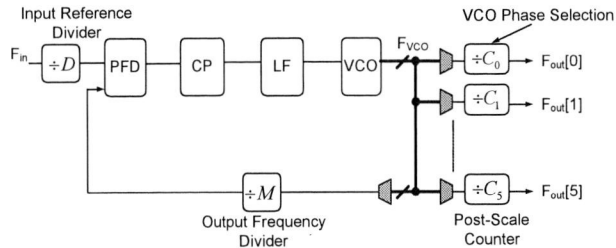

Fig. 5 Simplified block Diagram of PLL in Altera FPGA.

Table I
PLL SPECIFICATIONS FOR ALTERA STRATIX II GX FPGA

Fabrication Process	90nm
Input frequency (F_{in})	2~500MHz
Output frequency (F_{out})	2~550MHz
Input reference divisor (D)	1~512
Output frequency divisor (M)	1~512
Post-scale counter divisor (C)	1~512

down counter with a reloading cycle equal to one half of the desired output period followed by a divide-by-two counter to generate the period signal T_{per} with 50% duty cycle. The circuit of output pulse generator is depicted in Fig. 4. Before the rise edge of T_{per} signal, the preset counting value β (or $\alpha+\beta$) is loaded into CNT_f (or CNT_s). After 0 count is reached, the output D-type flip flop DFF_o will be triggered to set Start/Stop signal to 1 as required in Fig. 2(b) to make the output width equivalent to $\alpha T_s + \beta \Delta T$ exactly.

Since both coarse delay αT_s and fine delay $\beta \Delta T$ are generated by the same phase-locked loops, the coarse to fine resolution ratio $T_s/\Delta T$ will be accurately maintained by the negative feedback mechanism of phase-locked loops. It assures the performance of the proposed DTC insensitive to PVT variations. Assume the divisors of PLL_f and PLL_s frequency dividers (or prescalers) in Fig. 3 are designed to be A and B respectively, the periods T_f and T_s can be derived as

$$T_f = \frac{T_{ref}}{A}, \ T_s = \frac{T_{ref}}{B}. \quad (6)$$

With A larger than B, the effective resolution becomes

$$\Delta T = T_s - T_f = \frac{A-B}{AB} T_{ref}. \quad (7)$$

The coarse to fine resolution ratio $T_s/\Delta T$ equals

$$\frac{T_s}{\Delta T} = \frac{A}{A-B}. \quad (8)$$

For a given output width $n\Delta T$, the corresponding values of α and β can be derived from (5) to be

$$\alpha = \left\lfloor \frac{n}{A/(A-B)} \right\rfloor, \ \beta = n \ mod \ \frac{A}{A-B}. \quad (9)$$

To ease the decomposition calculation of α and β in (9), the divisors of PLL_f and PLL_s frequency dividers had better be designed as 2^K and 2^K-1 respectively. T_f and T_s become $T_{ref}/2^K$ and $T_{ref}/(2^K-1)$. By (7), the effective resolution can be derived as

$$\Delta T = T_{ref} / 2^K (2^K - 1), \quad (10)$$

which can be made extremely fine with large K. The coarse to fine resolution ratio $T_s/\Delta T$ becomes 2^K exactly. By (9), α and β are recalculated as

$$\alpha = \left\lfloor \frac{n}{2^K} \right\rfloor, \ \beta = n \ mod \ 2^K. \quad (11)$$

For an N-bit DTC, α and β can be simplified as the (N-K)-bit MSB value $[D_{N-1}:D_K]$ and the K-bit LSB value $[D_{K-1}:D_0]$ correspondingly. The complicated division hardware illustrated in (9) is no longer required.

IV. FPGA IMPLEMENTATION AND MEASUREMENT RESULTS

Except for phase-locked loops, all sub-circuits in Fig. 3 only utilize standard digital logics which can be readily implemented with FPGA chips. Fortunately, there are usually several embedded phase-locked loops which can be used as PLL_s and PLL_f in nowadays FPGA chips. By full FPGA realization, the design effort and cost of the proposed DTC can be significantly reduced. For function verification and performance evaluation, Altera Stratix II GX FPGA is adopted for circuit implementation. Since the DTC accuracy is dominated by the performance of phase-locked loops, the simplified PLL block diagram of the Stratix II GX FPGA is re-plotted in Fig. 5 along with the important design parameters listed in Table I for easy reference. To achieve the finest resolution, the divisors (M) of PLL output frequency dividers should be designed the closest to each other.

For Altera Stratix II GX series, the PLL output frequency F_{out} is limited to 550MHz. The finest resolution can be gotten by setting the input frequency F_{in} to 2.148MHz, the output frequency devisor M of PLL_s/PLL_f to 511/512, the input reference divisor D to 2 and the post-scale counter divisor C to 1. The output frequency can be derived as

$$F_{out} = F_{in} \div D \times M \div C. \quad (12)$$

We have

$$T_f = \frac{1}{2.148 \div 2 \times 512 MHz} = 1.81818ns$$

978-1-4244-2018-6/08 $25.00 © 2008 IEEE   229

$$T_s = \frac{1}{2.148 \div 2 \times 511 MHz} = 1.82174ns$$

The effective resolution reaches

$$\Delta T = T_s - T_f = 3.56ps \cdot \qquad (13)$$

For such an extraordinary fine resolution, the DTC output width was still measured from $1\Delta T$ to $2T_s$ for every input value to validate both coarse and fine resolutions and reveal the excellence of the propose circuit. The reference clock was generated by Agilent 81130A 400/660MHz Pulse/Pattern Generator. The delay difference between Start and Stop signals was accurately measured by Tektronix DPO 70404 digital oscilloscope with 25GS/s real time sample rate. Fig. 6 illustrates the output interval width versus the digital input value. Unlike the conventional versions, the proposed DTC has no device mismatch problem and thus possesses excellent linearity. The integral nonlinearity is verified to be merely -0.23~0.2LSB as depicted in Fig. 7. It ensures that every input bit is valid. The DTC functions well with 49 input bits and achieves the widest operation range of 33.4 minutes among its predecessors.

V. CONCLUSION

The proposed DTC utilizes a single vernier delay stage to realize the digital-to-time conversion function. Since the DTC fully adopts close-loop operation which is stabilized by two PLLs, both of the coarse and fine resolutions are promised to be insensitive to PVT variations. The finest effective resolution is verified to be 3.56ps for Altera Stratix II GX FPGA chips. Moreover, the programmable delay range is as large as 33.4 minutes with 49 functioning input bits. The measured INL is -0.23~+0.2LSB and thus every input bit of the DTC is ensured to be valid. The performance is even better than those of some commercial digital pulse generators with list prices over tens of thousand US dollars [12]. It makes the proposed DTC excellent for low cost but high accuracy instrumentation or testing applications.

ACKNOWLEDGMENTS

The authors would like to thank National Chip Implementation Center (CIC) for the support of design tools and National Science Council for the financial support under Grant NSC 96-2221-E-011-151. They would also want to thank GALAXY Taiwan for the useful design discusses and valuable help in FPGA implementations.

REFERENCES

[1] S. Katsu, T. Ueda, M. Kazumura and G. Kano , "A GaAs programmable timer with 125-ps delay time resolution," *IEEE ISSCC*, pp. 16~17, Feb 1988.

[2] C.-W. Branson, "Integrated Pin Electronics for a VLSI test system," *IEEE Transaction on Industrial Electronics*, vol. 36, pp. 23-27, MAY.1989.

[3] T.-I. Otsuji and N. Narumi, "A 10-ps resolution , process-Insensitive timing generator IC," *IEEE J. Solid-State Circuits*, vol. 24, No. 5, OCT. 1989.

[4] T. Otsuji, N. Narumi, "A 3-ns range, 8-ps resolution, timing generator

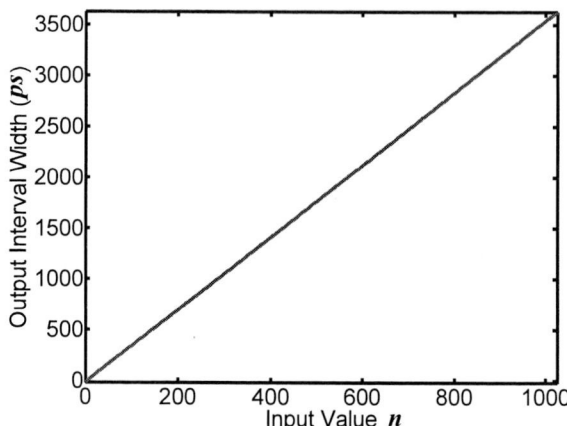

Fig. 6 Measurement result for output interval width vs. digital input value.

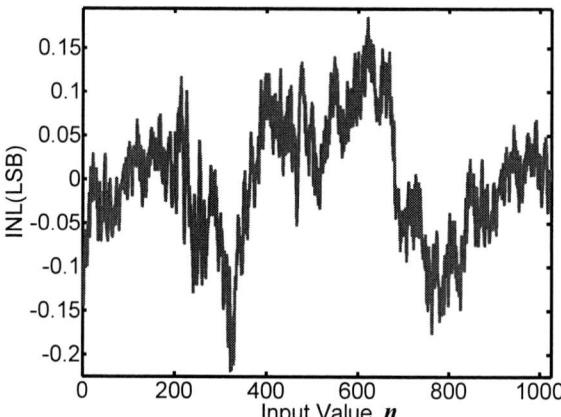

Fig. 7 Measured INL error.

LSI utilizing Si bipolar gate array," *IEEE J. Solid-State Circuits*, vol. 26, pp. 806 – 811, May 1991.

[5] T.-Y. Wang, S.-M. Lin and H.-W. Tsao, "Multiple Channel Programmable Timing Generators With Single Cyclic Delay Line," *IEEE Trans. Instrum. Meas.*, vol. 53, pp.1295-1303, Aug 2004.

[6] B. Arkin, "Realizing a Production ATE Custom Processor and Timing IC Containing 400 Independent Low-Power and High- Linearity Timing Verniers," *IEEE ISSCC*, pp.348-349, Feb 2004.

[7] T. Okayasu, M. Suda, K. Yamamoto, S. Kantake, S. Sudou and D. Watanabe, "1.83ps-Resolution CMOS Dynamic Arbitrary Timing Generator for >4GHz ATE Applications," *IEEE ISSCC*, pp. 2122- 2131, Feb 2006.

[8] T.-I. Otsuji, "A picosecond-accuracy, 700-MHz range, Si bipolar time interval counter LSI," *IEEE J. Solid-State Circuits*, vol. 28, pp. 941-947, Sept. 1993.

[9] T.-E. Rahkonen and J.-T. Kostamovaara, "The use of stabilized CMOS delay line for the digitization of short time intervals," *IEEE J. Solid-State Circuits*, vol. 28, pp. 887–894, Aug. 1993.

[10] P. Chen, C.-C. Chen, J.-C. Zheng and Y.-S. Shen, "A PVT Insensitive Vernier-Based Time-to-Digital Converter with Extended Input Range and High Accuracy," *IEEE Trans. on Nuclear Science*, vol. 54, pp. 294-302, Apr 2007.

[11] N. Weste and K. Eshraghian, "Principles of CMOS VLSI design," *Addison Wesley Longman*, 2nd Edition, 1994.

[12] 81110A 165/330MHz Pulse/Data Generator, available from the Agilent Corporation, www.agilent.com.

IEEE 2008 Custom Intergrated Circuits Conference (CICC)

RASP 2.8: A New Generation of Floating-gate based Field Programmable Analog Array

Arindam Basu*, Christopher M. Twigg†, Stephen Brink*, Paul Hasler*, Csaba Petre*, Shubha Ramakrishnan*, Scott Koziol*
and Craig Schlottmann*

*School of Electrical and Computer Engineering, Georgia Institute of Technology, Atlanta, Georgia 30332
†Department of Electrical and Computer Engineering, Binghamton University, Binghamton, New York 13902
Email: {arindamb,phasler}@ece.gatech.edu, ctwigg@binghamton.edu

Abstract—The RASP 2.8 is a very powerful reconfigurable analog computing platform with thirty-two computational analog blocks(CABs). Each CAB has a wide variety of sub-circuits ranging in granularity from multipliers and programmable offset wide linear range Gm blocks to NMOS and PMOS transistors. The programmable interconnects and circuit elements in the CAB are implemented using floating gate transistors. This system exhibits significant performance enhancements over its predecessor in terms of achievable signal bandwidth(> 50 MHz), accuracy(> 9 bits), dynamic range(> 7 decades of current), speed of floating-gate programming(> 200 gates/sec) and isolation between ON and OFF switches. The improved bandwidth is primarily due to an improved routing fabric that includes nearest neighbor connections. Programming performance improved drastically by implementing the entire algorithm on-chip with an SPI digital interface. Several complex system examples are presented.

I. RASP 2.8: OVERVIEW

The RASP 2.8 FPAA is a powerful system comprising over fifty thousand programmable floating gate elements which can be utilized as programmable interconnect as well as adaptive-computational elements. This leads to a platform capable of performing signal processing and computational tasks beyond a typical digital signal processor but at a fraction of its power. Similar advantages in computation have been demonstrated in [1], but this also marks a paradigm shift in the concept of analog design since its superior performance compared to earlier designs enables this to be used not only as a prototyping tool, but also as an attractive option for the final implementation.

The RASP 2.8 has thirty-two CABs connected by multi-level routing. The CABs are of two major types as shown in fig. 1(c). The first one has three operational transconductance amplifiers(OTA), three floating capacitors (500 fF each), two multi-input floating gates which can be used for constructing translinear circuits using MITE architectures, a voltage buffer, a transmission-gate with dummy switch for switched-capacitor applications and nMOS/pMOS transistor arrays with two common terminals for easily constructing source-follower or current-mirror topologies. All the OTAs are biased using floating gate transistors giving the user the option to tradeoff bandwidth, noise and power. Cascode biasing circuits valid for inversion levels are included as well. Two of the OTAs have floating-gate differential pairs which enable programming

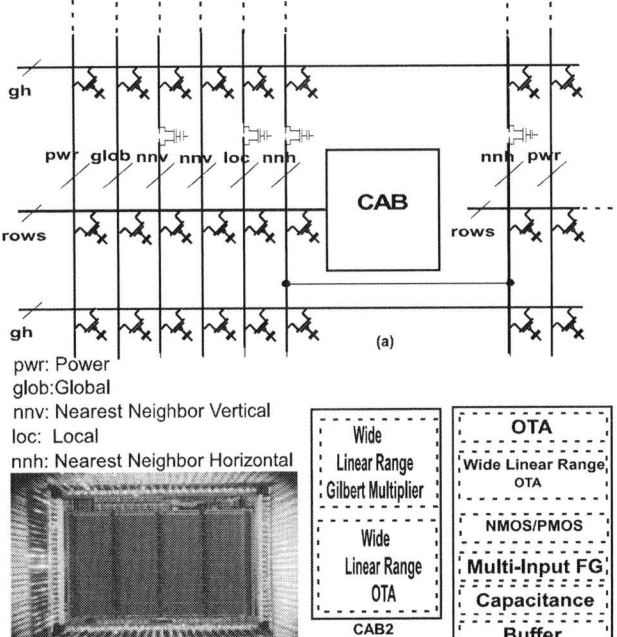

pwr: Power
glob: Global
nnv: Nearest Neighbor Vertical
loc: Local
nnh: Nearest Neighbor Horizontal

Fig. 1: **Chip level architecture:** (a) Routing architecture showing multi-level routing lines with different capacitances for improved bandwidth and connectivity. (b) Die Photo of the fabricated IC. (c) Two basic CAB types with internal components. These are complemented by circuits compiled in the switch fabric.

the offset of the amplifier as well as provide wide input linear range that is essential to reduce distortion in Gm-C filters and oscillators. The second type of CAB has a folded Gilbert multiplier in addition to a wide linear range OTA. The multiplier also has floating-gate differential pairs to reduce distortion. These CAB components can be connected using the switch-matrix consisting of floating-gate switches, which unlike other digital switch implementations, can be used for analog computations.

II. RASP 2.8: ARCHITECTURE

The present generation of FPAA exhibits significantly improved performance over its predecessor [2] primarily because of several architectural modifications that are described next.

978-1-4244-2018-6/08 $25.00 © 2008 IEEE

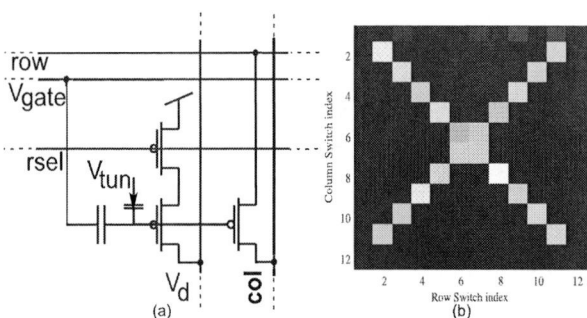

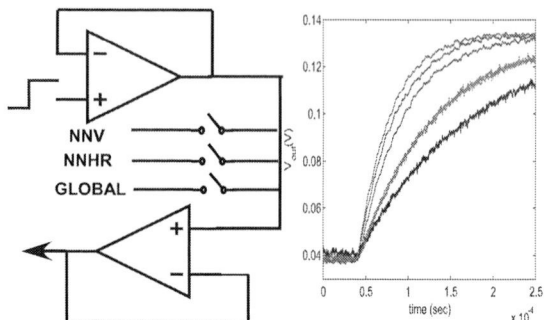

Fig. 2: **Improved Isolation:** (a) Source side selection together with indirect programming of the floating gate switches allows the RASP 2.8 to display impressive isolation between ON and OFF transistors. (b) Grayscale values in the figure correspond to logarithm of measured current from the array after the switches were programmed in this pattern.

A. Switch Isolation and Programming

The programmable switch matrix used in the earlier FPAA used the application of high gate voltage or high drain voltage as the method of isolation while the selected device had a low voltage at both the gate and drain terminals. However this method has a number of disadvantages, the primary one being over-injection of devices beyond the isolation point [3]. This IC employs source side selection [3] coupled with indirect programming as shown in fig. 2. The signal 'rsel' selects the desired row and removes the source current of other rows thus prohibiting injection. This leads to significantly better isolation compared to [2] which used 'V_{gate}' to control isolation. Hence, it is also possible to measure devices programmed to accurate currents located on the same column as an ON switch. Fig. 2 shows the current levels programmed into a pattern of 12x12 switches with the grayscale values representing measured currents. All the ON switches conduct more than 20 uA while the OFF black devices are at levels less than 40pA.

B. Routing

The routing architecture of the IC shown in 1(a) demonstrates the four major types of interconnections - local, nearest neighbor vertical and horizontal (nnv and nnh) and global. This granularity allows for high speed interconnects to be routed on low-capacitance lines like local or nearest neighbors while global connections are used only for I/O after the internal processing is complete. This results in huge power savings and facilitates low-power adaptive designs. Fig. 3 shows the configuration used to estimate the capacitances. The wide linear range OTA is biased at a Gm of 32 nS. Different routing lines corresponding to various capacitive loads are added successively and the step responses are measured. A voltage buffer is used to isolate the pad capacitance from the Gm-C element. Thus routing between CABs can be accomplished with relatively lower parasitic as compared to the earlier version and can achieve bandwidths of approximately 6 Mhz at around 100 nA of current. The achievable bandwidth within a CAB should be an order of magnitude higher. In addition to bandwidth, this characterization allows one to use the routing

Fig. 3: **Capacitance estimation:** Step responses are measured after adding one and three nearest neighbor vertical(nnv), three nearest neighbor horizontal(nnh) and one global line respectively as capacitive load to the Gm-C filter. Capacitances estimated from the resulting time constants are 151 fF for nnv, 228 fF for nnh and 763 fF for global lines.

to reduce kT/C noise as desired. The other feature of the routing scheme is bridge transistors that allow local lines to be bridged between CABs facilitating variable length connections without incurring the capacitance penalty of global lines.

C. On-chip Programming

Earlier generations of the FPAA used off-chip current measurement circuits which led to inaccuracy in the measurement due to noise and increased the minimum measurable current to the ESD leakage. Also, serial communication with a picoammeter is a time consuming operation resulting in large programming times. Fig. 4(a) shows the architecture of the current programming scheme which does all measurement operations on-chip and provides a digital SPI interface to a microprocessor (uP). Binary scaled current mode 7-bit DACs are used to supply the gate and drain voltages during both programming and operational modes. The drain selection circuit sets the drain to one of four choices depending on the current state in the programming algorithm. When the source of the floating-gate is being ramped-up, the drain is set to Vdd to prevent injection. For switch injection or accurate bias programming, the drain is then set to GND or a DAC voltage respectively for a fixed amount of time. Lastly, for measuring the currently programmed charge on the floating gate, the drain is switched to the current measurement circuitry.

The huge improvement in measuring accuracy, speed and dynamic range is obtained by using a logarithmic transimpedance amplifier described in [4]. Fig. 4(b) shows measured current from an off-chip picoammeter which saturates at the ESD leakage level of around 100pA while the inferred current from the logamp goes below 100 fA. The logamp is followed by a lowpass filter to limit bandwidth and improve noise performance [4] which has been measured to be around 9 bits.

The output voltage of the logarithmic amplifier is quantized by a ramp ADC as shown in the figure. The clock to the ADC is currently generated by a microprocessor and is limited to 25 Mhz resulting in an average conversion time of $500\mu s$ which should decrease in proportion to clock frequency. The digital

978-1-4244-2018-6/08 $25.00 © 2008 IEEE

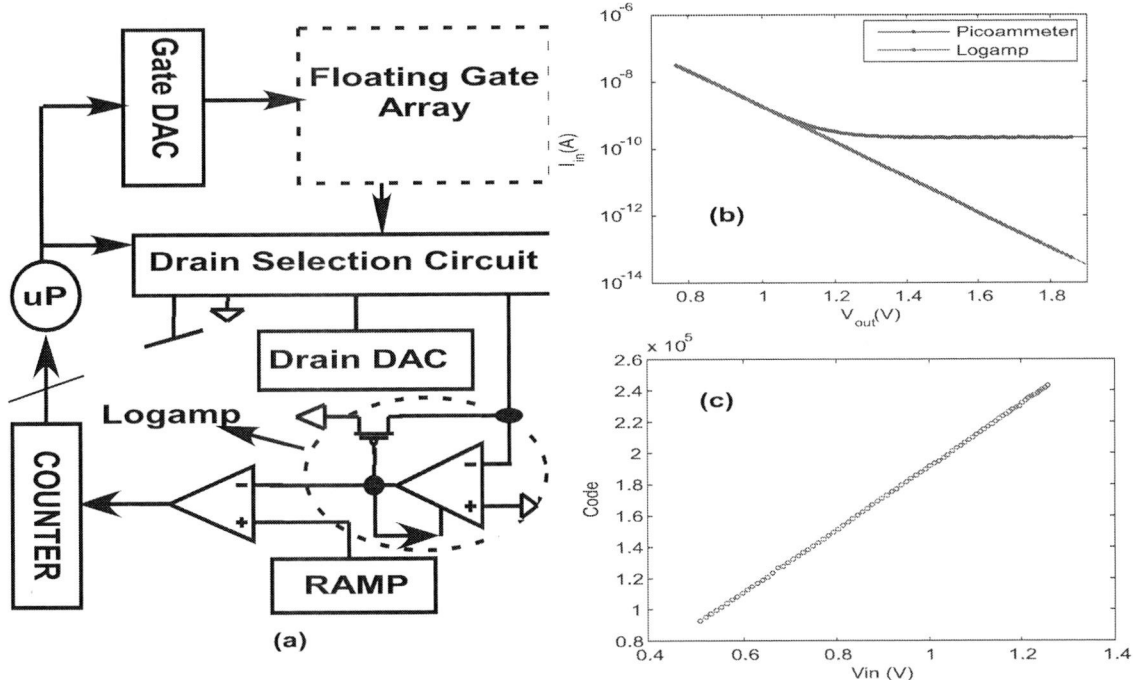

Fig. 4: **On-chip Programming:** (a) The scheme for on-chip programming is shown. DACs supply voltages to drain and gate. The programmed current is measured using a logarithmic I-V converter followed by a linear ramp ADC. (b) Measured I-V characteristic of logarithmic TIA showing improved accuracy over off-chip ammeter. (c) Ramp ADC characteristic displaying good linearity.

output word is sent to a microprocessor that implements the programming algorithm over a SPI interface.

Another important improvement is introduction of row-parallel programming for switches. The rows of the floating-gate array are selected by a decoder but the columns are selected using a shift register which enables selecting multiple columns per row. This leads to switch programming time given by $N_{rows} \times 100\mu s$.

D. I/O pad and Scanner Shift Register

Special bidirectional I/O pads have been incorporated in this IC which have buffer amplifiers capable of driving high capacitive loads when enabled. Their bandwidth is determined by a programmable floating gate device. Also an analog 16 bit shift register is available to scan through and observe different lines allowing the user an option to debug their circuit almost in a SPICE-like fashion.

III. CIRCUIT AND SYSTEM EXAMPLES

In this section we describe a few of the systems that have been implemented on the RASP 2.8. Fig. 5(a) demonstrates a wide linear range(WLR) OTA that is part of the CABs. To demonstrate the programmable nature of the offsets, the OTA was used as a comparator and linearly spaced charge differences were programmed into the differential pair. In the experiment, one terminal was fixed at 1V while the other was swept from 0 to 2.4V. The resulting curves are plotted in fig. 5(b) and show the threshold shifts. Fig. 5(c)

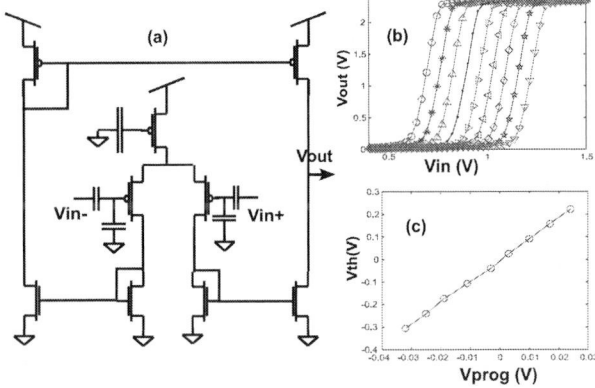

Fig. 5: **Programmable Offset OTA:** (a) The circuit for a wide input linear range OTA with a floating-gate differential pair and bias. The floating gates allow programming bias currents and offsets. (b) The OTA is used as a voltage comparator to demonstrate the different programmed offsets. (c) The measured threshold offsets are linearly related to the programmed charge difference.

plots the measured threshold against the programmed charge difference. The deviations from linearity are primarily because of inaccurate extraction of the threshold from the high-gain characteristics. This set of curves directly exhibit the feasibility of implementing a flash ADC in the FPAA accurate to the programming accuracy of 9 bits.

The next system is a second order current mode delta sigma converter. Fig. 6(a) shows the implementation in the

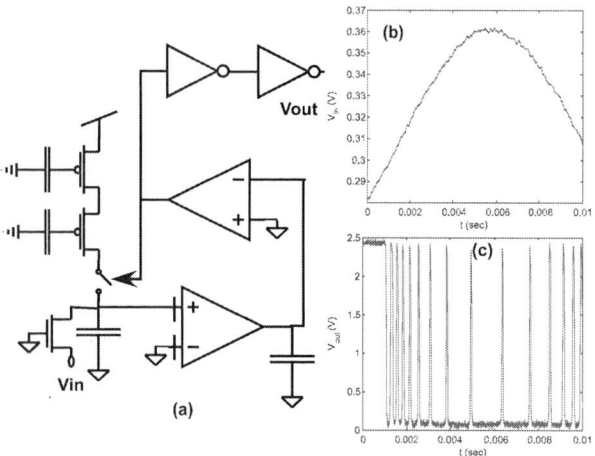

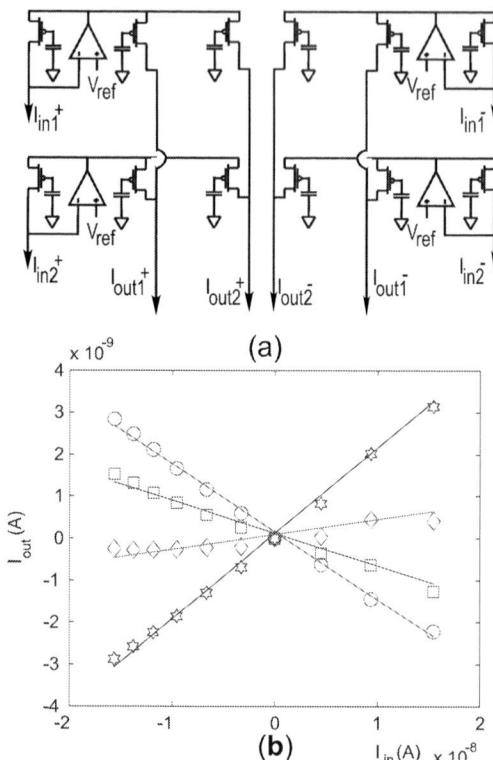

Fig. 6: **Current Mode Delta Sigma:** (a) The circuit for a second-order current mode delta sigma modulator. The first integration is performed by integrating current on capacitor and the second one is implemented using a WLR OTA. (b) Input sinusoidal signal to the modulator. (c) Output digital waveform from the modulator.

FPAA. The first integration is performed using a cascoded floating gate current source integrating charge on a capacitor. These floating gates are elements of the switch matrix and demonstrate the computational power of the analog switches. The second integration is performed by the WLR OTA on an explicit drawn capacitor. Fig 6(b) and (c) show the input sinusoidal signal and the modulator output respectively. A better implementation where the current source is switched using a differential pair instead of being turned off can be easily done and is not shown due to lack of space.

The third circuit fully exploits the computational ability of the switches [5]. It is a four quadrant vector matrix multiplier where the matrix of the weights is stored as charge difference on the floating gates. Fig. 7(a) shows the fully differential version of the circuit which can perform four quadrant multiplication. All the floating gates shown are part of the switch fabric and thus a really computation intensive operation can be obtained without having dedicated hardware in the CAB. The power of this architecture lies in its low power and scalability. Fig. 7(b) shows measured four quadrant multiplication on a linear scale.

Many other systems including amplitude and frequency modulators, oscillators, winner-take-all can also be easily implemented.

TABLE I: Table of Parameters

Process	$0.35\mu m$
Die Size	$3mm \times 3mm$
Power Supply	2.4V
Injection Vdd	$5.6V$
Number of CABs	32
Switch programming time	$N_{rows} \times 100\mu s$
Bias programming time	5 ms/element
Programming accuracy and range	9 bits over 100fA to 10 μA

Fig. 7: **Vector Matrix Multiplier:** (a) The circuit for a four quadrant vector matrix multiplier implemented in current mode. (b) Measured four quadrant multiplication showing different programmed weights.

IV. CONCLUSION

The RASP 2.8 generation of FPAA devices provide a powerful platform for prototyping and implementing large-scale signal processing applications. The programmable switch matrix composed of floating gate devices shows excellent isolation and can be readily utilized in computation. On chip programming interface allows current measurements below 100fA. The fully digital interface allows easy integration with a microprocessor. Programming times are around 5 ms for accurate biases and 100 μs per row of switches. Different levels of routing allow implementation of high performance circuits while allowing for fast turn-around times.

REFERENCES

[1] R. Melville G. Cowan and Y. Tsividis, "A vlsi analog computer/digital computer accelerator," *IEEE Journal of Solid-State Circuits*, vol. 41, no. 1, pp. 42–53, Jan 2006.

[2] C.M. Twigg and P.E. Hasler, "A large-scale reconfigurable analog signal processor ic," in *Proceedings of the IEEE Custom Integrated Circuits Conference*, Sept 2006, pp. 5–8.

[3] C.M. Twigg and P.E. Hasler, "Programmable conductance switches for fpaas," in *Proceedings of the International Symposium on Circuits and Systems*, May 2007, pp. 173–76.

[4] R. Robucci A. Basu and P. Hasler, "A low-power, compact, adaptive logarithmic transimpedance amplifier operating over seven decades of current," *IEEE Transactions on Circuits and Systems I*, vol. 49, no. 1, pp. 2167–77, Oct 2007.

[5] C.M. Twigg J Gray and P. Hasler, "Programmable floating gate fpaa switches are not dead weight," in *Proceedings of the International Symposium on Circuits and Systems*, May 2007, pp. 169–72.

IEEE 2008 Custom Intergrated Circuits Conference (CICC)

Modeling of Triple-Well Isolation and the Loading Effects On Circuits up to 50 GHz

Piljae Park and C. Patrick Yue

High-Speed Silicon Lab, University of California, Santa Barbara, CA 93106-9560

Abstract — This paper presents the noise isolation characteristics and substrate loading effects of NMOS devices in triple-well (TW) at frequencies up to 50 GHz. The importance of the series resistance in well bias path is investigated. Our study reveals that using large well bias resistors, which create a high substrate impedance, are beneficial to circuit performance in both saturation and triode bias regimes. However, small bias resistances are advantageous for better substrate noise isolation. By using test circuits designed in a 0.13-μm CMOS technology, a compact model for TW substrate impedance is developed to facilitate quantitative analyses. The dependence of TW isolation effectiveness on the bias resistance, well capacitance, and noise frequency are evaluated. To measure the impact of TW on circuit performances, the output impedance of a saturation-mode device and the insertion loss (IL) between source and drain of a triode-mode device are examined under different biasing conditions.

Index Terms — Triple-well model, Substrate noise isolation, Triple-well bias.

I. INTRODUCTION

Scaled CMOS technologies have facilitated multi-function, mixed-mode integration on a single chip. The ever increasing circuit density and integration has provided significant cost benefits but also raises new technical challenges. In particular, to accommodate these densely packed devices in a single substrate, the isolation of multiple signal sources must be handled with care, especially when the signal dynamic range is large. Figure 1(a) shows the cross-section of a triple-well (TW), which consists of a p-well (PW) embedded within a deep-N-well (DNW) to provide a body for the NMOS isolated from the P-substrate (P-sub). Compared to twin-well CMOS processes, it is well known that TW technology offers better substrate noise isolation and body bias control [1][2].

In this paper, substrate isolation utilizing TW and the loading effects on the circuits are examined for different well bias conditions. Two biasing schemes – AC floating (high-impedance) and AC short (low-impedance) – for the PW and DNW are evaluated at various bias voltages. By using a set of circuit test structures fabricated in a 0.13-μm CMOS technology, a compact model of the TW substrate impedance is first developed based on measured s-parameters up to 50 GHz. The TW model is then used to predict the measured substrate noise at different separation distances. The effect of well bias resistance at different noise frequencies is considered. By appending the TW model to the body node of a standard NMOS device, the

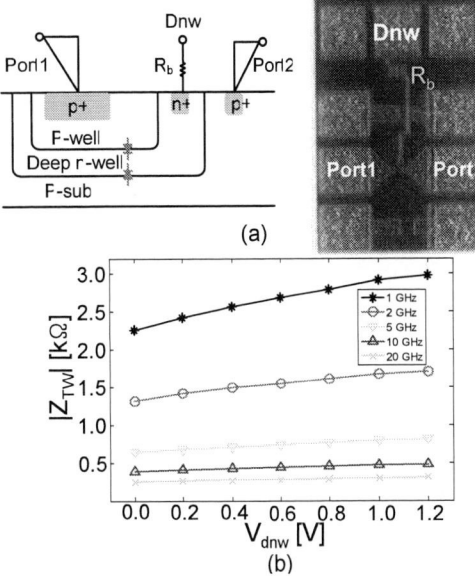

Figure 1. (a) Cross-sectional view and die photo of triple-well test structure. (b) Measured impedance characteristic.

loading effects of TW substrate impedance on the device characteristics are simulated and compared to measurement results.

II. EXPERIMENT RESULTS AND DISCUSSION

A. *Model of TW Substrate Impedance*

Typically, for the bias of a NMOS in TW, the lowest and highest voltages in design are applied to the PW and DNW, respectively. Due to the capacitive nature of the two back-to-back reverse-biased junctions, the TW substrate impedance (Z_{TW}) exhibits dependence on frequency and bias voltage. To analyze the TW impedance characteristics, the 2-port s-parameters are measured between a PW contact and a P-sub contact, as illustrated in Fig. 1(a). The test structure has a PW/DNW area of 11x11 μm^2 and a DNW/P-sub area of 14x14 μm^2. The Z_{TW}, under different DNW bias voltages, are extracted from the measured s-parameters. The bias for the PW and P-sub are connected to ground through off-chip bias-T's. The DNW bias, V_{dnw}, is varied from 0.0 to 1.2 V (the nominal V_{DD}) via a large on-chip resistor, R_b, of 70 kΩ (see Fig. 1(a)). R_b is employed to avoid shorting out the junction impedance, such that Z_{TW} can be observed. Figure 1 (b) plots the magnitude of the measured Z_{TW} as a function of V_{dnw}, at frequencies ranging from 1 to 20 GHz. The

978-1-4244-2018-6/08 $25.00 © 2008 IEEE

highest substrate impedance is achieved when V_{dnw} is at V_{DD}, which causes both the PW/DNW and DNW/P-sub

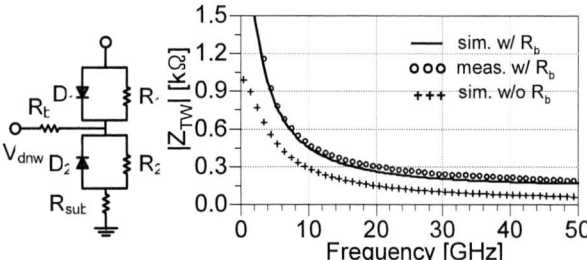

Figure 2. Triple-well substrate model and its impedance characteristics versus frequencies with and without R_b.

junctions to have the largest depletion region and hence the lowest capacitance. As frequency increases, the variation of $|Z_{TW}|$ with DNW bias reduces. At 1 GHz, when V_{dnw} is changed from 0 V to 1.2 V, $|Z_{TW}|$ is increased by 32%, while at 10 GHz, only a 18% increase is measured.

Based on the measured data, an equivalent circuit model of the TW is developed, as shown in Figure 2. The PW/DNW (53 fF) and DNW/P-sub (66 fF) diodes are modeled with parallel resistors, R_1 (1 kΩ) and R_2 (1.2 kΩ), respectively. Diode quality factors are set by the parallel resistors to match the measurement results. A series substrate resistor, R_{sub} of 140 Ω, is also included in the model. The measured and simulated $|Z_{TW}|$ frequency responses (with V_{dnw} at 1.2 V) are in excellent agreement as shown in Fig. 2. When the DNW is biased without R_b, which essentially provides an AC short path at DNW, $|Z_{TW}|$ decreases. The AC short path effectively eliminates the series impedance due to the DNW/P-sub diode (D_2), and consequently results in a lower $|Z_{TW}|$. The Z_{TW} model is used next for simulations of the TW noise isolation.

B. Substrate Noise of NMOS in Triple-well

To reduce substrate noise, various isolation structures, such as guard ring and pocket type structures, have been reported [1-3]. Typically, the isolation is measured between two standalone diffusion areas. However, it is more realistic to evaluate isolation directly from the source/drain junctions of a device under proper biasing. Figure 3(a) shows the cross-section schematic and the die photo of the TW test structure. The forward transmission, S_{21}, is measured with Port 1 at the source terminal of the NMOS and Port 2 at the P-sub contact. The NMOS device (100 μm/0.12 μm) is biased in saturation region with V_{gs} at 0.5 V and V_{ds} at 0.8 V. The AC floating bias resistors, R_{b1} and R_{b2} (both 70 kΩ), are used for PW (0 V) and DNW (1.2V), respectively, to avoid AC short path. P-sub contacts at two different distances, 350 μm and 650 μm, away from the NMOS source node are measured. In the simulation, the NMOS body node is connected to the Z_{TW}

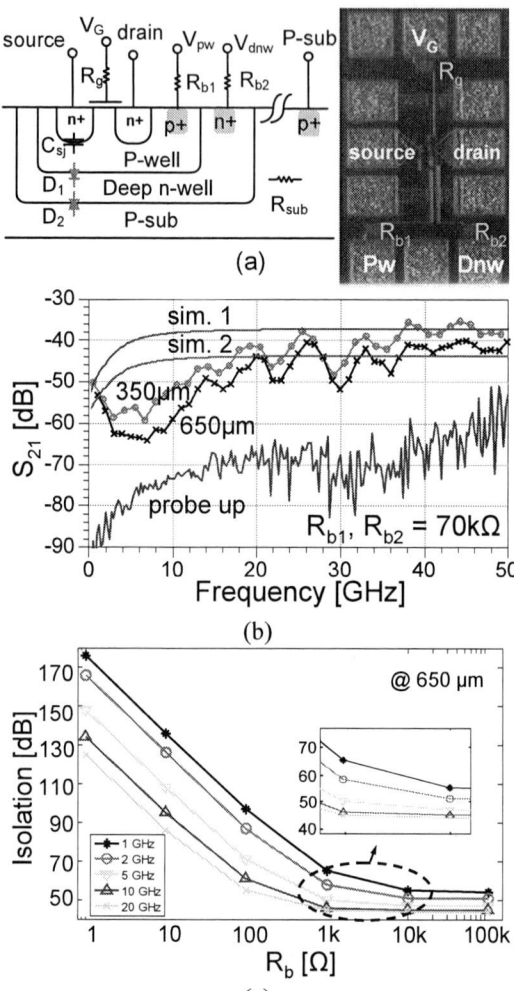

Figure 3. (a) TW isolation test structure. (b) Isolation between the source node of NMOS to the P-sub contact at 2 different distances. (c) Isolation dependence on bias resistance (R_b).

model. The substrate resistor, R_{sub}, in the Z_{TW} model is set to 7 kΩ and 15 kΩ, respectively, to model the 350-μm and 650-μm separation. In Figure 3(b), the measured S_{21} are plotted along with the simulation results. The isolation worsens as frequency increases because the impedance of the junction capacitances is dropping. With 300 μm increase in separation, 5-dB better isolation is observed up to 50 GHz, which is well predicted by our model.

The substrate noise coupling path from the source of the NMOS to the P-sub contact includes R_{sub} and three junction capacitors: source/PW, PW/DNW, and DNW/P-sub. As can be seen in Fig. 3(a), the substrate coupling path is also in parallel with the two bias paths for PW and DNW. Based on simulations with different R_b (=R_{b1}=R_{b2}) values, the effect of the well bias resistors on TW isolation is demonstrated in Fig. 3(c). For better isolation, a low-impedance bias path of PW and DNW is advantageous.

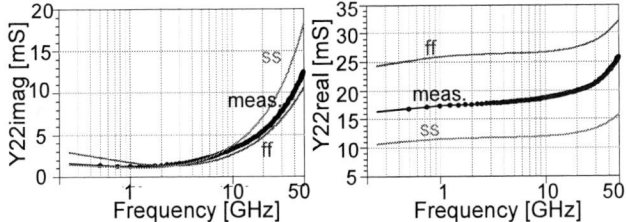

Figure 4. Y-parameter comparison of TW NMOS device in saturation region.

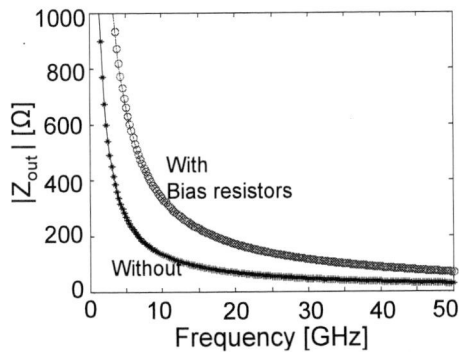

Figure 5. Simulated output impedance, $|Z_{out}|$, of TW NMOS with and without bias resistors.

Smaller R_b provides more effective AC short paths to shunt away the noise. At low frequencies, the impedances due to the junction capacitance in the substrate noise path are higher, therefore better isolation is achieved. The critical value for R_b, below which isolation can be improved, depends on the noise frequency. For example, with R_b equal to 100 Ω, 90-dB isolation can be achieved at 2 GHz. However, if the frequency increases to 20 GHz, R_b must be reduced by a decade to 10 Ω to maintain the same level of isolation. In general, the ratio between the impedance of the TW junction capacitance and R_b determined whether the well bias path can provide an effective AC ground path for the noise. When the noise coupling path impedances due to the TW junctions are high, for small TW area and at low frequencies, the requirement to minimize R_b to create a better AC short is relaxed. Therefore, TW isolation is more effective when it is applied separately for sensitive devices, rather than merging the TW together to enclose the entire analog section of a chip, because of the lower TW capacitance.

C. *TW Loading Effects on Saturation-Mode Devices*

To evaluate TW loading effect on a saturation-mode NMOS device, the test structure shown in Fig. 3(a) is used. Port 1 is still applied at the source node as in the case for the isolation measurements, but Port 2 is now connected to the drain node of the NMOS. The saturation bias condition is the same as in the previous section: V_{ds} at 0.8 V, V_{gs} at 0.5 V, V_{pw} at 0 V, and V_{dnw} at 1.2 V. At

the source and drain node, 10-pH series inductance are added to accurately model the residual parasitic that is not fully de-embedded using open pad structures. The simulation of the device characteristics is based on the Z_{TW} model attached to the NMOS body node. Figure 4 shows the output admittance (Y_{22}) comparisons between measurement and simulations. Slow-slow (ss) and fast-fast (ff) corners are simulated to account for the process variation. The measurement results are well within the range predicted by the corner models with the added Z_{TW} model. For saturation-mode devices operating as a voltage amplifier, high output impedance is desirable for better gain. PW and DNW biasing with and without bias resistors (70 KΩ) are considered. The simulated output impedance at the drain, $|Z_{out}|$, which is obtained form $|1/Y_{22}|$, is plotted in Fig. 5. Due to the capacitance associated with the drain node of the device, $|Z_{out}|$ decreases with frequency. At frequencies up to 50 GHz, AC floating TW bias (i.e. PW, DNW are biased with large R_{b1}, R_{b2}) helps to increase $|Z_{out}|$. In other words, to achieve higher intrinsic voltage gain for a TW NMOS device, it is preferred to using large well resistors. The tradeoff between good noise isolation and high device gain must be considered when determining the value of the well bias resistance.

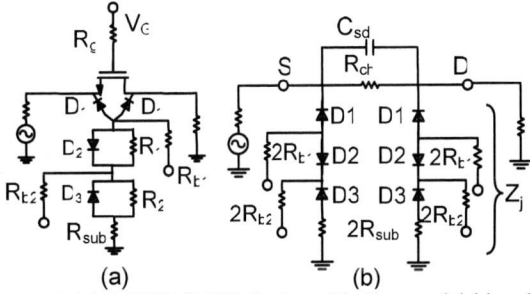

Figure 6. (a) A TW NMOS device with Z_{TW} model biased in triode mode for analyzing the TW loading effects on the insertion loss and off-state isolation of a switch. (b) Simplified equivalent circuit model to highlight the parasitics at the source and drain nodes.

D. *TW Loading Effects on Triode-Mode Devices*

When a NMOS device is utilized as a high-frequency switch, the device operates in either triode or cut-off regions. In triode mode, the power loss to the TW substrate impedance directly affects the insertion loss (IL) between source and drain. Similarly, in the cut-off mode, the isolation between source and train depends on Z_{TW}. Simulations are performed using the TW NMOS circuit model (shown Fig. 6) to study the effect of biasing with and without R_b (70 kΩ) for PW and DNW. As shown in Fig. 7, without R_b, at above 10 GHz, IL is higher than with a large R_b. This can be attributed to AC short path at the

978-1-4244-2018-6/08 $25.00 © 2008 IEEE   237

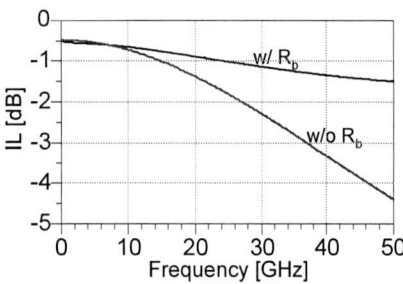

Figure 7. Simulated insertion loss with and without bias resistors R_b at P-well, and deep-n-well.

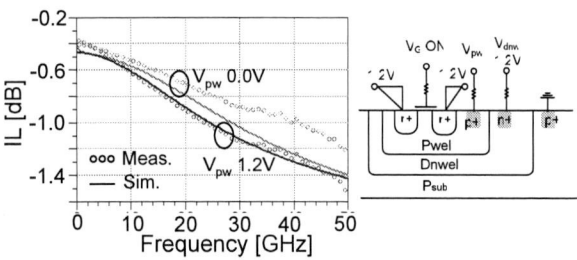

Figure 8. Measured and simulated insertion loss of triode-region TW NMOS device at different V_{pw} bias conditions.

PW and DNW. With R_{b1} and R_{b2} equal to 0 Ω, the source/drain substrate impedance, Z_j, is smaller and thus causes higher substrate loss. In turn, this leads to higher IL because more power is dissipated to the substrate instead of transferring from source to drain. Using large R_b for AC floating well bias has been reported in antenna switch design to reduce IL and increase linearity [4][5]. Without large R_b, both source/drain junction diodes, D_1 and D'_1, in Fig. 6(b) will turn on when a large negative signal swing (<−0.7 V) is applied to the switch.

The IL and isolation between the source and drain for the TW NMOS are compared between measurement and simulation using our model in Fig. 8 and Fig. 9, respectively. The device is biased in triode mode with V_{gs} and V_{gd} equal to 1.2 V, while the PW bias, V_{pw}, is altered to vary the substrate loading. The lowest IL is measured with V_{pw} at 0 V, which puts the highest reverse bias (1.2 V) across the source/drain-to-PW junction as well as the PW/DNW junction. With lower substrate impedance, degradation in IL of more than 0.5 dB is observed above 10 GHz. When the NMOS is in cut-off(V_{gs} and V_{gd} of − 1.2 V), the isolation between source and drain shows less than 1 dB difference with alteration of V_{pw} from 0 V to 1.2 V. Since the $|Z_{ch}|$ of off-state NMOS is comparable with $|Z_j|$ (refer to Fig. 6(b)), the change of substrate isolation (i.e. $|Z_j|$) is less significant to the isolation.

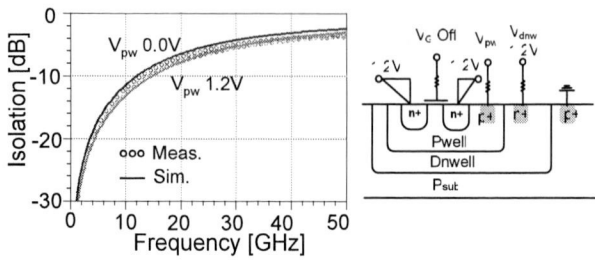

Figure 9. Measured and simulated isolation of off-state TW NMOS device at different V_{pw} bias conditions.

III. CONCLUSION

The TW isolation characteristic and its loading effects are investigated up to 50 GHz. A compact model for the TW substrate impedance has been developed to examine the effect of different TW bias schemes: AC floating and short bias. For better substrate noise isolation, an AC short bias, i.e. small well bias resistance, is desirable. However, an AC floating bias is advantageous for circuit performance: higher voltage gain in saturation mode and lower insertion loss in triode mode. For saturation mode device, higher output impedance is obtained with AC floating bias. Without AC floating bias resistors, substrate loss in triode-mode device worsens significantly above 10 GHz.

ACKNOWLEDGEMENT

This work was sponsored by SRC/DARPA FCRP C2S2. The authors would like to thank the United Microelectronics Corp. (UMC) for chip fabrication and Dong Hun Shin for his valuable inputs.

REFERENCES

[1] Vinella, R.M. et al, "Substrate Noise Isolation Experiments in a 0.18-μm 1P6M Triple-well CMOS process on a Lightly Doped Substrate," 2007 *IEEE Instrumentation and Measurement Technology Conference,* May 2007, pp. 1-6

[2] Barbier-Petot, C., Bardy, S., Biard, C., Descamps, P., "Substrate isolation in 90nm RF-CMOS technology," *European Microwave Conference*, Oct. 2005, pp. 4-7.

[3] Tung-Sheng Chen et al, "An efficient noise isolation technique for SOC application," *IEEE Transaction on Electron Devices*, Feb. 2004, pp. 255-260

[4] Mei-Chao Yeh. et al, "Design and analysis for a miniature CMOS SPDT switch using body-floating technique to improve power performance," *IEEE Transactions on Microwave Theory and Techniques*, Volume 54, Issue 1, Jan. 2006, pp. 31-39.

[5] Piljae Park, Dong Hun Shin, John J. Pekarik, Mark Rodwell, C. Patrick Yue, "A High-linearity, LC-tuned, 24-GHz T/R switch in 90-nm CMOS," to be presented at *IEEE Radio Frequency Integrated Circuits Symposium*, June 2008.

978-1-4244-2018-6/08 $25.00 © 2008 IEEE

A general weak nonlinearity model for LNAs

Wei Cheng, Anne Johan Annema, Jeroen A. Croon*, Dirk B.M. Klaassen* and Bram Nauta
IC Design Group, CTIT Research Institute, University of Twente
PO Box 217, 7500 AE Enschede, The Netherlands
e-mail: w.cheng@utwente.nl
*with NXP-TSMC Research Center, High Tech Campus 37, Eindhoven, The Netherlands

Abstract- **This paper presents a general weak nonlinearity model that can be used to model, analyze and describe the distortion behavior of various low noise amplifier topologies in both narrowband and wideband applications. Represented by compact closed-form expressions our model can be easily utilized by both circuit designers and LNA design automation algorithms. Simulations for three LNA topologies at different operating conditions show that the model describes IM components with an error lower than 0.1% and a one order of magnitude faster response time. The model also indicates that for narrowband IM2@w_1-w_2 all the nonlinear capacitances can be neglected while for narrowband IM3 the nonlinear capacitances at the drain terminal can be neglected.**

I. INTRODUCTION

The low noise amplifier (LNA) is a critical building block in the RF front-end. One of the important design specifications of the LNA is its distortion performance, typically specified in terms of IIP2 and IIP3. A number of papers present nonlinearity analyses for LNAs to provide design guidelines [1-2] or for LNA design automation purposes [3-4]. Volterra series theory is widely used as the major nonlinearity analysis approach [5]. Trying to avoid the complex calculation involved in Volterra series, other approaches include:

A. Per-nonlinearity distortion analysis [6]

B. Combined multisine and Volterra analysis [7]

C. Harmonic distortion analysis in feedback amplifiers [8]

Approach *A* treats a MOS transistor as a linear component with a nonlinear drain current source, which allows identifying the transistors that contribute most to the output nonlinearity. Although all the drain-source nonlinearities can be included, no information is given about which nonlinearity of the drain current has more effects [7]. Approach *B* splits the nonlinear behavior in similar contributions as in approach *A* while better insights on the nonlinearity contribution are achieved by using the selective Volterra analysis. Both approach *A* and *B* demand distortion simulations of the circuit to provide the data for post-processing, which is not meant for hand-calculation analysis. Alternatively approach *C* only uses conventional algebra to analyze harmonic distortion of feedback amplifiers but can't provide solutions for intercept 1-dB compression point and intermodulation distortions. Despite the complexity difference in all aforementioned methods the nonlinearity analysis must be redone for each new topology.

In this paper we introduce a general weak nonlinearity model that is independent of the LNA topologies. Theoretically, for any LNA topology both in narrowband or wideband applications, this model calculates output IM2 and IM3 of the circuit using simple closed-form expressions. The compact closed-form expression is just a linear combination of nonlinear coefficients of each MOS transistor and terminal AC gains. Therefore, the result of our model can be easily utilized by both circuit designers and LNA design automation algorithms without involving any complex nonlinearity analysis. In section II the weak nonlinearity model for MOS transistor valid from DC to RF frequencies is introduced. Using this MOS nonlinearity model a generalized distortion analysis for weakly nonlinear circuits is discussed in Section III, which presents the approach to obtain the general weak nonlinearity model. Section IV shows the benchmark on accuracy for our model using three different LNAs operating in different load condition and at different frequencies. Conclusions are drawn in section V.

II. MOS TRANSISTOR NONLINEARITY MODELING

In this paper it is assumed that MOS transistors' nonlinearities are the (main) cause for distortions in RF-circuits. Therefore, for analyzing the nonlinear behavior of RF circuits, modeling and describing transistor nonlinearity is essential. Taylor series are successfully and dominantly used to describe the weakly nonlinear behavior of the MOS transistor [1-5]. However, most of these descriptions simplify the MOS nonlinearity model to the following extent:

- Only consider transconductance nonlinearity [1-3].
- Consider all resistive drain current nonlinearity but neglect the charge-storage nonlinearity [5], [7- 8].

Aiming for validity from DC to the RF frequency range, we introduce a complete weakly nonlinear model taking into account both resistive and charge-storage nonlinearity.

A MOS transistor is a four-terminal device, in which all currents into and charges at nodes are nonlinear functions of the voltages across any two terminals. Mathematically the transistor can be modeled as a three-port network with the gate-source, drain-source and bulk-source voltage as the inputs and gate current, drain current and bulk current as outputs for any given DC bias. Therefore the transistor's weakly nonlinear behavior in the close vicinity of any DC bias point can be described by the multi-dimensional Taylor series up to (here) the third-order, which is given by

$$
\begin{aligned}
i_k(t) = & G_{100}^k v_{gs} + G_{200}^k v_{gs}^2 + G_{300}^k v_{gs}^3 + G_{010}^k v_{ds} + G_{020}^k v_{ds}^2 + G_{030}^k v_{ds}^3 + G_{001}^k v_{bs} \\
& + G_{002}^k v_{bs}^2 + G_{003}^k v_{bs}^3 + G_{110}^k v_{gs} v_{ds} + G_{120}^k v_{gs} v_{ds}^2 + G_{210}^k v_{gs}^2 v_{ds} + G_{101}^k v_{gs} v_{bs} \\
& + G_{102}^k v_{gs} v_{bs}^2 + G_{201}^k v_{gs}^2 v_{bs} + G_{011}^k v_{ds} v_{bs} + G_{012}^k v_{ds} v_{bs}^2 + G_{021}^k v_{ds}^2 v_{bs} + G_{111}^k v_{gs} v_{ds} v_{bs} \\
& + C_{100}^k \frac{dv_{gs}}{dt} + C_{200}^k \frac{dv_{gs}^2}{dt} + C_{300}^k \frac{dv_{gs}^3}{dt} + C_{010}^k \frac{dv_{ds}}{dt} + C_{020}^k \frac{dv_{ds}^2}{dt} + C_{030}^k \frac{dv_{ds}^3}{dt} + C_{001}^k \frac{dv_{bs}}{dt}
\end{aligned}
\tag{1}
$$

$$+C_{002}^k \frac{dv_{bs}^2}{dt} + C_{003}^k \frac{dv_{bs}^3}{dt} + C_{110}^k \frac{dv_{gs}v_{ds}}{dt} + C_{120}^k \frac{dv_{gs}v_{ds}^2}{dt} + C_{210}^k \frac{dv_{gs}^2 v_{ds}}{dt} + C_{101}^k \frac{dv_{gs}v_{bs}}{dt}$$

$$+C_{102}^k \frac{dv_{gs}v_{bs}^2}{dt} + C_{201}^k \frac{dv_{gs}^2 v_{bs}}{dt} + C_{011}^k \frac{dv_{ds}v_{bs}}{dt} + C_{012}^k \frac{dv_{ds}v_{bs}^2}{dt} + C_{021}^k \frac{dv_{ds}^2 v_{bs}}{dt} + C_{111}^k \frac{dv_{gs}v_{ds}v_{bs}}{dt}$$

where

$$C_{mnl}^k = \frac{1}{m!}\frac{1}{n!}\frac{1}{l!}\frac{\partial^{(m+n+l)}Q_X}{\partial V_{gs}^m \partial V_{ds}^n \partial V_{bs}^l} \quad ; G_{mnl}^k = \frac{1}{m!}\frac{1}{n!}\frac{1}{l!}\frac{\partial^{(m+n+l)}I_k}{\partial V_{gs}^m \partial V_{ds}^n \partial V_{bs}^l} \quad k \in \{g,d,b\}$$

are respectively the nonlinear capacitive and resistive coefficients with Q_k for charge at and I_k for current into terminal k. For simplicity reasons, in this paper the source and bulk are assumed to be connected, effectively reducing the MOS-transistor to a three-terminal device. As a result only the currents into and charges stored at the gate, drain and source are of concern and hence coefficients C_{mnl}^k and G_{mnl}^k with $l\neq 0$ are zero. Furthermore, the DC gate current is neglected and only the capacitive gate current is taken into account, which is a valid simplification for RF operation [9].

The weak nonlinearity model is shown in Fig. 1a, where the linear current sources ($i_{g,lin}$ and $i_{d,lin}$) and nonlinear current sources ($i_{g,nonlin}$ and $i_{d,nonlin}$) are separated for the circuit distortion analysis discussed in next section. Other capacitances that may be present in the MOS transistor structure can be included in this representation. For example the interconnect capacitance between the gate and drain may be added explicitly across the terminals, but can also be embedded in (1) by using e.g. the Blakesley transform.

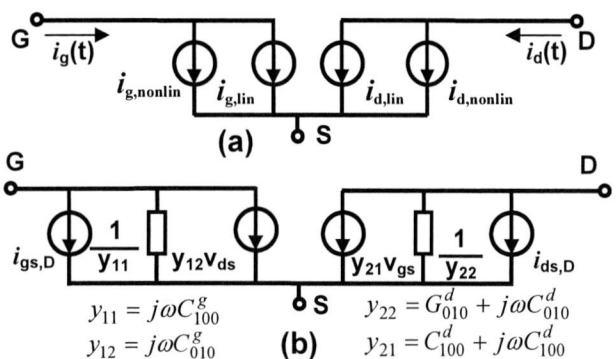

(a)

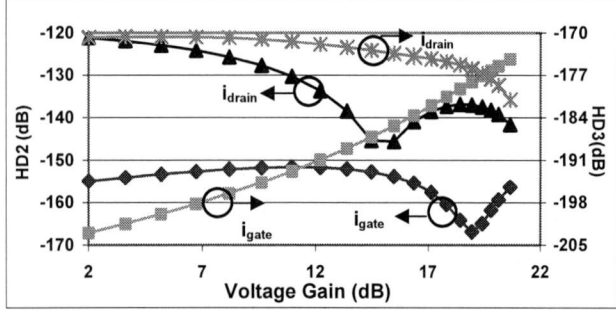

$$y_{11} = j\omega C_{100}^g \qquad y_{22} = G_{010}^d + j\omega C_{010}^d$$
$$y_{12} = j\omega C_{010}^g \qquad \textbf{(b)} \qquad y_{21} = C_{100}^d + j\omega C_{100}^d$$

Fig. 1. Weak nonlinearity model for the MOS transistor in (a) time-domain and (b) frequency domain.

Fig. 2. Comparison on HD2@2GHz and HD3@3GHz in gate and drain current between ADS [13] simulation and the MOS nonlinearity model for different voltage gains (v_{ds}/v_{gs}). Symbols represent model prediction, lines transistor-level circuit simulation using a commercial 90nm CMOS process (f_T=110GHz)

The nonlinear coefficients are extracted directly from state-of-art MOS model, namely the PSP model [14], which ensures excellent accuracy. Very good agreement with the simulated HD2 and HD3 in gate and drain current is observed for transistors with different dimensions and bias, one of which is shown in Fig. 2 with the transistor under different voltage gain (v_{ds}/v_{gs}) conditions.

III. GENERALIZED DISTORTION ANALYSIS

In this section a general circuit with N transistors is analyzed. A two-tone input signal $V_{IN}e^{j\omega_1 t} + V_{IN}e^{j\omega_2 t}$ is applied assuming that amplitude V_{IN} is small to ensure the circuit operates in the weakly nonlinear region. The MOS transistors are assumed to be the only nonlinearity sources in the circuit, although this assumption is not necessary. Since no topology information is involved the analysis result is valid for all topologies.

A. Dependent relation

The frequency-domain MOS nonlinearity model shown in Fig.1b is used. It consists of first-order y-parameters (directly converted from the linear coefficients in (1)) and two distortion current sources (gate distortion current source $i_{gs,D}$ and drain distortion current source $i_{ds,D}$) that contain both harmonic and intermodulation distortion elements. In the frequency domain $i_{gs,D}$ and $i_{ds,D}$ can be regarded as (dependent) small-signal multi-tone stimuli, therefore in any circuit with N transistors the distortion in the circuit output is a linear combination of $i_{gs,D}$ and $i_{ds,D}$ from each transistor. Moreover, the terminal voltages of each transistor (v_{gs} and v_{ds}) are linear combination of $i_{gs,D}$, $i_{ds,D}$ from each transistor and the two-tone input signal, which in total yield

$$v_{out,D} = \sum_{k=1}^{N} (H_{igsk} \circ i_{gsk,D} + H_{idsk} \circ i_{dsk,D}) \tag{2}$$

$$v_{dsj} = \sum_{k=1}^{N} (F_{igsk}^{dsj} \circ i_{gsk,D} + F_{idsk}^{dsj} \circ i_{dsk,D} + F_{vin}^{dsj}(\omega_1) \cdot V_{IN}e^{j\omega_1 t} + F_{vin}^{dsj}(\omega_2) \cdot V_{IN}e^{j\omega_2 t}) \tag{3}$$

$$v_{gsj} = \sum_{k=1}^{N} (F_{igsk}^{gsj} \circ i_{gsk,D} + F_{idsk}^{gsj} \circ i_{dsk,D} + F_{vin}^{gsj}(\omega_1) \cdot V_{IN}e^{j\omega_1 t} + F_{vin}^{gsj}(\omega_2) \cdot V_{IN}e^{j\omega_2 t}) \tag{4}$$

where H_{ixsk} $x \in \{g,d\}$ is the AC gain to the output voltage when an AC current source applied between the terminal x and source in transistor k; F_{ixsk}^{ysj} $x \in \{g,d\}$, $y \in \{g,d\}$, $x \neq y$ is the voltage gain from the current source attached between the terminal x and source in transistor k to the terminal voltage v_{ys} in transistor j; F_{vin}^{ysj} $y \in \{g,d\}$ is the AC gain from voltage input to the terminal voltage v_{ys} in transistor j.

In summary (2) suggests that the output voltage distortion can be easily calculated once all the distortion current sources are known. (1), (3) and (4) indicate a dependent relation between the controlling voltage v_{gs} and v_{ds} and the distortion current sources of each transistor as illustrated in Fig. 3.

B. Solving for the general model

The dependent relation shown in Fig 3 generates a linear

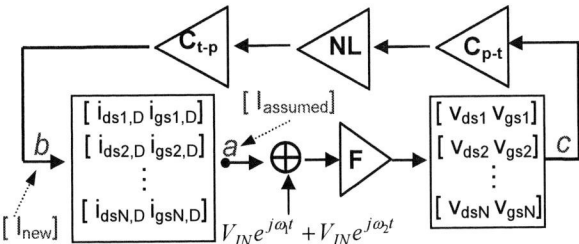

Fig. 3. Illustration of the dependent relation between the distortion current sources and the controlling voltages in any circuit with N transistors. Block "F" and "NL" represent the calculation conducted in (3)-(4) and (1) respectively. Block "C_{p-t}" represents the conversion from phasors to the according time-domain expressions and "C_{t-p}" for vice-versa.

equation matrix for solving the closed-form expressions of each distortion current sources ($i_{gsk,D}$, $i_{dsk,D}$, k=1,···N). At first the assumed expression matrix $[I_{assumed}]$ for $i_{gsk,D}$ and $i_{dsk,D}$ are input to node a. Given the weak nonlinear assumption, $[I_{assumed}]$ contains distortion elements up to the third-order, where the magnitude of second-order distortions should be proportional to V_{IN}^2 and the magnitude of the third-order distortions to V_{IN}^3, any terms with higher-order of magnitude is small enough to be neglected. Following the loop counterclockwise until node b we can obtain a new expression matrix $[I_{new}]$. Up to the third-order $[I_{assumed}]$ should be equal to $[I_{new}]$, which provides the solution for the expressions of $i_{gsk,D}$, $i_{dsk,D}$. Then the circuit output distortion can be easily obtained from (2).

Meanwhile the matrix solving is significantly simplified due to the observation in (3) and (4) that only the terms with fundamental tones contribute to the second-order distortion; only the terms with fundamental tones and second-order tones contribute to the third-order distortion. Therefore other terms in (3) and (4) can be neglected for the calculation of second and third order distortions.

As a result the closed-form expressions for the output harmonic and intermodulation distortions are obtained. Due to the space limitation only their compact forms are given

$$v_{out,\omega_{IM}} = \sum_{k=1}^{N} [P_{mn0,k}^d \cdot (G_{nm0,k}^d + j\omega_{IM} C_{nm0,k}^d) + P_{mn0,k}^g \cdot j\omega_{IM} C_{nm0,k}^g] \quad (5)$$

where $P_{mn0,k}^j$ is a function of according AC terminal gains; ω_{IM} is the intermodulation frequency; $G_{mn0,k}$ and $C_{mn0,k}$ are the nonlinear conductances and capacitances in transistor k at terminal gate/drain. We've observed that in a 90nm CMOS process up to 5 GHz $G_{mn0,k}^d$ is more than one order of magnitude larger than both $\omega C_{mn0,k}^d$ and $\omega C_{nm0,k}^g$ and thus for lower RF frequency (5) can be simplified to

$$v_{out,\omega_{IM}} \approx \sum_{k=1}^{N} (P_{mn0,k}^d \cdot G_{nm0,k}^d + P_{mn0,k}^g \cdot j\omega_{IM} C_{nm0,k}^g) \quad (6)$$

or $\quad v_{out,\omega_{IM}} \approx \sum_{k=1}^{N} (P_{mn0,k}^d \cdot G_{nm0,k}^d)$, if $P_{mn0,k}^g \approx P_{mn0,k}^d \quad (7)$

which provides the following circuit design insights for any LNA topology in a CMOS technology:

a. For narrowband IM2 (ω_{IM}@low frequency) all the nonlinear capacitances can be neglected and only three second-order conductances of each transistor are important, namely, nonlinear transconductance

(G_{200}), output conductance (G_{020}) and cross-modulation conductance (G_{110}).

b. For IM3 (ω_{IM}@ RF tone) the nonlinear capacitances at the drain terminal can be neglected and if $P_{mn0,k}^g$ is not much larger than $P_{mn0,k}^d$ the nonlinear capacitances at the gate terminal can also be neglected.

In summary this generalized distortion analysis shares an approach that has similarities with harmonic balance analysis in approximating all node voltages and branch currents by truncated Fourier series. By utilizing the diagram shown in Fig. 3 and cutting redundant frequency elements our analysis method simplifies complex weakly nonlinear analyses into relatively simple calculations. The resulting general weakly nonlinear model can be applied for different LNA topologies as will be shown next. The model uses simple closed-form expressions to describe the total output distortion for both narrowband and wideband application, which is just linear combination of the nonlinear coefficients of each transistor and terminal AC gains. In fact this general model can also be applied to [10] where a state-space approach is used to get low complexity analytic expressions of distortions for different fully balanced band Gm-C filters

IV. BENCHMARK ON ACCURACY

In order to evaluate the accuracy of the proposed general nonlinearity model, a common source (CS) amplifier, a narrowband cascode LNA [11] and a broadband noise-canceling LNA [12] are simulated in ADS using a commercial 90nm CMOS process. The topologies are shown in Fig. 4. The general LNA nonlinearity model is integrated in parametric cells (P-cells) [4] to provide distortion prediction.

The CS amplifier is an example verifying the two design insights given in the previous section. Fig. 5 shows that for narrowband IM2 three nonlinear conductances are the key factors, which verifies the design insight a. For the CS amplifier $P_{mn0,k}^g/P_{mn0,k}^d \approx g_m(Z_{Cgs}//R_s)$ is typically around 1~2 and thus insight b is applied, which is verified in Fig. 6 that all nonlinear capacitances at gate and drain terminals can be neglected for narrowband IM3. For the other two LNAs the 0simulated results by ADS and the results of our model are compared on output IM2 and IM3. The accuracy is defined by $\left|1 - \left|\dfrac{V_{ADS} - V_{model}}{V_{ADS}}\right|\right|$, where V_{ADS} and V_{model} are the ADS-simulated and model results for voltage magnitude of IM respectively. For the narrowband LNA the two-tone signal is at 5 GHz with 1 MHz spacing. By sweeping the load (Z_{Load}) the model is benchmarked for different gains as shown in Fig. 7. For the wideband LNA one tone is fixed at 1 GHz and the other tone is swept from 2.15 GHz to 10 GHz resulting in a perfect description of IM2 and IM3 at different frequency shown in Fig. 8.

In summary the benchmark shows the proposed general nonlinearity model predicts the output distortion for different topologies with very good accuracy. The simple closed-expressions with the model enable a very fast response time for calculating IM distortions (within 40 ms per one sweep)

providing over one order of magnitude advantage versus the transistor-level simulation in ADS, thus making it very suitable for implementation in a design automation loop for optimizing the circuit within short time.

V. CONCLUSION

A general weak nonlinearity model for different low-noise amplifier topologies was presented, which is achieved by our generalized weak nonlinearity analysis. Implemented by simple closed-form expressions this model provides a potential solution for LNA design automation with different topology candidates to improve the response time and for the circuit designers to gain insightful guidelines on LNA nonlinearity. Very good accuracy and short response time of this model is shown on intermodulation distortion calculation for different LNAs in both narrowband and wideband applications.

REFERENCES

[1] R.A.Baki, T.K.K. Tsang and M.N.El-Gamal, "Distortion in RF CMOS Short-Channel Low-Noise Amplifiers, *IEEE Trans. Microw. Theory Tech.*, VOL. 54, NO.1, pp. 46-56, Jan. 2006.

[2] W.Chen, G.liu, B.Zdravko and A.M.Niknejad, "A Highly Linear Broadband CMOS LNA Employing Noise and Distortion Cancellation," in Proceedings of *IEEE RFIC Symposium*, Hawaii, June 3-5, 2007.

[3] G. Tulunay, and S. Balkir, "Automatic synthesis of CMOS RF front-ends," in Proceedings of *IEEE ISCAS*, Greece, May 21-24, 2006.

[4] W.Cheng, A. J. Annema and B.Nauta , "A multi-step P-cell for LNA design automation," in Proceedings of *IEEE ISCAS*, May 2008, in press.

[5] P.Wambacq and W.Sansen, *Distortion Analysis of Analog Integrated Circuits*, Norwell, MA: Kluwer, 1998.

[6] P.Li and L.T.Pileggi, "Efficient per-nonlinearity distortion analysis for analog and RF circuits,"*IEEE Trans. CAD Des. Integr. Circuits Syst.*, vol. 22, pp. 1297-1309, 2003.

[7] J.Borremans, L.D.Locht, P.Wambacq and Y.Rolain, "Nonlinearity Analysis of Analog/RF Circuits using Combined Multisine and Volterra Analysis," in Proceedings of *IEEE DATE*, France, April 16-20, 2007.

[8] G.Palumbo and S.Pennisi, "High-frequency harmonic distortion in feedback amplifiers: Analysis and applications," *IEEE Trans. Circuit Syst. I, Reg. Papers*, vol. 50, no. 3, pp. 328-340, mar. 2003.

[9] A.J. Annema, et al,, "Analog Circuits in Ultra-Deep Sub-Micron CMOS," *IEEE J. Solid-State circuits*, vol. 402, No. 2, Jan. 2005

[10] A.Celik, Z.Zhang and P.P.Sotiriadis, "A state-space approach to intermodulation distortion estimation in fully balanced bandpass Gm-C Filters with weak nonlinearities" *IEEE Trans. Circuit Syst. I, Reg. Papers*, vol. 54, Issue 4, pp. 829-844, April 2007.

[11] T.Lee, *The design of CMOS radio-frequency integrated circuits*, 2nd ed., Cambridge: Cambridge University Press, 2004.

[12] C.–F. Liao and S.-I. Liu, "A Broadband Noise-Canceling CMOS LNA for 3.1-10.6-GHz UWB Receivers", *IEEE J. Solid-State circuits*, vol. 42, Issue. 2, Feb. 2007

[13] http://eesof.tm.agilent.com/products/ads_main.html

[14] http://www.nxp.com/Philips_Models/ mos_models/index.html.

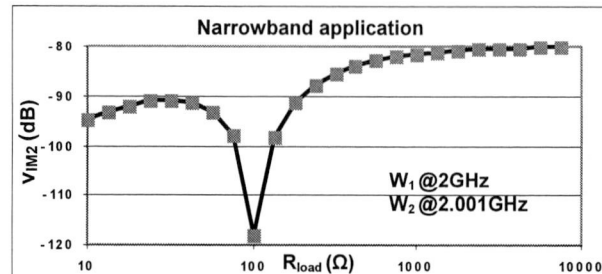

Fig. 5 Comparison between simulated (line) and model (symbol) results on output voltage IM2@1MHz for a CS amplifier as a function of load resistance. Squares represent the model (only contains three nonlinear conductances and neglects nonlinear capacitances).

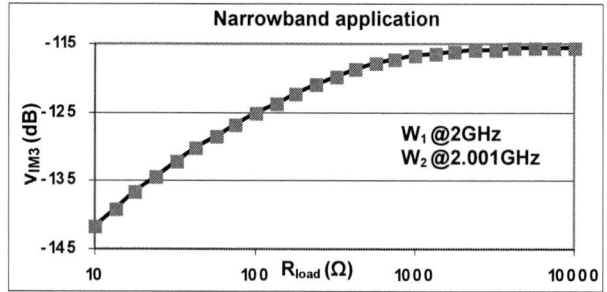

Fig. 6 Comparison between simulated (line) and model (symbol) result on output voltage IM3@1MHz for a CS amplifier as a function of load resistance. Squares represent the model (neglecting nonlinear capacitances at gate and drain terminals)

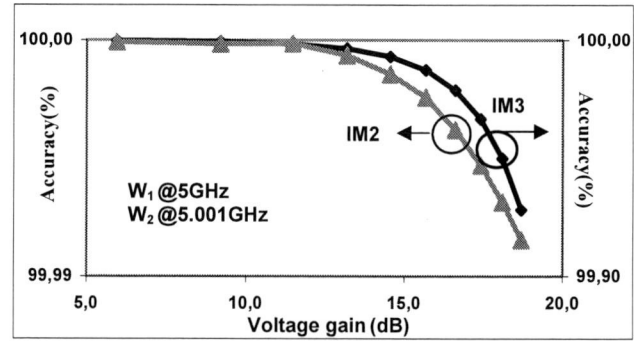

Fig. 7 Accuracy of our model on IM2 and IM3 for narrowband cascode LNAs over different voltage gain (v_{out}/v_{in})

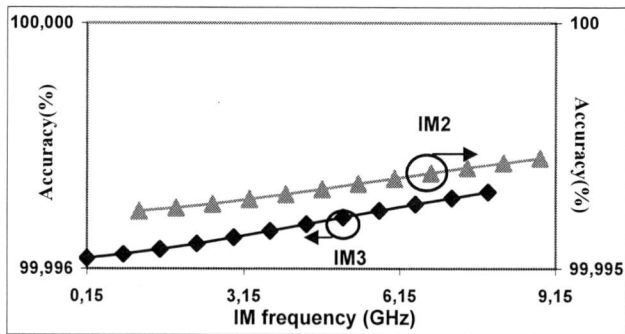

Fig. 8 Accuracy of our model on IM2 and IM3 prediction for a wideband LNA (IM2 @w_1-w_2 ranging from 1.15 GHz to 8.85GH; IM3@w_2-2w_1 ranging from 0.15GHz to7.85GHz)

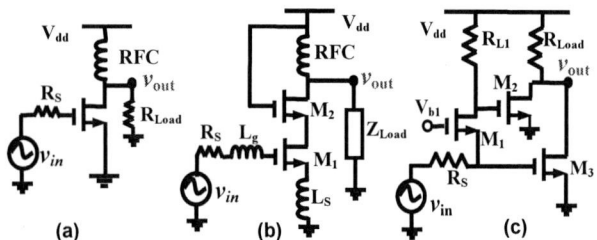

Fig. 4. Schematics of (a) CS amplifier, (b) narrowband cascode LNA and (c) wideband noise-cancelling LNA

IEEE 2008 Custom Intergrated Circuits Conference (CICC)

Modeling and Synthesis of Wide-Band Switched-Resonators for VCOs

Bodhisatwa Sadhu, Umaikhe E. Omole, and Ramesh Harjani

University of Minnesota

Abstract-**A two-level switched resonator model for synthesizing a wide-tuning range LC VCO is described. Viable resonator design styles are selected for achieving broadband frequency synthesis. A simple, generalized model is developed for design space exploration. Using this model, based on the performance specifications, a resonator style is selected. Also, the performance specifications are translated to individual resonator components. At the second level, detailed models for the resonator components are developed. The resonator is then designed and simulated based on these models. As an example, a wideband VCO using a new inductor switching technique is explored. Simulations performed in a 0.18 μm CMOS process achieved a tuning range from 3.35 to 8.34 GHz while exhibiting phase noise between -107 and -119 dBc/Hz at 1 MHz offset over the entire range. The VCO core dissipated 1.9 to 8.3 mW power using a 1.8-V supply.**

I. INTRODUCTION

The increased congestion of the wireless spectrum has made cognitive programmable radios highly desirable. To be able to operate over a wide range of frequencies, such programmable radios require very wide tuning range VCOs. Additionally, modern multi-carrier modulation schemes such as OFDM impose stringent phase noise specifications. In this paper we develop a two-level switched resonator model for synthesizing low phase noise, wide tuning range VCOs. The phase noise requirements make LC resonators the preferred choice. However, LC VCOs have traditionally been targeted for narrowband applications. Though switched inductor resonators provide a viable solution, very limited literature is available on the analysis and modeling of such systems. Therefore, the technique remains ill-exploited.

The general model developed in this paper provides design insight into the trade-offs involved in switched resonator circuits. More detailed models are then developed for circuit-level simulation. A method to synthesize wide-band, low phase noise LC VCOs based on these models is demonstrated.

The paper is organized as follows. In Section II, viable LC resonator styles for wideband VCO solutions are discussed. In Section III, generalized switched resonator model is developed and analyzed. Based on this model, a procedure for performing topology selection and translating the block-performance specifications into individual component specifications is developed. Detailed component modeling is then described in Section IV, and a design methodology for an optimized resonator based on these models is summarized in Section V. Simulation results from an example design used to validate the modeling and design framework are discussed in Section VI.

II. LC RESONATOR STYLES FOR WIDE-TUNING RANGE

LC VCOs can be tuned by changing the inductance, the capacitance or both [1]. In the absence of an effective method

to continuously tune the inductor, the inductance value can be altered by switching to a new inductor [1], or by tapping a portion of a single inductor [2][3]. The first technique is area inefficient and is not considered in this paper. Capacitors can be varied stepwise by using switched capacitor banks, and continuously using varactors. A general model for a switched resonator is shown in Fig. 1.

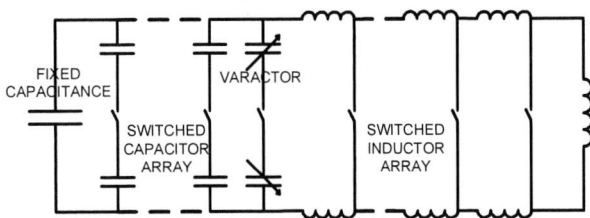

Fig. 1. A general switched resonator model

III. A SIMPLE MODEL FOR DESIGN SPACE EXPLORATION

A simple model for the switched resonator is shown in Fig. 3. Only one inductor switch is analyzed here, though the analysis can easily be extended to multiple switches. The frequency of oscillation is given by $f = 1/(2\pi\sqrt{LC})$.

The two frequency banks obtained from the one switch inductor is shown in Fig. 2.

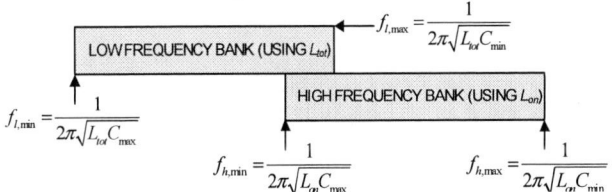

Fig. 2. Frequency synthesis scheme using the single switch resonator of Fig. 3

Here L_{tot} and L_{on} are the effective inductances when the switch is off and on respectively, and C_{min} and C_{max} are the minimum and maximum capacitances attainable. We also define a *switching ratio* $k = L_{off}/L_{tot}$, where L_{off} is the fraction of the inductor switched out. In the ideal case, $L_{off} = L_{tot} - L_{on}$. In order to obtain a continuous frequency range, we will make these frequency bands contiguous.

Without loss of generality, we assume that the switch is implemented by an nMOS transistor. We begin with simple models for the switch, depending on its region of operation. When on (transistor in triode), we can model the switch as a resistor in parallel with the switched out part of the inductor (Fig. 6). When off, it can be modeled (Fig. 4) as a capacitor (C ~ (C_{gd}+C_{db})/ 2). Using parallel to series transformations, we can absorb the switch parasitic into the inductor branch. An equivalent parallel RLC tank can then be constructed using series to parallel transformation of the combined inductor

S: Switch

Fig. 3. Simple model for switched resonator

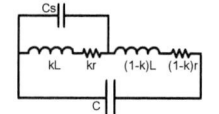

Cs: Effective transistor capacitance when off

r: Total parasitic resistance of inductor

Fig. 4. Resonator when the switch is off

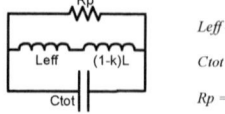

$Leff \sim kL$

$Ctot \sim C + kCs$

$Rp = (1+Q^2)r$

Fig. 5. Parallel RLC equivalent with switch off

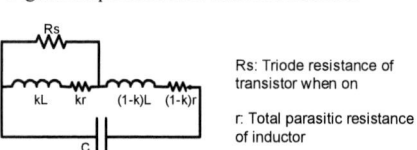

Rs: Triode resistance of transistor when on

r: Total parasitic resistance of inductor

Fig.6. Resonator when the switch is on

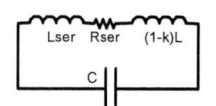

$$R_{ser} = \frac{R^2(R+kr)}{(R+kr)^2+(\omega kL)^2}$$

$$L_{ser} = \frac{R^2kL}{(R+kr)^2+(\omega kL)^2}$$

Fig. 7. All parasitic resistances absorbed into Rs when the switch is on

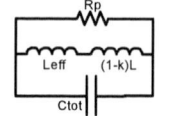

$Leff \sim Ls$

$Ctot \sim C$

$Rp = (1+Q^2)Rs$

Fig. 8. Parallel RLC equivalent with switch on

branch parasitics (Fig. 4. to Fig. 8). As is evident, the inductor switch introduces parasitics that impact the performance. This comes as a price paid for the wider frequency range obtained, and as we will see, the technique is not beneficial for narrowband tuning.

IV. DESIGN INSIGHT AND RESONATOR SYNTHESIS

Based on the simple model developed, we now present a design example to illustrate quantitatively the trade-offs involved in switched resonator design. In the model described, the inductor and capacitor designs, switch size, and the switching ratio k, are all design variables. The choice of inductors and capacitors for LC VCOs has already been explored in detail [4], and is fairly independent of the other variables. Given the finite length of this paper, and the mathematical clutter introduced by too many dimensions in our design space, we will focus only on choosing the switching ratio, and the switch size. For this example design, we use a frequency floor plan spanning 3-9 GHz, and a 1 nH spiral inductor based on criteria developed in [4] and will target the TSMC 0.18 μm technology.

From an exploration of the design space using this simple model, we will determine: 1) when switching the inductor is beneficial, and if so, 2) what is the optimal switching ratio (k), and 3) the optimal switch size.

We begin by deriving power and phase noise expressions based on this resonator model. Assuming a constant voltage swing at the output, the power dissipated can be derived as:

$$Power = \frac{R_S V_{sw} V_{DD}}{(\omega L)^2} \qquad (1)$$

Note that the worst case value occurs either at $f_{l,min}$ or $f_{h,min}$. As the inductance was chosen with due consideration to the power specification at $f_{l,min}$, we monitor $f_{h,min}$ for the worst case power performance. The power dissipation at $f_{h,min}$ is plotted Fig. 11.

Deriving the phase noise expression from Leeson's model, we obtain:

$$L(\Delta\omega) \cong 10.\log\left[\frac{kTFR_S}{V_{rms}}\left(\frac{\omega_0}{\Delta\omega}\right)^2\right] \qquad (2)$$

Due to the ω^2 term in expression (2), the worst case phase noise occurs at $f_{h,max}$. The phase noise at $f_{h,max}$ is plotted in Fig. 10. Observe that for resonators dominated by the inductor Q, the phase noise reduces with a reduction in the effective series resistance, R_{ser} (Fig. 7). Fig. 9 shows a plot of R_{ser} vs. k in the higher frequency bank. R_{ser} is lower than the effective parasitic

resistance of an un-switched inductor ($r=2.4\Omega$) only for higher values of k. This provides a lower limit for k (condition 1) for obtaining phase noise benefit from switching.

The start-up condition for oscillation is given by: $g_m R_p = (g_m L_s)/(C_p R_s) > 1$. Interestingly, pure capacitive tuning by using a low inductance value violates this condition at some point, showing a basic limitation of capacitive tuning for broadband purposes. Even for the switched inductor resonator described here, this capacitive tuning limitation provides an upper bound for k (condition 2) if contiguous banks are desired.

Summarizing these two effects, from condition 1, switched inductor tuning is not beneficial for low k values where an alternative capacitive tuning solution is feasible (in narowband applications). However, for broadband requirements, switched inductors not only provide a wideband solution where pure capacitive tuning fails (start-up condition), but also provides phase noise benefits as compared to pure capacitive tuning.

From Fig. 10 and Fig. 11, we note that the worst case power dissipation and phase noise increase drastically for high k values. These provide two more upper bounds for the value of k (conditions 3 and 4).

Also, from the plots in Fig. 9, Fig. 10 and Fig. 11, we note that the performance benefits from increasing the transistor width diminish beyond 600 μm. However, the parasitic capacitance continues to degrade linearly with increasing width, reducing the maximum frequency obtained. Based on this, we identify a 600 - 900 μm range for the optimum transistor width. Using this range, from the conditions 1, 3 and 4 derived, we find $0.25 < k < 0.65$. Within these limits, we want to maximize k for a higher frequency range. We consider condition 3 later to ensure contiguous frequency banks.

V. DETAILED COMPONENT DESIGN

Using the simple analysis above, ranges for the switch size and switching ratio k were obtained. In this section, we will develop detailed models for the individual resonator components to capture and analyze several higher order effects. This will facilitate schematic level simulations and help complete and fine-tune the resonator design.

A. Switched Inductor Design And Modeling

The switched spiral inductor is shown in Fig. 12. Inductor tapping has been a recent research trend [2][3], and a detailed model of the switched on-chip inductor has not been developed. With the technique shown in Fig. 12, the relatively

978-1-4244-2018-6/08 $25.00 © 2008 IEEE

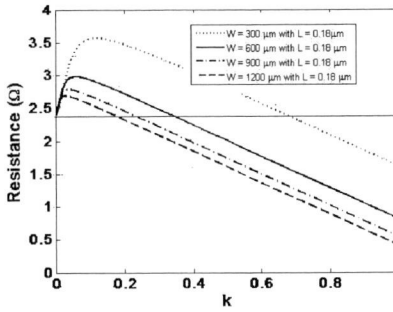

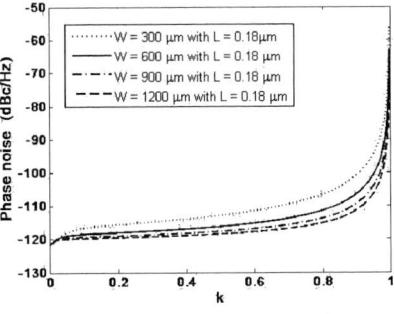

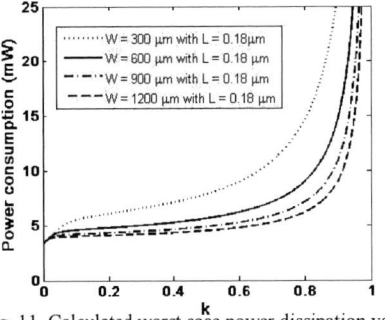

Fig. 9. Rser (see Fig. 6) vs k; here r (Fig. 5) = 2.4Ω   Fig. 10. Calculated worst case phase noise vs k Fig. 11. Calculated worst case power dissipation vs k

low quality inner part of the inductor is switched out, as compared to previous designs [3]. The nMOS switch was carefully positioned to reduce the effect of its parasitics. Recall that the switch is within the substrate that is physically lower than the inductor coils.

A circuit model for the switched spiral inductor (Fig. 13) was developed in ADS. It is an extension of the frequency independent physical $2\pi\psi$ model proposed in [5]. The spiral is modeled as a distributed structure consisting of an inner $2\pi\psi$ network with series/ladder RL elements $\{L_b, r_b, L_b', r_b'\}$. These are nested within an outer $2\pi\psi$ network with series/ladder elements $\{L_a, r_a, L_a', r_a'\}$. The inner section in Fig. 13 represents the portion of the spiral that has been switched out. A comparison of the simulated S-parameter components for the physical inductor (using Momentum) and the circuit model (using SpectreRF) shows excellent agreement over a wide frequency range (Fig. 14).

B. Switch Considerations

The trade-off between the switch on-resistance and off-capacitance was analyzed in the previous section. Here, we examine the frequency dependent substrate characteristics of the nMOS transistor in relation to the impact of the substrate response of the spiral. For this, the nMOS transistor model with resistive substrate network in [6] is converted via Y-transformation to an analogous model with frequency dependent shunt elements R_P' and C_P'

$$R_P' = R_{sub} + \frac{1}{(\omega C_{db})^2 R_{sub}}; \quad C_P' = \frac{C_{db} + C_\tau + (\omega R_{sub} C_{db})^2 C_\tau}{1 + (\omega R_{sub} C_{db})^2}$$

$$\text{where} \quad C_\tau = C_{gd} + \frac{C_{gb}}{2}, off \quad \text{and} \quad C_\tau = \frac{C_{gd}}{2}, on$$

similar to the shunt elements of the inductor π-model of [7]. The shunt elements of the switch and the inductor are then compared for different switch widths. From an analysis of the increase in the parasitic interaction between the spiral and the switch with increasing switch size, and exact width is selected. For our example design, the switch is constructed using three 0.18 µm parallel MOS devices each with sixty-four 4 µm wide fingers (overall width of 768 µm).

C. The Capacitance Bank

For this design, a capacitor bank was constructed using two 5-bit differential switched capacitor (using MIM capacitors) arrays, and pMOS inversion mode varactors. From simulations, the start-up condition was found to limit the freq-

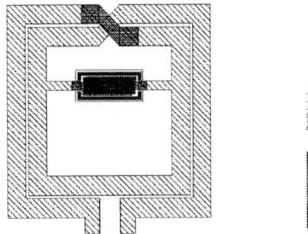
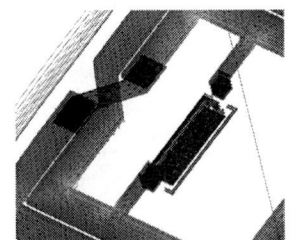

Fig. 12. The proposed switched inductor and its EM fill simulation(Momentum)

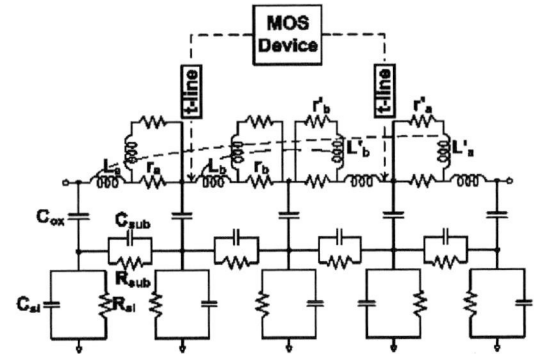

Fig. 13. Frequency independent model for the switched inductor

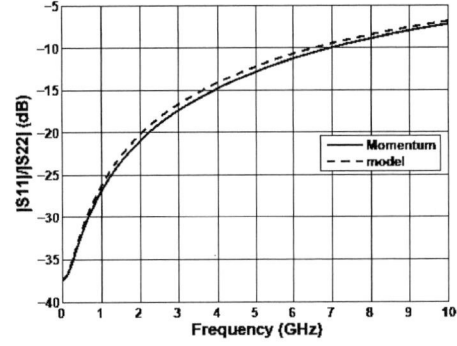

Fig. 14. Comparison of S-parameter magnitude ratios

uency tuning range to only 3.1 GHz.

For a higher tuning range, switching the inductor is necessary. In order to ensure contiguous bands, condition 2 needs to be satisfied. This limits the value of k to ~ 0.45. Since this value is greater than that given by condition 1, we expect a lowering of the phase noise in the upper bank by using this technique.

978-1-4244-2018-6/08 $25.00 © 2008 IEEE   245

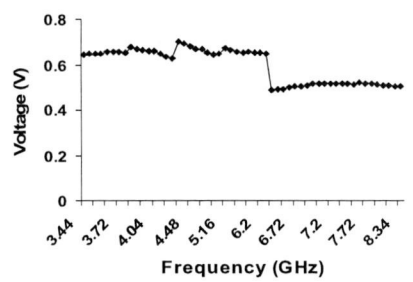

Fig. 15. Voltage swing vs. frequency

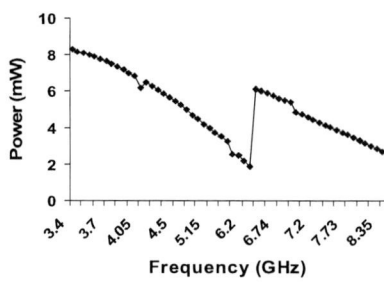

Fig. 16. Power consumption vs. frequency

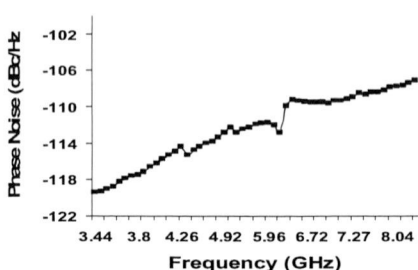

Fig. 17. Phase noise vs. frequency

VI. OVERALL DESIGN FLOW

The generalized design methodology developed and followed, based on the two-level model is summarized below.

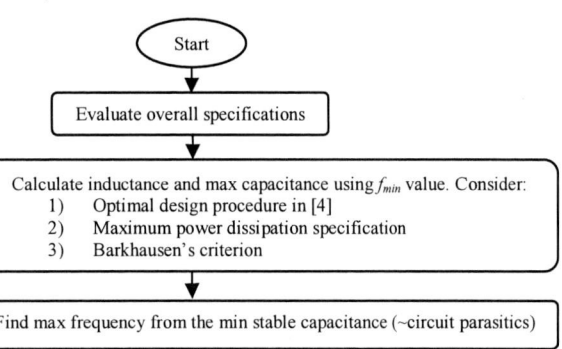

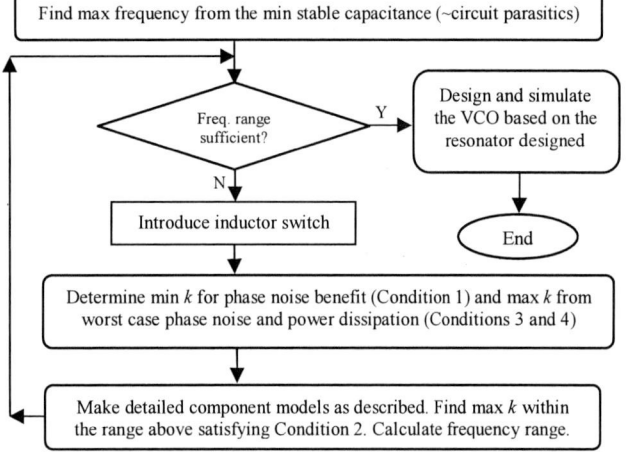

Fig. 18. Flowchart summary of design methodology

VI. SIMULATION RESULTS FOR A DESIGN EXAMPLE

A VCO was synthesized using the switched resonator flowchart above. Simulation results are discussed below. Table 1 gives a performance comparison of this design against recent work. The accurate modeling and optimized design strategy developed has helped us achieve an unprecedented wide tuning range (85%) with a good phase noise and reasonable power dissipation in simulation.

Tuning: Using the combination of discrete and continuous tuning described above, the VCO achieves a tuning range of 85 %, from 3.35 to 8.34 GHz. The nominal VCO frequencies in the off and on switched inductor modes are 6.37 and 8.34 GHz. Considerable overlap of the frequency banks is ensured to account for process variations. The *output voltage swing*

was kept approximately constant at 0.65V in the lower bank, and 0.5V in the upper bank (Fig. 15) by employing feed-forward current sources.

The *power dissipation* reduces with increasing frequency in each individual bank (Fig. 16) as predicted. The power dissipation varies from 1.9 mW to 8.3 mW.

The simulated *phase noise* is shown in Fig. 17. The plot does not show the predicted decrease in the phase noise when switching to the upper frequency bank due to a lower voltage swing here (0.5V compared to 0.65V in the low bank).

TABLE I: VCO PERFORMANCE COMPARISON					
Ref	f_0(GHz)	Tuning Range (%)	Phase Noise (dBc/Hz)@1MHz	Power (mW)	CMOS (μm)
[8]	3.07-5.61	58.7	-114.6 to -120.8	2.0 - 3.0	0.13 SOI
[9]	1.8	73	-126.5	2.6 - 10	0.18
This work	3.37-8.34	85	-107 to -119.4	1.9 - 8.3	0.18

VI. CONCLUSION

A switched resonator has been modeled at two levels. Based on the high-level model, the number of switching segments, switch sizes and optimal switching points to meet particular specifications can be quickly obtained. The detailed model developed enable further design space exploration through circuit simulation, rather than protracted EM simulations. An example design based on this framework attains a tuning range of 85% with good phase noise and low power dissipation. The framework described fits the design of a variety of broadband LC VCOs, including multiple inductor switching topologies. An algorithm based on this framework is easily developed for automating the resonator design process.

REFERENCES

[1] A. Kral, F. Behbahani, and A. A. Abidi, "RF-CMOS oscillators with switched tuning," *CICC,* pp. 555–558, 1998.
[2] S. M. Yim and K. K. O, "Demonstration of a switched resonator concept in a dual-band monolithic CMOS *LC* tuned VCO," *CICC,* pp. 205–208, 2001.
[3] J. Borremans *et al.*, "A single-inductor dual-band VCO in 0.06mm² 5.6 GHz multiband front-end in 90nm digital CMOS," *ISSCC,* pp. 324-325, 2008.
[4] D. Ham and A. Hajimiri, "Concepts and methods of optimization of integrated *LC* VCOs," *JSSC,* pp. 896–909, June 2001.
[5] Y. Cao *et al.*, "Frequency-independent equivalent-circuit model for on-chip spiral inductors," *JSSC,* pp. 419–426, March 2003.
[6] Y. Cheng and M. Matloubian, "On the high-frequency characteristics of substrate resistance in RF MOSFETs," *IEEE EDL,* pp. 604–606, Dec. 2000.
[7] C. P. Yue and S. S. Wong, "On-chip spiral inductors with patterned ground shields for Si-based RF ICs," *JSSC,* pp. 743–752, May 1998.
[8] N. H. W. Fong *et al.*, "Design of wideband CMOS VCO for multiband WLAN applications," *JSSC,* pp. 1333–1342, Aug. 2003.
[9] A. D. Berny *et al.*, "A 1.8-GHz *LC* VCO with 1.3 GHz tuning range and digital amplitude calibration," *JSSC,* pp. 909–917, April 2005.

978-1-4244-2018-6/08 $25.00 © 2008 IEEE

IEEE 2008 Custom Intergrated Circuits Conference (CICC)

Faster Statistical Cell Characterization using Adjoint Sensitivity Analysis

Ben Gu, Kiran Gullapalli, Yun Zhang, and Savithri Sundareswaran

Design Technology Organization, Freescale Semiconductor

7700 Parmer Lane, Austin, Texas 78729

Abstract—With the adoption of statistical static timing analysis (SSTA), the characterization of standard cell libraries for delay variations and output transition time (output slew) variations, referred to as statistical characterization, is becoming essential. Statistical characterization of intra-cell mismatch variations as well as inter-chip variations need to be performed efficiently with acceptable accuracy as a function of process parameter variations. The conventional approach to this problem is to model these mismatch variations by characterizing each device variation separately. However, this entails a cost that is proportional to the product of the number of devices (n_d) in the cell and the number of local statistical parameters (n_p), and characterization becomes infeasible. In this work, we propose an improved transient sensitivity analysis to accelerate statistical characterization. We compute sensitivities of node voltages with respect to any process/design parameters. These sensitivities are used to extract the sensitivities of delays and transition times. It is more critical to note the sparsity of the circuit's dependence on the statistical parameters (i.e., any given parameter directly impacts only a small portion of the circuit, sometimes only one device). By exploiting this sparsity we obtain a method that is $O(n_p)$, compared to $O(n_p \times n_d)$ of the conventional approach. As an example, for an AOI cell with 40 devices, the sensitivity analysis, compared to the standard approach using multiple simulations, results in more than 18X runtime improvements with better accuracy.

I. INTRODUCTION

Recently there has been significant interest in including process variations into static timing analysis [1]–[4]. Process variations of circuits are often described by device parameter variations, which can be classified into two broad categories: (a) global variations and (b) local variations. Typically, all chip-to-chip, across-wafer and wafer-to-wafer variations are combined as a global variation (also, commonly referred to as inter-chip variation). Variability across-chip (intra-die) is termed as local variation. Each parameter that has significant impact on the device characteristics has both global and local components of variation. The global component ΔX varies from chip-to-chip, but for a given chip this value is the same for all devices in the design. The local component, ΔR is the across-chip component, which can vary from device-to-device and captures both location or geometry dependent variations and mismatch or random variations.

With the adoption of statistical static timing analysis (SSTA), characterization of standard cell libraries for delay variations and output transition time (output slew) variations, referred to as statistical characterization, is becoming essential. When the intra-cell variations are considered in SSTA, the delay of a timing arc, D can be represented as follows:

$$D = D_0 + \sum_{i=1}^{m} d_i \Delta X_i + \sum_{j=1}^{n_d} \sum_{k=1}^{n_p} d_{jk} \Delta R_{jk} \qquad (1)$$

where D_0 is the nominal delay value characterized by setting variations ΔX_i, and ΔR_{jk} to zero, d_i and d_{jk} are direct sensitivities of the delay with respect to the global variations, ΔX_i and the local variations, ΔR_{jk} respectively. These sensitivities are deterministic quantities obtained from characterization results. The problem of statistical characterization becomes that of characterizing for these d_i and d_{jk} quantities, which are delay sensitivities to the global and local parameter variations. The most straightforward approach (referred to

as the brute–force method in this paper) to computing the intra-cell sensitivity is to run extra transient simulations (perturbed simulation) for each source of the local variations, R_{jk} in (1), after a nominal simulation with all the variations set to 0, and to evaluate the delay sensitivities, d_{jk}, by using the finite difference between the perturbed and nominal simulations. The runtime complexity of this approach is given by,

$$C_{bruteforce} = (n_p \times n_d + 1) \times C_{tran} \qquad (2)$$

where n_p is the number of intra-cell statistical parameters, n_d is the number of devices impacted by these parameters, and C_{tran} is the complexity of a transient simulation for the delay evaluation. This approach will quickly become infeasible as either n_d or n_p in a cell increases. In order to overcome this complexity, recent work has used empirical observations to cluster groups of devices that are varied together, instead of treating them as uncorrelated [5]. The method requires considerable insight into circuit behavior and is applicable mainly to sequential digital logic. For this class of circuits, the clustering approach leads to computational cost of $O(n_p)$.

In this work, we develop a more efficient and general method for the statistical cell characterization that does not require knowledge of the circuit. We improve the well–known adjoint transient sensitivity algorithm [6] to compute the delay sensitivities. This approach only requires one nominal transient simulation with all the local parameters set to their nominal values, and an extra sensitivity analysis to evaluate sensitivities of delay of the cell with respect of all the local parameters of all the devices in the cell. We show that by exploiting sparsity of device dependence on parameters and ignoring time points having insignificant contributions, we can reduce the cost of the extra sensitivity analysis to a insignificant level compared to the cost of the nominal transient simulation. This dramatically reduces the runtime of characterizations required to capture intra-cell mismatch variations. As an example, for statistical characterization of a AOI32x20 cell in Freescale 65nm technology, our approach is more than 18X faster than the brute–force method. More importantly, we show that the simulation complexity of our method asymptotically approaches $O(n_p)$ as the number of devices and parameters increases. The techniques we developed in this work to accelerate the statistical characterizations of standard cells can be generalized to improve the adjoint sensitivity algorithm to solve other sensitivity problems.

The paper is organized as follows: Section II provides an overview of transient sensitivity computation using the adjoint method. Section III presents the techniques we use to accelerate the adjoint method for statistical characterization. Experiments and accuracy analysis of several digital standard cells for delay variations are presented in Section IV. The results are illustrated for delay variations, but these techniques can be easily extended using a similar approach for output slew.

II. TRANSIENT SENSITIVITY ANALYSIS

The approach we take to speed up the statistical characterizations of standard cells is to use the transient sensitivity algorithm to evaluate d_{jk} (1). The transient sensitivity algorithm is well established

978-1-4244-2018-6/08 $25.00 © 2008 IEEE

and has been applied in different areas of circuit simulation [6]–[11]. In general, there are two well known methods for computing transient sensitivities: the direct and the adjoint methods [6], [7]. The advantages and disadvantages of these methods for sensitivity computation have been discussed in detail. The direct method is advantageous when the sensitivities of a large number of circuit responses with respect to a few circuit parameters are desired. On the other hand, the adjoint method is more efficient when the sensitivities of a few responses with respect to a large number of circuit parameters are required, which is the case for the statistical characterization of intra-cell variations.

A. Adjoint Sensitivity Algorithm

We start with a nonlinear circuit which can be described by the differential algebraic equation

$$\frac{d}{dt}q(x,t) + i(x,t) = 0 \tag{3}$$

where x is the node voltage or branch current in the circuit, and $q(x,t)$ and $i(x,t)$ are the charge/flux and current terms respectively. In (3), it is assumed that all q, i, and x have an explicit dependence on a local statistical parameter, p. In order to compute the sensitivity of x on p, dx/dp, we differentiate (3) with respect to p

$$\frac{d}{dt}\left(\frac{dq}{dp} + \frac{dq}{dx}\frac{dx}{dp}\right) + \frac{di}{dp} + \frac{di}{dx}\frac{dx}{dp} = 0 \tag{4}$$

$$\frac{d}{dt}\left(C\frac{dx}{dp}\right) + G\frac{dx}{dp} = -\frac{d}{dt}\left(\frac{dq}{dp}\right) - \frac{di}{dp} \tag{5}$$

where $G = di/dx$ and $C = dq/dx$ evaluated when $p = p_0$. Equation (5) can be written as, assuming, for example, the use of the backward Euler formula to discretize the time derivative,

$$dx/dp = (J)^{-1}(df/dp) \tag{6}$$

where

$$J = \begin{bmatrix} G_0 & 0 & \cdots & 0 \\ \vdots & \ddots & \ddots & \vdots \\ \cdots & -\frac{C_{n-2}}{h_{n-1}} & G_{n-1} + \frac{C_{n-1}}{h_{n-1}} & 0 \\ \cdots & 0 & -\frac{C_{n-1}}{h_n} & G_n + \frac{C_n}{h_n} \end{bmatrix} \tag{7}$$

and

$$\frac{df}{dp} = -\left[\frac{d}{dt}\left(\frac{dq}{dp}\right) + \frac{di}{dp}\right] \tag{8}$$

After a transient simulation of (3) with $p = p_0$, each sub-block of J is available from the Jacobian matrix of the Newton-Raphson iterations at each time point. The right hand side of (6), df/dp, can be evaluated using the finite difference method

$$\frac{df}{dp} = -\frac{1}{\Delta p}\begin{bmatrix} i_0(p_1) - i_0(p_0) \\ i_1(p_1) - i_1(p_0) + \frac{q_1(p_1) - q_1(p_0) - q_0(p_1) + q_0(p_0)}{h_1} \\ \vdots \\ i_n(p_1) - i_n(p_0) + \frac{q_n(p_1) - q_n(p_0) - q_{n-1}(p_1) + q_{n-1}(p_0)}{h_n} \end{bmatrix} \tag{9}$$

where $p_1 = p_0 + \Delta p$, and i_i and q_i are the current and charge at ith time point of the transient simulation of (3). The sensitivity, dx/dp, of the jth node of the circuit at the nth time point can be obtained by multiplying both sides of (7) by a unit vector $e^T = [0 \cdots 1 \cdots 0]$ with 1 on the location corresponding to the jth node and nth time point.

$$\frac{dx}{dp}\Big|_{t=t_n} = e^T(J)^{-1}\frac{df}{dp} \tag{10}$$

$$= \left(J^{-T}e\right)^T\frac{df}{dp} \tag{11}$$

So the flow of computing sensitivities using the adjoint method is to solve the transposed system

$$u = \left(J^{-T}e\right) \tag{12}$$

and then compute inner product of u and df/dp to obtain the sensitivity.

$$dx/dp = u^T(df/dp) \tag{13}$$

B. Computation of Delay Sensitivities

After dx/dp is computed, the sensitivity of delay or transition time of the cell, $d\tau/dp$, can be readily computed as outlined in [7]. For example, the sensitivity of the delay at the output node of a cell when its voltage (x) is equal to half of the supply voltage (V_{dd}) is given as:

$$\frac{d\tau}{dp}\Big|_{x=0.5V_{dd}} = -\frac{dx/dp}{dx/dt}\Big|_{x=0.5V_{dd}} \tag{14}$$

We compare the results of the adjoint sensitivity analysis with those of the brute–force approach, using a NAND2x2 cell in Freescale's 65nm technology, which is illustrated in Figure 1. Figure 2 compares sensitivity waveforms of the output node Z with respect to the local variations of the threshold voltage of each MOSFET in the cell at the last transition. The results given by the adjoint sensitivity method (lines) have a close agreement with the results given by the brute–force approach (circles).

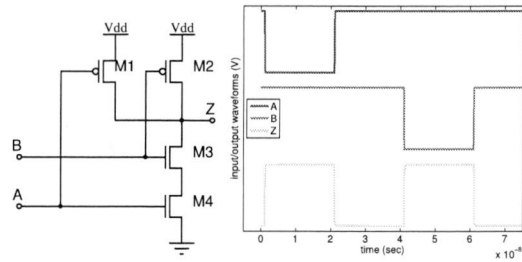

Fig. 1. A NAND2x2 cell in Freescale 65nm technology with all the parasitics ignored and its input/output switching activities.

III. ACCELERATING SENSITIVITY ANALYSIS

In this section, we provide a detailed analysis of computational complexity of the adjoint sensitivity algorithm compared to the brute–force approach. Using the adjoint sensitivity algorithm outlined in the last section, in order to compute a cell's delay sensitivities with respect to a local parameter we need to run a nominal transient simulation to evaluate J, solve the adjoint system to compute u in (12), compute df/dp and the inner product of df/dp and u by n_d times. Therefore, the cost of the adjoint sensitivity analysis is roughly given by,

$$C_{sensitivity} = C_{tran} + C_{adjoint} + n_p \times n_d \times C_{device} \tag{15}$$

978-1-4244-2018-6/08 $25.00 © 2008 IEEE   248

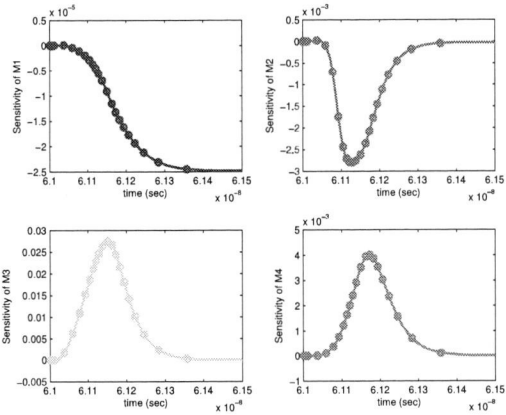

Fig. 2. Sensitivity waveforms of each MOSFET at the last transition, where lines are the results given by the sensitivity analysis and circles are those by the brute–force method.

where $C_{adjoint}$ is the cost to form and solve the adjoint system, and C_{device} is the cost of evaluating all the devices in the cell at all time points.

$$\frac{C_{bruteforce}}{C_{sensitivity}} = \frac{(n_p \times n_d + 1) \times C_{tran}}{C_{tran} + C_{adjoint} + n_p \times n_d \times C_{device}} \quad (16)$$

As n_p and n_d increase, the ratio asymptotically approaches C_{tran}/C_{device} given that C_{tran} and $C_{adjoint}$ in the denominator are not proportional to n_p and n_d. The complexity of a transient simulation in SPICE simulators can be approximated as

$$C_{tran} \approx C_{matrix} + n_{iter} \times C_{device} \quad (17)$$

where C_{matrix} is the cost of factorizing and solving the Jacobian matrix at every iteration of every time point of the simulation, and n_{iter} is the average number of the Newton-Raphson iterations at each time point. Typically, C_{matrix} is negligible compared to C_{device} for the standard cell circuits. Therefore, the maximum speed up we can achieve from using the adjoint method is roughly equal to n_{iter} (typically between 3 and 4 in SPICE simulators), compared to using the brute–force approach. This is apparently less than ideal considering that n_d can be fairly large for a complicated cell.

A. Sparsity of Device Evaluations

It is important to note that a local statistical parameter directly impacts only one device (sensitive device) in the cell. So instead of evaluating all the devices to compute df/dp vector, for a given parameter, we only need to evaluate the single sensitive device. Taking this into account, the ratio in (16) becomes:

$$\frac{C_{bruteforce}}{C_{sensitivity}} = \frac{(n_p \times n_d + 1) \times C_{tran}}{C_{tran} + C_{adjoint} + n_p \times C_{device}} \quad (18)$$

which is roughly equal to $(n_d + 1)/(1 + n_p/n_{iter})$ assuming that $C_{adjoint}$ can be neglected compared with C_{tran}. In other words, by exploiting the sparsity of dependence of devices on parameters, the overhead of the statistical cell characterization for each local parameter is only the cost of an extra device evaluation, C_{device}, which is about $1/n_{iter} \approx 1/3$ of the non-statistical characterization. In practical cell characterizations, very often we need to compute delay sensitivities at different transitions and different thresholds, i.e. the delay between falling of A and rising of Z ($A_f \rightarrow Z_r$), and the delay

between falling of B and rising of Z ($B_f \rightarrow Z_r$), *etc.*, in the NAND2x2 cell. This requires solving the adjoint system multiple times, so that the actual speed up achieved by using the adjoint sensitivity algorithm also depends on the ratio between $C_{adjoint}$ and C_{tran}.

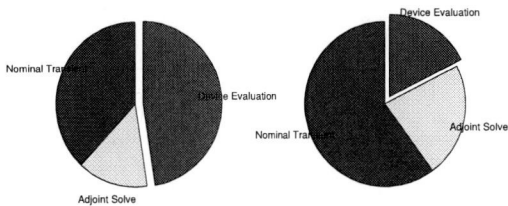

Fig. 3. Ratio of C_{tran}, $C_{adjoint}$ and C_{device} before and after exploiting device sparsity.

Figure 3 presents comparisons between C_{tran}, $C_{adjoint}$ and C_{device} for the NAND2x2 cell before and after we exploit the device sparsity with the adjoint system solved 16 times.

B. Ignoring Time Points with Insignificant Contributions

In (13) the sensitivity is evaluated as the cumulative summation of udf/dp of each time point.

$$dx/dp = \sum_{j=0}^{n} u_j (df/dp)_j \quad (19)$$

where $u_j(df/dp)_j$ is the contribution from the jth time point to the total sensitivity at the time point we are interested. Figure 4 plots the magnitude of udf/dp of each MOSFET in the NAND2x2 cell as function of time. An important observation is that the magnitude drops exponentially in reverse time, which means that the contributions from most of the prior time points are not significant to the total sensitivity except a few very close to the time point we are interested in. Another way to present this is to plot the cumulative summation of $u_j(df/dp)_j$ in (19) in reverse time. As shown in Figure 5, the cumulative sum saturates very rapidly as we travel through the time points backwards. Among the total 1110 time points in the simulation, the contributions from time points [0 1000] do not have impact on the accuracy of the sensitivity at the last time point up to four digits after the decimal point. Exploiting this behavior can further speed up the device evaluation time, which makes C_{device} become insignificant compared to C_{tran} and $C_{adjoint}$.

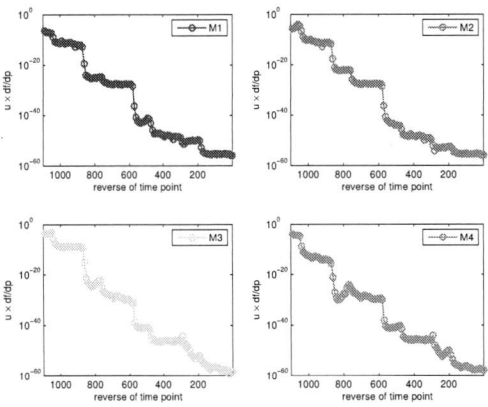

Fig. 4. $u_j(df/dp)_j$ as a function of reverse time.

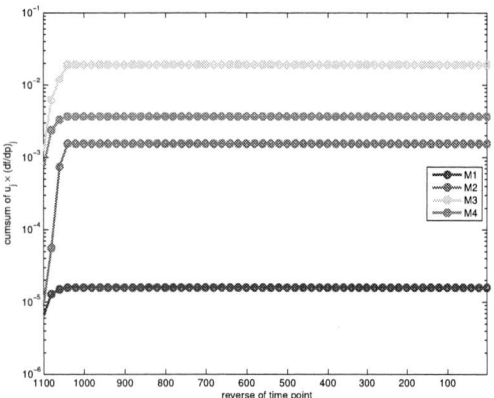

Fig. 5. Cumulative sum of $u_j (df/dp)_j$ as a function of reverse time.

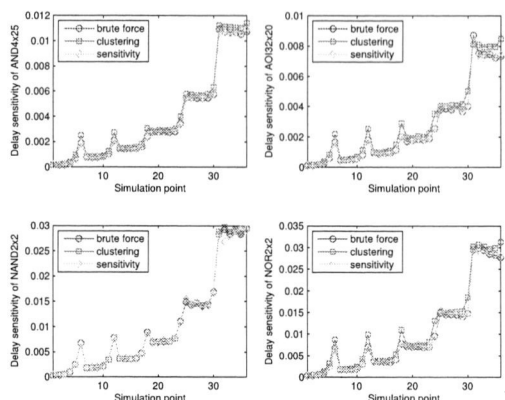

Fig. 6. Comparison of delay sensitivities computed using different methods of some standard cells in Freescale 65nm technology.

IV. BENCHMARK RESULTS

The algorithm we developed in this work has been implemented into Mica, the in–house circuit simulator of Freescale Semiconductor, and has been used in the statistical cell characterizations for 65nm technology. In order to validate the sensitivity algorithm developed in this work, we present characterization results of four different standard cells, AND4x25, AOI32x20, NAND2x2, and NOR2x2 using the brute–force method, the clustering method [5], and the sensitivity algorithm respectively. Figure 6 shows that the results given by the new algorithm at 36 different simulation points (different input slew and output load) have a close agreement with those given by the brute–force and the clustering method. Table I summarizes the CPU time improvements given by different methods on these four cells. Generally, the new algorithm provides significant runtime improvement over the brute–force method, and as outlined in the last section the improvement is more impressive for cells with larger n_d. Compared to the clustering method, our method can be considerably more efficient for most of cells without using any knowledge of the cell. Note on the AOI32x20 cell in Table I, the speed up of our algorithm over the clustering method, less than 2X, is not as significant as other cells. In this cell there are only few devices (cluster) contributing to the output transition, so most of the devices can be ignored in computing the overall delay sensitivities of the cell. The same idea has been employed to speed up our sensitivity algorithm further, and the results will be reported elsewhere.

TABLE I
SPEEDUP RATIO COMPARED TO THE BRUTE–FORCE METHOD

Cell	n_d	Brute-force	Clustering	Sensitivity
AND4x25	32	1.0	4.02	14.22
AOI32x20	40	1.0	13.20	18.86
NAND2x2	4	1.0	1.69	4.35
NOR2x2	4	1.0	1.51	4.15

V. CONCLUSIONS

In summary, an improved adjoint sensitivity algorithm has been developed to compute the sensitivities of delay, transition time, and *etc.*, with respect to the local statistical variations required by SSTA. We show that by exploiting the sparsity of device evaluations and ignoring time points with insignificant contributions, the computational efficiency of the adjoint sensitivity algorithm can be significantly better than the brute–force method. We demonstrate the superiority of the sensitivity analysis using results from some cells in Freescale 65nm library.

VI. ACKNOWLEDGMENT

The authors are grateful to all the members of the Freescale Mica team for being wonderful to work with. They also want to thank Rajendran Panda and Patrick McGuinness for several helpful discussions.

REFERENCES

[1] M. Orshansky and K. Keutzer. A general probabilistic framework for worst case timing analysis. In *Proceedings of Design Automation Conference 2002*, pages 556–561, 2002.

[2] A. Devgan and C. Kashyap. Block-based static timing analysis with uncertainty. In *Proceedings of IEEE ICCAD 2003*, 2003.

[3] J.A.G. Jess, K. Kalafala, S.R. Naidu, R.H.J.M. Otten, and C. Visweswariah. Statistical timing for parametric yield prediction of digital integrated circuits. In *Proceedings of Design Automation Conference 2003*, pages 17–24, 2003.

[4] V. Khandelwal and A. Srivastava. A General Framework for Accurate Statistical Timing Analysis considering Correlations. In *Proceedings of Design Automation Conference 2005*, 2005.

[5] Savithri Sundareswaran, Jacob Abraham, Alexandre Ardelea, and Rajendran Panda. Characterization of Standard Cells for Intra-Cell Mismatch Variation. In *Proceedings of International Symposium on Quality Electronic Design*, March 2008.

[6] S.W. Director and R.A.Rohrer. Generalized adjoint network and network sensitivities. *IEEE Trans. on Circuit Theory*, 16(8):318–323, August 1969.

[7] D.A. Hocevar, P. Yang, T.N. Trick, and B.D. Epler. Transient sensitivity computation for mosfet circuits. *IEEE Trans. on Computer-Aided Design of Integrated Circuits and Systems*, 4(10):609–620, October 1985.

[8] C.J. Chen and W.S. Feng. Relaxation-based transient sensitivity computations for MOSFET circuits. *IEEE Trans. on Computer-Aided Design of Integrated Circuits and Systems*, 14(2):173–185, August 1995.

[9] A.R. Conn, P.K. Coulman, R.A. Haring, G.L. Morrill, C. Visweswariah, and W.W. Chai. JiffyTune: circuit optimization using time-domain sensitivities. *IEEE Trans. on Computer-Aided Design of Integrated Circuits and Systems*, 17(12):1292–1309, December 1998.

[10] T.V. Nguyen, P. O'Brien, and D. Winston. Transient sensitivity computation for transistor level analysis and tuning. In *Proceedings of IEEE ICCAD 1999*, pages 120–123, 1999.

[11] Z. Wang and J. Roychowdhury. Obtaining Frequency Sensitivities to Variations Analytically from Parameterized Nonlinear Oscillator Phase Macromodels. In *Proceedings of the Custom Integrated Circuits Conference 2007*, pages 611–614, 2007.

An ESD-Protected 5-GHz Differential Low-Noise Amplifier in a 130-nm CMOS Process

Yuan-Wen Hsiao and Ming-Dou Ker

Nanoelectronics and Gigascale Systems Laboratory
Institute of Electronics, National Chiao-Tung University, Hsinchu, Taiwan

Abstract- **A novel ESD protection design for radio-frequency (RF) differential input/output (I/O) pads is proposed and successfully applied to a 5-GHz differential low-noise amplifier (LNA) in a 130-nm CMOS process. In the proposed ESD protection design, an ESD bus and a local ESD clamp device are added between the differential input pads to quickly bypass ESD current, especially under the pin-to-pin ESD-stress condition. With 10.3-mW power consumption under 1.2-V power supply, the differential LNA with the proposed ESD protection design has the human-body-model (HBM) ESD robustness of 3 kV, and exhibits 18-dB power gain and 2.62-dB noise figure at 5 GHz. Experimental results have demonstrated that the proposed ESD protection circuit can be co-designed with the input matching network of LNA to simultaneously achieve excellent RF performance and high ESD robustness.**

I. INTRODUCTION

Radio-frequency integrated circuits (RF ICs) are used in a lot of portable electronics devices for wireless communications. Recently, the feature size of MOS transistor in CMOS processes has been continuously scaled down to improve its high-frequency characteristics, which makes them more attractive to implement RF ICs. With the advantages of high integration capability and low cost for mass production, RF ICs operating in giga-Hz frequency bands have been fabricated in CMOS technology. In an RF receiver, low-noise amplifier (LNA) plays a very important role because it is the first stage in the RF receiver. Differential configuration is popular for LNA design because differential LNA has the advantages of common-mode noise rejection, less sensitivity to substrate noise/supply noise, and less bond-wire inductance variation [1]. In addition, the differential output signals of the differential LNA can be directly connected to the differential inputs of the double balanced mixer.

Electrostatic discharge (ESD), which has been the most important reliability issue in IC industry, is getting more attention in nanoscale CMOS technology. With the advances of CMOS processes, ESD robustness of CMOS ICs becomes worse because of the thinner gate oxide of MOS transistors. Since the input of LNA is usually connected to the external node of RF receiver chip such as the off-chip antenna, on-chip ESD protection circuits are needed for all input pads of LNA. However, applying ESD protection circuits at the input pads inevitably introduce impacts to RF performance. Therefore, LNA and ESD protection circuit have to be co-designed to simultaneously optimize RF performance and ESD robustness.

Recently, several ESD-protected LNAs in CMOS processes have been reported [2]-[6]. The typical ESD protection design with diodes at the input pad and the power-rail ESD clamp circuit had been verified [2]. Besides diodes, the inductor-based ESD-protected LNA had been proposed [3]. The inductor serves as part of the input matching network and provides ESD current path. With the parallel LC network resonating at the RF operation frequency, the shunt impedance of the ESD protection circuit becomes very large to mitigate the RF performance degradation [4]. Moreover, the LC-tanks, which resonate at the RF operation frequency, can be used between input pad and VDD, as well as between input pad and VSS, respectively [5]. In addition to single-ended LNA, ESD protection design for differential LNA had also been reported, recently [6].

In this paper, a novel ESD protection design for differential input/output (I/O) pads is proposed. Applied to a 5-GHz differential LNA, the ESD-protected LNA has been designed and successfully verified in a 130-nm CMOS process. The reference differential LNA without ESD protection was also designed and fabricated in the same wafer for comparison. Moreover, the pin-to-pin ESD stress on differential LNAs, which was never mentioned in the previous works, is first studied in this work.

II. DESIGN OF DIFFERENTIAL LNA

The circuit schematic of the differential LNA without ESD protection is shown in Fig. 1. The architecture of common-source with inductive degeneration is applied to match the source impedance (50 Ω) at the resonant frequency of 5 GHz. Cascode configuration provides good stability and reduces the Miller effect [7]. With the input matching network resonating at the RF operation frequency, the input impedance is purely real and proportional to the inductance of source inductor L_{S1} (L_{S2}). Therefore, input impedance matching can be achieved by choosing appropriate inductance for L_{S1} and L_{S2}. After the inductance of L_{S1} (L_{S2}) is determined, the remaining capacitive impedance needs to be cancelled by the inductance of gate inductor L_{G1} (L_{G2}). The drain inductor L_{D1} (L_{D2}) and drain capacitor C_{D1} (C_{D2}) form the output matching network. The LC-tank consisting of L_{TANK} and C_{TANK} is used to enhance the common-mode rejection. NMOS transistors M_1 - M_4 are fully silicided. All devices in Fig. 1 except the bias tee were fully integrated in the experimental test chip.

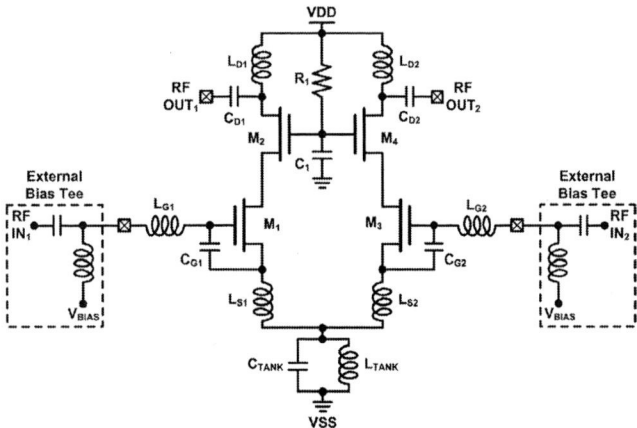

Fig. 1. Circuit schematic of differential LNA without ESD protection.

III. ESD PROTECTION DESIGN FOR DIFFERENTIAL LNA

A. ESD Protection Device

Silicon-controlled rectifier (SCR) is suitable for ESD protection design for RF ICs, because it has the features of high ESD robustness and low parasitic capacitance under a small layout area [8]. Besides, SCR had been demonstrated to be the optimum ESD protection device for high-speed differential I/O pads [9]. The P-type substrate-triggered SCR (P-STSCR) is utilized in this work to protect the LNA against ESD stresses. The cross-sectional view of the P-STSCR is shown in Fig. 2(a). The SCR path exists among the P+ diffusion (anode), N-well, P-well, and N+ diffusion (cathode). The equivalent circuit of the P-STSCR is shown in Fig. 2(b), which consists of a parasitic vertical PNP BJT and a parasitic lateral NPN BJT. The PNP BJT Q_{PNP} is formed by the P+ diffusion (anode), N-well, and P-well. The NPN BJT Q_{NPN} is formed by the N-well, P-well, and N+ diffusion (cathode).

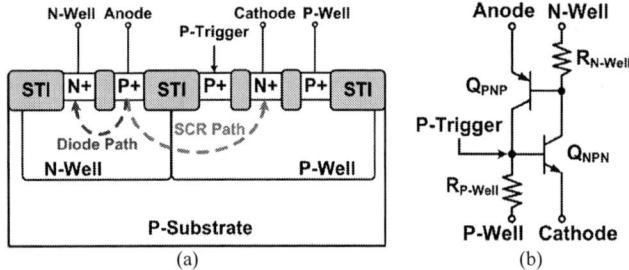

Fig. 2. (a) Cross-sectional view of P-STSCR. (b) Equivalent circuit of P-STSCR.

To quickly turn on the P-STSCR during ESD stresses, the P+ trigger diffusion (in the P-well region) was added between the anode and cathode. An ESD detection circuit was designed to inject trigger current to the P-trigger node during ESD stresses. After the trigger current is injected into the P-trigger node, the SCR is turned on to bypass ESD current. Since the holding voltage (2.58 V at 85 °C) of the SCR in this work is larger than the supply voltage of 1.2 V, the SCR can be safely used without latchup issue.

B. Differential LNA With Proposed ESD Protection Design

The schematic of differential LNA with the proposed ESD protection design is shown in Fig. 3. An ESD bus is inserted between the differential input pads and VDD. The anode of P-STSCR₁ is connected to the ESD bus with its cathode grounded. A P-diode is connected between each input pad and the anode of P-STSCR₁, whereas an N-diode is connected between VSS and each input pad. With the N-well of P-STSCR₁ directly connected to VDD, ESD current paths from the input pads to VDD can be established. Besides, a power-rail ESD clamp circuit was also designed to provide ESD current path between VDD and VSS to achieve comprehensive whole-chip ESD protection [10]. The power-rail ESD clamp circuit consists of an RC timer, an inverter, and an ESD clamp device realized by another SCR (P-STSCR₂). During ESD stresses, when ESD voltage is coupled to VDD, RC delay causes the PMOS in the inverter to be turned on and therefore to inject trigger current to turn on P-STSCR₁ and P-STSCR₂. Under normal circuit operating conditions, the output of inverter is kept at low (0 V) to turn off P-STSCR₁ and P-STSCR₂.

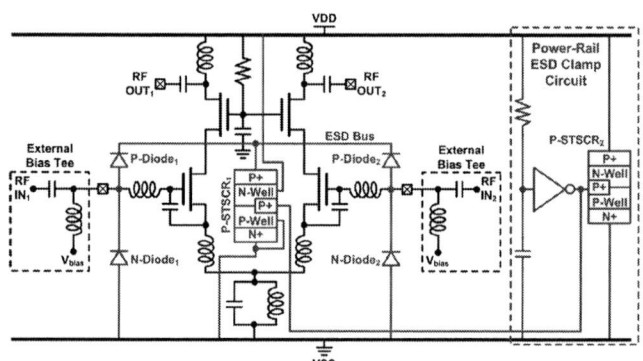

Fig. 3. Circuit schematic of differential LNA with the proposed ESD protection design.

IV. EXPERIMENTAL RESULTS

The differential LNAs without ESD protection and with the proposed ESD protection design have been fabricated in the same wafer in a 130-nm CMOS process to compare their performance. The chip microphotographs of the differential LNAs without ESD protection and with the proposed ESD protection design are shown in Figs. 4(a) and 4(b), respectively. The differential LNA without ESD protection occupies the area of 1070 μm × 630 μm, and the circuit area of the differential LNA with the proposed ESD protection design is 1090 μm × 750 μm. Both LNA consume 10.3 mW under 1.2-V power supply because the ESD protection circuit does not consume any DC power.

A. RF Performance

Fig. 5(a) shows the measured S21-parameters of the fabricated LNAs. Both LNAs have their best S-parameters around 5 GHz. The power gain of the differential LNA without ESD protection is 16.2 dB at 5 GHz, while the

978-1-4244-2018-6/08 $25.00 © 2008 IEEE

differential LNA with the proposed ESD protection design exhibits 18-dB power gain. The measured S11-parameters are compared in Fig. 5(b). Both of the fabricated LNAs have the S11-parameters of less than -25 dB at 5 GHz. The differential LNAs without ESD protection and with the proposed ESD protection design achieve best input matching at the same frequency, which demonstrates that no shift in the center operation frequency can be achieved after adding ESD protection circuits as long as the parasitic effects caused by ESD protection devices are well-characterized and included in the input matching network. As shown in Fig. 5(c), the differential LNA without ESD protection exhibits the S22-parameter of -9.3 dB at 5 GHz, whereas the differential LNA with the proposed ESD protection design has the S22-parameter of -11.4 dB. Fig. 5(d) compares the S12-parameters of these two LNAs. The S12-parameters of these two LNAs are better than -28 dB because good reverse isolation is one of the attributes in the cascode configuration.

The measured noise figures are shown in Fig. 6. The differential LNA without ESD protection has better noise figure, which is 2.16 dB at 5 GHz, whereas the differential LNA with the proposed ESD protection design exhibits the noise figure of 2.62 dB. The increase in noise figure is due to the addition of ESD protection devices connected to the RF input pads. The measured input-referred third-order intercept (IIP3) value is -12.5 dB for both LNAs. There is no significant difference on RF performances except noise figure between these two fabricated LNAs, which demonstrates the co-design effectiveness of LNA with ESD protection.

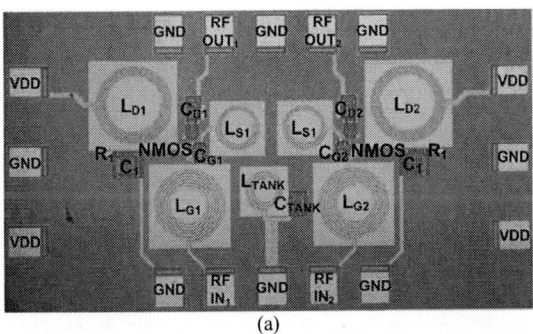

(a)

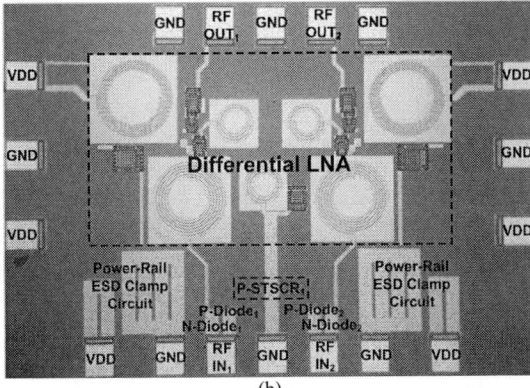

(b)

Fig. 4. Chip microphotographs of (a) differential LNA without ESD protection, and (b) differential LNA with the proposed ESD protection design.

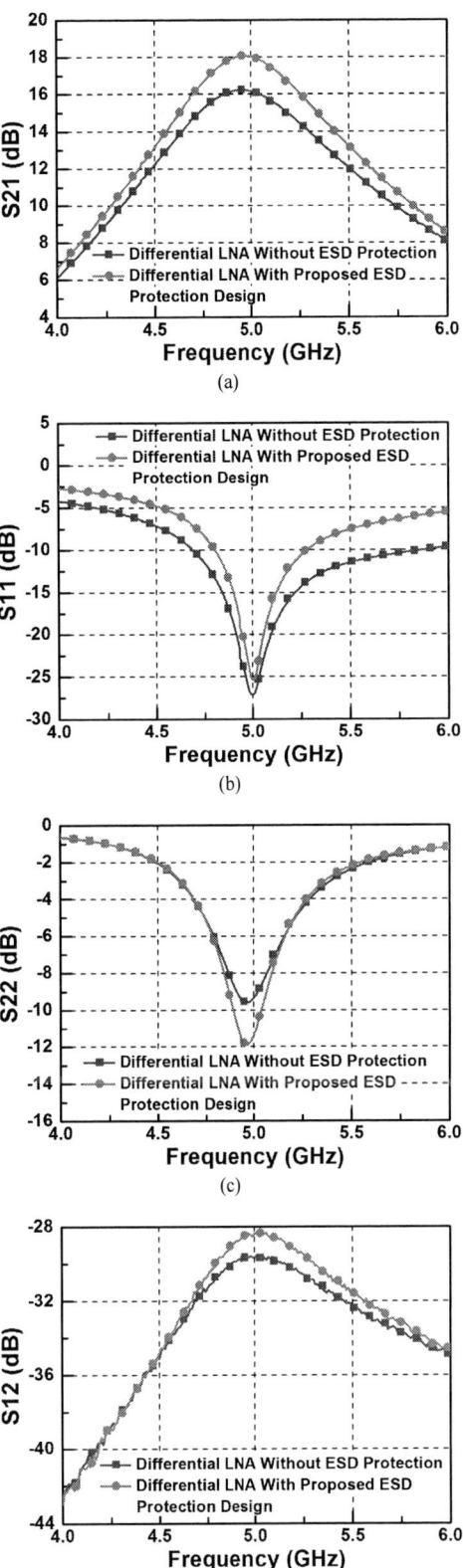

Fig. 5. Measured (a) S21-, (b) S11-, (c) S22-, and (d) S12-parameters of the differential LNAs without ESD protection and with the proposed ESD protection design.

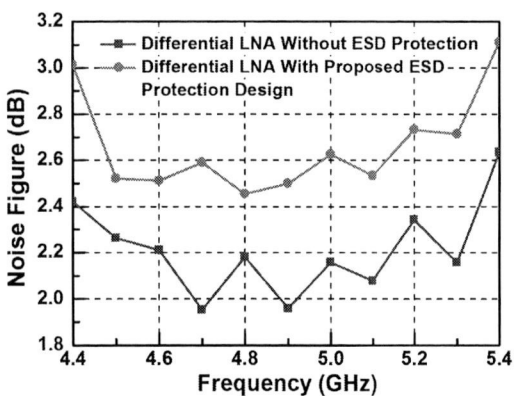

Fig. 6. Measured noise figures of the differential LNAs without ESD protection and with the proposed ESD protection design.

B. ESD Robustness

The human-body-model (HBM) ESD levels of these two fabricated LNAs are measured and listed in Table I. The differential LNA without ESD protection is very vulnerable to ESD. The differential LNA with the proposed ESD protection design has HBM ESD level of 3 kV, which meets the typical ESD specification of 2 kV for commercial ICs. During pin-to-pin ESD test, one input pad is zapped by ESD with the other input pad grounded, while VDD and VSS pads are floating. The ESD path during pin-to-pin ESD stresses is through P-diode$_1$, P-STSCR$_1$, and N-diode$_2$. Since P-STSCR$_1$ is placed near the differential input pads which can be quickly turned on during ESD stresses, the pin-to-pin ESD robustness is quite high of 6.5 kV. The differential LNA with the proposed ESD protection design can sustain VDD-to-VSS ESD stress of over 8 kV. Failure analysis has been also performed to investigate the failure mechanism. The scanning-electron-microscope (SEM) picture at the failure points is shown in Fig. 7. The failure points locate at the gate oxide of the input NMOS whose gate terminal is connected to the input pad.

TABLE I
HBM ESD LEVELS OF FABRICATED DIFFERENTIAL LNAs

HBM ESD Level	LNA Without ESD Protection	LNA With Proposed ESD Protection Design
Positive to VSS	< 50 V	3 kV
Negative to VDD	< 50 V	7 kV
VDD to VSS	0.5 kV	> 8 kV
Pin to Pin	< 50 V	6.5 kV

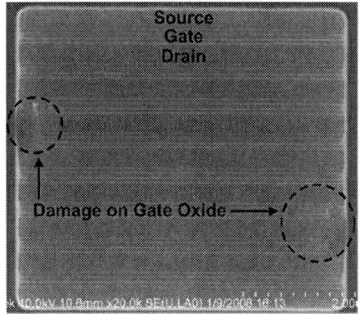

Fig. 7. SEM failure picture of differential LNA with the proposed ESD protection design after 3.5-kV PS-mode ESD test.

Because the ESD-clamping voltage between the input pad and VSS under 3.5-kV PS-mode ESD test is larger than the gate-oxide breakdown voltage, the gate oxide is damaged before the ESD protection devices damaged. To further enhance the ESD robustness, the turn-on resistance of the ESD protection devices and parasitic resistance along the ESD current path need to be further reduced.

V. CONCLUSION

A novel ESD protection design for differential I/O pads is proposed, especially considering the pin-to-pin ESD stress. Applied to a 5-GHz differential LNA, the proposed ESD protection design has been successfully verified in a 130-nm CMOS process. Once the parasitic effects caused by ESD protection devices can be accurately characterized, they can be matched without causing degradation on RF performance. Thus, the operation frequency of LNA will not be shifted and the RF performance can be still maintained after the ESD protection circuit is added into the chip. Excellent RF performance and ESD robustness can be simultaneously achieved by co-designing the LNA with ESD protection circuit.

ACKNOWLEDGMENTS

The authors would like to thank United Microelectronics Corporation for the support of chip fabrication. The authors would also like to thank Ansoft Corporation for the support of simulation tool Designer/Nexxim. This work was partially supported by National Science Council (NSC), Taiwan, under Contract of NSC96-2221-E-009-182.

REFERENCES

[1] D. Cassan and J. Long, "A 1-V transformer-feedback low-noise amplifier for 5-GHz wireless LAN in 0.18-μm CMOS," *IEEE J. Solid-State Circuits*, vol. 38, no. 3, pp. 427-435, Mar. 2003.

[2] M.-D. Ker, W.-Y. Lo, C.-M. Lee, C.-P. Chen, and H.-S. Kao, "ESD protection design for 900-MHz RF receiver with 8-kV HBM ESD robustness," in *Proc. IEEE Radio Freq. Integrated Circuit Symp.*, 2002, pp. 427-430.

[3] D. Linten, S. Thijs, M. Natarajan, P. Wambacq, W. Jeamsaksiri, J. Ramos, A. Mercha, S. Jenei, S. Donnay, and S. Decoutere, "A 5-GHz fully integrated ESD-protected low-noise amplifier in 90-nm RF CMOS," *IEEE J. Solid-State Circuits*, vol. 40, no. 7, pp. 1434-1442, Jul. 2005.

[4] S. Hyvonen and E. Rosenbaum, "Diode-based tuned ESD protection for 5.25 GHz CMOS LNAs," in *Proc. EOS/ESD Symp.*, 2005, pp. 9-17.

[5] M.-D. Ker, C.-I. Chou, and C.-M. Lee, "A novel LC-tank ESD protection design for giga-Hz RF circuits," in *Proc. IEEE Radio Freq. Integrated Circuit Symp.*, 2003, pp. 115-118.

[6] Y. Cao, V. Issakov, and M. Tiebout, "A 2kV ESD-protected 18GHz LNA with 4dB NF in 0.13μm CMOS," in *IEEE Int. Solid-State Circuits Conf. Dig. Tech. papers*, 2008, pp. 194-195.

[7] D. Shaeffer and T. Lee, "A 1.5-V, 1.5-GHz CMOS low noise amplifier," *IEEE J. Solid-State Circuits*, vol. 32, no. 5, pp. 745-759, May 1997.

[8] M.-D. Ker and K.-C. Hsu, "Overview of on-chip electrostatic discharge protection design with SCR-based devices in CMOS integrated circuits," *IEEE Trans. Device Mater. Reliab.*, vol. 5, no. 2, pp. 235–249, Jun. 2003.

[9] H. Sarbishaei, O. Semenov, and M. Sachdev, "Optimizing circuit performance and ESD protection for high-speed differential I/Os," in *Proc. Custom Integrated Circuits Conf.*, 2007, pp. 149-152.

[10] M.-D. Ker, "Whole-chip ESD protection design with efficient VDD-to-VSS ESD clamp circuits for submicron CMOS VLSI," *IEEE Trans. Electron Devices*, vol. 46, no. 1, pp. 173-183, Jan. 1999.

IEEE 2008 Custom Intergrated Circuits Conference (CICC)

A 65nm 3.4Gbps HDMI TX PHY with Supply-regulated Dual-tuning PLL and Blending Multiplexer

Jongshin Shin, Jaehyun Park , Bongjin Kim, Jongjae Ryu, Chiwon Kim , JiYoung Kim,
Seung-Hee Yang, Hyungoo Kim, and Jaewhui Kim

System LSI, Samsung Electronics, Korea

Abstract – **A 65nm HDMI TX PHY was designed with supply-regulated dual-tuning PLL and blending multiplexer. The proposed PLL uses a new dual-tuning scheme for small capacitor and low-jitter while keeping the supply regulation capability. A fractional-N operation for non-integer pixel clock generation was implemented with a blending multiplexer which enables seamless switching of high-speed multiphase clock. The fabricated PHY gives maximum 3.4Gbps data rate per channel and shows 34ps peak-to-peak data jitter.**

I. Overall Architecture

As demanding data rate of display devices increases with finer resolution and deeper color depth, more requirements are added on the TX PHY of a digital data transmission system like HDMI (high definition multimedia interface). To cover more video formats, wider tuning range PLL is needed and stringent jitter control is required for higher data rate. Together, along with a process migration an embedded environment of TX PHY should be considered.

Fig. 1 shows the block diagram of designed HDMI TX PHY for deep sub-micron SOC embedment. It has two PLL's, one is pixel clock generator (PCG) and the other is clock multiplication unit (CMU). The PCG makes a pixel clock required for a video processor and CMU multiplies that clock to deliver serial data through a physical media. The PCG and the CMU use the same PLL architecture shown in Fig. 2, except the CMU does not include the sigma delta modulator (SDM). For wide tuning range and low-jitter, a new dual-tuning technique was used with supply regulated VCO. A seamless switching of multiphase clock was used for fractional-N operations of PCG. Besides, the PCG was integrated within PHY area sharing well-established analog circumstance to fully utilize SOC environment. Since the clock jitter of PCG is controlled in the PHY layer, there is no need for a jitter-filter PLL used to guarantee a clean input clock for CMU in a discrete chip environment.

II. Supply-regulated Dual-tuning PLL

Supply regulation and dual-tuning control are useful PLL design techniques, respectively. Supply regulation increases power noise immunity and dual-tuning gives a capacitance

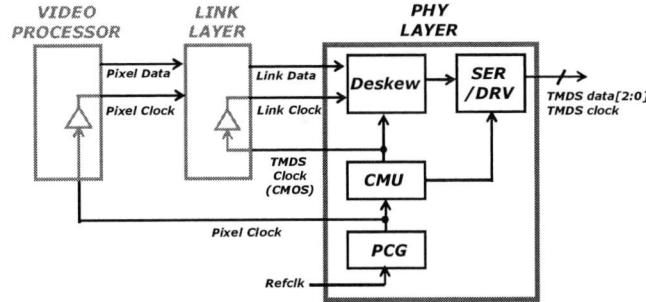

Fig.1. Block diagram of the proposed HDMI TX PHY with clocking scheme of link layer and video processor in a SOC

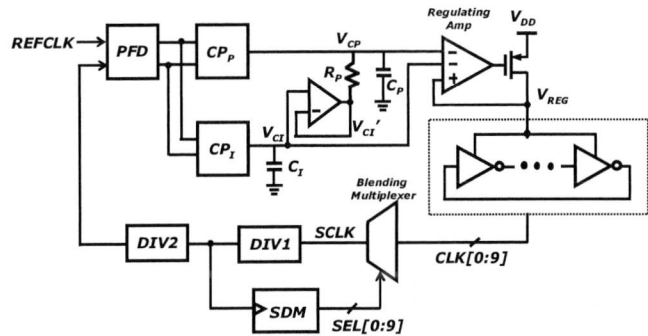

Fig.2 The proposed PLL architecture

multiplication [1] or jitter reduction with independent setting of proportional path and integral path gain of a VCO [2]. However, using both techniques at the same time is not straightforward. The difficulty of dual-tuning with supply-regulated VCO is that the control port of supply-regulated VCO is its supply but the supply node is only one. To use dual-tuning control, two control voltages from each path should be added first then applied to VCO supply node. Voltage mode summation can be done at the output of unit gain buffer taking output resistance of the buffer as a loop filter resistor as shown in Fig. 3(a) [3]. The summed voltage is applied to conventional regulating amp as shown in Fig. 3(b). But the VCO gain is same for both paths so that the constraint between VCO tuning range and VCO gain is not resolved, which leads to a trade-off of tuning range and jitter

978-1-4244-2018-6/08 $25.00 © 2008 IEEE 255

performance. Besides, the noise of OPamp_R is added in summed control voltage. In the proposed PLL, the voltage summation of two control path is achieved within the regulating amp as shown in Fig. 4. The weighted input branches in the amp make a weighted voltage summation and it gives separate VCO gain for each path. The weightings are set by the current ratio of M1 and M2, which are $1/(1+N)$ and $N/(1+N)$ for proportional (V_{CP}) and integral (V_{CI}) path, respectively. The resulted regulated voltage is

$$V_{reg} = \frac{1}{N+1} V_{CP} + \frac{N}{N+1} V_{CI} \qquad (1)$$

Therefore, wide tuning range is obtained through large VCO gain in the integral path and low-jitter is obtained by less noise sensitivity in the proportional path. In a PLL with replica VCO [4], similar summation was used but its dual-tuning quantity is V_{reg} and replica V_{reg} from the VCO and replica VCO in the feedback path and it was used for bandwidth extension to supply noise. In the proposed PLL, the DC voltage of proportional path is maintained to V_{CI} by unit gain buffer between V_{CI} and $V_{CI'}$ as shown in Fig2. Although a noise of unit gain buffer is also added to proportional path, the reduced VCO gain can mitigate the effect of buffer noise.

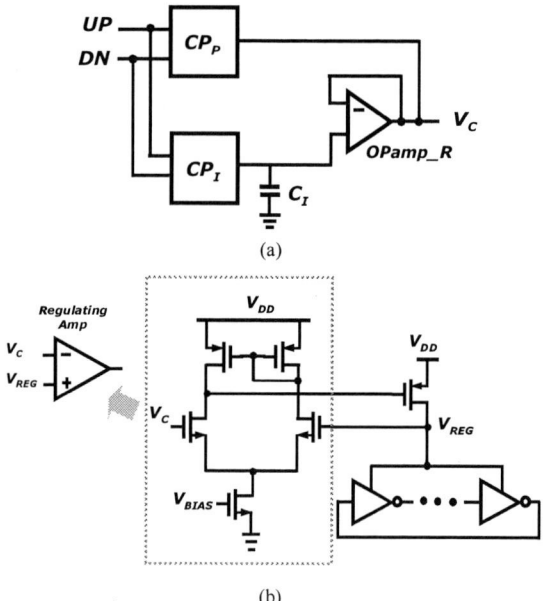

(a)

(b)

Fig3. Adding of proportional path and integral path voltage with a unit gain buffer [3] (a) for using a conventional regulating amp (b)

III. Fractional-N Frequency Synthesizing with a Blending Multiplexer

As shown in Fig. 2, the integrated pixel clock generator uses split divider [5] and multi-phase clock for fractional-N operations.

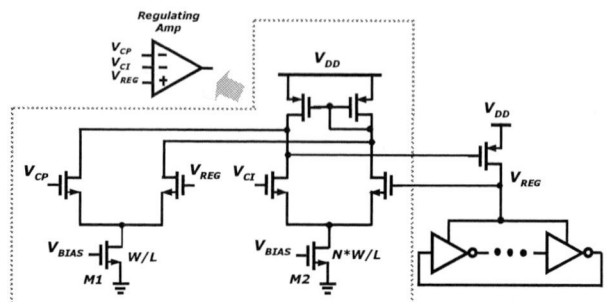

Fig.4. The proposed supply-regulated dual-tuning VCO

Compared with divider modulation method, phase selection method shows better jitter performance due to reduced quantization noise [6]. Increased speed of SDM allows unit phase shift at a time [5]. However, a phase selection through a multiplexer needs a caution since it is difficult to switch every phase in its settled level, especially for high frequency clock. The 10 phases of the designed VCO covers all 360 degree and selection signals (SEL[0:9]) arise at fixed clock timing of sigma delta modulator, where a simultaneous transition of input clock and selection signal is inevitable. As shown in Fig. 5, if SEL[k] and SEL[k+1] signals have a timing skew comparable with the phase distance of CLK[k] and CLK[k+1], none of 10 phase clocks may be selected with a missed edge. A timing controller for multiplexer switching [7][8] or sampling the SEL signal by appropriate clock phase [9] were reported to avoid simultaneous transition of SEL and CLK. But timing controller increases design complexity and sampling SEL signal itself may bring the same issue of edge missing.

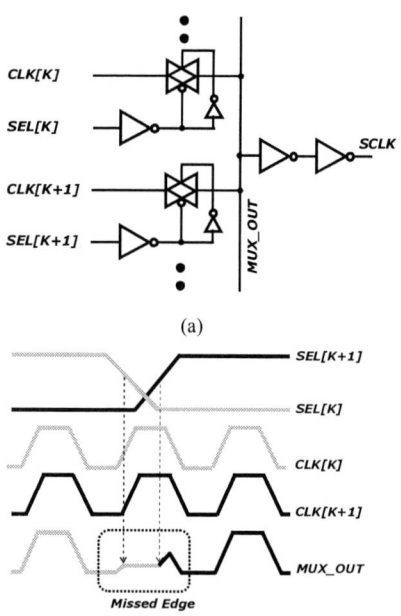

(a)

(b)

Fig.5. Conventional multiplexer (a) and edge missing due to simultaneous switching of input clock and selection signal (b)

In the proposed design, instead of trying to align selection signal with multi-phase clock signal, a blending procedure was inserted during a switching from a phase to the next phase. The multiplexer with blending function, blending multiplexer, was implemented using a topological analogy of a phase blender and a multiplexer as shown in Fig. 6. When both paths conduct input signals, a topology of 2:1 multiplexer is same with a phase blender. Thus, if the input signals are clocks with different phase and there is some time overlap in two selection signals during which both switches are closed, the multiplexer can work as a phase blender. Therefore, at any moment the output node has at least one low impedance path to the input nodes. It prevents edge missing at the MUX_OUT node in Fig.4 and keeps the phase of output clock between the phases of two input clocks when the simultaneous transitions occur.

To make timing overlap between adjacent selection signals, a half schmitt trigger was used. As shown in Fig. 7, the half schmitt trigger makes it difficult to change output state from HIGH to LOW. Thus it delays HIGH to LOW transition position, but does not change LOW to HIGH transition position. It makes overlap of HIGH state of two switching SEL signals. Fig.8 shows the proposed blending multiplexer and its operation when the CLK and SEL transit at the same time.

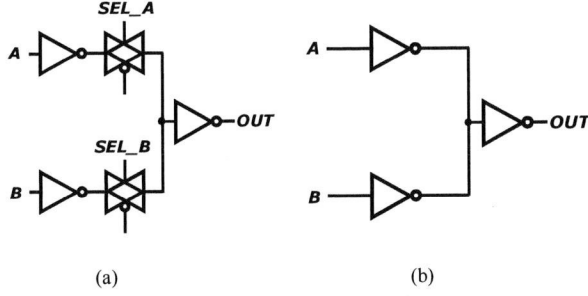

(a)   (b)

Fig.6 A topological analogy of a 2:1 multiplexer (a) and a phase blender (b)

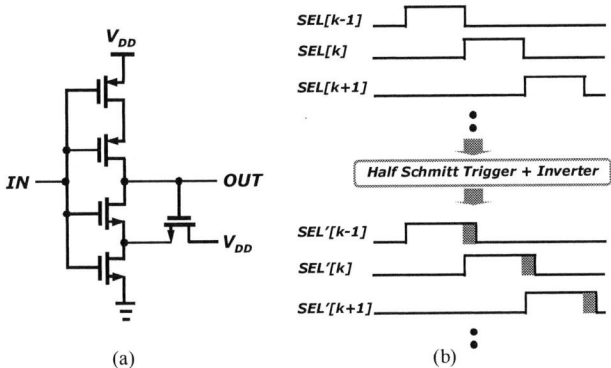

Fig.7 Half schmitt trigger (a) and its operation with multiple SEL signals (b)

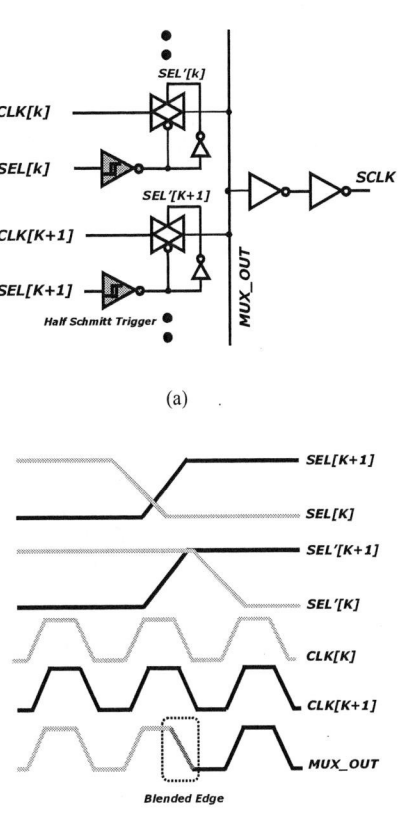

Fig.8 The proposed blending multiplexer (a) and its operation with simultaneous transit of input clocks and selection signals (b). The SEL'[k] and SEL'[k+1] are the results of half schmitt trigger.

IV. Experimental Results

The proposed PHY was fabricated with 65nm standard CMOS process. Measurements were helped by automated HDMI testing software with 1.3b spec. Fig.9 shows jitter of TMDS clock and data at 3.4Gbps transmission rate. They show pk-to-pk 28ps and pk-to-pk 34ps jitter, respectively. The eye diagram overlaid with compliance mask shows no mask touching. CMU tuning range is 500MHz~1700MHz. Fig. 10(a) shows an eye-diagram of 1080p/16bit data in 59.94Hz mode, where a fractional-N operation of the PCG is activated. Fig. 10(b) shows the synthesized TMDS clock frequency, which is 1000/1001 of 60Hz mode frequency (297MHz). The correct frequency confirms no missing edge with the proposed blending multiplexer. Fig.11 shows the micro photograph of the proposed PHY. A bias block and a control logic were shared for the PCG and the rest of PHY blocks. If an external PCG is used, the integrated PCG can be configured as jitter-filter for lower jitter or by-passed for lower power. Table I shows the performance summary of the fabricated PHY.

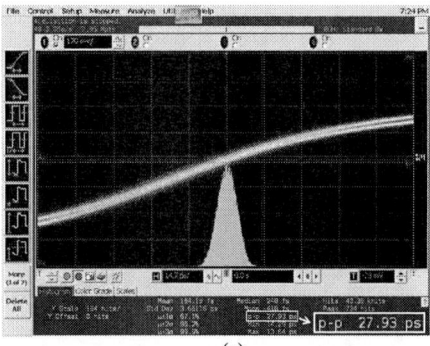

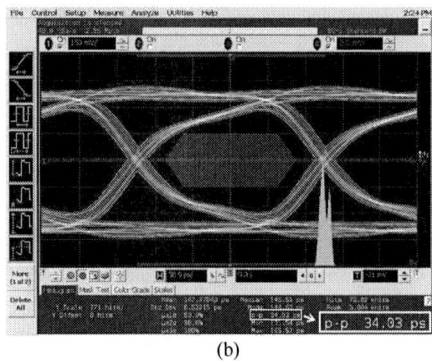

Fig.9. TMDS clock jitter (a) and data jitter (b) at 3.4Gbps.

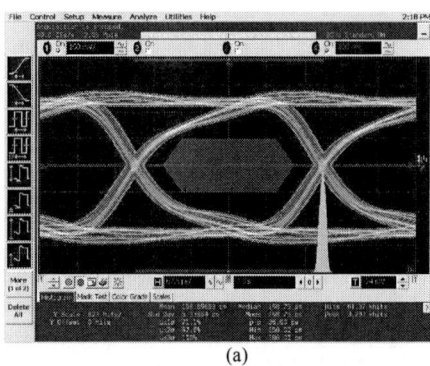

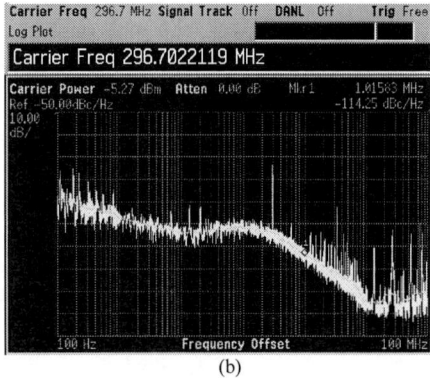

Fig.10. An eye-diagram of data at 1080p/16bit in 59.94Hz mode (a) and its TMDS clock frequency with phase noise profile (b)

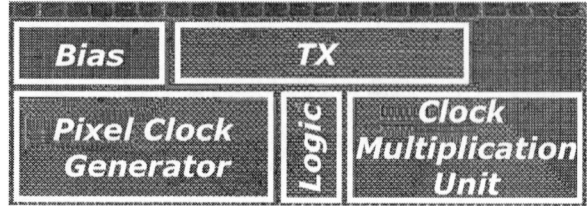

Fig.11 Micro photograph of the proposed PHY.

Table I. Performance Summary

Process	65nm CMOS
Power	60mW @720p/8bit 85mW@1080p/16bit
Supporting Data Rate	250Mbps~3.4Gbps
Supply	1.2/3.3V
Clock Jitter (@3.4Gbps)	28ps p-p
Data Jitter (@3.4Gbps)	34ps p-p
Size	0.56mm²

ACKNOWLEDGEMENT

The authors thank Jeonghan Kim of EMSC for project support and H. Okamura for technical review

References

[1] Jan Craninckx and Michel. S. J. Steyaert, " A Fully Integrated CMOS DCS-1800 Frequency Synthesizer", *IEEE J. Solid-State Circuits*, vol. 33, pp. 2054-2065, Dec. 1998.

[2] Chi-Wa Lo and Howard Cam Long, " A 1.5-V 900-MHz Monolithic CMOS Fast-Switching Frequency Synthesizer for Wireless Applications", *IEEE J. Solid-State Circuits*, vol.37, pp.459-470, April. 2002.

[3] K. Ken Yang *et al.,*"A 0.4-4-Gb/s CMOS Quad Transceiver Cell Using On-Chip Regulated Dual-Loop PLLs", *IEEE J. Solid-State Circuits*, vol.38, pp.747-754, May. 2003.

[4] Elad Alon et al., "Replica Compensated Linear Regulators for Supply-Regulated Phase-Lock Loops," *IEEE J. Solid-State Circuits*, vol.41, pp.413-424, Feb. 2006.

[5] J.Shin, et al.," A Low-Jitter Added SSCG with Seamless Phase Selection and Fast AFC for 3rd Generation Serial-ATA," *IEEE Custom Integrated Circuit Design*, pp. 409-412, 2006.

[6] C. Heng and B. Song, "A 1.8GHz CMOS fractional-N frequency synthesizer with randomized multi-phase VCO," *IEEE Custom Integrated Circuit Design*, pp. 427-430, 2002.

[7] H. R. Lee, O. Kim, G. Ahn, and D. K. Jeong, "A low-jitter 5000ppm spread spectrum clock generator for multi-channel SATA transceiver in 0.18um CMOS," *IEEE International Solide-State Circuit Conference*, pp. 162-163, 2005.

[8] Y. Moon, G. Ahn, H. Choi, N. Kim, and D. Shim, "A Quad 6Gb/s Multi-rate CMOS Transceiver with TX Rise/Fall-Time Control," *ISSCC Dig.Tech.Papers*,pp.84-85,Feb,2006.

[9] K. Omoto, K. Shimizu, and J. Okamura, "A Vernier Over Sampling and Alignment technique for Gb/s Serial Communication," *Asia Solid-State Circuit Conference*, pp.29-32,2005.

978-1-4244-2018-6/08 $25.00 © 2008 IEEE

IEEE 2008 Custom Intergrated Circuits Conference (CICC)

A Scalable Digitalized Buffer for Gigabit I/O

HungWen Lu[1], ChauChin Su[2] and Chien-Nan Liu[1]

[1]National Central University, Chung Li, Taiwan, [2]National Chiao-Tung University, Hsin Chu, Taiwan

[1]s9521011@cc.ncu.edu.tw, [2]ccsu@cn.nctu.edu.tw

Abstract—A serial I/O composed of inverters and transmission gates only is proposed to achieve high supply voltage scalability and low area overhead. The inverter with an inductive biasing circuit can extend bandwidth, and reduce the SSN simultaneously. With a TSMC 0.18μm CMOS process, the I/O occupies an area of 0.014mm² and operates from 4Gbps@1.9V to 1.5Gbps@1.1V.

Index Terms— Buffer, I/O, SSN

I. INTRODUCTION

Several low power and high bandwidth chip-to-chip buffers have been reported in [1]-[5]. When the supply voltage is reduced with the technology, the performance of the current mode and pulse mode buffers will be affected by small noise margins and overdrive voltages. In order to take the full advantage of the scaled down technology, [6] adopts the wide-bandwidth filter in [7] to design a I/O buffer that is only made of the standard inverter, as shown in Fig. 1(a). It can operate at low supply voltage and reduce the hardware. However, its bandwidth is limited. This work proposes an inductive biasing circuit to solve the lack of bandwidth. And this proposed buffer can achieve a low area and good power efficiency. In addition, it has a high scalability with different cell libraries, technologies, and supply voltages.

This work is organized as follows. Section II reviews the original digital buffer and introduces the proposed buffer. Section III describes the differences between the original and proposed buffers through comparing their bandwidths and *Simultaneous Switching Noise* (SSN) by SPICE simulations. Section IV discusses the implementation and measurement of the test chip. Finally, Section V concludes of this work.

II. ARCHITECTURE

A. Original Digital Buffer

Fig. 1(a) shows the original digital buffer. The driver enhances amplitudes and driving abilities. The gain of the inverter driver is highly related to its input common mode levels. The optimum common mode level is at the threshold voltage. Therefore, a biasing circuit designed by shorting the input and output of another inverter is inserted.

In addition to set the common mode levels, the biasing circuit has several benefits. First, the gain is lowered to achieve a higher input common mode range and a higher 3dB bandwidth. Additionally, the biasing circuit adds two diode resistors of $1/g_m$ between outputs and power supplies. For the RLC model from the package and the buffer, its damping factor is increased by adding $1/g_m$. Hence, it significantly

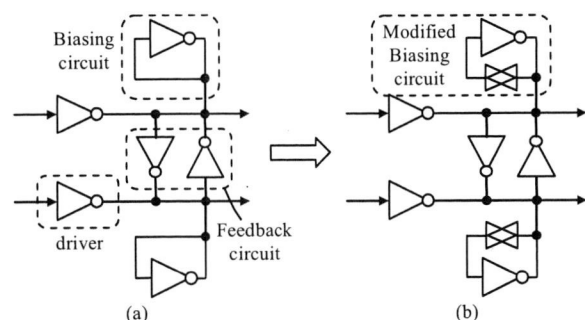

Fig. 1.(a) Reported digital buffer and (b) the proposed buffer

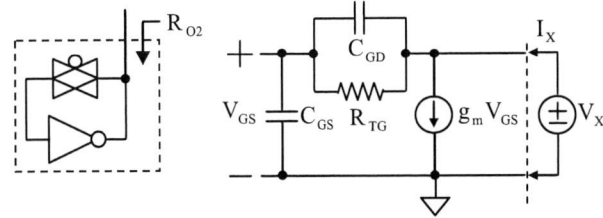

Fig. 2. The small signal model of the inductive biasing circuit.

reduces the SSN, but it consumes some static power.

For the inverter, the common mode gain is the same as the differential mode gain. It results in serious timing jitter caused by the common mode noise. Thus, a feedback circuit made of back to back inverters is added not only to suppress the common mode noise, but also to enlarge the input differential signal. However, the time constant at output is increased with the negative resistance of the feedback circuit, and the bandwidth is reduced.

B. Proposed Digital Buffer

To extend the bandwidth of the digital buffer, a transmission gate served as a resistor R_{TG} is inserted at the biasing circuit, as shown in Fig. 1(b). The transmission gate is turned on by digital signals. Therefore, the biasing circuit becomes an inductive load. By analyzing the modified biasing circuit, Fig. 2 shows a small signal model. And C_{GS} is the capacitance between the inputs of the inverter and the ground. C_{GD} is the capacitance between the inputs and outputs. g_m is the trans-conductance of the inverter. The impedance of the biasing circuit is obtained as:

$$R_{O2}(s) = \frac{V_X}{I_X} = \frac{R_{TG}(C_{GS}+C_{GD})S+1}{(R_{TG}C_{GD}S+1)(C_{GS}S+g_m)} \quad (1)$$

The poles and zero in (1) are

$$Z = \frac{1}{R_{TG}(C_{GS}+C_{GD})}, P_1 = \frac{1}{R_{TG}C_{GD}}, P_2 = \frac{g_m}{C_{GS}} \quad (2)$$

In (2), P_1 and Z both are inverse proportion to R_{TG}, and P_2 is independent to R_{TG}. The ratio (P_1/Z) equals to $(1+C_{GS}/C_{GD})$, it results in a fixed distance between Z and P_1 in the Bode plot.

978-1-4244-2018-6/08 $25.00 © 2008 IEEE

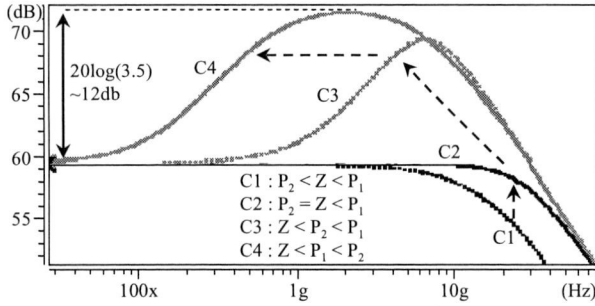

Fig. 3. Simulated frequency response of the inductive biasing circuit

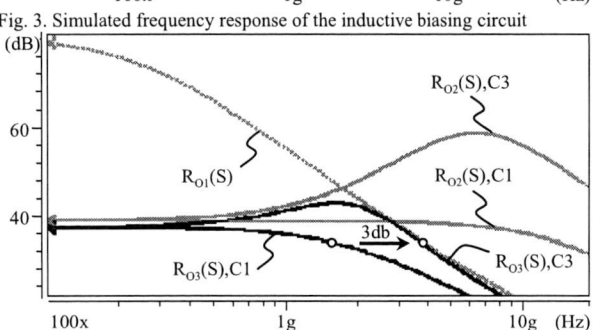

Fig. 4. Simulated frequency response of the output impedance of the reported buffer and the proposed one.

Fig. 3 is the relations of the poles and the zero for different R_{TG}. If R_{TG} is not added or R_{TG} is small, it results in $P_2 \ll Z < P_1$. $R_{O2}(S)$ is a single-pole system with a 3dB frequency at P_2. If R_{TG} is raised to have $Z < P_2 < P_1$ or $Z < P_1 < P_2$, some frequency peak is induced by Z. The frequency peak is saturated by the ratio of (P_1/Z). If (C_{GS}/C_{GD}) is 3.5, the maximum peak is 12 dB.

Fig. 4 shows how the bandwidth is extended by the inductive biasing circuit. $R_{O3}(s)$ is total output impedance which is a parallel connection of the output impedance of the driver ($R_{O1}(s)$) and the biasing circuit ($R_{O2}(s)$). r_O is DC value of $R_{O1}(s)$. And C_L is the loading capacitance at output. $1/R_{O3}(s)$ is dominated by $R_{O2}(s)$ at low frequency and by $R_{O1}(s)$ at high frequency. For $R_{TG}=0$, the 3dB frequency of $R_{O3}(s)$ is the product of the effective resistance $((1/g_m)//r_O)$ and the effective capacitance (C_L+C_{GS}). Assume that a zero is produced in $R_{O2}(s)$ to be the case of C3 in Fig. 3, and is placed in the crossing region of $R_{O2}(s)$ and $R_{O1}(s)$. At the frequency above Z, $R_{O2}(s)$ is far higher than $R_{O1}(s)$. Therefore, $R_{O3}(s)$ is close to $R_{O1}(s)$. The maximum 3dB frequency improvement is about

$$\omega_{3db} : \frac{1}{[(1/g_m)//r_O] \cdot (C_{GS} + C_L)} \Rightarrow \omega_{3db}' : \frac{1}{[(1/g_m)//r_O] \cdot C_L} \quad (3)$$

Note that if the R_{TG} is made of the poly resistor, it achieves higher bandwidth owing to less parasitic effect. But it results in a high area and a low circuit integrality with the other digital cells. Another issue is that R_{TG} needs to be designed in a specific range because over-peaking may cause signal overshoot and additional timing jitter in the eye diagram.

III. COMPARISON SIMULATION

By employing standard cells, the inverter buffer (B1), the conventional digital buffer (B2, B3) and the proposed ones (B4, B5) are designed to be a four-stage buffer, as shown in Fig. 5. M and S represent parallel and series connected

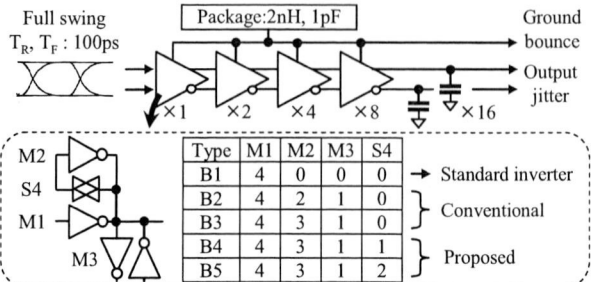

Fig. 5. Buffer comparison simulation setting

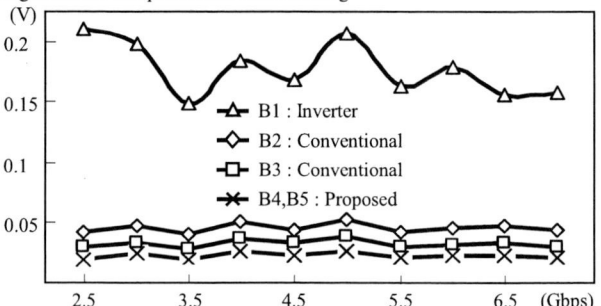

Fig. 6. RMS ground noise V.S. data rate

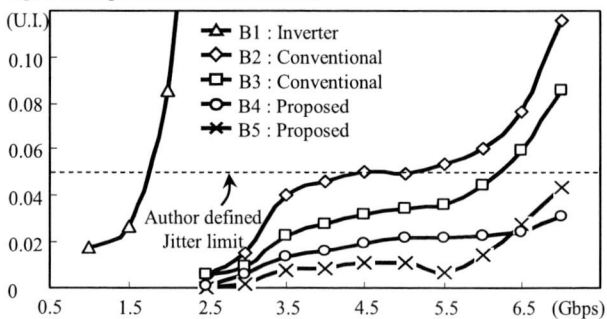

Fig. 7. Simulated Peak to peak output jitter V.S. data rate

numbers. For example, B1 consists of only inverters with 4 minimum devices in parallel. B2 and B3 consist of driving inverters, biasing inverters and feedback inverters [7]. B4 and B5 are the proposed drivers with an inductive biasing circuit (S4 ≠ 0). NMOS and PMOS of the logic gates are uniformly 0.45um/ 0.18um and 1.87um/0.18um.

Fig. 6 and Fig. 7 reveal the RMS ground bounce value and output jitter in *Unit Interval* (UI) at different data rates. The data rate of B1 is limited to 1.75Gbps owing to a serious SSN of 175mV. With biasing and feedback circuits, B2 and B3 reduce the ground bounce to 50mV and to 35mV, respectively. Without having inductive biasing circuit, the transition time of B2 and B3 is about 200ps, as shown in Fig. 8. The jitter of B2 and B3 increases significantly since 5Gbps due to the transition time is higher than the data period. Hence, the data rates of B2 and B3 are limited to the jitter caused by ISI effect rather than by the ground noise. When the bandwidth is compensated by the inductive biasing circuit, the transition time in B4 is reduced to 100ps, and the data rate is raised from 6Gbps to 7.5Gbps. For B5, the over inductive peaking result in an overshoot waveform. If the bit period is longer than the overshoot effects, the increase in jitter will be limited. Otherwise, it would cause more ISI and jitter. The driver and

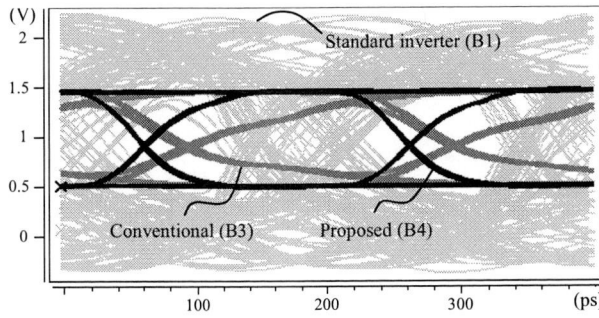

Fig. 8 Simulated eye diagram (5Gbps)

Fig. 9. Input common mode range simulation

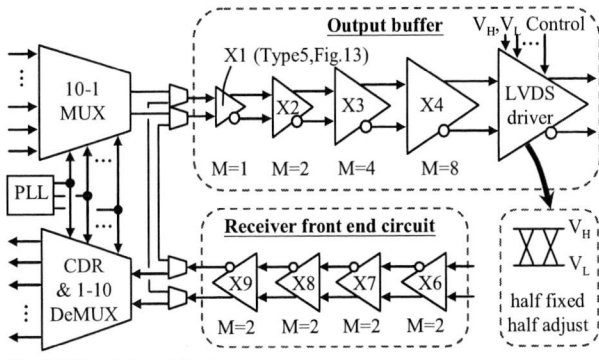

Fig. 10. Test chip architecture

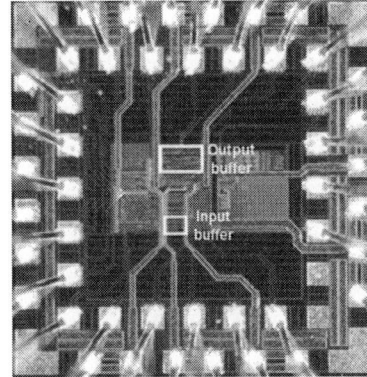

Fig. 11. Test chip photo

the package form a LC type filter. The filter produces power noise located at specific band. Hence, the increase of jitter cause a smaller slope occurred in all the buffers at the bit rate from 3.5Gbps to 5.5Gbps. Because the extra inductor made of biasing circuit is connected to power, it reduced the effective inductance at supply to further lower the SSN to 25mV.

Applying inverters as a buffer might raise questions in noise immunity and the allowable input common mode range. Regarding the noise immunity, Fig. 9 implies that the differential mode and the common mode gain at different common mode levels and different feedback circuit sizes. M1, M2 and S4 adopt the same setting as B4 does in Fig. 5, whereas M3 is a design variable. If the allowable gain variation is 3dB, for the case of M3=2 and M3=1, the usable common mode level is from 0.68V to 1.13V, and from 0.62V to 1.08V, respectively. In addition, the CMRR is varied from 16.5dB to 21.5dB and from 5.2dB to 8dB.

IV. EXPERIMENTAL RESULTS

Fig. 10 shows the system architecture being implemented. Both the output pre-driver and input pre-amplifier are designed by cascading several stages of the proposed digital buffers, except the last stage of the output driver. Each stage is composed of different numbers of digital buffer cells in parallel. Although the optimized case in simulation is B4 in Fig. 5, it becomes B5 instead when considering the parasitic capacitance caused by layout. For the transmitter, the circuit size is doubled stage by stage. For the receiver, the circuit sizes in each stage remain the same because of the on-chip driving.

The last stage of the output buffer is a LVDS driver. It is composed of many inverters as presented in [8], and it has an on-chip termination resistance of either 50Ω single-ended, or

100Ω differentially. The 50Ω termination is matched by a parallel connection of an 112Ω on-chip ploy resistor and a 90Ω turn-on resistor of data switches.

Fig. 11 shows the chip photo of the transceiver, which is fabricated in TSMC 0.18μm CMOS process. The output driver and the receiver occupy 0.0108mm² and 0.0032mm² of areas.

The test is performed on a PCB in Roger material. Four AC coupling capacitors were inserted to isolate the DC level from instrument and I/O. A Bit-Error-Rate-Tester (N4901B) generated a random data with a differential swing of 300mV. An on-chip bypass path is inserted to redirect the receiver front end output to the pre-driver input. A Wide-Bandwidth-Oscilloscope (11801C) measured the eye diagram. The test was focused on verifying a scaleable bandwidth and power consumption with supply voltage. Hence, the peak jitter and the total power consumption are measured under different supply voltages and data rates. We have to note that the measured results are accumulated value, including overall I/O.

TABLE I notes that the peak jitter normalized to UI when the supply voltage is swept from 1.9V to 1.1V, and when the data rate is swept from 1.5Gbps to 4Gbps. A general peak jitter limitation of 0.25UI is used to decide the maximum data rate. The gray region shows that the maximum data rate is 4Gbps/3Gbps/2.5Gbps/2Gbps/1.5Gbps when the supply voltage is 1.9V/1.7V/1.5V/1.3V/1.1V. TABLE II indicates that the consumed power per bit is obtained by dividing the power consumption by data rates. When the settings are adjusted from 1.9V and 4Gbps to 1.1V and 1.5Gbps, the efficiency is increased from 7.81pJ/bit to 2.77pJ/bit. The minimum supply voltage is about two times of threshold

TABLE I. OUTPUT PEAK JITTER AT DIFFERENT VDD AND DATA RATES

VDD	Data rate (Gbps)						
	1	1.5	2	2.5	3	4	5
1.9	-	-	0.08	0.16	0.2	0.25	0.43
1.8	-	-	0.09	0.17	0.21	0.27	0.44
1.7	-	-	0.11	0.2	0.22	0.3	0.47
1.6	-	-	0.13	0.22	0.27	0.34	0.54
1.5	-	-	0.17	0.24	0.31	0.4	0.64
1.4	-	-	0.2	0.3	0.38	0.48	0.75
1.3	-	-	0.24	0.34	0.45	-	-
1.2	0.14	0.23	0.31	0.47	-	-	-
1.1	0.15	0.25	0.39	0.67	-	-	-

TABLE II. POWER EFFICIENCY AT DIFFERENT VDD AND DATA RATES

VDD	Data rate (Gbps)						
	1	1.5	2	2.5	3	4	5
1.9	-	-	12.81	10.69	9.28	7.81	6.82
1.8	-	-	10.91	9.16	7.94	6.72	5.81
1.7	-	-	9.18	7.75	6.75	5.74	4.86
1.6	-	-	7.64	6.46	5.66	4.66	4.00
1.5	-	-	6.29	5.35	4.70	3.71	3.16
1.4	-	-	5.06	4.32	3.83	3.06	2.72
1.3	-	-	3.37	2.81	2.71	-	-
1.2	4.84	3.70	2.99	2.44	-	-	-
1.1	3.58	2.77	2.93	3.06	-	-	-

TABLE III. REPORTED I/O BUFFER COMPARISON

Reference	JSSC 05[2]	JSSC 06 [3]	JSSC 05 [4]	JSSC 05 [5]		This work		
Method	Pulse mode	Pulse mode	Current mode	Current mode		Voltage mode		
Technology (um)	0.1	0.18	0.13	0.35		0.18		
Supply Voltage (V)	1.8	1.8	1.8	1.8	1.8	1.1	1.5	1.9
Data rate (Gbps)	1	3	12	1.4	1.2	1.5	2.5	4
Power (mW)	5.6	15	78.9	23 (w/o pre-amp)	12.8 (w/o pre-amp)	4.2	13.4	31.2
Amplitude (mV)	100	-	> 500	340	340	130	335	700
Area (mm²)	0.028 (w/o Cap)	0.0014 (w/o Cap)	< 0.12	0.11	0.14	0.014		
Power efficiency (pJ/bit)	5.6	5	6.58	19.1	10.6	2.77	5.35	7.81

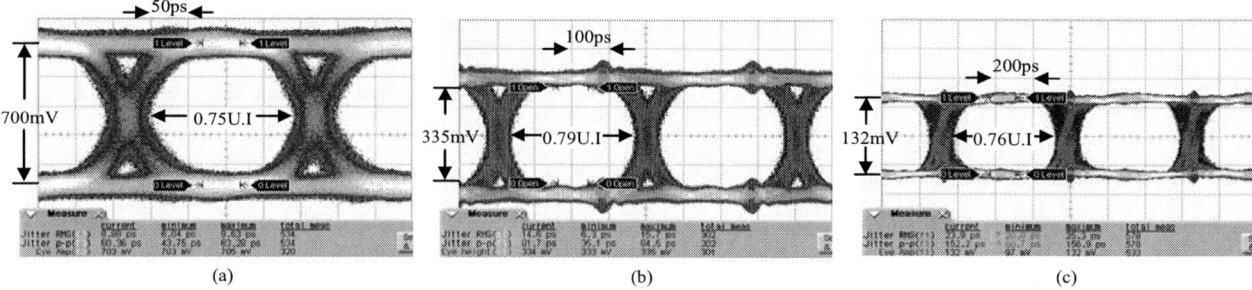

(a) (b) (c)

Fig. 12. Measurement eye diagrams at different data rates and supply voltages, (a) 4Gbps and 1.9V (b) 2.5Gbps and 1.5V (c) 1.5Gbps and 1.1V.

voltage, and it is 1.1V in 0.18um technology. Above 1.1V, the transmission gate in the biasing circuit will turn off. It induces the failure of the biasing circuit, and lowers the bandwidth. Fig. 12 is the eye diagram at different supply voltages and data rates. The output swing from a supply voltage of 1.9V to 1.1V varies from 700mV to 132mV when the data transition time is 60ps.

TABLE III compares this work to reported I/O designs. Pulse mode I/O [2] [3] reaches the lowest area, and has good power efficiency. Current mode I/O [4] achieves high bandwidth, but it needs high supply voltage. This current mode I/O is worse in power efficiency and area overhead. An improved current mode I/O [5] can operate at low supply voltage (0.35um, 1.8V), but its bandwidth and power efficiency become worse. This work benefits on a wide range of supply voltages, and has good power efficiency at low voltages.

V. CONCLUSION

Comparing the current mode and pulse mode buffer to a digitalized I/O made only of inverter and transmission gates, the digitalized I/O can achieve a high scalability with different cell libraries, technologies, and supply voltages. The proposed inductive biasing circuit raises the bandwidth and lowers the

ground noise. Through TSMC 0.18μm CMOS process, the proposed I/O occupies an area of 0.014mm². Measurement proves that the proposed I/O can work at low supply voltages, and it has good power efficiency. Additionally, the bandwidth has a high scalability with supply voltages.

REFERENCE

[1] Mick S., Wilson J., Franzon P., "4Gbps high-density AC coupled interconnection," *in Proc. IEEE CICC*, pp.133 – 140, May 2002.

[2] Jongsun Kim; Verbauwhede, I.; Chang, M.-C.F.; "A 5.6-mW 1-Gb/s/pair pulsed signaling transceiver for a fully AC coupled bus," *IEEE J. Solid-State Circuits*, Vol. 40, pp.1331 – 1340, June 2005.

[3] Luo *et al*,"3 Gb/s AC coupled chip-to-chip communication using a low swing pulse receiver," *IEEE J. Solid-State Circuits*, Vol. 41, pp.287-296, Jan. 2006.

[4] Jong K. Kim and T. S. Kalkur, "High-Speed Current-Mode Logic Amplifier Using Positive Feedback and Feed-Forward Source-Follower Techniques for High-Speed CMOS I/O Buffer," *IEEE J. Solid-State Circuits*, Vol. 40, pp.796-802, Mar. 2005.

[5] Mingdeng Chen, Silva-Martinez, J.; Nix, M.; Robinson, M.E., "Low-voltage low-power LVDS drivers," *IEEE J. Solid-State Circuits*, Vol. 40, pp. 472-479, Feb. 2005.

[6] Jacob Baker, "High speed digital signal buffer and method," U.S. Patent 6483347, Nov. 19, 2002.

[7] B. Nauta, "A CMOS transconductance-C filter technique for very high frequencies," *IEEE J. Solid-State Circuits*, pp. 142 - 153, Feb. 1992.

[8] Hsin-Wen Wang, Hung-Wen Lu and Chau-Chin Su, "A digitized LVDS driver with simultaneous switching noise rejection," *in Proc. IEEE AP-ASIC*, pp.240 – 243, Aug. 2004.

IEEE 2008 Custom Intergrated Circuits Conference (CICC)

Broadband Transimpedance Amplifier in 0.35-μm SiGe BiCMOS Technology for 10-Gb/s Optical Receiver Analog Front-End Application

J. C. Huang, Y. S. Lai, and K. Y. J. Hsu

Institute of Electronics Engineering, National Tsing Hua University, Hsinchu, Taiwan

Abstract-A 10-Gb/s fully integrated optical receiver analog front-end was realized in a commercial 0.35-μm 50-GHz f_T SiGe BiCMOS technology. An input capacitance immunization technique was used to diminish the large input parasitic capacitance of the transimpedance amplifier to achieve a high-gain, wide-bandwidth, low-noise performance. The overall transimpedance gain is 107.3-dBΩ, and the −3 dB bandwidth is 8.85-GHz with 1.1 pF input parasitic capacitances. The measured results demonstrate a differential output swing of 842 mV$_{p-p}$ with 50-Ω output load, a peak-to-peak jitter (jitter$_{p-p}$) of 13.22 ps, and a sensitivity of −16.4 dBm at a bit error ratio (BER) of 10^{-12} with a 2^{31}−1 PRBS test pattern. The total circuit, including output buffer, dissipates 195 mW under 3-V supply and occupies 0.7 x 0.7 mm² chip area.

I. INTRODUCTION

Traditionally, optical receiver analog front-end (AFE) and photodetectors were heavily dependent on high-cost III/V semiconductor technologies for implementation because the resultant devices and circuits exhibited excellent performance in speed and noise. In recent years, the SiGe BiCMOS technology provided a cost-effective alternative for 10-Gb/s communications ICs and for the development of photodetectors [1]. In this paper, the 0.35-μm 50-GHz f_T SiGe BiCMOS technology was chosen to realize an inductor-less, area-efficient, fully integrated optical receiver analog front-end for 10-Gb/s applications.

The whole bandwidth of the optical receiver AFE is limited by the total input capacitance of the transimpedance amplifier (TIA). To the authors' knowledge, the typical total input capacitance (C_{PD}), which includes the capacitances from the photodetector, the bonding pad, and the ESD protection circuit, is between 0.4 pF and 1 pF. In such case, high-gain, low-noise, and wide-bandwidth are difficult to achieve at the same time in a TIA for 10-Gb/s applications. To resolve these issues, we used an active feedback impedance to provide an equivalent negative capacitor and a low input impedance to diminish the large input capacitance so as to enhance the bandwidth performance. In addition, the two-stage TIA were properly designed to achieve high transimpedance gain and low noise performance. In the design, we directly coupled the TIA with the limiting amplifier (LA) on the chip to eliminate the inter-stage matching network. The Miller amplification capacitors were used to replace the large off-chip capacitors that are needed in the auto-threshold control circuit and the dc-offset cancellation (DOC) circuit. In the paper, we present a fully integrated 10-Gb/s optical receiver AFE with emphasis on the design of the broadband TIA without using the

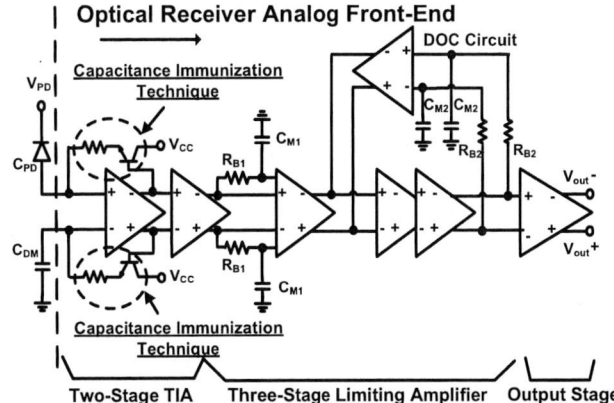

Fig. 1   The proposed optical receiver analog front-end architecture

conventional, area-consuming inductor peaking technique.

II. OPTICAL ANALOG FRONT-END ARCHITECTURE AND CIRCUITS IMPLEMENTATION

Fig. 1 shows the function blocks of the proposed optical receiver analog front-end which integrates a two-stage TIA, an auto-threshold control (ATC) circuit, a three-stage LA, a dc-offset cancellation (DOC) circuit, and an output stage in a single chip. The two-stage TIA converts and amplifies the single ended current signal to pseudo-differential ended voltage signal. The LA consists of a chain of three-stage Cherry-Hopper gain cells [2] that provide active inductors to enhance bandwidth, achieve high gain design, and deliver a certain output voltage swing to a logic level for data recovery. The first stage gain cell of LA uses the ATC circuit to be a trans-admittance stage which consists of a low pass filter (LPF) to convert the pseudo-differential signal to truly-differential one and provide a precise crossing percentage of the output swing. Two 121 pF equivalent capacitors (C_{M1}) are needed in this circuit block. The DOC circuit consists of a LPF to eliminate the dc offset voltage due to the process variation. It needs two 216 pF equivalent capacitors (C_{M2}).

A. Two-Stage Active Feedback Transimpedance Amplifier

As shown in Fig. 2(a), the first stage is a TIA with active feedback impedance to enhancement the bandwidth performance. Due to the active feedback impedance, we obtain equivalent low input impedance and negative shunt to ground capacitance (C_{NEG}). The C_{NEG} can strongly cancel the large input capacitance (C_{PD}) of TIA. That is, the resultant equivalent low input impedance can be used to sustain large C_{PD}. According to small signal analysis, the negative

978-1-4244-2018-6/08 $25.00 © 2008 IEEE   263

capacitance ($C_{NEG}(0)$), the input impedance ($Z_{in}(0)$), and the mid-band transimpedance gain ($A_Z(0)$) of the active feedback TIA can be expressed as

$$C_{NEG}(0) \cong \frac{-[C_{be3}g_{m3}R_{C1} + C_{bc1}g_{m3}^2 R_{F1}(R_{C1} - R_{F1})]}{(1 + g_{m3}R_{F1})^2} \quad (1)$$

$$Z_{in}(0) \cong \frac{(1 + g_{m3}R_{F1})}{g_{m3}(1 + g_{m1}R_{C1})} \quad (2)$$

$$A_Z(0) = \frac{v_{c1}}{i_{in}}(0) \cong \frac{g_{m1}R_{C1}(1 + g_{m3}R_{F1})}{g_{m3}(1 + g_{m1}R_{C1})} \quad (3)$$

where the C_{bc1} and g_{m1} are the base-collector capacitance and the transconductance of Q_1, the C_{be3} and g_{m3} are the base-emitter capacitance and the transconductance of Q_3, respectively. The analysis results of the diminished total input capacitance (C_{IN1}) of 0.674 pF ($\sim 0.61*C_{PD}$) and the negative capacitance ($C_{IN2}=C_{NEG}$) of −0.427 pF are shown in Fig. 2(b). The proposed capacitance immunization technique can enhance the −3 dB bandwidth by a factor of 1.63 without using large-area inductor peaking.

The second stage of The TIA is composed of a trans-admittance stage (TAS) and transimpedance stage (TIS), which achieves high gain-bandwidth product (GBW) and pseudo-differential signal output, and an emitter follower for level shifting [3]. The transimpedance gain and the bandwidth performance of the two-stage active feedback TIA can achieve 63.8-dBΩ and 8.85-GHz under a large C_{PD} of 1.1 pF.

B. Design Strategy of Limiting Amplifier

Conventional limiting amplifier (LA) is composed of identical cascaded gain cells, which is difficult to achieve high gain-bandwidth product. In our design, we used Cherry-Hooper gain cells to achieve high GBW performance. The first stage of LA that is composed of ATC input stage, TIS output stage, and DOC circuit is shown in Fig. 3. The Q_{L3} and R_{D3} provide equivalent active inductor load to improve bandwidth, and the R_{D4} and R_{F3} can achieve higher gain performance.

In the LA design, the number of gain stages was determined from gain bandwidth product performance [4]. Assume each gain cell is identical and approximated by a two-pole amplifier. The transfer function of the gain cell can be expressed as

$$A(s) = \frac{A_s \cdot \omega_n^2}{s^2 + 2\zeta\omega_n s + \omega_n^2} \quad (4)$$

where A_S denotes the dc gain, ω_n is the natural frequency, and ζ is the corresponding damping factor. Let the targeted −3 dB bandwidth of an n-stage LA be ω_{LA-3dB}, the voltage gain be A_{VLA}, and each gain cell be approximated by a two-pole amplifier. For a critical damping, the required GBW per stage can be expressed as

$$GBW = \omega_{LA-3dB} \times \left(\frac{1}{2^{1/n} - 1}\right)^{1/4} \times A_{VLA}^{1/n} \quad (5)$$

To provide a −3 dB bandwidth of 10-GHz and a voltage gain of 40-dB in the LA, a three-stage LA demands a GBW of 65-GHz per stage. To achieve the GBW performance, every

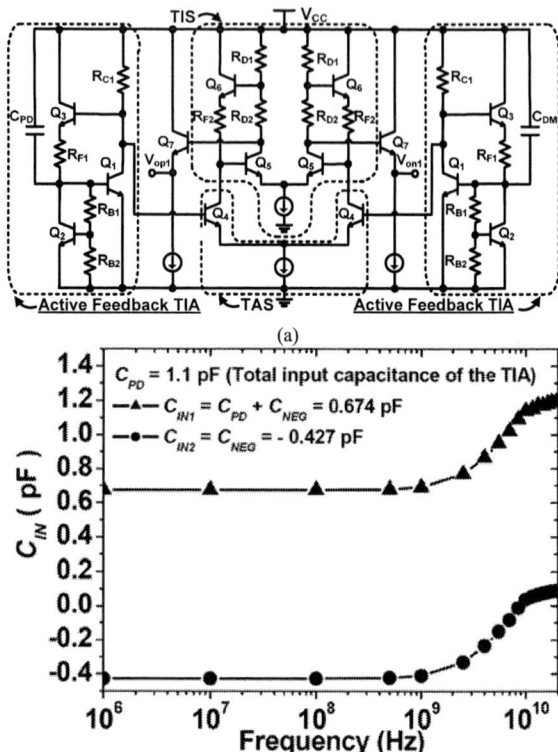

(a)

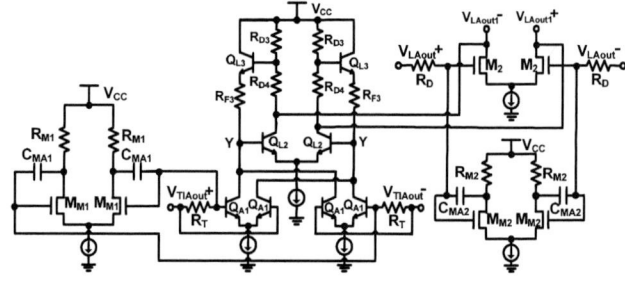

Fig. 2　(a) The two-stage TIA architecture. (b) The analysis results of the input parasitic capacitance.

Fig. 3　The input stage of limiting amplifier.

stage of the LA was designed to have 13.3-dB gain and 14.1-GHz bandwidth. The whole gain and bandwidth performance of the LA with output stage were designed to be 43.5-dB and 9.9-GHz. The output stage is in large-signal operation and the bandwidth degradation caused by the output stage is almost negligible.

The ATC and DOC circuits need large off-chip capacitors to provide low pass function, as shown in Fig. 1. As can be seen in Fig. 3, the Miller amplification capacitors [5] are used to realize the large equivalent capacitors of 121 pF and 216 pF, respectively. The real capacitors C_{MA1} and C_{MA2} are amplified by the differential amplifier gain factor of $(1+A_{D1})$ and $(1+A_{D2})$ due to the Miller effect. Both the capacitance values of the real capacitors (C_{MA}) in the ATC and the DOC circuits are only 14.8 pF, which can be integrated in a single chip to avoid off-chip interference and to reduce cost.

In this optical receiver AFE design, the whole circuit

blocks are directly coupled on the chip to avoid off-chip noise interference. Due to the direct coupling, conventional matching networks at the output of the TIA and the input of the LA are eliminated. The high gain-bandwidth product, low power consumption, and small chip area optical receiver AFE can be achieved.

III. EXPERIMENTAL RESULTS

The proposed optical AFE has been fabricated in a commercial 0.35-μm 50-GHz f_T SiGe BiCMOS technology. Fig. 4 shows the die photo whose area is 0.7 x 0.7 mm², including test pads. To accurately demonstrate the capability of capacitance immunization technology, two 1.1 pF n^+p diode capacitors, which include test pads, have been integrated on this chip to act as the large C_{PD}.

A. Frequency Domain Response

The post-layout simulated small-signal frequency responses of the optical receiver AFE are shown in Fig. 5. The solid line shows the transimpedance gain of 107.3-dBΩ, and the f_{H-3dB} of 8.85-GHz, for a 3-V supply. The dash line shows the transimpedance gain of 99.6-dBΩ, and the f_{H-3dB} of 6.7-GHz, for a 2.5-V supply. The frequency responses show the capacitance immunization technique can provide high bandwidth performance between 2.5-V to 3-V supply.

B. Time Domain Response

The Agilent 4906B serial BERT and 86100A DCA wide-bandwidth oscilloscope were used to measure the time domain response of the optical receiver AFE. In the measurement, $2^{31}-1$ pseudorandom bit sequence (PRBS) pattern was used. Fig. 6 illustrates the measured eye diagrams under 3-V and 2.5-V supplies, respectively. The eye diagrams were defined at a bit-error ratio (BER) of 10^{-12} using the $2^{31}-1$ PRBS test pattern for a power level of −10 dBm. For 10-Gb/s, the peak-to-peak jitter is about 13.22 ps (15.86 ps) under a 3-V (2.5-V) supply. For 12.5-Gb/s, the peak-to-peak jitter is about 17.87 ps under a 3-V supply. For 10-Gb/s, the differential output swing is 842 mV$_{p-p}$ (616 mV$_{p-p}$) under a 3-V (2.5-V) power supply.

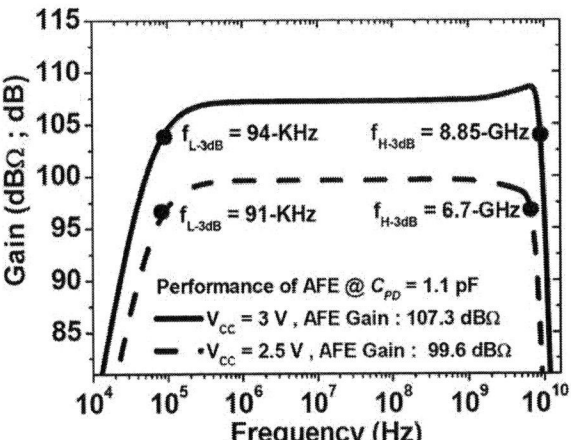

Fig. 5   Frequency response of the analog front-end

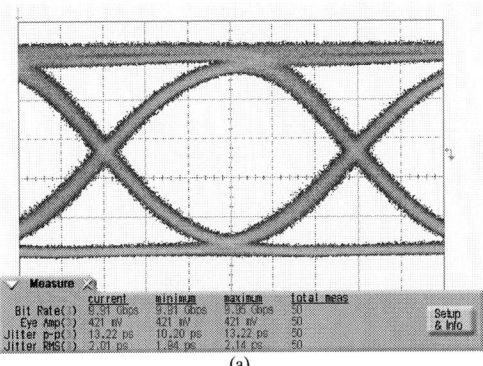

(a)

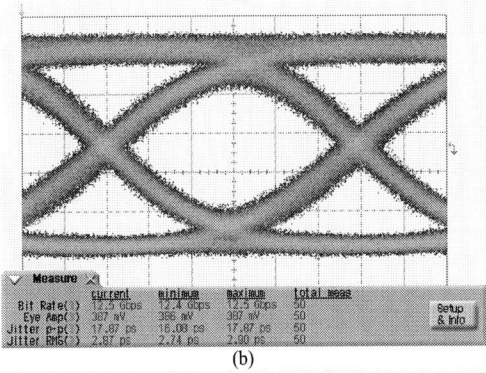

(b)

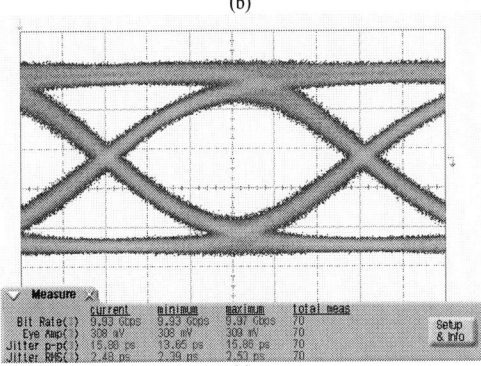

(c)

Fig. 6   The output eye diagrams with −10 dBm input using $2^{31}-1$ PRBS at BER=10^{-12}. (a) 10-Gb/s, X-scale: 17 ps/div, Y-scale: 90 mV/div, at V_{CC}=3-V; (b) 12.5-Gb/s, X-scale: 13.4 ps/div, Y-scale: 90mV/div, at V_{CC}=3-V; (c) 10-Gb/s, X-scale: 16.6 ps/div, Y-scale: 80mV/div, at V_{CC}=2.5-V.

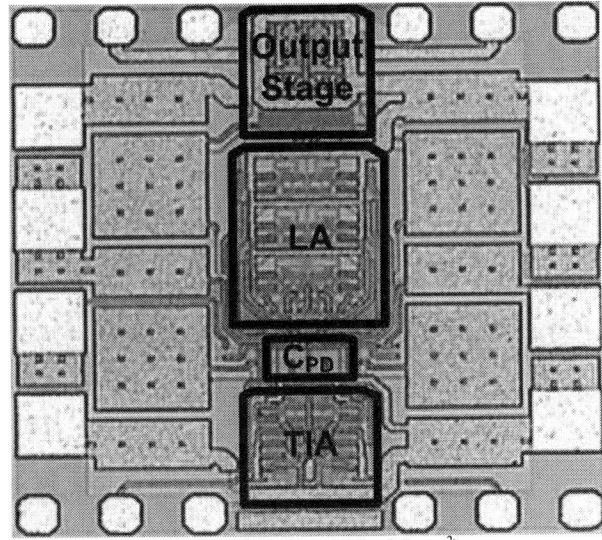

Fig. 4   Die photo (0.7 x 0.7 mm²)

Table I
PERFORMANCE COMPARISON of 10 Gb/s OPTICAL RECEIVER ANALOG FRONT-END

	Technology	PD capacitance (C_{PD})	PD Responsivity	Power dissipation(P_{DC})	Chip area	Output swing (differential)	Peak-to-Peak jitter	Sensitivity	GBW/P_{DC} (GHz-Ω/mW)
This work (3V supply)	0.35 μm SiGe BiCMOS	1.1 pF	0.9 A/W	195 mW (3V supply)	0.7 × 0.7 mm²	842 mV$_{p-p}$	13.22 ps	-16.4 dBm	10517.4 GHz-Ω/mW (post-sim.)
This work (2.5V supply)	0.35 μm SiGe BiCMOS	1.1 pF	0.9 A/W	106.3 mW (2.5V supply)	0.7 × 0.7 mm²	616 mV$_{p-p}$	15.86 ps	-12.7 dBm	6019.2 GHz-Ω/mW (post-sim.)
[6] 07'JSSC	0.2 μm SiGe BiCMOS	N.A.	N.A.	300 mW (3.3V supply)	1.8 × 1.8 mm²	300 mV$_{p-p}$	15 ps	16 uA	2061.1 GHz-Ω/mW (measurement)
[7] 07'VLSI System	0.18 μm CMOS	0.15 pF	1 A/W	199 mW (1.8V supply)	1.3 × 1.796 mm²	900 mV$_{p-p}$	30.96 ps	-13 dBm	1249 GHz-Ω/mW (post-sim.)
[8] 05'JSSC	0.18 μm CMOS	0.15 pF	0.85 A/W	210mW (1.8V supply)	1.028 × 1.796 mm²	400 mV$_{p-p}$	29.88 ps	-12 dBm	810.2 GHz-Ω/mW (post-sim.)

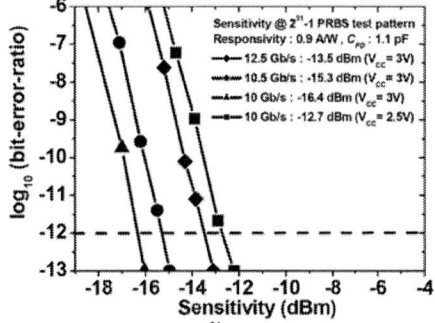

Fig. 7   Measured sensitivity for $2^{31}-1$ PRBS test pattern at 10-Gb/s, 10.5-Gb/s, and 12.5-Gb/s.

For 12.5-Gb/s, the differential output swing is 774 mV$_{p-p}$ under a 3-V supply. The eye diagrams show good eye-open condition and small jitter$_{p-p}$. These measured results indicate that the capacitance immunization technique can be used to enhance bandwidth performance. Fig.7 shows the bit error ratio (BER) performance of the receiver with a photodetector whose C_{PD} is 1.1 pF and whose responsivity is 0.9 A/W. For a BER less than 10^{-12}, the measured sensitivity of the analog front-end under 3-V operation is −16.4 dBm for 10-Gb/s, −15.3 dBm for 10.5-Gb/s, and −13.5 dBm for 12.5-Gb/s, respectively. For a BER less than 10^{-12}, the measured sensitivity of the analog front-end under 2.5-V supply is −12.7 dBm for 10-Gb/s. The input-referred noise current (I_N) of the optical receiver AFE is derived from its sensitivity performance which can be expressed as [7]

$$Sensitivity \cong 10\log\left[\frac{14.1I_N(r_e+1)}{2\rho(r_e-1)}1000\right] \text{ dBm} \qquad (6)$$

where ρ is the responsivity of the photodetector, and r_e is the extinction ration. The corresponding I_N is approximately 2.81 μA$_{rms}$ and 6.6 μA$_{rms}$ for 10-Gb/s with 3-V and 2.5-V supply voltages, respectively.

Table I summarizes the performance for some 10-Gb/s optical receiver analog front-ends that are based on Si-related technologies. Under 3-V (2.5-V) supply, the power dissipation of the AFE in this work is 195 mW (106.3 mW) in which 27.4 mW (19.7 mW) is consumed by the output buffer to drive a 50-Ω load. The chip area of this AFE is also noticeably smaller than others.

IV. CONCLUSION

This paper presents the implementation of a fully integrated optical receiver analog front-end in a 0.35-μm 50-GHz f_T SiGe BiCMOS technology. The design uses a two-stage TIA with a capacitance immunization technology to enhance gain-bandwidth performance of the TIA. In addition, the area-efficient optical receiver AFE provides an excellent gain-bandwidth product per DC power figure-of-merit of 10517.4 GHz-Ω/mW (GBW/P_{DC}), under a 1.1 pF input capacitance and a 3-V supply. With lower supply voltage (2.5-V, 106.3 mW), the optical receiver AFE still can achieve high GBW/P_{DC} of 6019.2 GHz-Ω/mW.

ACKNOWLEDGMENTS

The financial support from the National Science Council of Taiwan, R.O.C. is acknowledged. The authors would like to thank the Chip Implementation Center of the National Science Council of Taiwan, R.O.C. for the support of fabricating samples.

REFERENCES

[1] K.-S. Lai, J.-C. Huang, and K.Y.-J. Hsu, "Design and properties of phototransistor photodetector in standard 0.35-μm SiGe BiCMOS technology," *IEEE Trans. Electron Devices*, vol. 55, no. 3, pp. 774–781, Mar. 2008.

[2] C. D. Holdenried, J. W. Haslett, and M. W. Lynch, "Analysis and design of HBT Cherry-Hooper amplifiers with emitter-follower feedback for optical communications," *IEEE J. Solid-State Circuits*, vol. 39, no. 11, pp. 1959–1967, Nov. 2004.

[3] J. S. Weiner, *et al.*, "An InGaAs-InP HBT differential transimpedance amplifier with 47-GHz bandwidth," *IEEE J. Solid-State Circuits*, vol. 39, no. 10, pp. 1720–1723, Oct. 2004.

[4] S. Galal and B. Razavi, "10 Gb/s limiting amplifier and laser/modulator driver in 0.18 μm CMOS technology," *IEEE J. Solid-State Circuits*, vol. 38, no. 12, pp. 2138–2146, Dec. 2003.

[5] H. Tran, *et al.*, "6 KΩ 43 Gb/s differential transimpedance limiting amplifier with auto-zero feedback," *IEEE J. Solid-State Circuits*, vol. 39, no. 10, pp. 1680–1689, Oct. 2004.

[6] A. Maxim, "A 54 dBΩ + 42 dB 10 Gb/s SiGe transimpedance-limiting amplifier using bootstrap photodiode capacitance neutralization and vertical threshold adjustment," *IEEE J. Solid-State Circuits*, vol. 42, no. 9, pp. 1851–1864, Sep. 2007.

[7] W. Z. Chen and D. S. Lin, "A 90-dBW 10-Gb/s optical receiver analog front-end in a 0.18-mm CMOS technology," *IEEE Trans. Very Large Scale Integr (VLSI). Syst.*, vol. 15, no. 3, pp. 358–365, Mar. 2007.

[8] W. Z. Chen and D. S. Lin, "A 1.8-V 10-Gb/s fully integrated CMOS optical receiver analog front-end," *IEEE J. Solid-State Circuits*, vol. 40, no. 6, pp. 1388–1395, Jun. 2005.

IEEE 2008 Custom Intergrated Circuits Conference (CICC)

A Multifunction Transceiver RFIC for 802.11a/b/g WLAN and DVB-H Applications

Yin Shi, Fa Foster Dai*, Jun Yan, Xueqing Hu, Hua Xu, Ming Gu, Xuelian Zhang, Qiming Xu, Bei Chen, Fangxiong Chen, Peng Yu, Heping Ma, Fang Yuan and Richard C. Jaeger*

Smart Chip Integration Inc., CAS Semiconductor Integration Technology Research Center, Suzhou, China
* Department of Electrical and Computer Engineering, Auburn University, Auburn, AL 36849-5201, USA

Abstract-We present a multifunction wireless transceiver RFIC for WLAN and DVB-H applications. The RFIC includes a super-heterodyne transceiver for IEEE 802.11a/b/g WLAN applications and a zero-IF receiver for DVB-H application. The WLAN and DVB-H transceiver share the down-conversion mixers, the baseband VGAs, filters, and the PLL synthesizer, resulting in a small silicon area of 20mm^2. Using a 3.3V supply, the power consumption of the WLAN transceiver and the DVB-H receiver are 462mW and 396mW, respectively.

I. INTRODUCTION

The boom of wireless and mobile networks leads to an ever-increasing request for multi-function, low power, and low cost RFICs. With wireless local network (WLAN) and digital-video-broadcasting (DVB) standards operating in very different frequency bands [1-5], future trends point to integrated wireless terminals that can support variety of modes such as TV and Internet browser. Wireless transceivers that can support multi-standards have not been reported so far. This work presents a low-cost hybrid wireless transceiver RFIC for WLAN and DVB coexistence. Combining multi-standards into a single wireless transceiver achieves high performance, low cost, low power, and small form-factor. It thus allows multi-mode interoperability with transparent worldwide usage.

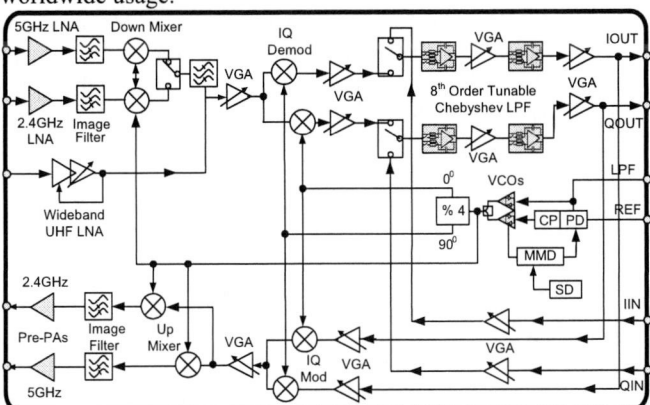

Fig. 1 Block diagram of the hybrid transceiver RFIC for WLAN 802.11a/b/g and DVB-H tuner applications.systems.

II. HYBRID TRANSCEIVER RFIC DESIGN

As illustrated in Fig. 1, the implemented multifunction hybrid transceiver RFIC comprises multi-band radio front-ends that can cover the WLAN 802.11a/b/g at 2.4GHz and 5.2GHz bands and the DVB-H UHF band from 470MHz to 862MHz. The illustrated multi-band transceiver is a direct-conversion receiver for DVB-H and a super-heterodyne transceiver for WLAN. The WLAN transceiver utilizes a walking IF, where the RF LO is 4 times of the IF LO. Thus both LOs can be derived from a single synthesizer. The IF chooses the upper-side band of the RF mixer for 2.4GHz band and the lower side-band of the RF mixer for 5.2GHz band. Thus, the walking-IF topology not only saves a synthesizer, but also moves the IF frequencies of the 2.4GHz and the 5.2GHz bands close to each other, allowing sharing the IF filter and the IF mixer with the IF ranging from 804MHz to 1161.2MHz. This frequency plan also allows the DVB-H receiver to share the IF-mixer of the WLAN receiver with its UHF input frequency from 470MHz to 860MHz. The baseband filter is an 8th order tunable Chebyshev low pass filter (LPF), which is shared by WLAN and DVB-H receivers. The WLAN transmitter also uses the same baseband filter with a loop-back scheme shown in Fig. 1.

The multi-standard wireless receiver comprises multiple LNAs in order to cover the entire WLAN and DVB-H bands. Both 5GHz and 2.4GHz LNAs are designed with 25dB gain at high-gain mode and about 0dB gain at low-gain mode.

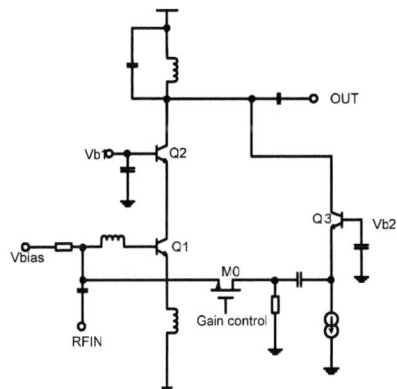

Fig. 2 WLAN 802.11a/b/g LNA circuit with high-gain and low-gain mode

The WLAN LNAs with tuned single-ended cascade common-emitter structure is shown in Fig.2, which has high gain and low gain modes for a wider dynamic range. Inductive emitter degeneration is used for better noise performance with high linearity. For the high gain mode, the cascode amplifier (Q1 and Q2) is powered on while the common base amplifier (Q3) is off. For the low gain mode, the NMOS switch (M0) and common base amplifier (Q3) is turned on while the cascode amplifier (Q1 and Q2) is turned off. The two gain modes step the gain by 25dB.

978-1-4244-2018-6/08 $25.00 © 2008 IEEE   267

Two image filters with more than 20 dB image rejection are inserted between the WLAN LNA and down-mixer to eliminate the images. Mixer output buffers are designed to be open-collector, sharing the same load to switch the WLAN and tuner functions. An on-chip bandpass filter with 300MHz bandwidth and 15dB rejection is employed to replace the off-chip SAW filters. In order to increase the dynamic range of the receiver, a variable gain amplifier (VGA) with selectable gain of 16/11.5/6/1.6 dB gain is inserted in the receiver RF path.

For wide band reception over the DVB-H UHF band, inductorless wide band LNAs is used. Fig. 3 shows the tuner front-end with variable gain LNA circuits. The fixed gain LNA is differential cascade amplifier with shunt feedback. Variable gain LNA supplies a large dynamic range for the system with the gain tuning range of from 5dB to 25dB with automatic gain control (AGC) based on the input power level. An adaptive biasing circuitry is used for automatic linearity compensation. The input referred 3rd-order intercept-point (IIP3) of the tuner front-end increases automatically for a strong input signal. Following the LNA is a VGA that is designed to have 0dB to 17dB gain control.

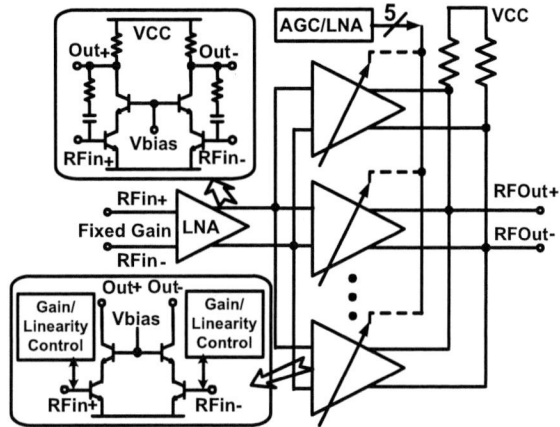

Fig. 3 DVB-H tuner front-end circuits with fixed gain LNA and AGCs.

The WLAN transmitter baseband VGA provides negative gain from -32dB to -26dB with a step size of 2dB. The receiver baseband variable gain range is 48.8dB from 11.4dB to 60.2dB with a step size of 1dB. Transmitter IQ modulator includes two mixers similar to the receiver IF mixers with image rejection.

The baseband filter is designed as an 8th order Chebyshev filter with temperature compensated programmable corner frequency. The cut-off frequency of baseband filter for WLAN can be programmed from 6.3MHz to 18.2MHz, while that for DVB-H tuner can be adjusted from 2.6MHz to 7.4MHz.

The power amplifier driver consists of a differential cascode amplifier, an emitter follower, and a common-emitter amplifier as shown in Fig.4. The cascode amplifier stage acts as a differential to single-ended converter. The cascode structure employed here can supply good inverse-isolation and wide bandwidth. A tank is used as the load to provide narrow-band gain with good out-of-band filtering. The common-emitter amplifier provides additional gain with good linearity

and sufficient voltage headroom. In a two-tone test, the IIP3 of the power amplifier driver is +10.2dBm.

The power amplifier driver is designed to drive the 50 Ω input of the off-chip 5.2GHz/2.4GHz power amplifier. The tank is resonated at 5.2GHz/2.4GHz, which also provides image rejection for out-of-band tones. The bypass Cb and Rb provide an ac path for the signals. The current dissipation of the power amplifier driver is 28mA.

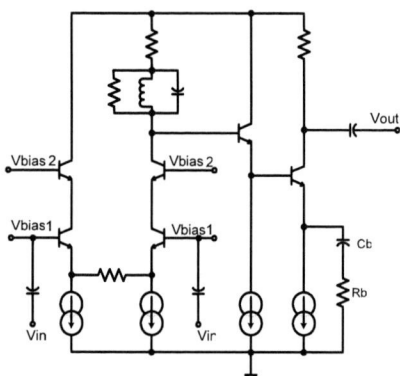

Fig. 4 Schematic of the PA driver with image rejection.

The pre-PA driver is designed to have 8dB gain at 5GHz and 11dB gain at 2.4GHz. An odd order harmonic filter is designed to remove the harmonics after the IQ modulator. With the pre-PA tanks, the filters in the Tx chain provides more than 40dB emission rejections for harmonics and spurs rejections, with the IIP3 of +10.2dBm.

The multi-band phase-locked-loop (PLL) comprises a 9-bit programmable multi-modulus divider (MMD) constructed with 8 cascaded divided-by-2/3 cells. With an additional control bit, the divide ratio can be programmed from 128 to 511. A divided by 1/2/6/8 programmable divider is employed at the output of PLL to generate the LO frequencies from 470MHz to 4.8GHz. Five wide-band VCOs are employed to cover the bands from 2900MHz to 4800MHz bands. Wide-tuning range PMOS VCOs are designed to allow the tank referenced to the ground, which leads to lower phase noise.

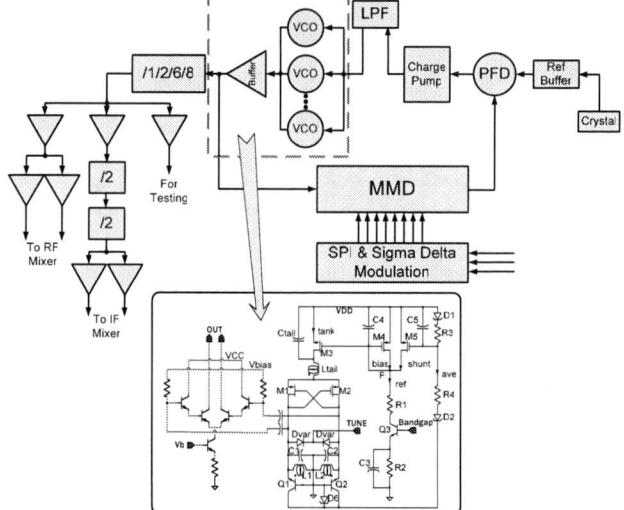

Fig.5 Wide-band fractional-N PLL synthesizer with a PMOS VCO

978-1-4244-2018-6/08 $25.00 © 2008 IEEE   268

The VCO design also employs an automatic amplitude control (AAC) circuitry that can keep the VCO in current limit region and alleviate the AM and AM-to-PM noise. AAC also provides perfect ambient-proof characteristics over process, temperature and frequency variations. VCO buffer utilizes the Darlington structure in order to sustain the large VCO output swing. VCO buffer has an open-collector with the load resistors laid out close to the mixer. Porting the LO in current mode through long distance avoids the interference and voltage decay. The phase frequency detector (PFD) employs a totally differential structure for symmetric and noise rejection. The charge pump (CP) is also differential and its current can be programmed from 140uA to 1190uA for adjusting the loop gain and bandwidth for different applications.

III. MEASURED RESULTS

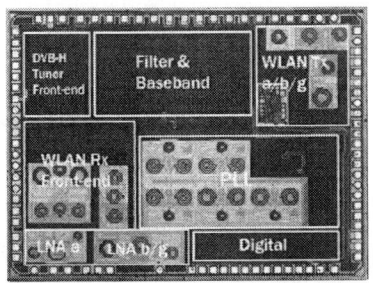

Fig. 6 Die photo of the WLAN/DVB-H hybrid transceiver RFIC.

The die photo of the WLAN/DVB-H hybrid transceiver RFIC is shown in Fig. 6. To our best knowledge, this design is the first multifunction hybrid transceiver RFIC with a compact die size of 4×5mm².

Fig. 7 shows the measured WLAN LNA S11 in the 802.11a and 802.11bg mode separately. The measured baseband filter V~F characteristics are shown in the Fig. 8. It shows that the filter in WLAN mode has a cut-off frequency at 9.8MHz with stopband rejection larger than 31dB at 12MHz. In DVB-H tuner mode, cut-off frequency is measured as 4MHz with rejection larger than 36dB at 4.8MHz. Passband ripple are all less than 1dB.

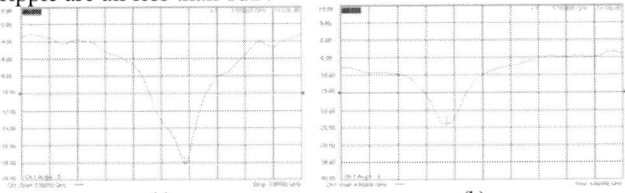

(a)   (b)

Fig. 7 Measured WLAN LNA S11 for 802.11a (a) and 802.11bg (b)

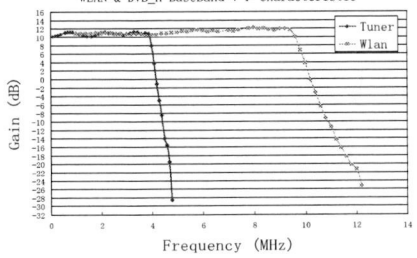

Fig. 8 Measured WLAN and tuner baseband filter V-F characteristics.

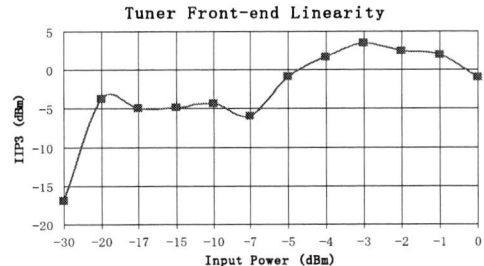

Fig. 9 Measured DVB-H front-end linearity boosting vs. input power.

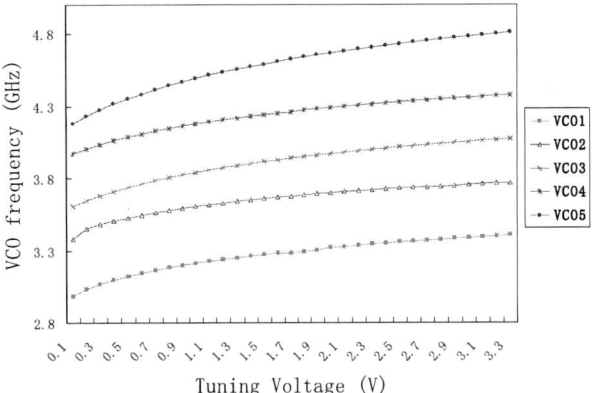

Fig. 10 Measured VCO tuning curves for 5 VCOs, i.e., VCO1 to VCO5

The measured tuner front-end linearity is shown in Fig.9. When the input power increases, the linearity compensation circuitry automatically boosts the front-end IIP3. When the power reaches a saturation level (around -3dBm), the IIP3 begins to drops. The measured IIP3 result agrees with the simulation well.

Fig. 10 shows the VCO tuning curves for VCO1 to VCO5 when the tuning voltage is changed from 100mV to 3.3V. The five VCOs can cover the 2.9GHz to 4.8GHz range with tuning voltage from 500mV to 3.0V. Fig.11(a) shows the measured PLL settling times with VCO frequency is changed from 3504MHz to 4056MHz, the settling time is about 250us with charge pump output current set as 170uA. Fig.11(b) shows the typical PLL phase noise measurement using VCO5 in the WLAN mode. The carrier frequency is 4608MHz and the VCO output power is -24.68dBm, the measured phase noise in locked state is: -93 dBc/Hz @ 1kHz; -100 dBc/Hz @ 100kHz; -112 dBc/Hz @ 1MHz.

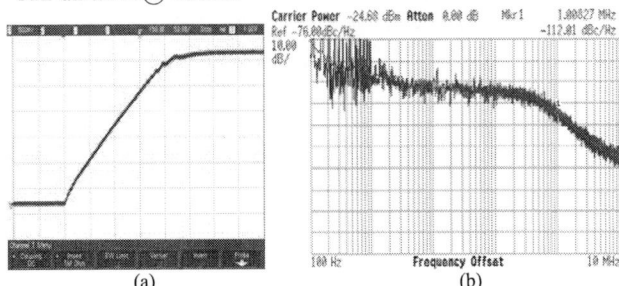

(a)   (b)

Fig. 11 Measured PLL settling time(50us/div) (a) and PLL phase noise (b) for WLAN mode

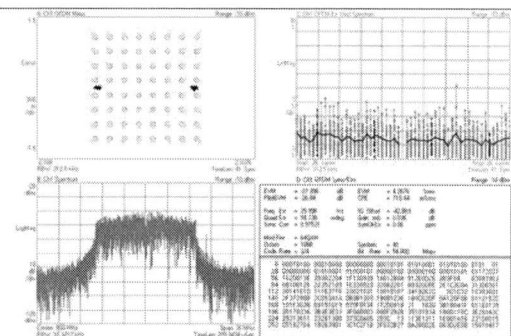

Fig. 12 Measure DVB-H tuner EVM

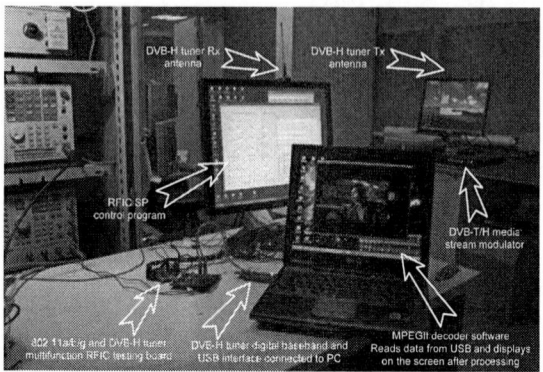

Fig. 14 WLAN/DVB-H hybrid transceiver RFIC DVB-H mode demo system

Fig. 12 shows the measured EVM of 4.26% rms for DVB-H tuner Rx. Fig.13 shows the multifunction RFIC testing board and Fig. 14 gives a complete DVB-H tuner demo test configurations. A DVB-T/H media stream modulator is placed in the back to modulate MPEG2 stream into DVB-H format and transmit the modulated signals at about 0dBm power level. In the front-end, apart about 4 meters from the back-end transmitter antenna, there is a 75ohm receiver antenna which is connected to the 802.11a/b/g and DVB-H multifunction RFIC board. The output signals of the RFIC transceiver IC are fed into the DVB-H digital baseband which is connected to a PC through the USB interface IO port. The PC receives the raw media stream signals and performs software processing to de-code the MPEG2 streams, and then the application program can play the video and audio on the PC. The RFIC is controlled by another serial-port-interface (SPI) control program which loads the control data bits into RFIC through RS232 interface to a MCU and then to the RFIC in SPI timing.

Table 1 summarizes the transceiver performances and compares them to previous work. This work presents a high performance, lower power, and smaller size multi-function transceiver RFIC that covers WLAN a/b/g and DVB-H bands. The prior solution for multi-standard radio implementation is to use multiple single-band transceivers with large die size and form factor. By integrating multi-standard radio functions into a single RFIC, this work gains great advantages over previous designs. The present work allows sharing of radio functional blocks such as the frequency synthesizer, mixers, VGAs and baseband filters. In addition to a complete WLAN a/b/g transceiver, a DVB-H tuner front-end with only $0.9 \times 1.2mm^2$ size was added to form a multi-function hybrid transceiver RFIC with $20mm^2$ total die size.

Table.1 Measured WLAN/DVB-H performance and comparison with others. Test conditions: (1)Simulated LNA NF/Measured Front-end NF, DVB-H: max gain, WLAN Rx front-end: 22dB gain; (2) Rx IIP3, WLAN: 39dB gain, DVB-H: max gain; (3) Tx OIP3, -10 dB gain; (4) 64QAM, 54Mbps; (5) @66 dB gain; (6) @-20dB gain; (7) @64 dB gain.

Parameters	802.11a		802.11b/g		DVB-H	Reference Work		
	Rx	Tx	Rx	Tx		WLAN	DVB-H	
	This work					[1]	[2]	[3]
	0.5um SiGe BiCMOS					0.18um SiGe	0.5um SiGe	0.18um CMOS
Size [mm²]: 20 (WLAN a/b/g + DVB-H)						12	16	7.84
Supply [V]: 3.3						1.8	2.8	2.8
G max dB	89	-20	95	-20	100	76	94~100	95.2
G min dB	5.4	-32	5.4	-32	18	6/20	/	2
NF dB(1)	3.8/5.6	33.7	3.5/5.2	32	5.5	5.6	3.1~4.6	4.5
IIP3 dBm(2)	13.2	/	14	/	-29.3	-11(5)	12(6)	-5(7)
OIP3 dBm(3)	/	18	/	16	/	17.3	/	/
I_{diss} .mA	95	101	80	87	62	112/134	65.7	66
PN dBc/Hz	-100, fc=4.6GHz @100kH				-98	/	/	-98.1
	-112, fc=4.6GHz @1MHz				-121	/	/	/
EVM %rms (4)	5.42	5.24	5.15	4.83	4.26	3	/	/

IV. CONCLUSION

In conclude, this paper presents a single-chip multi-function, multi-standard hybrid transceiver RFIC for WLAN 802.11a/b/g and DVB-H applications that has not been reported before. The bybrid transceiver RFIC was implemented in a 0.5um SiGe BiCMOS technology with $20mm^2$ total die size. The power consumption is only 462mW and 369mW for WLAN and DVB-H operating modes, respectively.

REFERENCES

[1] R. Ahola, et al., "A single-chip CMOS transceiver for 802.11a/b/g wireless LANs", IEEE J. Solid-State Circuits, vol. 39, no. 12, pp. 2250-2258, 2004.

[2] K. Iizuka et al., "A 184 mW Fully Integrated DVB-H Tuner With a Linearized Variable Gain LNA and Quadrature Mixers Using Cross-Coupled Transconductor", IEEE J. Solid-State Circuits, vol. 42, no. 4, pp. 862-871, 2007.

[3] K. Young-jin, et al. "A Multi-Band Multi-Mode CMOS Direct Conversion DVB-H Tuner", ISSCC. pp. 608-609, 2006.

[4] O.Charlon, et al., "A low-power high-performance SiGe BiCMOS 802.11a/b/g transceiver IC for cellular and bluetooth Co-existence applications", IEEE Journal of Solid-State Circuits, 2006. 41(7): pp. 1503-1512.

[5] Dave G. Rahn, et al., "A Fully Integrated Multi-Band MIMO WLAN Transceiver RFIC," IEEE Journal of Solid-State Circuits, vol. 40, no. 8, pp. 1629-1641, August, 2005.

Fig. 13 WLAN/DVB-H hybrid transceiver RFIC testing board

978-1-4244-2018-6/08 $25.00 © 2008 IEEE

IEEE 2008 Custom Intergrated Circuits Conference (CICC)

A Fully Integrated Zero-IF Mobile TV Tuner RFIC for S-band CMMB Application

Yin Shi, Fa Foster Dai*, Jun Yan, Hua Xu, Xuelian Zhang,
Heping Ma, Fang Yuan, Xin Guan, and Richard C. Jaeger*

Smart Chip Integration Inc., CAS Semiconductor Integration Technology Research Center, Suzhou, China
* Department of Electrical and Computer Engineering, Auburn University, Auburn, AL 36849-5201, USA

Abstract- **We present a single-chip tuner RFIC for the newly established Chinese Mobile Multimedia Broadcasting (CMMB) standard. This mobile digital TV tuner covers the CMMB frequencies from 2.635GHz to 2.660GHz. Implemented in a 0.35um SiGe technology with 4mm^2 die size, this tuner IC achieves noise figure of 2dB, input sensitivity of -100dBm, IIP3 of 17dBm, and total power of 162mW under a 2.8V supply.**

I. INTRODUCTION

The mobile TV has been evolved into a wide variety of standards in different countries. With DVB-H established in Europe and US, ISDB-T in Japan and DMB in Korea, China recently announced the adoption of her own mobile TV standard CMMB based on Satellite-Terrestrial Interactive Multi-service Infrastructure (STiMi). As Fig. 1 shows, the STiMi system is a mixed satellite and terrestrial wireless broadcasting system designed to provide audio, video and data service for handheld receivers with less than 7 inch wide LCD display, such as PMP, mobile phone, PDA and UMPC. CMMB utilizes two S-band satellites to cover the digital video broadcasting (DVB) over the wide area. In the populated metropolitan areas, CMMB utilizes the cellural base stations to enhance the digital video transmission in order to allow DVB reception with low-cost terminals. The satellite and terrestrial complementary network is combined to create a Single Frequency Network (SFN) using the 2635-2660 MHz band. The CMMB service operates in the S-band frequency with 25MHz bandwidth in order to offer 25 video and 30 radio channels.

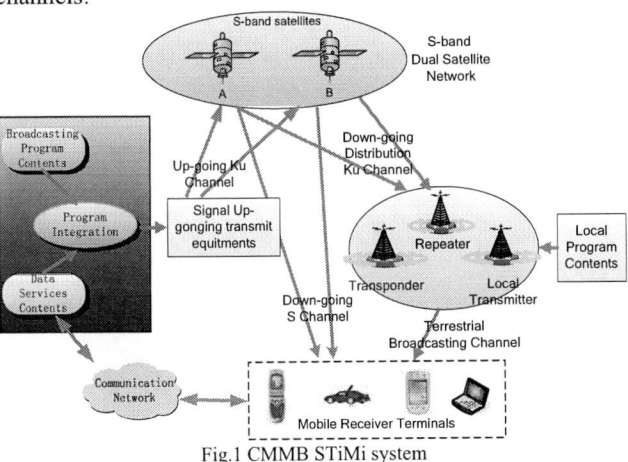

Fig.1 CMMB STiMi system

This paper presents the first reported CMMB tuner RFIC. In order to handle both satellite and terrestrial receptions, the CMMB tuner requires very high input sensitive and wide dynamic range, which requires very low noise and high linearity; small form factor and low cost, which requires high degree of integration; and low power consumption.

This CMMB tuner RFIC achieves high input sensitivity of -100dBm and high linearity of 17dBm IIP3. The power consumption of 162mW is optimized for longer mobile usage, and the chip is design to be compact with only 4mm^2 die size.

In Section II, the CMMB tuner RFIC design will be discussed in detail. Section III presents the measurement results of the critical blocks and the whole receiver system.

II. CMMB TUNER RFIC DESIGN

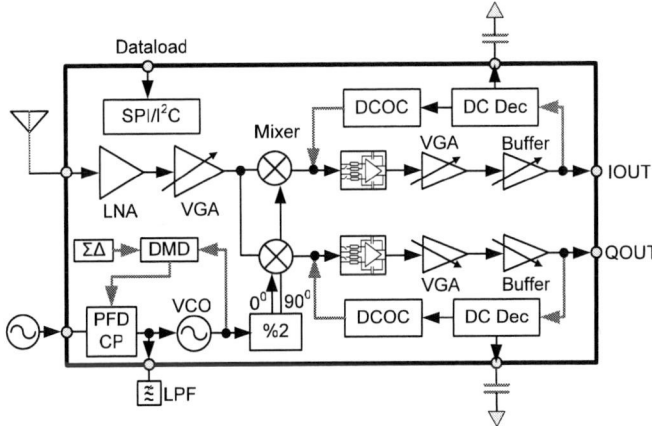

Fig. 2 Block diagram of the CMMB tuner RFIC.

As illustrated in Fig. 2, the fully integrated CMMB tuner includes all RF receiver functions without using any off-chip SAW filters. The front-end of the CMMB tuner consists of a single-ended low-noise amplifier (LNA) with dual gain mode (HG/LG), an RF VGA with continuously controlled variable gain, and a degenerated double-balanced Gilbert cell down-conversion I/Q mixer.

The LNA has a high-gain mode and a low-gain mode as shown in Fig. 3. For the high gain mode, an inductively degenerated cascode structure is used to achieve 15dB power gain and 1.6dB NF while drawing only 4mA. For the low gain mode, the MOS switch M0 is turned on, and the attenuation resistor R1 are used for -7dB power loss. The capacitor C1 is used for ac coupling. The IIP3 in the low gain mode can achieve more than +20dBm. In the high gain mode, the output capacitor C2 is designed to conjugate-match the output impedance of the LNA with that of the following RFVGA for

978-1-4244-2018-6/08 $25.00 © 2008 IEEE   271

best power transmission. In the low gain mode, the value of the C2 may not be on the power matched value, but it is not important since the gain is attenuated in this mode.

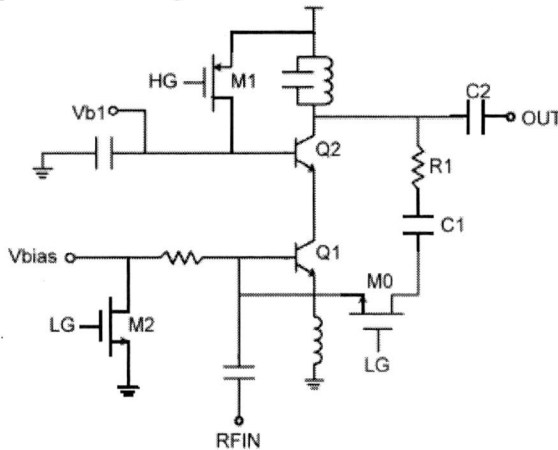

Fig. 3 Simplified schematics of the two-mode LNA.

In Fig. 3, the MOSFET switch M0, M1 and M2 are used to switch the gain of the LNA. In the high gain mode, all the switches are turned off, while in the low gain mode, all the switches will be turned on. In our design, we added the switch M1 for keeping the cascode transistor Q2 working in the safe area in the low gain mode. When the input signal is large and the circuit is in the low gain mode, the transistor Q1 is shut down by switch M2, the collector voltage of Q1 will be low without the switch M1. As a result, almost the entire supply voltage adding the large input signal swing will be added to the collector-emitter junction of the transistor Q2, potentially causing break down. However, with M1 turned on in the low gain mode, the base and emitter voltage of Q2 will be pulled to high and the collector-emitter junction voltage will be greatly reduced, which helped to keep Q2 working in the safe region in the low gain mode.

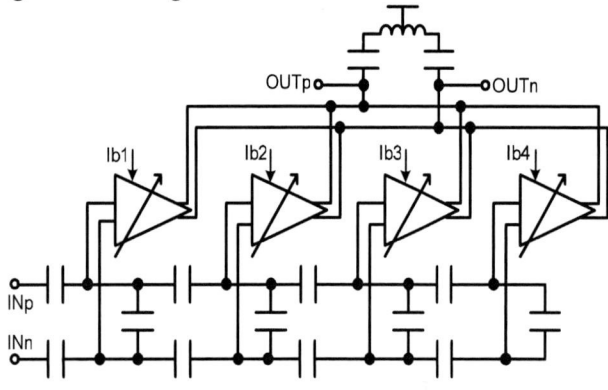

Fig.4 Simplified RFVGA schematic

The RF VGA consists of a three-stage capacitive attenuator with 13dB per stage attenuation and four amplifiers which smoothly switch outputs from the ladder attenuator, as shown in Fig. 4. The common load of the four core amplifiers is constructed by L and C1, C2, which act as a LC tank to

filter out the signal out of the band. This LC load also greatly released the critical headroom requirement, thus improved the gain and the linearity performance of the circuit.

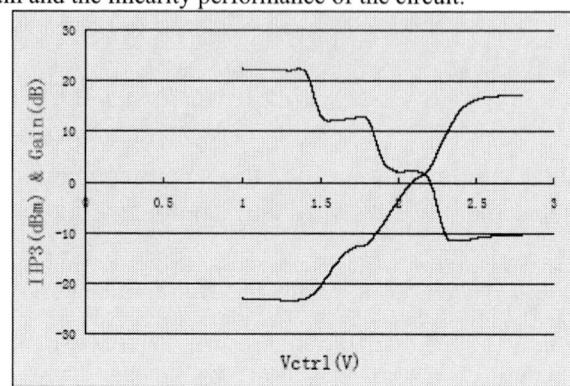

Fig. 5 Simulated gain and IIP3 dependence on control voltage of RFVGA

Fig. 5 shows the simulated IIP3 and gain of the RFVGA as control voltage changes. Comparing to the VGA used in [1], the proposed VGA greatly improves linearity and gain control range using the same amplifier structure with alternate current biasing.

A quadrature mixer shown in Fig. 6 has been developed to provide improved image rejection ratio (IRR) and reduced phase error [2]-[4]. It is basically a combination of two Gilbert mixers with shared transconductance stages. The total signals turn on in the order: LOQp, LOIp, LOQn, LOIn. For example, when the RF signal on transistor Q1 is high and the LOQp signal on Q7 is high, the voltage of the collector terminal of Q1 is pulled high and the transistors Q3, Q4 and Q8 are shut off. In this way, the total available current must flow through only a selected transistor according to the local signal sequence. This mechanism is called Q-I mutual interference and is useful for phase error suppression. In Fig. 6, the capacitors C1 to C4 are used to filter out the RF and LO frequency and their harmonics.

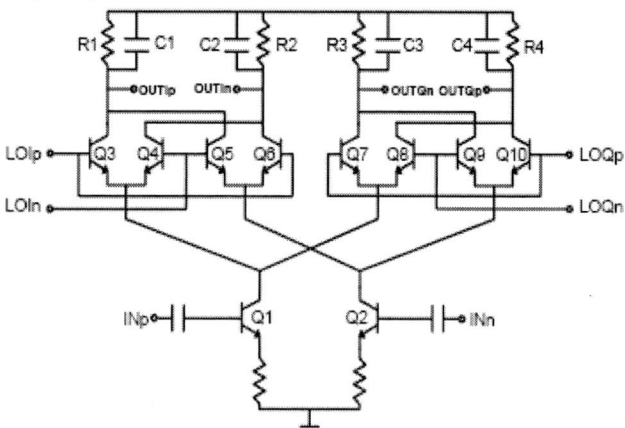

Fig. 6 Quadrature down-conversion I/Q mixer

The CMMB tuner IC comprises a baseband VGA, an 8th order Chebyshev filter and a DC offset compensation loop. The VGA is a cascode amplifier with current shunt from the load for gain tuning. The VGA achieves more than 40dB gain tuning and 18dBm IIP3 with temperature compensation. Fig. 7 shows a detailed structure of the baseband filter.

The baseband output is filtered to extract its DC components used for DC offset compensation for the down mixer. The DC offset at the mixer output is measured as 56 mV without compensation and it reduces to 6mV when the compensation loop is closed. The DC offset compensation loop consumes 412uA. The 8th order Chebyshev filter is implemented using Op-amp-RC integrator in a leapfrog configuration as shown in Fig. 7. It is less sensitive to the variation of component values, comparing to cascaded biquad. The filter provides 0.5dB passband ripple and -35dB attenuation at 6MHz with the cutoff frequency at 4MHz. An on-chip frequency auto-calibration is accomplished using a capacitor switching technique. The 5-bit capacitor arrays are controlled by the frequency auto-tuning circuits shown in Fig. 7. After calibration, the auto-tuning circuits can be turned off without additional power consumption.

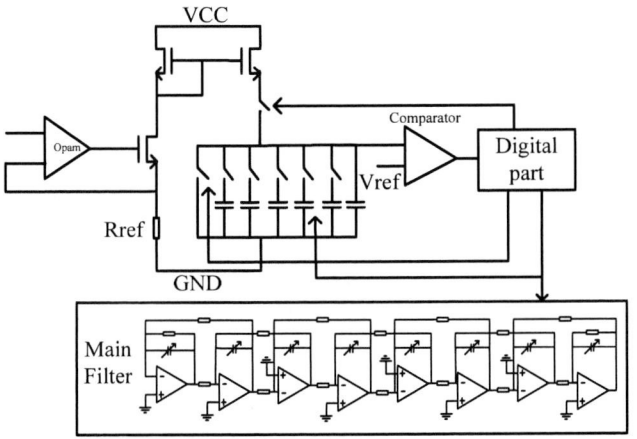

Fig. 7 Simplified 8th order Chebyshev filter with corner frequency auto-calibration.

The CMMB tuner includes an Fractional-N PLL comprising low noise PFD, current variable chare pump, a low phase noise VCO, P/P+1 dual modulus divider (DMD), and a 3 order MASH sigma-delta noise shaping digital block. DMD input is a high speed prescaler with 32 or 33 divide ratios. DMD provides a divide range from 352 to 354. The VCO output is divided by 2 as LO signal which can cover the 2635MHz to 2660MHz. This divider also provides the I/Q phase LO signal for I/Q down-conversion mixer. The VCO is designed to be a wide frequency tuning range from 5.1GHz to 5.75GHz. Wide-tuning range PMOS VCOs are designed to allow the tank referenced to the ground, which leads to lower phase noise. The VCO design provides good ambient-proof characteristics over process, temperature and frequency variations. The in-band phase noise of the phase-locked loop is measured as -90dBc/Hz at 100kHz offset.

III. CMMB TUNER RFIC MEASUREMENT RESULTS

The die photo of CMMB tuner RFIC with a compact die size of 4mm^2 is shown in Fig. 8. It is the first CMMB tuner RFIC reported so far.

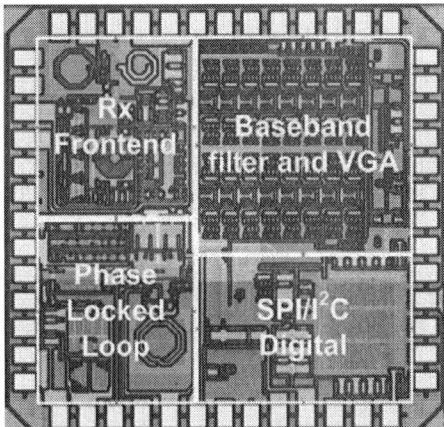

Fig. 8 Die photo of the CMMB tuner RFIC.

Fig. 9 shows the measured LNA S11 and NF plots. It can be seen from the NF curve that in the operation band from 2.635GHz to 2.660GHz, the NF is measured below 2.25dB.

Fig. 10 shows the measured RFVGA gain versus control voltage. The gain varies from -21dB to 19dB when the control voltage changes from 1.1V to 2.5V, achieving more than 40 dB variable gain. The measured curve fits the simulated one well. Fig. 11 shows the measured baseband VGA gain versus control voltage. The gain varies from about 7 dB to 48dB in 85 Degree Celsius ambient temperature. In the room temperature about 25 Degree, the measured curve fits the simulated curve well.

The measured baseband filter transfer characteristics are shown in the Fig. 12. The filter shows a cut-off frequency of 4.13MHz at -10 Degree Celsius and 4.41MHz at 85 Degree when the corner frequency auto-calibration circuit is turned on.

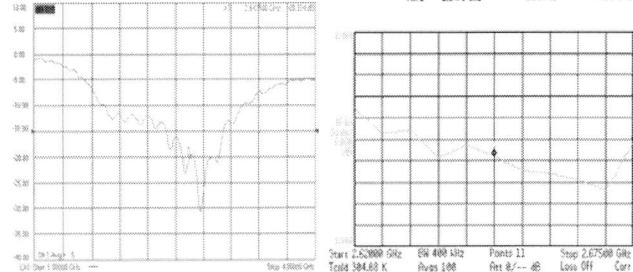

Fig. 9 Measured CMMB tuner LNA S11 and NF.

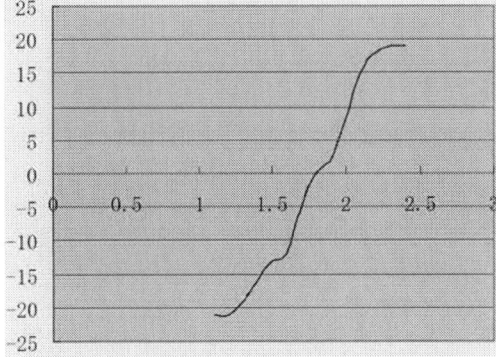

Fig. 10 Measured gain versus control voltage of RFVGA.

978-1-4244-2018-6/08 $25.00 © 2008 IEEE

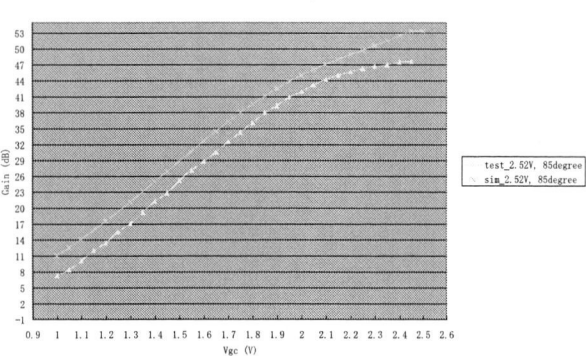

Fig. 11 Measured gain versus control voltage of baseband VGA in 85 Degree Celsius ambient temperature.

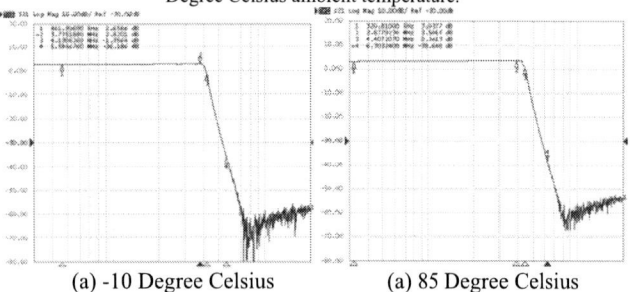

(a) -10 Degree Celsius   (a) 85 Degree Celsius

Fig. 12 Measured tuner baseband filter transfer function at different temperature

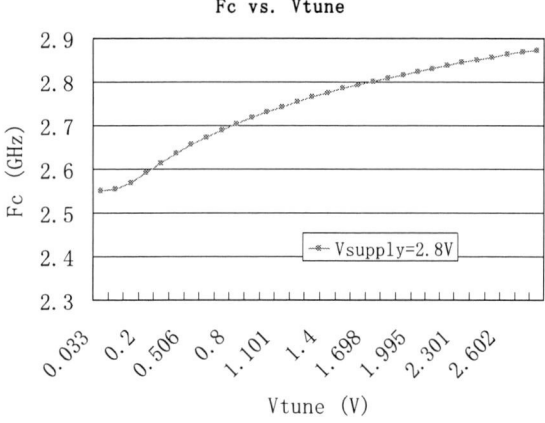

Fig. 13 Measured LO frequency tuning curve

Fig. 13 shows the measured LO frequency tuning curve. When VCO tuning VCO is varied from 0.5V to 2.5V, the LO frequency can covers the frequency band from 2.6GHz to 2.85GHz. The VCO frequency is twice the LO frequency from 5.2GHz to 5.7GHz.

Fig. 14 shows the measured LO phase noise. The carrier frequency is 5295MHz, the LO output power is -19.6dBm. The measured integrated rms phase noise is about 1.68 Degree.

Table 1 summarizes the tuner performances in which tuner sensitivity is measured when the output signal amplitude is beyond -20dBv and the waveform looks without obvious distortion. The tuner IC achieves low noise figure, high input sensitivity and high linearity, while consuming low power.

Table.1 Summary of CMMB tuner RFIC performance.

Parameters	Conditions	Meas.	Unit
Input signal power	LNA input	-100~0	dBm
IIP3	@0dB power gain	17	dBm
IIP2	@0dB power gain	44	dB
Noise figure	High gain	2.2	dB
System gain	Power gain	-15~89	dB
LO phase noise	@10kHz	-84	dBc/Hz
	@100kHz	-90	
	@1MHz	-113	
VCO tuning range		5.1~5.75	GHz
IQ amplitude mismatch		-32	dBc
Pass band		3.8	MHz
Pass band ripple		<1	dB
Stop band rejection		>50	dB
Settling time	DCOC	<100	µs
	PLL	200	µs
Power consumption	Power supply: 2.5~3.1V Temp range: -40~85 ^{0}C	46~58	mA
Current consumption	Sleeping mode	97	µA
Die size		2x2	mm^2

IV. CONCLUSION

In conclude, this paper presents a single-chip tuner RFIC for CMMB S-band applications which is never reported before. It was realized in 0.35um SiGe BiCMOS technology with 4mm^2 total die size. The power consumption is only 162mW under a 2.8V supply. This mobile digital TV tuner covers the CMMB frequencies from 2.635GHz to 2.660GHz., it achieves noise figure of 2dB, input sensitivity of -100dBm, IIP3 of 17dBm. This ensures the tuner RFIC has high sensitivity, wide dynamic range, fast switching and ambient-proof performance with low power consumption and low cost.

REFERENCES

[1] B. Gilbert, "A low-Noise Wideband Variable-Gain Amplifier Using an Interpolated Ladder Attenuator," *ISSCC Dig. Tech. Papers,* pp. 280-281, Feb. 1991.

[2] J. Bojer, "Quadrature mixer method and apparatus," U.S. Patent 6,029,059, Jun. 30, 1997.

[3] J. Harvey and R. Harjani, "Analysis and design of an integrated quadrature mixer with improved noise, gain, and image rejection," in *Proc. IEEE Int. Symp. Circuits and Systems,* May 2001, vol. IV, pp. 786-789.

[4] K. Taniguchi et al., "A study on a GPS dual-band image-reject mixer." IEICE, Tech. Rep. IEICE ICD2003-100.

[5] S. Azuma, et al., "A Digital Terrestrial Television(ISDB-T) Tuner for Mobile Application," *ISSCC Dig. Tech. Papers,* pp. 278-279, 2004.

[6] K. Iizuka, *et al.,* "A 184 mW Fully Integrated DVB-H Tuner With a Linearized Variable Gain LNA and Quadrature Mixers Using Cross-Coupled Transconductor", *IEEE J. Solid-State Circuits,* vol. 42, no. 4, pp. 862-871, 2007.

[7] K. Young-jin, *et al.* "A Multi-Band Multi-Mode CMOS Direct Conversion DVB-H Tuner", *ISSCC Dig. Tech. Papers,* pp. 608-609, 2006.

[8] Myung-Woon Hwang, *et al.*,"A 1.8dB NF 112mW Single-Chip Diversity Tuner for 2.6GHz S-DMB Applications", *ISSCC Dig. Tech. Papers,* pp. 616-617, 2006.

Compact Modeling   Session 11

Chair: Gennady Gildenblat, Arizona State University
Co-Chair: Brian Chen, Advanced Micro Devices

As CMOS technology continues its relentless scaling, new materials are being introduced and novel device structures are being considered. Adequate compact models and simulation techniques are required to account for their impact on circuit characteristics. This session showcases several examples in compact model development, enhancement, and application in connection with the latest technology advances.

The first, invited paper from University of California, San Diego presents a generalized analytical model for multi-gate MOSFETs of various configurations. Fully equipped with quantum- and short-channel effects and terminal charges expressions, the new model has been calibrated and validated by FinFET hardware. The paper also provides a brief review of the overall research activities on multi-gate MOSFET modeling.

The second, invited paper is a joint work from AMD, IBM and Freescale on partially depleted (PD) silicon-on-insulator (SOI) MOSFET model. It presents SOI-specific challenges in SOI MOSFET modeling, parameter extraction, and circuit simulation. The current status of the next-generation SOI model standardization process by Compact Model Council is briefly reviewed.

The third, invited paper from Arizona State University presents an analytical model of ionizing radiation effects on bulk CMOS transistors and integrated circuits that takes into account non-uniform defect distributions through surface potential equations. The new model leads to identification of sidewall doping as the primary technology feature to drive increased radiation tolerance in advanced deep-submicron bulk CMOS.

The fourth paper from IBM describes a new approach to characterize and model FET source/drain diffusion resistance by incorporating electric current trajectory information. The new model is verified with field solver simulations over various parameter values and validated by hardware data from a 65 nm SOI technology.

The final paper from Georgia Tech develops a hybrid technique using hardware-based mobility measurement and device electrostatic simulations to analyze the impact of interfacial oxide thickness variations on metal-gate high-K circuits. It is shown that the interfacial oxide thickness variation needs to be effectively controlled to reduce circuit variability and fully exploit the advantage of metal gate high-k technologies.

978-1-4244-2018-6/08 $25.00 © 2008 IEEE

Notes

978-1-4244-2018-6/08 $25.00 © 2008 IEEE

Compact Modeling of Multiple-Gate MOSFETs

Yuan Taur, Jooyoung Song, Bo Yu

The Department of Electrical and Computer Engineering, University of California - San Diego, La Jolla, CA 92093, USA
E-mail: taur@ece.ucsd.edu, jooyoungsong@ucsd.edu

Abstract **This paper reviews recent development on compact modeling of multiple-gate MOSFETs. Starting with a core model based on the analytic potential solutions for the highly symmetric double-gate (DG) and surrounding-gate (SG) MOSFETs, an explicit solution to the implicit algebraic equations with high accuracy has been developed. With the addition of quantum, short-channel effects, and capacitance formulations, the core model for DG MOSFETs has been expanded into a full-blown compact model which has subsequently been calibrated and validated by FinFET hardware. In view of the various types of experimental multiple-gate MOSFETs developed, the DG and SG MOSFET models have been generalized to other less symmetric structures, including quadruple-gate, triple-gate, Ω-gate, and Π-gate devices. Finally, research activities in other groups on multiple-gate MOSFETs are briefly summarized.**

I. INTRODUCTION

Multiple-gate (MG) MOSFETs have been proposed to extend the scaling limit of conventional bulk MOSFETs to 10-nm gate length and beyond [1], [2], owing to their excellent control of the short-channel effect and high current driving ability. Since there are many different MG structures investigated (Fig. 1), it is attractive to develop a unified, accurate and efficient compact model for all those MG MOSFETs. An analytic potential model for symmetric double-gate (DG) MOSFETs was first derived by solving Poisson and current continuity equations with the gradual channel approximation [3], [4]. It is significant to note that the analytic potential DG model is able to account for volume inversion effects because the charge-sheet approximation [5], commonly employed in all surface-potential based bulk MOSFET models, is not invoked. A similar approach has been applied to surrounding-gate (SG) MOSFETs as well [6]. These are possible because of the inherent symmetry of DG and SG MOSFETs which allows the Poisson equation to be reduced to 1-D ordinary differential equation in Cartesian and cylindrical coordinates respectively. However, Poisson's equations in other MG MOSFETs such as quadruple-gate (QG) MOSFETs, triple-gate (TG) MOSFETs, and triple-plus-gate (TPG) MOSFETs (Fig. 1), cannot be solved analytically due to lack of symmetry. Additional approximations are necessary to develop compact models for those MG MOSFETs.

In this paper, a unified compact model for MG MOSFETs is presented. Section II starts with the basic analytic potential current models for DG and SG MOSFETs and their explicit solutions. Section III describes the charge and capacitance models. Section IV covers the quantum and short-channel additions to the model. Calibration and validation of the completed model to FinFET hardware are discussed in Section V. Section VI extends the core models to quadruple-gate, triple-gate, and other multiple-gate MOSFETs. Section VII

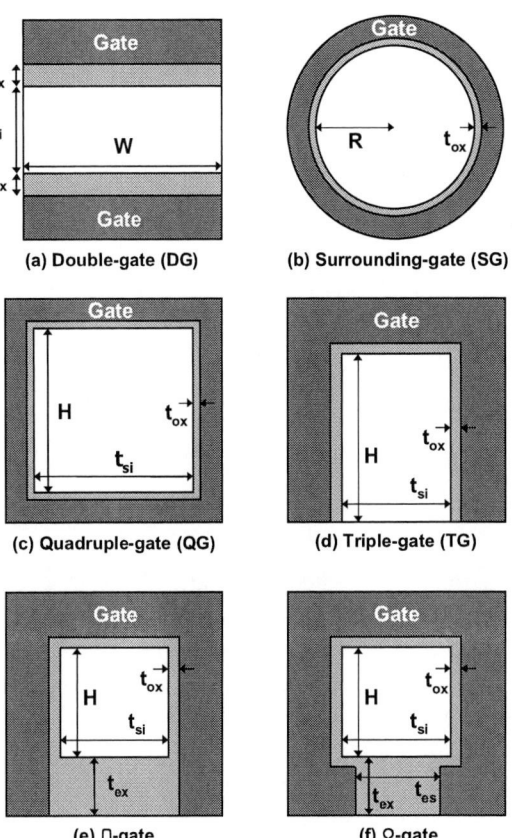

Figure 1 Schematic diagrams of MG MOSFET cross sections.

reviews other group's approaches and Section VIII concludes the paper.

II. CORE MODELS

A. Symmetric Double-Gate MOSFET (Fig. 1a)

For an undoped (or lightly-doped) silicon film, the potential at any point in the device is governed by Poisson's equation,

$$\frac{d^2\psi}{dx^2} = \frac{q}{\varepsilon_{si}} n_i e^{q(\psi-V)/kT} \qquad (1)$$

where V is the quasi-Fermi potential that varies laterally from the source to the drain, and is independent of x. The general solution takes the form

$$\psi(x) = V + \frac{2kT}{q}\ln\left[\frac{2\beta L_{Di}}{t_{si}}\right] - \frac{2kT}{q}\ln[\cos(2\beta x/t_{si})] \qquad (2)$$

where the intermediate parameter β is given by the boundary condition,

$$\frac{q(V_g - \Delta\phi - V)}{2kT} - \ln\left[\frac{2L_{Di}}{t_{si}}\right] = \ln\beta - \ln[\cos\beta] + 2r\beta\tan\beta \cdot \qquad (3)$$

978-1-4244-2018-6/08 $25.00 © 2008 IEEE

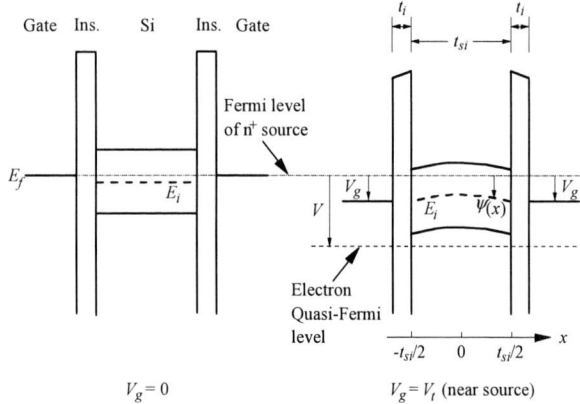

$V_g = 0$ $V_g = V_t$ (near source)

Figure 2 Electric Potential of SDG MOSFET with no bias and bias V_g.

Here the structural parameter for DG MOSFETs is defined as $r \equiv \varepsilon_{si} t_{ox} / \varepsilon_{ox} t_{si}$. The inversion charge density is given by $Q_m = 8\varepsilon_{si} (W/t_{si})(kT/q)\beta \tan\beta$. Integrating the current continuity equation yields the drain current for symmetric DG MOSFETs,

$$I_{ds} = \mu \frac{W}{L} \frac{8\varepsilon_{si}}{t_{si}} \left(\frac{kT}{q}\right)^2 \left[p(\beta_d) - p(\beta_s) \right] \qquad (4)$$

where the intermediary function p is defined as $p(\beta) = -2\beta\tan\beta + \beta^2 - 2r\beta^2\tan^2\beta$, and β_s and β_d are solved from the boundary condition (3) with V being the values at the source and drain respectively.

B. Surrounding-Gate MOSFET (Fig. 1b)

The compact model for SG MOSFET has been obtained in a similar way as DG MOSFETs. With cylindrical symmetry, Poisson's equation takes the form

$$\frac{d^2\psi}{d\rho^2} + \frac{1}{\rho}\frac{d\psi}{d\rho} = \frac{q}{\varepsilon_{si}} n_i e^{q(\psi-V)/kT}. \qquad (5)$$

Solving the equation yields the potential as below.

$$\psi(\rho) = V - \frac{2kT}{q} \ln\left[\frac{R}{L_{Di}\sqrt{1-\alpha}} \left(1 - \frac{(1-\alpha)\rho^2}{R^2} \right) \right] \qquad (6)$$

where α is given by

$$\frac{q(V_g - \Delta\phi - V)}{2kT} - \ln\left(\frac{2L_{Di}}{R} \right) = \ln\sqrt{1-\alpha} - \ln\alpha + s\frac{1-\alpha}{\alpha}. \qquad (7)$$

Here the structural parameter for SG MOSFET is $s \equiv 2\varepsilon_{si}\ln(1+t_{ox}/R)/\varepsilon_{ox}$. The inversion charge density is given by $Q_m = 8\pi\varepsilon_{si}(kT/q)(1-\alpha)/\alpha$. And the drain current for SG MOSFETs

$$I_{ds} = \mu \frac{8\pi\varepsilon_{si}}{L} \left(\frac{2kT}{q}\right)^2 \left[f(\alpha_d) - f(\alpha_s) \right] \qquad (8)$$

where the intermediary function f is defined as $f(\alpha) \equiv (-2/\alpha - \ln\alpha) + s(-1/\alpha^2 + 2/\alpha)$, with α_s and α_d being the values from the boundary conditions at the source and the drain.

C. Explicit Solutions for SDG MOSFET and SG MOSFET
1) General Method

Similar to the analytical approximation for the surface potential (PSP) compact model, the explicit analytical solution can be divided into three steps [7]. The first step starts with an initial estimate as an explicit continuous function of gate voltage, structural parameters, etc. Continuity through all regions is important to avoid discontinuity of high-order derivatives.

The second step is to modify the initial guess with high-order correction terms. Assuming that the implicit function to be solved is $f(x; a, b, c) = 0$, and x is approximated by $g(a, b, c)$, f can be expanded into Taylor series near g as

$$\sum_{n=0}^{\infty} \frac{1}{n!} \frac{\partial^n f(x; a, b, c)}{\partial x^n}\bigg|_{x=g(a,b,c)} \left[x - g(a, b, c) \right]^n. \qquad (9)$$

By taking the expansion to the third order, the correction term $h = x - g(a, b, c)$ is given by

$$h = -\frac{f_{g0}}{f_{g1}} \left(1 + \frac{f_{g0} f_{g2}}{2 f_{g1}^2} + \frac{f_{g0}^2 (3 f_{g2}^2 - f_{g1} f_{g3})}{6 f_{g1}^4} \right) \qquad (10)$$

where $f_{gn} = \dfrac{\partial^n f(x; a, b, c)}{\partial x^n}\bigg|_{x=g(a,b,c)}$.

If the accuracy of the solution after the second-step correction is not sufficient, another high-order correction can be applied as the third step to further modify the solution to obtain a highly accurate approximation for x

$$v(a, b, c) = g(a, b, c) + h(a, b, c) + w(a, b, c) \qquad (11)$$

where $w(a, b, c)$ is the new correction term in the last step.

2) Explicit Solution for Symmetric Double-Gate MOSFET

Using an intermediary variable $z \equiv \arctan\beta$ instead of β yields a better initial guess due to the relatively simple behavior of the variable z. Following the general approach above, an appropriate initial guess in this case is

$$z_1 = \sqrt{\left(\frac{2}{\pi^2 r^2}\right)^2 + \left(\frac{2}{\pi r}\right)^2 \ln^2(1 + e^{F/2})} - \frac{2}{\pi^2 r^2}. \qquad (12)$$

The final result z_3 is calculated by applying the above procedure adding two correction terms to z_1. The final error $|f(z_3(r, F); r, F)|$ is a criterion for measuring the accuracy of the explicit solution. For the functions above, the worst case error is found to be $|f| \cong 1.1 \times 10^{-11}$, indicating extremely high accuracy.

3) Explicit Solution for Surrounding-Gate MOSFET

The general approach can be applied to SG MOSFET as well with an initial guess

$$z_1 = \sqrt{\left(\frac{1}{2s^2}\right)^2 + \left(\frac{1}{s}\right)^2 \ln(1 + e^G)} - \frac{1}{2s^2}. \qquad (13)$$

The final result z_3 is calculated from the same procedure with two correction terms. The worst case error is found to be $|g| \cong 3.2 \times 10^{-13}$, which is extremely small.

III. CHARGES AND CAPACITANCES

A. Analytical Charge Model for double-gate MOSFETs

978-1-4244-2018-6/08 $25.00 © 2008 IEEE

The drain current model previously presented is sufficient for DC simulation. For transient simulation, an analytic and continuous expression of charges associated with each terminal needs to be developed [8]. Unlike the four-terminal bulk MOSFET, DG MOSFETs have only three terminals because the two gates are connected together and switched simultaneously. Therefore, only three terminal charges, i.e., 1) Q_g, 2) Q_d and 3) Q_s, associated with the gate, drain, and source, respectively, are required for circuit simulation. For the drain and source charges, we adopt Ward and Dutton's linear charge partition method [9].

$$Q_g = \frac{\mu_{eff}W}{I_{ds}}\int_0^L Q_i^2(V)dV = \frac{\mu_{eff}W}{I_{ds}}\int_{\beta_s}^{\beta_d} Q_i^2(\beta)\frac{dV}{d\beta}d\beta \quad (14)$$

Substitute $Q_i(\beta)$ and $dV/d\beta$ developed previously into Q_g yields

$$Q_g = \frac{\mu_{eff}W^2}{I_{ds}}\frac{2kT}{q}\left(\frac{8kT\varepsilon_{si}}{qt_{si}}\right)^2$$
$$\times \int_{\beta_s}^{\beta_d}\beta^2\tan^2\beta\left(\frac{1}{\beta}+2r\frac{\beta}{\cos^2\beta}+\tan\beta+2r\tan\beta\right)d\beta \quad (15)$$

The above integral cannot be carried out in closed-form. Approximations are made to obtain a continuous, analytical charge model with high accuracy. Designate $q_i = \beta\tan\beta = Q_i/(8\varepsilon_{si}kT/qt_{si})$ as the normalized charge sheet density, the modified boundary condition yields

$$dV = \frac{t_{si}}{4\varepsilon_{si}}\left(\frac{1}{2q_i}+2r\right)dq_i \cdot \quad (16)$$

Substituting the charge expression as a function of β, dV, and I_{ds} leads to

$$Q_g = 8WL\frac{kT}{q}\frac{\varepsilon_{si}}{t_{si}}\frac{0.25(q_{is}+q_{id})+\frac{2r}{3}(q_{is}^2+q_{is}q_{id}+q_{id}^2)}{0.5+r(q_{is}+q_{id})} \quad (17)$$

where $q_{is} = \beta_s\tan\beta_s$, $q_{id} = \beta_d\tan\beta_d$ are the normalized charge density at the source and drain.

Based on the modified current expression above, the drain current can be expressed by

$$I_{ds} \approx \mu\frac{W}{L}\frac{4\varepsilon_{si}}{t_{si}}\left(\frac{2kT}{q}\right)^2\left[\frac{q_i}{2}+rq_i^2\right]_{q_{id}}^{q_{is}} \cdot \quad (18)$$

Substituting y/L into the integral of Q_d and performing similar approximations as in the gate charge derivation, one can eventually obtain

$$Q_d = -8WL\frac{kT}{q}\frac{\varepsilon_{si}}{t_{si}}\frac{(k_1+k_2+k_3)}{[0.5+r(q_{is}+q_{id})]^2} \quad (19)$$

where
$$k_1 = (q_{is}+2q_{id})/24$$
$$k_2 = (5q_{is}^2+10q_{is}q_{id}+9q_{id}^2)r/24$$
$$k_3 = (2q_{is}^3+4q_{is}^2q_{id}+6q_{is}q_{id}^2+3q_{id}^3)2r^2/15$$

The resulting equations for the gate charge and drain charge are continuous throughout all operation regions. Once the normalized charge sheet density at the source and drain is given, terminal charges can be directly computed from the analytical model of Q_g and Q_d. The source charge is then given by $Q_s = -Q_g - Q_d$.

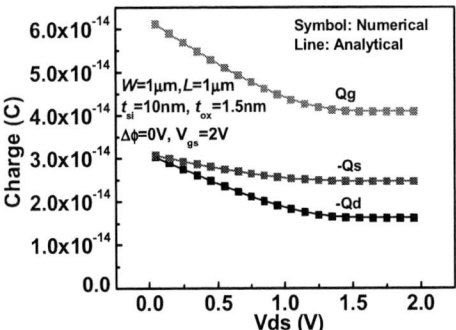

Figure 3 Terminal charges of a symmetric DG MOSFET from the analytical charge model in comparison with the numerical integration as a function of drain voltage.

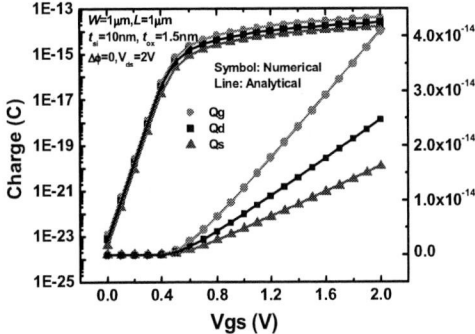

Figure 4 Terminal charges of a symmetric DG MOSFET from the analytic charge model compared with the numerical integration. Both log (left) and linear (right) scales are displayed.

B. Capacitance Model for double-gate MOSFETs

From the charge expressions developed previously, the capacitance model is developed by adopting widely-accepted Ward *et al*'s linear charge partition method. It is worthy to note that the developed capacitance model is not only valid for DG MOSFETs, but also for all three-terminal devices when the charge expression associated with each terminal is known.

Generally, a three-terminal device needs 9 non-reciprocal capacitances for small signal simulation. Charge conservation law yields several relationships that leave only four capacitances independent of each other, e.g., C_{gd}, C_{dd}, C_{dg}, and C_{gg}.

$$C_{gd} = -\frac{L^2g_{ds}^2}{\mu_{eff}I_{ds}}+\frac{Q_g}{I_{ds}}g_{ds}$$

$$C_{dd} = -\frac{L^2g_{ds}^2}{\mu_{eff}I_{ds}}-\frac{2Q_d}{I_{ds}}g_{ds}$$

$$C_{dg} = -\frac{L^2g_{ds}^2}{\mu_{eff}I_{ds}}+\frac{Q_g}{I_{ds}}(g_m+g_{ds})+\frac{2Q_d}{I_{ds}}g_m$$

$$C_{gg} = \frac{L^2[(g_m+g_{ds})^2-g_{ds}^2]}{\mu_{eff}I_{ds}}-\frac{Q_g}{I_{ds}}g_m$$

(20)

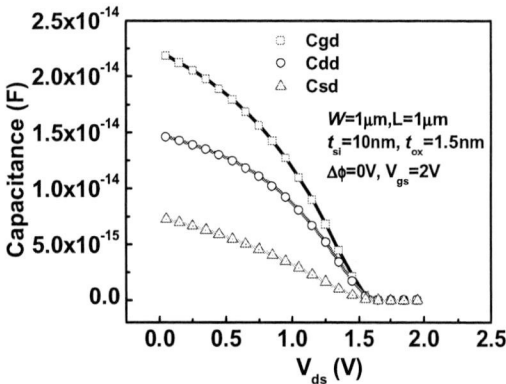

Figure 5 Capacitances of a symmetric DG MOSFET calculated by the analytical model (line) and numerical integration (symbol) as a function of drain voltage.

With the analytical charge model previously developed in the unified drain current model, the capacitances can be calculated analytically once β_s and β_d are solved from the boundary conditions. Thus, all the intrinsic parameters of DG MOSFETs required for DC, AC and large signal analysis are analytically derived. Comparison between the numerical simulations and the analytical values of the charges and capacitance in SDG case is given in Figs. 3, 4 and Fig. 5.

IV. SHORT-CHANNEL EFFECTS AND QUANTUM-MECHANICAL EFFECTS

A. Short-Channel Effects of Symmetric Double-Gate MOSFET

The essence of MOSFET 2-D geometry effect is well captured by the scale length theory which has been proven a powerful guideline to the minimum channel length design of bulk and DG MOSFETs. To fully quantify the short-channel V_t roll-off, DIBL, and subthreshold slope for compact modeling purposes, the pre-exponential factor of the 2-D potential term needs to be determined in addition to the exponential scale length. In the subthreshold region, the mobile charge and fixed charge are negligible. Using the superposition method neglecting high order terms, the 2-D potential expression in the subthreshold region has been solved [10].

$$\psi(x,y) = \frac{\Delta\phi_1 - \Delta\phi_2}{t_{si} + 2t_i\varepsilon_{si}/\varepsilon_{ox}}x + V_g - \frac{\Delta\phi_1 + \Delta\phi_2}{2} + \frac{b_1 \sinh[\pi(L-y)/\lambda_1] + c_1 \sinh(\pi y/\lambda_1)}{\sinh(\pi y/\lambda_1)}\cos(\pi x/\lambda_1) \quad (21)$$

where λ_1 is the scale length that can be solved from the following equation.

$$\tan(\pi t_{ox}/\lambda_1)\tan(\pi t_{si}/2\lambda_1) = \varepsilon_{ox}/\varepsilon_{si} \quad (22)$$

Once the potential is solved, I_{ds}-V_g expression in the subthreshold region can be derived from current continuity. The result includes all V_t roll-off, DIBL, and subthreshold slope degradation. Comparison of the V_t roll-off between analytical values and numerical simulations is presented in Fig. 6 for DG MOSFETs.

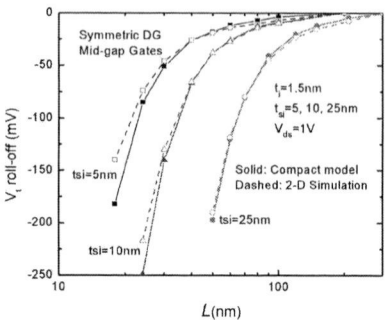

Figure 6 Threshold voltage roll-offs for symmetric DG MOSFETs with t_i=1.5nm and t_{si}=5,10 and 25nm compared with 2-D numerical simulation.

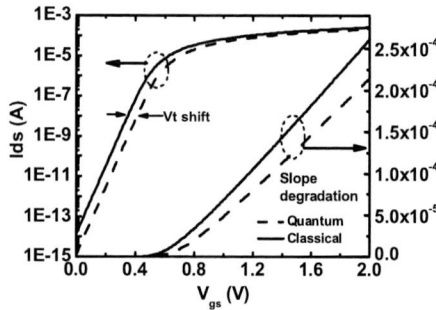

Figure 7 Drain current calculated by quantum mechanical solution and classical solution.

B. Quantum-Mechanical Effects

As the silicon thickness in MG MOSFETs decreases to 10 nm and below, quantum mechanical effect has become significant and must be taken into account in compact modeling of nano-scale devices. Quantum mechanical effects manifest themselves in thin film MOSFETs in two ways. First, the threshold voltage is shifted to a higher value due to the higher electron ground-state energy. The confinement of motion in a thin silicon film leads to discrete subbands and quantized electron energy. As a result, the electron concentration peaks away from the surface in contrast to the classical solution in which the carriers peaks at the surface. Numerical solutions from the Poisson's and Schrödinger equations in Fig. 7 show that the quantum mechanical effect causes the shift of the threshold voltage and the degradation of the slope of the I_{ds}-V_g curve, i.e., the gate capacitance C_g. Both effects have been implemented in compact model.

Due to volume inversion, the potential in the subthreshold region is essentially flat throughout the silicon film regardless of the oxide thickness. Neglecting the higher energy level and considering the lowest energy level only, the threshold voltage shift is given by

$$\Delta V_t = \frac{E_1}{q} = \frac{h^2}{8qm^* t_{si}^2} \quad (23)$$

where $m* \cong 0.91m_0$ is the larger effective mass in the two valleys, m_0 is the free electron mass.

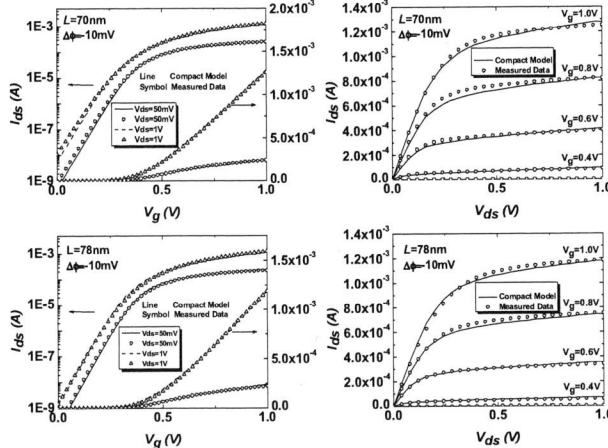

Figure 8 I_{ds} vs. V_g data with low and high drain voltages and I_{ds} vs. V_{ds} data for two short channel devices. Threshold voltage shift by DIBL effect and subthreshold slope degradation are adequately modeled.

Due to shifting of the peak carrier concentration, the inversion layer thickness increases and the gate capacitance is degraded. The inversion layer thickness can be modeled as

$$\delta_{inv} = \left(\frac{21\varepsilon_{si}\hbar^2}{2m^*qQ_i}\right)^{1/3}. \tag{24}$$

This can be lumped into an effective increase of the gate oxide thickness.

V. CALIBRATION TO FINFET HARDWARE

A complete compact model of FinFETs with the DG MOSFET model including short-channel and quantum mechanical effects has been implemented in SPICE. The compact model is shown to be in excellent agreement with experimental FinFET data [11]. The FinFETs have 30nm thick undoped body, and TiSiN metal gate with midgap workfunction. The height of the fins is 60nm, the gate length 10um to 90nm, and the number of fins 20.

First step of the systematic compact modeling is to extract the effective device width and effective oxide thickness from the long channel device data with given dimension parameters, considering quantum mechanical effects. The effective device width and oxide thickness are extracted from the capacitance vs. gate voltage data of 10um device. The next step is to extract the mobility from the drain current vs. gate voltage plot at low drain bias. A modified universal mobility model is used.

As the channel length decreases, the effect of the source-drain series resistance becomes significant. A linear source-drain series resistance model is assumed.

With the inclusion of short channel effects, channel length modulation, and velocity saturation, the calibrated results are presented in Fig. 8. As the figures show, the compact model successfully describes measured FinFET capacitances and currents covering all devices for a wide range of channel lengths.

VI. COMPACT MODELING OF OTHER MG MOSFETs

MG MOSFETs (Fig. 1 c,d,e,f) other than DG and SG

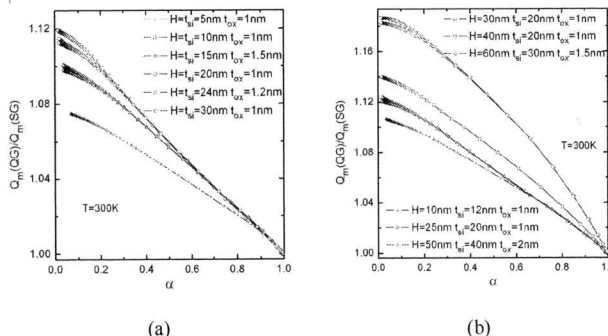

(a)   (b)

Figure 9 $Q_m(QG)/Q_m(SG)$ as a function of a. The result of $Q_m(QG)$ are obtained from 2-D TCAD numerical simulations, whereas the result of $Q_m(SG)$ are simply obtained from the analytic model, which has been validated.

MOSFETs do not have exact analytic solutions due to lack of symmetry. However, a complete compact model for MG MOSFETs can be obtained by extending the core models of DG and SG MOSFETs to the other multiple-gate MOSFETs based on the similarity in their asymptotic behavior in the subthreshold region and the above threshold region [12]. The basic idea is as follow: QG and SG MOSFETs both have gate-all-around structures, so that it is possible to modify the compact model for SG MOSFETs to that of QG MOSFETs; TG MOSFETs can be approximated by the average of QG and DG MOSFETS; the obtained the compact model for TG MOSFET can be modified to describe the TPG MOSFETs such as Π-gate and Ω-gate MOSFETs.

By applying Pao-Sah's integral, the drain current of MOSFETs can be obtained from the integration of inversion charge density of MOS capacitors as a function of the quasi-Fermi potential.

$$I_{ds} = \frac{\mu}{L}\int_{V_s}^{V_d} Q_m(V)dV \tag{25}$$

In subthreshold region, Q_m is proportional to the area of the cross-section of the silicon body because of volume inversion. In the above threshold region, Q_m is proportional to the perimeter of the cross-section regardless of corner effects.

A. Model for Quadruple-Gate MOSFETs (Fig. 1c)

QG MOSFET resembles SG MOSFET in many aspects. One difference is the corner effect of QG MOSFETs. The charge function of a QG MOS capacitor can be expressed as a multiplier of that of the SG MOS capacitor asymptotically.

In subthreshold, the drain current is proportional to the cross-sectional area of the device. A SG MOS capacitor has the same cross-sectional area as a QG MOS capacitor, if the radius of the SG capacitor is chosen to be $R = \sqrt{t_{si}H/\pi}$.

However, their well-above threshold behaviors are not matched, since the perimeters of the two devices are not identical. The ratio between the charges of SG and QG MOS capacitors, $Q_m(QG)/Q_m(SG)$ can be expressed as a function of α as follows:

$$Q_m(QG)/Q_m(SG) \cong 1 + C_1(1-\alpha) \tag{26}$$

978-1-4244-2018-6/08 $25.00 © 2008 IEEE   281

The range of α is $0 < \alpha < 1$. $\alpha \sim 1$ means the subthreshold region, where the charge ratio is one because of the chosen equal areas of the two devices. $\alpha \sim 0$ corresponds to the well-above threshold region where the charge ratio is $1 + C_1$ to account for the corner effects in QG MOSFETs. The results in Fig. 9 indicate that C_1 depends on the structure dimension ratios, t_{si}/t_{ox} and H/t_{ox}. Therefore, the charge of a QG MOS capacitor is expressed as

$$Q_m(QG) = 8\pi\varepsilon_{si}\frac{kT}{q}\frac{1-\alpha}{\alpha}\left[1 + C_1(1-\alpha)\right]. \tag{27}$$

Fig. 10 shows that $Q_m(V_g)$ calculated from the above model closely agrees with numerical simulation results.

B. Model for Triple-Gate MOSFETs (Fig. 1d)

For TG MOS capacitors, we need to approximate

$$Q_m \cong \frac{Q_m(DG) + Q_m(QG)}{2}. \tag{28}$$

Here DG and QG devices have the same structure parameters as the TG device. From the approximation, three different aspects of device behaviors are found. First, both sides have identical subthreshold behavior due to the same cross-sectional area of silicon body. Second, they have the identical inversion layer, which is about $t_{si} + 2H$. Finally, they have the same number of corners. These explain why both sides have identical well-above threshold behavior. The charge of the TG MOS capacitors can therefore be obtained as below.

$$Q_m(TG) = 4\pi\varepsilon_{si}\frac{kT}{q}\frac{1-\alpha}{\alpha}\left[1 + C_1(1-\alpha)\right]$$
$$+ 4\varepsilon_{si}\frac{H}{t_{si}}\frac{kT}{q}\beta\tan\beta \tag{29}$$

where α is from the boundary condition of SG MOSFETs, and β is from the boundary condition of DG MOSFETs.

C. Model for Tri-Plus-Gate MOSFETs (Fig. 1e,f)

TPG MOSFETs such as π-gate and Ω-gate MOSFETs can be approximated as the devices somewhere between TG and QG devices. Therefore, the charge of TPG MOS capacitor can be expressed as

$$Q_m \cong \frac{1-C_2}{2}Q_m(DG) + \frac{1+C_2}{2}Q_m(QG) \tag{30}$$

where C_2 is a parameter extracted from numerical simulations. Substituting the expression for DG MOS and QG MOS capacitors yields the charge expression of the TPG MOS capacitor as below.

$$Q_m(TPG) = (1+C_2)4\pi\varepsilon_{si}\frac{kT}{q}\frac{1-\alpha}{\alpha}\left[1 + C_1(1-\alpha)\right]$$
$$+ (1-C_2)4\varepsilon_{si}\frac{H}{t_{si}}\frac{kT}{q}\beta\tan\beta \tag{31}$$

where α is from the boundary condition of SG MOSFETs, and β is from the boundary condition of DG MOSFETs. Fig. 11 shows the excellent agreement between the analytical model and numerical simulations.

D. Unified Drain Current Model for multiple-gate MOSFETs

The expression for $Q_m(TPG)$ is general enough to cover any multiple-gate MOSFETs by appropriate choices of C_1 and

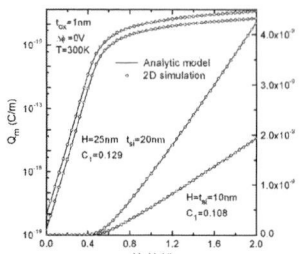 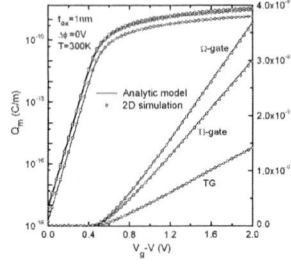

Figure 10 Mobile charge $Q_m(QG)$ as a function of V_g -V obtained from analytical model (solid lines) in comparison with the 2-D numerical simulation results (open circles).

Figure 11 Mobile charge $Q_m(TG)$ and $Q_m(TPG)$ as functions of V_g -V obtained from the analytical model (solid lines) in comparison with the 2-D numerical simulation results (open circles).

C_2. Therefore, a unified drain current model can be formulated. The final drain current expression is given by

$$I_{ds} = \mu\frac{4\pi\varepsilon_{si}}{L}\left(\frac{kT}{q}\right)^2(1+C_2)\left[f(\alpha) + C_1 d(\alpha)\right]_{\alpha_s}^{\alpha_d}$$
$$+ \mu\frac{4\varepsilon_{si}}{L}\frac{H}{t_{si}}\left(\frac{kT}{q}\right)^2(1-C_2)\left[p(\beta)\right]_{\beta_s}^{\beta_d} \tag{32}$$

where the functions f and p have been defined previously, and d is defined as

$$d(\alpha) \equiv \left(-\frac{2}{\alpha} - 3\ln\alpha + \alpha\right) + s\left(-\frac{1}{\alpha^2} + \frac{4}{\alpha} + 2\ln\alpha\right). \tag{33}$$

The table for the dimensionless parameters, C_1 and C_2 for MG MOSFETs is given below.

Device	H	t_{si}	t_{ox}	C_1	C_2
DG	W	t_{si}	t_{ox}	0	-1
SG	$\sqrt{\pi}R$	$\sqrt{\pi}R$	t_{ox}	0	1
QG	H	t_{si}	t_{ox}	C_1	1
TG	H	t_{si}	t_{ox}	C_1	0
TPG	H	t_{si}	t_{ox}	C_1	C_2

Table 1 Structural parameters, C_1 and C_2 values for MG MOSFETs.

VII. OTHER RESEARCH ACTIVITIES ON COMPACT MODELING OF MG MOSFETS

A. BSIM-CMG

A key assumption made in the analytic potential model described above is that the silicon film is either undoped or lightly doped. This allows the depletion charge term to be neglected from the Poisson equation that made the analytic solution possible. For not too heavily doped devices, the first-order effect of the depletion charge can be included in the analytical potential model by shifting the gate voltage V_g by $qN_aW/2C_{ox}$.

To describe the effect of heavy doping, the depletion charge term would have to be added to the Poisson's equation which has no rigorous analytical solution. A perturbation method has been introduced by Dunga et al [13] to obtain the solution to Poisson's equation. First, the potential with undoped body is obtained ignoring the body doping effect. By first-order approximation, the potential due to the perturbation is expressed as a function of the potential at the center of the body. The potential in the silicon body is obtained by the boundary condition at the silicon-oxide interface and the center of the body, which leads to an implicit equation with one unknown, the potential at the center of the body. While this model is applicable to both fully-depleted and partially-depleted cases and has been shown to agree with 2-D simulation results, no explicit solution has been published to date.

Furthermore, the charge-sheet approximation, commonly employed in conventional bulk MOSFET models, has been applied in BSIM-CMG [13] for drain current calculations, namely,

$$I_{ds,CSA} = -\mu W \left(Q \frac{d\psi_s}{dy} - \frac{kT}{q} \frac{dQ}{dy} \right), \qquad (34)$$

where Q is the mobile sheet charge density. This led to the final expression,

$$I_{ds,CSA} = 2\mu \frac{1}{C_{ox}} \frac{W}{L} \left| Q^2 + C_{ox} \left(\frac{2kT}{q} Q \right) \right|_{Q_s}^{Q_d}. \qquad (35)$$

The same result has been obtained in other publications (see below) even though charge-sheet approximation was not specifically mentioned.

It should be noted that the drift and diffusion current density,

$$J_y = -qn\mu \left(\frac{d\psi}{dy} - \frac{kT}{qn} \frac{dn}{dy} \right) = -qn\mu \frac{dV}{dy}, \qquad (36)$$

where n is the electron volume density, is often integrated to yield

$$I_{ds} = -\mu W Q \frac{dV}{dy} \qquad (37)$$

under the gradual channel approximation since the quasi-Fermi potential V and thus dV/dy is independent of x in the depth direction. Eq. (34), however, cannot be rigorously obtained by integrating Eq. (36) because the potential ψ is a function of both x and y.

Although the asymptotic behavior of the approximated drain current expression, Eq. (34), is same as that obtained from the rigorous Eq. (37) in the analytic potential model, there are significant errors in the intermediate region. Using the analytic potential model, the drain current with the charge-sheet approximation, Eq. (34), can be shown to be

$$I_{ds,CSA} = \mu \frac{W}{L} \frac{8\varepsilon_{si}}{t_{si}} \left(\frac{kT}{q} \right)^2 \left| p_{CSA}(\beta) \right|_{\beta_s}^{\beta_d} \qquad (38)$$

where $p_{CSA}(\beta) = -\beta \tan \beta - 2r\beta^2 \tan^2 \beta$. The difference to the rigorous drain current, Eq. (4), is clear. The error between two functions is shown in Fig. 12. As the geometry parameter r decreases, the maximum error in the intermediate region increases.

The drain current models with the charge-sheet approximation have been verified by numerical simulations and experiments including both common-gate operation [14] and independent-gate operation [15]. The physics and the

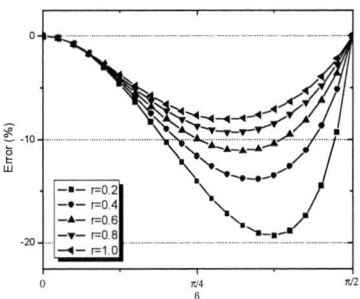

Figure 12 Error between $p(\beta)$ and $p_{CSA}(\beta)$ vs. β. In the subthreshold region, $\beta_s \sim 0$, $\beta_d \sim 0$. In the intermediate region, $\beta_s \sim 0$, $\beta_d \sim \pi/2$. In the strong-inversion region, $\beta_s \sim \pi/2$, $\beta_d \sim \pi/2$. The maximum drain current error occurs in the intermediate region, where the values of β_s and β_d are not close to each other.

behavior of the DG MOSFET under independent-gate operation are similar to those of conventional bulk MOSFETs, appropriate for the charge-sheet approximation.

To complete the compact model for DG MOSFETs, short-channel effects such as drain-induced barrier lowering, threshold voltage roll-off, and subthreshold current slope have been added to BSIM-CMG [16]. Quantum-mechanical effects including the increase of the effective oxide thickness due to the finite inversion layer thickness, and the threshold voltage shift due to electron confinement are also implemented. The quantum-mechanical model assumes an ideal triangular or square potential well depending on the specific geometry and bias voltages of the device. The correction terms can be added either to the potential [17], or the oxide thickness and threshold voltage.

To extend BSIM-CMG to triple-gate MOSFETs, corner effect is added to the compact model for DG MOSFETs. Since no physical features of the SG MOSFET model are implemented, it is not straightforward to further extend the model to other multiple-gate MOSFETs in Fig. 1 c, e, f.

B. PSP-based FinFET Model

A compact model for FinFETs has been presented by Smit *et al* [18] based on the compact model of DG MOSFETs with undoped body [3], employing various techniques developed in the bulk PSP-model. Using an approximated expression of inversion charge, a drain current expression identical to Eq. (35) is derived. This indicates that the approximation used is equivalent to the charge-sheet approximation in nature. The amount of error introduced by the approximation has been presented as a function of an intermediary parameter θ. Short-channel effects and quantum-mechanical effects have been integrated into the model in almost the same way as in the PSP model, due to the similarity of the current and charge expressions. The compact model has been verified by 2-D numerical simulations and experimental data [19].

C. European Group

A compact model with normalizations to make use of the EKV model has been presented by Sallese *et al* [20]. By re-writing Poisson's equation as a function of gate charge and applying the boundary conditions, an expression for gate charge is obtained containing an integration coefficient. The integration coefficient is approximated based on its asymptotic behavior to derive a drain current expression

$$I_{ds} \approx \left| -q_m^2 + q_m \right|_{q_{mS}}^{q_{mD}} \tag{39}$$

where $q_m \equiv Q_m \left(kT/q/4C_{ox} \right)$ is the normalized inversion charge. This again is identical to Eq. (35) obtained with the charge-sheet approximation. Thus the approximation of the integration coefficient is equivalent to the charge-sheet approximation and is subject to the same errors discussed in Fig. 12. By normalizing the currents and voltages with appropriate constants, existing pieces of the bulk EKV model can be incorporated into the DG model.

VIII. CONCLUSIONS

A unified strategy for compact modeling of various forms of multiple-gate MOSFETs has been presented. It is based on the closed-form, analytic potential solutions to Poisson's and current continuity equations for DG and SG MOSFETs with highly accurate explicit functions continuous throughout all ranges of voltage. The DG model has been expanded into a complete compact model by adding short-channel and quantum-mechanical effects to the core model. Charge and capacitance expressions for DG and SG MOSFETs have also been developed for transient simulations. The DG compact model has been validated by hardware calibration with experimental FinFET data. The core model has been extended to the less symmetric QG, TG and TPG MOSFETs based on the asymptotic behavior of inversion charge in the subthreshold and above-threshold regions.

REFERENCES

[1] *International Technology Roadmap for Semiconductors: 2006 Edition.* [Online] http://www.itrs.net.

[2] J.-P. Colinge, "Multiple-gate SOI MOSFETs," *Solid-State Electron.*, vol. 48, pp. 897-905, 2004.

[3] Y. Taur, "An Analytical Solution to a Double-Gate MOSFET with Undoped Body," *IEEE Electron Device Lett.*, vol. 21, pp. 245-247, 2000.

[4] Y. Taur, X. Liang, W. Wang, and H. Lu, "A Continuous, Analytic Drain-Current Model for DG MOSFETs," *IEEE Electron Device Lett.*, vol. 25, pp. 107-109, 2004.

[5] J. R. Brews, "A charge sheet model of the MOSFET," *Solid-State Electron.*, vol. 21, pp. 345-355, 1978.

[6] D. Jimenez, B. Iniguez, J. Sune, L. F. Marsal, J. Pallares, J Roig, and D. Flores, "Continuous Analytic I-V Model for Surrounding-Gate MOSFETs," *IEEE Electron Device Lett.*, vol. 25, pp. 571-573, 2004.

[7] B. Yu, H. Lu, M. Liu, and Y. Taur, "Explicit Continuous Models for Double-Gate and Surrounding-Gate MOSFETs," *Electron Devices, IEEE Trans. on*, vol. 54, pp. 2715-2722, 2007.

[8] H. Lu and Y. Taur, "An Analytic Potential Model for Symmetric and Asymmetric DG MOSFETs," *Electron Devices, IEEE Trans. on*, vol. 53, pp. 1161-1168, 2006.

[9] D. E. Ward and R. W. Dutton, "A charge-oriented model for MOS transistor capacitances," *Solid-State Circuits, IEEE Journal of*, vol. 13, pp. 703-708, 1978.

[10] X. Liang and Y. Taur, "A 2-D Analytical Solution for SCEs in DG MOSFETs," *Electron Devices, IEEE Trans. on*, vol. 51, pp. 1385-1391, 2004.

[11] J. Song, B. Yu, W. Xiong, C. H. Hsu, C. R. Cleavelin, M. Ma, P. Patruno, and Y. Taur, "Experimental Hardware Calibrated Compact Models for 50nm n-channel FinFETs," *SOI, 2007 IEEE Conf. of*, pp. 131–132, 2007.

[12] B. Yu, J. Song, Y. Yuan, and Y. Taur, "A Unified Analytic Drain Current Model for Multiple-Gate MOSFETs", unpublished.

[13] M. V. Dunga, C. -H. Lin, X. Xi, D.D. Lu, A. M. Niknejad, and C. Hu, "Modeling Advanced FET Technology in a Compact Model," *Electron Devices, IEEE Trans. on*, Vol. 53, No. 9, pp. 1971-1978, 2006.

[14] M. V. Dunga, C. –H. Lin, D. D. Lu, W. Xiong, C. R. Cleavelin, P. Patruno, J. –R. Hwang, F. –L. Yang, A. M. Niknejad, and C. Hu, "BSIM-MG: A Versatile Multi-Gate FET Model for Mixed-Signal Design," *VLSI Tech., 2007 IEEE Symposium on*, pp. 60-61, 2007.

[15] D. D. Lu, M. V. Dunga, C. –H. Lin, A. M. Niknejad, and C. Hu, "A Multi-Gate MOSFET Compact Model Featuring Independent-Gate Operation," *IEDM 2007*, pp. 565-568, 2007.

[16] C. Hu, C. –H. Lin, M. Dunga, D. Lu, and A. Niknejad, "A Versatile Multi-Gate MOSFET Compact Model: BSIM-MG," *NSTI-Nanotech 2007*, vol. 3, pp. 512-514, 2007.

[17] C.-H. Lin, M. V. Dunga, A. M. Niknejad, and C. Hu, "A Compact Quantum-Mechanical Model for Double-Gate MOSFET," *ICSICT 2006*, pp. 1272-1274, 2006.

[18] G. D. J. Smit, A. J. Scholten, G. Curatola, R. van Langevelde, G. Gildenblat, and D. B. M. Klaassen, "PSP-based Scalable Compact FinFET Model," *NTSI-Nanotech 2007*, Vol. 3, pp. 520-525, 2007.

[19] G. D. J. Smit, A. J. Scholten, N. Serra, R. M. T. Pijper, R. van Langevelde, A. Mercha, G. Gildenblat, and D. B. M. Klaassen, "PSP-based Compact FinFET Model Describing DC and RF Measurements," *IEDM 2006*, pp 175-178, 2006.

[20] J. –M. Sallese, F. Krummenacher, F. Pregaldiny, C. Lallement, A. Roy, and C. Enz, "A design oriented charge-based current model for symmetric DG MOSFET and its correlation with the EKV formalism," *Solid-State Electron.*, Vol.49, No. 3, pp. 485-489, 2005.

IEEE 2008 Custom Intergrated Circuits Conference (CICC)

Compact Modeling and Simulation of PD-SOI MOSFETs: Current Status and Challenges

(Invited Paper)

Jung-Suk Goo[1], Richard Q. Williams[2], Glenn O. Workman[3], Qiang Chen[1], Sungjae Lee[2], and Edward J. Nowak[2]

[1]Technology Development Group, Advanced Micro Devices Inc., Sunnyvale, CA 94088
[2]SOI Compact Modeling Group, IBM Corporation, Essex Junction, VT 05452
[3]Freescale Semiconductor Inc., 3501 Ed Bluestein Boulevard, Austin, TX 78721

Abstract- **This paper reviews the status and challenges of the modeling Partially-Depleted Silicon-On-Insulator transistors. Many challenges stem from the floating-body potential, which offers advantages in terms of performance and leakage, but presents complex electrical behavior. Circuit simulator considerations and the importance of model standardization are also highlighted.**

I. INTRODUCTION

Compact modeling, defined as numerical modeling using electrical elements to approximate semiconductor physics, is an essential part of chip design in both analog and digital applications. In this work, the compact modeling focus is on partially depleted silicon-on-insulator (PD-SOI) transistors that have been successfully manufactured in ULSI CMOS from the 225nm [1] through the 45nm nodes [2]. PD-SOI provides multiple benefits for CMOS scaling, including threshold (V_t) lowering due to the floating-body effect and reduced V_t versus L_{poly} sensitivity, reduced capacitive loading because of the buried insulator and the floating-body, reduced body effects in stack transistor circuits [3], and reduced soft-error rates [4]. The buried oxide further benefits SOI designs through the elimination of latch-up, well-implant proximity effects (WPE), and natural isolation of auxiliary device elements such as embedded DRAM, passives, high-voltage, and RF devices.

Compact models for PD-SOI transistors are necessarily more complex than bulk transistors due to the presence of the floating-body (FB). In reality, the floating-body is a complex 3D potential distribution that is influenced by many factors. In PD-SOI compact models, a single body-potential describes the electrical effects of the floating-body since local potential variations within the body charge are small compared to other non-uniformities under ordinary circumstances.

This paper is organized as follows. Section II describes general strategies for assembling a PD-SOI compact model. Section III discusses the complexities and challenges of modeling the FB potential, while section IV introduces the SOI body-contacted (BC) structure and its modeling and challenges. Section V discusses some of the simulator considerations for SOI circuit analyses. Section VI reviews current work towards standardizing compact model best practices and section VII concludes the paper.

II. STRATEGIES FOR PD-SOI COMPACT MODEL CALIBRATION

A. PD-SOI Compact Model Topology

The element topology guides the procedure for calibrating compact model parameters. Figure 1 shows a typical circuit element topology for a PD-SOI transistor. This topology follows the BSIMSOI model [5] and another similar construction is given in [6] with an excellent overview. A key feature differentiating PD-SOI from conventional bulk models is that the parasitic currents that drive charge into and out of the isolated body region must be explicitly included: gate-to-body, source and drain junction diodes, Gate Induced Drain Leakage (GIDL), and impact ionization. A simplified bipolar transistor model is used to model the source-body-drain regions (analogous to a lateral bipolar structure) to capture bipolar-driven base charge dump. A single-pole thermal sub-circuit is employed to model power dissipation out of the channel region through the buried oxide and other structures such as contact studs.

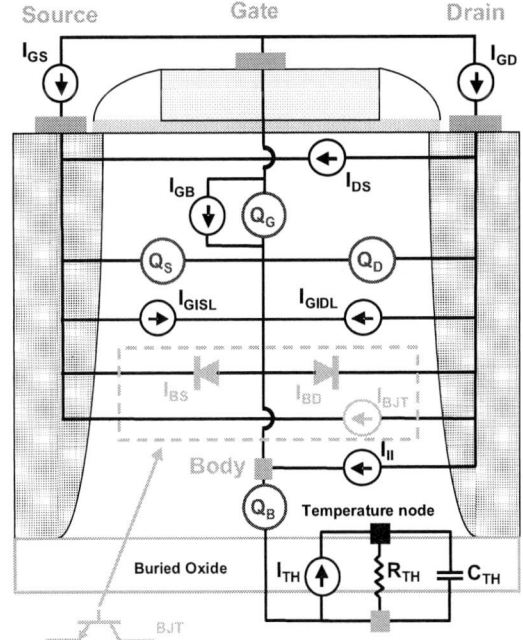

Fig. 1. PD-SOI compact model electrical element topology.

978-1-4244-2018-6/08 $25.00 © 2008 IEEE
285

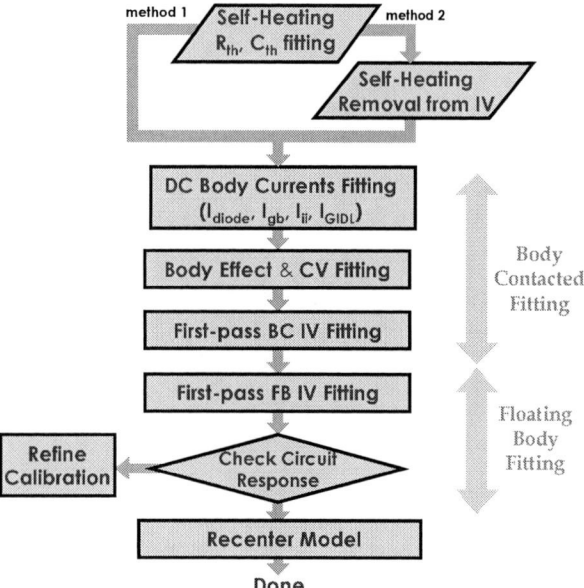

Fig. 2. Model parameter calibration flowchart for PD-SOI MOSFETs.

B. Overview of Model Calibration Process

Figure 2 summarizes the major steps used to calibrate PD-SOI model parameters. One of the unique steps in PD-SOI modeling is the treatment of self-heating that, like other parasitic effects, must be analyzed before calibrating the intrinsic MOSFET. One approach is to characterize the model thermal resistance and capacitance using MOSFET power versus temperature trends under both DC and transient conditions. Subsequent model calibration proceeds using these derived quantities (method 1 in Fig. 2). Figure 3 illustrates the DC extraction of thermal resistance.

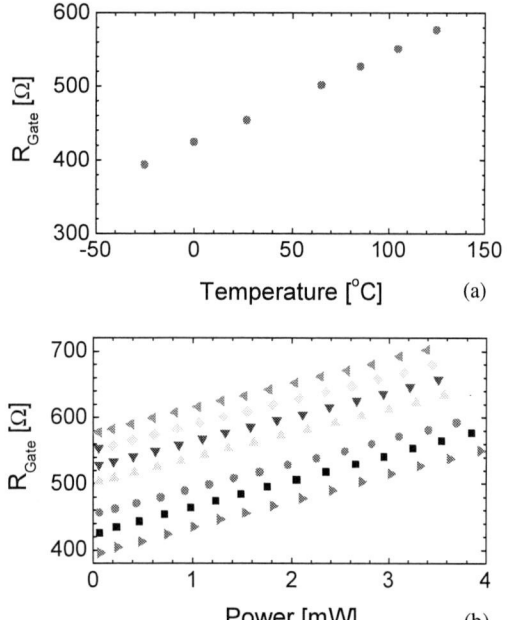

Fig. 3. (a) Calibration of temperature response of the gate resistance. (b) Measurment of the gate resistance as a function of FET power dissipation. $R_{th} = (dT/dR_g) \times (dR_g/dPower) = dT/dPower$.

Another approach (method 2) is to directly compensate the measured model I-V data for self-heating effects. This method can reduce the need to simultaneously tune temperature-dependent and temperature-independent parameters in the previous method. Hence, it is desirable to generate "self-heating free" data before parameter calibration. It should be mentioned that self-heating removal is not only implemented for the channel current but also for other parasitic currents such as the impact ionization current [7]. Figure 4 compares self-heated measurements to self-heating-free post-processed I-V curves assuming that a linear temperature interpretation is appropriate. Note also that some currents in some regions will have an exponential dependence on temperature, which requires a different algorithm for interpolation to remove self-heating.

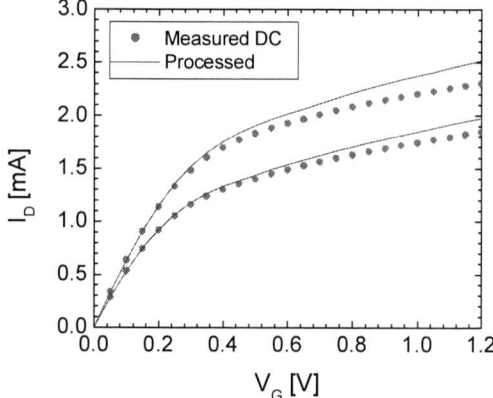

Fig. 4. Data is processed (solid line) to remove self-heating from DC measured data (circle). Shown for ambient of 27°C.

Model parameter calibration proceeds by using BC devices to characterize capacitance and parasitic DC currents because parameters related to several fundamental components such as MOS channel body effect, impact ionization, and junction diode characteristics cannot be directly measured on floating-body devices. Initial I-V fits for both BC and FB configurations are performed. Another unique PD-SOI consideration is that multiple parameters are shared between the FB and BC calibrations. This can introduce the need to refine the BC calibration by verifying parameter consistency for FB effects such as the SOI history-effect. It is not uncommon to iteratively fine-tune some model aspects to improve the fitting due to the increasingly challenging characterization tasks of these components.

While adjustment of MOSFET channel current characteristics has very limited impact on parasitic currents, adjustment of parasitic currents leads to noticeable change in channel current. Therefore, it is strongly recommended to begin the calibration flow with parasitic currents and then to evaluate high-level effects like history before refining the overall MOSFET model characteristics [8].

Once satisfactory model-to-hardware comparison results are achieved, the model is re-centered (i.e., tuned slightly) to match desired technology targets as needed. When this base model is completed, other features like statistical models and systematic variation (*e.g.*, layout-dependent stress [9]) can be incorporated by expanding the basic calibration flow in Fig. 2.

C. Challenges

Going forward from one node to another can introduce modeling difficulties. Some issues and solutions that have proven effective in past work are as follows:

1) Small body currents not easily measurable because of larger BC-specific parasitic currents (such as the bridge region I_{gb} which can overwhelm both channel-region I_{gb} in inversion and reverse-biased diode leakage, particularly at lower temp). Solution: choose bias regions carefully to highlight the intended signal, and to implicitly fit with history-effect and/or DC FB Drain Induced Barrier Lowering (DIBL) from I-V data.

2) Limited range for reverse-biased diode DC leakage due to the onset of full depletion in the body region. Solution: fit implicitly with history-effect and/or DC floating-body DIBL.

3) Limited range of DC body effect (V_t vs. V_{bs}) measurements due to large (non-linear bias-dependent) body resistance. Solution: choose data range with care.

4) Limited range of AC junction capacitance measurements due to large junction leakage. Solution: choose data range to emphasize intrinsic capacitive response.

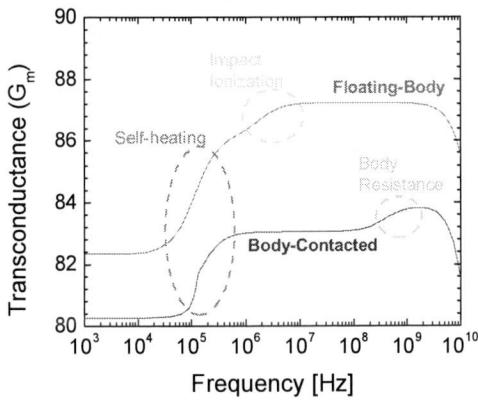

Fig. 5. Simulated G_m frequency response of a PD-SOI nMOSFET. These curves can be used to study detailed model response.

Figure 5 illustrates implicit calibration using RF data. In this figure, MOSFET Gm is plotted against frequency for a BC device and an FB device. This figure identifies the time constants associated with the transition of the body contact to floating-body mode; operation faster than the self-heating time constant; and, impact ionization currents interacting with the body charge.

III. PD-SOI FLOATING-BODY EFFECTS MODELING

The threshold voltage of the MOSFET is a function of the body potential, which affects the amount of the channel depletion charge. While multiple bodies are tied to a fixed potential in the conventional bulk CMOS, each thin-film body region of the PD-SOI CMOS is isolated; thus, its potential is determined by numerous components. For simplicity, this paper will only use nMOSFETs in the rest of discussion.

A. DC Floating-Body Effects

It has been known that a high drain bias can cause a kink in I_D vs. V_D characteristics due to an abrupt turn-on of majority carrier injection into the floating-body generated by the impact ionization [10]. This kink can occur even for drain voltages below the band-gap voltage owing to various energy gain mechanisms [11],[12].

The abrupt turn-on in inversion of gate-to-body tunneling of the ultra-thin gate oxide in the state-of-the-art PD-SOI technologies adds another charging mechanism to the body, leading to a second peak in linear transconductance [13].

Other mechanisms lead to more subtle changes in DC characteristics. At lower drain voltages, diode recombination and band-to-band junction tunneling currents will increase DIBL and off-current due to various highly non-linear components that control the floating-body charge. Clearly, these dependencies make it imperative to fit parasitic body effects prior to fitting the channel current in an FB SOI MOSFET.

B. Transient Floating-body Effects (History-Effect)

The body potential is determined by majority carrier density in the floating-body. Depending on the switching conditions, the majority carrier density increases or decreases by various DC and AC coupling effects. Some charging and discharging mechanisms are reversible, which means if the device returns to its initial state, the body potential on return will be the same as the initial one. On the other hand, some mechanisms are irreversible, making each individual device subject to their switching history; thus, this behavior is called the history-effect [14]. The history-effect can be characterized by measuring propagation delays in the following three extreme conditions [14]:

- 1st switch: This condition refers to the first transition after holding the circuit's input constant for a long time so that the body potentials can reach DC equilibrium. Depending on the direction of the input signal switching, the history-effect can be further distinguished as "1st switch 0→1" or "1st switch 1→0." This paper will use only the 0→1 condition for sake of simplicity .

- 2nd switch: This refers to the transition immediately after the 1st switch transition.

- DSS: Most circuits switch frequently thus may not return to the DC equilibrium. If a circuit switches constantly with a 50% duty cycle like an oscillator, the body potentials of the PD-SOI devices gradually converge to a steady-state condition (see Fig. 7); in turn, the rise-to-rise and fall-to-fall delays become the same. This state is often called DSS (Dynamic Steady State) [15] or SSS (Switching Steady State) [16].

Input Clock Shape

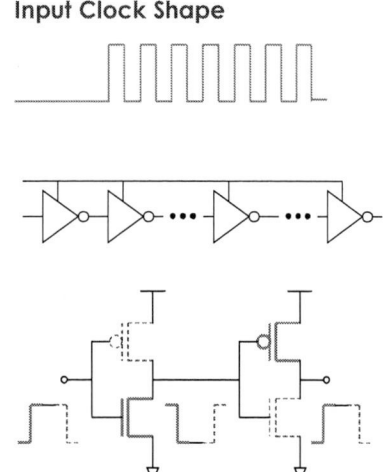

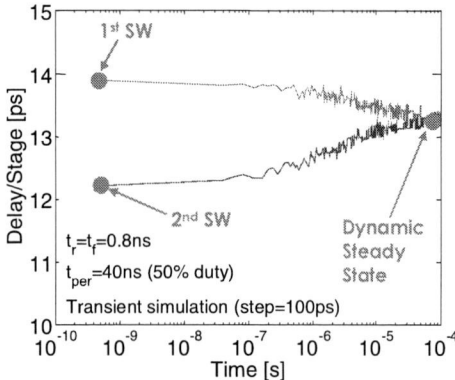

Fig. 6. Inverter delay-chain and its input pulse for characterizing the history-effect. Monitoring the stage delay in alternating stages gives 1st and 2nd switch delays.

Fig. 7. Evolution of switching delay in a floating-body PD-SOI inverter chain.

Most circuits neither switch every cycle nor sit at DC equilibrium for long periods of time. Nevertheless, these three conditions well represent the extreme situations that most circuits encounter. Generally, the DSS condition tends to be somewhere between the 1st and 2nd switch delays but, in some MOSFET device designs under high voltage conditions, it is possible for the DSS condition to result in the fastest delay as high impact ionization rates prevalent at high voltages drive body voltages higher than equilibrium condition values. As a metric for the worst-case delay offset, this paper quantifies the history-effect (H) as the relative difference of the propagation delay (τ) between the 1st and 2nd switches:

$$H = (\tau_{1st} - \tau_{2nd}) / \tau_{2nd}$$

C. What Determines the History-Effect?

The mechanisms that influence history can be roughly categorized by their response time [17]:

- Very fast capacitive couplings from the gate, source, and drain (reversible unless it triggers fast discharging process);
- Fast discharging through forward-biased source and drain junctions (irreversible); and,

- Slow charging through reverse-biased junctions, impact ionization, and gate tunneling current (irreversible).

In CMOS digital circuits, the nMOSFET typically determines the pull-down delay. As illustrated by the inverter body potentials in Fig. 8, $V_{body,1st}$ (the body potential of the 1st switch) is purely determined by the initial DC conditions and a capacitive coupling is added to the initial DC condition for $V_{body,2nd}$.

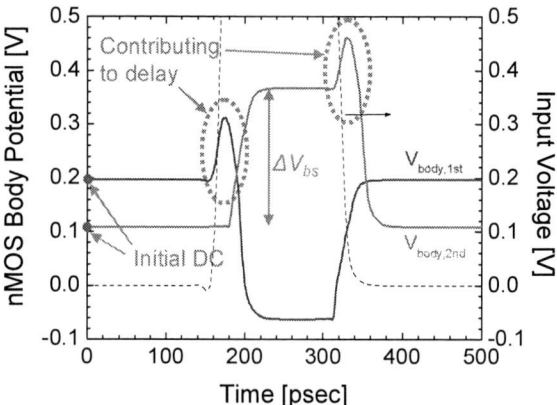

Fig. 8. Evolution of the nMOSFET floating-body potential in an inverter delay chain during a switching event. An extra boost in the body potential is introduced by gate coupling during switching.

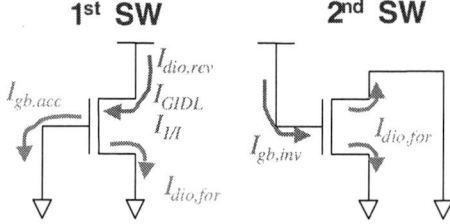

Fig. 9. Key current components to determine the initial DC body potential.

Figure 9 illustrates the current components that interact to generate the initial DC V_{body}. Down to 130nm technology, the initial DC condition was determined by the reverse and forward diode currents (consisting of diffusion, generation-recombination, band-to-band tunneling components) and GIDL. From 90nm onward, the gate-to-body valence band tunneling current (I_{gb}) has shown a visible impact, mostly on the 2nd switch [18]-[20]. GIDL and impact ionization currents are becoming essential factors in determining the body potential [21].

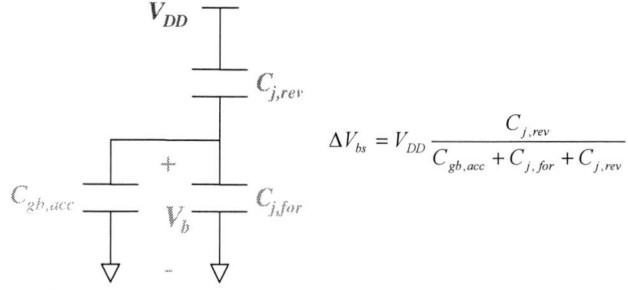

$$\Delta V_{bs} = V_{DD} \frac{C_{j,rev}}{C_{gb,acc} + C_{j,for} + C_{j,rev}}$$

Fig. 10. Key current components to determine the capacitive coupling of body potential.

978-1-4244-2018-6/08 $25.00 © 2008 IEEE

As illustrated in Fig. 10, the capacitive coupling amount in the 2nd switch condition can be simply understood as a capacitive voltage divider, consisting of gate-to-body capacitance and forward/reverse junction capacitance.

Finally, the body-effect of the MOSFET (V_t vs. V_{bs}) is the main transfer function of V_{bs} into propagation delay; thus, its fitting is critical in history-effect modeling. Figure 11 demonstrates that the measured history-effect can be successfully reproduced by simulation across a wide range of conditions when all the key components are properly modeled.

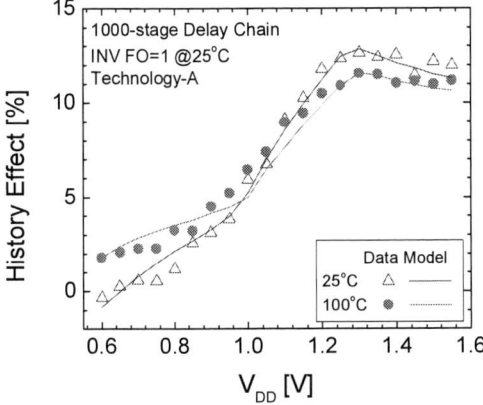

Fig. 11. History-effect of an inverter chain at different bias and temperature conditions, comparing measured data with simulation results.

D. Other Floating-Body Effects

The positive body potential in the floating-body PD-SOI tends to stay in the positive side for most of the time, which can result in elevated FET leakage following a switching event and activate the parasitic bipolar transistor. Combined with capacitive coupling, the body potential can exceed 0.7V in logic circuits and therefore approach the operating regime in which bipolar effects can be significant. The pass-gate logic operation is particularly vulnerable to bipolar effects [22]. Fortunately, the parasitic bipolar problem is partially alleviated by the trend toward lower power supply voltages.

E. Challenges

Statistical modeling of the history-effect, including corner model definitions, is a challenge due to the need for accurate floating-body effect characterization. Monte Carlo simulations can be used to investigate statistical history-effects, in which the variation of key statistical sensitivities include the model parameters for the parasitic currents and capacitances that drive floating-body effects.

Direct measurement of the floating-body effect using a single-stage logic gate requires very high accuracy test equipment, and test throughput is extremely low because of the duration of the measurement. One of the most widely used structures for characterizing the history-effect is a delay chain (as shown in Fig. 6) consisting of hundreds or thousands of stages of logic gates that allows characterization in the MHz range. However, the measured history-effect is averaged over all stages; thus, measuring the switching details of the history-effect is virtually impossible. The test structure also occupies

a fairly large area. Figure 12 shows statistical data for the history-effect and its comparison with the simulation results including corners.

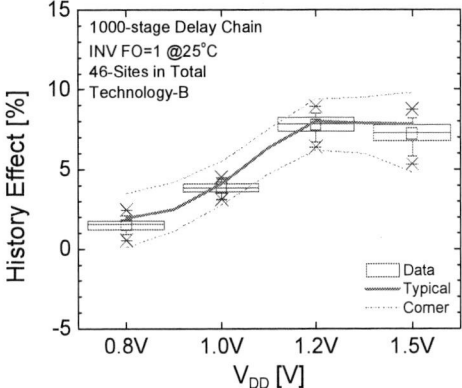

Fig. 12. Statistical history-effect data collected from all sites across a wafer and compared with corner model simulation results.

Recently, a novel in-line test circuit in a relatively compact macro has been proposed that enables fully automated test for quantifying the history-effect using only simple DC equipment [23]. This design trades the ability to monitor history in-line using digital test equipment for approximate precision in the measurement.

IV. PD-SOI BODY-CONTACTED DEVICE MODELING

In principle, a BC device can help eliminate or reduce the floating-body variation in sensitive circuits. It also offers an additional degree of freedom in circuit design.

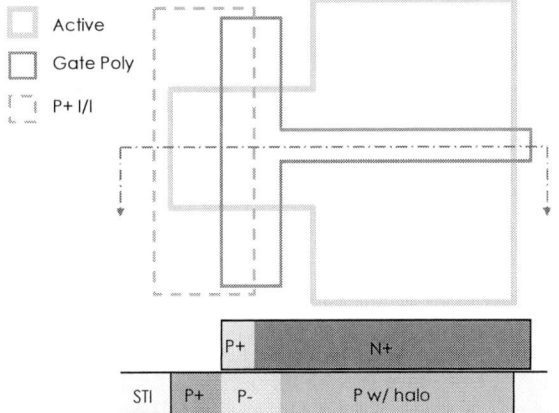

Fig. 13. An example of body-contacted PD-SOI nMOSFET layout and its cross-section.

A. Extra Parasitic Capacitance of the Extended-Gate

As illustrated in Fig. 13, the body-contacted PD-SOI MOSFET has an extended gate to bridge the channel implants to body contact diffusion (P- region in Fig. 13)[24]. The body contact structure adds extra overlap, gate-to-body, and body-to-substrate capacitance that introduces a substantial penalty in switching speed. Importantly, the gate-to-body capacitance of the bridge-region needs to be carefully modeled, as it is the most significant capacitive component since the P+ over P (or

N+ over N for a BC pMOSFET) region (as shown in the left portion of Fig. 13) is in accumulation.

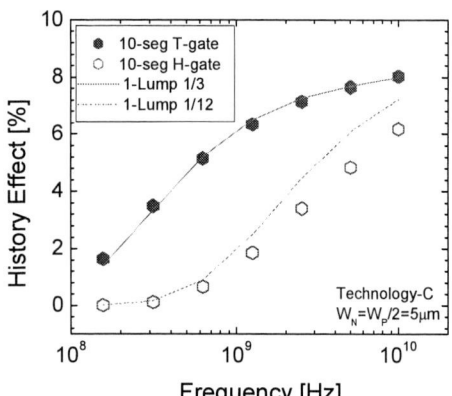

Fig. 14. History-effect of a body-contacted PD-SOI CMOS, comparing partitioned (into 10 segments) case with a single MOSFET case (with scaled body resistance).

B. Body Resistance

The combination of thin Si body regions and low channel doping can create a body RC time constant that is substantially higher than other circuit elements; thus, the electrical element network for the body has to be carefully modeled [25],[26]. The most accurate body resistance consists of several parallel resistances (well and halo implants) and series resistances (distributed or segmented body resistance) [24] and it is desirable to capture the bias-dependence as well [6]. A switching event causes a strong capacitive coupling to the body potential even in BC devices, and this leads to the history-effect similar to the one present in a floating-body device, i.e., no body contact structure is completely ideal. The distributed effect of the body resistance is an essential element that is needed to capture the BC history-effect correctly. In principle, the MOSFET can be partitioned into smaller segments along the width direction then connected in series [6],[24], but the cost is a significant penalty in simulation time. Alternatively, as demonstrated in Fig. 14, a single MOSFET approximation can reasonably reproduce a multi-partition channel by scaling the body resistance with mathematically-derived R equivalence factors: 1/3 for one-side body-contact (T-gate) and 1/12 for two-side body-contacts (H-gate). The same equivalence factors are also found in the BJT base resistance model [27] and MOSFET gate resistance model [5],[28].

V. PD-SOI FLOATING-BODY EFFECTS SIMULATION

A. Accuracy Options in SPICE

Since the evolution of the floating-body potential is on a much smaller voltage scale than the supply voltage level, the accuracy options in circuit simulations are crucial. First, the PD-SOI circuit simulations require higher accuracy in voltage convergence criteria than bulk circuit simulation. Otherwise, the change of the body potential can be incorrect, especially during fast switching at high supply voltage, leading to a significant error in propagation delay prediction, as demonstrated in Fig. 15.

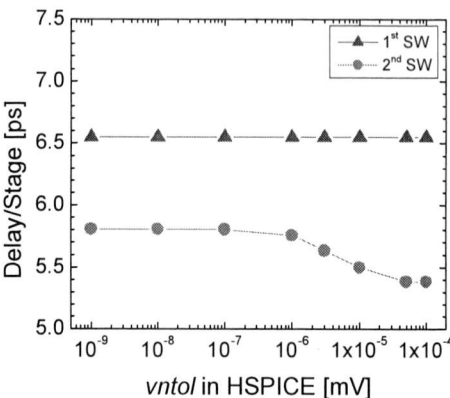

Fig. 15. Impact of SPICE transient voltage tolerance on switching delay accuracy of a floating-body PD-SOI circuit.

Common techniques to help simulation convergence also need to be carefully reviewed. One example is that SPICE adds a conductance across all capacitors, including junction diodes and gate oxides of MOSFET devices, during initial DC operating point calculations. In bulk circuits, this conductance causes virtually no impact in solutions. However, this conductance sometimes can be larger than parasitic currents in the PD-SOI device such as the gate-to-body tunneling current or the reverse diode current.

In this case, the initial DC operating point solutions from SPICE are incorrect. Special attention should be paid to numerical convergence criteria at low temperature because diode currents show much stronger thermal voltage sensitivity at lower temperatures.

B. Dynamic-Steady-State Simulation

One of the challenges in PD-SOI circuit simulations is computational efficiency in the steady-state assessment that is particularly critical for large multiple-input circuit macros, SRAMs, clock drivers, I/O devices, and analog circuits such as PLLs. Reaching steady-state generally takes an order of micro- (or even milli-) seconds due to the slow nature of charging and discharging components. This is an impractically long time period to run a straightforward SPICE simulation for large scale circuits. Several techniques have been proposed for steady-state simulation to achieve a reasonable computational cost.

The harmonic balance method directly finds the steady-state solution by solving the circuit equations in the frequency domain using a Fourier series [29]. Generally, this method is efficient but requires over-sampling and more than six harmonics to achieve consistent results with transient simulations [30] in some simulators. A more recently proposed technique tracks the net body charges for a single cycle of the specified periodic input pattern, then projects its evolution using the Newton method until the change of the net body charges becomes zero [31],[32],[33].

In case the simulation tool does not support such techniques mentioned before, there are approximate methods such as [33] in which it has been empirically found that the body potential can be initialized close to the steady-state solution for ring oscillators by setting all switching gate nodes to $V_{DD}/2$.

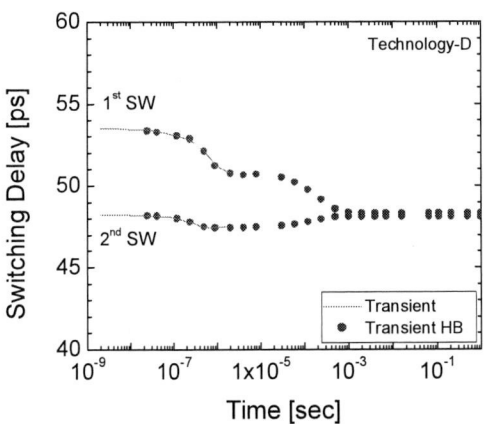

Fig. 16. History-effect in ten-stage inverter chain. Complete transient harmonic balance results are obtained in under 10min, while (incomplete) transient results run for more than a day. These results demonstrate the ability of harmonic balance to efficiently capture complex body effects.

C. Charging and Discharging Time

Many modern circuits experience sleep and wake-up modes to minimize power consumption. In such a case, it is crucial to estimate the charging time required to restore the steady-state and the discharging time to reset the potentials to DC 1st and 2nd switch values. The transient harmonic balance method enables running such simulations within affordable runtime by solving slow transients in the time domain (transient analysis), simultaneously with fast transients in the frequency domain (harmonic balance analysis) [34],[35]. Figure 16 shows the history-effect simulated using transient harmonic balance methods.

D. Considerations for Self-Heating Simulations

As previously mentioned, self-heating is modeled using an auxiliary node inside the MOSFET model and the solved potential on that node is interpreted inside the model as an increase in operating temperature. A side effect of this modeling approach is that the matrix stamp for each MOSFET grows due to the addition of the temperature node and simulation runtimes increase correspondingly. Fortunately, in high-performance logic simulations, switching occurs faster than the thermal time constant, so approximately the same simulation results can be obtained with self-heating disabled. However, self-heating should be enabled in circuits with lower operating frequencies or circuits with mixed fast and slow time constants.

VI. STANDARDS FOR SOI MOSFET MODELS

An important facet of compact modeling that spans academia, ULSI foundries, and chip design shops is standardization. Standardization helps ensure consistency in model implementation across circuit simulator platforms [36]. In addition, model developers receive recognition and academic funding for their contributions and circuit designers

benefit from improved model accuracy and features that are honed through detailed review during in the standardization process [37]. Today the compact model standards-setting body is the Compact Model Council (CMC) [38], which is hosted by the Government Electronics and Information Association (GEIA). The CMC evaluates a candidate model's ability to reproduce relevant fundamental phenomena and its numerical properties such as symmetry, continuity, convergence, and runtime.

The current CMC-standard model is BSIMSOI from the University of California at Berkeley. In 2006, the CMC began a process to develop a standard for next-generation SOI MOSFET models. The CMC has published documents outlining model requirements [39] that are applicable to PD- and FD-SOI technologies, as well as the procedure for standardization [40]. The CMC is currently soliciting candidate models.

VII. CONCLUSION

The modeling, characterization, and simulation of PD-SOI MOSFETs have been reviewed while highlighting the current and future challenges in building PD-SOI compact models. PD-SOI circuit simulation and model standardization efforts have also been summarized. Much of the complexity in PD-SOI compact modeling originates from the floating-body, which has attributes that can be difficult to measure. Despite these challenges, compact model structures, calibration methods, and circuit simulation approaches have been developed and implemented for accurate simulation of PD-SOI circuit behavior.

ACKNOWLEDGMENTS

This paper is distilled from Research Alliance Team collaborations and discussions at various ASTA (Advanced Semiconductor Technology Alliance) Research and Development Facilities over many years with many colleagues. Specifically, the authors would like to acknowledge the following for their contributions to this work: K. Bernstein at IBM and B. Rice at Freescale. Also the authors would like to acknowledge management support from A. Icel and N. Kepler at AMD; S. Springer and R. Wachnik at IBM; and S. Jallepalli and M. Zunino at Freescale.

Finally the authors would like to recognize the contributions to SOI compact modeling from academic research groups, in particular, those of Chenming Hu, Gennady Gildenblat, Jerry G. Fossum, and Mitiko Miura-Mattausch.

REFERENCES

[1] D. H. Allen, A. G. Aipperspach, D. T. Cox, N. V. Phan, and S. N. Storino, "A 0.2-um 1.8V SOI 550 MHz 65-b PowerPC microprocessor with copper interconnects," *Solid-State Circuits,* Vol. 34, No. 11, pp. 1430-1435, Nov. 1999.

[2] S. Narasimha *et al.,* "High Performance 45-nm SOI Technology with Enhanced Strain, Porous Low-k BEOL, and Immersion Lithography," Technical Digest of *Int. Electron Devices Meeting,* pp. 689–692, Dec. 2006.

[3] C. T. Chuang, P. F. Lu, and C. J. Anderson, "SOI for digital CMOS VLSI: Design considerations and advances," *Proc. IEEE,* vol. 86, no. 4, pp. 689–720, Apr. 1998.

[4] C. T. Chuang, K. Bernstein, R. V. Joshi, R. Puri, K. Kim, E. J. Nowak, T. Ludwig, I. Aller, "Scaling Planar Silicon Devices," *IEEE Circuits and Devices Magazine*, vol. 20, no. 1, pp. 6-19, Jan-Feb. 2004.

[5] BSIM Group, University of California, Berkeley, *BSIMSOI4.0 MOSFET Model Users' Manual*, http://www-device.eecs.berkeley.edu/~bsimsoi/

[6] W. Wu, X. Li, G. Gildenblat, G. Workman, S. Veeraraghavan, C. C. McAndrew, R. van Langevelde, G. D. J. Smit, A. J. Scholten, D. B. M. Klaassen and J. Watts, " PSP-SOI: A Surface Potential Based Compact Model of Partially Depleted SOI MOSFETS," Proceedings of the IEEE *Custom Integrated Circuits Conference*, pp. 41-48, September 2007.

[7] Q. Chen, Z.-Y. Wu, R. Y. K. Su, J.-S. Goo, C. Thuruthiyil, M. Radwin, N. Subba, S. Suryagandh, T. Ly, V. Wason, J. X. An, and A. B. Icel, "Extraction of Self-Heating Free I-V Curves Including the Substrate Current of PD SOI MOSFETs," Proceedings of *International Conference on Microelectronic Test Structures*, pp. 272-275, March 2007.

[8] Q. Chen, J.-S. Goo, N. Subba, X. Cai, J. X. An, T. Ly, Z.-Y. Wu, S. Suryagandh, C. Thuruthiyil, M. Radwin, L. Zamudio, J. Yonemura, F. Assad, M. M. Pelella, and A. B. Icel, "An A Priori Hysteresis Modeling Methodology for Improved Efficiency and Model Accuracy in Advanced PD SOI Technologies," Proceedings of *Nano Science and Technology Institute Nanotechnology Conference & Trade Shows*, pp. 159-162, May 2005.

[9] D. Chidabarrao, D. Jordan, J. McCullen, D. Onsongo, T. Wagner, R. Q. Williams, *Methodology for Layout-based Modulation and Optimization of Nitride Liner Stress Effects in Compact Models*, US Patent US7337420, Feb. 26, 2008.

[10] T. Skotnicki, G. Merckel, and A. Merrachi, "A re-examination of the physics of multiplication-induced breakdown in MOSFET's," Technical Digest of *Int. Electron Devices Meeting*, Dec. 1989, pp. 87–90.

[11] M. V. Fischetti and S. E. Laux, "Monte carlo study of sub-band-gap impact ionization in small silicon field-effect transistors," Technical Digest of *Int. Electron Devices Meeting*, Dec. 1995, pp. 305–308.

[12] K. G. Anil, S. Mahapatra, and I. Eisele, "Role of inversion layer quantization on sub-bandgap impact ionization in deep-sub-micron n-channel MOSFET's," Technical Digest of *Int. Electron Devices Meeting*, Dec. 2000, pp. 675–678.

[13] A. Mercha, J. M. Rafí, E. Simoen, E. Augendre, and C. Claeys, "Linear Kink Effect Induced by Electron Valence Band Tunneling in Ultrathin Gate Oxide Bulk and SOI MOSFETs," *IEEE Transactions on Electron Devices*, Vol. 50 No. 7 pp. 1675-1682, Jul. 2003.

[14] K. Bernstein and N. J. Rohrer, *SOI Circuit Design Concepts*, Chapters 3-4, Kluwer Academic Publishers, 1st Edition, 2000.

[15] M. M. Pelella, J. G. Fossum, M.-H. Chiang, G. O. Workman, and C. R. Tretz, "Analysis and Control of Hysteresis in PD/SOI CMOS," Technical Digest of *International Electron Device Meeting*, pp. 831-834, December 1999.

[16] A. Wei and D. A. Antoniadis, "Bounding the Severity of Hysteretic Transient Effects in Partially-Depleted SOI CMOS," Proceedings of *IEEE International SOI Conference*, pp. 74-75, October 1996.

[17] R. Puri and C.-T. Chuang, "Hysteresis Effect in Pass-Transistor-Based, Partially Depleted SOI CMOS Circuits," *IEEE Journal of Solid-State Circuits*, Vol. 35, No. 4, pp. 625-631, April 2000.

[18] P. Su, S. K. H. Fung, W. Liu, and C. Hu, "Studying the Impact of Gate Tunneling on Dynamic Behaviors of Partially-Depleted SOI CMOS Using BSIMPD," Proceedings of the *International Symposium on Quality Electronic Design*, pp. 487-491, March 2002.

[19] M. H. Na, J. S. Watts, E. J. Nowak, R. Q. Williams, W. F. Clark, "Predicting the SOI history-effect using compact models," Proceedings of *International Conference on Modeling and Simulation of Microsystems*, Vol, 2, pp. 183-186, May 2004.

[20] J.-S. Goo, J. X. An, C. Thuruthiyil, T. Ly, Q. Chen, M. Radwin, Z.-Y. Wu, M. S. L. Lee, L. Zamudio, J. Yonemura, F. Assad, M. M. Pelella, and A. B. Icel, "History-Effect-Conscious SPICE Model Extraction for PD-SOI Technology," Proceeding of the IEEE *International SOI Conference*, pp. 156-158, October 2004.

[21] Q. Chen, S. Suryagandh, J.-S. Goo, J. X. An, C. Thuruthiyil, and A. B. Icel, "Impact of Gate Induced Drain Leakage and Impact Ionization Currents on Hysteresis Modeling of PD SOI Circuits," Proceedings of *Workshop on Compact Modeling*, Vol, 3, pp. 570-573, May 2007.

[22] P.-F. Su, C.-T. Chuang, J. Ji, L. F. Wagner, C.-M Hsieh, J. B. Kuang, L.-C. Hsu, M. M. Pelella, S.-F. S. Chu, C. J. Anderson, "Floating-Body

[23] Effects in Partially Depleted SOI CMOS Circuits," *IEEE Journal of Solid-State Circuits* , Vol. 32 No. 8 pp. 1241-1253, Aug. 1997.

[23] D. J. Pearson, M. B. Ketchen, M. Bhushan, "Technique for Rapid, In-Line Characterization of Switching History in Partially Depleted SOI Technologies," Proceeding of the IEEE *International SOI Conference*, pp. 148-150, October 2004.

[24] P. Su, S. K. H. Fung, F. Assaderaghi, and C. Hu, "A Body-Contact SOI MOSFET Model for Circuit Simulation," Proceeding of the IEEE *International SOI Conference*, pp. 50-51, October 1999.

[25] C.-T. Chuang, P.-F. Lu, and C. J. Anderson, "SOI for Digital CMOS VLSI: Design Considerations and Advances," *Proceedings of the IEEE*, Vol. 86, No. 4, pp.689-720, Apr. 1999.

[26] G. O. Workman and J. G. Fossum, "A Comparative Analysis of the Dynamic Behavior of BTG/SOI MOSFET's and Circuits with Distributed Body Resistance," *IEEE Transactions on Electron Devices*, Vol. 45, No. 10, pp. 2138-2145, Oct. 1998.

[27] A. B. Phillips, *Transistor Engineering and Introduction to Integrated Semiconductor Circuits*, pp. 211-216, McGraw-Hill, 1st Edition, 1962.

[28] R. P. Jindal, "Noise Associated with Distributed Resistance of MOSFET Gate Structures in Integrated Circuits," *IEEE Transactions on Electron Devices*, Vol. 50 No. 7 pp. 1505-1509, Jul. 1984.

[29] K. S. Kundert, "Introduction to RF simulation and its application," *IEEE Journal of Solid-State Circuits*, Vol. 31, No. 10, pp. 1298-1319, October 1999.

[30] D. Anderson, "Comparison of SPICE VS. Harmonic Balance Simulations," Digest of the *Automatic RF Techniques Group Conference*, pp. 1-7, June 2000.

[31] V. Liot and P. Flatresse, "A New and Fast Method to Compute Steady State in PD-SOI Circuits and Its Application to Standard Cells Library Characterization," Proceeding of the IEEE *International SOI Conference*, pp. 170-171, October 2003.

[32] R. V. Joshi, K. Kroell, and C. T. Chuang, "A Novel Technique for Steady State Analysis for VLSI Circuits in Partially Depleted SOI," Proceedings of the *International Conference on VLSI Design*, pp. 832-836, June 2004.

[33] J.-O. Plouchart, J. Kim, B. J. Gross, K. Wu, R. Trzcinski, V. Karam, P. Hyde, R. Williams, M.-H. Na, J. McCullen, and W. Clark, "SOI 90-nm Ring Oscillator Sub-ps Model-Hardware Correlation and Parasitic-aware Optimization Leading to 1.94-ps Switching Delay," Technical Digest of *Int. Electron Devices Meeting*, Dec. 2005, pp. 1053-1056.

[34] D. Sharrit, "New Method of Analysis of Communication Systems," in *MTT-S Nonlinear CAD Workshop*, Jun. 1996.

[35] J. Roychowdhury, "Analyzing Circuits with Widely Separated Time Scales Using Numerical PDE Methods," *IEEE Transactions on Circuits and Systems I: Fundamental Theory and Applications*, Vol. 48, No. 5, pp. 578–594, May 2001.

[36] B. Brooks, K.Green, J. Krick, T. Vrostos, D. Weiser, "Standardization and Validation of Compact Models", Technical Proceedings of the *International Conference on Modeling and Simulation of Microsystems*, Vol, 1, pp. 653-656, May 2002.

[37] J. Watts, "Enhancing Productivity by Continuously Improving Standard Compact Models," Proceedings of the IEEE *Custom Integrated Circuits Conference*, pp. 623-639. Sep 2006.

[38] See http://www.geia.org/index.asp?bid=597

[39] CMC Subcommittee on SOI Next Generation SOI MOSFET Model Standardization, "Model Requirements," http://www.geia.org/index.asp?bid=746

[40] CMC Subcommittee on SOI Next Generation SOI MOSFET Model Standardization, "Procedure for Standardization," http://www.geia.org/index.asp?bid=746

IEEE 2008 Custom Intergrated Circuits Conference (CICC)

Modeling Ionizing Radiation Effects in Solid State Materials and CMOS Devices

(invited paper)

H. J. Barnaby, *Senior Member, IEEE*, M. L. Mclain, *Student Member, IEEE*,
I. S. Esqueda, *Student Member, IEEE*, X. J. Chen, *Student Member, IEEE*
Dept. of Electrical Engineering, Arizona State University, Tempe, AZ, 85287

Abstract—A comprehensive model is presented which enables the effects of ionizing radiation on bulk CMOS devices and integrated circuits to be simulated with closed form functions. The model adapts general equations for defect formation in uniform SiO_2 films to facilitate analytical calculations of trapped charge and interface trap buildup in structurally irregular and radiation sensitive shallow trench isolation (STI) oxides. A new approach whereby non-uniform defect distributions along the STI sidewall are calculated, integrated into implicit surface potential equations, and ultimately used to model radiation-induced "edge" leakage currents in n-channel MOSFETs is described. The results of the modeling approach are compared to experimental data obtained on 130 nm and 90 nm devices. The features having the greatest impact on the increased radiation tolerance of advanced deep-submicron bulk CMOS technologies are also discussed. These features include increased doping levels along the STI sidewall.

I. INTRODUCTION

Aggressive semiconductor scaling has reduced the susceptibility of most deep-submicron CMOS technologies to ionizing radiation damage. This is due primarily to the fact that, to first order, radiation-induced defect buildup in gate oxides scales with the oxide film thickness (t_{ox}) [1]. Radiation tolerance is further enhanced by the increased likelihood of positive charge annihilation or compensation by tunneling elections into the thin oxides from the adjacent materials [2]. Increased channel and body doping, required for short channel effects (SCE) and latchup suppression, has also improved the inherent radiation hardness of most deep-submicron CMOS processes [3]. Despite these positive trends, recent studies on commercial 130 nm and 90 nm bulk CMOS technologies suggest that damage from ionizing radiation exposure continues to have a measurable impact on many key integrated circuit (IC) specifications [3-5]. Fig 1 plots the normalized increase in SRAM standby current (I_{SB}) as a function of total ionizing dose [rad(Si)]. The SRAM array used in this experiment was fabricated in a commercial 90 nm bulk CMOS process [4]. After 2 Mrad(Si), I_{SB} increases between 40X to 75X depending on the power supply voltage. This not only increases the static power

This work was supported in part by the Defense Threat Reduction Agency and the U.S. Defense Advanced Research Projects Agency (Radiation Hardened by Design Program) and the Air Force Multidisciplinary Research Program of the University Research Initiative (AFMURI).

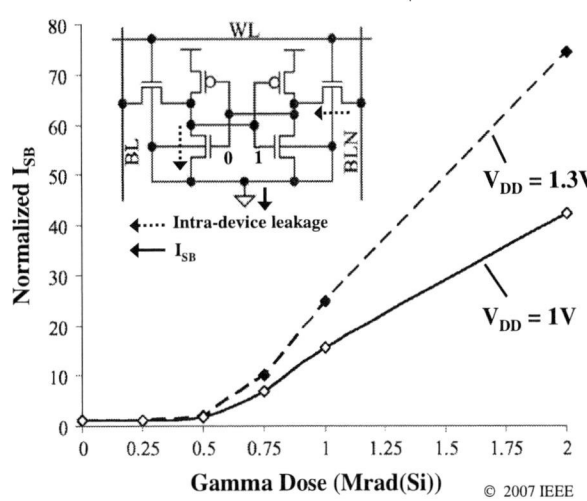

Fig. 1. Chip level measurement of SRAM I_{SB} vs. total ionizing (gamma) dose normalized to pre-irradiation level. Inset shows schematic of 6T SRAM indicating I_{SB} and intra-device leakage references [4].

consumption of embedded memory but may also cause a loss of circuit functionality by interfering with proper pre-charging and degrading read stability [4]. The mechanisms for increased standby current have been identified as leakage paths created as a result of radiation-induced defect buildup in isolation oxides [6-8]. These paths include: 1) leakage between the drain and source of an individual nFET along the isolation sidewall (edge leakage) and 2) leakage underneath isolation oxides between the drains and/or sources of adjacent nFETs or between nFET drain/source diffusions and n-well layers [3].

The deleterious impact of radiation environments on ICs can be mitigated through the use of well-established process hardening techniques. While radiation-hardening-by-process (RHBP) has the advantage of being an extremely reliable means of achieving hardened components, it is susceptible to low volume concerns such as yield, process instability, and high manufacturing costs. In order to leverage the economy-of-scale provided by the commercial electronics industry, developers of rad-hard electronics are beginning to adopt radiation-hardening-by-design (RHBD). In RHBD, electronic components are manufactured in un-hardened commercial

978-1-4244-2018-6/08 $25.00 © 2008 IEEE

processes but are designed with specialized hardening approaches in order to meet specified radiation performance criteria. This type of environment-specific design approach requires accurate models for radiation effects (e.g., edge leakage) that can be implemented in commercial circuit simulators. In this paper, we present analytical models for ionizing radiation damage in semiconductor materials and demonstrate how these models can be used to simulate radiation response in CMOS devices.

II. IONIZATION DAMAGE IN SILICON DIOXIDE

A. Radiation Damage Processes

Ionization damage is initiated when electron-hole-pairs (ehps) in a target material are generated from the energy deposited by ionizing radiation. The density of ehps generated is proportional to the energy transferred [9]. The total amount of energy deposited by a particle that results in ehp production *is* total ionizing dose (TID). The typical unit of TID is the rad, which denotes the energy absorbed per unit mass of the target material (1 rad = 100 ergs absorbed per gram).

The physical processes that occur, from the initial deposition of energy to the creation of ionization defects, are: 1) the generation of ehps, 2) the prompt recombination of a fraction of the generated ehps, 3) the transport of free carriers remaining in the oxide, and either 4a) the formation of positive trapped charge via hole trapping in defect precursor sites or 4b) the formation of interface traps via reactions involving hydrogen [10]. These processes are summarized graphically in Fig. 2 [11].

The conversion factor, g_0, that relates ehps generated to dose (process 1) is obtained using the following formula:

$$g_0 \left[\frac{\#ehp}{cm^3 rad} \right] = 100 \left[\frac{erg}{g} \right] \left[\frac{1}{rad} \right] \bullet \frac{1}{1.6x10^{-12}} \left[\frac{eV}{erg} \right] \bullet \frac{1}{E_p} \left[\frac{\#ehp}{eV} \right] \bullet \rho \left[\frac{g}{cm^3} \right], (1)$$

where E_p is the mean energy needed to ionize (~ 2x the bandgap) and ρ is the density of the target material. For SiO_2, which, with respect to ionization damage, is the material of greatest interest, $g_0 = 8.1 \times 10^{12}$ ehps/cm^{3}rad.

Once generated, a fraction of the ehps are annihilated through either columnar or geminate recombination (process 2) [9]. The remaining fraction of ehps that escape this initial recombination process is the charge yield, denoted by f_y. Charge yield is dependent on the local electric field in the material [11, 12] and can be approximated with the function

$$f_y \approx \frac{\varepsilon}{\varepsilon + \varepsilon_0}, \quad (2)$$

where ε is the magnitude of the local field and ε_0 is threshold field constant (< 10^6 V/cm).

In SiO_2, electrons, having a much higher mobility than holes, are rapidly swept out of the dielectric [13]. The surviving holes undergo polaron hopping transport via shallow traps in the SiO_2 (process 3) [13]. A fraction of these transporting holes may fall into trapping sites primarily located near

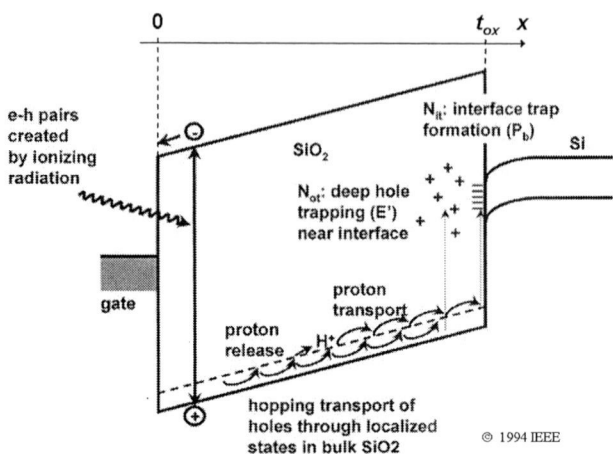

Fig. 2. Illustration of the main processes in TID damage in SiO_2 [11].

heterogeneous dielectric [14] or SiO_2-Si interfaces, thereby forming fixed trapped positive charge (process 4a) [13]. These trapping sites are generally characterized as neutral oxygen vacancies (E' centers) [15-20], however several studies have suggested that hydrogen containing defects in the oxide may also trap holes [21, 22]. Reactions between holes and hydrogen-containing defects or dopant complexes can also lead to the formation of a second type of ionization defect: the interface trap (process 4b) [23, 24].

B. Model for Oxide Trapped Charge Formation

In a simple one-dimensional MOS system, during radiation exposure, the continuity equation for holes in the oxide is

$$\frac{\partial p}{\partial t} = -\frac{\partial f_p}{\partial x} + \dot{D}g_0 f_y - R_p, \quad (3)$$

where p is the hole density (cm^{-3}), f_p is the hole flux (cm^{-2}s^{-1}), $\dot{D}$ is the radiation dose rate (rad/s), and R_p is the delayed hole recombination rate (cm^{-3}s^{-1}) [23, 25]. Assuming steady state conditions and a negligible recombination rate, the hole flux near the interface can be approximated as

$$f_p = \dot{D}g_0 f_y t_{ox}. \quad (4)$$

It should be noted that for high dose rates and low oxide fields, hole recombination cannot be neglected and must be retained in the model. Delayed carrier recombination is now thought to be a factor in the phenomenon of enhanced-low-dose-rate-sensitivity (ELDRS) [25].

Using (4), the rate of hole trapping near the SiO_2-semiconductor interface can be expressed as

$$\frac{\partial N_{ot}}{\partial t} = (N_t - N_{ot})\sigma f_p - \frac{N_{ot}}{\tau}, \quad (5)$$

where N_{ot} and N_t are the densities of trapped holes and hole trapping sites, respectively, σ is the hole capture cross-section, and τ is the trapped hole lifetime (s). The model in (5) assumes N_{ot} and N_t to be sheet densities at the interface. The lifetime factor relates to trapped hole annealing processes. If $N_t \gg N_{ot}$ and τ is large, the buildup of N_{ot} in an oxide can be

approximated as [26, 27]

$$\Delta N_{ot} = D g_0 f_y N_t \sigma t_{ox}, \qquad (6)$$

where D is the total ionizing dose deposited. Fig. 3 plots ΔN_{ot} as a function of radiation bias and exposure level for a 130 nm bulk CMOS technology [3]. These data were obtained from measurements on field oxide capacitors. As the figure indicates, trapped charge builds up roughly linearly with dose, which is consistent with (6). Moreover ΔN_{ot} is greater in parts irradiated with larger biases, i.e., 1.32V vs. 0V. This bias dependence is due to the dependence of charge yield on oxide field (refer to (2)).

C. Model for Interface Trap Formation

The initial processes that lead to N_{it} formation are similar to those that lead to the formation of oxide trapped charge (i.e., ehp generation, recombination, and hole transport). However, the formation of dangling bonds relies on several other reactions. The first reaction is between transporting holes and hydrogen containing defects (DH centers), which releases protons (H+) [23]. Similar to (3), the continuity equation for protons in the oxide is

$$\frac{\partial N_{H^+}}{\partial t} = N_{DH} \sigma_{DH} f_p - \frac{\partial f_{H^+}}{\partial x}, \qquad (7)$$

where N_{H+} is the proton concentration (cm^{-3}), f_{H+} is the proton flux ($cm^{-2}s^{-1}$), N_{DH} is the concentration of DH centers, and σ_{DH} is the capture cross-section of holes reacting with the DH centers (cm^2). Released protons diffusing or driven by an electric field to the SiO_2 interface can remove hydrogen atoms from H-passivated dangling bonds (SiH) via the simple reaction [23, 28]

$$SiH + H^+ \Rightarrow Si^+ + H_2. \qquad (8)$$

The resulting defect (Si^+) is the interface trap.

The formation *and* annealing of N_{it} can be described by the rate equation,

$$\frac{\partial N_{it}(t)}{\partial t} = \sigma_{gen} N_{SiH} f_{H^+} - \sigma_{pass} N_{it}(t) f_{H_2}, \qquad (9)$$

where f_{H2} is the molecular hydrogen flux ($cm^{-2}s^{-1}$) and σ_{gen} and σ_{pass} are capture cross sections related to the generation and passivation of silicon dangling bonds by hydrogen, respectively [28]. The positive term on the right side of (9) describes the N_{it} formation reaction in (8), and the negative term describes the passivation of N_{it} by H_2, the reverse reaction of (8). By combining (3), (7), and (9), N_{it} build-up can be modeled with a t_{ox}^n dependence (where $1 < n < 2$). The models also show that N_{it} buildup, like N_{ot}, is proportional to the TID level, and is greater when larger oxide fields are present in the oxide (or when larger radiation bias conditions are applied). These dependencies are shown experimentally in Fig. 3.

D. Impact of Hydrogen

Hydrogen is ubiquitous in today's semiconductor IC fabrication and packaging processes. During IC fabrication, hydrogen is used in wafer cleaning procedures, film

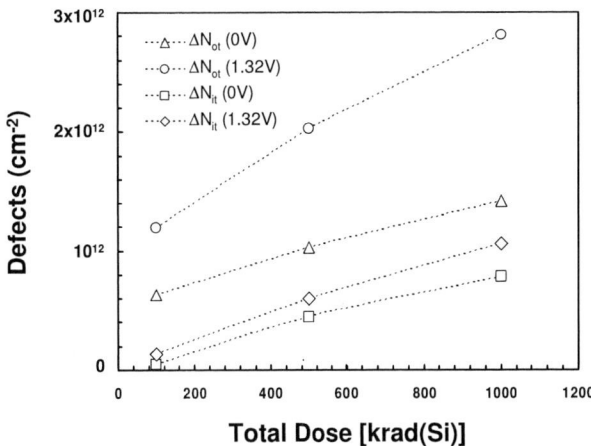

Fig. 3. Buildup of N_{ot} and N_{it} as a function of radiation bias and exposure level in STI field oxides capacitors fabricated in 130 nm bulk CMOS.

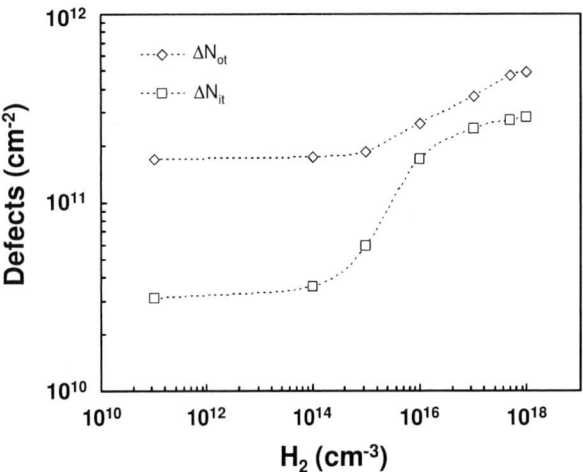

Fig. 4. Radiation-induced N_{it} and N_{ot} (after 30 krad(Si)) as a function of H_2 concentration in the oxide.

depositions, etches, high and low temperature forming gas anneals, etc. [29, 30]. During IC packaging, hydrogen can be introduced during die attach and in some applications during pre-seal forming gas baking [31].

Several studies have reported that excess hydrogen can greatly enhance a device's radiation susceptibility [21, 22, 32]. Fig. 4 shows the enhancement of radiation-induced interface traps and oxide trapped charge in bipolar base oxides (after 30 krad(Si)) as a function of in-package molecular hydrogen concentration [21]. Descriptions of the analytical models that relate H_2 concentration to N_{ot} and N_{it} buildup during radiation exposure are beyond the scope of this paper but can be found in [21, 22]. Nevertheless, the demonstrated impact of hydrogen on the radiation response strongly suggests the need to incorporate its effects when developing accurate radiation effects models for circuit simulators.

III. MODELING RADIATION EFFECTS ON DEVICES

A. Effects on Surface Potential

The effect of oxide and interfacial defects on surface potential can be modeled analytically with the equation

$$(V_{gb} - \phi_{ms} + \phi_{nt} - \psi_s)^2 = \gamma^2 \cdot \phi_t H(u). \quad (10)$$

Equation (10) represents a modified form of the implicit equation for surface potential [33, 34]. In this equation, ψ_s denotes surface potential, V_{gb} is the applied gate-to-body bias, ϕ_{ms} is the gate-to-body workfunction difference, γ is the body factor, ϕ_t is the thermal voltage, and $u = \psi_s/\phi_t$. The parameter ϕ_{nt} represents a defect potential which is added to model the effects of N_{ot} and N_{it}. The function $H(u)$ in (10) captures the charge contributions of both fixed charge and free carriers in the Si and is expressed as

$$H(u, \phi_n) = e^{-u} + u - 1 + e^{-\beta(2\phi_b + \phi_n)}(e^u - u - 1), \quad (11)$$

where ϕ_n is the channel voltage otherwise known as imref splitting, $\beta = 1/\phi_t$, and ϕ_b is the bulk potential [33, 34].

In MOS devices, the primary distinction between N_{ot} and N_{it} is that, to first order, the charge contributed by oxide trapped charge is fixed while the charge contributed by interface traps will vary with surface potential [35, 36]. If the energy distribution of interface traps, D_{it}, is assumed uniform, then the defect potential can be expressed analytically as,

$$\phi_{nt} = \frac{q}{C_{ox}}(N_{ot} - D_{it} \bullet (\psi_s - \phi_B)), \quad (12)$$

where q is the absolute value of the electron charge and C_{ox} is the oxide capacitance per unit channel area [35, 36].

Equation (10), through its inclusion of the defect potential, enables surface potential across a MOSFET channel to be calculated analytically not only as a function of terminal voltages and device parameters (e.g., oxide capacitance, workfunction difference, and doping), but also as a function of oxide and interfacial defects. This is critical when modeling radiation effects in CMOS devices.

B. Modeling Radiation-induced Edge Leakage

1) Edge leakage overview

Fig. 5 plots the drain current (I_d) vs. gate voltage (V_g) response as a function of total ionizing dose for a single stripe n-channel MOSFET ($W = 280$ nm, $L = 120$nm) manufactured in a 130 nm commercial bulk CMOS process [37]. For these experiments, during ionizing radiation exposure (^{60}Co gamma-rays), the gate was biased at 1.32 V while all other terminals were grounded (worst case conditions). During the measurements, the drain voltage was fixed at 50 mV while the source and body terminals were grounded. The isolation oxide for this technology is shallow trench (STI).

The data in Fig. 5 show a significant increase in the subthreshold current, which is indicative of defect buildup (primarily positive oxide trapped charge) in the STI. As stated in the introduction, damage to isolation oxides is typically the dominant cause of radiation-induced degradation in sub-micron

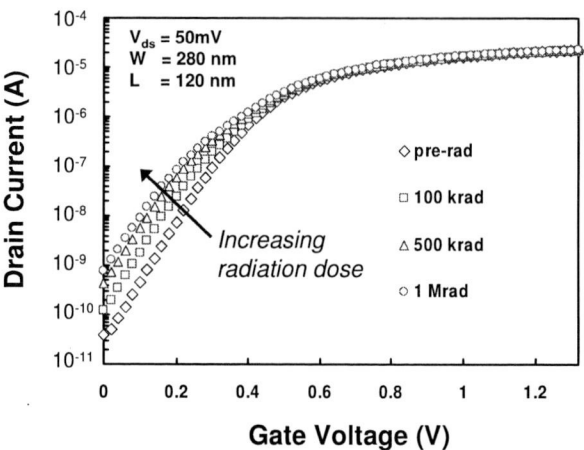

Fig. 5. I_d vs. V_g response of 130 nm single stripe n-channel MOSFET ($W = 0.28$ µm, $L = 0.12$ µm) prior to and after ionizing radiation exposure to ^{60}Co gamma rays.

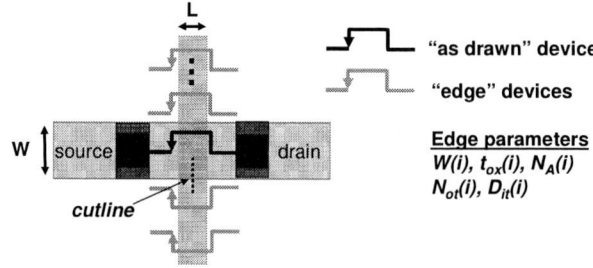

Fig. 6. Top view of a standard two edge nFET transistor showing the "as drawn" device in parallel with parasitic "edge" devices (elementary transistors).

CMOS technologies. For this 130 nm technology, charge buildup in the STI increases the defect potential (see (12)) along the trench sidewall. This increases current flow (leakage current) between the drain and source of the parasitic "edge" structures that exist in parallel with an "as drawn", single stripe device (Fig. 6).

2) Modeling defect buildup along STI sidewall

The "edge" devices (elementary transistors) represent distinct subdivisions of the conducting sidewall, extending from the "as drawn" gate oxide interface to the drain/source diffusion depth, as shown in Fig. 7 [6]. This figure represents the nFET cross-section along the cutline indicated in Fig. 6. In order to model the nFET radiation response in Fig. 5, defect generation and the effects of these defects on surface potential (and ultimately on the I_d vs. V_g characteristics) are modeled for each elementary transistor at a given total dose. The number (n) of "edge" devices chosen for the model must balance the need for simulation accuracy with computational efficiency. In the current model, n = 6. The ith elementary nFET ($1 \leq i \leq n$) is characterized by its edge parameters which include: oxide thickness [$t_{ox}(i)$], body doping concentration

$[N_A(i)]$, defect density $[N_{ot}(i), D_{it}(i)]$, and width $[W(i)]$. $W(i)$ is typically fixed to a constant value, W_s, for all i. As shown in Fig. 7, $t_{ox}(i)$ is calculated along the median electric field line intersecting the STI/p-type body interface at $z(i) = (i-1)W_s + W_s/2$.

For the model described here the STI profile is recessed and has corner rounding in order to reduce narrow channel effects and increase device reliability [38]. As such, the oxide thickness of the 1st elementary transistor is determined by the engineering of the STI corner. The precise structure of the corner oxide may be obtained from SEM cross-sections. If the STI aspect ratio is large, then

$$t_{ox}(i) \approx \frac{\pi}{2} z(i) \quad 2 \le i \le n. \quad (13)$$

These estimates of thickness can be used to approximate the local electric field in the elementary device's oxide during radiation exposure where

$$\varepsilon_{ox}(i) \approx \frac{V_{gb}}{t_{ox}(i)}. \quad (14)$$

Equations (2) and (14) are then used to find the charge yield within each elementary transistor. $N_{ot}(i)$ can subsequently be calculated with (6), under the specified assumptions. The product of N_t and σ in (6) is typically used as a fitting parameter, but is generally close to 1 for most advanced technologies.

Fig. 8 plots the calculated values for $N_{ot}(i)$ normalized to dose using the modeling methodology described above. Also included on Fig. 8 is the continuous, dose-normalized N_{ot} distribution numerically obtained through TCAD simulations. The TCAD modeling was performed with the Radiation Effects Module in the SILVACO ATLAS device simulator. As the figure indicates, the analytical results compare well to those obtained through numerical simulation.

Analytical computations of D_{IT} buildup are considerably more complex, but similar methods can be employed to obtain reasonable estimates of these distributions.

3) Modeling transistor electrical response.

Analytically modeling the impact of radiation-induced defects on the surface potential of each elementary transistor not only requires calculations of defect densities, but, as (10) implies, accurate estimates of effective oxide thickness and body doping concentration. A methodology for obtaining t_{ox} was presented in the previous subsection. Obtaining the doping distribution along the sidewall is difficult if not impossible to measure precisely. However a non-uniform profile with two peaks can generally be assumed and adjusted to fit to a given technology's requirements and scaling rules [39]. Fig. 9 plots a representative discrete doping profile, $N_A(i)$, as a function of depth along the STI sidewall. The first peak in the distribution (nearest to the surface, W = 0) is used for threshold voltage adjustment and sub-threshold swing control in the "as drawn" device. The second peak located further down the sidewall

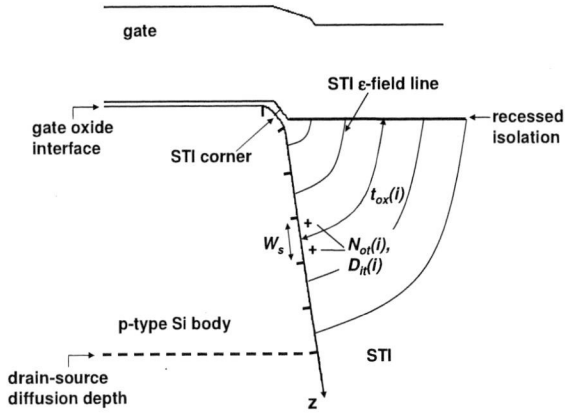

Fig. 7. nFET cross-section along cutline indicated in Fig. 6.

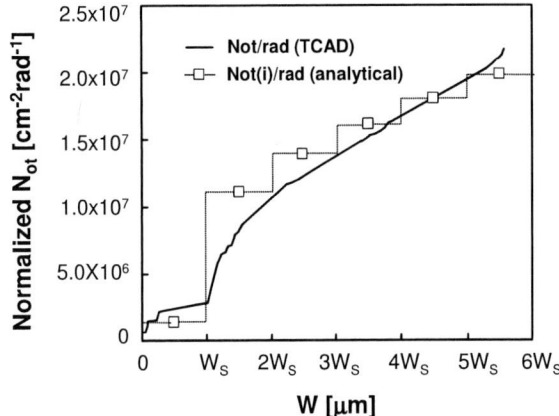

Fig. 8. $N_{ot}(i)$ normalized to dose vs. depth along the STI sidewall vs.. continuous, dose-normalized distributions of trapped charge computed numerically with TCAD simulations.

represents the punchthrough (PT) implant, with additional dopant species potentially added from sidewall implantation. Fig. 9 also plots in log scale the $t_{ox}(i)$ vs. sidewall depth, computed previously.

Once the parameters for each of the elementary transistors have been determined, surface potential responses can be calculated iteratively as a function of V_{gb} using (10). The resulting surface potentials at the source and drain ends of the channel are inserted into the closed form expressions for diffusion and drift current to compute the total drain current contributed by each "edge" device [40]. The sum of all of the "edge" drain currents therefore represents the total edge leakage current. When this sum is added to the "as-drawn" pre-irradiated characteristics, the degraded I_d vs. V_g curves can be modeled analytically for various levels of radiation exposure.

The approach described above was validated by comparing results of the model, computed in MATLAB, with radiation response data from measurements on two bulk CMOS

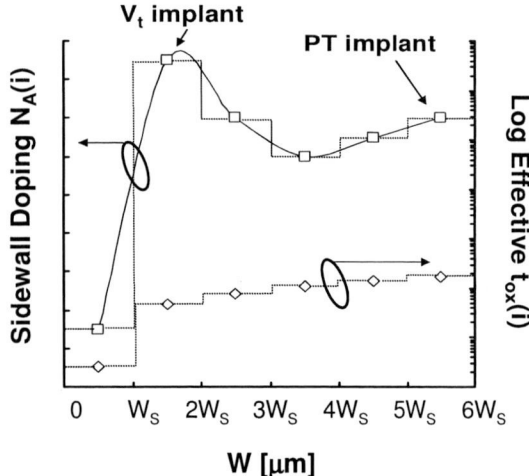

Fig. 9. Discretized doping profile, $N_A(i)$ and log of effective $t_{ox}(i)$ vs. depth along the STI sidewall. Doping assumed to have two peaks for Vt-adjust and punch-through implants.

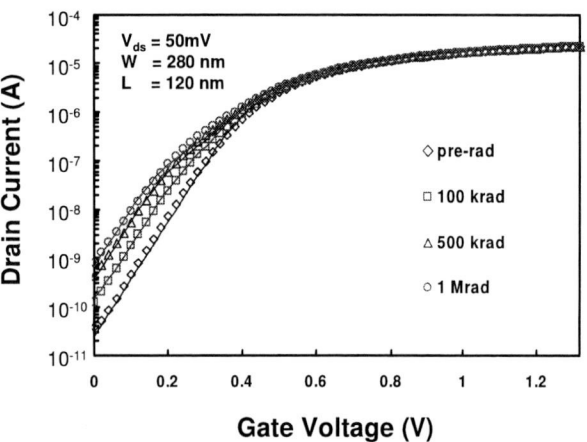

Fig. 10. Comparison of measured pre- and post-irradiation data (symbols) with modeled radiation response characteristics (solid lines) for single stripe nFETs in 130 nm technology (W= 280 nm, L = 120nm).

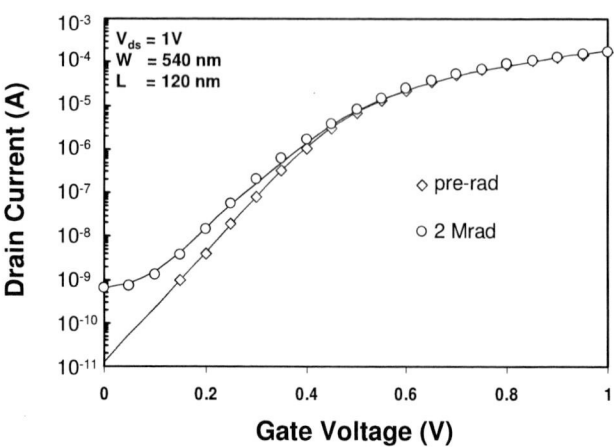

Fig. 11. Comparison of measured pre- and post-irradiation data (symbols) with modeled radiation response characteristics (solid lines) for single stripe nFETs in 90 nm technology (W= 540 nm, L = 120nm).

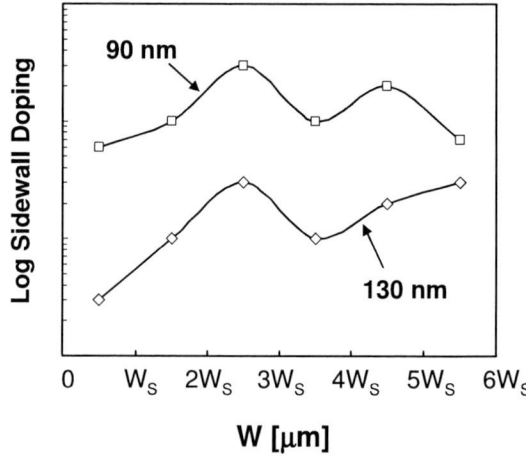

Fig. 12. Comparison of sidewall doping profiles extracted from models fits to the post-irradiation data in the 130 nm and 90 nm technologies.

technologies: 130 nm and 90 nm. In Fig. 10, the pre- and post-irradiation response characteristics obtained from the model are plotted (solid lines) along with the 130 nm experimental data (symbols), shown previously in Fig. 5. The figure demonstrates that the model gives an excellent fit to the experimental data. Fig 11 shows that the model can also be successfully applied to fit the response characteristics of a bulk CMOS 90 nm technology. For the 90 nm experiments, the gate was biased at 1 V while all other terminals were grounded during exposure to ^{60}Co gamma-rays. During the I_d vs. V_g measurements, the single stripe 90 nm nFETs (W = 540 nm, L = 120nm) were biased with a drain voltage of 1 V while the source and body terminals were grounded. It should be noted that accurate simulation of shorter length devices in the 90 nm technology requires the incorporation of additional short

channel effects models such as velocity saturation.

The 130 nm and 90 nm data reveal distinct differences in the radiation responses of these two technologies. Indeed, while the 130 nm results show measurable radiation-induced shifts below 500 krad(Si), after 2 Mrad(Si) the 90 nm devices remain fairly radiation tolerant. Differences in radiation responses between successive technology nodes has been observed in almost all sub-micron generations, dating back to 0.5 μm [41].

In addition to producing good fits to post-irradiated electrical characteristics, the proposed modeling approach is an effective tool for identifying the critical features that affect radiation response behavior from one technology generation to another. With respect to the differences observed in the 130 nm and 90 nm nodes, our model indicates that changes in

978-1-4244-2018-6/08 $25.00 © 2008 IEEE

doping profiles along the STI sidewall are having the greatest impact on radiation tolerance. Fig. 12 plots the sidewall doping profiles used to obtain the fits in Figs. 10 and 11. As the figure shows, the doping in the 90 nm technology structure is significantly larger. This leads to a suppression of the effects of radiation-induced defects on the STI sidewall surface potential, thereby reducing the edge leakage effects at the 90 nm node. Recent applications of this modeling approach have revealed sidewall doping and not gate oxide thickness to be the primary factor affecting radiation response in deep-submicron bulk CMOS technologies [37, 39, 42].

IV. CONCLUSIONS

The increased use of commercial deep-submicron technologies for use in integrated circuits operating in harsh radiation environments is leading to a greater demand for accurate models for radiation effects (e.g., edge leakage) that can be implemented in circuit simulators. The comprehensive radiation-enabled model presented in this paper suggests how these types of models may be realized. Through the use of closed form functions and reasonable estimates of critical technology variables that affect radiation response (e.g. STI geometry and sidewall doping), the effects of ionizing radiation on bulk CMOS devices are able to be modeled analytically, i.e., without numerical simulations. The model is based on a new technique which calculates non-uniform defect distributions and surface potential responses along the STI sidewall to model radiation-induced "edge" leakage currents in n-channel MOSFETs. Simulated results using the model compare well to experimental data obtained on 130 nm and 90 nm devices. The model is also used to identify sidewall doping as the primary technology feature driving increased radiation tolerance in advanced deep-submicron bulk CMOS.

ACKNOWLEDGMENT

The authors would like to thank Lew Cohn, of DTRA, Ken LaBel of NASA GSFC, Michael Fritze of DARPA, Warren Snapp of the Boeing Company, and Kitt Reinhardt of AFOSR for their support of this work and Larry Clark and Gennady Gildenblat of ASU for technical discussions.

REFERENCES

[1] N. S. Saks and M. G. Ancona, "Generation of interface states by ionizing radiation at 80K measured by charge pumping and subthreshold slope techniques," *IEEE Trans Nucl. Sci.*, vol. 34, pp. 1348-1354, 1987.

[2] A. Cester and A. Paccagnella, "Switching oxide ionizing radiation effects on ultra-thin oxide MOS structures," in *Radiation Effects and Soft Errors in Integrated Circuits and Electronic Devices*, R. D. Schrimpf and D. M. Fleetwood, Eds. New Jersey: World Sci., 2004.

[3] H. J. Barnaby, "Total-ionizing-dose effects in modern CMOS technologies," *IEEE Trans Nucl. Sci.*, vol. 53, pp. 3103-3121, 2006.

[4] L. T. Clark, K. C. Mohr, K. E. Holbert, X. Yao, J. Knudsen, and H. Shah, "Optimizing Radiation Hard by Design SRAM Cells," *IEEE Trans Nucl. Sci.*, vol. 54, pp. 2028-2036, 2007.

[5] W. Snapp, "DARPA rad-hard-by-design program results," in *GOMAC Tech 2006*. San Diego, CA, 2006.

[6] C. Brisset, V. F. Cavrois, O. Musseau, J. L. Leray, J. L. Pelloie, R. Escoffer, A. Michez, C. Cirba, and G. Bordure, "Two-Dimensional Simulation of Total Dose Effects on NMOSFET with Lateral Parasitic Transistor," *IEEE Trans. Nucl. Sci.*, vol. 43, pp. 2651-2658, 1996.

[7] I. S. Esqueda, H. J. Barnaby, and M. L. Alles, "Two-dimensional methodology for modeling radiation-induced off-state leakage in CMOS technologies," *IEEE Trans. Nucl. Sci.*, vol. 52, pp. 2259-2264, 2005.

[8] M. R. Shaneyfelt, P. E. Dodd, B. L. Draper, and R. S. Flores, "Challenges in Hardening Technologies Using Shallow-Trench Isolation," *IEEE Trans Nucl. Sci.*, vol. 45, pp. 2584-2592, 1998.

[9] T. R. Oldham, "Analysis of damage in MOS devices in several radiation environments," *IEEE Trans. Nucl. Sci.*, vol. 31, pp. 1236-1241, 1984.

[10] P. V. Dressendorfer, "Basic mechanisms for the new millennium," *1999 IEEE NSREC Short Course*, 1998.

[11] F. B. McLean and T. R. Oldham, "Basic mechanisms of radiation effects in electronic materials and devices," *Harry Diamond Laboratories Technical Report*, vol. HDL-TR, pp. 2129, 1987.

[12] M. R. Shaneyfelt, D. M. Fleetwood, J. R. Schwank, and K. L. Hughes, "Charge yield for cobalt-60 and 10-keV x-ray irradiations," *IEEE Trans. Nucl. Sci.*, vol. 38, pp. 1187-1194, 1991.

[13] T. R. Oldham, "Switching oxide traps," in *Radiation Effects and Soft Errors in Integrated Circuits and Electronic Devices*, R. D. Schrimpf and D. M. Fleetwood, Eds. New Jersey: World Sci., 2004.

[14] S. Lee, A. Raparla, Y. F. Li, G. Gasiot, R. D. Schrimpf, D. M. Fleetwood, K. F. Galloway, M. Featherby, and D. Johnson, "Total dose effects in composite nitride-oxide films," *IEEE Trans Nucl. Sci.*, vol. 47, pp. 2297-2304, 2000.

[15] J. R. Chavez, S. P. Karna, K. Vanheusden, C. P. Brothers, R. D. Pugh, B. K. Singaraju, W. L.Warren, and R. A. B. Devine, "Microscopic structure of the E' center in amorphous SiO₂," *IEEE Trans Nucl. Sci.*, vol. 44, pp. 1799-1803, 1997.

[16] D. M. Fleetwood, H. D. Xiong, Z. Y. Lu, C. J. Nicklaw, J. A. Felix, D. Schrimpf, and S. T. Pantelides, "Unified model of hole trapping, 1/f noise, and thermally stimulated current in MOS devices," *IEEE Trans Nucl. Sci.*, vol. 49, pp. 2674-2683, 2002.

[17] S. P. Karna, A. C. Pineda, R. D. Pugh, W. M. Shedd, and T. R. Oldham, "Electronic structure theory and mechanisms of the oxide trapped hole annealing process," *IEEE Trans Nucl. Sci.*, vol. 47, pp. 2316-2321, 2000.

[18] C. J. Nicklaw, Z. Y. Lu, D. M. Fleetwood, R. D. Schrimpf, and S. T. Pantelides, "The structure, properties, and dynamics of oxygen vacancies in amorphous SiO2," *IEEE Trans Nucl. Sci.*, vol. 49, pp. 2667-2673, 2002.

[19] S. T. Pantelides, S. N. Rashkeev, R. Buczko, D. M. Fleetwood, and R. D. Schrimpf, "Reactions of hydrogen with Si-SiO₂ interfaces," *IEEE Trans Nucl. Sci.*, vol. 47, pp. 2262-2268, 2000.

[20] J. K. Rudra and W. B. Fowler, "Oxygen vacancy and the E' center in crystalline SiO₂," *Phys. Rev. B, Condens. Matter*, vol. 35, pp. 8223-8230, 1987.

[21] X. J. Chen, H. J. Barnaby, B. Vermeire, K. Holbert, D. Wright, R. L. Pease, R. D. Platteter, G. Dunham, J. Seiler, S. McClure, and P. Adell, "Mechanisms of enhanced radiation-induced degradation due to excess molecular hydrogen in bipolar oxides," *IEEE Trans Nucl. Sci.*, vol. 54, pp. 1913-1919, 2007.

[22] J. R. Schwank, D. M. Fleetwood, P. S. Winokur, P. V. Dressendorfer, D. C. Turpin, and D. T. Sanders, "The role of hydrogen in radiation-induced defect formation in polysilicon gate MOS devices.," *IEEE Trans Nucl. Sci.*, vol. 34, pp. 1152-1158, 1987.

[23] S. N. Rashkeev, C. R. Cirba, D. M. Fleetwood, R. D. Schrimpf, S. C. Witczak, A. Michez, and S. T. Pantelides, "Physical model for enhanced interface-trap formation at low dose rates," *IEEE Trans Nucl. Sci.*, vol. 49, pp. 2650-2655, 2002.

[24] L. Tsetseris, R. D. Schrimpf, D. M. Fleetwood, R. L. Pease, and S. T. Pantelides, "Common origin for enhanced low-dose-rate sensitivity and bias temperature instability under negative bias," *IEEE Trans Nucl. Sci.*, vol. 52, pp. 2265-2271, 2005.

[25] J. Boch, F. Saigné, R. D. Schrimpf, J.-R. Vaillé, L. Dusseau, and E. Lorfèvre, "Physical model for the low-dose-rate effects in bipolar devices," *IEEE Trans Nucl. Sci.*, vol. 53, pp. 3655-3660, 2006.

[26] D. M. Fleetwood, T. L. Meisenheimer, and J. H. Scofield, "1/f noise and radiation effects in MOS devices," *IEEE Trans. on Electron Devices*, vol. 41, pp. 1953-1964, 1994.

[27] R. C. Lacoe, J. V. Osborn, R. Koga, S. Brown, and J. Gambles, "Total dose tolerance of the commercial TSMC 0.35µm CMOS process," *2001 IEEE Radiation Effects Data Workshop*, vol. IEEE cat no. 01TH8588, pp. 72-77, 2001.

[28] S. N. Rashkeev, D. M. Fleetwood, R. D. Schrimpf, and S. T. Pantelides, "Effects of hydrogen motion on interface trap formation and annealing," *IEEE Trans Nucl. Sci.*, vol. 51, pp. 3158-3165, 2004.

[29] D. M. Fleetwood, "Effects of hydrogen transport and reactions on microelectronics radiation response and reliability," *Microelectronics Reliability*, vol. 42, pp. 523-541, 2002.

[30] S. Matsuda, T. Sato, H. Yoshimura, Y. Takegawa, A. Sudo, I. Mizushima, Y. Tsunashima, and Y. Toyoshima, "Novel corner rounding rrocess for shallow trench isolation utilizing MSTS (Micro-Structure Transformation of Silicon)," *International Electron Devices Meeting, IEDM '98 Technical Digest*, pp. 6.2.1-6.2.4, 1998.

[31] R. K. Lowry, "Sources of volatile gases hazardous to hermetic electronic enclosures," *Ieee Trans. Elec. Packaging Man.*, vol. 22, pp. 319-323, 1999.

[32] R. A. Kohler, R. A. Kushner, and K. H. Lee, "Total dose hardness of MOS devices in hermetic ceramic packages," *IEEE Trans Nucl. Sci.*, vol. 35, pp. 1492, 1988.

[33] C. C. McAndrew and J. J. Victory, "Accuracy of approximations in MOSFET charge models," *IEEE Trans. Electron Devices*, vol. 49, pp. 72-81, 2002.

[34] W. Wu, T. Chen, G. Gildenblat, and C. C. McAndrew, "Physics-Based mathematical conditioning of the MOSFET surface potential equation," *IEEE Trans. Elec. Dev.*, vol. 51, 2004.

[35] E. Nicollian and J. Brews, *Metal Oxide Semiconductor Physics and Technology*. NY: John Wiley & Sons, 1982.

[36] P. S. Winokur, J. R. Schwank, P. J. McWhorter, P. V. Dressendorfer, and D. C. Turpin, "Correlating the radiation response of MOS capacitors and transistors," *IEEE Trans. Nucl. Sci.*, vol. 31, pp. 1453-1460, 1984.

[37] I. S. Esqueda, H. J. Barnaby, M. McLain, K. E. Holbert, M. Baze, J. D. Black, L. W. Massengill, B. Bhuva, R. D. Schrimpf, and F. Faccio, "Characterization of the Radiation Response of 0.13 µm n-Channel MOSFETs," in *IEEE Nuclear and Space Radiation Effects Conf.* Ponte Vedra Beach, FL, 2006.

[38] F. T. Brady, J. D. Maimon, and M. J. Hurt, "A Scaleable, Radiation Hardened Shallow Trench Isolation1," *IEEE Trans Nucl. Sci.*, vol. 46, pp. 1836-1840, 1999.

[39] M. McLain, H. J. Barnaby, K. E. Holbert, R. D. Schrimpf, H. Shah, A. Amort, M. Baze, and J. Wert, "Enhanced TID Susceptibility in sub-100 nm Bulk CMOS I/O Transistors and Circuits," *IEEE Trans Nucl. Sci.*, vol. 54, pp. 2210 - 2217, 2007.

[40] Y. Tsividis, *Operation and Modeling of the MOS Transistor*, 2nd ed. New York: McGraw-Hill, 1999.

[41] R. Lacoe, "CMOS Scaling Design Principles and Hardening-by-Design Methodologies," *IEEE NSREC Short Course*, 2003.

[42] M. McLain, H. J. Barnaby, K. Holbert, M. Baze, A. Amort, and J. Wert, "Gate Width Effects on the Radiation Response of sub-100 nm Bulk CMOS Two-Edge Transistors," in *HEART Conference*. Colorado Springs, CO, 2008.

IEEE 2008 Custom Intergrated Circuits Conference (CICC)

Characterization, Simulation, and Modeling of FET Source/Drain Diffusion Resistance

Ning Lu and Bill Dewey[1]

IBM Semiconductor Research and Development Center, Essex Junction, VT 05452 USA   lun@us.ibm.com

[1]IBM Electronic Design Automation, Fishkill, NY 12533 USA   bdewey@us.ibm.com

Abstract- **We present an innovative and comprehensive approach to characterize and model FET source/drain diffusion resistance. We present a set of new SPICE models for the parasitic resistance in FET source and drain regions. Our FET source/drain diffusion resistance model has been verified with field solver simulation results, and is found to be very accurate over a wide range of parameter values. We also present a set of micro testing structures to measure and characterize diffusion resistance. This approach has been validated using hardware data from a 65 nm SOI technology.**

I. INTRODUCTION

Parasitic resistance in CMOS FETs plays an important role in determining device and circuit performance. FET source and drain diffusion resistance has been modeled using a transmission line model [1]. The industry standard organization Compact Model Council has also put a simple source/drain diffusion resistance model into industry standard FET models, like BSIM3/4 models [2] and PSP FET model [3]. Due to various contact configurations and wide/narrow channel width situations, BSIM3/4 models and SPICE simulators also support user supplied number of squares and sheet resistance in a source or drain region [2, 4]. Many layout extraction tools also have capabilities of extracting FET source/drain diffusion resistance [5]. However, there are many areas in which the accuracy of FET source/drain diffusion resistance can be improved, both in layout extraction tools and in pre-layout/schematic level FET transistor models.

The problem of accurately modeling FET source/drain diffusion resistance is not a simple problem, since the electric current coming out of an FET channel will make a turn (more or less) in the silicide diffusion region before it reaches one or more contacts in the source (or drain) region. A good FET source/drain resistance model needs to cover various cases, including the case of electric current making a turn in the silicide diffusion region. The problem of modeling FET source/drain diffusion resistance is further complicated by the issue of how to characterize diffusion resistance using measured silicon hardware data, since any attempt to directly use FET structures will involve the characterization of FET current, which in itself is already a complex task.

In this paper, we present our approach of accurately characterizing and modeling FET source/drain diffusion resistance for advanced leading edge semiconductor technologies. We decompose the problem of modeling FET source and drain resistance into two parts. One part focuses on the characterization of diffusion resistance for electric current that **flows along a straight line** (across the channel). The result of this part of work is silicon validated diffusion sheet resistance and a width bias for resistance calculation. The other part focuses on accurately modeling the electric current in the FET source/drain region, which more or less **makes a turn** in the silicide region, with a known/given sheet resistance value and two known/given geometric bias values for the FET source/drain regions. Combining both analyses, we get an accurate and silicon validated FET source and drain resistance model for IBM advanced semiconductor technologies. The paper is organized in the following way. In Sec. II, we give our method of characterizing silicided (or non-silicided) diffusion resistance. In Sec. III, we show our simulation results on source and drain resistance, including FET channel resistance. In Sec. IV, we present our SPICE model for source and drain resistance. We summarize our results in Sec. V.

II. CHARACTERIZATION OF DIFFUSION RESISTANCE

Using the Kelvin (i.e., four-point) structures of long and straight diffusion strips of multiple design widths (W_{des}) (see Fig. 1), we first measure hardware data of silicide diffusion resistance, and then we fit measured data to the following resistance model:

$$R = \frac{L\,R_s}{W_{des} + 2\Delta_w}, \quad \text{or} \quad \frac{1}{R} = \frac{W_{des} + 2\Delta_w}{L\,R_s}, \tag{1}$$

where L is the length of a long diffusion strip (Fig. 1), R_s (in $\Omega/\square$) is (silicided) sheet resistance in the source/drain diffusion region, and Δ_W is the electric width bias per edge. Plotting measured conductance $1/R$ vs. the design width W_{des} of Kelvin structures of the same length L, we see that a linear relationship holds, as predicted by Eq. (1) (see Fig. 2). Then, we do a least-squares fit to match the data to a straight line, and extract a pair of sheet resistance R_s and electric width bias Δ_W for resistance calculation. We do this for 3 cases of N+ diffusion associated with a FET device: i) isolation bounded on both sides of a long diffusion structure (will call it a diffusion wire in the following), ii) polysilicon bounded on both ides of a long diffusion structure, and iii) isolation bounded on one side of a lone diffusion region and polysilicon bounded on the other side of the diffusion region. We will get 3 pairs of (R_s, Δ_W). When extracting sheet resistance from measured resistance data, sometimes a zero

978-1-4244-2018-6/08 $25.00 © 2008 IEEE   301

width bias, $\Delta_W = 0$, is effectively used. This will yield a different sheet resistance value. Worse, when hardware data is collected from a second width structure, a different sheet resistance value will be generated. This is especially true for polysilicon bounded diffusion case, since the FET spacer will dramatically reduce the width of silicide portion of a narrow diffusion strip. On the other hand, measured hardware data from a 65 nm SOI technology [6] has shown that, with the use of a non-zero width bias Δ_W, diffusion resistance for a wide range of diffusion width can be accurately represented by Eq. (1) (see Fig. 2).

As expected, the three sheet resistance values are very close to each other, based on hardware data from a 65 nm IBM SOI technology. We choose a middle $\dot{R}_s$ value, typically the R_s value of isolation and poly bounded case, as a unified R_s value for N+ diffusion. Also, as expected, the following relation is approximately true,

$$2\Delta_W(\text{poly-isol}) \approx \Delta_W(\text{poly-poly}) + \Delta_W(\text{isol-isol}) . \quad (2)$$

Then, a hardware based single sheet resistance value, R_s, and two hardware based width bias values, $\Delta_W(\text{poly-poly})$ and $\Delta_W(\text{isol-isol})$, are used in NFET source/drain resistance modeling (Fig. 3). Similarly, for P+ diffusion, we extract R_s, $\Delta_W(\text{poly-poly})$, $\Delta_W(\text{isol-isol})$ from hardware data, and then use them to model PFET source/drain resistance.

III. Simulation of Source/Drain Diffusion Resistance

Two IBM field solvers are used to obtain accurate simulation results on FET source/drain diffusion resistance. One of them, RESCAL [5], uses the finite-element method to solve the Laplace equation on a dense triangular grid, with an automatic clustering capability in region when electric current density is high. When the total resistance of (source + channel + drain) is computed, the RESCAL tool also plots the equal potential lines (see Fig. 4), which provides invaluable insight on how the electric current flows in both channel region and source/drain region. To help understand how a higher sheet resistance in the channel region modifies both equal potential lines and electric current lines in the whole (source + channel + drain) region, the ratio of channel region sheet resistance $R_s(\text{ch})$ to diffusion region sheet resistance R_s is increased from 1 to 1000 in Fig. 4. The other field solver, Cosmic [7], uses the boundary element method [8] and Green's functions to compute the capacitance (and resistance). In our benchmark testing, both field solvers lead to the same resistance values (see Table I).

TABLE I. Field Solver Results. $R_{tot} = R(\text{ch}) + 2\,R$.

$R_s(\text{ch}) / R_s$	R_{tot} / R_s		$R(\text{ch}) / R_s$	$2\,R / R_s$
	Cosmic	RESCAL		
1	0.7140	0.713	0.04	0.673
10	2.091	2.089	0.4	1.689
100	8.024	8.025	4	4.025
1000	45.12	45.22	40	5.22

IV. Modeling Source/Drain Diffusion Resistance

The center line of the FET channel along the width direction is an equal potential line. Thus, we can split the total (source + channel + drain) resistance into the sum of (source + half channel) resistance and (half channel + drain) resistance. Then, for (source + half channel) resistance, we divide the whole channel width into multiple sub-regions. When the number of contacts in a source (drain) region is N ($N \geq 1$), the number of sub-regions in the source (drain) is $M = 3N$ in an edge-contacted case (Fig. 5). Similarly, for (half channel + drain) region, we divide the whole channel width into multiple sub-regions. Since the sheet resistance in the channel region is much larger than the diffusion sheet resistance in the source/drain region, the electric current inside the channel region flows mostly along the channel length direction. As an approximation, we also regard that the electric current in each sub-width region flows along the sub-region boundary lines; namely, the total conductance for the whole width W is the sum of the conductance of all sub-width regions. Using the assumption that the channel resistance is large compared to source/drain diffusion resistance, the source/drain diffusion resistance is found to be

$$R = \frac{1}{I_{tot}^2} \sum_{m=1}^{M} I_m^2 R_m , \quad (3)$$

where R_m is the resistance of the mth sub-region, I_m is the electric current in the mth sub-region, and $I_{tot} = \Sigma I_m$ is the total electric current.

For a sub-region which directly faces a contact, its electric current is proportional to its width, i.e., the contact width w_c,

$$I_{2+3n} \propto w_c , \quad (4)$$

and its resistance is simply

$$R_{2+3n} = R_s s_1 / w_c , \quad (5)$$

where s_1 is the poly-contact spacing (always after accounting for poly edge bias).

For a sub-region which does not directly face a contact (call it a side region), its electric current can be found using a transmission line model,

$$I_m \propto w_0 \tanh \frac{w_m}{w_0}, \quad m = 1+3n, 3+3n, \quad w_0 = \sqrt{\frac{R_{ch} D}{2 R_s}}, \quad (6)$$

where w_0 is a characteristic width value, D is the electric dimension of, say, poly to isolation (i.e., after adjusting poly and isolation edge biases), and R_{ch} (in Ω-μm) is the FET channel resistance for a unit width, which can be estimated using $R_{ch} = V_{ds}/I_{ds} = V_{dd}/I_{on}$ or $R_{ch} = 0.5 V_{dd}/I_{ds} (V_{ds} = 0.5 V_{dd}, V_{gs} = V_{dd})$. When all $w_m \ll w_0$, I_m in Eq. (6) simplifies to an expression similar to Eq. (4), and Eq. (3) reduces to

$$R = \frac{1}{W^2} \sum_{m=1}^{M} w_m^2 R_m , \quad (7)$$

where $W = \Sigma w_m$. The resistance of a side region is calculated using two resistance extraction formulas, and then take the larger of two resistance values,

$$R_m = \max(R_{m,cf}, R_{m,tr}).\qquad(8)$$

For a narrower width sub-region (Fig. 6), its resistance is [9],

$$R_{m,cf} = \frac{DR_s}{w_m} - \frac{2R_s}{\pi}\ln\left(\sinh\frac{\pi(D-s_1)}{2w_m}\right),\quad w_m \leq \eta D.\quad(9)$$

The exact value of η is not needed due to the max operation in Eq. (8). When $(D - s_1) \gg w_m$, it approaches a value that is independent of both D and $(D - s_1)$,

$$R_{m,cf} = \frac{s_1 R_s}{w_m} + \frac{2R_s \ln 2}{\pi},\quad D-s_1 \gg w_m.\qquad(10)$$

For a wider width sub-region, its resistance is found by using a transmission line model,

$$R_{m,tr} = \frac{w_0 R_s}{D\tanh(w_m/w_0)} - \frac{R_{ch}}{w_m} + r_a,\quad w_m \geq \eta D.\quad(11)$$

Here, an adder resistance r_a accounts for the fact that the contact size w_c is actually smaller than the diffusion region dimension D (Fig. 7). Using Eq. (7) and Ref. [9], we get

$$r_a = -\frac{R_s}{2\pi D^2}\sum_{i=1}^{2}(w_c + 2s_i)^2 \ln\left(\cos\frac{\pi s_i}{w_c + 2s_i}\right).\qquad(12)$$

In a symmetric case of $s_1 = s_2$, the adder r_a reduces to

$$r_a = -\frac{1}{\pi}\ln\left(\cos\frac{\pi s_1}{D}\right),\quad\text{when } s_1 = s_2.\qquad(13)$$

When a sub-region width is not too short and not too long, the resistance expression simplifies to

$$R_{m,tr} \approx \frac{w_m R_s}{3D} + r_a,\quad \eta D \leq w_m < w_0,\qquad(14)$$

which is independent of FET parameters, like FET types, V_{dd}, on-current or an effective/high current, etc. In Eq. (14), the ratio of w_m/D is "the number of squares by eye". Figure 8 compares the extraction formula with the field solver RESCAL simulated values. As a comparison, we also plot in Fig. 8 the resistance given by the simple source/drain diffusion resistance model in BSIM3/4,

$$R_{BSIM} = \frac{R_s W}{3D},\quad\text{point contact, isolated;}\qquad(15a)$$

$$R_{BSIM} = \frac{R_s s_1}{W},\quad\text{wide contact.}\qquad(15b)$$

V. Summary

We have presented a new extraction method and a set of extraction formulas for accurately modeling FET source/drain diffusion resistance. Our extraction method has been verified with field solver simulation results. Application of our method to various other source/drain layouts can also be done.

Acknowledgment

The authors would like to thank IBM management for support.

References

[1] H. H. Berger, *Solid-State Electron*, **15**, 145–158 (1972). D. B. Scott, W. R. Hunter, and H. Shichijo, *IEEE Trans. Electron Devices*, **29**, 651–661 (1982).

[2] BSIM4.3.0 MOSFET Model, User's Manual, UC Berkeley, 2003.

[3] G. Gildenblat, et al., *IEEE Trans. Electron Devices*, **53**, 1979 (2006).

[4] W. Wong, et al., *Proc. International Solid-State and Integrated Circuit Technology Conference (ICSICT)*, 2006, pp. 1243–1247.

[5] P. Habitz et al., "ERIE, A new parasitic model extraction tool," IBM Micronews, vol. 7, no. 1, pp. 32–36, 2001.

[6] E. Leobandung, et al., *Proc. Symposium on VLSI Technology*, 2005, pp. 126–127.

[7] Cosmic User's Guide, Publication No. 0220-5535-00, 1993, IBM.

[8] A. E. Ruehli and P. A. Brennan, *IEEE Trans. Microwave Theory Tech.* MTT-**21**, pp. 76–82 (1973).

[9] P. M. Hall, *Thin Solid Films*, **1**, 277–295 (1967/68).

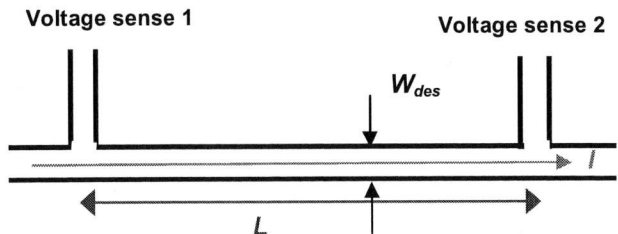

Fig. 1. Four-point (Kelvin) measurement structure for characterizing diffusion resistance. Electric current flows along a straight line.

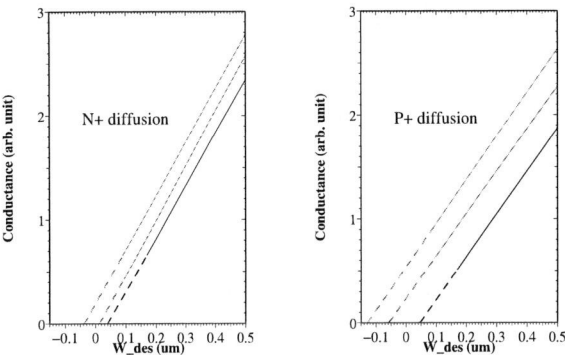

Fig. 2. Measured conductance $(1/R)$ of long and straight diffusion strips vs. the design width of strips. The top blue lines are poly bounded case, the middle red lines are poly and isolation bounded case, and the bottom black lines are isolation bounded case. The solid portion of lines is the data-fitting result, and the intersection of a dashed line with the X axis is $-2\Delta_w$.

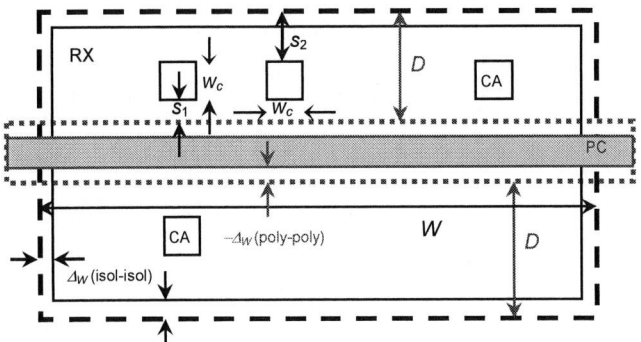

Fig. 3. FET source and drain regions. Solid lines are design dimensions. Dashed lines are electric dimensions for resistance calculation.

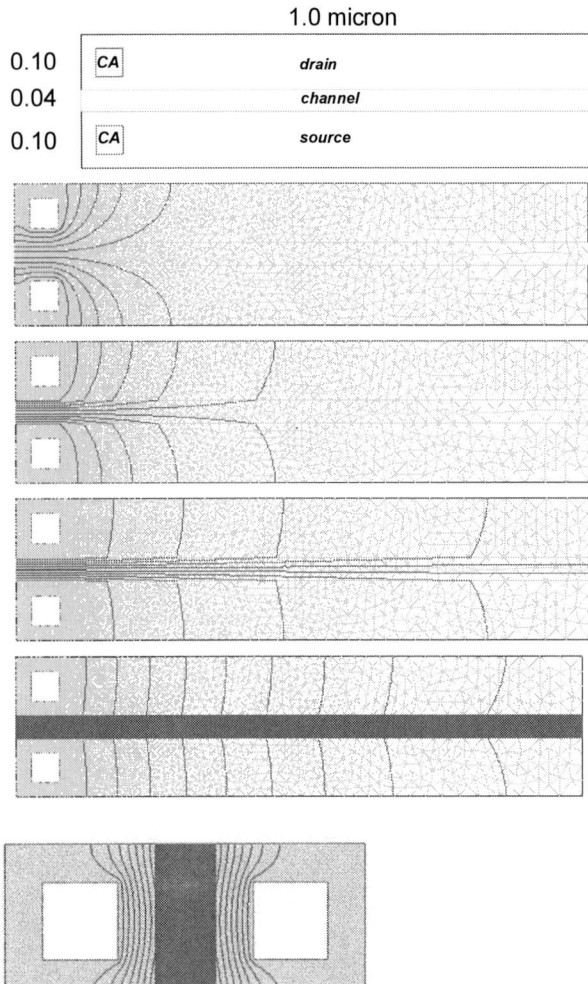

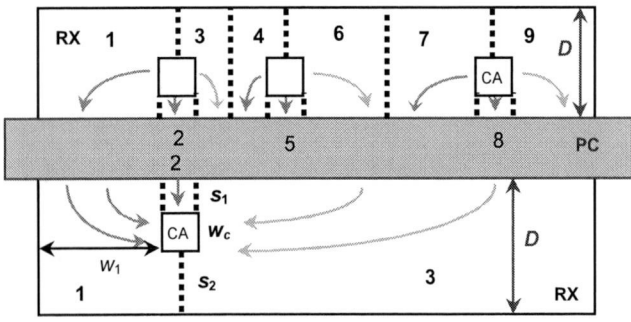

Fig. 4. Simulated FET source and drain structure (very top) and obtained equal potential lines from an IBM FEM field solver called RESCAL. In the bottom diagram (rotated 90 degrees), width is reduced to 0.10 μm. From top to bottom respectively, (i) the ratio of channel region sheet resistance to diffusion region sheet resistance, $R_s(ch) / R_s$, is 1, 10, 100, 1000, and 1000, (ii) the number of FEM elements is 10,000 for each case, and (iii) the number of equal potential lines is 10, 10, 10, 100, and 10,000.

Fig. 5. Method of decomposing a source or drain region into multiple sub-regions for calculating source/drain diffusion resistance.

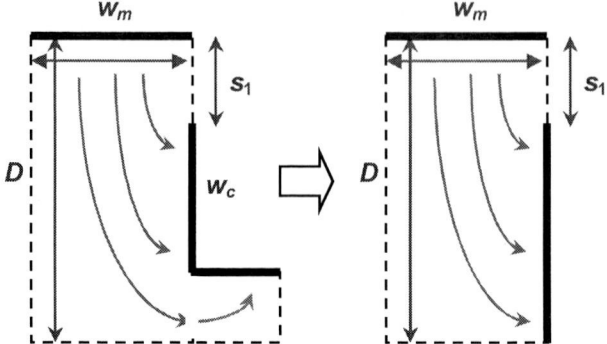

Fig. 6. The layout approximation used when obtaining Eq. (9). Thick solid lines are equal potential lines.

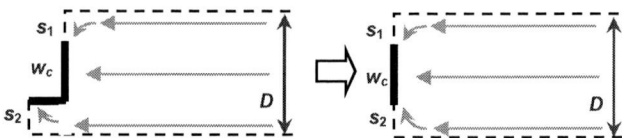

Fig. 7. The layout approximation used when obtaining Eqs. (12) and (13). Thick solid lines are equal potential lines.

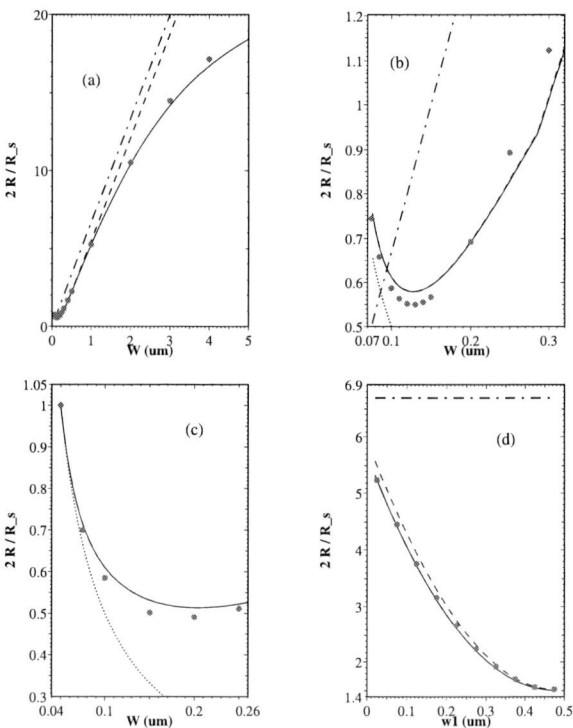

Fig. 8. The sum of source and drain diffusion resistance $2R$ (normalized to diffusion sheet resistance R_s) vs. total FET width W in (a), (b), and (c) and vs. the width w_1 of the left sub-region in (d). In (a) and (b), the width of the left sub-region w_1 and contact width w_C are fixed. Fig. (b) is an enlarged view of the lower-left corner in (a). In (c), the contact width w_C is fixed and the contact is always at the center. In (d), the total FET width ($W = 1$ μm) and the contact width w_C are fixed. Red symbols are the results of FEM simulations (50,000 elements are used in each simulation). The ratio of channel region sheet resistance to diffusion sheet resistance is 1000. Solid curves use Eqs. (3), (6), and (11), but dashed curves use Eqs. (7) and (14). Dash-dotted straight lines are Eq. (15a), and dotted curves are Eq. (15b).

IEEE 2008 Custom Intergrated Circuits Conference (CICC)

Analysis of the Impact of Interfacial Oxide Thickness Variation on Metal-Gate High-K Circuits

Minki Cho[1], Kingsuk Maitra[2], and Saibal Mukhopadhyay[1]

[1]School of ECE, Georgia Institute of Technology, Atlanta, GA; [2]Advanced Micro Devices, Albany, NY

mcho8@gatech.edu, kingsuk.maitra@amd.com, saibal@ece.gatech.edu

ABSTRACT

In this paper we analyze the impact of interfacial oxide thickness variations on metal-gate high-K circuits. A simulation methodology is developed which considers a holistic model of the interaction of interfacial oxide layer thickness variation with the transistor electrostatic and transport properties, The unique nature of the high-k devices is thus captured and its influence on circuit parameters is investigated. It is shown that, the interfacial oxide thickness variation need to be effectively controlled to reduce circuit variability and fully exploit the advantage of metal gate high-k technologies.

1. Introduction

The introduction of high-K gate stacks into MOSFET is a key driver for technology scaling in sub-45nm nodes [1]. High-K gate stack provides the advantages of much less gate leakage, resulting in better scalability. In sub-45nm nodes, the analysis and mitigation of variability is an integral part of circuit design. Therefore, it is important to identify the 'high-K specific' variation sources, and analyze their effects in CMOS circuits.

The 'interfacial oxide layer' is necessary to reduce the mobility degradation in high-K devices [2]. However, it acts as an important new variation source which impacts both device electrostatics (gate capacitance (C_g)) and threshold voltage (V_{th})) and transport (namely,

mobility). This effect is considerably different from the effect of oxide thickness variation in conventional poly-Si/SiON transistor [3]. Further, from a pure processing perspective, interfacial oxide thickness (t_{IL}) variation is unique to high-k/metal gate transistors. It is relatively easier to control the oxide thickness variation of a conventional poly-Si/SiON transistor compared to controlling t_{IL} of a high-k/metal gate transistor. This is because, t_{IL} can be on the order of a few atomic mono-layers (~5-6 ^{0}A) for high-k/metal gate transistors [4]. In this paper, we analyze the impact of interfacial oxide thickness variations on metal-gate high-K circuits. A methodology is developed to capture the effects of t_{IL} variation on device transport and electrostatics through measurement and device simulations, respectively. The effects on device propoerties are next used to analyze the circuit performances using simple analytical models. Our analysis suggests that, random variation in t_{IL} can result in a considerable variation in circuit properties.

2. Effects of Interfacial Oxide Thickness

In this section, we discuss how interfacial oxide thickness affects device electrostatics and transport.

2.1 Effect of t_{IL} variation on device electrostatics

Clearly, t_{IL} variation contributes to a variation in total gate capacitance (C_g) (through a variation in equivalent oxide thickness, t_{eq}, see (1)). The variation in t_{eq} directly impacts threshold voltage (V_{th}) of transistor (namely, an in increase in t_{eq} tends to increase V_{th}). Also, for short channel devices, a higher t_{IL} (and hence, higher t_{eq}) worsens the short-channel effect (SCE), resulting in higher drain induced barrier lowering (DIBL) and Vth roll-off. Therefore, for better scalability of a technology generation, a thinner t_{IL} is beneficial.

2.2 Effect of t_{IL} variation on device transport

Interfacial oxide thickness variation contributes to a variation in effective field (E_{eff}) which translates into a mobility variation in the channel [5]. However, apart from this well-understood field effect, additional scattering mechanisms are introduced and/or modified by presence of high-k gate stack. These are, (i) remote phonon scattering (RPS) dominant in low to medium E_{eff} or N_{inv} (inversion channel carrier density, which determines E_{eff}) [6-7], which gets worse with reduction in t_{IL} (ii) remote Coulomb scattering (RCS) also dominant in low to intermediate E_{eff} (may be present in poly-Si/SiON gate stacks as well, but could be controlled by optimized processing), which also worsens with reduction in t_{IL} and (iii) surface roughness scattering (SR) which dominates at high E_{eff} and behaves very differently from poly-Si/SiON gate stacks [8]. With increasing t_{IL}, the SR component becomes more poly-Si/SiON like and vice versa. All these scattering mechanisms are very sensitive to the t_{IL} variation, and results in a unique sensitivity of high-K transistor mobility to t_{IL} variation.

3. Analysis Methodology

In this section we present the analysis methodology to consider both the electrostatic effect (section 2.1) and the mobility effect (section 2.2). This provides a more realistic view of the interaction of high-k device parameters and circuit performance.

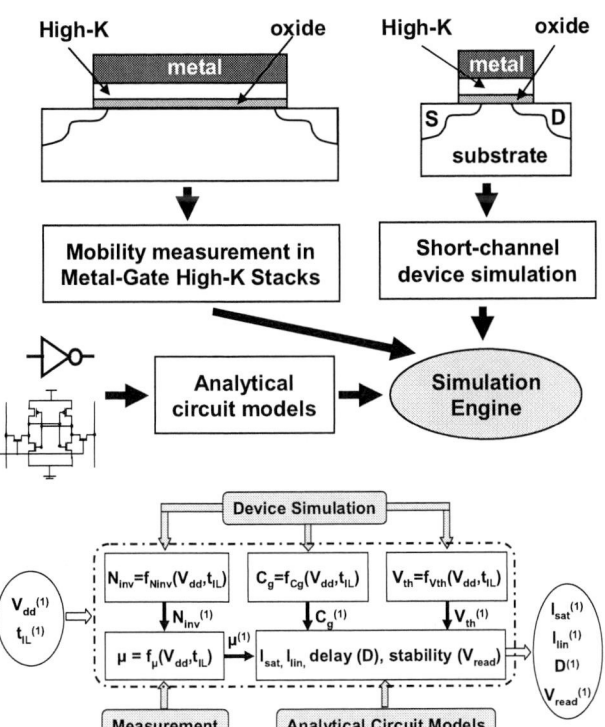

Fig. 1: Analysis framework (top) and mathematical engine (bottom)

978-1-4244-2018-6/08 $25.00 © 2008 IEEE

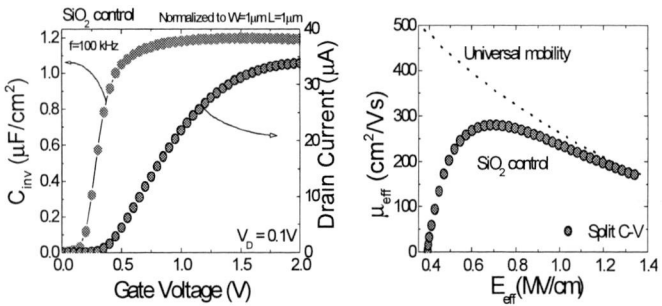

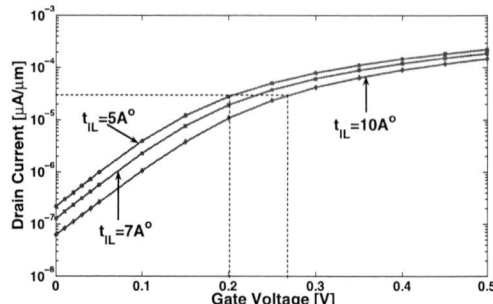

Fig. 2: Mobility measurement using the split CV and dc I_d-V_g technique [5, 10]

Fig. 3: Id-Vg characteristics of simulated 65nm device

3.1. Analysis Engine: Mixed-Mode-Circuit Analyzer

We develop a physics and measurement driven 'Mixed-Mode-Circuit Analyzer (MMCA)' to integrate the measurement data, device simulation, and circuit models into a single analysis engine. This helps perform circuit analysis at early cycles of technology development. While the effect of t_{IL} on C_g and V_{th} of a short-channel device can be captured in a drift-diffusion device simulator using a technologically relevant short-channel transistor, physically modeling the impact on mobility can be very challenging. On the other hand, the mobility measurements [i.e. $\mu_{eff} = f_\mu(N_{inv}, t_{IL})$] are performed at the very early stages of technology development while designing and characterizing gate stack. Therefore, MMCA simultaneously considers device *simulation* of short-channel MOSFET and *measurement data* on metal-gate High-K stacks to predict the impact of t_{IL} variation on circuit functionality. The overall structure of MMCA is shown in Fig. 1. First, drift-diffusion device simulations are performed ([9]) to obtain Cg [= $f_{Cg}(V_{dd}, t_{IL})$], V_{th} [=$f_{Vth}(V_{dd}, t_{IL})$], and N_{inv} [=$f_{Ninv}(V_{dd}, t_{IL})$] for different V_{dd} and t_{IL}. Next, the simulated values and measured data are used to predict mobility μ_{eff} and different V_{dd} and t_{IL}. Finally, the device parameters are used to predict the impact of t_{IL} variation on circuit functionality. The mathematical framework of predicting circuit parameter for a given V_{DD} and t_{IL} is summarized in Fig. 1. The Monte-Carlo simulation using this framework can predict the circuit variability for a given random t_{IL} variation.

3.2. Mobility Measurement

To study the effect of interfacial oxide thickness, HfO_2 gate stack was prepared with TiN, and TaN gate electrodes by gate-first approach with poly-Si capping on top [3]. 2.5nm HfO_2 thickness was used with interfacial oxide thickness varying between 0.5 to 1.0 nm respectively. The inversion oxide thickness (T_{inv}/EOT) was observed to vary between 1.4nm/1.0nm to 2.4nm/2.0nm. The room temperature (T=300K) mobility measurement is carried out by the well known split CV/dc I_d-V_g technique. As depicted in Fig. 2, a dc split-CV measurement is conducted on long channel transistors (5-10 μm) and the resulting N_{inv} is computed by integrating the split CV curve [5, 10]. An independent linear dc I_d-V_g measurement is carried out on the same transistor and the resulting mobility is carried out using: $\mu_{eff} = (L/W)(I_d/V_{ds}N_{inv}q)$.

3.3. Device Simulation

Device simulations were performed for a 65nm poly length bulk-CMOS device with metal-gate high-K stack. 2.5nm thick HfO_2 high-K layer was used and interfacial oxide thickness was varied from $5A^o$ to $10A^o$. The super-halo doping profile was used to obtain ~200nA/μm subthreshold leakage for t_{IL}=$5A^o$. Quantum correction was used in the simulation. The I_d-V_g simulations (Fig. 3) were performed in the subthreshold region to obtain the threshold voltage (defined as V_{gs} required for Id = 30μA). Next, N_{inv} was obtained by integrating (along the depth) electron concentration in the channel. The quantum correction required for T_{inv} was estimated from the

location of peak of the inversion charge distribution (say, t_{QM}). The effective gate capacitance (C_g) was computed as:

$$t_{eq} = \frac{\varepsilon_{SiO_2}}{\varepsilon_{HfO_2}}t_{HfO_2} + t_{IL} + \frac{\varepsilon_{SiO_2}}{\varepsilon_{Si}}t_{QM} ; C_g = \frac{\varepsilon_{SiO_2}}{t_{eq}} \qquad (1)$$

3.4. Circuit Models

To analyze the impact of interfacial oxide thickness on circuits we have considered simple short-channel models for transistor currents. The short-channel linear (I_{lin}) and saturation current (I_{sat}) was computed as [11]:

$$I_{lin} = \frac{\mu_{eff}C_g(W/L)[(V_{gs}-V_{th})V_{ds} - (m/2)V_{ds}^2]}{1 + (\mu_{eff}V_{ds}/\upsilon_{sat}L)}$$

$$I_{sat} = C_gW\upsilon_{sat}(V_{gs}-V_{th})\frac{\sqrt{1+2\mu_{eff}(V_{gs}-V_t)/(m\upsilon_{sat}L)}-1}{\sqrt{1+2\mu_{eff}(V_{gs}-V_t)/(m\upsilon_{sat}L)}+1} \qquad (2)$$

where, υ_{sat} is the saturation velocity. The current models are used to compute the high-to-low delay of an inverter (assuming a step input). During the transition, NMOS device remains in saturation for V_{out}=V_{dd} to V_{out}=V_{dd}-V_{th} (current is a weaker function of mobility). In the region, V_{out}=V_{dd}-V_{th} to V_{out}=0.5V_{dd}, NMOS operates in the linear region (current is a stronger function of mobility). Using the two operating regions delay is computed as:

$$T = C_L\int_{V_{DD}}^{0.5V_{DD}-V_t}\frac{dv_{out}}{I_{sat}} + C_L\int_{V_{DD}-V_t}^{0.5V_{DD}}\frac{dV_{out}}{I_{lin}} \qquad (3)$$

The above model can capture the effects of μ, C_g, and V_{th} on delay considering the different sensitivities of I_{lin} and I_{sat} to mobility.

We also analyze the effect of t_{IL} on SRAM. In particular, we consider the read disturb voltage (V_{read}, the voltage rise at the node storing '0' while reading an SRAM cell) of an SRAM cell. V_{read} is computed by equating current through pull-down (operating in linear mode) and access (operating in saturation) devices of the cell. The circuit model is obtained by solving the following [12]:

$$I_{lin\_pulldown}(V_{gs}=V_{dd}, V_{ds}=V_{read}) \qquad (4)$$
$$= I_{sat\_access}(V_{gs}=V_{ds}=V_{dd}-V_{read})$$

A lower V_{read} improves the cell stability.

4. Results and Discussions

In this section, we analyze the effect of t_{IL} variation on circuit properties. Note, the primarily aim of this paper is to analyze the effect of t_{IL} variation on a 'designed' metal-gate high-K device/circuit. Therefore, during our analysis we have considered same doping for all t_{IL} values.

4.1. Mobility Measurement

The measured mobility for a scaled dielectric stack shows that mobility reduces at a higher inversion charge density (or a higher gate electric field) (Fig. 4). Further, strong dependence of mobility on t_{IL} can also be observed. At moderate N_{inv}, thicker t_{IL} results in better mobility due to lower RCS and RPS. However, at higher N_{inv}, lower SR scattering in thinner t_{IL} can improve the mobility ('high-K

978-1-4244-2018-6/08 $25.00 © 2008 IEEE

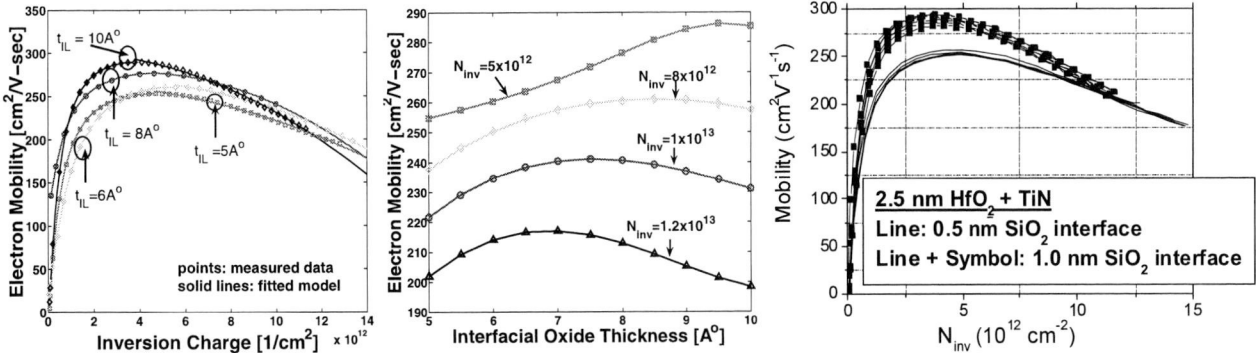

Fig. 4: Measurement results and model calibration: (a) Room temperature measured and fitted mobility at different N_{inv} and t_{IL}, (b) mobility vs t_{IL}, and (c) measured mobility showing larger variation at thicker SiO_2

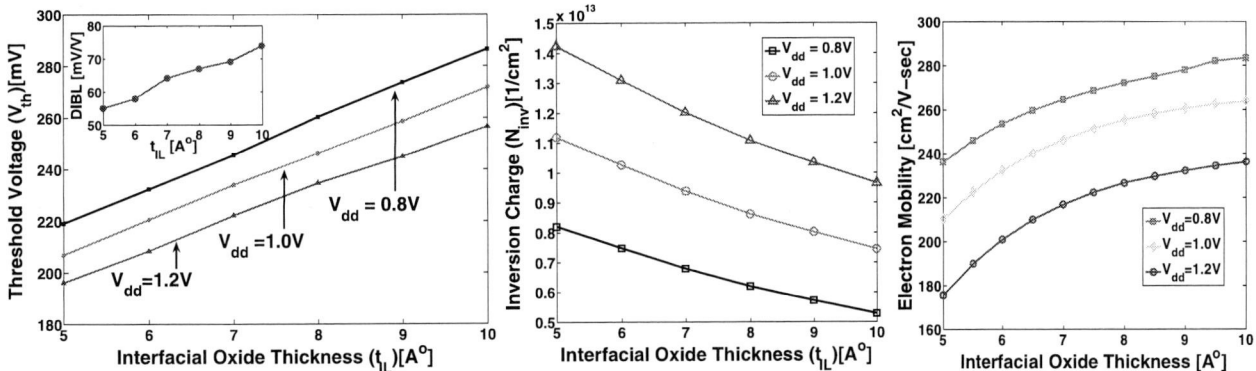

Fig. 5: Effects of t_{IL} on (a) threshold voltage, (b) inversion charge, and (b) mobility at a constant V_{dd}.

mobility crossover') (Fig. 4). A wafer level measurement also demonstrates that, mobility variation at a given N_{inv} is higher for thicker t_{IL} which is in accordance with the exponential dependence of high-K mobility on t_{IL} as predicted in [7]. For our analysis, we calibrated the measured mobility as a function of N_{inv} and t_{IL} and used the calibrated model in MMCA.

4.2. Device Simulations

The simulation of the short-channel MOSFET shows the effects of t_{IL} on device electrostatics. An increase in t_{IL} reduces C_g which in turn increases V_{th} (for constant doping) (Fig. 5a). The simulated values of DIBL, estimated as change in Vth due to 1V change in V_{ds}, shows that DIBL increases at an increased t_{IL} (Fig. 5a, inset). This confirms that SCE degrades at higher t_{IL}. Since use of super halo doping profile suppresses the SCE, we observed that V_{th} increases for a nominal increase in t_{IL}. Lower C_g and higher V_{th} reduce N_{inv} at a higher t_{IL} (Fig. 5b). Therefore, at a given V_{dd}, a change in t_{IL} modifies N_{inv}, which in turn modifies the mobility. Further, at a given N_{inv} mobility also depends on t_{IL}. Using MMCA, we observed that, at a given V_{dd}, mobility tends to improve with an increase in t_{IL} mainly due to lower N_{inv} (Fig 5c). This is important for circuit analysis, as digital circuits are designed to operate at a given V_{dd}.

4.3. Circuit Analysis: Sensitivity

The measured results of mobility and simulated results of C_g, N_{inv}, and V_{th} are used by MMCA to predict the effects of t_{IL} on circuits. For a given V_{dd}, lower C_g, and higher V_{th} reduces both linear and saturation current (Fig. 6). For linear current, the degradation is less pronounced due to improvement of mobility at a lower N_{inv} (for a higher t_{IL}). The degradation is stronger for saturation current, as it is a weaker function of mobility. At a higher V_{dd}, a small increase in t_{IL} can also marginally improve the

linear current as mobility improvement overshadows the C_g and V_{th} degradation. The effect of high-field mobility crossover was less pronounced in the currents when compared at equal V_{dd}.

A higher linear and saturation current at thinner t_{IL} improves the inverter delay (Fig. 7). The delay improvement is smaller when load is dominated by the gate capacitances as lower C_g at higher t_{IL} reduces both current and load (Fig. 7). When no parasitic capacitance is considered, effect of C_g is eliminated. Under this condition improved mobility at thicker t_{IL} can compensate for the V_{th} increase resulting in a delay reduction. At higher V_{DD}, the marginal improvement in I_{lin} at a thicker t_{IL} can result in a delay improvement (Fig. 7).

To analyze the read disturb voltage, we first consider a global change in t_{IL} (i.e. t_{IL} for both access and pull-down device changes by same amount). A 'global' change eliminates the C_g effect [see (2) and (4)]. Higher mobility and V_{th} at thicker t_{IL} (at a given V_{dd})

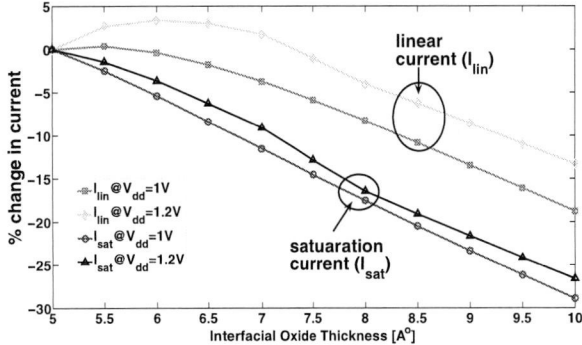

Fig. 6: Effect of t_{IL} on linear and saturation current.

978-1-4244-2018-6/08 $25.00 © 2008 IEEE

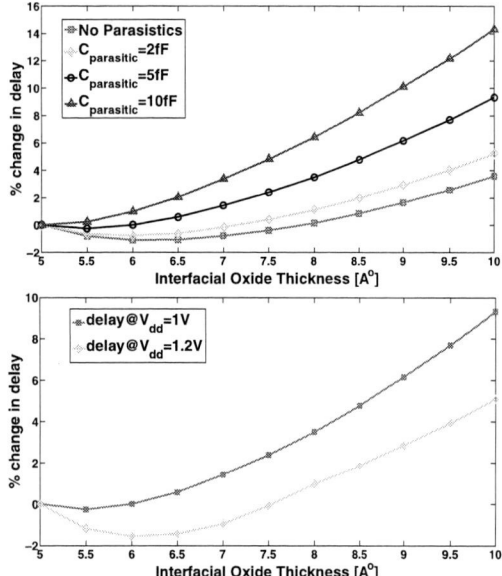

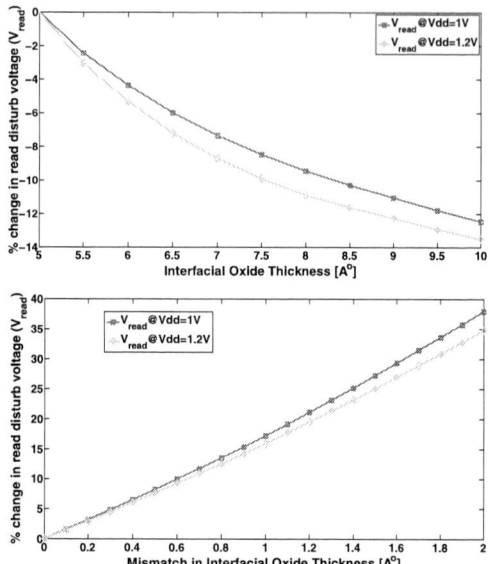

Fig. 7: Effect of t_{IL} on delay at different parasitic capacitances (@V_{dd}=1.0V) (top) and different V_{dd} @$C_{parasitic}$=5fF (bottom)

Fig. 8: Effect of t_{IL} on V_{read} for global variation (top) and mismatch (bottom)

increases the linear to saturation current ratio thereby lowering V_{read} (Fig. 8, note, lower V_{read} implies better cell stability) [12]. Next, we perform worst case mismatch analysis [higher t_{IL} (lower I_{lin}) of pull-down and lower t_{IL} (higher I_{sat}) of access, nominal t_{IL} = 8A°]. Mismatch in t_{IL} causes V_{th} and C_g mismatch, resulting in a significant increase in V_{read} (Fig. 8). The better mobility at a lower N_{inv} helps reduce the V_{read} increase to a certain extent.

4.4. Circuit Analysis: Variability

Monte-Carlo simulations were performed for randomly generated t_{IL} values (mean = 8A° with different standard deviation) using MMCA. It was observed that, even a small variability in t_{IL} can result in considerable delay variation. Further, due to the non-linear dependence of delay on t_{IL} (Fig. 7), Normal t_{IL} variation results in skewed delay distribution (Fig. 9a). Global random variations of t_{IL} has a marginal impact on V_{read} distribution (Fig. 9b). However, the local random variation (which results in mismatch) can have a much larger impact (Fig 9c). The distributions are also observed to be skewed towards a higher V_{read} values which is detrimental for read stability.

5. Conclusions

We have analyzed the effect of interfacial oxide thickness variation on performance and robustness of metal gate high-K

circuits using mobility measurement, device simulation, and circuit models. The sensitivity of circuit parameters, such as inverter delay and SRAM read disturb voltage, to interfacial oxide thickness variation have been studied. Our analysis suggests that, t_{IL} variation can have a strong impact on the circuit operation and acts as a new source of variations for metal-gate high-K technologies. Therefore, we conclude that, to fully exploit the advantages of metal-gate high-K technologies, interfacial oxide thickness variation needs to be well-controlled and its impacts on circuits need to be considered.

6. References

[1]. M. P Chudzik et. al, *VLSI Tech. Symp. 2007*
[2]. V. Narayanan et. al., *IEEE EDL*, vol. 27, p. 591, 2006
[3]. K. Maitra et. al., *JAP*, vol. 102, p. 114507, 2007
[4]. W. Tsai et. al., *IEDM 2003*, pp. 311-314
[5]. A. Kerber, doctoral dissertation, TU Darmstadt., 2003
[6]. M. V Fischetti et. al., *JAP*, vol. 90, p. 4587, 2001
[7]. M. Casse et. al., *IEEE TED* , vol. 53, p. 759, 2006
[8]. S. Jin et. al., IEEE TED, vol. 54, no. 9, Sept. 2007, pp. 2191-2203
[9]. MEDICI: 2-D device simulation tool: www.synopsys.com
[10]. K. Maitra, doctoral dissertation, NCSU 2005
[11]. Y. Taur and T. H Ning, Fundamentals of Modern VLSI Devices, 1998
[12]. S. Mukhopadhyay, *et. al*, *IEEE TED*, vol. 55, no. 1, Jan. 2008, pp. 152-162.

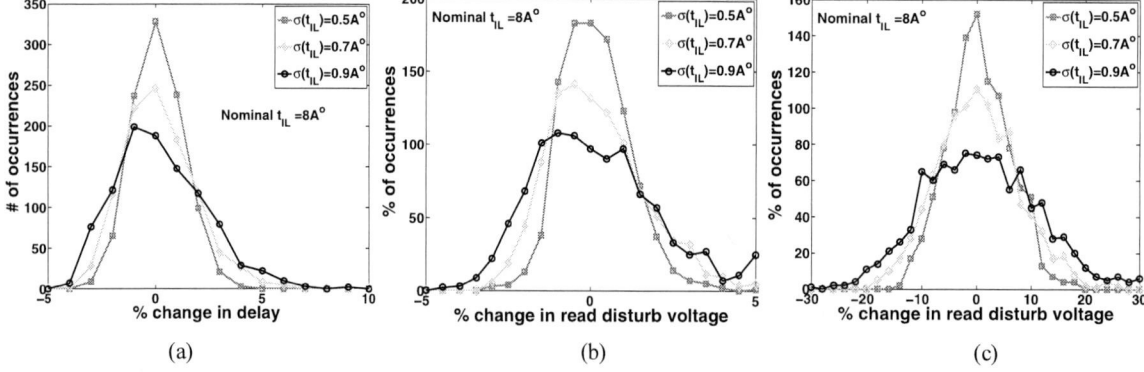

Fig. 9: Effect of random t_{IL} variation on (a) delay, and read disturb voltage considering (b) global, and (c) local random variation

High Speed A/D Converters

Session 12

Chair: Jennifer Lloyd, Analog Devices
Co-Chair: Takahiro Miki, Renesas

In this session several papers are presented which advance the state of the art in high-speed A/D conversion techniques. Included are pipelined, time-interleaved and flash conversion architectures, implemented in conventional CMOS (65nm to 0.18μm CMOS) processes. The design in last paper, which presents the highest speed converter in this session, is implemented in a SiGe BiCMOS process.

The first, invited paper from the University of Waterloo, presents a tutorial overview of time-interleaved A/D converters. The effects of timing errors and several sources of mismatch are discussed in detail, as well as a summary of current solutions and resulting state-of-the-art performance for this choice of converter architecture.

The second paper from National Semiconductor, presents a 12b 50 MSPS pipelined ADC which achieves excellent performance (0.27pJ/step FOM, 11.3 ENOB at 10 MHz input frequency). This design uses internal reference buffers, as well as op-amp sharing techniques and the exclusion of the S/H stage, to achieve low power dissipation.

The third paper from the University of California, Berkeley, UIUC, Bosch and Texas A&M, describes a 100 MSPS pipelined ADC with adaptive digital background calibration. The calibration technique improves the low-frequency SNDR and SFDR to reasonable levels allowing the use of power efficient analog components in the signal path.

The fourth paper from STM, India, describes an 11b 100 MSPS ADC based on a 2.5b pipeline followed by a 9b time-interleaved SAR converter. The architecture, described in detail, is aimed at providing lower power dissipation over a conventional pipelined ADC, and performance of 0.2 pJ/conversion step is demonstrated.

The fifth paper from Qualcomm and the University of Texas at Austin, describes a 10b 210 MSPS two-step ADC for a digital-IF receiver. This paper describes use of a capacitor network instead of a resistor ladder, as well as offset-canceling comparators, to achieve 52 dB SNDR at 100 MHz input frequency.

The final paper from Alcatel-Lucent, describes a 24 GSPS 5b flash ADC with a closed-loop THA for use in high throughput communications systems. The paper describes the architecture, as well as the implementation of the THA and comparator circuits, in detail. The converter is shown to provide 28 dB of SNDR at 1 GHz (sampled at 16 GSPS).

978-1-4244-2018-6/08 $25.00 © 2008 IEEE

Notes

IEEE 2008 Custom Intergrated Circuits Conference (CICC)

Time-Interleaved Analog-to-Digital Converters

David G. Nairn

Department of Electrical & Computer Engineering
University of Waterloo
Waterloo, Ontario
Canada N2L 3G1
nairn@uwaterloo.ca

Abstract-**This paper provides a tutorial review of time-interleaved analog-to-digital converters. After explaining the impact of offset, gain, timing and other mismatches on converter performance, current solutions to the mismatch problems are presented. The paper concludes with a summary of the current-state-of-the art for time-interleaved analog-to-digital converters.**

I. INTRODUCTION

Time-interleaved Analog-to-Digital Converters (ADCs) combine multiple sub-ADCs into a single ADC system that can achieve significantly faster sample rates than that of the sub-ADCs alone [1]. Historically, the primary drawback to time-interleaved ADCs, has been the need to tightly match the characterics of the sub-ADCs [2]. Recently, applications requiring the digitization of very wide-band signals, the availability of power efficient sub-ADCs and trimming techniques made possible by deep sub-micron Complementary Metal Oxide Silicon (CMOS) technologies, has lead to significant advances in the performance of time-interleaved ADCs.

Typical very wide-band applications include high-bandwidth oscilloscopes [3] and optical communication systems [4]. Each of these applications require signal bandwidths in the range of 10GHz. Unfortunately, currently available non-time-interleaved ADCs are incapable of sampling the input signals at the 20 to 24GSample/s rates that are required. Consequently it is necessary to use time-interleaved ADCs in these applications, simply to meet the required sample rates.

In other applications, the power dissipation of existing high-speed ADCs is too high for a practical system. In these cases, slow low-power ADCs can be time-interleaved to achieve the required sample rates with lower power dissipations [5]. A particularly attractive use for this approach is in ultra-wideband receivers, where it is desirable to embed the ADC into a single-chip system [6].

In addition, advances in digitally assisted analog circuits have greatly relieved the mismatch problem for time-interleaved ADCs. With the use of Digital Signal Processing (DSP), it is now possible to detect and correct the mismatches between sub-ADCs that have historically limited the resolution of time-interleaved ADCs [7-27]. Consequently, time-interleaved ADCs are now an attractive approach to implementing, high-speed/high-resolution, low-power ADCs.

To facilitate a better understanding of time-interleaved ADCs, this paper provides an introduction to time-interleaved

ADCs, their key limitations and current solutions to these limitations. In section II, the operation and terminology of time-interleaved ADCs is discussed along with an alternate approach to interleaving. Then in section III, the effects of mismatches between the sub-ADCs are reviewed. Section IV looks at current approaches for reducing and compensating the various mismatches between the sub-ADCs. In Section V, the current state-of-the-art is reviewed, followed by conclusions in section VI.

II. INTERLEAVING ADCS FOR HIGH-SPEED SAMPLING

Multiple sub-ADCs can be interleaved in either the time domain, as shown in Fig. 1 [2] or in the frequency domain, as shown in Fig. 2 [26]. While the approaches are similar, the time-interleaved approach is generally simpler to implement in current processing technologies.

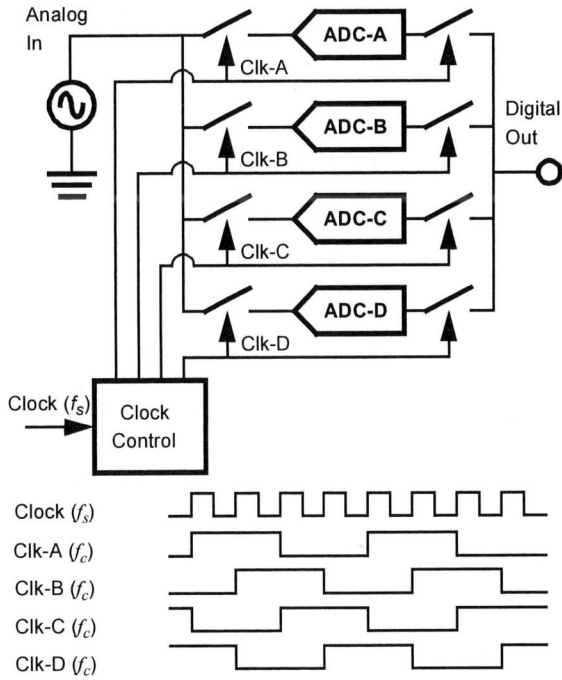

Figure 1) A time-interleaved ADC with $M = 4$.

When interleaved in the time domain, M sub-ADCs, each with a conversion rate of f_c are combined to achieve a system sampling rate of f_s. By appropriately shifting each sub-ADC's

978-1-4244-2018-6/08 $25.00 © 2008 IEEE

clock, as shown in Fig. 1, the analog input is effectively sampled at f_s which is given by:

$$f_s = M \cdot f_c. \tag{1}$$

M, the interleaving factor, directly indicates the speed improvement. When $M = 2$, time-interleaved ADCs are often called "ping-ponged" ADCs. Time-interleaved ADCs are also commonly referred to as parallel ADCs.[1] Note that if Nyquist rate performance is required, each sub-ADC's bandwidth must be at least $f_s/2$. In addition, as discussed in Section III, even for moderate resolutions, it is imperative that the sub-ADCs be well matched.

When interleaving in the frequency domain, each sub-ADC is preceded by a frequency selective filter, as shown in Fig. 2 [26]. In this case, each sub-ADC handles a much narrower band of frequencies. Nevertheless if the sub-ADCs are identical, they still require an input bandwidth of at least $f_s/2$, to achieve Nyquist rate performance. A attractive feature of this approach is that all of the sub-ADCs can be operated from the same low-speed clock. Unfortunately, to achieve good performance, the analog filters must be well matched in their transition bands.

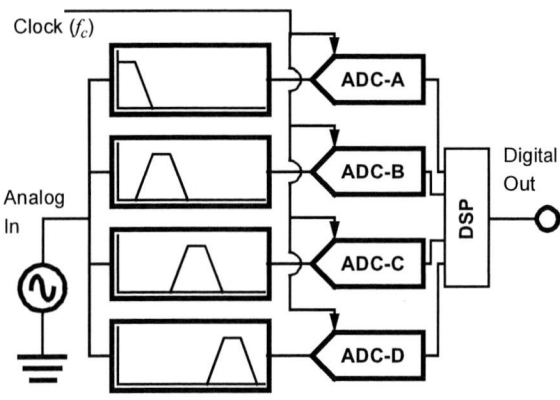

Figure 2) A band-splitting ADC.

Most current fabrication technologies are better suited to implementing digital circuits than to implementing well-matched analog circuits. In this design environment, it is preferable to minimize the analog circuitry and exploit digital circuits to enhance the performance of the remaining analog circuits. Consequently, the time-interleaved approach is generally preferred over the band-splitting approach for implementing high-speed ADCs.

III. Effects of Mismatches

In time-interleaved ADCs, each output sample comes from a single sub-ADC. Consequently, if the sub-ADCs are not perfectly matched, spurious tones appear in the interleaved output that are not present in the output of any of the sub-ADCs. Some mismatches, such as offsets and noise coupling produce

---

1. Unfortunately, flash ADCs are also often called parallel ADCs.

spurious tones in the interleaved ADC's output, even in the absence of an input signal. Other mismatches, such as gain mismatches and timing skews, only produce spurious tones when an input signal is present. In most untrimmed time-interleaved ADCs, the dominant spurious tones are due to offset, gain and timing skew mismatches and have been extensively analyzed by early researchers in the field [2], [7], [27]. Later researchers have also addressed the effects of bandwidth mismatches [28] and linearity mismatches [29]. Since each of these authors provides a good analysis of the mismatch effects, the purpose of the following discussion is to provide some understanding of the nature of the matching problem particularly for offset, gain and timing skew mismatches.

Offset mismatches

When two ADCs, with different offsets, V_{OS1} and V_{OS2}, are time-interleaved and there is no input signal, the output oscillates between the two offset levels, as shown in Fig. 3. The resulting square-wave has a frequency of $f_s/2$. The square wave's amplitude (and root mean squared (RMS) level) is given by

$$\Delta V_{OS} = \frac{|V_{OS1} - V_{OS2}|}{2} \tag{2}$$

To lower the offset tone's RMS level to the quantization noise level (*i.e* $0.289V_{LSB}$, where V_{LSB} is the least significant bit size), the offsets of the converters must match to about $V_{LSB}/4$.

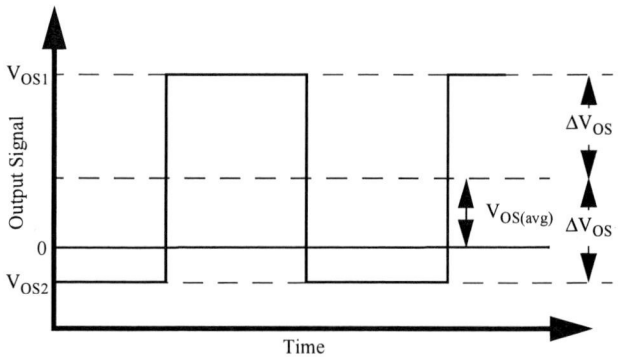

Figure 3) Output of two time-interleaved ADCs with different offsets and zero input ($f = f_s/2$, $V_{RMS} = \Delta V_{OS}$).

Often spurious tones at $f_s/2$ are not a problem. Unfortunately when M is greater than two, the spurious tones appear at $f = i f_s/M$, where $i = 0, 1, \dots M-1$. In this case, the tones due to offset mismatches appear in-band. The offset tones due to time-interleaving four ($M = 4$) ideal ADCs with random offsets between ± 1 LSB at the 8-bit level are illustrated in Fig. 4. Note that the tone at $f_s/4$ is only down 42dB from full scale. Consequently, to achieve a spurious free dynamic range (SFDR) at least equal to the ideal SNR, the sub-ADCs' offsets must match to about 2 bits better than the resolution of the sub-ADCs.

978-1-4244-2018-6/08 $25.00 © 2008 IEEE

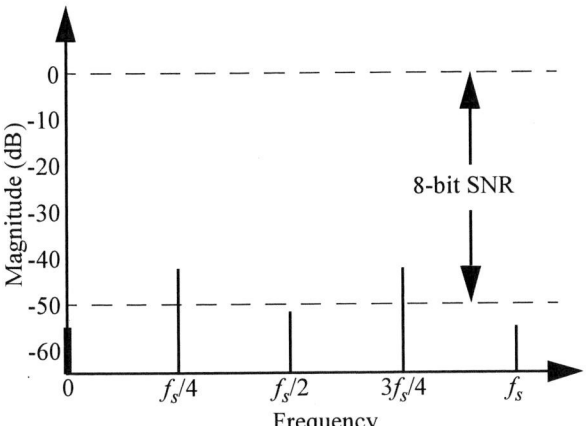

Figure 4) An FFT of a time-interleaved ADC's output with $M = 4$ and $V_{OS} \leq \pm 1/2^8$ Full Scale.

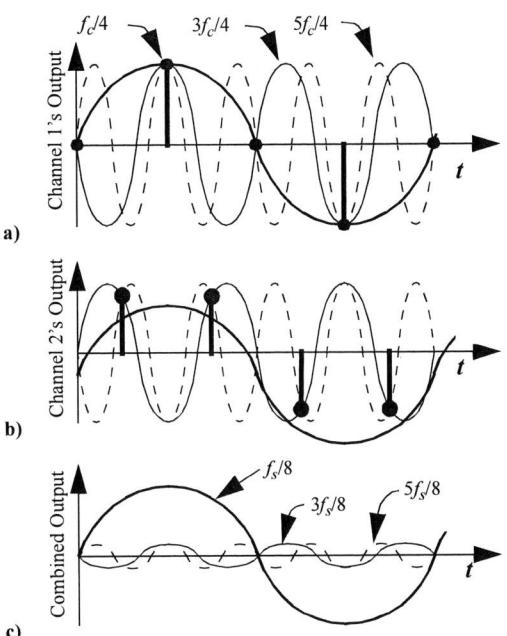

Figure 5) Effects of gain mismatches when $M = 1$, a) channel 1's output, b) channel 2's output, c) combined output.

Gain mismatches

The effects of a gain mismatch in a two channel ($M = 2$) time-interleaved ADC are illustrated in Fig. 5. The input to the system is a single tone at $f_s/8$, the two sub-ADCs have sample rates of $f_c = f_s/2$. So each sub-ADC sees an input at $f_c/4$, as shown in Fig. 5a and b. In the sampled time domain, these samples also represent sine-waves at $kf_c \pm f_{in}$ where k is any integer. The sine waves for $f_c - f_{in} = 3f_c/4$ and for $f_c + f_{in} = 5f_c/4$ are shown in Fig. 5. Note that the signal ($f_c/4$) is in phase between the two channels while the other components ($3f_c/4$ and $5f_c/4$) are $180°$ out of phase. In a single ADC these other signals are clearly outside the Nyquist band and can be neglected. When the two channels combine, the undesired spectral components cancel, if the two sub-ADCs have the same gain. If the gains do not match, imperfect cancellation results in residual tones at $kf_c \pm f_{in}$ or as they are more commonly expressed at $if_s/M \pm f_{in}$, where $i = 1, 2, \dots M - 1$. Note that the spurious tones are proportional to the amplitude of the input signal and to the relative mismatch. For a single tone input, the SFDR due to gain mismatches, in a two channel system, is given by:

$$SFDR_G = 20\log(\Delta A) \qquad (3)$$

Where ΔA is the gain mismatch. To obtain an SFDR equal to an ideal N-bit ADC's SNR, the sub-ADCs' gains must match to the N-bit level. This gain matching is twice the required matching of the MSB component in each sub-ADC [2].

Timing skews

The third dominant mismatch effect is due to sample time mismatches, these mismatches are often referred to as timing skews. Timing skews arise because the clock or signal delay to each sub-ADC's sampler differs from that of the other samplers. In the case of timing skews, the plots illustrated in Fig. 5 are modified such that the outputs from the two channels have the same amplitudes but different phase shifts. Consequently when the two channels are combined, there is incomplete cancellation of the undesired tones. The relative size of the tones depends on both the timing mismatches, ΔT and the frequency of the input signal, f_{in}. For a two channel system, the SFDR due to timing skews is given by;

$$SFDR_T = 20\log(2\pi f_{in}\Delta T) \qquad (4)$$

Similar to the gain mismatch tones, these tones appear at $if_s/M \pm f_{in}$, where $i = 1, 2, \dots M - 1$. A distinguishing feature of the timing skew tones is that they are frequency dependent because timing errors are more significant at higher frequencies, as illustrated in Fig. 6. To obtain an 8-bit dynamic range (50dB), at Nyquist for a 1GS/s time-interleaved ADC, ΔT must be less than 1ps.

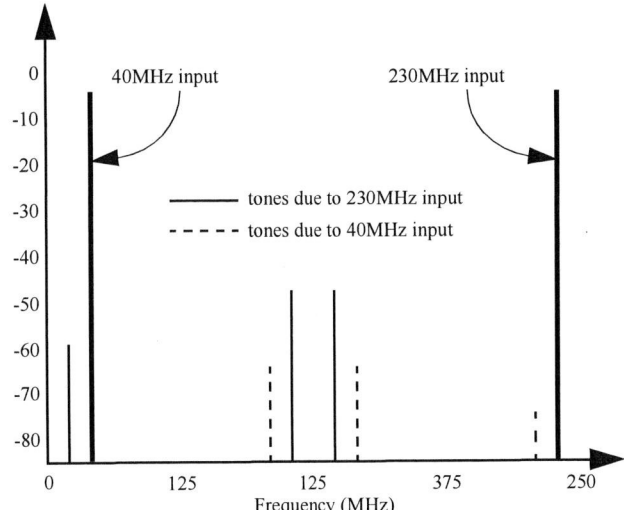

Figure 6) Timing skew spurious tones for a 500MSample/s interleaved ADC with $M = 4$ and $|\Delta T| <$ 10ps.

Other mismatches

Unlike offset, gain and timing skew mismatches most other mismatches are signal dependent. Bandwidth mismatches cause gain and timing skew mismatches that are frequency dependent [28]. To illustrate the problem, if the input to each sub-ADC of a 1GS/s ADC can be characterized as a first-order low-pass filter with a cutoff frequency of twice the Nyquist band $\pm5\%$, the resulting gain and timing skew mismatches, for $M = 2$, are as shown in Fig. 7. Note that the mismatches get worse at higher frequencies, such that the effective timing skew limits the SFDR to only 34dB. INL mismatches give rise to tones dependent on both frequency and signal size. Fortunately these tones are no worse than the spurious tones generated in a sub-ADC acting alone, but at different spectral locations [8]. As one goes to higher resolutions and speeds these additional sources of mismatch become more significant.

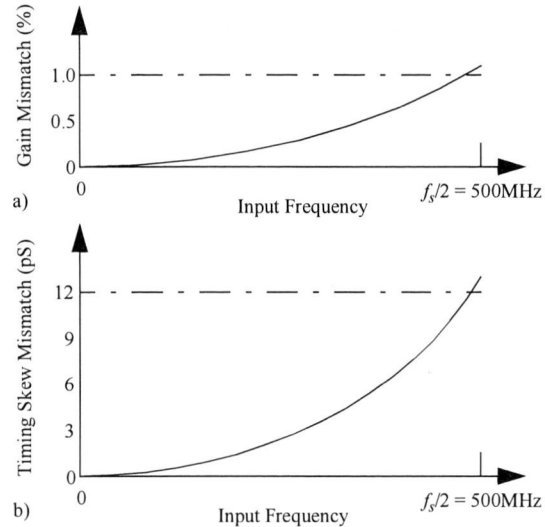

a)

b)

Figure 7) Mismatches for a 5% bandwidth mismatch a) Gain mismatch b) Timing skew.

Up to this point, the discussion has largely focused on determining the spurious tones when two sub-ADCs are time-interleaved. When M is greater than 2, the number of spurious components increases and their amplitudes tend to decrease. Nevertheless, the SNR degradation due to the spurious tones remains independent of M [27]. Consequently, the various formulae given above can still be used to estimate the total degradation due to mismatches and time-interleaving.

IV. DEALING WITH MISMATCHES

As discussed above, mismatches between the sub-ADCs lead to spurious tones a in time-interleaved ADC's output. For most systems, the offset gain and timing skew mismatches are the most significant. These mismatches are characterized by a constant or global mismatch for all signals, making them both relatively easy to detect and to correct. Other mismatches such as bandwidth differences and INL differences are both more difficult to characterize and to correct. Approaches for dealing with mismatches are discussed below.

Offset mismatch

Offset mismatches are one of the most significant sources of mismatch tones in time-interleaved ADCs. As a result, numerous circuit techniques and trimming methods have been developed to deal with offset mismatches.

Circuit techniques can be used to both reduce the sources of offsets and to reduce the magnitude of the spurious tones due to any remaining offset mismatches. One particularly attractive option for offset mismatch reduction is to share potential sources of error and thereby eliminate the mismatch. For example, if $M = 2$, the comparators [30] and the op amps [31] [18] can be shared in pipelined ADCs. Now when $M = 4$, the offset tones only appear at $f_s/2$, which is not a problem in many applications. Another common solution to offset problems in analog circuits is the use of chopping. Chopping effectively shifts the offset error to a higher, ideally out-of-band frequency. Unfortunately, if regular chopping is applied in time-interleaving applications, the offset is shifted to $f_c/2$ which only causes more problems. An alternative is to use a random chopping sequence [15]. In this case the offset energy is splattered fairly uniformly across the noise floor. Unfortunately, the increased noise may be undesirable and the offsets still need to be trimmed to achieve the desired level of performance [15].

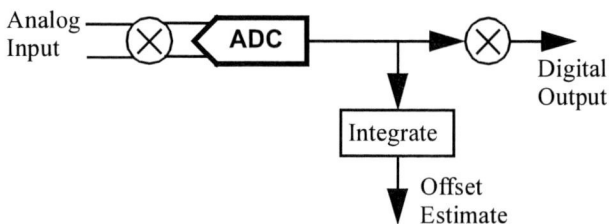

Figure 8) Random chopping within a sub-ADC channel.

Offset mismatches can be trimmed in a variety of ways. The first part of any trimming operation is to obtain an estimate of the offset. While a simple offset can be estimated by shorting the input and reading the digital output, this method is usually inappropriate for time-interleaved ADCs. As discussed in Section III, the offsets must match to better than the resolution of the sub-ADC. Provided the sub-ADC is sufficiently noisy, averaging a number of offset samples will yield sub-LSB offset estimates. Unfortunately, if the sub-ADCs have INL errors, it is not the offset at zero but the average offset of the sub-ADC's that is needed. The desired offset can be obtained during normal operation of the time-interleaved ADC, by averaging the output of the ADC provided there is no input signal at dc or that is correlated with the chopping frequency [11]. Fortunately, randomly chopping, as shown in Fig. 8 solves both of these problems [15]. The measured offset can then be applied to correct the offset at the input of the ADC (after chopping) or by simply subtracting the offset from the digital output (before un-chopping). The analog solution maximizes the sub-ADC's useful input range, while the digital solution is relatively trivial to implement. Note that the offset correction resolution must exceed the desired resolution of the ADC. Fortunately, these sub-LSB correction bits can be relatively noisy without significantly impacting the overall performance of the time-interleaved ADC. Consequently, offset mismatches are no longer a significant problem in time-interleaved ADCs

Gain mismatch

The second easiest mismatches to deal with are gain mismatches. Similar to offset mismatches, a variety of circuit techniques and trimming techniques have been developed to solve the gain mismatch problem.

Appropriate circuit techniques for avoiding gain mismatches depend on the architecture of the sub-ADCs. Since most ADCs' full-scale range, or gain, is set by the reference, a shared reference eliminates this source of mismatch [32]. Unfortunately, a shared reference provides cross talk between the sub-ADCs. For switched-capacitor based successive approximation ADCs (SARs) and pipelined ADCs, the capacitor arrays in the sub-ADCs should be laid out identically and oriented the same way [32]. For pipelined ADCs, another potential gain mismatch is the finite op amp gain. This source of gain mismatch can be avoided by increasing the op amp gain [32]. Unfortunately, since the op amp gain required for good matching in an interleaved ADC exceeds that required by the basic accuracy of the sub-ADCs, an increased gain may lead to a reduced speed for the sub-ADCs.

An alternative circuit technique to reducing spurious tones due to mismatches is to randomly interleave the sub-ADCs [33], [34]. As shown in Fig. 9, the use of a pseudo random, or some other sequence that avoids a repetitive use of the sub-ADCs effectively spreads the mismatch energy (from all sources of mismatch) over the ADC's Nyquist band. Alternatively, the sequence can be designed to spread the energy from the mismatch tones away from a particular band of interest. To achieve effective randomization, only one additional sub-ADC is required [34]. While randomization improves the SFDR, the overall Signal-to-Noise-plus-Distortion Ratio, SNDR (or SINAD), ideally remains constant. Consequently for high SNDR applications, it is still necessary to trim the sub-ADCs.

To trim the gain mismatch, it is first necessary to measure or estimate the gain mismatches. While the mismatches can be determined during manufacturing or during a calibration interval [3], due to parameter drift, it is preferable to estimate the mismatches during normal operation. If the characteristics of the input signal are known, this information can be used to greatly simplify the parameter estimation process [12]. Unfortunately, in most applications, the input signal's characteristics are unknown. In this case there are three options for estimating the mismatches; Add a calibration signal to the input [11], use a redundant ADC [13], perform a "blind estimation" of the signal [19]. When calibration signals are added, the input signal is filtered from the sub-ADCs' outputs and the calibration signal is extracted with correlation techniques. While good results can be achieved, the calibration signal typically reduces the allowable signal swing for the input signal. When redundant sub-ADCs are used, one sub-ADC is taken offline and calibrated at a time. If extreme care is not taken in this approach, additional

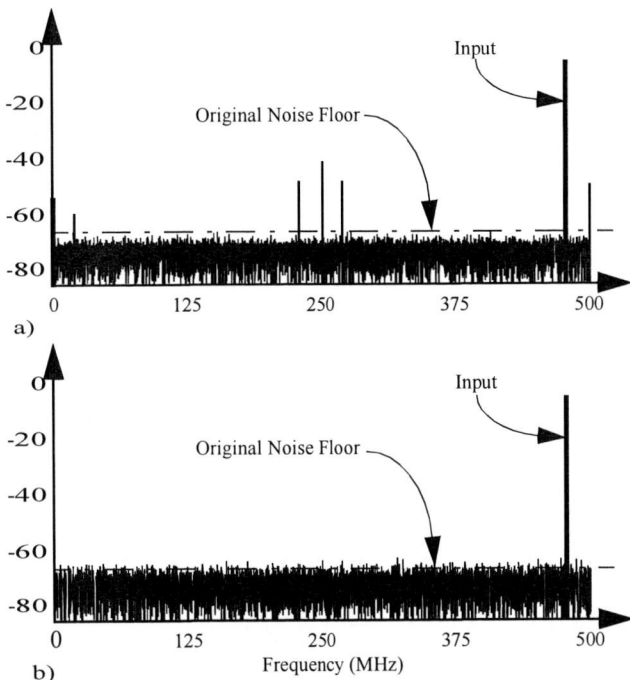

Figure 9) Different interleaving sequences; a) sequential, b) pseudo random.

spurious tones can be created due to changing loads on the system's analog input and the periodic nature of the calibration cycle. A "blind" gain mismatch can be estimated without knowledge of the input signal by comparing the relative power levels of the various channels. The power levels can be easily determined by summing the square of the signal in each channel [19]. Problems can arise if the input signal is correlated with f_c. To avoid this problem, a random interleaving sequence can be used to de-correlate the sub-ADC's sample rate from the input signal [21].

Once the gain mismatch has been estimated, the gain mismatches can be trimmed in either the analog domain or in the digital domain. The simplest way to correct the gain in the analog domain is to vary the reference. If one chooses to vary the reference, it will no longer be possible to share the same reference between all sub-ADCs. In the digital domain, the gain can be corrected with a single multiplier running at the channel rate, f_c. Similar to offset correction, the gain correction must be done at levels comparable to the resolution of the ADC. Consequently, the correction provided to the digital multiplier typically has two or three more bits than the desired resolution of the ADC [11].

Timing Skews

Of the three main sources of mismatch, timing skews have been the most difficult to estimate and generated the largest number of different solutions. Unlike the other mismatches, circuit techniques can reduce or even eliminate timing skew mismatches in some application. Nevertheless, many applications remain where it is necessary to trim or correct the timing skews.

The timing skew mismatches can be avoided by using a two-rank sample-and-hold (S/H) [35], as shown in Fig. 10. The input S/H runs at f_s while the sub-ADCs sample the S/H's output at f_c. This solution can be used when either a sufficiently fast process is available to implement the S/H [35], or if passive sampling is used to achieve sampling rates greater than would be possible with active samplers in the same technology [32]. Passive two-rank sampling can be implemented with either bottom plate sampling [36], [37], as shown in Fig. 11a, or with top-plate sampling [22], as shown in Fig. 11b. When bottom-plate sampling is used, the parasitic capacitance, C_P limits the SFDR improvement to about 15dB [37]. When top-plate sampling is used, it is essential to bootstrap switch S_1 to minimze charge injection and signal dependent clock-feed-through. In cases where a two-rank S/H solution can be used, it is often the best solution to the timing skew problem.

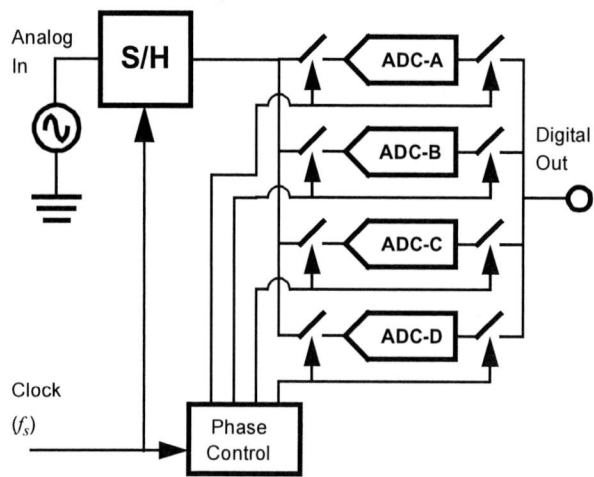

Figure 10) Eliminating timing skews with a two-rank S/H.

When separate input samplers are used, precise control of the sampling times are necessary to avoid timing skews. While delay-locked loops (DLLs) can be used to generate the sample clocks directly [32], it is generally much better to gate the master clock to avoid timing skews [31]. More recently, this latter method has been called clock-edge reassignment [14]. Using these techniques and careful layout, it is possible to reduce the timing skews to a few picoseconds which yields an SFDR around 50dB for a 100MHz input signal.

When it is necessary to trim or correct the timing skews, the first step is to obtain an estimate of the timing skews. A calibration cycle with sinusoidal test tones and the use of a discrete Fourier transform can be used to estimate the timing skews very accurately [7]. Alternatively, a calibration ramp can be added to the input signal, such that calibration can be performed during normal operation [16]. Unfortunately, these two approaches limit the usefulness of the time-interleaved ADC. A better approach is to estimate the timing skews directly from the input signal [17] [19] [25]. These algorithms are typically based on Fourier methods and correlation techniques. Conceptually, in the absence of offset and gain mismatches, timing

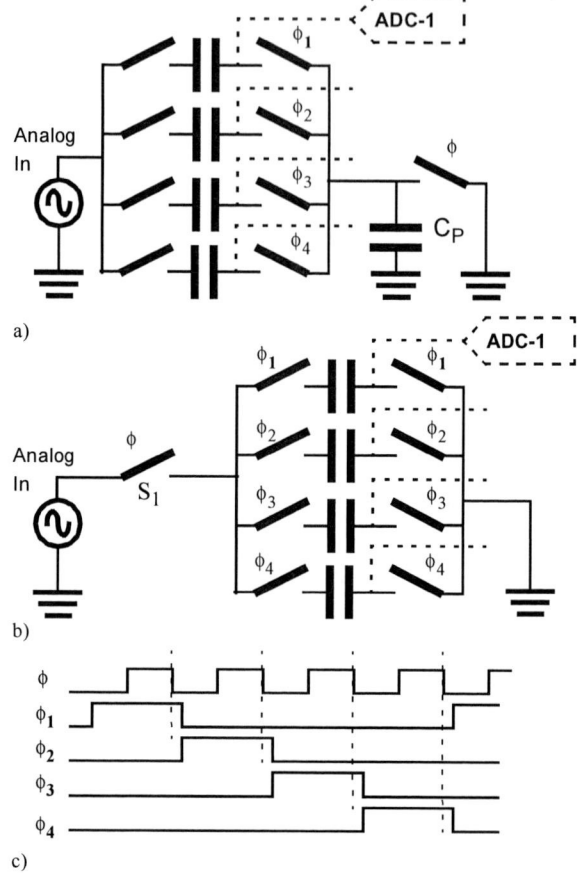

Figure 11) Passive two-rank S/H circuits, a) bottom-plate sampling b) top-plate sampling c) switch timing.

skews lead to phase differences between the sub-ADCs, as shown in Fig. 12. By squaring the difference between the two channels' outputs, a signal proportional to the timing skew can be generated [19]. Unfortunately, for robust timing skew estimates, high-speed multipliers are required.

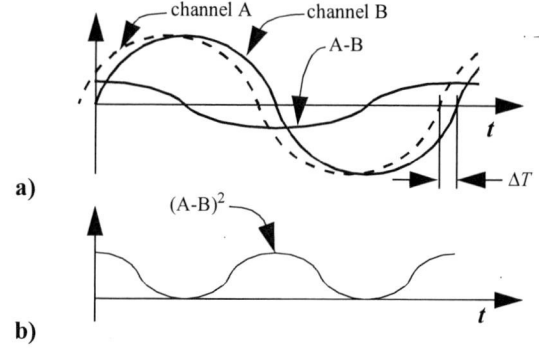

Figure 12) Detecting timing skews a) channel outputs b) squared difference between channel outputs.

Once the timing skew errors have been determined, they can be corrected either in the analog domain [10], which unfortunately significantly increases the jitter [25] or in the digital domain [9], [16], [19], [25]. In the digital domain, the timing skew correction problem can be though of as either a fractional delay problem or as an interpolation problem. In either case, the solutions require the use of a separate Finite Impulse Response (FIR) filter for each sub-ADC. These solutions require a significant amount of digital signal processing, which has now become a relatively simple matter to include on-chip due to the recent advances in digital CMOS processes.

Local mismatches

Once the global sources of sub-ADC mismatch are dealt with, other sources of mismatch become apparent. These other sources include bandwidth, INL and coupling mismatches. Since these mismatches depend on the input frequency and the signal strength, it is both more difficult to estimate the mismatches and to correct or compensate for them.

To date, the general approach to determining local mismatches has been to characterize the sub-ADCs over the complete range of interest (*i.e.* frequency, signal strength etc.) [8] [20]. Even when a simple first order bandwidth mismatch is being measured, in the presence of unknown gain and timing skews, the mismatches must be measured at two different frequencies to fully characterize the mismatches. For a practical time-interleaved ADC, 249 different test frequencies, spaced 0.8MHz apart where used to characterize the mismatches in a 400MS/s time-interleaved ADC [20]. The characterization data is then be used to model the channel mismatches.

Once the mismatches are characterized, appropriate filters [20], [23] are added after each sub-ADC to correct the mismatches. The filters are typically FIR filters, similar to those used to correct timing mismatches, with numerous taps. The array of filters is often referred to as a "filter bank". Unfortunately, at this time, it does not appear that techniques currently exist to continually monitor the long-term drift in the local mismatches and recalculate the coefficients for the filter bank in an efficient manner.

Summary

Solutions for detecting and correcting offset, gain and timing skew mismatches exist and can be readily implemented on-chip with the time-interleaved ADCs, using currently available CMOS processes. Signal dependent mismatches such as bandwidth and INL mismatches require significantly more computational overhead. Nevertheless, if the mismatches can be determined, digital filter banks can be used to provide the necessary corrections in modern CMOS processes.

V. CURRENT-STATE-OF-THE-ART

Time-interleaved ADCs have been reported in the available literature for almost thirty years. During this time, there have been significant increases in the sample rates and resolutions while the power dissipation has continued to decrease.

To illustrate the performance improvement, a Figure-Of-Merit (FOM) for ADCs has been plotted versus the year of publication in Fig. 13.[1] For the plot, the FOM is given by

$$FOM = Power/(2^{2 \cdot ENOB} \cdot f_s), \qquad (5)$$

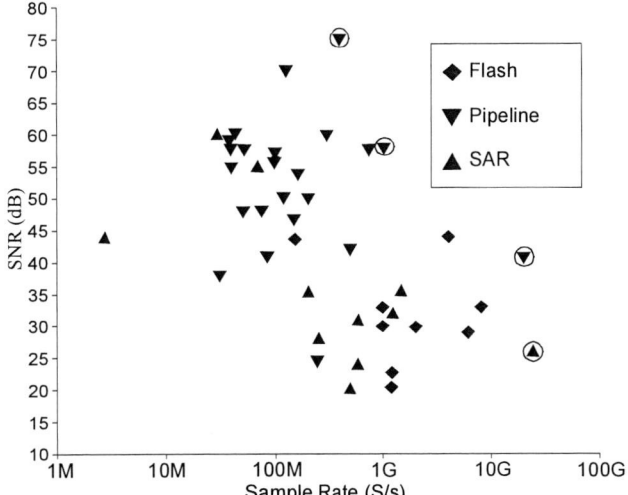

Figure 14) Reported SNR (or SNDR) *versus* sample rate.

where ENOB is the effective number of bits. The results indicate that the FOM has improved by more than an order of magnitude every decade. The chart also shows that while flash-based sub-ADCs may be attractive for achieving very high sample rates, they are not particularly efficient. Consequently there has been a renewed interest in the use for SAR ADCs to implement more efficient time-interleaved ADCs.

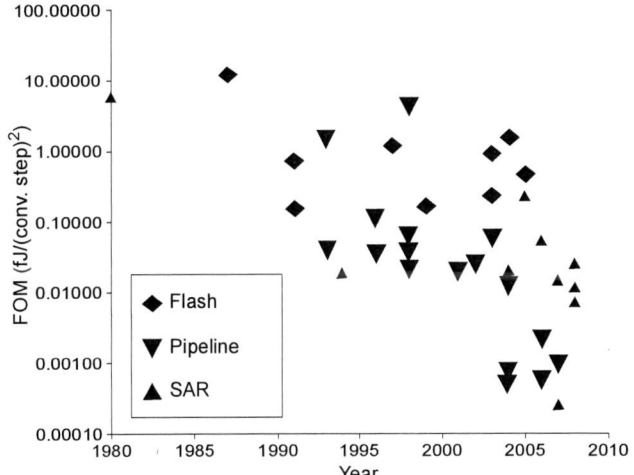

Figure 13) FOM *versus* Year.

Fig. 14 is a plot of the reported SNR *versus* sample rate. In the plot it can be seen that flash-based sub-ADCs tend to dominate the highest sample-rate time-interleaved ADCs. Nevertheless, SAR and pipelined ADCs appear to be competitive solutions at even the highest sample rates. To illustrate the leading edge, further details on the four circled data points are summarized in Table I below. Note that while the power levels seem high (*i.e.* measured in Watts), on a mW per MSample basis, they are all below 1mW per MSample, making them power competitive with lower speed ADCs.

---

1. Only ADCs based on sub-ADC architectures with more than two reported examples are plotted to avoid potentially misleading data points.

978-1-4244-2018-6/08 $25.00 © 2008 IEEE   317

TABLE I
RECENT REPORTED PERFORMANCE OF TIME-INTERLEVED ADCs

Sample Rate (f_s) (GS/s)	Resolution (N) (bits)	Power (W)	Reference
24	6	1.2	[4]
20	8	10	[3]
1	11	0.35	[22]
0.4	14	-	[20]

VI. CONCLUSIONS

Time-interleaving continues to be an attractive option for implementing high-speed ADCs. The performance of the earliest time-interleaved ADCs were limited by offset, gain and timing mismatches. Due to recent advances in both calibration techniques and digital CMOS processes, it is now possible to implement single-chip time-interleaved ADCs that self-calibrate the offset, gain and timing mismatches. These advances have lead to the development time-interleaved ADCs with sample rates of 1GS/s or more and resolutions at the 8 to 10-bit level. The remaining sources of mismatch, such as bandwidth mismatches, can be detected and corrected. Unfortunately, algorithms for continuous detection and correction of these remaining mismatches remains problematic. Nevertheless, if the recent progress in time-interleaved ADC performance is any guide, these remaining problems will also be solved in the near future.

REFERENCES

1. A.G.F. Dingwall, "Monolithic expandable 6-bit 20 MHz CMOS/SOS A/D converter," *IEEE J. Solid-State Circuits,* Vol. SC-14, pp. 926-932, Dec. 1979.
2. W.C. Black and D.A. Hodges, "Time interleaved converter arrays," *IEEE JSSC,* Vol. SC-15, No. 6, pp. 1022-1029, December 1980.
3. K. Poulton, R. Neff, B. Setterberg, B. Wuppermann, T. Kopley, R. Jewett, J. Pernillo, C. Tan and A. Montijo, "A 20GS/s 8b ADC with a 1MB Memory in 0.18um CMOS," *ISSCC,* pp. 18.1, February 2003.
4. P. Schvan, J. Bach, C. Falt, P. Flemke, R. Gibbons, Y. Greshischev, N. Ben-Hamida, D. Pollex, J. Sich, S.-C. Wang, J. Wolczanski, "A 24GS/s 6b ADC in 90nm CMOS," *ISSCC,* pp. 544-545, Feb. 2008.
5. D. Draxelmayr, "A 6b 600MHz, 10mW ADC Array in Digital 90nm CMOS," *ISSCC,* pp. 14.7, Feb. 2004.
6. B. Ginsburg and A. Chandrakasan, "Dual Scalable 500MS/s, 5b Time-Interleaved SAR ADCs for UWB," *IEEE CICC,* pp. 403-406, Sept. 2005.
7. Y-C. Jenq, "Digital Spectra of Nonuniformly Sampled Signals: A Robust Sampling Time Offset Estimation Algorithm for Ultra High-Speed Waveform Digitizers Using Interleaving," *IEEE Trans. Instrumentation and Measurement,* Vol. 39, No. 1, pp. 71-75, February 1990.
8. D.M. Hummels, J.J. McDonald and F.H. Irons, "Distortion compensation for time-interleaved analog-to-digital converters" *IEEE Inst. and Measurement Tech. Conf.,* pp 728-731, June 1996.
9. Y.C. Jenq, "Perfect Reconstruction of Digital Spectrum from non-Uniformly sampled signals" *IEEE Inst. and Measurement Tech. Conf.,* pp. 624-627, May 1997.
10. K. Poulton, K. Knudsen, J. Kerley, J. Kang, J. Tani, E. Cornish and M. VanGrouw, "An 8-GS/s 8-bit ADC System," *VLSI Conf.,* pp. 23-24, June 1997.
11. D. Fu, K. Dyer, S. Lewis and P. Hurst, "Digital Background Calibration of a 10b 40MSample/s Parallel Pipelined ADC," *ISSCC,* pp. 140-141, Feb. 1998.
12. T. Matsuura, T. Nara, T. Komatsu, E. Imaizumi, T. Matsutsuru, R. Horita, H. Katsu, S. Suzumura and K. Sato, "A 240-Mbps, 1-W CMOS EPRML Read-Channel LSI Chip Using an Interleaved Subranging Pipeline A/D Converter," *JSSC,* pp. 1840-1850, Nov. 1998.
13. K. Dyer, D. Fu, S. Lewis and P. Hurst, "An Analog Background Calibration Technique for Time-Interleaved Analog-to-Digital Converters," *JSSC,* pp. 1912-1919, Dec. 1998.
14. Y. Wang and B. Razavi, "An 8-bit 150MHz CMOS A/D Converter," *CICC,* pp. 7.3.1-7.3.4, May 1999.

15. J.K. Eklund and F. Gustafsson, "Digital Offset Compensation of Time-Interleaved ADC Using Random Chopper Sampling," *ISCAS,* pp. III-447-III-450, May 2000.
16. H. Jin and E.K.F. Lee, "A Digital-Background Calibration Technique for Minimizing Timing-Error Effects in Time-Interleaved ADCs," *TCAS-II,* pp. 603-613, July 2000.
17. J. Elbornsson and J.E. Eklund, "Blind Estimation of Timing Errors in Interleaved AD Converters," *Int'l Conf. on Accoustics, Speech and Signal Prosc.,* pp 3913-3916, June 2001.
18. L. Sumanen, M. Waltari and K. Halonen, "A 10-bit 200MS/s CMOS Parallel Pipeline A/D Converter," *JSSC,* pp. 1048-1055, July 2001.
19. S.M. Jamal, D. Fu, N.C-J. Chang, P.J. Hurst and S.H. Lewis, "A 10-b 120-Msample/s Time-Interleaved Analog-to-Digital Converter with Digital Background Calibration," *IEEE JSSC,* pp. 1618-1627, December 2002.
20. M. Seo, M.J.W. Rodwell and U. Madhow, "Comprehensive digital correction of mismatch errors for a 400-msamples/s 80-dB SFDR time-interleaved analog-to-digital converter," *IEEE Trans. on Microwave Theory and Techniques,* Vol. 53, Issue 3, Part 2, pp. 1072-1082, March 2005.
21. D.G. Nairn, "Signal conditioning system with adjustable gain and offset mismatches", United States Patent No. 6,900,750, May 31, 2005.
22. S. Gupta, M. Choi, M. Inerfield and J. Wang, "A 1GS/s 11b Time-Interleaved ADC in 0.13um CMOS," *ISSCC,* pp. 576-577. Feb. 2006.
23. T.-H. Tsai, P.J. Hurst and S.H. Lewis, "Bandwidth Mismatch and Its Correction in Time-Interleaved Analog-to-Digital Converters," *IEEE Trans. on Circuits and Systems II: Express Briefs,* Vol. 53, Issue 10, pp. 1133 - 1137, Oct. 2006.
24. C.C. Hsu, F.C. Huang, C.Y. Shih, C.C. Huang, Y.H. Lin, C.C. Lee and B. Razavi, "An 11b 800 MS/s Time-interleaved ADC with Digital Background Calibration," *ISSCC,* pp. 464-467, February 2007.
25. S. Huang and B.C. Levy, "Blind Estimation of Timing Offsets for Four-Channel Time-Interleaved ADCs," *IEEE Trans. Circuits and Systems Pt. I Regular papers,* Vol. 54, No. 4, pp. 863-876, April 2007.
26. G. Ding, C. Dehollain, M. Declercq and K. Azadet, "Frequency Interleaving Technique for High-Speed A/D Conversion," *ISCAS,* pp. I-857 - 860, May 2003.
27. A. Petraglia and S.K. Mitra, "Analysis of Mismatch Effects Among A/D Converters in a Time Interleaved Waveform Digitizer," *IEEE Trans. Instrumentation and Measurement,* Vol. 40, No. 5, pp. 831-835, Oct. 1991.
28. N. Kurosawa, H. Kobayashi, K. Maruyama, H. Sugawara and K. Kobayashi, "Explicit Analysis of Channel Mismatch Effects in Time-Interleaved ADC Systems", *TCAS-I,* pp. 261-271, March 2001.
29. J. Simoes, J. Landeck and C. Correia, "Nonlinearity of a Data-Acquisition System with Interleaving/Multiplexing," *IEEE Trans. Test and Measurement,* pp. 1274-1279, Dec. 1997.
30. M. Yotsuyanagi, T. Etoh and K. Hirata, "A 10-b 50-MHz Pipelined CMOS A/D Converter with S/H," *IEEE JSSC,* Vol. 28, No. 3, pp. 292-300, March 1993.
31. K. Nagaraj, H.S. Fetterman, J. Anidjar, S.H. Lewis, J. Alsayegh and R.G. Renninger, "A 250-mW, 8-b, 52-Msample/s Parallel-Pipelined A/D Converter with Reduced Number of Amplifiers," *IEEE JSSC,* pp. 312-320, March 1997.
32. C.S.G. Conroy, D.W. Cline and P.R. Gray, "An 8-b 85-MS/s Parallel Pipeline A/D Converter in 1-um CMOS," *IEEE JSSC,* Vol. 28, No. 4, pp. 447-454, April 1993.
33. H. Jin, E. Lee and M. Hassoun, "Time-Interleaved A/D Converter with Channel Randomization,", *IEEE ISCAS,* pp. 425-428, June 1997.
34. M. Tamba, A. Shimizu, H. Munakata and T. Komuro, "A Method to Improve SFDR with Random Interleaved Sampling Method," *ITC International Test Conf.* pp. 512-520, June 2001.
35. K. Poulton, J.J. Corcoran and T. Hornak, "A 1-GHz 6-bit ADC system," *IEEE JSSC,* Vol. Sc-22, No. 6, pp. 962-970, December 1987.
36. M. Waltari and K. Halonen, "Timing Skew Insensitive Switching for Double Sampled Circuits," *ISCAS,* pp. II-61-II-64, June 1999.
37. M. Gustavsson and N.N. Tan, "A Global Passive Sampling Technique for High-Speed Switched-Capacitor Time-Interleaved ADCs," *TCAS-II.* pp. 821-831, September 2000.

IEEE 2008 Custom Intergrated Circuits Conference (CICC)

A 12b 50MSPS 34mW Pipelined ADC

H. Yu, S. W. Chin, B. C. Wong

National Semiconductor

2900 Semiconductor Drive

Santa Clara, CA 95051 USA

Abstract- **A 12bit 50MSPS pipelined ADC is fabricated in 0.18µm CMOS process. Internal reference buffers without off-chip capacitors are implemented under 1.8V power supply voltage for 2Vp-p input signal swing. Opamp sharing and removal of explicit S/H stage are utilized for low power dissipation. Occupying 1.81x0.76 mm², ADC achieves SNR of 70.4 dBFS, SFDR of 86 dBFS and ENOB of 11.3b at Fin of 10MHz. It consumes 34mW with FOM of 0.27pJ/Step.**

I. INTRODUCTION

With the recent development in data conversion systems, higher speed and higher accuracy with lower power consumption are highly desired features for commercial portable devices. It is well known that trade-off exists between speed, accuracy and power consumption. Higher speed requires internal Opamps have wider bandwidth to settle to certain accuracy level, while accuracy imposes minimum requirement on Opamps' gain. Given a certain power budget, it is a deliberate job to satisfy both speed and accuracy goals at the same time. Many calibration techniques have been developed to relax the above trade-off and improve the power efficiency of the data converters, such as compensating low gain, low bandwidth and incomplete settling of Opamps, etc [1-5]. The disadvantage of calibration is complicated algorithm, additional digital circuitry and extra calibration cycles. This paper mainly utilizes the techniques of Opamp sharing and removal of explicit S/H stage to reduce power consumption [6-7]. In addition, extra comparators are included in the subADCs to lower residue voltage swing to improve linearity performance.

System portability also requires multiple ADCs to be integrated on the same chip and the removal of external components. Therefore a special fast-settling internal reference buffer without external bypass capacitor is designed in this work to handle 2Vp-p input signal swing. Since high input swing is adopted and no explicit S/H stage is present, the relaxed thermal noise requirement leads to smaller sampling capacitor sizes or smaller load capacitance for the Opamps, and this directly reduces the power consumption of the ADC. Eight low power ADCs are integrated on a single chip with an LVDS serializing interface for sending out digital data. Since no shared bypass capacitor for voltage reference is present, channel crosstalk is minimized. Other auxiliary blocks such as bandgap reference, self-tracking PLL, duty cycle correction and voltage boosters are implemented to achieve system performance, but we will focus on ADC design in this paper. Section II will give brief description of the ADC architecture, and section III will discuss the detailed circuit design. Then

the measurement results are presented in section IV and the conclusion is drawn in section V.

II. ACHITECTURE

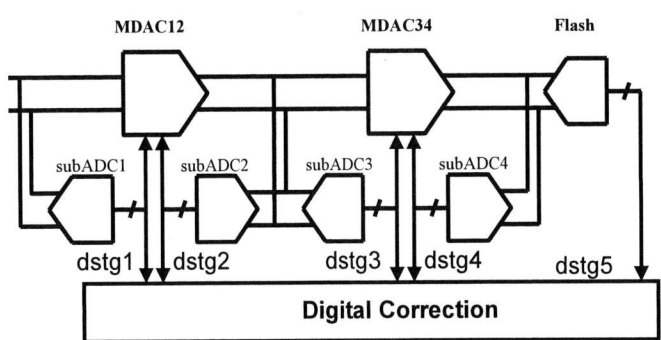

Fig. 1. 12-bit ADC architecture

The ADC architecture is shown in Fig. 1. There are totally five stages in this ADC, each one from the first to the fourth stage gives two effective bits and last flash ADC gives four effective bits. Removal of explicit S/H stage [6] and Opamp sharing [7] are utilized for low power dissipation. The first Opamp is shared by the first and the second MDAC stages, denoted as MDAC12 in the figure, and the second Opamp is shared by the third and fourth MDAC stages, denoted as MDAC34. Because of the memory effect of Opamp sharing, higher Opamp DC gain is required to reduce this effect, but the resulting additional power consumption is much lower than the power saved by sharing.

The choice of signal swing plays a critical role in power estimation and circuit topology selection. Under the drive of Moore's law, integrated circuit feature size and supply voltage decrease continuously and will follow this trend in the years coming. The lower supply voltage imposes many difficulties on the design of accurate analog circuitry. One important effect is that lower supply voltage limits the internal signal swing, which demands bigger capacitor size to maintain the dynamic range. Heavier capacitive loads are then added to the stages and higher power consumption is demanded from the Opamps to keep the same speed. Therefore, the input signal swing is better to be kept as high as possible for low power design. However, higher swing may drive the devices in the output stage of the amplifier to triode region, causing degradation in linearity. As a compromise, 2Vp-p is chosen for this 1.8V design. The choice of positive and negative reference voltages in ADC is also affected by the signal swing selection. Given a 10% design margin, the supply voltage may go down to 1.6V. If MDAC output common mode voltage is

978-1-4244-2018-6/08 $25.00 © 2008 IEEE   319

set at 0.8V, positive and negative reference voltages need to be 1.3V and 0.3V respectively, which makes the design of internal reference buffers a big challenge, especially when no external reference capacitors are used.

High internal signal swing can degrade linearity as discussed above. To have a better SFDR performance, extra comparators can be added to subADCs to give two extra bits for digital correction and hence reduce subADCs' quantization error. The resulting residue from MDACs is then reduced by half, which helps in keeping higher ADC linearity and tolerating sampling path mismatch in the first stage. The extra power consumption of these extra comparators is relatively low and can put under control when the subADCs' effective bits are around 1-2.

The design choice of capacitor sizes, Opamp noise specification are mainly derived from the ADC SNR performance. Opamp gain and bandwidth specifications can be derived from the residue voltage accuracy requirement of each stage. To have a more sophisticated approach, system level simulation is used to optimize power allocation. The effect of nonidealities such as finite Opamp gain and bandwidth, capacitor mismatch and thermal noises are analyzed and modeled mathematically. Matlab is used to simulate the ADC model and help to generate the required specifications for main circuit blocks and components, such as sampling capacitor sizes, comparator offsets, Opamp minimal gain and bandwidth etc. This model can give a very good estimation of the system power consumption.

III. CIRCUIT DESIGN

As discussed above, the internal reference buffer is a challenge for this design handling 2Vp-p input signal swing without external bypass capacitors. Fig. 2 shows the proposed schematic of a fast settling reference buffer. In this voltage reference buffer, transistors M1 and M2 share the same current to generate voltages Vrp and Vrn for high power efficiency. The gate voltages of M1 and M2 are biased by a replica circuit so that Vrp is set at 1.3V and Vrn is set at 0.3V. The quiescent output impedances at nodes Vrp and Vrn are 1/gm1 and 1/gm2 correspondingly. The resistor R1 is adjusted to establish a current Id1 as (Vrp-Vrn)/R1, where Id1 is directly related to the noise performance of the reference buffer. The current source topology is very effective in the dynamic charging of sampling capacitors in ADCs. In operation, the transient voltage at node Vrp will go down at the moment when the reference voltage buffer is switched to the sampling capacitors. At this time, the current going through transistor M1 will increase significantly, so that the equivalent output impedance drops to help bring Vrp up. The mechanism for negative reference voltage Vrn is the same as for voltage Vrp, except the swing direction is opposite. However, the gate voltagesVg1 and Vg2 may need to go beyond power supply voltages because of Vrp and Vrn voltage levels and the threshold voltages of M1 and M2. To overcome this problem, positive and negative boosted voltages Vdd2 and Vss2 are generated to supply minimum required currents to set up the desired voltages Vg1 and Vg2. The currents needed to

bias M5 or M6 are very small compared to the main current Id1. Consequently, the power cost to generate boosted voltages does not affect the overall power efficiency of the reference voltage buffer.

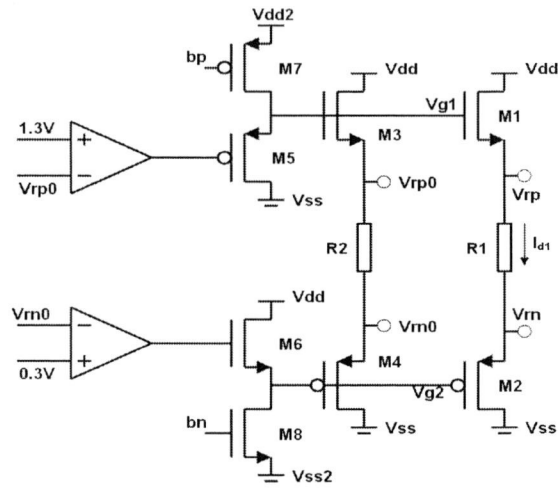

Fig. 2. Reference buffer schematic.

Another effort towards low power consumption is to remove the explicit S/H stage [6]. The detailed sampling circuitry topology is shown in Fig. 3. Both MDAC path and subADC1 should have similar sampling networks. Although the circuitry is drawn in a single-end mode, the real circuit is implemented in a fully differential mode.

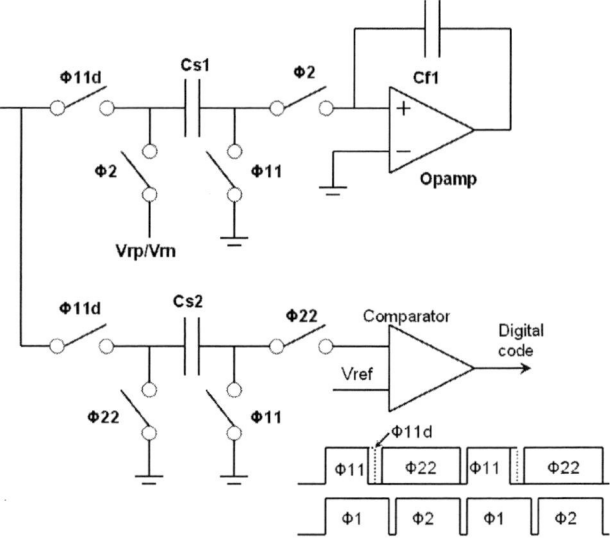

Fig. 3 ADC sampling paths and timing.

The Φ11 falling edge will sample the input signal and the time between the falling edges of Φ11d and Φ1 is reserved for subADC to digitize the input signal. The pulse width ratio of Φ11 to Φ1 is set to be 2:3. However, the mismatch between the MDAC sampling path and the subADC path can cause a sampling timing difference, which can be regarded as extra offsets of the comparators in subADC. When input signal frequency goes higher, the equivalent extra offset becomes

978-1-4244-2018-6/08 $25.00 © 2008 IEEE

bigger and it can cause the MDAC1 output residue voltage to have wider swing range, degrading SFDR. If a sinusoid signal is applied, and sampling path delay mismatch is Δt, the maximum extra residue range would be

$$G \times 2\pi \cdot f_{in} \cdot \Delta t$$

Where G is the MDAC gain, fin is the input frequency. For example, if 10MHz 1Vp-p sinusoid signal is applied and MDAC gain is 4, the extra residue voltage range is 25mV. The disadvantage of removing explicit S/H stage is that it could reduce the ADC's full power bandwidth (FPBW). To alleviate this effect, the time constants of the two signal paths are matched with each other in design and layout. For the situations where input signal frequency is under 100MHz, this is still a good approach for lower power dissipation. The sizes of switches and capacitors in the subADC1 path can be scaled down to keep extra loading capacitance ignorable for preceding driving circuits.

A two-stage operational amplifier is designed for MDAC12, illustrated in Fig. 4. The first stage is a telescopic topology which is a power efficient structure for a specific noise requirement. The first stage output swing is small, so the cascade devices are almost free in terms of power consumption to give more gain. The second stage is a simple differential stage to provide high output swing. The compensation capacitors are connected from the output nodes to the sources of the NMOS cascode devices instead of the first stage output nodes, which eliminate the feed forward path and the zero at the right hand plane for better phase margin. PMOS devices are used as input devices for their good flicker noise performance. Although not shown in Fig. 4, common mode feedback circuitry is used to establish stable common mode voltages at the first and second stages' output nodes. In the layout, circuit symmetry is a key point and the symmetry helps to reduce the second order harmonic, which is important for this application.

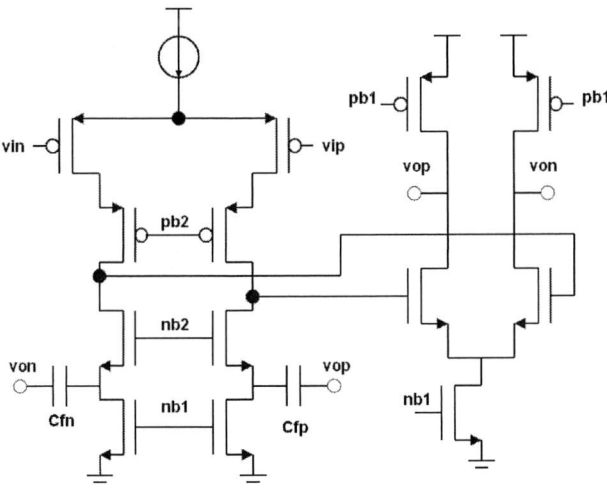

Fig. 4. Two stage Opamp for MDAC.

An over-range protection scheme is also implemented in the input and output of MDAC12 to make sure the circuit

recovers from input overdrive in one clock cycle. SubADC1 and subADC2 can trigger this protection operation when detecting overdriving situations.

IV. MEASUREMENT

Fig. 5 shows a measured 32768 point FFT of ADC output at nominal condition, where the sampling frequency Fs is 50MHz and input signal frequency Fin is 10MHz. The measured ENOB is 11.3bit, SNR is 70.4dBFS and SFDR is 86.0dBFS.

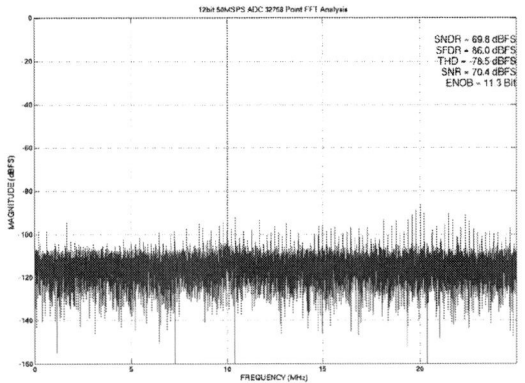

Fig. 5. FFT plot of ADC output at Fin=10M, Fs=50M.

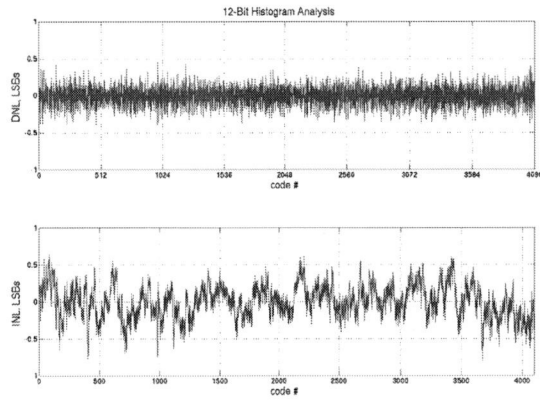

Fig. 6. DNL and INL plot of measured ADC output at Fin=10M, Fs=50M.

Shown in Fig. 6, the measured DNL is ±0.4LSB and INL is -0.8/+0.6LSB. ADC dynamic performance is measured as the function of input frequency and sampling frequency. Fig. 7 shows that with the help of sampling path matching, the full power bandwidth can go to 100MHz. When the input signal frequency goes up further, the 3rd order harmonic becomes dominant, which is consistent with the expectation that sampling mismatch can cause the MDAC to have higher residue voltage range, degrading harmonic performance. Fig. 8 is the measured dynamic performance when sweeping sampling frequency at a fixed input frequency of 10MHz.

978-1-4244-2018-6/08 $25.00 © 2008 IEEE

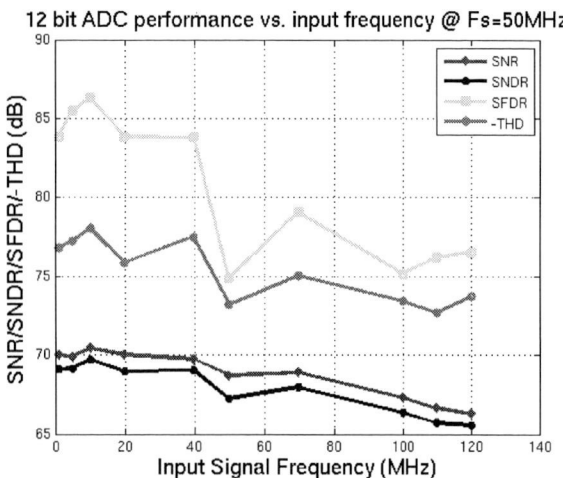

Fig. 7. ADC dynamic performance vs. input frequency.

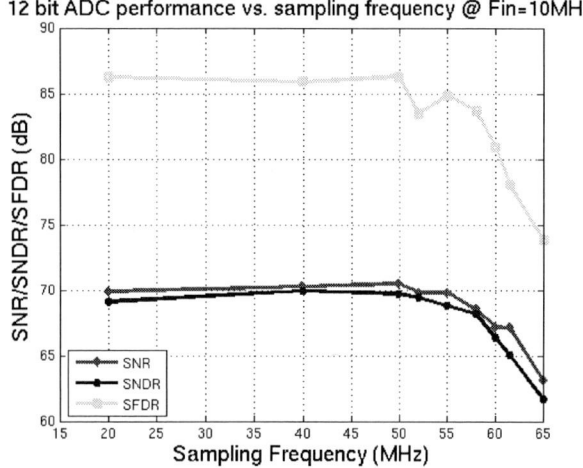

Fig. 8. ADC dynamic performance vs. sampling frequency.

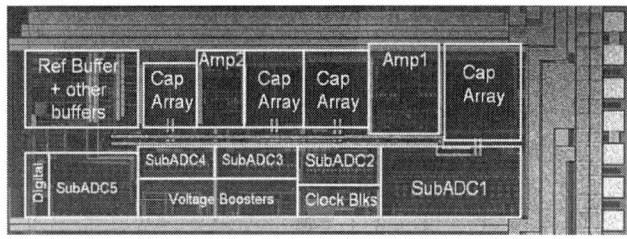

Fig. 9. One ADC 's die micrograph.

Measured power consumption is 34mW for sampling frequency Fs of 50MHz and input signal frequency Fin of 10MHz, leading to FOM (Power/(Fs*2^{ENOB})) of 0.27pJ/Conv step. The prototype chip is fabricated in a 2P6M 0.18μm CMOS process and Fig. 9 shows a micrograph of one ADC, where the active area of the ADC is 1.81x0.76 mm^2.

The circuit performance is summarized in Table I.

TABLE I
ADC PERFORMANCE SUMMARY

Process	0.18μm 2P6M 1.8V CMOS
Resolution/Sampling rate	12bit / 50 MSPS
Input Signal Range	2.0Vpp
SNR/SFDR	70.4dBFS
SFDR	86.0dBFS
THD	-78.5dBFS
SNDR/ENOB	69.8dB / 11.3bit
DNL	±0.4 LSB
INL	-0.8LSB ~ +0.6 LSB
Power Consumption	34.0mW
FOM1 [Power/(Fs*2^{ENOB})]	0.27 pJ/conv step
FOM2 [Power/(ERBW*2^{ENOB})]	0.13pJ/conv step
Area	1.81 x 0.76 mm^2

V. CONCLUSION

This paper presents a 12bit 50MSPS pipelined ADC for low power applications. Opamp sharing and removal of explicit S/H stage have been utilized for lower power consumption. More extra comparators are implemented in subADCs to lower the residue voltage swing in each stage, and this improves linearity performance and lower the memory effect caused by Opamp sharing. The main sampling path in the first MDAC is matched to the first subADC sampling path for higher FPBW. The prototype is fabricated in 0.18um 2P6M CMOS process. The measurement results show SNR of 70.4dBFS, SFDR of 86 dBFS, SNDR of 69.8 dBFS at the input frequency of 10MHz. The differential nonlinearity is ±0.4LSB and integral nonlinearity is -0.8~0.6LSB. The FPBW is measured to be 100MHz. The ADC consumes 34mW with FOM of 0.27pJ/Step.

ACKNOWLEDGMENTS

The authors gratefully acknowledge the support of Eric, Eugene and Khoa for their timely support for this work. We would also like to thank Don and George for reviewing ESD, and the layout team for productive collaborations.

REFERENCES

[1] E. Iroaga, B. Murmann, "A 12-Bit 75-MS/s Pipelined ADC Using Incomplete Settling" IEEE JSSC, pp. 748-756, Apr., 2007.
[2] C. R. Grace, P. J. Hurst and S. H. Lewis, "A 12-bit 80-Msample/s pipelined ADC with bootstrapped digital calibration", IEEE JSSC, vol. 40, no. 5, pp1038-1047, May 2005.
[3] D. Y. Chang, G. C. Ahn, U. K. Moon, "Sub-1V design techniques for high-linearity multistage/pipelined analog-to-digital converters", IEEE Transactions on Circuits and Systems-I, vol. 52, no. 1, pp 1-12, January 2005.
[4] C. R. Grace, P. J. Hurst and S. H. Lewis, "A 12-bit 80-Msample/s pipelined ADC with bootstrapped digital calibration", IEEE JSSC, vol. 40, no. 5, pp1038-1047, May 2005.
[5] B. Murmann, B. E. Boser, "Digitally Assisted Pipeline ADCs", Kluwer Academic Publishers, May 2004.
[6] Y. D. Jeon, S.-C. Lee, et al., "A 4.7mW 0.32 mm^2 10b 30MS/s Pipelined ADC without a Front-End S/H in 90nm CMOS", ISSCC Dig. Tech. Papers, pp. 456-457, Feb., 2007.
[7] K. Nagaraj, et al., "A 250-mW, 8-b, 52MS/s Parallel-Pipelined A/D Converter with Reduced Number of Amplifiers", IEEE JSSCC, pp. 312-320, Vol.32, No. 3, Mar., 1997.

IEEE 2008 Custom Intergrated Circuits Conference (CICC)

Background ADC Calibration in Digital Domain

Cheongyuen Tsang[1], Yun Chiu[2], Johan Vanderhaegen[3], Sebastian Hoyos[4], Charles Chen[1],
Robert Brodersen[1], Borivoje Nikolić[1]

[1]University of California, Berkeley, CA 94704
[2]University of Illinois at Urbana-Champaign,Urbana, IL 61801
[3]The Bosch Research and Technology Center, Palo Alto, CA 94304
[4]Texas A&M University College Station, TX 77843

Abstract— **A 100MS/s pipelined ADC is digitally calibrated by a slow $\Sigma\Delta$ ADC using a least-mean-square (LMS) algorithm. Both linear and nonlinear memoryless residue gain errors of the pipeline stages are adaptively corrected. With a 411kHz sinusoidal input, the peak SNDR improves from 28dB to 59dB and the SFDR improves from 29dB to 68dB. The complete 0.13μm ADC SoC occupies a die size of 3.7mm×4.7mm, and consumes a total power of 448mW.**

I. INTRODUCTION

Recent digital CMOS technology advancement has spurred a flurry of research activities in seeking digital adaptive techniques to address the inherent problem of analog impairments in a genre of mixed-mode and RF integrated circuits. In the context of pipelined ADCs, such approaches often result in relaxed matching and gain requirements of the switched-capacitor residue gain stages [1]–[4]; and in return, power efficiency and/or conversion speed can be improved. The latest effort in this regime seems to have been focusing on various nonlinear calibration schemes, as revealed by the recently reported works in the literature [5]–[7].

The general approaches of the reported calibration techniques thus far can be categorized into mainly three types: the statistics-based approach [5], the correlation-based approach [1]–[3], and the equalization-based ones [8], [9]. The former two share a common drawback of long calibration times, and thus are not suitable for time-variant operations [10]. In this work, we present a fast adaptive digital nonlinear calibration technique that is an extension of the work reported in [9]. We show that, with the aid of a slow $\Sigma\Delta$ ADC, the nonlinear residue transfer function of a pipeline stage can be fully reversed in a backend digital filter, enabling the use of simple cascode inverters as the residue amplifiers in the ADC. The resulting simplicity of the analog circuits potentially leads to a much relaxed design tradeoff between the circuit speed and accuracy, one of the most difficult aspects of analog design.

II. CODE DOMAIN FILTERING APPROACH

A detailed mathematical formulation of the proposed calibration technique is described in this section. First, we express the nonlinear transfer function of a 1.5-b/stage residue amplifier in the (digital) code domain as an FIR filter. It follows that the digital equivalence of the input analog voltage can be represented as a function of the output code from each pipeline stage and some circuit parameters related to matching, op-amp

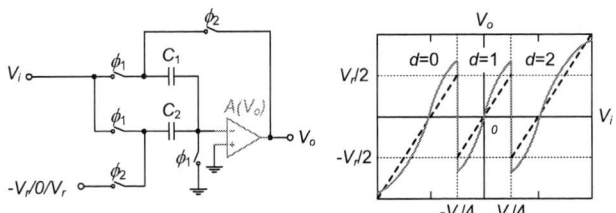

Fig. 1. Residue amplifier of a 1.5-b/stage pipelined ADC and its ideal (dashed line) and nonideal (solid line) voltage transfer function.

gain, and offset. We will then demonstrate that a properly designed nonlinear digital filter is sufficient to correct the analog errors using a known reference signal (obtained from the $\Sigma\Delta$ ADC).

A. Code Domain Formulation of 1.5-b Pipeline Stage

Fig. 1 shows the residue amplifier of a typical switched-capacitor 1.5-b pipeline ADC stage and its voltage transfer curve. The residue voltage can be expressed as [9]

$$V_o = \frac{V_i(C_1 + C_2) - (d-1)V_r C_2 + V_{os}(C_1 + C_2 + C_x)}{C_1(1 + \frac{C_1 + C_2 + C_x}{A(V_o)C_1})},$$
(1)

where, V_r is the reference voltage, C_x is the summing-node parasitic capacitance, V_{os} is the lumped offset voltage, $A(V_o)$ is the signal-dependent op-amp gain, and d is the digital decision of the current stage that assumes a value of 0, 1, or 2. The term $\frac{A(V_o)C_1}{C_1+C_2+C_x}$ is the loop gain of the residue amplifier. After dividing both sides of (1) by V_r, a digital representation is obtained,

$$D_i(\frac{C_1 + C_2}{C_1}) = D_o(1 + \frac{C_1 + C_2 + C_x}{C_1}\frac{1}{A(D_o)})$$
$$+ (d-1)(\frac{C_2}{C_1}) - D_{os}(\frac{C_1 + C_2 + C_x}{C_1}),$$
(2)

where, $D_i = \frac{V_i}{V_r}$, $D_o = \frac{V_o}{V_r}$, and $D_{os} = \frac{V_{os}}{V_r}$. Equivalently, (2) can be approximated by a power series expansion [11]:

$$D_i = D_o\alpha_1 + D_o^2\alpha_2 + D_o^3\alpha_3 + D_o^4\alpha_4 + ...$$
$$+ (d-1)\beta - D_{os}\gamma,$$
(3)

where, $\alpha_k = f_k(C_1, C_2, C_x, A(D_o))$, $\beta = (\frac{C_2}{C_1+C_2})$, $\gamma = (\frac{C_1+C_2+C_x}{C_1+C_2})$. Note that the truncated version of (3) resembles

978-1-4244-2018-6/08 $25.00 © 2008 IEEE   323

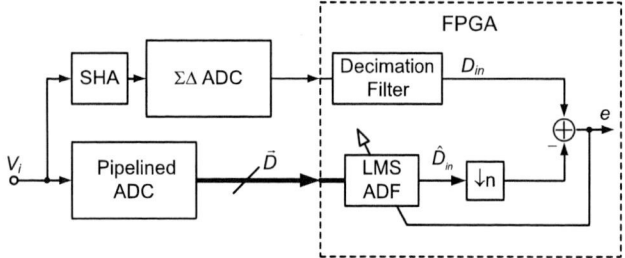

Fig. 2. Nonlinear adaptive digital calibration of pipelined ADC.

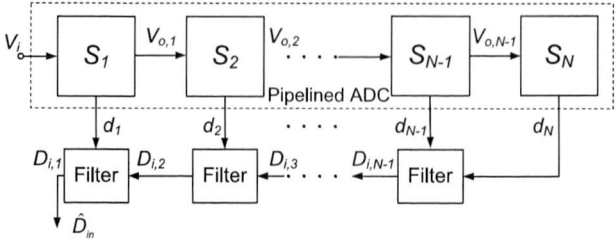

Fig. 3. Reverse pipelining of the multistage adaptive digital filter.

an FIR filter with coefficients α_k, β, and γ. For an ideal pipeline stage, the values of α_1 and β are 0.5, the values of $\alpha_2 - \alpha_k$ are 0, and the value of γ is 1. In practice, these coefficients are generally unknown due to PVT (process, voltage, and temperature) variations.

B. Code Domain Multistage Formulation

After deriving the input-output (D_i-D_o) relationship of a single-stage residue amplifier, we now extend this treatment to a complete multistage pipelined ADC. Using (3) and again assuming a 1.5-b/stage architecture, we arrive at the following expressions for D_i's from all stages [9]:

$$
\begin{aligned}
D_{i,1} =& D_{o,1}\alpha_{1,1} + D_{o,1}^2\alpha_{1,2} + D_{o,1}^3\alpha_{1,3} + \dots \\
& + (d_1 - 1)\beta_1 - D_{os,1}\gamma_1 \\
D_{i,2} =& D_{o,2}\alpha_{2,1} + D_{o,2}^2\alpha_{2,2} + D_{o,2}^3\alpha_{2,3} + \dots \\
& + (d_2 - 1)\beta_2 - D_{os,2}\gamma_2 \\
D_{i,3} =& D_{o,3}\alpha_{3,1} + D_{o,3}^2\alpha_{3,2} + D_{o,3}^3\alpha_{3,3} + \dots \\
& + (d_3 - 1)\beta_3 - D_{os,3}\gamma_3 \\
&\vdots \\
D_{i,N} =& D_{o,N}\alpha_{N,1} + D_{o,N}^3\alpha_{N,2} + D_{o,N}^3\alpha_{N,3} + \\
& (d_N - 1)\beta_N - D_{os,N}\gamma_N.
\end{aligned}
\tag{4}
$$

Since $D_{o,N} = D_{i,N+1}$, a digital representation of the input signal (D_{in}) can be obtained from the above equations (recursively):

$$
\begin{aligned}
D_{in} = D_{i,1} =& D_{i,2}\alpha_{1,1} + D_{i,2}^2\alpha_{1,2} + D_{i,2}^3\alpha_{1,3} + \dots \\
& + (d_1 - 1)\beta_1 - D_{os,1}\gamma_1.
\end{aligned}
\tag{5}
$$

Eq. (5) formulates a nonlinear FIR filter in the code domain. In order to obtain D_{in}, we need to find the coefficients β_i, γ_i, and $\alpha_{i,k}$. Note that the treatment in (5) can be generalized to an x.5-b/stage pipeline architecture.

C. Calibration Algorithm and System Architecture

As the filter tap values of (5) are unknown a priori, adaptive learning can be applied to obtain them on the fly. This leads to a background calibration approach using a gradient-descent algorithm. As shown in Fig. 2, the output digital codes ($\vec{D}$) from all pipeline stages of an inaccurate, high-speed pipelined ADC is decimated and applied to an LMS adaptive digital filter (ADF); while a slow-and-accurate $\Sigma\Delta$ converter is used

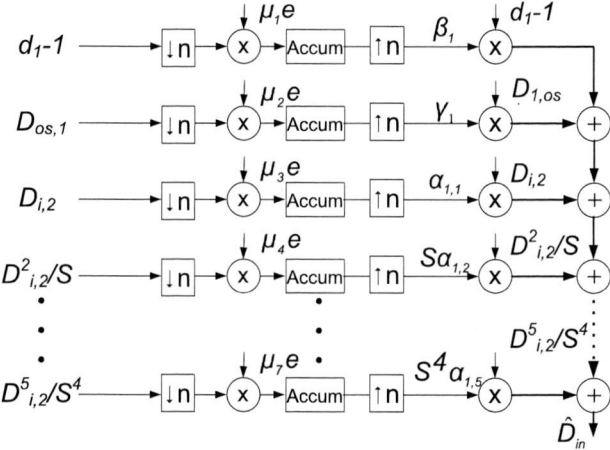

Fig. 4. Simplified coefficient update diagram for the first pipeline stage. Parameter S is a constant scaling factor.

to obtain an accurate version of D_{in} to compare with those procured from the pipeline. An LMS algorithm is used to update the filter coefficients at the sample rate of the $\Sigma\Delta$ converter. The recursive formulation of the adaptive digital filter as explained before is implemented as a reverse pipeline (Fig. 3). The detailed coefficient adaptation scheme is shown in Fig. 4 for the first stage. Note that a sampling clock skew may exist at the front-end between the SHA of the $\Sigma\Delta$ ADC and the pipelined ADC. As long as the resulting sampling error exhibits a symmetrical probability distribution in the long term, the effect is removed by the accumulator of the LMS loop.

III. CMOS PROTOTYPE DESIGN

In the prototype, the reference ADC consists of a SHA followed by a 2-1 MASH $\Sigma\Delta$ ADC [12]. A traditional flip-over switched-capacitor S/H architecture with a two-stage amplifier is used. As the calibration performance tends to be limited by the linearity of the reference path instead of noise, extra attention was paid in the SHA and $\Sigma\Delta$ design to ensure an almost 14-bit accuracy. The clock frequencies of the SHA and the $\Sigma\Delta$ ADC are $1/2^{14}$ and $1/2^4$ of that of the pipelined ADC, respectively. The sample rate of the SHA determines the update frequency of the ADF and its convergence time.

The pipelined ADC uses a SHA-less 1.5-b front-end stage, followed by 2.5-b stages from stage 2 to 6, and ending in a 3-

978-1-4244-2018-6/08 $25.00 © 2008 IEEE

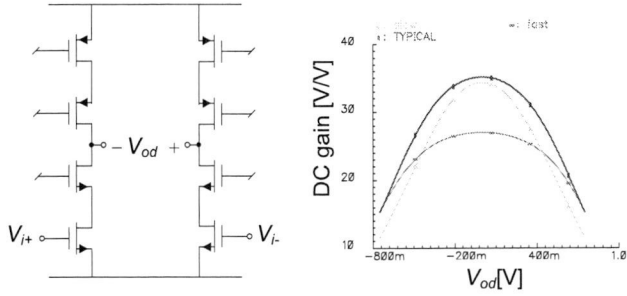

Fig. 5. Second-stage residue amplifier and its gain curves.

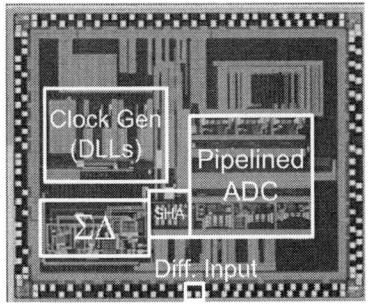

Fig. 6. Die photo of the prototype ADC (3.7mm×4.7mm).

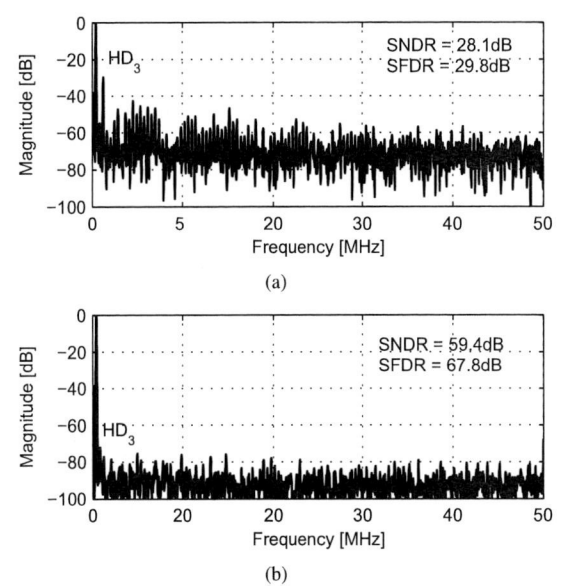

(a)

(b)

Fig. 7. Measured ADC spectra for f_s=100MHz, V_{in}=0dBFS, f_{in}=411kHz: (a) before calibration, (b) after calibration.

TABLE I
EXPERIMENTAL RESULTS

	Before cal.	After cal.
Process	0.13μm digital CMOS	
Supplies: Analog/Digital	1.35V/1.2V	
Full-scale Input(pk-pk)	±0.5V	
Sampling Rate	100MHz	
Chip Area	3.7mm×4.7mm	
$\Sigma\Delta$ Power	10mW	
SHA + Pip. ADC Power	332mW	
Clock Gen. Power (DLLs+Buffer)	106mW	
Total Power	448mW	
SNDR(411kHz)	28.1dB	59.4dB
SNDR(49MHz)	29.4dB	54.8dB
SFDR(411kHz)	29.8dB	67.8dB
SFDR(49MHz)	31.7dB	63.4dB
HD$_3$(411kHz)	-29.8dB	-72.2dB
HD$_3$(49MHz)	-31.7dB	-68.0dB

b flash stage. Altogether 14 raw bits are collected. Simple cascode pseudo-differential inverters (Fig. 5) are used as residue amplifiers to enable high conversion speed and power efficiency. A similar common-mode feedback/feed-forward circuit as in [13] is devised to set the common-mode levels along the pipeline. The (simulated) open-loop gain curves of the 2nd-stage amplifier are shown in Fig. 5. Relative sizes of the amplifiers are 7:4:1 in stage 1, 2, and 3, respectively. The sampling capacitors of the pipelined ADC and the SHA are 2pF and 0.5pF, respectively. To ensure good matching between the reference and the main signal paths, a replica switched-capacitor S/H structure was used by the SHA to mimic that of the pipeline. This also minimizes the systematic sampling clock skew existent in between the paths. They further share the same analog supply.

Clock generation/distribution of the ADC SoC is facilitated by two DLLs producing synchronized multi-phase clocks for the SHA, the $\Sigma\Delta$ and the pipelined ADCs. The analog portion of the prototype was fabricated in a triple-well 0.13μm digital CMOS process without analog options. Capacitors are implemented using fingered metal capacitor (MOM). A micrograph of the 3.7mm×4.7mm chip is shown in Fig. 6. The LMS ADF is implemented in FPGA.

IV. MEASURED RESULTS

In experiments, the LMS ADF converges within a few seconds after power-up. Nonlinear residue gain errors up to the 5th-order in stage 1 and 2 are calibrated, while only linear gain errors are filtered for the remaining stages. Fig.

7 shows the power spectral density (PSD) of the ADC before (a) and after (b) calibration. The input frequency is 411kHz and the sample rate of the pipelined ADC is 100.12MS/s. Before calibration, the SFDR is limited by HD$_3$ at 29.8dB; after calibration, the HD$_3$ drops to -72.2dB, and the SFDR is limited to 67.8dB by a spur at f_s/2. The SNDR improves from 28.1dB to 59.4dB before and after calibration, respectively. The results of a similar experiment performed on a 49MHz input are plotted in Fig. 8, wherein the SNDR improves from 29.4dB to 54.8dB and the SFDR improves from 31.7dB to 63.4dB. Fig. 9 plots the measured SNDR vs. the sampling frequency with a 10MHz input. Fig. 10 shows the measured SNDR and SFDR vs. the input frequency. Table I summarizes the measured performance of the prototype.

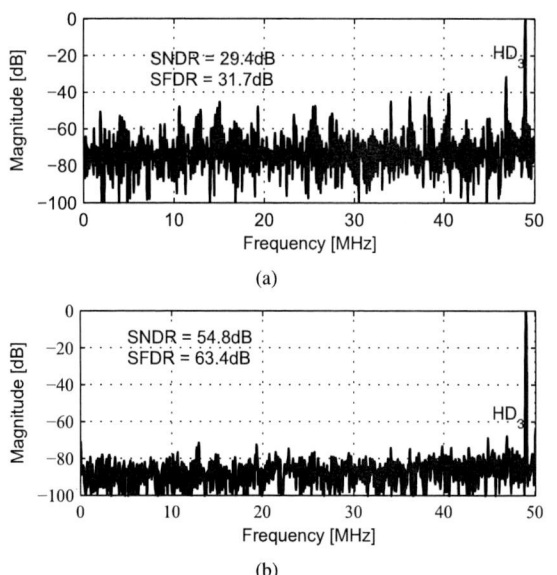

(a)

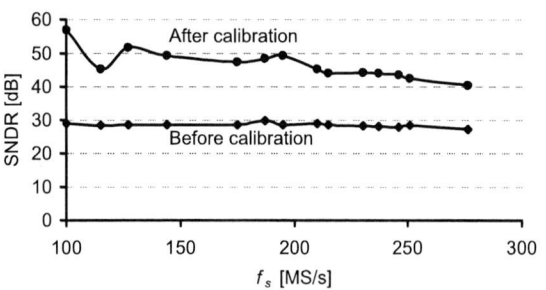

(b)

Fig. 8. Measured ADC spectra for f_s=100MHz, V_{in}=0dBFS, f_{in}=49MHz: (a) before calibration, (b) after calibration.

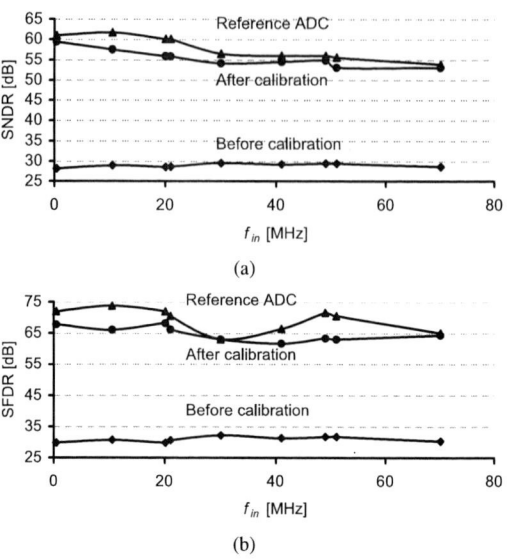

(a)

(b)

Fig. 10. Measured performance at 100MS/s: (a) SNDR, (b) SFDR.

Fig. 9. Measured SNDR versus f_s with 10MHz input.

V. CONCLUSION

A fully digital, adaptive, background calibration algorithm, capable of correcting linear as well as nonlinear residue gain errors in pipelined ADCs, has been reported. The technique allows the use of power-efficient, broadband, and low-gain cascode inverters as precision residue amplifiers. Experimental results show that the calibration improves the SNDR and SFDR of the prototype ADC by >25dB across the Nyquist band at a sample rate of 100MS/s.

ACKNOWLEDGMENT

The authors would like to acknowledge the contributions of the students, especially D. Stepanovic, faculty, and sponsors of the Berkeley Wireless Research Center, funding support from the National Science Foundation Infrastructure Grant No. 0403427, wafer fabrication donation of STMicroelectronics, and the support of the Center for Circuit & System Solutions (C2S2) Focus Center, one of the five research centers funded under the Focus Center Research Program, a Semiconductor Research Corporation program.

REFERENCES

[1] E. Siragusa and I. Galton, "A digitally enhanced 1.8-V 15-bit 40-MSample/s CMOS pipelined ADC," *Journal of Solid-State Circuits*, vol. 39, pp. 2126-2138, Dec. 2004.

[2] H.-C. Liu, Z.-M. Lee, and J.-T. Wu, "A 15 b 20 MS/s CMOS pipelined ADC with digital background calibration," *International Solid-State Circuits Conference*, 2004, pp. 454-455, 539.

[3] K. Nair and R. Harjani, "A 96 dB SFDR 50 MS/s digitally enhanced CMOS pipeline A/D converter," *International Solid-State Circuits Conference*, 2004, pp. 456-457, 539.

[4] S.-T. Ryu, S. Ray, B.-S. Song, G.-H. Cho, and K. Bacrania, "A 14 b-linear capacitor self-trimming pipelined ADC," *International Solid-State Circuits Conference*, 2004, pp. 464-465, 540.

[5] B. Murmann and B.E. Boser, "A 12-bit 75-MS/s pipelined ADC using open-loop residue amplification," *Journal of Solid-State Circuits*, vol.38, pp. 2040-2050, Dec. 2003.

[6] C. R. Grace, P. J. Hurst, and S. H. Lewis, "A 12 b 80 MS/s pipelined ADC with bootstrapped digital calibration," *International Solid-State Circuits Conference*, 2004, pp. 460-461, 539.

[7] A. Panigada and I. Galton, "Digital background correction of harmonic distortion in pipelined ADCs," *Transactions on Circuits and Systems I: Regular Papers*, vol. 53, pp. 1885-1895, Sept. 2006.

[8] X. Wang, P. J. Hurst, and S. H. Lewis, "A 12-bit 20-MS/s pipelined ADC with nested digital background calibration," *Custom Integrated Circuits Conference*, 2003, pp. 409-412.

[9] Y. Chiu, C. W. Tsang, B. Nikolic, and P. R. Gray, "Least mean square adaptive digital background calibration of pipelined analog-to-digital converters," *Transactions on Circuits and Systems I: Regular Papers*, vol. 51, pp. 38-46, Jan. 2004.

[10] J. McNeill, M.C.W. Coln, B. J. Larivee, ""Split ADC" architecture for deterministic digital background calibration of a 16-bit 1-MS/s ADC," *Journal of Solid-State Circuits*, vol. 40, pp. 2437-2445, Dec. 2005.

[11] Y. Chiu, "An adaptive filtering platform for digitally calibrated A/D conversion," Berkeley Wireless Research Center (BWRC) retreat presentation, Jan. 2004.

[12] C. Tsang, Y. Chiu, B. Nikolic, "A 1.2V, 10.8mW, 500kHz Sigma-Delta Modulator with 84dB SNDR and 96dB SFDR," *Symposium on VLSI Circuits*, 2006, pp.162-163.

[13] Y. Chiu, P. R. Gray, and B. Nikolic, "A 14-b 12-MS/s CMOS pipeline ADC with over 100-dB SFDR," *Journal of Solid-State Circuits*, vol. 39, pp. 2139-2151, Dec. 2004.

978-1-4244-2018-6/08 $25.00 © 2008 IEEE

A 1.2v 11b 100Msps 15mW ADC realized using 2.5b pipelined stage followed by time interleaved SAR in 65nm digital CMOS process

Pratap Narayan Singh, Ashish Kumar, Chandrajit Debnath, Rakesh Malik

STMicroelectronics India

Abstract- **This paper describes an 11b ADC realized using a 2.5b pipelined stage followed by 9b time interleaved SAR. Presented ADC designed in 65nm CMOS process occupies 0.3mm² area, achieves 59.1dB SINAD at 100Ms/s sampling frequency while dissipating 15mW power from 1.2V supply and resulting FOM is 0.20 pJ/step.**

I. INTRODUCTION

Pipelined architecture is considered to be best suited for more than 8b resolution with conversion speed greater than 50Msps. The overall power efficiency of pipelined ADC is governed by power consumption of opamps inside pipelined stages. Time-interleaved SAR architectures [1] offer high speed ADCs with lower power consumption than pipelined counterparts. These ADCs require S/H circuit when more than 8b resolution is targeted for analog input frequency till nyquist to satisfy sampling skew requirement. The proposed ADC utilizes an approach to implement the S/H circuitry in front of time-interleaved SAR sub-ADC in the form of a 2.5b pipelined stage and thus enhancing the resolution by 2 bits which results in resolution improvement of ADC. The hybrid architecture eliminates sampling skew problems of time-interleaved architecture since the pipelined stage delivers held output to sub-ADC. The proposed architecture also improvises the power consumption of a conventional pipelined architecture since the lower order bits are implemented by a time-interleaved SAR which consumes lesser power as compared to complete pipelined embodiment. The SoCs targeted for wireline, medical instrumentation and other applications require multiple instances of ADC with minimum device pins. To satisfy this requirement the proposed ADC incorporates a complete internal reference driver without external/internal decoupling capacitor.

Section II describes about implementation issues and advantages, section III describes about ADC architecture and implementation, section IV describes about MDAC amplifier and complete internal reference driver, section V describes about measurement results and section VI gives the conclusion.

II. IMPLEMENTATION ISSUES AND ADVANTAGES

In the present design time interleaved SAR stages are connected to the MDAC of the pipelined stage. In this configuration only one SAR stage is connected to MDAC by sampling switches at a time. So connecting a time interleaved SAR is similar as connecting one SAR at time except some additional load because of sampling switches of un-switched

stages. So the effective load at the output of the MDAC is the load of one SAR stage. In case of complete pipelined implementation, each MDAC stage is loaded by the sampling capacitor of next stage and these sampling capacitors cant be reduced after certain limit since reduction of the sampling capacitors results in increase of feedback factor of MDAC opamps and subsequently degradation of amplifier bandwidth. Moreover SAR ADCs inherently low power consuming as they do not require any opamp.

However time-interleaved ADCs have got limitation of sampling skew when input sampling frequency till nyquist frequency is required. But connecting them in front of the MDAC of the pipelined stage helps in eliminating it because of already held output from the residue calculation phase of the MDAC is applied as input to time interleaved SAR ADC.

Since these two architectures are entirely different so there is a difficulty of sharing the reference voltage source for both parts of the ADC. This is done by carefully synchronizing clock phase generators of the pipelined and SAR stage.

III. ADC ARCHIRECTURE

Figure1 shows the overall ADC architecture where input signal is sampled by a 2.5b pipelined stage and residue output is applied to 9b time-interleaved SAR (PSAR) sub-ADC. 3MSB outputs of pipelined stage are combined with 9b output of PSAR by digital error correction method to produce 11b digital output.

Pipelined stage is standard 2.5b implementation [2][6] with a Flash ADC composed of 6 comparators and each comparator is a dynamic latch preceded by a pre-amplifier to ensure the comparator offset remains inside tolerance limit for 2.5b stage with 1Vp-p input swing at 1.2v supply voltage. Multiplying DAC amplifier shown in figure2 is fully differential gain boosted folded cascode amplifier [3][6] with boosting amplifiers also boosted by fully differential gain boosters to ensure more than 70dB gain since output impedance of the standard digital MOS transistors is low and transistor lengths are also kept close to minimum to maximize bandwidth . The analog input is passively sampled by a bootstrapped switch [3] allowing input signal frequency beyond nyquist rate. Input sampling capacitors are 500fF each for KT/C noise criteria and also satisfies matching requirement. All the capacitors inside the ADC are fringe capacitors or metal sandwich capacitors and implemented in standard process option.

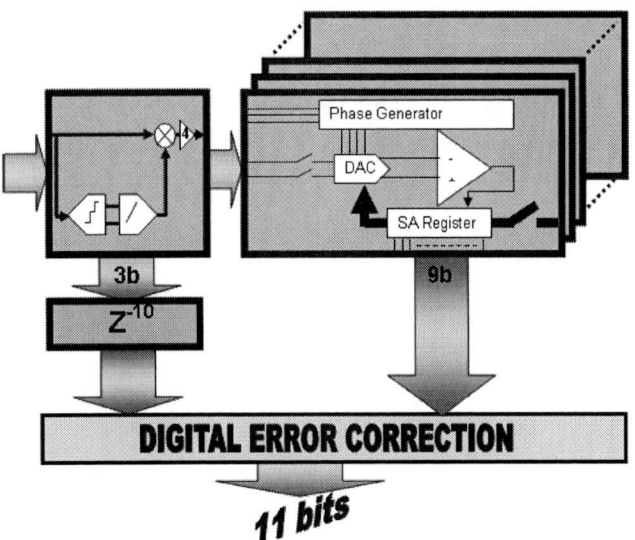

Figure 1: ADC Architecture

The 9b PSAR is implemented by connecting in parallel 12 SAR ADCs with a redundant bit. Each individual stage uses 2 clock cycles for sampling the residue of pipelined stage and 10 cycles for conversion and 12 stages in parallel makes conversion rate of PSAR equal to input clock frequency. Each slice in PSAR implements a 9b DAC with 5b MSB-DAC is realized by a 32 segment resistive string producing 32 reference voltages and 4b LSB-DAC also utilizes the same reference voltages. The MSB-DAC output voltage is combined with LSB-DAC output by 2 capacitors of ratio 32:1 to produce final DAC output. These 2 capacitors are also used as input sampling capacitors for the SAR ADC. The 32 reference voltages are shared in all 12 slices to eliminate inter-stage gain-error.

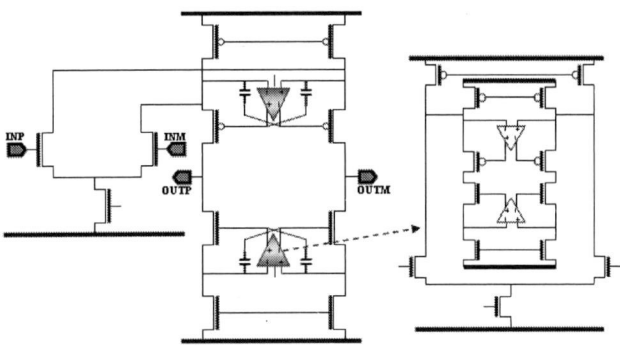

Figure 2: OPAMP Architecture

Each slice of PSAR uses an offset compensated comparator which ensures less than 0.5LSB offset in each stage and thus minimizes inter-stage offset variation and hence performance of the PSAR is governed by resistance matching of MSB DAC. A phase generator block is incorporated to manage the clock phases between pipelined stage and PSAR. It is worthwhile to mention at this point that the MSB DAC of PSAR can be implemented by capacitance array and since

capacitor exhibits much superior matching performance to resistors hence PSAR resolution can be increased to 10b and overall ADC to 12b or more. More pipelined stages can be cascaded at the front of PSAR to further increase resolution of the ADC keeping the power advantage intact over complete pipelined architecture and resolution enhancement over time interleaved SAR.

IV. REFERENCE DRIVER

A high drive super class-AB opamp shown in figure3 has been designed as a reference buffer to drive 320ohm resistance used to generate reference voltages for SAR ADC and capacitive load of pipelined and SAR stages. Comparing with super class-AB buffer [4], the proposed reference driver uses an improved common mode loop control and implements a pseudo differential output stage to minimize the dc power consumption keeping the high current driving capability intact and thus improving the overall efficiency.

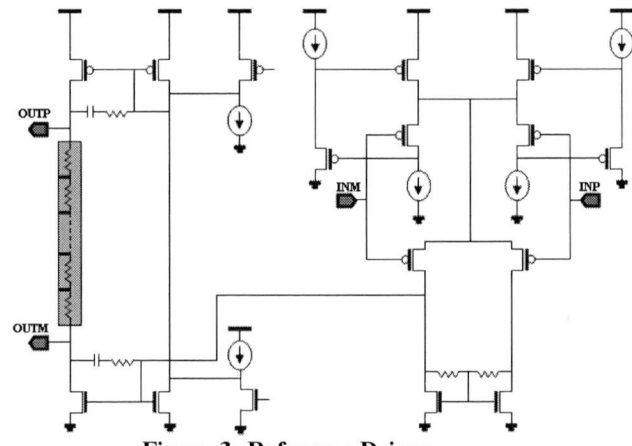

Figure 3: Reference Driver

Level shifted slew rate enhancement block is incorporated for reduced current variation across process corner, and improved input common mode range. The differential output of the driver used as high and low reference voltage for the ADC, successfully drives the resistive and capacitive load of the complete ADC up to 130Ms/s conversion rate. Thus the ADC integrates a complete internal reference without any need for internal/external decoupling capacitor and saves core area or device pin count as well as number of external components.

V. MEASUREMENT RESULTS

The prototype of the ADC is fabricated in 65nm digital CMOS technology. The ADC occupies 0.3mm^2 silicon area. Figure4 shows the measurement results of the ADC at 100Ms/s 1.2v supply. The measured INL, DNL are within +/- 1.5LSB and +/- 0.94LSB respectively. From the plot it is clearly visible that DNL of the ADC is dominated by resistive mismatch of time-interleaved SAR. The peak SNDR reaches 59.1dB (after full-scale correction) and SFDR 74.7dB with 10MHz input. Figure5, Figure6 and Figure7 demonstrate dynamic parameters with sweep of sampling frequency, input signal frequency and supply voltage respectively. Figure8 explores

978-1-4244-2018-6/08 $25.00 © 2008 IEEE   328

the robustness of static parameters with +/-10% supply variation and sampling frequency till 130MHz. The ADC dissipates only 15mW power from supply voltage 1.2v at 100Ms/s conversion speed. The FOM achieved by the ADC is 0.20pj/step. Computed figure of merit is defined by the following formula [5].

$$FOM = \frac{Power}{2^{ENOB} \times 2 \times F_{BW}}$$

The chip micrograph of the ADC is shown in Figure7, where core dimension is 500umx620um. Proposed architecture helps in extending capabilities of time-interleaved SAR ADC's to higher resolution, eliminating all fundamental limitations of interleaving and with their power efficiency preserved.

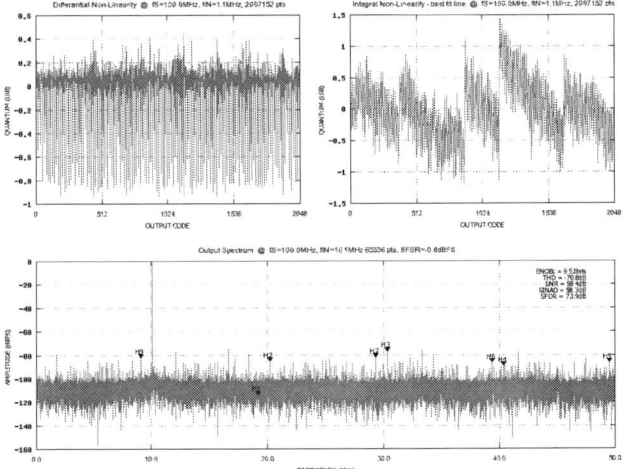

Figure 4: Static Performance and FFT

VI. CONCLUSION

An ADC implemented by a 2.5b pipelined stage followed by 9b time interleaved SAR ADC is proposed. The 9b SAR ADC helps in reducing power consumption and pipelined stage implements the MSBs and also ensures a held signal to be applied at SAR stage and thus eliminates sampling skew problem associated with time interleaved SAR ADCs. The test results of the ADC shows total power consumption of 15mW at 100 Msps with 9.53b ENOB resulting in FOM of 0.20pJ/Step.The ADC also integrates a fully internal reference which obviates any need for external capacitor thus reducing package pin as well as external components. And also this ADC architecture can be further extended to 12b or more resolution by preserving its power efficiency.

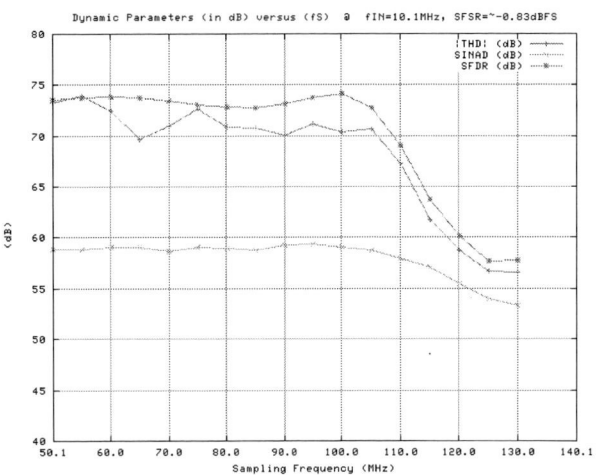

Figure 5: Dynamic Performance Vs Sampling Frequency

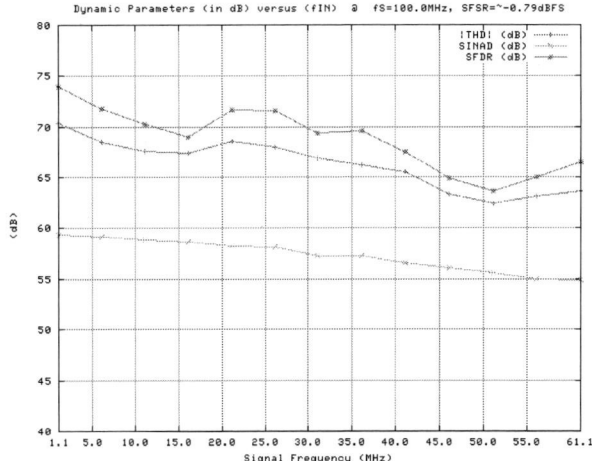

Figure 6: Dynamic Performance Vs Input Frequency

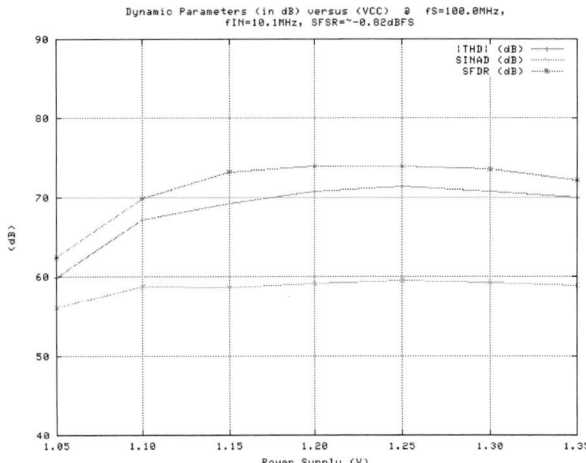

Figure 7: Dynamic Performance vs Supply

978-1-4244-2018-6/08 $25.00 © 2008 IEEE

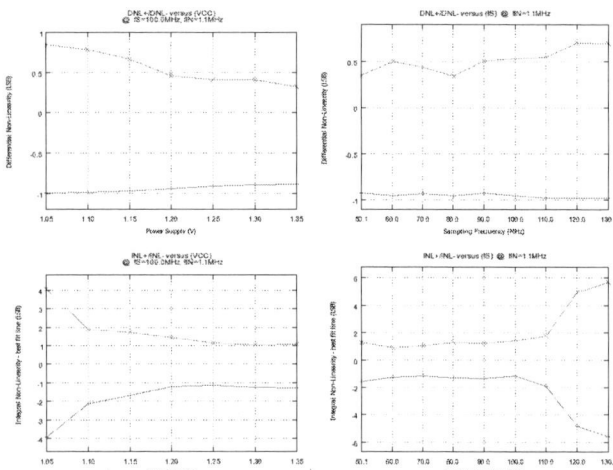

Figure 8: Static Performance

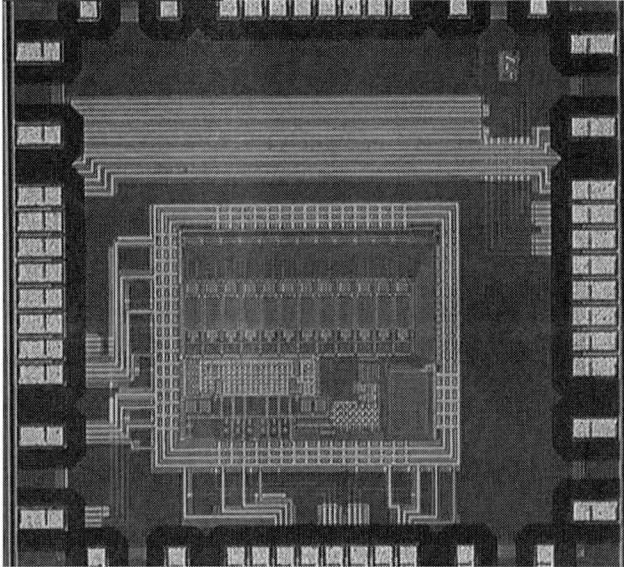

Figure 9: Chip Micrograph

Table 1: ADC Summary

Power Consumption ADC	15mW
Power Consumption Reference Driver	8mW
SINAD @ Fin 10Mhz	59.1dB
ENOB @ Fin 10Mhz	9.53
INL	+/-1.5 LSB
DNL	+/-0.94 LSB
FOM	0.20 pJ/Step
Power Supply	1.2V
Sampling Frequency	100Msps
Area	0.3mm^2

REFERENCES

[1] Dieter Draxelmayr, "A 6b 600MHz 10mW ADC Array in Digital 90nm CMOS", *ISSCC Dig. Tech. Papers,* pp. 264 - 527, Feb. 2004.

[2] Gulati, K. Peng, M. Pulincherry, A. Munoz, C. Lugin, M. Bugeja, A. Li, J. Chandrakasan, A. "A highly-integrated CMOS analog baseband transceiver with 180MSPS 13b pipelined CMOS ADC and dual 12b DACs" IEEE Proceedings of the Custom Integrated Circuits Conference 2005,pp. 515 – 518, Sept. 2005

[3]Yun Chiu Gray, P.R. Nikolic, B., "A 1.8 V 14 b 10 MS/s pipelined ADC in 0.18 /spl mu/m CMOS with 99 dB SFDR" *ISSCC Dig. Tech. Papers,* pp. 458 – 539, Feb. 2004

[4] Lopez-Martin, A.J. Baswa, S. Jaime Ramirez-Angulo Carvajal, R.G.Klipsch "Low-Voltage Super class AB CMOS OTA cells with very high slew rate and power efficiency" , *IEEE Journal of Solid-State Circuits*, Vol. 40 , No. 5, pp. 1068-1077, May 2005

[5] Robert Wang, Ken Martin, David Johns, Gangadhar Burra " A 3.3mW 12MS/s 10b Pipelined ADC in 90nm Digital CMOS " *ISSCC Dig. Tech. Papers,* pp. 278 – 279, Feb. 2005

[6] Singh, Pratap Narayan; Kumar, Ashish; Debnath, Chandrajit; Malik, Rakesh "20mW, 125 Msps, 10 bit Pipelined ADC in 65nm Standard Digital CMOS Process" *Custom Integrated Circuits Conference, 2007. CICC '07. IEEE, pp 189-192, 16-19 Sept. 2007*

IEEE 2008 Custom Intergrated Circuits Conference (CICC)

A 52mW 10b 210MS/s Two-Step ADC for Digital-IF Receivers in 0.13μm CMOS

Zhiheng Cao[1], Shouli Yan[2]

[1]Qualcomm, San Diego, CA 92121
[2]University of Texas at Austin, Austin, TX 78712
zcao@qualcomm.com, slyan@ece.utexas.edu

Abstract- **A 10b 210MS/s two-step ADC has been implemented in 0.13μm digital CMOS with an active area of 0.38mm². Using a proposed capacitor network implemented with small value interconnect capacitors which replaces the resistor ladder/multiplexer in conventional sub-ranging ADCs, and proposed offset canceling comparators, it achieves 74dB SFDR/55dB SNDR for 10MHz and 71dB SFDR/52dB SNDR for 100MHz inputs at 210MS/s while consuming 52mW from a 1.2V supply.**

I. INTRODUCTION

The digital-IF receiver architecture eliminates DC offsets and flicker noise problem in the direct-conversion architecture, and simplifies the frequency synthesizer and analog channel selection filter since fine channel tuning and filtering can be done in digital domain. As a result, wide-bandwidth integer-N PLLs can be used, reducing the channel switching time and over-all complexity of the system [1].

Digital-IF receivers require ADCs with a high sampling frequency and high SFDR at high input signal frequencies such that digital filtering can be used to extract desired signals in the presence of large interferers. If such ADCs can be integrated with the DSP in standard deep-submicron digital CMOS processes with low power consumption and small die area, the overhead of implementing the digital-IF architecture can be much reduced.

The pipeline architecture, which has been widely used for high sampling frequency and ≥8b resolution ADCs, is inefficient in these processes due to reduced transistor linearity, voltage gain and headroom. Because accurate residue amplification is necessary, techniques such as multi-stage amplifiers and nested gain boosting must be used, which limit the sampling frequency. On the other hand, two-step ADCs do not require closed-loop amplifiers, because the coarse flash ADC (CDAC) with relatively high resolution (5b in this design) much reduces the fine ADC (FADC) input swing, significantly relaxing its linearity requirement. Combined with well-matched small value interconnect capacitors [2] in nanometer CMOS processes to realize low-power highly linear residue generation, and offset cancellation techniques to realize compact and low-power flash ADCs, better linearity, higher speed and lower power consumption are achieved at 10b resolution.

II. ARCHITECTURE AND CIRCUITS

Fig. 1 shows a block diagram of the proposed ADC. The difference from a conventional sub-ranging ADC is that the 6b FADC, which determines the performance of the entire converter, does not use resistor reference ladder and that there is no multiplexer that chooses a sub-range in the reference ladder to be used by the FADC. Instead, two SAR-like 5-bit feedback capacitor DACs (CDAC) that work also as passive S/Hs are used to generate *vin*-(CADC output)-1 and *vin*-(CADC output)+1, where *vin* is the analog input in unit of CADC LSB. The FADC uses further capacitive interpolation to resolve 64 levels out of these two input boundary voltages and find the zero-crossing, converting the 6 LSB (Fig. 2).

These boundary voltages are generated by the appropriate DAC input codes and an additional LSB capacitor in the CDAC which shifts the boundary in such a way that in the presence of smaller than ±0.5LSB comparator offset in the 5-bit CADC, the two boundaries still have opposite sign, i.e. the FADC output does not clip, enabling correct 10-bit conversion after adding the CDAC input with the FADC output.

Obvious advantages compared with conventional resistor ladder + multiplexer sub-ranging ADC are 1) faster setting due to high turn-on voltages of the switches and 2) no reference resistor ladder which drives the FADC. Since the kT/C noise of each sampling cap of FADC may handle the zero-crossing and determine the noise floor of the entire ADC, the FADC could present significant total input capacitance (5pF in [3]) especially when more than two sampling caps are used. As a result, much power must be used in the resistor ladder to satisfy settling speed requirement. In the proposed scheme, two reference voltages, i.e. Vref+ and Vref- are used to drive the FADC through capacitive interpolation by the CDAC. These reference voltages are generated using gain-boosted source followers with large MOS capacitors connected at the output. The gain boosting allows orders of magnitude smaller current than what would be required in a resistor ladder to generate ~3Ω impedance below the gain-boosting loop bandwidth and above through the MOS capacitor [2]. Since there are only two voltage levels, die area consumed by these large capacitors can be tolerated.

To achieve high SFDR at high input frequencies, it is best to use all-passive sampling network, i.e. the preamp should not be involved during the tracking phase. This means that the CADC needs an extra half clock-cycle for the reference ladder to settle before it can latch and starts to regenerate (Fig. 3). To allow time for the CADC regeneration and thermometer-to-binary converter settling, the FADC input is held for 1 clock cycle (Fig. 4). This is accomplished by time-interleaving two S/H (P and Q), each one samples input at the same time when CADC samples the input, and is connected to FADC input only after the other one finishes sampling input (1 clock cycle

978-1-4244-2018-6/08 $25.00 © 2008 IEEE

later). A 6-bit (full scale of ±14ps) digitally controlled delay line illustrated in Fig. 5 calibrates the clock skew to each S/H, which is measured at power-up by applying an arbitrary periodic signal (with any frequency not in harmonic relation with the sampling frequency) and performing FFT of the ADC odd/even outputs to find the phase difference of the applied tone in each output. The FFT also finds the frequency of the applied tone hence the ideal phase difference when there is no sampling clock skew. Offset and gain errors are corrected by digital post-processing (through equalizing the mean and power of the odd/even outputs.)

The preamps in the FADC shown in Fig. 6 are differential pairs with load resistance adjusted considering the trade-off between gain, settling-speed and linearity (important because of the 2-to-5 interpolation, even with the much reduced swing residue input to the FADC). Stage-1 preamplifier decreases/increases its load resistance in reset/amplification phase to speed up settling/increase gain. While stage 1 uses output offset storage, stage 2~3 uses input/output offset storage [4], which is more power efficient than output offset storage. During reset, when the interpolation capacitors are charged, the impedance is low (1/gm of the input device) to satisfy settling speed requirement. During amplification, when the next stages' input capacitances, which is much smaller than the interpolation capacitor are charged, the impedance can be higher, hence less current/larger load resistance can be used to reduce power. Because the output capacitors, which also function as interpolation capacitors, store the error due to finite preamp gain, complete offset cancellation is possible [4].

Because the CDAC has no gain, comparator offset referred at the FADC input must be smaller than the 10b LSB of the entire ADC. The comparators in the FADC, as well as in the CADC, use a proposed offset canceling comparator (Fig. 7) to significantly reduce the required preamplifier stages and hence power consumption and silicon area. When "rst" is high, the offset of the preamp is stored, while the comparator regenerates. When "rst" goes low, the preamp starts to amplify the signal and settles before "latch" goes high, at which point the signal is applied to the regenerative comparator. While "latch" was low, the offset of the comparator is also stored and subsequently cancelled when "latch" goes high. The signal input to the regenerative comparator is a pulse rather than a step. The regeneration time domain waveform is given by $(1-\exp(-T/\tau))\exp(t/\tau) u(t)$, where T is the pulse width and τ is the regeneration time constant. We can see $T \sim \tau$ is enough not to slow down the regeneration much compared with conventional comparators where the input is a step.

III. Experimental Results

The ADC is fabricated in a 1P8M digital 0.13μm CMOS process without MIM capacitors and uses only 1.2V devices. It occupies 0.38mm^2 (Fig. 8) and draws 40mA from a 1.2V AVDD and 3mA at 210MS/s from a 1.2V DVDD, including reference buffers, bias and clock buffers, but excluding output drivers. Fig. 8 shows measured ADC output FFT (16384 point with hanning window) spectra at 210MS/s with a -3dBFS

100.3MHz single tone applied at the input. The 1st plot does not have any post-processing except DC removal and the input code to the clock delay line (Fig. 5) is 0. The tone at fs/2-fin is about -55dBc. The FFT after correcting offset and gain is shown in the 2nd plot. The tone at fs/2-fin reduces about 2dB to-57dBc. With timing skew correction alone, the tone reduces 6dB to -61dBc (3rd plot). This result indicates that timing skew dominates over gain/offset mismatch with this fin. With gain/offset and timing correction (last plot) all channel mismatch related tones become less than harmonic distortion terms and the SFDR becomes 71.7dBc. SFDR changes with measurement but is typically better than 71dBc for 100MHz signal input at 210MS/s. Fig. 10 shows measured maximum SFDR/SNDR for different signal frequencies and sampling frequencies (after correction of offset/gain/timing error). Without using an integrated signal buffer, >70dBc SFDR is maintained up to 100MHz signal frequency at 210MS/s. Fig. 11 shows SNDR/SFDR vs. input signal amplitude, from which we see that noise floor increases for >-40dBFS inputs that trigger change of CADC output. This is most likely caused by interference from CADC's binary encoder to analog input/reference bus. At 300MS/s there is less time for the analog bus to settle from this disturbance, resulting in big drop in performance above -40dBFS, even though good SFDR and SNDR is achieved at and below -40dBFS. Fig. 12 shows the INL and DNL. Due to the interference, the DNL at the codes where the CADC switches is much larger than what would otherwise be. This results in reduced SNDR. Nevertheless, the INL is still quite linear, confirming the high measured SFDR and demonstrating the linearity advantage of the proposed two-step architecture over the conventional pipeline architecture. Table 1 summarizes measured performance. Table 2 provides a comparison with other high speed 10b ADCs published in recent years based on the conventional pipeline architecture. Even with the interference problem, this ADC provides comparable SNDR and better SFDR, with less silicon area and power consumption.

Acknowledgments

The authors acknowledge Semiconductor Research Corporation (SRC) for partial funding, Europractice for their mini@sic program and Texas Instruments for help with measurement.

References

[1] A. Maxim et al., "A single chip DBS tuner-demodulator SoC using discrete AGC, continuous I/Q correction and 200MS/s pipeline ADCs," *RFIC Symposium,* pp. 511-514, June 2007.

[2] Z. Cao et al., "A 32mW 1.25GS/s 6b 2b/step SAR ADC in 0.13μm CMOS," *ISSCC Dig. Tech. Papers,* 2008.

[3] M. Clara, et al., "A 1.8V fully embedded 10b 160MS/s two-step ADC in 0.18μm CMOS," *CICC,* 2002.

[4] C. Sandner, M. Clara, A. Santner, T. Hartig, and F. Kuttner, "A 6bit, 1.2GSps low-power fash-ADC in 0.13μm digital CMOS," *IEEE J. Solid-State Circuits,* vol. 40, pp.1499-1505, July 2005.

[5] B. Hernes et al., "A 1.2V 220MS/s 10b Pipeline ADC Implemented in 0.13μm Digital CMOS," *ISSCC Dig. Tech. Papers,* 2004, pp. 256-257.

[6] D. Kurose et al., "55-mW 200MS/s 10-bit Pipeline ADCs for Wireless Receivers," *IEEE J. Solid-State Circuits*, vol. 41, no. 7, pp. 1589-1595, July 2006.

[7] S.-C. Lee et al., "A 10-bit 400MS/s 160-mW 0.13μm CMOS Dual-Channel Pipeline ADC Without Channel Mismatch Calibration," *IEEE J. Solid-State Circuits*, vol. 41, no. 7, pp. 1596-1605, July 2006.

[8] S.-T. Ryu et al., "A 10-bit 50MS/s Pipelined ADC With Opamp Current Reuse," *IEEE J. Solid-State Circuits*, vol. 42, no. 3, pp. 475-485, March 2007.

[9] S.-C. Lee et al., "A 10-bit 205MS/s 1.0-mm2 90-nm CMOS Pipeline ADC for Flat Panel Display Applications," *IEEE J. Solid-State Circuits*, vol. 42, no. 12, pp. 2688-2695, Dec. 2007.

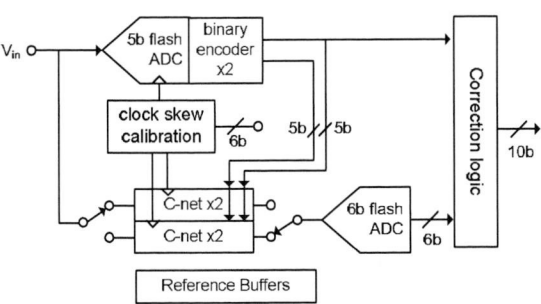

Fig. 1. Overall block diagram of the proposed ADC.

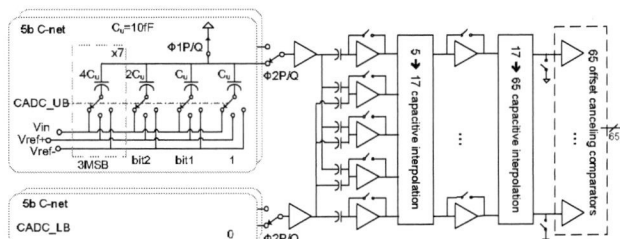

Fig. 2. 5b capacitor network and the 6b fine ADC (fully differential, one side shown).

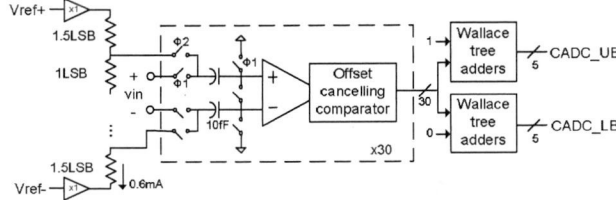

Fig. 3. 5b CADC (fully differential, one side shown).

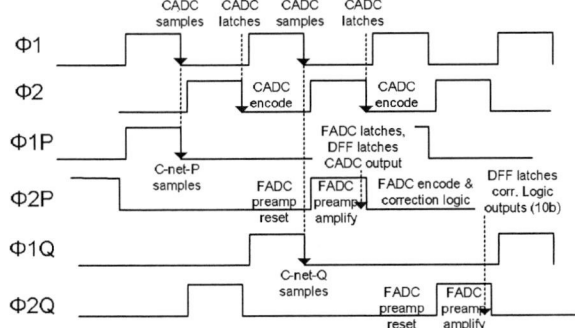

Fig. 4. Timing diagram of the two-step ADC.

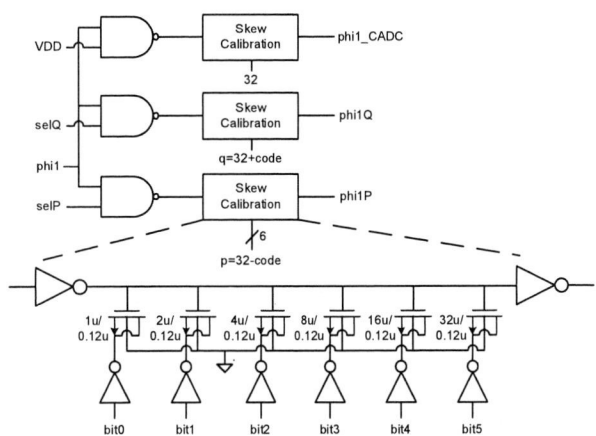

Fig. 5. Clock skew calibration circuits.

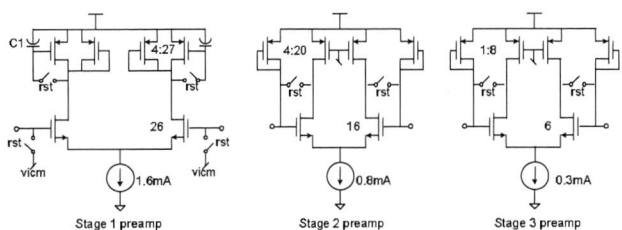

Fig. 6. Preamplifiers used in FADC.

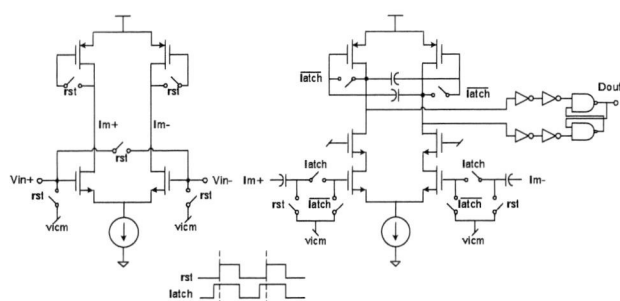

Fig. 7. Schematic of the proposed offset canceling comparator used by both FADC and CADC.

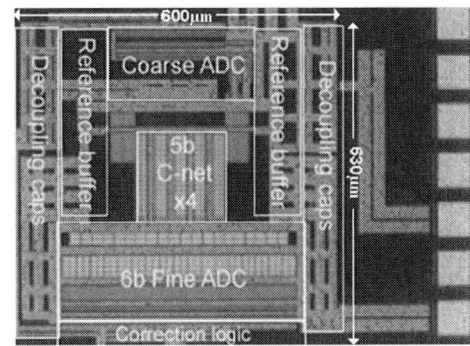

Fig. 8. Die photograph.

978-1-4244-2018-6/08 $25.00 © 2008 IEEE

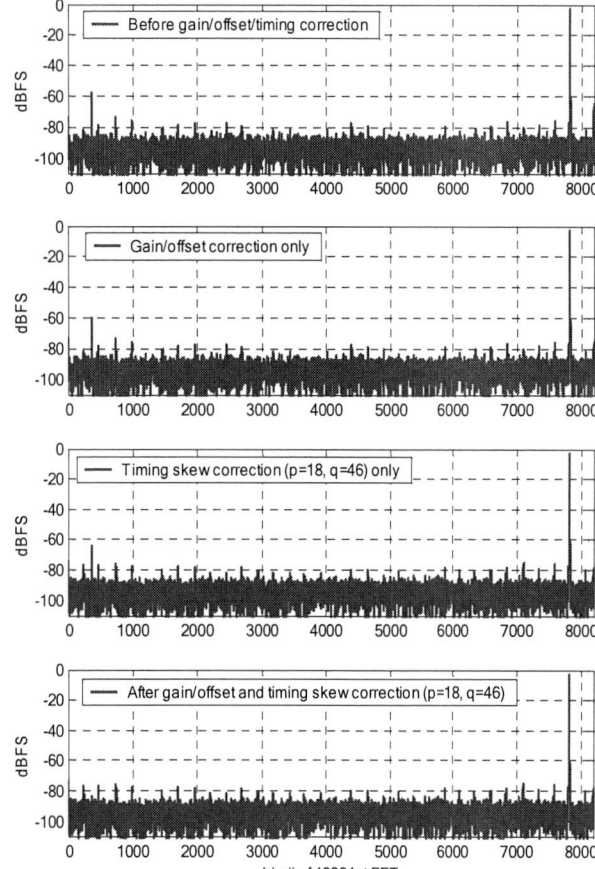

Fig. 9. Output FFT of 100.3MHz -3dBFS signal input at 210MS/s with and without gain/offset/timing corrections.

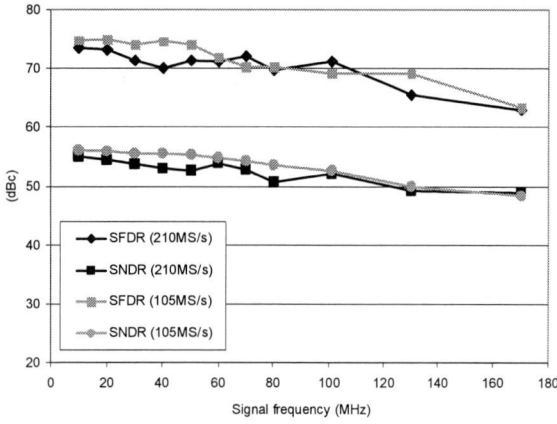

Fig. 10. Maximum SFDR/SNDR vs. input signal frequency.

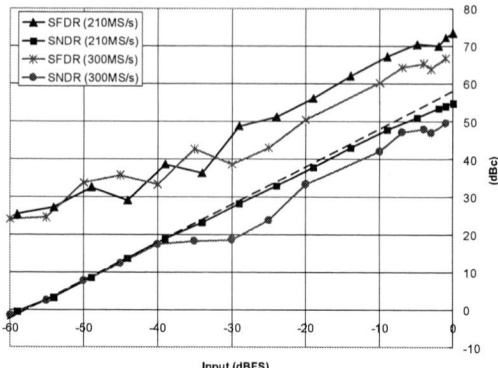

Fig. 11. SFDR and SNDR vs. input amplitude for 10MHz signal.

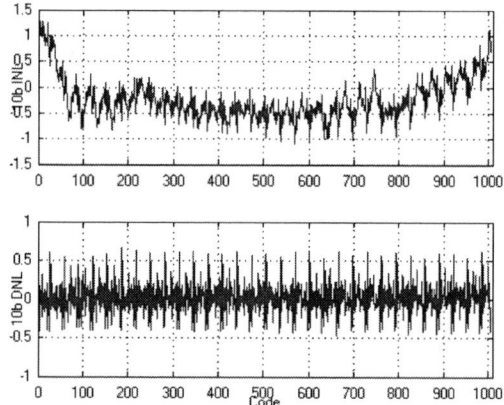

Fig. 12. INL/DNL measured with histogram method.

Table 1: Performance summary

Resolution	10 bits
Conversion rate	210MS/s
Input full-scale	0.9Vpp differential
Supply voltages	1.2V analog, 1.2V digital
Power dissipation	48mW analog, 4mW digital
SFDR/SNDR	74dB/55dB (10MHz) 71dB/52dB (100MHz)
Area	0.38mm^2 including decoupling caps

Table 2: Comparison with other 10b pipeline ADCs

	[5]	[6]	[7]	[8]	[9]	This work
Fs (MS/s)	220	200	400	50	205	210
SFDR (DC)	62dBc	66.5dBc	70.5dBc	70dBc	73dBc	74dBc
SFDR (100MHz)	N/A	N/A	60dBc	N/A	71dBc	71dBc
SNDR	54.3dB	54.4dB	56dB	56dB	55dB	55dB
Power	135mW	55mW	160mW	18mW	61mW	52mW
Area (mm^2)	1.3	1.26	4.2	1.43	1.0	0.38
Process	130nm	90nm	130nm	180nm	90nm	130nm
FOM (pJ/conv)	1.44	0.63	0.77	0.8	0.64	0.56

IEEE 2008 Custom Intergrated Circuits Conference (CICC)

A 24GS/s 5-b ADC with Closed-Loop THA in 0.18μm SiGe BiCMOS

Jaesik Lee, Joe Weiner, Pascal Roux, Andreas Leven, Young-Kai Chen

Alcatel-Lucent, Murray Hill, New Jersey, USA

Abstract-A 5-b flash ADC with a closed-loop THA is implemented in 0.18-μm SiGe BiCMOS. A global shunt feedback THA and a current-weighted comparator allow the ADC to achieve wide resolution bandwidth of 6.5GHz and high sampling rate up to 24GS/s. The ADC shows an SNDR of 28dB and an SFDR of 36dB with a 1GHz input sampled at 16GS/s. It consumes 3.3W from 3.6/3-V supplies and occupies 8.68mm² silicon area.

I. INTRODUCTION

The demand of high-speed ADCs is rapidly increasing with the advent of higher data capability communication systems like a 40Gb/s optical coherent system. The recent interest of coherent detection is mainly motivated by the ability to receive complex modulation formats as well as the ability to access the full information of optical field in the electrical domain [1]. In a 40-Gb/s coherent system (Fig. 1), four bits of information can be encoded per symbol when DQPSK is used in conjunction with double polarization multiplexing. Thus, DSP-enhanced receiver requires four 5-b 20GS/s ADCs so that the ADC should be ultra-fast and low-power. Moreover, the resolution bandwidth of the ADC must exceed 0.6*symbol rate (of 10GBaud) > 6 GHz to take place smaller OSNR penalty.

CMOS time-interleaved ADC [2], SiGe flash ADCs [3, 4, 5], or InP flash ADC [6] are recently developed ultra-high-speed ADCs. However, most state-of-the-art ADCs have failed to achieve enough resolution bandwidth for coherent application because of the lack of a high-speed, high-dynamic range (DR) sampling circuit. In this paper, a 24GS/s 5b flash ADC embedded a closed-loop track and hold amplifier (THA) in 0.18-μm SiGe is presented. The design achieves a linearity of 4.4 effective bits for a 1GHz input signal and the effective resolution bandwidth up to 6.5GHz at 16GS/s.

II. ARCHITECTURE

The architecture of the ADC is shown in Fig. 2. The input circuit of the ADC is constructed with a dedicated THA. The differential THA provides high-speed sampling operation and drives the flash quantizer with a full-scale range of 1-V_{p-p} differential. The quantizer consists of a resistive ladder followed by 9 first-stage preamplifiers (PA1), 17 second-stage preamplifiers (PA2), and 33 master (C1) - slave (C2) - master (C3) comparators. This design applies amplification and interpolation per each PA stage in order to reduce large parasitic capacitance at the output of the THA and to relieve comparator offset requirements. The 31 thermometer code

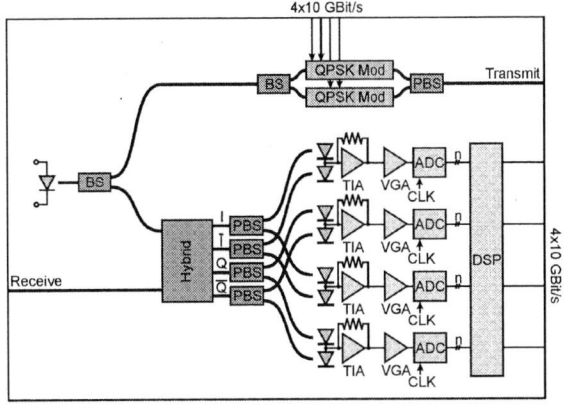

Fig. 1. Block diagram of a 10GBaud 40Gb/s coherent receiver [1].

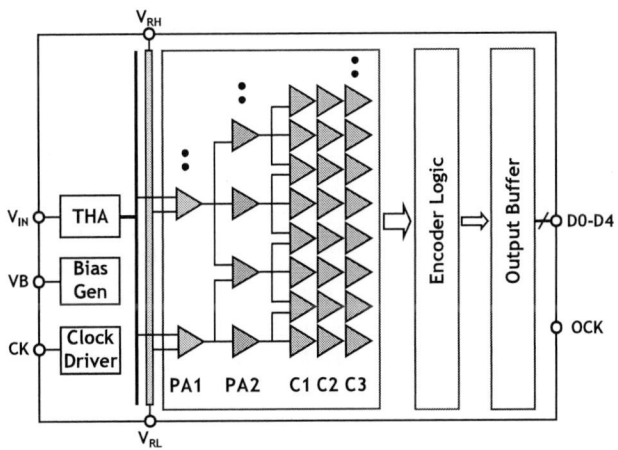

Fig. 2. Architecture of the ADC.

outputs are delivered to the encoder logic through differential microstrip transmission lines to suppress signal reflections. The encoder converts thermometer code into Gray code, and finally 5-b binary code.

III. Circuits Design

A. Global Shunt Feedback THA

A schematic of the THA is depicted in Fig. 3. At sampling frequencies of multiple tens of gigahertz, the wideband THA is hard to design, and critical for achieving good dynamic performance over broadband input signal. A compact wide-band THA is proposed using an active feedback

978-1-4244-2018-6/08 $25.00 © 2008 IEEE

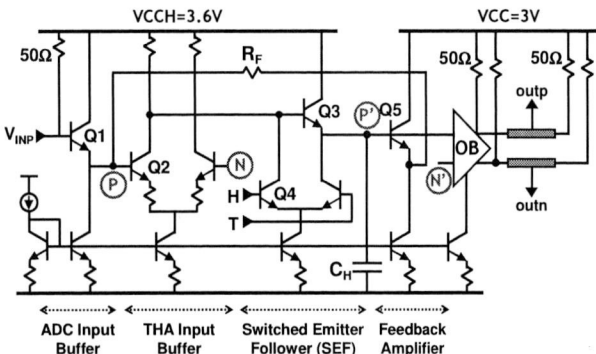

Fig. 3. Front-end THA with a closed-loop feedback.

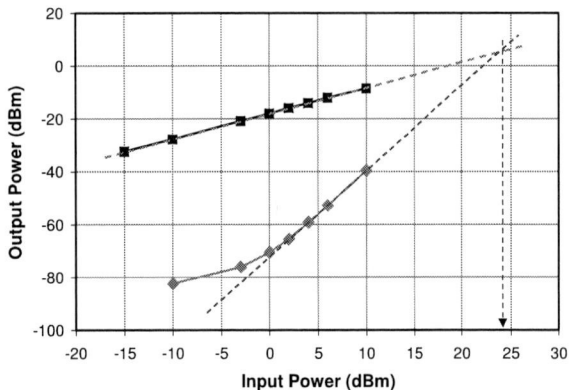

Fig.4. Measured input IP3 of the front-end THA.

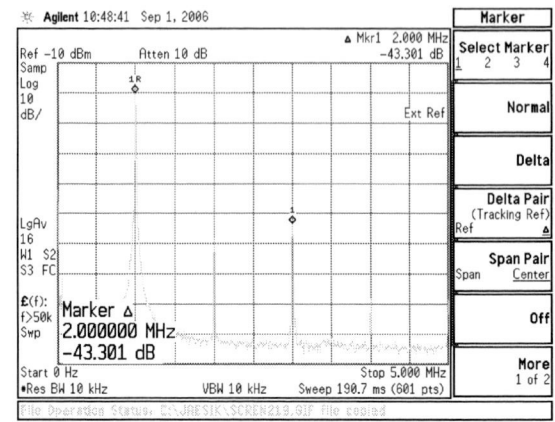

Fig.5. THA beat frequency test for track-to-hold mode with a 20GHz input signal and a beat frequency of 1MHz (Single-ended measurement).

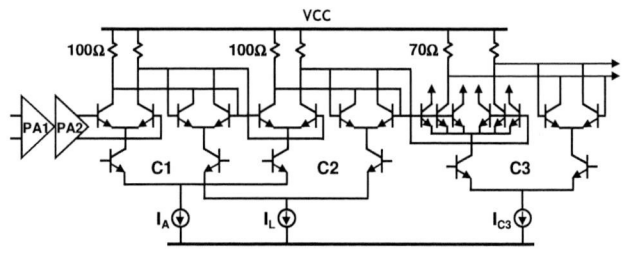

Fig.6. Schematic of M-S-M comparator array with bubble error correction.

topology. This global shunt feedback is employed at the output of the switched emitter follower (SEF) to the base of the THA input buffer to suppress the error from nonlinear parasitic and track-to-hold transition. The feedback is realized with a series combination of a unilateral feedback amplifier (Q5) and a shunt feedback resistor R_F. The feedback amplifier guarantees the isolation of a hold node from the input signal through R_F and compensates for the first-order hold-mode feedthrough caused by the C_{BE} of the SEF (Q3). Amount of feedback has been determined to make sure that the overall THA stability has been confirmed by S-parameter (S11 and S22) and K-factor. The ADC input emitter follower (Q1) that precedes the THA is necessary to decouple the input signal from the sampling circuit and ease the off-chip driving requirement for the THA. The input transistor (Q1) utilizes a large emitter length of 4x10-μm which is sufficient to drive the subsequent closed-loop THA with broadband input impedance of approximately 200Ω. In order to reduce the signal reflections the differential output of the THA is distributed to the PA1 through a balanced microstrip transmission line which achieves a characteristic impedance of 50Ω using a 2.8-μm-thick top metal layer (metal width of 10-μm) over M1 ground plane.

Measurements of the stand-alone THA show the advantages of the global shunt feedback THA which obtains 5.6dB gain of track mode dynamic range and 7.2dB gain of track-to-hold transition mode dynamic range, in comparison with the conventional open-loop THA. The track-mode bandwidth (S21) exceeds 26GHz in S-parameter measurements. This THA exhibits an input IP3 of 24dBm from two-tone intermodulation test with 10GHz and 10.1GHz input signals (Fig. 4), so that the SFDR is better than 48dBc with an input-referred noise power of -48dBm. The spectral characteristic of the beat frequency test at track-to-hold mode with $f_{IN} = f_S + \Delta f$, $f_S = 20$ GHz, and $\Delta f = 1$ MHz, at an input power of 0 dBm is obtained with the third-order harmonic distortion better than -43 dB (Fig. 5). The power dissipation of the THA including a clock buffer is 310 mW from 3.0/3.6 V supplies.

B. Current-Weighted Comparator

Two-stage PA array in conjunction with two-time interpolation allows reducing the number of preamplifiers from 33 to 9, thus reducing the nonlinear output capacitance of the THA. This topology also provides enough PA gain of 17 dB (= PA1 of 8dB + PA2 of 9dB) for the comparator's offset voltage as well as very small kickback noise. The overall bandwidth of the two-stage PA exceeds 20GHz. The comparator schematic shown in Fig. 6 consists of current-weighted master-slave (C1-C2) comparators followed by a

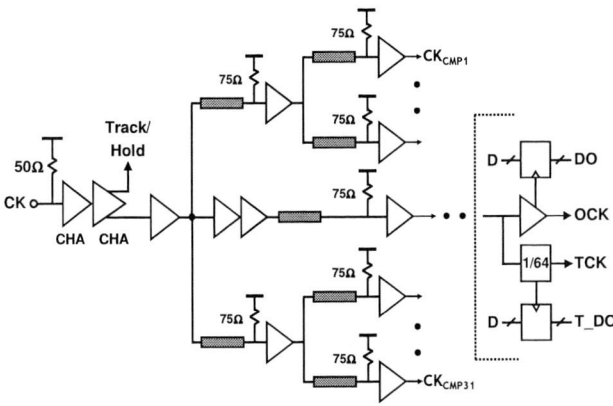

Fig.7. Clock distribution. The CHA denotes a Cherry-Hopper amplifier.

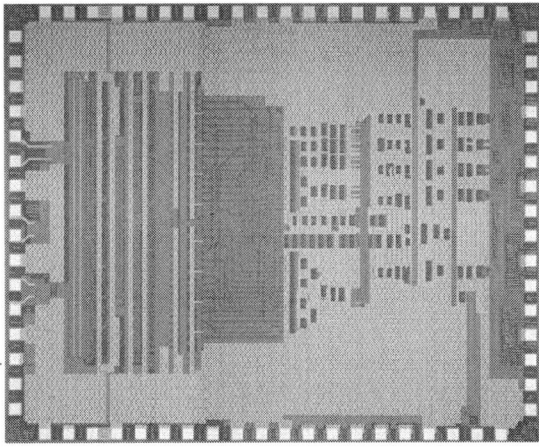

Fig.8. Die photograph.

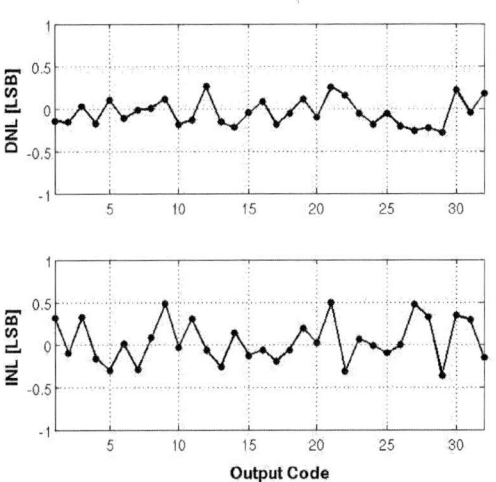

Fig.9. Static performance of the ADC.

third comparator stage (C3) with a bubble error correction to force a clean decision. The input BJTs in the comparator are chosen to guarantee 3σ offset voltage less than 0.2LSB (emitter length of 2.5-μm), while the transistors in the rest are chosen small (emitter length of 0.8-μm) for high-speed operation. The regeneration time constant is estimated about 12ps, mainly dependent on the device cut-off frequency and bias current. In the current-weighted M-S comparator, the bias current of the amplification (I_A=2.2mA) is emphasized over that of the latch (I_L=1.6mA) to suppress random offset and probability metastability, as well as reducing the overall comparator power dissipation. This 3-stage comparator guarantees a metastable error rate of less than 10^{-10} at 20GS/s.

C. *Circuits Implementations*

The routing of 31 differential thermometer traces is the most complicated and needs careful layout design. The balanced differential microstrip lines are used for interconnect with M6 (width of 3-μm for 75Ω) over M1 ground plan and M5 for routing bridge. This microstrip line is also used for clock distribution to reduce aperture jitter (Fig. 7). The encoded binary output can be switched to a test mode for decimated by-64 to enable easy acquisition by a logic analyzer.

IV. Measurement Results

The ADC has been implemented in a 0.18-μm SiGe BiCMOS. The chip occupies 2.63 x 3.3 mm^2 silicon area. Multiple power supplies are used: 3.6V for THA, 3V for analog part, and 3V for digital part. At the maximum clock frequency of 24 GHz, power dissipation is about 3.3W. Digital part including encoder logic and output buffers consumes around 1.6W, whereas the analog part including PA stages, comparator arrays, and line drivers for 31 thermometer codes consumes about 1.2W. The circuit board easily manages several watts power dissipation with simple stacked structure assembly. The ADC was mounted on the ceramic substrate via epoxy glue, which was then recessed into a circuit board. The substrate was carefully mounted on heat sink to minimize thermal resistance between the substrate and heat sink.

At 1 GS/s, measured DNL and INL are below 0.32 LSB and 0.53 LSB, respectively, as shown in Fig. 9. Figure 10 shows a spectrum measured for 1008.8 MHz input signal frequency at f_S = 16GS/s with 1/64 decimation (f_O =250 MHz). Thus, the fundamental spectrum is located at $f_{IN} - 4*f_O$ = 8.8 MHz. The SFDR is 35.8 dBc and effective number of bit (ENOB) is 4.4-bit. Fig. 11 shows the measured SFDR, SNDR, and SNR at the sampling rate of 16GS/s. While the SNR is better than 24dB up to Nyquist frequency, the SNDR is degraded to 21dB at Nyquist. The SNDR degradation at high input frequencies is mainly determined by signal-frequency-dependent spurious tones, rather than aperture jitter of 500fs. The effective resolution bandwidth (ERBW) is 6.5 GHz, and the figure of merit (FoM) of energy per conversion step is 11pJ. The die micrograph of the ADC is shown in Figure 8.

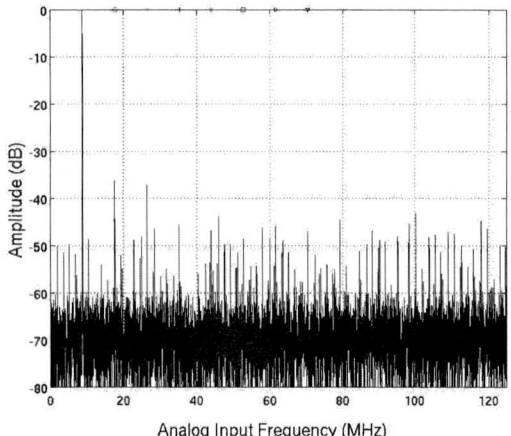

Fig.10. Output spectrum measured at 1008.8MHz input signal frequency. The sample rate is 16GS/s with 1/64 decimation (16384 FFT).

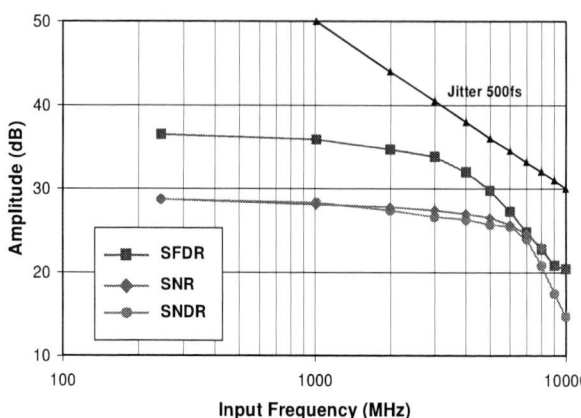

Fig.11. Measured SNDR and SFDR versus analog input frequency at f_S=16GS/s.

V. Conclusion

We report a 5-b 24GS/s ADC with enough resolution bandwidth for 10GBaud 40-Gb/s optical receivers. In order to achieve the best resolution bandwidth of 6.5GHz, a global shunt feedback THA has been embedded in the ADC. ENOB is > 3.5-bit up to Nyquist at 16GS/s. The FoM of 11pJ/conversion step, which is the best in comparison with prior arts, has been obtained with 3.3W power consumption.

Acknowledgements

This research was supported in part by DARPA under contract No. HR0011-05-C-0153 supervised by Drs. Jay Lowell.

	5-bit 20GS/s ADC
Input Frequency (f_{in})	< 10 GHz
Sampling Frequency (f_s)	> 20 GHz
DNL/INL	+/- 0.32 / 0.53 LSB
SNDR @ fin = 1GHz	> 28 dB
SNDR @ fin = 3GHz	> 26.5 dB
ERBW *	> 6.5 GHz
Timing Jitter (RMS)	< 500 fs
Power Dissipation @3/3.6V	~ 3.3 W
FoM **	11 pJ
Technology	SiGe 0.18-µm BiCMOS

Table 1. Summary of the ADC characteristics.

References

[1]. Leven et al., "Coherent Receivers for Practical Optical Communication Systems," OFC, 2007
[2]. K. Poulton et al., "20GS/s 8 Bit ADC with a 1MB Memory in 0.18um CMOS," *ISSCC Dig. Tech. Papers*, pp. 318-319, Feb. 2003.
[3]. J. Lee et al., "A 5-b 10-GS/s A/D Converter for 10-Gb/s Optical Receivers," *IEEE J. Solid-State Circuits*, vol. 39, pp. 1671-1679, Oct. 2004.
[4]. W. Cheng et al., "A 3b 40GS/s ADC-DAC in 0.12um SiGe," *ISSCC Dig. Tech. Papers*, pp. 262-263, Feb. 2004.
[5]. P. Schvan et al., "A 22GS/s 5b ADC in 0.13um SiGe BiCMOS," *ISSCC Dig. Tech. Papers*, pp. 572-573, Feb. 2006.
[6]. H. Nosaka et al., "A 24-Gsps 3-bit Nyquist ADC using InP HBTs for Electronic Dispersion Compensation," *MTT-S Digest*, vol. 1, pp. 101-104, June 2004.

978-1-4244-2018-6/08 $25.00 © 2008 IEEE

Biomedical, Sensors and MEMS

Session 13

Chair: Makoto Nagata, Kobe University
Co-Chair: Steve Garverick, Case Western University

Unique microsystems have emerged from the union of advanced CMOS circuits, sensors, and MEMS technologies. Key circuits includes front-end sensor signal processing and data acquisition, back-end electrical stimulation for biological tissue, and wireless communication between the implantable medical device and an external host transceiver, potentially using MEMS-based frequency references and electromechanical filters.

This session begins with implantable stimulators for biomedical applications. The first paper gives design details of a micro-stimulator that features magnetically rechargeable energy storage, exponential digital-to-analog conversion for stimulation, and wireless communication. The second paper describes a multi-chip, flexible retinal stimulator featuring adjustable photo-sensing to support its stimulation function.

The session continues with a high-resolution sensing techniques realized in CMOS technology. The third paper of the session discusses a digital Hall sensor that employs a MAGFET for magnetic field sensing and TDC for low-power digitization. The integrated sensor/circuit provides a resolution of 0.8 uT for fields as strong as 0.8 T, with a temperature variation of just 250 nT from -40 to +80 degree Celsius. The forth paper describes sensing circuitry for action potentials in extracellular neural recordings, featuring an on-chip successive approximation ADC for selective digitization of the minimum and maximum values of detected spikes. Power consumption is 1 uW from a 1.0-V supply.

The fifth and final paper of the session presents a digitally programmable oscillator that combines a CMOS fractional-N PLL circuit with a MEMS resonator in a single package. An oscillation frequency of 1.8 GHz is achieved with an offset of 600 kHz and a phase noise of -116 dBc/Hz. Frequency resolution is 220 Hz.

978-1-4244-2018-6/08 $25.00 © 2008 IEEE

Notes

978-1-4244-2018-6/08 $25.00 © 2008 IEEE

IEEE 2008 Custom Intergrated Circuits Conference (CICC)

A Biomedical Implantable FES Battery-Powered Micro-Stimulator

Eusebiu Matei, Edward Lee, John Gord, Patrick Nercessian, Phil Hess, Howard Stover, Taihu Li and James Wolfe
Alfred Mann Foundation, 25134 Rye Canyon Loop, Suite 200, Santa Clarita, CA, USA

ABSTRACT

The IC designs of an implantable functional electrical stimulation (FES) battery-powered micro-stimulator are presented in this paper. The battery is recharged by receiving power magnetically through a coil under supervision of a power management unit. To limit the heat generation on the stimulator when a large magnetic field is present, an on-chip rectifier is capable of reducing the Q factor of the coil using a time-varying method. The same coil is also used for generating and receiving magnetic fields such that the distances among 16 different stimulator pairs can be estimated simultaneously. The stimulation current is produced from an 8-bit exponential DAC that has a range between ~2.5µA and ~10mA. The stimulator communicates with an external master control unit through a 400MHz wireless link that can support a bit rate of ~2.5MB/s in high speed mode. The total number of stimulators that the communication system can support is 852.

1. Introduction

Functional electrical stimulation (FES) is a rehabilitation technique for the restoration of lost neurological function that may result from conditions such as stroke and spinal cord or head injury. FES utilizes electrical current pulses applied in programmable patterns to different nerves or reflex centers in the central or peripheral nervous system to produce functional movements. In this paper, an implantable micro-stimulator including different sensing functions such as bio-potential signal sensing, goniometry sensing (measurement of distances among different stimulators inside the body), etc. is presented. The availability of different sensing functions as well as the controllability of multiple micro-stimulators (up to 852) provides clinicians many opportunities to restore neurological functions such as, for example, arm rehabilitation after stroke. In this case, the stimulator can be used to augment muscle strength and exercise the muscles to speed up recovery [1]. It can also be used for reanimating paralyzed limbs [2], bladder control and respiration [3], blocking pain [4], etc.

Most current micro-stimulator designs are powered from an external magnetic source by inductively coupling to a coil inside the micro-stimulator [5]. As a result, a bulky coil outside the body is usually required for operating the stimulators. Furthermore, communications between the implanted stimulators and an external unit are often achieved via the same inductive link. In such case, the communication distances for these systems are often limited to about a foot and only a few stimulators at a low data rate are usually supported (e.g. [6]).

Our research effort is mainly focused on the development of an implantable FES micro-stimulator powered from a 3.6V, 3mAhr lithium ion battery, which is only required to be recharged inductively from an external magnetic source for a few hours once every 3 days. The cross section of the micro-stimulator is illustrated in Fig. 1. Besides the battery, it consists of 16 discrete capacitors, 2 discrete inductors, 2 ICs, a miniature crystal and a coil for recharging the battery inductively and for goniometry sensing. All the components as well as the device case are connected through a flex circuit board (FCB) as shown in Fig. 2. The ICs are mounted on the back of the FCB, which will be folded in half before fitting it inside the coil and then the device case. The cylindrical device case is a hermetically sealed, brazed ceramic case package suitable for protecting the electronics from harmful in vivo environmental conditions such as moisture and salt. The case is designed for a minimum life span of 80 years. One end of the case package is an iridium disc that is used as the stimulation cathode electrode. The opposite end is the stimulation anode electrode, made of platinum-iridium. The two electrodes are also used as an antenna for wireless communication.

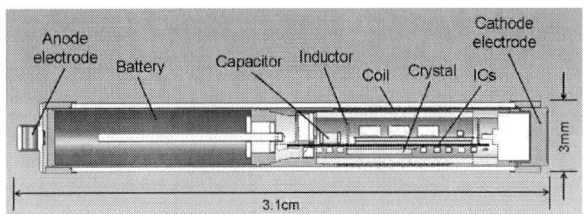

Fig. 1: Micro-stimulator cross section

One of the two ICs inside the stimulator is a high voltage IC that realizes the stimulation and power management function and also controls recharging of the battery. It also provides the capability of sensing bio-potentials and implements the analog front-end circuits for goniometry sensing. The other IC is a low voltage IC that implements the wireless RF transceiver for data communication with an external unit. Digital signal processing functions required for goniometry sensing are also realized in this IC. Due to the requirement of continuous operations without recharging the battery for 3 days, the total current consumption for both ICs is limited to about 40µA on average. As a result, power consumption is one of the main concerns in designing the ICs.

2. High Voltage IC Design

The block diagram of the high voltage IC based on a 0.5µm CMOS technology is shown in Fig. 3. A serial interface is used for communications between the two ICs. The die photo is shown in Fig. 4 and the die area is ~6 × 1.4mm². Realization of each block is described below.

Rectifier with Large Magnetic Field Protection

Reception of a magnetic field for charging the battery is achieved using the coil L_E and the capacitor C_E in Fig. 3.

978-1-4244-2018-6/08 $25.00 © 2008 IEEE

Their values are chosen to be tuned to 125kHz, which is the frequency of the external magnetic source. The induced AC voltage is rectified by an on-chip rectifier as shown in Fig. 5. In charging operations, the PMOSs are normally connected as cross couple pairs acting as a latch, which switches the node that has the higher induced voltage to the rectifier output V_{CHG}. N-diffusion/substrate diodes are used for rectifying the negative peak voltages and supplying the ground current. Since the same magnetic source is used for charging a number of implanted stimulators at the same time, some of the stimulators may be close to the magnetic source and this may result in a very large induced voltage. The rectifier must therefore be equipped with a protection circuit to reduce the induced voltage as well as power such that the temperature of the entire stimulator will not increase more than 2°C per FDA regulations. Although varying capacitance to achieve this is a possible solution, the required capacitors will become too large to be realized on chip. Alternatively, a parallel resistor can be applied to reduce the Q-factor of the resonated LC tank when a large magnetic field is detected. However, heat may still be generated from this resistor. Instead, we propose a time-varying method such that a low resistance path is created by turning on both PMOSs in Fig. 5 only when the differential input voltage is detected by comparators to be near the zero crossings. Since the drain-to-source voltages as well as the currents of the PMOSs are minimal at the zero crossings, heat dissipation is minimized. At the same time, the effective Q-factor of the LC tank is reduced. The low resistance path created by the PMOSs is activated asynchronously when V_{CHG} is detected by a comparator to be larger than 5V. Although this may create random ripples on V_{CHG}, this scheme does not affect charging of the battery and it minimizes the power requirement in the control loop without any compromises during low magnetic field conditions. An on-chip temperature sensor is used for detecting if the stimulator increases more than 2°C. If this happens, the stimulator will be deactivated. For a magnetic field strength of 1 gauss, the rectifier produces ~1.5mA, which is sufficient to charge the battery. Table 1 summarizes the measured increase in temperature when the stimulator is inside different media with different magnetic field strengths.

Battery Charger and Monitor

A simplified diagram for the battery charger and monitor is shown in Fig. 6. The charging process and profile are programmable such that a battery with different charging characteristic can be accommodated. When the charging field is present and the battery is discharged, the battery is charged by M_2 with a pre-charging current of ~100µA until the battery voltage V_{BAT} reaches ~2.5V (this value can be set by R_{CS}). Then the battery will be charged in a constant-current mode with current settings between 1mA and 4mA (step size of 200µA). A power-on-reset signal will be generated and the stimulator starts to operate in normal condition when V_{BAT} reaches ~3.2V and the signal BAT_L turns low. When V_{BAT} is charged to ~3.6V (set by R_{VS}), the battery is considered to be fully charged. The battery will continue to charge but the charging current will decrease gradually (charging in constant-voltage mode) through the feedback loop consisted of A_{CC} and M_5. An end-of-charge (EOC) signal will go high when the charging current drops to <10% of its nominal

setting. EOC will drop low again if the charging current increases to >20% of its nominal constant-current setting and the battery will be recharged again under these conditions. Fig. 7 illustrates the charging process experimentally with a capacitor used to emulate the battery.

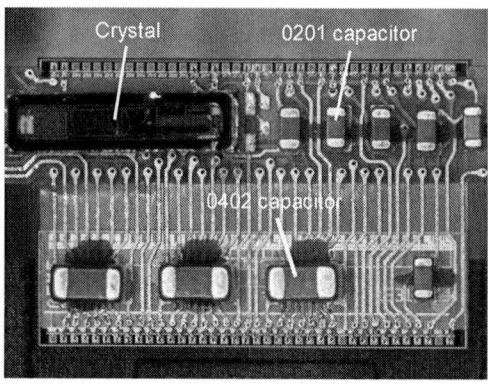

Fig. 2: Flex circuit board with different components

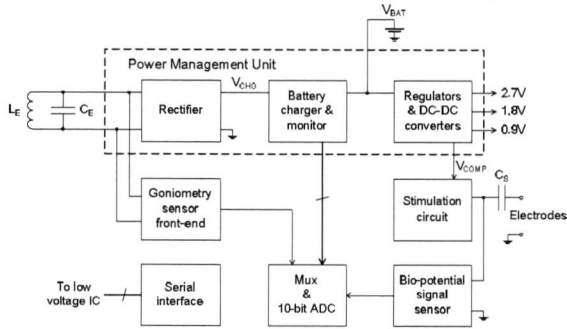

Fig. 3: Block diagram of the high voltage IC

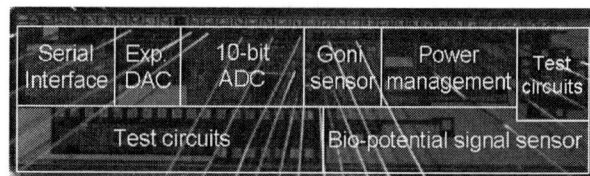

Fig. 4: Die photo of the high voltage IC

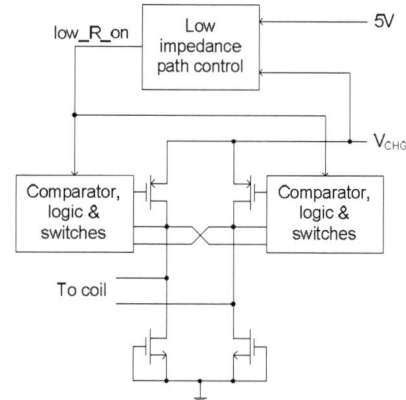

Fig. 5: Rectifier with large magnetic field protection

978-1-4244-2018-6/08 $25.00 © 2008 IEEE   342

Table 1: Increase in stimulator temperature

Magnetic field Strength	Increase in Temperature (°C)			
	Water	Olive oil	Phantom	Pure Fat
5 Gauss	0.4	1.35	0.65	0.75
11 Gauss	–	–	2.0	–
13 Gauss	–	–	–	2.05
20 Gauss	2.3	7.75	–	–

*52.4% water, 1.5% salt, 45% of sugar, 1% of HEC and 0.1% of Dowicil75

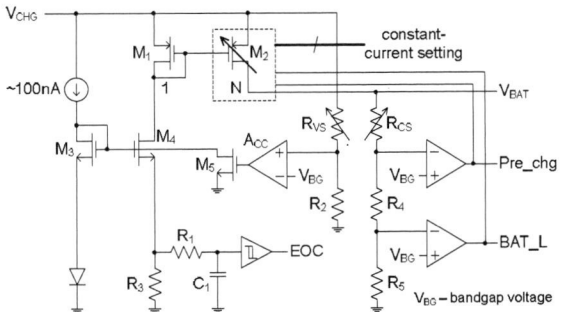

Fig. 6: Diagram for battery charger and monitor

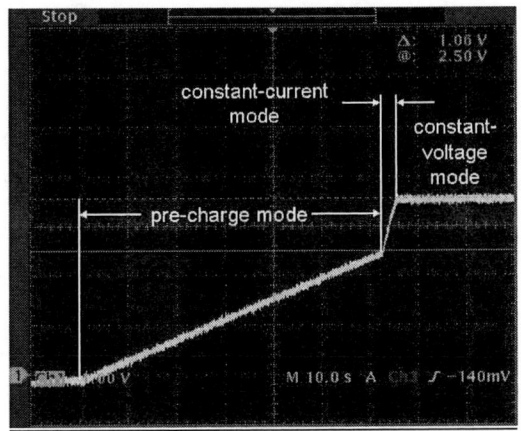

Fig. 7: Illustration of charging process

<u>Voltage Regulators</u>

In the stimulator, the digital circuits in the low voltage IC are powered from a 1.8V supply while most of the analog/RF circuits are supplied from a 2.7V supply, with a minimum V_{BAT} of ~3.2V. A number of linear regulators are used for providing these supply voltages as well as reference voltages for a 10-bit ADC inside the high voltage IC. As discussed later, the RF circuits and the associated digital signal processing circuits only turn on for ~11μs, including a startup time of 3 – 4μs for a communication time slot of 6.4 μs in a 11.2ms time frame. The current drawn from both 1.8V and 2.7V supplies may rise from a few hundred nA to ~5mA during the ~11μs time interval. To minimize power dissipation, the current consumptions of the regulators must be minimal during the off time and must settle before the communication time slot begins.

To obtain fast settling responses with very low bias currents, the regulators are designed with unity gain feedback and fixed loading capacitors. Appropriate reference voltages

are generated (using a bandgap reference) as inputs to the regulators. Fig. 8 shows the proposed voltage regulator structure. Typically, a buffer is used between the first differential stage and the output stage such that it provides a low capacitance load to the first stage and a low impedance to the output stage for fast settling. A buffer with some transient enhancement is also preferable [7]. To minimize the power dissipation, no buffer is used in this design. When the regulator output V_{REG} is reduced due to a step change in load current, both drain currents of M_5 and M_6 will increase. As a result, the tail current of the first stage will increase. Hence, the bandwidth as well as the slew rate will also be increased such that the entire regulator has a fast settling response. The W/L ratio between M_5 and M_6 is 1:40. The quiescent currents of the first stage and the second stage are both ~100nA when the load current is ~0A. Fig. 9 shows the transient response for a step change in load current from 10μA to 5mA. The voltage drop on V_{REG} is < 60mV. The entire power management unit dissipates ~1μA when the stimulator is idle.

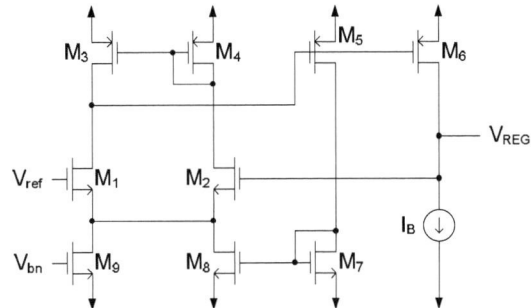

Fig. 8: Proposed voltage regulator structure

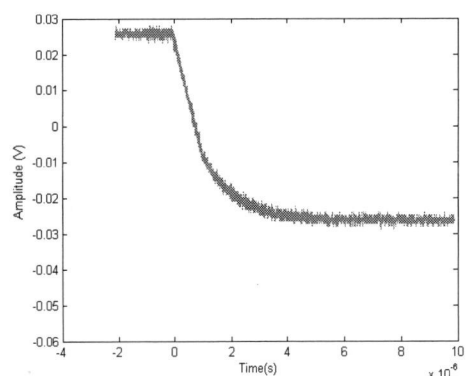

Fig. 9: Transient response of the voltage regulator

<u>Goniometry sensor</u>

Goniometry is the measurement of angles, which in limb implants translates to measurement of the distances among different implanted stimulators. It is based on generating a low frequency (~125kHz) magnetic field from one stimulator and measuring the field strength in other stimulators. The field strength drops with distance rapidly and is inversely proportional to a cube law. Transmitting and receiving magnetic fields are achieved using the charging coil L_E as well as the tuning capacitor C_E shown in Fig. 3. For monitoring multiple movements, like the fingers of one hand, one fixed transmitter can work with several receivers. If more

978-1-4244-2018-6/08 $25.00 © 2008 IEEE

than two different measurements must be obtained simultaneously (e.g., one system for each hand), different measurements must not interfere with each other. To achieve this goal, different measurements may be made utilizing slightly different frequencies around 125kHz with each channel ~90Hz apart. These different frequencies are generated using a direct digital frequency synthesizer inside the low voltage IC. The system can support different measurements using 16 different frequencies simultaneously.

The goniometry transmitter shown in Fig. 10 is powered from a 0.9V supply derived from the 2.7V supply using a switched-capacitor (SC) DC-DC converter in order to reduce power consumption. The MOSFET switches are only turned on for ~5% of each period to compensate the losses in each cycle and sustain the oscillation of the LC tank. Due to the Q factor (~20) of the coil, higher harmonics are suppressed. The output amplitude is ~0.5V.

The goniometry receiver is based on a super-heterodyne architecture. The receiver front-end illustrated in Fig. 11 consists of a low noise, programmable gain pre-amplifier and a switch mixer that down converts the signal to 4kHz. After buffering, the signal is then passed through a 2nd order active RC band-pass filter with programmable center frequency before digitizing with a 10-bit charge redistributed successive approximation ADC sampling at ~16kS/s. Afterwards, the digitized signal is down converted and separated into quadrature I and Q DC components in the digital domain inside the low voltage IC. Cascaded integrator-comb filters with sin(x)/x responses that produce nulls at multiples of ~90Hz are used for filtering out high frequency noise as well as interference from other measurement frequencies. The actual distance is then computed from the amplitudes of the I and Q components followed by a log-to-linear conversion for inverting the cube law characteristic. Both the goniometry transmitter and receiver consume <40 µA from the battery for a single measurement. Fig. 12 shows the measured distances and the corresponding measurement errors. The goniometry sensor can sense a distance up to ~15cm with an error of <1cm that is sufficient for most clinical applications. It can be observed from the figure that closer distances have lower measurement errors.

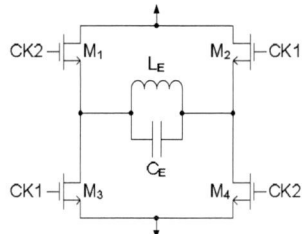

Fig. 10: Goniometry transmitter

Bio-potential signal sensor

The stimulator can also be used to sense bio-potential signals through the stimulator electrodes during the time between stimulation current pulses. Typical bio-potential signals such as central or peripheral nerve signals have amplitudes in the range between 10µV and a few mV with a frequency range between 100Hz and ~2kHz. Capacitors are usually required to block the buildup of DC potential on the electrodes as well as for safety issues. The high-pass pole

associated with the blocking capacitors must be low enough to meet the frequency range requirement. To achieve these requirements, a front-end amplifier design technique that utilizes a close loop configuration is used popularly [8]. In our design, an open-loop front-end amplifier illustrated in Fig. 13 is chosen such that noise of the front-end amplifier is minimized for a given current. Large input PMOSs, M_1 and M_2, are used for minimizing 1/f noise. The gate voltages of these PMOSs are usually pre-set at ground potential. Nevertheless, large PMOSs result in large input capacitances that attenuate the input signal because of the voltage division effect associated with the blocking capacitors. The input capacitances are dominated by C_{GD}'s and given as $A_o \cdot C_{GD}$, due to Miller effects, where A_o is the amplifier gain. These capacitances can be canceled by using capacitors of similar size to C_{GD} in positive feedback. The C_{GD}'s of M_3 and M_4 are used for this purpose. As a result, small blocking capacitors in the range of 50pF can be used. Since M_3 and M_4 are always off, they do not affect the bias condition of the amplifier. The amplifier second stage shown in Fig. 14 has a programmable high-pass pole between 10Hz and 300Hz that eliminates the DC offset of the front-end amplifier. If S_1 is opened after resetting the front-end amplifier, the offset of the entire amplifier will be stored on C_{H2} and the bio-potential sensor will have a very low high frequency pole at 0.035Hz during the sensing period. The output of the amplifier can be digitized for further processing. Furthermore, the bio-potential signal sensor also consists of a threshold detector circuit, which consists of 2 DACs, 2 comparators and digital circuitry for detecting the occurrence of neural spikes. Table 2 summarizes the measured performance of the bio-potential signal sensor.

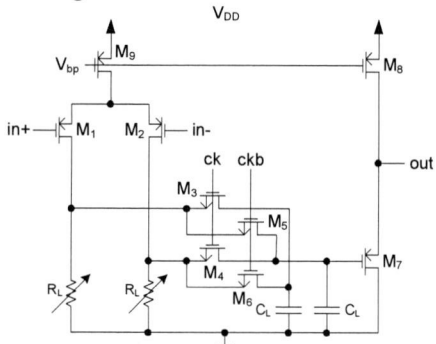

Fig. 11: Goniometry receiver pre-amplifier & mixer

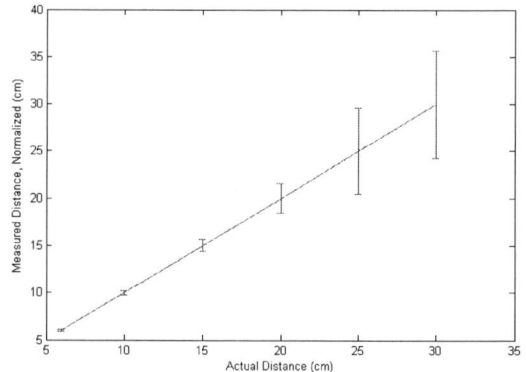

Fig. 12: Measured distances and errors

978-1-4244-2018-6/08 $25.00 © 2008 IEEE

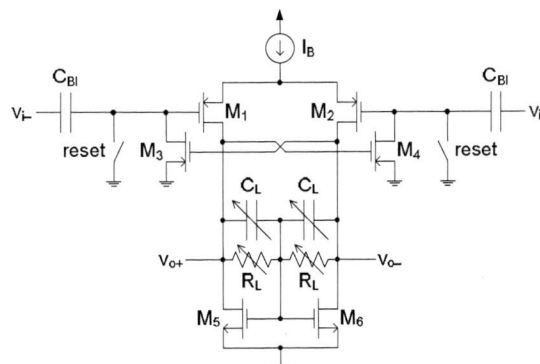

Fig. 13: Bio-potential front-end amplifier

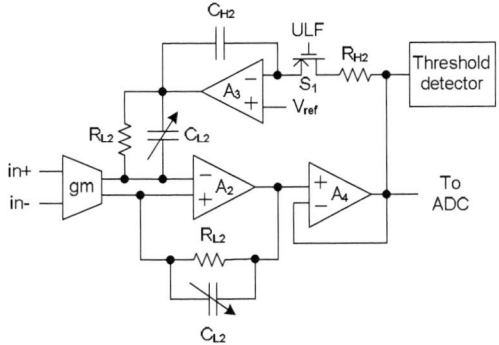

Fig. 14: Bio-potential second stage amplifier

Table 2: Performance summary of the bio-potential signal sensor

Max. power consumption	<33μW
Gain range	4 – 4000
Input referred noise	5.6μV$_{RMS}$ for max. BW (10Hz – 7kHz)
Overall input dynamic range	~82dB for max. BW (10Hz – 7kHz)
Programmable LP corner	300Hz – 7kHz
Programmable HP corner	10 – 300Hz

Stimulation circuit

The stimulation circuit consists of an 8-bit exponential DAC (EDAC) and a recharge circuit. Each incremental step in the EDAC increases the output current I_{out} by 3.3%. The stimulation capacitor C_S shown in Fig. 3 is assumed to be pre-charged to the compliance voltage V_{COMP}, which is generated from a SC DC-DC converter and has a value of either 1 or 2 V_{BAT}'s as necessary to drive current through the load impedance. The electrodes in Fig. 3 are in contact with the tissue, which can be modeled as a resistor in a range between a few hundred ohms to a few tens of kΩs. When stimulation is applied, the EDAC discharges C_S, producing a mono-phasic current pulse in the tissue according to the EDAC setting and the programmed duration (between 5.2μs and 5.2ms). After the stimulation pulse has been applied, C_S is recharged back to V_{COMP} by a current source that has a range between 0 and 500μA with a step size of 7.8μA. A

segmented approach as illustrated in Fig. 15 is proposed for the design of the EDAC. The entire EDAC can be viewed as a current mirror. The 4 LSBs control the equivalent W/L ratio of the diode-connected NMOS (M_L) with a reduction factor of 3.3% for every LSB step change. This small change is accomplished by a parallel connection of M_L with a series chain of NMOSs ($M_{L0} - M_{L15}$). By switching off S_{L1} to S_{L15} progressively according to each LSB step change, the equivalent length of M_{L0} increases and hence, the W/L ratio of the equivalent diode-connected transistor reduces by 3.3%. The V_{GS} of the equivalent diode-connected transistor is used for biasing the current sources in the MSB section. The 4 MSBs activate 1 to 16 of the 16 current sources using a thermometer decoder. The NMOS (i.e. M_{M2} ... or M_{M16}) in each current source increases its W/L ratio by a factor of $(1.033^{16}-1) = 0.681$ relative to the sum of the W/L ratios in the preceding current sources. High voltage transistors ($M_{C1} - M_{C16}$) are used as cascode devices at the stimulation circuit output since the output goes up to V_{COMP}, which has a maximum value of 2 max$\{V_{BAT}\}$'s (i.e., 8V). This segmented approach reduces the area requirement of the EDAC significantly when comparing to a direct transistor scaling approach. The output currents of the EDAC and their corresponding errors when compared to the ideal case are measured and shown in Fig. 16. For typical stimulation profiles, the stimulation circuit consumes ~10μA on average from the battery.

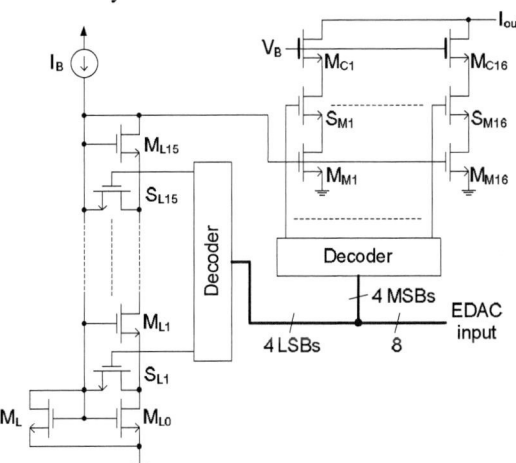

Fig. 15: Proposed exponential DAC design

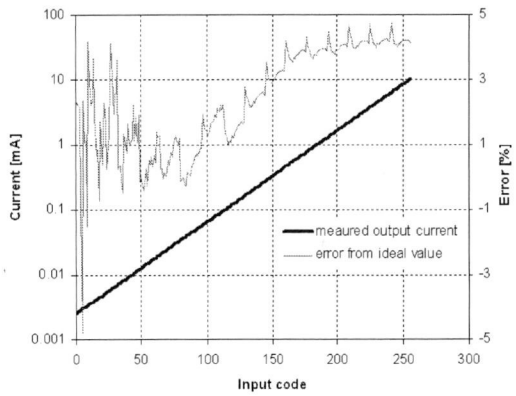

Fig. 16: Output and error of EDAC vs. input code

3. Low Voltage IC Design

The low voltage IC is implemented in a 0.18μm CMOS technology. The die photo is shown in Fig. 17. The die area is ~6 × 1.6 mm^2. The low voltage IC realizes the wireless transceiver that communicates with the external master control unit (MCU) as discussed in [9]. The communication system utilizes a QPSK modulation scheme and can communicate on one of 4 different channels to avoid interference. Each channel has a bandwidth of less than 10MHz. The communication distance is a few meters. Within each ~11.2ms data frame, one or more time slots of 6.4μs can be assigned to each stimulator that is being controlled by one MCU. If only one slot is used for receive or transmit, the effective bit rate is 1.36kB/s, which is adequate for our applications. The low duty cycle in a frame allows the system to support up to 852 stimulators with low average power consumption for each stimulator. To minimize power dissipation, the circuits are only turned on for 3 − 4μs before the assigned time slot. When one stimulator sends out data continuously in streaming (high speed) mode, the transmit bit rate is ~2.5MB/s. Due to signal attenuation within the body at very high frequencies, the carrier frequencies have been selected in the range between 405MHz and 440.2MHz. For this frequency range, attenuation is between 20dB and 30dB, depending on the depth of the stimulator within the body. Since the carrier frequencies are located not far from the TV broadcasting range, blocking due to large TV signals as well as adjacent channel interferences has to be considered.

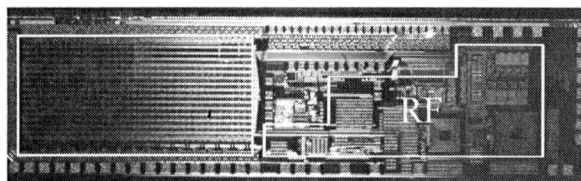

Fig. 17: Die photo of the low voltage IC

Wireless transceiver front-end

The wireless transceiver front-end architecture is shown in Fig. 18. The antenna using the two stimulator electrodes has impedance measured to be 33 − j3.8Ω at 420MHz in a phantom solution, which mimics body tissue. A discrete LC network is connected between the antenna and the RF input/output. It provides a voltage gain of ~13dB for the received signal such that relatively low bias current can be used for the LNA to achieve a low noise figure. The LC network is also used for transforming the impedances between the transmitter and the antenna to achieve good transmitted power efficiency. The design of the LNA is similar to a common source amplifier with an inductor and a programmable capacitor as the load. Gain adjustment of the LNA is achieved using a programmable degenerated source resistor. If the stimulator is only ~10km array from a high power TV, the strong TV signal (< −34dBm) may block the received signal although it is at least 40MHz away from the band of interest. Furthermore, the inter-modulation products due to two strong TV signals may fall within the received band. In this case, the single-ended LNA in combination with the mixer needs to have an IIP3 of ~−5dBm when it is

referred to the discrete LC network input. This is achieved by adjusting the source degeneration of the LNA. The current consumption of the LNA including a source follower output stage for driving the following mixer is ~0.9mA. A sampling mixer structure shown in Fig. 18 is used for achieving high linearity. The mixer is also used for converting the single-ended LNA output signal into differential I and Q signals for the following poly-phase filter/amplifier (PPF) stages as shown in Fig. 19. In this case, the required amplification and poly-phase filtering are merged into one block to achieve low power consumption. The RC poly-phase network provides the required phase shift for the IF current signals from M_1 − M_4, which convert the voltage output signals from the mixer into current signals. The output currents of the poly-phase network are then converted back to voltages by the regulated cascode transistors M_5 − M_8 and resistors R_5 − R_{10}. Each amplifier connected to the cascode transistors is a single stage amplifier with a simple common-mode feedback at the input stage. To achieve an image rejection ratio of >25dB for the IF bandwidth (3.75 − 8.75MHz), the RC poly-phase network has to be at least second order. Since the resistors in the poly-phase network also pass through DC bias currents, the DC voltages across these resistors limit the order in the poly-phase network due to voltage headroom issues. Nevertheless, high voltage gain is required for the IF signals. Hence, two first-order PPF stages in cascade are used. Each PPF stage provides a programmable gain between ~1dB and ~10dB with a gain step of ~3dB by adjusting R_1 − R_6.

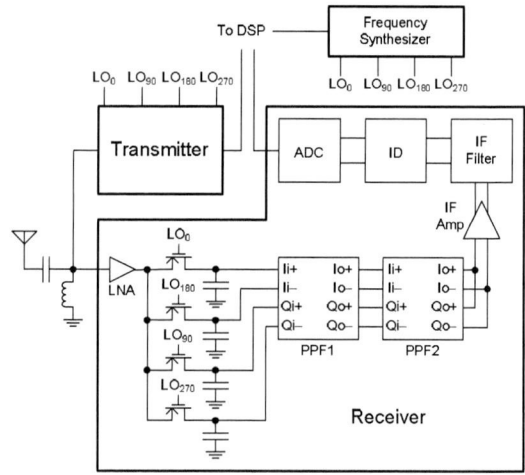

Fig. 18: Wireless transceiver front-end architecture

After the PPF stages, the IF signal is then amplified and low-pass filtered by the variable gain IF amplifier and the IF filter. The IF filter is a 4th order low-pass filter realized using two differential Sallen-Key biquads in cascade. The filter has a cutoff frequency at ~10MHz and provides an attenuation of ~20dB at ~15MHz. A current feedback amplifier structure was chosen for realizing the IF amplifier such that it can provide a relatively constant bandwidth for different gain settings with relatively low current consumption. The same amplifier structure is also used for realizing the differential constant gain amplifiers required in the biquads. Different amplifier gain requirements are achieved by adjusting the resistor values inside each amplifier.

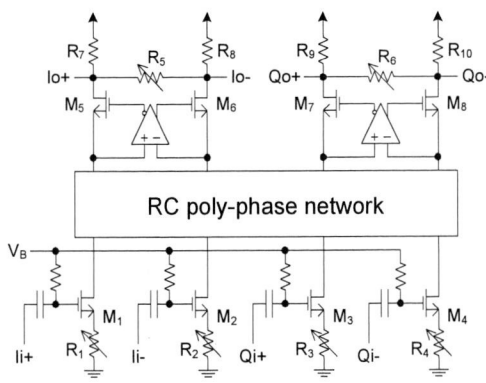

Fig. 19: Poly-phase amplifier/filter stage

The integrate-and-dump (ID) consists of a front-end transconductance (gm) stage and a switched-capacitor (SC) network. The gm stage converts the differential IF input voltage signal into a single-ended current signal, which is integrated, sampled-and-held by a capacitor. The sampled voltage on the capacitor is output to the ADC while another capacitor is receiving the next sample, then the capacitor is reset for another sample. As a result, the SC network provides additional low-pass filtering for the IF signal. The ID output is sampled by a 6-bit, 25MS/s interpolated ADC. The design of the interpolating amplifiers in the ADC is similar to a CMOS inverter with the source voltage of the PMOS controlled by a biasing circuit such that each amplifier operates as a class AB amplifier for low power consumption. The entire ADC dissipates 230µA at 1.8V.

The LO signals required in the receiver and the transmitter are obtained through a divide-by-4 circuit from a PLL based frequency synthesizer, which has an LC VCO operated in the range between 1.6GHz and 1.8GHz. To minimize power dissipation, the receiver (or the transmitter) and the frequency synthesizer are turned on only 3 − 4µs before each assigned time slot. Relatively fast phase/frequency locking is obtained by having a relatively high reference frequency at 24.86MHz and by changing the charge pump current from high to low within the ~3µs. In addition, the charge of the filter capacitor in the frequency synthesizer is held constant when the synthesizer is turned off. The 24.86MHz reference input is obtained from an on-chip oscillator that includes a 12-bit capacitor DAC and an off-chip crystal. The reference input frequency has an adjustment range of about 0.017%.

The transmitter was designed to satisfy the FCC regulation requirements for the transmit spectrum. The baseband ~2.5MB/s I and Q transmit data are passed through two 21-tap, ~25MS/s transmit filters before carrier modulation. Since there are only 3 different I or Q input transmit data within the 21 filter taps and since the transmit data are the same for every 10 filter sample periods, each filter is realized using a ROM lookup table with a simple DAC as shown in Fig. 20 for design simplification as well as low power consumption. For the 3 most recent I or Q input data, each state machine addresses the ROM that outputs the corresponding 10 filter output values. These output values are converted to analog envelope levels of the modulated I and Q signals using two DACs. Since there are only 7 possible non-zero envelope values, the design of the DACs is greatly simplified. After mixing with the LO signals at the current

mirror output stages, the modulated I and Q current signals are summed together at the output node of the transmitter. The signal is then transmitted through the antenna after going through the discrete LC network shown in Fig. 18.

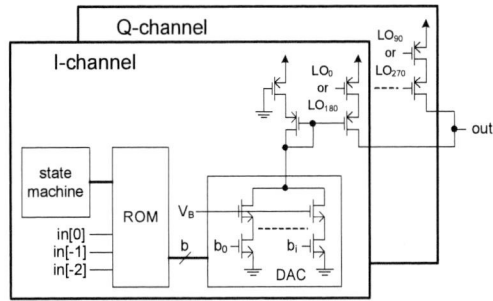

Fig. 20: Simplified diagram for the transmitter

<u>Digital signal processor</u>

During the 6.4µs receive time slot, the IF data from the 6-bit ADC is collected into a 3kB RAM and the digital signal processor (DSP) is turned off. This is to minimize noise coupling from the DSP to the transceiver front-end circuits. After the receive time slot, the DSP clocks are turned on. Fig. 21 is a simplified block diagram for the DSP. The RAM output is fed to a heterodyning band-pass filter followed by a poly-phase low-pass filter to produce I and Q outputs at early, late and center taps. The early and late taps are spaced ±5 taps for the header sequence and are spaced ±3 taps for the normal time slot data. To establish initial synchronization between the stimulator and the external unit, an 11-symbol BPSK Barker sequence is continually transmitted by the external unit in the data frame header slot. After synchronizing the data frame timing and settling of the AGC loop (both determined by the Barker correlation outputs), the DSP will stop capturing the header slot and read only from its assigned time slot. Packets received are encoded in a Reed-Muller(30,15) code, which is related to the BCH(31,16) code. Reference symbols included in the 17 received QPSK symbols of each packet are used for removing phase quadrant ambiguity after phase rotation. Since the packet is relatively short, the mean phase of all the symbols in a packet is sufficient to determine the rotation of the packet. The demodulated, rotated, decoded data are passed to a micro-code engine along with the detected error information to recode the data into codewords, which are correlated against the stored filter outputs to produce estimations on timing error. Based on this error, a digital PLL is used to track the phase of the data frame, and adjusts slowly the capacitor DAC for pulling the 24.86MHz reference frequency. The AGC loop is also updated continuously using normal time slot data. Transmission from the stimulator to the external unit uses a separate code that only requires 12 QPSK symbols, leaving 2 symbols of silence on either side of the transmitted packet as guard bands for other stimulators on the adjacent time slots. The micro-code engine inside the DSP also allows users to download a small program for customizing some operations for the stimulator. In addition, there is a 384-bit non-volatile memory, which stores some unique parameters for a specified stimulator such as the stimulator ID number and the parameters for the wireless transceiver.

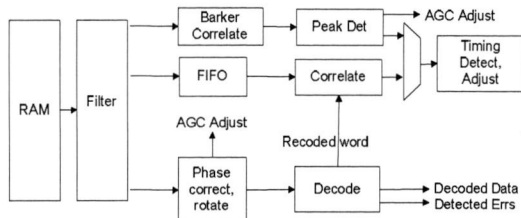

Fig. 21: Block diagram for the DSP

Wireless transceiver performance

When the stimulator is immersed ~20cm deep inside a phantom solution, the communication distance with the external unit is about 2m. The distance increases to over 50 feet in air. Communication between 16 stimulators and one MCU simultaneously has also been verified. The maximum transmit power was measured to be –2.3dBm. For this output power level, the peak powers of the side lobes are less than the total transmit power by 32dB, which satisfies the FCC requirement of <30dB. The total average current consumptions from the 3.6V battery for transmitting data using only 1 time slot is ~8μA, and is <7mA when using continuous streaming mode at ~2.5MB/s. The entire receiver has a minimum noise figure of 6.2dB and the minimum detectable signal level (MDS) with no interference was measured to be <–95dBm with 8dB SNR. The bit error rates for different input SNRs are shown in Fig. 22. The overall degradation when compared to the ideal SNR requirements for demodulating QPSK symbols is ~0.8dB for this code rate. When only 1 transmit time slot and 1 receive time slot are used in a data frame, a total average current of ~23μA is consumed from the battery. The transceiver performance is summarized in Table 3.

4. Summary

The IC implementations of a biomedical implantable FES battery-powered micro-stimulator are described in this paper. The battery is recharged from an external magnetic source through an on-chip rectifier with large field protection. The micro-stimulator consists of a bio-potential signal sensor, a goniometry sensor, and a stimulation circuit that can provide a wide range of stimulation current pulses for various clinical applications. Communication with the external control unit is accomplished using a 400MHz wireless transceiver, which has a maximum transmit bit rate of ~2.5MB/s. The wireless receiver has a MDS of <–95dBm for no interference. A maximum of 852 stimulators can be supported by the communication system. For typical clinical applications, the average current consumption of the stimulator from the battery is <40μA, such that it is only required to be recharged for a few hours every 3 days. The stimulator is currently undergoing animal trials.

REFERENCES

[1] L. Baker, C. Wederich, D. McNeal, C. Newsam, and R. Waters, NeuroMuscular Electrical Stimulation, 4th ed. Downey, CA: Los Amigos Research & Education Institute, 2000.

[2] T. Cameron, G. Loeb, R. Peck, J. Schulman, P. Strojnik, and P. Troyk, "Micromodular implants to provide electrical stimulation of paralyzed muscles and limbs," IEEE Trans. Biomed. Eng., vol. 44, no. 9, pp. 781–790, Sept. 1997.

[3] A. Talalla, "Electrical stimulation for bladder control,", Pacing and Clinical Electrophysiology, vol. 9, pp. 164–170, Mar. 1986.

[4] R. North and F. Wetzel, "Spinal cord stimulation for chronic pain of spinal origin: a valuable long-term solution," Spine, vol. 27, pp. 2584–2591, Nov. 2002.

[5] W. Heetderks, "RF powering of millimeter and submillimeter-sized neural prosthetic implants," IEEE Trans. Biomed. Eng., vol. 35, pp. 323–327, May 1988.

[6] P. Troyk and G. DeMichele, "Inductively-Coupled Power and Data Link for Neural Prostheses Using A Class-E Oscillator and FSK Modulation," Proc. IEEE 25th EMBS Conf., pp. 3376–3379, Sept. 2003.

[7] M. Al-Shyoukh, R. Perez, and H. Lee, "A transient-enhanced 20-μA-quiescent 200 mA-load low-dropout regulator with buffer impedance attenuation," Proc. of IEEE 2006 CICC, pp. 615–618, Sep. 2006.

[8] R. Harrison, "A Versatile Integrated Circuit for the Acquisition of Biopotentials," Proc. of IEEE 2007 CICC, pp. 115–122, Sep. 2007.

[9] E. Lee, P. Hess, J. Gord, H. Stover and P. Nercessian, "A 400MHz RF transceiver for implantable biomedical micro-stimulators," Proc. of IEEE 2007 CICC, pp. 173–176, Sep. 2007.

Table 3: Transceiver performance Summary

MDS	< –95dBm (no interference) –81.2dBm (2 TV interferences @ –34dBm) –87dBm (2 adjacent channel signals @ –54dBm)
Image rejection	> 30dB (2.9MHz – 9.5MHz)
RX gain range	33dB – 78dB (~3dB gain step size)
TX power range	–2.3dBm to –9dBm
Current consumptions	Freq. synthesizer: 440μA @ 2.7V & 1mA @ 1.8V RF receiver: 2.5mA @ 2.7V RF transmitter: 3mA @ 1.8V Total avg. for 1 TX and 1 RX slot: ~23μA @ 3.6V Total for streaming out @ 2.5MB/s: <7mA @ 3.6V

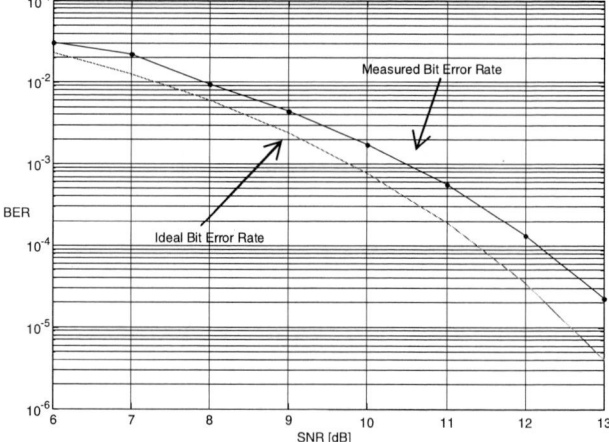

Fig. 22: Wireless receiver bit error rate plot

IEEE 2008 Custom Intergrated Circuits Conference (CICC)

CMOS LSI-based multi-chip flexible retinal prosthesis device for subretinal implantation

T. Tokuda[1], S. Sawamura[1], Y. Terasawa[2], Y. Tano[3], and J. Ohta[1]

[1]Graduate School of Materials Science, Nara Institute of Science and Technology (NAIST)
Takayama-cho 8916-5, Ikoma, Nara, 630-0192, JAPAN
[2]Vision Institute, R&D Div., NIDEK Co., Ltd., Hama 73-1, Gamagori, Aichi, JAPAN
[3]Graduate School of Medicine, Osaka University, 2-15 Yamadaoka, Suita, Osaka 565-0871, JAPAN

Abstract— A CMOS-based multi-chip flexible retinal stimulator with light-sensing function was designed. The light sensing function is implemented for light-controlled retinal stimulation in subretinal configuration. The basic functionality and feasibility of the multi-chip flexible retinal stimulator was confirmed. A photo-activating operation with adjustable photosensitivity was successfully implemented on the retinal stimulator without any additional input line. We demonstrated and characterized the photosensing function.

I. INTRODUCTION

LSI-based neural prosthesis / rehabilitation technology is attracting a lot of interest as the new therapeutic platform for various neural disabilities [1-8]. Retinal prosthesis is the neural prosthesis in which LSI-based device will play an essential role [3-8].

Considering that the complexity of visual information, retinal stimulator for retinal prosthesis must have many channels (1000 or more). It is no more realistic to design a multi-channel retinal stimulator without LSI multiplexer.

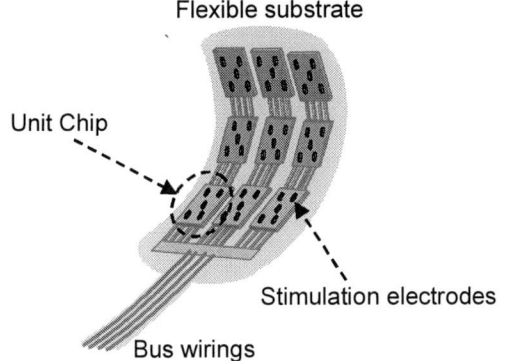

Figure 1. Concept of multi-chip flexible retinal stimulator.

We proposed an architecture for retinal stimulator named "multi-chip flexible stimulator" [9-12]. Fig. 1 shows the concept of the multi-chip flexible retinal stimulator. The stimulator consists of an array of small-sized neural stimulator named "unit chip" mounted on a flexible substrate. The unit chips are connected to the four-wire bus line and can be operated cooperatively. Appropriate spacing between the unit chips provide flexibility and the multi-chip stimulator can be bent to fit the curve of an eye.

In this paper, we present design, packaging and functional demonstration of the multi-chip flexible retinal stimulator. We also implement photosensing function onto the unit chip and demonstrate its function.

II. DESIGN AND PACKAGING OF MULTI-CHIP FLEXIBLE RETINAL STIMULATOR

We designed the unit chip using 0.35μm 2-poly, 4-metal standard CMOS fabrication technology. The operating voltage of the unit chip is 5 V. Fig. 2 shows block diagram and layout of the unit chip. Size of the unit chip is 600μm x 600μm. The unit chip has four inputs and nine stimulation electrodes. The input lines are; GND, VDD, CONT, and STIM. The basic circuit design is almost compatible with previously reported unit chips [11, 12]. In this work, we have implemented additional circuitry for photo-activation of the stimulation electrode, as shown in Fig. 2.

CONT is used for addressing and STIM is used for stimulation input. Each unit chip has 10-bit asynchronous counter as an address buffer. The unit chip counts the digital pulses applied on CONT line and interprets it as the address of stimulation electrode on specific unit chip assembled on the multi-chip retinal stimulator. The higher six bits are used to identify the unit chip to be used, and the lower four bits are used to define the stimulation electrode to be used on the selected unit chip. Only the selected electrode on the selected unit chip is connected to the STIM input via a transmission-gate switch circuitry. In the current unit chip with photosensing circuitry, the connection between STIM input and the selected stimulation electrode is established only in the case light is detected by the photosensing circuit of the corresponding stimulation electrode. Configuration of the

978-1-4244-2018-6/08 $25.00 © 2008 IEEE
349

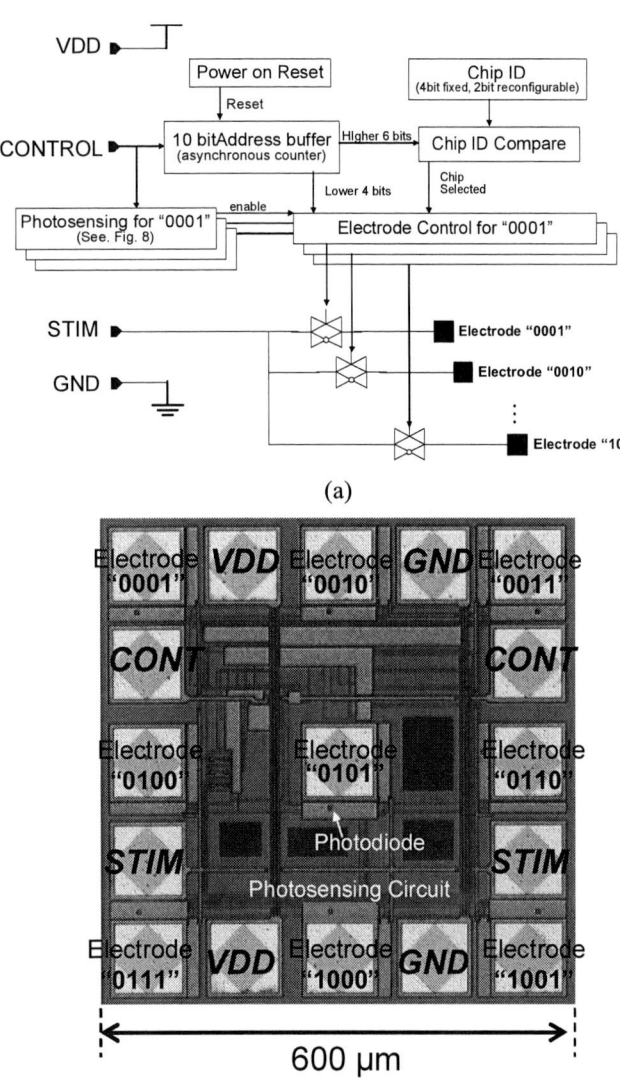

(a)

(b)

Figure 2. (a) Block diagram and (b) layout of the unit chip.

photosensing circuitry and operating scheme of photo-activation are described in section IV. The on-resistance of the switch is smaller then 200 Ohm in whole voltage region. We can perform retinal stimulation using any stimulation electrode on the multi-chip flexible retinal stimulator. In the current architecture, the stimulation signal is generated by an off-chip stimulator and the stimulation is performed in single-site stimulation mode.

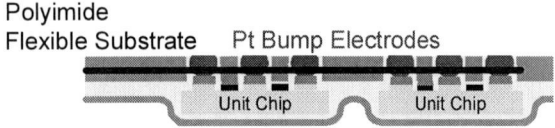

Figure 3. Structure of the multi-chip flexible retinal sitmulator.

The packaging is an essential issue for the CMOS-based retinal stimulator, too. Fig. 3 shows the current structure of the multi-chip retinal stimulator. We designed a CMOS die on which 4 x 4 unit chip circuits aligned in 1100µm pitch. We form Au bumps on the input and stimulation pads on the unit chip circuitry and assemble the die onto a polyimide flexible substrate using flip-chip bonding technique. Then, we separate the unit chip circuits using Deep RIE (reactive ion etching) process, and seal the backside of the unit chips with Teflon© film and epoxy molding.

We form Pt bump stimulation electrodes on the pre-formed Au bumps on the stimulation pads using conventional ball bonder. The outline of the stimulator was formed with excimer laser cutting process.

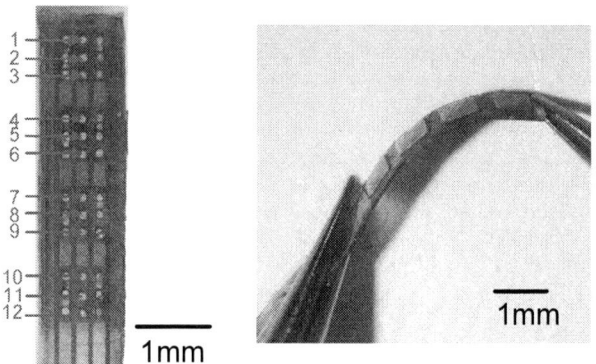

Figure 4. Multi-chip retinal stimulator with 1 x 4 configuration

Fig. 4 shows typical shape of fabricated retinal stimulator. The stimulator was designed for animal experiments using rabbits, and has a configuration of 1 x 4 unit chips. As shown in Fig. 4, the multi-chip stimulator has sufficient flexibility to fit the curvature of rabbit's eye.

III. EXPERIMENTAL CONFIRMATION OF THE FLEXIBILITY AND FUNCTIONALITY OF ADDRESSING AND CURRENT INJECTION

We have performed animal experiments to demonstrate the surgical compatibility and basic function (addressing and current injection) of the multi-chip flexible retinal stimulator without the photosensing function. Dutch belted rabbits with normal sight were used for the experiments. All the experiments were carried out following Institutional Guidelines of Osaka University and the ARVO Resolution on the Use of Animals in Ophthalmic Research.

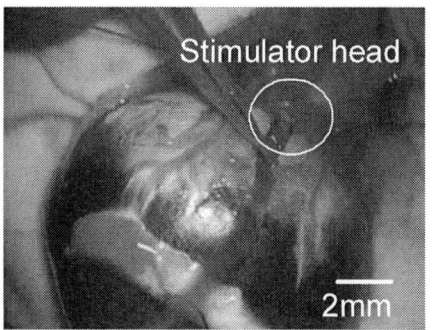

978-1-4244-2018-6/08 $25.00 © 2008 IEEE

Figure 5. Retinal stimulator inserted on rabbit's sclera

Prior to the retinal stimulation experiments, a rabbit was anesthetized and the retinal stimulator was implanted on the eye. The stimulator was implanted in a configuration for suprachoroidal transretinal stimulation (STS), in which the stimulator head was inserted in a scleral pocket. Fig. 5 shows a photograph of the retinal stimulator implanted into the scleral pocket during the experiments. A Pt counter electrode was inserted into vitreous.

The response in the rabbit's brain evoked by the retinal stimulation was observed by two screw electrodes inserted into the skull. One of the electrodes was configured to sense the field potential in visual cortex of the rabbit's brain, and the other electrode was used as a reference electrode. The detailed experimental configuration and procedure are described in ref. 12.

Figs. 6 and 7 show traces of electrically evoked potential (EEP) observed in the visual cortex measured via the screw electrodes. All the retinal stimulation was performed in anodic monophasic, current-controlled injection.

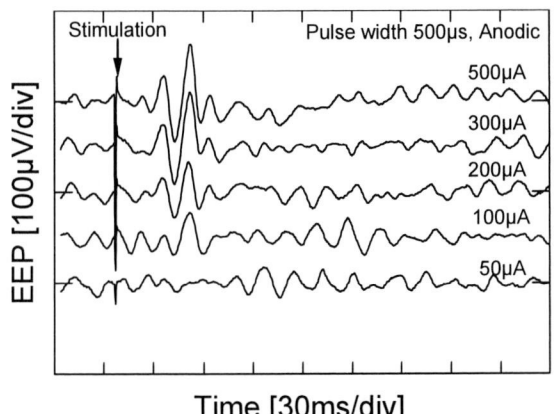

Figure 6. EEP signals evoked by the stimulation with the multi-chip stimulator. The electrode 2 (see Fig. 4) was used for stimulation.

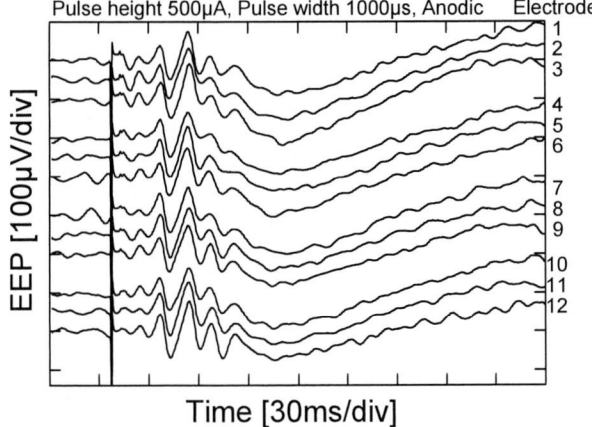

Figure 7. EEP signals evoked by the stimulation with the multi-chip stimulator. Electrodes 1-12 (see Fig. 4) were used.

Fig. 6 shows the EEP responses evoked by the stimulus with different injection current. The pulse duration was fixed

as 500µs, and the electrode 2 shown in Fig. 4 was used. Clear responses were observed in EEP traces by the stimulation with 100µA or larger. This threshold condition is reasonable for this type of retinal stimulation experiments. Therefore, we can conclude the fabricated multi-chip retinal stimulator is applicable for the animal experiments toward retinal prosthesis.

Fig. 7 shows the EEP traces observed by the stimulation trials using different electrodes. The stimulation condition was fixed as 500µA/1000µs. Clear responses were observed in EEP traces for the stimulation trials from either electrode. These results indicate that the multi-chip stimulator was correctly working, and the retinal stimulation was effectively performed in multi-chip operation mode.

IV. IMPLEMENTATION OF PHOTOSENSITIVITY FOR SUBRETINAL IMPLANTATION

As the photosensing circuitry, a simple active pixel sensor (APS) circuitry with digitized output was configured onto each stimulation electrode (Fig. 8). Reset transistor of the APS circuitry is closed when L (0 V) is applied on the CONT input, and opened with H (5 V). The level of photodiode (PD) is digitized by an inverter and latch circuitry. The output is latched during L is applied on CONT input. Thus, the APS circuitry implemented on each electrode accumulate photocarrier during H level is applied on CONT, and keep the digitized result during L is applied. We use this photosensing result to determine the stimulation electrode is activated or not. Threshold light intensity for electrode activation depends on the pulse length. It increases for the shorter pulse, and decreases for longer pulse. The number of pulses applied on CONT was used to identify the address of the stimulation electrode, and length of the last pulse for addressing was used to define the accumulation time of the APS circuitry. Using this control scheme, we succeeded to implement an adjustable photosensitivity without any additional control input.

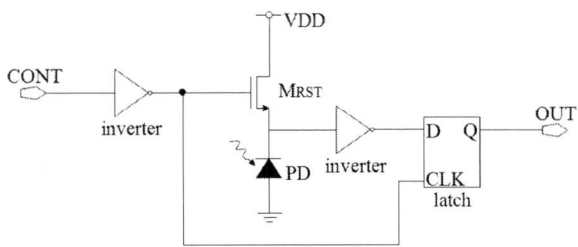

Figure 8. Photosensing circuitry implemened on the retinal stimulator

Fig. 9 shows the results obtained in an experimental demonstration. Alternative digital pulses are applied on CONT and STIM, and the output signals were monitored on the electrodes with addresses of "0000000111" and "0000001001". When the unit chip was illuminated, pulse applied on STIM just after the seventh (or ninth) pulse on CONT pulse were transferred to the electrode "0000000111" (or "0000001001"). However, the pulses were not transferred in dark situation. These results indicate that the implemented photosensing function is working correctly.

978-1-4244-2018-6/08 $25.00 © 2008 IEEE

Fig. 10 shows the threshold light intensity as a function of accumulation time (= length of the CONT pulse). The adjustability of the sensitivity in light sensing was clearly demonstrated by this result. As is expected from the implemented measurement scheme, the threshold light intensity is nearly inversely proportional to the accumulation time.

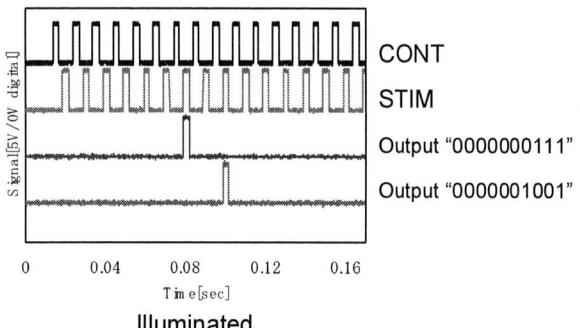

Illuminated

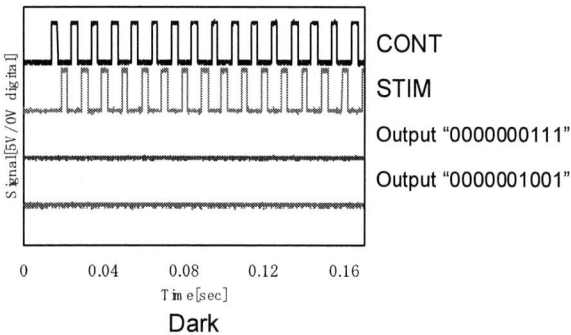

Dark

Figure 9. Sitnal tranfer demonstrations in illuminated / dark situations

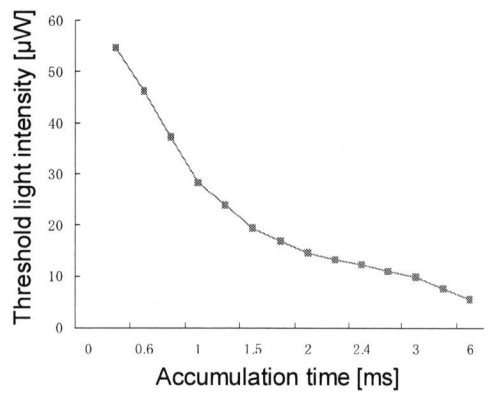

Figure 10. Threshold light intensity of the photosensing function

V. CONCLUSIONS

A CMOS-based multi-chip flexible retinal stimulator with photosensing function was designed. The photosensing function was adopted to apply the stimulator for light-controlled retinal stimulation in subretinal configuration. The flexibility, surgical compatibility, addressing and current injection functions of the stimulator were demonstrated. The photo-activation of the stimulation electrode provided by the photosensing circuit was demonstrated, too. Using the control (addressing) pulse to reset the APS circuitry, we have successfully implemented the photosensing function with adjustable sensitivity without any additional input.

ACKNOWLEDGMENTS

This work was supported by a Grant for Practical Application of Next-Generation Strategic Technology from The New Energy and Industrial Technology Development Organization (NEDO), Japan, and by Health and Labor Sciences Research Grants from the Ministry of Health, Labour, and Welfare of Japan.

REFERENCES

[1] K. Najafi, K.D. Wise, "An implantable multielectrode array with onchip signal processing," IEEE J. Solid-State Circuits, vol. sc-21, pp. 1035–1986 ,1986.

[2] C. Kim, K.D. Wise, "A 64-site multishank CMOS low-profile neural stimulating probe," IEEE J. Solid-State Circuits, vol. 31, pp. 1230–1238, 1996.

[3] M. S. Humayun, E. Jr. de Juan, J. D. Weiland, G. Dagnelie, S. Katona, R. Greenberg, and S. Suzuki, "Pattern electrical stimulation of the human retina," Vision Researchm vol. 39, pp. 2569-2576, 1999.

[4] S. C. DeMarco, W. Liu, P. R. Singh, G. Lazzi, M. S. Humayun, and J. D. Weiland, "An arbitrary waveform stimulus circuit for visual prostheses using a low-area multibias DAC," IEEE J. Solid-State Circuits, vol. 38, pp.1679-1690, 2003.

[5] E. Zrenner, A. Stett, S. Weiss, R. B. Aramant, E. Guenther, K. Kohler, K. D. Miliczek, M. J. Seiler, and H. Haemmerle, "Can subretinal microphotodiodes successfully replace degenerated photoreceptors?," Vision Res., vol. 39 pp. 2555-2567, 1999.

[6] T. Watanabe, R. Kobayashi, K. Komiya, T. Fukushima, H. Tomita, E. Sugano, H. Kurino, T. Tanaka, M. Tamai, and M. Koyanagi, "Evaluation of Platinum-Black Stimulus Electrode Array for Electrical Stimulation of Retinal Cells in Retinal Prosthesis System," Jpn. J. Appl. Phys, vol. 46, pp. 2785-2791, 2007.

[7] J. F. Rizzo III, J. Wyatt, J. Loewenstein, S. Kelly, and D. Shire, "Methods and Perceptual Thresholds for Short-Term Electrical Stimulation of Human Retina with Microelectrode Arrays," Invest. Ophthalmol. Vis. Sci., vol. 44, pp. 5355-5361, 2003.

[8] J. Ohta, N. Yoshida, K. Kagawa and M. Nunoshita, "Proposal of application of pulsed vision chip for retinal prosthesis," J. Jpn. Appl. Phys., vol. 41 pp. 2322-2325, 2002.

[9] T. Tokuda, Yi-Li Pan, A. Uehara, K. Kagawa, M. Nunoshita, J.Ohta, "Flexible and extendible neural interface device based on cooperative multi-chip CMOS LSI architecture," Sensors & Actuators: A, vol. 122, pp. 88-98, 2005.

[10] J. Ohta, T. Tokuda, K. Kagawa, T. Furumiya, A. Uehara, Y. Terasawa, M. Ozawa, T. Fujikado, and Y. Tano: IEEE Engineering in Medicine and Biology Magazine, vol. 25, pp. 47-59, 2006.

[11] T. Tokuda, S. Sugitani, M. Taniyama, A. Uehara, Y. Terasawa, K. Kagawa, M. Nunoshita, Y. Tano, and J. Ohta, "Fabrication and validation of a multi-chip neural stimulator for in vivo experiments toward retinal prosthesis", Jpn. J. Appl. Phys, vol. 46, pp. 2792-2798, 2007.

[12] T. Tokuda; R. Asano; S. Sugitani; M. Taniyama; Y. Terasawa; M. Nunoshita; K. Nakauchi; T. Fujikado; Y. Tano; J. Ohta, "Retinal stimulation on rabbit using CMOS-based multi-chip flexible stimulator toward retinal prosthesis," Jpn. J. Appl. Phys. 47 (4B), 2008, in press.

IEEE 2008 Custom Intergrated Circuits Conference (CICC)

A CMOS TDC-Based Digital Magnetic Hall Sensor Using the Self Temperature Compensation

Young-Jae Min, Soo-Won Kim

Department of Electronics and Computer Engineering

Korea University, Seoul, Korea

Abstract-A CMOS TDC-based digital magnetic Hall sensor using the self temperature compensation schemes is proposed. The proposed sensor consists of the sensor device, its bias and signal-processing circuit which is fully compatible with a standard CMOS technology. For the high magnetic sensitivity, the MAGFET is implemented with proper geometric parameters. The TDC-based digital circuit is proposed for the low power consumption and easy design. The self temperature compensation schemes of mobility and threshold voltage temperature effects are utilized for the low temperature variation. The proposed sensor is implemented with a standard 0.18μm CMOS technology and shows the good detectable resolution by 0.8μT under magnetic fields ranging from 0 to 0.8T and temperature variations by 250nT ranging from -40°C to +80°C. And the very low power consumption by 360μW at 1.8V supply voltage is measured.

I. INTRODUCTION

Recently published and commercial integrated micro-sensors have on-chip bias and signal-processing circuits replacing both discrete sensor devices and circuits. Low cost, easy design and high reliability are reached when the sensor device fabrication is compatible with integrated circuit technologies such as CMOS. The Magnetic Hall sensor is a transducer which converts a magnetic field into a corresponding electrical signal. Magnetic Hall sensors have many applications in fields as magnetic compass applications, DC brushless motor controls, computer storage devices, automotive applications, medical or process controls, etc [1]-[3]. The magnetic MOSFET (MAGFET) with the split drain structure is one kind of magnetic Hall sensor. The major advantage of the MAGFET is a fully compatibility with a standard CMOS technology [4], [5].

In conventional CMOS magnetic Hall micro-sensors, analog-to-digital converters (ADCs) were integrated to convert physical quantities of a magnetic field into corresponding digital output signals [4]-[8]. The Fig. 1 shows the typical block diagram of conventional magnetic Hall sensors. Because the magnetic Hall sensor devices give a weak output signal of the order of a few millivolts, its amplification is required. However, the magnetic Hall sensor devices suffer from imperfections that limit their precision. In particular, The CMOS magnetic Hall sensor devices such as MAGFET have the magnetic sensitivity drift due to temperature variations, mechanical stresses and ageing. In addition, operational amplifiers (OPAMPs) made in a CMOS technology have a large offset that is also subject to temperature variations, mechanical stresses and ageing.

Recent researches of magnetic field sensors focused on the reduction of these measurement errors. The offset cancellation of OPAMPs and the variation calibration of magnetic Hall sensor devices would become larger size and more complicated system.

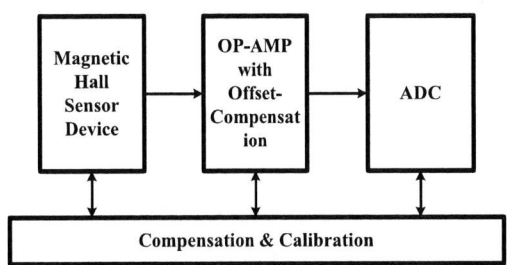

Fig. 1. Block diagram of conventional magnetic Hall sensor.

In this paper, a new CMOS time-to-digital converter (TDC) based digital magnetic Hall sensor to reduce temperature variations without additional compensation circuitry and calibration system is proposed. This paper is organized as follows. In Section II, the operation principle of the MAGFET is presented. Section III gives the architectures and techniques of the proposed magnetic Hall sensor circuits for the low temperature variation and power consumption. Section IV shows implementations and experimental results of the proposed design. And finally conclusions are drawn in Section V.

II. SENSOR DESCRIPTION

The Hall effect due to the Lorentz force on charge carriers is the basic principle of the MAGFET operations. The MAGFET with the split drain structure shown in Fig. 2 senses the magnetic field which is perpendicular to the channel of the device.

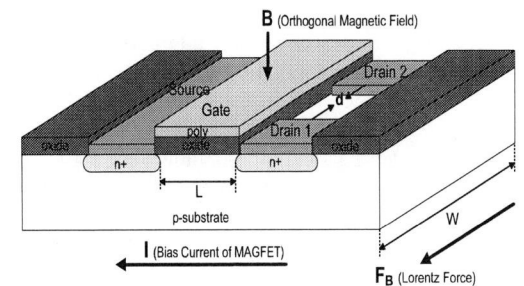

Fig. 2. Physical structure of n-channel MAGFET.

978-1-4244-2018-6/08 $25.00 © 2008 IEEE

In the presence of a magnetic field B perpendicular to the device surface, the current in the channel travelling from source to drain will be deflected by the Lorentz force F_B. By a current imbalance, differential current ΔI between the two drains generates as follow

$$\Delta I = I_{d1} - I_{d2} , \qquad (1)$$

where I_{d1} and I_{d2} represent each currents at Drain 1 and Drain 2. The magnetic sensitivity S of the MAGFET is defined by the following equation [9],[10]

$$S = \frac{\Delta I}{I \cdot B} , \qquad (2)$$

where $I = I_{d1} + I_{d2}$. Applying large gate or drain voltage will only increase the bias current of the MAGFET but not the differential current. Thus to enlarge bias current which is a denominator of (2) is not possible solution to obtain the high magnetic sensitivity.

As mentioned, the magnitude of the current imbalance is related to current deflection by Lorenz force. The current deflection ΔZ in the magnetic field B with the channel mobility μ_n of the MAGFET can be expressed as [11]

$$\Delta Z = L \cdot \mu_n \cdot B , \qquad (3)$$

where L is the length of the MAGFET. From (3), the proper large length and high channel mobility will increase the magnetic sensitivity.

The cascade of two or more MAGFETs can improve the sensitivity [12]. But the cascade structure increases the power consumption. In this paper, in order to achieve high magnetic sensitivity with low power consumption, only one MAGFET with the proper geometric parameters is proposed. The geometric parameters are W/L = 80μm / 40μm and d = 4μm.

III. CIRCUIT DESCRIPTION

The block diagram of the proposed circuit is depicted in Fig. 3. The proposed circuit consists of the bias of the MAGFET with diode connected transistors, magnetic field-to-pulse generator (MPG) by the function with two NOT gates and one XOR gate and TDC with 8-bit ripple carry counter. If the bias and MAGFET outputs $V_{bias}+$, $V_{bias}-$, which are corresponding differential voltage signals of a magnetic field, is applied in the MPG, the pulse P_{out} of the delay difference generates. The pulse-width corresponds to the magnetic flux density. The TDC converts this pulse-width into the 8-bit digital signals D_{out}.

In this proposed circuit, no additional temperature compensation circuitry and calibration system is implemented. For the compensation of the temperature variations and low power consumption, the self compensation schemes of the mobility and threshold voltage temperature effects [14], [15] are utilized in the bias and TDC. And a differential architecture of the TPG is proposed for the low temperature variation. The details circuits are discussed as following.

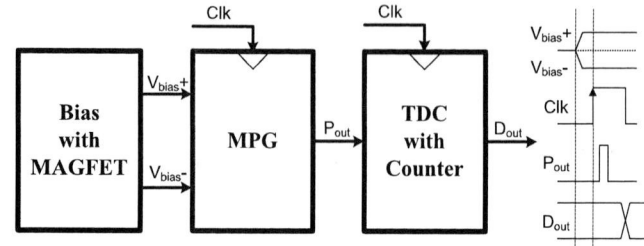

Fig. 3. Block diagram of proposed magnetic Hall sensor.

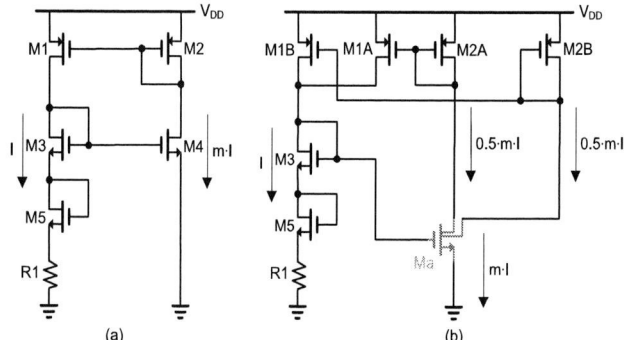

Fig. 4. CMOS current references. (a) First-order temperature-compensated current reference of [16]. (b) Proposed first-order temperature-compensated current reference with MAGFET.

A. Bias with self temperature compensation

The modified first-order temperature compensated current reference [16] is shown in Fig. 4(a). In the two cross-connected current mirrors which are Widlar current mirror M1–M3–R1 and current mirror M2–M4, the diode-connected NMOS transistor M5 is added. The derivative dI/dT of the reference current I with respect to the absolute temperature T can be obtained

$$\frac{dI}{dT} = \frac{\left(2k_{V_{Tn}} + k_{\mu_n}\right)V_{Tn} + \left(2k_{R_1} + k_{\mu_n}\right)m I R_1}{V_{Tn}/I + m R_1} , \qquad (4)$$

where μ_n, V_{Tn}, and m are the mobility of the electrons, the threshold voltage in NMOS transistors, and the current ratio between two branches. And the notation for the relative temperature coefficient $k_X = (1 / X) / (dX / dT) = d \ln X / dT$ is also adopted. From (4), the first-order temperature compensation can be achieved by sizing the resistance R_1 according following equations (5).

$$R_1 = -\frac{V_{Tn}}{mI} \cdot \frac{2k_{V_{Tn}} + k_{\mu_n}}{2k_{R_1} + k_{\mu_n}} . \qquad (5)$$

In this work, the MAGFET Ma replaces the transistor M4 as shown in Fig. 4(b) for the self temperature compensated bias. The flowing current of the MAGFET $m \cdot I$, which has nominal value of 15.28μA at 27°C and a residual temperature drift of about 600nA in the temperature range between -40°C to +80°C have been designed and simulated.

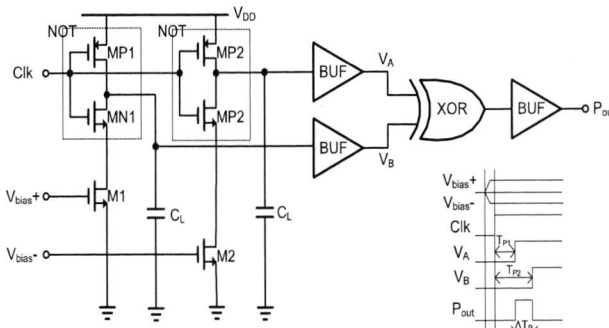

Fig. 5. Proposed magnetic field-to-pulse generator.

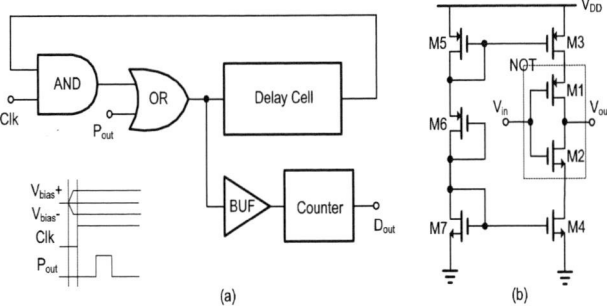

(a)   (b)

Fig. 6. (a) Block diagram of cyclic TDC. (b) Basic delay cell with temperature compensation of [19].

B. MPG with differential architecture

As shown in Fig. 5, the MPG consists of two NOT gates, one XOR gate, and three BUFFER gates. The two NOT gates with the bias of magnetic Hall sensor outputs $V_{in}+$, $V_{in}-$ generate delayed pulses V_A, V_B with clock input according to the quantity of the magnetic field. The function of the XOR gate is similar to the function of the phase detector in frequency synthesizers. The XOR gate generates the pulse of the delay difference P_{out}. The BUFFER gates are added for the invariable loading condition.

Load capacitors C_L are added at the output of the MPG. The dominant capacitor C_L among the parasitic capacitors according to the lumped capacitance model reduced the variations of the propagation delay by the process and electric conditions. The propagation delay of the NOT gate with C_L can be derived as [18]

$$T_P = \frac{(L/W)C_L}{\mu C_{OX}(V_{DD} - V_T)} \cdot \ln\left(\frac{1.5V_{DD} - 2V_T}{0.5V_{DD}}\right) \quad (6)$$

where

$$\mu = \mu_0(T_0)\left(\frac{T}{T_0}\right)^{\alpha_\mu}, \quad \alpha_\mu = -1.2 \sim -2.0, \quad (7)$$

$$V_T = V_T(T_0) + \alpha_{VT}(T - T_0), \quad \alpha_{VT} = -1 \sim -4mV/K. \quad (8)$$

Because of μ, V_T, C_L are the temperature dependent terms in (6), the propagation delay varies with the temperature function of mobility, threshold voltage and capacitance. As the temperature changes, the propagation delays T_{P1}, T_{P2} of two

Fig. 7. Die photo of proposed magnetic Hall sensor.

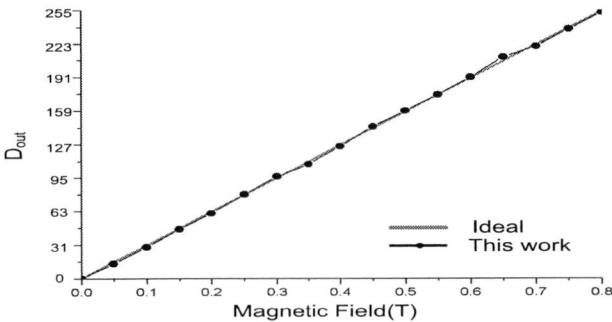

Fig. 8. Measured output of magnetic Hall sensor as a function of magnetic field at 27°C.

NOT gates change with a same fashion. The XOR gate output has only temperature-independent pulse of the difference ΔT_P between two propagation delays. Thus, the temperature-independent operation is achieved by this differential architecture.

C. Cyclic TDC with 8-bit ripple carry counter

The cyclic TDC with cyclic delay line has the greatly enhanced linearity and resolution compared with other TDCs with pulse-shrink delay elements. But the resolution variation can be high for about 100°C temperature range. Thus in this work, the cyclic TDC of [19] with self compensation of the mobility and threshold voltage temperature effects is designed. Because of the low temperature dependency of the counter, the semi-custom design with standard logic cells is used for the easy design of the 8-bit ripple carry counter.

IV. EXPERIMENTAL RESULTS

The proposed magnetic Hall sensor has been implemented in a standard 0.18μm single-poly four-metal (1P4M) CMOS process and occupied 180μm × 400μm without pads. The die photo of the proposed magnetic Hall sensor is shown in Fig. 7. The 8-bits ripple carry counter in the TDC has been implemented by using standard logic cells.

The performances of the implemented magnetic Hall sensor are measured by applying the magnetic field from 0 to 0.8mT with Lakeshore Electromagnet Model EM4-HVA, Gauss-Meter Model 420, and Hall-Prove Model 475. Fig. 8 shows the detectable resolution by 0.8μT and the temperature variation by 3% is measured during the thermal cycles of the temperature range between -40°C to +80°C. And the power consumption by 360μW at 1.8V supply voltage is measured.

TABLE I

PERFORMANCE SUMMARY AND COMPARISONS OF PUBLISHED MAGNETIC HALL SENSORS.

	This Work	[4]	[5]	[6]	[7]	[8]
Technology	CMOS 0.18μm	CMOS 0.7μm	CMOS 0.5μm	CMOS 0.5μm	BiCMOS 0.6μm	BiCMOS 0.6μm
Sensor Device	MAGFET	MAGFET	MAGFET	Hall Plate	Hall Plate	Hall Plate
Range	0 ~ 800mT	0 ~ 800mT	±100mT	±10.8mT	N/A	0 ~ 200mT
Detectable Resolution	0.8μT	12.5mT	1mT	0.33μT	120μT	N/A
Gain Error	<3%	< 3%	3%	N/A	N/A	N/A
Operation Temperature	-40 ~ +80°C	N/A	N/A	-30 ~ +95°C	-40 ~ +175°C	-50 ~ +150°C
Temperature Variation	250nT	N/A	N/A	<250nT	<2%(240nT)	50μT
Power Consumption	360μW	N/A	32.4mW	21mW	>18.81mW	NA
Area	0.072mm^2	N/A	0.9324mm^2	2.9mm^2	4.9mm^2	6mm^2

The magnetic and electric performance summary of the proposed magnetic Hall sensor is given in Table I. And performance comparisons with published magnetic Hall sensors are also listed. The proposed sensor has very low power consumption and small area with low temperature variation by 250nT of the temperature range between -40°C to +80°C. This work can give the improvement in sensitivity, temperature variation and power consumption without additional compensation circuitry and calibration system.

V. CONCLUSION

A highly sensitive and temperature independent digital magnetic Hall sensor is presented. The MAGFET as the sensor device and the bias, MPG and TDC circuits with the self temperature compensation and differential architecture are proposed. Compared with published works, better electric and magnetic performances are shown by experimental results. Because of the realization with digital circuits, this sensor can provide the simple and easy-design solution and the further compensation and calibration for process and temperature variation.

ACKNOWLEDGMENTS

The work is financially supported by Korea University and IC Design Education Center (IDEC) and financially supported by the Ministry of Education and Human Resource Development (MOE), the Ministry of Commerce, Industry and Energy (MOCIE) and the Ministry of Labor (MOLAB) through the fostering project of the Lab of Excellency.

REFERENCES

[1] C. L. Chien and C. R. Westgate, *The Hall Effect and Its Applications.* New York, NY: Plenum, 1980.
[2] J. E Lenz, "A Review of Magnetic Sensors," *Proc. IEEE,* vol. 78, no. 6, pp. 973-989, Jun. 1990.
[3] M. J. Caruso, T. Bratland, C. H. Smith, and R. Schneider, " A New Perspective on Magnetic Field Sensing," *Sensors Magazine,* vol. 15, no. 12, pp. 34-46, Dec. 1998.
[4] C. Rubio, S. Bota, J. G. Macias, J. Samitier, "Modelling, Design and Test of a Monolithic Integrated Magnetic Sensor in a Digital CMOS Technology Using a Switched Current Interface System," *Analog*

Integrated Circuits and Signal Processing, vol. 29, no. 1, pp. 115-126, 2001.
[5] C. H. Kuo, S. L. Chen, L. A. Ho, and S. I. Liu, "CMOS Oversampling ΔΣ Magnetic-to-Digital Converters," *IEEE J. Solid-State Circuits,* vol. 36, no. 10, pp. 1582-1586, Oct. 2001.
[6] J. C van der Meer, F. R. Riedijk, E. van Kampen, K. A. A. Makinwa, and J. H. Huijsing, " A fully integrated CMOS Hall sensor with a 3.65μT 3σ offset for compass applications," in *IEEE ISSCC, Dig. Tech. Papers,* vol. 1, pp. 246-247, Feb. 2005.
[7] M. Motz, D. Draxelmayr, T. Werth, and B. Forster, "A Chopped Hall Sensor With Small Jitter and Programmable "True Power-On" Function," *IEEE J. Solid-State Circuits,* vol. 40, no. 7, pp. 1533-1540, Jul. 2005.
[8] M. Motz, U. Ausserlechner, W. Scherr, and B. Schaffer, " An Integrated Magnetic Sensor with Two Continuous-Time ΔΣ–Converters and Stress Compensation Capability," in *IEEE ISSCC, Dig. Tech. Papers,* pp. 1151-1169, Feb. 2006.
[9] H. P. Baltes, and R. S. Popovic, "Integrated Semiconductor Magnetic Field Sensors," in *Proc. of the IEEE,* vol. 74, pp. 1107-1132, Aug. 1986.
[10] C. S. Roumenin, *Handbook of Sensors and Actuators 2, Solid State Magnetic Sensors,* Elsevier Science B. V., 1994.
[11] R. Rodriguez-Torres, E. A. Gutierrez-Dominguz, R. Klima, and S. Selberherr, "Analysis of Split-Drain MAGFETs," *IEEE Trans. Electron Devices,* vol. 51, no. 12, pp. 2237-2245, Dec. 2004.
[12] F. Ning, and E. Bruun, "An offset-trimmable array of magnetic-field-sensitive MOS transistors(MAGFETS)," *Sensor and Actuators A,* vol. 58, no. 2, pp. 109-112, Feb. 1997.
[13] S. I. Liu, J. F. Wei, and G. M. Sung, "SPICE Macro Model for MAGFET and its Applications," *IEEE Trans. Circuits and Syst. II: Analog and Digital Signal Processing,* vol. 46, no. 4, pp. 370-375, Apr. 1999.
[14] Y. Tsividis, *Operation and Modeling of the MOS Transistor,* Second Edition, McGraw-Hill, NY, 1999.
[15] I. M. Filanovsky, and A. Allam, "Mutual Compensation of Mobility and Threshold Voltage Temperature Effects with Applications in CMOS Circuits," *IEEE Trans. Circuits and Syst. I: Fundam. Theory Appl.,* vol. 48, no. 7, pp. 876-884, Jul. 2005.
[16] P. Fiori, and P. S. Crovetti, "A New Compact Temperature-Compensated CMOS Current Reference," *IEEE Trans. Circuits and Syst. II: Analog and Digital Signal Processing,* vol. 52, no. 11, pp. 724-728, Nov. 2005.
[17] J. M. Rabaey, A. Chandrakasan, and B. Nikolic, *Digital Integrated Circuits, A Design Perspective,* Second Edition, Prentice Hall, Upper Saddle River, NJ, 2003.
[18] T. A. Demassa and Z. Ciccone, *Digital Integrated Circuits,* Wiley, NY, 1996.
[19] P. Chen, C.C. Chen, C. C. Tsai, and W. F. Lu, "A Time-to-Digital-Converter-Based CMOS Smart Temperature Sensor," *IEEE J. Solid-State Circuits,* vol. 40, no. 8, pp. 1642-1648, Aug. 2005.
[20] S. I. Liu, J. F. Wei, and G. M. Sung, "SPICE Macro Model for MAGFET and its Applications," *IEEE Trans. Circuits and Syst. II: Analog and Digital Signal Processing,* vol. 46, no. 4, pp. 370-375, Apr. 1999.

978-1-4244-2018-6/08 $25.00 © 2008 IEEE

IEEE 2008 Custom Intergrated Circuits Conference (CICC)

A Micro-Power Neural Spike Detector and Feature Extractor in .13μm CMOS

Jeremy Holleman, Apurva Mishra, Chris Diorio, and Brian Otis

Department of Electrical Engineering
University of Washington
Seattle, Washington, 98195–2500

Abstract—We present a fully-integrated system for the detection and characterization of action potentials observed in extracellular neural recordings. The circuit includes an analog implementation of the nonlinear energy operator for spike detection. The minimum and maximum value of detected spikes are captured by peak detectors and digitized by an on-chip successive approximation ADC to provide a compact description of the spike waveform. The circuit is implemented in a .13 μm CMOS process. It occupies .17 mm^2 of chip area and consumes 1 μW from a 1 V supply.

I. Introduction

Many neural recording applications focus on action potentials, spikes generated by individual neurons. Action potentials occupy the frequency range roughly between 100 Hz and 5 kHz, and occur at a rate up to about 100 per second. Adequate digitization would require a neural signal to be sampled at 10 kS/s, even though around 90% of the digitized samples would not be part of an action potential. These "empty" samples must be processed using local computer cycles or transmitted via a wireless link for off-chip processing. Either choice results in unnecessary power dissipation and would be prohibitive for an implanted multi-channel system. Additionally, measurement of non-linear features such as spike amplitude and width requires that the signal be either oversampled or digitally interpolated, further increasing demands on the analog-digital converter (ADC) or local processor.

In this paper we present a circuit to perform spike detection in the analog domain, precluding the need to digitize the entire waveform. After a spike is detected, the maximum and minimum values are digitized with an 8-bit successive approximation ADC. By extracting the most important features of the signal in the analog domain, the power required to digitize the entire waveform is greatly reduced. Compared to a simple thresholding scheme, our architecture provides additional information by capturing the maximum and minimum values of the action potentials, which can be used for further processing, including spike sorting. Additionally, the nonlinear energy operator (NEO), which we use to implement our spike detector, has superior discriminatory ability to a threshold-based detector.

II. System Design

A. Architecture

The neural interface circuit, shown in Fig. 1(a), comprises a spike detector for distinguishing action potentials from

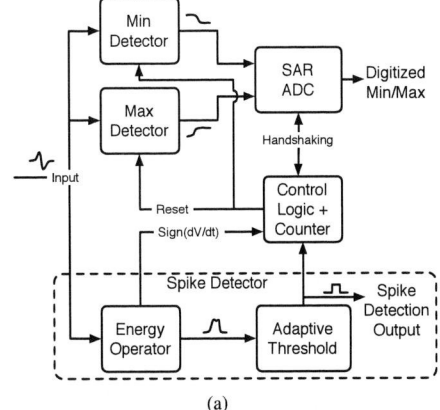

(a)

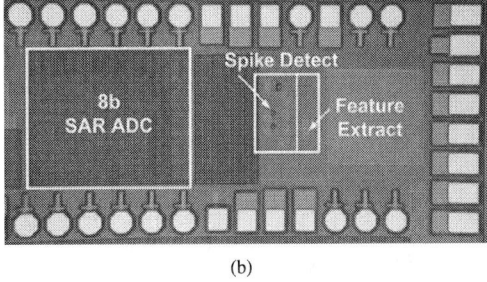

(b)

Fig. 1. (a) System architecture. (b) Die photo.

noise, positive and negative peak detectors to characterize the detected spike, and an ADC to digitize the spike maximum and minimum. The proposed circuit is intended to follow a pre-amplifier that amplifies the neural signal to an amplitude of approximately 400mV peak-peak. The spike detector is the first component in the signal chain. When a spike is detected, a counter is triggered to provide a delay equal to twice the width of the spike. The delay ensures that the maximum and minimum occur and are captured before the ADC is triggered. After the delay has elapsed, the "Ready" signal is asserted, which causes the ADC to digitize the captured minimum and maximum values. The digitized values are then read through a serial interface. After both conversions are complete, the ADC asserts the "Done" signal, which triggers a reset of the peak detectors and control logic, preparing the system for the next spike detection.

978-1-4244-2018-6/08 $25.00 © 2008 IEEE

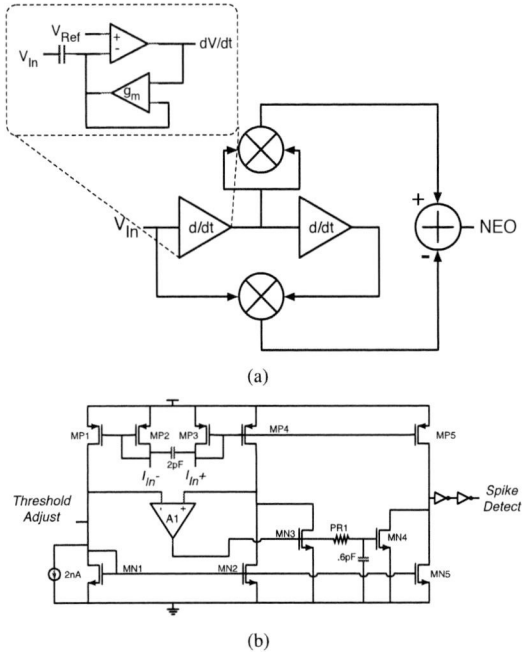

(a)

(b)

Fig. 2. Spike detector. (a) Nonlinear energy operator. (b) Adaptive threshold circuit. The NEO output in (a) is a differential current which is is the input to the thresholding circuit in (b).

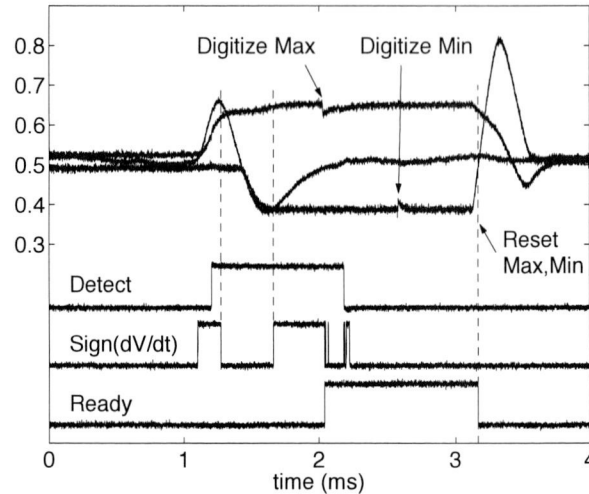

Fig. 3. Spike detection timing. Input, V_{Max}, and V_{Min} are shown at the top, followed by the digital timing signals involved in the detection sequence.

The detection sequence is shown in Fig. 3. The first two edges of $\text{sign}(dV/dt)$ after the spike detection, marked with dashed lines, indicate the time of the positive and negative peaks. The time between the two edges is the measured spike width. The rising edge of Ready initiates the A/D conversion. A handshaking signal from the ADC (not shown) indicates that the conversion is complete and resets the peak detectors and timing logic, marked by the third dashed line.

B. Spike Detector

A schematic of the spike detector is shown in Fig. 2. An analog implementation of the nonlinear energy operator (NEO) provides a differential output current which indicates the amount of activity in the input signal. The nonlinear energy operator (NEO), defined as

$$NEO(x) = \dot{x}^2 - \ddot{x}x,$$

has been found to discriminate between spikes and noise better than a simple thresholding detector, particularly when the signal-noise ratio (SNR) is low [1]. The spike detection operation is traditionally performed offline using a PC.

The two differentiations are performed by g_m-C differentiators. The multiplications are performed by Gilbert cells. The differential current outputs are connected to perform the subtraction. The multiplier inputs are differential, with the positive inputs taken from the single-ended outputs of the differentiators. The DC levels of the positive multiplier inputs are computed by low-pass filters (not shown) using a pseudo-resistor realized from anti-parallel diodes [2] and connected to the negative multiplier inputs. This arrangement,

made possible because there is no useful DC information in any of the signals, prevents offsets in the differentiators from corrupting the NEO output.

The adaptive thresholding circuit shown in Fig. 2(b) converts the NEO output into a binary spike detection signal. Any activity in the input signal, including noise, will result in a positive NEO output. In order to minimize false detections, the threshold must be set above the background noise level. The feedback loop formed by A1 and MN3 sets $I_{D,MN3}$ equal to the differential NEO input current. This quantity is then low-pass filtered and doubled through the current mirror formed by MN3 and MN4. The low-pass corner frequency is set to around 1-2 Hz by realizing PR1 as a pseudo-resistor formed from anti-parallel diodes. Thus the NEO input required to cause a detection is set at twice the average background activity. The current source in parallel with MN1 ensures that current is flowing through MN3 even when the differential input is zero. A threshold adjustment current can be injected to vary the sensitivity of the spike detector. The current mirrors in the thresholding circuit all have cascodes, which are omitted from the figure for clarity. Because the current mode signaling allows for small voltage swings, cascodes are possible even with a supply voltage of 1V.

C. Feature Extraction

Positive and negative peak detectors capture the extreme values of the signal. The positive peak detector is shown in Fig.4, and the negative detector uses the same topology with opposite polarity. The use of a differential pair to charge the storage capacitor allows $V_{SG,MP1}$ to be made less than 0 V, minimizing subthreshold current in MP1, which could cause the peak detector output to drift to V_{DD} during periods with little activity. When the Ready signal is issued to the ADC, Hold is simultaneously asserted in the peak detector. The peak detector input is forced to 0 V, preventing the output

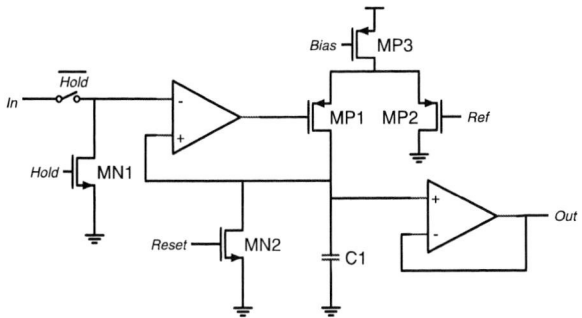

Fig. 4. Peak detector.

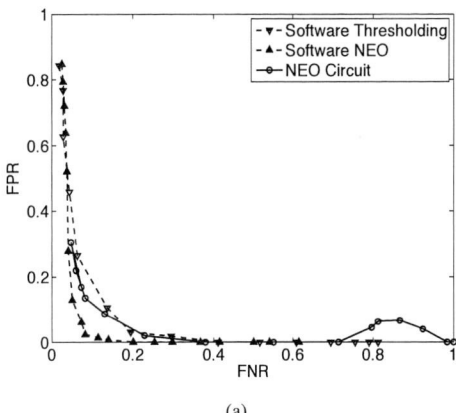

(a)

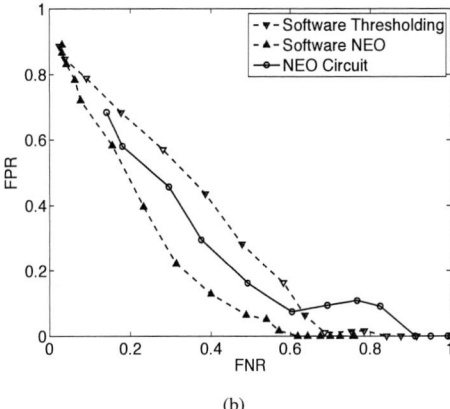

(b)

Fig. 5. False positive rate (FPR) versus false negative rate (FNR) for the threshold-based software spike detector, the proposed analog NEO detector, and a software implementation of the NEO operator. FPR = False detections / Total detections. FNR = Undetected spikes / Total true spikes. (a) SNR=10 dB. (b) SNR=6 dB.

from changing during the analog-digital conversion. After the conversion is complete, Hold is released, and Reset briefly forces the output to 0 V.

A digital counter is used in conjunction with the differentiators from the NEO to measure the width of the spike. The first differentiator has an auxiliary sign output. A change in the sign of the first derivative indicates a minimum or maximum in the input signal. After a spike is detected, the next change in the derivative sign starts the counter. The second change in the sign output causes the counter value to be registered for readout and the counter to count back down to zero. The additional delay allows time for the extreme values of the spike occur and be sampled by the peak detectors. When the counter returns to 0, the Ready signal is asserted to initiate conversions of the maximum and minimum voltages. The counter is also intended to provide a measurement of the spike width, defined as the time between the maximum and minimum of the spike.

D. Analog-Digital Converter

The 8-bit ADC was designed to operate from 10 kS/s to 100 kS/s. A successive approximation register (SAR) architecture was chosen for the ADC to minimize power consumption [3][4]. The digital ADC output is read serially from the comparator output. A synchronization signal, which is used internally to purge the DAC capacitor array and SAR logic after each conversion, also synchronizes the serial output.

III. RESULTS

The system was implemented in a .13μm CMOS process. The spike detector and feature extractor occupy a die area of 200μm x 220μm. The ADC occupies 295μm x 430μm, of which about 85% is consumed by the DAC capacitors.

The ADC is functional for sampling frequencies from 100 S/s to 1MS/s. At a 1.2 V power supply, with the positive and negative reference voltages taken from the supplies, and -0.25 dBFS input, SNDR=39.23 dB, and SFDR=46.5 dB, leading to an ENOB of 6.22 bits. The non-ideal ENOB arises from time-varying noise as well as non-linearity. The worst DNL is 0.31 LSBs. The worst INL is 1.42 LSBs.

To test the sensitivity of the spike detector, we used an artificial neural recording [5]. An artificial recording allowed variation of the noise level and spike rate, and provided a ground truth reference against which to compare spike detector accuracy. With an actual recording, there is no guaranteed correct reference, since a neural recording is subject to differences in interpretation, even among expert neurophysiologists [6].

Spike detections from the circuit were compared with labels from the generating software to determine the sensitivity and selectivity. For comparison, we also applied a threshold-based software spike detector to the same signal. The software detector indicated a spike whenever the absolute value of the input exceeded a specified threshold. We tested both detectors with several different threshold values to build the curves shown in Fig. 5. The y axis shows the false positive rate (FPR), the fraction of detections determined to be false. The x axis shows the false negative rate (FNR), the fraction of true spikes that were not detected. With a 10 dB SNR, shown in Fig. 5(a), the threshold-based software detector has good discriminative abilities. Fig. 5(b) shows the same curves measured with an SNR of 6 dB. With the noisier signal, the discriminative power of the NEO yields a superior detector at most threshold levels.

TABLE I

PERFORMANCE SUMMARY

Process	.13 μm CMOS
Active Area	.17 mm^2
V$_{DD}$	1.0 V
Power	.95 μW

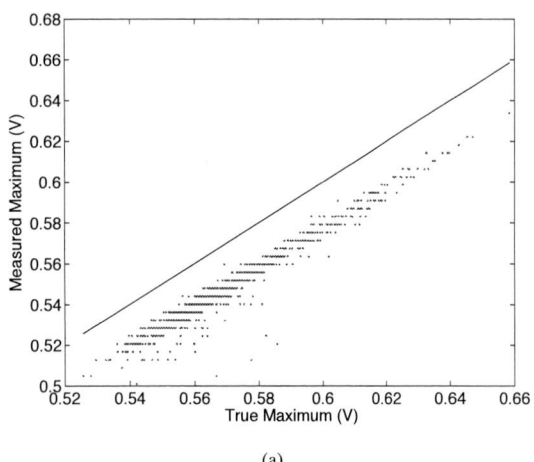

(a)

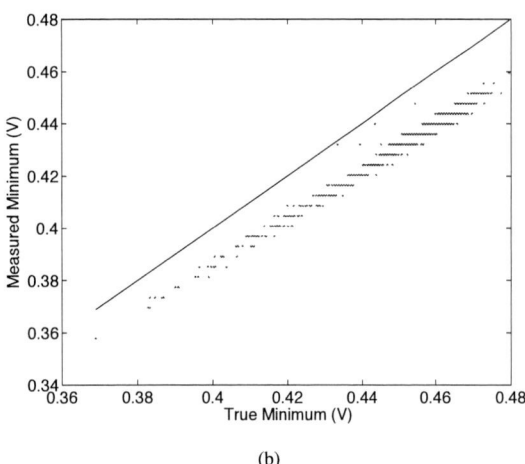

(b)

Fig. 6. Accuracy for capture and digitization of the spike maximum and minimum values. The solid lines indicate equality of true and measured values.

such as gain error and offset are less important than random variations due to noise. Offsets due to the peak-detecting circuits have been subtracted based on a DC measurement. The remaining offset is due to the ADC. The RMS error relative to a linear fit is 4.5 mV (1.2 LSB) and 2.0 mV (0.5 LSB) for the maximum and minimum, respectively.

Previous work which has implemented similar processing entirely in the digital domain [7], consumed approximately 1 μW/channel to perform spike detection and calculate the maximum, minimum, and width of detected spikes, in addition to the power required for the ADC.

IV. CONCLUSION

We have presented a low-power circuit which performs detection and feature extraction on extra-cellular neural signals. Our system demonstrates that complex digital logic can be replaced by compact low-power analog circuits. By performing simple computation in the analog domain, demands on analog-digital conversion and on local processing or communications bandwidth are reduced, enabling a reduction in overall system power.

The proposed system can easily be extended to multiple channels. Simple digital multiplexing would allow the ADC, the most area-intensive component, to be shared by many channels. By greatly reducing the required conversion rate, the analog processing we have described also facilitates the sharing of one ADC among many channels.

REFERENCES

[1] S. Mukhopadhyay and G. Ray, "A new interpretation of nonlinear energy operator and its efficacy in spike detection," *Biomedical Engineering, IEEE Transactions on*, vol. 45, no. 2, pp. 180–187, 1998.

[2] R. Harrison and C. Charles, "A low-power low-noise CMOS amplifier for neural recording applications," *Solid-State Circuits, IEEE Journal of*, vol. 38, no. 6, pp. 958–965, 2003.

[3] M. Scott, B. Boser, and K. Pister, "An ultralow-energy ADC for Smart Dust," *Solid-State Circuits, IEEE Journal of*, vol. 38, no. 7, pp. 1123–1129, 2003.

[4] P. Allen and D. Holberg, *CMOS Analog Circuit Design*. Oxford University Press New York, 2002.

[5] R. Vogelstein *et al.*, "Spike sorting with support vector machines," *Engineering in Medicine and Biology Society, 2004. EMBC 2004. Conference Proceedings. 26th Annual International Conference of the*, vol. 1, pp. 546–549, 2004.

[6] F. Wood, M. Black, C. Vargas-Irwin, M. Fellows, and J. Donoghue, "On the variability of manual spike sorting," *Biomedical Engineering, IEEE Transactions on*, vol. 51, no. 6, pp. 912–918, 2004.

[7] M. Chae *et al.*, "A 128-channel 6mw wireless neural recording IC with on-the-fly spike worting and UWB transmitter," in *Solid-State Circuits, 2006 IEEE International Conference Digest of Technical Papers*, 2008, pp. 146–147.

A digital counter is included to measure the width of the spike, defined as the time elapsed between the positive and negative peaks. However, due to variation in the relative timing between the rising edge of the detection signal and the first peak of the spike, the timing measurement is not reliable.

To test the accuracy of the peak detection and digitization, we simultaneously recorded the digital output of the ADC, the timing signals, and the input waveform. We then compared the ADC output to the actual minimum or maximum value that should have been digitized. The comparisons are shown in Fig. 6. At the end of each pair of conversions, the ADC handshaking signal resets the two peak detectors to allow a new peak to be captured. The true value, plotted on the x axis, is computed from the recorded input signal as the maximum or minimum value in the time interval between the beginning of a given digitization and the end of the last digitization. The results shown in Fig. 6 are for spikes detected when at least 2 ms has occurred since the most recent conversion-reset cycle. For applications such as spike sorting, deterministic errors

IEEE 2008 Custom Intergrated Circuits Conference (CICC)

A Compact and Programmable High-Frequency Oscillator Based on a MEMS Resonator

F. Nabki, F. Ahmad, K. Allidina, and M. N. El-Gamal

McGill University, Montreal, Canada

Abstract- **This paper presents a 1.65-2.0GHz digitally programmable oscillator. The oscillator is based on a MEMS resonator combined with a high-resolution 0.18µm CMOS fractional-N PLL. Due to the dimensions of the MEMS resonator (350µm x 130µm), the size of the entire system is ~6.05mm^2, and can be integrated into a single small form factor package. The phase noise for an oscillation frequency of 1.8GHz is −116dBc/Hz at a 600kHz offset, and the entire system consumes 50mW from a 2V supply. The PLL employs a 3rd-order 20-bit delta-sigma modulator to deliver an output resolution of ~220Hz, i.e. enabling a controlled frequency stability of better than 0.125ppm.**

I. INTRODUCTION

As data rates in communications systems continue to rise, there is an increased need for low-jitter high-frequency oscillators that are stable over both temperature and time. This is especially evident in optical networks and LANs operating at data rates close to 10Gbps. Currently, oscillators based on high-frequency fundamental (HFF) crystals or surface acoustic wave (SAW) resonators are used to fill this need. However, these devices are large, can only operate at a single frequency, and are complicated to manufacture with high frequency stability [1]. They also require off-chip connections to interface with the electronics, increasing the size, complexity, and cost of the overall system.

This paper presents a stable high-frequency oscillator based on a micro electro-mechanical (MEMS) resonator combined with a fractional-N phase-locked loop (PLL). In the context of this work, stability will always refer to the ability to produce an accurate frequency over both temperature and time. The oscillator presented here could also be realized using a stable low-frequency crystal; however, the small size of the MEMS resonator enables the integration of the entire oscillator into a single package. This considerably reduces the form factor of the system, and the shorter connections between the electronics and the MEMS device result in higher levels of performance, due to the reduced parasitics and losses.

II. THE OSCILLATOR

The system diagram of the oscillator is shown in Fig. 1. It includes a 3rd order 20-bit delta-sigma modulator which allows the PLL to achieve a very high output resolution, on the order of 220Hz. When combined with an automatic feedback loop to dynamically control the delta-sigma modulator, this oscillator could achieve a stability of less than 0.125ppm, while operating in the GHz range. This would enable the fabrication of tunable high-frequency oscillators possessing stability characteristics which are much better than those of currently available commercial HFF or SAW based oscillators.

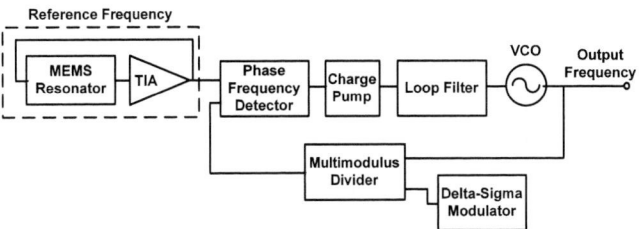

Fig. 1 – Simplified oscillator system diagram.

This paper will first provide a brief description of the MEMS resonator, along with its relevant characteristics. The circuits that make up the frequency synthesizer will then be presented, followed by measurement results of the entire system.

III. BRIEF DESCRIPTION OF THE MEMS RESONATOR

The MEMS resonators are fabricated in an in-house metalized, CMOS compatible, amorphous silicon carbide (*a*-SiC) surface micromachining process, described in detail in [2]. The specific device used in this work exhibits a motional resistance of ~26kΩ, a quality factor (Q) of 1040, and a resonant frequency of 8.3MHz. Various pictures depicting the fabricated clamped-clamped beam resonator and its characteristics are shown in Fig. 2. The entire structure with pads measures 350µm by 130µm, and the resonator's dimensions are 25µm by 45µm, with a 200nm air gap.

IV. THE FREQUENCY SYNTHESIZER

This section presents the circuits used in the frequency synthesizer shown in Fig. 1.

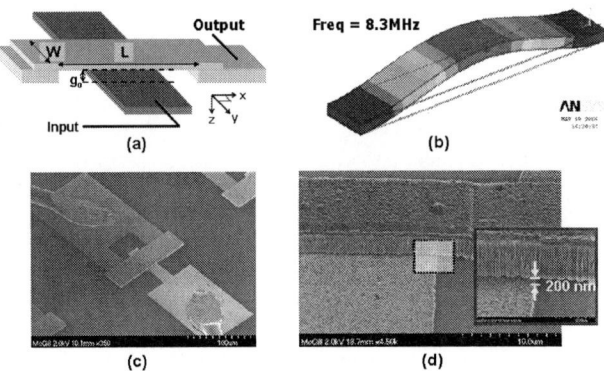

Fig. 2 – (a) 3-D model of the resonator.
(b) Simulated mode shape of the resonator.
(c) Micrograph of the resonator with wire bonds.
(d) Close-up SEM picture of the resonator showing a 200nm gap spacing.

978-1-4244-2018-6/08 $25.00 © 2008 IEEE

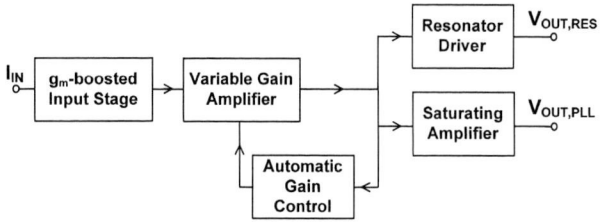

Fig. 3 – Simplified TIA system diagram.

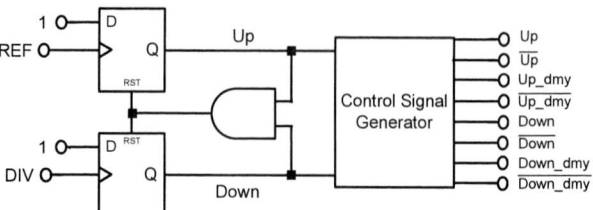

Fig. 4 – Phase frequency detector.

A. The Transimpedance Amplifier (TIA)

Due to the large motional resistance of a MEMS resonator, a very high-gain TIA is needed. At the same time, a large bandwidth is required to ensure that the phase shift around the loop is as close to zero as possible. From the resonator's characteristics, it can be determined that the necessary gain-bandwidth product of the TIA is 2.16THz.

A TIA was custom-designed for this application, and its block diagram is shown in Fig. 3. The input stage is based on a g_m-boosted common-gate amplifier, and is followed by a variable gain amplifier controlled by an automatic gain control (AGC) loop. This prevents the oscillation voltage from exceeding values which could exert non-linearities in the MEMS resonator and deteriorate noise performance [3]. The TIA has two voltage outputs: $V_{OUT,RES}$ is fed back into the resonator to form the positive feedback loop, and $V_{OUT,PLL}$ is a digital signal that serves as the reference frequency for the PLL. Both the input and output stages are designed to have very low impedances to avoid loading the resonator's quality factor.

The gain of the TIA can be varied from 17kΩ to 290kΩ, i.e. a tuning range of 25dB. The 3-dB bandwidths corresponding to these gains are 253MHz and 103MHz, respectively. Since these bandwidths are greater than ten times the resonant frequency of the MEMS device, the phase shift around the loop is ensured to be close to zero degrees.

B. The Phase Frequency Detector (PFD) and Charge Pump

The PFD used is shown in Fig. 4. It is based on D flip-flops, with outputs depending on the time differences between the rising edges of the reference and those of the divided oscillator waveforms. A delay is introduced into the reset signal path to eliminate the dead zone associated with this PFD. An output stage (control signal generator) is used to convert the up and down signals generated by the PFD into eight separate control signals [4]. The purpose of these signals is to mitigate non-idealities in the charge pump, as will be described in the next paragraph.

The charge pump schematic is shown in Fig. 5. A dummy branch (M7–M10) was added to preserve the voltages at the drains of M1 and M2, thus avoiding current spikes due to charge leakage. To eliminate changes in these voltages during transitions, the signals from the PFD are timed such that there is a small overlap between the time when the output branch turns on and the time the dummy branch turns off, and vice-versa. A simple differential amplifier in a unity-gain configuration is used to ensure that the voltage on the dummy branch matches that of the charge pump output [5].

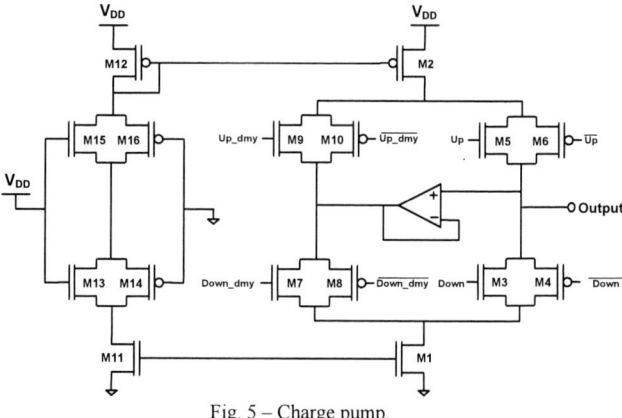

Fig. 5 – Charge pump.

The eight PFD control signals are needed in order to allow the use of complementary switches in the charge pump, which drastically reduces charge injection and charge feedthrough errors, provided a careful layout is used to match the transistors. This reduces the ripples on the control line, and thus the level of spurs created by the VCO.

Biasing is set by M11–M16, where M13–M16 are present to reduce the mismatch resulting from channel length modulation, when mirroring currents to the active branches of the charge pump.

C. The Loop Filter

The low frequency of the reference MEMS resonator (8.3MHz), combined with the 3rd order delta-sigma modulator necessitates a very small loop bandwidth, on the order of 25kHz. A simple passive or opamp-based loop filter would require a large capacitance to achieve this bandwidth, and would not be amenable to on-chip integration. Instead, the dual-path filter shown in Fig. 6 is used [4]. This type of architecture reduces the capacitance needed by a factor equivalent to the current multiplication between the two loop filter paths, i.e. corresponding to the variable B in Fig. 6. It however mandates the use of two charge pumps. It should also be noted that the inputs to this filter are negative, which means that the "Up" and "Down" signals from the charge pumps must be reversed to obtain the correct output. The low-pass filter R_1 and C_1 forms an additional pole at the output of the loop filter.

In this design, the charge pump currents were set to 60µA and 5µA. This reduces the sizes of the capacitances needed by a factor of twelve. It should also be noted that the output voltages of both charge pump branches are fixed at a voltage

978-1-4244-2018-6/08 $25.00 © 2008 IEEE

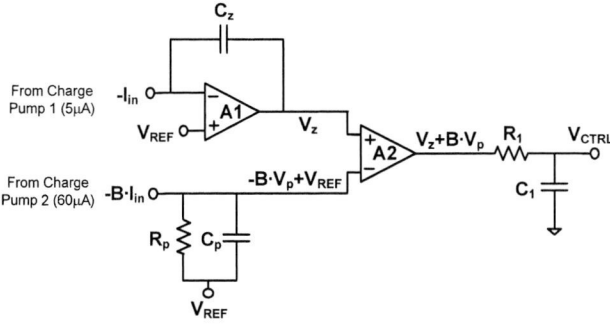

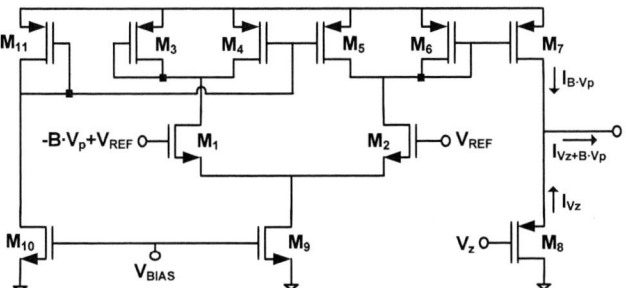

Fig. 6 – Dual-path loop filter.

Fig. 7 – Schematic of amplifier A2.

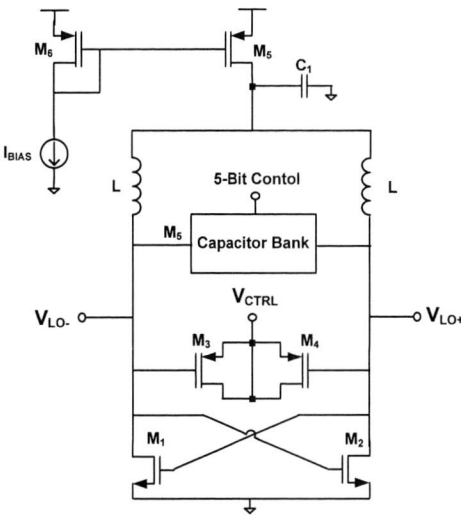

Fig. 8 – VCO schematic.

V_{REF}, when the loop is in lock. This helps ensure that the up and down currents do not suffer from mismatches due to channel length modulation. The amplifier used in the integrator (A1) is a simple one-stage differential pair with an active load. Amplifier A2 (detailed in Fig. 7) removes the DC offset (V_{REF}) at the output of the low-pass filter formed by C_p and R_p. Also, M_4 and M_5 source bias currents equivalent to those of the differential pair M_1 and M_2, so that any differences between V_{REF} and the low-pass filter voltage is mirrored to M_7 through M_6. At the same time, the common-drain stage of M_8 converts the integrated voltage (V_z) into a current and adds it to that of M_7. The body of M8 is tied to the source to eliminate the body effect and increase the available voltage swing.

D. The VCO

A top-fed cross-coupled VCO is used in this design. The schematic is shown in Fig. 8. The current mirror M_5 and M_6 serves to reduce sensitivity to power-supply variations, and capacitor C_1 reduces the drain current at the zero-crossings of the tank voltage. This decreases the phase noise by reducing the current injected during sensitive portions of the cycle [6].

Accumulation mode PMOS varactors are used to provide fine tuning of the VCO, and a bank of five digitally switched capacitors is used to provide coarse tuning. With these two mechanisms, the tuning range of the VCO is 1.65GHz to 2GHz. The gain of the VCO has an average value of 150MHz/V. The inductors are intertwined to decrease their areas.

One output of the differential VCO is fed to the divider through an inverter chain, while the other terminal is fed to an output buffer to drive the 50Ω load of the measurement equipment.

E. Multimodulus Divider and Delta-Sigma Modulator

A 6-bit programmable pulse-swallow divider is used in this design for its simplicity and ability to interface well with the delta-sigma modulator's 4-bit output without overflow. True signal phase clock (TSPC) logic was used for the programmable counters and the prescaler in the divider to increase the speed of the circuits.

The delta-sigma modulator is a single-loop 3rd order modulator with multiple feedforward. It is important that the delta-sigma calculation is performed on the opposite edge of the clock as the edge which activates the PFD. This ensures that the switching noise from the delta-sigma modulator does not affect the PLL during lock. Dithering is also implemented to suppress the quantization noise idle tones. The modulator has a 20-bit input, which results in a PLL output resolution of:

$$resolution = \frac{f_{ref}}{2^{20}} \times N , \qquad (1)$$

where f_{ref} is the reference frequency and N is the divider ratio. This provides a high output resolution for the frequency synthesizer, on the order of 220Hz. By dynamically controlling the delta-sigma modulator through an automatic frequency control loop, this high resolution could be translated into a very high oscillator stability, i.e. better than 0.125ppm.

V. EXPERIMENTAL RESULTS

The frequency synthesizer and the TIA are fabricated in a 0.18μm CMOS process. A micrograph of the chip is shown in Fig. 9, alongside the MEMS resonator. The resonator is significantly smaller than an on-chip inductor. The entire design occupies an area of ~6.05mm² and consumes 50mW from a 2V supply.

The synthesizer was tested both with the 8.3MHz MEMS resonator and a 10MHz temperature controlled crystal oscillator (TCXO). Fig. 10 shows a comparison of the phase noises obtained with each reference at an output frequency of 1.8GHz. The phase noise profiles are clearly similar, while the

978-1-4244-2018-6/08 $25.00 © 2008 IEEE   363

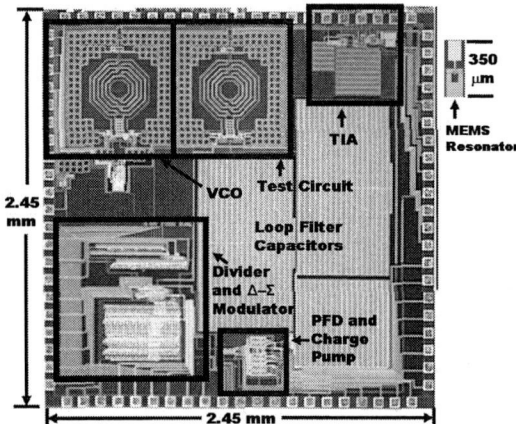

Fig. 9 – Chip micrograph, along with a MEMS resonator (to scale).

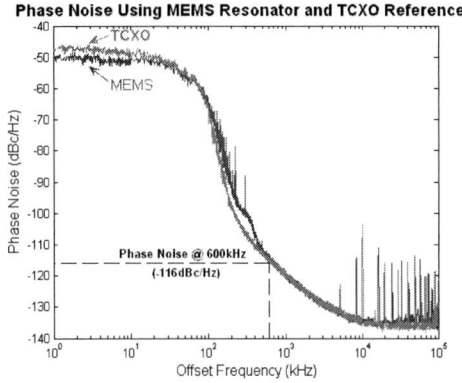

Fig. 10 – Phase noise comparison using different references.

in-band phase noise of the PLL is actually slightly better with the MEMS resonator. The phase noise at a 600kHz offset is -116dBc/Hz. The optimization of the on-chip RF VCO is currently underway to reduce this value to below -121dBc/Hz, which will expand the potential applications of this oscillator to the wireless DCS-1800 standard.

A plot of the spectrum with three output frequencies is shown in Fig. 11. The delta-sigma modulator was used to tune the output frequency in steps of 2.5MHz. This is far from the minimum resolution of the PLL, but it illustrates the oscillator's ability to synthesize frequencies which are not multiples of the reference. The smallest measurable step size around the frequency of 1.8GHz is 220Hz, which could yield a frequency reference with stability on the order of 0.125ppm. Fig. 11 also shows that the power of the reference spurs is 56dB below that of the carrier.

Fig. 12 shows the effect of dithering on the fractional spurs created by the delta-sigma modulator. The centre frequency for both curves is 1.7GHz. It can be seen that dithering reduces the power of the spurs by at least 11dB. Once the fractional spurs are suppressed close to the carrier by dithering, the noise shaping of the delta-sigma modulator becomes more apparent.

It should be noted that the system reported here can be used to implement a highly stable oscillator, in terms of initial frequency accuracy and drift compensation, in any frequency

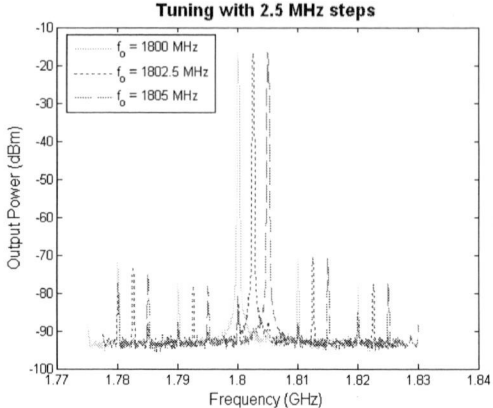

Fig. 11 – Fractional tuning of the output frequency.

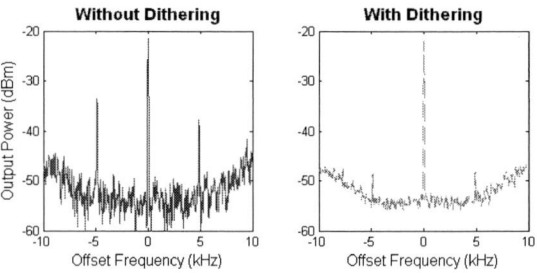

Fig. 12 – Effect of dithering on the fractional spurs.
(Centre frequency = 1.7GHz)

range simply by changing the tuning range of the RF VCO. For instance, a VCO in the range of 600MHz would make this design applicable to the SONET standards of optical networks.

VI. CONCLUSION

This paper presented a compact and programmable high-frequency oscillator based on a MEMS resonator and a high-resolution PLL. Thanks to the MEMS-based reference signal, the entire system can be integrated into a single small form factor package. The oscillator tuning range is from 1.65GHz to 2GHz, and the high-resolution makes it possible to obtain a stability of less than 0.125ppm.

REFERENCES

[1] R. Clark, "Programmable Crystal Oscillators with Sub-ps Jitter and Multiple Frequency Capability," white paper, Silicon Laboratories Inc., Austin, TX.

[2] F. Nabki, T. A. Dusatko, S. Vengallatore, and M. N. El-Gamal, "Low-Temperature (<300°C) Low-Stress Silicon Carbide Surface Micromachining Fabrication Technology," *Hilton Head Workshop 2008: A Solid-State Sensors, Actuators and Microsystems Workshop,* Accepted, 2008.

[3] F. Nabki and M. N. El-Gamal, "Modeling and Simulation of Micro Electromechanical (MEM) Beam Resonator-Based Oscillators," *IEEE International Symposium on Circuits and Systems,* Accepted, 2008.

[4] J. Craninckx and M. S. J. Steyaert, "A Fully Integrated CMOS DCS-1800 Frequency Synthesizer," *IEEE Journal of Solid-State Circuits,* vol. 33, pp. 2054-2065, December 1998.

[5] W. Rhee, "Design of High-Performance CMOS Charge Pumps in Phase-Locked Loops," *IEEE International Symposium on Circuits and Systems,* vol. 2, pp. 545-548, July 1999.

[6] A. Hajimiri and T. H. Lee, "Design Issues in CMOS Differential LC Oscillators," *IEEE Journal of Solid-State Circuits,* vol. 34, pp. 717-724, May 1999.

Advanced SoCs – Techniques and Applications   Session 14

Chair: Steve Wilton, University of British Columbia
Co-Chair: Arif Rahman, Xilinx Research Labs

The complexity of SoC designs is increasing. Not only must SoC designers integrate novel algorithmic and architectural innovations, but do so while addressing variability, reliability, and power constraints. This session addresses some of these techniques, and describes novel architectures for complex SoCs.

The first paper gives an overview of device variability, defects, aging, and noise. Experimental characterizations on the chip and wafer level are presented. Exploiting regularity is key; the paper shows how this can be addressed through a design methodology that makes use of Restricted Design Rules. The concepts described in this paper will become increasingly important to all SoC designers.

The second paper describes a 9Gbit/s serial transceiver for on-chip signaling. As frequencies rise, distributing global signals across a die is increasingly challenging. The transceiver described in this paper addresses this problem. The transmitter can send 8 bits at 1.125 GByte/s over a distance of 5.8mm. A bit error rate of less than 10^{-10} is demonstrated in a 0.13μm CMOS process.

The third paper describes the implementation of a low power baseband receiver for DVB-T/H using OFDM to tolerate noisy multipath environments. The power consumption of this chip is 28mW which may make it suitable for portable devices.

The fourth paper describes a non-volatile magnetic flip-flop cell for SoC designs. Such a non-volatile cell is important in building low-power systems in which parts of the chip must be powered-down frequently. The performance of the magnetic flip-flop was shown to be comparable to a normal CMOS D flop-flop.

The fifth paper presents a monolithic pulsed time-of-flight measurement system with 12ps precision. The system is integrated on a single chip, meaning low noise performance is especially important.

The sixth paper describes a wireless real-time monitoring system that detects force and other conditions in orthopedic implants. A particular challenge of this application is that all data communications and power delivery must be done wirelessly. The chip can supply 400μW across 20cm.

The final paper presents a tessellation-enabled shader for use in a graphics engine. A novel architecture reduces the memory bandwidth by a factor of 1/250, and costs only 6.2% additional logic. The power consumption is low enough to make it feasible for portable devices.

978-1-4244-2018-6/08 $25.00 © 2008 IEEE

Notes

Characterization and Design for Variability and Reliability

Kevin Nowka, Sani Nassif, Kanak Agarwal

IBM Austin Research Laboratory, Austin, Texas 78758

Abstract- **Device variability due to sub-wavelength lithography, layout complexity, and random effects is impacting manufacturable design. Defects, aging, and noise must also be accounted for in design and manufacturing. Characterization structures to quantify these effects, measured behaviors, the resulting models, and design and tools mitigation actions are presented in this paper.**

I. INTRODUCTION

The IC industry is built on a foundation of simulation and prediction. VLSI design flows rely on models for frequency, power, area, yield, and cost which are derived from characterization conducted at multiple levels: manufacturing processes, devices, gates, units, cores, and chips. Technology complexity beyond the 90nm node is making this characterization and representation increasingly difficult. As a result, the model predictive capability is severely degraded.

Technology has become so complex that it is not well represented by design rules. The design/technology interface information has increased substantially, resulting in substantial increase in the number of design rules as shown in Fig 1. The complex non-linear relations necessary to achieve yieldable designs via ground rules is becoming difficult.

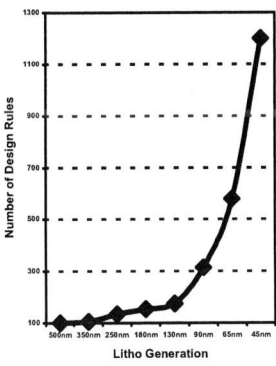

Fig. 1. Number of design rules by generation

Semiconductor manufacturing is increasingly challenging as scaling exposes greater device non-idealities. Performance gain per lithography generation is reducing. Gain has increasingly come from non-scaling-related innovations such as Cu-interconnect, SOI technology [1], and device stress [2, 3]. These innovations, along with demanding lithography, are increasing the technology cost with scaling. The resulting device variability necessitates larger margins, reducing performance benefits. Fig. 2 shows the range of variability in

fabricated device performance and power consumption for a 65nm SOI technology.

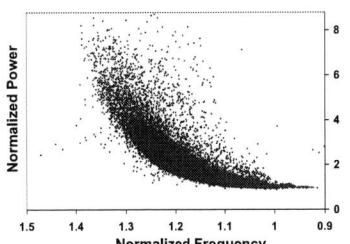

Fig. 2. Processor frequency vs. power consumption

II. VARIABILITY

Variability fundamentally is a lack of uniformity: Individual devices on a chip are NOT identical and individual chips are NOT identical. Fluctuations imply unpredictability, which ultimately reduces design profitability. Variability, for a variety of reasons, is increasing with technology scaling.

A. Physical Variability vs. Operational/Environmental

Variability results from many sources: (1) Physical variation in the characteristics of devices and wires caused by IC manufacturing processes, device aging, and wear-out. These effects occur on the time scale of months to years. (2) Environmental changes in power supply, temperature, and local noise coupling caused by the specifics of the design implementation. These effects occur on the time scale of the operating clock period (10^{-6} to 10^{-9} sec).

B. Physical Extent of Variation

In addition to physical vs. operational variability, an important character of variability is the physical extent of the resulting variation. Design characteristics can exhibit "chip mean" behavior. This variation can be attributed to factors which are most pronounced in context of die-to-die across a wafer, between wafers in a lot, lot-to-lot, or even fab-facility to fab-facility. The behavior results in a shift in the mean chip electrical characteristics and thus VLSI circuit performance. Finally, localized device-to-device variation results from random-dopant fluctuation, line-edge roughness, and localized non-catastrophic defects.

C. Systematic vs. Random Variation

Variability can be systematic or random. Systematic variation is one with a known quantitative relationship to a source which can readily be modeled and simulated. A model may be physics-based or empirical. Given a model, a designer may null out the impact. For random variation, the sources are unknown or are too difficult or costly to model, generate, or simulate. The designer must use worst-case analysis, thereby increasing design margin. Some systematic phenomena are so complex that tractable models are infeasible. These must be treated as random.

In current and next generation silicon CMOS device-based design, the most crucial random variability effect is threshold variation from random dopant fluctuation resulting from ion implantation [4]. The variation in device threshold results from variation in the number and placement of implanted ions. For any given implanted particle, a small perturbation in the initial direction or velocity can cause a large change in the eventual resting location. While the location of every dopant atom cannot be predicted individually, the statistical probability that an atom will reach a given depth can be. Given such a prediction, estimates of the dopant atom density vs. depth can be generated. With enough dopant atoms in the channel, the ensemble behavior dominates. In earlier technologies, the channel of a MOSFET contained thousands of dopant atoms; with this large number of implanted ions, individual ion placement was less critical. In modern devices, however the number of dopant atoms is countable (~100). With so few atoms, the averaging effect of large number of ions is not exhibited; thus both the number and the location of individual atoms matter. Device threshold variation is represented by Pelgrom's model [5]. Here WL represents gate area and A_{VT0} and S_{VT0} are process-related constants modeling the dependence of V_T mismatch on gate area and physical spacing between devices.

$$\sigma_{V_{T0}}^2 = \frac{A_{V_{T0}}^2}{WL} + S_{V_{T0}}^2 D^2 \qquad [1]$$

Line-edge-roughness (LER) is also a significant cause of random device performance variation [6]. LER is due to variations in the number of photons incident on a local area during exposure and the characteristics of the photoresist [7].

Another significant random variability is due to gate oxide thickness variations. With oxides as thin as 1nm, variations of one to two atomic-spacing [8] result in substantial random variation in gate leakage currents.

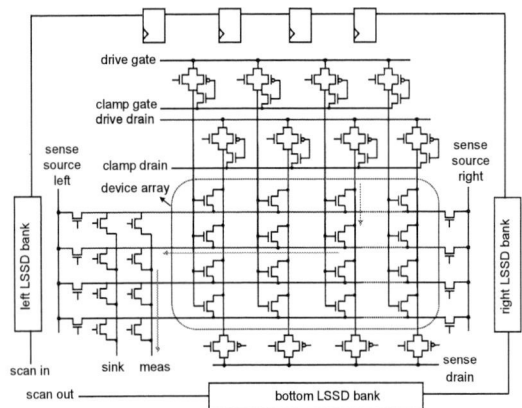

Fig. 3. Addressable characterization array structure

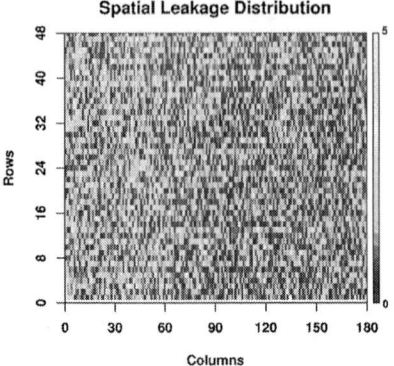

Fig. 4. Spatial device leakage map

To experimentally characterize the affect of local variability on device electrical characteristics, several array-based characterization structures were developed [9, 10]. Small devices were arranged in an addressable array as shown in Fig 3 to allow the current-voltage (IV) measurement of 96,000 individual devices [9]. The electrical behavior of the devices was measured by sweeping drain and gate voltages. Parameter statistics were extracted from the measurements. Fig. 4 shows the leakage exhibited little spatial correlation. The effects of random dopant fluctuation (RDF) are observed in the V_T variability of the small devices (Fig. 5). To isolate and characterize the local variability associated with gate oxide variation, a similar characterization structure was developed. An addressable array of SRAM sized devices were developed and characterized for gate leakage current [11]. Fig 6 shows the variation characteristics for these devices.

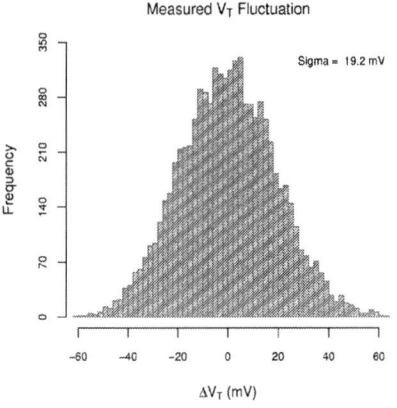

Fig. 5. Threshold variation for 65nm SOI process

Systematic variations can be classified as across field (dependent on position in reticle) systematic and layout dependent systematic variations. Across field systematic variations are caused by photolithographic and etching sources such as dose, focus, and exposure variations [12], lens aberrations [13], mask errors [14], and variations in etch loading [15]. These variations exhibit strong spatial correlation. Across-field variations can cause identical devices

978-1-4244-2018-6/08 $25.00 © 2008 IEEE

at different locations in the reticle to behave differently. These effects are characterized by embedding test structures at several positions in the reticle.

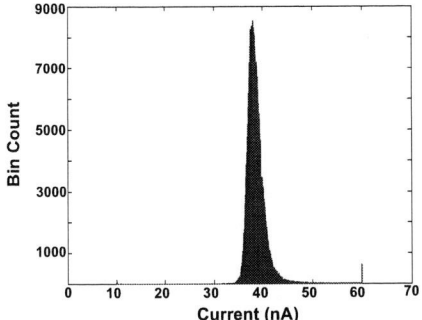

Fig. 6. Gate leakage current distribution

The layout dependent variations, however, can cause two different layouts of the same device to have different characteristics even when the two instances of the device are located close to each other in the reticle. Layout dependent variations differ from random variations because they are predictable and can be modeled as a function of deterministic factors such as layout structure and surrounding topological environment [16]. Layout diversity characteristics can affect active and poly density, active/poly/metal interconnect pitches, device orientation, device sizes, contact and via numbers and placement, and other significant non-regularities.

Advanced technologies exhibit large variability due to lithography. Sub-wavelength lithography, such as the use of 193nm light sources to fabricate devices with feature sizes of 45nm, is a significant source of device variability. Optical Proximity Correction (OPC) is applied to improve fidelity. Resolution enhancement techniques introduce local "context dependent" layout correction. The resulting layouts are dependent on neighborhood; the same mask shapes can print differently in different areas, resulting in additional variability. Implementation diversity, traditionally used for performance, area, and power optimization, results in local context dependence which increases variability.

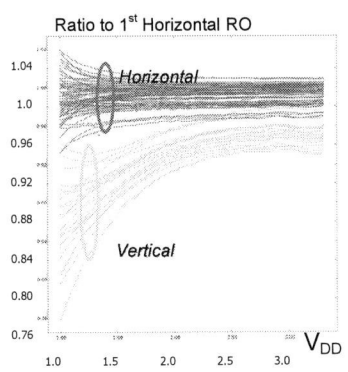

Fig. 7. Poly orientation performance effect vs. supply voltage

An example within-die layout dependence results from differences in poly orientation. A characterization structure to assess poly orientation [42] demonstrated an approximately 5% mean performance difference as shown in Fig. 7. The sensitivity to supply voltage indicates a device threshold variation. Poly pitch has been shown to have significant impact on critical dimension (CD) variation.

Many of the layout dependent effects can be minimized by restrictive design rules (RDRs) [17, 18] and optical proximity correction (OPC) [19, 20] techniques. However, in spite of RDRs and OPC, a large set of layout configurations can still cause a significant systematic shift in the device parameters. For example, the spacing of polysilicon contact landing pad to the diffusion (active) region shows approximately 3% device performance impact in 65nm SOI [21].

Besides lithographic variations, there are other sources of variation that can cause a device parameter to vary with its local layout environment. Well-proximity effect is a well known cause of local layout dependent variation [22]. In deep well implants, dopant ions can scatter from the edge of the well masking photoresist layer and increase the surface implant dose in devices located near the edge of the well, resulting in a layout dependent V_T offset. Stress is another source of layout dependent variation. The introduction of uniaxial stress in channel areas for carrier mobility enhancement via dual stress liners (DSL) [2] and e-SiGe [3] results in layout dependent stress variation.

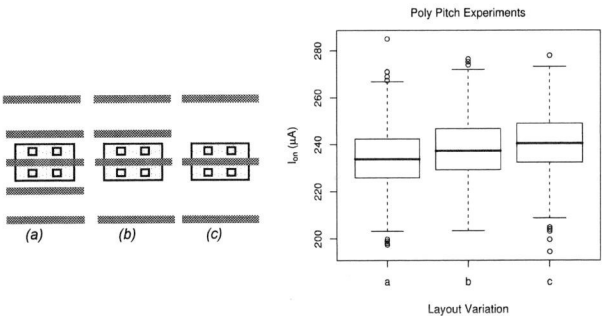

Fig. 8. Poly density induced current variability

Array-based structures were fabricated in a 65 nm bulk technology characterize layout dependent variations. Fig. 8 shows the measured results for the layout experiment where the impact of neighboring poly configurations on a device performance was investigated [21]. The measurements show a clear layout based systematic variation in the on-currents.

There are other sources of variation that depend on layout density in a larger window than the local environment dependencies discussed above. The well known example is pattern dependent Cu dishing and oxide erosion during Cu chemical mechanical polishing (CMP) process [23]. Rapid thermal anneal (RTA) process may also induce layout density dependent intra-die variations [24]. During RTA, local anneal temperature can vary across the die due to differences in radiation reflectivity of the layout patterns. The die regions with higher local anneal temperature result in better thermal absorption of junctions resulting in lower threshold voltage and lower extrinsic resistance of the device [25].

978-1-4244-2018-6/08 $25.00 © 2008 IEEE

Cross-wafer variation of performance is apparent in the wafer map in Fig 9. Die to die performance variation within a single lot of one in-situ performance monitor has been shown to exhibit max-to-min delays of more than 1.5X. Cross-die systematic performance variation of identical structures has been shown to exhibit delay variability of up to 1.4X [26, 41].

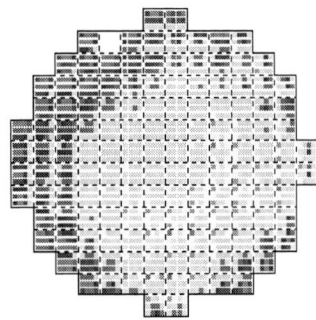

Fast ■■■ ■□□□ □□□ ■ Slow

Fig. 9. Wafer performance map

III. RELIABILITY

Device reliability loss can be characterized as extrinsic failures (Infant mortality), operational failures (Transients), and intrinsic failures (Wearout).

Infant mortality, caused by various defect mechanisms, can be accelerated via burn-in. Semiconductors fail due to a variety of mechanisms: (1) defects, (2) high field effects including Hot Carrier Injection (HCI), Negative Bias Temperature Instability (NBTI) and metal Electro-Migration (EM), and (3) noise phenomena either self-generated or externally induced.

Extrinsic failures are caused predominantly by contamination during manufacture. These result in "topological" changes in circuit due to electrical shorts, opens, or bridges. Extra conducting material can result from photolithographic printing errors, particulate contamination, incomplete etch, or incomplete metal polish. In ultra-thin gate oxides, pinholes can form due to surface imperfections, crystal defects, nitride cracking or oxygen deficiency at Si-SiO$_2$ interface. Fig. 10 shows a gate-oxide pinhole identified and isolated using a parametric gate current characterization macro.

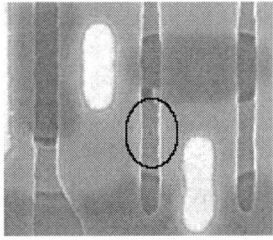

Fig. 10. Non-catastrophic gate-oxide pinhole

Operational failures can result from self-generated or external noise sources On-chip noise coupling is an issue as scaling increases the density wires and the number of wiring levels. For interconnect performance wire resistance is minimized resulting in increased line-to-line coupling. Coupling noise is results when the switching of one wire is capacitively-coupled to neighbor lines. Numerous noise analysis and reduction techniques exist. Off-chip noise can be caused by energetic particles (alpha particles or cosmic rays) that cause ionization in Si [27, 28]. In both cases, small circuits are more sensitive than large circuits. Both phenomena effectively inject extra charge which, if greater than a critical charge threshold, will cause upset failures in SRAMs, latches, or dynamic circuit storage nodes.

Intrinsic failures in device or wires can result from high-field or high-current wear-out phenomena. In devices these phenomena include NBTI [29-31], Hot-ē [32], and gate-oxide breakdown [33, 34]. The primary wear-out phenomenon for wires is electro-migration [35]. Electro-migration occurs as energetic electrons transfer momentum to metal atoms at grain boundaries. As the atoms move, the conductors thin, concentrating the carriers and speeding up further wear-out. Eventually, continuity is lost, resulting in an open circuit.

This mechanism is modeled with Black's Model [36]:

Mean Time To Failure $MTTF = A \times J_{avg}^{-n} \times e^{\frac{C}{kT}}$ [2]

Where, A=area, J=avg current density, T=temperature, n~2, and C~8000.

Conductors which carry large currents are particularly subject to catastrophic failure from electro-migration (EM). Failure is accelerated in conductors where the current flow is unidirectional. High temperature further exacerbates the problem. Typical location of this in VLSI designs is in the power delivery network or at localized metal/via or metal/contact interfaces of devices connected to the power distribution. EM is getting worse with scaling because of various interface scattering phenomena.

Hot-ē injection is caused by horizontal high fields and NBTI by vertical fields. The high field/temperature breaks contaminant H bonds, H+ ions diffuse, creating charge at the Si/SiO2 interface. NBTI increases the magnitude of the Vt of the device. This change is characterized by a relatively large increase in threshold on a small time scale and a smaller but additive increase on a longer time scale. In current technologies this effect is most significant in PFETs. The net device impact is that devices slow down as they age. Devices which are "stressed" by maintaining a lateral field across the device wear out faster due to hot-e and devices which are "stressed" by maintaining a verticle field across the device wear out faster due to NBTI.

Aging of device oxides is usually modeled by the following Weibull distribution of timing-dependent dielectic breakdown [34]. Here F is the cumulative failure probability and T_{BD} is the time-to-breakdown.

$$F(T_{BD}) = 1 - \exp\left[-\left(\frac{T_{BD}}{\alpha}\right)^{\beta}\right]$$ [3]

978-1-4244-2018-6/08 $25.00 © 2008 IEEE

Dealing with reliability defects involves avoiding manufacturing defects due to contaminants. Dealing with on-chip noise involves managing wire spacing, signal-edge-rates, and switching patterns often through CAD tools. Avoiding off-chip noise includes shielding techniques, make devices bigger, or adding redundancy/coding/checking. Avoiding reliability issues involves minimizing device stress by limiting the fields placed across the devices, the temperature, or the time the devices are exposed to stress.

IV. CIRCUIT IMPACT

Trends for substantial variation blur the distinction between failure modes due to structural (topological), and parametric (variability) defects. The impact of variability on delay and power is widely established and addressed by design tools. A secondary impact on reliability is well established, but with less automation support. The impact of variability on system resilience is emerging at the 45nm node; few tools exist to address this.

Introduction of process and environmental sensitivity into VLSI design tools is necessary to manage variability. Examples are statistical timing analysis [43-45] and function-based OPC. Manufacturing awareness has clearly entered the "lower" levels of design. Current physical implementations are highly influenced by OPC/RET at the front-end, and by CMP at the back-end levels. Variation-tolerance supported at the system or at the circuit levels is possible.

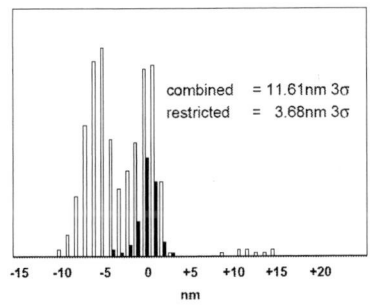

Fig 11: Reduction in linewidth variation through RDR [17]

Much of the variability in current designs is due to layout diversity. Rigid regularity through Restricted Design Rules (RDRs) or regular fabrics can reduce variability at the cost of physical design constraint. Diversity can be limited to a smaller subset of predictable and manufacturable constructs by aggressive use of RDRs. Fig 11 shows the benefit of RDRs in reducing linewidth variation [17]. The two histograms present the 1) 180nm, 200nm, 250nm, and 1000nm pitches with both horizontal and vertical orientations (combined) and 2) only 250nm pitch with horizontal orientation (restricted).

A design methodology called L3GO (*Layout using Gridded Glyph Geometry Objects*) [18] was developed to implement RDRs and exploit regularity. In the L3GO approach, layout is performed using *glyphs* instead of shapes. Three basic glyph types are used: 1) *Line glyph* (poly, metal wiring, etc.) 2) *Rectangle glyph* (active, n-well, etc.), and 3)

Point glyph (contacts, vias, etc.). Design restrictions can be enforced on each layer such as layer specific X/Y grids, legal placement points, preferred orientation, and layer-specific allowed wire widths, etc. A glyph based layout is converted into manufacturable target wafer shapes through a process called *elaboration*. Fig 12 shows an example of a glyph based layout and corresponding target wafer shapes. The elaborator can absorb minor changes in the design rules without requiring a re-design of the L3GO layout, thereby improving design predictability, reducing systematic yield loss, and simplifying design rules.

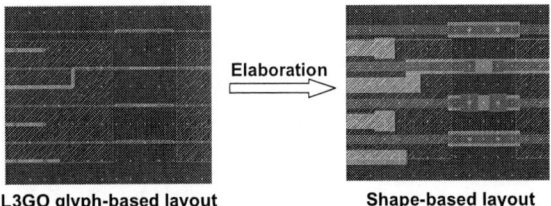

Fig 12: L3GO design methodology for implementing RDRs [18]

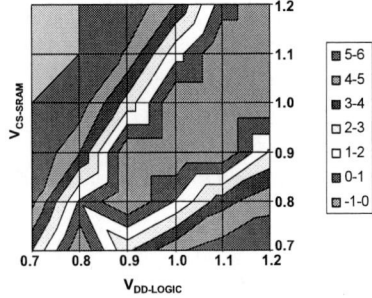

Fig 13: 65-nm PD/SOI SRAM cell stability vs supply voltages [40]

The most significant circuit impact of process variation is observed in SRAM arrays [37, 38]. In order to achieve high yield in large arrays, SRAM cell failure probabilities must be extremely low; to ensure that a 4 MB cache has at most one failure, the cell must operate correctly up to the 5.44 sigma portion of the distribution [39]. Such levels of cell stability are difficult at low supply voltages, Fig 13 shows the measured data for 65 nm PD/SOI SRAM cell in logic ($V_{DD-LOGIC}$) and SRAM ($V_{CS-SRAM}$) supply voltage space [40]. Here different regions represent the stability in sigma with zero being worst and six being the best. Yield modeling tools linked to empirical data, used to generate the data shown in Fig. 13, allow circuit and layout design and manufacturing to accommodate parametric variation in the development of robust array designs.

V. CONCLUSION

Technology scaling and design complexity result in increasing variability and reliability challenges. The characterization, modeling, and mitigating of these effects is key to manufacturable design. Variability characterization structures have been developed to understand the characteristics of these effects. Deriving empirical results for systematic variability enables more aggressive design.

The interface between design and manufacturing is becoming ever more complex. One of the most effective ways to improve design/technology coupling is to explicitly limit implementation diversity through the use of restricted design rules or implicitly through regular-fabrics. Uniformity improves predictability, reduces variability, and increases yield. Variability and reliability awareness must increasingly be reflected in designs and design tools.

REFERENCES

[1] G.G. Shahidi, "SOI Technology for the GHz Era", *IBM J. of Res. And Dev.*, vol. 46, pp. 121-131, 2002.

[2] H.S. Yang *et al.*, "Dual Stress Liner for High Performance Sub-45nm Gate Length SOI CMOS Manufacturing", *IEDM*, pp. 1075-1077, 2004.

[3] T. Ghani *et al.*, "A 90nm High Volume Manufacturing Logic Technology Featuring Novel 45nm Gate Length Strained Silicon CMOS Transistors", *IEDM*, pp. 11.6.1-11.6.3, 2003.

[4] R.W. Keyes, "The Impact of Randomness in the Distribution of Impurity Atoms on FET Threshold", *J. of App. Phys.*, pp. 251-259, 1975.

[5] M. Pelgrom, A. Duinmaijer, A. Welbers, "Matching Properties of MOS Transistors", *IEEE J. of Solid State Circuits*, vol. 24, 1989, pp. 1433-1440.

[6] P. Oldiges *et al.*, "Modeling Line Edge Roughness Effects in sub 100 Nanometer Gate Length Devices", *SISPAD*, pp. 131-134, 2000.

[7] T. Brunner, "Why Optical Lithography Will Live Forever", *J. of Vac. Sci. Technology*, B 21, no. 6, pp. 2632–2637, 2003.

[8] K. Bowman, *et al.*, "Impact of Extrinsic and Intrinsic Parameter Variations on CMOS System on a Chip Perforamnce", *IEEE Intln. ASIC/SOC Conference*, pp. 267-271, 1999.

[9] K. Agarwal *et al.*, "A Test Structure for Characterizing Local Device Mismatches", *Symp. on VLSI Circuits*, pp. 67-68, 2006.

[10] K. Agarwal, S. Nassif, F. Liu, J. Hayes and K. Nowka, "Rapid Characterization of Threshold Voltage Fluctuation in MOS Devices", *Intl. Conf. on Microelectronic Test Structures*, pp. 74-77, 2007.

[11] R. Kanj *et al.*, "Design Considerations for PD/SOI SRAM: Impact of Gate Leakage and Threshold Voltage Variation", *IEEE Trans. on Semiconductor Manuf.*, vol. 21, no. 1, pp. 33-40, Feb 2008.

[12] Y. Borodovsky, "Impact of Local Partial Coherence Variations on Exposure Tool Performance", *SPIE*, vol. 2440, pp. 750-770, 1995.

[13] T. Brunner, "Impact of Lens Aberrations on Optical Lithography", *IBM J. of Res. and Dev.*, vol. 41, pp. 57-67, 1997.

[14] A.K. Wong, R. Ferguson and S. Mansfield, "Mask Error Factor in Optical Lithography", *IEEE Trans. on Semiconductor Manuf.*, vol. 13, pp. 235–242, May 2000.

[15] C. Hedlund, H. Blom, and S. Berg, "Microloading Effect in Reactive Ion Etching," *J. of Vac. Sci. Tech.*, vol. 12, pp. 1962–1965, 1994.

[16] J. Watts, N. Lu, C. Bittner, S. Grundon, J. Oppold, "Modeling FET Variation within a Chip as a Function of Circuit Design and Layout Choices", *Nanotech Workshop on Compact Modeling*, pp. 87-92, 2005.

[17] L. Liebmann *et al.*, "High Performance Circuit Design for the RET-Enabled 65-nm Technology Node", *SPIE*, vol. 5379, pp. 20-29, 2004.

[18] M. Lavin, F. Heng and G. Northrop, "Backend CAD Flows for Restrictive Design Rules", *Intl. Conf. on Computer-Aided Design*, pp. 739-746, 2004.

[19] L. Liebmann *et al.*, "TCAD Development for Lithography Resolution Enhancement", *IBM J. of Res. and Dev.*, vol. 45, pp. 651-665, Sep 2001.

[20] L. Liebmann, "Resolution Enhancement Techniques in Optical Lithography, It's not Just a Mask Problem", *SPIE*, v. 4409, pp. 23-32, 2001.

[21] K. Agarwal and S. Nassif, "Characterizing Process Variation in Nanometer CMOS", *Design Automation Conf.*, pp. 396-399, 2007.

[22] T.B. Hook *et al.*, 'Lateral Ion Implant Straggle and Mask Proximity Effect", *IEEE Trans on Elec. Devices*, vol. 50, pp. 1946-1951, Sep 2003.

[23] J.-Y. Lai, N. Saka and J.-H. Chun, "Evolution of Copper-Oxide Damascene Structures in Chemical Mechanical Polishing", *J. of Electrochem. Soc.*, pp. G31-G40, 2002.

[24] P.J. Timans *et al.*, "Challenges for Ultra-Shallow Junction Formation Technologies beyond the 90nm Node", *Intl. Conf. on Adv. Thermal Processing of Semiconductors*, pp. 17-33, 2003.

[25] I. Ahsan *et al.*, "RTA-Driven Intra-Die Variations in Stage Delay, and Parametric Sensitivities for 65nm Technology," *Symp. on VLSI Technology*, pp. 170-171, 2006.

[26] A. Gattiker *et al.*, "Data Analysis Techniques for CMOS Technology Characterization and Product Impact Assessment," *International Test Conference*, 2006.

[27] J. F. Ziegler, "Terrestial Cosmic Ray Intensities", *IBM Journal of Res. and Dev.*, vol. 42, Jan 1998.

[28] R. Baumann, "Radiation-Induced Soft-Error in Advanced Semiconductor Technologies", *IEEE Trans. on Device and Materials Reliability*, vol. 5, no. 3, pp. 305-316, Sep 2005.

[29] B. Deal, M. Sklar, A.S. Grove and E.H. Snow, "Characteristics of the Surface-State Charge (Qss) of Thermally Oxidized Silicon", *J. Electrochem. Soc.*, 114:266-74, 1967.

[30] C.E. Blat, E.H. Nicollian and E.H. Poindexter, "Mechanism of Negative-Bias-Temperature Instability", *J. Appl. Phys.*, vol. 69, no. 3, pp. 1712-1720, 1991.

[31] G. Massey, "NBTI: What We Know and What We Need to Know: A Tutorial Addressing the Current Understanding and Challenges for the Future", *Integrated Reliability Workshop*, pp. 199-211, Oct. 2004.

[32] T. H. Ning, "Hot-Electron Emission from Silicon into Silicon Dioxide," *Solid-State Electron.*, vol. 21, pp. 273–282, 1973.

[33] S.-H. Lee, H.-J. Cho, J.-C. Kim and S.-H. Choi, "Quasi-breakdown in Ultrathin Gate Oxides under High Field Stress," *IEDM*, pp. 605–608, 1994.

[34] E.Y. Wu *et al.*, "CMOS Scaling beyond the 100-nm Node with Silicon-dioxide-based Gate Dielectrics", *IBM J. of Res. And Dev.*, vol. 46, pp. 287-398, 2002.

[35] I.A. Belch and J. Sello, "The Failure of Thin Aluminum Current-Carrying Strips in Oxidized Silicon", *Physics of Failure in Electronics*, vol. 5, pp. 496-505, 1966.

[36] J.R. Black, "Mass Transport of Aluminum by Momentum Exchange with Conduction Electrons", *IEEE Intl. Reliability Physics Symp.*, pp. 148-159, 1967.

[37] D. Burnett, K. Erington, C. Subramanian, K. Baker, "Implications of Fundamental Threshold Voltage Variations for High-Density SRAM and Logic Circuits", *Symp. on VLSI Technology*, pp. 15-16, 1994.

[38] A. Bhavanagarwala, X. Tang, J. Meindl, "The Impact of Intrinsic Device Fluctuations on CMOS SRAM Cell Stability, *J. Solid State Circuits*, pp. 658-665, 2001.

[39] R. Heald and P. Wang, 'Variability in sub-100nm SRAM designs", *ACM/IEEE Intl. Conf. on Computer-Aided Design*, pp. 347-352, 2004.

[40] R. Joshi *et al.*, "Statistical Exploration of the Dual Supply Voltage Space of a 65 nm PD/SOI CMOS SRAM Cell", *European Solid-State Dev. Res. Conf.*, pp. 315-318, 2006.

[41] M. Ketchen and M. Bhushan, "Product-representative at-speed test structures for CMOS characterization", *IBM Journal of Research and Development*, pp. 451-468, Jul/Sept 2006.

[42] D. Boning *et al.*, "Test Structures for Delay Variability", *TAU Workshop*, 2002.

[43] A. Devgan and C. Kashyap, "Block-Based Static Timing Analysis with Uncertainty", *Intl. Conf. on Computer-Aided Design*, pp. 607-614, 2003.

[44] H. Chang and S. S. Sapatnekar, "Statistical Timing Analysis Considering Spatial Correlations Using a Single PERT-like Traversal," *Intl. Conf. on Computer-Aided Design*, pp. 621-625, 2003.

[45] C. Visweswariah, *et al.*, "First-order Incremental Block-Based Statistical Timing Analysis" *Design Automation Conf.*, pp. 331-336, 2004.

A 9Gbit/s Serial Transceiver for On-chip Global Signaling over Lossy Transmission Lines

JunYoung Park[1,2], Joshua Kang[1], Sunghyun Park[1,3], Michael P. Flynn[1]

[1]University of Michigan, Ann Arbor, MI
[2]Qualcomm, San Diego, CA
[3]Qualcomm, Campbell, CA

Abstract-A 9Gbit/s serial link transceiver for on-chip global signaling is presented. A transmitter serializes 8b 1.125Gbyte/s parallel data and transmits over 5.8mm of lossy on-chip transmission line. The receiver de-serializes the data with the help of a digitally-tuned interpolator. An error checking block verifies the recovered and de-serialized data against the original data and counts the number of discrepancies. The prototype transceiver, implemented in 0.13μm 8 metal CMOS, achieves 9Gbit/s with four pre-defined data patterns and a measured BER is less than 10^{-10}.

I. INTRODUCTION

With the increase in clock frequencies to multi-GHz rates, it has become impossible to move data across a die in a single clock cycle with conventional parallel bus-based communication. There are also reliability problems due to timing errors, skew, and jitter in fully synchronous systems. Noise, coupling, and inductive effects become significant for both intermediate length and global routing. Global signaling is responsible for an ever increasing portion of total power consumption. Buses are consuming too much area, yet interconnect is reverse scaling [1], while the required communication bandwidth on an IC is growing exponentially. Repeaters are commonly used to improve long distance signaling by breaking up long lines into shorter sections. Although this technique is very effective, the optimum number of repeaters can be large. The repeaters have an adverse effect on power consumption and chip area, and also generate significant supply noise. The performance of repeaters is also sensitive to edge rates. Furthermore, repeaters are not always optimally located; for example, it may not be possible to optimally place repeaters when routing over a critical logic block.

This work presents techniques for global serial signaling over lossy on-chip transmission lines, as an alternative to global parallel buses with repeaters. Unlike some other schemes for high-speed on-chip signaling, there is no requirement for up-conversion, modulation [2], equalization [3], or special metal processing. An optimum resistive termination scheme, first presented by the authors in [4], allows large bandwidth (10's GHz) communication over lossy transmission lines (>200Ω) without Inter-Symbol-Interference (ISI). An interleaved pulse driver scheme efficiently drives the transmission line. At the receive-end a digitally-tuned sampler samples the received signal at the optimum phase, regardless of clock and signal delays. The prototype transceiver achieves 9Gbit/s over a 5.8mm on-chip transmission line with a measured BER of less than 10^{-10}.

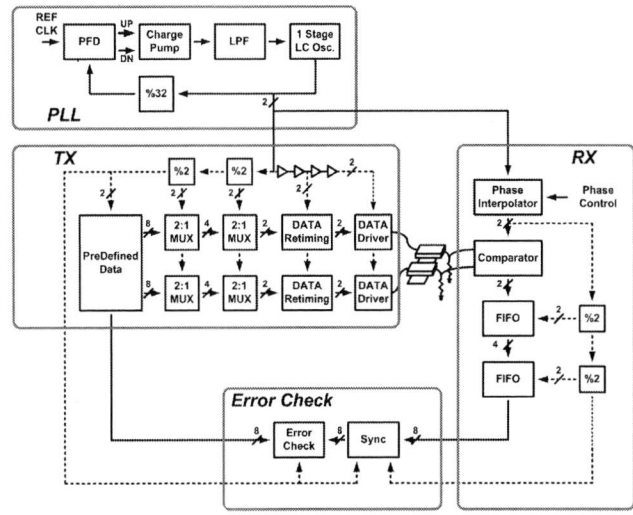

Fig. 1. Block diagram of the serial transceiver for on-chip global signaling.

II. TRANSCEIVER FOR ON-CHIP GLOBAL SIGNALING

Fig. 1 shows a block diagram of the transceiver. The prototype device consists of a PLL, a transmitter, a receiver, and error checking logic for self-test. An LC oscillator based PLL generates a 4.5GHz differential clock signal, which is routed to both the transmitter and the receiver. The transmitter serializes 8b wide 1.125Gbyte/s parallel data and drives the serialized data down the lossy on-chip transmission line.

Optimum resistive termination at the receive-end minimizes dispersion and ISI [4]. A phase-tuned receiver samples and de-serializes the received signal. Since the sampling instant is tuned to match the received signal eye, there is no requirement to match the clock and signal routing or clock and data signal delays. The 4.5GHz PLL output clock is divided by two (2.25GHz) and four (1.125GHz) for use in serializing the parallel 8b 1.125Gbyte/s data at the transmitter [5] and de-serializing the serial 9Gbit/s data at the receiver, respectively. An error checking block verifies the recovered data from the receiver against the original data and counts the number of discrepancies.

A. Lossy on-chip transmission line

A lossy on-chip differential transmission line, as shown in Fig. 2, is used for a global signaling. Metal interconnect resistivity in conventional CMOS poses significant challenges to implementing long (~10mm) transmission lines for global on-chip, digital communication. Series resistance causes

978-1-4244-2018-6/08 $25.00 © 2008 IEEE

dispersion, leading to considerable ISI. Optimum termination and appropriate selection of the characteristic impedance of the transmission line minimizes the dispersion caused by the low frequency slow-wave RC effect [1]. The signal propagates along the transmission line at the speed of light of the dielectric. A shielded coplanar transmission line is formed with two 6µm wide metal-5 wires separated by 3µm, over a 21µm wide metal-2 ground plane. The simulated characteristic impedance is 30.6Ω and the line is terminated at the receiver with a 30.6Ω poly-silicon resistor.

B. Transmitter

A low-output-resistance 9GHz line driver is impractical in 0.13µm CMOS technology, so instead an interleaved transmit architecture is adopted. Two interleaved parallel streams of 4b data at 1.125Gbit/s are serialized to 4.5Gbit/s. The final 9Gbps serialization is achieved using a pair of interleaved line drivers. Fig. 3 shows one of the two identical 4-bit serializers. In order to properly serialize the data, the original parallel data at 1.125GHz are distributed in the bit sequence *D1, D3, D5,* and *D7* to one module, and *D2, D4, D6,* and *D8* to the other identical module. The serialized outputs are buffered in order to drive the next stage.

The final 9Gbit/s serialization is achieved using interleaved 4.5GHz line drivers formed by *M1, M2, M3,* and *M4* as shown in Fig. 4. Two 4.5GHz 1b data patterns drive *DS_1* and *DS_2*. Only one driver is active during each half cycle of a 4.5GHz clock, facilitating 9Gbit/s serialization at the node *OUT*. Separate dynamic pre-drivers drive the NMOS and PMOS line driver devices, with the help of two differential 4.5GHz clock phases, *CK* and *CKB*. Data arrives at the pre-drivers ahead of the clock signals so that the internal pre-driver nodes (*xp1, xn1, xp2,* and *xn2*) are pre-evaluated. During the clock periods (*CK* and *CKB*) the driver signals (*cp1, cn1, cp2,* and *cn2*) are generated activating only one of four line driver devices (*M1, M2, M3,* or *M4*).

C. Receiver

The receiver samples the received signal with interleaved samplers operating at 4.5GHz (the same frequency as the transmitter). The sampled data is de-serialized and down-sampled to 1.125GHz. Due to the signal delay over the long transmission-line, the phase of the sample clock signals is tuned for proper data recovery. With appropriate phase control, we can employ only two comparators sampling at 4.5GHz. The comparator clock phases are adjusted to sample the input data at the center of data eye.

An RC-CR filter and a phase interpolator block, shown in Fig. 5, allow control of the sampling phase. The differential outputs of the PLL, *PLL_out+* and *PLL_out-*, are the inputs to a pair of RC-CR filters, which generates four equally-spaced 4.5GHz clock signals. The digitally phase controlled interpolator takes these four signals and generates four phase controlled clock signals for the two sampling comparators. Although the two signals from an RC-CR filter have a 90-degree phase difference at all frequencies, the magnitudes of these two signals are only the same at 1/(RC). Therefore, gain

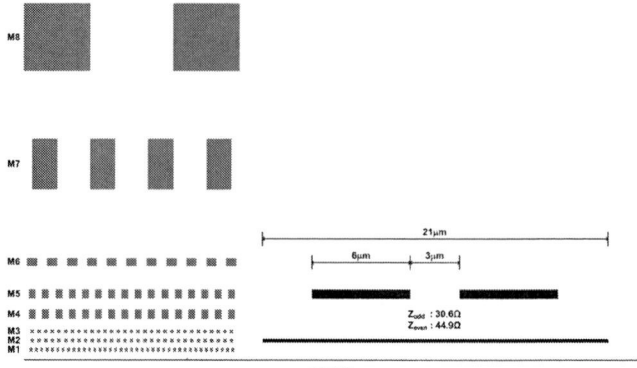

Fig. 2. Cross-section of the on-chip transmission line.

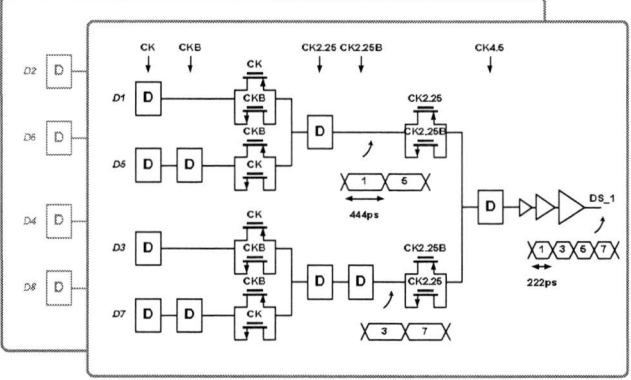

Fig. 3. Schematic of the data serializer.

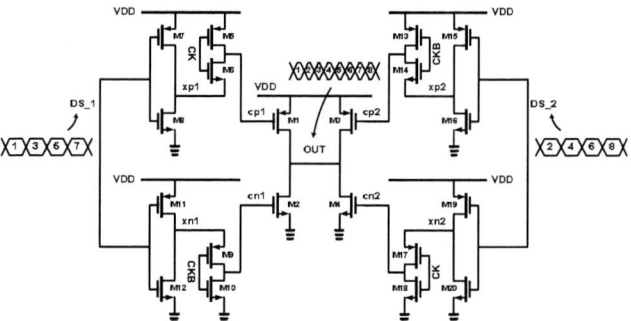

Fig. 4. Schematic of the interleaved line drivers and pre-drivers.

stages follow the RC-CR filters to compensate for amplitude mismatches [6].

The phase of the output signals of the interpolator is changed by controlling the tail bias currents. Two sets of differential clock signals (or four phases) with a phase difference of 90-degrees, (i.e. 56ps at 4.5GHz), drive the inputs. Since there are eight differential control-switches, turning one switch on or off causes approximately 7ps of advancement or delay, respectively. The gain of the interpolator also helps suppress amplitude mismatches of the signals from the RC-CR filters.

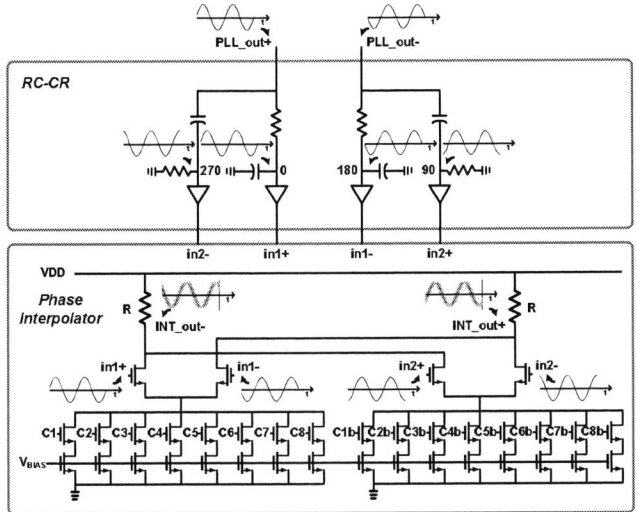

Fig. 5. Schematic of the RC-CR and phase interpolator.

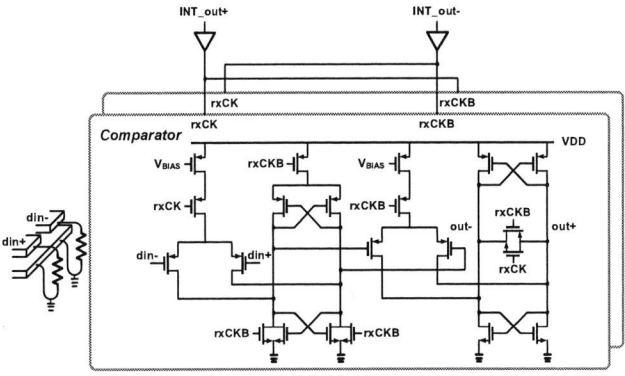

Fig. 6. Schematic of the comparator.

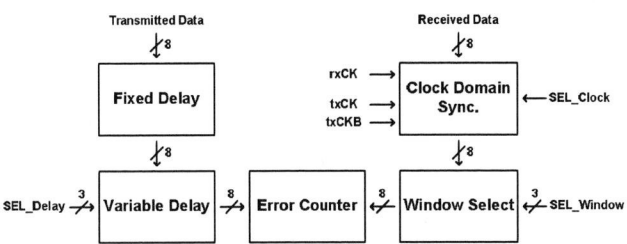

Fig. 7. Block diagram of the self-test error-check block.

Comparators are used at the first stage of the receiver to sample the received signal and convert a small input signal to CMOS voltage levels. Most high-speed comparators consist of a preamplifier followed by a regenerative latching stage, with each stage driven by complementary clock signals. While on, the preamplifier operates as a differential amplifier and the latching stage, in turn, amplifies the signal.

The voltage signal level at the long end of the lossy transmission line (from the transmitter) can be as small as 200mV, due to resistive attenuation along the transmission line. Since the data rate is fast (9Gbps), the comparator is formed as a cascade of two pairs of preamps and latching stages, as shown in Fig. 6, in order to have sufficient gain to avoid a metastability. Since the sampling instances for the comparators differ in phase by 180-degrees, the outputs of both comparators are synchronized to a single clock phase before being de-serialized.

D. Error Check

An error-check block compares data patterns from the transmitter and from the receiver, and generates an error signal when there is a mismatch. The error counter receives and compares two 8-bit data words every 1.125GHz clock cycle. As shown in Fig. 7, the original 8-bit data sent to the transmitter is stored in registers in the *Fixed Delay* and *Variable Delay* blocks until the serialized data is sampled and recovered at the receiver, and sent to the error-check block. In order to decide whether the recovered data is same as the original data sent by the transmitter, the data from the receiver is first transferred to the clock domain of the transmitter. The clock synchronization block performs clock domain alignment. A *WINDOW SELECT* block stores recovered data for two 1.125GHz clock cycles and selects an 8-bit window from the 16 stored data-bits for comparison.

Every clock cycle, the error counter shown in Fig. 8 compares two 8-bit input data patterns and increments the error count if a mismatch is found. By performing a bit-wise XOR of the two 8-bit data words and ORing the result, an

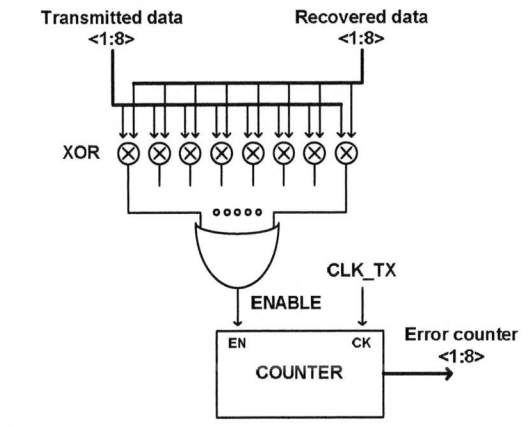

Fig. 8. Error counter.

ENABLE signal increments the error count whenever a mismatch is found.

III. EXPERIMENTAL RESULTS

The prototype is fabricated in a 0.13µm 8M CMOS process, and measures 1.7mm×2.1mm including pads. The transmitter and receiver occupy 0.15mm^2 and 0.56mm^2 respectively. The chip micrograph is shown in Fig. 9. The 5.8mm transmission line is routed from the transmitter to the receiver outside the pads. Measured data waveforms are shown in Fig. 10 and Fig. 11. Fig. 10 shows the pre-defined serialized patterns, 10101010, 11001100, 1110000, and 01001110, at the output of the drivers, along with the 4.5GHz PLL clock, all of which are measured directly using a GSG probe on a probe station. The received de-serialized data are monitored by the on-chip error checking logic and the error count output is recorded with a mixed-signal oscilloscope.

978-1-4244-2018-6/08 $25.00 © 2008 IEEE

When a deliberate mismatch is introduced between transmitted and recovered data through a deliberate timing error, the error counter accumulates as shown on the left side of Fig. 11. When the receiver correctly recovers data, the error counter does not increment as shown on the right side of Fig. 11. The measured BER is less than 10^{-12} with a 10101010 pattern, and is less than 10^{-10} with a 01001110 pattern. The prototype operates with 1.5V supply and the total currents for the analog circuits (i.e. charge pump and LC oscillator in the PLL, comparator, and phase interpolator), transmitter, receiver, and error checking logic are 70mA, 280mA, 120mA, and 160mA, respectively. The performance of the chip is summarized in Table I.

IV. CONCLUSION

A complete on-chip transceiver communicating over a 5.8mm on-chip transmission line is described. Since the signal propagates along the line at the speed of light of dielectric, the latency of a long link is reduced significantly. Therefore, serial signaling promises low power consumption for high-bandwidth low-latency on-chip data links.

Fig. 9. Chip micrograph.

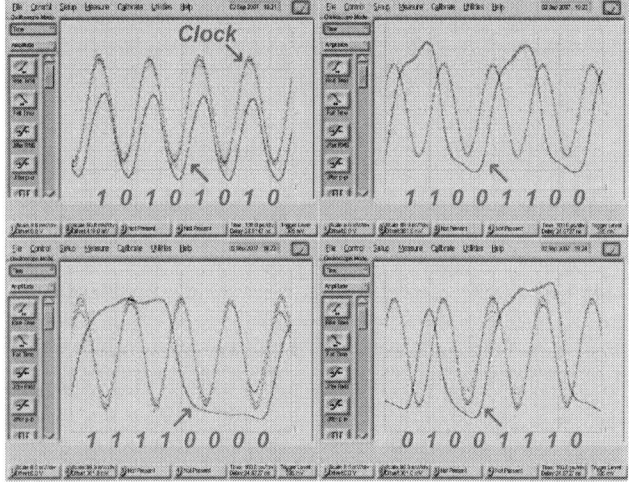

Fig. 10. Measured clock and serialized data at the output of the drivers.

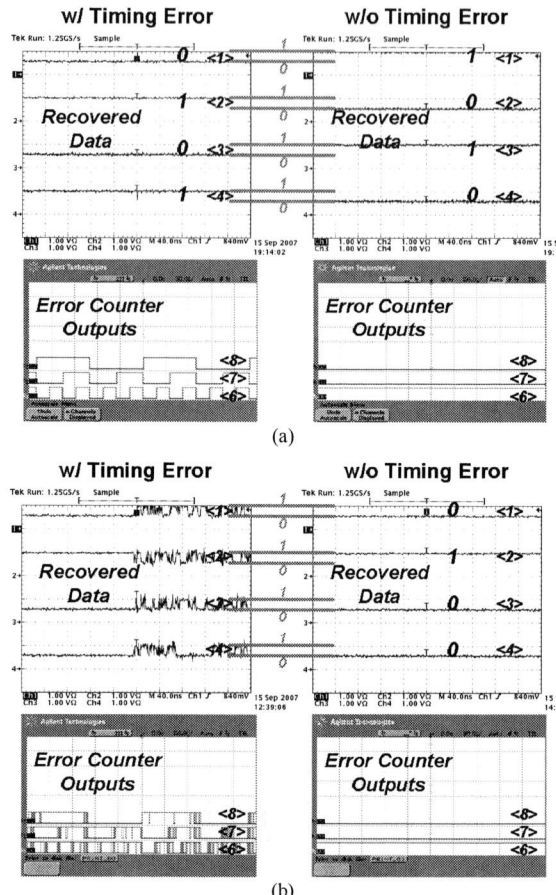

Fig. 11. Output of the self checking logic with (left) and without (right) deliberate timing error for the (a) 10101010 and (b) 01001110 patterns.

TABLE I
PERFORMANCE SUMMARY

Area	1.7mm × 2.1mm 0.5mm × 0.3mm (TX) 1.4 mm × 0.4mm (RX)	
Power (Ref. CLK : 140MHz, VDD: 1.5V)	Analog	105mW
	TX	420mW
	RX	180mW
	Error Check	240mW
BER	10^{-12} (10101010) 10^{-10} (01001110)	
Technology	0.13μm 8M CMOS	

REFERENCES

[1] M. Flynn, J. Kang, "Global signaling over lossy transmission lines," *ICCAD*, Nov. 2005.
[2] R. Chang, et al., "Near speed-of-light signaling over on-chip electrical interconnects," *J. Solid-State Circuits*, pp. 834-838, May. 2003.
[3] R. Ho, K. Mai, M. Horowitz, "Efficient on-chip global interconnects," *Dig. Symp. VLSI Circuits*, Jun. 2003.
[4] J. Kang, J. Park, M. Flynn, "Global High-Speed Signaling in Nanometer CMOS," *ASSCC Dig. Tech. Papers*, Nov. 2005.
[5] P. Chiang, et al., "A 20-Gb/s 0.13-μm CMOS serial link transmitter using an LC-PLL to directly drive the output multiplexer," *J. Solid-State Circuits*, pp. 1004-1011, Apr. 2005.
[6] B. Razavi, RF Microelectronics, 1998.

IEEE 2008 Custom Intergrated Circuits Conference (CICC)

A 28mW OFDM Baseband Receiver Chip for DVB-T/H with All Digital Synchronization

Ting-Chen Wei[1], Wei-Chang Liu[1], Chi-Yao Tseng[2], Syu-Siang Long[2], Shyh-Jye Jou[1], and Muh-Tian Shiue[2]

Department of Electronics Engineering, National Chiao Tung University, Taiwan R.O.C.[1]
Department of Electrical Engineering, National Central University, Taiwan R.O.C.[2]

Abstract—An OFDM baseband receiver chip for DVB-T/H application is proposed in this paper. With all-digital jointed detection/synchronization loops and channel estimation, the proposed receiver chip can compensate 200ppm sampling clock offset (SCO) and ± 50 subcarrier spacing carrier frequency offset (CFO) in multipath environment. The total memory requirement of this chip is 102.8KB and the total equivalent gate count (including memory) is about 806,800 gates. By using 0.18μm CMOS process, the power consumption is 28mW at 1.45 V, 40MHz and core size of this chip is 3600μm × 3600μm.

I. INTRODUCTION

DVB-T [1] is the digital TV broadcasting standard proposed by ETSI in 1997. Additional features for handheld devices; DVB-H [2] is also released in 2004. The DVB-T/H standard adopts OFDM to provide high data rate and spectrum efficient. The length of an OFDM symbol in DVB-T/H can be either 2048 (2K mode), 4096(4K mode) or 8192 (8K mode) and four GI lengths, namely 1/32, 1/16, 1/8 and 1/4, are selected to conquer different inter symbol interference (ISI) effects. Different modulations schemes like, QPSK, 16 QAM and 64QAM, are also provided for different data rate transmissions. In addition, three types of pilots, continual pilots, scattered pilots and transmission parameters signaling (TPS) pilots, are inserted in frequency domain for synchronization, channel estimation and transmitting system parameters.

In this paper, a low power baseband receiver architecture for DVB-T/H and its VLSI implementation are presented. The proposed receiver chip contains a Mode/GI/Symbol

detection to detect the transmission mode and locate the symbol position, a 2K/4K/8K FFT to demodulate the OFDM symbols, a carrier synchronization loop to compensate the carrier frequency offset (CFO), a sampling clock synchronization to compensate the sampling clock offset (SCO) and a channel estimation (CE) to calculate the channel response. In addition, to reduce the power consumption, several low power schemes and hardware sharing methods are adopted in this chip.

The organization of the paper is as follows. In section II the baseband receiver architecture and demodulation flow are introduced. Section III shows the system simulation results and the chip implementation and comparison. Then, Section IV is the conclusion.

II. RECEIVER ARCHITECTURE AND DEMODULATION FLOW

Fig.1 shows the block diagram of the proposed baseband receiver architecture. The demodulation flow is divided into acquisition and tracking [3][4]. In the acquisition stage, the receiver blindly detects the transmission mode and length of GI, finds the location of the OFDM symbol and estimates the fractional CFO (FCFO) and the integer CFO (ICFO). The Mode/GI/Symbol detection adopts our previous work [5]. A division free (replaced with subtraction) algorithm of a normalized maximum correlation is used in Mode/GI detection. This algorithm combining twisted memory access does not need to refill or replenish the input data samples. Thus, the longest latency of the proposed detection scheme is 17280 samples and is comparable with Mode/GI detection in parallel which needs large memory storages and operations. Besides, to share the memory elements with channel

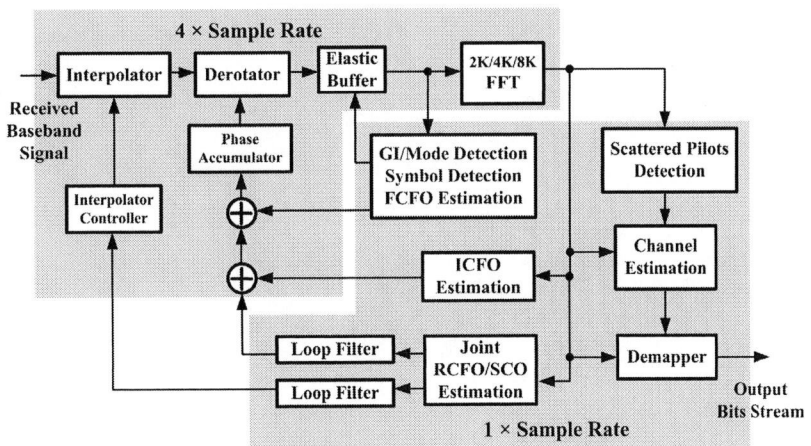

Fig.1 The proposed DVB-T/H baseband receiver architecture

978-1-4244-2018-6/08 $25.00 © 2008 IEEE   377

estimation (CE), the delay line in Mode/GI/Symbol detection comprises several 1K single-port SRAM blocks to increase the read-write ability. After completing these two detections, the memory bank is released for channel estimation to reduce the area cost. The total required storage size is 21KB SRAM.

Then, demodulation flow enters into tracking stage and residual CFO (RCFO) and SCO are separately compensated by a Cubic-Langrage interpolator [6][7] and a CORDIC based derotator [8]. An elastic buffer is used to control the data rate. Both carrier and sampling clock synchronization work in digital domain without feedbacks to the analog front end devices. Our previous work [9][10] proposes an jointed architecture for ICFO, residual CFO (RCFO) and SCO estimation as shown in Fig.2. In ICFO estimation; this architecture uses Series-In-Parallel-Out (SIPO) to temporarily store sign bit of the samples from FFT; hence, it is not necessary to store each received data that comes from FFT. As a result, the usage of memory is reduced by 90 %, the access number is reduced by 94% with SIPO length equal to twelve. Besides, a differential encoding method is used to record the continual pilot positions and the distribution of differential encoding positions is periodic. Therefore, storage requirement of recording continual plots positions is reduced by 77% in implementation. The design overhead is an accumulator and control unit to accumulate the difference values. Besides, The ICFO detection and RCFO/SCO estimation share the memory to reduce hardware cost. The carrier synchronization can compensates ±50 shifted subcarrier spacing (equivalent to ±220kHz at 2K mode and ±55kHz at 8K mode) and the clock synchronization can compensate 200ppm sampling clock offset.

12 cycles, only one SRAM block is used to store the newly arrived scatted pilots, and the other memory blocks are shut down for power saving.

Finally, the received signals ($R_{re} + jR_{im}$) in the data subcarriers are equalized in frequency domain by multiplying the reciprocal of the channel response ($C_{re} + jC_{im}$) in corresponding subcarriers and generates the transmitted signal ($D_{re} + jD_{im}$). For hard decision, the transmitted signals are translated into bit stream (y_0, y_1...,y_5); hence, the exact values of the transmitted signals are not necessary. For example, for a decision boundary 'B' of the hard decision demapper, Equation (1) can be modified into (2); then, the demapper only judges (2) to generate the bit stream. This work proposes a architecture of division free channel equalizer combining with hard demapper as shown in Fig.4. This architecture reduces the cost of a divider which is replace with a complex multiplier (3 multipliers and 5 adders/subtractors). In addition, the multiplication of the several constant normalized power gain (B) are implemented by Canonical Signed Digits [12] (CSD) multipliers which are composed of wire shifting connection, MUX and adders. As a result, the division operation is replaced with the multiplication in channel equalization.

$$D_{re} + jD_{im} = \left(R_r + jR_i\right) \times \frac{C_r - jC_i}{C_r^2 + C_i^2} \begin{array}{c} > \\ < \end{array} B$$

(1)

$$\underbrace{\left(R_r + jR_i\right) \times \left(C_r - jC_i\right)}_{Complex\ Multiplier} - \underbrace{B \times \left(C_r^2 + C_i^2\right)}_{CSD\ Multiplier} \begin{array}{c} > \\ < \end{array} 0$$

(2)

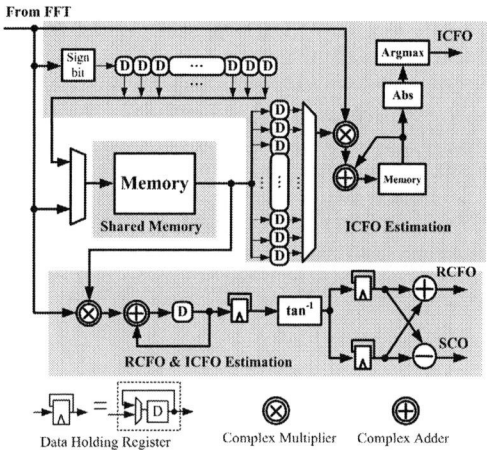

Fig.2 Jointed architecture for ICFO, residual CFO (RCFO) and SCO estimation

Fig.3 Channel estimation architecture

After coming into steady stage, the receiver detects the scattered pilot mode, estimates the channel response, equalizes and demaps the constellation into bits stream. This work adopts the 2-D predictive channel estimation [11]. The channel estimation architecture shown in Fig.3 is modified from [11] and includes seven SRAM blocks which separately store scatted pilots of previous received seven symbols with each composed of two 1K SRAM blocks. The stored scattered pilots of the previous received seven symbols are read every 12 cycles into data-holding registers. Within these

978-1-4244-2018-6/08 $25.00 © 2008 IEEE   378

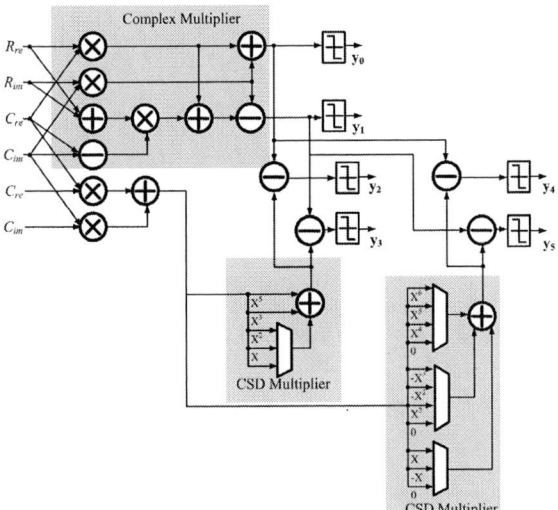

Fig.4 Architecture of division free channel equalizer combining with hard demapper

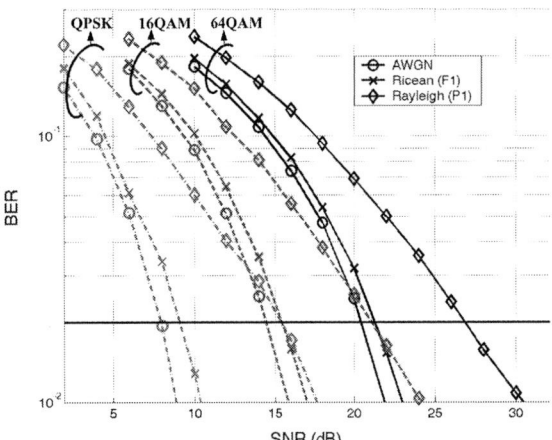

Fig.5 Simulated RTL uncoded BER performance at 8K mode, 1/4 GI, RCFO = 0.05 subcarrier spacing and SCO = 200ppm

III. HARDWARE IMPLEMENTATION AND COMPARISON

The baseband receiver chip contains two clock rate domains. The CORDIC based derotator, the Cubic Langrage interpolator and the 2K/4K/8 K multi-modes FFT operate at 4×sampling rate and the Mode/GI/symbol detection, the channel estimation and the jointed SCO and RCFO estimation work at 1×sampling rate. For the 8MHz channel bandwidth, the required sample rate is 9.14MHz, and the 4×sampling rate is 36.56MHz. A pipeline-based FFT is composed of Radix-2 and Radix-2/4/8 butterfly and uses single-path delay feedback (SDF) [13] to realize 2K/4K/8K points FFT. The delay line in SDF architecture uses single-port memory to reduce the area cost and unit can achieve 40dB SQNR.

The total memory requirement of this work is 102.8KB (99KB SRAM and 3.8KB ROM): 76KB SRAM and 3.7KB ROM for FFT, 21KB SRAM for channel estimation and Mode/GI/Symbol detection, and 2KB SRAM and 0.1KB ROM for synchronization. This chip does not include the channel decoder. For a hard Viterbi decoder, the uncoded BER (before Viterbi hard decoder) of 2×10^{-2} can archive the required coded BER for a rate 1/2 convolution code [14]. Fig.5 shows simulated RTL uncoded BER performance under different noise level, modulations and channel models.

Using 0.18μm, 1.8 V digital CMOS process, the core area of the proposed DVB-T/H receiver is 12.96 mm^2. The die photo of this receiver is shown in Fig.6 which includes scan-chain and memory built-in self-test (BIST) for testing. All the blocks were designed by cell-based approach. The chip can operate under supply voltage range from 1.4V to 1.8V. The maximum operating frequency is 60MHz at 1.8V. When running at 40 MHz, this chip consumes 43 mW at 1.8V and 28mW at 1.45V in the 1/4 GI mode. A comparison with previous reported DVB receiver [15] is listed in TABLE I. The Chen's chip [15] contains OFDM demodulation and channel decoding. To make a fair comparison, the area and power consumption of the Chen's OFDM demodulation is calculated according to its die photo and power profiling. In summary, the proposed chip provides a low power and low area solution to the DVB-T/H baseband receiver.

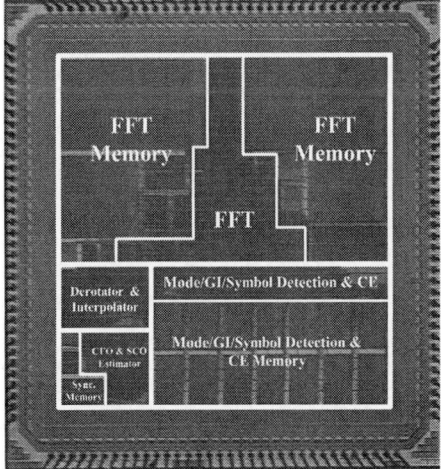

Fig.6 Die photo of the proposed DVB-T/H baseband receiver IC

TABLE I
COMPARISONS BETWEEN PREVIOUS REPORTED DVB-T/H BASEBAND RECEIVER WITH THIS WORK

	Chen [15]	Ours
Technology	0.18μm	0.18μm
Supply Voltage	1.8 V Core, 3.3V I/O	1.8 V Core, 3.3V I/O
Core Size	≒6.5 mm × 5.4 mm = 35.10 mm²	3.6 mm × 3.6 mm = 12.96 mm²
Input Clock Frequency	109.71 MHz	40 MHz (36.56MHz)
Power Consumption	250mW (1/32 GI Mode)	43mW (1/4 GI Mode)
Memory of OFDM Demodulation	SRAM: 118KB	SRAM: 99KB, ROM: 3.8KB
Area of OFDM Demodulation	≒ 25 mm²	12.96 mm²
Operation Frequency of OFDM Demodulation	36.56 MHz	40 MHz (36.56MHz)
Power Consumption of OFDM Demodulation	≒200mW(1/32 GI Mode)	43mW (Measured Power at 1/4GI mode) 55mW (Normalized power at 1/32 mode)
Power Consumption @ Low Supply Voltage	NA	28mW (Measured Power at 1/4GI mode,40MHz, 1.45V)

IV. CONCLUSION

This paper proposed an OFDM baseband receiver chip for DVB-T/H. The chip integrates Mode/GI detection, symbol detection, multimode FFT, channel estimation, carrier frequency synchronization and sampling clock synchronization loop. Both synchronization loops work in digital domain to relax the specifications of the front end devices. Mode/GI detection, symbol detection and channel estimation share the same memory bank to reduce the hardware cost. A division free channel equalizer combining with hard demapper reduces the cost of a divider. Total memory requirement is 102.8KB and the total equivalent gate count (including memory) is about 806,800 gates. This receiver chip was fabricated in a 0.18μm 1P6M technology and its core size is 12.96 mm². Measurement results indicates that the power consumption is 28mW running at 40MHz from 1.45V supply voltage.

ACKNOWLEDGMENT

This project is supported by NSC. The authors greatly appreciate the Chip Implementation Center (CIC), R.O.C., for the fabrication and measurement of the chip.

REFERENCES

[1] ETSI, "Digital Video Broadcasting: Framing Structure, Channel Coding and Modulation for Digital Terrestrial Television, *European Telecommunication Standard* EN 300 744 V1.5, Nov. 2004.

[2] ETSI, "Transmission System for Handheld Terminals (DVB-H)," *European Telecommunication Standard* EN 302 304 V1.1.1 Nov. 2004.

[3] M. Speth, S. A. Fechtel, G. Fock and H. Meyr, "Optimum receiver design for wireless broadband systems using OFDM—part I ," *IEEE Trans. Commun.,* vol. 47, no. 11, pp. 1668–1677, Nov. 1999.

[4] M. Speth, S. Fechtel, G. Fock and H. Meyr, "Optimum receiver design for OFDM-based broadband transmission part II: A case study," *IEEE Trans. Commun.,* vol. 49, no. 4, pp. 571–578, Apr. 2001.

[5] W. C. Liu, T.C. Wei and S. J. Jou, "Blind Mode/GI detection and coarse symbol synchronization for DVB-T/H," in *Proc. ISCAS 2007,* New Orleans, May 2007, pp. 2092–2095.

[6] M. Gardner, "Interpolation in digital modems-part I: Fundamentals," *IEEE Trans. Commun.,* vol. 41, no. 3, pp. 501–507, Mar. 1993.

[7] L. Erup, M. F. Gardner and R.A. Harris, "Interpolation in digital modems-part II: Implementation and performance," *IEEE Trans. Commun.,* vol. 41, no. 6, pp. 998–1008, June 1993.

[8] Y. Ahn, S. Nahm and W. Sung, "VLSI Design of a CORDIC-based Derotator," in *Proc. ISCAS 1998,* May 1998, pp. 449–452.

[9] T. Z. Wei, S. J. Jou and M. T. Shieu, "Memory reduction ICFO estimation architecture for DVB-T," in *Proc. ISCAS 2006,* Greece, May 2006, pp. 3406–3409.

[10] C. Y. Tseng, T.C. Wei, W.C. Liu and S. J. Jou, "Low power and power aware design for DVB-T/H baseband inner receiver," *in Proc. IEEE VLSI-DAT 2007,* Apr. 2007, pp. 1−4.

[11] T. A. Lin, C. Y. Lee, "Predictive equalizer design for DVB-T system," in *Proc. ISCAS 2005,* vol. 2, May 2005, pp. 940–943.

[12] K. K. Parhi, *VLSI Digital Signal Processing Systems: Design and Implementation,* New York: Wiley-Interscience Publication, 1999, pp. 505–528.

[13] E. H. Wold and A. M. Despain, "Pipeline an parallel-pipeline FFT processors for VLSI implementation," *IEEE Trans. Comput.,* vol. 33, no. 5, pp. 414–426, May 1984.

[14] P. Banelli, "Theoretical analysis and performance of OFDM signals in nonlinear fading channels," *IEEE Trans. Wireless Commun.,* vol. 2, no. 2 pp. 284–293, Mar. 2003.

[15] L. F. Chen, Y. Chen, L. C. Chien, Y. H. Ma, C. H. Lee, Y. W. Lin, C. C. Lin, H. Yu. Liu, T. Y. Hsu and C. Y. Lee, "A 1.8V 250mW COFDM baseband receiver for DVB-T/H applications," in *Proc. IEEE ISSCC 2006,* Feb. 2006, pp. 1002–1011.

IEEE 2008 Custom Intergrated Circuits Conference (CICC)

Nonvolatile Magnetic Flip-Flop for Standby-power-free SoCs

Noboru Sakimura, Tadahiko Sugibayashi, Ryusuke Nebashi and Naoki Kasai

Device Platforms Laboratories, NEC Corporation,

1120 Shimokuzawa, Sagamihara, Kanagawa, 229-1198 Japan

Abstract— A nonvolatile Magnetic Flip-Flop (MFF) primitive cell for SoC design libraries has been developed using a unique MRAM process. It has high design compatibility with conventional CMOS LSI designs. MFF maximum frequency was estimated to be 3.5 GHz, which is comparable to that of a normal CMOS DFF. An MFF test chip was fabricated with the process. The chip's functional performance was sufficiently high to demonstrate the potential of MFFs, which helps to reduce the power dissipation of SoCs dramatically.

I. INTRODUCTION

CMOS process scaling has enabled clock frequencies in most SoCs to be over 200 MHz. However, because it has also caused increased power consumption, many low-power circuit schemes have been utilized in SoCs since the development of the 65nm generation process. In such schemes, various means of effectively reducing the charging component of power consumption are employed, among them a gated clock or dynamic voltage-and-frequency scaling[1]. To reduce the leaking component, means utilized include multiple V_{TH} levels or back-bias control.

One well-known low-power circuit scheme is an on-chip power-isolation switch (OPS) scheme[2]. This scheme is very effective for reducing leakage when an LSI's function block is in the standby state. With this scheme, however, the OPS of the function block cannot be activated very often because the frequent switching degrades performance (e.g. processing speed) of the LSI.

If all the logic gates connected to the OPS are nonvolatile, the degradation can be avoided. In particular, if the function block circuitry is clock-synchronized, it is only necessary for all the flip flops (F/F) to be nonvolatile. Therefore, it is essential to develop a nonvolatile F/F to achieve higher performance LSIs with lower power consumption. In particular, the employing nonvolatile F/Fs can enhance a more efficient use of energy in SoCs for stanby-power-critical and quick-startup applications such as intelligent remote controls, medical sensor systems, etc, in the future.

Although nonvolatile F/Fs have been investigated for many years using the FeRAM process due to the merits mentioned above[3], the market for them has not been growing. Because this problem seems to have occurred as a result of high operational voltage, write endurance limitations, and low compatibility with the CMOS process of the FeRAM process, discussing on MRAM-based nonvolatile F/Fs were initiated[4].

We have already developed an MRAM process for a 2T1MTJ MRAM cell as well as a 1-Mbit MRAM macro[5-6]. This process is unique and more suitable for high speed (e.g.

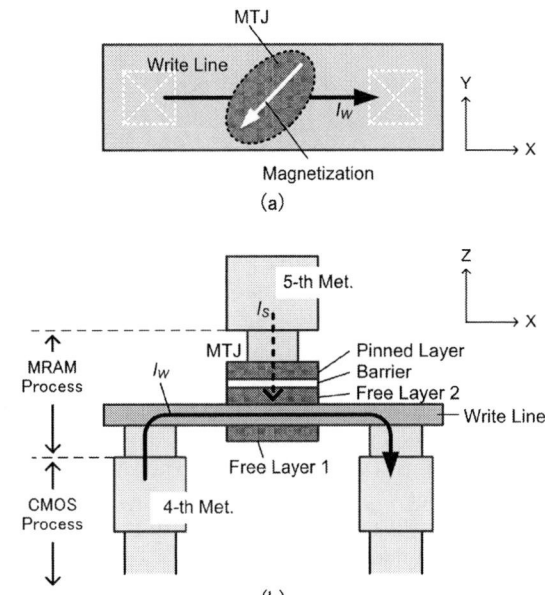

Fig. 1. Our developed MRAM technology, (a) top view and (b) cross section.

over 200-MHz) operation than any other MRAM process. This process has unlimited write endurance, low operational voltage and high compatibility with the CMOS process. In this paper, we will describe our development of a nonvolatile MFF primitive cell for SoC design libraries and our measurements of an MFF test chip made using our process.

II. MRAM TECHNOLOGY FOR SYSTEM LSI

Magnetic Tunneling Junctions (MTJs) are composed of two magnetic layers separated by barrier: a free magnetic layer and a pinned magnetic layer. The magnetization of the free layer is used for information storage. The MTJ resistance is either low or high depending on the relative magnetization, parallel or anti-parallel, of the free layer with respect to the pinned layer whose magnetization is fixed. The free layer's magnetization is switched by a magnetic field, which is generated by a bi-directional writing current (I_W) through the conductors.

Figure 1 shows the top view and cross section of our previously proposed MRAM technology[5-6]. In this technology, a thin base electrode of the MTJ is used as the write

978-1-4244-2018-6/08 $25.00 © 2008 IEEE

line, and the MTJ is directly placed on it to effectively enhance the magnetic field for the switching of the free layer's magnetization. We reported that our MRAM technology helps to reduce the I_W down to less than 1 mA[5], which is much smaller than that in conventional MRAMs. The technology is also superior to other MRAM technologies in terms of high-speed operation and compatibility with standard logic processes. The MTJ structure eliminates read disturbance because the sensing current (I_S) path is different from the I_W path. This helps to achieve nano-second read/write operation[6]. The optional MRAM process is added only in the insulator between the 4-th and 5-th metals, this ensures that the manufacturing process does not become complicated.

III. MAGNETIC FLIP-FLOP

Figure 2 shows an example of a 6-transistor SRAM cell-based nonvolatile latch circuit employing our MRAM technology. A bit of data and its complement are stored in a pair of MTJs to expand the sensing signal. Each MTJ is connected to the source of the transistor M3, which is a part of the cross-coupled inverter. The base electrode of each MTJ is connected to the write circuit, which supplies bi-directional I_W according to the bit of data. As can be seen from Fig. 2, the I_W paths are laid out so that a writing magnetic field can be supplied to each MTJ in opposite directions.

When the write clock (WCKb) is activated at low level, the write circuit backs up the held data in the latch circuit to the MTJ pair. Then, the face-to-face NOR gates (NR and NRb) provide I_W into the write line of each MTJ. When the power-on signal (PON) is activated at low level, a bit of stored data in the MTJs is recalled into the latch circuit. The M4 activated by PON makes the latch circuit balanced and a clamp voltage of around 100 mV is supplied to each MTJ. Then, the sensing currents I_S and I_{SB} penetrate through MTJ elements J and Jb respectively. When the M4 is turned off, the difference between I_S and I_{SB} is amplified by the cross-coupled inverters, and soon the latch circuit outputs the stored data in MTJs with logic amplitude. Our proposed nonvolatile latch makes it possible to

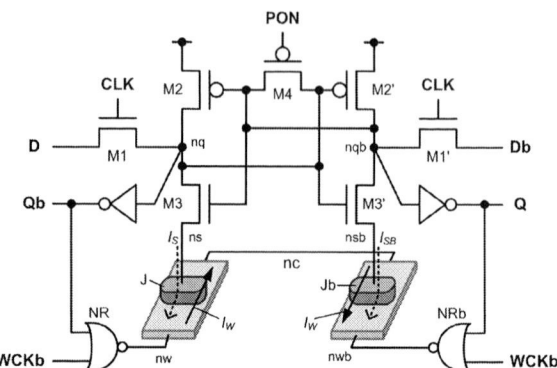

Fig. 2. Circuit diagram of our proposed MRAM-based latch.

carry out both latch and store operations simultaneously because all of the lower electrodes of MTJs are grounded all the time.

Figure 3 shows a circuit diagram of the MFF, in which the MRAM-based latch mentioned above is adopted as a slave latch. The recall and store operation-timing chart of the MFF is shown in Fig. 4. When the power is supplied, the recall operation starts immediately on the condition that the clock (CLK) is at low-level. Then, the 1-bit information saved in the MTJs is loaded to the MRAM-based slave latch, and the MFF output is determined.

One problem is that the additional circuit for non-volatilization may degrade the F/F performance. In particular, the parasitic capacitance of node nq (and nqb) shown in Fig. 2 degrades the performance because of the large-size NOR gates for 1-mA I_W. To prevent the capacitance from increasing, the master-latch output is used as the write circuit input data and I_W is conducted by newly added transistors M5 and M6, though it is necessary for the store-operation timing to be synchronized with the rise-edge of the CLK.

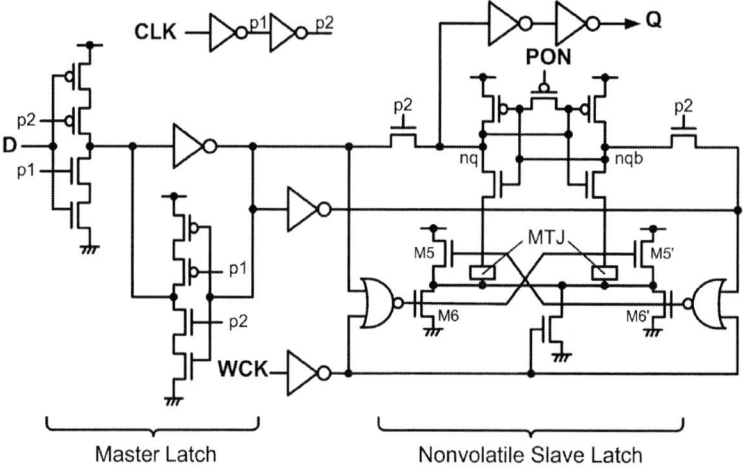

Fig. 3. MFF circuit employing MRAM-based slave latch

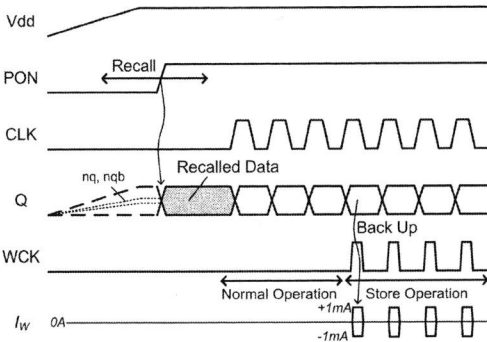

Fig. 4. Recall and store operation-timing chart of MFF.

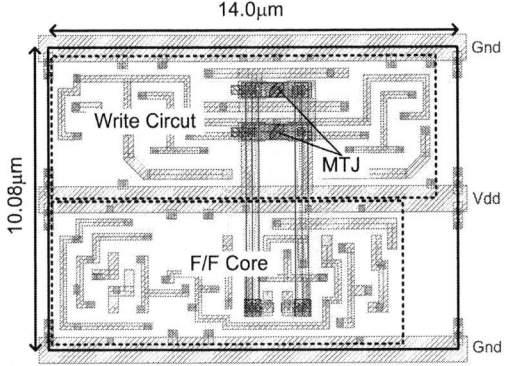

Fig. 5. MFF layout.

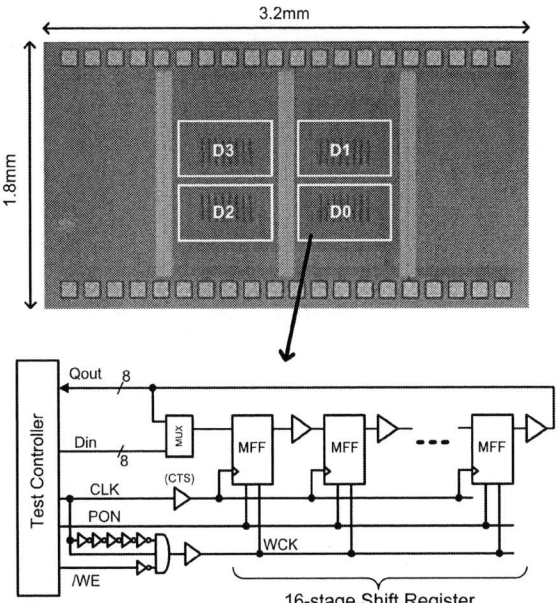

Fig. 6. Microphotograph of MFF test chip, which is divided into four domains. A 16-stage, 8-bit shift register using various types of MFFs is fabricated in each domain.

Figure 5 shows the MFF's primitive-cell layout designed under 0.15-μm standard CMOS rules. The cell does not include any special power lines for data retention. Although the cell area of the MFF was twice as large as that of a standard DFF, a maximum-toggle frequency of 3.5 GHz was achieved by SPICE simulation. The power dissipation per bit was 23μW in normal mode and 296μW in store mode at 200-MHz clock frequency with 1-ns I_W width. The simulated performance of the MFF was almost the same as that of a normal DFF. The results encourage us to apply our MFF into mainstream SoCs as high-speed non-volatile F/Fs.

IV. TEST CHIP DESIGN

Figure 6 shows a microphotograph of an MFF test chip fabricated using 0.15-μm, 1.5V CMOS and our MRAM technology. The chip consists of four domains for evaluating various types of MFFs. A 16-stage, 8-bit shift register using MFF is implemented in each domain. The domain is selected by external pins and 8-bit output data bus is shared in each domain. As can be seen in the figure, a 1-ns-width WCK is generated from CLK and /WE to prevent increased overhead of power consumption in store operation. The test chip has a double-rate test mode for achieving high-speed evaluation even in probing

tests. In the test mode, internal CLK is generated from external quadrature-phase-shifted clocks (CLK0-1) as shown in Fig. 7(a).

The test chip was designed using a conventional design flow and CAD tools. The core layout, including MFFs, can easily be designed merely by adding an MFF primitive cell into the ASIC design library because the MFF has no special power line. However, when the MFF cell is X-mirror or Y-mirror arranged, the easy-axis direction of MTJ magnetization and the I_W direction are reversed from these of normal arrangement. To solve the problem, we designed four kinds of MFF layouts for normal, X-mirror, Y-mirror and XY-mirror arrangement.

V. EXPERIMENTAL RESULTS

Our MFF's useful characteristics, which are that it is nonvolatile, high-speed, and easy to use, were demonstrated by evaluating the test chip. Figure 7(a) shows measured store-operation waveforms at 400-MHz clock frequency. In the operation, 8-bit data in each register was shifted to the next-stage register while being backed up at each clock cycle. The set data in each register was output one after another in normal-operation mode as shown in Fig. 7(b). When the power was re-supplied after it had been shut down once, the saved data in all registers were recalled (Fig. 7(c)). The recalled data in each register was re-output; the same result was achieved for the set data before the power was shut down (Fig. 7(d)). The MTJ resistance was 5 kΩ and its magneto-resistive (MR) ratio was around 70% at 100-mV bias in the chip.

Figure 8 shows a SHMOO plot of store operation clock

978-1-4244-2018-6/08 $25.00 © 2008 IEEE   383

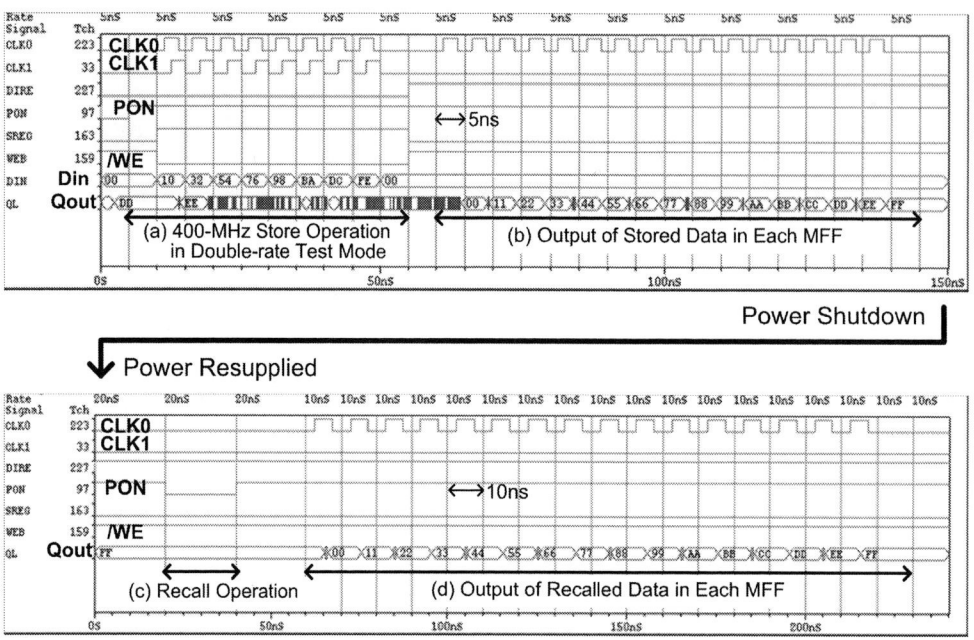

Fig. 7. Measured waveforms of 400-MHz store operation of MFF used shift register, and recall operation when power was re-supplied.

frequency. The test chip was measured with a 333-MHz store-and-recall operation at over 2.6V, and no failed bits were discovered. The lower limit voltage of the pass area was higher than that estimated in the circuit design process. If the MRAM process and MTJ layout are optimized, measurements with over 400-MHz operation at the logic-supply voltage can be expected.

VI. CONCLUSION

An MFF primitive cell for SoC design libraries has been developed. Its performance was estimated to be comparable to that of a normal CMOS DFF. An MFF test chip was measured and no failed bits were discovered. We are confident that MFFs can be incorporated into normal SoC designs in the near future to provide much lower power LSIs.

REFERENCES

[1] J. Lee, et al.," Dynamic Voltage and Frequency Scaling (DVFS) Scheme for Multi-Domains Power Management," *A-SSCC 2007*, Nov. 2007, pp. 360-363.

[2] K. Fukuoka, et al., "A 1.92µs-wake-up time thick-gate-oxide power switch technique for ultra low-power single-chip mobile processors," *Symp. VLSI Circuits*, Jun. 2007, pp. 128-129.

[3] S. Masui, et al.,"Design and applications of ferroelectric nonvolatile SRAM and flip-flop with unlimited read/program cycles and stable recall," *CICC 2003*, Sep. 2003, pp. 403-406.

[4] W. Zhao, et al., "Integration of Spin-RAM technology in FPGA circuits," *ICSICT '06*, Oct. 2006, pp. 799-802.

[5] H. Honjo, et al.,"Performance of write-line inserted MTJ for low-write-current MRAM cell," *Journal of Applied Physics* 103, 07A711, 2008.

[6] N. Sakimura, et al.," A 250-MHz 1-Mbit embedded MRAM macro using 2T1MTJ cell with bitline separation and half-pitch shift architecture," *A-SSCC 2007*, Nov. 2007, pp. 216-219.

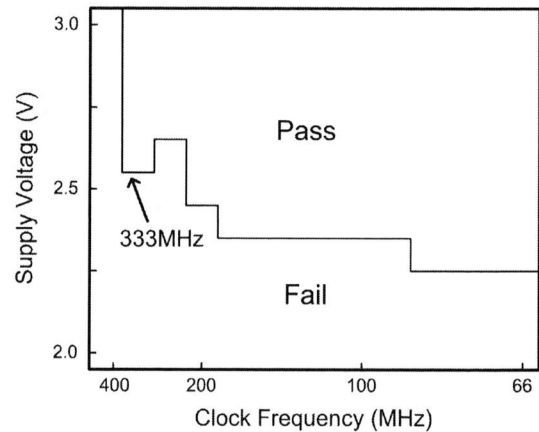

Fig. 8. MFF measured SHMOO plot in store operation.

IEEE 2008 Custom Intergrated Circuits Conference (CICC)

A Fully Integrated Pulsed-LASER Time-of-Flight Measurement System with 12ps Single-Shot Precision

T. Copani, B. Vermeire, A. Jain, H. Karaki, K. Chandrashekar, S. Goswami, J. Kitchen, H. H. Chung, I. Deligoz,
B. Bakkaloglu, H. Barnaby, and S. Kiaei

Arizona State University

Abstract- **A monolithic pulsed time-of-flight measurement system is presented in this paper. The circuit consists of an optical receiver front-end and time-to-digital converter. Received pulse amplitude detectors are also included to correct measurement for walk error. The IC is fabricated in a 0.18-μm BiCMOS process. The receiver front-end achieves a 2.3GHz bandwidth and NF lower than 5dB. The system shows a single-shot precision of 12ps and a measurement range of 188μs. The measurement rate can be as high as 700kHz. The power consumption is 148mA from a 3.5V supply for analog and E²CL circuitry, and a 1.8V supply for CMOS circuitry.**

I. INTRODUCTION

Fully integrated optical time-of-flight measuring systems have numerous applications such as measuring liquid levels in large tanks, height and distance measurement for buildings, collision detectors/avoiders, and 3-D imaging of objects. Basically, the instrument measures the time interval between a transmitted LASER pulse and the return pulse reflected by the target. The LASER beam approach is preferred over pulsed RADAR because of its natural capability to focus on small targets. In contrast, the RADAR approach requires complex antennas and phase arrays to achieve same beam forming capability. Also noteworthy, time-of-flight systems often require a measurement accuracy within few tens of ps, which translates into a length resolution of few millimeters. However, up until now such systems have not exploited a high degree of circuit integration to achieve the required performance.

Usually, pulsed time-of flight measuring instruments consist of two optical receiver channel ICs (i.e., start and stop pulse channels) to convert the optical signals into accurate electrical timing pulses, followed by a time-to-digital converter IC [1,2]. However, the integration of the complete architecture onto a single chip is highly desirable as it reduces power consumption, size, weight and the cost of the device. On the other hand, the IC approach introduces different challenges, which include crosstalk, low voltage operation (decreased dynamic rage in the front-end), higher noise factors compared to discrete design technologies (e.g. GaAs).

A fully integrated pulsed time-of-flight measurement system with a single-shot resolution of 12ps and a measurement range of 188μs is presented in this paper. The IC has been implemented in a 0.18μm BiCMOS process, which features 5 metal layers and MIM capacitors. The paper is organized as follows, system overview is provided in Section II.

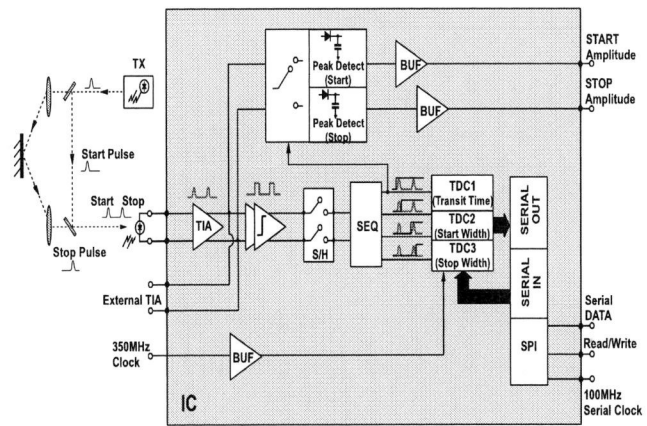

Fig. 1. Block diagram of the time-of-flight measurement system.

Circuit design criteria and trade-offs are analyzed in Section III. Section IV presents and discusses main experimental results. Conclusions are finally drawn in Section V.

II. SYSTEM OVERVIEW

The block diagram of the proposed architecture is shown in Fig. 1. The system is comprised of an optical receiver front-end block, a pulse amplitude indicator (peak detector), a pulse edge discriminator (SEQ), and a time-to-digital converter (TDC). Moreover, a Serial Peripheral Interface (SPI) is included to configure the resolution of time digitizers and for measurement read-out. The receiver front-end consists of an external photo detector (e.g., avalanche photodiode), a low-noise transimpedance amplifier (TIA), a limiting amplifier with threshold and an E²CL switch. The TDC block includes three configurable time digitizers to measure the transit time between the start and stop pulse edges and the widths of the received pulses, respectively. The pulse width measurements along with the amplitude information are used to compensate for the walk error, which is the error introduced by crossing a fixed threshold with different pulse amplitudes, as depicted in Fig. 2.

978-1-4244-2018-6/08 $25.00 © 2008 IEEE

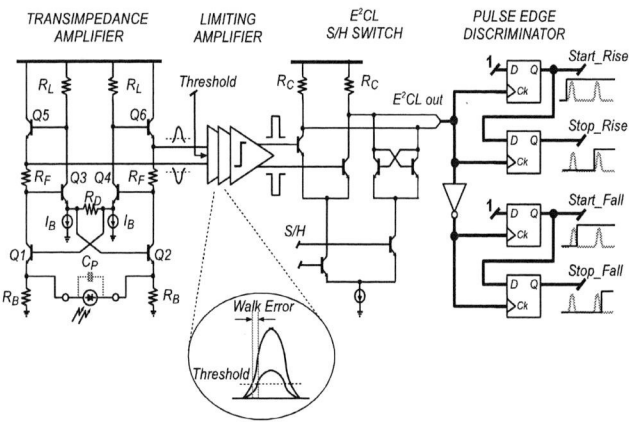

Fig. 2. Receiver front-end and pulse edge discriminator.

II. CIRCUIT DESIGN

The simplified schematic of the receiver front-end is shown in Fig. 2. The TIA is implemented combining a common base input stage (Q_1-Q_2) with a modified Cherry-Hooper amplifier (Q_3-Q_6) to simultaneously achieve high input dynamic range and wide bandwidth for detection of pulses as narrow as 1ns. The differential amplifier also implements a feedback path to reduce the input impedance of the common base input stage. Furthermore, the cross-coupled input stage, Q_1-Q_2, shunts the degeneration resistor R_D, thus providing zero peaking in the frequency response.

The real part of the input impedance of the TIA of Fig. 2 can be approximated by

$$R_i = \Re(Z_i) \approx 2\left(\frac{1}{g_{m1,2}} - \frac{R_F}{1 + R_L/R_D}\right) \quad , \qquad (1)$$

where $g_{m1,2}$ is the trasconductance of the input transistors Q_1 and Q_2. Therefore, for a fixed bias current of the input BJTs, the input resistance R_{in} is reduced by the feedback loop through Q_3, Q_4, R_D and R_F. Equation (1) also gives the design criteria for the resistor values to avoid circuit instability, that is negative input resistance. The simulated behavior of R_{in} for different values of the degeneration resistor, R_D, is shown in Fig. 3.

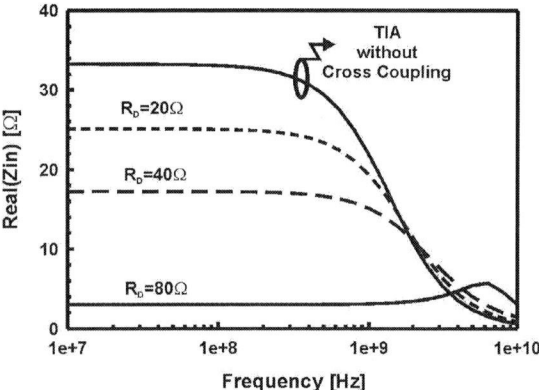

Fig. 3. Characteristic of the TIA input resistance.

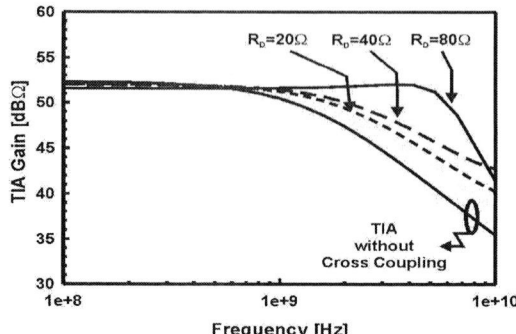

Fig. 4. Frequency characteristic of the TIA gain.

The input pole due to photodiode capacitance, C_P, is moved to higher frequencies by reducing the input resistance of the TIA. Moreover, the local feedback through the cross-coupled BJTs produces a zero in the frequency response. The resulting transimpedance gain can be approximated by

$$T(j\omega) \approx \frac{R_F}{1 + j\omega C_P R_i} \cdot \left(1 + \frac{j\omega C_{\pi 1,2} R_D}{g_{m1,2} R_F}\right) \quad , \qquad (2)$$

where $C_{\pi1,2}$ is the base-emitter capacitance of transistors Q_1 and Q_2. The characteristic of the TIA gain is plotted against frequency for different values of the degeneration resistor, R_D, in Fig. 4.

The differential output of the TIA is fed to a three-stage limiting amplifier that provides 75dBV of gain. The receiver front-end can be alternatively switched to an external transimpedance amplifier when the application has more stringent noise requirements. Therefore, the limiting amplifier adopts an adjustable input threshold voltage to discard unwanted pulses and ringing due to bonding wire inductances. The output of the limiting amplifier drives an E²CL S/H switch, to enable or disable the receiver channel. The start/stop logic pulses are fed to sequencer logic circuitry which generates the timing signals *Start_Rise*, *Start_Fall*, *Stop_Rise* and *Stop_Fall* used for time-to-digital conversion as shown in the timing diagram of Fig. 5. Furthermore, the timing signals generated by the edge discriminator are also used to control the peak detectors.

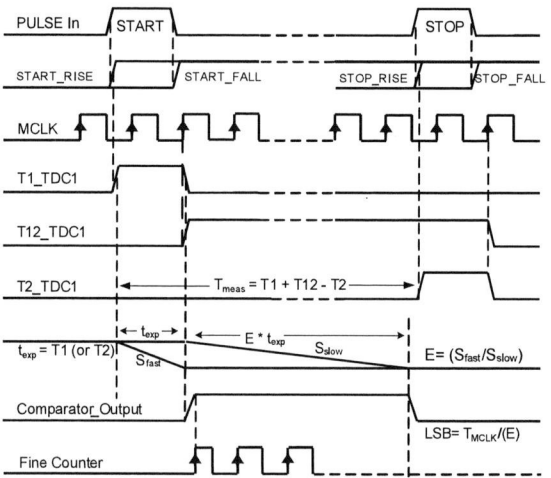

Fig. 5. Timing diagram of the time-to-digital conversion.

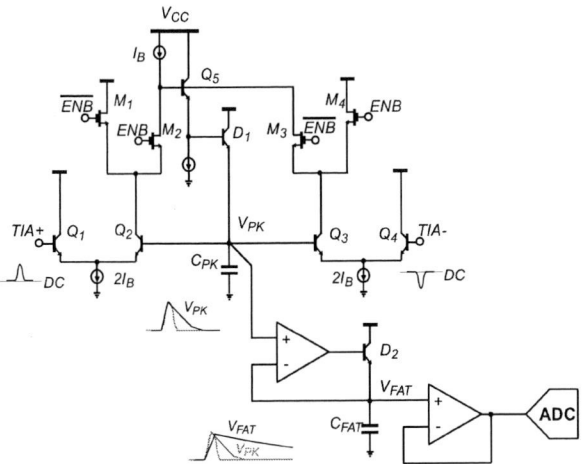

Fig. 6. Schematic of the multi-stage peak detector.

The peak detectors (PD) provide the amplitude of both start and stop pulses coming from the TIA, respectively. The simplified schematic of the adopted peak detector is shown in Fig. 6. The topology consists of a dual stage precision rectifier followed by a voltage buffer to drive external ADCs. The input super diode enables the capture of fast pulses by charging a relatively small capacitor, C_{PK}, through the Q_1-Q_2 differential pair, Q_5 follower and D_1 diode. The input rectifier uses a topology to both interface to differential TIA as well as implement a sample/hold through the Q_3-Q_4 reference differential pair, M_1-M_4 switching pairs and ENB signal. After a pulse has been detected, output node voltage V_{PK} is charged with respect to the TIA DC voltage and bias current of the Q_3-Q_4 pair is steered to Q_3. When ENB signal is low, I_B current source is steered through M_3 and Q_3 while bias current of the Q_1-Q_2 pair is steered to V_{CC} through M_1. Therefore, the input rectifier holds the detected amplitude across C_{PK} while being insensitive to incoming pulses as long as ENB is low. The second super diode (D_2, C_{FAT}) holds the sampled amplitude for the following A/D conversion stage.

Each of the three configurable time digitizers are based on analog dual slope interpolation for pulse stretching [2,3]. For each time interval measurement, timing signals $T1$, $T2$ and $T12$ are generated and measured as shown in Fig. 5. $T1$ and $T2$ are measured relative to the second clock pulse (to avoid metastability) following the start/end of the time interval. The coarse measurement of the time interval is taken using the synchronous $T12$ signal. Fig. 7 shows the block diagram of the TDC. Signals $T1\_TDC1$, $T2\_TDC1$ and $T12\_TDC1$ are used to measure the transit time between the start and stop pulse. The fine measurement is taken by using a 12b synchronous counter on the expanded $T1$ and $T2$ signal. The simplified schematic of the analog interpolator used for expansion is shown in Fig. 7. The ratio of discharge rates of the two charge-pumps (i.e., S_{fast} and S_{slow}) is equal to the expansion ratio used. The charge pumps are implemented using differential current-steering topologies where the tail current determines the expansion ratio in a range between 250 to 1000. The comparator consists of four preamplifiers and a latch.

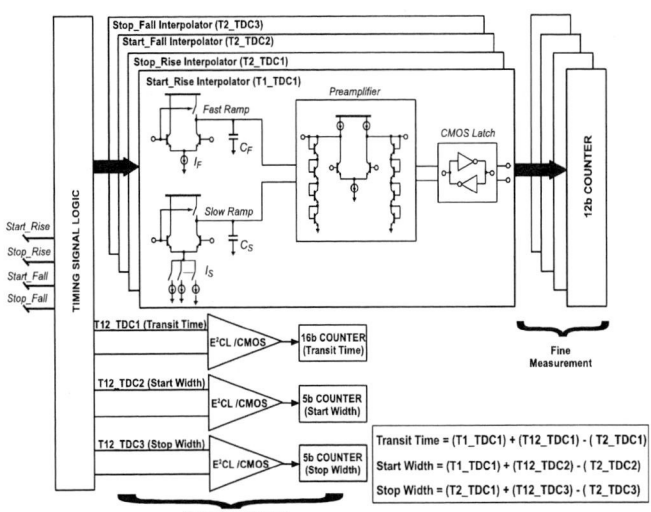

Fig. 7. Time-to-digital converter diagram.

The preamplifiers are NPN differential pairs with folded diode connected NPNs as a load to provide accurate voltage gain matching between the stages. The coarse measurement for $T12\_TDC1$ is taken using a 16b counter. Signals $T1\_TDC1$, $T2\_TDC2$ and $T12\_TDC2$ are used for width measurement of the start pulse. Similarly, $T2\_TDC1$, $T2\_TDC3$ and $T12\_TDC3$ are used to sample the width of the stop pulse. Fractional $T2\_TDC2$ and $T2\_TDC3$ are fine counted using same dual slope interpolation, whereas synchronous $T12\_TDC2$ and $T12\_TDC3$ are counted using 5b counters, as shown in Fig. 5. All the critical circuits in TDC are implemented using E²CL for jitter and speed consideration.

The width and amplitude of the received pulse are used to calibrate for walk error, easing out any requirements on gain control at the receiver input. Upon completion of measurement, the values of coarse and fine counters are stored in the on-chip hold registers and are provided to the on-board DSP through a 100MHz clock frequency SPI.

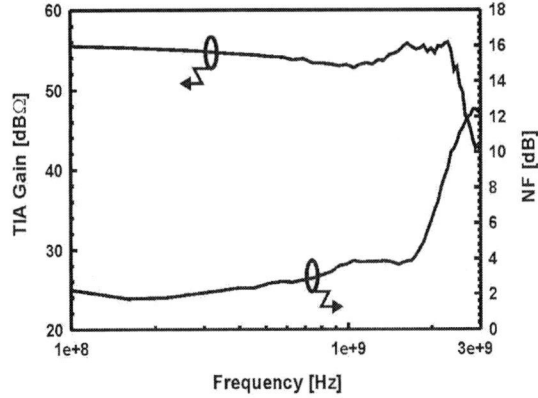

Fig. 8. Measured TIA gain and noise figure.

978-1-4244-2018-6/08 $25.00 © 2008 IEEE

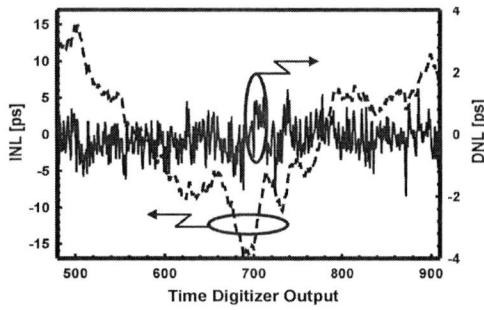

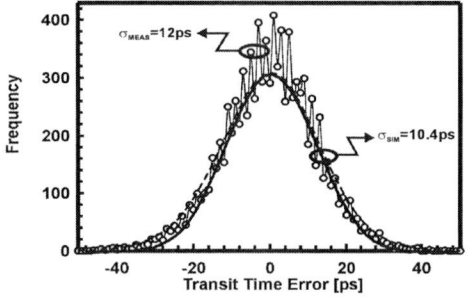

Fig. 9. a) Measured INL and DNL. b) Single-shot noise histogram.

IV. MEASUREMENTS

The proposed system is fabricated in a 0.18-μm BiCMOS process. The circuit uses a 350MHz reference clock for the time digitizers achieving a measurement range of 188μs. The power consumption is 148mA from a 3.5V supply for analog and E²CL circuitry and a 1.8V supply for CMOS circuitry. The measurement rate can be as high as 700kHz. The measured TIA bandwidth is as high as 2.3GHz while gain is 56dBΩ. The measured noise figure is below 5dB across a 2-GHz bandwidth, as reported in Fig. 8. Figs 9a,b show both the measured linearities (INL, DNL) as well as the noise histogram for a fixed 75-ns delay. The measured single-shot precision is as low as 12ps. The measured performance agrees well with system level equations and simulations. Indeed, estimated noise can be modeled by

$$\sigma_{sim}^2 = \sigma_{quant}^2 + \sigma_{CLK}^2 + \sigma_{Rx}^2 + \sigma_{TDC}^2 \quad , \tag{3}$$

where the first term is the quantization noise, the second term is the noise from jittery clock and, finally the last two terms are the noise due to the front-end and time-to digital converter, respectively. The noise contributors can be approximated by the following expressions

$$\sigma_{CLK}^2 = (N_{coarse} + 2N_{fine} / E)\sigma_P^2 \tag{4}$$

$$\sigma_{Rx}^2 = \frac{\sigma_G^2}{2A^2 \ln(A/V_{th})}\sigma_A^2 + 2V_{n,LA}^2 / S + \sigma_{TIA}^2 \tag{5}$$

$$\sigma_{TDC}^2 = 2\left(\frac{V_{n,CP}^2 + V_{n,Comp}^2}{S_{fast}^2} + \frac{V_{n,CP}^2}{S_{TI}^2}\right) \quad . \tag{6}$$

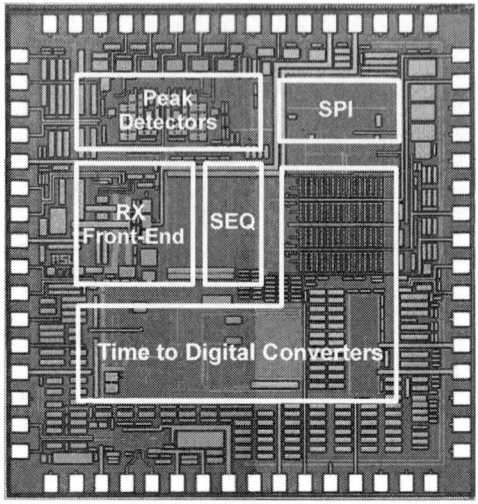

Fig. 10. Die micrograph.

In the above expressions: N_{coarse} is the length of the coarse counters after time expansion; σ_P^2 is the period jitter of the 350MHz clock; σ_G and A are the standrd deviation and amplitude of the received pulses, assuming a gaussian shape approximation; σ_A^2 is the noise at the peak detector output; V_{th} and $V_{n,LA}$ are the threshold and noise of the limiting amplifier, respectively; σ_{TIA}^2 is the noise contribution of the TIA; $V_{n,CP}$ and $V_{n,Comp}$ are the noise contribution of charge pumps and comparators; S_{fast} is the slope of the fast capacitor discharge and S_{TI} is the slew rate of the TI_TDC timing signal of Fig. 5. Using circuit simulations and (3)-(6) the estimated noise results 10.4ps against a measured noise of 12ps.

The die micrograph is shown in Fig. 10. The chip occupies a 2.8x2.8mm² area and is assembled on a 56-pin low-profile lead-less package.

V. CONCLUSION

A monolithic time-to-digital converter with optical receiver front-end has been presented in this paper. Novel circuit topologies as well as different design approaches have been adopted to achieve low noise performance.

ACKNOWLEDGEMENTS

The authors gratefully acknowledge BAE Systems for support throughout the project.

REFERENCES

[1] Palojarvi, P.; Ruotsalainen, T.; Kostamovaara, J., "A 250-MHz BiCMOS receiver channel with leading edge timing discriminator for a pulsed time-of-flight laser rangefinder," *Solid-State Circuits, IEEE Journal of*, vol.40, no.6, pp. 1341-1349, June 2005.

[2] Chen, P.; Chen, C.-C.; Shen, Y.-S., "A Low-Cost Low-Power CMOS Time-to-Digital Converter Based on Pulse Stretching," *Nuclear Science, IEEE Transactions on*, vol.53, no.4, pp. 2215-2220, Aug. 2006.

[3] Raisanen-Ruotsalainen, E.; Rahkonen, T.; Kostamovaara, J., "An integrated time-to-digital converter with 30-ps single-shot precision," *Solid-State Circuits, IEEE Journal of*, vol.35, no.10, pp.1507-1510, Oct 2000

978-1-4244-2018-6/08 $25.00 © 2008 IEEE

IEEE 2008 Custom Intergrated Circuits Conference (CICC)

A Low-Power IC Design for the Wireless Monitoring System of the Orthopedic Implants

Hong Chen, Chen Jia, Yi Chen, Ming Liu, Chun Zhang, Zihua Wang, *Senior Member, IEEE*

Tsinghua University

Abstract-**This paper proposes an architecture of the wireless monitoring system for the real-time monitoring of the orthopedic implants. The system monitors the implant duty cycle, detects abnormal high amounts of force, and other conditions of the orthopedic implants. Data for diagnosis is communicated wirelessly by Radio Frequency (RF) signal between the embedded chip (inside body) and the remote circuit (outside body). In different working modes the system can be powered by the RF signal or stiff lead zirconate titanate (PZT) ceramics which are able to convert mechanical energy inside the orthopedic implant into electrical energy. The Radio Frequency (RF) circuits with the working frequency of 2.4GHz have been taped out with 0.18μm CMOS technology with 50μW. It can supply 400μW power over a distance of 20cm between the two transceivers. The power circuits have been taped out with 0.35μm CMOS technology. The circuits including RF circuits, Analog Digital Converter (ADC), and Micro control Unit (MCU) have been implemented in 0.18μm CMOS process.**

I. INTRODUCTION

Total Knee Replacement (TKR) can release patients with degenerative joint disease from severe pain and immobility due to osteoarthritis. However, TKR implants will fail because of wear, loosening, misalignment, etc. As a result, revision surgery will be conducted, bringing more pain to the patients. Therefore, early diagnosis of abnormalities is critical for avoiding injury. Embedded implant sensors could provide new in-vivo diagnostic capabilities that reduce these clinical complications and lead to improved implant materials and designs [1].

All the electronic sensors need power. Battery has been the power source of most electric-driven devices. However, the limited lifetime and physical dimension have rendered traditional batteries unacceptable for some power-critical or maintenance-free real-time embedded applications such as the wireless sensor, orthopedic implants etc. In [2], an analysis on the power generation characteristics of the stiff PZT ceramics and its equivalent circuit is put forward. It is then verified by simulation and experimental results. It was found that the maximum power (about 1.2mW) is generated when the four PZTs receive uniform and maximum force from the implant. Compared to [1], smaller PZT elements are used in the work, which specifically addresses the problem of limited space available inside an implant.

In this paper, we consider a low power IC design of the wireless monitoring system of orthopedic implants. The sensors are encapsulated within the implants, and provide in-vivo diagnostic data (i.e., force, pressure, etc.) to monitor the primary responsibility of the implant.

II. SYSTEM ARCHITECTURE

The proposed system architecture shown in figure 1 consists of two parts: one embedded in the orthopedic implants, and the other outside the body, both of which are analog-digital mixed-mode circuits. Inside the embedded part, the sensors are applied to obtain in-vivo data which will be then saved in EEPROM. The embedded part can be powered by RF signal as well as the PZT elements [2]. The low power MCU is the center control unit and is in charge of power management, control of data writing process, and other logic control operations. Outside the patient's body, there is a RF front-end module connected to a computer or a portable recording device. The two parts of the system communicate with each other by the RF signal. The data from the EEPROM in the embedded part can be stored in the recording device, which can be mounted on the leg of the patient, and will be used by the doctor afterwards for analysis and diagnosis. In the opposite direction, the control information can be wirelessly transmitted to the embedded chip when necessary.

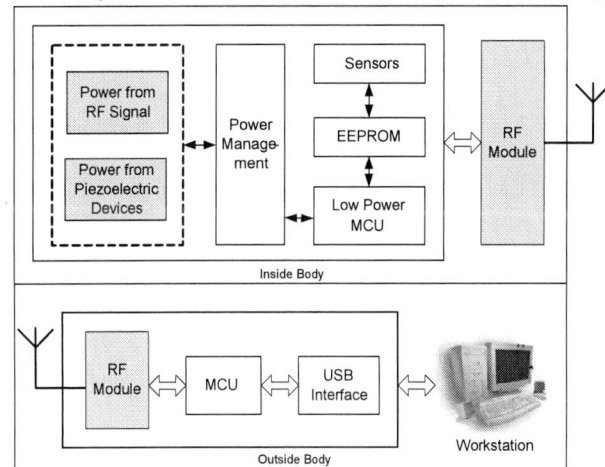

Fig. 1. System architecture

Obviously, the chip embedded is the main part of the system. It includes RF circuits, micro antenna, power circuits, piezoelectric elements, memory, and a low power MCU. The wireless transceiver in the RF circuits works in half-duplex mode and transmits or receives data using the micro antenna which should be small enough to be placed in the orthopedic implants. The power circuit should transfer the power from PZT elements into useful power for other circuits. The piezoelectric elements located in the orthopedic implants can generate about 1.2 mW of regulated power [2]. The A/D

978-1-4244-2018-6/08 $25.00 © 2008 IEEE   389

converter is designed to transform the biological information into digital data to be stored in the memory.

III. CIRCUIT DESIGN

A. RF Circuits Design

Figure 2 illustrates the block diagram of the RF circuits. The "RF_Limiter" circuit keeps the amplitude of the input RF signal within a limited range to avoid damage due to excessive voltage. The "Rectifier" circuit uses Schottky diodes to rectify the current, and the "Regulator" circuit provides steady voltage when the input voltage is fluctuating. The "Oscillator" circuit supplies clock for the EEPROM, and the "Bias" circuit provides the bias current for other circuits. Obviously, the "Modulator" circuit and "Demodulator" circuit modulate and demodulate the signal respectively. The "RST" circuit gives the reset signal to the entire system. Figure 3 is the reset circuit, figure 4 is the demodulator circuit, and figure 5 is the power recovery circuit.

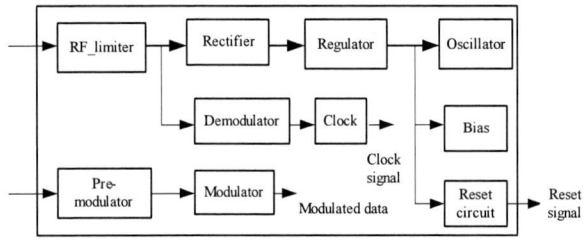

Fig. 2. Wireless analog front-end circuit block diagram

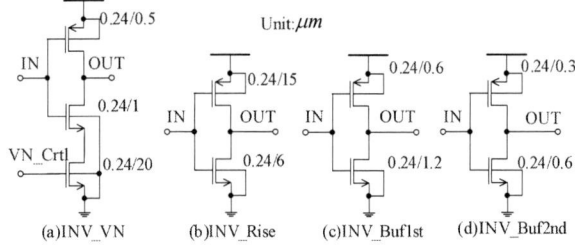

Fig. 3. Reset circuit

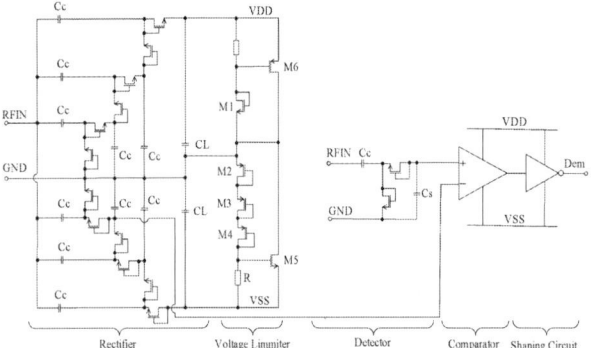

Fig. 4. Demodulator circuit

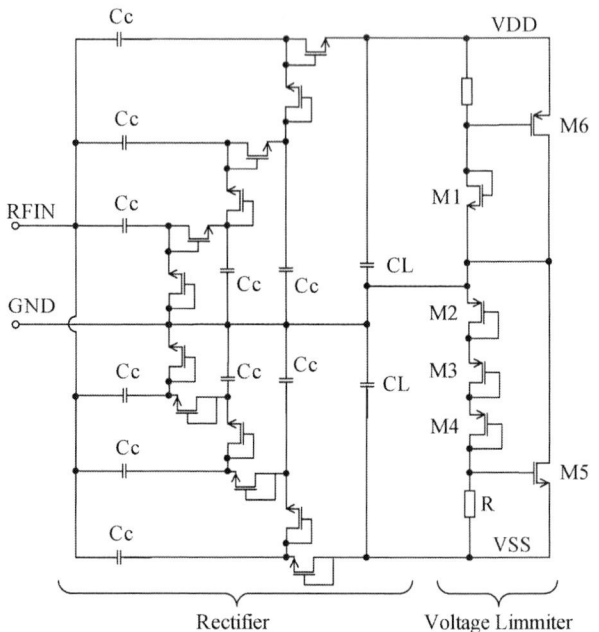

Fig. 5. Power recovery circuit

The circuit diagram in Fig.2 has been taped out with 0.18μm CMOS technology as shown in Fig. 6. The size of the chip is 0.5mm*1mm. It can supply 400μW power over a distance of 20cm between the two transceivers, and the working frequency is 2.4GHz. The power dissipation is about 50μW.

Fig. 6. Layout of the RF circuit

The test result is shown in Fig. 7(a), in which the upper is the demodulated signal which has some glitch caused by the capacitor of the testing circuit, and the lower is the clock signal extracted from the demodulated signal. The reset signal, which is used to reset the digital circuit, is shown in Fig.7 (b). Its high value is equal to the voltage values extracted from the received signal. The test result verifies that the RF circuit can work well.

978-1-4244-2018-6/08 $25.00 © 2008 IEEE

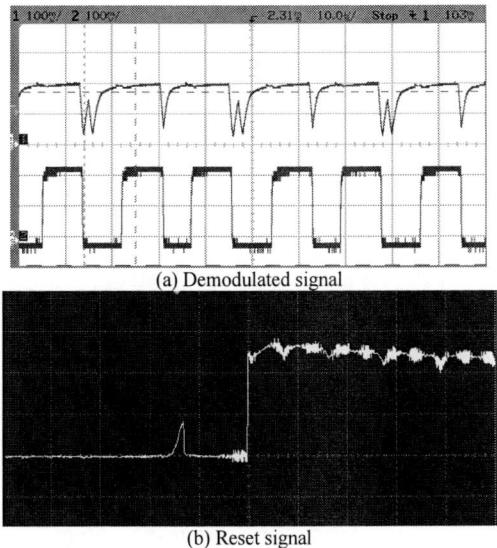

(a) Demodulated signal

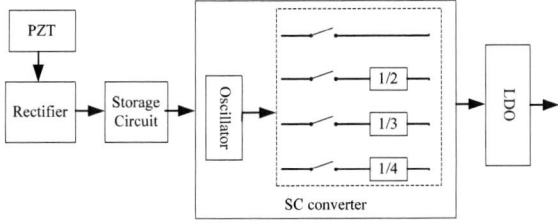

(b) Reset signal

Fig. 7. Test result of the chip

B. Power Circuit Design

The output of PZT varies with the force applied on the TKR implant. In [2] it can be seen that the piezoelectric signal consists of mainly the frequencies between 1Hz and 4Hz. At 1Hz, the power is maximized (about 0.5mW); for frequencies higher than 4 Hz, the power approaches zero. Knowing the frequency characteristic of the power, the proposed power regulation circuit is shown in Fig. 8. .

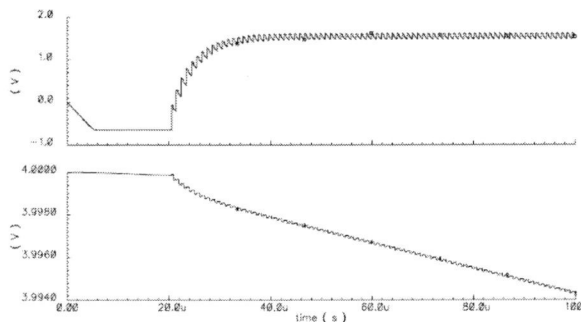

Fig. 8 Schematic power circuit

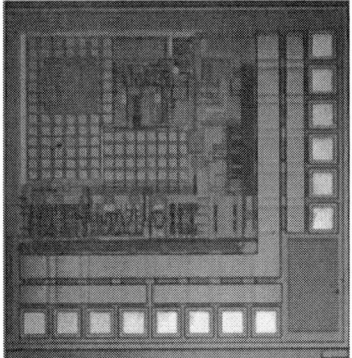

Fig. 9. Simulation result of LDO.

In Fig. 8, the power generated by PZT is first rectified by a full-wave bridge rectifier to convert the bipolar piezoelectric output to a unipolar output. Then, a capacitor is used as storage element in parallel with the load circuit. An oscillation circuit is embedded to generate clock signals for an "SC converter" (Switching Capacitor based on DC-DC Converter), which will lower the input voltage from approximately 10V to 2V. The input voltage is controlled by four programmable

switches, as shown in the dashed block in Fig. 8. Therefore, the power circuit can provide four different voltages to other circuits. Finally, a Low Drop-Out (LDO) voltage regulator reduces the voltage further to a steady value around 1.5V.

The HSPICE simulation result of the LDO circuit is shown in the Fig. 9: the input voltage in the circuit is 4V and the output voltage is approximately 1.5V.

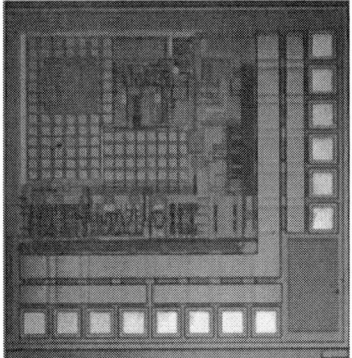

Fig. 10. Die microphotograph of the power circuit.

The SC converter circuits have been taped out at the process of 0.35μm CMOS HV technology. The microphotograph of the chip is illustrated in Fig. 10. The test results of the chip are shown in Table I. It can be seen that the SC converter can transfer the input voltage (9.5V) from the PZT elements into 2.9V which will be dealt with by LDO circuit in the future work. The total efficiency of the SC converter is 28.1% at fulltime working mode.

TABLE I
TEST RESULT OF THE CHIP

Input Voltage (9.5V) Load resistance (5.6 KOhms)	Results
Bandgap reference	60uA
Step-down charge pump	298uA
Clock driver	20uA
Resistive voltage divider	7uA
Test buffer	180uA
Output voltage detector	
Oscillator	
Summary consumed current	565uA
Output voltage	2.906V
Output current	519uA
Efficiency of charge pump	53.27%
Total efficiency	28.1%

C. ADC design

The analog-to-digital convertor (ADC) converts the analog signal from the sensors into digital signal. According to system requirement, the ADC works at a sample frequency of 400Hz and has the resolution of 8 bits. The two-step cyclic ADC is adopted.

The ADC has four analog input channels. At one time only one channel will be chosen to be converted into digital signal. The ADC has two working modes, the normal mode and the sleep mode. When the ADC is in normal mode, it converts the analog inputs into digital outputs. When the ADC is in sleep mode, all the circuits will not work and almost no power will be consumed. This is achieved by turning off the bias circuit

978-1-4244-2018-6/08 $25.00 © 2008 IEEE

of the analog part and clock-gating the digital circuits.

The whole circuit of the ADC with the 1.5-bit correction circuit is shown in Fig.12. The digital logical circuit consists of a shift register and an end adder.

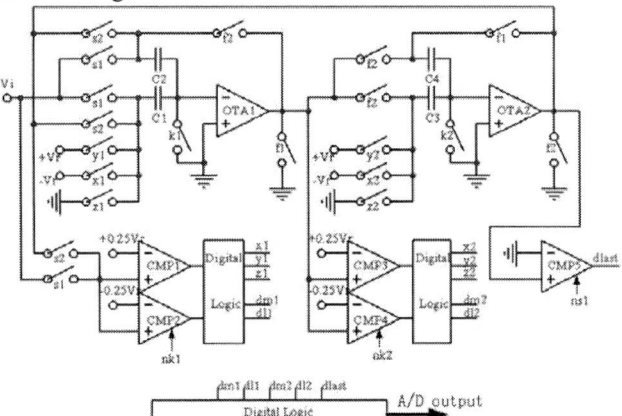

Fig.11 The top circuit of the ADC

The convertor is fabricated at 0.18μm standard CMOS process with single-poly, six-layer metal. All the simulation results were obtained using the software of Spectre as follown. The frequency of the input signal and the clock frequency were set to full scale, 25.765Hz, and 1MHz, respectively. A typical low-frequency fast Fourier transform (FFT) output spectrum is shown in Fig.13. The spurious free dynamic range (SFDR) is approximately 61dB. The signal-to-noise-plus-distortion ratio (SNDR) of the ADC is 48dB. Then, from formula (1), the effective numbers of bits (ENOB) is about 8bits.

$$ENOB = \frac{SNDR(dB) - 1.76}{6.02} \tag{1}$$

In the normal mode, the ADC core (not including I/O) consumes 14μW at the 1.8V supply with the sample frequency of 400Hz. In sleep mode, the ADC core consumes less than 100nW at 1.8V supply.

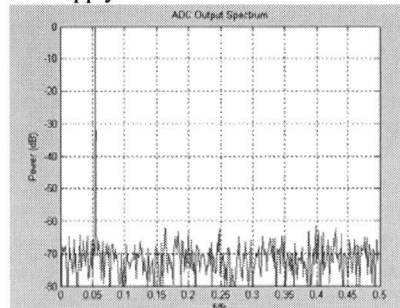

Fig.12 ADC output spectrum for full scale input sine wave

IV. ASIC DESIGN

The 0.18-μm CMOS process technology is used for the digital IC inside the human body. The layout of the chip embedded is shown in Fig. 13. It will be verified in the testing system in the next work.

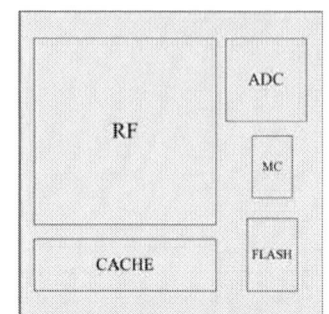

Fig. 13. The layout of the circuit embedded

V. CONCLUSION

This paper proposes a novel architecture of the wireless monitoring system for the real-time monitoring of the orthopedic implants. The analog-digital mix-mode system monitors the implant duty cycle, detects high amounts of force, and other conditions of the orthopedic implants. In different working modes the system can be powered by the RF signal or PZT elements. The Radio Frequency (RF) circuits with the working frequency of 2.4GHz have been taped out with 0.18μm CMOS technology with 50μW. It can supply 400μW power over a distance of 20cm between the two transceivers. The SC converter can transfer the input voltage (9.5V) from the PZT elements into 2.9V which will be dealt with by LDO circuit in the future work. The total efficiency of the SC converter is 28.1% at fulltime working mode. The ADC circuit is also designed to meet the requirement of the system. The circuits including RF circuits, Analog Digital Converter (ADC) and Micro control Unit (MCU) have been implemented in 0.18μm CMOS process. The low power and small size chip make it possible to put the chip inside the implant to be checked, Future work includes some clinical experiments test in the application where PZT elements are used for power generation in a TKR implants.

REFERENCES

[1] Stephen R. Platt, Shane Farritor, and Hani Haider, "On Low-Frequency Electric Power Generation With PZT Ceramics," ITTT/ASME Transactions on mechatronics, Vol.10, No.2, pp. 240-252, April 2005.

[2] Hong Chen, Chen Jia, Chunsheng Liu, "Power Harvesting Using PZT Ceramics", Proceeding of 2007 IEEE International Symposium on Circuit and Systems (ISCAS2007), pp. 557-560, New Orleans, USA, May 2007

[3] Triet T. Le, Jifeng Han, Annette von Jouanne, et al. "Piezoelectric Micro-Power Generation Interface Circuits", IEEE journal of solid-state circuits, vol. 41, no. 6, June 2006, pp. 1411-1420

[4] Nizar Lajnef, Shantanu Chakrabartty, Niell Elvin,Alex Elvin. "Piezo-powered floating gate injector for self-powered fatigue monitoring in biomechanical implants" [C]. Proceedings of the 2007 IEEE International Symposium on Circuits and Systems (ISCAS'07), May 2007, Page(s):89 – 92

[5] Paul Gerrish, Erik Herrmann, Larry Tyler, Kevin Walsh. "Challenges and Constraints in Designing Implantable Medical ICs" [J]. IEEE Transactions on Device and Materials Reliability, Vol. 5, No. 3, September 2005, Page(s): 435 - 444

[6] Tae Song Kiml, Si Young Song, Han Jung, etc.. "Micro Capsule Endoscope for Gastro Intestinal Tract" [C]. Proceedings of the 29th Annual International Conference of the IEEE EMBS (EMBC'07), Lyon, France, August 23-26, 2007. Page(s):2823 – 2826

IEEE 2008 Custom Intergrated Circuits Conference (CICC)

Tessellation-Enabled Shader for a Bandwidth-Limited 3D Graphics Engine

Kyusik Chung, Chang-Hyo Yu, Donghyun Kim, and Lee-Sup Kim

Dept. of EECS, KAIST, 373-1 Guseong-dong, Yuseong-gu, Daejeon, 305-701, Republic of Korea

Abstract-A tessellation-enabled shader (TES), 1/250 memory bandwidth saving geometry processor, is proposed for a mobile 3D graphics engine. By utilizing operational characteristic of tessellation, the TES is implemented with 6.2% additional logic of a dedicated unit based on a conventional vertex shader. With optimization of key components of a shader, a symmetric dual-core TES is fabricated using 0.18um CMOS technology and processes 120Mvertices/s at 100MHz while consuming 272mW of power.

I. INTRODUCTION

Multimedia system-on-a-chip (SoC) is widely adopted in mobile applications such as personal multimedia players, PDAs, mobile phones, and etc. and 3D graphics is one of key features in it to improve user interface or extend applications. A 3D graphics engine for a mobile multimedia SoC should be implemented within limited area and power specifications. Expanded VLIW architecture [1], fixed point arithmetic datapath [2], DVFS (dynamic voltage and frequency scaling) [4], and other techniques have been proposed to improve performance or performance per watt. However, their processing ability cannot be handled within limited memory bandwidth of shared bus architecture as described in TABLE I. Moreover, a number of multimedia IPs including audio and video accelerators as well as a host processor compete with each other to utilize the bus in an SoC. Efficient use of memory bandwidth is a critical issue for the performance of the system.

For a bandwidth-efficient vertex processor, vertex cache is proposed to reduce the number of vertex fetch by removing redundant operations for the same vertex [6]. Tessellation is another effective method to resolve the bandwidth problem of a vertex processor [5][8]. As shown in Fig. 2, internal generation of vertices by tessellation reduces the considerable amount of data transfer for vertex fetch. Despite the obvious effectiveness, only a few high performance gaming systems integrated dedicated tessellators which consisted of floating point datapath array and complex control logic [5][7]. The implementation overhead prevents a mobile 3D graphics engine from embedding a dedicated tessellator.

In this paper, we propose a programmable tessellator, named a tessellation-enabled shader (TES), which is based on a conventional vertex shader with 6.2% additional silicon area. Since the TES also supports full functionality of vertex shading, it can replace a conventional vertex shader as a geometry processor without the implementation overhead. It is especially advantageous for a mobile 3D graphics engine which suffers from the limitations of area and memory bandwidth. At the expense of shading performance, the TES saves the memory bandwidth consumed for geometry data

transmission up to 1/250. For this architecture, we analyze the operations of tessellation and divide them into two parts: intensive arithmetic operations are performed by the floating point datapath of a shader and tessellation-specific control logic is processed by a dedicated unit. In addition to the functionality of tessellation, optimized key components of a shader such as 4D vector dot product unit (DP4 unit), special function unit (SFU), and data fetch unit (DFU) increase area-efficiency of a shader while sustaining 120Mvertices/s transformation rate of the previous work [1]. A symmetric dual-core of the TES is fabricated using 0.18um one-poly four-metal standard CMOS technology on 4.95x5.75mm^2 die size. And the effectiveness of tessellation by the TES is investigated in the view of memory bandwidth saving and shading performance degradation.

The remainder of this paper is organized as follows: Section II explains operational characteristics of tessellation and detailed architecture of the proposed TES. System architecture and implementation results of a dual-core TES are described in Section III. Section IV illustrates and analyzes experimental results. Finally, conclusion is summarized in Section V.

II. TESSELLATION-ENABLED SHADER

A. Operational Characteristics of Tessellation

Tessellation in 3D graphics is a technique to divide polygonal or parametric primitves into a number of triangles on the primitives [8]. Fig. 2 illustrates a cubic Bézier curve as a representative parametric primitive. It is specified with four control points and a higher order polynomial instead of many vertices. Tessellation consists of sequential evaluation of $\mathbf{P}(t)$ for a number of parametric coordinates and connection

TABLE I
BANDWIDTH REQUIREMENT OF PREVIOUS VERTEX PROCESSORS

Prev. works	[1]	[2]	[3]	[4]	**
Required bandwidth (Gbps)*	61.4	4.7	6.4	72.2	3.2

* Required memory bandwidth is computed by: 128bits x (4 attributes) x (shading performance)
** Supported memory bandwidth is computed by: 32bits x 100MHz

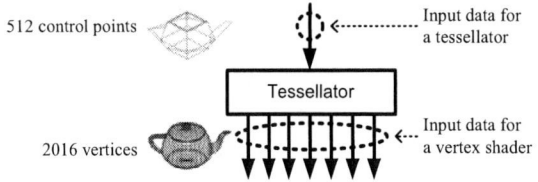

Fig. 1. The effectiveness of tessellation to reduce the amount of data transfer for vertex fetch.

978-1-4244-2018-6/08 $25.00 © 2008 IEEE

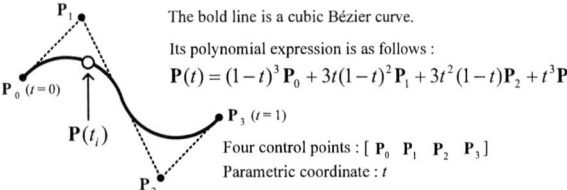

The bold line is a cubic Bézier curve.

Its polynomial expression is as follows :

$$\mathbf{P}(t) = (1-t)^3\mathbf{P}_0 + 3t(1-t)^2\mathbf{P}_1 + 3t^2(1-t)\mathbf{P}_2 + t^3\mathbf{P}_3$$

Four control points : $[\ \mathbf{P}_0\quad\mathbf{P}_1\quad\mathbf{P}_2\quad\mathbf{P}_3\ ]$
Parametric coordinate : t

Fig. 2. A cubic Bézier curve and its polynomial expression.

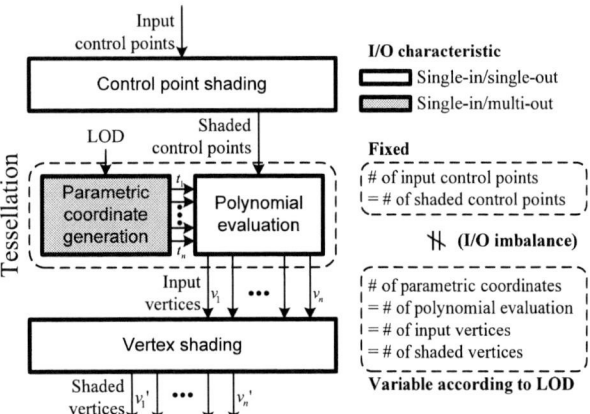

Fig. 3. Relationship between input and output of a geometry pipeline embedding tessellation.

of the evaluated results to generate lines or triangles. The former part is called *polynomial evaluation* and the latter is called *parametric coordinate generation* in our architecture.

A geometry pipeline embedding tessellation is illustrated in Fig. 3. Unlike two shading operations which have single-in/single-out characteristic, tessellation generates multiple vertices from a fixed number of control points. How many vertices are generated is dynamically determined by level-of-detail (LOD) which is set by a host processor. Although the I/O imbalance is an important property of tessellation to save memory bandwidth, a conventional vertex shader cannot handle it appropriately. The vertex shader is inherently designed to process single-in/single-out stream data. Therefore, dedicated tessellators implemented in high performance 3D graphics accelerators perform both control point shading and tessellation-specific operations with powerful floating point datapath and complex control logic [7].

For shader-based tessellation, we divide operations of tessellation into two parts according to the I/O characteristic as shown in Fig. 3. *Polynomial evaluation* performs a number of floating point arithmetic to evaluate higher order polynomial for each parametric coordinate. Its computational and I/O characteristics are very similar to a vertex shader: massive floating point arithmetic and single-in/single-out I/O. On the other side, *parametric coordinate generation* controls order of operations to generate line or triangle strips which are crack-free [8] and efficient for vertex cache. And it also controls quality of tessellated results according to dynamically varying LOD, which is called adaptive tessellation [8]. For the crack prevention and adaptive tessellation, *parametric coordinate*

generation should perform dynamic flow control frequently which is supported but not efficiently processed by a vertex shader. In addition, it has single-in/multi-out I/O characteristic. Therefore, we propose a dedicated unit for *parametric coordinate generation*. It is tightly coupled with a conventional vertex shader and executes tessellation-specific control logic. Detailed procedure of tessellation by the TES is described in the following subsections.

B. Tessellation with a Vertex Shader

Since *polynomial evaluation* requires a number of floating point arithmetic, it can be efficiently processed by utilizing SIMD floating point datapath of a vertex shader. To implement a shader program, the polynomial expression of a cubic Bézier curve in Fig. 2 is rearranged to matrix-vector form as follows:

$$
\begin{aligned}
\mathbf{P}(t) &= (1-t)^3\mathbf{P}_0 + 3t(1-t)^2\mathbf{P}_1 + 3t^2(1-t)\mathbf{P}_2 + t^3\mathbf{P}_3 \\[2mm]
&= \begin{bmatrix}(1-t)^3 & 3t(1-t)^2 & 3t^2(1-t) & t^3\end{bmatrix}\begin{bmatrix}\mathbf{P}_0\\\mathbf{P}_1\\\mathbf{P}_2\\\mathbf{P}_3\end{bmatrix} \\[2mm]
&= \begin{bmatrix}1 & t & t^2 & t^3\end{bmatrix}\begin{bmatrix}1 & 0 & 0 & 0\\-3 & 3 & 0 & 0\\3 & -6 & 3 & 0\\-1 & 3 & -3 & 1\end{bmatrix}\begin{bmatrix}\mathbf{P}_0\\\mathbf{P}_1\\\mathbf{P}_2\\\mathbf{P}_3\end{bmatrix}
\end{aligned}
\tag{1}
$$

4-by-4 matrix in the middle of (1) is called a characteristic matrix and specifies a type of a parametric surface such as Bézier, Hermite, B-spline, NURBS, and so on [8]. The TES can support various types of parametric surfaces by exchanging the characteristic matrix or implementing different shader programs. This is an advantageous feature because the inusability of a tessellator in previous works often results from the limited support for the types of parametric surfaces. Besides, SIMD floating point datapath of a vertex shader is optimized to calculate matrix-vector multiplications; twenty 2-input multiplications and fifteen 2-input additions of (1) are computed by only five 4D dot product instructions of the TES. Therefore, it is reasonable to process *polynomial evaluation* with a vertex shader at the expense of shading performance.

C. Parametric Coordinate Generator

A parametric coordinate generator (PCG) is fixed pipeline that controls tessellation-specific operations. Fig. 4 illustrates a control method of the PCG to support crack-free and adaptive tessellation. When a parametric coordinate patch is generated for each control point patch, it is divided into four segments which hide irregular patterns caused by fractional part of LOD at the center of the patch [5]. In a segment, parametric coordinates are computed in a regular manner like a rasterizer of a 3D graphics pipeline. And power of parametric coordinate such as t^2 or t^3 is needed to evaluate a higher order polynomial and to reduce computational burden of a shader. Forward differencing (FD) is an effective computing method in this case; it evaluates higher order polynomial with only additions for a fixed step [8]. The PCG implements FD for the computation of parametric coordinates

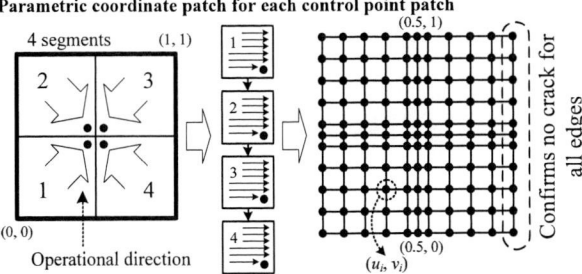

Fig. 4. Segmentation and control method of the PCG to support continuous LOD and crack-free tessellation.

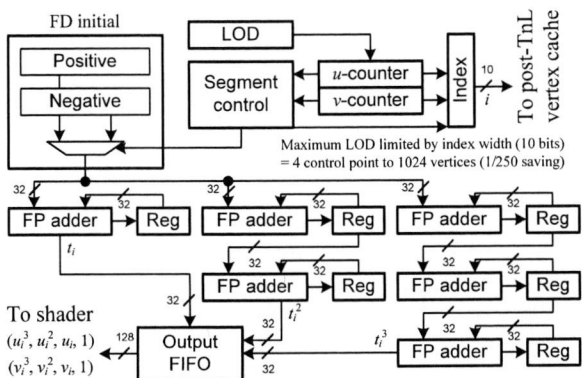

Fig. 5. Architecture of the PCG with forward differencing datapath.

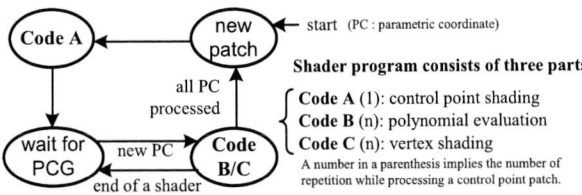

Fig. 6. Structure of a shader program for tessellation and its operating sequence.

as shown in Fig. 5. With six floating point adders, $(t^3, t^2, t, 1)$ is calculated at every clock cycle and stored into an output FIFO which is connected to a shader. For surface tessellation, two-dimensional parametric coordinate (u, v) is generated with its powers. If an LOD changes, initial values of FD are changed so that the tessellated result can be adjusted adaptively. Since the TES with the PCG can produce 1024 vertices from four control points, memory bandwidth is saved up to 1/250. The proposed PCG is implemented with 6.2% additional logic gates to a conventional vertex shader.

C. Structure of a Single Shader Program for Tessellation

Since API state changes degrade the performance of a 3D graphics engine [9], the TES performs tessellation with a single shader program which consists of three parts of operations as shown in Fig. 6. Code A and B are added in front of vertex shading codes for tessellation. When a control point patch is fetch to the TES, Code A processes control point shading such as transformation. And then the shader waits for

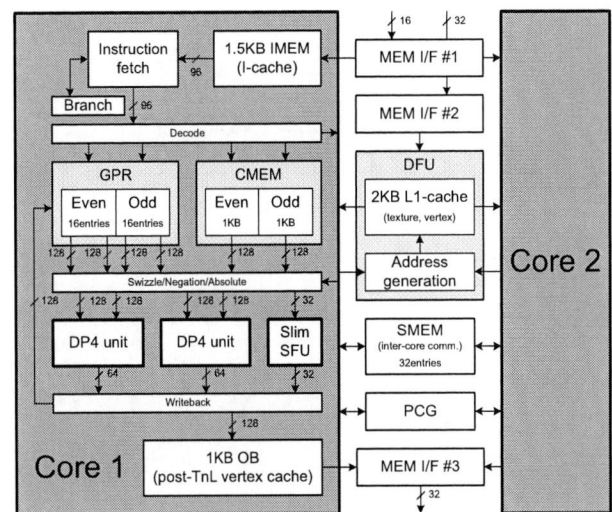

Fig. 7. System architecture of a dual-core TES.

the PCG to generate a parametric coordinate which is required for polynomial evaluation. Finally, shaded vertex is produced after processing Code B and C. Shaded control points, results of Code A, are common for all vertices generated from the current control point patch. Therefore, Code A is performed only one time for each control point patch. Only Code B and C are repeated for each parametric coordinate. If many vertices are generated by tessellation, overhead of implementing Code A would become negligible. Overhead of Code A and B for tessellation is analyzed in detail in Section IV.

III. SYSTEM ARCHITECTURE AND IMPLEMENTATION RESULTS

Fig. 7 shows the system architecture of a dual-core TES which can operate as a programmable tessellator and a vertex shader. Each core is multi-threaded two-issue VLIW architecture with two 4D dot product units (DP4 units) and a slim special function unit (SFU). The DP4 unit adopts newly proposed architecture which especially optimized to 3D graphics applications [10]. And the slim SFU is area-optimized version of the previous work [1] by extracting DP4-like part of internal datapath. Shading performance of each core is 60Mvertices/s and the dual-core performs twice better.

For optimal implementation, redundant and infrequently used modules such as a data fetch unit (DFU) and a PCG are shared by two cores. The DFU consists of a L1 cache and an address generator. It fetches two type of input data used in a vertex shader: vertex and texture. Since vertex texture is not used frequently, merging two fetch units does not degrade the performance of a vertex shader. And the L1 cache is utilized as a pre-TnL vertex cache and a vertex texture cache at the same time. The PCG is also shared because it is fast enough to feed parametric coordinate to two cores (See Fig. 8). A dual-port shared memory (SMEM) is implemented to shading the results of Code A such as transformed control points.

The dual-core TES is fabricated using 0.18um 1P4M standard CMOS technology on 4.95x5.75mm² die area and it consumes 516K logic gates and 15KB on-chip memory. Chip specification and microphotograph are shown in Fig. 9.

978-1-4244-2018-6/08 $25.00 © 2008 IEEE

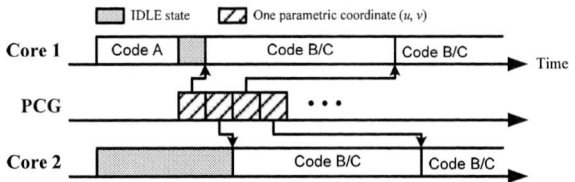

Fig. 8. System architecture of a dual-core TES.

Process technology	0.18μm 1P4M CMOS
Power supply	1.8V (core), 3.3V (I/O)
Transistor counts	1.04M gates (0.5M logic gates, 15KB SRAM)
Die size	4.95mm x 5.75mm
Performance	120Mvertices/s @100MHz 1/250 BW saving
Power consumption	< 272mW
Features	OpenGL|ES 2.0, Shader Model 3.0 support Programmable tessellation
Package	208 pin QFP

Fig. 9. Chip specification and die micrograph.

IV. EXPERIMENTAL RESULTS

The RTL implementation of the TES is used for the simulation to verify the effectiveness of tessellation. Fig. 10 illustrates four test objects which are constructed with cubic Bézier surfaces. Each Bézier sufaces is defined with 16 control points and a cubic polynomial. For varying LOD, the number of generated vertices from control points is counted. Regardless of the number of generated vertices, the amount of data transfer is determined by the number of control points. That results in memory bandwidth saving. In Fig. 10, the change of the relative memory bandwidth for varying LOD is illustrated. If the relative memory bandwidth is less than 1.0, the bandwidth is saved by tessellation. Vertex processors introduced in the previous works [1-4] have a fixed relative memory bandwidth of 1.0. For our test objects, the relative memory bandwidth becomes less than 1.0 when LOD becomes larger than 2. 67.4% of memory bandwidth are saved in this simulation and up to 1/250 of saving can be achieved by increasing LOD to the maximum value.

As described in Section II, *Code A* and *B*, which includes control point shading and polynomial evaluation, are appended in front of a conventional shader program to support tessellation. It means that the memory bandwidth saving is achieved at the expense of the shading performance with the PCG. The overhead can be analyzed by comparing the number of instructions that are executed for conventional vertex shading and shader-based tessellation. If $L(C_{conv})$ is code length of conventional vertex shading (*Code C_{conv}*), $L(C_{tess})$ could be expressed as $L(C_{conv}) - L(A)/k$, where k is the number of control points. That is, *Code C* of tessellation is shortened because *Code A* performs a part of *Code C* of vertex shading such as transformation. Therefore, the overhead to shading performance compared with conventional vertex shading can be expressed as the following equation:

$$\frac{L(A) + n\{L(B) - L(A)/k\}}{nL(C_{conv})} \times 100(\%) \qquad (2)$$

where n is the number of generated vertices by tessellation. According to (2), overhead of *Code A* becomes negligible as n

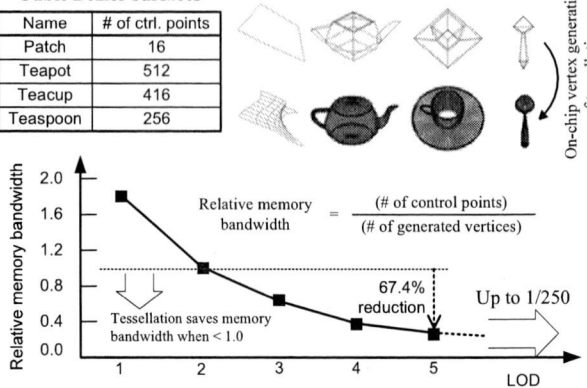

Fig. 10. System architecture of a dual-core TES.

increases. If n is large enough, the overhead would be $L(B)/L(C)$.

V. CONCLUSION

In this paper, we propose a programmable tessellator, TES, for a mobile 3D graphics engine. By utilizing the operational characteristics of tessellation, the TES is implemented with 6.2% additional logic gate to a conventional vertex shader and saves memory bandwidth consumed to transfer vertex data up to 1/250. An area-efficient dual-core TES is fabricated using 0.18um one-poly four-metal standard CMOS technology and each core adopts two-issue VLIW architecture with optimized datapath such as DP4 unit and a slim SFU. It processes 120Mvertices/s at 100MHz operating frequency while consuming 272mW of power.

ACKNOWLEDGMENTS

Chip fabrication was supported by IDEC at KAIST, Republic of Korea.

REFERENCES

[1] C.-H. Yu, K. Chung, D. Kim, and L.-S. Kim, "An energy-efficient mobile vertex processor with multithread expanded VLIW architecture and vertex cache," *IEEE JSSC*, vol. 32, no. 10, pp. 2257-2269, Oct 2007.

[2] J.-H. Woo, et al., "A 152mW mobile multimedia SoC with fully programmable 3D graphics and MPEG4/H.264/JPEG," *SOVC Dig. Tech. Papers*, pp. 220-221, Jun 2007.

[3] Y.-M. Tsao, et al., "An 8.6mW 12.5Mvertices/s 800MOPS 8.91mm² Stream Processor Core for Mobile Graphics and Video Applications," *SOVC Dig. Tech. Papers*, pp. 218-219, Jun 2007.

[4] B.-G. Nam, J. Lee, K. Kim, S. J. Lee, and H.-J. Yoo, "A 52.4mW 3D graphics processor with 141Mvertices/s vertex shader and 3 power domains of dynamic voltage and frequency scaling," *ISSCC Dig. Tech. Papers*, pp. 278-279, Feb 2007.

[5] H. Moreton, "Watertight tessellation using forward differencing," *Proc. of the ACM SIGGRAPH/EUROGRAPHICS Workshop on Graphics Hardware*, pp. 25-32, Jul 2001.

[6] K.Chung, C.-H. Yu, and L.-S. Kim, "Vertex cache of programmable geometry processor for mobile multimedia application," *Proc. of ISCAS*, May 2006.

[7] J. Andrews and N. Baker, " Xbox 360 system architecture," *IEEE Micro*, vol.26, no.2, pp.25-37, Mar 2006.

[8] T. Akennine-Moller and E. Haines, Real-time rendering second edition, AK Peters, 2002.

[9] D. Blythe, "The direct3D 10 system," *Proc. of ACM SIGGRAPH*, vol 25, pp. 724-734, Jul 2006.

[10] D. Kim and L.-S. Kim, "A floating-point unit for 3D vector inner product with reduced latency," unpublished

IC Technology – More Moore and More Than Moore

Session 15

Chair: Alvin Loke, Advanced Micro Devices
Co-Chair: Jordan Lai, Taiwan Semiconductor Manufacturing Company

This session of exclusively invited papers covers a selection of key developments extending CMOS scaling as well as several non-CMOS technologies with pervasive applications. Included are an overview of lithography options beyond 45nm, high-K/metal-gate technology, high-speed BiCMOS, and opportunities orthogonal to conventional scaling.

The first paper, from IMEC, surveys the three major lithography technology options remaining for CMOS manufacturing at 32nm and beyond: 193nm immersion lithography with high refractive index media, 193nm double patterning, and 13.5nm extreme ultraviolet lithography. Merits and current challenges of each approach are highlighted along with the importance of resist materials development.

The second paper introduces Intel's 45nm production technology with metal gate and high-K gate dielectric. The paper reveals a "gate-last" process integration scheme that incorporates dual-band-edge work function metal gates with multiple uniaxial channel straining techniques for mobility enhancement.

The third paper summarizes STMicroelectronics' state-of-the-art SiGe BiCMOS technology for millimeter-wave and high-speed applications. The paper details a 230-GHz f_T, 280-GHz f_{max} process and provides circuit examples, including a 77-GHz receiver, enabled by such intrinsic device performance.

The final paper, from ON Semiconductor, discusses new classes of application-specific products that provide additional value based on functional innovation and diversification rather than on lithographic scaling. Such "More-than-Moore" products exploit cost-effective mature technologies to incorporate digital and non-digital functionality into compact systems at package and chip levels.

978-1-4244-2018-6/08 $25.00 © 2008 IEEE

Notes

Lithography Options for the 32nm Half Pitch Node and Beyond

K. Ronse, Ph. Jansen, R. Gronheid, E. Hendrickx, M. Maenhoudt, M. Goethals, G. Vandenberghe

IMEC, Kapeldreef 75, B-3001 Leuven, Belgium

Abstract-There are still three major technological lithography options for high volume manufacturing at the 32nm half pitch node: 193nm immersion lithography with high index materials, enabling NA>1.6; 193nm double patterning and EUV lithography. In this paper the pros and cons of these three options are discussed. The extendibility of these options beyond 32nm half pitch is important for the final choices to be made. High index 193nm immersion lithography requires high index resist materials, which are under development but still far removed from the target refractive index and absorbance specifications and not to mention lithographical performance. For double patterning the pitch may be relaxed, but the resists still need to be able to print very narrow lines and/or trenches. Moreover, it is preferred for the resists to support pattern or image freezing techniques in order to step away from the litho-etch-litho-etch approach and make double patterning more cost effective. For EUV, besides the high power light source, the resist materials need to meet very aggressive sensitivity specifications. In itself this is possible, but it is difficult to simultaneously maintain performance in terms of resolution and line width roughness.

I. INTRODUCTION

With the production of the ASML XT:1900i scanner, water-based immersion lithography has proven to be able to afford extension of numerical aperture (NA) up to 1.35. This affords printing half pitches down to ~45nm with acceptable k_1 (>0.3). Further increase beyond this NA is not possible with a water-based 193nm immersion approach and a decision is needed on the technological choice for patterning at the 32nm half pitch node and beyond. According to the ITRS roadmap the 32nm node should go into production in 2011, meaning that development should start in 2009; the 22nm node should go in production in 2014, requiring development to start in 2012. The semiconductor industry is taking a lot of time to reach this decision. Currently, three contenders are being pursued, each with their own benefits and drawbacks: high-index 193nm immersion lithography, enabling high NA (>1.6); double patterning, enabling low k_1 (<0.25); or reducing the imaging wavelength by moving to the Extreme Ultra-Violet (EUV, λ=13.5nm) wavelength. In this paper the technological advantages and challenges of these three options will be discussed.

II. HIGH-INDEX 193NM IMMERSION LITHOGRAPHY

High-index 193nm immersion allows scanners to have NA beyond the limit of water-based systems (NA = 1.35). The target is to achieve NA >1.6 which is required to resolve 32nm half pitch in a single lithography step. The advantage of this option is that much of the current infrastructure can be used for this technology. On the downside, this technology requires new high-index glass material for the final lens element as well as high-index immersion liquid with refractive index (RI) >1.8 and both with high transparency. Moreover, to efficiently couple the light into the photoactive material, high-index polymers (RI >1.9, preferably 2.0)[1] for resist formulation are required. Therefore, three independent breakthroughs in material development are required for this route to succeed.

From a materials properties perspective, the main requirements on the high-index fluids are refractive index, optical transparency at 193nm and dn/dT (sensitivity of refractive index to temperature). As an initial step, focus has been on the so-called second generation immersion fluids. Target RI for these materials is 1.65 and several materials have emerged as candidates.[2] A noteworthy characteristic of these materials is that they have a low contact angle (typically 20-40°) on most resist and top-coat surfaces (Figure 1) compared to water (70-90°). This requires adaptations in the design of the fluid control in the scanner, since the current design would result in significant quantities of fluid being left on the wafer upon high-speed scanning.

Using immersion interference lithography, decent patterning and acceptable chemical compatibility with resist and top-coat materials has been demonstrated for these liquids. Typical current (dry or immersion) 193nm resists exhibit similar exposure latitude and profiles when printed with water or a second generation high-index liquid (Figure 2). This demonstrates that the chemical interaction between high-index liquid and resist (or top coat) can be kept under control. Using second generation high-index liquids the feasibility of this approach to print the 32nm half pitch has been demonstrated (Figure 2). The effective NA of 1.51 (which is not accessible using water with RI=1.44 as immersion liquid) for this pattern demonstrates the working principle of immersion lithography beyond water.

Transmission targets are met for several of the second generation high-index liquids, but the fluid absorbance increases upon use. Transmission of the fluids decreases with the presence of any contaminants. (absorbed gasses, leached components from resists...) From a cost perspective the fluids cannot be used in a single-pass concept (as used for water-based immersion lithography). Active fluid regeneration systems have been demonstrated and are claimed to deliver acceptable levels for fluid purification, resulting in satisfactorily low total fluid consumption.[3,4] Another issue is in the laser-induced lens contamination that has been observed with these materials. Lens transmission has been shown to decrease upon exposure in contact with typical second-generation high-index fluids.

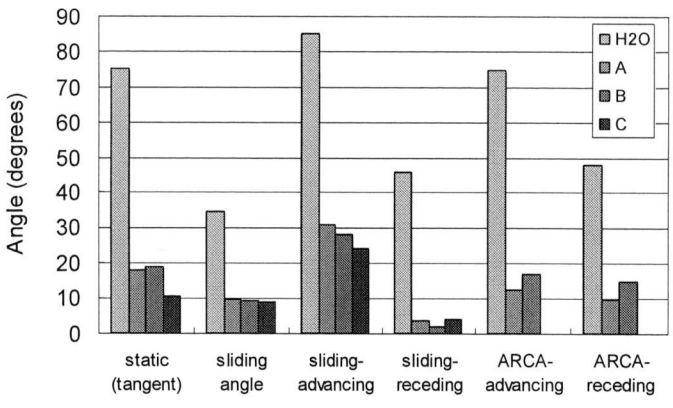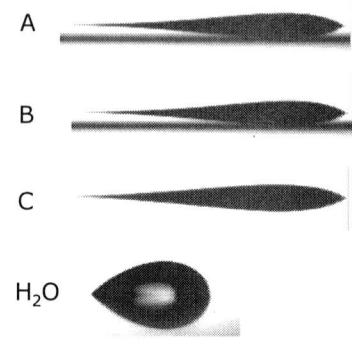

Figure 1. Left: Static contact angle, sliding angle, sliding advancing and receding angle, captive advancing and receding contact angles of second generation high index liquids A, B and C compared to water. Substrate is 'dry' 193nm resist Sumitomo PAR-817 on bare Si. Right: pictures of sliding high index liquid droplets on Sumitomo PAR-817.

Although preliminary results indicate that this can be cleaned, it makes integration of the fluids in a lithography tool more complex. However, the current main issue for the high-index fluids is that RI >1.8 is required for 32nm half-pitch imaging. Such so-called third generation immersion fluids are being explored, but simultaneously meeting target specifications for RI and transmission has so far not been successful. No solution appears to exist for chemically pure fluids. Therefore attempts are ongoing to increase RI using nano-particle dispersions in second generation high index

immersion fluids[5]. In theory this provides a path to meeting target specifications for RI and transmission simultaneously, but formulation of such fluids is far from straightforward and still needs to be proven in practice.

The limited reports that have thus far emerged on high-index resists demonstrate the material challenges for increasing RI of the resist polymers while maintaining good patterning characteristics.[6,7] The most straightforward method to increase photoresist RI is by increasing the heteroatom (N, S, P) content of the photoresist polymer. Simulation and

HIL	EL water (%)	EL HIL (%)	Picture HIL (72nm pitch)			HIL B (64nm pitch)
			37nm	32nm	24nm	
A	15	18				
			37nm	32nm	24nm	
B	21	22				
			37nm	32nm	24nm	
C	21	21				

Figure 2. Exposure latitude at 72nm pitch obtained with water and different high-index liquids (HIL) by interferometric exposure. The images in the middle are top-down SEM images of the 72nm pitch lines printed at different exposure dose. The images on right are of 64nm pitch obtained using high-index liquid B.

experimental work have demonstrated that such increase in heteroatom content is in almost all cases accompanied by an increase in polymer absorbance. Moreover, the obtained increase in polymer RI is rather modest. Using this approach patternable polymers have been reported with n=1.78, compared to ~1.72 for current resist formulations. The reduction of the RI in the resist formulation compared to that of the base polymer is partially due to the additives in resists that are required for patterning (photo-acid generator-PAG, quencher, casting solvent, etc.) Efforts to increase resist RI are ongoing, but are still to a large extent in an exploratory phase and taking place at universities, rather than at resist vendors. Similar as for high index liquids, also in case of resists the addition of high index nano-particles is being considered[8]. If a promising route can be identified at one of these institutes, it will still take significant effort from the resist vendors to translate this into production-worthy materials. Commercial high index resists should therefore not be expected to come to market within the next few years.

Finally, the progress on the side of the high-index lens material is slower than originally anticipated. Several materials with high-refractive indices (compared to the current 1.55 for quartz or CaF_2) have been identified, but none of them has been proven to simultaneously meet the specifications on RI, transmission, intrinsic and stress birefringence. The most prominent contenders in this respect are $BaLiF_3$, ceramic spinel and LuAG (Lutetium-Aluminum Garnet). $BaLiF_3$ has the advantage that it was proven to meet the transmission specifications and no technological showstoppers appear for incorporating this material in a lens design. However the RI-increase for this material (1.65) is relatively small and would only allow an increase in NA to 1.45. Such increase is too low to support the investments that are required to realize the technology. Ceramic spinel has a sufficient high RI (1.92), but is a poly-crystalline material. This characteristic makes light scattering critical and also affects material transmission. For high-index glass material most focus is on LuAG, which also has a high RI (2.1). It does however have some intrinsic birefringence, but modeling studies indicate that this could be compensated for in the lens design. The major problem for LuAG is that despite significant efforts to grow pure crystals of this material, it was so far not proven to meet the transmission specifications[9]. This proof of concept is required in the very near future for high index immersion lithography to keep the focus of attention. In terms of timing for the 32nm half pitch node, high-index immersion is probably already late for the applications that require the most aggressive timing, such as memory, but may still meet the timing for logic device makers.

Given the material developments that are still required for high-index immersion lithography, it is no longer considered to be a viable option for patterning at the 32nm half pitch node, certainly not for memory. The earliest feasible timing for a full-field high-index liquid scanner is by early 2011. However, with 193nm double patterning on the horizon high-index immersion may be an interesting option to be combined with double patterning at the 22nm node.

III. 193NM DOUBLE PATTERNING

Double patterning is a technique that comes in many flavors. The basic idea is that if a pitch of interest is not achievable in a single lithography step, the design is split over

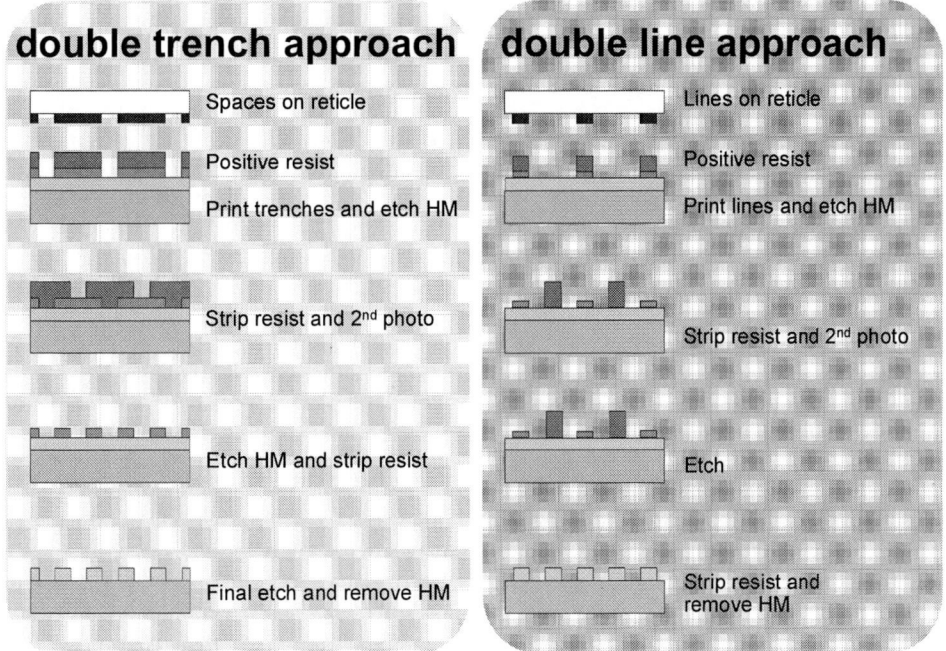

Figure 3. Schematic representation of double trench and double line approaches for double patterning.

978-1-4244-2018-6/08 $25.00 © 2008 IEEE   401

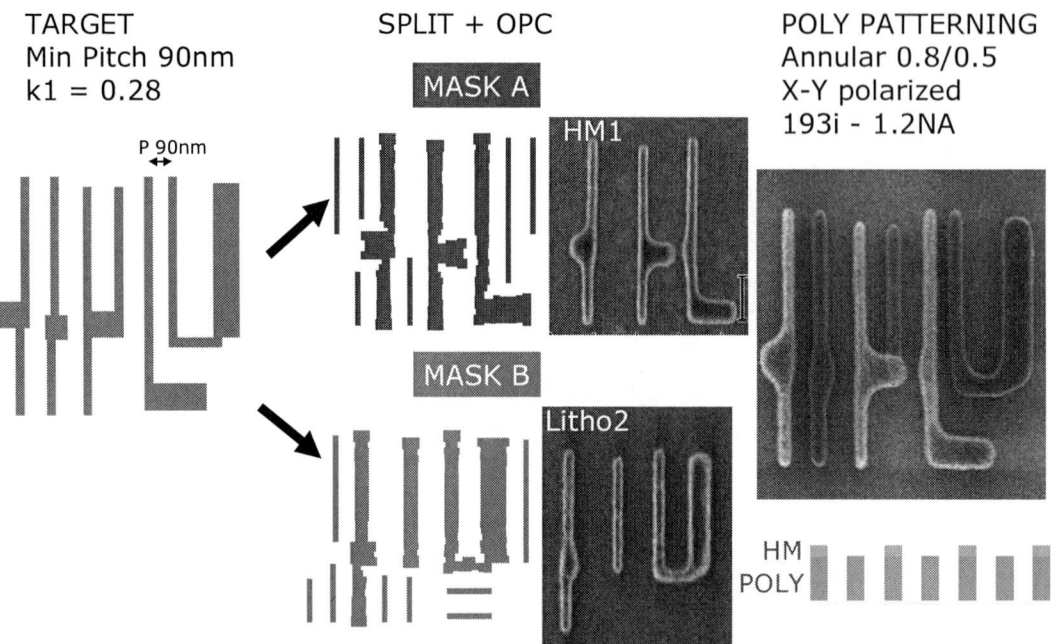

Figure 4. Approach and results for splitting a logic gate cell, with 45nm minimal half-pitch for double patterning at 1.2NA (k_1=0.28). A target design (left) is split and corrected for optical proximity, resulting in Mask A and B (and their corresponding individual wafer prints). The combined result before removing the hard-mask (resulting in different contrast in the SEM) is shown on the right, demonstrating the good match with the target design.

two lithography layers in a way that the minimum pitch is relaxed (and preferably doubled) with respect to the target pitch. In this way the effective k_1 of the total process (i.e. the combination of the two lithography steps) can drop below the theoretical limit of 0.25 for a single patterning process. From a processing standpoint, the easiest way to implement this is by transferring the first litho step into a hard-mask layer by etch and subsequent imaging and etching of a second photoresist layer. This Litho-Etch-Litho-Etch (or LELE) approach can for instance be achieved either by double line or double space patterning as shown in Figure 3.

A big advantage of double patterning 193nm lithography for the 32nm node is that it builds on existing platforms. Early development of the technology is already possible with the tools that are currently available on the market. Out of the three technological options that are discussed in this paper, it is the most mature and most likely to be in time for patterning at the 32nm node. From a scanner perspective the main technological challenge is in meeting the overlay requirements. Alignment between the first and second exposure pass is critical for meeting the line or space critical dimension (CD) uniformity target (depending on which process is chosen). Next to this, software is needed to automatically and robustly split designs over the two patterning steps. This is especially challenging for logic designs, which are relatively random by nature. Using some example clips, a proof of principle has been demonstrated using LELE for processing (Figure 4).[10]

From a processing point of view, the LELE approach requires 'just' normal scaling from the approaches for the larger nodes. However, the drawback of the LELE approach is that the total lithography cost essentially doubles. The full lithography process is done twice per layer, with even an extra intermediate etch step in between (not to mention extra hard mask materials that may be required). This problem can, however, be circumvented by novel material developments. The most attractive approach is the use of non-linear optical materials, since this would allow double exposure rather than double patterning. The two exposures that are needed to print a single layer can, in this case, be performed without removing the wafer from the exposure chuck which is advantageous for overlay reasons. However, it does require materials that have 'forgotten' the dose in underexposed areas from the first exposure, by the time they see the second exposure such that, $f(I_{Pass1} + I_{Pass2}) \neq f(I_{Pass1}) + f(I_{Pass2})$.[11] To achieve such a behavior, reversible contrast enhancement layers and optical threshold materials are being considered. A lot of work, however, is still focusing on theoretical analysis and it is not straightforward to come up with appropriate materials that possess all the required characteristics.

An intermediate option from a processing point of view is to freeze the pattern from the first image through some additional processing (for instance in the litho track) before coating and exposing a second layer of resist over the first (Litho-Process-Litho-Etch). The principal requirement is that the process not only enables that the resist from the first image withstands the solvent that is applied when coating the second layer of resist, but also that the first layer is not further developed after the second exposure (mainly of concern for positive-tone resists in the double line approach). A method to

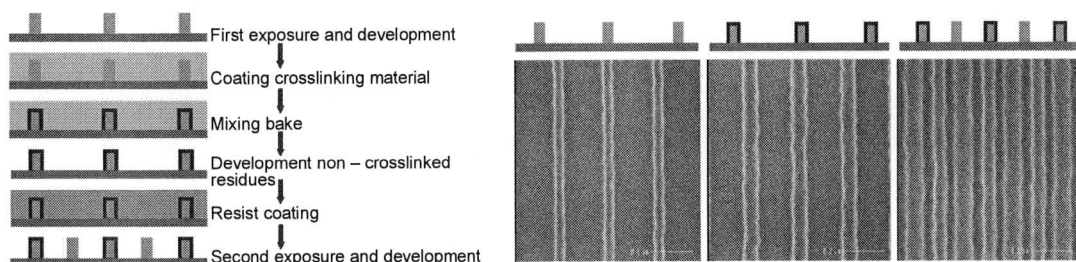

Figure 5. Approach and results for pattern freezing through overcoat and cross-link. During the mixing bake, the residual acid from the patterns of the first litho step cross-link the overcoat material. The cross-linked layer protects the first pattern during spin-coat of the second resist layer. The right SEM picture shows the final result for 60nm half pitch lines/spaces at 0.75NA.

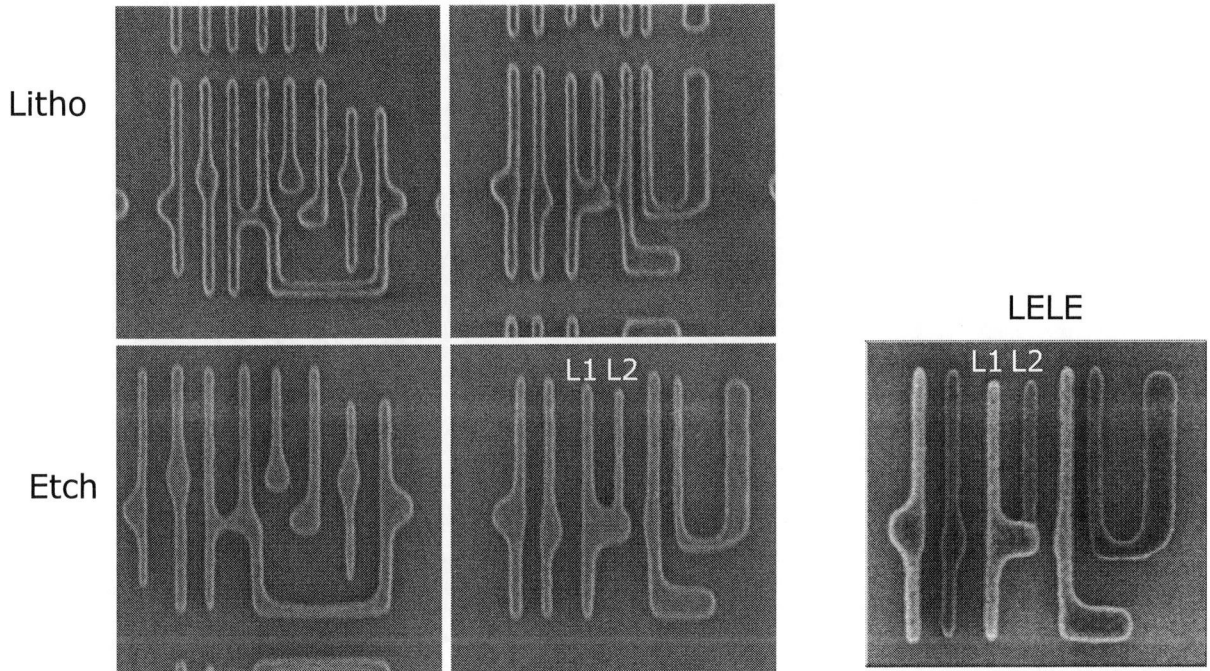

Figure 6. 2-dimensional patterning results after litho and after etch for a pattern freezing technique (showing also the comparison to litho-etch-litho-etch).

achieve this is, for instance, by depositing an overcoat after the first layer is imaged and developed, that cross-links in a subsequent bake step under the influence of residual acid and PAG. The result is the formation of a thin crust on the resist pattern that protects the resist from the solvent of the second layer and the second development step (Figure 5). A drawback of this method is that the lines are widened by the cross-linking step and thus affects the requirement of printing even narrower lines with good CD control for double patterning.

Very promising results have been recently reported using this approach[12]. High resolution, good CD uniformity and excellent 2D patterning properties could be demonstrated after

litho and etch, making this approach comparable to the much more extensive and costly Litho-Etch-Litho-Etch approach (figure 6).

An alternative method is to use two different resists whose chemistries are sufficiently different such that the solvent of the second resist does not react with the polymer of the first resist (Figure 7). For this approach materials of opposite tonality need to be used. This is because in areas where there are patterns of the first layer, the second layer needs to be removed. In a first litho step the positive-tone resist is patterned. The undeveloped polymer of this resist does not dissolve in the solvent of the second, negative-tone resist.

978-1-4244-2018-6/08 $25.00 © 2008 IEEE

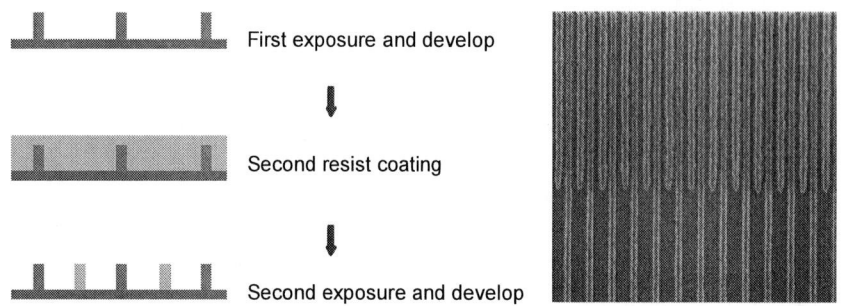

Figure 7. Approach and results for double patterning using a combined positive/negative tone resist process. The SEM image on the right shows 50nm half pitch results at 0.85NA for the negative (short lines) and positive (long lines) images. Note that opposite tone materials are required to avoid that the first layer is developed after the second litho step.

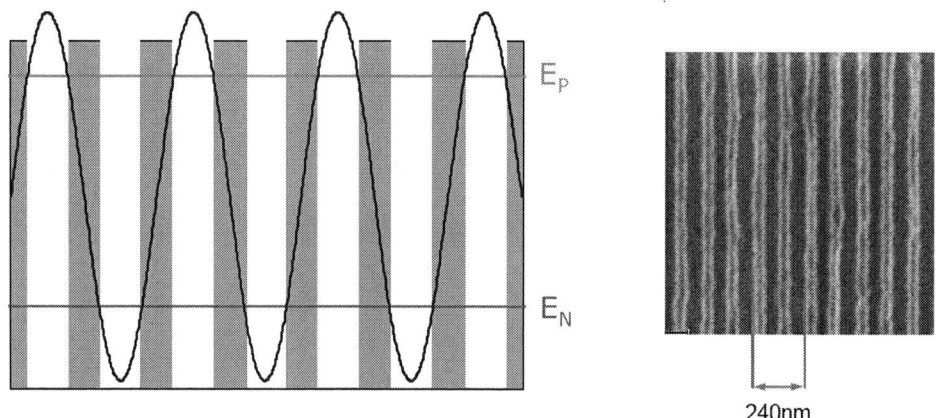

Figure 8. Approach and results for patterning using double development. The SEM image on the right shows results for a 240nm pitch (1X) mask feature printed at 0.75NA. This results after a single exposure in 60nm half pitch resist lines on the wafer.

A third example of an alternative process for making double patterning more cost-effective is in dual tone development.[13] In this approach only a single exposure step is used for printing the dense pattern. A sinusoid aerial image is used and two different chemistries are used in separate development steps. In the positive development process, those parts of the resist that have received a dose higher than E_P are developed. In the negative development process, parts of the resist are developed that have received a dose below E_N. This approach results in frequency doubling in a single exposure step and effectively the slopes of the aerial image are printed (Figure 8). It should be emphasized that this self-aligned approach requires an extra trim exposure to remove conflict areas (such as line ends, similar to using alternating PSM). However this second exposure is less critical in terms of CD and overlay. The dual development process is an attractive solution for very regular designs (such as memory), but will be more difficult to apply in logic designs.

A fourth example is self-aligned double patterning (SADP), a spacer doubling technique, which has emerged as an option for 32nm NAND flash devices. Scalability to 32nm half pitch has been demonstrated.[14] SADP avoids the overlay challenges of double exposure because the patterns are created from a single exposure and doubling is achieved by spacer formation and bottom hard mask etch.

IV. EXTREME UV LITHOGRAPHY

The main benefit of Extreme UV Lithography is in its potential extendibility. The enormous jump (from 193 to 13.5nm) that is made in imaging wavelength allows reduction in NA and increase in k_1, while maintaining the improved resolution. For instance for 32nm half pitch imaging at 0.25NA (target NA for the early full-field systems) k_1 will be 0.59. The higher k_1 removes a lot of the optical proximity effects and therefore avoids the need for their corrections and for assist feature placement for mask making. Additionally, EUV lithography will be a single exposure technique and should therefore be more cost-effective than some of the techniques discussed above, especially for high volume products. IMEC has recently installed one of the very first EUV full-field tools. This ASML Alpha Demo Tool (EUV ADT) is providing important learning to the industry on the implementation of EUV lithography in silicon processing. One of the biggest technological hurdles for introducing EUV lithography is the availability of EUV light sources with sufficient power for high-throughput imaging. EUV requires all-reflective optics since the transmission of all materials to EUV light is low. Mo/Si multilayer coated mirrors are used for this purpose, but their maximum reflectivity is ~70%. This means that even with a high power source, the photoresist

978-1-4244-2018-6/08 $25.00 © 2008 IEEE

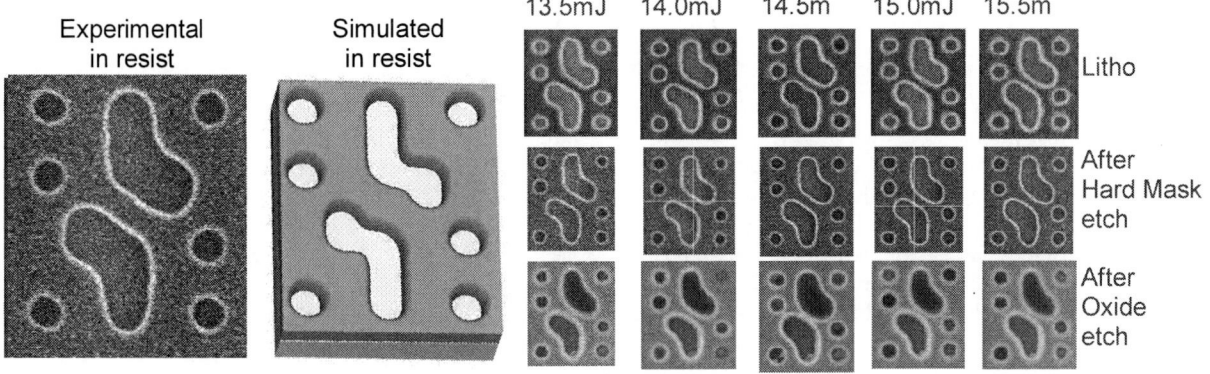

Figure 9. Results of implementation of EUV in a 32nm SRAM cell at contact level

needs to be very sensitive to make efficient use of the expensive photons that reach the wafer surface.

Recently quite some progress has been reported on the exposure results of the EUV ADT[15,16]. Besides imaging results, the first implementations of EUV lithography as a technology to pattern the critical levels of the next technology nodes are being investigated (Figure 9).

The current target for EUV photoresist sensitivity is 10mJ/cm^2. The feasibility for reaching this sensitivity target has already been demonstrated today. However, the challenge is to reach the target while maintaining performance for resolution and line-width roughness (LWR)[17]. The interdependence of LWR, Sensitivity and Resolution has been described in literature and termed the 'Triangle of Death' or 'Lithographic Uncertainty Principle' (Figure 10). Simultaneous optimization of resist formulations for these three performance parameters has been selected as one of the top three critical issues for successful introduction of EUV lithography for three years in a row at the annual EUVL Symposium.

The main area of concern for chemically amplified resists at the 32nm node is how to meet the very tight LWR specifications. Although the problem may be somewhat more severe for EUV than for the other technologies (due to the unfavorable shot-noise statistics), it is of concern for all optical patterning technologies that target this node. Currently there are no known chemically amplified resists (for EUV or 193) that meet these specifications. Besides this, metrology of such low LWR values is by far not trivial. Various novel resist concepts are being proposed to tackle several of the issues that are thought to limit current LWR values. These include non-chemically amplified resists, molecular glass resists and many others.

Other critical issues for EUVL are defect free EUV blank

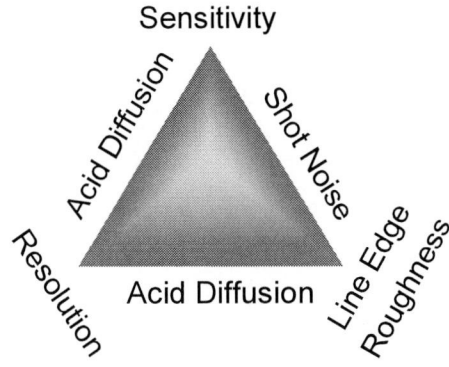

Figure 10. Graphical representation of the 'Lithographic Uncertainty Principle', which demonstrates the interdependence of Line-Width Roughness (LWR), Sensitivity and Resolution and the physical processes through which they are linked.

and mask manufacturing and low defect addition during shipping and use of the reticles in the fab; and finally optics manufacturing specifications and lifetime.

V. CONCLUSIONS

For the 32nm half pitch node, three imaging techniques are under consideration: 193nm high-index immersion, 193nm double patterning and Extreme UV lithography. This node should go into pilot development in 2009, meaning that production-worthy tools and infrastructure should be in place by that time.

Both 193nm high-index immersion and EUV lithography have suffered delays in their development so that it has become unlikely that they will meet this target. However, both techniques (high-index immersion in combination with double patterning) are still contenders for imaging at the 22nm half-pitch node.

For the 32nm half pitch node, it is likely that double patterning will be the technology of choice. However, in order to make the technique more cost-effective and avoid the Litho-Etch-Litho-Etch approach, significant material development is still required. This may be achieved either through the use of materials with a non-linear optical response (double exposure), or through process-induced freezing of the first image. Whatever the final choice for patterning is going to be, it is clear that significant resist material challenges lie ahead of us.

ACKNOWLEDGEMENTS

The authors are indebted to the entire IMEC lithography team for their input and helpful discussions on the materials presented in this paper. Special thanks to Roel Gronheid for his contribution to this manuscript.

REFERENCES

[1] W. Conley "The Application of High Refractive Index Photoresist for 32nm Device Level Imaging" Proc. SPIE, *2007*, **6519**, 65190Q.

[2] E. Hendrickx, S. Postnikov, P. Foubert, R. Gronheid, B.-S. Kim "Screening of Second-Generation High-Index Liquids" Proc. SPIE, *2007*, **6519**, 65190A.

[3] H. Sewell, J. Mulkens, P. Graeupner, D. McCafferty, L. Markoya, S. Donders, N. Samarakone, R. Duesing "Extending Immersion Lithography with High Index Materials - Results of a feasibility study" Proc. SPIE, *2007*, **6520**, 65201M.

[4] R. H. French, V. Liberman, H. V. Tran, J. Feldman, D. J. Adelman, R. C. Wheland, W. Qiu, S. J. McLain, O. Nagao, M. Kaku, M. Mocella, M. K. Yang, M. F. Lemon, L. Brubaker, A. L. Shoe, B. Fones, B. E. Fischel, K. Krohn, D. Hardy, C. Y. Chen "High Index Immersion Lithography with Second Generation Immersion Fluids to Enable Numerical Apertures of 1.55 for Cost Effective 32 nm Half Pitches" Proc. SPIE, *2007*, **6520**, 65201O.

[5] P. Zimmerman et al, "Development and evaluation of a 193-nm immersion generation-three fluid", Proc. SPIE, Vol.6923-09, to be published (2008).

[6] I. Blakey, L. Chen, B. Dargaville, H. Liu, A. Whittaker, W. Conley, E. Piscani, G. Rich, A. Williams, P. Zimmerman "Novel High-Index Resists for 193 nm Immersion Lithography and Beyond" Proc. SPIE, *2007*, **6519**, 651909.

[7] Y. Nishimura, T. Kawakami, K. Hoshiko, T. Tominaga, M. Shima, S. Kusumoto, T. Shimokawa, K. Hieda "Novel Photoresist Materials Development with High Refractive Index at 193nm for Next Generation Immersion Lithography" 3rd International Symposium on Immersion Lithography, Kyoto, Japan, 2006.

[8] P. Zimmerman et al, "Development of an operational high refractive index resist for 193-nm immersion Lithography, Proc. SPIE, Vol. 6923-05, to be published (2008).

[9] L. Parthier et al, "High-Index Lens Material LuAG: Development Status and Progress", Proc. SPIE, Vol. 6924-69, to be published (2008).

[10] V. Wiaux, E. Hendrickx, J. Bekaert, S. Verhaegen, G. Vandenberghe, S. Locorotondo, S. Beckx, J. Finders, M. Dusa, J. Quaedackers, B. Vleeming "193nm Immersion Lithography Towards 32nm Half-Pitch Using Double Patterning" 3rd International Symposium on Immersion Lithography, Kyoto, Japan, 2006.

[11] J. Byers, S. Lee, K. Jen, P. Zimmerman, N. J. Turro, C. G. Willson "Double Exposure Materials: Simulation Study of Feasibility" *J. Photopolym. Sci. Technol.* **2007**, *20*, 707-717.

[12] M. Maenhoudt et al, "Alternative process schemes for double patterning that eliminate the intermediate etch step", to be published in Proceedings SPIE, Vol. 6924-24, 2008.

[13] S. Tarutani, H. Tsubaki, M. Yoshidome, K. Wada, W. Hoshino, S. Kanna, N. Nishikawa, K. Mizutani, N. Ohshima, K. Shitabatake "Development of Materials and Processes for 32nm Node Immersion Lithography Process" 4th International Symposium on Immersion Lithography, Keystone, Colorado, 2007.

[14] A. Khan, "Enabling Etch Technology for Patterning Beyond 32nm", Proc. SPIE, to be published

[15] H. Meiling et al, "Performance of the full field EUV systems", Proc. SPIE, Vol. 6921-21, to be published (2008).

[16] G. Lorusso et al, "Imaging performance of the EUV alpha demo tool at IMEC", Proc. SPIE, Vol. 6921-24, to be published 2008.

[17] R. Gronheid et al, "Lithography Options for the 32nm Half Pitch Node and Their Implications on Resist and Material Technology", Proc. of SPIE Vol. 6827, p. 68271V 2007.

AUTHOR INDEX

Agarwal, Kanak 367
Agrawal, Ankur 491
Ahmed, Haseeb 433
Ahn, Hyung Ki 801
Ahn, Sunghoon 487
Akopyan, Filipp 701
Al-Hashimi, Bashir 571
Allee, David 199
Allstot, David 611
Amrutur, Bharadwaj 145
Annema, Anne Johan 239
Aparin, Vladimir 777
Arakali, Abhijith 475
Arimoto, Kazutami 717
Aroca, Ricardo A. 505
Arslan, Umut 445
Arvind, N. V. 145
Asbeck, Peter 789
Auth, Chris 407
Avanzo, Flavio 615
Ayers, James 629
Bakir, Muhannad 705
Bakkaloglu, Bertan 385
Balakrishnan, Varsha 13
Balamurugan, Ganesh 743
Barale, Francesco 81
Barnaby, Hugh 293, 385
Bashirullah, Rizwan 23
Basu, Arindam 231
Bertholet, Nicole 509
Bhargava, Mudit 445
Bhatia, Deepak 23
Bichan, Mike 651
Blaauw, David 453, 543, 579
Borremans, Jonathan 53
Bottarel, Valeria 591

Brama, Riccardo 595
Brink, Stephen 231
Brodersen, Robert 323
Burnell, David 771
Butzen, Paulo 133
Cai, Liuchun 727
Cao, Yu 13
Cao, Zhiheng 331
Carlson, Andrew 441
Carnes, Josh 739
Carusone, Anthony Chan 479, 651, 675
Casper, Bryan 743
Chabert, Laurent 509
Chandrashekar, Kailash 385
Chang, Hsiang-Hui 797
Chang, Ken 467
Chantre, Alain 415
Chen, Bei 267, 271
Chen, Charles 323
Chen, Fangxiong 267, 271
Chen, Hong 389
Chen, Jia 389
Chen, Ke- Horng 27, 35
Chen, Poki 227
Chen, Po-Yu 227
Chen, Qiang 285
Chen, X. J. 293
Chen, Yanfei 531
Chen, Yi 389
Chen, Young-Kai 335
Cheng, Horace 675
Cheng, Wei 239
Chevalier, Pascal 415
Chiang, Ed 761
Chin, Sing 319
Chin, Tj 467

AUTHOR INDEX

Chiu, Yun 323
Cho, Choongyeun 223, 667
Cho, Min-Hyung 539
Cho, Minki 305
Cho, Sangyeun 223
Choi, Minseok 721
Choi, Moo-Yeol 99
Chu, Hyunho 587
Chu, Kun-Da 513
Chun, Jung-Hoon 467
Chung, Hayun 599
Chung, Hoon Hee 385
Chung, Kyusik 393
Cicek, Paul-Vahé 207
Ciftcioglu, Berkehan 603
Clark, Lawrence 199
Cole Jr., Edward 57
Collados, Manel 781
Cong, Peng 559
Copani, Tino 385
Cressler, John 63
Croon, Jeroen 239
Dai, Fa Foster 267, 271, 525
Dallago, Enrico 591
Dang, Bing 705
Daniel, Erik 663
Das, Bishnu Prasad 145
Davis, W. Rhett 697
Dawn, Debasis 81
De Nicola, Paolo 713
De Paola, Francesco M. 607, 615
Debnath, Chandrajit 327
Dehos, Cedric 509
Deligoz, Ilker 385
Depaoli, Emanuele 595
Dewey, Bill 301

Dhanasekaran, Vijay 749
Diorio, Chris 357
Doi, Yoshiyasu 505
Dong, Qiaohua 457
Dong, Yunzhi 625
Dusatko, Tomas A. 203, 361
El-Gamal, Mourad 207, 619
El-Gamal, Mourad N. 203, 361
Elkhatib, Tamer 551
English, George 463
Erba, Simone 595
Esqueda, Ivan 293
Eum, Nakwoong 721
Evans, William 771
Fang, David 701
Feng, Guoyou 457
Fiez, Terri 629, 633
Flynn, Michael 373, 757
Foo, Zhiyoong 579
Frattini, Giovanni 591
Friedman, Daniel 463
Friedman, Eby 693
Fritz, Karl 663
Fujita, Hiroyuki 93
Fukuzawa, Fumitaka 717
Funaki, Hideyuki 93
Gambhir, Manisha 749
Garcia, Patrice 415, 509
Garnier, Christophe 415
Genesi, Raffaella 607
Gerosa, Gian 433
Gerrits, John F. M. 625
Gierkink, Sander 471
Gilbert, Barry 663
Goethals, Mieke 399
Gondi, Srikanth 475

AUTHOR INDEX

Goo, Jung-Suk 285

Gord, John 341

Goswami, Sushmit 385

Greyling, Braam 713

Groeseneken, Guido 53

Gronheid, Roel 399

Gu, Ben 247

Gu, Ming 267, 271

Gubbins, David 187

Gullapalli, Kiran 247

Guo, Lu 457

Guo, Song 547

Guo, Zheng 441

Hamada, Mototsugu 31

Hamada, Takayuki 531

Hamoui, Anas 555

Hanson, Scott 453

Hanumolu, Pavan 739

Hanumolu, Pavan Kumar 187, 475, 491, 753

Hara, Hiroyuki 31

Harjani, Ramesh 243, 727

Hasan, Mohammad 433

Hashemi, Hossein 75

Hashimoto, Eiri 689

Hasler, Paul 231

Hendrickx, Eric 399

Hess, Phil 341

Heydari, Payam 521

Holleman, Jeremy 357

Honda, Kazutaka 127

Horowitz, Mark 219

Hossain, Masum 479

Hoyos, Sebastian 323

Hsiao, Yuan-Wen 251

Hsieh, Bing-Yu 797

Hsieh, Ping-Hsuan 567

Hsieh, Yao-Jen 767

Hsu, Klaus Yune-Jane 263

Hu, Xueqing 267, 271

Huang, Gang 705

Huang, Hong-Wei 27, 35

Huang, Ji-Chen 263

Huang, Juin-Wei 513

Huang, Tian-Wei 497

Humble, James 663

Hung, B. 761

Hwang, Myeong-Eun 449

Inaoka, Kazuhiro 717

Irwin, J. David 525

Ishida, Koichi 149

Ishimi, Koichi 717

Itaya, Kazuhiko 93

Ito, Minoru 41

Itoh, Shinya 127

Jackson, Sandra 701

Jaeger, Richard 267, 271

Jaeger, Richard C 525

Jain, Anuj 385

Jain, Vipul 521

Jamadagni, H. S. 145

Jansen, Philippe 53, 399

Jaussi, James 743

Javid, Babak 521

Jee, Dong-Woo 535

Jenkal, Ravi 697

Jou, Shyh-Jye 377

Jun, Young-Hyun 679

Jung, Inhwa 487

Kajita, Mikihiro 731

Kang, Joshua 373, 757

Karaki, Habib 385

Kasai, Naoki 381

AUTHOR INDEX

Kaviani, Kambiz 467

Kawahito, Shoji 127

Keane, John 133

Kennedy, Joseph 743

Kent, John 423

Ker, Ming-Dou 251, 767

Kertis, Robert 663

Keskin, Gökçe 9

Khaligh, Alireza 195

Kiaei, Sayfe 385

Kibune, Masaya 531

Kim, Bongjin 255

Kim, Chiwon 255

Kim, Chris 133

Kim, Chris H. 437

Kim, Chulwoo...........................487, 587

Kim, Daeik 223, 667

Kim, Dong Woon 579

Kim, Donghyun 393, 575

Kim, Duckhwan 721

Kim, Hyungoo 255

Kim, Jaeha 735

Kim, Jaewhui...............99, 123, 183, 255

Kim, Jinkyu 721

Kim, Jiyoung 255

Kim, Jongdae 539

Kim, Jonghae 223, 667

Kim, Jongsik 641

Kim, Ju-Hwa 123

Kim, Jungmoon 587

Kim, Kihong 641

Kim, Kinam 679

Kim, Kwi-Dong 539

Kim, Lee-Sup...................393, 575, 679

Kim, Namsoo 777

Kim, Seon Wook 587

Kim, Seungsoo 641

Kim, Soo-Won 353

Kim, Sungdo 721

Kim, Tae-Hyoung 437

Kim, Yi-Gyeong 539

Kim, Yontae 487

Kim, Youngcho 641

Kimball, Donald 789

King, Calvin 705

Kinget, Peter 761

Kitahara, Takeshi 31

Kitao, Masaya 717

Kitchen, Jennifer 385

Klaassen, Dick 239

Ko, Wen 559

Kondo, Hiroyuki 717

Koo, Bontae 721

Koziol, Scott 231

Krishnan, Srikanth 13

Kristan, Ernst 713

Kulkarni, Jaydeep P. 137

Kumar, Ashish 327

Kuo, Chung-Hung 767

Kuo, Tzu-Chieh 563

Kurahashi, Peter 753

Kuroda, Tadahiro 531

Kwon, Jong-Kee 539

Kwon, Sunwoo 111

Kwon, Youngsu 721

Kyung, Moon- Jung........................... 175

Lai, Juan-Shan 227

Lai, Yu-Sheng 263

Larson, Larry 789

Larson, Lawrence E. 777

Laskar, Joy 81

Le Tual, Stephane............................. 509

AUTHOR INDEX

Lee, Bumha 187
Lee, Edward 341
Lee, Haechang 467
Lee, Hae-Seung 99, 123
Lee, Hoi 547
Lee, Ho-Young 123
Lee, Hyuckjae 721
Lee, Jaesik 335
Lee, Joohyun 721
Lee, Kang-Jin 183
Lee, Kyehyung 103, 107
Lee, Sungjae 285
Lee, Sung-No 99
Lee, Wei-Ming 175
Lei, Yu 457
Leibowitz, Brian 735
Lemoine, Dominique 207
Levacq, David 583
Leven, Andreas 335
Li, Chun-Huai 767
Li, Jing 211
Li, Li 757
Li, Pengfei 23
Li, T. L. 761
Li, Taihu 341
Li, Xin 9, 445
Lim, Daihyun 223
Lin, Chia-Hsiang 27, 35
Lin, David 757
Lin, Kuan-Yu 619
Lin, Tsung-Hsien 637
Lin, Xue 23
Lin, Yu- Shiang 543
Linten, Dimitri 53
Liu, Andrew 599
Liu, Chien-Nan 259

Liu, Chun-Ting 767
Liu, Haixin 211
Liu, Jason 437
Liu, Ming 389
Liu, Tsu-Jae King 441
Liu, Wei-Chang 377
Liu, Yao- Hong 637
Liu, Zheng 127
Loizou, Philipos 547
Long, John R. 625
Long, Syu-Siang 377
Lou, Shuzuo 71
Lu, Hung Wen 259
Lu, Minhua 89
Lu, Ning 301
Lu, Ting-Chou 767
Luo, Xian 215
Luong, Howard 71
Ma, Heping 267, 271
Ma, Jirong 215
Ma, Kwang-Sung 183
Madden, Christopher 735
Maebashi, Takanore 689
Maenhoudt, Mireille 399
Maghari, Nima 111
Mai, Ken 9, 445
Maitra, Kingsuk 305
Malik, Rakesh 327
Manohar, Rajit 701
Manstretta, Danilo 607, 615
Mansuri, Mozhgan 743
Marchetti, Maxime 509
Masui, Norio 717
Matei, Eusebiu 341
Matsumoto, Shuuji 31
Matsushita, Hiroaki 41

AUTHOR INDEX

Maxey, Jay 567
Mayaram, Kartikeya 629, 633, 739
McCartney, Mark P. 445
McLain, Michael 293
McNally, Iain 571
Meindl, James 705
Melamed, Samson 697
Mellon, Carnegie 9
Meterelliyoz, Mesut 137
Miatton, Daniele 591
Miller, Matt 611
Miller, Matthew R. 103
Min, Junki 641
Min, Young- Jae 353
Minakawa, Takuya 583
Mineo, Christopher 697
Mishra, Apurva 357
Mita, Makoto 93
Miyakawa, Nobuaki 689
Moon, Kyoung-Ho 123, 183
Moon, Kyoung-Jun 183
Moon, Un-Ku 111, 187, 739, 753
Mooney, Randy 743
Moquillon, Laurence 509
Morche, Domique 509
Mukhopadhyay, Saibal 305
Murmann, Boris 115, 179
Nabki, Frederic 203, 207, 361
Naeemi, Azad 705
Nairn, David 311
Nakajima, Masami 717
Nakamoto, Satoshi 731
Nakamura, Natsuo 689
Nakamura, Yasumi 583
Nakayama, Naoya 731
Nakayama, Shigeto 689

Nassif, Sani 1, 367
Nauta, Bram 239
Nebshi, Ryusuke 381
Neihart, Nathan 611
Nercessian, Patrick 341
Nho, H. Henry 219
Nicolson, Sean 415
Nicolson, Sean T. 505
Niiyama, Taro 149, 583
Nikaeen, Parastoo 179
Nikolic, Borivoje 141, 323, 441
Nomura, Shuou 31
Nowak, Edward 285
Nowka, Kevin 367
O, Kenneth 163
Oh, Jaegeun 587
Oh, Kwang-Il 679
Oh, Kyujin 163
Oh, Tae-Hwan 123
Ohta, Jun 349
Ok, Sunghwa 587
Okpisz, Alex 433
Okumura, Naoto 717
Okushima, Mototsugu 53
O'Mahony, Frank 743
Omole, Umaikhe 243
Otani, Sugako 717
Otero, Carlos Tadeo Ortega 701
Otis, Brian 357
Pain, Bedabrata 175
Palaskas, Yorgos 155
Pan, Liyang 215
Pang, Liang-Teck 141, 441
Panitantum, Napong 633
Pareschi, Fabio 483
Park, Byeong-Ha 801

AUTHOR INDEX

Park, Ho-Jin 99, 123

Park, Hong-June 535

Park, Jaehyun.............................255

Park, Junyoung 373

Park, Kihyuk 721

Park, Kwang-Il 679

Park, Min 579

Park, Piljae 235

Park, Seung- Jin 535

Park, Sunghyun 373

Pavlidis, Vasilis.............................693

Pavlovic, Nenad 781

Pellerano, Stefano 155

Perego, Rich.............................467

Perfecto, Eric.............................89

Perumana, Bevin 81

Petre, Csaba.............................231

Philpott, Rick.............................663

Piazzese, Nicolo Ivan.............................509

Piccolella, Salvatore 713

Pileggi, L. 9, 171, 445

Pinel, Stephane 81

Pisati, Matteo.............................595

Plouchart, Jean-Olivier 667

Pozzoni, Massimo.............................595

Prasad, Jagdish 423

Prete, Edoardo 45

Proesel, Jon.............................9

Proesel, Jonathan 171

Pruvost, Sebastien 415, 509

Puma, Giuseppe Li.............................713

Rakers, Pat 611

Ramakrishnan, Shubha 231

Ravi, Ashoke.............................155

Razavi, Behzad 655

Reddy, Vijay.............................13

Redouté, Jean-Michel 683

Ren, Jihong 735

Rhee, Woogeun 801

Ricotti, Giulio 591

Riley, John 433

Ronse, Kurt.............................399

Roux, Pascal.............................335

Rovatti, Riccardo.............................483

Roy, David.............................415

Roy, Kaushik 137, 211, 449

Rylyakov, Alexander 463

Ryu, Jongjae 255

Sacho, Yutaka 689

Sadhu, Bodhisatwa 243

Safarian, Zahra 75

Saito, Yoshihiro 717

Sakimura, Noboru 381

Sakugawa, Mamoru 717

Sakurai, Takayasu.............................149, 583

Salahuddin, Sayeef 211

Salama, Khaled 551

Sanchez-Sinencio, Edgar.............................749

Sanders, Seth R.............................785

Sanduleanu, Mihai A. T..............................625

Sankaran, Swaminathan 163

Sanzogni, Davide 595

Sato, Hironori 31

Sato, Masayuki 717

Savidis, Ioannis 693

Sawamura, Shigeki 349

Scaccianoce, Salvatore 509

Schipani, Monica 591

Schlottman, Craig.............................231

Scholz, Mirko 53

Scuderi, Angelo 509

Sekar, Deepak 705

AUTHOR INDEX

Seki, Keiko 31

Sen, Padmanava 81

Seo, Jae-Sun 453

Seok, Mingoo 453

Serratore, Alberto 509

Setti, Gianluca 483

Shekhar, Sudip 743

Shen, Jie 467

Shi, Xudong 467

Shi, Yin 267, 271

Shimizu, Toru 717

Shin, Hyunchol 641

Shin, Jaewook 641

Shin, Jongshin 255

Shiue, Muh-Tian 377

Shringarpure, Rahul 199

Shu, Yun-Shiang 175

Silva-Martinez, Jose 749

Sim, Jae-Yoon 535

Singh, Pratap Narayan 327

Sode, Mikiki 731

Song, Bang-Sup 175

Song, Jooyoung 277

Song, Minyoung 487

Song, Peilin 137

Sonkusale, Sameer 191

Sperling, Michael 463

Srivastava, Kamalesh 89

Stauth, Jason T 785

Stellari, Franco 137

Steyaert, Michiel 19, 683

Stover, Howard 341

Su, Chau Chin 259

Su, David 645

Sugibayashi, Tadahiko 381

Sun, Yuanfeng 801

Sundareswaran, Savithri 247

Sundlof, Brian 89

Suzuki, Kazuhiro 93

Svelto, Francesco 595

Swabey, Matthew 571

Sylvester, Dennis 453, 543, 579

Tachibana, Fumihiko 31

Takahashi, Kazuhiro 93

Takamiya, Makoto 149, 583

Takeuchi, Kan 41

Talbot, Gerry 45

Tamura, Hirotaka 531

Tanaka, Genichi 41

Tano, Yasuo 349

Taufique, Mohammed 433

Taur, Yuan 277

Tecpoyotl-Torres, Margarita 555

Temes, Gabor 107

Temes, Gabor C 103

Terasawa, Yasuo 349

Thacker, Hiren 705

Thijs, Steven 53

Tierno, Jose 463

Tokuda, Takashi 349

Tokunaga, Carlos 579

Tomita, Yasumoto 531

Tomkins, Alexander 505

Toshiyoshi, Hiroshi 93

Toyoda, Shinjiro 689

Trakimas, Michael 191

Tsang, Cheongyuen 323

Tseng, Chi-Yao 377

Tsuboi, Yoshiro 31

Tsukamoto, Sanroku 531

Twigg, Christopher M 231

Uppili, Shrinivas 199

AUTHOR INDEX

Van Veenendaal, Gerrit 625

Van Zeijl, Paul T. M. 781

Vandenberghe, Geert 399

Vanderhaegen, Johan 323

Vannier, Cyril 713

Varona, Jorge 555

Venchi, Giuseppe 591

Venkatraman, Shrinivas 133

Venugopal, Sameer 199

Vermeire, Bert 385

Vezyrtzis, Christos 761

Viola, Paolo 595

Visvanathan, V. 145

Voinigescu, Sorin 415

Voinigescu, Sorin P. 505

Vytyaz, Igor 739

Wambacq, Piet 53

Wang, Chao-Shiun 513, 517

Wang, Chorng-Kuang 513, 517

Wang, Huei 497

Wang, Nan 457

Wang, Ping-Ying 797

Wang, Wenping 13

Wang, Yan 107

Wang, Zhihua 801

Wang, Zi 457

Wang, Zihua 389

Watanabe, Yoshinori 31

Wei, Gu-Yeon 491, 599

Wei, Ting-Chen 377

Weiner, Joseph 335

Wen, Shon-Hang 517

Wens, Mike 19

Wilcock, Reuben 571

Williams, Richard 285

Willson, Alan 563

Wilson, Peter 571

Wiser, Robert 645

Wolfe, James 341

Wong, Bill 319

Wong, S. Simon 219

Wooley, Bruce 645

Workman, Glenn 285

Wu, Dong 215

Wu, Hsinta 163

Wu, Hui 603

Wu, Ting 467, 739

Wu, Wanghua 625

Wu, Xiaochun 195

Xiao, Limin 583

Xu, Hua 267, 271

Xu, Jun 215

Xu, Liang 457

Xu, Qiming 267, 271

Xu, Yang 195

Yamamoto, Osamu 717

Yamamoto, Takuji 505

Yamane, Fumiyuki 31

Yamashita, Takahiro 31

Yan, Jun 267, 271

Yan, Shouli 331

Yan, Yaru 215

Yang, Bo 13

Yang, Chih-Kong Ken 567

Yang, Dayu 525

Yang, Seung- Hee 255

Yao, Xiang 457

Yasutomi, Keita 127

Yeh, David 81

Yeum, Wang-Seup 99

Yoon, Gilwon 587

Yoon, Jae-Sung 575

AUTHOR INDEX

You, Seung-Bin 99

Young, Darrin 559

Young, Michael Wieckowski 579

Yu, Bo 277

Yu, Chang-Hyo 393, 575

Yu, Hao 319

Yu, Peng 267, 271

Yu, Xuefeng 525

Yu, Xueyi 801

Yuan, Fang 267, 271

Yue, C. Patrick 235

Zan, Hsiao-Wen 767

Zargari, Masoud 645

Zhan, Jing-Hong 797

Zhang, Chun 389

Zhang, Lin 603

Zhang, Xuelian 267, 271

Zhang, Yun 247

Zhao, Yi 625

Zhu, Xiaolei 531

2008 IEEE Custom Integrated Circuits Conference

(CICC 2008)

San Jose, California
21-24, September 2008

Pages 407-804

IEEE Catalog Number: CFP08CIC-PRT
ISBN: 978-1-4244-2018-6

**Copyright © 2008 by the Institute of Electrical and Electronic Engineers, Inc
All Rights Reserved**

Copyright and Reprint Permissions: Abstracting is permitted with credit to the source. Libraries are permitted to photocopy beyond the limit of U.S. copyright law for private use of patrons those articles in this volume that carry a code at the bottom of the first page, provided the per-copy fee indicated in the code is paid through Copyright Clearance Center, 222 Rosewood Drive, Danvers, MA 01923.

For other copying, reprint or republication permission, write to IEEE Copyrights Manager, IEEE Service Center, 445 Hoes Lane, Piscataway, NJ 08854. All rights reserved.

***This publication is a representation of what appears in the IEEE Digital Libraries. Some format issues inherent in the e-media version may also appear in this print version.**

IEEE Catalog Number:	CFP08CIC-PRT
ISBN 13:	978-1-4244-2018-6
Library of Congress No:	85-653738

Additional Copies of This Publication Are Available From:

Curran Associates, Inc
57 Morehouse Lane
Red Hook, NY 12571 USA
Phone: (845) 758-0400
Fax: (845) 758-2633
E-mail: curran@proceedings.com

TABLE OF CONTENTS

FRONT MATTER

SESSION 1- KEYNOTE PRESENTATION

KEYNOTE PRESENTATION
More than Moore
Dave Bergeron

SESSION 2- STATISTICAL MODELING

Process Variability at the 65nm Node and Beyond (INVITED)...1
Sani Nassif

Mismatch Analysis and Statistical Design at 65 nm and Below (INVITED)...9
Larry Pileggi, Gökçe Keskin, Xin Li, Ken Mai, Jon Proesel, Carnegie Mellon

Statistical Prediction of Circuit Aging under Process Variations...13
Wenping Wang, Vijay Reddy, Bo Yang, Varsha Balakrishnan, Srikanth Krishnan, Yu Cao

SESSION 3 – POWER MANAGEMENT

A Fully-Integrated 0.18μm CMOS DC-DC Step- Down Converter, Using a Bondwire Spiral Inductor 19
Mike Wens and Michiel Steyaert

A 90-240MHz Hysteretic Controlled DC-DC Buck Converter with Digital PLL Frequency Locking...23
Pengfei Li, Deepak Bhatia, Xue Lin and Rizwan Bashirullah

Fast Transient Technique (FTT) in Buck Current- Mode DC-DC Converters for Low-Voltage SoCv Systems...27
Chia-Hsiang Lin, Hong-Wei Huang, Ke- Horng Chen

A Process Variation Compensation Scheme Using Cell-Based Forward Body-biasing Circuits Usable for 1.2V Design...31
Fumihiko Tachibana, Hironori Sato, Takahiro Yamashita, Hiroyuki Hara, Takeshi Kitahara, Shuou Nomura, Fumiyuki Yamane, Yoshiro Tsuboi, Keiko Seki, Shuuji Matsumoto, Yoshinori Watanabe, Mototsugu Hamada

Built-in Resistance Compensation (BRC) Technique for Fast Charging Li-Ion Battery Charger 35
Chia-Hsiang Lin, Hong-Wei Huang, Ke-Horng Chen

SESSION 4- HIGH-SPEED TEST, CHARACTERIZATION, AND DEBUG

A Voltage Drop Aware Crosstalk Measurement with Multi-Aggressors in 65 nm Process 41
Genichi Tanaka, Kan Takeuchi, Minoru Ito, Hiroaki Matsushita

Measurements of the Silicon Die Characteristics of Packaged Drivers for High-Speed I/O (INVITED) 45
Gerry Talbot and Edoardo Prete

Inductor-Based ESD Protection under CDM-like ESD Stress Conditions for RF Applications 53
Steven Thijs, Mototsugu Okushima, Jonathan Borremans, Philippe Jansen, Dimitri Linten, Mirko Scholz, Piet Wambacq and Guido Groeseneken

Non-Destructive IC Defect Localization Using Optical Beam-Based Imaging (INVITED)........................57
Edward Cole Jr.

SESSION 5- BROADBAND CIRCUIT TECHNIQUES FOR EMERGING WIRELESS COMMUNICATIONS

Emerging Application Opportunities for SiGe Technology (INVITED)...............................63
John Cressler

A 0.8GHz-10.6GHz SDR Low-Noise Amplifier in 0.13-μm CMOS 71
Shuzuo Lou and Howard Luong

A 1.3-6 GHz Triple-Mode CMOS VCO Using Coupled Inductors................................75
Zahra Safarian and Hossein Hashemi

SESSION 6- ADVANCED SoC/SiP INTEGRATION AND CO-DESIGN

A SOC/SOP Co-design Approach for mmW CMOS in QFN Technology (INVITED) 81
Joy Laskar, Stephane Pinel, Padmanava Sen, Bevin Perumana, Debasis Dawn, David Yeh and Francesco Barale

Chip to Carrier C4 Technology Challenges with Pb-free Solders (INVITED)............................89
Eric Perfecto, Brian Sundlof, Kamalesh Srivastava and Minhua Lu

A Study on Process-compatibility in CMOS-first MEMS-last Integration................................93
Kazuhiro Takahashi, Makoto Mita, Hiroyuki Fujita, Kazuhiro Suzuki, Hideyuki Funaki, Kazuhiko Itaya, Hiroshi Toshiyoshi

SESSION 7- HIGH RESOLUTION CONVERTERS

A 101-dB SNR Hybrid Delta-Sigma Audio ADC using Post Integration Time Control 99
Moo-Yeol Choi, Sung-No Lee, Seung-Bin You, Wang-Seup Yeum, Ho-Jin Park, Jae-Whui Kim and Hae-Seung Lee

An 8.1 mW, 82 dB Delta-Sigma ADC with 1.9 MHz BW and -98 dB THD 103
Kyehyung Lee, Matthew R. Miller and Gabor C. Temes

A 2.5mhz Bw and 78db Sndr Delta-sigma Modulator Using Dynamically Biased Amplifiers...............107
Yan Wang, KyeHyung Lee and Gabor Temes

74dB SNDR Multi-Loop Sturdy-MASH Delta-Sigma Modulator Using 35dB Opamp Gain 111
Nima Maghari, Sunwoo Kwon and Un-Ku Moon

A/D Converter Trends: Power Dissipation, Scaling and Digitally Assisted Architectures (INVITED) 115
Boris Murmann

A 16b 10MS/s Digitally Self-Calibrated ADC with Time Constant Control 123
Tae-Hwan Oh, Ho-Young Lee, Ju-Hwa Kim, Ho-Jin Park, Kyoung-Ho Moon, Jae-Whui Kim and
Hae-Seung Lee

A 15b Power-Efficient Pipeline A/D Converter Using Non-Slewing Closed-Loop Amplifiers 127
Shoji Kawahito, Kazutaka Honda, Zheng Liu, Keita Yasutomi and Shinya Itoh

SESSION 8- CHARACTERIZATION AND TEST METHODS FOR DEVICE VARIABILITY IN NANOSCALE TECHNOLOGIES

**An Array-Based Test Circuit for Fully Automated Gate Dielectric Breakdown
Characterization** 133
John Keane, Shrinivas Venkatraman, Paulo Butzen, Chris Kim

A High Sensitivity Process Variation Sensor Utilizing Sub-threshold Operation 137
Mesut Meterelliyoz, Peilin Song, Franco Stellari, Jaydeep P. Kulkarni and Kaushik Roy

Measurement and Analysis of Variability in 45nm Strained-Si CMOS Technology 141
Liang-Teck Pang, Borivoje Nikolic

Within-Die Gate Delay Variability Measurement using Re-configurable Ring Oscillator 145
Bishnu Prasad Das, Bharadwaj Amrutur, H.S. Jamadagni, N.V. Arvind and V. Visvanathan

**Expected Vectorless Teacher-Student Swap (TSS) Test Method with Dual Power Supply
Voltages for 0.3V Homogeneous Multi-core LSI's** 149
Taro Niiyama, Koichi Ishida, Makoto Takamiya and Takayasu Sakurai

SESSION 9- BROADBAND CIRCUIT TECHNIQUES FOR EMERGING WIRELESS COMMUNICATIONS

MIMO Techniques for High Data Rate Radio Communications (INVITED) 155
Yorgos Palaskas, Ashoke Ravi and Stefano Pellerano

Wireless Interconnection within a Hybrid Engine Controller Board 163
Swaminathan Sankaran, Kyujin Oh, Hsinta Wu and Kenneth O

SESSION 10- PANEL DISCUSSION
Sure, Moore's Law Can Continue, But Should It:

A 0.6-to-1V Inverter-Based 5-bit Flash ADC in 90nm Digital CMOS 171
Jonathan Proesel and Lawrence Pileggi

A 10~15b 60MS/s Floating-Point ADC with Digital Gain and Offset Calibration 175
Yun-Shiang Shu, Moon- Jung Kyung, Wei-Ming Lee, Bang-Sup Song and Bedabrata Pain

**Digital Correction of Dynamic Track-and-Hold Errors Providing SFDR > 83 dB up to fin
= 470 MHz** 179
Parastoo Nikaeen and Boris Murmann

A 65nm CMOS 1.2V 12b 30MS/s ADC with Capacitive Reference Scaling 183
Kang-Jin Lee, Kyoung-Jun Moon, Kwang-Sung Ma, Kyoung-Ho Moon and Jae-Whui Kim

A Continuous-time Input Pipeline ADC 187
David Gubbins, Bumha Lee, Pavan Kumar Hanumolu and Un-Ku Moon

A 0.8 V Asynchronous ADC for Energy Constrained Sensing Applications 191
Michael Trakimas and Sameer Sonkusale

Modeling, Design and Optimization of Hybrid Electromagnetic and Piezoelectric MEMS Energy Scavengers 195
Xiaochun Wu, Alireza Khaligh and Yang Xu

Amorphous Silicon Logic Circuits on Flexible Substrates 199
Rahul Shringarpure, Lawrence Clark, Sameer Venugopal, David Allee and Shrinivas Uppili

Frequency Tunable Silicon Carbide Resonators for MEMS Above IC 203
Frederic Nabki, Tomas A. Dusatko, Mourad N. El-Gamal

MEMS Wafer-Level Vacuum Packaging with Transverse Interconnects for CMOS Integration 207
Dominique Lemoine, Paul-Vahé Cicek, Frederic Nabki, Mourad El-Gamal

Variation-Tolerant Spin-Torque Transfer (STT) MRAM Array for Yield Enhancement 211
Jing Li, Haixin Liu, Sayeef Salahuddin and Kaushik Roy

Pure Logic CMOS Based Embedded Non-Volatile Random Access Memory for Low Power RFID Application 215
Liyang Pan, Xian Luo, Yaru Yan, Jirong Ma, Dong Wu and Jun Xu

A High-speed, Low-power 3D-SRAM Architecture 219
H. Henry Nho, Mark Horowitz and S. Simon Wong

Early Prediction of Product Performance and Yield Via Technology Benchmark 223
Choongyeun Cho, Daeik Kim, Jonghae Kim, Daihyun Lim and Sangyeun Cho

A FPGA Vernier Digital-to-Time Converter with 3.56ps Resolution and -0.23~+0.2LSB Inaccuracy 227
Poki Chen, Juan-Shan Lai and Po-Yu Chen

RASP 2.8: A New Generation of Floating-gate based Field Programmable Analog Array 231
Arindam Basu, Christopher M. Twigg, Stephen Brink, Paul Hasler, Csaba Petre, Shubha Ramakrishnan, Scott Koziol and Craig Schlottman

Modeling of Triple-Well Isolation and the Loading Effects On Circuits up to 50 GHz 235
Piljae Park and C. Patrick Yue

A General Weak Nonlinearity Model for LNAs 239
Wei Cheng, Anne Johan Annema, Jeroen Croon, Dick Klaassen and Bram Nauta

Modeling and Synthesis of Wide-B, Switched- Resonators for VCOs 243
Bodhisatwa Sadhu, Umaikhe Omole and Ramesh Harjani

Faster Statistical Cell Characterization using Adjoint Sensitivity Analysis 247
Ben Gu, Kiran Gullapalli, Yun Zhang and Savithri Sundareswaran

An ESD-Protected 5-GHz Differential Low-Noise Amplifier in a 130-nm CMOS Process 251
Yuan-Wen Hsiao and Ming-Dou Ker

A 65nm 3.4Gbps HDMI TX PHY with Supplyregulated Dual-tuning PLL and Blending Multiplexer 255
Jongshin Shin, Jaehyun Park, Bongjin Kim, Jongjae Ryu, Chiwon Kim, JiYoung Kim, Seung- Hee Yang, Hyungoo Kim and Jaewhui Kim,

A Scalable Digitalized Buffer for Gigabit I/O, ...259
Hung Wen Lu, Chau Chin Su, Chien-Nan Liu

Broadband, Transimpedance Amplifier in 0.35-um SiGe BiCMOS Technology for 10-Gb/s Optical Receiver Analog Front-End Application 263
Ji-Chen Huang, Yu-Sheng Lai and Klaus Yune-Jane Hsu

A Multifunction Transceiver RFIC for 802.11abg WLAN and DVB-H Applications 267
Yin Shi, Fa Foster Dai, Jun Yan, Xueqing Hu, Hua Xu, Ming Gu, Xuelian Zhang, Qiming Xu, Bei Chen, Fangxiong Chen, Peng Yu, Heping Ma, Fang Yuan and Richard Jaeger

A Fully Integrated Zero-IF Mobile TV Tuner RFIC for S-b, CMMB Application 271
Yin Shi, Fa Foster Dai, Jun Yan, Xueqing Hu, Hua Xu, Ming Gu, Xuelian Zhang, Qiming Xu, Bei Chen, Fangxiong Chen, Peng Yu, Heping Ma, Fang Yuan and Richard Jaeger

SESSION 11- COMPACT MODELING

Compact Modeling of Multiple-Gate MOSFETs (INVITED) 277
Yuan Taur, Jooyoung Song and Bo Yu

Compact Modeling and Simulation of PD-SOI MOSFETs: Current Status and Challenges (INVITED) 285
Jung-Suk Goo, Richard Williams, Glenn Workman, Qiang Chen, Sungjae Lee, Edward Nowak

Modeling Ionizing Radiation Effects in Solid State Materials and CMOS Devices (INVITED) 293
Hugh Barnaby, Michael Mclain, Ivan Esqueda and X. J. Chen

Characterization, Simulation,Modeling of FET Source/Drain Diffusion Resistance 301
Ning Lu and Bill Dewey

Analysis of the Impact of Interfacial Oxide Thickness Variation on Metal-Gate High-K Circuits 305
Minki Cho, Kingsuk Maitra and Saibal Mukhopadhyay

SESSION 12- HIGH SPEED A/D CONVERTERS

Time-Interleaved Analog-to-Digital Converters (INVITED) 311
David Nairn

A 12b 50MSPS 34mW Pipelined ADC 319
Hao Yu, Sing Chin and Bill Wong

Background ADC Calibration in Digital Domain, ...323
Cheongyuen Tsang, Yun Chiu, Johan Vanderhaegen Sebastian Hoyos, Charles Chen, Robert Brodersen and Borivoje Nikolic

A 1.2v 11b 100Msps 15mW ADC Realized using 2.5b Pipelined Stage Followed by Time Interleaved SAR in 65nm Digital CMOS Process 327
Pratap Narayan Singh, Ashish Kumar, Chandrajit Debnath and Rakesh Malik

A 52mW 10b 210MS/s Two-Step ADC for Digital-IF Receivers in 0.13μm CMOS 331
Zhiheng Cao and Shouli Yan

A 24GS/s 5-b ADC with Closed-Loop THA in 0.18μm SiGe BiCMOS 335
Jaesik Lee, Joseph Weiner, Pascal Roux, Andreas Leven and Young-Kai Chen, Alcatel-Lucent

SESSION 13: BIOMEDICAL, SENSORS AND MEMS

A Biomedical Implantable FES Battery-Powered Micro-Stimulator (INVITED) 341
Eusebiu Matei, Edward Lee, John Gord, Patrick Nercessian, Phil Hess, Howard Stover, Taihu Li and James Wolfe

CMOS LSI-based Multi-chip Flexible Retinal Prosthesis Device for Subretinal Implantation 349
Takashi Tokuda, Shigeki Sawamura, Yasuo Terasawa, Yasuo Tano and Jun Ohta,

A CMOS TDC-Based Digital Magnetic Hall Sensor Using the Self Temperature Compensation 353
Young- Jae Min and Soo-Won Kim

A Micro-Power Neural Spike Detector and Feature Extractor in .13μm CMOS...................357
Jeremy Holleman, Apurva Mishra, Chris Diorio and Brian Otis

A Compact and Programmable High-Frequency Oscillator Based on a MEMS Resonator 361
Frederic Nabki, Tomas A. Dusatko, Mourad N. El-Gamal

SESSION 14- ADVANCED SoCs- TECHNIQUES AND APPLICATIONS

Characterization and Design for Variability and Reliability (INVITED)...................367
Kevin Nowka, Sani Nassif and Kanak Agarwal

A 9Gbit/s Serial Transceiver for On-chip Global Signaling over Lossy Transmission Lines...................373
JunYoung Park, Joshua Kang, Sunghyun Park, Michael Flynn

A 28mW OFDM Baseb, Receiver Chip for DVBT/ H with All Digital Synchronization...................377
Ting-Chen Wei, Wei-Chang Liu, Chi-Yao Tseng, Syu-Siang Long, Shyh-Jye Jou and Muh-Tian Shiue

Nonvolatile Magnetic Flip-Flop for Stanby-powerfree SoCs...................381
Noboru Sakimura, Tadahiko sugibayashi, Ryusuke Nebshi, Naoki Kasai

A Fully Integrated Pulsed-LASER Time-of-Flight Measurement System with 12ps Single-Shot Precision...................385
Tino Copani, Bert Vermeire, Anuj Jain, Habib Karaki, Kailash Chandrashekar, Sushmit Goswami, Jennifer Kitchen, Hoon Hee Chung, Ilker Deligoz, Bertan Bakkaloglu, Hugh Barnaby and Sayfe Kiaei

A Low-Power IC Design for the Wireless Monitoring System of the Orthopedic Implants...................389
Hong Chen, Jia Chen, Yi Chen, Ming Liu, Chun Zhang and Zihua Wang

Tessellation-Enabled Shader for a Bandwidth- Limited 3D Graphics Engine 393
Kyusik Chung, Chang- Hyo Yu, Donghyun Kim and Lee-Sup Kim

SESSION 15- IC TECHNOLOGY – MORE MOORE AND MORE THAN MOORE

Lithography Options for the 32nm Half Pitch Node and Beyond (INVITED)...................399
Kurt Ronse, Philippe Jansen, Roel Gronheid, Eric Hendrickx, Mireille Maenhoudt, Mieke Goethals and Geert Vandenberghe

45nm High-k + Metal Gate Strain-Enhanced CMOS Transistors (INVITED) 407
Chris Auth

Will BiCMOS Stay Competitive for mmW Applications (INVITED)...................415
Patrice Garcia, Alain Chantre, Sebastien Pruvost, Pascal Chevalier, Sean Nicolson, David Roy, Sorin Voinigescu, Christophe Garnier

Microelectronics for the Real World: "Moore" versus "More than Moore" (INVITED) 423
John Kent and Jagdish Prasad

SESSION 16- EMBEDDED MEMORY

A 512-KB Level-2 Cache Design in 45 nm for sub- 2W Low Power IA Processor Silverthorne 433
Mohammed Taufique, Alex Okpisz, Haseeb Ahmed, John Riley, Mohammad Hasan, Gian Gerosa

A Voltage Scalable 0.26V, 64kb 8T SRAM with Vmin Lowering Techniques and Deep Sleep Mode 437

Tae-Hyoung Kim, Jason Liu and Chris H. Kim

Compensation of Systematic Variations Through Optimal Biasing of SRAM Wordlines 441

Andrew Carlson, Zheng Guo, Liang-Teck Pang, Tsu-Jae King Liu and Borivoje Nikolic

Variation-Tolerant SRAM Sense-Amplifier Timing Using Configurable Replica Bitlines 445

Umut Arslan, Mark P. McCartney, Mudit Bhargava, Xin Li, Ken Mai,Lawrence T. Pileggi

A 135mV 0.13µW Process Tolerant 6T Subthreshold DTMOS SRAM in 90nm Technology 449

Myeong-Eun Hwang and Kaushik Roy

Robust Ultra-Low Voltage ROM Design, ... 453

Mingoo Seok, Scott Hanson, Jae-Sun Seo, Dennis Sylvester,David Blaauw

A Million Cycle 0.13µm 1Mb Embedded SONOS Flash Memory Using Successive Approximated Read Calibration 457

Nan Wang, Xiang Yao, Yu Lei, Guoyou Feng, Qiaohua Dong, Liang Xu, Lu Guo and Zi Wang

SESSION 17- CLOCKING CIRCUITS

A Wide Tuning Range (1 GHz-to-15 GHz) Fractional-N All-Digital PLL in 45nm SOI 463

Alexander Rylyakov, Jose Tierno, George English, Michael Sperling and Daniel Friedman

Clocking Circuits for a 16Gb/s Memory Interface, ... 467

Ting Wu, Xudong Shi, Kambiz Kaviani, Haechang Lee, Jung-Hoon Chun, TJ Chin, Jie Shen, Rich Perego and Ken Chang

A 1V 15.6mW 1-2GHz -119dBc/Hz @ 200kHz clock multiplying DLL 471

Sander Gierkink

A 0.5-to-2.5GHz Supply-Regulated PLL with Noise Sensitivity of -28dB 475

Abhijith Arakali, Srikanth Gondi,Pavan Kumar Hanumolu

20 GHz Low Power QVCO and De-skew Techniques in 0.13µm Digital CMOS 479

Masum Hossain and Anthony Chan Carusone

A 3 GHz Spread Spectrum Clock Generator for SATA Applications Using Chaotic PAM Modulation 483

Fabio Pareschi, Gianluca Setti and Riccardo Rovatti

A 1.5 GHz Spread Spectrum Clock Generator with 5000ppm Piecewise Linear Modulation 487

Minyoung Song, Sunghoon Ahn, Inhwa Jung, Yontae Kim and Chulwoo Kim

A 8x5 Gb/s Source-Synchronous Receiver with Clock Generator Phase Error Correction 491

Ankur Agrawal, Pavan Kumar Hanumolu and Gu-Yeon Wei

SESSION 18- MILLIMETER-WAVE CIRCUIT TECHNIQUES

Millimeter-wave CMOS Integrated Circuits for Gigabit WPAN Applications (INVITED) 497

Tian-Wei, Huang and Huei Wang

A Zero-IF 60GHz Transceiver in 65nm CMOS with > 3.5Gb/s Links 505

Alexander Tomkins, Ricardo A. Aroca, Takuji Yamamoto, Sean T. Nicolson, Yoshiyasu Doi and Sorin P. Voinigescu

Low-Cost Fully Integrated BiCMOS Transceiver for Pulsed 24-GHz Automotive Radar Sensors..509
Laurence Moquillon, Patrice Garcia, Sebastien Pruvost, Stephane Le Tual, Maxime Marchetti,
Laurent Chabert, Nicole Bertholet, Angelo Scuderi, Salvatore Scaccianoce, Alberto Serratore,
Nicolo Ivan Piazzese, Cedric Dehos and Domique Morche

A 0.13um CMOS Fully Differential Receiver with On-Chip Baluns for 60GHz Broadb, Wireless Communications..513
Chao-Shiun Wang, Juin-Wei Huang, Kun-Da Chu and Chorng-Kuang Wang

A Low power 20 GHz 1.5 Gb/s CMOS Injection- Pulling FSK Modulator and Frequency Discriminator for 60GHz Links...517
Shon-Hang Wen, Chao-Shiun Wang and Chorng-Kuang Wang

A 24/77GHz Dual-B, BiCMOS Frequency Synthesizer 521
Vipul Jain, Babak Javid and Payam Heydari

An X/Ku-B, Frequency Synthesizer Using A 9- Bit Quadrature DDS 525
Xuefeng Yu, Fa Foster Dai, Dayu Yang, J. David Irwin, Richard C. Jaeger,

A Dynamic Offset Control Technique for Comparator Design in Scaled CMOS Technology.................531
Xiaolei Zhu, Yanfei Chen, Masaya Kibune, Yasumoto Tomita, Takayuki Hamada, Hirotaka Tamura,
Sanroku Tsukamoto, Tadahiro Kuroda,

A Low-Voltage OP Amp with Digitally Controlled Algorithmic Approximation ...535
Dong-Woo Jee, Seung- Jin Park, Hong-June Park and Jae-Yoon Sim,

A 105.5 dB, 0.49 mm2 Audio Sigma Delta Modulator using Chopper Stabilization and Fully Randomized DWA..539
Yi-Gyeong Kim, Min-Hyung Cho, Kwi-Dong Kim, Jong-Kee Kwon and Jongdae Kim

An Ultra Low Power 1V, 220nW Temperature Sensor for Passive Wireless Applications......................543
Yu- Shiang Lin, Dennis Sylvester and David Blaauw

A 9-Bit Configurable Current Source with Enhanced Output Resistance for Cochlear Stimulators...547
Song Guo, Hoi Lee and Philipos Loizou

Super-Resolution: Imaging beyond the Pixel Size Limit 551
Tamer Elkhatib and Khaled Salama

Polysilicon Vertical Actuator Powered with Waste Heat 555
Jorge Varona, Margarita Tecpoyotl-Torres and Anas Hamoui

Low Noise uWatt Interface Circuits for Wireless Implantable Real-Time Digital Blood Pressure Monitoring 559
Peng Cong, Wen Ko and Darrin Young

A Flexible Decoder IC for WiMAX QC-LDPC Codes...563
Tzu-Chieh Kuo and Alan Willson

Minimizing the Supply Sensitivity of CMOS Ring Oscillator by Jointly Biasing the Supply Control Voltage 567
Ping-Hsuan Hsieh, Jay Maxey, Chih-Kong Ken Yang

The Superchip: Innovative Teaching of IC Design and Manufacture....................571
Peter Wilson, Reuben Wilcock, Matthew Swabey, Iain McNally and Bashir Al-Hashimi

A 3D Graphics Processor with Fast 4D Vector Inner Product Units and Power Aware Texture Cache 575
Jae-Sung Yoon, Donghyun Kim, Chang-Hyo Yu and Lee-Sup Kim

Timing Yield Enhancement Through Soft Edge Flip-Flop Based Design 579
Michael Wieckowski Young, Min Park, Carlos Tokunaga, Dong Woon Kim, Zhiyoong Foo, Dennis Sylvester and David Blaauw

1/5 Power Reduction by Global Optimization based on Fine-GrainedBody Biasing 583
Yasumi Nakamura, David Levacq, Limin Xiao, Takuya Minakawa, Taro Niiyama, Makoto Takamiya and Takayasu Sakurai

A DC-DC Converter with a Dual VCDL-based ADC and a Self-Calibrated DLL-based Clock Generator for an Energy-Aware EISC Processor 587
Sunghwa Ok, Jungmoon Kim, Gilwon Yoon, Hyunho Chu, Jaegeun Oh, Seon Wook Kim and Chulwoo Kim

Active Autonomous AC-DC Converter for Piezoelectric Energy Scavenging Systems 591
Enrico Dallago, Daniele Miatton, Giuseppe Venchi, Valeria Bottarel, Giovanni Frattini, Giulio Ricotti, Monica Schipani

A 10Gb/s Receiver with Linear Backplane Equalization and Mixer-Based Self-Aligned CDR 595
Simone Erba, Massimo Pozzoni, Matteo Pisati, Riccardo Brama, Davide Sanzogni, Emanuele Depaoli, Paolo Viola and Francesco Svelto

A 12.5-Gbps, 7-bit Transmit DAC with 4-Tap LUTbased Equalization in 0.13µm CMOS 599
Hayun Chung, Andrew Liu and Gu-Yeon Wei

Active Deskew in Injection-Locked Clocking 603
Lin Zhang, Berkehan Ciftcioglu and Hui Wu

A 53 GHz DCO for mm-Wave WPAN, ...607
Raffaella Genesi, Francesco M. De Paola and Danilo Manstretta

Twisted Inductors for Low Coupling Mixed-Signal and RF Applications 611
Nathan Neihart, David Allstot, Matt Miller, Pat Rakers

A Common-Base Linear RF Power Amplifier for 3G Cellular Applications 615
Flavio Avanzo, Francesco M De Paola and Danilo Manstretta

Design of Low Power CMOS Ultra-Wideb, 3.1- 10.6 GHz Pulse-Based Transmitters 619
Kuan-Yu Lin and Mourad El-Gamal

SESSION 19- LOW POWER AND NON-TRADITIONAL RF TRANCEIVERS

Energy-efficient Wireless Front-end Concepts for Ultra Low Power Radio (INVITED) 625
John R. Long, Wanghua Wu, Yunzhi Dong, Yi Zhao, Mihai A.T. Sanduleanu, John F.M. Gerrits and Gerrit van Veenendaal

A 0.4 nJ/b 900MHz CMOS BFSK Super- Regenerative Receiver 629
James Ayers, Kartikeya Mayaram and Terri Fiez

A 900-MHz Low-Power Transmitter With Fast Frequency Calibration For Wireless Sensor Networks 633
Napong Panitantum, Kartikeya Mayaram, Terri Fiez

A 3.5-mW 15-Mbps O-QPSK Transmitter for Realtime Wireless Medical Imaging Applications 637
Yao- Hong Liu and Tsung-Hsien Lin

A CMOS Direct Conversion Transmitter With IEEE 802.22 Cognitive Radio Applications 641
Jongsik Kim, Seungsoo Kim, Jaewook Shin, Youngcho Kim, Junki Min, Kihong Kim and Hyunchol Shin,

A 5-GHz Wireless LAN Transmitter with Integrated Tunable High-Q RF Filter 645
Robert Wiser, Masoud Zargari, David Su and Bruce Wooley

SESSION 20- ADVANCED WIRELINE TECHNIQUES

A 6.5 Gb/s Backplane Transmitter with 6-tap FIR Equalizer and Variable Tap Spacing 651
Mike Bichan and Anthony Chan Carusone

Phase-Locking in Wireline Systems: Present and Future (INVITED) 655
Behzad Razavi

A 20Gb/s SerDes Transmitter with Adjustable Source Impedance and 4-tap Feed-Forward Equalization in 65nm Bulk CMOS 663
Rick Philpott, James Humble, Robert Kertis, Karl Fritz, Barry Gilbert,Erik Daniel

Wideband, mmWave CML Static Divider in 65nm SOI CMOS Technology (INVITED) 667
Daeik Kim, Choongyeun Cho, Jonghae Kim,Jean-Olivier Plouchart

A 32/16 Gb/s 4/2-PAM Transmitter with PWM Pre- Emphasis and 1.2 Vpp per side Output Swing in 0.13-µm CMOS 675
Horace Cheng and Anthony Chan Carusone

A 5-Gb/s/pin Transceiver for DDR Memory Interface with a Crosstalk Suppression Scheme 679
Kwang-Il Oh, Lee-Sup Kim, Kwang-Il Park, Young- Hyun Jun and Kinam Kim

EMI Resisting Smart-power Integrated LIN Driver with Reduced Slope Pumping 683
Jean-Michel Redouté and Michiel Steyaert

SESSION 21- LEVERAGING THE THIRD DIMENSION

Stacking Technology Based on 8-inch Wafers using Direct Connection between TSV and Microbump 689
Nobuaki Miyakawa, Eiri Hashimoto, Takanore Maebashi, Natsuo Nakamura, Yutaka Sacho, Shigeto Nakayama,Shinjiro Toyoda

Clock Distribution Networks for 3-D Integrated Circuits 693
Vasilis Pavlidis, Ioannis Savidis and Eby Friedman

Inter-Die Signaling in Three Dimensional Integrated Circuits 697
Christopher Mineo, Ravi Jenkal, Samson Melamed and W. Rhett Davis

Variability in 3-D Integrated Circuits, ...701
Filipp Akopyan, Carlos Tadeo Ortega Otero, David Fang, Sandra Jackson and Rajit Manohar

3D Heterogeneous Integrated Systems: Liquid Cooling, Power Delivery,Implementation (INVITED) 705
Muhannad Bakir, Calvin King, Deepak Sekar, Hiren Thacker, Bing Dang, Gang Huang, Azad Naeemi and James Meindl

SiP for GSM/EDGE in CMOS Technology, ...713
Giuseppe Li Puma, Ernst Kristan, Paolo De Nicola, Cyril Vannier, Braam Greyling and Salvatore Piccolella

Heterogeneous Multicore SoC for Secure Multimedia Applications 717
Hiroyuki Kondo, Masami Nakajima, Sugako Otani, Osamu Yamamoto, Norio Masui, Naoto Okumura, Mamoru Sakugawa, Masaya Kitao, Koichi Ishimi, Masayuki Sato, Fumitaka Fukuzawa, Kazuhiro Inaoka, Yoshihiro Saito, Kazutami Arimoto and Toru Shimizu

A 159.2mW SoC Implementation of T-DMB Receiver including Stacked Memories............................721
Joohyun Lee, Sungdo Kim, Jinkyu Kim, Duckhwan Kim, Youngsu Kwon, Minseok Choi, Kihyuk Park, Bontae Koo, Nakwoong Eum and Hyuckjae Lee

SESSION 22- NOISE AND OSCILLATOR SIMULATION

Modeling, Measurement and Mitigation of Crosstalk Noise Coupling in 3D-Ics........................727
Liuchun Cai,Ramesh Harjani

A Method Using Circuit/Substrate Macro Modeling to Analyze Substrate Noise in a 3.2-GHz 350Mtransistor Microprocessor..731
Mikiki Sode, Mikihiro Kajita, Naoya Nakayama and Satoshi Nakamoto,

Characterization of Random Decision Errors in Clocked Comparators............................735
Brian Leibowitz, Jaeha Kim, Jihong Ren and Christopher Madden

Noise Tolerant Oscillator Design Using Perturbation Projection Vector Analysis............739
Igor Vytyaz, Josh Carnes, Ting Wu, Pavan Hanumolu, Un-Ku Moon and Kartikeya Mayaram

Strong Injection Locking of Low-Q LC Oscillators,..743
Mozhgan Mansuri, Frank O'Mahony, Ganesh Balamurugan, James Jaussi, Joseph Kennedy, Sudip Shekhar, Randy Mooney and Bryan Casper

SESSION 23- ANALOG TECHNIQUES

A Low Power 1.3GHz Dual-Path Current Mode Gm-C Filter......................................749
Manisha Gambhir, Vijay Dhanasekaran, Jose Silva-Martinez and Edgar Sanchez-Sinencio

A 1V Downconversion Filter Using Duty-cycle Controlled Bandwidth Tuning................753
Peter Kurahashi, Pavan Kumar Hanumolu and Un-Ku Moon

A Reconfigurable FIR Filter Embedded in a 9b Successive Approximation ADC............757
Joshua Kang, David Lin, Li Li and Michael Flynn

Voltage References for Ultra-Low Supply Voltages (INVITED)..................................761
Peter Kinget, Christos Vezyrtzis, Ed Chiang, B. Hung and T.L. Li

Design of Bandgap Voltage Reference Circuit with all TFT Devices on Glass Substrate in a 3-μm LTPS Process...767
Ting-Chou Lu, Ming-Dou Ker, Hsiao-Wen Zan, Chung-Hung Kuo, Chun-Huai Li, Yao-Jen Hsieh and Chun-Ting Liu

Deep Submicron Effects on Data Converter Building Blocks (INVITED)......................771
William Evans and David Burnell

SESSION 24- ADVANCED SUBSYSTEMS FOR CONNECTIVITY AND CELLULAR RADIO

A Highly Linear SAW-less CMOS Receiver Using a Mixer with Embedded Tx Filtering for CDMA..777
Namsoo Kim, Lawrence E. Larson and Vladimir Aparin

High-power Digital Envelope Modulator for a Polar Transmitter in 65nm CMOS............781
Manel Collados, Paul T.M. van Zeijl and Nenad Pavlovic

A 2.4GHz, 20dBm Class-D PA with Single-Bit Digital Polar Modulation in 90nm CMOS......785
Jason T. Stauth and Seth R. Sanders

Linearity and Efficiency Enhancement Strategies for 4G Wireless Power Amplifier Designs (INVITED) 789
Larry Larson, Donald Kimball and Peter Asbeck

An Analog Enhanced All Digtial RF Fractional-N PLL With Self-Calibrated Capability 797
Ping-Ying Wang, Jing-Hong Zhan, Hsiang-Hui Chang and Bing- Yu Hsieh

A Delta-Sigma Fractional-N Synthesizer with Customized Noise Shaping for WCDMA/HSDPA Applications 801
Xueyi Yu, Yuanfeng Sun, Woogeun Rhee, Zhihua Wang, Hyung Ki Ahn, Byeong-Ha Park

Author Index

Welcome from the CICC Committee

Welcome to CICC 2008, the 30th annual IEEE Custom Integrated Circuits Conference and leading international conference for integrated circuit development at the DoubleTree Hotel in San Jose, California. In addition to the core technical lecture and poster presentations, CICC offers attendees a total educational experience with educational sessions, a keynote address, a special luncheon lecture, exhibits, panels, tutorials, and stimulating networking events. The conference begins with a keynote address entitled "More Than Moore" by Dave Bergeron, CEO of SVTC. He will describe directions in semiconductor innovation that leverage older technology nodes to develop novel functions in passive devices, high- and low-voltage transistors and a host of other products. Our conference luncheon guest speaker, Dr. Alberto Sangiovanni-Vincentelli will speak on the use of IC's in automotive design.

Ann Marie Rincon
General Chair

The conference begins on Sunday, September 21 with educational sessions taught by practicing experts working at the leading edge of their fields. The technical session themes are; The Fundamentals of Analog Design, High-Speed Serial IO Design, and Coping with Technology Scaling. This year we also added a special professional development session on Effective Presentations and Technical Writing.

The Custom Integrated Circuits Conference has grown and changed with the industry and continues to showcase technical papers describing the most advanced analog and digital circuits and their applications. This year's program will start with the keynote address on "More Than Moore" by Dave Bergeron, CEO of SVTC. For over thirty years productivity has been driven primarily by Moore's Law, providing twice the number of circuits on a piece of silicon every 18-24 months. Progress has been measured by increased density and performance. While it's true that scaling continues and 22nm IC's are not far away, progress today is often measured by low power consumption and 3-D integration for portable applications, green technologies that preserve the environment, the ability to operate in hostile high temperature/high vibration environments, and high reliability solutions for life-critical applications. The evolution of technology to solve these issues has been dubbed "More than Moore." A panel discussion on Monday afternoon titled "Sure, Moore's Law Can Continue, But Should It?" will further debate issues around this growth area.

David Nairn
Conference Chair

Jacqueline Snyder
Technical Program
Chair

The paper submissions in 2008 allow us to bring you a conference of the highest technical caliber. This year 120 papers were selected from 364 submissions and organized into 24 sessions. The topics addressed by these high quality papers include the latest innovations in 3-D circuits, ADC's, sensors and displays, SOC's, wireless circuits, power management, high performance wired interfaces, PLLs, embedded memories, simulation and modeling, design test and debug issues, and manufacturing developments. Highlights include invited and tutorial papers from leading experts in industry and academia.

Our Monday evening Welcome Reception and Tuesday Conference Reception are professional networking at its very best! Our exhibits area will include booths from prominent industry suppliers and Poster Sessions. The poster session is a unique forum for in-depth discussions with authors.

CICC is co-located with the IEEE Behavior, Modeling, and Simulation Conference 2008. BMAS will take place September 25– 26 at the DoubleTree Hotel, San Jose, California. Visit the BMAS website at www.bmas-conf.org for complete conference information.

Steering Committee

Henry Chang
Designer's Guide Consulting

David Nairn
University of Waterloo

Ann Rincon
ON Semiconductor

Jacqueline Snyder
Marvell Semiconductor

Trudy Stetzler
Texas Instruments

Larry Wissel
IBM

Organizing Committee

General Chair
Ann Rincon, ON Semiconductor

Conference Chair
David Nairn, University of Waterloo

Technical Program Chair
Jacqueline Snyder, Marvell Semiconductor

Educational Sessions Chair
Shahriar Mirabbasi, University of British
Columbia

Exhibits Chair
Tom Andre, EverSpin Technologies

Panel Chair
Aurangzeb Khan, Consultant

Publicity Chair
Arif Rahman, Xilinx Research Labs

Sponsorship Chair
Eric Naviasky, Cadence Design

Best Paper Awards
Jennifer Lloyd, Analog Devices

Treasurer
Trudy Stetzler, Texas Instruments

Technical Program Committee

Analog Circuit Design
Seated left to right: George LaRue, Washington State University, David Nairn, University of Waterloo, Donald Thelen, ON Semiconductor, Jennifer Lloyd, Analog Devices, *Standing left to right:* Takahiro Miki, Renesas, Ken Suyama, Epoch Microelectronics, University, Yusuf Haque, Consultant, *Not Pictured :* Yun Chiu, University of Illinois, Eric Naviasky, Cadence

Biomedical, Sensors, Displays, and MEMS
Standing left to right: Ken Szajda, LSI, , Dawn Fitzgerald, Aurora Enterprises, Steve Garverick, Case Western University, *Standing left to right:*, Makoto Nagata, Kobe University, Mourad El Gamal, McGill University, Edward Lee, Alfred Mann Foundation, *Not Pictured:* Sang-Soo Lee, Pixelplus Semiconductor

Characterization, Debug and Test
Seated left to right: Gordon Roberts, McGill University, R Hamid Mahmoodi, San Francisco State University, Jeanne Trinko Mechler, IBM, Mike Li, Altera,

Digital Circuits and SoC-SIP Designs and Methodology
Seated left to right: Charles Thomas, NICTA, Arif Rahman, Xilinx, Rakesh Patel, Altera Corp., Raj Amirtharajah, University of California, Davis, Henry Chang, Designer's Guide Consulting, *Standing left to right:* Ann Marie Rincon, ON Semiconductor, Aurangzeb Khan, Consultant, Paul Billig, Cavendish Kinetics, Osamu Takahashi, IBM, Steve Wilton, University of British Columbia, Ric Williams, Sun Microsystems, Mike Seningen, Intrinsity, *Not Pictured:* Ram Krishnamurthy, Intel

Embedded Memory
Seated left to right: Takashi Akioka, Renesas Technology, Tom Andre, EverSpin Technologies, Kenji Noda, NScore, Jean-Christophe Vial, Infineon, *Not Pictured:* Subramani Kengeri, TSMC, Larry Wissel, IBM

Manufacturing
Seated left to right: Philipe Jansen, Infineaon, Jordan Lai, TSMC, David Sunderland, Boeing Satellite Systems, *Not Pictured:* Rich Liu, Macronix, Alvin Loke, AMD

Power Management
Seated left to right: Jerry Zheng, Iwatt, Makoto Takamiya, University of Tokyo, Gordon Lee, Qualcomm, Vikas Chandra, ARM, Lawrence Clark, Arizona State University

Simulation and Modeling
Seated left to right: Gennady Gildenblat, Arizona State University, Hidetoshi Onodera, Kyoto University, Hong-Ha Vuong, LSI, *Standing left to right:* Colin McAndrew, Freescale Semiconductor, Inc., Rob Jones, IBM, Brian Qiang Chen, AMD, *Not Pictured:* Larry Nagel, Omega Enterprises

Wired Communications
Seated left to right: Ed van Tuijl, Philips Research, Tony Chan Carusone, University of Toronto, Kimo Tam, Analog Devices, *Standing left to right:* Ken Chang, Rambus, Shahriar Mirabbasi, University of British Columbia, Dennis Fischette, AMD, *Not Pictured:* Jin Liu, University of Texas, Dallas, Cormac O'Connell, TSMC

Wireless Designs
Seated left to right: Nobuyuki Itoh, Toshiba, Ramesh Harjani, University of Minnesota, Trudy Stetzler, Texas Instruments, John Rogers, Carleton University, Andrea Mazzanti, Universita di Modena, *Standing left to right:* Earl McCune, Panasonic, Ranjit Gharpurey, University of Texas, Howard Luong, Hong Kong University of Science and Technology, Fa Foster Dai, Auburn University, Stefan Drude, NXP Semiconductors, *Not Pictured:* Payam Heydari, University of California, Irvine, Cicero Vaucher, NXP Semiconductors

CONFERENCE OVERVIEW

	OAK BALLROOM	FIR BALLROOM	PINE BALLROOM	CEDAR BALLROOM	OTHER ROOMS	BAYSHORE FOYER	DONNER BALLROOM
SUNDAY, SEPTEMBER 21							
8:00 am - 5:00 pm	Ed Session 1 - High Speed I/O	Ed Session 2 - Coping with Technology Scaling	Ed Session 3 - Fundamentals of Analog Design		Silicon Valley Room 2:00 PM – 5:00 PM Effective Technical Writing and Presentations Session	Ed Session Registration 7:30 am – 2:00 pm Technical Session Registration 2:00 pm - 5:00 pm	
MONDAY, SEPTEMBER 22							
7:30 am - 5:00 pm						Technical Session Registration 7:30 am - 5:00 pm	
8:15 am - 9:30 am	1. Keynote Presentation						Exhibits Open
10:00 am - 12:00 pm	2. Statistical Modeling	3. Power Management	4. High-Speed Test, Characterization, & Debug.	5. Broadband Circuit Techniques for Emerging Wireless Communications I			4:00 – 8:00 pm
1:30 - 5:30 pm	6. Advanced SoC/SiP Integration & Co-Design	7. High Resolution Converters	8. Characterization & Test Methods for Device Variability in Nanoscale Technologies 10. Panel Discussion 4:00 pm – 5:30 pm	9. Broadband Circuit Techniques for Emerging Wireless Communications II			
5:00 pm - 7:00 pm							Poster Session
5:30 pm - 8:00 pm							Welcome Reception
TUESDAY, SEPTEMBER 23							
8:00 am - 5:00 pm						Technical Session Registration 8:00 am - 5:00 pm	Exhibits Open
8:25 am - 12:00 pm	11. Compact Modeling	12. High Speed A/D Converters	13. Biomedical, Sensors and MEMS	14. Advanced SoCs – Techniques and Applications			4:00 – 8:00 pm
12:00 pm – 1:50 pm					Cedar Ballroom - CICC Luncheon		
2:00 pm –5:00 pm	15. IC technology – More Moore and More Than Moore	16. Embedded Memory	17. Clocking Circuits	18. Millimeter-Wave Circuit Techniques			
5:00 pm – 7:00 pm							Poster Session
5:30 pm – 8:00 pm							Conference Reception
WEDNESDAY, SEPTEMBER 24							
8:00 am - 3:00 pm						Technical Session Registration 8:00 am - 3:00 pm	
8:25 am - 12:00 pm	19. Low Power and Non-traditional RF Tranceivers	20. Advanced Wireline Techniques	21. Leveraging the Third Dimension				
1:30 pm – 5:00 pm	22. Noise and Oscillator Simulation.	23. Analog Techniques	24. Advanced Subsystems for connectivity and Cellular Radio				

EDUCATIONAL SESSION 1   Oak Ballroom

Chairperson: Shahriar Mirabbasi, University of British Columbia

High-Speed I/O

Organizer: Tony Chan Carusone, University of Toronto
Co-Organizer: George LaRue, Washington State University

9:00 am - 10:50 am
E1-1 Multi-Gigabit I/O Design for Microprocessor Platforms
Randy Mooney (Intel)
A discussion of the constraints of the design space for microprocessor platforms, and a look at the state of the technologies required to deliver bandwidth in these platforms. These technologies include analysis tools, interconnect components, modulation and equalization choices that fit the constraints, and the various circuits required in silicon. This is followed by a look at future platform requirements, and the potential solution space to meet those needs.

11:10 am - 1:00 pm
E1-2 Jitter and Signal Integrity at 10 Gbps
Mike Li (Altera Corp.)
In this tutorial presentation, we will first review where the technology is heading to for the multiple Gbps high-speed links and I/O buses for devices and systems in networks and computers. Second, we will discuss why jitter and signal integrity have become the major challenges, as well as limiting factors for developing those high-speed, high performance, high volume, and low-cost I/O devices and systems as the data rate approaches 10 Gbps and beyond. Third, we will discuss the jitter and signal integrity modeling, simulation, verification and characterization methodologies within the context of a serial link. We will cover these ever evolving cutting edge topics from generic perspective, as well as practical application perspective, with real-world examples from multiple Gbps link technologies such as Giga Bit Ethernet (GBE), PCI Express (PCIe), Fibre Channel (FC), with emphasis on their latest generations operating at single lane data rates in the vicinity of 10 Gbps. Emerging challenges such as jitter amplification and mitigation, equalization optimization and verification, on-chip jitter de-embedding will also be covered.

2:00 pm - 3:50 pm
E1-3 Equalization & High-Speed Transceiver Design
Jared Zerbe (Rambus)
Equalization is an ever-critical aspect of serial data systems and is even beginning to expand into high-speed parallel systems. This tutorial provides a basic overview of the serial data transmission problem and the goals of equalization, along with some of the practical challenges at high speeds and some vision of its future. The architecture of equalized systems is explained, with detail on key types of equalizers such as linear receive equalizers, transmitter pre-emphasis, and DFE, along with the pros and cons of each type. The tutorial will also teach how various equalizer components can be used together to mitigate each other's weaknesses. Equalizers will be presented from various viewpoints, including effectiveness and practical circuit design as well as future trends. Effectiveness of different approaches on different practical environments will be compared. Simulation approaches for equalization will be discussed. Finally, as an alternate to equalization, certain modulation approaches such as 4-PAM & duo-binary will be covered and pros and cons reviewed.

4:10 pm - 6:00 pm
E1-4 Clocking and CDRs
Jafar Savoj (Qualcomm)
High-purity clock generation enables longer reach in wired communication systems. Optical and copper standards set an upper bound on the maximum noise and distortion added to the signal at the source. This tutorial describes means of efficient clock generation and distribution in high-performance chips to satisfy requirements imposed by the standards. Clock and data recovery (CDR) circuits are an integral part of wired communication systems. With the accelerated rate of device scaling in recent decades, CDR architectures have transitioned from fully analog into mixed-mode and digital implementations. The tutorial later addresses the evolution of CDR architectures, as well as the design of their building blocks.

EDUCATIONAL SESSION 2   Fir Ballroom

Coping with Technology Scaling

Organizer: Colin McAndrew, Freescale Semiconductor
Co-Organizer: Foster Dai, Auburn University

9:00 am – 10:50 am
E2-1 Technology and Reliability
Paul Packan (Intel)
Continued technology scaling drives not only developments and changes in device designs, characteristics, and performance, but also involves an ever expanding complexity of constraints that must be applied and phenomena, like variability and stress, that must be taken into account to enable the design of billion+ transistor ICs. This tutorial will review nMOS and pMOS scaling trends and their impact on circuit performance benchmarks, including power. Sources of variability and their impact on circuit performance at scaled supply voltages will be reviewed, as will circuit level techniques to mitigate problems caused by variability. The impact of both device and circuit architectures will be discussed, as related to memories, RF circuits, power consumption, and performance. Scaling has also lead to issues with reliability, and these will be reviewed. Finally, because of the growing complexity of restrictions that must be applied to make functions ICs, lithography, layout, and design rule issues that affect the manufacturability of designs will be discussed.

11:10 am – 1:00 pm
E2-2 Logic and Memory Scaling Challenges
Bora Nikolic (University of California, Berkeley)
Digital logic and memory are expected to scale down in area by 50% with each new technology node. This is the only key benefit of technology scaling as the active and leakage power limit the rate of further logic speed increase. This tutorial will address the main challenges and known solutions for keeping the expected scaling rate: increased cost of design and manufacturing, design under power limitation, impact of technology variability, and design with added technology features.

2:00 pm – 3:50 pm
E2-3 CAD and Modeling Issues
Sani Nassif (IBM)
Technology scaling is not just a problem for the manufacturing engineers; it presents unique challenges for those who must use this same technology to produce working high performance chips in volumes that can lead to profit. Activities like OPC and DFM have become common place terms for designers and EDA engineers, and are all part of the response to the increasing complexity of the design/manufacturing interface. This interface has historically been defined by layout design rules and so-called corner models. Both of these representations are unraveling as we enter the 45nm node with thousands of design rules, and with overall manufacturing variability becoming the most significant challenge faced by design. In this tutorial, we will review the design/manufacturing interface and show current trends, explain how technology characterization and modeling leads to specific challenges for the representation of technology in simulation tools, and finally review some of the design responses to technology scaling that leverage adaptivity and regularity.

4:10 pm – 6:00 pm
E2-4 Analog and RF Design Issues in Deep Submicron CMOS Technology
Behzad Razavi (University of California, Los Angeles)
This tutorial presents the challenging issues in analog and RF design as technology nodes go beyond 65 nm and 45 nm. Noise-power-speed and mismatch-power-speed trade-offs resulting from supply scaling are quantified and the effect of switch nonlinearities in sampling circuits is formulated. Phenomena such as output resistance nonlinearity and the gate leakage current are studied and their impact on circuits such as PLLs and op amps is summarized. Noise-linearity trade-offs in passive and active RF mixers and various deep-submicron effects in LC oscillators are also presented and low-voltage circuit techniques are described.

EDUCATIONAL SESSION 3   Pine Ballroom

Fundamentals of Analog Design

Organizer: Sang Soo Lee, Hynix Semiconductor

9:00 am - 10:50 am
E3-1 Amplifiers
Boris Murmann (Stanford University)
This lecture covers a systematic methodology for the design of high performance operational transconductance amplifiers (OTAs) in deep sub-micron CMOS technology. The first part of this presentation reviews the basic design equations and power/speed/noise tradeoffs in OTAs using a two-stage Miller-compensated design as an example. In the second part, Spice-generated look-up tables are introduced as a means to bridge the gap between simulation, hand analysis and Matlab optimization. Using tabulated device data that captures the fundamental tradeoff between speed (gm/Cgg) and transconductance efficiency (gm/ID), the proposed method yields near-optimal designs without the need for iterative Spice simulations or expensive CAD tools.

11:10 am - 1:00 pm
E3-2 References
Wing-Hung Ki (Hong Kong University of Science and Technology)
In this lecture, the treatment of voltage references that is systematic and coherent, rigorous but not excessive is attempted. The talk starts with fundamentals of voltage references. Popular bandgap references (BGRs) are then discussed, with emphasis on CMOS bandgap references using parasitic BJTs in a CMOS process. Performance parameters such as temperature coefficient, power supply rejection, line and load regulation, and loop gain are introduced. For BGR with simple structures, analytic results on loop gain and power supply rejection are presented. The development of op-amp based BGR for reducing effect due to op-amp input offset voltage, folded resistor for lowering power supply voltage (Vdd) requirement, and folded resistor divider for further lowering Vdd requirement, are traced. Non-op-amp based BGRs are discussed, starting with the 4T current-voltage-mirror (CVM) scheme in replacing the op-amp. The principle of symmetrical matching is then introduced to minimize systematic errors due to channel length modulation, and an 8T symmetrically matched CVM is used to realize a BGR with improved power supply rejection. The BGRs discussed are designed using a 0.18 procedure and simulation results are presented. Design issues such as trimming, resistor strings and organization of voltage references in an IC system are also sketched.

2:00 pm - 3:50 pm
E3-3 PLL
Behzad Razavi (UCLA)
This tutorial deals with the analysis and design fundamentals of PLLs. Various voltage-controlled ring oscillator topologies are described that can be used in timing applications up to several gigahertz. Next, type I PLLs and their shortcomings are studied, leading to type II (charge-pump) PLLs as a superior choice. The dynamics of the PLLs are derived, the effect of various charge pump nonidealities is presented, and circuit techniques for alleviating these effects are summarized. Lastly, a design procedure for PLLs is outlined and demonstrated by a transistor-level implementation.

4:10 pm - 6:00 pm
E3-4 DAC
Doug Mercer (Analog Devices)
Modern communication systems have spawned a growing interest in high performance, high speed Digital to Analog Converter designs which can be easily embedded into larger mixed signal systems. Implementing larger systems in addition require peripheral support D/A functions outside the main signal path in applications such as tuning and calibration. The tutorial will concentrate on D/A converter design in MOS process technologies and cover these three topics.

    1) A brief look Digital to Analog conversion first principles including a description of the D/A function and the key specifications that define the performance of a D/A.
    2) Common D/A architectures will be explored with these first principles in mind. The advantages and disadvantages of each will discussed.
    3) Case studies of example CMOS implementations will be included.

SPECIAL AFTERNOON WORKSHOP   Silicon Valley Room

Effective Technical Writing and Presentations

2:00 PM – 5:00 PM
Ann Marie Rincon (ON Semiconductor)

My technical work is outstanding - why didn't my paper get accepted? I thought my description was very clear - why was my thesis misunderstood?

This class will provide answers to these questions and help engineers and programmers write clear, concise technical papers. The writing do's and don'ts covered in this class can be applied to other technical documents such as application notes, product specifications and emails.

The class will provide:
       A standard technical paper outline and a description of each section
       Tips for submitting a paper to an external conference
       General writing tips including do's and don'ts
       Tips for translating your technical paper into an effective presentation
       Several lucky attendees will receive a copy of "The Elements of Style" by William Strunk Jr. and E.B. White.

Keynote Presentation   Session 1

8:15 am   **Welcome and Opening Remarks**
                 Awards Presentations
                 Keynote Speaker Introduction
                 Ann Marie Rincon, General Chairman

8:30 am   **Keynote Presentation**

"More Than Moore"
Dave Bergeron, CEO, SVTC Technologies

There's plenty of life left in Moore's Law, as today's leading-edge semiconductor developments will attest. However, there is a new direction in semiconductor innovation that takes advantage of old technology nodes. It's called "More than Moore."

This approach leverages the CMOS backbone established processes and technologies to develop novel function in passive devices, high- and low-voltage transistors, MEMS and a host of other products.

In this keynote, Dave Bergeron describes the evolving infrastructure and expanding marketplace, and lays out options and strategies for "More than Moore" players in today's semiconductor industry.

Dave Bergeron is the Chief Executive Officer, SVTC Technologies. Dave recently served as Executive-in-Residence at Tallwood Venture Capital, where he evaluated semiconductor chip products and equipment opportunities. Prior to joining Tallwood, Dave held senior management positions at Applied Materials, Candescent Technologies and IBM Microelectronics. These positions included VP and General Management responsibilities for a semiconductor equipment product line; semiconductor fab operation management responsibilities, including CMOS DRAM and CMOS LOGIC product development; and multiple technology product integration responsibilities supporting an advanced display opportunity. Dave has authored 17 U.S. patents grants and has published more than 30 technical bulletins and papers. Dave received a B.S. in Physics and an M.S. in Applied Mathematics from Georgetown University.

Notes

Statistical Modeling   Session 2

Chair: Hidetoshi Onodera, Kyoto University
Co-Chair: Hong-Ha Vuong, LSI

With device dimensions in the nanometer regime, variability is now a serious concern in LSI design. Aggressive scaling, along with ever increasing technology complexity, leads to an explosion in the magnitude of variability while also introducing new sources of variability that need to be characterized and modeled. The topic of this session, statistical modeling of performance variability and circuit reliability, is therefore one of key challenges for achieving robust design of LSIs.

Our first paper reviews in a first part the current status of performance variability. It examines the sources of variability and explains how they can be characterized using test structures. Examples of measured variabilities at 65nm bulk and SOI processes are disclosed, which include the distribution of threshold voltages within a die and a breakdown of spatial variability into each component of "Lot-to-Lot", "Wafer-to-Wafer", "Within Wafer", "Within Die", etc. This paper then examines variability trends for future scaled technologies. It also shows that the impact of variability is changing its character such that the parametric performance variability is moving closer to the region of catastrophic faults.

Our second paper, taking an SRAM sense amplifier at a 65nm bulk CMOS process as a test vehicle, analyzes mismatches in NMOS and PMOS threshold voltages and discusses a statistical design method using transistor sizing. A linear response surface model is derived that relates input offset voltage of the amplifier to threshold voltages variations. A successful application of the model in the statistical sizing is verified by the measured variability of fabricated test circuits.

The last paper discusses statistical modeling of circuit performance degradation over time. Analytical solutions are developed that efficiently predict the statistics of circuit timing and the leakage due to NBTI and process variations. It is shown that the degradation rate and its variance can be predicted from the characteristics of transistor degradation and circuit performance sensitivity to aged parameters, which are independent of the type and the amount of process variations. An aging model is implemented into SPICE and verified by simulation results and 65nm silicon data.

Notes

IEEE 2008 Custom Intergrated Circuits Conference (CICC)

45nm High-k + Metal Gate Strain-Enhanced CMOS Transistors

Chris Auth

Logic Technology Development, Intel Corp., Hillsboro, OR, U.S.A.

Abstract

At the 45nm technology node, high-k + metal gate transistors were introduced for the first time on a high-volume manufacturing process [1]. The introduction of a high-k gate dielectric enabled transistors with a 0.7x reduction in Tox (electrical gate oxide thickness) while reducing gate leakage 1000x for the PMOS and 25x for the NMOS transistors. Dual-band edge workfunction metal gates were introduced, eliminating polysilicon gate depletion and providing compatibility with the high-k gate dielectric. High-k + Metal gates have also been shown to have improved variability at the 45nm node [2].

In addition to the high-k + metal gate, the 35nm gate length CMOS transistors have been integrated with a third generation of strained silicon and have demonstrated the highest drive currents to date for both NMOS and PMOS. An SRAM cell size of $0.346\mu m^2$ has been achieved while using 193nm dry lithography. High yield and reliability has been demonstrated on multiple single, dual-, quad- and six-core microprocessors.

Introduction

One of the key methods to enable transistor gate length scaling over the past several generations has been to scale the gate oxide. This improves the control of the gate electrode over the channel enabling both shorter channel lengths and higher performance. As the gate oxide was scaled the gate leakage increased; this increase in gate leakage was insignificant until the 90nm technology node (Fig. 1). At the 90nm and 65nm nodes, the scaling of the gate oxide slowed as a result of the power limitations from the increase in gate leakage. In order to overcome this at the 45nm technology, a gate dielectric with a higher dielectric constant (high-k) has been introduced. This enabled a >25x gate leakage reduction while scaling the gate oxide thickness (Tox) by 0.7x.

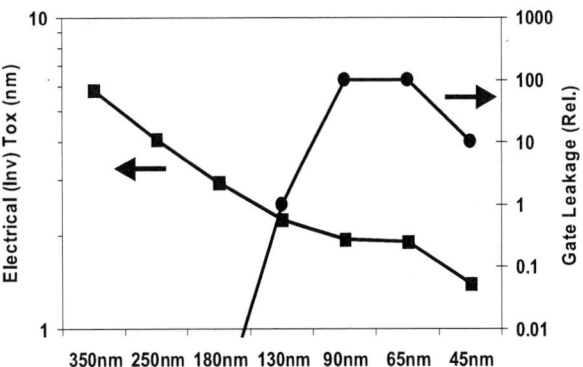

Figure 1: Trend of inversion Tox and gate leakage vs. technology node (source: Intel)

The introduction of high-k gate dielectrics has been slowed by several issues [3-5]. The first was the interaction by the high-k material with the existing polysilicon gates. This interaction led to high trap densities at the interface that pinned the Vt of the transistor to an undesirable value. The second was the degradation of the channel mobility in the presence of high-k dielectrics. The third issue was the poor reliability of the high-k dielectric.

The gate electrode effectiveness has also been increasingly impacted by poly depletion effects. This has led to lower drive currents when the transistor is turned on. By selecting a compatible metal gate electrode with the high-k gate dielectric, both the poly depletion effects and the Vt pinning at the high-k/polysilicon interface can be eliminated while providing higher channel mobilities [6].

In introducing high-k + metal gate transistors for the 45nm generation, these significant challenges needed to be overcome. First, both the materials to use for the high-k dielectric and the dual-band edge metals needed to be determined. Second, an integrated CMOS flow needed to be developed which optimized the channel mobility of while meeting the reliability requirements for the technology. The development of this CMOS flow was complicated by the need to mesh the process requirements of the metal gate process with the thermal limitations of the junction formation steps and with the uniaxial strain-

978-1-4244-2018-6/08 $25.00 © 2008 IEEE

inducing steps that have become central to the transistor architecture.

Along with the above mentioned improvements in performance and gate leakage with high-k + metal gate, a key requirement of the technology node was an increased packing density for the transistors. For each node, an ~50% area scaling is expected, and this technology continues that trend. A key challenge to overcome in this scaling is the loss of performance due to scaling of the stress-inducing features of the technology. Use of 193nm dry lithography for critical layers at the 45nm technology node was preferred over moving to 193nm immersion lithography due to lower cost and greater maturity of the toolset. In order to achieve the tight 160nm gate and contact pitch requirements, unique gate and contact patterning process flows were developed and implemented.

Transistor Process Flow

The two common methods for introducing a metal gate to the standard CMOS flow include either a "gate-first" or "gate-last" process. Most comparisons of these two process flows focus on the ability to select the appropriate workfunction metals, the ease of integration, or the ability to scale; however, these comparisons typically fail to comprehend the interaction of the process flows with the strain-inducing techniques. By use of a high-k first and metal gate-last flow, it is possible to maximize the benefits of the stress-inducing steps and high temperature junction formation, while minimizing the thermal processing of the workfunction metals.

In the metal gate-first flow (Fig. 2), the high-k dielectric and dual-metal processing are completed prior to the polysilicon gate deposition. The dual metal gates are then subtractively etched along with the poly gates prior to S/D formation. In contrast, for the high-k first and metal gate-last flow, a standard polysilicon gate is deposited after the hafnium-based high-k gate dielectric deposition. This is followed by a standard polysilicon processing flow through the salicide formation steps.

Gate-First	High-k First, Gate-Last
-Isolation	-Isolation
-Hi-k gate deposition	-Hi-k gate deposition
-Dual Metal-Gate Dep	-Poly-Si gate dep/patterning
-Poly-Silicon deposition	-S/D formation
-Poly-Si/metal etch	-Salicide/Contact etch stop
-S/D formation	-1st ILD dep/polish
-Salicide/Contact etch stop	**-Poly Si gate removal**
-1st ILD dep/polish	**-Dual-Metal Gate dep**
-Contact formation	-Contact formation

Figure 2: Comparison of unique steps in gate-first and high-k first, metal gate-last process flows. Key differences are highlighted in bold.

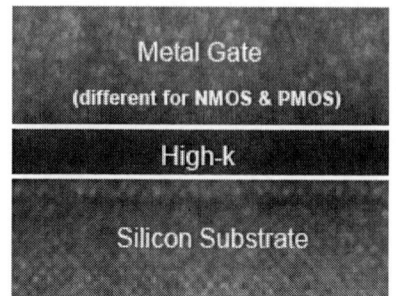

Figure 3: TEM of High-k/Metal gate stack

After deposition of the contact etch stop and the first interlayer dielectric (ILD0) films, a polish step is used to expose the poly gates and enable removal of the dummy poly. The PMOS workfunction metal is then deposited. A patterning step removes the PMOS metal from the NMOS area. The NMOS workfunction metal is deposited and the gate trenches are filled with Al for low gate resistance (Fig. 3). By using novel gapfill techniques, robust gate resistance is enabled to sub-30nm gate lengths (Fig. 4). A metal polish step is used to remove the excess metal and planarize the gate trenches. The flow then continues with the contact and interconnect processing steps.

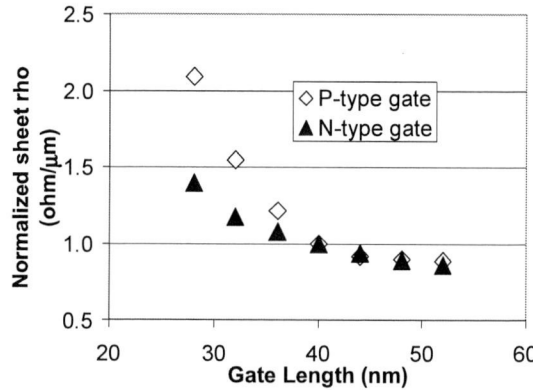

Figure 4: Gate sheet rho versus gate length showing scalability of gate fill process

Fig. 5 shows a TEM of the high-k/metal gate NMOS and PMOS transistors with the embedded SiGe S/D strain layer on the PMOS and Ni salicide. The strained silicon techniques that were first introduced at the 90nm and 65nm nodes were further enhanced in this generation. The Ge concentration of the embedded SiGe S/D was increased to 30% from the previous generations of 23% at the 65nm technology node [7] and 17% at the 90nm technology node [8].

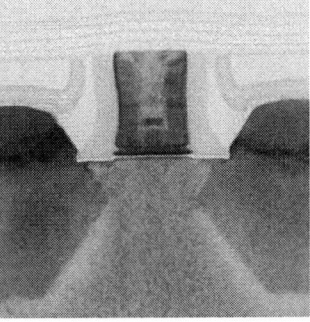

Figure 5: TEMs of High-k + Metal Gate NMOS (left) and PMOS (right) transistors

Design rules & 193nm Dry Patterning

Contacted gate pitch (CGP) is a key measure of front end density, and the scaling to 160nm maintains the 0.7x scaling trend (Fig. 6). This is the most aggressive CGP reported to date for a 45nm high performance logic technology. The contact process has also been modified with trench contacts replacing conventional circular contacts for lower series resistance. Trench contact based local routing improves layout density, especially for cross-coupled inverter pairs that are very common in microprocessor SRAM and register file arrays. Tight pitches and trench contacts allow SRAM cell size to be scaled to 0.346µm² (Fig. 7).

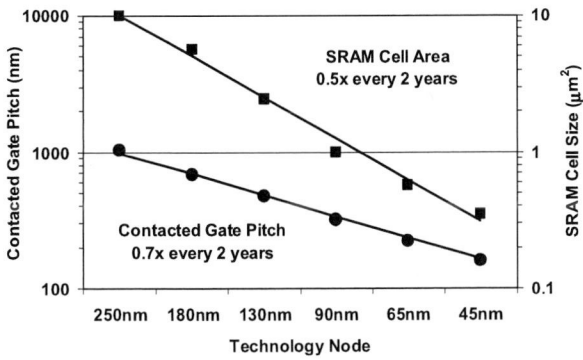

Figure 6: Contacted gate pitch and SRAM cell size scaling trend for Intel's technology nodes

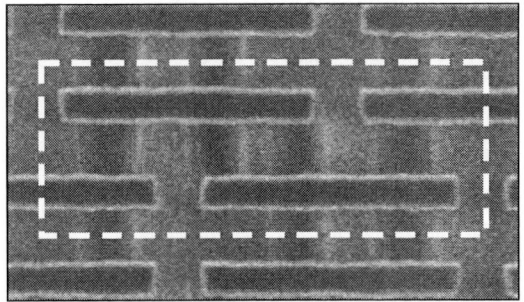

Figure 7: Diffusion and poly layers for 0.346 •µm² 6-T SRAM cell

In order to enable these tight pitches by use of low cost 0.92NA 193nm dry patterning, innovative processes were developed to produce robust patterning. This is demonstrated by the fidelity of the poly lines in Fig. 7. The gate patterning process uses a double patterning scheme. Initially the gate stack is deposited including the polysilicon and hardmask deposition. The first lithography step patterns a series of parallel, continuous lines (Fig. 8). Only discrete pitches are allowed with a horizontal orientation, with the smallest at 160nm, to assist in the patterning. A second masking step is then used to define the cuts in the lines. The two-step process enables abrupt poly endcap regions, devoid of rounding which allows for tight contact-to-gate (CTG) design rules (Fig. 9). There are no additional masking steps from this process since the 65nm generation also used two reticles for poly patterning.

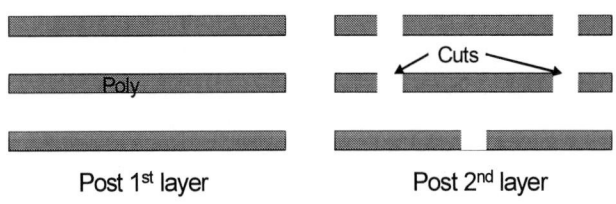

Figure 8: Illustration of double Poly patterning showing parallel, continuous lines post first patterning step (left) and with cuts in the lines post second patterning step (right)

The contact patterning process uses a similar pitch restriction to facilitate lithography but not the common orientation requirement. Trench diffusion contacts run parallel to the gates with discrete pitches, while trench gate contacts run orthogonal to the gates. Use of trench contacts has the added benefits of lowering the contact resistance by >50% and allowing use as a local interconnect, which improves SRAM/logic density by up to 10%.

978-1-4244-2018-6/08 $25.00 © 2008 IEEE

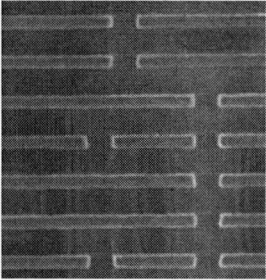

Figure 9: Top-down SEM post poly patterning process showing 160nm poly pitch with minimum gate length lines. Note the square poly ends, devoid of rounding.

Transistor Results

The introduction of the high-k gate dielectric delivered a dramatic gate leakage reduction relative to 65nm transistors of >25X for NMOS and 1000X for PMOS (Fig. 10).

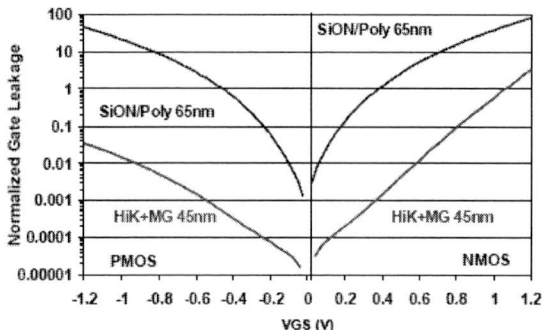

Figure 10: Gate leakage reduction of 25-1000x with use of high-k + metal gate relative to 65nm technology with SiON gate dielectric + polysilicon gates

The high-k + metal gate transistors exhibit excellent NMOS and PMOS short channel effects (SCE) and drain-induced barrier-lowering (DIBL) due to the combination of Tox scaling and the optimal workfunction metal gates (Fig. 11, 12). The excellent gate control is also illustrated in the well-behaved subthreshold characteristics with <90mV/dec subthreshold slopes (Fig. 13).

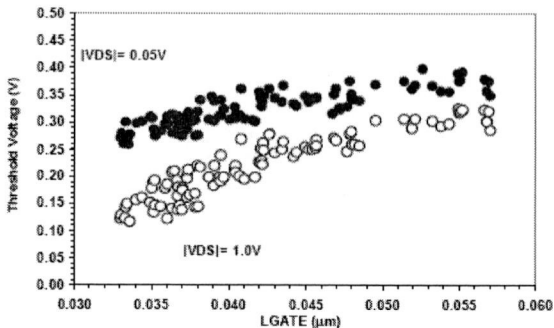

Figure 11: NMOS Vt vs. Lgate shows excellent SCE and DIBL

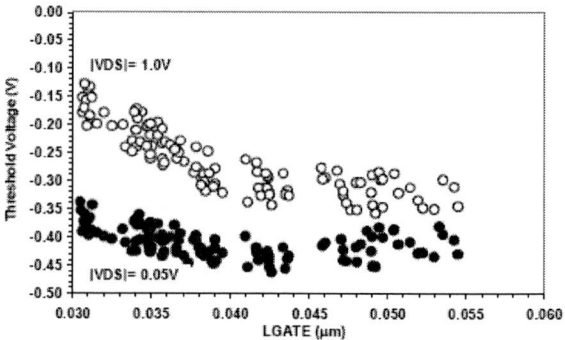

Figure 12: PMOS Vt vs. Lgate shows excellent SCE and DIBL

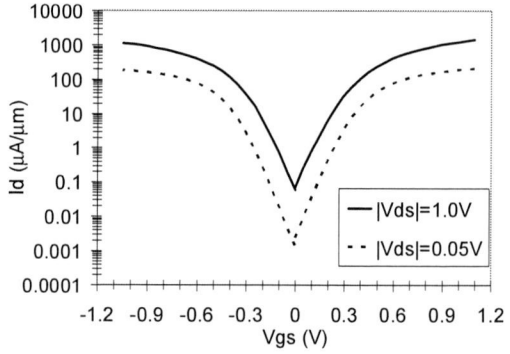

Figure 13: Subthreshold Id-Vgs for both NMOS and PMOS transistors

PMOS performance is improved by using high-k+metal gate as well as by the enhancements to the embedded SiGe processing. The PMOS drive current (Fig. 14) of 1.07 mA/µm is a marked 51% improvement over 65nm [9]. NMOS drive current (Fig. 15) is 1.36mA/µm, 12% better than the previous generation, 65nm transistors. The average drive current improvement versus 65nm is 32% at the same voltage and Ioff despite scaled transistor pitch. The linear drive currents show similar enhancements with PMOS (Fig. 16) at 0.178mA/µm and NMOS (Fig. 17) at 0.192mA/µm. These drive currents are benchmarked at

978-1-4244-2018-6/08 $25.00 © 2008 IEEE   410

1.0V, a low 100nA/μm Ioff and at 160nm contacted gate pitch. Both the saturated and the linear drive currents represent the best drive currents reported to date for a 45nm technology at low Ioff. Figure 18 shows the transistor performance vs. gate pitch for this generation illustrating that both density and performance are improved with this transistor flow.

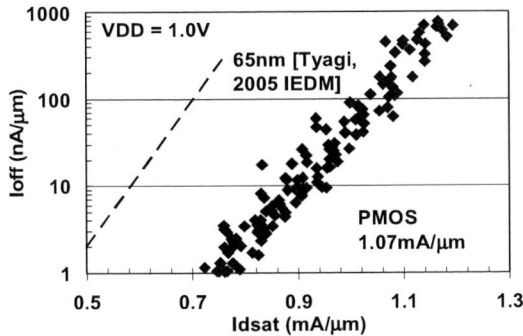

Figure 14: PMOS Idsat of 1.07mA/μm at 100nA/μm Ioff and Vdd =1.0V

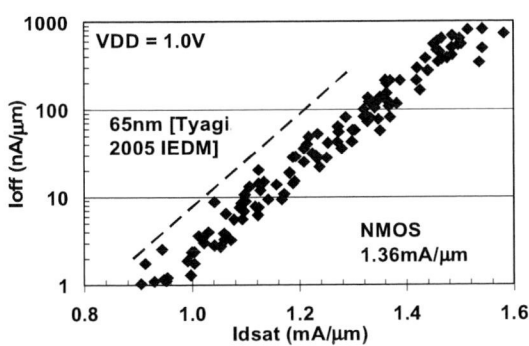

Figure 15: NMOS Idsat of 1.36mA/μm at 100nA/μm Ioff and Vdd =1.0V

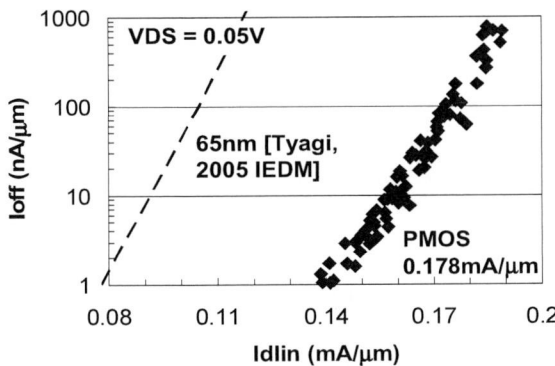

Figure 16: PMOS Idlin of 0.178mA/μm at 100nA/μm Ioff and Vds=0.05V.

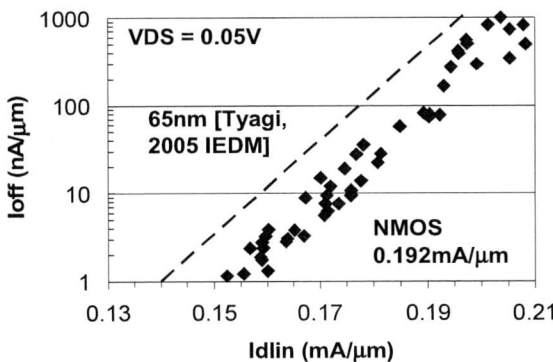

Figure 17: NMOS Idlin of 0.192mA/μm at 100nA/μm Ioff and Vds=0.05V

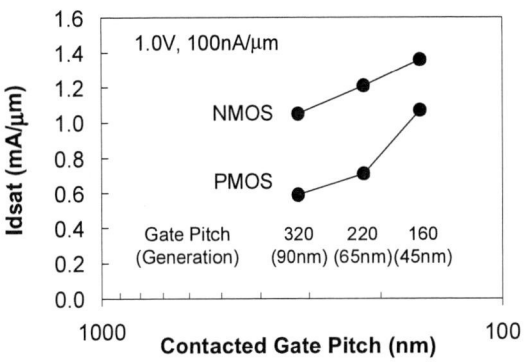

Figure 18: Performance vs. Gate pitch for 90, 65, 45nm generations

Stress enhancement in a Metal gate flow

Since its introduction at the 90nm node, strain has become a central performance enhancement element for the standard CMOS flow. The most commonly used techniques for implementing strain in the transistors include embedded SiGe in the PMOS S/D, stress memorization for the NMOS, and a nitride stress-capping layer for NMOS and PMOS devices (Fig. 19).

65nm Method	45nm Method
PMOS	**PMOS**
Embedded SiGe S/D	Embedded SiGe S/D **with higher %Ge**
	Poly Gate Removal Enhancement
NMOS	**NMOS**
Tensile Nitride Cap	**Tensile Trench Contacts**
Gate Stress Memorization + S/D Stress Memorization	**Metal Gate Stress (MGS)** + S/D Stress Memorization

Figure 19: Comparison of stress enhancement methods for 65nm and 45nm nodes. New features highlighted in bold.

A significant benefit from using a gate-last flow comes from removing the poly gate from the transistor after the stress-enhancement techniques are in place. It has been shown that the stress benefit from the embedded S/D SiGe process is enhanced through this removal of the poly gate since the poly gate acts as a buffer counteracting the effect of the embedded S/D SiGe [10]. This benefit can be illustrated in simulation with an estimated 50% increase in lateral compressive stress by removal of the polysilicon gate (Fig. 20). The combined impact of the increased Ge fraction in the embedded S/D and the strain enhancement from the gate-last process allow for 1.5x higher hole mobility compared to 65nm, despite the scaling of the transistor pitch from 220nm to 160nm.

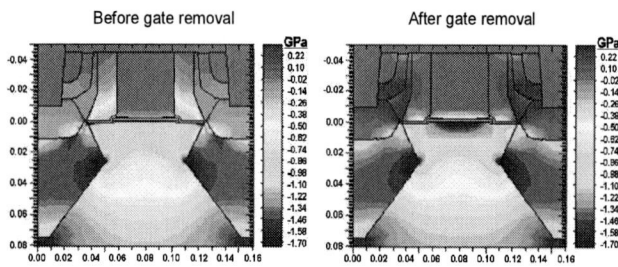

Figure 20: Stress contours in the PMOS transistor before and after the removal of the polysilicon dummy gate. Stress in the channel is shown to increase 50% from 0.8GPa to >1.2 GPa.

For the NMOS device, two methods of stress enhancement have been employed in this technology. First, the loss of the nitride stress layer benefit, due to scaling the poly pitch from the 65nm technology node, has been overcome by the introduction of trench contacts and by tailoring the contact fill material of the contacts to induce a tensile stress in the channel. The NMOS response to tensile (control) vs. compressive contact fill materials is shown in figure 21. The stress impact of the trench contact fill material on the PMOS device is mitigated by use of the raised S/D inherent in the embedded SiGe S/D process.

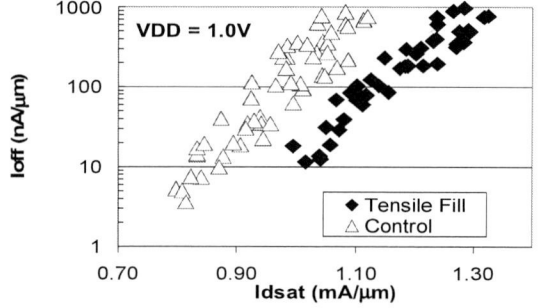

Figure 21: Ion-Ioff benefit of tensile Contact Fill showing a 10% NMOS Idsat benefit. Contact resistance is matched for the two fill materials.

For NMOS stress memorization, there are two primary methods commonly used, one is memorization of stress in the Source/Drain (S/D) of the device and the other is memorization in the poly gate [11]. The metal gate-last flow is compatible with the S/D method while the poly gate component would be compromised. To compensate for this, the poly gate method is replaced by Metal Gate Stress (MGS): modifying the metal-gate fill material to directly induce stress in the channel [12]. By introducing a gate fill material with a compressive stress, the performance of the NMOS device is enhanced and additive to the contact fill technique (Fig. 22). By use of a dual-metal process with PMOS first, the stress of the NMOS gate is decoupled from the PMOS gate through optimization of the PMOS gate stack to buffer the stress.

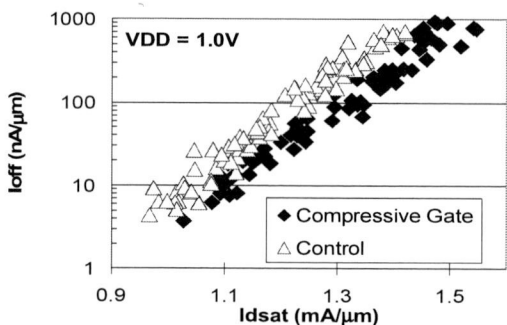

Figure 22: Ion-Ioff benefit of compressive gate stress showing a 6% NMOS Idsat gain. Tensile Contact Fill is used on both sets of data.

Ring Oscillators

The transistor performance gains are reflected in the ring oscillator (RO) performance data. Gate delay data from ring oscillators with a fanout of 2 is benchmarked at 100nA/μm Ioff each for NMOS and PMOS, at 1.2V for 65nm and a lower 1.1V for 45nm. The ring oscillators use the minimum contacted gate pitch (220nm and 160nm) for each technology. Despite the scaling of both voltage and contacted gate pitch, FO=2 gate delay is reduced from 6.65pS (65nm) to 5.1pS (45nm), for a gain of 23% (Fig. 23). The table in fig. 24 breaks out the RO gains between NMOS/PMOS Idsat, Idlin, and the gate and junction capacitances illustrating the marked impact of the PMOS performance gains on the ring oscillators.

978-1-4244-2018-6/08 $25.00 © 2008 IEEE   412

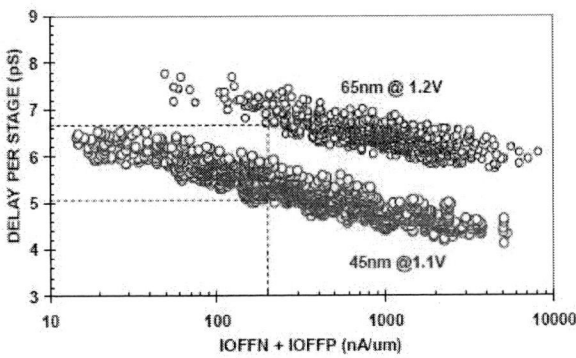

Figure 23: Ring Oscillator delay vs. leakage for fanout=2. Comparison of delay for 65nm vs. 45nm is at 1.2 and 1.1V, respectively.

Component	Benefit (%)
PMOS Idsat	+13
PMOS Idlin	+18
NMOS Idsat	+3
NMOS Idlin	+2
Cjunction	+2
Cgate/Cov	-8
Voltage Scaling	-7
Total	+23

Figure 24: Breakdown of RO gains vs. 65nm results. The voltage scaling term accounts for the reduction in VDD from 1.2V (65nm) to 1.1V (45nm).

Reliability

TDDB and bias-temperature reliability (BTI) of high-k + metal transistors have been a key concern [5]. The reliability concerns have been overcome through optimization of the gate stack. NMOS TDDB (Fig. 25) is improved; supporting a 30% higher electric-field compared to 65nm SiON transistors. PMOS NBTI (Fig. 26) is improved at the same electric-field and is matched at 50% higher electric-field, relative to 65nm SiON transistors.

NMOS PBTI (Fig. 27) is better than 65nm and supports 15% higher electric-field; the net BT shift for NMOS and PMOS is matched to 65nm at 30% higher field. Note that while NMOS transistors are considered stable for SiON/Poly, they actually show PBTI at very high electric-fields.

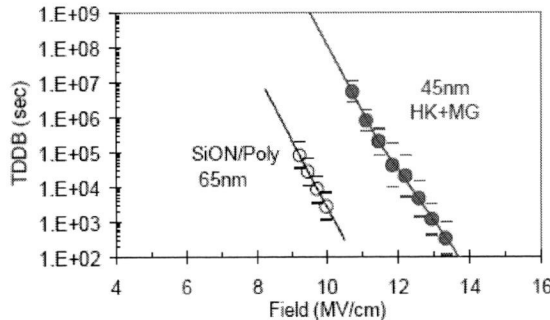

Figure 25: NMOS TDDB time to fail vs. electric field

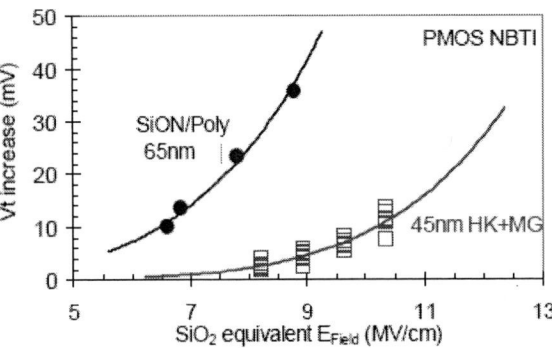

Figure 26: PMOS NBTI Vt shift vs. electric field

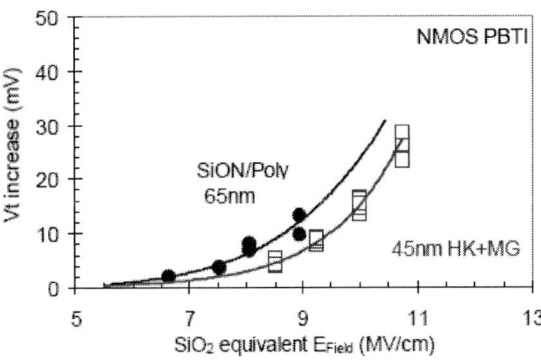

Figure 27: NMOS PBTI Vt shift vs. electric field

Conclusion

High-k + metal gate transistors have been integrated into a manufacturable 45nm process for the first time. Selection of the metal-gate flow (high-k first, metal-gate last) was made to maximize the benefit from the strained silicon steps. Novel stress techniques were also developed to replace the stress methods that are compromised due to scaling and the metal gate flow. The scaling of the transistor density was achieved through development of new poly and contact patterning schemes The resultant transistors provide record drive current at low leakage and at tight contacted gate pitch achieving both performance

978-1-4244-2018-6/08 $25.00 © 2008 IEEE

and density benefits. This is demonstrated in the ring oscillators with a 23% gate delay reduction compared to 65nm at the same Ioff and 10% lower VDD. Reliability concerns have been overcome through engineering of the gate stack.

Acknowledgments

The author gratefully acknowledges the collaborative efforts of his colleagues in the Portland Technology Development Group, the Technology Computer Aided Design Group, and in the Corporate Quality and Reliability Group.

References

1. K. Mistry, et al, "A 45nm Logic Technology with High-k+Metal Gate Transistors, Strained Silicon, 9 Cu Interconnect Layers", 193nm Dry Patterning, and 100% Pb-free Packaging, IEDM Tech. Dig. pp 247-250, 2007
2. K. Kuhn, "Reducing Variation in Advanced Logic Technologies: Approaches to Process and Design for Manufacturability of Nanoscale CMOS", IEDM Tech. Dig. pp. 471-474, 2007
3. V. Misra, G. Lucovsky,and G. Parsons, "Issues in High-k Gate Stack Interfaces," MRS Bull., vol. 27, no. 3, pp. 212-216, 2001.
4. C. Hobbs et al., "Fermi Level Pinning at the Poly-Si/Metal Oxide Interface," in Symp. VLSI Tech. Dig., pp. 9-10, 2003.
5. G. Ribes et al., "Review on High-k Dielectrics Reliability," IEEE Trans. on Device and Materials Rel., vol. 5, no. 1, pp. 5-19, 2005.
6. R. Kotlyar et al., "Inversion Mobility and Gate Leakage in High-k/Metal Gate MOSFETs," IEDM Tech. Dig., p. 391, 2004.
7. P. Bai et al., "A 65nm Logic Technology Featuring 35nm Gate Lengths, Enhanced Channel Strain, 8 Cu Interconnect Layers, Low-k ILD and 0.57 um^2 SRAM Cell", IEDM Tech. Dig., pp. 657-660, 2004
8. T. Ghani et al., "A 90nm high volume manufacturing logic technology featuring novel 45nm gate length strained silicon CMOS transistors", IEDM Tech. Dig., pp. 978-980, 2003.
9. S. Tyagi, et al, "An advanced low power, high performance, strained channel 65nm technology", IEDM Tech. Dig. pp 1070-1072, 2005
10. J. Wang, et al, "Novel Channel-Stress Enhancement Technology with eSiGe S/D and Recessed Channel on Damascene Gate Process", VLSI Sym. Tech. Dig. pp 46-47, 2007
11. A. Wei, et al, "Multiple Stress Memorization In Advanced SOI CMOS Technologies", VLSI Sym. Tech. Dig. pp 216-217, 2007
12. C. Kang, et al, "A Novel Electrode-Induced Strain Engineering for High Performance SOI FinFET utilizing Si (110) Channel for Both N and PMOSFETs", IEDM Tech. Dig. pp885-888, 2006.

IEEE 2008 Custom Intergrated Circuits Conference (CICC)

Will BiCMOS stay competitive for mmW applications ?

Patrice Garcia[1], Alain. Chantre[1], Sébastien Pruvost[1], Pascal Chevalier[1], Sean T. Nicolson[2], David Roy[1],
Sorin P. Voinigescu[2], Christophe Garnier[1]

[1] STMicroelectronics, 850 rue Jean Monnet, F-38926 Crolles, France

[2] Edward S. Rogers, Sr., Dept. of Electrical & Comp. Eng., University of Toronto, Toronto, ON M5S 3G4, Canada

Abstract- **This work summarizes upcoming millimeter wave and high speed applications which will benefit from advanced SiGe BiCMOS process. The performance of a 230GHz f_T 280GHz f_{max} process is detailed and future improvements are discussed.**

Intrinsic transistor performance for millimeter wave design has been compared with that of advanced 65nm Low-Power CMOS. To help process comparison, design examples are also given and circuit optimizations to reach optimum noise figure are discussed. Recent realizations at 24GHz and 77GHz in SiGe BiCMOS are presented, demonstrating state of the art results on both receivers.

I. INTRODUCTION

A decade ago, major applications demanding stringent performance were GSM, GPS and DECT; most of them were made of external components such as off-chip Low Noise Amplifiers. These standards were primarily not defined to be compatible with silicon monolithic integration. In the mean time, higher frequency or high power applications were dedicated to GaAs. In the golden age of Analog/RF design, designers faced to the paradigm of trying to integrate new products using new architectures with limited silicon performances; much effort went into fulfilling integration constraints with the targeted performance and cost. Today, advanced CMOS processes offer extensive performances to address new applications (WCDMA, Wimax, UWB) and the finality is not to prove the ability to integrate but mainly to focus on cost reduction while offering more and more features (multi-standards, higher bandwidth). These existing applications are probably a small step towards a new age of radio communications offering real time features and services by exploiting new frequency bands and higher bandwidth.

Full HD-video, security, and high-speed data transport open a new horizon for many indoor and outdoor applications and challenges for RF/Analog designers demanding even higher silicon process performances.

In this paper a BiCMOS process dedicated to millimeter wave will be detailed and linked with applications requirements.

The paper is organized as follows: Section II emphasizes the main driving applications demanding state of the art processes. Theoretical results are reported and discussed in Section V, where the performance of BiCMOS and CMOS is compared. An advanced BiCMOS process is detailed Section IV. Experimental results are reported and discussed in Section VI. Finally, conclusions are drawn in Section VII.

II. UPCOMING APLICATIONS

A. Wireless Video

Wireless HD and Wireless HDMI have great potential of market growth for upcoming indoor applications. Wireless HD (WiHD) was already demonstrated at the last CES showing video transmission between two transceivers. HDMI may first start with HDTV, DVDR, STB, Game console, then portable PCs & WPAN, followed by Video cameras, handheld devices (Ipods, MP3), and digital cameras. The requirements based on IEEE 802-15.3c suggest high bandwidth for video stream transmission located in the new 60GHz frequency band. 1600x1200 pixels at 24bits per point with 60 Hz refresh rates lead to 5Gbit/s uncompressed data-rate. This bandwidth is compatible with the newly allocated UWB frequency band from 57 GHz to 65GHz.

Modulation to support video streaming is not fully defined yet: Orthogonal Frequency Modulation (OFDM) or Quadrature Phase Shift Keying (QPSK) will be selected soon by the consortium. A transmission range up to 10m is requested allowing video to be received in meeting rooms. Line of sight is the first application, but Multiple Input Multiple Output (MiMO) transceivers or beam forming will be necessary to reach the targeted range while reducing the emitted power. The outcome of these applications will be discussed in the next section.

B. Automotive Radar

Automotive radar is the second application which will benefit from advanced silicon technology. Although automotive radar has existed sine 1990 [1], new frequency bands and cost reduction are the two main drivers pushing for monochip integration. Pulsed Ultra Wide Band (UWB) and Frequency Modulation (FMCW) radars will probably coexist in the near future depending on the application. FMCW radars are already shipped in the high end automotive market (Daimler Chrysler, Toyota, BMW, …) for Automotive Cruise Control (ACC). Mostly dedicated for highway driving, ACC allows target detection up to 300m distance with one sensor. Inversely, pulsed UWB short range radar (SRR) is foreseen for near and midrange detection leading to pre-crash, parking aids, stop and go applications using multiple sensors [2]. Nowadays, two frequency bands are allowed by regulation laws for automotive radar: 77GHz is preferred for FMCW thanks to the good directivity of the radiated beam, while 24GHz is preferred for UWB. However, the 24GHz band will be progressively abandoned in 2013 to the benefit of the 79GHz band.

978-1-4244-2018-6/08 $25.00 © 2008 IEEE

Performance specifications and robustness of radar sensors dedicated to the automotive market lead to stringent requirements today. Maximum range detection, distance accuracy, target discrimination, angular resolution and response time are related to architecture definition and lastly to process performance.

C. Wireless Imager

An extension of automotive radar is high frequency imaging applications [3].Imaging at radio frequencies is a more recent innovation, although the basic phenomenon has been understood for decades. Passive imaging can be used with the use of several radar sensors to build a precise radar image of an object using natural emissions from scenes at radio, microwave, and millimeter wavelengths. All natural objects whose temperatures are above absolute zero emit passive millimeter-wave radiation occurring at 35 GHz, 94 GHz, 140 GHz, and 220 GHz called blackbody radiation [4]. Many applications could be covered by microwave imaging such as airport safety or all-weather vision.
On the contrary, active imaging based on millimeter wave transceivers can be used for medical diagnosis, security imaging and in-situ non destructive testing. The signal generated by each detection element(s) represents the intensity for a single pixel in the image. When these intensities are displayed on a video monitor, a visual-like, image appears which is easily interpreted by an untrained observer.

D. Point to Point Communications

Point to Point communication is another application that will take advantage of BiCMOS process capabilities. Frequency bands have been recently allocated to the unlicensed spectrum resources: 24.25GHz to 24.45GHz and 25.05-25.25 GHz bands are currently used for communications between Base Station (BTS) and network nodes. But 72.5- 82.5GHz ISM bands are going to be used to extend LAN backbones between buildings or to support high-speed access services for first/last miles where links up to 30 / 50km are possible.
The 24GHz (and especially 80GHz) operation allows the use of small antennas. There are several key features of 24/80 GHz operation that are attractive for point-to-point applications. The fact that, such signals do not readily pass through building materials or foliage, enables additional rejection of hostile interferers. The combination of these attributes allows highly robust, dependable operation of point to point outdoor links.
On the contrary, 60GHz unlicensed communications are subject to natural atmospheric absorption phenomenon. The atmospheric absorption at 60 GHz gives the advantage to naturally increase the interferers as does multi-path attenuation even if this affects the desired signals. Generally, this results in broadband wireless links of a kilometer or less which are well suited for indoor communications.

E. Gigabit Ethernet

In 2010, the gigabit Ethernet standard (802.3ba) will be extended up to 100 Gbits/s following the decision of the IEEE HSSG (High Speed Study Group of the Institute of Electrical and Electronics Engineers) working group, allowing maximum bandwidth of 12 Go/s. At this data rate, it will be possible to transfer the content of a full DVD via a local network in less than a second. Unfortunately, this speed should be only dedicated to physical layers from 10km to 40km, while lower data rate (40Gbits) will be used in the meantime for 100m distances. Nowadays, Fast Ethernet (100 Mbit/s) is becoming a common standard widely used in the industry for networking. To answer an exponential demand in high-speed data transfers, telecom operators are asking to double their available bandwidth every 18 months, while the growth rate is slower in the industry (double every 2 years). 100 Gbits/s backbones and routers should be available between 6 European countries.

III. PERFORMANCES REQUIREMENTS

All these future applications have different requirements to fulfill. The aim of this paper is not to enter into details in the system specifications, but it is interesting to give a general overview of the requirements of future millimeter wave applications. The table below shows a short comparison between wireless specifications (Table. 1). 100Gbits Ethernet is not added in the table as output power requirement is not comparable with other standards.

TABLE 1
WIRELESS REQUIREMENTS SUMMARY

	SRR	P.to P.	WiHD
Freq. (GHz)	79	80	60
LNA NF (dB)	6.5	5	9
PA Pout (dBm)	20		10
Sφ (dBc/Hz) @1MHz	-105		
Throughput	N.A	1Gbs	2.1Gbs
BW. (GHz)	4	1.4	5
distance	25m	800m	10m
modulation	Pulsed	BFSK	OQPSK

To transmit HD video, WiHD requires more bandwidth and uses Offset Quadrature Phase Shift Keying (OQPSK), a complex modulation which leads to stringent I/Q phase errors at the transmit path. Moreover, heavy digital processing is necessary to perform the 2.1Gbits/s demodulation. This high data rate requires highly directional beam forming. However, the Low Noise Amplifier (LNA) noise figure and the Power Amplifier (PA) power constraints are relaxed compared with SRR or Point to Point communications. It can be assumed that the WiHD transceivers are foreseen to be advantageously integrated using advanced CMOS process including digital. However, this partitioning assumes that the power dissipation of the digital part can be kept adequately low.
Regarding 77GHz SRR, LNA noise figure and output power requirements are more stringent. In the UWB radar, a short pulse is radiated to be backscattered by the target.

To be compatible with the power density mask, pulse peak power remains quite low (10dBm). Taking into account distance attenuation and target cross section, the received pulse is below the noise floor of the RX path, in such a way that several integrations are needed to recover the pulse position. The LNA noise figure is a tradeoff between target distance, and integration time. The 4GHz equivalent bandwidth is due to the 0.3ns pulse duration. Duration is directly related to the range-detection accuracy as the delay between the emitted and received signal is due to the speed of light; shorter pulses improve radar resolution. In principle no modulation is required from the signal but coding bits could be added to improve robustness to interferers. VCO phase noise is such impairment that which seriously affects the performance. This phase noise can be seen as an equivalent jitter which degrades the coherent detection by limiting the Signal to Noise Ratio [5].

All these applications call for devices with high mmW performances that only advanced BiCMOS process could achieve today with sufficient margin.

IV. BiCMOS TECHNOLOGY – STATUS AND TRENDS

MOSFETs frequency performances largely benefit from the downscaling of gate lengths below 100nm as one moves along the CMOS roadmap. Nevertheless, for a given speed, SiGe heterojunction bipolar transistors (HBTs) still present many advantages over CMOS devices, such as lower $1/f$ noise, higher output resistance and higher voltage capability. Moreover, the given device speed can be obtained in a BiCMOS process relying on n-3 CMOS node as compared to pure CMOS technology. This implies considerable cost advantage, and explains why SiGe BiCMOS remains the technology of choice for industrial companies willing to enter the emerging millimeter-wave market, while the tremendous challenges presented by pure CMOS millimeter-wave circuit design attract more academic efforts.

A. State-of-the-art production 0.13μm BiCMOS technology

At STMicroelectronics, the most advanced high-speed SiGe BiCMOS technology in production today is the so-called BiCMOS9 [6]. This technology, which is based on a 0.13μm CMOS core process, offers a quasi self-aligned SiGe HBT with 160GHz f_T / f_{max} and 1.8V BV_{CEO}, dual V_T (high performance / low leakage) and dual gate oxide (1.2V / 2.5V) CMOS devices, passives and a 6-level copper back-end. A high-voltage HBT (BV_{CEO} = 3V) is also available.

BiCMOS9 has been designed to address optical networking and wireless applications up to 40Gb/s - 40GHz. Numerous circuits have been fabricated in this technology, demonstrating its suitability for the targeted applications space [7]. The largest circuit in production today has a surface area close to 200 mm². A yield as large as 55%, which is a record for such a big ASIC, has been reached thanks to a $D0$ as low as 50 defects / m² / level.

B. Millimeter-wave dedicated 0.13μm BiCMOS technology

Applications such as 60GHz WLAN, 77GHz automotive radars and 100Gb/s optical communications cannot be addressed with BiCMOS9 due to two main limitations.

The first limitation comes from the HBT performances, which must be increased well above 200GHz f_T and f_{max}. A Fully Self-Aligned (FSA) HBT architecture is then mandatory to minimize parasitic elements. Several FSA HBT structures based on non selective SiGe:C base epitaxy have been described in literature, but they are generally quite complex and difficult to integrate in a BiCMOS flow. Instead, we have favoured the so-called FSA/SEG structure, where the self-alignment of emitter and base electrodes is provided by the Selective Epitaxial Growth (SEG) of the SiGe:C base through the emitter window. This HBT construction is much easier to integrate with CMOS devices, and also uniquely enables the self-alignment of the base/collector junction to the emitter opening, which further minimizes transistor parasitics. The main technological challenge is placed on the control of the SEG process, and in particular on the efficient incorporation of carbon in the SiGe to minimize base widening during processing [8].

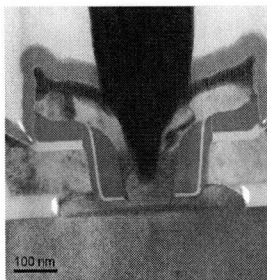

Fig.1 TEM cross-section of the main region of the FSA/SEG SiGe HBT

The TEM cross-section in Fig.1 shows the main region of the FSA/SEG HBT developed for very high frequency applications. The picture outlines the 20nm-thick SiGe:C film grown selectively to top and sidewall nitride layers in the 270nm-wide emitter opening, and the 130nm-wide effective monocrystalline emitter. Fig.2 summarizes the high frequency performances achieved using this structure in a bipolar-only process which includes the main constraints of the later BiCMOS integration. 230GHz f_T and 280GHz f_{max} are obtained, together with a record low minimum noise figure NF_{min} < 2dB at 60GHz. The industrial capability of this process has been checked through uniformity and yield investigations [7]. The suitability of this technology to address emerging applications such as 77GHz automotive radars and millimeter-wave imaging has been tested through the fabrication of a variety of critical circuit blocks [7]

978-1-4244-2018-6/08 $25.00 © 2008 IEEE

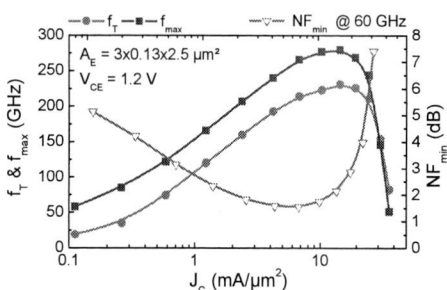

Fig.2 f_T, f_{max} and NF_{min} @ 60GHz vs. current density of a high speed FSA/SEG SiGe HBT

The second limitation of BiCMOS9 comes from the metal interconnects. This technology uses the same 6-level "digital" copper back-end as the core CMOS process, which is not adequate to support high performance transmission lines (TL's) at millimeter-wave frequencies. Indeed, attenuation constants α of 1.2dB/mm at 80GHz are obtained for 50Ω microstrip TL's in this technology, while ~0.5dB/mm would be needed.

Development of a second generation 0.13µm SiGe BiCMOS technology has therefore been initiated to overcome these two limitations. BiCMOS9MW adds to BiCMOS9 the high performance FSA/SEG SiGe HBT with 230GHz f_T and 280GHz f_{max} discussed above, as well as a millimeter-wave dedicated copper back-end, with two thick inter-metal dielectrics (1.5µm) and copper lines (3µm). The last copper level is therefore located at 8.3µm above the second metal level, which drastically reduces the attenuation constant of microstrip TL's.

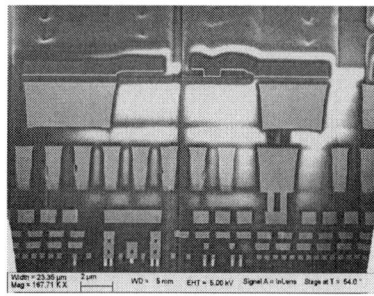

Fig.3 SEM cross-section showing the main features of the BiCMOS9MW technology (HBT, FETs, thick-Cu back-end, MIM)

Parameter	Unit	HS	MV	Comments
C_{BC}	fF	14.4	8.7	V_{CB}=0V
V_{BE} @ f_{Tmax}	V	0.91	0.84	
β @ f_{Tmax}		235	442	
BV_{CEO}	V	1.56	2.08	V_{BE}=0.7V
f_T	GHz	240	120	V_{CB}=0.5V
f_{MAX}	GHz	270	280	from f_{P20dB}

Table 2: Measured static and dynamic characteristics for BiCMOS9MW HS and MV SiGe HBTs (0.12×4.85µm² emitter area)

Fig. 3 shows a cross-sectional SEM picture which outlines the main features of this technology. The successful integration of

new front-end and back-end features is further demonstrated by measured parameters for high speed (HS) and medium voltage (MV) SiGe HBTs (Table 2), and passive devices (Table 3) [9].

Device	Value	Parameter	Comment
Standard T Line	50Ω	0.40dB/mm	at 40GHz
		0.50dB/mm	at 60GHz
		0.58dB/mm	at 77GHz
Induct. 1	74pH	$Q = 25.2$	at 60GHz
Induct. 2	107pH	$Q = 24.0$	at 77GHz
MIM	39pF	$C_0 = 1.96fF/µm^2$	at 25°C
		$I_{leak} <1nA/cm^2$	at -5V
		$BV > 28V$	at 25°C

Table 3: Measured BiCMOS9MW back-end devices characteristics

C. Future trends

Contrary to CMOS, there is no general consensus on a roadmap for BiCMOS technology evolution, despite recent efforts from the ITRS Radio Frequency and Analog/Mixed-Signal Wireless Technology Working Group [10]. As mentioned above, high speed BiCMOS developments are driven by the increase of optical communications data rate and by the emergence of millimeter-wave applications. This results in a step-by-step roadmap aiming at the best combination between CMOS density and HBT performance. This roadmap is expected to continue as long as SiGe HBT performances can be increased and new applications emerge, because it is unlikely that pure CMOS will be able to compete with BiCMOS on such aspects as output power, reliability and design margin.

While SiGe HBT performances are not directly linked to a CMOS node, the move from one node to the next is generally favorable. Indeed, this usually implies a reduced overall thermal budget, and the possibility to engineer more abrupt vertical doping profiles in the device, which is needed to increase the transit frequency f_T. At the same time, the better lithography offered by the new CMOS node allows tighter horizontal dimensions and smaller overlays, which is beneficial to the maximum oscillation frequency f_{max}. In fact, the main obstacle to the development of a first 90nm or 65nm SiGe BiCMOS technology appears to be the cost associated with the use of 300mm processing, which must be justified by large digital density. 60GHz wireless applications discussed in Section II, which require a lot of digital processing, may provide the products needed to justify such advanced 300mm SiGe BiCMOS technology development. In any event, it should be clear that 32nm or even 45nm CMOS millimeter wave circuits are not likely to be any cheaper, while their capability to move from research to production is still very uncertain.

In the mean time, advanced developments are on-going at STMicroelectronics to explore the scaling of SiGe HBT performances to 0.5THz. This work is part of a more global European effort - the so-called DOTFIVE Project [11] - aiming to establish a leadership position for the European semiconductor industry in the area of SiGe HBTs for millimeter-wave applications. Preliminary results, including the demonstration of a SiGe HBT with record f_T values of 0.41THz and 0.64THz at room and cryogenic temperatures respectively, will be presented at BCTM 2008 [12].

V. SiGe Transistor Performances

It is commonly agreed that HBT transistors offer superior performances for mm-wave design. Numerous publications pointed out the higher performance of BiCMOS over bulk CMOS [13]. However, advanced CMOS processes, too, provide also high speed capabilities, such as f_T which is of the same order of magnitude as that of SiGe HBTs. It is important to remind circuit designers why, for a given f_T performance, SiGe HBTs are better than MOSFETs in the millimeter-wave range and how wide the gap between the two devices is.

At the process level, only few parameters are necessary to compare BiCMOS and CMOS device performances. The most used parameter, f_T is extracted using the current gain formula h21=|Y21/Y11| as a function of operating frequency [15]. In advanced CMOS, the current density at which f_T reaches a maximum, is constant over different technology nodes [16]. But, for SiGe HBTs, current densities offering maximum f_T are not constant [13]. Figure 4 compares f_T performances of SiGe HBTs with different CMOS nodes.

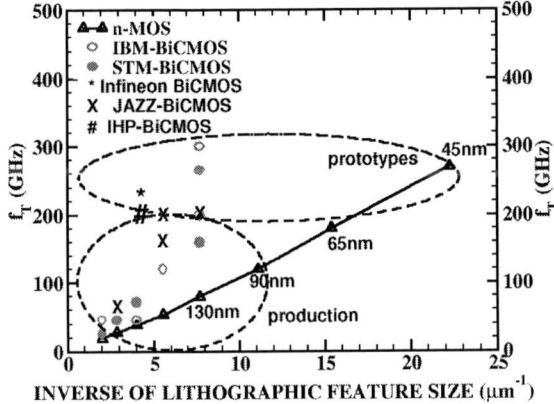

Fig.4. fT scaling in CMOS and SiGe BiCMOS technology nodes.

Reported 45nm bulk CMOS results and STM-BiCMOS are comparable in terms of pure intrinsic speed, but 65nm CMOS also exhibits also interesting performance [17][18].

Minimum Noise figure (NFmin) follows the same trend as shown in figure 5 [19]. For this comparison, a 20GHz frequency has been chosen as measurements uncertainties are reduced at this frequency (Fig. 5). From 20GHz measurements, extrapolated NF_{min} 60GHz should be close to 3dB; not so far from SiGe transistor. Although CMOS speed and noise figure are not far from the BiCMOS ones, power

gain should be also investigated. To address power gain of active device, several definitions coexist in the literature: Mason unilateral gain (U), Maximum Stable Gain (MSG), and associated gain Gass [20].

U provides power gain when an active device has been unilateralized with external circuitry. This ideal case is practically difficult to achieve at millimeter wave frequency without stability issue. MSG is a more realistic absolute limit where device is perfectly matched at its input and output ports.

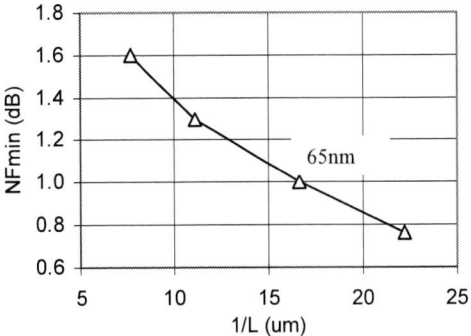

Fig.5. NFmin scaling in CMOS versus technology nodes.

With respect to MSG, the SiGe HBT exhibits higher power gain than 65nm or 45nm CMOS at 60GHz. This can be further analyzed using the MSG formula: MSG=|S21/S12|.

By developing S21 and S12 respectively and taking into account simplifications occurring at frequency of interest, it can be demonstrated that MSG can be rewritten in a more convenient expression: MSG = $gm/\omega C_{BC}$ with a good accuracy. gm and C_{BC} (or C_{GD} for mosfet transistor) are respectively the transconductance and base-collector capacitance; ω the operating pulsation.

TABLE 4
CMOS- 0.13um BiCMOS devices comparison

	C65nm LP	C45nm LP	B9MW	B9MW
J (mA/um)	0.36	0.39	0.45	0.94
Lg / We (um)	0.060	0.050	0.13	0.13
W / Le (um)	1	0.7	4x8	4x8
Ids, Ic (mA)	14.5	19.2	14.4	30
Vgs, Vbe (V)	0.82	0.8	0.87	0.97
Gm (mS)	40	40	277	292
fT (GHz)	166	160	150	230
C_{GS}, C_{π} (fF)	28	44	227	169
C_{GD}, C_{BC} (fF)	10.7	14	60	31
MSG (dB)	10	8.3	10.9	13.9
Gds, Ro (Ω)	171	147	22K	2800
U (dB)	13.2	12.5	16.4	18
k	0.38	0.41	0.46	0.55
NFmin (dB)	2.7	2.1	2.2	3.3
OIP1 (dBm)	5.7			9.1
Swing (V)	1.77			2

978-1-4244-2018-6/08 $25.00 © 2008 IEEE   419

It is interesting to notice that C_π or C_{GS} are not the relevant parameter for MSG in such a way that even with high f_T, 45nm CMOS cannot compete with SiGe in providing mmW power gain. The following table summarizes this comparison.

Table 4 shows MSG, NFmin, at 60GHz of C65nm, C45nm and SiGe BiCMOS9MW devices (B9MW) biased at current density leading to both NFmin and peak f_T. C45nm results have been obtained from L=50nm measurements. Since f_T scales inversely with mosfet's gate length L [21], 200GHz peak f_T can be extrapolated from measurements. Load Pull simulations have been performed to evaluate 1dB Output Compression Point (OIP1) capabilities of single devices in stable condition. These values have been simulated with transistor models showing good MSG correlation with silicon measurements. However, noise figures are difficult to extract at 60GHz, and the accuracy is limited to +/- 0.5dB for CMOS [22]. The noise measurement uncertainty is mainly due to the noise source impedance which is not perfectly matched to 50Ω and also to the noise figure meter sensitivity when device exhibits low gain in the measurement frequency band.

In this example, at 14.5mA of current consumption, SiGe HBTs offer higher performance (MSG, OIP1 and NFmin) than 65nm CMOS. Moreover, the MOSFET model does not take into account for interconnect capacitances. Due to the interdigited layout of mosfet, this will lead to 1dB MSG reduction in 45nm and only 0.4dB reduction in SiGe. In addition, as mosfet transistors exhibits 6 times lower C_{GD} than HBT, layout interconnects will seriously limit the MSG with respect to the HBT degradation. CMOS might be an interesting alternative when power gain and noise are not an issue. Moreover, the microstrip line -or matching network-losses have to be taken into account to evaluate the final performances.

Figure 6 shows the comparison between microstrip line constant attenuation using the Split-Thru deembedding methodology [14] and the equivalent model. Thanks to the thick upper copper layers, the transmission losses are only 0.5dB/mm at 60GHz when characteristic impedance Z_C is chosen to be 50Ω. Low loss networks can be integrated for noise and impedance matching without performance penalty.

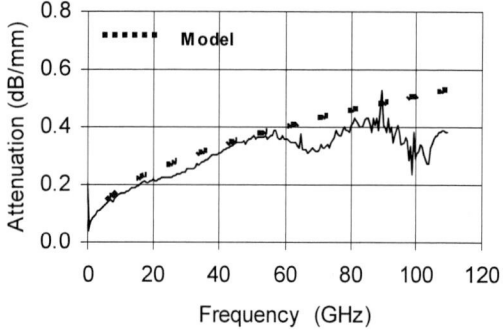

Fig.6. BiCMOS9MW optimized transmission line attenuation

For the previous devices, it is shown that input matching losses of C65nm, C45nm and SiGe are respectively 2.2dB, 2.8dB and only 0.8dB at 60GHz. Taking into account all these

degradations, the achievable power gain can be estimated to 5dB and 2.5dB below the MSG for 65nm CMOS and BiCMOS9MW respectively.

The last advantage of SiGe is its competitive reliability behavior by comparison with the CMOS. In 65nm CMOS technology, Hot carrier injection takes place at nominal voltage i.e. for Vds=Vgs=1.2V [23]. This mechanism results in CMOS parameter degradation (gm, linear and saturation current…) during its operation. For digital applications, where the degradation only occurs during transient, device lifetime criterion commonly required in the Semi-conductor industry is 0.2 year for 10% parameter shift (corresponding to more than 10 years AC). For analogue/RF application where DC voltage can be applied all along the product's life, some precautions must be taken to minimize or manage HCI impact. Designer can either reduce the operating voltage, use cascode topology to reduce lateral field or increase the CMOS channel length since the HCI lifetime is improved for larger L.

.Unfortunately, at millimeter frequencies, both the cascode topology and the larger gate length will result in worse NFmin performances [22][24].

VI. CIRCUITS REALIZATIONS

0.13um SiGe BiCMOS offers different flavors which greatly simplify the integration of millimeter wave systems. The first example is the 24GHz automotive SRR sensor designed in the previous generation of SiGe process targeted for 40Gb/s High Speed data rates [6]. The block diagram of a zero-IF transceiver (Fig. 7) was realized and tested including the following functions: LNA, Quadrature Down-mixer, PLL, TX pulse generation and VGA Low-Pass filter. First version of stand alone IPs have been already discussed previously [25] and optimized to ease the integration. This transceiver was intended to fulfill 24GHz radar requirements: 30m maximum detection range with 25cm resolution range. To perform early transceiver characterization in pulsed mode, the IC has been wire bounded on a daughter board (Fig. 8). This allows the emitted pulse at the TX output to be injected at the LNA differential inputs RF+, RF- using an external balun. Great care has been taken during the layout phase to minimize TX to RX leakage using a fully differential LNA

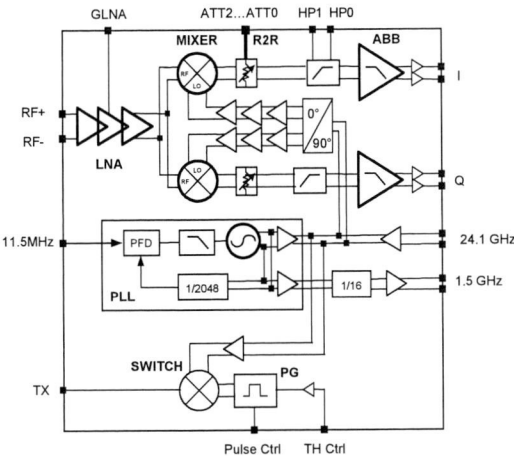

Fig. 7. Block diagram 24GHz RF sensor

. On-wafer measurements have also been performed to provide accurate performances evaluation. The overall power consumption is 640mW for NF=2.7dB DSB with 700MHz bandwidth and 1ns pulse at 0dBm peak power in 5.9mm2.

Fig. 8. Chip on board sensor and microphotograph

Using the state of the art BiCMOS process available at STMicroelectronics [9], similar integration was started to address the 77GHz automotive radar. Stand alone IPs and a receiver chain whose block diagram is shown in Fig. 9 have been transferred from the previous SiGe process to take advantage of the thick copper back-end dedicated to mmW. A die photo of the receiver, including LNA, VCO, and mixer with on-chip balun and IF amplifier, is shown in Fig.10.

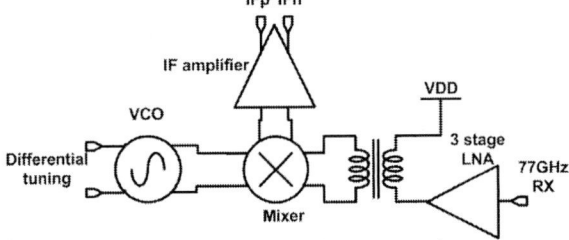

Fig. 9. Block diagram of a low-power receiver.

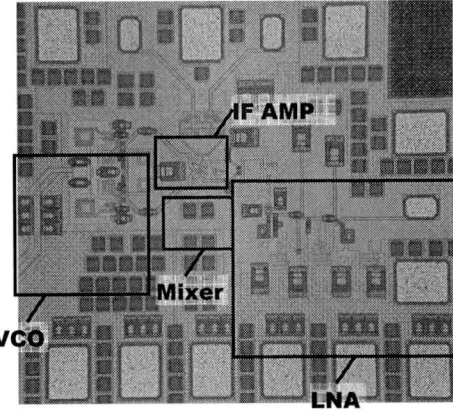

Fig. 10. Die photo of the receiver (LNA, VCO, mixer, and IF amplifier). (515um×460um).

The direct-conversion receiver consumes 123mW from 1.5V and 1.8V supplies, has a record low (for silicon) noise figure of 4.8dB (Fig.11), 30dB gain and the phase noise at 1MHz offset remains lower than -98dBc/Hz up to 100 °C. [26]. By comparison, a receiver has been fabricated using STMicroelectronics 65nm GP CMOS [27]. Reported noise figure at 77GHz [17] is 7dB which corroborates our analysis.

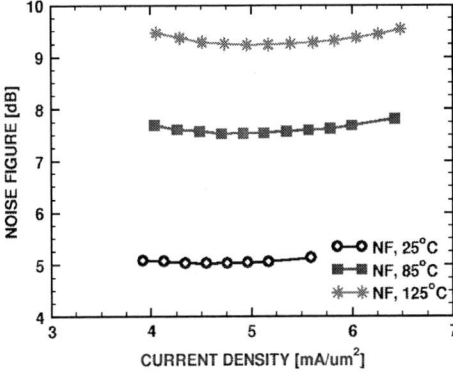

Fig.11. Measured receiver DSB noise figure versus current density in the LNA and temperature for 75GHz LO and 77GHz RF.

Single-stage stand-alone LNA has also been designed at 77GHz. LNA measured gain at 79GHz is 6.8dB as the simulation is about 8dB (Fig. 12).

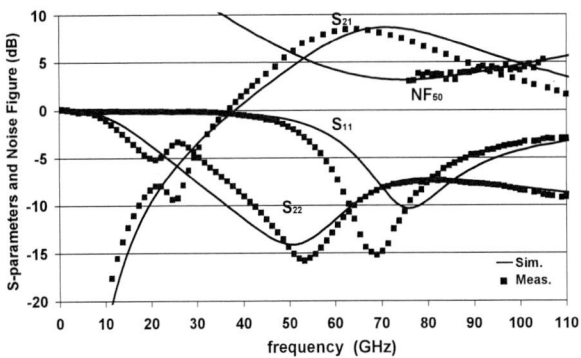

Fig.12. Measured LNA noise figure versus frequency at room temperature

978-1-4244-2018-6/08 $25.00 © 2008 IEEE

Then, the measured Noise Figure is 3.2dB at 79GHz and showing good accuracy with simulations. Moreover the minimum noise figure simulation is 3.1dB at this frequency. This LNA has been optimized at 6mA of current consumption under 1.5V.

VII. CONCLUSION

Although published realizations in advanced CMOS show interesting results, SiGe HBTs continue to offer state of the art performance for millimeter wave design. Thanks to its higher achievable power gain, lower noise figure, dedicated back-end and intrinsic robustness to hot carrier effects, SiGe BiCMOS processes will continue to be the best choice for future mmW and high-speed products.

ACKNOWLEDGMENTS

The authors thank the many members of STMicroelectronics, Crolles, France: Olivier Richard, Thomas Quemerais, Patrick Scheer, Clement Charbuillet involved in the various aspects of this work. As well as people involved in the system studies at the CEA-LETI, Grenoble, France: Dominique Morche and Cédric Dehos.

REFERENCES

[1] B.-E. Tullsson, "Alternative applications for a 77 GHz automotive radar", IEEE International Radar Conference, 2000, pp. 273 - 277

[2] H. Ogawa, et al, "Technology development of short range ultrawide-band radar system", International Workshop on Ultra Wideband Systems, 2004, pp. 351-355

[3] E. Laskin, et all, "80/160-GHz Transceiver and 140-GHz Amplifier in SiGe Technology", Radio Frequency Integrated Circuits (RFIC) Symposium, 2007, pp. 153-156

[4] Christopher D. Haworth et all, "Image Analysis for Object Detection in Millimetre-wave Images", SPIE Conference Proceedings Vol. 5619, 2004

[5] Poh Boon Hor Ko, C.C. Wanjun Zhi, "BER performance of pulsed UWB system in the presence of colored timing jitter", International Workshop on Ultra Wideband Systems, 2004, pp. 293- 297

[6] M. Laurens et al., "A 150GHz fT / fmax 0.13μm SiGe:C BiCMOS technology", in Proc. BCTM, 2003, pp 199-202.

[7] P. Chevalier et al, "High-Speed SiGe BiCMOS Technologies: 120-nm Status and End-of-Roadmap Challenges", in Proc. SiRF, 2007, pp. 18-24.

[8] D. Dutartre et al, "Si/SiGe epitaxy: a ubiquitous process for advanced electronics", in IEDM Tech. Dig., 2007, pp 689-692.

[9] G. Avenier et al, "0.13μm SiGe BiCMOS technology for mm-wave applications", submitted to BCTM.

[10] W.M. Huang et al, "RF, Analog and Mixed Signal Technologies for Communication Ics - An ITRS Perspective", BiCMOS Circuits and Technology Meeting, 2006, pp 1-7.

[11] www.dotfive.eu

[12] B. Geynet et al, "SiGe HBTs featuring >400GHz at room temperature", submitted to BCTM.

[13] S.P. Voinigescu et all, "SiGe BiCMOS for Analog, High-Speed Digital and Millimetre-Wave Applications Beyond 50 GHz", Bipolar/BiCMOS Circuits and Technology Meeting, 2006, pp. 1-8.

[14] A.M. Mangan et all, "De-embedding transmission line measurements for accurate modeling of IC designs", IEEE Transactions on Electron Devices, 2006, pp. 235 - 241

[15] T. Manku, "Microwave CMOS-device physics and design", IEEE Journal of Solid-State Circuits, Volume 34, Issue 3, 1999, pp. 277 – 285

[16] T.O. Dickson, "The Invariance of Characteristic Current Densities in Nanoscale MOSFETs and Its Impact on Algorithmic Design Methodologies and Design Porting of Si(Ge) (Bi)CMOS High-Speed Building Blocks", IEEE Journal of Solid-State Circuits, 2006, Volume 41, Issue 8, pp. 1830 - 1845

[17] E. Laskin et all, "95GHz Receiver with Fundamental Frequency VCO and Static Frequency Divider in 65nm Digital CMOS" ISSCC, 2008, pp. 180-181

[18] A. Mazzanti et al., "A 24 GHz sub-harmonic receiver front-end with integrated multi-phase LO generation in 65nm cmos", ISSCC, 2008, pp216-217.

[19] C. H. Doan, S. Emami, A. M. Niknejad, R.W. Brodersen, "Millimeter-Wave CMOS Device Modeling and Simulation, ", International Symposium on Circuits and Systems, 2004, pp. 524-527

[20] M.S. Gupta, "Power gain in feedback amplifiers, a classic revisited", IEEE Transactions on Microwave Theory and Techniques, Volume 40, Issue 5, 1992, pp. 864 - 879

[21] A.J. Joseph et all, "Status and Direction of Communication Technologies - SiGe BiCMOS and RFCMOS", Proceedings of the IEEE, Vol. 93, No. 9, 200", pp. 1539-1557

[22] B. Martineau et all, "80 GHz low noise amplifiers in 65nm CMOS SOI", 33rd European Solid State Circuits Conference, 2007, pp. 348 – 351

[23] G. Groeseneken et all, "Recent trends in reliability assessment of advanced CMOS technologies", International Conference on Microelectronic Test Structures, 2005, pp. 81 - 88

[24] S. T. Nicolson, S. P. Voinigescu, "Methodology for Simultaneous Noise and Impedance Matching in W-Band LNAs", IEEE Compound Semiconductor Integrated Circuit Symposium, 2006, pp. 279-282

[25] S. Pruvost et al, "Low Noise Low Cost Solutions for Pulsed Automotive Radar sensors", RFIC-IMS, RTUI1B-3, Honolulu, Hawaii, USA, 3-5 June 2007, pp 387-390.

[26] S. Nicolson et al, "A Low-Voltage SiGe BiCMOS 77-GHz Automotive Radar Chipset", IEEE Trans. On MTT, in press, 2008.

[27] F. Arnaud et all, "Low cost 65nm CMOS platform for Low Power & General Purpose applications", Symposium on VLSI Technology, 2004, pp.10 - 11

IEEE 2008 Custom Intergrated Circuits Conference (CICC)

Microelectronics for the Real World: "Moore" versus "More than Moore"

John P. Kent, IEEE, Senior Member and Jagdish Prasad, IEEE, Senior Member
ON Semiconductor
2300 Buckskin Road
Pocatello, Idaho 83201 USA

Abstract- Memories and microprocessors improvements rely on the continued scaling in silicon based CMOS technologies known as "Moore's" law. However, new classes of products are emerging that provide additional value based on functional innovation and diversification instead of scaling. This functional diversification is being called "More-than-Moore". Product innovation in "More-than-Moore" technologies is differentiated by circuit design, architecture, embedded software and unique process technology. These innovations enable the use of older, proven technologies in highly reliable products. This approach allows for non-digital functions such as RF, power control, passive components, sensors and actuators to migrate from the system board level into a package level (SiP) or chip level (SoC) implementation. The objective of "More-than-Moore" is to incorporate digital and non-digital functionality into compact systems. "More-than-Moore" technologies are application specific and focus on the interface between the "Analog" and the digital world. In this paper we will discuss the applications of "More-than-Moore" concept and its utility in real life.

I. INTRODUCTION

For more than 40 years, the semiconductor industry has continuously realized a rapid pace of improvements in integration level, cost, speed, power, and function in its devices. Most of these improvements are the result of exponential decreases in the minimum feature size used to fabricate integrated circuits (ICs). This trend has been dictated by Moore's law which states that the number of functions per chip doubles every 24 months. This trend is shown in Figure 1 below:

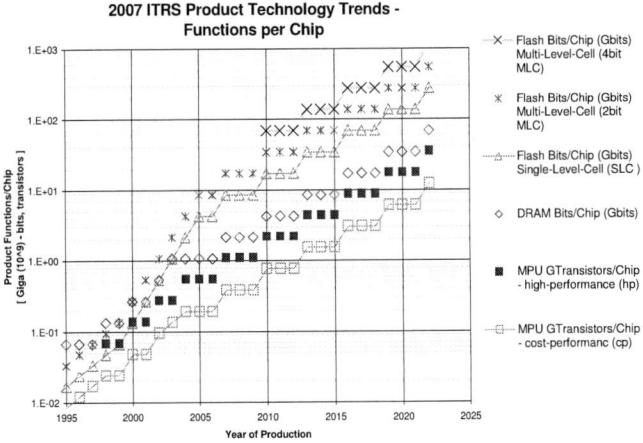

Figure 1: Technology trends over last 13 years and projection for next 17 years following Moore's law (courtesy: ITRS 2007)

Another significant part of Moore's law is the observation that the cost will decrease by 30% per year per function which has led to significant improvements in economic productivity and over all quality of life through the availability of affordable computers, communication, medical and consumer electronics in most areas of the world.

The improvements provided by Moore's law (also called "scaling") are enabled by large R&D investments. Total R&D spending by semiconductor companies exceeded $45 billion in 2007, an increase of nearly 12% from 2006. Overall R&D spending by semiconductor companies worldwide increased at an annual average rate of 10% between 2003 and 2007. This growing size of the required investment going forward can not be afforded by a single IC manufacturer and therefore has motivated industry collaboration as R&D partnerships and consortia such as IMEC in Europe and Sematech in the US. In addition, the cost to build a fab has increased from $10M in 1970 to about $3 billion in 2006 and is projected to approach $10 billion by the year 2012, thus making scaling a cost prohibitive proposition. This trend is presented in Figure 2 below.

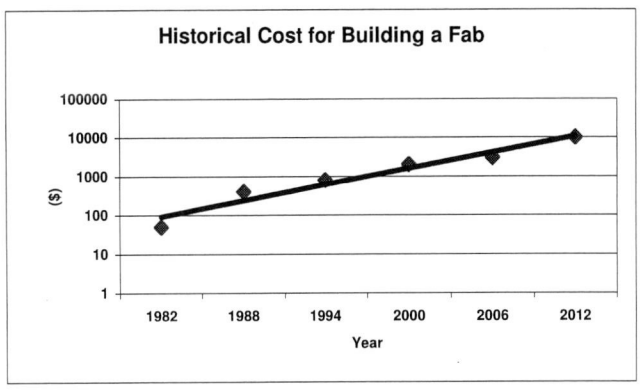

Figure 2: Historical cost for building IC manufacturing facility. Data is collected from various published sources.

In this paper, we will discuss a new trend of functional diversification also known as "More-than-Moore" that provides additional value to the customer via product innovation and less cost to the developer by leveraging mature technologies. We will focus especially on emerging markets such as power over Ethernet (POE), sensors, actuators, bio sensors (Lab-on-Chip), and system-in-package (SiP) that will

978-1-4244-2018-6/08 $25.00 © 2008 IEEE

use new innovations in mature technologies for high yields, low cost, long life time, and high reliability.

II. MORE-THAN-MOORE: FUNCTIONAL DIVERSIFICATION

A comparison between technology applications representing the realm of "Moore" versus "More-than-Moore" is shown in Figure 3.

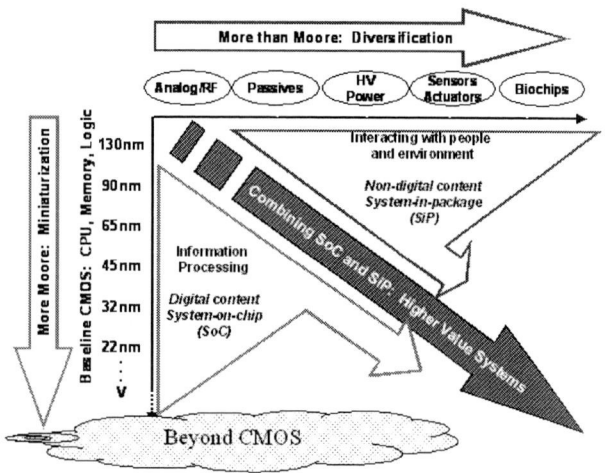

Figure 3: Scaling (Moore's law) on vertical axis and functional diversification (More than Moore) on Horizontal axis. (ITRS2007).

The "More-than-Moore" approach allows for non-digital functionalities such as RF, power control, passive components, sensors and actuators to migrate from the system board level into a package level (SiP) or chip level (SoC) implementation. The objective of "More-than-Moore" is to incorporate digital and non-digital functionality into compact systems. "More-than-Moore" technologies are application specific and focus on the interface between "Analog" and the "Digital" world. Example of this interface is shown in Figure 4. Interfacing the digital and analog world will require high voltage, high power, RF technologies to enable "More-than-Moore" diversification by integrating MEMs, sensors and actuators either by SoC or SiP.

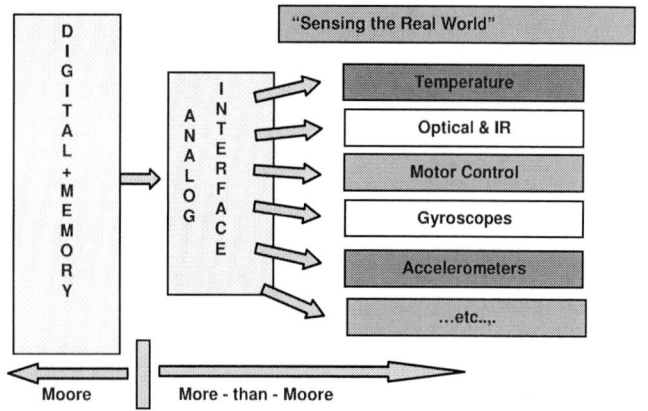

Figure 4: Interface between "Real World" and Application specific technologies enabled by functional diversification "More-than-Moore"

A. "More-than-Moore" Innovation Driven Technology Roadmap

Product innovation in "More-than-Moore" technologies is differentiated by circuit design, architecture, embedded software and unique process technology. These innovations enable the use of proven technologies in highly reliable products. The innovation driven technology roadmap is shown in Figure 5 along with Moore's law drivers DRAM and CMOS.

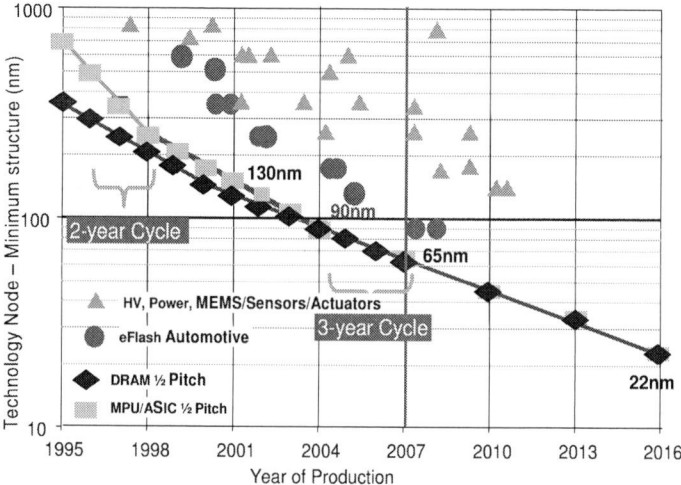

Figure 5: "More-than-Moore" technology Roadmap with "Moore's" drivers DRAM and CMOS. "More-than-Moore" products are manufactured in proven technologies for high reliability and thus are far behind "Moore's" technologies. Source: ITRS.

Some of the "More-than-Moore" products enabled by innovation and functional diversification are listed in table 1.

Table 1 summarizes several of these product innovations that leveraged mature technologies (0.35u node) and enabled new products.

Market	Product	Challenge	MTM Innovation
Consumer Industrial	POE	High voltage low R_{on}, Cable discharges, EOS	High Voltage device architecture, Robust ESD protection
Automotive	Motor Drivers, Sensors, Actuators	Harsh Environment	Thick power metal to distribute heat Bond 8mm wire to address vibration
Automotive, Industrial, Consumer	Micro-bolometer	Operation at ambient temperature with low dark noise	New sensing material integrated into CMOS
Medical	Bio Sensors Lab-on-a-chip	Integration of Bio sensing materials in CMOS	Converging micro-electronics with bio-sensing materials
Consumer Medical	Demand for dense memories	Dense memory integration in CMOS	SiP

978-1-4244-2018-6/08 $25.00 © 2008 IEEE

B. Power over Ethernet: XtreMOS

Power over Ethernet (PoE) is a rapidly emerging market based on device innovations that are capable of handling high voltages, cable discharges and other electrical over stress (EOS) events. This innovation will be able to provide power over Ethernet cables allowing the "Ethernet plug" form factor to become "iconic" for DC power supply, similar to the 12V "car cigarette lighter". The same wires that carry Ethernet data are also able to carry high voltage (68V or more) DC power. Two high voltage devices are needed to manage the power. One device is used at the source end to distribute the power from the Switch to the cable and the other at the application end to control the voltage and the power level required by the application. These kinds of applications require switching of large currents while withstanding high voltages. Thus one of the important requirements of these devices is low on-state resistance (R_{on}) for a given breakdown voltage V_{bd} to minimize the energy loss.

The "More Than Moore" innovation that enables POE is the optimization of device on-resistance for power transistors. Resurf [1] and super-junction (SJ) [2] concepts are to achieve low R_{on}. However, there is a 1D silicon limit that prevents further improvements. Improvements in the transistor performance beyond 1D silicon limit in the voltage range < 120V has been demonstrated using the concept of depletion of the drift region in vertical MOS capacitors [3]. These power transistors (called "XtreMOS") are integrated in a 0.35 μm CMOS process flow and are compatible with standard CMOS processing. A cross section of the "XtreMOS" is shown in Figure 6 along with the isolation scheme. The operating voltages of "XtreMOS" devices are in the range of 50V-120V.

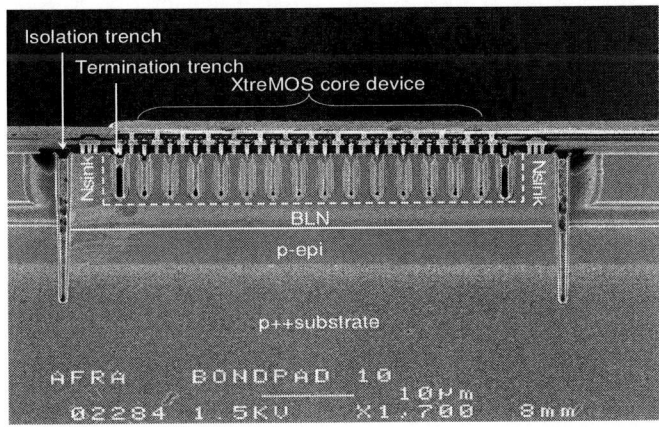

Figure 6: XtreMOS cross section showing the device core, termination and isolation trench and buried layer (BLN).

This XtreMOS transistor is bench marked against the published data for various device concepts which is presented in Figure 7. Figure 7 indicates that R_{on} (area) can be improved at least by a factor of 3 using the XtreMOS concept for devices in the range of 50-120V.

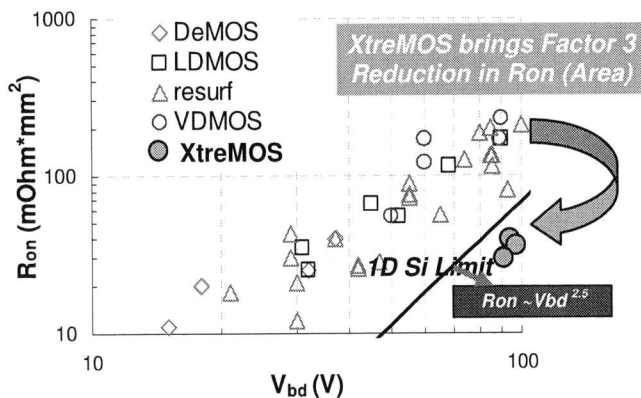

Figure 7: Benchmarking against published data. This data shows that the "XtreMOS" R_{on} is better than 1D Si and comparable to the SJ limit. Source: ON Semiconductor internal data.

One important requirement of this application is the protection from cable discharge. XtreMOS devices meet these requirements since these are robust, have low Ron, are small in size and have low device capacitance for fast switching.

C. "More-than-Moore" in Automotive Applications

Electronics are used throughout the automobile for non-critical as well as critical applications as shown in Figure 8.

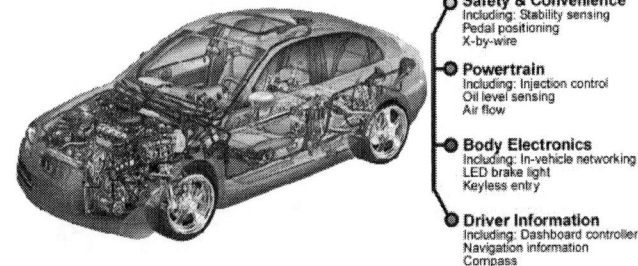

Figure 8: Showing the applications of electronics in a car. Source: Gartner, April 2007.

Semiconductor electronics contents in a car are more than 50% of the total electronics by value in 2008. Growth trend in automotive electronics is shown in Figure 9.

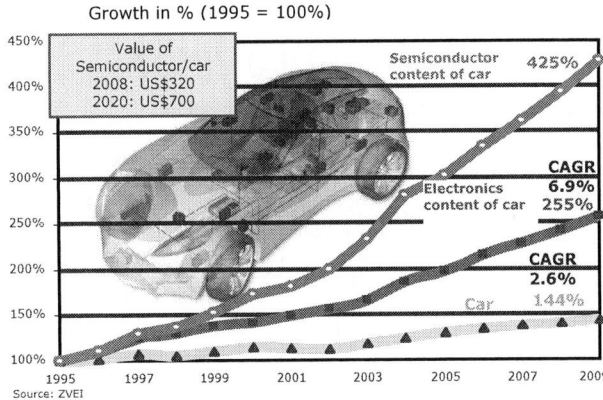

Figure 9: Automotive Microelectronics Trends – More Than 50% of $-value of Electronics is Semiconductors. Source: ZVEI

These applications have very challenging requirements: 1) high voltages to drive motors, relays, communication interfaces, sensors, actuators and optical drivers, 2) extreme mechanical stress such as vibrations, 3) integrated technologies such as non-volatile memory and significant analog and digital integration, 4) robust environment that requires operation at high temperatures to 200°C and, 5) high reliability (<1 part per million failure rate) which requires proven characterized technology and robust devices.

One example of 'next generation' micro-electronics for automotive applications is the starter/alternator that will be packaged in a single module. The semiconductor chip will be mounted directly onto this module in form of a bare die. This chip has to work under extreme mechanical stress (vibrations) and high temperatures to 200°C. The "More Than Moore" innovation that meets this challenge is processing thick Cu/Au (10-15μm) as a top metal to enable thick wire bonding. The auto industry uses 2mil Al wire bonding as a standard and will require 8 mil wire bonding for next generation devices. This is shown in Figure 10. Thick top metal has many advantages: 1) reduces the resistance important for switches; 2) helps reduce the hot spots and, 3) allows bonding on active circuit thus resulting in die area reduction.

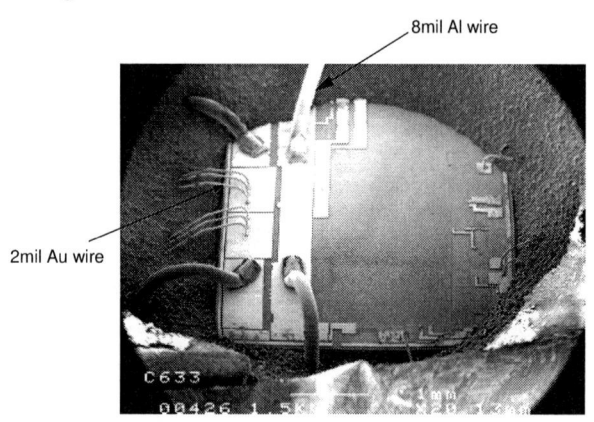

Figure 10: Shows 2 mil Au and 8mil Al wire bonding to a die for next generation auto applications

Another challenge is to meet demanding electromagnetic compatibility (EMC) requirements. As automotive electronics contents grow and complex electronics modules are distributed throughout the vehicle, the issue of EMC becomes a challenge. These challenges are met by using the design practices that minimizes: 1) electromagnetic susceptibility (EMS), 2) protecting from large supply transients caused by switching of inductive loads such as starters and, 3) minimize electromagnetic emission (EME) that has impact on other components. Designers can use number of design techniques available to them to meet these harsh environment challenges.

D. Microbolometer: IR Sensors for Automotive, Military, Industrial and Consumer Applications

So far infrared imaging has been primarily developed for military applications and has been based on quantum detection that operates at liquid nitrogen temperature thus restricting the use of this technology. An innovation in microbolometer technology that incorporates new materials in the CMOS process allows infrared thermal detection at room temperature. This has opened the door for low cost infrared imaging systems for automotive, industrial and commercial applications.

A bolometer works on the principle of temperature rise due to infrared radiation absorbed by an infrared absorbing material. The radiation absorbed by an absorbing material is dependent on the wavelength λ of the incident radiation. One of the materials used to detect temperature changes in bolometer is vanadium oxide (VOx). The relative power spectral response of VOx is presented in Figure 11.

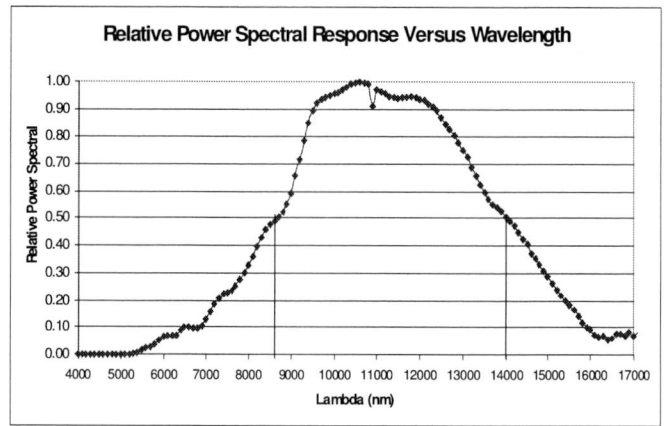

Figure 11: Power spectral response of VOx used in a bolometer. VOx based bolometer works in infrared region. Maximum efficiency is obtained in the wavelength range of 9-14μm.

The vanadium oxide (VOx) based bolometer works in the wavelength range of 9-14 μm. Striking radiation on a VOx pixel changes its temperature thus changing the electrical resistance of the pixel. This change in resistance can be detected by passing a current through the pixel. As pixel resistance changes so does the current. This change in temperature can be read out as change in current by a read-out integrated circuit (ROIC) and is used to produce an image.

A block diagram of a complete sensor circuit is shown in Figure 12. This circuit uses CMOS process technology as a ROIC and an integrated micro-electro-mechanical system (MEMS) process to build a sensor.

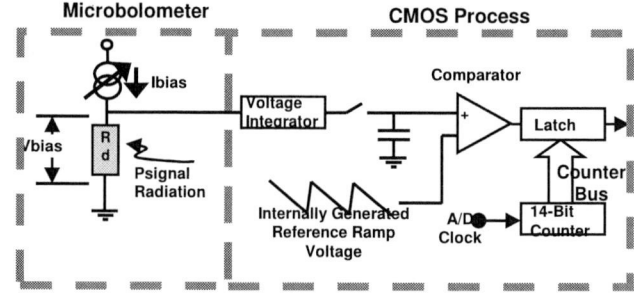

Figure 12: Showing ROIC (CMOS) and added micro bolometer circuit.

A combined MEMS/CMOS process flow is used to manufacture a bolometer. A typical fabrication process for bolometer sensor is shown in Figure 13.

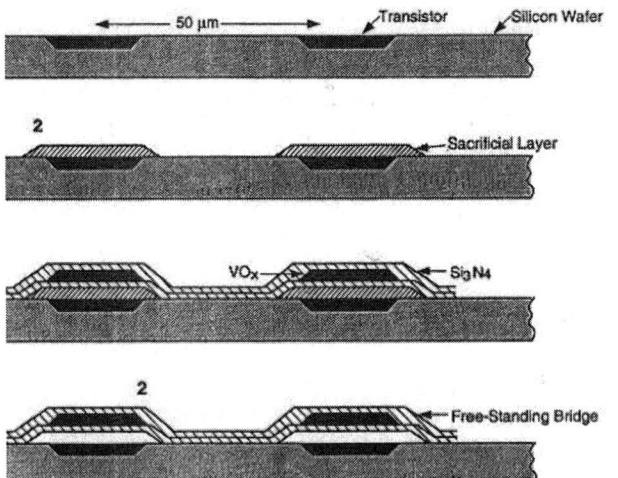

Figure 13: Figure 13: Presents a bolometer process flow showing various process steps. Courtesy: Yang Zhao, Ph.D. Thesis, University of California, Berkeley, 2002

The SEM picture of a microbolometer pixel manufactured using this process flow is shown in Figure 14. Figure 14 show various interconnect elements of the pixel such as readout contact, leg and a microbolometer bridge.

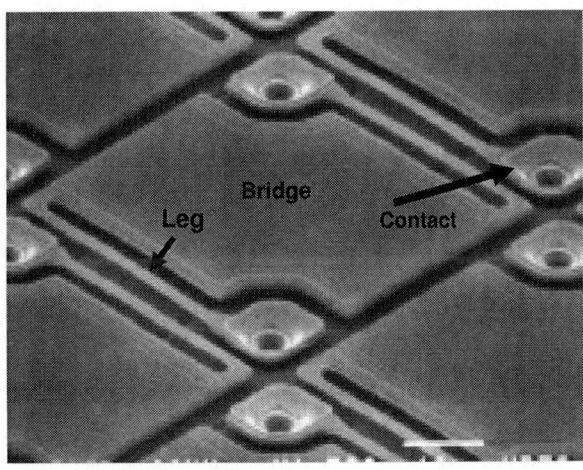

Figure 14: SEM picture of a bolometer pixel showing interconnect elements like bridge, leg and contact.

Infrared bolometer applications are increasing beyond military in the areas such as automotive safety, security, and law enforcement and consumer electronics where object detection in darkness or low visibility due to fog and other unfavorable climatic conditions. One of the examples of microbolometer application is shown in Figure 15. In this picture a camera display is shown of the road ahead during night time driving.

Figure 15: Picture of a display camera mounted in the dash board of a car displaying the road hazards during night time driving.

E. Bio-Field Effect Transistors (FETs) for Medical Applications (Lab – on- Chip)

Almost all methods for bio-materials detection, monitoring and analysis are used in laboratory setting and are not portable. However, recent innovations in microelectronics have made it possible to integrate various kinds of bio sensors for bio-material detection and analysis. These sensors include, Ion sensing Field Effect Transistor (ISFET) [4], microbial-FET [5] and other membrane based sensors. These are inexpensive, small portable devices that can be used at the point of need in situations where quick diagnostic is needed. One such innovation is portable Lab-on-Chip developed by NASA that is shown in Figure 16.

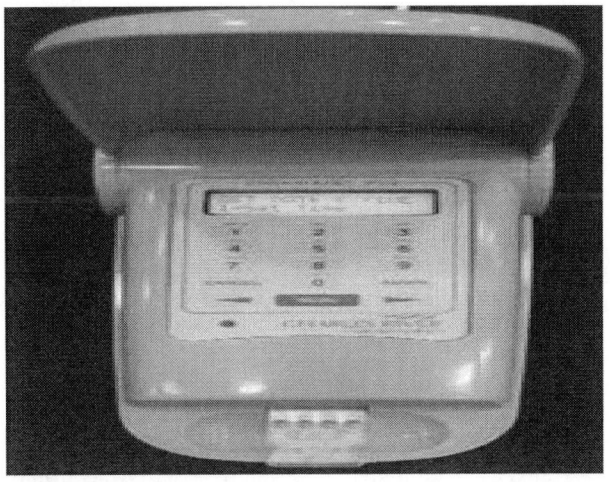

Figure 16: Lab-on-Chip developed by NASA showing analyzing unit. Courtesy: NASA, Marshal Space Center website.

The convergence of microelectronics and various sensing materials have made micro-sensors possible. In general all biosensors have a Field Effect Transistor (FET) that works as a transducer integrated with different ion sensitive materials depending on the application. One example of this application is ion sensing field effect transistor (ISFET) [6, 7]. ISFET is a field effect transistor in which standard polysilicon gate is replaced by a structure that is sensitive to ion concentration. This ISFET structure is shown in Figure 17.

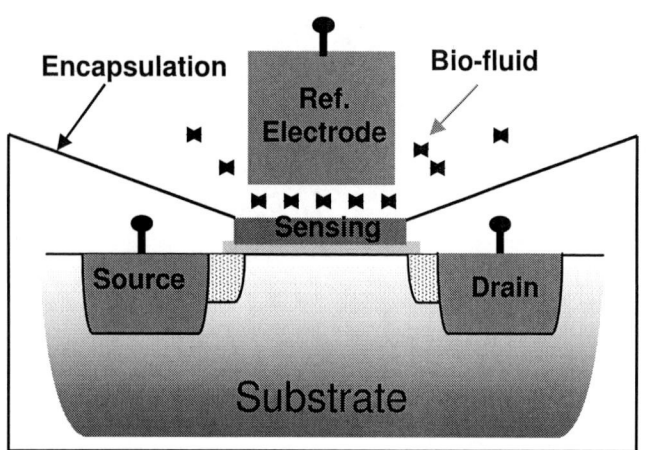

Figure 17: A cross section of an ISFET showing encapsulation and gate structure built on a CMOS process flow. Gate material depends upon the application but is usually an ion sensitive material.

ISFET works on the principle that the change in ion concentration will cause a change in the transistor threshold voltage (V_t) and thus the change in transistor's current-voltage (I-V) characteristics [8] that can be measured and correlated to the ion concentration in fluid.

Further diversification of ISFET can be achieved by integrating other sensors like oxygen, temperature and light sensors on a single chip that can be used as Cell Monitoring System (CMS) [8]. These cell monitors based on ISFET allow on-line and non invasive measurements of cellular parameters such as changes in extra cellular acidification due to addition of drugs. These on-line and non-invasive cell monitors will be very helpful in analyzing early drug uptake and reversing its effect if needed. The other applications of these biosensors could be toxicological monitoring of the environment [9].

F. System-in-Package (SiP) and System-on-Chip (SoC) Solutions to Conventional Scaling

Functional density trends can be maintained by innovations in packaging such as system-in-package (SiP) and by integrating multiple circuits into a single chip (SoC). SiP can be defined as "a combination of multiple active electronic components of different functionality, assembled in a single unit that provides multiple functions associated with a system. SiPs may contain passives, MEMS, optical components and other packages and devices [ITRS 2007]." One such example is shown in Figure 18.

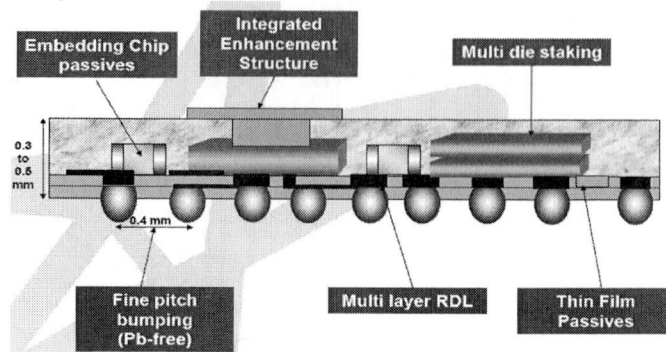

Figure 18: Showing circuit components in a 3D SiP. New interconnect technologies are needed to enable the migration of micro-systems to 3D SiP. Courtesy: IME, Singapore.

These technologies provide low cost solutions for continued improvement in density, performance and size at the system level as an alternative to conventional scaling.

Use of SiP vs. SoC will depend on the cost/function, time to market, system complexity and application. This relationship for SiP and SoC is shown in Figure 20 on a relative scale.

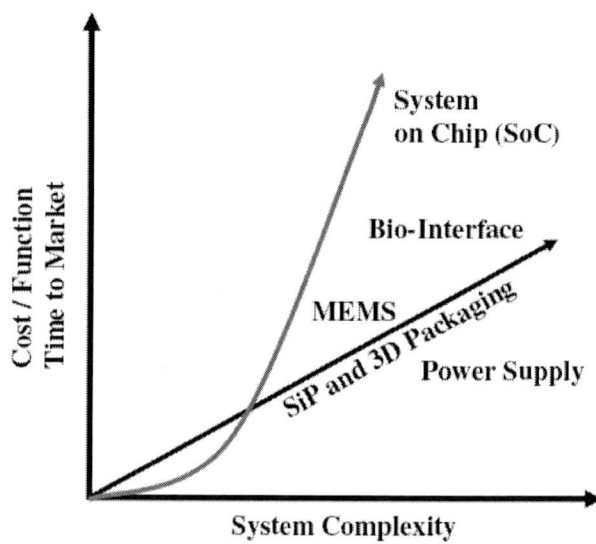

Figure 19: Showing the relative relationship among cost/function, time to market and system complexity on a relative scale for SiP and SoC. Source: ITRS meeting, July 2007.

Figure 19 shows that SiP provides advantages over SoC in most markets especially in markets that are dominated by consumer electronics. However, there will be some applications where SoC will provide a better alternative. One such application is implantable medical devices where reliability, size, and form factors are important selection criteria.

SiP technology builds on the integration of new elements in the base single-chip packaging. Emerging technologies that will be integrated in this base technology include wafer-level packaging, die stacking, through-silicon vias (TSV), and

978-1-4244-2018-6/08 $25.00 © 2008 IEEE   428

embedded actives and passives. Integrating these technologies into SiP provides a solution to achieve cost effective functional diversification (More-than-Moore).

1. SiP Requirements

There are many general requirements for SiP and can vary with the application. Some of these requirements are: (a). small form factors, (b). high functional density, (c). large memory capacity, (d). high reliability, (e). low package cost, (f). rapid time-to-market, (g). wireless connectivity. Extensive requirements are listed in reference [10]. Many types of SiP packages are already in production or are in development to meet these requirements. These are horizontal placement, stacked, and embedded structures. Examples of these major categories are shown in Figure 20.

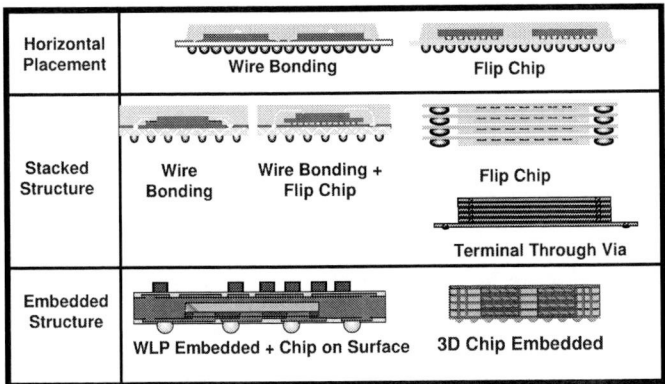

Figure 20: Showing innovation in SiP architectures to realize "More-than-Moore" diversification. Source: 2007 ITRS, Assembly and packaging, page 37.

The highest level of integration is achieved through 3-D packaging. Die stacking has been used for stacked memory/logic devices because of the high cost of embedding large memories in a chip. Consumer products such as cell phones, ipods and other consumer electronics that require large memories are driving the need of high level integration that can be achieved by SiP technologies. In die stacking wire bonding is used to connect the stacks to the package substrate and therefore is not a very efficient process. Therefore, new technologies are needed that are more efficient. One such emerging technology is TSV that allows more efficient die stacking and high level of integration.

2. SiP Challenges

There are number of challenges that SiP technology has to address to meet its potential. Following are some primary challenges.

(a). Wafer Thinning

SiP requirement for wafer thinning will reach 8 μm by the year 2015 [10]. This will pose a number of technical challenges for wafer thinning process. The challenges associated with handling the thinned wafers and thinned singulated die will be even more difficult.

(b) Reliability

Reliability, quality and manufacturing yield are key requirements of SiP. There are number of SiP failure modes and failure mechanisms. Some of these are: 1. coherent crack formation due to thermal mechanical mismatch, 2. interfacial delamination due to interfacial reactions causing loss of adhesion, 3. voids and pore formation due to mechanical creep, electro-migration and thermo-migration, and 4. material decomposition and bulk reactions. SiP technology challenges and potential solutions are discussed extensively in reference [10].

(c). Thermal dissipation

SiP has to provide the same thermal dissipation for each component as in a single chip package. SiP will have the most thermal variations due to different types of active components that have different power levels and incorporation of passive components that have very little heat generation. This temperature variation causes hot spots and damaging stress levels and will be worse in TSV structures due to number of stacked dies and reduced surface area available for heat dissipation. Potential solutions to resolve thermal dissipation issue require incorporating an improved thermal interface material for efficiently spreading heat and removing thermal energy from the heat source to the heat sink. Methods of forming CMOS compatible thermo-fluidic heat sinks and the use of micro-pipe I/O interconnects have been recently reported [11, 12]. One such example is shown in Figure 21.

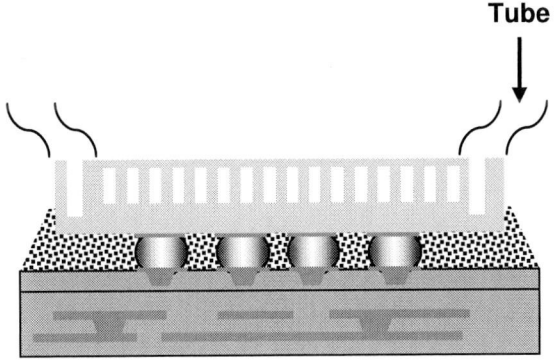

Figure 21: Showing (a) back-side integrated fluidic heat sink and back and front-side inlets/outlets.

(d) Signal and Power Integrity and Shielding

Signal integrity for high density interconnects is a primary challenge for 3D SiP integration. Cross talk, impedance discontinuities and timing skew are the major concerns. Furthermore, voltage drop for high speed signals may limit power delivery. Radiated noise within the package at high frequency and external electromagnetic noise sources should be shielded to make high density SiP integration successful.

978-1-4244-2018-6/08 $25.00 © 2008 IEEE

One example of noise if not shielded properly is shown in Figure 22.

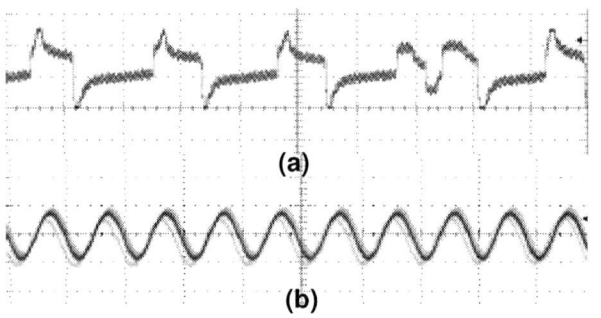

Figure 22: Showing (a) noise introduced in the signal by components in SiP and (b) signal with ground metallic shield integrated in SiP package.

(e). Testing of SiP

There are many challenges to address for SiP testing such as system level test, fine pitch capabilities, low cost of ownership and no damage while testing interconnects to name the few.

System level test solutions for SiP should provide similar test to those of standard packaging. Standard packaging process flow involves (a) pre-packaging test, (b) electrical test of the package substrate for opens and shorts, (c) assembly of chip, and (d) functional test of the packaged chip. 3D SiP testing involves many more steps depending on the level of integration thus making it much more challenging. Interconnection and functional tests must be performed at each level of a SiP. For example, first step in 3D SiP packaging is testing of all chips to be embedded followed by assembly of the 1st layer. Interconnection test at this point is same as that of standard packaging. However, it becomes increasingly challenging with each additional layer. Next step in SiP process is 2nd layer stack up with interconnects and micro-via process, assembly of the second layer and interconnection test of the second layer and so on. This increasing complexity of SiP testing will require custom socket designs, failure analysis for 3D SiP with TSV and probe access to the internally embedded die in the case of failure of a SiP component during final test.

Cost effective solutions of SiP testing are required since most SiP today are used in consumer electronics. SiPs will require both conventional and BIST testing due to access limitations in high component density, high speed of RF and digital circuits and the system level test requirements.

III. RELIABILITY

Automotive customers demand "0 ppm" failure rate over the product life. However, in practice the failure rate is 100 ppb – 1 ppm. Most cars have long life (10-15 years) but the good news is that they have limited driving time (10,000 hrs). High reliability requirements are for two reasons: 1) low early failure rate, and 2) safety and long lifetime - failure rate. Implications of these requirements are: 6-sigma design and high test coverage at different temperatures for I_{ddq} and V_{stress} as a screen for early life time failures. This requires ownership and commitment by design, manufacturing, test, and application levels. SiP and 3D integration will add more challenges to reliability as we add more components and integration levels.

IV. SUMMARY

Memories and microprocessor roadmaps require silicon based CMOS technologies that follow "Moore's" law. Enabling Moore's law requires large R&D investments. Total R&D spending by semiconductor companies exceeded $45 billion in 2007, an increase of nearly 12% from 2006. Overall R&D spending by semiconductor companies worldwide increased at an annual average rate of 10% between 2003 and 2007. Thus keeping up with scaling is becoming expensive and unaffordable.

However, a new trend of functional diversification is emerging that does not necessarily scale but provides additional value to the customer. This functional diversification is being called "More-than-Moore". The "More-than-Moore" approach allows the development of new products through innovation in circuit design, process modules architecture, embedded software, unique process technology and packaging solutions. These innovations enable the use of older, proven technologies in highly reliable products. Packaging advances (SiP and SOC) allow non-digital functions such as RF, power control, passive components, sensors and actuators to migrate from the system board level into a package level implementation. However, 3D level SiP integration has many challenges such as power dissipation, noise shielding, reliability and testing. These challenges have to be resolved before SiP technology can meet its potential.

V. REFERENCES

[1] J. Appels and V. Vaes, "HV thin layer devices (resurf devices," IEDM Technical Dig., 1979, p238.
[2] T. Fujihira, "Theory of semiconductor superjunction devices", Jpn. J. Appl. Phys., 36, 1997 p6254.
[3] P. Moens et al, "XtreMOS, the first integrated power device breaking the silicon limit", IEDM, 2006, p919.
[4] A.P. Soldatkin et al, Sensors and Actuators, 1985, no 8, p91
[5] F. Scheller, F. Schubert, Biosensors, 1992, p. 92
[6] Young-Chul Lee and Byung-Ki Sohn, J. Korean Phys. Soc. 2002, Vol.40, p.601.
[7] M. Kollar, Measurement Science Review, Vol. 6, 2006, p39.
[8] G.F Blackburn, Biosensors, 1987, p. 481.
[9] M. Kraus et al, Bioscope I, 1993 p. 24.
[10] ITRS, Assembly and Packaging, 2007, p.40
[11] B.Dang, M.S. Bakir and J. Meindl, IEEE, EDL, 2006, vol.27, p. 117.
[12] M.S. Bakir, B. Dang and J. Meindl, IEEE, CICC, 2007

Embedded Memory

Session 16

Chair: Kenji Noda, NScore
Co-Chair: Jean-Christophe Vial, Infineon

Penetrating into sub-tenth micron regime, variations in transistor characteristics are having a more critical impact on the design of high-density memories, such as SRAM, Flash and ROM. In this session, two papers from industry and five papers from academia will be presented.

The first paper form Intel presents an extremely high-speed and low-power level-2 cache SRAM design on a micro-processor in 45nm process generation.

The second paper from University of Minnesota describes unique techniques to operate 8T-SRAM at only 0.26 V. They use reverse short channel effect to improve cell write margin and read performance.

The third paper from U. C. Berkeley develops a sensor circuit to compensate for systematic variation affecting SRAM performance and stability. The sensors control word-lines at the optimum voltage level.

The forth paper from Carnegie Melon University presents a configurable replica bitline technique, which selects a subset of replica bitline driver cells from a pool of cells to optimize sense-amplifier timing.

The fifth paper from Purdue University presents a DTMOS based 6T-SRAM operating in the subthreshold region.

The sixth paper from University of Michigan describes a methodology for designing ROM to operate at ultra-low-voltage.

The last paper in this session from Shanghai Hua Hong NEC Electronics presents a 1-Mb embedded SONOS Flash memory having a calibration mechanism to compensate for variations in cell transistor characteristics.

Notes

978-1-4244-2018-6/08 $25.00 © 2008 IEEE

IEEE 2008 Custom Intergrated Circuits Conference (CICC)

A 512-KB Level-2 Cache Design in 45-nm for Low Power IA Processor Silverthorne

Mohammed H. Taufique, Alex Okpisz, Haseeb N. Ahmed, John R. Riley, Mohammad M. Hasan, Gian Gerosa

Intel Corporation

1501 S. Mopac, Suite 400. Austin, TX 78746

(512) 314-0597; mohammed.h.taufique@intel.com

Abstract

A self-timed, phase based 512KB L2 cache design in 45nm CMOS process for a low power IA core is presented. The design supports back to back access every core clock by performing sense amplifier (SA) evaluation and SA precharge (SAPCH) in one phase. Dynamic latch is used in the data-out path to enable the use of narrowest SA and latch pulse (LATCK) width during high volume manufacturing (HVM). Power gated (PG) WL driver, sleep and deep sleep for the SRAM array along with floating the bit lines and tri-state write driver schemes are implemented to reduce cache leakage power. The design meets the performance requirement of 2.0 GHz and 1.0 GHz at 1.0V and 0.75V respectively at 90C.

Introduction

This paper describes a self-timed, phase-based 8-way 1024-entry 512-KB L2 cache design in 45nm CMOS process designed for Intel's low power IA Silverthorne processor [1]. L2 cache line size is 64-bytes performing back to back (B2B) accesses on any given 32-byte chunk every core clock. The design is fuse-able to 4-WAY 256KB and extensible to 1MB with same latency. It has in-line Single Error Correction Double Error Detection (SECDED) mechanism for the data and tag arrays for enhanced reliability. Column redundancy is implemented on the data array to improve die yield. Total area of the cache including the clock bays is 3.05mm^2.

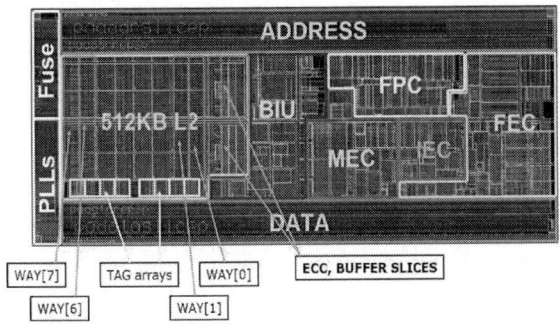

Figure 1: Silverthorne L2 cache Floor plan

Cache Sub array Organization

Figure 1 illustrates the processor floor plan consisting of the L2 data bank, L2 tag bank and BBL blocks. L2 DATA and TAG consist of 32 17.5-KByte data and 8 4.5-Kbyte tag sub arrays, respectively. L2 data sub-array has 256 cells on the bit line and 136 cells on the WL to achieve 70% array efficiency. TAG, STATE, ECNT and the LRU bits are combined into the one

TAG sub array to reduce area overhead and leakage power. L2 sub arrays use 0.3816um^2 6-T SRAM cell that is capable of retaining data below 650mV.

In the floor plan, TAG blocks are placed at the bottom of the DATA arrays to make the design expandable to a 1-MB L2 design by placing additional 512KB module at the bottom. Furthermore, this alleviates the timing requirement for the HIT signals going to the corresponding data array to "WAKE" up the sub arrays. The sub array output bus is connected in an 8-wide domino NOR configuration and only one of eight WAY-slices are accessed during each transaction. OR-ing of the 256-bit data-out bus is done in structure data path (SDP) blocks placed in between two data sub arrays. This allows for track sharing of the output bus and reduces device count over a conventional mux-type implementation.

Architectural Features for Power Savings

Several micro-architectural features are incorporated to reduce average power. During low performance operation, micro-code uses control registers in the BBL control unit to disable a pair of ways at a time. The disablement starts from the higher numbered WAYs such as 7/6 first, 5/4 second, 3/2 third and 1/0 last. A given WAY is flushed before disablement so that the entire way ends up invalid. The Tag array stays above retention value to maintain its invalid state while data drops below retention. During Intel® Deep Power down Mode, the voltage plane for the entire L2 including TAG, DATA and ECC logic are cut-off.

Data array remains at its retention level until a qualified ECC corrected HIT signal is received from the TAG array. This generates the SRAMWAKE signal for the desired WAY (only 4 out of 32) to reach up to full voltage from the retention level. Data array decode, access, in-line ECC and transmit to the L1 occur in 4 cycles for the 1st chunk. During low performance operation, micro-code uses control registers in the BBL control unit to disable a pair of ways at a time. The Tag stays above retention value to maintain its invalid state while data drops below retention. During Intel® Deep Power down Mode, the voltage plane for the entire L2 is cut-off. Additionally, phase based design with serialized chunk access reduces total leakage power as peripheral blocks i.e. repeater and ECC blocks are shared. This also reduces sequential count in the sub-array by 2X and noise associated with accessing all 64-byte in one cycle.

Self-timed Timer Design and Implementation

L2 cache is a phase based self-timed design for the read data-out and fully synchronous for the writes. L2 cache design is a phase based self-timed design to allow for a single-cycle throughput on a chunk basis, reduce overall latency to the level 1 cache (L1) and reduce sequential count. Recently many

978-1-4244-2018-6/08 $25.00 © 2008 IEEE

433

designs have moved away from the phase based self timed design in favor of cycle based [2] or decouple sense amplifier based fully-synchronous phase-based [3] design. Although cycle based design is acceptable for L3 caches with higher latency, it is not suitable for low latency caches. Similarly, decouple sense amplifier differential requirement is almost 2x to consider for faster cache using low leakage process.

Figure 2 shows the timing relationship and interlocking of the different clocks during the read operation. As shown in the diagram, word line (WL) is ON for the complete phase to allow for writing into the array, developing the Sense amplifier (SA) differential and pre-charging the bit lines (BLPCH). SA enable

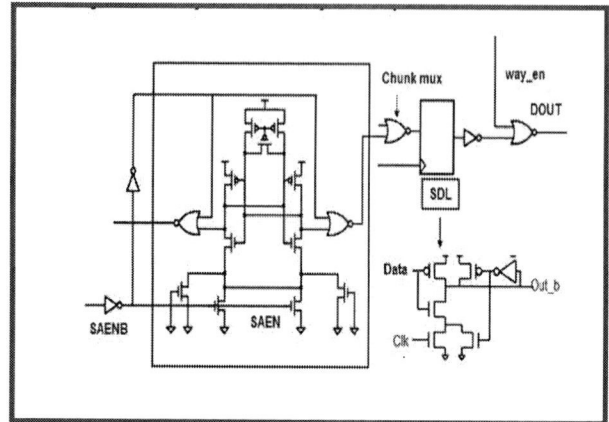

Figure 3: Cache data-out Implementation

Figure 4 shows the simulated SA pulse width requirement for the SDL and latch across various voltages using statistical variation tool. SDL relaxes the pulse-width requirement by 2X and 3X at 1.1V and 0.75V respectively. Since the SA evaluated

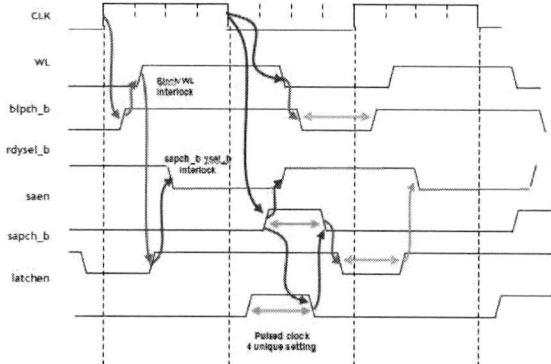

Figure 2: Read Timing Waveforms for Self timed Scheme

(SAEN), SA evaluation, DATA capture by the latch clock (LATEN) and sense amplifier precharge (SAPCH) are all done in the next phase. This require SA and LAT clock to be self-timed. A key challenge in the self-timed implementation is controlling pulse-width in HVM due to process variations. Narrower pulse-width affects data capture into the latch and wider pulse-width (PW) hurt SAPCH time impacting the frequency. Current design mitigates the challenge by replacing the conventional latch with a set-domino latch (SDL). Additionally, SAEN and LATCHEN clocks are fuse programmable to delay the launch and control the PW. A NOR gate is added at the SA output to avoid false discharge of the sense nodes and to ensure SDL evaluation occurs only during the access. Figure 3 shows the sense amplifier and the SDL used for the data array. This circuit requires and gated-NOR to ensure robustness during low frequency operation.

Additionally, bit line precharge window is adjusted dynamically for both read and write operations in the sub array Timer block to ensure adequate precharge time at all SA fuse setting and at all operating condition. In this scheme, BLPCH clock is muxed between the read and write operation to ensure adequate precharge time at all SA fuse settings and at all operating conditions.

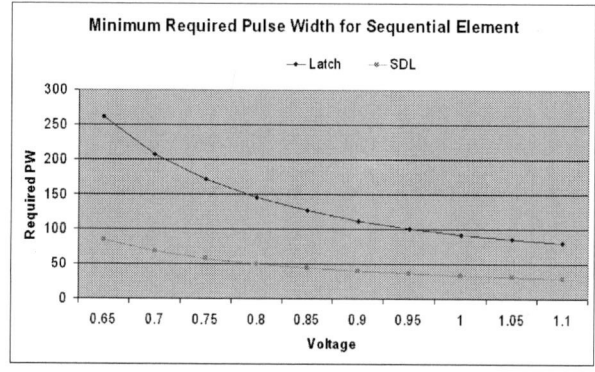

Figure 4: Self-timed PW Requirement (LATCH vs. SDL)

data can flow transparently through the SDL and its reset time is faster than the latch capture time; current design can tolerate narrower SA pulse over convention design process.

Cache write is also phased based and it is fully synchronous. WL, column selects (YSEL) and bit line pre-charge (BLPCH) are controlled by clock edges.

Leakage Power Reduction Techniques

Several circuit techniques are used to reduce the stand-by power of the L2 while maintaining its frequency target. Figure 5 shows the schematic for the power gated (PG) WL driver implementation. Unlike conventional designs [2], this scheme uses one of the pre-decoded bits as the WL power gating devices and is shared among other 8 (out of 256) drivers. This avoids the requirement for any early pre-decoded signal and reduces cache dynamic power by limiting the numbers of WL that are awaked.

This enables up to 8X reduction in leakage for the WLDRV circuits.

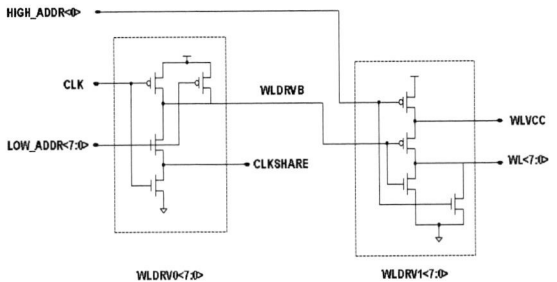

Figure 5: Power gated Word line driver scheme

Figure 6 shows single stack tri-state able write driver to reduce leakage power of the cache. Single stack write drivers enable

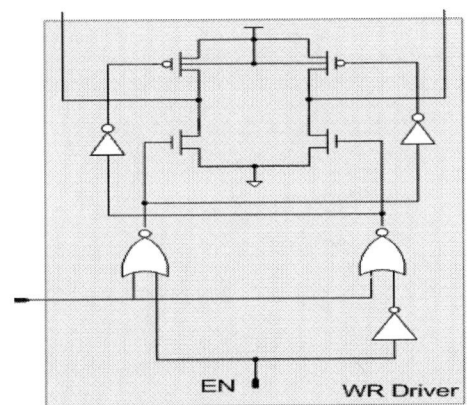

Figure 6: Power Gated (PG) WL Driver scheme

better scalability for performance at lower supply.

Figure 7 shows the L2 data array implementation of drowsy and shut-off [3] modes using sleep PMOS transistors (ST) and PG. In the drowsy mode, PG transistors are OFF; 4 levels of digital bias settings are used to hit the retention value (650mV) and SRAMVCC between 1.1V to 0.80V resulting in a 2.5X reduction in leakage power. During the shut-off mode, both PG and ST are off to provide additional 10X reduction in leakage power. Additionally, bit line and write drivers are kept floating for the WAYS that are not being accessed. Total leakage power is 20mW at 1.0V, 90C typical process corner.

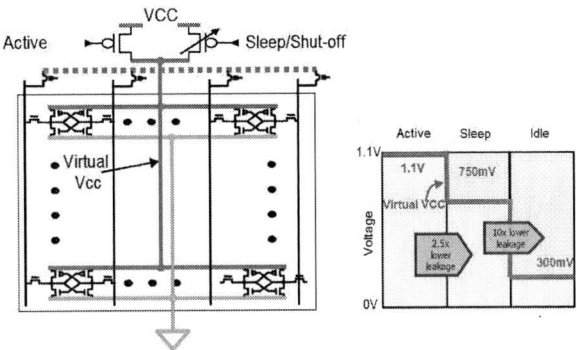

Figure 7: L2 cache sleep circuit and shut-off mode

Cache Debug Features

Both Data and Tag modules have PBIST to debug the arrays at speed. Additionally, self timed SA pulse width and launch time are being modulated using 4-bit [3:0] programmable fuse settings. First two bits control the pulse-width of the SA and LATCH clocks to enable the SA evaluation and data capture into the latch. The last two bits control the launch time of the clocks to provide more SA differentials. So [0011] would refer to SA/ LATCH narrowest pulse width (60ps @ 1.05V TZST 0C) setting with the largest delay of the clock launch time. One the other hand, [1100] would refer to the widest pulse width and fastest launch time for the SA and LATCH clocks. Each setting provides 30ps incremental delay at 1.05V in nominal process corner. As shown in figure 8, current design yields the highest frequency distribution at [0000] setting, as desired for maximum frequency. This also shows that design is sensitive to delaying the launch of the SA [0001], as would be the case for decouple SA, as opposed to the wider SA pulse width [0100].

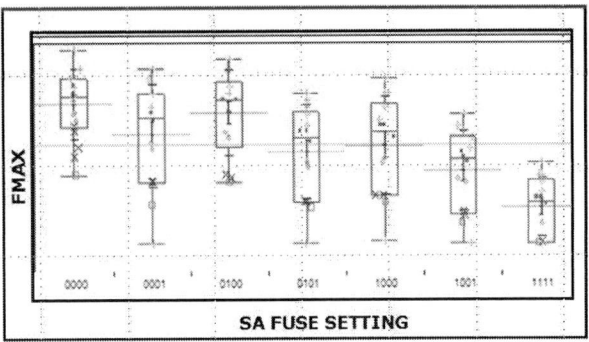

Figure 8: Fmax vs. SA Fuse Settings

In addition to fuse settings for Fmax characterization, cache also has stability test mode (STM) and low yield analysis (LYA) to characterize the bit cells for yield improvement.

As shown in Figure 8, the cache is characterized to operate

978-1-4244-2018-6/08 $25.00 © 2008 IEEE

above 2 GHz at 1.0V and 1 GHz at 0.75V at 90C for a typical unit.

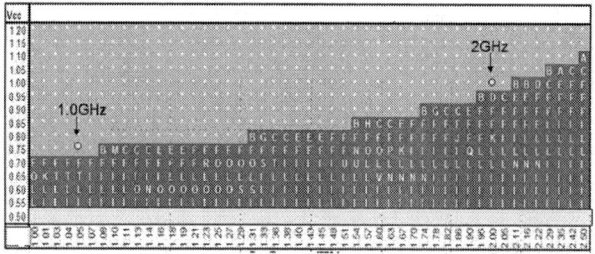

Figure 8: Cache voltage (V) vs. Frequency (GHz) SHMOO

Summary

A solution to address self-timed pulse issue for cache design with single cycle throughput and back to back access is demonstrated. Dynamic latch allowed narrowest pulse setting for the Cache design. Additionally, all leakage saving techniques is demonstrated to work in silicon at all operating conditions. Cache meets 2.0GHz and 1.0GHz frequency at 1.0V and 0.75V respectively.

ACKNOWLEDGMENTS:

The authors gratefully acknowledge the technical guidance from Kevin Zhang, and Eric Wang. Special thanks to Mike Tapia and Eng Chin Beh for providing the Silicon Data.

References

[1] G. Gerosa, et al., "A Sub 1W to 2W IA Processor for Mobile Internet Devices in 45nm Hi-K Metal Gate CMOS", ISSCC Digest of Technical Papers, pp.256-257, Feb. 2008.
[2] K. Zhang, "A 153Mb SRAM Design with Dynamic Stability Enhancement and Leakage Reduction in 45nm Hi-K Metal Gate CMOS Technology", ISSCC Digest of Technical Papers, pp. 376-377. Feb. 2008.
[3] Yih Wang, "A 1.1Ghz 12uA/Mb-leakage SRAM design in 65nm Ultra-Low-Power CMOS Technology with Integrated Leakage Reduction for Mobile Applications, ISSCC, Feb 2007

IEEE 2008 Custom Intergrated Circuits Conference (CICC)

A Voltage Scalable 0.26V, 64kb 8T SRAM with V_{min} Lowering Techniques and Deep Sleep Mode

Tae-Hyoung Kim, Jason Liu and Chris H. Kim

Department of Electrical and Computer Engineering
University of Minnesota, Minneapolis, MN 55455, USA

Abstract- **A voltage scalable 0.26V, 64kb 8T SRAM with 512 cells per bitline is implemented in a 130nm CMOS process. Reverse short channel effect was utilized to improve cell write margin and read performance. A marginal bitline leakage compensation scheme was used during read operation to lower V_{min} down to 0.26V. Floating write bitline and read bitline, auto wordline pulse width control, and a deep sleep mode minimize the active and standby leakage power consumption.**

I. INTRODUCTION

Subthreshold logic circuits are becoming increasingly popular in ultra-low power applications where minimal power consumption is the primary design constraint [1][2][3][4]. Subthreshold static CMOS logic can operate while consuming roughly an order of magnitude less power than in the normal strong-inversion region. However the MOS current becomes an exponential function of gate and threshold voltage in the subthreshold regime. This leads to an exponential increase in MOS current variability under Process-Voltage-Temperature (PVT) fluctuations.

SRAMs that can operate under a wide range of supply voltages are necessary for achieving high-performance during normal modes while minimizing the power consumption during low voltage modes [5]. For reliable operation from the super-threshold region down to the deep sub-threshold region, key memory design metrics such as noise margin, speed, and power consumption need to be examined at different supply voltages and accounted for as such. In the subthreshold region, conventional 6-T SRAMs fail to deliver the density and yield requirements due to the reduced Static Noise Margin (SNM), poor writability, limited number of cells per bitline, and reduced bitline sensing margin. Decoupled SRAM cells have been proposed to make the read mode SNM equal to the hold mode SNM by isolating the SRAM cell nodes from the bitline [6][7][8][9]. Writability has been improved in prior designs by using a higher supply voltage for the write access transistors at the cost of generating and routing the extra supply voltage [6].

In this work, we demonstrate a 64kb SRAM with several circuit techniques that can be activated at ultra low voltages to expand the operating range: (i) 8T SRAM cell utilizing Reverse Short Channel Effect (RSCE) for improved writability and read performance, (ii) Marginal Bitline Leakage Compensation (MBLC) scheme for improved read sensitivity and precharge elimination, (iii) floating Read BitLines (RBL) and Write BitLines (WBL) to minimize bitline leakage, (iv) deep sleep mode, and (v) automatic read wordline pulse width control for power reduction and improved bitline sensing margin.

II. V_{min} LOWERING CIRCUIT TECHNIQUES

A. 8T SRAM Cell Utilizing Reverse Short Channel Effect

Reverse Short Channel Effect (RSCE) becomes pronounced at lower supply voltages due to the significantly reduced Drain Induced Barrier Lowering (DIBL) effect [9]. This increases the current drivability per width as the channel length increases at sub-0.6V. The proposed 8T SRAM cell uses 3X longer channel length in the write access devices and 2X longer channel length in read path devices (Fig. 1). This improves the write margin from -90mV to 70mV and the read performance by 52% at 0.2V without increasing bitline capacitance. The proposed 8T SRAM cell utilizing RSCE has an area overhead of 20% compared to conventional 8T cells [10].

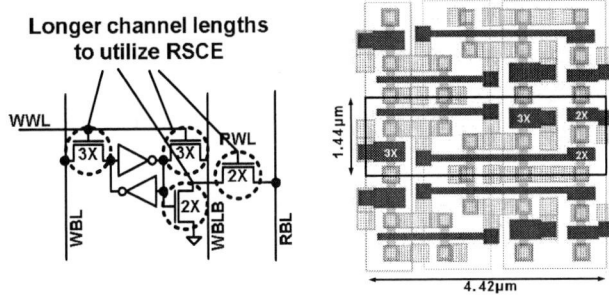

Figure 1. Schematic and layout of the proposed 8T SRAM cell utilizing Reverse Short Channel Effect (RSCE).

B. Marginal Bitline Leakage Compensation (MBLC) Scheme

A Marginal Bitline Leakage Compensation (MBLC) scheme shown in Fig. 2 compensates for the RBL leakage in the unaccessed cells. The RBL voltage is tuned to settle just above the Sense Amplifier (SA) trip point by progressively turning on the marginal compensation devices, which is based on a replica bitline circuit. When a logic '0' is read, only a small swing is required to change the SA output, which is beneficial when a cell current is comparable to the bitline leakage current. The logic level of RBL during read operation is decided by the balance between the cell read current, the pull-down leakage current, and this marginal compensation current. The marginal compensation current should be large enough to produce logic '1' for the worst case pull-down leakage current, while still being small enough to produce logic '0' for the pull-down cell current and the smallest bitline leakage. The optimal compensation current is generated by the replica bitline. A feedback loop charges RBL_REPLICA up to a point where the SA output switches to '1'. The worst case data pattern is hardwired in the replica bitline. Initially, the SA output is '0' because PCHG<3:0> is initialized with '1's, turning off all marginal precharge devices. An increasing

978-1-4244-2018-6/08 $25.00 © 2008 IEEE

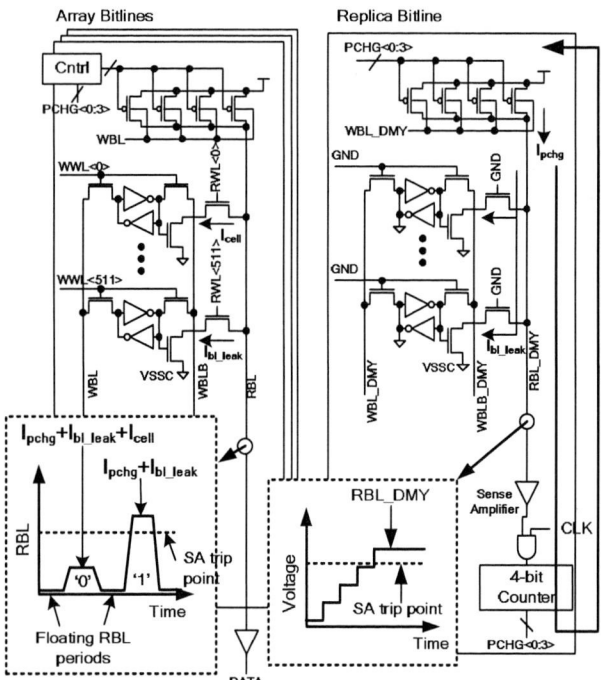

Figure 2. Marginal Bitline Leakage Compensation (MBLC) scheme.

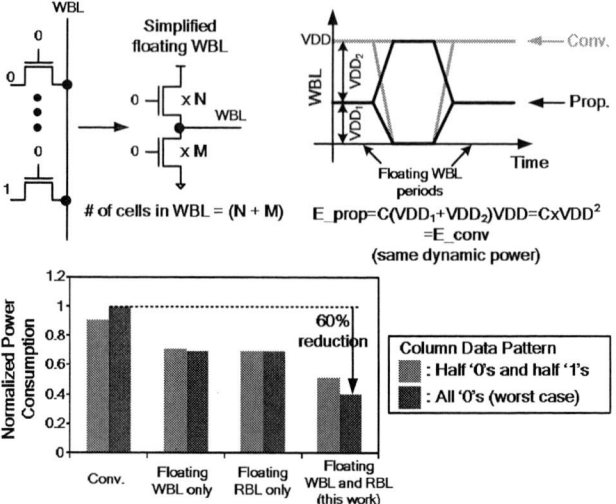

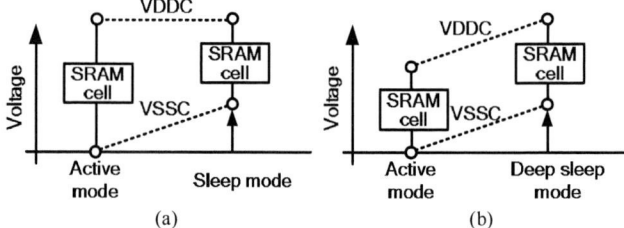

Figure 3. Power reduction using floating read and write bitlines.

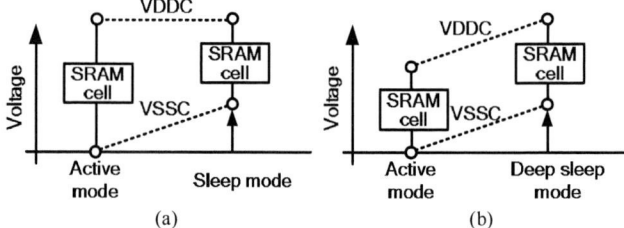

Figure 4. (a) Conventional sleep mode. (b) Proposed deep sleep mode.

number of pull-up devices are then turned on until the SA output switches to '1'. Additional margin can be given by providing added compensation current through selectively turning on extra devices in the accessed bitline. The optimal compensation current depends on the data pattern in a column, since the amount of bitline leakage is related to the column data. This data dependency was accounted for by connecting the body of the compensating PMOS devices to the floating WBL voltage, which is also determined by the data pattern stored in the SRAM column. Increasing the number of cells storing '0' decreases the floating WBL voltage, which in turn increases the amount of marginal compensation current by forward body biasing the PMOS devices.

III. ACTIVE LEAKAGE REDUCTION AND DEEP SLEEP MODE

A. Floating Read/Write Bitlines

Leakage current in inactive memory cells account for most of the SRAM power consumption. Circuit techniques for leakage control are critical for reducing the total power consumption, especially in the sub-threshold region. RBL leakage is one of the most dominant leakage components and is inevitable in conventional memories where bitlines have to be precharged to VDD. In our design, RBL is left floating without being precharged whenever RWL is low. During the read operation, the MBLC scheme provides the compensation pull-up current to generate logic high or low levels in the RBL with large sensing margin. Write bitline (WBL) is also left floating when WWL is low so that it will automatically settle to a level which minimizes the leakage current as shown in Fig. 3. The proposed scheme has no energy overhead during the write operation compared to the conventional scheme. A total SRAM leakage reduction of 44% to 60% can be obtained by using floating RBLs and WBLs.

B. Deep Sleep Mode

Sleep transistors are popular for reducing SRAM leakage current in standby mode by collapsing the virtual supply rails. However, due to the fact that the voltage margin is already close to the functionality limit, it is difficult to use conventional footer sleep transistors for subthreshold designs. In this work, we propose a deep sleep mode illustrated in Fig. 4 to reduce the standby leakage in subthreshold memory designs. During sleep mode, the proposed scheme raises both VDDC and VSSC while keeping the cell voltage, VDDC-VSSC, constant to reduce leakage while maintaining the same cell stability in deep sleep mode. SRAM cell leakage is reduced due to the negative VGS in the write access transistors and the increased threshold voltage of the pull-down NMOS devices due to the reverse body bias. Raising the VDDC and VSSC too high may increase the WBL and WBLB levels because they are decided by the column data pattern and the SRAM cell node voltages. If both VDDC and VSSC are raised excessively, the pull-up path in the interfacing circuit becomes leaky and current starts to flow from WBL and WBLB to VDDP. Fig. 5 demonstrates the leaky current path and the normalized SRAM leakage current reduction obtained by using the proposed deep sleep mode. By applying an optimal supply voltage (VDDC=0.83V, VSSC=0.60V), 87% reduction in cell leakage was obtained during a deep sleep mode.

C. Automatic Wordline Pulse Width Control

The pulse width of wordline signals should be carefully controlled to avoid access failures while minimizing the delay and power overhead. Due to the fact that read speed is highly dependent upon the PVT conditions, a scheme to

978-1-4244-2018-6/08 $25.00 © 2008 IEEE

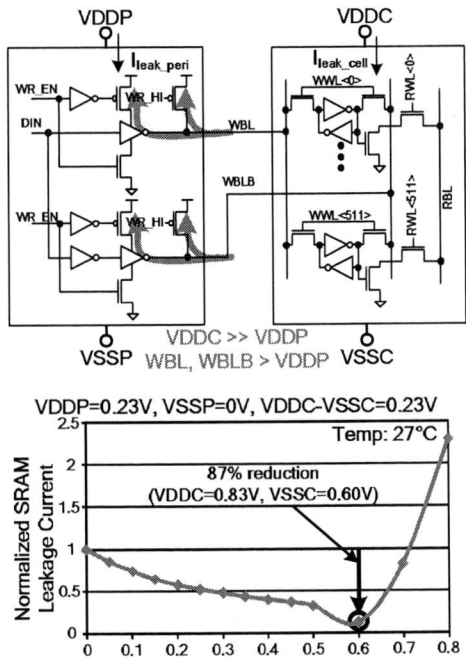

Figure 5. Deep sleep mode and simulated leakage reduction.

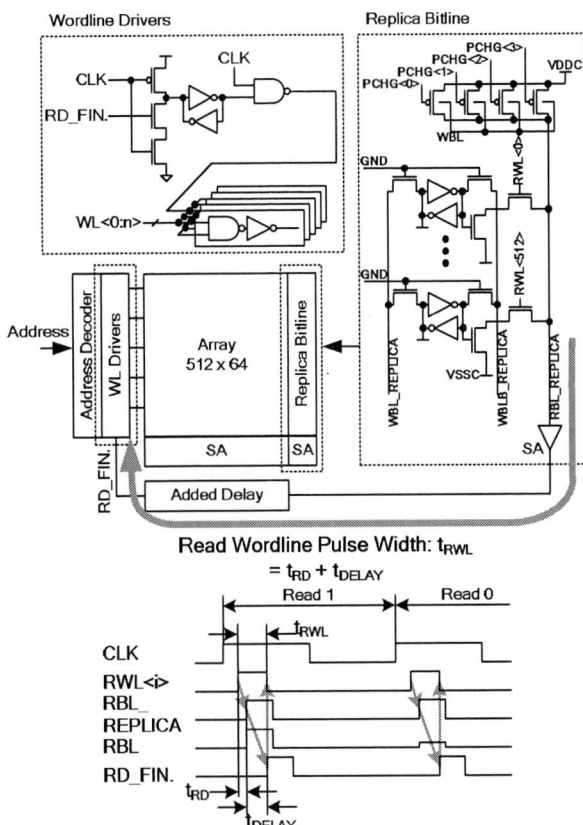

Figure 6. Read wordline pulse width control for PVT tracking.

automatically adjust the read wordline pulse width based on PVT variations is proposed (Fig. 6). A replica bitline is used to generate the amount of delay necessary for the SA to capture the read data. The delayed SA output RD_FIN from the replica bitline disables the read wordline and shuts off the marginal precharge devices. Added delay, which can be trimmed, takes care of the within die variations. The proposed wordline pulse width control scheme also reduces the unnecessary read power consumption by minimizing the wordline enable time.

IV. EXPERIMENTAL RESULTS

A 64kb SRAM was fabricated in a 130nm CMOS technology with a nominal supply voltage of 1.2V. Fig. 7 shows the measured power consumption and leakage current. We observed SRAM cells functional down to 0.23V running at 100kHz and consuming 4.3μW (Fig. 7 (a)). At 0.4V, the operation frequency was 6.7MHz with a power consumption of 10.8μW. The measured SRAM leakage currents from different dies are shown in Fig. 7 (b). Variation in leakage current was 2.0X at 0.3V due to its exponential dependency on device threshold voltage. The normalized leakage current measured at different temperatures is shown in Fig. 7 (c). The leakage current at 110°C is 3.4X larger than that at 27°C when the supply voltage is 0.23V. Fig. 7 (d) illustrates the normalized leakage current reduction achieved using the proposed deep sleep mode. The total SRAM leakage including the array and peripheral components was reduced by 69% in deep sleep mode by raising the VSSC to 0.45V while maintaining a cell voltage of 0.23V. The initial leakage reduction is large when raising VSSC due to the strong negative Vgs effect in conjunct with the reverse body biasing

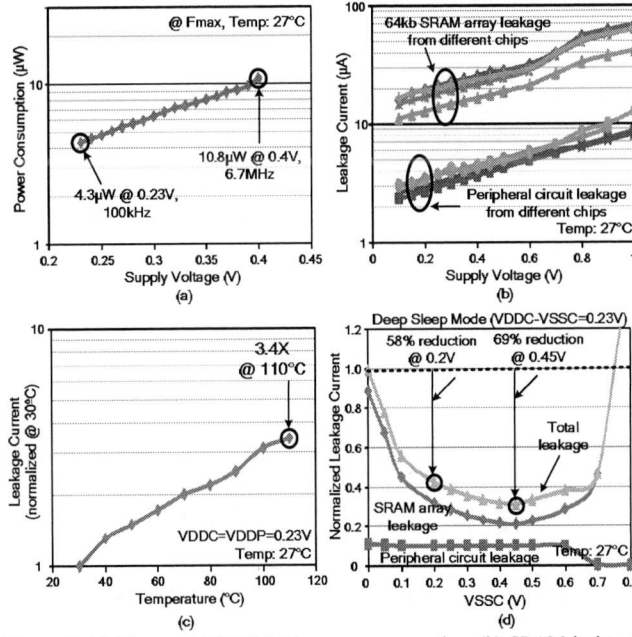

Figure 7. (a) Measured SRAM total power consumption. (b) SRAM leakage current varying supply voltage. (c) Normalized leakage current at different temperature. (d) Leakage current reduction in deep sleep mode.

effect. 58% leakage reduction was achievable using a VSSC of 0.2V during deep sleep mode. The smaller offset in VDDC and VSSC improves the efficiency and area overhead of the

978-1-4244-2018-6/08 $25.00 © 2008 IEEE

charge pumps that generate the boosted voltages. In this test chip, we used an external supply for the higher supply voltages needed during the deep sleep mode.

Fig. 8 (a) shows the shmoo plot when the proposed MBLC scheme is on and off. When the MBLC scheme is off, a conventional fixed precharge device is used. The V_{min} of the implemented SRAM is reduced from 0.28V to 0.23V by activating the MBLC scheme. V_{min} map for read and write operations for an 8-by-8 subarray is shown in Fig. 8 (b). We have also tested the feedback control circuit for the MBLC scheme which compensates the bitline leakage on-the-fly. The 4 bit counter used in the MBLC requires up to 16 clock cycles to generate the optimal precharge strength. Fig. 9 shows SA

outputs with two different trip points. It is shown that a SA with a higher trip point requires additional cycles to turn on more number of compensation devices. Similarly, more devices should be turned on for a larger bitline leakage current due to process variations. The die photo and chip performance summary are given in Fig. 10. The proposed MBLC and read wordline pulse width control scheme incur an area overhead of 1.3%.

V. CONCLUSIONS

A 64kb SRAM was fabricated in a 130nm CMOS technology with 512 cells per bitline. The SRAM was functional down to 0.23V running at 100kHz and consuming 4.3μW. RSCE was utilized in the read and write port of the SRAM cell to improve write margin and read performance. The MBLC scheme was proposed to expand V_{min} down to 0.23V. Floating read and write bitline scheme was proposed to eliminate RBL leakage current during non-read operation and minimize write bitline leakage current during non-write operation. Deep sleep mode for subthreshold memories was proposed to further reduce the leakage current during standby mode. The total SRAM leakage including the array and peripheral components was reduced by 69% in deep sleep mode by raising the VSSC to 0.45V while maintaining a constant cell voltage. An automatic read wordline pulse width control scheme improves readability and reduces wasted read power by tracking the PVT variations.

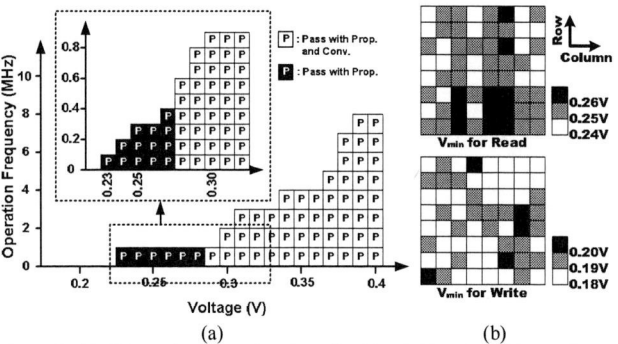

Figure 8. (a) Shmoo plot for an SRAM cell with a 0.23V V_{min}. (b) V_{min} for read and write from an 8-by-8 subarray.

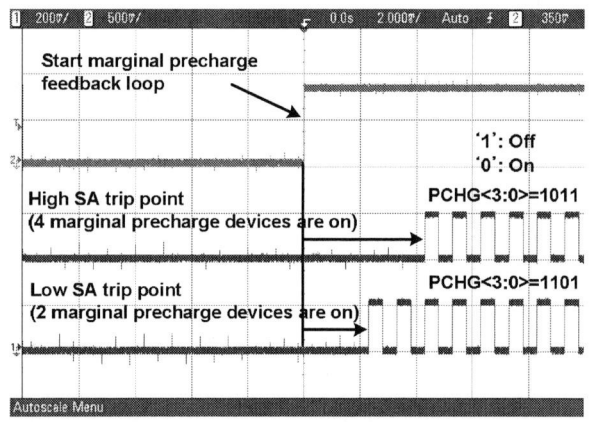

Figure 9. Output waveforms from marginal bitline leakage compensation control circuit.

REFERENCES

[1] H. Kim, H. Soeleman, K. Roy, "An ultra-low power DLMS filter for hearing aid applications", IEEE Trans. VLSI Systems, Volume 11, , pp. 1058-1067, Dec. 2003.

[2] A. Bryant, J. Brown, P. Cottrell, et al., "Low-power CMOS at Vdd=4kT/q", Device Research Conference, pp. 22-23, 2001.

[3] B. Zhai, S. Hanson, D. Blaauw, D. Sylvester, "Analysis and mitigation of variability in subthreshold design", International Symposium on Low Power Electronics and Design, pp. 20-25, Aug. 2005.

[4] A. Wang, A.P. Chandrakasan, "A 180-mV subthreshold FFT processor using a minimum energy design methodology", IEEE J. of Solid-State Circuits, Volume 40, pp. 310-319, Jan. 2005.

[5] L. Chang, Y. Nakamura, R.K. Montoye, J. Sawada, et al., "A 5.3GHz 8T-SRAM with operation down to 0.41V in 65nm CMOS", VLSI Circuits Symposium, pp. 252-253, June 2007.

[6] B.H. Calhoun, A. Chandrakasan, "A 256kb Sub-threshold SRAM using 65 nm CMOS," International Solid-State Circuits Conference, pp. 628-629, Feb. 2006.

[7] L. Chang, D.M. Fried, J. Hergenrother, J.W. Sleight, et al., "Stable SRAM cell design for the 32 nm node and beyond," Symposium on VLSI Technology, pp. 128-129, June 2005.

[8] J. Chen, L. T. Clark, T. Chen, "An Ultra-Low-Power Memory with a Subthreshold Power Supply Voltage," IEEE J. of Solid-State Circuits, Volume 41, pp. 2344-2353, Oct. 2006.

[9] T. Kim, J. Liu, J. Keane, C. Kim, "A high-density subthreshold SRAM with data-independent bitline leakage and virtual ground replica scheme", International Solid-State Circuits Conference, pp. 330-331, Feb. 2007.

[10] T. Kim, J. Liu, C. Kim, "An 8T Subthreshold SRAM Cell Utilizing Reverse Short Channel Effect for Write Margin and Read Performance Improvement", Custom Integrated Circuits Conference, pp. 241-244, Oct. 2007.

Technology	130nm 8-metal CMOS
SRAM Area	0.72x0.85mm^2
VCC	≥0.26V
Size	64kb
Performance	100kHz @ 0.23V, 27°C
	15MHz @ 0.60V, 27°C
Power Consumption	4.3μA @ 0.23V, 27°C
	34μA @ 0.60V, 27°C
Deep Sleep Mode Leak.	69% reduction

Figure 10. Chip microphotograph and performance summary.

IEEE 2008 Custom Intergrated Circuits Conference (CICC)

Compensation of Systematic Variations Through Optimal Biasing of SRAM Wordlines

Andrew Carlson, Zheng Guo, Liang-Teck Pang, Tsu-Jae King Liu, and Borivoje Nikolić

Dept. of Electrical Engineering and Computer Sciences, University of California, Berkeley 94720

E-mail: acarlson@eecs.berkeley.edu, Tel.: (510) 643-2558

Abstract – Increasing process variability is slowing SRAM scaling by reducing both read and write margins. Existing techniques to compensate for systematic variations optimize cell stability with excessive penalty to writeability. To maximize overall yield, a sensor circuit is presented that optimizes the read / write tradeoff in the presence of process, voltage, and temperature variations. Sensors implemented in a low-power 45nm test chip adjust the wordline voltage to track changes in the optimal value within 30mV over the entire range of operation.

I. INTRODUCTION

Process variations represent a major challenge for continued SRAM scaling. By altering the relative device strengths, variations in parameters such as threshold voltage (V_T), gate length (L_G), or active width can upset the cell balance and cause read or write failures. Additionally, SRAM cells (Fig. 1) are vulnerable during a half-select state, when the wordline (WL) is active but the cell is not being written. Reducing the supply voltage adversely affects the stability and even more so the writeability of SRAM, as shown in Fig. 2. Balancing the noise read/write margins at the nominal supply voltage does not guarantee that they will be balanced under all operating conditions and for all supply voltages in the operating range.

Process variations can be systematic or random in nature. Systematic variations, such as those caused by changes in critical-dimension bias or alignment, are spatially correlated, affecting multiple cells in close proximity. The spatial correlation can extend over several microns (within die) to several millimeters (die-to-die), or from wafer to wafer. In order to detect spatially correlated variations, small sensor circuits can be placed on the periphery of an SRAM array [1, 2], using partial cell layouts to replicate the environment of the actual SRAM cells.

The information from the sensors can be used to compensate average read or write margins of the arrays through V_{DD} adjustments, well-, bitline- or WL-biasing. Ohbayashi *et al.* use several replica access transistors with an identical layout to PG devices in the cell [1]. These transistors oppose a wide PMOS driver to set the WL bias voltage (V_{WL}) for cell access. The circuit thereby compensates for systematic NMOS variations by adjusting V_{GS} on the PG transistors, which increases robustness to read upset but degrades writeability. Additional circuitry is used to recover the writeability to a desired minimum threshold. Takeda *et al.* use a "monitoring circuit" of one-and-a-half cells to determine the minimum V_{WL} or maximum V_{DD} to ensure writeability [2]: the internal nodes of a monitor cell are connected to an op-amp, which sets V_{DD} - V_{WL} to force the cell into a meta-stable state (with equal internal node voltages), which is the minimum DC condition for a successful write. This approach can increase robustness during read well beyond what is practically required, with an excessive penalty to write access time.

Cells in the same row must simultaneously be both writeable and robust to half-select upset. There is a well-understood tradeoff between the write access time and stability during a half-select condition, which at DC can be measured by the read static noise margin (RSNM) [3]. Increasing the strength of the PG devices improves writeability but degrades RSNM. Increasing the strength of

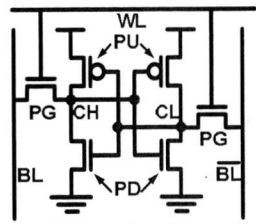

Fig. 1. Device and node naming convention for the 6-T SRAM cell

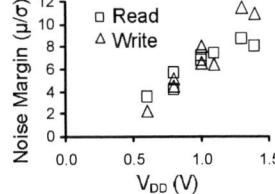

Fig. 2. Measured noise margins. Write margins decrease faster with decreasing supply voltage than do read margins.

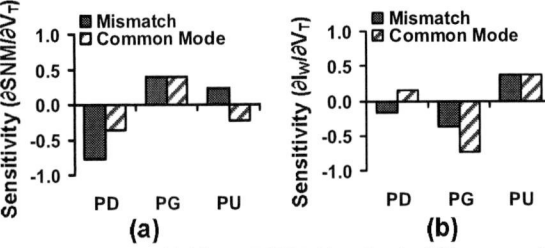

Fig. 3. Normalized sensitivities to RSNM (a) and writeability current [5] (b), illustrating both mismatch and common mode variations affect SRAM.

the PU devices tends to produce the opposite effect. Due to this read / write tradeoff, optimizing for writeability impairs the half-select stability, and vice-versa. An improved approach to correction for systematic variations that optimizes this tradeoff can maximize overall SRAM yield.

In this work, a new circuit is presented to optimize the read / write tradeoff. It uses multiple sensor cells with gate and active SRAM layouts to generate the optimal V_{WL} in the presence of systematic process, voltage, and temperature variations. A test chip in an industrial 45nm bulk CMOS process is used to demonstrate operation of the variation sensor.

II. SYSTEMATIC CELL VARIABILITY

The variability of any particular SRAM cell is determined by a combination of many random and systematic variations at the device level. Both types of variation can affect device pairs in a common or differential (i.e., mismatch) mode. The relative importance of common-mode and mismatch variations can be illustrated with simulated, normalized sensitivities of read stability and writeability metrics to V_T variation (Fig. 3). RSNM is much more sensitive to mismatch variations than common-mode variations in the PD devices, whereas it is equally sensitive to mismatch and common-mode variations in the PG devices, since each of these affects only one half of a cell (Fig. 3a). For writeability, the sensitivities are more distributed, and are accentuated for common-mode variations in the PG devices (Fig. 3b) due to the complementary pull-down / pull-up behaviors of these devices on each side of the cell. Note that common-mode variations generally degrade either stability or writeability via the read / write tradeoff, whereas mismatch variations always degrade one half of a cell. Control of both common-mode and mismatch variations is therefore important for SRAM robustness.

978-1-4244-2018-6/08 $25.00 © 2008 IEEE

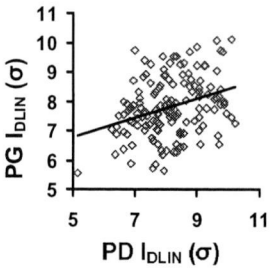

Fig. 4. I_{DLIN} measurements of adjacent PG and PD devices (normalized by their standard deviations) are weakly correlated ($R^2 = 0.11$). Systematic variations account for about 11% of total PG I_{DLIN} variability.

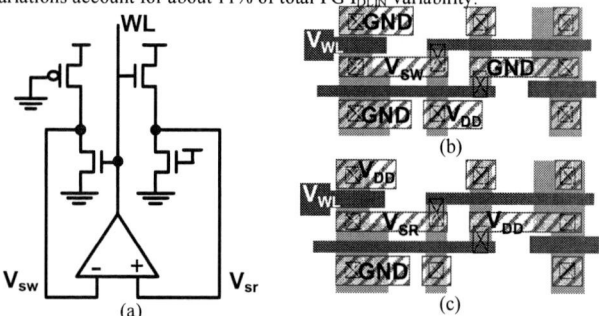

Fig. 5. Multiple copies of cells with identical SRAM layouts through the first metal layer are configured for worst-case write (a, left side; layout: b) or worst case read (a, right side; layout: c). The op-amp changes V_{WL} to optimize the read / write tradeoff for systematic variations.

Variation analyses have traditionally focused on the random component for several reasons. The magnitude of random variations has been increasing with continued technology scaling and is beginning to reach prohibitive levels. Unlike systematic variations, random variations cannot be compensated by a process adjustment. Systematic variations are harder to model, since they are not independent and can arise from complex sources.

Systematic variations can cause both common-mode and differential variations in the cell performance because they can arise from multiple, opposing sources. For example, a systematic bias in transistor gate lengths would be a source of common-mode variation, whereas a gate misalignment would cause a mismatch. The total variation can be decomposed into a net systematic component (σ_s) and a random component (σ_r). Fig. 4 illustrates the correlation between adjacent PD and PG I_{DLIN} (I_{DS} at $V_{DS} = 0.1V$, $V_{GS} = 1.0V$) measured from 144 cells in an early industrial 45nm process. The correlation of the currents is $R^2 = 0.11$, suggesting that approximately 11% of the total variation is due to spatially correlated sources. Variations from a perfectly random source alone would be expected to give a correlation less than this with 91% confidence. Among all device pairings, the correlation between PG and PD I_{DLIN} was the largest, indicating that the process has a very low within-die systematic variability. Greater systematic variations are expected at the die-to-die and wafer-to-wafer levels, however.

III. VARIATION SENSOR DESIGN

The systematic process variation can be expected to degrade either writeability or cell stability during a half-select. Fig. 5 illustrates a variation sensor circuit that restores the balance using the read / write tradeoff. The sensor comprises half-cells configured for a worst-case read or a worst-case write. Actual SRAM bitcell layouts are used up through the first metal layer to ensure maximum sensitivity to layout-sensitive

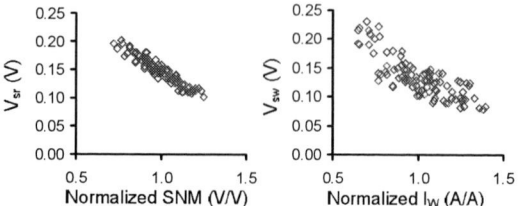

Fig. 6. Simulated outputs of the worst-case read and write sensor cells, V_{sr} and V_{sw}, are negatively correlated with RSNM and I_w, respectively. The read / write tradeoff for a set of cells can be estimated with minimal area overhead using these metrics.

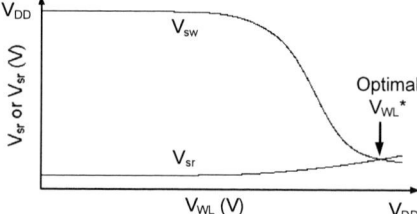

Fig. 7. The optimal $V_{WL}*$ is that which minimizes the maximum of V_{sr} and V_{sw}. Because of the read / write tradeoff, this point is where $V_{sr} = V_{sw}$.

variations. The worst-case condition for writing a cell consists of a voltage divider with the PG device contesting a fully-on PU device to bring the internal node voltage, V_{sw}, low. In the layout, this can be achieved (for example, on the CH side) by connecting BL and CL to GND, with V_{DD} at the source of the appropriate PU device, and the other nodes floating (Fig. 5b). Similar connections can be made to configure the cell for a worst-case write on the CL side. The worst-case read condition for a cell consists of a resistive voltage divider with a fully-on PD contesting the adjacent PG device to bring the internal node voltage, V_{sr}, to a low value. In the layout, this can be achieved by connecting BL and CL to V_{DD} and by connecting V_{SS} to GND on the CH side of the cell (Fig. 5c). The rest of the nodes are left floating. In all cases, the gate of the PG is connected to WL, and each sensor half-cell is the size of one SRAM bitcell.

The internal node voltages, V_{sr} and V_{sw}, can be used to estimate the read stability and writeability of a set of cells. Lower V_{sr} corresponds to less read disturb and therefore higher read stability. Lower V_{sw} corresponds to an easier and faster write. Fig. 6 illustrates good correlation between these voltages and RSNM or I_W (a writeability metric [4]), as determined by Monte Carlo simulation. The correlations in Fig. 6 can be improved by including secondary contributions from the other four transistors in the SRAM cell, but this requires additional connections that increase overall sensor area.

In both read and write cases, the strength of the PG transistor is determined by V_{WL}. Wordline biasing was chosen due to its relatively large impact on the write / read (half-select) tradeoff; however, other biasing or timing approaches could also be used. Raising V_{WL} increases the PG strength, lowering V_{sw} but raising V_{sr}. An op-amp is used to find the optimal wordline bias $V_{WL}*$ for which the maximum of V_{sr} and V_{sw} is minimized (Fig. 7). Because of the read / write tradeoff, the minimum is achieved when $V_{sr} = V_{sw}$. Depending on the requirements of the application, $V_{WL}*$ may be chosen to be a fixed offset away from this point without affecting the sensor's response to systematic variations.

In the presence of variations, the sensor modifies V_{WL} to maintain $V_{sr} = V_{sw}$. Fig. 8 illustrates the simulated response of

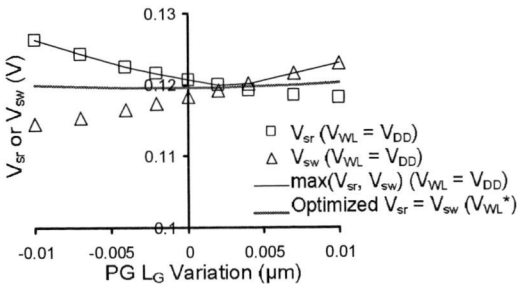

Fig. 8. The optimal $V_{WL}{}^*$ is adjusted to minimize the maximum of V_{sr} and V_{sw} in the presence of simulated systematic variations, such as on PG L_G.

a sensor to variations in the PG gate length. Increasing gate length weakens the PG transistor, improving read stability, and raising V_{sw}. V_{sr} and V_{sw} are plotted with the op-amp output connected to WL in feedback vs. with V_{DD} connected to WL. The sensor correctly lowers the maximum of V_{sr} and V_{sw} at all points but one, where the optimal $V_{WL}{}^* = V_{DD}$.

Each sensor can comprise multiple, parallel copies of half-cells. These are connected in parallel with the outputs V_{sr} tied together for the read half-cells, and V_{sw} tied together for the write half-cells. Increasing the number of parallel half-cells reduces the variability due to random sources by a factor of $1/\sqrt{N}$. The half-cells can optionally include different cell layout orientations to control for systematic variations associated therewith.

IV. EXPERIMENTAL RESULTS

The variation sensor is implemented in silicon in an early industrial 45nm process, as illustrated in Fig. 9. Six sensors are placed near SRAM arrays across a 2mm × 2mm die. Each sensor consists of sixteen half-cells: eight for the worst-case read and eight for the worst-case write. All cell orientations are included. The input and outputs of each sensor (V_{WL}, V_{sr}, and V_{sw}) are connected via a transmission multiplexer to one shared PMOS-input amplifier. In this proof-of-concept design, a simple 1-stage folded-cascode amplifier with a gain of more than 60dB is implemented in a 9000 μm² area. Large 1.8V, thick-oxide transistors were used to ensure a high gain and small 3σ offset of 2.4mV. In a mature process, when the matching data is known, a much smaller thin-oxide amplifier could be used. V_{WL} is also connected to an external pin for data collection.

Fig. 10 illustrates the average response of several SRAM cells to different V_{WL}. Because only the bitlines and wordlines are externally accessible in these cells, the bitline read margin (BLRM) and bitline write margin (BLWM) are used to measure read stability and writeability. BLRM is defined as V_{DD} minus the smallest cell supply voltage that retains the cell state with V_{BL}, V_{BLB}, and $V_{WL} = V_{DD}$. BLWM is defined as the highest V_{BL} that flips the cell state with $V_{WL} = V_{BLB} = V_{DD}$. These metrics have been shown to correlate well with RSNM and I_W, respectively [5]. With increasing V_{WL}, BLRM decreases and BLWM increases, similar to the expected trend from Fig. 5. The $V_{WL}{}^*$ indicated by the variation sensor is near the crossover point where BLRM = BLWM. A small offset can be expected since the metrics are different from the V_{sr} and V_{sw} equalized by the op-amp.

V_{DD} and substrate biases can also be used to skew the read / write tradeoff. With $V_{WL} = V_{DD}$, increasing V_{DD} improves writeability faster than read stability (Fig. 11a). A lower V_{WL} can restore the balance, so the variation sensor output

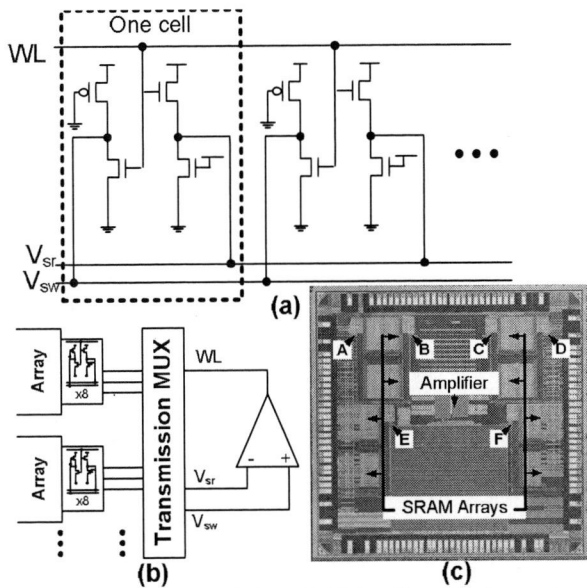

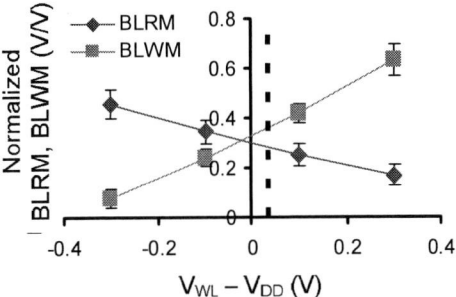

Fig. 9. Variation sensors consisting of 16 half cells are implemented on an SRAM testchip in an industrial 45nm process (a). The sensors are situated in close proximity to SRAM arrays and share a single amplifier, connected through a transmission multiplexer (b). Die photo of the chip, indicating six distributed variation sensors (lettered) and nearby SRAM arrays (c).

Fig. 10. BLRM, a measure of read stability, decreases with increasing WL voltage, while BLWM, a measure of writeability, increases. Data are taken for several SRAM cells in a dense array. Error bars indicate +/- one standard deviation. The optimal WL bias generated by the nearest variation sensor is indicated by the dotted line.

generates a decreasing $V_{WL}{}^* - V_{DD}$. The read / write tradeoff is highly sensitive to V_{DD} variations, with an average of 40 mV of $V_{WL}{}^*$ required to compensate for a 100 mV change in V_{DD}. With increasing n-well bias, the PMOS devices become weaker, improving the writeability. This effect is similar to that produced by burn-in phenomena such as negative bias temperature instability (NBTI), which increases PMOS V_T over time and has been shown to lower SRAM yield [6, 7]. The variation sensor appropriately decreases $V_{WL}{}^*$ in response to the rise in PMOS V_T (Fig. 11b). With increasing p-well bias, the NMOS devices become forward-biased, increasing their drive currents, and improving the writeability. $V_{WL}{}^*$ is observed to decrease accordingly (Fig. 11c). An average decrease in $V_{WL}{}^*$ with temperature of $\partial V_{WL}{}^* / \partial T = -0.2$ mV/K was also observed (Fig. 11d).

In order to evaluate across-chip variations, 512 cells from each of six arrays at different locations on the die are measured. BLRM and BLWM are measured for each cell at $V_{WL} = V_{DD} - 0.1$ V and $V_{DD} + 0.2$ V. A linear relationship between BLRM, BLWM, and V_{WL} is assumed over this interval based on the measurements illustrated in Fig. 10. The V_{WL} where BLRM = BLWM is then calculated for each cell. Because of variability, there is a distribution of V_{WL} from each

978-1-4244-2018-6/08 $25.00 © 2008 IEEE

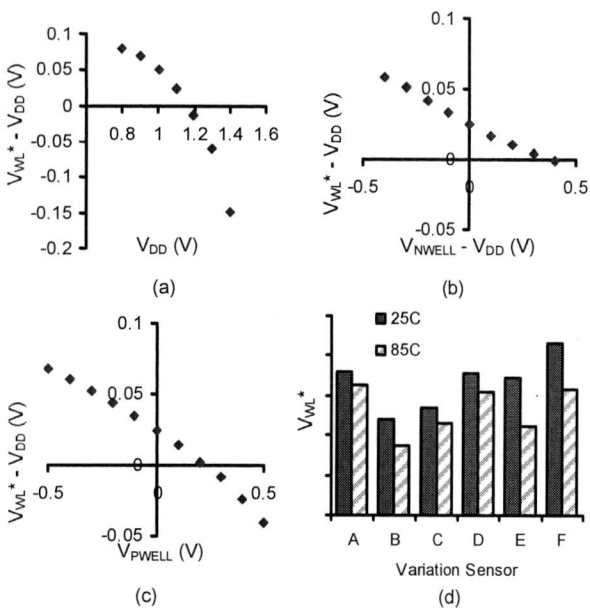

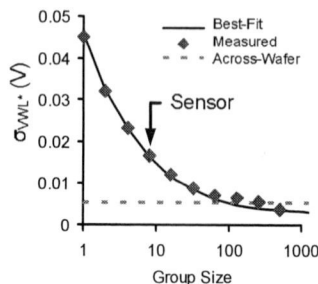

Fig. 12. Measurements of groups of N SRAM cells show a decreasing random variation component as N increases. The least squares fit from theory (eqn. 1) indicates very low within-die systematic variation of σ_s = 3 mV, which is below the detection limit of the 16 half-cell sensors.

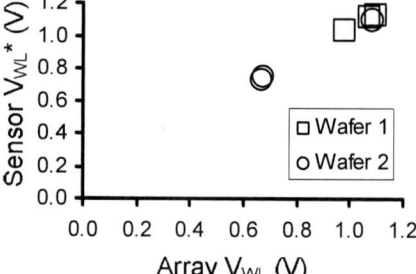

Fig. 11. Increasing V_{DD} (a), n-well bias (b), p-well bias (c), and temperature (d) all have the effect of favoring writeability. The variation sensor generates a lower $V_{WL}*$ to restore the balance between writeability and read stability.

Fig. 13. Chip-to-chip correlation of sensor $V_{WL}*$ and optimal V_{WL} as measured from arrays. Each data point represents an average of six variation sensors and over 256 array cells. A range of chips, V_{DD}, substrate biases, and temperatures were used to verify operation under various operating conditions.

array. The V_{WL} are normally distributed with a standard deviation of 47 mV that comprises a random (σ_r) and systematic (σ_s) component. To isolate the systematic variation, groups of N cells can be averaged. The calculated V_{WL} are then expected to be normally distributed with a standard deviation of:

$$\sigma_{VWL} = \sqrt{\frac{\sigma_r^2}{N} + \sigma_s^2} \qquad (1)$$

Fig. 12 illustrates the measured σ_{VWL} for different N among all six arrays. The least squares fit curve of (1) is also shown, with σ_r = 46 mV and σ_s = 3 mV. This corresponds to a very low level of across-chip variation, far below the 17 mV that the implemented sensors are able to measure with only N = 8 cells. The measured value of $\sigma_{VWL}*$ from 42 sensors on seven chips (with die-to-die and wafer-to-wafer systematic variations subtracted out) is 14 mV, close to the expected value. A large sensor of N > 512 cells would be required to detect systematic variations at σ_s = 3 mV; however, such a sensor would be impractical since this level is too small to affect yield.

With simultaneous process, voltage, and temperature variations, however, a larger spread in the data is measured. At each point in Fig. 13, the average of all six variation sensors on a chip is plotted against the average optimal V_{WL} (at BLRM = BLWM) from the arrays. To reduce noise, only chips with over 256 measured cells are plotted. The sensor $V_{WL}*$ has a 50mV offset, but otherwise tracks the optimal value within 30mV over a 500mV range. Since each chip has its own op-amp, an additional source of variability in the sensor output is expected.

The variation sensor can compensate for high levels of systematic variation, where a reduction in σ_s is expected to improve yield. The sensors should contain at least N > σ_r^2 / σ_s^2 cells to ensure a strong signal. The separation between multiple sensors on a chip determines the minimum distance over which spatially correlated variations can be detected. Sensors placed 400 µm apart, for example, will be unable to

detect variations that are correlated within 40 µm, even if they are large in magnitude. In the test arrays, measurements for cells within ranges of 10 - 200µm exhibited a very low σ_s in this process, however, so only one sensor is required for these ranges.

V. CONCLUSION

Systematic variations can upset the read / write balance of an SRAM cell and contribute to yield loss. A variation sensor circuit is presented to restore the balance by co-optimizing writeability and cell stability during half-select. The sensor can be used to increase SRAM robustness not only to process variations, but also to bias and temperature fluctuations over time, and works well over a wide range of V_{DD}.

Acknowledgement

This work was supported by the Center for Circuit & System Solutions (C2S2) Focus Center, one of five research centers funded under the Focus Center Research Program, a Semiconductor Research Corporation program; the SRC Mahboob Khan Fellowship; and the National Science Foundation Infrastructure Grant No. 040342. Chip fabrication was donated by STMicroelectronics. The authors acknowledge the support of Ernesto Perea and Jean-Pierre Schoelkopf.

References

[1] S. Ohbayashi et al., "A 65 nm SoC embedded 6T-SRAM design for manufacturing with read and write cell stabilizing circuits," *VLSI Circ. Dig.*, 2006, pp.

[2] K. Takeda et al., "Redefinition of write margin for next-generation SRAM and write-margin monitoring circuit," *ISSCC*, 2006, pp. 34.5.

[3] E. Seevinck et al., "Static noise margin analysis of MOS SRAM cells," *IEEE JSSC*, vol. SC-22, 1987, pp. 748-754.

[4] C. Wann et al., "SRAM cell design for stability methodology," *IEEE VLSI-TSA*, 2005, pp. 21-22.

[5] Z. Guo et al., "Large-scale read / write margin measurement in 45nm CMOS SRAM arrays," *VLSI Circ. Dig.*, 2008, in press.

[6] M. Ball et al., "A screening methodology for VMIN drift in SRAM arrays with application to sub-65nm nodes," *IEDM Tech. Dig.*, 2006, pp. 705-708.

[7] A. Carlson, "Mechanism of increase in SRAM Vmin due to negative bias temperature instability," *IEEE TDMR*, 2007, pp. 479-487.

Variation-Tolerant SRAM Sense-Amplifier Timing Using Configurable Replica Bitlines

Umut Arslan, Mark P. McCartney, Mudit Bhargava, Xin Li, Ken Mai, and Lawrence T. Pileggi
Electrical and Computer Engineering Department, Carnegie Mellon University
5000 Forbes Ave., Pittsburgh PA 15213
{uarslan,mmccartn,mbhargav,xinli,kenmai,pileggi}@ece.cmu.edu

Abstract - A configurable replica bitline (cRBL) technique for controlling sense-amplifier enable (SAE) timing for small-swing bitline SRAMs is described. Post-silicon selection of a subset of replica bitline driver cells from a statistically designed pool of cells facilitates precise SAE timing. An exponential reduction in timing variation is enabled by statistical selection of driver cells, which can provide 14x reduction in SAE timing uncertainty with 200x less area and power than a conventional RBL with equivalent variation control. We describe the post-silicon test and configuration methodology necessary for cRBLs. To demonstrate the efficacy of the proposed cRBL technique, we present measured results from a 90nm bulk CMOS 64kb SRAM testchip.

I. INTRODUCTION

To achieve fast, low-power read operations, SRAMs use small-swing bitlines (BL) and clocked sense amplifiers (sense-amps). For reliable operation at high speed, however, the sense-amp enable (SAE) signal must track the small-swing BL delay across global and local process, voltage, and temperature (PVT) variations. If the SAE signal fires before the differential BL signal exceeds the sense-amp offset, a read failure may occur at the sense-amp output. Conversely, if the SAE fires too late, then the access time and power increase unnecessarily. The SAE signal is usually self-timed using replica bitlines (RBL) (Fig. 1), since they track BL delays better than simple buffer chains over global PVT skews [1]. However, the susceptibility of the RBL delay to local transistor mismatch (which is becoming the dominant source of variation with CMOS scaling) has become a major concern [2].

A conventional RBL column uses replica memory cells that are essentially identical to the core cells (Fig. 3a). During a read, the replica wordline (RWL) signal is asserted turning on a fixed number of replica driver cells that discharge the RBL. These driver cells are hardwired to store "0" while the rest of the replica cells function as dummy loads on the RBL. The full-swing RBL signal is then inverted and buffered to generate the SAE signal. The RBL signal is also used to turn off the active wordline to limit the bitline swing and save power. Global PVT variations cause memory cells on the same die to have correlated read current variations and thereby good tracking between RBL and BL delays. Local transistor mismatch (increasingly dominated by threshold voltage (Vt) mismatch due to random dopant fluctuations), however, causes uncorrelated current variations between memory cells on a chip. This degrades tracking between BL and RBL delays, increasing the possibility of read failures or unnecessary

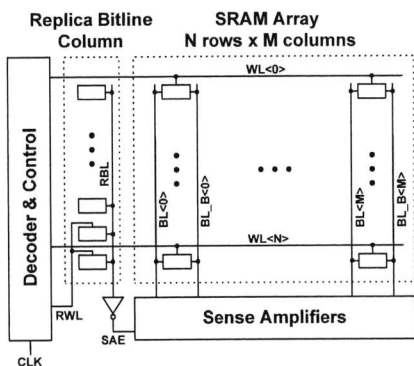

Fig. 1 – Block diagram of an SRAM array using a replica bitline column for sense-amp enable timing.

latency and power. Fig. 2 illustrates the two-sided SAE timing constraint and how a large variability in the SAE timing would cause failures.

RBL delay variation due to mismatch can be reduced by increasing the number of driver cells in the RBL column [3]. Using this method, however, the number of driver cells needed to cancel mismatch grows rapidly with scaling and ultimately necessitates use of multiple RBL columns. At advanced process nodes, the size and power of the RBL necessary for sufficiently controlled SAE timing variation becomes impractical. As an alternative to this brute force approach to RBL delay variation, we propose a configurable RBL (cRBL) column that selects a subset of a pool of potential driver cells (configurable cells) (Fig. 3b). A post-silicon characterization and configuration step can select the subset of configurable cells that offers the best cancellation of local mismatch.

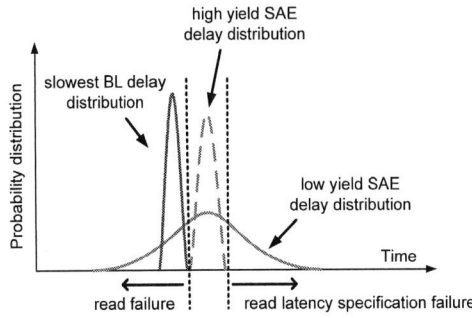

Fig. 2 – Illustration of probability distribution of slowest BL delay and SAE delay. SAE timing suffers from two-sided timing constraint.

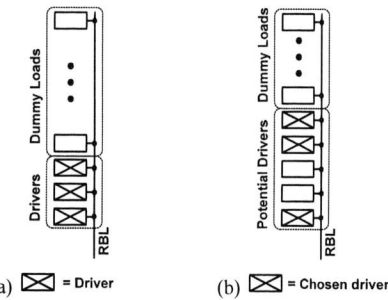

(a) ⊠ = Driver (b) ⊠ = Chosen driver

Fig. 3 – Example RBL columns: (a) Conventional RBL with 3 fixed driver cells. (b) Configurable RBL with 5 potential driver cells.

Statistical analyses show that a small number of configurable cells will suffice to provide excellent mismatch cancellation. Thus, the replica timing path can remain a single column of cells and does not incur any additional power or area penalty for delay variation control. The cRBL is superior to a delay-adjustable buffer chain, since it tracks BLs more robustly over voltage and temperature variations after post-silicon configuration.

II. CONFIGURABLE REPLICA BITLINE DESIGN AND ANALYSIS

A key cRBL design decision involves choosing the number of configurable, driver, and dummy cells for the given performance and yield specifications. As the cRBL essentially exploits device randomness, we use statistical models and simulations to determine those design parameters.

For initial design exploration, we have built a Gaussian distribution model of the SRAM cell current that captures local process variations at a global PVT corner. The parameters of the distribution are obtained from transistor-level Monte Carlo (MC) simulations using statistical device models. Next, we have performed MC simulations using a numerical solver (MATLAB) for a varying number of driver cells selected from a given number of configurable cells. For each MC sample, the configuration (i.e. subset) of driver cells that best cancel current variation is selected. Fig. 4 demonstrates standard deviation of current as a function of the number of driver cells,

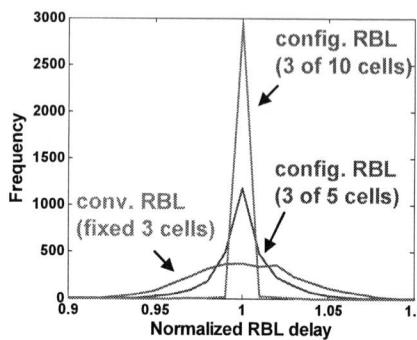

Fig. 5 – Normalized delay distribution of cRBL that selects 3 drivers that best cancels mismatch compared to the conventional RBL with 3 fixed drivers. Results are from 3,000-sample Monte Carlo simulations that vary local process parameters at a global process corner.

sweeping the number of configurable cells from 2 to 10. The analysis shows that the optimal number of drivers (minimum point of the curves in Fig. 4) increases gradually with the number of configurable cells. Using subsets of 2 or 3 driver cells offers the best cancellation of mismatch with 10 or fewer configurable cells.

Based on the results from initial analysis, we simulate a transistor-level model of a cRBL using an industrial 90nm bulk CMOS technology for accurate timing analysis. The cRBL uses a single column and selects a subset of 3 driver cells that best cancels the RBL delay variation. The number of dummy cells is chosen as 256 to obtain an appropriate RBL delay with respect to the slowest BL delay. The cRBL delay distributions are obtained from MC simulations for both 5 and 10 configurable cells (Fig. 5). The results show that the standard deviation of cRBL delay decreases exponentially with the number of configurable cells while the conventional RBL delay variation decreases sub-linearly with the number of active drivers (Fig. 6). This exponential scaling enabled by statistical driver selection is the reason for the cRBL's superior efficiency. For example, the RBL delay variation (as measured by σ/μ) can be reduced by ~14x by using 10 configurable cells. To achieve equal variation control, a conventional RBL would require

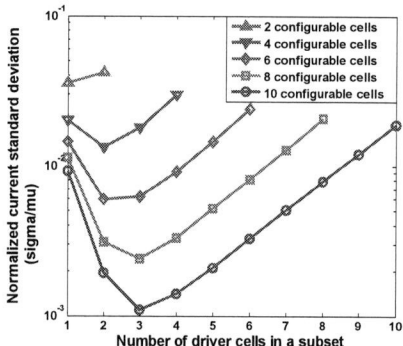

Fig. 4 – cRBL normalized current variation as a function of number of driver cells in a subset for a given number of configurable cells. Results are from 10,000-sample Monte Carlo simulations that use Gaussian model for local cell current variations at a global PVT corner.

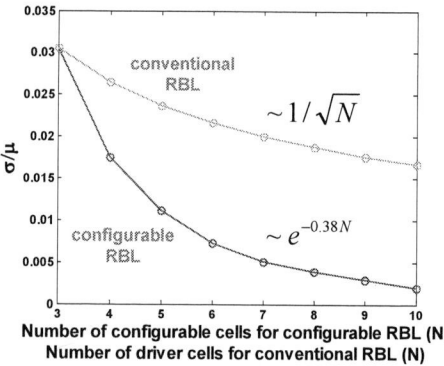

Fig. 6 – Scaling of RBL delay variation (σ/μ) as a function of the number of configurable cells for cRBL and the number of active driver cells for conventional RBL.

978-1-4244-2018-6/08 $25.00 © 2008 IEEE

~600 active driver cells using ~200 columns and thus increase the power and area overhead of the self-timing path by ~200x.

III. CONFIGURABLE REPLICA BITLINE TEST AND CONFIGURATION

Although the cRBL is superior to the conventional RBL, it relies on post-silicon test and configuration. Therefore, the development of efficient test and configuration techniques is required to make the cRBL practically viable. The primary goal of the cRBL test/configuration is to select a configuration of replica cells that provides the appropriate SAE timing so that the memory can be read correctly at the specified clock speed. For a given cRBL configuration, the tester (either on-chip built-in self-test logic or automated test equipment), which can read and write data from every address in the memory at-speed (e.g. by applying a specific marching pattern [4]), can detect a read failure and thereby can report a pass/fail evaluation. If the tester finds a configuration that evaluates as pass for the whole array, the configuration can be used to control SAE timing during normal memory operation. To avoid failures due to change in environmental conditions, an additional delay margin can be added to the BL path during test to force marginally fast configurations to fail. To further increase read speed and/or reduce bitline power, the above test scheme can be improved to enable speed binning of configurations. Repeating the array read/write tests for varying additional delays on the BL path, configurations can be classified based on their speed at the expense of test time and complexity.

Simulating the proposed post-silicon test/configuration scheme to quantitatively analyze yield recovery enabled by the cRBL necessitates a timing model of the memory that includes the uncertainty of both BL and RBL delays. We have used the same modeling approach as the initial design explorations. SRAM cell currents are modeled as Gaussian random variables and a first-order timing model is used to compute the BL and RBL delays. MC simulations are performed to obtain the statistics of yield recovery enabled by the cRBL. We have applied the analysis technique to a 64kb SRAM prototype. The results given in Table I demonstrate that using cRBLs the parametric yield due to self-timing failures can be increased from ~83% to over 99.99% at a specified clock frequency.

Although the test overhead is not considered in our analysis and a built-in self-test (BIST) engine was not implemented in the test chip, we believe that it would be of a low overhead cost relative to the benefit obtained from cRBLs.

TABLE I
YIELD RECOVERY BY CONFIGURABLE RBL

# of configurable cells	Self-timing yield
3 (conventional RBL)	83.083%
4	95.686%
6	99.667%
8	99.975%
10	99.998%

Results from 1M-sample MC simulations assuming no memory redundancy. The selection criteria for a passing configuration is that it should be at least 40ps (~1 F04 inverter delay) slower than the slowest BL and faster than a specified limit. The resulting SAE timing uncertainty is smaller than ~50ps.

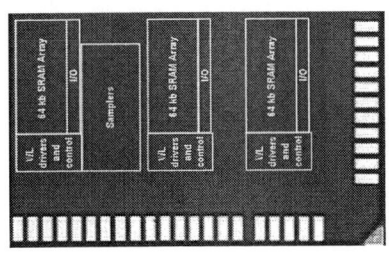

Fig. 7 – Die photo of the 2mm x 1.2mm test chip with three 64kb SRAM macros one of which is instrumented with on-die samplers in 90nm bulk CMOS.

IV. TEST CHIP AND MEASUREMENT RESULTS

To demonstrate the efficacy of the proposed cRBL technique, a 64kb SRAM prototype has been designed and manufactured in 90nm bulk CMOS (Fig. 7). The implemented cRBL uses a single column with 10 configurable cells and 256 dummy load cells (Fig. 8). A scan chain sets the configurable cell control bits determining which driver cells are used. The test chip is instrumented with on-chip samplers that enable precise read-out of internal node waveforms and accurate measurement of internal delays [5].

Fig. 9 shows the sampled waveforms for internal signals in the bitline and the self-timing paths during a read at 1.0V supply and 100 MHz external clock (CLK). CLK to SAE delay of each cRBL configuration is measured by sampling clock and the SAE signal with 10 ps resolution. Fig. 10 shows the measured CLK to SAE delays from one test chip for conventional RBL mode using 3 fixed driver cells and the cRBL mode using 10 configurable cells. While the conventional RBL provides a fixed delay of 1.68ns, the cRBL provides a delay tunable between 1.66-1.75ns (90ps tuning range). The measurements from 18 test chips show that the tuning range of the cRBL varies between 60-130ps. These results agree closely with the expected tuning range obtained from transistor-level MC simulations (Fig. 11).

We have also processed the measured delay data to emulate a post-silicon configuration step. A cRBL with 10 configurable cells can emulate a cRBL with fewer configurable cells by

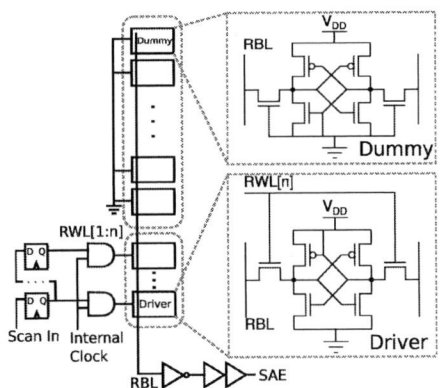

Fig. 8 – Circuit schematic of the implemented cRBL.

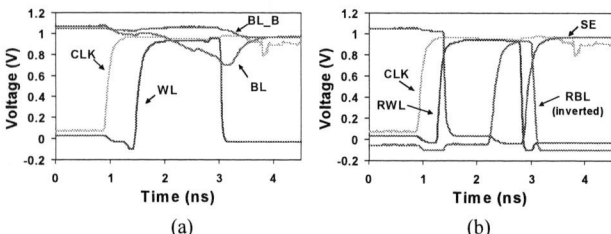

(a)   (b)

Fig. 9 – Measured waveforms during read operation for internal signals in the (a) bitline path and (b) self-timing path.

simply limiting the set of configurable cells we choose from to be fewer than the full 10. For a cRBL mode using N configurable cells, the configuration providing the least delay difference from a reference delay is selected over all 3 of N configurations. We have obtained the variation in the selected configuration's delay for each cRBL mode by taking measurements from 18 test chips. The reference delay is estimated as the mean delay of all statistically independent 3-driver subsets across all tested chips. Ideally, the reference delay should represent the no-mismatch RBL delay for each chip at its own global process corner, but this information is unavailable from our experimental chip. Fig. 12 demonstrates the variation in SAE timing after post-silicon configuration for the conventional RBL mode and the cRBL mode using 5 and 10 configurable cells. The standard deviation of the SAE delay is reduced by ~60% (i.e. ~2.5x) using 10 configurable cells. However, it should be noted that the measured improvement in variation control is pessimistic due to the unavailability of perfect reference delays.

V. CONCLUSIONS

Post-silicon configuration of RBLs can be used to cancel SAE timing variations due to local mismatch. Due to the exponential decay of variation enabled by statistical selection of driver cells, the cRBL can provide ~14x reduction in SAE timing uncertainty equivalent to a conventional RBL with ~200x the area and power. With the improvement of conventional memory test techniques to enable cRBL test/configuration, over 10% parametric yield loss due to self-timing failures can be recovered. Such configurable designs offer promising solutions for self-timing of small-swing bitline SRAMs in future process technologies.

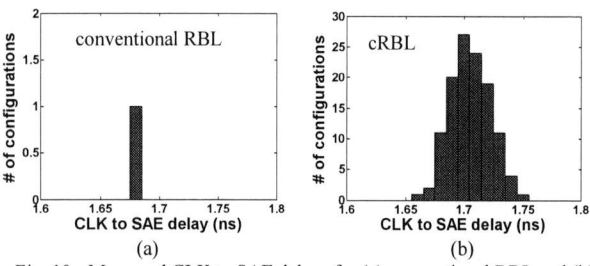

(a)   (b)

Fig. 10 - Measured CLK to SAE delays for (a) conventional RBL and (b) configurable RBL using 10 configurable cells (120 selectable configurations) from one test chip. Using cRBL, SAE delay is tunable within a 90ps range.

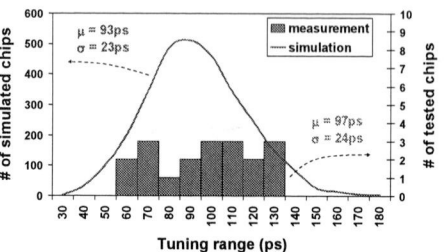

Fig. 11 – Measured cRBL tuning range from 18 test chips.

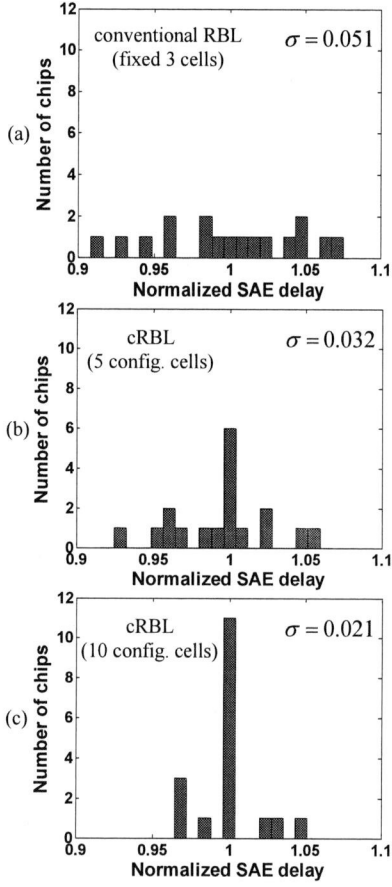

Fig. 12 - Measured normalized SAE delay histograms for (a) conventional RBL using 3 fixed driver cells (b) cRBL using 3 of 5 configurable cells and (c) cRBL using 3 of 10 configurable cells.

REFERENCES

[1] B. Amrutur and M. Horowitz, "A replica technique for wordline and sense control in low-power SRAM's," *JSSC*, Aug. 1998.

[2] R. Heald and P. Wang, "Variability in sub-100nm SRAM designs," *ICCAD*, Nov. 2004.

[3] K. Osada, *et al.*, "Universal-Vdd 0.65-2.0-V 32-kB cache using a voltage-adapted timing-generation scheme and a lithographically symmetrical cell," *JSSC*, Nov. 2001.

[4] R.D. Adams, *High performance memory testing*, Kluwer Academic Publishers, 2003.

[5] R. Ho, *et al.*, "Applications of on-chip samplers for test and measurement of integrated circuits," *Symposium on VLSI Circuits*, 1998.

IEEE 2008 Custom Intergrated Circuits Conference (CICC)

A 135mV 0.13µW Process Tolerant 6T Subthreshold DTMOS SRAM in 90nm Technology

Myeong-Eun Hwang and Kaushik Roy

Purdue University, West Lafayette, IN 47907, USA

Abstract

Cell stability and tolerance to process variation are of primary importance in subthreshold SRAMs. We propose a DTMOS based 6T SRAM suitable for subthreshold operation. For variation tolerant memory peripheral circuitry, we apply β-ratio modulation technique. DTMOS SRAM array fabricated in 90nm technology operates down to 135mV consuming 0.13µW at 750Hz. The proposed SRAM achieves 200% improvement in read static noise margin at iso-area compared to the conventional 6T SRAM at a supply voltage of 200mV.

Introduction

Efficient power management is becoming increasingly important with the rapid growth of portable, wireless, and battery-operated applications, such as laptop computers, personal digital assistants (PDAs), and portable communication devices. One of the promising approaches to achieve ultralow power dissipation in such applications is to use sub-threshold (sub-V_T) logic [1][2]. In sub-V_T logic, circuits operate with a supply voltage (V_{dd}) lower than the transistor threshold voltage (V_T) and utilize the subthreshold leakage current as the operating current.

Lowering the supply voltage reduces the dynamic power quadratically and leakage power exponentially. However, supply voltage scaling also limits signal swing and thus reduces noise margin. Further, aggressive technology scaling in the sub-100nm region increases the sensitivity of the circuit electrical parameters to process variation (PV) [3][4]. This may result in device mismatch between the adjacent transistors and limits the circuit operation in the sub-V_T region, particularly for memories [5]. Embedded cache memories are expected to occupy 90% of the total die area of a system-on-a-chip [6]. Nano-scaled SRAM cells having minimum-sized transistors are vulnerable to inter-die as well as intra-die PVs. In addition to PVs, such as random fluctuations in electrical circuit parameters, low voltage operation results in reduced noise margin and various memory failures, such as read/hold/access/writes failure [7]. This seriously degrades robustness in sub-V_T memory systems.

For robust sub-V_T SRAMs, various types of SRAMs have been proposed (Fig. 1). To maintain the clarity of the discussion, the conventional 6 transistor CMOS cell is referred to as "6T", and the 5, 7, 8 and 10 transistor SRAM cells are referred to as the 5T [8], 7T [9], 8T [10 and 10T [11], respectively. Adaptive circuit techniques such as source biasing, dynamic V_{dd} have been also proposed to improve PV tolerance [12]. The authors in [13] used reverse short-channel effect to improve writability and to decrease circuit variability. While 5T, 8T, and 10T cells employ single ended reading 6T and 7T cells utilize differential read operation. 8T and 10T cells use an extra sensing circuit for reading the cell contents, achieving improved read SNM. Recently, a memory cell operating at the supply voltage of 103mV has been reported with single ended read operation [14].

*This work was sponsored in part by SRC and Boeing.

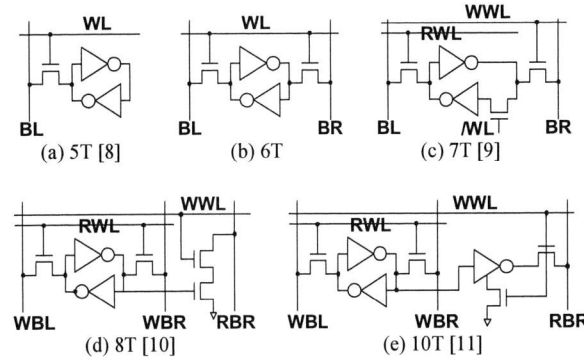

Fig. 1. Various SRAM cells.

This paper describes a dynamic threshold MOS (DTMOS) based 6T SRAM suitable for robust sub-V_T operations using positive feedback in the cells and body biasing in peripherals. The proposed SRAM requires no architectural change compared to the conventional 6T SRAM. Some highlights of our design are:

- Considerable improvement in read and hold stabilities for sub-V_T operation;
- Full functionality at low voltages, all the way to 135mV;
- Variation tolerant peripheral circuitry for sub-V_T memory.

DTMOS Technique

Fig. 2(a) shows a basic DTMOS inverter where the gate and the body of the transistor are tied together. Because of body effect, the threshold voltage (V_T) of the device changes dynamically based on input data. When the input 'in' is LOW, PMOS is ON with low-V_T and NMOS is OFF with normal-V_T (high-V_T). When 'in' is HIGH, the situation is the opposite. In other words, during the ON state, DTMOS scheme lowers V_T, increasing the overdrive voltage ($V_{gs}-V_T$).

Fig. 2(b) compares the subthreshold current-voltage characteristic (I_d-V_{gs}) of DTMOS and standard CMOS transistors obtained from HSPICE simulations at V_{dd}=300mV in 90nm technology. During the OFF state, DTMOS scheme raises V_T, decreasing the leakage current. As a result, the trip voltage (V_M) of the DTMOS inverter increases or decreases depending on the direction of input transition. In the active mode, the inverter switches from LOW to HIGH with a higher speed because of the low-V_T PMOS. In the standby mode, the static leakage current is determined by the subthreshold current of the high-V_T NMOS.

In sub-V_T DTMOS, the body terminal provides the additional transconductance (g_{mb}=$\delta I_d/\delta V_{bs}$), resulting in higher operational current through the device. This improves voltage transfer characteristics (VTC) of sub-V_T DTMOS inverter as shown in Fig. 2(c), where we refer to the full (PMOS-only) DTMOS scheme as "DT" ("pDT") and the conventional CMOS scheme as "CMOS".

978-1-4244-2018-6/08 $25.00 © 2008 IEEE

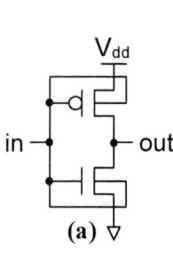

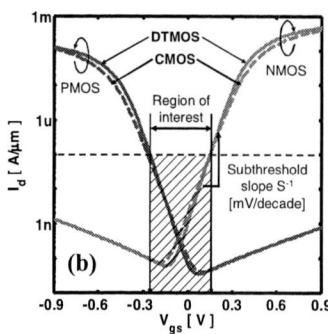

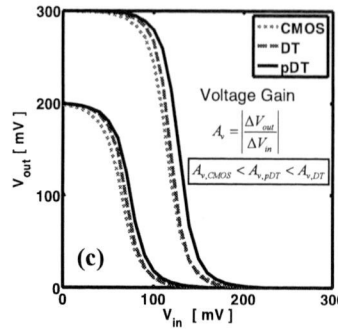

Fig. 2. DTMOS inverter. (a) Schematic, (b) I_d-V_{gs} curve and (c) VTC (simulation in 90nm technology). CMOS stands for the conventional CMOS scheme. "DT" means DTMOS scheme applied to both PMOS and NMOS transistors and "pDT" means DTMOS scheme applied to the PMOS transistor only.

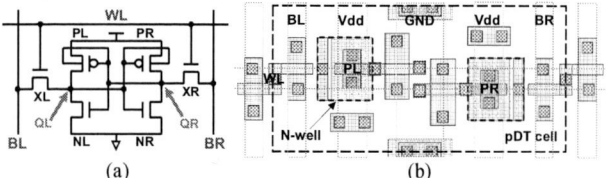

Fig. 3. Proposed 6T pDT SRAM cell. (a) Schematic. (b) Layout.

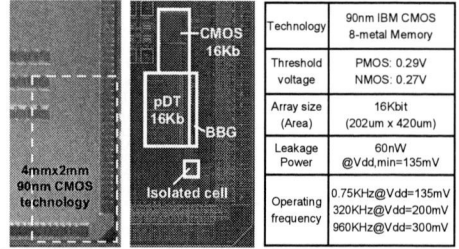

Fig. 4. Chip view and summary of pDT SRAM measurements.

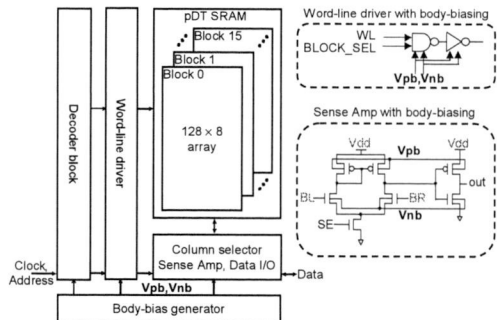

Fig. 5. pDT SRAM. Four columns share one sense amplifier.

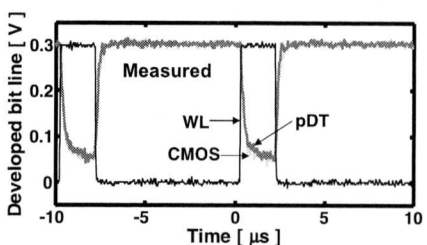

Fig. 6. Read operation @V_{dd}=300mV and Freq=100KHz.

The improved VTC results in an improved noise margin and tolerance to process variation. Note that the voltage gain (Av) of an inverter is the product of the transconductance and output resistance.

Proposed 6T DTMOS-based SRAM

Fig. 3(a) shows the proposed pDT 6T cell. At low V_{dd}'s, degradation in the gain of the cross-coupled inverter is of concern. To improve the inverter characteristics, DTMOS scheme is used. Transistors PL-NL form one pDT inverter while PR-NR form the other inverter. XL and XR are the access transistors. Positive feedback from the pDT inverter changes the inverter trip voltage dynamically depending on the direction of state transition. This gives a near ideal inverter characteristics in the cell essential for robust memory operation. The pDT cell utilizes differential operation, giving better noise immunity. Fig. 3(b) shows one possible layout of the pDT cell. Due to separation of body contacts for PMOS devices, the pDT cell requires 78% larger area than the standard CMOS cell in 90nm process.

Fig. 4 shows the test chip view and summary of the 16Kb pDT SRAM fabricated in 90nm CMOS memory technology. To measure the static noise margin of pDT and conventional SRAMs, their isolated cells are implemented with direct probing capability. Fig. 5 illustrates the architecture of pDT SRAM operating with one single supply voltage. DTMOS scheme (pDT) is used to the core array to increase cell sta-

bilities while body bias is applied to the peripheral circuits to improve variation tolerance in logic. The body bias generator (BBG) provides body bias voltages to the transistors in peripheral circuits (V_{pb} and V_{nb} for the pull-up and the pull-down networks, respectively).

Memory Operations

During a read operation (with QL=0 and QR=1 say), the voltage of the node QL rises due to voltage division between the access transistor XL and the pull-down transistor NL. If this voltage is greater than the trip voltage of the other inverter PR-NR, the content of the cell can get flipped resulting in a read failure. To avoid a read failure, it is necessary to increase the trip voltage of the inverter PR-NR. In a pDT cell, Transistors PR and NR have lower-V_T and higher-V_T, respectively, and thus increase the trip voltage of the inverter which is storing HIGH. DTMOS action helps preserve the logic '1' value of the cell. During a read operation with QL/QR=1/0, the situation is the opposite and DTMOS action helps preserve the logic '0'. Fig. 6 shows the waveforms when the bit-line is discharged during a read operation at V_{dd}=300mV.

978-1-4244-2018-6/08 $25.00 © 2008 IEEE

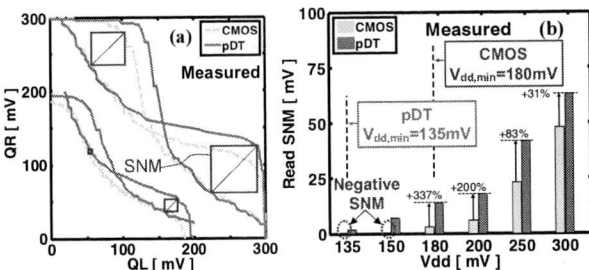

Fig. 7. Read SNM. (a) V_{dd}=300mV and 200mV. (b) The average of 32 measured data.

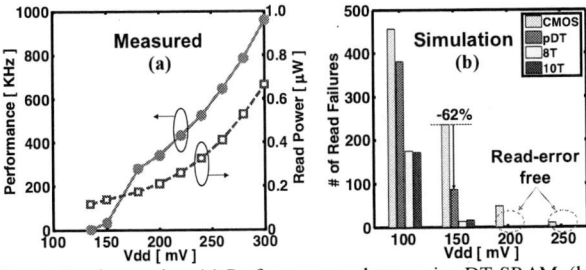

Fig. 8. Read operation. (a) Performance and power in pDT SRAM. (b) Read failure.

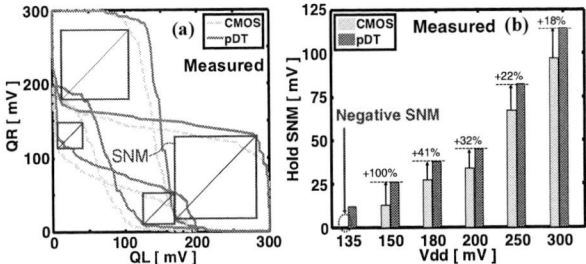

Fig. 9. Hold SNM. (a) V_{dd}=300mV and 200mV. (b) The average of 32 measured data.

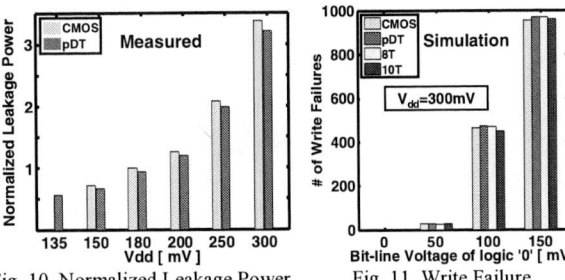

Fig. 10. Normalized Leakage Power in hold mode.

Fig. 11. Write Failure.

Fig. 7(a) shows the read static noise margin (SNM) of the pDT cell and its conventional counterpart (referred to as CMOS) measured at low V_{dd}'s. Fig. 7(b) shows the average SNM computed out of 32 measured data samples. The cell has two cross-coupled pDT inverters. pDT scheme augments the PMOS transistor (i.e., lowering PMOS V_T) and thus makes one inverter hold '1' stronger. This restrains the other inverter (involved in discharging) from flipping the cell value. As a result, the pDT cell improves the read SNM by 31% (200%) at V_{dd}=300mV (200mV) compared to the conventional cell and the degree of improvement in read SNM is considerable as V_{dd} is lowered. pDT SRAM shows correct functionality at a minimum V_{dd} of 135mV, read power of 0.13μW, and frequency of 750Hz.

One may consider upsizing would improve the read SNM in the conventional cell. At high V_{dd}'s (i.e., super-V_T region), the drain current varies linearly with the gate voltage. Transistor sizing increases the SNM considerably. However in the sub-V_T region, the drain current depends exponentially on the gate voltage. Any device upsizing will result in marginal change in the drain current. Thus in the sub-V_T region, SNM is relatively independent of device sizing [7]. For example, compared to the '5X-increased' conventional cell, the proposed 'minimum sized' pDT cell shows 160% improvement in read SNM at V_{dd}=200mV. This means that for a stable SRAM cell operating in the sub-V_T region, DTMOS technique can be more efficient than simple transistor upsizing in the conventional 6T cell.

Fig. 8(a) shows performance and power consumption during read operations in pDT SRAM. The minimum supply voltage of 135mV in the pDT SRAM provides a wider range of operable supply voltage whereas the conventional SRAM fails to operate below 180mV.

Fig. 8(b) shows the read failure obtained from 1000 Monte-Carlo simulations at iso-area condition in the sub-V_T

region. pDT SRAM exhibits 62% improvement compared to the conventional SRAM at 150mV (the measured minimum supply voltage was 180mV). To achieve iso-read error probability with pDT SRAM (V_{dd}=200mV), conventional SRAM requires the supply voltage of 330mV. Note that at a supply voltage of >200mV, pDT SRAM provides read-error free operation as do 8T and 10T SRAMs.

In the standby mode, supply voltage can be lowered to save leakage power and SRAMs enter the hold state while preserving the cell values. Hold SNM measured at low V_{dd}'s is shown in Fig. 9(a). The improved VTC in a pDT inverter pair also increases the cell stability, introducing 18% (32%) improvement in hold SNM over the CMOS cell at 300mV (200mV) as shown in Fig. 9(b). Higher hold SNM clearly increases immunity to bit-line disturbance from environmental noise. Further, higher hold SNM provides more room for V_{dd} scaling, therefore, reduction in leakage power. Normalized leakage power of the pDT and the CMOS SRAMs measured during hold operations is shown in Fig. 10.

Write operation is performed externally to the cell and its analysis is important from environmental noise point view. Fig. 11 shows the write failure at iso-area. 1000 Monte-Carlo simulations have been performed at V_{dd}=300mV and various voltages of a bit-line holding '0'. pDT SRAM shows comparable write failure with other SRAMs.

BBG for Variation-Tolerant Peripheral Circuits

Variation tolerant peripherals are also essential for robust sub-V_T memory operations since noise margin is seriously limited in the sub-V_T region. The relative strength of PMOS and NMOS transistors changes due to process variation. If the β-ratio (pull-up to pull-down ratio) between PMOS and NMOS transistors is correctly chosen, the gate trip voltage (V_M) can be $0.5V_{dd}$, allowing maximum noise margin for peripheral circuits. The β-ratio given by $β=W_p/W_n$ needs to be designed such that $β \times I_{sub,p}/I_{sub,n}$ is close to unity where

978-1-4244-2018-6/08 $25.00 © 2008 IEEE

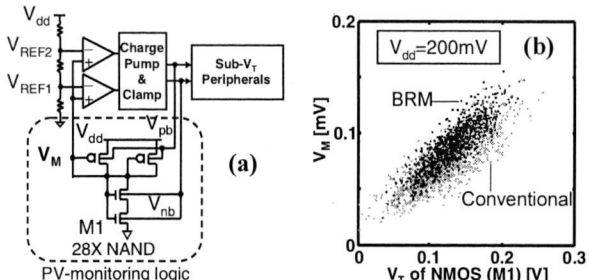

Fig. 12. BBG for subthreshold peripherals with β-ratio modulation. (a) Schematic. (b) Distributions of NAND trip voltage V_M and NMOS V_T at V_{dd}=200mV (simulation).

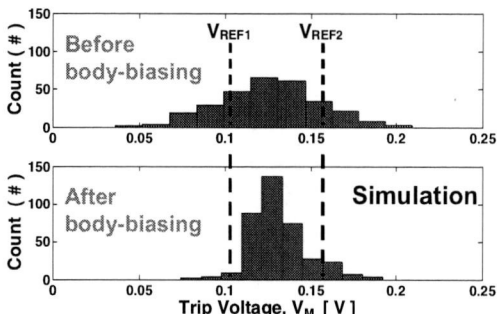

Fig. 13. V_M distribution before and after body biasing.

$I_{sub,p/n}$ is the PMOS/NMOS subthreshold current. This guarantees equal strength of the pull-up network and the pull-down network, which in turn reduces short circuit power and the impact of process variation (PV). However, the ratio of $I_{sub,p}/I_{sub,n}$ exponentially depends on V_T fluctuation and thus the impact of PV is more serious in the sub-V_T region.

In this work we use body biasing (BB) as a method to compensate device mismatches and to maximize circuit robustness under PV [4]. Note that body biasing is more effective in the sub-V_T region, since a little change in the bias can affect current in an exponential way. Fig. 12(a) shows the implemented body bias generator (BBG) that supplies BB voltages for the peripherals and performs adaptive β-ratio modulation (BRM). The NAND V_M in PV-monitoring logic is compared against two reference potentials of V_{REF1} and V_{REF2}. If V_M is below a predetermined reference potential (V_{REF1}), indicating that the pull-up network (PUN) is stronger than the pull-down network (PDN), we apply forward BB to the PUN to make the PUN and the PDN equally strong. If V_M is above V_{REF2}, the PDN is forward body biased. The generated body biases are again fed to the NAND gate, and the updated V_M is compared against the reverence voltages. Under PV, any mismatch between the PUN and the PDN leads to skewed NAND VTCs. 1000 Monte-Carlo simulations are done and the distributions of NAND V_M and its NMOS V_T at V_{dd}=200mV are plotted in Fig. 12(b). Fig. 13 illustrates other Monte-Carlo simulations of V_M before and after applying BB. Note that the application of BB can successfully change the relative strength of two networks and reduce the spread of both V_M and V_T, leading to increased robustness in sub-V_T peripheral circuits. This illustrates the effectiveness of β-ratio modulation (BRM) technique for sub-V_T operations.

To avoid failure due to N/P mismatches in the sub-V_T region it is essential to modulate the β-ratio adaptively such that V_M is close to $0.5V_{dd}$. BBG equalizes the transistors in strength and shifts V_M to $0.5V_{dd}$ while narrowing variability ($\sim\sigma/\mu$) in V_M distribution. Note that the circuit performance is also maximized when the PMOS and NMOS are of equal strength, i.e., V_M=$0.5V_{dd}$. Hence, for both maximum performance and correct functionality under PV, we need to achieve the proper β-ratio in sub-V_T memory peripherals. Impact of PV in BBG can be reduced by increasing the size of PV-monitoring logic (e.g., 28X NAND in Fig. 12(a)).

Conclusions

This paper proposes a DTMOS based 6T SRAM suitable for sub-V_T operation. Cell stability and variation resilience are most important in sub-V_T SRAMs. Positive feedback mechanism in the cross-coupled DTMOS inverters in the cell provides considerable improvement in stability at the expense of area whereas, just increasing size of the cell does not improve cell stability. To increase tolerance to process variation in memory peripherals, we used body bias technique, adaptively adjusting the relative strength between pull-up and pull-down networks. Measurements from a test chip fabricated in 90nm technology confirm the validity of the proposed SRAM for robust sub-V_T operation. Our 16Kb DTMOS SRAM achieves 200% (32%) improvement in read (hold) stability at a supply voltage of 200mV and operates down to 135m, consuming 0.13uW at a frequency of 750Hz.

References

[1] H. Soeleman and K. Roy, "Ultra-low power digital subthreshold logic circuits", *IEEE ISLPED*, pp.94-96, Aug. 1999.

[2] A. Wang and A. Chandrakasan, "A 180mV FFT processor using subthreshold circuit techniques", *IEEE ISSCC*, pp.292-529, 2004.

[3] S. Borkar, T. Karnik, S. Narendra, J. Tschanz, A. Keshavarzi, and V. De, "Parameter Variations and Impact on Circuits and Microarchitecture", *IEEE/ACM DAC*, pp.338-342, June 2003.

[4] M-E.Hwang, A.Raychowdhury, K.Kim, and K.Roy, "A 85mV 40nW Process Tolerant 8x8 FIR Filter with Ultra-Dynamic Voltage Scaling", *IEEE VLSI Circuit Symp.*,pp.145-155, Japan, June 2007.

[5] A.J. Bhavnagarwala., T. Xinghai, and J.D. Meindl, "The impact of intrinsic device fluctuations on CMOS SRAM cell stability", *IEEE Journal of Solid-State Circuits*, Vol.36, No.4, pp.658-665, April 2001.

[6] N. Yoshinobu, H. Masahi, K. Takayuki and K. Itoh, "Review and future prospects of low-voltage RAM circuits", *IBM Journal of Research and development*, Vol.47, No. 5/6, pp. 525-552, 2003.

[7] S. Mukhopadhyay, H. Mahmoodi, K. Roy, "Modeling of failure probability and statistical design of SRAM array for yield enhancement in nanoscaled CMOS",*IEEE TCAD*,Vol.24,No.12, pp.1859-1880,Dec.2005.

[8] I. Carlson, S. Andersson, S. Natarajan, and A. Alvandpour, "A high density, low leakage, 5T SRAM for embedded caches", *ESSCIRC*, pp. 215-218, Sept. 2004.

[9] K. Takeda, Y. Hagihara, Y. Aimoto, M. Nomura, Y. Nakazawa, T. Ishii, and H. Kobatake, "A read-static-noise-margin-free SRAM cell for low-V_{dd} and high-speed applications", *IEEE ISSCC*, pp.478-479, Feb. 2005.

[10] L. Chang, D.M. Fried, J. Hergenrother, J.W. Sleight, R.H. Dennard, R.K. Montoye, L. Sekaric, S.J. McNab, A.W. Topol, C.D. Adams, K.W. Guarini,W. Haensch, "Stable SRAM cell design for the 32nm node and beyond",*IEEE Symp. on VLSI Tech.*,pp.128-129,Feb.2005.

[11] B.H. Calhoun and A. Chandrakasan, "A 256kb Sub-threshold SRAM in 65nm CMOS", *IEEE ISSCC*, pp. 628-629, Feb. 2006.

[12] H. Kawaguchi, Y. Itaka, and T. Sakurai, "Dynamic leakage cut-off scheme for low-voltage SRAM's", *IEEE VLSI Circuit Symp.*, pp.140-141, June 1998.

[13] T. Kim,J. Liu,J. Keane,and C. H. Kim,"A High-Density Subthreshold SRAM with Data-Independent Bitline Leakage and Virtual Ground Replica Scheme",*JSSCC Tech. Digest*, pp.330-331, Feb.2006.

[14] J. Chen, L.T. Clark, T.-H. Chen, "An Ultra-Low-Power MemoryWith a Subthreshold Power Supply Voltage", *IEEE JSSC*, Vol. 41, No. 10, pp. 2344-2353, Oct. 2006.

IEEE 2008 Custom Intergrated Circuits Conference (CICC)

Robust Ultra-Low Voltage ROM Design

Mingoo Seok, Scott Hanson, Jae-Sun Seo, Dennis Sylvester, David Blaauw

Electrical Engineering and Computer Science
University of Michigan, Ann Arbor
{mgseok@ hansons@ jseo@ dmcs@ blaauw@} umich.edu

Abstract— SRAM dominates standby power consumption in many systems since the power supply cannot be gated as in logic blocks. The use of ROM for parts of instruction memory can alleviate this power bottleneck in mobile sensing applications such as implantable biomedical and environmental sensing systems, which can spend up to 99% of their lifetimes in standby mode. However, robust ROM design becomes challenging as the supply voltage is reduced aggressively. In this paper, three different ROM topologies are investigated and compared for ultra-low voltage operation. A simple method to estimate the theoretical robustness at low voltage is proposed and applied to the ROM topologies. A test circuit fabricated in a carefully-selected 0.18μm CMOS technology reveals that our proposed static NAND ROM structure improves performance by 26X, energy by 3.8X and lowest functional supply voltage by 100mV over a conventional dynamic NAND ROM.

I. INTRODUCTION

Applications that depend on energy scavenging or on-die batteries for a power source require ultra–low power consumption to guarantee long lifetime. Such applications range from implantable biomedical systems to environmental sensing systems among others. Scaling supply voltage near or below the device threshold voltage (V_{th}) has recently emerged as an attractive method to save switching energy [1][2]. In addition to minimizing switching energy, standby energy reduction is particularly critical since sensing systems can spend >99% of their lifetimes in standby mode [3].

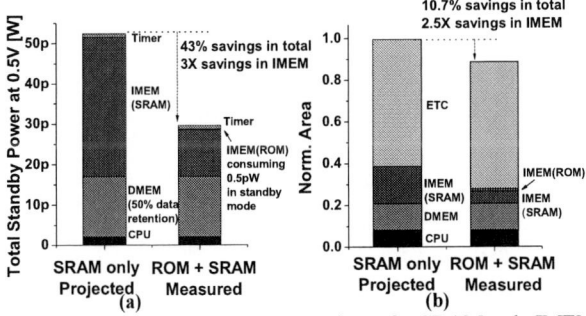

Figure 1. (a) Power and (b) area comparisons for SRAM-only IMEM (projected) and an SRAM/ROM hybrid IMEM (measured).

The authors of [4] observe that SRAM (Static Random Access Memory) for both data and instructions is the dominant source of total standby energy consumption in a typical sensing platform. It is therefore paramount to minimize the standby power consumption of the memories. Although most data memory (DMEM) must be both read and written, instruction memory (IMEM) can be re-optimized to take advantage of its read-only nature. For example, by storing common procedures in ROM (Read Only Memory) with a power gating switch designed for ultra-low voltage operation, both standby power and area can be reduced. Figure 1 shows that standby power can be reduced by 43% and area reduced by 10.7% in a sensing platform by replacing 128 out of 192 SRAM words with power-gated ROM.

However, there are four key challenges for designing robust ROM at ultra-low voltages: 1) The reduced on-current to off-current ratio causes robustness problems, 2) there is potentially a large skew in

beta ratio (relative strength between NFET and PFET) at low voltage, 3) for dynamic ROM styles, conventional keepers (half-latches) are likely to lose state and 4) significant variability further complicates each of the previous three issues.

In this paper, we explore the design of ultra-low voltage ROM. First we investigate the challenges of designing conventional dynamic NAND ROM at ultra-low voltages and propose circuit techniques to overcome these challenges. We also propose a back-of-the-envelope method, referred to as a *current margin plot*, which estimates the theoretical functionality of ROM at ultra-low voltages and provides guidelines for design decisions. We then propose two alternative ROM topologies, static NAND ROM and static NAND-NOR ROM, that improve robustness, performance, and energy-per-operation compared to dynamic NAND ROM. The current margin plot is used to estimate robustness for the two static ROM topologies. We conclude by describing a 0.18μm test chip that includes structures for each of the three ROM topologies discussed. The 0.18μm technology is chosen due to a superior balance between switching and leakage energy relative to more recent technologies. Measurements show that the static NAND ROM improves performance by 26X, energy by 3.8X, and minimum functional supply voltage by 100mV over a conventional dynamic NAND ROM.

II. DYNAMIC NAND ROM DESIGN

This section first investigates the challenges of ultra-low voltage ROM design with a particular focus on the dynamic NAND ROM topology (Figure 2(a)), which is commonly used in superthreshold operation. Although dynamic NOR is also commonly used in superthreshold regime, it is not considered in this study due to the reason discussed in the section II.C. Then a method called a current margin plot is proposed to show theoretical robustness at ultra-low voltages, which can be applied to any ROM topology. Using this method we describe the design of a dynamic NAND ROM targeting ultra-low voltage operation.

A. Challenges of dynamic NAND ROM design at low voltages

The dynamic NAND ROM (Figure 3(a)) operates in two phases: precharge and evaluation. In precharge when clock is low, the dynamic node is charged up to V_{dd}. In evaluation when the clock is high, the dynamic node is either discharged to V_{ss} by stacked NFETs or held at V_{dd} by a half-latch keeper depending on the read word line signals. Having an NFET for a specific read word line means a high output value since the NFET for the word line is turned off.

Operation becomes less robust in the ultra-low voltage regime for several reasons. First, the on- to off-current ratio is reduced (Figure 3(a)), resulting in robustness problems since on-current becomes less distinguishable from off-current. This problem is exacerbated in ROM design since ROM usually has a large number of FETs in series for NAND, or in parallel for NOR styles [5]. The FETs in series limit on-current while FETs in parallel increase the worst-case leakage current. As shown in Figure 3(b), the on-current decreases super-linearly as more FETs are connected in series, resulting in a worse on- to off-current ratio. In the technology used in this work, the on- to off-current ratio of 32-stacked NFETs is only ~152X at 0.5V, which is several orders of magnitude smaller than at nominal voltage.

978-1-4244-2018-6/08 $25.00 © 2008 IEEE

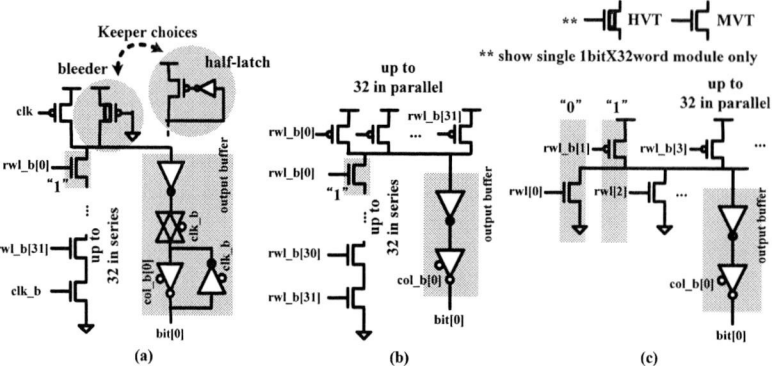

Figure 2. Schematics of three ROMs for ultra-low voltage: (a) dynamic NAND, (b) static NAND, (c) static NAND-NOR

Additionally, the large skew in beta ratio can further aggravate this on- to off-current ratio problem. Beta ratio can be dramatically different between subthreshold and superthreshold since on-current is exponentially dependent on V_{th} and devices are typically optimized for superthreshold operation. In this technology, the min-sized NFET is ~20X stronger than the min-sized PFET at V_{dd}=0.5V, compared to 2.7X at V_{dd}=1.8V, as shown in Figure 3(a). Therefore the ratio of PFET on-current to NFET off-current is reduced by roughly 20X in addition to the already-reduced on-current to off-current ratio of ultra-low voltage operation.

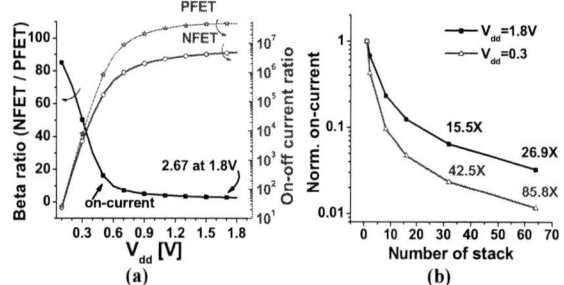

Figure 3. (a) Beta ratio and on- to off-current ratio, (b) on-current reduction over number of stack (for minimum-sized FET)

The functionality of the half-latch keeper (Figure 2(a)) in the ultra-low voltage regime is another problem for dynamic ROM design. The half-latch becomes more important at low voltages since the charge on dynamic nodes is reduced linearly with scaled supply voltage while leakage current stays almost constant. The half-latch is not able to maintain its state for two reasons. First, its retention ability is reduced. When we view the half-latch keeper as broken back-to-back inverters in SRAM, its static noise margin is known to degrade in ultra-low voltage regime. Second, the same amount of charge sharing has a more destructive effect at low voltages. Finally, large variability further complicates low voltage dynamic NAND ROM design. Simulations show that if NFET off-current is larger than nominal by 1σ and PFET on-current smaller than nominal by 1σ due to process variations, the total on- to off-current ratio is reduced by 4.8X at 0.5V.

B. On-current to off-current plot

In this section a back-of-the envelope method, a "current margin plot", is described to estimate the theoretical robustness of ROM in the ultra-low voltage regime. All the factors mentioned in the previous section are accounted for in the method. Here we apply it to a 32-stack dynamic NAND ROM with a high-V_{th} PFET bleeder with length of 0.45μm and width of 0.33μm (Figure 2(a)).

Figure 4 shows the margin of on- to off-current ratio in the 32-stack dynamic NAND ROM with bleeder for two different operations at different voltages: 1) evaluation of a one (eval-1) and 2) evaluation of

a zero (eval-0). The eval-1 (left side) is the case where the output is maintained by the bleeder. Here the worst case off-current through the NFET stack should be smaller than the current that the bleeder provides to guarantee functionality. A guardband equal to the standard deviation of off-current is included to estimate worst-case off-current and is denoted as VAR in Figure 4. On the other hand, complete discharge through stacked NFETs is required for eval-0 (right side of Figure 4). Here the discharging current through NFETs should be larger than the bleeder current. However the discharging current is reduced by the series connection (STK) and guardband for process variation (VAR), resulting in a range of just 20X between the stack and the bleeder at 0.5V.

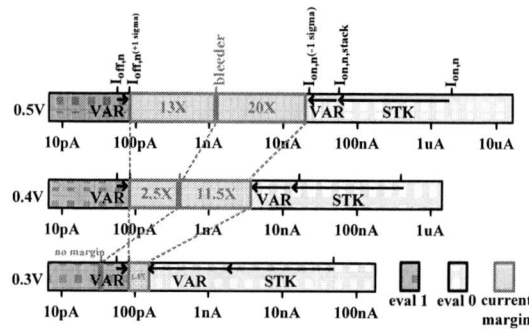

Figure 4. Current margin plot for 32-stack dynamic NAND ROM with HVT bleeder

The magnitude of the guardbands for FET current due to process variation (VAR) is determined based on the standard deviation taken from 1000 Monte Carlo simulations. Here, the log of current is assumed as a Gaussian distribution, which is based on the fact that subthreshold current is an exponential function of normally distributed V_{th}, which is the dominant factor for current variation [6] at ultra-low voltage. One thing to note here is that variation in off-current is almost constant while on-current variation is increased with smaller supply voltage, which is considered in the current margin plot. We set VAR based on the current at $\mu \pm 1\sigma$ of V_{th}.[1] The current reduction due to the large stack (STK) is based on simulation results (Figure 3(b)).

The margin of on- to off-current ratio provides information about two circuit metrics: robustness and performance. Clearly larger margins offer more robustness in light of substantial process variation. In addition to robustness, the margin dictates circuit performance. For instance, a large margin for the eval-1 case implies fast recovery from signal degradation of dynamic nodes. A large margin between dis-

---

[1] A single standard deviation is used here due to the assumption of small designs at ultra-low voltages (e.g., sensor processors) but a more conservative value could also be employed.

charging current and bleeder current is preferred for reduced contention.

The margin of the on- to off-current ratio is diminished as the supply voltage is scaled down. Since process variation in the bleeder current can further degrade the margin, robustness at ultra-low voltage for this ROM topology is questionable.

C. A 32-stack dynamic NAND ROM with HVT bleeder

In the previous section, a 32-stack dynamic NAND ROM with a bleeder is used to illustrate the current margin plot. This section discusses the design decisions regarding ROM topology, stack height and keeper style, which collectively point to a 32-stack dynamic NAND ROM with bleeder as a reasonable design choice.

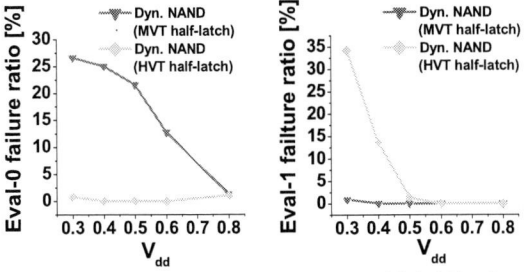

Figure 5. Failure rate for the dynamic NAND with half-latches. (1000 Monte Carlo iterations with die-to-die and mismatch variations)

First, dynamic NAND is chosen over dynamic NOR ROM due to the large skew in beta ratio. Consider a 32-stack dynamic NOR ROM. The off-current through 31 parallel-connected NFETs is only ~20X smaller than the on-current of a single PFET serving as keeper at V_{dd}=0.5V, even without considering process variation. In addition, this reduced on- to off-current ratio degrades performance, which is one of the primary advantages of the NOR topology. Larger leakage power and larger footprint compared to the NAND topology are other drawbacks. Therefore the dynamic NAND structure is chosen over a NOR topology in this study. Note that this decision is motivated primarily by technology limitations. A different technology may make the NOR structure more attractive.

Second, given a dynamic NAND structure, we select a stack height of 32. A taller stack reduces area but can cause robustness problems due to small discharging currents and significant charge sharing. A stack of 64 devices would reduce the margin of on- to off-current ratio between the bleeder current and the worst case on-current by 2X compared to a stack of 32 devices.

Finally, a HVT bleeder is chosen over the half-latch keeper configuration. Monte Carlo simulation with die-to-die and mismatch variations shows that two half-latches with different strengths, (a medium V_{th} (MVT) device with W/L=0.33 μm/1.8μm and an HVT device with W/L= 0.33μm/0.45μm) fail to discharge (eval-0) or hold (eval-1) dynamic nodes in the ultra-low voltage regime as shown in Figure 5. The MVT half-latch often becomes so strong that series-connected NFETs are unable to discharge the dynamic node while the HVT half-latch is often unable to supply enough charge to overcome charge sharing. However the HVT bleeder operates more robustly than the half-latch in the ultra-low voltage regime. An HVT bleeder provides the same on-current as the HVT half-latch for eval-0, so the series connected NFETs are able to discharge the dynamic node. Additionally, the bleeder constantly provides current even after the dynamic node accidentally pulls low, so the correct value will eventually be restored in contrast to a half-latch.

Setting the appropriate strength of the bleeder is important due to the tradeoff between recovery and contention. The margin of on- to off-current in Figure 4 specifies the available strength that the bleeder can have. If the bleeder strength resides outside the margin, it can cause incomplete discharge for eval-0 or poor recovery for eval-1. However setting bleeder strength is a non-trivial task. While keeper strength is adjusted at nominal V_{dd} through gate sizing, it cannot be applied in an area efficient manner in the ultra-low voltage regime due to the exponential variability in current. As shown in Figure 4, if a MVT device is used as a bleeder, the on-current is reduced by ~3 orders of magnitude to reside in the allowed margin, requiring infeasible length biasing. Therefore other knobs such as using different V_{th} devices or applying body bias should be considered. We use a different V_{th} in this work to avoid generating an extra body bias.

III. STATIC NAND AND NAND-NOR ROM

Although we have shown that dynamic NAND ROM can operate at very low voltages, the performance, energy-per-operation and minimum functional voltage are unsatisfactory due to the small current margin. The bleeder is among the most important components of this design style, which gives rise to a challenging sizing tradeoff. Therefore, new topologies without a bleeder are worth investigating. In this section, we describe the design of 2 full-static ROMs: 32-stack NAND and 32-leg NAND-NOR as shown in Figures 2(b) and 2(c). Since these structures have no bleeder, a larger on- to off-current margin is expected.

A. Investigating static ROM topologies

This section applies the current margin plot to the two static ROM topologies to investigate theoretical robustness at ultra-low voltages. Figure 6 shows the margin of on- to off-current ratio for the 32-stack static NAND ROM. Only the eval-0 case is considered here since it is the most stringent for large NFET stacks. Larger margin is still observed after incorporating the effect of parallel connection of PFETs (denoted as PAR), series connection of NFETs, and guardbands for process variation. The avoidance of the bleeder also helps increase the margin and ease design. Overall a 17X margin is maintained at a very aggressive V_{dd} of 0.3V, compared to zero margin for the 32-stack dynamic NAND ROM described earlier.

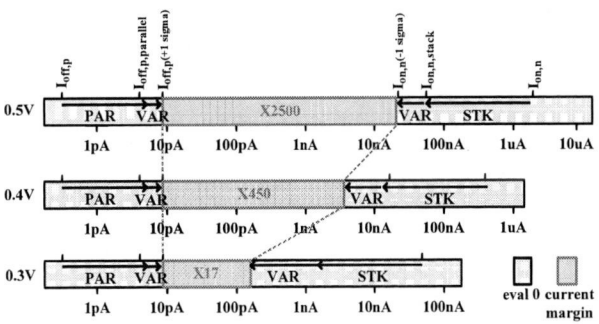

Figure 6. Current margin plot for 32-stack static NAND ROM

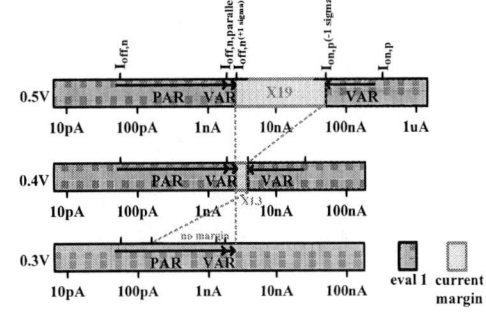

Figure 7. Current margin plot for 32-leg static NAND-NOR ROM

If NFET and PFET strengths are balanced, the static NAND ROM can be improved by replacing the long stack with parallel legs using inverted input signals as shown in Figure 2(c). However since the technology used in this study has a large beta ratio at low voltages, this topology is less robust than the static NAND ROM. Figure 7 shows the much reduced on-off margin in the eval-1 case where a single PFET contends with 31 NFETs, which is same as in a dynamic NOR topology. The current margin disappears at 0.3V, as in the dynamic NAND ROM. However, better robustness is expected over dynamic NAND ROM since no bleeder is present. Although the NAND-NOR ROM topology is not ideal in the technology used in this study, it may be a good candidate for technologies with more balanced beta ratios at low voltage.

B. Static NAND ROM Monte Carlo Analysis

Since the current margin plot is a first-order method to estimate robustness, Monte Carlo simulations considering all sources of variation are performed to investigate the effectiveness of the plot as well as the robustness of the ROM topologies. As shown in Figure 8, the static NAND ROM starts to fail at 0.3V in the eval-0 case due to the large NFET stack. The failure voltage is higher than that estimated by the current margin plot since the latter considers only one standard deviation of variation. In comparing topologies, the minimum functional voltage for the static NAND ROM is larger than that for the dynamic NAND ROM by nearly 200mV, confirming that the current margin plot is able to track the trends as well as the static NAND ROM is more robust.

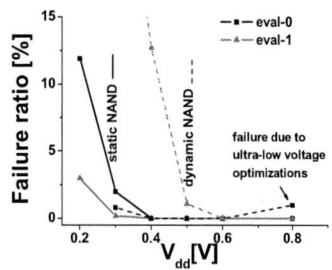

Figure 8. 1000 Monte Carlo SPICE simulations for two ROM topologies considering mismatch and die-to-die

IV. MEASUREMENT RESULTS

A 10x128bit dynamic NAND ROM with HVT bleeder, a 10x128bit static NAND ROM, and a 10x128bit static NAND-NOR ROM were fabricated in a 0.18μm CMOS technology. Each ROM contains an identical set of random data patterns as well as patterns causing worst case charge sharing. The worst case pattern for the dynamic NAND ROM and the static NAND ROM is 31 series-connected NFETs while the worst case for the static NAND-NOR structure is a single PFET connected to 31 parallel-connected NFETs. Relevant silicon measurements and dimensions are shown in Figure 10.

The two static topologies show dramatically improved maximum operating frequency, energy-per-operation, and minimum functional voltage compared to the dynamic NAND ROM. The small on- to off-

current ratio degrades performance in the dynamic NAND ROM, leading to substantially lower performance. The static NAND ROM and the static NAND-NOR ROM show similar energy, performance and minimum operating voltage numbers, though the static NAND ROM has a small advantage over the static NAND-NOR ROM, as predicted in previous sections.

Figure 9 shows the effect of variability on the performance of the static NAND ROM. The variation in maximum operating frequency (σ/μ) across 20 dies at 0.35V is just 18%. This number falls between the bounds set by a Monte Carlo simulation that includes die-to-die and mismatch variation and another Monte Carlo simulation considering mismatch only. Since all the 20 dies come from a single wafer, it is not surprising that the relatively small measured variability is much closer to the variability predicted by the mismatch-only simulation.

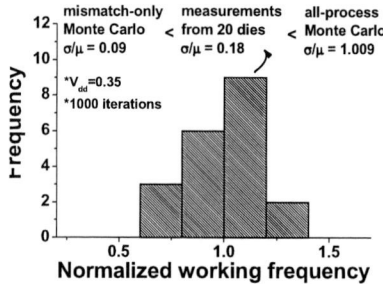

Figure 9. Histogram of operating frequency of static NAND ROM

V. CONCLUSIONS

Three different ROM topologies for ultra-low voltage operation are investigated with the test chip fabrication in an industrial 0.18μm CMOS technology. The challenges in ultra-low voltage design are analyzed and incorporated in the current margin plot which is devised for estimating theoretical low voltage robustness. Silicon measurements shows that the static NAND ROM shows 26X faster performance, 3.8X smaller energy-per-operation and 100mV smaller minimum working voltage than the dynamic NAND ROM with 33% area penalty. The static NAND-NOR ROM is also studied as a potential candidate for other technologies.

REFERENCES

[1] A. Wang, *et al*, "A 180mV FFT Processor Using Subthreshold Circuit Techniques", ISSCC, 2004

[2] M. Hwang, *et al*, "A 85mV 40nW Process-Tolerant Subthreshold 8x8 FIR Filter in 130nm Technology", Symposium on VLSI Circuits, 2007

[3] L. Nazhadili, *et al*, "SenseBench: toward on accurate evaluation of sensor network processors", Workload Characterization Symposium, 2005

[4] M. Seok, *et al*, "The Phoenix Processor: A 30pW Platform for Sensor Applications", Symposium on VLSI Circuits, to be published in 2008

[5] S. Hsu, *et al*, "A 9GHz 320x80bit Low Leakage Microcode Read Only Memories in 65nm CMOS", ESSCIRC, 2006

[6] N. Verma, et al, "Nanometer MOSFET Variation in Minimum Energy Subthreshold Circuits", TED, vol. 55, no. 1, Jan, 2008

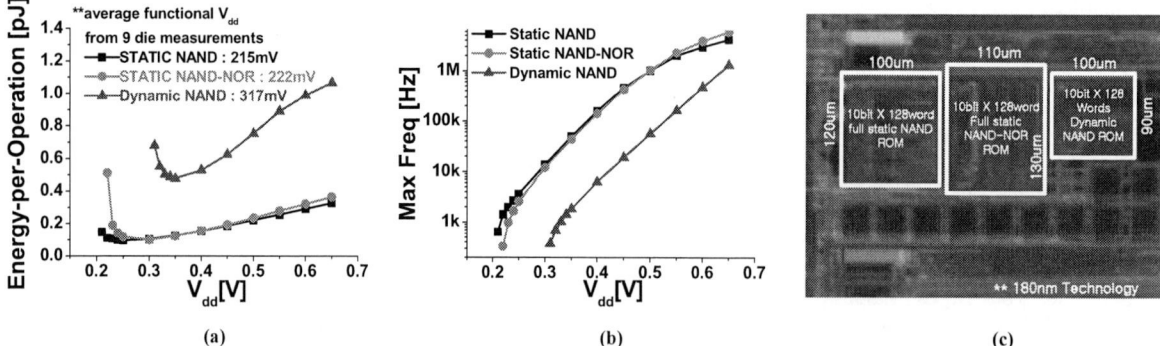

Figure 10. (a) Measured energy-per-operation (b) measured frequency measurement (c) die photo and dimensions

IEEE 2008 Custom Intergrated Circuits Conference (CICC)

A Million Cycle 0.13um 1Mb Embedded SONOS Flash Memory Using Successive Approximated Read Calibration

N. Wang, X. Yao, Y. Lei, G.Y. Feng, Q.H. Dong, L. Xu, L. Guo, Z. Wang, T.S. Tang

Shanghai Hua Hong NEC Electronics Company, Limited

Abstract- **A 1Mb embedded 2T-SONOS Flash macro is implemented in 0.13um logic compatible process. The Flash macro has improved reliability and yield with a power-on Successive Approximated Read Calibration (SARC). Word-line decoder area is greatly reduced using 1.8V transistors to tolerate high voltage. Source degenerated compensation is implemented to enhance read margin. The Flash macro consumes 1.0mA at 50ns 1.8V access and 0.5uA in standby mode, and achieves one million cycling and 20-year data retention.**

I. INTRODUCTION

Embedded Flash memory technology is growing fast with the requirements from MCU and other SoC applications. Meanwhile, Flash based embedded nonvolatile memories are adopted to replace EEPROM in smart card and security applications for lower cost. These applications increasingly request higher endurance performance and longer retention time. Scaling down to 0.13um and below, SONOS technology [1][2] is getting more competitive for its compatibility to logic process, good reliability and scalability to next generation.

This paper is to present a 1Mb SONOS Flash macro with low power, low leakage and higher reliability, which is ideal for smart card and SoC products with high security requirement.

Successive Approximated Read Calibration (SARC) is introduced for reliability and yield. To cover all SONOS cell corners and the entire lifetime, specific parameters for sensing reference current need to be programmed to special area in the Flash macro as trimming data. SARC works during power-on procedure to get trimming bits, solving the deadlock that reading out trimming bits need the trimmed parameters set by themselves.

Word-line decoder is traditionally large using high voltage transistors to sustain programming voltage. A novel Voltage Divided High Voltage Tolerance (VDHVT) method is proposed to allow implementation with all low voltage transistors, which results in 5% area reduction for the whole macro.

To compromise process and temperature extreme of cell selector transistor, a self-compensated loading is implemented in the sense amplifier to gain margin for decision circuit and in turn the read speed. Moreover, a dynamic comparator is used for fast sensing.

II. ARCHITECTURE

The 0.62um2 memory cell is a 2T single-ended stack structure with select gate at the source side and is programmed and erased with Fowler-Nordheim (FN) tunneling. It is not necessary to have either redundancy or complex algorithm for erase control as the memory is not suffered from over-erase

and is relatively relaxed from disturb. Figure 1 shows the architecture for this Flash macro and Table 1 gives the voltage setting for each operation mode. The page size is 128 bytes. It takes 5ms to program a page or the whole chip and less than 10ms to erase a page or the chip. After erase, the threshold voltage is reduced and the state is defined as zero. The output bus can be configured to x8, x16 or x32 to accommodate 8-b, 16-b and 32-b MCU.

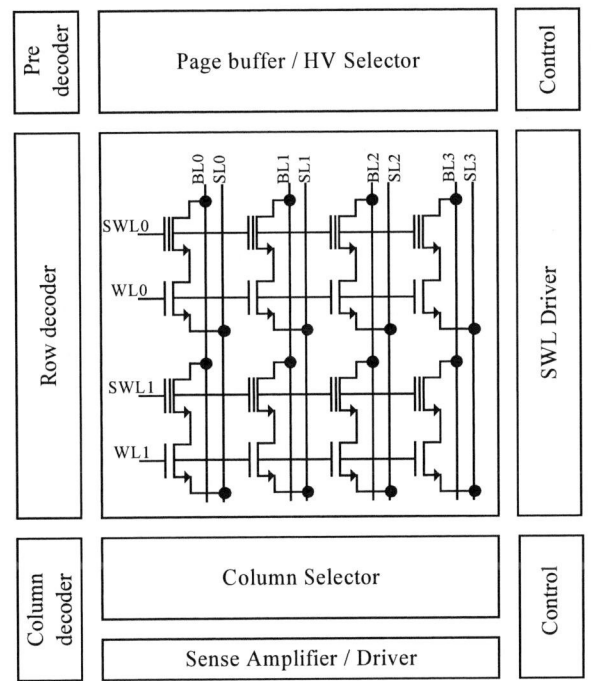

Fig. 1. Flash Macro Architecture

Table 1. Voltage condition of each operation mode

Mode	WL	SWL	BL	SL	Bulk
Read	VCC	GND	VBL	GND	GND
Erase	VCC	VNEG	VPOS	Open	VPOS
Program	VNEG	VPOS	VNEG	Open	VNEG

Note: VNEG=-4V, VPOS=7V, VBL~1V, VCC=1.5-2.0V

III. READ PATH AND CALIBRATION

During read mode, the SONOS word line (SWL) is fixed to ground while selected word line (WL) is set high by row decoder. Bit line voltage VBL is set to around 1V for adequate cell current.

The current sense amplifier, shown in Figure 2, is divided to three parts: loading compensation, current to voltage converting stage and dynamic decision stage. IREF is from REFGEN block as the reference current to compare with the cell current. IREF is also used to bias PMOS loading in the first stage to its saturation region to obtain high gain. As the word line is not boosted in this macro, the memory cell have source degenerated effect by the select transistor especially at minimum supply voltage. To compensate for the degradation, the PMOS loading is adjusted by a similar source degenerated scheme. Npass is the same as the select transistor in the cell. By doing so, the split voltage difference after equalization is increased. The bit line and reference line voltages are limited by the two native MOS with gates connected to VBLG. To improve the read access time, the second stage of the sense amplifier is high-speed comparators with dynamic operation. The latching time for 1mV difference is less than 2ns from post simulation result at worst case. To minimize offset voltage, the comparator is optimized for layout matching. In house measurement shows that the offset is within +/-15mV for 48 dies across industry temperature range.

For embedded applications, read power reduction is emphasized. In addition to low power design for every building block, the consumption at read mode is further reduced for 8-bit output data-bus width as all other sense amplifiers are turned off.

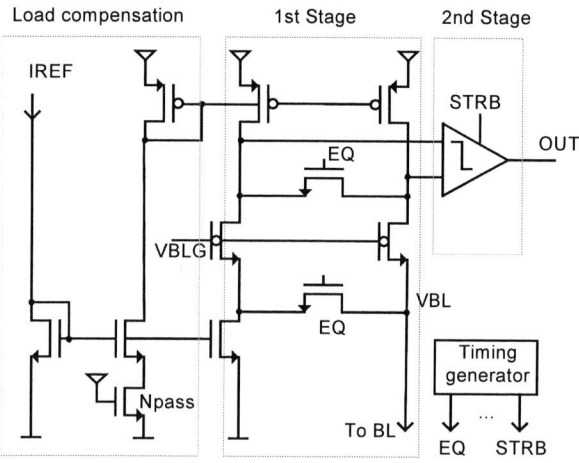

Fig. 2. Sense Amplifier Block

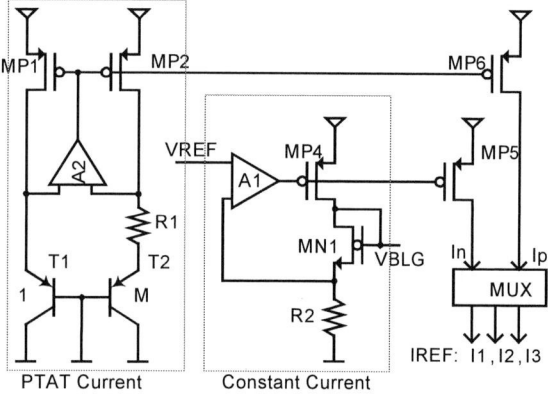

Fig. 3. REFGEN block

REFGEN block, shown in Figure 3, generates reference current which can be trimmed for both absolute value and temperature coefficient (TC). Temperature compensation and TC adjustment is archived by mixing PTAT current and constant current with different ratio. The constant current part also generates VBLG as the limiter voltage for bit line.

Though the cell threshold voltage within one die or one wafer is well controlled, there are extreme corners in which the cell current changes greatly from fast corner lots to slow corner lots. It is possible that the current of fast corner programmed cell at the end of its life, i.e. after data retention, is higher than the current of slow corner erased cell at the end of its life. In Figure 4, the phenomenon is presented as the overlapping of FPC and SEC at EOL. It is difficult to trace the cell lifetime using dummy cell, so the reference current is trimmed to proper value to guarantee the good yield. However, although the trimmed data can be written to a special area in the macro, it is not safe to read out the data directly without the trimmed reference current. This is a deadlock. It can be solved using fuse but it will both increase process and testing cost, and enlarge the chip area. Nonvolatile-SRAM (NV-SRAM) can be introduced for its robust differential architecture, but will also increase chip complexity, die size and testing cost.

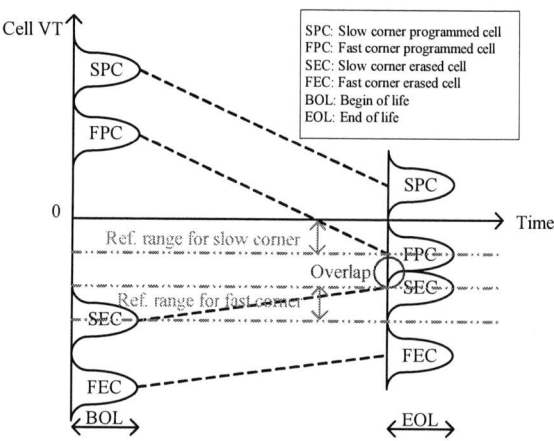

Fig. 4. Cell VT distribution for fast corner and slow corner

In this macro, SARC, a novel calibration method, is proposed to overcome the problem. 5 bits are used to trim the current DAC for the magnitude of the reference current. Unlike traditional method to write single 5 bits directly to special area, in this macro, each bit is written with its complementary to form 5 pair bits. During the power-on step, the macro will enter into a calibration procedure. Internal state machine behaves as an approximating approaching operation to read out 5 pair bits and will determine one bit of the trimmed data every cycle by judging the number of '1's against the number of '0's. After 5 cycles, the 5 bits are correctly decided and consequently loaded back to set the current DAC. After the calibration, the read path is ready for normal read operation. Figure 5 shows the diagram of the SARC.

978-1-4244-2018-6/08 $25.00 © 2008 IEEE   458

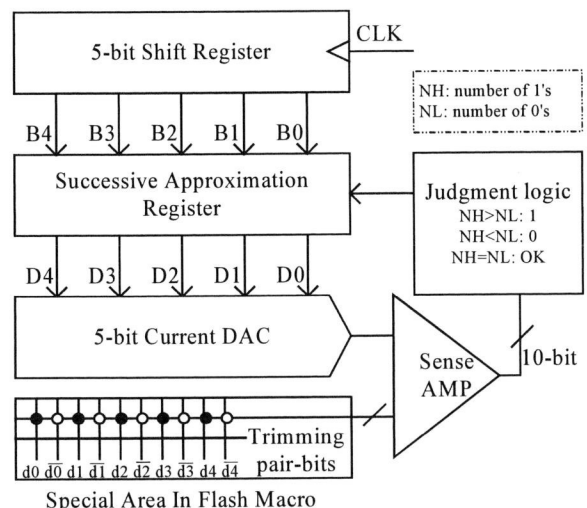

Fig. 5. SARC Diagram

IV. HIGH VOLTAGE PATH

For normal application, the Flash macro works as a page mode program and erase with a 1024-bit buffer. Before programming, there need a procedure to write all bytes into the buffer. After erase, Flash cells of the selected page are cleared to '0', with lower threshold voltage. Then after programming, Flash cells that have '1' in their relating bit in the buffer will be programmed to '1', with higher threshold voltage.

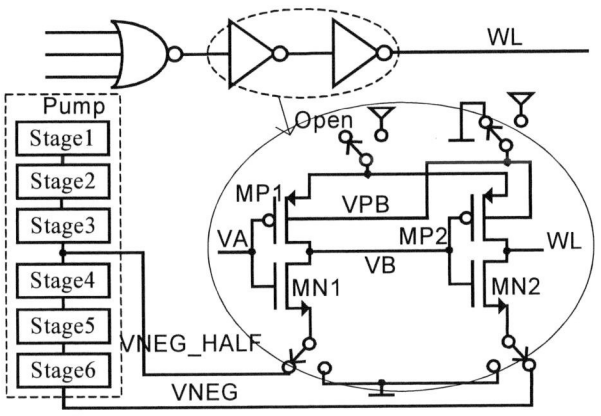

Fig. 6. VDHVT schematic

Table 1 illustrates the setting of cell nodes in erase and program mode. Both erase and program are using F-N tunneling effect with 5 to 10ms for erase and 5ms for program.

The high voltage path is optimized for compact layout area. The word line control is to generate decoding signals in read mode and is forced to negative high voltage (VNEG) during program mode. Traditional implementation will need to use thick oxide transistor to tolerate high voltage stress. However, these transistors are slow and need to be designed to large

dimensions for speed in read mode. Bootstrapping is an alternative solution but still consumes large layout area. This design explores a novel design technique, Voltage Divided High Voltage Tolerance (VDHVT), to replace high voltage transistors with all low voltage MOS. Shown in Fig 7, for the final two inverting stages, the PMOS bulks and sources are switched to ground and high impedance respectively. Thanks to the deep N-well, the NMOS (MN1, MN2) sources and bulks can be switched to VNEG_HALF and VNEG for 1st stage and 2nd stage respectively. And the input of the inverter (VA) is set to zero. As VNEG is no higher than junction breakdown voltage and VNEG_HALF is around half of VNEG generated from the pump circuit, there is no over stress to any nodes in the row decoder. Without loss of generality, it is a method to divide the high voltage by N to limit the stress within device specification. Though it is at the cost of more complexity for decoder and VNEG/N generation, for this case, it is acceptable and effective. The result is more than 5% improvement in total macro area.

V. MEASUREMENT RESULT AND CONCLUSION

Implemented in 0.13um low leakage CMOS compatible process, this embedded SONOS Flash macro is working under supply voltage ranged from 1.5 to 2.0V with very low power consumption. At the access time of 50ns and 8bit output, the current is 1.0mA and standby leakage is less than 1uA at room temperature. Program and erase current is as low as 550uA.

For reliability, the macro is cycled for 200k cycles with all 1Mb cells and the result is extrapolated to one million cycles. Both function check and margin test show that there is still adequate window of threshold voltage after the stress. Meanwhile, data retention with preconditioning of 10k cycles is verified and at least 20 years are guaranteed within industrial level temperature range. For preconditioning of 100k cycles, there is at least 10 years' retention time. The reliability data are shown in Figure 7 and Figure 8, which proves that the read path and the calibration is effective.

Table 2 summarizes the key parameters of the SONOS Flash macro.

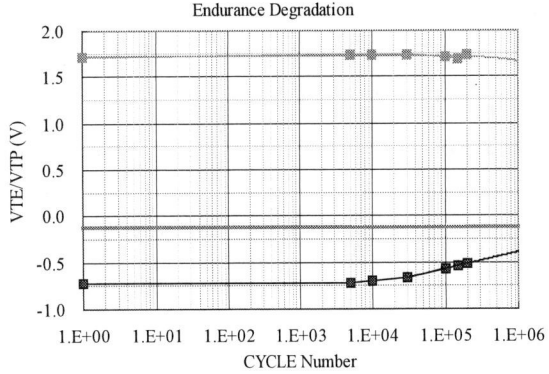

Fig. 7. Endurance result at 55C

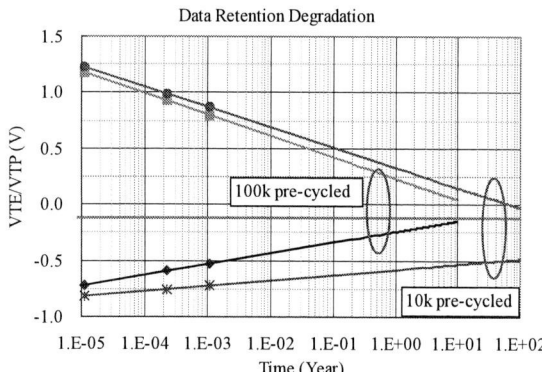

Fig. 8. Data retention result at 95C

Table 2. Flash Macro Key Performance

Process	0.13um 1P3M Low Leakage CMOS
Cell Size	0.62um2 (2T)
Macro Size	936um X 1224um = 1.14mm2
Supply Voltage (VCC)	1.5 to 2.0V
Read Access Time	47.5ns
Bulk/Page Program Time	5ms
Bulk/Page Erase Time	5 to 10ms
IDD at Standby Mode	<0.5uA
IDD at Program/Erase	0.55mA
IDD at Read Mode	1.0mA at 50ns access
Endurance (Industrial level)	>1M Cycles
Data Retention (Industrial level)	>20 Years (10k pre-cycled)

Figure 9 shows the die photo of the SONOS Flash macro with 1.14mm2 area.

ACKNOWLEDGMENTS

The authors would like to thank Vijay Raghavan, Ryan Hirose, Paul Ruths and Sean Mulholland from Cypress Semiconductor for valuable discussion.

REFERENCES

[1] J. H. Kim, I. W. Cho, G. J. Bae, et al., "Highly manufacturable SONOS non-volatile memory for the embedded SoC solution," 2003 Symposium on VLSI Technology Dig. Tech. Papers, pp. 31-32, 2003.
[2] M. K. Seo, S. H. Sim, M. H. Oh, et al., " A 130-nm 0.9-V 66-MHz 8-Mb (256Kx32) local SONOS Embedded Flash EEPROM," IEEE J. Solid-State Circuits, Vol. 40, No. 4, pp. 877-883, 2005.

Fig. 9. Flash macro die photo

Clocking Circuits

Session 17

Chair: Dennis Fischette, AMD
Co-Chair: Kimo Tam, Analog Devices

The first paper reports an all digital fractional-N PLL in state-of-the-art 45 nm SOI CMOS. The authors study the impact of various orders of feedback divider modulator on the period jitter.

The second paper describes 8 GHz clocking circuits for a 16Gb/s per pin memory interface. This work combines an LC-PLL and a ring-PLL in a 65 nm CMOS process to achieve sub-ps rms jitter.

The third paper presents a reconfigurable PLL/DLL based clock multiplier implemented in a 90 nm CMOS process. Starting in PLL-mode, a lock-detect circuit switches the circuit to DLL and simultaneously reconfigures the loop filter.

The fourth paper reports a supply-regulated phase-locked loop employing a split-tuned architecture to decouple the tradeoff between supply noise rejection and power consumption.

The fifth paper describes a quadrature VCO topology that combines the low power of –gm oscillators with the inherent buffering of Colpitts oscillators. The prototype achieves 20 GHz oscillation in 0.13 μm CMOS and provides linear phase control for deskew applications.

The sixth paper describes a spread spectrum clock generator for 3 GHz SATA applications that incorporates a chaotic PAM modulator to achieve a >14 dB peak reduction in EMI.

The seventh paper describes a spread-spectrum clock generator that employs a piecewise-linear modulation profile to emulate a "Hershey kiss" modulation profile with lower implementation complexity.

The final paper describes an 8x5 Gb/s source-synchronous receiver that employs a cascaded-DLL architecture that avoids the filtering of the jitter on the received clock to enhance jitter tolerance bandwidth. It also reduces phase spacing mismatch in the DLLs by 40%.

978-1-4244-2018-6/08 $25.00 © 2008 IEEE

Notes

IEEE 2008 Custom Intergrated Circuits Conference (CICC)

A Wide Tuning Range (1 GHz-to-15 GHz) Fractional-N All-Digital PLL in 45nm SOI

Alexander Rylyakov[1], Jose Tierno[1], George English[2], Michael Sperling[2], Daniel Friedman[1]

[1]IBM T.J. Watson Research Center, Yorktown Heights, NY
[2]IBM , Poughkeepsie, NY

Abstract-An all static CMOS (45nm SOI) all-digital fractional-N PLL has a wide tuning range (from 0.84 GHz to 13.3 GHz, at 1.0V, 65°C) and supports a broad range of multiplication factors (up to 1,000x) and reference clock speeds (from 2 MHz to 1 GHz). At 125°C the period jitter of the 4.12 GHz clock (206 MHz reference) is 1.1ps rms (11.4ps pp) at 1.3V (52.4mW), and 2.2ps rms, (22.7ps pp) at 0.7V (9.7mW). The area of the PLL is 175µm x 160µm.

I. INTRODUCTION

An all-digital PLL is an ideal candidate for microprocessor and ASIC clocking applications. It uses the same exact devices and circuit families as all other logic components on the chip, enabling it to match their performance and yield. An all-digital implementation significantly reduces the PLL area and enhances noise immunity, programmability and testability. For a recent review and examples of digital PLL techniques see, e.g., [1] and [2]. This work is a further development of the base architecture variants reported in [3] and [4], with the main emphasis on the fractional-N operation. Fractional-N capability increases the frequency resolution of the PLL which makes it a highly desirable feature for an ASIC or microprocessor clock generator. We study the impact of various orders of the feedback divider modulator on the period jitter, both rms and peak-to-peak. These values define the shortest clock cycle at a given frequency and are the main figures of merit for the target applications.

II. CIRCUIT DESIGN

The PLL top level block diagram is shown in Fig. 1. The proportional-integral loop filter (LF) processes the 1-bit "early/late" output of the bang-bang PFD. The two most recent PFD outputs are used to select one of four signed 7-bit constants and the result is integrated in an 8-bit accumulator. The overflow-underflow ("inc/dec") signals from the accumulator are applied to the digitally controlled oscillator (DCO). The remaining 8-bit "frac" output of the LF accumulator is sampled by a 2nd-order delta-sigma modulator (DSM) and converted into a 3-bit "dither" signal for the DCO. Fig. 2 shows the timing relationships between the PLL output, the global internal clock ("clkg") and the "phold" signal.

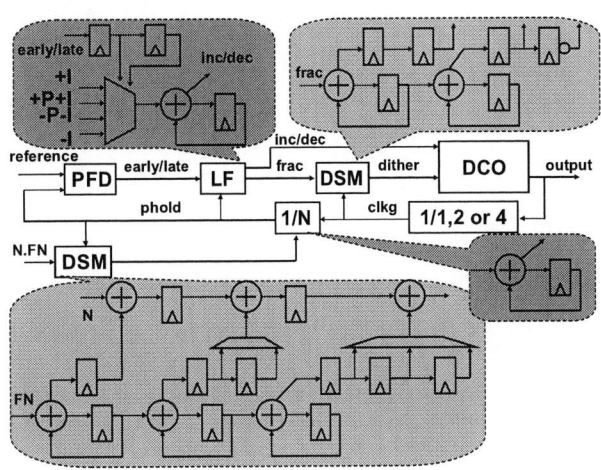

Fig. 1 Top level block diagram.

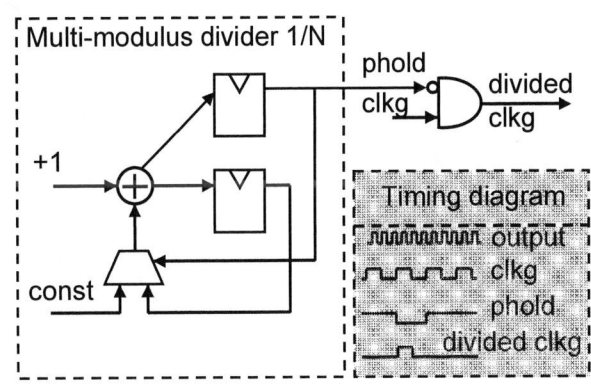

Fig. 2 DPLL internal clocking.

Division by N is achieved by generating the "phold" as the overflow signal from a reloadable counter (1/N in Fig. 1) and then using it as the mask for "clkg", enabling output of one "clkg" pulse out of N, as shown in Fig. 2. The 1/N counter can be loaded with a different 8-bit starting value ("const" in Fig. 2) every reference clock cycle, which makes it a multimodulus divider, ready for fractional-N implementation. The output clock can have the same frequency as "clkg", or it

978-1-4244-2018-6/08 $25.00 © 2008 IEEE   463

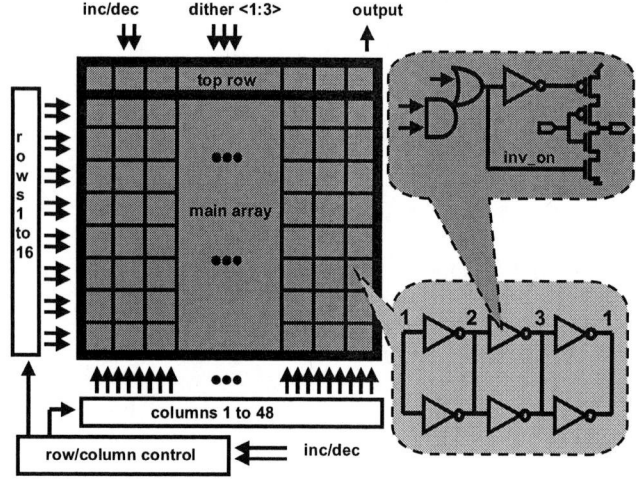

Fig. 3 DCO schematic.

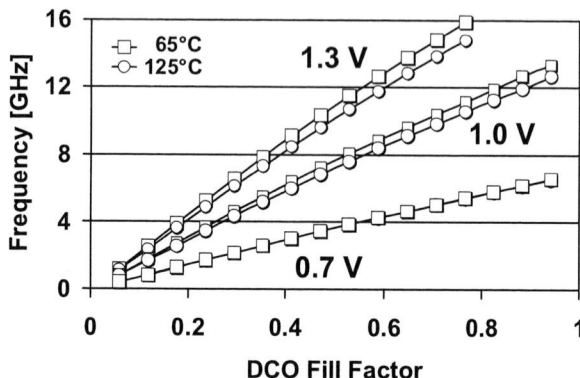

Fig. 4 DCO tuning curves at V_{DDA}=0.7, 1.0 and 1.3V. Each curve was measured at 65°C (square marker) and at 125°C (circle marker). At V_{DDA}=0.7V the 65°C data overlaps the 125°C points.

can be divided by 2 or 4. A 3rd-order DSM is used to convert the static 16-bit frequency command word (8-bit integer part "N" and 8-bit fractional part "FN") into the 8-bit dithering signal for the 1/N counter. All 10 adders shown in Fig.1 are instances of the same 8-bit Kogge-Stone adder. The 3-stage ring DCO schematic is shown in Fig. 3. The main array is an 8x16 matrix of 2x3 blocks of tri-state inverters with local row- column logic, scaled to 45nm from the 65nm design described in [3]. The tri-state inverters in the main array are individually controlled by the local "inv_on" signals. The tri-state inverters in the top row are always on, except the 3 "dither" inverters, and the 2 "inc/dec" inverters. These last two inverters act to reduce loop latency by applying the output of the LF one clock cycle ahead of the row-column control block [4]. The DCO power supply (V_{DDA}) is separate from that of the digital logic (V_{DD}). Unlike earlier designs, the signals going in and out of the DCO are converted from one voltage domain to another by dedicated voltage level shifters.

III. EXPERIMENTAL RESULTS

The PLL was fabricated in a 45 nm SOI CMOS [5] and occupies 175μm x 160μm. The open-loop DCO tuning curves at constant power supply voltages are shown in Fig. 4. The DCO fill factor is defined as the ratio of on-state to total tri-state inverters in the array. The DCO gain K_{DCO} (minimum frequency step), shown in Fig. 5, is measured at constant voltage and at constant current through the array. Clearly, the constant current condition can significantly reduce K_{DCO}, while also limiting the tuning range. A smaller K_{DCO} is expected to lower the output jitter due to the reduced quantization noise of the DCO. The integer-N (both DSMs off) PLL period jitter and power dissipation numbers at output frequencies close to 4GHz are summarized in Table I. For the data, summarized in Fig.6 and in Table II, a 3.8 GHz clock was synthesized from a 200 MHz reference as either an integer or a fractional-N multiplier at different orders of the

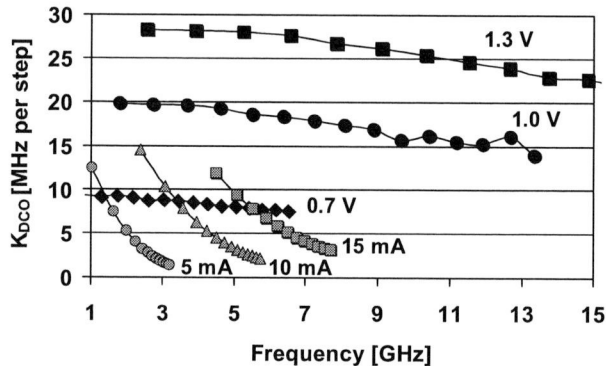

Fig. 5 DCO gain at V_{DDA}=0.7, 1.0 and 1.3V and at I_{DDA}=5, 10 and 15mA. The curves were measured at 65°C.

divider DSM. All experiments reported in this work have been carried out with the DCO-DSM turned off. The peak to peak period jitter numbers in Tables I and II were measured on more than 3 million clock samples each.

TABLE I
INTEGER-N DPLL PERFORMANCE SUMMARY

4.12 GHz output (206 MHz × 4 × 5) at constant V_{DDA}						
V_{DDA}, V	I_{DDA}, mA	V_{DD}, V	I_{DD}, mA	T, °C	Jitter, ps rms	Jitter, ps pp
0.7	7.6	0.7	3.8	65	2.1	22.1
1.0	11.7	1.0	6.9	65	1.2	13.0
0.7	8.2	0.7	5.7	125	2.2	22.7
0.7	8.2	1.0	11.1	125	1.5	19.2
1.0	12.8	1.0	10.7	125	1.3	14.8
1.3	19.5	1.3	20.8	125	1.1	11.4
3.8 GHz output (200 MHz × 1 × 19) at constant I_{DDA}						
0.76	7.5	1.0	11.1	65	1.3	17.8
0.97	10	1.0	11.1	65	1.0	12.8
1.14	12.5	1.0	11.1	65	1.1	12.2
1.27	15	1.0	11.1	65	1.2	11.9

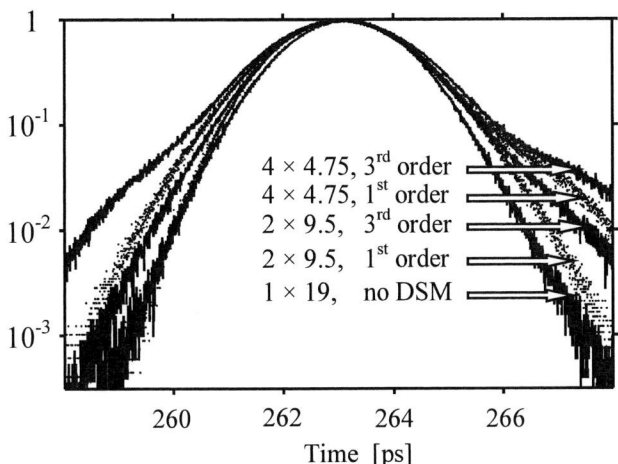

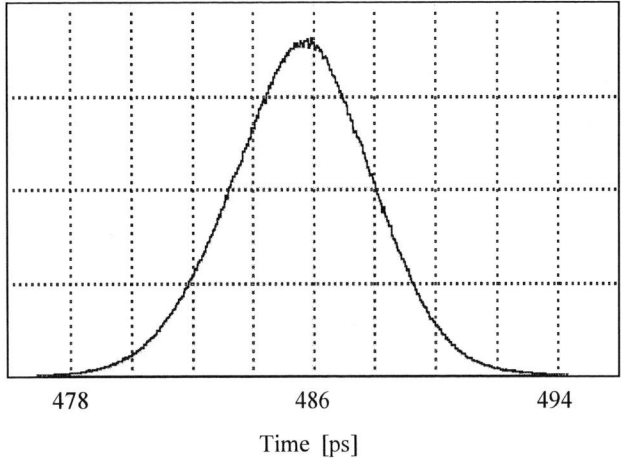

Fig. 6 Semilog scale normalized period jitter histograms at 3.8 GHz (200 MHz reference). The curves (from the inside to the outside) are: 1 × 19 - no divider DSM, 2 × 9.5 - 1st order divider DSM, , 2 × 9.5 - 3rd order, 4 × 4.75 - 1st order, 4 × 4.75 - 3rd order.

Fig. 7 Linear scale period jitter histogram of the 2.06 GHz output at 2.06 MHz reference (multiplication factor: 1,000 = 4 × 250). The period jitter is 2.5 ps rms, 29 ps peak-to-peak (3.1 million clock cycles); $V_{DDA} = V_{DD}$=0.75V, 65°C.

TABLE II
FRACTIONAL-N DPLL PERFORMANCE SUMMARY

3.8 GHz (200 MHz reference) at $V_{DDA} = V_{DD}$=1.0V, 25°C				
Multiplication factor	DSM order	Power, mW	Jitter, ps rms	Jitter, ps pp
1 × 19	-	20.1	1.0	11.8
2 × 9.5	1	17.5	1.1	12.2
2 × 9.5	2	17.6	1.1	13.2
2 × 9.5	3	17.9	1.2	14.6
4 × 4.75	1	16	1.3	15.2
4 × 4.75	2	16.2	1.4	16.5
4 × 4.75	3	16.5	1.5	18.3

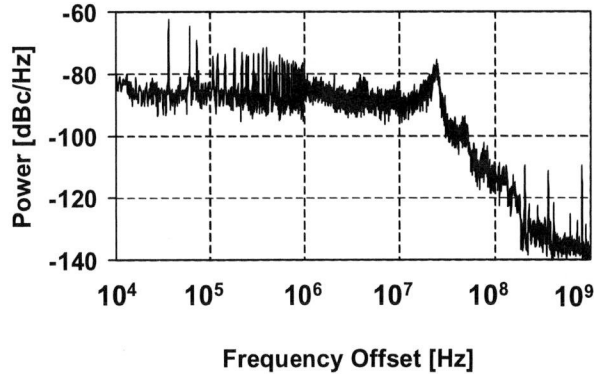

Fig. 8 Phase noise plot at 4.12 GHz output (206 MHz reference), $V_{DDA} = V_{DD}$=1.3V, 65°C.

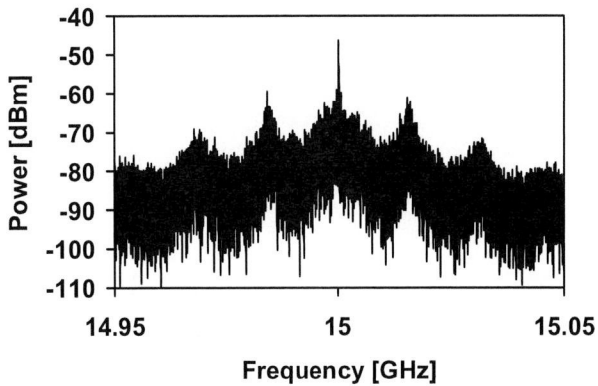

Fig. 9 Spectrum of 15 GHz output (250 MHz reference), $V_{DDA} = V_{DD}$=1.3V, 65°C. Span: 100 MHz, resolution bandwidth: 10 kHz.

    Fractional-N operation was also verified for more than 50 different multiplication factors and DSM orders, with the overall peak to peak frequency deviation from ideal being less than 0.12ppm, limited by the test equipment.
    In a separate study, a 2 GHz clock was synthesized from references ranging from 2 MHz to 1 GHz, at power supplies from 0.75V to 1.3V and temperatures from 25°C to 125°C. The observed period jitter of the 2 GHz clock was better than 3.5ps rms, at all settings. Period jitter histogram of the 2.06 GHz output clock synthesized from a 2.06 MHz reference is shown in Fig. 7.
    Fig. 8 shows the measured phase noise plot of the 4.12 GHz signal. The broad spur around 20 MHz offset from the carrier is the result of the limit cycle behavior, typical of digital PLLs using bang-bang PFDs [6],[7]. Correct PLL locking behavior was verified for frequencies up to 15GHz, limited by the bandwidth of the output driver. The measured spectrum of the 15 GHz output is shown in Fig. 9. The details of the DPLL physical design are shown in Fig. 10.

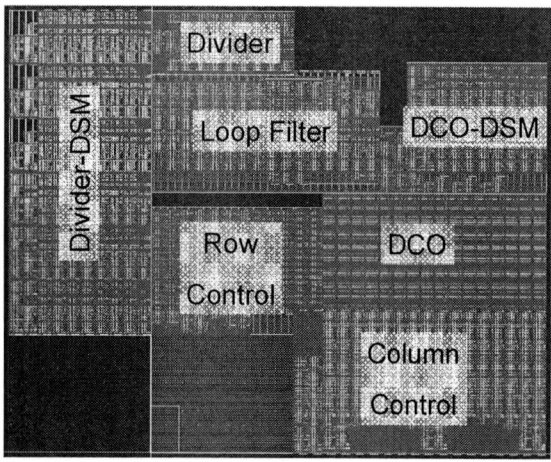

Fig. 10 DPLL layout. Area: 175 μm x 160 μm. Total number of transistors in the design: ~25,000.

ACKNOWLEDGEMENTS

The authors would like to thank B. Parker, D. Kuchta, P. Muench, G. Smith, R. Dussault, S. Gowda and M. Soyuer. This work was supported by MPO contract #HR0011-07-9-0002.

REFERENCES

[1] P. Hanomolu, G. Wei, U. Moon, *et. al.*, "Digitally-Enhanced Phase-Locking Circuits," *Proc. of CICC,* pp. 361-368, Sept. 2007

[2] R.B. Staszewski, J.L. Wallberg, S. Rezeq, et al., "All-Digital PLL and Transmitter for Mobile Phones", *IEEE J. Solid State Circuits*, vol. 40, no. 12, pp. 2469-2482, Dec. 2005.

[3] J.A. Tierno, A.V. Rylyakov, D. Friedman, "A Wide Power Supply Range, Wide Tuning Range, All Static CMOS All Digital PLL in 65 nm SOI", *IEEE J. Solid State Circuits* , vol. 43, no. 1, pp. 42-51, Jan. 2008.

[4] A. Rylyakov, J. Tierno, D. Turker, *et. al.*, "A Modular All Digital PLL Architecture Enabling Both 1-to-2GHz and 24-to-32GHz Operation in 65nm CMOS", *ISSCC Dig. Tech. Papers,* pp. 516-517, Feb. 2008.

[5] S. Lee, B. Jagannathan, S. Narasimha, *et.al.*, "Record RF performance of 45-nm SOI CMOS Technology", *2007 IEEE International Electron Devices Meeting (IEDM)*, pp. 255-258, Dec. 2007.

[6] N. Da Dalt, "A Design-Oriented Study of the Nonlinear Dynamics of Digital Bang-Bang PLLs," *IEEE Trans. on Circ. Syst. Part I*, vol. 52, pp. 21–31, January 2005.

[7] V. Kratyuk, P. Hanumolu, K. Ok, K. Mayaram, and U. Moon, "A digital PLL with a stochastic time-to-digital converter*," IEEE Symp. VLSI Circuits*, pp. 38-39, Jun. 2006.

IEEE 2008 Custom Intergrated Circuits Conference (CICC)

Clocking Circuits for a 16Gb/s Memory Interface

Ting Wu, Xudong Shi, Kambiz Kaviani, Haechang Lee, Jung-Hoon Chun,
TJ Chin, Jie Shen, Rich Perego, and Ken Chang
Rambus Inc. 4440 El Camino Real, Los Altos, CA 94022, USA

Abstract - **8GHz clocking circuits for a 16Gb/s/pin asymmetric memory interface [1] are described. A combination of an LC-PLL and a ring-PLL achieves improved jitter performance for multiple phase outputs with a wide frequency range. A direct phase mixer and a digitally controlled duty-cycle corrector (DCC) are time-multiplexed between transmitter (TX) and receiver (RX), thereby reducing area and power. The prototype chip implemented in a 65nm CMOS technology has measured 734fs RJ (rms) at the TX output when operating at 16Gb/s.**

I. Introduction

Driven by today's advances in personal computing, game consoles, mobile phones, and HDTV, the demand for high bandwidth memory interfaces has been steadily growing. As the performance demands of memory interfaces increase, clock generation and distribution techniques are among the most important design considerations [2]. To meet the market requirements, clocking circuits must achieve both low jitter and low power at an increasingly higher data rate.

In this paper, we present a high-performance clocking architecture featuring dual-loop PLLs on the controller side of an asymmetric memory interface [1]. An LC oscillator based PLL produces a low jitter high frequency clock as the reference clock of a ring oscillator based PLL. The ring-PLL then generates quadrature phase outputs, which drive phase mixers for clock phase adjustments. As a consequence, the overall architecture achieves improved jitter performance for multiple phase outputs with a wide frequency range.

To simplify the interface circuitry of the associated DRAM devices, the controller interface must ensure a ~50% duty-cycle with 360-degree phase adjustment at both TX and RX circuits. The phase adjustment and duty-cycle adjustment are achieved by using a direct phase mixer and a duty-cycle corrector (DCC), respectively. In this work, one phase mixer and one DCC are time-multiplexed between TX and RX on a bi-directional I/O. This reduces the area overhead significantly. However, to support minimal overhead memory read/write bus turnaround, a conventional analog DCC loop [3] is not adequate. In this work, a new digitally controlled DCC is employed to meet the requirement. As a result, both the phase mixer and the DCC are digitally controlled. Each uses a MUX to select the corresponding control code for TX and RX operations. The digital phase and DCC codes are determined during initial and periodic timing and duty-cycle calibrations, respectively.

This paper is organized as follows. In Section II, the architectural choices are considered and our clocking scheme is described. Section III presents the circuit techniques for area and power reduction. This is followed by measurement results in Section IV. Finally, Section V concludes this paper.

II. Clocking Architecture

There are several conflicting requirements for our clock generator: low jitter, a wide frequency range, and 360 degrees phase adjustment. Although ring oscillators have great flexibility to generate multiple phase outputs that are necessary for phase adjustment, their poor phase noise performance would not meet the timing budget for 16Gb/s operation. On the contrary, LC-tank oscillators have superior noise performance compared to ring oscillators. Unfortunately, they only produce two phases with a much narrower frequency range.

In order to provide a wide frequency range, an LC-VCO requires a large varactor or a bank of switched capacitors, resulting in reduced tank quality factor and increased AM-to-FM noise conversion. Quadrature phases can be achieved by either using a quadrature LC oscillator or dividing a 2x faster LC-VCO with a divide-by-2 divider. However, both approaches make the design of LC-VCO and PLL more challenging. A quadrature LC-VCO typically impairs phase noise. A 2x faster LC-VCO requires a 16GHz divider. It is also difficult to acquire accurate models for both active and passive devices operating at 16GHz frequency.

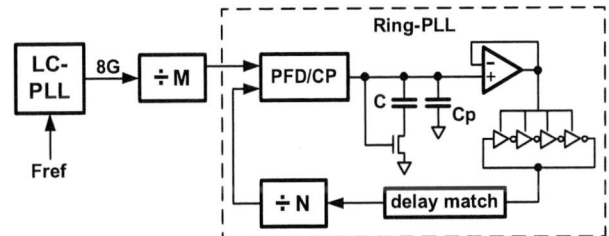

Fig. 1: Clocking architecture consists of an LC-PLL and a ring-PLL.

To regain the benefits of the LC-VCO PLL, we developed our dual-PLLs architecture. As shown in Fig. 1, an LC-PLL and a ring-PLL are cascaded. From the external reference clock, the LC-PLL generates a fixed output frequency, 8GHz, with a fixed divider ratio. This 8GHz output is fed to a divide-by-M divider which provides the reference clock for the ring-PLL. The ring-PLL has another programmable divider, divide-by-N, in the feedback loop to generate an 8GHz × (N/M) frequency under lock condition. The overall output frequency range is limited by the tuning range of the ring VCO rather than that of the LC-VCO. In our work, M is set to be either 4 or 5, and another integer number N is programmable from 1 to 6. With this configuration, the clock generator is able to provide a frequency range from 1.6GHz up to 9.6GHz that is only limited by the ring oscillator.

Fig. 1 also shows the supply regulated ring-PLL, where the ring-VCO consists of four differential inverter stages. A larger loop bandwidth is allowed for the ring-PLL due to its

978-1-4244-2018-6/08 $25.00 © 2008 IEEE

increased reference frequency. Consequently, the noise from the ring VCO is greatly attenuated. To optimize the noise performance of the whole clock architecture, the bandwidths of both LC-PLL and ring-PLL are carefully chosen. In this work, the LC and the ring PLLs have a loop bandwidth of approximately 3MHz and 80MHz, respectively. The supply regulator for the ring VCO has an approximately 1GHz bandwidth; hence, it has a negligible impact on the PLL dynamic. A replica compensated regulator [4] is adopted to achieve both high power supply rejection and low power consumption.

One drawback of our clocking architecture is the increased die area due to the additional ring-PLL. Although the increased ring-PLL bandwidth leads to a smaller loop filter capacitor, we still need to alleviate the area constraints. In previous designs, a second Charge Pump (CP) driving an opamp [3] was used to obtain an active loop filter resistor and an adaptive PLL bandwidth at the expense of area and power overhead. In this work, a MOS transistor operating in the linear region is employed as the loop filter resistor (Fig. 1). It can be shown that, when both the MOS transistor and the CP current are controlled by PLL control voltage, the PLL bandwidth is tolerant to process, voltage, and temperature (PVT) variations. Therefore, the ring-PLL area and power are reduced without compromising the PLL performance.

Only one LC-PLL is used for a whole data-byte, while each two data-bit bi-directional I/O circuits share one ring-PLL [1]. Other clocking circuits, including phase mixers, DCCs, and clock buffers, are placed between the two I/O circuits to shorten the high-speed clock wires. Reducing the power and area overhead of these circuits that are close to the I/O circuits are important. In the next Section, we describe our circuit techniques for power and area reduction.

III. Circuit Design

A. Time-multiplexing technique

Time-multiplexing is an efficient way to improve area efficiency by sharing common components among different functional blocks. On a bi-directional I/O link, the TX and RX operate at different times. Therefore, we can employ the time-multiplexing technique for each pair of TX/RX. In Fig. 2, the quadrature outputs from the ring VCO are fed to a phase mixer, whose differential outputs drive a duty-cycle corrector (DCC). The phase mixer and the DCC are time-multiplexed between the output clocks tclk and rclk for TX and RX, respectively. Each uses a MUX to select the corresponding control codes for TX and RX operations. Two separate clock buffers are used due to the different loading requirements for the TX and RX. With this architecture, the number of phase mixers and DCCs are halved, leading to a significant amount of area reduction. Further power reduction can be achieved by disabling the clock buffers that are not in operation.

Fig. 2 also shows our duty-cycle error correction loop consisting of the DCC and the clock buffers in the forward path and an m-to-1 MUX, a duty-cycle error detector, and a FSM in the feedback path. The duty-cycle error detector is essentially an analog integrator followed by a comparator. The integrator is a fully-differential charge pump with a common-

mode feedback circuit [3]. To reduce the additional duty-cycle distortion due to the detector, its input-referred offset should be minimized, leading to the use of large transistor size and an area penalty. Since the duty-cycle detection and calibration only happen periodically, the duty-cycle error detector is time-multiplexed among multiple clock signals. A novel duty-cycle error free MUX is used to select one clock signal during each individual duty-cycle calibration phase.

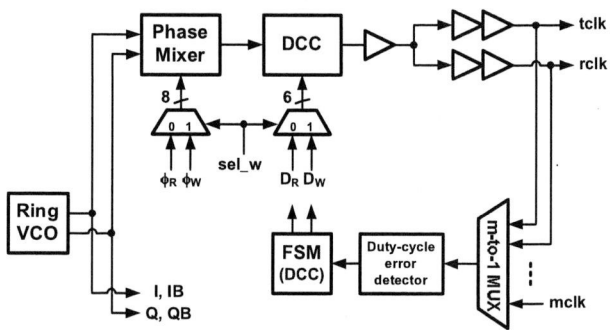

Fig. 2: Time-multiplexed phase mixers and DCCs between TX and RX, and time-multiplexed duty-cycle error detector among multiple clocks.

B. Direct phase mixer

The phase mixer is used to adjust the output clock phase with respect to the input over 360 degrees. Traditional phase mixer [3] consists of a MUX that selects two phases from the quadrature phases and a phase interpolator that interpolates between the chosen phases. However, this type of the phase mixer suffers from charge-injection and clock feed-through from the input devices, resulting in degraded phase linearity.

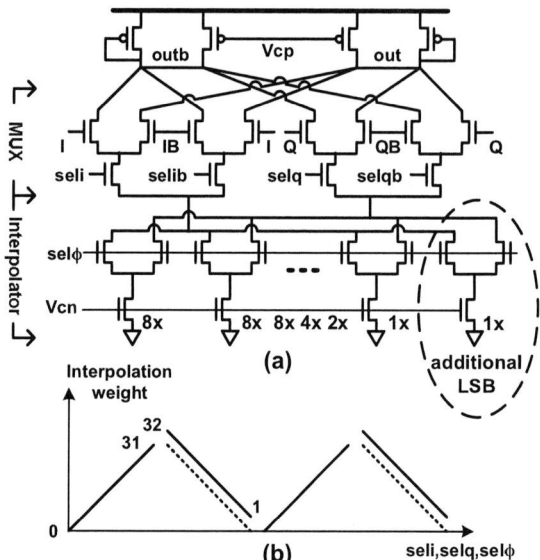

Fig. 3: (a) Simplified schematic of direct phase mixer. (b) Conceptual phase transfer of the phase mixer in (a) in comparison with the traditional results (shown in the dashed lines).

To address this issue, a direct phase mixer depicted in Fig. 3(a) is employed in this work. The MUX and the phase interpolator are cascoded rather than being cascaded, similar to [5]. By doing so, each output node is connected to four transistors that are driven by the quadrature phases. Therefore, the overall injected charge at the output remains the same and is no longer dependent on the chosen phases, resulting in improved phase linearity. Additionally, the cascoded circuits reduce total power consumption. Since the cascoded switches operate in the deep-linear region, the resulting supply voltage headroom reduction is very small.

To avoid having the same interpolation weights across each quadrant boundary, in Fig. 3(a), we added one LSB current cell. The control bit for this additional LSB is only varied at each phase boundary. Fig. 3(b) shows the resulting phase transfer curve (shown in the solid lines) in comparison with the traditional one (shown in the dashed lines). The interpolation weights vary from 0:31 LSBs to 32:1 LSBs rather than from 0:31 LSBs to 31:0 LSBs. As a result, two consecutive phase outputs at the phase boundaries are made different by one LSB. In this work, one LSB is defined as 0.5UI/32, or 976fs at 16Gb/s.

In order to guarantee phase linearity across frequency range, both the phase mixer and its input buffers are biased based on the ring-PLL control voltage.

C. Digitally controlled Duty-Cycle Corrector (DCC)

The digitally controlled DCC is shown in Fig. 4. A small swing differential input signal from the phase mixer goes through an AC-coupled capacitor and drives an inverter with a feedback resistor [6]. The feedback resistor DC biases the inverter input to its switching point. Therefore, the input signal is amplified to a full CMOS output after 3 inverter stages. In addition, any duty-cycle error of the input signal will be corrected by the first inverter stage via the feedback resistor [7]. Consequently, this circuitry combines both functions of low-to-high swing amplification and duty-cycle error correction, resulting in a significant area reduction.

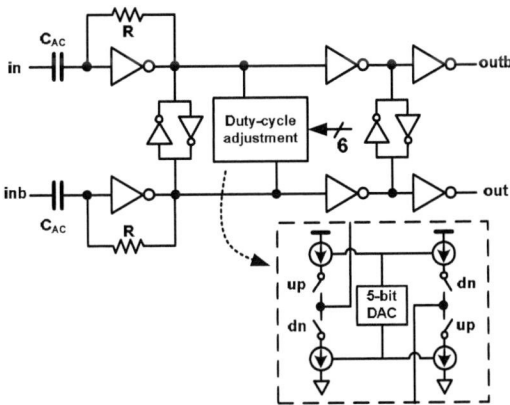

Fig. 4: Digitally controlled duty-cycle corrector (DCC).

Given an input with ±10% duty-cycle error, simulation results show that the resulting duty-cycle error at the output is reduced to ±3% over PVT variations. This unfortunately still falls short of our target of ±1% duty-cycle error for 8GHz clocks. To mitigate this problem, a digitally controlled duty-cycle adjustment circuitry is added at the output of the first inverter stage. The adjustment circuitry consists of two charge-pump circuits with one bit polarity control and a 5-bit

current-steering DAC. Depending on the polarity, the adjustment circuitry sources or sinks current at the first-stage output nodes, thereby applying a differential offset to the signal. As a result, the duty-cycle of the output is varied. The duty-cycle adjustment range is approximately ±4% with an LSB of 0.125%.

D. Duty-cycle error correction

The digital codes for the duty-cycle adjustment circuitry are determined by the duty-cycle error correction loop (Fig. 2). During one DCC calibration cycle, the duty-cycle error of a chosen clock signal is detected by the duty-cycle error detector which consists of an integrator and a comparator. The integrator accumulates the duty-cycle error over a number of clock phases. According to the integrator output, the comparator determines the polarity of the present duty-cycle error and feeds it to a Finite State Machine (FSM), which correspondingly increments or decrements the control value for the duty-cycle adjustment circuitry. This finishes one duty-cycle calibration cycle. In the next cycle, these updated control values are loaded to the adjustment circuitry, which varies the output duty-cycle toward its desired 50% value.

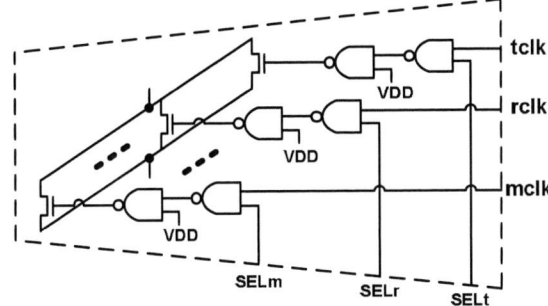

Fig. 5: Simplified schematic of the m-to-1 MUX used in the time-multiplexed duty-cycle error correction loop (Fig. 2).

To relax the area constraint, the duty-cycle detector is time-multiplexed by multiple clock signals with an m-to-1 MUX. In this work, m is 5 with two pairs of tclk/rclk and one feedback clock as the MUX input. Care should be taken to minimize both extra duty-cycle error (introduced by the MUX and the detection circuit) and coupling between the input clocks. To mitigate this extra distortion, we proposed a duty-cycle error free MUX as depicted in Fig. 5. It consists of multiple pairs of NAND gates for multiple input signals. In each pair, two identical NAND gates are cascaded. The first stage NAND gate is driven by a clock signal and the corresponding select signal, while the second stage driven by the first stage output and VDD. Through the two cascaded identical stages, any introduced systematic duty-cycle error from the first stage is cancelled at the second stage output. Therefore, the additional duty-cycle error due to the MUX is minimized. The outputs from the MUX drive the gate nodes of NMOS transistors connected in parallel in a CML buffer stage. Since there is only one active signal at the first stage output for all pairs of NAND gates, the coupling from other clock signals to the selected clock path which occurs at the CML buffer output is also minimized.

IV. Measurement Results

The prototype chip was fabricated in TSMC 65nm G+ CMOS technology with flip-chip package assembly. Fig. 6 shows the die photo of one data-byte and two data bits. The LC-PLL and ring-PLL occupy $0.048mm^2$ ($220\mu m \times 220\mu m$) and $0.014mm^2$ ($280\mu m \times 50\mu m$) die area, respectively. An approximately $0.01mm^2$ area reduction was achieved by using the time-multiplexed components in the design.

Fig. 7 shows the measured TX output jitter histogram with a clock pattern running at 16Gb/s. The measured RJ (rms) and TJ (BER=10^{-12}) at TX output are 734fs and 13.0ps, respectively. The power (from simulations) of the LC-PLL and one ring-PLL are 36mW and 21mW from a 1.1V supply voltage, respectively.

By changing the TX phase codes manually, the output phases are obtained and used to generate the INL of the phase transfer curve. As shown in Fig. 8, the direct phase mixer achieves a measured INL better than 4ps.

Fig. 9 shows the measured duty-cycle at the TX output when the DCC control code is varied manually. Without duty-cycle error calibration, the output duty-cycle is 51% (correspondingly at code 32). With the help of the calibration, the duty-cycle error is reduced to 0.2%. Across chips and measuring conditions, the calibrated duty-cycle measures from 48.5% to 51%.

Table 1 summarizes the performance of the clocking circuits in our prototype chip.

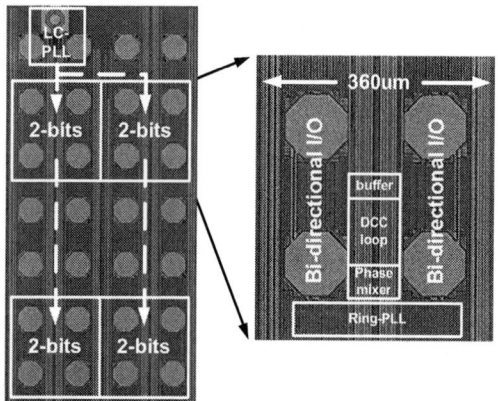

Fig. 6: Die photograph showing one data byte and two data bits.

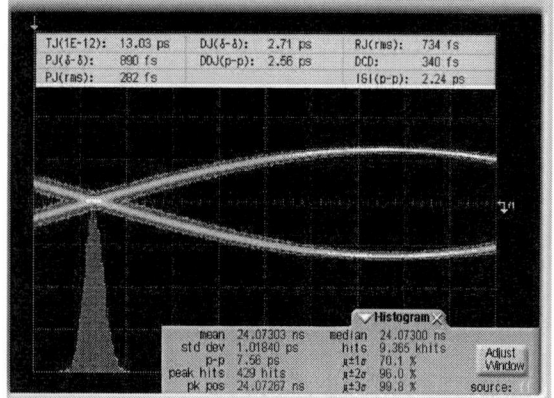

Fig. 7: Measured TX output jitter histogram when operating at 16Gb/s.

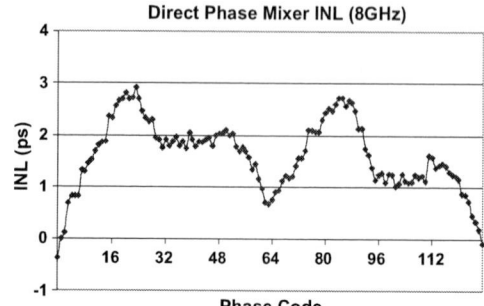

Fig. 8: Measured INL of the direct phase mixer phase transfer curve.

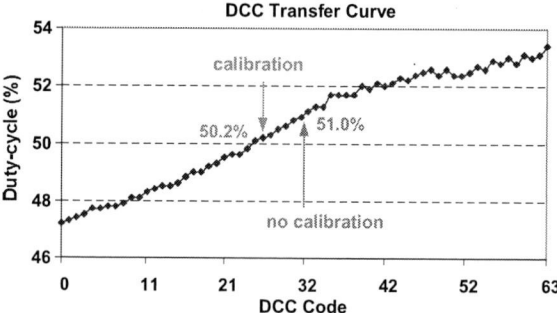

Fig. 9: Measured output duty-cycle vs. control code of the DCC. Duty-cycle calibration improved the duty-cycle error from 1% to 0.2%.

Table 1: Performance summary

Technology	TSMC 65nm G+
Supply voltage of clocking circuits	1.1V
Locking range of LC-PLL / Ring-PLL	8GHz ±5% / 1.6-9.6GHz
Power of LC-PLL / Ring-PLL	36mW / 21mW
TX RJ (rms) / TJ (BER=10^{-12})	734fs / 13.0ps
Phase linearity	INL < 4ps
Duty-cycle across chips	48.5% - 51%

V. Conclusions

We have presented 8GHz clocking circuits for a 16Gb/s/pin asymmetric link memory interface. The ring-PLL noise performance is improved by using an LC-PLL as a reference clock generator. The area constraints have been greatly alleviated by using time-multiplexed building blocks. Operating at 16Gb/s, the prototype chip has measured 0.73ps RJ (rms) jitter performance.

Acknowledgments

The authors gratefully thank Wendem Beyene and Chris Madden for testing assistance.

References

[1] K. Chang, et al., "A 16Gb/s/link, 64GB/s bidirectional asymmetric memory interface cell," to be appeared in *IEEE Symp. VLSI Circuits*, Jun. 2008.

[2] D. Pham, et al., "The design methodology and implementation of a first-generation CELL processor: a multi-core SOC," *IEEE Custom Integrated Circuits Conf.*, pp. 41-45, Sep. 2005.

[3] K. Chang, et al., "A 0.4-4Gb/s CMOS Quad transceiver cell using on-chip regulated dual-loop PLLs," *IEEE Symp. VLSI Circuits*, pp.88-91, Jun. 2002.

[4] E. Alon, et al., "Replica compensated linear regulators for supply regulated phase-locked loops," *IEEE JSSC*, vol. 41, pp. 413-424, Feb. 2006.

[5] M. Meghelli, et al., "A 10Gb/s 5-tap-DFE/4-tap-FFE transceiver in 90nm CMOS," *ISSCC Dig. Tech. Papers*, pp.213-214, Feb. 2006.

[6] J. Savoj and B. Razavi, "A CMOS interface circuit for detection of 1.2Gb/s RZ data," *ISSCC Dig. Tech. Papers*, pp.278-279, Feb. 1999.

[7] C. Menolfi, et al., "A 16Gb/s source-series terminated transmitter in 65nm CMOS SOI," *ISSCC Dig. Tech. Papers*, pp.446-447, Feb. 2007.

IEEE 2008 Custom Intergrated Circuits Conference (CICC)

A 1V 15.6mW 1-2GHz -119dBc/Hz @ 200kHz clock multiplying DLL

Sander L.J. Gierkink

Conexant Systems, 100 Schulz Drive, Red Bank, NJ 07701

*Abstract-*A low-phase-noise 1-2GHz clock multiplier is configurable as PLL or DLL. Starting in PLL-mode, a lock-detect circuit switches the circuit to DLL and simultaneously reconfigures the loop filter. Upon loss of lock the circuit automatically falls back to PLL-mode. The number of stages in the VCO/delay line is programmable to 8-10-12-14-16, implementing band selection with modest tuning gain K_{VCO}. Multiplication is selectable from 2 to 64. The 0.11mm^2 90nm CMOS chip also includes charge pump and divider, both programmable. With 80MHz reference, phase noise is essentially flat and measures -125dBc/Hz and -119dBc/Hz @ 200kHz offset for 0.92 and 1.92GHz output respectively. Total maximum power consumption is 15.6mW from a 1V supply.

I. INTRODUCTION

A PLL, with minimum loop order two, suppresses VCO phase noise over a bandwidth typically less than $1/20^{th}$ of crystal reference frequency, due to loop stability requirements. A DLL, with one order less, suppresses delay line noise up to approximately half the crystal frequency, provided the periodic phase realignment is sufficiently strong. On the other hand, PLL static phase offset does not compromise frequency accuracy and a resultant reference spur can in principle be filtered by the loop filter. DLL static phase offset affects frequency accuracy and results in large reference spurs. Literature provides techniques to reduce DLL static phase offset [4,5] This paper focuses on minimizing DLL random phase noise, assuming a sufficiently stable low-frequency reference clock is available. A frequency domain model is explored [1], then circuitry is shown to maximize realignment strength, followed by measurements of a complete 1V 90nm 1-2GHz clock multiplier.

II. PHASE NOISE TRANSFER FUNCTIONS

Figure 1 top left shows the DLL frequency-domain model. The transfer functions $H_{up}(j\omega)$ and $H_{rl}(j\omega)$ model reference noise up-conversion and phase realignment respectively [1]. The factor β, valued between <0-1>, models the strength of phase realignment by the reference. The model utilizes the VCO voltage–to-phase transfer $2\pi K_{VCO}/s$ with K_{VCO} in Hz/V. $H_{rl}(j\omega)$ transforms the VCO into a delay line by adding a first-order high-pass, resulting from the periodic realignment by reference *REF*. The table lists relevant transfer functions and approximations for $\omega < 0.5 \cdot \omega_{REF}$, with $H_{ref}(j\omega) = \phi_{out}/\phi_{ref}$ and

$H_{vco}(j\omega) = \phi_{out}/\phi_{vco}$. Extensive time-domain simulations of a DLL in *Matlab* utilizing random number generators to model noise and zero crossing detection to track phase deviation, followed by Fourier transforms to plot phase PSD show good agreement with the expressions.

A simple, incorrect DLL model widely applied in literature uses $H_{up}(j\omega)=0$ and $H_{rl}(j\omega)=1$ and models the delay line's voltage-to-phase transfer as $2\pi K_{VCO}T_d$ with T_d the time of one delay-line circulation. However, it underestimates REF noise up-conversion and delay line noise suppression. It also fails to predict potential jitter-peaking in $H_{ref}(j\omega)$ [2].

Figure 1 top right shows $H_{up}(j\omega)$ and $H_{rl}(j\omega)$ while the bottom plot shows the closed-loop transfer functions $H_{ref}(j\omega) = \phi_{out}/\phi_{ref}$ and $H_{vco}(j\omega) = \phi_{out}/\phi_{vco}$ for β=0.25,0.5,1, all with f_{REF}=20MHz, N=100, I_{CP}=400uA and C=2pF. Larger β (better realignment) gives more VCO noise suppression, but also more HF reference noise upconversion. Small β results in peaking in both reference upconversion and VCO noise. Lowering loop gain by increasing C/lowering I_{CP} removes this peaking. Note that for β=1 $H_{vco}(j\omega)$ has +40dB/dec slope for $\omega<<\omega_{REF}$, changing to +20dB/dec for larger $\omega<\omega_{REF}$. The +40dB/dec region extends with larger loop gain, useful for suppressing delay line 1/f noise as shown in section IV.

III. CIRCUITRY

Figure 2 shows the delay stage. Inputs *A* and *B* each drive a tri-state inverter, (de)activated by *enB*. Both inverters drive the same input of a NAND structure. The other NAND input is driven by an inverter, connected to output P_n. This forms a latch that secures P_n when, after a short delay, it resets the active input *A* or *B*. As a result, waveform duty cycle is low, enabling easy pulse removal/reinsertion as shown later.

The PLL/DLL schematic in figure 4 is comprised of an 8/10/12/14/16-stage VCO/delay line of programmable length with single-stage "exit branch" plus an 8-stage reference delay line. All stages are identical and tuned by the PLL/DLL. Division factor N is programmable in the range 2-64. At startup, all stages are briefly reset and lock is low. This configures the loop filter as 2nd-order for the PLL and disables *eninj*, preventing pulse updates. In both PLL and DLL mode, rising *REF*-edges inject a pulse in the reference delay line. Pulses at stages *M4* and *P16* are phase locked by the PLL. Upon phase lock, signal *lock* goes high on a falling *REF*-edge, reconfiguring the loop filter for DLL-mode and enabling pulse removal/reinsertion by subsequent rising *REF*-edges.

978-1-4244-2018-6/08 $25.00 © 2008 IEEE

Figure 3 shows the DLL pulse removal/reinsertion process. Pulses at $M4$ and $P16$ are locked. A rising REF-edge injects a pulse in the reference delay line $M1..8$. While this pulse travels down the delay line, $eninj$ goes high. This enables $M4$ to inject a pulse into stage $P1$ and redirects the pulse arriving at stage $P16$ to the exit branch, where it terminates. It can be shown that noise of the reference delay line is non-dominant in the final DLL output phase noise.

Charge sharing between injected- and removed pulse is minimized, maximizing β and thus suppression by $H_{rl}(j\omega)$ (see figure 1). The tri-state inverters in figure 2 and the exit branch in figure 4 isolate the injected/removed pulse. Separate low-output-impedance buffers control the tune node of oscillator- and reference delay line, suppressing spikes on the tune node that would otherwise arise during resets of a delay stage.

The loop filter in figure 4 is programmable independently for PLL- and DLL. In DLL the resistor can be shorted and the capacitance reduced in 6 steps of 10pF. The lock-detect circuit automatically switches between programmed configurations.

Figure 5 shows the lock-detect circuit. At startup, the multiplier acts a PLL. When rising edges at the detector's inputs $in_{1,2}$ line up, latch SR_1 is set. This condition is alternately retimed in latches $SR_{2,3}$. If lock is maintained over 32 consecutive REF-cycles, $lock$ goes high on a falling REF-edge, switching the multiplier to DLL-mode. At the very first non-lock event, $lock$ goes low, resetting the detector's counter and switching the multiplier back to PLL-mode, until lock is re-established. This prevents false-lock problems normally encountered in DLLs. Setting $enDLL$ in figure 4 low forces the multiplier to work solely in PLL-mode.

IV. EXPERIMENTAL RESULTS

The PLL/DLL is realized in 1V 90nm standard CMOS. Figure 6 shows a die photograph. Active area is $0.11mm^2$.

Figure 7 shows the output spectrum with f_{out}=1.92GHz, f_{ref}=80MHz over 300MHz span, using 8 stages in the ring. Trace 3 is the PLL, traces 1,2 are the DLL with HP8662A signal generator and Pletronics SQ33 crystal clock oscillator as reference respectively. Trace 1 clearly shows the sinc-shaped sideband spectrum due to upconversion of reference noise, as seen in the plot of $H_{ref}(j\omega)$ with β=1 in figure 1. Trace 2 has lower phase noise than trace 1 due to the lower phase noise of the crystal reference compared to the HP8662A. Even when using a crystal reference, the DLL phase noise is larger than that of the PLL at offsets larger than approx. 20MHz, showing that upconversion of reference noise is dominant at these offsets. Reference spur is -35.2dBc, comparable to [3].

Figure 8 shows the spectrum at f_{out}=1.92GHz over 20MHz span. Trace 1 is the PLL with f_{ref}=80MHz, traces 2,3 are the DLL with f_{ref}=40MHz and 80MHz respectively (N=48,24), all measured with a crystal clock reference. The DLL with 80MHz reference has 27dB better phase noise than the PLL at 1MHz offset. The difference between traces 2 and 3 is close to 6dB, accounting for a factor 2 difference in division factor N.

Figure 9 shows the DLL spectrum at f_{out}=1.92GHz, f_{ref}=80MHz, N=24 at various capacitor C settings of the DLL loop filter and charge pump current I_{CP}. With increased loop gain (larger I_{CP} and lower C) the close-carrier phase noise decreases, as it increases the transition frequency between the +40/+20dB/dec region of $H_{VCO}(j\omega)$ with β=1 in figure 1.

The table in figure 10 summarizes DLL phase noise at offsets of 200kHz and 1MHz for various division factors N, measured with 40,60,80MHz crystal clock reference oscillators. The number of stages used in the ring is 16,12,8 for f_{out}=0.96,1.44,1.92GHz respectively. Taking into account the DLL "flat" phase noise and normalizing for total power P [mW], division factor N and output frequency f_{out} a DLL figure-of-merit can be defined as:

$$\text{FOM}_{DLL} = 10 \cdot \log\left(\frac{L \cdot P}{f_{out}^2 \cdot N^2}\right)$$

with phase noise L taken here at 200kHz offset. This work shows a minimum 9dB improvement over state-of-the-art.

Figure 11 shows accumulated random RMS jitter vs. time interval τ for PLL (loop bandwidth BW=0.2MHz) and DLL at f_{out}=1.92GHz, measured with a sampling scope. DLL jitter is τ-independent, showing that β=1. It is about 1.26ps and 1.70ps for N=24,48 respectively. The jitter ratio for N=48/N=24 is approximately $\sqrt{2}$. PLL jitter is proportional to $\sqrt{\tau}$ for small τ, and flat for $\tau>1/2/\pi/BW$, as expected.

V. CONCLUSIONS

A reconfigurable 1-2GHz PLL/DLL clock multiplier is presented with maximized realignment strength β=1 for maximum delay line noise suppression. Ring length, loop filter, charge-pump current and division factor are all programmable. A lock-detect circuit continuously monitors lock and automatically switches to PLL mode to prevent false lock in DLL mode. The effects of loop gain and reference noise upconversion on DLL phase noise are discussed and measured. Compared to DLLs in literature this work presents a minimum 9dB improvement in random phase noise, normalized for multiplication factor, power and frequency.

REFERENCES

[1] S. Ye and I. Galton, "Techniques for Phase Noise Suppression in Recirculating DLLs", IEEE J. Solid-State Circuits, vol. 39, no. 8, pp. 1222-1230, Aug 2004.

[2] M.-J. Lee, W.J. Dally et. al., "Jitter Transfer Characteristics of Delay-Locked Loops – Theories and Design Techniques", IEEE JSSC, vol. 38, no. 4, pp. 614-621, Apr 2003.

[3] R. Farjad-Rad et. al. "A Low-Power Multiplying DLL for Low-Jitter Multigigahertz Clock Generation in Highly Integrated Digital Chips", IEEE JSSC, vol. 37, no. 12, pp. 1804-1812, Dec 2002.

[4] P.C. Maulik and D.A. Mercer, "A DLL-Based Programmable Clock Multiplier in 0.18-μm CMOS With -70 dBc Reference Spur", IEEE JSSC, vol. 42, no. 8, pp. 1642-1648, Aug 2007.

[5] S. Gierkink, "An 800MHz -122dBc/Hz-at-200kHz Clock Multiplier based on a Combination of PLL and Recirculating DLL", Proc. ISSCC 2008, vol. 51, pp. 454-455, Feb. 2008.

978-1-4244-2018-6/08 $25.00 © 2008 IEEE

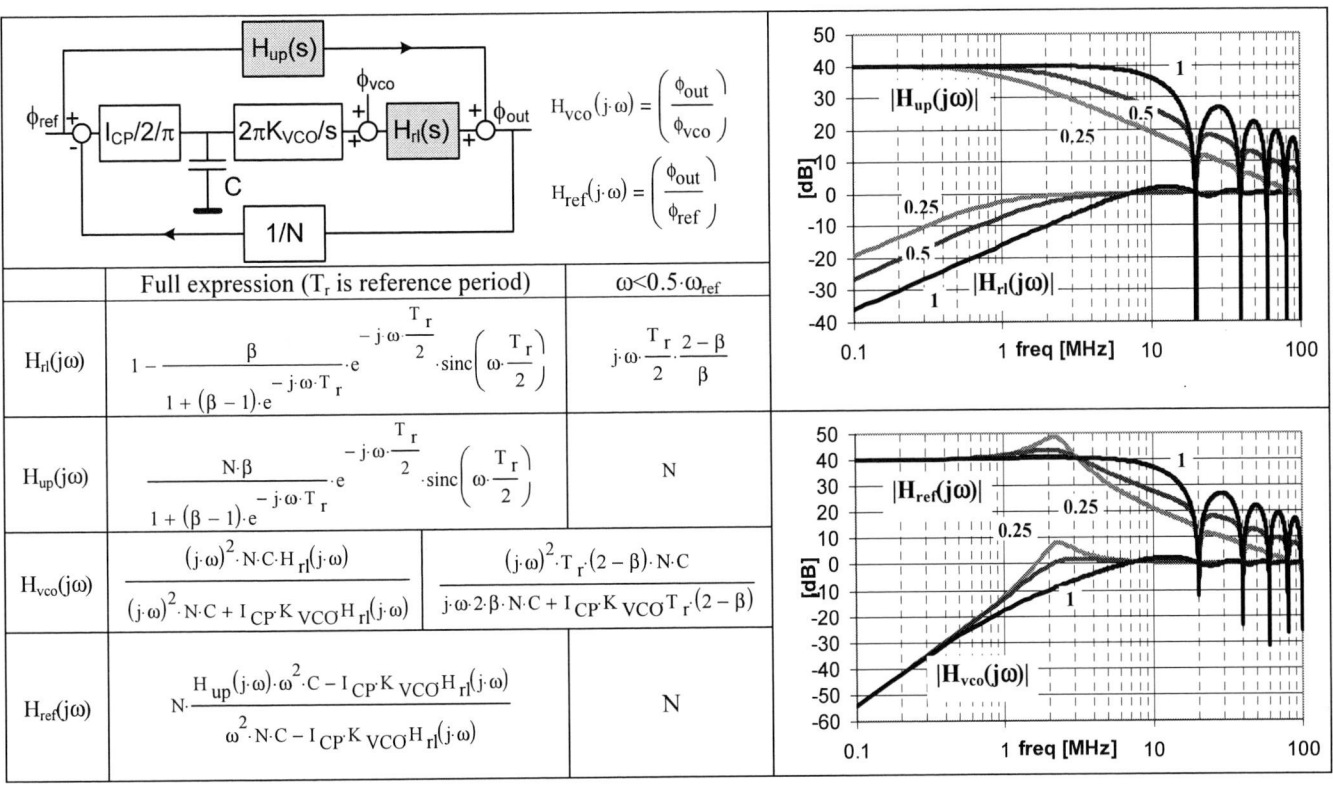

Fig. 1. Freq.-domain DLL model [1] (top left); transfer functions and low-freq. approximations (left); plot of transfer functions for β=0.25,0.5,1 (right).

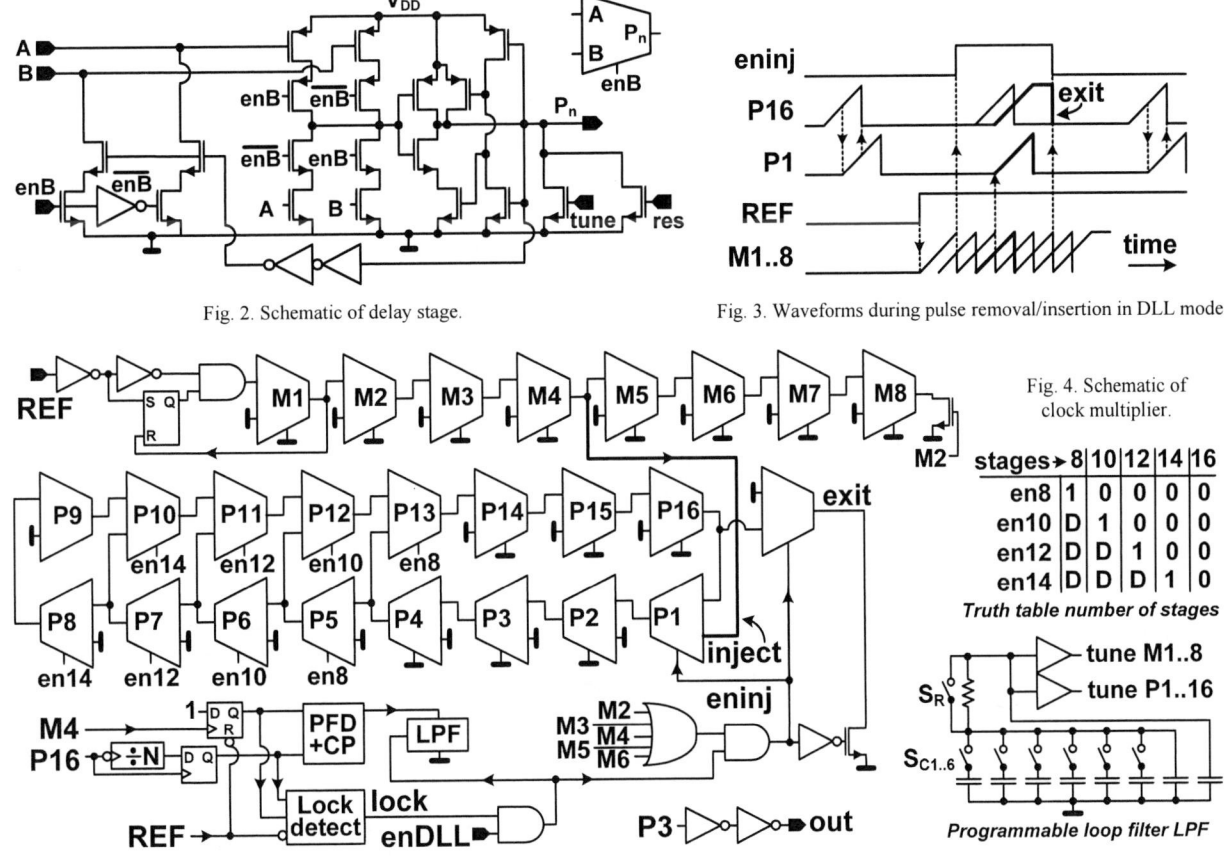

Fig. 2. Schematic of delay stage.

Fig. 3. Waveforms during pulse removal/insertion in DLL mode.

Fig. 4. Schematic of clock multiplier.

978-1-4244-2018-6/08 $25.00 © 2008 IEEE

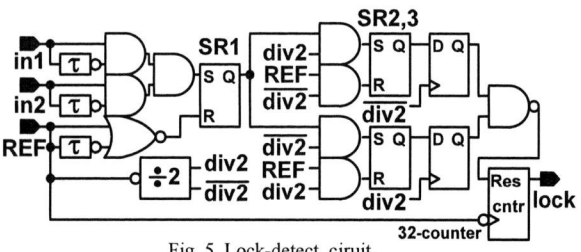

Fig. 5. Lock-detect ciruit.

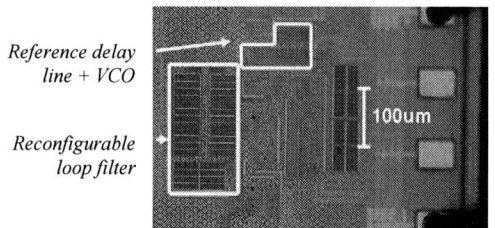

Fig. 6. Die photograph

Reference delay line + VCO

Reconfigurable loop filter

100um

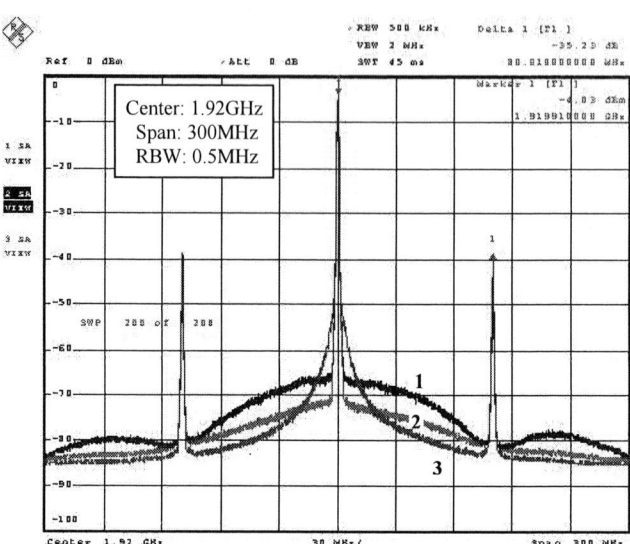

Fig. 7. PLL/DLL output spectrum; the latter with two different REF sources.

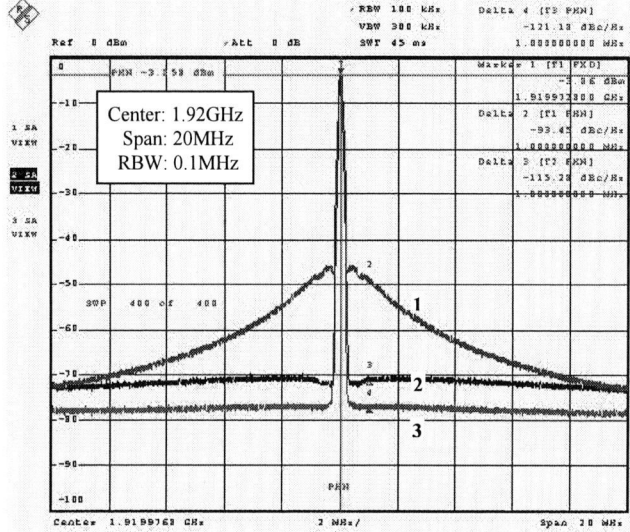

Fig. 8. PLL/DLL output spectrum; the latter with 40/80MHz reference.

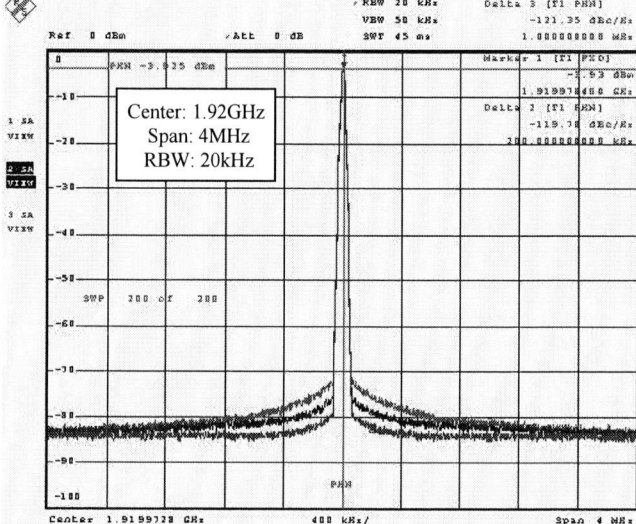

Fig. 9. DLL output spectrum at different loop gains.

f_{REF} MHz	N	f_{OUT} GHz	P_{tot} mW	L(200kHz) dBc/Hz	L(1MHz) dBc/Hz	FOM dB
40	24	0.96	11.6	-122.5	-124.4	-199.1
60	16	0.96	12.4	-122.6	-123.9	-195.6
80	12	0.96	11.9	-125.4	-128.9	-195.9
40	36	1.44	15.3	-118.9	-119.3	-201.3
60	24	1.44	15.4	-121.3	-122.1	-200.2
80	18	1.44	15.6	-122.7	-124.7	-199.0
40	48	1.92	14.4	-115.5	-115.7	-203.2
60	32	1.92	14.6	-118.2	-118.5	-202.3
80	24	1.92	14.9	-119.7	-121.2	-201.2
4	25	0.1	8.6	-117	[1]	-175.6
250	8	2.0	12	-110	[3]	-183.3
30	8	0.24	16	-116	[4]	-169.6
100	8	0.8	15	-122	[5]	-186.4

Fig. 10. Performance summary and comparison.

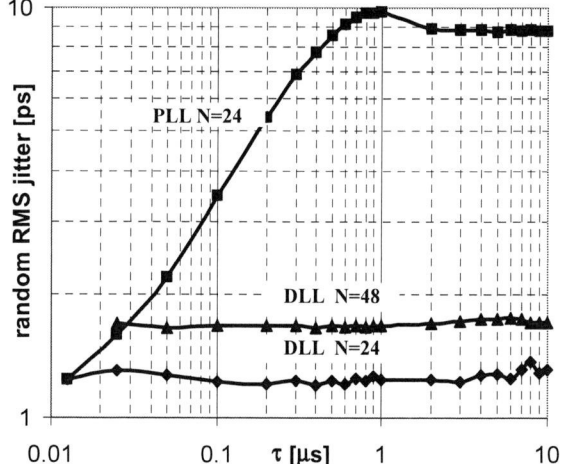

Fig. 11. Random RMS jitter vs. time interval τ; f_{out}=1.92GHz.

IEEE 2008 Custom Intergrated Circuits Conference (CICC)

A 0.5-to-2.5GHz Supply-Regulated PLL with Noise Sensitivity of -28dB

Abhijith Arakali, Srikanth Gondi[1], and Pavan Kumar Hanumolu

School of EECS, Oregon State University, Corvallis, OR 97331

[1]Kawasaki Microelectronics America, Inc., R&D Division, San Jose, CA 95131

Abstract—A supply-regulated phase-locked loop (PLL) employs a split-tuned architecture to decouple the tradeoff between supply-noise rejection performance and power consumption. The prototype PLL, incorporating a novel regulator, is fabricated in a $0.18\mu m$ digital CMOS process and operates from 0.5 to 2.5GHz. At 1.5GHz, the proposed PLL achieves a worst-case noise sensitivity of -28dB (0.5rad/V), an improvement of 20dB over conventional solutions, while consuming 2.2mA from a 1.8V supply.

I. INTRODUCTION

Power-supply noise is a major performance limiting factor in PLLs used for clock generation in large digital integrated circuits. As processes scale, the impact of power-supply noise on PLLs is further exacerbated. To alleviate the effect of supply noise on the jitter of PLLs, various supply regulation techniques are employed [1]-[4]. Figure 1 shows a general representation of a supply-regulated PLL, wherein the control voltage for the VCO is applied to its supply through a low drop-out regulator. A low drop-out voltage is needed in the

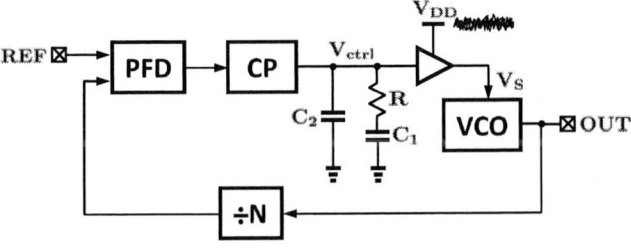

Fig. 1. Conventional supply-regulated PLL.

regulator to maximize the available supply and hence improve the frequency range of the VCO. As is generally observed, the most significant effect of supply noise is on the VCO and hence only the VCO supply is regulated.

Conventional supply regulation techniques used in [1]-[4], all suffer from one fundamental limitation— since the regulator is inside the PLL loop, the poles in the regulator degrade the stability of the loop dictating the use of a much wider regulator bandwidth (compared to the PLL bandwidth), hence leading to excessive power consumption. Additionally, regulators used in [1], [2] also suffer from poor supply noise rejection performance.

A commonly used regulator schematic is shown in Fig. 2, where the poles associated with the amplifier and the output node are denoted as ω_a and ω_o, respectively. The transfer

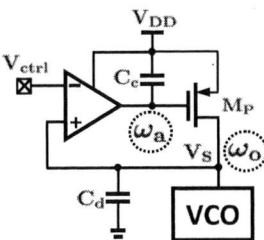

Fig. 2. Conventional regulator used in a supply-regulated PLL.

function from supply V_{DD} to V_s ($|\frac{V_s}{V_{DD}}|$) is shown in Fig. 3 for two conditions: (i) $\omega_a < \omega_o$, and (ii) $\omega_a > \omega_o$. Also shown is

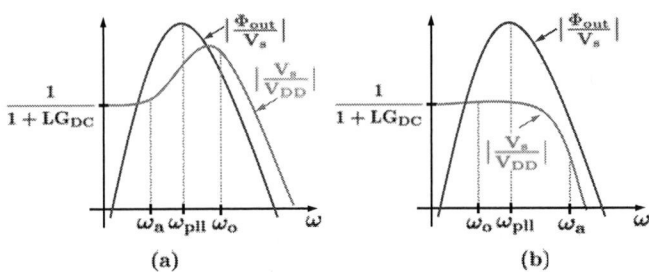

Fig. 3. Conventional regulator supply-noise rejection profile: (a) $\omega_a < \omega_o$ and (b) $\omega_a > \omega_o$ (LG$_{DC}$: regulator DC loop gain).

the VCO's sensitivity to supply-noise on the PLL output phase Φ_{out} ($|\frac{\Phi_{out}}{V_s}|$). Considering the feedback loop of the PLL, this transfer curve has a bandpass characteristic with the center frequency approximately at ω_{PLL}, the PLL bandwidth. It is evident from Fig. 3 that ω_o must be made the dominant pole, to maximize the supply rejection. This in turn dictates that the amplifier pole, ω_a, be very large— a requirement that is difficult to achieve with low power dissipation. In [4], peaking in the supply noise transfer function is reduced by introducing a local feedback using a replica branch. Even though this technique enables superior supply-noise rejection performance compared to that achieved in [1], [2], it still suffers from the same tradeoff between supply noise rejection performance and power consumption.

In this work, a novel supply-regulated PLL is presented that decouples the tradeoff between supply-noise rejection performance and power consumption. Also, a new regulation scheme is presented that is conducive to the proposed PLL architecture. The combined solution greatly improves the

978-1-4244-2018-6/08 $25.00 © 2008 IEEE

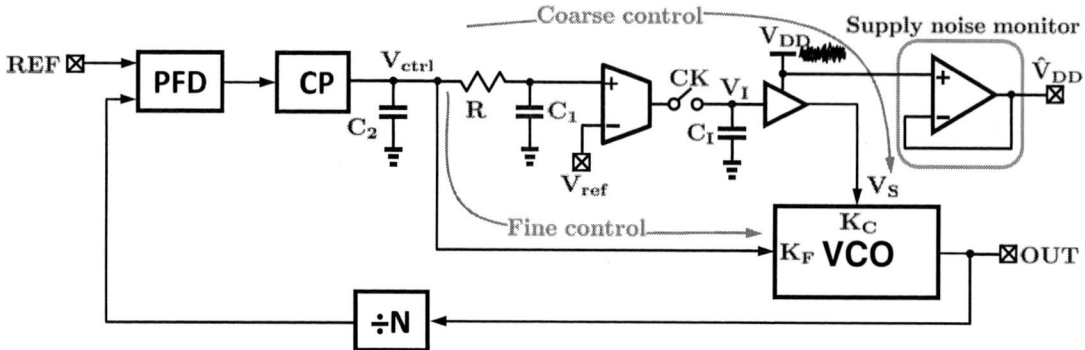

Fig. 4. Proposed supply-regulated PLL.

supply-noise rejection properties with minimal power penalty and without compromising on any of the other performance metrics of the PLL itself. The rest of the paper is organized as follows. Section II describes the proposed architecture while Section III discusses the issues with regulator design. Section IV and Section V present the experimental results and conclusions, respectively.

II. PROPOSED ARCHITECTURE

As described in the previous section, the conventional supply-regulation techniques had to contend with the regulator in the main-loop of the PLL resulting in unfavorable tradeoffs between supply-rejection and power dissipation. This tradeoff can be decoupled if the supply regulator is not included in the main loop of the PLL. Following this line of thought, Fig. 4 shows the proposed split-tuned supply-regulated PLL. The PLL consists of a dual-input VCO which has a high-gain path (K_C), that typically has a high sensitivity to supply noise, and a low-gain path (K_F) that is relatively insensitive to supply noise. The two input paths to the VCO are part of a low-bandwidth frequency-tracking loop (shown as coarse control), and a wide-bandwidth phase-tracking loop (shown as fine control). Since the dynamics of the PLL are primarily dictated by the fine control path, the placement of the regulator in the coarse control path helps decouple the tradeoff between regulator bandwidth (and hence power dissipation), and supply-noise rejection properties, a fundamental improvement over existing solutions [1]-[4]. A wide-bandwidth voltage follower circuit operating from a clean supply is also incorporated to directly monitor the on-chip supply.

The coarse control path in Fig. 4 integrates the voltage across capacitor C_I, and generates the corresponding coarse control voltage V_I. This control voltage is buffered through a low-dropout regulator to generate the supply voltage of the VCO, much like the traditional PLL shown in Fig. 1. The effective size of C_I is reduced by using a switched-G_m-C integrator that is clocked by a 0.1% duty cycle clock source.

Figure 5 compares the simulated performance of a traditional supply-regulated PLL with the proposed split-tuned supply-regulated PLL. In this comparison, the traditional regulator circuit, shown in Fig. 2, is used in both architectures

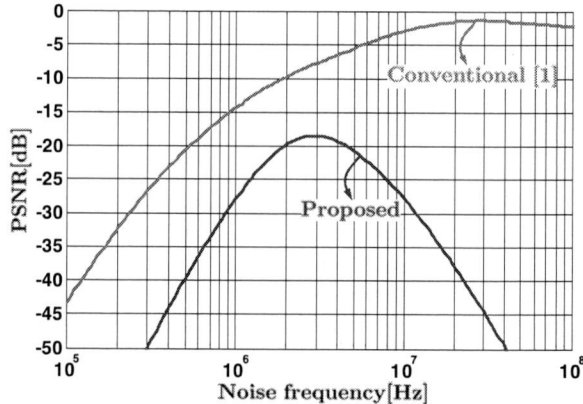

Fig. 5. Simulated PLL's power supply noise rejection (PSNR) performance for conventional [1] and proposed PLL.

with similar overall power dissipation. The proposed architecture shows a 12dB improvement in supply-noise rejection performance over conventional solutions, confirming that the proposed architecture is inherently well suited for supply regulation in PLLs.

In the implemented prototype, fine tuning in the VCO is done by varying the delay cell output time constant [5]. As a result, fine tuning control input is completely decoupled from the supply and hence, supply noise does not appear in this path. Any small amount of supply noise coupled through circuit parasitics has minimal impact on the PLL output jitter because of the low fine-VCO-gain (K_F).

III. REGULATOR DESIGN

The design of the regulator requires considerations that specifically exploit the benefits offered by the proposed PLL architecture. It is important to note that even though the regulator is placed in the coarse control path, which has minimal impact on the dynamics of the PLL, its supply-noise rejection performance depends on the loop dynamics. Consequently, the regulator should be designed in conjunction with the PLL loop response to supply noise. Since the proposed PLL architecture does not impose any constraints on the regulator, the conventional regulator shown in Fig. 2 can be used in

978-1-4244-2018-6/08 $25.00 © 2008 IEEE

the proposed PLL. The design techniques to improve the combined regulator's and overall PLL's supply-noise rejection performance are discussed next.

Let us first consider the scenario where the amplifier pole is dominant ($\omega_o > \omega_a$). It is clear from Fig. 3(a) that supply regulation performance can be improved by misaligning the peaks of the two curves so as to obtain a lower overall peak of the product of the two curves. This can be achieved by having a regulator bandwidth that is either much higher or much lower compared to the PLL bandwidth. Note that in both cases a large DC loop gain is needed to achieve good supply-noise rejection. Misaligning the peaks by having a very wide regulator bandwidth, as done in conventional supply-regulated PLLs, increases the power dissipation of the amplifier. Reducing the regulator bandwidth, a design choice allowed only in the proposed architecture incurs severe area penalty due to large capacitors, C_c and C_d.

On the other hand, if the regulator output pole is dominant ($\omega_a > \omega_o$), then the peaking in the supply-sensitivity curve is reduced (Fig. 3(b)). This design choice however leads to: (a) increased power dissipation due to large ω_a, and (b) large chip area required to implement C_d. Note that the output impedance is reduced by the large feedback loop gain thereby mandating a very large capacitor C_d to reduce ω_o.

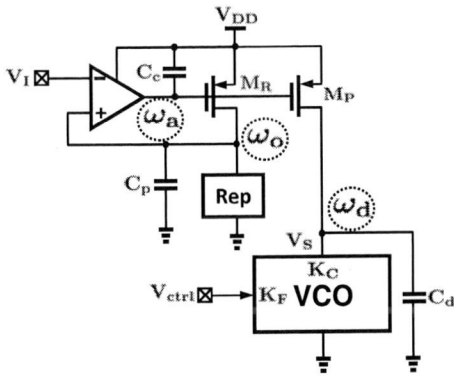

Fig. 6. Proposed regulator for the split-tuned PLL.

To circumvent these issues, we propose the regulator circuit shown in Fig. 6 in which a replica branch (M_R + Rep) is used to mimic the supply noise on the main branch (M_P + VCO). Now with the feedback eliminated in the main branch, a reasonably small decoupling capacitor C_d can be used to introduce the dominant pole at ω_d in the supply noise transfer function. The supply-sensitivity performance for this regulator is shown in Fig. 7, where the dominant pole is at ω_d. Intuitively, the regulator's loop gain provides low-frequency rejection, while the decoupling capacitor, C_d, further suppresses noise at frequencies beyond ω_d leading to a 20dB/decade roll-off of the sensitivity profile. The loop gain degrades beyond ω_a, resulting in a flattened power-supply rejection response. In other words, the pole of the amplifier contributes a zero to the overall transfer function and this degrades the suppression around ω_a. Beyond the regulator

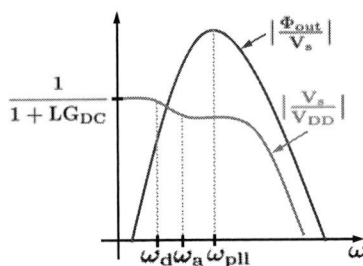

Fig. 7. Proposed regulator's supply-noise rejection profile.

bandwidth, the supply-noise rejection curve starts to roll-off again.

Although in [4], a replica branch was used, it was for an entirely different purpose— to improve the low-frequency supply-noise suppression. In this solution, however, the loop consisting of the replica branch and amplifier is used to suppress the noise across a wide frequency band. The absence of a global loop around V_s significantly improves performance by avoiding the supply-noise degradation due to amplifier pole, ω_a. In the proposed regulator, only the amplifier pole at ω_a and the replica pole at ω_o are part of the regulated feedback loop but, the pole at ω_d is not part of the regulated feedback loop and hence can be made sufficiently small with a reasonably small decoupling capacitor, C_d. The freedom to have $\omega_a < \omega_o$ leads to a regulator with the least power dissipation. In fact, ω_a can be arbitrarily small (but higher than ω_d), independent of the PLL dynamics. Figure 8 compares

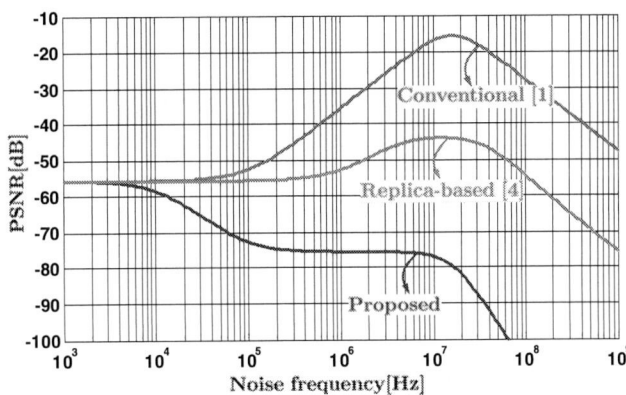

Fig. 8. Simulated power supply noise rejection (PSNR) performance of different regulators.

the simulated performance of the proposed regulator with the conventional [1] and replica-based regulators [4], indicating that using the replica branch as proposed provides greatly improved supply-noise rejection performance.

IV. EXPERIMENTAL RESULTS

A prototype was developed using the proposed split-tuned architecture that included the proposed regulator circuit. Fig. 9 shows the die photo of the prototype that is realized in a $0.18\mu m$ digital CMOS process, and occupies an active die area of 0.093mm^2 ($300\mu\text{m} \times 310\mu\text{m}$). The supply-regulation test was performed by modulating the regulator supply with a sinusoidal tone. Package parasitics along with the decoupling capacitance and other stray parasitics dampen the injected noise on-chip. To ensure the reliability of the test, the actual noise level reaching the regulator is measured by a wide-bandwidth voltage follower circuit that operates from a clean supply, shown as supply-noise monitor in Fig. 4.

Figure 10 shows the measured supply-noise rejection performance for this prototype, indicating an improvement of at least 20dB over the conventional supply-regulated PLL and 15dB over the replica-based regulator PLL [4]. The equivalent time domain peak-to-peak jitter values are also shown in Fig.10 for a $200\text{mV}_{\text{pk-pk}}$ sinusoidal input. A worst case peak-to-peak jitter of only 10ps is added to the output which translates to a sensitivity of 50fs/mV (0.5rad/V). The jitter histogram, under this worst-case condition, is shown in Fig. 11 indicating that the standard deviation of the jitter increases by only 3ps. Table I summarizes the key measurement results for this prototype. The power consumption of the PLL operating at 1.5GHz is 3.9mW from a 1.8V supply, of which 1.2mW is consumed in the VCO and only 0.27mW in the regulator.

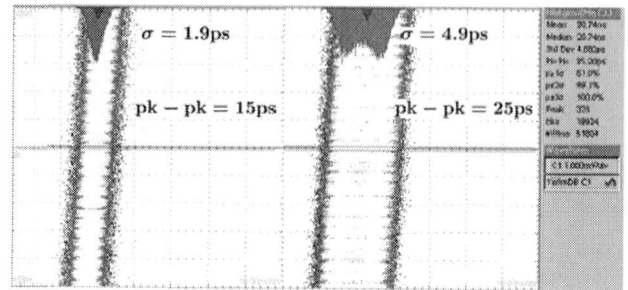

Fig. 11. Jitter histograms: (a) Without supply noise. (b) With supply-noise amplitude of $200\text{mV}_{\text{pk-pk}}$ at 8.85MHz.

TABLE I

PERFORMANCE SUMMARY

Technology	$0.18\mu m$ CMOS
Supply Voltage	1.8V
Operating Frequency	0.5-2.5GHz
Jitter(rms/pk-pk) @ 1.5GHz	1.9ps/15ps (no noise)
	4.9ps/25ps (200mV noise)
Power Consumption @ 1.5GHz	VCO: 1.2mW
	Regulator : 0.27mW
	Rest of PLL : 2.5mW
	Total : 3.9mW
Active Die Area	$0.093mm^2$

V. Conclusion

The design of a regulator in conjunction with the proposed supply-regulated PLL offers the advantage of the regulator being designed independent of the PLL loop, giving tremendous flexibility in the design and leading to a low-power solution with significantly improved supply noise rejection properties.

VI. Acknowledgements

The authors would like to thank Satoru Suenaga, Masa Konishi, Kouichi Abe, Yoshinori Nishi, and Suyama Takashi of Kawasaki Microelectronics for fabrication support. This work was partly supported by SRC under contract 2007-HJ-1597.

References

[1] V. von Kaenel *et al.*, "A 320 MHz, 1.5 mW @ 1.35 V CMOS PLL for microprocessor clock generation," *IEEE J. Solid-State Circuits*, vol. 31, pp. 1715 - 1722, Nov. 1996.

[2] K. Chang *et al.*, "A 0.44-Gb/s CMOS quad transceiver cell using on-chip regulated dual-loop PLLs," *IEEE J. Solid-State Circuits*, vol. 38, pp. 747 - 754, May 2003.

[3] S. Sidiropoulos *et al.*, "Adaptive bandwidth DLLs and PLLs using regulated supply CMOS Buffers," *Symp. VLSI Circuits Dig.*, pp. 124 - 127, June 2000.

[4] E. Alon *et al.*, "Replica compensated linear regulators for supply-regulated phase-locked loops," *IEEE J. Solid-State Circuits*, vol. 41, pp. 413 - 424, Feb. 2006.

[5] M. Johnson and E. Hudson, "A variable delay line PLL for CPU-coprocessor synchronization," *IEEE J. Solid-State Circuits*, vol. 23, pp. 1218 - 1223, October 1988.

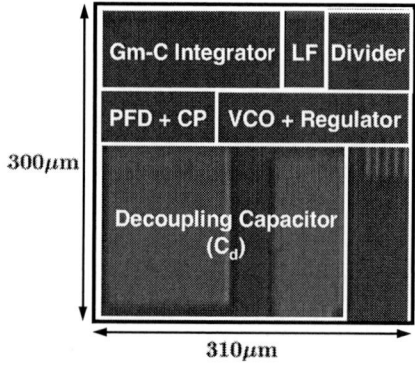

Fig. 9. Die photograph.

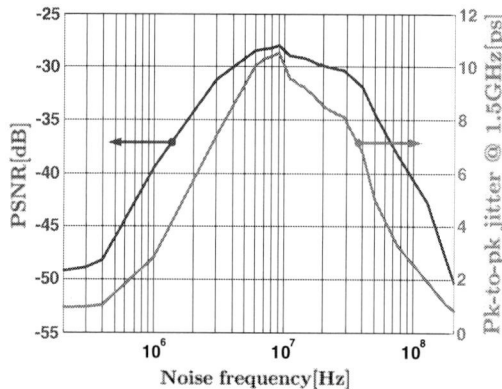

Fig. 10. Measured power supply noise rejection (PSNR) performance and jitter with a supply-noise amplitude of $200\text{mV}_{\text{pk-pk}}$.

IEEE 2008 Custom Intergrated Circuits Conference (CICC)

20 GHz Low Power QVCO and De-skew Techniques in 0.13μm Digital CMOS

Masum Hossain and Anthony Chan Carusone
Dept. of Electrical and Computer Engineering, University of Toronto

Abstract- **A novel VCO topology is proposed that combines the low power of $-g_m$ oscillators with the inherent buffering of Colpitts oscillators. Using this topology, a quadrature VCO (QVCO) was implemented in 0.13 μm digital CMOS consuming 32 mW at 20 GHz with just over 10% tuning range. The measured phase noise of the QVCO at 20.17 GHz is -102.41 dBc/Hz at 1 MHz offset. Because the load is isolated from the tank, the QVCO can directly drive 50-Ohm impedances or large capacitive loads with no additional buffering. A technique to use the QVCO to deskew clocks is also presented whereby the QVCO accepts a small forwarded clock amplitude of 20 mV, and provides a 200 mV peak-to-peak differential clock output with linear control of the phase over the complete range, 0-360°.**

I. INTRODUCTION

Clock generation and distribution consumes significant power and area in high-speed I/Os. To reduce power consumption per link, a shared clock source may be used. This shared clock may be generated either in the receiver [1] or at the transmitter and then forwarded to the receiver [2]. A phase interpolator must then be included in each link's receiver to compensate for skew (fig.1) [1,3,4].

A common approach to the problem of clock generation and distribution is to employ a low-jitter VCO, within a phase-locked loop (PLL), then buffer the output with several CML and CMOS stages to distribute the clock [1,5]. In this work we propose a VCO with an inherent buffer that re-uses the VCO bias current and provides large driving capacity without additional power consumption.

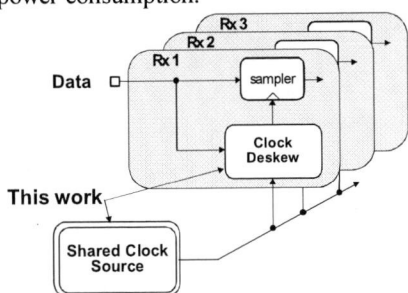

Fig. 1. Shared clocking for high density I/O [1-4].

An efficient technique for deskewing is to use an injection locked oscillator (ILO) whose free-running frequency is detuned away from the input frequency [6]. A problem with this approach has been that for large phase shifts considerable variation is observed in the jitter tracking bandwidth and output clock amplitude [7]. In this work, by selectively injecting either one or the other side of a quadrature VCO (QVCO), the required phase adjustment range is cut in half.

II. LOW POWER VCO ARCHITECTURE

Before introducing the proposed topology, two popular VCO topologies are reviewed: Colpitts and cross-coupled. The proposed topology, introduced at the end of this section, combines the advantages of both.

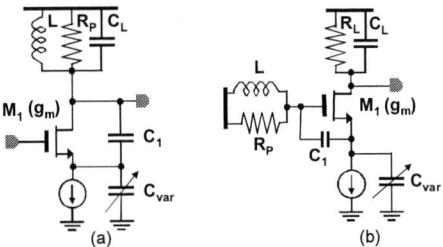

Fig. 2 Colpitts VCO (a) Conventional (b) CMOS version of modified Colpitts

A. Colpitts

Fig. 2 shows the equivalent half circuits two variants of the Colpitts VCO: fig. 2(a) is the well known conventional Colpitts and fig. 2(b) is a CMOS implementation of the bipolar microwave oscillator discussed in [8]. Although both VCO topologies have the same start up and oscillation conditions, the implementation in fig. 2(b) provides inherent buffering [8]: the tank is coupled to the load only through C_{GD} whereas in 3(a) the load capacitance(C_L) is directly across the tank. This is the main advantage of this modified Colpitts VCO.

If g_m is the small-signal transconductance of M_2 and R_P models tank losses, the condition to ensure oscillation of the colpitts VCO is:

$$g_m \geq \frac{4}{R_P} \quad (1)$$

Providing sufficient bias current to meet this requirement can lead to very high power consumption in CMOS Colpitts VCOs. For example, in this design tank inductance (L) and capacitance (C) are chosen to be 350 pH and 150 fF respectively to achieve the targeted 20 GHz oscillation. In the given digital process, this inductance was realized with a single loop. EM simulation of the optimized structure (including metal fill inside the loop required by the design rules) shows a Q of 5 which is typical for digital CMOS processes [5,7]. Capacitors C_1 and C_{var} are chosen to be 400 fF and 250 fF respectively. For the given value of Q, the required transconductance to meet the oscillation condition was found to be approximately $G_m=\omega^2 C_1 C_{var} R_s=25mS$ [9]. Here R_s is the series loss of the inductor. For the given CMOS 0.13 um process, a bias current density of 0.3mA/um provides a transconductance of 0.75 mS/um. Thus a single ended VCO

978-1-4244-2018-6/08 $25.00 © 2008 IEEE

consumes 10 mA of current, which leads to 20 mA of total current consumption for the differential Colpitts VCO.

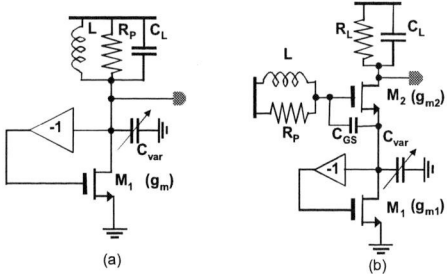

Fig. 3 (a) Cross-coupled VCO (b) Proposed topology

B. Cross-coupled

The equivalent half circuit for the cross-coupled VCO is shown in fig. 3(a). Here, the ideal gain of '-1' is used to represent the positive feedback resulting from the cross-coupling in a fully-differential circuit. Compared to the Colpitts VCO, the cross-coupled VCO has a relaxed oscillation condition:

$$g_m \geq \frac{1}{R_P} \approx \sqrt{\frac{C_{eq}}{L}} \left(\frac{Q_L Q_C}{Q_L + Q_C} \right) = \sqrt{\frac{C_L + C_{var}}{L}} \left(\frac{Q_L Q_C}{Q_L + Q_C} \right) \quad (2)$$

where g_m is the transconductance of M_1 and R_p models the tank loss. Q_L and Q_C are the quality factors of inductor and varactor respectively. The capacitance C_{var} is the varactor and C_L takes into account additional load and parasitic capacitors connected to this node. Considering the same tank circuit as in the previous section, the g_m required to meet the oscillation condition is found to be 6 mS. Assuming the same current density (0.3 mA/um), a differential cross coupled VCO consumes only 5 mA to provide the required g_m, one quarter the current of the Colpitts VCO. However, if used as a clock generator for a wide parallel bus, the output node is heavily loaded by C_L. Hence, to achieve the target oscillation frequency, the varactor C_{var} must be made small resulting in poor tuning range [10]. To avoid the loading effect CML buffers are used in [5] and [9], but these consume additional power negating the benefit of cross-coupled oscillators.

In summary, the Colpitts topology provides sufficient tuning range and output power but consumes large power. On the other hand although cross-coupled VCOs consume less power, they require additional buffer and they are more susceptible to load parasitics.

C. Proposed topology

To address these issues, we propose the architecture in fig. 3(b) which combines the useful properties of both cross-coupled and Colpitts architectures: The inherent buffering of the modified Colpitts topology (2(b)), and the low power oscillation condition of the cross-coupled VCO (fig. 3(a)). In this architecture the transistor M_2 is introduced in the tank of the cross-coupled VCO to isolate the output node from the tank similar to the modified Colpitts VCO. Effectively, M_2 serves as a buffer which can directly drive 50-ohm or large capacitive load. Since it uses the same VCO bias current, there is no additional DC power consumption.

There are two sources of negative resistance in this topology: (i) Due to the cross-coupling, M_1 provides a negative resistance of $-1/g_{m1}$. (ii) M_2 also provides a negative resistance of approximately $-g_{m2}/(C_{GS} C_{var} \omega^2)$. Now potentially there are two modes of operation: (i) A Colpitts VCO, where the negative resistance provided by M_2 is the dominating one and is large enough to compensate the tank loss (ii) A cross coupled VCO, where the oscillation occurs due to the negative resistance provided by M_1. Between these two modes, cross-coupled mode of oscillation requires less transconductance hence lower power consumption.

Table-I
Summary of VCO topology comparison

	Frequency of Oscillation	Minimum Required g_m
Cross-Coupled	$f_{osc} = \left(2\pi \sqrt{L (C_{var} + C_L)} \right)^{-1}$	$\geq \dfrac{1}{R_P}$
Colpitts	$f_{osc} = \left(2\pi \sqrt{L \dfrac{C_1 C_{var}}{C_1 + C_{var}}} \right)^{-1}$	$\geq \dfrac{4}{R_P}$
Proposed VCO	$f_{osc} = \left(2\pi \sqrt{L \dfrac{C_{GS} C_{var}}{C_{GS} + C_{var}}} \right)^{-1}$	$\geq \dfrac{1}{R_P} - \dfrac{C_{var}}{C_{GS}} g_{m2}$

The derived oscillation condition and oscillation frequency for the proposed cross coupled oscillator are given in table-I. These results are in good agreement with qualitative description given above: f_{osc} is independent of C_L and required minimum transconductance is slightly less than $1/R_P$. Thus M_1 is sized to meet the oscillation condition and M_2 is used as a buffer only. The total current consumption for a differential VCO is 8 mA which results in 60% power reduction compared to the Colpitts implementation.

D. Qudrature VCO (QVCO)

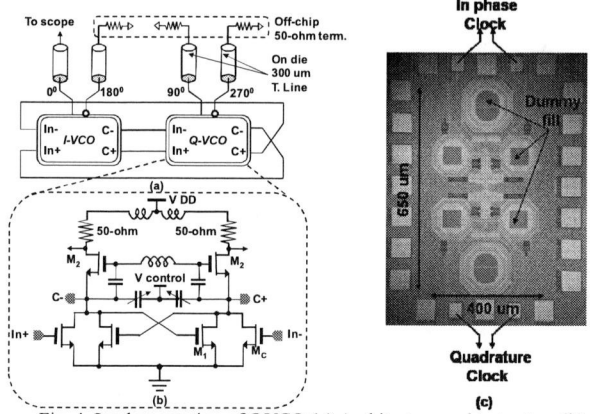

Fig. 4. Implementation of QVCO (a) Architecture and test set up (b) Detail schematic of QVCO (c) Die photo of QVCO in 0.13 um CMOS

A quadrature version of the proposed VCO topology was implemented by coupling 2 differential VCOs operating at the same frequency[11,12]. A schematic is shown in figures 4a and b. The coupling is provided by active devices, Mc.

Quadrature (4-phase) VCOs in general have several disadvantages compared to their differential (2-phase) counterparts: a) Due to the additional DC power consumption in the coupling devices, the power consumption of the quadrature VCO is usually more than twice the power consumption of their differential version. b) In the quadrature implementation both tanks operate slightly off resonance which results in higher phase noise and reduced tank impedance compared to the differential version.

In cross-coupled topology, the coupling devices load the coupling node with additional parasitic capacitance which further reduces the tuning range of the cross-coupled QVCOs. To ensure 90° phase locking, the quadrature coupling transistors M_C are one-half the size of the cross-coupled transistors M_1, which results in additional 8 mA of current consumption. Thus the total current consumption was 24 mA from 1.2 V supply. A die photo of the implemented quadrature VCO is shown in fig. 4(c). For testing, the QVCO directly drives 0.3mm on die transmission line and 50-ohm off-chip termination without additional buffer.

E. Experimental Results

Measured results of the QVCO are summarized in fig. 5. The VCO has a tuning range of 2 GHz. Measured single ended output power driving a 50-ohm load varies from -12 dBm to -14 dBm over the tuning range. A captured phase noise plot at 20.17 GHz is shown in fig. 6.

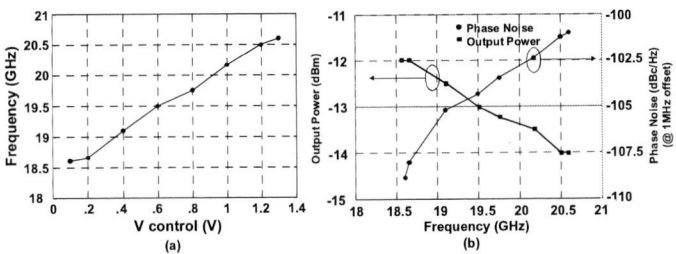

Fig. 5. QVCO performance summary (a) Tuning curve (b) Phase noise and output power as a function of frequency

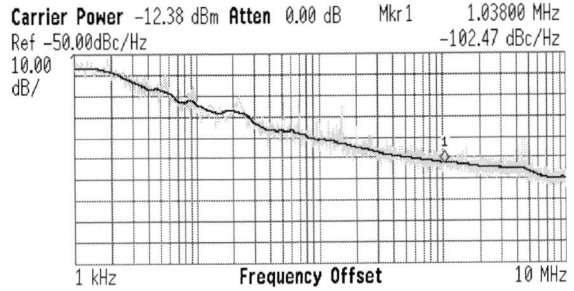

Fig. 6. Measured phase noise of the QVCO at 20.17 GHz

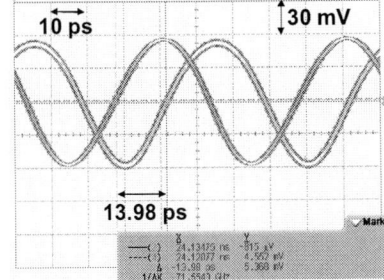

Fig. 7 Time domain QVCO output at 20 GHz showing 0° and 90° phase

To verify the quadrature operation, 0° and 90° outputs are captured on an oscilloscope where any mismatch in the length of measurement cables has been calibrated out (fig.7). For comparison, key performance metrics for different VCO topologies are summarized in table-II. According to the ITRS 2003[13], the figure-of-merit for VCOs is:

$$FoM = 10\log_{10}\left(\left(\frac{f_{osc}}{\Delta f}\right)^2 \frac{1}{L(\Delta f)P_{diss}(\text{mW})}\right) \quad (3)$$

Our earlier conclusion regarding Colpitts and cross-coupled VCOs are in good agreement with the measured results from [9]: cross-coupled VCO can achieve a significant advantage over Colpitts for low power applications. However, this advantage is significantly compromised when the buffer is included in the performance metric. In addition as pointed out in the previous section, there is significant performance degradation in cross coupled QVCOs compared to their differential counterparts [11,12] . Although the tank Q in this VCO is much lower compared to the other VCOs listed in the table, this VCO topology is still has a FoM better than other QVCOs in CMOS. The differential 10GHz Colpitts VCO designed in [9] consumes more power than the 20GHz QVCO designed in this work, which demonstrates the low power advantage of the proposed topology.

Table II
Comparison for state-of-art CMOS VCOs

	[9] CSICS'06	[9] CSICS'06	[10] JSSC'07	[11] JSSC'04	[12] VLSI'05	This work
Technology	90-nm CMOS	90-nm CMOS	0.13-um CMOS	0.13-um CMOS	90-nm SOI	0.13-um CMOS
Frequency	10 GHz	10 GHz	26 GHz	10 GHz	40 GHz	20 GHz
Topology	Colpitts	Cross-Coupled	G_m Tuned	Cross-Coupled	Cross-Coupled	Cross-Coupled
Diff./Quadrature	Diff.	Diff.	Diff.	Quad.	Quad.	Quad.
Tuning Range	12.2 %	15.8 %	23.6 %	15%	12.5 %	10.2 %
Inductor Q/ Transformer Q	10	10	18	-------	18	5
Phase Noise (dBc/Hz@1 MHz)	-117.5	-109.2	-92.6	-95	-87 @3 MHz	-102.41
VCO power VCO+ Buffer	36 mW	7.5 mW 17.5 mW	43.6 mW 50 mW	14.4 mW -------	------- 81 mW	------- 32mW
FOM (VCO) (dB) (VCO+ Buffer)	181.9 -------	180.4 176.5	163.9 163.3	163.4 -------	------- 150.4	------- 173.45

III. DE-SKEW TECHNIQUES WITH JITTER FILTERING

Injection locking was introduced in [14] as an effective method to filter out jitter and duty cycle distortion from a high frequency reference clock. Recently in [6], an ILO is used as a local clock generator which provides several advantages: (i) Due to its high sensitivity, ILOs can operate with very small input amplitude. The ratio of the input clock amplitude to the VCO output amplitude is known as injection strength. Thus the reference clock can be distributed with low power which translates into large power savings. (ii) Since an ILO behaves as a 1st order PLL, it rejects high frequency jitter and is less susceptible to power supply noise. (iii) The clock can be deskewed by detuning the free running frequency of the ILO. For small injection strengths, the deskew range is smaller than 360° [7]. With large injection strength, it is possible to extend the deskew range but this requires a wide tuning range in the ILO. Furthermore, providing skews near ± 180 degree results in considerable variation in the jitter tracking bandwidth and output clock amplitude [7]. To address these issues, we propose a deskew technique utilizing the QVCO as shown in fig.8(a). This proposed scheme allows us to selectively inject either of the differential VCOs in the QVCO. The measured skew versus control voltage is shown in fig. 8(b). Two deskew curves (AB and CD) are shown due to injection in *I-VCO* and *Q-VCO* respectively. Since *I-VCO* and *Q-VCO* are oscillating in quadrature, they maintain 90° phase difference between each other.

In [6,7] a single differential VCO is used as an ILO which has a deskew curve very similar to each of those in fig. 8(a). To obtain the $0\text{-}360^\circ$ phase selection capability, full length of the curve is utilized. Notice the nonlinear compression observed close to the edges of the lock range. Fig. 9(a) shows the captured deskewed clocks for this portion of the curve. Variation in output amplitude is observed, and clock phases are nonlinearly spaced.

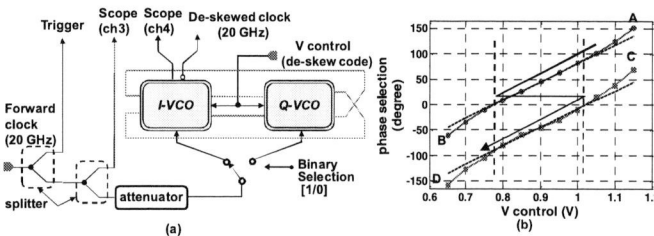

Fig. 8 (a) Proposed deskew technique (b) Corresponding measured deskew curves at 20 GHz for I-VCO and Q-VCO injection

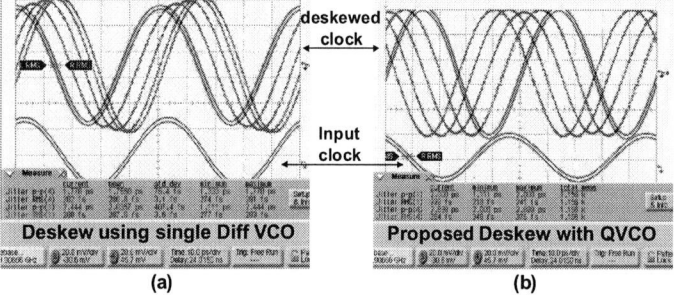

Fig.9 (a) Measured deskewed clock using I injection only (d) Measured deskew using proposed technique

In the proposed technique, only the linear portions of the deskew curves are used. Hence, the ILO can provide linear control of the phase shift, relatively constant output amplitude and relatively little variation of the jitter transfer bandwidth. Now the forwarded clock is injected to the in-phase VCO to achieve $0\text{-}180^\circ$ phase shift only. For $180^\circ\text{-}360^\circ$, we shift the injection to *Q-VCO* and use linear portion of its deskew curve. Thus the proposed technique allows us to accomplish $0\text{-}360^\circ$ phase selection with linear phase steps and negligible amplitude variation, as shown in fig. 9(b). In this experiment the forwarded clock amplitude was 20 mV and the deskewed differential peak to peak clock output was 200 mV for an injection strength of 0.1. Due to the additional VCO in

quadrature, this technique will consume more power compared to [6] and [7]. However, this 20 GHz deskew scheme can be implemented using total 35 mW only, which still compares favorably with using a complete DLL for deskewing, as in for example [14], which requires many buffers to delay the clock and perform phase selection.

IV. CONCLUSION

In this work we have introduced a novel VCO topology capable of driving large capacitive loads without a buffer and with lower power than Colpitts VCOs. Using this topology, a QVCO is designed with a FoM comparable to state-of-art solutions in spite of a much lower tank Q. Its inherent buffering makes it useful for clock generation and distribution to the large capacitive loads in high-speed I/Os. It can also be used as an ILO to deskew a forwarded clock. It provides more linear skew control with less variation in output amplitude than previous solutions.

ACKNOWLEDGMENTS

The authors would like to acknowledge F. O'Mahony, M. Mansuri and B. Casper of CRL (Intel Circuit Research Lab at Hillsboro,OR) for their contribution in clock deskew technique presented in this work. This work is supported by Intel and fabrication facilities were provided by Gennum Corporation.

REFERENCES

[1] H. Takauchi *et. al.* , "A CMOS Multichannel 10-Gb/s Transceiver," *IEEE J. Solid-State Circuits*, vol. 38, no. 12, pp. 2094–2100, Dec, 2003.

[2] B. Casper *et. al.* "A 20 Gb/s forwarded clock transceiver in 90-nm CMOS,"in *IEEE ISSCC Dig. Tech. Papers*, Feb, 2006, pp. 90-91

[3] R. Kreienkamp *et. al.*, "A 10-Gb/s CMOS Clock and Data Recovery Circuit With an Analog Phase Interpolator," *IEEE J. Solid-State Circuits*, vol. 40,no. 3, pp. 736–743, Mar, 2005.

[4] C. Kromer *et. al.* , "A 25-Gb/s CDR in 90-nm CMOS for High-Density Interconnects," *IEEE J. Solid-State Circuits*, vol. 41,no. 12, pp. 2921–2929, Dec, 2006.

[5] F. O'mahony *et. al.* , "A low-jitter PLL and repeaterless network for a 20Gb/s link," in *IEEE Symp. On VLSI Circuits Dig. of Tech. Papers*, 2006.

[6] L. Zhang, B. Ciftcioglu, M. Huang, H. Wu, "Injection-locked clocking: a new GHz clock distribution scheme," *Custom Integrated Circuits Conference*, San Jose, California, September 2006

[7] F. O'Mahony *et. al.* , "A 27Gb/s Forwarded Clock I/O Receiver using an Injection-Locked LC-DCO in 45nm CMOS", *IEEE International Solid-State Circuits Conference*, Feb. 2008.

[8] N. Nguyen, R.G. Meyer , "Start up and frequency stability in high-frequency oscillators," *IEEE J. Solid-State Circuits*, vol. 27,no. 5, pp. 810–820, May, 1992.

[9] K.W. Tang et. al. "Frequency Scaling and Topology Comparison of mm-wave CMOS VCOs," *IEEE CSICS*, pp.55-58, Nov,2006.

[10] K. Kwok, J. Long, "A 23-to-29 GHz Transconductor-Tuned VCO MMIC in 0.13µm CMOS," *IEEE J. Solid-State Circuits*, vol. 42,no. 12, pp. 2878–3997, Dec, 2007.

[11] S. Li, I. Kipnis, M. Ismail, "A 10-GHz CMOS Quadrature LC-VCO for Multicore Optical Applications," *IEEE J. Solid-State Circuits*, vol. 38,no. 10, pp. 1626–1634, Oct, 2003.

[12] F. Ellinger and H. Jäckel, "38-43 GHz Quadrature VCO on 90nm VLSI CMOS with Feedback Frequency Tuning," *VLSI Circuits Symposium*, Kyoto, Japan, June , 2005.

[13] International technology roadmap of semiconductor. www.itrs.org, 2003.

[14] H.Ng et. al., "A Second-Order Semidigital Recovery Circuit Based on Injection Locking," *IEEE J. Solid-State Circuits*, vol. 38,no. 12, pp. 2101–2110, Dec, 2003

978-1-4244-2018-6/08 $25.00 © 2008 IEEE

IEEE 2008 Custom Intergrated Circuits Conference (CICC)

A 3 GHz Spread Spectrum Clock Generator for SATA Applications Using Chaotic PAM Modulation

Fabio Pareschi[*‡], Gianluca Setti[*‡] and Riccardo Rovatti[†‡]

[*]ENDIF - University of Ferrara, via Saragat 1, 44100 Ferrara - ITALY
[†]DEIS - University of Bologna, viale risorgimento 2, 40136 Bologna - ITALY
[‡]ARCES - University of Bologna, via Toffano 2/2, 40125 Bologna - ITALY
Email: {fabio.pareschi,gianluca.setti}@unife.it, rrovatti@arces.unibo.it

Abstract—This paper proposes a prototype of a Spread Spectrum Clock Generator which is the first known specifically meant for 3 GHz Serial ATA-II applications. The modulation is obtained from a fractional PLL which employs a Delta-Sigma modulator. A further innovative aspect of our work is that our prototype takes advantage of a *chaotic* PAM as driving signal, instead a triangular signal as in all spread spectrum generators proposed in literature for SATA-II. In this way we avoid the periodicity of the modulated clock, completely flattening the peaks in the power spectral density. The circuit prototype has been designed in 0.13 μm CMOS technology and achieves a peak reduction greater than 14 dB measured at RBW = 100 kHz. The chip active area is 0.27×0.78 mm^2 and the power consumption is as low as 14.7 mW.

I. INTRODUCTION

The reduction of Electro-Magnetic Interference (EMI) in electronic devices is an issue of increasing interest for designers. Digital signals, due to their sharp edges and their synchronization with a periodic clock signal, are preeminent sources of interference. In particular, several problems may arise with high–speed serial interfaces connecting signal processing units and peripheral devices. Interestingly, many of the protocols employed to this purpose explicitly mention EMI-related problems, as well as possible solutions. Among them, we can include the Serial AT Attachment (*SATA*) protocol [1].

This computer bus technology is designed for fast data transmission to and from Hard Disk Drives, according to the most recent specifications (known as SATA-II), at a clock speed of 1.5 GHz or 3.0 GHz. The protocol also allows the introduction of *spread spectrum* techniques on its clock signal, in order to perform an intrinsic EMI reduction.

The spread spectrum clock technique consists of introducing a continuous and slight *delay* or *anticipation* in the reference clock edges, thus avoiding a perfect periodicity. The result is the introduction of new components in the power spectra of all synchronous signals, but also a reduction in their peak level. This point of view is perfectly coherent with FCC and EC regulations [2] that link EMI compliance with the ability of fitting the interfering power spectrum below a predefined mask.

The most common way to get a spread spectrum clock is through a *frequency modulation* of the clock signal [3]:

$$s(t) = \text{sgn}\left(\cos\left(2\pi f_0 t + 2\pi \Delta f \int_{-\infty}^{t} \xi(\tau)d\tau \right) \right) \quad (1)$$

where f_0 is known as carrier frequency, Δf is the frequency deviation and $-1 < \xi(\tau) < 1$ the driving signal. If we refer to the first harmonic, its energy is spread, according to Carson's rule, into the frequency band $[f_0 - \Delta f, f_0 + \Delta f]$. Of course, this may impair the synchronization process between transmitter and receiver; for this reason it is necessary that the receiver device is designed to be compatible with the introduced modulation. This is particularly important in asynchronous applications, like the Serial ATA, to allow a correct clock recovery. To this purpose, the SATA-II protocol fixes the modulation parameters, allowing only a 30–33 KHz driving signal $\xi(t)$ and a 5000 ppm *down-spreading* frequency deviation [1]. Intuitively assuming that f_0 is the non-modulated clock frequency, a down-spreading is an asymmetric modulation where the output frequency cannot exceed f_0. In this way the allowed spreading band is $[f_0 - 2\Delta f, f_0]$. If we recast equation (1) to fit SATA-II specifications, we get $\Delta f = 0.005 \cdot f_0/2$

Under the above constraints, the performance of the system in terms of EMI reduction (measured as the reduction of the peak level in the power spectrum) depends only on the driving signal $\xi(t)$ employed, which is the only degree of freedom. Poor results are given by a simple sinusoidal $\xi(t)$ signal; better performances can be achieved with a triangular waveform; while we can get even better performances with a more complicated periodic driving signal [3]. Recently, many papers in literature proposed a 1.5 GHz spread spectrum clock generator for SATA-II [4], [5], [6], [7] employing a triangular waveform as modulating signal. Almost all of them are based on a fractional PLL employing a $\Delta\Sigma$ modulator.

In this paper we present a prototype of a spread spectrum clock generator which is the first one running at 3 GHz that is SATA compatible. A further fundamental difference between the prototype presented here and others proposed in literature is that our generator does not exploit a triangular driving signal, but a uniform chaotic PAM modulated driving signal, i.e. where $\xi(t)$ is a sequence of impulses whose amplitude is given by a chaotic system. The advantages of a chaotic modulation are known in literature [8] and can be summarized as a complete lack of peaks in the power spectrum when using the proper parameters.

In fact the modulation index m, defined as the ratio between the frequency deviation Δf and the frequency f_m of the modulating signal $\xi(t)$ (i.e. $m = \Delta f / f_m$) is very high in the case of serial ATA-II. We get $m \simeq 110$ for a 1.5 GHz SATA clock and $m \simeq 220$ for a 3.0 GHz SATA clock. It is shown in [8] that under this assumption the uniform chaotic PAM modulation outperforms any periodic modulation, including the patented modulation proposed in [3]. A sketch comparison between the power spectra of a triangular modulation, the proposed chaotic modulation and of an unmodulated clock can be observed in Figure 1. The advantage in terms of EMI reduction of the proposed modulation with respect to the triangular one depends on the resolution bandwidth used in evaluating the spectrum (which may represent the sensitivity of the EMI victim); the narrower the bandwidth, the larger the difference. As an example, when using the parameters allowed by the 3 GHz SATA-II spread-spectrum clock and a resolution bandwidth of 10 kHz the proposed chaotic modulation achieves a theoretical additional EMI reduction of about 7 dB.

The paper is organized as follows. Section II explains the working principle of the prototype and presents a few simulation results. The measurement results are presented in

978-1-4244-2018-6/08 $25.00 © 2008 IEEE

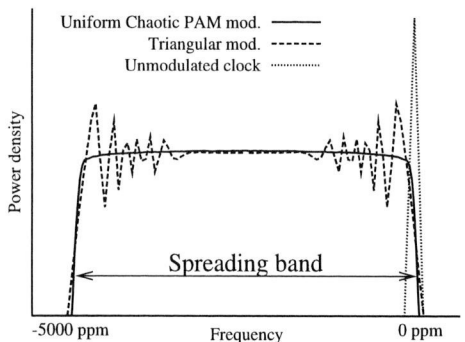

Fig. 1. Sketch comparison between the power spectra of an unmodulated clock (dotted line) and of a clock with a down-spreading modulation employing a triangular signal (dashed line) and a chaotic PAM signal (continuous line).

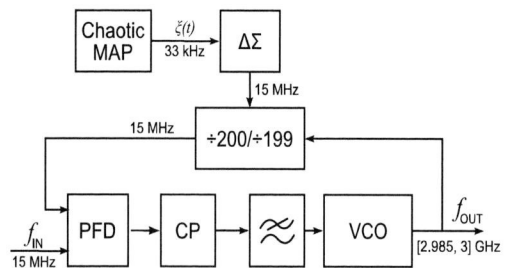

Fig. 2. Block diagram of the proposed spread spectrum clock generator.

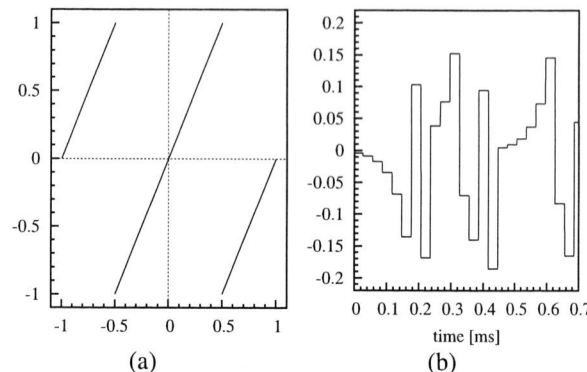

Fig. 3. (a) ideal M implemented in the prototype; and (b) output signal $\xi(t)$ of the chaotic map from circuital simulation (differential voltage).

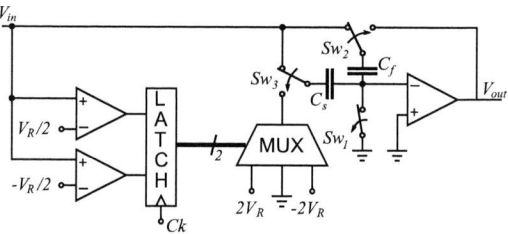

Fig. 4. Schematic of the circuit used for implementing the chaotic map of Figure 3-a.

section III, whereas conclusions are given in the last section.

II. PROPOSED SPREAD SPECTRUM CLOCK GENERATOR

The block diagram of the proposed prototype is shown in Figure 2. This circuit is very similar to a classic fractional PLL [9] employing a phase/frequency detector, a charge pump, a VCO and an integer divider on the feedback path. The divider ratio is controlled by a first-order $\Delta\Sigma$ modulator, thus exploiting a fractional divider under the assumption of a low frequency $\xi(t)$ signal and of a high $\Delta\Sigma$ clock frequency (with respect to the PLL closed-loop bandwidth). In this way, with an input frequency of 15 MHz, the PLL can generate any output frequency between 15 MHz × 200 = 3 GHz and 15 MHz × 199 = 2.985 GHz (that is 3 GHz − 5000 ppm) with a linear dependence on $\xi(t)$.

As mentioned before, in our prototype the driving signal $\xi(t)$ is the PAM signal

$$\xi(t) = \sum_k x_k g(t - kT) \qquad (2)$$

where $g(t)$ is the impulse shape and where the coefficients x_k are given by the discrete-time autonomous system

$$x_{k+1} = M(x_k) \qquad (3)$$

where $x_k \in X$ and the mapping function M is a proper non-linear function $M : X \mapsto X$. Systems such as (3) are usually referred to as *chaotic maps*; starting from an initial state x_0 (which is set in real systems by noise at circuit startup), they generate a succession $\{x_k\}$ that has very peculiar features. Roughly speaking, this is a bounded, non-periodic, deterministic sequence that, due to the increasing unpredictability of the system evolution under the assumption

of noise perturbations, can be handled only as a random sequence.

The main property of these systems is that, under certain assumptions on M, they can be modeled as a stochastic process where the state x_k at time step k is a random variable with a probability density function ρ that depends only on M.

More details on a frequency modulation exploiting (2) as driving signal (*chaotic PAM modulation*), as well as a complete and extensive analysis on chaotic maps, can be found in [8]; for the purposes of this paper, it is sufficient to recall that, under the assumption of a large modulation index (defined in this case as $m = \Delta f \cdot T$) and of a rectangular-shaped impulse $g(t)$, the power spectrum of the modulated signal around the carrier frequency takes the shape of ρ. As a result, we can optimize the power spectrum using a map whose ρ does not present any peak, i.e. an *uniform* ρ. The example of Figure 1 has been obtained with a chaotic map generating symbols whose density is uniform in X.

The chaotic map we used to get the optimal symbol density is the following

$$M(x_k) = \begin{cases} 2x_k + 2 & \text{if } x_k \leq -\frac{1}{2} \\ 2x_k & \text{if } -\frac{1}{2} < x_k \leq \frac{1}{2} \\ 2x_k - 2 & \text{if } x_k > \frac{1}{2} \end{cases} \qquad (4)$$

where we have assumed that X is the normalized interval $X = [-1, 1]$. This map can be found drawn in Figure 3-a. Figure 3-b instead shows a simulation of the PAM signal $\xi(t)$ representing the output of the chaotic map.

The schematic of the circuit implementing function (4) is depicted in Figure 4, and it is taken from [11], where the authors showed that this switched capacitors circuit, commonly used in 1.5 bit/stage pipeline ADC converters [10], can be conveniently used to implement the chaotic map (4) with $X = [-V_R, V_R]$ when setting $C_s = C_f$

978-1-4244-2018-6/08 $25.00 © 2008 IEEE

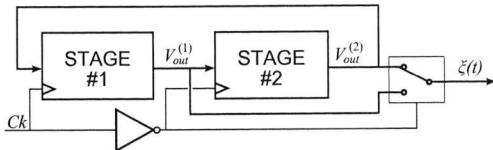

Fig. 5. Pipeline structure used to get the desired dynamical behaviour.

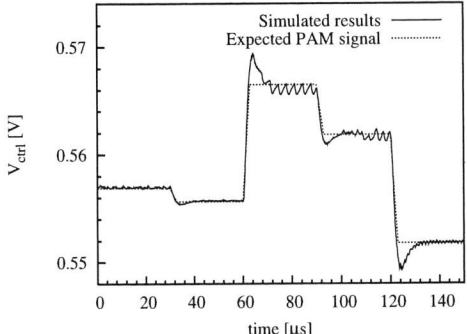

Fig. 6. VCO control voltage during a post-layout simulation (solid line), compared with the expected PAM signal (dotted line).

Note that this circuit introduces a delay of half clock cycle between output and input. In order to achieve a dynamic behaviour as in (3) it is necessary to close this circuit into a loop after a delay of full clock cycle. To get the additional delay required, we follow [11] and add another identical block working on the opposite clock phase, as in Figure 5. With this arrangement, assuming $2T$ is the period of the chaotic map clock, the symbols x_k are available alternately at the outputs of the two stages, at the rate of one symbol every time step T; to get the $\xi(t)$ as in (2) it is enough to add two simple pass transistors as in figure.

It is important to stress that the additional cost in terms of area and power consumption with respect to using a simpler delay circuit (e.g. a sample/hold) is negligible, since only the two comparators and few pass transistors are added.

Accordingly to serial-ATA specifications, the chaotic map generates the PAM signal $\xi(t)$ with a symbol rate of $1/T = 33$ kHz, while the $\Delta\Sigma$ modulator clock has been set to 15 MHz, corresponding to an oversampling ratio of about 450.

The PLL bandwidth is set to an intermediate value between the two above frequencies. In this way, the low-frequency variations of the PAM signal $\xi(t)$ can modulate the output clock, while the effect of the high-frequency variations of the $\Delta\Sigma$ modulated signal are strongly attenuated. Note that any residual modulation given by the $\Delta\Sigma$ modulator gives rise to an undesired *deterministic* jitter in the output clock. By setting the PLL closed loop bandwidth to 63 KHz, the VCO control voltage has the shape depicted in Figure 6, which compares it with the output of the chaotic map. Two remarks need here to be made: *i.* the worst case deterministic jitter introduced in the output clock by the residual modulation of the $\Delta\Sigma$ converter is experimentally evaluated in 0.4 ps rms; *ii.* due to the low frequency components of $\xi(t)$ and to the low PLL cut–off frequency, a VCO control voltage as in Figure 6 can easily be rebuilt by a clock recovery circuit. High–level simulations show that the additional jitter of the recovered clock is much lower with respect to the deterministic jitter of the transmitter clock.

As a final remark, it is necessary to stress that the chaotic map requires an analog implementation, since no digital circuit

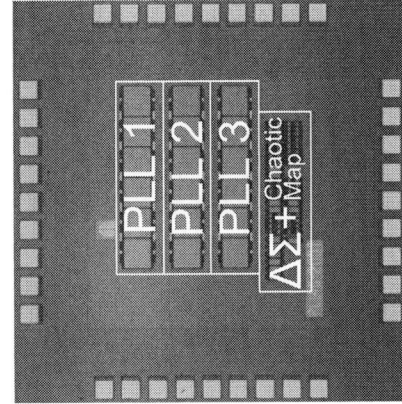

Fig. 7. Microphotograph of the chip prototype.

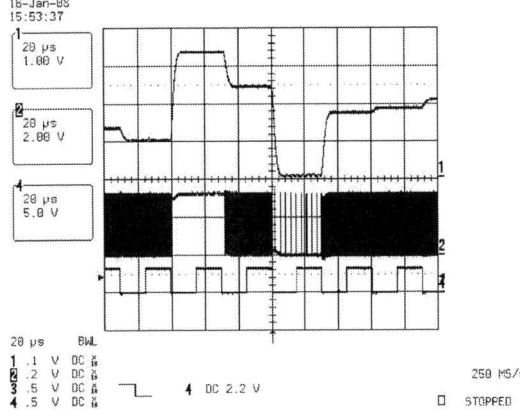

Fig. 8. Waveforms from the chaotic map/$\Delta\Sigma$ block. From bottom to top: clock of the chaotic map; driving signal $\xi(t)$ modulated by the $\Delta\Sigma$ converter; and the same signal demodulated by an external low-pass filter.

can exhibit chaotic behaviour. However, the low working frequency and the possibility of sharing the voltage references between the chaotic maps and the $\Delta\Sigma$ converter allow us a low power design. The overall power consumption of both circuits is estimated in 3.8 mW, much lower than the power consumption of the 3 GHz PLL.

III. MEASUREMENT RESULTS

We fabricated the 3 GHz spread spectrum clock generator SATA-II compatible in 130 nm CMOS technology. Figure 7 is a microphotograph of the chip prototype. Note that the prototype includes three different PLLs; this solution has been adopted to extend the chip tunable range from 2.5 GHZ to 3.5 GHz.

Due to this hardware replication, the power consumption of the entire chip is quite high, and it has been measured in 37 mW on the 1.2 V core power supply line. By a comparison with simulation results, we can say that the power consumption of a single PLL and of the chaotic map/$\Delta\Sigma$ block is as low as 14.7 mW.

The behavior of the block constituted by chaotic map and the $\Delta\Sigma$ modulator can be observed in Figure 8. In this figure we can found the chaotic signal $\xi(t)$ modulated by the $\Delta\Sigma$ converter (center trace) and demodulated by an external low-pass filter (upper trace), as well as the chaotic map clock (lower trace).

978-1-4244-2018-6/08 $25.00 © 2008 IEEE

	This work	[4]	[5]	[6]	spec.
Mod. Method	$\Delta\Sigma$, chaotic	$\Delta\Sigma$, triang.	$\Delta\Sigma$, triang.	$\Delta\Sigma$, triang.	N/A
Working Frequency [GHz]	**3.0**	1.5	1.5	1.5	1.5/3.0
Mod. Freq. [kHz]	33.0	31.25	31.1	30.0	30.0-33.0
EMI Reduction [dB, RBW = 100 KHz]	**14.5**[1]	*9.6* [2]	10	9.8	>7
Spreading Ratio [ppm]	-5000/+0	5070[3]	-5000/+350	-5000/+0	-5000/+0
PLL Random jitter [ps rms]	5.9 [4]	N/A[5]	8.1	3.2	<12
PLL BW [kHz]	63	100	300	N/A	N/A
CMOS Technology [μm]	0.13	0.18	0.15	0.13	N/A
Power consumption [mW]	**14.7**	27	54	77	N/A
Chip area [mm^2]	**0.27 x 0.78**	0.44 x 0.48	0.88 x 0.48	1.75 x 0.94	N/A

(1) Since the frequency deviation is doubled with respect to the other solutions, there is a +3 dB EMI reduction since the spreading ratio is constant. In any case, this solution achieves the best reduction. (2) Paper [4] report a EMI reduction of 19.6 dB at RBW = 10 Khz; the value reported here is an estimation. (3) The exact spreading range is not indicated. (4) Estimated from phase noise. (5) A peak to peak jitter 30 ps is indicated.

TABLE I

MEASUREMENT SUMMARY AND COMPARISON WITH PREVIOUS WORKS.

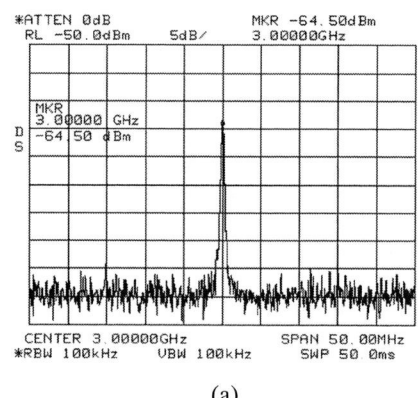 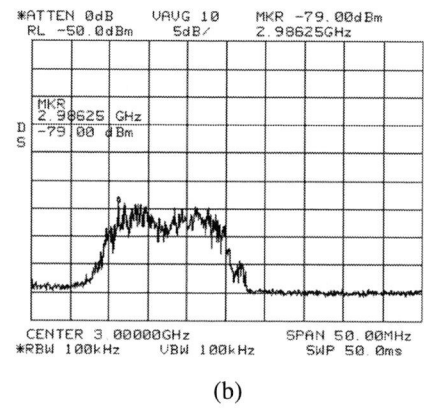

(a)   (b)

Fig. 9. Comparison between power spectrum of the output clock in non-spread spectrum (a) and spread spectrum (b) mode. Measurement conditions: RBW = 100 KHz, VBW = 100 KHz, Center Frequency = 3.0 GHz, Span = 50 MHz, positive peak detector mode. EMI reduction can be evaluated in 14.5 dB.

Finally, Figure 9 shows the measurements results of the output spectrum both in spread spectrum and in non-spread spectrum mode. The comparison is done at a resolution bandwidth of 100 KHz (suggested in [1]), and shows a flat spectrum in the range of interest with a peak reduction of 14.5 dB. From the measure of the phase noise in the non-spread spectrum, and accordingly to [12], we can assume a random jitter of 5.9 ps. Note also that the modulated clock spectrum presents the expected down-spreading modulation shape as expected by the SATA-II standard.

Measurement results are summarized in Table I. In the same table we also compared results from our prototype with results from other solutions in literature. All prototypes in [4], [5], [6] exploit a triangular modulation. Our prototype achieves the best performances in terms of power consumption and EMI reduction. Furthermore, it is the only one working at a 3 GHz frequency, while all other prototypes work at 1.5 GHz.

Note also that some prototypes use a spreading ratio larger than 5000 ppm. SATA-II protocol allows a ± 350 long term deviation from the main frequency, and some designers uses this additional allowed deviation for spreading purposes. However [1] clearly states that the accepted spreading ratio is only $-5000/+0$ ppm; larger ratios give rise to additional EMI reduction (up to +0.3 dB) making the comparison unfair.

IV. CONCLUSIONS

We developed a PLL-based spread spectrum clock generator compatible with 3 GHz serial ATA specifications. The prototype exploits a chaotic PAM modulation, thus flattening all peaks in the clock power spectrum. In fact, the EMI reduction can be measured in 14.5 dB peak reduction, that

is the highest reduction among all prototypes presented in literature. Furthermore, the generator presented here achieves the lowest power consumption, which is only 14.7 mW.

REFERENCES

[1] Serial ATA International Organization, "Serial ATA Revision 2.6", Feb. 2007.

[2] Federal Communication Commission "FCC methods of measurement of radio noise emission from computing devices", *FCC/OST MP-4*, 1987

[3] K.B. Hardin, J.T. Fessler, D.R. Bush, "Spread spectrum clock generation for the reduction of radiated emission", in *Proc. Int. Symp. Electromagnetic Compatibility*, pp. 227-231. Aug. 1994.

[4] Yi-Bin Hsieh, Yao-Huang Kao, "A New Spread Spectrum Clock Generator for SATA Using Double Modulation Schemes" in *Proc. IEEE Custom Integrated Circuits Conference CICC '07*, pp. 297-300. Sept. 2007.

[5] M. Kokubo, *et al.*, "Spread-Spectrum Clock Generator for Serial ATA with Multi-Bit $\Sigma\Delta$ Modulator-controlled Fractional PLL" in *IEICE Tarns. Electron.*, vol E89-C, no.11. Nov. 2006.

[6] H.R. Lee, Ook Kim, Gijung Ahn, D.K. Jeong, "A low-jitter 5000ppm spread spectrum clock generator for multi-channel SATA transceiver in 0.18μm CMOS" in *IEEE Solid-State Circuits Conf. Dig. of Tech. Papers ISSCC '05*, vol 162-163. Feb. 2005.

[7] W.T. Chen, J.C. Hsu, H.W. Lune, C.C. Su "A spread spectrum clock generator for SATA-II" in *Proc. IEEE Int. Symp. Circuits and Systems ISCAS '05*, vol. 3, pp. 2643-2646, May 2005.

[8] G. Setti, G. Mazzini, R. Rovatti, and S. Callegari, "Statistical modeling of discrete time chaotic processes: Basic finite dimensional tools and applications", in *Proc. of IEEE*, vol. 90, no. 5, pp. 662-690. May 2002.

[9] R.E. Best, "Phase-locked Loops: Design, Simulation and Applications", 5th edition, McGraw-Hill. 2003.

[10] B. Razavi, *Principles of Data Conversion System Design*, Wiley-IEEE Press, November 1994.

[11] F. Pareschi, G. Setti, R. Rovatti, "A Fast Chaos-based True Random Number Generator for Cryptographic Applications", in *Proc. IEEE European Solid-State circ. conf. ESSCIRC '06*, pp 130-133. Sept. 2006.

[12] A. Zanchi, A. Bonfanti, S. Levantino, C. Samori, "General SSCR vs. Cycle-to-Cycle Jitter Relationship with Application to Phase Noise in PLL", in *Proc. IEEE Southwest Symph. Mixed Signal Design SSMSD '01*, pp 130-133. Feb. 2001.

A 1.5 GHz Spread Spectrum Clock Generator with a 5000ppm Piecewise Linear Modulation

Minyoung Song, Sunghoon Ahn, Inhwa Jung, Yongtae Kim, Chulwoo Kim

Korea University

Abstract- **A spread spectrum clock generator is implemented in a 0.18μm CMOS process employing the proposed piecewise linear modulation profile to significantly reduce EMI with a simple implementation. A high resolution fractional divider to reduce quantization noise from the modulation is proposed as well. A peak power reduction level of 14.2dB with 5000ppm down spreading and 27.88ps$_{pp}$ of jitter in the SSCG without modulation are measured.**

I. INTRODUCTION

Reducing Electromagnetic Interface (EMI) in digital chips has become one of the most important issues as the degree of integration and the clock speed increase to enhance chip performance. External filters can be used to increase the rise and fall times of the clock but this is not cost effective. Low voltage differential clocking can be applied but at the cost of level conversion and complex routing. A simple and cost effective solution to reducing EMI is to use a spread spectrum clock generator (SSCG), which is a PLL-based clock generator with an output frequency modulation. The amount of EMI reduction depends on the frequency modulation profile. The Hershey-Kiss modulation achieves high EMI reduction, but implementation is complicated due to the modulating signal's non-linearity [1]. On the other hand, triangular modulation has the advantage of simple implementation, but it is not as effective as Hershey-Kiss modulation in EMI reduction. The proposed piecewise linear (PWL) modulation profile consists of linear signals for simple implementation and it maintains high EMI reduction. One solution for frequency modulation in SSCGs is to directly modulate the VCO control voltage using a programmable charge pump [2]. However, this technique suffers from PVT variations and it generates additional jitter. Another scheme is to modulate the division ratio of the programmable feedback divider using a ΔΣ modulator [3-5]. However, the PLL bandwidth must be narrow enough to filter out the jitter generated by the ΔΣ modulator. This paper describes a piecewise-linear modulated SSCG with the proposed fractional divider (FDIV) for accurate frequency control.

II. CIRCUIT DESIGN

A. The Piecewise Linear (PWL) Modulation

Fig. 1(left) shows the conventional triangular modulation profile that is widely used due to its simplicity in implementation. This modulation profile can be generated with a few standard digital cells such as accumulators. The disadvantage of the triangular modulation profile is the loss in EMI reduction. Large peaks occur at the edges of the spectrum within the modulated bandwidth. The well-known Hershey-Kiss modulation profile and its spectrum are shown in Fig.

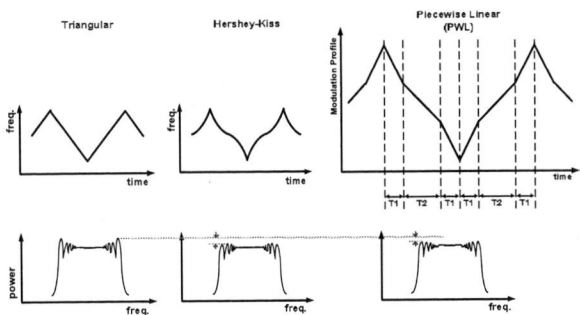

Fig.1. Conventional (triangular, Hershey-Kiss) and proposed modulation profiles.

1(center). The spectrum peaks are reduced due to the increased signal slopes at the maximum and minimum points of the modulation profile as well as the decreased signal slope at the zero crossing point of the modulation profile [1]. However, the modulation profile itself is hard to implement and complicated circuits such as read only memories (ROMs) are required.

To overcome the EMI reduction loss in triangular modulation while maintaining its simple implementation, the piecewise linear modulation profile is proposed. As shown in Fig. 1(right), the proposed PWL modulation profile consists of linear signals that have different slopes. By increasing the slopes at the maximum and minimum points, higher EMI reduction can be achieved. Implementation is also simple since the PWL modulation profile can be generated using an accumulator with a simple α-controller as shown in Fig. 2. There is a trade-off between EMI reduction and simplicity. If the number of piecewise linear sections is increased, higher EMI reduction can be achieved at the cost of increased α-controller complexity.

Fig. 2 shows the block diagram of the proposed PWL

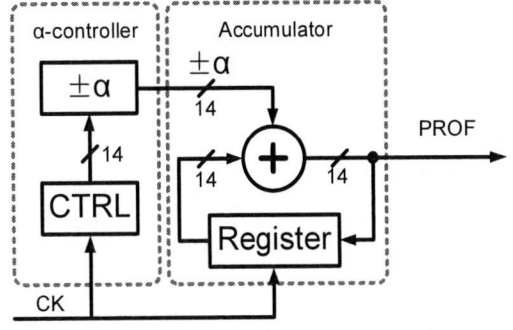

Fig.2. Block diagram of the proposed piecewise linear (PWL) modulation profile generator.

978-1-4244-2018-6/08 $25.00 © 2008 IEEE

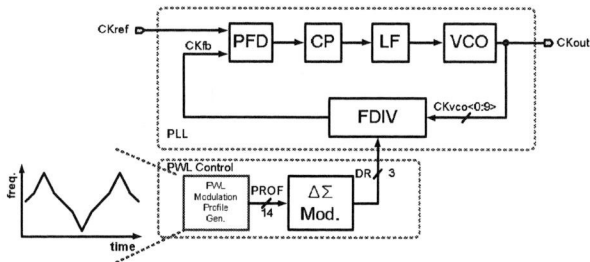

Fig.3. Block diagram of the proposed SSCG.

modulation profile generator. It consists of an α-controller and an accumulator (ACC). The α-controller is used for generating an add value (±α) for the ACC, which varies the slope of the modulation profile. The α-controller is synchronized with the reference clock of the PLL to easily generate the modulation frequency. The add value is generated using a reference clock that is divided several times. This value is changed periodically for the proposed modulation profile. The ACC fed α generates the proposed modulation profile.

B. Overall Architecture

Fig. 3 shows the block diagram of the proposed SSCG. It consists of a PLL loop with a 25MHz reference clock for frequency synthesis and a frequency modulator for PWL modulation generation.

The profile generator generates the proposed PWL modulation profile and it is modulated by a ΔΣ modulator. By using a ΔΣ modulator, the number of control bits can be reduced while the quantization noise is pushed into the high frequency band. There is a trade-off between its order and the stability of the ΔΣ modulator. A higher order ΔΣ modulator shapes much more of the quantization noise, but it is hard to compose a stable loop. A MASH type ΔΣ modulator can be used to overcome this trade-off [6]. It is a cascade system that consists of several low order ΔΣ modulators. An M•N order MASH ΔΣ Modulator is obtained by using M cascaded Nth order ΔΣ modulators. In analog design such as data converters, it is hard to cancel the quantization noise of each stage, which degrades the noise shaping of the ΔΣ modulator. However, in a digital ΔΣ modulator, the quantization noise is canceled by simple digital logic. Fig. 4 shows the block diagram of a 3rd-order MASH 1-1-1 ΔΣ Modulator that is used in the proposed SSCG. The division ratio of the feedback divider is controlled

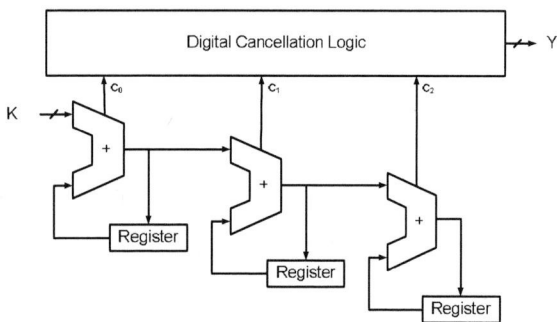

Fig.4. Block diagram of 3rd-order MASH 1-1-1 ΔΣ Modulator.

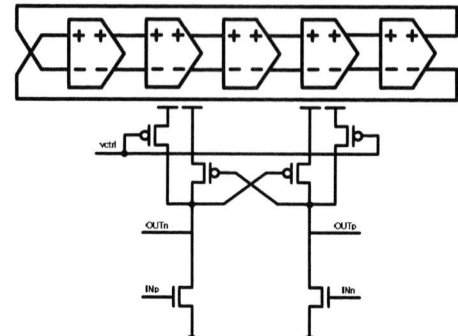

Fig.5. Ring type VCO (top) and differential positive feedback delay cell (bottom) [8].

from 60 to 59.7 and the output clock frequency varies at a rate of 33 kHz from 1.5 GHz to 1.4925 GHz correspondingly to produce 5000ppm of down spreading.

C. The Low-jitter PLL

5000ppm is an appropriate range for the modulating frequency. A low-jitter PLL is required for accurate clock spectrum spreading and the PLL is shown in Fig. 3. It comprises a phase-frequency detector (PFD), a charge pump (CP), a loop filter (LF), a voltage-controlled oscillator (VCO), and a fractional divider (FDIV). Jitter in the PLL can be generated by PFD static phase error, VCO noise, CP current mismatch, leakage current in the LF, and skew in the FDIV. The PFD generates a phase difference with a static error that increases the PLL jitter. To reduce the static phase error, a PFD that has a short reset path delay such as a TSPC-based PFD can be applied [7]. VCO noise is the main source of the PLL jitter. The VCO generates the PLL output directly, so VCO noise is also passed through the PLL output. Hence, an LC type of VCO can obtain a significant phase noise reduction, but it is hard to generate multi-phase outputs when using this kind of oscillator.

A ring type of VCO generates multi-phase outputs by adding delay cells [8]. However, it has a great deal of phase noise compared to an LC oscillator, so a low noise-sensitive delay cell using the ring type of VCO should be considered. Fig. 5 shows a differential delay cell with positive feedback that has low sensitivity to noise. The current mismatch between the charging and discharging current in the CP causes the phase mismatch to increase. To overcome this, an offset-cancelled charge pump [9] can be applied as shown in Fig. 6.

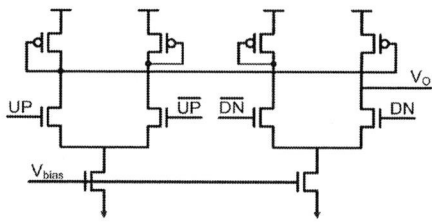

Fig.6. Offset-cancelled charge pump (CP) schematic [9].

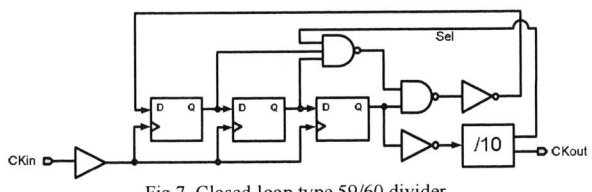

Fig.7. Closed-loop type 59/60 divider.

It has only NMOS transistors for equal current switching. Both charging and discharging currents are the same by virtue of current mirroring. Leakage current in the LF also contributes to this mismatch especially if the update period is quite long. Either using low-leakage capacitors or updating the state of the PLL faster can reduce the effect of the leakage current in the LF. MIM capacitors instead of MOS capacitors are used in the proposed SSCG due to their low leakage current although MIM capacitors occupy a large area and have lower capacitances than MOS capacitors. The reference clock's 25MHz update rate is high enough to reduce the leakage effect.

The high speed reference clock allows a higher PLL bandwidth and the capacitance in the LF can be decreased, so the LF occupies a smaller area even when using a MIM capacitor. Skew in the FDIV also generates phase errors to the PFD because skew in the FDIV generates phase offset and thus static jitter. Furthermore, because the FDIV has a high division ratio, the VCO phase error accumulates in proportion to the division ratio and this increases the phase error.

D. The Fractional Divider

Because the input signals of the FDIV are high-speed multi-phase signals, the main clock path should use a simple division ratio that is easily controlled. Feedback control systems are commonly used to design this feedback divider for simplicity as shown in Fig. 7. However, the main clock path is relatively long and the division ratio controlling it is hard to achieve in this closed-loop system. As shown in Fig. 7, the main clock path is composed of several gates, which limits the input clock speed.

To overcome this problem, an open-loop fractional divider is introduced. The division ratio of the divider can be controlled with ease since the main clock is decoupled from the division procedure. The open-loop fractional divider can

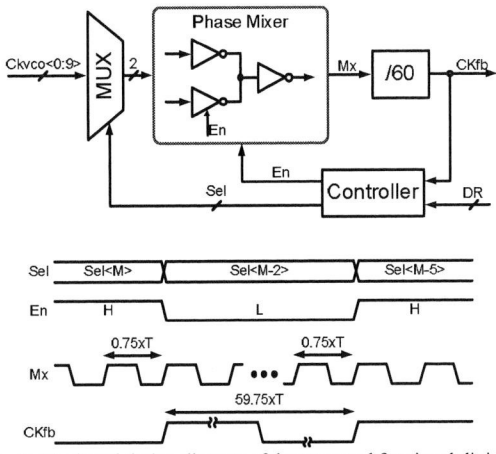

Fig.9. Block and timing diagram of the proposed fractional divider.

be implemented with a phase selecting division technique [10]. The main clock path with multiplexing can be shorter than the closed-loop type. In the proposed SSCG, a fractional divider with a phase selecting technique is used.

A fine resolution feedback divider is also required for low jitter in the PLL loop. Low jitter is achieved through reduction in the magnitude of the quantization error. As more $\Delta\Sigma$ modulator output bits are used, more quantization noise is shaped. To increase the number of $\Delta\Sigma$ modulator output bits, a high-resolution fractional divider is required. Fine resolution also increases the code efficiency of the modulation profile. To control the division ratio from 60 to 59.7 using a 60-59 dual modulus divider, the modulation profile uses only 30% of the full code length. On the other hand, the code efficiency can be increased to 100% using a 60-59.7 fractional divider. A division ratio of 59.7 can be achieved by shifting 3 of the 10 multiphase clocks [11] as shown in Fig. 8.

Counting 59 successive input clock periods and one with a 0.7 input clock period (3 phase shifted) leads to a division ratio of 59.7. The resolution can be reduced to 0.1 with 10 multiphase clocks. The number of multiphase clocks must be increased to achieve a finer resolution, but the VCO frequency is a limiting factor. The proposed fractional divider (FDIV) that is used as the feedback divider in the PLL loop is shown in the top of Fig. 9. A phase mixer is introduced in the proposed FDIV. By mixing two clock phases, an intermediate

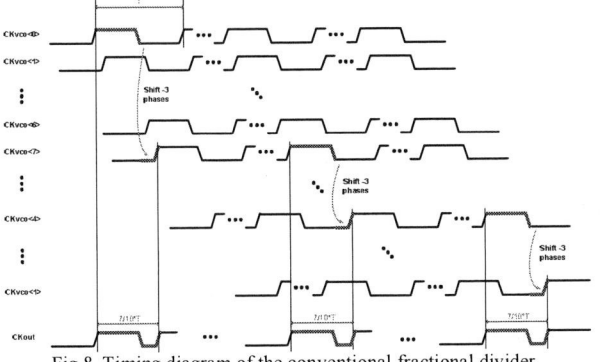

Fig.8. Timing diagram of the conventional fractional divider.

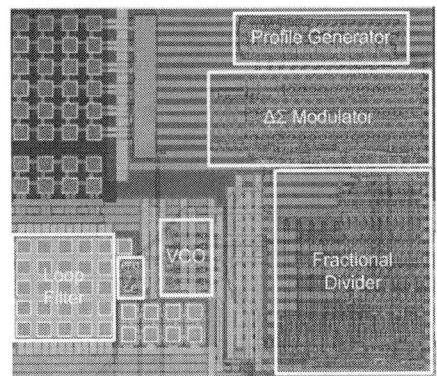

Fig.10. Chip microphotograph.

978-1-4244-2018-6/08 $25.00 © 2008 IEEE

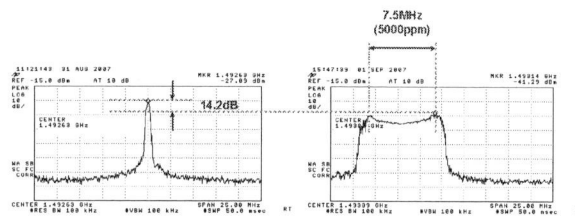

Fig.11. The measured spectrum of SSCG w/ and w/o modulation.

clock phase can be generated. Using 20 effective clock phases generated by the VCO and the phase mixer, a fine resolution of 0.05 can be achieved. For example, periodically jumping 5 of the 20 multiphase clocks every 60 successive clock cycles leads to division ratio of 59.75 as shown in the bottom of Fig. 9. The division ratio of the proposed FDIV can be controlled from 60 to 59.7 with a resolution of 0.05.

III. EXPERIMENTAL RESULTS

Fig. 10 shows the micrograph of the 0.49mm^2 chip fabricated in a 0.18μm CMOS process. Fig. 11 shows the spectrum results of the proposed SSCG. The resolution bandwidth and video bandwidth are 100 kHz, while the frequency span is 25MHz. When it is not in the spread-spectrum mode, we can see a single 1.5GHz tone. In the spread-spectrum mode, the single tone is spread with the piecewise linear frequency modulation. A peak power reduction level of 14.2dB and 5000ppm (7.5MHz) of spreading are measured. The Peak-to-peak and rms jitters of the SSCG without modulation are 27.88ps$_{pp}$ and 3.77ps$_{rms}$, respectively as shown in Fig. 12. A performance comparison of the proposed SSCG and conventional schemes is summarized in Table I. The supply voltage is 1.8V and the power consumption of the SSCG is 40mW with a 1.5GHz output.

IV. CONCLUSION

By virtue of the proposed piecewise modulation profile, a significant area savings is achieved without using the ROM that is required for the Hershey-Kiss modulation profile, and a dramatic EMI reduction is achieved compared to the triangular modulation profile. Furthermore, the proposed fractional

TABLE I
PERFORMANCE COMPARISON OF SSCGS FOR SERIAL ATA

	Sugawara[3] SOVC 02	Aoyama[4] SOVC 03	Kokubo[5] ISSCC 05	Lee[11] ISSCC 05	This Work
Modulation Profile	Triangular	Triangular	Triangular	Triangular	Piecewise Linear
Modulation Bandwidth	-5150~350ppm 30.3 ~ 33.3us	-5000 ~ 0ppm 32us	-5000 ~ 0ppm 32us	-5000 ~ 0ppm 33us	-5000 ~ 0ppm 32us
Modulation Method	Phase Interpolation	Phase Interpolation	Delta-Sigma	Delta-Sigma	Delta-Sigma
Peak Reduction	7dB	5.43dB	10dB[12]	9.8dB	14.2dB
Output Frequency	1.5GHz	1.5GHz	1.5GHz	1.5GHz	1.5GHz
Peak to peak jitter*	57.2ps	-	-	41.01ps	27.88ps
RMS jitter*	9.3ps	-	-	3.07ps	3.77ps
Process	0.13um	0.15um	0.15um	0.18um	0.18um
Area	-	-	0.42mm^2	0.31mm^2	0.49mm^2
Power	-	-	54mW	-	40mW

*A jitter w/o spread-spectrum mode

divider increased the division resolution without increasing the number of VCO delay cells, which reduces the quantization noise and in turn is useful for piecewise modulation. Using the proposed scheme, a 1.5GHz 5000ppm down spread SSCG with 14.2dB EMI reduction is achieved for serial ATA and the proposed scheme can be easily applied to other applications.

ACKNOWLEDGMENTS

This work was supported by "System IC 2010" project of Korea Ministry of Knowledge Economy and the fabrication was supported by IDEC.

REFERENCES

[1] K.B. Hardin, J. T. Fessler, D.R. Bush, "Spread Spectrum Clock Generation for the Reduction of Radiated Emissions," *IEEE Intl. Symp. on Electromagnetic Compatibility*, pp. 227-231, Aug., 1994.

[2] H.H. Chang *et al.*, "A Spread-Spectrum Clock Generator with Triangular Modulation," *IEEE J. Solid-State Circuits*, vol.38, no.4, pp. 673-676, Apr. 2003.

[3] M. Sugawara *et al.*, "1.5 Gbps, 5150ppm Spread Spectrum SerDes PHY with a 0.3mW, 1.5 Gbps Level Detector for Serial ATA," *IEEE Symp. VLSI Circuits*, pp. 60-63, 2002.

[4] M. Aoyama *et al.*, "3Gbps, 5000ppm Spread Spectrum SerDes PHY with frequency tracking Phase Interpolator for Serial ATA," *IEEE Symp. VLSI Circuits*, pp. 107-110, 2003.

[5] M. Kokubo *et al.*, "Spread-Spectrum Clock Generator for Serial ATA using Fractional PLL Controlled by ΔΣ Modulator with Level Shifter," *ISSCC Dig. Of Tech. Papers*, pp. 160-161, 2005.

[6] T. Hayashi, Y. Inabe, K. Uchimura and A. Iwata, "A Multistage Delta-sigma Modulator without Double Integration Loop," *ISSCC Dig. Of Tech. Papers*, pp. 182-183, 1986.

[7] S. Kim, K. Lee, Y. Moon, D.K. Jeong and H.K. Lim, "A 960-Mb/s/pin Interface for Skew-Tolerant Bus Using Low Jitter PLL," *IEEE J. Solid-State Circuits*, vol.32, no.5, pp. 691-700, May. 1997.

[8] J. Lee, B. Kim, "A 960-Mb/s/pin Interface for Skew-Tolerant Bus Using Low Jitter PLL," *IEEE J. Solid-State Circuits*, vol.35, no.8, pp. 1137-1145, Aug. 2000.

[9] J. Maneatis, "Low-Jitter Process-Independent DLL and PLL Based on Self-Biased Techniques," *IEEE J. Solid-State Circuits*, vol.31, no.11, pp. 1723-1732, Nov. 1996.

[10] J. Craninckx *et al*, "A 1.75-GHz/3-V Dual-Modulus Divide-by-128/129 Prescaler in 0.7-μm CMOS," *IEEE J. Solid-State Circuits*, vol.31, no.7, pp. 890-897, Jul. 1996.

[11] H. R. Lee O. Kim, G. Ahn, D.-K. Jung, "A Low-Jitter 5000ppm Spread Spectrum Clock Generator for Multi-Channel SATA Transceiver in 0.18 μ m," *ISSCC Dig. Of Tech. Papers*, pp. 162-163, 2005.

[12] M. Kokubo *et al.*, "Spread-Spectrum Clock Generator for Serial ATA with Multi-Bit ΔΣ Modulator-Controlled Fractional PLL," *IEICE Trans. Electron*, vol.E89-C, no.11, pp. 1682-1687, Nov. 2006.

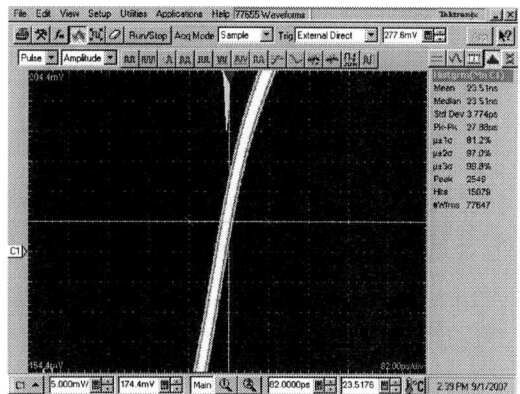

Fig.12. Jitter histogram of the SSCG without modulation.

IEEE 2008 Custom Intergrated Circuits Conference (CICC)

A 8x5 Gb/s Source-Synchronous Receiver with Clock Generator Phase Error Correction

Ankur Agrawal, Pavan Kumar Hanumolu* and Gu-Yeon Wei

Harvard University, Cambridge, MA 02138, *Oregon State University, Corvallis, OR 97331

Abstract— **This paper describes the design and implementation of a 8x5Gb/s source-synchronous receiver in a 0.13μm CMOS technology. The receiver employs a cascaded-DLL architecture that avoids filtering of the jitter on the received clock to enhance jitter tolerance bandwidth. A technique is proposed to correct phase spacing mimatch in DLLs that reduces the error standard deviations by more than 40% and improves receiver timing margins.**

I. INTRODUCTION

The need for high I/O bandwidth in multi-chip digital systems has led to the widespread use of parallel links. These links are generally source synchronous, with a clock sent along with the data signals for receiver timing recovery. As data rates increase, successful data recovery in the presence of jitter requires precise positioning of the sampling clock. Receivers need to perform per-pin skew compensation [1] while preserving the correlation in the jitter between the transmitted clock and data.

Source-synchronous receivers often use multi-phase clock generators to drive phase interpolators [2]. Multiple clock phases are also required when interleaved samplers are employed to easily accomodate high off-chip data-rates. Phase locked loops (PLL) or delay locked loops (DLL) can be used to generate multi-phase clocks. While the phase filtering action of a PLL reduces the jitter correlation between the incoming clock and data, DLLs are susceptible to systematic and random phase offsets and mismatch that can significantly reduce timing margins and degrade achievable data rates. If these phase errors can be corrected, DLLs are a better choice than PLLs for multi-phase clock generation in source-synchronous receivers.

This paper presents a cascaded-DLL architecture for receivers that avoids any phase filtering in the path of the received clock and incorporates techniques to correct for phase spacing errors in DLLs. It requires neither phase interpolators nor the distribution of multi-phase clocks over long on-chip wires. The next section describes the trade-offs between using DLLs and PLLs in source-synchronous receivers.

II. SOURCE SYNCHRONOUS RECEIVER DESIGN CONSIDERATIONS

Fig. 1 shows the architecture of a general source-synchronous transceiver. The transmitter sends a parallel word of data along with a clock to the receiver. To save clock power and avoid jitter amplification, often the frequency of the transmitted clock is stepped down and a multiplying PLL

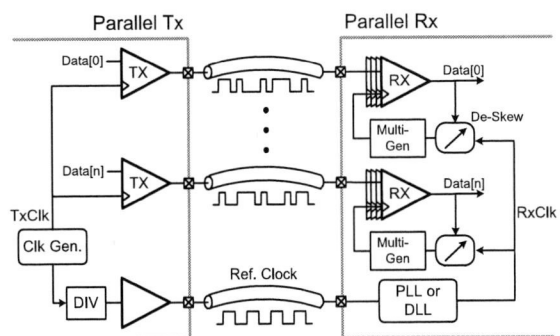

Fig. 1. General Source Synchronous Transceiver

or DLL is used in the receiver to step the frequency back up. This clock is then distributed to each of the receiver slices using either clock buffers or passive distribution [2], [3]. The receiver slices need to perform skew compensation to correct for flight time variations over the PCB traces. If multiple data bits are transmitted in each RxClk cycle, the receivers also need to generate multiple clock phases to sample the incoming data.

A single PLL can be used for multi-phase clock generation and clock de-skew [4]. Alternatively, a combination of a DLL and a phase interpolator can be used to perform the two tasks independent of each other. PLLs have a low-pass transfer function from the phase of the reference clock to the output clock and, thus, filter out the middle and higher frequency jitter on the received clock. On the other hand, DLLs have a nearly all-pass phase transfer function, and are able to preserve the correlation between the jitter on the incoming data and received clock, resulting in good jitter tolerance over a wide frequency range. Recently reported designs [2], [5] have avoided the use of PLLs in the path of the received clock to achieve wide jitter tracking bandwidth.

Our test chip, the details of which shall be discussed in the following sections, enables a direct comparison between the jitter tracking bandwidths of PLL and DLL based timing recovery. Fig. 2 plots the jitter tolerance curves for BER < 10^{-9} for 2 different configurations of the test chip. The "DLL-only" case is the typical configuration of the test chip (as described in Section III), where a quarter rate clock is directly fed to the local recievers. In the "PLL/DLL" case, a sub-rate clock is first multiplied on-chip using a PLL to the desired frequency. The jitter tolerance bandwidth for the DLL-only case is >100 MHz and is limited by the difference in the on-chip path length between the data and clock signals. In

978-1-4244-2018-6/08 $25.00 © 2008 IEEE

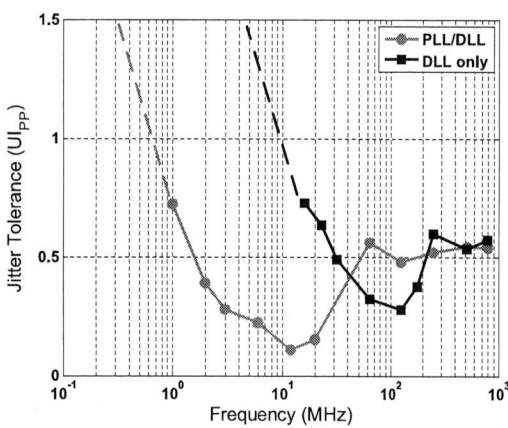

Fig. 2. Jitter Tolerance Plots for the DLL only case and PLL/DLL case measured at 3.2 Gb/s.

constrast, a jitter tolerance bandwidth of around 10 MHz is obtained for the PLL/DLL case. The PLL bandwidth limits the maximum achievable jitter tolerance bandwidth in this case. Due to test equipment limitations, a maximum jitter of 0.72 UI can be introduced. These measurements confirm that phase filtering should be avoided in the path of the received clock in source-synchronous receivers to improve jitter tracking.

However, there are some caveats to using DLLs: (1) A small difference between the charge pump up and down currents can result in static phase spacing error at the end of the delay line in a DLL, while only causing a slight increase in clock jitter in a PLL. (2) DLLs that lock the delay of the delay line to half a reference clock cycle are very sensitive to the duty cycle of the reference clock signal. (3) The shape of the reference clock entering the delay line can exacerbate phase spacing mismatches in DLLs. By carefully addressing each of these issues, DLL-based timing recovery can be made better suited to high-speed source synchronous receivers.

The first two issues listed above are well known and solutions have been proposed in literature [6]. The last one is more subtle. The shape of the clock, characterized by its voltage swing and the slope of its edges, entering the first delay cell can be different from those entering subsequent stages. This results in adjacent delay stages having different delays. Preshaping the clock signal by adding identical delay cells before the delay line helps, but does not eliminate the problem. The optimal clock shape is one that would be produced by a voltage-controlled oscillator composed of the same delay cells and having the same control voltage as the delay line. Fig. 3 shows the simulated nonlinearity in the phase spacings for a DLL consisting of a 4 stage differential delay line after preshaping using an identical delay line. It also shows the simulated non-linearity achieved after a technique to shape the reference clock signal is used. This technique, which we call phase spacing error correction (PSEC), is described in greater detail in Section IV. In the next section the receiver architecture is described that is able to easily incorporate techniques to correct for the phase spacing errors mentioned above.

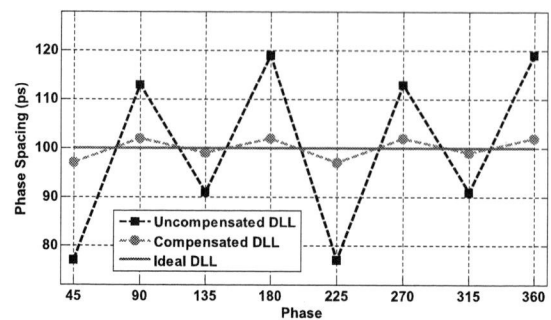

Fig. 3. Phase-spacing error caused by the shape of reference clock.

III. SYSTEM ARCHITECTURE

Fig. 4 presents details of the receiver architecture. A quarter rate clock is boosted to on-chip voltage levels and distributed to individual receiver slices using simple clock buffers. Each receiver slice consists of 2 DLLs, samplers, skew-compensation logic and phase spacing error correction circuits. The global RxClk first goes through a duty cycle corrector (DCC) to compensate for any distortion from the clock distribution buffers. The first DLL, called the phase-deskew DLL, adds a variable amount of delay to the clock path such that its output clock is phase aligned to the incoming data. The second DLL produces the 8 clock phases that drive 8 interleaved edge and data samplers. The phase detector generates early/late information that is forwarded to a digital filter. The digital filter controls a DAC that adds offsets between the up and down currents of the charge pump of the phase-deskew DLL that results in skewing in output clock. The tuning range of the de-skew circuit is greater than ± 0.5 UI. Also shown in the figure is the phase spacing error correction loop that balances the delays of the delay cells by adjusting the slope of the clock edges. Additional details of the receiver slice can be found in [7].

The cascaded DLL architecture is chosen for a number of reasons. It provides a simple mechanism to de-skew the global RxClk. No extra preshaping delay buffers are required in the clock path before the DLL used for multiphase clock generation. It also provides a convenient location for placing the duty cycle correction circuit. The duty cycle of the clock entering the second DLL is sensed and tuned to 50%, but the correction circuit is placed before the first DLL in order to not disturb the shape of the clock signal for the second DLL. The offsets in the phase-deskew DLL do not matter as they get absorbed into the overall delay of its delay line.

IV. CIRCUIT DESIGN

A. DLL

Fig. 5(a) shows the details of the DLL. The use of an active loop filter offers two advantages. It provides a 2nd-order low pass transfer function from I_{CP} to Φ_{out} that is able to filter out the quantization noise in the control bits to the DAC. The feedback amplifier also biases the output node of the charge pump to V_{REF}, irrespective of the delay of the delay line. By

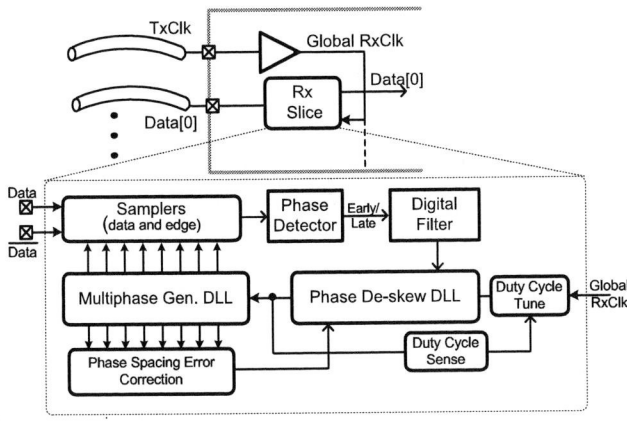

Fig. 4. Proposed receiver block diagram

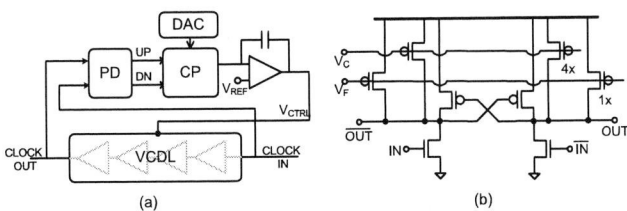

Fig. 5. (a) DLL block diagram, (b) Delay cell schematic

adjusting V_{REF}, small mismatches in the charge pump up and down currents can be compensated. The delay cell, that uses pseudo differential inverters with rail-to-rail swing, is shown in Fig. 5(b). An external coarse voltage (V_C) is used to bring the delay line to within the lock range of the DLL.

B. Phase Spacing Error Corrector (PSEC)

As described in Section II, the shape of the clock signal entering the DLL affects the delays through the delay cells. To correct for this effect, the rise and fall times of the input clock to the second DLL are adjusted by sinking or sourcing small amounts of current from the penultimate delay stage of the first DLL. A feed-back loop, shown in Fig. 6, consists of XOR based delay detectors, charge pump, loop filter and a V/I converter. As the error is such that all odd delay stages are either faster or slower than all even delay stages, the charge pump dumps or removes charge proportional to the difference between the odd and even delays. Multiple error detectors are used to average out the effects of random mismatches between the adjacent stages. The charge pump employs feedback biasing to ensure that static and dynamic mismatches in the up and down currents do not introduce offsets that can result in residual delay mismatches. The bandwidth of this loop is set such that there is no conflict with the de-skew loop.

Simulation results (Fig. 3) indicate that the RMS DNL reduces from 16.9 ps to 2.1 ps. The efficacy of this method is limited by random mismatches in the delay cells and XOR gates.

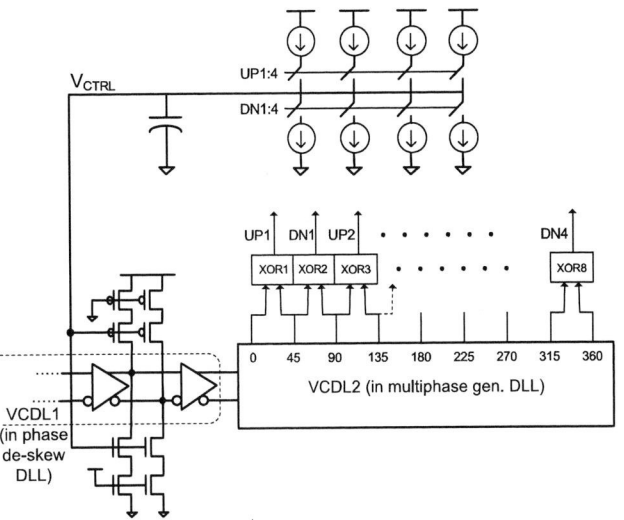

Fig. 6. Phase Spacing Error Corrector Loop

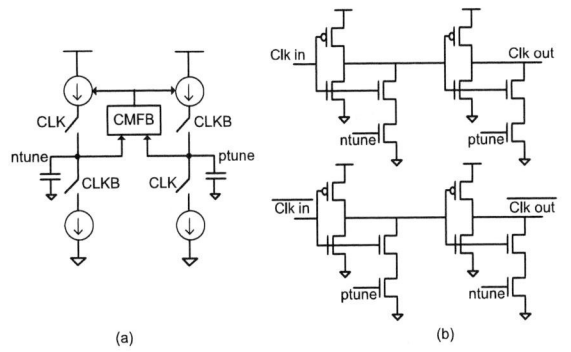

Fig. 7. (a) Duty-cycle sensor (DC_sense), (b) Duty-cycle corrector (DC_tune)

C. Duty Cycle Corrector

The duty cycle corrector (DCC) shown in Fig. 7, consists of a duty cycle sensing block (DC_sense) placed before the second DLL, and a duty cycle tuning block (DC_tune) placed before the first DLL.

The DC_sense circuit is a differential charge-pump that produces two voltages, *ntune* and *ptune* depending on the magnitude of duty cycle distortion in the clock signal. Common-mode feedback ensures *ntune* and *ptune* are differntial with a common-mode of $V_{DD}/2$. The DC_tune circuit, similar to that in [6], uses these voltages to slow down or speed up one of the clock edges. It consists of 2 tuning stages, where each stage consists of an inverter with weak pull-down and a parallel pull-down path whose strength depends on *ntune* or *ptune*. Consequently, each stage can adjust the falling transition of the clock signal. Simulation results show that the circuit suppresses duty cycle distortion by more than 10x.

V. EXPERIMENTAL RESULTS

A test chip fabricated in a $0.13\mu m$ CMOS logic process operates off a 1.2V supply voltage. To verify the efficacy of the peripheral correction loops, Fig. 8 plots the measured

978-1-4244-2018-6/08 $25.00 © 2008 IEEE

phase spacings with the PSEC loop turned on and off for 2 representative DLLs in our parallel receiver. The reference clock frequency is 1.25GHz and the nominal delay of each stage is 100ps. A reduction in the differential nonlinearity (DNL) is observed. The residual error can be attributed to random delay cell mismatches that our correction loops cannot correct. In Fig. 9 we plot the rms value (standard deviation) of the DNL in phase spacings for 7 DLLs from 7 receiver slices, with the PSEC loop enabled and disabled. The 8th channel is inaccessible due to insufficient connector spacing on the board. On average, the RMS DNL reduces by 42%.

Fig. 10 presents the integral nonlinearity (INL) in the phase spacings with the DCC loop turned on and off. INL is defined as the deviation from the ideal location of the clock edge. We observe a reduction in the INL in phase spacings from 25.8ps to 7.8ps. The dotted line shows the theoretical INL for a 10% duty cycle error. Phase spacing measurements were made by sweeping a 1010... data pattern at 5 Gb/s across half a reference clock cycle and observing sampler outputs.

The 8x5Gb/s receiver consumes a total power of 258mW (including all output drivers), which translates to an efficiency better than 6.45mW/Gbps. Each receiver slice occupies an area of 350μm x 450μm. A die micrograph is shown in Fig. 11 with floorplan overlays.

VI. CONCLUSION

This paper describes a 8x5Gb/s source-synchronous receiver that avoids any phase filtering in the path of the received clock. Although source-synchronous receivers prefer DLLs over PLLs to enhance jitter tolerance bandwidth, DLL based receivers can suffer from timing margin degradation due to phase spacing mismatches. Techniques to reclaim the lost timing margin are proposed and experimentally verified.

VII. ACKNOWLEDGMENTS

The authors would like to thank UMC for chip fabrication.

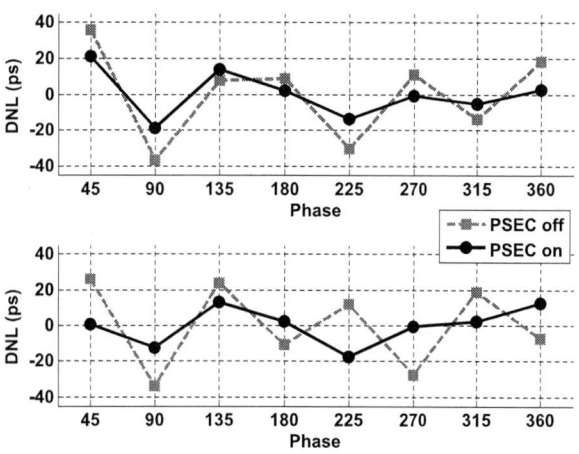

Fig. 8. Differential non-linearity in phase spacings with the PSEC loop disabled and enabled for 2 representative DLLs

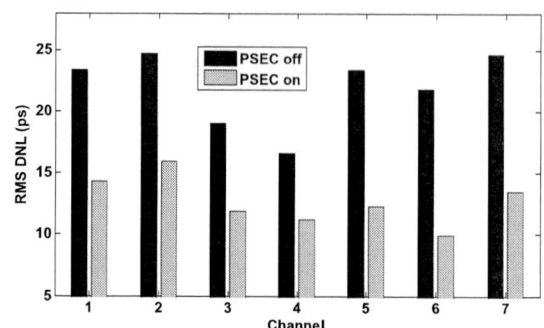

Fig. 9. Comparison of the RMS DNLs for 7 DLLs with the PSEC loop disabled and enabled

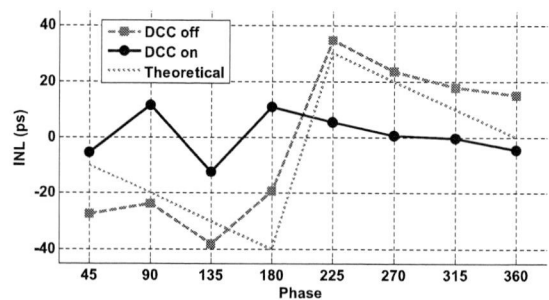

Fig. 10. Integral non-linearity for the phase spacings with the duty-cycle correction loop disabled and enabled

REFERENCES

[1] E. Yeung and MA Horowitz, "A 2.4 Gb/s/pin simultaneous bidirectional parallel link with per-pin skew compensation," *Solid-State Circuits, IEEE Journal of*, vol. 35, no. 11, pp. 1619–1628, 2000.

[2] B. Casper et al., "A 20Gb/s Forward Clock Transceiver in 90nm CMOS," *ISSCC Dig. Tech. Papers*, pp. 90–1, 2006.

[3] K. Chang, S. Pamarti, K. Kaviani, E. Alon, X. Shi, TJ Chin, J. Shen, G. Yip, C. Madden, R. Schmitt, et al., "Clocking and circuit design for a parallel I/O on a first-generation CELL processor," *ISSCC Dig. Tech. Papers*, pp. 526–615, 2005.

[4] P. Larsson, "A 2-1600-MHz CMOS clock recovery PLL with low-Vdd capability," *Solid-State Circuits, IEEE Journal of*, vol. 34, no. 12, pp. 1951–1960, 1999.

[5] F. O'Mahony et al., "A 27Gb/s Forwarded-Clock I/O Receiver using an Injection-Locked in 90nm CMOS," *ISSCC Dig. Tech. Papers*, pp. 452–453, 2008.

[6] A.H. Tan and G-Y. Wei, "Phase Mismatch Detection and Compensation for PLL/DLL Based Multi-Phase Clock Generator," *Conference 2006, IEEE Custom Integrated Circuits*, pp. 417–420, 2006.

[7] A. Agrawal, P.K. Hanumolu, and G-Y. Wei, "A 8x3.2Gb/s Parallel Receiver with Collaborative Timing Recovery," *ISSCC Dig. Tech. Papers*, pp. 468–469, 2008.

Fig. 11. Chip micrograph with floorplan overlays

Millimeter-Wave Circuit Techniques

Session 18

Chair: Payam Heydari, University of California, Irvine
Co-Chair: Nobuyuki Itoh, Toshiba Corporation

The availability of vastly unutilized spectra in millimeter-wave (MMW) frequency range together with huge interest in high data rate wireless communications and radar sensors has led to research efforts in designing millimeter-wave integrated circuits (MMW-ICs). On the other hand, the aggressive scaling of silicon technologies has created the possibility of developing MMW-ICs using these technologies. Designing ICs at such high frequencies close to unity-gain frequency of the transistor, however, face new challenges from device characterization and circuit design to measurement complexity.

The first paper in this session gives an overview of CMOS MMW-IC for Gigabit WPAN Applications. This invited paper compares the RF performance of the reported 60-GHz CMOS transceivers, and also discusses the architecture design trade-offs between performances and dc power consumption.

The second paper presents a direct conversion 60GHz transceiver in a 65nm CMOS process. The integrated transceiver employs direct BPSK modulation, 60GHz LO tree, static divider, and Gilbert-cell mixer. A transmit-receive link without ADCs and IQ-mixer was established and achieved 3.5Gb/s data-rate over 2m distance.

The third paper demonstrates design of a 24-GHz transceiver for automotive short-range radar. Designed and fabricated in a 130nm BiCMOS SiGe process, the protype chip achieves a 47-dB gain with 3-dB DSB noise figure, and a dynamic range of 31-dB in its receive path. In the transmit side, 1-ns pulses modulated at 24 GHz are obtained using an RF switch, which is able to deliver a continuous wave output power of 0dBm.

The fourth paper presents a fully differential receiver with on-chip baluns for 60GHz broadband wireless communications. This design consists of on-chip baluns, gm-boosted current-reuse LNA, sub-harmonic dual-gate down conversion mixer, second IF mixer and baseband gain stage. The 60GHz receiver front-end has been implemented in a standard 130nm RF CMOS technology.

The fifth paper presents a low power FSK MODEM in 130nm CMOS that can support over than 1 Gb/s data rate for high-speed multimedia data communications. Compared with open-loop direct modulation of a voltage-controlled oscillator (VCO), the direct injection-pulling of a VCO enables the high speed FSK modulation without suffering frequency drift problem.

The sixth paper presents the design and implemnetation of the first MMW dual-band 24/77GHz frequency synthesizer in a 180nm SiGe BiCMOS technology. All circuits except the VCOs are shared between the two bands. The synthesizer chip exhibits a locking range of 23.7-26.9GHz/75.65-78.6GHz with a power consumption of 95mW from 2.5/1.8V supplies. The phase noise at 1MHz offset from the carrier is less than 100dBc/Hz in both bands.

Finally, the seventh paper presents an X/Ku-band fine-tuning frequency synthesizer using a quadrature DDS implemented in a 180nm SiGe BiCMOS technology. The frequency synthesizer comprises a 9-bit quadrature DDS, an 11.7GHz quadrature VCO and image rejection mixers. The outputs of the quadrature DDS are downconverted to 9.4-11.7GHz and up-converted to 11.7-14.0GHz, respectively. The die area of the synthesizer is $3.0 \times 3.0 \text{mm}^2$ and the power consumption is 2.6W under a 3.3V supply.

978-1-4244-2018-6/08 $25.00 © 2008 IEEE

Notes

IEEE 2008 Custom Intergrated Circuits Conference (CICC)

Millimeter-wave CMOS Integrated Circuits for Gigabit WPAN Applications

Tian-Wei Huang and Huei Wang

(Invited Paper)

Dept. of Electrical Engineering and Graduate Institute of Communication Engineering
National Taiwan University, Taipei, Taiwan, 106, R.O.C

(E-mail: twhuang@cc.ntu.edu.tw, hueiwang@ntu.edu.tw)

Abstract — The 60-GHz gigabit wireless personal area network (WPAN) is an attractive application for next-generation dual-mode broadband wireless network. The CMOS technologies enable the low-cost manufacturability and system-on-chip (SoC) integration for 60-GHz low-power mobile devices. This paper compares the RF performance of the reported 60-GHz CMOS transceivers, and also discusses the architecture design trade-offs between performances and dc power consumption.

Index Terms — CMOS, Power Amplifier, Modulator, Demodulator, Millimeter-wave Transceiver

I. INTRODUCTION

Recently, the usage of dual-mode wireless broadband technologies, 3G and WLAN, is booming among the enterprise customers. 3G provides broadband download speeds in domestic metropolitan areas. On the other hand, to avoid high international 3G roaming charges, WLAN hotspots provide fast bi-directional connections everywhere else. The next wave of dual-mode broadband wireless is suggested in [1] as the combination of WLAN and gigabit wireless personal area network (WPAN). Gigabit fast file transfer or uncompressed HDTV video streaming need millimeter-wave (MMW) gigabit WPAN [2] as ad-hoc short-links, but the WLAN provides the infrastructure-mode with medium range coverage, as shown in Fig. 1. The IEEE 802.15 WPAN standard organization has started to draft its 60-GHz 2/4-Gbps WPAN standard within the task group 3c.

According to the usage model of MMW WPAN [3], there are two types of 60-GHz radio under development: i) One-way radio for video streaming (10 meter) or download applications (1 meter), and ii) Two-way radio for multi-gigabyte multi-media files transfer between portable devices (less than 1 meter), as shown in Fig. 2. Especially for the two-way radio applications, the ultra-low power consumption in RF/base-band technologies is required. Therefore, this paper will focus on the discussion of the trade-offs between RF performance and dc-power consumption in MMW transceiver designs.

Many 60-GHz transceiver building blocks have been implemented in silicon CMOS technologies [4]-[20] due to the low manufacturing cost. Currently, most existing 60-GHz CMOS RFICs are designed for one-way radio with two separate chips, one for transmitter and one for receiver. If a two-way TDD WPAN radio is desired, a 60-GHz low-loss switch is required between the transceiver and antenna. For higher level of integration, a single-chip 60-GHz 4-Gbps low-power transceiver with an on-chip SPDT switch has been successfully demonstrated in [7].

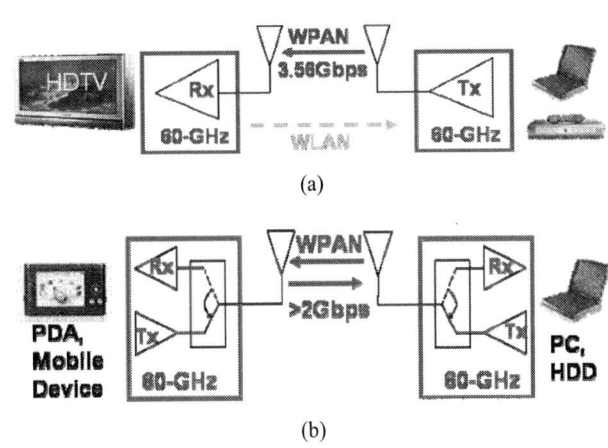

(a)

(b)

Fig. 2. (a) One-way 60-GHz radio for uncompressed HDTV video streaming, and (b) two-way 60-GHz radio for gigabyte multi-media file transfer.

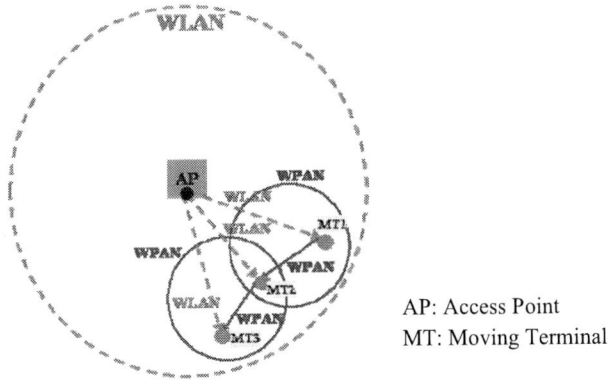

AP: Access Point
MT: Moving Terminal

Fig. 1. Dual-mode broadband and wireless network (DMBWN)

978-1-4244-2018-6/08 $25.00 © 2008 IEEE

II. LOW-POWER TRANSCEIVER BUILDING BLOCKS

To design a low-power MMW WPAN, the power-budget and gain-budget for different transceiver architectures are critical. For example, we need a power amplifier with adequate power gain, a modulator with high OP_{1dB}, and a demodulator with low LO power drive. Therefore, we will discuss the power-budget and gain-budget for the following building blocks in the transceiver.

A. Power Amplifier

Because of the low breakdown voltage of CMOS devices, and large metal/substrate loss at MMW frequencies, the CMOS power amplifier is the most challenging task among all the MMW transceiver circuits. Currently, the 90-nm CMOS is the most popular technology for MMW CMOS power amplifier at 60-GHz [4]-[5], [8]-[11]. According to the world-wide 60-GHz WPAN regulations, the maximum system output power limit is +10 dBm. To provide adequate output power for the 60-GHz WPAN system, the maximum PA power design goal needs to include the switch insertion loss or bond-wire assembly loss. Another concern is the linearity in wireless gigabit high-speed systems since the operation power level of PA is suggested to be either near P_{1dB} point or with power back-off from P_{1dB} [21]. Most published 60-GHz CMOS PAs have output P_{1dB} under 10 dBm, in Table I. Because of the low output power of the upconverter (or modulator), the linear gain of PA needs to be 10 dB or higher to achieve better PAE.

To achieve the maximum system output power of +10 dBm, we need a 60-GHz CMOS PA with +13 dBm output power or higher in linear operation range to overcome the switch and packaging losses. In order to increase the linear output power or output P_{1dB} of millimeter-wave CMOS power amplifier, several power combining topologies were proposed [11]-[12]. Balanced power combining in Fig. 3 have been widely used in traditional power amplifier design, especially for binary combining cases. The schematic of the two-stage individual broadband amplifier is shown in Fig. 4, where the transistors are 16 fingers (4-μm/finger) in each stage with total gate periphery of 64 μm.

Table I. 60-GHz 90-nm CMOS Power Amplifiers

Ref	OP_{1dB}(dBm) [PAE]	Psat(dBm) [PAE]	Gain (dB)	V	P_{dc} (mW)
[4]	+5.1 [5.9%]	+8.4	17	N/A	54
[5]	+10.5	+11.5 [8.5%]	16.3	1.0	150
[5]	+5.2	+8.5 [7.0%]	15.2	0.7	89
[8]	+8.2	+10.6	8.3	1.2	229
[9]	+9.0	+12.3 [8.8%]	5.6	1.0	113
[10]	+6.7[20%]	N/A	9.8	1.0	N/A

Nevertheless, the size and high frequency loss of CMOS broadside coupler (Fig. 5) makes the CMOS balanced topology not practical for four-element or eight-element power combining. In [12], a 60-GHz four-element CMOS power combining, in Fig. 6, using distributed active transformer (DAT), is successfully demonstrated within a miniature chip area (0.78 mm x 0.82mm). The chip size of the four-element CMOS DAT PA is even smaller than the two-element balanced PA.

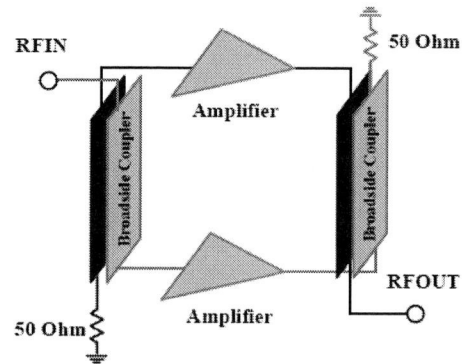

Fig. 3. The CMOS balanced amplifier with two-element combining [11].

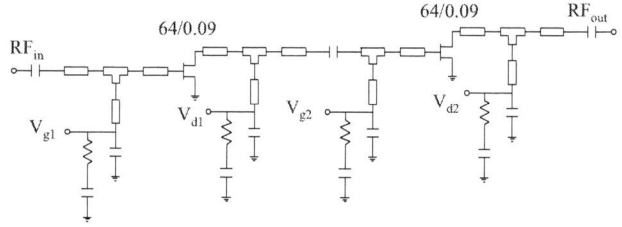

Fig. 4. Schematic of individual broadband amplifier in Fig. 3.

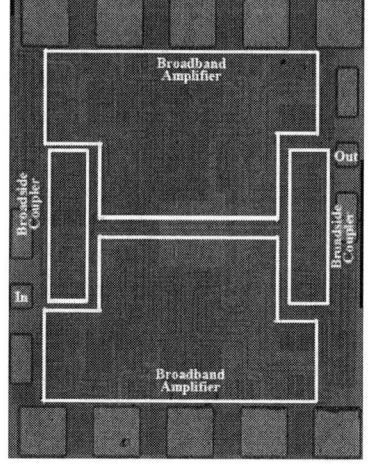

Fig. 5. Chip photograph of the balanced amplifier in 90-nm CMOS with chip size of 0.78 mm x 0.92 mm [11].

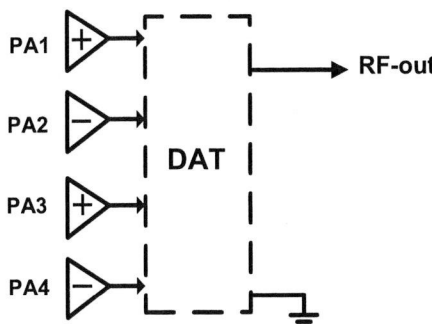

Fig. 6. CMOS Distributed Active Transformer (DAT) power amplifier with four-element combining [12].

B. T/R Switch

For WPAN TDD system, a T/R switch is required between transceiver and antenna. To minimize the assembly and packaging loss, an integrated low-loss CMOS T/R switch is required to enhance the transmitter output power or receiver noise figure. A 50-94-GHz 90-nm CMOS traveling-wave switch in [13] has 3.3 dB insertion loss and 27-dB isolation in Fig. 7. To optimize the narrow band performance, a 57-67-GHz 130-nm CMOS SPDT (Fig. 8), with 2.5-dB insertion loss and 30-dB isolation was implemented in a 60-GHz CMOS transceiver [7]. The input P_{1dB} of the traveling-wave switch, +15 dBm, is adequate for the 60-GHz WPAN system applications.

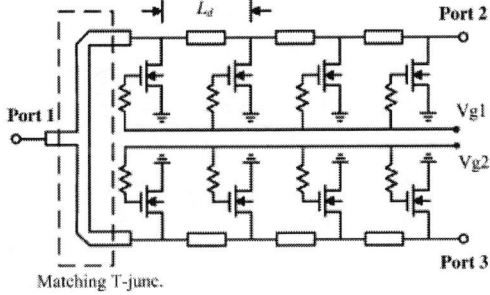

Fig. 7. Schematic of the 50-94-GHz 90-nm CMOS SPDT switch [13].

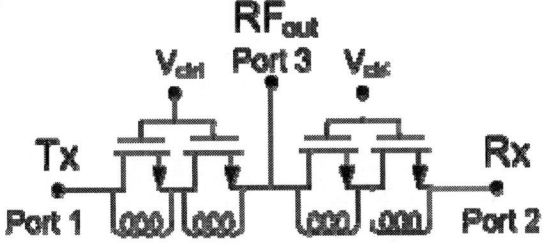

Fig. 8. Schematic of the 57-67-GHz 130-nm CMOS SPDT [7].

C. Modulator/Up-converter

To achieve higher spectral efficiency and data rate, the IQ modulation is essential in the 60-GHz 802.15.TG3c WPAN standard draft. There are four types of 60-GHz IQ modulator topologies in Table II: i) Heterodyne transmitter using resistive mixer for RF up-conversion [4]. The IF frequency Gilbert-cell IQ modulator has better IQ balance than the active modulator in RF frequency; ii) Direct conversion modulator using passive double balanced resistive mixer [5]. The passive mixer consumes no dc power, but the high conversion loss (-17 dB) and high LO power (5-7 dBm) require addition dc power for RF/LO driver, with output P_{1dB} of -17 dBm; iii) Direct conversion reflection type modulator [7], [14] (Fig. 9). This type of reflection-type modulator only needs ultra-low LO power, less than -12 dBm, and therefore the VCO can be used as LO input without additional driver. iv) Sub-harmonic direct conversion transmitter using active Gilbert-cell mixer with half of LO frequency [15] in Fig. 10. The schematic of the sub-harmonic Gilbert-cell mixer is shown in Fig. 11. Each transistors of the LO switching quad turn on and off alternatively during one period of the applied LO signal, which creates a signal of twice of the LO frequency. The active sub-harmonic modulator needs 7-dBm LO at 30 GHz, and consumes 75.9-mW dc power with output P_{1dB} of -19dBm.

Since the LO power requires additional dc power to amplify the VCO output signal, especially at MMW regime, it is observed that the reflection-type modulator in [7] demonstrates lowest dc power consumption among these four-type modulators.

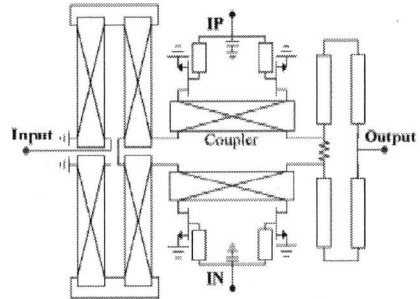

Fig. 9. Schematic of the 35-110-GHz reflection-type modulator in 65-nm CMOS [14].

Table II. 60-GHz CMOS Modulator Comparison

Ref.	OP_{1dB} (dBm)	Conversion Gain (dB)	LO power (dBm)	LO Driver	P_{dc} (mW)
[4]	+3 IF:7-13GHz	N/A	N/A	Yes	17
[5]	-17	-17	+5	Yes	0
[7]	N/A	-13	< -12	No	0
[15]	-19	-6	+7 30-GHz LO	Yes	79.9

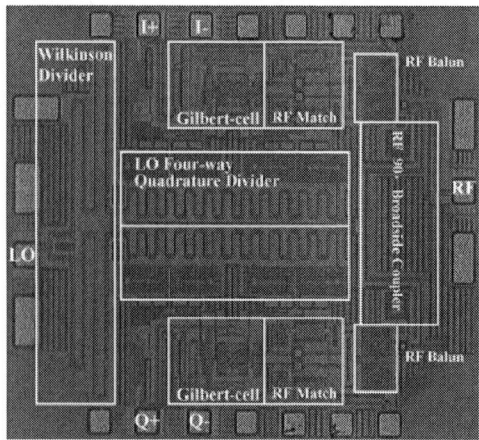

Fig. 10. Chip photograph of the 130-nm CMOS 35-65 GHz sub-harmonic modulator (0.98 mm x 0.8 mm) [15].

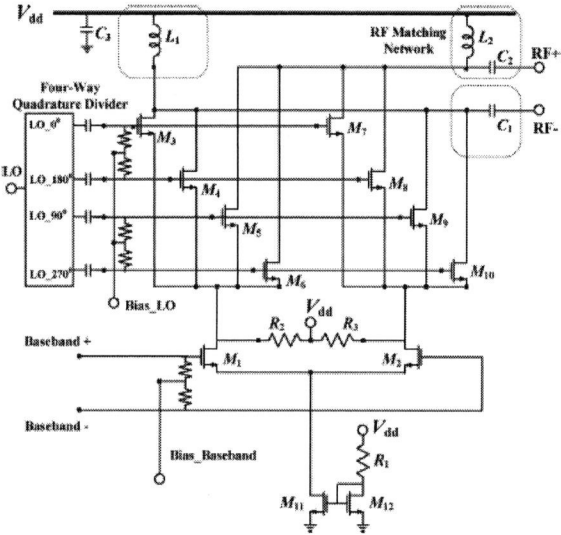

Fig. 11. Schematic of the 35–65-GHz CMOS sub-harmonic Gilbert-cell mixer in Fig. 10 [15].

D. De-Modulator/Down-converter

There are five types of 60-GHz IQ de-modulator topology in Table III: i) Heterodyne receiver using resistive mixer for RF down-conversion [4]; ii) Direct conversion de-modulator using passive double balanced resistive mixer [5]; iii) Sub-harmonic direct conversion receiver using active Gilbert-cell mixer and poly-phase RF filter with half of LO frequency [6]; iv) Direct conversion six-port de-modulator [7] (Fig. 12), which requires ultra-low LO power of lower than -12 dBm, therefore CMOS MMW VCO can be used as LO drive without additional driver;

v) Sub-harmonic direct conversion receiver using active Gilbert-cell mixer with half of LO frequency [15]. The active sub-harmonic de-modulator (Fig. 13) needs 8-dBm LO at 30 GHz, and consumes 90.8-mW dc power with input P_{1dB} of -5 dBm. The schematic of down-conversion Gilbert-cell mixer has similar LO circuits as shown in Fig. 11, with the RF and IF ports switched.

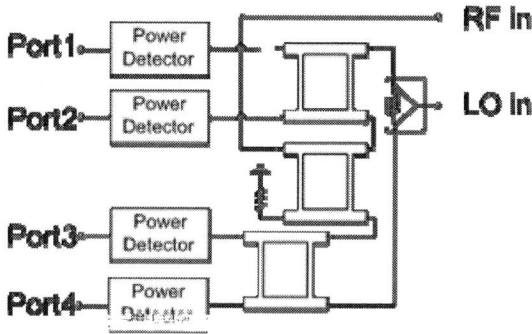

Fig. 12. Block diagram of the six-port reflectometer de-modulator [7].

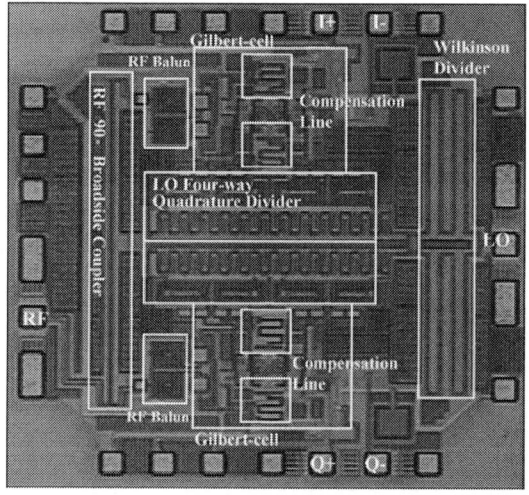

Fig. 13. Chip photograph of the 130-nm CMOS 35-65 GHz sub-harmonic demodulator with chip size of 1.0 mm x 1.0 mm. [15].

Table III. 60-GHz CMOS De-Modulator

Ref.	Conversion Gain (dB)	LO power (dBm)	LO Driver	P_{dc} (mW)
[4]	+19 (IF:7-13GHz)	N/A	Yes	19
[5]	-17	+5	Yes	0
[6]	N/A	N/A	No	23
[7]	N/A	< -12	No	0
[15]	-7.5	+8	Yes	90.8

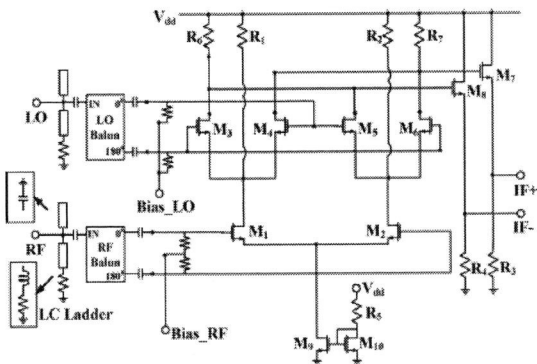

Fig. 14. Schematic of the 25-75-GHz 90-nm CMOS Gilber-cell down-conversion mixer [16].

Compared with the fundamental Gilber-cell down-converter [16] shown in Fig. 14, the 30-GHz LO power of sub-harmonic de-modulator is easier to be generated from CMOS drivers than the 60-GHz LO in fundamental converter. The six-port de-modulator needs special four-port DSP to convert the data back to regular IQ base-band data, which means additional dc power for the DSP circuits is needed. The power detectors in six-port de-modulator have limited dynamic range, so additional variable gain amplifier (VGA) needs to be inserted in front of RF input port of de-modulator to adjust the receiving signal level within the dynamic range of the power detectors.

E. Frequency Divider

The architectures of high-speed frequency divider design can be categorized into three groups: static (flip-flop-based) frequency divider, Miller frequency divider, and injection-locked frequency divider. The flip-flop-based static frequency divider has wide locking range, but limited maximum locking frequency due to the parasitic capacitance of transistors. A Miller frequency divider demonstrates high locking frequency, but it suffers from narrow locking range problem.

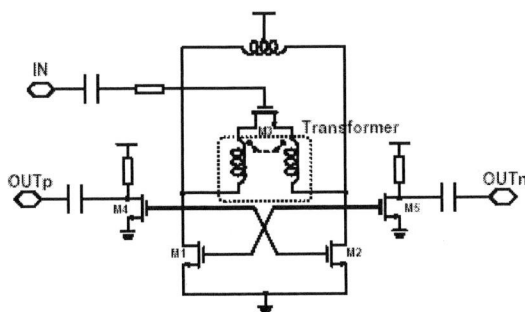

Fig. 15. Schematic of the 130-nm CMOS 50-62 GHz injection

locked frequency divider with transformer feedback [17].

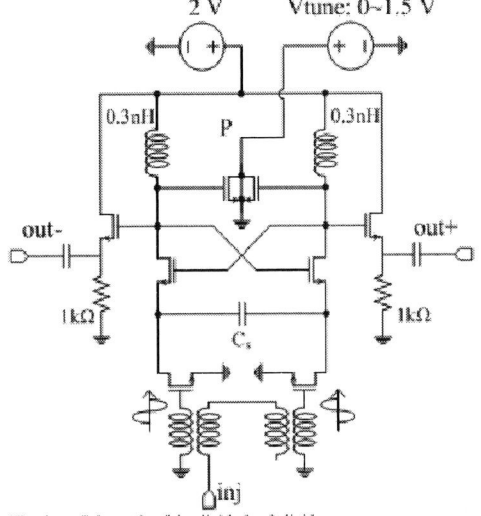

Fig. 16. Schematic of the 66-GHz CMOS divided by 3 divider.

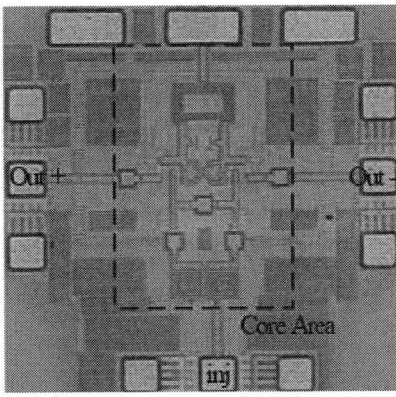

Fig. 17. Chip photograph of the 66-GHz CMOS divided by 3 divider (0.65 mm x 0.64 mm) [18].

Compared with others, injection-locked frequency divider has the highest operating frequency. However, conventional injection-locked frequency dividers suffer from narrow locking range nature. A new type of injection-locked frequency dividers is presented [17] (Fig. 15), using a series inductive-peaking technique to increase the locking range. For frequency division by an odd integer, the injection signal needs to be differential, as shown at the input port of Fig. 16. Fig. 17 is a chip photo of a 66-GHz divided by three divider, which generates lower divided frequencies for PLL system integration [18].

F. VCO/PLL

For practical WPAN high data rate links, PLL and VCO are essential. Two single-chip MMW PLLs in 90-nm CMOS have been reported with 80-mW [19], and 88-mW [20]. The PLLs of the heterodyne system in [4] need 60 mW for the RF PLL/VCO and 40 mW for the IF PLL/QVCO. The phase noise of the WPAN is also important in order to achieve low bit-error-rate (BER) [21], for example, -93 dBc/Hz at 1MHz offset for 10^{-6} BER.

III. MMW CMOS TRANSCEIVER

Based on the previous discussions, an example of link budget for 60-GHz gigabit wireless systems is given in Table IV. The bandwidth is defined in the channellization description of IEEE 802.15.TG3c proposal [22]. The 7-dB antenna gain is suggested according to the user friendly half-power beam-width (HPBW) of $60°$ without any special alignment instructions. The transmitter power level could be increased from 6 to 10 dBm, but the linear operation range is reqired for the gigabit links. To increase the distance of gigabit links, one can increase the transmitter power level or use high-gain directional antenna with special alignment procedures. For the video streaming applications with a 10-meter range requirement, we need 20-dB or more receiving signal power to maintain the SNR level as in one-meter case, then the transmitter power needs to be close to the maximum limit, +10 dBm, and the antenna gain needs to be increased from 7 dB to 15-20-dB gain. Nevertheless all above calculations are based on limited level of multi-path fading, which needs to be confirmed through extensive 60-GHz short-link channel measurements in different home/office environment.

Table IV. An Example of Link Budget for 60-GHz Systems

Bandwidth (GHz)	2.16
Ambient Noise (dBm)	-80.6
Noise Figure (dB)	10
Other losses (dB)	10
Tx Antenna Gain (dBi)	7 (HPBW ~ $60°$)
Rx Antenna Gain (dBi)	7 (HPBW ~ $60°$)
Path Loss @1m (dB)	68 (Ideal Free Space)
Tx Signal Power (dBm)	6 (Linear Power)
Rx Signal Power (dBm)	-48
SNR (dB)	12.6

To implement a 60-GHz uncompressed HDTV video streaming system, the transmitter and receiver chips can be separated in a one-way 60-GHz radio. If the gigabyte multi-media file transfer is required in 60 GHz, a 60-GHz single-chip transceiver chip is preferred with an on-die TDD switch, as shown in Fig. 18, for a two-way radio system. The block diagram of 60-GHz transceiver (Fig. 19) shows that a CMOS MMW VCO provides sufficient LO power for both 60-GHz modulator and de-modulator. The single chip solution can provide a low-loss switch and a loop-back self-test function.

Table V compares the reported 60-GHz CMOS tranmitters. After including adequate power amplifier and MMW PLL in

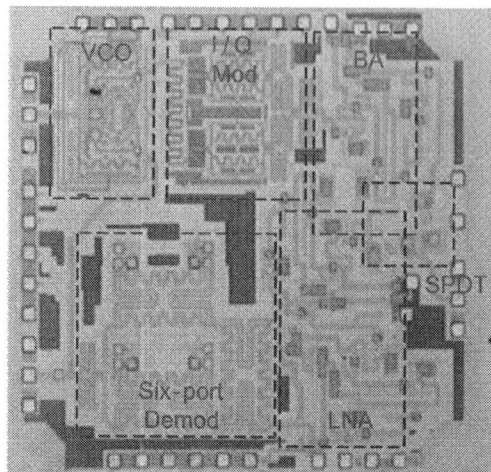

Fig. 18. Chip photograph of the 130-nm CMOS 60-GHz single-chip transceiver (1.65 mm x 1.5 mm) [7].

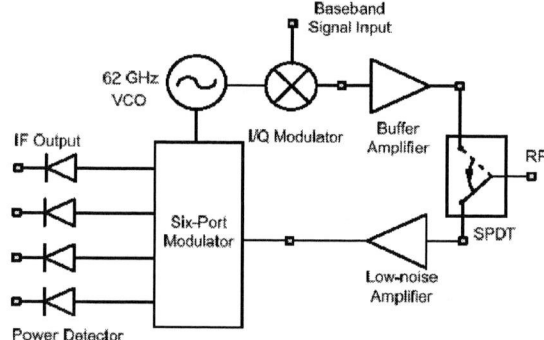

Fig. 19. Block diagram of the 60-GHz transceiver in Fig. 18.

transmitter, it is possible to implement a gigabit CMOS transmitter with dc power consumption less than 200 mW. Based on the 200-mW dc power budget, the output P_{1dB} of the transmitter is difficult to reach the maximum system output power limit, +10 dBm, unless we can design a high efficiency 60-GHz CMOS PA with 20% [10] or higher PAE or a MMW linearizer [23] for better linearity near saturation power output. The dc power consumption of transmitter blocks is plotted in Fig. 20. The transmitter in [7] can provide output power around 0 dBm without the usage of power amplifier. For a practical WPAN application, it is desired to include the PLL and PA with the IQ modulator/VGA/VCO, as mentioned in [7].

Table VI shows the comparison of the reported 60-GHz CMOS receivers, which indicates the possibility of fabricate a gigabit MMW receiver with dc power consumption less than 200 mW. The receiver in [6] demonstrates excellent dc power consumption and receiver noise figure, as shown in Fig. 21, but the VCO phase noise can be improved. In order to realize a gigabit high-data rate receiver with adequate phase noise in the 60-GHz source [21], it is desired to combine the LNA/de-modulator/VCO [6] with a low-phase-noise PLL.

IV. Conclusions

It is concluded that CMOS technologies indeed enable a low-cost and low-power 60-GHz transceiver within the power budget of a hand-held mobile device, such as mobile phone, iPod, digital camera, or digital HDTV camcorder. Hence the 60-GHz gigabit WPAN technology will be a candidate for the next-generation multi-mode broadband wireless network, especially for the short-range high-speed wireless applications.

Acknowledgement

This work was supported by the National Science Council and Ministry of Education of Taiwan, R.O.C. (contract no. NSC 96-2219-E-002-021, NSC 95-2219-E-002-011, NSC 95-2221-E-002-084-MY2, and 95R0062-AE00-01).

60-GHz Low-Power Tx

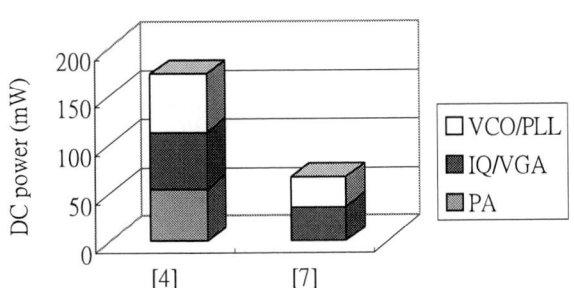

60-GHz Low-Power Rx

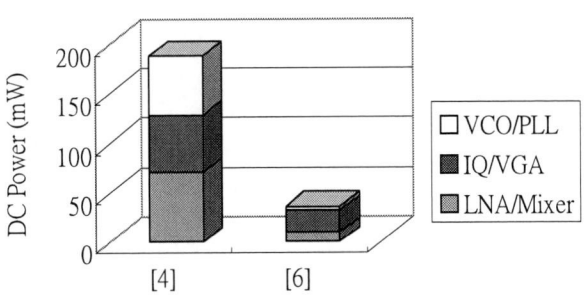

Fig. 20. DC power consumption of 60-GHz low-power transmitters.

Fig. 21. DC power consumption of 60-GHz low-power receivers.

Table V. Comparison of the reported 60-GHz Low-power CMOS Transmitters

Ref.	CMOS Tech.	RF Freq. (GHz)	Type	PA OP_{1dB} (dBm)	PA P_{dc} (mW)	IQ Mod./VGA P_{dc} (mW)	VCO/PLL P_{dc} (mW)	Total DC power, mW, (V)
[4]	90nm	57-65	Heterodyne (IF:7-13GHz)	+5.1	54	Active IF IQ Passive RF mixer 59	30 (VCO) +30 (PLL)	173, (1.8 V)
[5]	90nm	60-64	Direct Conv.	+5.2	89	Passive RF IQ (with LO driver)	No VCO	133, (0.7 V)
[7]	130nm	57-70	Direct Conv.	~+0	No PA	Passive RF IQ 36.9	30 (VCO)	66.9

Table VI. Comparison of the reported 60-GHz Low-power CMOS Receivers

Ref.	CMOS Tech.	RF Freq. (GHz)	Data Rate (Gbps)	Type	LNA P_{dc} (mW) [Gain,dB]	NF (dB)	IQ Mod. and VGA P_{dc} (mW)	VCO/PLL P_{dc} (mW)	PN (dBc/Hz@ 1MHz)	Total DC power mW, (V)
[4]	90nm	57-65	15.0	Heterodyne (IF:7-13GHz)	29 [16]	9 (Rx)	Active IF IQ Passive RF mixer	30(VCO) +30(PLL)	-95	189 (1.8 V)
[5]	90nm	60-64	2.6	Direct Conv.	N/A [18.7]	8.3 (LNA)	Passive RF IQ (with LO driver)	No VCO	No VCO	206 (1.0 V)
[7]	130nm	57-70	4.0	Direct Conv.	31 [20]	5.7-8.8	Passive RF IQ 36.9	30 (VCO)	-92	61
[6]	90nm	57-61	N/A	Direct Conv. (SHP)	9.6 [13]	8 (LNA)	Active RF IQ Active IF IQ	30GHz VCO	-87	36 (1.2V)

978-1-4244-2018-6/08 $25.00 © 2008 IEEE

REFERENCES

[1] C. K. Tzung *et al.*, "Dual-mode broadband and wireless network (DMBWN)", *IEEE 802.15-07/0692r0*, May 2007.

[2] http://www.ieee802.org/15/pub/TG3c.html

[3] A. Sadri "Summary of the usage models for 802.15.3c", *IEEE 802.15-06/0369r09*, 2006

[4] S. Pinel, *et al.*, "A 90nm CMOS 60GHz radio," *2008 ISSCC*, pp. 130-131, February 2008.

[5] M. Tanomura, *et al.*, "TX and RX front-ends for 60GHz band in 90nm standard bulk CMOS," *2008 ISSCC*, pp.558-559, February 2008.

[6] A. Parsa, *et al.*, "A 60GHz CMOS receiver using a 30GHz LO," *2008 ISSCC*, pp.190-191, February 2008.

[7] C-H. Wang, *et al.*, "A 60GHz low-power six-port transceiver for gigabit software-defined transceiver applications," *2007 ISSCC*, Febbruary 2007.

[8] T. Suzuki, *et al.*, "60 and 77GHz power amplifiers in standard 90nm CMOS," *2008 ISSCC*, February 2008.

[9] D. Chowdhury, *et al.*, "A 60GHz 1V +12.3dBm transformer-coupled wideband PA in 90nm CMOS," *2008 ISSCC*, pp.560-561, February 2008.

[10] B. Heydari, *et al.*, "A 60 GHz power amplifier in 90-nm CMOS technology," *IEEE Custom Integrated Circuits Conference (CICC)*, pp. 769-772, Sept. 2007.

[11] Y-N. Jen, J-H. Tsai, T-W. Huang, and H. Wang, "A V-band fully-Integrated CMOS distributed active transformer power amplifier for 802.15.TG3c wireless personal area network applications" submitted to *IEEE Compound Semiconductor IC Symposium (CSICS)*, Oct. 2008.

[12] J-H. Tsai, Y-L. Lee, T-W. Huang, C-M. Yu, and John G. J. Chern, "A 90-nm CMOS broadband and miniature Q-band balanced medium power amplifier," *IEEE MTT-S Int. Microwave Symp.*, June 2007.

[13] S-F. Chao, *et al.*, "A 50-94 GHz CMOS SPDT switch using traveling-wave concept," *IEEE Microwave and Wireless Component Letters*, vol. 17, no. 2, pp. 130-132, Feb. 2007

[14] H-Y. Chang, H. Wang, and W. Lin, "A miniature 35-110-GHz modified reflection-type BPSK modulator using 65-nm CMOS technology," *IEEE MTT-S Int. Microwave Symp.*, June 2007.

[15] J-H. Tsai, and T-W. Huang, "35–65-GHz CMOS broadband modulator and demodulator with sub-harmonic pumping for MMW wireless gigabit applications," *IEEE Trans. Microwave Theory Tech.*, Vol. 55, No. 10, pp.2075-2085 Oct. 2007.

[16] J-H. Tsai, P-S. Wu, C-S. Lin, T-W. Huang, John G.J. Chern, and W-C. Huang, "A 25-75-GHz broadband Gilbert-cell mixer using 90-nm CMOS technology," *IEEE Microwave and Guided Wave Letters*. Vol. 17, No. 4, pp. 247-249, April 2007

[17] Y-H. Wong, W-H. Lin, J-H. Tsai, and T-W. Huang, "A 50-to-62GHz wide-locking-range CMOS injection-locked frequency divider with transformer feedback," to be presented in *IEEE Radio Frequency Integrated Circuits (RFIC) Symp. Digest*, June 2008.

[18] Chung-Chun Chen, Chi-Hsueh Wang, Ming-Fong Lei, Mei-Chen Chuang, and Huei Wang, "66-GHz divide-by-3 injection-locked frequency divider in 0.13-μm CMOS technology," in *Asia Solid-State Circuits Conference (ASSCC) Proceeding*, Jeju, South Korea, Nov. 2007.

[19] C. Lee, *et al.*, "A 58-to-60.4GHz frequency synthesizer in 90nm CMOS," *2007 ISSCC*, pp. 196-197, February 2007.

[20] Jri Lee, "A 75-GHz PLL in 90-nm CMOS technology," *2007 ISSCC*, pp. 432-433, February 2007.

[21] C.-S. Choi, "RF impairment models for 60GHz-band SYS/PHY simulation", *IEEE 802.15-06/0477r0*, 2006

[22] H. Harada *et al.*, "Unified and flexible millimeter wave WPAN systems supported by common mode," *IEEE 802.15-07/0761r10*, Sept. 2007.

[23] J-H. Tsai, H-Y. Chang, P-S. Wu, Y-L. Lee, T-W. Huang, and H. Wang, "Design and analysis of a 44-GHz MMIC low-loss built-in linearizer for high-linearity medium power amplifiers," *IEEE Trans. Microwave Theory Tech.* vol. 54, No. 6, pp. 2478-2496, June 2006.

IEEE 2008 Custom Intergrated Circuits Conference (CICC)

A Zero-IF 60GHz Transceiver in 65nm CMOS with > 3.5Gb/s Links

A. Tomkins[1], R. A. Aroca[1], T. Yamamoto[2], S. T. Nicolson[1], Y. Doi[2], S. P. Voinigescu[1]

[1] Edward S. Rogers Sr. Department of Electrical and Computer Engineering, University of Toronto, Canada
[2] Fujitsu Laboratories LTD., Kawasaki, Japan

Abstract—This paper presents a 1.2V 60GHz zero-IF transceiver fabricated in a 65nm CMOS process with a digital back-end. The chip includes a receiver with 14.7dB gain, a low 5.6dB noise figure, a 60GHz LO distribution tree, a 64GHz static frequency divider, and a direct BPSK modulator operating over the 55-65GHz band at data rates exceeding 3.5Gb/s. The chip consumes 374mW (232mW) from 1.2V (1.0V) and occupies 1.28x0.81mm^2. The transceiver was characterized over temperature up to 85°C and for power supplies down to 1V. A manufacturability study of 60GHz radio circuits is presented with measurements of transistors, the low-noise amplifier, and the receiver on typical and fast process splits. The transceiver performance is demonstrated using a 3.5Gb/s 2-meter wireless transmit-receive link over the 55-64 GHz range.

I. INTRODUCTION

Some of the first mass-consumer products operating in the mm-wave spectrum are likely to be targeted at the emerging 60GHz-radio market. These products take advantage of the 7GHz of unlicensed spectrum available from 57 to 64GHz to provide data-rates that far exceed those of alternative standards such as 802.11n or Ultra Wide-Band. Multi-Gb/s wireless point-to-point transmission is possible over short-ranges using basic modulation techniques.

Unlike earlier 60GHz radio chip-sets reported in SiGe BiCMOS [1] or CMOS [2]–[9], this paper presents the first fundamental frequency, zero-IF 60GHz wireless transceiver. It employs the simplest architecture, as needed in rapid file-transfer applications, utilizing direct BPSK modulation at 60GHz, a fundamental frequency static divider, and which does not require image rejection or ADCs in the receiver. The transmitter data input accepts baseband NRZ data at rates beyond 4Gb/s. Measurement results over temperature, power supply, and process corners are presented.

II. TRANSCEIVER ARCHITECTURE

A block diagram of the direct modulation, zero-IF radio transceiver is presented in Fig 1. With the exception of a fundamental frequency voltage-controlled oscillator, this chip integrates all of the critical 60-GHz blocks, including an LNA, mixer, direct modulation BPSK transmitter, fundamental frequency static divider, and 60-GHz LO tree. Since, due to time constraints, no IF amplifier was included, all the receive-path gain is at 60 GHz. Baseband NRZ data is provided from off-chip to the transmitter and is recovered at the IF output of the receiver, without any digital signal processing or analog-to-digital conversion.

III. CIRCUIT BUILDING BLOCKS

All circuits are differential, except the LNA, and were designed using the constant current density biasing technique described in [10] which renders them immune to V_T and bias current variation. Furthermore, by employing transformer-coupling and AC-folded

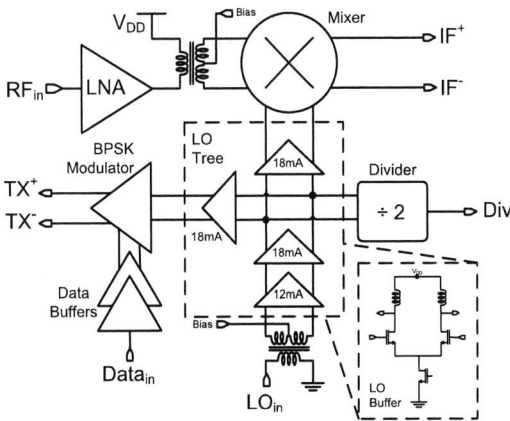

Fig. 1.   60GHz direct modulation BPSK transceiver architecture.

cascode topologies in all blocks, no more than a single high-speed transistor is ever stacked above a current source. This guarantees operation from 1.2 or 1.0V supplies, with the largest possible V_{DS}, which maximizes the power gain and minimizes the noise figure of the transistor. Unlike in a conventional cascode, in an AC-folded cascode, the V_{DS} of the common-source transistor is insensitive to V_T variations of the common-gate transistor. All of these techniques combine to allow the circuits presented here to be operable in LP, GP, or HS flavors of a 65nm CMOS manufacturing process. The penalty for using AC-folded cascodes is a doubling of current consumption compared to the traditional cascode, which is only partially compensated by the reduction in power supply voltage.

A. Low-Noise Amplifier

The single-ended LNA schematic is shown in Fig 2. It consists of three cascaded CS-CG stages, with the transistor sizes increasing from stage to stage. The input stage is noise and impedance

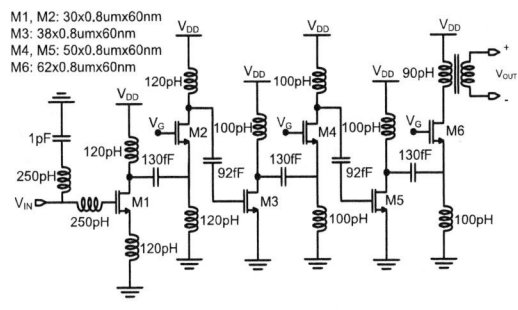

Fig. 2.   1.2V, 3-stage cascaded CS-CG, 60GHz low-noise amplifier schematic.

978-1-4244-2018-6/08 $25.00 © 2008 IEEE   505

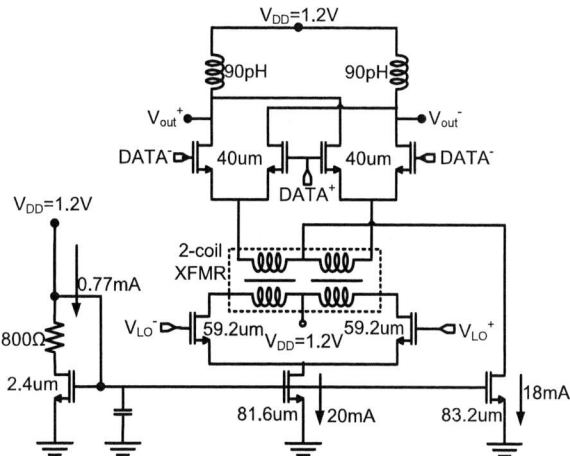

Fig. 3. BPSK modulator schematic.

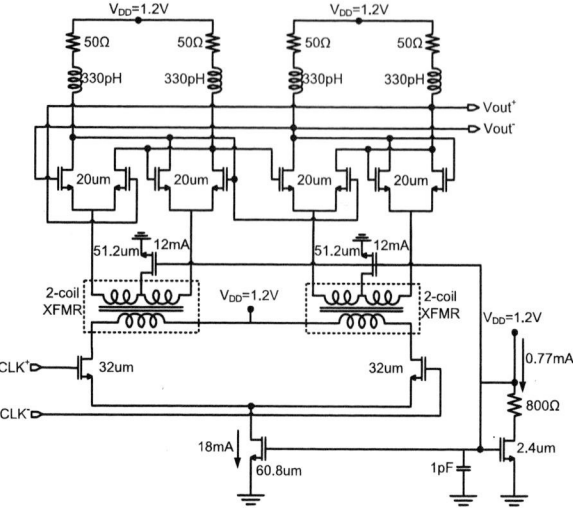

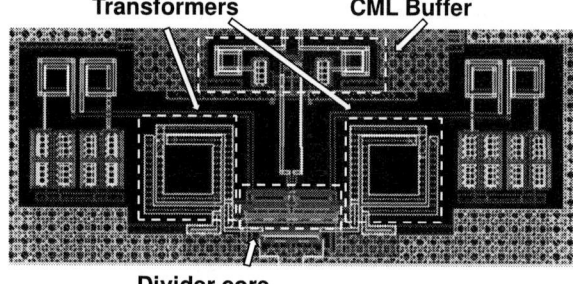

Fig. 4. Static frequency divider schematic (top), and layout details (bottom).

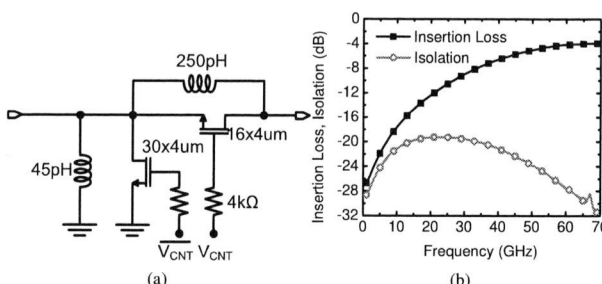

Fig. 5. (a) Schematic and (b) measured insertion loss and isolation of series-shunt switch.

matched to 50Ω and is biased at the receiver minimum noise figure current density of 0.3mA/μm at V_{DS}=1.2V. All transistors feature 0.8μm finger width with double-sided gate contacts. The last stage of the LNA is terminated in a transformer which acts as single-ended to differential converter between the LNA and the double-balanced mixer. The LNA, which can also act as a low-noise power amplifier, consumes 80mA (60mA) from a 1.2V (1.0V) power supply and has a measured saturated output power of 7.5dBm (4.6dBm).

B. BPSK Modulator and Mixer

Fig 3 illustrates the BPSK modulator. The LO differential pair at the bottom and the data quad at the top are coupled through a transformer that allows the maximization of the V_{DS} across each transistor as well as the independent biasing of the two sides of the circuit. The data-path is formed from a cascade of three current-scaled CML buffers with a 7GHz bandwidth.

The mixer topology is identical to that of the BPSK modulator and is scalable to 140 GHz [11]. The mixer output is matched to 50Ω and drives the signal directly off chip.

C. Static Frequency Divider

The schematic of the 60GHz static frequency divider is shown in Fig 4 (top). It employs a new topology that features a single differential pair at the clock input which drives the two latches through two transformers. Eliminating one of the two clock differential pairs results in reduced area and power consumption. The transformers were realized with two vertically stacked coils with inputs and outputs aligned along a diagonal line of symmetry. Fig 4 (bottom) illustrates the layout detail including the first of three CML buffers at the divider output.

D. Tuned Clock Tree

The 53-65GHz off-chip LO signal is distributed to the critical circuit blocks using a tuned LO distribution tree designed to have 25% bandwidth. The latter consists of cascaded differential buffers with inductive loads and a fanout of three, as shown in Fig 1, with both 12mA and 18mA buffers employed.

E. Tuned mm-Wave Switch

While not included in the aforementioned transceiver, a stand-alone 50-70GHz series-shunt switch as in Fig 5a was manufactured on the same dies and characterized separately. S-parameter

measurements on the stand-alone mm-wave switch in Fig 5b show a nominal insertion loss of 3.9dB and an isolation of 28dB at 60GHz.

IV. IMPLEMENTATION AND MEASUREMENTS

The transceiver was fabricated in a 65nm CMOS process with a 7-metal digital back-end and MIM capacitors. For a 80x60nmx1μm NFET device with double-sided gate contacts, peak f_{MAX} and f_T values of 320GHz and 220GHz were measured at current densities of approximately 0.3 and 0.4 mA/μm respectively.

A micro-photograph of the transceiver is shown in Fig 6. It occupies 1.28x0.81mm^2. The LO and RF signals are distributed along μ-strip lines formed in metal 7 over a shunted metal 1 and metal 2 ground plane. A side-wall consisting of all metals

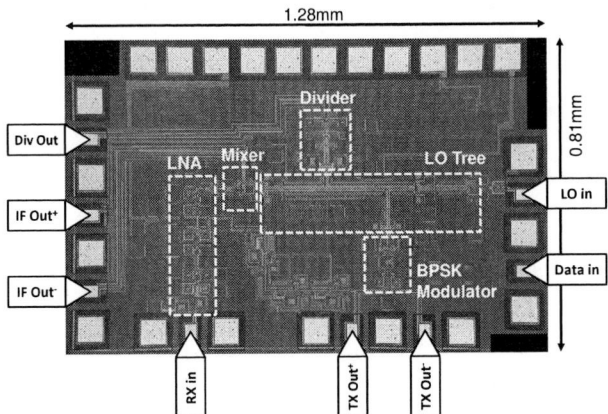

Fig. 6. Die photograph of the transceiver. The total die area is 1.28x0.81mm².

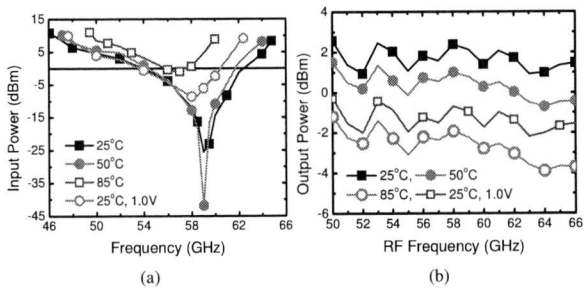

Fig. 7. Measured (a) divider and LO-tree sensitivity, and (b) transmitter output power all over temperature.

shunted together forms a Faraday cage-like structure around each transmission line and between circuit blocks, improving isolation.

The static frequency divider sensitivity was measured in the transceiver, including the LO distribution network. The self-oscillation frequency is 59.2GHz at room-temperature with a 1.2V supply. As illustrated in Fig 7a, the divider operates over a frequency range of 46-65GHz at 1.2V, and 47-62GHz at 1.0V.

Measured total integrated power at the output of the transmitter is plotted in Fig 7b versus frequency. Measurements are shown over temperature up to 85°C, with peak output power above +2.4dBm at 25°C and 58GHz.

The manufacturability of the 60GHz transceiver was studied by measuring transistors, the low-noise amplifier, and the stand-alone receiver over fast and typical process corner lots. Compared to

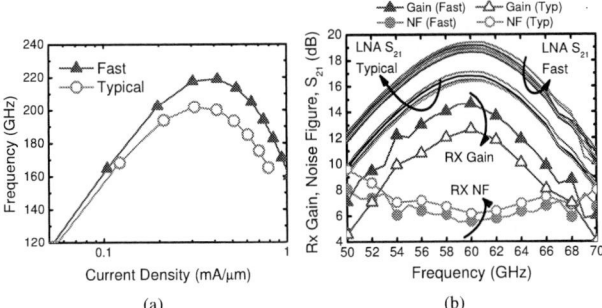

Fig. 8. Measured (a) f_T, and (b) RX gain, noise figure, and LNA S_{21} over fast and typical process corners.

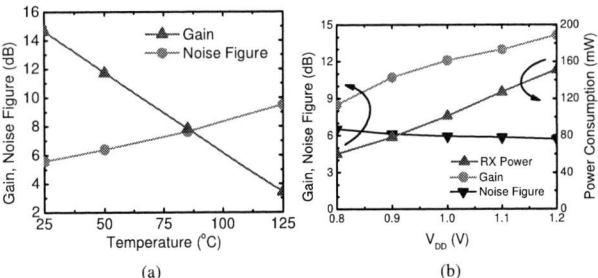

Fig. 9. Measured receiver gain and noise figure over (a) operating temperature, and (b) power supply.

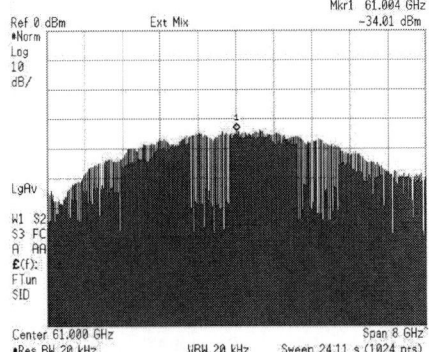

Fig. 10. 3.5Gb/s PRBS transmitter spectrum at 61GHz LO.

the standalone RX breakout measurements, the receiver gain in the transceiver was degraded by 3dB because of insufficient LO power at the mixer LO port. The LO-tree fanout in the transceiver is 3, whereas it is only 1 in the RX breakout. Performance results for both versions are summarized in Table 1 at the end of this paper.

Results in Fig 8a indicate a 10% drop in peak f_T from the fast to typical corner splits. It should be noted that although there is significant V_T variation across corners, the measured peak f_T current density is essentially constant. LNA s-parameter measurements are plotted in Fig 8b, along with receiver gain and noise figure, all for the typical and fast corners. The LNA displays a peak gain of 19.2dB at 60GHz and a 3dB bandwidth extending from 54GHz to 66GHz. A peak receiver conversion gain of 14.7dB and a 50Ω noise figure of 5.6dB are noted, both occuring at 60GHz. The DSB noise figure remains below 6dB over an RF bandwidth of 58 to 63GHz. The receiver measurements indicate that the impact of process variation on the receiver can be attributed solely to transistor g_m and f_T degradation due to an increase in gate length. Note that the peak gain frequency of the LNA and of the receiver and the minimum noise figure frequency do not shift with process variation.

Measurements of the receiver over temperature in Fig 9a show an approximate 7dB gain degradation and 2dB noise increase at 85°C. Fig 9b illustrates that the gain and noise figure do not seriously degrade as V_{DD} is reduced to 1.0V, with a gain drop of 2.5dB, and a noise figure increase of only 0.3dB. The gain degradation over temperature and supply can be easily compensated with a VGA at IF. The measured input return loss of the receiver is less than -10dB, and the input compression point is -22dBm.

Finally, a transmit-receive link was demonstrated in the 55-64GHz range by employing one transceiver chip in transmit-mode on a probe station, and another transceiver chip in receive-mode

978-1-4244-2018-6/08 $25.00 © 2008 IEEE

TABLE I

PERFORMANCE COMPARISON FOR 60GHz TRANSCEIVER/RECEIVER CHIP-SETS. BRACKETS SHOW MEASURED RESULTS FROM 1.0V.

Ref.	Gain (dB)	NF (dB)	IP_{1dB} (dBm)	S_{11} (dB)	TX Power (dBm)	Modulation, Data Rate	V_{DD} (V)	P_{DC} (mW)	Chip(s) Area (mm^2)	Process and Integration
[1]	72	5.5-6.5	-36	-15	17.0	MSK 2.0Gb/s	2.7, 4.0	996	12.2	0.13 SiGe, TX, RX, 20GHz PLL, IQ mod/demod
[2]	16-21	5.5-7	-21	-10	n/a	n/a	1.2-1.5	60	0.3	90nm, LNA, mixer, IF buffer, off-chip LO
[5]	51	9	-30	-15	8.4	16-QAM 15Gb/s	1.8	362	6.5	90nm, TX, RX, PLL, IQ mod/demod
[6]	55.5	6.2	-26	-10	n/a	n/a	1.0	24	1.6	90nm, LNA, mixer, VGA, buffer, off-chip LO
[7]	18.3-22	5.7-8.8	-27.5	-	n/a	n/a	1.2	36	0.2	90nm, LNA, 30GHz LO, RF mixer, IF mixer
[8]	19.7	11.7	-17	-	6.0	QPSK 2.6Gb/s	0.7, 1.0	339	4.5	90nm, TX, RX, IQ mod/demod, off-chip LO
This work (RX)	14.7 (12.1)	5.6 (5.7)	-22	-10	n/a	n/a	1.2 (1.0)	151 (101)	0.5	65nm, single-chip RX, off-chip LO
This work (TXRX)	11.3 (8.9)	5.6 (5.8)	-22	-10	2.4 (-0.7)	BPSK 3.5Gb/s+	1.2 (1.0)	374 (232)	1.0	65nm, single-chip TXRX, LO tree, divider, off-chip LO

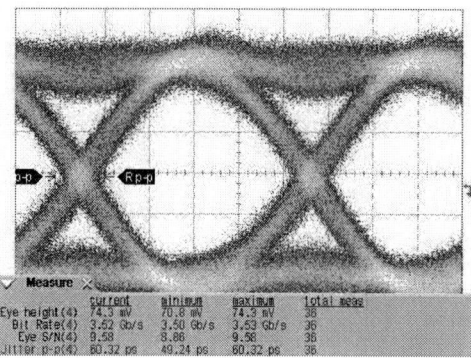

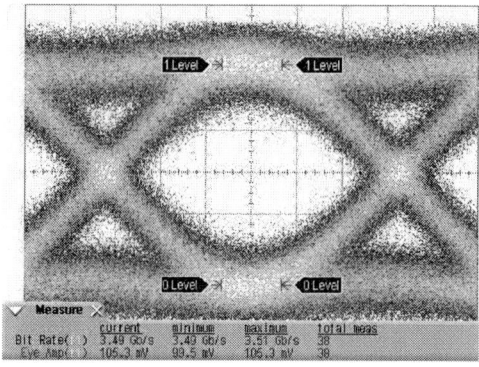

Fig. 11. 3.5Gb/s PRBS eye diagram at the output of the receiver for wired (top) and wireless (bottom) setups.

on a second probe station, approximately 2 meters away. Both wired and wireless demonstrations were conducted, with high-speed signal probes and 67GHz cables connecting to either horn antennas with 25dBi gain, or 2 meters of cable. Both channels are estimated to have approximately 35-40dB of loss at 60GHz. An external amplifier with 30dB gain and 4GHz bandwidth was connected between the IF output of the receiver and a sampling oscilloscope. The transmitter output spectrum is shown in Fig 10 for a 3.5Gb/s 2^7 - 1 PRBS data signal. In Fig 11, the wired (top) and wireless (bottom) received eye diagrams at the receiver IF output are shown. The bandwidth of the external IF amplifier limited the data rate experiments to 5Gb/s. These transceiver link

experiments demonstrate for the first time that a simple zero-IF radio architecture, without ADCs and IQ mixer, can be realized in CMOS at 60 GHz.

V. CONCLUSION

A 1.2V 60GHz zero-IF transceiver has been implemented in a 65nm CMOS technology occupying only 1.28x0.81mm^2. Employing direct BPSK modulation, a 60GHz LO distribution tree, a fundamental frequency static divider, and zero-IF down-conversion, this transceiver represents the simplest architecture appropriate for high-frequency, high-bandwidth data-transfer applications. A 2-meter wireless transmit-receive demonstration between two probe stations acts as a proof-of-concept, achieving data-transfer rates in excess of 3.5Gb/s over the 55-64GHz range.

ACKNOWLEDGMENTS

This work was funded by Fujitsu Limited. The authors would like to thank Jaro Pristupa for CAD support, CFI, OIT, and ECTI for test equipment. We would also like to thank Dr. W. Walker of Fujitsu Laboratories of America Inc. for his support.

REFERENCES

[1] B. Floyd et al., "Short course: SiGe BiCMOS transceivers for millimeter-wave," in BCTM, Sep. 2007.

[2] D. Alldred et al., "A 1.2V, 60GHz radio receiver with on-chip transformers and inductors in 90nm CMOS," in IEEE CSICS, Nov. 2006, pp. 51–54.

[3] S. Emami et al., "A highly integrated 60GHz CMOS front-end receiver," in ISSCC Dig. Tech., Feb. 2007, pp. 190–191.

[4] C.-H. Wang et al., "A 60GHz low-power six-port transceiver for gigabit software-defined transceiver applications," in ISSCC Dig. Tech., Feb. 2007, pp. 192–193.

[5] S. Pinel et al., "A 90nm CMOS 60GHz radio," in ISSCC Dig. Tech., Feb. 2008, pp. 130–131.

[6] B. Afshar et al., "A robust 24mW 60GHz receiver in 90nm standard CMOS," in ISSCC Dig. Tech., Feb. 2008, pp. 182–183.

[7] A. Parsa et al., "A 60GHz CMOS receiver using a 30GHz LO," in ISSCC Dig. Tech., Feb. 2008, pp. 190–191.

[8] M. Tanomura et al., "TX and RX front-ends for 60GHz band in 90nm standard bulk CMOS," in ISSCC Dig. Tech., Feb. 2008, pp. 558–559.

[9] T. Mitomo et al., "A 60-GHzCMOS receiver front-end with frequency synthesizer," IEEE JSSC, vol. 43, no. 4, pp. 1030–1037, 2008.

[10] T. Dickson et al., "The invariance of characteristic current densities in nanoscale MOSFETs and its impact on algorithmic design methodologies and design porting of Si(Ge) (Bi)CMOS high-speed building blocks," IEEE JSSC, vol. 41, no. 8, pp. 1830–1845, 2006.

[11] S. Nicolson et al., "A 1.2V, 140GHz receiver with on-die antenna in 65nm CMOS," in RFIC Symp., paper RMO3C-2, Jun. 2008.

978-1-4244-2018-6/08 $25.00 © 2008 IEEE

IEEE 2008 Custom Intergrated Circuits Conference (CICC)

Low-Cost Fully Integrated BiCMOS Transceiver for Pulsed 24-GHz Automotive Radar Sensors

L. Moquillon[1], P. Garcia[1], S. Pruvost[1], S. Le Tual[1], M. Marchetti[1], L. Chabert[1], N. Bertholet[1], A. Scuderi[2],
S. Scaccianoce[2], A. Serratore[2], N. Piazzese[2], C. Dehos[3], D. Morche[3]
STMicroelectronics, [1]Crolles - France and [2]Catania - Italy
[3]CEA-LETI, Grenoble, France

Abstract- **This work presents the performance of a 24-GHz transceiver for automotive short-range radar.**

Complying with the pulsed radar sensor requirements, this single-chip is the first silicon realization with this high-level of integration. The single-chip results were obtained at 24 GHz using a standard 0.13μm BiCMOS SiGe 170GHz f_T featuring 1.7V BVceo. This allows a receive path of 47 dB gain with 3 dB$_{DSB}$ noise figure providing a dynamic range of 31 dB. In the transmission section, 1-ns pulses modulated at 24 GHz are also obtained using an RF switch, which is able to deliver a continuous wave output power of 0dBm.

I. INTRODUCTION

The 24GHz pulsed wave automotive radar sensors give the opportunity for silicon technology to access the automotive market and other millimeter applications. Previous works [1] already present the benefit of silicon in terms of performance like cut-off frequency, cost, and integration.

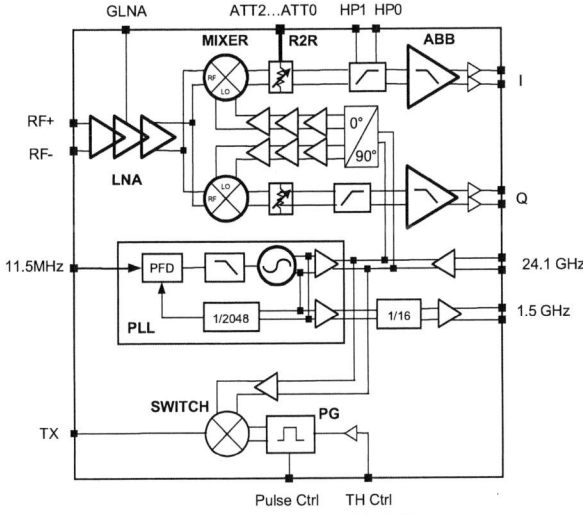

Fig. 1. Block diagram of the RF sensor.

The 24-GHz UWB transceiver block diagram of the short-range radar (SRR) is presented in Fig. 1. The RF architecture is based on a zero-IF transceiver and it is composed of three main sections: the receiver, the 24-GHz synthesizer, and the transmitter.

The paper is organized as follows: Section II covers the design of the main circuit blocks composing the transceiver. Experimental results are reported and commented upon in Section III. Finally, conclusions are drawn in Section IV.

II. TRANCEIVER CIRCUIT DESIGN

A. Receiver

The three stage LNA is a true differential common emitter topology (Fig. 2) with a tail current source for each stage. This topology exhibits best noise/linearity trade-off.

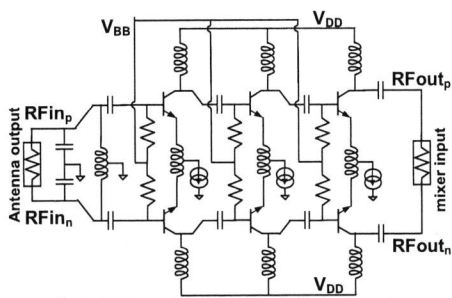

Fig. 2. Differential LNA simplified schematic

The LNA has two states of gain allowing a good management of the input RF signal regarding the distortion. The input pad capacitances are symbolized by the capacitors connected to the ground at RFin$_p$ and RFn$_n$. Electro-Static Discharge protection at the LNA input is performed by the symmetric inductance shorted to the ground in the middle plane of this component. For high gain mode, the LNA provides, at 24GHz, S_{21}=24dB and NF=2.9dB for 54mA of current consumption.

The LNA is followed by a pair of Zero IF Gilbert cell mixers in quadrature with 2.5GHz bandwidth designed in order to simplify the pulse down-conversion. The mixer is integrated twice to perform quadrature I/Q channels. The load of the mixer is low enough (70Ω) for linearity and bandwidth purposes. A polyphase filter is used to generate quadrature LO signals. The mixer exhibits a voltage conversion gain of 12.8dB for 31.5mA of current.

The analog baseband (see Fig. 1) is designed to filter signals and high frequency noise that can be aliased back during off-chip Analog to Digital conversion. It is thus made of:

- a filtering and amplifying stage, which provides 24dB gain and 1GHz second order low-pass filtering through a very compact Gm-R Sallen-Key topology,

- a 1st order high-pass filter which removes 24GHz FMCW interferers that are also shifted to baseband during mixing operation,

- a programmable attenuator which provides constant amplitude signal to the gain stage,

- a 50Ω output buffer.

978-1-4244-2018-6/08 $25.00 © 2008 IEEE

B. Frequency Synthesizer Design

This work uses a 24-GHz PLL, whose block diagram is depicted in Fig. 1. The PLL adopts a well-established architecture composed of phase/frequency detector (PFD), charge pump, external loop filter, VCO, and *N*-integer divider. Using a division ratio *N* of 2048, a 24.125-GHz carrier is achieved with a PLL reference frequency f_{REF} of 11.78 MHz.

The PFD adopts a tri-state topology with a delay unit on the reset path, which eliminates the dead zone of the charge pump characteristic, thus improving linearity.

In-band PLL noise contributions are minimized adopting a symmetric topology for the charge pump, whose noise analysis and design description are reported in [2]. An integrated second order loop filter is used to set the PLL bandwidth at 200 kHz.

The VCO (Fig. 3) uses a bipolar core with a high-Q *LC* resonator. An important point is the design of both capacitors and inductors in the *LC* tank. This resonator exploits accumulation MOS (A-MOS) devices [3], which provide the better compromise between quality factor and tuning capability with respect to diode varactors available in BiCMOS technology. The resonator is completed using a single-turn inductor whose value at 24 GHz is 230 pH. The VCO has a simulated phase noise of -104.5dBc/Hz at 1MHz offset from the carrier with a tuning range of 20%.

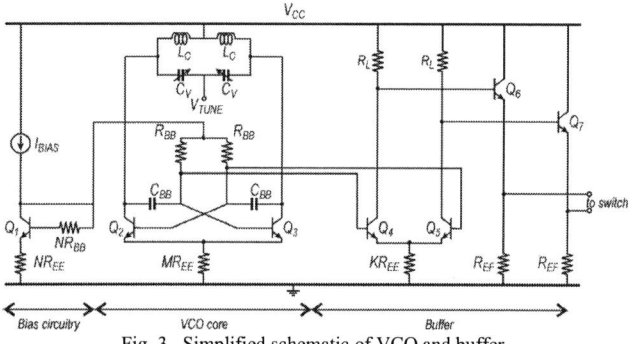

Fig. 3. Simplified schematic of VCO and buffer.

A buffer (emitter-follower) is used to connect the oscillator core with other active stages.

Finally, the *N*-integer divider is designed using a chain of 11 divide-by-two stages. Each stage uses flip-flops in closed-loop master-slave configuration. High-speed flip-flops are designed using current-mode-logic D-latches.

C. Pulsed Transmitter Design

Pulses are easily generated by the Pulse Generator shown in Fig. 4. It is composed of a single-to-differential converter, two parallel inverter chains, A and B, and an AND gate, whose output directly drives the switch. After the differential conversion, the input signal runs on the two parallel paths, A and B. In the path B this signal is inverted and properly delayed by means of a tunable-delay buffer. The introduced delay sets the pulse width T_{pulse} at the output of the AND gate.

The tunable-delay buffer is designed using a variable *RC* load, which is implemented by means of a MOS varactor controlled by the *Pulse Ctrl* signal.

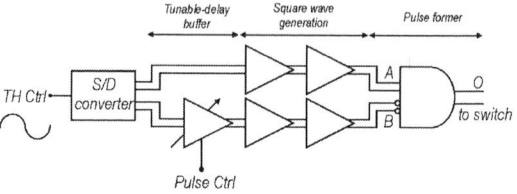

Fig. 4. Pulse generator circuitry

The simplified schematic of the switch is shown Fig. 5. A current steering approach is adopted. The 24-GHz signal drives the *RF* port, while the pulse signal coming from the pulse generator steers the RF signal into the output port to generate the 24-GHz modulated pulse. To be compliant with the emission limits of the ETSI EN 302-288 standard [6] -a 24GHz 0dBm of EIRP (Effective Isotropic Radiated Power) - the switch is properly designed to deliver a 0-dBm peak output power. A target distance accuracy of 15cm is performed using a single rectangular pulse with a 1ns duration, modulating a sinusoidal carrier at frequency f_c=24.125GHz with a repetition period T_{PRF} = 250ns. A resonant load composed of MIM capacitor, C_L, and integrated transformer, T_{LOAD}, is adopted. The design of the integrated transformer is an issue of great concern. First a simplified lumped model [4] was exploited to define the transformer geometrical parameters and then a commercial 3D EM simulator, Ansoft HFSS, was used to simulate the layout structure, taking all the connection paths into account. As well the monolithic transformer provides both differential-to-single-ended conversion of the output signal and ESD protection. Thanks to the secondary inductor T_{LOAD}, the current generated by electrostatic discharge is shorted to the ground without any ESD protection diode.

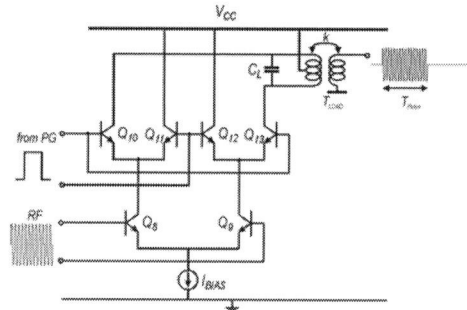

Fig. 5. Simplified schematic of TX output switch

III. EXPERIMENTAL RESULTS

The die micrograph of the fabricated transceiver is shown Fig. 6. The differential topology is used to reduce the substrate noise and to get rid of the ground signal loop issues. Floorplan and pinout are compatible with both on-wafer measurements and flip-chip assembly. ESD standard protection has been included. RX is located on the upper part of the chip with

LNA (a), mixer (b), and analog baseband (d). Local Oscillator (e) and TX switch Power (f) are designed on the lower. Great care has been taken to ensure proper LO signal interconnect between the VCO driver (e), the quadratic generator (c) and the mixer.

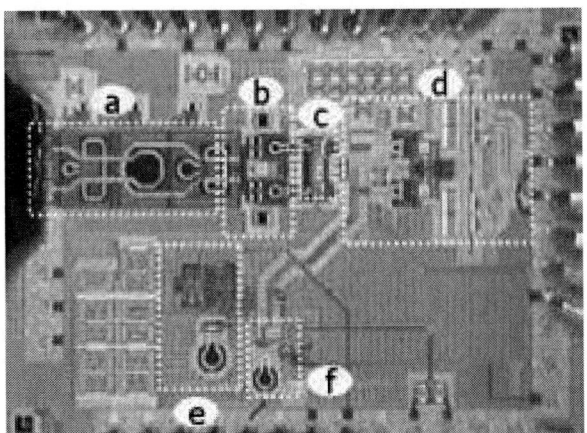

Fig. 6. Die photo of the single-chip integration
die size (all inclusive): 5.9mm²

A. PLL and TX Results

To evaluate the continuous wave performance of the PLL, VCO, and switch, a transmitter without the pulse generator was also integrated and characterized.

Using an Agilent 4440A spectrum analyzer, the output spectrum is centered at 24.125 GHz carrier. The carrier of PLL presents two tones at a frequency offset of ±11.78 MHz (PLL reference frequency f_{REF}) whose levels are 50 dBc lower. The measured output power is about -4.65 dBm, which corresponds to a delivered power of -0.8 dBm considering a cable loss of 3.8 dB at 24 GHz. The maximum power is 1.5 dBm measured at 24.9 GHz.

The VCO exhibits a tuning range of 4.7 GHz from 20.4 GHz up to 25.1 GHz, when f_{REF} sweeps from 9.78 MHz up to 12.2 MHz respectively. The measured phase noise is -104.3 dBc/Hz at 1 MHz offset from the carrier at 24.125 GHz, as shown Fig. 7. The VCO core consumes 5.8 mA from a 2.5 V power supply.

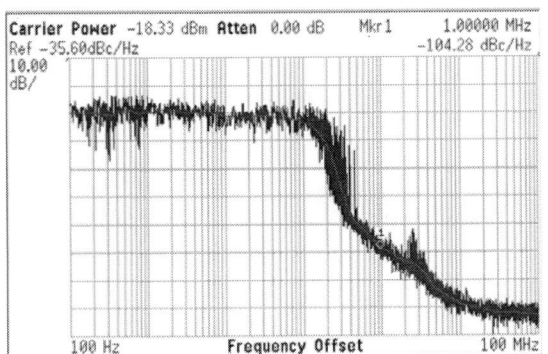

Fig. 7. Measured phase-noise at 1 MHz offset from the 24.125 GHz carrier.

The UWB transmitter was tested using the setup described by the ETSI standard [8]. Fig. 8 shows the measured spectrum of the UWB transceiver with *TH Ctrl* fixed at 10 MHz. The spectrum presents a main lobe centered at 24.125 GHz with two nulls at 23.25 and 25.08 GHz respectively, which indicate a pulse transmission duration of 1.1 ns.

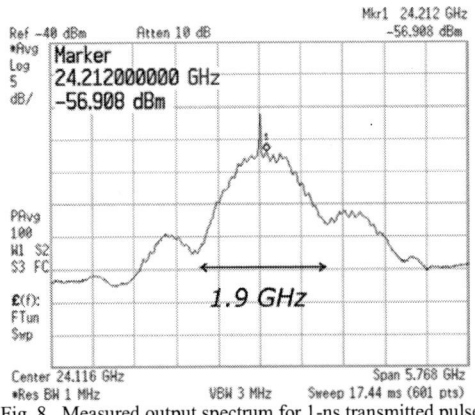

Fig. 8. Measured output spectrum for 1-ns transmitted pulse
(RBW = 1MHz)

B. On Wafer RX Results with LO Signal from PLL

The on-wafer measurements and simulations of RX power conversion gain and noise figure are shown in Fig.9 for the high gain mode. The noise has been measured with spectrum analyzer. The noise figure has been extracted by C/N method integrated on a bandwidth of 100MHz to 1GHz. The -3dB output IF bandwidth is 900 MHz with second order roll-off.

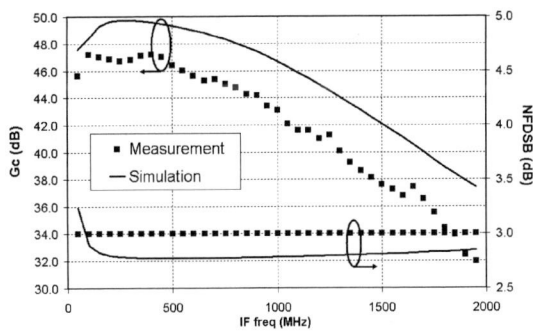

Fig. 9. Simulation and experimental RX conversion gain and noise figure versus IF frequency (LO=24.0GHz)

TABLE 1
UWB TRANSCEIVER MEASURED AND SIMULATED PERFORMANCES SUMMARY

	Measurements	Simulations
Current consumption (mA)	190 (Rx) 66(PLL)	186 (Rx) 68 (PLL)
Rx Power Gain (dB)	47	49
IF Bandwidth (MHz)	900	980
NF$_{DSB}$ (dB)	3.0	2.7
1dB Input compression (dBm)	-47	-46
LO leakage on RF (dBm)	-75	n.a.

Table 1 presents the on wafer measurements and simulations of the whole receiver. Good agreement between experimental results and simulations is obtained for conversion gain, noise figure and linearity.

C. On Board 24GHz Pulsed Results on RX

An aggressive pitch, a high number of pads, and a soft RF substrate are 3 difficulties that reduce the chance of success for a Flip-Chip assembly. To provide early transceiver measurements in pulsed mode, the chip was assembled on board using wire bonding. For the Rx-Tx test, a 1ns Tx output pulse is injected at the LNA input with cable as shown in Fig. 10.

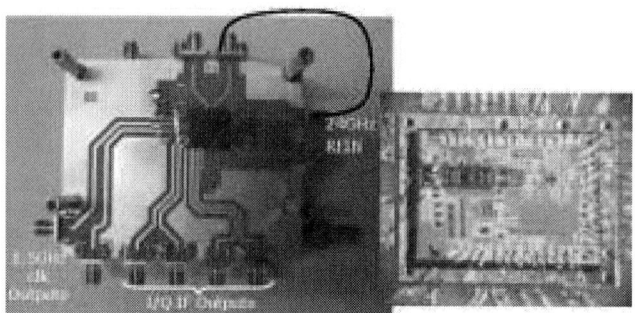

Fig. 10. Chip on board PCB and micrograph

The experimental results are shown in Fig.11 for 2 different cable lengths (in order to delay the received impulsions). The TX pulse is generated at the negative slope of the TH Ctrl signal (input signal of pulse generator). The TX leakage on RX shows the emission of pulse.

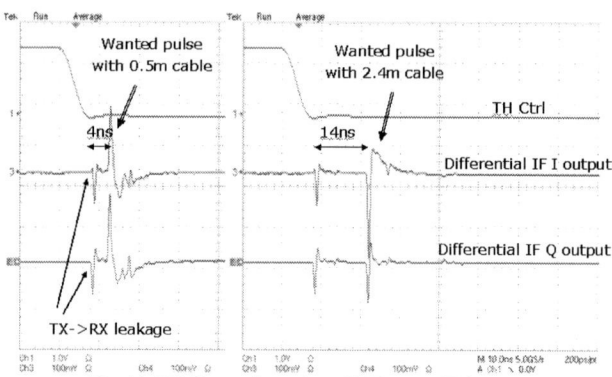

Fig. 11. RX Conversion gain and NF versus IF frequency

After 4ns for 0.5m of cable and 14ns for 2.4m of cable, the pulse is perfectly detected in the delayed time range. This board gives the opportunity to perform a true detection including external A/D and FPGA.

D. Performance Comparison

This work has been compared with other published transceivers in table 2. This work obtains good experimental results especially for noise figure.

IV. CONCLUSION

This work presents an UWB single-chip transceiver for 24GHz pulsed radar sensor. The comparison of the measurements and the simulations reports a high accuracy level which complies with the ETSI specifications. However a larger characterization including temperature and process variation will be necessary to assess the achievement of the single-chip integration. Finally the flip-chip assembly will give the opportunity to improve the main sensor performance (also regarding cost and integration level) allowing the Silicon technology to access to the automotive radar sensor market.

TABLE 2
COMPARISON WITH PUBLISHED RECEIVER

	This work (on wafer)	[7] (on wafer)	[8] (in QFN)	[9] (on wafer)
Freq(GHz)	22-26	21.3-29	22-26	23-25
Integration	Rx, Tx, Pusle former, PLL	Rx, VCO, Pulse former	Rx, Integretor	Rx, VCO, Dividers
Process	0.13μm BiCMOS	0.18μm CMOS	HBT SiGe	65nm CMOS
Max gain (dB)	47	38.1	45	31.5
NF $_{DSB}$(dB)	3.0	5.6	7.8	6.5
Power (mW)	640	131	1082	92.4
Area (mm²)	5.9	3	n.a.	2.1

ACKNOWLEDGMENTS

The authors thank the many members of STMicroelectronics Crolles: E. Ó hAnnaidh and L. David involved in the various aspects of this work. As well as peoples involved in the system studies at the CEA-LETI: L. Dussopt, and at STMicroelectronics Catania: E. Messina and A. Galluzzo.

REFERENCES

[1] S. Pruvost, et al, "Low Noise Low Cost Solutions for Pulsed Automotive Radar sensors", RFIC-IMS, RTUI1B-3, Honolulu, Hawaii, USA, 3-5 June 2007, pp 387-390.

[2] T. Copani, S. A. Smerzi, G. Girlando, and G. Palmisano, "A 12-GHz silicon bipolar dual-conversion receiver for digital satellite applications," IEEE J. Solid-State Circuits, vol. 40, pp. 1278-1287, Jun. 2005.

[3] P. Andreani and S. Mattisson, "On the use of MOS varactors in RF VCO's," IEEE J. Solid-State Circuits, vol. 35, pp. 905-910, Jun. 2000.

[4] M. Tiebout, "A CMOS direct injection-locked oscillator topology as high-frequency low-power frequency divider," IEEE J. Solid-State Circuits, vol. 39, pp. 1170-1174, Jul. 2004.

[5] T. Biondi, A. Scuderi, E. Ragonese, and G. Palmisano, "Analysis and modeling of layout scaling in silicon integrated stacked transformers," IEEE Trans. Microwave Theory Tech., pp. 2203-2210, Apr. 2006.

[6] ETSI EN 302-288-1 v1.1.1 (2005-01).

[7] V. Jain, S. Sundararaman, and P. Heydari, "A cmos 22-29Ghz front-end for UWB Automotive Pusle-Radars," CICC. 2007, pp 757-760.

[8] I. Gresham, et al, "A Fully Integrated 24 Ghz SiGe receiver chip in a low-cost QFN plastic package", Radio Frequency Integrated Circuits Symposium, 11-13 June 2006.

[9] A. Mazzanti et al., "A 24 GHz sub-harmonic receiver front-end with integrated multi-phase LO generation in 65nm cmos", ISSCC 2008, pp216-217.

978-1-4244-2018-6/08 $25.00 © 2008 IEEE

IEEE 2008 Custom Intergrated Circuits Conference (CICC)

A 0.13 μ m CMOS Fully Differential Receiver with On-Chip Baluns for 60GHz Broadband Wireless Communications

Chao-Shiun Wang, Juin-Wei Huang, Kun-Da Chu and Chorng-Kuang Wang

Graduate Institute of Electronics Engineering & Department of Electrical Engineering

National Taiwan University, Taipei, Taiwan

Abstract — **This paper presents a fully differential receiver with on-chip baluns for 60GHz broadband wireless applications. This design consists of on-chip baluns, gm-boosted current-reuse low-noise amplifier (LNA), sub-harmonic dual-gate down conversion mixer, second IF mixer and baseband gain stage. Fully differential circuit technique is adopted to obtain good common mode performance. The gm-boosted current-reuse differential LNA mitigates the noise, gain, robustness, stability, and integration issues associated with previous solutions. The sub-harmonic dual-gate down conversion mixer will prevents the issue of the third harmonic of the LO as well. The measured conversion gain and input P1dB of the proposed receiver are 30 dB and -27 dBm, respectively. The measured noise figure at 100 MHz baseband output is around 10 dB. The proposed 60GHz receiver dissipates 44 mW with a 1.2 V supply voltage. The fully differential receiver with the on-chip baluns is implemented in a standard 0.13μm 1P8M+ RF CMOS technology.**

Index Terms — Receiver, LNA, MIXER, Current-reuse, Gm-boosting, Sub-harmonic, Dual-gate

I. INTRODUCTION

With the 7 GHz unlicensed spectrum around 60 GHz being available, increasing interests are drawn in this band for high data rate wireless consumer applications. The application of 60 GHz nowadays contains high-speed wireless LAN, and wireless home video and data link that need lots of bandwidth. The scaling down of CMOS technology, the unity current gain frequency, f_T, in MOSFET can be comparable to that in III/V technologies. CMOS technology provides advantage of high integration with the digital baseband processing circuitry, which make 60-GHz CMOS transceiver a lowest-cost option.[1]

This paper investigates the issues regarding to the design of the fully differential receiver architecture. On-chip baluns, gm-boosted current-reuse low-noise amplifier (LNA), sub-harmonic dual-gate down conversion mixer, second IF mixer and baseband gain stages are realized. The proposed receiver performs differential separation in the RF input and down converts the signal to 20GHz and subsequently to baseband. Fully differential circuit technique is adopted to obtain good common mode performance. The gm-boosted current-reuse differential LNA mitigates the noise, gain, robustness, stability, and integration issues associated with previous solutions. The sub-harmonic dual-gate down conversion mixer will prevents the issue of the third harmonic of the LO as well. This receiver

architecture lent itself better to integration and signal integrity while consuming lower power.

The paper is organized as follows: section II presents the fully differential receiver architecture. The detail circuit diagrams and design considerations of the proposed 60GHz receiver blocks are discussed in section III. Section IV demonstrates the 60GHz receiver with on-chip baluns measurement results and the performance comparisons. Finally, the conclusion is given in section V.

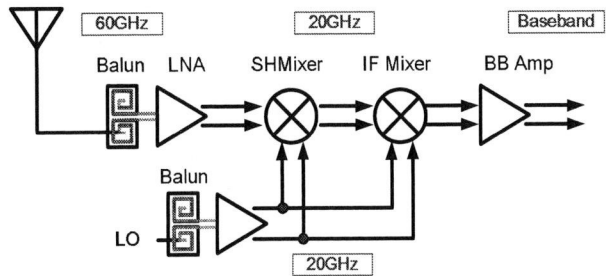

Fig. 1. The proposed 60GHz fully differential receiver with on chip baluns

II. FULLY DIFFERENTIAL RECEIVER ARCHITECTURE

Fig. 1 shows the fully differential receiver architecture. The circuit consists of on-chip baluns, gm-boosted current-reuse low-noise amplifier (LNA), sub-harmonic dual-gate down conversion mixer, second IF mixer and baseband gain stages. This architecture describes a heterodyne receiver architecture that only incorporates a 20GHz LO. The proposed receiver performs differential separation in the RF input and down converts the signal to 20GHz and subsequently to baseband. It prevents the issue of the third harmonic of the LO as well. The proposed configuration advances itself better to integration and signal integrity while consuming lower power.

Fully differential circuit can provide good common mode noise rejection and power supply variation. Since it is extremely difficult to externally generate the differential 60GHz RF signals, a single-ended-to-differential (S/D) converter (balun) is realized on the chip. The Marchand balun structure is employed for signal coupling and impedance matching in the RF input and the local oscillator (LO) signals. Whereas the reactive component is too small in millimeter-wave application, transmission lines are used for bias network, impedance matching, and interstage connection. In order to obtain the high quality factor (Q) at 60GHz, Thin-film

---

This program is supported by the National Science Council (*NSC*), Taiwan, R.O.C.

978-1-4244-2018-6/08 $25.00 © 2008 IEEE   513

microstrip (TFMS) configuration is chosen for transmission lines. It can be easily meandered and save the chip area. The layout of the whole TFMS transmission lines are verified by EM simulation to guarantee the performance at 60GHz.

III. RECEIVER BUILDING BLOCK DESIGN

A. On-Chip Marchand Balun

On-chip balun offers high levels of integration for cost savings and develops less parasitic due to shorter interconnections to devices. In addition, the balun should occupy a small area and maintain design simplicity. For high design error tolerance, a wideband, passive Marchand-type balun is used. [2]

Fig. 2 shows the design of the balun. The shape of the balun is optimized to reduce occupied area and minimizes the distance between the differential ports. The desired distance between the differential ports is closely affected by the layout requirements of the circuit that uses the balun. This Marchand type balun uses the next highest metal layer strip of half wavelength. The single-ended port is at one end while the other end is left open-circuited. Two metal strips are located immediately above the next highest metal layer strip in the top metal. One end of each top metal strip is connected to ground while the other end connects to each of the differential ports.

Port 1 refers to the single-ended port. Ports 2 and 3 refer to the differential balanced ports. |S21| and |S31| are the measured attenuations through ports 2 and 3 respectively. The theoretical values of |S21| and |S31| of an ideal balun are −3 dB, since each differential port divides to half of the power at the single ended port. Around 60 GHz, the measured values of |S21| and |S31| are −6.5dB and −6.2 dB respectively. This large bandwidth indicates a good broadband amplitude balance. The measured phase difference at the differential output is 187◦. This calculates to a phase imbalance of 7.0◦. Within an acceptable phase imbalance of ±10◦, the frequency range is between 50GHz and 70GHz. Therefore, the on-chip Marchand balun operates well between 50GHz and 70GHz, confirming its broadband characteristics.

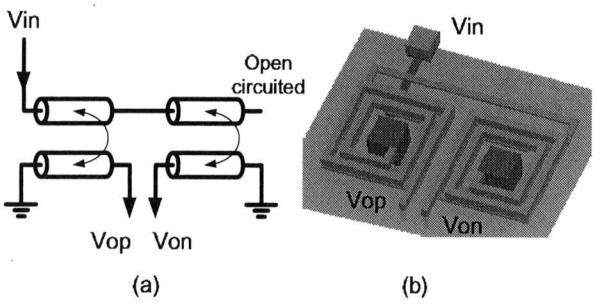

Fig. 2 On-chip Marchand balun

B. Gm-boosted and Current-reuse LNA

Gm-boosted is a well-established technique that uses an amplifier in place of A(s) to increase the dc gain of a cascode stage. [3] With negative feedback as shown in Fig. 3(a), the

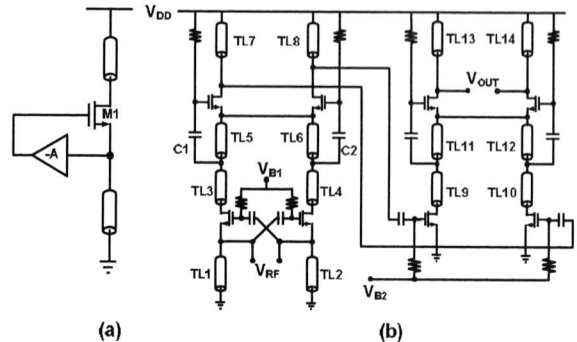

Fig. 3 Gm-boosted current-reuse low noise amplifier

effective transconductance of M1 is increased by a factor of [1 + A(s)] and the noise factor F is reduced to

$$F = 1 + \frac{\gamma}{\alpha(1 + A(s))}$$

where α and γ are empirical process- and bias-dependent parameters. The noise performance can be improved as the transconductance is increased. In a fully-differential circuit, a cross-connection can be used to provide the negative polarity of the feedback inherently. Another advantage of the common gate differential input stage with gm-boosted is that the input effective transconductance is doubled due to the capacitive cross-coupling. Thus, the current consumption can be reduced.

In order to reduce the hungry power consumption of the 60GHz LNA circuit, the current-reuse technique is used to improve the gain performance. The complete gm-boosted current-reuse differential low noise amplifier comprises the input pre-amplifier plus one additional gain stage. The noise performance is improved through the use of a gm-boosted technique, while the gain performance is improved using current-reuse techniques. [4]

The detail schematic of the proposed LNA is shown in Fig.3(b). The inverting gain is naturally available (with A = 1) in differential configuration, wherein TL1 and TL2 resonates with the input node capacitance at the operating frequency. The implementation of the gain between source and gate is mainly determined by noise considerations. In Fig. 3(b), a common source second stage is cascoded upon a gm-boosted to reuse the bias current. The output of the first stage is connected to the input of the second through a coupling-capacitor (C1,C2); another large bypass-capacitor is used at the source of M2 to provide an ac ground. TL3, TL4, TL5 and TL6 form a matching network between the two cascode stages. There is more power efficient and stable than a conventional single-stage common source LNA (CS-LNA) with comparable high gain.

The post-amplifier gain cell employs a common source and common source (CS-CS) cascoded LNA topology, where the second stage shares the bias current of the first stage to save power. The transmission line values were chosen to impedance matching at each interstage. TL7, TL8, TL13 and TL14 construct the amplifier load to resonate with the parasitic capacitances. The whole LNA circuit only draws a supply current of 6 mA at 1.2V supply voltage. Fig. 4 shows the floor plan of the differential LNA. The folded TSMS transmission lines allow placement of the active devices in close proximity and symmetry. The lowest metal ground plane is maximized under the microstrips to provide a good quality factor for the TFMS transmission line.

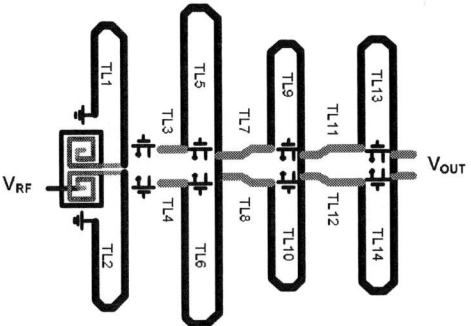

Fig. 4 Floorplan of the proposed gm-boosted current-reuse low noise amplifier

C. Sub-Harmonic Dual-Gate Down Conversion Mixer

In the integrated receiver design, DC offsets and LO-RF feed through of the mixers is the most important design issues since it falls within the bandwidth. To overcome these issues, a passive sub-harmonic mixer is chosen. The LO is a sub-harmonic of the RF frequency alleviating the DC offset due the LO-RF feed through and the third harmonic conversion of the LO signals.

The conventional dual-gate down conversion mixer is implemented as shown in Fig. 5(a).[5] In the dual gate mixer, the LO signal is applied at the gate of the upper MOS as demonstrated in Fig. 5(a). The large swing LO signal varies the dc bias of V_{G2} and also the drain-to-source voltage of the bottom MOS since the upper MOS acts as a source follower. Hence, the time-varying drain-to-source voltage of bottom MOS varies the transconductance (g_m) and also modulates the power gain to provide frequency conversion. Due to the inherent frequency doubling is achieved at the common-source of these transistors, sub-harmonic mixing with one-half the LO frequency can be obtained, as shown in Fig. 5(b).

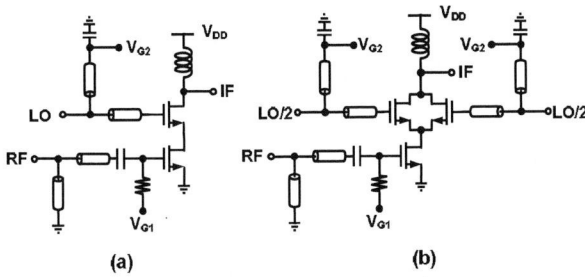

Fig. 5 Dual-gate down conversion Mixer

The detail circuit diagram of the sub-harmonic dual-gate mixer is shown in Fig. 6. The core of the first down converter in this design is the sub-harmonic dual-gate mixer. The 60 GHz RF is mixed with an LO at 20 GHz to down convert to 20 GHz IF frequency. The size of the transistors determines the noise and conversion gain performance of the mixer.

Large transistors have smaller on resistance and hence have better noise performance.

The IF down converter is also depicted in Fig.6. The amplifier output inductor L1 and L2 resonate with the total capacitance seen at the node X and Y. Most of the 20GHz IF signal is commutated with the same LO at 20 GHz to baseband by M1, M2, M3 and M4.

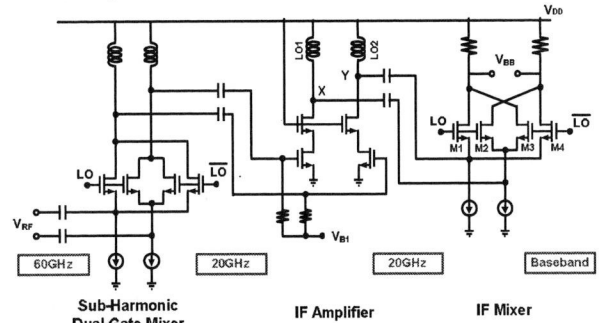

Fig.6 The schematic of the proposed sub-harmonic dual-gate mixer

C. LO and IF Buffer

The LO buffers constitute a differential pair with inductive loads. The Marchand balun is also implemented for signal coupling and impedance matching in the LO signals.

The baseband output buffer consists of a simple differential pair followed by open-drain common-source devices that can drive 50Ω instrumentation. The targeted voltage gain is about 10 dB to ensure the receiver output noise overwhelms the noise floor of spectrum analyzers and noise figure meters. To deliver a high voltage swing to a 50Ω load, each of the open-drain common-source devices consumes 5mA DC current.

IV. EXPERIMENTAL RESULTSE

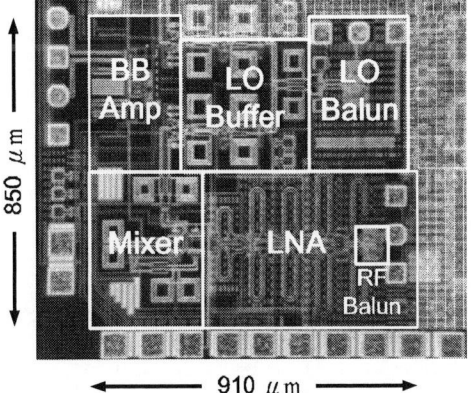

Fig.7. Die micro photographic of the proposed 60GHz receiver

The 60GHz fully differential receiver with on-chip baluns is implemented in a standard 0.13μm RF CMOS technology. The 15die micro photographic of the proposed receiver as shown in Fig.7 occupies a 910 x 850 μm² active area. The folded TSMS transmission lines are chosen for bias

978-1-4244-2018-6/08 $25.00 © 2008 IEEE   515

network, impedance matching, and interstage connection in the LNA. The lowest metal ground plane is maximized under the microstrips to provide a good quality factor for the TFMS transmission line. The current of the 60GHz fully differential receiver is about 37 mA from a 1.2 V supply voltage. The proposed gm-boosted current-reuse LNA circuit only consumes 7.2 mW.

The input referred compression point measured at the peak frequency, 61GHz, is -27dBm, as shown in Fig.8. Fig. 9 demonstrates the measured conversion gain with a peak of 30dB at 60GHz and a 3dB bandwidth of 4GHz. The noise figure is around 10dB, which is measured by the noise meter at the 100MHz baseband output and de-embedded the on-chip Machand balun loss. Table.1 compares the measured performance of the receiver with that of the published receivers.

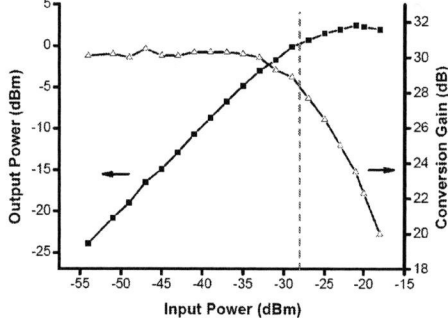

Fig.8. Measured output power and conversion gain

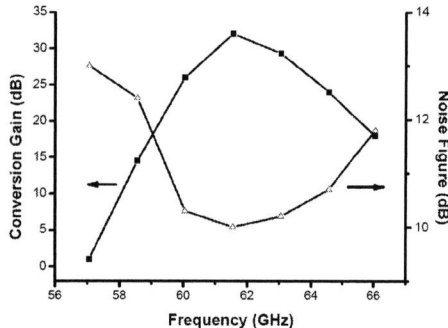

Fig.9. Measured conversion gain and noise figure vs. frequency

Table 1. Performance comparison of the 60GHz receiver

	[6]	[7]	This Work
Frequency Range	57~63GHz	49~53GHz	60~65GHz
Noise Figure (dB)	10.4 dB	7~8 dB	10~12dB
Conversion Gain (dB)	11.8dB	26~31dB	25~30dB
P_{1dB} (dB)	-15dB	-25.5dB	-27dB
Power Dissipation	77mW	80mW	44mW
Active Chip Area	~3mm^2	0.15mm^2	0.765mm^2
Technology	0.13μm CMOS	90nm CMOS	0.13μm CMOS

V. CONCLUSION

A CMOS fully differential receiver with on-chip baluns for 60GHz broadband wireless applications is proposed in this paper. This chip consists of on-chip baluns, gm-boosted current-reuse LNA, sub-harmonic dual-gate down conversion mixer, second IF mixer and baseband gain stage. Fully differential circuit technique is adopted to obtain good common mode performance. The gm-boosted current-reuse differential LNA mitigates the noise, gain, robustness, stability, and integration issues associated with previous solutions. The noise performance is improved through the use of a gm-boosting technique, while the gain performance is improved using current-reuse techniques. The sub-harmonic dual-gate down conversion mixer will prevents the issue of the third harmonic of the LO as well. The measured conversion gain and input P1dB of the proposed receiver are 30 dB and -27 dBm, respectively. The measured noise figure at 100 MHz baseband output is around 10 dB. The proposed 60GHz receiver dissipates 44 mW with a 1.2 V supply voltage. By fabricating in the 0.13μm standard CMOS RF process, all circuits including the on-chip baluns can be integrated on a single chip. The proposed receiver configuration advances itself better to integration and signal integrity while consuming lower power.

ACKNOWLEDGMENTS

The authors would like to acknowledge NTU-MediaTek Lab. (MediaTek Inc.) and NSC R.O.C. for supporting this research and Chip Implementation Center (CIC) for chip fabrication. They are also grateful to High Frequency Testing Center (HFTC) NDL on chip measurement.

REFERENCES

[1] B. Razavi, "Gadgets Gab at GHz," IEEE Spectrum, vol. 45, pp. 46-58, Feb. 2008.

[2] S. A. Maas, The RF and Microwave Circuit Design Cookbook. Boston, MA: Artech House, 1998.

[3] W. Zhuo, S. Embabi, J. Pineda de Gyvez, and E. Sanchez-Sinencio, " Using capacitive cross-coupling technique in RF low-noise amplifiers and down-conversion mixer design," Proc. Eur. Solid-State Circuits Conf. (ESSCIRC), pp. 116–119, 2000.

[4] Jeffrey S. Walling, et al., "A gm-Boosted Current-Reuse LNA in 0.18μm CMOS," in Radio Frequency Integrated Circuits (RFIC) Symposium, 2007 IEEE , vol., no., pp.613-616, 3-5 June 2007

[5] Chao-Shiun Wang, et al., "A CMOS RF front-end with on-chip antenna for V-band broadband wireless communications," 33rd European Solid State Circuits Conference, 2007. ESSCIRC , vol., no., pp.143-146, 11-13 Sept. 2007

[6] S. Emami, C. H. Doan, A. M. Niknejad and R. W. Brodersen, "A HighlyIntegrated 60GHz CMOS Frond-End Receiver," ISSCC Dig. Tech. Papers, pp. 190-191, Feb. 2007.

[7] B. Razavi, "A mm-Wave CMOS Heterodyne Receiver with On-Chip LO and Divider," ISSCC Dig. Tech. Papers, pp. 188-189, Feb. 2007.

IEEE 2008 Custom Intergrated Circuits Conference (CICC)

A Low power 20 GHz 1.5 Gb/s CMOS Injection-Pulling FSK Modulator and Frequency Discriminator for 60GHz Links

Shon-Hang Wen, Chao-Shiun Wang and Chorng-Kuang Wang

Graduate Institute of Electronics Engineering and Department of Electrical Engineering,
National Taiwan University, Taipei, Taiwan

Abstract-**This paper presents a low power frequency shift keying (FSK) modulator and demodulator (MODEM) that can support over than 1 Gb/s data rate for high-speed multimedia data communications. Compared with open-loop direct modulation of a voltage-controlled oscillator (VCO), the direct injection-pulling of a VCO enables the high speed FSK modulation without suffering frequency drift problem. The FSK discriminator uses injection-locked technique to perform frequency-to-phase transformation with low power consumption. Due to the lack of delay-dependent term in this new transformation, the accurate delay control circuits can be removed. Realized in the 0.13-μm CMOS technology, the FSK MODEM can achieve maximum data rate of 1.5 Gb/s. The power consumption of core modulator and demodulator is 14.4 mW.**

I. INTRODUCTION

With 7-GHz unlicensed bandwidth at 60 GHz being available, many kinds of transceivers have been implemented for high data rate wireless communications [1]-[6]. For the incorporation of high-speed components into portable devices, the power consumption becomes a critical issue in the next-generation high-data-rate transceiver design. In modern communication systems, orthogonal frequency division multiplexing (OFDM) scheme is popular due to its robust and anti-fading ability. However, the stringent performance of power amplifier and analog-to-digital converter restricts the low cost and low power transceiver implementation [3]. Different from it, frequency shift keying (FSK) modulation eliminates these power-hungry components, such as high-speed ADC/DAC and power amplifier, and permits the low power and high speed transceiver design. Fig. 1 shows the block diagram of the low power 60GHz transceiver.

This paper presents a 20 GHz 1.5 Gb/s FSK modulator and demodulator (MODEM). Section II and III present the FSK modulator and demodulator details respectively, and Section IV summarizes the experimental results.

II. INJECTION-PULLING FSK MODULATOR

To achieve high-speed FSK modulation, the open-loop direct modulation of a voltage-controlled oscillator (VCO) is always used for both power saving and wide bandwidth properties. However, the frequency drift problem limits it in burst-mode operation. Fig. 2 shows the injection-pulling FSK modulator architecture. The direct injection-pulling of a VCO by two fixed tones solves the frequency drift problem. An injection-locked oscillator (ILO) can track the input weak signal and amplify it with non-linear gain. To implement two fixed tones without two PLLs, we use a single sideband mixer to generate these two wanted tones in terms of input data. The integrated modulator is composed of two injection-locked

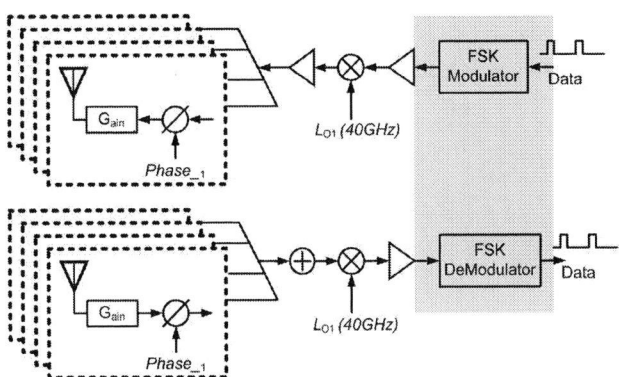

Fig. 1. High data rate wireless transceiver

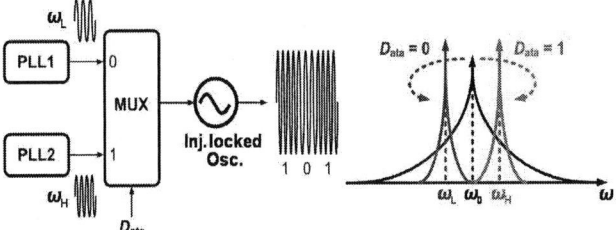

(a)   Conceptual illustration

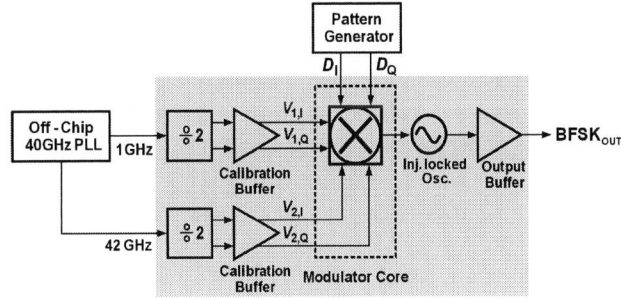

(b)   Modulator architecture

Fig. 2. The Injection-pulling FSK modulator

frequency dividers, two calibration buffers, a modulator core and ILO.

Fig. 3 shows the FSK modulator core circuit. By stacking two double-sideband (DSB) mixers and a selector at each side, the power consumption of modulator can be reduced. If D_Q and D_I are both logic low, the modulator generates a lower sideband signal due to the summation of two in-phase currents from each-side DSB mixers and vice versa. The low frequency signals ($V_{1,I}$ and $V_{1,Q}$) are injected from M_9-M_{16} for less signal phase shift to achieve precise in-phase multiplication. This arrangement also reduces the bias current of M_1-M_8 for complete switching.

978-1-4244-2018-6/08 $25.00 © 2008 IEEE   517

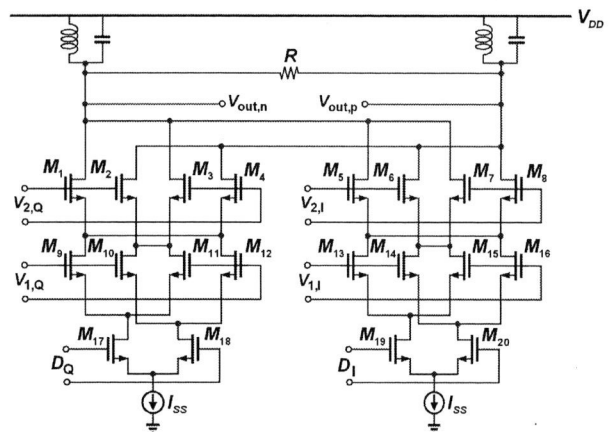

Fig. 3. Low power FSK modulator core

III. INJECTION-LOCKED FSK DEMODULATOR

Classic Bilotti's quadrature demodulator is widely employed in integrated circuits for FM/FSK signals, which comprises a passive phase shift network and a phase detector [5]-[6]. If a FSK signal passes through a phase shift network that acts as a delay cell, the FSK signal turns into

$$y(t) = A_C \cos\big((\omega_C \pm \Delta\omega)t - \omega_C D - \theta(t)\big) \quad (3)$$

, where A_C is constant amplitude, D is a delay period and $\theta(t)$ is output phase that depends on input frequencies. To guarantee a phase shift ($\omega_C D$) of 90° for normal operation of phase detector, the accurate delay control circuits are necessary. In comparison with delay cell, the ILO also performs frequency-to-phase transformation but does not require accurate delay controller. The FSK signal after ILO becomes [7]

$$y(t) = A_C \cos\big((\omega_C \pm \Delta\omega)t - \theta(t)\big) \quad (4)$$

Since there is no delay-dependent term ($\omega_C D$), the accurate delay controller is not required in this architecture. Fig. 4 shows the FSK demodulator architecture.

A. Quadrature injection-locked oscillator

The demodulator utilizes a quadrature injection-locked oscillator (QILO), as shown in Fig. 5, to perform the frequency-to-phase transformation. If the input frequency is larger than free-running frequency of QILO, the phase shift of LC tank increases from θ_0 to θ_{inj} and vice versa. This phase shift is required for the oscillator to sustain oscillation. In other words, the phase of QILO is proportional to the input frequency. Since the fixed phase difference between input and output is 26° while the input frequency is identical to the free-running frequency of QILO, the simple phase detector can be realized by a DSB mixer without losing much gain.

B. Phase detector

There are many design issues in a quadrature DSB mixer. First, since the mixer should have enough and stable conversion gain, the large signal loss at drain of M_{1-2} should be avoided. Second, the phase shift caused by the RC network at drain of M_{1-2} must be considered. Third, the V_{RF} signal delay

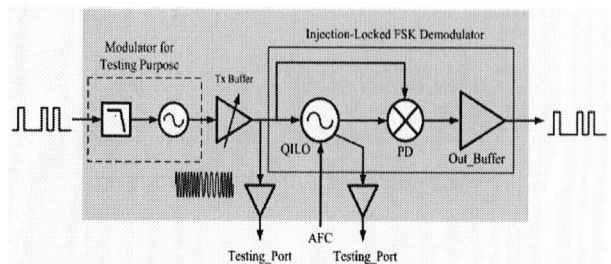

Fig. 4. Injection-locked FSK demodulator

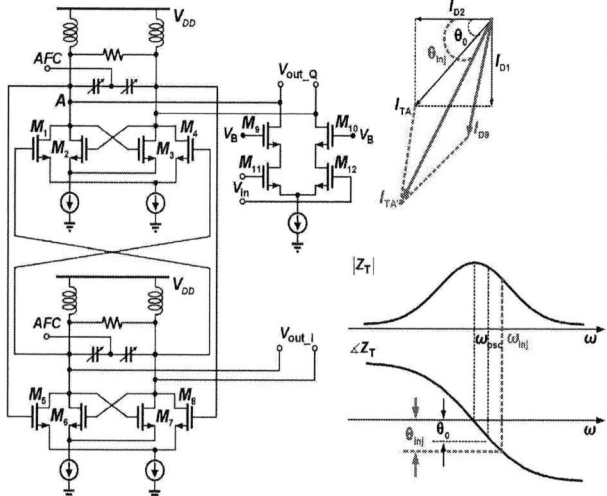

Fig. 5. Quadrature injection-locked oscillator

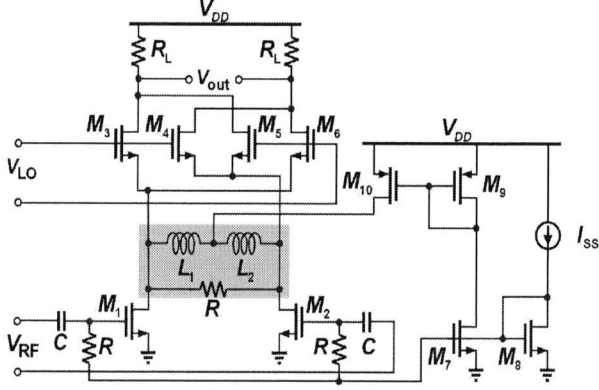

Fig. 6. Phase detector with inductive peaking

through M_{1-2} should be the same as the V_{LO} signal delay caused by the injection-locked cell. The phase detector uses the inductive peaking technique to solve these critical problems, as shown in Fig. 6. The inductors L_1 and L_2 can resonate with the excess parasitic capacitances from transistors and wires for less signal loss. Second, since there is no phase shift provided by LC tank at resonance frequency, the quadrature multiplier can operate normally. Finally, the delay of RF signal can be designed adequately through adjusting quality factor of LC tank. The current-splitting technique is also used here to increase mixer's gain. First, due to the low bias current in switching pairs M_3-M_6, the complete switching can be achieved even with small switching signal V_{LO}.

978-1-4244-2018-6/08 $25.00 © 2008 IEEE   518

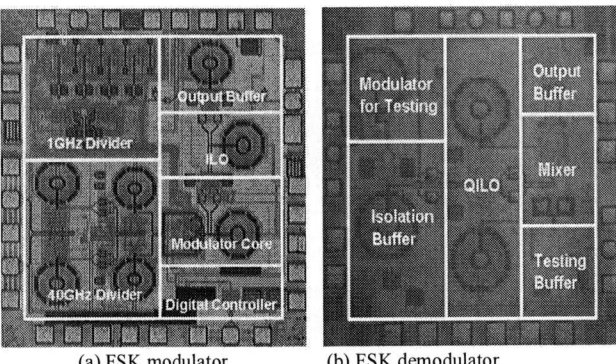

(a) FSK modulator   (b) FSK demodulator

Fig. 7. Die photo

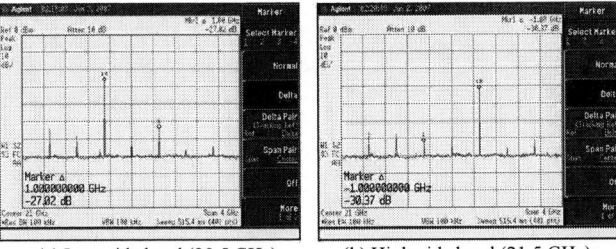

(a) Low-side band (20.5 GHz)   (b) High-side band (21.5 GHz)

Fig. 8. Output FSK spectrum at (a) Data is logic high
and (b) Data is logic low

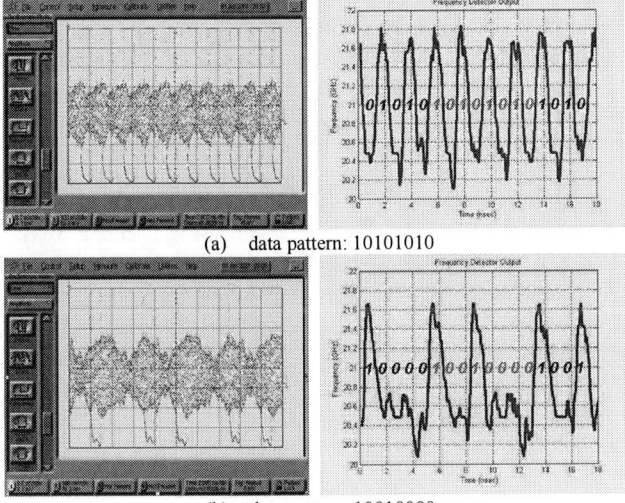

(a) data pattern: 10101010

(b) data pattern: 10010000

Fig. 9. 1 Gb/s FSK output waveform (left panel)
and corresponding demodulated data (right panel)

Second, the large load resistance can be used without suffering headroom limit.

IV. MEASUREMENTS

The FSK modulator and demodulator are fabricated in a 0.13-μm CMOS technology and tested on a high-frequency probe station. Fig. 7 shows the micrographs of modulator and demodulator which occupy 1.56 mm^2 and 1 mm^2 respectively. The core size of modulator and demodulator is 0.22 mm^2 and 0.18 mm^2 only.

The output spectrum of FSK modulator at different codes is shown in Fig. 8. If the transmitted data is logic low ($D_Q = 0$),

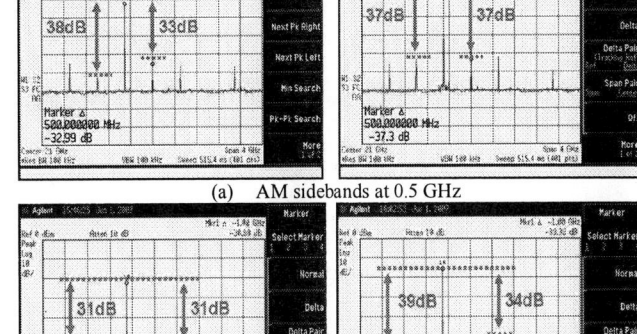

(a) AM sidebands at 0.5 GHz

(b) AM sidebands at 1 GHz

Fig. 10. Output spectrum before (left panel)
and after enabling (right panel) ILO

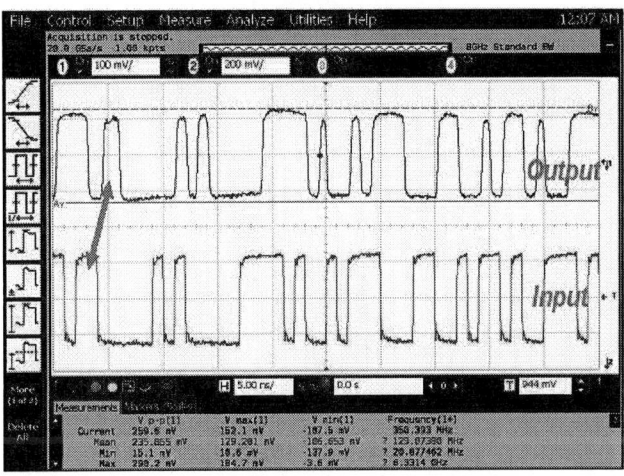

Fig. 11. Output spectrum before (left panel)
and after enabling (right panel) ILO

the output frequency is 20.5 GHz and vice versa. The estimated signal bandwidth is 2.2 GHz. To verify 1 Gb/s FSK modulation, we extract waveform data by a sampling oscilloscope (Agilent 86100C) and calculate its instantaneous frequency through a MATLAB program. Fig. 9 shows the FSK output waveforms and corresponding demodulated data under two periodic data patterns of 10010000 and 10101010. Fig. 10(a) shows the spectrum of a FSK signal before enabling ILO. Due to nonlinearity of modulator, the more than two AM sidebands are around the signal. After enabling ILO, these AM sidebands are suppressed. In addition to signal tracking, the ILO rejects the maximum spur at 0.5 GHz from 33 dB to 37 dB and spur of 1 GHz from 31 dB to 34 dB. The ILO improves the signal to spurs ratio (SSR) from 21 dB to 25 dB.

The FSK demodulator function is measured by an on-chip VCO. Due to the frequency drift of a VCO, the continuous-mode PRBS of length 2^7-1 is used in our measurement only to prevent long-run case. Fig. 11(a) shows that the FSK

978-1-4244-2018-6/08 $25.00 © 2008 IEEE   519

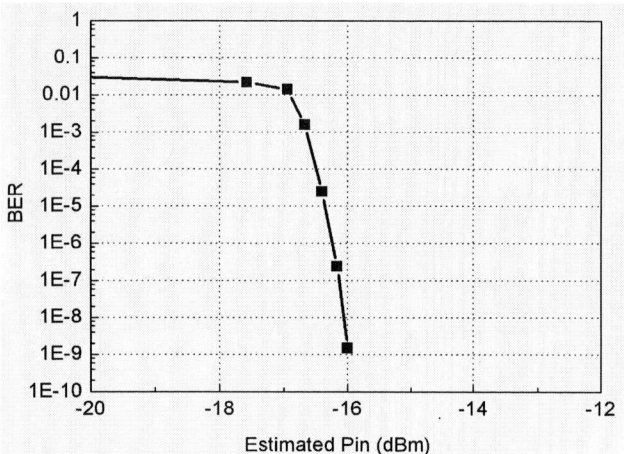

Fig. 12. BER

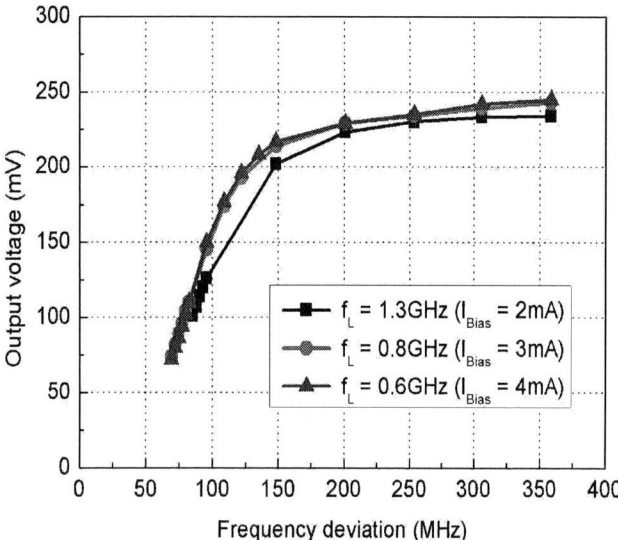

Fig. 13. Demodulation gain

Table. I. Performance summary of modulator

Reference	This work	[6]
Process	0.13-μm CMOS	0.13-μm SiGe
Modulation	BFSK	MSK
Intermediate frequency	20 GHz	9 GHz
Data Rate	1.5 Gbps	2 Gbps
Image Rejection Ratio	34 dB	N/A
LO Rejection Ratio	37 dB	N/A
Spur Rejection Ratio	37 dB	N/A
Settling Time	380 ps	N/A
Supply	1.8 V	2.7 V
Power Dissipation (core)	7.2 mW	10.8 mW
Chip area (core)	0.22 mm^2	0.03 mm^2

Table. II. Performance summary of demodulator

Reference	This work	[6]
Process	0.13-μm CMOS	0.13-μm SiGe
Modulation	FM/FSK	MSK
Intermediate frequency	20GHz	9GHz
Data Rate	1.5Gbps	2Gbps
Operation range	±0.8GHz	±1GHz
Demodulation gain	1.96V/GHz	-
Supply	1.2-V	2.7-V
Power Dissipation (core)	7.2mW	32.4mW*[1]
Chip area (core)	0.18mm^2	0.02mm^2
*[1] [6] contains the limiter power.		

achieves the high-speed operation and solves the frequency drift problem which emerges from direct modulation of a VCO. The injection-locked FSK demodulator removes the critical delay control circuits of a traditional quadrature demodulator and reduces the power consumption. The power consumption of core modulator and demodulator is 14.4mW only. These techniques provide low power solution for next-generation high-data-rate wireless communications.

ACKNOWLEDGMENTS

The authors would like to thank MediaTek, National Science Council (NSC) and Chip Implementation Center (CIC) for support and chip implementation.

REFERENCES

[1] K. Ohata, et al., "Wireless 1.25Gb/s Transceiver Module at 60GHz-Band," *ISSCC Dig. Tech. Papers*, pp. 236-237, Feb., 2002.

[2] C. -H. Wang, et al., "A 60GHz Low-Power Six-Port Transceiver for Gigabit Software-Defined Transceiver Applications," *ISSCC Dig. Tech. Papers*, pp. 192-193, Feb., 2006.

[3] A. Jerng and C.G. Sodini, "A Wideband ΔΣ Digital-RF Modulator for High Data Rate Transmitter," *IEEE J. Solid-State Circuits*, vol. 42, no. 8, pp. 1710-1722, August, 2007.

[4] Troy J. Beukema, B. Manor and Scott K. Reynolds, "Receiver and integrated AM-FM/IQ demodulators for gigabit-rate data detection," *U.S Patent 2007/0178866 A1*, Aug. 2, 2007

[5] Nakagawa, Tadao; Nishikawa, Kenjiro; Piernas, Belinda; Seki, Tomohiro;
Araki, Katsuhiko, "60-GHz Antenna and 5-GHz Demodulator MMICs for More Than 1-Gbps FSK Transceivers," *European Microwave Conference, 2002. 32nd*, pp. 1-4, Oct. 2002

[6] A. Valdes-Garcia, T. Beukema and S. Reynolds, "Multi-Mode Modulator and Frequency Demodulator Circuits for Gb/s Data Rate 60 GHz Wireless Transceivers", *IEEE Custom Integrated Circuits Conference (CICC)*, pp. 639-642, Sept. 2007.

[7] H. R. Rategh and T. H. Lee, "Superharmonic injection-locked frequency dividers," *IEEE J. Solid-State Circuits*, vol. 34, no. 6, pp. 813–821, June, 1999.

demodulator output waveform (top) is identical to the delayed input waveform (bottom). The maximum data rate of FSK demodulator is 1.5 Gb/s. Fig. 12 shows the relationship between BER and input power of FSK demodulator. The input power of FSK demodulator can be adjusted by varying the transmitter buffer's gain. The estimated sensitivity of FSK demodulator is -16 dBm while the BER is less than 10^{-9}. Fig. 13 shows that the demodulation gain of demodulator is 1.96 V/GHz at one-side locking range (f_L) of 0.8 GHz. The different bias conditions are also measured. If the frequency deviation of signal is less than operation range of ILO, the demodulation gain keeps constant almost. In other words, the BER of demodulator is not sensitive to the bias variation. The table I and II summarize the performance of modulator and demodulator.

V. CONCLUSION

This paper presents the FSK modulator and demodulator which achieve maximum data rate of 1.5 Gb/s in a 0.13-μm CMOS technology. The injection-pulling FSK modulator

978-1-4244-2018-6/08 $25.00 © 2008 IEEE

IEEE 2008 Custom Intergrated Circuits Conference (CICC)

A 24/77GHz Dual-Band BiCMOS Frequency Synthesizer

Vipul Jain, Babak Javid, and Payam Heydari

Nanoscale Communication Integrated Circuits Lab, University of California, Irvine, CA 92697

Abstract — The design of a millimeter-wave dual-band phase-locked frequency synthesizer in a 0.18μm SiGe BiCMOS technology is presented. All circuits except the voltage controlled oscillators (VCOs) are shared between the two bands. A W-band divide-by-3 frequency divider is used inside the loop after the VCOs to simplify division-ratio reconfiguration. The 0.9mm² synthesizer chip exhibits a locking range of 23.8-26.95/75.67-78.5GHz with a low power consumption of 50-75mW from a 2.5V supply. The closed-loop phase noise at 1MHz offset from the carrier is less than −100dBc/Hz in both bands. The proposed frequency synthesizer is suitable for integration in direct-conversion transceivers for K/W-band automotive radars and heterodyne receivers for 94GHz imaging applications.

Index Terms — BiCMOS integrated circuits, frequency synthesizer, phase-locked loop, automotive radar, VCO, injection-locked divider, dual-band, millimeter-wave.

I. INTRODUCTION

The proliferation of various millimeter-wave (MMW) frequency bands allocated by the Federal Communications Commission (FCC) has enabled new generations of short- and long-range automotive radar sensors. MMW automotive radars have mostly been implemented in fast, expensive III-V compound semiconductor technologies as monolithic microwave integrated circuits (MMICs). This is primarily because of the extremely high frequency of operation mandated by these radar transceivers. Recent advances in silicon (Si) and silicon-germanium (SiGe) technologies have made it possible to design silicon-based MMW integrated circuits [1]-[3].

Low cost requirements will necessitate multiband operation with lower component count in future generations of radar sensors. For instance, long and short-range detection can potentially be combined by integration of 24GHz and 77GHz radars on a single chip. This paper addresses the design of a dual-band frequency source for such systems.

W-band VCOs and frequency synthesizers have been reported in both SiGe [2], [4], and CMOS [5]-[6] technologies recently. Most of the reported synthesizers suffer from either high power dissipation, limited tuning range or high reference frequency. This paper presents a low-power K/W-band frequency synthesizer in a 0.18μm SiGe BiCMOS process with 200GHz f_T/f_{max} heterojunction bipolar transistors (HBTs). All components except the loop filter are integrated on-chip. The synthesizer has been designed for integration with a dual-band automotive radar transceiver. It can potentially be used in a 94GHz heterodyne receiver by utilizing the 77GHz VCO

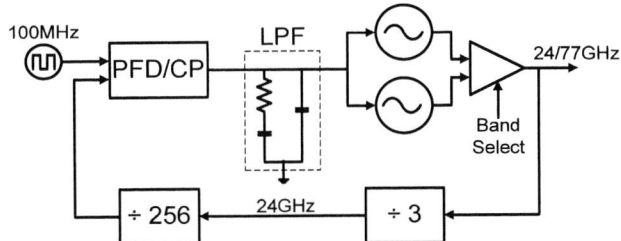

Fig. 1: Block diagram of the dual-band frequency synthesizer

output as the first local oscillator (LO) and the divide-by-6 output as the second LO.

The remainder of this paper is organized as follows: Section II discusses the synthesizer architecture and circuit design of the constituent circuits. Measurement results are presented in Section III. Finally, Section IV provides concluding remarks.

II. DESIGN OF DUAL-BAND MMW FREQUENCY SYNTHESIZER

The block diagram of the charge pump-based phase-locked frequency synthesizer is shown in Fig. 1. It consists of two LC VCOs, a divide-by-3 injection-locked frequency divider (ILFD), a divide-by-32 emitter-coupled logic (ECL) frequency divider, a divide-by-8 static CMOS frequency divider, a CMOS phase frequency detector (PFD), a CMOS charge pump (CP) and an off-chip low pass filter (LPF). In the W-band mode, the 77GHz VCO is enabled and the divide ratio is 768. The ILFD is locked to the 77GHz VCO output. In the K-band mode, the 24GHz VCO is enabled. In this mode, the ILFD is locked to the 24GHz VCO output and thus acts as a buffer, resulting in a divide ratio of 256. Although the ILFD could be used as the VCO for the K-band mode (with the 77GHz VCO disabled), the ILFD phase noise is inadequate for this purpose. This is because the ILFD in this work incorporates a tank with a low quality factor (Q) in order to achieve maximum injection-locking range. The proposed scheme allows the use of the same low phase-noise reference in both operating bands of the synthesizer. The reference frequency of the frequency synthesizer is 92-105MHz. Details of the circuit design of the synthesizer building blocks are discussed next.

A. 24GHz and 77GHz Voltage Controlled Oscillators

Fig. 2(a) shows the circuit schematic of the 77GHz VCO. It is based on a differential Colpitts topology. Microstrip transmission lines $T_{1,2}$ are used at the HBT base terminals to realize small inductance values (~25pH) with a high Q (~20).

978-1-4244-2018-6/08 $25.00 © 2008 IEEE

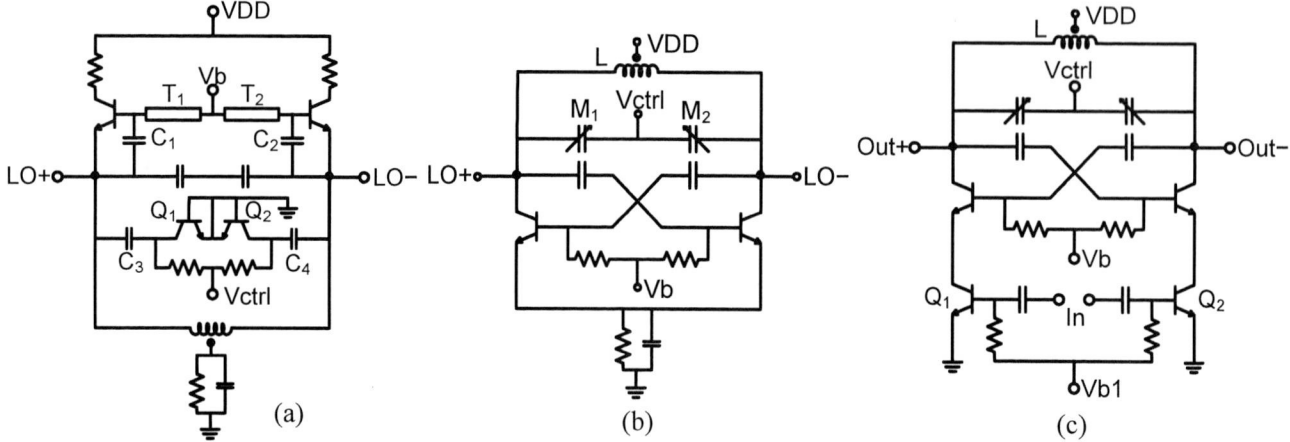

Fig. 2: Schematics of (a) 77GHz VCO, (b) 24GHz VCO, and (c) 77GHz divide-by-3 injection-locked frequency divider.

Inductive degeneration is used to improve phase noise [4] and frequency-voltage linear range. Resistive biasing is used to avoid additional noise contributions due to transistors in a conventional active tail current source. MIM capacitors $C_{1,2}$ connected between the transistor base and emitter terminals reduce the effect of voltage-variant base-to-emitter device capacitance non-linearity on VCO phase noise. HBT varactors $Q_{1,2}$ with variable base-collector junction capacitance ($Q \approx 10@77GHz$) are used for frequency tuning as the Q of 0.18μm MOS varactors is inadequate at 77GHz. Differential operation is achieved by connecting two MIM capacitors $C_{3,4}$ across the emitters of the two HBTs. The VCO has been designed for a center frequency of 78GHz with a tuning range of 4GHz.

A differential cross-coupled LC oscillator topology is used for the 24GHz VCO, as shown in Fig. 2(b). The center-tapped inductor L and accumulation-mode MOS varactors $M_{1,2}$ form the VCO tank. Similar to the 77GHz VCO, resistive biasing is used to avoid phase noise degradation. The simulated tuning range of the VCO is from 24GHz to 28.5GHz.

Both VCOs are followed by two emitter-follower buffer stages, to provide sufficient isolation from the output load, and an open-collector differential amplifier stage. The open-collector outputs of the 24GHz and 77GHz differential buffer chains are tied together and then connected to the load resistors. A digital signal is used to switch between the two bands by turning on or off the NMOS tail current sources in the two differential pairs.

B. 77GHz Injection-Locked Divide-by-3 Frequency Divider

Harmonic injection-locked frequency dividers are attractive at millimeter-wave frequencies as they have low power consumption and provide flexibility in the choice of division ratio. Static frequency dividers can be used at these frequencies [2] but at the cost of higher power dissipation and higher phase noise. In this paper, a cascode HBT-based injection-locked LC oscillator, based on the work reported in [7]-[8], has been used to realize a division ratio of three. As shown in Fig. 2(c), it consists of a cross-coupled LC VCO with the tail current source replaced by a pair of input

common-emitter HBT amplifiers $Q_{1,2}$. Without an input signal, the circuit operates as a free-running oscillator at 24GHz. When a 77GHz differential input signal is applied, it modulates the free-running state of the LC tank. Due to the non-linearity of the cross-coupled pair, several inter-modulation products result from the multiplication of the input signal and the tank oscillation. For a sufficiently large input signal, the output is locked to the intermodulation product at one-third of the input frequency.

ILFD circuits typically suffer from a limited locking range. In this work, the tank Q, the varactor C_{max}/C_{min} ratio, and the input amplifier gain have all been optimized in order to maximize the locking range and the free-running tuning range. The tank inductance has a Q of 9@25GHz and MOS varactors have been used to provide a tuning range from 24.5GHz to 28.3GHz. The divider power consumption is 15mW in the 77GHz mode.

In the 24GHz mode, the divider acts as a tuned buffer and locks to the VCO output frequency. Since the amplifier gain is higher at 24GHz, the current consumption can be decreased to obtain the same locking range as that in 77GHz mode. The minimum power dissipation in the 24GHz mode is 5mW.

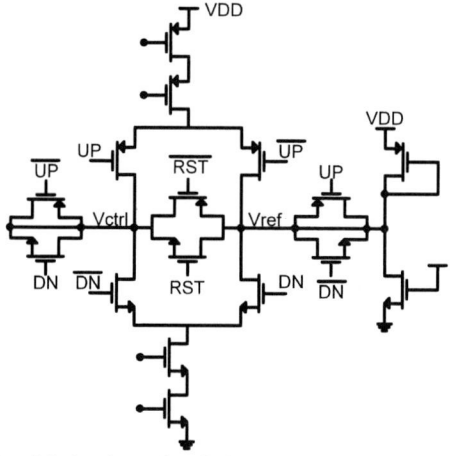

Fig. 3: Simplified schematic of the charge pump. Bias details not shown.

978-1-4244-2018-6/08 $25.00 © 2008 IEEE   522

Fig. 4: Die micrograph of the dual-band frequency synthesizer. Chip size is 1 x 0.8mm².

C. Divider Chain, PFD/CP and Loop Filter

A chain of five static ECL dividers follows the ILFD and consumes only 15mW. Three static flip-flop based CMOS dividers further divide the frequency down to the reference frequency and provide a rail-to-rail signal at the input of a standard tri-state PFD. Gradual increment of the signal amplitude through the ECL divider chain, optimized for low power-consumption, efficiently eliminates the need for an ECL-to-CMOS converter prior to the CMOS divider chain.

The charge pump schematic, inspired by the topology in [9], is shown in Fig. 3. Cascode current sources reduce the effect of the VCO control voltage variation on the charge pump UP/DOWN currents until V_{ctrl} comes within $2V_{dsat}$ of the supply rails, which in turn broadens the linearity of the PLL loop. Also it reduces the UP/DOWN current mismatch. The use of a dummy branch to steer the charge-pump current for the time when V_{ctrl} is not integrating any charge, plus the charge injection and clock feed-through cancellation provided by the dummy switches, reduce the non-idealities of the charge pump circuit. The loop filter is placed off-chip to compensate for any frequency modeling errors in the MMW circuits. A Spectre-RF/Verilog-A co-simulation methodology

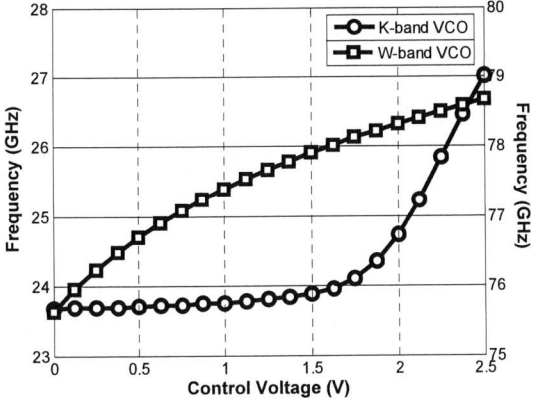

Fig. 5: Measured tuning range of the two VCOs.

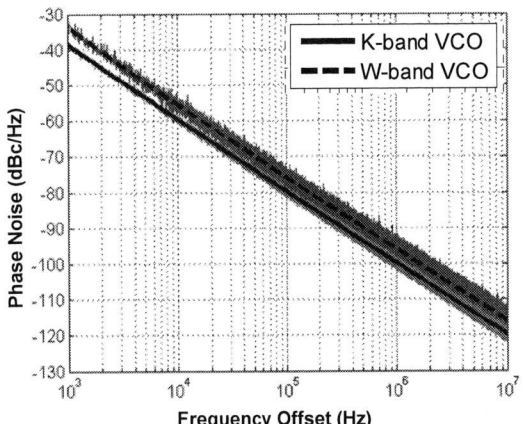

Fig. 6: Measured phase noise of the free-running VCOs.

was adopted for closed-loop simulations of the frequency synthesizer. The synthesizer has been optimized for a target loop bandwidth of 1MHz.

III. MEASUREMENT RESULTS

The dual-band frequency synthesizer was fabricated in a 0.18μm 200GHz f_T/f_{max} SiGe BiCMOS process with six metal layers. The 2.8μm-thick top metal was used to realize inductors and transmission lines in the VCOs and ILFD. Signal distribution between building blocks was done in the 1.6μm-thick penultimate metal to minimize coupling to the oscillator tanks. Fig. 4 depicts the micrograph of the 1 x 0.8mm² chip.

On-wafer measurements were carried out to characterize the synthesizer performance. A WR-10 waveguide-based setup was used for the 77GHz mode, including an Agilent 11970W harmonic mixer. The reference signal is provided by a 50-125MHz voltage controlled crystal oscillator. The divider chain was disabled to measure the free-running VCOs. The K-band VCO achieves a tuning range from 23.68GHz to 27GHz while the W-band VCO can be tuned from 75.6GHz to 78.6GHz, as shown in Fig. 5. The phase noise of the two VCOs is shown in Fig. 6 and is better than −95dBc/Hz at 1MHz offset from the carrier.

The output spectrum of the locked synthesizer in the two bands is shown in Fig. 7. The reference spurs at the output are 47-50dB below carrier. The synthesizer can be locked from 23.8GHz to 26.95GHz and 75.67GHz to 78.5GHz in the two bands. The output power of the synthesizer is −9.5dBm at 25.6GHz and −17.8dBm at 76.8GHz after de-embedding the losses of the measurement assembly. Fig. 8 shows the closed-loop phase noise performance of the synthesizer along with the phase noise of the reference input. The locked 24GHz VCO output shows a phase noise of −112dBc/Hz, −114dBc/Hz and −117dBc/Hz, at 100kHz, 1MHz and 10MHz offsets from the carrier, respectively. The corresponding phase noise of the locked 77GHz VCO output is −102dBc/Hz, −103.5dBc/Hz and −116dBc/Hz, respectively.

The frequency synthesizer consumes 50mW in the 24GHz mode and 75mW in the 77GHz mode. The 77GHz and 24GHz VCOs require 10mA and 4mA, respectively. The ILFD consumes a maximum of 6mA.

978-1-4244-2018-6/08 $25.00 © 2008 IEEE

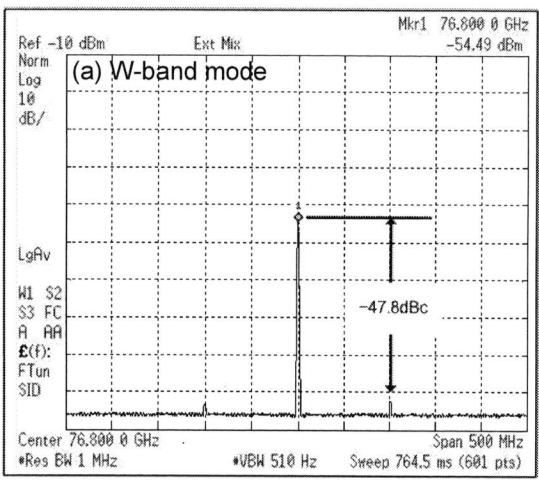

Fig. 7: Measured output spectrum of the synthesizer in (a) the W-band mode and (b) the K-band mode. Measurement setup losses have not been de-embedded.

The measured performance of the dual-band synthesizer is summarized in Table I.

IV. CONCLUSION

Design and implementation of a dual-band MMW frequency synthesizer in a 0.18μm BiCMOS technology has been demonstrated. The highly-integrated synthesizer targets 24/77GHz automotive radars, and is also suitable for 94GHz imaging applications. Measurements of the fabricated prototype demonstrate excellent results. The locking range of the synthesizer is from 23.8GHz/75.67GHz to 26.95GHz/78.5GHz. The phase noise performance is better than −100dBc/Hz at 1MHz offset from the carrier. Use of a divide-by-3 frequency divider results in a low power consumption of 50-75mW. This work represents the first step towards the realization of fully-integrated dual-band MMW radar transceivers.

ACKNOWLEDGMENTS

The authors thank Jazz Semiconductor for chip fabrication. They acknowledge Dr. Magnus Wiklund of Fujitsu Laboratories of America for valuable technical discussions.

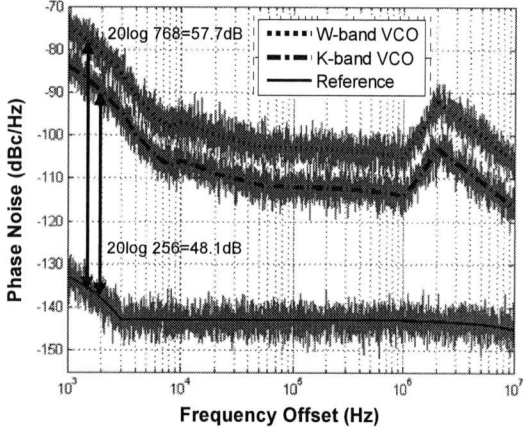

Fig. 8: Measured phase noise of the locked synthesizer output. Reference phase noise is limited by noise floor of spectrum analyzer.

TABLE I
SUMMARY OF THE MEASURED PERFORMANCE

	K-band	W-band
Locking Range	23.8-26.95GHz	75.67-78.5GHz
Phase Noise	−114dBc/Hz@1MHz	−103.5dBc/Hz@1MHz
Spurs	−49.5dBc	−47.8dBc
Output Power	−9.5dBm	−17.8dBm
Settling Time	<25μs	<25μs
Power Dissipation	50mW	75mW
−VCO	10mW	25mW
−ILFD	5mW	15mW
−Static Divider	15mW	15mW
−PFD/CP	5mW	5mW
Technology	0.18μm BiCMOS	
Die Area	1x0.8mm²	

REFERENCES

[1] A. Natarajan, *et al.*, "A 77-GHz Phased-Array Transceiver With On-Chip Antennas in Silicon: Transmitter and Local LO-Path Phase Shifting," *IEEE J. Solid State Circuits,* vol. 41, pp. 2807-2819, Dec., 2006.

[2] S. Trotta, *et al.*, "A 79GHz SiGe-Bipolar Spread-Spectrum TX for Automotive Radar," *ISSCC Dig. Tech. Papers,* pp. 430-613, Feb., 2007.

[3] I. Gresham, *et al.*, "A fully integrated 24 GHz SiGe receiver chip in a low-cost QFN plastic package," *Radio Frequency Integrated Circuits Symp.,* June 2006.

[4] H. Li and H.-M. Rein, "Millimeter-Wave VCOs With Wide Tuning Range and Low Phase Noise, Fully Integrated in a SiGe Bipolar Production Technology," *IEEE J. Solid State Circuits,* vol. 38, pp. 184-191, Feb., 2003.

[5] E. Laskin, *et al.*, "95GHz Receiver with Fundamental Frequency VCO and Static Frequency Divider in 65nm Digital CMOS," *ISSCC Dig. Tech. Papers,* pp. 180-181, Feb., 2008.

[6] J. Lee, "A 75-GHz PLL in 90-nm CMOS Technology," *ISSCC Dig. Tech. Papers,* pp. 432-613, Feb., 2007.

[7] H. Wu and L. Zhang, "A 16-to-18GHz 0.18-μm Epi-CMOS Divide-by-3 Injection-Locked Frequency Divider," *ISSCC Dig. Tech. Papers,* pp. 2482-2491, Feb., 2006.

[8] J. Jeong and Y. Kwon, "A Fully Integrated V-Band PLL MMIC Using 0.15-μm GaAs pHEMT Technology," *IEEE J. Solid State Circuits,* vol. 41, pp. 1042-1050, May, 2006.

[9] A. L. S. Loke, *et al.*, "A Versatile 90-nm CMOS Charge-Pump PLL for SerDes Transmitter Clocking," *IEEE J. Solid State Circuits,* vol. 41, pp. 1894-1907, Aug., 2006.

IEEE 2008 Custom Intergrated Circuits Conference (CICC)

An X/Ku-Band Frequency Synthesizer Using A 9-Bit Quadrature DDS

Xuefeng Yu, Fa Foster Dai, Senior Member, IEEE
Dayu Yang, J. David Irwin, Life Fellow, IEEE, Richard C. Jaeger, Fellow, IEEE)
Dept of Electrical and Computer Engineering, Auburn University, Auburn, AL 36849-5201

Abstract-This paper presents an X/Ku-band fine-tuning frequency synthesizer using a quadrature DDS implemented in a 0.18μm SiGe BiCMOS technology. The frequency synthesizer comprises a 9-bit quadrature DDS, an 11.7GHz quadrature VCO and image rejection mixers. The outputs of the quadrature DDS are down-converted to 9.4~11.7GHz and up-converted to 11.7~14.0GHz, respectively. The die area of the synthesizer is 3.0x3.0mm^2 and the power consumption is 2.6W under a 3.3V supply. The chip is measured with a 48-pin leadless free ceramic package and external cooling.

I. INTRODUCTION

In the next generation radar system, there are emerging trends toward digitization in radar receiver designs by applying direct intermediate frequency-to-digital conversion (IF sampling) and direct digital synthesis (DDS). The digital radar receivers can obtain much higher precision, low noise, low power and better stability than analog counterparts. Moreover, it can retain the flexibility of digital techniques such as direct digital modulation and waveform generation. A DDS generates a digitized waveform of a given frequency by accumulating phase changes at a higher clock frequency. Microwave range DDS has been developed in both InP and SiGe technologies [1-3] with output frequency up to 10GHz. It's highly desirable to develop frequency synthesis means for X/Ku-band applications. By mixing the outputs of a quadrature DDS (QDDS) and a quadrature VCO, X/Ku-band waveform generation can be achieved.

This paper presents a 0.18μm SiGe BiCMOS X/Ku-Band frequency synthesizer, which consists a 9-bit quadrature DDS, an 11.7GHz quadrature VCO and image rejection mixers. The Nyquist output of the quadrature DDS for 4.6GHz clock input is 2.3GHz. With up-convert and down-convert mixers, 9.4~14.0GHz range can be covered. Packaged in a 48-pin ceramic leadless package, the MMIC occupies 3.0x3.0mm^2 and consumes 2.6W.

II. ARCHITECTURE AND CIRCUIT DESIGN

The conceptual diagram of the frequency synthesizer is shown in Fig. 1. The quadrature outputs from the local oscillator are mixed with the outputs of a quadrature DDS and the mixers outputs are summed and subtracted with each other, so the up-converted and down-converted sine waveforms are derived [4]. The local oscillator generates quadrature outputs with relatively fixed output frequency ω_0, which are mixed

with the outputs of a quadrature DDS. Then the mixer outputs are summed and subtracted with each other, so the up-converted cosine waveforms with a frequency of $\omega_0 + \omega$ or $\omega_0 - \omega$ are derived. Assuming the local oscillator frequency is higher than the output frequency of the quadrature DDS, the above mixing scheme can be used to up convert the DDS output frequency to a higher frequency band.

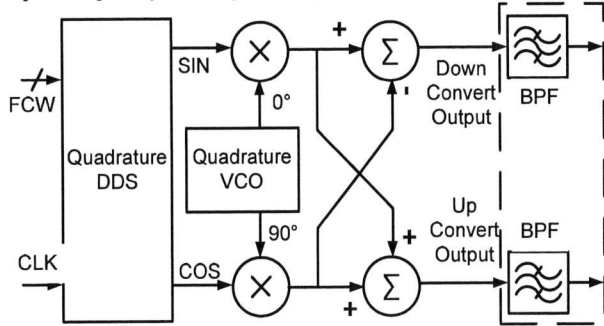

Fig. 1 Concept diagram of the frequency synthesizer.

Theoretically the output should be clean of alias images. However, in practice the DDS output contains harmonics and spurs that significantly deteriorate the purity of desired output waveforms. The imperfections of the mixers due to leakage and second order effects will introduce spurs and harmonics to the output signals. In the multiple GHz DDS design, this will be more complicated due to the impact post by large scale circuit and huge power dissipation.

The frequency synthesizer contains three major parts, the quadrature DDS, quadrature VCO and mixers. DDS can provide quadrature signals with accurate I/Q matching, as shown in Fig. 2. The quadrature DDS is formed by merging two sine-weighted current steering DACs and a 9-bit pipeline accumulator. The nonlinear DAC approach is still attractive for the microwave DDS design because it provides drastically speed improvement to the ROM based or algorithm based DDS design. To reduce the effect of the amplitude error introduced spurs in an ultra high speed DDS needs to be taking into account during the design. The phase truncation error introduced spurs have already minimized because only one bit of the phase accumulator output has been truncated.

978-1-4244-2018-6/08 $25.00 © 2008 IEEE

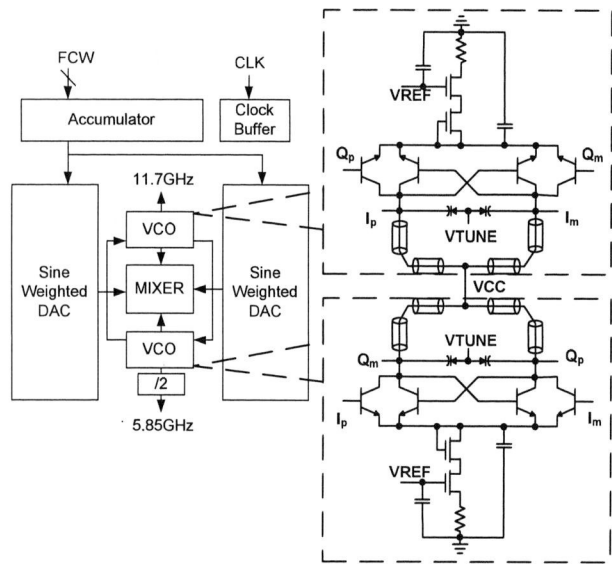

Fig. 2 Block diagram with the circuit of quadrature VCO.

The input frequency control word (FCW) specifies the output frequency of the quadrature DDS. The output of the quadrature VCO is tuned to 11.7GHz and is also divided by 2 to generate 5.85GHz for potential use as the DDS clock. The quadrature VCO design adopts a standard cross-coupled LC-VCO topology. The center tapped inductor in the LC-tank has been replaced by 4 transmission line inductors to facilitate a symmetrical and compact layout. However, the Q factor is relatively lower than typical spiral inductors, which needs to be accounted for in the design. To reduce the losses of the inductors, thick analog metals are used for the connections between the transmission line inductors. To produce the 90 degree phase word, binary number of '01' need to be added to the two most significant bits of the DAC input. Translating the add function into gate level, the output of the MSB is the results of an Exclusive-OR (XOR) of the first two MSB inputs

and the output of the 2nd MSB is the inversion of the 2nd MSB input. Because all the digital logics have differential outputs, only one XOR gate is needed to be inserted at the inputs of the sine-weighted DAC to converter it to DAC with 90 degree output phase difference.

The essential building block of the nonlinear DAC is the sine weighted current source matrix. The smallest unit current of each current source is 0.1mA, which should provide the current switches with enough switching speed when toggling. The largest current in the current source is 0.7mA, which is composed of 7 identical current sources. The current switch contains two differential pairs, with minimal sized transistors, and a cascade transistor, to isolate the current sources from the switches, and improve the bandwidth of the entire group of switching circuits. In this ultra-high speed DDS design, the ROMless structure with two nonlinear current steering DACs is employed, and the sine/cosine mapping function is performed by a sine-weighted DAC instead of using the traditional ROM-based sine waveform look-up-table. By eliminating the ROM, speed of the DDS is improved and the power consumption is reduced. This quadrature DDS comprises a 9-bit pipeline accumulator and two 8-bit sine-weighted current-steering DACs. To produce the 90 degree phase, an XOR gate is inserted into the inputs of one sine-weighted DAC. Since the out frequency cannot exceed the Nyquist rate, an 8-bit frequency control word (FCW) is fed into a 9-bit pipeline accumulator with the MSB of the accumulator input tied to zero. The LSB of the 9-bit phase word is truncated, and its MSB is used to provide the proper mirroring of the sine waveform about the π phase point. Its 2nd MSB is used to invert the remaining 6-bits for the 2nd and 4th quadrants of the sine wave prior to the decoding logic. The outputs of 3:8 column-row decoders go to the switch matrix to control the switches in each DAC cell. The latch and switch matrices contain 64 cells.

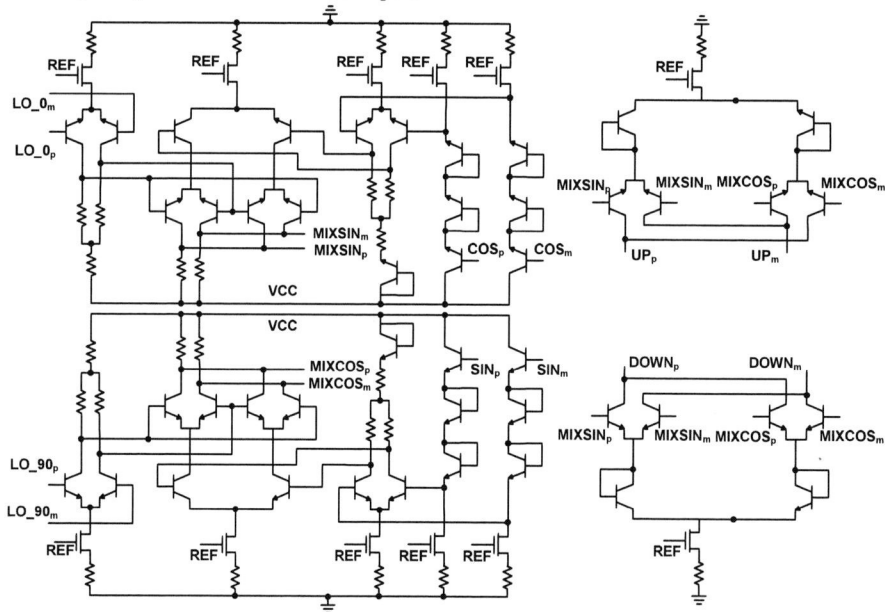

Fig. 3 Circuits of up-convert and down-convert mixers.

978-1-4244-2018-6/08 $25.00 © 2008 IEEE

The implemented ultra-high speed DDS presents the first mm-wave quadrature DDS design reported so far. When compared with other single-phase mm-wave DDSs [1-3], it's more complex, yet more compact and has lower power, as shown in Table 1. The minimum size of the InP transistor is much larger than that of the SiGe transistor. Although the current density needed to achieve peak f_T frequency in InP and SiGe technologies are similar, the current required to operate the minimum size SiGe transistor is much less. It is for this reason that the SiGe DDS leads to a superior power efficiency performance.

TABLE.1 ULTRA-HIGH SPEED DDS PERFORMANCE COMPARISON. THIS WORK IS QUOTED FOR SINGLE-PHASE / QUADRATURE-PHASE.

Technology f_T/f_{max} [GHz]	InP 137/267 [1]	InP 300/300 [2]	SiGe 120/100 [5-6]
Emitter area of min npn [μm^2]	1.5x4	0.4x2	0.2x0.64
Current density at peak f_T [mA/μm^2]	1~1.2	5	6
Peak f_T current of min npn [mA]	7.2	4	0.77
Break down voltage B$_{vceo}$ [V]	8	4	1.8
Accumulator size [bit]	8	8	9
DAC resolution [bit]	7	7	8
Max clock frequency [GHz]	9.2	13	9.6/6.3
SFDR [dBc]	30	26.67	30/26
Power consumption [W]	15	5.42	1.9/2.5
Number of Transistors	3000	1646	9600 /13500
Die size [mm^2]	8x5	2.7x1.45	2.3x0.7 /2.3x2.5
FOM[GHz/W/Phase]	0.5	2.4	5.1/5.04

To shorten the connections to the mixers and make the layout as symmetrical as possible, the mixers are placed in the center of the chip, and two VCOs and two sine-weighted DACs are placed at the opposite sides of the mixers. The die photo of the frequency synthesizer is shown in Fig. 4. The active area is approximately 2.5x2.5mm^2.

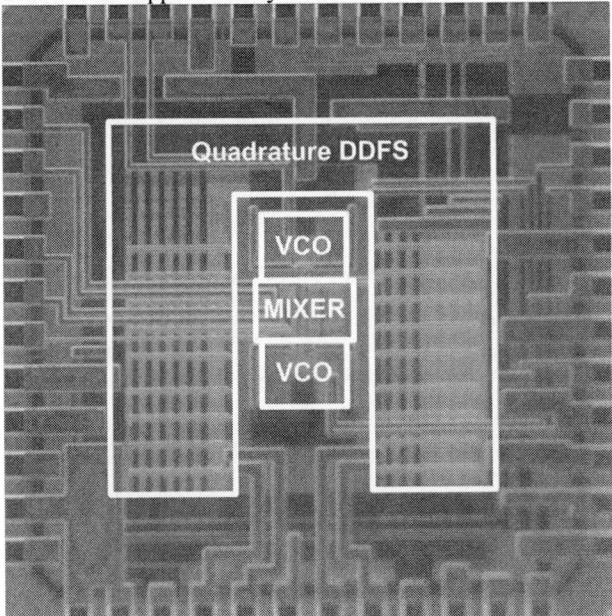

Fig. 4 Frequency synthesizer die photo.

III. MEASURED RESULTS

The test is performed on ceramic leadless free packaged chips. The test board was built using Rogers RO4003 laminate board, which has a loss tangent of less than 0.003 and good temperature stability. To convert the single-ended signal to differential clock inputs, a 180 degree 3dB hybrid coupler is employed at the clock input. For the differential outputs, a second hybrid coupler is inserted into the output path to covert them into single-end for testing. To ensure the chips working in the safe range, external air cooling is used. The measured I/Q waveforms with a digital oscilloscope confirm the 90 degree phase difference of the outputs of the quadrature DDS, as shown in Fig. 5. The measured amplitude imbalance is 5% and the phase imbalance is 2 degree.

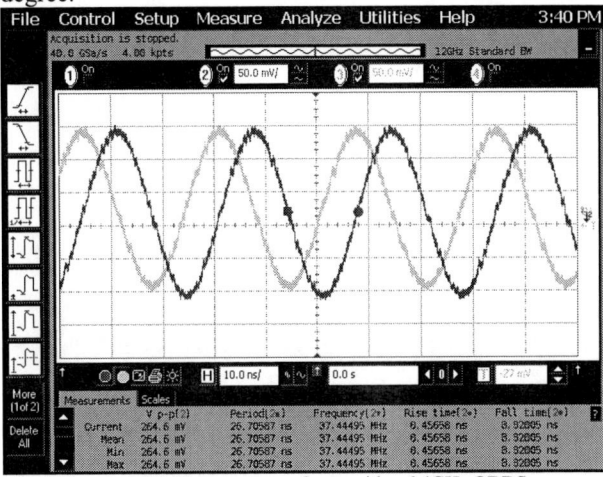

Fig. 5 Measured 37MHz output waveforms with a 6.4GHz QDDS

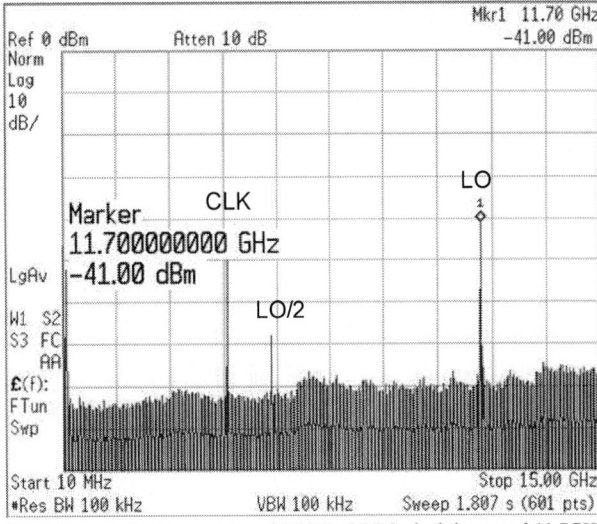

Fig. 6 Measured output spectra of 4.6GHz QDDS clock input and 11.7GHz LO output.

The spectra of the frequency synthesizer outputs shown in Fig. 6-8 are taken at the down-convert output side without calibrate the attenuation. Fig. 6 is the output spectrum of 11.7GHz VCO output and 4.6GHz DDS clock input when

DDS has been turned off. The leaked clock power to the mixer output is -50dBm and the power of leaked local oscillator is -41dBm. The spur at 5.85GHz is purely due to the leakage of the divide-by-2 output of the local oscillator, which contains built in divider for test purpose. The divided output of the local oscillator is attenuated by 27dB.

Fig. 7 shows the Nyquist output spectrum of the DDS with a 4.6GHz clock input when the local oscillator has been turned off. The output of the DDS is located close to 2.31GHz. Since the measurement is taken at the mixer output side, the DDS output power has been significantly reduced. The output power of the quadrature DDS with single output is approximately -53.75dBm.

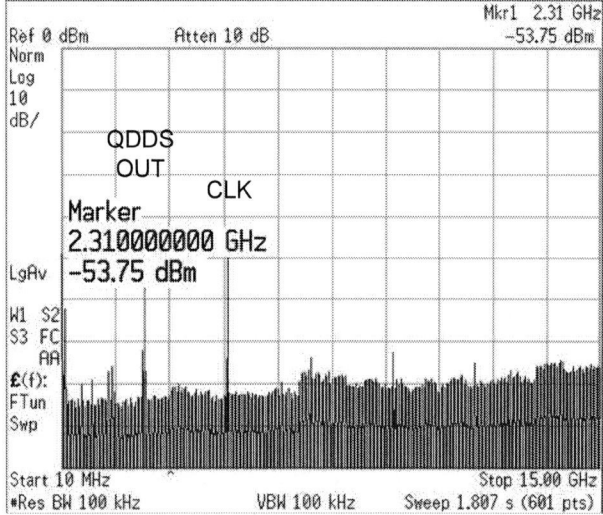

Fig. 7 Measured output spectra of 4.6GHz QDDS clock input and 2.3GHz QDDS output.

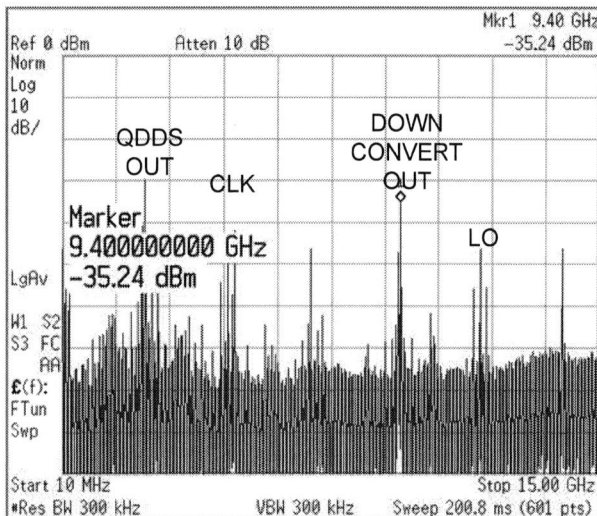

Fig. 8 Measured output down-converted 9.4GHz output.

In Fig. 8, the local oscillator and DDS are switched on and the output spectra of mixed outputs are shown. The frequency of the down-converted signal is 9.4GHz with a power of -35.24dBm and the up-converted 14.0GHz signal can also be noticed which has a power of -46dBm. During

the measurement, one of the quadrature outputs of the quadrature VCO shows a strong distortion. One of the reason cause it can be explained as the interconnection wires connecting the transmission line inductors used in the LC tanks are possibly induce inductive peaking and drive some of the transistors into saturation. The single side band suppression has been also affect by the imbalance of the integrated quadrature VCO, which will be improved in the next version.

IV. CONCLUSION

In this work, a 0.18µm SiGe BiCMOS X/Ku-band frequency synthesizer has been implemented and tested. Containing a 9-bit quadrature DDS, an 11.7GHz quadrature VCO and image rejection mixers, this frequency synthesizer provides a solution for digital tuned and modulated signal generator over 10GHz area. Combined with single side band mixers, the design can cover 9.4~14.0GHz output range. The chip is packaged in a 48-pin ceramic leadless package and occupies $3.0x3.0mm^2$ area. With a 3.3V power supply and 4.6GHz external clock input, the total power consumption of the chip is 2.6W.

ACKNOWLEDGEMENT

The authors would like to acknowledge Vasanth Kakani for the VCO design. The authors would like to thank Eric Adler and Geoffrey Goldman at the Army Research Lab for funding this project, Nat Albritton and Bill Fieselman at Amtec Corporation for business management.

REFERENCES

[1] A. Gutierrez-Aitken, et al, "Ultrahigh-speed direct digital synthesizer using InP DHBT technology," *IEEE J. Solid State Circuits*, vol. 37, no. 9, pp. 1115-1121, Sept. 2002.

[2] S. E. Turner and Kotecki, "Direct digital synthesizer with ROM-Less architecture at 13-GHz clock frequency in InP DHBT technology," *IEEE Microwave and Wireless Components Letters*, vol. 16, no. 5, pp. 296-298, 2006.

[3] S. E. Turner, D. E. Kotecki, "Direct Digital Synthesizer With Sine-Weighted DAC at 32-GHz Clock Frequency in InP DHBT Technology," *IEEE J. Solid State Circuits*, vol. 41, no. 10, pp. 2284-2290, Oct. 2006.

[4] Cushing R., (2000). "Single-Sideband Upconversion of Quadrature DDS Signals to the 800-to-2500-MHz Band," *Analog Dialog*, v. 34, no. 03, May 2000.

[5] Xuefeng Yu, Foster Dai, Dayu Yang, Vasanth Kakani, J. David Irwin and Richard C. Jaeger, "A 9-Bit 6.3GHz 2.5W Quadrature Direct Digital Synthesizer MMIC," *IEEE VLSI Symposium*, pp. 52-53, Kyoto, Japan, June 2007.

[6] Xuefeng Yu, Foster Dai, J. David Irwin and Richard C. Jaeger, "A 9-bit 9.6GHz 1.9W direct digital synthesizer implemented in 0.18µm SiGe BiCMOS technology," to appear in *IEEE Journal of Solid-State Circuits*, June 2008.

Tuesday Poster Session

Notes

978-1-4244-2018-6/08 $25.00 © 2008 IEEE

IEEE 2008 Custom Intergrated Circuits Conference (CICC)

A Dynamic Offset Control Technique for Comparator Design in Scaled CMOS Technology

Xiaolei Zhu[1], Yanfei Chen[1], Masaya Kibune[2], Yasumoto Tomita[2], Takayuki Hamada[2],

Hirotaka Tamura[2], Sanroku Tsukamoto[2] and Tadahiro Kuroda[1]

[1] *Department of Electronics and Electrical Engineering, Keio University, 3-14-1, Hiyoshi, Kohoku, Yokohama, Japan*

Phone/Fax: +81-45-566-1779, E-mail: zhuxl@ kuro.elec.keio.ac.jp

[2] *Fujitsu Laboratories Limited, Kawasaki, Japan*

Abstract-A principle of charge compensation approach for comparator offset control is analyzed. A dynamic offset control technique that employs charge compensation by timing control is proposed for comparator design in scaled CMOS technology. The analysis has been verified by fabricating a 65 nm CMOS 1.2 V 1 GHz comparator that occupies 25 x 65 μm^2 and consumes 380 μW. Circuits for offset control occupies 21% of the areas and 12% of the power consumption of the whole comparator chip.

I. INTRODUCTION

Converter architectures that incorporate a large number of comparators in parallel to obtain a high throughput rate impose stringent constraints on the delay, resolution, power dissipation, input voltage range, input impedance, and area of those circuits [1]. To lower the power consumption and increase the speed, the channel lengths of the transistors are minimized at the expense of an increased level of mismatch.

For higher speed and resolutions in deep sub-micron CMOS technology it is often necessary to cancel or calibrate for these effects by means of circuit or algorithmic techniques [2]-[6]. Among recent comparator offset cancellation techniques, an approach using capacitors with dynamic correction to adjust the output loads of comparators [5] degrades the response. Another approach using capacitors to control current for offset cancellation [6] requires refreshing for the capacitors.

This work proposes a dynamic offset control technique based on charge compensation by timing adjustment. The analysis shows that the offset in the comparator can be controlled with appropriate design. We verify the analysis by fabricating a 65 nm CMOS 1.2 V 1 GHz comparator that occupies 25 x 65 μm^2 and consumes 380 μW.

II. DYNAMIC OFFSET CONTROL TECHNIQUE

A. Principle of Charge Compensation Approach for Comparator Offset Control

Fig. 1 shows a comparator consists of preamplifier and cross-coupled inverters, I_1 and I_2, which work as a regenerative latch. The differential input signal is amplified by the preamplifier and converted into current signals by PMOS transistors M_1 and M_2. The resulting current signals, I (M_1) and I (M_2), are introduced into the regenerative stage through the inverter PMOS source nodes, a and b. Logic levels are

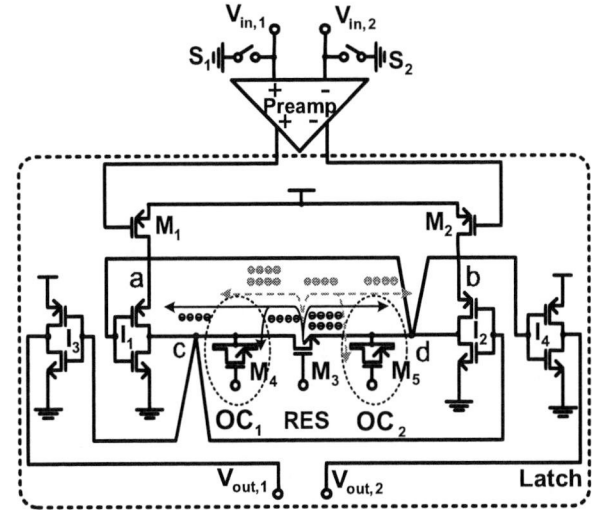

Fig. 1. Comparator with dynamic offset control technique.

produced at V_{out} when the latch stage is enabled. The regenerative nodes, c and d, are connected through reset transistor M_3. Two source-drain shorted transistors M_4 and M_5 are implemented on each regenerative node respectively. M_4 and M_5 have the same channel length with that of M_3 while have the half channel width of M_3.

The principle of the dynamic offset control is explained in detail as follows.

During the reset period, the differential inputs pair of the preamplifier is biased to ground through switches S_1 and S_2. Assuming that the offset of the preamplifier is positive and the latch stage is ideally symmetric, then the voltage on node a is lower than node b since the current I (M_2) > I (M_1). During this period, the M_3 is on by keeping its gate voltage, RES, at high while M_4 and M_5 are off by applying their gate voltages, OC_1 and OC_2, at low. Thus the regenerative nodes are shorted together through M_3, resulting in $V_c \approx V_d$. The total charge Q_{ch} in the inversion layer of M_3 is

$$Q_{ch} = LWC_{ox}(V_{RES} - V_c - V_{TH}) = LWC_{ox}V_{od} \qquad (1)$$

where L denotes the effective channel length, C_{ox} multiplied by W presents the total capacitance per unit length, and V_{TH} denotes the threshold voltage of M_3. Equation (1) indicates that Q_{ch} depends on the overdrive voltage V_{od}. Because the

978-1-4244-2018-6/08 $25.00 © 2008 IEEE

531

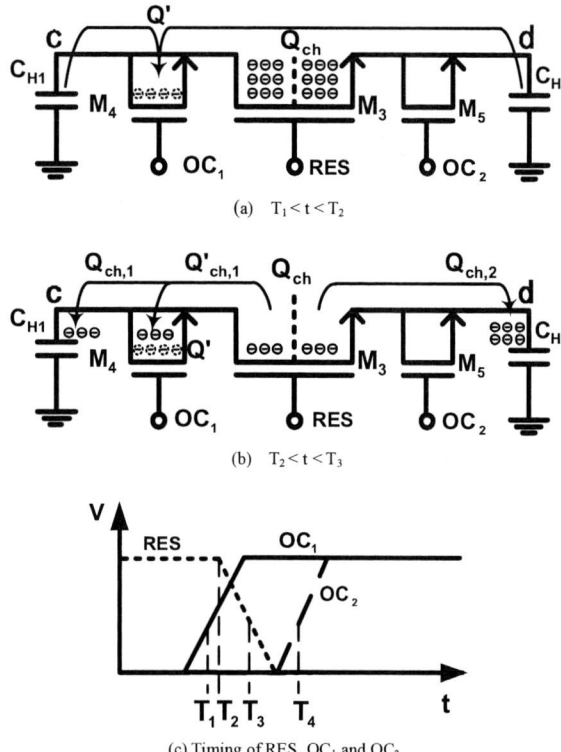

(a) $T_1 < t < T_2$

(b) $T_2 < t < T_3$

(c) Timing of RES, OC_1 and OC_2.

Fig. 2. Mechanism of offset control by charge absorption.

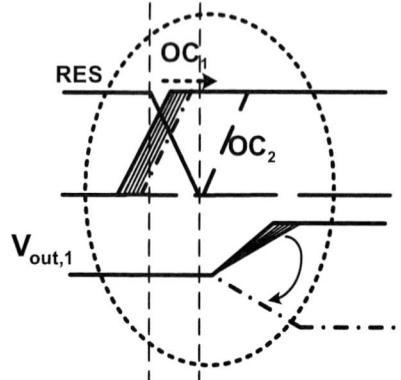

Fig. 3. Timing diagram of clock, offset control and output.

charges absorbed by M_4 and M_5 also have a dependency on the overdrive voltage, the same as M_3 as explained in (1), the voltage at node c and d will be influenced by the overdrive voltage of M_4 and M_5 at the moment M_3 turns off.

To perform the offset compensation, the channel charge is absorbed by M_4 and M_5 with appropriate timings. Fig. 2 shows the mechanism of the proposed timing-based offset control. Here, we explain the case where a positive input-referred offset is compensated for.

Let T_1, T_3 and T_4 be the moments at which the overdrive voltages reach the thresh levels of M_4, M_3 and M_5, T_2 is the moment at which M_3 starts switching from on to off. The timings are chosen to be $T_1 < T_2 < T_3 < T_4$. This means M_3 starts to turn off while the charge absorption by M_4 is still continuing.

During the period between T_1 and T_2, M_4 absorbs charges Q', illustrated in Fig. 2 (a), from parasitic capacitors C_{H1} and $C_{H2,}$ on the nodes c and d. Thus the charge absorption has approximately the same contribution to the potentials on nodes c and d. Assuming the channel transit time for M_4 is large compared to $T_2 - T_1$, the channel of M_4 will be partially depleted during this period.

During the period between T_2 and T_3, the latching operation starts as reset transistor M_3 is being turned off by RES. Meanwhile, charge injection and clock feedthrough [7] appear at the latch stage, however the charge injection dominates the effects on the potentials of the regenerative nodes c and d. Since M_4 is turned on while M_5 remains off in this period, as illustrated in Fig. 2 (b), some of the charge Q'$_{ch,1}$ injected from

the drain of M_3 is absorbed by M_4, and the remaining charge $Q_{ch,1}$ is deposited by C_{H1} on node c. The charge $Q_{ch,2}$ injected from the source of M_3 is not absorbed by M_5 but deposited by C_{H2} on node d. Thus the charge absorption by M_4 in this period contributes to generating a compensation voltage at the initial state of the latch regeneration. Adjusting the phase between T_1 and T_2 controls the amount of the compensation voltage.

The value of $T_4 - T_3$ is chosen to be large enough comparing to the regeneration time constant. This ensures that the input signal along with the compensation voltage grows enough by the time when M_5 starts to absorb charge from C_{H2} on node d. Thus the charge absorption by M_5 will affect little on the latch output.

Analysis above shows the threshold level can be controlled by timing adjustment of OC_1 and OC_2.

B. Analysis of Timing Control

Based on the aforesaid mechanism, the timing of signals OC_1 and OC_2 for offset control should be adjusted in time domain in order to let two transistors M_4 and M_5 switch an appropriate manner to allocate the charge injected by M_3 between nodes c and d so as to control the offset. As shown in Fig. 3, the timing diagram of clock, offset control and output shows how the dynamic offset control technique works. Here, assuming that the comparator has a positive input offset, OC_2 lags OC_1. To train the circuit, inputs are shorted to ground (S1 and S2 are on in Fig. 1) and the latching operation is performed over a series of phase steps on signal OC_1 with fixed control signals OC_2 and RES. When the phase offset is perfectly cancelled, in theory, the output will remain metastable forever. In practice, it latches to opposite states on each side of the phase offset that most closely cancels the offset. The phase offset is then fixed to one of these two values. In the case of negative offset the training operation is the same, but OC_1 lags OC_2.

C. Building Blocks

Based on the timing analysis in section II.*B*, the dynamic offset control needs to control the timing of OC_1 and OC_2. Fig. 4 shows an analog-to-timing controller (ACT). Two delay elements (I_2, I_3), two tunable delay elements (I_7, I_8) and four buffers (I_1, I_4, I_5, I_6) are implemented to control the timing of

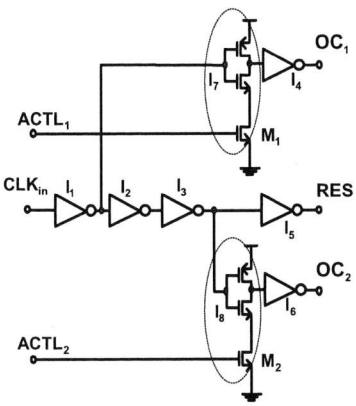

Fig. 4. Analog-to-Timing Convertor.

OC_1, OC_2 and RES. In the case of negative offset, OC_1 and OC_2 are swapped.

Although the timing among RES, OC_2 and OC_1 is controlled by implementing analog voltage, it will be feasibly controlled by digital, replacing the M_1 and M_2 in Fig. 4 with small digitally controlled switches to regulate the current and adjust timing. Such a digital to timing controller could be useful for the real applications.

D. Simulation Results

Fig. 5 shows simulated wave forms with the comparator in Fig. 1 and the timing controller in Fig. 4, giving 50 mV of input offset to the preamplifier. Gain of the preamplifier is 3dB. $ACTL_1$ is set 550 mV to ensure M_5 turns on after M_3 completely turns off. $ACTL_1$ is swept from 0.7 V to 1.2 V in 10 mV steps. The transition appears at the latch output when $ACTL_1$ changes from 850 mV to 840 mV.

Fig. 6 shows the relationship between $ACTL_1$ and the input offset, by changing the input offset to the preamplifier. According to this simulated result, 500 mV of $ACTL_1$ variation causes OC_1 delay to change by approximately 50 ps and 75 mV input offset is controlled. The control ratio ($\Delta V_{off} /\Delta ACTL_1$) is 0.167.

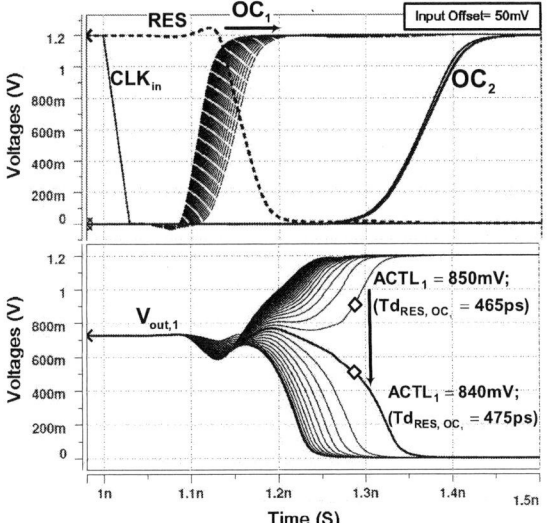

Fig. 5. Simulation results.

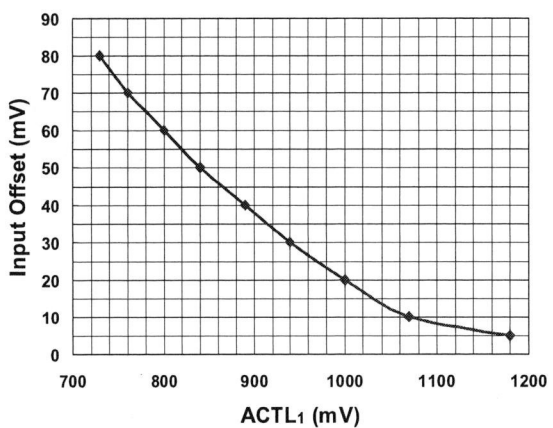

Fig. 6. Relationship between analog control signal and input offset.

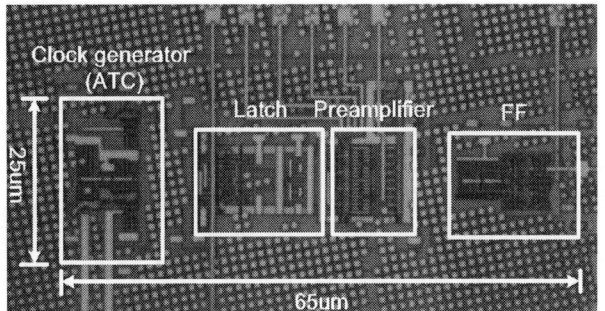

Fig. 7. Photograph of the comparator test die.

III. EXPERIMENTAL RESULTS

Fig. 7 is a chip photograph of the comparator, including blocks of Fig. 1, Fig. 4 and an output flip-flop, fabricated in a 65 nm CMOS process. It occupies 25 x 65 μm^2 and consumes 380 μW.

Fig. 8 (a) shows the threshold level of the comparator response at 1 GS/s with 1.2 V power supply. $ACTL_1$ is controlled with a 100 mV gap. Analog signals are acquired at the rising edge of the clock. An offset of 10 mV is controlled by 100 mV of $ACTL_1$ variation. Fig. 8 (b) also shows a response with a 500 MHz analog input frequency, synchronized to the 1 GHz clock. The analog signal is acquired at the top and bottom voltage and their outputs are compared. 100 mV of $ACTL_1$ variation controls the threshold level of the comparator to be higher than the top of analog signal, so after the transition of $ACTL_1$ to be lower level the comparator output is fixed to low. This means the analog signal is lower than reference revel.

Fig. 9 shows the relationship between $ACTL_1$ and the input offset. The control ratio is 0.104 which means 1mV offset can be controlled by 10 mV of $ACTL_1$. It is well balanced between the controllable range and resolution to control several tens of mV offset with mV order accuracy. There is 60% mismatching between simulation and the actual performance which will be caused by the parasitic on nodes c and d, transition rate of RES signal [8] and modeling of the elements.

978-1-4244-2018-6/08 $25.00 © 2008 IEEE   533

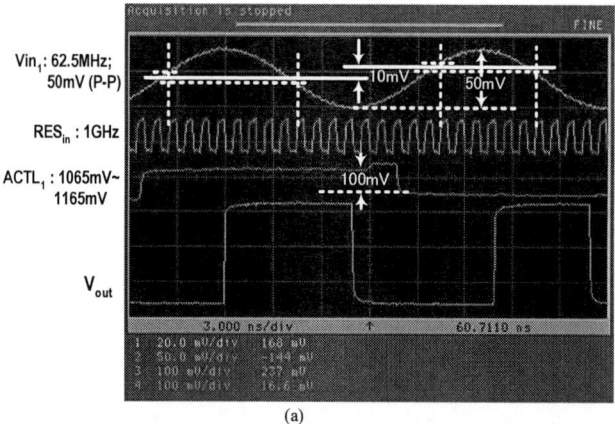

(a)

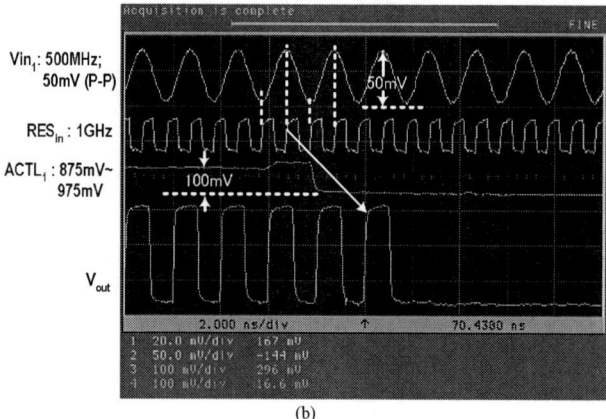

(b)

Fig. 8. Oscilloscope photographs of typical waveforms
for input offset control test.

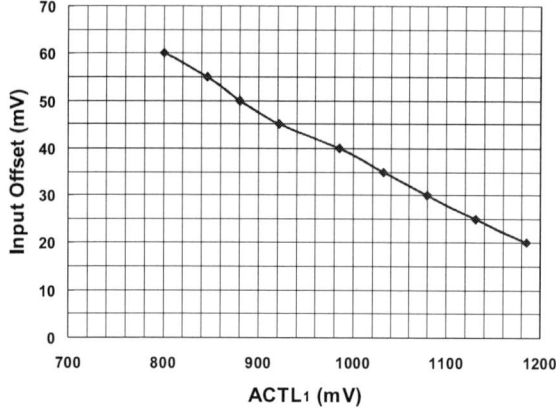

Fig. 9. Relationship between control signal and input offset
based on dc measurement results.

IV. CONCLUSION

A new offset control technique based on charge compensation by timing control is confirmed by a 1 GHz comparator fabricated in 65 nm CMOS with 1.2 V power supply. The comparator occupies 25 x 65 μm^2 and consumes 380 μW.

ACKNOWLEDGMENT

The authors would like to thank Junji Ogawa of Fujitsu Laboratories Ltd. for his encouragement and valuable discussions, and Yuki Yamamoto of Fujitsu Microelectronics Ltd for her generous support.

REFERENCES

[1] B. Razavi and B. A. Wooley, "Design Techniques for High-Speed, High-Resolution Comparators," *IEEE Journal of Solid-State Circuits*, vol. 27, No. 12, pp. 1916-1926, Dec. 1992.

[2] J. Atherton et al., "An Offset Reduction Technique for Use with CMOS Integrated Comparators and Amplifiers," *IEEE Journal of Solid-State Circuits*, vol. 27, No. 8, pp. 1168-1175, Aug. 1992.

[3] D. Weinlader et al., "An eight channel 36G Sample/s CMOS timing analyzer," in *IEEE ISSCC Dig. Tech. Papers*, pp.170-171, Feb. 2000,.

[4] K. L. J. Wong and C. K. K. Yang, "Offset Compensation in Comparators with Minimum Input-Referred Supply Noise," *IEEE Journal of Solid-State Circuits*, vol. 39, No. 5, pp. 837-840, May 2004.

[5] G. V. der Plas, S. Decoutere, and S. Donnay, "A 0.16 pJ/Conversion-Step 2.5 mW 1.25 GS/s 4b ADC in a 90 nm digital CMOS process," in *ISSCC Dig. Tech. Papers*, pp.2308–2309, Feb. 2006

[6] P. M. Figueiredo et al., "A 90nm CMOS 1.2V 6b 1GS/s Two-Step Subranging ADC," in *ISSCC Dig. Tech. Papers*, pp. 568-569, Feb. 2006.

[7] A. M. Abo, P. R. Gray, "A 1.5-V, 10-bit, 14.3-MS/s CMOS pipeline analog-to-digitalconverter," *IEEE Journal of Solid-State Circuits*, vol. 34, No. 5, pp. 599-567, May 1999.

[8] F. Maloberti, "Data Converters," Springer 2007.

978-1-4244-2018-6/08 $25.00 © 2008 IEEE

IEEE 2008 Custom Intergrated Circuits Conference (CICC)

A Low-Voltage OP Amp with Digitally Controlled Algorithmic Approximation

Dong-Woo Jee, Seung-Jin Park, Hong-June Park, and Jae-Yoon Sim

Pohang University of Science and Technology(POSTECH),
San 31, Hyojadong, Namgu, Pohang, Kyungbuk, 790-784, Korea
Tel : +82-54-279-2378, Fax : +82-54-279-2930, E-mail : minuano@postech.ac.kr

Abstract - This paper presents a new architecture of digitally controlled algorithmic OP amp suitable for scaled CMOS technologies. With inverter-based gain stages and digitally-assisted damping control, the amplifier achieves high-gain and wide input/output ranges even at the minimally allowable supply voltage by digital circuits. The amplifier, implemented in a standard 0.18 μm CMOS, shows a DC gain of 73 dB and 95 % settling time of 41 ns at 0.5 V step input.

I. INTRODUCTION

As CMOS technologies have been scaled down to nanometer regime, analog circuit design is becoming increasingly difficult. One of the most difficult analog building blocks is operational amplifier (OP amp). To design a high-gain OP amp, cascode structures have been conventionally used. The threshold voltage, however, is not scaled down proportionally to the supply voltage due to the sub-threshold leakage problem. So the multiple stacks of transistors for the cascoding result in significant reduction in output swing. The difficulty in designing a high-gain OP amp is even exacerbated by the low output resistance of scaled CMOS devices.

A number of different approaches have been taken to address these challenges. As an alternative to cascoding for high-gain amplifier, the cascading of amplifier stages has been researched with complicated frequency compensation techniques such as nested Miller[1] and nested Gm-C[2,3] schemes. However, the frequency compensation for more than three stages is very difficult to achieve and hardly considered. For approaches to avoid explicit OP amp, switching techniques such as dynamic amplifiers[4] and comparator-based switched capacitor circuit[5] have been researched to replace the use of OP amp.

This paper proposes a new architecture of digitally-assisted algorithmic OP amp. With inverter-based cascaded gain stages, the amplifier achieves high gain even at the minimally

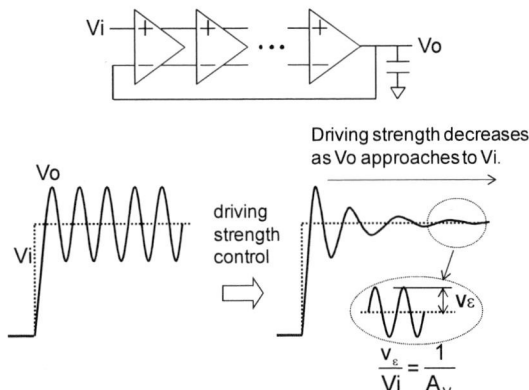

Fig. 1. Concept of the proposed high-gain OP amp by damping control.

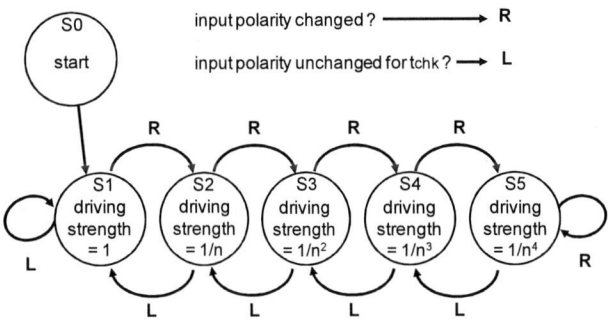

Fig. 2. State diagram for the algorithmic approximation.

allowable supply voltage of digital circuits. Instead of applying complex frequency compensation techniques, a new damping control technique is introduced to solve the stability problem.

II. ALGORITHMIC APPROXIMATION

Fig. 1 shows the concept of the proposed approach for high-gain with the multiple gain stages. Since each gain stage generates a low-frequency pole, the closed loop configuration causes oscillation if the number of stages is more than 3. In this unity-gain feedback example, the steady state response on

978-1-4244-2018-6/08 $25.00 © 2008 IEEE   535

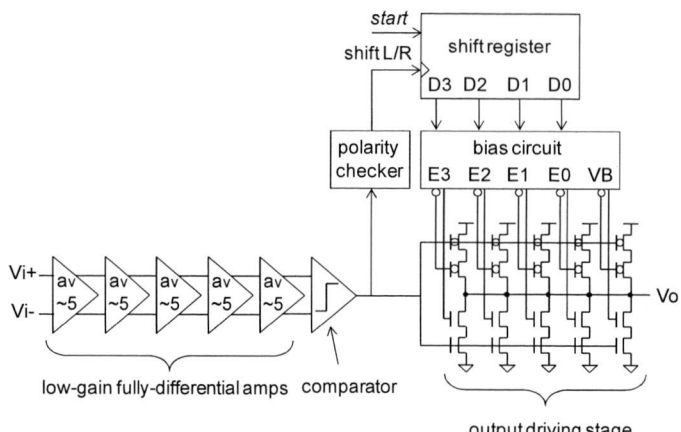

Fig. 3. Circuit diagram of the proposed algorithmic OP amp.

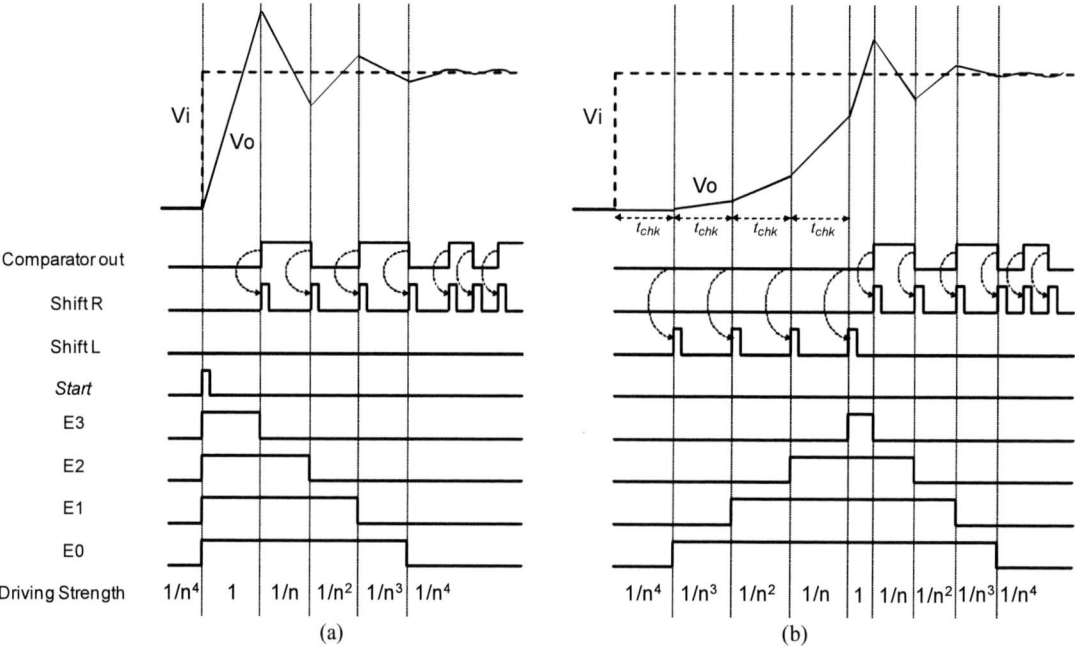

Fig. 4. Operations of the driving strength control in unity-gain feedback configuration with *start* signal (a), and without *start* signal (b).

Vo shows an oscillation around Vi with the amplitude and period determined by the loop delay and driving strength.

Since the driving strength of the last stage determines the slew, decreasing the output current reduces the oscillation amplitude. If the oscillation amplitude is settled to within a certain range by controlling the driving strength of output stage, Vo can be considered to be sufficiently converged to Vi. Since the oscillation is the steady state response, there should be some finite ripple(v_ε) which can be regarded as a gain error. The effective voltage gain(A_V) of the amplifier can be defined as shown in Fig. 1.

Fig. 2 shows the state diagram for the proposed digitally controlled algorithmic approximation. S0 state is for the initial state and S1~S5 states are for the driving strength control. The state diagram begins with S0 state whenever *start* signal is applied. The state transition is performed by checking the polarity of the differential input so that the driving strength would be smaller if the polarity toggles, or larger if the polarity remains unchanged for a certain period of time(t_{chk}). The state transition performs the change in the output driving strength. The strength scaling factor, n, should be optimized considering fast settling time and small ripple error. In this work, n was designed to be 5.

III. CIRCUIT DESCRIPTION

Fig. 3 shows the circuit diagram of the proposed algorithmic OP amp. In the gain stage, to achieve high gain as well as low supply voltage operation, multi-stage amplifier architecture is applied: five-stages of low-gain(~5) fully-differential amplifiers followed by a single-ended

978-1-4244-2018-6/08 $25.00 © 2008 IEEE

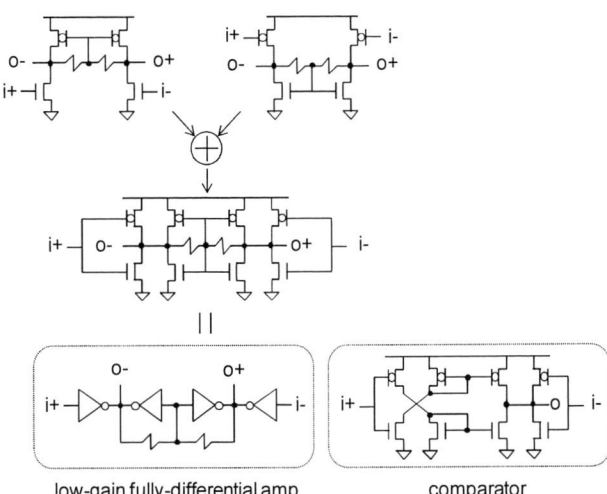

Fig. 5. Amplifier schematics for rail-to-rail input/output range.

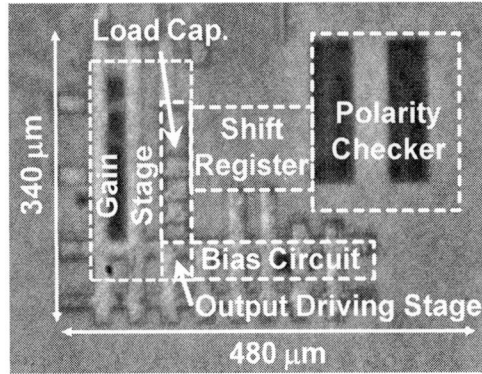

Fig. 6. Photomicrograph of the OP amp.

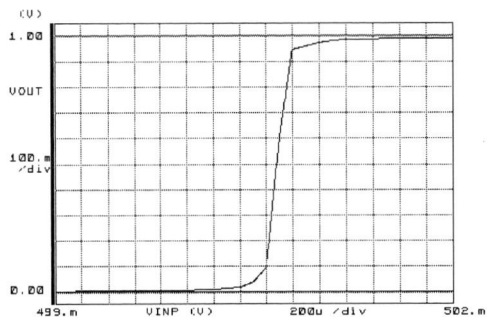

Fig. 7. Measured open-loop DC transfer curve

amplifier as a comparator. The polarity checker block monitors the comparator output and provides the input to the shift register for the state transition. The four outputs of the shift register block enable the bias voltages for the output driving stage. Since the comparator output changes with large swing, to reduce the gate-to-drain feed-through noise on Vo, the transistors for current bias are placed nearer to Vo.

Since the comparator output toggles whenever the polarity of the input difference changes, the polarity checker monitors the comparator output. If the comparator output toggles, which means Vo is oscillating, the polarity checker generates *shift R* signal to reduce the ripple by decreasing output driving strength. On the other hand, if the comparator output remains unchanged for a specified period t_{chk}, the polarity checker generates *shift L* signal. Since the unchanged comparator output indicates that Vo has not reached its target yet, the driving strength increases for larger slew rate.

For faster settling time, the *start* signal can be simultaneously applied with the step input to begin output driving with the maximum strength. This is recommended when the OP amp is used in discrete-time applications with a large step input such as Nyquist-rate ADC design. Fig. 4(a) and (b) show the difference in step response illustrating the processes of the driving strength change with and without applying the *start* signal, respectively.

Fig. 5 shows the amplifier schematics for the gain stages. The low-gain fully differential amplifier is designed with simplified common-mode gain suppression. For a wide input range, two amplifiers with NMOS and PMOS input stages are combined. Since the resistors can be shared by the two amplifiers, the resultant amplifier can be formed with only four inverters and two resistors. It can be understood that the two inverters in the middle and two registers perform the common-mode operation. For the comparator circuit, a higher-gain single-ended amplifier is used with complementary input stages. Since the amplifier circuits are implemented with only one PMOS and one NMOS between

power rails, almost rail-to-rail input and output ranges are achieved even at the lowest supply voltage required for digital circuit operation.

IV. MEASUREMENTS

For verification, the proposed OP amp was implemented in a standard 0.18 μm CMOS process typically used with supply voltage of 1.8V. Fig. 6 shows the photomicrograph of the OP amp with a chip size of $340 \times 480 \ \mu m^2$ including on-chip 3 pF load capacitance. The minimum operating voltage was 0.7 V which was limited by the operation of digital circuits. Fig. 7 shows open-loop DC transfer curve. The DC gain of 73dB obtained in Fig. 7 indicates the practical limit of the maximum achievable gain which is different from the ripple-defined gain definition shown in Fig. 1. The ripple-defined gain approaches to the maximum achievable gain as the output capacitance increases at the cost of settling time.

Fig. 8 shows measured large signal transient response when unity-gain feedback configuration is formed. With the supply voltage of 1 V, a step input of 0.25 V-to-0.75 V was applied. The results show 95% settling time of 41ns with *start* input and 614ns without *start* input. Due to parasitic capacitance associated with pin loading and PCB trace, the total load capacitance was increased and expected to be 7pF by correlating with simulation. The simulation shows 99.9 % settling time of 113 ns with the criterion of the ripple error of less than 0.5 mV. Fig. 9 shows the fastest simulated transient

Table. I. Performance summary

Operating Supply Voltage	≥ 0.7 V
Power consumption@1 V-Vdd	2.78 mW
DC gain	73 dB
Input CMR	0.1 V ~ 0.9 V
CMRR	40 dB
Simulated 99.9 % settling time (0.5 V-step, 1.8 pF)	90 ns(w. *start* input)
	504 ns(w.o. *start* input)
Measured 95 % settling time (0.5 V-step, 7 pF)	41 ns(w. *start* input)
	464 ns(w.o. *start* input)

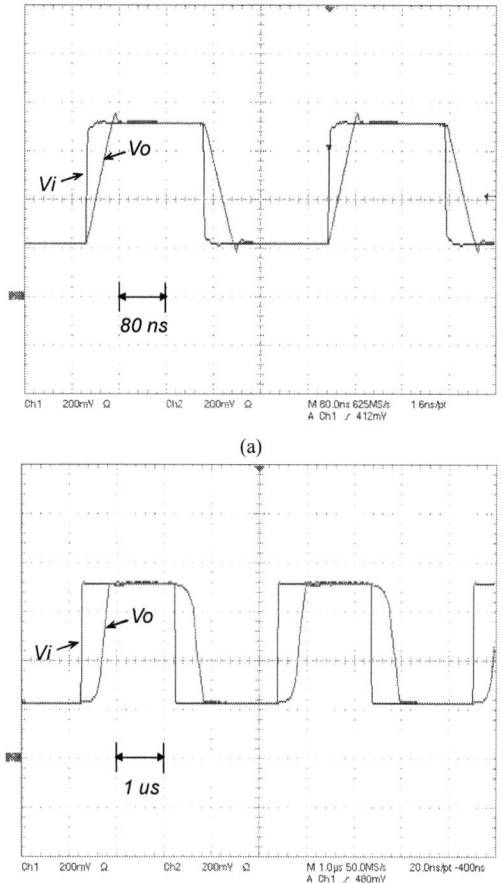

(a)

(b)

Fig.8. Measured transient step response in closed loop configuration with *start* input (a) and without *start* input (b). Blue lines represent step input and red lines represent output.

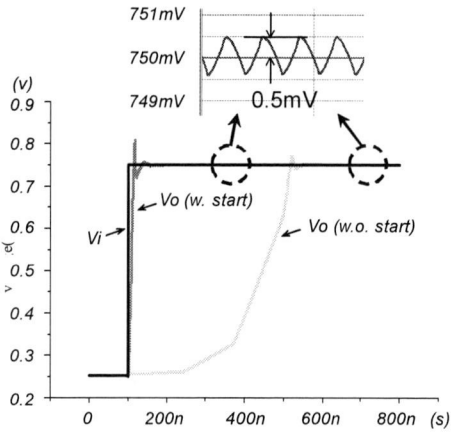

Fig.9. Simulation result of step response in closed loop configuration with 1.8pF load capacitance.

response with 1.8 pF output load capacitance which is the minimum value to be able to control the ripple error less than 0.5 mV, 0.1 % of input. The result shows 99.9 % settling time of 90 ns. As shown in Fig. 9, the finite ripple error is well controlled within the desired range by the proposed output

driving strength control. The settling time is mainly determined by circuit delay time since the polarity checker detects input information after the gain stage circuit delay time. Therefore, the transient response can be improved as the process scales down. Table. I summarizes the performance.

V. CONCLUSION

A new architecture of digitally controlled algorithmic OP amp is proposed to achieve high gain and wide input/output ranges even at the minimally allowable supply voltage by digital circuits. The proposed algorithmic damping control eliminates stability problem of a multiple gain stage amplifier. The OP amp was fabricated with a standard 0.18 μm CMOS, and measurements show a DC gain of 73 dB and 95 % settling time of 41 ns for 0.5 V step response at the supply voltage of 1V. Since the proposed amplifier employs only inverter-based gain stages and digital circuits, it is highly promising in scaled CMOS processes where low-voltage and high-speed digital circuits are available.

ACKNOWLEDGEMENT

This work was supported by the IDEC and BK21 programs of Korea.

REFERENCES

[1] G. Palumbo, et. al., "Design methodology and advances in nested-miller compensation," *IEEE Transactions on Circuit and Systems-I*, vol. 49, pp. 893-903, Jul. 2002

[2] F. You, et. al., "Multistage amplifier topologies with nested Gm-C compensation," *IEEE Journal of Solid-State Circuits.*, vol. 32, pp. 2000-2011, Dec. 1997

[3] J. S. Lee, et. al., "A design guide of 3-stage CMOS operational amplifier with nested-Gm-C frequency compensation," *Journal of Semiconductor Technology and Science*, vol. 7, pp. 20-27, Mar. 2007

[4] M. A. Copeland, et. al., "Dynamic amplifier for m.o.s. technology," *Electronics Letters,* vol. 15, pp. 301-302, May, 1979

[5] T. Sepke, et. al., "Comparator based switched capacitor circuits for scaled CMOS technologies," *IEEE International Conference of Solid-State Circuit,* pp. 220-221, 2006

978-1-4244-2018-6/08 $25.00 © 2008 IEEE

IEEE 2008 Custom Intergrated Circuits Conference (CICC)

A 105.5 dB, 0.49 mm^2 Audio ΣΔ Modulator using Chopper Stabilization and Fully Randomized DWA

Yi-Gyeong Kim, Min-Hyung Cho, Kwi-Dong Kim, Jong-Kee Kwon, and Jongdae Kim

Electronics and Telecommunications Research Institute (ETRI), 138 Gajeongno, Yuseong-gu, Daejeon, 305-700, Korea

E-mail: kimyig@etri.re.kr

Abstract-An audio ΣΔ modulator achieves 105.5 dB dynamic range over 20 kHz audio bandwidth. A chopper stabilization technique is used in both the first integrator and the reference buffer to prevent degradation of the dynamic range and the peak signal-to-noise-plus-distortion-ratio due to flicker noise. A fully randomized data weighted averaging is used as a dynamic element matching technique to suppress the generation of spurious tones with a negligible increase in the in-band noise compared to conventional data weighted averaging. The chip was fabricated in 0.13 μm CMOS technology (I/O devices) and occupies a small chip area of 0.49 mm^2. The total power consumption is 9.9 mW from a 3.3 V supply.

I. INTRODUCTION

The recent growth of digital consumer applications such as digital TV has increased the demand for the technology known as the system-on-chip (SoC), which includes not only digital signal processors but also mixed-mode circuits. An audio codec is an essential item among various mixed-mode circuits. The audio codec requires low-cost audio analog-to-digital converters (ADC) that are capable of high-performance. These audio ADCs must achieve a dynamic range (DR) and a peak signal-to-noise-plus-distortion-ratio (SNDR) of more than 100 dB and 95 dB, respectively, at low power. Moreover, the chip area must be minimized for the low-cost.

The ΣΔ modulator is a key block of the audio ADC. The multi-bit architecture is a possible route for high-performance. Multi-bit modulators have several advantages, including enhanced modulator stability, relaxed settling requirements of the loop filter, and reduced quantization noise compared to single-bit modulators. However, they have mismatch problem of a digital-to-analog converter (DAC) in the feedback path. The DAC has inherently nonlinear characteristics due to a device mismatch. In order to resolve the nonlinearity of the internal DAC in the feedback path, a dynamic element matching (DEM) technique is required. In addition, the generation of unwanted in-band tones and an increase in in-band noise by DEM are minimized to achieve the target performance.

In order to satisfy the target DR and the peak SNDR, the flicker noise has to be minimized. The flicker noise generated in the first integrator and the reference buffer mainly affects the performance of the modulator. The noise generated in the first integrator degrades the DR of the modulator. The noise generated in the reference buffer degrades the peak signal-to-noise-ratio (SNR) and peak SNDR. Therefore, a technique is required to reduce the flicker noise effect in a ΣΔ modulator.

This paper presents a high-performance low-cost audio ΣΔ modulator with a reference buffer. In order to minimize the contribution of the flicker noise, chopper stabilization

technique is used in both the first integrator and the reference buffer. The fully randomized data weighted averaging (DWA) is used as a DEM for suppression of the generation of spurious tones with a negligible increase of in-band noise.

II. ARCHITECTURE

Fig. 1 shows the architecture of the modulator. The modulator is a second-order single loop design with a feed-forward path to the input of the second integrator and a 17-level quantizer. The feed-forward path makes the first integrator process a second-order high-pass filtered signal component, thereby reducing the signal swing at the first integrator output and improving the distortion performance. In addition, the reduced swing range enables the utilization of a folded cascode OTA, thus reducing the power consumption. In order to remove the effect of the flicker noise, the first integrator and the reference buffer adopt a CHS technique. A fully randomized DWA is used as a dynamic element matching technique. The loop filter is implemented with fully differentially switched capacitor circuitry. The modulator is operated with a clock speed of 6.144 MHz which corresponds to an OSR of 128. The input signal range of the modulator is 2 V$_{RMS}$ and reference voltages are 3.0 V and 0 V.

III. CIRCUIT TECHNIQUES AND IMPLEMENTATION

A. Chopper Stabilization

Flicker noise occurring in the first integrator goes through the loop filter with the characteristics of a signal transfer function (STF). The flicker noise will then appear at the output of the modulator with little attenuation. Therefore, a technique for the removal of the flicker noise is needed to achieve a DR of more than 100 dB. CHS is a suitable method for the removal of the flicker noise effect in a ΣΔ modulator due to the low-pass filter characteristic of the integrator [2]. Fig. 2 shows the first integrator and the timing diagram of the CHS. The frequency of the CHS is set at a sampling frequency/2 (Fs/2). The CHS makes it possible to increase the

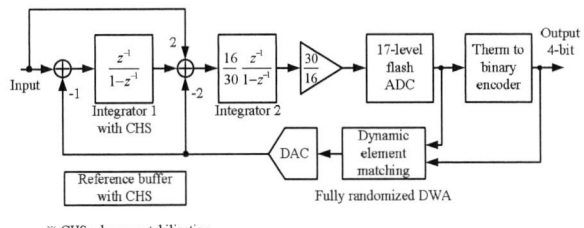

※ CHS: chopper stabilization

Fig. 1. ΣΔ modulator architecture.

978-1-4244-2018-6/08 $25.00 © 2008 IEEE

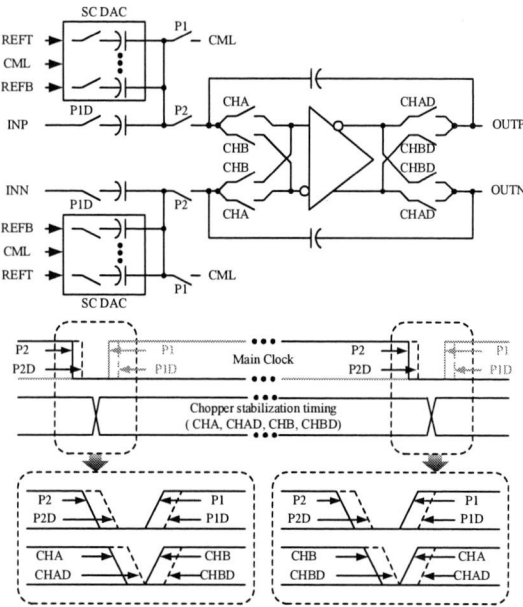

Fig. 2. First integrator and chopping timing diagram.

noise in the signal bandwidth, as spectrally shaped high-frequency (Fs/2) quantization noise can be modulated down to the signal bandwidth [3].

To reduce the coupling effect of the shaped high-frequency quantization noise, chopping is performed at a range between P2 and P1D. In addition, to minimize the signal-dependent charge injection, chopping is performed with non-overlap timing. That is, after the P2 phase switches shown in Fig. 2 are opened, CHA (or CHB) switches are opened. Following this, CHAD (or CHBD) switches are opened. Next, CHB (or CHA) switches are closed and CHBD (or CHAD) switches are subsequently closed.

Fig. 3 shows the effect of flicker noise in the reference buffer. In the time domain, DAC output is generated from the multiplication of the DAC input signal and the reference voltage. In the frequency domain, the output is the result of convolution between the input spectrum and the reference voltage spectrum. Thus, if the reference voltage contains flicker noise, the noise appears around the input signal frequency in the frequency domain. Due to the input signal dependency, although there is only a slight noise effect in case of a small input signal, there is an increase in the level of noise around the input signal frequency in case of a large input signal. Therefore, the noise will degrade the peak SNR and the peak SNDR.

Fig. 4 shows the reference buffer schematic and the timing diagram of the CHS. REFT and REFB are set to 3.0 V and 0 V, respectively. The flicker noise that appears in the reference voltage spectrum is mainly generated at the error amplifier. Thus, the error amplifier adopts a CHS technique. The chopping is performed at the frequency of Fs/2. The flicker noise generated at the error amplifier is modulated up to Fs/2 via CHS technique. The modulated noise is attenuated by an off-chip capacitor. Therefore, the DAC is supplied with clean

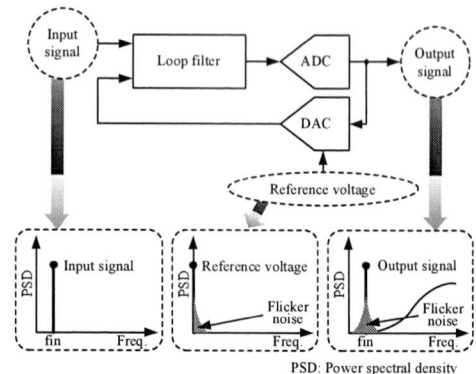

Fig. 3. Effect of flicker noise in reference buffer.

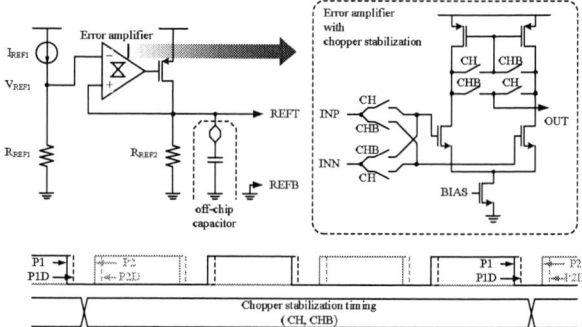

Fig. 4. Reference buffer and chopping timing diagram.

DC voltage and generates the DAC output signal without additive noise.

B. Fully Randomized DWA

The DWA technique, which has a first-order mismatch shaping property, is a practical and effective DEM technique. However, DWA experiences what is known as a tonal problem, which involves the generation of unwanted in-band tones due to the periodic property of DWA. Thus, this problem limits the dynamic performance of a ΣΔ modulator. To overcome the tonal problem of DWA, the periodic property of DWA must be removed. In this design, the randomized DWA [1] is used as the DEM. The randomized DWA can be implemented with either partial randomization [1] or full randomization. In the partially randomized DWA, the origin pointer jumps to a new random pointer only in the case of complete rotation, which implies that the start pointer equals the origin pointer. Although this implementation suppresses the generation of tones, this technique increases the in-band noise, degrading the SNR. To minimize SNR degradation, a fully randomized DWA is used here. In the fully randomized DWA, the selection of DAC elements is similar to that of the conventional DWA except that the origin pointer jumps to a new random pointer at every full turn, as shown in T3 code of Fig. 6. The one full turn implies that all elements are selected 1 time. Fig. 5 (a) shows the simulated spectrum of the modulator using the conventional DWA, the partially randomized DWA, and the fully randomized DWA for an input amplitude of -51 dB with a random DAC element mismatch less than 0.2 %. The SNDRs for the conventional

978-1-4244-2018-6/08 $25.00 © 2008 IEEE   540

DWA, the partially randomized DWA, and the fully randomized DWA are 67.1 dB, 66.7 dB, and 68.9 dB, respectively. Although degradation of the SNDR occurs in the case of the conventional DWA and the partially randomized DWA due to spurious tones and an increase in the in-band noise, respectively, there is little degradation of the SNDR in the case of the fully randomized DWA. Fig. 5 (b) shows the SNDR versus the input level. The amplitude level of the input signal is increased in 5 dB steps from -60 dB to 0 dB. The SNDR at each input level is determined by averaging 10 trials of simulation results, performed with a random DAC element mismatch less than 0.2 %. Although the average SNDR for the partially randomized DWA are degraded, the fully randomized DWA results show little degradation of the average SNDR compared to the conventional DWA. Some simulation results of the conventional DWA show the generation of spurious tones. Therefore, these results show that the fully randomized DWA is effective for the suppression of the generation of spurious tones with only a negligible increase in in-band noise.

Fig. 6 shows a hardware implementation block diagram of fully randomized DWA, shown for 8-bit input of a thermometer code for simplicity. For a simple description, the values of the input code, T and B, as well as that of a random number R are set to 5, 5 and 1, respectively. The element marked with slanted lines at the T3 code in Fig. 5 is the origin pointer, which shifts according to a random number, SH2. This random number, SH2, is inserted at every full turn. The hardware implementation of the fully randomized DWA is comprised of three shifters and some logic gates. The first shifter rotates the T code in the same manner as DWA. The second shifter shifts only the left side elements (dashed border line) among the selected elements to the right by SH2 in case of a full turn of T1. SH2 is generated using a subtractor, AND gates, and OR gates, and are hence a random number so as not to overlap with the selected elements. The third shifter rotates the code T2 by the SH3, which is an accumulation of SH2, to correct the start pointer. These operations are performed by logarithmic shifters and a modified shifter, named a local element shifter in Fig. 6. The local element shifter is constructed from switch units that consist of two NMOS transistors, one NAND gate, and one inverter, shown in Fig. 6.

C. Integrators and Flash ADC

The integrators employ a fully differential folded-cascode OTA. This OTA does not require an additional compensation capacitor and achieves high bandwidth with a capacitor load. Therefore, it has a high level of power efficiency. A fully differential gain boosting technique is used to achieve high DC gain as low DC gain causes a leakage of quantization noise. A gain boosting amplifier dissipates less than 30 % of the total amplifier power. The unity gain bandwidth of the first OTA is 60 MHz and the open loop DC gain is 113 dB. The value of the first sampling capacitor and reference capacitor is 4.2 pF and 4.9 pF, respectively, which satisfy a thermal noise target of -108 dB. The second integrator is scaled down to reduce the power consumption. The flash ADC consists of 16 comparators composed of a preamp and a regenerative latch.

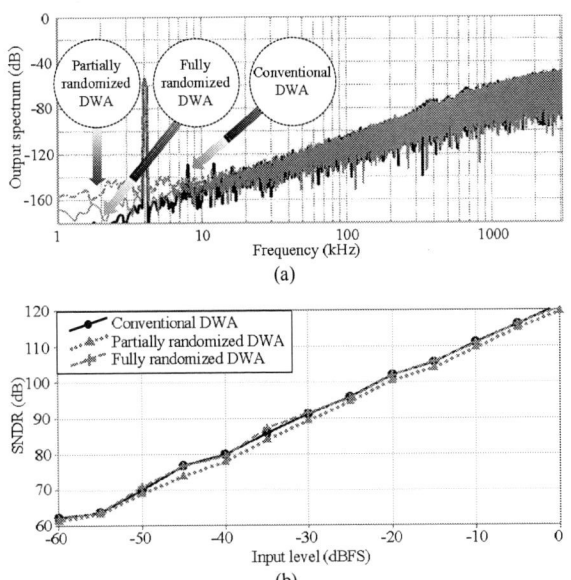

Fig. 5. Comparison of DWA, partially randomized DWA, and fully randomized DWA with a random DAC element mismatch less than 0.2 %.
(a) Output spectrum.
(b) SNDR versus input level.

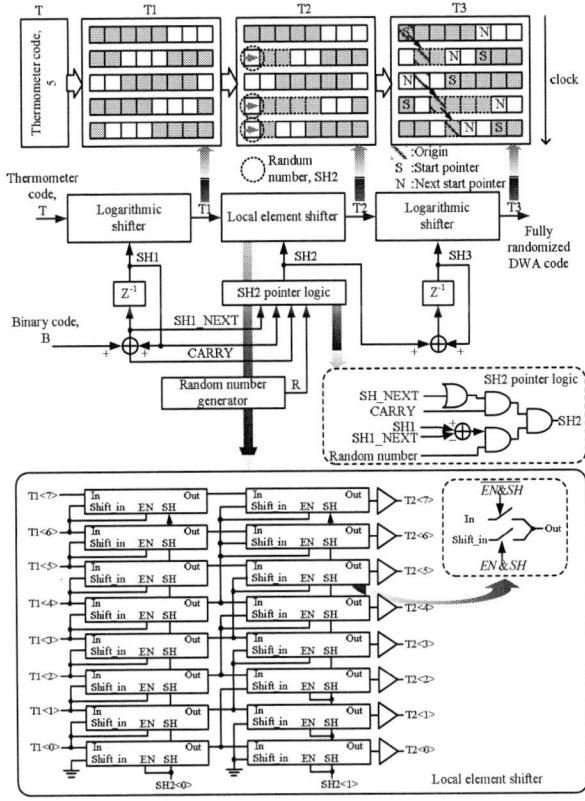

Fig. 6. Block diagram of fully randomized DWA and example codes.

The reference voltages of the flash ADC and common mode voltages of the integrators are generated by resistive dividers.

IV. MEASUREMENT RESULTS

The prototype modulator was fabricated in 0.13 μm CMOS technology (I/O Devices) and occupies a 0.49 mm² (0.97 mm x 0.5 mm) active area, as illustrated in Fig. 7. The analog and digital section operates on a 3.3 V supply with a total power consumption of 9.9 mW. The modulator achieves a 97.5 dB peak SNDR (A-weighted) and a 105.5 dB dynamic range (A-weighted) over a bandwidth of 20 kHz.

Fig. 8 (a) shows the measured output spectrum of a 1 kHz and 0 dBFS sinusoidal signal. The SNR (A-weighted) values of the reference buffer CHS ON and OFF are 101.9 dB and 99.1 dB, respectively. The SNDR (A-weighted) values of the reference buffer CHS ON and OFF are 97.2 dB and 96.5 dB, respectively. Fig. 8 (b) shows the measured output spectrum of a 1 kHz and -20 dBFS sinusoidal signal. The SNDR (A-weighted) values are 85.5 dB and 76.0 dB with the integrator CHS ON and OFF, respectively. These results show that peak SNR/SNDR and DR are enhanced by the CHS technique in the reference buffer and the integrator. A plot of the SNR and SNDR values versus input level is shown in Fig. 9. TABLE I gives a summary of the measured performance.

V. CONCLUSIONS

The audio ΣΔ modulator with the reference buffer is presented in this paper. The CHS is applied to the first integrator and to the reference buffer to minimize the contribution of the flicker noise. The fully randomized DWA is used to reduce the effect of the nonlinear DAC. The modulator achieves a DR (A-weighted) of 105.5 dB, a peak SNDR of 97.5 dB over an audio bandwidth of 20 kHz with power consumption of 9.9 mW power consumption. The design occupies an active area of 0.49 mm².

ACKNOWLEDGMENTS

This work was supported by the IT R&D program of Ministry of Knowledge Economy, Rep. of Korea [2006-S-006-02, components/Module Technology for Ubiquitous Terminals].

REFERENCES

[1] M. Vadipour, "Techniques for preventing tonal behavior of data weighted averaging algorithm in Σ-Δ Modulators," IEEE Transactions on Circuits and Systems II: Analog and Digital Signal Processing, Vol. 47, No. 11, pp. 1137-1144, November 2000.

[2] A. B. Early, "Chopper stabilized delta-sigma analog-to-digital converter," U.S. Patent 4 939 516, July 3, 1990.

[3] Y. Yang, A. Chokhawala, M. Alexander, J. Melanon, and D. Hester, "A 114-dB 68-mW chopper-stabilized stereo multibit audio ADC in 5.62 mm²," IEEE Journal of Solid-State Circuits, Vol. 38, No. 12, pp. 2061–2068, December 2003.

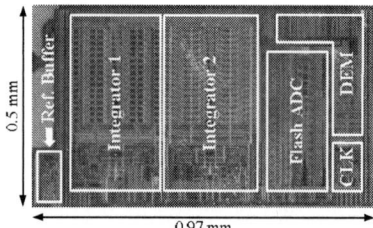

Fig. 7. Photograph of the chip.

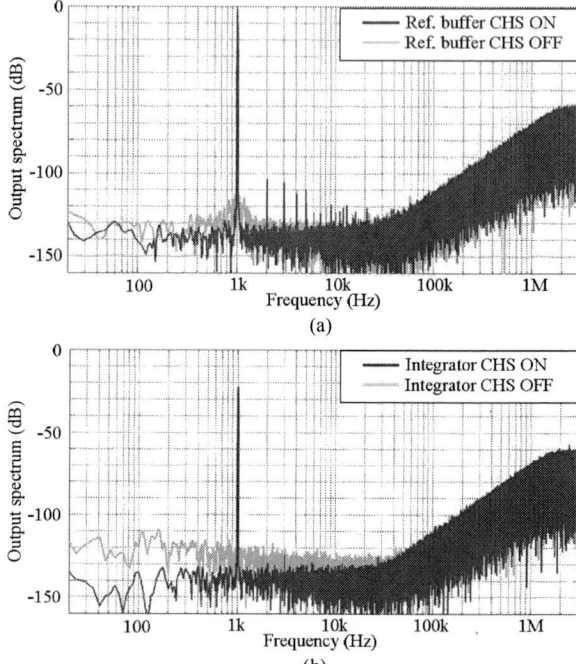

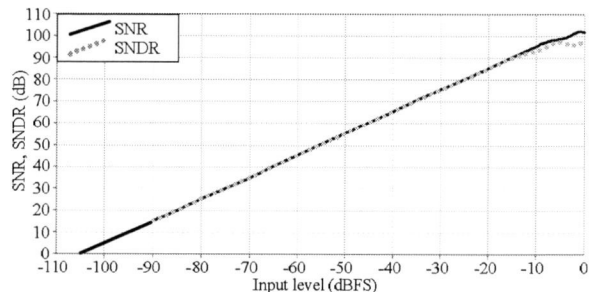

Fig. 8. 1 M FFT plot of (a) 0 dB input and (b) -20 dB input, 1 kHz sine wave with a sampling frequency of 6.144 MHz.

Fig. 9. SNR and SNDR versus the input level.

TABLE I

PERFORMANCE SUMMARY

Power supply voltage	3.3 V
Signal bandwidth	20 kHz
Sampling frequency	6.144 MHz
Peak SNDR	97.5 dB (A-weighted)
Dynamic range	105.5 dB (A-weighted)
Full scale input	2 V_{RMS} differential
Total power consumption	9.9 mW (Analog:5.28 mW, Digital: 4.62 mW)
Chip area	0.49 mm² (0.97 mm x 0.5 mm)
Technology	0.13 μm CMOS technology, 3.3 V I/O Devices

IEEE 2008 Custom Intergrated Circuits Conference (CICC)

An Ultra Low Power 1V, 220nW Temperature Sensor for Passive Wireless Applications

Yu-Shiang Lin, Dennis Sylvester and David Blaauw

Dept of EECS, University of Michigan

1301 Beal Ave, Ann Arbor, MI 48109-2122

{yushiang,dennis,blaauw}@eecs.umich.edu

Abstract- This work presents a low power temperature sensor that is suitable for passive wireless systems. The test chip is fabricated with a 0.18μm CMOS technology and the total area is 0.05mm^2. With temperature inaccuracy of $-1.6°C/+3°C$ from $0°C$ to $100°C$, the temperature sensor consumes only 220nW at 1V under room temperature. The data conversion rate is 100 sample/s with an output resolution of $0.3°C$, which is sufficient for most sensor applications.

I. INTRODUCTION

Since the last decade, smart temperature sensors have growing demands on VLSI, automotive, and wireless sensing applications due to their low cost. Monitoring VLSI chip temperature plays a key role on long-term system level reliability and performance. Rapidly increasing transistor numbers require embedded sensors with small area and low power that can be spread over the chip for temperature management [1]. Sensors that produce low power consumption not only helps with power grid integrity but also alleviates self-heating issues. Recently, growing interests in building monitoring systems with wireless telemetry or RFID cards demand even more stringent power consumption [2, 3]. The energy range which is defined as the distance from the transponder and reader that is just enough to operate the transponder can be extended by cutting down the power dissipation [4]. In the work reported in [5], the temperature sensor consumes 10μW compared to 2μW by the reader for their passive RFID transponder. This means that the power consumption of the temperature sensor is highly related to the working distance of such wireless systems.

Smart temperature sensor ICs were first developed using bandgap reference and analog-to-digital converters (ADCs) [6, 7]. Such sensors typically are able to achieve better than $\pm1°C$ accuracy with calibration. Combining with offset cancellation, dynamic element matching and room-temperature calibration, accuracy of $\pm0.1°C$ with 247.5μW power consumption was reported [8]. Time-to-digital converter (TDC) was also proposed to measure the temperature by tracking a pulsed signal along a delay line [9]. In this work, our

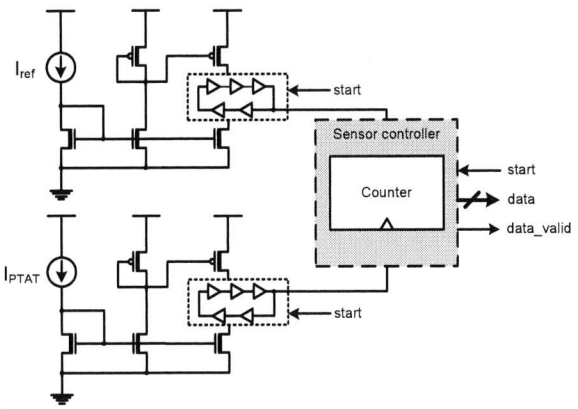

Figure 1: Temperature sensor block diagram.

goal is to implement a temperature sensor with sub-μW power dissipation with acceptable accuracy for ultra low power passive wireless sensor applications. In the section II, the architecture of our proposed circuits will be discussed and the power consumption will be analyzed. Measurement results will be shown in section III and is followed by the conclusion in section IV.

II. LOW POWER TEMPERATURE SENSOR DESIGN

Fig. 1 shows the block diagram of the temperature sensor. Temperature insensitive current source I_{ref} and proportional to absolute temperature (PTAT) current source I_{PTAT} are generated seperately. Each current source is mirrored and fed into the current-starved ring oscillator to translate the temperature information into frequency. Afterwards, the clock signals are fed into an UP-counter that is triggered by a *start* signal in order to produce a digitized output. The sensor controller decides when the conversion should start and responds by a *data_valid* signal when the data is available. The key blocks of this work is to generate current sources I_{ref} and I_{PTAT} with low power dissipation and is still able to maintain reasonable temperature characteristics.

Generating I_{PTAT} is a commonly used technique in bandgap reference design for compensating the complementary to absolute temperature (CTAT) current sources. Fig. 2 shows the schematic for such purpose that was originally implemented with bipolar circuits

978-1-4244-2018-6/08 $25.00 © 2008 IEEE   543

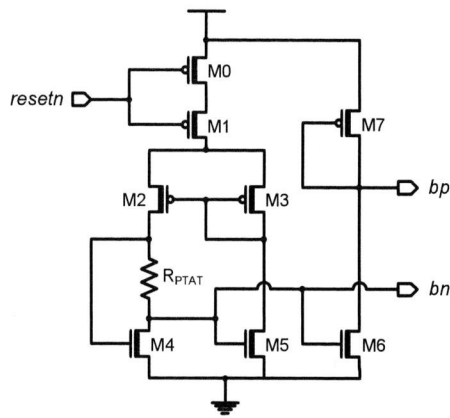

Figure 2: Schematic for I_{PTAT} generation.

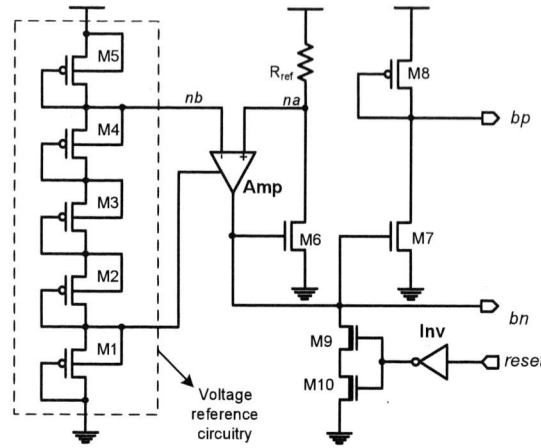

Figure 3: Schematic for I_{ref} generation.

[10]. CMOS transistors can be used in place of bipolar transistors when operating in the subthreshold region. In this way, we can reduce the power consumption of this block significantly, which accounts for roughly 30% of the total power dissipation. When V_{gs} is less than V_{th} and V_{ds} is larger than three V_T, the drain current of transistor M4 and M5 can be approximately written by:

$$I_{sub} = \mu C_{OX} \cdot \frac{W}{L} \cdot V_T^2 \cdot \exp\left[\frac{V_{gs} - V_{th}}{nV_T}\right] \qquad (1)$$

where V_T and V_{th} are the thermal voltage represented by kT/q and V_{th} is the threshold voltage of the transistor, respectively. Through current mirror transistors M2 and M3, the current through resistor R_{PTAT} can be expressed as

$$I_{R_{PTAT}} = \frac{nV_T}{R_{PTAT}} \ln\left[\frac{W_5 W_3 L_4 L_2}{W_4 W_2 L_5 L_3}\right] \qquad (2)$$

Assuming that V_{th} mismatch is ignored. By properly biasing the circuit, the output current is proportional to V_T. The sensitivity to the geometric variations can be minimized by designing a large value in the log function. Large transistor sizes also help to reduce the impact on threshold voltage due to random doping fluctuations.

The temperature insensitive current source is generated by a self-biasing technique. The circuit diagram is shown in Fig. 3. M1 through M5 are diode-connected transistors used to provide bias voltages that are proportional to the supply voltage. The voltage of nb is replicated to node na through negative feedback loop consisting of transistor M6, resistor R1 and the amplifier. Therefore, the drain current of M6 can be defined by $(V_{dd} - V_{na} - V_{os})/R_{ref}$, where V_{os} is the input offset voltage of the amplifier. The fractional temperature

coefficient (TC_F) of I_{d6} is

$$TC_F(I_{d6}) = \frac{1}{I_{d6}} \cdot \frac{dI_{d6}}{dT} \qquad (3)$$

$$= \frac{1}{V_{dd} - V_{na} - V_{os}} \cdot \left(\frac{dV_{na}}{dT} - \frac{1}{R_{ref}} \cdot \frac{dR_{ref}}{dT}\right) \qquad (4)$$

To reduce the non-ideal temperature effect on the sensor we do the following: 1) the resistor is chosen so that the second-order temperature coefficient (TC2) is minimized; and 2) transistors M1-M5 should be identically sized to eliminate the first term of Eq. 4.

It is noted that in this work, the voltage reference circuitry in Fig. 3 was implemented as a voltage divider. Thus, I_{ref} is inversely proportional to the supply voltages and lead to changing output value with power supply noises. To fix this issue in the future, the voltage reference should be re-designed to have constant output regardless of the supply voltage.

Both I_{ref} and I_{PTAT} blocks generate analog voltages bn and bp to provide the starving voltage for the ring oscillator. Temperature information in I_{ref} and I_{PTAT} are translated into frequency for the signals clk_i and clk_l. In Fig. 4, the sensor controller is shown as well as the timing diagram. clk_i and clk_l are used to clock the q-counter and the d-counter, respectively. When $start$ is 0, both counter outputs are cleared. Triggered by input signal $start$, the controller asserts output $data\_valid$ after the q-counter gets overflowed 2^{10} cycles later. $data\_valid$ immediately stops both counters from changing their content until $start$ goes to 0 again to reset the states. The temperature sensor including the I_{ref} and I_{PTAT} blocks are implemented so that they can be deactivated during sleep state by asserting $reset$ signal. When high conversion rate in not required, the

978-1-4244-2018-6/08 $25.00 © 2008 IEEE

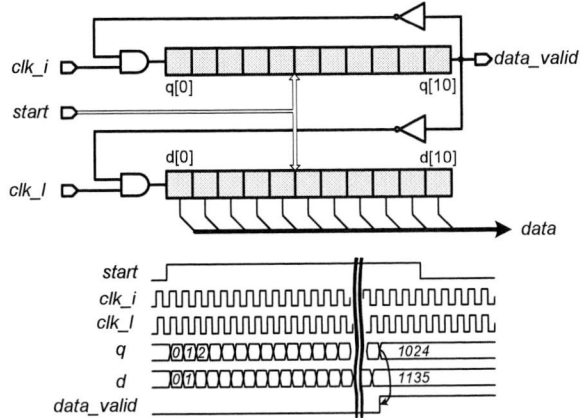

Figure 4: Block diagram and timing diagram of the sensor controller.

temperature sensor can be periodically deactivated to save power.

The total power of our proposed temperature sensor can be written as follow:

$$P_{tot} = V_{dd} \cdot [(n+1)I_{PTAT} + (m+1)I_{ref}] + P_{ctrl} \quad (5)$$

where n, m are the multiplication constants of current mirrors. P_{ctrl} is the power consumption of the sensor controller. For simplification, static power consumption is neglected in this first order analysis. Therefore, P_{ctrl} can be expressed as $\alpha C_c V_{dd}^2 f_{clk}$ given the total capacitance C_c, effective activity factor α and clock frequency f_{clk}. Considering f_{clk} as a function of I_{ref}, I_{PTAT} and V_{dd}, Eq. 5 can be re-written as

$$P_{tot} = k1 \cdot \frac{V_{dd}}{R_{PTAT}} \cdot V_T + k2 \cdot \frac{V_{dd}^2}{R_{ref}} \quad (6)$$

where k1 and k2 are geometry and process related constants. It is shown that 1) the power consumption of the sensor is a linear function of temperature; and 2) power consumption can be proportionally reduced by using large resistors. The size of resistors are determined by the target current consumption of 200nA and by matching I_{ref} and I_{PTAT} at room temperature. In this work, 6.2MΩ and 3.2MΩ P+ poly resistors are chosen for R_{ref} and R_{PTAT}, respectively.

III. MEASUREMENT RESULTS

The chip was implemented in a 0.18μm 1P6M digital CMOS process. The total area of the temperature sensor module is 0.05mm². The die photo is shown in Fig. 5. In this test chip, 85% of the area is dominated by the resistor for biasing the current sources.

The measurement is setup inside a TestEquity environment chamber TE-105A. The power consumption is

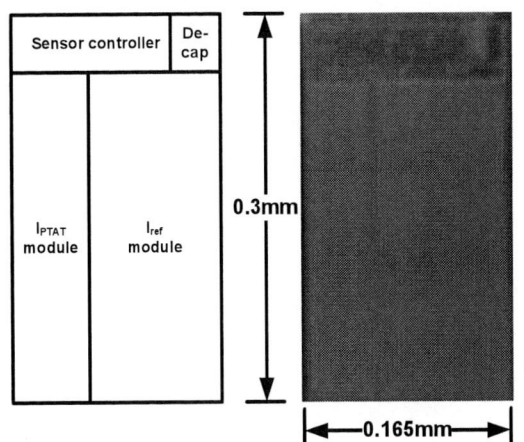

Figure 5: Die photo of the temperature sensor.

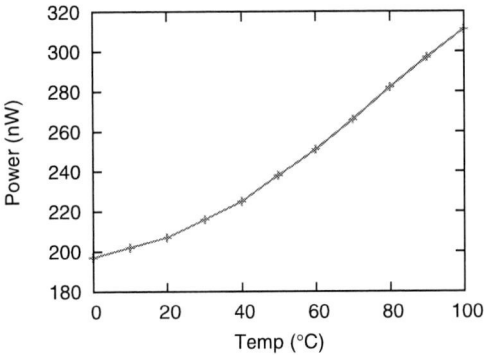

Figure 6: Power consumption of the temperature sensor.

measured by a Keithley electrometer 6517A and the results are shown in Fig. 6. The supply voltage is set to 1V while the nominal supply voltage for this technology is 1.8V. The power consumption increases from 200nW to 310nW from 0°C to 100°C, which matches the expected trend from Eq. 6. The slope at higher temperature is larger mainly because leakage components become non-negligible in this region. It is noted that there is a trade-off between power consumption and area. While most area are dominated by the resistors, reducing the resistance by half also reduce total area by 43%. In the same time, the conversion rate is also doubled because of the boost in ring oscillator's starving current. In this test chip, clk_i is running at 100kHz for an equivalent of 100 samples/s. This is sufficiently fast for most applications, and in fact we can lower the conversion rate to lower the reading noise as will be shown in Fig. 8.

The temperature inaccuracy of 5 test samples after two-point calibration are shown in Fig. 7. The temperature error is ranging from -1.6°C to +3°C over the

Table 1: Comparison of temperature sensors.

Sensor	Inaccuracy ($^\circ C$)	Power Consumption	Technology	Area (mm^2)	Temperature range ($^\circ C$)	Conversion rate (samples/s)
[6]	±1	7μW	2μm	1.5	-40∼120	50
[7]	±1	1mW	0.6μm	3.32	-55∼125	40k
[8]	±0.1	247.5μW	0.7μm	4.5	-55∼125	1∼10
[9]	-0.7/+0.9	10μW	0.35μm	0.175	0∼100	10k
[5]	-1.8/+2.2	10μW	N/A	N/A	0∼100	∼2
[3]	±1	0.9μW	0.18μm	0.2	27∼47	N/A
This work	-1.6/+3	0.2∼0.31μW	0.18μm	0.05	0∼100	100

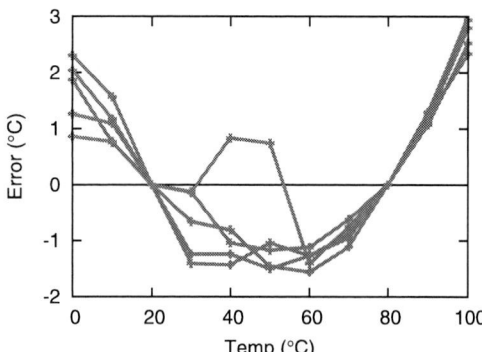

Figure 7: Temperature inaccuracy of the temperature sensor with two-point calibration at 20°C and 80°C.

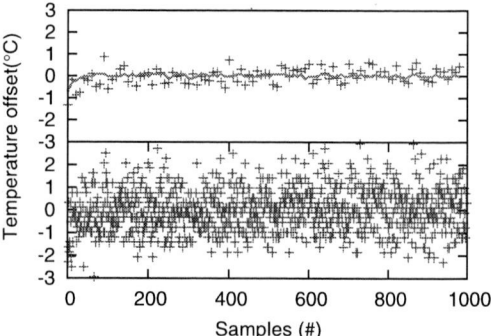

Figure 8: Temperature inaccuracy over samples (top: 10 samples/s; bottom: 100 samples/s; solid line: actual temperature).

sweeping range from 0°C to 100°C. With 11 bits output from the sensor controller, the temperature resolution is 0.3°C.

Table 1 lists the previous works on smart temperature sensors and compares the key circuits parameters to this work. It can be seen that our proposed temperature sensor adopts an approach that is favorable for low power operation at the expense in terms of temperature inaccuracy. The total area of our test chip is comparable or even smaller than other works after considering the translation of different technologies.

Fig. 8 shows the long term characteristics of the sensor by setting up the chip in the temperature chamber (top: 10 samples/s; bottom: 100 samples/s). After taking 1000 samples successively, the 3σ inaccuracy value over the samples is 2.5°C. By lowering the conversion rate to 10 samples/s, the 3σ inaccuracy is reduced to 0.28°C by averaging the samples. The actual temperature is also shown in solid line in Fig. 8.

IV. Conclusion

In this work, we implemented an ultra low power temperature sensor for passive wireless applications. At room temperature, it consumes merely 220nW while continuously running. By utilizing a temperature independent current source I_{ref} and PTAT current source I_{PTAT}, the temperature information can be synthe-

sized and translated into digital output in a conversion rate of 100 samples/s. Measured data shows that the temperature inaccuracy of the temperature sensor is -1.6°C/+3°C from 0°C to 100°C.

References

[1] D. Duarte et al. Temperature sensor design in a high volume manufacturing 65nm cmos digital process. *CICC*, pages 221–224, Sept 2007.

[2] F. Kocer and M. Flynn. An RF-powered, wireless CMOS temperature sensor. *IEEE Sensors Journal*, 6(3):557, 2006.

[3] Zhou S. and Wu N. A novel ultra low power temperature sensor for UHF RFID tag chip. *ASSCC*, pages 464–467, Nov 2007.

[4] K. Finkenzeller. *RFID handbook*. John Wiley & Sons, 2003.

[5] K. Opasjumruskit et al. Self-powered wireless temperature sensors exploit rfid technology. *Pervasive Computing, IEEE*, 5(1):54–61, Jan.-March 2006.

[6] A. Bakker and J. Huijsing. Micropower CMOS temperature sensor with digital output. *JSSC*, 31(7):933–937, July 1996.

[7] M. Tuthill. A switched-current, switched-capacitor temperature sensor in 0.6-μm CMOS. *JSSC*, pages 1117–1122, 1998.

[8] M. Pertijs, K. Makinwa, and J. Huijsing. A CMOS smart temperature sensor with a 3σ inaccuracy of ±0.1°C from -55°C to 125°C. *JSSC*, 40(12):2805–2815, Dec. 2005.

[9] P. Chen et al. A time-to-digital-converter-based CMOS smart temperature sensor. *JSSC*, 40(8):1642, 2005.

[10] K. Kimura. Low voltage techniques for bias circuits. *IEEE Tran. on Circuits and Systems*, 44(5):459, May 1997.

978-1-4244-2018-6/08 $25.00 © 2008 IEEE

A 9-Bit Configurable Current Source with Enhanced Output Resistance for Cochlear Stimulators

Song Guo, Hoi Lee, and Philipos Loizou
Department of Electrical Engineering
University of Texas at Dallas
Richardson, TX 75080-3021, USA

Abstract - **This paper presents a configurable current source for cochlear stimulators. A switchable multi-bias active-cascode architecture is developed to provide a 9-bit output current in a small implementation area. A stacking MOS structure enables the current source to achieve high output resistance and large voltage compliance. Implemented in a standard 0.35μm CMOS process, the current source can source a maximum 1mA current, provide ≥4.77V voltage compliance under a 5V supply and achieve >50MΩ output resistance in 0.26mm^2.**

I. INTRODUCTION

Cochlear implants are well-accepted prosthetic devices to restore hearing to profoundly deaf people by delivering electrical stimuli to auditory nerves. Fig. 1 shows a structure of a cochlear implant [1], which consists of an external module and an implanted system. In the implant, a current stimulator consisting of current source(s) is used to generate biphasic charge-balanced current pulses with a suitable amplitude, period and pulse width configured by the decoder to stimulate auditory nerves through an electrode array such that hearing can be restored. Since hearing restoration is determined by the quality of the current pulses, the performances of the current source in the stimulator are crucial. Studies have found that large amplitude of 1mA anodic or cathodic stimulus current generated by the current source is needed to elicit auditory percept under variations in stimulation types (bipolar or monopolar), placement of electrodes, number of remaining ganglion cells, etc [2]. In addition, the output current of the current source should have large amplitude resolution such that the cochlear implant could be more flexible to be adjusted under variability in patient thresholds for comfortable stimulation levels. However, with the increase in the amplitude resolution, the implementation area of the current source greatly increases. Large implementation area can lead to errors in gradients, edge effects, etc during the chip fabrication, which degrades the accuracy of the output current. Moreover, the current source should have high output resistance with large voltage compliance to ensure accuracy of the output current under variations of the electrode impedance. Unfortunately, the above performance requirements of the current source are very challenging to achieve simultaneously.

Previous reported current sources either for retinal [3] or cochlear [4], [5] stimulators are based on a wide-swing

This research is partially supported by National Institute of Health (NIH) under contract NIH-NIDCD-N01-DC-6-0002.

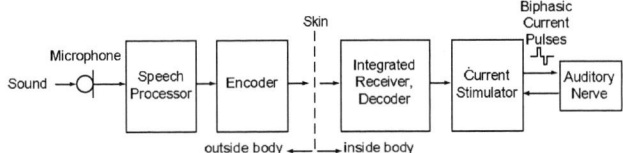

Fig. 1. Structure of a cochlear implant system.

cascode topology [3], [4] or voltage-controlled resistors (VCR) [5]. To achieve small chip area, the output resistance of the reported wide-swing cascoded current source is only about hundreds of kΩ due to the use of small-size transistors and the resulting channel-length modulation effect [3]. On the other hand, the output resistance of the reported VCR current source can achieve over 10MΩ [5]. In the VCR current source, different current amplitudes is generated by varying the gate-to-source voltage of a triode-region transistor. However, the intrinsic non-linear behavior between current and gate-to-source voltage of the triode-region transistor limits the current amplitude resolution in the VCR current source.

This paper presents a current source using a switchable multi-bias active-cascode digital-to-analog converter (DAC), which can source/sink a 1mA full-scale output current with 9-bit configurable amplitudes in a small implementation area. A stacking MOS structure is also developed to allow the current source to achieve large output resistance with a wide voltage compliance range. Table I provides performance comparisons of different current sources for current stimulators. Compared to all reported current sources, the proposed current source achieves the largest output current, amplitude resolution, voltage compliance and output resistance, while small implementation area can still be maintained. Details of the proposed current source will be discussed in later sections of this paper.

II. PROPOSED CURRENT SOURCE

The proposed current source is the core of a 8-channel bipolar current stimulator in our portable cochlear implant research platform for animal studies [6]. Fig. 2 shows a conceptual diagram of a single-channel bipolar current stimulator, in which two control signals are used to determine on/off of switches $SW_{s1} - SW_{s4}$ for generating biphasic current pulses between two adjacent electrode sites A and B. In particular, during anodic (cathodic)-pulse phase, SW_{s1} (SW_{s2}) and SW_{s4} (SW_{s3}) allow the current sinking from site A (B) to site B (A) through the auditory nerve. Since the biphasic

TABLE I
PERFORMANCE COMPARISONS WITH PREVIOUSLY REPORTED CURRENT SOURCES

	JSSC 03 [3]	JSSC 06 [4]	TBME 05 [5]	This Work
Supply Voltage	±5	±2.5	5	+5
Maximum Output Current (mA)	0.4	0.5	0.21	1
Resolution	8 bits	8 bits	5 bits	9 bits
Maximum INL (LSB)	-3.11	N.A.	N.A.	2.9
Maximum DNL (LSB)	2.15	N.A.	N.A.	0.8
Voltage Compliance (V)	3.6V (cathodic current) 3.35V (anodic current)	±2.0	4.25	4.77
Output Resistance (MΩ)	0.443	N.A.	>10	>50
Area (mm²)	0.227	N.A.	0.05	0.26
Technology	standard 1.2μm CMOS	High-voltage 0.5μm CMOS with thick gate oxide	standard 1.5μm CMOS	standard 0.35μm CMOS

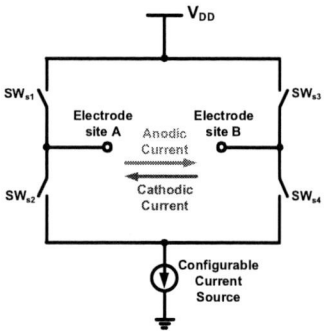

Fig. 2.   Conceptual diagram of a bipolar current stimulator.

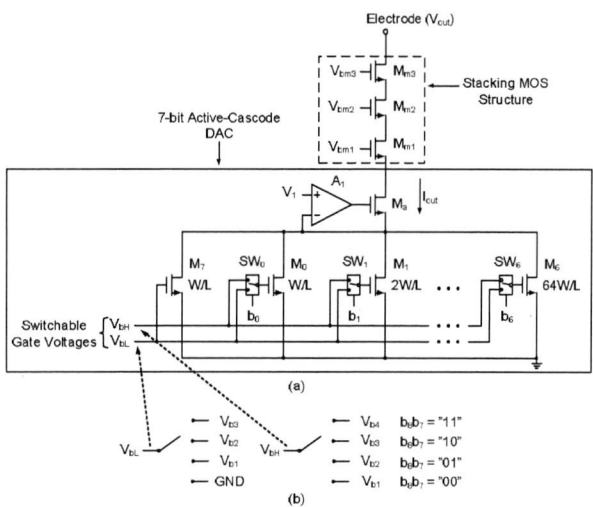

Fig. 3.   (a) Structure of the proposed configurable current source with its (b) 2-bit switchable bias scheme.

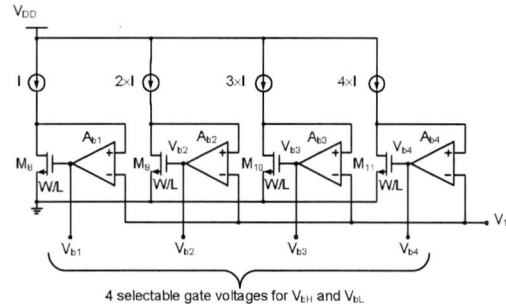

Fig. 4.   Bias voltage generation circuit.

current pulses are provided by the same current source, charge balance can be achieved and the current source determines the performances of the current stimulator.

A. Structure and Operational Principle

Fig. 3(a) shows the structure of the proposed current source, which consists of an active-cascode 7-bit current-mode DAC realized by transistors M_a, M_0 - M_6 and an amplifier A_1. The active-cascode architecture can provide high resistance at the drain of M_a. Input control code of the DAC $b_0 - b_6$ is used to connect gate voltages of M_0 - M_6 to either V_{bH} or V_{bL} through switches $SW_0 - SW_6$. For example, the gate voltage of M_n is connected to V_{bL} (V_{bH}) through SW_n when $b_n = 0$ (1), where n is from 0 to 6. Fig. 3(b) shows a 2-bit switchable bias scheme, which allows gate voltages V_{bL} and V_{bH} of $M_0 - M_7$ to select from different bias voltages $V_{b1} - V_{b4}$ based on an input control code b_8b_7. Voltages $V_{b1} - V_{b4}$ are provided by the bias voltage generation circuit shown in Fig. 4, in which current I is the least significant bit of the output current I_{out}.

To generate the first 7-bit output current I_{out}, $b_8b_7 = 00$ and thus V_{bL} and V_{bH} are connected to 0 and V_{b1}, respectively. With high-gain amplifiers A_1 and A_{b1}, the drain voltages of $M_0 - M_8$ are maintained to be the same. Since all $M_0 - M_8$ are designed to operate in the saturation region, the current from M_8 can be accurately mirrored to transistors M_n when $b_n = 1$ as

$$I_n = 2^n I \qquad . \qquad (1)$$

Therefore, the active-cascode DAC is binary-weighted to realize a 7-bit output current I_{out} as

$$I_{out} = I(b_0 + 2b_1 + 2^2 b_2 + 2^3 b_3 + 2^4 b_4 + 2^5 b_5 + 2^6 b_6) \quad . \quad (2)$$

By controlling $b_0 - b_6$ to be either 1 or 0, the output-current ranging from 0 to 127I can be generated.

In order to achieve 9-bit resolution, the 2-bit switchable multi-bias scheme selects appropriate gate voltages for $M_0 - M_7$ based on the input control code $b_8b_7 = 00, 01, 10,$ or 11. As described above, when $b_8b_7 = 00$, the dynamic range of I_{out} is from 0 to 127I. When the input code increases from 127 to 128, b_8b_7 changes from 00 to 01 to cause $V_{bL} = V_{b1}$. It allows the gate voltage of $M_0 - M_6$ to remain at V_{b1} when $b_0 - b_6 = 0$. I_{out} due to $M_0 - M_6$ is thus still kept at 127I. The compensation transistor M_7 provides an additional I to allow the total I_{out} of the current source smoothly increasing from 127I to 128I. Therefore, with $b_8b_7 = 01$, the dynamic range of I_{out} from 128I to 255I can be realized by setting different

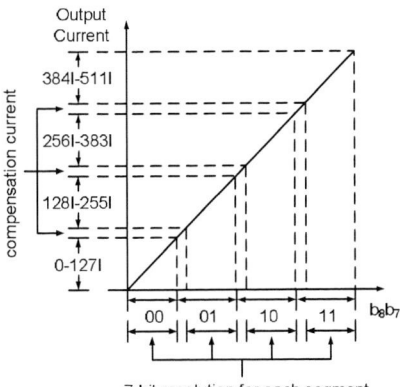

Fig. 5.   Relationship between output current and 2 additional bits.

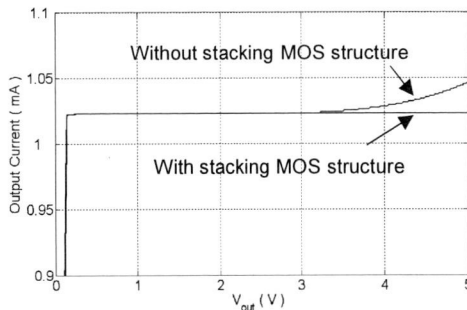

Fig. 6.   Simulated output current under different output voltages.

values of $b_0 - b_6$. Similarly, the output current from 256I to 383I and from 384I to 511I can also be generated by the proposed 2-bit switchable multi-bias strategy and 7-bit current-mode DAC with monotonic increasing behavior shown in Fig. 5. The compensation transistor M_7 is important to ensure smooth transitions between two different 7-bit segments of the output current.

The implementation area of the proposed current source is greatly reduced by the proposed 2-bit switchable bias generation scheme. In the 7-bit current-mode binary-weighted DAC shown in Fig. 3(a), the large-size transistor M_6 occupies almost half of the total implementation area of transistors $M_0 - M_6$. Obviously, if a conventional 9-bit binary-weighted DAC structure is adopted by using two additional large transistors with sizes of 128(W/L) and 256(W/L), the total implementation area would be about 4 times as that of the existing 7-bit DAC structure. It should be noted that the area occupied by the 2-bit switchable bias generation scheme is compact due to the use of small-size transistors. The proposed current source based on switchable multi-bias 7-bit current-mode DAC architecture is thus an area-efficient way to realize 9-bit current amplitude resolution. The compact design of the proposed current source also provides good current matching and linearity of the output current.

B. Output Resistance & Voltage Compliance Considerations

The output resistance of the active-cascode DAC is given by

$$R_o = A \cdot g_{ma} r_{o1} r_{oa} \qquad (3)$$

where A is the gain of amplifier A_1, r_{o1} is the equivalent output resistance of transistors M_0 - M_7, and g_{ma} and r_{oa} are the transconductance and output resistance of M_a, respectively. Since V_1 is large enough (~0.18V) to ensure all transistors $M_0 - M_7$ in the saturation region, the output resistance R_o is A times larger than that of wide-swing or fully cascode current sources, where $A \gg 1$. Although the output resistance R_o of VCR current source in [5] has a similar expression as (3), its R_o value is limited by the resistance of the triode-region transistor. As a result, the proposed active-cascode DAC structure can provide the largest output resistance.

However, as the input supply of the current source as high as 5V is needed in the stimulator, the drain voltage of M_a can

be close to 5V. When drain-to-source voltage $V_{ds,Ma}$ of M_a is much larger than the overdrive voltage $V_{ov,Ma}$, transistor M_a will suffer from the hot carrier effects [7]. The drain-to-substrate current $I_{DB,Ma}$ of M_a will then be greatly increased, thereby decreasing the output resistance of M_a (r_{oa}) and then the value of R_o. As shown in Fig. 6, the simulated output current of the current source (without stacking MOS structure) increases when the output voltage is beyond 3V, which indicates the decrease in the output resistance of the current source under hot carrier effects. In fact, hot-carrier effects have more serious impact on a transistor implemented in a standard sub-micron CMOS process than a transistor realized either by a high-voltage or a long-channel process due to larger electric field at the drain of a standard sub-micron CMOS transistor. It implies that it is difficult for the current source to maintain at high output resistance over a wide range of the voltage compliance under a high supply voltage and using a standard sub-micron CMOS process.

In the proposed current source, a stacking MOS structure shown in Fig. 3(a) using transistor $M_{m1} - M_{m3}$ on top of the active-cascode DAC is developed to address the above technical challenge. The purpose of three transistors $M_{m1} - M_{m3}$ is to limit drain-to-source voltage V_{ds} across each transistor to a reasonably small value for eliminating hot carrier effects and minimizing substrate current of each transistor when the output voltage V_{out} of the current source is close to the supply rail. It should be noted that when V_{out} is small, M_{m1} - M_{m3} will be driven to the deep linear region with a very small voltage drop across the stacking MOS structure. The output resistance of the current source is guaranteed to achieve at least R_o given in (3), as transistors M_a and $M_0 - M_{11}$ are all in the saturation region. When V_{out} is large, transistors $M_{m1} - M_{m3}$ would operate in the saturation region to further increase the output resistance of the current source. From Fig. 6, the simulated output current of the proposed current source is maintained constant under a wide range of V_{out}. High output resistance with an extended voltage compliance is thus achieved in the proposed current source under a high supply voltage of 5V and a standard 0.35μm CMOS process by using the proposed stacking MOS structure.

C. Circuit Implementations

In the proposed current source, amplifiers A_1, $A_{b1} - A_{b4}$ are crucial to its output resistance and the accuracy of mirroring current. A single-stage folded-cascode structure with

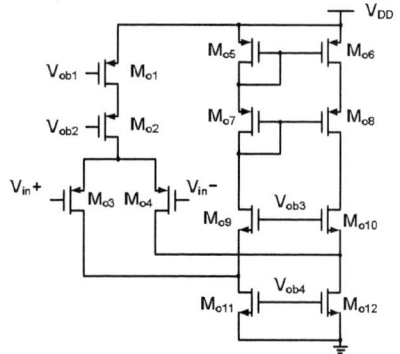

Fig. 7.   Operational amplifier in the proposed current source.

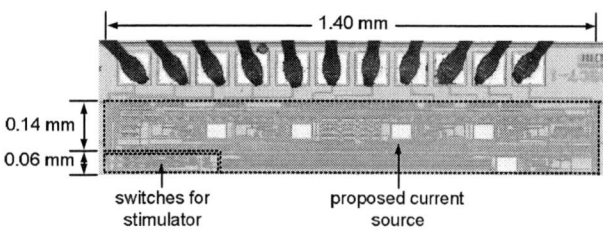

Fig. 8.   Micrograph of a single-channel current stimulator with the proposed current source.

transistors $M_{o1} - M_{o12}$ shown in Fig. 7 is used to implement A_1, $A_{b1} - A_{b4}$ for achieving high low-frequency gain. Since V_1 at the positive terminal of A_1 is a low voltage for minimizing the drain-to-source voltage of transistors $M_0 - M_{11}$ and maximizing the voltage compliance of the current source, the input differential pair of A_1 is realized by pMOS transistors M_{o3} and M_{o4}. With a 5V input supply, a cascode current mirror with transistors $M_{o5} - M_{o8}$ can provide high output resistance at the amplifier output across its output swing.

III.   MEASUREMENT RESULTS

In order to verify the performances of the proposed current source, a single-channel bipolar current stimulator is realized in a standard 5-V 0.35μm process. With the same concept shown in Fig. 2, the single-channel stimulator can generate biphasic current pulses to the auditory nerve by using the proposed current source. Fig. 7 shows the micrograph of the current stimulator, in which the chip area of the proposed current source is 0.26mm^2.

Table I provides the detailed measurement results. The proposed current source can source a maximum 1mA output current with 9-bit resolution. In particular, Fig. 9 shows the measured output currents of the proposed current source under different output voltages and input digital codes. The measured output resistance is at least 50MΩ for the output voltage between 0.23V and 5V. It justifies that both high output resistance and wide voltage compliance are achieved simultaneously in the proposed current source.

Fig. 10 shows the measured current pulses generated by the current stimulator. With two control signals to define on/off of switches in the stimulator, charge-balanced biphasic current pulses with the pulse amplitude of 1022μA are generated.

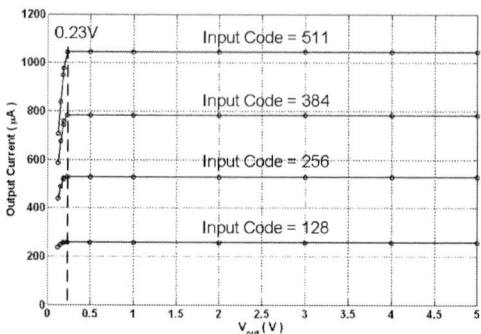

Fig. 9.   Measured output current under different output voltages.

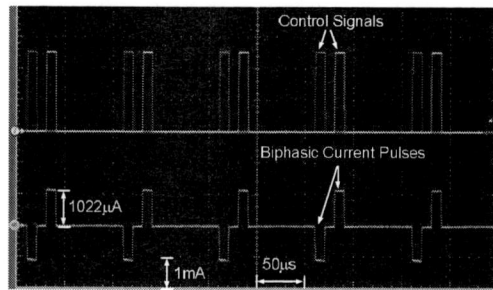

Fig. 10.   Measured charge-balanced biphasic current pulses using a single-channel bipolar current stimulator.

IV.   CONCLUSION

A 9-bit configurable current source using the proposed active-cascode 7-bit binary-weighted DAC and 2-bit switchable multi-bias scheme has been introduced and verified on silicon. The active-cascode DAC increases the output resistance, whereas 2-bit switchable multi-bias strategy helps to decrease the implementation area of the current source. Additionally, a stacking MOS structure on top of the DAC allows the current source to maintain high resistance over a wide voltage compliance range. The proposed current source is suitable for the current stimulation in cochlear implants.

REFERENCES

[1] P. Loizou, "Mimicking the human ear," *IEEE Signal Process. Mag.,* vol. 15, pp. 101 - 130, Sep. 1998.

[2] M. Chatterjee, "Effects of stimulation mode on threshold and loudness growth in multielectrode cochlear implants," *J. Acoust. Soc. Am.* vol. 105(2), pp. 850 - 860, Feb. 1999.

[3] S. C. DeMarco, W. Liu, P. R. Singh, G. Lazzi, M. S. Humayun, and J. D. Weiland, "An arbitrary waveform stimulus circuit for visual prosthesis using a low-area multibias DAC," *IEEE J. Solid-State Circuits,* vol. 38, pp. 1679 - 1690, Oct. 2003.

[4] P. T. Bhatti and K. D. Wise, "A 32-site 4-channel high-density electrode array for a cochlear prosthesis," *IEEE J. Solid-State Circuits,* vol. 41, pp. 2965 - 2973, Dec. 2006.

[5] M. Ghovanloo and K. Najafi, "A compact large voltage-compliance high output-impedance programmable current source for implantable microstimulators," *IEEE Trans. Biomed. Eng.,* vol. 52, pp. 97 - 105, Jan. 2005.

[6] A. Lobo, P. Loizou, N. Kehtarnavaz, M. Torlak, H. Lee, A. Sharma, P. Gilley, V. Peddigari, and L. Ramanna, "A PDA-based research platform for cochlear implants," in *Proc. International IEEE/EMBS Conference on Neural Engineering,* May 2007, pp. 28 - 31.

[7] Y. Tsividis, *Operation and modeling of the MOS transistor,* 2nd ed. Singapore: McGraw-Hill, 1999.

IEEE 2008 Custom Intergrated Circuits Conference (CICC)

Super-Resolution: Imaging beyond the Pixel Size Limit

Tamer A Elkhatib and Khaled N Salama
Electrical, Computers and Systems Engineering
Rensselaer Polytechnic Institute
Troy, NY
Email: elkhat@rpi.edu, khaled_salama@ieee.org

Abstract—We have implemented a new high resolution imaging system independent of the image sensor's pixel size. This super-resolution is achieved by integrating a nano-aperture patterned in the first metal layer within the pixel using a standard CMOS process. The image sensor's focal plane is scanned with a sub-micron step to obtain the super-resolution image. To experimentally verify the operation of our technique, we have fabricated a standard 3-Transistors (3T) active pixel sensors with integrated nano-apertures in a 0.13 m CMOS technology. Here,we describe the concept of our super-resolution imaging and elaborate on our fabricated design and experimental setup.

I. INTRODUCTION

The Pixel size of an image sensor ultimately determines the resolution of digital imaging. Recent advance in sub-micron technology with the demanding market of higher megapixels imagers have boosted the successive reduction in pixel size of commercial image sensors from 2.8 m till finally 1.4 m [1]. On the other hand, the first sub-micron pixel sizes have been reported recently [2] in a frame-transfer CCDs. However these miniaturized pixels meet great challenges, the most urgent ones are low light sensitivity and high crosstalk between adjacent pixels that may affect the quality of obtained image badly [3],[4]. It is clear that the pixel size of image sensors is approaching the diffraction limit of visible light.

To enhance the resolution of digital imaging beyond the pixel size, an active research field in image signal processing, referred to as super-resolution or sub-pixel imaging [5],[6], has successfully increased the effective resolution of images by capturing multiple low-resolution images through many sub-pixel dimension movements between an object and the image sensor, then a single high resolution image is obtained by applying image processing algorithms which are beyond the scope of this paper. However, these methods require an extensive image processing computations which in turn requires more digital processing circuitry and are not power efficient. In addition, there are some limitations on the hardware used to achieve the sub-pixel movements [7].

This paper presents a simple modification in the current pixel structures by integrating a nano-aperture inside each pixel. This integration will limit the detection area of the pixel to the size of the nano-aperture, then the sensor will be shifted with a submicron steps in a standard scanning fashion to sample the image, and finally a super-resolution image is formed by direct combinations of the detected frames

without any need for image processing. Such improvement will enable very high resolution imaging with good quality for many applications such as near-field in vivo imaging and digital microscopy.

Thanks to the recent development in sub-micron (0.18 m and below) CMOS technology, the realization of a small aperture in a thin metal layer in a standard CMOS process is feasible. To verify this super-resolution imaging concept experimentally, we have fabricated several standard 3T active pixels with different nano-apertures in a 0.13 m CMOS technology.

II. SUPER-RESOLUTION CONCEPT

If an optical lens is free from optical aberrations, the maximum image resolution R obtained using such a lens is determined mainly by its numerical aperture NA according to the following relation:

$$ R \quad 0.61 \times \frac{\lambda}{NA} \qquad (1) $$

where λ is the illumination wavelength. This resolution represents the radius of the airy disc, which is simply the diffraction pattern representing a point source at the object plane after its projection through the optical lens at the detector plane as shown in Fig.1(a). To achieve the best imaging resolution, two pixels in the image sensor should sample the airy disc pattern [3]. For an optical lens with moderate NA, the radius of the airy disc is in the submicron range.

To enable digital imaging with resolutions that break the limit of pixel size, we completely cover the photodetector in the pixel with the first metal layer and then pattern a nano-aperture in this metal layer with dimensions comparable to the required imaging resolution. This nano-aperture, located at the focal plane of the used optical lens, will limit the light to pass to the pixels photodetector as shown in Fig. 1(b); in such a way each pixel can sample portion of the image with the maximum possible resolution of the optical lens, and the complete image can be obtained by shifting the sensor with submicron steps that are equal to the required resolution in a standard scanning fashion. By this method, digital imaging resolution is completely independent of the pixel size and depends only on the nano-apertures dimension and the sensor scanning step.

978-1-4244-2018-6/08 $25.00 © 2008 IEEE

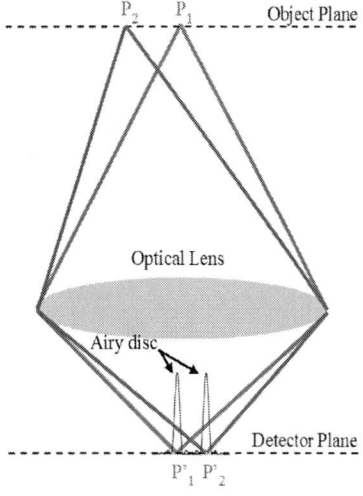

Fig.1(a)

Fig. 1(b)

Fig. 1. Schematic showing our super-resolution imaging concept (a) Cross-section of airy disc formed by projection of point source through an optical lens (b) Cross-section of a typical pixel with integrated nano-aperture that can only resolve a single airy disc pattern

A. Challenges

The main challenge for utilizing a nano-aperture in imaging is the limited amount of light reaching photodetector inside the pixel and a poor quality of image is traditionally expected. This is not entirely accurate because the direct integration of the nano-aperture inside the pixel and the small separation distance (few hundreds of nanometers) between the aperture and the photodetector. This allows the best detection of all transmitted light through the nano-aperture. Thus its extremely important to pattern the nano-aperture in the first metal layer in a CMOS technology. On contrary, if the separation distance between the aperture and the photodecter is just few microns or more, very limited amount of light can be detected through the aperture and the image quality will deteriorate.

In addition, our method doesn't suffer from the image sensors performance challenges of miniaturizing the pixel size. In fact, the pixel size and the photodetector area in our method can remain large enough (few microns) independent of the acquired resolution. This enables the imager to achieve good dynamic range, high well capacity and minimize crosstalks between adjacent pixels. The nano-aperture mainly limits the

incident light intensity on the photodetector surface. The image sensor can just be considered to operate under low light conditions which will have an effect on the detector's Signal to Noise Ratio SNR. However, an acceptable SNR can be achieved by increasing the illumination intensity or increasing the integration time of the sensor as discussed in the following section.

B. Advantages

Our novel super-resolution imaging concept offers many advantages. Obviously the key advantage is breaking the pixel size resolution tradeoff. In addition, very high spatial resolution images can be obtained using a few megapixels imager. For example, an image sensor with one megapixels array and pixel size of 5 m including a 0.5 m nano-aperture can achieve a resolution of 0.5 m and a total spatial resolution of 100 megapixels by scanning the image just in few seconds. Another important feature is combining the multi-captured frames without any need for image signal processing unlike the current sub-pixel imaging techniques. Finally our system doesn't require any extra fabrication steps and can be fabricated in any standard CMOS technology.

III. NANO-APERTURE SNR

There is no analytical method to investigate the light transmission through our suggested integrated nano-aperture and so numerical simulations are essential. We have modeled the problem in three dimensions (as shown in Fig. 2) and used finite element method to solve for the total power that reaches the surface of a 1.5×1.5 m photodetector through different square apertures sizes.

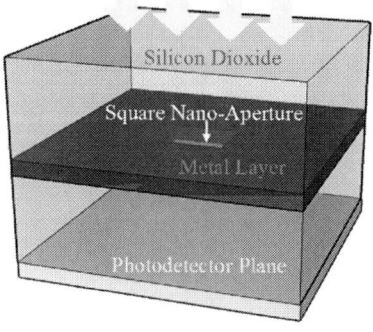

Fig. 2. A schematic of our 3D geometrical model in COMSOL Multiphysics. Photodetector plane is 0.5 m underneath the metal layer of thickness 0.3 m. These dimensions match the used 0.13 m CMOS technology.

For simplicity, we considered a plane wave with wavelength of $550nm$ to represent the average wavelength of a white light. Figure 3 summarizes our simulation results for light transmission through different nano-apertures sizes. To compare the efficiency of photodetection through these nano-apertures, Fig. 4 shows the total amount of light detected through different aperture sizes normalized to the total amount

978-1-4244-2018-6/08 $25.00 © 2008 IEEE

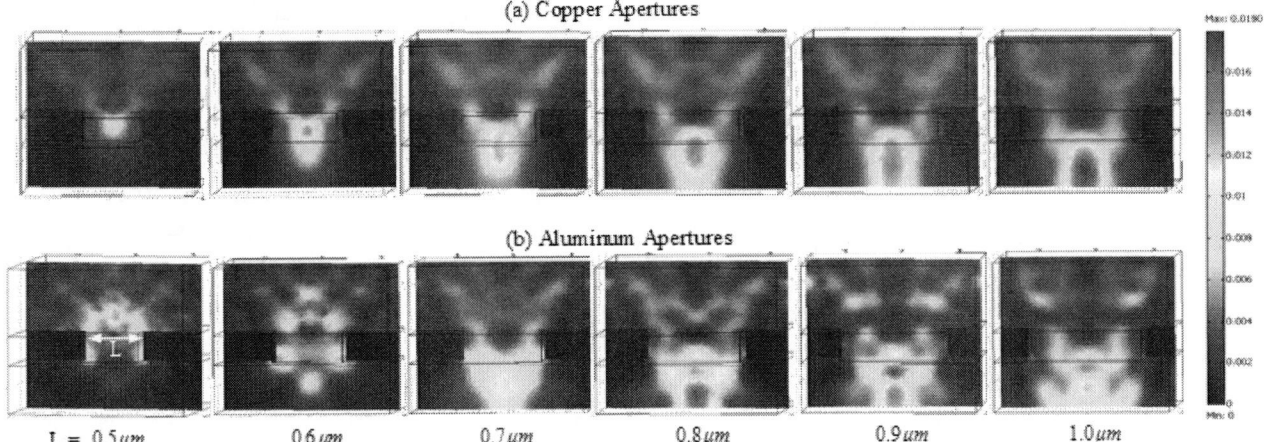

(a) Copper Apertures

(b) Aluminum Apertures

L = 0.5 um   0.6 um   0.7 um   0.8 um   0.9 um   1.0 um

Fig. 3. 3D optical simulation results showing a cross section plots of the time average power flow (a.u) through nano-apertures with different sizes that are patterned in (a) aluminum or (b) copper metal layers. The effect of surface plasmons on light transmission through aluminum apertures is obvious[8].

of light detected with no integrated nano-aperture. we also compared these results to the detection through sub-micron photodetector sizes without integrated aperture assuming a uniform illumination intensity on the pixel.

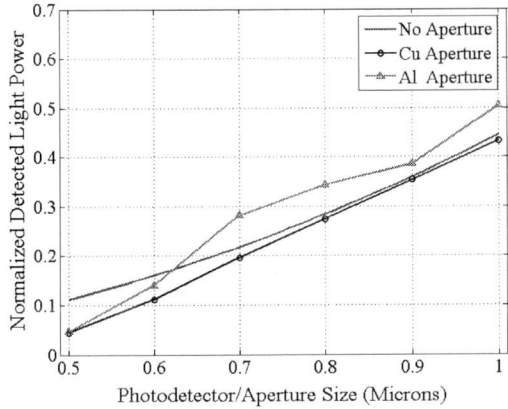

Fig. 4. Normalized total light power detected by a 1.5×1.5 m^2 photodetector through different integrated square sub-micron aperture sizes patterned in aluminum or copper metal layer, and compared to that detected by sub-micron photodetector sizes without integrated aperture

These simulation results reveal three important aspects:

For large submicron apertures, the detected light through aluminum apertures is slightly higher than that from copper apertures or even equivalent photodetector sizes with no aperture; this can be explained by the enhanced transmission due to surface plasmons in the aluminum metal layer.

For large submicron apertures, the detected light through copper apertures is close to that of equivalent photodetector sizes with no aperture; this is due to the efficient detection of the near field transmitted light that leads to an effective area of the photodetector that is slightly

equivalent to aperture's area.

For small submicron apertures, the detected light through both aluminum and copper apertures is much lower than that of equivalent photodetector sizes with no aperture; this is due to the rapid decay of the near field transmission through the aperture with the distance leading to inefficient detection.

The SNR of the integrated nano-aperture pixels can be estimated using the following formulas:

$$SNR = 10log\frac{(i_{ph} \times t_{int})}{(q(i_{ph} + i_{dc}) \times t_{int} + \quad_r)} \qquad (2)$$

$$i_{ph} = q \times \phi \times QE \times A_{eff} \qquad (3)$$

where i_{ph} and i_{dc} are the photocurrent and dark current respectively, t_{int} is the integration time, $\quad_r$ is the average noise power, ϕ is the incident photon flux, QE is the imager quantum efficiency, A_{eff} is the effective area of photodetector, and q is the electron charge.

To achieve a good quality imaging, a minimum SNR of $30dB$ is required [9]. This can be achieved by increasing the illumination intensity or the exposure time or both provided that the imager does not saturate. We estimated in Fig. 5 the minimum photometric exposure (illumination intensity × exposure time) required to achieve a SNR of $30dB$ as a function of nano-aperture sizes based on the results of our numerical simulations. These results show that it is feasible to achieve a good quality imaging using our super-resolution method with illumination intensity of 300 ux and integration time of $10ms$ using an aperture size of a 0.6 m.

IV. EXPERIMENTAL SETUP

In order to experimentally verify our super-resolution imaging technique, we designed and fabricated a standard 3T active pixel sensor (APS) with integrated nano-apertures in a 0.13 m 1.2V 7M2P CMOS process. A schematic of the 3T APS is

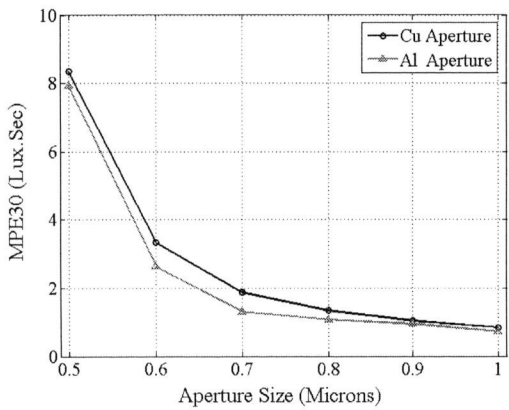

Fig. 5. The minimum photometric exposure to produce a SNR of $30dB$ (MPE30) [10] as function of different integrated nano-aperture sizes

shown in Fig. 6. The metallization layers of this process are made of copper. The thickness of metal 1 layer is 0.29 m and the separation distance to the silicon surface is 0.5 m. The minimum nano-aperture size is determined by the minimum allowable opening area in metal 1 (0.366 m) which is equivalent to a square nano-aperture with size of 0.6 m. We selected an $n+/p_{sub}$ junction as the pixels photodiode because it is the nearest junction to the silicon surface. The designed area of this photodiode is 3×3 m .

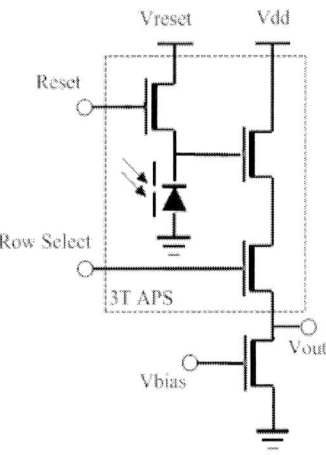

Fig. 6. A schematic of the fabricated photodiode type standard 3T active pixel sensor

We have just received our fabricated chip shown in Fig. 7. We will obtain the high resolution images from this sensor using a 3D nano-positioning stage from ThorLab that offer a minimum scanning step of $25nm$ and range of $4mm$ in the three dimensions. A LabView Program has been already designed to control this stage from computer and obtain images. We also use a single aspherical lens with NA of 0.68 to minimize the optical aberration and enable optical resolution in sub-micron scale.

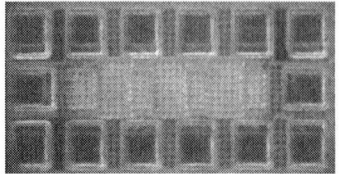

Fig. 7. Die photo of the Fabricated Sensor

V. CONCLUSION

We introduced a new method to achieve a super-resolution digital imaging independent of the pixel size by integrating a nano-aperture in each pixel of the image sensor. This integration can be implemented in any standard submicron CMOS process. We have shown that the transmission through this nano-apertures can achieve good image qualities by optical nummerical simulations and SNR calculations. We have designed and fabricated a testing chip of a standard 3T APS integrated with different nano-aperture sizes in a 0.13 m technology to test this method experimentally.

VI. ACKNOWLEDGMENT

The authors would like to thank George Soliman at Rensselaer Polytechnic Institute for his valuable discussions and support in numerical optical simulations that helped to improve the quality of this paper.

REFERENCES

[1] G. Agranov, R. Mauritzson, S. Barna, et al., "Super Small, Sub 2 m Pixels for Novel CMOS Image Sensors," In 2007 International Image SensorWorkshop, pp. 307-310, 2007.

[2] K. Fife, A. El Gamal and H.-S. P. Wong, "A 0.5 m Pixel Frame-Transfer CCD Image Sensor in 110nm CMOS," IEDM Technical Digest, pp. 1003-1006, 2007.

[3] P. Catrysse and B. Wandell, "Roadmap for CMOS image sensors: Moore meets Planck and Sommerfeld," Proceedings of the SPIE,Vol. 5678, pp. 1-13, 2005.

[4] H. Rhodes, G. Agranov, C. Hong, et al., "CMOS imager technology shrinks and image performance," In 2004 IEEE Workshop on Microelectronics and Electron Devices, pp. 7-18, 2004.

[5] M. Ben-Ezra, A. Zomet, and S. K. Nayar, "Video super-resolution using controlled subpixel detector shifts," IEEE Transactions on Pattern Analysis and Machine Intelligence, Vol. 27, No. 6, pp. 977-987, 2005.

[6] K. Yu, N. Park, D. Lee, and O. Solgaard, "Superresolution digital image enhancement by subpixel image translation with a scanning micromirror," IEEE Journal of Selected Topics in Quantum Electronics, Vol. 13, No. 2,pp. 304-311, 2007.

[7] S. Farsiu, D. Robinson, M. Elad, and P. Milanfar, "Advances and challenges in Super-Resolution," International Journal of Imaging Systems and Technology, Vol. 14, pp. 47 57, 2004.

[8] T. Thio, K. M. Pellerin, R. A. Linke, H. J. Lezec, , and T. W. Ebbesen, "Enhanced light transmission through a single subwavelength aperture," Optics Letters, Vol. 26, No. 24, pp. 1972-1974, 2001.

[9] F. Xiao, J. E. Farrell, B. Wandell, "Psychophysical thresholds and digital camera sensitivity: The thousand photon limit," Proceedings of the SPIE , Vol. 5678, pp. 75-84, 2005.

[10] J. E. Farrell, F. Xiao, S. Kavusi, "Resolution and Light Sensitivity Tradeoff with Pixel Size," Proceedings of the SPIE, Vol. 6069, 2006.

IEEE 2008 Custom Intergrated Circuits Conference (CICC)

Polysilicon Vertical Actuator Powered with Waste Heat

J. Varona[1], M. Tecpoyotl-Torres[1], and A. A. Hamoui[2]

[1]Universidad Autónoma del Estado de Morelos, Mexico; and [2]McGill University, Canada

Abstract-This paper presents a new micro thermal actuator with bi-directional vertical motion. Traditional thermal actuators require an electric current to generate heat by Joule effect and obtain motion via thermal expansion. This work offers a simplified polysilicon thermal actuator designed to operate with an external heat source and, for example, scavenge heat from the surrounding medium. The actuator was fabricated using a standard surface micromachining process (PolyMUMPs). The thermal actuation characteristics, analysis, and experimental results for this novel micro-mechanical device are provided to illustrate performance and potential applications.

I. INTRODUCTION

Thermal actuation has been extensively used in Micro-Electro-Mechanical Systems (MEMS). Thermal actuators typically exhibit larger displacements and forces than electrostatic actuators and their thermal expansion is linearly related to the applied heat.

Deflection of thermal actuators strongly depends on their geometry. The principle of operation is to make an electric current flow through the device structure and induce Joule heating. The corresponding increase in temperature causes the structural material to expand and generate deflection when the device is mechanically constrained.

As Joule heating consumes considerable electrical energy, typical thermal microactuators also require an external battery that is several times the size of the microsystem itself. It is the power supply unit that severely limits portability and practicability of microsystems technology.

The idea of exploiting energy resources from the environment presents an opportunity to develop autonomous microsystems that do not rely on a bulky battery with fairly limited energy storage capacity. Through the application of energy harvesting techniques, micro-devices could scavenge power from ambient heat, light, vibrations, EM waves, etc.

Microactuator technologies find applications in a great variety of fields where a reduction in size and weight is important. Particularly compelling is the possibility of implementing MEMS and electronic circuitry together on the same substrate that can be manufactured in high volumes and low cost due to the mass production nature of semiconductor wafers.

As an alternative to the classical electrically driven thermal actuator, this work presents the design and characterization of a vertical actuator that operates by exploiting thermal energy already present in the surrounding environment. Such conditions in which a high heat density is available are very common in practical situations where MEMS are usually found, for example in the automotive industry and in every application using electronic circuitry that dissipates heat.

II. DESIGN CONCEPT

Historically, the analysis of thermal actuators began with the development of the horizontal thermal actuator [1, 2]. The basis for the operation of these devices consists in obtaining asymmetrical thermal expansion between two adjacent and physically joined structural microbeams (the 'hot' and 'cold' arms respectively). Numerous works on the subject of lateral thermal actuators have been previously published [1-4]. The actuation principle of a vertical thermal actuator (VTA) is based on that of a horizontal actuator.

Variations of the vertical actuator are well documented [5-7]. One of the preferred architectures for implementing a vertical thermal actuator is illustrated in Fig. 1. This design comprises two 'U' shape structures, one on top of the other, that are connected at the tip of the actuator. If a voltage is applied between the anchor pads of the top (or bottom) level structure, electrical current will flow solely through this layer thereby increasing its temperature and causing a correspondent thermal expansion that will deflect the tip of the actuator downwards (or upwards).

This paper reports a customized actuator design that generates vertical deflection as a result of external heating of the device. A simplified geometry that does not entail providing a return path to ground for a driving current on the same layer is introduced in Fig. 2.

The structure is formed by two cantilever-type beams fabricated in two different layers one on top of the other and separated by an air gap. The beams are independently anchored on the substrate at one end and linked to each other at the other end by means of a via. The width of the upper cantilever is made smaller than that in the lower level so that the beams are asymmetric. Also, the top layer is about 25% thinner than the bottom one due to characteristics imposed by the MUMPs fabrication process.

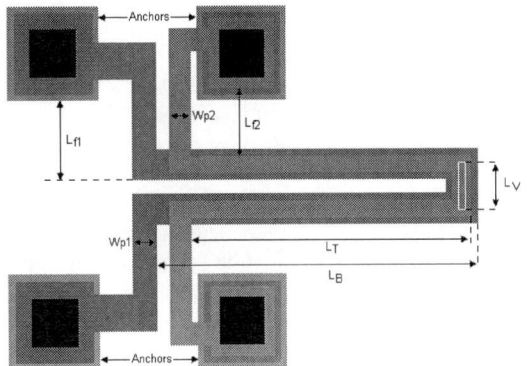

Fig. 1. The classical 'U' shape electrothermal vertical actuator.

978-1-4244-2018-6/08 $25.00 © 2008 IEEE

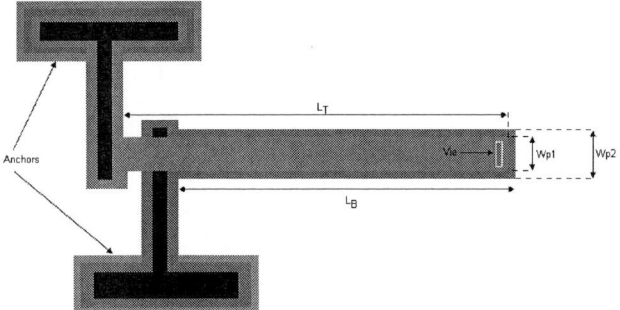

Fig. 2. Schematic view of the proposed double single- beam vertical thermal actuator.

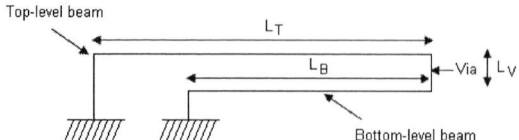

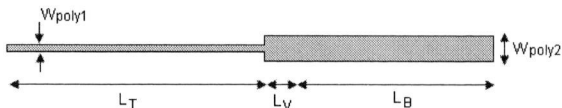

Fig. 3. Simplified coordinate system and diagram of the 'unfolded' actuator of Figure 2 showing the two microbeams connected in series.

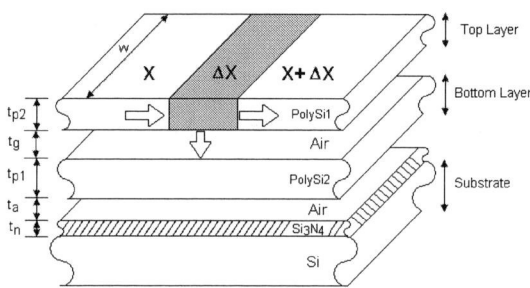

Fig. 4. Cross-sectional view of the actuator's structure.

In the presence of a heat source, energy is transferred to the structure of the actuator which in turns increases its temperature. The different geometry of the two microbeams leads to a net expansion of the top layer structure that generates vertical deflection in the direction of the substrate.

Bidirectional displacement may be achieved if heat is applied separately through one of the anchors. In such case, the temperature along the actuator will decrease as one moves in the direction of heat flow as dictated by heat transfer theory. Accordingly, one of the beams will expand more than the other, driving the tip of the actuator against the colder beam.

Among several options that can be applied to more easily drive the actuator independently at either anchor pad is the use of flip-chip transfer technology [8]. Flip-chip techniques allow releasing the actuator from its substrate and liberating the driving pads that may be transferred onto another substrate to integrate a system on a chip.

Design concept relies on the availability of a sufficiently high heat source. External heat sources are fairly common in environments where MEMS find practical use. Examples include automotive and aerospace industry where high heat transfer rates and temperatures beyond 500°C are commonly available [9]. Another potential widespread source of heat for MEMS is found in modern electronic VLSI integrated circuits that may have power densities in excess of 40W/cm^2 [10, 11].

III. THERMAL ANALYSIS AND MODELING

A. Temperature Distribution

The thermal analysis of the vertical thermal actuator illustrated in Fig. 2, treats the two microbeams as elements that are connected in series as depicted in schematic of Fig. 3. The total length of the device is $L_T+L_V+L_B$ with the segment L_B being thicker than L_T as indicated before. The following analysis also assumes that heat is applied at the anchor pad of the upper beam while the anchor of the lower beam is kept in contact with the substrate at room temperature.

The heat losses by radiation may be ignored based on the fact that radiation becomes significant only at high temperatures (>1000°C) [12] while operating temperatures for this kind of actuators are relatively low. The steady-state spatial thermal distribution can be derived from examining a differential element of the hot-arm structure of width w, thickness t and length Δx as illustrated in Fig. 4.

Following the First Law of Thermodynamics, under steady-state, the energy brought into the element at point x is equal to the energy that comes out at point x+Δx plus the energy which is transferred to the ambient:

$$-kwt\frac{dT}{dx}\bigg|_x = -kwt\frac{dT}{dx}\bigg|_{x+\Delta x} + hw(T-T_a)\Delta x \qquad (1)$$

where k is the thermal conductivity of the structural material, T is the operating temperature, T_a is the ambient temperature, and h is the convection coefficient that accounts for the heat losses by conduction through the air.

After re-arranging equation (1) and taking the limit as $\Delta x \rightarrow 0$, the following second-order differential equation is obtained:

$$\frac{d^2T}{dx^2} - \frac{h}{kt}(T-T_a) = 0 \qquad (2)$$

If changing variables in the form of $\theta = T-T_a$ and $B = \sqrt{\frac{h}{kt}}$, the general solution for equation (2) is:

$$T_{(x)} = T_s + C_1 e^{Bx} + C_2 e^{-Bx} \qquad (3)$$

A particular solution for boundary conditions $\theta_{(0)}=\theta_2$ and $\theta_{(L)}=0$ yields:

$$C_1 = \frac{\theta_2}{1-e^{2BL}} , \quad C_2 = \theta_2 - C_1 \qquad (4)$$

Based on equations (3) and (4), Fig. 5 presents an example for the thermal distribution along the beams of the vertical actuator with a total length of 395μm.

978-1-4244-2018-6/08 $25.00 © 2008 IEEE   556

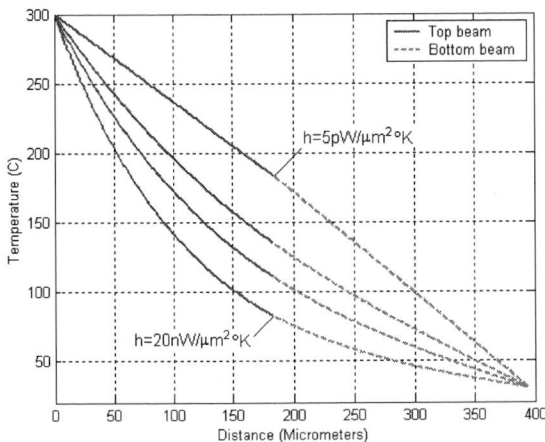

Fig. 5. Solution for temperature distribution along the actuator beams for different values of the convection coefficient h.

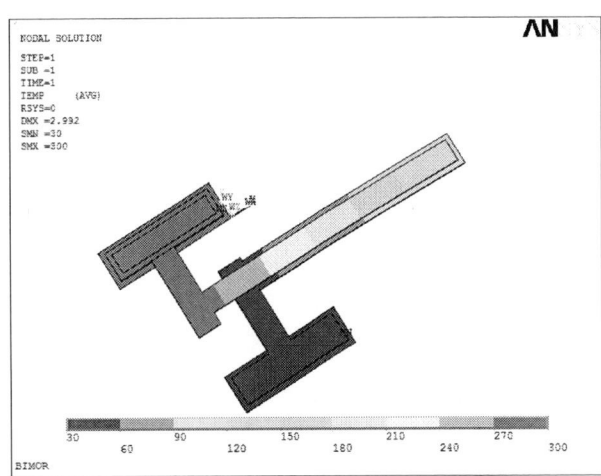

Fig. 6. Numerical results for the temperature distribution of the VTA.

Estimation of the heat convection coefficient is a non-trivial task as it can change dramatically from one situation to the next. The effect of the exact value of the convection coefficient h on the temperature profile can be observed as the distribution changes from a linear to exponential shape.

B. Thermal Expansion

Thermo-mechanical actuation has been comprehensively studied in MEMS [2, 3, 7]. The linear thermal expansion ΔL is given by the following set of equations:

$$\Delta L = K^{-1}F_{thermal} \,, \quad F_{thermal} = A\sigma \,, \quad \sigma = E\alpha\Delta T \qquad (5)$$

where K is the stiffness coefficient, A is the cross-sectional area, E corresponds to the Young's Modulus, $\Delta T=(T-T_a)$, σ represents the thermal stress and α is the thermal expansion coefficient of the material.

The thermal expansion for each of the two arms is obtained by integrating along the beam structure:

$$\Delta L = \alpha \int_0^L (T - T_a)dx \qquad (6)$$

and substituting the appropriate temperature distribution from equation (3) yields:

$$\Delta L = \alpha \int_0^L (C_1 e^{Bx} + C_2 e^{-Bx})dx \qquad (7)$$

Finally, the mechanical deflection of the actuator can be estimated based on the thermal expansion by using some of the well established structural engineering methods presented in [3, 7] that analyze the bending moments acting on the hot-arm by solving a set of simultaneous equations.

C. Finite Element Simulation

To simulate, finite element analysis (FEA) was performed using the commercial software ANSYS™. Multi-physics models including mechanical and thermal domains were used while imposing boundary conditions as mentioned above. In modeling, material properties and characteristic parameters of the selected fabrication process were considered.

Fig. 6 shows the temperature distribution of the thermal actuator when the anchor pad of the upper level microbeam is brought to 300°C. Simulation results for the out-of-plane tip deflection of the actuator as a function of temperature behave as predicted by the analytic model in [2].

IV. FABRICATION AND EXPERIMENTAL RESULTS

The proposed vertical thermal actuator was fabricated using the commercial Multi-User MEMS process (MUMPs), comprising three standard layers of surface micromachined polysilicon. The actuator was then released removing the sacrificial PSG layer by a buffered hydrofluoric (HF) etchant. Scanning electron microscope (SEM) images of the fabricated actuator are shown in Fig. 7.

Initially, Joule heating was used to warm up one of the anchor pads and let the heat being transferred to the actuator by solid thermal conduction mainly. To do so, an electrical circuit was formed across the anchor pad alone without having any current flowing through the structure of the actuator. This setup permitted to electrically control the temperature and the amount of energy applied at the anchor pad according to the Lenz-Joule law.

Fig. 7. SEM micrograph of one of the thermal actuators fabricated for testing.

978-1-4244-2018-6/08 $25.00 © 2008 IEEE

A power supply unit and a source meter were used to provide the current and monitor the voltage simultaneously. A second experimental setup using an electric heating resistor connected to a thermometer was employed to raise the temperature of the substrate of the whole microchip and actuate the devices by pure heat conduction.

Fig. 8 shows the experimental results for displacement as a function of pad temperature. It is worth mentioning that the maximum displacement in the direction of the substrate is limited by the height of the actuator from the substrate which in this case is equal to 2-µm as can be appreciated in the bottom-right inset of Fig. 7.

It can be observed that sufficiently large displacements, useful for many practical applications, are achieved by direct heating of the actuator pad. This is an indication of the feasibility to implement these thermal actuators in many situations where an external heat source is readily available.

The frequency response of the actuator was determined by alternately applying and ceasing to apply the heat source to one of the anchor pads. The main heat dissipation mechanisms are convective heat transfer from the top surface and heat conduction from the bottom surface through the air to the substrate. The contribution of radiative heat transfer may be neglected due to its minimal contribution as stated in section III. Experimentally, it was found that the thermal actuator can operate at a frequency of up to 22-Hz while still displaying full-swing deflection. Thermo-mechanical efficiency and frequency response strongly depend on how the heat is applied to the actuator and on the conditions for heat dissipation. While in some cases the substrate can act as a heat sink, in some others heat dissipation and cyclical operation could be improved by releasing the actuator from its original substrate (e.g. by means of flip-chip, etc.) and transferring it onto a better heat sink that may be part of a more elaborated microsystem involving several mechanisms to transfer the energy from one component to another and obtain useful work.

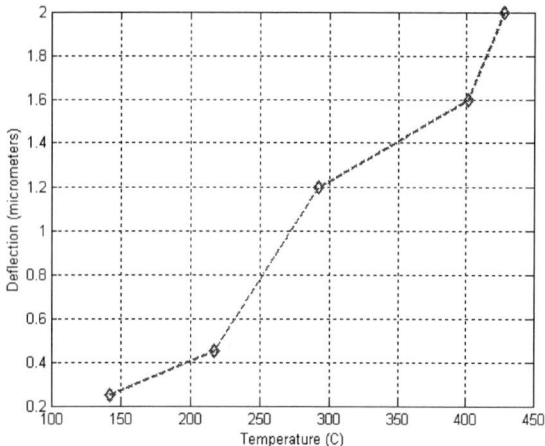

Fig. 8. Displacement of the actuator tip as a function of pad temperature in the direction to the substrate.

V. CONCLUSIONS

A novel vertical thermal actuator device which can be manufactured using a cost-effective standard fabrication process has been introduced. As an alternative to the electrically driven thermal actuators, this design has been optimized for its usage with an external heat source that opens the possibility for energy scavenging.

Actuation of the proposed VTA can be achieved through a variety of means, from the typical Joule heating process to direct bulk heating of the actuator structure. Practically, any source of heat producing a temperature change in the material can be used; heat may be the result of combustion, solar radiation, laser, hot air, geothermal or nuclear energy, direct conduction from a hot surface, etc.

Temperature distribution and thermo-mechanical analysis and modeling have been undertaken. Experiments show that fairly large and useful displacements can be achieved at relatively low operating temperatures. Thermal to mechanical energy conversion has a broad range of practical applications including sensors and actuators for complex electronic systems and MEMS devices.

ACKNOWLEDGMENTS

The authors wish to thank Prof. Antonio Ramírez-Treviño from CINVESTAV-GDL and Prof. José Mireles Jr. from UACJ for providing access to test equipment and laboratory facilities. The first author gratefully acknowledges financial support from CONACYT scholarship and the Doctoral Scholars Loan from SAE.

REFERENCES

[1] H. Guckel, J. Klein, T. Christenson, K. Shrobis, M. Landon and E.G. Lovell, "Thermo-magnetic metal flexure actuators" Proc. of IEEE Solid-State Sensor and Actuator Workshop pp 73–5, 1992.

[2] Q. Huang and N. Lee, Analysis and design of polysilicon thermal flexure actuator, J. Micromech. Microeng. 9, 64-70, 1999.

[3] N.D. Mankame and G.K. Ananthasuresh, Comprehensive thermal modeling and characterization of an electro-thermal-compliant microactuator, J. Micromech. Microeng. 11 452-62, 2001.

[4] D. Yan, A. Khajepour and R. Mansour, Modeling of two-hot-arm horizontal thermal actuator, J. Micromech. Microeng. 13 312–322, 2003.

[5] J.H. Comtois and V. M. Bright, "Applications for surface-micromachined polysilicon thermal actuators and arrays", Sensors and actuators, vol. A (58), pp. 19—25, 1997.

[6] W. D. Cowan and V. M. Bright, "Vertical thermal actuators for micro-opto-electro-mechanical systems", Proc. SPIE, vol. 3226, pp. 137-146, 1997.

[7] D. Yan, A. Khajepour and R. Mansour, Design and Modeling of a MEMS bidirectional vertical thermal actuator, J. Micromech. Microeng. 14 841–850, 2004.

[8] J.H. Lau, Flip chip technology, McGraw-Hill, New York, 1995.

[9] H. Fu et al., A One-Dimensional Model for Heat Transfer in Engine Exhaust Systems, SAE Technical Paper 2005-01-0696.

[10] Y. Cheng et al., Thermal Analysis for Indirect Liquid Cooled MultiChip Module Using Computational Fluid Dynamic Simulation and Response Surface Methodology, IEEE Trans. Comp. Packag. Technol., vol. 29, no. 1, pp. 39-46, March 2006.

[11] C.JM. Lasance and R.E. Simons, Advances In High-Performance Cooling For Electronics, Electronics Cooling, November 2005. Available on-line at:
www.electronics-cooling.com/articles/2005/2005_nov_article2.php

[12] R. Hickey, M. Kujath and T. Hubbart, Heat Transfer Analysis and Optimization of Two-beam Microelectromechanical Thermal Actuators, J. Vac. Sci. Technol. pp. 971, 2002.

IEEE 2008 Custom Intergrated Circuits Conference (CICC)

Low Noise μWatt Interface Circuits for Wireless Implantable Real-Time Digital Blood Pressure Monitoring

Peng Cong, Wen H. Ko, and Darrin J. Young
Department of Electrical Engineering and Computer Science
Case Western Reserve University
Cleveland, Ohio 44106

Abstract-A wireless implantable blood pressure monitoring microsystem employs an MEMS capacitive pressure sensor, which converts pressure signal to a capacitance change with a sensitivity of approximately 0.7 fF/mmHg. Low-noise μWatt correlated-double-sampling capacitance-to-voltage interface converter followed by an 11-bit cyclic ADC further converts the capacitive signal to a digital output for data processing and telemetry. The prototype microsystem achieves a resolution of 0.1 mmHg over a 1 kHz bandwidth with a dynamic range of 59 dB and consumes 18 μA from a 2V supply.

I. INTRODUCTION

DNA sequencing of small laboratory animals together with *in vivo* real-time long-term biological information, such as blood pressure, core body temperature, biopotential signals, etc., is ultimately crucial for various advanced biomedical and system biology research to identify genetic variation susceptibility to diseases and to potentially develop new treatment methods [1]. Common blood pressure monitoring techniques for small animals employ an invasive catheter-tip transducer inserted into a blood vessel or a tail cuff device. Both approaches are inadequate for long-term monitoring [2]. Recently, a wireless less-invasive long-term implantable blood pressure monitoring microsystem with RF powering and data telemetry capability is developed [3] as shown in Fig. 1. The system employs an instrumented elastic circular cuff wrapped around a blood vessel to sense real-time blood pressure waveforms. The elastic circular cuff is made of bio-compatible elastomer and is filled with low viscosity bio-compatible insulating fluid with an immersed pressure sensing microsystem, consisting of an MEMS pressure sensor and a custom-designed IC. The system measures the pressure waveform inside the cuff coupled from the expansion and contraction of the vessel. Because the measured waveform represents a down-scaled version of the vessel blood pressure with a typical scaling factor of 0.2, it is essential to have a pressure sensing system with high resolution. Miniaturization is critical for the implantable system to measure pressure from a blood vessel with a diameter in a range of sub-millimeter. Therefore, an MEMS capacitive pressure sensor together with μWatt interface electronics is developed and reported in this paper. A correlated-double-sampling (CDS) capacitance-to-voltage (C/V) converter is designed to implement a precision sensor interface. An 11-bit cyclic ADC, counting for the necessary pressure variation range and environmental effects, is used to digitize the pressure information with a bandwidth of 1 kHz, which is sufficient for the proposed biomedical application, for data processing and telemetry. Furthermore, it is difficult to provide a high supply voltage to power electronics in a wireless implantable battery-less microsystem. Therefore, low voltage design is critical for a such system. For 1.5 μm CMOS process selected for this design with V_{TP} and V_{TN} approximately 1V and 0.7V, respectively, a 2V power supply is chosen as a trade-off between circuit design complexity and radio-frequency powering constraints.

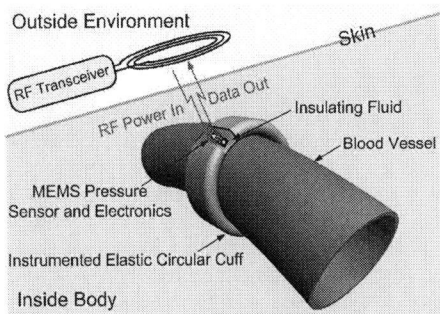

Fig. 1. Proposed wireless implantable blood pressure monitoring system

II. MEMS CAPACITIVE PRESSURE SENSOR

MEMS capacitive pressure sensor is chosen for the proposed blood pressure monitoring due to its miniature size, adequate sensitivity, zero DC power dissipation, and time stability. Fig. 2 presents a cross-sectional view of a sensor, which consists of an edge-clamped silicon diaphragm over a vacuum cavity. The diaphragm deflects under an increased external pressure; hence, an increased capacitance value. An MEMS capacitive pressure sensor is designed and fabricated using a similar wafer bonding and etch-back process as described in [4]. Fig. 3 shows SEMs of a fabricated pressure sensor exhibiting a dimension of 0.4 x 0.5 x 0.4 mm³ with a measured nominal capacitance value of 2pF and a sensitivity of approximately 0.7 fF/mmHg.

III. CDS CAPACITANCE TO VOLTAGE CONVERTER

Given a typical scaling factor of 0.2 and the sensor sensitivity, to achieve 1 mmHg resolution of blood pressure inside a vessel, the sensing system calls for a capacitance resolution of 140 aF over 1 kHz bandwidth. A switched-capacitor correlated-double-sampling interface scheme with a 2 kHz sampling frequency is used to suppress DC offset and $1/f$ noise to achieve a precision MEMS sensor interface. Fig. 4 shows a simplified schematic of the C/V converter.

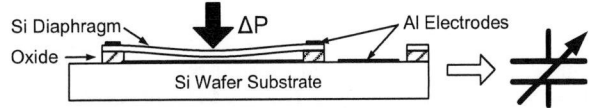

Fig. 2. Capacitive pressure sensor cross-sectional view

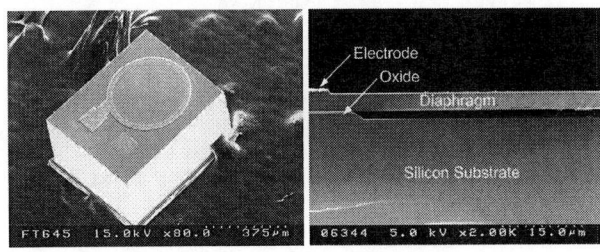

a. 3-D view   b. Cross-sectional view

Fig. 3. SEMs of fabricated MEMS capacitive pressure sensor

978-1-4244-2018-6/08 $25.00 © 2008 IEEE

An input common-mode feedback (ICMFB) circuit is incorporated with the converter to minimize the input common-mode shift caused by the driving clock (V_S); hence minimizing any offset due to mismatching of parasitic capacitances and drift over time. C_{FB} of 5 pF is chosen to sufficiently compensate the input common-mode shift for sensors with nominal capacitance up to 5 pF.

A 2.5-gain stage is designed to connect the pre-amplifier to the input of ADC. There are three major functions for the gain stage. i). Boost up signal level, as the input range of the ADC and the output range of the pre-amplifier are $2V_{pp}$ and $0.8V_{pp}$, respectively. ii). Achieve subtraction between two samples of the pre-amplifier for CDS. During the first phase, Φ_1 is "Low" and Φ_2 is "High," and the thermal noise, DC offset, and $1/f$ noise of the pre-amplifier are sampled onto C_H at the end of this phase. During the second phase, Φ_1 is "High" and Φ_2 is "Low," the signal gain is defined by the ratio of C_H and C_{II}, which is designed to be 2.5. The sensing information, which is $(\Delta C_S/C_I)V_S$, with thermal noise, DC offset, and $1/f$ noise of the pre-amplifier will be added to C_H, and the difference between two phases will be amplified and then sampled by ADC. The DC offset and $1/f$ noise closely track each other between the two phases, thus will be greatly suppressed. However, the thermal noise becomes doubled. iii) The bandwidth of the pre-amplifier is limited by the thermal noise requirement. The 2.5-gain stage is, therefore, designed with a sufficiently high bandwidth to interface with the ADC.

As shown in Fig. 4, a reference capacitor, C_R, closely matched to sensor capacitance is required. The sensor can be designed with a differential architecture, however, requiring a much involved microfabrication process [5]. Alternatively, an automatic capacitance offset cancellation scheme can be readily implemented to obtain a closely-matched reference capacitor for the MEMS sensor, as shown in Fig. 5. During the initial phase of the circuit operation, a digitally controlled reference capacitor array at the amplifier input will be cycled through to find a reference capacitance closely matching the sensor nominal capacitance. The digital reference capacitor thus can greatly suppress the output offset voltage, ensuring a proper circuit operation and dynamic range requirement. C_I is designed to be 300 fF to achieve a C/V converter sensitivity of 10 μV/aF.

A fully differential telescopic architecture with PMOS input transistors is chosen for its low noise and low power dissipation compared to other architectures, such as folded-cascode amplifier. The amplifier employs a conventional switched-capacitor common-mode feedback (CMFB) circuit. The input pair operates in sub-threshold region to maximize g_m for a given bias current with a reduced circuit speed, which is, however, sufficient for biomedical applications, where typical signal bandwidth is from a few Hz to kHz. The transconductances of input devices are selected based on the thermal noise requirement. Class A/AB amplifier, similar to the amplifier used in ADC, is chosen to implement the 2.5-gain stage for its low power dissipation as well as large output signal range. The circuit details will be described in Section IV. The output noise of this gain stage is designed to be comparable to the pre-amplifier to achieve a total C/V converter output noise of 1 mV$_{rms}$ over 1kHz bandwidth, equivalent to a 100 aF resolution.

IV. CYCLIC ADC

An 11-bit ADC is required to digitize the analog output of C/V converter. Cyclic (algorithmic) ADC is selected for its low power dissipation, adequate resolution, small silicon area, and design simplicity. Fig. 6 shows simplified schematic of cyclic ADC using redundant signed digit (RSD) and ratio-independent multiply-by-two schemes [6].

A. Redundant signed digit (RSD) scheme

The advantages of redundant signed digit (RSD) scheme are its immunity to linearity errors caused by offset voltages and comparator accuracy. It can be shown that the decision levels, V_L and V_H, can vary over a wide range without introducing missing or redundant codes, or corresponding degradation in linearity.

B. Ratio-independent multiply-by-two algorithm

To avoid capacitors matching limitation for precise multiply-by-two operation in the multiplying DAC (MDAC), a ratio-independent multiply-by-two algorithm is used. The input voltage is sampled twice followed by an exchange of sampling capacitor and integrating capacitor [6]. In addition, the amplifier resets in each clock cycle, which suppresses the amplifier DC offset and $1/f$ noise. Although the precise closed-loop gain is not limited by capacitor matching, it is limited by the amplifier finite open-loop gain. Table 1 shows simulated amplifier open loop gain requirement versus ADC linearity. It can be seen that a gain greater than 72 dB is required to achieve an INL less than 1 LSB. Therefore, a two-stage amplifier topology is desirable.

C. Noise Analysis

ADC resolution is ultimately limited by the ADC sampling switch noise and amplifier noise. The total noise referred to the ADC input can be expressed as

$$\frac{4}{3}\overline{V_{n,1}^2} + \frac{1}{3}\overline{V_{n,2}^2} \ , \tag{1}$$

where $\overline{V_{n,1}^2}$ is noise power generated in S/H and $\overline{V_{n,2}^2}$ is noise power generated in MDAC.

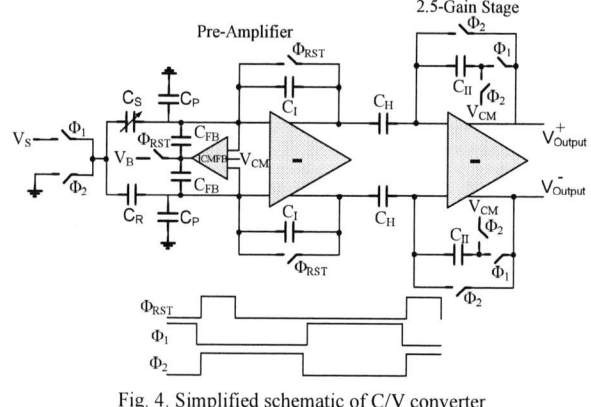

Fig. 4. Simplified schematic of C/V converter

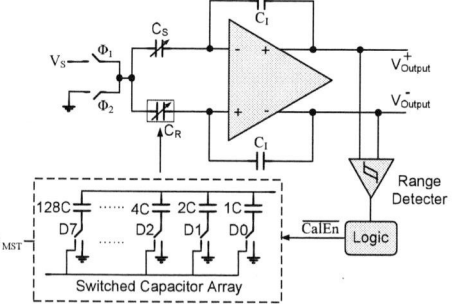

Fig. 5. C/V converter with automatic capacitance offset cancellation

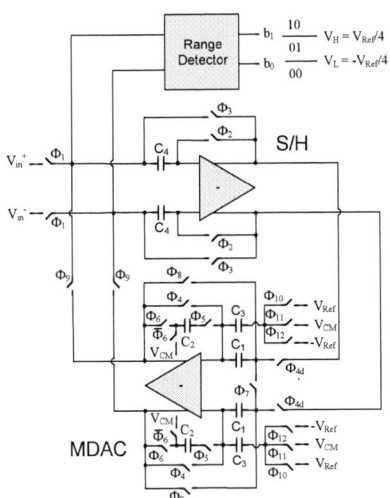

Fig. 6. Simplified schematic of cyclic ADC

Table 1. Simulated amplifier open loop gain requirement for certain linearity

Loop Gain Error (%)	DNL (LSB)	INL (LSB)	Required Amp Gain (dB)
0.02	0.12	0.4	86
0.1	0.5	1	72
0.2	0.7	1.8	66
0.4	Missing Code	4	60

Assume all capacitors in Fig. 6 have a same value of C, and the compensation capacitor in two-stage amplifiers used in S/H and MDAC is C_C. By using the analysis described in [7], noise can be estimated in (2) and (3)

$$\overline{V_{n,1}^2} = 2\frac{kT}{C} + \frac{8}{3}\frac{kT}{C_C}, \quad (2)$$

$$\overline{V_{n,2}^2} = 20\frac{kT}{C} + \frac{40}{3}\frac{kT}{C_C}, \quad (3)$$

Combining (1) to (3), the total noise can be expressed as:

$$9.5\frac{kT}{C} + 9\frac{kT}{C_C}, \quad (4)$$

The total noise is designed to be comparable to the quantization noise as a trade-off between ADC resolution and power consumption. The optimal choice of C and C_C will be presented in the following section.

D. Circuit implementation

Two-stage Class A/AB amplifier is chosen for its low power dissipation compared to two-stage Class A counterpart. In a typical two-stage amplifier, the lowest non-dominant pole, which is determined by the transconductance of the second stage and its load capacitance, needs to be designed higher than the unit gain frequency to ensure stability. For a given second stage transconductance requirement, due to the push and pull function in the Class AB stage, the second stage bias current of Class A/AB amplifier can be approximately half of that in the two-stage Class A amplifier. Fig. 7 shows the detailed schematic of Class A/AB amplifier used in MDAC with a maximum loading of 2C. The input transistors and transistors in the second stage are biased in sub-threshold region for low power dissipation. Different current values of I_1 and I_2 thus different transconductances in the two stages, call for different values of C and C_C to meet the noise and settling requirements. Simulation shows that for C = 1pF, C_C = 1.7pF, and $I_1 = 0.4\mu A$, $I_2 = 0.8\mu A$, the lowest power dissipation can be achieved while satisfying all other design requirements. M_{10} and M_{14} are controlled by the first stage

output, and M_{11} and M_{15} are controlled by the corresponding complementary output of the first stage through a current mirror formed by $M_{12, 13}$ and $M_{16, 17}$ to achieve the push and pull function. The resistor, R_Z, is implemented by a N type MOSFET operated in triode region for stability. It can be noticed that the common-mode output of the first stage influences the bias currents in the second stage but does not affect the second-stage output common-mode voltage, thus common-mode voltages of the two stages are controlled by two independent common-mode feedback loops [8]. Furthermore, the closed-loop gain in S/H is less than that in MDAC Class A/AB amplifier. Therefore, the unit gain frequency of the S/H stage can be lowered to achieve the same settling accuracy. Class A/AB amplifier topology is also used in S/H with all the transistors size and bias currents being scaled down by factor of two.

V. MEASUREMENT RESULT

The integrated circuit has been fabricated by AMI 1.5 μm CMOS two-ploy two-metal process with a chip size of 2.2 mm x 2.2 mm, as shown in Fig 8. The chip also includes a number of other building blocks for testing purpose, which are not described in the paper. ADC static testing is performed by standard histogram method, as results shown in Fig. 9. Both DNL and INL are below +/- 0.1 LSB indicating an adequate linearity. Fig. 10 shows a normalized power spectral density with a -0.1dBFS input, showing a flat noise floor with -90 dB 3rd harmonic distortion. SNDR is calculated to be 65 dB with 1 kHz bandwidth, equivalent to an ENOB of 10.5. The ENOB is constant for any input frequency up to $f_S/2$.

The MEMS pressure sensor is then directly placed over the IC and wire bonded to form electrical connections for the prototype system without any off-chip component required as shown in Fig. 11. The unit is tested inside a pressure chamber for system characterization.

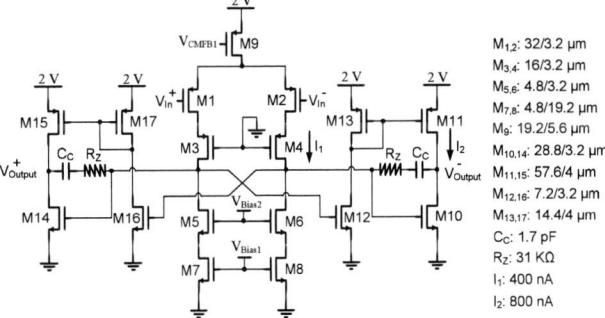

Fig. 7. Detailed schematic of Class A/AB amplifier in MDAC

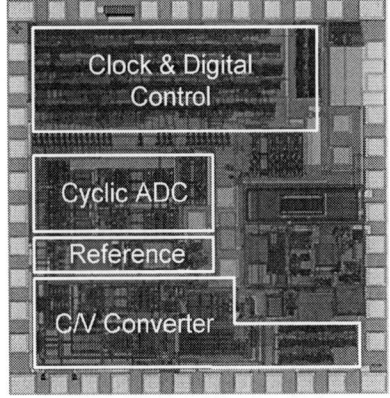

Fig. 8. Fabricated IC micrograph

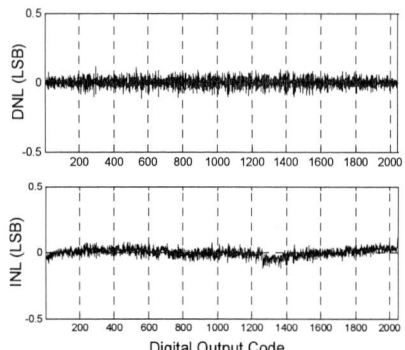

Fig. 9. DNL and INL measurement

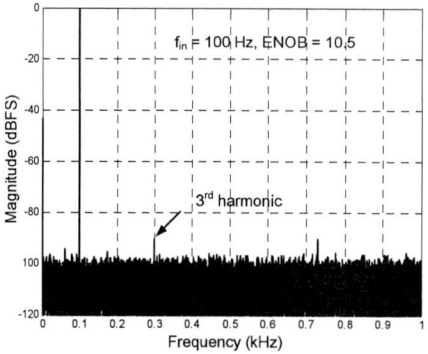

Fig. 10. Normalized power spectral density with -0.1dBFS 100 Hz input signal

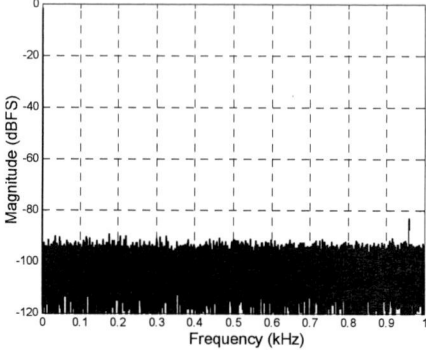

Fig. 13. ADC power spectral density with pressure sensor connected to C/V converter

Fig. 12 shows the measured output digital code versus a 200 mmHg pressure variation, indicating a system sensitivity of approximately 7 LSB/mmHg with a hysteresis below 1% full scale and nonlinearity of 2.5% full scale. The nonlinearity is due to the MEMS sensor operating in non-touch mode, which is also responsible for the low hysteresis. Fig. 13 presents ADC power spectral density with the pressure sensor connected to C/V converter, showing an approximately 6 dB increase in noise floor compared to ADC tested alone due to the noise of the C/V converter of approximately 700 μV_{RMS}, which is close to the designed value of 1 mV_{RMS}. The total noise is calculated to be 800 μV_{RMS} referred to the ADC input, which is equivalent to 80 aF or 0.1 mmHg resolution with 59 dB dynamic range. Table 2 summarizes system performance. The spurious tones in Figs. 10 and 13 are likely coupled from the on-chip digital electronics.

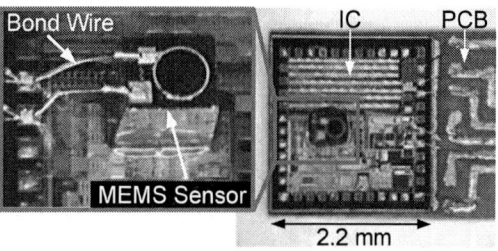

Fig. 11. Prototype system with pressure sensor directly wire bonded to IC

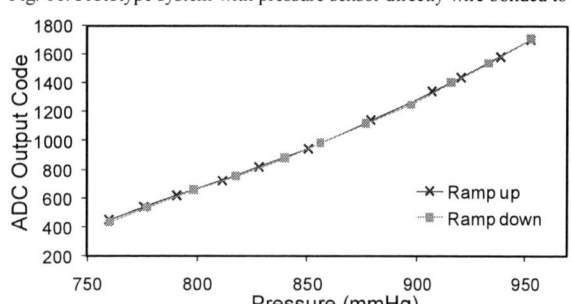

Fig. 12. ADC output digital code versus pressure variation

Table 2. Performance summary

System		Cyclic ADC	
Technology	1.5-µm CMOS	DNL	< +/- 0.1 LSB
System Sensitivity	7 LSB/mmHg	INL	< +/- 0.1 LSB
Noise over 1 kHz	80 aF or 0.1 mmHg	Clock Frequency	48 kHz
		Conversion Rate	2 KS/s
C/V Converter		THD	-90 dB
Power	12 µA @ 2V	ENOB	10.5
Sensitivity	1 mV/100 aF	Input Range	2 V_{PP}
f_s	2 kHz	Power	6 µA @ 2V

VI. CONCLUSION

A low-noise µWatt pressure sensing microsystem is developed for wireless implantable real-time blood pressure monitoring of small laboratory animals. An MEMS capacitive pressure sensor and integrated electronics are designed to achieve small size and low power dissipation as well as high resolution. The custom-designed electronics chip is much larger than the MEMS sensor in the prototype design, thus allowing more advanced IC technology to be used for further system miniaturization. The prototype microsystem achieves a resolution of 0.1 mmHg over a 1 kHz bandwidth, which is equivalent to a blood vessel pressure resolution of 0.5 mmHg, with a dynamic range of 59 dB and consumes 18 µA from a 2V supply. The demonstrated performance is superior compared to any commercial implantable blood pressure sensing technology, thus enabling state-of-the-art bioimplant system to be realized.

ACKNOWLEDGMENT

This work is supported by National Science Foundation under grant # EIA-0329811.

REFERENCES

[1] B. Hoit, S. Kiatchoosakun, J. Restivo, D. Kirkpatrick, K. Olszens, H. Shao, Y. Pao, and J. Nadeau, "Naturally Occurring Variation in Cardiovascular Traits among Inbred Mouse Strains," *Genomics*, Vol. 79, no. 5, May 2002, pp. 679-685.

[2] P. Cong, K. Olszens, D. J. Young, and W. H. Ko, "Implantable Blood Pressure Monitoring of Small Animal for Advanced Biological Research," *the 13th International Conference on Solid-State Sensors, Actuators and Microsystems*, June 2006, pp. 2002-2005.

[3] P. Cong, D. J. Young, and W. H. Ko, "Wireless Less-Invasive Blood Pressure Sensing Microsystem for Small Laboratory Animal In Vivo Real-Time Monitoring," *the Fifth International Conference on Networked Sensing Systems*, Kanazawa, Japan, June, 2008, in press.

[4] D. J. Young, J. Du, C. A. Zorman, W. H. Ko, "High-Temperature Single Crystal 3C-SiC Capacitive Pressure Sensor," *IEEE Sensors Journal, Special Issue on Microsensors and Microactuators*, August, pp. 464-470, 2004.

[5] C.H. Mastrangelo, X. Zhang, and W.C. Tang, "Surface-micromachined capacitive differential pressure sensor with lithographically defined silicon diaphragm," *Journal of Microelectromechanical Systems*, Vol. 5, Issue 2, pp.98 – 105, June 1996.

[6] D. Johns and K. Martin, *Analog Integrated Circuit Design*, John Wiley & Sons, New York, 1996.

[7] R. Schreier, J. Silva, J. Steensgaard and G. C. Temes, "Design-oriented estimation of thermal noise in switched-capacitor circuits," *IEEE Trans. Circuits Syst. I*, vol. 52, no.11, pp. 2358- 2368, Nov. 2005.

[8] S. Rabii and B. Wooley, "A 1.8-V digital-audio sigma-delta modulator in 0.8-µm CMOS," *IEEE J. Solid-State Circuits*, vol. 32, Issue 6, pp. 783-796, June 1997.

IEEE 2008 Custom Intergrated Circuits Conference (CICC)

A Flexible Decoder IC for WiMAX QC-LDPC Codes

Tzu-Chieh Kuo and Alan N. Willson, Jr.

Electrical Engineering Department

University of California, Los Angeles, CA 90095-1594, USA

Abstract-A programmable and power-efficient decoder IC employing the layered-decoding message-passing algorithm and the low-complexity offset-based Min-Sum check algorithm for irregular QC-LDPC codes is presented. The iterative decoder can be reconfigured to decode all the QC-LDPC codes defined in the Mobile WiMAX standard. Specifically, the decoder achieves a throughput of 68 Mbps at a 100-MHz clock rate for the rate-1/2 length-2304 code with ten iterations, and occupies a core area of 3.4 mm^2 in 0.18-μm CMOS technology and dissipates estimated 165 mW from a 1.8-V supply. It is 53% smaller, 86% lower in complexity, and has better energy efficiency than other published WiMAX LDPC decoder ASICs, to the best of our knowledge.

I. INTRODUCTION

Low-density parity-check (LDPC) codes have been included as optional codes in new standards such as in the IEEE 802.11n (Wi-Fi) and 802.16e (Mobile WiMAX) standards [1] due to their excellent coding gains. These LDPC codes are based on structured irregular quasi-cyclic (QC) codes with great levels of scalability for supporting multiple code structures of various code rates and code lengths. Therefore, a decoder for these applications must be very flexible and reconfigurable. Additionally, the hardware implementation must be energy-efficient and have low complexity in order to be practically integrated into VLSI chips for portable devices. The decoder chip presented here sufficiently addresses these important design considerations by incorporating many optimization techniques at the algorithm and the architecture levels, and consequently our decoder is significantly smaller than other WiMAX LDPC decoders and exhibits better energy efficiency.

First, the decoder is based on the layered-decoding message-passing algorithm in [2] and the offset-based Min-Sum check algorithm in [3]. Fig. 1 illustrates the operating principle of the decoding algorithms, where the columns and the rows of a parity-check matrix **H** are shown as bit nodes and check nodes, respectively. A bit node and a check node are connected if they share a common entry in **H**. The matrix is divided into blocks (layers) of non-overlapping parity-check rows (check nodes), and check messages are indirectly exchanged between these block rows through a Bit-Sum memory that contains the *a posteriori* log-likelihood ratio (AP-LLR) of the coded bits (bit nodes). A check message represents the AP-LLR estimate for a coded bit by a check node, based on the information (bit messages) provided by its neighboring bit nodes. The estimate of AP-LLRs improves with each iteration of message exchange.

In general, the layered decoding algorithm provides approximately a factor of two improvement in decoding convergence speed. The update of check messages by a check

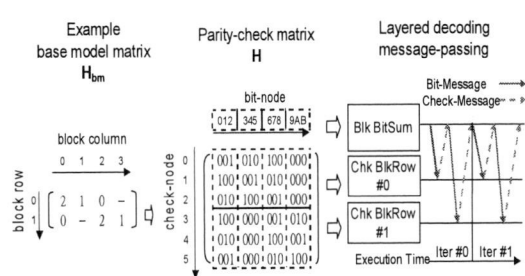

Fig. 1. Layered-decoding message passing for QC-LDPC codes.

node is performed in a check-function unit (CFU) implementing the Min-Sum check algorithm, in which the magnitudes of the check messages are obtained from the two smallest magnitudes of the bit messages it receives, rather than complicated computations over all its bit messages. This simplifies the computation complexity, and the decoding performance loss of this sub-optimal algorithm is compensated using an offset value to reduce the overestimated check message magnitude.

At the architecture level, the decoder is based on a partial-parallel architecture, in which multiple CFUs are implemented to process blocks of non-overlapping check nodes in parallel [4]. The computation in a CFU is performed sequentially; therefore, the CFUs can operate with different code structures of irregular check degrees. Although the maximum parallelism factor of WiMAX codes is 96, our decoder chip employs only a subset of 24 CFUs because the PHY data rate of current Mobile WiMAX is lower than 40 Mbps [5]. This achieves a significant complexity reduction. In addition, our initial implementation shows that memories in a low-complexity decoder could consume more than 60% of its total power dissipation. Therefore, we have developed consolidated memory architectures for power reduction, and this also simplifies the design tasks in place-and-route with a reduced number of memory instances. Finally, the power dissipation has been further reduced using the clock gating technique. The chip architecture is described in the following section.

II. CHIP ARCHITECTURE

Fig. 2 shows the architecture of the decoder chip. It consists of a code-configuration memory with interface, a central control block, a multi-bank Bit-Sum memory (BM), a message-routing network using cyclic shifters, CFUs, and a decision output memory. The decoder has 24 CFUs available, and the number of CFUs enabled (*P*) depends on the number of

978-1-4244-2018-6/08 $25.00 © 2008 IEEE

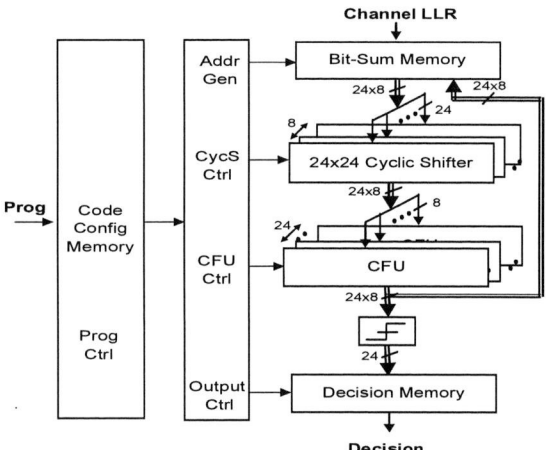

Fig. 2. Chip block diagram.

check nodes (M) in **H**. P must be a factor of M so that AP-LLRs can be effectively organized in the BM. The 5-bit LLRs of coded bits received from the channel are first stored in the BM as the initial estimates of the 8-bit AP-LLRs, and then the AP-LLRs are iteratively updated by the decoder. When processing a block of P consecutive check nodes in the CFUs, the code parameters, including the check degree (dc), the cyclic-shift value and the block column number, are retrieved from the configuration memory to drive miscellaneous controllers. The cyclic shift value and the block column number are used to identify the specific bit nodes connecting to the check nodes in the process. The AP-LLRs of these bit nodes are sequentially retrieved from the BM and routed to their destination CFUs through the cyclic shifters. In turn, the new AP-LLRs obtained from the CFUs are then put back in the BM while their signs are also stored in the output memory as hard decisions. The operation continues to work on the next block of P check nodes. When all check nodes in **H** have been processed by the CFUs, a new iteration starts over from the first block of check nodes. The decoding operation stops when a programmable iteration number is reached or when all the parity checks are satisfied, and the decoded bits are ready to be sent out through an output interface.

A. Check Function Unit

The implementation of a CFU is shown in Figs. 3 and 4. In check-node processing, the CFU sequentially subtracts the old

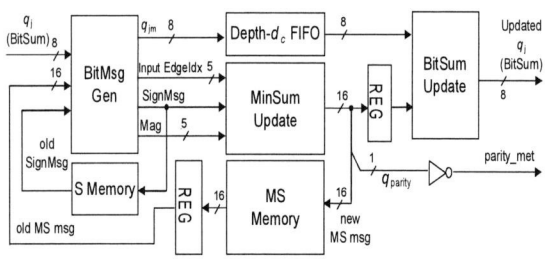

Fig. 3. CFU block diagram.

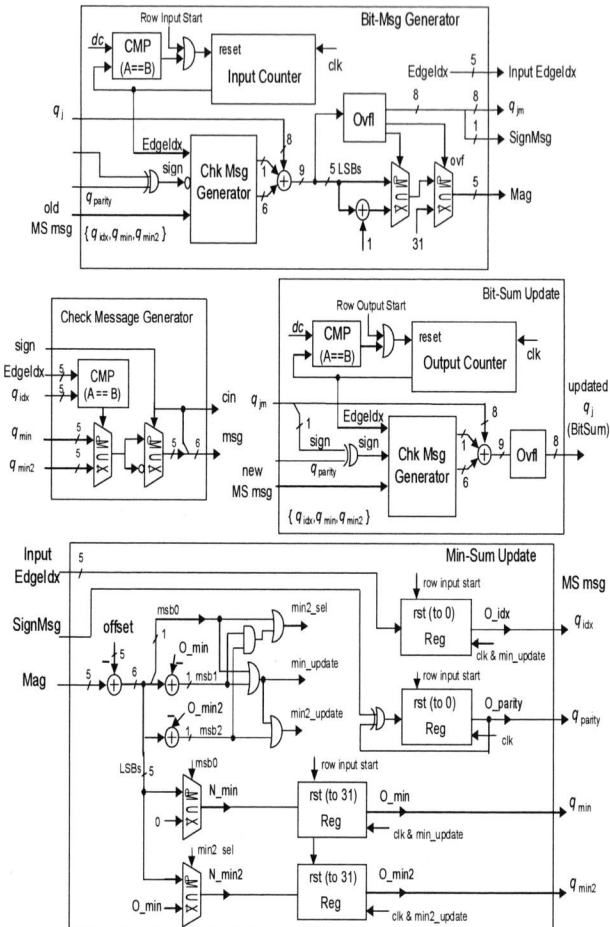

Fig. 4. CFU block diagram, continued.

check messages from the incoming AP-LLRs q_j's to generate the bit messages q_{jm}'s. The magnitudes of the q_{jm}'s are compared in the Min-Sum update block to obtain the new MS message, which is then used to produce the new check messages to combine with the corresponding q_{jm}'s in the Bit-Sum update block to produce the new AP-LLRs. An MS message of a check node contains its tentative parity value (q_{parity}), the two smallest magnitudes ($\{q_{min}, q_{min2}\}$) of its bit messages, and the index (q_{idx}) of the bit node producing q_{min}. It is created by a sequential search over the bit messages as shown in the schematic of the Min-Sum update block.

All the check messages of a check node are produced from its MS message and the individual sign values of its bit messages, as shown in the schematic of the check-message generator block. Notice that a negative check message is represented as a 1s complement number with an additional carry-in signal that is used by the adders in the succeeding stages; therefore the addition required for 2's complement negation has been removed. The subtraction in the bit-message generation block has been simplified to an addition by inverting the sign signal of the check message. Furthermore, the reduction of check message magnitudes by a programmable offset value is effectively implemented on bit messages in the

Min-Sum update block, in which the delay path is less critical than in the Bit-Sum update block. The data q_{jm}'s and q_j's are saturated to eight bits, while the magnitudes are saturated to five bits for the reduction of computation complexity with very little performance degradation. The sign values and the MS message, rather than the individual check messages, are put into memories for producing the old check messages in the next iteration. This significantly reduces the number of memory bits required, especially for high check-degree codes. The clock signals of the registers are gated off when there are no updates available. Furthermore, the small memories in the 24 CFUs are grouped together whenever a centralized memory consumes less power than the distributed ones.

The CFUs work on new check nodes while their Bit-Sum update blocks are still generating the updated AP-LLRs for current check nodes. However, when the operation arrives at the end of the current layer, the decoder is put into wait cycles for the CFUs to complete if the next layer overlaps the current layer. The numbers of wait cycles required depend on the given parity-check matrix, therefore they are also obtained from the code configuration memory. A parity check is assumed to be satisfied if its tentative parity value q_{parity} is zero. If all parity checks in the current iteration have been satisfied, the decoding operation can be early terminated for saving power if this feature is enabled.

B. Bit-Sum Memory

A partial-parallel LDPC decoder typically contains many small memory instances that cause inefficiency in area and power dissipation due to the duplicated row and column decoders in the memory instances. In order to minimize the power consumption, the BM is implemented with six memory banks as shown in Fig. 5, rather than 24 memory banks to supply AP-LLRs for the 24 independent CFUs as in a conventional partial-parallel architecture. Each memory bank has 96 words with each word containing the AP-LLRs of four adjacent bit nodes. The maximum code length supported is 2304 as required in WiMAX codes. However, four of the P AP-LLRs requested by the P CFUs might be stored in two separate locations in a memory bank as a consequence. Therefore, the read ports of the memory banks operate at twice the system clock rate in order to retrieve the desired AP-LLRs from two separate locations during a system clock cycle, when required. The desired contents obtained in the first read access are stored in the registers in the auxiliary circuits, followed by a second read access to retrieve the rest of the AP-LLRs. The multiplexers direct the right AP-LLRs to the cyclic shifters, which in turn align the AP-LLRs in the right order for the CFUs. In contrast, the write ports of the memory banks operate at the system clock rate with common control signals to facilitate the write-back of AP-LLRs from the CFUs without a backward routing network. The computation latency in the CFUs ensures that the old memory contents have been retrieved before they are being replaced. Consequently, the location of an AP-LLR in the BM will change from layer to layer and from iteration to iteration. However, the mapping rule is fully deterministic, and the read/write addresses of the BMs and the revised cyclic shift values for the shifters can be computed on the fly. In a system clock cycle, only at most one out of the six BMs requires double accesses, and at most three of the 24 register-sets in the auxiliary circuits are active, so a clock-gating technique has been applied to reduce the power consumption again. This consolidated memory structure saves about 40 mW from a 24-bank implementation. This is a 20% savings from our initial implementation.

C. Cyclic Shifter

The cyclic shifters are implemented with barrel shifters having programmable wrap-around positions to support the 19 different code lengths in WiMAX. Our decoder implements only the forward routing of AP-LLRs as in [4] and [6]. When only parts of the CFUs are enabled, the shifters force zero values on the unused data buses to reduce switching activity.

III. IMPLEMENTATION RESULTS AND COMPARISONS

A prototype decoder ASIC operating at a 100-MHz clock frequency has been implemented in TSMC 0.18-μm technology and is currently under fabrication. Its layout plot is shown in Fig. 6. The core size is 2.23 mm × 1.52 mm and the

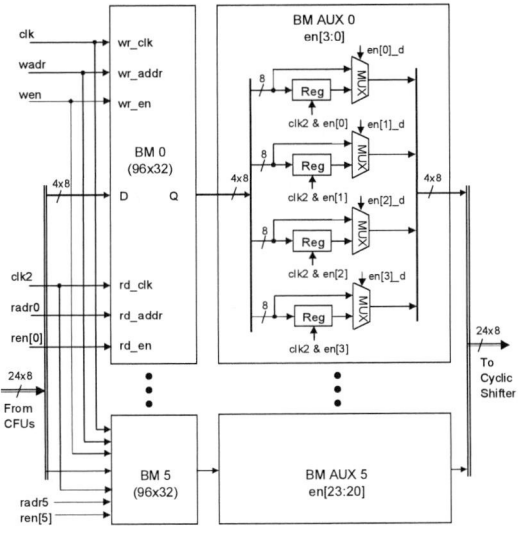

Fig. 5. Bit-Sum memory architecture.

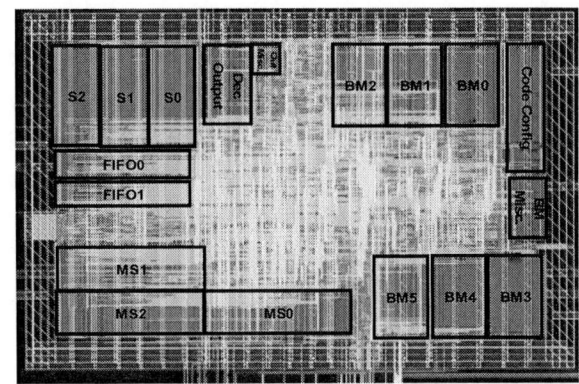

Fig. 6. Plot of LDPC decoder core.

post-layout estimated power consumption is 165 mW at 1.8 V. The decoding performance in an AWGN channel for the length-2304 codes is shown in Fig. 7. Similar performance is achieved for the other code sizes and there are no noticeable error floors resulting from the finite word-lengths. Fig. 8 presents the decoding throughput for ten iterations. The decoder has lower throughput at certain code block lengths that can use just a limited number of CFUs. The decoder is more efficient for long codes as more CFUs can be enabled; therefore, more iterations could be performed to further enhance the error-rate performance for these long codes.

The comparison of the decoder with other representative WiMAX LDPC decoders is summarized in Table I, where the throughput and power dissipation are characterized using the rate-1/2, length-2304 code, and technology normalization is done with respect to the 0.18-μm feature size and 1.8-V supply voltage. Our decoder is 53% smaller and employs fewer memory bits and significantly fewer logic gates than the other decoders. Our design complexity is 86% lower in terms of logic gates. The multi-code decoder proposed here has energy efficiency (energy consumption per decoded bit per iteration [8]) that is 27% better than the fixed-code decoder in [7]. The effective energy performance is 45% better, as the convergence

TABLE I. COMPARISON OF DECODER IMPLEMENTATIONS

Design	[7]	[4]	[6]	This Work
Technology (μm)	0.13	0.13	FPGA	0.18
Code Rate	1/2	1/2, 2/3A, 2/3B, 3/4A, 3/4B, 5/6		
Code Length	576 : 96 : 2304			
No. of Processors	4C/8B	96	24	24
Clock Rate (MHz)	83.3	333	110	100
Decoding Throughput (Mbps)	60.6	666	72	68
No. of Iterations	8	15	10	10
Power (mW)	52	N/A	N/A	165
Norm. Core Area (mm^2)	8.53	7.34	N/A	3.39
Normalized Energy Efficiency (pJ/bit/iter.)	334	N/A	N/A	243
Memory Bits	76800	N/A	65760	55576
Logic Gate Count	420 K	~500 K	N/A	55 K

rate of the layered-decoding message-passing algorithm is about 1.7 times faster than the two-phase message-passing decoding algorithm in [7]. Using the consolidated memory structure, our decoder ASIC only requires a total of 20 memory instances, compared to 38 memory instances in [6] and 100 memory instances in [7].

IV. CONCLUSION

A programmable iterative decoder for irregular WiMAX LDPC codes has been presented. Using efficient low-complexity decoding algorithms and optimizations at the architecture and circuit levels, this decoder has demonstrated significant improvements over other published designs, with respect to complexity, area, and energy efficiency. The decoder can also be programmed to decode other QC-LDPC codes of appropriate sizes, limited only by the on-chip memories.

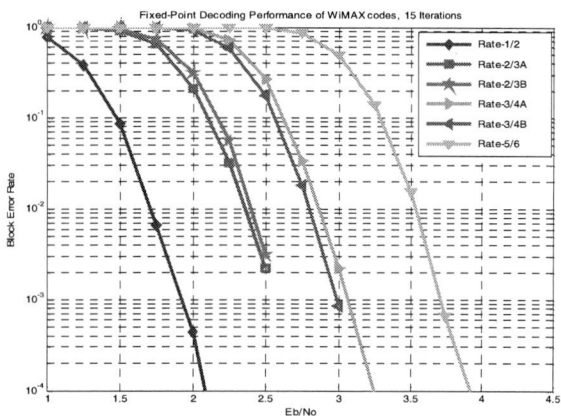

Fig. 7. Decoding performance in AWGN channels.

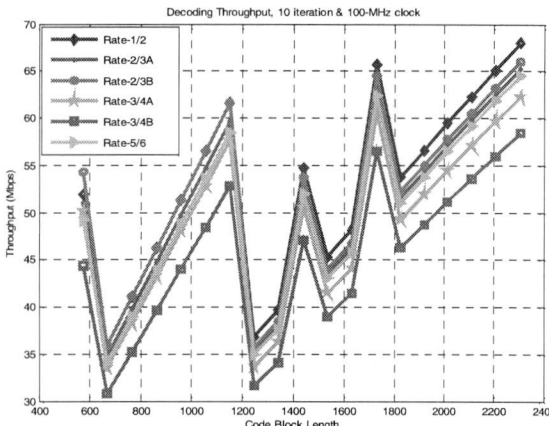

Fig. 8. Decoder throughput performance.

REFERENCES

[1] IEEE 802.16e, Part 16: Air interface for fixed and mobile broadband wireless access systems, IEEE std 802.16e-2005, Feb. 2006.

[2] D. Hocevar, "A reduced complexity decoder architecture via layered decoding of LDPC codes," in *IEEE Workshop on Signal Processing Systems, SIPS 2004*, pp. 107-112, 2004.

[3] J. Chen, A. Dholakia, E. Eleftheriou, M. Fossorier, and X. Hu, "Reduced-complexity decoding of LDPC codes," *IEEE Trans. Communications*, vol. 53, pp. 1288-1299, Aug. 2005.

[4] T. Brack, M. Alles, F. Kienle, and N. When, "A synthesizable IP core for WiMAX 802.16e LDPC code decoding," in *Proc. 2006 Personal Indoor and Radio Communications Conference (PIMRC'06)*, Helsinki, Sept. 2006.

[5] WiMAX Forum, Mobile WiMAX – Part I: A technical overview and performance evaluation, Aug. 2006.

[6] K. Gunnam, G. Choi, M. Yeary, and M. Atiquzzaman, "VLSI architectures for layered decoding of irregular LDPC codes of WiMAX," in *Proc. IEEE International Conference on Communications (ICC)*, pp. 4542-4547, June 2007.

[7] X. Shih, C. Zhan, C. Lin, and A. Wu, "A 19-mode 8.29 mm^2 52-mW LDPC decoder chip for IEEE 802.16e system," in *Symposium on VLSI Circuits, Dig. Tech. Papers*, pp. 16-17, June 2007.

[8] M. Mansour and N. Shanbhag, "A 640-Mb/s 2048-bit programmable LDPC decoder chip," *IEEE Journal of Solid-State Circuits*, vol. 41, no. 3, pp. 684-698, March 2006.

IEEE 2008 Custom Intergrated Circuits Conference (CICC)

Minimizing the Supply Sensitivity of CMOS Ring Oscillator by Jointly Biasing the Supply and Control Voltage

Ping-Hsuan Hsieh, Jay Maxey[*], and Chih-Kong Ken Yang

University of California, Los Angeles, 56-147A Engineering IV Building, Los Angeles, CA 90095
*Texas Instruments, Dallas, TX 75243
Tel: 310.206.3665 Fax: 310.206.8495 E-mail: abin@ee.ucla.edu

Abstract—A method to minimize the supply sensitivity of a CMOS ring oscillator is proposed through joint biasing of the supply and the control voltage. The technique can supplement a number of common supply rejection techniques. The proposed CMOS ring oscillator is designed and implemented with a charge-pump based phase-locked loop in 65-nm technology to demonstrate the robustness against the supply fluctuation. Taking advantage of the negative static supply sensitivity of the ring oscillator with proper combination of the bias voltages, the rms jitter of the 4-GHz output clock is reduced from 10.66-ps to 5.04-ps while subject to switching noise with magnitude of 2.5% of the supply voltage at 150-MHz. Furthermore, more than 4.5× of reduction in the power consumption is achieved.

I. INTRODUCTION

Voltage-controlled oscillators (VCOs) have been extensively used as a key part of phase-locked loops (PLLs) in high-performance microprocessors and high-speed digital communication systems for on-chip clock generators and clock recovery. The VCOs' jitter performance in these applications can impact the output clock's timing jitter which often limits the system performance. As more functional blocks are integrated onto a single IC, VCOs experience large supply noise because of the digital switching activities. The resulting timing jitter can be much larger than the jitter caused by the inherent device noise of the oscillator [1, 2]. Many methods have been adopted to suppress the supply noise seen by a VCO with the use of differential structures [3] and voltage regulators [4]. Other approaches compensate the VCO's intrinsic positive supply sensitivity with additional circuitry which has negative supply sensitivity [3, 5, 6].

This work proposes a technique that utilizes an inherent characteristic of many VCOs to reduce supply sensitivity. It has been observed that a VCO with the Lee-Kim delay cell [7] has positive supply sensitivity for most of the operating frequencies [6]. The VCO's supply sensitivity decreases at low supply voltage and even becomes negative at high operating frequencies. Thus, a bias point with zero static supply sensitivity exists for each operating frequency. The mechanism

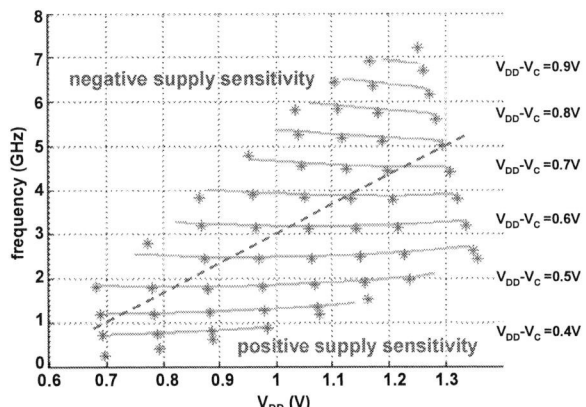

Fig. 2. Measured operating frequency vs. supply voltage with different V_{DD}-V_C.

is discussed in the first section of the paper. This optimal biasing can be adopted in conjunction with many other supply noise rejection techniques to further lower the supply sensitivity. This paper describes a possible implementation and measurement results. The results show additional benefits in power consumption.

II. THE VCO DESIGN

Fig. 1 shows the design of the VCO embedded in an analog PLL with the detailed schematic of the delay cell. NMOS transistors are used as the delay stage with PMOS transistors serving as current source loads. Cross-coupled pair is used to guarantee oscillation with differential outputs. The VCO's operating frequency is controlled by adjusting the PMOS transistors' current through the gate-to-source voltage V_{GS}. The delay cell is similar to the Lee-Kim delay cell. The NMOS-transistor cross-coupled pair is adopted for higher operating frequency [8].

Fig. 2 illustrates a variant of the typical K_{VCO} plot by showing how the oscillator's operating frequency is a function of the supply voltage (V_{DD}) for various control voltage bias of the PMOS transistor (V_{DD}-V_C). The measured results are plotted in the figure and match well the prediction from simulation. From this figure we notice that the VCO's supply sensitivity can be either positive or negative depending on different bias of the VCO's supply voltage and control voltage. The dotted line in Fig. 2 marks the transition (points of zero supply sensitivity).

On the right-hand side of the dotted line in Fig. 2, with a fixed V_{GS} of the PMOS transistors, as the supply voltage increases the operating frequency increases which corresponds to positive supply sensitivity. Fig. 3 shows the simulated oscillation waveform of the VCO running at 4.2-GHz with the supply

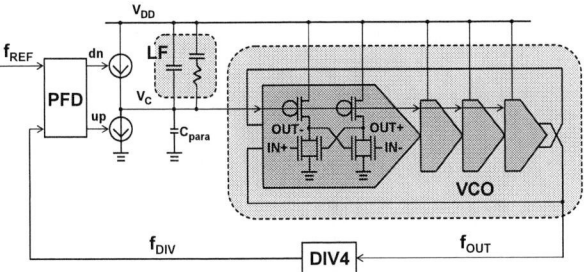

Fig. 1. Block diagram of the designed PLL.

978-1-4244-2018-6/08 $25.00 © 2008 IEEE

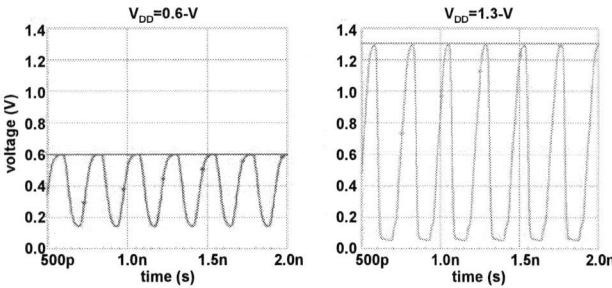

Fig. 3. Simulated oscillation waveform of the VCO running at 4.2-GHz with the supply voltages of 0.6-V and 1.3-V.

voltages of 0.6-V and 1.3-V. Note that as the supply voltage increases, the voltage swing also increases. The increase in frequency with supply is due to an increase in the NMOS current (larger V_{GS}) causing a faster falling edge. The PMOS current increases only slightly due to the finite output impedance. Because of the larger voltage swing, the rising transition is slower. The net impact on the frequency is the combination of the rising and falling edges. The increase in the current dominates over the increase in the oscillation amplitude and, as a result, causes positive supply sensitivity when the supply voltage is high.

On the left-hand side of the dotted line, where the supply voltage is low, the oscillation waveform is heavily clamped by the supply and reduced as the supply voltage decreases. Moreover, the lower bound of the oscillation waveform rises in order to increase the loop gain for sustaining the oscillation. With a smaller voltage swing, a constant PMOS current enables faster rising transitions. The higher lower-bound of the oscillation waveform leads to a smaller decrease in the average pull-down current. The net impact on frequency is again an increase with decreasing supply voltage.

Based on the previous paragraphs, the major reason for the oscillator to exhibit the negative supply sensitivity is the asymmetry in the pull-up and pull-down mechanism. The existence of the zero supply sensitivity and the magnitude of both the positive and negative supply sensitivity depend on the design parameters such as the P/N size ratio of the delay-cell transistors and the relative strength of the cross-coupled pairs. Fig. 4 shows the simulated operating frequency vs. supply voltage with different V_{GS} of the PMOS transistors in two cases with different P/N transistors size ratio. Fig. 4(a) shows the

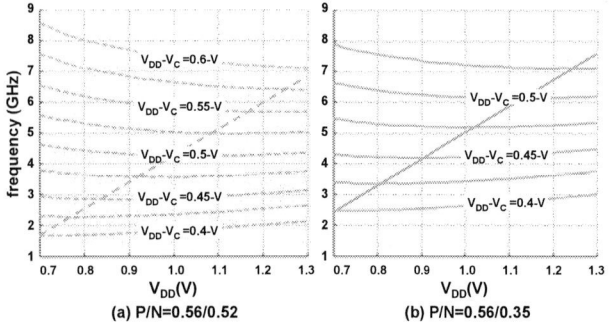

Fig. 4. The simulated operating frequency vs. supply voltage with different V_{DD}-V_C.

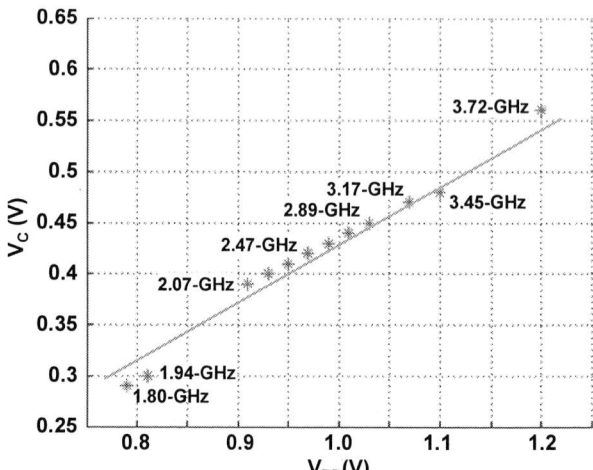

Fig. 5. The control vs. the supply voltage with zero supply sensitivity.

result with the P/N ratio of 0.56/0.52 while Fig. 4(b) shows the result with the P/N ratio of 0.56/0.35. The figures show that larger NMOS transistors shift the zero supply sensitivity towards the right and the resulting oscillator exhibits larger negative supply sensitivity. Note that increasing the NMOS transistors size has a tradeoff with the K_{VCO} because of the reduced frequency range. Zero supply sensitivity over a target operating frequency range can be achieved with proper sizing of the transistors.

III. SYSTEM IMPLEMENTATION

Either open-loop or closed-loop method can be employed to properly bias the oscillator's supply and control voltage in order to achieve low supply sensitivity. Fig. 5 shows the measured relationship between the supply voltage and the control voltage with the notation of operating frequency on top of the curve when the oscillator exhibits zero supply sensitivity. The result in Fig. 5 shows a roughly linear relationship. Therefore, a simple linear amplifier as shown in Fig. 6 can be adopted to change the supply voltage according to the control voltage while the control voltage is used as a port to change the oscillation frequency (or vice versa). On-chip timing jitter measurement circuitry [9] can be further adopted to measure the output clock's timing jitter and adjust the supply voltage to minimize jitter even in the presence of environmental change or dynamic supply noise.

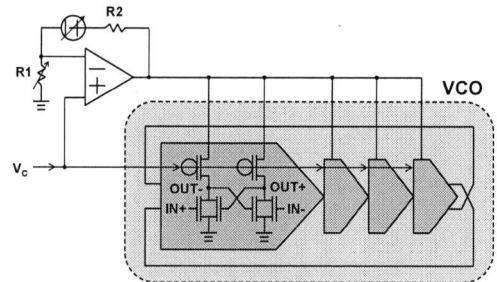

Fig. 6. Example of voltage biasing circuitry.

978-1-4244-2018-6/08 $25.00 © 2008 IEEE

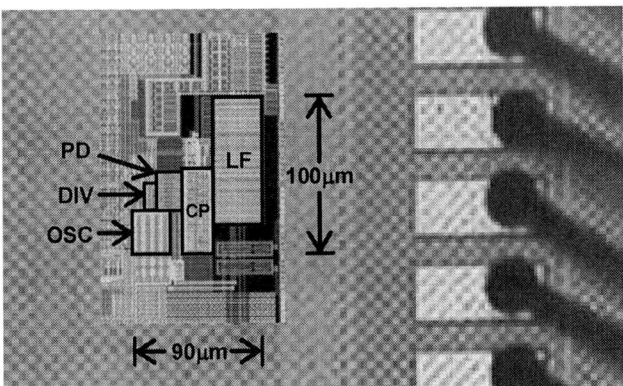

Fig. 7. Micrograph of the test-chip.

For an oscillator placed in a phase-locked loop as shown in Fig. 1, the V_{GS} of the PMOS transistors is controlled by the negative feedback loop such that the output frequency and phase track with the reference. Ideally, the control voltage follows the supply voltage in the presence of dynamic supply noise to maintain a constant V_{DD}-V_C. In practice, however, the V_{GS} of the PMOS is changed because of the parasitic capacitance between the PMOS gate and the ground as shown in Fig. 1. In the case shown, the change in the V_{GS} of the PMOS causes the operating frequency to increase as supply voltage increases. This positive supply sensitivity is limited often by careful layout. The proposed technique can further compensate the positive sensitivity by properly biasing the VCO in the region of negative supply sensitivity. Larger NMOS transistors can be used when designing the VCO to increase the region of negative supply sensitivity.

IV. MEASUREMENT RESULTS

The VCO with a charge-pump PLL is fabricated in 65-nm CMOS technology with nominal V_{DD} of 1.1-V. Fig. 7 shows the die micrograph overlaid with the layout picture. The entire PLL occupies 90-μm×100-μm with the VCO occupying

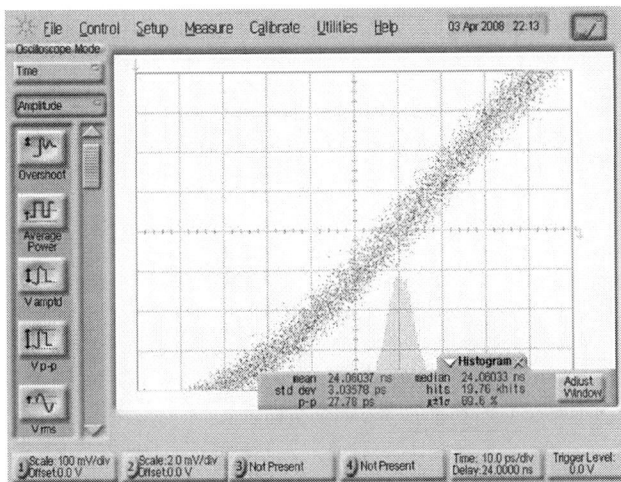

Fig. 8. PLL output histogram at 4-GHz with supply voltage of 1.04-V and V_{DD}-V_C of 0.66-V.

TABLE I
SUMMARY OF THE TEST-CHIP PERFORMANCE

	reference frequency	1-GHz				
	output frequency	4-GHz				
	loop bandwidth	150±5-MHz				
	supply voltage (V)	0.75	0.77	1.04	1.27	1.33
	supply current (mA)	10.29	11.15	18.92	25.20	26.65
	power (mW)	7.72	8.59	19.68	32.00	35.44
quiet supply	rms jitter (ps)	3.20	3.10	3.04	3.11	3.21
100-MHz	rms jitter (ps)	3.43	3.62	4.02	4.56	5.99
130-MHz	rms jitter (ps)	4.23	5.71	6.54	7.17	8.06
150-MHz	rms jitter (ps)	5.04	6.00	6.60	8.59	10.66
200-MHz	rms jitter (ps)	3.73	3.64	3.78	4.57	4.99

30-μm×30-μm. Digital calibrations are built into the charge pump and the loop filter for adjustable loop characteristics. The PLL has an input reference frequency of 1-GHz, and an output clock at 4-GHz. A jitter histogram of the output clock from an oscilloscope is shown in Fig. 8. The VCO operates with the supply voltage of 1.04-V and V_{DD}-V_C equal to 0.66-V, and the output clock has the rms jitter of 3.04-ps and the p2p jitter of 27.78-ps. However, the inherent noise of the oscilloscope is 2.5-ps rms and contributes a substantial amount of the measured noise.

The measurement results are summarized in Table 1. With quiet supply, the timing jitter is mostly due to the thermal noise in the devices. The jitter performance is worse when the supply voltage is high because of the poor symmetry between the pull-up and pull-down slope [10].

With the noise injection device switching on and off current out from the supply with the configuration as shown in Fig. 9 [11], noise with estimated amplitude of the 2.5% of the supply at different frequencies is injected onto the supply. Fig. 10 shows the measured output clock jitter when the circuit is biased at different supply voltages. The loop filter parameters such as the charge pump current, the loop filter resistance and the loop filter capacitance are adjusted such that the loop bandwidth is near 150±5-MHz under different supply voltages. The noise from the supply experiences a band-pass filter to the clock output phase. As we can see, although the intrinsic noise performance tends to be better when the supply voltage is high,

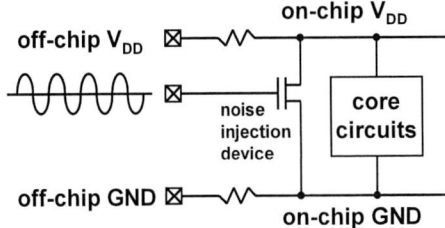

Fig. 9. Noise injection configuration.

978-1-4244-2018-6/08 $25.00 © 2008 IEEE

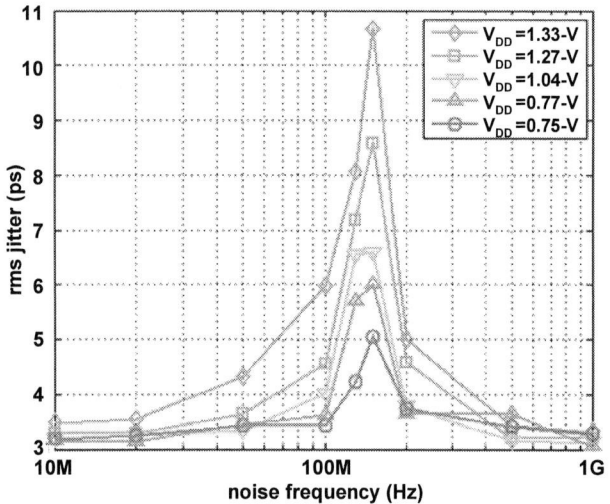

Fig. 10. Measured output rms jitter vs. switching noise frequency with different supply voltages.

the performance can be easily overwhelmed by the jitter due to the switching activities with the frequencies close to the loop bandwidth. By operating the VCO at lower supply voltage, the VCO is more robust to the supply fluctuation. The performance is improved as the oscillator is biased towards the optimum. In our design, while the supply sensitivity is substantially reduced, some residual sensitivity to the supply noise is consistently measured due to the large parasitic capacitance in our design. It is important to also note that the power consumption is dramatically reduced using the smaller supply voltage. As shown in Table 1, more than 4.5× of reduction in the power consumption is achieved when the supply voltage is decreased from 1.33V to 0.75V.

V. CONCLUSIONS

A ring oscillator is designed and implemented with a charge-pump based PLL in 65-nm CMOS technology. The supply sensitivity of the oscillator can be either positive or negative depending on the supply voltage and the operating frequency. For each operating frequency, there exists a set of bias voltages at which the VCO exhibits zero static supply sensitivity. The negative static supply sensitivity with low supply voltage can be utilized to compensate any positive supply sensitivity that may result from parasitic capacitance at the presence of dynamic supply noise. The measurement results show the oscillator's robustness against the supply fluctuation in the presence of switching noise.

The author would like to thank Texas Instruments for chip fabrication and technical support.

REFERENCES

[1] F. Herzel and B. Razavi, "A study of oscillator jitter due to supply and substrate noise," in *IEEE Trans. Circuits Syst. II*, vol. 46, pp. 56-62, Jan. 1999.

[2] P. Heydari, "Analysis of the PLL Jitter Due To Power/Ground and Substrate Noise," in *IEEE Trans. Circuits Syst. I*, vol. 51, pp. 2404-2616, Dec. 2004.

[3] I.-C. Hwang and S.-M. Kang, "A self-regulating VCO with supply sensitivity of <0.15%-delay/1% supply," in *IEEE ISSCC Digest of Technical Papers*, Feb. 2002.

[4] M. Brownlee, P. Hanumolu, K. Mayaram, and U.-K. Moon, "A 0.5 to 2.5GHz PLL with Fully Differential Supply-Regulated Tuning," in *IEEE ISSCC Digest of Technical Papers*, Feb. 2006.

[5] M. Mansuri, and C.-K. K. Yang, "A Low-Power Adaptive Bandwidth PLL and Clock Buffer with Supply-Noise Compensation," in *IEEE J. Solid-State Circuits*, vol. 38, pp. 1804-1812, Nov. 2003.

[6] T. Wu and U.-K. Moon, "An On-chip Calibration Technique for Reducing Supply Voltage Sensitivity in Ring Oscillators," in *IEEE J. Solid-State Circuits*, vol. 42, pp. 775-783, Apr. 2007.

[7] J. Lee and B. Kim, "A Low-Noise Fast-Lock Phase-Locked Loop with Adaptive Bandwidth Control," in *IEEE J. Solid-State Circuits*, vol. 35, pp. 1137-1145, Aug. 2002.

[8] B. Razavi, K. Lee, and R. Yan, "Design of High-Speed, Low-Power Frequency Dividers and Phase-Locked Loops in Deep Submicron CMOS," in *IEEE J. Solid-State Circuits*, vol. 30, pp. 101-109, Feb. 1995.

[9] T. Xia and J.-C. Lo, "Time-to-voltage converter for on-chip jitter measurement," in *IEEE Trans. Instrum. Meas.*, vol. 52, pp. 1738-1748, Dec. 2003.

[10] A. Hajimiri, S. Limotyrakis, and T. H. Lee, "Jitter and Phase Noise in Ring Oscillators," in *IEEE J. Solid-State Circuits*, vol. 34, pp. 790-804, Jun. 1999.

[11] S. Sidiropoulos and M. A. Horowitz, "A Semidigital Dual Delay-Locked Loop," in *IEEE J. Solid-State Circuits*, vol. 32, pp. 1683-1692, Nov. 1997.

The Superchip: Innovative Teaching of IC Design and Manufacture

Peter R. Wilson, Reuben Wilcock, Iain McNally, Matthew Swabey and Bashir Al-Hashimi

{prw, rw3, bim, mas, bmah} @ecs.soton.ac.uk

University of Southampton

Abstract- **In this paper we describe how through intelligent chip architecture, a large cohort (~100 students) of undergraduates can be given effective practical insight into IC design by designing and manufacturing their own individual ICs. To achieve this, the "Superchip" has been developed, which allows (without excessive cost in terms of time or resources) multiple student designs to be fabricated on a single IC, and encapsulated in a standard package. We demonstrate how the practical process has been tightly coupled with theoretical aspects of the degree course and how transferable skills are incorporated into the design exercise. The paper provides details of the chip architecture, test regime, test vectors, and an example design.**

I. INTRODUCTION

A. Background

Recent advances in IC CMOS process technology have forced Electronics departments world-wide to adapt their educational programs to equip students with the right skills and knowledge needed by industry. In addition, design cycle times (the time it takes to get from product specification to delivery to the market) are being driven ever shorter. The skills that are required to support this level of design are rapidly changing, as are the software and hardware tools required for engineers. In addition to the technical skills, we have taken an approach of team organization that is enabling for students and provides invaluable skills in terms of time management, team working, collaboration and interpersonal skills. The Electronics Engineering undergraduate program at the University of Southampton has run successfully for many years and provides a good grounding in hardware design. This paper demonstrates Southampton's adaptation to modern industry requirements, the educational rationale for this exercise and results are presented of its implementation and delivery.

B Learning Strategy

A key part of the strategy for learning has been to provide a solid experiential learning platform based on the Kolb learning cycle [1] and in particular by using small groups [2]. The strength of this approach is clearly the tutorial style with students able to progress at their own pace with a structured work plan to facilitate learning. While this is desirable in itself, there is the added benefit of empowering students by allowing them to organize their groups into whichever structure suits them best. This was an interesting step to take, as the intuitive assumption is often made that students must be given a tight framework within which to work, with very clear operating instructions. Our experience in this design exercise

has however been that the students welcome the responsibility and enjoy the fact that there is a "real" deadline for chip manufacture to meet, not just an artificial deadline, typical for most coursework at undergraduate level. Providing literature prior to the session [3] enables students to take a less linear approach to the design process and enable much more iteration and creativity to take place. This is intentionally in place to help develop students that can do design – not just rote type learning. Biggs [4] provides a useful framework to assist in the strategy of preparation and we apply different methods of delivery and assessment [6] to engage large classes more directly. A key aspect of this approach is to use student-oriented learning [5]. In our courses collaborative group work and peer review prove effective and useful in this context, and we incorporate this aspect into the design exercise.

C. Learning Outcomes, Key Skills and Assessment

In order to ensure that the individual student's experience is satisfactory, learning outcomes have been designed in the context of an integrated process of teaching, learning and assessment. This is essential to provide the student with a high quality of learning in a rapidly changing field. In this course we have taken a view of learning that considers the academic aspects of the work and links this to the industrially oriented aspects. Fourteen specific learning outcomes were devised for this course. The integration of key skills for industry is critical for engineering students in general as discussed by Woods et al [7], but in this particular field it is even more acute. In order to ensure that the learning outcomes in relation to the proposed course structure were appropriate, a matrix based approach was employed as described by Felder and Brent [8] to analyze the exercise structure in relation to learning outcomes. In this particular design exercise, we have identified relevant key skills and tied them into specific learning outcomes in a coherent manner that will provide the basic framework for the students to achieve a successful outcome and assessment has been considered in the context of the variety of skills, platforms and learning outcomes required, described by Felder and Brent in [5]. The approach we have taken to ensure this with a relatively inexperienced group of undergraduates is to provide design freedom, within a tightly constrained design tool framework.

D Design of the Exercise

The design exercise is divided into two main areas. In the first semester, the design and implementation stages take place. Teams will undertake the following activities, paper design, schematic capture, design verification and simulation, layout, post layout verification and simulation, and finally design package production. These steps will be described in more detail later in this paper. During the design stage there is an emphasis on best design practice, design for testability and fault tolerance.

Since integrated circuit designs must in principle be right first time, CAD tools are of course used extensively in this exercise for design entry, verification and simulation. Exposure to such CAD tools and techniques is considered by us to be a vital part of this exercise. After the design package has been completed by each team (A to P), the individual designs are incorporated onto the single Superchip layout and final checks undertaken. The complete layout package is delivered to AMS for fabrication. This takes around three months, and when the chips return, they can be tested during Semester two. In Semester two, the student teams re-convene to develop test vectors to enable automatic testing of their designs, carry out simulations to validate their test vectors using their original designs in simulation, and finally test their individual ICs.

II. The Superchip
A Introduction
A crucial aspect of the program is the ability to effectively support a large number of individual IC designs without excessive cost in terms of time or resources. In order to achieve this, the "Superchip" has been developed as shown in figure 1, which allows multiple student designs to be fabricated on a single IC, and encapsulated in a standard package. This has been achieved through innovative design techniques, some of which are discussed in the following sections of this paper. There are 16 separate design slots within this single chip, and the cohort is divided into teams of around 6 students enabling a large number of students to develop separate designs as a group.

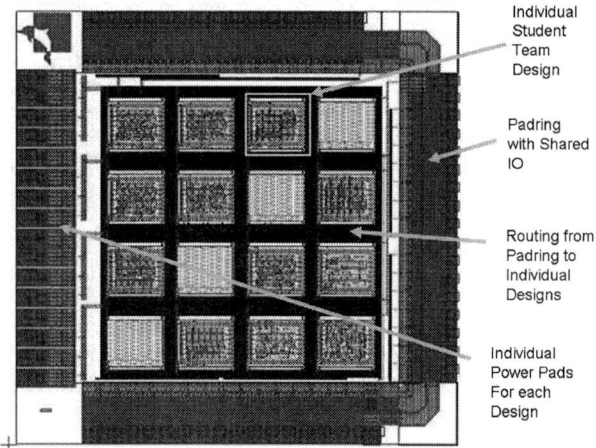

Figure 1: Typical Southampton Superchip IC layout

B Details of the Superchip Layout
The chip is designed using the Austria Microsystems C35B4 (0.35μm) CMOS process, with 4 metal layers available

through multi project wafers (MPW). The Chip infrastructure consists of a padring which has 24 digital inputs, 24 digital outputs, 16 individual site power supplies (VDD) operating at 3.3V, a global VDD at 3.3V and a global Ground. This gives a total of 66 pins. The chip is packaged in a JLCC68 (68pin) package, with two spare pins, for use in general laboratory situations.

The individual student design sites are buffered and selected using separate power connections (VDD), so when a site is powered, then its inputs and outputs are also enabled. Within the ring of buffers and power for each site, a miniature pad ring has been created which is the interface to the Superchip that the students see.

III. The Design Process
A Introduction
In this section of the paper, we will introduce the key stages in the design process, particularly with reference to the students backgrounds from their first year in terms of knowledge, how this relates to the theoretical program of study and also the context of the skills looking forward to later on in their degree course. The design exercise is schedule as early as possible in the second year of the undergraduate program to provide as much time as possible to ensure that the ICs can be made in adequate time for testing the second semester. This has the obvious implication that we must assume the students only have the knowledge obtained in their first year by this stage. This has the effect of defining the type of designs that can be undertaken (simple digital synchronous or combinatorial logic design). Typical applications have therefore included sequence recognition, counter design, ALU design and oscillator design. In addition, the students have limited analogue electronics experience (basic CMOS transistor knowledge) and basic knowledge of electronic design tools.

B The Student Design Kit
As in any IC design, we provide the students with a complete design kit. This includes a library of schematic symbols, layout abstract cell views, simulation files, design rule check files and design extraction files. This is not however to be confused with the standard AMS design kit. In this case the design kit has a much reduced number of digital gates for the students to work with (shown in table 1). This greatly simplifies the scale of the design kit and it becomes markedly less intimidating psychologically. For example, a typical gate layout is shown in figure 2, where the inverter is simplified to the power rails (VDD and VSS) in Metal 1 as horizontal tracks, and the vertical signal tracks (A, Y) in Metal 2. In comparison with the full layout cell, the complexity is hugely reduced, thus enabling student versions of software to be easily used, and also to minimise the design complexity. For more advanced students, the full layout cell views could certainly be used instead of the limited abstract views.

The routing is constrained, so that the students are not allowed to route over the cells, and are restricted to Metal 1 and Metal 2. They use a standard "routing channel" strategy to manually connect up the cells. For each abstract cell, there is an equivalent Spice model for analog simulation, and a VHDL model for digital simulation. During the initial design phase, the students are restricted to Spice simulation, so it is important for them to ensure that not only are the cells connected correctly, but that the VDD and VSS are also connected.

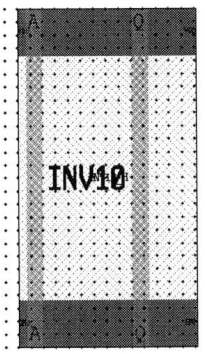 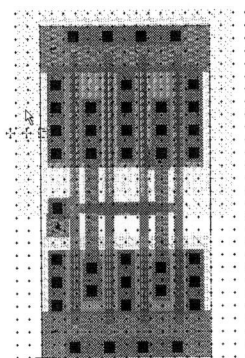

Figure 2: Typical Cell Abstract and Full Layout -Inverter

Cell Name	Cell Decription
inv10	Inverter
nand2	Two Input Nand gate
nand3	Three Input nand Gate
nand4	Four Input Nand Gate
nor2	Two Input Nor gate
nor3	Three Input Nor Gate
nor4	Four Input Nor Gate
xor2	Two Input XOR gate
xnor2	Two Input XNOR gate
dff	D-Type Flip Flop with reset
Tie1	Tie to VDD
Tie0	Tie to GND
MUX21	Two Input Multiplexer

Table 1: Design Kit Cells

C Design Tools and Methods

Given the limited knowledge of the students, we use the same PC based schematic design and simulation software they are familiar with in their first year studies. While these are necessarily optimum from an IC design perspective, as an introduction, they work well due to the existing familiarity of the students with the software. The overall process is shown in figure 2. Each stage of the process is discussed in the following sections of this paper.

D Design Specification

The design specification is published for all the teams and an introductory lecture is given to explain the detailed concepts, deadlines, tools and methods in detail. This is also an opportunity for the students to meet up with their other team members. As discussed previously, a typical design specification may be an 8 bit ALU or similar level of functionality. In recent years, a ring oscillator has also been added as a specific item which can be used to test the process

operation in a more "analog" function and enable the students to carry out some probing of high speed digital signals.

E Design process

A diagram of the standard design flow is shown in figure 3. The initial design phase is a typical "paper" design, where the team will discuss the options both for the functionality of the design, but also the implications for its fabrication.

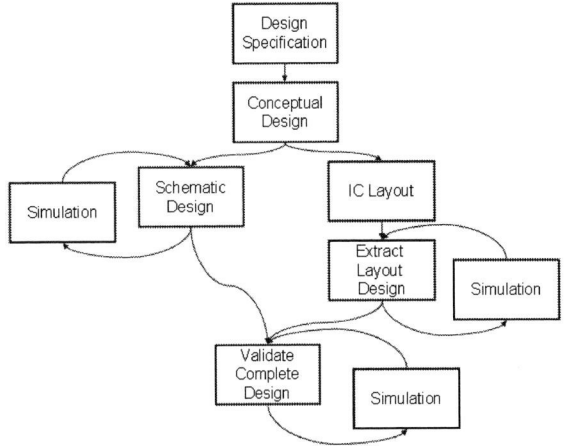

Figure 3: Design Process Flow Diagram

For example, in the ALU design one option is to design 8 bit functions in turn and link together, whereas an alternative approach would be to create a one bit "slice" and then simply replicate this 8 times. The student teams create a schematic of their design using the Orcad schematic capture software, from which they can simulate their design in Spice, or extract a VHDL model for digital simulation. We use an analog approach in the initial stages of the design to familiarize the students with the concepts of power consumption, realistic rise and fall times, overshoot, ground-bounce and device characteristics impacting on fan-out and loading.

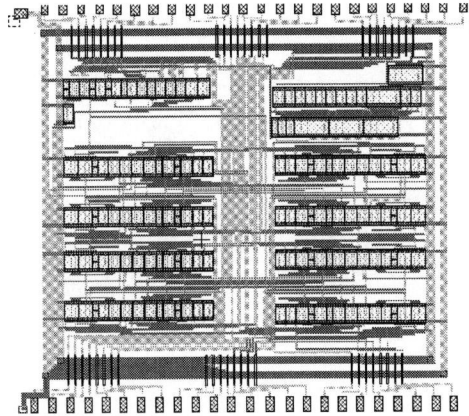

Figure 4: Student Design – 8-bit ALU

The IC Layout is carried out using the L-Edit software from Tanner EDA and the same Spice test benches are used to validate the extracted Spice model from the layout to ensure the designs are consistent. LVS (Layout Versus Schematic) is

978-1-4244-2018-6/08 $25.00 © 2008 IEEE   573

also possible within L-Edit, and we introduce the students to the use of Design Rule Checking (DRC) at this stage also. The student designs must pass DRC prior to completion of the lab, and the functionality of the design (schematic and layout) has to be fully demonstrated to the lab supervisor prior to "sign-off". A typical example design is an 8-bit ALU, which has a completed layout as shown in figure 4.

As can be seen from the layout, we use a standard cell based approach of manually laying out rows of abstract cells, with routing channels between the rows in two metal layers(M1 and M4). Although this is a standard 4 Metal Layer process, and it is possible to route over the top of the standard cells using M3 and M4, at this stage of the programme, it is useful to illustrate the concepts in channel based routing, and by manually routing, the students also have to think about the physical design, and implications of poor choice of cell placement.

V. IC Validation and Verification

A IC Test Board

When the ICs return from fabrication, the key task is to test them to ensure the basic functionality is correct, and also to carry out some basic performance measurements (timing, power consumption, oscillator frequency) to verify the design criteria have been satisfied. While it is straightforward in principle to carry out this type of testing by building a prototype test board, we take the view that it is more productive to provide a basic test infrastructure and to introduce the students to the concept of test vectors at this stage. To this end, we have developed a standard chip tester board, with a USB interface and chip socket, so semi automatic testing can be easily and quickly carried out on the individual designs. The credit card sized board is shown in figure 5, and uses a standard USB interface chip (CP2102) and a PIC to manage the interface between the PC and the test board, and the students also have full access to every pin via probe points directly next to the package.

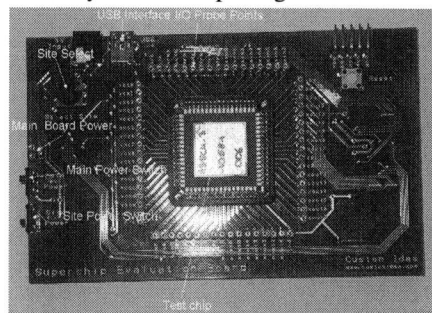

Figure 5: Superchip Test Board

B Test Vector Validation

In order to develop the test vectors efficiently, the same schematic used to design the layout, can also be used to extract a VHDL model of the design. Using this digital model, the test software used to connect to the test board can also export a VHDL test bench, so the test vectors can be tested using the model of the design, in this case in the Modelsim simulator, thus validating the test vectors prior to testing the IC itself. The test vectors are created in an XML style format,

making editing and validation simple, with an example file shown below:

```
# Test Vector File
<PinDef>
clk,nrst,tclk,trst,tdi,freein0,freein1,freein2,ai
n0,ain1,ain2,ain3,ain4,ain5,ain6,ain7,aout0,aout1
,aout2,aout3,aout4,aout5,aout6,aout7,bout0,bout1,
bout2,bout3,bout4,bout5,bout6,bout7
</PinDef>
<TestVector>
#  cin      ain       aout     bout
10000000 00000000 10000000 00000000
01000000 00000000 01000000 00000000
00100000 00000000 00100000 00000000
</TestVector>
```

The basic format is divided into two sections: **PinDef**, which is the pin definitions, and **TestVector**, which gives all the individual test stages. The test vectors can be either static 0, static 1 or a clock pulse denoted by C. The tester is not designed for high speed testing, but for the type of designs implemented (usually something like a frame decoder, ALU or sequence detector) these are perfectly adequate. (Lines beginning "#" are comments and ignored).

VI. Conclusions

This design exercise is unique in that the a cohort of second year undergraduates will have experienced a complete CMOS IC design process flow during their 4-year degree programme including making their own ICs. This is the most recent innovation in a long history of CMOS design and fabrication undertaken by undergraduates at Southampton and since 2004 over 400 students have produced their own designs on Silicon using this approach. The benefits to industry are clear, as the students leave the University with not only the theoretical and design skills, but also a practical knowledge of real design deadlines, team-working and the achievement of designing, making and testing their own ICs. The paper has described the architecture of the Superchip, the test board and the test vector approach used. We conclude that this demonstrates how large numbers of undergraduate or postgraduate students can be taught the essentials of IC design in a practical and cost-effective manner.

VII. REFERENCES

[1] Kolb, D.A., "Experiental Learning: Experience at the source of learning and development", Prentice-Hall, Englewood Cliffs, NJ, 1984

[2] Brown S., "The art of teaching small groups", New Academic, Spring 1997, pp3-6

[3] Atman C.& Bursic K., "How Effective are Textbooks in Teaching the Engineering Design Process", Frontiers in Education Conference, Nov 1995, Georgia.

[4] Biggs J., "Teaching for Quality Learning at University", Open University Press, 2002.

[5] Felder, R.M., and R. Brent. 2001. "Effective strategies for cooperative learning," J. Cooperation & Collaboration in College Teaching, 10(2), 63–69.

[6] Qualters D., "Using classroom assessment data to improve student learning", Penn State University, USA.

[7] Woods, D.R., R.M. Felder, A. Rugarcia, and J.E. Stice. 2000. "The Future of Engineering Education. 3. Developing Critical Skills", Chem. Engr. Education, 34(2), 108–117.

[8] Felder, R.M., and R. Brent, "Designing and Teaching Courses to Satisfy the ABET Engineering Criteria", J. Engr. Education, 92(1), 7-25 (2003)

[12] W. Wolf, J. Madsen, "Embedded System Education for the Future", Proceedings of the IEEE, vol. 18(1), pp. 23-30, 2000

A 3D Graphics Processor with Fast 4D Vector Inner Product Units and Power Aware Texture Cache

Jae-Sung Yoon, Donghyun Kim, Chang-Hyo Yu, and Lee-Sup Kim

Dept. of EECS, KAIST, Guseong-dong, Yuseong-gu, Daejeon, KOREA

cucujs@mvlsi.kaist.ac.kr

Abstract-**A 3D graphics system integrating two symmetric unified shader cores for mobile application is presented. To utilize instruction, data, and task level parallelism, a dual-core, dual-issue VLIW and multi-threading method is adopted. For efficient processing, an IEEE-754 compliant fast 4D vector inner product arithmetic unit for matrix multiplication, an internal bus system and a configurable texture cache technique to reduce power consumption in texture unit are proposed. By these methods, the proposed processor achieves 143Mvertices/s and 2.3Gtexels/s consuming the power of 367mW. Also, 45% performance improvement and 26% increase in performance per power ratio are achieved.**

I. INTRODUCTION

Modern GPUs and mobile 3D graphics engines integrate programmable processor cores, known as shaders, for their numerical computation of 3D graphics pipeline. Shaders are used as vertex and pixel processing components of the entire processor, known as vertex shader and pixel shader. Also a unified shader was presented for high efficiency by performing both vertex and pixel processing. In the mobile multimedia system using these shaders, a design challenge of the 3D graphics processor is sufficient processing capability within limited power consumption. There are several researches [1-7] about the embedded 3D graphics system using high performance and low power shader.

Vertex shaders [2-6] are usually focused on the ability of matrix multiplication because the most frequent geometry operation is vertex transformation. A SIMD datapath [2][3][5] was presented to evaluate an inner product of 4-dimensional vectors in one instruction. To utilize instruction level parallelism, a VLIW architecture [4] was also adopted in the vertex shader. It enables the processor to perform a geometry transformation in a single cycle. It is maximally four times faster than SIMD architecture, but functional units cannot be fully utilized because task level parallelism is not supported.

A recently introduced pixel shader[7] was focused on the texture operation which is the most important operation in the pixel shading. Texture unit fetches various texture images from the external texture memory. Since this operation consumes most power in the processor, power management scheme should be considered.

A unified shader should consider both characteristics of vertex shader and pixel shader. In a unified shader, the performance is proportional to the matrix processing power in vertex shading and the texture fetching ability in pixel shading. Since it uses both vertex and pixel data, an efficient input and output buffer system is also needed. But a recently introduced unified shader[8] has only SIMD architecture with conventional vertex and pixel buffer.

In the datapath of shaders, a fixed point arithmetic unit [2] was used to take advantage of fast and small core area. Also, a logarithmic number system [5] for more fast and power efficient processing was presented. It achieves the fastest result of 141Mvertices/s of vertex processing performance. Since most geometric operations in 3D graphics require a numeric system with wide dynamic range and moderate precision, a floating point datapath [3] was introduced. However a conventional floating point datapath is slower than fixed point or logarithmic datapath. Consequently, the speed of floating-point computation is one of the most important factors in overall performance of 3D graphics processors.

In this paper, we propose a 143Mvertices/s and 2.3Gtexels/s vertex and pixel unified processor supporting up-to-date embedded 3D graphics API, OpenGL ES 2.0. To achieve both high performance and low power in this processor, we used following approaches. In the system level, a two-core and two-issue system for high datapath utilization using task level parallelism is proposed. Also, an internal bus system for efficient and flexible data communication between shaders and I/O bus is used. To optimize shader core, an IEEE 754 compliant floating point arithmetic unit for fast matrix multiplication is introduced. It can reduce total processing time. Also, a configurable texture cache decreases the power consumption of a texture unit. The proposed processor achieves 45% improvement in performance and 26% increase in performance per power ratio. The implemented processor integrates 3.4M transistors and 14KB SRAM in $4.5x4.5mm^2$ die size using 1.8V 0.18μm CMOS technology.

The rest of the paper is organized as follows. In section 2, details of the proposed processor architecture is described. Then, section 3 shows the optimization methods of shader core. Test and implementation results will be shown in section 4. Finally, section 5 will conclude this paper.

II. Overall Processor Architecture

Figure 1 shows the overall block diagram of the proposed 3D graphics processor. A host interface unit fetches vertex/pixel data from external memory into an internal bus system(IBS). The IBS stores vertex/pixel data, and distributes them to two shaders by control signals from scheduler. Unified shaders execute geometry and rendering operations, and write back results to the IBS.

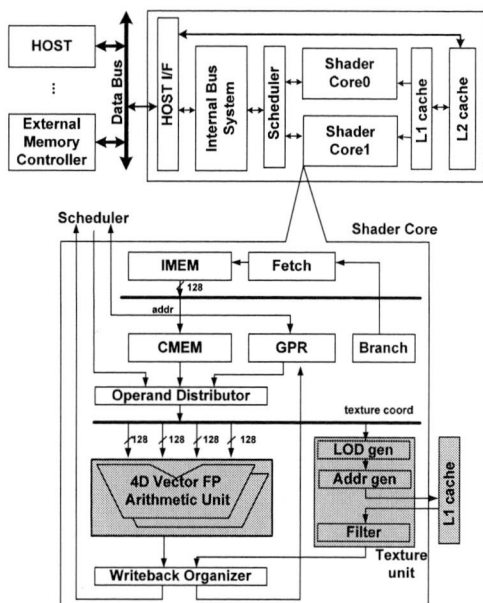

Fig. 1. Overall architecture of the proposed processor

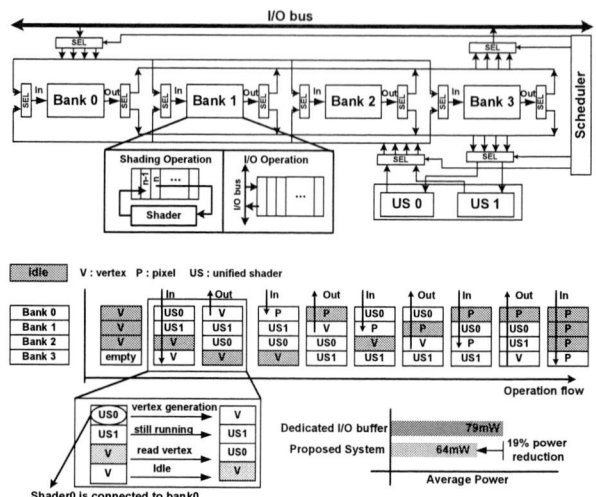

Fig. 3. Internal bus system and operation flow

A. Unified Shader

When the shader acts as a vertex shader, it performs geometry operations from matrix and vertices. The matrix is fetched from the constants memory (CMEM), and vertices are fetched from an internal bus system. According to the fetched instructions, the arithmetic unit performs the matrix multiplication and lighting operations for vertices. A pixel rendering and texture operations are performed when the shader acts as a pixel shader. To perform the trilinear filtering in a single cycle in the texture unit of both shaders, the L1 texture cache is shared and provides 16 ports to shader cores.

To utilize the instruction level parallelism(ILP), a four-issue VLIW architecture[4] is effective. However, it cannot utilize task level parallelism(TLP) because functional units are tightly coupled with a single shader core. When we perform only one task which is a vertex or pixel program, the ILP has limitation because complex functions such as NRM(normalize), transcendental functions, and LIT(lighting) are composed of sequence of primitive instructions which are dependent on each other. In this case, functional units cannot be fully occupied by four instructions, and should wait until the dependencies are removed. This problem is more serious when the shader acts as a pixel shader. The proposed architecture

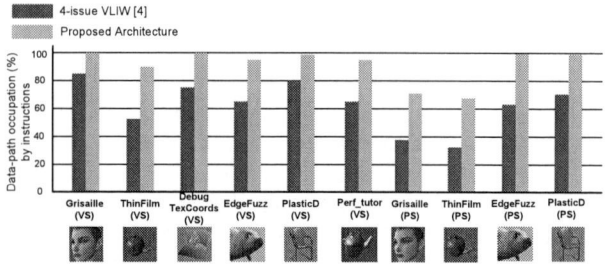

Fig. 2. Datapath occupation rate by instructions

separates the processing core into two shaders to perform independent tasks simultaneously maintaining matrix operation performance. Therefore, functional units are nearly in full operation, so that they achieve average 30% decrease of processing time as shown in the Fig. 2.

To utilize the vertex/pixel data level parallelism, multi-threading is adopted which can hide the latency of datapath, and thus four threads are running at the same time.

B. Internal Bus System

The proposed internal bus system (IBS) is implemented to provide efficient data transmission among shaders. A traditional unified shader has its own input and output buffer for vertex and pixel data respectively [8]. Since each shader cannot access buffers of another shader in the two shader system, there are two limitations: When a shader performs vertex shading, buffers for pixel data should be idle, and vice versa. Also, when the input buffer receives input data from I/O bus, the shader should stop until the transmission is finished. The IBS alternates the dedicated vertex/pixel input and output buffer of shaders. The IBS is composed of four-banked SRAMs and a scheduler. To satisfy the data input and output bandwidth requirement between shaders and I/O bus, each bank has two operating modes which are shading mode and I/O mode. The vertex/pixel data is transferred between I/O bus and IBS in the shading mode, or transferred between IBS and shaders in I/O mode. When the shading is in operation with some banks, remained banks transfer data with I/O bus. Therefore, the shader operation time can hide the latency of the I/O access.

In this system, the scheduling obeys the following rules: The shader can access any bank if all banks are filled with either vertex or pixel. If some banks are filled with pixels, they have higher priority to be accessed by the shader. From this scheduling, at least one of four banks is in an idle state as shown in the Figure 3. On average 19% power reduction can be achieved compared to the shader with the dedicated I/O buffer.

978-1-4244-2018-6/08 $25.00 © 2008 IEEE   576

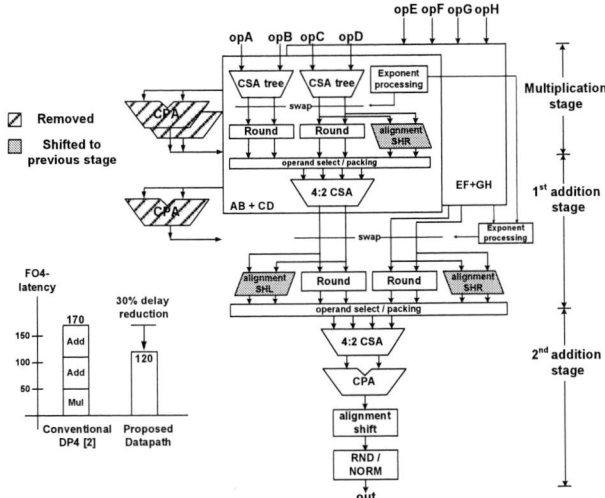

Fig. 4 Floating-point datapath for 4-D Vector Inner Product with reduced delay

II. Shader Core Implementation

In the unified shader core, processing units can be classified into arithmetic unit and texture unit as shown in Fig. 1. The arithmetic unit is a quad floating-point SIMD which processes four 32-bit floating point scalars such as *RGBA* and *XYZW*. Texture unit receives texture coordinate from an operand distributor, and fetches texture images from the external texture memory. In this system, the performance is proportional to the matrix processing power in vertex shading and the texture fetching ability in pixel shading. The following subsections describe details of the proposed optimization methods.

A. 4D Vector Inner Product Unit

In shader programs, 4D vector inner product (DP4: dot product for 4D vector) is the most important and frequent operation [4][5] for matrix multiplication. The proposed arithmetic unit [13] of Fig. 4 supports IEEE754 single-precision format, and produces the results of four multiplications, two additions, and one DP4. The DP4 operation requires 3 fundamental instructions, which are 4-way SIMD multiplication and two SIMD additions for $Z = (AB + CD) + (EF + GH)$. To reduce critical delay of the arithmetic unit, we used the following techniques: The effective reduction of latency is achieved by merging multiple operations into one, and by configuring parallel paths for independent or exclusive operations. We removed two CPAs of multiplication stage and first addition stage, and retained intermediate roundings after each step. The alignment shifts of addition stages are performed in parallel with the rounding of previous stages. Therefore, in cycle 1, the partial products of multiplier tree is generated. In cycle 2, the rounding of multiplication product is performed in parallel of the alignment shift of the first adder. In cycle 3, the rounding of the sum from the first adder is performed in parallel of the alignment shift of the second adder. In cycle 4, finally, the real

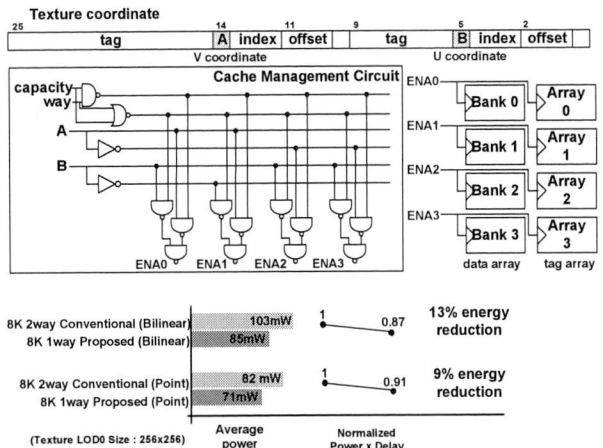

Fig. 5 Dynamic Configurable Texture Cache

addition occurs in the CPA of the last stage. By using this technique, the critical path of the proposed DP4 unit is reduced by 50 FO4-inverter delay compared to previous DP4 implementation [4] using discrete multipliers and adders.

B. Configurable Texture Cache

A texture cache is the most massive processing unit in the pixel shading operation, and the share of power consumption is about 20% in the processor. The unified shader supports both vertex and pixel texture fetch. The difference of these two textures is that the pixel texture needs tri-linear filtering for image interpolation, but the vertex texture needs bilinear or point filtering for normal map and other geometry related algorithms. For tri-linear filtering, texture cache should access the separated regions of memory at once. Therefore, in this case, two-way set associative cache is effective. But, for bilinear or point filtering, the direct-mapped cache is effective for energy saving in spite of its miss penalty. To exploit this feature, we used a configurable cache [9]. The proposed texture cache in Fig. 5 can be configured dynamically by host command to be either direct-mapped or two-way set associative, and also can change the cache capacity to 4KB or 8KB. To change the configuration, an index field of the texture coordinate needs more bits which are marked as A and B. It enables the index field to select more lines across banks. A cache management circuit (CMC) generates turn-off signals from the address bits A and B according to the capacity and way selection signals. Since the index selects only one or two banks, at least two of four banks can be turned off from the bits A and B. By this technique, the proposed texture cache achieves 30% power reduction and 13% energy reduction for selected configurations.

VI. Implementation and Test Results

To evaluate performance, we used shader programs from Nvidia FX Composer[10]. Fig. 6 illustrates cycle count requirement for vertex processing and performance per power comparison with previous works. Shader programs can be classified into matrix operations and arithmetic operations. In

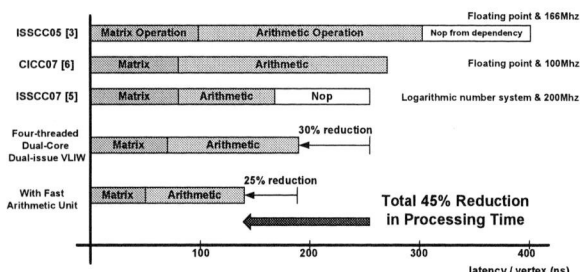

Fig. 6 Processing time evaluation

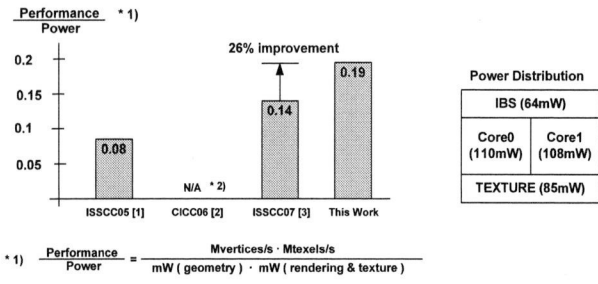

Fig. 7 Comparison with the previous works

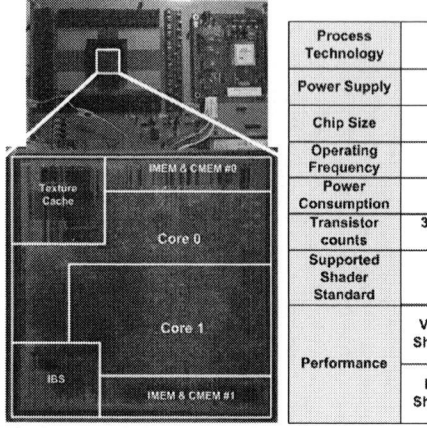

Fig. 8 test results

Process Technology	0.18um CMOS	
Power Supply	1.8V	
Chip Size	4.5mm x 4.5mm	
Operating Frequency	143 Mhz	
Power Consumption	367mW	
Transistor counts	3.4M logic, 14KB on-chip SRAM	
Supported Shader Standard	OpenGL ES 2.0 Shader Model 3.0	
Performance	Vertex Shading	143Mvertices/s (Geometry Transformation)
	Pixel Shading	2.3Gtexels/s (Trilinear Filtering)

the conventional SIMD datapath [3][5], no operation penalty occurs from instruction dependency.

The proposed two core system for task level parallelism of shaders can reduce the arithmetic operation time about 30% compared to floating point processor[6]. Also, the fast floating-point 4D-vector inner product unit reduces the cycle time of arithmetic operation to the logarithmic number system[5], and reduces the matrix operation compared to previous works. By these techniques, total cycle time is reduced about 45%. As a result, the proposed shader is faster than both conventional floating-point and logarithmic arithmetic functional units. Total power consumption in Fig. 7 is larger than previous works because additional registers and control logics are needed for two shader cores. However, the proposed processor achieved 26% improvement in the performance per power metric. Fig. 8 summarizes the features of the chip, and shows the chip micrograph and evaluation board for verification. The implemented processor integrates 3.4M transistors and 14KB SRAM in 4.5x4.5mm^2 die size using 1.8V 0.18μm CMOS technology.

VII. Conclusion

In this paper, we proposed a 3D graphics processor based on dual-core dual-issue VLIW architecture and shader core optimization techniques. To achieve both high performance and low power in this processor, we used four key approaches: a two-core and two-issue system for high datapath utilization, an internal bus system for scalability and power reduction, an IEEE 754 compliant floating point (FP) arithmetic unit with reduced delay for fast matrix multiplication, and a configurable texture cache to decrease the power consumption.

The instruction, data, and task level parallelism was exploited using dual-core dual-issue VLIW architecture which gives about 30% performance improvement. Also, fast 4D-vector arithmetic unit achieves additional 25% performance increase. A configurable texture unit reduced 30% power consumption in texture cache. By these techniques, the proposed processor achieves 45% improvement in performance and 26% increase in performance per power ratio.

REFERENCES

[1] F. Arakawa, et al., "An Embedded processor Core for Consumer Applications with 2.8GFLOPS and 36M Polygons/s FPU," Proc. of ISSCC Dig. Tech. Papers, pp.334-335, Feb 2004.

[2] J. Sohn, et al., "A 50Mvertices/s Graphics Processor with Fixed-Point Programmable Vertex Shader for Mobile Applications," Proc. of ISSCC Dig. Tech. Papers, pp.192-193, Feb 2005.

[3] D. Kim, et al., "An SoC with 1.3Gtexels/s 3D Graphics Full Pipeline Engine for Consumer Applications," Proc. of ISSCC Dig. Tech. Papers, pp.190-191, Feb 2005.

[4] C-H Yu, et al., "A 120Mvertices/s Multi-threaded VLIW Vertex Processor for Mobile Multimedia Applications," Proc. of ISSCC Dig. Tech. Papers, pp.408-409, Feb 2006.

[5] B-G Nam, et al., "A 52.4mW 3D Graphics Processor with 141Mvertices/s Vertex Shader and 3 Power Domains of Dynamic Voltage and Frequency Scaling," Proc. of ISSCC Dig. Tech. Papers, pp.278-279, Feb 2007.

[6] C-H. Yu, et al., "A 186Mvertices/s 161mW Floating-Point Vertex Processor for Mobile Graphics Systems," Proceedings of the IEEE Custom Integrated Circuits Conf.,pp579-582,2007.

[7] S-H. Kim, et al., "A 36fps SXGA 3D Display Processor with a Programmable 3D Graphics Rendering Engine," Proc. of ISSCC Dig. Tech. Papers, pp.276-277, Feb 2007.

[8] J-H. Woo, et al., "152mW Mobile Multimedia SoC with Fully Programmable 3D Graphics and MPEG4/H.264/JPEG," Proceedings of the IEEE Symposium on VLIS Circuits Conf.,pp.220-221,2007.

[9] Chuanjun Zhang, et al., "A Highly Configurable Cache Architecture for Embedded Systems," Proc. of International Symposium on Computer Architecture, pp. 136-146, Jun., 2003.

[10] Nvidia Corp., FX Composer 2.0, http://www.nvidia.com

[11] OpenGL ES 2.0, http://khronos.org

[12] Shader model 3.0, http://microsoft.com

[13] D. Kim, et al., "A Floating-Point Unit for 4D Vector Inner Product with Reduced Latency," IEEE Transactions on Computers, submitted.

IEEE 2008 Custom Intergrated Circuits Conference (CICC)

Timing Yield Enhancement Through Soft Edge Flip-Flop Based Design

Michael Wieckowski, Young Min Park, Carlos Tokunaga, Dong Woon Kim,
Zhiyoong Foo, Dennis Sylvester, David Blaauw
University of Michigan, Ann Arbor, MI

Abstract- **The first evaluation of a soft-edge flip-flop is presented as an alternative to useful-skew and latch-based designs for variation compensation in a 16-bit 8-tap FIR filter in 0.13μm CMOS. An 11.2% performance improvement was achieved over a standard hard edge data flip-flop (9.2% when post-silicon useful-skew is applied).**

I. INTRODUCTION

Increasing process variation in advanced technology nodes due to random sources, such as dopant fluctuation and line-edge roughness, along with complexity in CAD analysis for timing closure, have become limiting factors in the performance and yield of ASIC designs. Traditionally, two-phase latch-based circuits have been used to address these issues due to their relatively high tolerance to variation and their inherent ability to borrow time and pass slack between pipeline stages [1]. While these qualities are attractive from a performance standpoint, latch based systems incur significant power overhead in clock distribution along with an overall increase in design complexity. As an alternative, the combination of hard edge data flip-flops (DFF) with useful clock skew have been used to similarly enhance performance [2]. While effective at mitigating stage delay imbalance and clock tree variation, it is less effective when compensating for individual path delay variation. This is due to the fact that the useful skew can only be applied to groups of flip-flops, among which path delay may vary substantially. This limitation in achievable performance gains is shown in silicon measurements later in this work.

To address timing yield issues without the overhead and penalties seen in other techniques, we have chosen to investigate a soft edge flip-flop (SFF) that maintains synchronization at a clock edge, but also has a transparency window, or "softness", around it. This flip-flop offers similar variation tolerance to that seen in latch based designs along with performance improvements due to time borrowing. Furthermore, the soft flip-flop provides these gains with minimal overhead in complexity, area, and power.

II. BACKGROUND

It is well known that introducing a transparency window around the clock edge provides benefits in performance by reducing sensitivity to clock skew and jitter and by allowing some degree of time borrowing between pipelined stages [3]. As a demonstration of this effect, we simulated a simple two-stage pipeline composed of balanced inverter chains with an equivalent 10 FO4 stage delay in a commercial 45 nm technology using the setup shown in Fig. 1.

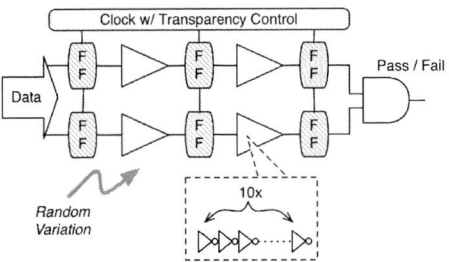

Fig. 1 - Simulation setup for testing transparency windows

Monte Carlo analysis was performed considering only random mismatch among all of the transistors in the pipeline. This mismatch induces delay imbalance among the stages, forcing one path to dictate the maximum attainable clock frequency [4]. As seen in Fig. 2, adding a small window of transparency around the clock edge directly mitigates this effect by allowing some degree of time borrowing, and in turn providing an improvement in overall performance. This improvement would be greater once global variations, clock skew, and clock jitter were considered.

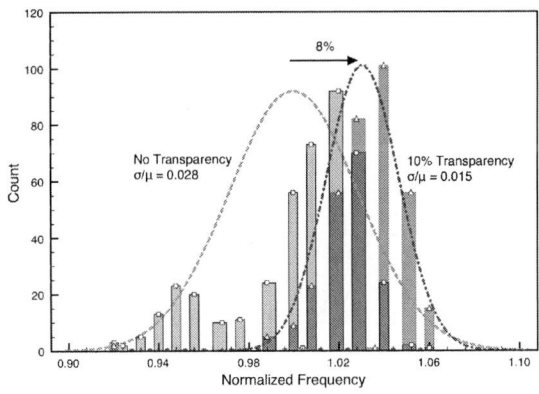

Fig. 2 – The effect of small transparency windows at nominal supply.

A soft edge flip-flop implementation to create such a transparency window has been previously presented in [7], but its application space was focused on time borrowing in on-chip interconnects. Other implementations with similar transparency windows have been based on pulsed latches or hybrid latch-flip-flops [3,6-10]. These designs require the generation of a precisely timed pulse that is triggered from the arriving clock edge. Robust generation of this pulse under process and environmental variation is becoming increasingly difficult and has led to limited use of this technique in only the most timing critical circuits. The transparency window of the

978-1-4244-2018-6/08 $25.00 © 2008 IEEE

design presented in this work makes use of a standard master-slave flip-flop with a delay element driving the clock from the slave stage to the master stage. This delay element can be implemented using a variety of circuits and is inherently more tolerant to variation than a pulse generator. We discuss our proposed implementation in the following section.

III. SOFT EDGE FLIP-FLOPS

In order to measure the effectiveness of the proposed soft edge technique, a common data flip-flop (DFF) design was used allowing us to mask out any circuit specific effects and to specifically isolate the influence of the transparency window. As shown in Fig. 3, a traditional master-slave flip-flop was modified to perform as a soft flip-flop (SFF) by separating the clocks to the master and slave latching stages.

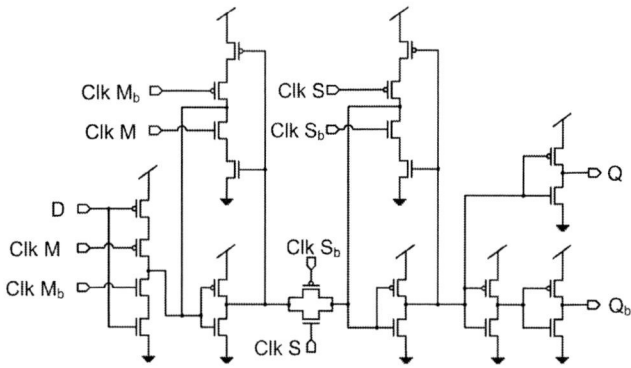

Fig. 3 – Soft edge flip-flop schematic.

To generate a transparency window and create the proposed soft clocking edge, the master stage clock is delayed with respect to the slave stage. Since the master stage is transparent low and the slave stage is transparent high, this delay results in a short window each cycle where both stages are transparent as shown in Fig. 4.

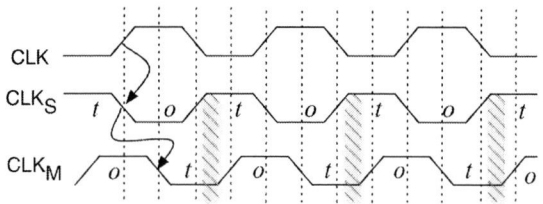

Fig. 4 - Delaying the master clock (CLK_M) relative to the slave clock (CLK_S) creates a window of transparency (t = transparent, o = opaque).

As discussed in Section II, this transparency window provides a direct improvement in overall performance. Creating this window by delaying the master stage clock is much simpler and more robust than generating a local timing pulse as in pulsed registers. This follows from the fact that even under extreme variation where the delay approaches zero, the soft flip-flop converges to the same functionality as a standard data flip-flop. This is in contrast to pulsed registers where a reduction in the pulse width can result in functional failure. In this work, we adopted a simple approach to generating the master stage delay that allowed flexibility in

choosing the size of the transparency window, for the purpose of experimentation. As shown in Fig. 5, tunability was implemented by using a series of different length inverter chains and selecting one as the master delay element using a scan controlled multiplexer.

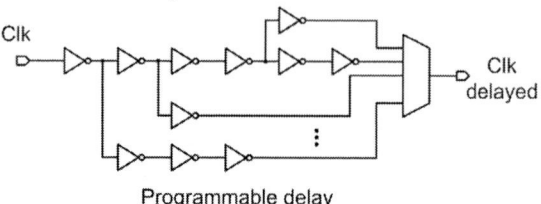

Fig. 5 - Soft flip-flop delay generator.

This delay generator allowed tuning of the transparency window in 10 ps steps from 0 to 250 ps. In a practical implementation, the flip-flop and the delay generator would be specifically designed and optimized for a particular performance target. In addition, it is important to point out that the power overhead of the delay generated can be easily amortized through sharing among groups of flip-flops. Similar techniques are used to share pulse generators in other methodologies and have proven effective [10].

IV. EXPERIMENTAL SETUP

An experimental test chip was designed containing one core with a 16-bit 8-tap FIR filter and a second core with only inverter chains. To determine the effects of the proposed soft edge technique, the pipeline of each core was segmented using both soft flip-flops and standard data flip-flops. This was accomplished using the scheme shown in Fig. 6 where data is fed to an SFF and a DFF simultaneously and their outputs are multiplexed to the succeeding stage of combinational logic. In addition, the aforementioned delay generator was included and shared among all SFFs in each stage bank. The area overhead of this arrangement was only 7.2% and would be dramatically lower given a more intelligent delay generation scheme. A second delay generator, also shown in Fig. 6, was included to assign tunable useful skew to each bank.

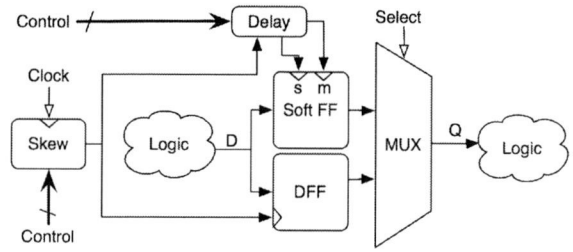

Fig. 6 - Method for selecting between SFF and DFF.

The four-stage FIR filter was designed as shown in Fig. 7 with particular attention paid to balancing all of the pipeline stages. The multipliers were divided into three stages and the final adder took only a single stage. The launching and receiving synchronization elements were chosen as hard edge DFFs in order to close any time borrowing that might occur due to transparency. In order to directly compare the

978-1-4244-2018-6/08 $25.00 © 2008 IEEE

performance of the SFF to a traditional DFF under the same process and environmental variations, each synchronization element contained both flip-flops and received the same clock from the local distribution network.

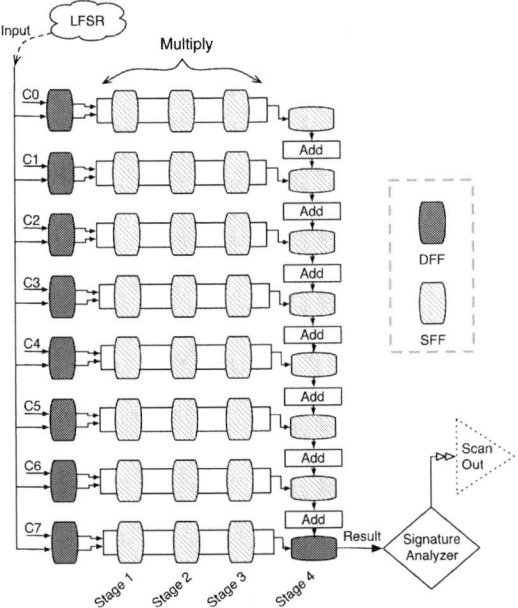

Fig. 7 – Four-stage FIR filter implementation.

The arrangement of the second test core containing four stages of only inverter chains is shown in Fig. 8. Just as in the FIR filter, SFF and DFF elements were included at each stage boundary with DFF elements on both ends to provide time borrowing closure.

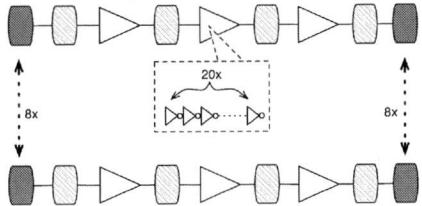

Fig. 8 - Four stages of balanced inverter chains.

Each stage was exactly balanced in the nominal case and contained a chain of inverters equaling a 20 FO4 stage delay. This structure was chosen as a test bench for the proposed soft edge technique at reduced supply voltages. Since the relative sensitivity to the threshold voltage increased at lower supplies, any local mismatch due to random sources, such as random dopant fluctuations (RDF), would express themselves as imbalance in the inverter path delays. As a result, the effect of the transparency window on random variation induced mismatch could be isolated from its systematic counterpart and quantified with respect to performance.

V. MEASURED RESULTS

We fabricated the described FIR filter and balanced inverter pipeline in a 0.13 μm CMOS technology with areas of 0.48 mm^2 and 0.1 mm^2, respectively as shown in Fig. 9.

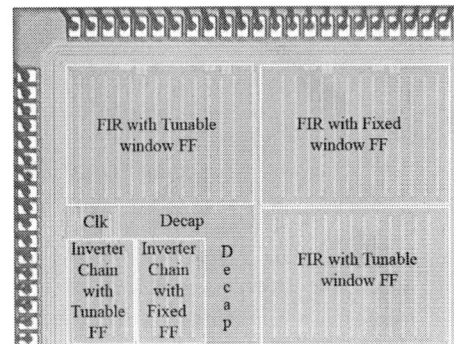

Fig. 9 - 0.13 μm test chip micrograph.

Performance measurements at nominal supply voltage of the FIR (1.2V) for 33 die showed an 11.7% improvement in the maximum operating frequency of the SFF circuits when compared to the DFF, and 9.2% improvement when optimal post-silicon skew was applied as seen in Fig. 10.

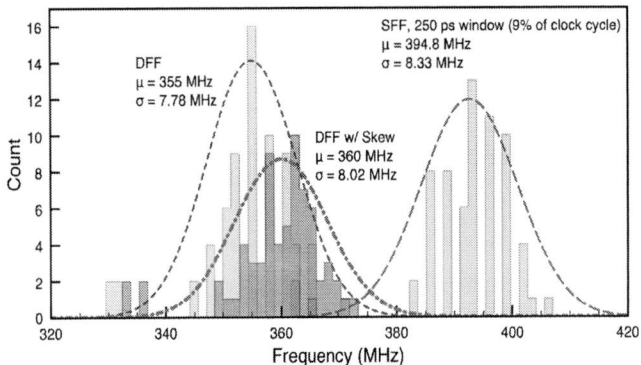

Fig. 10 - Test chip FIR filter frequency distributions for DFF, DFF with useful skew, and SFF.

In addition, all of the FIR filters were measured as a function of transparency window size using the tunable delay generators in each bank. Fig. 11 shows that an increasing window size results in a proportional increase in performance and saturates when the window size approaches 10% of the cycle time.

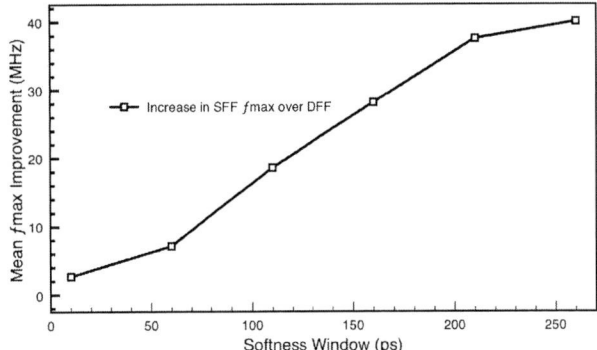

Fig. 11 - Frequency improvement of SFF over DFF FIR filter as a function of softness.

As the window sized is increased, a tradeoff occurs due to hold time constraints becoming more stringent. In addition, the larger delay introduced to the master clock will lead to higher power consumption. It is therefore desirable to use the minimum amount of softness necessary to achieve a particular performance target. Since separate delay generators were used for each bank of the FIR filter, we could determine the effects of softness to each stage of the pipeline independently. As seen in Fig. 12, the performance improvement of each stage saturates at different values of softness. Therefore, choosing the smallest window size that results in performance saturation will minimize hold time constraints and reduce overall system power. This window size assignment can be performed at design time, and a CAD algorithm for such softness assignment was recently proposed in [13] and showed a total power overhead of 2% on average with comparable performance gains as experimentally shown in this work. Alternatively, window sizes could be dynamically programmed during post-silicon testing.

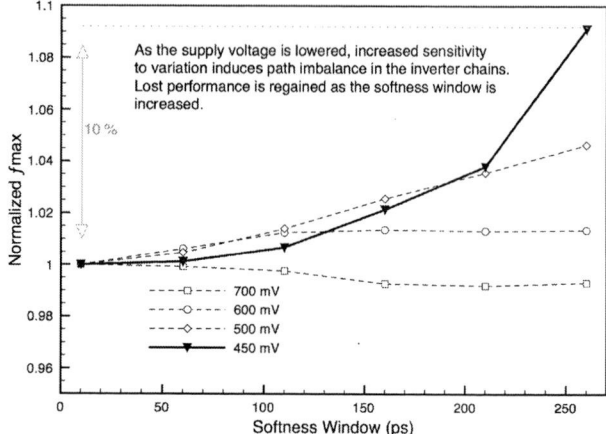

Fig. 13 - Frequency improvement of SFF over DFF inverter chains as a function of softness at low voltage

SFF can achieve gains in performance over 10% compared to traditional designs even under harsh low-voltage conditions. These gains come at minimal area and power penalties and can readily be incorporated into a standard design flow.

REFERENCES

[1] D. Harris, *Skew-Tolerant Circuit Design*, Stanford University, 1999.

[2] V. Nawale and T.W. Chen, "Optimal useful clock skew scheduling in the presence of variations using robust ILP formulations," *International conference on Computer-aided design*, 2006, pp. 27-32.

[3] H. Partovi et al., "Flow-through latch and edge-triggered flip-flop hybrid elements," *IEEE International Solid-State Circuits Conference*, 1996, pp. 138-139.

[4] K.A. Bowman, S.G. Duvall, and J.D. Meindl, "Impact of die-to-die and within-die parameter fluctuations on the maximum clock frequency distribution for gigascale integration," *IEEE Journal of Solid-State Circuits*, vol. 37, 2002, pp. 183-190.

[5] K. Bowman et al., "Time-borrowing multi-cycle on-chip interconnects for delay variation tolerance," *International symposium on Low power electronics and design*, 2006, pp. 79-84.

[6] A. Scherer et al., "An out-of-order three-way superscalar multimedia floating-point unit," *IEEE International Solid-State Circuits Conference*, 1999, pp. 94-95.

[7] J. Tschanz et al., "Comparative delay and energy of single edge-triggered & dual edge-triggered pulsed flip-flops for high-performance microprocessors," *Proceedings of the 2001 international symposium on Low power electronics and design*, 2001, pp. 147-152.

[8] F. Klass et al., "A new family of semidynamic and dynamic flip-flops with embedded logic for high-performance processors," *IEEE Journal of Solid-State Circuits*, vol. 34, 1999, pp. 712-716.

[9] H. Partovi, "Clocked Storage Elements," *Design of High-Performance Microprocessor Circuits*, A.P. Chandrakasan, W.J. Bowhill, and F. Fox, eds., Wiley-IEEE Press, 2000.

[10] D. Krueger, E. Francom, and J. Langsdorf, "Circuit Design for Voltage Scaling and SER Immunity on a Quad-Core Itanium Processor," *IEEE Internation Solid-State Circuits Conference*, 2008, pp. 94-95.

[11] V. Joshi, D. Blaauw, and D. Sylvester, "Soft-Edge Flip-Flops for Improved Timing Yield: Design and Optimization," *International Conference on Computer-Aided Design*, 2007, pp. 667-673.

[12] A. Datta et al., "A Statistical Approach to Area-Constrained Yield Enhancement for Pipelined Circuits under Parameter Variations," *Asian Test Symposium*, 2005, pp. 170-175.

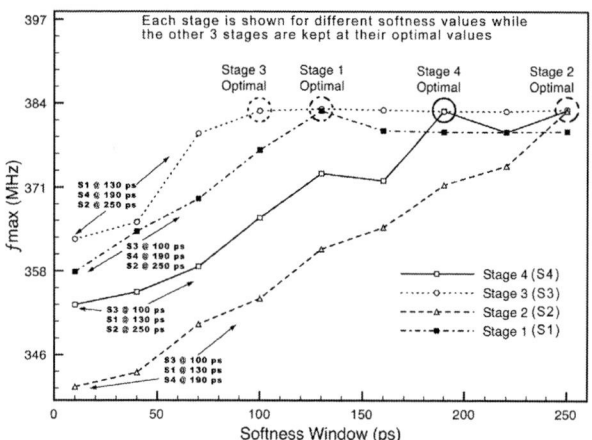

Fig. 12 - Optimization of softness windows for each FIR stage

As explained in Section IV, low voltage operation of pipelined circuits incurs an additional performance penalty due to the heightened sensitivity to threshold voltage variation. The end result is greater imbalance in the pipeline stages, translating to larger timing margins and design area [12]. The proposed soft edge technique helps to reclaim these losses by reducing the sensitivity of the system frequency to critical path delay variation. This improvement was measured for the balanced inverter core over a typical supply voltage scaling range, and is demonstrated in Fig. 13. At higher voltages near the nominal supply of 1.2V, softness has negligible effect on performance since the stages are perfectly balanced. As the voltage is lowered toward the threshold voltage, soft edge clocking reclaims over 10% of the performance lost to variation-induced stage delay imbalance.

VI. CONCLUSION

Soft edge clocking based on latch transparency windows is a well-understood technique that has been shown in this work to provide significant gains in performance for modern process technologies plagued by random sources of variation. Since high performance pulsed techniques are intolerant to variation, and hence unsuitable for variation compensation, we

1/5 Power Reduction by Global Optimization based on Fine-Grained Body Biasing

Yasumi Nakamura, David Levacq, Limin Xiao, Takuya Minakawa, Taro Niiyama, Makoto Takamiya, and Takayasu Sakurai

The University of Tokyo, 4-6-1 Komaba, Meguro-ku, Tokyo 153-8505, Japan

*Abstract-*A fine-grained body bias control to compensate both the process and design variations is proposed. A test chip was fabricated in 90nm CMOS process. The proposed global optimization scheme reduced power by 23% compared with an as-fabricated chip power and by 11% compared with the power optimized by the conventional local optimization approach. Also, the proposed global optimization scheme reduced power by 19% compared with an as-fabricated chip power within 20 test iterations with simulated annealing algorithm.

I. INTRODUCTION

The adaptive substrate bias control and the post-fabrication tuning of parameters [1-4] are the recent trend in designing power-efficient LSI's to cope with the increasing random device variability. The systematic design variation, which is the error between the simulation results at a design stage and the measured results of a fabricated chip, is also getting an important issue. This is partly because the chip is getting more and more complicated and delay estimation gets inherently more difficult and partly because the recent process introduces systematic delay deviation from designed value by new phenomena such as stress-induced drain current variation and imperfect optical proximity correction.

In the post-fabrication tuning, the parameters can be either locally or globally optimized. [4] shows a local optimization example of the body bias by monitoring critical path replicas in the 21 domains. Local optimization means that the parameter is locally optimized by looking at the local value of the parameter and modifying it to the designed value. This approach, however, cannot compensate the systematic variations and has the mismatch between the critical path replica and the real critical path. This approach may not work

for cases where short-range correlated variation is small as is shown in Fig.1 which shows that there is no specific peak in slow spatial frequencies.

In contrast, [1-3] show the global optimization of the clock skew in the 52 clock domains [1]. In this approach, however, the required tunable skew circuits consume power and the skew tuning does not reduces the leakage of the circuits.

In this paper, a fine-grained body bias control to compensate both the random (process) and systematic (design) variations is proposed and the globally minimized power at a constant performance is investigated through measurements.

II. LOCAL OPTIMIZATION VS GLOBAL OPTIMIZATION

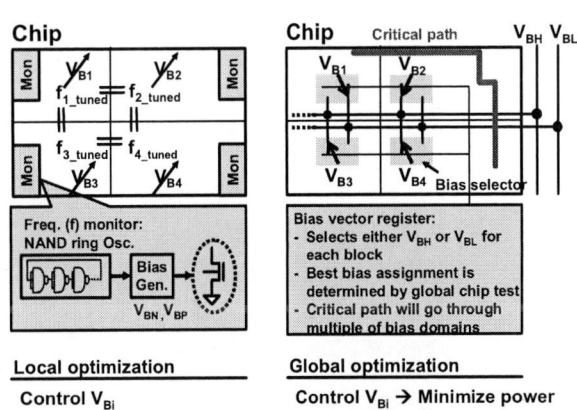

Fig.2. Local optimization (Left) and global optimization (Right).

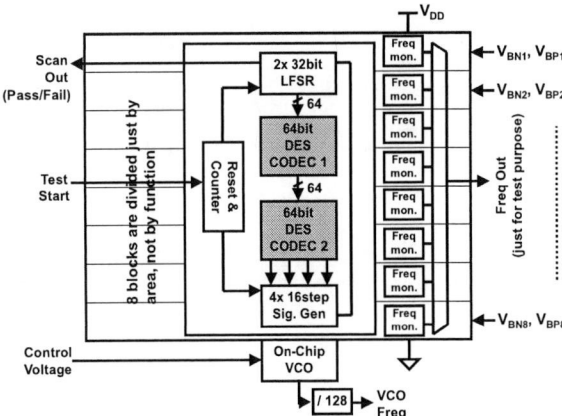

Fig. 3. Block diagram of the fabricated test chip.

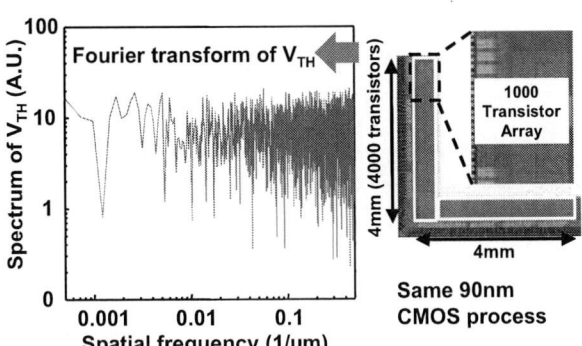

Fig. 1. Spectrum of within-die process variation

978-1-4244-2018-6/08 $25.00 © 2008 IEEE

There are two approaches to determine the body bias values for each biasing region (Fig.2). One is the local optimization scheme, which aims at making V_{TH} of all the regions is equal to compensate within-die V_{TH} variations [4].

The other is the proposed global optimization scheme, which aims at achieving the lowest power consumption while the real critical path can operate at the desired frequency. Real

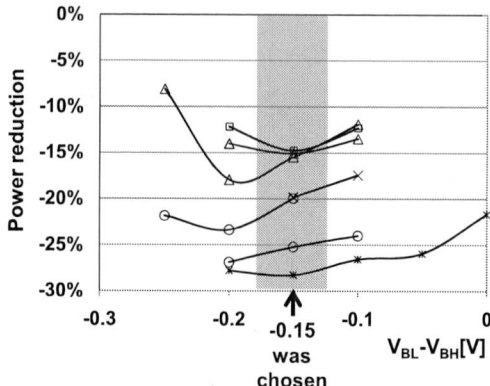

Fig. 6. Measured V_{BH}-V_{BL} – power dependence for 6 chips. This is for PMOS well bias and similar optimization is done for NMOS.

critical paths are tested at a chip level. The global optimization can compensate not only the process variation but also the systematic design variation. This can be used as a method for post-fabrication clock skew tuning (time borrowing) without introducing parametric delay component which is large.

To evaluate the effect of the fine-grain global body bias optimization, a test chip was fabricated in 1V 90nm CMOS and measured. Fig. 3 shows the block diagram of the test chip. The circuit under test is two series-connected 64bit DES CODECs, which is driven by 2 32bit LFSR input vector generator, and 4 out of 64bit output are compacted with 16step signature generator. The chip has 8 body bias domains and each region has a 31-stage FO3 NAND ring oscillator divided by 64 as a frequency monitor. Test is carried out with a PC with 16ch D/A board which generate the body bias voltages.

The design flow is shown in Fig.3. The only difference between a normal digital circuit and the proposed circuit is to divide the chip into multiple body biasing domains. In the test chip, this division is done just by area and not by function. This means that any division can be made without considering the functional borders and can be applied to any chip.

The micrograph and layout of the fabricated chip is shown in Fig.5. As shown in Table 1, it occupies 2400um x 2400um in area.

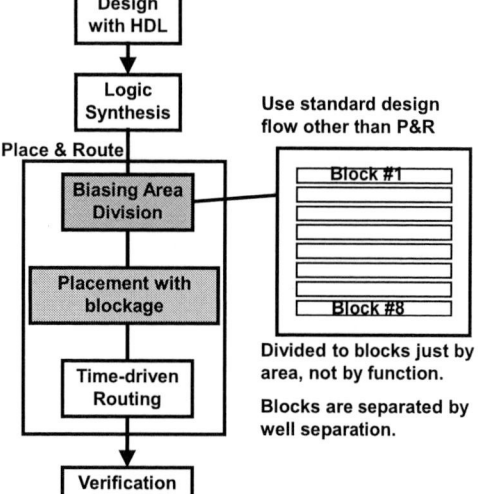

Fig. 4. Design flow of the test chip

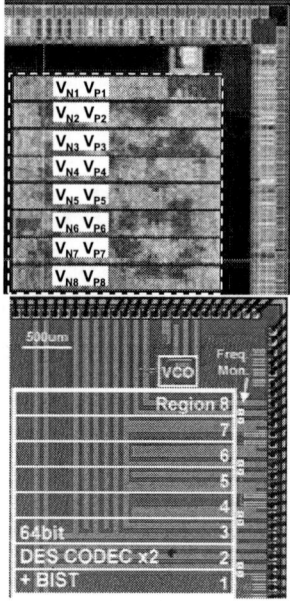

Fig.5. Chip layout (up) and micrograph of fabricated chip (down)

Table 1 Summary of the fabricated chip

Process	1V 90nm CMOS
Area	2400um x 2400um
# of gates	1M gates
Test	BIST (Built-In Self-Test) with 4-bit 16step signature generator

III. METHODS FOR DETERMING VB VECTOR

Since each body bias voltage ($V_{BP1\sim8}$ for PMOS and $V_{BN1\sim8}$ for NMOS) can take any analog value, finding the best value to obtain the lowest power out of millions of patterns is difficult. In this paper, each body bias voltage is limited to "high" or "low", namely V_{BH} and V_{BL}. V_{BH} is set to the lowest value where the chip can operate when $V_{BN1}=V_{BN2}=\ldots=V_{BN8}=V_{BP1}=\ldots=V_{BP8}=V_{BH}$.

V_{BL} is then set to V_{BH}-0.15V. Fig. 5 shows the reason, the V_{BL}-V_{BH} – Power reduction ratio dependence for 6 chips. Here, power reduction ratio = 0 when the power consumption is same as with the worst case V_{BH} or the V_{BH} where all the chips can operate. Power reduction ratio shows a gentle minimum between -0.2V and -0.1V for various chips.

Three methods are tried to determine the V_B vector, exhaustive search, best vector LUT and simulated annealing.

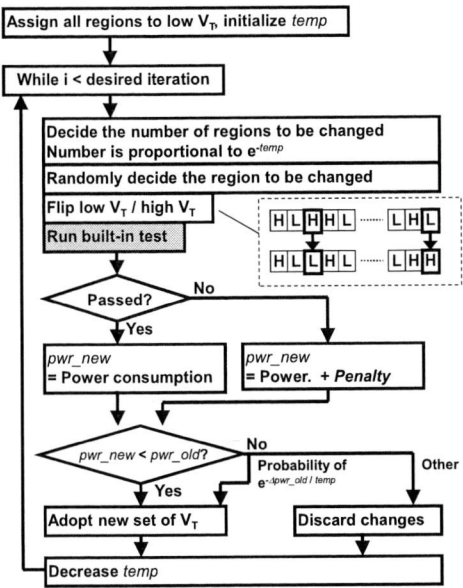

Fig. 6. Simulated annealing algorithm applied to the test chip

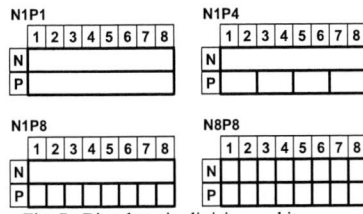

Fig. 7. Bias domain division and its name

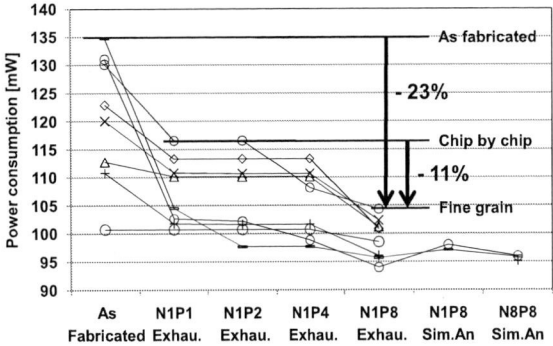

Fig. 8. Measured relationship between grain size and power consumption

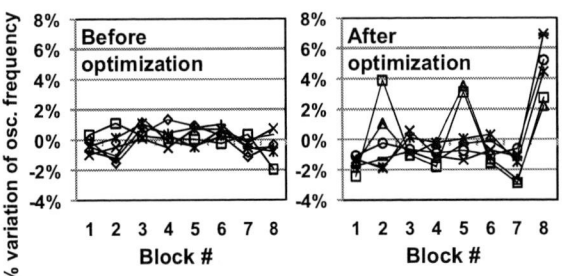

Fig. 9. NAND ring monitor frequency for each bias block. Each line corresponds to a chip

A. Exhaustive Enumeration

The simplest way to find the best vector to reduce power is to test all possible vectors. It is possible for small number of regions, though, even for 8 regions or 16 parameters (V_{BN}, V_{BP} x 8), 65536 tests are required, which is not practical if the number of domains are large.

B. Best Vector LUT

An alternative method is to apply the exhaustive test (or long-series simulated annealing) in the development stage and find out the best vector look-up table (LUT) for each V_{TN}, V_{TP}. This will take time in the development stage but once the LUT is established, there is not test time overhead in each die test.

C. Simulated Annealing

Simulated annealing is an algorithm to find a global minimum value in large search space by stochastic approach. To apply simulated annealing to this bias vector optimization, the algorithm shown in Fig. 6 is adopted. The key point here is to add a "penalty constant" to the consumed power if the circuit fails since if the circuit fails to operate at the desired frequency, the bias vector is "very bad" even if the power is low. With the introduction of the penalty, the search becomes a simple bound-free minimum search.

The difference among there three methods are shown in the following section.

IV. MEASUREMENT RESULTS

In this section, the word "NiPj" denotes that there are i bias domains for NMOS and j bias domains for PMOS as shown in Fig. 7.

First, the relationship between grain size and power consumption is shown in Fig. 8(a). Since the within-die process variation is small and random (see Fig. 9), the same bias is happed to be set to all the bias domains. From the figure, it is clear that the finer the grain size is, the lower the power consumption becomes. N1P8 bias control is shown to reduce 23% of the power consumption compared with as-fabricated chip without post-tuning.

NAND ring oscillation frequency at the lowest power consumption in N1P8 exhaustive search is shown in Fig. 9. From this, it is seen that within-die variations are below 3% while design variations goes up to 10% which is seen from the figure at right hand side.

The measured relationship between number of test iterations and the power reduction in the proposed simulated annealing method for one chip is shown in Fig. 10. Within 20 iterations, more than 19% power reduction is achieved.

Fig. 11 shows the comparison among the three methods for N1P8 parameter set. The power reduction efficiency of one certain chip is compared the methods themselves. Exhautive search of course shows the best result of 29% power reduction but it may not be practical in cases because the test time increase is considerably increased. On the other hand, by using

978-1-4244-2018-6/08 $25.00 © 2008 IEEE

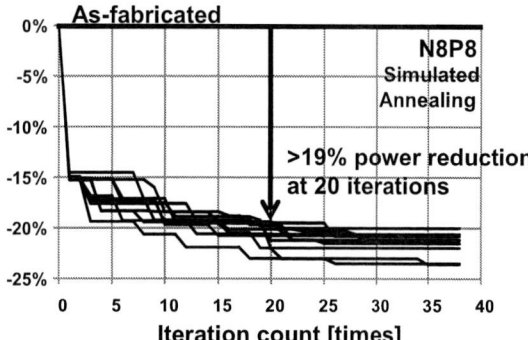

Fig. 10. Measured iteration – power characteristics in simulated annealing

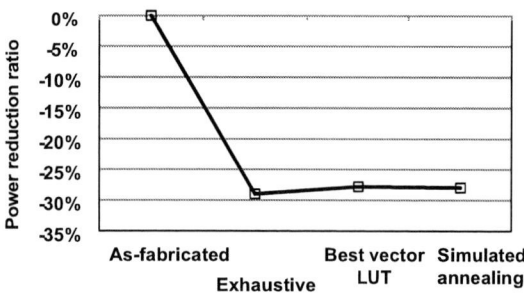

Fig. 11. Comparison among exhaustive search, LUT, and simulated annealing (N1P8)

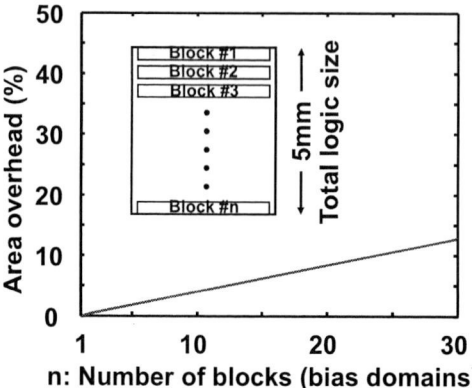

Fig. 12. Number of blocks vs area overhead (calculated)

simulated annealing, the shorter test time is expected and the power reduction ratio stays the same as 28%. As to the best vector LUT approach, the power reduction ratio is 28% and in this look-up table approach, no test time overhead is needed since the time-consuming optimization efforts are made only once in the development stage. In measuring this value, the best vector is not the best vector for the specific chip but it is the best vector for the chip set.

IV. DISCUSSIONS

In order to divide a logic circuit into multiple bias blocks, extra area for well separation is required. Fig. 12 shows the relationship between the number of division and the area overhead caused by this well separation. In the fabricated process, this overhead does not exceed 5% until 12 divisions for 5mm x 5mm chip.

V. CONCLUSION

The fine-grained body bias control to compensate both the random (process) and systematic (design) variations was proposed and the effectiveness was demonstrated with the 90 nm CMOS test chips. The proposed scheme compensates die-to-die V_{TH} variation and the systematic design variations. Undesired inequality of critical path delay among pipeline stages are compensated in this scheme. Compared with as-fabricated chip, proposed global optimization approach reduces the power by 23% and by 11% compared with the

power optimized by the conventional local optimization approach. Also, the proposed global optimization scheme reduced power by 19% compared with as-fabricated chip power within 20 test iterations with simulated annealing algorithm. The best vector LUT approach is also practical. The proposed schemes are considered to be promising for achieving the power-efficient LSI's in scaled devices with reasonable area and test overhead.

ACKNOWLEDGMENTS

This work is partially supported by MEXT and Hitachi, Ltd. The VLSI chips were fabricated through the chip fabrication program of VLSI Design and Education Center (VDEC), the University of Tokyo, with the collaboration by STARC, Fujitsu Limited, Matsushita Electric Industrial Company Limited., NEC Electronics Corporation, Renesas Technology Corporation, and Toshiba Corporation.

REFERENCES

[1] D. Deleganes, J. Douglas, B. Kommandur, and M. Patyra, " Designing a 3GHz, 130nm, Pentium® 4 processor," IEEE Symp. on VLSI Circuits, pp. 130-133, June 2002.
[2] E. Takahashi, Y. Kasai, M. Murakawa, and T. Higuchi, "A post-Silicon clock timing adjustment using genetic algorithms," IEEE Symp. on VLSI Circuits, pp. 13-16, June 2003.
[3] J. Friedrich, B. McCredie, N. James, B. Huott, B. Curran, E. Fluhr, G. Mittal, E. Chan, Y. Chan, D. Plass, S. Chu, H. Le, L. Clark, J. Ripley, S. Taylor, J. Dilullo, and M. Lanzerotti, "Design of the Power6 microprocessor," ISSCC, pp. 96-97, Feb. 2007.
[4] J. Tschanz, J. Kao, S. Narendra, R. Nair, D. Antoniadis, A. Chandrakasan, and V. De, "Adaptive body bias for reducing impacts of die-to-die and within-die parameter variations on microprocessor frequency and leakage," ISSCC, pp. 422-423, Feb. 2002.

978-1-4244-2018-6/08 $25.00 © 2008 IEEE

IEEE 2008 Custom Intergrated Circuits Conference (CICC)

A DC-DC Converter with a Dual VCDL-based ADC and a Self-Calibrated DLL-based Clock Generator for an Energy-Aware EISC Processor

Sunghwa Ok, Jungmoon Kim, Gilwon Yoon, Hyunho Chu,
Jaegeun Oh, Seon Wook Kim, and Chulwoo Kim

Department of Electrical Engineering, Korea University, Seoul, Korea

Abstract—This paper describes a dynamic voltage and frequency scaling (DVFS) scheme for the dynamic power management (DPM) of the extendable instruction set computing processor. The DVFS circuit comprises a digitally-controlled DC-DC buck converter with a dual VCDL-based ADC and a low-power and low-jitter DLL-based clock generator with self-calibration. The prototype is fabricated in a 0.18-μm CMOS process. The implemented DVS circuit provides a supply voltage from 1.4V to 1.8V and the DFS circuit dynamically generates the system clock from 7.5MHz to 120MHz according to the workload of the embedded processor. The DVS and DFS circuits occupy 2.72 mm² and 0.27 mm² active areas, respectively.

I. INTRODUCTION

The extendable instruction set computing (EISC) processor has been used for several portable multimedia applications [1]. They should be realized with a low-power design because the battery has a limited energy. A dynamic voltage and frequency scaling (DVFS) scheme can be applied to provide both V_{DD} and clock that varies with the workload of the EISC processor.

Also, applications with the embedded processor are likely to be exposed to various and severe noisy environments. Among the many noise and variation sources, temperature fluctuations and timing uncertainties can degrade the performance of the processor severely.

Our techniques dramatically reduce the sensitivity to these variations. Some efforts were made to make a DC-DC converter less sensitive to noise and supply variations in [2-3]. Also, the fast tracking response for dynamic voltage scaling (DVS) has been researched with various approaches [4-5]. So as to be compatible with the digital interface of the embedded system, an efficient DVS scheme was achieved by an adaptive digital DC-DC converter with fast response owing to coarse and fine controls [4]. Improvements in this paper are further accomplished through a DC-DC converter using a dual VCDL-based ADC with temperature compensation, which contributes to making the dynamic power management (DPM) tolerate temperature variations better than other converters.

Conventional dynamic frequency scaling (DFS) circuits usually use a PLL for the frequency scaling circuit. However, the PLL-based frequency scaling scheme is not attractive

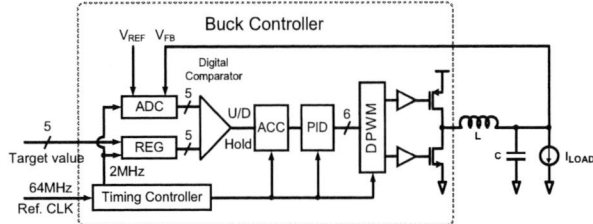

Fig. 1. Digital DC-DC buck converter for DVS.

when noise causes sudden changes in the supply voltage due to the limited bandwidth of the PLL, which causes jitter accumulation. In this work, a DLL-based clock generator that inherits the superior characteristics of a DLL is used to implement DFS. Through the proposed self-calibration technique, the clock generator can provide the multiplied low-jitter clock to the processor to save energy.

II. DVFS CIRCUIT IMPLEMENTATION

A. Digital DC-DC Converter with a dual VCDL-based ADC

To control the adaptive digital DC-DC converter, an ADC is required, which constitutes an overhead. The digital pulse-width modulator (DPWM) is also needed to control the duty cycle of the power FETs. Figure 1 shows the block diagram of the digital DC-DC buck converter for DVS. The main input signals include a 5-bit binary code as the target value for regulating the output voltage, a 64MHz external clock, and the output feedback voltage, V_{FB}. By comparing the feedback voltage with the target value, the digital comparator produces either an *Up*, *Down*, or *Hold* signal. As the timing controller triggers the accumulator, it determines the 6-bit binary values for the DPWM. The duty cycle controlled by the DPWM drives the power FETs. The control loop is assured to be stable by a PID controller [6]. The 1.4V to 1.8V supply voltage is provided to the CPU through a power FET switch and an output inductor.

In order to use an ADC in the DC-DC converter, the ADC must be capable of high speed operation. To detect the instantaneous output voltage, the ADC's operating speed should be ten times higher than the switching frequency. In addition, the power consumption of the ADC must be

978-1-4244-2018-6/08 $25.00 © 2008 IEEE   587

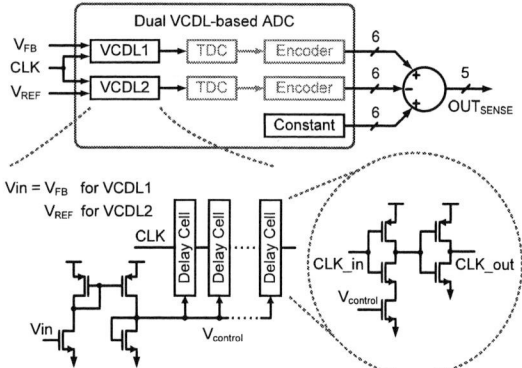

Fig. 2. Proposed temperature-compensated VCDL-based ADC.

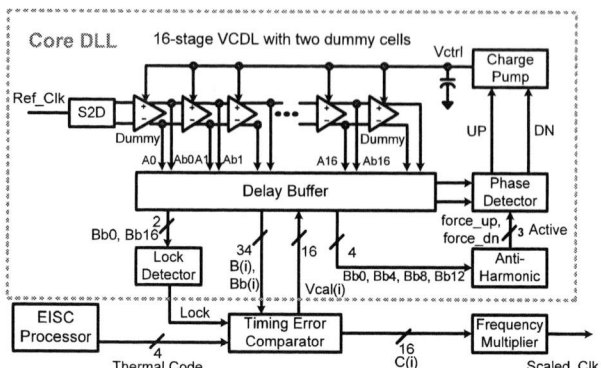

Fig. 3. Proposed self-calibrated DLL-based clock generator.

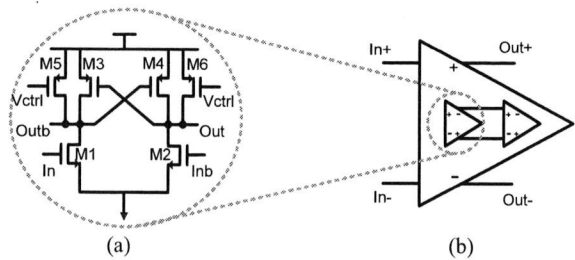

Fig. 4. (a) Differential delay unit cell and (b) delay cell of VCDL.

minimized to below tens of μA so that it can be a part of the controller for the DC-DC converter. The resolution of the ADC does not need to be high. With high resolution ADCs, it would be even more difficult to design a PID circuit. Hence, it is desirable to have an ADC with an input range of 1.4V-1.8V and 5-bit resolution. The optimal ADC with these characteristics for the DC-DC converter has not been reported yet. The sigma-delta ADC has both low-power and good reliability. Its speed, however, is too low to be attractive. The flash ADC has a simple architecture featuring fast operation and low resolution, but its power consumption in the resistor arrays and comparators is so large that it becomes prohibitive for the low-power system.

As mentioned in [7], synthesizable ADCs are good candidates for an optimal ADC in the DC-DC converter. Thus, we propose an ADC for the DC-DC converter with a dual voltage-controlled delay line (VCDL), as shown in Fig. 2. A feedback voltage (V_{FB}) controls the amount of the delay of the VCDL. A time-to-digital converter (TDC) then converts the delay into the thermal codes that constitute the output of the A/D conversion. The power consumption of the ADC is substantially decreased because the conversion occurs only when the output pulse of the VCDL triggers the TDC, and the A/D conversion speed is fast enough because the conversion time is just less than one clock cycle.

To mitigate the effects of temperature variation, a dual VCDL is employed. The feedback output voltage (V_{FB}) is applied to one of the VCDLs, while the reference voltage (V_{REF}) is applied to another, which should be in the input range of the VCDL. The output word (OUT_{SENSE}) can be negative from the subtraction. To solve this problem, a constant value is added. Moreover, the result of the subtraction may have a fixed offset due to the PVT variations. In this regard, the reference voltage level is controlled so that the variation can be compensated.

The ADC using a dual VCDL has an input offset and its INL and DNL may seem improper. However, a fine control is achievable by adjusting the reference voltage (V_{REF}) because the ADC can obviously detect the difference of voltages and there is no hysteresis that affects the ADC's operation. The conversion time is simply equal to the amount of delay that occurs when the clock passes through the VCDL. The total

delay of the VCDL is less than 24ns. In other words, the operating speed of the ADC can be higher than 40MHz. The power consumed by a conversion depends on the input voltage. The higher the reference, the greater the power consumption, most of which dissipates in the VCDLs and flip-flops.

B. Self-calibrated DLL-based Clock Generator

The DLL-based clock generator is proposed to provide a clean clock signal to the processor. It takes a 4-bit thermal code from the EISC processor and generates the multiplied clock signal for DFS. By selecting each multi-phase signal from the delay buffer according to the thermal code, the frequency multiplication ratio is determined [8]. The entire architecture of the proposed clock generator is depicted in Fig. 3.

The 16-stage VCDL uses a differential clock as an input signal from the single-to-differential converter (S2D). Compared to the conventional single-ended delay cell with current-starved inverters, differential delay cells guarantee that the clock generator will have better immunity to supply and substrate noises which could possibly cause delay mismatches between the delay cells. Each delay cell consists of two delay unit cells as shown in Fig. 4 (a). The cross-coupled transistors M3 and M4 operate as a latch to ensure that a fully differential clock appears at every delay cell's output. M5 and M6 control the delay of the delay cell by adjusting the drain currents. This type of delay unit cell achieves full swing output and does not need a tail current source. Hence, it results in an area reduction and saves power by removing a level conversion circuit and a bias circuit [9]. The VCDL generates 16 pairs of evenly spaced multi-phase

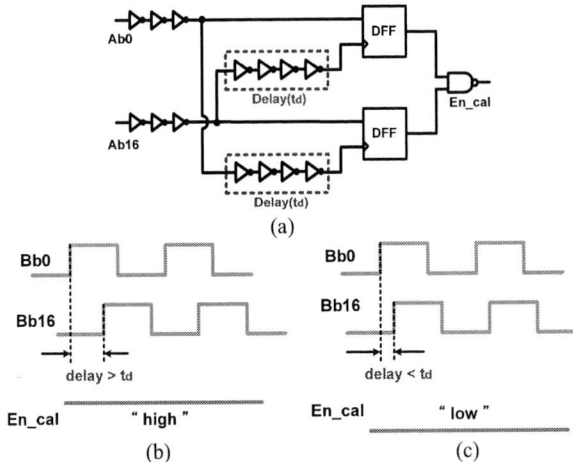

(a)

(b) (c)

Fig. 5. (a) Proposed lock detector and (b), (c) its timing diagram.

clock signals for frequency multiplication. Also, dummy delay cells are inserted in each terminal of the VCDL for load balancing and transistor matching.

The proposed lock detector is shown in Fig. 5. It takes two signals (Bb0 and Bb16) from the delay buffer. If the delay between the two signals is smaller than the predetermined delay (td), then the lock detector generates the *Lock* signal and it starts both the frequency multiplier and the calibration blocks. Because the frequency multiplier and calibration blocks operate only after the core DLL locks, the initial error and power consumption can be reduced.

An anti-harmonic block detects four signals (Bb0, Bb4, Bb8, and Bb12) from the delay buffer and creates three output signals that indicate the lock state. It forces either the *UP* or *DN* signal of the phase detector (PD) to be high when the output rising edge of the last delay cell is outside of the correct locking range. Only the *Active* signal is high when the output of the VCDL is within the proper frequency range [10] and it allows the PD to conduct normal operation. The anti-harmonic block enables the clock generator to operate over a wide frequency range while avoiding harmonic lock, which is undesirable for accurate frequency multiplication.

The proposed self-calibration block consists of the delay buffer and the timing error comparator. The delay buffer consists of 16 delay buffer cells. Each delay buffer cell consists of a 2-stage VCDL with small delay coverage. Figure 6 shows the timing diagram of the delay buffer output clocks and a schematic of the timing error comparator. The timing error comparator compares the relative pulse widths of *Cal_dn* and *Cal_up* and charges or discharges the MOS capacitor according to the difference in the pulse widths. The control voltage Vcal(i+1) feeds back to the delay buffer to adjust the delay of B(i+1) in order to place B(i+1) right in the middle of the two neighbor signals.

By doing so, the proposed self-calibration scheme can reduce the delay mismatch without affecting the core DLL loop behavior unlike in [11]. In addition, since the proposed scheme uses an analog calibration method instead of a digital method, the circuit does not suffer from the finite digital

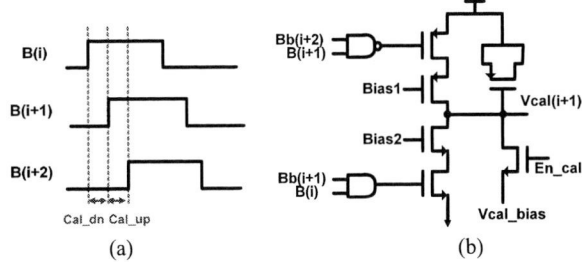

(a) (b)

Fig. 6. (a) Timing diagram of delay buffer output and (b) proposed timing error comparator.

Fig. 7. Chip micrograph.

quantization error. Because of these advantages over the conventional calibration methods, the fixed pattern jitter and spur can be effectively reduced.

III. MEASUREMENT RESULTS

The circuit is implemented in a 0.18-μm CMOS process. Figure 7 shows the chip photo of the DVFS circuit with the embedded EISC processor. The fabricated digital DC-DC converter and DLL-based clock generator occupy active areas of 2.72 mm² and 0.27 mm², respectively.

The temperature characteristics of the ADC with a dual VCDL in DC-DC converter are shown in Fig. 8 by testing at various temperatures (5℃, 45℃, and 85℃). Little deviation is observed in the worst case condition. When the ADC circuits are operating at 2MHz, about 230μW of power is dissipated when the reference voltage is 1.8V and the temperature is 45℃.

Figure 9 shows the frequency multiplication with the reference frequency. The measurement result verifies that the output clock frequency of the clock generator is scaled along with the digital thermal code. Figure 10 shows the jitter performance of the DFS circuit without and with the proposed self-calibration. With the proposed self-calibration, the rms jitter of the output clock is reduced from 11.1 ps to 9.7 ps at 120MHz, which shows a considerable reduction of the jitter.

Figure 11 shows the normalized power consumption of the EISC processor in several DVFS modes. The result confirms that the proposed DVFS circuit can effectively save the battery energy by reducing the dynamic power consumption of the EISC processor.

978-1-4244-2018-6/08 $25.00 © 2008 IEEE

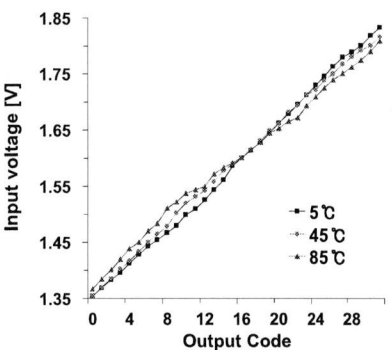

Fig. 8. Temperature variation insensitivity of the ADC.

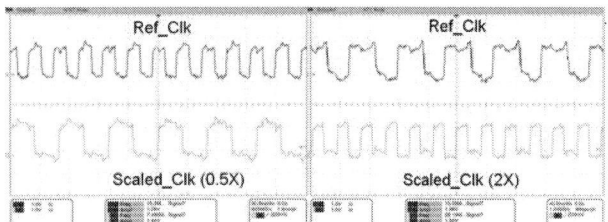

Fig. 9. Dynamic frequency scaling with the reference clock.

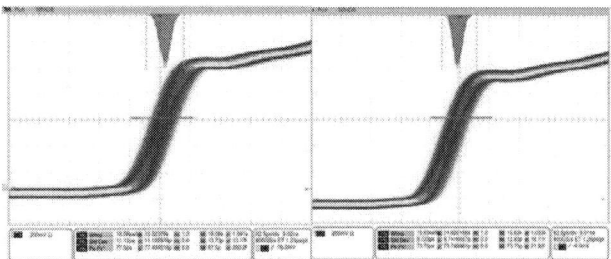

Fig. 10. Jitter characteristics without and with the self-calibration.

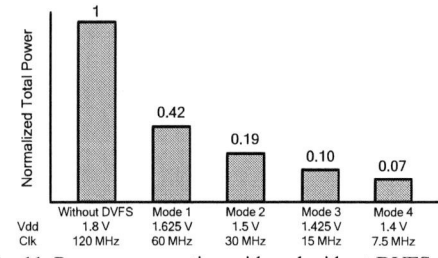

Fig. 11. Power consumption with and without DVFS.

Table I Performance summary

Process	0.18-µm CMOS
Digital DC-DC buck converter	
Input / Output Voltage	3.3 V / 1.4 ~ 1.8 V with 12.5 mV/step
Switching frequency	1 MHz
Max. output current	400 mA
Maximum efficiency	91 %
Tracking time	42 µs
Voltage ripple	4.8 mV
Active area	0.8 mm x 3.4 mm
DLL-based clock generator	
Output clock frequency	7.5, 15, 30, 60, and 120 MHz
RMS jitter	9.7 ps @ 120 MHz
Peak-to-Peak jitter	73.7 ps @ 120 MHz
Power consumption	28 mW @ 120 MHz, 1.8 V_{DD}
Active area	0.51 mm x 0.53 mm

Table I summarizes the performance of the proposed DVFS circuit.

IV. CONCLUSION

A digitally-controlled DC-DC buck converter with a dual VCDL-based ADC and a low-power/low-jitter DLL-based clock generator with self-calibration for the DPM of the embedded processor have been presented.

The DC-DC buck converter provides an adaptive supply voltage from 1.4V to 1.8V with a resolution of 12.5mV/step. The DC-DC converter achieves temperature tolerance through the proposed dual VCDL-based ADC.

The DLL-based clock generator can produce five levels of multiplied frequencies from 7.5MHz to 120MHz with a reference frequency of 15MHz. The analog self-calibration with the delay buffer can noticeably mitigate the delay mismatch, leading to excellent jitter performance.

The proposed DVFS circuit can provide a clean supply voltage and a system clock to the EISC processor even in the presence of severe variation sources, while achieving the energy-efficient DPM.

ACKNOWLEDGMENT

This work was supported by the Korea Science and Engineering Foundation (KOSEF) grant funded by the Korea government (MOST) (No. R0A-2007-000-20059-0) and IDEC.

REFERENCES

[1] Alex K. Jones et al., "An automated, reconfigurable, low-power RFID tag," in *Design Automation Conf.,* in *Proc.* Jul. 2006, pp. 131–136.

[2] Masakatsu Nakai et al., "Dynamic voltage and frequency management for a low-power embedded microprocessor," *IEEE J. Solid-State Circuits*, vol. 40, no. 1, pp. 28-35, Jan. 2005.

[3] Marnie Wong et al., "A low noise buck converter with a fully integrated continuous time ΣΔ modulated feedback controller," in *IEEE Custom Integrated Circuits Conf.*, Sep. 2007, pp. 377-380.

[4] Janghoon Song et al., "An efficient adaptive digital DC-DC converter with dual loop controls for fast dynamic voltage scaling," in *IEEE Custom Integrated Circuits Conf.*, Sep. 2006, pp. 253-256.

[5] Kohei Onizuka et al., "V_{DD}-hopping accelerators for on-chip power supply circuit to achieve nanosecond-order transient time," *IEEE J. Solid-State Circuits*, vol. 41, no. 11, pp. 2382-2389, Nov. 2006.

[6] Gu-Yeon Wei et al., "A fully digital, energy-efficient, adaptive power-supply regulator," *IEEE J. Solid-State Circuits*, vol. 34, no. 4, pp. 520-528, Apr. 1999.

[7] Jinwen Xiao et al., "A 4 µA quiescent-current dual-mode digitally controlled buck converter IC for cellular phone applications," *IEEE J. Solid-State Circuits*, vol. 39, no. 12, pp. 2342-2348, Dec. 2004.

[8] Jin-Han Kim et al., "A 120-MHz–1.8-GHz CMOS DLL-based clock generator for dynamic frequency scaling," *IEEE J. Solid-State Circuits*, vol. 41, no. 9, pp. 2077-2082, Sep. 2006.

[9] Joonsuk Lee et al., "A low-noise fast-lock phase-locked loop with adaptive bandwidth control," *IEEE J. Solid-State Circuits*, vol. 35, no. 8, pp. 1137-1145, Aug. 2000.

[10] Kyunghoon Chung et al., "An anti-harmonic, programmable DLL-based frequency multiplier for dynamic frequency scaling," in *Proc. IEEE Asian Solid-State Circuits Conf.*, Nov. 2007, pp. 276-279.

[11] Chan-Hong Park et al., "A 1.8-GHz self-calibrated phase-locked loop with precise I/Q matching," *IEEE J. Solid-State Circuits*, vol. 36, no. 5, pp. 777-783, May. 2001.

[12] Hsiang-Hui Chang et al., "A 0.7–2-GHz self-calibrated multiphase delay-locked loop," *IEEE J. Solid-State Circuits*, vol. 41, no. 5, pp. 1051-1061, May. 2006.

IEEE 2008 Custom Intergrated Circuits Conference (CICC)

Active Autonomous AC-DC Converter for Piezoelectric Energy Scavenging Systems

E. Dallago, D. Miatton, G. Venchi
Department of Electrical Engineering
University of Pavia, 27100 Pavia, Italy

V. Bottarel, G. Frattini*, G. Ricotti, M. Schipani
STMicroelectronics
20010 Cornaredo, Milan, Italy

Abstract- **The paper focuses on an electronic interface which can be used into Piezoelectric Energy Scavenging Systems (PESS). These systems convert the energy of mechanical vibrations into electrical energy using a piezoelectric transducer to realize a power supply for low power electronic systems. To obtain a suitable supply source an AC-DC conversion of the output signal of these transducers is needed and, since the output power level of the energy scavenger can be very low, the conversion should be as efficient as possible.**
This paper shows an active voltage doubler AC-DC converter for PESS. A novel driving circuitry topology is presented; it has the advantage to be tolerant with respect to the process variations. The converter uses exclusively a fraction of the harvested energy to supply itself and a bias circuit has been designed to make the total current consumption supply independent.
A test chip was diffused in 5V CMOS STMicroelectronics technology. Experimental results show the effectiveness of this solution and efficiencies higher than 90% have been obtained for different load values.

I. INTRODUCTION

In recent years a lot of studies focused on energy scavenging systems which are used to harvest the normally lost environmental energy and to convert it into electrical energy. This solution can be attractive where batteries are a bottleneck for the whole system (i.e. they have a finite life time and their replacement or recharge is not feasible or too expensive). An energy scavenging system, instead, is a theoretically endless energy source. Energy scavenging systems could supersede batteries thanks to two driving forces: 1) reduction of the power consumption of the supplied electronic system, 2) optimization of the harvesting system. This last point can be further subdivided into optimization of the energy transducer and optimization of the electronic interface which has to manage and store the harvested energy.
In literature many papers can be found which describe methodologies to realize the energy-scavenger [1], [6], [8], [10]. A lot of these works are focused on the conversion of the energy associated to mechanical vibrations since they can be easily found in many environments [1], [7].
This paper describes a system based on a piezoelectric transducer since it is one of the more efficient which can be used [1]-[2].
The electrical energy at the output of this transducer is a

strong and irregular function of time [1]-[4], [9], hence, to realize a DC supply source, an AC-DC conversion is needed. The paper focuses on the optimization of the electronic interface. In particular an active voltage doubler AC-DC converter, which uses only a fraction of the harvested energy to supply its active devices, is presented.
A possible way to realize the driving circuitry of an active voltage doubler is to use comparators [3]-[4], [9] which sense the voltage drop across the switches and drive them. In this paper a different solution is proposed where comparators are replaced with operational amplifiers. A capacitance of 1 µF is used to store the harvested energy which is partially used to supply the active circuitry of the converter so that no external power source is required. Since the voltage across the storage capacitance is variable the converter is also enriched with a circuitry which, differently from [4], makes the bias current supply independent.
A test chip was diffused in 5V CMOS STMicroelectronics technology. Experimental results show the effectiveness of the AC-DC converter in terms of its efficiency, which is higher than 90%.

II. PIEZOELECTRIC ENERGY SCAVENGING SYSTEM

A Mechanical Aspects

The considered piezoelectric transducer is a cantilever which works in 31-mode when it is excited by the mechanical vibrations. To have a maximally efficient conversion of the mechanical vibrations into electrical energy the cantilever should be excited at its resonant frequency which can be varied adding a mass on its free end [1]. This allows the energy transducer to be tuned with the vibrations which are present in a specific environment. Some experimental measurements [1] showed that the frequency range of mechanical vibrations existing into civil environments is approximately (60-380) Hz.
The piezoelectric transducer can be modeled at resonance by the equivalent circuit [1]-[2] shown in Fig. 1. V_{PO} is a sinusoidal voltage source whose frequency is equal to the transducer resonance frequency and whose amplitude is equal to the open circuit output voltage, while C_P is the electrical capacitance of the piezoelectric cantilever.
The results presented in the paper are based on a piezoelectric cantilever whose geometrical dimensions are 30x15x0.1 mm (LxWxH). The piezoelectric material used is a soft Lead Titanate Zirconate (PZT). In this case C_P is equal to 36 nF, according to experimental measurements.

---

* Now he is with National Semiconductor, Strada 7, R3, 20089, Rozzano (Mi), Italy

978-1-4244-2018-6/08 $25.00 © 2008 IEEE

B. Front End Circuitry

Fig. 1 shows the active topology of the proposed AC-DC converter: switches $M_{SW}1$ and $M_{SW}2$ are realized with p-channel and n-channel MOSFET respectively while the driving circuitry is composed of operational amplifiers OP1 and OP2, and by the biasing circuitry. The DC sources indicated in Fig. 1 as V_{OS} are the equivalent input offset voltages of the operational amplifiers. The energy stored into capacitance C_S is also used to supply this circuitry. Since the voltage across the storage capacitance is variable, a supply independent bias circuitry is used to prevent a non linear power consumption of the driving circuitry.

During the start-up the energy stored into C_S is not enough to supply the active devices. In this case the operation of the converter is guaranteed by the body-source diode of MOSFETs $M_{SW}1$ and $M_{SW}2$. Consequently the proposed rectifier can be seen as a parallel of two AC-DC converters: an active one with an higher efficiency and a passive one with a lower efficiency, the latter working only during start-up. The performances of the proposed solution have been simulated with a resistive load R_L. Since the effectiveness of the energy transfer from the piezoelectric transducer to the load depends on the value of the load resistance, it was varied between 50 kΩ to 1 MΩ seeking the optimum value.

To introduce the working principle let us consider Fig. 2 which shows the case of OP1 (similar considerations apply to OP2). It is possible to apply Kirchhoff Voltage Law to the external mesh:

$$V_{IN} + V_{SD} - V_{OS} = 0 \qquad (1)$$

If the operational amplifier has a DC gain equal to A the voltage on the gate of the $M_{SW}1$ is:

$$V_G = A \cdot V_{IN} \qquad (2)$$

Equations (1) and (2) can be solved as a function of V_G:

$$V_G = A \cdot (V_{OS} - V_{SD}) \qquad (3)$$

In the ideal case the DC gain A of the operational amplifier is infinite; as a consequence the difference in equation (3) has to vanish in order to have a finite value of the gate voltage. In this way, a regulation loop is obtained which modulates the gate voltage V_G to keep the source to drain voltage of the MOSFET equal to offset voltage for each value of the drain current.

Fig. 3 gives a graphical representation of this working principle: in the ideal case when drain current is positive the regulation loop sets the working point of the MOSFET at the intersection between its characteristics and the offset voltage. As the current decreases, the loop moves the working point of the MOSFET at lower values of its source to gate voltage, until the current is equal to zero. At this point the regulation loop turns the MOSFET off: since there is no intersection between the MOSFET characteristics and the offset voltage for negative values of the current, they are not allowed and regulation loop holds the transistor off. This principle guarantees that no oscillations of the driving signal will take place.

Furthermore, being a negative current impeded, any unwanted discharge of both C_P and C_S is prevented. Symmetrical considerations apply to OP2. The value of the offset is not critical because it has simply to be far enough from zero so that the process mismatches will not change its sign. While it is true that an higher offset gives higher losses on the switch, they are still negligible for practical offset values. This makes the circuit quite tolerant to process mismatches.

The real operational amplifier has a finite DC gain; this means that the value of the voltage across the MOSFET is slightly different from the theoretical one. Nevertheless it can be demonstrated that, with the previously described choice of the offset, this will not affect the operation of the circuit.

III. DESIGN OF THE AC-DC CONVERTER

A. Supply Independent Bias Circuitry

Since the driving circuitry should be supplied only by the harvested energy its power dissipation should be as low as possible, in particular the proposed circuit was designed to have an average current consumption of about 500nA.

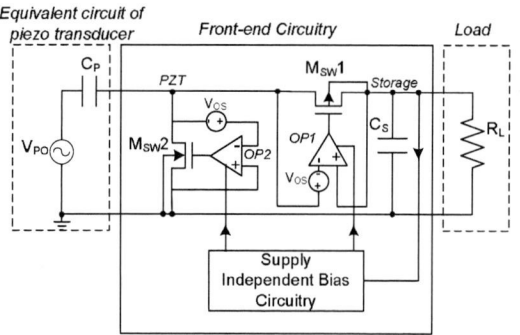

Fig. 1 Schematic of the proposed ESS. Equivalent circuit of piezoelectric transducer when it is excited at its resonant frequency, front-end circuitry and load can be identified.

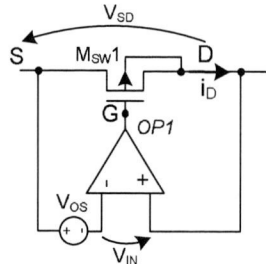

Fig. 2 Regulation loop composed by the operational amplifier OP1 and MOSFET MSW1.

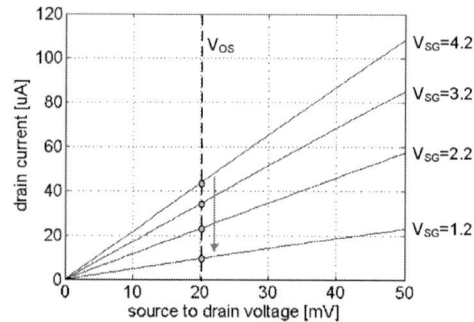

Fig. 3 Graphical solution of the regulation loop composed by the operational amplifier and MOSFET $M_{SW}1/2$.

A dedicated circuitry was added to keep this current consumption constant as the supply voltage increases, while it would have naturally increased as well: as a consequence the power consumption depends on the supply voltage in a linear way. The proposed scheme, shown in Fig. 4, is modified with respect to [5] by adding diode connected MOSFET (M4). The system, in fact, requires supply currents in the range of tens of nA: in the scheme without M4, resistance R1 would have been in the order of tens of MΩ, which is too area expensive for an integrated solution. The effect of M4 is to reduce the voltage drop across R1 and this lowers its value for a given current I1. Furthermore, a start up function is needed, which was obtained with dummy MOSFETs ML1 and ML2. In particular, the leakage of their body-drain diode was exploited: its effect is to inject a current into node A and B so to have the start-up aid. This solution allows us to avoid additional start-up circuitry, reducing total power consumption.

Simulations shows that the circuitry starts to regulate when the supply voltage is higher than about 680 mV. Above this value the voltages V_{BiasP} and V_{BiasN} can be used to mirror a supply independent current.

The total current consumption of the bias circuitry is about 100nA.

B. Operational Amplifiers

The schematics of the proposed operational amplifiers are shown in Fig. 5a. Because of the level of their input voltage ranges, OP1 and OP2 have to be supply compatible and ground compatible respectively.

The bias of the operational amplifiers is given by the supply independent bias circuitry. The operational amplifiers were designed so that they are able to work with the minimum possible value of the supply voltage. In this way the active rectifier takes over the passive one as soon as a very low voltage has been stored into capacitance C_S.

A 5pF capacitance was introduced to compensate the regulation loop.

The offset voltage was obtained mismatching the aspect ratio of the input MOSFETs M_A and M_B: the values of the target offset voltages are equal to 26mV and 21mV for OP1 and OP2 respectively. Fig. 7 presents the simulation results of 500 MonteCarlo iterations, showing the possible spread of these voltages. It is possible to see that this spread is sufficiently small to have the correct sign also in case of process mismatches.

The aspect ratio of the MOSFETs $M_{SW}1$ and $M_{SW}2$ has to be chosen so to avoid the loop saturation. The expression of the drain current when the MOSFET is turned on is, in a first approximation, is equal to:

$$i_D = k\frac{W}{L}(V_{GS}-V_{th})V_{DS} \qquad (4)$$

where V_{th} is the threshold voltage of the MOSFET, W/L is its aspect ratio and k is its characteristic constant.

The term into the parenthesis is the overdrive voltage: its value is modulated by the regulation loop which, for a each drain current, varies the gate voltage. If the overdrive voltage is enough, the source to drain voltage is equal to V_{OS}, that is:

$$i_D = k\frac{W}{L}(V_{GS}-V_{th})V_{OS} \qquad (5)$$

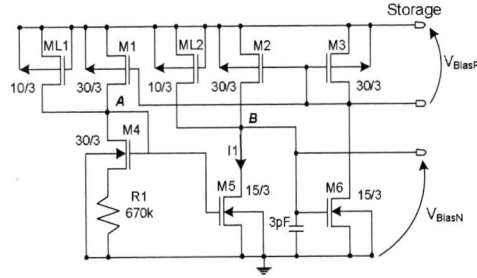

Fig. 4 Schematic of the proposed supply independent bias circuitry.

On the other hand if, for a given current, the overdrive is not enough (which means that the op-amp is saturated), the regulation loop does not work and the source to drain voltage of $M_{SW}1$ or $M_{SW}2$ is not equal to offset voltage. The aspect ratio has to be designed to accommodate the maximum expected current.

Fig. 5b shows a picture of the diffused AC-DC converter.

Tab. 1 resumes the characteristics obtained for the amplifiers: in particular the current consumption is about 200 nA for each amplifier. V_{dd-min} is the minimum supply voltage required by the operational amplifiers to work.

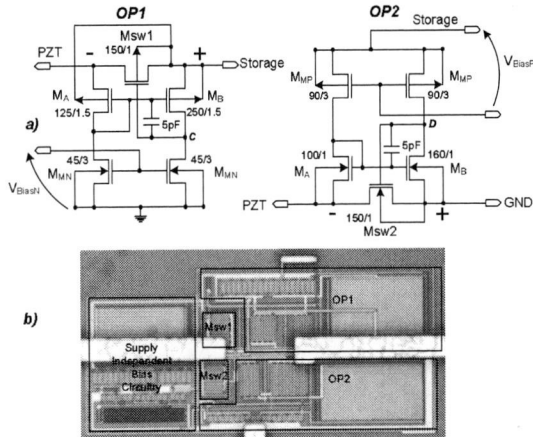

Fig. 5 a)Schematic of the proposed operational amplifiers; b) picture of the diffused AC-DC converter.

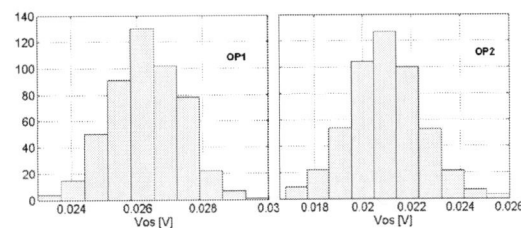

Fig. 6 MonteCarlo simulation of the offset voltages of the designed operational amplifiers.

TABLE I.   CHARACTERISTICHS OF THE DESIGNED OPERATIONAL AMPLIFIERS

	GAIN-DC [dB]	Band-Width [Hz]	Vdd-min [V]	Voffset [V]	Current Consumption [nA]
OP1	50.37	2500	0.73	26e-3	200
OP2	49.78	2610	0.675	21e-3	200

978-1-4244-2018-6/08 $25.00 © 2008 IEEE

IV. EXPERIMENTAL RESULTS

An experimental characterization of the supply independent bias circuit was performed using a stand alone version of the circuit depicted in Fig. 7b: three MOSFETs Ma, Mb and Mc were added to amplify the bias current by a factor 30: this current is then converted into a voltage drop by means of an external resistance, equal to 2 MΩ. Experimental characterization was done suppling the bias circuit with a triangular waveform in the range 0-4.8 V. Fig. 7a shows that, as soon as the supply voltage is higher than 680 mV, the circuit generates a constant voltage drop across the 2 MΩ resistance which corresponds to a constant bias current, as it was predicted by the simulations.

Fig. 8 shows the efficiency of the AC-DC converter as a function of the value of the load resistance R_L. These results were obtained exciting the piezoelectric transducer with a B&K 8410 shaker at a frequency equal to 200 Hz. The amplitude of the vibrations has been regulated so to have a peak to peak output voltage of the transducer, at no load condition, equal to 3 V. It is possible to see that experimental values are similar to the simulated ones: a measured efficiency higher than 90% has been obtained.

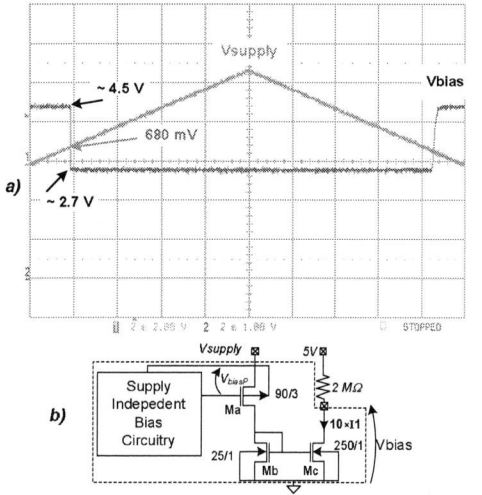

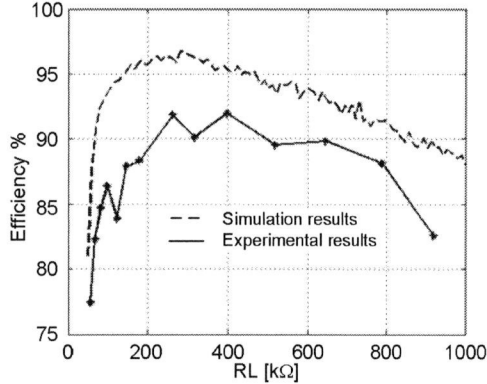

Fig. 7 a) Experimental results of the supply independent bias circuitry; b) circuitry used to do the characterization.

Fig. 8 Efficiency of the proposed AC-DC converter.

V. CONCLUSIONS

This paper presents a piezoelectric energy scavenging system where the AC-DC conversion is realized with an active voltage doubler which uses exclusively a fraction of the harvested energy to supply itself.

A novel solution was presented where comparators have been replaced with operational amplifiers. They sense the voltage across the switching MOSFETs and drive their gate terminals. A regulation loop fixes the drop voltage across the switch at a value equal to the equivalent input offset voltage of the operational amplifier itself. This solution is tolerant to the process variations since the obtained offset value is not a critical parameter, only its sign is important.

Furthermore a supply independent bias circuitry was implemented in order to make the current consumption of the driving circuitry independent on the supply voltage. Average current consumption of the whole active elements is about 500nA.

A test chip has been diffused in CMOS 5V STMicroelectronics technology and experimental results are presented.

An efficiency of about 90% was obtained in a wide range of load values.

VI. References

[1] S. Roundy, "Energy Scavenging for Wireless Sensor Nodes with a focus on Vibration to Electricity Conversion", PhD Thesis, The University of California Berkeley, Spring 2003.

[2] S. Roundy et al., "Improving Power Output for Vibration-Based Energy Scavengers", Pervasive Computing January-March 2005, pp. 28-36, Published by the IEEE and IEEE ComSoc.

[3] T. T. Le, Jifeng Han, A. von Jouanne, K. Mayaram, T. S. Fiez, "Piezoelectric Micro-Power Generation Interface Circuits" IEEE Journal of Solid State Circuits, pp. 1411-1420, Vol. 41, NO. 6, June 2006.

[4] E. Dallago, G. Frattini, D. Miatton, G. Ricotti, G. Venchi, "Self-Supplied Integrable High Efficiency AC-DC Converter for Piezoelectric Energy Scavenging Systems", International Sympoium on Circuits and Systems ISCAS 2007, New Orleans LO, 27-30 May 2007, pp. 1633-1636.

[5] Z. Dong, P. E. Allen, "Low-Voltage, Supply Independent CMOS Bias Circuit", The 2002 45th Midwest Symposium on Circuits and Systems, 4-7 Aug 2002, pp. 568-570, Vol. 3.

[6] F. Peano, T. Tambosso, "Design and Optimization of a MEMS Electret-Based Capacitive Energy-Scavenger", Journal of Microelectromechanical Systems, pp. 429-435Vol. 14, NO. 3, June 2005.

[7] M. Renaud, T. Sterken, P. Fiorini, R. Puers, K. Baert, C. van Hoof, "Scavenging Energy from Human Body, Design of a Piezoelectric Transducer", The 13th International Conference on Solid-State Sensors, Actuators and Mycrosystems, Seoul, Korea, June 5-9, 2005, pp. 784-787.

[8] I. Stark, "Thermal Energy Harvesting with Thermo Life®", Proceedings of the International Workshop on Wearable and Implantable Body Sensor Networks, BSN 2006, pp. 19-22, IEEE Computer Society.

[9] J. Han, A. von Jouanne, T. Le, K. Mayaram, T. S. Fiez, "Novel power conditioning Circuits for Piezoelectric Micro Power Generators", APEC 2004, , February 2004, Vol. 3, pp. 1541-1546.

IEEE 2008 Custom Intergrated Circuits Conference (CICC)

A 10Gb/s Receiver with Linear Backplane Equalization and Mixer-Based Self-Aligned CDR

S. Erba[1], M. Pozzoni[1], M. Pisati[1], R. Brama[3], D. Sanzogni[1],
E. Depaoli[1], P. Viola[1], F. Svelto[2]

[1]STMicrolectronics, Pavia, Italy.
[2]Universitá degli Studi di Pavia, Pavia, Italy.
[3]Universitá di Modena e Reggio Emilia, Reggio Emilia, Italy.

Abstract – A 65nm CMOS receiver including a tapered chain linear equalization and a mixer based clock recovery circuit capable of SSC tracking is presented. The proposed architecture works up to 10 Gb/s with transmission channels with more than 20dB loss at Nyquist, while consuming 110mA and occupying 0.25mm².

I. INTRODUCTION

In the last years, serial communication has faced new challenges, due to the increase of the transmission speed while assuring a reliable interconnection through lossy cables and backplanes. In this environment, one of the main impairment is represented by the inter symbol interference (ISI), mainly caused by the limited bandwidth of the interconnecting channel. Two common techniques are widely used to overcome the ISI limitation: linear equalization, and decision feedback equalization (DFE) [1].

Despite the advantages of DFE in presence of transmission channels with severe losses and significant crosstalk interferers, performing the timing recovery over DFE equalized data requires various impairments to be overcome.

In case of direct feedback DFE topology, the loop delay has to be less than 1/2 bit period [2], whereas in case of loop unrolling the CDR front-end complexity increases with the number of DFE thresholds [3].

On the other hand, for moderate channel losses and crosstalk interference, a receiver topology based on simplified equalization and clock recovery, as depicted in Fig. 1, represents an effective solution for low power high frequency links operation.

This work presents a 10Gb/s receiver based on linear equalization and on a novel digital CDR technique, minimizing the hardware requirement while being capable of high jitter tolerance performances for a reliable serial interconnection.

II. CLOCK RECOVERY TECHNIQUE

To date high speed serial interfaces are communication systems employing non-return-to-zero modulation (NRZ) to maximize the transmission data throughput.

As shown in Fig. 2, no power at the transmission rate is present in the spectrum of a NRZ coded random data stream.

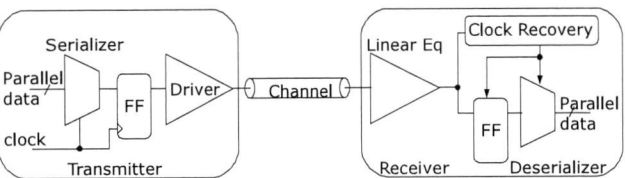

Fig. 1. Link employing linear equalization at the receiver.

This prevents the use of a phase locked loop (PLL) to recover the signal clock. For this reason, clock recoverers based on spectral line methods can be employed. They apply a second order non-linearity to generate a data-rate tone, thus allowing the usage of a PLL to recover the clock. A possible implementation of a spectral line clock recovery is represented in the block diagram of Fig. 3, where the PLL phase detection is accomplished by a mixer.

In this architecture the recovered clock is not necessarily aligned to the incoming data and a delay in the sampling path is required to match the intrinsic delay of the clock recovery loop.

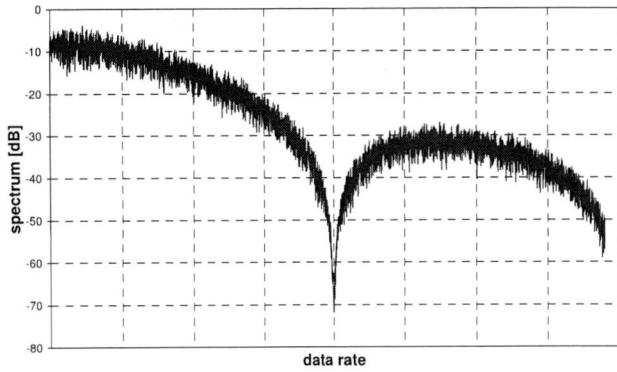

Fig. 2. NRZ coded random data spectrum.

This limitation of spectral line timing has historically limited application to low frequency CDRs, while early/late solutions have been commonly preferred in high speed applications, for their self-aligned nature.

978-1-4244-2018-6/08 $25.00 © 2008 IEEE

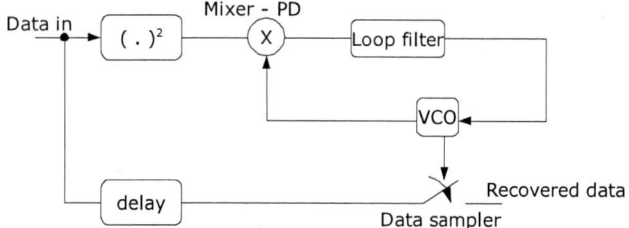

Fig. 3. Timing loop based on the spectral line method.

Early-late techniques usually sample the incoming data in the center and in the edge, and the samples are de-multiplexed to a digital clock recovery engine, implementing the phase detection and the adequate early-late filter [4], as indicated in Fig. 4. On the other hand, de-multiplexing of the sampled data required by the early-late technique introduces a serious hardware complexity over the spectral line, increasing the overall latency of the loop and the pattern dependent noise injection into the analog supply.

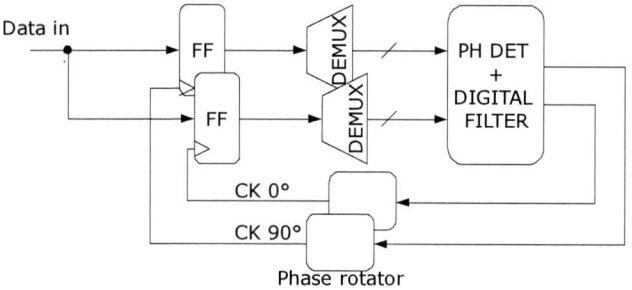

Fig. 4. Early/late digital CDR.

For this reason, as discussed in the following sections, a specific spectral line based CDR has been developed simplifying the hardware implementation, when compared to alternative early-late solutions, while preserving the clock-data self alignment.

III. IMPLEMENTED ARCHITECTURE

Considering the timing loop of Fig. 3, the output of the phase detector mixer is given by:

$$PD_{out}(\phi) = s(t)^2 \cdot \sin\left[\omega_{VCO}t + \phi\right] \qquad (1)$$

where $s(t)$ represents the random incoming data and ϕ is the phase difference between data and clock.

Because the spectrum of $s(t)^2$ presents a tone at the data rate, the desired phase information is down-converted to baseband.

The normalized average of PD_{out} is plotted in Fig. 5 as a function of the phase difference ϕ.

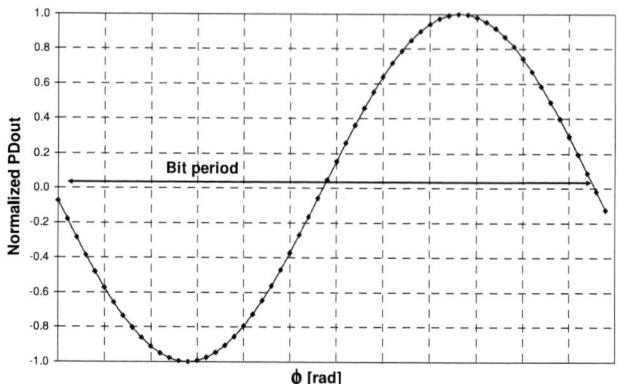

Fig. 5. Phase detector characteristic function.

The phase detector characteristic function is odd and one unit interval (UI) periodic. Its sign, therefore, carries the timing information required by the timing loop to lock to the data phase.

The block diagram of the proposed CDR is shown in Fig. 6. To minimize the power consumption, the architecture employs two I/Q half rate clocks, achieving the same phase detection characteristic of the full rate implementation of Fig. 3.

This signal is accumulated over 8 bit symbols (12 for 10Gb/s) and its sign is sent as a binary phase information to a digital clock recovery implementing both the loop filter, by a proportional–integral (PI) controller, and the clock phase update.

The CDR loop is closed by two phase rotators, synthesizing programmable phases from two I/Q half rate clocks.

The single-bit interface between the accumulator and the digital PI controller represents a significant hardware simplification with respect to the conventional early-late topology of Fig. 4, where de-multiplexing of data samples at the center and at the transitions is required.

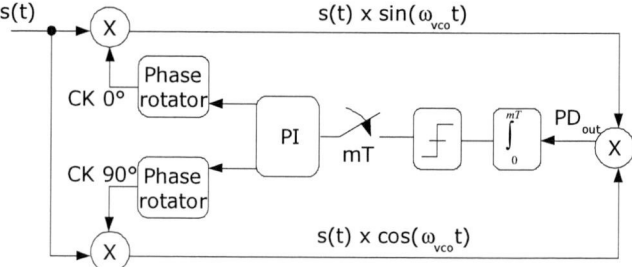

Fig. 6. Block diagram of the implemented clock recovery.

Fig. 7 shows the timing diagram of the main signals in lock condition. In case of a phase error between *CK90°* and *s(t)*, the average of the signal PD_{out} changes its sign, thus supplying the phase information to the PI controller.

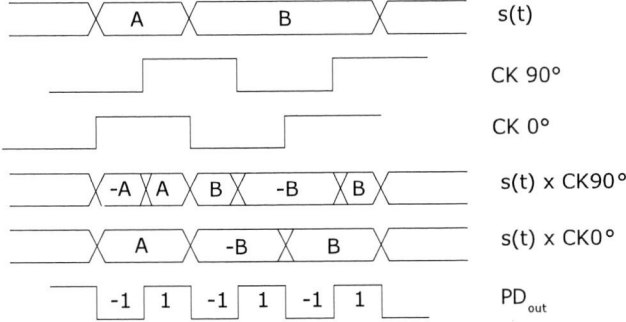

Fig. 7. Lock condition timing diagram.

The above time diagram shows that self-alignment can be assured, if rising (or falling) edges of the clock $CK90°$ are used to sample the data in the data recovery path.

As shown in the next section, custom techniques have been applied to match the CDR mixer and the data sampler clock delays, thus allowing a self-aligned implementation.

IV. CIRCUIT DESCRIPTION

The described clock recovery has been integrated in a complete receiver operating from 6 to 10Gb/s.

As detailed in Fig. 8, the half rate data recovery is driven by a Programmable Gain Amplifier (PGA) and a Feed Forward Equalizer (FFE).

The recovered data are de-multiplexed by a factor of ten and sent to an internal PRBS checker. The PGA allows the linear operation of the equalizer, regulating the internal data swings to 0.6Vppd, from an input data swing between 0.4V and 1.6Vppd. The FFE block applies a programmable frequency boost to compensate for the low pass behavior of the transmission channels.

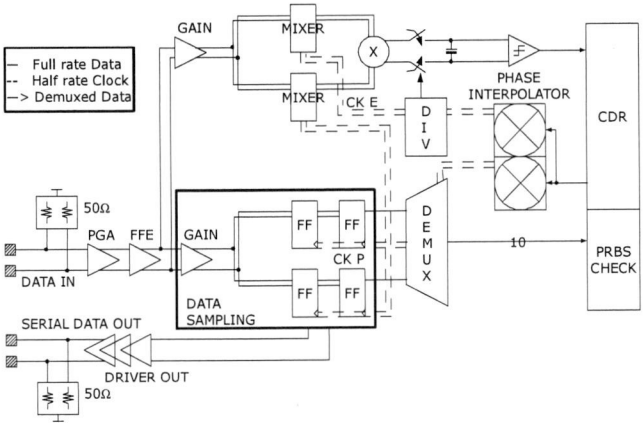

Fig. 8. Receiver's architecture.

As depicted in Fig. 9, four RC source-degenerated differential pairs have been sized as a tapered chain to optimize both the bandwidth and the current consumption.

The widths of the MOS differential pairs and the bias current are increased, while the load and source degenerating resistance are scaled down through the chain.

Each stage can be programmed in four steps by changing both the capacitance and the degenerating resistance, synthesizing a boost programmable from 0 to 4dB. The peaking frequency of each stage is scaled between 3GHz and 5GHz inside the chain, thus allowing a better matching of the channel slopes.

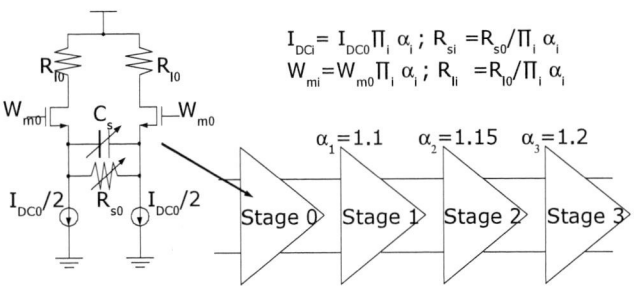

Fig. 9. Equalizer stage.

The linear input chain drives both the data and the clock recovery: the parasitic delays of the two paths are matched by designing a replica gain stage and by matching the same design of the data samplers inside the mixers, as in Fig. 10.

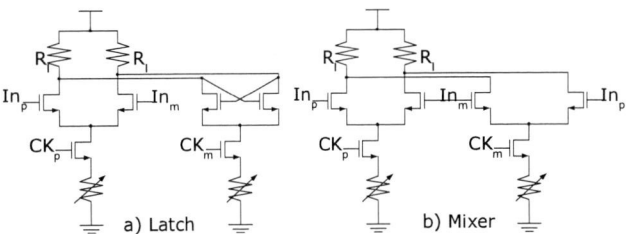

Fig. 10. Front end latch and mixer.

To operate up to 10Gb/s while keeping margins for jitter tolerance, data samplers and matched mixers employ pseudo CML, driven by half-rate differential CMOS clock. The current consumption of the stage is reduced according to the data rate by programming the tail current through programmable MOS resistances.

The mixers output are amplified and sent to the double balanced multiplier shown in Fig. 11. Two Gilbert cells (M1 – M4 and M5 – M8) are shunted with swapped inputs, to assure symmetry between the phase and quadrature signal paths. The multiplier's output current is filtered by a second order low pass (made of R_L, R_F and C_F), sampled each eight bits and then reset for a bit period. The sign of this phase error information is sent to the digital PI controller.

To minimize the latency of the loop, allowing a bandwidth capable of 5000ppm spread spectrum clock (SSC) tracking, the PI is operated at 750MHz for 6Gb/s and it has been implemented in only 2 stages pipeline, as shown in Fig. 12.

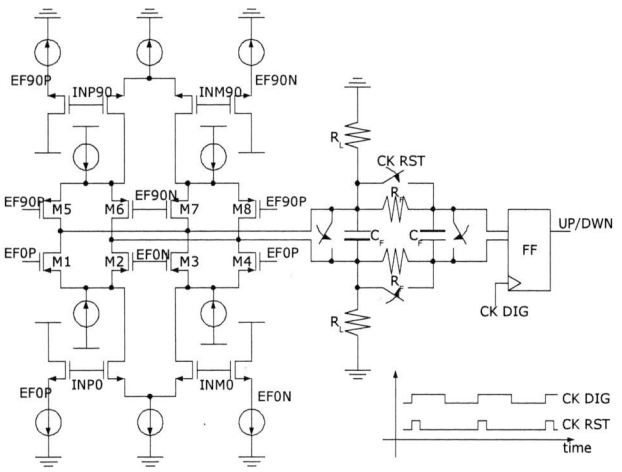

Fig. 11. Double balanced multiplier and reset integrator.

This digital core includes a double-step phase update, allowing +/-3900ppm 6Gb/s SSC tracking with 32 phases per UI or +/- 7800ppm with 16 phases per UI. Moreover, a programmable decimation has been inserted to optimize the integral path frequency resolution versus data rate and ppm spread (from +/- 200ppm to -5000ppm in SSC tracking).

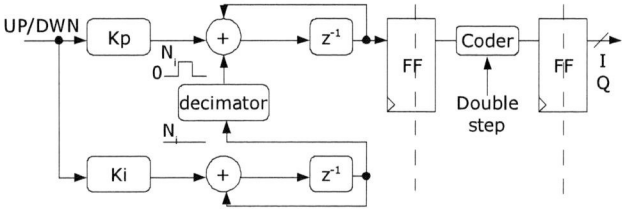

Fig. 12. Digital CDR architecture

The CDR feedback is closed by the two phase interpolators synthesizing the P and E CDR clocks, with a maximum resolution of 32 phases per unit interval (UI). A serial data output is included to check the operation before de-multiplexing while the parallel PRBS checker allows verifying the whole de-serializing path.

V. EXPERIMENTAL RESULTS

The receiver operates from 6Gb/s to 10Gb/s, consuming 110mA at maximum speed from 1.2V supply voltage, including the 10 bits data de-multiplexer and the digital core.

It supports more than 20" FR4 at 10Gb/s (20dB loss at Nyquist) in presence of 200ppm frequency shift, while keeping additional 0.4 UI of sinusoidal jitter tolerance on top of the data and clock sources random jitter, as indicated in Fig. 13. The tests have been performed both with PRBS7 and with the jitter tolerance compliance pattern (CJTPAT).

Fig. 14 shows the receiver performance with +/- 2500ppm center spreading SSC modulating the data in 6Gb/s operation. The sinusoidal jitter tolerance exceeds 0.4 UI up to 36" FR4 backplanes (20dB loss).

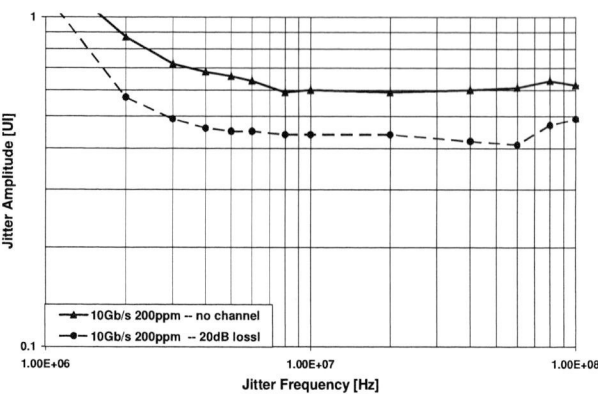

Fig. 13. Sinusoidal jitter tolerance performance at 10Gb/s.

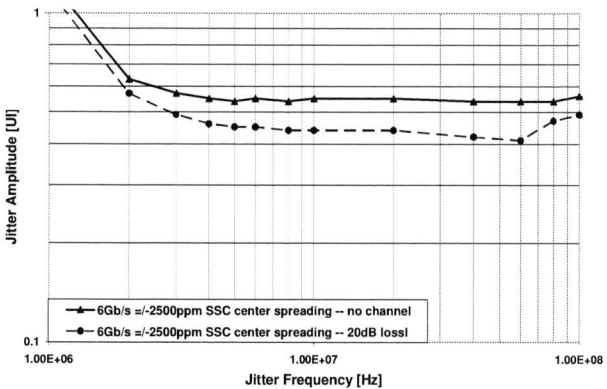

Fig. 14 Sinusoidal jitter tolerance in presence of SSC modulation.

VI. CONCLUSIONS

Measured results show the capability of the proposed architecture to allow a safe operation up to 10Gb/s over transmission channels with 20dB losses. Moreover SSC tracking has been demonstrated.

Despite its simplicity, this solution proves to address high data rate operation and we believe it is attractive even for future higher bit rate transmissions.

REFERENCES

[1] M. Pozzoni, S. Erba, P. Viola, E. Depaoli, D. Sanzogni, R. Brama, D. Baldi, M. Repossi, F. Svelto, "Clock Recovery and Equalization Techniques for Lossy Channels in Multi GB/S Serial Links", invited speech at AACD 2008, Apr. 2008.

[2] M. Pozzoni, S. Erba et al., "A Multi Standard 1.5 to 10Gb/s Latch-Based 3-Tap DFE Receiver with a SSC Tolerant CDR for Serial Backplane Communication", accepted for presentation at 2008 Symp. on VLSI Circuits, June 2008.

[3] B.S. Leibowitz et al., "A 7.5Gb/s 10-Tap DFE Receiver with First Tap Partial Response, Spectrally Gated Adaptation, and 2nd Order Data Filtered CDR", ISSCC Dig. of Tech. Papers, pp 228-229, Feb. 2007.

[4] M. Meghelli et al., "A 10Gb/s 5-Tap-DFE/4-Tap-FFE Transceiver in 90nm CMOS", ISSCC Dig. of Tech. Papers, pp 80-81, Feb. 2006.

IEEE 2008 Custom Intergrated Circuits Conference (CICC)

A 12.5-Gbps, 7-bit Transmit DAC with 4-Tap LUT-based Equalization in 0.13μm CMOS

Hayun Chung, Andrew Liu and Gu-Yeon Wei

School of Engineering and Applied Sciences, Harvard University

Abstract—This paper presents a 12.5-Gbps transmitter that uses a lookup table (LUT)-based equalizer to compensate for within-die imperfections. An equalization technique with 2x sampling is proposed to accommodate timing offsets in the multiphase clocks used for 8:1 serialization. LUT code remapping is also demonstrated to compensate for mismatch effects that introduce nonlinearity in the transmit DAC. Experimental results of a 7-bit resolution transmitter with 4-tap equalization, implemented in 0.13μm CMOS, show the LUT-based equalizer can significantly improve the signal integrity of an otherwise closed eye for data transmitted at 12.5-Gbps.

I. INTRODUCTION

Random and systematic parameter fluctuations can degrade circuit performance and signal integrity in high-speed transmitters. For example, when multiphase clocks are used for data serialization, uneven clock phase spacing due to random and systematic variations can lead to inter-symbol timing mismatch. Also, DAC current mismatch introduces nonlinear characteristics to the transmitter output that cannot be easily addressed with conventional linear equalization techniques. As transmitter jitter is amplified by channel bandwidth limitations, especially at high data rates [1,2], inter-symbol timing mismatch can further degrade performance. Thus, for high data rate transceivers, compensation of mismatch effects in the transmitter is desirable. Consequently, digitally-assisted analog circuits are commonly used to compensate for within-die imperfections [3].

This paper demonstrates a 12.5-Gbps transmit DAC with 7-bit resolution and 4-tap equalization using look-up tables (LUT). The LUT-based equalizer supports both linear and nonlinear forms of equalization and also offers flexibility to easily reprogram equalization tap settings [4,5]. In addition to compensating for channel losses, the proposed equalizer can be used to mitigate various mismatch effects. The transmitter employs an 8-way interleaved structure using 8 multiphase clocks to alleviate on-die speed limitations. Unfortunately, these clock phases are susceptible to variations that lead to inter-symbol timing mismatch and cause eye closure. While phase adjusters can be employed to reduce phase mismatch [6,7], we can leverage the flexibility of the existing transmit equalizer to improve signal integrity by treating the mismatch effects as inter-symbol interference (ISI). Other within-die mismatch effects, such as DAC current mismatch and nonlinearity, and deviations in pulse peak positions, are also addressed by remapping LUT codes. While this paper relies on an external sampling scope and offline analysis, an on-chip sampling scope can be used to measure transmitter characteristics and drive an internal self-calibration engine [8].

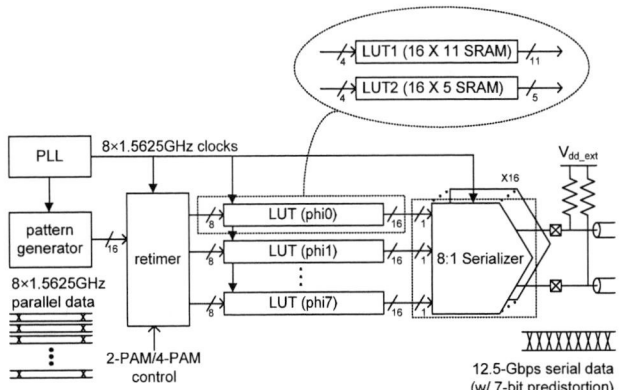

Fig. 1. Overall transmitter block diagram

The remainder of the paper is organized as follows. The next section describes the proposed transmitter architecture and circuits. Section III then provides an analysis of how within-die mismatch affects transmitter performance. Lastly, a description of transmit-side equalization techniques to compensate for various mismatch effects and their measured results are discussed in Sections IV and V, respectively.

II. CIRCUIT DESIGN

Fig. 1 presents the block diagram of an 8-way interleaved transmitter architecture. It uses a ring-oscillator-based phase-locked loop (PLL) to generate 8 evenly-spaced 1.5625-GHz clock phases for 8:1 serialization. The retimer block samples and rearranges parallel data from a pseudo-random pattern generator (or input data memory) to generate the addresses for 8 parallel sets of LUTs, each running off of a different clock phase. The LUT uses a set of data (current, next, and two previous bits) as addresses that correspond to 4-tap mappings of transmit equalizer outputs. The parallel implementation was originally chosen to accommodate speed limitations of the automatically-generated memories, but the interleaved structure confers the benefit of supporting individual equalizer settings for each serializer path to compensate for phase mismatch effects. As shown in the inset of Fig.1, one LUT set comprises two 16-word SRAMs, instead of one 256-word SRAM, to further overcome speed limitations and to reduce power and area overheads. LUT1 uses a set of current and previous bits as an address to generate a portion of the transmitter output that consists of the main-cursor and first post-cursor components for two adjacent bits in each symbol output. The 11-bit code out of each LUT1 ultimately maps to one of 128 possible output levels (7-bit resolution) with 4-bit binary coding and thermometer coding for the remaining

978-1-4244-2018-6/08 $25.00 © 2008 IEEE

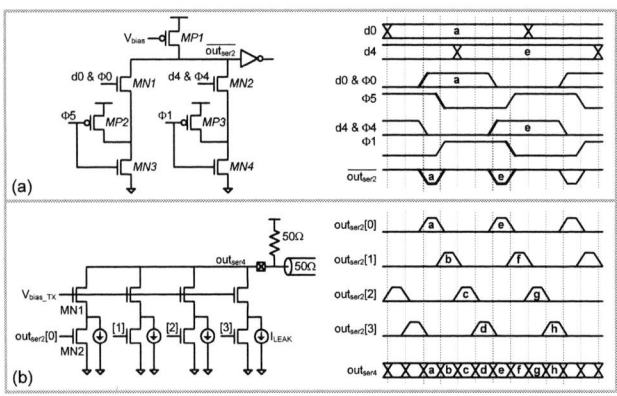

Fig. 2. Two-stage 8:1 serializer schematics and timing diagrams: (a) 2:1 multiplexer and (b) 4:1 multiplexer + current DAC.

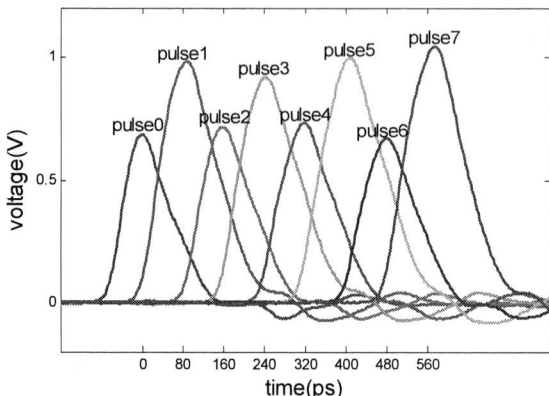

Fig. 3. Transmitter pulse responses through 8 interleaved paths.

higher-order bits. LUT2 uses a set of next and second previous bits as an address that maps to the pre-cursor and second-post cursor compensation components of the transmitter output. Since the pre-cursor and second post-cursor components typically exhibit smaller amplitudes in transmit-side equalization, the LUT2 output requires fewer bits (5-bit binary). Each LUT generates data for 1 out of 8 interleaved symbols combined together via an 8:1 serializer. The 11- and 5-bit codes out of each LUT are connected to 16 parallel serializers.

Fig. 2 illustrates the 8:1 serializer circuitry. To reduce capacitive loading on the output node, serialization occurs in two stages—four parallel 2:1 multiplexers followed by a 4:1 multiplexer combined with the output DAC. Fig. 2(a) shows the circuit schematic and timing diagrams of the 2:1 multiplexing stage. A pseudo-nMOS structure enables high-speed transitions, where pMOS transistors MP2 and MP3 speed up rising transitions and reduce static current in MP1. To eliminate one level of transistor stacking in the final 4:1 multiplexer, otherwise required for delineating each symbol, the 2:1 multiplexer generates pulsed data patterns. Fig. 2(b) shows the circuit schematic and timing diagrams of the subsequent 4:1 multiplexing stage. Leaker devices enable fast switching in each pull-down stack. The 16 parallel sets of 4:1 multiplexer blocks introduce large capacitive loading and, hence, come at the end to leverage the low-impedance output node. A drawback of this highly-parallel structure is its susceptibility to various forms of mismatch and uneven phase spacing between the 8 clocks. On the other hand, the resulting imperfections can be viewed as circuit-induced ISI that the LUT-based equalizer can compensate.

III. WITHIN-DIE MISMATCH EFFECTS

Phase spacing in multiphase clocks are susceptible to random and systematic parameter fluctuations along the clock paths, e.g., inter-stage mismatch in the multi-stage ring oscillator, unbalanced clock distribution network, etc. Unevenly-spaced multiphase clocks can degrade performance in time-division multiplexed (serializing) transmitters due to inter-symbol period mismatch. In the proposed structure, where the 8:1 serializer and transmit DAC are merged together, the effects

of uneven phase spacing becomes more serious—both symbol time and voltage swing can vary. Fig. 3 shows the pulse responses of each interleaved transmit path combined to operate at 12.5-Gbps. As shown in the figure, clock phase mismatch leads to irregular widths, heights, and peak positions for the 8 pulses, severely degrading the signal integrity of the transmitter output.

In addition to the phase mismatch across the multiphase clocks, phase mismatch can also be present in the different paths corresponding to the same clock phase. This is mainly due to unbalanced clock loads and wire length mismatch. For example, DAC branches that have smaller load and/or are closer to the clock buffer turn on/off earlier than branches with larger load and/or are farther away from the clock buffer. Such imbalances translate to deviations in the pulse peaks away from ideal positions. As these pulse responses with different peak positions add up at the transmitter output based on different DAC codes, the output pulse peak position also deviates. Fig. 4 presents an overlay of the differential transmitter output of *pulse1* for DAC codes 0 to 127. The X's correspond to the measured peak positions for each code, which exhibits a slight tilt between positive and negative outputs. The vertical line corresponds to the nominal peak position of *pulse1*. Fig. 4(b) presents the INL (error) of the differential DAC output pulses with respect to the peak of each pulse and to a fixed sampling point. The S shape of the INL error curve can be attributed to saturation effects resulting from on-chip bandwidth limitations. Note that these INL results corresponding to high-speed pulses include the effects of mismatch in clock paths and DAC currents. Since we assume symbols are aligned to nominal sampling points, we later calibrate the INL error via LUT code remapping based on the DAC output at the fixed sample point rather than at the peaks of each pulse.

It is difficult to isolate the effects of DAC current mismatch while operating at high frequencies, because clock phase mismatch also affects pulse heights. Hence, we also measured the DAC at a lower frequency (~5 Gbps). Fig. 5 presents the low-frequency DAC INL (error) plot of *pulse1*. Note that the INL error due DAC current mismatch alone is much smaller, within 1 LSB, by avoiding bandwidth limitations from the channel and transmitter circuitry.

978-1-4244-2018-6/08 $25.00 © 2008 IEEE

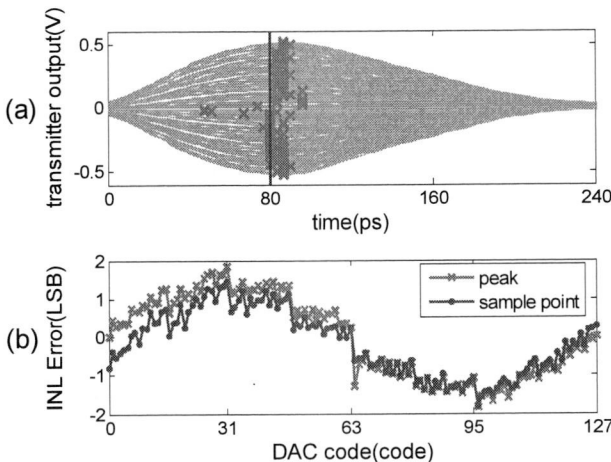

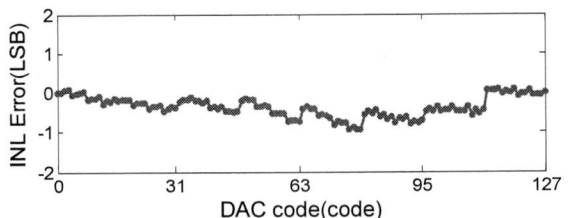

Fig. 4. Effects of within-die mismatch and DAC nonlinearity on pulse1 at 12.5-Gbps: (a) deviations in pulse peak position and (b) 7-bit DAC INL error.

Fig. 5. Effects of DAC current mismatch and nonlinearity on pulse1 at low frequency (~5 Gbps).

IV. EQUALIZATION

While the LUT-based equalizer can also compensate for channel loss, the focus of this section is on the compensation of within-die variations—improving the worst-case eye by compensating for phase mismatch and DAC current mismatch. Uneven spacing of the multiphase clocks is the dominant cause of signal integrity degradation, creating the worst-case eye. While this issue can also be addressed with phase adjusters [6,7], we investigate how the proposed LUT-based equalizer can also alone compensate for the clock phase mismatch effects.

As shown in Fig. 3, both random and systematic variations can lead to mismatch in the 8 pulse responses corresponding to different serializer paths that present different characteristics and require separate equalization settings for each path. Using 8x8 Multi-Input-Multi-Output (MIMO) mapping [6], each individual pulse response can be treated separately to facilitate finding the appropriate equalization settings. Eq. 1 shows an example of a process to find tap coefficients for a particular pulse (*pulse3*). Note that scaled versions of pulse1, pulse2, and pulse4 are used to cancel out the two post-cursor and one pre-cursor components of *pulse3*, respectively. This conventional equalization technique uses 1x sampling of each cursor and, thus, referred to as 1xEQ.

When the cursors are sampled with evenly spaced time periods, forcing a pre-cursor and two post-cursors to zero can compensate for the effect of uneven phase spacing. However, to compensate for phase mismatch effects even further, Eq. 2

$$
\begin{bmatrix}
pulse1 & pulse2 & pulse3 & pulse4 \\
(post2- & (post1- & (main- & (pre- \\
cursor) & cursor) & cursor) & cursor) \\
& & & \\
(1x\ sampled\ pulse\ responses)
\end{bmatrix}
\begin{bmatrix}
w_{post2} \\
w_{post1} \\
w_{main} \\
w_{pre}
\end{bmatrix}
=
\begin{bmatrix}
\vdots \\
0 \\
0 \\
1 \\
0 \\
0 \\
\vdots
\end{bmatrix}. \quad (1)
$$

uses mid points between cursors for equalization and requires 2x sampling (2xEQ). In other words, 2xEQ forces inter-symbol data transitions to specific levels.

$$
\begin{bmatrix}
pulse1 & pulse2 & pulse3 & pulse4 \\
(post2- & (post1- & (main- & (pre- \\
cursor) & cursor) & cursor) & cursor) \\
& & & \\
(2x\ sampled\ pulse\ responses)
\end{bmatrix}
\begin{bmatrix}
w_{post2} \\
w_{post1} \\
w_{main} \\
w_{pre}
\end{bmatrix}
=
\begin{bmatrix}
\vdots \\
0 \\
0.5 \\
1 \\
0.5 \\
0 \\
\vdots
\end{bmatrix}. \quad (2)
$$

To compensate for other within-die mismatch effects, such as DAC nonlinearity and deviations in pulse peak positions, the LUT codes are remapped for each of the 8 individual serializer paths based on DAC INL plots shown in Fig. 4(b).

V. MEASURED RESULTS

A test-chip prototype of the transmitter with LUT-based equalization, fabricated in 0.13µm CMOS, utilizes external termination resistors and voltage to facilitate testing across a wide range of current settings for the DAC at the expense of larger ringing due to parasitic bond wire inductance. In order to reduce ripples caused by bond wire inductance, two 5 ohm resistors were added in series after the differential bond wires and before connecting to the termination resistors.

Fig. 6 shows the eye diagrams of the transmitter output after propagating through a pair of 3-inch coaxial cables and into a high-speed sampling scope while operating at 12.5-Gbps. As shown in Fig. 6(a), 4 out of the 8 eyes are completely closed with any compensation. By treating the internal circuit mismatch as ISI, Figs. 6(b), (c), and (d) show LUT-based equalization can salvage this otherwise unusable transmitter and compensate various within-die mismatch effects. 2xEQ with LUT code remapping provides the best jitter performance as well as the most evenly distributed phase spacing by forcing the data transition position to the mid-point between cursors.

To verify the benefits of forcing data transitions to the mid-point between symbols in 2xEQ, mean and rms jitter of zero crossings while transmitting random data are measured. Fig. 7 shows one of the histograms used for the measurement.

Fig. 8(a) plots the INL (error) of mean data transition positions across 8 symbols when 1xEQ and 2xEQ are applied. Note that results without equalization are not included in the plot, because half of the eyes are completely closed making it impossible to measure zero crossing points. While both 1xEQ and 2xEQ can compensate for phase mismatch effects, 2xEQ

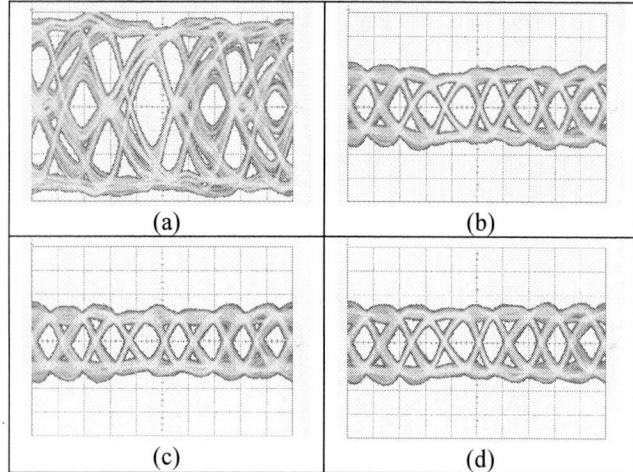

Fig. 6. Eye diagrams of transmitter at 12.5-Gbps:
(a) without equalization, (b) with 2xEQ before LUT code remapping,
(c) with 1xEQ, and (d) with 2xEQ after LUT code remapping.

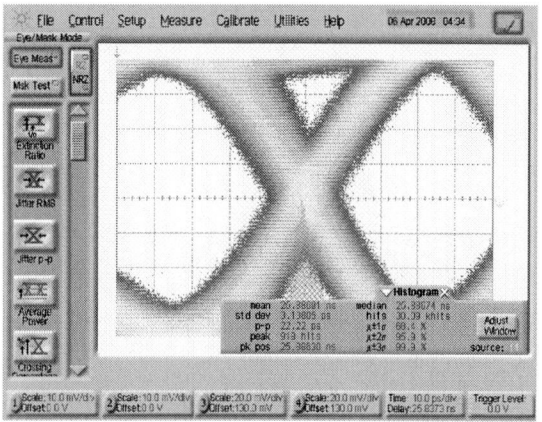

Fig. 7. Histogram of zero crossings for mean and rms jitter measurement.

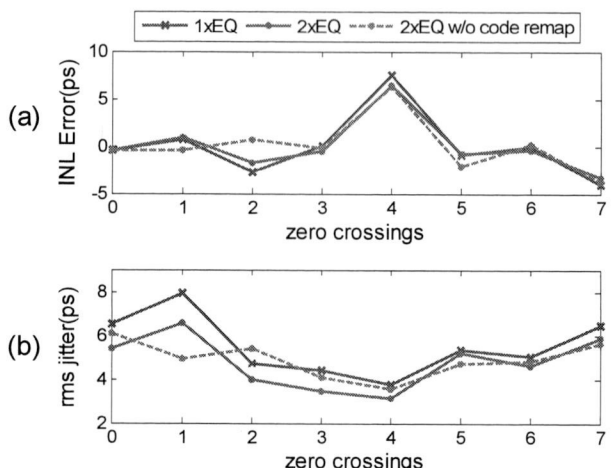

Fig. 8. Data transition point measurements: (a) mean INL error, (b) rms jitter.

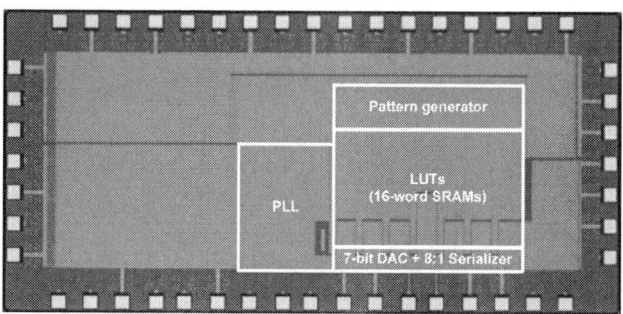

Fig. 9. Die photo and overlay of basic blocks.

enables lower average error across the 8 crossings. The standard deviations of INL error for 1xEQ and 2xEQ are 3.43ps and 2.89ps, respectively. The results for 2xEQ before LUT code remapping are also included in Fig. 8 to show that DAC nonlinearity can degrade performance due to inaccurate code settings. The standard deviation of INL error for 2xEQ before LUT code remapping is 2.93ps.

Fig. 8(b) plots the rms jitter at the data crossing across 8 symbols when 1xEQ and 2xEQ are applied. For all the 8 zero crossing points 2xEQ provides better jitter performance compared to 1xEQ. The average RMS jitter for 1xEQ, 2xEQ, and 2xEQ before LUT code remapping are 5.51ps, 4.77ps, and 4.90ps with standard deviation of 1.36ps, 1.19ps, and 0.80ps, respectively. Figs. 8(a) and 8(b) show that 2xEQ with LUT code remapping provides the best equalization setting.

Fig. 9 presents the die photo of the test chip. The overall chip area, including I/O pads, 16 interleaved 16-word SRAMs, and PLL, is 3.08×1.47 mm^2, with 1.03 mm^2 of active area. While operating at 12.5-Gbps off of a 1.2-V supply, the chip consumes 425mW where 36% of this power is consumed in test-related blocks (e.g., pattern generator and retimer).

ACKNOWLEDGMENTS

This work was supported in part by Samsung and the NSF CCF0325228. The authors also thank UMC for test-chip prototype fabrication.

REFERENCES

[1] P. K. Hanumolu et al., "Jitter in High-Speed Serial and Parallel Link," *IEEE ISCAS*, 2004.

[2] B. Casper et al., "Future Microprocessor Interfaces: Analysis, Design and Optimization," *IEEE CICC*, Sep., 2007.

[3] B. Murmann, "Digitally Assisted Analog Circuits," *IEEE Micro*, 2006.

[4] W. Dally and J. Poulton, "Transmitter Equalization for 4Gb/s Signaling," *Hot Interconnects*, 1996.

[5] B. Casper et al., "A 20Gb/s Forwarded Clock Transceiver in 90nm CMOS," *IEEE ISSCC*, Feb., 2006.

[6] V. Stojanovic et al., "Transmit Pre-emphasis for High-Speed Time-Division-Multiplexed Serial-Link Transceiver," *IEEE ICC*, 2002.

[7] A. H.-Y. Tan and G.-Y. Wei, "Phase Mismatch Detection and Compensation for PLL/DLL Based Multi-Phase Clock Generator," *IEEE CICC*, Sep., 2006.

[8] V. Stojanovic et al., "Autonomous Dual-Mode (PAM2/4) Serial Link Transceiver with Adaptive Equalization and Data Recovery," *IEEE JSSC*, Apr., 2005.

978-1-4244-2018-6/08 $25.00 © 2008 IEEE

IEEE 2008 Custom Intergrated Circuits Conference (CICC)

Active Deskew in Injection-Locked Clocking

Lin Zhang, Berkehan Ciftcioglu, and Hui Wu
Department of Electrical and Computer Engineering, University of Rochester, Rochester, NY

Abstract— This paper presents an injection-locked clock (ILC) distribution system with a new active deskew mechanism based on the built-in phase tuning of injection-locked oscillators (ILO). The proposed technique removes the required deskew delay lines and associated power dissipation, clock latency and jitter accumulation in conventional active deskew schemes. A test chip was fabricated in a standard $0.18\mu m$ digital CMOS process to demonstrate this new technique. Working at 3.5GHz clock frequency, the ILOs in the ILC achieved 40ps deskew range with a step size of 1.25ps. The deskew loop successfully achieved a skew reduction from the preset value of 16ps to 2ps. The cycle-to-cycle jitter degradation from clock input to clock output is measured only 0.04ps.

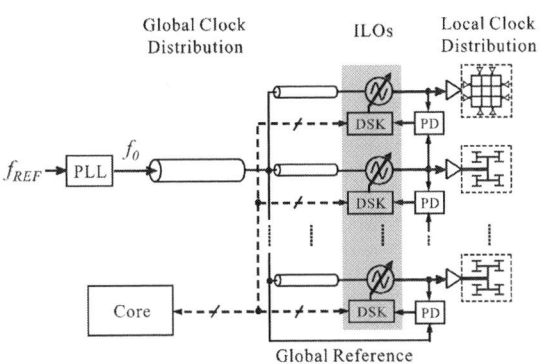

Fig. 1. ILC with active deskew based on the ILO delay tunings.

I. INTRODUCTION

Injection-locked clocking (ILC) has been proposed for multi-GHz clock distribution [1], which uses injection-locked oscillators (ILO) as local clock generators. Compared to conventional clocking based on buffered trees and grids, it can achieve lower jitter while consuming less power [2]. Compared to other resonance-based clocking schemes proposed recently [3], [4], ILC's jitter performance is not limited by the quality factor Q of on-chip resonator, which explains why injection locking has recently been adopted for resonant clocking [5]–[7].

Skew, however, remains a major challenge to be addressed in these new clocking schemes including ILC. Balanced H-tree and grid structure have been used in conventional clock distribution to reduce skew [8]. Due to the load mismatch, and process, voltage and temperature (PVT) variations, it becomes increasingly difficult to use this method to effectively control the skew in a multi-GHz system. More seriously, as the load capacitance and the temperature profile of the microprocessor vary when running different tasks, induced skew changes with time. Passive skew reduction techniques cannot predict and control such dynamical skew variations, and active deskew is needed.

Conventional active deskew methods [9], [10] compensate the skew by adding tunable delays to different clock paths. They are designed to reduce the clock skew after the chip fabrication, and capable of tracking the skew variations dynamically. The tunable delay is typically implemented by active delay lines which are loaded with switched-capacitor arrays [9], or built with current starved buffers [10]. These approaches proved effective in conventional clocking and have been applied to resonant clocking [7]. However, adding active delay lines has several disadvantages. First, it consumes extra power; second, it increases the clock latency substantially due to the delay tuning requirement; most importantly, the extra active delay line tends to degrade the clock jitter significantly. This is because power supply noise coupled through clock

buffers is the main contributor to jitter accumulation in the conventional clock distribution [11], and adding active delay lines for deskew further increases the length of the buffer chain in the clock signal path.

In a multi-GHz clock distribution system, jitter degradation due to deskew circuitry must be minimized considering the ever-decreasing timing margin when clock frequency increases [13]. Therefore, we propose a new active deskew scheme that utilizes the phase tuning capability of ILOs to compensate the skew [1], [15]. This deskew scheme does not require the additional deskew delay lines and hence does not suffer in clock latency and jitter while consuming minimal extra power. In this paper, we will for the first time demonstrate such an active deskew method in an ILC system.

II. ILO-BASED ACTIVE DESKEW

In order to avoid the jitter degradation of the conventional active deskew scheme, the active deskew scheme for ILC is based on the ILO's phase tuning capability, as shown in Fig. 1. The ILO at the input of each local clock domain works as both a local clock regenerator and a deskew buffer. Clock skews between different clock domains, or between each clock domain and a global reference, are measured by phase detectors and sent to deskew control logics (DSKs) associated with each ILO. The DSKs, under the coordination of the microprocessor core, then control the phase shift of each ILO to compensate the skew and accomplish the deskew function. It is worth noting that, skew can also be intentionally inserted for skew scheduling with the desired skew information loaded to the microprocessor core.

The key for such an active deskew scheme is the phase tuning capability of ILOs, which has been demonstrated in our previous work [1], [14], [15]. As shown in Fig. 2a, in the locked state, the output phase, $\phi_i + \Delta\phi$, of an ILO is

978-1-4244-2018-6/08 $25.00 © 2008 IEEE

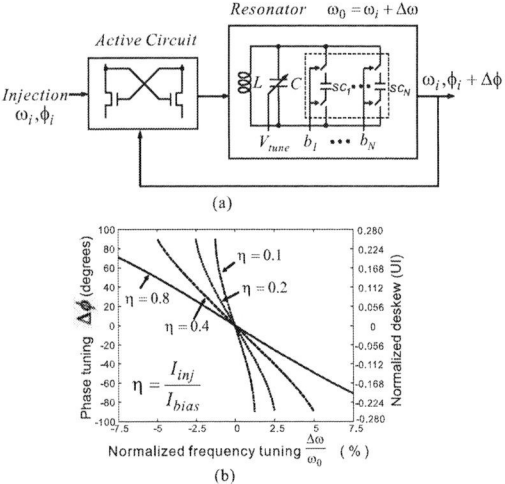

(a)

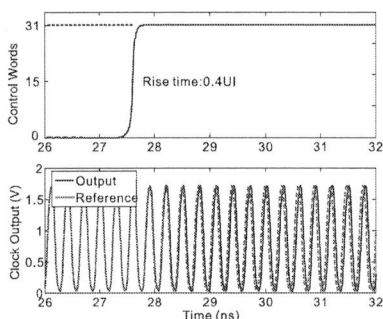

(b)

Fig. 2. (a) A generic ILO model, showing frequency tuning by a varactor or N-bit switched capacitor array. and (b) the phase shift characteristics with respect to frequency tuning, assuming the resonator quality factor of 4.

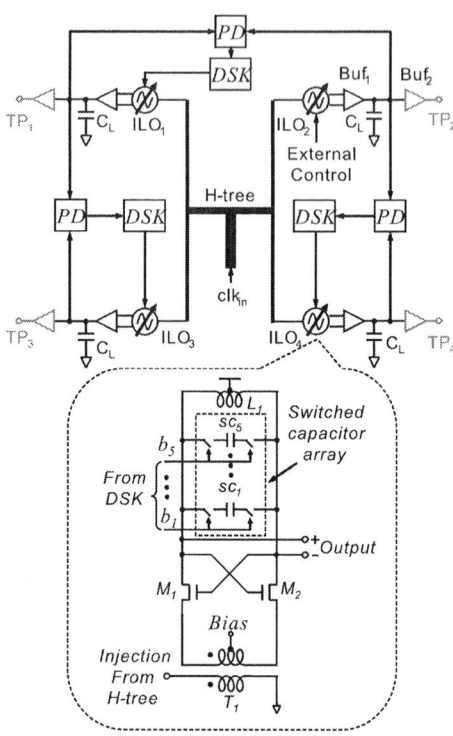

Fig. 4. An injection-locked clocking system with ILO-based active deskew. Four ILOs are driven by the input clock through an H-tree. Each ILO is buffered by Buf_1 to drive 2pF of on-chip load capacitor (C_L), which also converts the ILO differential output to a single-ended signal. Output buffers Buf_2 drive the test ports (TP_x).

Fig. 3. Simulated transient response of an ILO with full-scale deskew step.

determined by the difference between the input frequency ω_i, and the free-running oscillation frequency ω_0, of the ILO, which means that the ILO phase can be tuned by varying its free-running oscillation frequency ω_0 determined by its resonator (Fig. 2a). This phase tuning characteristics of an ILO is plotted in Fig. 2b, which shows the phase shift with respect to frequency tuning at different injection ratios [15]. It can be seen that the phase shift tuning characteristics is pretty linear in the middle of the tuning range for all injection ratios, which is good for linear deskew implementation. Also since the delay tuning of ILOs are centered around zero phase shift, there is no clock latency increase as in conventional active deskew approaches.

Frequency tuning of an ILO can be implemented like in a generic voltage controlled oscillator by varactors [1] or switched capacitor array [15] in its resonator. The switched capacitor array approach is more suitable for digital deskew control. Thanks to the digitized control signals, it exhibits better immunity to power supply and substrate noise, which is a serious threat to analog circuits in a noisy microprocessor environment.

Unlike conventional deskew delay lines, an ILO has less

paths to pick up power supply noise. Even when noise manages to couple into the ILO, it is largely high-pass filtered, considering that an ILO behaves like an PLL with large loop bandwidth. Thus the jitter degradation of this ILO based deskew scheme is reduced compared to conventional deskew schemes. Interestingly, in [7], it was proposed that injection locking would reduce the jitter introduced by conventional deskew delay lines, which exhibits an low-pass filtering characteristics.

Since the deskew control logic is digital, it is critical to verify the ILO's dynamic behavior, and particularly, how an ILO responds to deskew control steps. As shown in Fig. 3, the ILO transient response to a full-scale step of the control word (from 0 to 31) is simulated in time domain, which means all the switched capacitors in the ILO resonator are turned on. From the waveform, we can see the transient time is within several clock cycles even with a pessimistic rise time on the control signal. Also the transition dynamics does not disrupt the ILO output phase.

III. TEST CHIP IMPLEMENTATION

Fig. 4 shows the schematic of the test chip to demonstrate the proposed ILO-based active deskew in ILC. The input clock signal is distributed by a passive H-tree to each clock domain, and injection-locked to an ILO. Each ILO drives a 2pF clock load, which models the local clock load in real processors,

978-1-4244-2018-6/08 $25.00 © 2008 IEEE

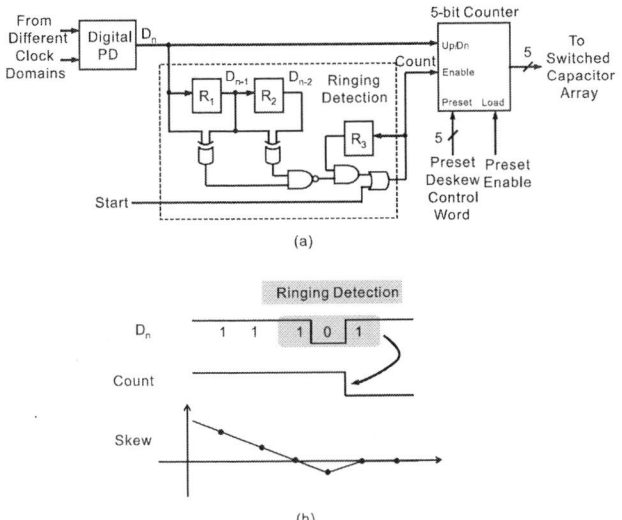

(a)

Ringing Detection

(b)

Fig. 5. (a) Deskew logic algorithm, and (b) an example of the deskew sequence which shows the design for ringing prevention.

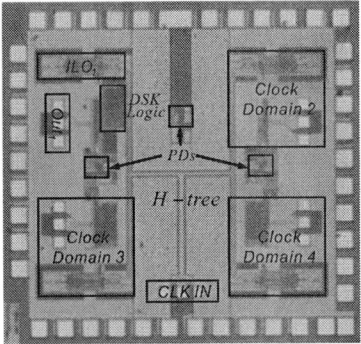

Fig. 6. Test chip die photo.

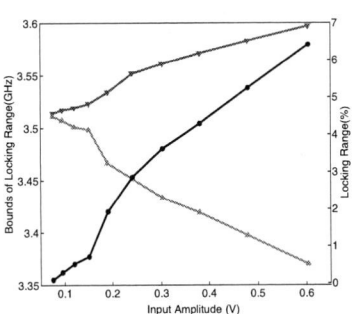

Fig. 7. Measured locking range for the ILC network.

through a differential to single-ended buffer (Buf_1). The ILOs use a newly developed transformer-direct-injection topology [15], as shown in the blow-up of Fig. 4. A 5-bit binary-weighted switched capacitor array is implemented in the LC resonator of each ILO for phase tuning. The 5-bit binary coded digital control is generated from a 5-bit bi-directional counter built inside the deskew control logics (DSKs). The DSK algorithm and an example of the deskew sequence is shown in Fig. 5a and Fig. 5b. When the counter counts UP, the ILO free-running frequency decreases, and the ILO phase tuning increases, and vice versa. The counter value can also be preset from external, as shown in Fig. 5a This enables a manual adjustment of the ILO delays for test purpose. The clock phases from adjacent clock domains are compared by a digital phase detector, and the 1-bit skew information is fed into the deskew control logic, which is D_n in Fig. 5. The skew information for two previous cycles are also stored by two registers R_1 and R_2 to implement the ringing detection and prevention algorithm in the deskew control logic. Once a ringing happens, the DSK forces the counter to enter a stop state, until a start from external to restart the deskew control logic.

The test chip was fabricated in a standard $0.18\mu m$ digital CMOS technology. The clock frequency is set at 3.5GHz, representing the state-of-the-art processor speed. The transformer and inductors in the ILOs are all built with the $0.35\mu m$-thick top metal layer. The transformers have a k factor of 0.77 at 3.5GHz. The symmetric spiral inductor have an inductance value of 2.8nH and quality factor of 4.1 at 3.5GHz. The die photo of the test chip is shown in Fig. 6, which measures 2mm by 2mm.

IV. MEASUREMENT RESULTS

The test chip is measured on a probe station. Locking range of the ILC network is measured up to 6.5% with input amplitude of 0.6V (Fig. 7). Time domain clock waveforms from each clock domain are measured on 50GHz sampling oscilloscope to study the clock timing. When comparing the timing of different clock domains, connector and cable delay mismatch is first characterized and then used to calibrate the measured results.

The free-running frequency tuning and locked state phase tuning of ILOs are first characterized to show the deskew capability of the ILC network, as shown in Fig. 8. The measured ILO free-running frequency tuning is pretty linear with a step size of 2.5MHz. This linear frequency tuning generated a linear phase tuning for ILOs in the locked state, with range of 40ps and a step size of 1.25ps Then the dynamics of the deskew loop is measured as shown in Fig. 9. An initial skew of -16ps is preset between the two clock domains before the deskew loop starts. The deskew loop reduces the skew to a final residual value of 2ps within 15 cycles of the deskew clock with a little overshoot. The residual skew can be attributed to the deskew step size limitation and phase detector offset.

The phase noise of ILC output clock is measured and compared with that of the input clock source and free-running ILO (Fig. 10a). The ILC output tracks the phase noise of the source clock up to 10MHz, and shows up to 60dB improvement over free-running case near the offset of 10KHz. Cycle-to-cycle jitters for both ILC output and input clock are measured with a self-triggered method, and compared in Fig. 10b. The jitter accumulated in ILC is largely negligible (0.04ps).

Each ILO in the test chip consumes 12mA from a 1V

978-1-4244-2018-6/08 $25.00 © 2008 IEEE

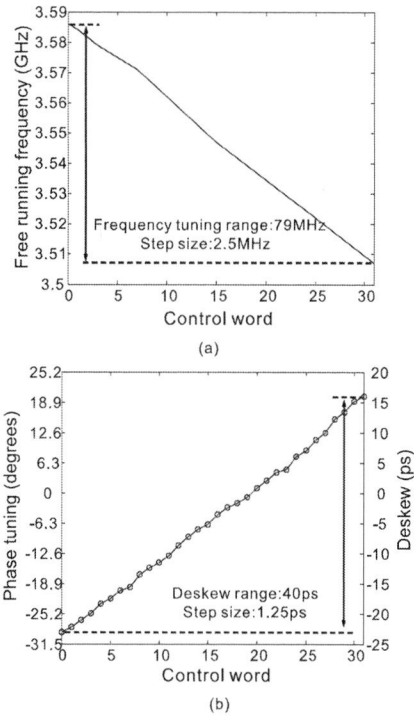

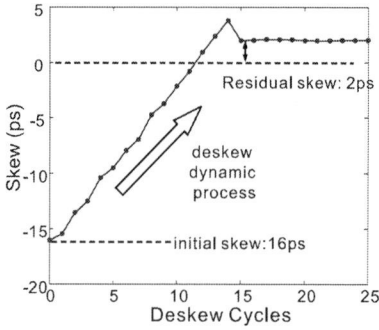

Fig. 8. (a) Measured free-running frequency tuning, and (b) delay tuning under locked state by 5-bit switched capacitor array.

Fig. 9. Deskew dynamics of the deskew loop.

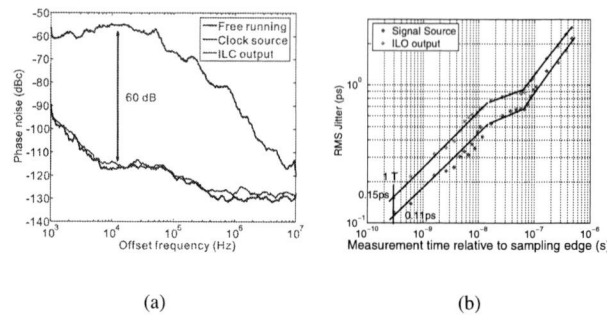

(a)   (b)

Fig. 10. (a) Phase noise of the ILC output in comparison with input clock and free-running ILO. (b) Cycle-to-cycle jitter test for ILC output and input clock. The degradation is only 0.04ps.

supply, and each Buf_1 consumes 40mA from 1.8V. Each PD and DSK consumes 6.8mA from 1.8V.

ACKNOWLEDGMENT

The authors would like to thank B. Chatterjee, A. Bahai, P. Holloway, M. Bohsali, J. Yu, A. Shah, V. Abellera, P. Misich, and J. Wan of National Semiconductor for their support in chip fabrication. We also thank P.J. Restle of IBM for helpful discussions. We acknowledge Sonnet Software for EM simulation tools support.

REFERENCES

[1] L. Zhang, B. Ciftcioglu, M. Huang, and H. Wu. Injection-Locked Clocking: A New GHz Clock Distribution Scheme. *IEEE Custom Integrated Circuits Conf. Dig. Tech. Papers*, pages 785–788, 2006.

[2] L. Zhang, A. Carpenter, B. Ciftcioglu, A. Garg, M. Huang, and H. Wu. Injection-Locked Clocking: A Low-Power Clock Distribution Scheme for High-Performance Microprocessors. *to appear in TVLSI*, 2008.

[3] F. O'Mahony, C.P. Yue, M.A. Horowitz, and S.S. Wong. A 10-GHz Global Clock Distribution Using Coupled Standing-Wave Oscillators. *IEEE J. Solid-State Circuits*, 38(11):1813–1820, Nov. 2003.

[4] S.C. Chang, P.J. Restle, K.L. Shepard, N.K. James, and R.L. Franch. A 4.6GHz Resonant Global Clock Distribution Network. In *IEEE Int. Solid-State Circuits Conf. Dig. Tech. Papers*, pages 342–343, 2004.

[5] B. Mesgarzadeh, M. Hansson, and A. Alvandpour. Jitter Characteristic in Resonant Clock Distribution. In *Solid-State Circuits Conference, ESSCIRC 2006. Proceedings of the 32nd European*, pages 464–467, 2006.

[6] L. Lee and C.K. Yang. An Adaptive Low-Jitter LC-Based Clock Distribution. In *IEEE Int. Solid-State Circuits Conf. Dig. Tech. Papers*, pages 182–183, 2007.

[7] Z. Xu and K. Shepard. Low-Jitter Active Deskewing Through Injection-Locked Resonant Clocking. *IEEE Custom Integrated Circuits Conf. Dig. Tech. Papers*, pages 9–12, 2007.

[8] P.J. Restle et al. A Clock Distribution Network for Microprocessors. *IEEE J. Solid-State Circuits*, 36(5):792–799, May 2001.

[9] S. Tam, S. Rusu, U.N. Desai, R. Kim, J. Zhang, and I. Young. Clock Generation and Distribution for the First IA-64 Microprocessor. *IEEE J. Solid-State Circuits*, pages 1545–1552, Nov. 2000.

[10] S. Tam, R.D. Limaye, and U.N. Desai. Clock Generation and Distribution for the 130-nm Itanium 2 Processor With 6-MB On-Die L3 Cache. *IEEE J. Solid-State Circuits*, 39(4):636–642, April 2004.

[11] N.A. Kurd, J.S. Barkatullah, R.O. Dizon, T.D. Fletcher, and P.D. Madland. A Multigigahertz Clocking Scheme for the Pentium 4 Microprocessor. *IEEE J. Solid-State Circuits*, 36(11):1647–1653, Nov. 2001.

[12] S. Verma, H. Rategh, and T. Lee. A Unified Model for Injection-Locked Frequency Dividers. *IEEE J. Solid-State Circuits*, 38(6):1015–1027, Jun. 2003.

[13] A.V. Mule, E.N. Glytsis, T.K. Gaylord, and J.D. Meindl. Electrical and Optical Clock Distribution Networks For Gigascale Microprocessors. *IEEE Transactions on VLSI*, 10(5):582–594, Oct. 2002.

[14] L. Zhang and H. Wu. A Double-Balanced Injection-Locked Frequency Divider for Tunable Dual-Phase Signal Generation. *IEEE Radio-Frequency Integrated Circuits (RFIC) Symposium Digest of Papers, pp.137-140*, 2006.

[15] L. Zhang, B. Ciftcioglu, and H. Wu. A 1V, 1mW, 4GHz Injection-Locked Oscillator for High-Performance Clocking. *IEEE Custom Integrated Circuits Conf. Dig. Tech. Papers*, pages 309–312, 2007.

A 53 GHz DCO for mm-Wave WPAN

Raffaella Genesi, Francesco M. De Paola, and Danilo Manstretta

Dipartimento di Elettronica, Università degli Studi di Pavia
via Ferrata 1, 27100, Pavia, Italy

Abstract— **A digitally-controlled oscillator (DCO) for mm-wave wireless applications is demonstrated for the first time. Low phase noise is achieved adopting an efficient complementary topology and optimizing the programmable-capacitor array. A fine frequency resolution of 1.8MHz is obtained using switched metal-oxide-metal capacitors while coarse frequency tuning is achieved using switched accumulation-mode MOS varactors. The DCO, implemented in 90nm CMOS technology, has a power consumption of 2.34 mW, an oscillation frequency ranging from 51.3 to 53.3 GHz, and a measured phase noise of -116.5 dBc/Hz at 10MHz from the carrier, with a resulting figure-of-merit equal to -187.2 dBc/Hz.**

I. INTRODUCTION

The ongoing progress in CMOS technology and circuit design research is opening the way to highly-integrated low-cost wireless systems working in the millimeter-wave frequency range [1][2]. One of the possible applications are wireless networks, that, thanks to the widespread success of WLAN systems and the ever growing demand for higher data-rate communications, are leading to increasing congestion in the frequency bands below 10 GHz. The 7-GHz unlicensed band around 60 GHz offers the possibility of proving short-range wireless data communications at rates of several gigabits per second and is therefore being subject of intensive research. Thanks to the reduced wave-length, the antenna has a reasonable size that allows it to be integrated with a complete radio on the same chip [3], reaching unprecedented levels of integration. Such an ambitious goal presents many design challenges mainly because of the high operative frequencies and the low analog qualities of MOS transistors in deep-submicron CMOS technologies, supporting the use of a digitally intensive design approach. In this paper, a low phase noise digitally-controlled oscillator (DCO) for high performance mm-wave wireless personal area network transceivers is presented. The oscillator was designed in a 90nm CMOS process without advanced analog features and adopts a fully-digital control technique, making it compatible with all-digital frequency synthesis. In section II the oscillator digital tuning circuitry is described; section III and IV report on the oscillator design and experimental characterization; section V draws some conclusion.

II. DIGITALLY CONTROLLED CAPACITOR BANK DESIGN

In a DCO the variable capacitance consists in an array of digitally-controlled capacitors where the total variable capacitance and the minimum capacitance step are determined by the desired tuning range and by phase noise considerations.

Modeling the DCO frequency quantization as a uniformly-distributed random variable [4], its effect on phase noise, or frequency quantization noise (FQN), is theoretically given by:

$$PN(f) = \left(\frac{f_{LSB}}{f}\right)^2 \frac{T_{REF}}{12} (\sin cf T_{REF})^2 \qquad (1)$$

where T_{REF} is the PLL reference frequency, f_{LSB} is the frequency resolution of the DCO and f is the frequency offset from the carrier. Assuming f_{LSB} equal to 2MHz and 20 ns T_{REF}, a FQN of -102.3 dBc/Hz at 10MHz offset from the carrier results, which is higher than what can be achieved in terms of thermal-induced phase noise at reasonable power levels. The FQN can be lowered further using $\Sigma\Delta$ noise-shaping techniques as proposed in [4]. For instance, using a 2nd-order $\Sigma\Delta$ modulation at 500MHz, FQN for the same f_{LSB} would be lowered to -147.8 dBc/Hz at 10MHz from the carrier. A target f_{LSB} of 2MHz has been assumed in this work.

One of the major shortcomings in the implementation of LC-tank oscillators above 10GHz is the low quality factor (Q) of the capacitors, that usually dominates the overall Q of the resonator. This is even more true for variable capacitors, both MOS varactors and switched Metal-oxide-Metal (MoM) capacitors, where additional losses are brought about by active device limitations (gate resistance, finite channel conductance and parasitic drain/source capacitance). To better cope with technology limitations at high frequencies, as opposite to a previous implementation at 10-GHz [5], the capacitor bank was segmented into three sections, implemented with different structures. Tuning range and Q are the primary drivers for the coarse tuning bank, while small minimum capacitance step is the main requirement for the fine tuning bank. Segmentation also allows to reduce the overall size of the capacitor arrays compared to a uniform array of equal resolution, minimizing the parasitic inductances that would hinder oscillation at 50GHz. The coarse and intermediate sections are intended to be used in the initial phase of the locking algorithm, similarly to what is done in analog PLLs when the VCOs uses switched tuning, while the fine tuning bank is used in the normal operation of the PLL. For the coarse tuning section, two different tuning devices have been considered in this work: the NMOS in n-well accumulation-mode (a-MOS) varactor, switched between the extremes of its C-V characteristic, and the NMOS transistor operated as a switch [6]. To compare the two devices, a variable-capacitance figure of merit (FoM_C) has been used:

$$FoM_C = \frac{1}{\omega_0 R_{ON} C_{OFF}} \qquad (2)$$

where ω_0 is the oscillation frequency, R_{ON} is the resistance of the device when it exhibits the maximum value of capacitance

---

This work has been carried out in the framework of the Italian National Research Program FIRB (contract nr. RBIP063L4L).

978-1-4244-2018-6/08 $25.00 © 2008 IEEE

(corresponding to the accumulation region for the varactor) while C_{OFF} is the minimum capacitance. In this both quality factor and tuning capability (C_{max}/C_{min}) are taken into account. At the time of design, a scalable model of the a-MOS varactors was not available in the technology design kit. Based on the device model of an a-MOS varactor with fixed device size (finger width of 1.6µm, channel length of 400nm), a custom *Verilog-A* scalable model has been developed using simplified device equations [7]. Based on this approximate varactor model and on the switch device model, a comparison of the two structures has been carried out for different device sizes. Depending on the chosen channel length (L_f) and finger width (W_f), the a-MOS varactor or the switched-MoM capacitor features an higher FoM$_C$. At W_f=1µm and L_f=100nm the a-MOS varactor reaches the maximum FOM_C, equal to 128 at 50 GHz, that is approximately three times greater then the MOS switch's maximum one, even neglecting the bottom-plate capacitance of the latter. Hence, an a-MOS varactor was used for the coarse tuning section, while custom-designed MoM capacitors were used to achieve a finer frequency resolution than would be attainable using a minimum-sized a-MOS varactor. The coarse tuning bank is formed by 31 identical a-MOS, controlled by a 5-bit binary word. According to the approximate model this should cover a frequency range of about 3.2 GHz with a frequency resolution equal to 104 MHz. The fine tuning bank is formed by 31 MoM capacitors with series NMOS switches controlled by a 5-bit thermometer code to ensure monotonicity. The physical implementation of two adjacent elements is shown in Fig. 1. A single top layer (M6) is used for all capacitors while the bottom layer of each capacitor unit is implemented as a metal stack (M_4-M_1) connecting to the NMOS transistor switch.

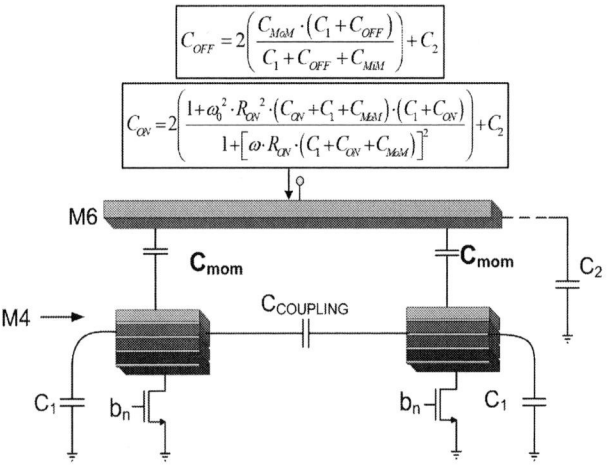

Fig. 1 Layout of the two adjacent fine tuning elements

In order to obtain the desired frequency resolution of about 2 MHz, an LSB capacitance step in the order of 10 aF is required. This has been achieved using a small MoM capacitor with a top-bottom capacitance in the order of 60 aF, a bottom-ground capacitance of about 80 aF and a minimum-sized switch in series. The actual capacitance variation, taking into account the finite on-resistance and the parasitic capacitance introduced by the switch is given by the expression in Fig. 1. The unavoidable coupling between adjacent cells determines a non-linearity in the code-to-frequency characteristic. The sizing of each plate and the distance between two adjacent plates have been optimized using EMX electromagnetic simulator. According to electromagnetic simulations, the error caused by the capacitive coupling between adjacent elements of the bank is lower than 1/9 *LSB*. Extending the tuning range of the fine-tuning section up to the frequency resolution of the coarse-tuning section with some margin would require more than 100 elements, resulting in an exceedingly large structure. To bridge the gap between the coarse and fine tuning elements, an intermediate tuning bank has been added. The intermediate tuning bank is made of 7 MoM capacitors with series NMOS switches controlled by a 3-bit binary word.

III. OSCILLATOR DESIGN

In Fig. 2 the DCO schematic is illustrated. The core consists of a complementary NMOS-PMOS cross-coupled pair that generates the negative resistance required to sustain oscillation. The use of a complementary topology allows to achieve near double current efficiency as compared to the single-pair topology, enhancing the oscillator efficiency while keeping the low-voltage transistors, including the a-MOS varactors, always below V_{DD}. Furthermore, with the a-MOS gates biased around V_{DD}/2 and a rail-to-rail voltage swing, close to maximum tuning of the varactors is achieved without requiring the control voltage, connected to the a-MOS source/drain terminals, to swing above V_{DD}.

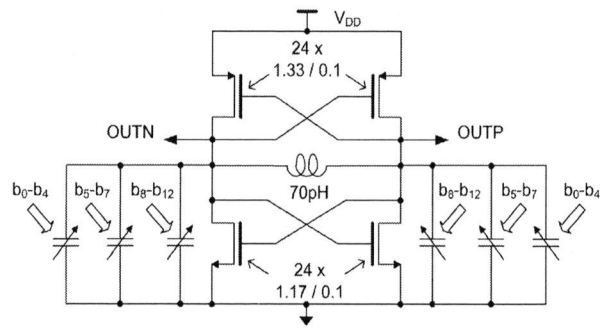

Fig. 2 Digitally-controlled oscillator schematic.

The core transistors have been sized according to the main targets of minimizing phase noise and complying with the oscillation start-up condition with some margin (g_mR_{TANK}=3) while taking into account layout-related parasitics. For a given total device width, reducing the finger width helps lowering gate resistance but increases device perimeter and interconnect parasitic capacitances. The use of a complementary pair introduces an additional degree of freedom in the design and layout of the transistors. In order to minimize layout parasitics, PMOS and NMOS transistors were laid out with the same number of fingers. The finger widths have been chosen in order to optimize the f$_{MAX}$ of the pair. The expected overall fixed capacitance contributed by the active devices, including interconnect parasitics, is approximately 90 fF. In order to minimize parasitic series inductances, the coarse tuning varactors were laid out right beside the active devices, as shown in Fig. 3. The single-turn inductor, physically implemented on the top metal layer (*M6*), has been optimized for maximum Q and to facilitate the placement of active core and capacitor arrays. An accurate inductor circuit model, which is of primary importance for accurate simulations, has been derived from Agilent-Momentum EM simulations. The inductor layout and its circuital double-π model [8], valid in the range 50-70GHz, are reported in Fig. 3. Inductance value is about 70 pH, with a differential Q of 19 at 60 GHz.

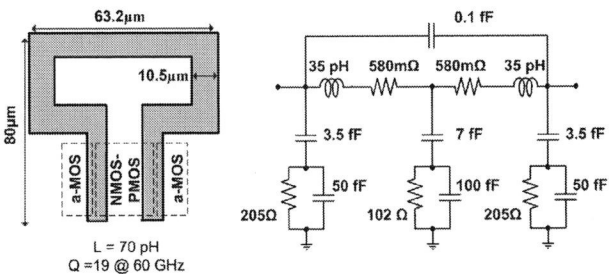

Fig. 3 Layout and equivalent circuit model of the inductor

IV. EXPERIMENTAL RESULTS

The test chip has been designed and implemented in TSMC 90nm CMOS process with 6 levels of metal and ultra thick top metal. A chip micrograph is reported in Fig. 4. Bias and control pads are located on the top (not showed) while the two outputs, suitable for on-chip measurement with GSG probes, are in the lower portion. The chip, including all pads, measures 1078 x 760 μm², but most of the chip area is occupied by the output buffer, which is only required for testing purposes. The core circuit area is only 106 x 83 μm².

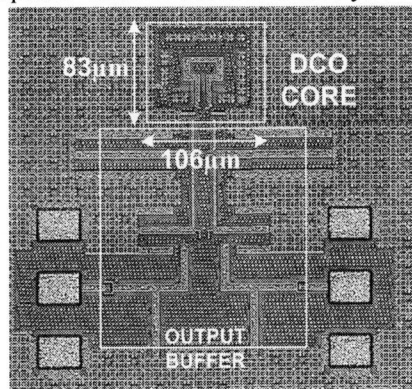

Fig. 4 Die microphotograph.

The measurement setup for the DCO characterization is reported in Fig. 5. The oscillator was characterized on-chip by directly probing its output in single-ended configuration. The outgoing signal from the probes is down-converted by means of a Wisewave V-band mixer, driven by a 51-53GHz local-oscillator provided by an Agilent E8527D generator. In order to improve phase noise measurement accuracy, an intermediate-frequency amplifier on a custom-made board is used to boost the signal level. Phase noise is finally measured using an R&S FSUP signal analyzer using the cross-correlation method. The lack of the tail current generator in this design makes the oscillator more sensitive to disturbances on the power supply. This has been addressed during testing using two separate boards: one small-sized microwave board where the chip is down-bonded and large bypass capacitors are used to heavily filter the supply and control lines; one larger control board that is used to generate and process the control signals.

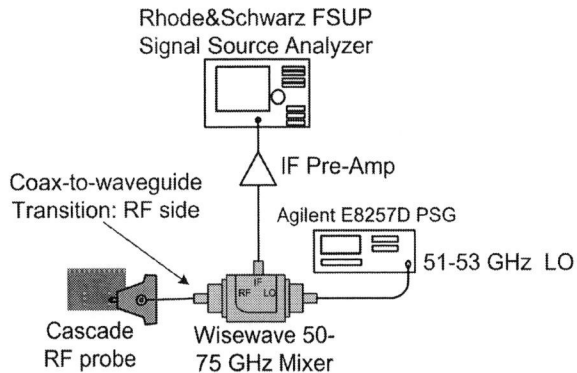

Fig. 5 Experimental setup for the DCO characterization

The oscillation frequency as a function of the coarse tuning 5-bit binary control word is given in Fig. 6. The measured tuning range of 2 GHz, is lower than predicted probably due to underestimation of the minimum depletion capacitance of the a-MOS varactor by the approximate model used in this work. The average frequency step is 64 MHz, with less than 50-MHz differential non-linearity (DNL). The intermediate tuning section shows a tuning range of 102 MHz, an average frequency step of 14 MHz and a DNL of less than 3 MHz. The frequency range covered by the fine tuning section is 54 MHz, with an average step of 1.8 MHz. Accurate direct measurements of the individual fine frequency steps were not possible using the aforementioned setup. Therefore, an indirect measurement has been carried out taking advantage of the thermometer-code control to toggle a single capacitor at a time, performing a narrow-band digital frequency modulation. This results in a pair of sidebands around the carrier with relative amplitudes proportional to the frequency step that can be easily measured using the spectrum analyzer. The modulating frequency must be chosen carefully in order to minimize unwanted effects due to on-board parasitics on the ground and control lines, and such that the narrow-band assumption is satisfied. A modulating frequency between 30-40MHz has been chosen. Due to an implementation mistake in the control board only a sub-set of the fine-tuning capacitors could be tested. The measured frequency step is equal on average to 1.78 MHz, consistently with the direct measurement carried out between the extremes of the tuning code, with a maximum deviation of 200 kHz.

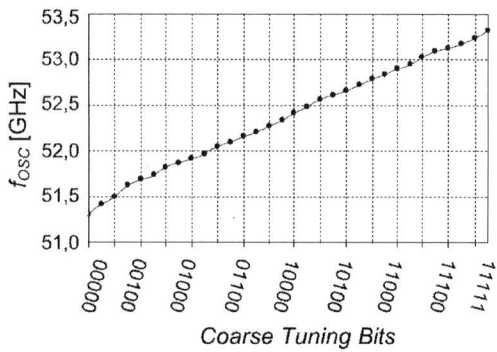

Fig. 6 Measured oscillation frequency versus coarse tuning control word.

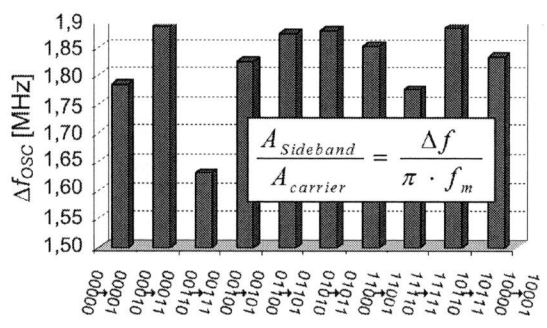

$$\frac{A_{Sideband}}{A_{carrier}} = \frac{\Delta f}{\pi \cdot f_m}$$

Fig. 7 Fine-tuning frequency step measurements results.

The measured phase noise at the center of the tuning range is reported in Fig. 8. The phase noise at 10-MHz offset from the carrier results -116.5 dBc/Hz and remains substantially flat over the tuning range. The DCO core consumes 2.34 mW, when operated with a 1.2 V supply and presents a figure-of-merit (FoM) of -187.2 dBc/Hz, which compares favorably with other LC oscillators operating in this frequency range. In Fig. 9 the FoM of the proposed DCO is compared with other published LC-tank and standing-wave oscillators (SWO) operating in the mm-wave frequency range.

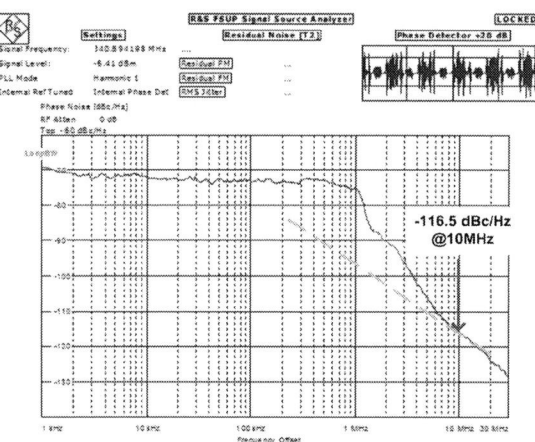

Fig. 8 Measured phase noise at 52.45GHz oscillation frequency.

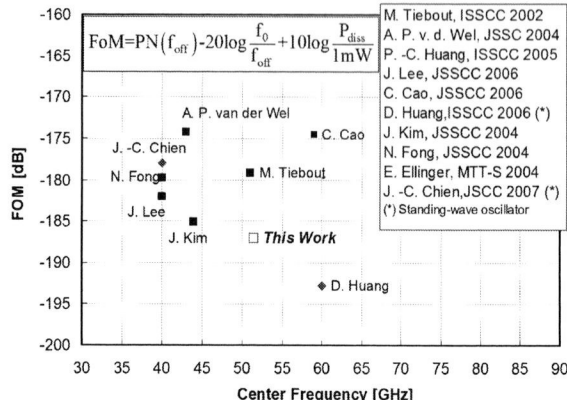

Fig. 9 Millimeter-wave oscillators FoM comparison.

V. CONCLUSIONS

A low phase-noise digitally-controlled oscillator suitable for millimeter-wave wireless applications has been proposed and demonstrated for the first time in this paper. The circuit, implemented in a digital 90nm CMOS process, shows very good efficiency and enables to employ fully digital frequency control in the frequency synthesizer.

ACKNOWLEDGMENT

The authors wish to thank R. Castello and Marvell for providing technology access and S. Shia (TSMC) for design kit support. The authors gratefully acknowledge Integrand Software and Agilent for providing simulation tools and Rhode-Schwarz for phase noise measurement equipment.

REFERENCES

[1] B. A. Floyd, et al., "SiGe Bipolar Transceiver Circuits Operating at 60 GHz", in *IEEE Journal of Solid-State Circuits*, vol. 40, No. 1, pp. 156-167, Jan 2005

[2] C. H. Doan, S. Emami, A. M. Niknejad, R.W. Brodersen, "Millimeter-Wave CMOS Design", in *IEEE Journal of Solid-State Circuits*, vol. 40, No. 1, pp. 144-155, Jan 2005

[3] C.-H. Wang et al., "A 60GHz Transmitter with Integrated Antenna in 0.18μm SiGe BiCMOS Technology", in *IEEE International Solid-State Circuits Conference, Digest of Technical Papers (ISSCC 2006)*, pp 659 - 668, Feb. 6-9 2006

[4] R.B. Staszewski, P. T. Balsara, "All-Digital Frequency Synthesizer in Deep-Submicron CMOS", *Wiley and Sons, 2006*

[5] N. Da Dalt, C. Knopf, M. Burian, T. Hartig, H. Eul, "A 10b 10GHz digitally controlled LC oscillator in 65nm CMOS", in *IEEE International Solid-State Circuits Conference, Digest of Technical Papers (ISSCC 2006)*, pp 188 – 189, Feb. 6-9 2006

[6] H. Sjoland, "Improved Switched Tuning of Differential CMOS VCOs", *IEEE Transactions on Circuits and Systems-II: Analog and Digital Signal Processing*, Vol.49, No.5, pp 352-355, May 2002

[7] J.Victory, Y. Zhixin, G. Gildenblat, C. McAndrew, A. J. Zheng, "A Physically Based, Scalable MOS Varactor Model and Extraction Methodology for RF Applications", *IEEE Transactions on Electron Devices*, Vol.52, No.7, pp. 1343-1353, July 2005

[8] T.O. Dickson, M. A. LaCroix, S. Boret, D. Gloria, R. Beerkens, S.P. Voinigescu ,"30-100-GHz Inductors and Transformers for Millimeter-Wave (Bi)CMOS Integrated Circuits", *IEEE Transaction on Microwave Theory and Techniques*, Vol.53, No.1, pp 123-133, January 2005.

IEEE 2008 Custom Intergrated Circuits Conference (CICC)

Twisted Inductors for Low Coupling Mixed-Signal and RF Applications

Nathan M. Neihart[1], David J. Allstot[1], Matt Miller[2] and Pat Rakers[2]

[1]University of Washington, Seattle, WA, 98195
[2]Freescale Semiconductor Inc, Lake Zurich, IL, 60047

Abstract—Parasitic magnetic coupling is a major design challenge for integrated circuit designers. Fundamentally, it originates in conventional spiral inductors because the magnetic field is not localized, extending far beyond the perimeter. This paper introduces a twisted winding scheme for inductors that increases the localization of the magnetic field, reducing parasitic magnetic coupling by as much as 3100X and the edge-to-edge spacing of inductors by 10X. These results are validated in a 0.18μm CMOS process.

I. INTRODUCTION

Parasitic magnetic coupling is a major design challenge in integrated systems employing integrated inductors, which are found in virtually all modern integrated circuits from analog-intensive FIR filters to multiple-input multiple-output (MIMO) transceivers. Moreover, in the face of continued scaling in CMOS technologies, fully-differential designs requiring differential to single-ended conversion are becoming more common. In these systems, integrated transformers are often employed to perform differential to single-ended conversion and vice-versa, further complicating the issue of magnetic coupling. Fundamentally, parasitic coupling originates in conventional spiral inductors because the magnetic field is not localized; it extends far beyond the perimeter of the inductor causing parasitic currents to be induced in nearby inductors.

Fully integrated single- and multiple-antenna transmitters are especially sensitive to parasitic coupling. A prominent source of parasitic coupling in these systems is magnetic coupling between the inductive load of a power amplifier (PA) and the tank inductor in the local oscillator (LO) generator. This results in frequency pulling, as well as frequency modulation, of the LO signal by the PA signal through injection locking [1]. Additionally, parasitic magnetic coupling in multiple-antenna transceivers increases the correlation between ideally independent RF chains, consequently reducing the effectiveness of MIMO techniques targeting high data rate applications such as spatial multiplexing. In the case of severe coupling, the MIMO algorithm can fail completely, reducing the data rate to zero.

This point is underscored in recent 2-antenna spatial multiplexing transceivers that have been realized in silicon wherein either critical inductors are implemented off-chip in the package substrate to avoid parasitic magnetic coupling concerns [2],[3] or substantial calibration is necessary [4].

Traditional methods for reducing the effects of parasitic coupling include using a relatively large separation distance between ideally unrelated inductors and, in fully integrated transmitters, offsetting the carrier frequency from the LO frequency. Separating the inductors (e.g., by as much as 100μm) wastes silicon area and increases cost. Offsetting the carrier frequency, f_C, from the LO frequency, f_{LO}, has been used successfully, but due to physical limitations of the process, it becomes impractical at mm-wave frequencies and beyond. Clearly, there is a need for area and frequency efficient methods that better contain the magnetic fields of integrated inductors.

This paper introduces a simple winding scheme for integrated inductors that reduces the amount of parasitic magnetically coupled power by as much as 3100X (35dB). Furthermore, the edge-to-edge spacing between inductors using this new winding scheme has been reduced by 10X (e.g., from 100μm to 10μm). In recent work, similar 'figure-8' structures were independently developed for use in power combining applications but no investigation into the design of inductors using the new layout was reported [5]. Section II introduces the new winding scheme for integrated inductors. Section III presents measured results and compares the twisted inductors to traditional spiral inductors in terms of area, Q, self-resonant frequency, etc. Finally, conclusions are made in Section IV.

II. TWISTED INDUCTORS

In a traditional integrated spiral inductor the magnetic field is oriented in one of two directions, depending on the direction of current flow, and extends far beyond the perimeter of the structure. As the time-varying magnetic field passes through a nearby inductor, parasitic current is induced resulting in coupling. In order to reduce the level of magnetic

978-1-4244-2018-6/08 $25.00 © 2008 IEEE

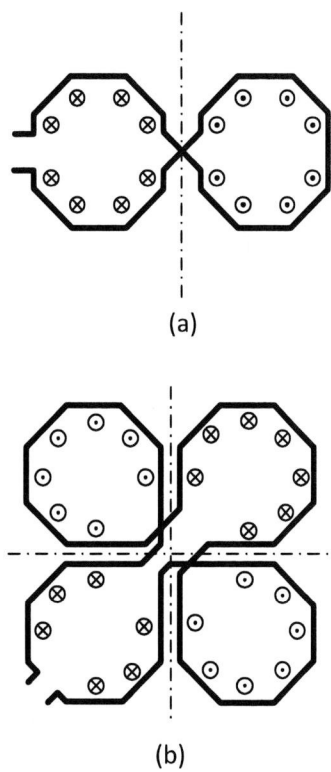

(a)

(b)

Figure 1: Twisted winding scheme showing the polarities of the magnetic field in each lobe for a (a) 2-lobed and a (b) 4-lobed inductor. The dashed line represents the axes of minimum coupling.

coupling, the localization of the magnetic field of the inductor should be increased. Improved localization of the magnetic field can be achieved by using a winding scheme that resembles a section of a twisted pair, Fig. 1(a). By winding the inductor in this manner, the magnetic fields in each lobe are equal in magnitude but opposite in polarity. Thus, beyond the perimeter of the inductor the magnetic fields cancel each other. Another way of saying this is that the flux lines that leave one 'lobe' ideally enter the other lobe and thus do not couple to unrelated inductors. The level of magnetic field cancellation is maximum along the dashed lines in Fig. 1 (referred to as the axes of minimum coupling), and minimum along the orthogonal axis, for example, in Fig. 1(a).

Two different inductors are designed: a 2-turn, 1nH 2-lobe inductor similar to Fig. 1(a) and a 2-turn, 1.5nH 4-lobe inductor similar to Fig. 1(b). A 3D electromagnetic simulator is used to simulate the level of magnetic coupling at 5GHz as a function of distance for the following three test cases: the 2-lobe inductor and a traditional spiral inductor (2-lobe/spiral pair), the 4-lobe inductor and a traditional spiral inductor (4-lobe/spiral pair) and two equal inductance spiral inductors (spiral/spiral pair). Fig. 2 shows that the level of simulated coupling in the 2-lobe/spiral pair is reduced by approximately

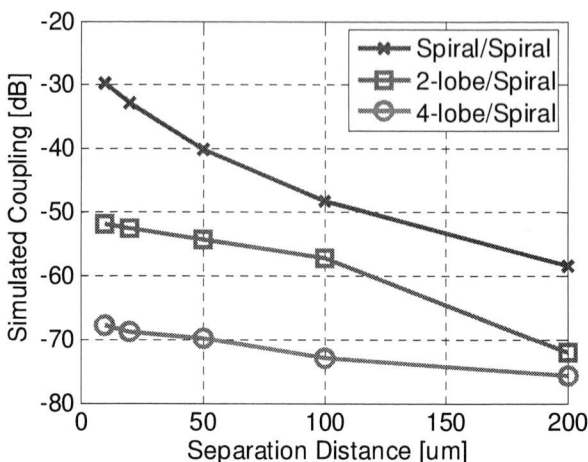

Figure 2: Simulation results of the magnetic coupling versus separation distance at 5GHz between the spiral/spiral, 2-lobe/spiral and 4-lobe/spiral pairs.

300X (25dB) at a spacing of 10µm compared to the spiral/spiral pair. Further improvement is realized in the 4-lobe/spiral pair where there is an approximate reduction of 3100X (35dB) at 10µm separation distance. Interestingly, no further benefit is gained by using a 2-lobe/2-lobe or 4-lobe/4-lobe pair. Fig. 2 also shows that two conventional spiral inductors must be separated by approximately 100µm in order to achieve the same level of coupling as the 2-lobe/spiral pair with only 10µm of separation. The separation must be even greater to match the level of coupling of the 4-lobe/spiral pair. In the competitive market, it is desired to have closely-spaced inductors to minimize silicon area and cost. The coupling advantages of the twisted structure are greatest at small separation distances between inductors; hence, this solution is attractive for dense circuit layouts.

One drawback to the twisted inductors is the large variation in magnetic coupling shown in Fig. 3(a), at 5GHz, for different positions of the neighboring spiral inductor. Fig. 3(b) illustrates how the spiral inductor is positioned in order to obtain the results shown in Fig. 3(a). The spacing between the twisted and spiral inductors was maintained at 10µm. Because of this large variation, careful layout is needed to ensure that critical inductors are placed to minimize parasitic coupling. The 4-lobe inductor gives more freedom in placement by virtue of its two axes of minimum coupling. Moreover, the level of coupling is at least 10dB less than that of the 2-lobe inductor for all angles and even at maximum coupling it is still 16dB lower that a standard spiral inductor.

III. MEASURED RESULTS

To demonstrate the dramatic reduction in both parasitic coupling and edge-to-edge inductor spacing, a test chip was fabricated in a 0.18µm CMOS process. Fig. 4 shows a

978-1-4244-2018-6/08 $25.00 © 2008 IEEE

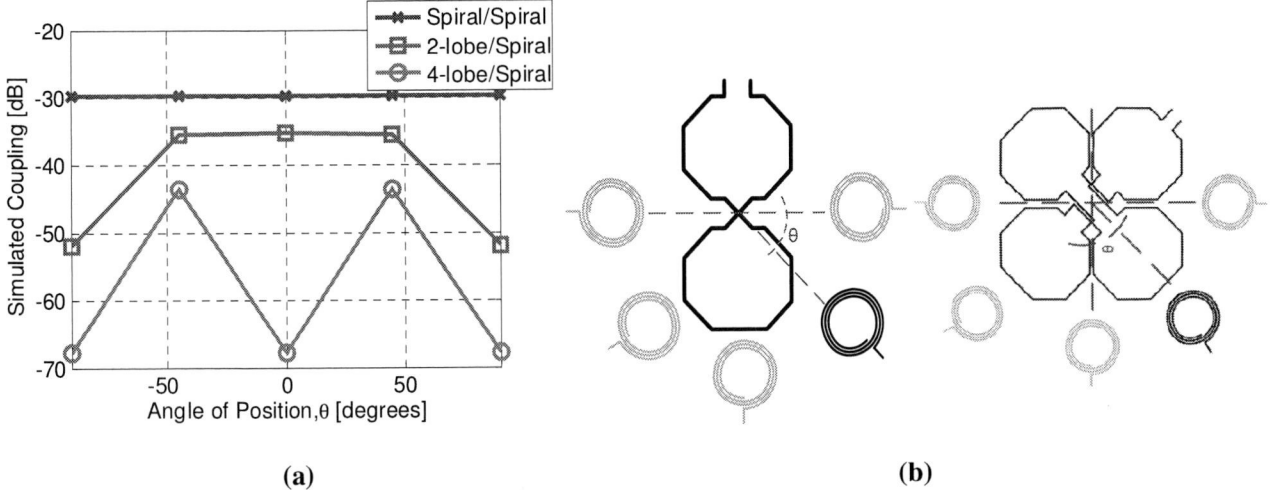

(a)　　　　　　　　　　　　　　　　(b)

Figure 3: (a) Simulation results of magnetic coupling versus inductor position at 5GHz for three different structures: spiral/spiral, 2-lobe/spiral and 4-lobe/spiral pairs. (b) Illustration of test setup showing how the spiral inductor is positioned in order to obtain the results shown in (a). Inductor spacing is maintained at 10μm.

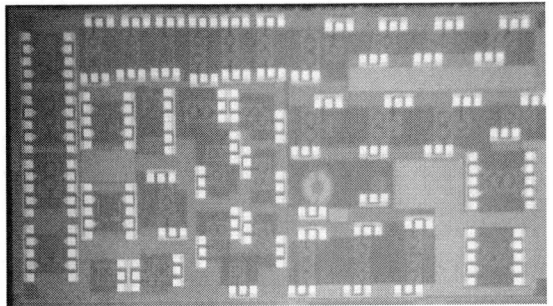

Figure 4: Test chip for validation of twisted inductors. Fabricated in a 0.18μm CMOS process the chip measures 5mm X 2.5mm.

microphotograph of the test chip, which consists of 2-lobe/spiral and 4-lobe/spiral pairs with separation distances of 10μm, 20μm and 50μm. The inductance of the 2-lobe, 4-lobe, and spiral inductors is 1.1nH, 1.4nH and 2.2nH, respectively. All of the measurements are performed at 5GHz unless otherwise noted. Moreover, for all measurement results, the pads and extraneous routing are de-embedded.

A. Coupling Results:

Fig. 5 shows the measured coupling versus separation distance for the 2-lobe/spiral, 4-lobe/spiral and spiral/spiral pairs. Plotted over the measured data points in Fig. 5 are the results obtained from simulation, showing a good agreement between the two. Simulation results of similar structures developed independently also confirm our measured results [6]. The spiral inductors have been placed along the axes of minimum coupling for both the 2- and 4-lobe inductors. Although they were not used in this test chip, guard rings added around each of the inductors can further reduce the parasitic magnetic coupling at the expense of area [7].

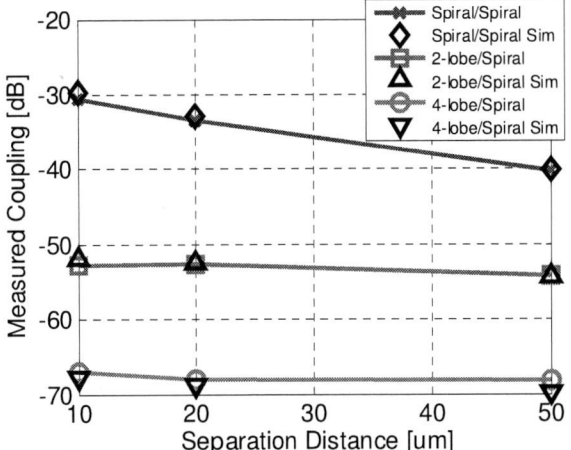

Figure 5: Measured coupling versus separation distance for three different structures at 5GHz: spiral/spiral, 2-lobe spiral and 4-lobe/spiral pairs. Simulated values are plotted over the measured values showing good agreement between simulations and measurements.

The coupling was also measured as a function of inductor position for the three different pairs. The edge-to-edge spacing remained constant at 10μm for all positions. The results are plotted in Fig. 6. As expected, the magnetic coupling for the spiral inductor is the same for all angles. The level of coupling of the 2- and 4-lobe inductors is closely predicted by simulation results. For angles of approximately ±45° the magnetic coupling is only 6dB less than that of the standard spiral. The 4-lobe coupling shows at least a 10dB improvement over the 2-lobe inductor for all angles. Furthermore the presence of an extra axis of minimum coupling greatly reduces the level of coupling for θ=0°.

If the number of lobes is greater than four, the level of coupling is further reduced due to an increasing number of

978-1-4244-2018-6/08 $25.00 © 2008 IEEE　　　613

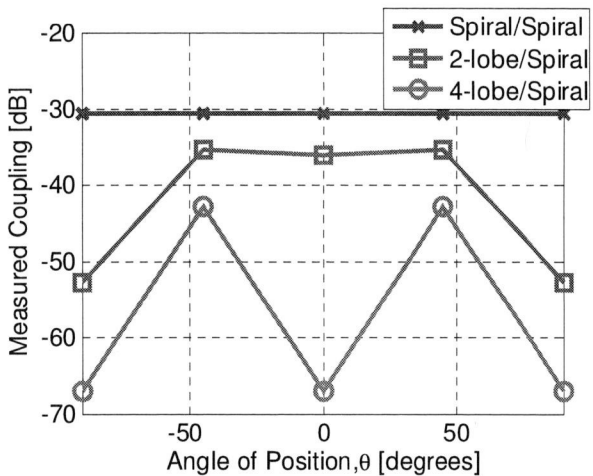

Figure 6: Measured coupling versus inductor position for three different structures. The inductor spacing in all cases is 10µm and frequency of operation is 5GHz.

axes of minimum coupling. Currently the requirement for a layout to utilize angles that are integer multiples of 45° limits the number of lobes that can be easily fabricated.

B. Comparison of Twisted Inductors to Spiral Inductors

Finally the twisted inductors are compared with standard spiral inductors in terms of inductance, area, Q, and self-resonant frequency. Two 4-lobe inductors were characterized: a 1-turn, 500pH inductor and a 2-turn, 1.5nH inductor. Three 2-lobe inductors were characterized: a 1-turn 300pH inductor, a 2-turn 1nH inductor and a 4-turn, 8.5nH inductor. These twisted inductors are compared to two different spiral inductors: a 2-turn 500pH inductor and a 4-turn 2.25nH inductor. Again all measurements are performed at 5GHz.

Table I summarizes the relevant parameters for these test inductors. Due to the increased sheet resistivity resulting from the extra metal vias needed to realize the twists in the twisted inductors, the Q is slightly degraded when compared to the traditional spiral inductor. The self-resonant frequency, f_{SR}, of the twisted inductors is slightly lower due to the larger area needed, but it is still high enough for usage in the 1-10GHz band for all but the 8.5nH 2-lobe inductor. A spiral inductor of similar inductance would have the same limitation, however.

One trade-off worth noting is that the minimum realizable inductance of a twisted inductor is going to be slightly higher than that of a standard spiral. This is because, for a given minimum sized loop, a 2-lobe inductor will use two such loops

TABLE I
MEASURED PARAMETER COMPARISON OF TWISTED INDUCTORS TO SPIRAL
INDUCTORS

Inductor	L [nH]	R_s [Ω]	Q@5GHz	F_{SR} [GHz]	Area[µm²]
1 Turn 4-lobe	0.587	4.55	4.1	>40	116 x 116
2 Turn 4-lobe	1.43	13.5	3.3	18	116 x 116
1 Turn 2-lobe	0.290	2	4.6	>40	118 x 56
2 Turn 2-lobe	1.1	7.31	4.7	20	163 x 80
4 Turn 2-lobe	8.4	45.35	5.8	9	247 x 127
2 Turn Spiral	0.452	2.1	6.8	>40	75 x 75
4 Turn Spiral	2.27	8.35	8.5	17	100 x 100

and a 4-lobe inductor will require four loops thus increasing the amount of inductance over a single loop spiral inductor. This is not expected to be a problem, however, for inductances greater than approximately 200pH.

IV. CONCLUSIONS

We have demonstrated that using a simple twisted winding scheme for integrated inductors reduces the level of parasitic magnetic coupling by as much as 3100X (35dB) whereas the edge-to-edge spacing is reduced by 10X. This 10X reduction in inductor spacing, and the resulting savings in overall circuit area, offsets the slight increase in area of a twisted inductor over a spiral inductor of equal inductance. Moreover these benefits are realized using a standard CMOS process with no added processing steps so this technique is applicable to a wide range of technologies.

V. REFERENCES

[1] B. Razavi, "RF IC design challenges," *Design Automation Conference*, pp. 408-413, June 1998.

[2] Y. Palaskas, et al. "A 5GHz 108Mb/s 2x2 MIMO transceiver with fully-integrated +16dBm Pas in 90nm CMOS," *ISSCC*, pp. 362-363, 659, Feb. 2006.3

[3] Y Palaskas, et al. "A 5-GHz 108-Mb/s 2x2 MIMO transceiver RFIC with fully-integrated 20.5dBm P_{1dB} power amplifiers in 90-nm CMOS," *IEEE J. Solid-State Circuits*, vol. 41, pp. 2746-2755, Dec. 2006.

[4] A. Behzad, et al. "A fully integrated MIMO multiband direct conversion CMOS transceiver for WLAN applications (802.11n)," *IEEE J. Solid-State Circuits*, vol. 42, pp. 2795-2808, Dec. 2007.

[5] G. Liu, P. Haldi, T. K. Liu and A. M. Niknejad, "Fully integrated CMOS power amplifier with efficiency enhancement at power back-off," *IEEE J. of Solid-State Circuits*, vol. 43, no. 3, pp. 600-609, March 2008.

[6] T. Mattsson, "Method of and inductor layout for reduced VCO coupling," US Patent No. 7,151,430, Dec. 2006.

[7] J.H. Mikkelsen, O.K. Jensen, and T. Larsen "Crosstalk coupling effects of CMOS co-planar spiral inductors," *Custom Integrated Circuits Conference*, pp. 371-374, Oct. 2004.

978-1-4244-2018-6/08 $25.00 © 2008 IEEE

IEEE 2008 Custom Intergrated Circuits Conference (CICC)

A Common-Base Linear RF Power Amplifier for 3G Cellular Applications

Flavio Avanzo, Francesco M. De Paola, and Danilo Manstretta

Dipartimento di Elettronica, Università degli Studi di Pavia
via Ferrata 1, 27100, Pavia, Italy

Abstract— A linear power amplifier for 3G cellular applications is presented. The amplifier operates in common-base configuration and can sustain output voltages in excess of BV_{CEO}. The chip, implemented in a 0.25μm SiGe:C technology, occupies 2.76 mm². When operated from a 4.5 V supply, the amplifier has a measured power gain of 20 dB at 1.85 GHz. At 1dB Compression Point, the amplifier delivers 27 dBm with a power-added efficiency of 33%. Saturated output power is 28.2 dBm with 37% power-added efficiency.

I. INTRODUCTION

Third-generation cellular [1] and modern wireless LAN standards use modulation schemes with high spectral efficiency but non-constant envelope and therefore require linear amplification. The high peak-to-average power ratio typical of these signals forces the power amplifier (PA) to work several dBs below its peak output power, even when transmitting at maximum average output power. As a result, for a given average output power, the optimum load impedance and the amplifier average efficiency are significantly reduced compared to constant envelope operation. This problem is even more difficult to address using the most advanced silicon processes. In fact, the tremendous improvements in f_T and f_{MAX} of modern SiGe and SiGe:C technologies have been partly achieved exploiting the tradeoff between transit time and breakdown voltage: i.e. higher f_T and f_{MAX} have been exchanged for a reduction in breakdown voltage [2][3]. The consequent reduction in supply voltage pushes the optimum load impedance further down. As the load impedance level is reduced, the impedance transformation ratio from the PA output to the antenna 50Ω load increases, making the matching network more sensitive to components variations and parasitic elements that ultimately impact operative bandwidth and efficiency. To address these issues, a common-base topology that is able to sustain output voltages well in excess of the collector-emitter breakdown voltage (BV_{CEO}) has been adopted in this work. The circuit topology and its voltage capability are discussed in Section II. Design and experimental results are reported in Section III. Section IV draws some conclusion.

This work has been carried out in the framework of the Italian National Research Program FIRB (contract nr. RBAP06L4S5).

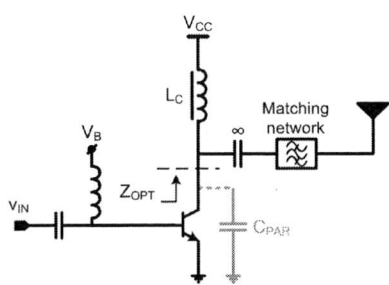

Fig. 1 Classic common-emitter amplifier power stage.

II. COMMON-BASE POWER AMPLIFIER OPERATION

The classical solution adopted in the power stage of linear power amplifiers [4] is a common-emitter amplifier such as the one reported in Fig. 1. The optimum load is given by a parallel inductance that cancels out the total output capacitance seen at the transistor collector and a parallel resistance that is set essentially by the supply voltage and the desired power level:

$$R_{OPT} \cong \frac{(V_{CC} - V_{SAT})^2}{2P_{SAT}} \qquad (1)$$

where V_{CC} is the supply voltage, V_{SAT} is the minimum collector-emitter voltage that keeps the device in the linear operating region and P_{SAT} is the maximum output power. An high supply voltage is desirable since it allows to deliver a large output power with a relatively high optimum load impedance, simplifying the output impedance transformation network. High speed SiGe technologies can sustain higher voltages compared to CMOS and are therefore usually preferred in RF power amplifier designs. As the technology evolves toward higher frequency capabilities, the contextual decrease in breakdown voltage pushes the supply voltage down, significantly lowering the optimum load impedance. At the technological level this has been addressed by providing several transistor designs, optimized for high power or high frequency operation [3]. Hence, using high-voltage transistors biased well below BV_{CEO} is a viable, albeit inefficient, solution under optimum load conditions

978-1-4244-2018-6/08 $25.00 © 2008 IEEE   615

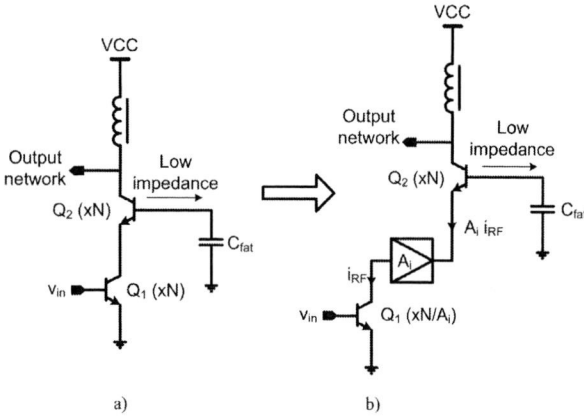

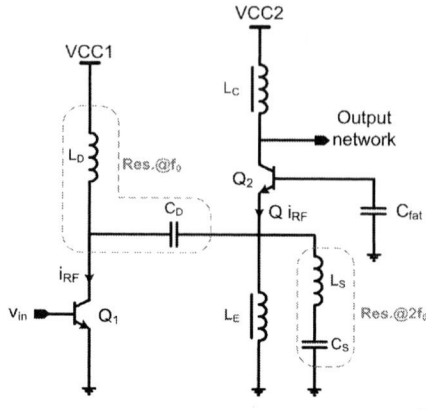

Fig. 2 Alternative PA configurations: a) cascode amplifier; b) cascode amplifier with inter-stage impedance-transformation network.

Fig. 3 Common-base amplifier with LC impedance-transformation network and second harmonic short.

Nonetheless, ensuring reliable operation of a power amplifier in real-world conditions is less straightforward. In fact, load mismatch conditions, e.g. due to antenna impedance variations, can lead to instantaneous over-voltages that, for a voltage standing-wave ratio (VSWR) of 10:1, can go up to four times the supply voltage [5]. Under worst-case conditions, such as during battery recharge, an output voltage as high as 20 V may be reached [5]. This means that, even using high-voltage devices, special protection countermeasures need to be taken to avoid device failure. A possible solution comes from the device bias configuration. When the output voltage exceeds BV_{CEO}, avalanche current multiplication occurs in the collector-base junction and an additional current in the base terminal is generated, with opposite sign compared to the base bias current [3][6]. An avalanche current as small as the base bias current could then lead to device breakdown. On the other hand, if the bias network presents a sufficiently low impedance at the base terminal and has sufficient compliance to absorb the avalanche-generated current, the collector is allowed to rise significantly above BV_{CEO} before reaching a breakdown condition, ideally up to collector-base breakdown (BV_{CBO}), which is much larger than BV_{CEO}. In order to approach this theoretical limit, the base impedance must be kept low at DC and all (even and odd) harmonics of the RF signal frequency. In practice, achieving this condition without introducing undue losses in the RF path is not trivial in a common-emitter amplifier, where the input RF signal and the bias network are connected through the same terminal (i.e the base) to the power transistor. The most straightforward and reliable solution would be to use a cascode amplifier, as shown in Fig. 2.a. This solution makes it easy to provide a low impedance to the base terminal of the output (common-base) transistor, where no RF signal is present, by simply using a large capacitor, while the common-emitter device is protected from high voltages by the common-base device. On the other end,

such a configuration is highly inefficient because the voltage stacking of the two devices, both working at high current levels, significantly increases power dissipation. A more efficient solution, conceptually similar to the one proposed in [7], is shown in Fig. 2.b. The inter-stage passive impedance transformation network scales down the impedance level going from the common-emitter to the common-base device. This amounts to a passive current amplification and, as a result, the common-emitter device signal current level is significantly lower compared to the common-base device. Assuming that the inter-stage network is AC-coupled, the common-emitter device can be biased at a reduced current level, in principle proportionally to the current amplification contributed by the inter-stage network. A simplified version of the circuit implementation proposed in this work is reported in Fig. 3. The circuit consists of a power stage (Q_2), a driver stage (Q_1) and an inter-stage LC matching network. The power stage provides voltage and power amplification and has nominally unity current gain. Hence, its power gain is given by the ratio of the load and input (emitter) impedance. The LC resonator formed by L_D and C_D resonates at the fundamental frequency and, together with Q_2, provides current amplification. In fact, ignoring the parasitic capacitance at the output of the driver, the RF signal current flowing in L_D and C_D (hence also in Q_2) is equal to the driver's output current multiplied by the loaded quality factor (Q), given by

$$Q = \left(\frac{1}{Q_D} + \sqrt{\frac{C_D}{L_D}} \frac{1}{g_{m2}} \right)^{-1} \cong \sqrt{\frac{L_D}{C_D}} g_{m2} \qquad (2)$$

where g_{m2} is the transconductance of the power transistor and Q_D is the quality factor of inductor L_D. The use of an LC network to perform the current amplification allows for a certain degree of flexibility in the choice of the current gain, the upper bound to the gain being the inductor

978-1-4244-2018-6/08 $25.00 © 2008 IEEE   616

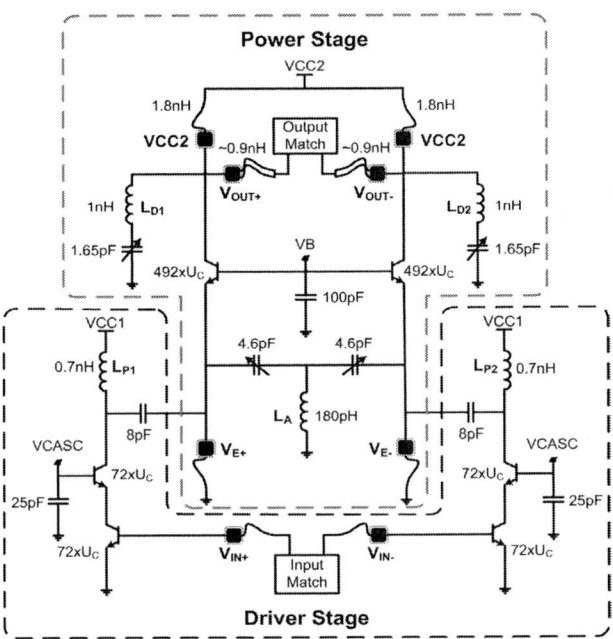

Fig. 4 Complete circuit schematic of the differential common-base power amplifier with the driver stage.

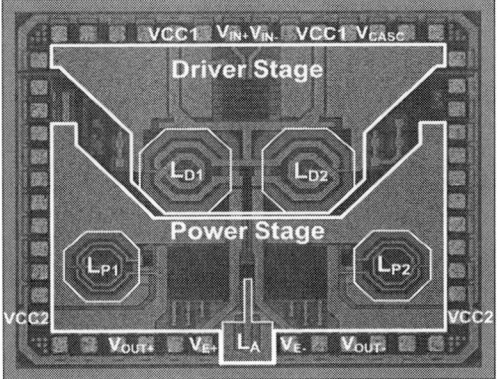

Fig. 5 Die microphotograph

quality factor. A large capacitance C_D, hence a low inductance L_D, is desirable in order to minimize the contribution of the inductor quality factor Q_D to the loaded Q and the relative weight of the driver output capacitance compared with C_D, minimizing the sensitivity to parasitic elements. In this way most of the power supplied by the driver stage is delivered to the power stage and the loaded Q is less dependent on process variations. On the other hand, as C_D is increased, current gain decreases. The choice of C_D also impact the driver efficiency. From this point of view, C_D should be set according to the optimum loading condition:

$$Q\sqrt{\frac{L_D}{C_D}} = R_{OPT,DRV} \cong \frac{(V_{CC1} - V_{SAT})A_i}{I_{RF}} \qquad (3)$$

where V_{CC1} is the driver voltage supply, A_i is the current gain and I_{RF} the power stage maximum RF current. Control of the impedance at the source of the power transistor at the harmonics of the signal is key to determine the desired amplification class. Usually power transistors are operated in class AB mode, that provides the best trade-off between linearity and efficiency. In this operating mode, a large second harmonic is present in the power transistor emitter current [4]. If a large impedance is present at the emitter of the power transistor around the second harmonic of the signal frequency, the current at this frequency is rejected and the transistor can not work properly in class-AB, leading to lower efficiency and early compression. A second harmonic short is implemented in Fig. 2.b using L_S and C_S. A parallel large inductor L_E provides a low impedance path for the DC current.

III. CIRCUIT DESIGN AND EXPERIMENTAL RESULTS

The complete schematic of the common-base power amplifier is shown in Fig. 4. The PA was designed in a low cost 0.25μm SiGe:C technology from ST-Microelectronics. The technology features high break-down voltage transistors with BV_{CEO} of 6 V and BV_{CEO} of 19 V, typical f_T equal to 30 GHz and f_{MAX} of 60 GHz. A differential solution was chosen to further boost the optimum impedance load. In fact, using a differential topology, the optimum (differential) load is increased by a factor of four compared to the expression in (1). The PA was optimized to operate according to 3GPP/UMTS specifications [1] in power class 3, with a center frequency of 1.95 GHz, average output power of 24 dBm and peak output power around 27 dBm. The voltage supply for the power stage was set to 4.5 V, to ensure safe operation even in the worst load mismatch conditions and avoid thermal runaway [8]. Estimating a V_{SAT} of about 1V, with a supply voltage V_{CC} set to 4.5 V, the optimum differential load resistance is approximately 50 Ω or 25 Ω per side. In practice, due to parasitic elements and non-linearity, a lower impedance is required, or equivalently an higher signal current. The power stage transistors were designed in order to reach the desired power level with maximum efficiency and linearity. In fact when the output impedance is fixed, there is only one transistor size (in terms of total emitter area and device unit size) which maximizes linearity and efficiency for the desired output power. If the transistor is made larger than the optimum size voltage compression will dominate the output power compression; if the transistor's size is lower than the optimum, current compression dominates [1]. The transistor optimum size has been found by sweeping the transistor size with fixed optimum load and by simulating the output power and efficiency. At the power stage's output, a common-mode series resonant network is used to shunt second harmonic components in the output current in order extend the linear output range . Since second harmonics appear as a common mode, only one common-mode small-sized .

978-1-4244-2018-6/08 $25.00 © 2008 IEEE 617

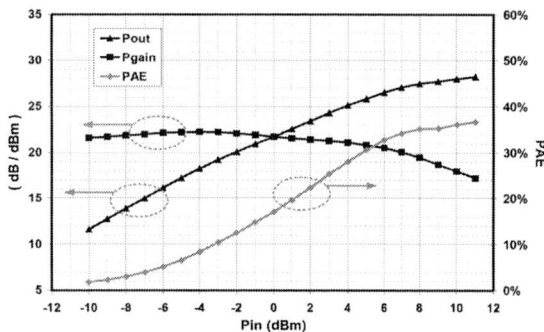

Fig. 6 Measured power gain, output power and PAE at 1.8 GHz.

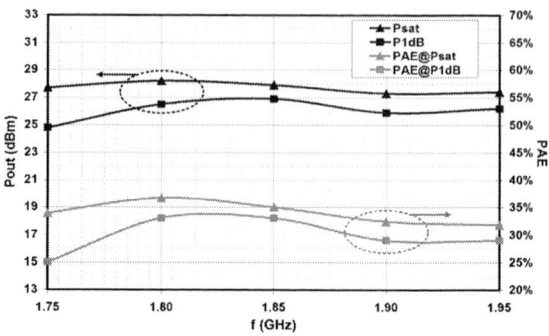

Fig. 7 Measured P-1dB and power added efficiency versus frequency.

inductor (L_A in Fig. 4) is needed, saving area. The series capacitance is made controllable using MOS switches, allowing to accommodate for process variations. A bond-wire is used to supply the DC current at the power stage emitter. This bond-wire does not carry RF signal current and, due to the low impedance present at the emitter, the exact value of its inductance is not important as long as it is sufficiently high. The driver stage provides additional stable power gain and supplies the necessary current drive to the power stage. The inter-stage matching network was sized to obtain a current gain of about 4. According to the inductor model and circuit simulations the loss due to the finite inductor Q is less than 1 dB. Since the power level is much lower than in the power stage, a cascode amplifier can be used in the driver stage, improving reverse isolation and with limited impact on the overall efficiency. The driver was designed such that power compression occurs at a power level greater than the maximum input power needed by the output stage. The overall power compression is then dominated by the power stage. The chip microphotograph is reported in Fig. 5. The die area is 2.76 mm². The chip was tested in a chip-on-board configuration. Three bond-wires are used for each output: one connected to the power supply and two connected to the signal output. A 4 pF capacitor toward ground on each output implements a close-to-optimum impedance transformation. Optimum loading was achieved using manual impedance tuners, whose outputs are measured using a dual-input power meter. The tuners are individually adjusted to compensate for the inevitable mismatch in bond-wire inductances when using non-automated bonding techniques. The measured tuner loss (0.2 dB), cable and connectors loss (0.3 dB) are de-embedded from the measurements. Precise input impedance matching was not implemented, hence exact power gain measurements were not possible. The measured power gain, output power and power-added-efficiency at 1.8 GHz are shown in Fig. 6. The measured 1dB compression point, saturated output power and related efficiencies as a function of frequency are shown in Fig. 7.

IV. CONCLUSIONS

Linear power amplifier design issues associated with the lowering breakdown voltages in advanced SiGe processes are addressed in this work. A common-base linear power amplifier which can sustain output voltages in excess of BV_{CEO} is presented. The chip, implemented in a 0.25µm SiGe:C technology, when operated from a 4.5 V supply, has a measured power gain of 20 dB at 1.85 GHz. At 1dB Compression Point, the amplifier delivers 27 dBm with a power-added efficiency of 33%. Saturated output power is 28.2 dBm with 37% power-added efficiency.

ACKNOWLEDGMENT

The authors gratefully acknowledge A. Liscidini and F. Ramaioli for help with testing. The authors wish to thank K. Torki (CMP) for design kit support.

REFERENCES

[1] "3rd Generation Partnership Project; Technical Specification Group Radio Access Networks; UE Radio Transmission and Reception (FDD)", Release 7, TS 25.101, V7.1.0 (2005-09).

[2] K. K. Ng, M. R. Frei, and C. A. King, "Reevaluation of the ft BVceo limit on Si bipolar transistors," *IEEE Trans. Electron Devices*, vol. 45, no. 8, pp. 1854–1855, Aug. 1998.

[3] H. Veenstra, G. A. M. Hurkx, D. van Goor, H. Brekelmans, and J. R. Long, "Analyses and Design of Bias Circuits Tolerating Output Voltages Above BVCEO", IEEE Journal of Solid-State Circuits, vol. 40, NO. 10, pp2008-2018, Oct 2005

[4] S. C. Cripps, "RF Power Amplifiers for Wireless Communications", Norwood, MA, Artech House, 1999

[5] A. Scuderi, L. La Paglia, A. Scuderi, F Carrara, and G. Palmisano, "A VSWR-Protected Silicon Bipolar RF Power Amplifier with Soft-Slope Power Control", IEEE Journal of Solid-State Circuits, vol. 40, NO. 3, pp 611-621, Mar 2005

[6] M. Rickelt, H.-M. Rein, and E. Rose, "Influence of impact-ionization induced instabilities on the maximum usable output voltage of Si-bipolar transistors," *IEEE Trans. Electron Devices*, vol. 48, no. 4, pp. 774–783, Apr. 2001

[7] D. T. S. Cheung, and J. R. Long, "A 21-26 GHz SiGe Bipolr Power Amplifier MMIC", IEEE Journal of Solid-State Circuits, vol. 40, No. 12, pp 2583-2597, Dec 2005

[8] A. Inoue, S. Nakatsuka, R. Hattori, and Y. Matsuda, "The maximum operating region in SiGe HBTs for RF power amplifiers," in *IEEE MTT-S Int. Microwave Symp. Dig.*, vol. 2, pp. 1023–1026, Jun. 2002

IEEE 2008 Custom Intergrated Circuits Conference (CICC)

Design of Low Power CMOS Ultra-Wideband 3.1-10.6 GHz Pulse-Based Transmitters

Kuan-Yu Lin and Mourad N. El-Gamal

Department of Electrical and Computer Engineering, McGill University
3480 University Street, Montreal, Quebec, Canada H3A 2A7
E-mail: kuan-yu.lin@mail.mcgill.ca, mourad.el-gamal@mcgill.ca

Abstract - The design of two low power CMOS ultra-wideband (UWB) pulse-based transmitters is reported. The goal is to propose simple, low power, and tunable topologies for full-band and sub-band UWB transmitters. The first transmitter utilizes a gated ring oscillator, an NMOS switch, and a passive pulse shaping filter to generate a 3.1-10.6 GHz full-band UWB signal. The second transmitter multiplies a carrier and a triangular signal to up-convert and generate a low side-band UWB signal for sub-band applications. We propose the use of two NMOS switches in series to perform this multiplication while consuming minimum power. Control voltages incorporated in both designs are used to adjust the shapes of the pulses to compensate for mismatch and process variations. Both the full-band and sub-band transmitters, fabricated in a 0.18 μm CMOS process, generate FCC compliant UWB signals with a supply voltage of 1.8 V and power consumptions of 237.4 and 254.9 μW, respectively.

I. INTRODUCTION

With the continuous expansion of the wireless market, the demand for better performance, lower cost, and longer battery life devices constantly increases. New and existing standards are constantly evolving to meet these demands. In 2002, the Federal Communications Commission (FCC) amended the Part 15 rules to approve the use of ultra-wideband (UWB) frequencies ranging from 3.1 to 10.6 GHz for communication systems. The outdoor FCC spectral mask for an UWB signal is illustrated in Fig. 1. The UWB signal's Equivalent Isotropically Radiated Power (EIRP) is limited to a maximum of -41.3 dBm/MHz, in order to protect existing wireless technologies operating within this frequency band. The UWB signal must also have a bandwidth of greater than 500 MHz or a fractional bandwidth (i.e. the ratio of its bandwidth over its center frequency) larger than 20%. Therefore, depending on the standard, an UWB transceiver can either use the full-band of 3.1 to 10.6 GHz, or divide it into multiple sub-bands similar to the WiMedia Alliance standard, where multiple 524 MHz bands are used to communicate.

The UWB technology offers many advantages compared to traditional narrowband technologies. The wide frequency bandwidth nature of this technology offers a large channel capacity for high data rate communication (e.g. >100 Mbps), an excellent multi-path immunity, and a low susceptibility to interference from other wireless systems. In an UWB pulse-based system, the transceiver architecture can be significantly simplified, since the UWB signal is generated using a carrierless pulse signal. Due to its low EIRP, the UWB signal has a low probability of interception and detection.

Furthermore, additional power saving is achieved, since the

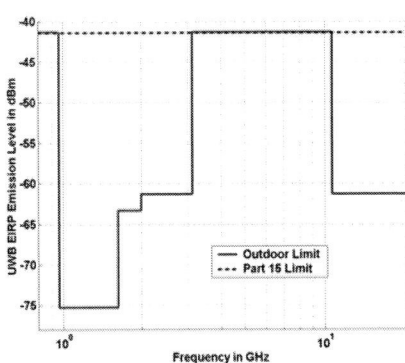

Fig. 1. FCC outdoor spectral mask for the UWB technology.

total power transmitted is very low compared to that of a traditional narrowband technology.

This paper presents FCC compliant full-band and sub-band UWB transmitters for low power applications. The designs are implemented in a standard CMOS 0.18 μm process.

The following section surveys existing pulse-based UWB transmitters in the literature. In Section 3, the system overview and the design of the UWB transmitters are examined in detail. The paper concludes with experimental results and discussions.

II. REVIEW OF UWB PULSE-BASED TRANSMITTER DESIGNS

A pulse-based transmitter produces modulated pulses with precise frequency characteristics that satisfy the FCC UWB spectral mask and they are generally categorized into analog and digital implementations [1]-[4].

The transmitter in [1], based on an analog approach, generates UWB pulses using a multiplier circuit. A triangular signal is multiplied with a carrier oscillating signal to produce an up-converted triangular pulse. By controlling the width of the triangular signal and the frequency of the carrier signal, the center frequency and bandwidth of the UWB pulse can be tuned to satisfy the FCC UWB spectral mask. This design consumes 2 mW @ 1.8 V for a pulse repetition frequency (PRF) of 40 MHz in a 0.18 μm CMOS process.

An alternative analog-based approach utilizes a passive on-chip filter to shape the spectrum of the pulse into the FCC UWB spectral mask [2]. The resulting pulse is then driven by a wideband amplifier to the 50Ω antenna. In this approach, the driver amplifier consumes constant biasing current. Therefore, the transmitter consumes moderate power of 10.8 mW @ 1.8 V in a 0.18 μm CMOS process with a large silicon area of 0.96 mm².

978-1-4244-2018-6/08 $25.00 © 2008 IEEE   619

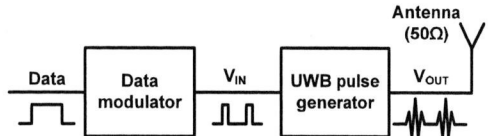

Fig. 2. Block diagram of the UWB transmitter architecture used here.

For low power applications, the transmitter is better implemented using a digital approach, due to the low duty cycles involved. The transmitter in [3] utilizes a NOR gate and inverters to generate a 1st order pulse. This pulse is then shaped into a 2nd order Gaussian pulse by a switching amplifier with a passive pulse shaping network. This technique can significantly reduce the power consumption to below 100 μW @ 1.8 V for a PRF of 100 MHz. However, this transmitter generates only a 2nd order Gaussian pulse, which theoretically cannot efficiently satisfy the FCC UWB spectral mask [5].

The transmitter in [4], also based on a digital approach, utilizes a set of digital triangular pulse generators. Each generator creates a triangular pulse with a different pre-determined delay time. All the pulses are combined in a specific order and driven to the 50Ω antenna. This technique generates digitally specific pulse shapes, which eliminates the need for analog filtering. However, this design is sensitive to component mismatches, which can disrupt the pulse shapes. Since the power is only consumed during switching transients, this transmitter, fabricated in a 0.18 μm CMOS technology, consumes only 675 μW @ 1.8 V for a PRF of 1 MHz.

III. LOW POWER UWB TRANSMITTER TOPOLOGY

The design of UWB pulse-based transmitters presents many challenges. For example, for a full-band UWB transmitter, the output power spectral density (PSD) should cover as much area possible under the FCC UWB spectral mask. This maximizes the total energy transmitted and leads to a higher probability of detection in the receiver. For sub-band UWB transmitters, the PSD of the UWB pulse should have low sidelobes to minimize the interference with adjacent sub-bands. To compensate for mismatch and process variations, the transmitter should also be able to control the pulse shape and, consequently, the output PSD.

In this work, all the above requirements are addressed while minimizing the power consumption to less than 300 μW. The UWB transmitter architecture used here is illustrated in Fig. 2. Both the modulator and UWB pulse generator are implemented using a digital approach, in order to eliminate continuous power consuming analog circuits. An on-off keying (OOK) modulation scheme is utilized here for simplicity and low power consumption. When the baseband data signal is high, the modulator block generates a square wave signal with a pre-determined pulse repetition frequency. The modulator is turned off to save power, when the data signal is low. On every rising edge of V_{IN}, the UWB pulse generator generates an UWB pulse at its output V_{OUT}.

Full-band UWB pulse generator – The schematic of the full-band UWB pulse generator is shown in Fig. 3(a). First, a voltage-controlled current-starved inverter and a NOR gate produce a variable width rectangular pulse, V_{REC}, to power on

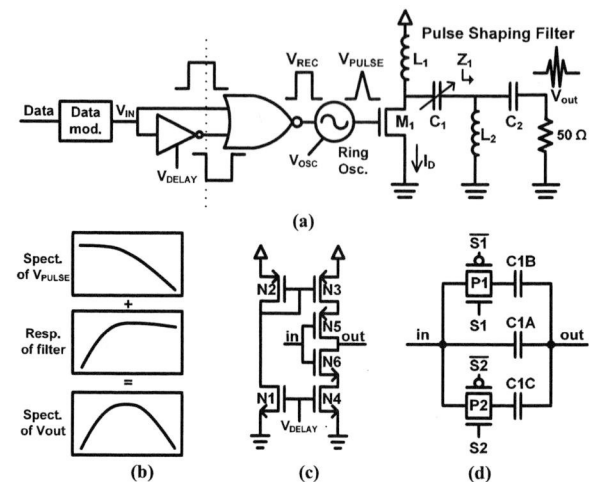

Fig. 3. Schematic of (a) the full 3.1-10.6GHz UWB transmitter, (b) the conceptual view of the operation of the UWB pulse generator, (c) the voltage-controlled current-starving inverter, and (d) the tunable capacitor C_1.

the fast start-up tunable ring oscillator. The delay of the inverter is controlled by the amount of current supplied to it, which is set by the voltage V_{DELAY} in a current mirror, as shown in Fig. 3(c). This delay establishes the pulse width of the signal V_{REC}, which then sets the number of the oscillation cycles of V_{PULSE} and, consequently, the width of the UWB pulse. The voltage V_{OSC}, applied to the ring oscillator, controls its frequency of oscillation. Therefore, the bandwidth and the center frequency of the UWB output pulse can be tuned by the voltages V_{OSC} and V_{DELAY} to compensate for mismatch and process variations.

The V_{PULSE} signal generated by the ring oscillator however does not meet the FCC spectral mask in the low frequency range, as described in Fig. 3(b). To attenuate these frequencies, a high-order high pass filter is needed. The pulse shaping high pass filter in this work is shown on the right hand side of Fig. 3(a). An NMOS switch M1 acts as a switching transconductance amplifier, whose current (I_D) is filtered by the pulse shaping filter and sent to the 50Ω antenna. The pulse shaping filter consists of a fourth order high pass network, implemented by cascading two second order LC high pass filters (i.e. a total of two inductors and two capacitors). Since the filter is passive, no power is consumed. MIM capacitors and standard on-chip inductors are used. The transfer function of the pulse shaping filter in the S-domain is written as:

$$H(s) = \frac{V_{OUT}}{I_D} = \left(\frac{s^2 L_1 C_1}{s^2 L_1 C_1 + s Z_1 C_1 + 1} \right) \left(\frac{s^2 L_2 C_2}{s^2 L_2 C_2 + s R_L C_2 + 1} \right). \quad (1)$$

The fourth order high pass filter offers abrupt roll-off due to the four zeros at s=0. The LC values (i.e. pole locations) are chosen to meet the stop band edge at 3.1 GHz, as shown in Fig. 1. To control the corner frequency of the filter for spectral mask fitting purposes, the design utilizes a variable capacitor C1, consisting of three capacitors and two pass transistors, as shown in Fig. 3(d). This variable capacitor C1 achieves four different capacitances: 60, 120, 140, and 200 fF.

978-1-4244-2018-6/08 $25.00 © 2008 IEEE

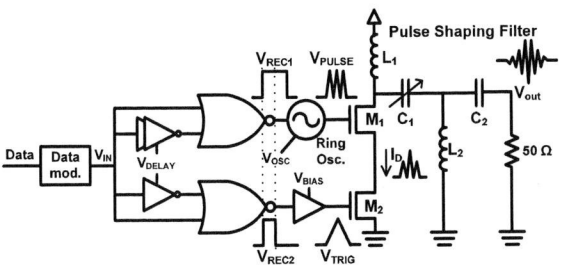

Fig. 4. Schematic of the low sidelobe sub-band UWB transmitter.

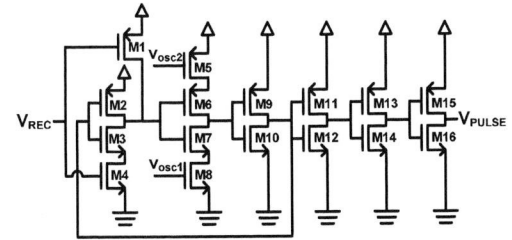

Fig. 5. Schematic of the fast start-up tunable ring oscillator.

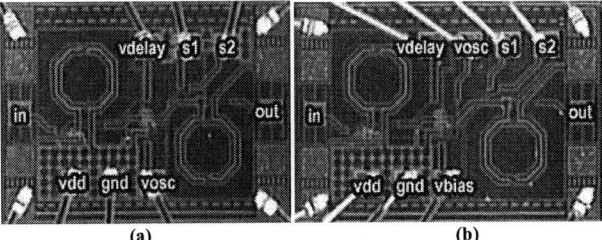

Fig. 6. Photographs of (a) the full-band, and (b) the sub-band 3.1-10.6GHz UWB transmitters.

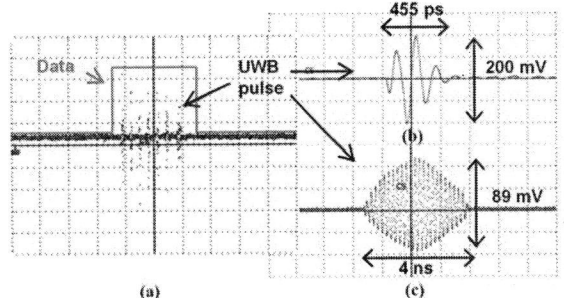

Fig. 7. (a) Measured modulated UWB pulses in the time domain: (b) the full-band UWB pulse, and (c) the 528MHz bandwidth UWB pulse.

Low sidelobe sub-band UWB transmitter – Low bandwidth UWB pulses can easily be generated using gated sine wave generators [6]. The full-band UWB transmitter here (Fig. 3(a)) can also produce lower bandwidth UWB pulses by changing the control voltage V_{DELAY} to lengthen its oscillation time. However, the generated gated sine wave signal possesses high sidelobes, which can cause interference. To suppress these sidelobes, the gated sine wave can be multiplied by a triangular signal, forming an up-converted triangular pulse. This multiplication technique has been explored in [5]. However, in this design, instead of using a power consuming differential multiplier circuit, two NMOS switches, M1 and M2, are used. This significantly reduces its power consumption and complexity.

Fig. 4 shows the schematic of the low sidelobe sub-band UWB transmitter. Similar to the first design, a high pass pulse shaping filter is used to suppress the low frequency content of the pulses. A fast start-up tunable ring oscillator generates the carrier signal, V_{PULSE}, which is connected to the top NMOS switch, M1. The triangular signal, V_{TRIG}, is generated by charging and discharging the gate capacitance of M2. The charging and discharging current is controlled by a voltage-controlled current-starved buffer. It charges when V_{REC2} is high, and discharges when V_{REC2} is low. Since the NMOS switches M1 and M2 are in series, the resulting drain current, I_D, has the same shape as the voltage multiplication of the inputs.

To synchronize the multiplication of the carrier and the triangular pulse, two rectangular pulse generators are used. Both rectangular pulses, V_{REC1} and V_{REC2}, are generated at the same starting time. Since the triangular pulse consists of charging and discharging periods, the pulse width of the bottom signal V_{REC2} is two times shorter than that of the top signal, V_{REC1}, to make sure that both signals V_{TRIG} and V_{PULSE} overlap in time.

Fast start-up tunable ring oscillator – A three stage ring oscillator, shown in Fig. 5, is used in this work. The frequency of oscillation is tuned with transistors M5 and M8. The

biasing voltages v_{OSC1} and v_{OSC2} are set by a current mirror circuit with the control voltage v_{OSC}. Transistors M1 and M4 force initial oscillation conditions when turned off, therefore, allowing almost instantaneous oscillation startup time. At the output, a buffer, consisting of inverter stages, is used to drive the NMOS switch.

Data modulator – The OOK modulator generates a square wave signal when the baseband data signal is high. This modulator is implemented using ON/OFF transistor switches and a clock generator. The clock generator must operate at a pre-determined PRF. In this work, the PRF varies between 10 and 200 MHz. The clock generator is implemented using a low power current starving scheme [3], consisting of two current starved inverters and a low power Schmitt trigger in a three stage ring oscillator configuration.

IV. TRANSMITTERS PERFORMANCE

The output time domain waveform and the PSD of the UWB transmitter are measured using a Tektronix TDS8000 50 GHz digital sampling scope and an Agilent 4440A 26.5 GHz spectrum analyzer with a resolution bandwidth of 1 MHz, respectively. The photograph of the two transmitter designs are shown in Fig. 6. The full-band and sub-band transmitters occupy chip areas of 0.49 and 0.52 mm^2, respectively. With a voltage supply of 1.8 V, the power consumption of the full-band and sub-band designs are 237.4 and 254.9 μW, respectively.

Fig. 7 shows the measured OOK modulated UWB pulses of the full-band and the sub-band UWB transmitters. The FCC compliant full-band pulse has a peak to peak voltage of 200 mV and a pulse width of 455 ps, as shown in Fig. 7(b). The sub-band 528 MHz bandwidth pulse has a peak to peak voltage of 89 mV and pulse width of 4 ns, as shown in Fig. 7(c). A higher peak voltage is obtained from the first transmitter because the single NMOS switch pulls more current (I_D) than the two NMOS switches in series, due to the

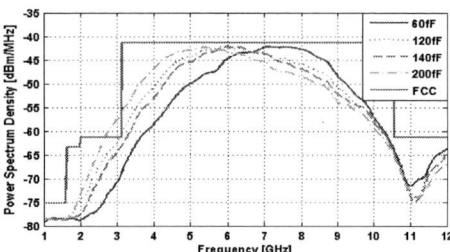

Fig. 8. Measured PSD of the full-band transmitter for different capacitances C1.

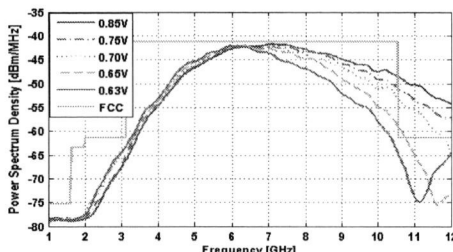

Fig. 9. Measured PSD of the full-band transmitter for different voltages v_{OSC}.

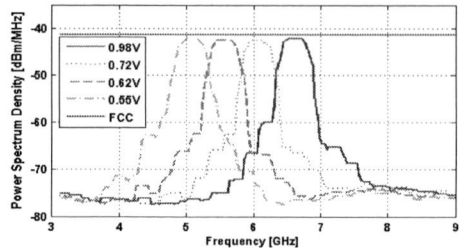

Fig. 10. Measured PSD of the 528 MHz bandwidth sub-band transmitter for different voltages v_{OSC}.

single switch inherent lower resistance path.

Fig. 8 shows the envelope of the measured output PSD for different capacitance values C1. The outputs are generated by the full-band UWB transmitter at a PRF of 110 MHz. As shown, a smaller capacitance C1 shifts the lower corner frequency of the filter to a higher frequency. This corner frequency shifting is used to compensate for variations. In Fig. 8, the signal meets the FCC UWB spectral mask when the 120 fF capacitance C1 is selected. This UWB signal has a -10 dB bandwidth of 5 GHz and achieves a spectral power efficiency of 30%, which is the ratio between the pulse radiation power to the maximum FCC radiation power limit of 556 µW, when the full 7.5 GHz bandwidth is considered.

Fig. 9 shows the envelope of the measured output PSD at different pulse center frequencies. Lowering the center frequency reduces the upper corner frequency of the UWB signal for spectral mask fitting purposes. To lower the center frequency, the voltage v_{OSC} is decreased. In Fig. 9, the UWB pulse meets the FCC UWB mask by setting the voltage v_{OSC} to 0.63 V. The performance of the full-band transmitter design is summarized in Table I.

Fig. 10 shows four 528 MHz bandwidth pulses at different

center frequencies generated by the second transmitter at a PRF of 30 MHz. In Fig. 10, the bandwidth is set to 528 MHz by biasing the voltage v_{DELAY} to 0.51. The voltage v_{DELAY} can change the pulse bandwidth from 500 to 5000MHz. Similarly, the voltage v_{OSC} can tune the center frequency from 4.8 to 6.8 GHz, as shown in Fig. 10. This up-converted triangular pulse transmitter provides a sidelobe rejection of more than 20 dB, as shown in Fig. 10. Its performance is summarized in Table I. A comparison to a recent transmitter in the literature [1] is included in Table I.

V. CONCLUSION

Full-band and sub-band UWB transmitters are presented in this work. Both designs employ an OOK modulation scheme and a digital pulse generation approach to minimize power consumption. A simple, tunable, and low power topology for the generation of full-band UWB pulses is presented for the first transmitter. It uses a gated tunable ring oscillator and a pulse shaping filter to generate full-band FCC compliant UWB pulses. The second transmitter multiplies a carrier and triangular signal to produce low sidelobe sub-band UWB pulses. The feasibility of using two NMOS switches to perform the multiplication while consuming minimum power was demonstrated. Control voltages incorporated in both designs adjust the shapes of the pulses to compensate for mismatch and process variations. The full-band and sub-band transmitters consume only 237.4 and 254.9 µW, respectively, making them among the lowest power FCC compliant tunable 3.1-10.6 GHz UWB transmitters reported to date.

ACKNOWLEDGMENT

The authors would like to acknowledge the Canadian Microelectronics Corporation (CMC) for fabricating the test chips, and NSERC and ReSMiQ for financial support.

REFERENCES

[1] J. Ryckaert, et al. "Ultra-wide-band transmitter for low-power wireless body area networks: design and evaluation", *IEEE Trans. on Circuits and Syst. I*, pp.2515-2525, Dec. 2005.

[2] K. W. Wong, S. R. Karri, and Y. Zheng, "Low-power full-band UWB active pulse shaping circuit using 0.18-µm CMOS technology", in *Proc. Radio Freq. Integr. Circuits Symp.*, June 2006.

[3] T. K. K. Tsang and M. N. El-Gamal, "Fully integrated sub-microWatt CMOS ultra wideband pulse-based transmitter for wireless sensors networks", in *Proc. IEEE Int. Symp. on Circuits and Syst.*, pp. 670-673, May 2006.

[4] T. Norimatsu, et al., "A novel UWB impulse-radio transmitter with all-digitally-controlled pulse generator", in *Proc. IEEE Int. Solid-state Circuits Conf.*, pp. 267-270, Sept. 2005.

[5] H. Sheng, et al. "On the spectral and power requirements for ultra-wideband transmission", in *Proc. IEEE Conf. on Communications*, pp.738-742, May 2003.

[6] Y.-H. Choi, "Gated UWB pulse signal generation", *Int. Workshop on Ultra Wideband Syst.*, pp. 122- 124, May 2004.

TABLE I
PERFORMANCE SUMMARY OF THE TWO CMOS UWB TRANSMITTERS IN THIS WORK, AND A COMPARISON WITH [1].

	Design 1	Design 2	TCAS'05 [1]
Technology	CMOS 0.18 µm	CMOS 0.18 µm	CMOS 0.18 µm
Topology	Pulse generator and pulse shaping filter	Multiplier and pulse shaping filter.	Multiplier for triangular pulse, and carrier.
Vdd	1.8 V	1.8 V	1.8 V
Modulation	OOK	OOK	PPM
Pulse repetition freq. (PRF)	110 MHz	30 MHz	40 MHz
Pulse bandwidth	500 – 5000 MHz	500 – 5000 MHz	200 – 2000 MHz
Power	237.4 µW	254.9 µW	2 mW
Area	0.49 mm²	0.52 mm²	0.36 mm²
FCC compliant	Yes	Yes	Yes
Other feature	Low power	Low sidelobes	Low sidelobes

978-1-4244-2018-6/08 $25.00 © 2008 IEEE

Low Power and Non-Traditional RF Tranceivers

Session 19

Chair: Ramesh Harjani, University of Minnesota
Co-Chair: Andrea Mazzanti, Universita di Modena

This session explores non-traditional techniques to develop energy efficient radio transceivers. Ultra low power radios are one of the key enabling technologies necessary for ubiquitous untethered computing systems. As wireless circuits gain popularity it has become increasingly necessary to consider alternate architectures and designs that provide added systems benefits. Some of the applications that will be discussed in this session include sensor networks, medical imaging, cognitive radios and wireless LANs.

The first three papers focus on different aspects of low power radio design for low, medium and high data rate applications. The first paper discusses energy efficient design concepts that focus on ultra low power radio designs. It provides an overview of some of the newer circuit design techniques for ultra low power and focuses on energy per bit a figure of merit to compare designs. In particular, the paper provides two low power wireless designs, one for high data rate and one for low to medium data rate. The second and third papers focus on wireless sensor network applications. The second paper discusses an ultra low power super-regenerative receiver design for binary FSK signals. The authors show that in comparison to traditional on-off keying the use of BPSK provides an additional 3dB of link margin and allows the receiver to be operated at higher data rates. The third paper focuses on a low power transmitter design that supports both OOK and BPSK signals.

The fourth, fifth and final papers describe transmitter circuit designs for different applications and utilize a multitude of architectures to increase integration. In particular, the fourth paper discusses the design for a high data rate O-QPSK transmitter circuit for wireless medical imaging applications. The transmitter design is primarily focused on implantable/swallowable miniature biomedical devices that can support a 10 frames per second transmission rate. The fifth paper discusses a fully programmable direct conversion radio transmitter for potential IEEE 802.22 use. The transmitter design incorporates in-band filters to suppress harmonics within the 54-862MHz TV band. The final paper incorporates tunable high-Q RF filters for wireless LAN applications in standard CMOS technologies. The 200MHz bandwidth filter operates at 5GHz 802.11a and provides spectral mask and EVM compatible OFDM signals.

Notes

IEEE 2008 Custom Intergrated Circuits Conference (CICC)

Energy-efficient Wireless Front-end Concepts for Ultra Lower Power Radio

John R. Long, W. Wu, Y. Dong, Y. Zhao, M.A.T. Sanduleanu[1], J.F.M. Gerrits[2] and G. van Veenendaal[3]

Electronics Research Laboratory/DIMES, Delft University of Technology
Mekelweg 4, 2628CD Delft, The Netherlands
1. Philips Research, Eindhoven, The Netherlands
2. CSEM, Neuchatel, Switzerland
3. IC Lab, NXP Semiconductors, Eindhoven, The Netherlands

Abstract — **Two ultra low power wireless concepts are described in this paper. A high data rate receiver demonstrator consisting of LNA, subharmonic I/Q mixer and transimpedance IF amplifiers realizes an energy consumption of 1.75 nJ/bit at 10Mbit/s in the 17GHz band. A high-band FM-UWB receiver demonstrator, which achieves a measured sensitivity of -84.3dBm at 100kbit/s while consuming just 6mW is also described.**

Introduction

Ultra low power radio interfaces are poised to enable short-range wireless applications such as remote health and environmental monitoring, logistics and inventory control, and home or office automation. Due to cost and size considerations, a low-complexity approach is required that allows one or multiple users to share the same RF bandwidth while offering robustness to potential sources of interference, multi-path effects and antenna mismatch.

State-of-the-art radios are compared in Fig. 1 with respect to data rate, power consumption and their resultant energy/bit. Ultra low power (ULP) operation, as characterized by energy/bit, can be realized in many ways. For example, energy consumption may be traded for reduced RF performance by simplifying the transceiver architecture in order to reduce circuit complexity. Activating the radio only when necessary conserves energy at the cost of transceiver 'up time'. Energy consumption is related to the format selected, as circuit complexity is a function of the modulation and detection method, however, reliability and link throughput must also be considered. In addition, energy/bit performance is proportional to operating frequency and RF front-end bandwidth. Two recently reported examples of ULP radios shown in Fig. 1 are the wake-up receiver [1] and impulse-UWB transceiver [2]. The 2GHz wake-up receiver uses on/off keying (OOK) and envelope detection in a simple tuned-RF receiver to achieve -72dBm sensitivity, 50μW power consumption and throughput in the 10-100kbit/s range.

The super-regenerative (SR) and tuned-RF (TRF) receiver architectures shown in Fig. 2 are commonly used low-power implementations due to their simplicity (e.g., in keyless entry and RFID applications). Periodic quenching of a self-oscillating mixer in the SR receiver converts the OOK-RF signal to baseband. SR receiver performance is typically limited by stability and selectivity. An RF local oscillator, PLL synthesizer and downconverting mixer are not required in a TRF receiver, which drastically reduces power consumption. Envelope detection in the receive path demodulates the OOK-RF signal. Sensitivity of a TRF radio is limited by the Q-factor of RF band selection filters, which are required for channel selectivity and to minimize out of band interference.

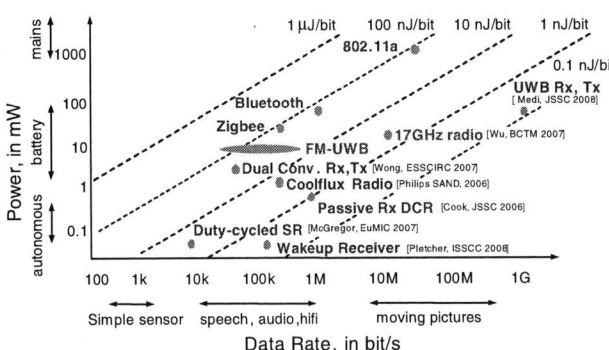

Fig. 1: Energy consumption space for low power radio.

Two ULP radios which utilize more robust modulation schemes than OOK and operate in RF bands where interference is minimal for enhanced link reliability are described in this paper.

A sub-2nJ/bit, low duty cycle 17GHz radio aimed at indoor wireless sensor networks (WSNs) is described in the first section of this paper [3]. Low energy consumption and small form factor are achieved by operating in the 17GHz unlicensed band at data rates up to 10Mbit/s.

Ultrawideband FM (FM-UWB), which is intended for 1-1000 kbit/s data rate WPAN applications [5] is described in the second half of the paper. Advantages of FM-UWB are its relatively simple receiver and transmitter architectures and robust low-power, low-cost IC implementations.

17GHz Low Energy/bit Receiver

The 17GHz-band radio receiver of Fig. 3 is aimed at ultra low power sensor network applications. Power consumption will be higher at 17GHz (an unlicensed band in Europe), but the

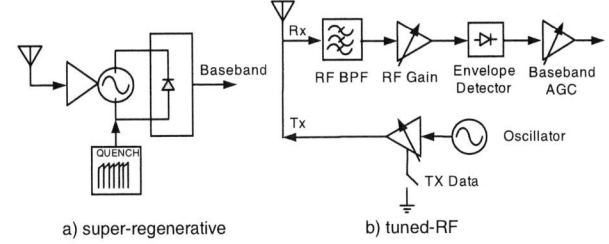

Fig. 2: Ultra low power receiver examples.

978-1-4244-2018-6/08 $25.00 © 2008 IEEE   625

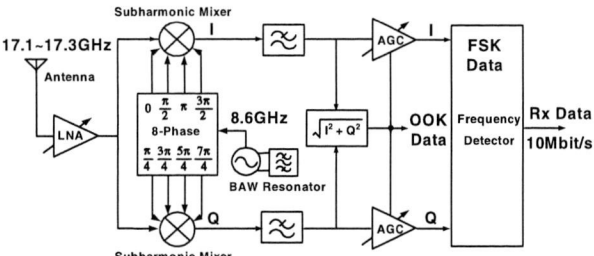

Fig. 3: Block diagram of 17GHz high data rate receiver.

smaller antenna size, reduced interference and 200MHz of available bandwidth are distinct advantages for a WSN transceiver. This low duty cycle, high bit rate direct conversion receiver can realize energy consumption on the order of 1nJ/ bit and a turn-on time of 2μs because a PLL synthesizer is not required. The simple direct-conversion architecture without an ADC also reduces overall power consumption. An 8.6GHz BAW resonator VCO generates a low phase noise frequency reference, which is used as the LO signal. The VCO runs at one-half the RF input frequency because of current limitations on the operating frequency of BAW technology. Sub-harmonic mixers driven by multi-phase LO inputs are employed for I/Q downconversion. On/off keying (OOK) and frequency shift keying (FSK) demodulation are both implemented, but in the analog domain in order to conserve power. Square root detection of the I/Q signal amplitudes demodulates an OOK signal, while an analog frequency discriminator realizes FSK demodulation. The down-converted signal lies above the band where flicker noise from the bipolar devices (1/f corner of approx. 200kHz) degrades the receiver bit-error rate.

DC offsets arising from self-mixing are avoided by using a subharmonic mixer operating at one-half of the RF input frequency. The schematic of the single-balanced I/Q mixer is shown in Fig. 4. Quadrature LOs drive each mixer pair (e.g., Q_2-Q_5 and Q_6-Q_9 in Fig. 4), so a total of 8 LO phases are required. The multi-phase LO is generated by an on-chip filter and interpolation network. The mixers share the same inductively degenerated common-emitter input stage in the single-ended RF input path. Each transistor in the mixing quad

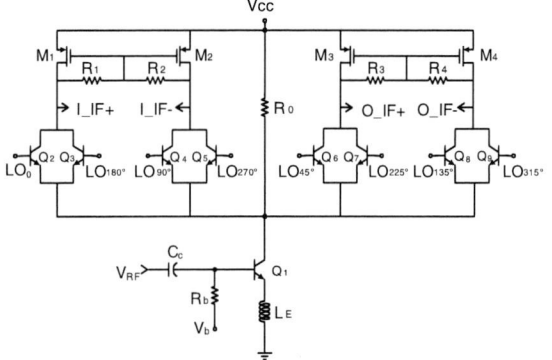

Fig. 4: Subharmonic mixer schematic.

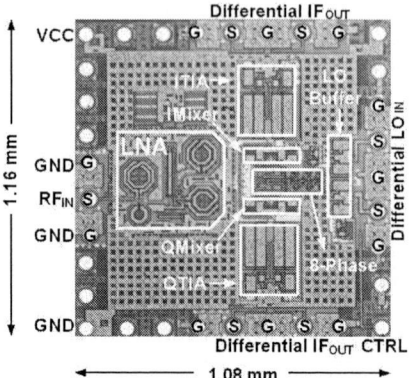

Fig. 5: 17GHz receiver testchip photomicrograph.

is biased at 150μA giving a total of 1.5mA flowing through Q_1.

Table 1: Measured 17GHz Rx performance.

Specification	Measurement
Data rate	10Mbits/s
-1dB input compression point	-16dBm
Power consumption (active)	17.5mW

The prototype 17GHz transceiver shown in Fig. 5 is fabricated in NXP Semiconductors QUBIC4X 0.25μm SiGe:C BiCMOS technology. The receiver front-end testchip consists of the LNA, sub-harmonic I/Q downconverting mixer and transimpedance IF amplifiers (from Fig. 3). Data from experimental measurements are listed in Table 1 and plotted in Fig. 6. The measured conversion gain is 25-30dB into a 50Ω load at 10MHz IF, and noise figure as a function of IF is approximately 12dB across the band. The 1.25mm² testchip has a measured IIP$_3$ of -26.3dBm and consumes 17.5 mW in active mode. The resultant energy consumption for a 10Mbit/s data rate is 1.75 nJ/bit for the receiver, which is approximately one-third of the energy consumption of conventional designs.

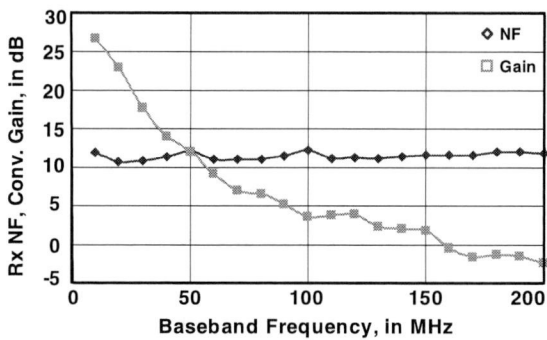

Fig. 6: Measured noise figure and conversion gain for the 17GHz Rx.

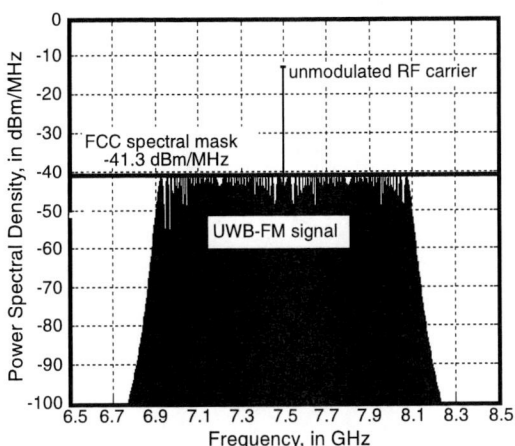

Fig. 7: FM-UWB signal spectrum (Δf=600MHz, β=600).

Ultrawideband FM (FM-UWB)

Frequency modulation (FM) has the unique property that RF bandwidth B_{RF} is defined by the modulating signal bandwidth (f_m) and modulation index β, that can be chosen freely. The approximate bandwidth of an FM-UWB signal is given by Carson's rule

$$B_{RF} = 2(\beta + 1)f_m = 2(\Delta f + f_m). \qquad (1)$$

Choosing ($\beta \gg 1$) creates an ultrawideband signal that can occupy a bandwidth which can be adjusted by modifying the deviation (Δf) of the wideband FM signal.

To create the constant-envelope FM-UWB signal shown in Fig. 7, data is first modulated onto a low-frequency sub-carrier using binary FSK (e.g., 1MHz subcarrier for 100kbit/s data with β_{SUB}=2 via direct digital synthesis or DDS), as shown in the block diagram of Fig. 8. The RF carrier amplitude drops by a factor $10\log_{10}(\beta)$=28dB under modulation (see Fig. 7), making this FM-UWB example FCC compliant. The RF-voltage-controlled oscillator (RF-VCO in Fig. 8) is the only RF circuit required for carrier modulation. FDMA techniques may be employed at the sub-carrier level to accommodate more than one user.

The FM-UWB receiver (see Fig. 8) consists of a wideband LNA, wideband FM demodulator, and one (or several) low-frequency filtering, frequency conversion, amplification, and subcarrier demodulation stages. A wideband (e.g., 500MHz) demodulator and not a frequency converting mixer divides the front-end and baseband sections of the receiver. Being wideband, the front-end allows rapid synchronization (e.g., <1ms) in ad-hoc networks. Its simplicity, with no local oscillator or carrier synchronization leads to the potential for lower power consumption hardware compared to other system implementations, without sacrificing robustness. The scheme is well-suited to low and medium data rate applications (e.g., up to 1Mbit/s) where battery lifetime, weight and size are important design considerations.

A multi-user system requires multiple baseband sections (e.g.,

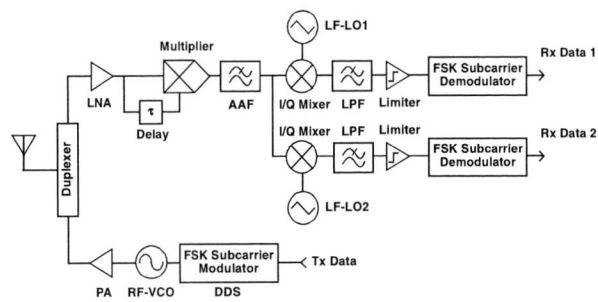

Fig. 8: FM-UWB system block diagram.

as shown for 2 users in Fig. 8). All users share the same RF carrier and distinguish themselves by using different sub-carrier frequencies (i.e., FDMA).

The key receiver building block is the wideband FM demodulator [4]. A fully-balanced, fixed-time-delay design is shown in Fig. 9. This circuit was developed as a proof of the FM-UWB concept, and consists of an LC delay (all-pass filter, APF), gain stage and Gilbert multiplier. The differential input signal is applied to both the APF and multiplier stages, and it can be shown that the demodulator's dynamic range is optimized by selection of the time delay in the APF path [5]. The APF circuit produces a delay of 100ps over the band from 7.2–7.7GHz. The differential pair and tuned load of the gain stage increases the delay time to 500ps, while also developing a signal amplitude sufficient to drive the quad of the multiplier and yield an FM demodulated output at terminals V_{IFout}. Bias current in the APF and gain stages is forced to flow via L_{AC} through input pair Q_9-Q_{10} of the multiplier, thereby increasing the transconductance of the multiplier input pair and the demodulator gain. Noise generated by the multiplier quad is suppressed by the preceding gain stage and also by operating quad transistors Q_{13}-Q_{16} at a low bias current (i.e., 320µA). The Miller effect in each diff pair in the demodulator is neutralized by diode-connected transistors that feedback

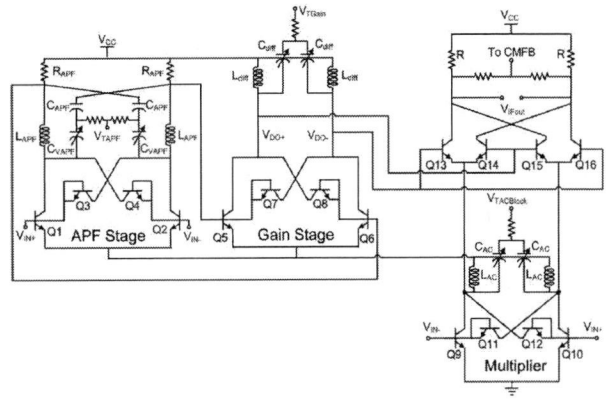

Fig. 9: UWB-FM wideband demodulator schematic.

978-1-4244-2018-6/08 $25.00 © 2008 IEEE

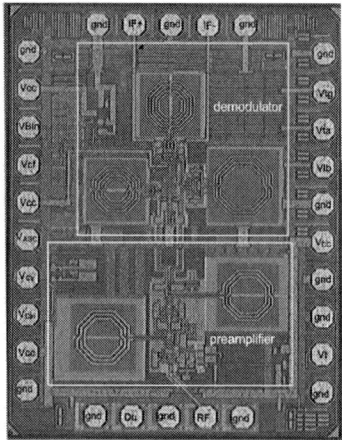

Fig. 10: FM-UWB receiver IC prototype photomicrograph.

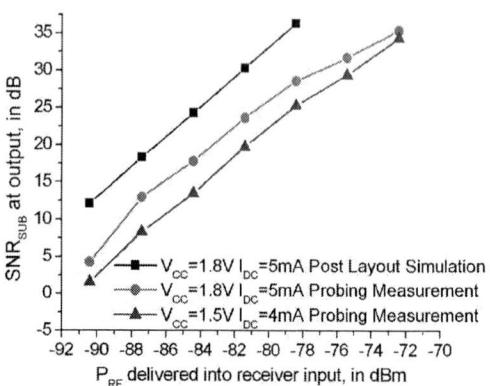

Fig. 11: Sensitivity curves for the prototype FM-UWB receiver.

compensating currents from output back to the input.

The photomicrograph of a prototype FM-UWB receiver front-end is shown in Fig. 10. The IC is fabricated in NXP Semiconductors 0.25μm QUBIC4X SiGe:C BiCMOS technology. The circuit consists of a 30dB gain, two-stage RF preamplifier followed by the wideband FM demodulator of Fig. 9. The receiver is designed to operate in the upper band for UWB devices allocated in Europe and has an RF bandwidth of 500MHz centered at 7.45GHz when packaged in a 16-pin HVQFN package. The measured results obtained from on-wafer probing of the IC are summarized in Table 2. When operating from a (nominal) 1.8V supply, the measured receiver sensitivity is -86.8dBm. This RF power level gives a minimum SNR of 14dB at baseband, which is sufficient to realize a bit-error rate of 10^{-6} at a data rate of 100kBit/s. Power consumption is 9.1mW from a 1.8V supply, 6mW from a 1.5V supply (with a slightly reduced sensitivity of -84.3dBm). Tests also confirmed that the receiver functions properly at supply voltages as low as 1.4V. The circuit was designed to operate with a RF input sensitivity better than -80dBm (14dB SNR at baseband) over the anticipated variation in processing, -40 to +85°C operating temperature range, and 10% variation from nominal supply voltage. The 1.13x1.43mm^2 chip has an active area of 0.88mm^2.

Table 2: Measured FM-UWB Receiver Performance.

Specification	Measurement	
Rx Sensitivity	-86.8dBm	-84.3dBm
Supply Voltage	1.8V	1.5V
Power Consumption	9.1mW	6mW

The signal to noise ratio at the demodulator output (i.e., at baseband, SNR_{SUB}) as a function of RF input power is plotted in Fig. 11. The upper curve shows the post-layout simulation result obtained using Cadence Spectre-RF™ for the receiver biased at 5mA from 1.8V. The lower curves illustrate the variation in measured sensitivity as the supply voltage and current vary from nominal (1.8V) down to 1.5V. The trend predicted by simulation agrees well with the measured data, although the measured data is 5-9dB below the simulated SNR at nominal supply. Nevertheless, the measured results are sufficient to realize a link span greater than 10m at a data rate of 100kBit/s, with the potential for further improvements in power consumption, integration level, data rate and chip area required.

Conclusions

RF receiver front-ends for a high band FM-UWB system and a 17GHz high data rate radio were described in this paper. High data rate radio interfaces are advantageous for burst-mode connections. FM-UWB offers simple, compact implementations for short-range, continuous-streaming applications. Both schemes have the potential for realizing even lower energy consumption in future implementations.

Acknowledgments

IC fabrication and support for the 17GHz project was provided by Philips Research. The FM-UWB work is supported by the 6th Framework European Magnet-Beyond project, and the IC Lab of NXP Semiconductors in collaboration with CSEM.

References

[1] N. Pletcher, S. Gambini, J. Rabaey, "A 2GHz 52μW Wake-up Receiver with -72dBm Sensitivity Using Uncertain-IF Architecture," *ISSCC 2008 Tech. Papers,* pp. 524-525, Feb. 2008.

[2] A. Medi, W. Namgoong, "A High Data-Rate Energy-Efficient Interference-Tolerant Fully Integrated CMOS Frequency Channelized UWB Transceiver for Impulse Radio," *IEEE Journal of Solid-State Circuits,* vol. 43, April 2008, pp. 974-980.

[3] W. Wu, M.A.T. Sanduleanu, X. Li, J.R. Long, "17GHz RF Front-Ends for Low-Power Wireless Sensor Networks", *Proceedings of the 2007 IEEE-BCTM,* pp. 164–167, Oct. 2007.

[4] Y. Dong, Design of a Low-power FM-UWB Demodulator, M.Sc. thesis, Delft University of Technology, Feb. 2008.

[5] J.F.M. Gerrits, M.H.L. Kouwenhoven, P.R. van der Meer, J.R. Farserotu, J.R. Long, "Principles and Limitations of UWB-FM Communications Systems," *EURASIP Journal of Applied Signal Processing,* vol. 2005, no. 3, pp. 382-396.

IEEE 2008 Custom Intergrated Circuits Conference (CICC)

A 0.4 nJ/b 900MHz CMOS BFSK Super-Regenerative Receiver

James Ayers, Kartikeya Mayaram, and Terri S. Fiez
School of Electrical Engineering and Computer Science, Oregon State University
Corvallis, Oregon 97331, USA

Abstract— An ultra low-power super-regenerative receiver for BFSK signals has been designed and fabricated in a standard 0.18 μm CMOS process. The use of BFSK allows the receiver to operate at higher data rates and also gives an approximate 3 dB SNR performance increase over the more traditional OOK modulation. At 1Mb/s the receiver consumes 0.4 nJ/b making it the lowest energy integrated super-regenerative receiver to date.

I. INTRODUCTION

As the field of ultra low-power electronics progresses, wireless sensor networks (WSNs) are becoming more common in our daily lives. The amount of data these networks need to collect, process, and communicate continues to increase while the amount of power they consume needs to remain minimal. In addition, many networks need to operate autonomously for long lengths of time from a single battery, or energy scavenged from the surrounding environment [1].

The super-regenerative architecture has long been known to provide a good solution for low-power receivers. To date, several designs [2–5] have demonstrated very low power operation, but typically have low data rates or rely on special micro electromechanical systems (MEMS) processing. This is partially due to that fact that the designs all employ On / Off Keying (OOK) which suffers from low data rates due to issues both at the circuit and system levels.

On the system level, binary frequency-shift keying (BFSK) signaling is generally preferred over OOK because FSK is less susceptible to fading and noise. An OOK signal needs about 3 dB more signal-to-noise ratio (SNR) than a BFSK signal to achieve the same theoretical bit-error rate (BER) [6]. On the circuit level, the maximum data rate in an OOK system is determined by the amount of time the transmit oscillator needs to build oscillations when a '1' is being sent and also by the amount of time needed for the oscillations to die out when a '0' is being sent. In cases when a high-Q MEMS resonator is used to minimize power consumption, the turn-on and turn-off times of the oscillator increase which further reduces the data rate.

This paper describes a super-regenerative receiver for BFSK signals. The use of BFSK signaling allows this receiver to increase data rates and reliability over the traditional OOK approach while still maintaining extremely low-power operation with only a single external inductor. Section II describes the BFSK receiver operation and low power design techniques for the oscillator and baseband circuits. The measured results

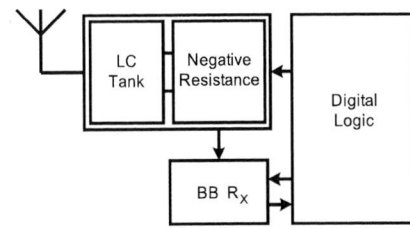

Fig. 1. Simplified block diagram for BFSK super-regenerative receiver.

for a fabricated test chip are reported in Section III and finally Section IV provides conclusions.

II. CIRCUIT DESIGN

A simplified block diagram for the BFSK super-regenerative receiver is shown in Figure 1. The BFSK super-regenerative receiver operates by comparing the amplified signal from two different frequencies rather than a single frequency as in the classic design. During each incoming bit period the oscillator is quenched twice, once when the tank is tuned to f_1 and again when the tank is tuned to f_2. Oscillations will build at both frequencies, but they will build much faster when the tank is tuned to the same frequency as the input signal. This is because the LC tank will provide voltage gain to the incoming signal when both the tank and the signal are at the same frequency. A complete system level analysis of the BFSK super-regenerative receiver can be found in [7].

The oscillations from both tank frequencies are then rectified and compared to determine the incoming data. The tank frequency with the largest rectified voltage is the frequency of the incoming signal. The receiver is made up of the core oscillator, baseband circuitry which consists of rectifiers and a comparator, and the digital control circuitry that provides the timing for all the circuit blocks and data alignment. The following subsections will describe the implementation of each of these blocks in detail.

A. Oscillator

A schematic of the core oscillator circuit and quench switch is shown in Figure 2. A differential input signal enters the oscillator through the two series capacitors that serve as part of the input matching network while also preserving the DC operating point of the oscillator.

The oscillator core is constructed using an LC tank architecture with both NMOS and PMOS cross coupled pairs to maximize the total g_m. The LC tank is made up of a single

978-1-4244-2018-6/08 $25.00 © 2008 IEEE

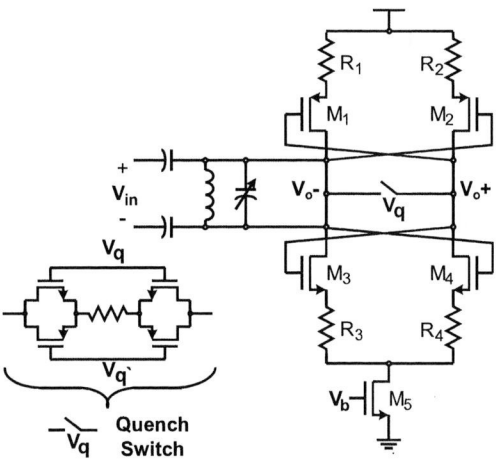

Fig. 2. Schematic of the core oscillator.

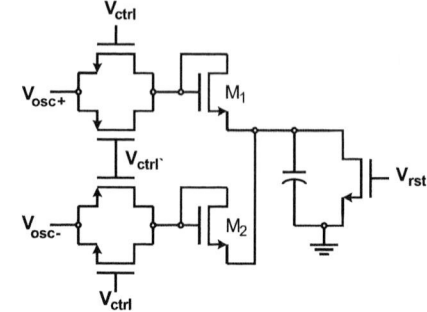

Fig. 3. Schematic of the passive rectifier.

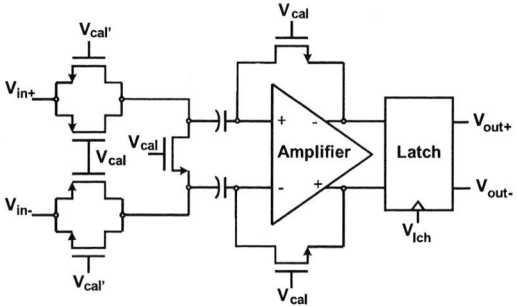

Fig. 4. System level schematic of the comparator.

external inductor and two integrated varactors, 1 large and 1 small. The large varactor is controlled by an off chip analog control voltage to allow the center frequency of the oscillator to be tuned anywhere in the 900 MHz ISM band. The smaller varactor is controlled by a digital signal that is switched continuously during operation to re-tune the oscillator between f_1 and f_2. Depending on the desired bandwidth and power requirements of the system the Δf (f_2-f_1) can be easily controlled by varying the amplitude of the digital control signal.

The negative resistance of the oscillator is made up of two sets of source degenerated cross-coupled pairs (M_1 - M_4). The source degeneration helps to regulate the g_m of the transistors while also providing a greater immunity to the g_m drift over temperature. A triple-well process is used which allows the substrate connection of both sets of transistors to their sources in order to minimize the threshold voltages and hence, the amount of voltage headroom needed.

Transmission gates are used in the quench switch in order to minimize the variance in switch conductance across the entire range of oscillation. The resistor in the quench switch provides the signal that allows the oscillations to build when the switch is opened. When the oscillator is quenched (i.e., the switches are closed), the quench switch resistor dissipates all of the power in the input signal creating a small voltage across the tank. When the switch opens, the small voltage generated by the resistor determines the amount of time the oscillator takes to build oscillations. The resistor value also determines how fast the oscillations die out when the circuit is quenched. Therefore, the resistor value must be optimized for both speed and sensitivity. If its value is large, the resulting voltage drop across the resistor will also be larger which will increase the receiver's sensitivity. If it is made too large the oscillator will take longer to quench, which will reduce the maximum quench rate, and hence the maximum data bandwidth.

The quench switch is driven digitally by the digital control logic. The differential control signal is generated by two capacitively loaded inverters. The capacitive loading creates a ramp function that reduces the amount of charge injection

created by the quench switch in order to eliminate false triggering of the oscillator.

B. Baseband Circuitry

In order to determine the incoming signal frequency, the oscillations generated by the oscillator at f_1 and f_2 need to be "down-converted" and compared. This is accomplished by the use of two passive rectifiers and a comparator. The rectifier schematic is shown in Figure 3.

As in the quench switch, the rectifier switches are also realized using transmission gates to maximize switch conductance across the entire swing of the input oscillations. The rectification is performed by the diode-tied NMOS transistors M_1 and M_2 whose substrate connections are tied to their sources to maximize their efficiency. The charge that is collected from the oscillations is stored on a large capacitor. For the duration of the rectification the capacitor charging remains in the linear (constant current) region and does not charge completely. This ensures a roughly linear relationship between the duration of the oscillations and the voltage stored on the capacitor which maximizes the voltage difference between the two rectified voltages.

Two rectifiers store the voltages for f_1 and f_2 so they can be compared. The v_{ctrl} signal for f_1 first closes the switches to allow the f_1 signal to be obtained, then the switches open and the signal is held while the signal for f_2 is obtained. Once the comparison has been made the reset signal is applied simultaneously to both of the reset switches and the capacitor is reset.

The comparison between the two rectified signals is performed by the latched comparator shown in Figure 4. The amount of offset that is present in the comparator will limit the overall sensitivity of the receiver as it will determine the smallest signal from the rectifier that can be resolved. To improve the overall sensitivity of the system, offset correction in the comparator is used. During the calibration stage the input of the comparator is disconnected from the rectifier and the calibration switches store the amplifier offset on the input capacitors where it will be subtracted from the input signal when it is sampled.

The comparator's preamplifier is shown in Figure 5. PMOS inputs are used in the amplifier to accommodate the low common mode voltage of the rectified signal. The impedance of the diode connected loads (M_4 and M_5) is boosted by the addition of the cross-coupled devices (M_6 and M_7). These cross-coupled devices are sized to be slightly smaller than the diode connected load to reduce their effective g_m (and increase the amplifier gain) while still maintaining amplifier stability.

The cross-coupled latch is shown in Figure 6. When the V_{lch} signal is high the NMOS and PMOS cross-coupled pairs are disconnected and the outputs are both pulled low. When the V_{lch} signal goes low, the pull-down switches on the output open and the two cross-coupled pairs use positive feedback to amplify the signal from the preamplifier to force a binary decision. When the comparators decision is made the top two PMOS switches turn off to eliminate the DC current that flows while the output is valid.

III. MEASUREMENT RESULTS

The receiver was fabricated in a 0.18 μm triple-well CMOS process and is shown in Figure 7. The digital logic is not shown in the photo because it was spaced away from the analog circuity (about 300 μm to the right) in order to minimize substrate noise coupling. The total active area of the receiver including the testing buffers and digital logic (excluding pads) is less than 0.15 mm^2.

Figure 8 shows the measured signals when the receiver is running at 500 kb/s (1 MHz quench). The bottom of Figure 8 shows the output of the two rectifiers while the top shows the resulting data. At 1 μS both of the rectifiers are reset and the rectifier corresponding to f_2 acquires the signal from the oscillations generated by the incoming signal. The voltage on the f_2 rectifier is held while the signal for f_1 is acquired. Since the input to the receiver is at f_1 the rectifier corresponding to

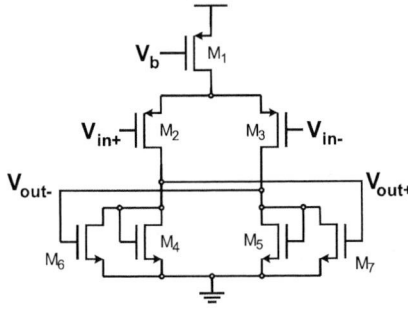

Fig. 5.   Preamplifier schematic for the latched comparator.

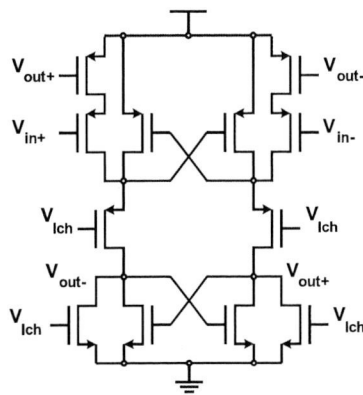

Fig. 6.   Latch schematic for the latched comparator.

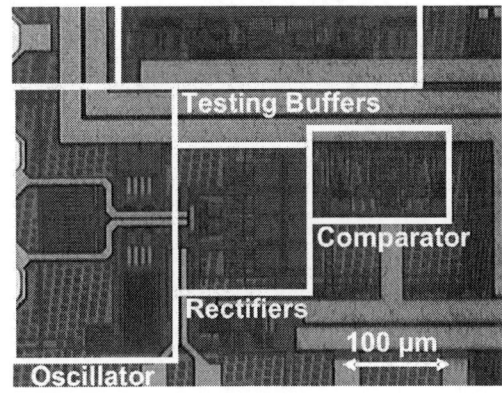

Fig. 7.   Chip photograph.

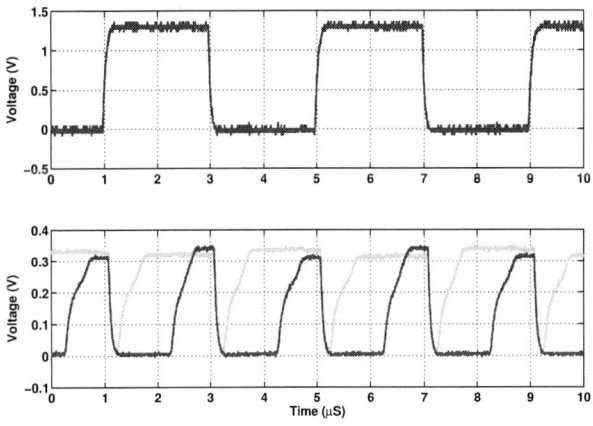

Fig. 8.   Measured receiver operation for a '1010' pattern at 500 kb/s.

f_1 builds a larger voltage and when the decision is made (at 3 μS) the data signal goes low.

The bit error rate (BER) measurements for the receiver are shown in Figure 9. For each point on Figure 9 a sequence of over 4 million normally distributed random points were collected. The receiver is most sensitive at low data rates as it has more time to build oscillations from small input signals. The minimum measured BER for the receiver is just under 5×10^{-7} and is achieved at all data rates for input power levels greater than or equal to -68 dBm. A maximum sensitivity of

978-1-4244-2018-6/08 $25.00 © 2008 IEEE

TABLE I

BFSK SUPER-REGENERATIVE RECEIVER COMPARED TO OTHER RECENT SUPER-REGENERATIVE DESIGNS.

	Vouilloz '01	Otis '05	Chen '07	M.-Geniz '07	This Work
Frequency	1 GHz	1.9 GHz	2.4 GHz	2.4 GHz	900 MHz
Modulation	OOK	OOK	OOK	OOK	BFSK
Data Rate	100 kb/s	5 kb/s	500 kb/s	11 Mb/s	1 Mb/s
Power	1.2 mW	400 μW	2.8 mW	2.1 mW	$\geq$ 244 μW
Energy	12 nJ/b	80 nJ/b	5.6 nJ/b	0.19 nJ/b	0.4 nJ/b
Sens. (BER$\leq 10^{-3}$)	-107.5 dBm	-100.5 dBm	-80 dBm	-80 dBm	-82 dBm
Technology	0.35μm CMOS	0.13μm CMOS	0.35μm CMOS	Discrete BJTs	0.18μm CMOS
Tank	External LC	BAW Resonator	Internal LC	Microstrip	External L

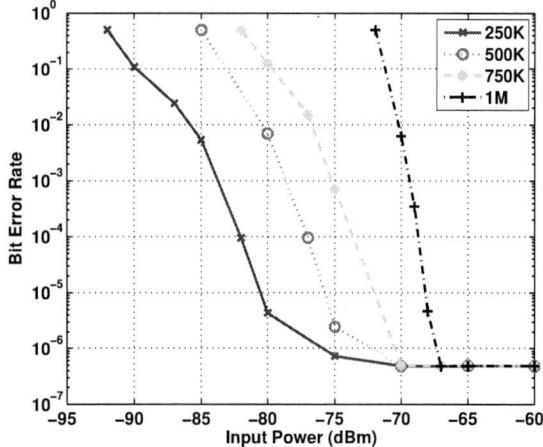

Fig. 9. Measured bit error rate versus input power for different data rates.

TABLE II

SUMMARY OF MEASURED RESULTS.

Min. Power	244 μW
Supply Voltage	1.3 V
Max. Data Rate	1 Mb/s
Min. Energy	400 pJ/b
Min. Sensitivity	-90 dBm
Frequency	900MHz ISM
Technology	0.18μm CMOS
Area	< 0.15 mm^2

-90 dBm is achieved when operating at 250 kb/s.

The receiver consumes 244 μW, 300 μW, 340 μW, and 400 μW from a 1.3V supply while operating at 250 kb/s, 500 kb/s, 750 kb/s, and 1 Mb/s, respectively. The highest efficiency is achieved at 1 Mb/s where the receiver consumes 400 pJ/b.

Table I compares the receiver to recently published super-regenerative receivers and Table II summarizes the measured results. The design presented here is the lowest power super-regenerative receiver reported. It also has the highest data

rate and the lowest energy per bit of any integrated super-regenerative receiver.

IV. CONCLUSION

An ultra low-power BFSK super-regenerative receiver for wireless sensor networks has been presented. By quenching the receiver at twice the data rate and modulating the tank frequency BFSK can be used instead of the traditional OOK. This allows an increase in performance while still maintaining the ultra low-power design. In the oscillator, a single external inductor along with simple digital control enables low power operation. A completely passive rectifier and low power design techniques minimize the power consumed in the baseband circuitry. The prototype circuit demonstrates the smallest and lowest power integrated super-regenerative reported to date.

V. ACKNOWLEDGMENT

This work was supported in part by NSF grant DBI-0529223 and a grant from the US Army Research Laboratory through Oregon Nanoscience and Microtechnologies Institute (ONAMI).

REFERENCES

[1] J.M. Rabaey, J. Ammer, T. Karalar, S. Li, M.S. Otis, and T. Tuan, "Picoradios for wireless sensor networks: the next challenge in ultra-low power design," in *International Solid-State Circuits Conference*, Feb. 2002, pp. 200–201.

[2] A. Vouilloz, M. Declercq, and C. Dehollain, "A low-power CMOS super-regenerative receiver at 1 GHz," *IEEE Journal of Solid-State Circuits*, vol. 36, no. 3, pp. 440–451, Mar. 2001.

[3] B. Otis, Y.H. Chee, and J. Rabaey, "A 400μW-RX, 1.6mW-TX super-regenerative transceiver for wireless sensor networks," in *IEEE International Solid-State Circuits Conference*, Feb. 2005, pp. 396–397.

[4] J.-Y. Chen, M. P. Flynn, and J. P. Hayes, "A fully integrated auto-calibrated super-regenerative receiver in 0.13-μm CMOS," *IEEE Journal of Solid-State Circuits*, vol. 42, no. 9, pp. 1976–1985, Sept. 2007.

[5] F.X. Moncunill-Geniz, P. Palà-Schönwälder, C. Dehollain, N. Joehl, and M. Declercq, "An 11-Mb/s 2.1-mW synchronous superregenerative receiver at 2.4 GHz," *IEEE Transactions on Microwave Theory and Techniques*, vol. 55, no. 6, pp. 1355–1362, June 2007.

[6] B.P. Lathi, *Modern Digital and Analog Communication Systems*, Oxford University Press, Oxford, New York, 1998.

[7] J. Ayers, K. Mayaram, and T. Fiez, "A low power BFSK super-regenerative transceiver," in *International Symposium on Circuits and Systems*, May 2007, pp. 3099–3102.

978-1-4244-2018-6/08 $25.00 © 2008 IEEE

IEEE 2008 Custom Intergrated Circuits Conference (CICC)

A 900-MHz Low-Power Transmitter With Fast Frequency Calibration for Wireless Sensor Networks

Napong Panitantum, Kartikeya Mayaram, and Terri S. Fiez

School of EECS, Oregon State University, Corvallis, OR 97331

Abstract-**A low-power transmitter designed for wireless sensor networks operating in a 900-MHz ISM band has been fabricated in 0.18-μm CMOS technology. The transmitter has a power efficiency as high as 26% while radiating an output signal of -2.9 dBm. It supports both OOK and BFSK modulation. A fast frequency calibration is developed to reduce the RF tuning time down to 36 μs and thus further minimize the overall energy consumption.**

I. INTRODUCTION

Emerging wireless sensor networks (WSNs) enable us to collect data autonomously even in remote areas such as a deep forest or a desert [1]. A WSN node must operate with low energy consumption in order to operate as long as possible with a limited supply of energy. The highest power consuming part of a WSN node is the transceiver and in particular the transmitter. Unlike the traditional homo- or hetero-dyne transmitters, a direct modulation architecture has advantages in WSN applications as it simplifies the transmitter circuit to just an oscillator and an amplifier as shown in Fig. 1. The direct modulation transmitter normally uses less power but it is suitable only for simple OOK or FSK modulation schemes [2]-[6].

In addition to the use of a low power architecture, the energy dissipated when not transmitting data has to be minimized such as the frequency tuning of the oscillator. This is particularly important for a WSN transmitter as only a few hundreds of bits are transmitted. As an example, a data transmission of 250 bits at 250 kbps requires only 1 ms. Thus a conventional frequency tuning that usually lasts longer than 100 μs consumes additional energy of 10% or more. A faster approach is implemented and integrated into the transmitter in this work to minimize this energy overhead.

Frequency tuning is a necessary operation so that the carrier frequency of a transmitter can be matched with a receiver. This operation is always required before a WSN node starts communication since the oscillator frequency may change due to variations in its environmental conditions, e.g. temperature, humidity, etc. A calibration must be carried out to ensure that the calibrated frequency will be within the input bandwidth of a receiver. The target receiver for this transmitter is a super-regenerative receiver as it is by far the lowest energy-per-bit receiver to date [7]-[8].

This paper presents a high-efficiency transmitter with a low latency frequency calibration operating in the 900-MHz ISM band. Section II describes the detailed operation of the frequency calibration technique and its frequency accuracy. The transmitter circuit design is presented in Section III. Measurement results of a prototype transmitter are given in Section IV and the conclusions are summarized in Section V.

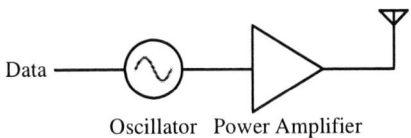

Oscillator   Power Amplifier

Fig. 1. Direct modulation architecture.

II. FREQUENCY CALIBRATION

In this work the oscillator frequency is controlled by a digital signal. To calibrate this digitally controlled oscillator (DCO), the digital code has to be selected for the oscillator frequency to be within the input bandwidth of a receiver. For most oscillators, frequency calibration is done through a phase-locked loop [1], [9] or a frequency-locked loop [2]. While these closed-loop approaches have high accuracy, they are too slow since the calibration time is inversely proportional to the loop bandwidth. In general, the loop bandwidth is at least ten times lower than the reference frequency. In [10] an open-loop system has been developed that needs 4 cycles of reference frequency for a single frequency comparison. The open-loop frequency calibration described in this paper further reduces the comparison time to only 2 cycles.

A. Basic Operation

The open-loop system of Fig. 2, excluding grayed boxes, comprises a programmable divider, a rising-edge comparator, and a controller. The divider acts as a programmable counter. At the beginning, it is reset and then starts counting when the controller deactivates a reset signal. The reset signal is pulled down at the same time as F_{ref} rises as shown in the timing diagram of Fig 3(a). The counter generates a rising-edge output $F_{osc/N}$ as an overflow signal after the Nth rising-edge of the input arrives where N is a dividing number. Ideally $F_{osc/N}$ and the next rising-edge of F_{ref} will rise simultaneously if both signals have the same frequency. The rising-edge comparator compares the frequency of F_{ref} and $F_{osc/N}$ by detecting the rising-edge of these two signals and latches the one that rises first. This latch output indicates the signal that has a higher frequency.

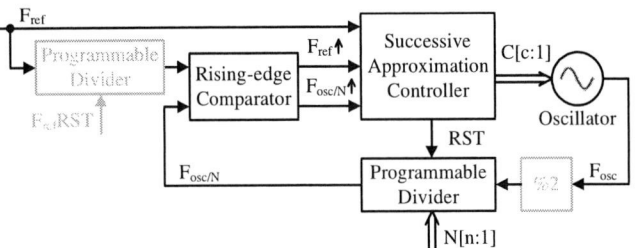

Fig. 2. Block diagram of the proposed open-loop frequency calibration.

978-1-4244-2018-6/08 $25.00 © 2008 IEEE

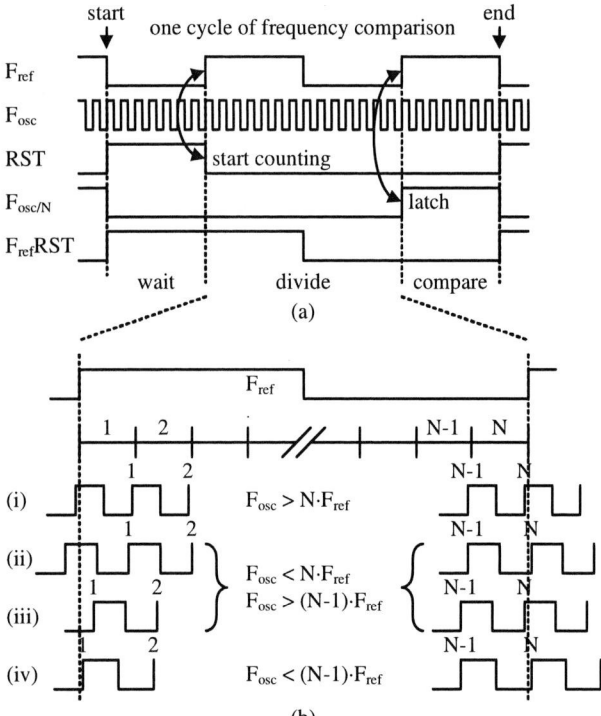

Fig. 3. Timing diagram for (a) one cycle of frequency comparison, and (b) the divided period of the frequency comparison.

The controller uses a successive approximation algorithm to determine the frequency control code. First, all bits are set to '0' corresponding to the lowest frequency and then a comparison of the frequencies is performed. If the comparator output indicates that the divided oscillator frequency is lower than the reference frequency then the controller changes the MSB to '1'. This frequency comparison repeats in order from the MSB to the LSB.

As seen in Fig. 3(a), one cycle of frequency comparison takes 2 cycles of F_{ref}; half a cycle waiting for the oscillator frequency after a change, a cycle for the dividing process, and another half a cycle for comparison. Thus the total calibration time is $2 \cdot c$ cycles of F_{ref} where c is the number of frequency control bits.

B. Circuit Design

Fig. 4 shows the structure of a programmable frequency divider [11]. It is composed of a cascade of 2/3-divider cells. For n cells, the range of the divider number N varies from 2^n to $2^{n+1}-1$. In order to cover the entire 900-MHz ISM band (902-928 MHz) with a 1-MHz resolution, n = 9. However, only 6 programmable bits are necessary. The first 3 MSBs are fixed to a constant of '110' and N = 896-959. Conventionally the divider does not require a reset signal, but one is inserted here to align the divider counting start point with the rising-edge of F_{ref}. The functionally equivalent standard logic of the 2/3-divider cell is shown in Fig. 5 with a reset signal. The actual circuit is implemented by using a true-single phase clock (TSPC) [12] design to minimize the number of transistors and power consumption.

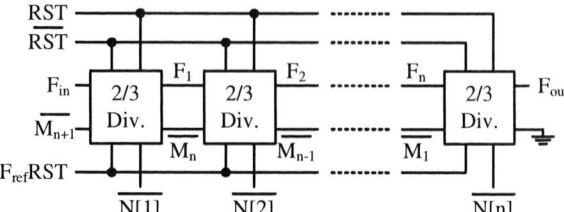

Fig. 4. Programmable frequency divider structure.

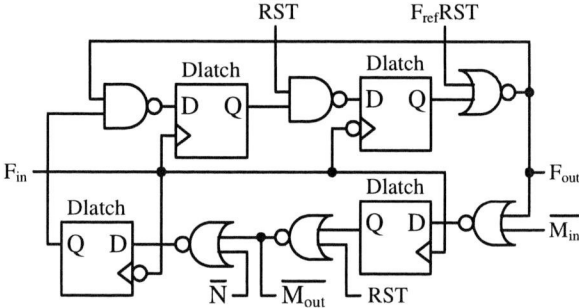

Fig. 5. Functionally equivalent standard logic of the 2/3-divider cell.

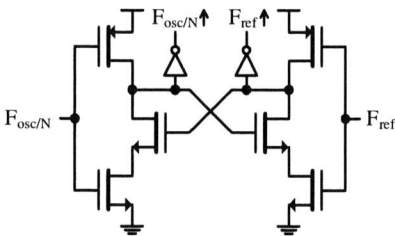

Fig. 6. Rising-edge comparator schematic.

The schematic of a rising-edge comparator is depicted in Fig. 6. When both inputs are low, the two outputs are low as well. If one of the inputs goes high, the output corresponding to that signal rises and latches up preventing the other output to rise.

In Fig. 2, a programmable divider is inserted between F_{ref} and the rising-edge comparator. This divider is reset by the $F_{ref}RST$ signal displayed in Fig. 3(a). The purpose of this divider is not to divide F_{ref} but to mimic the delay between the Nth rising-edge of F_{osc} and the rising-edge of $F_{osc/N}$. A fixed divide-by-2 circuit is also added between the oscillator and the programmable divider as a buffer, pre-divider and differential-to-single-ended converter. Thus a 0.5-MHz F_{ref} is used for a resolution of 1-MHz. With a F_{ref} of 0.5-MHz and the 9-bit DCO explained in the Section III, the system needs only 36 μs to complete one frequency calibration cycle.

C. Calibration Accuracy

Ideally, the frequency error after calibration should be equal to the resolution of the divider. However, the initial phase uncertainty of the oscillator frequency leads to an increase in the worst-case frequency error. As shown in Fig. 3(b), in cases (i) and (iv) the system has no problem in determining that F_{osc} is faster or slower than $N \cdot F_{ref}$, respectively. But if F_{osc} is in between $(N-1) \cdot F_{ref}$ and $N \cdot F_{ref}$, the system can interpret it as being faster or slower than $N \cdot F_{ref}$ depending on the initial phase as illustrated in cases (ii) and (iii). Both cases have the

978-1-4244-2018-6/08 $25.00 © 2008 IEEE

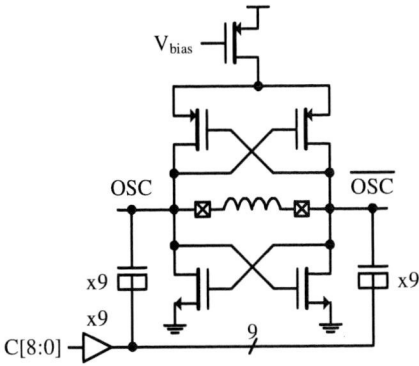

Fig. 7. Oscillator schematic.

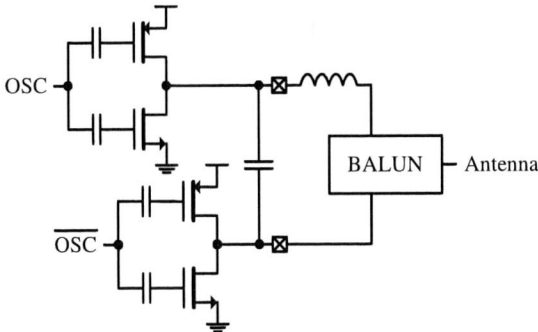

Fig. 8. Power amplifier schematic (bias circuit is not shown).

same oscillator frequency but the initial phase in case (ii) is larger than in case (iii). Then the Nth rising-edge in case (iii) arrives earlier than the rising-edge of F_{ref} and the system misinterprets it as F_{osc} being faster than $N \cdot F_{ref}$. This results in an increase in the frequency error to about 1.5 times the resolution of the divider or about 1.5 MHz for this transmitter. Therefore, this transmitter can communicate with receivers that have an input bandwidth that exceeds 1.5 MHz such as 3 MHz and 1.8 MHz of [7] and [8], respectively.

III. TRANSMITTER DESIGN

The direct modulation transmitter consists of two key blocks, an oscillator and a power amplifier. The role of the oscillator is to generate a designated carrier frequency directly. This architecture supports both OOK and BFSK modulation by turning on/off the oscillator and by switching the oscillator frequency, respectively. The power amplifier is utilized to buffer the oscillator and drive an antenna with high efficiency.

A. Oscillator

An LC oscillator has been chosen because of its lower power consumption and better phase noise. The schematic of the LC oscillator is depicted in Fig. 7. The start-up gain is increased by using both NMOS and PMOS cross-coupled pairs. This is important in reducing the bias current to lower the power consumption.

The oscillator frequency is tuned by MOS varactors. The values of C_{min}, C_{max} are designed to cover the frequency range about ±10% from the ISM band or approximately 190 MHz in order to tolerate variations in the inductor, bondwires and other parasitics. The varactor is split into 9 bits, 2 thermometer-coded MSBs and 7 binary-weighted LSBs. Thus 384 unique frequencies can be realized with an approximate resolution of 0.5 MHz which is smaller than the divider resolution.

The inductor is a 10.4 nH off-chip inductor since the inductance of the LC oscillator at 900 MHz is too high to be economically integrated on the chip. The off-chip inductor also offers a higher Q which leads to lower power consumption than with the use of an on-chip inductor.

B. Power Amplifier

The power amplifier schematic is illustrated in Fig. 8. The amplifier is similar to an inverter but the transistors are sized

and biased so that it operates as a push-pull amplifier. The amplifier is driven directly by the oscillator and the input capacitance of the amplifier is a part of the oscillator resonant capacitance.

The matching network is made up of an on-chip capacitor and an off-chip inductor. The 50-ohm antenna is transformed into a higher impedance at the output of the power amplifier for a high efficiency radiation at low power output levels.

IV. MEASUREMENT RESULTS

The transmitter has been fabricated in a 0.18-μm standard CMOS technology and occupies an active area of only 0.25 mm². The die photograph of the transmitter is shown in Fig. 9.

The open-loop frequency calibration proposed in this paper is fast. In this transmitter (9-bit DCO and 0.5-MHz F_{ref}), the calibration process lasts only 36 μs for OOK modulation and 72 μs for BFSK modulation. The oscillator frequency after calibration is plotted in Fig. 10. All of the calibrated frequencies lie within 1.5-MHz boundaries (displayed by dashed lines) as described in Section II. This accuracy is sufficient for a receiver that has a high input bandwidth such as the super-regenerative receiver [7]-[8].

The oscillator has a frequency tuning range from 864 MHz to 1.06 GHz. The power on/off time for the oscillator is 120 ns and 70 ns, respectively. Assuming the on/off time is 10% of a bit period in OOK modulation, the maximum data rate of the transmitter will be 1 Mbps. The transmitter consumes only 1.94 mW at 0.9-V supply, 0.42 mW in the oscillator and 1.52 mW in the power amplifier. It can transmit an output of -2.9 dBm equivalent to an efficiency of 26%. This output level is adequate for data communication over 10-20 meters. Fig. 11 shows the output spectrum at 915 MHz. It has a low phase noise of -123 dBc/Hz at 1 MHz offset. Therefore, the transmitted signal generates very low interference to other devices occupying the same 900-MHz ISM band.

Table I summarizes the key performance parameters of recent WSN transmitters, comparing them to this work. The proposed transmitter has among the best efficiency. Those with higher efficiency were fabricated in a more advanced technology. This transmitter has more flexibility in terms of an RF programmability and a modulation scheme than [4]-[5] since those transmitters can generate only one frequency fixed by the physical dimensions of a MEMS device and are unsuitable for FSK modulation. The calibration time of this

978-1-4244-2018-6/08 $25.00 © 2008 IEEE

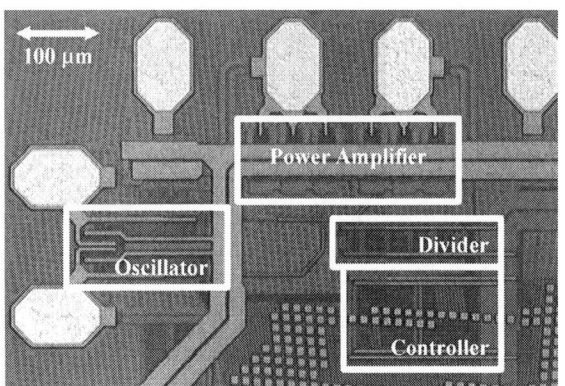

Fig. 9. Die photograph of the transmitter.

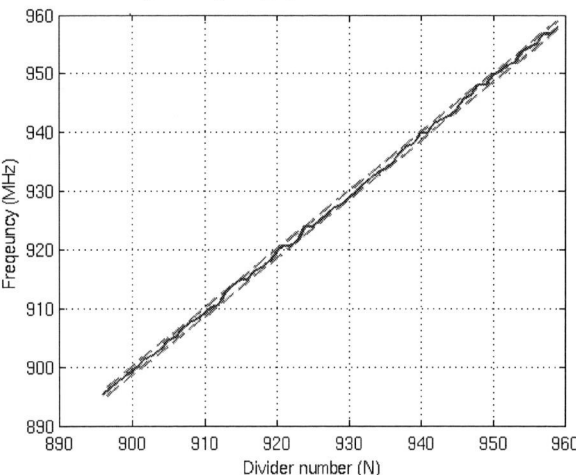

Fig. 10. Measured calibrated frequency vs. divider number.

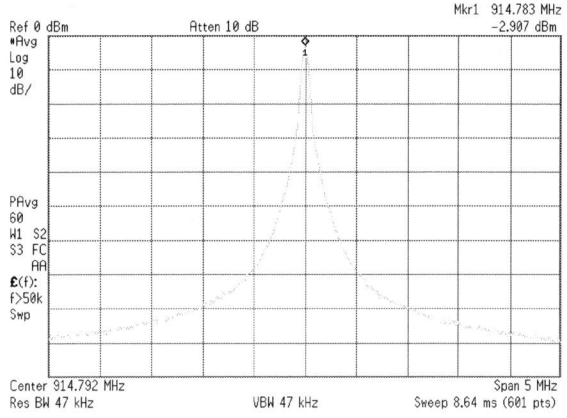

Fig. 11. Measured output spectrum of the transmitter at 915 MHz.

transmitter is almost 3 times faster than that of a closed-loop calibration in [1] even with a slower F_{ref}.

V. CONCLUSION

The design of a low-power WSN transmitter is presented. The transmitter can be configured to communicate at any frequency of the 900-MHz ISM band with either OOK or BFSK modulation. An open-loop frequency calibration feature is integrated in this transmitter. This improves the calibration time and the energy dissipation for short data transmission times of WSNs. The transmitter is designed to achieve high efficiency by utilizing an LC oscillator and a push-pull power amplifier. The prototype chip was fabricated in a 0.18-μm CMOS technology and the transmitter demonstrates an efficiency of 26% at -2.9 dBm output radiated power.

ACKNOWLEDGMENT

This work was supported in part by NSF grant DBI-05292230 and a grant from the US Army Research Laboratory through Oregon Nanoscience and Microtechnologies Institute (ONAMI).

REFERENCES

[1] J. Porter *et al.*, "Wireless sensor networks for ecology," *BioScience Mag.*, vol. 55, no. 7, pp. 561-572, July 2005.
[2] N. Boom, W. Rens, and J. Crols "A 5.0mW 0dBm FSK transmitter for 315/433 MHz ISM applications in 0.25μm CMOS," *European Solid-State Circuits Conf.*, Sep. 2004, pp. 199-202.
[3] A. Molnar, B. Lu, S. Lanzisera, B. W. Cook, and K. S. J. Pister, "An ultralow power 900 MHz RF transceiver for wireless sensor networks," *IEEE Custom Integrated Circuits Conf.*, Oct. 2004, pp. 401-404.
[4] B. P. Otis, Y. H. Chee, and J. M. Rabaey, "A 400μW-RX, 1.6mW-TX super-regenerative transceiver for wireless sensor networks," *Int. Solid-State Circuits Conf.*, Feb. 2005, pp. 396-397,606.
[5] Y. H. Chee, A. M. Niknejad, and J. M. Rabaey, "An ultra-low-power injection locked transmitter for wireless sensor networks," *IEEE J. Solid-State Circuits*, vol. 41, no. 8, pp. 1740-1748, Aug. 2006.
[6] B. W. Cook, A. Berny, A. Molnar, S. Lanzisera, and K. S. J. Pister, "Low-power 2.4-GHz transceiver with passive RX front-end and 400-mV supply," *IEEE J. Solid-State Circuits*, pp. 370-371,659, Dec. 2006.
[7] F. X. Moncunill-Geniz, P. Pala-Schonwalder, C. Dehollain, N. Joehl, and M. Declercq, "An 11-Mb/s 2.1-mW synchronous superregenerative receiver at 2.4 GHz," *IEEE Trans. on Microwave Theory and Techniques*, vol. 55, no. 6, pp. 1355-1362, June 2007.
[8] J.-Y. Chen, M. P. Flynn, and J. P. Hayes, "A fully integrated auto-calibrated super-regenerative receiver in 0.13-μm CMOS," *IEEE J. Solid-State Circuits*, vol. 42, no. 9, pp. 1976-1985, Sep. 2007.
[9] A. Aktas and M. Ismail, "CMOS PLL calibration techniques," *IEEE Circuits and Devices Mag.*, vol. 20, no. 5, pp. 6-11, Mar., 2001.
[10] T.-H. Lin and Y.-J. Lai, "An agile VCO frequency calibration technique for a 10-GHz CMOS PLL," *IEEE J. Solid-State Circuits*, vol. 42, no. 2, pp. 340-349, Feb. 2007.
[11] C. S. Vaucher *et al.*, "A family of low-power truly modular programmable dividers in standard 0.35-μm CMOS technology," *IEEE J. Solid-State Circuits*, vol. 35, no. 7, pp. 1035-1045, July 2000.
[12] J. Yuan and C. Svensson, "High-speed CMOS circuit technique," *IEEE J. Solid-State Circuits*, vol. 24, no. 1, pp. 62-70, Feb. 1989.

TABLE I

PERFORMANCE COMPARISON OF RECENT WSN TRANSMITTERS.

	Tech. (μm)	Freq (GHz)	Vdd (V)	Prad (mW)	P (mW)	Eff. (%)	Ext. component	Calibration time (μs)
[2]	0.25	0.43	1.3	1.00	5.0	20	Inductor	200 @ 8-MHz F_{ref}
[3]	0.25	0.9	1.2, 1.4	0.25	1.3	19	Inductor	NA @ 32-kHz F_{ref}
[4]	-	1.9	1.0	0.38	1.7	22	BAW	-
[5]	0.13	1.9	1.0	1.00	3.6	28	FBAR	-
[6]	0.13	2.4	0.4	0.30	1.0	30	-	not included
This work	0.18	0.9	0.9	0.51	1.94	26	Inductor	36, 72 @ 0.5-MHz F_{ref}

A 3.5-mW 15-Mbps O-QPSK Transmitter for Real-time Wireless Medical Imaging Applications

Yao-Hong Liu and Tsung-Hsien Lin

Graduate Institute of Electronics Engineering and Department of Electrical Engineering
National Taiwan University, Taipei, Taiwan

Abstract—A 400-MHz phase-mux-based O-QPSK transmitter (TX) for medical imaging applications is presented in this paper. The modulation signal is generated by selecting one of the four quadrature phases to the TX output via a proposed Phase MUX. An inverter-type power amplifier is utilized for it is compatible with the quasi constant-envelope nature of the O-QPSK modulation. Fabricated in a 0.18-μm CMOS process, the whole TX draws 2.9 mA from a 1.2-V supply. With a maximum 15-Mbps data rate, the TX achieves an energy efficiency of 0.23 nJ/bit and delivers an output power up to -7 dBm.

I. INTRODUCTION

Implantable/swallowable miniature bio-medical devices are playing increasingly important roles in medical applications, e.g. in a wireless ultrasound capsule endoscope system [1]. For real-time high-resolution images with a maximum frame rate of 10-frames/sec, a transmission data rate up to 10 Mbps is required. In these systems, the high-data-rate transmitter (TX) often dominates the battery energy consumption. In this paper, an energy-efficient phase-mux-based TX architecture is proposed, which can support various B/QPSK modulation schemes.

The proposed TX is designed to operate at 400 MHz and may be applicable to devices conform to the Medical Implantable Communications Service standard [3]. The 400-MHz band is chosen primarily because the signal propagation characteristic in a human body at this frequency is better than other available frequency bands (e.g. 900-MHz or 2.4-GHz ISM bands) [4]. The TX output power is constrained to 250 μW (i.e. -6 dBm), for this is adequate for a 2-meter link. The O-QPSK modulation is adopted in this work. PSK type modulation is preferred here because of its better noise immunity compared with ASK and FSK modulations. Furthermore, the O-QPSK modulation has a smaller peak-to-average power ratio than that of the QPSK, which allows an efficient switch-type power amplifier (PA) to be utilized to facilitate low-power consumption.

This paper is organized as follow. In Section II, the proposed TX architecture is introduced. Key building blocks are detailed in Section III. Finally, Section IV and V present the experimental results and conclusion, respectively.

II. PROPOSED O-QPSK TRANSMITTER

A conventional mixer-based direct up-conversion TX is depicted in Fig. 1. This architecture is well-studied and is adopted in many phase modulation systems. It typically contains two digital-to-analog converters (DACs), two filters, a summed quandrature Gilbert-cell mixer, and a PA. However, higher transmission data rate usually leads to higher power consumption of the DACs and filters. Therefore, power-bandwidth trade-off is inevitable in this architecture. In addition, this architecture is sensitive to process, supply voltage, and temperature (PVT) variations. Large device sizes and higher power consumption are mandated for the analog circuits to overcome various non-idealities, such as offset, IQ-mismatch, and non-linearity effects, etc.

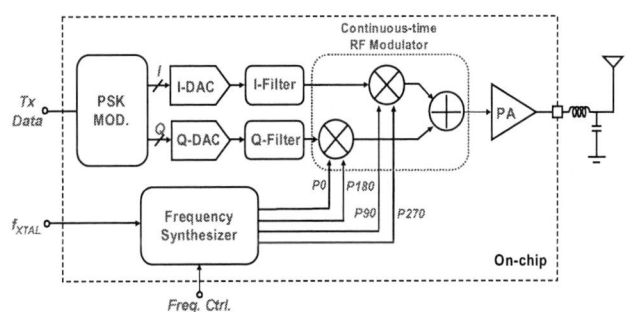

Fig. 1. Block diagram of a conventional direct-conversion transmitter.

To address the issues mentioned above, a phase-mux-based TX architecture is proposed in this work, as shown in Fig. 2. In this architecture, the quadrature mixer in the conventional TX is largely simplified to the proposed Phase MUX to reduce the power consumption as well as the chip area. The O-QPSK modulator implements a compact digital baseband. The digital baseband data are directly sent to the phase-mux-based RF modulator; thus eliminating two high-speed DACs and wide-bandwidth filters. The external output matching network essentially also functions as a filter to limit the signal bandwidth.

A PLL-based frequency synthesizer operating at twice the TX carrier frequency is implemented. Such frequency plan avoids the PA pulling effect. The VCO output is divided by 2 to generate four phases in quadrature. According to the baseband TX data, one of the phases is then selected via the Phase MUX and sent to the PA; thus, creating phase modulation signals.

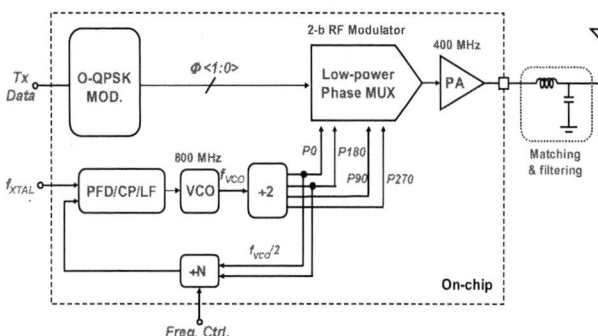

Fig. 2. Block diagram of the proposed phase-mux-based TX.

III. CIRCUIT IMPLEMENTATIONS

A. Digital O-QPSK Modulator

Offset-QPSK is a variation of the QPSK modulation. As illustrated in Fig. 3, the O-QPSK data is generated by offsetting one of the data stream by half the symbol period. Along the process, a transition phase is introduced, which prevents a direct transition between two diagonally opposite points (e.g. [0 1] and [1 0]) in the constellation. This ensures that the signal power will not drop to zero during data transition; therefore, peak-to-average power ratio is reduced and the variation of signal envelope is smaller.

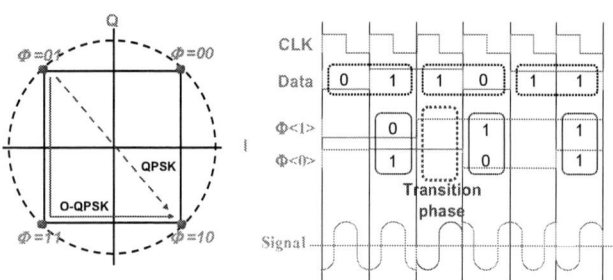

Fig. 3. O-QPSK modulation.

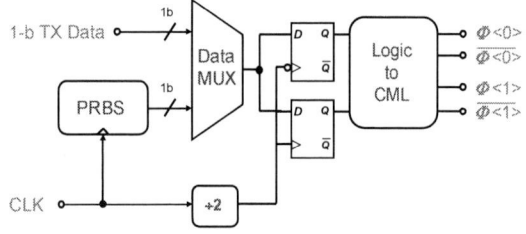

Fig. 4. Block diagram of the digital O-QPSK modulator.

Fig. 4 shows the digital O-QPSK data generator. The input 1-b TX data is sampled by two DFFs with inverted clock phases at half of the clock rate. This circuit produces a 2-b data stream, with one delayed by half symbol period with respect to the other. The data stream is then converted to the differential form for the Phase MUX. The input data can be selected either from the pseudo-random bit stream (PRBS) generator or from an external data source by the Data MUX. The PRBS generator is implemented primarily for evaluation purpose. Logic-to-CML converter converts the rail-to-rail data to differential CML signals before being sent to the high-speed CML Phase MUX.

B. Phase MUX

The proposed Phase MUX circuit is shown in Fig. 5. The circuit selects a pair of complementary phases according to the symbol $\Phi<1:0>$. The selected pair is then converted to a single-ended signal to drive the following PA. As the baseband data $\Phi<1:0>$ have large swings, the circuit can easily realize a high-data-rate PSK modulator with low current consumption (less than 0.4 mA). Adequate quadrature accuracy is ensured by well-balanced LC-VCO and divide-by-2 frequency divider designs.

The proposed Phase MUX is free from the LO leakage problem that exists in the conventional mixers, since the effects of offset voltage in M1~M6 are eliminated by fully switching of these differential pairs. This circuit is also less sensitive to PVT variations, since there is no transconductance dependency in such a digital-like implementation. Furthermore, no analog filtering is employed in this TX architecture. Therefore, it generates BPSK/QPSK type modulations with low power consumption.

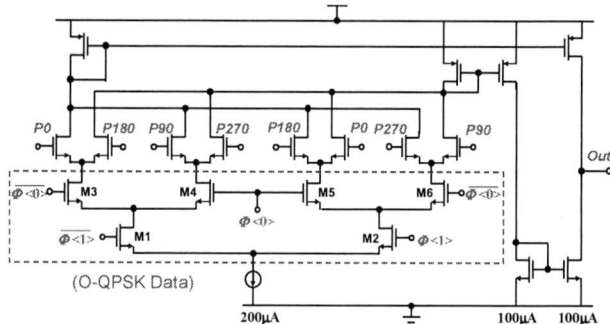

Fig. 5. Simplified schematic of the proposed Phase MUX.

C. Power Amplifier

Fig. 6 depicts the inverter-type PA adopted in this work. It is compatible with the quasi constant-envelope nature of the O-QPSK modulation signal and has a better power efficiency. The 1st-stage of the PA is self-biased to reduce the 2nd-order harmonic. The output of the PA drives an external matching network which also serves as a filter to remove other harmonic components. PA efficiency is defined as output power divided by average DC power. The efficiency here is mainly limited by the requirement of keeping a low average DC power consumption, as expressed in (1) and illustrated in Fig. 6, where k is the duty cycle of the current, and I_{Peak} is determined by load impedance, device sizes, and the input driving signal V_X.

$$PA\ efficiency \equiv \frac{P_{OUT}}{DC\ Power} \propto \frac{I_{Peak}^2 R_L}{V_{DD} \cdot I_{Avg}} \propto I_{Avg} \qquad (1)$$

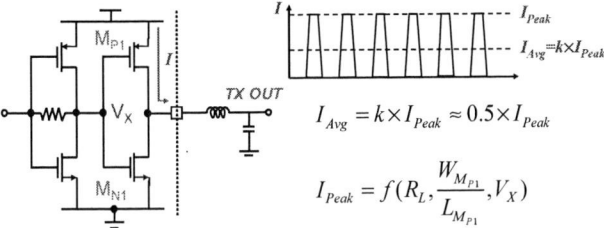

$$I_{Avg} = k \times I_{Peak} \approx 0.5 \times I_{Peak}$$

$$I_{Peak} = f(R_L, \frac{W_{M_{P1}}}{L_{M_{P1}}}, V_X)$$

Fig. 6 The inverter-type power amplifier.

D. VCO and Frequency Synthesizer

A frequency synthesizer (FS) must have a low phase noise to achieve a high-quality communication. In addition, a fast-locking FS is required to minimize the energy wasted during the start-up period. This is especially important when the energy source is limited. In this work, the FS is based on a type-II integer-N phase-locked loop (PLL). A loop bandwidth of 250 kHz is selected for it allows fast locking, optimum phase noise, and integration of an area-efficient 2nd-order passive loop filter. This 800-MHz FS consumes only 1.9 mA (including the VCO) from a 1.2-V supply.

Fig. 7(a) shows the LC-VCO circuit which utilizes both cross-coupled NMOS and PMOS pairs. The VCO is operated at twice the carrier frequency. In addition to avoiding the pulling effect from PA, higher operating frequency also allows a smaller inductor with a better quality factor. Moreover, the LC-tank impedance is increased at higher frequency, enabling a better oscillation condition. The VCO covers frequencies ranging from 710 to 870 MHz, which supports TX output from 355 to 435 MHz. To cover the 20% tuning range while keeping a small VCO tuning gain (K_{VCO}), a 2-bit switched capacitor array is incorporated in the VCO tank circuit. The measured frequency tuning curves are shown in Fig. 7(b), where the peak K_{VCO} is around 100 MHz/V. The VCO consumes only 0.8 mA.

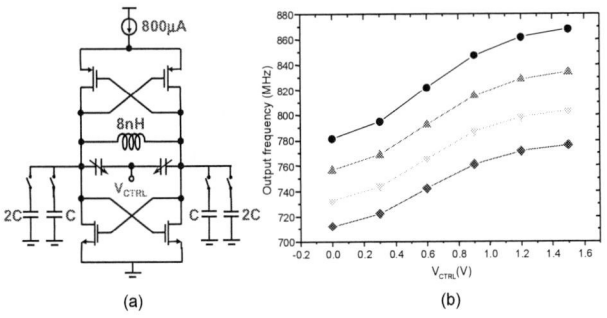

Fig. 7 (a) Simplified schematic of the LC-VCO, (b) measured VCO tuning curves.

IV. EXPERIMENTAL RESULTS

This chip is fabricated in the TSMC 0.18-μm 1p6m CMOS technology. The PA consumes only 0.6 mA and delivers an output power up to -7 dBm, which is equivalent to 22% power efficiency. According to (1), the efficiency will be better if the PA delivers a higher output power. Fig. 8 shows the measured TX output spectrum (at 403.2 MHz) and the eye diagram at a data rate of 15 Mbps. The modulated signal achieves an EVM of 14%, which is equivalent to 17-dB SNR. The EVM is mainly dictated by the non-linear inverter-type PA. However, such signal quality is still more than adequate for typical QPSK demodulators to satisfy a BER of better than 10^{-3}. Fig. 9 plots the overall O-QPSK modulation quality (in measured EVM and SNR) versus various transmission data rate (from 300 kbps to 15 Mbps).

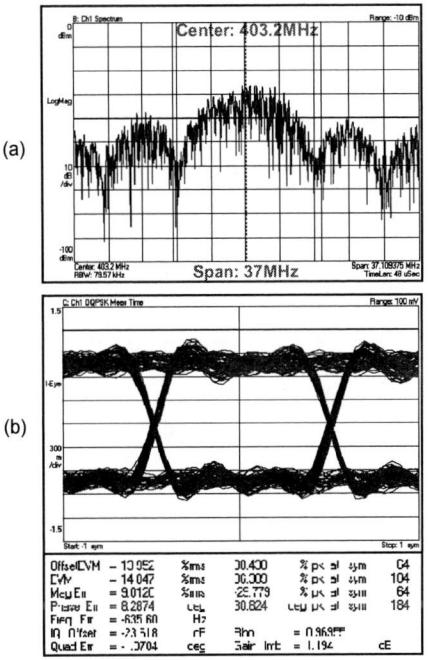

Fig. 8. Measured 15-Mpbs O-QPSK modulation characteristics. (a) Modulation spectrum, (b) eye digram.

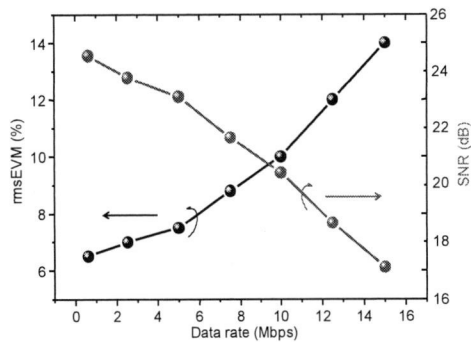

Fig. 9. Measured TX EVM and SNR versus various data rate.

Fig. 10 shows the experimental results of the FS. The VCO frequency (f_{VCO}) is locked to 806.4 MHz. The PLL phase noise (measured at TX output) is -107.9 dBc/Hz and -

110 dBc/Hz at 10-kHz and 100-KHz offset, respectively (Fig. 10(a)). The integrated phase error (from 1 kHz to 30 MHz) is only 0.31 degree. The PLL loop bandwidth is chosen to be 250 kHz to achieve a fast locking within 16 µs. The loop settling transient behavior (shown in Fig. 10(b)) is monitored by observing the voltage on the loop filter (V_{CTRL}).

The complete O-QPSK TX consumes only 2.9 mA from a 1.2-V supply voltage. It achieves an energy efficiency of 0.23 nJ/bit at a data rate of 15 Mbps. While the PA is usually the most power hungry component in a long-range communication system, it is the VCO and PLL that dictates the power consumption in a short-range application as in this work. Fig. 10 shows the chip photo. The whole chip occupies an area of 1.2 mm^2; while the core circuits occupy 0.7 mm^2. The proposed Phase MUX and PA only utilize very small portion of the chip area.

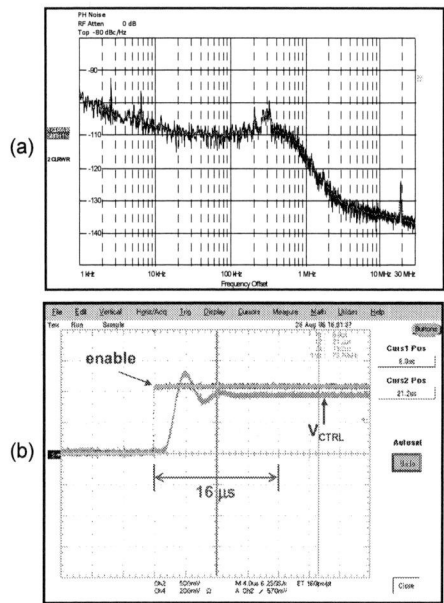

Fig. 10. Measured FS performance: (a) phase noise (measured at TX output); (b) locking behavior.

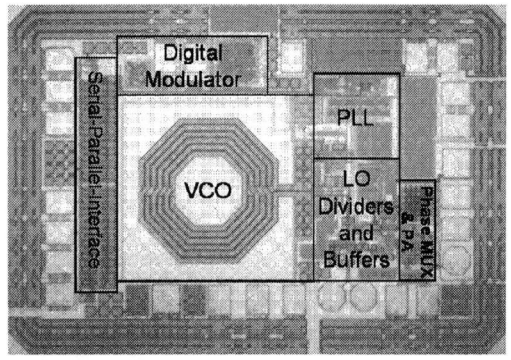

Fig. 11. Chip photo.

Table I summaries the experimental results of the proposed TX and compares its performance with other related works. The proposed TX achieves a high data rate with very low power consumption, resulting in good energy/bit performance. To take into account the TX output power, which reflects the transmission range, a FOM defined in (2) is adopted for comparison purpose.

$$FOM \equiv \frac{P_{OUT} \times Data\ rate}{Power\ consumption} (\mu W \cdot bit/nJ) \qquad (2)$$

The proposed O-QPSK TX achieves the best FOM of 870 µW×bit/nJ at maximum data rate.

TABLE I. SUMMARY AND COMPARISON WITH OTHER WORKS

	This work	[2]	[3]
Process (CMOS)	0.18 µm	0.13 µm	0.18 µm
Supply voltage	1.2 V	360 ~ 600 mV	2 ~ 3 V
Modulations	O-QPSK	FSK	G/FSK
Architecture	Phase MUX	Direct-VCO	IQ mixers
Operating freq.	355~435 MHz	1.95~2.38GHz	400/433 MHz
Max. data rate	**15 Mbps**	0.5 Mbps	0.8 Mbps
Max. P_{OUT}	-7 dBm	-7 ~ -5 dBm	0 dBm
TX current	**2.9 mA**	3.2 mA	5 mA
Energy/bit	**0.23 nJ/bit**	2.6 nJ/bit	12.5 nJ/bit
FOM (µW×bit/nJ)	**870**	121	80

V. CONCLUSION

In this work, an ultra-low-power phase-mux-based TX has been demonstrated to support O-QPSK modulation with a wide data rate range. To the authors' knowledge, this TX achieves the best energy/bit efficiency so far. The proposed TX with the Phase MUX is inherently a digital design, and can be easily ported to advanced digital CMOS processes, and benefit from technology scaling.

ACKNOWLEDGMENT

The authors wish to thank the Chip Implementation Center (CIC), Taiwan, for fabricating this chip. This work is supported by the National Science Council, Taiwan; project contract number: NSC95-2218-E-002-036.

REFERENCES

[1] Y.-H. Liu and T.-H. Lin, "An Energy-Efficient 1.5-Mbps Wireless FSK Transmitter with A ΣΔ-Modulated Phase Rotator, " *Proc. ESSCIRC*, pp. 488-491, Sep. 2007.
[2] B. C. Cook, A. D. Berny, A. Molnar, S. Lanzisera, and K. Pister, "An Ultra-Low Power 2.4GHz RF Transceiver for Wireless Sensor Network in 0.13-µm CMOS with 400-mV Supply and Integrated Passive RX Front-End, " *ISSCC Dig. Tech Papers*, pp. 370-371, Feb. 2004.
[3] Zarlink Corp.: Datasheet ZL70100, May 2005.
[4] J. Ryckaert, et al., "Channel model for wireless communication around human body" *Electronics Letters*, vol. 40, pp. 543-544, Apr. 2004.

978-1-4244-2018-6/08 $25.00 © 2008 IEEE

IEEE 2008 Custom Intergrated Circuits Conference (CICC)

A CMOS Direct Conversion Transmitter With Integrated In-Band Harmonic Suppression for IEEE 802.22 Cognitive Radio Applications

Jongsik Kim, Seungsoo Kim, Jaewook Shin, Youngcho Kim, Junki Min*, Kihong Kim*, Hyunchol Shin

High-Speed Integrated Circuits and Systems Lab., Kwangwoon University, Seoul 139-701, Korea
*Samsung Electro-Mechanics Co., Suwon 443-803, Korea

Abstract- **A CMOS direct conversion transmitter for IEEE 802.22 cognitive radio applications is presented. In-band harmonic distortions are effectively suppressed across the full TV band by exploiting single-conversion dual-path architecture with integrated harmonic rejecting mixers and RF tunable filters. A fractional-N synthesizer with a single *LC* VCO and a wideband muti-modulus (2/3/4/6/8/12/16/24) divider block provide multiphase LO signals. Implemented in 0.18 μm CMOS, the transmitter delivers 0-dBm with in-band distortions less than -42 dBc across 54 - 862 MHz band *without off-chip filters*. Image and LO leakage components are also suppressed below -45 dBc and -36 dBc, respectively, through calibration circuitry. Measured P_{1dB} and OIP_3 are +9 dBm and +20 dBm, respectively.**

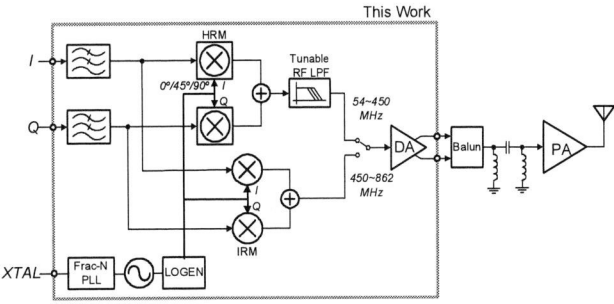

Figure 1. Cognitive radio transmitter architecture.

I. INTRODUCTION

Cognitive radio (CR) is considered as an enabling technology to overcome the current situation of the limited spectrum resources. It is to intelligently detect the spectrum usage at a given location and utilize temporarily unoccupied spectrum without interfering incumbent users. In 2004, FCC opened TV band for the CR applications. Afterwards, IEEE 802.22 working group initiated the first worldwide standardization activity for the CR-based wireless communication in the TV band between 54 and 862 MHz [1].

CR requires sensing, receiving, and transmitting functions preferably in a single CMOS chip. Recently a spectrum sensing device integrated into a CMOS receiver was reported for the first time, but only covered UHF band [2]. We anticipate that more advanced CR receivers with sensing functions will appear further in future by adopting various existing advanced TV tuner arts. On the other hand, however, CMOS transmitters that are applicable to CR are very limited in literature. A wideband transmitter covering 100MHz – 2.5GHz was reported for SDR applications [3]. But their Cartesian feedback structure was not suitable for a wideband signal and their DDS was not desirable either due to the relatively high current dissipation and high spurious tones. A wideband cable TV transmitter reported in [4] provided very limited output power (-16dBm) and employed high phase-noise ring oscillator and no PLL.

In this paper, we present a TV-band direct-conversion transmitter and a frequency synthesizer in CMOS for IEEE 802.22-based CR applications. Effective suppression of in-band distortions is achieved through integrated harmonic suppressing architecture and highly linear building block designs. Wideband LO generation is achieved through a

wideband fractional-N frequency synthesizer with a single *LC* VCO and a wideband multi-modulus divider block.

II. ARCHITECTURE

For the CR transmitter, two architectural candidates are considered from the prior TV tuner arts. The most conventional architecture in wideband tuners is the dual conversion structure [5]. It suffers from high power dissipation and requires external SAW filters, thus not suitable for a single chip. For single chip realization, the single conversion architecture with low-IF [6] or zero-IF [7] is more suitable. Maxim *et al.* [7] demonstrated that a single-conversion zero-IF structure with harmonic rejection mixer and on-chip filtering could eliminate an otherwise bulky off-chip *LC* tracking filters that was used in [6]. But their two PLL's and two VCO's to cover the wideband were still a circuit overhead in [7].

In this work, we adopt single-conversion zero-IF transmitter architecture. Critical in-band distortions are effectively suppressed by integrated on-chip harmonic rejection mixers and accompanying RF tunable filters. Also, we adopt wideband fractional-N frequency synthesizer architecture with a single *LC* VCO and a wideband LO generation (LOGEN) block.

Fig. 1 shows the transmitter architecture. Baseband I/Q signal is fed to the baseband filter. The following up-conversion mixers are formed in dual paths. Harmonic rejection mixer (HRM) covers the lower half band from 54 to 450 MHz. The HRM rejects 3rd and 5th harmonic distortions that always fall in the band. The following RF tunable low pass filter (LPF) further suppresses the 7th and higher harmonics. The upper half band from 450 to 862 MHz imposes less severe in-band distortions. Thus a regular image

978-1-4244-2018-6/08 $25.00 © 2008 IEEE   641

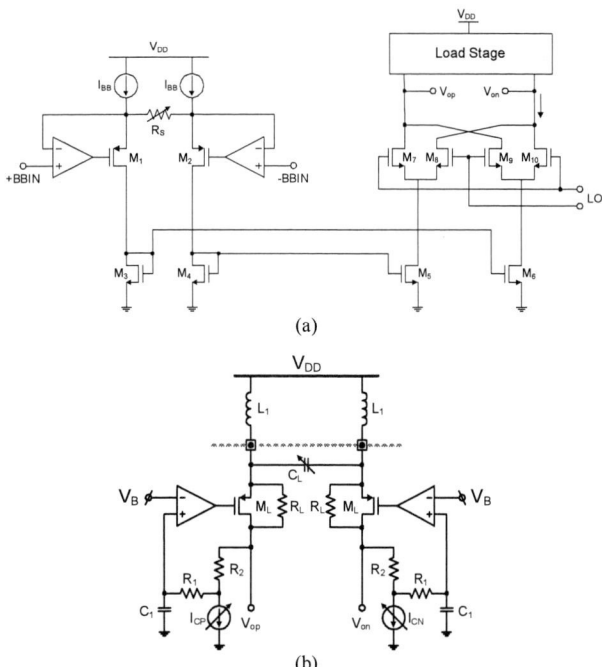

(a)

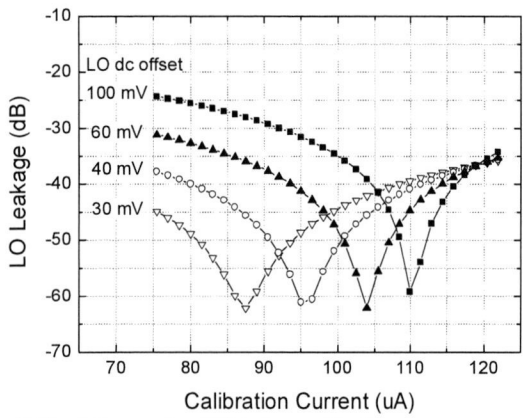

(b)

Figure 2. (a) Mixer core schematic. (b) Mixer load stage schematic.

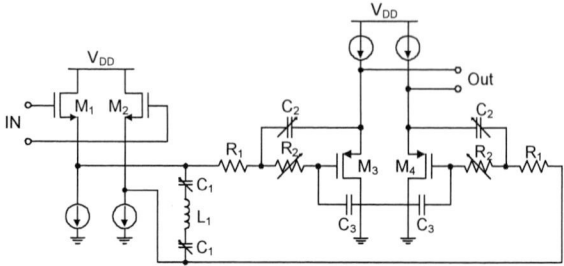

Figure 4. Tunable RF filter.

[Figure 3 graph]

Figure 3. Simulation results of LO leakage calibration.

rejection mixer (IRM) is adopted for the sake of low current dissipation. A differential cascode driver amplifier delivers 0-dBm output power, whose P_{1dB} is set higher to cope with 10-dB PAPR of OFDM signals. An external power amplifier is used to boost the output power up to 30 dBm. A fractional-N PLL with a single VCO generates a 1.0 ~ 1.8 GHz signal and subsequent multi-modulus LOGEN delivers appropriate multi-phase LO signals to the mixers. This direct-conversion dual-path architecture is suitable for a single chip implementation across the full band without requiring external filters

III. CIRCUIT DESIGN

A. Wideband Transmitter

The 110-mV baseband I/Q signal is fed to the baseband filter to remove the residual quantization noise created by preceding DAC. The baseband filter is a 6th-order Chebyshev II biquad. In order to support all the possible bandwidths of 5,

6, 7, and 8 MHz, the bandwidth is tunable from 2.1 to 6.5 MHz in 64 steps. The gain is also tunable from -15 to +15 dB. DC offset cancellation loop is implemented as a feedback loop, for which the lower cutoff frequency is designed to be tunable from 630 Hz to 9.4 kHz to cope with the various subcarrier spacings of OFDM signals of CR.

Fig. 2(a) shows the core schematic of the mixer that is used for both HRM and IRM. Note that the HRM has three of these in parallel performing multiphase ($0^{\circ}/45^{\circ}/90^{\circ}$) mixing with $1:\sqrt{2}:1$ g_m-ratio. This circuit structure is chosen for its high linearity [8]. The input stage is linearized through g_m-boosting with opamp and degeneration resistor R_S, in which R_S is variable to adjust the mixer gain. Current-mode amplification between the input and switching stages is employed to attain high linearity.

Load stage of the mixer is crucial for broad bandwidth and high linearity. Fig. 2(b) details the load schematic. R_L and M_L provide load resistance. Off-chip inductors (82 nH for HRM and 6.2 nH for IRM) are used for shunt peaking to compensate the gain drop at high frequency and thus ensure a broad bandwidth. A 4-bit switched capacitor array C_L is also used to maintain the gain flatness across an almost octave bandwidth. LO leakage calibration circuit is realized in the load stage. A dc level at the mixer output nodes V_{op} and V_{on} are adjusted by varying I_{CP} and I_{CN} in order to counteract any process-induced dc offset at LO signals, which in turn remove the LO leakage at the RF output. The tuning resolution of the V_{op} and V_{on} is 4.45 mV. The simulation results in Fig. 3 shows that the LO leakage is successfully removed against LO dc offsets by adjusting the calibration current I_{CP}. In addition, a image leakage calibration circuit, though not shown here, is also added in LO buffers. All of these calibration circuitry allow us to keep the LO and image leakage components below -40dBc across the full TV band under hostile PVT variations.

Fig. 4 shows the RF tunable LPF schematic. It consists of input buffer, notch filter (L_1 and C_1), and Sallen-Key LPF ($M_{3,4}$, $R_{1,2}$, $C_{2,3}$). The source follower input buffer $M_{1,2}$ is designed in order not to degrade the gain and linearity at the interface with the preceding mixer. The corner frequency of the Sallen-Key is tuned from 100 to 500 MHz by using C_2 and R_2. The transmission zero is designed to be around the 7th harmonic of the corner frequency, but the far-out attenuation is found to be not sufficient in this case [9]. Thus, we add an additional LC notch filter (L_1, C_1) such that the resonant frequency falls on the 9th harmonic of the Sallen-Key's corner frequency. Moreover, they are designed to track each other by

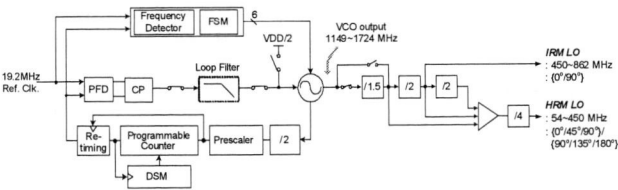

Figure 5. Frequency synthesizer and LO generation.

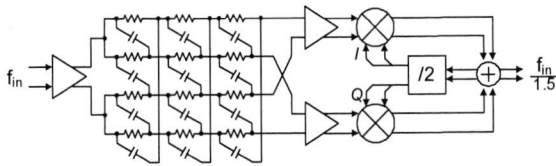

Figure 6. Wideband semidynamic divide-by-1.5 circuit.

using variable C_1, C_2, and R_2. By doing so, the attenuation for the 3rd and higher in-band harmonics is always kept below -20 dBc across the whole band.

Driver amplifier uses a differential cascode type. An external balun is used to deliver a single-ended output signal to the next power amplifier. To achieve high linearity over a wide band, resistive source degeneration is adopted. Degeneration resistor is optimized for high linearity with acceptable gain and noise figure. The simulated OIP_3 and P_{1dB} are at least +24 dBm and +10 dBm across the whole band.

B. Wideband Fractional-N Synthesizer

A fractional-N synthesizer is required for the CR transmitter because it must handle the variable channel bandwidths of 5, 6, 7, and 8 MHz. Fig. 5 depicts the fractional-N synthesizer and its accompanying LOGEN. A single LC VCO with a wideband LOGEN effectively covers the full band. The VCO generates 1149 - 1724 MHz, and the LOGEN converts it to 8 sub-banded signals through the variable division ratio 2/3/4/6/8/12/16/24. The $0°/45°/90°$ I/Q signals for HRM are generated by the divide-by-4 placed at the end while simple I/Q signals are generated by the divide-by-2.

In LOGEN, all the dividers are conventional current-mode logic type for its noise superiority, except for the divide-by-1.5. The divide-by-1.5 whose structure is shown in Fig. 6 is designed as a semidynamic type. To achieve almost an octave bandwidth without requiring a quadrature VCO, a wideband single-sideband mixer is realized by employing a three stage RC polyphase filter for the quadrature generation at the input. The peak phase mismatch and attenuation of the polyphase filter is 0.25 degree and -4.5 dB, respectively for octave bandwidth.

A 3-bit-level 20-bit-resolution third-order single-loop delta-sigma modulator is employed to achieve low phase noise. In OFDM system, it is known that the integrated phase noise is more critical than the spot phase noise for SNR degradation [10]. We have optimized the loop bandwidth and parameters in order to attain the integrated phase noise of less than 1 RMS degree or equivalently -35 dBc.

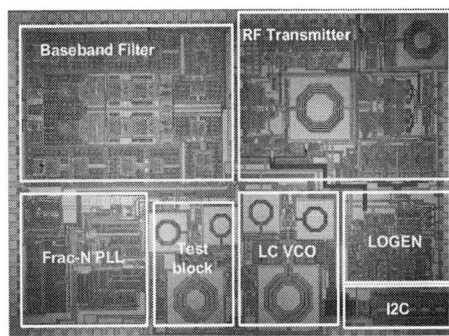

Figure 7. Chip micrograph.

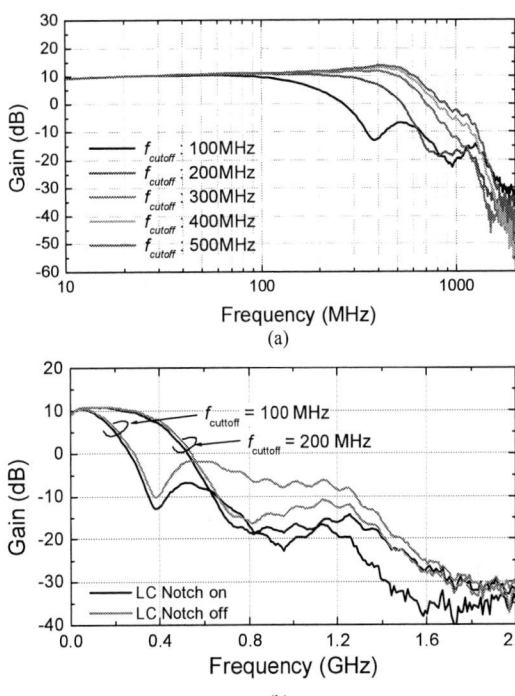

Figure 8. Measured characteristics of RF tunable filter. (a) Tunable bandwidth from 100 to 500 MHz. (b) Effect of LC notch filter for the corner frequencies of 100 and 200 MHz.

IV. MEASUREMENT RESULTS

The transmitter is fabricated in a 0.18-μm RF CMOS technology. Fig. 7 is a micrograph of the fabricated chip. The silicon area is 2.4×3.4 mm^2 including a test block. The chip is packaged in a 48-pin LPCC and tested on an evaluation board. An I^{2}C interface is used for digital programming and tuning. The full chip including synthesizer draws 84 ~ 119 mA from a 1.8-V supply depending on the operation modes.

Fig. 8(a) shows the RF tunable filter characteristics. As can be seen, the corner frequency is tuned from 100 to 500 MHz in 5 steps. Fig. 8(b) illustrates the effects of the LC notch filter. It shows that the far-out stop-band attenuation is significantly improved by using the LC notch filter. Fig. 9 shows the measured output spectrum of the transmitter *without any off-chip filter*, where 3-MHz single-tone baseband input is used. The worst in-band distortion occurs when LO is the lowest 54 MHz. As clearly compared in Fig. 9(a), the in-band distortions

978-1-4244-2018-6/08 $25.00 © 2008 IEEE

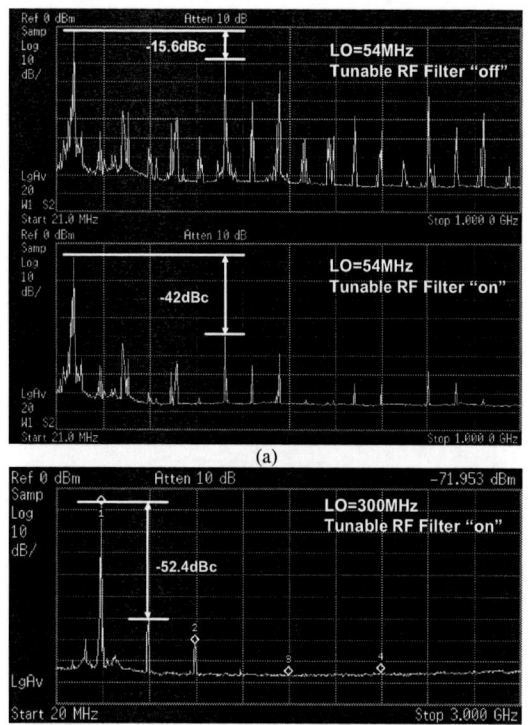

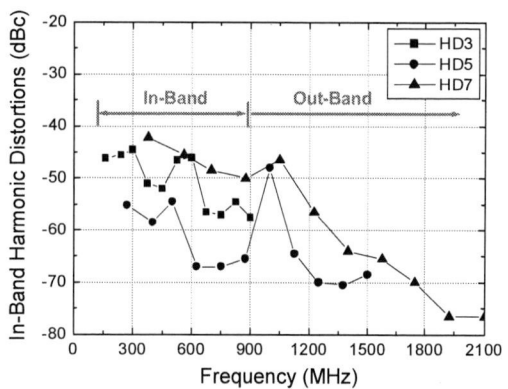

Figure 9. Measured output spectrum with (a) LO = 54MHz and tunable RF filter on and off, and (b) LO = 300MHz and tunable RF filter on.

Figure 10. Measured harmonic distortions against RF frequency.

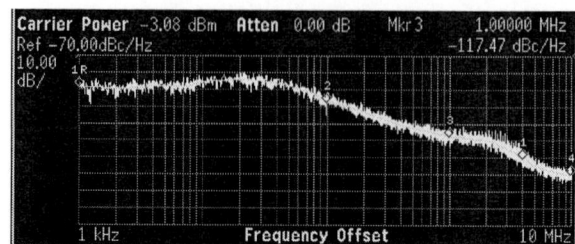

Figure 11. Measured phase noise of fractional-N PLL at 1.6 GHz.

TABLE I. MEASURED PERFORMANCE SUMMARY

Technology	0.18 μm CMOS
Operating Frequency	43.75 ~ 900 MHz
Output P_{1dB}	+9 dBm
Output IP_3	+20 ~ +25 dBm
In-band harmonic distortions	< -42 dBc
LO Leakage	< -36 dBc
Image Rejection	< -45 dBc
Integrated Phase Noise	-34.6 dBc @ 1.6 GHz
Supply Voltage	1.8 V
Current Consumption	84 ~ 119 mA

V. CONCLUSION

A zero-IF single-conversion transmitter IC for IEEE 802.22 Cognitive Radio applications has been presented. Realized in 0.18-μm CMOS, it covers the full TV band from 54 to 862 MHz. The in-band harmonic distortions are effectively suppressed by adopting dual-path upconversion structure with integrated harmonic rejecting mixers and tunable filters. A fractional-N frequency synthesizer with a single VCO and a wideband LOGEN is used for providing LO signals across the full band. Measurement results have shown acceptable performances for CR applications in future.

ACKNOWLEDGMENTS

This work has been supported in part by MIC ITRC program IITA-2008-C1090-0801-0038, and Seoul R&BD Program GR070039, and SEMCO.

REFERENCES

[1] IEEE 802.22 working group, http://www.ieee802.org/22
[2] J. Park et al., "A Fully-Integrated UHF Receiver with Multi-Resolution Spectrum Sensing (MRSS) Functionality for IEEE 802.22 Cognitive Radio Applications," IEEE ISSCC, Feb. 2008
[3] G. Cafaro et al., "A 100 MHz – 2.5 GHz Direct Conversion CMOS Transceiver for SDR Applications," IEEE RFIC Symp., Jun. 2007
[4] M. Borremans et al., "A CMOS Dual-Channel 100-MHz to 1.1-GHz Transmitter for Cable Applications," IEEE JSSC, Dec. 1999
[5] J. –M. Stevenson et al., "A Multi-Standard Analog and Digital TV Tuner for Cable and Terrestrial Applications," IEEE ISSCC, Feb. 2007
[6] V. Fillatre et al., "A SiP Tuner with Integrated LC Tracking Filter for Both Cable and Terrestrial TV Reception," IEEE ISSCC, Feb. 2007
[7] A. Maxim et al., "0.13u CMOS Hybrid TV Tuner Using a Calibrated Image and Harmonic Rejection Mixer," Symp. VLSI Cir., Jun. 2007
[8] H. Yoo et al., "A 97.2 mW 1.8 GHz Low Power CMOS Transmitter for Mobile WiBro and WiMAX," IEEE RFIC Symp., Jun. 2007
[9] P. Su, "A Dual-Band Enhanced Harmonic Rejection Filter for Modulators in GSM and DCS Transmitters," ESSCIRC, Sep. 2003
[10] H. Steendam et al., "The Effect of Carrier Phase Jitter on the Performance of OFDMA Systems," IEEE T. Comm., Apr. 1998

are dramatically diminished by using the tunable RF filter. Fig. 9(b) shows the in-band distortion characteristics when LO=300MHz. Fig. 10 summarizes the measured in- and out-band harmonics. As can be seen, the in-band harmonic distortion improves as the frequency increases and is always maintained below -42 dBc. The image and LO leakages are also measured less than -45 and -36 dBc, respectively, across the full band. The P_{1dB} and OIP_3 of the full transmitter are measured to be about +9 dBm and +20 ~ +25 dBm, respectively, across the full band. Fig. 11 is the PLL phase noise directly measured at the VCO output frequency of 1.6 GHz. The spot phase noises are -89 and -117 dBc/Hz at 1 kHz and 1 MHz offsets, respectively, and the integrated phase noise from 1 kHz to 4 MHz is -34.6 dBc. Table I summarizes the measured performance of the transmitter chip.

IEEE 2008 Custom Intergrated Circuits Conference (CICC)

A 5-GHz Wireless LAN Transmitter with Integrated Tunable High-Q RF Filter

Robert F. Wiser, Masoud Zargari, David Su, and Bruce A. Wooley

Center for Integrated Systems, Stanford University, Stanford, CA 94305

Abstract

A tunable high-Q RF filter suitable for wireless transmitters has been implemented in standard CMOS technology. Two stagger-tuned Q-enhanced resonators form a filter that can be tuned across frequency and bandwidth. A 5.145-GHz filter with 200-MHz bandwidth and 0.8-dB of ripple is demonstrated in an 802.11a sliding-IF transmitter. The transmitter provides spectral mask and EVM-compliant output power of –8.26 dBm for a 54-Mb/s OFDM signal. At lower output power, an EVM of –32.3 dB is achieved.

Introduction

The availability of wireless local area networks (WLAN) using protocols such as 802.11a/b/g/n is rapidly expanding. Unlicensed frequency bands, spectrally efficient modulation, and interoperable equipment have promoted the widespread adoption of these systems. The cost of WLAN transceivers has been lowered by leveraging system-on-a-chip (SoC) implementations, where the baseband digital processing is integrated with RF front-end circuitry. In such implementations unwanted spurs can be generated by various sources, such as mixing harmonics, synthesizer spurs, wideband circuit noise, and digital clocking harmonics [1]. In a transmitter, the spectral mask requirements can be violated by spurs appearing in the output spectrum. Spurs also risk violating the FCC emission limit in designated restricted bands, where, instead of being limited relative to the peak transmit power, there is an absolute limit of –41.25 dBm in a 1-MHz bandwidth [2].

Band-select RF filters are commonly employed to attenuate spurs due to factors such as LO feedthrough and harmonics, whereas architectural innovations have eliminated the need for intermediate frequency (IF) filters. The selectivity provided by surface acoustic wave (SAW), or similar, discrete filters at gigahertz frequencies has proven difficult to achieve in integrated implementations. Q-enhancement techniques have emerged as one means of enabling the integration of highly selective filters; however, heretofore their application has been limited by low dynamic range (DR) and difficulty in synthesizing wide-bandwidth filters [3].

This paper introduces an integrated, tunable, high-Q, RF filter. Active circuits are used to enhance the Q of individual parallel LC resonators. A cascade of two independent stagger-tuned Q-enhanced resonators is used to create a low-ripple, wide-bandwidth filter. An experimental prototype of such a filter has been implemented within a sliding-IF 802.11a WLAN transmitter in a 0.18-μm CMOS technology.

Transmitter Architecture

Multiple benefits derive from the integration of the RF band-select filter in the transmitter. Monolithic integration of the transceiver becomes possible as the transmit/receive switch can be brought on chip without incorporating the transmit band-select filter in the receive path, which would directly affect the noise figure. Placing an RF transmit filter before the power amplifier (PA) stages avoids the power loss of an external filter. The attenuation of spurs and wideband noise by the filter allows for the use of larger antenna gains without violating FCC restricted band emission limits.

A WLAN transmitter using 64-QAM modulation at a data rate of 54 Mb/s requires a signal-to-noise ratio (SNR) of approximately 30 dB [1]. Combined with a power backoff between 10 and 17 dB, the required DR for the transmitter is 40-47 dB, which is within the achievable DR of a Q-enhanced filter. The transmit chain must be designed to maintain the requisite SNR while keeping the signal within the filter DR. For example, if the filter can handle a maximum signal power of –10 dBm and has an integrated noise floor of –60 dBm, the signal power at the input of the filter must be between –30 and –20 dBm. The preceding and subsequent circuits of the transmitter must be designed around this power level.

As an example application, Fig. 1 shows the proposed architecture for a sliding-IF 802.11a WLAN transmitter. The signal path is fully differential. During transmission, the in-phase (I) and quadrature (Q) baseband inputs are up-converted to RF and then passed through a stagger-tuned filter consisting of two tunable Q-enhanced resonators. Gain control is provided after filtering by a variable-gain amplifier (VGA), which would drive the PA followed by the antenna. The outputs of the mixer and the two Q-enhanced resonators are multiplexed at the input of the VGA. The input of the second resonator can be taken as either the output of the mixer or the output of the first resonator. In this way, the signal path is programmable and can select either the baseline transmitter path with no Q-enhanced resonators, or a path with either the first or second resonator individually or both resonators in cascade. The frequency synthesizer signal is generated off-chip

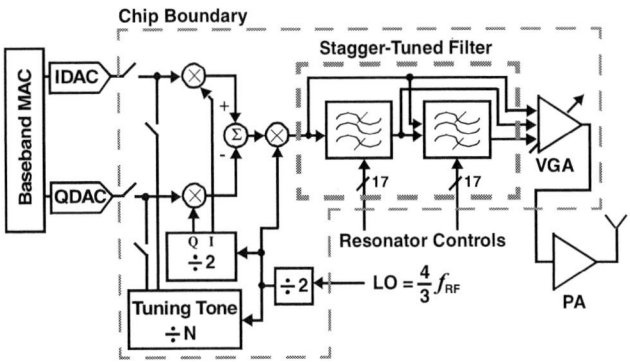

Fig. 1. Transmitter architecture.

and divided by two on chip to produce the LO signal for the double-balanced mixer; it is then divided again by two to produce the I and Q LO signals for the quadrature mixer. The first and second LO signals are at 1/3 and 2/3 the transmit frequency, respectively. The image generated by the second up-conversion lies at 1/3 the transmit frequency and is far enough out-of-band to be adequately filtered by the subsequent stages.

Stagger-Tuned Filter

A unique feature of the proposed transmitter is the band-limiting filter formed by two independent stagger-tuned Q-enhanced resonators. As shown in Fig. 2, the Q-enhanced resonator circuit consists of a variable input transconductance, an LC tank, and a resistively-degenerated cross-coupled transistor pair. The cross-coupled transistor pair, M_1 and M_2, presents a negative conductance in parallel with the LC tank that acts to partially cancel the conductive loss of the resonator and thus enhance its Q. The negative conductance is digitally controlled by adjusting the bias current, $I_{Q,bias}$, supplied to M_1 and M_2. Multiple binary-weighted current-mirror pairs are selected by a 7-bit control word. A fixed degeneration resistance acts to linearize the negative conductance. The tunable resonator capacitance, C_{tank}, is realized using multiple binary-weighted poly-poly capacitors, with an LSB of 15 fF, selected using series PMOS switches and a 6-bit control word. Resonator gain is tuned by applying a 4-bit control word to the gates of the cascode transistors of multiple binary-weighted input differential pairs. Cascoding reduces interaction between the resonators. By controlling the negative conductance, capacitance, and input transconductance, the resonator can be tuned to achieve a specific Q, resonant frequency, and gain. The minimum ripple of the filter is limited by the frequency tuning step size, which manifests itself as an error in the resonator center frequency.

Since a significant portion of the loss in the LC tank can be modeled as a resistor in series with the inductor, the equivalent parallel loss is frequency dependent and can only be completely cancelled by the negative conductance at a single frequency. Individual Q-enhanced resonators do not exhibit significant asymmetry in their frequency response. However, asymmetry appears when Q-enhanced resonators are coupled

to form wide bandwidth filters [3]. An additional compensation mechanism is then needed to re-balance the filter, or a wideband loss compensation technique needs to be employed [4]. These approaches preclude a simple automatic tuning scheme for adjusting the wide-bandwidth filter frequency response. Stagger-tuning multiple independent resonators allows a wide bandwidth filter to be formed without introducing asymmetry in the passband. The filter is designed by the placement of the individual second-order resonator poles so that the pole locations for the overall filter result in the desired frequency response. Design equations for a stagger-tuned filter are presented in [5].

To create the desired filter shape, the stagger-tuned resonators must be tuned to specific center frequencies and Q's. They should be tuned individually before being cascaded to form the filter. Because there is no resonator coupling, the overall filter response does not require additional tuning. The simplest tuning method uses a replica resonator as a VCO in a PLL. The frequency and bias current control signal are then applied to the main resonator. While this properly tunes the resonant frequency, it produces large but unknown Q's. When tuning to a high but specific Q, the matching requirements between the two resonators become too severe. One simple method of directly tuning to a specific resonator frequency and Q was introduced in [6]. The resonator magnitude is measured at three specific frequencies: the desired center frequency and the –6-dB bandwidth frequencies. The resonator is then tuned to fit the desired profile at these frequencies.

The frequency tuning range for a stagger-tuned filter will be similar to that of a coupled resonator filter, as both are governed by the resonator frequency tuning range. However, the bandwidth tuning range of a stagger-tuned filter will be larger than that of a similar coupled resonator filter. The Q of an individual resonator increases exponentially with the negative conductance, leading to a large Q tuning range. In a stagger-tuned filter, the bandwidth tuning range is determined by the resonator frequency tuning range, Q tuning range, and the frequency tuning step size. This contrasts with a coupled resonator filter, where, for example in a series-C coupled filter, the filter Q is set by the ratio of the coupling capacitance to the resonator capacitance.

Implementation

An experimental prototype of the proposed transmitter has been integrated in a 0.18-μm 2P5M CMOS technology and occupies an active area of 1.25mm^2. A reference LO is generated off-chip as a single-ended signal and brought on chip at twice the desired LO frequency in order to avoid generating a close-in spur via direct bondwire coupling to the output. A unity-gain output buffer is used to drive the 50-Ω load. The chip is packaged in a 44-lead gull-wing QFP. Control signals are provided through a low-speed serial-to-parallel interface.

The stagger-tuned filter is formed with two tunable resonators. The resonators use bondwire inductor loads in order to broaden their frequency tuning range. Different lengths of bondwires were tested to arrange for a wide range of induc-

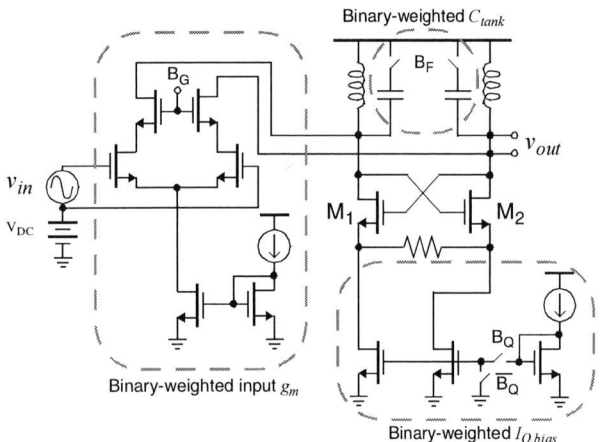

Fig. 2. Schematic for a tunable Q-enhanced resonator circuit.

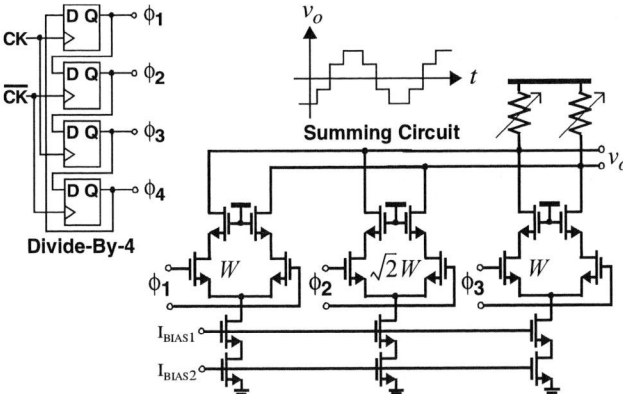

Fig. 3. Schematic of harmonic-rejection summer.

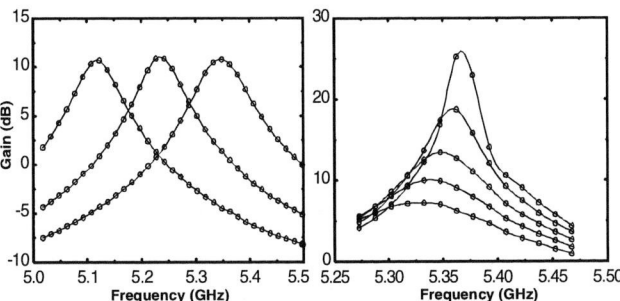

Fig. 4. Measured resonator tuned across frequency and Q.

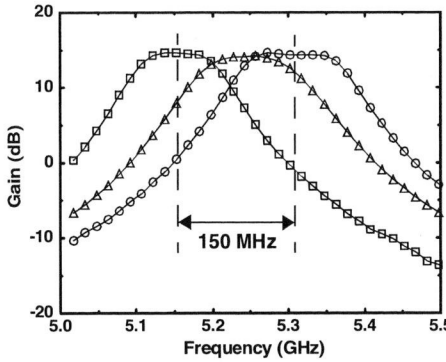

Fig. 5. Measured filter response tuned to various center frequencies.

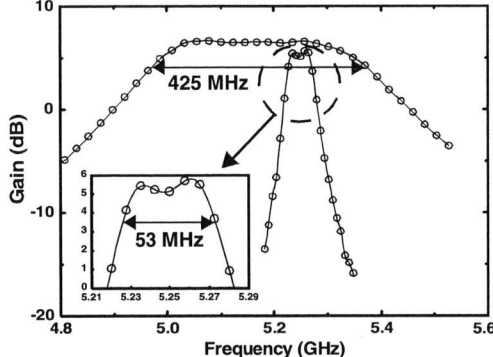

Fig. 6. Measured filter response tuned to different bandwidths.

tance. In a manufacturing environment, the bondwire lengths can vary by 30%. This variation can be compensated with an on-chip array of digitally selectable capacitors. The resonators are located at one corner of the chip, with the bondwires oriented orthogonally to reduce resonator crosstalk. The differential bondwire loads are connected to a shared SMD capacitor located in the package for each resonator, before connecting to the supply pin. This enables shorter bondwires and hence smaller inductance values.

The programmable signal path of the prototype allows each resonator response to be extracted individually. The VGA and second resonator have multiplexed inputs, formed by identical selectable cascoded input differential pairs. A programmable frequency divider is used to create baseband tuning tones from the RF LO signal. High-speed current-mode logic circuits are used above 1 GHz, while standard digital blocks form the rest of the divider chain. The final stage of the divider is an analog summing circuit driven by a four-phase generating divide-by-4 circuit, shown in Fig. 3. The summing circuit adds different signal phases to create a small-signal output tone with rejected 5th and 7th harmonics [7]. A programmable low-pass filter can be used to further suppress harmonics. The up-conversion chain converts these baseband tones to the desired tuning tones at RF. The response is amplified by the VGA to obtain a large-swing signal for amplitude measurement using a peak detector. On chip digital circuitry could be used to control the tuning process. By using a programmable divider to generate the RF tuning tones, the need for a frequency synthesizer is avoided. Besides reducing power and area, the divider allows faster switching and a wider frequency range of tuning tones.

Measurement Results

For tuning the experimental prototype, a signal generator creates a fixed baseband tone while the external LO is swept in frequency, and the response is measured with a spectrum analyzer. In this way, the individual resonator responses are extracted from the baseline transmitter response. The resonators are manually tuned to various center frequencies and Q's across the allowable tuning range, as shown in Fig. 4. Measured frequency tuning for the individual resonators spans 5.120 to 5.345 GHz for a 4.3% tuning range, while the bandwidth is maintained between 75 and 90 MHz. Unforeseen additional parasitic capacitance limits the frequency tuning range. The Q can be tuned from 39 up to oscillation (infinite Q). Inherent resonator Q with no enhancement was measured to be 5, which is half the value used in design and simulation. This significantly increases the implemented filter power consumption and noise floor, but demonstrates that typical spiral inductor Q's will be sufficient for use in stagger-tuned filters.

Two independent resonators form the stagger-tuned filter, which is tunable across the range of the individual resonators. Fig. 5 shows a fixed bandwidth filter tuned to different center frequencies. The filter center frequency varies from 5.16 to 5.31 GHz, for a tuning range of 150 MHz. During frequency tuning, the bandwidth and peak-to-peak ripple are maintained between 120 and 152 MHz and below 0.6 dB, respectively. Fig. 6 shows the filter tuned to different bandwidths, which range from 53 MHz to 425 MHz while the peak-to-peak ripple is less than 0.6 dB. The resonator gains are adjusted to keep the filter gain approximately constant.

978-1-4244-2018-6/08 $25.00 © 2008 IEEE   647

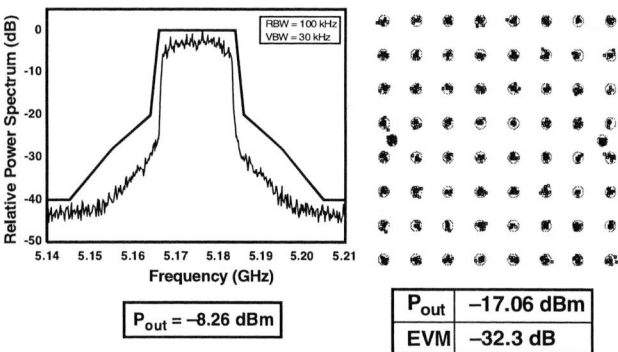

Fig. 7. Measured transmitter spectral response and constellation.

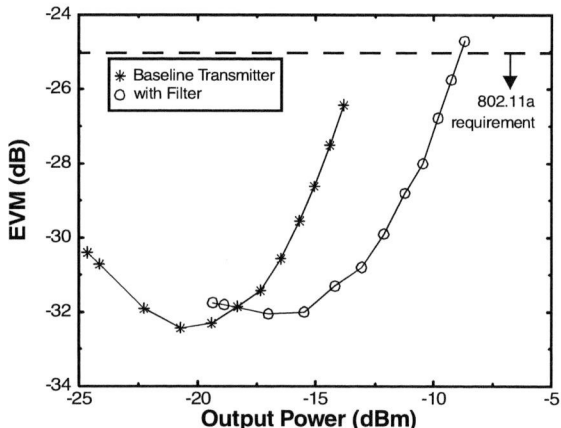

Fig. 8. Measured EVM versus output power.

For characterizing the experimental prototype transmitter, an FPGA and external DAC's are used in place of the MAC and I/QDAC's shown in Fig. 1. The FPGA and DAC's are clocked at 80 MHz, and generate a 64-QAM OFDM signal coded at 54 Mb/s data rate. The power level is controlled by the gain of the external DAC's. The chip output buffer is matched using a sliding capacitor and converted to a single-ended signal by a discrete balun. The test board directly drives a spectrum analyzer. A power meter is used to determine the total integrated output power. When measuring the transmitter EVM, an external active mixer down-converts the 5-GHz output to a 1-GHz signal, which is then digitized and analyzed on a computer. The mixer conversion loss was measured to be 4 dB, and integrated power levels were confirmed using the power meter.

The filter used in the transmitter prototype for system-level measurements is centered at 5.145 GHz with 200-MHz bandwidth and 0.8-dB of ripple. The transmitter with stagger-tuned filtering meets the 802.11a specifications for a 54 Mb/s OFDM signal at an output power of −8.26 dBm. The constellation and spectral response are shown in Fig. 7. The maximum output power that is EVM-compliant is limited by the transmitter linearity, as can be seen from Fig. 8, in which the power levels are adjusted for mixer conversion loss. The minimum achievable EVM of −32.3 dB is limited by the transmitter noise

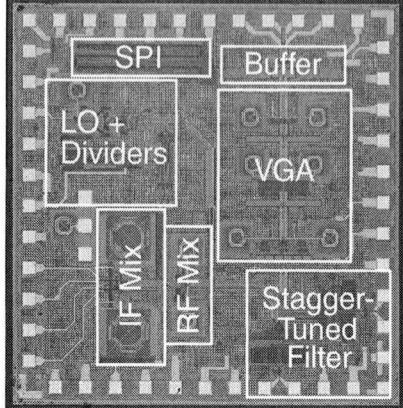

Fig. 9. Chip micrograph.

floor. For lower output powers, the EVM degrades due to the impact of the decreasing SNR on the signal constellation. The signal path of the transmitter, excluding the output buffer, dissipates 190 mW from a 1.8-V supply, of which the filter dissipates 39.1 mW. Fig. 9 shows the chip micrograph.

Conclusion

This paper demonstrates the use of a wide bandwidth Q-enhanced filter in an 802.11a WLAN transmitter. Stagger-tuning creates a wide bandwidth filter using simply tuned Q-enhanced resonators. An experimental transmitter meets 802.11a spectral mask and EVM requirements.

Acknowledgments

This work was supported in part by an Anne T. and Robert M. Bass Stanford Graduate Fellowship, and by Texas Instruments, Inc. and National Semiconductor, Inc.

References

[1] M. Zargari, S. Mehta, and D. Su, "Challenges in the design of CMOS transceivers for the IEEE 802.11 wireless LANs: past, present and future," *Radio Frequency Integrated Circuits Symposium*, pp. 353-356, June 2005.
[2] CFR 47, Federal Communications Commission, "Part 15 - Radio Frequency Devices," May 2007.
[3] D. Li and Y. Tsividis, "Design techniques for automatically tuned integrated gigahertz-range active LC filters," *IEEE J. Solid-State Circuits*, vol. 37, pp. 967-977, Aug. 2002.
[4] J. Kulyk and J. Haslett, "A monolithic CMOS 2368 +/- 30 MHz transformer based Q-enhanced series-C coupled resonator band-pass filter," *IEEE J. Solid-State Circuits*, vol. 41, pp. 362-374, Feb. 2006.
[5] M. Dishal, "Design of dissipative band-pass filters producing desired exact amplitude-frequency characteristics," *Proceedings of the IRE*, pp. 1050-1069, Sept. 1949.
[6] H. Liu and A.I. Karsilayan, "An accurate automatic tuning scheme for high-Q continuous-time bandpass filters based on amplitude comparison," *IEEE Trans. on Circuits and Systems II*, vol. 50, pp. 415-423, Aug. 2003.
[7] J.A. Weldon, et al., "A 1.75-GHz highly integrated narrow-band CMOS transmitter with harmonic-rejection mixers," *IEEE J. Solid-State Circuits*, vol. 36, pp. 2003-2015, Dec. 2001.

Advanced Wireline Techniques

Session 20

Chair: Ed van Tuijl, Axiom-IC Twente
Co-Chair: Ken Chang, Rambus

This session shows present and future mult-gigabit transceivers in backplane, memory, and broadband wired applications, and a EMI resistant line driver for automobile design.

The first paper describes a 6-tap FIR 6.5Gb/s backplane transmitter. The equalizer uses variable tap spacing to optimize its tap weights as opposed to the conventional fixed bit rate tap spacing.

The second paper presents the challenges and future trends in the design of phase locked loops and clock and data recovery circuits and systems for 80-100Gb/s applications.

The third paper describes a 20Gb/s SerDes transmitter in 65nm Bulk CMOS. The transmitter uses multi-slice technique to realize adjustable source impedance and 4-tap feed-forward equalization.

The fourth paper describes the design and measurement of a wideband mmWave CML static divider in 65nm SOI CMOS. The wideband divider is robust against process variability in sub-100nm CMOS technology and is useful as the building block for very high-speed PLL front-end.

The fifth paper is a dual mode 4/2-PAM transmitter providing 32/16 Gb/s bandwidth in 0.13um CMOS technology. The transmitter uses Pulse-Width Modulation pre-emphasis technique and provides 1.2V output swing to compensate for heavily attenuated channels.

The sixth paper presents a 5Gb/s/pin transceiver for DDR memory interface in 0.18um CMOS technology. The proposed transceiver uses a staggered memory bus topology and a glitch canceller to suppress crosstalk-induced distortions in a memory channel.

The last paper presents a smart-power driver for local interconnect network (LIN) bus that is commonly used in automotive applications. This integrated driver exhibits reduced slope pumping and a high degree of immunity to electromagnetic interference.

978-1-4244-2018-6/08 $25.00 © 2008 IEEE

Notes

978-1-4244-2018-6/08 $25.00 © 2008 IEEE

IEEE 2008 Custom Intergrated Circuits Conference (CICC)

A 6.5 Gb/s Backplane Transmitter with 6-tap FIR Equalizer and Variable Tap Spacing

Mike Bichan, Anthony Chan Carusone

University of Toronto

10 King's College Road

Toronto, ON M5S 3G4 CANADA

Abstract–This paper presents a 6.5 Gb/s transmitter for use in backplane links. This transmitter incorporates a finite impulse response filter with programmable tap spacing in the output driver to compensate for intersymbol interference. Using jitter-minimizing tap weights computed using a behavioral model of the transmitter, it is shown that at 6.5 Gb/s peak-to-peak data-dependent jitter is reduced by over 50% by using a tap spacing of 0.53 unit intervals (UI) instead of the usual 1 UI.

I. INTRODUCTION

In high-speed backplane communication links, the limited channel bandwidth introduces inter-symbol interference (ISI) into the received signal. Equalizers in the transmitter and receiver have been widely used to compensate for loss at high frequency and allow data to be sent at rates higher than the bandwidth of the channel. A common configuration involves a decision feedback equalizer (DFE) in the receiver and a feed-forward equalizer (FFE) in the transmitter [1]. These equalizers consist of a delay line that generates phase-shifted versions of the input signal and an output driver for each tap. The taps are typically baud-spaced, and are generated by a cascade of flip-flops clocked at the bit rate. The useful number of taps is determined by the length of the impulse response of the channel. For high-loss channels with long impulse responses, up to five baud-spaced taps have been used at multi- Gb/s data rates [2, 3].

Fractionally-spaced equalizers can increase the useful number of taps by inserting additional taps between the baud-spaced ones. The precise delays between taps can be generated by lumped LC structures that absorb the parasitic capacitance of the output driver [4]. Alternatively, the taps can be generated using a variable delay line that references its delay to the period of the data signal, such as in the receiver equalizer proposed in [5]. A variable delay line dissipates more power than the cascade of flip-flops typically used to generate baud-spaced taps, but it has the advantage of being easily reconfigured to provide fractional tap spacings. This reconfigurability can permit lower jitter when equalizing backplane links and is also desirable because of the need for bandwidth and power scalability [6].

Using an equalizer with a variable tap spacing allows us to investigate the effect of varying the tap spacing on the jitter performance of a backplane communication link. We show that the optimal tap spacing is not necessarily one unit interval (UI) or a simple fraction thereof. For the 24" backplane channel con-

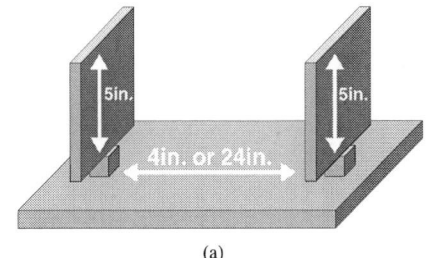

(a)

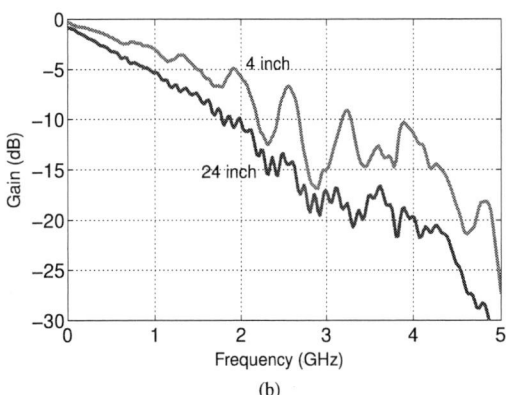

(b)

Fig. 1. (a) Backplane channel, (b) Measured frequency response of the 4" and 24" Tyco backplane channels.

sidered here, choosing the optimal tap spacing can significantly reduce received jitter.

II. CIRCUIT DESCRIPTION

This paper describes a transmit-side equalizer for communication over backplane channels such as the one pictured in Fig. 1(a). This channel is typical of legacy backplanes in use today. The proposed finite impulse response (FIR) equalizer, shown in Fig. 2, consists of a six-tap delay line with a variable gain stage for each tap. The delay line is tuneable to allow the use of different tap spacings and different bit rates. Each tap has adjustable gain with 4-bit resolution.

The variable gain stage is broken into six slices, one for each tap of the delay line. The currents from the six slices are summed in 50 Ω load resistors to produce the output voltage. As shown in Fig. 3, each slice of the gain stage consists of a sign selection switch followed by a preamplifier and a 3-bit adjustable output driver, for a total of 4 bits. When a given filter tap is not in use,

978-1-4244-2018-6/08 $25.00 © 2008 IEEE   651

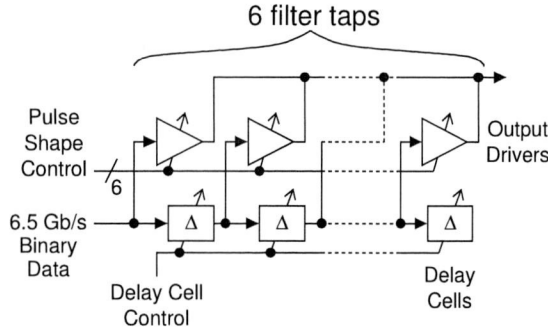

Fig. 2. Transmitter block diagram.

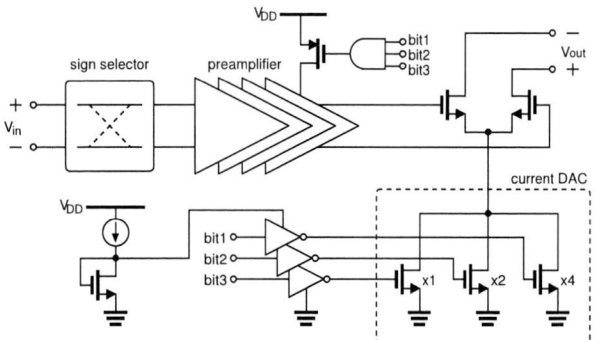

Fig. 3. Schematic diagram of one output driver slice.

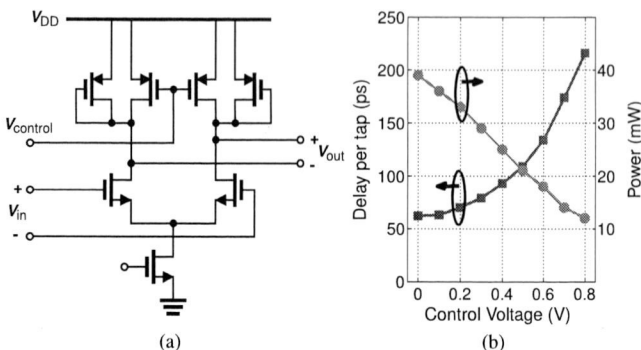

Fig. 4. (a) Delay stage. Each delay cell block in Fig. 2 contains four such stages, (b) measured tap delay and total power of the delay line.

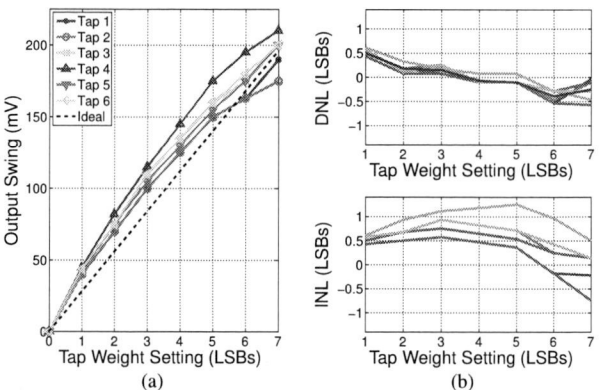

Fig. 5. Linearity of the output driver DAC. (a) Output swing for each digital code, (b) DNL and INL.

the corresponding gain stage slice is automatically shut down in order to save the power that would have been burned in the preamplifier. Unfortunately, the linearity of the current-mode D/A converter (DAC) used in this stage is not very good. However, as long as we can characterize the nonlinearity we can take the nonlinearity into account when optimizing the pulse shape for equalization of a given channel.

The delay line generates six phase-shifted versions of the input data using five delay cells. Each delay cell consists of four stages like the one in Fig. 4(a) with a differential pair driving symmetric loads. This type of load acts as a resistance with an adjustable value. As the resistance increases, the delay of the stage increases while the voltage swing is kept constant by a replica bias feedback loop that reduces the tail current. Four stages are required in order for each cell to generate a significant delay while still maintaining the required bandwidth. With fewer stages the delay per cell would be too small.

The delay cell is tuneable from 62–216 ps, as shown in Fig. 4(b). Increasing the delay reduces the power dissipation from 40 down to 12 mW, providing power scaling for slower data rates. At a delay of 120 ps the bandwidth of the delay line is sufficient for data rates only up to 5 Gb/s. To achieve higher delay while maintaining bandwidth, multiple delay cells can be used as one by simply turning off the intervening taps. For example, to generate a 125 ps tap spacing, we can use either one delay cell with a delay of 125 ps or two cascaded delay cells each with a delay of 62.5 ps. The latter option will result in higher bandwidth.

III. CHANNEL AND CIRCUIT MODELING

The intended channel for this transmitter is a backplane with two 5" daughtercards connected to a 16-layer non-backdrilled motherboard. The frequency response of the channel is shown in Fig. 1(b). The 24" channel has more attenuation but the 4" channel suffers more from reflections. These reflections are caused by impedance discontinuites at the vias and connectors and they show up as ripples in the frequency response.

or a 6-tap filter with 4-bit resolution for each tap weight, there are $(2^4)^6 \approx 1.6$ million possible filter configurations. To help with tap weight selection a behavioral model of the transmitter and channel was created. To start, we first measure the nonlinearity of the output driver DAC. The nonlinearity was characterized by measuring the output swing for all tap weight settings with only one tap operational. The linearity of the output driver DAC is shown in Fig. 5. While the linearity is far from ideal, we compensate for this deficiency by choosing the transmitted pulse shape appropriately. This flexibility loosens the requirements on the DAC which means it can be designed to consume less voltage headroom and area.

FThe slew rate in simulation is then limited to the value that is observed in measurement. With six output driver slices whose currents are summed together, there is significant parasitic capacitance at the output node that limits the slew rate.

978-1-4244-2018-6/08 $25.00 © 2008 IEEE

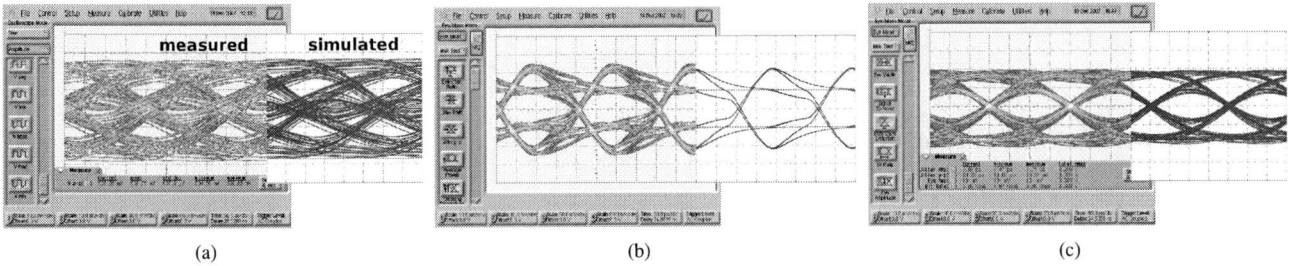

Fig. 6. Measured and simulated behavioral model 5 Gb/s eye diagrams for a PRBS7 pattern: (a) Output of unequalized 24" backplane channel, (b) transmitter output for half-baud-spaced jitter-minimizing pulse shape, (c) equalized 24" backplane channel.

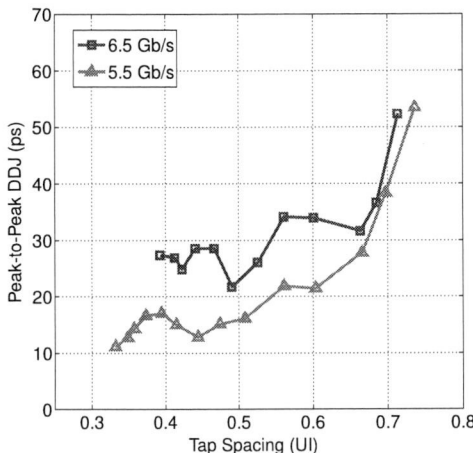

Fig. 7. Transmit-side DDJ vs. delay per cell of the delay line.

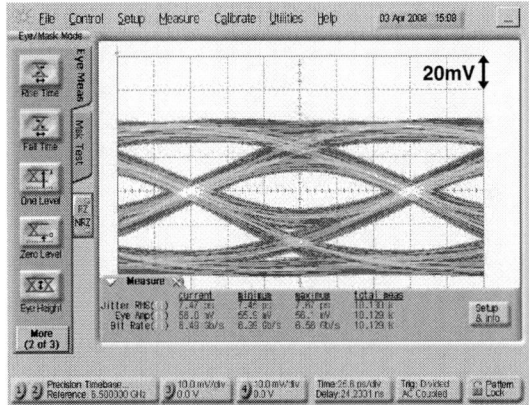

Fig. 8. Received 6.5 Gb/s equalized eye diagram for the 24" backplane channel. Tap spacing is 0.53 UI and the optimal tap weights are [+2 -1]. Total transmitter power for this configuration is 42 mW.

This slew-rate-limited signal is then put through an RLC circuit that models the parasitics of the QFN package. Finally the measured s-parameters of the backplane channel give us the receive-side signal that results.

Using this model, behavioral simulations were performed to evaluate all possible tap weight settings. For each configuration (channel length, number of taps, and tap spacing) the tap weights resulting in the lowest simulated jitter were selected.

Fig. 6 shows the match between model and measurement for three sample eye diagrams at 5 Gb/s. Fig. 6(a) shows the unequalized signal received over a 24" backplane, while Figs. 6(b) and (c) show the transmit- and receive-side signals for the optimal transmitted pulse shape. The tap weights for the optimal pulse in this case are [+1 +5 -3] with a tap spacing of 100 ps.

IV. MEASUREMENT RESULTS

We first consider the bandwidth of the delay line. As the delay generated by the line increases the bandwidth of the line decreases. The increase in transmit-side jitter as the tap spacing increases can be seen in Fig. 7. When the delay becomes 0.7 of a UI the delay line can no longer operate as the jitter has increased too much. This limitation forces us to use multiple delay cells in cascade when implementing longer delays.

An eye diagram of the received signal over 24" backplane for optimal tap weights and tap spacing at 6.5 Gb/s is shown in Fig. 8. In this case, the optimal tap spacing was 0.53 UI, much less than the typical tap spacing of 1 UI.

The tap weights were chosen only to minimize data-dependent jitter (DDJ), so we use the jitter decomposition feature of the oscilloscope to examine only the part of the jitter that is due to ISI. Random jitter remains roughly constant across all transmitter configurations around 1 ps rms.

Fig. 9 shows the improvement in DDJ as the tap spacing is varied for a data rate of 6.5 Gb/s. Since DDJ, unlike random jitter, is bounded, peak-to-peak DDJ is plotted. At 6.5 Gb/s the equalizer benefits significantly from using a smaller tap spacing; DDJ can be halved compared with conventional baud-rate tap spacing. As shown in Fig. 10, even at 5.5 Gb/s the jitter varies by almost 10 ps as the tap spacing is changed, underlining the importance of choosing the optimal tap spacing. The total power of the delay line is also plotted for both of these graphs. Note that the jump in the middle of these graphs is caused by the limited bandwidth of the delay line. Once the desired tap spacing is greater than twice the minimum delay of the line (62.5 ps), two delay cells are combined with their individual delays halved.

A die photo of the transmitter, implemented in 90 nm CMOS, is shown in Fig. 11. The power varies from 40–80 mW depending on the tap spacing and output swing. At 5 Gb/s with half-baud tap spacing and a 500 mV output swing, 63 mW is consumed. Of this, 23 mW is consumed in the delay line and 40 mW is consumed in the output drivers. This is comparable to the power reported in [1, 7] considering that those baud-spaced equalizers have a smaller number of taps as seen in Table I.

978-1-4244-2018-6/08 $25.00 © 2008 IEEE

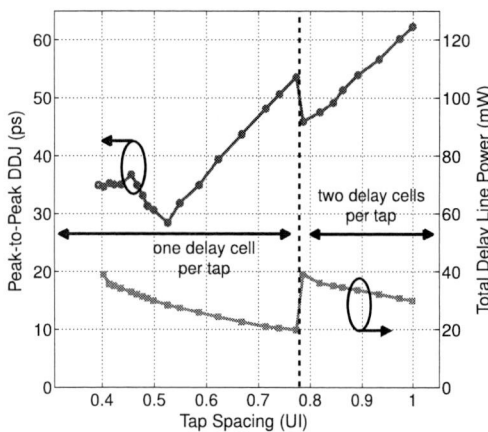

Fig. 9. Measured receive-side DDJ for the 24" backplane channel at 6.5 Gb/s. Optimal tap weights are recalculated for each tap spacing individually.

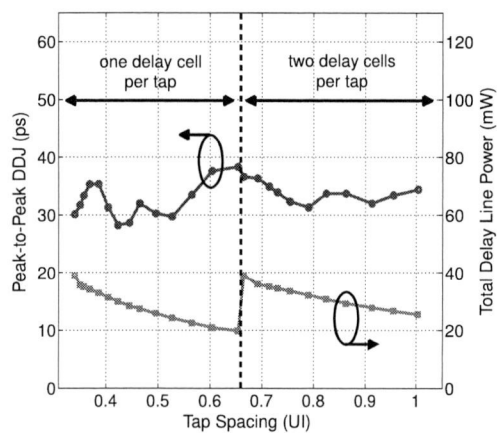

Fig. 10. Measured receive-side DDJ for the 24" backplane channel at 5.5 Gb/s. Optimal tap weights are recalculated for each tap spacing individually.

V. CONCLUSION

A 6.5 Gb/s transmit-side equalizer with the flexibility to use variable fractional tap spacings was presented. While the output driver DAC used here is slightly nonlinear, this nonideality is taken into account in the transmitter model. This transmitter model is then used with a channel model to choose the tap weights that minimize received jitter. The spacing between filter taps was shown to strongly influence the received peak-to-peak DDJ. For example, at 6.5 Gb/s over a 24" backplane channel, the DDJ could be cut in half, from 62.3 ps to 28.4 ps, by decreasing the tap spacing from the typical 1 UI to 0.53 UI.

ACKNOWLEDGMENT

The authors thank CMC Microsystems and Intel Corporation for their help and support.

TABLE I
TRANSMIT-SIDE EQUALIZER PERFORMANCE COMPARISON.

	This work	[1]	[7]
Process	90 nm	90 nm	90 nm
Area	0.105 mm^2	0.16 mm^2	0.036 mm^2
Voltage	1.2 V	1.0 V	1.0 V
Power	40–80 mW	70 mW	24 mW
Data Rate	6.5 Gb/s	10 Gb/s	6 Gb/s
No. of taps	6	4	2
Tap Spacing	variable	baud	baud

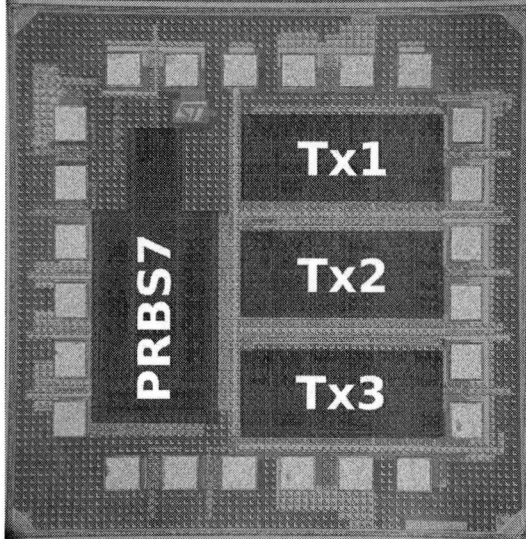

Fig. 11. The transmitter prototype was implemented in 90 nm CMOS. Die size is 1 mm × 1 mm.

REFERENCES

[1] J. F. Bulzacchelli, *et al.*, "A 10-Gb/s 5-tap DFE/4-tap FFE transceiver in 90-nm CMOS technology," *IEEE J. Solid-State Circuits*, vol. 41, no. 12, pp. 2885–2900, Dec. 2006.

[2] R. Payne, *et al.*, "A 6.25-Gb/s binary transceiver in 0.13-μm CMOS for serial data transmission across high loss legacy backplane channels," *IEEE J. Solid-State Circuits*, vol. 40, no. 12, pp. 2646–2657, Dec. 2005.

[3] W. Gai, *et al.*, "A 4-channel 3.125Gb/s/ch CMOS transceiver with 30dB equalization," in *Symp. on VLSI Circuits Dig. Tech. Papers*, 2004, pp. 138–141.

[4] C. Pelard, *et al.*, "Realization of multigigabit channel equalization and crosstalk cancellation integrated circuits," *IEEE J. Solid-State Circuits*, pp. 1659–1670, Oct. 2004.

[5] X. Lin, H. Lee, and J. Liu, "A continuous-time adaptive FIR equalizer with INV-AIL delay line for 2.5Gb/s data communication," in *IEEE Custom Integrated Circuits Conf. (CICC) Dig. Tech. Papers*, 2005, pp. 413–416.

[6] G. Balamurugan, *et al.*, "A scalable 5-15Gbps 14-75mW low power I/O transceiver in 65nm CMOS," in *Symp. on VLSI Circuits Dig. Tech. Papers*, 2007, pp. 270–271.

[7] J. F. Buckwalter, M. Meghelli, D. J. Friedman, and A. Hajimiri, "Phase and amplitude pre-emphasis techniques for low-power serial links," *IEEE J. Solid-State Circuits*, vol. 41, no. 6, pp. 1391–1399, June 2006.

IEEE 2008 Custom Intergrated Circuits Conference (CICC)

Phase-Locking in Wireline Systems: Present and Future[1]

Behzad Razavi

Electrical Engineering Department

University of California, Los Angeles

Abstract

This paper describes the challenges in the design of phase-locked loops and clock and data recovery circuits as speeds approach 80-100 Gb/s. Skew and jitter issues are presented and the effect of reference phase noise, charge pump noise, reference spurs, and loop filter leakage is quantified. The phase noise performance of cascaded loops is analyzed and two new architectures are proposed.

I. INTRODUCTION

Our appetite for higher data rates remains unabated. The ever-growing volume of data in wireless and wireline links demands a proportional increase in the aggregate or serial throughput rates. At the physical layer, other challenges such as distortion of data, power dissipation, and packaging also manifest themselves as higher rates are sought.

This paper deals with phase-locking functions in high-speed wireline transceivers, emphasizing new developments and predicting possible trends for future systems. While modern transceivers employ many other functions as well, phase-locking has not only proved a difficult bottleneck but extended its reach into other building blocks, thus claiming its own place in the top design challenges.

Section II reviews present trends in the field. Section III describes the challenges in transmit (TX) phase-locked loop (PLL) design, including speed and jitter issues and the use of cascaded loops. Section IV deals with clock and data recovery (CDR) circuits, and Section V points to future trends.

II. PRESENT TRENDS

Recent work in phase-locking for wireline systems has entailed a number of interesting trends that are likely to intensify in the future.

(1) Use of RF Design Concepts. The vast research into the problem of RF synthesis has greatly benefited wireline transceivers as well. Examples include the use of on-chip inductors, the design of low-noise oscillators and frequency dividers, methods of phase noise and spur reduction in phase-locked loops, and the use of injection-locked oscillators for clock distribution [1].

[1]This work was supported by Kawasaki Microelectronics, Realtek Semiconductor, and Skyworks, Inc. Chip fabrication was provided by TSMC.

(2) Convergence of Functions. The need for global optimization of the performance has discouraged the design of building blocks as independent modules. For example, the transmit PLL and the receive (RX) CDR circuits are now viewed as one entity [2], particularly to avoid mutual injection pulling. Also, the equalization, eye opening, and CDR functions have become intertwined [3, 4].

(3) Convergence of PLLs and DLLs. The potentially lower jitter of delay-locked loops and the synthesis capabilities of PLLs have motivated work on combining the two [5, 6].

(4) Use of Phase Interpolation. The desire to move toward "digital" PLLs and to avoid oscillators within the CDR circuit has translated to extensive phase interpolation, albeit at the cost of routing, mismatch, and jitter issues.

III. TRANSMIT PLL

The TX PLL generates a full-rate clock whose integer sub-multiples drive the multiplexer (MUX) chain. More importantly, it drives a retiming flip flop (FF) in the data path [Fig. 1(a)] so as to remove the output jitter of the MUX. This jitter arises from two sources, namely, the propagation mismatches within the MUX and the duty cycle error at the output of the $\div 2$ stage. Note that duty cycle errors in the full-rate clock are unimportant. It is the need for this retiming that makes the design of the TX PLL and its surrounding circuitry difficult.

A critical issue in the architecture of Fig. 1(a) stems from the delay of the $\div 2$ circuit, ΔT. As shown in Fig. 1(b), this delay displaces the edges of the MUX output, causing sampling closer to the data zero crossings. This effect can be suppressed by inserting an equal delay in the clock path of the FF, but the required full-rate bandwidth complicates the design of such a delay stage. We return to this issue in Section III.D.

Another important issue in the TX of Fig. 1(a) originates from the coupling of the data transitions from the D input of the FF to its clock input. Exemplified by the capacitive path depicted in Fig. 2, this effect heavily corrupts the phase of the oscillator unless a buffer with high reverse isolation is interposed between the PLL and the FF. Unfortunately, such a buffer must incorporate inductors at high frequencies, thereby complicating the routing of the signals.

In high-speed transmitters, the PLL and data path designs are intimately related. As illustrated in Fig. 3, the TX design begins with the output driver (possibly including equalization),

978-1-4244-2018-6/08 $25.00 © 2008 IEEE

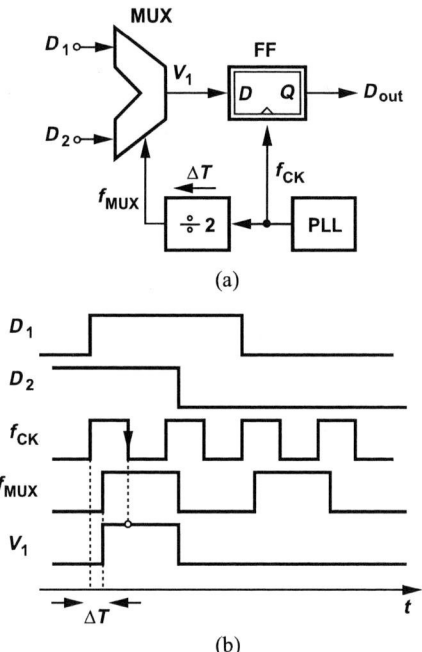

(a)

(b)

Fig. 1. (a) Interface between PLL and data path, (b) effect of divider skew.

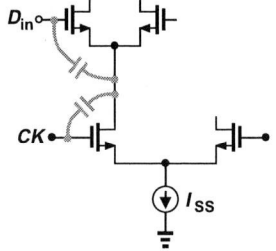

Fig. 2. Coupling of data to clock through FF capacitances.

aiming to deliver the necessary output swing to the load (a laser or a transmission line) and an on-chip back-termination resistor. Since bandwidth requirements limit the number of stages used in the driver [7], its input capacitance, C_{dr}, tends to be large, thus dictating a flip flop with high drive capability and hence high currents. To accommodate such currents with a limited voltage headroom, the FF itself employs wide transistors and exhibits large input capacitances, C_D and C_{CK}. This in turn necessitates high currents in the MUX and the VCO

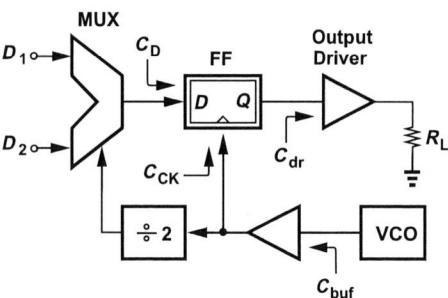

Fig. 3. Propagation of driver capacitance to VCO.

buffer, creating large capacitive loads for the VCO and the ÷2 stage. In other words, the output driver's input capacitance "propagates" to the VCO.

Due to the propagation of the output driver's input capacitance, the VCO must typically drive a large capacitance itself even if the path through its buffer and the FF may employ some tapering. At frequencies approaching 100 GHz, the design of such a VCO becomes exceedingly difficult as it requires very small inductors, e.g., around 50 pH.

Figure 4(a) shows an oscillator topology that exhibits poten-

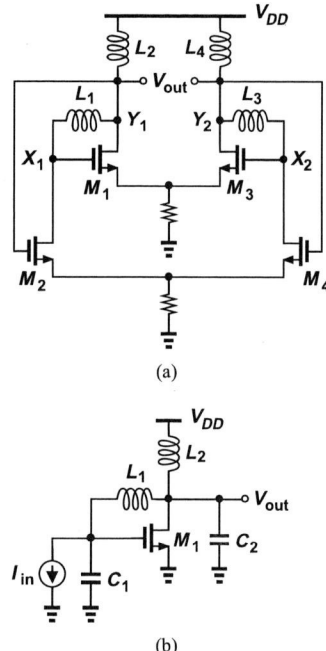

(a)

(b)

Fig. 4. (a) Oscillator based on inductive feedback, (b) basic amplifier circuit.

tial for high-speed operation [8]. The circuit is derived from the single-ended transimpedance amplifier illustrated in Fig. 4(b), whose transfer function exhibits two imaginary poles at

$$\omega_{p2,p2'} = \pm j\sqrt{\frac{3+\sqrt{5}}{2LC}}. \quad (1)$$

The magnitude of this pole is about 0.62% higher than that of a second-order LC tank, allowing operation at high speeds. For example, oscillation frequencies in the range of 80 GHz to 128 GHz have been achieved in 90-nm CMOS technology [8]. To tune the frequency, varactors must be tied from nodes X_1, X_2, Y_1, and Y_2 to the control voltage. By virtue of inductive feedback, the circuit can drive heavy capacitive loads and operate from a low supply voltage.

Figure 5 compares the simulated phase noise of this oscillator with that of a standard cross-coupled topology at 80 GHz, assuming a given inductor design, a given power consumption, and a given buffer. Interestingly, the inductive feedback suppresses the flicker noise contribution of M_1 and M_3 to the phase noise because a low-frequency voltage perturbation in

978-1-4244-2018-6/08 $25.00 © 2008 IEEE

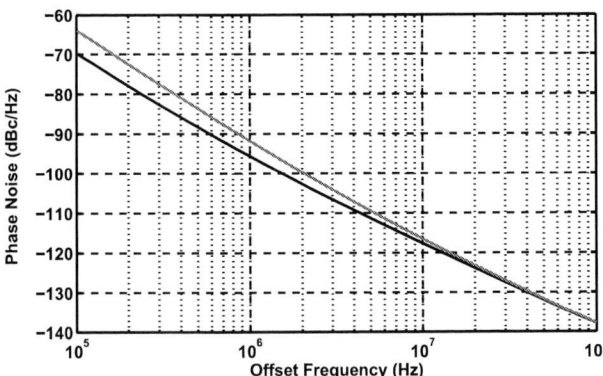

Fig. 5. Simulated phase noise of the inductive-feedback oscillator (black line) and cross-coupled oscillator (gray line).

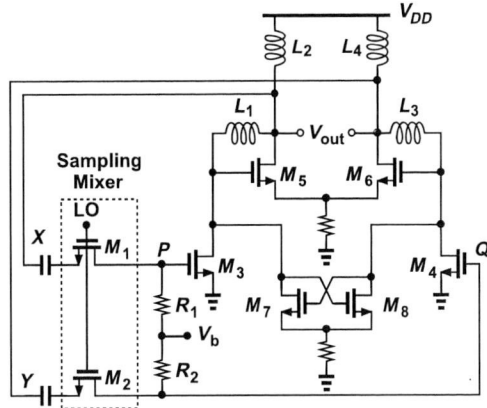

Fig. 6. Miller divider based on inductive-feedback amplifier.

series with the gate of, say, M_1, cannot change the phase difference between V_{X1} and V_{Y1}. Note that the oscillator of Fig. 4(a) is loaded with much greater capacitance so that it operates at the same frequency as the cross-coupled topology.

Even with the topology of Fig. 4(a), the design of VCOs at frequencies approaching 100 GHz faces other critical challenges. First, the quality factor, Q, of inductors does not scale linearly with frequency, beginning to saturate above 50 GHz. For example, [10] reports a Q of 12 for 180-pH inductors at 60 GHz, and [11] a Q of 17 for 400-pH inductors at 50 GHz. Second, the Q of varactors is likely to fall *below* the Q of inductors at these frequencies. Both effects exacerbate the trade-offs among phase noise, power dissipation, tuning range, and output swings.

A. Frequency Dividers

High-speed frequency dividers pose another serious challenge to the design of PLLs. In addition to speed, other important parameters of dividers include the minimum required input voltage swing, the input capacitance, the output drive capability, and the complexity, especially the number of inductors required in each topology. Static (flip-flop-based) topologies suffer from a limited speed and injection-locked dividers (ILDs) exhibit a limited lock range, placing the overall PLL design at risk. This is because, due to modeling inaccuracies and process variations, the lock range of ILDs may not enclose the desired frequency, thus causing lock failure or false lock [9]. Two other topologies, namely, the Miller divider and the "heterodyne PLL" [9] can achieve a wider lock range at the cost of greater complexity.

The circuit technique illustrated in Fig. 4(b) can also improve the speed of frequency dividers. Shown in Fig. 6 is a Miller topology employing the inductive feedback configuration [8]. The cross-coupled pair increases the loop gain and hence the lock range. Also, M_1 and M_2 form a differential "sampling mixer," which presents less loading to the amplifier than conventional double-balanced passive mixers. Specifically, the capacitance at node P switches periodically between X and Y in a conventional mixer, thereby introducing a re-

sistance between these two nodes and lowering the gain of the amplifier. Here, on the other hand, the voltage is simply stored on the capacitance for a half cycle (if R_1 and R_2 are sufficiently large).

The circuit of Fig. 6 achieves high speeds even in 90-nm CMOS technology. For example, one choice of inductor values provides a lock range of 88 to 104 GHz.

Heterodyne phase-locking is another candidate for high-speed dividers. Depicted in Fig. 7 in its simplest form, a

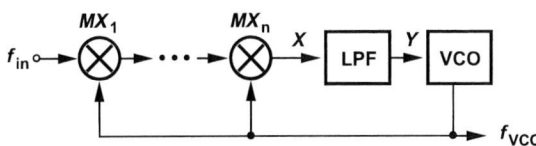

Fig. 7. Basic heterodyne PLL.

heterodyne PLL mixes the input with the VCO output n times, generating a frequency component at X given by $f_{in} - nf_{VCO}$. If the circuit locks, this component must be equal to zero, and f_{VCO} equal to f_{in}/n. Other divide ratios can be realized by inserting dividers in the feedback loop and/or at the input ports of the mixer [9]. A prototype realized in 0.13-μm CMOS technology operates from 64 to 70 GHz while consuming 6.5 mW.

The use of consecutive mixers in a heterodyne PLL raises the possibility of false lock due to unwanted mixing products. However, it can be shown that for divide ratios up to 4, the limited VCO tuning range prohibits false lock.

B. Jitter Issues

The TX PLL produces the dominant jitter in the transmitted data if the flip flop and output driver in the data path exhibit sufficient bandwidth. Jitter becomes much more pronounced as we approach 80-100 GHz because (1) the Q of inductors begins to saturate and the Q of varactors is likely to be even lower; (2) the very large frequency multiplication factor realized by the PLL dramatically amplifies the reference phase noise, S_{REF}. The PLL loop bandwidth must therefore be chosen carefully.

978-1-4244-2018-6/08 $25.00 © 2008 IEEE

Reference Phase Noise The choice of the loop bandwidth is governed by the available frequencies, phase noise, and cost of crystal oscillators. Since low-noise, low-cost crystal oscillators typically operate at frequencies no higher than 100 MHz, we assume a PLL multiplication factor, N, of roughly 1000 and hence a 60-dB amplification of the reference phase noise within the loop bandwidth. As shown in Fig. 8, a natural

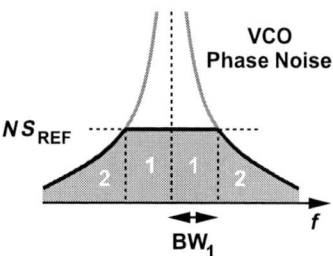

Fig. 8. Effect of reference phase noise.

choice for the loop bandwidth, BW_1, is given by the intersection of the amplified reference phase noise and that of the free-running VCO. For example, a 100-MHz crystal oscillator displays a constant phase noise of about -150 dBc/Hz beyond 100-kHz offset, suggesting that we seek the offset frequency at which the VCO phase noise falls to -90 dBc/Hz. Assuming a $1/f^2$ roll-off beyond the loop bandwidth, noting that regions 1 and 2 have equal areas, and integrating the phase noise we obtain the rms jitter as

$$\text{Jitter} = \frac{\sqrt{4NS_{REF} \cdot BW_1}}{2\pi} T_{CK}, \qquad (2)$$

where the factor of 4 accounts for the areas in regions 1 and 2 on both sides of the carrier, and T_{CK} denotes the carrier period. If $NS_{REF} = -90$ dBc/Hz and $BW_1 = 2$ MHz, then the rms jitter is equal to 1.42% of the clock period, a marginally acceptable value.

An interesting point that arises here relates to the frequency of jitter. As observed in the above calculations, most of the jitter originates from offsets up to a few tens of megahertz. For date rates of 80 to 100 Gb/s, this low-frequency jitter is readily removed by the RX CDR circuit and, therefore, proves unimportant. However, the transmit data mask may still be violated.

Charge Pump Noise The large multiplication factor also raises concern regarding the charge pump (CP) noise. We introduce an analysis here to determine the CP-induced phase noise at the output within the loop bandwidth.

First, suppose the Up and Down currents in the charge pump exhibit a mean value of I_P and a static mismatch of ΔI. It can be shown that such a mismatch gives rise to an input static phase error of

$$\Delta T = \frac{\Delta I}{I_P} T_{RST}, \qquad (3)$$

where T_{RST} denotes the width of the Up and Down pulses in the locked condition (approximately equal to five gate delays).

If expressed in picoseconds—rather than in radians—this error appears at the PLL output without multiplication. To convert the output error to a phase quantity, we normalize ΔT to T_{CK} and multiply the result by 2π:

$$\Delta\phi_{out} = 2\pi \frac{\Delta I}{I_P} \cdot \frac{T_{RST}}{T_{CK}}. \qquad (4)$$

Next, let us consider the noise of each current source, $\overline{I_n^2}$, as a mismatch between the two. Since the noise powers of the two current sources add, we have, within the loop bandwidth,

$$\overline{\Delta\phi_{out}^2} = 4\pi^2 \frac{2\overline{I_n^2}}{I_P^2} \frac{T_{RST}^2}{T_{CK}^2}. \qquad (5)$$

In addition, the harmonics of the Up and Down pulses downconvert high-frequency noise to baseband. Since the harmonics have a sinc envelope that crosses zero at $1/T_{RST}$, we assume that the number of harmonics is equal to $(1/T_{RST})/f_{REF}$ and they have roughly equal amplitudes. Since each harmonic downconverts two noise sidebands and since all of the sidebands are uncorrelated, we must multiply (5) by $2(1/T_{RST})/f_{REF}$:

$$\overline{\Delta\phi_{out,tot}^2} = 4\pi^2 \frac{2\overline{I_n^2}}{I_P^2} \frac{2NT_{RST}}{T_{CK}}, \qquad (6)$$

where N denotes the PLL multiplication factor. For thermal noise of a MOSFET, $\overline{I_n^2} = 4kT\gamma g_m = 4kT\gamma(2I_P)/(V_{GS} - V_{TH})$, yielding

$$\overline{\Delta\phi_{out,tot}^2} = 4\pi^2 \frac{16kT\gamma}{(V_{GS} - V_{TH})I_P} \frac{2NT_{RST}}{T_{CK}}. \qquad (7)$$

For example, with $\gamma = 1$, $I_P = 1$ mA, $V_{GS} - V_{TH} = 100$ mV, $N = 1000$, $T_{RST} = 30$ ps, and $T_{CK} = 10$ ps, we obtain $\overline{\Delta\phi_{out}^2} = -98$ dBc/Hz.

Effect of Reference Spurs The trade-off between the loop bandwidth and the level of output reference spurs creates another constraint in the design. Assuming the first harmonic of the ripple on the control voltage is expressed as $V_m \cos\omega_{REF}t$, we write the output as

$$V_{out} = V_0 \cos\left(\omega_{CK}t + \frac{V_m K_{VCO}}{\omega_{REF}}\sin\omega_{REF}t\right). \qquad (8)$$

The zero crossings therefore deviate from their ideal points by a maximum of $\pm V_m K_{VCO}/\omega_{REF}$ radians, exhibiting a peak-to-peak jitter equal to

$$J_{pp} = \frac{1}{2\pi} \frac{2V_m K_{VCO}}{\omega_{REF}} T_{CK}. \qquad (9)$$

Since the relative magnitude of the sidebands in the output spectrum is given by $V_m K_{VCO}/(2\omega_{REF})$, we conclude that the relative jitter, J_{pp}/T_{CK}, and the relative sideband magnitude are nearly equal. For J_{pp}/T_{CK} to remain below 1%, the sidebands must fall to −40 dBc, a relatively relaxed requirement.

Effect of Capacitor Leakage The gate leakage current has reached significant values in 45-nm technology. Plotted in Fig. 9 is the simulated leakage for a 10 μm/0.5 μm device

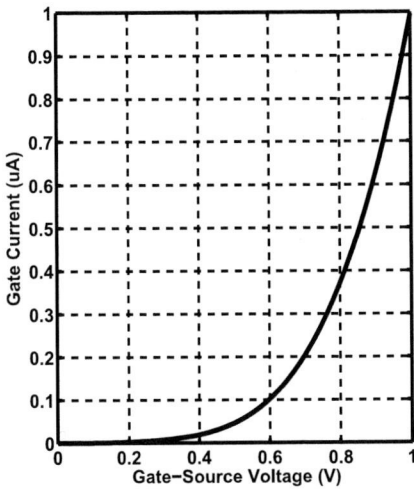

Fig. 9. Gate leakage in 45-nm technology.

with a gate dielectric thickness of 20 $\overset{\circ}{A}$. (The source and drains are grounded.) Note that the strong dependence of the leakage on the gate-source voltage makes cancellation difficult.

The gate leakage readily manifests itself if the loop filter incorporates MOS capacitors. As illustrated in Fig. 10 [12], the leakage current I_G discharges the loop filter while the

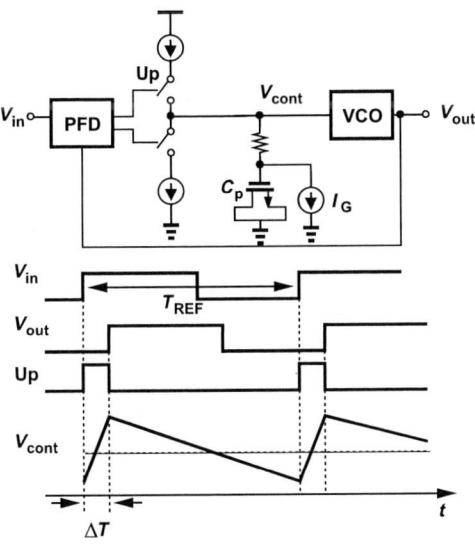

Fig. 10. Effect of gate leakage on PLL performance.

charge pump is off. In the steady stage, the PLL develops a phase offset, ΔT, during which the CP replenishes the charge drained by I_G. Thus, the peak-to-peak ripple on the control voltage is given by $(I_G/C_P)T_{REF}$, where it is assumed $\Delta T \ll T_{REF}$.

Interestingly, the "self-droop" rate I_G/C_P is independent of the MOS lateral dimensions and hence a constant of the technology, reaching 1.2 mV/ns for 45-nm devices at $V_{GS} = 0.6$ V. For example, if $T_{REF} = 10$ ns, then a ripple of 12 mV$_{pp}$ appears on the control voltage, yielding large sidebands at the VCO output. If the sidebands are to remain 40 dB below the carrier (as shown above), then $K_{VCO} < 520$ MHz, a very difficult condition to meet for a VCO running at 80-100 GHz.

C. Cascaded PLLs

The large multiplication factor required to translate f_{REF} to f_{CK} makes cascaded PLLs [13] a plausible alternative. Specifically, we seek the conditions under which such a cascade exhibits less jitter than a single PLL.

Consider the cascade shown in Fig. 11, where we assume the following are constant: f_{REF}, $N_1 N_2$, the (free-running)

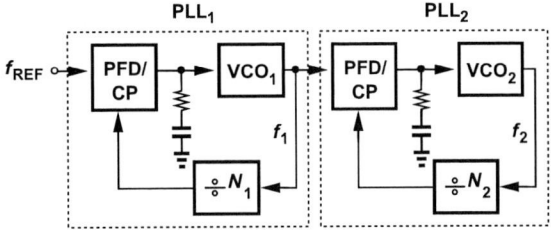

Fig. 11. Cascaded PLLs.

phase noise of VCO$_2$ (S_2), and the phase noise of the reference (S_{REF}). We seek the optimum choice of N_1, N_2, and the loop bandwidths of the two PLLs, BW$_1$ and BW$_2$. As we will see, the utility of cascaded PLLs directly depends on the phase noise of VCO$_1$ (S_1) relative to that of VCO$_2$. We denote the Q's of the two oscillators by Q_1 and Q_2.

We analyze three scenarios for the two VCOs. (1) The phase noise (at a given frequency offset) directly scales with frequency, $S_2 = N_2 S_1$. This occurs if the Q remains relatively *constant* from f_1 to f_2. (2) The phase noise is constant, $S_2 = S_1$, requiring that $Q_2 = N_2 Q_1$. (3) Due to tuning range limitations at f_2, VCO$_2$ is much more difficult to design than VCO$_1$, and $S_2 > N_2 S_1$.

Figure 12(a) plots on a log scale the single-sideband (SSB) profiles of S_1 and S_2 for the first scenario. If BW$_1$ is chosen at the intersection of S_1 and $N_1 S_{REF}$ and BW$_1$ =BW$_2$, then the amplified phase noise of PLL$_1$ adds to the shaped phase noise of VCO$_2$, resulting in the "humps" shown in S_{out} and hence higher jitter than that of a single PLL (the gray curve). It can be shown that BW$_2$ >BW$_1$ or BW$_2$ <BW$_1$ yield even higher jitter. In other words, this scenario provides no jitter advantage over a single loop.

Figure 12(b) illustrates the second scenario, where the amplified phase noise of PLL$_1$ dominates, thereby producing a larger jitter than does the first scenario. This holds regardless of the choice of BW$_2$.

Shown in Fig. 12(c) is the third scenario. Here, the amplified phase noise of PLL$_1$ reaches $N_1 N_2 S_{REF}$ but extends only to BW$_1$. Thus, if BW$_2$ is maximized (subject to conditions such

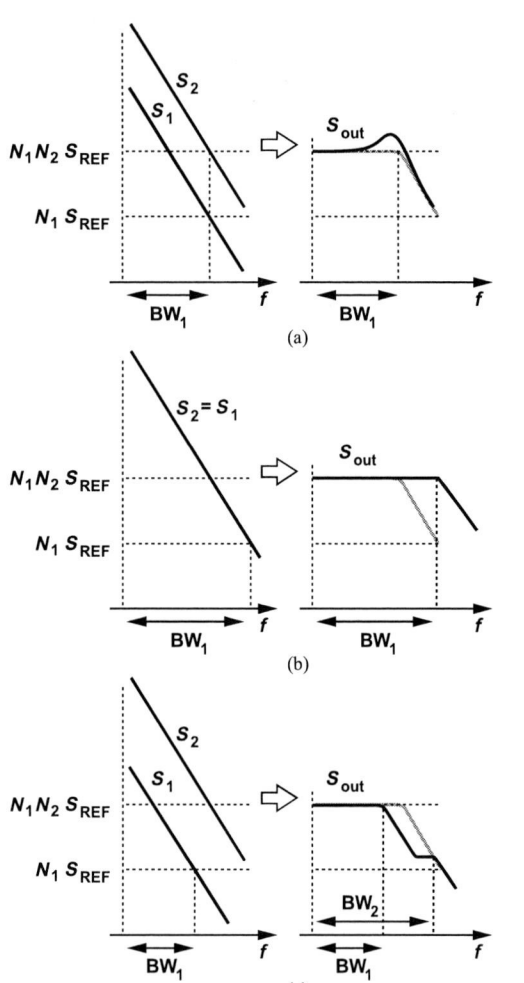

Fig. 12. Output phase noise of cascaded PLLs for (a) $S_2 = N_2 S_1$, (b) $S_2 = S_1$, and (c) $S_2 > N_2 S_1$.

as $BW_2 < 0.1 N_1 f_{REF}$), then the cascade exhibits less jitter than a single PLL.

D. A New Approach

We propose a TX/PLL interface that can relax many of the issues described above. Illustrated in Fig. 13(a), the idea is to employ a half-rate PLL and a frequency doubler to generate the full-rate clock. The twofold reduction in the required PLL speed greatly eases the design of the VCO and frequency dividers. More importantly, the architecture eliminates the troublesome divider delay depicted in Fig. 1. Instead, the delay through the doubler must match that through the MUX, a simpler task because they have the same polarity.

The proposed approach nonetheless entails two important issues. First, the doubler must avoid attenuation so that it only doubles the phase noise and generates large output swings that can directly clock the FF. This can be accomplished by means of the inductively-loaded symmetric XOR gate shown in Fig. 13(b) [14]. Second, duty-cycle distortion in the PLL

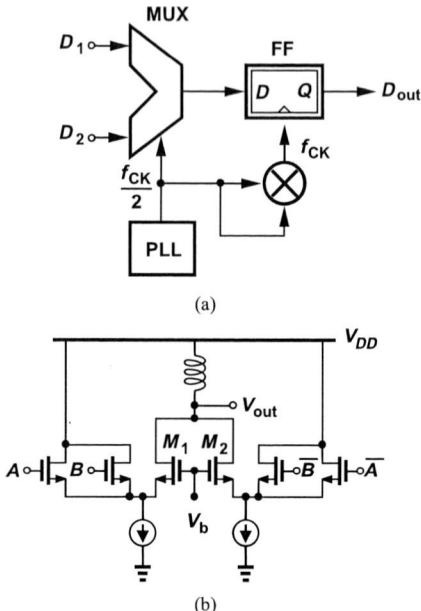

Fig. 13. (a) Proposed TX PLL interface, (b) doubler implementation as a low-voltage symmetric XOR .

output translates to displacement of every other falling (or rising) edge at the doubler output. Fortunately, the inductive (resonant) load in Fig. 13(b) reduces this effect by about 12 dB. Note also that the common-gate transistors M_1 and M_2 provide a high reverse isolation, suppressing the coupling of data transitions to the oscillator.

IV. RX CDR CIRCUIT

The CDR circuit presents its own challenges in receiver design, especially if it is integrated along with the transmitter. In addition to recovering the clock with a small jitter and retiming the data optimally, the circuit must remain immune to both the noise generated by the data edges at the TX output and the injection-pulling effected by the TX PLL. Arising from coupling through the substrate and the package, these two types of corruption directly determine the choice of the transceiver architecture.

A. Phase-Interpolating CDR Circuits

A class of CDR circuits dealing with these issues is based on shared PLLs with phase interpolation [2]. Illustrated in Fig. 14 [2], the idea is to perform phase comparison with the input data through the use of interpolated phases obtained from the TX PLL. The phase detector (PD) selects the interpolated edges so as to optimally sample the data (in the middle and at the zero crossings of the eye). In the presence of an offset between f_{VCO} and the data rate, the interpolated phases rotate at a rate equal to the offset. Thus, if the CDR loop bandwidth remains sufficiently greater than the frequency offset, the data is sampled properly.

The above architecture can employ discrete or continuous phase interpolation. With the former, the phase rotation in

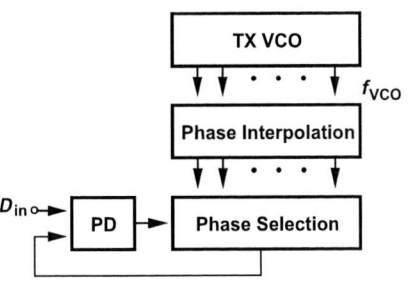

Fig. 14. CDR loop using interpolated phases of TX VCO.

the presence of a frequency offset occurs in discrete steps, failing to retime the data optimally. For this reason, a large number of finely-spaced edges must be produced, leading to a complex layout and mismatches and unwanted couplings in the routing. Continuous interpolation is therefore better suited to high-speed design as it requires only quadrature phases.

Figure 15 shows an example of CDR with continuous phase

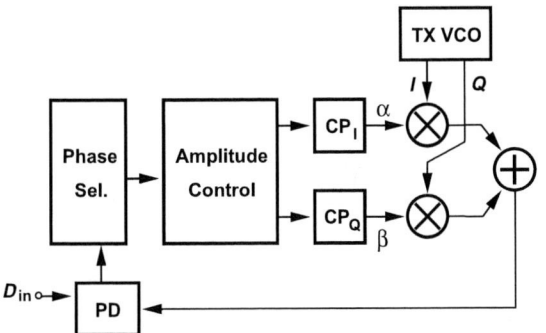

Fig. 15. CDR loop using continuous phase interpolation.

interpolation [2]. Here, quadrature phases provided by the TX PLL are summed with weights α and β and the result is applied to a bang-bang phase detector. From the phase error, the PD commands the two charge pumps CP_I and CP_Q to adjust α and β, respectively. The loop locks when every other edge of the interpolated clock samples the data at its zero crossings.

The CDR loop of Fig. 15 employs an amplitude control circuit to avoid a degenerate state. Since the phase of the interpolated clock only depends on α/β, these weights may diminish toward zero while, in principle, maintaining the proper phase. As a result, the interpolated clock amplitude continues to decline, eventually causing lock failure. The amplitude control circuit monitors α and β individually, adjusting their value if they fall or rise excessively.

PLL sharing with phase interpolation suffers from two critical drawbacks at very high speeds. First, the required quadrature or multiphase VCO topology inevitably degrades the phase noise of the TX PLL—where jitter is most important. Second, with the large layout dimensions of the TX and the RX —especially if many peaking inductors are used—it becomes exceedingly difficult to route the TX PLL outputs to the RX CDR circuit. The issue of sharing a PLL between TX and RX paths has already manifested itself in ultra-wideband RF transceivers operating at 10 GHz and in new RF designs targeting 60 GHz. In these cases, it is advantageous to employ two

independent synthesizers so as to avoid long interconnects.

B. VCO-Based CDR Circuits

With the foregoing issues plaguing phase-interpolating CDR circuits, the conventional, VCO-based architectures appear a more feasible approach at data rates of 80-100 Gb/s. Except for regenerator units required in long-haul optical communication, most applications make it desirable to employ a half-rate or quarter-rate CDR topology. This is because (a) mutual injection pulling [15] prohibits equal nominal frequencies for the TX VCO and the RX VCO; and (b) data demultiplexing in the receive path is greatly simplified if it is realized within the CDR circuit.

The choice between half-rate and quarter-rate architectures is determined by a number of factors: (1) if the TX PLL operates at half rate [e.g., as in the architecture of Fig. 13(a)], then the CDR circuit must run at quarter rate or lower; (2) the VCO design and layout becomes more complex as the number of required phases increases [16].

C. A New Approach

In order to avoid injection pulling between the TX PLL and the RX CDR circuit, it is desirable to choose a *non-integer* relationship between their VCO frequencies. We propose a CDR architecture that can coexist with a full-rate or half-rate TX PLL with minimal pulling. Shown in Fig. 16(a), the loop incorporates a VCO running at $f_{CK}/3$ (f_{CK} denotes the

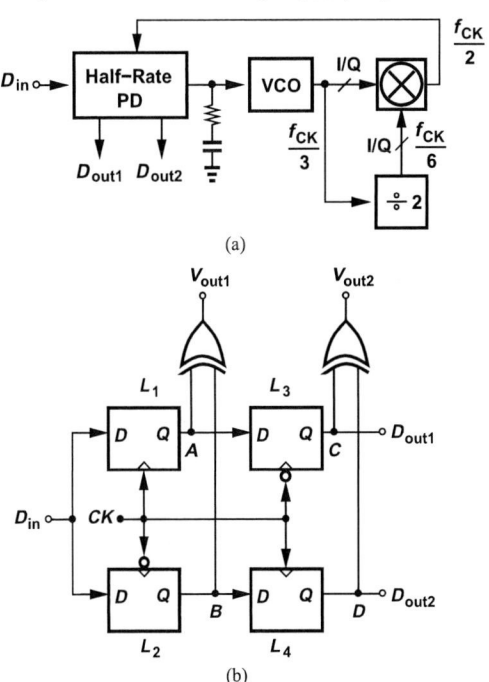

Fig. 16. (a) Proposed CDR architecture, (b) realization of the PD.

full-rate clock) and a $\div 2$ circuit generating $f_{CK}/6$. These two frequencies are applied to a single-sideband mixer so as to yield a half-rate clock. A half-rate PD performs phase comparison with the input data while producing demultiplexed outputs D_{out1} and D_{out2}.

978-1-4244-2018-6/08 $25.00 © 2008 IEEE

The design of the VCO for operation at $f_{CK}/3$ is relatively relaxed even though it must provide quadrature outputs for SSB mixing. However, since the mixer does not generate quadrature phases of the half-rate clock, the PD topology must be chosen accordingly. A half-rate PD that does not require quadrature clock phases is reported in [17] and shown in Fig. 16(b). Here, latches L_1-L_4 serve as the phase detector, and V_{out1} and V_{out2} are applied to a charge pump or V/I converter.

In the SSB mixer of Fig. 16(a), mismatches produce a fraction of the unwanted sideband at $f_{CK}/3 - f_{CK}/6 = f_{CK}/6$, which is 25 to 30 dB below the wanted component. Fortunately, an inductively-loaded mixer such as that in Fig. 13(b) suppresses this component by about 20 dB. Note that third-order nonlinearity at the input ports of the mixer is benign because it results in a component given by $3(f_{CK}/3) - 3(f_{CK}/6) = f_{CK}/2$.

V. FUTURE TRENDS

As wireline transceivers target speeds greater than 40 Gb/s, a number of trends are likely to emerge.

(1) Millimeter-wave device modeling and circuit techniques will find increasingly broader usage in wireline designs.

(2) While conventional transmission standards typically do not distinguish among different jitter frequencies in the TX mask, future standards may add a spectral mask to reveal components that are more difficult to remove in the receiver.

(3) Future "circuit-friendly" standards may allow a greater bandwidth for CDR circuits so that the recovered clock simply tracks the data edges rather than ignores the input jitter. This larger bandwidth will allow suppressing the phase noise of the CDR VCO to a greater extent.

(4) Layout and packaging of high-speed transceivers will also draw upon the techniques developed in millimeter-wave systems.

VI. CONCLUSION

High-speed transceivers continue to present interesting challenges, requiring more than ever TX data path and PLL co-design, TX PLL and RX CDR co-design, and the use of RF and millimeter-wave techniques. It appears that the reference phase noise will play a major role in the jitter of PLLs, and cascaded loops will offer marginal improvement. A TX/PLL interface and a CDR architecture are proposed as a means of relaxing some of these issues.

REFERENCES

[1] M. Sasaki, "A 9.5GHz 6ps-Skew Space-Filling-Curve Clock Distribution with 1.8V Fill-Swing Standing-Wave oscillators," *ISSCC Dig. Tech. Papers,* pp. 518-519, Feb. 2008.

[2] F. Yang et al, "A CMOS low-power multiple 2.5-3.125-Gb/s serial link macrocell for high IO bandwidth network

ICs," *IEEE J. Solid-State Circuits*, vol. 37, pp. 1813-1821, Dec. 2002.

[3] S. Gondi and B. Razavi, "Equalization and Clock and Data Recovery Techniques for 10-Gb/s CMOS Serial Links," *IEEE J. Solid-State Circuits*, vol. 42, pp. 1999-2011, Sept. 2007.

[4] H. Noguchi et al, "A 40Gb/s CDR Cicruit with Adaptive Decision-Point Control Using Eye-Opening Monitor Feedback," *ISSCC Dig. Tech. Papers,* pp. 228-229, Feb. 2008.

[5] R. Farjad-Rad et al, "A low-power multiplying DLL for low-jitter multigigahertz clock generation in highly integrated digital chips," *IEEE J. Solid-State Circuits*, vol. 37, pp. 1804-1812, Dec. 2002.

[6] S. Gierkink, "An 800MHz -122dBc/Hz-at-200kHz Clock Multiplier Based on a Combination of PLL and Recirculating DLL," *ISSCC Dig. Tech. Papers,* pp. 454-455, Feb. 2008.

[7] S. Galal and B. Razavi, "10-Gb/s Limiting Amplifier and Laser/Modulator Driver in 0.18um CMOS Technology," *IEEE J. Solid-State Circuits*, vol. 38, pp. 2138-2146, Dec. 2003.

[8] B. Razavi, "A Millimeter-Wave Circuit Technique," to appear in *IEEE J. Solid-State Circuits*, Sept. 2008.

[9] B. Razavi, "Heterodyne Phase Locking: A Technique for High-Speed Frequency Division," *IEEE J. Solid-State Circuits*, vol. 42, pp. 2877-2892, Dec. 2007.

[10] K. Scheir et al, "Design and Analysis of Inductors for 60 GHz Applications in a Digital CMOS Technology," *Proc. 69th ARFTG Microwave Measurement Conference,* June 2007.

[11] T. Dickson et al, "30-100 GHz Inductors and Transformers for Millimeter-Wave (BI)CMOS Integrated Circuits," *IEEE Trans. Microwave Theory and Techniques*, vol. 53, pp. 123-133, Jan. 2005.

[12] B. Razavi, "Design Considerations for Future RF Circuits," *Proc. International Conference on Circuits and Systems*, pp. 741-744, May 2007, New Orleans.

[13] M. Kossel et al, "A low-jitter wideband multiphase PLL in 90nm SOI CMOS technology," *ISSCC Dig. Tech. Papers,* pp. 414-415, and slide supplement, Feb. 2005.

[14] B. Razavi, K. F. Lee, and R. H. Yan, "Design of High-Speed Low-Power Frequency Dividers and Phase-Locked Loops in Deep Submicron CMOS," *IEEE J. Solid-State Circuits*, vol. 30, pp. 101-109, Feb. 1995.

[15] B. Razavi, "Mutual Injection Pulling Between Oscillators," *Proc. IEEE Custom Integrated Circuits Conference*, pp. 675-678, Sept. 2006.

[16] J. Lee and B. Razavi, "A 40-Gb/s clock and data recovery circuit in 0.18um CMOS technology," *IEEE J. Solid-State Circuits*, vol. 38, pp. 2181-2190, Dec. 2003.

[17] J. Savoj and B. Razavi, "A 10-Gb/s CMOS Clock and Data Recovery Circuit with a Half-Rate Linear Phase Detector," *IEEE J. Solid-State Circuits*, vol. 36, pp. 761-768, May 2001.

IEEE 2008 Custom Intergrated Circuits Conference (CICC)

A 20Gb/s SerDes Transmitter with Adjustable Source Impedance and 4-tap Feed-Forward Equalization in 65nm Bulk CMOS

R. A. Philpott, *Member, IEEE*, J. S. Humble, *Member, IEEE*, R. A. Kertis, *Member, IEEE*, K. E. Fritz, B. K. Gilbert, *Fellow, IEEE*, and E. S. Daniel, *Member, IEEE*

Special Purpose Processor Development Group, Mayo Clinic, Rochester, MN 55905, U.S.A.

Abstract-The design and wafer probe test results of a 20Gb/s Source-Series Terminated SerDes transmitter are presented. The integrated circuit, fabricated in a 65nm bulk CMOS technology, transmits pre-emphasized data through the use of a 4-tap feed-forward equalizer. Transmitter output impedance is adjustable from 45 to 55 ohms. A power consumption of 167mW at 1.1V was measured at a transmit rate of 20Gb/s.

I. INTRODUCTION

Source-Series Terminated (SST) transmitters have been known to offer a lower power solution for SerDes signal transmission while maintaining the ability to handle a large range of termination voltages [1,2]. This makes this type of transmitter compatible with multiple termination standards compared to their CML counterparts. This paper will describe an SST transmitter with 4-tap feed-forward equalization (FFE) operating at data rates up to 20Gb/s. Key features of this work include adjustable source impedance for optimized matching to variations in transmission media impedances and the achievement of 20Gb/s data rate using a standard bulk CMOS technology. Also included is an on-chip programmable PRBS test pattern generator with an enhanced parallel architecture to meet the testing speed requirements. The output stage contains T-coils which enable broad-band impedance matching with a return loss of -11dB over 20GHz while still providing ESD protection for the macro [3].

II. Architecture

Figure 1 shows a high level block diagram of the SerDes transmitter. In this design, a programmable pattern generator is used to generate 8-bit wide pseudo-random data patterns. An 8:2 multiplexer serializes the pattern data into odd and even data streams. The half-rate interleaved transmitter passes both serial data streams through a 4-tap digital finite impulse response (FIR) filter. The FIR filter performs feed-forward equalization on the transmit data. The FIR filter is comprised of a four-delay shift register with tap sign select and 2:1 multiplexers integrated within the 64 driver slices. The tap sign select function allows each tap of the feed-forward equalizer to be assigned either a positive or negative value. The 2:1 multiplexers individually assign each of the 64 driver slices to the 'main' tap or to one of either the 'pre', 'post1' or 'post2' taps. Impedance adjustment can be used to offset output impedance variability across process variations as well

as to match the driver output impedance to the impedance of the desired channel fabric. The ESD device has been integrated into a T-coil which isolates the device capacitance of the ESD structure from the driver output. A configuration register is used to control the pattern generator, set FFE transfer function and adjust output impedance.

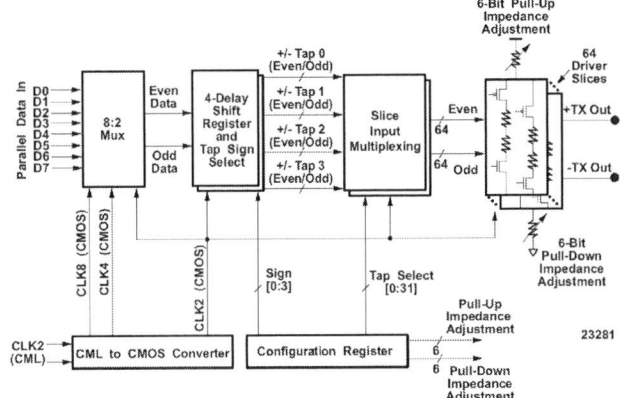

Fig. 1. SerDes Transmitter High-Level Block Diagram

The clock generation circuitry converts a reference current-mode logic (CML) half-rate clock signal into the full rail-to-rail CMOS clocks needed by the transmitter. Half-rate, quarter-rate and eighth-rate clocks are generated on-chip and distributed to the appropriate parts of the transmitter. Duty-cycle correction is incorporated on-chip to maximize the transmitter data rate while minimizing deterministic jitter [1].

III. Circuit Concepts

The transmitter accepts parallel input data via an 8-bit wide data bus at the 8:2 data multiplexer. The 8-bit data stream is generated by an on-chip programmable data generator described later in this section. A simplified schematic of the 8:2 data multiplexer is shown in Figure 2. The parallel input data is latched by eight flip-flops that are clocked at eighth-rate. Four latches, running on the opposite phase of the eighth-rate clock are used to delay four of the data bits by half a clock period. Four 2:1 multiplexers convert the 8-bit wide bus operating at eighth-rate into a 4-bit wide bus operating at quarter-rate. The process is repeated using two 2:1 multiplexers, two latches, and the quarter-rate clock to

978-1-4244-2018-6/08 $25.00 © 2008 IEEE   663

create a 2-bit wide bus operating at half-rate. The 2-bit wide data (i.e. interleaved data) is used throughout the transmitter until it reaches the SST output driver, where it is converted into full-rate data.

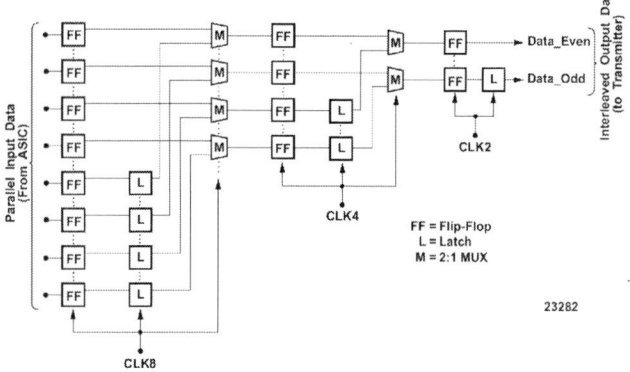

Fig. 2. Block Diagram of 8:2 data multiplexer

In order to implement a four-tap feed-forward equalizer, delayed versions of the interleaved transmitter data must be created. This is accomplished with the four-stage shift register shown in Figure 3. A four-tap FFE requires pre-cursor data, which occurs one bit before the main data, the main data, which has no time delay, post1 data, which is delayed by one bit relative to the main data, and post2 data, which is delayed by two bits. XOR gates are added into the data path for each of the four taps to allow the tap to have either a positive or negative sign.

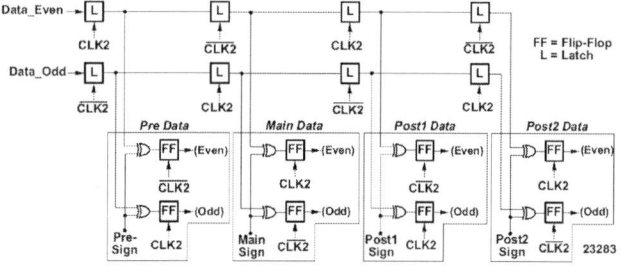

Fig. 3. Schematic of Four-Delay Shift Register With Tap Sign Select

The input to each of the 64 output driver slices has its own front-end logic to connect to the four-delay shift register. To allow every SST driver slice to be assigned to any tap weight, a 4:1 multiplexer would be required. However, in 65nm bulk CMOS, the propagation delay of a standard-cell 4:1 or 2:1 multiplexer could not meet the timing requirements for the period of a half-rate 20Gb/s clock. To achieve timing, a custom 2:1 multiplexer was used, which was implemented with four data capture latches and two transmission gates shown in Figure 4. The 2:1 multiplexer constraint on all driver slices limits equalization options. A choice was made to allow all 64 driver slices to act as main taps, but only a maximum of 8 can be programmed as pre taps, a maximum of 16 as post1 taps, a maximum of 8 as post2 taps and still have 32 assigned to the main tap. The SST differential driver slices require true

and complement data to be generated and present at their inputs with accurate timing placement between them. This was accomplished through the use of complementary pass-gate logic (CPL) latches, which accept the true and complement data (and associated skew) from the previous latch and inverter and aligns the complementary signals. In a similar transmitter design implemented in SOI [1], the driver slice logic used a standard cell 4:1 multiplexer and was able to meet timing requirements which resulted in significant power savings over this work.

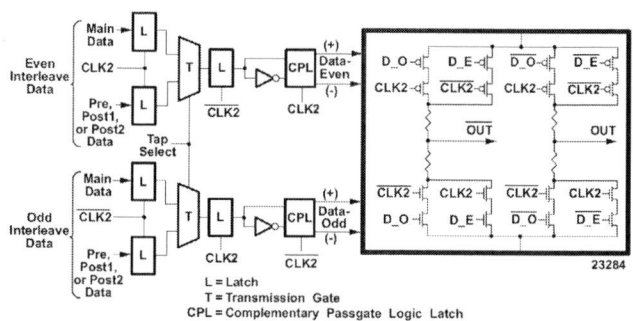

Fig. 4. Simplified Schematic of SST Output Driver Slice Logic

The method used for adjusting the source impedance of the transmitter outputs from 45 to 55 ohms is shown in Figure 5. An adjustable resistance is added in series with the pull-up and pull-down networks of all 64 transmitter slices. The selectable resistors compensate for the varying resistance of the FETs and resistors due to process variation. Additionally, the transmitter source impedance can be adjusted to match manufacturing variations in channel fabric impedances; minimizing reflections at the transmitter output.

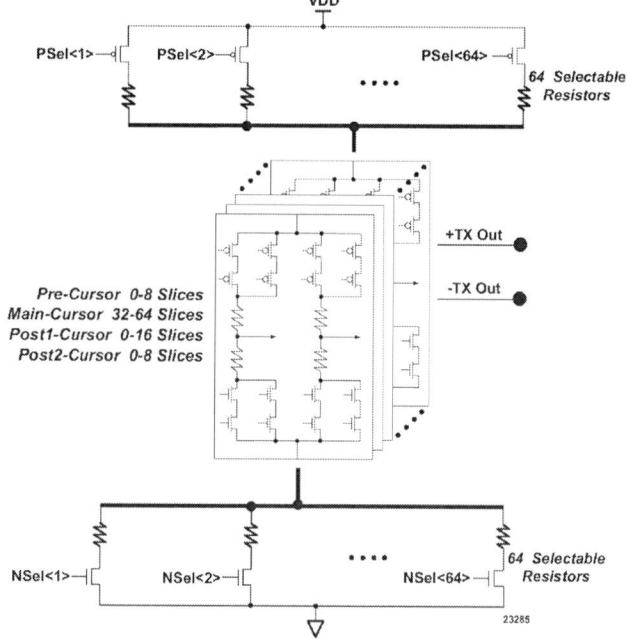

Fig. 5. Diagram of 64 Transmitter Slices With Source Impedance Adjustment

A photograph of the SST transmitter is shown in Figure 6. The die dimensions are 1.0mm x 1.0mm. Differential clock inputs enter from the top and are routed to the clock receiver with 50 ohm transmission lines. The differential outputs are routed through T-coil networks prior to exiting at the bottom. Shift register and pattern generation control occurs from the left and right sides. Embedded within the macro is 100pF of decoupling capacitance for supply noise suppression.

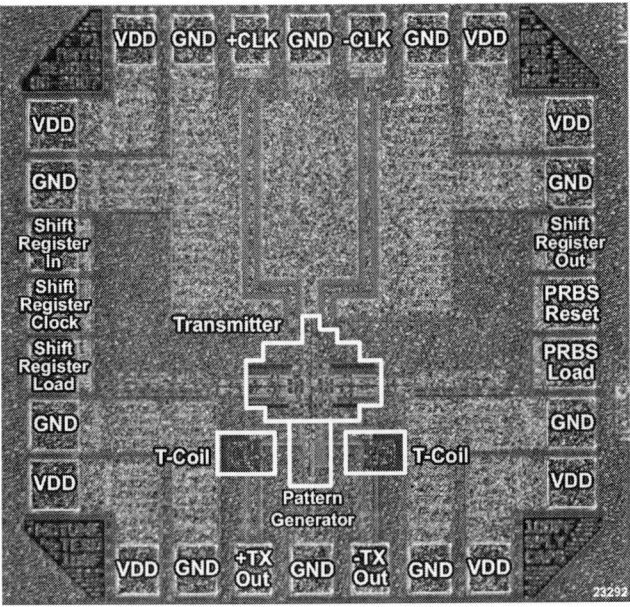

Fig. 6. Photograph of SST Transmitter Macro

The detailed circuit layout of the transmitter macro is shown in Figure 7. The 64 slices are partitioned into two halves. The impedance tuning circuitry is partitioned into four equal blocks located in each of the four layout quadrants. In the center of the macro are the data multiplexer, FFE tap control and CMOS clocking. The area of the macro is 0.025mm^2.

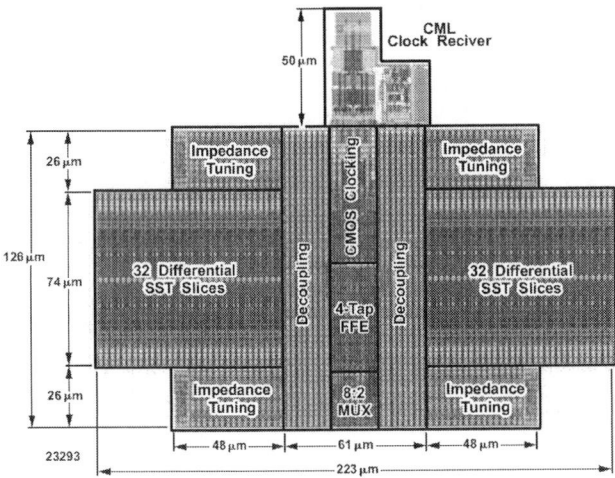

Fig. 7. Detailed Circuit Layout of Transmitter Macro

IV. Testing and Measured Performance

Testing of the SST transmitter was performed using a wafer probe station. An Agilent 8257D signal generator was used for the clock source due to its very low phase noise close-in to the carrier frequency. An Agilent Infiniium DCA-J 86100C scope with an Agilent 86107A precision time base was used to capture eye diagrams and measure jitter..

To simplify testing, a programmable pattern generator was integrated on-chip with the SST transmitter. A parallel design architecture, shown in Figure 8, was adopted to achieve input data speeds while minimizing power consumption [4]. Four pseudo-random bit sequence test patterns (2^7-1, 2^{15}-1, 2^{23}-1, 2^{31}-1) are available as well as a user-programmable 40-bit repeating pattern. An all-zero detect circuit is included to ensure the pseudo-random sequence generators start up in a valid state.

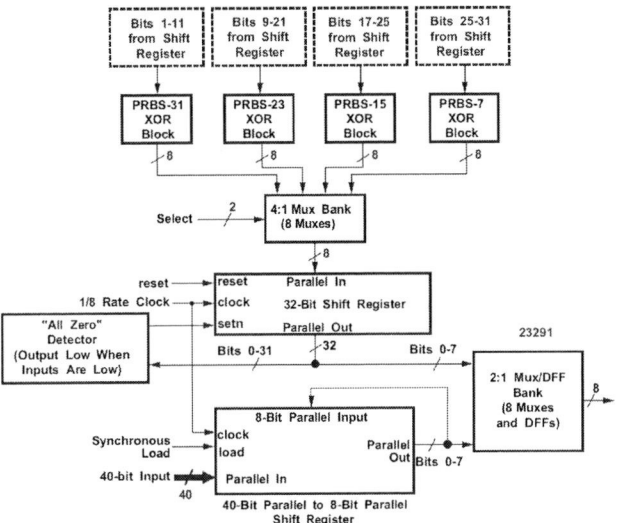

Fig. 8. Block Diagram of Programmable Pattern Generator

Figure 9 shows the measured eye diagram with FFE (pre=0, main=49, post1=11 and post2=4) applied to a 20Gb/s PRBS-15 signal at a supply voltage of 1.1V. The measured eye opening was 300mVpp. The measured random rms jitter was 239fs, with a duty cycle distortion error of 870fs (0.0174 UI). A deterministic jitter of 5.64ps was observed and a total jitter of 8.95ps was achieved at a BER of 1E-12. Figure 10 shows the return loss for three different transmitter output configurations: no ESD, with ESD and ESD embedded into a T-Coil. For each case, the output impedance was tuned to 50 ohms at DC. The best return loss of -11dB was achieved over broadband range up to 20GHz (supportive of a 40Gb/s data rate) with the use of a T-Coil. The measured power consumption of the transmitter including the CML clock receiver, but minus the pattern generator power, at different power supply voltages and data rates are shown in Figure 11. Table 1 compares our work to other transmitter examples in 65nm bulk CMOS and SOI. At 16Gb/s our design consumes twice the power of the reported SOI design due to the use of 64 driver slices versus 44 [1] and the custom 2:1 multiplexer

978-1-4244-2018-6/08 $25.00 © 2008 IEEE   665

function and CPL latches required in each driver slice to meet timing requirements at 16Gb/s and 20Gb/s. This is the first known reported example of an NRZ-based transmitter achieving 20Gb/s with adjustable source impedance in a 65nm bulk or SOI CMOS technology.

Specification	This Work	This Work	IBM Zurich [1]	TI [5]
Technology	65nm bulk	65nm bulk	65nm SOI	65nm bulk
Maximum Data Rate	20Gb/s	16Gb/s	16Gb/s	12.5Gb/s
Power Consumption	167mW @ 1.1V	115mW @ 1.0V	57.7mW @ 1.0V	~130mW
Power Efficiency	8.3 mW/Gbps	7.2 mW/Gbps	3.6 mW/Gbps	~10.4 mw/Gbps
Feed Forward Equalization	4-tap FIR	4-tap FIR	4-tap FIR	4-tap FIR
Active Area	0.025mm^2	0.025mm^2	~0.013mm^2	~0.08mm^2

Table 1. Comparison of Mayo SerDes with Other Recent Examples

V. CONCLUSION

A 20Gb/s Source-Series Terminated (SST) SerDes transmitter using 65nm bulk CMOS technology has been developed and tested. Tunable output impedance was implemented to achieve a good source match at DC. A T-Coil was used to improve the output impedance matching over a broad frequency range. A 4-tap feed-forward equalizer was demonstrated which minimizes the total output jitter to 8.95ps at an error rate of 1e-12. A measured power consumption of 167mW at 1.1V was observed which is comparable to other 65nm bulk CMOS transmitters but has significant power consumption and area penalties compared to what can be achieved using SOI technology.

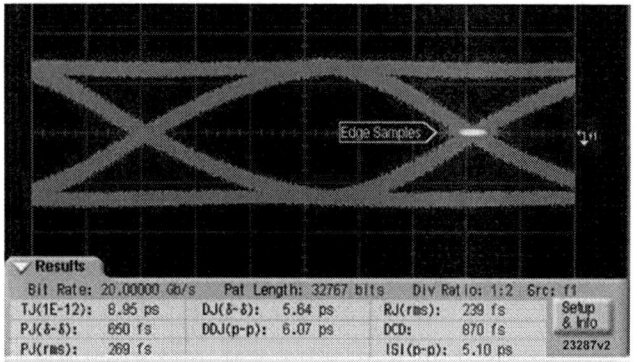

Fig. 9. Differential Eye Diagram of PRBS-15 at 20Gb/s After Equalization

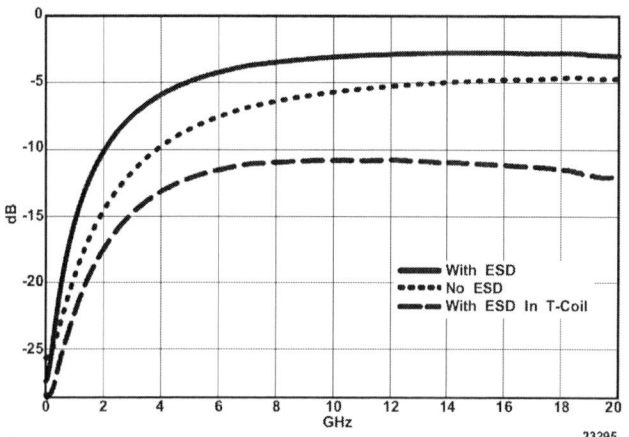

Fig. 10. Measured Return Loss of Transmitter Output

ACKNOWLEDGMENTS

The authors would like to thank Christian Menolfi, Thomas Toifl and Martin Schmatz from IBM-Zurich for education in Serdes transmitters. Additional thanks go to Victor Gammel and Jason Prairie for assistance with hardware characterization and Elaine Doherty, Teri Funk, Deanna Jensen and Steve Richardson for preparation of artwork.

REFERENCES

[1] C. Menolfi, T. Toifl, P. Buchman, et al., "A 16Gb/s Source-Series Terminated Transmitter in 65nm CMOS SOI," *ISSCC Digest of Technical Papers,* pp. 446-447, Feb. 2007.
[2] M. Kossel, C. Menolfi, J. Weiss, et al., "A T-Coil-Enhanced 8.5Gb/s High-Swing Source-Series-Terminated Transmitter in 65nm Bulk CMOS" *ISSCC Digest of Technical Papers,* pp. 110-111, Feb. 2008.
[3] S. Galal, B. Razavi, "Broadband ESD Protection Circuits in CMOS Technology," *ISSCC Digest of Technical Papers,* pp. 182-183, Feb. 2003.
[4] Wei-Zen Chen, Guan-Sheng Huang, "A Low Power Programmable PRBS Generator and a Clock Multiplier Unit for 10 Gbps Serdes Applications," *Proceedings of IEEE International Symposium on Circuits and Systems (ISCAS),* pp. 3273-3276, 21-24 May 2006.
[5] M. Harwood, et al., "A 12.5Gb/s SerDes in 65nm CMOS Using a Baud-Rate ADC with Digital RX Equalization and Clock Recovery," *ISSCC Digest of Technical Papers*, pp. 436-437, Feb. 2007.

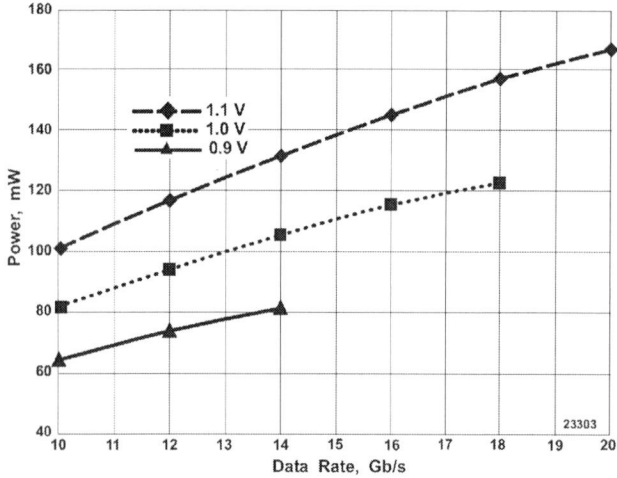

Fig. 11. Transmitter Power Consumption vs. Data Rate

IEEE 2008 Custom Intergrated Circuits Conference (CICC)

Wideband mmWave CML Static Divider in 65nm SOI CMOS Technology

Daeik D. Kim, Choongyeun Cho
IBM Semiconductor R&D Center
Hopewell Junction, NY
Email: {dkim, cycho}@us.ibm.com

Jonghae Kim
Qualcomm
San Diego, CA
jonghaek@qualcomm.com

Jean-Olivier Plouchart
IBM T. J. Watson Research
Yorktown Heights, NY
plouchar@us.ibm.com

Abstract—A wideband millimeter-wave (mmWave) CML static divider fabricated in 65nm SOI CMOS technology is presented. The mmWave system realization trend and engagement in sub-100nm CMOS technologies are summarized. CML static divider's circuit analysis, sensitivity curve, and simulations are explored. The input-locking hysteresis and divider DC bias tuning are employed to extend the divider operation range. The divider performance measurements are presented with hysteresis-assisted gain and figure-of-merits. A scalable statistical estimation is proposed, and it is validated with a full 300mm wafer measurements. The divider exhibits wideband mmWave performance to overcome the process variability in sub-100nm CMOS processes.

I. INTRODUCTION

The 60GHz range milli-meter wave (mmWave) physical links are emerging as a next generation short-range wide-bandwidth communications channel [1], [2]. The CMOS technology is becoming a strong candidate for the mmWave system design platform due to its manufacturing capability, system-on-chip (SoC) integration with baseband and digital intellectual property, and high-speed performance through technology scaling [3]. As CMOS FET's gate length scales aggressively, the device density has increased, and high-speed performance also has been improved down to 45nm. For example, 45nm SOI NFET's f_T is beyond 400GHz [4], and it provides enough design margin for mmWave analog system, though the device speed will reach the physical limit soon [5]. The adversaries of sub-100nm CMOS are the up-front costs for development and mask [6], and the aggravated defects and process variability [7], [8]. The variability affects the analog clock generation and transceiver front-ends of mmWave SoC more than the digital block. It is because of the small device dimension and parasitic capacitance contribution, which is relative to the total parasitic allowance. Especially the mmWave tranceiver and PLL shown in Fig. 1 are susceptible to the variation. The PLL front-end components - VCO and the pre-scaling frequency divider [9] - are potential bottlenecks for SoC chip-limited yield (CLY) due to the variability. It is essential to have a wideband tunable VCO and wideband divider to overcome the variation, while the technology stabilization is enforced.

This paper presents wideband CML static divider analysis, design, and measurements in 65nm SOI CMOS, as summarized in Fig. 1. The mmWave system implementation trend and the technology-to-circuit interaction between CMOS and

mmWave analog system design are discussed as backgrounds in II. The CML static divider design process is reviewed in III. The divider small-signal analysis, sensitivity curve, models for hysteresis between input-locked and self-oscillation modes, circuit design parameters, and simulation results are explored. In IV, divider test methodology is presented. The mmWave test setup, divider measurements, performance comparison, figure-of-merits (FoMs), process variation, and scalable statistical estimation are examined.

II. BACKGROUNDS

The mmWave channel realization trends are reviewed from technology and application perspectives in II-A. The increasing overlap and collaboration between CMOS technology and mmWave analog design are discussed in II-B.

A. Application

The high-definition multimedia contents are overloading existing channels with data bandwidth and the cumbersome cable connections. To meet the demands, the interests on mmWave system have been elevated for last several years [1], [10]–[12]. A survey on IEEE International Solid-State Circuits

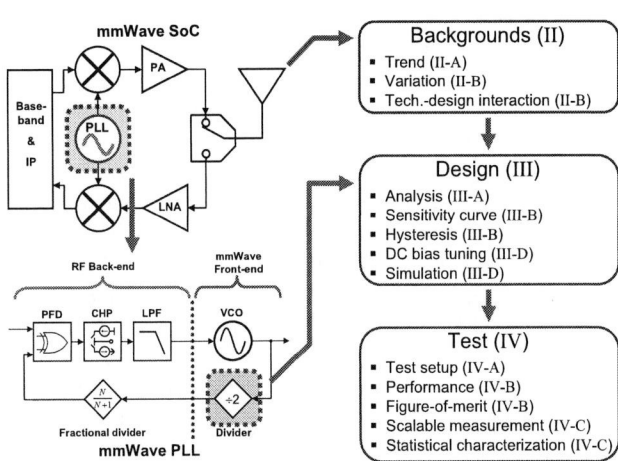

Fig. 1. Overview of the paper. The mmWave channel SoC implementation trend and CMOS technology-to-design interaction are reviewed (II). The PLL front-end frequency divider is the main concern in the paper, and it is susceptible to the process variability in sub-100nm along the VCO. The divider design (III) and measurements (IV) are presented.

978-1-4244-2018-6/08 $25.00 © 2008 IEEE

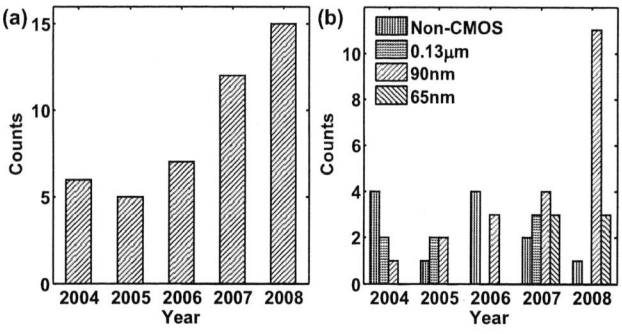

Fig. 2. Last 5-year trend of above-50GHz circuits in ISSCC. (a) The number of papers have been increasing steadily. Also more mature systems and higher-level of integrations have appeared. (b) Technology engagement trend. The 90nm CMOS has become the major platform in 2008, and 65nm implementations begin to appear. There are a few 0.25μm and 45nm entries not shown in the plot for simplicity.

TABLE I
CMOS mmWave Divider Performance

Type	f_{min}	f_{max}	BW (GHz)	CMOS	Ref.
Static	<64.7	94.4	>29.7	65nm	[17]
	73	100	27	65nm	[18]
	5	66	61	90nm	[19]
Dynamic	85.2	96.2	11	90nm	[13]
	82	94.1	12.1	65nm	[14]
	64	70	6	0.13μm	[15]
	62.9	71.6	8.7	90nm	[16]

Conference (ISSCC) for the last 5 years is given in Fig. 2. Circuits that operates above 50GHz are collected for the plot. The number of papers has increased in Fig. 2(a). In 2008, most of circuits are being implemented in 90nm CMOS, and the number has been increasing steadily. Also 65nm circuits are emerging, beginning in 2007. Once 0.13μm CMOS was considered enough for mmWave, but designers are moving to new scaled technologies for better performance. There are trade-offs among the fabrication cost, foundry access, and the design margin. It is yet to say that 90nm will be the main mmWave platform. It is likely that the popularity will shift toward 65nm and 45nm for the high-speed design margin and state-of-the-art performance.

As shown in Fig. 1, the prescaling frequency divider is an essential function block for mmWave signal generation. Many of mmWave PLL implementations use dynamic injection-locked designs to take advantage of the high-speed performance [13]–[16]. The CML-based static dividers operate at lower frequency, but they have wider operation range [17]–[19]. The dynamic and static classification is based on its operational concept and origin. Their performances are arranged in the Table I. The performances in the table are not necessarily accurate metrics, due to the uncertainties in the measurement setup. Still they are regarded as the best-effort results and demonstration. In general, the CML static dividers

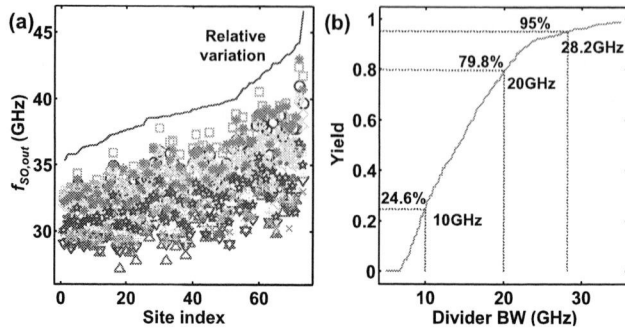

Fig. 3. Divider variability and yield. (a) Divider self-oscillation frequency f_{SO} sorted with the ascending site variation. High-speed circuits involve more variation. The relative variation has no unit and it is scaled to fit the plot. There are 73 site × 12 wafers = 876 data points. (b) Estimated divider yield with input-referred divider operation bandwidth. A divider meets the specification when it covers at 57-64GHz with a given bandwidth. The bandwidth is assumed symmetric around the self-oscillation frequency. About 28.2GHz operating range is required for 95% yield.

offer wider bandwidth than dynamic dividers. Also the CML dividers have reached higher $f_{max,div}$, but the different CMOS technologies make it difficult to compare.

When a mmWave system engages sub-100nm CMOS, the process variability should be considered for manufacturability. The variability of a CML divider is plotted in Fig. 3. One lot of 12 wafers in 65nm SOI CMOS was measured for the plot. It is using self-oscillation frequency f_{SO} to estimated the divider variability [18]–[20]. The variability experienced by injection-locked and CML static dividers is different. Nevertheless, the front-end of line (FEOL) component variation should be similar in both designs. Using the f_{SO} data and divider bandwidth, a CLY of a 57-64GHz mmWave PLL is calculated. In Fig. 3, 28.2GHz divider range is necessary to meet 95% divider yield. Most of dynamic dividers fall short of the bandwidth, and the CML static divider becomes an attractive alternative to overcome the process variation.

B. CMOS Technology and mmWave Design

As mmWave system designs adopt sub-100nm CMOS, the interaction between circuit and device technology becomes essential [21]. The technology complexity, process variability, and system integration challenge should be conveyed to designers. For example, a mmWave designer should be aware of not only technology native device offerings but also secondary options. Fig. 4 shows 65nm SOI CMOS FEOL and back-end of line (BEOL) stack diagram, and device options to make the most out of it. The analog FET performance is optimized with the adjusted gate-to-contact pitches and the number of contacts, as shown in Fig. 4(c). The pitch relaxation improves FET high-speed performance by reducing parasitic gate-to-contact capacitance and enhancing the stress liner efficiency and the carrier mobility [22], [23]. For passives, the technology options should be considered carefully. The different BEOL dielectric materials and air gap [24] could be offered as layer options. Inductor and BEOL vertical native capacitor (VNCAP) designs are determined along the

978-1-4244-2018-6/08 $25.00 © 2008 IEEE

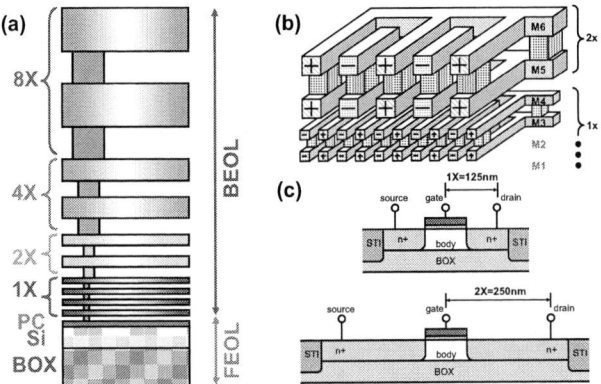

Fig. 4. Analog circuit design requires more knowledge on technology offering with scaling. (a) SOI CMOS technology vertical stack with FEOL and BEOL diagram. (b) BEOL vertical native capacitor (VNCAP) diagram with $1\times$ and $2\times$ metal levels are shown. VNCAP is formed without additional mask sets and scaled with technology. (c) A 65nm SOI NFET diagram with native $125\mu m$ pitch between source/drain contacts to poly-Si gate. Also relaxed $2\times$ ($250\mu m$) pitch is shown.

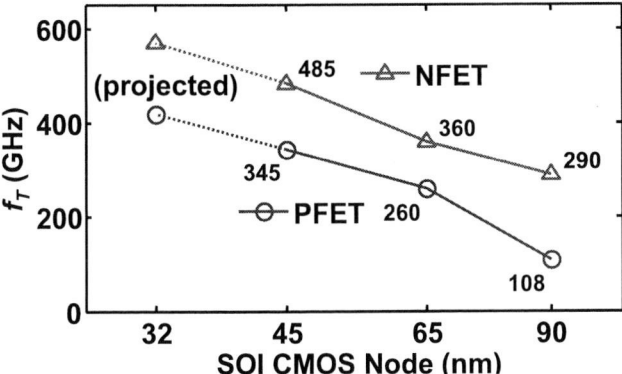

Fig. 5. The SOI CMOS technology current gain cut-off frequency f_T scaling trends in sub-100nm nodes. Not only the device density, but also high-speed performances have been scaled down to 45nm node.

technology options [25], [26]. After all, the CMOS platform itself becomes a menu to choose. CMOS performance has been scaled in high-speed performance so far as shown in Fig. 5 [4], [22], [27]. High-speed analog device performance will be improved along the novel digital device development, such as asymmetric FET and FinFET, and node scaling. The technology adoption trend in Fig. 2 shows that 90nm becomes the majority for mmWave circuits. Previously, 0.18 and 0.13μm CMOS technologies have been RF SoC design platform for several years. The mmWave CMOS platform adoption is limited by the technology accessibility, process variability, model uncertainty, manufacturing stability, and development and mask costs. It will take years to complete the technology performance exploration and mmWave system integration.

III. CML STATIC DIVIDER DESIGN

The CML static divider design process is described. The circuit is analyzed with small-signal model in III-A. The ana-

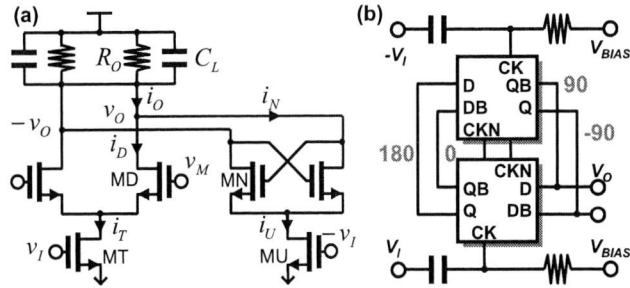

Fig. 6. CML static divider schematic diagrams. (a) A latch stage with a differential and a negative g_m pairs. (b) Divider as a master-slave flip-flop with AC coupled RF input and DC bias V_{BIAS} control. The phase relation is used to model the divider in steady state.

lytic results are used to derive sensitivity curve and hysteresis models in III-B. The circuit design is reviewed in III-C. The analysis is supported by circuit simulations in III-D.

A. Circuit Analysis

CML static divider topology is analyzed with a small-signal model and approximations. Fig. 6 shows diagrams of a latch stage and a divider as a master-slave flip-flop. The topology and its variants have been implemented in several CMOS generations [17]–[19], [28], [29] and non-CMOS technologies [30], [31]. They are referred to as static dividers, in contrast to the dynamic injection-locked dividers [13]–[16]. The topology also has been implemented as technology performance benchmark vehicles. In spite of the common use, the circuit has not been analyzed in detail. The use of sensitivity curve for performance characterization lacks a general model, whereas VCO has phase noise models [32].

The CML static divider is solved by approximating the differential stages as single-balanced mixer and with steady-state complex analysis at the output frequency. The circuit small-signal analysis is based on several assumptions. They are: 1) Small-signal approximation of differential pair g_m, 2) Input and output frequency locking in single-balanced mixer, 3) Steady-state operation at output frequency, and 4) Bi-stable divider status between input-locked (IL) mode and self-oscillation (SO) mode.

The differential pair small-signal gain $g_{M,D}$ is modulated by the tail input v_I and tail current, and it is approximated with a power series expansion assuming that tail DC current is much larger than small-signal current $i_t \ll I_T$. Then high-order terms are ignored.

$$g_{M,D} \approx \sqrt{\beta_D I_T}\left[1 + \frac{i_t(t)}{2I_T}\right] \qquad (1)$$

The differential pair output current $i_D = g_{M,D} \cdot v_m$.

$$i_D(t) = G_{M,D}v_m(t) + \sqrt{\frac{\beta_D\beta_T}{2}}v_i(t)v_m(t) \qquad (2)$$

The multiplication term $v_i(t)v_m(t)$ involves complicated harmonics as a mixer. For simplicity, the input v_i frequency f_{in} is twice of local v_m frequency f_{out}. The third harmonic at $3f_{out}$

978-1-4244-2018-6/08 $25.00 © 2008 IEEE

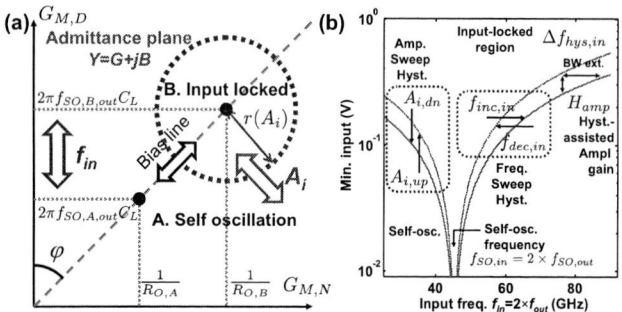

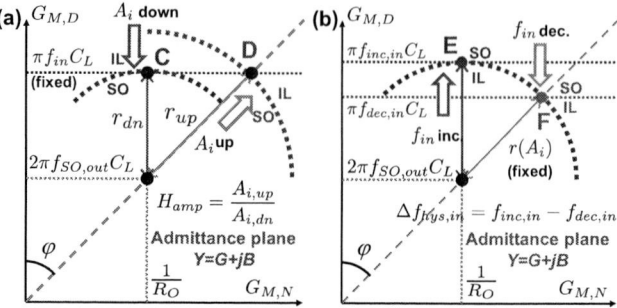

Fig. 7. CML static divider small-signal solution diagram. The approximations lead to an admittance plane circle or point equation. (a) Small-signal SO and IL modes diagram. The divider remains at the bias line for SO. For IL mode, the circuit condition deviates from the bias line to lock the divider to the input signal. The small-signal circuit solution stays on a circle, whose radius is a function of input signal amplitude A_i. (b) Input-referred frequency sensitivity curve with hysteresis. The small-signal solution suggests that there will be hysteresis in amplitude and frequency directions.

Fig. 8. CML static divider amplitude and frequency hysteresis models. (a) Amplitude sweep hysteresis model. The input signal amplitude A_i's downward sweep from IL mode experiences a step transition at point C to SO mode with a frequency jump. The A_i's upward sweep begins to lock when the self-oscillation frequency meets the desired output frequency at point D. (b) Frequency sweep hysteresis model. The input frequency f_{in} increase from IL mode sweep makes a frequency jump at point E to SO. The f_{in} decrease sweep locks to input when f_{SO} reaches the desired output frequency at F.

is ignored. It is because the entire loop works as a low-pass filter at high-frequency region [28].

$$v_i(t)v_m(t) \approx \frac{A_i A_o}{2} \cos\left[2\pi f_o t + \frac{\pi}{2} + \phi\right] \quad (3)$$

At the output frequency, the circuit node maintains steady-state phase noted in Fig. 6(b), including inversion from differential stage. The local input v_M and output v_O in Fig. 6(a) has 90-degree difference. The ϕ is the phase difference between RF input v_I and output v_O. The $-v_I$ in Fig. 6(a) has a 180-degree phase difference at the input frequency. The steady-state current summation at the output node is in (4). The R_O and C_L are the effective small-signal load resistance and capacitance at the output node. The equation is solved in real and imaginary terms, using trigonometric relations and the Pythagorean identity. Then it is arranged as an equation of circle in (5), for geometric visualization and analysis. The $G_{M,N}$ and $G_{M,D}$ are considered as both variables and constants. It denotes a circle on admittance plane $Y = G+jB$. It is centered at $(1/R_O, 2\pi f_{SO}C_L)$, where f_{SO} is the self-oscillation frequency. The radius is a linear function of input signal amplitude A_i. Considering angles θ and ϕ, not every point on the circle is a solution for the circuit, but it is useful for the circuit behavior visualization.

B. Sensitivity and Hysteresis

The analytic results are utilized to develop divider sensitivity curve and hysteresis models. The implications of the circle in (5) is explained in Fig. 7. There are two cases: A) self-oscillation (SO) and B) input-locked (IL) modes. We assume that the divider is bistable between the SO and IL modes. When $A_i = 0$, the radius of the circle is 0, and the solution remains on the bias line as a point. This corresponds to the SO mode. In the IL mode, input should be strong enough $A_i > 0$, so that the circuit solution moves away from the bias point to a point on the circle. By adjusting V_{BIAS}, the bias point moves along the bias line. For given a condition, a divider in IL mode will exit the mode and enters SO mode, when either

input frequency or A_i changes so that IL mode solution does not exist. Also a divider in SO mode will remain in the same mode, moving along the bias line until it finds an IL mode solution. Using the model, the divider sensitivity curve and hysteresis model in Fig. 7(b) are obtained.

The derivation begins with amplitude direction hysteresis at a fixed frequency in Fig. 8(a). When the divider input amplitude A_i goes down from an IL mode, there is more than one solution until point C. Passing C, the divider will make a step-like frequency transition to SO mode, since there is no solution for IL mode. When A_i goes up from a SO mode, the output frequency moves along the bias line till it meets the desired output frequency at D. The divider moves into the IL mode smoothly, by moving along the circle.

The definition of the divider sensitivity curve (SC) is the minimum input power that a divider operates at each frequency. Therefore, there are two different minimum power levels, whether it begins from IL or SO mode. The A_i downward sweep SC $A_{i,dn}(f)$ is more optimistic, and it is obtained by calculating the minimum A_i necessary to remain locked at the desired output and input-referred frequency $f_{in} = 2f_{out}$.

$$A_{i,dn}(f_{in}) \geq \frac{2\sqrt{2}}{\sqrt{\beta_D\beta_T + \beta_N\beta_U}}|G_{M,D} - \pi f_{in}C_L| \quad (6)$$

The A_i upward sweep SC is $A_{i,up} = A_{i,dn}/\cos\varphi$, using the trigonometric relation in Fig. 8(a). The curves in Fig. 7(b) are obtained with (6) and scaling by $1/\cos\varphi$. The SC slope is $2\sqrt{2} \cdot \pi C_L/\sqrt{\beta_D\beta_T + \beta_N\beta_U}$ (V/Hz), and it serves as a FoM. Fig. 8(a) predicts the frequency-direction hysteresis, whose model is described in Fig. 8(b). For frequency sweep, the A_i, or the radius of the circle is fixed. When input frequency increases from IL mode, there is more than one solution till point E. After E, the divider will make a sudden frequency change to SO mode, since there is no solution beyond E. As the input frequency decreases from SO mode, there is no solution on the bias line, until it reaches point F. Then the divider will be able to lock to input gradually without frequency change.

$$-\left[\frac{1}{R_O}+j2\pi f_o C_L\right] = -jG_{M,D}-G_{M,N}+\left[j\sqrt{\frac{\beta_D\beta_T}{2}}+\sqrt{\frac{\beta_N\beta_U}{2}}\right]\frac{A_i}{2}e^{j\phi} \tag{4}$$

$$\left(G_{M,N}-\frac{1}{R_O}\right)^2+\left(G_{M,D}-2\pi f_o C_L\right)^2=\left[\frac{A_i}{2\sqrt{2}}\sqrt{\frac{\beta_D\beta_T}{2}}+\frac{\beta_N\beta_U}{2}\right]^2 \tag{5}$$

C. Circuit Design

The CML divider was implemented in 65nm SOI CMOS. The circuit does not have waveguides, and relies on discrete active and passive components. In the early technology development stage, the device and parasitic models are far away from the reality. Though the design uses target models, the divider high-speed performance tends to improve over generation, while the topology ensures wideband operation against process variation. The model f_{SO} is about 40GHz, and the estimated sensitivity curve slope $|\partial A_i(f_{in})/\partial f_{in}|$ is about 11.6mV/GHz. For a 0.5V peak-to-peak input, the divider will operate up to $2 \times f_{SO}$+21.6GHz=101.6GHz.

To maximize the divider high-speed performance, relaxed $2\times$ gate-to-contact pitch FETs are used with dual gate connections. The g_m gain from the pitch relaxation is about 6.7%, while the g_{ds} also increases, so that the self gain is reduced. Still the NFET f_T is enhanced by 15.2% [23], and the layout optimization is useful to boost the circuit design margin. The pitch relaxation does increase circuit area by about 41%, but it is still small compared with passive components and interconnects. The FET sizes are W_D=10 and W_N=8μm, and the differential pair must be stronger than the negative g_m pair to overwrite latch.

A poly resistor of about 200Ω is used instead of a PFET load to reduce the output node v_O parasitic loading. An inductive load can be used to cancel out capacitance [18], [33], and to improve high-speed performance, which would increase circuit footprint. The BEOL VNCAP was used for AC coupling [25], [26]. The VNCAP implementation is highly compatible with digital CMOS technology, and requires no additional mask sets. Also capacitance density and qualify factor is comparable to MIM capacitors. To reduce substrate coupling, the lowest metal layer was avoided. It reduces the capacitance density, but the Q-factor should be higher slightly. The Q-factor is limited by the minimum feature metal layers (1$\times$) due to line resistance, and twice width (2$\times$) metal staggered capacitance has lower metal resistance. The VNCAP area is $10\times10\mu m^2$, and estimated capacitance is about 200fF. The VNCAPs were also used to decouple the power supply connections around the circuit. For testing, a pair of 150μm-pitch RF+DC wedge probe padsets were placed in the layout. The divider core area is about $40\times40\mu m^2$.

D. Simulation

Differential signals are directly applied to the tail gate without AC coupling and resistor. Signal DC level is used as V_{BIAS}. The output is terminated with a bias-tee so that

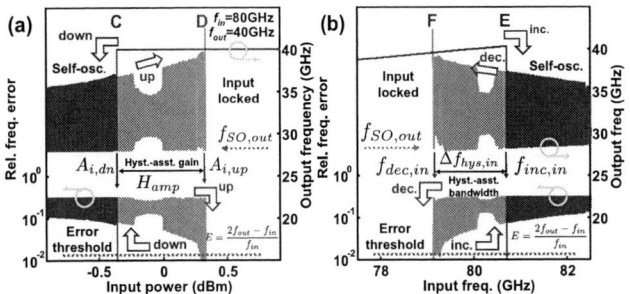

Fig. 9. CML static divider hysteresis simulation. (a) Amplitude sweep hysteresis simulation. At 80GHz, input signal amplitude is swept between IL and SO modes. The upper curves show the divider output frequency and the lower curves are relative frequency error. The divider is in IL mode when the error is below a preset 1% threshold level. As predicted in the model, the A_i-downward sweep makes a step-like frequency transition at C, while upward sweep smoothly moves to IL mode from SO at D. (b) Frequency sweep hysteresis simulation at 0dBm input power. The frequency-increasing sweep makes a sudden transition at E, as described in the model, and the frequency-decreasing sweep makes a gradual transition at F.

only AC components are visible at the output. After extracting the zero-crossing timing, and comparing the period with input frequency, the divider IL or SO mode is determined. Before obtaining divider SC, amplitude and frequency sweep hystereses are observed with in Fig. 9. The error threshold level is 1% of input frequency to determine the divider mode. The amplitude sweep was performed at 80GHz. As modeled in III-A, the input signal amplitude upward $A_{i,up}$ and downward $A_{i,dn}$ sweeps show different threshold levels. The $A_{i,up}$ needs more signal power to lock to input, and $A_{i,dn}$ maintains IL mode for lower input power. The ratio between the $A_{i,up}$ and $A_{i,dn}$ is considered as an amplitude-direction hysteresis gain H_{amp}, so that the divider can operate with lower input level when the initial condition is IL mode. Similar phenomenon is observed with frequency sweep in Fig. 9(b). The frequency operation range is effectively extended by $\Delta f_{hys,in}$, when the frequency and amplitude conditions begin from the IL mode.

As discussed, there are two SCs, one for upward $A_{i,up}$ and the other for downward $A_{i,dn}$ as plotted in Fig. 10(a). The amplitude hysteresis is noticeable at the high frequency, and therefore it extends the divider operation range. As expected, the hysteresis transition depends on the sweep speed, step size, and number of repetition. An example is plotted in Fig. 10(b). In the plot, the upward- and downward- amplitude and the frequency-increasing and decreasing sweeps are overlaid. The hysteresis-assisted operation extension range overlaps, but they are not exactly same. There are simulation errors since the hysteresis behavior is subject to simulation sweep setup. The

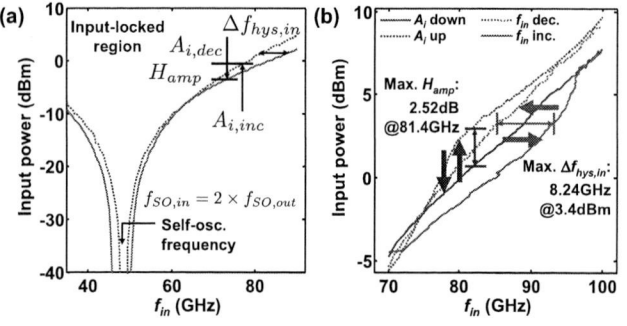

Fig. 10. CML static divider hysteresis simulation results. (a) Amplitude sweep SC and hysteresis over wide input frequency range. (b) A zoomed frequency and amplitude window was repeated with more detailed sweeps in frequency and amplitude. Two hysteresis extension ranges overlap, but they are not exactly same.

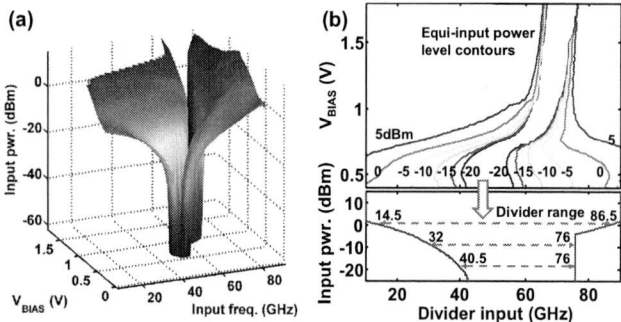

Fig. 11. CML static divider SC tuning with DC bias. (a) 3-D plot with V_{BIAS} tuning. (b) Effective dividable frequency with minimum input power at each frequency.

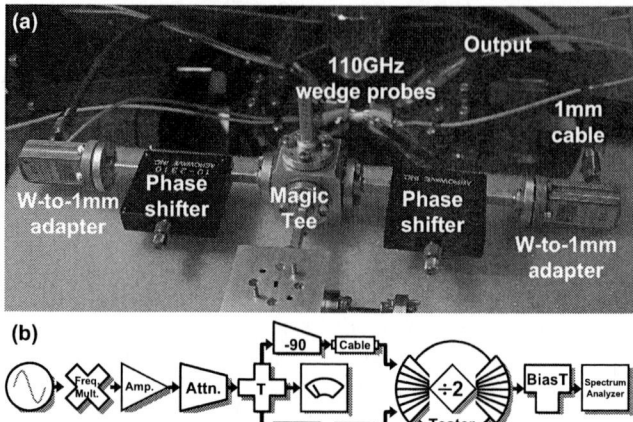

Fig. 12. (a) Divider test setup photo. The W-band waveguide limits below-65GHz low-frequency range access. (b) Test setup diagram. The input signal power is calibrated with a power meter at the magic tee input. Signal loss in the W-band connection makes it difficult to estimate the actual power that reaches the circuit input. Phase shifter needs to be adjusted at each frequency for better divider performance.

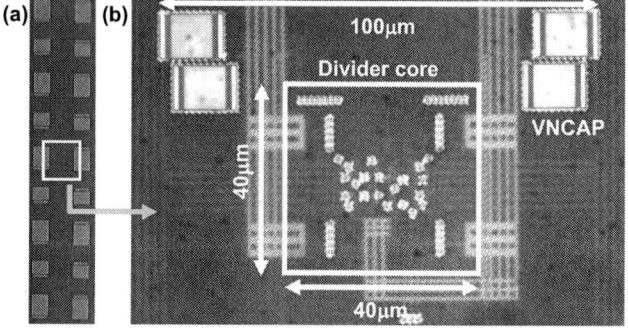

Fig. 13. Divider chip die photos taken at (a) top metal (M10) and (b) fourth metal (M4) levels. Two 150-μm pitch wedge probe padsets are used for RF and DC connections. Small dots are inter-metal vias, and bright squares are BEOL VNCAPs for AC coupling and power supply de-couplings.

frequency sweep gain is more important since a VCO makes frequency-direction sweeps for PLL locking.

Using separate DC tail control, the divider can adjust the f_{SO} and bandwidth. While the operation bandwidth decreases as V_{BIAS} increases, still it is a useful technique to broaden the effective operation range. The V_{BIAS} sweep with $A_{i,dn}$ SC is in Fig. 11. It shows the divider SC in 3-D. By obtaining the minimum input power level that the divider operates at each frequency across V_{BIAS} conditions, an effective operation range is obtained in Fig. 11(b). With V_{BIAS} tuning, the divider can accept low-level power over wide range. For example, it operates from 32 to 76GHz, or a 44GHz bandwidth at -10dBm power level.

IV. MEASUREMENTS

The divider measurements, variation, and statistical scalable characterization are discussed. The test setup is introduced in IV-A. The divider performances and FoM are compared in IV-B with hysteresis and bias tuning. Divider performance variation, benchmark, and statistical scalable divider characterization are discussed in IV-C.

A. Test Setup

A typical test setup photograph and a diagram are in Fig. 12. The difficulty of mmWave CML divider test originates

from high-frequency input, setup signal loss, and differential input. A 6× frequency multiplier is used to generate W-band frequency, from a 50GHz signal source. It is leveled greater than 10dBm by a W-band amplifier. The signal is attenuated for divider experiment, and it is split into two paths with a magic tee. Phase shifters adjust signal phases to make differential inputs. After exiting waveguides, differential signals go through 1mm cables and 110GHz wedge probes. The input power is calibrated at the output of the attenuator, and it is not trivial to estimate the real power at the circuit input. The setup is not appropriate for frequency sweep SC, since the phase shifter introduces uncertainty.

B. Measurement

The divider was fabricated in 65nm SOI CMOS as shown in the die photograph in Fig. 13. The divider output frequency is captured and compared with the input signal through spectrum analyzer. To determine the divider IL and SO modes

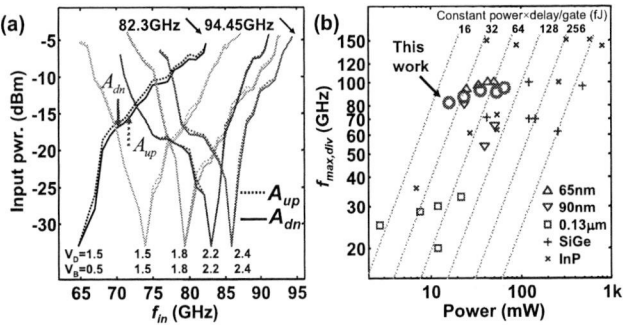

Fig. 14. CML static divider sensitivity curve measurements and FoM. (a) The solid line is a downward and the dotted line is an upward amplitude sweep sensitivity curves. An optimistic 3dB signal loss is assumed from the attenuator to the circuit input. The maximum hysteresis gain observed is about 0.74dB at V_{DD}=1.5 and V_{BIAS}=0.5V. (b) Power-delay product has been used as a divider FoM. The circuit power is normalized by the number of equivalent logic gates. CMOS dividers operate at V- and W-bands with minimal switching energy performances.

TABLE II
CML STATIC DIVIDER PERFORMANCE

V_{DD}	1.5	1.5	1.8	2.2	2.4
V_{BIAS}	0.5	1.5	1.8	2.2	2.4
Power (mW)	15.8	23.2	35.2	53.5	64.9
$f_{SO,out}$ (GHz)	32.3	37.0	39.7	41.5	43.0
Avg. H_{amp} (dB)	0.74	0.46	0.55	0.46	0.44
$f_{div,max}$ (GHz)	82.3	87.2	92.7	91.2	94.4

TABLE III
CML STATIC DIVIDER FoM

Ref.	Tech.	f_{max} (GHz)	$\frac{P}{8f_{max}}$ (pJ)	$|\partial A_i/\partial f|$ (mV/GHz)	$\Delta f/f_{SO}$ ratio
[31]	InP	143.6	78.3	79.8	1.392
[34]	SiGe	72.8	94.4	91.7	1.841
[19]	90nm	66.0	96.6	83.3	1.271
[18]	65nm	100.2	63.2	282.0	0.256
This work	65nm	82.3	24.0	85.0	0.545
	65nm	94.4	85.9	282.6	0.215

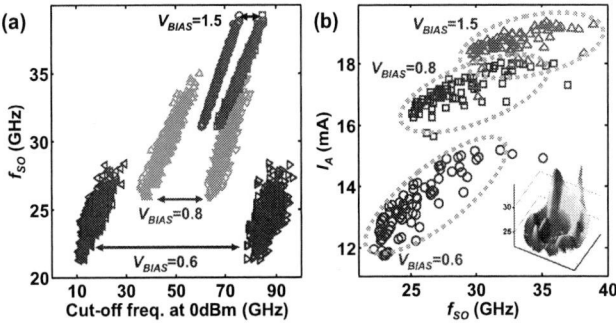

Fig. 15. Divider statistical simulation and measurement. (a) Simulation of divider cut-off frequency at 0dBm with different V_{BIAS}. Total 500 Monte Carlo points are used. The f_{in} increase and decrease sweeps are simulated, and the output is analyzed to obtain at which frequency a divider switches to SO from IL mode. There are strong correlations between f_{SO} and cut-off frequencies. (b) Divider active current I_A and f_{SO} at several V_{BIAS} settings. They show good correlations. The data were obtained from 73 chip sites in a 300mm 65nm SOI wafer, whose mappping is at the bottom right.

accurately, the phase noise of input and output should be compared. In practice, the spectral purity of the divider is observed. Three upward and downward amplitude sweeps are measured and averaged at each frequency, and the upward and downward SCs are plotted in Fig. 14. Several different V_{DD} and V_{BIAS} combinations were used to obtain the plot, and their performances are arranged in Table II. In case of V_{DD}=1.5V and V_{BIAS}=0.5V, the average hysteresis gain H_{amp} between upward and downward SCs is about 0.74dB.

The divider performance comparison has been done with the maximum divider operation frequency $f_{max,div}$, divider bandwidth, and the power-delay product [34] in Fig. 14(b). As described in the mmWave test setup, the actual power at the circuit input is not readily available, and it is obtained from test equipment signal loss estimations. The input power level in each paper is different depending on the setup and calibration. Still they are considered as best-effort measurements. Table III shows FoM comparison with f_{max}, power-delay product [34], SC slope (mV/GHz), and bandwidth over f_{SO} ratio [28]. When a divider has wide operation range, the SC slope $|\partial A_i/\partial f|$ tends to have lower value, and the bandwidth to f_{SO} ratio $\Delta f/f_{SO}$ becomes larger. While the table provides perspectives on the design, a new FoM that considers the setup power level uncertainty should be developed.

C. Scalable Statistical Characterization

The sub-100nm CMOS variation is one of the motivations for wideband CML static divider as discussed in II-A and Fig. 3. It is necessary to measure divider variation to estimate yield, while just one SC measurement is time consuming, and its automation is difficult due to the nonlinearity. The (6) states that a SC is defined when we have circuit design and DC parameters, and f_{SO} as a RF parameter. Therefore, the divider SC performance is reliably estimated with DC measurements and f_{SO}, assuming a good model-to-hardware correlation (MHC). With this estimation method, the statistical characterization of a divider becomes scalable.

The estimation begins with Monte Carlo simulations on the divider bandwidth as plotted in Fig. 15(a) and DC parameters, such as active and quiescent currents I_A and I_Q. The plot suggests that the f_{SO} reflects the frequency direction cut-off frequency well. Also I_A and I_Q are useful to enhance the estimation accuracy. An estimator is trained with the simulated cut-off frequency, I_A, I_Q, and nominal SC. Then hardware results, such as I_A and f_{SO} data in Fig. 15(b), are used to scale and to offset the estimator for MHC. The divider SC estimation from scalable DC and f_{SO} measurements is verified by sampling divider sensitivity thresholds at 65 and 70GHz with 73 dividers in a 300mm wafer as shown in Fig. 16(a).

978-1-4244-2018-6/08 $25.00 © 2008 IEEE

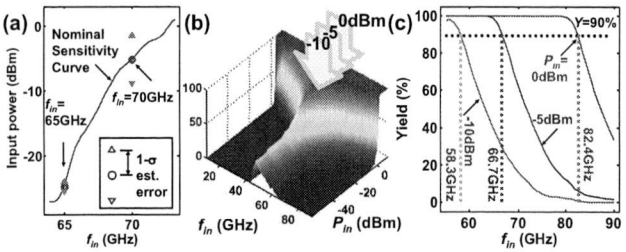

Fig. 16. Divider statistical measurement and estimation. (a) Divider estimation errors at 65 and 70GHz. Overall RMS error is 55% of standard deviation. (b) Divider yield calculation for input power and frequency. Yield is obtained from the scalable statistical measurements. (c) Yield plot slice at 0, -5, -10dBm inputs. The divider operates up to 82.4GHz at 0dBm with 90% yield.

Whole SC curve measurements on all chip sites are too time consuming, and do not add much information. The estimation RMS error is about -2.2dB of 1-σ standard deviation. Using the estimator, the dividers statistical yield for input power and frequency is calculated in Fig. 16(b). For example, the divider will have 90% yield for a 0dBm input at 82.4GHz, as plotted in Fig. 16(c).

V. CONCLUSION

The wideband CML static divider design and measurements in 65nm SOI CMOS were presented. Circuit analysis provided design equations, evaluation metrics, and scalable statistical method. The wideband divider overcomes the process variability in sub-100nm CMOS technology, and therefore useful as a prescaling frequency divider in mmWave PLL front-end. The scalable statistical measurements proved that divider's wideband capability enables high-yield operation.

ACKNOWLEDGMENT

The authors appreciate the support of K. Rim, S. Stiffler, P. Gilbert, and G. Patton at IBM SRDC, S. Reynolds and B. Floyd at IBM T. J. Watson Research, and D. Lim at Massachusetts Institute of Technology.

REFERENCES

[1] S. K. Reynolds, B. Floyd *et al.*, "A silicon 60-GHz receiver and transmitter chipset for broadband communications," *IEEE J. Solid-State Circuits*, vol. 41, no. 12, pp. 2820–2831.

[2] B. Razavi, "A millimeter-wave CMOS heterodyne receiver with on-chip LO and divider," *IEEE J. Solid-State Circuits*, vol. 43, no. 2, pp. 477–485.

[3] O. Takahashi, C. Adams *et al.*, "Migration of Cell Broadband Engine from 65nm SOI to 45nm SOI," in *ISSCC Dig. of Tech. Papers*, 2008, pp. 86–87, 597.

[4] S. Lee, B. Jagannathan *et al.*, "Record RF performance of 45-nm SOI CMOS technology," in *IEDM Tech. Dig.*, 2007, pp. 255–258.

[5] W. Sansen, "Analog design challenges in nanometer cmos technologies," in *IEEE ASSCC*, 2007, pp. 5–9.

[6] J. Rabaey, F. De Bernardinis *et al.*, "Embedding mixed-signal design in systems-on-chip," *Proc. IEEE*, vol. 94, no. 6, pp. 1070–1088.

[7] C. Cho, D. D. Kim *et al.*, "Decomposition and analysis of process variability using constrained principal component analysis," *IEEE Trans. Semicond. Manuf.*, vol. 21, no. 1, pp. 55–62.

[8] K. Agarwal and S. Nassif, "The impact of random device variation on SRAM cell stability in sub-90-nm CMOS technologies," *IEEE Trans. VLSI Syst.*, vol. 16, no. 1, pp. 86–97.

[9] D. Kim, J. Kim *et al.*, "A 75GHz PLL front-end integration in 65nm SOI CMOS technology," in *Symp. on VLSI Circuits Dig. of Tech. Papers*, 2007, pp. 174–175.

[10] B. Razavi, "Gadgets gab at 60 GHz," *IEEE Spectr.*, vol. 45, no. 2, pp. 46–49, 56–58.

[11] S. Pinel, S. Sarkar *et al.*, "A 90nm CMOS 60GHz radio," in *ISSCC Dig. of Tech. Papers*, 2008, pp. 130–131.

[12] M. Tanomura, Y. Hamada *et al.*, "TX and RX front-ends for 60GHz band in 90nm standard bulk CMOS," in *ISSCC Dig. of Tech. Papers*, 2008, pp. 558–559.

[13] K.-H. Tsai, L.-C. Cho *et al.*, "3.5mW W-band frequency divider with wide locking range in 90nm CMOS technology," in *ISSCC Dig. of Tech. Papers*, 2008, pp. 466–467, 628.

[14] P. Mayr, C. Weyers, and U. Langmann, "A 90GHz 65nm CMOS injection-locked frequency divider," in *ISSCC Dig. of Tech. Papers*, 2007, pp. 198–199, 596.

[15] B. Razavi, "Heterodyne phase locking: a technique for high-frequency division," in *ISSCC Dig. of Tech. Papers*, 2007, pp. 428–249.

[16] K. Yamamoto and M. Fujishima, "70GHz CMOS harmonic injection-locked divider," in *ISSCC Dig. of Tech. Papers*, 2006, pp. 472–481.

[17] D. D. Kim, J. Kim, and C. Cho, "A 94GHz locking hysteresis-assisted and tunable CML static divider in 65nm SOI CMOS," in *ISSCC Dig. of Tech. Papers*, 2008, pp. 461–462, 628.

[18] D. Lim, J. Kim *et al.*, "Performance variability of a 90GHz static CML frequency divider in 65nm SOI CMOS," in *ISSCC Dig. of Tech. Papers*, 2007, pp. 542–621.

[19] J.-O. Plouchart, J. Kim *et al.*, "Performance variations of a 66GHz static CML divider in 90nm CMOS," in *ISSCC Dig. of Tech. Papers*, 2006, pp. 2142–2151.

[20] D. Kim, C. Cho, and J. Kim, "Scalable statistical measurement and estimation of a mmWave CML static divider sensitivity in 65nm SOI CMOS," in *RFIC Symp.*, 2008.

[21] M. Chang, "Foundry future: Challenges in the 21st century," in *ISSCC Dig. of Tech. Papers*, 2007, pp. 18–23, 597.

[22] S. Lee, J. Kim *et al.*, "SOI CMOS technology with 360GHz f_t NFET, 260GHz f_t PFET, and record circuit performance for millimeter-wave digital and analog system-on-chip applications," in *Symp. on VLSI Technology Dig. of Tech. Papers*, 2007, pp. 54–55.

[23] D. Kim, J. Kim *et al.*, "Manufacturable parasitic-aware circuit-level FETs in 65-nm SOI CMOS," *IEEE Electron Device Lett.*, vol. 28, no. 6, pp. 520–522.

[24] S. Adee, "Winner: Semiconductors the ultimate dielectrics is ... nothing," *IEEE Spectr.*, vol. 45, no. 1, pp. 39–42.

[25] R. Aparicio and A. Hajimiri, "Capacity limits and matching properties of integrated capacitors," *IEEE J. Solid-State Circuits*, vol. 37, no. 3, pp. 384–393.

[26] D. Kim, J. Kim *et al.*, "Symmetric vertical parallel plate capacitors for on-chip RF circuits in 65-nm SOI technology," *IEEE Electron Device Lett.*, vol. 28, no. 7, pp. 616–618.

[27] J.-O. Plouchart, N. Zamdmer *et al.*, "A 243-GHz f_t and 208-GHz f_{max}, 90-nm SOI CMOS SoC technology with low-power mm-wave digital and RF circuit capability," *IEEE Trans. Electron Devices*, vol. 52, no. 7, pp. 1370–1375.

[28] U. Singh and M. Green, "High-frequency CML clock dividers in 0.13-μm CMOS operating up to 38 GHz," *IEEE J. Solid-State Circuits*, vol. 40, no. 8, pp. 1658–1661.

[29] J.-O. Plouchart, J. Kim *et al.*, "A 0.123 mW 7.25 GHz static frequency divider by 8 in a 120-nm SOI technology," in *Proc. ISLPED*, 2003, pp. 440–442.

[30] J.-S. Rieh, B. Jagannathan *et al.*, "SiGe heterojunction bipolar transistors and circuits toward terahertz communication applications," *IEEE Trans. Microw. Theory Tech.*, vol. 52, no. 10, pp. 39–42.

[31] D. Hitko, T. Hussain *et al.*, "A low power (45mW/latch) static 150GHz CML divider," in *IEEE CSICS*, 2004, pp. 167–170.

[32] A. Hajimiri and T. H. Lee, "A general theory of phase noise in electrical oscillators," *IEEE J. Solid-State Circuits*, vol. 33, no. 2, pp. 179–194.

[33] J.-O. Plouchart, D. Kim *et al.*, "A 1.2V 15.6mW 81GHz 2:1 static CML frequency divider with a band-pass load in a 90nm SOI CMOS technology," in *IEEE CSICS*, 2007.

[34] M. Sokolich, C. Fields *et al.*, "A low-power 72.8-GHz static frequency divider in AlInAs/InGaAs HBT technology," *IEEE J. Solid-State Circuits*, vol. 36, no. 9, pp. 1328–1334.

978-1-4244-2018-6/08 $25.00 © 2008 IEEE

IEEE 2008 Custom Intergrated Circuits Conference (CICC)

A 32/16 Gb/s 4/2-PAM Transmitter with PWM Pre-Emphasis and 1.2 Vpp per side Output Swing in 0.13-μm CMOS

Horace Cheng, and Anthony Chan Carusone Member, IEEE

Department of Electrical and Computer Engineering, University of Toronto [*]

Abstract– **A dual-mode 4/2-PAM transmitter is described that extends pulse-width modulated pre-emphasis to data rates of 16 Gb/s and 32 Gb/s in 2-PAM and 4-PAM modes respectively. Implemented in a 0.13-μm CMOS process to accommodate the wide output swing of 1.2 Vpp per side, the transmitter compensates for 30 dB and 9 dB of loss at one-half the symbol rate in 2-PAM and 4-PAM modes respectively.**

I. INTRODUCTION

This paper describes a CMOS transmitter for high-loss wireline channels at data rates exceeding 10-Gb/s. In general, these losses may be combated by equalization and, in severe cases, through the use of 4-level Pulse Amplitude Modulation (4-PAM). Ultimately, large output swing is also required for high-loss channels. This transmitter combines wide output swing (1.2 Vpp per side) with large loss compensation (up to 30-dB at one-half the symbol rate) and the option to operate in 4-PAM mode at 16 GSymbols/s. The state-of-the-art in transmit equalization for multi-Gb/s electrical wireline links is described in [1] where a Pulse-Width Modulation Pre-Emphasis (PWM-PE) technique is used to achieve low-BER operation at 5 Gb/s over a cable with 33-dB loss at 2.5 GHz. However, with an output swing of 600 mVpp only approximately 15 mV of eye opening remained at the receiver-end of the cable. In an attempt to increase the eye opening, the output swing was increased to 1.8 Vpp per side in [2] but the performance at 5 Gb/s suffered considerably. In this paper, a high-speed Current Mode Logic (CML) implementation extends PWM-PE to 16-Gb/s for binary (2-PAM) signals with an output swing of 1.2 Vpp per side providing approximately 30 mV of eye opening after a cable with over 30 dB of loss at one-half the symbol rate.

Furthermore, in this work the PWM-PE technique is applied for the first time to 4-PAM signals. 4-PAM transmitters require particularly accurate Inter-Symbol Interference (ISI)-cancelation due to their reduced level-spacing. To ensure ample eye opening at the receiver, 4-PAM transmitters generally require many Finite Impulse Response (FIR) pre-emphasis taps. For example, 4 taps of pre-emphasis with 500-mVpp swing per side were used in a 4-PAM 25-Gb/s transmitter to compensate for approximately 3 dB of loss at one-half the symbol rate [3]. To compensate for 14.5 dB of loss in a 4-PAM 24-Gb/s transmitter, the pre-emphasis filter length was increased to 13 taps and the output swing increased to 800 mVpp per side [4]. All of

these tap weights must be accurately optimized to ensure the received 4-PAM eye patterns are open. In this work, by adjusting only one pre-emphasis parameter, up to 9 dB of loss is compensated at one-half the symbol rate for a 32-Gb/s 4-PAM link.

II. PRE-EMPHASIS

In order to reduce ISI at the receiver input, the overall frequency response of the transmitter and channel should be as flat as possible up to at least one-half of the symbol rate. Pre-emphasis at the transmitter generally attenuates the low-frequency content of the signal, in order to flatten the overall response.

Pre-emphasis based on Pulse-Width Modulation (PWM), as proposed in [2], has the following pulse response:

$$
p_{\text{PWM}}(t) = \begin{cases} 0 & t < 0 \\ 1 & 0 \leq t < d \cdot T_s \\ -1 & d \cdot T_s \leq t < T_s \\ 0 & t \geq T_s \end{cases}
$$

The PWM-PE pulse response spans only one unit interval and its low-frequency attenuation is controlled by varying its duty cycle, the parameter d ($0.5 \leq d \leq 1$).

Similarly, a 2-tap Finite Impulse Response Pre-Emphasis (FIR-PE) transmitter has only one adjustable parameter (assuming the peak output level is constrained). Its output pulse response is,

$$
p_{\text{FIR}}(t) = \begin{cases} 0 & t < 0 \\ r & 0 \leq t < T_s \\ r - 1 & T_s \leq t < 2 \cdot T_s \\ 0 & t \geq 2 \cdot T_s \end{cases}
$$

The pulse response spans two Unit Interval (UI) with the amount of pre-emphasis controlled by the parameter r ($0.5 \leq r \leq 1$). However, PWM-PE provides a better match to the inverse of the channel response for electrical wireline links dominated by skin effect losses.

A channel comprising a total of 36 meters of coaxial cable and six connectors was characterized using a Vector Network Analyzer (VNA). The measured channel response is shown as triangular markers in Fig. 1. A model of the cable including skin-effect and dielectric losses was fitted to the measurements and is plotted as the dashed line in Fig. 1. Both the 2-tap FIR and PWM pulses were optimized and the resulting combined pre-emphasis pulse and channel responses are also plotted in Fig.

---

[*]This work was supported by Gennum Corp. and NSERC.

978-1-4244-2018-6/08 $25.00 © 2008 IEEE

1. The overall frequency response of the PWM-PE matches the inverse of the channel loss curve more closely, resulting in an overall flatter response and subsequently less ISI. However, note that the flattened response has an attenuation of roughly 30 dB from dc to 10 GHz. Hence, the transmitter requires a wide output swing of 1.2 Vpp to ensure reasonable eye amplitude. Fig. 2 shows the simulated eye diagram at the receiver for a PWM-PE transmitter with a duty-cycle of $d = 0.52$ and swing of 1.2 Vpp per side.

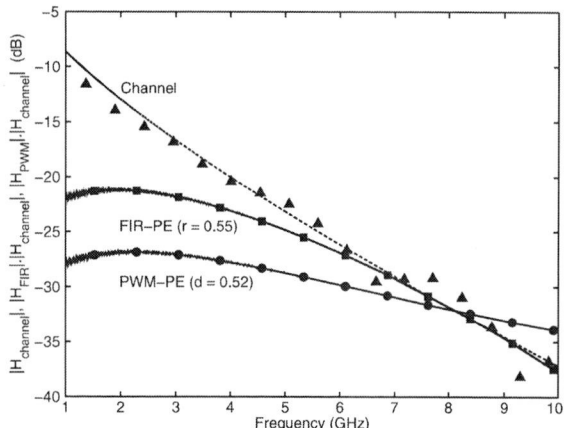

Fig. 1. Channel characterization and combined pre-emphasis pulses with channel frequency responses.

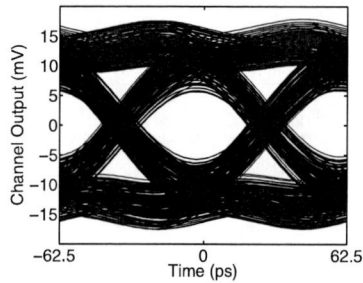

Fig. 2. System-level simulation eye diagram of channel output with PWM-PE (52% duty-cycle) and transmitter output 1.2 Vpp per side.

Although previously only applied to binary signals, PWM-PE is a linear operation and, hence, can be applied to a 4-PAM system. The corresponding 4-PAM symbol pulse shapes are shown in Fig. 3, assuming a Gray line code.

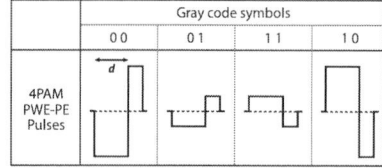

Fig. 3. Output pulses of the Gray-coded 4-PAM transmitter with PWM-PE.

III. SYSTEM-LEVEL DESIGN

A 0.13-μm CMOS process was chosen because its drain-source breakdown voltage of 1.6 V, can accommodate the 1.2

Vpp per side output swing. However, achieving the required speed in 0.13-μm CMOS was a significant challenge.

A. System Architecture

Fig. 4 illustrates a functional block diagram of the transmitter. When operating in 4-PAM mode at the maximum data rate of 32Gb/s, two single-ended data inputs at 16 Gb/s are translated into a three-bit thermometer code through a Gray encoder. The single-ended clock input at 16 GHz passes through several buffers and the Duty Cycle Control (DCC) circuit to generate a differential PWM clock. The PWM clock and the three encoded data streams combine at XOR gates to create three equally-weighted binary PWM data streams. These are combined at the output stage to create a 4-PAM output signal at 16 GSymbol/s. When operating in 2-PAM mode, the input Least Significant Bit (LSB) is connected to ground and a full-rate binary signal is applied to the Most Significant Bit (MSB) input. The Gray coding ensures a full-swing output. All logic was implemented with CML.

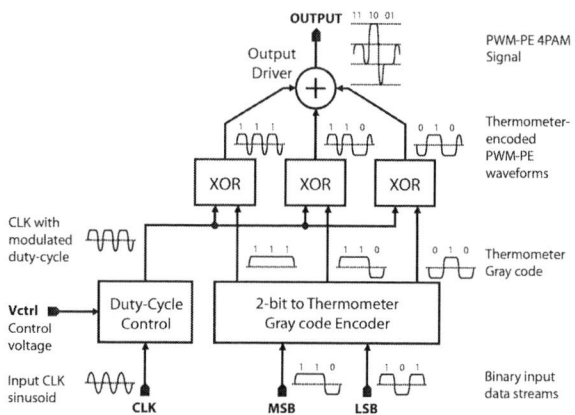

Fig. 4. Functional block diagram of transmitter system architecture.

IV. CIRCUIT DESIGN

A. Current Mode Logic (CML) Design

A good understanding of CMOS CML design was essential to achieving the required speed in 0.13-μm CMOS. It was found that by biasing the differential pair transistors with high current densities and relaxing the full-switching criterion usually enforced in CML design, higher operating speed can be achieved. Specifically, approximately 5% to 10% of the total tail current continues to flow through the "off" branch of all CML buffers, providing approximately a 48% increase in the logic's maximum switching speed. In addition, very low fanout had to be maintained to reduce the capacitive load on each CML stage. Moreover, in CML stages where high self-loading exists, such as the DCC circuit and XOR gates, inverse scaling (i.e. a fanout less than unity) is used [5]. All CML gates employ inductive peaking and operate from a 1.8-V supply.

B. Duty Cycle Control (DCC)

The DCC schematic is shown in Fig. 5. The differential offset inputs, $V_{offset+}$ and $V_{offset-}$, are derived from a single control voltage (not shown). The resulting dc offset current adjusts the duty-cycle of the differential clock. The output duty-cycle ranges from 50% to 75%. PWM-PE can be turned off for low loss channels by terminating the single-ended clock input and full-switching of dc offset differential pair resulting in standard Non-Return-to-Zero (NRZ) 4/2-PAM transmitted symbols. Note the use of inverse scaling at the DCC output due to its high capacitive self-load. A fanout ratio of only 0.36 is required to maintain sufficient bandwidth there, followed by several fanout-of-1 high-gain stages to generate a clean PWM clock.

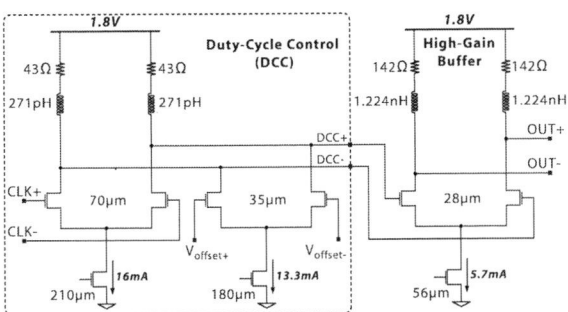

Fig. 5. Schematic of DCC stage.

C. Output Driver

A cascode topology is chosen for the output stage to provide wide swing. Fig. 6 illustrates the schematic. It operates from a separate 3-V power supply. The cascode devices shield the input differential pairs from excessive drain-source voltages and reduce the input Miller capacitance. Standard 0.13-μm NFETs with minimal gate lengths are used throughout. The output driver is designed to provide a return loss of better than -10 dB in either 75-Ω or 50-Ω environments. Bias-tees provide ac-coupling to test equipment while presenting the output with the expected dc load. It is possible to reduce the transmitter output swing for low loss channels to a minimum of 400 mVpp per side by adjusting the output stage tail currents.

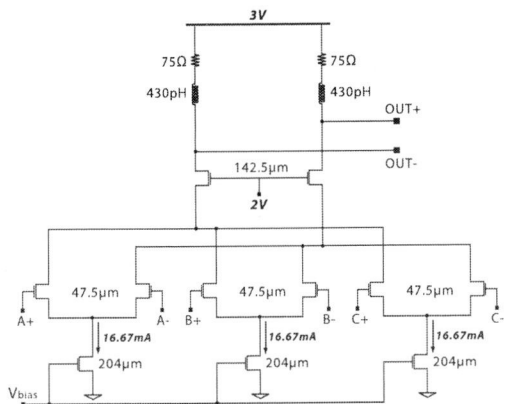

Fig. 6. Schematic of output driver.

V. EXPERIMENTAL RESULTS

The prototype transmitter integrates 165 spiral inductors and 80 CML gates into a die area of 1.71mm × 1.83mm in a 0.13-μm CMOS technology. Fig. 7 is a die photo of the transmitter. All measurements were performed on-die.

The transmitter draws a total power of 1.578 W from the two power supplies. The output driver consumes 147 mW of power

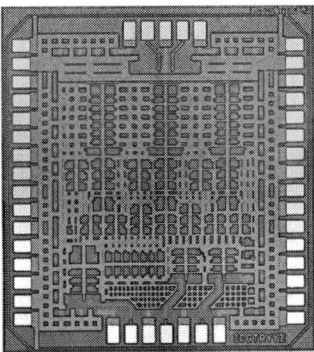

Fig. 7. Die photo of transmitter.

from 3-V power supply. The remaining 1.43 W is consumed from the 1.8-V supply, of which 35% is consumed by the clock distribution, 37% by the encoder, and the rest for the generation and buffering of the PWM-PE data. For applications that call for a lower output swing, the same architecture can be implemented in a more advanced CMOS technology (with lower drain-source breakdown voltages) permitting the use of fewer stages of higher fanout buffers and lower supply voltages, thus resulting in a large power savings.

To characterize the channel loss compensation in 2-PAM mode of operation, the same 36-meter coaxial cable channel (as shown in Fig. 1) is used. Fig. 8a illustrates the transmitter output with 53.2% duty-cycle. The corresponding channel output eye diagram is shown in Fig. 8b and has an eye amplitude of approximately 30 mV, which agrees with the simulated eye diagram shown in Fig. 2. After 2×10^7 measurements, the oscilloscope extrapolated a Bit-Error Rate (BER) of better than 10^{-12} with approximately 0.25 UI margin at the channel output.

In 4-PAM mode, its maximum data rate is 32 Gb/s. The test channel comprises six sections of 1-meter coaxial cables with associated connectors has a loss of 8.9 dB at half the symbol rate (8 GHz) and 17.2 dB at half the bit rate (16 GHz). By manually adjusting the duty-cycle of the transmit pulse, and thus the amount of pre-emphasis, it is possible to obtain an open eye at the output of the channel. Fig. 9a illustrates the transmitter 4-PAM output with 64% duty-cycle. The corresponding channel output eye diagram is shown in Fig. 9b. The 4-PAM eye opening is approximately 30 mV at the output of the channel.

VI. CONCLUSIONS

Table I compares this work with current state-of-the-art CMOS binary transmitters. It compares favorably in terms of loss compensation, output swing, and speed. A comparison with published CMOS 4-PAM transmitters is provided in Table II. This transmitter is the first of its kind to incorporate PWM-PE in 4-PAM in addition to being the fastest implementation in CMOS reported to date. The capability to switch between 4-PAM and 2-PAM, adjustable pre-emphasis (50%–75% duty-cycle, or NRZ), and adjustable output amplitude makes it suitable for use in a wide range of electrical wireline links.

REFERENCES

[1] J.-R. Schrader, E. A. M. Klumperink, J. L. Visschers, and B. Nauta, "Pulse-width modulation pre-emphasis applied in a wireline transmitter,

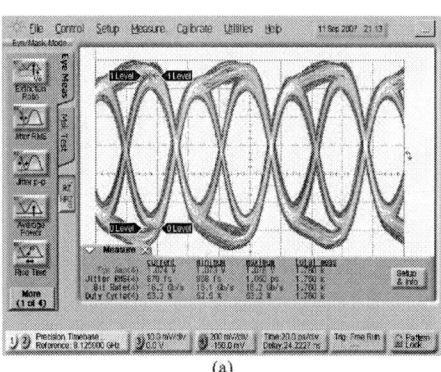

(a)

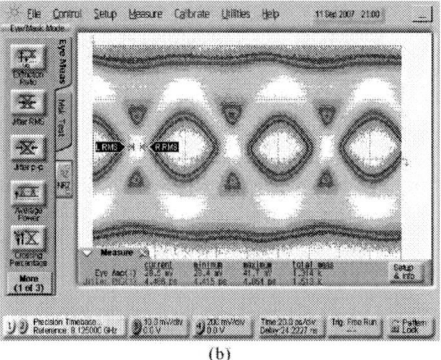

(b)

Fig. 8. Eye diagrams of 2-PAM mode pre-emphasis experiment: (a) Transmitter output; 53.2% duty-cycle and (b) Channel output.

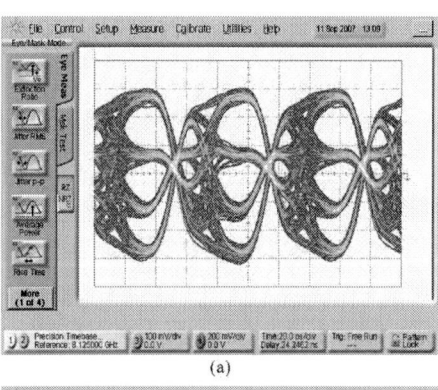

(a)

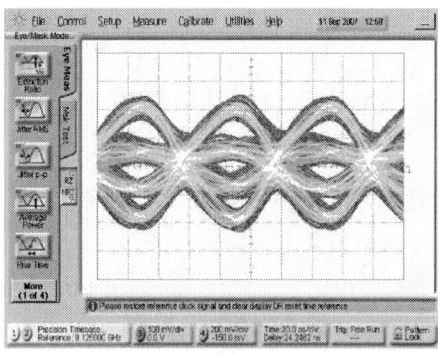

(b)

Fig. 9. Eye diagrams of 4-PAM mode pre-emphasis experiment: (a) Transmitter output; 64% duty-cycle and (b) Channel output.

TABLE I
COMPARSION OF 2-PAM TRANSMITTERS

	CMOS Process	Output Swing (mV)	Data Rate (Gb/s)	Loss Compensation (dB)	Power (mW)
[6]	0.13-μm	260	40	-	~2700
[7]	0.13-μm	350	30	-	150
[8]	0.18-μm	4000	13.6	-	600
[1]	0.13-μm	600	5	31 (@ 2.5GHz)	110
[2]	90-nm	700	4	22 (@ 1.25GHz)	-
This work	0.13-μm	1250	16	30.3 (@ 8 GHz)	1578

TABLE II
COMPARSION OF 4-PAM TRANSMITTERS

	CMOS Process	Output Swing (mV)	Data Rate (Gb/s)	Loss Compensation (dB)	Power (mW)
[3]	90-nm SOI	520	25	3 (@ 6.25GHz)	101.8
[4]	90-nm	800	24	14.5 (@ 6GHz)	510
[9]	0.18-μm	600	10	-	120
[10]	0.25-μm	600	10	3.7 (@ 2.5GHz)	222
[11]	0.4-μm	1100	10	-	1000
This work	0.13-μm	1250	32	8.9 (@ 8 GHz)	1578

achieving 33dB loss compensation at 5-Gb/s in 0.13-μm CMOS," *IEEE Journal of Solid-State Circuits*, vol. 41, pp. 990–999, April 2006.

[2] J. Schrader, E. Klumperink, and B. Nauta, "Wireline equalization using pulse-width modulation," *Proceedings of 2006 Custom Integrated Circuits Conference (CICC)*, pp. 591–598, 2006.

[3] C. Menolfi, T. Toifl, R. Reutemann, M. Ruegg, P. Buchmann, M. Kossel, T. Morf, and M. Schmatz, "A 25Gb/s PAM4 transmitter in 90nm CMOS SOI," *Proceedings of the 2005 International Solid State Circuits Conference (ISSCC)*, pp. 77–78, 2005.

[4] A. Amirkhany, A. Abbasfar, J. Savoj, M. Jeeradit, B. Garlepp, V. Stojanovic, and M. Horowitz, "A 24Gb/s software programmable multi-channel transmitter," *VLSI Circuits, 2007. Digest of Technical Papers. 2007 Symposium on*, pp. 38–39, 2007.

[5] E. Sackinger and W. Fischer, "A 3-GHz 32-dB CMOS limiting amplifier for SONET OC-48 receivers," *IEEE Journal of Solid-State Circuits*, vol. 35, pp. 1884–1888, December 2000.

[6] J. Kim, J.-K. Kim, B.-J. Lee, M.-S. Hwang, H.-R. Lee, S.-H. Lee, N. Kim, D.-K. Jeong, and W. Kim, "Circuit techniques for a 40-Gb/s transmitter in 0.13-μm CMOS," *Proceedings of the 2005 International Solid State Circuits Conference (ISSCC)*, pp. 150–151, 2005.

[7] P. Westergaard, T. O. Dickson, and S. P. Voinigescu, "A 1.5-V, 20/30-Gb/s CMOS backplane driver with digital pre-emphasis," *Proceedings of 2004 Custom Integrated Circuits Conference (CICC)*, pp. 23–26, 2004.

[8] D. Li and C. Tsai, "10-13.6 Gbit/s 0.18μm CMOS modulator drivers with 8 Vpp differential output swing," *Electronics Letters*, vol. 41, no. 11, pp. 643–644, 2005.

[9] K. Farzan and D. A. Johns, "A CMOS 10-Gb/s power-efficient 4-PAM transmitter," *IEEE Journal of Solid-State Circuits*, vol. 39, pp. 529–532, March 2004.

[10] C. LIN and C. TSAI, "Multi-gigabit pre-emphasis design and analysis for serial link," *IEICE Transactions on Electronics*, vol. E88-C, no. 10, pp. 2009–2019, 2005.

[11] R. Farjad-Rad, C.-K. K. Yang, M. A. Horowitz, and T. H. Lee, "A 0.4-μm CMOS 10-Gb/s 4-PAM pre-emphasis serial link transmitter," *IEEE Journal of Solid-State Circuits*, vol. 34, pp. 580–585, May 1999.

IEEE 2008 Custom Intergrated Circuits Conference (CICC)

A 5-Gb/s/pin Transceiver for DDR Memory Interface with a Crosstalk Suppression Scheme

Kwang-Il Oh[1], Lee-Sup Kim[1], Kwang-Il Park[2], Young-Hyun Jun[2], Kinam Kim[2]

[1]Dept. of EECS, KAIST, 373-1 Guseong-dong, Yuseong-gu, Daejeon, Republic of Korea

[2]Samsung Electronics, Gyeonggi-Do, Republic of Korea

Abstract-A 5-Gb/s/pin transceiver for DDR memory interface is proposed with a crosstalk suppression scheme. The proposed transceiver implements a staggered memory bus topology and a glitch canceller to suppress crosstalk-induced distortions in a memory channel. The transceiver is implemented using 0.18μm CMOS process and operates at 5-Gb/s. The results demonstrate widened eye diagram and lower bit error rate. The eye width and height of the proposed scheme increases 28.3% and 11.1% compared to the conventional memory transceiver, respectively. The p-p jitter of output data is 52.82-ps.

I. INTRODUCTION

High-bandwidth DRAMs have recently been important components for 3-D graphics applications such as graphic processing unit and game consoles. In order to maximize memory I/O bandwidth, a parallel off-chip bus topology is generally used in a memory interface. Due to the continuous increment of data rate, the increased number of channels made routings on PCB more complex and dense in parallel off-chip bus topology. As a result, channels have become more susceptible to the coupling by neighboring channels. The coupling of energy from one channel to another which is called crosstalk causes the serious impairment of data. Hence, the crosstalk impedes high speed data transmission above 5-Gb/s in a memory interface.

The crosstalk occurs whenever the electromagnetic fields from different structures such as channels interact. This phenomena cause crosstalk-induced timing distortion since the propagation time of a transmission line varies depending on the transition of adjacent channel [1]. Due to high data rate and small channel space, the distortion induced by crosstalk becomes so significant that it needs to be compensated for signal integrity.

Several attempts to eliminate near-end crosstalk noise using FIR filters have been implemented in [2], [3]. However, compensation schemes are limited to the crosstalk amplitude equalization. To cancel out crosstalk-induced timing distortion, an equalizer block that adjusts the time delay of data transition according to the state of data has been suggested [1]. However, in this scheme, a designer has a burden to know the exact amount of compensation time which also can be changed by the process variation. Crosstalk in the memory interface is also addressed in [4]. Although it limits the number of maximum transition channels, it does not compensate for the distorted signals.

This paper presents a DDR memory transceiver that compensates for both crosstalk-induced timing distortion and amplitude distortion to eliminate far-end crosstalk noise in a memory channel. To eliminate crosstalk-induced timing

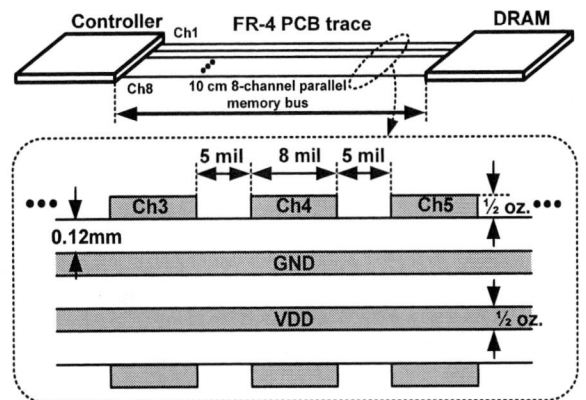

Fig. 1. Conventional memory interface configuration and cross sectional view of PCB traces.

distortions, we configure the memory bus as a staggered bus topology. Since large portion of the crosstalk is induced by the adjacent channel, the transition position of aggressing channel is moved to the middle of bit period of victim channel. As a result, the distortion is shifted from a time to an amplitude domain. The proposed glitch canceller removes crosstalk-induced amplitude distortion in the staggered bus. Based on the transition information of the adjacent channels at the transmitter, appropriate amplitude is added or subtracted to the victim channel.

An implementation of the DDR memory transceiver with a crosstalk suppression scheme is demonstrated that works to 5-Gb/s. We verify the operation of the proposed scheme by measuring the output waveform of the dense parallel memory channel. The improved performance is demonstrated in terms of eye-opening and bathtub curve results.

II. CROSSTALK IN MEMORY CHANNEL

Conventional parallel bus configurations and cross sectional views of PCB traces for the memory interface are shown in Fig. 1. In this configuration, high-speed signaling suffers from crosstalk-induced timing distortion because of close proximity between channels. In a conventional memory bus, each channel is operated by a single clock domain. If a data-transition occurs in a channel, timing deviation of the transition-edge is observed in the adjacent channel. Since the signal velocity of the transmission line depends on the signal patterns, the delay of data changes with the data transition [5]. The propagation velocity, *v*, of a signal in a transmission line with the presence of any magnetic or dielectric materials near the conductors is expressed as follows

978-1-4244-2018-6/08 $25.00 © 2008 IEEE

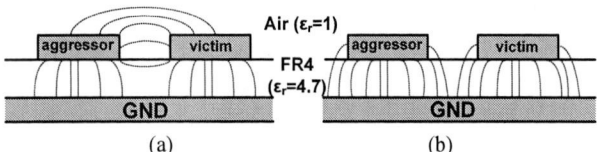

(a)   (b)

Fig. 2. Electric-field distribution around aggressor and victim line: (a) switching in the opposite direction, (b) switching in the same direction.

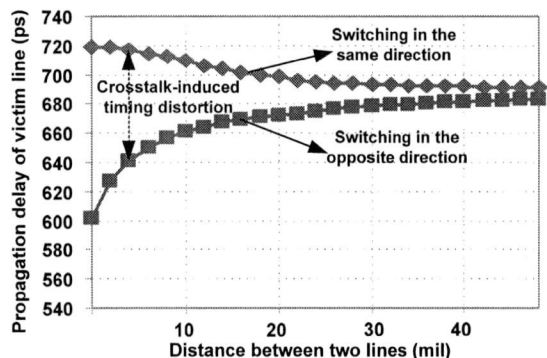

Fig. 3. Variation in propagation delay of victim line (10-cm).

$$v = \frac{c}{\sqrt{\varepsilon_r \mu_r}} \qquad (1)$$

where c is the velocity of light in vacuum, ε_r is a relative electric permittivity of the material surrounding the conductors, and μ_r is a relative magnetic permeability of the material surrounding the conductors. In PCB traces, nonmagnetic dielectric insulator is used; $\mu_r = 1$. In that case, (1) becomes

$$v = \frac{c}{\sqrt{\varepsilon_r}}. \qquad (2)$$

When the aggressor lines are switching in the opposite direction to the victim line, the effective electric permittivity (ε_{r_eff}) is reduced because many of electrical fields between victim and aggressor line are in air with low electric permittivity. When the aggressor lines are switching in the same direction as the victim line, ε_{r_eff} increases since many of electrical fields are in bulk material with high electric permittivity. This is illustrated in Fig. 2.

Based on (2) and deviations of effective electric permittivity, crosstalk-induced timing distortion (t_{dist}) is derived as follows

$$t_{dist} = \frac{s}{v_{opp}} - \frac{s}{v_{same}} = \frac{s}{c}\left(\sqrt{\varepsilon_{r\_opp}} - \sqrt{\varepsilon_{r\_same}}\right) \qquad (3)$$

where s is the length of the PCB microstrip line, ε_{r_opp} is a relative electric permittivity when two lines switch in the opposite directions, and ε_{r_same} is a relative electric permittivity when two lines switch in the same directions.

The impact of the crosstalk-induced timing distortion is severe when lines are very close like adjacent channels in memory interface. Fig. 3 shows the variation in propagation delay of victim line as the distance between lines is changed. The simulation is performed based on the structure shown in

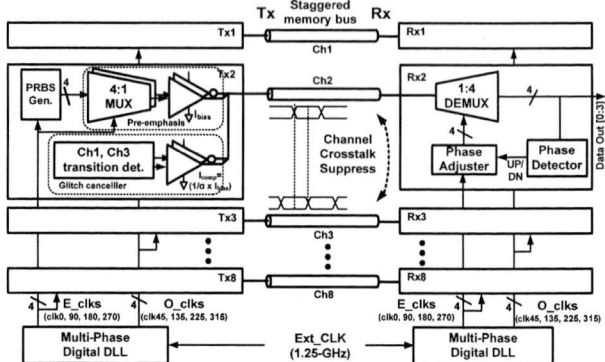

Fig. 4. Proposed memory transceiver block diagram with a crosstalk suppression scheme. A staggered memory bus and a glitch canceller are implemented.

Fig. 1. As the distance between lines increases, the coupling effect decreases significantly.

III. ARCHITECTURE

Fig. 4 shows the proposed memory transceiver block diagram with a crosstalk suppression scheme. An 8-channel 4-way interleaved scheme is employed. At the transmitter side, random data are generated from the PRBS generator and multiplexed into serialized data stream by a 4:1 MUX and a pre-emphasis circuit. At the receiver side, the transmitted data from transmitter is de-multiplexed into four 1.25-Gb/s data by a 1:4 DEMUX block. A multi-phase DLL is used to generate evenly-spaced 8-phase clocks (clk0, clk45, clk90, clk135, clk180, clk225, clk270, clk315). 8-phase clocks are divided into two groups as E_clks (clk0, clk90, clk180, clk270) and O_clks (clk45, clk135, clk225, clk315). The E_clks and O_clks from the multi-phase DLL are delivered to the even-channel transmitter and the odd-channel transmitter, respectively. Since the four clocks of each channel are evenly spaced as 90° phase difference, four data are serialized per a channel during a single period of the external clock.

IV. STAGGERED MEMORY BUS

To eliminate crosstalk-induced timing distortion on the memory channel, even channels and odd channels are synchronized by two different clock domains, respectively. Fig. 5 shows the concept of the proposed staggered memory bus. Each channel of the staggered memory bus is synchronized by E_clks or O_clks. A transmitter in the even-channel is synchronized by the E_clks and the odd-channel is synchronized by the O_clks. The receiver also uses the same clock strategy so that it recovers the received data based on the each clock domain. The delay difference of even channel and odd channel is $\tau/2$ which is half of the one bit period (τ). Since the transition point of the adjacent channel is shifted as $\tau/2$, there is no discrepancy of time delay due to the crosstalk. As shown in the Fig. 3, the crosstalk-induced timing distortion decreases significantly as the distance between lines increases. Therefore, a jitter induced by the inductive or capacitive coupling from the adjacent channel is minimized because the distance between two adjacent channels with the same transition is effectively doubled.

978-1-4244-2018-6/08 $25.00 © 2008 IEEE   680

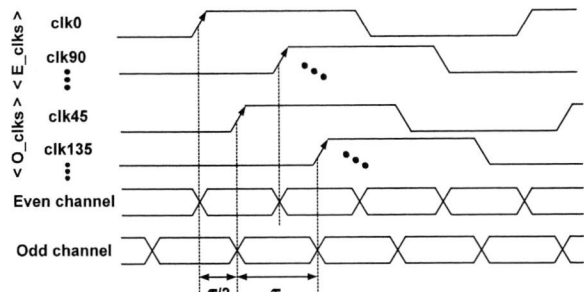

Fig. 5. Operation waveforms of the staggered memory bus.

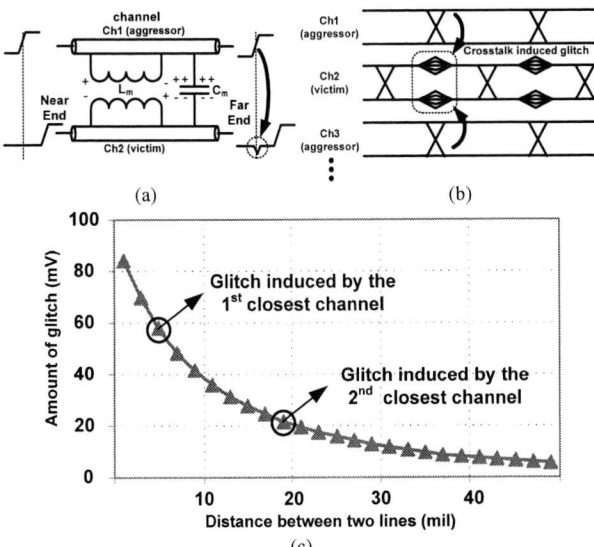

Fig. 6. (a) Opposite-direction voltage spike of the victim channel. (b) Crosstalk-induced glitch of staggered memory bus. (c) Amount of glitch with respect to the distance between two lines.

V. TRANSITION DETECTOR AND GLITCH CANCELLER

In the staggered memory bus, the crosstalk-induced glitch is observed since the transition edge of aggressing channel is moved to the middle of the bit period of victim channel. Fig. 6 shows the crosstalk-induced glitch. The opposite-direction voltage spike is induced at the victim channel due to inductive coupling. The amount of glitch induced by the second closest channel is about 30% of the amount of glitch by the first closest channel. To eliminate the crosstalk-induced glitch in the proposed staggered memory bus, a glitch canceller is proposed at the output driver of the transmitter. Since the amount of the induced glitch decreases as the distance of the two lines increases, the glitch canceller compensates for the glitch induced by transition on the 1st closest channels. Fig. 7 shows the schematic of the glitch canceller. The specific information about transition of adjacent channel, such as which channel acts as an aggressor and the direction of transition, is obtained from the transition detector as RISE, FALL. Current and previous data of the adjacent channels are fed to the transition detector to get transition information. In the case of Ch2, the glitches resulted from the transition of Ch1 or Ch3 should be compensated. If Ch1 or Ch3 falls, small pulse should be subtracted from Ch2 since the opposite

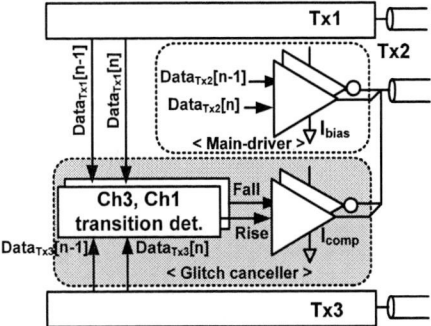

Fig. 7. Block diagram of glitch canceller.

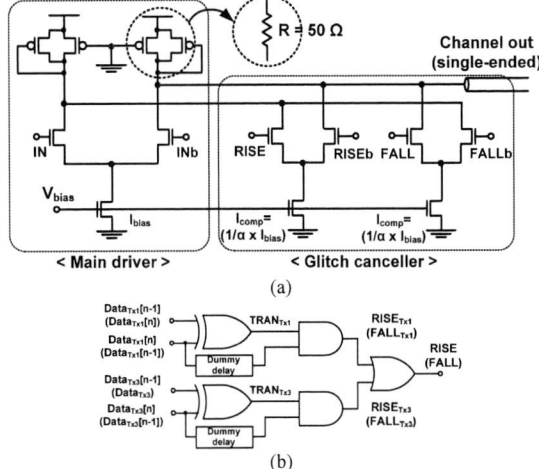

Fig. 8. (a) Schematic of the glitch canceller. (b) Transition detector.

voltage spike is generated on the victim channel. In the same way, if Ch1 or Ch3 rises, small pulse should be added to Ch2.

Fig. 8 (a) shows the glitch canceller. Fractional current ($I_{comp}= (1/\alpha)\times I_{bias}$, $\alpha=5$) is added to or subtracted from main driver according to the transition information from the transition detector. α is ratio of output swing to amount of glitch. The pre-emphasis scheme is also implemented but not shown in the figure. Fig. 8 (b) shows the single-ended version of the transition detector. After XOR of the previous and current data bit of the adjacent channels (TRAN), current data bit and TRAN are provided to AND gate. RISE signal is generated if one or both of two adjacent channels rise. Similarly, FALL signal is generated if one or both of two adjacent channels fall. If two adjacent channels transit in the opposite directions, no compensation is performed because both RISE and FALL are generated, and the compensation effect is cancelled out.

VI. MEASUREMENT RESULTS

The transceiver chip for DDR memory interface is implemented in a 0.18-μm CMOS process with 1.8V supply voltage. The sequence length of random data from PRBS generator is 2^{11}-1 and operates with 1.25-GHz external clock. An 8-channel 4-layer FR-4 test board is implemented for the measurement of far-end crosstalk of the channel. The cross sectional view of the coupled microstrip lines is the same as shown in Fig. 1. As the two adjacent lines are located at a

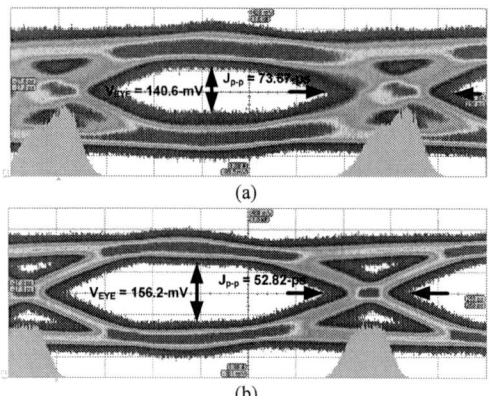

(a)

(b)

Fig. 9. Data eye-diagram (a) without and (b) with crosstalk suppression scheme.

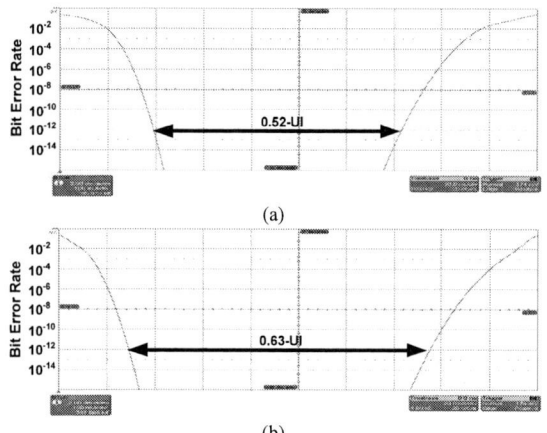

(a)

(b)

Fig. 10. Bathtub curve result (a) without and (b) with crosstalk suppression scheme.

distance of 5-mil, the coupling capacitance and the coupling inductance of two lines are 1.18-pF and 6.1-nH, respectively. The characteristic impedance of the PCB trace and the termination is designed as 50-Ω for minimizing effect of signal reflection.

Fig. 9 shows the measured eye diagram of the proposed memory transceiver. The output data waveform is measured through the 10-cm FR-4 PCB trace spaced 5-mil apart. The jitter histograms around a threshold from the data eye without and with crosstalk suppression scheme are also shown in Fig. 9. The measured p-p jitter (J_{p-p}) of output data is 73.67-ps and 52.82-ps without and with crosstalk suppression scheme, respectively. The height of eye (V_{EYE}) of conventional memory bus is measured as 140.6-mV due to the crosstalk-induced jitter caused by adjacent channel. The height of eye is widened as 156.2-mV with the staggered memory bus and the glitch canceller. 5-Gb/s data rate with BER less than 10^{-12} is measured through the bathtub curve as shown in Fig. 10. The bathtub curve measured without and with crosstalk suppression scheme shows an opening of 0.52-UI and 0.63-UI at BER of 10^{-12}, respectively. This indicates the substantial eye improvement is possible due to the proposed crosstalk suppression scheme. A summary of results is tabulated in Table I. The micrograph of the fabricated chip is shown in Fig.

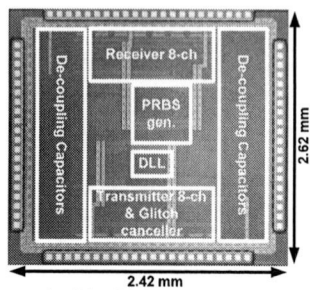

Fig. 11. Microphotograph of the fabricated chip.

11. The active area is $190 \times 90 \ \mu m^2$/Ch for transmitter, $175 \times 200 \ \mu m^2$/Ch for receiver, $70 \times 230 \ \mu m^2$ for DLL, respectively. The total power consumptions of transmitter, receiver, and DLL are 57mW/Ch, 30mW/Ch, and 15mW, respectively. The specification of test chip is summarized in Table II.

VII. CONCLUSION

A 5-Gb/s/pin transceiver for DDR memory interface is proposed with a crosstalk suppression scheme. The proposed transceiver implements a staggered memory bus topology and a glitch canceller to suppress crosstalk-induced distortions in a memory channel. The improved performance is demonstrated in terms of eye-opening and bathtub curve results of the dense parallel memory channel. The eye width and height of the proposed scheme increases 28.3% and 11.1% compared to the conventional memory transceiver, respectively.

TABLE I. PERFORMANCE IMPROVEMENT OF PROPOSED MEMORY TRANSCEIVER

	J_{p-p}	V_{EYE}	BER=10^{-12}
Without crosstalk suppression	73.67-ps	140.6 mV	0.52-UI
With crosstalk suppression	52.82-ps	156.2 mV	0.63-UI

TABLE II. SPECIFICATION OF TEST CHIP

Process	0.18μm CMOS
Supply voltage	1.8 V
Active area	$190 \times 90 \ \mu m^2$/Ch [Tx] $175 \times 200 \ \mu m^2$/Ch [Rx] $70 \times 230 \ \mu m^2$ [DLL]
Data rate	5-Gb/s
Measured BER	$< 10^{-12}$
Power consumption	57 mW/Ch [Tx] 30 mW/Ch [Rx] 15 mW [DLL]

ACKNOWLEDGEMENT

This work was supported by the IDEC at KAIST, Republic of Korea.

REFERENCES

[1] J. F. Buckwalter and A. Hajimiri, "Cancellation of Crosstalk-Induced Jitter," IEEE J. Solid-State Circuits, vol. 41, no. 3, pp. 621-631, March 2006.

[2] Y. Hur et al., "Equalization and Near-End Crosstalk (NEXT) Noise Cancellation for 20-Gb/s 4-PAM Backplane Serial I/O Interconnections," IEEE Trans. Microwave Theory Tech., vol. 53, no. 1, pp. 246-255, January 2005.

[3] C. Pelard et al., "Realization of multigigabit channel equalization and crosstalk cancellation integrated circuits," IEEE J. Solid-State Circuits, vol. 39, no. 10, pp. 1659–1669, October 2004.

[4] Seung-Jun Bae et al., "An 80nm 4Gb/s/pin 32bit 512Mb GDDR4 Graphics DRAM with Low Power an Low Noise Data Bus Inversion," IEEE J. Solid-State Circuits, vol. 43, no. 1, pp. 121-131, January 2008.

[5] Eric Bogatin, Signal Integrity – simplified. New Jersey: Prentice Hall, 2004.

IEEE 2008 Custom Intergrated Circuits Conference (CICC)

EMI Resisting Smart-power Integrated LIN Driver with Reduced Slope Pumping

Jean-Michel Redouté and Michiel Steyaert

Dept. ESAT - MICAS

Katholieke Universiteit Leuven

Heverlee, Belgium

Email: [jean-michel.redoute, michiel.steyaert]@esat.kuleuven.be

Abstract—This paper presents the design of a Local Interconnect Network (LIN) smart-power driver exhibiting a high degree of immunity against electromagnetic interference (EMI). Improving previous designed architectures, this circuit switches to a low power consumption mode when there is no EMI, while presenting an increased immunity to slope pumping. Measurements illustrate that this proposed LIN driver complies to the 5W direct power injection (DPI) specification.

I. INTRODUCTION

A local interconnect network (LIN) is a low speed (max. 20 kbit/s) serial communication protocol which is typically used in automotive applications [1]. Since the electromagnetic environment in automotive applications is strongly polluted by electromagnetic interference (EMI), this results in very stringent electromagnetic compatibility (EMC) requirements [2], [3]. In particular, to allow for a correct data transmission, the duty cycle of the signal on the LIN bus should not be corrupted by EMI which is injected into the LIN bus [1]. In [4], an integrated LIN driver has been presented with a high degree of immunity against EMI, but at the expense of a high and continuous bias current consumption. In addition, this circuit is very susceptible to slope pumping, which introduces significant deviations in the rising and the falling slopes of the LIN bus signal, and consequently in the propagation delay and in the duty cycle. Both points are addressed and solved in the proposed EMI resisting LIN driver. As illustrated in the measurement section, this LIN driver guarantees an excellent performance, and this up to the worst case 5W forward power direct power injection (DPI) measurement [5].

II. BASIC PRINCIPLE OF THE EMI RESISTING LIN DRIVER

The simplified schematic of the EMI resisting LIN driver is shown in Fig. 1 [4]. The EMI injection in the LIN bus is modeled by a sinusoidal voltage source V_{emi} injecting a forward power of maximally 5W through the bus capacitor C_b into the LIN bus as described in the DPI specification [5]. The slope control circuit block boosts the digital input signal (V_{TX}) to the battery voltage level (between 8V and 18V) and adds a constant slope of 5 μs in order to reduce electromagnetic emission (V_{in}) [1]. The copy and limiter circuit block copies the signal on the LIN bus (V_{LIN}) to the non-inverting opamp input while limiting this signal between predefined levels. There are, however, two major issues associated to this

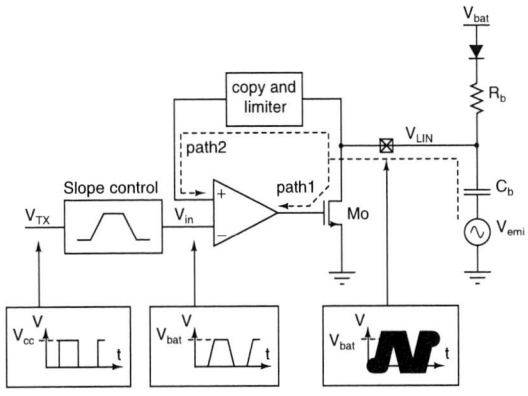

Fig. 1. Basic EMI resisting LIN driver circuit.

solution. Firstly, the opamp has to have a very low output impedance in order to minimize the EMI coupling through the parasitic drain-gate capacitance of Mo to its gate (labeled "path 1" in Fig. 1): this low opamp output impedance was achieved at the severe penalty of a heavy and continuous power consumption in [4]. Secondly, the EMI which is copied to the non-inverting opamp input and limited between specific levels introduces a considerable shift in the rising and falling slope of V_{LIN}: this shift has been identified as slope pumping in [4]. Slope pumping distorts the duty cycle and the propagation delay of the LIN bus signal. Both major issues are tackled in the presented LIN driver design.

III. SMART-POWER EMI RESISTING LIN DRIVER WITH REDUCED SLOPE PUMPING

A. Introduction

The proposed LIN driver simplified top schematic is depicted in Fig. 2. The LIN bus signal is copied to the non-inverting opamp input by means of M1-M2 and M3-M4. The signal on the non-inverting opamp input is limited between $V_{in} + |V_{GS6}|$ and $V_{in} - |V_{GS7}|$ by M6-M7. Diodes D1 and D2 prevent the EMI from discharging in the power supply rails through the parasitic drain-bulk diodes of Mo and M1.

B. Smart-power mode

The opamp output impedance must be low to decrease the high frequency coupling from the LIN bus to the gate

978-1-4244-2018-6/08 $25.00 © 2008 IEEE

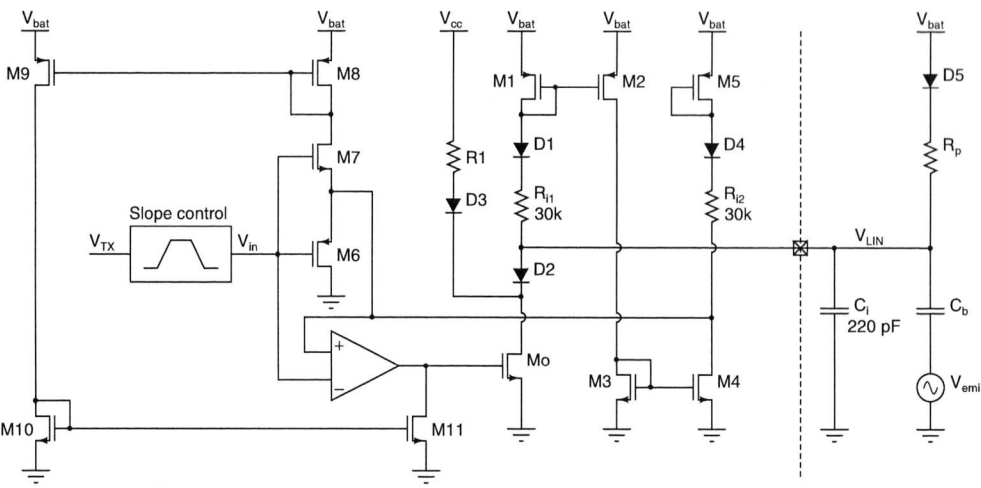

Fig. 2. EMI resisting smart-power LIN driver with reduced slope pumping.

of Mo through its parasitic drain-gate capacitance: if high frequency EMI couples to the gate of Mo, it will pump the net charge across the parasitic gate-source capacitance of Mo (C_{gso}) upwards, increasing the gate voltage of Mo and hereby progressively shorting the LIN bus to ground. This is a perfect example of the typical EMI induced charge pumping phenomenon as was described in [6]. The opamp used in [4] is depicted in Fig. 3.a. In this opamp, M4 and M5 keep V_{out} at a fixed DC level using local negative feedback, preventing charge pumping: in addition, the output impedance of this opamp is extremely low. Neglecting R, the opamp small signal output impedance is equal to:

$$Z_{out} = \frac{1}{g_{m4} + g_{m5} + r_o g_{m3} g_{m4}} \qquad (1)$$

where r_o is the parallel combination of the (high) output resistances of M1 and M3. Previous observations hold as long as M3, M4, M5 are biased in their saturation region: however, in the LIN driver circuit, the opamp is typically switched between the battery voltage and ground. Consequently, its output is mostly either high either low, in which case a small signal analysis is not always readily applicable. Let's consider what happens in both cases.

1) V_{LIN} is low: Mo is conducting drain current and the opamp output is high. Transistors M3, M4 and M5 in the opamp are biased in saturation. The opamp output impedance is therefore very low and equals (1). Transistor M4 sources the current which flows through M5 and which charges the parasitic gate-source capacitance of Mo (C_{gso}).

2) V_{LIN} is high: Mo is off and the opamp output is low. Transistors M3 and M4 are off. The diode connected transistor M5 sinks the charge which is stored in C_{gso}. The output impedance of the opamp is now much higher, since M3 and M4 no longer provide voltage gain to reduce the output impedance with the local negative feedback. The output impedance now solely depends on M5, and for small signals, is equal to $1/g_{m5}$. Consequently, for high EMI levels coupling

from the drain to the gate of Mo through its parasitic drain-gate capacitor, a situation might occur when the charge which is pumped into C_{gso} becomes larger than the charge which M5 is capable of sinking. If such a situation happens, the voltage on the gate of Mo increases, forcing more drain current to flow through Mo: this results in a distorted and collapsed output signal. The duty cycle of such output signal is no longer equal to its original value, generating an erroneous data transmission. This issue can be solved by increasing the $\frac{W}{L}$ of transistor pair M4 and M5. Unfortunately, this accounts for a higher static power consumption. In addition, owing to the nonlinear I-V characteristic of a MOS diode connected transistor, the parasitic C_{gso} of Mo is not discharged linearly, which makes this structure inherently susceptible to EMI induced charge pumping. In [4], the possibility of connecting a resistor between V_{out} and ground is suggested, as this provides a constant and linear discharge path for the parasitic C_{gso} (represented by R in Fig. 3.a). Not surprisingly, this resistor equally accounts for extra power consumption. Considering all the above reasons, M5 has been redesigned as a current source, as shown in Fig. 3.b. This current source discharges the parasitic C_{gso} of Mo linearly because it sinks a constant current, since M5 operates as a current source as long as V_{out} stays above V_{dssat5}. If V_{out} drops below V_{dssat5}, M5 operates in its triode region, shorting V_{out} to ground. In order to overcome the high static power consumption, a smart-power circuit has been implemented which switches on and shorts the gate of Mo to its source when the voltage V_{LIN} collapses owing to a large EMI which is injected into the LIN bus. This allows the continuous bias current through M4 and M5 to be small and barely sufficient to pull the gate of Mo to ground in the absence of EMI (2.3mA in this design). Refer to Fig. 2: the clipping structure detects when the LIN bus output collapses to ground owing to large EMI levels: M7 detects the sudden signal drop at the non-inverting opamp input, and turns on M8. This current is copied, scaled and sinked via M11 hereby pulling the gate of Mo to ground, which restores

978-1-4244-2018-6/08 $25.00 © 2008 IEEE   684

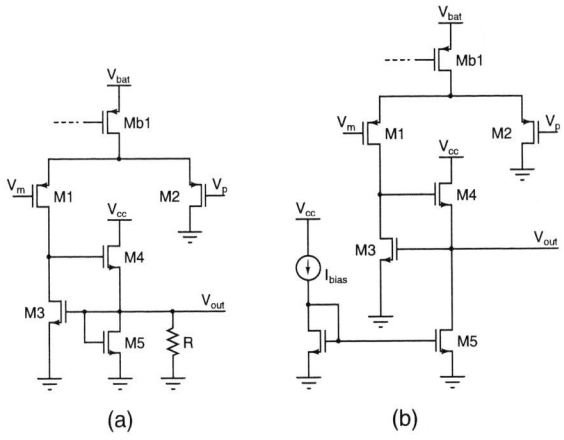

Fig. 3. (a) Opamp with diode connected M5 - (b) Opamp biased with a (small) constant bias current through M5.

TABLE I

OPAMP AVERAGE POWER CONSUMPTION

V_{bat} = 12V, V_{cc}=3.3V, R_p=1kΩ, C_b=1nF V_{TX} transmission speed = 20 kbit/s		
EMI	Opamp $P_{average}$ previous LIN driver [4] (mW)	Opamp $P_{average}$ present LIN driver (mW)
no EMI	28	11
EMI=10V@1 MHz	26	12
EMI=45V@1 MHz	28	14
EMI=10V@10 MHz	26	12
EMI=45V@10 MHz	28	15
EMI=10V@100 MHz	26	11
EMI=45V@100 MHz	25	13

V_{LIN}. As soon as the necessary output level is reached, or as soon as the EMI injection stops, M11 shuts off and the LIN driver resumes its low power operation. The average opamp power consumption is listed in table I. Observe that the average consumed power in the present LIN driver opamp is lower than in the LIN driver presented in [4]. More importantly, observe that the present LIN driver opamp switches to a higher power consumption for increasing EMI levels. This indicates that the smart-power principle is working properly, and that a much smaller continuous bias current through M5 may be used, reducing the static power consumption to a much lesser value.

C. Slope pumping reduction

Slope pumping distorts the rising and falling edges of the LIN bus signal and consequently corrupts the duty cycle and the propagation delay [4]. Consider that the LIN bus signal V_{LIN} is appropriately copied to the non-inverting opamp input: the (large) EMI which is superposed on that output signal is clipped between $V_{in} + |V_{GS6}|$ and $V_{in} - |V_{GS7}|$ by respectively M6 and M7 (refer to Fig. 2). Owing to this, the differential signal between the opamp inputs ($V_d = V_p - V_m$) can roughly be approximated by a square wave with V_A as upper and $-V_B$ as lower level. The duty cycle of this square wave is close to 50%, and the average drain current of Mo over

time ($\overline{I_{do}}$) can be calculated using basic formulas describing the MOS differential pair [7]:

$$\overline{I_{do}} = \lim_{T \to \infty} \frac{1}{T} \int_{-T/2}^{T/2} I_{do}(t).dt$$
$$= K.\frac{\mu C_{ox}}{2}.\frac{W_1}{L_1}.\overline{(V_{GS1} + V_d - V_t)^2} \quad (2)$$

Where K represents the scale factor of current mirror M3-Mo. Above equation can be expanded as follows:

$$\overline{I_{do}} = K.\frac{\mu C_{ox}}{2}.\frac{W_1}{L_1}. \left((V_{GS1} - V_t)^2 \right.$$
$$\left. +2.(V_{GS1} - V_t).\overline{V_d} + \overline{V_d^2} \right) \quad (3)$$

Substituting V_d by a square wave with V_A as upper and $-V_B$ as lower level yields:

$$\overline{I_{do}} = K.\frac{I_B}{2} + K.g_{m1}.\left(\frac{V_A - V_B}{2} \right) + K.\left(\frac{V_A^2 + V_B^2}{2} \right) \quad (4)$$

where I_B represents the DC tail current of the differential pair and g_{m1} the transconductance of M1. According to (4), the average value of the drain current flowing through Mo is pumped to a higher value owing to the large EMI which is superposed on the non-inverting opamp input. This effect is identified as slope pumping. Observe that the slope pumping phenomenon solely occurs during the slopes: once V_{LIN} reaches the high or the low state, the opamp drives the output transistor Mo as a switch, hereby forcing the output to the same level as input signal V_{in}. As can be seen in (4), slope pumping lengthens the rising slopes and shorts the falling slopes of V_{LIN}: clearly, the latter is not an issue, but the former may severely distort the duty cycle and perturb the propagation delay. It is therefore important to reduce the slope pumping on the rising slope as much as possible. Basically, there are three ways to achieve this.

1) Decrease V_A and V_B: the first possibility is to decrease V_A and V_B. However, both have to be larger than the steady-state error. As calculated in [4], the steady-state error is equal to 0.5 V, which defines the lower boundary of V_A and V_B. In this case, V_A and V_B are respectively equal to $V_{in} + |V_{GS6}|$ and $V_{in} - |V_{GS7}|$ (Fig. 2).

2) Ensure that $V_A=V_B$: if V_A is perfectly equal to V_B, the second (linear) term in (4) is canceled, which reduces the slope pumping. Unfortunately, V_A and V_B are directly affected by the steady-state error, which in turn depends on the load present on the LIN bus: moreover, this load is not a priori known. This varying steady-state error causes V_A and V_B to constantly differ from each other.

3) Decrease K: a third way to decrease slope pumping is to reduce the current mirror scaling factor K. This corresponds to decreasing the opamp gain: however, this equally increases the steady-state error and limits the maximum voltage level at the opamp output, which increases the low state of the LIN bus output signal. It is therefore important to reduce the opamp gain solely during the rising slopes and the high LIN bus output state, while reverting to a high opamp gain during the

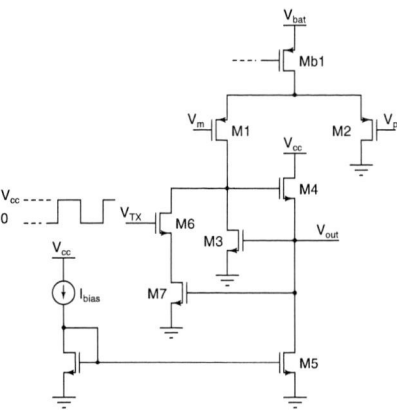

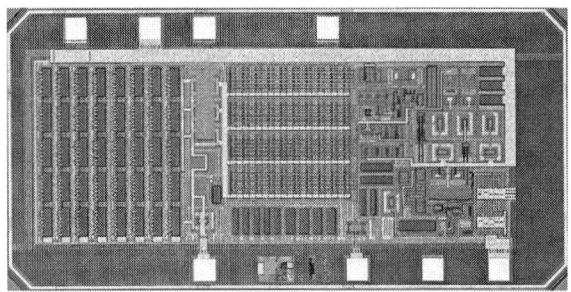

Fig. 4. Opamp biased with a (small) constant bias current. Transistor M6 is switched on when V_{TX} toggles to a high state, hereby lowering the voltage gain of the opamp and attenuating the slope pumping.

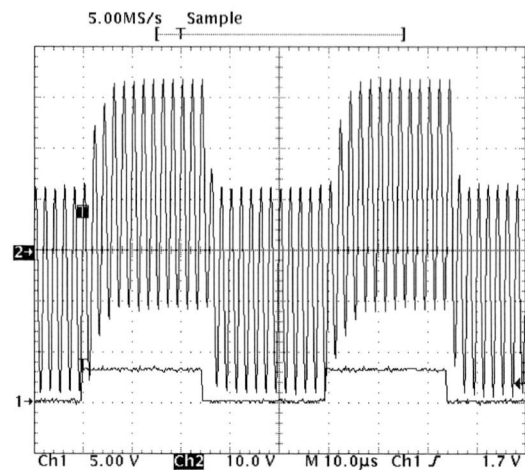

Fig. 6. 5W DPI measurement of the EMI resisting smart-power LIN driver: although the LIN bus is heavily distorted, the duty cycle is not altered.

TABLE II

DPI MEASUREMENT RESULTS

V_{bat}=12V, V_{cc}=3.3V, R_p=1kΩ , C_b=1nF				
V_{TX} transmission speed = 20 kbit/s				
EMI frequency (MHz)	Forward EMI power (W)	Duty cycle (%)	Prop. delay (rising edge) (μs)	Prop. delay (falling edge) (μs)
-	-	49.2	3	3
0.3	5.99	42.4	4.4	1
1	6.02	42.8	4.2	1
10	5.95	44	3.6	1
40	5.25	45.6	3.2	1.4
100	5.26	46	3	1.4
500	5.01	46	2.6	1
1000	5.05	49.2	3	3

Fig. 5. Microphotograph of the EMI resisting smart-power LIN driver with reduced slope pumping.

low output state. This has been implemented as follows (Fig. 4): M6 turns on as soon as the digital input signal (V_{TX}) is high. Transistor M7 is then in parallel with M3, hereby reducing the opamp gain. Once V_{TX} drops, M6 shuts off, the opamp output signal increases and forces a low output state on the LIN bus. This way, slope pumping is effectively countered.

IV. MEASUREMENTS

The proposed LIN driver (Fig. 2) has been integrated using the AMIS I3T80 high-voltage 0.35 μm CMOS technology (Fig. 5). A forward EMI power of maximally 5W has been injected through the external bus capacitance C_b, ranging from 150 kHz to 1 GHz according to the DPI specification [5]: this corresponds to an EMI source amplitude of 45V, which is of course a huge value. Measurements show that the proposed LIN driver withstands this 5W DPI test for the listed frequency range (Fig. 6): table II summarizes the results. Observe the almost constant propagation delay, and the minimal impact of EMI on the duty cycle and on the propagation delay: as specified in [1], the duty cycle is comprised between 39.6% and 58.1% at all times.

V. CONCLUSION

A robust EMI resisting integrated smart-power LIN driver with reduced slope pumping has been presented, and compared

against previously designed structures. DPI measurements corroborate the theoretical analyses, and prove that this circuit produces a duty cycle that is well within the specifications and this under the highest levels of EMI (5W of forward injected power) which are listed in the DPI specification.

ACKNOWLEDGMENT

The authors wish to thank AMI Semiconductor Belgium for their cooperation, and the IWT (Institute for the Promotion of Innovation by Science and Technology) for partial funding.

REFERENCES

[1] LIN Consortium, LIN specification package revision 2.0, http://www.lin-subbus.org, available online.
[2] H. Casier, "Electronic circuits in an automotive environment", International Solid-State Circuits Conference, tutorial, February 2004.
[3] K. Appeltans, "LIN transceivers: a case study of automotive design", International Solid-State Circuits Conference, ATAC design forum, Feb. 2005.
[4] J.-M. Redouté, M. Steyaert, "An EMI resisting LIN driver in 0.35-micron high-voltage CMOS", IEEE Journal of Solid-State Circuits, vol. 42, no. 7, pp. 1574 - 1582, July 2007.
[5] IEC 62132-4, Ed. 1: IC's Measurement of E/M immunity 150 kHz to 1GHz - Part 4: Direct RF power injection method.
[6] J.-M. Redouté, M. Steyaert, "Current mirror structure insensitive to conducted EMI", IEE Electronics Letters, vol. 41, no. 21, October 2005, pp. 1145-1146.
[7] W. M. C. Sansen, Analog design essentials, Springer, 2006.

978-1-4244-2018-6/08 $25.00 © 2008 IEEE

Leveraging the Third Dimension

Session 21

Chair: Rakesh Patel, Altera
Co-Chair: Michael Seningen, Intrinsity

Leveraging the 3rd Dimension by stacking silicon and assembling it into a single package is already underway in the miniaturization of consumer electronics and more. This session will address the mature package level wire-bond-3D stack, mature multi-chip package, and the silicon level 3D stack enabled by Through-Silicon-Via (TSV). The reasons for moving to TSV-3D are driven by some of the major challenges introduced with process scaling and/or complex SOCs such as leakage/power, interconnect delay, bandwidth/performance, non-optimum process technology, memory, and circuits resistant to scaling.

The first paper from Honda Research Institute, Japan addresses the yield associated with the TSV. Yield being one of the key barriers to adoption should make this a very interesting paper to attend. A wafer-to-wafer assembly process is described with almost 100% electrical connections and 60% functional yield.

The next three papers from University of Rochester, North Carolina State and Cornell University cover multi-tier design considerations with respect to clock architecture, network on chip interconnections, and process and thermal variability as the design stretches across different wafers. The traditional 2D design constraints are further enhanced in the third dimension.
The clocking in a synchronous system extends as defined in paper two extends into the 3D. A comparison of 3D clock network topologies is presented with clock skew and power dissipation measurements. Paper three discusses a three dimensional network on chip (NOC). A simple 3D mesh interconnection network enables the sharing of global resources for complex systems on silicon while consuming a low 2mW per transaction. Paper four tackles the challenges of process variability introduced as we scale the nanometer technologies.

Paper five is an invited paper from Georgia Institute of Technology, Atlanta provides solutions to tackle the thermal and power delivery concerns with TSV. Liquid cooling, TSV power delivery and implementation are addressed. In addition, optical interconnects are fabricated to enable signal routing.

The next three papers highlight the benefits of heterogeneous assembly, security, and memory density as enabled by wire-bond or multi-chip package assembly.
The heterogeneous assembly in paper six from Infineon Technologies is ideal for separating the analog from the digital, where the analog is now free to select an optimum process. The next paper from Renesas Corporation shows how the System-In-Package automatically allows a system design to ensure that sensitive access is secured. Lastly, paper eight describes how a T-DMB SOC with stacked memory allows for a 159.2mW total power by moving the memory technology from SRAM to DRAM.

978-1-4244-2018-6/08 $25.00 © 2008 IEEE

Notes

IEEE 2008 Custom Intergrated Circuits Conference (CICC)

Stacking Technology based on 8-inch Wafers using Direct Connection between TSV and Micro-bump

Nobuaki Miyakawa, Eiri Hashimoto, Takanori Maebashi, Natsuo Nakamura, Yutaka Sacho, Shigeto Nakayama and Shinjiro Toyoda

Honda Research Institute Japan Co., Ltd.

Abstract

We have developed a unique TSV structure and evaluated the connectivity between TSV and micro-bump. The connection resistances are less than 0.7 ohm and the capacitance of TSV is less than 3pF, respectively. The electrical connection between each wafer was almost 100% and the functional yield reached more than 60%.

Introduction

Many research organizations have been investigating stacking technology, both in the areas of assembly technology and in device technology (1)-(8). Due to underdevelopment in stacking device technology, this area is still far from the industrial manufacturing stage. The main reason is a lack of actual results with large-scale wafers, for example the relationship between yield data and stacked layers, and also the actual data effected through using stacking technology.

The connection yield between each chip or each wafer, that is, the connection between TSV and micro-bump, is one of the key technologies to increase the stacking-device yield. If the connection yield is almost 100%, that stacking-device yield depends on each stacked wafer's yield.

In this paper, we describe the structure of the connection portion, its characteristics, and the results of the trial manufacture using our proposed connection method between TSV and micro-bump.

Expected Technology

Fig. 1 shows the negative aspects of 2 dimensional LSI, and our remedial technology. With 2 dimensional LSI, usually long wires exist due to larger chip size, which makes critical paths. We should use conventional technology, not requiring large investment, to solve the problems of 2 dimensional LSI.

One of the expected technologies is 3 dimensional technology, as it can utilize short wires due to smaller chips, and shortened vertical wires between layers by using wafer-thinning technology. In fact, chip performance and power consumption can be improved by almost all wires becoming shorter.

This stacking technology will be useful as a future method for high performance, multi functions fabricated by different kinds of processes, for low power consumption, and downsizing.

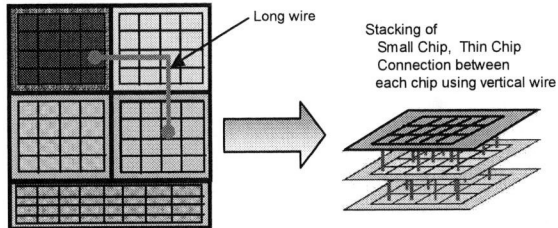

Fig. 1   Representative Negative Aspect of 2D LSI, and Expected Technology

TSV Structure and its Connectivity

Fig.2 compares the connection portion of many previous cases and ours.

In the previous connection portions shown at left, the number of factors for total resistance is five, and the resistance value dispersion between layer bumps increases for point-to-point connections.

Almost all used back-side metal for stacking between upper and lower layer, therefore the process needed an isolation layer and back side metal to avoid electrical short circuit between substrate and back-side metal. The total process was lengthened by adding the two processes.

We have decreased the process steps from wafer thinning to the back-side metal formation by developing a method that only uses wafer thinning.

The right side shows the cross section and total resistance value of our structure. The number of factors for the total resistance is three and the resistance value dispersion between each layer is small and stable for face-to-face connection.   Also, the connection process is shortened.

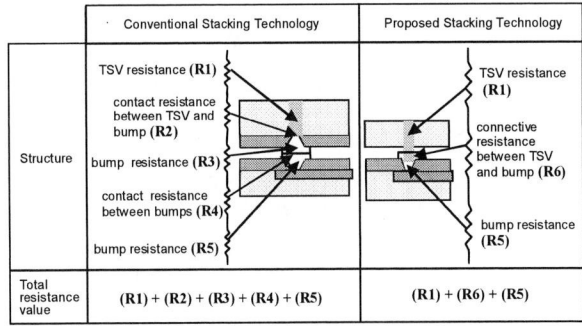

Fig. 2 Comparison between Conventional and Proposed Method

Trial Process Flow and Connective Characteristics

Fig.3 shows our trial process flow. This technology adds a wet etching process to the conventional grinding and polishing process of the upper wafer, and there, the TSV is stuck to the Si substrate. It is a direct stacked method wherein the TSV of the upper wafer is directly connected with a surface micro-bump of a lower wafer. The back-side micro-bumps are not needed, therefore the total process length is simplified and the reliability of the device is improved.

Our TSV structure has been fabricated before/after the MOS transistor fabrication process for a process temperature. The dielectric is fabricated by SiO_2/poly-si and the conductor is fabricated by burying tungsten, after the trench hole for TSV structure is opened by conventional deep-trench etching technology.

After that multi-wiring, the wafer is mounted on a support glass for thinning in two steps: back grinding and wet etching. The wet etching is an additional wafer thinning process.

For the lower wafer fabrication process, the wafer is also fabricated by multi wiring. However, the lower wafer has a bow due to this process (Fig.3 (b)). Some surface micro-bumps of the lower wafer are formed, after that the wafer bow is compensated for (Fig3(c) and Fig.4).

Next, the wafers are prepared for stacking, which process starts from alignment and bonding between each one (Fig3 (c)). There, TSV is stuck to the Si substrate, wherein the TSV of the upper wafer is directly connected with a surface micro-bump of the lower wafer.

And then an adhesive is injected between stacked wafers. The direction of the injection is from the wafer edge to the wafer center. Fig. 5 shows the injection results which were checked by using Scanning Acoustic Tomograph.

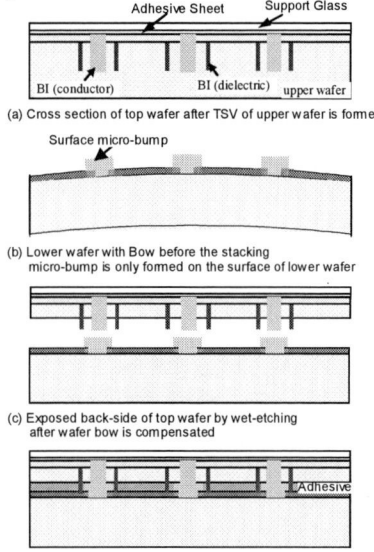

(a) Cross section of top wafer after TSV of upper wafer is formed

(b) Lower wafer with Bow before the stacking
micro-bump is only formed on the surface of lower wafer

(c) Exposed back-side of top wafer by wet-etching
after wafer bow is compensated

(d) Device is hardened by adhesive after each wafer is bonded

Fig.3 Stacked process flow for wafer-to-wafer

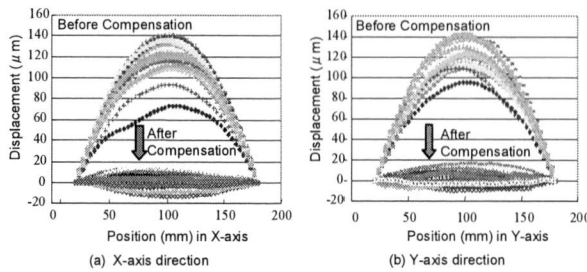

(a) X-axis direction   (b) Y-axis direction

Fig.4 Compensation for wafer bow

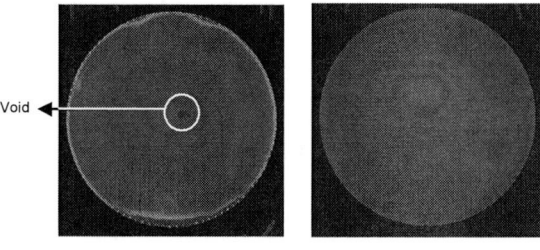

(a) Before Improvement (Void Exists)   (b) After Improvement (No Void)

Fig.5 Improvement of the Adhesive Injection

Fig. 6 shows the data of the electrical connectivity and frequency distribution between TSV and micro-bump using TEG chips for an 8-inch wafer after 2-layer stacking. The upper drawing is the measurement structure for the interconnection resistance value of the whole wafers between TSV and micro-bump of a stand-alone case and the frequency distribution.

Almost all values of the stand-alone case are less than 0.7 ohm and the frequency distribution curve is very sharp. The frequency distribution data is made by 6 stacking wafers.

Fig.7 shows the measurement structure for the interconnection resistance value (a), the photograph of the cross section (b) and the frequency distribution of 3-layer stacking from 1st to 3rd layers, using 6 stacking wafers (c). If too much force is added to the connection portion for bonding, the TSV will snap. But the TSV as shown here will not snap.

Fig.8 shows the electrical characteristics between TSV and micro-bump, the measurement results of the resistance value, and the capacitance of TSV using 300 chain cases. The mean of resistance value dispersion is about 1.5 ohm per chain by including the parasitic resistance values for the wiring, and the capacitance of TSV is less than 3pF per TSV from such measurement results.

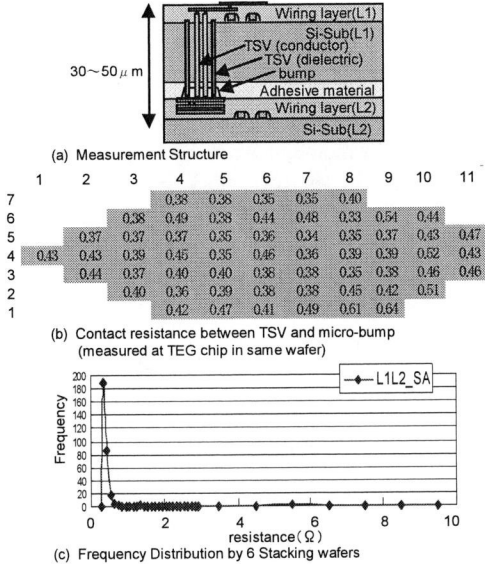

(a) Measurement Structure

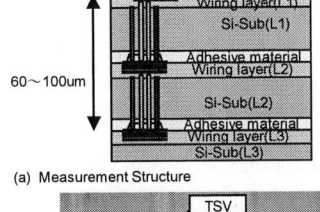

	1	2	3	4	5	6	7	8	9	10	11
7				0.38	0.38	0.35	0.35	0.40			
6			0.38	0.49	0.38	0.44	0.48	0.33	0.54	0.44	
5		0.37	0.37	0.37	0.35	0.36	0.34	0.35	0.37	0.43	0.47
4	0.43	0.43	0.39	0.45	0.35	0.46	0.36	0.39	0.39	0.52	0.43
3		0.44	0.37	0.40	0.40	0.38	0.38	0.35	0.38	0.46	0.46
2			0.40	0.36	0.39	0.38	0.38	0.45	0.42	0.51	
1				0.42	0.47	0.41	0.49	0.61	0.64		

(b) Contact resistance between TSV and micro-bump (measured at TEG chip in same wafer)

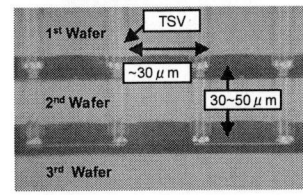

(c) Frequency Distribution by 6 Stacking wafers

Fig.6 Electrical Connectivity between Two Layers

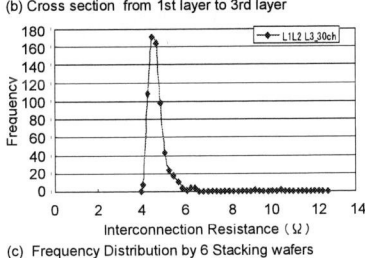

(a) Measurement Structure

(b) Cross section from 1st layer to 3rd layer

(c) Frequency Distribution by 6 Stacking wafers

Fig. 7 Connectivity from 1st layer to 3rd layer

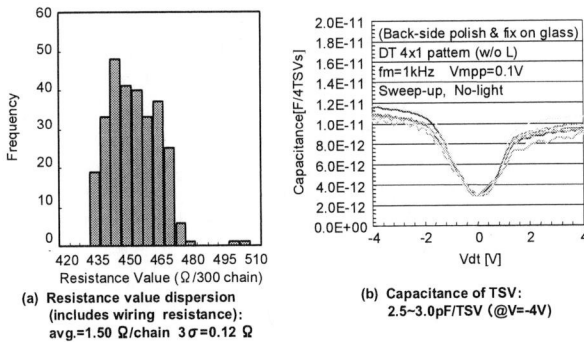

(a) Resistance value dispersion (includes wiring resistance): avg.=1.50 Ω/chain 3σ=0.12 Ω

(b) Capacitance of TSV: 2.5~3.0pF/TSV (@V=-4V)

Fig.8 Characteristics of (TSV + micro-bump)

Results of Trial Manufacture using 8-inch wafers

Fig. 9 shows the structure of a 3-wafer stack: each wafer is structured by a microprocessor (L1), custom circuits including analog circuits (L2), and 64M bit SDRAM (L3), respectively. The number of TSV between L1 wafer and L2 wafer becomes more than 650k, and that of TSV between L2 wafer and L3 wafer is more than 180k. It contains more than 2Mgates and 64Mbit SDRAM. The device is 8.44mm x 4.69mm and has more than 1400 TSV wires. Fig. 10 shows the photograph of the device and the cross section of the die after 3 layers are stacked.

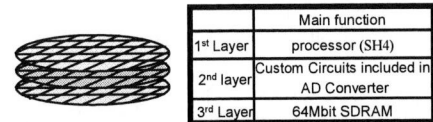

	Main function
1st Layer	processor (SH4)
2nd layer	Custom Circuits included in AD Converter
3rd Layer	64Mbit SDRAM

Fig.9 Structure of stacked device

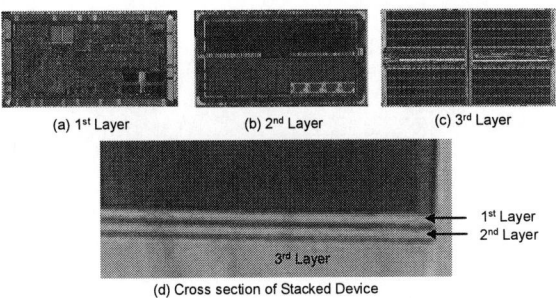

(a) 1st Layer   (b) 2nd Layer   (c) 3rd Layer

(d) Cross section of Stacked Device

Fig.10 Device photograph of Layers 1, 2 & 3, and cross section photograph of Stacked Device

Fig. 11 shows the evaluation results of the probe test after 3-wafer stacking in five cases. The resultant yields comprising all functional tests are over 60%. Such values suggest a high probability for industrial fabrication in the near future, although some further investigation remains.

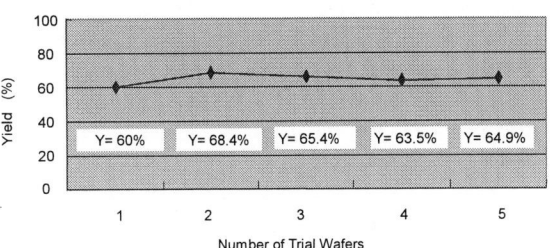

Fig. 11 Probe Evaluation Results of Trial Devices

Fig. 12 details operational frequency and power consumption. The red line shows the relationship of the trial manufacture and the blue line shows that of MCM device with same function. Comparatively, the trial device demonstrates 2 times the operational frequency but at less than one-third power consumption of MCM device under the same test conditions.

978-1-4244-2018-6/08 $25.00 © 2008 IEEE

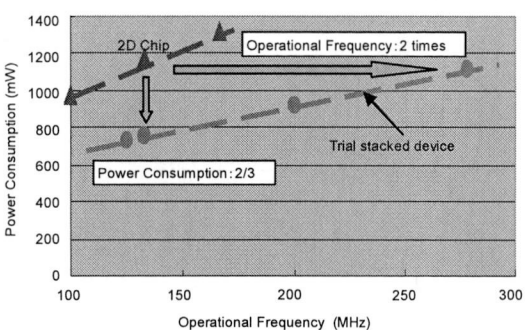

Fig. 12 Relationship between Operational Frequency and Power Consumption

Fig.13 indicates the results under the consumer-spec. standard. Our trial manufacture readily shows that there is no problem with the structure from the point of reliability.

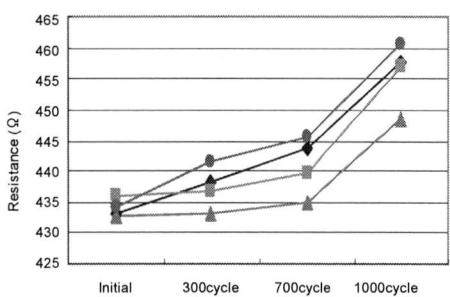

Fig. 13 Reliability Test Results (JEDEC Std. JESD22-A104)

Summary

1. We propose a new connection method between TSV and micro-bump.
   - Connection method:　　　　Direct Connection
   - Electrical connectivity :　　　　　　100 %
   - Resistance value:　　　　　　　　< 0.7 Ω
2. The trial manufacture of 3 stacked wafers using proposed TSV and micro-bump structure based on 8-inch wafers.
   - Yield of 3-layer stacked devices:　　over 60%
   - Operational Frequency:　　　　　2 times
     Compared with MCM device under same test condition
   - Power Consumption:　　　reduced over 30%
   - Reliability Test of thermal cycle :
     　　　　　　　more than 1000 cycles
     　　　　　　　(under consumer-specs)

Acknowledgements

The authors would like to thank Hitachi, Ltd. Micro Device Division for providing wafer fabrication using 0.18 um CMOS technology based on 8-inch wafers and for the evaluation of test wafer and test devices. And the authors would like to thank Soliton Systems K.K. for the stacked data conversion, Renesas Technology Corp. (formerly SuperH, Inc), Hitachi Information & Communication Engineering, Ltd. and Hitachi ULSI Systems Co., Ltd. for the design of SH4. The authors also would like to thank Mr. Nozomu Horino for his support.

References

(1)T. Matsumoto, Y. Kudoh, M. Tahara, K.. H. Yu, N. Miyakawa, H. Itani, T. Ichikizaki, A. Fujiwara, H. Tsukamoto, and M. Koyanagi, , "Three dimensional integration technology based on wafer bonding technique using micro-bumps," in Proc. Int. Conf. Solid State Devices and Mater, 1073-1074, 1995.

(2) T. Matsumoto, M. Satoh, K. Sakuma, H. Kurino, N. Miyakawa, H.Itani, and M.Koyanagi, "New Three-Dimensional Wafer BondingTechnology Using Adhesive Injection Method," in Proc. Int. Conf. Solid State Devices and Mater,460-461,1997.

(3) M. Koyanagi, H. Kurino, K. W. Lee, K. Sakuma, N. Miyakawa, and H.Itani, "Future system-on-silicon LSI chips," IEEE Micro, vol. 18, no.4, 17-22, 1998.

(4) A. Iwata, M. Sasaki, T. Kikkawa, S. Kameda, H. Ando, K. Kimoto, D. Arizono, H.Sunami, "A 3D Integration Scheme Utilizing Wireless Interconnections for Implementing Hyper Brains," ISSCC Dig. Tech. Papers, 262-263, 2005.

(5) T. Fukushima, Y. Yamada, H. Kikuchi and M. Koyanagi, "New Three-Dimensional Integration Technology Using Self-Assembly Technique," IEEE Int'l. Electron Devices Meeting Conference Digest Technology Papers, 359-362, 2005.

(6) T. Maebashi, N. Nakamura, N. Nakayama and N. Miyakawa, "New Fabrication Method for Multi-Layer Stacked Devices using Wafer-to-Wafer Stacked Technology based on 8-inch Wafers, ESSDERC, 251-254, 2007.

(7) N. Miyakawa, T. Maebashi, N. Nakamura, S. Nakayama, E. Hashimoto and S. Toyoda, "New Multi-Layer Stacking Technology and Trial Manufacture " 3D Architectures for Semiconductor Integration and Packaging, 2007

(8) N. Miyakawa, E. Hashimoto, T. Maebashi, N. Nakamura, Y. Sacho, S. Nakayama and S. Toyoda "3D Stacking Device Technology using Wafer-to-Wafer Stacked Method " IMAPS Device Packaging 2008, THA1.

Clock Distribution Networks for 3-D Integrated Circuits

Vasilis F. Pavlidis, Ioannis Savidis, and Eby G. Friedman
Department of Electrical and Computer Engineering
University of Rochester, Rochester, New York 14627
{pavlidis, iosavid, friedman}@ece.rochester.edu

Abstract - **Three-dimensional (3-D) integration is an important technology that addresses fundamental limitations of on-chip interconnects. Several design issues related to 3-D circuits, such as multi-plane synchronization, however, need to be addressed. A comparison of three 3-D clock distribution network topologies is presented in this paper. Experimental results of a 3-D test circuit manufactured by the MIT Lincoln Laboratories are also described. Successful operation of the 3-D test circuit at 1.4 GHz is demonstrated. Clock skew and power dissipation measurements for the different clock topologies are also provided.**

I. INTRODUCTION

An omnipresent and challenging issue for synchronous digital circuits is the reliable distribution of the clock signal to the many thousands of sequential elements distributed throughout a synchronous circuit [1], [2]. The complexity of this task is further increased in 3-D ICs as sequential elements belonging to the same clock domain (*i.e.*, synchronized by the same clock signal) can be located on multiple planes. Another important issue in the design of clock distribution networks is low power consumption, since the clock network dissipates a significant portion of the total power consumed by a synchronous circuit [3], [4]. This demand is stricter for 3-D ICs due to the increased power density and related thermal limitations.

In 2-D circuits, symmetric interconnect structures, such as H- and X-trees, are widely utilized to distribute the clock signal across a circuit [2]. The symmetry of these structures permits the clock signal to arrive at the leaves of the tree at the same time, resulting in synchronous data processing. Maintaining this symmetry within a 3-D circuit, however, is a difficult task.

An extension of an H-tree to three dimensions does not guarantee equidistant interconnect paths from the root to the leaves of the tree. The clock signal propagates through vertical interconnects, typically implemented by through silicon vias (TSVs) from the output of the clock driver to the center of the H-tree on the other planes. The impedance of the TSVs can increase the time for the clock signal to arrive

This research is supported in part by the National Science Foundation under Contract No. CCF-0541206, grants from the New York State Office of Science, Technology & Academic Research to the Center for Advanced Technology in Electronic Imaging Systems, by grants from Intel Corporation, Eastman Kodak Company, and Freescale Semiconductor Corporation, and foundry support from MIT Lincoln Laboratories.

at the leaves of the tree on these planes as compared to the time for the clock signal to arrive at the leaves of the tree located on the same plane as the clock driver. Furthermore, in a multi-plane 3-D circuit, three or four branches can emanate at each branch point. The third and fourth branches propagate the clock signal to the other planes of the 3-D circuit. Similar to a design methodology for a 2-D H-tree topology, the width of each branch is reduced by a third (or more) of the segment width preceding the branch point in order to match the impedance at that branch point. This requirement, however, is difficult to achieve as the third and fourth branches are connected by a TSV.

Global signaling issues in 3-D circuits, such as clock signal distribution, are essentially unexplored. Recent papers consider thermal effects on buffered 3-D clock trees [5] and H-tree topologies [6], [7]. No experimental characterization of 3-D clock distribution networks, however, has been presented. Measurements from a 3-D test circuit fabricated by the MIT Lincoln Laboratories (MITLL) [8] employing several clock distribution architectures are presented for the first time in this paper.

In the following section, the design of the 3-D test circuit is described. A brief discussion of the MITLL process is provided in Section III. Experimental results and a discussion of the characteristics of the three clock distribution networks are presented in Section IV. Some conclusions are offered in Section V.

II. DESIGN OF THE 3-D TEST CIRCUIT

The test circuit consists of three blocks. Each block includes the same logic circuit but implements a different clock distribution architecture. The total area of the test circuit is 3 mm × 3 mm, and each block occupies an approximate area of 1 mm^2. Each block contains about 30,000 transistors with a power supply voltage of 1.5 volts. The design kit used for the implementation has been provided by North Carolina State University [10]. The common logic circuitry used in each clock module is described in Section II-A, and the different clock distribution architectures are reviewed in Section II-B.

A. 3-D Circuit Architecture

The logic circuit common to the three blocks is described in this section. An overview of the logic circuitry is depicted in Fig. 1. The function of the logic is to emulate different switching patterns of the circuit and load conditions for the clock distribution networks under investigation. The logic is repeated in each plane and includes pseudorandom number generators (PNG), a six by six crossbar switch, control logic for the crossbar switch, several groups of four-bit counters and current loads, and RF output pads for probing.

The pseudorandom number generators use linear feedback shift registers and XOR operations to generate a random 16-bit word every clock cycle once the generators are initialized [9]. The data flow in this circuit can be described as follows. After resetting the circuit, the PNGs are initialized and the control logic connects each input port to the appropriate output port. Since the control logic includes an eight-bit counter, each input port of the crossbar switch is successively connected every 256 clock cycles to each output port.

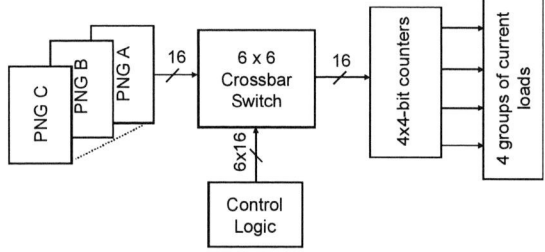

Fig. 1   Block diagram of the logic circuit included on each plane for each clock topology.

The current loads are implemented with cascode current mirrors, as shown in Fig. 2. In these cascode current mirrors, the output current I_{out} closely follows I_{ref} as compared to a simple current mirror. The reference current I_{ref} is externally provided to control the amount of current drawn from the circuit. The gate of transistor M5 is connected to the MSB of a four-bit counter, shown in Fig. 2 as the *sel* signal. This additional device is used to switch the current sinks. The width of the devices shown in Fig. 2 is $W_1 = W_2 = W_3 = W_4 = 600$ nm, and $W_5 = 2000$ nm.

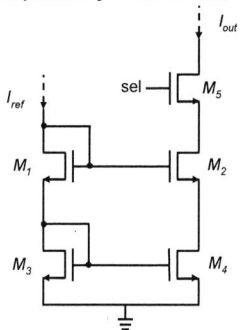

Fig. 2   Cascoded current mirror with an additional control transistor.

Several capacitors are included in each circuit block and serve as an extrinsic decoupling capacitance and are implemented by MIM capacitors. Additionally, each of the circuit blocks is supplied by separate power and ground pads (three pairs of power and ground pads per block) to ensure that

each block can be individually tested. Furthermore, one pair of power and ground pads is connected to the pad ring in order to provide protection from electrostatic discharge.

B. 3-D Clock Topologies

Several clock network topologies for 3-D ICs are described in this section. These architectures combine different topologies which are commonplace in 2-D circuits, such as H-trees, rings, and meshes [2]. Each of the three blocks includes a different clock distribution structure, which are schematically illustrated in Fig. 3. The dashed lines depict vertical interconnects implemented by through silicon vias.

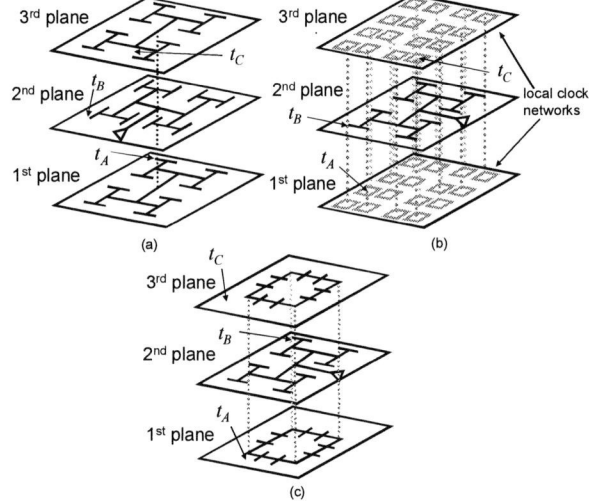

Fig. 3   Various 3-D clock distribution networks within the test circuit, (a) H-trees, (b) H-tree and local rings/meshes, (c) H-tree and global rings.

In each of the circuit blocks, the clock driver for the entire clock network is located on the second plane. The location of the clock driver is chosen to ensure that the clock signal propagates through identical vertical interconnect paths to the first and third planes, ideally resulting in the same delay when the clock signal reaches the first and third planes. The clock driver is implemented with a traditional chain of tapered buffers [11], [12]. Additionally, buffers are inserted at the leaves of each H-tree in all three topologies.

The architectures employed in the blocks are:

Block A: All of the planes contain a four level H-tree (*i.e.,* equivalent to 16 leaves) with identical interconnect characteristics. All of the H-trees are connected through a group of TSVs at the output of the clock driver. Note that the H-tree on the second plane is rotated by 90° with respect to the H-trees on the other two planes. This rotation eliminates inductive coupling between the H-trees. All of the H-trees are shielded with two parallel lines connected to ground.

Block B: A four level H-tree is included in the second plane. All of the leaves of this H-tree are connected by TSVs to small local rings on the first and third planes. As in Block A, the H-tree is shielded with two parallel lines connected to ground. Additional interconnect resources are used to form

978-1-4244-2018-6/08 $25.00 © 2008 IEEE

local meshes. Due to the limited interconnect resources, however, achieving a uniform mesh in each ring is difficult. Clock routing is constrained by the power and ground lines as only three metal layers are available on each plane [8].

Block C: The clock distribution network for the second plane is a shielded four level H-tree. Two global rings are utilized for the other two planes. Buffers are inserted to drive each ring, which are connected by TSVs to the four branch points on the second level of the H-tree. The registers in each plane are connected either directly to the ring or to buffers at the leaves of the tree on the second plane.

III. FABRICATION OF THE 3-D TEST CIRCUIT

The manufacturing process developed by MITLL for fully depleted silicon-on-insulator (FDSOI) 3-D circuits is summarized here [8]. The MITLL process is a wafer level 3-D integration technology with up to three FDSOI wafers bonded to form a 3-D circuit. The diameter of the wafers is 150 mm. The minimum feature size of the devices is 180 nm, with one polysilicon layer and three metal layers interconnecting the devices on each wafer. A backside metal layer also exists on the upper two planes, providing the pads for the TSVs and the I/O, power supply, and ground pads for the entire 3-D circuit. An attractive feature of this process is the high density TSVs. The dimensions of these vias are 1.75 μm × 1.75 μm, much smaller than the size of the through silicon via in many existing 3-D technologies [13], [14]. An intermediate step of the fabrication process is illustrated in Fig. 4, where some salient features of this technology are also depicted.

IV. EXPERIMENTAL RESULTS

The clock distribution network topologies of the 3-D test circuit are evaluated in this section. The fabricated circuit is depicted in Fig. 5, where the different blocks can be distinguished. Each block includes four RF pads for measuring the delay of the clock signal. The pad located at the center of each block provides the input clock signal. The clock input is a sinusoidal signal with a DC offset, which is converted to a square waveform at the output of the clock driver. The remaining three RF pads are used to measure the delay of the clock signal at specific points on the clock distribution network within each plane. A buffer is connected at each of these measurement points, and the output of this buffer drives the gate of an open drain transistor connected to the RF pad.

A clock waveform acquired from the topology combining an H-tree and global rings, shown in Fig. 3a, is illustrated in Fig. 6, demonstrating operation of the circuit at 1.4 GHz. The clock skew between the planes of each block is listed in Table I. For the H-tree topology, the clock signal delay is measured from the root to a leaf of the tree on each plane, with no other load connected to these leaves. The skew between the leaves of the H-trees on planes A and C (*i.e.* t_{AC})

is essentially the delay of a stacked TSV traversing between the three planes to transfer the clock signal from the target leaf to the RF pad on the third plane. The delay t_B is larger due to the additional capacitance coupled into that quadrant of the H-tree on the second plane. This capacitance is intentional on-chip decoupling capacitance placed under this quadrant, increasing the measured skew, t_{BC} and t_{BA}. This topology produces, on average, the lowest skew as compared to the other two topologies.

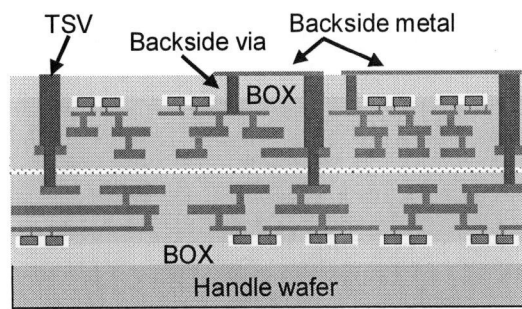

Fig. 4 Intermediate step of the MITLL process [8]. The second plane is flipped and bonded with the first plane. The backside metal layer and vias and the through silicon vias are also shown.

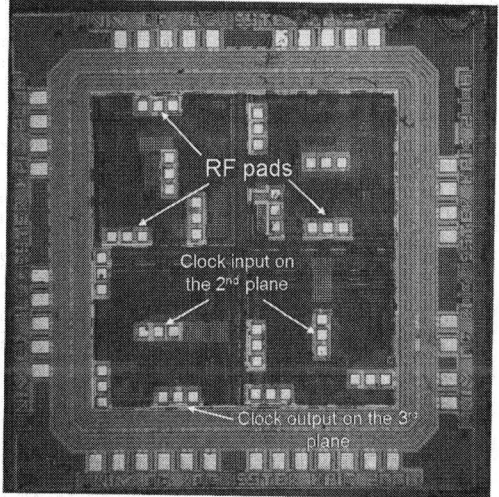

Fig. 5 Fabricated 3-D circuit. Some of the RF pads are also depicted.

Table I   Measured clock skew among the planes of each block.

Clock distribution network	Clock skew [ps]		
	$t_{BA} = t_B - t_A$	$t_{BC} = t_B - t_C$	$t_{AC} = t_A - t_C$
H-trees (Fig. 3a)	32.5	28.3	-4.2
Local meshes (Fig. 3b)	-68.4	-18.5	49.8
Global rings (Fig. 3c)	-112.0	-130.6	-18.6

The clock skew among the planes is greater for the local mesh topology as compared to the H-tree topology, primarily due to the unbalanced clock load for certain local meshes. The greatest difference in the load is between the measurement points on planes A and B, which also produces the largest skew for this topology. The increase in skew, however, is moderate. Additionally, the local meshes

978-1-4244-2018-6/08 $25.00 © 2008 IEEE

ever, is moderate. Additionally, the local meshes reduce the local skew for the load connected to each sink of the H-tree.

Alternatively, the clock distribution network that includes the global rings exhibits very low skew for planes A and C, those planes that include the global rings. Although the clock load on each ring is non-uniformly distributed, the load balancing characteristic of the ring yields a relatively low skew between these planes. Since the clock distribution network on the second plane is implemented with an H-tree, skew t_{BC} and t_{BA} is significantly larger than t_{AC}. Note that the leaf of the H-tree, where the clock signal delay for the second plane is measured, is located at a great distance from the rings on planes A and C (see Fig. 3c). A combination of the H-tree and global rings, consequently, is not a suitable approach for 3-D circuits due to the large difference in distance that the clock signal traverses on each plane.

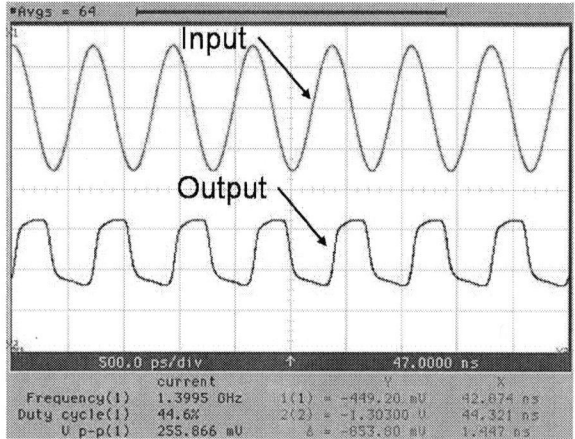

Fig. 6 Clock signal input and output waveform from the topology illustrated in Fig. 3c.

In Table II, the measured power consumption of the blocks operating at 1 GHz is reported. The local mesh topology demonstrates the lowest power consumption. This topology requires the least interconnect resources for the global clock network, since the local meshes are connected at the output of the buffers located on the last level of the H-tree on the second plane. In addition, this topology requires a moderate amount of local interconnect resources as compared to the H-tree and local mesh topologies. Alternatively, the power consumed by the H-tree topology is the highest as this topology requires three H-trees and additional wiring for local connections to the leaves of each tree. Finally, the global rings block consumes slightly less power than the H-tree topology due to the reduced amount of wiring resources used by the global clock network.

V. Conclusions

Three topologies to globally distribute a clock signal in 3-D circuits have been evaluated. A 3-D test circuit, based on the MITLL 3-D IC manufacturing process, has been designed, fabricated, and measured and is shown to operate at 1.4 GHz. Clock skew measurements indicate that a topology

that combines the symmetry of an H-tree on the second plane and local meshes on the other two planes will result in low clock skew for 3-D circuits while consuming the lowest power as compared to the other investigated topologies.

Table II   Measured power consumption of each block at 1 GHz.

Clock distribution network	Power consumption [mW]
H-trees (Fig. 3a)	260.3
Local meshes (Fig. 3b)	168.3
Global rings (Fig. 3c)	228.5

Acknowledgments

The authors would like to thank Yunliang Zhu, Lin Zhang, and Professor Hui Wu of the University of Rochester for their help during testing and the MIT Lincoln Laboratories for fabricating the 3-D circuit.

References

[1] E. G. Friedman (Ed.), *Clock Distribution Networks in VLSI Circuits and Systems,* IEEE Press, New Jersey, 1995.

[2] E. G. Friedman, "Clock Distribution Networks in Synchronous Digital Integrated Circuits," *Proceedings of the IEEE,* Vol. 89, No. 5, pp. 665-692, May 2001.

[3] D. W. Bailey and B. J. Benschneider, "Clocking Design and Analysis for a 600-MHz Alpha Microprocessor," *IEEE Journal of Solid-State Circuits,* Vol. 22, No. 11, pp. 1627-1633, November 1998.

[4] T. Xanthopoulos *et al.,* "The Design and Analysis of the Clock Distribution Network for a 1.2 GHz Alpha Microprocessor," *Proceedings of the IEEE International Solid-State Circuits Conference,* pp. 402-402, February 2001.

[5] J. Minz, X. Zhao, and S. K. Lim, "Buffered Clock Tree Synthesis for 3D ICs Under Thermal Variations," *Proceedings of the IEEE International Asia and South Pacific Design Automation Conference,* pp. 504-509, January 2008.

[6] M. Mondal *et al.,* "Thermally Robust Clocking Schemes for 3D Integrated Circuits," *Proceedings of the IEEE International Conference on Design, Automation and Test in Europe,* pp. 1206-1211, April 2007.

[7] V. Arunachalam and W. Burleson, "Low-Power Clock Distribution in a Multilayer Core 3D Microprocessor," *Proceedings of the ACM/IEEE International Great Lakes Symposium on VLSI,* May 2008.

[8] *"FDSOI Design Guide,"* MIT Lincoln Laboratories, Cambridge, 2006.

[9] W. Cui, H. Chen, and Y. Han, "VLSI Implementation of Universal Random Number Generator," *Proceedings of the IEEE Asia-Pacific Conference on Circuits and Systems,* Vol. 1, pp. 465-470, October 2002.

[10] Available on-line: http://www.ece.ncsu.edu/erl/3DIC/pub

[11] N. Hedenstierna and K. O. Jeppson, "CMOS Circuit Speed and Buffer Optimization," *IEEE Transactions on Computer-Aided Design of Integrated Circuits and Systems,* Vol. CAD-6, No. 2, pp. 270-281, March 1987.

[12] B. S. Cherkauer and E. G. Friedman, "A Unified Design Methodology for CMOS Tapered Buffers," *IEEE Transactions on Very Large Scale Integration (VLSI) Systems,* Vol. VLSI-3, No. 1, pp. 99-111, March 1995.

[13] M. W. Newman *et al.,* "Fabrication and Electrical Characterization of 3D Vertical Interconnects," *Proceedings of the IEEE International Electronic Components and Technology Conference,* pp. 394-398, June 2006.

[14] P. Dixit and J. Miao, "Fabrication of High Aspect Ratio 35 μm Pitch Interconnects for Next Generation 3-D Wafer Level Packaging by Through-Wafer Copper Electroplating," *Proceedings of the IEEE International Electronic Components and Technology Conference,* pp. 388-393, June 2006.

978-1-4244-2018-6/08 $25.00 © 2008 IEEE

IEEE 2008 Custom Intergrated Circuits Conference (CICC)

Inter-Die Signaling in Three Dimensional Integrated Circuits

Christopher Mineo, Ravi Jenkal, Samson Melamed, and W. Rhett Davis

Department of Electrical and Computer Engineering, North Carolina State University

{chris_mineo, rsjenkal, slmelame, rhett_davis}@ncsu.edu

Abstract—This work discusses a three dimensional network on chip (3D NoC) fabricated in the 0.18μm MIT Lincoln Laboratories 3D FDSOI 1.5V process. As a proof of concept, a three tier, 27 node, NoC test chip occupying 4 mm^2 per tier was designed and tested. It is the first of its kind to demonstrate successful inter-tier signaling in a complex three dimensional design, and validates the technology as a viable alternative to the continued scaling of conventional CMOS processes. Simulated results show that when implemented in this 3D process, simple 3D mesh interconnection networks allow for the sharing of global routing resources for complex systems while consuming an extremely low 2 mW of power per transaction. Using these results, we establish the need for a 3D network simulator to quantify the advantage 3D circuit implementations have over 2D.

I. INTRODUCTION

As the cost of continuing to scale down CMOS technologies increases, researchers have been considering three dimensional integrated circuits (3DICs) [1] as a path to continued performance and device density gains through process advancements. For a given design, 3DICs have the benefit of reduced average wire length over an equivalent traditional 2D implementation, and therefore typically will have shorter path delays and consume less power. Simultaneously, in recent technology nodes wire propagation delay has began to dominate gate delay; these trends necessitate more elegant global chip-level communication systems than those common today. Additionally, as digital systems continue to increase in complexity, parallel processors with an ever-increasing number of cores are becoming prevalent [2]. Networks on chip (NoC) have emerged as a way to enable processing elements to communicate efficiently on-chip. These interconnection networks provide the design reuse, well controlled interconnect electrical properties, and signal routing resource sharing that will make the massively parallel digital systems of the future possible.

The implementation of an NoC in a 3D technology is particularly attractive because of the additional degrees of connectivity stacked dice allow. Many high performance interconnection networks have complex topologies where their elements are of high node degree. Traditional 2D technologies only allow a large circuit macro to have four adjacent neighbors, and I/Os only around its perimeter. 3D technologies, however, allow a circuit an additional two nearest neighbors. Inputs and outputs to macros in other dice, or other tiers, can be placed throughout the area of the entire circuit.

For these reasons we have designed the three dimensional network on chip (3D NoC) test chip, fabricated in the 0.18μm 3D technology described in the following section. The 3D NoC is a 27 node 3D mesh interconnection network of nodes that send messages over the network just as independent processing elements would communicate with one another over a network. Primarily, this chip serves as a proof of concept to show that digital signaling between the tiers of a 3D processes is feasible and reliable. Secondly, it emulates the traffic in a 3D parallel processing system, so that we can characterize the network performance and power consumption for a high level network simulator that quantifies the tradeoffs in NoC design space. The 3D NoC is the first chip of which we are aware that demonstrates functionally correct inter-tier and intra-tier signaling in a complex synthesized fully digital circuit. We have verified, through test vectors, that all 162 top level NoC links in the 3D NoC have yielded.

In Sec. II we will discuss the details of the MIT 3D fabrication process. The specifics of the 3D NoC architecture will be explained in Sec. III-A, and the significance of our design methodology be discussed in Sec. III-B. In Sec. IV we present our simulated and measured results, and Sec. V gives an overview of the NoC simulator.

II. MIT LINCOLN LABORATORIES 3D PROCESS

The Massachusetts Institute of Technology Lincoln Laboratories (MITLL) 1.5V three dimensional fully-depleted silicon-on-insulator (3D FDSOI) 0.18μm process fabricates wafers of three dice independently. The wafers are stacked, and connections between the three dice, or inter-tier vias, are inserted. The chips are then scribed, so the result are 3DICs made up of three dice (or tiers) each. Each tier has three metal layers, in addition to a metal layer between the top two tiers, and a metal layer on top of the entire stack. The bottom tier, tier A, is face up in the stack, while the middle tier, tier B, and the top tier, tier C, are flipped before being stacked on top of tier A such that they are face down. All of the primary chip inputs and outputs connect to tier C. A cross section of the process can be seen in Fig. 1.

Along with the benefits, 3D technologies also expose a new set of potential problems. Perhaps the caveat most likely to create failing chips is the new source of timing variation. Present 2D digital designs are prone to timing variation across a single die. Random and systematic fluctuations in the manufacturing process can cause some devices on a die to switch faster than

978-1-4244-2018-6/08 $25.00 © 2008 IEEE

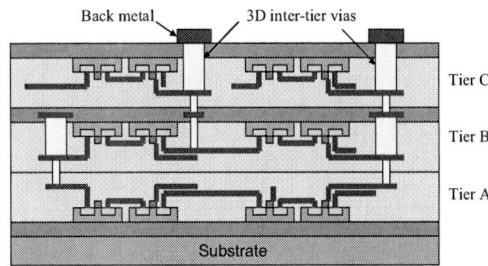

Fig. 1. MITLL 0.18μm 3D FDSOI chip stack [3]

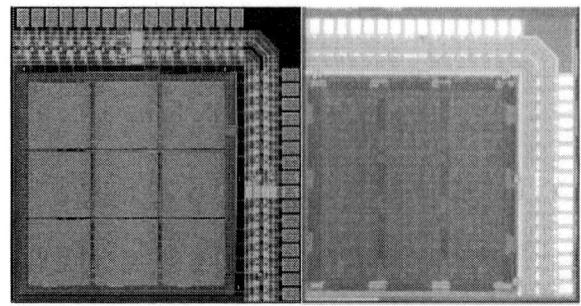

Fig. 2. Tier C layout of the 3D NoC (left) and die photograph (right)

others, as well as cause the rate of signal propagation down wires to vary across the chip. If not accounted for in timing calculations, this can cause unexpected setup or hold time violations. The timing variation from wafer to wafer in 3D is much larger than the timing variation across a single die, so we must account for this in our design methodology. However, 3D technologies enable the creation of high performance circuits and new types of systems, such as the 3D NoC. Interconnection networks with a high degree of connectivity truly exploit the advantages of vertical interconnect.

III. 3D NETWORK ON CHIP

Complex systems are becoming harder to design and verify, often with costly global interconnect which impose significant limitations on the performance and power consumption. Dally and Towles advocate system on chip design via a modular approach in which functional units do not communicate via global dedicated wires, but over a central interconnection network [4]. Using such a network brings uniformity to global interconnect through well controlled electrical parameters, simplifying chip timing. Furthermore, designs become very modular and can benefit from design reuse. Additionally, Heo and Asanović focus on the global interconnect issues of todays chips from a power consumption standpoint [5]. The authors suggest using dedicated wires for local interconnections that fall completely within a tile, and replacing global interconnections that cross tile boundaries with a centralized interconnection network. They aim to minimize power consumption through the sizing of the tiles. These results by other researchers motivate us to explore interconnection networks in the 3DIC design space.

A. Network on Chip Architecture

The 3D NoC is a 81k gate equivalent design (363,108 devices) occupying a 2mm by 2mm footprint on each of three tiers in the MITLL 0.18μm 3D FDSOI process described in Sec. II. The 3-ary 3-cube 27 node synchronous interconnection network consists of 9 nodes tiled on each tier. The top tier of the chip is shown in Fig. 2; the 9 nodes on this tier and the pads are clearly visible. As can be seen in Fig. 3, each node includes a 7 port single stage fixed priority router, one port of which connects to a functional unit internally within the node. The remaining ports allow the router exchange messages with its ±X or ±Y neighbors using links within its own tier,

and messages with the ±Z nodes through inter-tier vias. The 3D nature of this process allows us to implement a design such as a *k*-ary 3-cube, in which a unit has 6 theoretical nearest neighbors, with a physical design where all 6 of those neighbors are actually adjacent to the unit. This allows the links between the nodes to be very short and messages to pass through a node in a single clock cycle. Such a trivial placement would not exist for this topology in a 2D technology, leading to very long inter-node links which would make the network, perhaps prohibitively, slower and more power hungry.

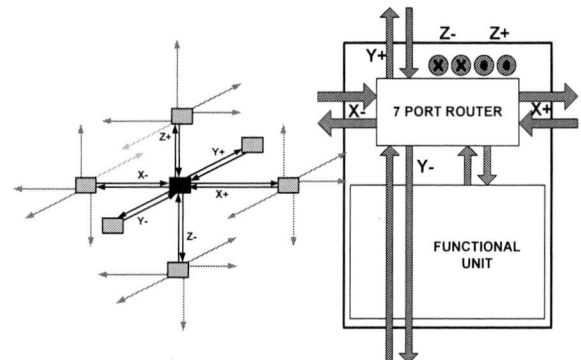

Fig. 3. 3D NoC node block diagram

A port is comprised of 2 unidirectional 16 bit links made up of address, data, valid, and acknowledge signals. The functional unit in each node makes use of a linear feedback shift register (LFSR) to receive incoming messages and compute a new pseudorandom address to forward that data to. Once injected into the network, single byte messages will travel around the network according to an adaptive version of the XYZ routing algorithm [6] until the chip is reset, behaving just as would parallel processing hardware communicating over a network.

B. Design Methodology

Studies of 3D technologies such as those in [7], [8], and many others have shown that there is great promise for reducing global interconnect length through 3DIC circuit implementations. The work of [7] shows the use of 3DICs to reduce power consumption, path delays, and average interconnect length for an example design. This is achieved via a complex

3D design methodology involving partitioning a gate level netlist of the design into n groups for an n-tier process. The tiers are placed and routed independently, and inter-tier vias are inserted and aligned to make the necessary connections between the tiers. Methodologies such as this should produce high performance circuits that almost certainly reduce power and increase performance over their 2D counterparts, yet at the time of writing we have no knowledge of any fully functioning fabricated digital chip synthesized and created completely with electronic design automation tools as large as the 3D NoC.

With such a design methodology, in order for a 3D chip in an n-tier process to yield, n defect-free dice must be combined. Since, presently, the wafers are stacked before the chips are scribed, we cannot yet test individual dice before assembling a 3D chip. Additionally, as mentioned in Sec. II most 3D design methodologies cannot tolerate inter-wafer wire or device speed variation. Cells are placed to minimize a metric, such as number of inter-tier vias, with little regard for actual timing paths. Timing paths are often allowed to cross tier boundaries as many times as is convenient for the physical design, such that timing violations are very likely if a "fast" die and "slow" die are combined in the same 3D chip. As 3D processes mature, inter-tier via yield is improving drastically, however chip timing could still be affected by significant inter-tier via parasitic RC variation.

Instead, in the case of the 3D NoC all tier crossings are carefully planned with inter-tier vias specially instantiated. The logic comprising the individual nodes is synthesized, placed, and routed using conventional 2D digital design methodology. The NoC is then assembled and connected such that only inter-node NoC links cross tier boundaries; the only inter-tier vias used are for power, ground, clock, and reset routing, and as the vertical links in the 3D mesh network. Since there are memory elements at the outputs of each node, a minimal number of timing paths cross tier boundaries. The timing arcs that do contain inter-tier vias, contain no more than one. Furthermore, the nodes use a valid/acknowledge handshaking interface to communicate, so there is handshaking at all inter-tier vias carrying data.

IV. EVALUATION OF ARCHITECTURE

The 3D NoC was designed and simulated with a post-place and route operating frequency of 145 MHz. At this speed, the test chip is expected to consume a maximum of 120.5 mW when operating with 17 messages simultaneously in transit, though as shown in Fig. 4, the power consumption varies with the number of messages in the network. The power increases linearly with the number of messages, showing that the power per inter-node transaction is a constant 2.013 mW on average. As we continue to increase the number of messages in the network, the power drops drastically because there begins to be significant contention for NoC links as the network approaches deadlock. With no messages in the network, 86.1 mW of dynamic power is consumed by clock power and the LFSRs in the functional units, the routers are not switching.

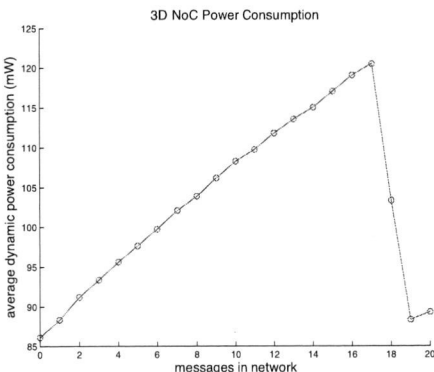

Fig. 4.   3D NoC dynamic power consumption varies with network traffic

The 3D NoC was mounted and wire bonded to a custom designed PCB as shown in Fig. 5 for testing. Input stimulus was applied by a digital pattern generator via SMA cables to header pins. Messages were preloaded into the digital pattern generator for injection into the 3D NoC through the +X port on the upper right node of tier C. The output pins allow us to monitor, using a logic analyzer, the -X output link on the upper right node of tier C to be sure the correct messages cross that link at the correct times.

Fig. 5.   3D NoC mounted on a test PC board

Functional correctness was verified by a test vector of 12 messages that utilize all 162 unidirectional links in the network, and use the functional unit in each of the 27 nodes. Furthermore, we observe that the routers correctly arbitrate between their input ports when there is contention for an output port. Messages predictably traverse the network both within a single tier and across tiers, thus validating both inter-tier and intra-tier signaling in 3DICs. However, at the time of review, the chip has been unable to operate functionally correct at frequencies higher than 2.25 MHz. We are working to resolve this and continue with our power analysis. The nominal process power supply voltage is 1.5V, but the 3D NoC operates at 2.25 MHz with supply voltages from 0.9V to 2.5V. These flat characteristics suggest an unidentified problem in the test setup, perhaps having to do with decoupling the power and ground supply rails. The 3D NoC demonstrates that reliable inter-tier signaling is feasible in a 3D process, but high level network simulation is necessary to see how much we can benefit from 3D architectures over 2D.

V. NoC SIMULATOR

A goal of creating the 3D NoC is to validate and provide information to a high level system simulator presently in development. As discussed in Sec. III, we are entering a time where global interconnect and communication networks in massively parallel systems are becoming of extreme importance. Considering this, we need an efficient method to make the many interconnection network design decisions, such as topology, level of parallelism, etc. The NoC simulator currently under development simulates low-density parity check decoding as a benchmark.

In recent years low-density parity check (LDPC) decoding has received much attention. The computationally intensive yet highly parallelizable message-passing based LDPC decoding algorithm for error correction coding (ECC) [9] is capable of achieving very low bit error rates, making it among the most powerful known ECC algorithms. Many researchers [10], [11] have published high performance LDPC architectures offering varying degrees of parallelism, spanning a wide range of power/performance/code re-configurability options. The wide range of possible tradeoffs in LDPC decoder hardware make it an attractive application for the NoC simulator.

For a task to be implemented over an interconnection network, such as LDPC decoding, we can use a cycle accurate, fast, high level simulator to simulate both the message passing and computation. We can then plot hardware performance versus level of parallelism (number of processing elements), as in Fig. 6. The plot shows that, for both 2D and 3D mesh networks, as you increase the number of parallel LDPC processing nodes, and hence increase area and power consumption, the number of cycles required to decode a single codeword decreases; decoding throughput increases. Furthermore, it quantifies the significant throughput advantage in using the 3D mesh topology, ranging from $2\times$ to $4\times$. This demonstrates the performance advantage of a topology that would suffer from routing congestion and inefficient top level interconnect if implemented in a 2D process, but would not in a 3D process. The simulator helps to quantify this tradeoff, as well as quantify the difference in using various network topologies, routing algorithms, etc. Data taken from the 3D NoC chip will validate the NoC simulator and provide information to characterize the power consumption of interconnection networks. Furthermore, the functional correctness of the 3D NoC shows that implementing an interconnection network in a 3D technology is a feasible option.

VI. CONCLUSION

In this work we have described the 81k G.E. 3D NoC test chip, the first functionally correct complex digital design to demonstrate successful inter-tier signaling in a 3D process. It validates the notion that 3DICs are a viable alternative to shrinking feature size; they allow us to continue to use a mature technology node while benefiting from reduced average

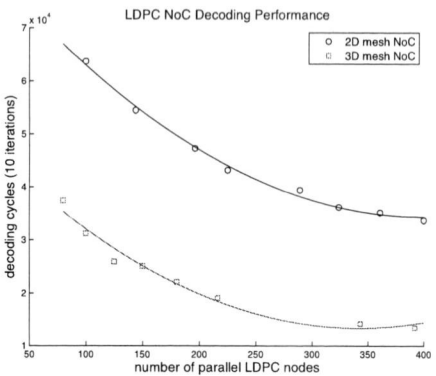

Fig. 6. Plots showing the decoding performance of various 2D and 3D mesh LDPC networks for a 16200 bit, rate $\frac{1}{4}$ code

interconnect length. As complex digital systems turn to interconnection networks to handle their global signal routing, we will undoubtably see the advantages of vertical interconnect.

REFERENCES

[1] V. Suntharalingam, R. Berger, J. Burns, C. Chen, C. Keast, J. Knecht, R. Lambert, K. Newcomb, D. O'Mara, D. Rathman, D.D. A10 Rathman, D. Shaver, D.C. A11 Shaver, A. Soares, A.M. A12 Soares, C. Stevenson, C.N. A13 Stevenson, B. Tyrrell, B.M. A14 Tyrrell, K. Warner, K. A15 Warner, B. Wheeler, B.D. A16 Wheeler, D.-R. Yost, D.-R.W. A17 Yost, and D. Young, D.J. A18 Young, "Megapixel CMOS image sensor fabricated in three-dimensional integrated circuit technology," in *Proc. Digest of Technical Papers Solid-State Circuits Conference ISSCC. 2005 IEEE International*, R. Berger, Ed., 2005, pp. 356–357 Vol. 1.

[2] S. Borkar, "Thousand Core ChipsA Technology Perspective," in *Proc. 44th ACM/IEEE Design Automation Conference DAC '07*, 2007, pp. 746–749.

[3] *MITLL Low-Power FDSOI CMOS Process Design Guide*, Revision 2006:7 ed., Massachusetts Institute of Technology Lincoln Labs, October 2006.

[4] W. Dally and B. Towles, "Route packets, not wires: on-chip interconnection networks," in *Proc. Design Automation Conference*, B. Towles, Ed., 2001, pp. 684–689.

[5] S. Heo and K. Asanovic, "Replacing global wires with an on-chip network: a power analysis," in *Proc. International Symposium on Low Power Electronics and Design ISLPED '05*, K. Asanovic, Ed., 2005, pp. 369–374.

[6] J. Duato, S. Yalamanchili, and N. Lionel, *Interconnection Networks: An Engineering Approach*. San Francisco, CA, USA: Morgan Kaufmann Publishers Inc., 2002.

[7] W. Davis, H. Hua, A. Sule, C. Mineo, S. Melamed, M. Steer, and P. Franzon, "Wire-delay reduction analysis of a 3-tier, 8-point Fast Fourier Transform 3D-IC," in *International VLSI Multilevel Interconnect Conference*, October 2005, pp. 474–479.

[8] W. Davis, J. Wilson, S. Mick, J. Xu, H. Hua, C. Mineo, A. Sule, M. Steer, and P. Franzon, "Demystifying 3D ICs: the pros and cons of going vertical," *IEEE Design & Test of Computers*, vol. 22, no. 6, pp. 498–510, 2005.

[9] R. Gallager, "Low-density parity-check codes," *IEEE Transactions on Information Theory*, vol. 8, no. 1, pp. 21–28, 1962.

[10] A. Blanksby and C. Howland, "A 690-mW 1-Gb/s 1024-b, rate-1/2 low-density parity-check code decoder," *IEEE Journal of Solid-State Circuits*, vol. 37, no. 3, pp. 404–412, 2002.

[11] P. Urard, E. Yeo, L. Paumier, P. Georgelin, T. Michel, V. Lebars, E. Lantreibecq, and B. Gupta, "A 135Mb/s DVB-S2 compliant codec based on 64800b LDPC and BCH codes," in *Proc. Digest of Technical Papers Solid-State Circuits Conference ISSCC. 2005 IEEE International*, E. Yeo, Ed., 2005, pp. 446–609 Vol. 1.

IEEE 2008 Custom Intergrated Circuits Conference (CICC)

Variability in 3-D Integrated Circuits

Filipp Akopyan, Carlos Tadeo Ortega Otero, David Fang, Sandra J. Jackson and Rajit Manohar
Computer Systems Laboratory
Cornell University, Ithaca, NY 14853
Email: {filipp, cto3, fang, sandra, rajit}@csl.cornell.edu

Abstract—**In recent years, there has been a trend among digital and analog circuit designers towards three-dimensional integration. There has been some debate regarding the applicability of 3-D technology to general logic circuits, especially with regard to thermal issues. We examine process variations on the same layer, across layers, and cross-chip variations. We show how the performance of each layer of the 3-D chip varies with temperature, and demonstrate the effect of heat pipes on circuit performance.**

I. INTRODUCTION

The constant trend toward increased logic density and higher complexity of integrated circuits faces challenges due to interconnect limitations. The constraint of mapping complicated devices and their interconnections onto a single device plane can result in artificial increases in wire length and delay, leading to degrading performance. Three-dimensional (3-D) integration has been proposed to remove the planar constraint from IC design [4], [8], [10], [13], [17]. The prospect of being able to place subcircuits physically closer to each other is an attractive option to shorten interconnect wires, and improve performance and density.

Currently deployed technologies for 3-D integration include: wire bonding, microbump, contactless, and through-vias that connect separately manufactured device layers (die stacking) [6], [7], [13], [15]. Structures that are likely to benefit from shortening wires (as a result of 3-D integration) are those that are sensitive to interconnect-latency, such as memory arrays, and FPGA interconnects [3], [10], [12]. Die stacking also allows designs to integrate Systems on Chip (SoC) using *different* process technologies (e.g. RF CMOS, SiGe) [5].

However, to exploit these benefits, designers must confront new challenges from 3-D integration: reduced yield, increased heat density (compounded with inferior thermal dissipation), and intra-chip process variations. Poor thermal properties threaten to increase local temperatures, which can degrade performance and increase leakage of transistors [1], [9], [11], [13], [17], [19]. Proposed solutions to reduce chip temperature include: better heat sinks, packaging solutions [19], and inter-tier vias to reduce inter-tier thermal resistance [1], [11], [14].

Process variation is a major threat to the viability of die-stacked integration [2]. Since device layers are manufactured separately and then assembled, variation is compounded. Variation can result in yield loss when a fault on a single layer breaks a design, and can result in performance loss when catering to worst case delays or in requiring generous timing margins. Thermal properties and performance mismatch between tiers makes designing 3-D synchronous circuits even more challenging. Circuit designers may resort to designing more conservatively, or employing multiple clock domains to isolate the impact of variation. Self-timed or asynchronous circuits can tolerate drastic variations, and may help pioneer large-scale, 3-D integrated designs even if the process has not been well-characterized [16]. We designed a 3-D test chip to study the impact of process variations in 3-D integration on performance mismatch between different locations on the same tier, across different tiers, and across different assembled chips. We measure how performance of various circuits respond to temperature, and profile how heat dissipates laterally within a tier and vertically through tiers. We also demonstrate how strategic placement of heat pipes (through-vias) improves local thermal conductivity (and hence, performance) at crucial locations within the chip.

Although our experiments are based on one particular process, we believe that the qualitative observations we make are relevant for similar technologies.

II. EXPERIMENTAL SETUP

Process. The MIT-LL 3-D through-via process [15] was used for our design. This process offers three die stacks (tiers) and three metal layers per tier. MIT-LL 3-D features $180nm$ devices on a fully depleted silicon-on-insulator (FDSOI) technology, with $1.5\mu m$ tungsten-filled through-vias.

During the fabrication, three 150-mm FDSOI wafers are integrated. Wafers are labeled as tiers A, B and C; with tier A being the base wafer. Handle substrates of tiers B and C are removed. Tier B is inverted, aligned and bonded to tier A using a low-temperature wafer-wafer oxide. 3-D vias are etched through the oxides, filled with tungsten and planarized using chemical-mechanical polishing. Tier C is deposited onto the structure using a similar process. During post-fabrication, bond pads and heat sink cuts are created.

Circuits. Our test structures were placed on all three-tiers of the process. All of the tiers have identical circuitry for comparison purposes. Each tier was divided into a 3x3 array of identical tiles (sites), and labeled as shown in Fig. 1(a).

In order to demonstrate performance degradation due to heating and to illustrate process variations of the 3-D design, we have chosen to use asynchronous (self-timed) circuits. Asynchronous circuits operate without a clock and use handshakes to communicate data. Asynchronous circuits have the capability of operating correctly at maximum local throughput

978-1-4244-2018-6/08 $25.00 © 2008 IEEE

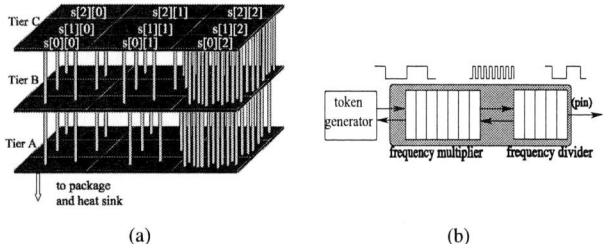

Fig. 1. (a) Floorplan; (b) Multiplier/divider chain

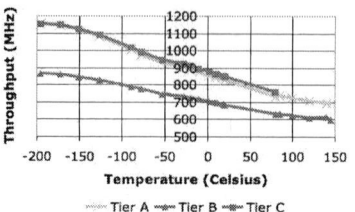

Fig. 3. External Cooling/Heating

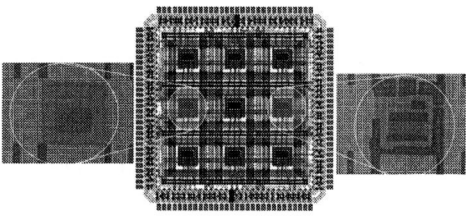

Fig. 2. Layout of Tier C, sites s[0][2], s[1][2] and s[2][2] contain extra heat pipes; Die size is $3.2mm$ x $3.2mm$, site size is $1mm^2$ approximately

in presence of continuous changes in delays through the digital logic. As opposed to synchronous designs, asynchronous circuits allow us to directly measure the maximum local throughput of each part of the circuit without making any modifications to them. Sources of delay differences include temperature, supply voltage, and process variations.

Each test site consists of a valid data (token) generator connected to several frequency multipliers to attain a reasonable internal frequency. The output of the last multiplier is connected to frequency dividers, as shown in Fig. 1(b). Dividers are used to reduce the switching rate of output signals. The output of the last divider indirectly measures the frequency of the multiplier-divider junction. We can calculate the maximum throughput by monitoring the output of the last divider and multiplying by a value based on the number of the divider stages. Even though the multiplier-divider circuit occupies some area, the junction point is most sensitive to temperature variation, since that signal switches much more often than the rest of the multiplier-divider signals. The high-frequency junction is site-centered for uniform heating.

At each site, a large number of wide nFET transistors surrounds the digital circuitry. These structures can act as local heat generators. Gate terminals of these heating transistors are controlled using $V_{thermal\_bias}$ signal. Drain terminals are connected to $V_{thermal\_vdd}$ signal. Both, $V_{thermal\_bias}$ and $V_{thermal\_vdd}$, are exposed to the pins and are controlled externally. Sources of all heating transistors are tied to ground.

Global signals are connected as a 3-D mesh using inter-tier vias. Sites s[0][2], s[1][2] and s[2][2] are provided with extra inter-tier vias (heat dissipation pipes) as shown in Fig. 1(a) and 2. These additional heat pipes create a temperature gradient when the circuit is generating large amount of heat.

III. RESULTS AND DISCUSSION

Calibration. Prior to testing the presented structure using on-chip heating transistors, the circuits were calibrated. To have a better approximation of the behavior-governing curve, we calibrated the circuit on a wide range of temperatures. The lowest temperature in our experiments was -196°C (77K), which is the boiling temperature of liquid nitrogen. We then slowly raised the circuit temperature to room temperature, while recording throughput of each tier at different chip temperatures. Throughput measurements were taken approximately every 20°C. After reaching room temperature, we placed the chip into an oven and continued taking temperature measurements and corresponding throughputs until all of the tiers on the chip stopped operating, which happened at approximately 150°C, as shown in Fig. 3. Before recording each data point, the chip was left in the cryogenic chamber or the oven for some time to achieve uniform temperature.

Fig. 3 demonstrates that tiers A and C behave almost identically, however, the throughput of tier B is lower on the entire range of temperatures. This emphasizes the wafer-to-wafer process variations, since all tiers were manufactured on different wafers. All three tiers have an almost linear throughput to temperature dependence. However, for higher accuracy, we have used polynomial approximations of these curves, while performing local heating experiments as described in the next subsection. As seen in Fig. 3, circuits on tiers A and B operate correctly up to 150°C, however, tier C structures stop working when temperature reaches approximately 85°C.

To demonstrate the correct operation of self-timed circuits in the presence of voltage supply variations, we performed voltage sweeps (all tiers) of the chip at -196°C (77K) and at room temperature, as shown in Fig. 4.

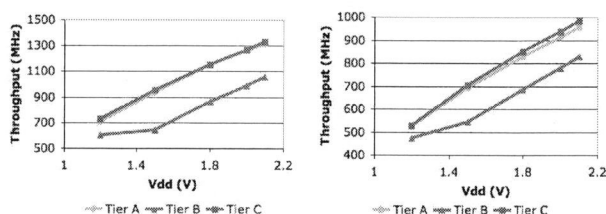

Fig. 4. V-Supply Sweeps at -196°C (77K) and at 21°C (294K) respectively

These two graphs confirm the previous observation that the throughput of tier B is less than the throughputs of the other two tiers throughout the range of tested supply voltages. The performance values of tiers A and C are almost identical during the voltage supply sweeps.

Process Variation. The standard deviation of the cycle time as a percentage of the mean for a single tier was found to be 2.2% (tier A), 3.5% (tier B), and 3.7% (tier C), with the overall variation within a single chip at 10.0%. The process variation for the *s[1][1]* site across chips was found to be 8.2% (tier A), 11.3% (tier B), and 7.6% (tier C), with the overall variation across tiers and chips at 10.9%. For our batch of chips, the tier-to-tier variations were similar to the chip-to-chip variations.

Local Heat Generation. We used the calibrated chip to perform local heating experiments. For this purpose, heating transistors were utilized. In the following experiments, at each heating scenario (setting), we measured throughput of the multiplier-divider structure on all sites. This corresponds to 9 measurements per tier (3x3 sites), and 27 total measurements per chip (3 tiers) for every setting. Circuit throughput measurements at the sites enable us to calculate the temperatures of each individual site on all tiers using the three calibration curves in Fig. 3. In all of the experiments, the digital part was powered by a separate constant 1.8V supply. During the experiments, the drains ($V_{thermal\_vdd}$) of all heating transistors were varied simultaneously, as well as the gates ($V_{thermal\_bias}$). Each measurement was taken after the throughput of all sites settled to some steady-state value.

The first setting we used was with all of the heating transistors turned off ($V_{thermal\_vdd} = 0V$, $V_{thermal\_bias} = 0V$). This is equivalent to a self-heating setting of our structures (all 27 sites contribute to self-heating), as shown in Fig. 5(a). The site layout and heat pipe locations in this figure match the floorplan of Fig. 1(a), except all tiers are plotted in one plane. For the next heating setting, $V_{thermal\_vdd}$ was raised. As the drain voltage increased (with transistors still off, $V_{thermal\_bias} = 0V$), the leakage current through heating transistors amplified severely. This setting increased the chip temperature, which decreased the throughputs of test structures on all sites, as shown in the first several columns of Fig. 6.

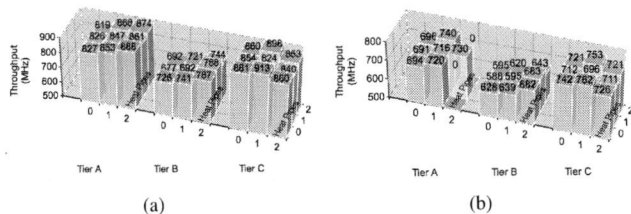

Fig. 5. (a) Self - Heating: $V_{thermal\_vdd} = 0V$, $V_{thermal\_bias} = 0V$; (b) Temperature Increase: $V_{thermal\_vdd} = 1.8V$, $V_{thermal\_bias} = 0.9V$

Fig. 6 is organized in the following manner. The leftmost column is a sequential list of sites for tiers A, B and C (corresponding to Fig. 1(a), 5(a) and 5(b)); sites with heat pipes are shaded. The rest of the white columns, labeled $f(MHz)$, represent throughputs of test structures on corresponding sites at various $V_{thermal\_vdd}$ and $V_{thermal\_bias}$ set-

tings. Grey columns, labeled *norm_T*, represent normalized site temperatures. Site temperatures were computed by taking the throughput of test structure on the site and obtaining the corresponding temperature using the curves in Fig. 3 (separate curves were used for each tier). Temperatures for all sites under all heating settings were then normalized to *site s[0][0] on tier A* at $V_{thermal\_vdd} = 0V$, $V_{thermal\_bias} = 0V$ (which has the value of 1 after normalization) on the Kelvin temperature scale. This normalization allows us to easily observe temperature trends across all tiers at a given setting (table columns), and also as chip temperature rises due to voltage increase on $V_{thermal\_vdd}$ and $V_{thermal\_bias}$ (table rows).

It is important to observe that cross-tier circuit throughput difference does not necessarily translate to proportional temperature difference between the two corresponding sites. The three tiers were manufactured on different wafers and initially had a substantial throughput mismatch.

| $V_{thermal\_vdd}$ | 0.0 V | | 1.8 V | | 1.8 V | | 1.8 V | | 2.0 V | |
$V_{thermal\_bias}$	0.0 V		0.0 V		0.9 V		1.8 V		2.2 V	
	f (MHz)	norm_T	f (MHz)	norm_T	f (MHz)	norm_T	f (MHz)	norm_T	f (MHz)	norm_T
Tier A										
s[0][0]	827	1.00	780	1.12	694	1.37	665	1.46	647	1.52
s[0][1]	853	0.94	807	1.05	720	1.29	~	~	~	~
s[0][2]	888	0.85	840	0.97	~	~	~	~	~	~
s[1][0]	826	1.00	781	1.12	691	1.38	669	1.45	~	~
s[1][1]	847	0.95	800	1.07	716	1.30	~	~	~	~
s[1][2]	861	0.92	814	1.03	730	1.26	~	~	~	~
s[2][0]	819	1.02	776	1.13	696	1.36	~	~	~	~
s[2][1]	868	0.90	825	1.00	740	1.23	~	~	~	~
s[2][2]	874	0.89	833	0.99	~	~	~	~	~	~
Tier B										
s[0][0]	726	0.85	695	0.97	628	1.27	605	1.37	594	1.42
s[0][1]	741	0.78	711	0.91	639	1.21	614	1.33	603	1.38
s[0][2]	787	0.61	752	0.74	682	1.03	656	1.14	639	1.21
s[1][0]	677	1.05	651	1.16	588	1.45	564	1.57	556	1.61
s[1][1]	692	0.98	660	1.12	595	1.42	651	1.16	569	1.55
s[1][2]	788	0.60	753	0.74	683	1.02	655	1.14	640	1.21
s[2][0]	692	0.98	660	1.12	595	1.42	575	1.51	569	1.54
s[2][1]	721	0.87	692	0.98	620	1.30	598	1.40	~	~
s[2][2]	744	0.77	711	0.91	643	1.20	621	1.30	608	1.36
Tier C										
s[0][0]	881	0.92	831	1.03	742	1.22	716	1.28	~	~
s[0][1]	913	0.85	853	0.96	762	1.18	736	1.23	718	1.27
s[0][2]	860	0.97	806	1.08	726	1.25	700	1.31	683	1.35
s[1][0]	854	0.98	797	1.10	712	1.28	686	1.34	668	1.38
s[1][1]	824	1.04	775	1.15	696	1.32	670	1.37	653	1.41
s[1][2]	840	1.01	787	1.12	711	1.29	687	1.34	664	1.39
s[2][0]	860	0.97	807	1.08	721	1.27	696	1.32	~	~
s[2][1]	898	0.89	838	1.02	753	1.20	730	1.25	711	1.29
s[2][2]	853	0.98	801	1.10	721	1.27	691	1.33	673	1.37

Fig. 6. Circuit Temperature Variation Using Heating Transistors

Besides varying the drain voltages of the heating transistors, we also varied the gate voltages, which allowed direct current flow from $V_{thermal\_vdd}$ to ground. A snapshot of the chip with $V_{thermal\_vdd} = 1.8V$, $V_{thermal\_bias} = 0.9V$ is presented in Fig. 5(b). The next setting that we used was $V_{thermal\_vdd} = 1.8V$, $V_{thermal\_bias} = 1.8V$. At this setting two sites on tier A stopped working correctly due to *overheating*. These sites are represented by '~' on the table. Our last setting was $V_{thermal\_vdd} = 2.0V$, $V_{thermal\_bias} = 2.2V$; at this setting almost all of the sites on tier A stopped working as well as some sites on tiers B and C (also represented by '~'). Throughput values for each site at various settings of the heating transistors are presented in Fig. 6.

As expected, temperature increases as we raise $V_{thermal\_vdd}$ and $V_{thermal\_bias}$. Also, temperature decreases towards the sites with more heat pipes, since they have lower effective thermal resistance to the heat sink.

The test structures' throughputs on tiers A, B, and C are quite different, since the tiers were manufactured on separate wafers. However, on tiers A and B there is a trend that shows increasing throughput from regions without heat pipes towards the regions with heat pipes (from left to right on Fig. 5(a) and 5(b)). Tier C does not follow this trend; we attribute this difference to the fact that tier C was the top-most tier and had an additional heat escape path through the package. Some of the chips that we have tested actually had the same throughput (and temperature) variation trend on tier C as on tiers A and B. The lids on those chips, however, were sealed with epoxy (unlike the tested chip) and there was no extra escape path for heat. Nonetheless, since all of those chips had *some* sites that were inoperable due to manufacturing and packaging issues, we do not report detailed results from those chips.

As for chip-to-chip variations, we have measured the throughput and power consumption on seven chips manufactured in the same process technology. The circuit throughput measurements for site s[1][1] (center site) with heating transistors off ($V_{thermal\_vdd} = 0V$, $V_{thermal\_bias} = 0V$) on each tier of each chip are presented in Fig. 7 and show a large variation between the chips. All earlier presented data was obtained using Chip2. The power consumption measurements for the entire digital circuitry (all 27 sites) for all seven chips with heating transistors off ($V_{thermal\_vdd} = 0V$, $V_{thermal\_bias} = 0V$) were normalized to the mean power and are shown in Fig. 7. The power consumption was rather consistent on five of the tested chips; however on the other two chips it was more than two times larger. Since some of the sites on the tested chips were not operational, they could have had undesired behavior that led to increased power consumption.

Presented measurements and results emphasize the issues of process variation and yield in this 3-D process technology.

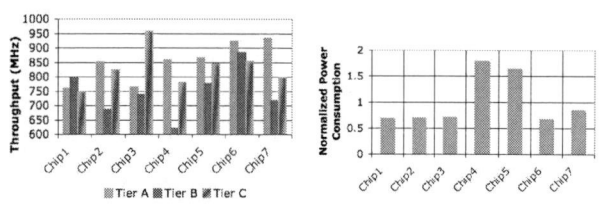

Fig. 7. Cross Chip Throughput Variation and Cross Chip Power Variation w/ $V_{thermal\_vdd} = 0V$, $V_{thermal\_bias} = 0V$ for the s[1][1] site

IV. CONCLUSION

In this paper, we examined the variations present in 3-D circuits. Thermal dissipation and process variations on-tier, between tiers and between chips were studied. Designers have to be aware of performance and reliability issues that may arise as a result of increased heat density on different tiers. We have also verified the effectiveness of inserting through-vias as heat dissipating pipes for managing the on-chip temperature. We have demonstrated that intra-chip process variations, that

are considered minor in 2-D designs at 180nm, are much more crucial in 3-D because designers have to deal not only with intra-die variations, but also with wafer-to-wafer variations.

Having different tiers of the same chip running at different frequencies presents a complication in circuit design especially when 3-D is used to partition logic circuits on multiple tiers. Asynchronous circuits offer an appealing alternative to battle the substantial process mismatch between different tiers.

Several chips manufactured in the described process technology had various faults due to defects. Due to lower yields of 3-D chips, fault tolerant circuits and repetitive structures could be used to increase effective yield. Designs have to be optimized not only for performance and power consumption but also for manufacturability.

Acknowledgments. We would like to thank DARPA and MIT Lincoln Labs for the opportunity to fabricate this project. We would also like to thank Christopher C. LaFrieda.

REFERENCES

[1] J. Cong and Y. Zhang. Thermal-driven multilevel routing for 3d ics. In *Design Automation Conference*, pages 121–126. ASP-DAC, 2005.

[2] C. Ferri et al. Strategies for improving the parametric yield and profits of 3d ics. In *ICCAD '07*, pages 220–226, Piscataway, NJ, USA, 2007.

[3] D. Fang et al. A three-tier asynchronous FPGA. In *International VLSI/ULSI Multilevel Interconnection Conference*, 2006.

[4] K. Banerjee et al. 3-D ICs: a novel chip design for improving deep-submicrometer interconnect performance and systems-on-chip integration. In *Proceedings of the IEEE*, 2001.

[5] L. Xue et al. Three-Dimensional Integration: Technology, Use, and Issues for Mixed-Signal Applications. In *IEEE Transactions on Electron Devices*, 2003.

[6] R.J. Drost et al. Proximity communication. *Solid-State Circuits, IEEE Journal of*, 39(9):1529–1535, Sept. 2004.

[7] R.M. Lea et al. A 3-d stacked chip packaging solution for miniaturized massively parallel processing. *IEEE Transactions on Advanced Packaging*, 22(3):424–432, Aug 1999.

[8] S. Das et al. Technology, performance, and computer-aided design of three-dimensional integrated circuits. In *ISPD '04*, pages 108–115, New York, NY, USA, 2004. ACM.

[9] S. Das et al. Timing, Energy, and Thermal Performance of Three-Dimensional Integrated Circuits. In *ACM Great Lakes symposium on VLSI*, 2004.

[10] T. Kgil et al. PicoServer: Using 3d stacking technology to enable a compact energy efficient chip multiprocessor. In *ASPLOS*, 2006.

[11] T. Y. Chiang et al. Thermal Analysis of Heterogeneous 3-D ICs with Various Integration Scenarios. In *IEEE Int. Electron Devices Meeting*, 2001.

[12] V. Suntharalingam et al. Megapixel CMOS image sensor fabricated in three-dimensional integrated circuit technology. In *IEEE Solid-State Circuits Conference*, volume 1, pages 356–357, 2005.

[13] W. R. Davis et al. Demistifying 3D ICs: The Pros and Cons of Going Vertical. In *IEEE Design and Test of Computers*, 2005.

[14] B. Goplen and S. Sapatnekar. Thermal via placement in 3d ics. In *ISPD '05*, pages 167–174, New York, NY, USA, 2005. ACM.

[15] MIT Lincoln Labs. *MITLL Low-Power FDSOI CMOS Process: Designer Guide*. MIT, Cambridge,MA, 2006.

[16] A. J. Martin. The limitations to delay-insensitivity in asynchronous circuits. In *ARVLSI*, pages 263–278. MIT, 1990.

[17] K. Puttaswamy and G. H. Loh. Thermal Analysis of a 3D Die-Stacked High-Performance Microprocessor. In *Great Lakes Symp. on VLSI*, 2006.

[18] A. Rahman and R. Reif. System-Level Performance Evaluation of Three-Dimensional Integrated Circuits. In *IEEE Transactioons on VLSI Systems*, 2000.

[19] A. Rahman and R. Reif. Thermal Analysis of Three-Dimensional (3-D) Integrated Circuits (ICs). In *IEEE Int'l Interconnect Technology Conf.*, IEEE Press, 2001.

IEEE 2008 Custom Intergrated Circuits Conference (CICC)

3D Heterogeneous Integrated Systems:
Liquid Cooling, Power Delivery, and Implementation

Muhannad S. Bakir, Calvin King, Deepak Sekar, Hiren Thacker, Bing Dang,
Gang Huang, Azad Naeemi, and James D. Meindl

Microelectronics Research Center; Georgia Institute of Technology
791 Atlantic Drive, N.W.; Atlanta, Georgia 30332-0269
muhannad.bakir@mirc.gatech.edu

Invited Paper

Abstract – **This paper describes a novel 3D integration technology that enables the integration of electrical, optical, and microfluidic interconnects in a 3D die stack. The electrical interconnects are used to provide power delivery and signaling, the optical interconnects are used to enable optical signal routing to all levels of the 3D stack, and the microfluidic interconnects are used to cool each level in the 3D stack and thus enable stacking of high-performance (high-power) dice. These interconnects are integrated in a 3D stack both as through-silicon vias (TSVs) and as input/output (I/O) interconnects. Design trade-offs (TSV density, power supply noise, thermal resistance, and pump size), fabrication, and assembly are reported.**

I. Introduction

Historically, advances in the field of packaging and system integration have not progressed at the same rate as ICs. In fact, today's silicon ancillary technologies have truly become a limiter to the performance gains possible from advances in semiconductor manufacturing, especially due to cooling, power delivery, and signaling [1, 2]. In recent years, the mainstream adoption of C4 bumps on organic substrates in the 1990's (although C4 was developed in the 1960's and used on ceramic substrate by IBM) was a significant transition for chip input/output (I/O) interconnect technology, which enabled higher density interconnection with low-cost package substrates. Today, it is widely accepted that three-dimensional (3D) system integration is a key enabling technology and has recently gained significant momentum in the semiconductor industry. Three-dimensional integration may be used either to partition a single chip into multiple strata to reduce on-chip global interconnect length [3] and/or used to stack chips that are homogenous or heterogeneous. An example of 3D stacking of homogenous chips is memory chips, while an example of heterogeneous chip stacking is memory and microprocessor chips. There are a number of interconnect challenges that need to be addressed to enable stacking of high-performance dice (see Figure 1). The origins of 3D integration date back to 1960 when James Early of Bell Laboratories discussed 3D stacking of electronic components and predicted that heat removal would be the primary challenge to its implementation [4]. This has indeed proven to be the case for today's high-performance integrated circuits. When two $100W/cm^2$ microprocessors are stacked on top of each other, for example, the net power

density becomes $200W/cm^2$, which is beyond the heat removal limits of air cooled heat sinks. Thus, cooling is the key limiter to the stacking of high-performance chips today. Power delivery to a 3D stack of high power chips also presents many challenges and requires careful and appropriate resource allocation at the package level, die level, and intrastratal interconnect level [5]. Finally, the prospects of photonic device integration (through monolithic or heterogeneous integration) with CMOS technology require the support of optical interconnect networks between 3D stacks and potentially within a stack.

This paper describes a 3D integration technology that provides fully compatible and wafer-level batch fabricated electrical, optical, and microfluidic interconnect networks. These interconnect networks consist of both through-silicon vias (TSVs) and I/O interconnects. The primary focus of the paper will be on the electrical and fluidic 3D interconnect networks. Section II provides an overview of current 3D integration technologies. The configuration of the 3D integration technology under consideration is discussed in Section III. Section IV discusses implementation of the 3D technology. Modeling of microchannel heat sink performance and intrastratal interconnect density are discussed in Section V. The impact of the latter on power supply noise in the 3D stack is also discussed. Finally, Section VI is the conclusion.

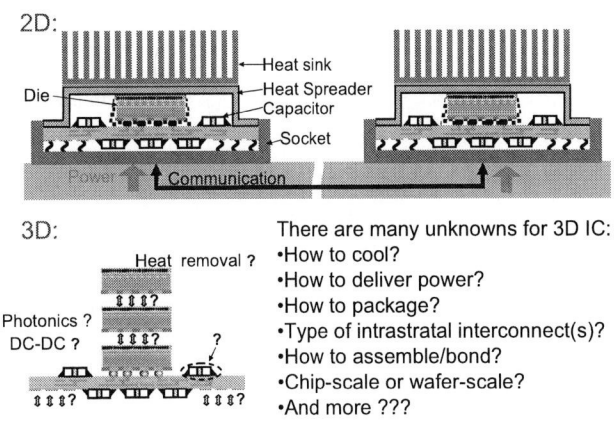

Figure 1: Schematic illustration of challenges associated with the stacking of high-performance dice.

978-1-4244-2018-6/08 $25.00 © 2008 IEEE

II. Existing Packaging and 3D Integration Technologies

Today, cooling of microprocessors is achieved using an air-cooled heat sink that is several-thousand times larger than the volume of the silicon chip [1]. Aside from the form-factor issue, an air-cooled heat sink (and heat spreader), at best, provides a junction-to-ambient thermal resistance of 0.5 °C/W. This is the key reason why stacking of high-performance (high-power) chips has not been demonstrated so far, because, simply put, it is hard enough to cool a single chip. Stacking N chips increases the power density by N and would most certainly require advanced cooling solutions and *integration* methodologies. Figure 2 is a representative schematic illustration of the 3D integration technologies that have been proposed to date and consists of three categories. The first category consists of 3D stacking technologies that do not utilize TSVs and are shown in Figure 2 (a)-(c). The second category consist of 3D integration technologies that require TSVs (Figure 2 (d)-(e)), and the third category consists of monolithic 3D systems that make use of semiconductor re-crystallization to form active levels that are vertically stacked (with on-chip interconnects possibly between). Of course, a combination of all these technologies is possible.

The non-TSV 3D systems span a wide range of different integration methodologies. Figure 2 (a) illustrates stacking of fully packaged dice. Although this may offer the advantages of being low cost, simplest to adopt, fastest to market, and modest form-factor reduction, the overhead in interconnect length and low-density interconnects between the two dice do not enable one to fully exploit the advantages of 3D integration. Figure 2 (b) illustrates the most common method to stacking memory dice, which is based on the use of wire bonds. Naturally, this 3D technology is suitable for low-power and low-frequency chips due to the adverse effect of wire bond length, low density, and peripheral limited pad location for signaling and power delivery. Figure 2 (c) illustrates the use of wireless signal interconnection between different levels using inductive coupling (capacitive coupling is also possible, but more limiting) [6]. This approach is quite elegant for low-power chips that require high-data rate signaling (without the need for TSVs). Power delivery, however, requires use of wire bonds for top dice in the stack, which are not applicable for high-performance/power chips. There are several derivatives to the topologies described above, such as the dice embedded in polymer approach [7]. This approach, although different from others discussed, makes use of a redistribution layer and vias through the polymer film, and thus is a hybrid die/package level solution. It is important to note that all non-TSV approaches rely on stacking at the die/package level (die-on-wafer possible for inductive coupling and wire bond) and thus do not utilize wafer-scale bonding. This may serve to impose limits on economic gains from 3D integration due to cost of the serial assembly process.

Figure 2 (d) and (e) illustrate 3D integration based on TSVs. The former figure illustrates bonding of dice with C4 bumps and TSVs. The short interconnect lengths and high density of interconnects that this approach offers are important

advantages. Compared to wire bonding, it is possible to have several orders of magnitude larger number of interconnects. Although it is possible to bond at the wafer level, this approach is most suitable for die-level bonding (using a flip-chip bonder) and thus faces some of the same economic issues described above. Figure 2 (e) illustrates 3D stacking based on thin-film bonding (metal-metal or dielectric-dielectric) [8-10]. Not only are solder bumps eliminated in this approach, but also increased interconnect density and tighter alignment accuracy can be achieved when compared to the previous approach due to the fact that these approaches are based on wafer-scale bonding. Thus, they utilize semiconductor based alignment and manufacturing techniques.

Finally, Figure 2 (f) illustrates a purely semiconductor-manufacturing (non-packaging) approach to 3D integration. The main enabler to this approach is the ability to deposit an amorphous semiconductor film (Si or Ge) on a wafer during the IC manufacturing process and re-crystallize to form a single-crystal film using a number of techniques [11, 12]. Ultimately, this approach may offer the most integrated system with least interconnects possible but may not provide chip-size areas for device fabrication in the stack.

It is important to note that none of the above described 3D integration technologies address the need for cooling in a 3D stack of high performance chips. This is a significant omission and imposes a constraint on the ability to fully utilize the benefits of 3D technology. As such, new 3D integration technologies are needed for such applications.

III. A Novel 3D Integration Technology

In this paper, we describe a novel 3D system that features low-cost and fully compatible electrical, optical, and fluidic (trimodal) TSVs and I/Os between strata, as illustrated in Figure 3, and is an extension of previous work [1, 13-16]. To our knowledge, this is the first such 3D integration technology and is fundamentally different from currently demonstrated 3D integration technologies. The electrical interconnect network is used for power delivery and signaling between strata, optical interconnects are used to enable the coupling of an optical signal to/from the substrate to a point within the 3D stack, and fluidic interconnects are used to enable the rejection of heat from each stratum in the 3D stack. One implementation of this approach is shown in Figure 4. Each silicon die in the 3D stack contains the following features: 1) a monolithically integrated microchannel heat sink; 2) through-silicon electrical (copper) vias (TSEVs), through-silicon optical vias (TSOVs) and through-silicon fluidic (hollow) vias (TSFVs); 3) solder bumps (electrical I/Os), optical pins [17], and microscale polymer pipes (fluidic I/Os) [18] on the side of the chip opposite to the microchannel heat sink. Microscale fluidic interconnection between strata is enabled by the combination of through-wafer fluidic vias and polymer pipe I/O interconnects. The chips are designed such that when they are stacked, each chip makes electrical, optical, and fluidic interconnection to the dice above and below. Consequently, power delivery and signaling can be

978-1-4244-2018-6/08 $25.00 © 2008 IEEE

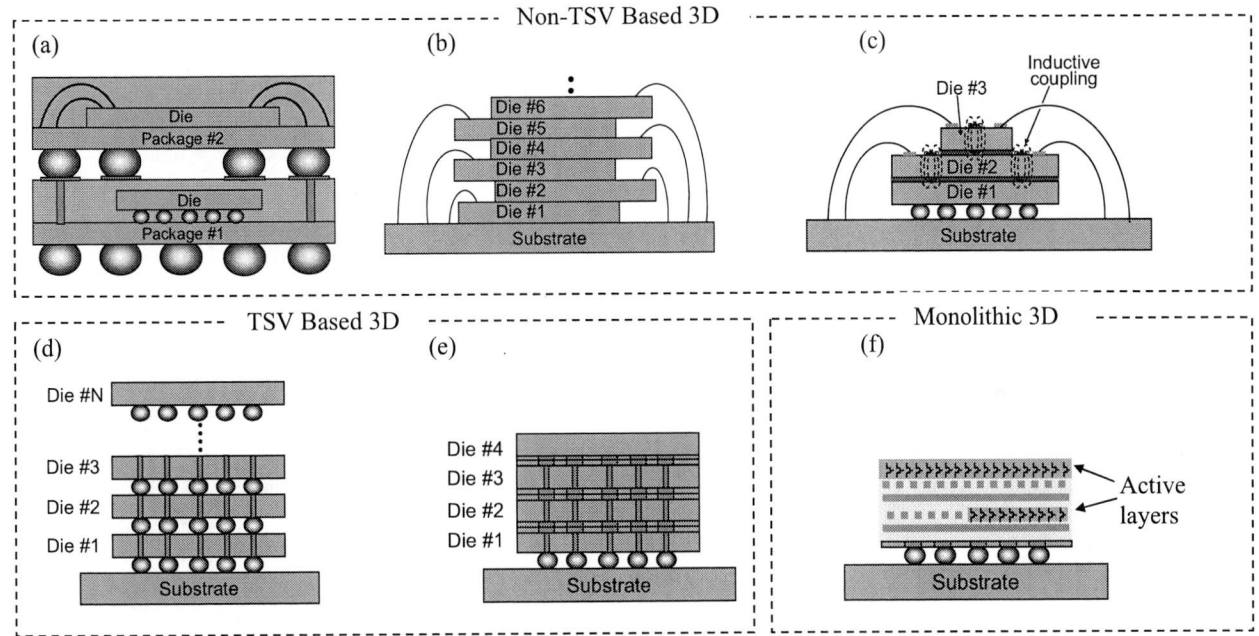

Figure 2: Schematic illustration of various 3D integration technologies.

supported by the electrical interconnects (solder bumps and copper TSVs), and heat removal for each stratum can be supported by the fluidic I/Os and microchannel heat sinks. The optical interconnects provide unusual flexibility to system integration.

filling TSVs with polymer have been devised with the spin-on process shown in the figure yielding the desired results. Using the measured polymer loss of 3.2 dB/cm, it is estimated that the optical power loss through the vias will be approximately 3%.

Figure 3: Schematic illustration of the goal of the research in this paper.

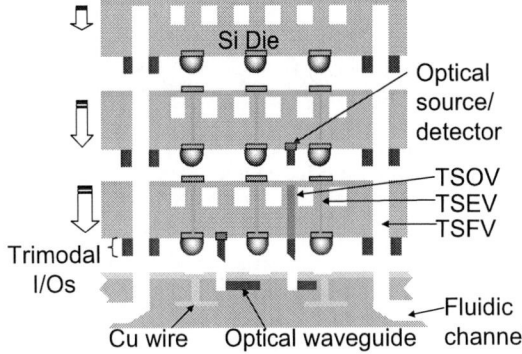

Figure 4: Schematic illustration of one possible implementation of the system shown in Figure 3.

IV. Fabrication and Assembly

The process used for fabricating the polymer (optical) through-silicon vias is shown in Figure 5 [16, 19]. It consists of through-silicon via etching, passivation of the via sidewalls with silicon-dioxide, via filling using a polynorbornene polymer, and polishing to remove any excess polymer. For guided wave optical transmission, the index of refraction of the selected polymer is greater than that of the passivation material, which serves as an optical cladding. Via etching is performed using the Bosch process in an ICP system. The silicon-dioxide also functions as electrical isolation for the electrical TSVs fabricated in subsequent process steps. Numerous processes for

In order to achieve high heat transfer, low thermal resistance, and low pressure drop, a relatively tall microchannel heat sink is needed (>250 μm, typically). As a result, this necessitates a thick silicon wafer and is different from other 3D integration technologies, which seek to polish the silicon wafer to as small a thickness as possible before wafer handling and mechanical strength become limiters. This possibly may impose a constraint on the density of TSVs achievable. Assuming a maximum aspect ratio of 20:1 (conservative estimate) for a copper filled via, a TSV diameter

978-1-4244-2018-6/08 $25.00 © 2008 IEEE   707

of 15 μm is needed for a 300 μm thick wafer (300/20). A trade-off analysis between TSV density and microchannel heat sink performance is presented in the next section. Cross-sectional optical images of fabricated electrical TSVs in a silicon wafer without ((a)) and with ((b) & (c)) a microchannel heat sink are shown in Figure 6. In Figure 6 (b), the microchannel is 200μm tall and 100μm wide, while in Figure 6 (c), the microchannel is 300μm tall and 100μm wide. The need for such a wide range of microchannel heights will be discussed in the next section. In both cases, the overall chip area is 1x1.2 cm^2 and the copper TSVs are 50μm in diameter. The thermal resistance of the microchannel heat sink for single chip was previously measured [1, 20]. The junction-to-ambient thermal resistance of the heat sink was 0.24 °C/W at a flow-rate (de-ionized water as coolant) of ~65 ml/min without TSVs.

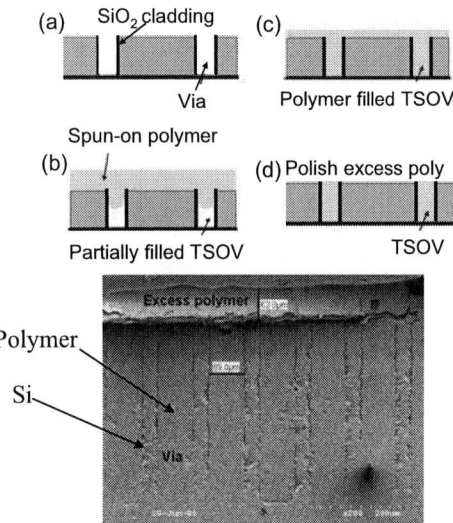

Figure 5: Schematic illustration of the fabrication process and SEM image of polymer-based through-silicon optical vias.

The process used to fabricate the dice is shown in Figure 7. The process begins by fabricating electrical TSVs (Figure 7 (a)) followed by the fabrication of trenches and microfluidic TSVs into the silicon wafer, as shown in Figure 7 (b). SEM images of the trenches and microfluidic TSVs are shown in Figure 8. Next, the trenches are encapsulated to form the microchannels, as shown in Figure 7 (c). For this experiment, the process of using a polymer overcoat described in [21] was used to encapsulate the trenches. Vias were next formed into the overcoat polymer to simultaneously expose the electrical TSVs and form fluidic vias that ultimately allow fluid flow to the upper and lower dice. Following this process step, copper pads are patterned above the electrical TSVs to facilitate solder bonding during assembly. Finally, solder bumps and microfluidic polymer micropipes (electrical and fluidic I/Os, respectively) are fabricated on the side of the wafer opposite to where the microchannel heat sink is located using processes

reported previously [1] (Figure 7(d)). SEM images of the I/Os are shown in Figure 9.

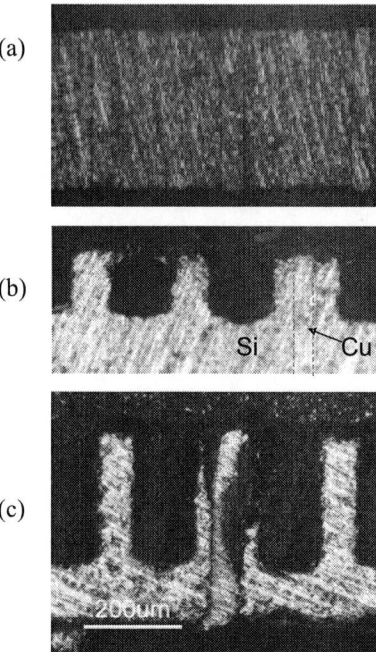

Figure 6: Optical images of a silicon wafer with through-silicon electrical vias (a) and the subsequent fabrication of two different aspect ratio microchannel heat sinks around the TSVs (b) & (c).

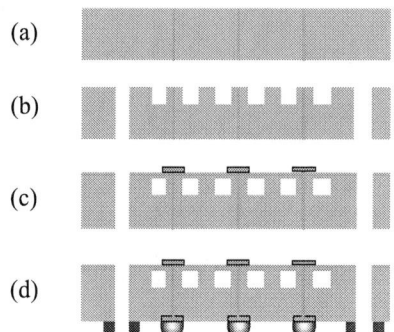

Figure 7: Schematic illustration of the process used to fabricate silicon dice, at the wafer level, that each contain electrical and microfluidic TSVs and I/Os.

The assembly process for microchannel cooled 3D ICs [14] is outlined in Figure 4. The bottommost die in the 3D stack is first assembled onto the substrate using a flip-chip bonder. Standard flip-chip assembly is used to perform this step (no unique processes are used), making our proposed technology fully compatible with current IC assembly infrastructure. The process used for assembly involves preheating the die and the substrate to 180°C and 140°C, respectively, bringing the two into contact with a compression force of 200g, and elevating the temperature of the chip and the substrate to 230°C and 150°C, respectively. Following the

978-1-4244-2018-6/08 $25.00 © 2008 IEEE   708

assembly of the bottommost die, subsequent dice in the stack are assembled in a sequential manner. This process continues until all dice are bonded. Next, an encapsulant is dispensed to seal the fluidic pipes, as demonstrated for single chips previously [18]. A two-die stack and a four-die stack using the above outlined assembly process are shown in Figure 10 (microchannel heat sink was not included to simplify the assembly experiment). A cross-sectional SEM image of the two-die stack highlighting the polymer micropipes is shown in Figure 11.

Figure 8: SEM image of an array of microchannels and through-silicon fluidic vias.

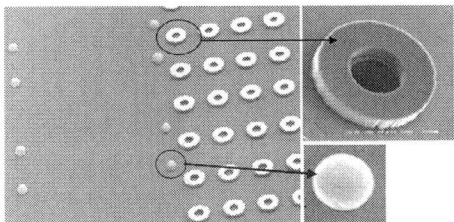

Figure 9: SEM images of electrical (solder) and microfluidic (polymer micropipe) chip I/O interconnects.

V. Modeling of Microchannel Heat Sink Performance and Power-Supply Noise in 3D Stack of Chips

In this section, we describe the modeling of microchannel heat sink performance (single-phase cooling) and TSV density. For this analysis, the width (W) and length (L) of the die are assumed to be 1 cm. Figure 12 illustrates a cross-sectional schematic of a microchannel heat sink containing an area-array distribution of TSVs per microchannel (silicon) wall. The width of the microchannel wall (W_W) can be expressed as a function of TSV parameters as

$$W_W = 20 \ \mu m + P_{TSV}\left(N_{TSVcolumn} - 1\right) + D_{TSV},$$

where P_{TSV}, D_{TSV}, and $N_{TSVcolumn}$ are the TSV pitch, TSV diameter, and the number of TSV columns in the microchannel wall ($N_{TSVcolumn} \geq 1$). The first term in the above equation (the constant 20 μm) is used to indicate the minimum clearance needed between the edges of the outermost TSVs and the microchannel wall (see Figure 12). It is assumed in this study that

$$P_{TSV} = 1.5 D_{TSV}.$$

The total number of TSVs ($N_{totalTSVs}$) that can be routed through the chip (only through the microchannel wall, in this study) can be expressed as

$$N_{totalTSVs} = N_{TSVcolumn}\left(n+1\right)\frac{L}{P_{TSV}},$$

where n is the total number of microchannels, which can be expressed as

$$n = \frac{W - W_W}{W_W + W_C}.$$

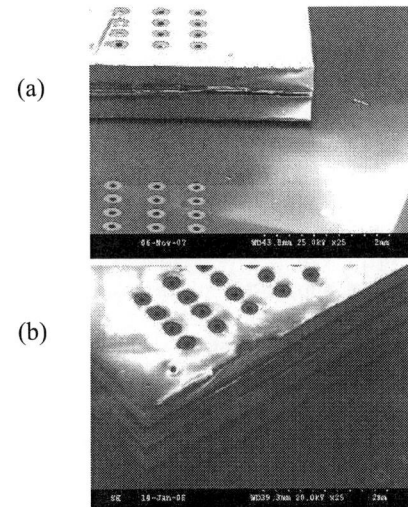

(a)

(b)

Figure 10: SEM images of two-die (a) and four-die (b) stacks that contain electrical and fluidic I/Os and TSVs. In this experiment, no microchannels were included to simplify the process and assembly experiments.

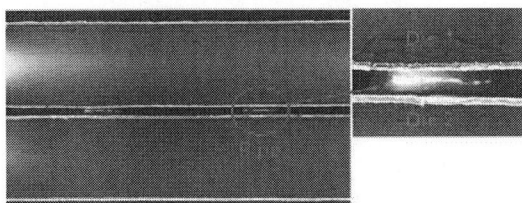

Figure 11: Cross-sectional SEM images of the two-die stack shown in Figure 10 (a).

The set of equations used to model the thermal resistance and pressure drop in a microchannel heat sink is shown in Figure 13, which models heat flow in the microchannel wall in one-dimension. Figure 14 illustrates microchannel heat sink thermal resistance and pressure drop as a function of TSV diameter, number of TSV columns per wall, and microchannel thickness. As the number of TSV columns per wall increases, the thermal resistance and pressure drop increase. This occurs because as the number of TSV columns increases, the wall width increases, which leads to a smaller number of total microchannels (for fixed channel width). This has the consequence of decreasing the overall surface area in contact with the fluid (increases thermal resistance) and decreases the total number of microchannels available (increases pressure drop because the variable n in the ΔP equation shown in Figure 13 decreases). Moreover, as the diameter of the TSV increases, both thermal resistance and pressure drop increase for the same arguments presented above. Moreover, similar trends are observed as channel thickness decreases. Nevertheless, these values of thermal resistance are much lower than for air-cooled heat sinks. In Figure 14, we assume that high aspect ratio vias can be formed into the silicon. For the case when TSV diameter is 5 μm and channel thickness is 200 μm, the via aspect ratio is 40:1. A copper filled trench with an aspect ratio of 49:1 has been demonstrated using a thinned Si wafer [22]. The total number of TSVs possible to route are also shown in the figure for each of the six curves.

The above analysis provides insight into the tradeoffs but does not provide an optimum solution. To this end, Matlab was used to numerically optimize the microchannel heat sink dimensions for various constraints. The objective function chosen for minimization was defined as W_C/W_W (yields the most area to route TSVs). The goal is to minimize the objective function while meeting an upper limit on thermal resistance and a certain pressure drop value. For constraints $R<0.2$ °C/W, $\Delta P=100$ kPa, and $Z=300$ μm, the optimized channel geometry dimensions are approximately $W_W=168$ μm and $W_C=80$ μm. When the constraints are modified such that $\Delta P=40$ kPa, the optimized channel geometry dimensions are $W_W=93$ μm and $W_C=102$ μm. Assuming a TSV diameter of 10 μm, the maximum number of TSVs possible for these two cases is ~170k ($\Delta P=40$ kPa) and ~265k ($\Delta P=100$ kPa). Thus, when the upper limit on pressure drop is increased from a value of 40 kPa to 100 kPa, the number of TSVs that can be fabricated through the wall increases (due to an increase in wall width).

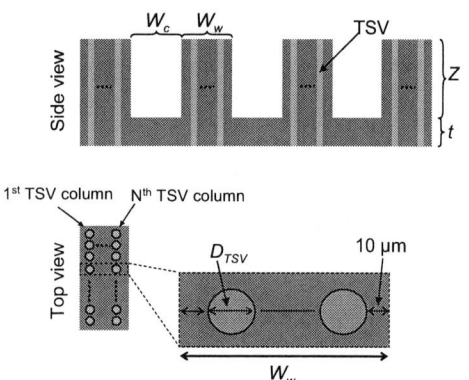

Figure 12: Schematic illustration of a microchannel heat sink with TSVs.

This analysis will ultimately be important to understand the potential trade-offs in microchannel cooling and signaling and power delivery in a 3D stack. The increasing functional complexities of digital systems results in a higher level of

integration and a larger number of dice to be stacked. Recently, compact physical models for power supply noise in a 3D stack of chips were derived and account for on-chip power delivery interconnect design, on-chip decoupling capacitor allocation, package-level inductance, and the total number of TSVs and chip power/ground pads [5]. The models are based on the simplified circuit model for the 3D power distribution network shown in Figure 15. The decoupling capacitance (including both the intentionally added decoupling capacitors and the equivalent capacitance of non-switching transistors) per unit area of die i is represented by C_{di}. The current density for an active block of die i is represented by $J_i(s)$ in the Laplace domain. L_p represents the per pad loop inductance associated with the package, which is connected to the bottommost die (level 1). Each through-via between two stacked levels is modeled by an inductor L_{via} and resistor R_{via} (including the parasitics of solder bumps when they are used between dice).

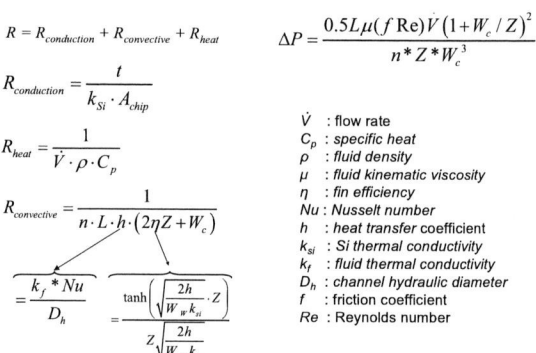

Figure 13: Set of equations used to calculate thermal resistance (R) and pressure drop (ΔP) in a microchannel heat sink [20, 23].

Normally, the switching activities of two circuit blocks under the same footprint in a 3D stack are highly correlated because an important purpose of 3D integration is to make shorter interconnects between communicating blocks when they are far apart in 2D (conventional) chip placement. Therefore, we must consider the worst-case scenario when all dice are switching simultaneously. In this analysis, it is assumed that each die in the stack consumes an on-current density of 100 A/cm². An empirical value of 0.5nH is assumed for the package inductance associated with each power/ground I/O to the package, and it is assumed that 20% area of each die is allocated for decoupling capacitance. The resistance of each segment in the power/ground global interconnect grid (R_{si}) in Figure 15 is calculated as 0.22 Ω [5], and the number of power (ground) wires between two power (ground) pads is calculated as 43.

Figure 16 illustrates the worst-case power supply noise in the topmost and bottommost die when all chips are switching as a function of the number of dice in the stack. If we increase the total number of dice and examine the power supply noise of the topmost and bottommost dice, we can observe that the noise condition in the 3D stack becomes unacceptable when compared to a single die case (left most region of the plot). This is especially true for the topmost level because the noise level changes from 180 mV for the single die case to 790 mV for the ten-die stack case (supply voltage assumed to be 1 V).

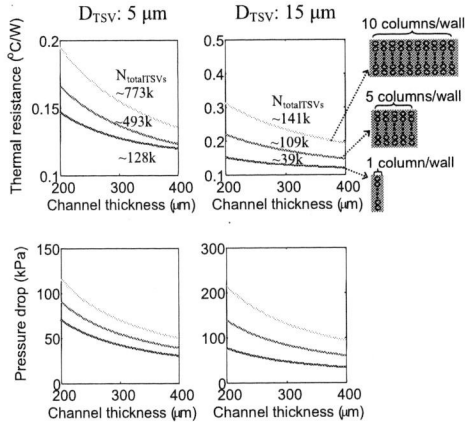

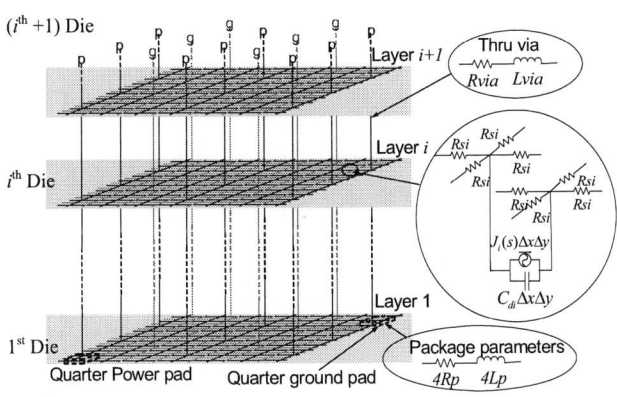

Figure 14: Thermal resistance and pressure drop in a microchannel heat sink (rectangular fins) as a function of TSV diameter, number of TSV columns per wall, and channel thickness. Assumptions: channel width (W_C) is 80μm, flow rate (V) is 5×10^{-6} m^3/s, base silicon thickness (t) is 100 μm, constant heat flux, de-ionized water as coolant, and fully developed laminar flow.

Figure 15: Simplified circuit model of the power distribution network in a 3D stack containing (i+1) dice [5].

Excessive power-supply noise can yield to profound performance degradation and logic failure, and thus, solutions are needed to tackle the power integrity problem brought by 3D integration. Traditionally, to suppress noise to a safe level, we can either adopt an on-chip solution (adding more on-chip decoupling capacitors) or a package level solution (adding more power/ground I/Os). In 3D systems, the power integrity problem arises from the three-dimensional topology, which may also lead to solutions in the third-dimension. To this end, it is important to model the impact of the number of TSVs and

I/Os on the power supply noise. Using compact physical modeling [5], it can be shown that increasing both the number of power/ground I/Os and TSVs in each level can greatly reduce the noise level to the single die case (Figure 17). When the number of TSVs and power/ground I/Os are both increased from approximately 5,000 to approximately 27,500, the worst-case power supply noise in the stack decreases from ~270 mV to ~100mV. This analysis highlights the importance of high-density TSVs in a 3D stack.

Another avenue to reducing the power supply noise in a 3D stack of chips is to dedicate entire levels in the chip stack to decoupling capacitors. This can be used to greatly offset the simultaneous switching noise due to the inductance of the package. The impact of such an implementation is shown in Figure 18. Compared to a four-die stack with no such decoupling capacitors (~400 mV worst-case power supply noise (Figure 16)), Figure 18 (a)-(c) illustrate the worst-case power supply noise when two complete 'decoupling dice' are added to a stack of four high-power dice (same assumptions as above). Depending on their location in the stack and proximity to the topmost functional die, the worst-case power supply noise can be reduced by up to 51 %. In fact, the worst-case power supply noise can be reduced to ~199 mV, which is comparable to the worst-case power supply noise of a single chip (182 mV (Figure 16)).

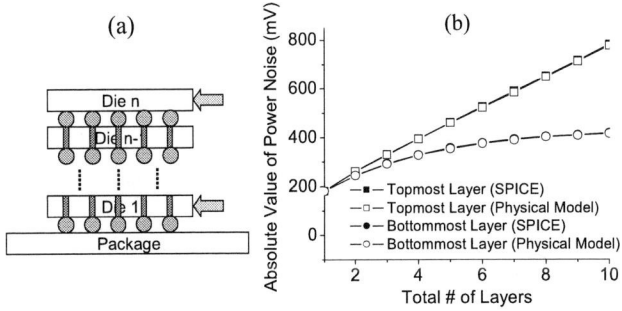

Figure 16: Schematic illustration of 3D stacked dice (a). Worst-case power supply noise in the top and bottom dice when all dice are switching simultaneously.

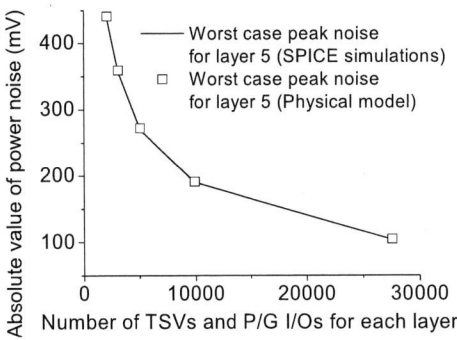

Figure 17: Power supply noise as a function of the number of power/ground TSVs and I/Os for a five-die stack.

978-1-4244-2018-6/08 $25.00 © 2008 IEEE

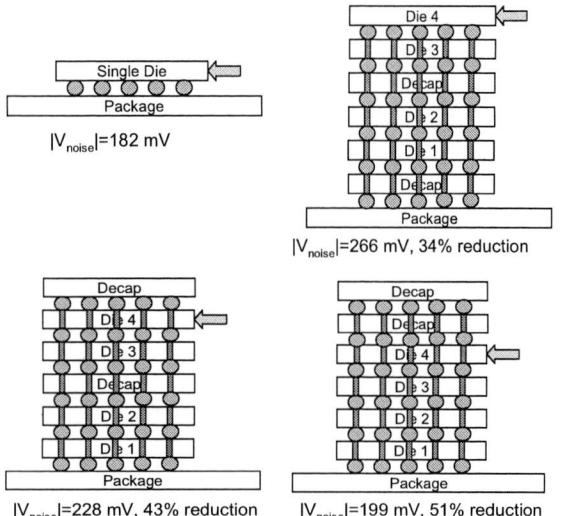

Figure 18: Impact of adding chip-scale decoupling capacitors on the worst-case power supply noise for the topmost die compared to a four-die stack with no such decoupling capacitors (Figure 16). For reference, the power supply noise in a single die is also shown.

VI. Conclusion

This paper describes a 3D technology that supports electrical, optical, and microfluidic interconnection networks (TSVs and I/Os) that is radically different from other 3D technologies. Such a 3D system will provide circuit and system designers a new avenue for improved performance due to the ability to stack high-performance chips and provides a new paradigm for system integration. Modeling of TSV density, microchannel heat sink performance, and power supply noise are discussed and reveal the importance of optimization.

Acknowledgements:

The authors acknowledge the support of the Interconnect Focus Center, one of five research centers funded under the Focus Center Research Program, a DARPA and Semiconductor Research Corporation program. This work is also in part based upon work supported by the National Science Foundation under Grant No. 0701560.

References:

[1] M. S. Bakir, B. Dang, and J. D. Meindl, "Revolutionary nanosilicon ancillary technologies for ultimate-performance gigascale systems," in *Proc. IEEE Custom Integrated Circuits Conf.*, 2007, pp. 421-428.

[2] G. G. Shahidi, "Evolution of CMOS technology at 32 nm and beyond," in *Proc. IEEE Custom Integrated Circuits Conf.*, 2007, pp. 413-416.

[3] J. W. Joyner, P. Zarkesh-Ha, and J. D. Meindl, "Global interconnect design in a three-dimensional system-on-a-chip," *IEEE Tran. Very Large Scale Integration (VLSI) Systems,* vol. 12, pp. 367-372, 2004.

[4] J. Early, "Speed, power and component density in multielement high-speed logic systems," in *Proc. IEEE Int. Solid-State Circuits Conf.*, 1960, pp. 78-79.

[5] G. Huang, M. Bakir, A. Naeemi, H. Chen, and J. D. Meindl, "Power delivery for 3D chip stacks: physical modeling and design implication," in *Proc. IEEE Conf. on Electrical Performance of Electronic Packaging,* 2007, pp. 205-208.

[6] H. Ishikuro, N. Miura, and T. Kuroda, "Wideband inductive-coupling interface for high-performance portable system," in *Proc. IEEE Custom Integrated Circuits Conf.*, 2007, pp. 13-20.

[7] P. D. Moor, W. Ruythooren, P. Soussan, B. Swinnen, K. Baert, C. Van Hoof, and E. Beyne, "Recent advances in 3D integration at IMEC," in *Proc. Materials Research Society Symp.*, 2007.

[8] J. Q. Lu, Y. Kwon, G. Rajagopalan, M. Gupta, J. McMahon, K. W. Lee, R. P. Kraft, J. F. McDonald, T. S. Cale, R. J. Gutmann, B. Xu, E. Eisenbraun, J. Castracane, and A. Kaloyeros, "A wafer-scale 3D IC technology platform using dielectric bonding glues and copper damascene patterned inter-wafer interconnects," in *Proc. IEEE Int. Interconnect Technol. Conf.*, 2002, pp. 78-80.

[9] C. S. Tan, K. N. Chen, A. Fan, and R. Reif, "A back-to-face silicon layer stacking for three-dimensional integration," in *Proc. IEEE Int. SOI Conference*, 2005, pp. 87-89.

[10] J. A. Burns, B. F. Aull, C. K. Chen, Chang-Lee Chen, C. L. Keast, J. M. Knecht, V. Suntharalingam, K. Warner, P. W. Wyatt, and D. R. W. Yost, "A wafer-scale 3-D circuit integration technology," *IEEE Trans. Electron Devices,* vol. 53, pp. 2507-2516, 2006.

[11] D. J. Witte, F. Crnogorac, D. S. Pickard, A. Mehta, Z. Liu, B. Rajendran, P. Pianetta, and R. F. W. Pease, "Lamellar crystallization of silicon for 3-dimensional integration," *Microelectronic Engineering,* vol. 84, pp. 1186–118, 2007.

[12] J. Feng, Y. Liu, P. B. Griffin, and J. D. Plummer, "Integration of Germanium-on-Insulator and Silicon MOSFETs on a Silicon Substrate," *IEEE Electron Device Lett.,* vol. 27, pp. 911-913, 2006.

[13] M. S. Bakir and J. D. Meindl, "Heat removal and power delivery for 3D SoC," *Design of 3D-Chipstacks Forum Presentation, Int. Solid-State Circuits Conf.,* 2007.

[14] C. K. King, D. Sekar, M. S. Bakir, B. Dang, J. Pikarsky, and J. D. Meindl, "3D Stacking of Chips with Electrical and Microfluidic I/O Interconnects," in *Proc. Electronics Components and Technol. Conf.*, 2008.

[15] D. Sekar, C. King, B. Dang, T. Spencer, H. Thacker, P. Joseph, M. S. Bakir, and J. D. Meindl, "A 3D-IC Technology with Integrated Microchannel Cooling," in *Proc. Int. Interconnect Technol. Conf.*, 2008.

[16] H. Thacker, O. Ogunsola, A. Carson, M. Bakir, and J. Meindl, "Optical through-wafer interconnects for 3D hyper-integration," in *Proc. IEEE Lasers & Electro-Optics Society Annual Meeting,* 2006, pp. 28-29.

[17] M. S. Bakir, T. K. Gaylord, K. P. Martin, and J. D. Meindl, "Sea of polymer pillars: compliant wafer-level electrical-optical chip I/O interconnections," *IEEE Photonics Technol. Lett.,* vol. 15, pp. 1567-1569, 2003.

[18] B. Dang, M. S. Bakir, and J. D. Meindl, "Integrated thermal-fluidic I/O interconnects for an on-chip microchannel heat sink," *IEEE Electron Device Lett.,* vol. 27, pp. 117-119, 2006.

[19] H. Thacker, "Probe Modules For Wafer-Level Testing Of Gigascale Chips With Electrical And Optical I/O Interconnects," Ph. D. Thesis, Georgia Institute of Technology, 2006.

[20] B. Dang, "Integrated Input/Output Interconnection and Packaging for GSI," Ph. D. Thesis, Georgia Institute of Technology, 2006.

[21] B. Dang, P. Joseph, M. Bakir, T. Spencer, P. Kohl, and J. Meindl, "Wafer-level microfluidic cooling interconnects for GSI," in *Proc. IEEE Int. Interconnect Technol. Conf.*, 2005, pp. 180-182.

[22] J. H. Wu, J. Scholvin, and J. A. del Alamo, "A through-wafer interconnect in silicon for RFICs," *IEEE Trans. Electron Devices,* vol. 51, pp. 1765-1771, 2004.

[23] D. B. Tuckerman and R. F. W. Pease, "High-performance heat sinking for VLSI," *IEEE Electron Dev. Lett.,* vol. 2, pp. 126-129, 1981.

IEEE 2008 Custom Intergrated Circuits Conference (CICC)

SiP for GSM/EDGE in CMOS Technology

G. Li Puma[1], E. Kristan[1], P. De Nicola[2], C. Vannier[3], B. Greyling[4], S. Piccolella[5]

Infineon Technologies, Duisburg, Germany[1], Sophia-Antipolis, France[2], Xi'an, China[3],
Villach, Austria[4], Padova, Italy[5]

Abstract- **The development in the field of RF and baseband (BB) integration in nanoscale CMOS technology for cellular systems over the last recent years has shown significant progress [1], [3], [4] The successful integration of the RF transceiver with digital baseband processor enables mobile phone manufacturer to build ultra-low cost phones for GSM/GPRS in CMOS technology [2]. This trend towards continuous system integration for mobile phones with an advanced feature set providing high data rate communication, multimedia and camera capabilities [4]. The support of various features requires a system solution including the power-management unit (PMU) with highly efficient DC-DC converters to reduce the overall power consumption. However, this imposes a major challenge for the integration of the RF due to crosstalk and thermal heating effects caused by the PMU and BB part.**

I. INTRODUCTION

A system in package (SiP) integration solution for GSM/EDGE consisting of two monolithic integrated dies is presented. The first die includes the BB and RF-transceiver part fabricated in a standard 0.13µm CMOS technology with 6 metal layers and MIM capacitors. A second die includes the power management unit and audio power amplifier in 0.25µm CMOS technology. This approach offers the advantage to choose the optimal technology for each die and collectively reduces the integration risks due to the separation of the PMU and the audio power amplifier from the sensitive RF-part. Both dies are flip-chipped side-by-side into a 10x10mm very thin BGA-293 package with a dual metal layer substrate for routing. Special care has been taken for pin layout and package interconnections since package crosstalk has a crucial impact on the RF-performance. Most connections from the PMU die to the BB power supplies are done in the package substrate. Critical high current and sensitive RF power supplies are connected on the PCB with the possibility to place external blocking capacitors close to the supply input at the RF side.

II. SYSTEM OVERVIEW

An overview of the complete SiP is given in Fig. 1. The digital signal processor (DSP) subsystem comprises the modem functionality and audio features. Hardware accelerators are integrated and connected to the DSP-bus to support the firmware (e.g. equalizer, channel decoder and ciphering unit). The CPU subsystem which is based on an ARM microcontroller architecture includes the external bus interface to connect the off-chip memories, general interfaces like USB, I2C and SPI to support multimedia functionality like camera and display and to connect an application processor for enhanced multimedia capabilities. The clock for

the BB is derived from the 26MHz generated by a digitally controlled crystal oscillator (DCXO) and a 32 kHz clock for standby modes. The CPU and the DSP (Teaklite) can run at clock frequencies of up to 260 MHz and 178 MHz respectively suitable for video encoding and streaming.

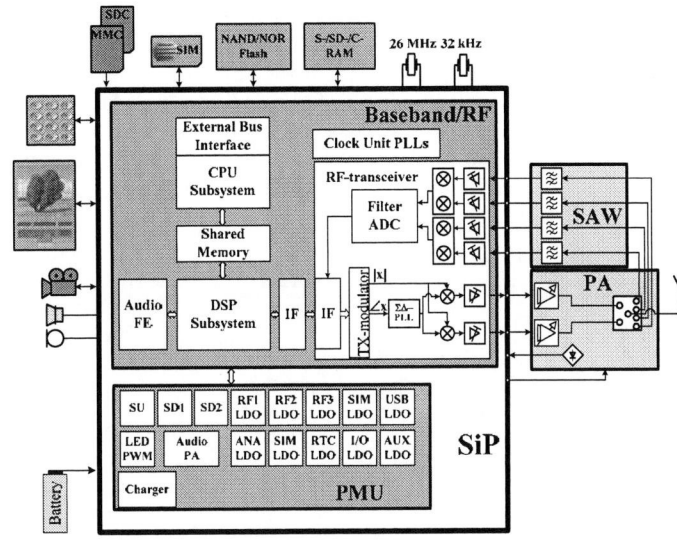

Fig. 1. Block Diagram of the SiP

The supply voltage for the BB core and external memories is provided by two step-down converters (SD1 and SD2) in the PMU. The clock frequency for both SDs is 900 kHz which is well above the critical 400 kHz GSM transmit modulation mask. In order to achieve maximum efficiency both step-down converters support low power modes with output voltages available at any time and under all load conditions. A step-up converter (SU) provides supply voltages higher than the battery voltage for serial connected LED's used in LCD, keypad backlight and photo-flash. Several low-drop regulators (LDOs) supply different power rails, among them three ultra-low noise LDOs for the RF-transceiver part.

The PMU includes a charger which can charge various types of batteries. Charging is software controlled and includes trickle charge (pre-charging controlled by PMU), fast charge (controlled by the Baseband), active trickle charge (controlled by Baseband), USB charge, reverse supply mode (supply power from battery back to charger pin), boot mode (supply SiP without battery connected). In addition, the PMU comprises pulsewidth modulation (PWM) vibrator drive and a 400mW audio amplifier for driving 6Ω to 8Ω loudspeakers.

978-1-4244-2018-6/08 $25.00 © 2008 IEEE

The audio amplifier has a THD less than 1% with 4.8V supply and 6Ω load and can be directly connected to the battery.

Communication to and control of the PMU is mainly done via the I2C interface. The PMU includes power-on reset and initialization circuitries as well as protection circuitries like watchdogs for over and under voltage level detection. For protection of the device a temperature measurement is implemented in the PMU. The temperature is polled every 1ms. In case an overtemperature is detected an interrupt is generated to prevent damage of the device.

II. RF ARCHITECTURE

The RF-transceiver shown in Fig. 2 is connected via an on-chip serial interface to the BB part avoiding the need of any off-chip interconnection between BB and RF which leads to reduction of power consumption and eliminates PCB radiation issues. The receiver is based on a zero-IF architecture with a second order anti-aliasing filter and a 4th order, 3 bit continuous-time $\Sigma\Delta$ analog-to-digital converter (ADC). The ADC achieves 90dB dynamic range at 240 kHz bandwidth which is sufficient to cope with the critical N+2 adjacent channel interferer.

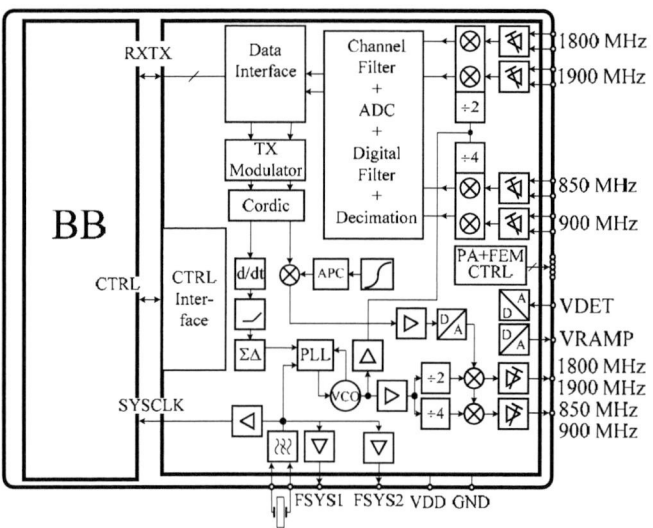

Fig. 2   Schematic of the RF-Subsystem

A source-follower feedback stage LNA is used with approximately 150Ω input impedance and 1.2 dB noise figure. In order to achieve a very low susceptibility against crosstalk and noise at supply and ground lines a fully differential coil-less circuit topology has been chosen.

For the transmitter a direct modulator concept for GPRS and a small signal polar modulator architecture for EDGE is used. The polar modulator is an attractive transmitter solution to generate a complex PAM-transmitter avoiding external RF filters that would likely be necessary with other transmitter topologies. The main advantage compared to cartesian

transmitter architectures is the inherent robustness against PA crosstalk. On the other the polar modulator requires precise control of the time delay between the amplitude and phase path. A delay mismatch of approximately 40ns which corresponds to roughly 1/100 of the symbol time is required to keep the spectral emissions below 54dBc at the critical 400kHz corner offset frequency from the carrier to meet the EDGE specification requirements.

A fractional-N synthesizer including the loop filter and the DCXO are fully integrated.

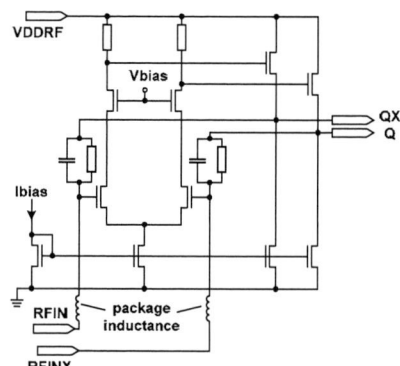

Fig. 3   Schematic of the LNA

For transmit operation the digital symbol stream enters either an 8PSK or GMSK modulator, depending on the desired mode. The phase information is differentiated to obtain a frequency signal and given to a pre-emphasis filter to compensate for the narrow PLL bandwidth [6]. Due to the increased bandwidth requirements for the EDGE phase modulation signal an adjustment of the loop filter bandwidth and a compensation of oscillator sensitivity variation by controlling the current of the charge-pump is done before each transmit slot. The envelope path performs a compensation of the dc-offset to less than 1% of the maximum signal swing to meet the spectral requirements.

The envelope signal and phase signal are combined by a mixer generating the EDGE modulated carrier signal. At the mixer output a band-pass filter with an integrated coil suppresses the critical harmonic components.

An automatic power control loop (APC) unit enables closed loop power ramping for accurate control of the output power level under VSWR mismatch conditions. Depending on the modulation scheme whether GMSK or 8PSK is used the PA output power is controlled either via the control signal VRAMP or by regulating the PA input power.

III. RF INTEGRATION CHALLENGES

The integration of both analog/RF and digital circuits on the same die not only offers many benefits, but also creates technical difficulties, especially in ultra-deep-submicron CMOS technologies. Since analog circuits are inherently vulnerable to noise and crosstalk, the higher levels of integration towards VLSI with over 1 million transistors per

978-1-4244-2018-6/08 $25.00 © 2008 IEEE   714

chip clocked at ever higher frequencies make the signal integrity problem increasingly challenging as technology scales down.

The most important prerequisite for the RF integration is a careful floorplan to avoid parasitic noise injection by the digital BB and PMU parts. A die micrograph of the SiP is shown in Fig. 3. The sensitive RF circuits like VCO and LNAs are located in the upper part in order to achieve maximum distance to the microprocessor and external bus unit interface which are located at the bottom part of the BB. Thereby digital noise coupling from the processor via the substrate to the sensitive RF is minimized. The substrate with 20Ohmcm is rather high resistive, but it is essential to get the required isolation between the noisy processors and the RF circuits. The technology allows to place bump-pads over active area which minimizes the length of signal and supply lines and their parasitics. No special process options are used like deep trench or triple well to meet the isolation requirements which certainly would be suitable to minimize crosstalk, but on the other would also add additional costs.

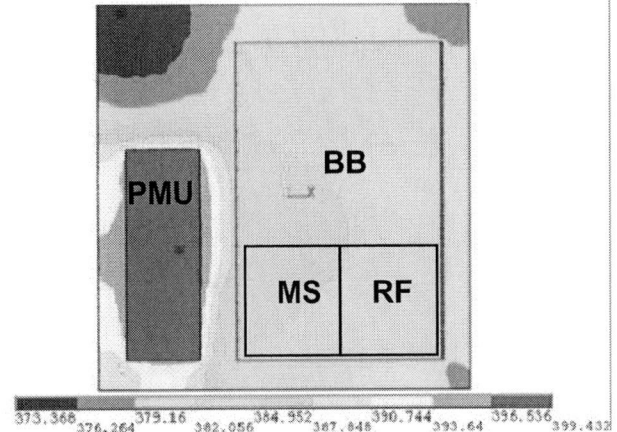

Fig. 5   Chip Temperature Distribution in Kelvin

IV. RF MEASUREMENT RESULTS

The RX reference sensitivity in Fig. 6 was measured in burst mode with running BB and PMU. The achieved sensitivity in the range of −112.5dBm in all four GSM bands demonstrates no degradation due to SiP integration.

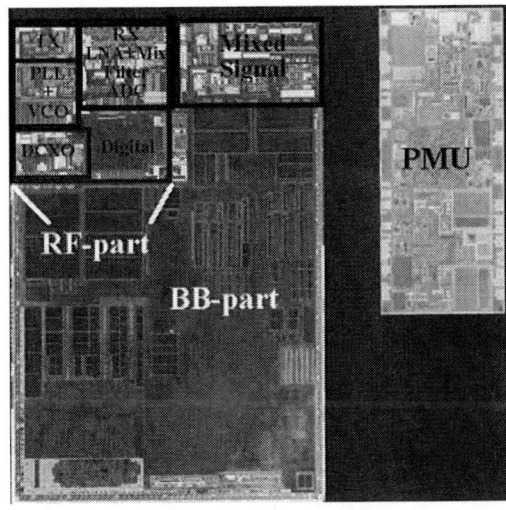

Fig. 4   Micrograph of the SiP

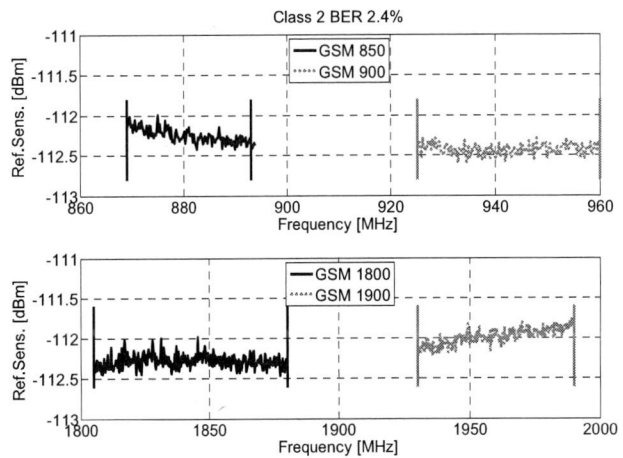

Fig. 6   Reference Sensitivity in Low- and High Bands

Thermal stress caused by the PMU is another important challenge for the RF integration, since thermal gradients may lead to frequency and voltage drifts during GSM receive and transmit bursts. The most sensitive part of the system is the high frequency oscillator due to its integrated LC resonator. It exhibits a high susceptibility to variation of the temperature resulting in a sliding settling behavior of the PLL. Fig. 5 shows a simulation result of the chip temperature distribution. The PMU can be clearly identified as the major heat source of the system as indicated by the red contours. Separation of the PMU device clearly reduces the thermal stress imposed onto the sensitive RF-part significantly and hence reduces the risk of performance degradation.

Important parameters for the RF integration are spurious emissions into the RX bands. Fig. 7 shows a TX spectrum in the 900MHz receive band with 914.8 MHz transmit carrier frequency measured with a notch filter. Only one out of maximal five exceptions occurring at a harmonic of the 26 MHz reference is taken which is similar to the performance of stand-alone transceivers.

978-1-4244-2018-6/08 $25.00 © 2008 IEEE

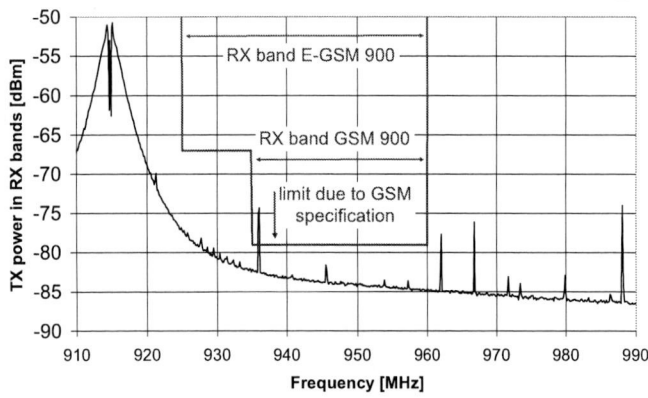

Fig. 7   Measured TX Spurious in RX bands

An important figure of merit which gives evidence for the quality of the EDGE transmit modulation signal is the error-vector magnitude (EVM). Measurements at extreme conditions shown in Fig. 8, achieve RMS-EVM smaller than 2% which is limited by the noise performance of the polar modulator itself.

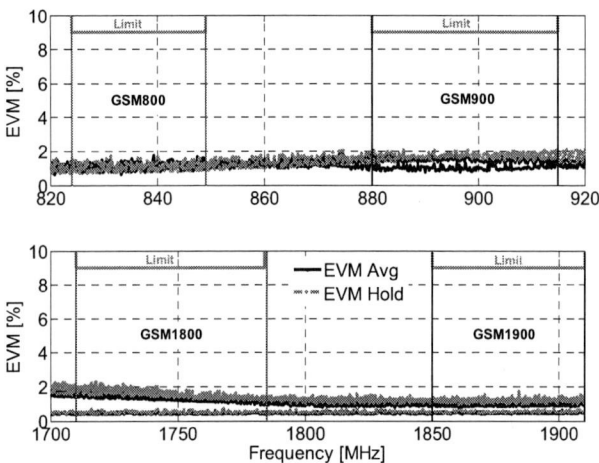

Fig. 8   Measured EVM at Low- and High bands

V. CONCLUSION

The successful integration of a SiP for GSM/EDGE in CMOS technology including the PMU with high efficient DC-DC converters has been presented. The achieved RF-performance is similar to stand-alone transceivers and is not deteriorated by the SiP integration.

ACKNOWLEDGMENTS

The authors gratefully acknowledge the complete engineering team.

REFERENCES

[1] P.-H. Bonnaud, M. Hammes, A. Hanke et al., "A Fully Integrated SoC for GSM/GPRS in 0.13μm CMOS", ISSCC Dig. Tech. Papers, pp. 482-484, Feb., 2006.

[2] M. Hammes, C. Kranz, J. Kissing et al., "A GSM Baseband Radio in 0.13μm CMOS with Fully Integrated Power-Managment" ISSCC Dig. Tech. Papers, pp. 264-602, Feb. 2007.

[3] S. Mehta, W. W. Si, H. Samavati et al., "A 1.9GHz Single-Chip CMOS PHS Cellphone," ISSCC Dig. Tech. Papers, pp. 484-485, Feb. 2006.

[4] R. B. Staszewski, J. Wallberg, S. Rezeq et al., "All-Digital PLL and GSM/EDGE Transmitter in 90nm CMOS," ISSCC Dig. Tech. Papers, pp. 316-317, Feb. 2005.

[5] H. Perrott, T. T. Tewksbury, C. G. Sodini, "A 27-mW CMOS Fractional-N synthesizer using digital compensation for 2.5-Mb/s GFSK modulation", IEEE Journal of Solid-State Circuits, Vol. 32, No. 12, December 1997, pp. 2048-2060.

[6] 3GPP TS 45.005: GSM Specification: "Radio Transmission and Reception", Version 7.2.0., 09.2005

Heterogeneous Multicore SoC for Secure Multimedia Applications

Hiroyuki Kondo, Masami Nakajima, Sugako Otani, Osamu Yamamoto, Norio Masui, Naoto Okumura,
Mamoru Sakugawa, Masaya Kitao, Koichi Ishimi, Masayuki Sato, Fumitaka Fukuzawa,
Kazuhiro Inaoka*, Yoshihiro Saito*, Kazutami Arimoto, and Toru Shimizu

Renesas Technology Corp., Itami, Hyogo, Japan
*Renesas Solutions Corp., Osaka, Japan

Abstract---A heterogeneous multicore SoC (System on a Chip) has been developed for HD (High-Definition) multimedia applications that require secure DRM (Digital Rights Management). The SoC integrates three types of processors: two specific-purpose accelerators for a cipher and a high-resolution video decoding; one general-purpose accelerator (MX: Matrix processor); and three CPUs. This is how our SoC achieves high performance and low power consumption with hardware customized for video processing applications that process a large amount of data. To achieve secure data control, hardware memory management and software system virtualization are adopted. The security of the system is the result of the cooperation between the hardware and software on the system. Furthermore, a highly tamper-resistant system is provided on our SiP (System in a Package), through DDR memories and Flash ROM that contain confidential information in one package. This secure multimedia processor provides a solution to protect contents and to safely deliver secure sensitive information when processing billing transactions that involve digital content delivery. The SoC was implemented in the 90nm generic CMOS technology.

I. INTRODUCTION

Recently, digital content protection standards such as DTCP-IP, Windows Media DRM (Janus) and Broadcast Flag have been established. However, in each case, decryption software is executed on non secure hardware. As a result, a vulnerability arises in which encryption key is disclosed or code is easily modified to access data without authorization. On the other hand, excess security sacrifices usability such as copy frequency and portability.

In this paper, we propose a novel secure system using our SoC with SiP technology and software solution.

1. Atomic operation of charging and viewing

The problem with a conventional system is that the partitions of charging, decryption and image processing are themselves large monolithic side-attack targets. Atomic operation of these processes deletes partitions and eliminates problems of charging omission and copyright infringement by illegal coping of data.

2. Heterogeneous core and SiP for faster communication and decryption

Faster communication between external devices and faster decryption are indispensable when dealing with digital contents including motion video format like MPEG. A multifunction motion video decoder is integrated on the chip to be compatible with MPEG-2/H.264/VC-1 on DTV (Digital Television) and DVD (Digital Video Disc). A symmetric-key cryptography accelerator for decoding

multimedia contents and a public key encryption IP for charging and user confirmation are also integrated. In addition, communication between DDR SDRAMs (Double-Data-Rate Synchronous Dynamic Random Access Memory) and flash memory cannot be tampered with because all communication routes are wired in the chip. So, cores, hardware accelerators and memories in the package provide strong security.

3. Hardware virtualization for strong software/ hardware security

To establish a strong security system, the system on this processor virtualizes hardware resources and an OS (Operating System) and applications are prohibited from accessing hardware resources directly. The most distinguishing feature of the processor is that the multimedia block and the secure block are isolated and communication between these blocks is executed on the virtualization layer. Therefore, this processor provides a secure and flexible system to satisfy users' demands.

This SoC realizes a good balance between protection of intellectual property and convenience when viewing digital contents.

II. ARCHITECTURE

A. Chip Overview

To decode high performance and low power cipher and high-definition video, and to be compliant with multi-standard specifications, we have developed a SoC that adopts heterogeneous multicore architecture. Fig. 1 and Table I show the system block diagram and the functional feature. The multicore SoC integrates three types of processors: two specific-purpose accelerators for a descriptor and a high-resolution multifunction video decoder; one general-purpose accelerator (MX) for image filtering; and three CPUs for data-flow control and data processing. The three CPUs are connected to a common pipelined bus with a cache coherence mechanism[1].

Each CPU is a 32-bit RISC architecture and a synthesizable processor, including a double floating-point processing unit, a memory management unit, three 8kB memories for level 1 instruction cache, level 1 data cache, local memory, and a debug module. The CPU is a 7-stage dual-issue pipelined processor.

The three CPUs are connected to the shared pipelined bus with MESI protocol, one of cache coherence mechanisms[1]. The CPU block is a three way conven-

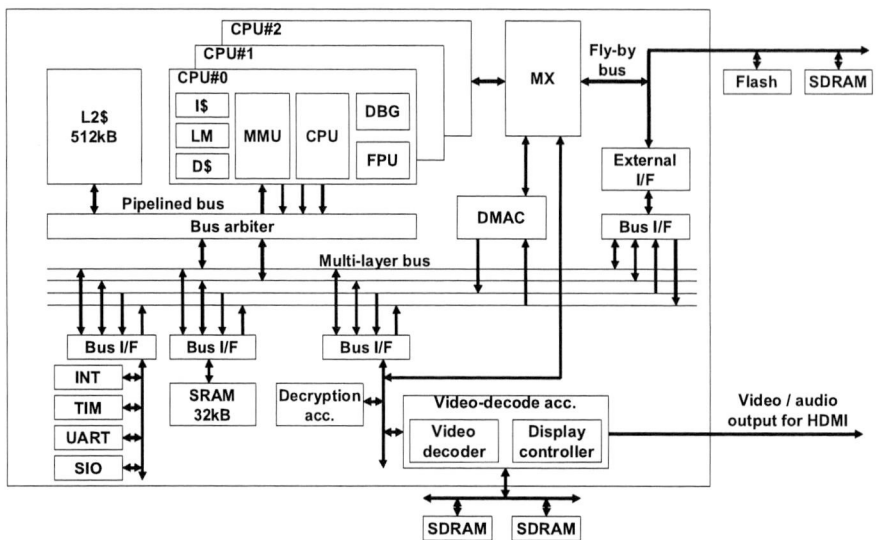

Fig. 1. Block diagram of the SoC.

tional SMP from a coherence perspective, and it supports a unique CPU grouping mode that divides CPUs into several groups. CPUs can only be snooped from other CPUs in the same group. The pipelined bus is connected to a 512kB L2 cache, which is directly connected to the multi-layer system bus.

These three types of processors are interconnected with a high-bandwidth multi-layer system bus on this chip. Three CPUs communicate through the L2-cache. An embedded SRAM, internal I/O, SDRAM and MX, are all connected by this multi-layer system bus. This four-layer system bus provides a high speed transfer rate by accessing these resources in parallel. The key to enhance MX performance is how much the bus transfer data to and from the MX. A specialized high bandwidth data transfer mechanism "fly-by bus" is used to improve data transfer rate between MX and external memories. The fly-by function of the fly-by port and the DMAC (Dynamic Memory Access Controller) achieves transferring a large amount of data from external memory to MX.

B. General-purpose Accelerator

The general purpose accelerator operates image filtering for image quality improvement. The MX is a SIMD (Single Instruction, Multiple Data) massively parallel processor[2], which is specialized to process a large amount of image data. The MX has 640 processing elements of 2bits each, a 32kB instruction memory, and an 80kB data memory with a four-bank configuration.

Because the MX is a massively parallel processor that requires artisan programming skills, it is a challenge to maximize the performance of the MX hardware, and we used a C-language parallel processing library (MX library) for that.

Programs including the MX library run on one CPU. When a library call is made, several parameters must be

TABLE I
FUNCTIONAL FEATURE OF THE SoC

CPU	32-bit RISC Processor (270MHz) x 3 SMP
	L1$:8kB(I)+8kB(D), LM:8kB, MMU, FPU
Memory	L2-cache : 512kB
	Internal SRAM : 32kB
General purpose accelerator	Matrix Processor : MX (270MHz)
	2b-PE x 640, I-SRAM : 32kB, D-SRAM : 80kB
Video-decode accelerator	Decoding feature :
	MPEG-2 MP@HL, MP@ML
	H.264/AVC (MPEG-4 AVC) HP@L4.1, MP@L4.1
	VC-1 AP@L3
	PES (Packetized Elementary Stream)
	Resolution : 1,920pixels x 1,080lines
Decryption accelerator	AES-CBC 128-bit
	AES-CTR 128-bit
	AES-CMAC 256-bit
Bus	Multi-layer bus (4-layer)
	Pipelined bus
	Fly-by bus

transferred from the CPU to the MX. In a conventional MX system, the MX is connected to the CPU with a system bus. In this system, it takes several cycles to transfer the parameters of a library call. To increase performance, the MX is directly connected to the CPU using the CPU's coprocessor interface. With this connection, parameter transfer takes only one cycle. In the case of an FIR (Finite Impulse Response) filter, it improves performance by 42% as shown in Fig. 2.

C. Specific-purpose Accelerator

To decode high performance and low power cipher and a high-definition video, two specific-purpose accelerators are integrated in the proposed SoC. A decryption accelerator is compliant with multi-standard specifications: AES-CBC 128-bit, AES-CTR 128-bit, and

978-1-4244-2018-6/08 $25.00 © 2008 IEEE

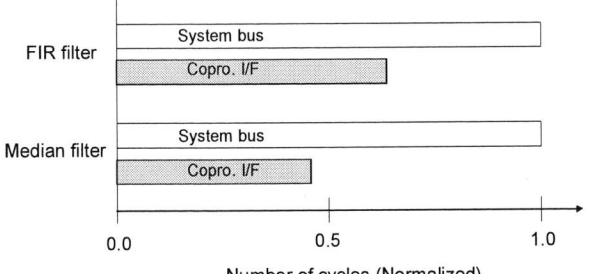

Fig. 2. Comparison of performance.

AES-CMAC 256-bit. A high-definition video-decode accelerator is compliant with multi-standard specifications as shown in Table I.

The block diagram of the video-decode accelerator is shown in Fig. 3. The accelerator consists of a video decoder, a display controller, an audio I/F, a bus I/F, and a DDR I/F.

The video-decode accelerator uses two DDR2 SDRAMs to store a large amount of temporary data. The accelerator is directly connected to the DDR2 SDRAMs by an exclusive bus to improve access latency and throughput performance.

Decrypted data of video stream from the decryption IP comes to the video-decode accelerator via the system bus. First, the video decoder transfers data to the DDR2 SDRAM via the input buffer. Each module of the video decoder and the display controller read data from the DDR2 SDRAM and decode and restore data to the DDR2 SDRAM. While storing temporary data, video stream processing proceeds from the VLD of the video decoder to the frame memory control module of the display controller. The decoded data is transferred from the SDRAM to the outside of the SoC by the display controller. Audio data goes to the audio I/F through the bus I/F. Audio data are sent to the outside of the SoC.

There is a security problem when this connection of DDR2 SDRAMs and the video decoder might allow a third party / adversary to observe the decrypted video data which is sent outside of SoC. To raise the security level, the SoC and the DDR2 SDRAMs are enclosed in an identical package (SiP). Moreover, decoded video data is directly transferred from the SDRAM to a display controller. The path from the accelerator to the display is localized, improves the tamper-resistance of the system.

III. SECURE MEDIA SYSTEM

Fig. 4 shows the software architecture and the clustering architecture. This heterogeneous multicore SoC is divided into two blocks. One is the secure media block and the other is the application block. The secure media block consists of a video decoder and two CPUs and other modules. Real Time OS runs and crypto, image decoding and display processes are executed atomically on this secure block. In the application block, a general purpose OS such as Linux or T-kernel handles the file management,

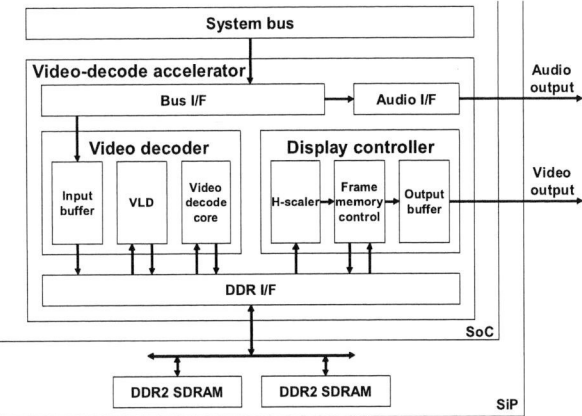

Fig. 3. Block diagram of video-decode accelerator.

GUI (Graphical User Interface) and network operations. User applications run on the general purpose OS. This technology is known as the Micro Clustering Model[3][4].

The software architecture consists of four layers. From the bottom up, they are the hardware layer, the hypervisor layer, OS layer and the user layer which includes the main task, the video task and the audio task.

The hypervisor layer is a thin software layer inserted between the hardware and the operating system. The multicore hypervisor is a software on the hypervisor layer which can provide the operating system with virtual hardware or limit its access to memory. It is based on a para-virtualization model, which is employed to improve performance by presenting each VM (Virtual Machine) with an abstraction of the hardware that is similar but not identical to the underlying physical hardware. Not fully emulating hardware results in low performance overhead.

In this system, communication between the secure OS and the application OS is realized by calling the OS communication API (Application Programming Interface) in the hypervisor.

The cooperation of the multicore hypervisor and the Micro Clustering Model improves tamper-resistant. The secure media block and the application block are effectively isolated by a firewall. Undetectable malware on the application block cannot tamper with the secure block's status or attack resources on the secure media block. In addition, the multicore hypervisor can allocate

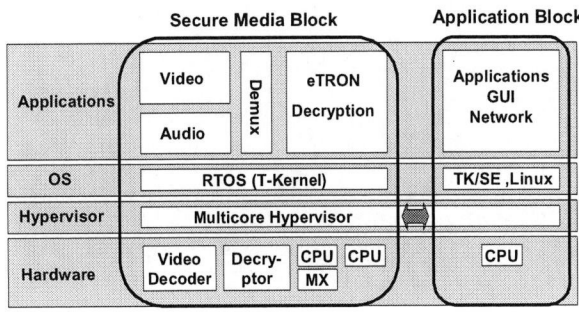

Fig. 4. Software layer configuration.

978-1-4244-2018-6/08 $25.00 © 2008 IEEE   719

hardware resources like memory between virtual blocks. Therefore, it provides the flexibility to adjust the system to the most suitable combination for an application's demands. This system succeeds in strengthening the security and being flexible enough to manage various hardware configurations.

IV. IMPLEMENTATION

Fig. 5 shows the micrograph of the SoC. All logic circuits on the chip were constructed using standard cells except for the memories, DDR-PHY, and PLL. The chip was fabricated in a 90nm generic CMOS-process with eight layers of copper interconnects. The chip integrates 4.35 million gates and 1.1MB memories in a 6.3mm x 6.3mm logic area.

Fig. 6 shows the micrograph of the SiP and Fig. 7 illustrates image of a cross section of the SiP. The die is packaged in an FCBGA (Flip-Chip Ball Grid Array) with 729 pins. Two DDR2 SDRAMs and a flash memory are enclosed in this package. Table II shows the physical feature of the SoC and the SiP.

V. CONCLUSION

A heterogeneous multicore SoC has been developed for high-definition multimedia applications that require secure DRM. The SoC integrates three types of processors: two specific-purpose accelerators for a cipher and a high-resolution video decode; one general-purpose accelerator; and three CPUs. By this means, our SoC accomplished high performance and low power consumption with integrating fitted to the applications processing a large amount of data on video processing.

To achieve secure data control, hardware memory

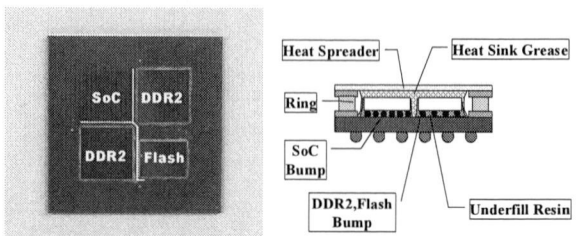

Fig. 6. Micrograph of the SiP.   Fig. 7. Cross section of the SiP.

Table II
PHYSICAL FEATURE OF THE SOC AND THE SIP

Technology	90nm Generic CMOS (8 layers)
Chip Size	$6.35 \times 6.35 mm^2$
Clock frequency	Internal: 270MHz max. External bus: 135MHz
Power supply	Core:1.0V,I/O:3.3V,DDR2:1.8V (Vref=0.9V)
Power consumption	2.0 W
SiP	$29 \times 29 mm^2$ 729pin FCBGA 4 chips in a package - Multicore SoC : $8.00 \times 8.00 mm^2$ Die size - DDR2 SDRAM : 256MBx2, $8.39 \times 8.58 mm^2$ Die size - Flash memory : 32MB, $5.74 \times 7.64 mm^2$ Die size

management and software system virtualization are adopted. The security of the system is the result of the cooperation between the hardware and software on the system.

Furthermore, highly tamper-resistant is provided on our SiP, because DDR memories and Flash ROM that contain confidential information into one package.

This secure multimedia processor provides a solution to protect contents and to make a safely deliver secure sensitive information when billing processing in digital contents delivery.

ACKNOWLEDGMENT

The authors thank Professor Koshizuka. This work is partially funded by the NEDO (New Energy and Industrial Technology Development Organization), via Grant #0628002.

REFERENCES

[1] S. Kaneko et al., "A 600MHz Single-Chip Multiprocessor with 4.8GB/s Internal Shared Pipelined Bus and 512kB Internal Memory", IEEE Journal of Solid-State Circuits, Vol. 39, No.1, pp. 184-193, January 2004

[2] H. Noda et al., "The Design and Implementation of the Massively Parallel Processor Based on the Matrix Architecture", IEEE Journal of Solid-State Circuits Vol.42, No.1, pp. 183-192, January 2007.

[3] M. Nakajima et al., "Design of a Multi-Core SoC with Configurable Heterogeneous 9 CPUs and 2 Matrix Processors", VLSI Symposium on Circuits, Dig. Tech. Papers, Paper 2.2, pp.14-15, June 2007.

[4] H. Kondo et al., "Design and Implementation of a Configurable Heterogeneous Multicore SoC With Nine CPUs and Two Matrix Processors", IEEE Journal of Solid-State Circuits Vol.43, No.4, pp.892-901, April 2008

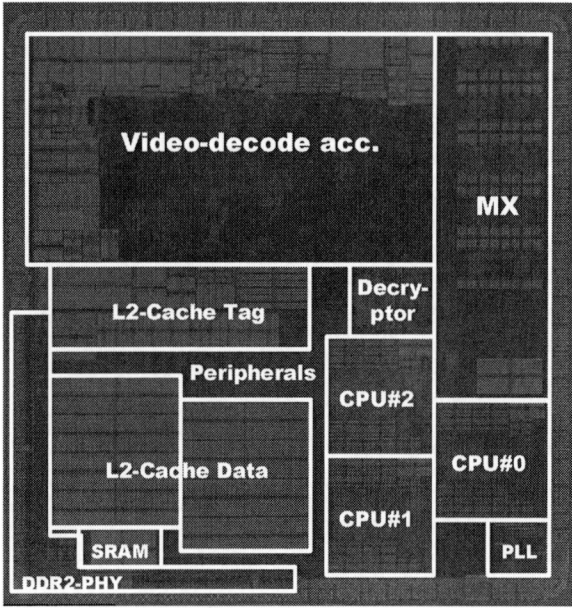

Fig. 5. Micrograph of the SoC.

A 159.2mW SoC Implementation of T-DMB Receiver including Stacked Memories

Joohyun Lee*, Sungdo Kim*, Jinkyu Kim*, Duckhwan Kim*,
Youngsu Kwon*, Minseok Choi*, Kihyuk Park*, Bontae Koo*, Nakwoong Eum*, Hyuckjae Lee**
* Electronics and Telecommunication Research Institute (ETRI), Korea
** Information and Communication University (ICU), Korea

Abstract – **This paper describes a system on chip (SoC) implementation of terrestrial digital multimedia broadcasting (T-DMB) receiver which integrates RF tuner, analog to digital converter (ADC), baseband processor, and multimedia processor in single silicon wafer. The pseudo-SRAM (PSRAM) and SDRAM are doubly stacked with method of silicon in package (SIP). A low-IF RF tuner and a 10bits pipelined ADC is used in this work as IP cores. Baseband processor contains Eureka-147 digital audio broadcasting (DAB) modem, MPEG1-Layer2 decoder, and outer decoder for T-DMB. Multimedia processor is consists of 32bit embedded micro processor, 24bit fixed-point DSP, and H.264/AVC hardware core. The T-DMB SoC was fabricated by using 0.13um 1poly 8metal (1P8M) CMOS process and it gives successful performance of 159.2mW total power dissipation including PSRAM and SDRAM at supply voltages of 1.2V, 2.5V for core and I/O respectively.**

I. INTRODUCTION

T-DMB system has been commercially launched in Korea. The T-DMB system is a broadcasting service which makes possible enjoying high quality audio and video at any time, at any place. Although T-DMB system is based on Eureka-147 DAB system [1], additional technology is also included to make this service possible in real world. Baseband system includes Eureka-147 receiver and outer decoder which consists of Reed-Solomon decoder and convolution de-interleaver [2]. Multimedia system is consists of bit-sliced-arithmetic-coding (BSAC) audio decoder and H.264/AVC video decoder [3][4].

Baseband processor needs large memory for time-deinterleaving and multimedia system also needs large memory for audio, video frame buffering. To complete T-DMB receiver we also need RF tuner and ADC functions.

T-DMB service mainly concerns on mobile device such as mobile-phone or hand-held equipments. In mobile application the power consumption and the area of chipset is important factor. In this work, all elements for T-DMB receiver are integrated in a single chip SoC. The implementation and its results is presented in this paper. The measured data of fabricated T-DMB SoC are also presented.

II. SoC OVERVIEW

The simplified block diagram for T-DMB SoC is shown in Fig. 1 and detailed block diagram is shown in Fig. 2. T-DMB SoC is consists of RF tuner, ADC, baseband processor, multimedia processor, and memories. There is two memories in Fig. 1. These are pseudo-SRAM(PSRAM) and SDRAM.

The PSRAM is used for time de-interleaving in baseband processor, SDRAM is used for audio and video frame buffering in multimedia processor. The capacities of each are 4Mb, 64Mb respectively and these memories are stacked on T-DMB SoC. The wire bonding length was approximately 2.5um and 1.0um for PSRAM and SDRAM respectively.

T-DMB broadcasting service use Band-III (174~216Mhz) frequency band. The RF tuner selects one ensemble and output intermediate frequency (IF) signal to ADC. IF signals are sample by ADC, and sampled data are processed by baseband processor. Baseband processor demodulates the sampled data and output MPEG transport-stream (TS) to multimedia processor. Multimedia processor parses TS and de-multiplexes it to video and audio elementary-stream (ES). Audio ES is decoded by DSP and output to external digital-to-analog-converter (DAC), and video ES is decoded by H.264/AVC hardware core and output to LCD via on chip LCD controller. RF tuner, ADC, and Memories are treated as IP cores and baseband processor and multimedia processor are designed under considerations for low power dissipation. And then, all these elements are integrated in single chip T-DMB SoC.

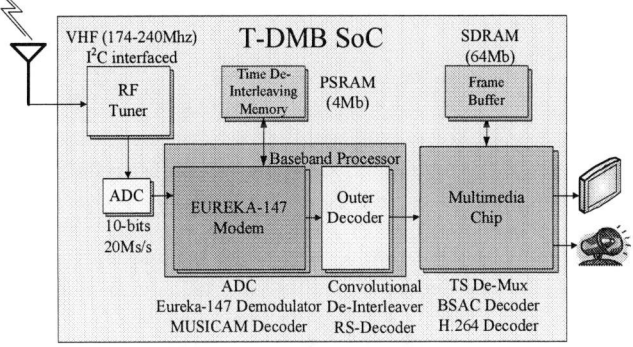

Fig. 1. Single Chip T-DMB Receiver Block Diagram

III. RF TUNER & ADC

RF tuner which used in this work is a low-IF RF tuner. A low-IF tuner does not have severe DC offset problem and there is no needs of discrete components for off-chip register and capacitor filter bank. Channel setting can be done via I²C bus.

RF tuner selects one T-DMB ensemble [1] and output the IF signals differentially to ADC block. The output signal level is $500mV_{p-p}$ and impedance is matched with ADC input buffer.

The IF output signal is also connected to PAD. This is for the purpose of debugging (Fig. 2).

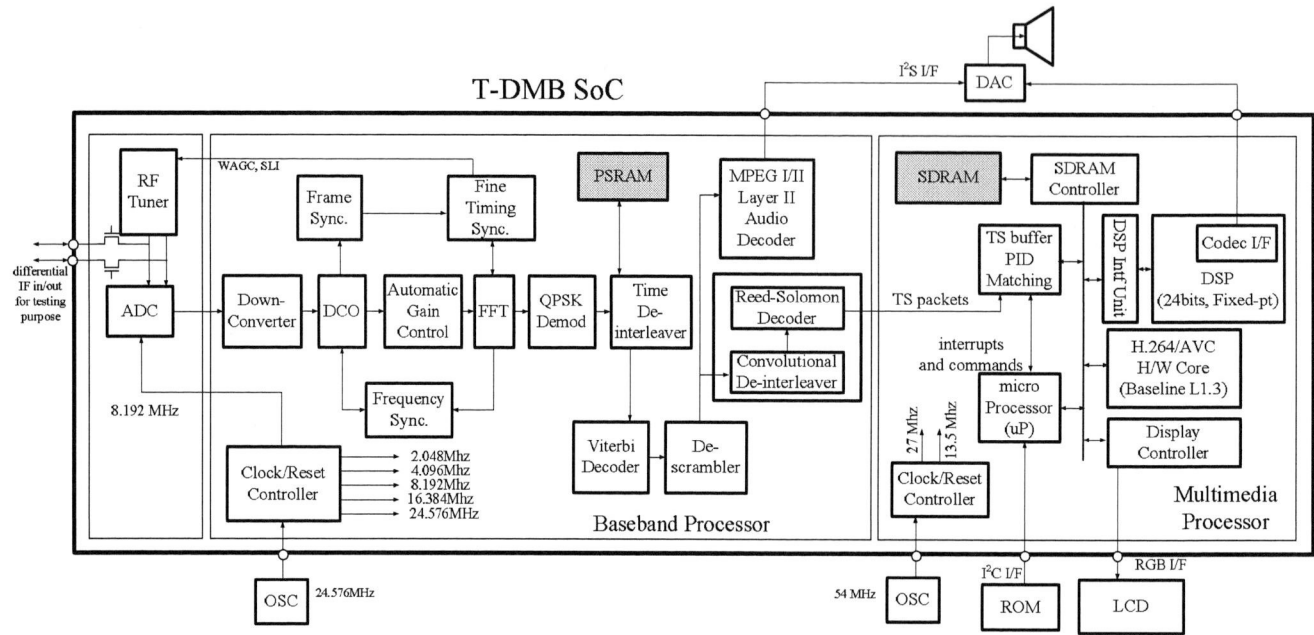

Fig. 2. Detailed T-DMB SoC Block Diagram

A 10-bits 20Msample/s pipelined ADC is used in this work, and used ADC is already silicon proven. The details of design and performances can be found in [5]. This ADC dissipates small power of 5mW and the area is only 0.26mm². These features and performance are very suitable for T-DMB SoC. For impedance and input signal level matching, buffer was inserted at input stage of ADC. For isolation from digital noise, a guard ring which has width of 113um was inserted around RF tuner.

IV. BASEBAND PROCESSOR

The ADC operated with sampling clock of 8.192Mhz and the sampling clock is generated from baseband processor. The down converter (Fig. 2) in baseband processor converts sampled IF signal to in-phase and quadrate-phase baseband signal. The sampling clock is 8.192Mhz and signal bandwidth is 2.048Mhz thus, the possible IF signals are given by

$$f_{if} = 2.048Mhz + n \cdot 4.096Mhz$$

where
$n = 0,1,2...$
f_{if} = frequency of IF signal

(1)

Most of conventional heterodyne tuners have 38.192Mhz IF output and the RF tuner in this work has 2.048Mhz IF output. Thus this sub-sampling technique is useful for both heterodyne tuner and low-IF tuner as well as attractive for several different IF frequencies.

Synchronization is performed in two steps. First one is initial acquisition (Fig. 3), and second is tracking.

During initial acquisition mode, the first thing to do is frame synchronization. Once frame timing is acquired, fine frequency synchronization is continuously performed for every symbol so, fractional portion of frequency offset is compensated. After that, coarse frequency synchronization estimates integral portion of frequency offset and compensates it. Fine timing synchronization is performed at the last of initial acquisition mode, and it estimates exact symbol timing and compensates it. This operation sequences are due to the dependencies among each synchronization algorithms.

In tracking mode, fine frequency synchronization is responsible for tracking of frequency-offset variation, and fine timing synchronization is responsible for timing drift.

The popular guard interval based (GIB) algorithm [6] is used for fine frequency synchronization algorithm and similar algorithm with [7] is used for coarse frequency synchronization.

Channel impulse response (CIR) is calculated in every frames to achieve fine timing synchronization. This can be done with received phase reference symbol (PRS) and original PRS. The timing offset can be derived as following

$$CIR = invFFT(r_k / z_k)$$
$$\delta = \underset{k}{Max}(CIR)$$

where

r_k = received PRS

z_k = original PRS

δ = timing offset

(2)

The frequency offset can be compensated with phase rotation operation. This is done via digital controlled oscillator (DCO) in Fig. 2. The T-DMB system use differential QPSK modulation. In this case, SNR degradations due to frequency offset is smaller than 0.1dB if frequency offset is smaller than 0.01 of subcarrier spacing. The designed DCO has resolutions of 0.001 of subcarrier spacing. Automatic gain control (AGC) is performed before FFT to prevent overflow in FFT operations (Fig. 2).

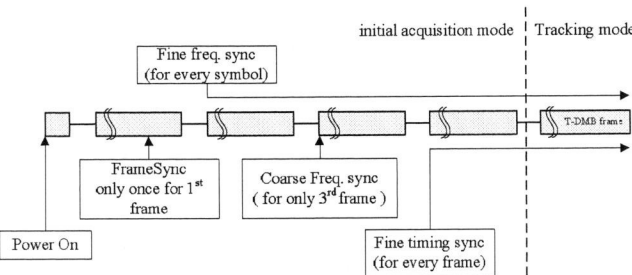

Fig. 3. Synchronization flow

After synchronization, OFDM demodulation is performed by FFT and QPSK demodulator block (Fig. 2). Time de-interleaver block performs de-interleaving of 384ms data. And then, forward error correction (FEC) is followed by de-puncturing of data. A viterbi decoder is used for FEC and it has the parameters of 4bit soft decision metric, truncation length of 128. T-DMB also includes the MPEG1-Layer2 audio service (This is also known as MUSICAM). A fully hardwired MUSICAM decoder is designed and included.

The outer decoder is adopted in T-DMB to meet the bit error rate (BER) performance for the multimedia service [2]. The outer decoder is consists of convolutional de-interleaver and Reed-Solomon decoder (RSD). RSD has parameters of (204,188,t=8) which is a shortened code from (255, 239) and generator polynomial and primitive polynomial are given by

$$g(x) = (x + \lambda^0) \cdot (x + \lambda^1) \cdot (x + \lambda^2) \cdot ... \cdot (x + \lambda^{15})$$
$$p(x) = x^8 + x^4 + x^3 + x^2 + 1 \tag{3}$$

where

$p(x) = primitive\ polynomial$

$g(x) = generator\ polynomial$

V. MULTIMEDIA PROCESSOR

The multimedia processor in this work is capable of TS demultiplexing, H.264 decoding, BSAC decoding and objects synchronization. T-DMB system use H.264/AVC and BSAC standards for video and audio application respectively.

Although T-DMB system use baseline profile level 3 of H.264/AVC, there is more restrictions to simplify hardware implementation [8]. H.264/AVC decoder in this work is conforms to T-DMB specification and designed with fully hardwired logic under low power considerations.

Baseband processor outputs TS to multimedia processor. TS buffer (Fig. 2) captures the TS packets and sends interrupts to micro processor (uP). The uP is responsible for scheduling of all video decoding procedures and parsing of incoming TS packets. After interrupt is acknowledged, uP parses the TS packets and extracts packets elementary stream (PES) packets, synchronization layer (SL) packets, and ES in consequence [8]. After parsing TS packets uP decides the T-DMB service channel and set the program identification (PID) information to PID matching unit in Fig. 2.

Once PID is set, the PID matching unit can select the correspondent packets and move the data to SDRAM independently so, the work load of uP can be reduced.

After TS packets are de-multiplexed, ES is stored in SDRAM which is the frame buffer of audio and video data. If the decoding start command is received from uP, the H.264/AVC H/W core (Fig. 2) patches the frames from SDRAM and decodes it. Once decoding procedures for a frame is completed, decoded frame is stored to decoded picture buffer (DBP) area in SDRAM.

Block diagram of H.264/AVC H/W core is shown Fig. 4. Command processor decodes the uP's commands to its internal command codes and sends to each blocks. Variable length decoder (VLD) decodes exp-golomb code and fixed length code (FLC) which is used for data other than video residual data such as syntax elements. VLD also decodes context-adaptive-variable-length-code (CAVLC) which is used for quantized transform coefficients i.e. video residual data. Transform block performs all types of H.264/AVC transforms such as 4x4 integer transform, 2x2 hadamard transforms and it is also incorporated with de-quantization.

After booting procedure, DSP continuously checks whether the audio frame data are ready in SDRAM. If audio ES is stored in SDRAM, uP sends information of audio ES to DSP interface unit (Fig. 2). After that, patching the audio ES and decoding procedures are performed by DSP. Decoded audio data are sent to external DAC via on chip I²S interface.

uP is also parses the PCR, OCR, CTS, DTS in formations from incoming data packets to synchronize the audio, video object [3].

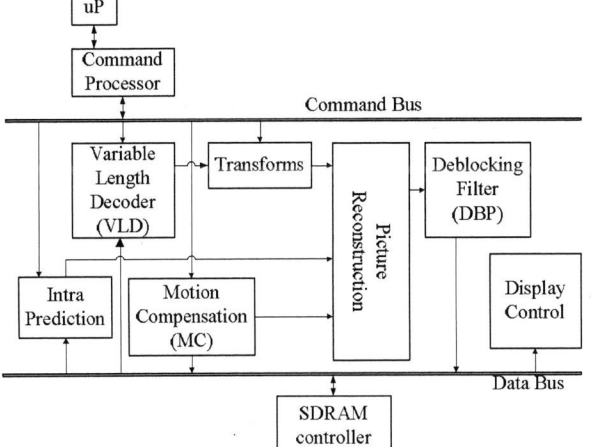

Fig. 4. H.264/AVC H/W core block diagram

VI. IMPLEMENTATION AND MEASUREMENTS

The clock domains of baseband processor and multimedia processor are carefully partitioned in 8 groups of 2.048Mhz, 4.096Mhz, 8.192Mhz, 16.384Mhz, 24.576Mhz, 13.5Mhz, 27Mhz, and 54Mhz. This optimization of clock domain is not just variations of operating frequency but the data processing algorithms for each functional block are also optimized for their clock domain. So, we can reduce the power and the size of T-DMB SoC. The detailed clock domain partitioning is

978-1-4244-2018-6/08 $25.00 © 2008 IEEE

summarized in Table. 1, and T-DMB SoC implementation results are summarized in Table. 2.

The fabricated T-DMB SoC photo and its stacked photo are shown in Fig. 5-Fig. 6 and measured RF tuner performances are summarized in Table. 3. ADC and RF tuner performances are same as that of their stand-alone specifications. The power dissipation measurements are summarized in Table. 4.

VII. CONCLUSION

All elements for T-DMB receiver are integrated in a single chip SoC and is fabricated with 0.13um 1P8M CMOS process. Memories (PSRAM, SDRAM) are stacked on it by SIP process. The fabricated chip is verified with wireless outdoor receiving test and it shows successful performance with total power dissipation of 159.2mW including PSRAM and SDRAM at supply voltages of 1.2V, 2.5V for core and I/O respectively.

Table. 1 Clock Domain Partitions

Clock Domain	Module Name
24.575 Mhz	FFT, FreqSync, RF tuner
16.384 Mhz	Viterbi, RSD
8.192 Mhz	ADC, Down Converter, Time De-interleaver
4.096 Mhz	QPSKdemod, TSbuffer, MP2decoder
2.048 Mhz	AGC, DCO, FrameSync FineTimingSync, De-scrambler
54 Mhz	DSP
27 Mhz	H.264core, uP, SDRAM controller
13.5 Mhz	Display Ctrl.

Table. 2. T-DMB SoC Implementation Results

Block	Area or Gate Counts	Descriptions
RF tuner	2.2 mm x 2.5 mm	
ADC	0.59 mm x 0.44 mm	
Baseband Proc.	636,972 gates	including memory (398,816 bits)
Multimedia Proc.	441,525 gates	Including memory (291,672 bits)
DSP	895,867 gates	including memory x/y-mem:16Kx24 for each. p-mem:16Kx24
Total	6.0mm x 6.0mm (1,974,368 gates)	

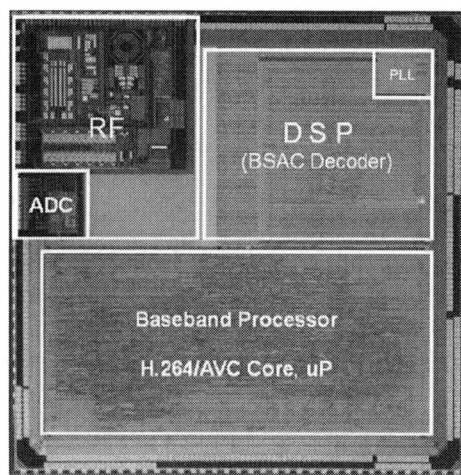

Fig. 5. T-DMB SoC Chip Photograph

Fig. 6. Memory Stacked T-DMB SoC Photograph

Table. 3. Measured RF tuner Performance

Parameters	Value	Descriptions
Sensitivity	-100 dBm	Gaussian channel
Receiver Noise Figure	3.75 dB	Cascade, max. gain
Receiver Linearity(IIP3)	+19 dBm	minimum gain condition
LO Phase noise	-88 dBc/Hz	100 KHz offset freq.
Adjacentb Ch selectivity	38 dB	1.712 Hz offset from center freq.
Far-off selectivity	48 dB	5.0 MHz offset from center freq.
Image Rejection Ratio	50 dB	Max. gain mode

Table. 4. T-DMB SoC Power Dissipation Measurements

Block	Current (mA)	Supply (V)	Power (mW)
RF tuner	22.0	1.2	26.4
ADC	4.0	1.2	4.8
Baseband+Multimedia	40.0	1.2	48
I/O+PSRAM+SDRAM	32.0	2.5	80
Total			159.2

VIII. ACKNOWLEDGMENT

This work is supported in part by the Ministry of Information and Communication of Korea.

Authors appreciate to Youngdeuk Jeon, who designs ADC and gives hand to make ADC as part of SoC.

IX. REFERENCES

[1]. "Radio Broadcasting Systems; Digital Audio Broadcasting to mobile, portable and fixed receivers", ETSI EN 300 401, 2001

[2]. "Digital Audio Broadcasting ; Data Broadcasting – MPEG-2 TS streaming", ETSI TS 102 427, 2005

[3]. "Information technology – Coding of audio-visual objects – part 10: Advanced video coding", ISO/IEC 14496-10

[4]. "Information technology – Coding of audio-visual objects – part 3:Audio", ISO/IEC 14496-3

[5]. Young-Deuk Jeon, Seung-Chul Lee, et al, "A 5-mW 0.26mm^2 10-bit 20-MS/s Pipelined CMOS ADC with Multi-Stage Amplifier Sharing Technique", IEEE ESSCIRC, pp.544-547, 2006

[6]. Moose, Paul H. "A Technique for Orthogonal Frequency Division Multiplexing Frequency Offset Correction." IEEE Transactions on Communications, vol. 42, No. 10, Oct. 1994, pp. 2908-2914

[7]. Keukjoon Bang , Namshin Cho "A Coarse Frequency Offset Estimation in an OFDM System Using the Concept of the Coherence Phase Bandwidth", IEEE Transactions on Communications, vol.49, no.8, Aug. 2001

[8]. "Digital Audio Broadcasting DMB video service; UserApplication Specification", ETSI TS 102 428, 2005

Noise and Oscillator Simulation

Session 22

Chair: Colin McAndrew, Freescale Semiconductor
Co-Chair: Larry Nagel, Omega Enterprises

Although many circuits can be efficiently analyzed by the general algorithms available in SPICE-like circuit simulators, there are certain classes of problems where brute-force simulation is not useful. One such problem is analyzing the effect of noise on switching circuits, where the probability of encountering an issue is so low that the simulation times needed to catch problems is prohibitively large. A second is for oscillator analysis, where conventional time-domain methods must simulate for a very large number of oscillation periods to capture details of slower superimposed variations or of noise. For these classes of problems domain-specific analysis algorithms have been developed that enable efficient modeling and simulation of the narrow problem of interest. Historically many of these developments have been reported at CICC, and this session continues that tradition.

The first paper addresses an approach to try to mitigate the crosstalk noise between circuit blocks by using a 3-dimensional Faraday cage to shield and isolate the circuit blocks. The proposed approach is verified both by measurement and by simulation.

The second paper presents a divide-and-conquer approach to simulating the effect of substrate noise on very large digital systems, with 100's of millions of transistors, where brute-force simulation is not feasible. Partitioning the chip into appropriate size domains, selected to balance current density variations, and then constructing macro-models for each domain, enables efficient and accurate full-chip modeling and simulation of substrate noise.

The third paper addresses the problem of the inability of normal noise simulation, which is based on linearized, static small-signal models, to analyze the effect of noise on mixed-signal blocks, which are nonlinear and time varying. The authors show that the periodic noise analysis algorithm can be used to characterize the bit-error-rate of clocked comparators.

Phase noise in oscillators has been studied using the perturbation projection vector (PPV) approach. The fourth paper in the session uses the PPV approach to analyze several voltage controlled oscillators and enable noise tolerant design.

The final paper in the session introduces a new approach to analyzing injection locked oscillators. This approach is applicable to low quality factor LC tanks as are typically found in digital CMOS processes used for microprocessors, and is verified against measurements from a 45nm industrial process.

978-1-4244-2018-6/08 $25.00 © 2008 IEEE

Notes

IEEE 2008 Custom Intergrated Circuits Conference (CICC)

Modeling, Measurement and Mitigation of Crosstalk Noise Coupling in 3D-ICs

Liuchun Cai and Ramesh Harjani

University of Minnesota, Minneapolis, MN 55455, Email: harjani@ece.umn.edu

Abstract—**Faraday cages have traditionally been used to provide isolation from electromagnetic fields. In this paper, we describe the use of Faraday cages for reducing crosstalk in 3D ICs. We validate our methodology with a combination of simulation and measurements from fabricated prototype designs. Measurement and simulation results show that the crosstalk between the transmitter and receiver reduces by about 75dB up to 10GHz by using a Faraday cage in combination with tier-to-tier isolation, which is one of best performance reported so far. We further develop a lumped equivalent model for crosstalk with and without a Faraday cage. There is good agreement between measurement, 3D electromagnetic simulation and lumped circuit simulation.**

I. INTRODUCTION

3D ICs provide an attractive alternative to traditional two dimensional integrated circuits (2D ICs). In the case of homogenous integration (i.e., same technology), they provide increased computational power with reduced wiring and additionally, provide the possibility of heterogenous integration (i.e., different technologies) that may be more suitable for RF and mixed-signal circuits. In this paper we focus on mixed-signal (RF/analog/digital) design in CMOS 3D ICs. Such 3D ICs allow for the integration of full systems that include RF circuits with significant digital signal processing capability. However, for such integration to be feasible it is critical to provide sufficient isolation between the sensitive analog/RF circuits and the digital circuits. Noise injected by digital circuits into the substrate can severely degrade the performance of RF/analog circuits.

Over the years, in 2D planar ICs a number of techniques for reducing crosstalk via the substrate has been developed. Guard rings around a noise source provide a low resistance path to AC ground and minimize the amount of noise injected into the substrate. Well-designed guard rings can suppress the amount of coupling at frequencies below 1GHz, but have proven to be quite ineffective above 3GHz [1]. Deep trench technologies have shown isolation that is better than 40dB for 0.5-20GHz, but result in fairly complex processes [2]. Patterned ground shields under inductors improves isolation for RF circuits by an additional 25dB [3]. Other noise isolation technologies that perform well at frequencies below 1GHz, include triple well, on-chip decoupling and buried oxide. The isolation frequency performance can be increased to approximated 10GHz by using SOI technology with a high resistivity substrate [4].

In this paper, we present and verify noise isolation in 3D ICs by using separate tiers and Faraday cages (FC). The design and implementation are based on a 0.18μm fully-depleted silicon-on-insulator (FDSOI) technology shown in Fig.1 [5],

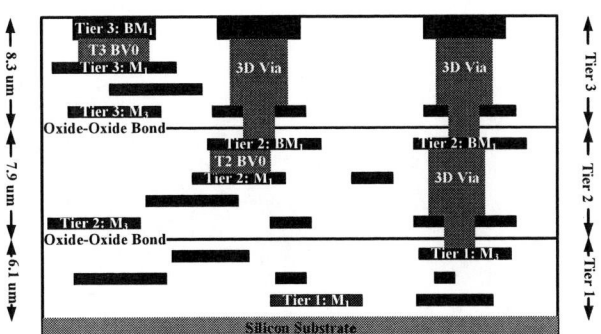

Fig. 1: A 3-tier 3D-IC layout including layer thickness information [5], [6]

[6]. The three active tiers, with three metal layers each, are first individually fabricated using a FDSOI CMOS process. The resistivity of each tier is about 2000 Ohms·cm. The handling wafers of the top two tiers are then removed and these tiers are flipped over and integrated with the bottom tier by using intertier vias. Faraday cages have traditionally been used to block-out external EM fields. We evaluate the efficacy of using such cages in 3D ICs. It is critical to note that Faraday cages are not possible in traditional 2D ICs because metal layers are only available above the active devices. In the MITLL 3D IC technology FCs can be created by using metal layers in any two tiers and using the inter-tier vias to complete the sides of the Faraday cage.

The paper is organized as follows. In Section II we describe the test chip and the experimental and simulation setup for the different coupling conditions in detail. Section III provides measurement and simulation results. Section IV presents a lumped model of the coupling effects based on physical parameters, simulations and calculations. Section V provides some concluding remarks.

II. EXPERIMENTAL & SIMULATION SETUP

3D ICs allow for significant digital signal processing to be available in close proximity to analog/RF circuits which is of great interest for future programmable and cognitive radios. However, the close proximity also introduces the potential problem of noise coupling. In the next subsections we explore an experimental setup that confirms the coupling problem and also identify techniques that are valuable for their mitigation.

A. Experimental Setup

Fig. 2 shows the microphotograph for a 5mm × 2.5mm test chip that was used to investigate noise coupling in 3D ICs. Test structures without Faraday cages on tier 3, displayed in Fig. 2

978-1-4244-2018-6/08 $25.00 © 2008 IEEE

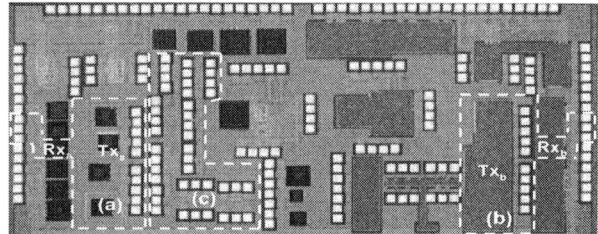

Fig. 2: Micrograph of test die (a) test structures without Faraday cage on tier 3 (b) test structures with Faraday cage on tier 3 (c) de-embedding structures

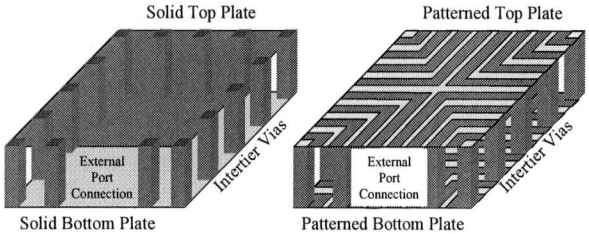

Fig. 3: Cross sectional view of Faraday cages; Left: solid top and bottom layers; Right: patterned top and bottom layers

(a), are used as a reference for test structures with Faraday cages on the same tier. The transmitter (TX) and receiver (RX) sections in this experiment use identical inductors. The transmitter consisted of four inductors at different distances from the receiver. The distances between transmitter inductor centers and receiver inductor centers are $710\mu m$, $510\mu m$, $435\mu m$ and $430\mu m$. The thickness of metal layers 1-3 on the tier 3 are all 630nm. All inductors are based on a square spiral structure where metal 3 is used to form the coil and metal 2 is used for the underpass. Inductors for test structures in Fig. 2 (a) have 3.25 turns with an inner diameter of $88\mu m$, and the width and the spacing of the spiral are $10\mu m$ and $4\mu m$ respectively. The inductors are configured as a one-port device with the other port connected to ground. The ground ring serves as a current return path, and is at a fixed distance of $100\mu m$ away from the outer diameter of the inductor.

The test structures with Faraday cages (FC) on tier 3 are shown in Fig. 2 (b). The test structures for the TX and RX are identical except for the Faraday cage. The two set of structures, with and without Faraday cage, are on opposite sides of the chip. We ensure that the distance between the TX center and RX center remained unchanged with and without the FC. For this experiment, the FC consists of the back metal (BM_1) on tier 3, the back metal (BM_1) on tier 2 and intertier vias between tier 3 and tier 2. It is important to note that creating this FC does not involve any additional fabrication steps. The two different kinds of Faraday cages are shown in Fig. 3.

For our experiment, we use the measured S_{21} as the figure of merit for isolation. Measurements were done from 50MHz to 10GHz with the help of an Agilent 8719Es two-port network analyzer for the test structures shown in Figs 2(a) and 2(b). The network analyzer is configured to perform broadband measurements. Probe calibration was done using CASCADE's 101-190 substrate using a Short-Open-Load-Thru technique. De-embedding structures shown in Fig. 2(c) were used to

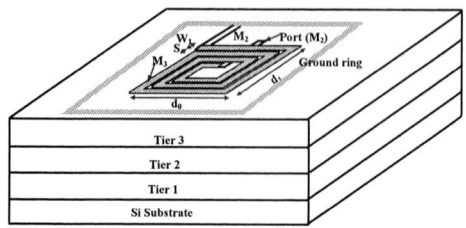

Fig. 4: Cross-sectional view of a spiral inductor on tier 3

eliminate the influence of the transition region between the probe, probe contact and the device under test.

B. Electromagnetic Simulation Methodology

Not all test cases could be verified via the test setup above. Therefore, to complete the picture we supplement the measurements from the experimental setup with full-wave electromagnetic (EM) simulations. EM simulations were validated via measurements and will be used to guide future isolation structure design. Ansoft HFSS was used for our EM simulations [7].

The solid model used within HFSS for the FDSOI CMOS 3D IC is shown in Fig. 4. The details of the structures used to simulate noise coupling correspond to the test structures shown in Fig. 2 (a) and (b). Two other simulation structures are also discussed; one is that of test inductors on the different tiers (tier 3 and tier 2) with/without Faraday cages; another one is the same structure as shown in Fig. 2 (b) except that the Faraday cage uses patterned top and bottom plates. All inductor design parameters are exactly the same as those shown in Fig. 2(a) and (b).

III. RESULTS AND DISCUSSIONS

Fig. 5 shows measured and simulation results for the reference structures in Fig. 2 (a). The graph plots the measured S_{21} with the probes in the air and measured and simulated values of S_{21} at different separations between the two structures. The separation between the structures is varied from $430\mu m$ to $710\mu m$. We note a good matching between measurement and simulations. Additionally, we note that increasing the distance between TX and RX by about $200\mu m$ reduces the crosstalk only 5dB at 1GHz. Clearly, distance alone is not the solution.

Fig. 6 shows the measurement and simulation results for the structures in Fig. 2 (b). We note that that magnitude of crosstalk is much lower (on average 30dB) than shown in Fig. 5. However, these results include the effects of the metal connection between the inductors and the pads that were used for testing purposes but were outside the Faraday cages, which unfortunately also contribute to crosstalk. In fact, the metal connection outside the Faraday cage is the primary contributor to the increase in crosstalk with frequency. Unfortunately, it is not possible to make measurements without these interconnections. We note that there is a good agreement with measurement and simulations at higher frequencies. The discrepancy at lower frequencies (below 2.5GHz) can be attributed to the crosstalk between the probes in the air. We also note that, unlike the results in Fig. 5, spacing between the structures has very limited impact on the crosstalk.

978-1-4244-2018-6/08 $25.00 © 2008 IEEE

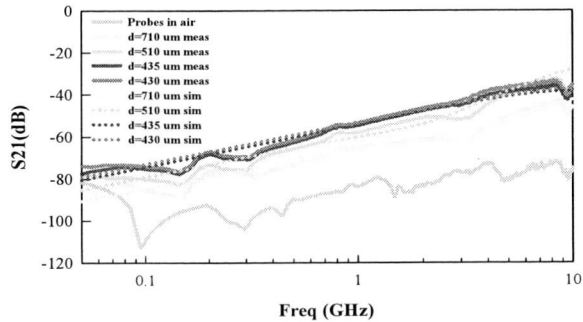

Fig. 5: Measurement and simulation results without Faraday cage [Fig. 2(a)]

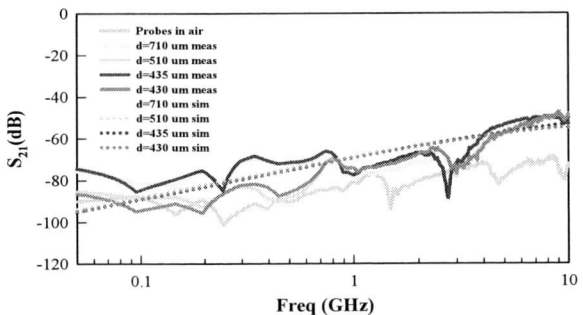

Fig. 6: Measurement and simulation results with Faraday cage [Fig. 2(b)]

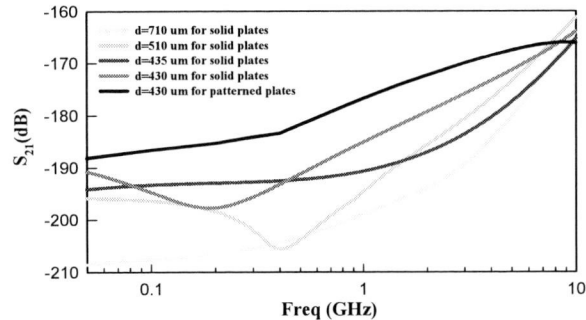

Fig. 7: The simulation performance of TX and RX both with Faraday cage

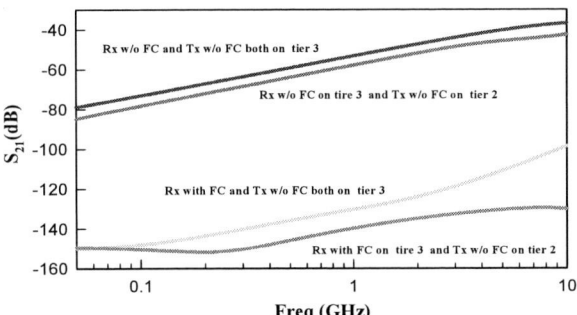

Fig. 8: Simulation results for different structures (d= 430μm)

As we are unable to eliminate the connection to the outside for our FC measurements and we have validated the close agreement between measurement and simulation results, we can now use HFSS as a tool to show the real impact of these cages. Fig. 7 shows the simulation results for inductors that are similar to the ones shown in Fig. 2(b) but are completely enclosed within a Faraday cage. In comparison to the measurement results shown in Fig. 5 the crosstalk is reduced by an additional 100dB, which is the highest value ever reported. The slight increase in crosstalk with frequency is due to a combination of numerical precision and port to port coupling during simulations.

A solid Faraday cage provides excellent isolation, however, the image current induced by the magnetic field in the solid surface of the Faraday cage flows in opposite direction and will reduce the total inductance and Q (i.e., Eddy current effect). Orthogonally slotting the top and bottom plates can be used to eliminate this effect without reducing the isolation properties [3]. The slots, which should be smaller than the operation wavelength, act as an open circuit to the induced current. To evaluate this property we pattern the top and bottom plates of the Faraday cages while maintaining everything else. Simulations for patterned FCs show a reduction in the isolation by 10dB (for 430μm separation) as shown in Fig. 7. Note, this is insignificant in comparison to the absolute values. However, both the inductance and inductance Q increase by about 40% at 5GHz.

Next, we evaluate tier-to-tier noise coupling with and without Faraday cages using the same structures shown in Fig. 2(a) and 2(b). For our first experiment, the TX and RX are placed without Faraday cages and are separated 430μm. The receiver

is kept on tier 3 while one transmitter each are placed on tier 2 and tier 3. The simulation results included in Fig. 8, show that using different tiers increases the isolation by an additional 10dB. For the second experiment, we repeat the procedure done in the previous experiment, but this time we include one of either the TX or RX in a Faraday cage. Here we see that using separate tiers increases isolation by 20dB. However, more importantly, using Faraday cages improves isolation by over 75dB.

IV. LUMPED MODELING OF NOISE COUPLING EFFECTS

Several equivalent circuit models have been developed to model crosstalk in integrated circuits [8]–[11]. In order to accurately understand the physical behavior of the Faraday cage, Fig. 9(a) shows a simplified equivalent circuit used for modeling each single spiral inductor [12]. The spiral inductor structure is represented by an inductance L_s, a series resistance R_s, the coupling capacitance C_c and resistance R_c. The parallel parasitics result from a combination of oxide capacitance $C_{oxeffect}$ representing the capacitance value of the oxide layer in each tier between the inductor and the top of silicon substrate. The substrate parasitics are represented by R_{sub} and C_{sub}. The value of each of the elements in the equivalent circuit is optimized to match the measured results. As seen in Fig. 9(b) there is there is extremely good matching between measured and modeled parameters.

To evaluate the coupling effects between two inductors with/without Faraday cages, a simple lumped element equivalent circuit is shown in Fig. 10. This noise coupling effect (crosstalk) is modeled by an RC path (R_{cp}//C_{cp}). Expressions of equivalent elements R_{cp} and C_{cp} have been derived for

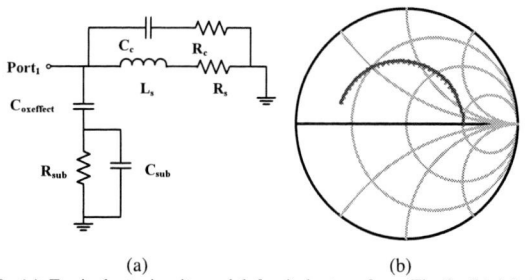

(a)   (b)

Fig. 9: (a) Equivalent circuit model for inductors from Fig 2; (b) Modeled data (dotted line) and measurement data (solid line) from 1GHz to 10GHz

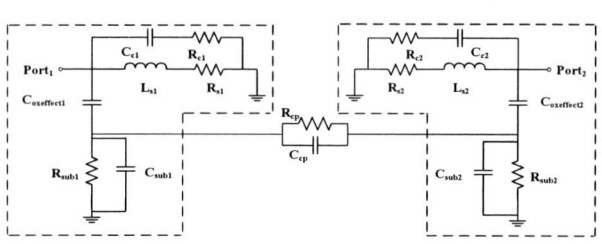

Fig. 10: The lumped equivalent circuit for the crosstalk between two inductors

inductors without Faraday cage in [9], [13], [14] and are included here for completeness.

$$R_{cp} = [K_1 \frac{\pi \varepsilon_0 \sigma_{si}}{4ln[\frac{\pi(d-W)}{W+t}+1]} W]^{-1} \qquad (1)$$

$$C_{cp} = [K_1 \frac{\pi \varepsilon_0 (\varepsilon_{si}+1)}{4ln[\frac{\pi(d-W)}{W+t}+1]} W] \qquad (2)$$

Here K_1 is the fringing factor. For our test structures without a Faraday cage K_1 is equal to 1.37 for d= 430μm. W is the average outer diameter of the inductor and t is the thickness of conductors. Fig. 11 shows very good agreement between the results of crosstalk simulation using the simple equivalent circuit presented above and the measurements for the test structure (d= 430μm) listed in Fig. 2 (a) without FC.

When two inductors are close to each other, their electromagnetic field patterns interact, a portion of the signal present at one inductor will be transferred to the other one by coupling through $C_{oxeffect}$. Simulations for structures with Faraday cages indicated that oxide capacitance $C_{oxeffect}$ can be treated as an effective coupling factor for the Faraday cage. For the receiver with Faraday cage and transmitter without Faraday cage, the $C_{oxeffect}$ for the receiver part is about $1/100^{th}$ of that for the receiver without a Faraday cage. The fringing factor K_1 is reduced to one third. For both receivers and transmitters with Faraday cages, the $C_{oxeffect}$ for both the receiver and the transmitter is $1/1000^{th}$ of that for receiver and transmitter without Faraday cages. The fringing factor K_1 is reduced to one tenth. This simple lumped model can be used to accurately predict the impact of coupling with and without Faraday cages by just modifying the value of $C_{oxeffect}$ as shown in Fig. 11.

V. CONCLUSIONS

In this paper, we model and measure noise coupling in 3D ICs. We also develop a new technology for reducing this

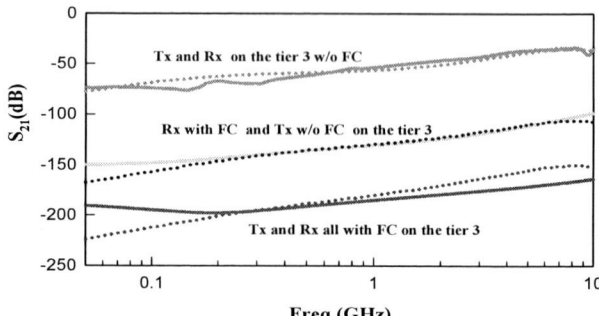

Fig. 11: Lumped circuit simulation results for different structures (d= 430μm). Good agreement between lumped circuit simulation data, measured data and HFSS simulation data. The doted line is lumped circuit simulation data

crosstalk. To validate our model and our design techniques we used a combination of simulation and measurement results from inductors (implemented on different tiers) with and without Faraday cages. We use the structure without a Faraday cage as our reference. Measurement and simulation results show that isolating either the TX or RX alone reduces the noise coupling by 75dB. However, enclosing both the TX and RX in Faraday cages reduces the noise coupling by an additional 25dB. We also developed and verified a simplified lumped equivalent circuit model for noise coupling. In 3D ICs the use of Faraday cages, combined with placing the RF/analog circuits and digital circuits on separate tiers, results in extremely good noise isolation. This allows us to fully integrate system that have both sensitive analog/RF circuits and require significant digital signal processing. Additionally, it is well known that noise isolation decreases at higher frequencies, therefore 3D ICs using Faraday cages may be the preferred process technology for high performance mixed-signal systems in the future.

ACKNOWLEDGMENT

This research was sponsored by a grant from DARPA. Chips were fabricated at MIT Lincoln Labs.

REFERENCES

[1] A. L. P. et.al., "Substrate noise coupling through planar spiral inductor," *IEEE JSSC*, pp. 877–884, June 1998.
[2] C. S. K. et.al., "Deep trench guard to suppress coupling between inductors in RF ICs," *IEEE MTT-S Digest*, pp. 1873–1876, 2001.
[3] C. P. Y. et.al., "On-chip spiral inductors with patterned ground shields for Si-based RF ICs," *VLSI Symposium*, pp. 85–86, 1997.
[4] T. B. et.al., "On-chip RF isolation techniques," *IEEE BCTM*, pp. 205–211, 2002.
[5] C. K. et.al., "3D integration for integrated circuits and advanced focal planes," 2007.
[6] "MITLL low-power FDSOI CMOS process: Design guide," 2006.
[7] "HFSS, Ansoft Inc," 2004.
[8] K. Joardar, "A simple approach to modeling cross-talk in integrated circuits," *IEEE JSSC*, pp. 1212–1219, Oct. 1994.
[9] J. P. R. et.al., "Substrate crosstalk reduction using SOI technology," *IEEE Trans. on Electron Devices*, pp. 2252–2261, December 1997.
[10] C. J. C. et.al., "Characterization and modeling of on-chip inductor substrate coupling effect," *IEEE RF IC*, pp. 311–314, 2002.
[11] J. H. M. et.al., "Crosstalk coupling effects of CMOS co-planar spiral inductors," *IEEE CICC*, pp. 371–374, 2004.
[12] C. L. Chen, "Effects of CMOS process fill patterns on spiral inductors," *Microwave and Optical Technology Letters*, pp. 462–465, March 2003.
[13] J. H. M. et.al., "Coupled microstrip parallel-gap model for improved filter and coupler design," *Electron Lett.*, pp. 377–379, 1983.
[14] M. Kirschning and R. H. Jansen, "Accurate wide-rang design equations for the frequency-dependent chararersitics of parallel coupled micristrip lines," *IEEE Trans. Microwave Theory Tech.*, pp. 83–90, 1984.

978-1-4244-2018-6/08 $25.00 © 2008 IEEE

IEEE 2008 Custom Intergrated Circuits Conference (CICC)

A Method Using Circuit/Substrate Macro Modeling to Analyze Substrate Noise in a 3.2-GHz 350M-transistor Microprocessor

Mikiko Sode, Mikihiro Kajita*, Naoya Nakayama, Satoshi Nakamoto

NEC Electronics Corporation, NEC Corporation*

1753, Shimonumabe, Nakahara-ku, Kawasaki, Kanagawa, Japan

ABSTRACT

Substrate noise analysis has become increasingly important in recent LSI design. This is because substrate noise, which affects PLLs, causes jitter that results in timing error. Conventional analysis techniques of substrate noise are, however, impractical for large-scale designs that have hundreds of millions of transistors because the computational complexity is too huge. To solve this problem, we have developed a fast substrate noise analysis technique for large-scale designs, one in which a chip is divided into multiple domains by using current density variation and the circuits of each domain are reduced into a macro model. Using this technique, we have designed a microprocessor chip for use in the supercomputer SX-9 (die size: 20mm x 21mm, frequency: 3.2GHz, transistor count: 350M). Computation time with this design is five times shorter than [2] that with a 1/3000 scale design using a conventional technique, while resulting discrepancy with measured period jitter is less than 15%.

1. Introduction

SX-9 is the world's highest performance vector supercomputer and contains a large number of 102-GFLOPS microprocessors, as well as very high-speed memory interfaces. The design of this microprocessor was very challenging because of its massive scale and high operational frequency (die size: 19.84mm × 21.04mm, frequency: 3.2GHz, transistor count: 350M, technology: 65nm CMOS). Since the constraints on timing are extremely tight for such a high-speed microprocessor, it is essential to estimate clock jitter precisely. Clock jitter is mainly caused by power supply noise in PLLs and clock distribution networks. Although the power supply lines of the PLLs are in general separated from those of noise sources to protect them from supply noise, supply noise propagates and affects the PLLs via the semiconductor substrate (i.e., creating substrate noise). Substrate noise affecting PLLs causes clock jitter, which degrades the timing margin of the logic circuits. In order to ensure design quality with respect to timing, then, it is essential to precisely analyze substrate noise as well as power supply noise.

In response to this need, a number of techniques for substrate noise analysis have been proposed [1-5]. The main technique to handle substrate noise is by using order reduction and gate level macro modeling to reduce the computational complexity[2][3]. However, it is difficult to handle digital circuit like microprocessor designs at practical CPU time. To solve this problem, domain-based macro modeling method is proposed. We analyze Domain size vs. CPU Time, Domain size vs. accuracy of substrate resistance and Domain size vs. accuracy of propagation noise. Trade-off problem is explained between accuracy and CPU time by changing the Domain size. We explain relationship between current density variation and accuracy which is the key idea to solve trade off problem and propose the partition method to achieve both low complexity and high accuracy. The proposed

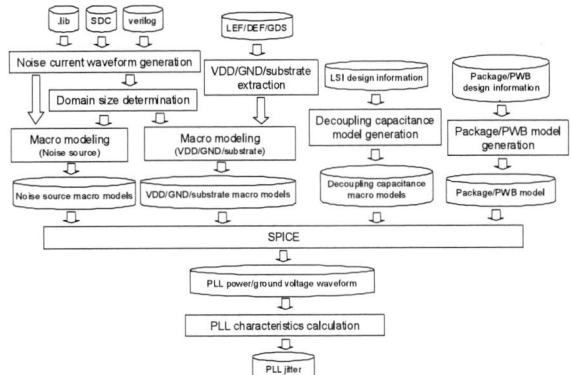

Figure 1. Proposed analysis flow

method decides analysis accuracy by using current density variation. Also we explain the macro modeling method that minimize the error of substrate resistance, even if the Domain size is changed. And we explain how to make the LSI model and the simulation method of the effects of the substrate noise on PLL performance.

Using this substrate noise analysis technique, we have designed a microprocessor for use in the supercomputer SX-9. In this design, because of the severity of our timing constraints, we estimated not only substrate noise but also clock jitter due to that substrate noise, and we verified the accuracy of our estimated clock jitter in comparison with measurement results. Our experimental results have shown that accuracy for estimated jitter is higher than 85%, and that analysis time is five times shorter than that with a 1/3000 scale design using a conventional technique [2]. That is, we estimate that the computation speed with our technique is roughly 15000 times faster than that with a conventional technique. Since our substrate noise analysis technique provides both accuracy and fast operations, it is well-suited to the design of microprocessors having hundreds of millions of transistors.

2. Analysis Flow

The proposed analysis flow is shown in Fig. 1 and described below.

STEP 1 Doman size determination

First, a power library (.lib) [7] based on switching activity ratios is generated for each primitive block. The current waveform for each instance is then generated, either using an SDC file [7] or a switching activity ratio, as well as net connectivity information and the power library. Next, the domain size for macro modeling is determined based on the current distribution in the chip layout.

STEP 2 Generation of noise source model

978-1-4244-2018-6/08 $25.00 © 2008 IEEE

A current waveform (noise source model) for each macro model is generated by superposition of individual-instance current-waveforms, as determined on the basis of domain size and layout design information.

STEP 3 Generation of power/ground/substrate model

First, an RC model of the LSI power/ground lines and substrate is extracted. Then, the RC model is reduced to a mesh network on the basis of domain size for macro modeling.

STEP 4 Generation of decoupling capacitance model

First, decoupling capacitance is calculated for each primitive block. Next, a decoupling capacitance model for each macro model is generated by summing up the decoupling capacitance of instances included in the macro model region.

STEP 5 Generation of package and PWB models

The power/ground of the package are modeled in the form of a lumped RLC model. And the PWB power/ground are also modeled in the form of a lumped RLC model, using an existing electromagnetic field solver.

STEP 6 Generation of SPICE netlist

The noise source model and the decoupling capacitance model are connected to the LSI power/ground/substrate model to make an LSI model. The package and PWB power/ground models are connected to the LSI model to generate a SPICE netlist.

STEP 7 Calculation of PLL power/ground voltage waveforms

SPICE simulation (transient analysis) is performed for the netlist to obtain PLL power and ground voltage waveforms.

STEP 8 Calculation of PLL characteristics

PLL simulations are performed using the waveforms to calculate the PLL jitter caused by the PLL itself and by the CTS circuits. We used a newly developed method which transforms PLL power/ground voltage waveforms into jitter. With this method, jitter is estimated with almost the same accuracy as with SPICE, but roughly 100k times faster than with SPICE.

Note that the flow is achieved with common design libraries, which means that the method can be applied to any kind of chip.

3. Method of substrate noise analysis for large-scale chips

In order to analyze substrate noise in large-scale chips, i.e., those having hundreds of millions of transistors, within a practical time-period, we propose an efficient method using macro-modeling (Fig. 2). The method is summarized below.

A chip is divided into multiple domains by using current density variation. Circuits, power supplies, grounds, and the substrate in each domain are reduced and integrated into a macro model. The substrate noise analysis is performed using a combination of all the created macro models.

A macro model is shown in Fig. 2(c). Substrate noise injection paths are generally thought to be categorizable into three types: (1) from ground contacts to the substrate through resistive connections, (2) from power contacts to the substrate through capacitive connections, and (3) from transistor gates to the substrate through capacitive connections. In ordinary digital circuit operations, since the amount of noise injected along the

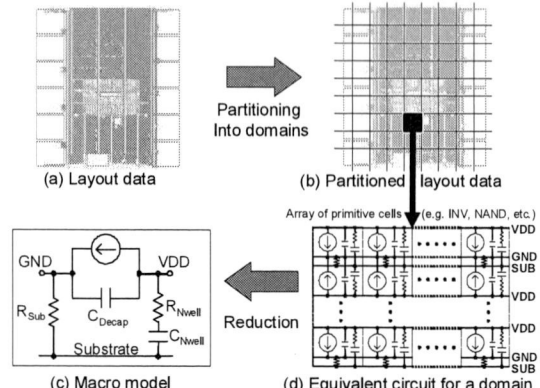

Figure 2. Complexity reduction in substrate noise analysis

third path (i.e., leak current [6]) is negligible, we did not consider the third path when building our macro model.

In Fig. 2(c), *Rsub* represents resistance for the ground contacts within the relevant domain, while *RsubN* represents the resistance for the power contacts. *Cwell* represents the capacitance for the domain's N-well, *Decap* represents its decoupling capacitance, and *Icurrent* represents the current source for the switching currents of the transistor gates.

In order to investigate the degree to which the domain division strategy is able to reduce computational time, we checked the relationship between CPU time and domain size, using data from a 20mm x 21mm chip. Results are shown in Fig. 3. As may be seen, after a certain point, reduced domain size results in an

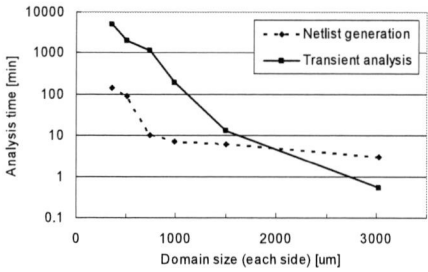

Figure 3. CPU time for substrate netlist generation and propagation waveform generation

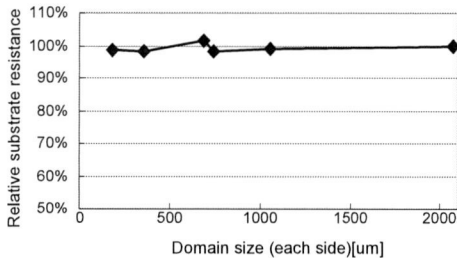

Figure 4. Accuracy of substrate resistance analysis vs. domain size in a digital circuit region

978-1-4244-2018-6/08 $25.00 © 2008 IEEE

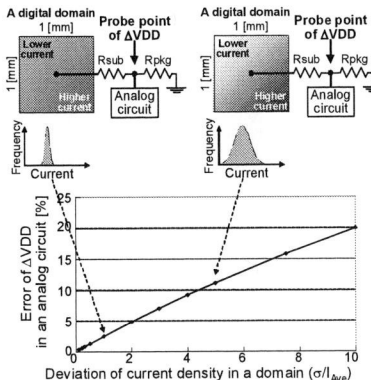

Figure 5. Relationship between standard deviation of current density and accuracy of substrate noise analysis

exponential rise in CPU time. It is clear that there are limits to the reduction of domain size, and that it is necessary to consider the tradeoff between CPU time and the analysis accuracy that results from reduced domain sizes.

We also investigated the relationship between domain size and analysis accuracy. We first created two domains, each representative of a part of a digital region, and placed them at a distance of 100 um from each other. We then calculated the substrate resistance between them for various domain sizes. We then created a macro model for each of the two domains, calculated the resistance between the two models for various domain sizes, and then compared the results with those obtained for the original, non-reduced domains. Maximum error was 0.96% up to a domain size of 9000 transistors, which verifies the accuracy of our macro-model. Next, we created two new domains using as their cells the previous macro models, whose accuracy had been experimentally verified. We then created a macro model for each of these two new domains and compared resistance calculation results as before. The results are shown in Fig. 4. As may be seen, the accuracy of substrate resistance calculation varies little with variation in domain size for the range of sizes tested (here, up to 2000um). This result shows that, for a uniform digital circuit domain, analysis accuracy depends little on domain size.

By way of contrast, in a digital circuit region within which there is high variation in current density (i.e. switching activity ratios), larger domain size results in significantly worse accuracy of substrate noise estimation. Figure 5 shows simulation results for the relationship between the standard deviation of current density in a domain and the error in estimated substrate noise. The results indicate that the error in estimated substrate noise is roughly proportional to the standard deviation of current density in the domain. For example, to suppress this error to less than 15% (i.e. to maintain an accuracy of higher than 85%), the standard deviation of current density must be suppressed to less than three times the average current density. To suppress deviation of current density, we iteratively divide any digital circuit domain with high deviation in current density until the standard deviation of current density in each divided domain reaches a desired value or less. Figure 6 shows the relationship between domain size and the accuracy of estimated substrate noise. By iteratively dividing domains, we are able to decrease the standard deviation of current density (dotted line, right axis), which improves the accuracy of substrate noise estimation (solid line, left axis).

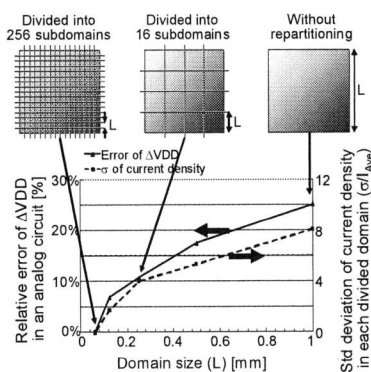

Figure 6. Relationship between domain size and accuracy of substrate noise analysis

In consideration of the above, we have adopted a macro-modeling policy in which:

(1) for digital circuit regions, models are reduced as much as possible while deviation value of current density is smaller than the desired value.

(2) for analog circuit regions, a smaller domain size is adopted since the contact pitch and current density are usually not uniform in those regions.

4. Application to the SX-9

We applied our macro modeling technique to a microprocessor for use in the supercomputer SX-9. To ensure the accuracy, we iteratively divide any digital circuit domain with high deviation in current density until the standard deviation (σ) of current density in each divided domain reaches three times the average current density ($Iave$). Table 1 shows the number of domains in which the standard deviation in current density satisfies a desired value or not. First, the sizes of all domains are set to 2688 um. In this first partioning, 40 domains satisfy the desired standard deviation in current density ($\sigma/Iave \le 3$), and another 10 domains does not satisfy it. The 10 domains are repartitioned into 40 subdomains, and two of them satisfy the desired standard deviation and another 13 do not. The domains which do not satisfy the desired standard deviation are iteratively repartitioned, and finally all domains satisfy it with a domain size of 168 um. Figure 7 shows the distribution of standard deviation in current density ($\sigma/Iave$) in each repartitioning step. This shows that domains with high deviation in current density are divided into subdomains with

Table 1. Process for determining domain sizes

	Domain size (side length) [um]				
	2688	1344	672	336	168
OK ($\sigma \le 3I_{ave}$)	40	2	5	15	12
NG ($\sigma > 3I_{ave}$)	10	13	11	4	0
No count (PLL, SerDes, etc.)	6	25	34	25	4

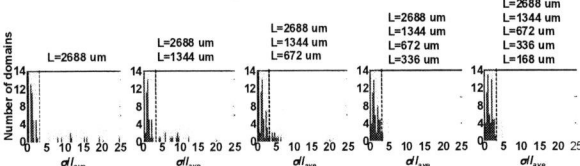

Figure7. Distributions of σ/I_{ave} in the process for determining domain sizes

lower deviation step by step. As a result, the microprocessor chip is divided into only 74 domains which consist of 40 domains of 2688um in size, 2 domains of 1344 um, 5 domains of 672 um, 15 domains of 336 um, and 12 domains of 168 um. By reducing netlist for each domain into a macro model, we are able to obtain both accuracy and fast operations.

5. Experimental Results

To verify the accuracy of our proposed analysis method, we fabricated a test chip and measured the period jitter in the PLL circuit caused by power supply noise. The test chip is the same in size and number of transistors as a CPU chip in an SX-9. A PLL circuit was placed at the center of the chip as a victim and surrounded by core logic circuits as aggressors which would generate power supply noise. The power supply and ground lines of the PLL circuit were completely separated from those of the core logic to prevent power supply noise from propagating via metal wiring, which meant that noise generated in the core logic region would be propagated only through the Si substrate. The core logic circuits offered the function of variable clock gating, and the gating cycle could be varied. Streamed data were toggled at each positive edge of each core clock cycle.

In this experiment, the gating cycle of the core logic was used as a parameter and varied from 31.25 MHz to 250 MHz. The measurement of period jitter was performed using an oscilloscope (Infinium 54855A by Agilent Technology). For each gating cycle, power supply noise (peak-to-peak) was measured using an on-die detector, and results are shown in Fig. 8. In our simulations, this power supply noise was modeled as a voltage source inserted between the power supply and ground of the core logic, and voltage fluctuation at the ground was assumed to be injected into the Si substrate via ground contacts.

We divided a digital portion of the test chip into main domains of 1mm x 1mm size, and we then performed macro modeling. The computation time required for noise analysis with this design was roughly 24 hours. In Fig. 9, the ratios of period jitter to clock cycle in the simulation are compared with those in the measurement. As may be seen, simulation results are consistent with measurement results, and the discrepancies between results obtained with our proposed method and measured period jitter were less than 15% (i.e., the accuracy of our proposed method was more than 85%) for all cases of clock gating. This accuracy is high enough even for the design of such high performance processors as the CPUs used in an SX-9.

6. Conclusions

We have developed a fast substrate noise analysis technique for large-scale designs, one in which a chip is divided into multiple domains and the circuits of each domain are reduced into a macro model. To maintain accuracy, we adaptively change the size of each domain according to circuit type or switching activity ratio. In this way, we are able to achieve both high accuracy and low complexity.

To verify the accuracy of our proposed analysis method, we fabricated a test chip and measured period jitter for its PLL circuit, and we compared period-jitter-to-clock-cycle ratios obtained in simulations with those obtained in measurements. The discrepancies between results obtained with our proposed analysis and measured period jitter were less than 15% for all cases of

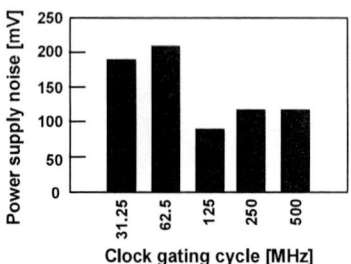

Figure 8. Dependency of power supply noise on clock gating cycle

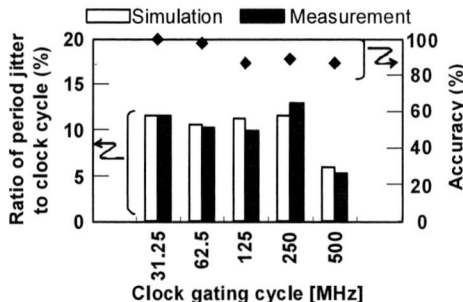

Figure 9. Comparison of period jitter obtained in simulations and measurements

clock gating. Computation time with this design was roughly 24 hours, including that required for jitter analysis.

7. Acknowledgment

The authors like to thank Mr. Shunichi kaeriyama for valuable discussions on substrate noise analysis.

8. References

[1] M. Nagata, "On-Chip Measurements Complementary to Design Flow for Integrity in SoCs," Proceedings of the 44th Design Automation Conference, pp. 400-403, 2007.

[2] M. Badaroglu, et. al., "SWAN:High-Level Simulation Methodology for Digital Substrate Noise Generation," IEEE Trans. VLSI Systems, Vol. 14, No. 1, pp. 23-33, January 2006.

[3] G. Van der Plas, M. Badsroglu, G. Vandersteen, P. Dobrovolny, P. Wambacq, S. Donnay, G. Gielen, H. De Man, "High-level Simulation of Substrate Noise in High-Ohmic Substrates with Interconnect and Supply Effects," Proceedings of the 41st Design Automation Conference, pp. 854-859, 2004.

[4] Jeong-Yeol Kim, Ho-Soon Shin, Jong-Bae Lee, Noon-Hyun Moon, Jeong-Taek Kong, "SilcVerify:An Efficient Substrate Coupling Noise Simulation Tool for High-Speed & Nano-Scaled Memory Design," Proceedings of the 8th International Symposium on Quality Electronic Design (ISQED '07), pp. 475-480, March 2007.

[5] R. Murgai, S. M. Reddy, T. Miyoshi, T. Horie, M. B. Tahoori, "Sensitivity-based modeling and methodology for full-chip substrate noise analysis," Proceedings of the Design Automation and Test in Europe Conference and Exhibition Designers' Forum (DATE '04), Vol. 1, pp. 610-615, Feb. 2004.

[6] K. Roy, S. Mukhopadhyay, H. Mahmoodi-meimand, "Leakage Current Mechanisms and Leakage reduction Techniques in Deep-Submicrometer CMOS Circuits," Proceedings of the IEEE, Vol. 91, No. 2, February 2003.

[7] SYNOPSYS design manual.

IEEE 2008 Custom Intergrated Circuits Conference (CICC)

Characterization of Random Decision Errors in Clocked Comparators

B. S. Leibowitz, J. Kim, J. Ren, C. J. Madden

Rambus, Inc.

Abstract-**Clocked comparators have found widespread use in noise sensitive applications such as wireline receivers, A/D converters, and memory bit-line detectors. However, their nonlinear, time-varying behavior and discrete output levels have discouraged the use of traditional small-signal noise simulation techniques such as .NOISE in SPICE. This paper asserts that the periodic noise analysis available from RF circuit simulators can provide insight into the intrinsic sampling and decision operations of clocked comparators and help develop a linear periodically time-varying (LPTV) noise model that accurately predicts the decision error probability. Two comparators are simulated and compared to laboratory measurements. A 90nm CMOS comparator is measured to have an equivalent input-referred random noise of 0.73 mVrms for DC inputs, matching simulation results with a short channel excess noise factor γ=2.**

I. INTRODUCTION

A clocked comparator is a circuit element that makes decision as to whether the input signal is high or low at every clock cycle. It has found widespread use in applications where digital information needs to be recovered from analog signals, such as A-to-D converters, wireline receivers, and memory bit-line detectors. To ensure correct detection on each comparison, the analog input must have sufficient magnitude to overcome deterministic errors such as offset and hysteresis, as well as random errors due to device thermal noise and 1/f noise. Therefore, it is important to estimate the probability of a decision error given these factors. This paper presents a way to simulate the decision error probability due to random noise.

This paper describes a method to simulate the random noise effects in clocked comparators using the linear periodically time-varying (LPTV) simulation techniques that have been used primarily for RF circuits such as LNAs and mixers [1]. While such small-signal linear analysis methods have been considered unsuitable for clocked comparators as they are strongly nonlinear in the sense that they produce hard ones and zeros, we show that the LPTV model for comparators can indeed provide many design insights including sampling aperture, regeneration gain, and noise contribution during the decision process. In addition, LPTV circuit simulation methods such as periodic steady-state (PSS) analysis and periodic noise (PNOISE) analysis can efficiently help us create those models [1,2,4,5].

Note that the decision error probability is different from the metastability failure probability, which has been extensively studied in literature. While metastability failure refers to a situation where the comparator cannot make a firm decision within a given period of time due to insufficient input swing, a random decision error in this paper refers to a situation where the comparator makes a decision, but the decision is incorrect due to random noise.

This paper is organized as follows. First, it describes the linear, periodically time-varying (LPTV) noise model for clocked comparators and discusses the method of estimating the decision error probability or equivalent input-referred noise. Second, it demonstrates the generation of the LPTV noise models for clocked comparator examples with the use of RF simulator analyses such as PSS and PNOISE. Finally, simulation results are compared to laboratory measurements to validate the methodology.

II. PERIODIC NOISE SIMULATION

A. Noisy Comparator Model

Fig. 1 shows a model of a clocked comparator that periodically samples an input signal $V_i(t)$ and compares the samples to an implicit zero reference to produce a sequence of decision results D_k. The nonlinear and noisy sampling process is represented by a nonlinear filter followed by an ideal sampler. The sampler is followed by an ideal slicer to produce the periodic decision results, so that the probability of a decision error is governed by the signal to noise ratio at the sampler output. While practical comparator circuits generally have no such explicit distinction between the sampling and decision elements, this mathematical model is useful for quantifying decision error probabilities.

If $V_i(t)$ is a DC or periodic signal harmonic with clock ϕ, we can model the nonlinear filter by its periodic steady state response, i.e. the output $V_o(t)$ that is also harmonic with clock ϕ, and its time-varying impulse response $h(t, \tau)$ that describes the small-signal output response $v_o(\tau)$ at observation time τ to an impulse at input $v_i(t)$ at time t [2]. The small-signal output at a specific output sampling time τ_{obs} can be more simply expressed as an impulse sensitivity function (ISF) (1). Written as an integral form in (2), the ISF is seen to represent a time

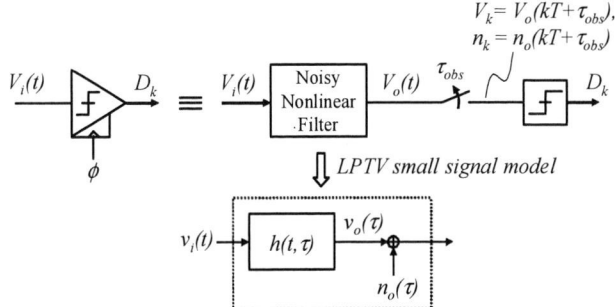

Fig 1. Clocked comparator model including a nonlinear filter, ideal sampler, and ideal slicer (top), and linearized PTV filter model with noise (bottom)

varying sensitivity profile of the output with respect to small-signal inputs. In other words, it is the shape of the comparator's sampling aperture and its Fourier transform yields the comparator's frequency response (aperture bandwidth) [8]. For the special case of a DC input, the area of the ISF is equivalent to the small-signal DC gain a_0 at output time τ_{obs} (i.e. the regeneration gain) (3).

$$ISF(t) = h(t, \tau_{obs}) \qquad (1)$$

$$v_o(\tau_{obs}) = \int_{-\infty}^{\infty} v_i(t) \cdot ISF(t) \cdot dt \qquad (2)$$

$$a_0 = v_o(\tau_{obs})/v_{i,dc} = \int_{-\infty}^{\infty} ISF(t) \cdot dt \qquad (3)$$

The LPTV filter model also includes an additive cyclostationary Gaussian noise process $n_o(\tau)$ at the output. The sampled output noise $n_k = n_o(kT + \tau_{obs})$ is characterized by a sampled power spectral density $PSD(f, \tau_{obs})$ and corresponding variance (4). This sampled noise adds directly to the sampled periodic steady state output $V_k = V_o(kT + \tau_{obs})$. Hence, the decision error probability can be expressed as in (5).

$$\sigma^2 = \overline{n_k^2} = \int_0^{1/2T} PSD(f, \tau_{obs}) \cdot df \qquad (4)$$

$$P(error_k) = Q(SNR) = \tfrac{1}{2} erfc\left(SNR/\sqrt{2}\right) \qquad (5)$$

$$SNR = \sqrt{V_o(\tau_{obs})/\sigma} \qquad (6)$$

For input frequencies well below the sampling bandwidth of the ISF, we can alternatively divide the rms output noise σ by the DC gain a_0 to predict an equivalent rms input noise σ_i.

B. Comparator Noise Simulation

In this section, we simulate a StrongARM style comparator [3] shown in Fig. 2 using the SpectreRF PSS and PNOISE analyses [1]. By considering the periodic steady state output and the sampled noise $PSD(f,t)$ across time for a small DC input, we can gain insight into the noise behavior of the comparator and how to apply the noise model described in the previous section.

Fig. 3 shows the periodic steady state response of the differential output near the time of the rising clock edge. Only a portion of the entire clock period of 625ps (1.6GHz) is shown, so the return to reset behavior after the comparison completes is not visible in Fig. 3. The upper plot shows the

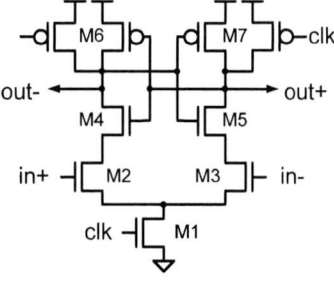

Fig. 2. StrongARM style comparator.

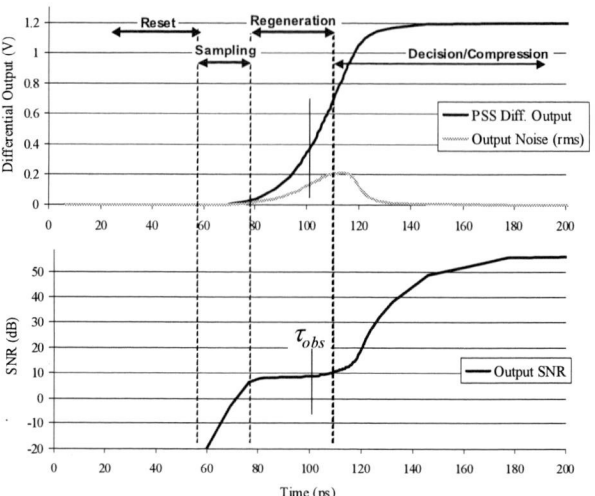

Fig. 3. Comparator periodic simulation results near the clock edge.

large signal PSS output as well as the rms value of the differential output noise, found by integrating the sampled noise $PSD(f,t)$ produced by PNOISE at each simulation time. The lower plot shows the differential output SNR versus time.

Four regions of operation are visible in Fig. 3. Initially the sampler is held in a reset state, and the PSS differential output is zero. During reset there is a small but non-zero rms output noise determined by the reset devices operating in the linear region and the output capacitance. Thus, the SNR approaches zero, or $-\infty$dB, during reset. In the second region, the input signal is sampled and transferred to the output nodes, resulting in a rapid rise in the output SNR. As time progresses, exponential regeneration of the output voltage by M4-M7 takes hold. This exponential regeneration in time means that signal and noise impulses injected to the output node earlier in the cycle have exponentially larger impact than equivalent injections at later times. In other words, the sensitivity of the output to new signal and noise components is decaying exponentially with time [8]. As these sensitivities approach zero, the circuit continues regenerating the output voltage resulting from the previous input and noise signals, but no longer accepts new signal and noise contributions. Thus, both the signal and noise components grow at the same rate resulting in an approximately constant SNR, as shown in the third region of Fig. 3. The fourth operation region begins when large signal output compression occurs, ultimately producing a logic-level decision output. The output voltage after saturation to a hard logic level is complete is insensitive to incremental signal or noise at any previous time, so the output noise power returns to a small value similar to the reset phase noise power in this region. While a decision error may occur, the decision output itself is essentially noise free. At this stage, the saturated outputs correspond to the ideal slicer output D_k in Fig.1 rather than to the sampled filter output V_k.

The observation time τ_{obs} in Fig. 3 represents a time at which large signal nonlinearity has not yet led to compression of the output noise power. Because the nonlinear decision response has not been excited yet, the behavior of the circuit can be accurately modeled as the LPTV small-signal model in

Fig. 1 up to this time point. Also, since future signal and noise events no longer impact the output, the decision outcome has already been determined. That is, the future operation of the circuit is to regenerate the already present output signal and noise to a logic output level, a process that can be modeled by the ideal sampler and slicer of Fig. 1. Thus, the decision error probability is given by (5) and (6), where $V_o(\tau_{obs})$ is the PSS output at the indicated τ_{obs}. Similarly, the ISF of the comparator can be derived by evaluating the PAC simulation result $h(t, \tau)$ at $\tau = \tau_{obs}$ [6].

The choice of evaluation time τ_{obs} is not unique, as a range of times within the regeneration phase of operation satisfy the criteria discussed above. However, the times within this range all have approximately the same SNR, and predict similar decision error probabilities.

III. SIMULATION AND MEASUREMENT RESULTS

The periodic noise simulation technique described above is applied to two different wireline receivers (Fig. 4) and the results are compared with the measured random noise performance in both cases to verify the results. Receiver A, fabricated in a 90nm CMOS process, uses 2× interleaved StrongARM comparators from Fig. 2 to directly sample a differential input. The differential input is terminated by poly-silicon resistors (not shown) to match the 100Ω differential channel impedance. Receiver B, fabricated in a 65nm CMOS process, has a similar differential termination and comparator design, but the input signal passes through a linear front-end circuit, consisting of a linear equalizer and preamplifiers, before the 4× interleaved comparators.

A. Receiver A – Direct Input Sampling

In the direct input sampling receiver, random noise includes contributions from the 2× interleaved comparators themselves, as well as thermal noise from the termination resistors. A differential input capacitance greater than 2pF (including pad metallization and ESD) limits the rms thermal noise contribution of the termination resistors to less than 100μV, which is found to be negligible compared to the rms equivalent input noise of the comparator itself.

The standalone comparator is simulated with a range of noise factors $\gamma=1,2,3$ to account for the excess noise seen in short channel MOS devices, since this information was not provided by the CMOS foundry [7]. For each case, the random decision error probability, or BER, is simulated for small DC inputs $V_{i,dc}$ according to (5) as described in Section II. The simulation data in Fig. 5 show the resulting $BER(V_{i,dc})$ for $\gamma=2$. The simulation results are then fit to an input referred additive Gaussian noise model (7), which incorporates parameters μ_i

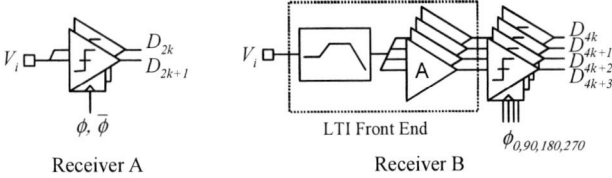

Fig. 4. Architecture of two simulated and measured differential receivers.

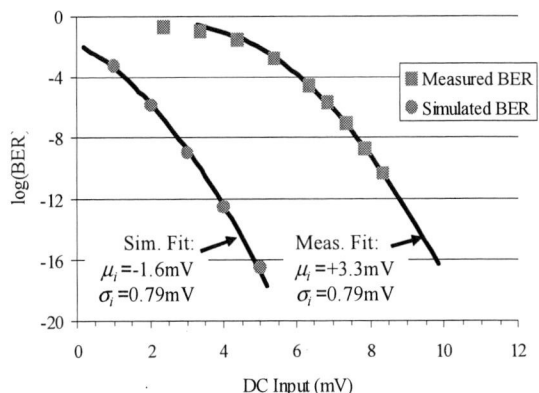

Fig. 5. Receiver A simulated ($\gamma=2$) and measured receiver BER.

TABLE I
SIMULATED AND MEASURED RMS INPUT NOISE

Receiver	Simulated (mV)				Measured (mV) (Pos. / Neg. / Avg)
	$\gamma=1$	$\gamma=2$	$\gamma=3$	$\gamma=4$	
(A) Direct Sampling					(0.79 / 0.65 / 0.72)
BER fit σ_i	0.59	0.79	0.94		
$\sigma_i = \sigma / a_0$	0.55	0.73	0.87		
(B) LTI Front-End					(0.87 / 0.83 / 0.85)
BER fit σ_i		0.62	0.73	0.83	
$\sigma_i = \sigma / a_0$		0.64	0.75	0.86	

and σ_i for the input referred offset and the rms sample noise, respectively. The model fit values for σ_i corresponding to excess noise factors $\gamma=1,2,3$ are listed in the first row of Table I. Alternatively, an equivalent input referred rms noise voltage may be estimated by dividing the comparator's rms output noise σ by its simulated small-signal gain a_0 at time τ_{obs}, yielding similar results listed in the second row of Table I (simulated at $V_{i,dc}=3$mV).

$$BER(V_{i,dc}) = Q\left(\frac{V_{i,dc} - \mu_i}{\sigma_i} \right) \qquad (7)$$

The equivalent input referred noise for Receiver A was measured in the laboratory by directly measuring *a posteriori* decision error probability $BER(V_{i,dc})$ for various DC inputs. The input stimulus was generated from two precision power supplies connected through high-ratio resistive voltage dividers to attenuate any possible external supply noise. The BER was detected by an external BERT via an on-chip loopback path from the comparator outputs. Fig. 5 shows the measured BER results for positive DC inputs with the BERT detecting errors against an "all ones" pattern. The measured data for BER below 2E-5 are fit to (7) to arrive at a measured rms input noise σ_i. The procedure is repeated for positive and negative input voltages, resulting in rms input noises of 0.79mV and 0.65mV, respectively, for an average rms value of 0.72mV. This measured rms input noise approximately matches the simulation results shown in Table I for $\gamma=2$.

Fig. 6 shows the simulated equivalent rms input referred noise and -3dB aperture bandwidth for Receiver A with $\gamma=2$ across a range of supply voltages. The aperture bandwidth is the -3dB magnitude response frequency of the Fourier

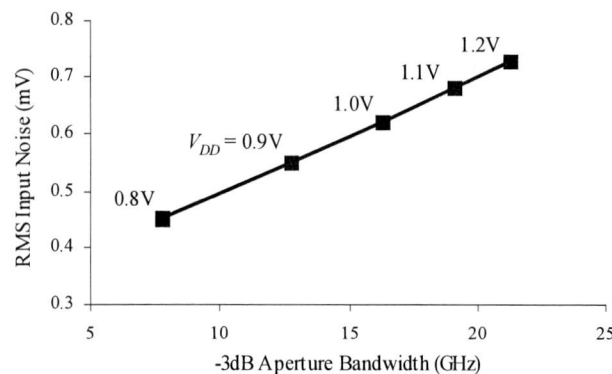

Fig. 6. Simulated input noise versus aperture bandwidth as V_{DD} is varied for Receiver A (CM input = V_{DD}-200mV, γ=2).

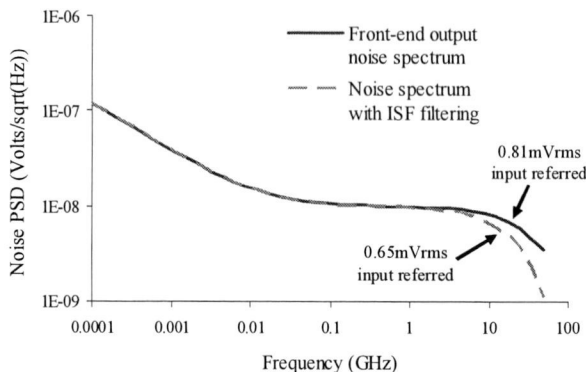

Fig. 7. Simulated output noise spectrum of front-end circuits in Receiver B.

transform of the simulated ISF. Increasing the supply voltage increases the aperture bandwidth of the comparator, consequently making it sensitive to external signal and noise inputs over a wider bandwidth, but also increases the impact of device noise within the comparator itself.

B. Receiver B – Linear Time-Invariant (LTI) Front End

In many applications comparators are preceded by LTI circuits such as equalizers and preamplifiers which also add noise to the input signal and contribute to the random decision error probability. While it is possible to separately simulate noise contributions from such LTI circuits with a linear AC noise simulation, it is important to note that the ISF of the comparator will filter the output noise PSD of these preceding stages, reducing the total noise power they contribute toward decision errors. This noise reduction is not free however; it comes at the expense of reducing signal bandwidth since the signal itself is also filtered by the ISF. Periodic simulation on the combined LTI front end plus comparator circuit properly accounts for these interactions.

The same simulation and measurement techniques used for Receiver A were applied to Receiver B incorporating LTI circuits and comparators. Again, foundry information is unavailable for the excess noise factor γ, so simulations were performed over the range γ=2,3,4. The simulated and measured random noise performance listed in Table I show that the measured rms input noise of 0.85mV is close to the simulation results for γ=4 in this 65nm CMOS process.

Separate periodic simulation of the comparator and linear AC simulation of the LTI front-end were also performed to examine the impact of comparator ISF filtering on the front-end output noise PSD. The simulated LTI front-end noise for γ=3 is plotted in Fig. 7. The solid line shows the noise PSD at the output of the LTI circuits. The dashed line shows the same noise PSD filtered by the normalized Fourier transform of the comparator ISF, showing less high frequency noise above the aperture bandwidth of approximately 20GHz. Referred to the input nodes by the DC gain of the front-end circuits, the total rms noise voltages for these two power spectra are 0.81mV and 0.65mV, respectively. Such impact on the total noise power is seen despite the relatively high aperture bandwidth of the comparator because the integrated noise power is relatively sensitive to the PSD at high frequencies. The

comparator's own rms noise contribution, referred to the input by the DC gain of the front-end circuits, is 0.47mV. With or without consideration of ISF noise filtering, the combined front-end and comparator rms equivalent input referred noise is calculated to be either 0.77mV or 0.94mV, respectively. Compared to the simulation results of 0.73 and 0.75mV for periodic simulations on the complete receiver shown in Table I for γ=3, we see that separate simulation of the LTI front-end and comparator noise contributions yields comparable results when the ISF filtering effect is included.

IV. DISCUSSION

An LPTV model of separate nonlinear sampling and decision operations has been shown to be applicable for simulating random decision error probability in clocked comparators using periodic simulation techniques. Comparators typically do not have separate sampling and decision circuits, but periodic simulation shows that for small input signals (when random decision errors are possible) these operations are temporally separated, allowing them to be modeled as consecutive operations to enable LPTV simulation. Periodic simulation results for two receivers, one with and one without linear circuits before the comparators, match the measured noise performance in both cases.

REFERENCES

[1] K. S. Kundert, "Introduction to RF Simulation and Its Application," IEEE JSSC, Vol. 34, No. 9, 1999.

[2] L. Zadeh, "Frequency Analysis of Variable Networks," Proc. I.R.E., Vol. 38, No. 3, pp. 291-299, Mar. 1950.

[3] J. Montanaro et al., "A 160MHz, 32b, 0.5W CMOS RISC Microprocessor," IEEE JSSC, Vol. 31, No. 11, 1996.

[4] M. Okumura, et al., "Numerical Noise Analysis for Nonlinear Circuits with a Periodic Large Signal Excitation Including Cyclostationary Noise Sources," IEEE Trans. on Circuits and Systems-I, Vol. 40, No. 9, pp. 581-590, Sept. 1993.

[5] A. Demir, et al., "Time-Domain Non-Monte Carlo Noise Simulation for Nonlinear Dynamic Circuits with Arbitrary Excitations," IEEE Trans. on Computer-Aided Design, Vol. 15, No. 5, pp. 493-505, May 1996.

[6] J. Kim, et al., "Impulse Sensitivity Function Analysis of Periodic Circuits," submitted to ICCAD 2008.

[7] R. P. Jindal, "Hot-Electron Effects on Channel Thermal Noise in Fine-Line NMOS Field-Effect Transistors," IEEE Trans. on Electron Devices, Vol. 33, No. 9, pp. 1395-1397, Sept. 1986.

[8] H. Johansson and C. Svensson, "Time Resolution of NMOS Sampling Switches Used on Low-Swing Signals," IEEE JSSC, Vol. 33, No. 2, 1998.

IEEE 2008 Custom Intergrated Circuits Conference (CICC)

Noise Tolerant Oscillator Design Using Perturbation Projection Vector Analysis

Igor Vytyaz, Josh Carnes, Ting Wu, Pavan Kumar Hanumolu, Un-Ku Moon, and Kartikeya Mayaram
School of Electrical Engineering and Computer Science
Oregon State University
Corvallis, OR 97331

Abstract– **The impact of an oscillator's intrinsic and extrinsic noise sources on its noise performance is evaluated using the perturbation projection vector (PPV) analysis. The projection of a perturbation into the phase deviation for white and flicker noise, as well as for deterministic perturbations is explained qualitatively. The PPV analysis is then applied to two ring-type voltage controlled oscillators (VCO). Comparisons with measured results demonstrate the usefulness of the PPV analysis for design of noise tolerant oscillators.**

I. INTRODUCTION

Voltage controlled oscillators (VCO) are critical blocks in the design of phase-locked loops for a variety of applications such as RF transceivers, digital communication systems and on-chip timing circuits, as well as for accurate signal sampling. The noise performance of the oscillator is an important specification in each of these applications. Therefore, an accurate understanding of the generation and reduction of noise due to both intrinsic and extrinsic noise sources is crucial for low noise oscillator design.

The analysis of phase noise in oscillators requires a description of the frequency content of the noise sources and the periodic sensitivity of the oscillator's phase to that noise [1], [2]. In [1], the impulse sensitivity function (ISF) was introduced to improve upon the inadequacies of a linear phase noise analysis and to provide design insights. The calculation of the ISF requires several transient simulations and post processing of the simulation data. A rigorous perturbation projection vector (PPV) based analysis was introduced in [2] for calculating phase noise using a circuit simulator. The PPV serves as a transfer function from a noise source to the output, and is similar to the ISF. Hence, it can be used to identify the sensitivity of a node to noise.

This paper describes the use of the PPV in analyzing common ring-based VCO structures [3], [4]. In Section II the PPV based analysis is described followed by practical implications in Section III. Section IV describes two oscillator cells and the insights gained from a PPV analysis. Comparisons with measured results are provided in Section V and conclusions are provided in Section VI.

II. PPV-BASED OSCILLATOR NOISE ANALYSIS

The oscillator periodic steady-state (PSS) solution $x_s(t)$ satisfies the periodicity constraint

$$x_s(t+T) = x_s(t) \tag{1}$$

where T is the oscillation period, and $f_0 = 1/T$ is the oscillation frequency. A perturbation $n(t)$, e.g., intrinsic or extrinsic noise, causes an uncertainty in zero crossings of the oscillator output, also known as the timing jitter.

The noisy oscillator solution $x_n(t)$ is expressed in terms of the noiseless PSS solution $x_s(t)$

$$x_n(t) = x_s(t + \alpha(t)) + a(t) \tag{2}$$

where $a(t)$ is an orbital deviation that remains small, $\alpha(t)$ is a phase deviation that can grow unbounded [2].

The phase deviation $\alpha(t)$ is the solution of the following nonlinear differential-algebraic equation (DAE)

$$\frac{d}{dt}\alpha(t) = v_1^T(t + \alpha(t))n(t) \tag{3}$$

where $v_1(t)$ is a T-periodic vector, known as the perturbation projection vector [2]. The time-dependent PPV quantitatively describes how additive noise $n(t)$ is being converted by the oscillator into phase deviation. Given a perturbation $n(t)$, a PPV $v_1(t)$, and an initial phase deviation $\alpha(0)$, the phase deviation $\alpha(t)$ can be computed as

$$\alpha(t) = \alpha(0) + \int_0^t v_1^T(\tau + \alpha(\tau))n(\tau)d\tau \tag{4}$$

In the frequency domain, this operation is equivalent to selecting the baseband component (time-domain integration) of the perturbation signal, downconverted by mixing with the PPV[1] (time-domain multiplication). Therefore, the phase deviation and the timing jitter depend not simply on the noise power, but rather on how much of its strong spectral components are downconverted by the PPV.

III. PRACTICAL IMPLICATIONS OF THE PPV

As an example, consider four voltage perturbation signals in Figure 1(a) and three PPVs in Figure 1(b). The phase deviation $\alpha(t)$ is computed based on (4) with $\alpha(0) = 0$ for all 12 combinations of perturbations and projection vectors, and is shown in Figure 1(c). As expected, the phase deviation is larger in those cases when the downconversion of a strong spectral component of a perturbation takes place.

---

[1] The phase deviation $\alpha(t)$ appears as the argument of $v_1^T(t + \alpha(t))$, which spreads the PPV spectrum before it takes part in the downconversion.

978-1-4244-2018-6/08 $25.00 © 2008 IEEE

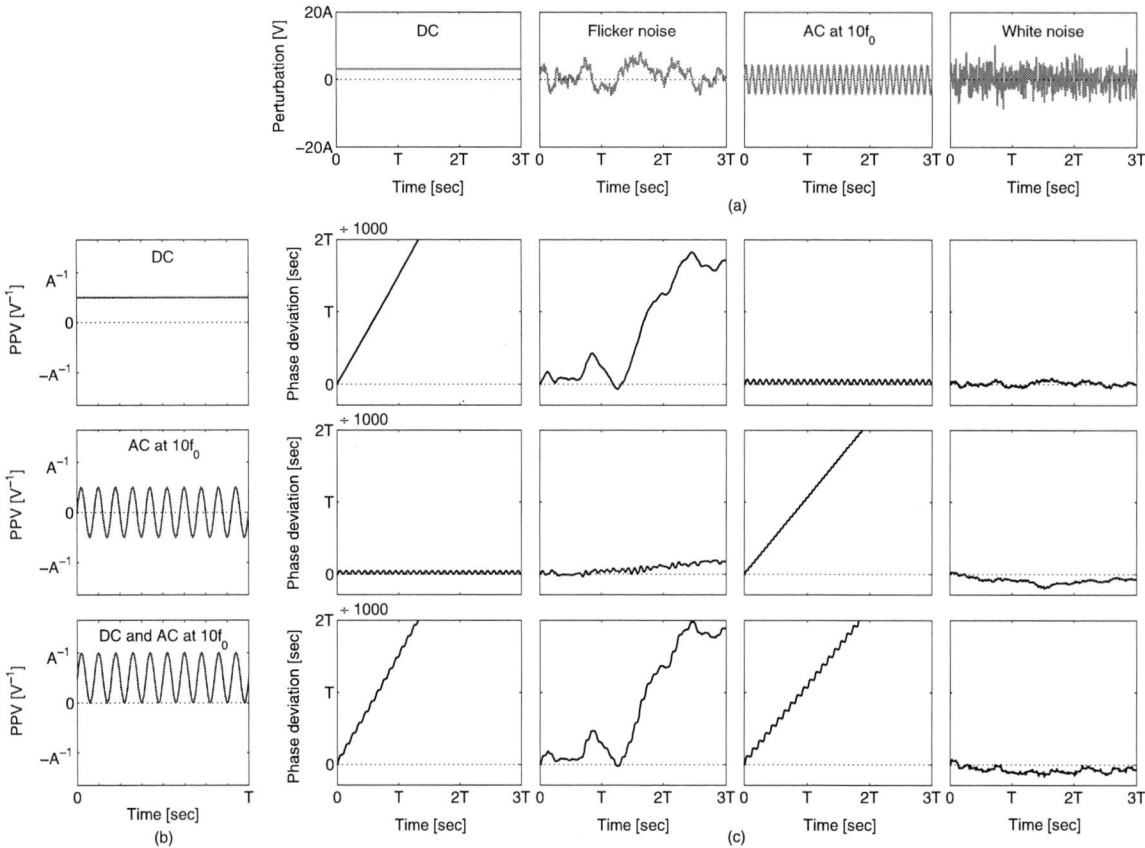

Figure 1: (a) Perturbation signals $n(t)$, (b) perturbation projection vectors $v_1(t)$, and (c) phase deviations $\alpha(t)$ corresponding to a combination of a perturbation signal above, and a PPV on the left. The phase deviations are computed based on (4) with a zero initial phase deviation $\alpha(0) = 0$.

Let us discuss some of the combinations and corresponding phase deviations in more detail.

- PPV: <u>DC</u>, perturbation: <u>DC</u>

  The phase deviation for this combination can be computed analytically based on (4) as $\alpha(t) = v_1 n t$. It grows in time at a constant rate, which is equivalent to a change in the oscillation frequency of $\Delta f_0 = d\alpha(t)/dt = v_1 n$. If the perturbation changes the control voltage V_{ctrl} of a VCO, the DC component of the corresponding PPV v_{1Vctrl} can be used to compute the VCO gain as[2]

$$K_{VCO} = -f_0 \cdot \mathrm{DC}(v_{1Vctrl}) \qquad (5)$$

- PPV: <u>DC</u>, perturbation: <u>Flicker (1/f) noise</u>

  The 1/f noise power dominates at low frequencies, and is projected into a large phase deviation by a constant, nonzero PPV.

- PPV: <u>AC at 10f_0</u>, perturbation: <u>Flicker noise</u>

  The flicker noise decreases with an increase in frequency, and therefore, its downconversion by a tone at $10f_0$ is

small. The flicker noise causes a small phase deviation if the PPV has no low-frequency components.

- PPV: <u>AC at 10f_0</u>, perturbation: <u>AC at 10f_0</u>

  If the perturbation and PPV spectral components are aligned in the frequency domain, the downconverted signal is strong at the baseband and the phase deviation grows at a nearly constant rate. The rate of growth depends on the phase alignment of the perturbation and PPV signals. The signals of Figure 1 are in-phase, which causes the phase deviation to grow quickly.

- PPV: <u>AC at 10f_0</u>, perturbation: <u>DC</u>

  In this case the PPV and perturbation signals are localized at different frequencies. The combination of these two results in a weak baseband signal, and therefore, in a small phase deviation.

- PPV: <u>any</u>, perturbation: <u>White Gaussian noise</u>

  The power of the white noise is spread along a wide range of frequencies. Thus, the white noise has a relatively small power content at any given frequency. Strong components of the PPV are localized in frequency and project a small amount of white noise into the phase deviation.

When the PPV is a combination of a number of components, e.g., PPV: <u>DC and AC at 10f_0</u>, it follows from (4) that the

---

[2] The voltage of a voltage source appears with a minus sign in the corresponding modified nodal analysis equation. Hence the minus sign in (5).

978-1-4244-2018-6/08 $25.00 © 2008 IEEE   740

resulting phase deviation is a combination of the phase deviations obtained with the particular PPV components.

While optimizing an oscillator design, the nature of the power supply noise must be taken into consideration, and the PPV for the power supply voltage must be minimized, or optimally shaped in the frequency domain to reject the strongest frequency components of the power supply noise.

IV. PPV BASED ANALYSIS OF RING VCOs

To determine the amount of phase noise that appears at the output of an oscillator, the contributing noise sources must be identified and the oscillator's periodic susceptibility to those sources must be described. The noise sources can be intrinsic to the oscillator such as thermal and flicker contributions from transistors, or they can be extrinsic and thus affect the phase noise through the supply or control of the VCO. The susceptibility of the oscillator to the noise sources can be described by the PPV which shows the sensitivity of the oscillator phase noise to perturbations over a period of oscillation for a particular node.

Designing oscillators with minimal susceptibility to extrinsic noise has become an important specification in many phase-locked loop (PLL) applications that are subjected to noisy digital supplies. Therefore, the PPV can be used as an effective tool for comparing various oscillator structures. Given that the PPV is a periodic waveform, it can be shown that reducing the DC component of the PPV makes the oscillator less susceptible to low frequency noise, and reducing the AC components makes the oscillator less susceptible to high frequency noise.

A. VCO with Maneatis Delay Cell

A ring VCO with the Maneatis delay cell [3] is used to demonstrate a PPV based analysis on a specific structure. Results from this analysis can be directly compared to results obtained for different structures simply by comparing the PPV frequency contents of each case through observations in the time domain or information in the frequency domain. All simulations are performed using a 5-stage ring VCO in a $0.18\mu m$ process, with an oscillation frequency of approximately 200 MHz.

Maneatis delay cell based VCOs rely on symmetric loads, dynamic biasing, and effective impedance boosting to achieve the superior supply noise rejection and wide tuning range for which they are known. Shown in Figure 2, a single delay cell of the VCO is composed of a differential pair and two symmetric loads. Within a specific output swing, the I-V curve of the loads has odd symmetry with nearly equal slopes on either side of an inflection point. Although the impedance characteristic of a single load is still non-linear and varies with the output swing, the output impedances on both sides of the differential circuit are equal when the common-mode lies on the inflection point of symmetry. Due to this impedance equality, any supply noise becomes common-mode noise at the cell output and is rejected by the common-mode rejection of the next delay cell. Full swing delay cells, or cells utilizing cross-coupled latches do not share this benefit.

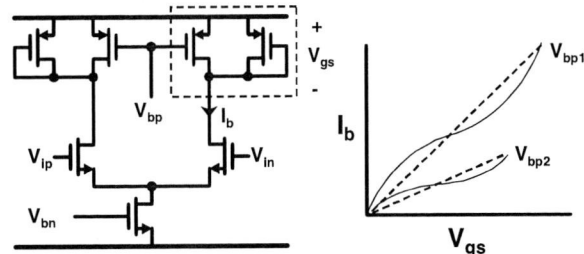

Figure 2: Maneatis delay cell with symmetric loads.

Using an appropriate bias network in a delay cell with symmetric loads, the control voltage can servo the other bias in the delay cell to ensure that the output swings between V_{DD} and V_{bp} where the loads are most symmetric. An improved biasing scheme (boosted dynamic biasing) uses an additional opamp to sense supply changes in the replica branch and make adjustments in the delay cells. The benefit of the improved biasing scheme as well as other differences are observed with a PPV based analysis in [5].

A comparison of the PPVs from the two biasing schemes is shown in Figure 3. The sensitivity to low frequency supply noise is depicted by the DC component of the PPV and is 300 times smaller when using the additional opamp in the bias circuit. This suggests that the opamp is effectively boosting the output impedance of the delay cell current sources by monitoring the replica bias thereby improving the supply rejection on the order of the opamp loop gain.

The PPV waveforms in Figure 3 also show the 10th harmonic of the oscillation frequency, f_0. This is an expected behavior from [6] as differential ring oscillators are sensitive to noise at $2N \times f_0$ where N is the number of delay cells.

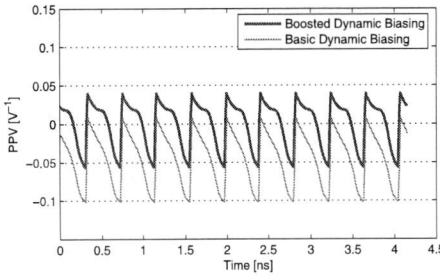

Figure 3: Supply PPV of VCO with basic dynamic bias and boosted dynamic bias.

B. VCO with Lee-Kim Delay Cell and Supply Compensation

Now, consider a four-stage differential ring oscillator based on the Lee-Kim delay cell [4] with or without power supply noise compensation. In [7] a new technique has been presented for compensating supply noise. Here we analyze the effectiveness of this approach using PPVs. A schematic of the delay cell is shown in Figure 4. The power supply noise compensation is enabled by changing the control current (not shown in the figure) and effectively changing the Vbn voltage.

Two sets of PPVs for the oscillator with and without power supply noise compensation are analyzed. Figure 5

978-1-4244-2018-6/08 $25.00 © 2008 IEEE   741

shows the PPVs for the power supply and ground. As expected, both power supply and ground rejection is improved when the compensation is enabled (DC component is reduced as well as the amplitude). As demonstrated in [7] through fabricated prototypes the compensation scheme is effective. Here, we have shown that the PPV analysis can be used to assess the suitability of a noise compensation approach prior to fabrication.

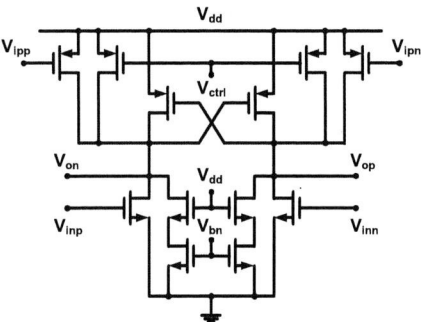

Figure 4: Delay cell of the four-stage differential ring oscillator.

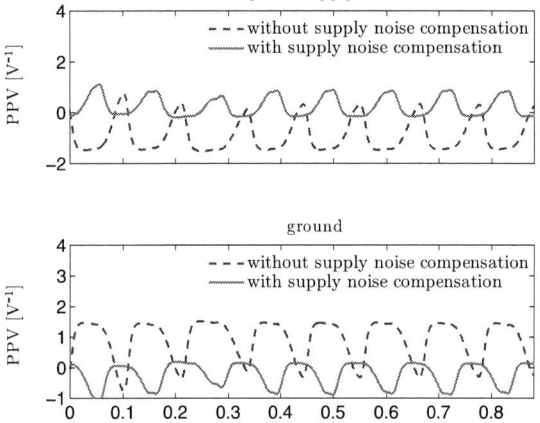

Figure 5: Power supply PPVs (top) and ground PPVs (bottom).

V. COMPARISON WITH MEASUREMENTS

The effect of the calibration current I_B on supply noise compensation in the oscillator of [7] is analyzed with the PPV analysis and compared with jitter measurements. The dependence of the dc value of the PPV, $DC(v_{1Vdd})$, on the calibration current code is shown in Figure 6. The dc component of the PPV is important as supply noise is a low frequency bandlimited noise due to the parasitics and routing. The higher harmonics of the PPV are not important for low frequency noise (Figure 1).

It is seen that the magnitude of the DC of the PPV is zero at the nominal value of current I_B. Also, shown in this figure is the measured jitter using supply compensation for the ring oscillator. Two different low frequency noise sources were injected at the supply. For a particular value of the calibration current the PLL jitter is minimized independent of the frequency of the supply noise. The simulated PPV result is in

agreement with measurements once again demonstrating the usefulness of the PPV analysis.

VI. CONCLUSIONS

The perturbation projection vector (PPV) is a useful characteristic for an oscillator and determines its sensitivity to different types of noise. The projection of a perturbation into the phase deviation for random and deterministic perturbations is described qualitatively. Analysis of two different ring VCOs demonstrates the effectiveness of the PPV technique in identifying delay cells and biasing arrangements that are tolerant to supply noise. In addition, supply compensation/regulation schemes can be evaluated with a PPV analysis. A ring VCO with supply compensation has been analyzed and the PPV predictions have been shown to correlate with measurements.

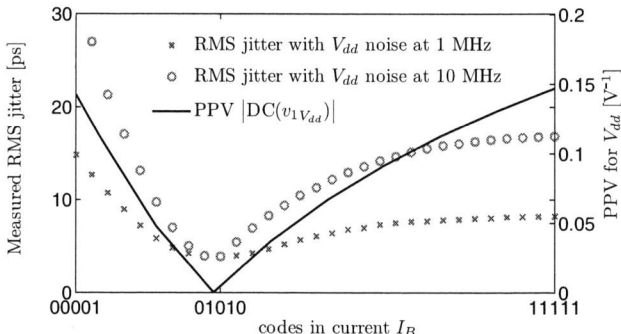

Figure 6: Magnitude of the DC component of the PPV for the power supply voltage, v_{1Vdd}, and measured RMS jitter of a 1.4GHz PLL using supply compensation for different values of the compensation current. The jitter is minimized by selecting a proper control current and is independent of the frequency of supply noise. The prediction from the PPV analysis is in good agreement with measurements.

ACKNOWLEDGMENT

This work is funded in part by SRC under contracts 2003-HJ-1076 and 2005-HJ-1326.

REFERENCES

[1] A. Hajimiri and T. Lee, "A general theory of phase noise in electrical oscillators," *IEEE J. Solid-State Circuits*, vol. 33, pp. 179-194, Feb. 1998.

[2] A. Demir, A. Mehrotra, and J. Roychowdhury, "Phase noise in oscillators: a unifying theory and numerical methods for characterization," *IEEE Trans. Circuits & Syst.-I*, vol. 47, pp. 655-674, May 2000.

[3] J. G. Maneatis and M. A. Horowitz, "Precise delay generation using coupled oscillators," *IEEE J. Solid-State Circuits*, vol. 28, pp. 1273-1282, December 1993.

[4] J. Lee and B. Kim, "A low-noise fast-lock phase-locked loop with adaptive bandwidth control," *IEEE J. Solid-State Circuits*, vol. 35, pp. 1137-1145, Aug. 2002.

[5] J. Carnes, I. Vytyaz, P. K. Hanumolu, K. Mayaram, and U. Moon, "Design and analysis of noise tolerant ring oscillators using Maneatis delay cells," *ICECS 2007*, Dec. 2007.

[6] V. Kratyuk, I. Vytyaz, U. Moon, and K. Mayaram, "Analysis of supply and ground noise sensitivity in ring and LC oscillators," *Proc. ISCAS 2005*, pp. 5986-5989, May 2005.

[7] T. Wu, K. Mayaram, and U. Moon, "An on-chip calibration technique for reducing supply noise sensitivity in ring oscillators," *IEEE J. Solid-State Circuits*, vol. 42, no. 4, pp. 775-783, April 2007.

IEEE 2008 Custom Intergrated Circuits Conference (CICC)

Strong Injection Locking of Low-Q LC Oscillators

Mozhgan Mansuri, Frank O'Mahony, Ganesh Balamurugan, James Jaussi, Joseph Kennedy, Sudip Shekhar*, Randy Mooney and Bryan Casper

Intel Corporation and *University of Washington, Seattle, WA
JF2-04, 2111 NE 25th Ave, Hillsboro, OR, 97124 USA

Abstract-This paper presents a new equation for injection-locked LC oscillators (ILOs) based on a *series* RL element in parallel with C. Unlike previous ILO equations based on a *parallel* RLC tank approximation, the proposed equation is valid for any tank Q and injection strength (K). The model reveals several important properties of low-Q and/or high-K ILOs such as asymmetry in the lock range, reduced phase shift, and higher bandwidth. Experimental results validate the model.

I. INTRODUCTION

Injection locked oscillators (ILOs) are widely used in wireless and wireline building blocks such as clock synthesizers, dividers or multipliers. Recently, ILOs that filter and phase shift the clock for microprocessors and I/Os have been demonstrated [1], [2]. Within its lock range, an ILO behaves as a Type I phase-locked loop (PLL), but has potential for less complexity, area, and power than a conventional PLL. LC oscillators are of great interest due to their superior phase noise performance and intrinsic supply noise rejection compared with CMOS ring oscillators. The behavior of injection-locked LC oscillators has been extensively studied in the past [3], [4], [5], [6]. The ILO model developed by Adler is valid for weak injection strength (K<<1) [3], [4]. Paciorek later extended Alder's equation to a general equation which models the ILO behavior for any injection strength [5]. However, all previously published ILO models are based on an *equivalent* parallel RLC tank, where the parallel resistor, R, accounts for all sources of tank loss. Given that tank Q is usually dominated by the series resistive loss of the inductor, L, the parallel RLC tank is an approximation that is only valid for a narrow frequency range around the free-running frequency of an LC oscillator and for a high tank Q (e.g. Q>10).

Digital CMOS processes are extensively used to implement VLSI systems that require precise clocks, such as microprocessors and memories. However, inductors implemented in digital processes with a low-resistivity substrate can achieve relatively modest Q, with typical values of 4-5. Accounting for other circuit losses, the overall tank Q can be as low as 2-3. Used in an ILO, these low-Q inductors result in large ILO lock range [5], so that the tank model must be accurate far from the tank's free-running frequency. In addition, there are applications that benefit from large injection strength in ILOs to maximize phase tuning range and tolerate frequency variations due to PVT [1], [2]. Stronger injection increases the lock range [5], which further increases the

frequency range over which the tank model must be accurate. Consequently, low-Q and/or high-K ILOs are not narrow band systems and thus, they can not be accurately modeled by an ILO with an equivalent parallel RLC tank.

In this paper, we develop an injection locking equation based on a series RL tank and calculate the ILO properties. Then, we analytically and experimentally demonstrate that, unlike the parallel RLC equation, our equation correctly predicts the behavior of an ILO for any Q and K values.

II. INJECTION LOCKING MODEL OF AN LC OSCILLATOR

Fig. 1a shows a conceptual LC oscillator with tuned tank. With no injection, the LC oscillator runs at its free-running frequency (ω_0) and the tuned tank contributes no phase shift ($\phi = 0$). Injecting the oscillator with a signal whose frequency is within the oscillator lock range forces the frequency of oscillator to be equal to the injected clock frequency (ω_{inj}). When $\omega_{inj} \neq \omega_0$ is injected, the tank contributes a non-zero phase shift, $\phi \neq 0$. For instance, if $\omega_{inj} > \omega_0$, the tank contributes a negative phase shift, $-\phi$ (Fig. 1a). To maintain oscillation at ω_{inj}, the total tank current (I_T) must have the same phase angle with an opposite polarity (ϕ). As a result, the ILO introduces a phase shift (θ) between the injected clock and its output (Fig. 1b). From Fig. 1b, the relationship between θ and ϕ is:

$$\tan\phi = \frac{I_{inj}\cdot\sin\theta}{I_{osc}+I_{inj}\cdot\cos\theta} = \frac{K\cdot\sin\theta}{1+K\cdot\cos\theta}, \quad K = \frac{I_{inj}}{I_{osc}} \quad (1)$$

Fig. 2 shows two models for a lossy LC tank, one with a series RL in parallel with C and one with its equivalent parallel RLC tank. The calculated phase and amplitude of both tuned tanks for Q=2.5 and Q=10 are shown in Fig. 3. Unlike the equivalent parallel tank, the phase and amplitude of a series RL tank are not symmetric with respect to ω_0. Based on Fig. 3, the asymmetry becomes worse as Q is reduced. The asymmetry within the ILO lock range ($\omega_0-\omega_L \leq \omega_0 \leq \omega_0+\omega_L$) should be modeled accurately because it impacts the ILO behavior. Also,

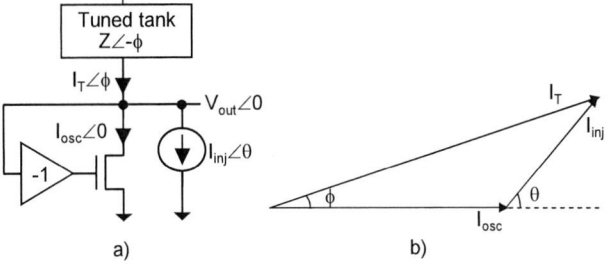

Fig. 1. a) An injection-locked LC oscillator, b) ILO vector diagram

978-1-4244-2018-6/08 $25.00 © 2008 IEEE   743

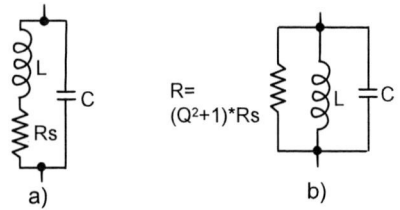

Fig. 2. Tuned tank, a) series RL in parallel with C tank, b) its equivalent parallel RLC tank

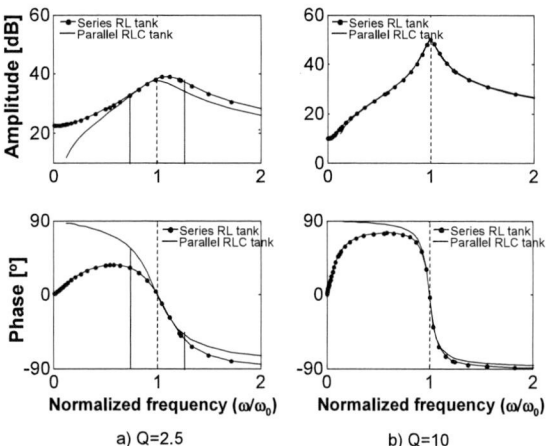

a) Q=2.5   b) Q=10

Fig. 3. Amplitude and phase of series vs. parallel tank for a) Q=2.5, b) Q=10

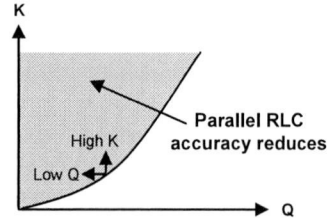

Fig. 4. Limitation of an ILO model based on a parallel RLC tank approximation

for high injection strength (K), the lock range increases. Therefore, for a low-Q and/or high-K ILO, an equivalent parallel RLC tank fails to accurately model a series RL tank because the narrow frequency range assumption does not hold. Fig. 4 qualitatively plots the range of Q and K values at which the state-of-the-art ILO equation fails to correctly model the ILO.

To analytically demonstrate the ranges of K and Q values at which a parallel tank model fails, we first derive the locking equation for a series RL tank, similar to the methodology used in [5]. Both phase and amplitude of the tank impact the ILO properties. The tank amplitude at different frequencies is not constant (Fig. 3); Thus, for constant injection signal amplitude, K varies as a function of the frequency. By making K a function of the frequency, the amplitude variation can be included into the model. For simplicity, however, we assume the oscillation amplitude and thus K, is constant (similar to [5]) and we only consider the tank phase. The phase produced by a series RL tank and its equivalent parallel RLC tank for a frequency ω is given in (2) and (3), respectively:

Series RL tank : $\phi = \arctan\left[\dfrac{-2Q(\omega-\omega_0)}{\omega_0}\cdot\left(\dfrac{1}{2}\cdot\left(1-1/Q^2\right)^{1.5}\cdot\left(\dfrac{\omega+\omega_0}{\omega_0^2}\cdot\omega\right)\right)\right]$ (2)

Parallel RLC tank : $\phi = \arctan\left[-\dfrac{2Q(\omega-\omega_0)}{\omega_0}\right]$ (3)

From (1) and (2) for a series RL tank (and similarly from (1) and (3) for a parallel RLC tank [5]):

Series RL tank : $\dfrac{2Q}{\omega_0}\cdot(\omega-\omega_0)\cdot\left[\dfrac{1}{2}\cdot\left(1-1/Q^2\right)^{1.5}\cdot\left(\dfrac{\omega+\omega_0}{\omega_0^2}\cdot\omega\right)\right]=-\dfrac{K\cdot\sin\theta}{1+K\cdot\cos\theta}$ (4)

Parallel RLC tank : $\dfrac{2Q}{\omega_0}(\omega-\omega_0)=-\dfrac{K\cdot\sin\theta}{1+K\cdot\cos\theta}$ (5)

where ω is the oscillator instantaneous frequency and is equal to $\omega_{inj}+d\theta/dt$. By replacing ω in (4) and (5), the locking equations for the two tanks are obtained:

Series RL tank : $\left(\dfrac{d\theta}{dt}+\omega_{inj}\right)^3/\omega_0^2-\dfrac{d\theta}{dt}=\omega_{inj}-\left(\dfrac{\omega_0}{Q\cdot\left(1-1/Q^2\right)^{1.5}}\right)\cdot\dfrac{K\cdot\sin\theta}{1+K\cdot\cos\theta}$ (6)

Parallel RLC tank : $\dfrac{d\theta}{dt}=\omega_0-\omega_{inj}-\left(\dfrac{\omega_0}{2Q}\right)\dfrac{K\cdot\sin\theta}{1+K\cdot\cos\theta}$ (7)

With the injection locked equations developed for a series RL tank, we next calculate and compare the ILO lock range, phase shift and bandwidth for both injection locking equations and for different K and Q values.

A. Lock Range

The lock range of an ILO is defined to be the maximum frequency difference ($|\omega_0-\omega_{inj}|$) for which locking occurs. To calculate the lock range, we set dθ/dt=0 in (6) and (7). Then, we differentiate ω_{inj} with respect to θ and solve for the maximum phase shift, θ_{ssmax}. Similar to the result shown in [5] for a parallel tank, $\theta_{ssmax}=\pi-\cos^{-1}(K)$ is calculated for both series and parallel tanks. By substituting θ_{ssmax} into (6) and (7), the lock range for both tank models is calculated:

Series RL tank : $(\omega_0\pm\omega_L)^3-(\omega_0\pm\omega_L)\cdot\omega_0^2=\pm\dfrac{\omega_0^3}{Q\cdot\left(1-1/Q^2\right)^{1.5}}\cdot\dfrac{K}{\sqrt{1-K^2}}$ (8)

Parallel RLC tank : $\omega_L=\omega_{L+}=|\omega_{L-}|=\dfrac{\omega_0}{2Q}\cdot\dfrac{K}{\sqrt{1-K^2}}$ (9)

Equation (8) combines two sets of equations: *minus* and *plus* signs are used to calculate the negative lock range, ω_{L-}, (i.e. $\omega_{inj}-\omega_0<0$), and positive lock range, ω_{L+}, (i.e. $\omega_{inj}-\omega_0>0$), respectively. Therefore, unlike a parallel RLC tank, the lock range for a series RL tank is not symmetric around the oscillator free-running frequency (ω_0). Table I summarizes the lock range calculated by (8) and (9) for different Q and K values. For a constant K (=0.25), as Q reduces, the parallel tank model underestimates the lock range and also fails to predict the asymmetry between positive and negative lock range. Increasing K with a constant low Q (=2.5) also results in more asymmetric lock range using the series tank model, while a parallel tank model fails to predict it.

B. Phase Shift

One application of an ILO is to shift / deskew the phase of an injected clock [1, 2]. By setting dθ/dt=0 in (6) and (7), we calculate θ_{ss} for series and parallel tank models, respectively. θ_{ss} depends on Q, K and the frequency difference ($\omega_0-\omega_{inj}$).

978-1-4244-2018-6/08 $25.00 © 2008 IEEE   744

TABLE I
CALCULATED LOCK RANGE FOR A SERIES RL AND PARALLEL RLC TANK

		Lock range calculated for *series* RL tank		Lock range calculated for *parallel* RLC tank
		$Q*|\omega_{L-}|/\omega_0$	$Q*\omega_{L+}/\omega_0$	$Q*(|\omega_{L-}|=\omega_{L+})/\omega_0$
K=0.25	Q=10	0.134	0.129	0.129
	Q=5	0.143	0.132	0.129
	Q=2.5	0.189	0.153	0.129
Q=2.5	K=0.125	0.086	0.078	0.063
	K=0.25	0.189	0.153	0.129
	K=0.5	0.534	0.314	0.289

By rearranging (7), it can be shown that for a parallel tank, calculating θ_{ss} as a function of $Q(\omega_0-\omega_{inj})/\omega_0$ makes the phase shift independent of Q and only dependent on K. Fig. 5 compares the calculated θ_{ss} for a series tank versus its equivalent parallel tank for different Q and K values. For a constant K (=0.25) and for high Q (>5), θ_{ss} approximated by the equivalent parallel tank closely matches θ_{ss} calculated by the series tank. However, as Q reduces, the error in θ_{ss} calculated from the equivalent parallel tank model increases relative to the series RL tank model. Therefore, at low Q (<5), the ILO equation for a parallel tank fails to predict the correct θ_{ss} (Fig. 5a). For a constant low Q (=2.5), the parallel tank model overestimates the phase shift for a given frequency offset from ω_0 and underestimates the oscillator frequency range required to provide $-90°<\theta_{ss}<90°$. As K increases, the error in approximating θ_{ss}, using a parallel tank, increases (Fig. 5b).

C. ILO Bandwidth

Similar to a PLL, an ILO tracks the input jitter within its bandwidth and rejects it outside the bandwidth. The ILO bandwidth for an ILO with a parallel RLC tank has previously been calculated [5], [6]. Instead of finding a closed-form equation for (6), we use a behavioral model to calculate the bandwidth for any ILO phase shift. The time-based behavioral model utilizes (6) to calculate the oscillator output phase at any time for any injected input phase (or frequency). By adding

sinusoidal jitter to the injection signal, we calculate the ratio of output to input peak-to-peak jitter. By sweeping the jitter frequency, the input to output jitter transfer function and bandwidth are calculated. For consistency, we repeat the same steps for (7) to calculate the bandwidth for an equivalent parallel tank model. (Results match well with the closed-form equation in [6]). As shown by [6], the bandwidth of an ILO using a parallel tank model is inversely proportional to Q. Thus, the product of normalized bandwidth (=BW/ω_{inj}) and Q for a parallel tank model is only dependent on K and θ_{ss}. Fig. 6 compares the calculated ILO bandwidth for the two tank models for different Q and K values. For a constant K and higher Q (>5), the bandwidth estimated from the parallel tank model matches well with series tank model. However, as Q reduces, the parallel tank model underestimates the bandwidth and fails to predict the asymmetry in bandwidth around $\theta_{ss}=0$ (Fig. 6a). For a constant low Q (=2.5), the parallel tank model again underestimates the bandwidth and fails to predict the asymmetry. As K increases, the error in approximating ILO bandwidth, using a parallel tank model, increases (Fig. 6b).

III. BEHAVIORAL AND EXPERIMENTAL RESULTS

In this section, we discuss and compare experimental results with the parallel RLC and series RL injection locking equations. The LC-ILO designed in [2] and fabricated in a 45nm digital CMOS process was used for measurements. The differential inductor has a Q of 3-4 across the frequency tuning range. Accounting for other circuit losses, the overall tank Q is estimated to be 2.5 at the injection frequency (tested around 12.5-13GHz). The oscillator free-running frequency and the injection strength are digitally controlled. For some measurements, we increased K from the maximum digital control of 0.25 to ~0.38 by increasing the injection amplitude.

Table II summarizes the measured and calculated lock range for both series and parallel tank ILO equations and for three different K values with Q=2.5. The lock range predicted with a series RL tank matches well with the measured lock range. At higher K, where the tank asymmetry is more pronounced, the

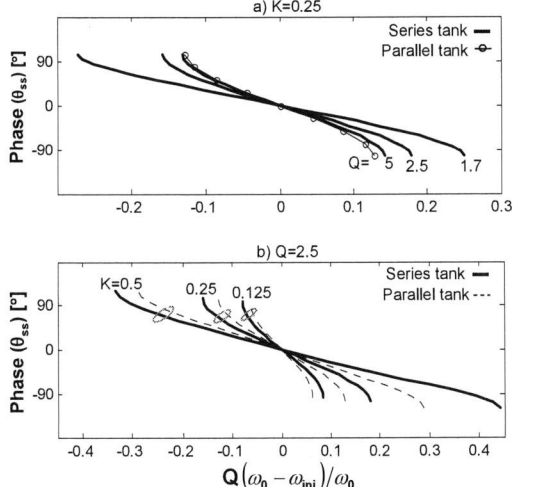

Fig. 5. Calculated ILO phase shift for series RL vs. parallel RLC tank for, a) K=0.25, and b) Q=2.5

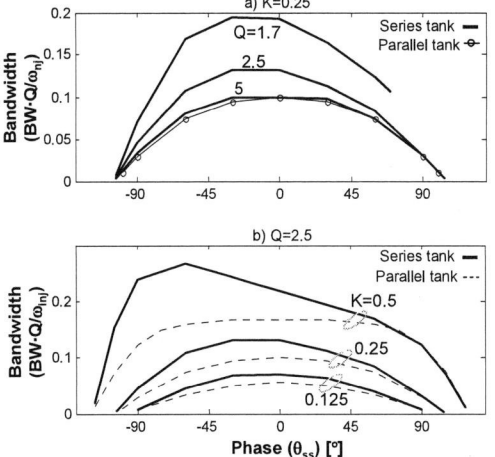

Fig. 6. Calculated ILO bandwidth for series RL vs. parallel RLC tank for, a) K=0.25, and b) Q=2.5

TABLE II
MEASURED AND CALCULATED LOCK RANGE

	Measured lock range		Lock range calculated for series RL tank		Lock range calculated for parallel RLC tank
	$|\omega_{L-}|/\omega_0$	ω_{L+}/ω_0	$|\omega_{L-}|/\omega_0$	ω_{L+}/ω_0	$(|\omega_{L-}|=\omega_{L+})/\omega_0$
K=0.125	0.042	0.039	0.034	0.031	0.025
K=0.25	0.083	0.060	0.077	0.062	0.052
K=0.38	0.119	0.076	0.117	0.086	0.075

ILO based on the parallel RLC tank results in over designing to achieve the desired lock range and fails to predict asymmetry at higher K.

Fig. 7 shows the measured and calculated phase shift as a function of the oscillator frequency for two K values, and for a tank Q of 2.5. The phase shift predicted with the series RL tank model closely matches with the measured result. However, the equivalent tank overestimates the phase sensitivity to frequency and thus, it underestimates the frequency range required to achieve $-90^\circ < \theta_{ss} < 90^\circ$ phase shift. At K=0.38, both measured and predicted θ_{ss} with the series RL tank model indicate that the oscillator frequency range required to achieve $\theta=-90^\circ$ is larger than that needed for $\theta=90^\circ$. The phase shift predicted by the equivalent parallel tank fails to predict this frequency range asymmetry at higher K=0.38 (Fig. 7b). Consequently, designing an ILO based on the equivalent tank can result in lower phase shift range than desired.

Fig. 8 summarizes the calculated and measured bandwidths for two injection strength values and Q=2.5. The parallel RLC tank model underestimates the ILO bandwidth at lower Q and fails to predict asymmetric bandwidth about the zero phase shift at higher K (Fig. 8b). However, the series RL tank successfully predicts the bandwidth with zero phase shift and closely approximates the bandwidth with phase shift between ~-70° and 90°. At phase shifts close to -90°, the error between the series model and measured result is most likely caused by K variation of the ILO across the tuning range. The ILO used for the measurement has no amplitude tracking mechanism between the oscillator and injection signal. Therefore, the frequency dependency of the oscillator amplitude (Fig. 3) causes K to change. By including the estimated K at any frequency (based on Fig. 3) into our model, the error significantly reduces. Another likely source of error for the series RL model comes from lumping all tank loss into a resistor in series with the inductor. Based on these measurements, however, the series-only loss approximation adequately predicts the filtering behavior of this low-Q ILO.

The measured properties of this low-Q ILO verify that our injection lock model based on a series RL tank predicts the system characteristics for *low-Q* and *low-Q, higher-K* ILOs while the commonly used ILO model based on an equivalent parallel RLC tank fails to predict them. The presented model can predict and optimize the ILO behavior for any Q and K.

REFERENCES

[1] L. Zhang, et al., "Injection-locked clocking: a new GHz clock distribution scheme, *Proc. IEEE CICC*, pp. 785-788, Sep. 2006.

[2] F. O'Mahony, S. Shekhar et al., "A 27Gb/s forwarded-clock I/O receiver using an injection-locked LC-DCO in 45nm CMOS," *ISSCC Dig. Tech. papers*, pp. 452-453, Feb. 2008

[3] R. Adler, "A study of locking phenomena in oscillators," reprinted in *Proc. IEEE*, vol. 61, pp. 1380-1385, Oct. 1973

[4] B. Razavi, "A study of injection locking and pulling in oscillators," *IEEE JSSC*, pp. 1415-1424, Sep. 2004

[5] L. J. Paciorek, "Injection locking of oscillators," *Proc. IEEE*, vol. 53, no. 11, pp. 1723-1728, Nov. 1965

[6] A. Mirzaei, et al., "The quadratue LC oscillator: a complete portrait based on injection locking," *IEEE JSSC*, pp. 1916-1932, Sep. 2007

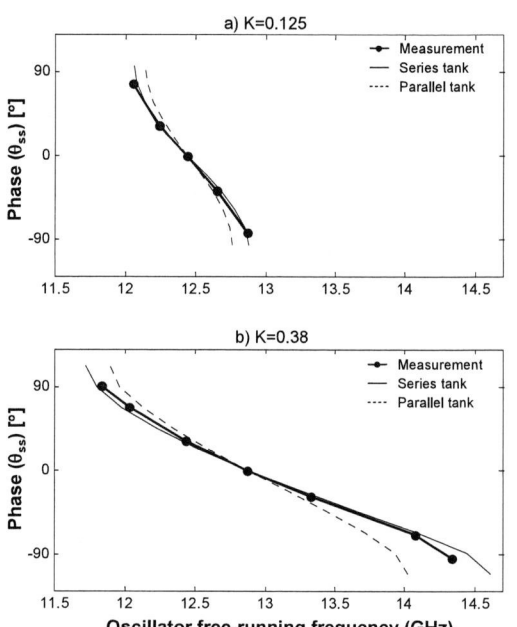

Fig. 7. Measured and calculated phase shift of ILO with Q=2.5 for a) K=0.125, b) K=0.38

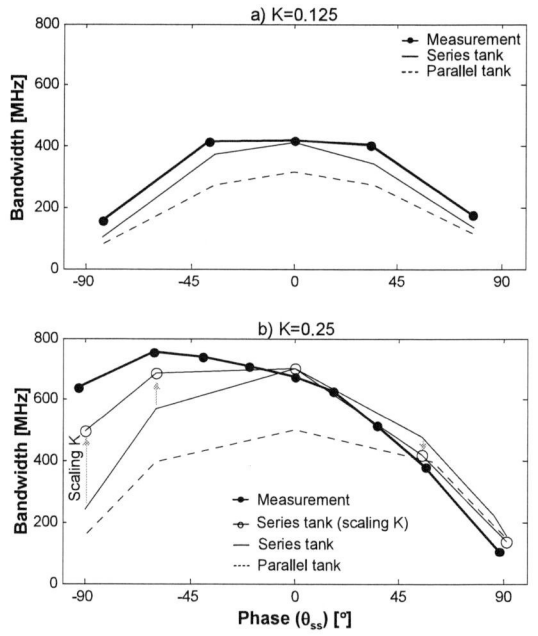

Fig. 8. Measured and calculated bandwidth of ILO with Q=2.5 for a) K=0.125, b) K=0.25

Analog Techniques

Session 23

Chair: Don Thelen, ON Semiconductor
Co-Chair: Ken Suyama, Epoch Microelectronics

This year's analog techniques session includes six papers that you will find interesting and informative if you design analog and mixed signal circuits. We have three papers describing frequency domain filters, two papers on voltage references (one is invited), and an invited paper on deep submicron effects.

Our first paper is about a Gm-C filter with a bandwidth of 1.3GHz. This fourth order Butterworth lowpass filter is built in a 0.13um process and consumes 24mW. An innovative class AB transconductor is used to achieve low distortion and good power supply rejection.

Next we have a downconversion filter based on the switched resistor technique. The switches that are used to tune RC time constants also perform the mixing function for the down converter. The circuit achieves a +/-50% tuning range and is suitable for low voltage applications.

The third paper presents an FIR filter embedded in a successive approximation ADC. The capacitor array in the ADC is used to realize a reconfigurable FIR low pass filter during the input sampling phase. One application of this ADC/filter combination is a digital radio receiver.

Next is an invited paper on voltage references for power supplies less than the voltage required to turn on a silicon P-N junction diode. Two techniques are described, one using Schottky diodes, and the other is an all MOSFET approach.

Our fifth paper presents a voltage reference realized in a polycrystalline thin film silicon process. This process has been used to realize high resolution displays. The bandgap reference is a step towards realizing system on glass applications.

The last paper is invited, and discusses deep submicron effects on data converter building blocks. Physical effects such as NBTI, STI, Nwell proximity and length of diffusion are presented and design practices to function reliably with these effects are discussed.

978-1-4244-2018-6/08 $25.00 © 2008 IEEE

Notes

978-1-4244-2018-6/08 $25.00 © 2008 IEEE

IEEE 2008 Custom Intergrated Circuits Conference (CICC)

A Low Power 1.3GHz Dual-Path Current Mode Gm-C Filter

Manisha Gambhir, Vijay Dhanasekaran, Jose Silva-Martinez, Edgar Sánchez-Sinencio

Department of Electrical and Computer Engineering, Texas A&M University

Abstract— **A class-AB, dual path building block for Gm-C filter that has high linearity and high PSRR is presented. The class-AB action provides good linearity and gm/Id while the dual path approach solves the PSRR problem associated with conventional class-AB OTA. A current-mode 4th order Butterworth filter designed using the proposed building block in 0.13μm CMOS technology provides 54.2dB IM3 and 52dB SNR in 1.3GHz bandwidth while consuming as low as 24mW. It occupies 0.1 mm^2.**

I. INTRODUCTION

Power-efficient high frequency continuous time filters are desirable to support multi-Gbps data communication in portable systems. Gm-C type filters are routinely used for channel selection and equalization [1-2] and their power-efficiency is limited by that of its main building block, the operational transconductance amplifier (OTA). The power-efficiency of an OTA is defined as the ratio of the signal-to-noise-power-ratio (SNR) to the power dissipation for a given transconductance (gm) and minimum distortion performance. Since the distortion performance heavily relies on the input signal amplitude, linearization of the OTA plays an important role in improving its power efficiency. For instance, for a certain third harmonic amplitude ratio (HD3) of a differential pair OTA, given by Vi^2/(32*V$_{GST}$^{2}) (where Vi is the input signal amplitude and V$_{GST}$ is the overdrive voltage of the transistor), Vi can be increased by a factor of 2 if the V$_{GST}$, and hence the power consumption (for same gm), is increased by the same factor. Thus, the SNR can be improved by 6dB if the power consumption is doubled, which represents a power efficiency improvement by a factor of 2. This is clearly an advantage over noise reduction through impedance scaling, where the SNR improves only by 3dB when the power consumption is doubled, which leaves power efficiency unaltered. A linearity improvement through mere increase in V$_{GST}$, however, is ultimately limited by voltage headroom and mobility degradation effects. This fact had spurred a lot of research in OTA linearization techniques over the past two decades [3-5].

Among several linear OTA schemes, the Nauta's OTA [4] provides superior power efficiency for the following reasons. A) the supply current is re-used between NMOS and PMOS devices, which results in twice the gm for the given power and the output noise compared to OTA schemes that use only NMOS or PMOS as transconductor. This corresponds to factor of 4 power efficiency

improvement (see Fig.1) B) linearization is achieved through class-AB action and it also works without reducing the gm. Note that the gm-vs-Vi curve of the Nauta's OTA in Fig.2 doesn't roll off sharply due to bias current limitation unlike a differential pair OTA. C) no additional noise due to linearization element or unequal division of bias current source noise between differential pair as seen in source degeneration and multi-tanh schemes. Simulation of best designed differential pair and Nauta's OTA shows the maximum input voltage swing for a given HD3 is about 33% higher in case of Nauta's OTA owing to its superior gm linearity. Thus Nauta's OTA provides about 7 times (4*1.33^2) power efficiency improvement over a differential pair OTA, which other linearization schemes don't achieve.

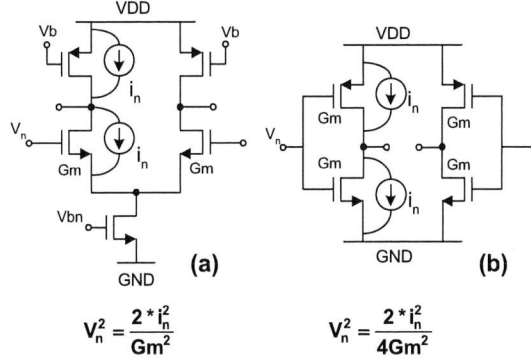

Fig.1 Input referred noise voltage for a given power
a) Differential pair OTA b) Nauta's OTA

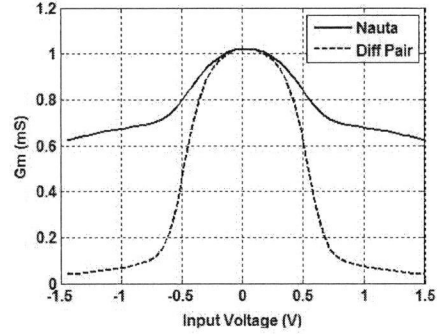

Fig.2 Gm variation as a function of input voltage

Although the Nauta's OTA is the best in terms of power efficiency, there are several practical problems associated with it. Firstly, the power supply rejection ratio of the OTA

978-1-4244-2018-6/08 $25.00 © 2008 IEEE   749

is paltry -6dB. This is due to the fact that a power supply voltage perturbation would generate half as much output current as the same voltage perturbation applied at the input would generate. Secondly, the bias current (and hence the gm) is a strong function of the threshold voltage (VT) of the transistors, the supply voltage and carrier mobility of the transistors. This results in large gm variation across temperature and process corners. Use of a low-dropout regulator as suggested in [4] is not effective in eliminating high frequency supply noise and would significantly reduce the power efficiency of the OTA.

In section II, we present a new building block that retains the linearity benefits of the Nauta's OTA while side-stepping the above-mentioned issues associated with it. Section III describes the design of a 1.3GHz, 4th order Butterworth filter in UMC's 0.13μm CMOS technology that uses the proposed building block. Section IV provides the experimental results. It is shown that the power efficiency of this filter is significantly better than the state-of-the-art Gm-C filters. Conclusions are drawn in section V.

II. BASIC BUILDING BLOCK

One possible way to generate signal currents independent of supply noise is by using current mirrors. This section describes a highly-Linear-supply-Robust-Current-Mirror (LRCM) which embodies the linearity property of the Nauta's transconductor while being impervious to supply noise and VT variations. Later in this paper, it will be shown that this element can easily be substituted for the traditional OTA for the design of OTA-C filters.

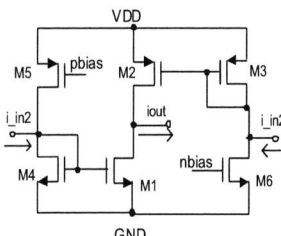

Fig. 3 Current biased complementary mirror

In order to make the output current of complementary transconductor (Fig. 1(b)) independent of supply noise, perturbations at VDD needs to be coupled onto the gate of the PMOS (M2) and the noise at ground to be coupled onto the gate of NMOS (M1). This necessitates that the NMOS and PMOS gates not to be tied together. Coupling of the noise present at the supply to the respective gates can be achieved if the main transconductor is biased through an independent current source and the input signal is applied in current mode instead of voltage. The modified structure in this case evolves to the one shown in Fig.3. Note that signal currents i1 and i2 are in-phase (and not differential) input currents. Transistors M5, M6 are biased to conduct a fixed DC current. Mirroring action between M4 and M1 and also between M3 and M2 ensures that the PMOS gates of M4 and M1 carry the noise voltage present at VDD and the NMOS gates of M2 and M3 carry the noise present at

the GND. In this case, the output current iout is free from supply noise and is directly related to input current (i_in1=i_in2) through the mirroring ratio. Note that signal commutation in this building block is in current mode. That is, in principle, the basic building block is analogous to a current mirror. It is instructive to examine the linearity property of the structure in Fig. 3. This structure can be viewed as a cascade of two functional stages. While M4 and M3 convert the incoming current to a signal voltage, M1 and M2 form an equivalent Nauta's transconductance stage that generate the output current. Linearity of the transconductor formed by M1-M2 is comparable to Nauta's due to complementary action. However, since M3 and M4 are biased through fixed current source (M5 and M6) linearity of current to voltage (I-V) conversion is limited to that of an ordinary diode load.

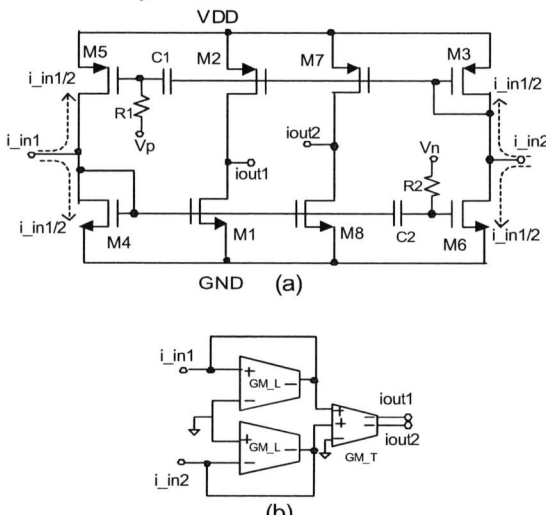

Fig. 4 (a) Proposed Highly Linear Supply Robust Current Mirror (b) Equivalent representation

In order to improve overall linearity of the structure in Fig. 3 close to that of complementary Nauta's structure, both PMOS and NMOS should participate in I-V conversion. That is, the gates of M6 and M5 should carry signal voltages equivalent to that at the gates of M3 and M4 respectively, without coupling the VDD noise onto the NMOS gates and vice-versa. To achieve this, the basic structure of Fig. 3 is modified to one that in Fig. 4(a). Here, the gate of M5 and M6 are capacitively connected to M3 and M4 respectively such that the signal swings at gates M3, M4, M5 and M6 are similar (as gm3=gm4=gm5=gm6); where gmi is the transconductance associated with transistors Mi and so on. With further analysis it can be shown that for correlated in-phase input currents (i_in1=i_in2) and equal transconductance (gm3=gm4=gm5=gm6), the input current i_in1(i_in2) partitions equally between M4 and M5 (M3 and M6) as shown in Fig. 4(a).

I-V conversion in the proposed structure of Fig. 4(a) is carried by M5-M4 and M3-M6 for respective arms. The signal voltages thus generated are applied to the inputs of the transconductor formed by M1-M2. The gates of all PMOS carry VDD noise while the NMOS gates carry the noise from GND. Thus, the output signal current generated has minimal supply noise. The DC bias for the structure is generated through resistors R1 and R2: vp and vn are generated through reference current bias. This ensures transconductances are independent of VT variations due to process and temperature. Since a dual path input current is essential for supply independent and yet linear output, an additional transconductor arm comprising of M7 and M8 is added to generate two copies of the output current: iout1 and iout2 that can be fed as input to a potential next stage. Note that the technique of processing signals using multiple paths does not affect the power efficiency as shown in [6]. The proposed LRCM block can also be represented as in Fig. 4(b). Note that this is a single ended representation with i_in1 and i_in2 being in-phase currents. Pursuing the formal small signal analysis it can also be shown that GM_L = 2*gm3 = 2*gm4 and Gm_T = 2gm1 = 2 gm2.

The salient features of the proposed LRCM block are: The linearity of this modified structure is similar to that of an equivalent Nauta's transconductor. However, the proposed building block has improved PSRR and is relatively robust to supply and VT variations. For example, LRCM can achieve HD3 as high as -70dB for peak to peak differential swing of 250mV. Supply rejection for this structure is found to be -25dB. This structure also maintains a low voltage operation with requirements of headroom even lower than that of inverter based Nauta's transconductor.

III. FILTER DESIGN

This section describes the application of LRCM for design of wideband current mode filters. OTAs, in particular current mirrors, have been used in previous works [4], [7] to realize the current-mode filters. Fig. 5 shows biquadratic section of a typical current mode filter. Transconductors Gm1-4 long with C determine filter's resonance frequency (ωo) while $\omega o/Q$ is determined by losses presented by Gm1-3 and C. The expression for the output current is given by:

$$i\_out = \left(\frac{Gm2Gm5/C^2}{s^2 + \frac{Gm1+Gm3}{C}s + \frac{Gm1Gm3+Gm2Gm4}{C^2}} \right) * i\_in \quad (1)$$

Fig. 5 A typical current-mode biquadratic section

The proposed LRCM which is equivalent to a load transconductor followed by an OTA and can replace the

structure inside the dotted box in Fig. 5. Cascaded sections of the two biquads are used to realize a fourth order Butterworth filter. Values of the transconductors are carefully chosen to satisfy butterworth transfer function and a uniform signal swing across all integrating nodes.

For practical implementation of the filter common-mode stability needs to be ensured. Filters with fully-differential OTAs usually employ Common-Mode-Feedback (CMFB) loops for each of the integrating nodes. However, when pseudo-differential OTAs or simple current mirrors are used, the common mode gain of each of these stages is relatively high. This necessitates stronger CMFBs, which are costly in terms of power. Past works [4], [7] have relied on introducing additional common mode losses in all nodes such that loop gain for common mode signals is less than unity. Additional losses are introduced by increasing Gm1 and Gm3. In order to preserve the transfer function negative impedances are shunted to Gm1 and Gm3 to compensate for these losses for the differential signals. This approach introduces additional circuit noise and increases power consumption. For example, for a biquad with Q=1.1 (Gm1 = Gm3 = 0.5*Gm2 = 0.5*Gm4), power consumption increases by 75% for the common mode losses as in [4].

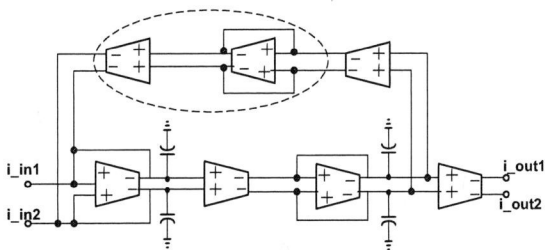

Fig. 6 Proposed Current mode Biquadratic section employing LRCM (Single ended version)

An alternate strategy is proposed here, wherein an additional mirroring stage is introduced in the feedback. Fig. 6 shows the realization of a current mode biquad using three LRCM elements (Fig. 4). The additional mirror (circled in Fig. 6) ensures a negative feedback for common mode signals. For a power increase of 20% the phase introduced by the additional mirror is minimal to cause any appreciable shift in filter's poles. Since all the integrating nodes are low impedance, common mode levels are self regulated and dedicated common mode feedback loops are not required.

IV. EXPERIMENTAL RESULTS

A prototype of the proposed filter was fabricated in UMC 0.13um CMOS technology. The micrograph of the chip is shown in Fig.7. In order to generate current inputs, a pair of V-I buffers were used at the input. The transfer function of the filter is found by calibrating the filter output with that of a copy V-I buffer.

978-1-4244-2018-6/08 $25.00 © 2008 IEEE

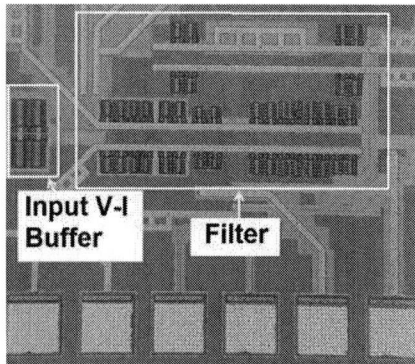

Fig. 7 Chip micrograph

Fig.8 shows the measured magnitude response of the filter. The ripple around 1GHz is due to mismatch in the peaking (due to package parasitics) between filter and calibration path.

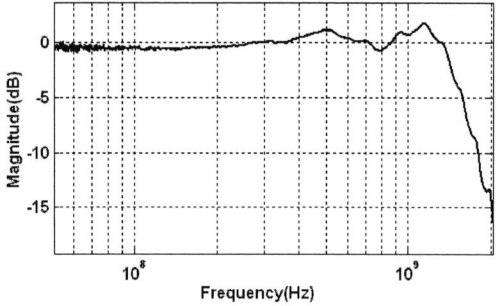

Fig. 8 Measured magnitude response

A two tone test performed with tones around 1GHz with 10MHz spacing shows IM3 of -54.2dB (see Fig.9). PSSR of the filter (using single-ended output) was measured to be 27dB at 100MHz. Key performance parameters of the proposed filter and other benchmark Gm-C filters are tabulated in Table.1.

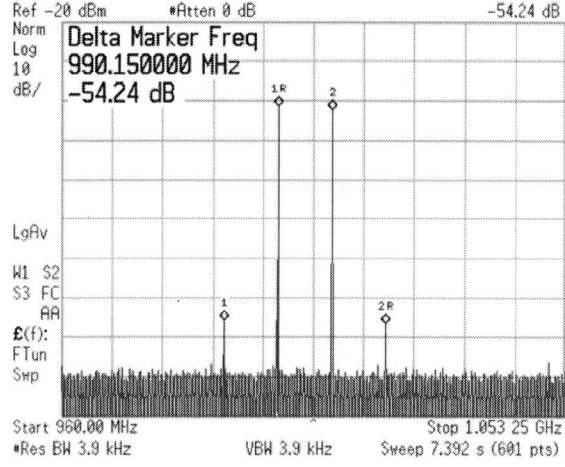

Fig. 9 Intermodulation test with input tones around 1GHz

TABLE 1
PERFORMANCE SUMMARY AND COMPARISONS

Reference	[8]	[9]	[10]	This work
Technology	0.35µm CMOS	0.18µm CMOS	0.13µm CMOS	0.13µm CMOS
Supply (V)	3.3	1.5	1.2	1.2
Filter Order	5	4	5	4
BW (MHz)	500	1000	240	1300
Noise (µVrms)	366	1389	117	266
Linearity	THD=-40dB at 0.5Vpp	IM3=-43dB at 0.35Vpp	OIP3=-18dBV	IM3=-54dB at 0.3Vpp
Power (mW)	100	175*	24	24

* Includes automatic tuning

V. CONCLUSIONS

A highly linear, dual-path, current mirror based element with excellent supply rejection is presented. A fourth order Butterworth filter with bandwidth of 1.3GHz is implemented using the proposed element. The power efficiency of the filter designed using the proposed technique compares favorably to that of similar class of filters.

ACKNOWLEDGMENT

Authors would like to thank UMC (United Microelectronic Corp) for the fabrication support.

REFERENCES

[1] M. Elmala, B. Carlton, R. Bishop, K Soumyanath, "A 1.4V, 13.5mW, 10/100MHz 6th order elliptic filter/VGA with DC-offset correction in 90nm CMOS," *IEEE Radio Frequency Integrated Circuits Symposium*, Jun. 2005, pp. 189-192

[2] D. Sun, A. Xotta, A. A. Abidi, "A 1 GHz CMOS Analog Front-End for a Generalized PRML Read Channel," *IEEE J. Solid-State Circuits*, vol. 40, no. 11, pp. 2275-2285, Jul. 2005.

[3] A. Nedungadi, T.R. Viswanathan, "Design of linear transconductance elements," *IEEE Trans Circuits Syst.*, vol CAS-31, pp. 891-894, Oct 1984.

[4] B. Nauta, "A CMOS Transconductance-C Filter Technique for Very High Frequencies," *IEEE J. Solid-State Circuits*, Vol. 27, no.2, pp. 142-153, Feb. 1992

[5] K. Salimi, E Krummenacher, C. Dehollain, M. Declercq, "Continuous-time CMOS circuits based on multi-tanh linearisation principle," *Electron. Lett.*, Vol.38, no. 3, pp. 103-104, Jan 2002.

[6] G. Efthivoulidis, L. Toth, Y.P. Tsividis, "Noise in Gm-C Filters," *IEEE Trans. Circuits Syst. II*, vol. 45, no. 3, pp. 295-302, Mar. 1998.

[7] S. L. Smith, E. Sanchez-Sinencio, "Low voltage integrators for high-frequency CMOS filters using current mode techniques," *IEEE Trans. Circuits Syst. II*, vol. 43, no. 1, pp. 39–48, Jan. 1996.

[8] S. Pavan, T. Laxminidhi, "A 70-500MHz Programmable CMOS Filter Compensated for MOS Nonquasistatic Effects," *Proc. IEEE ESSCIRC*, Sept. 2006, pp. 328-331.

[9] T. Y. Lo and C. C. Hung, "A 1GHz OTA-Based Low-Pass Filter with A High-Speed Automatic Tuning Scheme," *ASSCC Dig. Tech. Papers*, Nov. 2007, pp. 408-411.

[10] V. Saari, M. Kaltiokallio, S. Lindfors et al., "A 1.2V 240MHz CMOS Continuous-Time Low-Pass Filter for a UWB Radio Receiver," *ISSCC Dig. Tech. Papers*, Feb. 2007, pp. 122-123.

IEEE 2008 Custom Intergrated Circuits Conference (CICC)

A 1V Downconversion Filter Using Duty-cycle Controlled Bandwidth Tuning

Peter Kurahashi, Pavan Kumar Hanumolu and Un-Ku Moon

School of EECS, Oregon State University, Corvallis, OR 97331

Abstract— This paper describes a downconversion filter which uses variable delay clocks to simultaneously perform downconversion mixing and filter bandwidth tuning. This method of bandwidth tuning is highly linear and applicable to low supply voltages. The test chip fabricated in a $0.18\mu m$ CMOS process achieves 19.2dBV IIP3 at 1V and has a bandwidth that is tunable over a $\pm 50\%$ range. The downconversion filter mixes and filters an 830MHz input to a nominal 300kHz bandwidth at DC.

I. INTRODUCTION

The combination of a mixer followed by a low-pass filter is commonly used in downconversion receiver architectures [1][2]. Low-pass filtering after mixing suppresses interferers and reduces dynamic range requirements of later stages. A common implementation of the mixer/filter combination is for a passive current mixer to feed into an opamp RC transconductance stage. The passive current mixer provides high linearity while operating with low flicker noise [3]. The opamp RC transconductance stage filters the mixer output and provides to the mixer a virtual ground allowing it to operate in current mode.

The pole locations of integrated active RC filters are sensitive to process, voltage and temperature (PVT) variations. PVT variations in integrated resistors and capacitors can cause corner frequency variation by up to 50%. Duty-cycle based tuning has been shown to be an affective method of providing a high degree of low-distortion tuning at low-supply voltages [4].

In this work, we apply the idea of duty-cycle based tuning to the combination passive-current-mixer/transconductance-filter. The resulting architecture which we call a clock delay tunable downconversion filter, provides low-distortion tunability at low supply voltages. Allowing low-distortion tunability of the filter pole enables the filter to be used in ways that were not previously feasible. One application would be to incorporate this downconversion filter into the first stage of a highly-selective baseband filter where pole location is critical. Another application is to use the downconversion filter in an IF-to-baseband $\Delta\Sigma$ ADC [5] where the filter becomes part of the $\Delta\Sigma$ ADC loop filter. This paper is organized as follows. Section II highlights the overall architecture of the downconversion filter, Section III provides circuit details of the key building blocks, Section IV describes the filter conversion gain, and finally measurement results of the prototype IC are presented in Section V.

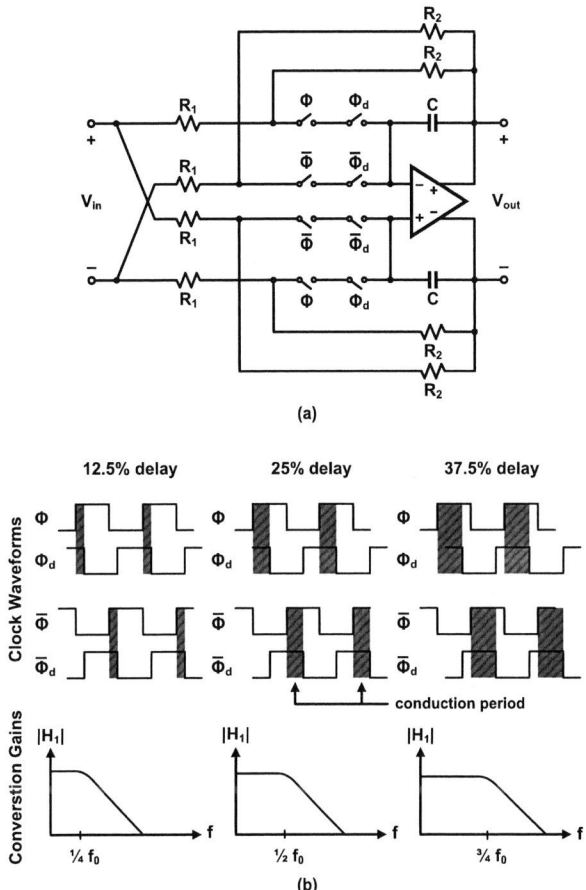

Fig. 1. (a) Ideal downconversion filter circuit. (b) Clock waveforms and 1st harmonic conversion gains for three different clock delays.

II. SYSTEM ARCHITECTURE

The ideal circuit diagram of the proposed downconversion filter is shown in Fig. 1. The downconversion filter is a modification of a passive current switching mixer with an active RC filter output stage. In this downconversion filter topology, the switching clock performs two functions: The first function is downconversion by commutating the input signal current, as is normally done in switching mixers. The second function, which has been added in this work, is the control of the filter bandwidth by tuning the effective duty-cycle of the clock. This method of bandwidth tuning is highly linear and can be used at low supply voltages.

978-1-4244-2018-6/08 $25.00 © 2008 IEEE   753

The top part of Fig. 1(b) shows the clock waveforms used in the circuit. All clocks have a 50% duty-cycle and are derived from the same local oscillator (LO). ϕ_d is a delayed version of $\bar{\phi}$ and $\bar{\phi}_d$ is a delayed version of ϕ. Current flows from the input and feedback resistor branches to the integrating capacitors C when both series switches are on. This conduction period is shown in grey on the clock waveforms. Input branch currents, through resistors R_1, are commutated while feedback branch currents, through resistors R_2, are not. Depending on the amount by which ϕ_d and $\bar{\phi}_d$ are delayed, the conduction period length can be changed. By controlling the conduction period, the average current for a given branch input voltage can be changed. This is equivalent to changing the time-averaged branch resistance and hence the filter time-constant.

The bottom part of Fig. 1(b) illustrates the downconversion filter 1st harmonic conversion gains for different clock delays. These conversion gains have a first-order response with a corner frequency dependent on clock delay. Because changing the clock delay changes the time-averaged branch resistance, an equivalent change in filter corner frequency is observed. f_0 is the cutoff frequency when the clock delay is 50%, or equivalently when conduction occurs over the entire clock period. Note that conduction occurs twice per clock period, once for the non-inverted clock and once for the inverted clock. Ignoring switch resistance, the filter cutoff frequency is

$$f_{bw} = 2 \cdot d \cdot f_0 = \frac{d}{\pi \cdot R_2 C} \quad for \quad 0 < d < 0.5 \quad (1)$$

where d is the clock delay as a fraction of the clock period.

Switching could have been performed with single series switches and a variable duty-cycle clock, but at high frequencies and low duty-cycles, finite rise/fall times degrade the clock signal integrity. For this reason, the tuning method of Fig. 1 is used to maintain all clocks at a 50% duty-cycle.

The filter bandwidth can be tuned precisely by using the master-slave tuning scheme shown in Fig. 2. The slave is the downconversion filter. The master contains scaled copies of the mixer branches and capacitors used in the filter. The time-constant created by the mixer branch and capacitor copies are compared to a reference time period T_{ref}. This reference time period is the period of an accurate reference clock ϕ_{ref}. The time-constant error between that of the elements in the master and the reference time period is output as a voltage. This voltage is then amplified and feed back using the clock delay cell to drive the time-constant error to zero. Because the delayed clocks are also applied to the downconversion filter, the filter will have a time constant proportional to T_{ref}.

III. CIRCUIT DESIGN

A. Downconversion Filter

Fig. 3 shows the circuit implementation of the downconversion filter. The series connected nMOS switches $M_1 - M_8$ perform the mixing and filter bandwidth tuning. Switches M_9 and M_{10} keep the current summing node at a low differential impedance while the series switches are off. Switches M_{11} and

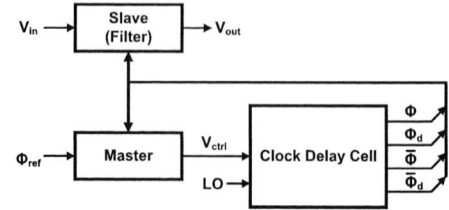

Fig. 2. Master-slave tuning.

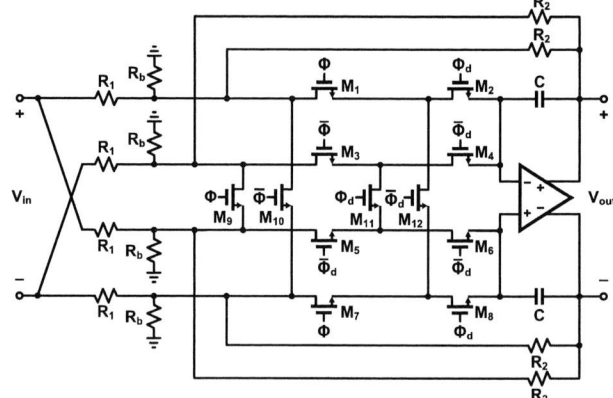

Fig. 3. Downconversion filter implementation.

M_{12} discharge the parasitic capacitance at the node between the series switches. Resistors R_b sink common-mode current such that internal nodes can be biased at a low voltage while the input and output common-modes are biased near mid-rail.

B. Filter Opamp

The filter differential opamp uses a two-stage folded cascode structure similar to that used in [4]. The opamp used in the common-mode feedback circuit has a single stage folded cascode structure with pMOS inputs. The input pair is biased near ground with resistive voltage dividers made up of common mode sense resistors R_{o1} and a pull-down resistor R_{o2}. This biasing is similar to that of the main opamp and allows low-voltage operation.

C. Master Circuit

The master is similar to that described in [4]. Two sets of branches feed into a differential integrator. The outer branches are SC resistors with resistance $T_{ref}/(2 \cdot C_{sc})$, where T_{ref} is the period of the reference clock ϕ_{ref}. Two parallel SC branches have been used instead of one branch to distribute charge transfer over both reference clock phases. For the same effective SC resistance, the peak charging ripple is lower than the single branch case. This helps reduce reference clock feedthrough. The inner branches are similar to the mixing branches in the downconversion filter. They have a resistance $R/(2 \cdot d)$, where d is the relative delay between mixing clocks ϕ and ϕ_d. Mismatch between branch resistances causes current to flow into the integrator, changing the control voltage V_{ctrl}.

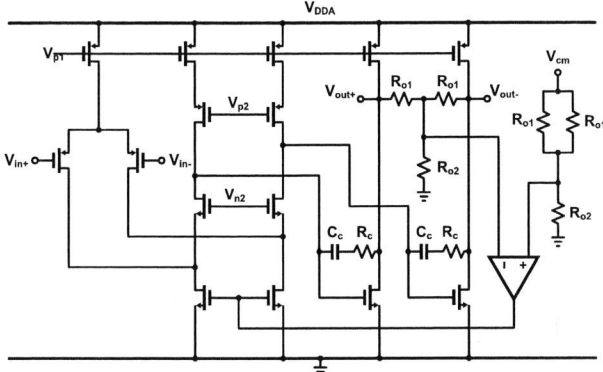

Fig. 4. Filter opamp.

The second opamp stage provides differential to single ended conversion and first order filtering of the mismatch current. Resistors R_{m2} bias the opamp input pair for low voltage operation. A second filter pole created with R_{m4} and C_{m4} further suppress clock reference feedthrough to the clock delay cell. A negative feedback loop is created through the delay cell, adjusting the relative clock delay between ϕ and ϕ_d and therefore the mixing branch resistance. Once the feedback loop reaches equilibrium, the branch resistances will be equal. Equating branch resistance gives

$$T_{ref} = C_{sc}R_{eq} \cong \frac{C_{sc}R}{d}. \qquad (2)$$

Equation (2) shows that the time constant $C_{sc}R_{eq}$ can be set precisely using an accurate reference clock. R_2 and C_{sc} are made from the same unit elements as the resistors and capacitors in the downconversion filter, therefore the time-constants between the filter and the master will track over PVT variation.

D. Delay Cell

The clock delay cell, shown in Fig. 6, creates relatively delayed clocks and their complements. The relative clock delay is adjustable via the input control voltage V_{ctrl}. The clock delay cell consists of three types of blocks: delay cells, alignment cells and output buffers. The delay cell contains two cascaded inverters with a capacitively loaded internal node. The capacitive load is adjustable by varying its resistance to ground through a triode MOSFET. By adjusting the internal capacitive load, the delay through the cell can be changed. The alignment cells preserve the 50% duty-cycle of the complementary clocks. Output buffers help drive the mixing switches.

IV. CONVERSION GAIN

The DC conversion gain of the downconversion filter is weakly dependent on the relative clock delay. The 1st harmonic conversion gain is found to be

$$H_1(s) = \frac{1}{2}\left(sinc(d) + \frac{1-d}{d}sinc(1-d)\right)\frac{R_2/R_1}{1+s\dfrac{R_2C}{2d}} \qquad (3)$$

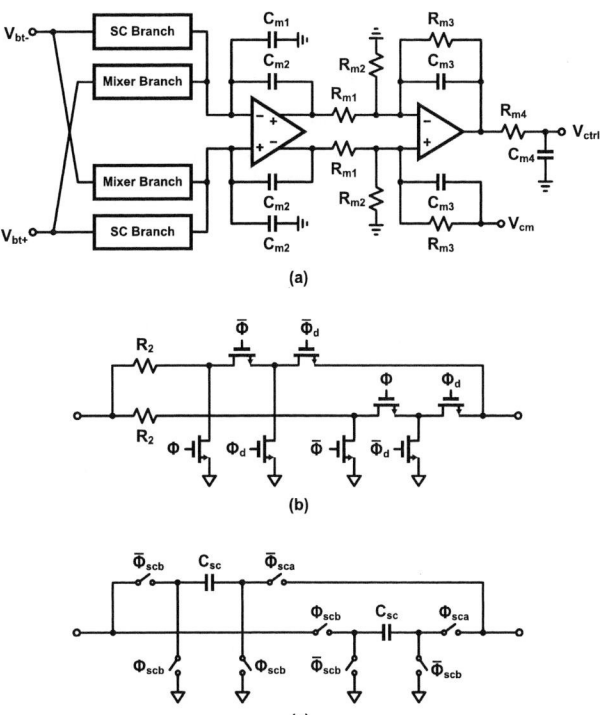

Fig. 5. (a) Master circuit. (b) Mixer Branch. (c) SC Branch.

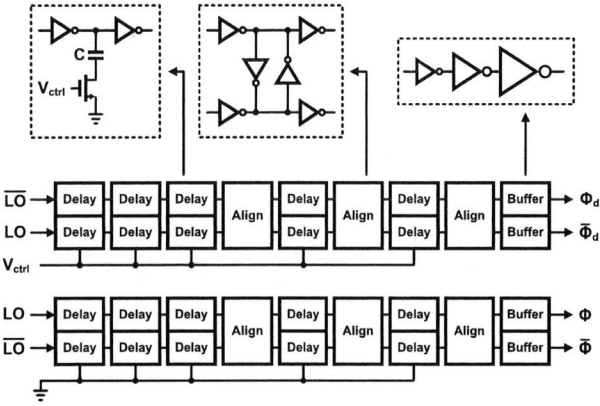

Fig. 6. Delay cell circuit.

where the relative clock delay d can vary from 0 to 0.5. For nominal delays less than 0.25, the conversion gain varies less than 2dB over a $\pm 50\%$ tuning range.

V. EXPERIMENTAL RESULTS

The prototype IC was fabricated in a 0.18μm CMOS process. The die micrograph is shown in Fig. 7. The active die area is 0.12mm^2. The measured filter conversion gains over a tuning range of $\pm 50\%$ are shown in Fig. 8. The input frequency band is located at 830MHz. The filter can tune from 150kHz to 450kHz with a nominal cutoff frequency of 300kHz. The nominal conversion gain is 9.6dB. Fig. 9 shows the measured intermodulation distortion. The downconversion

978-1-4244-2018-6/08 $25.00 © 2008 IEEE

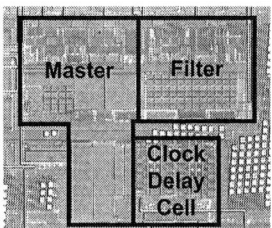

Fig. 7. Die photo.

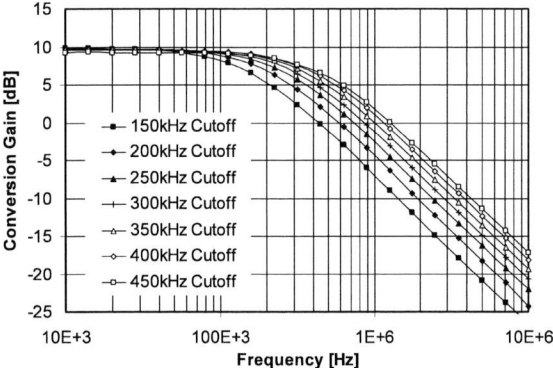

Fig. 8. Measured conversion gains over a ±50% tuning range.

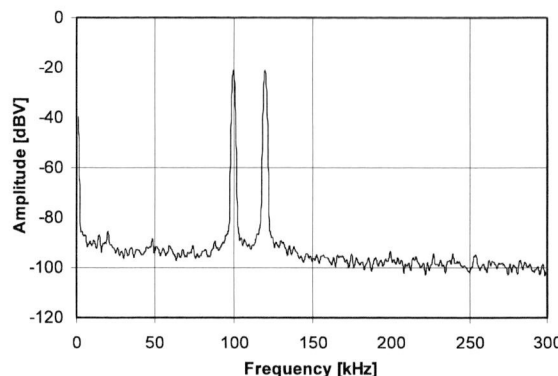

Fig. 10. Measured two tone output spectrum with -31dBV inputs, downconverted from an 830MHz carrier frequency.

TABLE I
PERFORMANCE SUMMARY

Supply Voltage	1V
IIP3 / OIP3	19.2dBV / 28.8dBV
1dB Compression Point (input/output)	-15dBV / -6.7dBV
Output SNR (1kHz-300kHz, Ref. to 1Vpp diff. output)	62dB
Input Frequency Band Center	830MHz
Nominal Cutoff frequency	300kHz
Nominal Conversion Gain	9.6dB
Power Consumption	3.0mW
Active Area	0.12mm^2
Technology	0.18μm CMOS

filter achieves 19.2dBV IIP3 while operating at 1V. The sharp rise in the 3rd order intermodulation component for inputs greater than -19dBV is caused by signal saturation at the opamp output. For an input amplitude of -18.7dBV when the IM3 is at -80dBV, each output tone swings 1Vpp differential on a 1V supply. The input referred 1dB compression point is -15dBV. Fig. 10 shows the measured output spectrum with a two tone -31dBV input. The performance of the test chip is summarized in Table I.

VI. CONCLUSION

The downconversion filter architecture provides downconversion mixing and bandwidth tuning using a variable delay clock. This method of bandwidth tuning is highly linear and applicable to low supply voltages.

ACKNOWLEDGMENT

The authors would like to thank Jazz Semiconductor for providing fabrication of the prototype IC. This work was supported by the Center for Design of Analog-Digital Integrated Circuits (CDADIC), NSF CAREER CCR-0133530, and Analog Devices, Inc.

REFERENCES

[1] S. Zhou and M. F. Chang, "A CMOS passive mixer with low flicker noise for low-power direct-conversion receiver," *IEEE J. Solid-State Circuits*, vol. 40, no. 5, pp. 1084-1093, May 2005.

[2] M. Valla *et al.*, "A 72-mW CMOS 802.11a direct conversion front-end with 3.5-dB NF and 200-kHz 1/f noise corner," *IEEE J. Solid-State Circuits*, vol. 40, no. 4, pp. 970-977, Apr. 2005.

[3] E. Sacchi *et al.*, "A 15 mW, 70 kHz 1/f corner direct conversion CMOS receiver," *Proc. IEEE Custom IC Conf.*, Sep. 2003, pp. 459-462.

[4] P. Kurahashi *et al.*, "Design of low-voltage highly linear switched-R-MOSFET-C filters," *IEEE J. Solid-State Circuits*, vol. 42, pp. 1699-1709, Aug. 2007.

[5] P. G. R. Silva *et al.*, "An IF-to-baseband $\Delta\Sigma$ modulator for AM/FM/IBOC radio receivers with a 118 dB dynamic range," *IEEE J. Solid-State Circuits*, vol. 42, no. 5, pp. 1076-1089, May 2007.

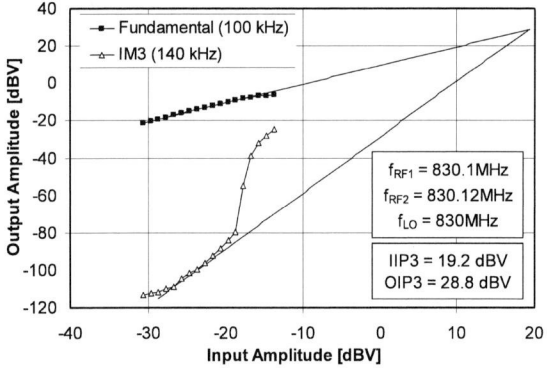

Fig. 9. Measured intermodulation distortion.

IEEE 2008 Custom Intergrated Circuits Conference (CICC)

A Reconfigurable FIR Filter Embedded in a 9b Successive Approximation ADC

Joshua Kang, David T. Lin, Li Li and Michael P. Flynn

University of Michigan
1301 Beal Avenue, EECS
Ann Arbor, MI 48109 USA

Abstract-A reconfigurable FIR filter and 9b SAR ADC combination in 0.13μm CMOS is presented. The filter does not require additional analog circuitry, but is implemented by using the SAR capacitor array with a modified tracking and sampling scheme. The prototype filter-ADC can be digitally configured as a 4-tap filter, as one of two different 12-tap filters, or without any filtering. The prototype occupies an active area of 0.68mm², achieves 45dB SNDR and dissipates 7.3mW power at 5MS/s. The lowest frequency notch of the embedded filter attenuates by as much as 30.5dB in 4-tap mode and 38.4dB in 12-tap mode.

I. INTRODUCTION

A key bottleneck in the development of the digital radio receiver is the difficulty in implementing integrated, low power, programmable filters. This paper introduces a filtering technique based on modification of sampling in a SAR ADC. The filter can be adapted on the fly to target blockers and interferers by changing its sampling configuration; yet, the filter-ADC combination adds minimally to power consumption, complexity and area when compared to a SAR ADC alone. This filter also supplements the anti-aliasing filter, thus allowing a reduction in ADC conversion rate and/or a reduction in the complexity of DSP or analog filtering. The fabricated prototype can be digitally configured as a 4-tap filter, as one of two different 12-tap filters, or without any filtering.

The proposed architecture achieves a higher level of integration when compared to recently demonstrated work on charge-domain filtering based digital radios. This is accomplished by integrating the filter into the ADC sampling process. Such integration can eliminate the explicit filtering stages found in traditional receiver front-ends, as shown in Fig. 1.

The receiver in [1] utilizes discrete-time (DT) current sub-sampling decimation filters and adjusts sampling rates and capacitor groupings in order to optimize the frequency response for each of 4 different GSM/GPRS bands. The authors of [2] demonstrate a direct conversion receiver for different communication standards by utilizing DT capacitive decimation anti-aliasing filters. The output of both [1] and [2] require resampling by an external ADC. The proposed design achieves DT filtering by modifying capacitive sampling in a SAR ADC and therefore eliminates resampling at the ADC. A key advantage of the proposed filtering technique is that the input signal is sampled only once by the filter-ADC combination, so that noise is not introduced both by DT filter sampling and ADC resampling of the filter output. Assuming that kT/C noise dominates, noise is halved. Furthermore, since there is only one set of capacitors, the total capacitance is reduced by a factor of 4.

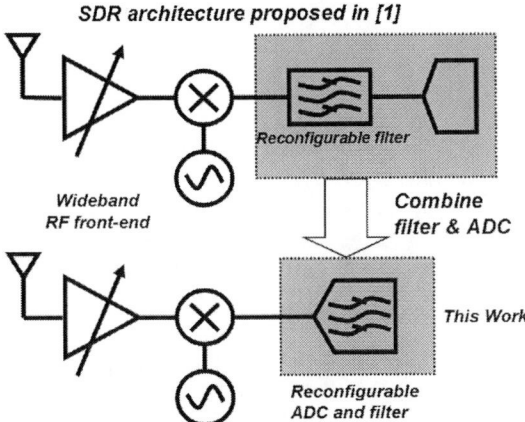

Fig. 1. Proposed reconfigurable digital radio receiver architecture.

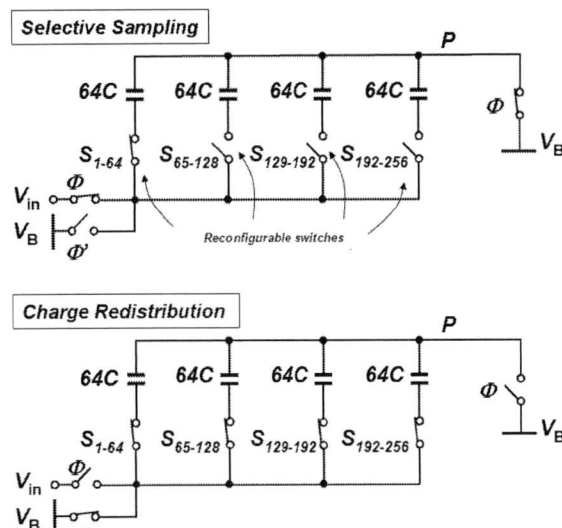

Fig. 2. Principle of filtering with charge sharing sampling.

The work was supported in part by Analog Devices Inc. and the Semiconductor Research Corporation.

978-1-4244-2018-6/08 $25.00 © 2008 IEEE

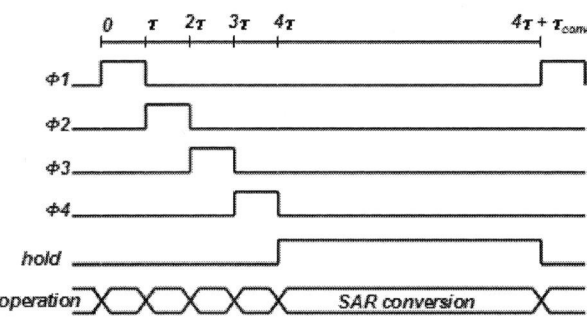

Fig. 3. Timing diagram of embedded FIR filter
and SAR ADC operation.

II. EMBEDDED FIR FILTERING SCHEME

The charge sharing DT FIR filter is embedded in the sampling process of a SAR ADC. The SAR ADC capacitors are arranged into groups that are proportional to the relative coefficient sizes of an FIR filter. For example, the implementation of a 4-tap filter with coefficients [.25 .25 .25 .25] requires 4 groups of capacitors, as shown in Fig. 2. Assuming this particular arrangement, the input voltage is sampled onto capacitor group 1, 2, 3 and 4 at time 0, τ, 2τ and 3τ, respectively. The timing diagram for the filtering and SAR conversion process described above is shown in Fig. 3. After acquiring these four samples, all of the capacitor groups are charge shared, as shown in lower diagram of Fig. 2. This averages the previously sampled voltage levels and implements the FIR transfer function, $H(z)=(z^{-1}+z^{-2}+z^{-3}+z^{-4})/4$. The successive approximation operation that follows is identical to that of a SAR ADC without filtering, but produces filtered output code since it operates on a filtered input. Once the conversion process completes, the process described above repeats.

The timing diagram in Fig. 3 shows that multiple samples are collected during the sampling phase and no samples are collected during the conversion phase. Nonetheless, this non-uniform sampling results in the desired FIR filtering response. Exactly enough samples are collected during each sampling phase to generate a single filtered output sample, each of which is generated at the ADC conversion rate.

The overall ADC conversion rate, F_S, is described in (1), where n_{tap} is the number of taps of the selected filter and t_{conv} is the time required for A/D conversion.

$$F_S = 1/(n_{tap} \cdot \tau + t_{conv}) \qquad (1)$$

The FIR filter is partly characterized by the input sampling rate, $F_{S,filter}$, which is the inverse of τ in (1). Since the ADC conversion rate, F_s, and the input sampling rate, $F_{S,filter}$ are not required to be the same, the embedded filter rate and response are highly reconfigurable.

As with any ADC, the conversion rate determines the Nyquist frequency, $F_S/2$, above which the input signal aliases. On the other hand, the input sampling rate determines the

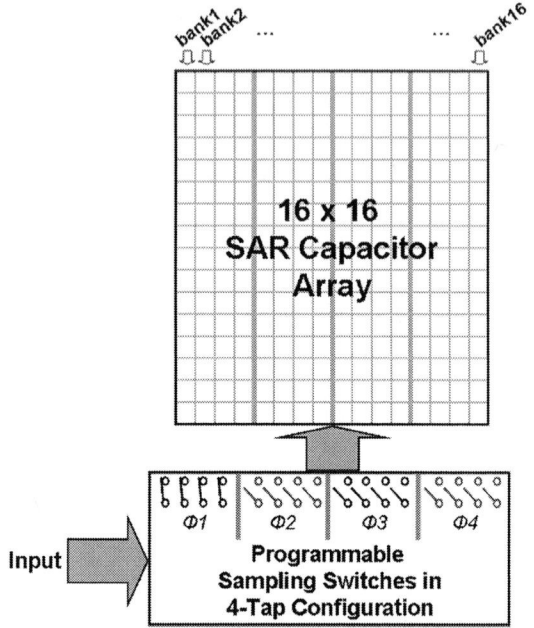

Fig. 4. High-level diagram of an embedded programmable filter in the DAC of a SAR ADC.

TABLE I
TESTED PROGRAMMABLE SAMPLING SWITCH CONFIGURATIONS

Filter Type	Filter Coefficient	-3dB frequency
No Filter	[16]/16	None
4-Tap	[4 4 4 4]/16	$2.0 \times F_S$
12-Tap #1	[1 1 1 2 1 2 2 1 2 1 1 1]/16	$1.4 \times F_S$
12-Tap #2	[4 0 1 1 1 1 1 1 1 1 0 4]/16	$1.0 \times F_S$

frequency span of the FIR filter response (i.e. $-\pi$ to π corresponds to $-F_{S,filter}/2$ to $F_{S,filter}/2$) and the frequency above which the filter response repeats. Inherent to the filter-ADC architecture, the input sampling rate is higher than the conversion rate, which allows the FIR filter to block multiple aliasing bands before the response repeats. The ideal and measured filter responses are discussed in section IV.

III. MODIFYING THE FILTER RESPONSE

The capacitive DAC in the prototype 9 bit SAR ADC contains 256 unit capacitors. These capacitors are grouped into 16 banks during the sampling phase, as shown in Fig 4. When the ADC operates without filtering, the analog input voltage is sampled onto all 16 capacitor banks simultaneously during the sampling period τ. When operating with the embedded FIR filter active, the 16 banks are arranged into multi-bank groups that are proportional to the relative coefficient sizes of an FIR filter with the desired frequency response.

The filter response can be modified by dividing the capacitor banks into different groupings and adjusting the sampling process appropriately. For example, a 12-tap filter with the coefficients [4 0 1 1 1 1 1 1 1 1 0 4]/16 can be realized by dividing the capacitor banks into 2 groups of 4 capacitor banks and 8 groups of 1 bank. Sampling occurs over 12 time periods (i.e. n_{tap} in (1) is 12). The fabricated prototype

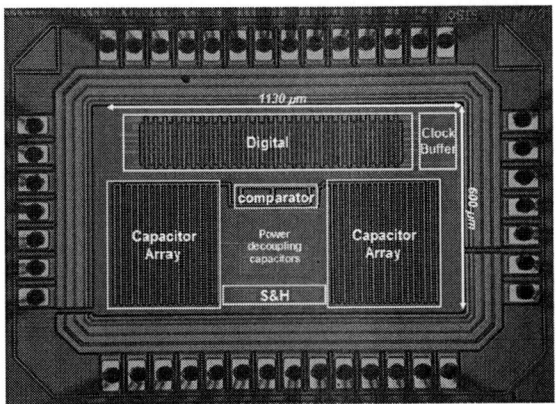

Fig. 5. Die microphotograph.

demonstrates the ability to customize the filter response by implementing 3 different arbitrarily chosen, externally controllable filtering responses. Table I summarizes the implemented filter configurations.

While three basic filtering modes are implemented to demonstrate the SAR ADC filtering technique, the architecture itself is highly adaptable to the dynamic filtering needs of a particular system. For example, the designer of a direct conversion receiver for a narrowband channel can divide the sampling capacitors into more banks, in order to enhance tap coefficient resolution. This permits precise placement of the filter nulls at multiples of F_S, thus allowing the filter to perform narrowband anti-aliasing. On the other hand, the designer of a receiver affected by unpredictable interferer frequencies and powers can program of the digital logic to support a variable number of tap coefficients, so that filter suppression can be enhanced or relaxed during system operation by increasing or decreasing the number of available filter zeros, respectively. Both of these changes require few modifications to the existing hardware. The first case typically requires only the addition of sampling related control lines and switches, since each capacitor bank of the DAC already consists of multiple unit-sized capacitors that can easily be regrouped. The second case typically requires little more than changing Verilog code.

IV. PROTOTYPE MEASUREMENTS

The reconfigurable FIR filter embedded in a SAR ADC is implemented in 0.13μm CMOS technology. The prototype occupies an active area of 0.68mm² including routing and power decoupling capacitors. Fig. 5 shows the die microphotograph of the prototype. The digital logic, including SAR logic and filter sampling control, is entirely synthesized from Verilog and auto-placed and routed. Simple changes to the Verilog code that implements SAR ADC control logic enable the filtering capability.

The measured attenuation at the lowest frequency notch is 30.5dB at $2.5 \times F_S$ for the 4-tap mode and 38.4dB at $1.7 \times F_S$ and 28dB at $1.1 \times F_S$ for the two 12-tap modes. Fig. 6 through Fig. 9 overlay the measured frequency response (in black) for all 4 modes of operation on top of the ideal FIR filter response (in grey).

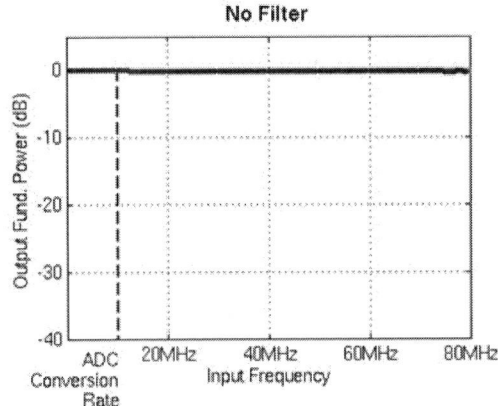

Fig. 6. Measured and ideal ADC response without filter turning on.

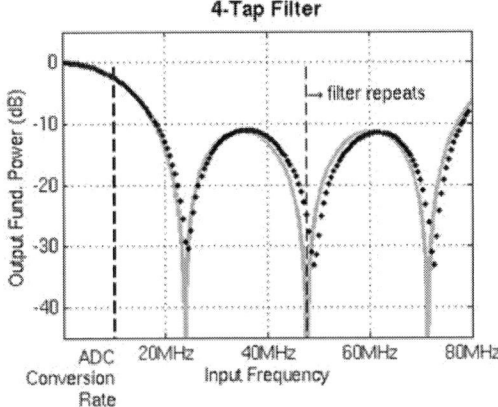

Fig. 7. Measured and ideal 4-tap [4 4 4 4]/16 filter response.

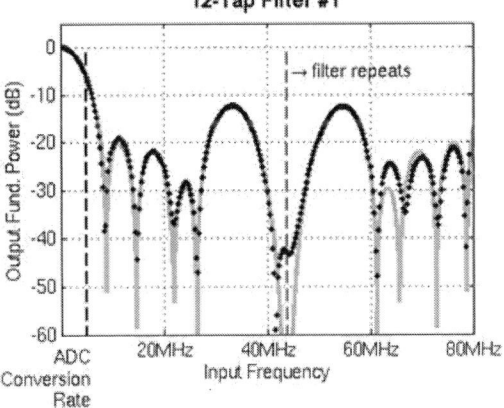

Fig. 8. Measured and ideal 12-tap
[1 1 1 2 1 2 2 1 2 1 1 1]/16 filter response.

978-1-4244-2018-6/08 $25.00 © 2008 IEEE   759

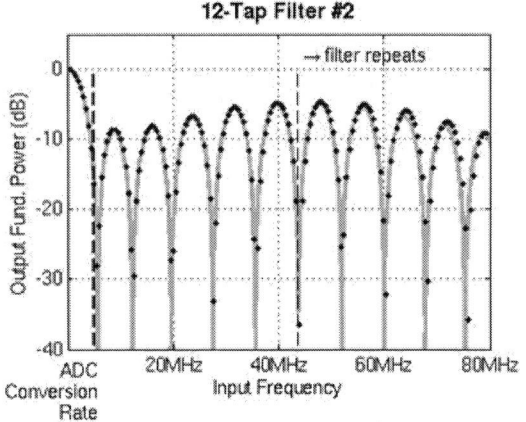

Fig. 9. Measured and ideal 12-tap
[4 0 1 1 1 1 1 1 1 1 0 4]/16 filter response.

TABLE II
PERFORMANCE SUMMARY

Technology	1P8M 0.13 μm CMOS with MIM capacitor			
Active Area	1.13×0.60 mm^2			
Resolution	9b			
DNL/INL	< 0.9LSB / < 1.2LSB			
Filter Modes	No filter	4-tap filter	12-tap filter #1	12-tap filter #2
First notch attenuation and freq.	N/A	-30.5dB @ $2.5 \times F_S$	-38.4dB @ $1.7 \times F_S$	-28dB @ $1.1 \times F_S$
ADC conversion rate	< 15MS/s	< 11MS/s	< 5.2MS/s	< 5.2MS/s
SNDR	48.9dB @ 12.5MS/s, 2MHz input	43.4dB @ 10MS/s, 0.5MHz input	45.0dB @ 5MS/s, 0.5MHz input	45.2dB @ 5MS/s, 0.5MHz input
Power @ 1MS/s	3.34mW	4.77mW	3.71mW	3.70mW
Power @ 5Ms/s	5.84mW	9.02mW	7.32mW	7.31mW

Table II summarizes the measured performance of the prototype. The ADC converts at up to 15MS/s without filtering enabled, 10MS/s with the 4-tap filter enabled and 5MS/s with the 12-tap filter enabled. Power consumption includes the power consumed by the internal clock buffer, digital, analog and excludes pad and external clock power.

Fig. 10 shows the measured output spectrum of a 2MHz input at 12.5MS/s conversion rate with no filtering. The measured SNDR and SFDR are 48.94dB and 54.36dB, respectively. Fig. 11 and Fig. 12 show DNL and INL plots of the 9b ADC. DNL and INL are within 0.9LSB and 1.2LSB, respectively.

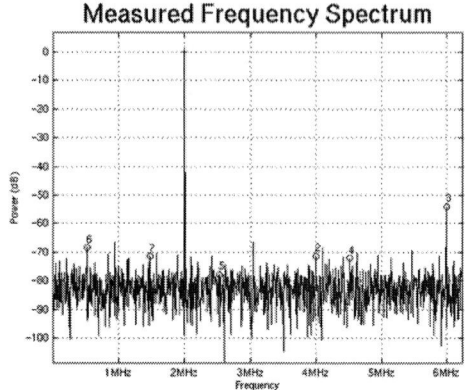

Fig. 10. Measured output spectrum at 12.5MS/s and 2MHz input.

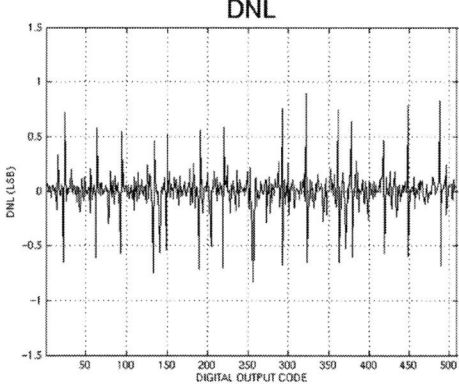

Fig. 11. Measured DNL plot.

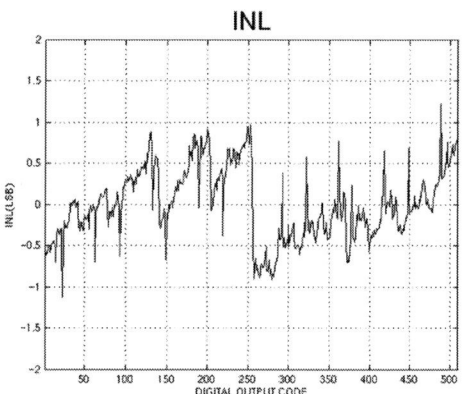

Fig. 12. Measured INL plot.

REFERENCES

[1] K. Muhammad, et al., "A Discrete Time Quad-band GSM/GPRS Receiver in a 90nm Digital CMOS Process," *Proc. CICC*, pp. 809–812, Sept. 2005.
[2] R. Bagheri, et al., "An 800-MHz–6-GHz Software-Defined Wireless Receiver in 90-nm CMOS," *IEEE J. Solid-State Circuits*, vol. 41, pp. 2860–2876, Dec. 2006.

978-1-4244-2018-6/08 $25.00 © 2008 IEEE

Voltage References for Ultra-Low Supply Voltages

P. Kinget, C. Vezyrtzis, E. Chiang, B. Hung[*] and T.L. Li[*]

Dept. of Electrical Engineering
Columbia University
New York, NY 10027, USA
[*]United Microelectronics, Taiwan, R.O.C.

Abstract—**The majority of integrated voltage references have, so far, been limited to a minimum supply voltage above 0.7-V often due to the voltage headroom required for the forward-biased operation of a PN-junction. This paper reviews design techniques for low voltage reference design and explores the feasibility of designing a voltage reference with a supply voltage below 0.7-V in a standard CMOS process. Two ultra-low-voltage solutions are explored in detail, a reference circuit based on CMOS compatible Schottky diodes and a MOS-only reference circuit.**

I. INTRODUCTION

Voltage references are indispensible building blocks of many analog and mixed-signal system-on-a-chip integrated circuits. They provide references for on-chip power management blocks, for signal conditioning and signal measurement, or for building blocks such as ADCs and DACs and further make possible the appropriate biasing of analog circuits so that they become much less unaffected by process, temperature and voltage (PVT) variations. Parts of reference circuits can further often be configured to serve as on-chip die temperature monitors, e.g. on large digital microprocessors.

The continued scaling of transistor dimensions in CMOS technologies in combination with the need to reduce power density in large digital ICs is driving down the supply voltage for nanoscale CMOS technologies. Supply voltages well below 1V and even down to 0.5V for low power applications are projected in the ITRS roadmap [1]. The most energy efficient operation of digital circuits and systems occurs in the supply voltage range of 0.3V to 0.5V in deeply scaled technologies [2][3]. Other applications such as self-powered sensor nodes to enable ambient intelligence also require ultra-low voltage mixed-signal ICs [4]. Recent research has demonstrated the feasibility of designing analog and RF circuits that can operate with ultra-low supply voltages down to 0.5V; this includes basic building blocks, analog filters, analog-to-digital converters, and even fully integrated RF receivers (see [5] and its references). In this paper, the realization of voltage references operating down to 0.5V is investigated. The widely used bandgap voltage reference [6] configurations with PN diodes cannot be operated with such low supply voltages [7]. We explore several alternatives that are available in any standard CMOS technology for the design of a voltage reference. They resemble the technique used in standard bandgap designs to create a temperature independent quantity, namely the addition of two quantities, one of which experiences a positive temperature dependence and the other a negative, which results in a temperature independent reference.

II. SUPPLY VOLTAGE REQUIREMENTS FOR CMOS VOLTAGE REFERENCES

The bandgap voltage reference [6] has been the most established on-chip voltage reference circuit for many years. It uses the complementary-to-absolute-temperature (CTAT) characteristic of the forward *voltage drop* across a PN junction and linearly combines it with the proportional-to-absolute-temperature (PTAT) characteristic of the *voltage difference* between two PN junctions with equal currents, but different current densities, to obtain a temperature independent output voltage. When these voltages are combined for a zero temperature coefficient, the resulting output reference voltage is around 1.205V, which is the bandgap voltage for Silicon. In bipolar ICs the PN junctions are typically formed using bipolar transistors. In CMOS processes, parasitic vertical bipolar PNP devices are generally used. These reference circuits typically require a supply voltage larger that 1.4V to properly function (see e.g., [7]).

Instead of linearly summing voltage drops, a CTAT *current*, derived using a forward-biased PN junction, and a PTAT *current* can be linearly combined, and converted back into a temperature-independent output voltage with arbitrary value using the circuit topology presented in [8]. This topology has a minimum supply voltage equal to the PN-diode turn on voltage (about 0.7V) plus a saturation voltage across a current source (about 0.2V) and can thus be operated below 1V, down to about 0.85V or 0.9V [9][10][11].

CMOS-only voltage references can operate with lower supply voltages and are not bound by the limitations due to the PN diode turn on voltage. Most designs exploit the negative temperature dependence of the threshold voltage of MOS devices (see e.g. [12]) and combine it with a PTAT characteristic which can be obtained with MOS devices in weak inversion; high performance references operating down to 0.85V V_{DD} have been demonstrated [13]. Using enhancement and depletion mode devices in a fully depleted MOS/SIMOX process, a reference operating from a supply down to 0.6V was demonstrated in [14]. However, such design is not compatible with standard digital CMOS technologies.

Specialized techniques for the realization of high precision reference voltages include floating-gate designs [15]; they require special devices and each reference needs to be programmed after fabrication. Additionally, increasing gate leakage effects make such structures less compatible with deeply-scaled CMOS processes.

In this paper we first explore the replacement of the PN diode in a low voltage bandgap structure with a standard-CMOS-compatible Schottky diode in Section III. We then investigate in Section IV the use of the CTAT characteristic of the forward drop across a diode-connected nMOS device with its body shorted to its gate, and thus with a forward-biased body-source junction and reduced threshold voltage V_T. For both designs a proof-of-principle IC prototype is presented with experimental data. Section V compares the results of various low voltage approaches.

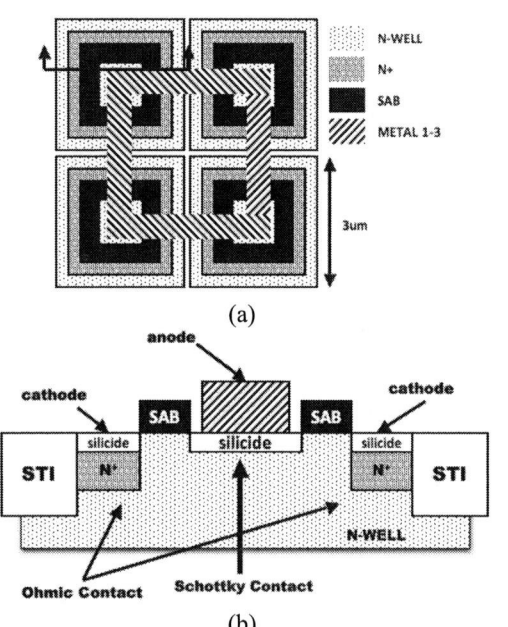

(a)

(b)

Figure 1   Layout (a) of 4 Schottky diode unit cells; (b) cross-section of a Schottky diode-unit in a standard CMOS process

III. Schottky Diodes And Their Use In A Voltage Reference

A. CMOS-Compatible Schottky Diodes

Schottky diodes can be realized in a standard CMOS process, by directly contacting an N-WELL region without an N+ implant. Figure 1 shows the layout and a cross section of the Schottky diodes laid out in a standard 90nm CMOS technology from UMC. In modern CMOS processes silicides are used at the Silicon-Metal interface and the Schottky barrier diode is formed where the silicide contacting the metal is in contact with the N-WELL; the N-WELL is then contacted with an ohmic contact. The metal is the diode anode and the well is the cathode. The current through the diode, I_D, is determined by thermionic emission across the barrier:

$$I_D = \left[AR^*T^2 \exp\left(\frac{-\Phi_B}{nkT/q}\right) \right] \cdot \left[\exp\left(\frac{V_D}{nkT/q}\right) - 1 \right] \quad (1)$$

where V_D is the voltage across the diode, kT/q is the thermal voltage, n is the diode ideality factor and is close to 1; Φ_B is the Schottky barrier height, A is the area of the diode, T is the absolute temperature and R^* is the effective Richardson constant [16]. This I-V dependence is very similar to the PN-

diode I-V relationship or the I_C-V_{BE} dependence for a bipolar transistor.

The difference of the forward voltage of two Schottky diodes operated with identical currents but different current densities can be used to generate a PTAT voltage. The dependence of the forward voltage drop to temperature for a fixed or temperature dependent current, is also very similar to the results obtained for a PN diode. (Note that the R^* term is only very weakly temperature dependent.) For the Schottky diode, the extrapolated forward voltage at zero Kelvin is equal to the Schottky barrier Φ_B. It depends on the work function of the metal and semiconductor used and is typically in the order of 0.2 to 0.6V.

These properties make the Schottky diode an excellent replacement for the traditional PN-junction diode for voltage references operating with ultra-low supply voltages. Although Schottky diodes can be readily designed in standard CMOS processes, they are typically not included in the standard designkit or simulation models. For the purpose of demonstrating an ultra-low V_{DD} reference, we characterized a number of diode test structures provided by the foundry. We further built a custom verilog-A based model to incorporate the diodes in the circuit design flow. I-V characteristics were measured over temperature to extract the temperature coefficient.

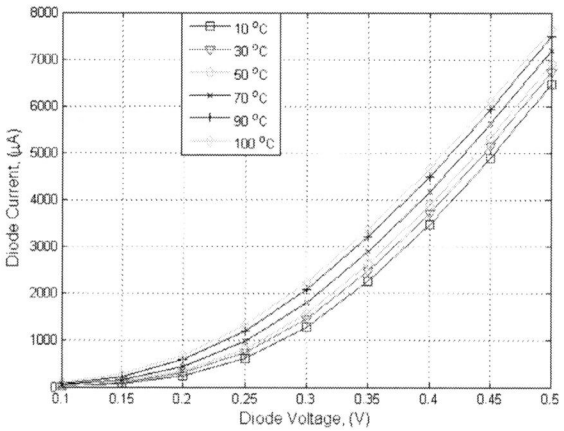

Figure 2   Measured I-V characteristics for various temperatures of a 32 unit Schottky diode with the layout of Fig. 1.

Figure 2 shows the measured I-V characteristics for various temperatures for a 32-unit diode test structure. The turn-on voltage of the Schottky diode is between 0.2 and 0.3V, which is significantly lower that the 0.6 to 0.7V typically needed to turn on a PN junction diode. The forward voltage across the Schottky diode had a measured temperature coefficient of -0.4mV/°C (from 10 to 100°C). The I-V characteristics also show a parasitic resistance in series with the diode, which is caused by the series resistance due to the NWELL used to contact the structure. A parasitic resistance of approximately 100 Ohms/unit-diode was extracted from the I-V characteristics.

978-1-4244-2018-6/08 $25.00 © 2008 IEEE

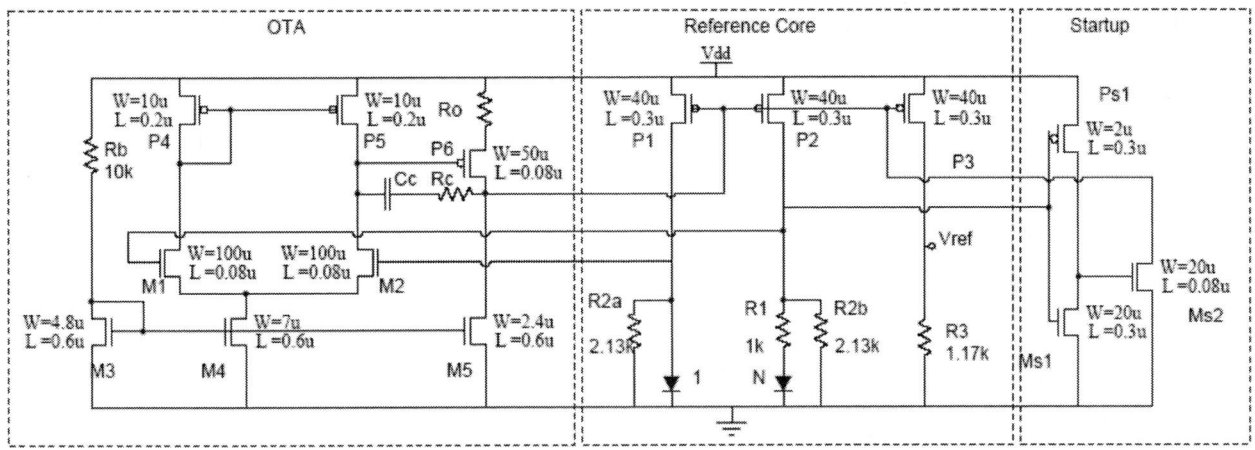

Figure 3   Schematic of the Schottky-diode-based reference.

B.   Voltage Reference Using Schottky Diodes

Schottky diodes have been used in the design of voltage reference operating from a 2.5-V supply voltage in [17]. Here, we demonstrate that they enable the realization of ultra-low V_{DD} references when used in the current-summing based bandgap topology from [8]. The schematic of the designed reference is shown in Figure 3. The reference circuit consists of two almost identical branches; one branch contains a single Schottky diode unit and the other branch has N diode units in parallel (N=6 in this design). An operational transconductance amplifier (OTA) is used to force the voltages across resistors R2a and R2b to be equal, with a nominal value of 250mV, which drops with a rate of 0.4mV/°C for increasing temperature. The difference of the forward-bias voltage of the diodes is dropped across resistor R1, creating the PTAT current, while the current through R2a and R2b is a CTAT current. The nominal value of the voltage across the resistor R1 increases with a rate of 0.22mV/°C. Resistors R1 and R2a (which is equal to R2b) are sized as 1kΩ and 2.13kΩ respectively, in order to obtain a zero temperature dependence. The sum of the PTAT and CTAT current is mirrored into resistor R3 to create the reference voltage. Devices Ms1-Ms2 and Ps1 form a startup circuit, which ensures that the circuit can not stay in the zero-current state.

An ultra-low voltage OTA is required in this circuit and was designed as a two stage Miller compensated amplifier. Its schematic is included in Figure 3. It employs a differential input pair, consisting of transistors M1 and M2, operating in weak inversion so that sufficient headroom remains available for the current source M4 while the input common mode level is about 250mV. The amplifier is self-biased from the supply (using the resistor Rb and transistor M3) and consumes 100uA of current, providing a DC gain of 40dB.

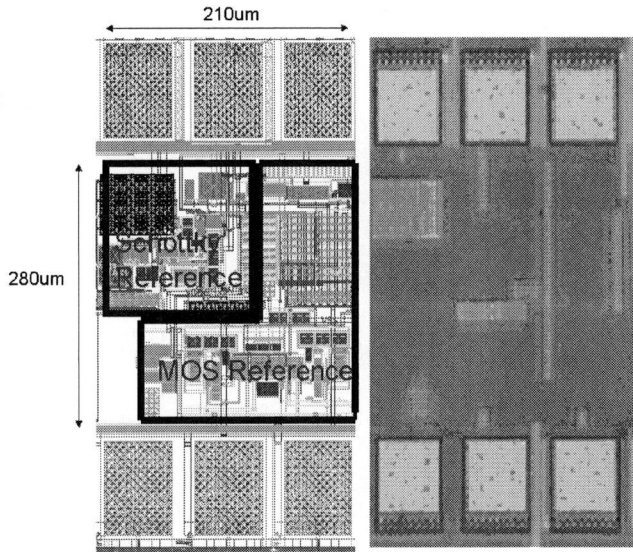

Figure 4   Die Microphotograph (right) and corresponding layouts (left) of the Schottky and MOS-only reference circuits.

C.   Experimental Results

The layout and die photograph of the experimental prototype in a standard 90nm CMOS technology is shown in Figure 4. The circuit is very compact and occupies only 0.019um². The circuit samples were measured using wafer probing. Figure 5 shows the measured reference output voltage w.r.t. the supply voltage V_{DD}; the reference is functional with supply voltages down to 0.5V. For the nominal V_{DD} of 0.6V the supply sensitivity is 3mV/100mV.

978-1-4244-2018-6/08 $25.00 © 2008 IEEE   763

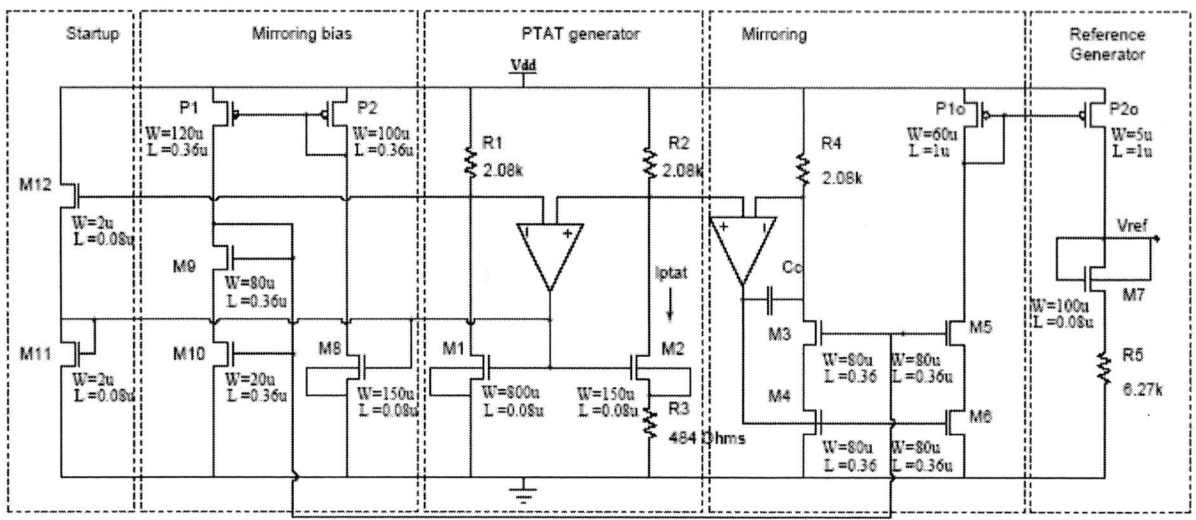

Figure 7   Schematic of the MOS-only reference .

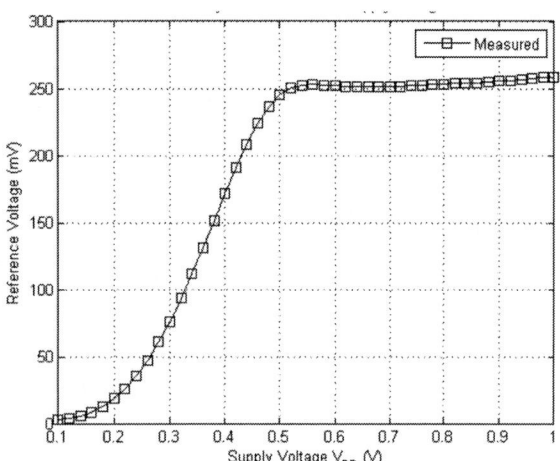

Figure 5   Measured Schottky reference supply voltage dependence

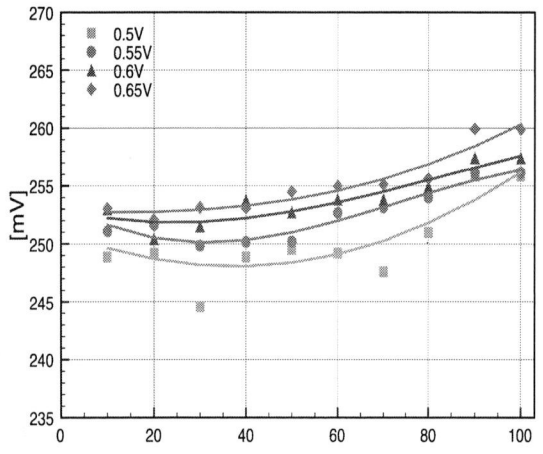

Figure 6   Measured Schottky reference temperature dependence between 10 and 100°C while operating from different supply voltages.

Measurements over temperature, plotted in Figure 6, show that the output voltage reference ranges from 250mV to 256mV when measured from 10 to 100°C with a supply voltage V_{DD} of 0.6V, which results in an effective temperature coefficient of 0.07mV/°C or 263ppm/°C. At the same V_{DD}, the reference voltage range was 251 ± 4mV for 18 samples. No trimming was performed for any of the measurements. The performance of the reference circuit for supply voltages of 0.55V, 0.6V and 0.65V is summarized in Table 1. This proof-of-principle circuit consumes 375uW at room temperature. For this design, limited diode data and models were available restricting the choice of bias points. As more characterization data and better device models become available to the designers, significant power savings can be expected. Also the power consumption of the OTA can be significantly reduced.

IV. VOLTAGE REFERENCE UTILIZING ONLY MOS DEVICES

Several MOS-only references have been demonstrated operating below 1V down to 0.85V [12][13]. They typically rely on the negative temperature coefficient of the threshold voltage to obtain a CTAT characteristic and use the gate-source voltage difference of weakly inverted MOS devices with different current densities for the PTAT characteristic [18]. To obtain low voltage operation, threshold reduction techniques such as forward body biasing is often applied. E.g. in [13] the body of a weakly inverted PMOS transistor is connected to the gate and used as the basis for a high precision, 0.85-V V_{DD} reference. Next we explore an ultra-low voltage, MOS-only design with an NMOS device with a forward-biased body-source connection using the commonly available deep n-well implants used in standard CMOS technologies.. Note that ultra-low supply voltages (<0.7V), the risk of latchup due to forward biasing is generally not present.

A.   CMOS PTAT

The replacement of the PTAT cell in the voltage reference can be achieved by replacing the PN diodes with MOS devices operating in the weak inversion region, exploring the I-V_{GS}

978-1-4244-2018-6/08 $25.00 © 2008 IEEE   764

characteristic behavior of weak inversion operation. In this design we used the topology shown in Figure 8 (b) [19] which greatly reduces the effect of OTA offset voltages compared to the conventional topology (Figure 8 (b)). The PTAT voltage is created as the difference between the V_{GS} voltages of transistors M1 and M2, creating a PTAT current of nominal value 150uA in this design. M1 and M2 are deep-n-well devices with source and body shorted. This circuit is incorporated in the MOS-only reference shown in Figure 7.

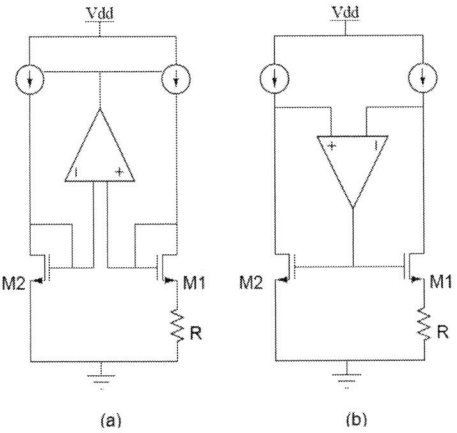

(a)   (b)

Figure 8   The conventional circuit (a) to create a PTAT voltage across resistor R with weakly inverted MOS devices is more sensitive than the circuit in (b) .

B. CMOS Ultra-Low Voltage CTAT

The CTAT part of the reference in Figure 7 is created as the drain-to-source voltage of an NMOS device, M7, operating in the weak inversion region, whose gate and body are tied to the drain. The body-source junction is forward biased effectively lowering the threshold voltage of the device. This configuration is called a dynamic-threshold MOS in [13]. Due to the CTAT dependence of the threshold voltage, the V_{GS} has a CTAT characteristic when the device is biased in weak inversion with a PTAT current. The temperature coefficient of this CTAT voltage will increase with increasing current through the device, as well as with an increased body effect coefficient and transistor W/L ratio.

C. CMOS-only Ultra-Low V_{DD} Reference

The complete schematic of the designed voltage reference is presented in Figure 7. The PTAT current, generated by devices M1 and M2 and resistors R1 and R2, is mirrored to resistor R5 and transistor M7, which create the PTAT and CTAT parts of the voltage reference respectively. Transistors M3-M6, along with M8-M10 and P1-P2, form a low-voltage cascade current mirror, to accurately copy the PTAT current. Transistors M11-M12 form the startup circuit of the reference, ensuring that the transistors M1 and M2 do not operate in zero-current state. Capacitor C_C is used for compensation. In this proof-of-principle design, the same OTAs as for the Schottky reference (Figure 3) have been reused.

D. Experimental Results

A circuit prototype was realized in a standard 90nm CMOS technology. The layout and die photo are shown in Figure 4; the circuit occupies only 0.07um^2.

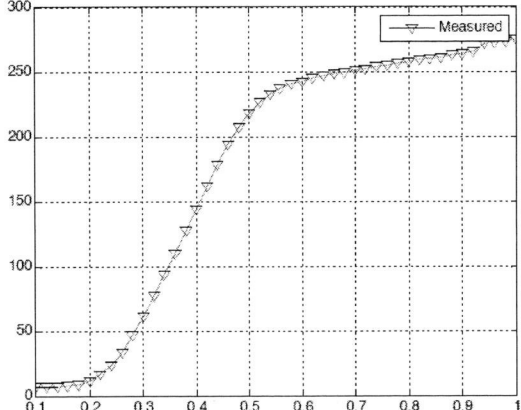

Figure 9   Measured reference voltage V_{REF} w.r.t. supply voltage V_{DD}

Measurements of the MOS-only reference circuit showed its functionality for supply voltages as low as 0.5V. The reference voltage supply dependence is shown in Figure 9. The reference voltage's temperature dependence, shown in Figure 10, is from 241.5 to 245mV when measured from 5 to 100°C, which corresponds toan effective temperature coefficient of 0.04mV/°C or 160ppm/°C, at a V_{DD} of 0.6-V. The nominal output voltage is 256 ± 4mV for 16 measured samples. No trimming was performed on any of the circuits. The total consumed power of this proof-of-principle prototype was 547uW at room temperature.

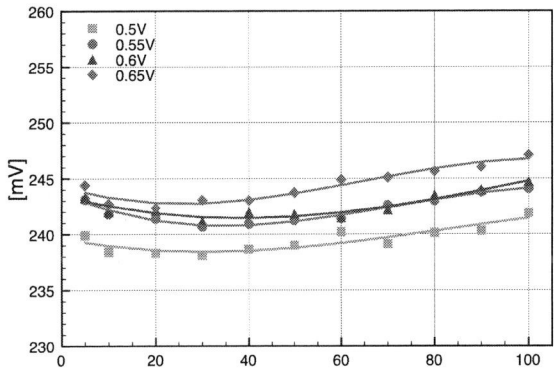

Figure 10   MOS-only reference temperature dependence

V. DISCUSSION AND CONCLUSIONS

Table 1 reviews the different performance characteristics of sub-1V reference circuits, both bandgap based as well as MOS-only based designs. As outlined in the paper several design techniques exist to design references that can operate with a supply voltage significantly below the Si bandgap voltage. In this paper we presented two solutions to further reduce the supply voltage for reference circuits to well below 0.7V and demonstrate functionality down to 0.5V.

Lower reference voltage designs intrinsically have a poorer relative temperature dependence performance (in ppm/°C) than higher reference voltage designs. Still, at this point the performance of the (untrimmed) ultra-low voltage designs cannot match the performance of designs based on more

978-1-4244-2018-6/08 $25.00 © 2008 IEEE   765

established design techniques and well characterized and modeled components. Schottky diodes offer a promising alternative to design ultra-low voltage references. An attractive feature of a Schottky based reference is that the reference voltage can be traced back to a physical parameters, the Schottky barrier, which is primarily dependent on material properties. MOS-only reference designs also offer interesting opportunities, in particular in combination with forward body biasing. Although MOS based designs are often more prone to process variations, experimental data in [13] suggests a better matching behavior for such MOS configuration which can be safely used in an ultra-low voltage context.

The performance of ultra-low voltage references can be expected to significantly improve as more reliable circuit models and device characterization data becomes available to the designer, and as design techniques are further developed and mature driven by the growing need for ultra-low voltage system-on-a-chip ICs.

VI. ACKNOWLEDGMENTS

The authors thank United Microelectronic Corp. (UMC) for chip fabrication and Y. Baeyens and Y.K. Chen of Bell Laboratories for temperature-sweep measurement support.

VII. REFERENCES

[1] "The International Technology Roadmap for Semiconductors (2006 edition)," ITRS, 2006, (Online), http://public.itrs.net.

[2] B. Calhoun, A. Wang, and A. Chandrakasan, "Modeling and sizing for minimum energy operation in subthreshold circuits," *IEEE J. of Solid-State Circuits*, Vol. 40, no. 9, pp. 1778 – 1786, Sept. 2005.

[3] S. Hanson et al., "Ultralow-voltage minimum-energy CMOS," *IBM Journal of Research and Development*, vol. 50, no. 4/5, pp. 469-489, July/Sept. 2006.

[4] J. Rabaey et al., "Ultra-low-power design, the roadmap to disappearing electronics and ambient intelligence," *IEEE Circuits and Devices Magazine*, vol.22, no.4, pp. 23-29, July-Aug. 2006

[5] P. Kinget, "Designing analog and RF circuits for ultra-low supply voltages," *Proceedings of the 33rd European Solid State Circuits Conference (ESSCIRC)*, pp.58-67, Sept. 2007.

[6] R.J. Widlar, "New Developments in IC Voltage Regulators," *IEEE J. of Solid-State Circuits*, vol. SC-16, pp. 2-7, Feb. 1971.

[7] P. K. T. Mok and K. N. Leung, "Design considerations of recent advanced low-voltage low-temperature-coefficient CMOS bandgap voltage reference," in *Proceedings of the IEEE Custom Integrated Circuits Conference*, pp. 635-642, 3-6 Oct. 2004.

[8] H. Banba et al. "A CMOS Bandgap Reference Circuit with Sub-1V Operation," *IEEE J. Solid-State Circuits*, vol. 34, pp. 670-674, May 1999.

[9] K.N. Leung et al., "A Sub-1-V 15-ppm/°C Bandgap Voltage Reference Without Requiring Low Threshold Voltage Device," *IEEE J. Solid-State Circuits*, vol. 37, pp. 526-530, April 2002.

[10] A. Boni, "Op-Amps and Startup Circuits for CMOS Bandgap Referecnces With Near 1-V Supply," *IEEE J. Solid-State Circuits*, vol. 37, pp. 1339-1342, October 2002.

[11] J. Doyle et al., "A CMOS Subbandgap Reference Circuit with 1-V Power Supply Voltage," *IEEE J. of Solid-State Circuits*, vol. 39, no. 1, pp. 252-255, Jan. 2004

[12] G. Giustolisi et. al., "A Low-Voltage Low-Power Voltage Reference Based on Subthreshold MOSFETs," *IEEE J. Solid-State Circuits*, vol. 38, pp. 151-154, Jan. 2003.

[13] A. Annema, "Low-power bandgap references featuring DTMOSTs," *IEEE J. of Solid-State Circuits*, vol. 34, no. 7, pp. 949-955, Jul 1999.

[14] M. Ugajin et. al., "A 0.6-V voltage reference circuit based on Σ-VTH architecture in CMOS/SIMOX," *Symp. VLSI Circuits Dig.*, pp. 141 - 142, June 2001.

[15] B. K. Ahuja et al., "A 0.5uA Precision CMOS Floating-Gate Analog Reference," *IEEE International Solid State Circuits Conference*, 2005, pp. 286-288.

[16] K. N. Ng, "Complete Guide to Semiconductor Devices," McGraw Hill, Inc. 1995.

[17] D. Butler et. Al., "Low-Voltage Bandgap Reference Utilizing Schottky Diodes", *Midwest Symposium on Circuits and Systems*, vol. 2, pp. 1794-1797, Aug. 2005.

[18] Y. Tsividis, R. Ulmer, "A CMOS voltage reference," *IEEE J. of Solid-State Circuits*, vol.13, no.6, pp. 774-778, Dec 1978.

[19] F. Serra-Graels, J. L. Huertas, "Sub-1V CMOS Proportional-to-Absolute Temperature References," *IEEE J. of Solid-State Circuits*, vol. 38, no. 1, Jan. 2003

TABLE I.   PERFORMANCE COMPARISON FOR SUB-1V V_{DD} VOLTAGE REFERENCES

	This work						Doyle [11]	Boni [10]	Leung [9]	Giustolisi [12]	Annema [13]	Ugajin [14]
	Schottky Diode based (untrimmed)			MOS-only (untrimmed)			Bandgap	Bandgap	Bandgap	MOS	MOS	SiMOX
Technology	Standard 90nm CMOS			Standard 90nm CMOS			0.5um	0.18um	0.6um	1.2um	0.35um	N/A
V_{DD} [V]	0.55	0.6	0.65	0.55	0.6	0.65	0.95	0.85	0.98	1.2	0.85	0.6
P [uW]	398*	450*	506*	482*	547*	614*	10	9	17.6	4.3	1.02	100
V_{REF} [mV]	247	251	253	241	243	245	631	500	603	293	650	530
$\Delta V_{REF}/\Delta V_{DD}$ [mV/100mV]	3	3	3	5	3	10	N/A	2.2	2.2	N/A		N/A
TC_{EFF} [mV/°C]	0.07	0.07	0.05	0.03	0.04	0.05	N/A	0.01	0.01	N/A	0.04	0.06
TC_{EFF} [ppm/°C]	270	263	200	150	160	227	17	33	15	119	57	N/A
# Samples	18			16				17	N/A	10	N/A	24
Min /Max V_{REF} [mV]	240 / 251	246 / 253	249 / 255	246 / 264	250 / 263	254 / 265		428.5 / 499.5	N/A	284.5 / 306.1	N/A	514 / 546
Spread [mV]	11	7	6	18	13	11		16	N/A	21.6	4.5	32

* not optimized

IEEE 2008 Custom Intergrated Circuits Conference (CICC)

Design of Bandgap Voltage Reference Circuit with all TFT Devices on Glass Substrate in a 3-μm LTPS Process

Ting-Chou Lu[1], Ming-Dou Ker[1], Hsiao-Wen Zan[2], Chung-Hung Kuo[3], Chun-Huai Li[3],

Yao-Jen Hsieh[3], and Chun-Ting Liu[3]

[1]Institute of Electronics, National Chiao-Tung University, Hsinchu, Taiwan.

[2] Display Institute, National Chiao-Tung University, Hsinchu, Taiwan.

[3]AU Optronics Corporation, Science-Based Industrial Park, Taiwan.

Abstract-A bandgap voltage reference (BGR) circuit designed with the low-temperature polycrystalline silicon (LTPS) thin-film transistors (TFTs) on glass substrate is proposed, which has been successfully verified in a 3-μm LTPS process. The experimental results have shown that the measured temperature coefficient of the new proposed bandgap voltage reference circuit is around 195 ppm/°C under the supply voltage of 10V. The proposed bandgap voltage reference circuit can be applied on precise analog circuits for System-on-Panel (SoP) or System-on-Glass (SoG) applications.

I. INTRODUCTION

Recently, low-temperature poly-Si (LTPS) thin-film transistors (TFTs) technology has been used to fabricate compact and high-resolution displays [1]. LTPS active-matrix liquid crystal displays (AMLCDs) integrated with driver and control circuits on glass substrate have been practically applied in some portable products, such as mobile phone, digital camera, and notebook, etc. The CPU, memory, timing controller, digital-to-analog converter (DAC), and driver buffer had been implemented on glass substrate with the LTPS TFT process [2], [3]. How to integrate more analog and digital circuits on panel is important for SoP (System-on-Panel) or SoG (System-on-Glass) applications.

Voltage reference generators are widely used in analog and digital circuits, such as DRAM, flash memory, analog-to-digital converter (ADC), and so on. The bandgap reference (BGR) circuit is the key design in analog circuits to provide a stable voltage reference with low sensitivity to temperature and supply voltage. So far, many techniques in CMOS silicon processes have been proposed to develop voltage or current references, which can be almost independent of temperature and power-supply voltage.

In CMOS silicon technology, the parasitic vertical bipolar junction transistors (BJTs) or the diodes had been commonly used in the BGR circuits [4], [5]. The main idea is to use the temperature-dependent voltage drop across the diode-connected BJTs (or the diodes) to modulate and stabilize the output voltage. With the negative temperature coefficients (TCs) of the diode-connected BJTs (or the

diodes), a positive TC can be generated by a proper circuit design to compensate the negative TC and to stabilize the output voltage.

The incorporation of BJTs into CMOS technology somehow makes the process control difficult. Therefore, it was also considered to use only MOSFETs in the BGR circuit to simplify the process and to reduce the operating voltage/power. The voltage across MOSFETs is sensitive to temperature only when the MOSFETs are biased in the sub-threshold region. The gate-to-source voltage of MOSFETs in sub-threshold region is strongly dependent on temperature and exhibits a negative temperature coefficient. Some successful demonstrations of BGR circuits realized with MOSFETs in sub-threshold region have been reported in CMOS technology [6], [7]. However, to precisely bias the devices in the sub-threshold region is quite difficult with consideration of process variation.

Although the BGR circuit has been reported to provide a stable output reference voltage in CMOS technology, the LTPS BGR circuit on glass substrate was never reported in the literature. The conventional BGR circuit incorporated with BJTs or diodes is a great challenge for LTPS process since the characteristics of the poly-Si BJTs or the poly-Si diodes are still unknown or lack of reliable control. The use of LTPS TFT devices biased in the sub-threshold region is also not practical because the poly-Si TFT devices often suffer from significant threshold voltage variation. However, the I-V characteristics of LTPS TFT devices have been found to be strongly dependent on temperature when the devices are operated in the saturation region [8], [9]. Therefore, the LTPS BGR circuits can be realized by using only LTPS TFT devices. The particular thermionic-emission characteristic of LTPS TFT devices is first used in this work to design BGR circuit on glass substrate.

In this paper, a method to realize the circuit of bandgap voltage reference in LTPS process is proposed. Without additional laser trimming after fabrication, the new proposed bandgap voltage reference circuit has been verified on the glass substrate with the output voltage V_{REF} of 6.87 V at the room temperature. The temperature coefficient of BGR

978-1-4244-2018-6/08 $25.00 © 2008 IEEE

output voltage is 195 ppm/°C under V_{DD} power supply of 10 V when the temperature varies from 25 °C to 125 °C.

II. TRADITIONAL BANDGAP REFERENCE CIRCUIT IN CMOS TECHNOLOGY

A traditional implementation of bandgap voltage reference circuit in CMOS technology is shown in Fig. 1 [10]. In this circuit, the output voltage (V_{REF}) is the sum of a base-emitter voltage (V_{EB}) of BJT Q_3 and the voltage drop across the upper resistor R_2. The BJTs (Q_1, Q_2, and Q_3) are typically implemented by the diode-connected parasitic vertical PNP bipolar junction transistors in CMOS process with the current proportional to exp (V_{EB}/V_T), where V_T (=kT/q) is the thermal voltage. Under constant current bias, V_{EB} is strongly dependent on V_T as well as temperature. The current mirror (formed by M_1, M_2, and M_3) is designed to bias Q_1, Q_2, and Q_3 with identical current. Then, the voltage drop on the resistor R_1 can be expressed by

$$V_{R1} = V_T \ln(\frac{A_1}{A_2}), \quad (1)$$

where A_1 and A_2 are the emitter areas of Q_1 and Q_2. It is noted that V_{R1} exhibits a positive temperature coefficient when A_1 is larger than A_2. Besides, since the current flows through R_1 is equal to the current flows through R_2, the output voltage of the traditional bandgap voltage reference circuit can be written as

$$V_{REF} = V_{EB3} + \frac{R_2}{R_1}V_T \ln(\frac{A_1}{A_2}). \quad (2)$$

The second item in Eq. (2) is proportional to the absolute temperature (PTAT), which is used to compensate the negative temperature coefficient of V_{EB3}. In general, the PTAT voltage comes from the thermal voltage V_T with a temperature coefficient about +0.085 mV/°C in CMOS technology, which is quite smaller than that of V_{EB}. After multiplying the PTAT voltage with an appropriate factor (R_2/R_1) and summing with V_{EB}, the output voltage V_{REF} of bandgap reference circuit can result in very low sensitivity to temperature.

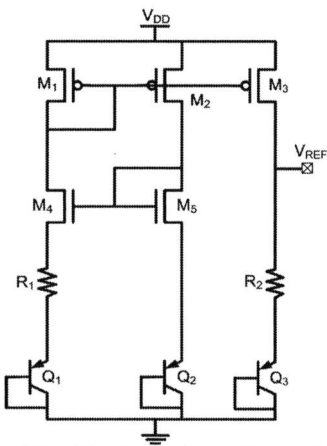

Figure 1. The traditional bandgap voltage reference circuit realized in CMOS technology with parasitic vertical PNP bipolar junction transistors.

III. NEW PROPOSED BANDGAP VOLTAGE REFERENCE CIRCUIT IN LTPS TECHNOLOGY

From the analysis on traditional BGR circuit, it has been known that the realization of BGR circuit strongly depends on the temperature coefficient of BJTs (Q_1, Q_2, and Q_3). In other words, the exponential term exp(V_{EB}/V_T) in the I-V relationship of BJTs makes it possible to obtain a PTAT voltage from the voltage difference between a large-area BJT and a small-area BJT.

For LTPS TFT devices, the drain current I_{DS} of devices operated in the saturation region can be expressed as [11]

$$I_{DS} = \frac{W}{2L}\mu_0 C_{ox}(V_{GS} - V_{TH})^2 \exp(-\frac{V_B}{V_T}), \quad (3)$$

where μ_0 is the carrier mobility within the grain, L denotes the effective channel length, W is the effective channel width, C_{ox} is the gate oxide capacitance per unit area, V_{TH} is the threshold voltage of TFT device, and V_{GS} is the gate-to-source voltage of TFT device. V_B is the potential barrier at grain boundaries which is associated with the crystallization quality of the poly-Si film. Under small V_{GS}, V_B is large. When the V_{GS} increases, V_B decreases rapidly. A typical relationship between V_B and V_{GS} is depicted in Fig. 2, which is exactly measured from an N-type TFT (NTFT) device. When the devices are operated under small V_{GS}, it is found that the drain current I_{DS} of devices is dominated by the exponential term and can be estimated by

$$I_{DS} = W\alpha \exp(-\frac{V_B}{V_T}), \quad (4)$$

where α is treated as a constant under small gate bias (V_{GS}). Then, the equation for V_B can be derived as

$$V_B = V_T \ln(\frac{W\alpha}{I_{DS}}) = \frac{kT}{q}\ln(\frac{W\alpha}{I_{DS}}). \quad (5)$$

When there is a variation of temperature ΔT, the corresponding variation of V_B is

$$\Delta V_B = \frac{k\Delta T}{q}\ln(\frac{W\alpha}{I_{DS}}). \quad (6)$$

From Eq. (6), it can be found that the temperature coefficient (TC) of V_B can be modulated by the channel width. The larger channel width gives rise to the larger TC of V_B.

From Fig. 2, the variation of V_B is related to the variation of V_{GS}. Assuming that the variation of V_{GS} (ΔV_{GS}) is very small, a negative linear approximation can be given between ΔV_B and ΔV_{GS} as

$$\Delta V_{GS} = -\frac{1}{m}\Delta V_B = -\frac{k\Delta T}{mq}\ln(\frac{W\alpha}{I_{DS}}) = -\frac{\Delta V_B \Delta T}{m\Delta T}, \quad (7)$$

where m is the absolute value of the slope under linear approximation as shown in the inset of Fig. 2. The devices biased at small V_{GS} can exhibit a large V_B if its channel width is enlarged. As a result, the LTPS TFT devices with larger channel width exhibit larger absolute value of TC. Such assumption has been verified by the following measurements

978-1-4244-2018-6/08 $25.00 © 2008 IEEE

on the LTPS TFT devices. All the devices are n-type poly-Si TFT devices fabricated in the same run with commercial excimer laser annealing process. The channel length of NTFT devices is fixed as 6 μm and the LDD length is 1.25 μm.

First, the measurement of the TC of diode-connected LTPS TFT devices with channel width of 6 μm is performed by changing the temperature from 25°C to 125°C. Under a constant driving current of 10 μA, the V_{GS} as a function of temperature is plotted in Fig. 3(a). It can be observed that when temperature increases from 25°C to 125°C, the V_{GS} decreases from 1.88 V to 1.66 V. In Fig. 3(a), the temperature coefficient of this TFT device with channel width of 6 μm is approximated as -2.15 mV/°C. Furthermore, the TC of diode-connected LTPS TFT devices with a wide channel width of 30 μm is measured. Under a constant driving current of 10 μA, the V_{GS} as a function of temperature is plotted in Fig. 3(b). As temperature changes from 25°C to 125°C, the V_{GS} decreases from 1.23V to 0.78V, significantly. In Fig. 3(b), the temperature coefficient of this TFT device with 30-μm channel width is approximated as -4.85 mV/°C. As predicted, the LTPS TFT device with a larger channel width exhibits a larger absolute value of TC.

After the evaluation on the TC of poly-Si TFT devices with different channel widths, the new LTPS BGR circuit can be implemented in Fig. 4. In this design, the TFT devices M_1, M_2, M_3, M_4, and M_5 are biased in saturation region. The diode-connected NTFT devices M_6, M_7, and M_8, which replace the diode-connected BJTs in the traditional CMOS BGR circuit (Fig.1) are also biased in saturation region. The nodes n_1 and n_2 are designed to have equal potential by the current mirror circuit.

The channel width of M_6 (W_6) is larger than the channel width of M_7 (W_7), so the TC of M_6 is more negative than the TC of M_7. The voltage drop across the resistor R_1 (V_{R1}) therefore exhibits a positive TC. If the dependence of m on V_{GS} is neglected, the variation of V_{R1} (ΔV_{R1}) as a function of ΔT can be expressed as

$$\Delta V_{R1} = \frac{k\Delta T}{mq}\ln\left(\frac{W_6}{W_7}\right) = \frac{k\Delta T}{mq}\ln N. \quad (8)$$

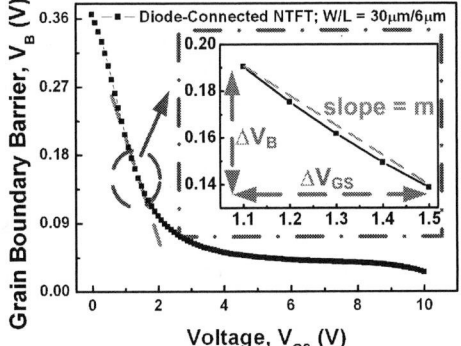

Figure 2. The dependence between potential barrier V_B and gate-to-source voltage V_{GS} of diode-connected N-type TFT (NTFT) device.

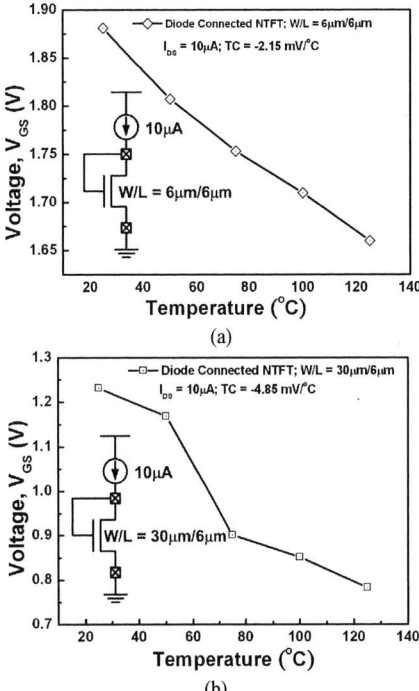

Figure 3. The measured temperature coefficients (TC) of diode-connected TFT devices under a constant drain current of 10μA with (a) channel width of 6μm, and (b) channel width of 30μm, in LTPS process.

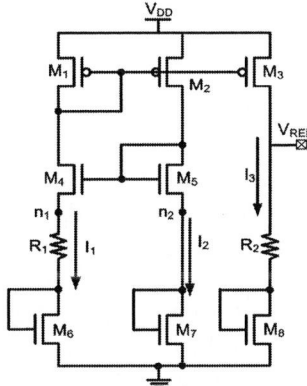

Figure 4. The implementation of the new proposed bandgap voltage reference circuit with all TFT devices in a LTPS process.

Obviously, ΔV_{R1} is proportional to the absolute temperature (PTAT). Hence, a PTAT loop is formed by M_6, M_7, and R_1. The PTAT current variation ΔI_1 can be written as

$$\Delta I_1 = \frac{k\Delta T}{mqR_1}\ln N, \quad (9)$$

where N ($=W_6/W_7$) is the channel width ratio of M_6 and M_7, and V_T is the thermal voltage. The current mirror, which is composed of M_1, M_2, and M_3, imposes equal currents in these three branches I_1, I_2, and I_3 of the circuit. The output voltage (V_{REF}) is the sum of a gate-source voltage of TFT M_8 (V_{GS8}) and the voltage drop across the upper resistor (V_{R2}). Therefore, the output voltage variation (ΔV_{REF}) of the new proposed bandgap reference circuit can be expressed as

$$\Delta V_{REF} = \Delta I_3 R_2 + \Delta V_{GS8} = \frac{R_2}{R_1}\frac{k\Delta T}{mq}\ln N + \Delta V_{GS8}, \qquad (10)$$

where R_1 and R_2 are the resistors shown in Fig. 4. The first item in Eq. (10) with positive TC is proportional to the absolute temperature (PTAT), which is used to compensate the negative temperature coefficient of ΔV_{GS8}. After multiplying the PTAT voltage with an appropriate factor (proper ratio of resistors) and summing with ΔV_{GS8}, the output voltage of bandgap reference circuit can result in very low sensitivity to temperature.

IV. EXPERIMENTAL RESULTS

The new proposed bandgap reference circuit has been fabricated in a 3-μm LTPS technology. The chip photo of the fabricated bandgap reference circuit on glass substrate is shown in Fig. 5. The threshold voltage of TFT devices is about $V_{thn} \approx V_{thp} \approx 1.25$ V at 25 °C. The total gate area of M_6 is 480 μm^2 and that of M_7 is 80 μm^2 in this fabrication. The resistors in this chip, formed by ploy resistors with minimum process variation, are used to improve the accuracy of resistance ratio. The chip size of the fabricated bandgap reference circuit is 400 × 380 μm^2. The power supply voltage V_{DD} is set to 10 V, and the total operating current is 8.97 μA. The measured results of the output voltage V_{REF} are shown in Fig. 6, where the R_2 is drawn with different values in the test chips. As R_2 is equal to 500 kΩ, the measured temperature coefficient of the fabricated bandgap reference circuit on glass substrate is around 195 ppm/°C (without laser trimming after fabrication) from 25 to 125 °C, whereas the output voltage (V_{REF}) is kept at 6.87 V.

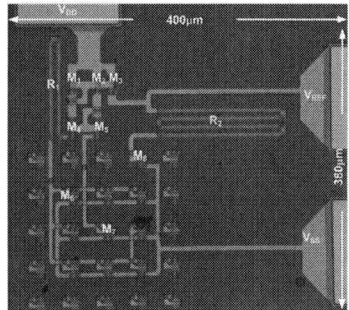

Figure 5. Chip photo with PAD of the new proposed bandgap reference circuit fabricated in a 3-μm LTPS process.

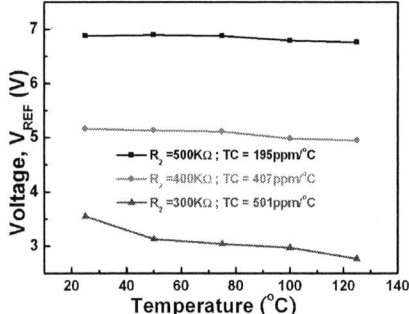

Figure 6. The measured output voltage (V_{REF}) of the fabricated bandgap reference circuit without laser trimming after fabrication.

V. CONCLUSION

The new proposed bandgap voltage reference circuit realized by all TFT devices has been successfully verified in a 3-μm LTPS process. The measurement results of the bandgap voltage reference are V_{REF} of 6.87 V with temperature coefficient of 195 ppm/°C, which consumes an operating current of 8.97 μA under supply voltage of 10 V on glass substrate. The new proposed bandgap voltage reference circuit can be used to realize the precise analog circuits in LTPS process for System-on-Panel (SoP) or System-on-Glass (SoG) applications.

ACKNOWLEDGEMENT

This work was supported by the project (NSC 96-2221-E-009-127-My2) from National Science Council, Taiwan; the project from AU Optronics Corporation, Taiwan; and the project (MOEA-96-EC-17-A-07-S1-046) from the Ministry of Economic Affairs, Taiwan.

REFERENCES

[1] T. Matsuo and T. Muramatsu, "CG silicon technology and development of system on panel," in *SID Tech. Dig.*, 2004, pp. 856–859.

[2] Y. Nakajima, Y. Kida, M. Murase, Y. Toyoshima, and Y. Maki, "Latest development of "System-on-Glass" display with low temperature poly-Si TFT," in *SID Tech. Dig.*, *2004*, vol. 21, No. 3, pp. 864–867.

[3] Y. Nakajima, "Ultra-low-power LTPS TFT-LCD technology using a multi-bit pixel memory circuit," in *SID Tech. Dig.*, 2006, pp. 1185–1188.

[4] K.-N. Leung and K.-T. Mok, "A Sub-1-V 15-ppm/°C CMOS bandgap voltage reference without requiring low threshold voltage device," *IEEE J. Solid-State Circuits*, vol.37, no. 4, pp.526-530, Apr. 2002.

[5] K.-N. Leung, K.-T. Mok, and C.-Y. Leung, "A 2-V 23- A 5.3-ppm/°C curvature-compensated CMOS bandgap voltage reference," *IEEE J. Solid-State Circuits*, vol. 38, no. 3, pp.561-564, Mar. 2003.

[6] G. Vita and G. Iannaccone, "A Sub-1-V, 10-ppm/°C, nanopower voltage reference generator," *IEEE J. Solid-State Circuits*, vol.42, no. 7, pp.1536-1542, July 2007.

[7] G. Giustolisi, G. Palumbo, M. Criscione, and F. Cutri, " A low-voltage low-power voltage reference based on subthreshold MOSFETs," *IEEE J. Solid-State Circuits*, vol. 38, no.1, pp. 151-154, Jan. 2003.

[8] M. Jacunski, M. Shur, A. Owusu, T. Ytterdal, M. Hack, and B. Iniguez, " A short-channel DC spice model for polysilicon thin-film transistors including temperature effects," *IEEE Trans. Electron Devices*, vol. 46, no. 6, pp.1158-1146, Jun. 1999.

[9] T.-C. Lu, H.-W. Zan, and M.-D. Ker, "Temperature coefficient of diode-connected LTPS poly-Si TFT and its application on the bandgap reference circuit," in *SID Tech. Dig.*, 2008, *in press*.

[10] G. Rinconmora, *Voltage Reference from Diodes to Precision High-Order Bandgap Circuits*, Wiley Publishers, pp. 26-28, 2002.

[11] A. Hatzopoulos, D. Tassis, N. Hastas, C. Dimitriadis, and G. Kamarinos, "On-state drain current modeling of large-grain poly-Si TFTs based on carrier transport through latitudinaland longitudinal grain boundaries," *IEEE Trans. Electron Devices*, vol. 52, no. 8. pp. 1727-1733, Aug. 2005.

IEEE 2008 Custom Intergrated Circuits Conference (CICC)

Deep Submicron Effects on Data Converter Building Blocks

W.P.Evans, D.Burnell
Cadence Design Systems
6210 Old Dobbin Lane, Suite 100
Columbia, MD 21045

Abstract–**Physical effects in deep submicron processes can affect reliability, performance, and even functionality in common circuit building blocks used in data converters. NBTI (Negative Bias Temperature Instability), STI (Shallow Trench Isolation) stress, and NWELL proximity effects will be reviewed and examples are given of circuit topologies and layout practices which can minimize the detrimental aspects these effects on commonly used blocks such as comparators, op-amps, and high speed flip-flops.**

I. INTRODUCTION

As technology processing nodes have scaled from sub micron to deep submicron, secondary effects are becoming increasingly significant. While the complexity of SPICE models has evolved to capture these, often the effects are dependent on the subtleties of layout which are not sufficiently anticipated at the time of simulation. Beyond layout, the aging of the device characteristics over lifetime is not included in SPICE models. This paper will discuss some of the more important effects and describe pitfalls in design of common circuit building blocks used in data converters. Circuits designed and laid out without taking these effects into account run the risk of reliability issues, reduced yield and in some cases, non-functional silicon – even when SPICE simulation predicts otherwise.

In this paper, NBTI (Negative Bias Temperature Instability) effects are discussed as they relate to comparator design and the long term reliability of comparators. STI (shallow Trench Isolation) stress effects are described and an example given for their impact on operational amplifier performance. Finally, the NWELL proximity effect is described along with an example that illustrates the vulnerability of a high speed flip-flop used in low power, high speed flash data converters.

II. NBTI AND COMPARATOR RELIABILITY

NBTI describes a reliability issue which is a consideration when designing comparators used in data converters. NBTI stress accelerates hole trapping and interface state generation at the gate-channel interface which increases Vt and degrades mobility. Fig. 1 is an illustration threshold shift after burn in for different oxide thicknesses. Sensitivity is process and foundry dependent. Higher gate-source voltages accelerate this threshold shift with time and in general, PMOS devices

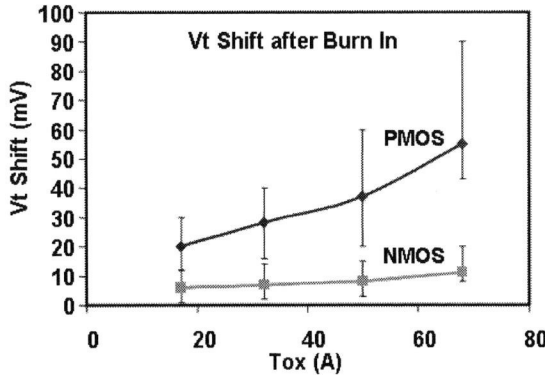

Fig. 1   NBTI Threshold Drift vs. Oxide Thickness for a typical 0.13um process

have a greater sensitivity than NMOS devices. Smaller technology nodes appear increasingly susceptible even for equivalent 2.5V or 3.3V I-O transistors.

The CMOS latches used in comparators are especially susceptible to this threshold drift effect since comparator latches are generally biased with V_{GS} of one transistor at VDD while V_{GS} of the matching transistor sits at 0 volts. Fig. 2 shows a simplified schematic of a basic CMOS latch commonly used in a comparator circuit. Note that in the latched state both NMOS and PMOS latching transistor pairs have different V_{GS} (either 0V or VDD). This bias condition is the worst possible situation for threshold drift with time.

When designing a pipeline data converter, the required accuracy of the comparator thresholds depends on the number of bits in a pipeline stage and the input range of the ADC. For example, a three bit pipeline stage on an ADC with a 1.0V peak-peak differential range would require a minimum comparator accuracy of 1.0V/16 or 62.5mV to avoid major linearity errors. In a similar manner, a four bit stage would require 31.25mV threshold accuracy. The comparator in general cannot consume this entire error budget so the actual comparator offset must be less.

At first glance, these offset requirements do not look challenging. However, if the ADC is burned in when set in a power down mode where the comparator is in the latched state, the offset of the comparator will drift considerably

978-1-4244-2018-6/08 $25.00 © 2008 IEEE

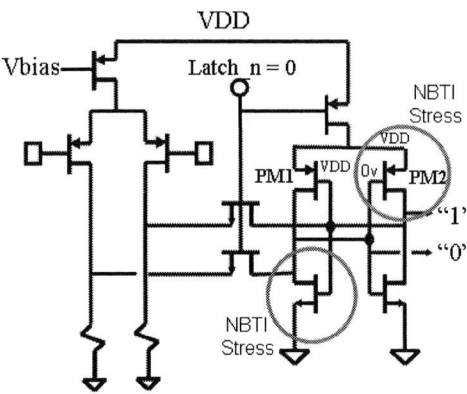

Fig. 2 CMOS latch illustrating possible NBTI offset problems

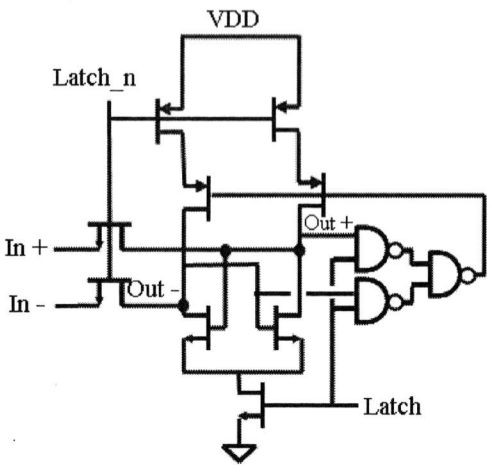

Fig. 3 CMOS dynamic latch with lower sensitivity to PMOS NBTI threshold drift

during a lifetime burn in. Observing the graph in Fig. 1, it isinteresting to note that the offset of the PMOS transistors in the latch could increase by as much as 90mV on some lots if the comparators were made using the 3.3V IO transistors.

This amount of drift would cause linearity degradation in an ADC made with these transitiors if the ADC is burned in with these comparators latched in a given state. Note that if the ADC is burned in when set in a power down mode with the comparator is in the latched state, the offset of the comparator will drift considerably during this lifetime burn in.

Fortunately, once aware of this phenomenon, one can use circuit topologies to minimize its effect. One can take advantage of the fact that NMOS transistors are less susceptible to NBTI shift than PMOS. For example the latch in Fig. 2 could be redesigned with an NMOS input preamplifier stage. If the common mode of the voltage leaving the pre-amp is high enough, the offset of the PMOS transistors in the output latch stage will have little effect on the overall comparator offset. Since the NMOS transistors often exhibit less NBTI drift, this change is often effective.

The latch shown in Fig. 3 was developed for a converter where the latch needed to function with a low input common mode. With this comparator topology, the out+ and out- nodes are precharged to a low common mode voltage in the unlatched state. When the latch command goes high, either out+ or out- will transition to a high state forcing the output of the final NAND gate high and the latched state will be held on the out+ and out- nodes. This latch architecture is less susceptible to NBTI PMOS threshold shift since the V_{GS} on the PMOS devices in the latch which must match are now approximately equal. This latch has been used successfully in a number of 12 bit pipeline ADCs with sample rates from 50 MHz to 100MHz in 0.13um, 0.11um, and 0.80um technologies.

There are several other options that will reduce comparator sensitivity to NBTI. Designing the latch using low voltage, core transistors will result in a smaller NBTI threshold drift.

Using fewer bits in the pipeline stages and adding gain in front of the comparator both reduce sensitivity to Vt drift.

Finally circuitry can be added to remove NBTI stress to critical matched components whenever a converter is in a powered down state. Appropriate switches and logic should be added when possible be sure V_{GS} is set to 0 volts in power down on pairs of transistors which must match.

The key point is comparator and other analog designs need to be analyzed for offset drift due to NBTI effects to avoid lifetime failures. As this requirement is becoming more popular with design customers, EDA tools that simulate NBTI and HCI (Hot Carrier Injection) aging on circuit performance are entering the design flow mainstream. To be useful, these reliability tools need to be calibrated specifically for the devices in each process node. Recently several of the major foundries have begun making calibration files available, particularly at the leading edge 45nm and 65nm nodes. Because aging effects depend critically on voltage stress history and temperature of each device, demonstration of aging margins is becoming a part of design verification.

III. STI STRESS EFFECTS AND AMPLIFIER DESIGN

The STI stress effect refers to the modulation of the mobility and Vt near the transistor edge from the increasing stress in the active transistor silicon upon approaching the dielectric interface. This is also often called the LOD (Length of Diffusion) effect referring to the SPICE parameter name for the length of space between the poly gate and the STI edge. For many years, IC designers have realized that layout has a large effect on critical device matching. Most design houses have developed best layout practices to improve matching and reduce temperature gradient effects. As dimensions have shrunk with deep submicron processes, layout asymmetries which would cause subtle offset errors in the past now can cause complete functional failure of circuits - even though they simulate and function without a problem at the schematic level.

The semiconductor industry began to realize the magnitude of this problem and at the 0.13um technology node as LOD parameters became more prevalent in foundry SPICE models, these effects could be simulated from an extracted layout. It is important for the designer to understand the magnitude and importance of these effects if final silicon is to match simulation.

Fig. 4 is a plot showing an example of how Ids varies as a function of distance from the STI boundary when biased at low current where Vgs is within 200mV of threshold. This has important ramifications in staple circuits of analog design like current mirrors and differential amplifiers. Fig. 5 shows a schematic and sample layout of a current mirror with a ratio of 1:5 which neglects LOD effect. The problem arises because many of the fingers in the 5x device are much further from the STI transistor edge than the fingers in the x1 transistor. From Fig. 4 we see that Ids mismatches as large as 30% can be possible so in extreme cases, our current mirror could be in error by as much as 30%.

An example of a functional failure caused by LOD effects is illustrated in Fig. 6 and Fig. 7 where the Fig. 6 shows the desired bias currents and Fig. 7 shows the actual currents after taking LOD layout effects into account. In this example, a target load current on each drain of the PMOS differential pair is 4mA and as long as the common mode loop can adjust the sum of the NMOS currents in N1 and N3 to this amount, the loop will set up correctly. With an improper layout, LOD effects cause the tail current of the differential pair to drop from 8mA to 7.2mA and the constant load current in N1 and N2 increases to 3.6mA. The common mode loop is now on the edge of failure since N3 and N4 can not drop below 0mA. If this circuit is simulated in a schematic view that doesn't accurately capture the final layout LOD geometry (assuming the SPICE model treats LOD effects), one would never see this problem. To catch this problem in simulation, the designer would have to monitor the bias currents in a post-layout extracted simulation.

As indispensable as SPICE simulations are in today's world,

the first line of defense is a robust layout methodology that takes care to make the environments of matching devices as equivalent as possible with respect to encroaching secondary effects. Catching problems in parasitic extraction is risky and expensive. This methodology can be enhanced by good process design kits (PDKs) allow the designer to specify transistors which need to match and provide links in LVS (layout vs. schematic) to make sure these assumptions are

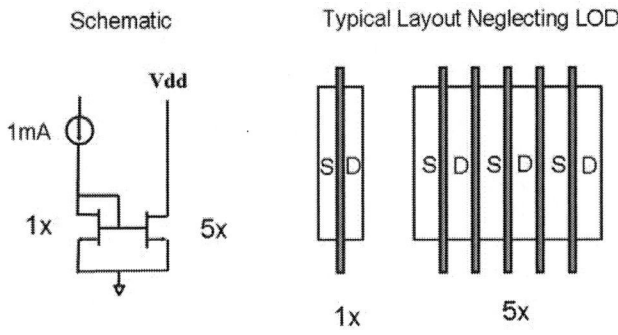

Fig. 5 Current mirror layout structure illustrating LOD mismatch

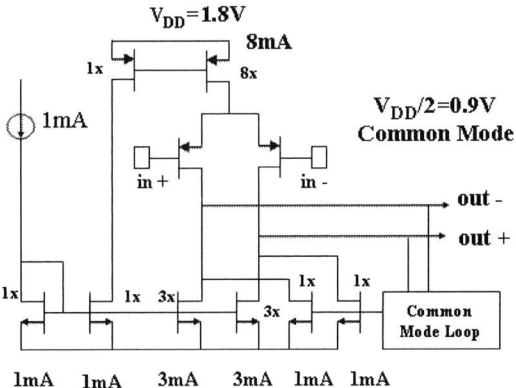

Fig. 6 Expected amplifier bias condition

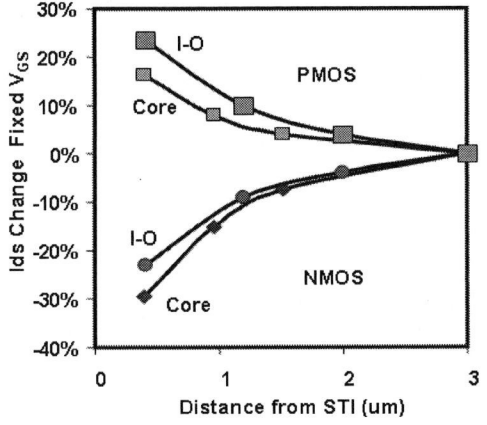

Fig. 4 Ids vs. Distance from STI boundary
Typical 0.13um process

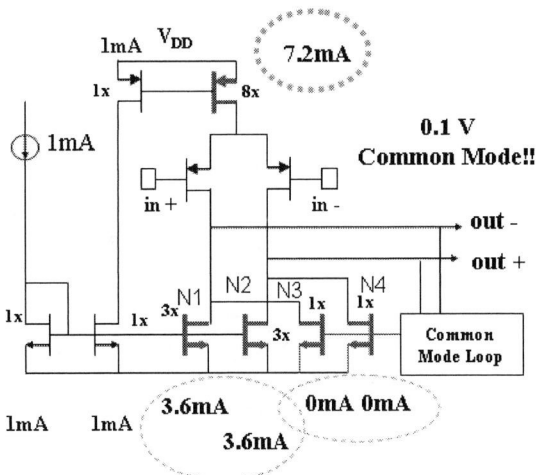

Fig 7 Bias condition after taking into account LOD effect

978-1-4244-2018-6/08 $25.00 © 2008 IEEE

realized in layout. A simple approach might be to check and assure LOD distances (Sa and Sb) are greater than a target spacing for matched transistors, typically 2um to 3um.

Efficient layouts can minimize asymmetries even further. Fig. 8 shows an example layout of the same current mirror in Fig. 6 which is has been optimized for matching in a minimum area. Here dummy fingers are added on each end of the arrayonly once for the transistor pair. Also the "master" transistor is embedded roughly one quarter of the distance into the slave mirror transistor to get closer to the average slave gate. Contributors to the variation of the STI stress effect that aren't captured in SPICE are the differences in the surrounding environment of matching transistors. Minimum STI spacing between neighboring transistors will create less stress than wider STI spaces, so good layout practices consider matching the local environment on all sides of matching transistors.

III. Nwell Proximity Effects

The Nwell proximity effect refers to a dopant pile up near the Nwell and Pwell boundary which shifts Vth (more enhancement) within ~2um of the well edge for both NMOS and PMOS devices. Foundry models at 90nm nodes and below generally include hooks for this effect, but capturing an accurate layout description of all the contributing NWell distances is difficult in most design flows for most PDKs.

Fig. 9 shows how this effect can change thresholds at distances as far as 2um from this boundary. Fig. 10 shows the layout and schematic of the input stage of a flip-flop used in a high speed flash converter which was an early victim of this effect before it was modeled correctly. In this example (similar to to flip-flop described in [1]) the close spacing between the Nwell surrounding the PMOS transistors caused a threshold shift in transistors N1 and P1 and P2. The Well proximity decreased the strength of the PMOS transistors much more than the NMOS transistors, which in turn caused the flip-flop to stop operating correctly low supply voltages contrary to SPICE simulation.

This problem was uncovered by comparing silicon results with simulations using modified BSIM4 models to take into account the Nwell proximity and LOD effects.

Once these effects and the parasitic capacitance were accounted for, the flip-flop simulation correctly predicted the fall out at low supply voltages.

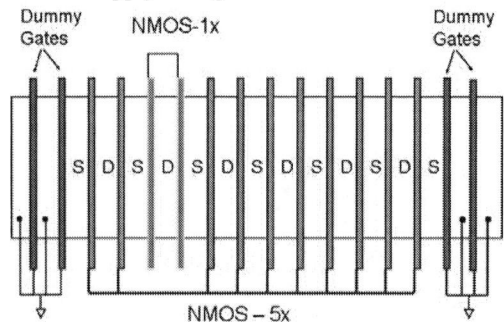

Fig. 8 Layout to minimize LOD mismatch errors

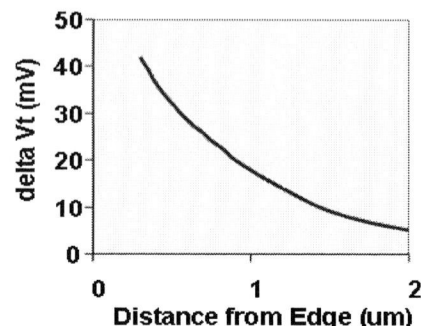

Fig. 9 Shift in threshold vs. distance from Nwell Pwell boundary

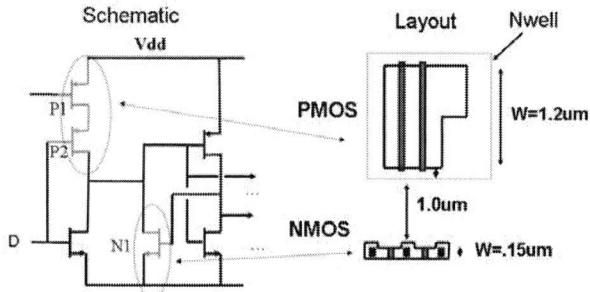

Fig. 10 Front stage of flip-flop illustrating Nwell layout induced failure

IV. Conclusion

STI stress, Nwell Proximity and NBTI are phenomena that can effectively change transistor performance by equivalent Vt shifts of 10-60mV depending on the severity of layout asymmetry and voltage stress. As feature sizes shrink, 2um buffers for Nwell and PWell enclosures and LOD lengths become increasingly onerous. Designing within the sphere of their influence is a practical necessity, and good layout methodology must accommodate that. Today's challenge for circuit designers and PDK (Process Design Kit) developers is to develop design flows and layout methodologies which allow designers to simulate circuits at the schematic level which anticipate the final layout and verify compliance. Additionally NBTI aging effects need to be considered along with process variations and wafer variations to assure long term reliability. Circuits can be designed to minimize the NBTI stress, or aging margins need to be determined and allocated.

References

[1] Q. Huang and R. Rogenmoser, "Speed Optimization of Edge-Triggered CMOS Circuits for Gigahertz Single-Phase Clocks," IEEE JSSC VOL 31 No.3 March 1996 pp456-465

[2] P. Drennan, M. Kniffin, D. Locascio, "Implications of Proximity Effects for Analog Design" *Conference 2006, IEEE Custom Integrated Circuits*, 10-13 Sept. 2006, pp169-176

[3] R. Vattikonda, W.Wang, Y Cao, "Modeling and minimization of PMOS NBTI effect for robust nanometer design", *Design Automation Conference, 2006 43rd ACM/IEEE*, 24-28 July 2006 pp 1047-1052

Advanced Subsystems for Connectivity and Cellular Radio

Session 24

Chair: Stefan Drude, NXP Semiconductors
Co-Chair: Earl McCune, Panasonic

With the continued pressure on cost and performance for mobile communication devices, research on new approaches for subsystems and building blocks continues. In this session a number of examples are given for latest developments in the areas of receiver design, power amplifiers, and frequency synthesis.

In CDMA systems the uplink and the downlink channels are active simultaneously. The first paper in this session addresses the associated challenge of transmitter power leakage into the local receiver front-end. A new solution is proposed that avoids the use of SAW band-roofing filters commonly found in CDMA receiver designs. A combination of a low noise amplifier with high linearity and a dedicated passive mixer structure with integrated filtering achieves the desired intermodulation performance. Test results presented for this 0.18 μm CMOS design demonstrate its advantages over a more conventional receiver structure.

In the second paper an 8-bit envelope modulator is presented which is part of a polar transmitter. The target application for this envelope modulator is Bluetooth and wireless LAN. Implemented in 0.65 μm CMOS technology this design operates primarily in the 2.45 GHz frequency range. Compared to earlier designs this proposal achieves a higher output power of 16.7 dBm and drain efficiency of 24% in WLAN mode, while maintaining a total EVM near 2.7%.

The polar transmitter architecture is also the topic of the next paper, in which a pulsed RF approach is presented to implement amplitude modulation. Noise shaping techniques are applied in the modulator to increase the resolution and to reduce out-of-band noise.

Ahead of the break, our invited paper in this session gives a summary of linearity and efficiency enhancement strategies for 4G wireless power amplifier designs. In order to meet cost and performance targets for 4G power amplifier designs, only highly integrated solutions should be considered as serious contenders. A number of strategies are reviewed including DC bias adjustment with required transmitter output power levels, envelope elimination and restoration, envelope tracking. Finally, a number of linearization techniques are covered in the presentation, which aim at increasing the achievable output power levels for given performance requirements.

Advanced designs for frequency synthesis is the topic after the break. First is a combination of an all digital phase locked loop frequency synthesizer with an analog feed-forward path. The proposed circuit aims at providing a phase modulation path which is insensitive to quantization and non-linearities of the digital controlled oscillator (DCO). Additionally a self-calibration function is proposed to improve robustness against variations in analog components, switching noise, and DCO gain.

The last paper presents a noise shaping technique for a frequency synthesizer, which employs $\Sigma\Delta$ modulation and fractional-N techniques to meet WCDMA and HSDPA requirements. A test chip in 180 μm CMOS is developed to demonstrate how a hybrid FIR filter can implement the required noise shaping, while operating under the constraints of low power requirements. The experimental results show that the proposed method is useful to meet the phase noise mask requirements for various RF applications including WCDMA/HSDPA standards.

978-1-4244-2018-6/08 $25.00 © 2008 IEEE

Notes

A Highly Linear SAW-less CMOS Receiver Using a Mixer with Embedded Tx Filtering for CDMA

Namsoo Kim, Lawrence E. Larson, Vladimir Aparin*

University of California at San Diego, 9500 Gilman Drive, La Jolla, CA 92093, USA

* Qualcomm, 5775 Morehouse Drive, San Diego, CA 92121, USA

Abstract- An embedded filtering passive receiver mixer is used to overcome transmitter power leakage without the use of a SAW filter. The receiver IC exhibits more than +60dBm of Rx IIP$_2$, 2.4dB Rx NF, and +77dB of Triple Beat (TB) with 45MHz transmit leakage at 900MHz Rx frequency while consuming only 18mA from a 2.1V supply.

I. INTRODUCTION

Zero-IF RF front-end architectures [1] are attractive for cellular systems due to lower cost and Bill-of-Material (BOM). But the external SAW filter after the LNA stage has been an essential component of cellular systems for several reasons. The two main reasons for using an external SAW filter in full duplex communication systems, like CDMA and WCDMA, are Triple Beat (TB) and IIP$_2$ performance requirements at the transmit (Tx) offset frequency. Tx power can leak to the LNA despite the duplexer isolation between the Rx and Tx band. The typical duplexer isolation is 55dB in the CDMA Cellular band, and the maximum Tx power can be as high as +27dBm, resulting in -28dBm of Tx power at the Rx input port. This strong Tx power can cause well-known cross-modulation distortion (XMD) [2].

This distortion problem is usually dominated by the LNA performance, because Tx rejection by an external SAW filter reduces the mixer TB requirement. More importantly, the reduced Tx power at the mixer input reduces the IIP$_2$ performance concern at the Tx offset frequency, since the second-order distortion at the Tx frequency offset can increase the noise floor of the receiver in a Zero-IF system.

A SAW–less receiver system is desirable since it eliminates the SAW filter as well as the external matching components. There have been several efforts to implement a SAW-less CDMA receiver. A Tx canceller was reported in [3], which used an LMS adaptive filter requiring up/down conversion mixers and a low-pass filter in the loop. This method suffers from several performance problems. First, the NF of the receiver is degraded due to the operation of the LMS loop. Second, the rejection varies depending on the group delay of the external matching network. Third, the TB performance of the overall system can be degraded due to the LMS loop.

An on-chip Tx reject band-pass filter using bond-wire inductors is reported for a WCDMA system in [4]. This method has benefits of saving area compare to an on-chip inductor and increasing the selectivity of the filter due to the high-Q of the bond-wire. But this method may have limited feasibility in real production due to bond wire variations

In this paper, an embedded filtering receiver is introduced.

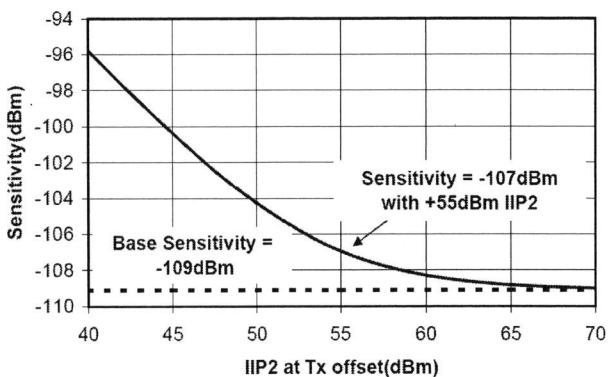

Fig. 1. IIP$_2$ performance vs sensitivity.

The system does not require a SAW filter and obtains an additional 15dB of Tx rejection; it shows more than +60dBm of receiver IIP$_2$ at Tx offset and +77dB TB. The double side-band (DSB) NF is 2.4dB and the total current consumption is 18mA from 2.1V supply.

II. REQUIREMENT OF SAW-LESS RCEIVER FOR CDMA

A. Sensitivity Specification

As explained in Section I, the IIP$_2$ performance at the Tx frequency offset degrades the sensitivity. The trade-off between IIP$_2$ and sensitivity with -28dBm of Tx leakage power is shown in Fig. 1. The relationship can be defined as

$$Sensitivity = 10 \cdot \log_{10} \left[10^{IM_{2,Tx}/10} + \kappa \cdot T \cdot B \cdot 10^3 \cdot 10^{NF_{ant}/10} \right] + CNR \tag{1}$$

where, κ is Boltzmann's Constant, T is absolute temperature, B is signal bandwidth (1.23MHz in CDMA), CNR is Carrier to Noise Ratio, the NF_{ant} is the noise figure referred to the antenna, and $IM_{2,Tx}$ is Tx IM$_2$ power in the Rx band. As can be seen from Fig. 1, the sensitivity is highly dependent on the Tx IIP$_2$ performance, and a +55dBm Tx IIP$_2$ is required for a 2dB sensitivity degradation. A +55dBm receiver IIP$_2$ - without some form of Tx rejection - is extremely challenging due to the high LNA gain at the Tx frequency.

B. Single Tone De-sense (STD) Specification

When the Tx power leaks to the Rx port, and a jammer is present in the vicinity of the Rx band, cross-modulation distortion (XMD) becomes the key determiner of the linearity and phase noise requirements of the receiver. The receiver needs to have +8dBm of effective IIP$_3$ to avoid corrupting the STD test defined in the CDMA standard [5].

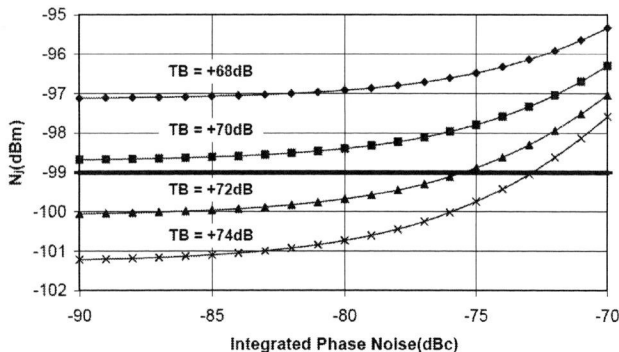

Fig. 2. Trade-off between phase noise and TB performance.

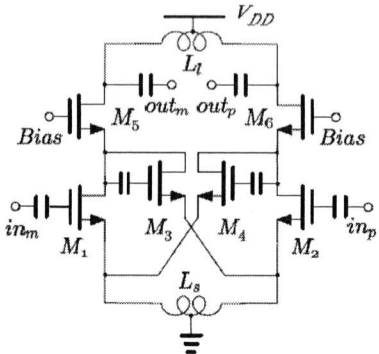

Fig. 3. Simplified schematic of differential APD LNA.

With a SAW filter between the LNA and mixer, the STD mainly depends on the linearity of the LNA, since the SAW filter will reject the Tx leakage by 35dB. On the other hand, a SAW-less receiver places an additional linearity burden on the mixer and the following stages. In addition, the phase noise requirement of the VCO at the jammer offset becomes extraordinarily difficult to meet because of reciprocal mixing. The receiver noise floor with jammer can be expressed as

$$N_j = 10 \cdot \log_{10} \left[10^{(P_j - TB)/10} + 10^{(P_j + P_{phase})/10} + \kappa \cdot T \cdot B \cdot 10^3 \cdot 10^{NF_{ant}/10} \right] \quad (2)$$

where, P_j is jammer power at antenna (in dBm), P_{phase} is phase noise integrated over signal bandwidth with the center frequency at jammer offset (in dBc), and N_j is noise floor with jammer. The TB can be measured by applying three tones, two Tx tones and one jammer tone. Then, the resulting XMD tone will appear at the base-band. The TB is defined as the difference between jammer tone and the XMD tone. The trade-off between phase noise and TB performance is shown in Fig. 2. It shows the phase noise requirement for -99dBm of N_j with various TB performances at -30dBm of jammer power [5]. The noise floor with jammer can never meet -99dBm if the TB is +68dB. But the phase noise can be relaxed to -75dBc with +72dB of TB.

III. SAW-LESS RECEIVER DESIGN

A. CMOS LNA Design

Due to the high IIP_3 requirement of the LNA, the conventional source degenerated LNA is not suitable. There are various ways of designing a highly linear LNA, such as the modified derivative super-position method (MDS) [6] and the active post-distortion method (APD) [7]. In this design, the APD method is chosen. Using this method, the complexity of the bias circuitry and the related input parasitic capacitance can be reduced. The simplified schematic of the LNA is shown in Fig. 3. M_1, M_2, M_5, and M_6 form the main signal path while M_3 and M_4 act as IM3 cancellers.

The LNA adopts a differential architecture. A differential LNA provides several advantages compared to its single-ended counterpart. First, there is no need for an active or passive balun to connect the doubly-balanced mixer after the LNA. An active balun will cause additional current consumption and linearity degradation and a passive balun

will create area and noise figure penalties due to the passive circuit losses. The differential design requires more external input matching components, but the overall reduction of external components is still significant. The detailed analysis of the linearity and NF of the APD method in a differential LNA is described in [8].

B. Embedded Filtering Passive Mixer Design

In a SAW-less receiver, the mixer becomes the critical component in terms of the linearity performance of the receiver.

The mixer is based on a doubly-balanced passive approach. A passive mixer gives better linearity and NF performance, especially in narrow-band communication systems [9]. The flicker noise (1/f) of the mixer can corrupt the integrated noise, but the passive mixer will not introduce significant flicker noise, since there is no dc current.

The impedance seen at the LNA output is quite high, due to the relatively high impedance of the inductor at the LNA output. The low impedance of the mixer input ensures that the RF current flows to the mixer input with minimal loss. The IM2 component generated in the LNA can be blocked by a coupling capacitor between the LNA output and the mixer input. Hence, there will be almost no voltage signal swing at the LNA output due to the low impedance provided by the mixer. The nonlinearity of the mixer plays a role after the Tx signal is down-converted.

The problem of using a conventional passive mixer with a transimpedance amplifier (TIA) at the output is illustrated in Fig. 4. The finite opamp gain-bandwidth causes a large signal swing at the TIA input and introduces nonlinearity from both

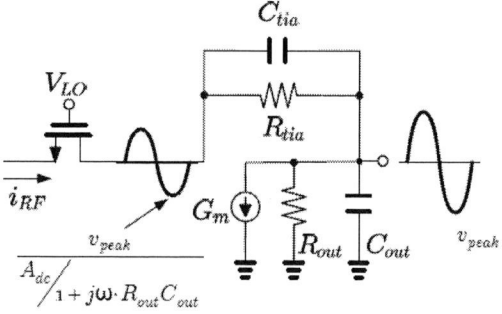

Fig. 4. The problem of using a conventional passive mixer and a TIA with a large Tx leakage signal.

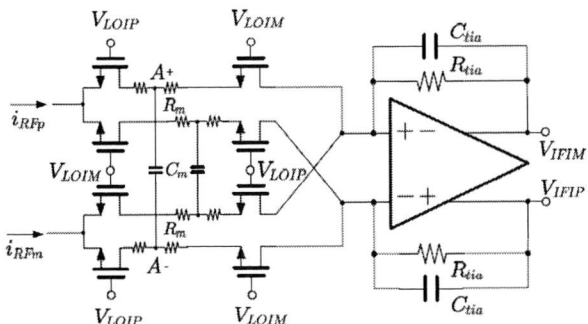

Fig. 5. Simplified schematic of embedded filtering passive mixer.

the Mixer and the TIA. Even if the TIA has a high gain-bandwidth product, it has to provide tremendous current into C_{tia} to avoid introducing nonlinearity caused by incomplete capacitor charging.

The simplified schematic of the embedded filtering passive Mixer (EFP Mixer), which avoids this problem, is shown in Fig. 5. The EFP Mixer has two switches connected in series, and each switch is controlled by opposite phases of the local oscillator (LO) signal. In between the switches is a series RC filter, which determines the bandwidth of the filtering. The RF current flows into the first set of switches, and is down-converted as a nominal passive mixer. The current at node A and the voltage across C_m, when the positive LO signal (V_{LOIP}) is high, can be expressed by

$$i_A(t) \approx g_{m,LNA} \cdot A_{rf} \cdot e^{-1/\tau \cdot 2 f_{LO}} \cdot \left[\cos(\omega_{rf}t) + \frac{2}{3\pi} \cdot \cos(2\omega_{rf}t) \right. \tag{3}$$
$$\left. - \frac{2}{15\pi} \cdot \cos(4\omega_{rf}t) \cdots \right] \cdot LO(t)$$

$$v_{C_m} = \frac{1}{C_m} \cdot \int i_A(t)dt \cdot \left[1 + \left(\frac{f_{offset}}{f_{3dB}} \right)^2 \right]^{-\frac{1}{2}} \tag{4}$$

where, $\tau = R_m C_m$, $f_{3dB} = (2\pi\tau)^{-1}$, A_{rf} is the amplitude of the RF input voltage, and $g_{m,LNA}$ is the transconductance of the LNA. This voltage generates a current into the transimpedance amplifier, inversely proportional to R_m, when the negative LO signal (V_{LOIM}) is high, and the integration in (4) provides a low-pass filtering action.

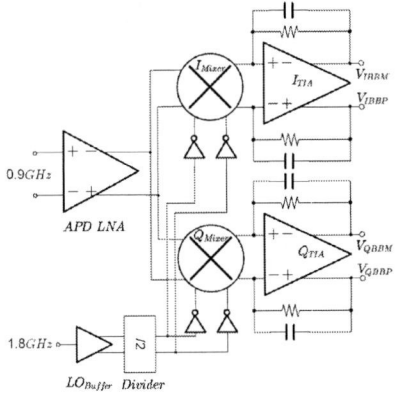

Fig. 6. Block Diagram of embedded filtering SAW-less receiver.

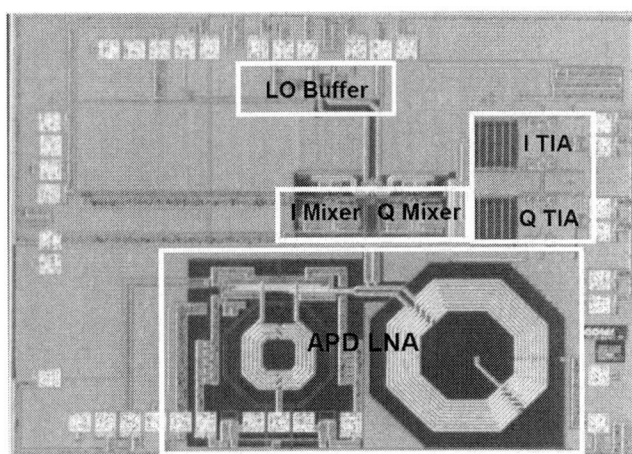

Fig. 7. Die-photograph of embedded filtering receiver.

IV. MEASUREMENT RESULTS AND DISCUSSION

The block diagram of the receiver is shown in Fig. 6 and a die photograph is shown in Fig. 7. It consists of a differential APD LNA, embedded filtering passive Mixer, TIA, LO buffer, divider, and input LO buffer. A conventional passive mixer version of the receiver was fabricated at the same time, in order to have a comparison with the EFP mixer.

The measured frequency responses for both receivers with the differing mixers are shown in Fig. 8. The measured gain performances are 42dB for EFP mixer and 44dB for conventional mixer, respectively. The rejection is normalized since there is 2dB of gain difference. The TIA provides 1.5MHz of 3dB cutoff frequency and the EFP mixer is designed to have an additional pole at 10MHz. As can be seen, the proposed mixer has 15dB more rejection at 45MHz offset, which is the Tx offset frequency in the CDMA CELL band. The performance of the conventional design illustrates the mixer limitation issue mentioned in Section III. Due to the finite opamp gain-bandwidth, the TIA cannot provide sufficient rejection at a high frequency offset.

The TB performance of the proposed receiver is shown in Fig. 9. Two Tx tones with -31dBm each at 45MHz offset and a jammer at 1MHz offset with -30dBm of power are applied. With the cancellation path on, the TB tone is at -65.8dBm, which implies 77.8dB of TB performance. With the APD cancellation path off, the TB is 65.3dB. This difference between cancellation ON and OFF implies two important facts. First, the APD method improves the TB by 12.5dB. Second, the TB variation with the cancellation ON and, then OFF means the mixer does not contribute *any* TB, otherwise the TB would not vary with the cancellation being turned ON and OFF, since it would be dominated by the mixer. Therefore, the additional rejection provided by EFP mixer dramatically improves system linearity performance.

The measured IIP$_2$ performance at the Tx offset frequency is shown in Fig. 10. The conventional passive Mixer starts showing a 4-to-1 nonlinearity, which is strongly nonlinear behavior, from -30dBm. The proposed EFP mixer, whereas, starts showing strongly nonlinear behavior above -24dBm of power.

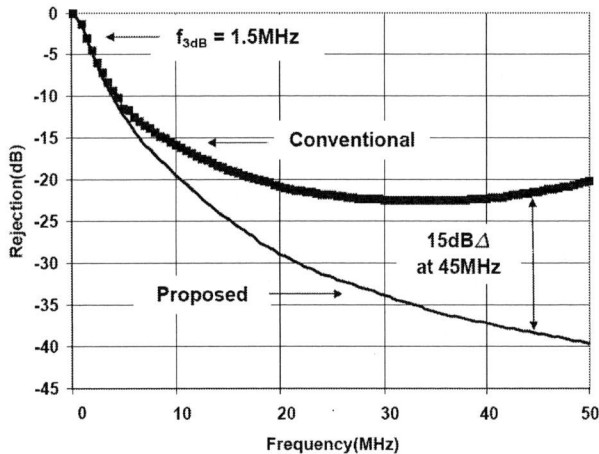

Fig. 8. Measured frequency response comparison.

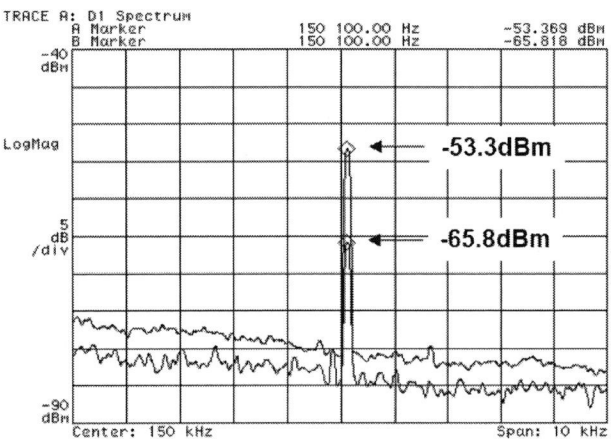

Fig. 9. Measured TB with and without cancellation path in APD LNA.

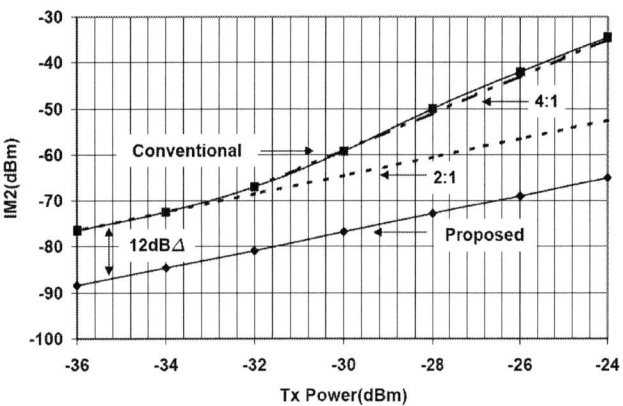

Fig. 10. Measured IIP2 performance comparison at Tx offset (45MHz).

The IIP_2 performance with the EFP mixer is +60dBm and +50dBm with the conventional passive mixer. The conventional mixer exhibits strongly (4:1) nonlinear behavior at -30dBm, which is a key power level in the specification. As a result, the IIP_2 performance of the conventional receiver is further degraded by 2dB.

The overall performance comparison is depicted in Table I. Each receiver is measured with the APD cancellation ON and

Table. I: Receiver Performance Comparison

	Proposed Rx	Conventional Rx
Vdd	2.1V	2.1V
Idd	18mA/17mA[1]	18mA/17mA
VSWR	<2:1	<2:1
Voltage Gain	42/44dB	44/46
Noise Figure	2.4/3.4dB	2.0/2.8dB
IIP_2 at 45MHz	+60/+65dBm	+50/+55dBm
Triple Beat	+77.8/+65.3dB	+47/+47dB
Tx Rejection	37dB	22dB

1 Cancellation ON/OFF

OFF. The Tx IIP2 performance is important when there is no jammer present. The APD cancellation path can be turned OFF with this situation. The cancellation path needs to be ON only when the jammer is present. The EFP mixer has less gain and higher NF than a conventional mixer due to additional loss in the second set of switches.

The total power consumption for the signal path is 18mA, 14mA for the APD LNA and 4mA for the I/Q TIAs. The chip is fabricated in a 0.18μm CMOS process with 5 metal and 1 poly layer (5M1P). The total area is 2.25mm^2 including all the related pads and ESD circuitries.

V. CONCLUSION

A highly linear SAW-less receiver is introduced. The system performance trade-off with and without a SAW filter is explained. The high IIP_2 of the receiver at the Tx offset is essential to avoid corrupting the system's sensitivity performance, and a high TB is required to avoid STD performance degradation due to transmitter leakage. Thanks to the embedded filtering in the mixer and the APD method in the LNA, the required high linearity is achieved with low noise figure and power consumption.

ACKNOWLEDGMENTS

The authors gratefully acknowledge the assistance and support of Qualcomm and UCSD Center for Wireless and Communications (CWC).

REFERENCES

[1] V. Aparin, et al., "A fully-integrated highly linear zero-IF cellular CDMA receiver," *IEEE ISSCC Dig. Tech. Papers*, pp. 324-325, Feb. 2005.

[2] V. Aparin, L.E. Larson, "Analysis and reduction of cross-modulation distortion in CDMA receivers," *IEEE Trans. Microw. Theory Tech.*, vol. 51, no. 5, pp. 1591-1602, May 2003.

[3] V. Aparin, G.J. Ballantyne, C.J. Persico, A. Cicalini, "An integrated LMS adaptive filter of TX leakage for CDMA receiver front ends," *IEEE J. Solid-State Circuits.*, vol. 41, no. 5, pp. 1171-1182, May 2006.

[4] H. Khatri, P.S. Gudem, L.E. Larson, "Integrated RF interference suppression filter design using bond-wire inductors," *IEEE Trans. Microw. Theory Tech.*, accepted for publication, 2008.

[5] CDMA 2000 Standard TIA/EIA IS-2000 Series Rev. A.

[6] V. Aparin, L.E. Larson, "Modified derivative superposition method for linearizing FET low-noise amplifiers," *IEEE Trans. Microw. Theory Tech.*, vol. 53, no. 2, pp. 571-581, Feb. 2005.

[7] N. Kim, V. Aparin, K. Barnett, C.J. Persico, "A cellular-band CDMA 0.25μm CMOS LNA linearized using active post-distortion," *IEEE J. Solid-State Circuits.*, vol. 41, no. 7, pp. 1530-1534, July 2006.

[8] N. Kim, V. Aparin, K. Barnett, "Differential Amplifier with active post-distortion linearization," *U.S. Patent 20070229154*

[9] B. Razavi, "Design considerations for direct-conversion receivers," *IEEE Trans. Circuits Syst. II, Analog Digit. Signal Procss.*, vol. 44, no. 6, pp. 428-435, Jun. 1997.

IEEE 2008 Custom Intergrated Circuits Conference (CICC)

High-power Digital Envelope Modulator for a Polar Transmitter in 65nm CMOS

Manel Collados
NXP Semiconductors Research
High Tech Campus 37
5656AE Eindhoven, The Netherlands

Paul T.M. van Zeijl
Philips Research
High Tech Campus 37
5656AE Eindhoven, The Netherlands

Nenad Pavlovic
NXP Semiconductors Research
High Tech Campus 37
5656AE Eindhoven, The Netherlands

Abstract—In this paper, an 8-bit envelope modulator as part of a digital polar transmitter for Bluetooth and WLAN is demonstrated. The modulator performs digital-to-analog conversion, up-mixing and power amplification, allowing for an area-efficient, fully-integrated transmitter architecture. The circuit delivers 24.8dBm peak-power and 16.7dBm WLAN OFDM power with a mean EVM of 2.7% at 2GHz. The measured peak-power drain efficiency is 51% while the efficiency for OFDM is 24%. In Bluetooth EDR mode a 19.7dBm signal with 5%-rms, 13%-peak EVM, and 26% drain efficiency has been measured. The circuit is fabricated in CMOS 65nm with 50Å thick-oxide devices and 2.5V power supply.

I. INTRODUCTION

The widespread use of CMOS technology in the design of Bluetooth, WLAN, and most recently cellular radios, has been accompanied by a digitization trend in traditionally analog RF blocks [1], and [2]. The adoption of CMOS has encouraged designers to embrace a more digital approach in or order to extract the best performance out of the technology.

Another trend in mobile communications has been the increase of standards and operation modes that have to be supported by the same device. This has triggered the concept of software defined radio (SDR). According to this principle, it is desirable to design radios which can cover as many standards as possible provided that performance is not compromised. In our case, we aim for a transmitter that covers both Bluetooth (BT) and WLAN in the 2.45GHz ISM band.

Cost reduction by means of integration and area reduction has always been a powerful driving force in the semiconductors industry. Area reduction not only comes from using a smaller technology, but also from combining as many functions as possible in the same block. This is in line with the transmitter concept presented here, which combines digital-to-analog converter, mixer, and power amplifier (PA) in one circuit.

Finally, battery life issues have brought a renewed interest for polar modulation techniques in the quest for better overall efficiency. Polar techniques based on PA supply modulation are theoretically very attractive, but fall short in the bandwidth which they can efficiently achieve, [3]. The approach here presented does not have these bandwidth limitations, and therefore can be used for wideband modulations like WLAN.

II. POLAR TRANSMITTER ARCHITECTURE

Most non-constant envelope wireless transmitters are based on the Cartesian decomposition of the baseband signal, like the one shown in Fig. 1. The upper digital-to-analog converter (DAC) processes the in-phase component $x[n]$, while the lower one processes the quadrature component $y[n]$. Each DAC is followed by a low-pass filter (LPF), which removes non-desired spectrum components. After filtering, the two signals are up-mixed and subsequently added, producing the RF modulated signal. The RF signal is then amplified by a pre-driver, before being amplified by a, mostly external, PA. The transmitted signal can be written in function of its quadrature components as

$$s(t) = A\left(x(t)\cos(\omega_c t) - y(t)\sin(\omega_c t)\right) . \quad (1)$$

Another increasingly popular transmitter architecture is based on the polar decomposition of the baseband signal. The phase $\phi(t)$ and envelope $r(t)$ can be calculated from the in-phase $x(t)$ and quadrature $y(t)$ components as

$$r(t) = \sqrt{x^2(t) + y^2(t)} \quad (2)$$

$$\phi(t) = \arctan\left(\frac{y(t)}{x(t)}\right), \quad (3)$$

and the transmitted signal can be written as

$$s(t) = A\,r(t)\cos\left(\omega_c t + \phi(t)\right) . \quad (4)$$

Polar transmitters modulate the phase of the carrier and subsequently change the amplitude of the phase-modulated carrier according to the envelope signal. This allows for the utilization of more efficient PAs. In the digital polar architecture sketched in Fig. 2, the phase-modulated carrier is obtained by changing the instantaneous frequency of a digitally-controlled oscillator (DCO). Instantaneous frequency deviations are calculated as the difference between consecutive phase samples divided by the sampling period. On the other hand, the amplitude is modulated using an envelope DAC (EnvDAC), [4]. The EnvDAC is an array of RF-current sources that are switched on or off depending on the amplitude value. Notice that $r[n]$ and $f_i[n]$ directly control the envelope and frequency modulators. Since there is no reconstruction filter in the envelope path, a bandpass filter might be required to attenuate the envelope aliases. The aliases coming from the phase path are lesser of a

978-1-4244-2018-6/08 $25.00 © 2008 IEEE 781

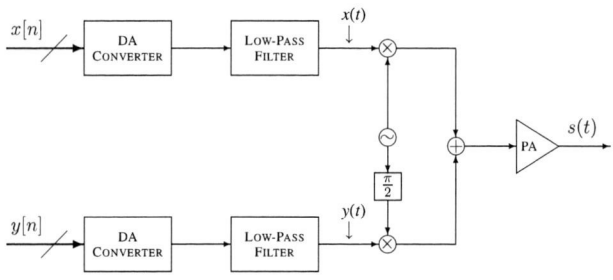

Fig. 1.   Cartesian transmitter.

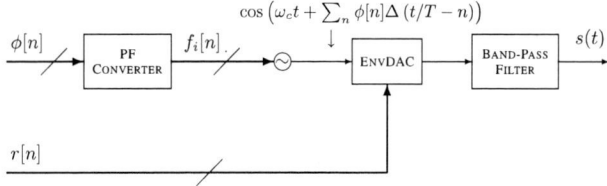

Fig. 2.   Digital polar transmitter.

problem, since phase samples are linearly interpolated. In any case, if the sampling frequency is high enough, the natural filtering of the sinc function can be enough to bring envelope aliases below acceptable levels.

The histogram of either the in-phase or quadrature OFDM signals corresponds pretty well with a Gaussian distribution

$$\text{pdf(x)} \approx \frac{1}{\sqrt{2\pi}\sigma} e^{\frac{-x^2}{2\sigma^2}}, \tag{5}$$

as a consequence, the histogram of the envelope signal can be well approximated by a Rayleigh distribution

$$\text{pdf(r)} \approx \frac{r}{\sigma^2} e^{\frac{-r^2}{2\sigma^2}} . \tag{6}$$

Fig. 3 shows the histogram of an envelope signal which power has been normalized to 1. The DC current consumption of the EnvDAC closely follows the distribution of the envelope.

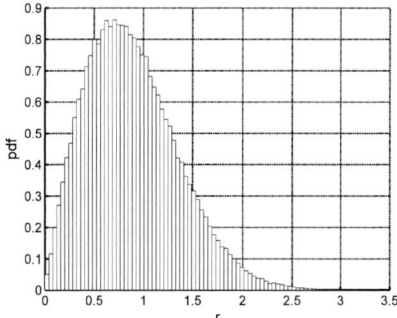

Fig. 3.   OFDM's envelope histogram.

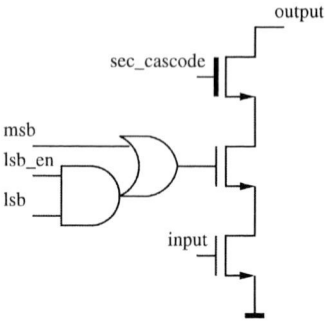

Fig. 4.   EnvDAC unit-cell.

III. ENVDAC CIRCUIT IMPLEMENTATION

The EnvDAC unit-cell consisting of a triple cascode stage and an AND-OR gate is shown in Fig. 4. The second cascode is a thick-oxide (50Å) device and its main function is to protect the lower transistors from the large voltage swing at the output (close to 5V for high output powers). Its gate is connected to a 2V external reference. The middle transistor is used as a switch and it determines whether the current of a particular unit-cell is added to the output. The gate voltage of the middle transistor is either 0 or 1.2V, depending on the values of the digital inputs *msb*, *lsb_en*, and *lsb*. The gate of the lower transistor, or input, is driven by the RF signal plus biasing voltage. Both input RF swing and biasing point can be adjusted to change output power and/or mode of operation. The AND-OR gate is part of the binary-to-thermometer decoding which translates the 8-bit envelope word to its equivalent 255-bit thermometer-encoded value. 256 unit-cells are arranged in a 16 x 16 matrix. The 8-bit input is split into 4 MSB's and 4 LSB's. The MSB's determine how many columns are ON, while the LSB's select how many unit-cells from the column with a valid *lsb_en* are ON. Two of these matrices are used to create a pseudo-differential structure with common *msb*, *lsb_en*, and *lsb* signals, but 180°-shifted RF inputs. Finally, 4 pseudo-differential matrices are combined to form the complete design. Each pseudo-differential matrix uses a different clock-phase for the decoding circuitry. The use of N clock-phases emulates an N-times higher clock, where the data released in $N-1$ out of N sampling periods is a linear interpolation of two accurate data samples. This topology significantly reduces the power of the first $N-1$ aliases [5], without the need for accurate data interpolation.

Fig. 5 shows the chip micrograph as implemented in a CMOS 65nm process. The envelope bits are converted from differential to single-ended before being fed to the binary-to-thermometer encoders. The four clock-phases are generated on-chip using a divider. The effective EnvDAC area amounts to $440\mu m$ x $220\mu m$.

The die has been packaged using an HVQFN48 and mounted on a FR4 epoxy PCB. To achieve the required output power an impedance transformation ratio of 8 is required. This is accomplished using an L-type transformation network comprising the package bondwire inductance and an on-PCB SMD capacitor. It is estimated that the bondwires add

978-1-4244-2018-6/08 $25.00 © 2008 IEEE

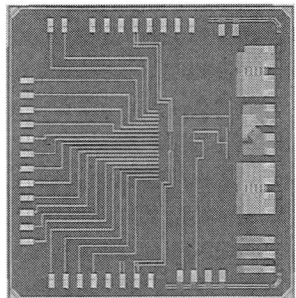

Fig. 5. Chip micrograph.

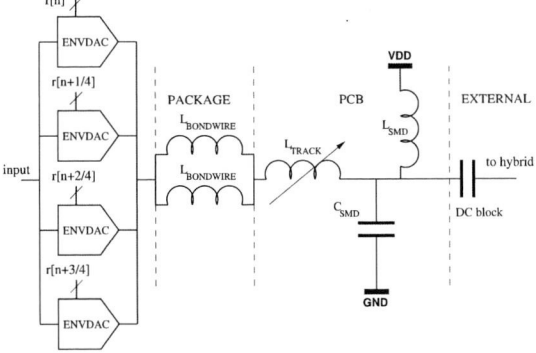

Fig. 6. Matching network of one of the output branches.

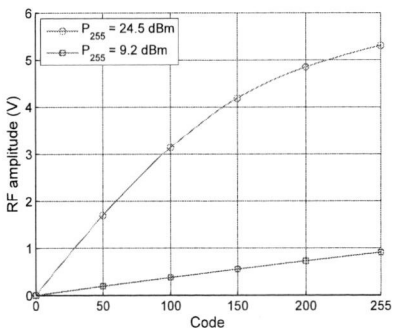

Fig. 7. Code-to-AM responses for two output powers settings.

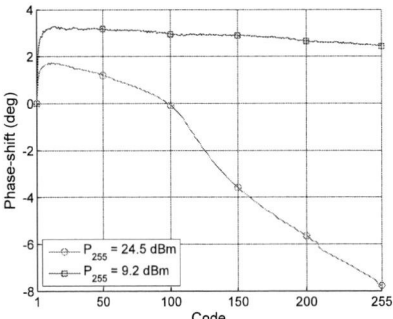

Fig. 8. Code-to-PM responses for two output powers settings.

0.8nH, a value slightly lower than the required 1nH, therefore extra inductance is added using a variable-length transmission line between the package and the SMD capacitor. After the matching network, the two outputs are routed using 50Ω transmission lines. The outputs are added off-PCB using an external hybrid. Fig. 6 shows a single-ended representation of the design including the matching network.

IV. MEASUREMENT RESULTS

The circuit was designed to operate around 2.45GHz but a slight underestimation of the circuit parasitics shifted the roll-off frequency to 2.1GHz. In the following, measurements are performed at 2GHz. During the dynamic behavior measurements the phase-modulated carrier is obtained using an external vector modulator.

A. Static behavior characterization

The static behavior of the EnvDAC can be described by its code-to-AM and its code-to-PM responses. Ideally, the code-to-AM response has a constant slope, but in practice the slope saturates for high codes. The amount of saturation depends on how close to the maximum output power the device is operating. Fig. 7 shows the code-to-AM response for a low and a high output power setting. For the high output power setting the biasing voltage is increased from 520mV to 700mV and the input power is increased by 15dBm. Fig. 8 shows the code-to-PM response for the two different output powers. Ideally, the phase-shift is constant for all codes, however in reality it decreases for increasing codes. Table I summarizes

the presented static behavior measurements. In Fig. 9 the output power in dBm and the drain efficiency versus code are shown for the two output power settings. It is clear that in order to have good efficiency predistortion needs to be applied. In the next subsection we use offline predistortion based on the measured code-to-AM/PM curves. In a complete solution, a calibration phase is required during which the code-to-AM could be estimated by measuring the code-to-DC transfer, and the code-to-PM response could be measured using a PLL. After that, envelope/phase predistortion can be performed using a look-up table.

B. Dynamic behavior characterization

The dynamic behavior of the envelope DAC is first verified with a 3Mb/s Bluetooth EDR signal using a sampling frequency of 300MHz. The input power and biasing point are adjusted to achieve an output power of 13dBm. A mean/peak EVM of 1.1/2.8% for a DC current of 66mA is measured. For this low output power, the supply voltage can be reduced to 2V

TABLE I

STATIC BEHAVIOR SUMMARY

setting	P_0 (dBm)	P_1 (dBm)	P_{255} (dBm)	DR (dB)	max. $\Delta\phi$ (^{o})
low power	-56.2	-37.1	9.2	46.3	3.3
high power	-41.8	-18.3	24.5	42.8	9.5

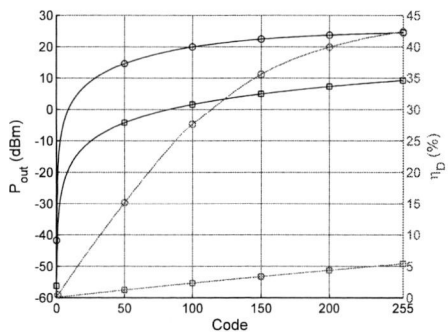

Fig. 9. Output power vs. efficiency for two output powers settings.

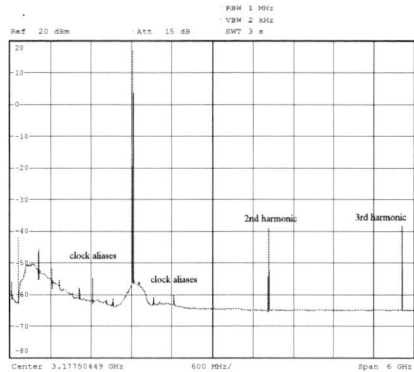

Fig. 10. Measured far-out Bluetooth EDR spectrum.

with hardly any change. In a second Bluetooth measurement, an output power of 19.7dBm is achieved for a current of 143mA, and a 5/13% mean/peak EVM. Fig. 10 shows a measured far-out spectrum. The noise at low frequencies is generated by an external pre-driver.

In Fig. 11 the output spectrum when processing a 54Mb/s WLAN OFDM signal is shown. The input power, biasing voltage, and relative envelope/phase path delay are optimized such that an average power of 16.7dBm with a mean EVM of 2.7% and a drain efficiency of 24% are obtained. The sampling frequency is 200MHz, but thanks to the multi-phase clocking the residual aliases 200MHz away from the carrier frequency are 20dB below the required IEEE 802.11g transmit spectrum mask. The first strong images are 800MHz away, and they have been measured to be -50dBc (left), and -55dBc (right).

C. Comparison with other work

A comparison between this digital envelope modulator and other implementations when used in WLAN OFDM mode is shown in Table II. The presented work shows the best output power, drain efficiency, and EVM.

V. CONCLUSION

In this paper the feasibility of a high-power envelope modulator as part of a digital polar transmitter for WLAN and BT is demonstrated. The modulator performs digital-to-analog conversion, up-mixing and power amplification, allowing for a

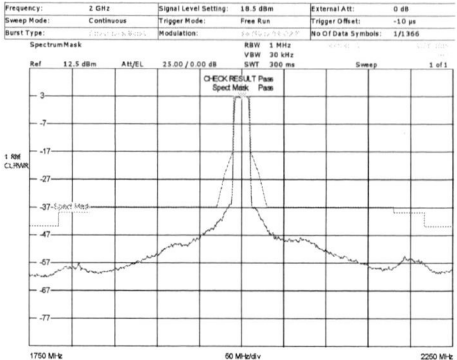

Fig. 11. Measured WLAN OFDM output spectrum.

TABLE II

COMPARISON WITH OTHER DIGITAL ENVELOPE MODULATORS

Reference	CMOS Technology	Supply voltage (V)	P_{out} (dBm)	η_D (%)	EVM (%)
[6]	0.18μm	1.7	13.6	9.2	4.6
[4]	90nm	3.0	9.4	8.1	4.9
this work	**65nm**	**2.5**	**16.7**	**24.0**	**2.7**

fully-integrated transmitter solution. It is shown how the use of multi-phase clocking reduces alias levels to acceptable values without the need for accurate envelope-data interpolation. The circuit delivers sufficient output power for most WLAN and BT implementations, and that with a comfortable EVM and transmit spurious emissions margin. The drain efficiency value in WLAN mode is a competitive 24%. The circuit uses 50Å thick-oxide devices and 2.5V power supply.

ACKNOWLEDGMENT

The authors would like to thank Joost Briaire and Pieter van Beek for their help during measurements. They would also like to thank Mark van der Heijden and Anton de Graauw for useful discussions. Finally, they thank Dennis Jeurissen, Pieter Prieshof, and Gerard van der Weide for their help during layout and packaging.

REFERENCES

[1] R.B. Staszewski, et al., "All-Digital TX Frequency Synthesizer and Discrete-Time Receiver for Bluetooth Radio in 130-nm CMOS," *IEEE J. of Solid-State Circuits*, vol. 39, issue 12, pp. 2278–2291, Dec. 2004.

[2] P. Eloranta, P. Seppinen, S. Kallioinen, T. Saarela, and A. Parssinen, "A Multimode Transmitter in 0.13 um CMOS Using Direct-Digital RF Modulator," *IEEE J. of Solid-State Circuits*, vol. 42, issue 12, pp. 2774–2784, Dec. 2007.

[3] A. Shameli, A. Safarian, A. Rofougaran, M. Rofougaran, and F. de Flaviis, "A Novel DAC Based Switching Power Amplifier for Polar Transmitter," *IEEE Custom Integrated Circuits Conf.*, pp. 137–140, Sept. 2006.

[4] P.T.M. van Zeijl, and M. Collados, "A Multi-Standard Digital Envelope Modulator for Polar Transmitters in 90nm CMOS" *IEEE Radio Frequency Integrated Circuits Symposium*, pp. 373–376, June 2007.

[5] Y. Zhou, and J. Yuan "A 10-bit wide-band CMOS direct digital RF amplitude modulator," *IEEE J. of Solid-State Circuits*, vol. 38, issue 7, pp. 1182–1188, July 2003.

[6] A. Kavousian, D.K. Su, and B.A. Wooley, "A Digitally Modulated Polar CMOS PA with 20MHz Signal BW," *IEEE Int. Solid-State Circuits Conf. Dig. Tech. Papers*, pp. 78–79, Feb. 2007.

978-1-4244-2018-6/08 $25.00 © 2008 IEEE

IEEE 2008 Custom Intergrated Circuits Conference (CICC)

A 2.4GHz, 20dBm Class-D PA with Single-Bit Digital Polar Modulation in 90nm CMOS

Jason T. Stauth and Seth R. Sanders

University of California, Berkeley, CA, 94720, USA

Abstract — **Polar transmitters are a promising alternative to traditional Cartesian architectures in terms of flexibility and power efficiency. Such systems are often difficult to implement due to wideband amplitude and phase signals and may require predistortion to meet EVM and spectral requirements. This work demonstrates a highly-linear digital-polar system implemented in 90nm CMOS. The amplitude is controlled with pulse-density modulation of the RF carrier. Phase information is provided with the RF clock. The class-D PA achieves peak efficiency of 38.5% at 2.4GHz, including power of the PA drivers and insertion loss of the bandpass filter. The system does not require predistortion, and achieves rms-EVM levels of 1.8-2.1% for π/4DQPSK and 8DPSK test vectors. The spectral mask for Bluetooth 2.1+EDR is satisfied under normal operating conditions.**

I. INTRODUCTION

Single-chip integration for low-power portable wireless systems is highly desirable to drive lower costs and smaller package footprints. In addition, research directions in cognitive and software-defined radio seek flexible architectures that can operate at multiple carrier frequencies and with different standards. This work presents a promising direction in both the system architecture and circuit implementation that may improve both the flexibility and integration level of the radio transmitter.

Polar transmitter architectures have gained interest in recent years due to improvements in flexibility and efficiency over traditional Cartesian architectures [1, 2]. In the Polar example, the complex RF signal is represented with amplitude and phase components instead of traditional I and Q vectors. This allows the power amplifier (PA) to operate with drain modulation, envelope restoration (ER), or envelope tracking (ET) modes [1-4]. If the power supply of the PA is modulated by an efficient DC-DC or hybrid switching-linear amplifier, the average efficiency of the system can be improved significantly [1, 4]. In true polar modulation there is no need for quadrature upconversion in the mixing stage. This can reduce passive components and make it easier to adjust the carrier frequency of the transmitter. Common difficulties in polar systems result from the fact that amplitude and phase are not bandlimited as with Cartesian I and Q signals. This drives the need for wideband amplitude and phase modulators to meet spectral masks and achieve low error-vector magnitude (EVM) levels. Also, supply modulated and envelope tracking PAs may experience significant AM-AM and AM-PM

distortion from the polar process. Many published solutions require complex predistortion algorithms to meet linearity requirements [1, 2].

This work presents a digital polar modulation transmitter with a class-D power amplifier. Instead of modulating the supply voltage, the amplitude is controlled with pulse-density modulation of the RF carrier. To achieve high resolution and reduce out-of-band noise, we use two stages of noise shaping. The first stage uses multi-bit ΔΣ modulation operating at 100MHz (up to 100 times oversampling). The second stage is a pre-programmed 1-bit pulse-density modulation (PDM) scheme operating at the RF carrier frequency.

The second-stage modulator and class-D PA are implemented in 90nm CMOS. To achieve higher output power (up to 20dBm) with standard oxide devices, the PA uses a cascoded CMOS output stage. A matching network and bandpass filter are used to help attenuate out-of-band noise. The class-D PA achieves efficiency of 38.5% at 2.4GHz including all driver power and insertion loss of the filter.

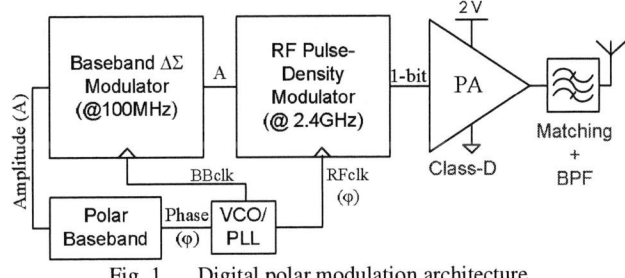

Fig. 1   Digital polar modulation architecture

II. ARCHITECTURE

The architecture, shown in Fig. 1, uses a polar digital baseband to control the amplitude and phase components of the transmitted RF signal. The amplitude information passes to the first stage ΔΣ modulator. The ΔΣ stage converts the signal to an oversampled representation with 10 quantization levels and a polarity bit. The polarity bit is used to smooth discontinuities in the phase signal when the amplitude is near zero. This is done by implementing phase inversions in the amplitude path instead of the phase path which significantly reduces the bandwidth of the *RFclk* signal. This also improves the linearity of the ΔΣ process when the amplitude is near zero by preventing saturation of the multi-bit output.

978-1-4244-2018-6/08 $25.00 © 2008 IEEE

The ΔΣ modulator controls a programmed RF pulse-density modulator that converts the signal to 1-bit quantization operating at 2.4GHz. The ΔΣ and PDM blocks are synchronized with clock signals that contain the polar phase information. In the case of the PDM stage, the clock signal operates at the RF carrier frequency of 2.4GHz. In a fully integrated implementation, the polar baseband would control a VCO to regulate the baseband and RF clock signals. In our implementation the RF clock with phase information is generated off chip with a Labview PXI system.

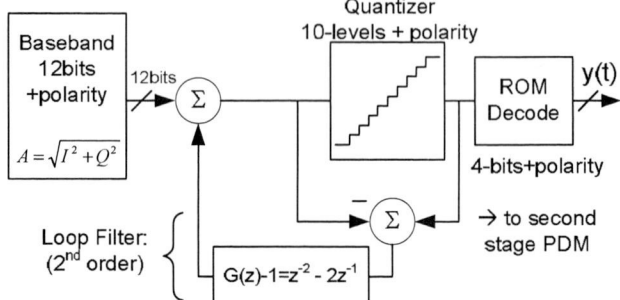

Fig. 4 Error-feedback digital ΔΣ modulator

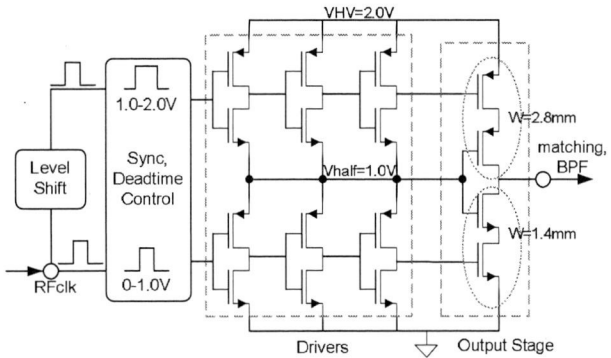

Fig. 2 CMOS Class-D cascode PA

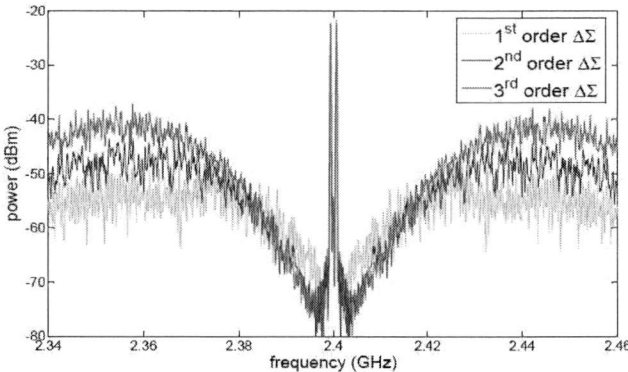

Fig. 5 ΔΣ Modulator: 1st, 2nd, 3rd-order comparison

IV. ΔΣ MODULATOR

The ΔΣ modulator is implemented with a digital error-feedback structure, as shown in fig. 4 [5]. In this example, the quantizer extracts 10 levels and the polarity of the 12-bit representation. A ROM decoder maps the bi-polar 10-level quantization to 4-bits with polarity for the next-stage PDM block. The system is clocked at the baseband frequency of 100MHz, so peaks in the noise spectrum occur roughly 50MHz centered at the carrier frequency.

Fig. 5 shows the simulated frequency response of the system with various loop filters for a suppressed carrier AM signal. The 3rd order system has the lowest noise at the center frequency, but highest peak out-of-band noise. The first-order system has lower peak out-of-band noise, but may have tones in the output spectrum for certain input signals [5]. A second-order filter achieves both favorable noise-shaping and reduced possibility of tones in the output spectrum. The second-order filter was used in the final design.

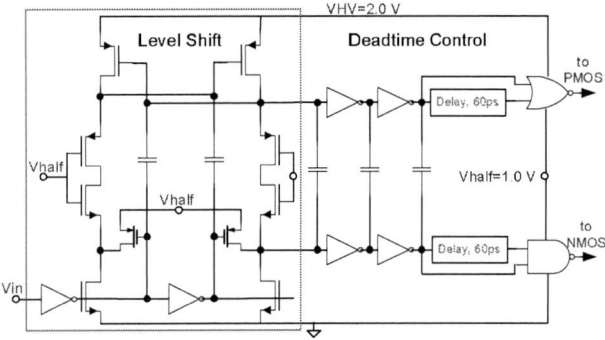

Fig. 3 Level shift and deadtime control

III. CLASS-D PA

Shown in fig. 2, the class-D PA operates with both NMOS and PMOS complementary devices. With PMOS f_t exceeding 40GHz in the 90nm process, the P-channel device does not significantly reduce efficiency. To achieve higher power, the output stage is cascoded. The PA provides power gain and interfaces with a matching network and high-order bandpass filter to attenuate out-of-band quantization noise. The level shift circuit, shown in fig. 3, interfaces the 1-bit polar signal to the cascode output stage. The PA normally operates with maximum power rail $VHV=2.0V$ and a mid power rail $Vhalf=1.0V$. This limits the maximum oxide stress to 1.0V in normal operation. The deadtime circuit prevents shoot-through current and synchronizes the output stage voltage waveforms. A nominal 60ps deadtime optimizes power efficiency while providing a reasonable buffer for process and temperature variation. A 100pF on-chip bypass capacitor supplies high-frequency current from the VHV node.

V. PULSE-DENSITY MODULATOR

The pulse-density modulator (PDM) block converts the output of the ΔΣ block to a 1-bit representation. The ΔΣ output is sampled with a synchronous 100MHz clock and then re-synchronized to a set of shift registers operating at 2.4GHz ($RFclk$ signal with phase information). The shift registers generate a PDM waveform from pre-programmed binary codes stored in an on-chip ROM. The codes represent bit sequences corresponding to 10-level amplitude quantization. Tones from the programmed bit sequences occur far from the carrier frequency to maximize filter attenuation. The clock recovery circuit inverts the phase of the clock depending on the polarity bit. This allows polarity information to be synchronized with the amplitude path, eliminating wideband phase inversions in the phase path. The PDM generation block mixes the PDM signal from the shift-registers with the RF clock. The RF clock contains phase information and the PDM signal contains amplitude information. The result is a polar modulated 1-bit output that is provided to the class-D PA.

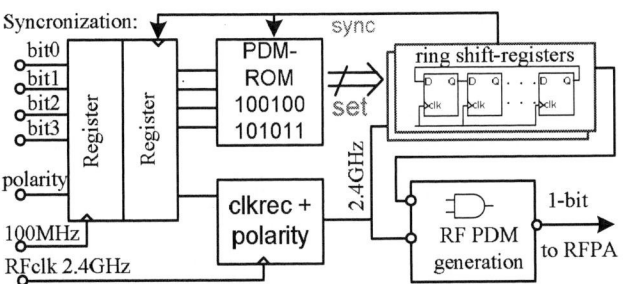

Fig. 6 Pulse-density modulator block

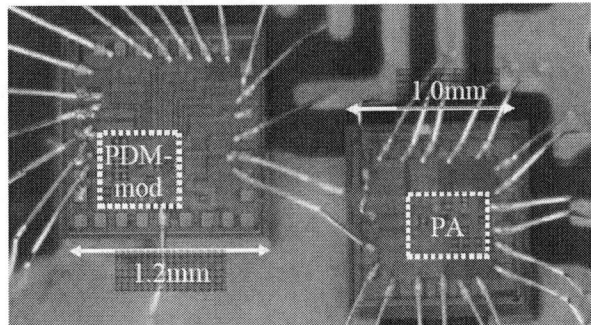

Fig. 7 Die Photo

VI. EXPERIMENTAL RESULTS

The system is implemented with two test chips in 90nm CMOS. A first chip contains the class-D PA and drivers (active area 0.15mm²). A second chip contains the PDM structure, synchronization, clock recovery and ROM (active area 0.2mm²). A die photo is shown in fig. 7. The baseband and $\Delta\Sigma$ modulator are implemented in an FPGA. The system is tested with a Labview PXI setup with RF upconverter and downconverter. The labview system and FPGA use a synchronous clock to perform time-alignment of the amplitude and phase signals. The output filter is implemented off-chip and consists of a traditional L-matching network and an additional surface-mount bandpass filter component. The bandpass filter is a Johanson 2450BP41D100B component for the 2.45GHz band with 1.3dB maximum insertion loss.

Fig. 8 shows the total output power and efficiency of the class-D PA across supply voltage. Here, efficiency includes power of the PA drivers and loss in the matching network, but does not include insertion loss of the bandpass filter. The peak output power of 20dBm occurs for the maximum supply voltage of 2.4V (oxide stress of 1.2V). The peak efficiency of 40.7% occurs for a supply voltage of 2.0V. With the bandpass

filter included, peak efficiency was measured at 38.5%. Fig. 9 shows linearity and efficiency at each of the 10-levels of amplitude quantization. Linearity for the PA is high with pulse-density modulation. Importantly, efficiency stays higher at lower power levels than with typical class A/AB power amplifiers. Efficiency of the PA, including driver power, stays above 25% for 10dB power backoff.

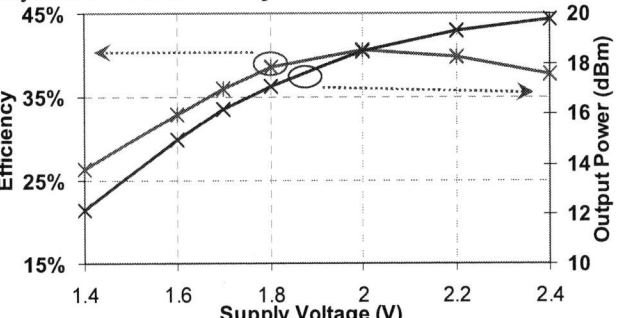

Fig. 8 Efficiency and output power vs supply voltage

VHV=2.0V, Vhalf=1.0V, fo=2.4GHz, Tamb=27°C

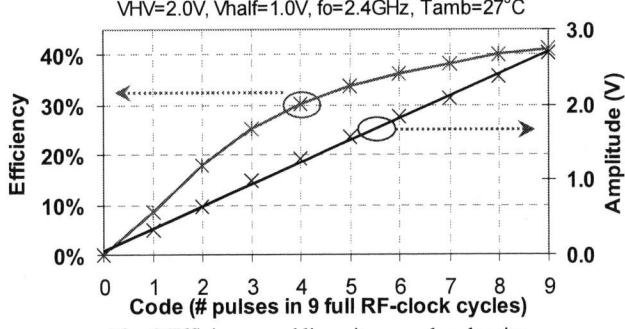

Fig. 9 Efficiency and linearity vs pulse-density

The average efficiency of the system is further improved with a current recycling scheme. Since the *Vhalf* node in the PA operates at 1.0V, this node can be shared with the 1.0V V_{DD} node. This allows excess current from driving the PMOS power device in the PA to be used to power the remaining 1.0V circuitry. At 2.4GHz this reduces the current drawn from the 1.0V supply from 11.7mA to 3.7mA. The current recycling scheme allows the entire system to operate with 30% average efficiency for π/4DQPSK and 8DPSK modulated signals with approximately 3dB PAPR.

Low digital processing power and high average efficiency make this scheme amenable to low-power portable wireless standards, such as Bluetooth. Comparable work published in the literature achieves higher spectral fidelity at the tradeoff of significantly higher power consumption. [6] demonstrates a system with $\Delta\Sigma$ DACs operating at 5.4GHz that meets cellular coexistence requirements for 802.11b/g and 802.16e. Reported current consumption is 83mA from a 1.2V supply for the digital block. RF current consumption is 128mA for rms output power of 2.6dBm. Our solution requires between 2-5mW for the digital PDM block and roughly 55mW RF power for 12-14dBm rms output power level.

978-1-4244-2018-6/08 $25.00 © 2008 IEEE

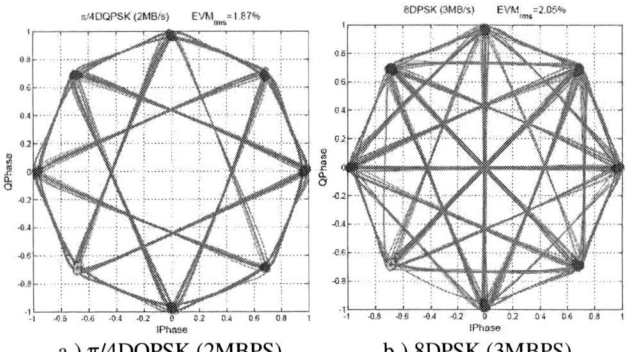

a.) π/4DQPSK (2MBPS)   b.) 8DPSK (3MBPS)

Fig. 10 Measured constellation diagrams

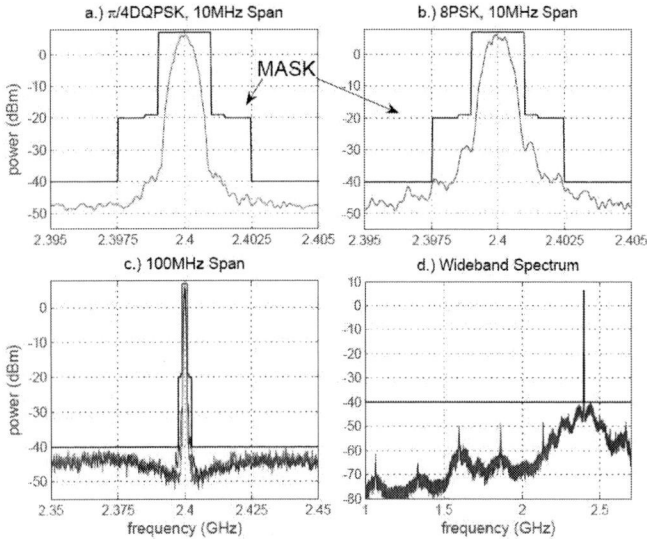

Fig. 11 Transmitter output spectrum, 100kHz RBW

Fig. 10 shows the measured constellation diagram for π/4DQPSK and 8DPSK test vectors. These test signals meet the requirements of the Bluetooth 2.1+EDR standard for the 2MBPS and 3MBPS datarates [7]. Measured EVM for π/4DQPSK is $EVM_{rms} = 1.8\%$ and $EVM_{peak} = 2.8\%$, and for 8DPSK, $EVM_{rms} = 2.1\%$ and $EVM_{peak} = 2.9\%$. As seen for the 8DPSK data in Fig. 10-b, the transmitter is capable of generating constellation trajectories that pass through the origin (infinite peak-minimum ratio). The system remains linear near zero amplitude, minimizing low amplitude distortion by synchronizing polarity shifts with the digital polarity bit in the amplitude path. This significantly reduces the bandwidth requirements of the phase modulator and improves performance for high PAPR signals.

Quantization noise of the digital modulator is a limiting factor for spectral performance. Fig. 11a-d shows that the nearband spectral requirements for Bluetooth 2.1+EDR are satisfied for both π/4DQPSK and 8DPSK test signals. Average power levels at the spectrum analyzer are between 11-14dBm. The wideband spectral mask is satisfied due to the addition of the bandpass filter at the output. However, it should be noted that in peak-hold mode, which is required for the Bluetooth standard, there is little margin for the -40dBm

power limit (100kHz RBW). Shown in fig. 11-c, the lowest margin occurs near peaks in the quantization noise spectrum 50MHz from the center frequency. Low margin in the spectral mask could make the system susceptible to load pull, power supply variation, or other scenarios which cause wideband noise levels to increase. This can be improved by operating the ΔΣ process at higher clock frequencies, or reducing the transmitter power level. In a fully-integrated version, it may be practical to operate the ΔΣ modulator at 250MHz or more, reducing the peak noise by 3-10dB, depending on the filter component. Due to limited margin in meeting the spectral-mask, we do not present this as a fully-functional Bluetooth transmitter, but rather a demonstration of a new and interesting topology worthy of further investigation. The high efficiency and excellent linearity show that this is a promising alternative to traditional Cartesian and polar transmitters.

VII. CONCLUSION

We have presented a new architecture for digital polar modulation using a class-D PA. The amplitude is controlled with pulse-density modulation through baseband ΔΣ modulation and an RF pulse-density modulation circuit. The PA achieves peak efficiency of 38.5% at 2.4GHz, including driver power and insertion loss of the filter. The system achieves high linearity across the full power range, achieving rms EVM of approximately 2%. The spectral masks for the Bluetooth 2.1+EDR standard are met for both 8DPSK and π/4DQPSK test vectors. Limited margin near peaks in the quantization noise spectrum motivates additional research to further reduce out-of-band noise levels.

REFERENCES

[1] F. Wang, D. Kimball, D. Y. Lie, P. Asbeck, and L. E. Larson, "A Monolithic High-Efficiency 2.4GHz 20dBm SiGe BiCMOS Envelope-Tracking OFDM Power Amplifier," *IEEE Journal of Solid State Circuits*, vol. 42, pp. 1271-1281, June 2007.

[2] P. Reynaert and M. Steyaert, "A 1.75-GHz polar modulated CMOS RF power amplifier for GSM-EDGE," *IEEE Journal of Solid State Circuits*, vol. 40, pp. 2598-2608, Dec. 2005.

[3] F. H. Raab, P. Asbeck, S. Cripps, P. B. Kenington, Z. B. Popovic, N. Pothecary, J. F. Sevic, and N. O. Sokal, "Power Amplifiers and Transmitters for RF and Microwave," *IEEE Transactions on Microwave Theory and Techniques*, vol. 50, March 2002.

[4] J. T. Stauth and S. R. Sanders, "Optimum Biasing for Parallel Hybrid Switching-Linear Regulators." *IEEE Transactions on Power Electronics*, vol. 22, pp. 1978-1985, Sept. 2007.

[5] S. R. Norsworthy, R. Schreier, G. C. Temes, and IEEE Circuits and Systems Society., *Delta-Sigma data converters : theory, design, and simulation*. New York: IEEE Press, 1996.

[6] A. Pozsgay, T. Zounes, R. Hossain, M. Boulemnakher, V. Knopik, and S. Grange, "A Fully Digital 65nm CMOS Transmitter for the 2.4-to-2.7GHz WiFi/WiMAX Bands using 5.4GHz DS RF DACs," *IEEE International Solid State Circuits Conference (ISSCC) Dig. Tech. Papers*, vol. 51, pp. 360-361, Feb. 2008.

[7] Bluetooth_Special_Interests_Group_(SIG), "Bluetooth technical specifications," in *www.bluetooth.org*, 2008.

IEEE 2008 Custom Intergrated Circuits Conference (CICC)

Linearity and Efficiency Enhancement Strategies for 4G Wireless Power Amplifier Designs

Larry Larson, Donald Kimball and Peter Asbeck
University of California, San Diego
larson@ece.ucsd.edu

Abstract

Next generation wireless transmitters will rely on highly integrated silicon-based solutions to realize the cost and performance goals of the 4G market. This will require increased use of digital compensation techniques and innovative circuit approaches to maximize power and efficiency and minimize linearity degradation. This paper summarizes the circuit and system strategies being developed to meet these aggressive performance goals.

1. Introduction

The linear microwave power amplifier/module industry has historically been dominated by non-standard semiconductor technologies (like GaAs and Si-LDMOS), relatively low levels of integration (hybrid assembly techniques are still common), and little interaction with baseband signal processing. In this regard, the typical power amplifier module today resembles the low-frequency analog/data converter module of forty years ago. This raises the question of whether microwave power amplifiers are poised to undergo a shift to highly integrated, richly digital, implementations — much the way analog and mixed-signal circuits (and indeed RF transceivers in general), transitioned to full CMOS within the last decade.

There are several factors that are pushing power amplifier technology towards higher levels of integration, and co-integration with other portions of the transceiver. The first is the relentless cost pressure of the modern consumer wireless industry, where every penny saved is important in a 1 Billion unit/year market. The second is the significant performance gains that can be realized if *baseband* information is used to improve the performance of the *RF* chain. This baseband information is readily available if the PA is co-located with the baseband digital processor, or if there is a common digital interface between the baseband processor and the RF transceiver.

At the same time, the performance of microwave power amplifiers (in terms of linearity and dc power consumption) is exquisitely sensitive to the characteristics of the device, and decades of III-V HBT/PHEMT and Si-LDMOS device optimization has made the performance and cost of these solutions very impressive. This continuing refinement is one of the reasons these technologies continue to dominate the power amplifier market while silicon-based technolo-

gies have taken over all remaining RF transceiver functions. In addition, the relatively low breakdown voltage of scaled silicon devices (SiGe HBTs or CMOS) makes it difficult to achieve the high power output required for many applications. Finally, the move towards wideband OFDM waveforms — with their high peak-to-average ratios — has reduced the power-added efficiency achievable using standard circuit techniques. For example, typical OFDM power amplifiers achieve PAE's of only 20-30%, which for a 27 dBm output power means that as much as 1-2W of waste heat is generated. Raising this efficiency to 40-50% would have an enormous impact on the battery life and capabilities of mobile devices.

This paper will examine these tradeoffs, with particular attention to the circuit and signal processing techniques being developed in highly integrated silicon-based wireless power amplifiers. The key challenges for silicon-based power amplifiers are improving efficiency, increasing output power, and enhancing linearity.

2. Signal Characteristics of Wireless Standards

Power amplifier performance is largely dictated by the baseband waveform characteristics, the center frequency, and the output power. As shown in Table 1, there are several ways to characterize the waveform characteristics. The Peak-to-Average Ratio (PAR) compares the peak modulated output power to its short-term average; as the industry move to more spectrally efficient modulation formats like OFDM this ratio is growing, creating additional linearity and efficiency challenges. Power amplifier efficiencies typically peak at the maximum saturated output power, but decline precipitously as the power is reduced, so a high PAR waveform implies that the power amplifier is spending most of its time in a low-efficiency "backed-off" mode of operation.

The Error Vector Magnitude (EVM) is a modulation quality metric widely used in digital RF communications systems, especially in 3G and OFDM-based modulation standards. It is essentially a measure of the in-band accuracy of the modulation of the *transmitted* waveform, i.e.

$$EVM = \frac{\sqrt{\sum_n |e(k)|^2}}{n} \qquad (1)$$

978-1-4244-2018-6/08 $25.00 © 2008 IEEE

Table 1. Signal Characteristics of Modern Wireless Standards.

System	PAR(dB)	PMR(dB)	PCDR(dB)	Bandwidth(MHz)	Access Type
GSM	0	0	0	0.2	TDMA
EDGE	3.2	17	30	0.2	TDMA
CDMA ONE	5.5-12	∞	60	1.25	CDMA
UMTS	3.5-7	∞	80	5	CDMA
CDMA 2000	4-9	∞	80	1.25	CDMA
802.11a/g	8-10	∞	25	20	TDMA
WiMax	8-10	∞	25	20	TDMA

where $e(k)$ is the normalized magnitude of the error vector at symbol time k, and n is the number of samples over which the measurement is made.

The EVM is essentially an in-band measure of the waveform quality, while spectral regrowth is an out-of-band measure of waveform quality. This is typically measured by the adjacent channel power regrowth (ACPR) and the alternate channel power regrowth (AltCPR). The ACPR is typically measured as the ratio of the signal power in the desired channel to the distortion power in an adjacent channel; the alternate channel power regrowth is a measure of the ratio of the signal power in the desired channel to the distortion power in the alternate channel. These two measures are shown in Figure 1.

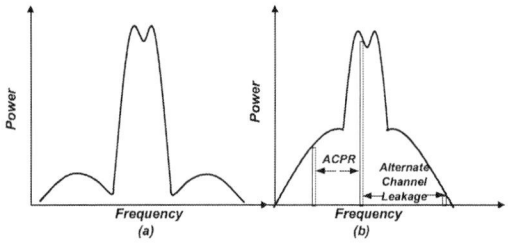

Figure 1. Spectrum of transmitted signal: (a) Spectrum of ideal transmitted modulated signal. (b) Spectrum of distorted signal illustrating ACPR and alternate channel regrowth.

The performance of the power amplifier can be dramatically improved if the peaks of the waveform can be reduced, lowering the PAR, while keeping the average power and bandwidth the same, and this is the basis for a variety of *crest-factor reduction* (CFR) algorithms [1–3]. These algorithms rely on tone reservation and carrier phasing techniques, or on selective clipping of the waveform. The clipping algorithms lower the peak by adding an "inverse" of the peak to the original waveform at the appropriate time. Viewed from the perspective of a typical OFDM waveform, these peaks occur at a time when all the individual carriers add coherently, so removing the peak adversely affects all of the carriers at the same time, with detrimental effects on the error vector magnitude (EVM) performance and resulting spectrum regrowth. In addition, the spectrum of this added inverse peak is often very broad, creating significant spectral regrowth outside the band of the original signal. A

variety of clever algorithms have been developed to minimize these limitations.

Sperlich [4] uses an interpolation technique to reduce CFR by selectively adding uncorrelated in-band noise to the signal. This improves spectral regrowth at the expense of EVM performance. Vaananen [5] adopts a peak windowing method to minimize adjacent channel power regrowth. Baxley [6] adopts a technique where in-band and out-of-band components are separately optimized, and in-band processing minimizes EVM and out-of-band processing minimizes regrowth.

A sophisticated signal processing engine is required to accomplish these improvements, and an example of a CFR engine is given by Wegener [7]. The circuit was fabricated using a 130 nm CMOS process, and required 1.8M gates at a power dissipation of 2.8 W when operating in two-channel transmit diversity mode. In this case, successive clipping of multiple peaks and filter stages were required to minimize the overall spectral regrowth and minimize EVM for CDMA, WCDMA, and OFDM signals. Typical reductions in PAR of 2-3 dB are achievable; this level of improvement seems modest, but can result in a 50% reduction in power amplifier dc power consumption in some cases. Clearly, as VLSI technology improves and the algorithms become more sophisticated, the overhead associated with the CFR reduction will drop dramatically.

A second method to characterize the waveform is by the Peak-to-Minimum Ratio (PMR), which is the ratio of the peak waveform power to the minimum waveform power, again measured over a short period of time. A high PMR waveform presents a challenge to the power amplifier from an efficiency perspective, and it can also make certain linearization techniques difficult to implement. In these cases, the output power reaches near zero for a brief period of time.

The Power Control Dynamic Range (PCDR) is a measure of the maximum variation in the average output power, measured over a long period of time (measured in days or even weeks). This variation is especially large in CDMA-based systems, because of the power control loop required to prevent near-far interference in spread-spectrum systems. A high PCDR presents problems for maximizing power amplifier efficiency, since the long term average output power is typically well below that where the peak efficiency occurs. In most CDMA mobile systems, the average output power is log-normally distributed, with a mean of roughly

978-1-4244-2018-6/08 $25.00 © 2008 IEEE

0 dBm and a peak of approximately 30 dBm. So, the PCDR is typically over 60 dB in most CDMA-based systems. OFDM-based systems like WiMax can have similar PCDR constraints.

3. Power Efficiency Enhancement Strategies

Silicon technology has an intrinsic disadvantage compared to III-V technology with respect to achievable efficiency (due to the lower f_T and breakdown voltage) so design-oriented approaches are required to realize the necessary dc power consumption goals. The use of varying classes of operation (like Class-AB, B, D, E, etc.) to achieve these improvements is of limited comparative value, since III-V devices can use them just as easily as Si-based devices. So, more sophisticated techniques are required for dc power reduction in silicon-based power amplifiers, and the two most popular are *DC Bias Variation* and *Load-Line Variation*. These techniques are illustrated in Fig. 2.

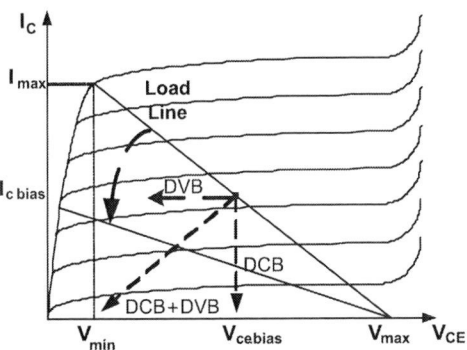

Figure 2. Illustration of efficiency enhancement techniques. Dynamic current biasing (DCB) alters the dc bias current in response to changes in the RF output power. Dynamic voltage biasing (DVB) alters the bias voltage in response to the RF power changes. The load-line presented to the transistor can also be changed in response to changes in power.

DC Bias variation techniques have been thoroughly investigated, and fall into several categories. Dynamic Current Bias (DCB) variation approaches have found their way into several commercial power amplifier products [8]. In their simplest embodiment, the dc bias current is reduced in response to reductions in the RF output power, as shown in Fig. 2. This simple approach can achieve reductions in dc power consumption, although a typical microwave power amplifier will exhibit significant gain reduction when its bias current is reduced, and this effect requires compensation. A better approach may be to switch parallel devices in and out with MOSFET switches to alter the dc bias current, which will maintain a constant power gain even as the dc current consumption drops [9]. In the simplest case, the

power amplifier can be switched from a "low-power" version to a "high-power version."

DC bias variation can also be applied to the drain/collector voltage of the transistors, with even greater effect, and these Dynamic Voltage Biasing (DVB) techniques are known as Envelope Tracking (ET) or Envelope Elimination and Restoration (EER). These techniques have a long pedigree — in fact, EER was developed initially in the 1930's as an efficiency enhancement technique for kilowatt AM transmitters [10]!

Dynamic power supply schemes are usually separated into two types: Envelope Elimination and Restoration (EER) and Envelope Tracking(ET). Figure 3 shows the principles of traditional EER and ET systems. EER uses a combination of a high efficiency switched-mode PA with an envelope re-modulation circuit [11–15]; ET utilizes a linear PA and a controlled supply voltage, which closely tracks the output envelope. When the supply voltage tracks the instantaneous output envelope, it is known as Wide Bandwidth ET (WBET) [16–19]; when the supply voltage tracks the long-term average of the output envelope, it is known as Average ET (AET) [20,21]. AET techniques are especially useful for power control schemes, such as the reverse link in CDMA, where the long-term variation in average power is much greater than 20dB [22]. However, they improve the efficiency only modestly for high PAR signals such as OFDM.

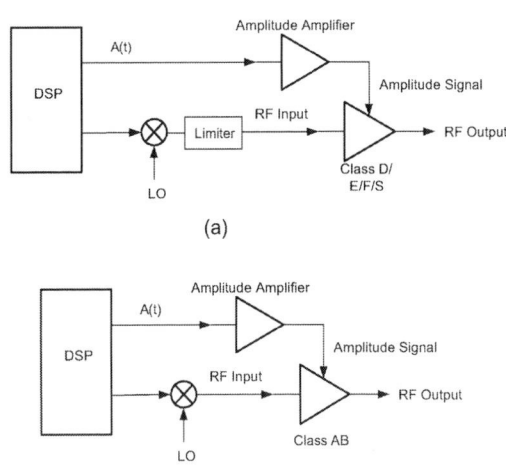

Figure 3. Block diagram of (a) Envelope Elimination and Restoration (EER) power amplifier and (b) Envelope Tracking (ET) amplifier

In both EER and wideband ET systems, the collector/drain supply of the RF power transistor dynamically changes with the output envelope, so the RF transistor operates with higher efficiency over a wide dynamic range of output power. Theoretically, EER is more efficient than ET, since the RF transistor is always operating in a switching mode. In the traditional EER system, the input RF signal is applied to a limiter (as shown in Figure 3); this is a problem for some wide dynamic range OFDM signals, where the peak-to-minimum ratio is essentially infinite [23]. By contrast, ET systems are better positioned to accommodate high peak-to-minimum signals because the amplifiers operate in a linear (if slightly compressed) mode at all output power levels, so the resulting gain variation is manageable.

Due to the nonlinear operations of the transformation of the I and Q signals to amplitude and phase, the bandwidths of the amplitude signal $A(t)$ and the phase signal are much wider than that of baseband signal. This imposes practical challenges to the traditional EER transmitter, and typically limits it to narrow-bandwidth applications [24, 25]. By contrast, the ET system requires a lower envelope amplifier bandwidth and less precise time-alignment between the envelope and RF paths [26]. Therefore, it is more easily applied to applications requiring a wide signal bandwidth such as OFDM.

The use of wideband ET techniques can boost the RF PA drain/collector average efficiency, but the total system efficiency is determined by the product of the envelope amplifier efficiency and the RF transistor drain/collector efficiency [27–29]. Thus, a high-efficiency envelope amplifier design is critical to the EER/ET system. This design is itself quite challenging, since the amplifier needs to provide a signal to a load (the power amplifier collector or drain) of several ohms or less, at a frequency of well over 20 MHz. Fortunately, efficient switching-mode dc-dc converters have been demonstrated with bandwidths sufficient for wideband OFDM modulation. In many cases, these converters can be integrated along with the power amplifier to realize a completely monolithic solution (except for the external inductor). A recent example of a monolithic SiGe BiCMOS envelope-tracking power amplifier (PA) for 802.11g OFDM applications was demonstrated at 2.4 GHz. The die included a high-efficiency high-precision envelope amplifier and a two-stage SiGe HBT PA. The two-stage amplifier exhibited a 12-dB gain, $< 5\%$ EVM, 20-dBm OFDM output power, and an overall efficiency (including the envelope amplifier) of 28% [30].

The output of the envelope amplifier has to be time-aligned to the output of the RF amplifier, so that extra distortion is not created by the resulting time mismatch between the two paths. Fortunately, the wideband ET system is less sensitive to this misalignment effect than the traditional EER amplifier, and so the effect of small misalignment on EVM is negligible.

Load-line variation techniques have been developed as an approach for reducing PA dc power consumption when the RF output power is reduced. A higher impedance is presented to the device as the power is reduced, maintaining a constant voltage swing a the drain/collector. There are several ways of accomplishing this variation.

The Doherty amplifier [31, 32] utilizes two amplifiers in parallel — the auxiliary amplifier performs a "load pull" on the main device so that the voltage swing across the main device stays roughly constant over the highest lev-

els of the output power, even though the RF current in the main device is increasing, as shown in Fig. 4. When the RF output power is reduced, the auxiliary amplifier turns off, reducing the overall dc current consumption considerably in the "backed-off" mode of operation. Multiple stages of the Doherty can be employed to realize even greater power saving benefits as the RF output power is reduced [33, 34]. Although the Doherty amplifier topology was invented decades ago, it has enjoyed renewed interest recently for base station applications [35–37] and handset applications [38–40].

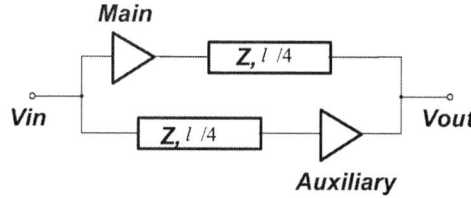

Figure 4. The Doherty amplifier uses a main amplifier and an auxiliary amplifier in parallel. The auxiliary amplifier only provides power for the highest output power conditions, and "load-pulls" the main power amplifier so that its voltage swing remains constant as the output power varies.

The load impedance presented to the transistor can be varied directly to adaptively modify the impedance presented to the transistor in response to the required output voltage. The impedance can be varied with pin-diodes, MEMs devices [41] or by varactor diode tuning of a transmission line impedance transformer, as shown in Fig. 5 [42]. A wide range of impedances can be achieved with this technique and the high efficiency of the power amplifier can be maintained over a broad range of output powers. One example of an adaptively tuned amplifier provided 13dB gain, 27-28 dBm output power at the 900, 1800, 1900 and 2100MHz bands [43]. For the communication bands above 1GHz optimum load adaptation resulted in efficiencies between 30-55% over a 10dB output power control range, a dramatic improvement compared to constant load impedance performance. Dynamic load line modulation techniques can also be employed to provide amplitude modulation of the output waveform, promising an alternative approach for non-constant amplitude modulation.

4. Power Enhancement Strategies

Scaled SiGe HBT and CMOS devices possess a relatively low breakdown voltage, which means that their dc operating voltage must also be low, and therefore their RF output is limited. This limitation is traditionally overcome in microwave circuit design by lossless power-combining techniques like the Wilkinson power combiner shown in Fig. 6(a). Unfortunately, these distributed approaches require a large die area, though this can be reduced through

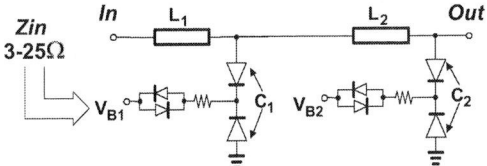

Figure 5. A variable impedance transmission line tuner can present a variety of impedance levels to the power amplifier, so that the voltage swing across the collector remains high even at low output power levels, maximizing efficiency [43].

the use of lumped equivalents of quarter-wave transmission lines.

Transformer-based power combining approaches are more popular for silicon power amplifiers, because of their compact implementation and low-loss. Series/parallel transformer approaches have been described [44–46] where the output is the sum of the output voltages from N parallel power amplifiers as shown in Fig. 6(b). These approaches also take advantage of the inherent symmetry of balanced operation to minimize the return currents flowing through the ground connections, a distinct advantage for monolithic high-power implementations. An example of the feasibility of this concept is a 2.4-GHz 2-W 2-V (450 mW at 1V) CMOS power amplifier with 50-Ω input and output match and power added efficiency (PAE) of 41% [47].

The voltage across the transistor can also be reduced by series operation of multiple devices. This approach was proposed by Sowlati [48] to realize a $0.18\mu m$ CMOS power amplifier that provided 23-dBm output power with a power-added efficiency (PAE) of 42% at 2.4 GHz. The amplifier operated reliably with a 2.4V supply even with short gate length devices. This approach is shown schematically in Fig. 6(c), and the RC-network connecting the upper gate device to the output is designed to equally distribute the total output voltage swing between the two devices.

5. Linearization Strategies

After efficiency and output power have been optimized, the final step is to improve the linearity of the amplifier. Traditional feedback techniques are of limited utility, because there is little excess gain at microwave frequencies to provide the necessary correction. However, if the high-frequency signal can be translated back down to complex baseband, then the higher available gain at baseband can be used to provide adequate feedback. This is the basis for *Cartesian Feedback*, which has been used for many years in relatively narrow bandwidth communication systems [49], and is shown in Fig. 7(a). The inherent matching of the I and Q channels in the Cartesian feedback approach is a distinct advantage compared to amplitude/phase feedback approaches.

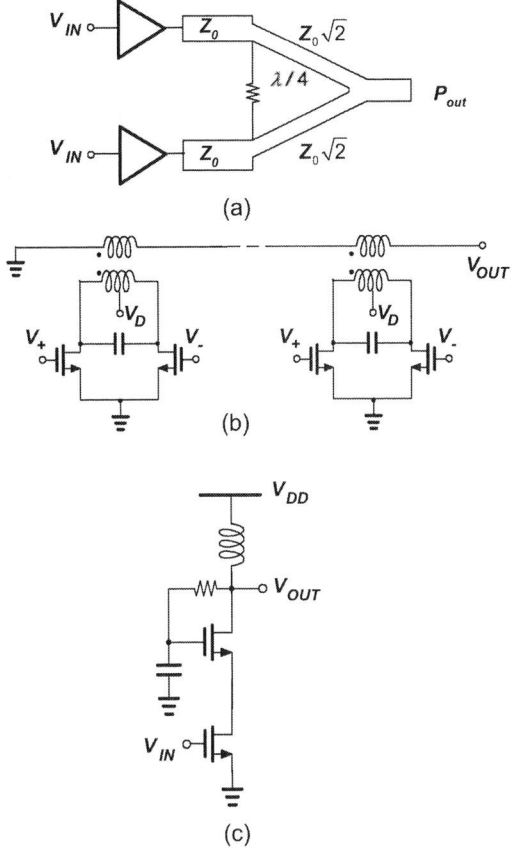

Figure 6. Lossless combiners for power amplifier applications (a) Wilkinson power combiner (b) Transformer combiners add the voltage output of several parallel amplifiers (c) Series connection of low-voltage FETs allows a high voltage to appear across the complete circuit.

However, the delay around the feedback loop, due to the high-Q output matching network, limits the achievable bandwidth of the system. Recent advances in phase compensation approaches, combined with fully integrated implementations of power amplifiers, have extended the bandwidth of Cartesian feedback techniques into the low-MHz range [50, 51]. However, the achievable bandwidth of any feedback system will always be limited, and broadband modulation approaches will require open-loop approaches. An open-loop analog pre-distortion approach using a polynomial pre-distorter was recently implemented to realize an improved bandwidth [52].

The most promising open-loop approach is to perform pre-distortion using digital techniques at baseband frequencies. This technique is illustrated in Fig. 7(c) and is known as *adaptive digital pre-distortion* [53]. In this case, the AM-AM and AM-PM distortion through the amplifier is determined and a digital signal processor provides the appropriate pre-distorted in-phase and quadrature-phase signals for the baseband upconverter. Several different versions of

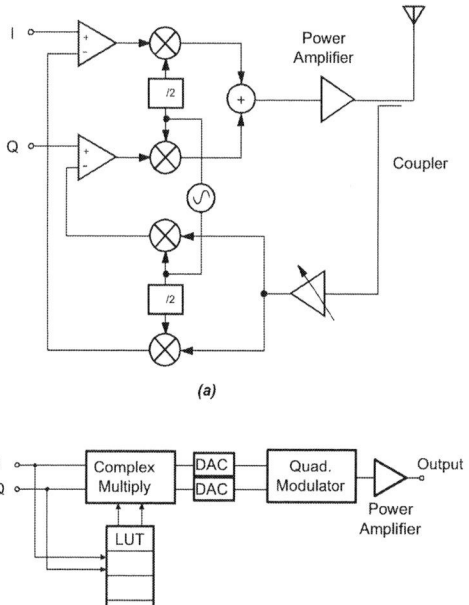

(a)

(b)

Figure 7. Linearization approaches for power amplifiers (a) Cartesian feedback provides complex baseband compensation of Power amplifier nonlinearities (b) Adaptive digital predistortion compensates the nonlinearities in the digital domain prior to baseband conversion.

adaptive predistortion have been developed [54, 55]. With the increasing sophistication of DSP techniques, as well as lower power high-speed DACs, adaptive digital predistortion has recently become possible for mobile terminals as well [56].

There are several limitations to adaptive digital predistortion: significant bandwidth expansion, a relatively high digital processing overhead, and a susceptibility to memory effects (long time constant variations in the PA gain and phase responses that are difficult to incorporate into the predistorter transfer function). In addition, the predistortion coefficients require constant updating, due to aging, temperature and power supply variation effects over long periods of time. Adaptive digital predistortion works best for standards requiring a high PAR (from Table I) and moderate-low PCDR, such as base stations, wireless LAN cards, and EDGE. It also works best in conjunction with efficiency enhancements schemes such as the Doherty amplifier or Envelope Tracking/EER, since it does little to enhance the overall efficiency of the basic amplifier, but it can have an enormous effect on linearity.

Another limitation of digital predistortion is the bandwidth expansion, which puts greater burden on the sampling rate of the DAC, as well as the bandwidth of the subsequent RF stages. The magnitude of the bandwidth expansion depends on the desired linearity improvement; if IM3 is the main source of the undesirable distortion, then a bandwidth expansion by a factor of three is required.

6. Conclusions

Upcoming 4G wireless systems will require innovations in all areas of power amplifier design and development. The migration of power amplifiers from III-V to silicon-based technologies and the use of digital techniques to correct the imperfections and limitations of RF power amplifiers represent the next stage of development for high-frequency wireless devices, especially since transistor technology is nearly "maxed-out" in performance. These new approaches will lead to dramatic improvements in the efficiency and bandwidth capabilities of multi-mode wireless transmitters.

7. Acknowledgments

The authors wish to thank Professors Ian Galton and Gabriel Rebeiz of UCSD and Professor Leo DeVreede of TU Delft for valuable discussions. We also acknowledge the generous support of Ericsson, Nokia, Conexant, Motorola, Freescale, Cree, Nitronex, ST Microelectronics, the UCSD Center for Wireless Communications and the California Institute for Telecommunications and Information Technology for their support.

8. Bibliography

[1] C. Zhao, R. J. Baxley, G. T. Zhou, D. Boppana, and J. S. Kenney, "Constrained clipping for crest factor reduction in multiple-user OFDM," in *Radio and Wireless Symposium, 2007 IEEE*, 9-11 Jan. 2007, pp. 341–344.

[2] P. Swaroop and K. G. Gard, "Crest factor reduction through in-band and out-of-band distortion optimization," in *Radio and Wireless Symposium, 2008 IEEE*, 22-24 Jan. 2008, pp. 759–762.

[3] J.-H. Chen and J. S. Kenney, "A crest factor reduction technique for WCDMA polar transmitters," in *Radio and Wireless Symposium, 2007 IEEE*, 9-11 Jan. 2007, pp. 345–348.

[4] R. Sperlich, Y. Park, G. Copeland, and J. Kenney, "Power amplifier linearization with digital pre-distortion and crest factor reduction," in *Microwave Symposium Digest, 2004 IEEE MTT-S International*, vol. 2, 6-11 June 2004, pp. 669–672Vol.2.

[5] O. Vaananen, J. Vankka, T. Viero, and K. Halonen, "Reducing the crest factor of CDMA downlink signal by adding unused channelization codes," in *Spread Spectrum Techniques and Applications, 2002 IEEE Seventh International Symposium on*, vol. 2, 2-5 Sept. 2002, pp. 455–459vol.2.

[6] R. Baxley, C. Zhao, and G. Zhou, "Magnitude-scaled selected mapping: A crest factor reduction scheme for OFDM without side-information transmission," in *Acoustics, Speech and Signal Processing, 2007. ICASSP 2007. IEEE International Conference on*, vol. 3, 15-20 April 2007, pp. III–373–III–376.

[7] A. Wegener, "High-performance crest factor reduction processor for WCDMA and OFDM applications," in *Radio Frequency Integrated Circuits (RFIC) Symposium, 2006 IEEE*, 11-13 June 2006, p. 4pp.

[8] J. Miller, "Reduce power consumption in the mobile handset's radio chain," *Mobile Handset Design Line*, 2006.

[9] J. Deng, P. Gudem, L. Larson, D. Kimball, and P. Asbeck, "A SiGe PA with dual dynamic bias control and memoryless digital predistortion for WCDMA handset applications," in *Radio Frequency integrated Circuits (RFIC) Symposium, 2005. Digest of Papers. 2005 IEEE*, 12-14 June 2005, pp. 247–250.

[10] H. Chireix, "High power outphasing modulation," *Proc. IRE*, vol. 23, no. 9, pp. 1370–1392, Nov. 1935.

[11] F. Raab, "Drive modulation in Kahn-technique transmitters," in *Microwave Symposium Digest, 1999 IEEE MTT-S International*, vol. 2, 13-19 June 1999, pp. 811–814vol.2.

[12] F. Raab and D. Rupp, "High-efficiency single-sideband HF/VHF transmitter based upon envelope elimination and restoration," in *HF Radio Systems and Techniques, 1994., Sixth International Conference on*, 4-7 Jul 1994, pp. 21–25.

[13] F. Raab, B. Sigmon, R. Myers, and R. Jackson, "High-efficiency L-band Kahn-technique transmitter," in *Microwave Symposium Digest, 1998 IEEE MTT-S International*, vol. 2, 7-12 June 1998, pp. 585–588vol.2.

[14] F. Raab and D. Rupp, "High-efficiency multimode HF/VHF transmitter for communication and jamming," in *Military Communications Conference, 1994. MILCOM '94. Conference Record, 1994 IEEE*, 2-5 Oct. 1994, pp. 880–884vol.3.

[15] D. Su and W. McFarland, "An IC for linearizing RF power amplifiers using envelope elimination and restoration," in *Solid-State Circuits Conference, 1998. Digest of Technical Papers. 45th ISSCC 1998 IEEE International*, 5-7 Feb. 1998, pp. 54–55,412.

[16] G. Hanington, P. Chen, V. Radisic, T. Itoh, and P. Asbeck, "Microwave power amplifier efficiency improvement with a 10 MHz HBT dc-dc converter," in *Microwave Symposium Digest, 1998 IEEE MTT-S International*, vol. 2, 7-12 June 1998, pp. 589–592vol.2.

[17] N. Schlumpf, M. Declercq, and C. Dehollain, "A fast modulator for dynamic supply linear RF power amplifier," in *Solid-State Circuits Conference, 2003. ESSCIRC '03. Proceedings of the 29th European*, 16-18 Sept. 2003, pp. 429–432.

[18] S. Cripps, *RF Power Amplifiers for Wireless Communications.* Artech House, 1999.

[19] F. Wang, A. Ojo, D. Kimball, P. Asbeck, and L. Larson, "Envelope tracking power amplifier with pre-distortion linearization for WLAN 802.11g," in *Microwave Symposium Digest, 2004 IEEE MTT-S International*, vol. 3, 6-11 June 2004, pp. 1543–1546Vol.3.

[20] J. Staudinger, B. Gilsdorf, D. Newman, G. Norris, G. Sadowniczak, R. Sherman, and T. Quach, "High efficiency CDMA RF power amplifier using dynamic envelope tracking technique," in *Microwave Symposium Digest., 2000 IEEE MTT-S International*, vol. 2, 11-16 June 2000, pp. 873–876vol.2.

[21] B. Sahu and G. Rincon-Mora, "A high-efficiency, dual-mode, dynamic, buck-boost power supply IC for portable applications," in *VLSI Design, 2005. 18th International Conference on*, 3-7 Jan. 2005, pp. 858–861.

[22] J. Groe and L. Larson, *CDMA Mobile Raiod Design.* Artech House, 2000.

[23] T. Sowlati, Y. Greshishchev, C. T. Salama, G. Rabjohn, and J. Sitch, "Linear transmitter design using high efficiency class E power amplifier," in *Personal, Indoor and Mobile Radio Communications, 1995. PIMRC'95. 'Wireless: Merging onto the Information Superhighway'., Sixth IEEE International Symposium on*, vol. 3, 27-29 Sept. 1995, p. 1233.

[24] T. Sowlati, D. Rozenblit, E. MacCarthy, M. Damgaard, R. Pullela, D. Koh, and D. Ripley, "Quad-band GSM/GPRS/EDGE polar loop transmitter," in *Solid-State Circuits Conference, 2004. Digest of Technical Papers. ISSCC. 2004 IEEE International*, 15-19 Feb. 2004, pp. 186–521Vol.1.

[25] A. Hietala, "A quad-band 8PSK/GMSK polar transceiver," in *Radio Frequency integrated Circuits (RFIC) Symposium, 2005. Digest of Papers. 2005 IEEE*, 12-14 June 2005, pp. 9–12.

[26] D. K. L. L. F. Wang, A. Yang and P. Asbeck, "Design of wide-bandwidth envelope tracking power aplifiers for OFDM applications," *IEEE Trans. Microwave Theory and Techniques*, vol. 53, pp. 1244–1255, April 2005.

[27] F. Wang, D. Kimball, J. Popp, A. Yang, D. Lie, P. Asbeck, and L. Larson, "Wideband envelope elimination and restoration power amplifier with high efficiency wideband envelope amplifier for WLAN 802.11g applications," in *Microwave Symposium Digest, 2005 IEEE MTT-S International*, 12-17 June 2005, p. 4pp.

[28] J.-H. Chen, K. U-yen, and J. Stevenson Kenney, "An envelope elimination and restoration power amplifier using a CMOS dynamic power supply circuit," in *Microwave Symposium Digest, 2004 IEEE MTT-S International*, vol. 3, 6-11 June 2004, pp. 1519–1522Vol.3.

[29] J. Popp, D. Lie, F. Wang, D. Kimball, and L. Larson, "Fully-integrated highly-efficient RF Class E SiGe power amplifier with an envelope-tracking technique for EDGE applications," in *Radio and Wireless Symposium, 2006 IEEE*, 17-19 Jan. 2006, pp. 231–234.

[30] F. Wang, D. F. Kimball, D. Y. Lie, P. M. Asbeck, and L. E. Larson, "A monolithic high-efficiency 2.4-GHz 20-dBm SiGe BiCMOS envelope-tracking OFDM power amplifier," *Solid-State Circuits, IEEE Journal of*, vol. 6, pp. 1271–1281, 2006.

[31] W. Doherty, "A new high efficiency power amplifier for modulated waves," *Proc. IRE*, vol. 24, no. 9, pp. 1163–1182, Sept. 1936.

[32] B. Kim, J. Kim, I. Kim, and J. Cha, "The Doherty power amplifier," *Microwave Magazine, IEEE*, vol. 7, no. 5, pp. 42–50, Oct. 2006.

[33] K.-J. Cho, W.-J. Kim, J.-Y. Kim, J.-H. Kim, and S. Stapleton, "N-way distributed Doherty amplifier with an extended efficiency range," in *Microwave Symposium, 2007. IEEE/MTT-S International*, 3-8 June 2007, pp. 1581–1584.

[34] B. Shin, J. Cha, J. Kim, Y. Woo, J. Yi, and B. Kim, "Linear power amplifier based on 3-way Doherty amplifier with predistorter," in *Microwave Symposium Digest, 2004 IEEE MTT-S International*, vol. 3, 6-11 June 2004, pp. 2027–2030Vol.3.

[35] J. Gajadharsing, O. Bosma, and P. van Westen, "Analysis and design of a 200 w LDMOS based Doherty amplifier for 3 G base stations," in *Microwave Symposium Digest, 2004*

IEEE MTT-S International, vol. 2, 6-11 June 2004, pp. 529–532Vol.2.

[36] T. Yamamoto, T. Kitahara, and S. Hiura, "High-linearity 60W doherty amplifier for 1.8ghz w-cdma," in *Microwave Symposium Digest, 2006. IEEE MTT-S International*, June 2006, pp. 1352–1355.

[37] N. Ui, H. Sano, and S. Sano, "A 80w 2-stage GaN HEMT Doherty amplifier with 50dBc ACLR, 42% efficiency 32dB gain with DPD for WCDMA base station," in *Microwave Symposium, 2007. IEEE/MTT-S International*, 3-8 June 2007, pp. 1259–1262.

[38] Y. Zhao, A. Metzger, P. Zampardi, M. Iwamoto, and P. Asbeck, "Linearity improvement of HBT-based Doherty power amplifiers based on a simple analytical model," in *Microwave Symposium Digest, 2006. IEEE MTT-S International*, June 2006, pp. 877–880.

[39] M. Iwamoto, A. Williams, P.-F. Chen, A. Metzger, C. Wang, L. Larson, and P. Asbeck, "An extended Doherty amplifier with high efficiency over a wide power range," in *Microwave Symposium Digest, 2001 IEEE MTT-S International*, vol. 2, 20-25 May 2001, pp. 931–934vol.2.

[40] J. Kang, D. Yu, K. Min, and B. Kim, "A ultra-high PAE Doherty amplifier based on 0.13-mum CMOS process," *Microwave and Wireless Components Letters, IEEE*, vol. 16, no. 9, pp. 505–507, Sept. 2006.

[41] D. Mercier, K. Van Caekenberghe, and G. Rebeiz, "Miniature RF MEMS switched capacitors," in *Microwave Symposium Digest, 2005 IEEE MTT-S International*, 12-17 June 2005, p. 4pp.

[42] K. Buisman, L. de Vreede, L. Larson, M. Spirito, A. Akhnoukh, T. Scholtes, and L. Nanver, "Distortion-free varactor diode topologies for RF adaptivity," in *Microwave Symposium Digest, 2005 IEEE MTT-S International*, 12-17 June 2005, p. 4pp.

[43] W. Neo, X. Liu, Y. Lin, L. de Vreede, L. Larson, S. Spirito, A. Akhnoukh, A. de Graauw, and L. Nanver, "Improved hybrid SiGe HBT class-AB power amplifier efficiency using varactor-based tunable matching networks," in *Bipolar/BiCMOS Circuits and Technology Meeting, 2005. Proceedings of the*, 9-11 Oct. 2005, pp. 108–111.

[44] A. Hajimiri, "Fully integrated RF CMOS power amplifiers - a prelude to full radio integration," in *Radio Frequency integrated Circuits (RFIC) Symposium, 2005. Digest of Papers. 2005 IEEE*, 12-14 June 2005, pp. 439–442.

[45] P. Haldi, D. Chowdhury, G. Liu, and A. Niknejad, "A 5.8 GHz linear power amplifier in a standard 90nm CMOS process using a 1V power supply," in *Radio Frequency Integrated Circuits (RFIC) Symposium, 2007 IEEE*, 3-5 June 2007, pp. 431–434.

[46] K. H. An, Y. Kim, O. Lee, K. S. Yang, H. Kim, W. Woo, J. J. Chang, C.-H. Lee, H. Kim, and J. Laskar, "A monolithic voltage-boosting parallel-primary transformer structures for fully integrated CMOS power amplifier design," in *Radio Frequency Integrated Circuits (RFIC) Symposium, 2007 IEEE*, 3-5 June 2007, pp. 419–422.

[47] I. Aoki, S. Kee, D. Rutledge, and A. Hajimiri, "A fully-integrated 1.8-V, 2.8-W, 1.9-GHz, CMOS power amplifier," in *Radio Frequency Integrated Circuits (RFIC) Symposium, 2003 IEEE*, 8-10 June 2003, pp. 199–202.

[48] T. Sowlati and D. Leenaerts, "A 2.4-GHz 0.18-um CMOS self-biased cascode power amplifier," *IEEE Journal of Solid-State circuits*, vol. 38, pp. 1318–1324, 2003.

[49] M. Briffa and M. Faulkner, "Dynamically biased Cartesian feedback linearization," in *Vehicular Technology Conference, 1993 IEEE 43rd*, 18-20 May 1993, pp. 672–675.

[50] J. Dawson and T. Lee, "Automatic phase alignment for a fully integrated cmos Cartesian feedback power amplifier system," in *Solid-State Circuits Conference, 2003. Digest of Technical Papers. ISSCC. 2003 IEEE International*, 2003, pp. 262–492vol.1.

[51] S. Chung, J. W. Holloway, and J. L. Dawson, "Open-loop digital predistortion using Cartesian feedback for adaptive rf power amplifier linearization," in *Microwave Symposium, 2007. IEEE/MTT-S International*, 3-8 June 2007, pp. 1449–1452.

[52] N. Mizusawa, S. Tsuda, T. Itagaki, and K. Takagi, "A polynomial-predistortion transmitter for WCDMA," *Solid-State Circuits Conference, 2007. ISSCC 2007. Digest of Technical Papers. IEEE International*, pp. 350–608, 11-15 Feb. 2007.

[53] Y. Park, W. Woo, R. Raich, J. Stevenson Kenney, and G. Zhou, "Adaptive predistortion linearization of RF power amplifiers using lookup tables generated from subsampled data," in *Radio and Wireless Conference, 2002. RAWCON 2002. IEEE*, 11-14 Aug. 2002, pp. 233–236.

[54] Y. Seto, S. Mizuta, K. Oosaki, and Y. Akaiwa, "An adaptive predistortion method for linear power amplifiers," in *Vehicular Technology Conference Proceedings, 2000. VTC 2000-Spring Tokyo. 2000 IEEE 51st*, vol. 3, 15-18 May 2000, pp. 1889–1893vol.3.

[55] E. Cottais, Y. Wang, and S. Toutain, "Experimental results of power amplifiers linearization using adaptive baseband digital predistortion," in *2005 European Microwave Conference*, vol. 3, 4-6 Oct. 2005, p. 4pp.

[56] J. Cavers, "The effect of quadrature modulator and demodulator errors on adaptive digital predistorters," in *Vehicular Technology Conference, 1996. 'Mobile Technology for the Human Race'., IEEE 46th*, vol. 2, 28 April-1 May 1996, pp. 1205–1209vol.2.

IEEE 2008 Custom Intergrated Circuits Conference (CICC)

An Analog Enhanced All Digtial RF Fractional-N PLL With Self-Calibrated Capability

Ping-Ying Wang, Jing-Hong Conan Zhan, Hsiang-Hui Chang and Bing-Yu Hsieh

MediaTek Inc., Hsin-Chu, Taiwan

ABSTRACT

In this paper, an analog enhanced all digital fractional-N PLL is proposed. An analog feed-forward circuits replace the time-to-digital converter used in conventional all digital PLL (ADPLLs) to provide a linear phase modulation path which is insensitive to quantization error and non-linearity of digital controlled oscillator (DCO). Its advantages include 1) Eliminating fractional spurs and noise induced by quantization error and the latency induced by the digital circuits in ADPLLs 2) Relaxing both digital controlled oscillator (DCO) and analog feed-forward circuit design requirements. 3) Providing a linear phase modulation path which can be self calibrated by using the digital loop filter. 4) Reducing loop filter area by using digital loop filter. The fractional spurs are 9 to 30dB lower than the latest reported ADPLLs. At 3.6GHz under fractional-N mode operation, the fractional spur is under -75 dBc, the phase noise is -115.6dBc/Hz @400KHz, -134.9dBc/Hz @3MHz. The performance satisfies GSM/GPRS/EDGE system requirements.

Keywords: ADPLL, Digital controlled Oscillator

I. INTRODUCTION

All-digital PLLs (ADPLLs) [1]][2][3][4] are of great research interest recently because their digital loop filters can be precisely controlled. Besides, capacitors traditionally required in the loop filter are not necessary, which reduces area and avoids the associated gate-leakage-induced noise. Despite all the benefits, such PLL is still limited to few applications because non-linearity of the loop dynamics and latency induced by quantization errors and digital circuits degrades the spur/noise performance when the fractional-N [5] or direct frequency modulation technique [6][7] is applied into ADPLLs. The problems are described below.

In ADPLLs, non-linearity of the loop could be introduced by quantization error, non-linearity and meta-stability of time-to-digital converter (TDC) and the DCO. The non-linearity of the loop will induce fractional spurs when a high order delta-sigma fractional-N technique is applied into PLLs [5]. To suppress the fractional spur induced by the quantization error and noise of DCO in ADPLLs, performance optimization for both spur and DCO's noise is more difficult than that of a conventional digital charge pump PLL thus the design of ADPLLs is complicated. Furthermore, the spurs caused by non-linearity cannot be eliminated by a digital loop filter because the resolution of the digital loop filter limited by DCO are finite and the quantization itself cannot be cancelled by a digital algorithm or a digital loop filter only.

Using a PLL as a modulator [6][7], the linearity of VCO/DCO gain is important to minimize phase error. It is harder to guarantee the linearity of DCO gain [2][3] without any additional circuits [8] compared to the conventional VCO design using an analog varactor, thus the design complexity is also increased.

Another problem is drawbacks induced by TDC used in digital phase frequency detector of ADPLL. For delta-sigma fractional-N operation, the TDC will require large dynamic range and fine resolution with minimum area overhead. A ring type TDC [2] [4] is used, however, it is more prone to supply noise because the jitter will be accumulated in ring type circuits. The inverter-based designs of TDC also induce large transient power to increase noise floor and a dedicated power pin is usually needed so the cost is increased.

Moreover, the TDC induce latency because the digital circuits need several clock cycles to process digitized phase errors. The latency will limit the highest reference frequency used in PFD, thus the conventional ADPLLs [1][2][3][4] can not be applied into high bandwidth system such as high speed clock data recovery applications [9][10]. Besides, the design of the high resolution TDC is still full custom as design of analog circuits so the design cost is not reduced compared with the digital charge pump PLL using a simple charge pump. In addition, the loop gain of the ADPLLs depends on resolution of the TDC which is also sensitive to temperature, process and voltage so the design is not robust compared to the design of digital charge pump PLLs.

To alleviate all problems mentioned above. This paper reports an analog enhanced all digital sigma-delta fractional-N synthesizer which replaces the noisy TDC by an analog feed-forward path. The analog feed-forward circuits eliminates the noise and fractional spur problems, has good PSRR, and produces lower transient noise. The digital integral approach maintained benefits of using digital capacitors. The analog feed-forward design parameter including the proportional gain of DCO can be calibrated by the digital integral path with 20 bit resolution, relaxing the analog feed-forward path circuit's design compared to the conventional analog PLL. The synthesizer is also insensitive to non-linearity and quantization error of DCO, relaxing the DCO design requirements compared to the conventional ADPLLs. Moreover, the architecture provides a linear phase modulation path and self calibrated capability so it is easy to be applied into advanced modulation scheme [6][7].

II. ARCHITECTURE

The block diagram of the proposed synthesizer is shown in Fig. 1. The proportional path consists of a conventional PFD and an analog feed-forward circuits (AFFC), which is AC coupled to the DCO. The analog feed-forward circuit corrects phase in the time domain using output generated from a conventional PFD. In the integral path, the frequency of the reference clock is sampled by a bang-bang PFD and is integrated digitally. The analog feed-forward circuit and the bang-bang operation eliminate the TDC which is usually required in conventional ADPLL. The associated transient and switching noise are therefore greatly reduced.

AFFC uses phase error signals generated from a conventional PFD, whose output pulse width is proportional to phase error to adjust the capacitance of the varactor. The relation between frequency, capacitance and control voltage is shown in Fig. 2(a).

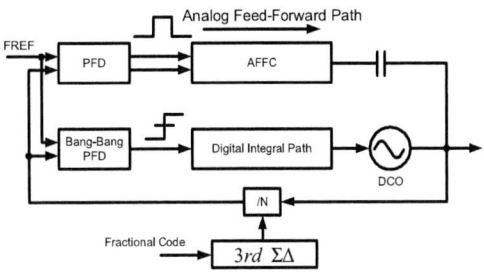

Fig. 1 The blocks diagram of the analog enhanced all digital PLL

978-1-4244-2018-6/08 $25.00 © 2008 IEEE   797

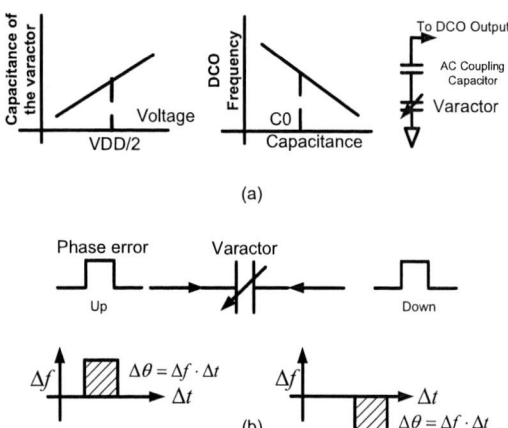

(a)

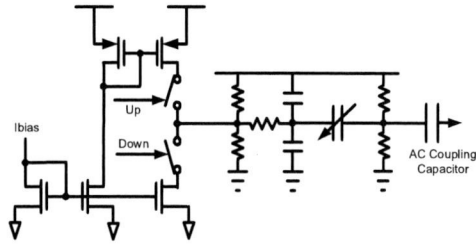

(b)

Fig. 2 Operation principle of AFFC

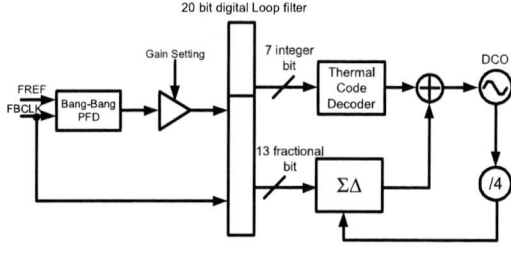

Fig. 3 The schematic of AFFC

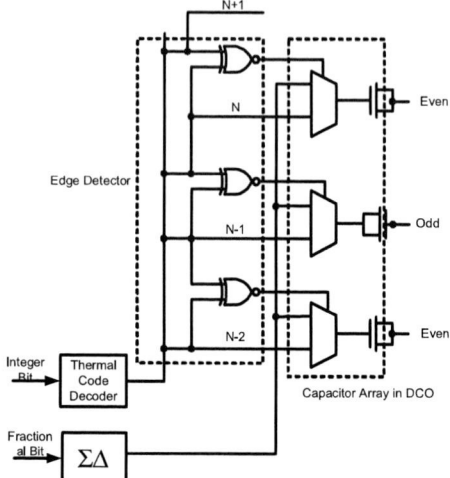

Fig. 5 The function blocks and schematic of our proposed architecture by the edge decoder as shown in Fig.5. Every capacitor can be switched to be either fractional bit or integer bit by control signal which is generated by detecting the transition bit of thermal code. Since fractional bit is turned on or off to be the integer bit when fractional code overflows or underflows, there is no discontinuity of capacitance. The monotonicity is maintained. The bang-bang operation also makes sure that only one unit capacitor in the DCO is switched when the phase error is update so the switching noise is also reduced. The gain of DCO in the analog feed-forward path is minimized to reduce the spur induced by the switching noise in AFFC The measured reference spur is about -86dBc, thus proving that the switching noise is reduced.

In next section, we will use s-domain linear model analysis to prove that the loop bandwidth is determined by the gain in the analog feed-forward path and the gain variation induced by bang-bang operation in the integral path has no impact on loop bandwidth.

III. ANALYTIC ANALYSIS

The linear model for the proposed PLL is shown in Fig. 6. The analog feed-forward path corrects phase error in the time domain. The integral path circuits track the frequency of the reference clock in the digital domain by a bang-bang PFD.

The PLL loop dynamics is analyzed as follows. The equivalent capacitance of the digital filter is derived by comparing the integral path digital filter with linear model of conventional digital charge pump PLL:

$$Hcon(s) = (Kp + \frac{Kz}{s \cdot C}) \cdot \frac{Kvco}{N \cdot s} \qquad (1)$$

The coefficient for the $1/s^2$ term represents the frequency gain, which is defined as frequency correction by the loop divided by phase error and divider N. In a conventional charge pump PLL, the coefficient is a constant:

$$\frac{Ich \cdot Kvco}{2\pi N \cdot C}$$

This expression shows that the frequency correction by the loop is linearly proportional to the phase error in a conventional charge pump-based PLL. For our bang-bang operation the frequency correction is not linearly proportional to the phase error. Instead, the frequency correction is constant for arbitrary phase error thus

---

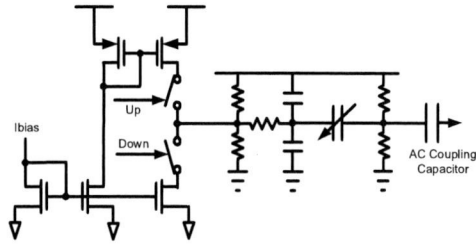

Fig. 4 The blocks diagram of the digital integral path's circuits

The correction of phase is linearly proportional to the phase error because the frequency / phase change of the DCO is linearly proportional to change of the capacitance as shown in Fig. 2(b).

A current driver is chosen over a voltage driver whose common mode is biased at VDD/2 by a resistive voltage divider to provide higher gain and good PSRR as shown in Fig. 3. A low pass filter (LPF) filters out transient ripple. This low pass filter can use thin oxide devices to reduce area because gate leakage is of no concern.

The digital integral path circuit is shown in Fig. 4. It has a 4-bit gain control, a 20-bit digital loop filter to replace the large zero capacitor in a conventional analog PLL, a 13-bit delta-sigma modulator (SDM), an edge detector and a 7-bit thermal code decoder. Five Hz effective frequency resolution is achieved by 900MHz SDM to dither the fractional capacitor which has the same capacitance as the integer bit has. The monotonicity of frequency is maintained because the integer bit is thermal-coded and fractional bit tracks the latest transient bit of the thermal code

978-1-4244-2018-6/08 $25.00 © 2008 IEEE

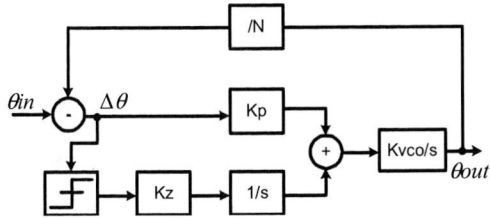

Fig. 6 Block diagram of the hybrid type I/II PLL

the coefficient will depend on the phase error which is

$$\frac{\Delta f}{2\pi \cdot \Delta te \cdot N},$$

where Δf is the step size of the frequency correction and Δte is the timing error produced by the PFD. The frequency gain will become infinity as the phase error approaches zero. Comparing these equations, the effective capacitance C_{eff} in the digital integral path is given by

$$Ceff = \frac{\Delta te \cdot Ich \cdot Kvco}{\Delta f}$$

In a conventional fractional-N PLL with a high order multi-bit delta-sigma modulator, Ich is 200uA and Kvco is 25MHz/V. The timing error is from -4Tvco to 4Tvco, or equivalently, -1.2ns to 1.2ns for 3.6GHz carrier frequency. The minimum frequency step Δf is 5 Hz in our design. Therefore the effective capacitance of the digital loop filter is on the μF order, which is three orders of magnitude higher than capacitance values traditionally integrated on chip for analog PLLs.

Large effective capacitance leads to a high damping factor and reduces jitter peaking. This is a type I loop filter characteristic. Moreover, because of the bang-bang operation in the integral path, the static phase error is kept zero due to bang-bang operation of the integral path, which is a typical characteristic in a type II loop filter.

The loop bandwidth (-3dB of closed loop transfer function) for a conventional charge pump PLL is [11]

$$\omega-3dB = \omega n \left\{ 1 + 2\zeta^2 + \left[\left(1 + 2\zeta^2\right)^2 + 1 \right]^{1/2} \right\}^{1/2} \quad (2)$$

where ωn is the nature frequency and ζ is the damping factor. The relation between ωn and ζ is given by

$$Kvco \cdot Ich \cdot R / 2\pi N = 2\zeta \cdot \omega n,$$

where Kvco is the effective DCO gain in the analog feed-forward path, Ich is the charge pump current used in the analog feed-forward path and R is resistance in the proportional path s shown in Fig. 7. Approximating ζ by infinity as reasoned in paragraphs above, we have

$$\omega-3dB = Kvco \cdot Ichp \cdot R / 2\pi N$$

The loop bandwidth is independent of effective capacitance in the integration path. Thus the gain variation induced by the bang-bang operation will not affect the loop bandwidth so the loop bandwidth is well controlled by the analog feed-forward circuits with the linear phase correction. In the next section, we will explain how the gain of the analog feed-forward circuits can be calibrated by the digital integral path to demonstrate the self-calibration capability of the design.

IV. SELF-CALIBRATION

The principle of the calibration is based on 1) the common mode voltage variation in AFFC will result in frequency drift which will be compensated by the digital loop filter in the digital

integral path. 2) The digital code of the loop filter can be calibrated by the fractional code of the SDM in the closed loop operation to obtain the exact relation between the fractional frequency and the digital code in the digital loop filter. The proposed digital method also can be used to calibrate analog components including current source mismatch, switching noise and DCO gain in the analog proportional path, thus the design is robust while keeping comparable performance of analog PLLs and the feasibility of ADPLLs.

The calibration current is shown in Fig. 7. A current mirrored from the charge pump current with a ratio β is injected into a resistor. The DCO gain in the analog feed-forward path is defined as the frequency variation of the DCO divided by the voltage variation in the varactor which is $\Delta f/(\beta$ Ichp*R), and the -3dB frequency is

$$\omega-3dB = \Delta f / \beta \cdot 2\pi N \quad (3)$$

where Δf is the frequency variation when the current is injected into the resistor. Since β is well defined by device geometry ratio, Equation (3) shows the -3dB frequency only depends on Δf, which can be calibrated in closed loop with high resolution as described below.

Fig. 7 shows the functional diagram for the proposed calibration technique. When the calibration current is injected into a resistor, the varactor value changes due to the voltage drop β*Ichp*R in the proportional path. This voltage in turn produces a frequency offset. The digital loop filter in the digital integral path will react to compensate for this intentionally injected frequency error, and output a corresponding digital code. This code is recorded by flip-flops as the target value. The calibration current is then turned off. The fractional code of the divider now varies depending on the output of the digital comparator, which flags if the digital code approach the target value or not. The calibration loop reads the flag to adjust fractional code of the PLL until the flag indicates that the digital code is equal to the target value. Fractional code multiplied by reference clock represents the frequency offset Δf, and can be used to calculate the close loop bandwidth according to (3). The low frequency noise of the DCO is tracked by the closed loop operation so the resolution of the frequency calibration can be up to KHz order which depends on the digital resolution of the SDM divider. In this design, the fractional divider is fifteen bits so that 1 KHz resolution is obtained. Since AFFC provide a linear phase modulation path, which can be self calibrated by the digital filter in the loop, the architecture is easy to be applied into the modulators [6][7].

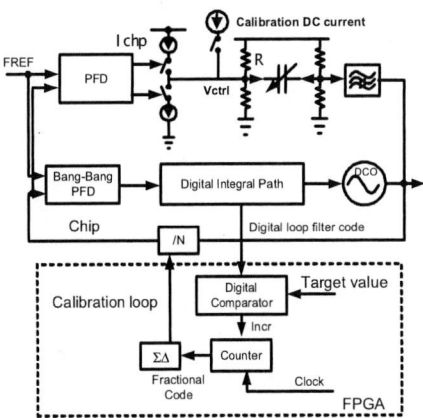

Fig. 7 Functional diagram of proposed loop bandwidth calibration

V. SILICON RESULTS

The phase noise in Fig. 8 is measured phase noise at 3.6GHz output with 26MHz reference clock when the fractional spurs are occurred at 400KHz harmonics, which is the worst case for GSM. No spur is observed and therefore the fractional spur is at least below the measured phase noise level, which is -75dBc. This is 10~30dB lower than reported in [2][3] as shown in Table 1. The worst fractional spur is 9dB lower than reported in [4] when operating under the same condition, where quantization error of the DCO is 20KHz and fractional code is 400KHz.

Fig. 9 shows the measured phase noise has low jitter peaking as a PLL with type I filter. The curve labeled as type I loop filter is obtained by disabling the digital integral path of our design. The two curves are overlapped to prove the prediction of the analytic analysis in section III, which proves that the loop dynamics of our design is the same as the PLL with type I loop filter. The static phase error is zero because the measured output of the bang-bang PFD shows equal possibility for phase lead/lag. With a zero static phase error and low jitter peaking, this architecture can be also applied into high bandwidth applications [9] and low jitter peaking application such as SONET [10].

The calibrated fractional code of the fractional divider is 39/2048 when the code of the digital loop filter is equal to the target value and the input frequency is 26MHz, so the frequency offset Δ f induced by the voltage offset β *Ichp*R is equal to 26MHz multiplied 39/2048. According to (3), the loop bandwidth is 96KHz with Δ f= 495.12KHz, β=1/168 and N=138. The calibrated DCO gain (which is Δ f/(β *Ichp*R)) in the proportional path is 11.90MHz/V. We used spectrum analyzer to measure the DCO gain to verify the calibrated value. Fig. 10 shows DCO gain is 11.88MHz/V. The close agreement between the measured and calibrated values validates the proposed calibration technique.

The prototype is fabricated in 130nm CMOS and occupies 0.85mm² with 40mA current consumption at 1.5V.

References

[1] J. Zhuang, Q. Du and T. Kwasniewski,"A 4GHz Low Complexity ADPLL-based Frequency Synthesizer in 90nm CMOS", *CICC Dig Tech Papers*, Sept. 2007

[2] Chun-Ming Hsu, M. Z. Straayer, M. H. Perrott., " A Low-Noise Wide-BW 3.6GHz digital sigma-delta Fractional-N Frequency Synthesizer with a Noise-Shaping Time-to-Digital Converter and Quantization Noise Cancellation," *ISSCC Dig. Tech. Papers.*, Feb 2008, pp.340–341.

[3] Colin Weltin-Wu, Enrico Temporiti, D. Baldi, F. Svelto., " A 3GHz Fractional-N All-Digital PLL with Precise Time-to-Digital Converter Calibration and Mismatch Correction," *ISSCC Dig. Tech. Papers.*,Feb 2008, pp.344–345

Table I A brief comparison of fractional spur performance for all digital fractional-N PLLs

	ISSCC 2008 [2]	ISSCC 2008[3]	ISSCC 2008[4]	Our Design
Fractional Spur	-53dBc @1MHz Channel Space	-45dBc@ 100KHz Channel Space	-66dBc@ 400KHz Channel Space	-75dBc@ 400KHz Channel Space
Frequency	3.6GHz	3GHz	3.6GHz	3.6GHz

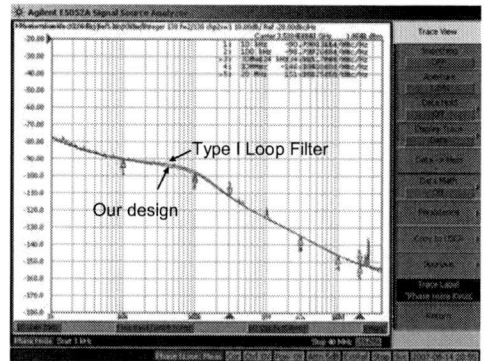

Fig. 9 Comparison of measured phase noise between type I and our proposed design

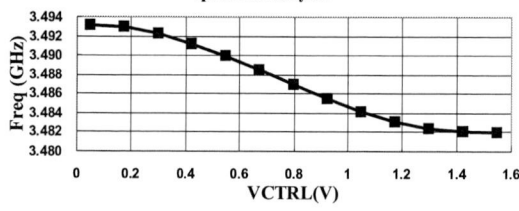

Fig. 10 Measured DCO gain in the proportional path by a spectrum analyzer to verify the technique proposed.

[4] Hsiang-Hui Chang, Ping-Ying Wang, Jing-Hong Zhan and Bing-Yu Hesieh "A Fractional Spur Free All-Digital PLL with Loop Gain Calibration and Phase Noise Cancellation for GSM/GPRS/EDGE," *ISSCC Dig. Tech. Papers.*Feb 2008, pp200~201

[5] Riley T.A.D., Copeland M.A., and Kwasniewski T.A., "Delta-sigma modulation in fractiona-N synthesizer," *IEEE J. Solid-States Circuits*, vol. 28, pp. 553 - 559, May 1993

[6] M. H. Perrot, T. T. Tewksbury, C. G. Sodini, "A 27-mW CMOS Fractional-N Synthesizer Using Digital Compensation for 2.5-Mbls GFSK Modulation," *IEEE Journal of Solid- State Circuits,* Vol. 32, No. 12, December 1997, pp. 2048- 2060

[7] Vuketich, M.Scholz, R. Weigel and J. Fenk, "GSM 900/DCS 1800 fractional-N modulator with two-point-modulation," *IEEE MTT-S International Microwave Symposium Digest*, vol. 1, June 2002, pp. 425-428.

[8] R. B. Staszewski et al., N. Barton, M. –C. Lee and D. Leipold, ''ALL-Digital PLL and Transmitter for Mobile Phone,'' *IEEE Journal of Solid-State Circuits*, vol. 401, No. 125, pp. 2469 - 2482, Dec 2005

[9] Nicola Da Dalt et al., ".A Compact Triple-Band Low-Jitter Digital LC PLL with Programmable Coil in 130-nm CMOS", *IEEE Journal of Solid-State Circuits*, vol. 40, No. 7, pp. 1482 -1490, July 2005.

[10] M-H. Perrott et al., "A 2.5Gb/s Mulit-Rate 0.25um CMOS CDR Utilizing a Hybrid Analog/Digital Loop Filter",. *ISSCC Digest of Technical papers*, pp. 328-329, Feb. 2006.

[11] J. Encinas "Phase Locked Loops", Chapman & Hall press, pp 83

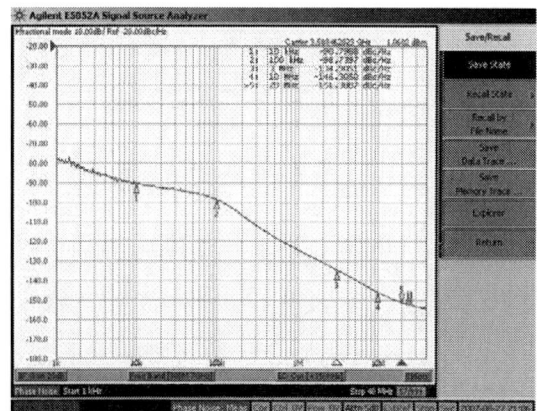

Fig. 8 The measured phase noise

IEEE 2008 Custom Intergrated Circuits Conference (CICC)

A ΔΣ Fractional-N Synthesizer with Customized Noise Shaping for WCDMA/HSDPA Applications

Xueyi Yu[2], Yuanfeng Sun[1], Woogeun Rhee[1], Zhihua Wang[1]
[1]*Institute of Microelectronics*
[2]*Department of Electronic Engineering*
Tsinghua University, Beijing 10084, China

Hyung Ki Ahn, Byeong-Ha Park
RF Development Team
Samsung Electronics, Co., Ltd.
Giheung-Gu, Yongin-City, Korea

Abstract-**This paper describes a quantization noise reduction method in ΔΣ fractional-N synthesizer design based on a semidigital approach. By employing a phase shifting technique, a low power hybrid finite impulse response (FIR) filtering is realized which is suitable for RF applications. A prototype fractional-N synthesizer is implemented in 180nm CMOS for WCDMA/HSDPA applications. Experimental results show that the proposed method can effectively suppress out-of-band phase noise to meet the phase noise mask requirements in various RF applications.**

I. INTRODUCTION

A fast growing 3G data service, High Speed Downlink Packet Access (HSDPA), needs to provide a peak data rate of up to 14.4Mb/s for the downlink as well as a high SNR at the output. Accordingly, HSDPA compliant transceivers have stringent requirements for the frequency synthesizer in terms of settling time and phase noise. For example, the settling time must be less than 150μs within 0.1ppm frequency accuracy or 50μs within 20ppm frequency accuracy. To have the EVM contribution of the PLL less than 2% of the overall transmitter, in-band phase noise should be less than –89dBc/Hz at 2GHz output [1].

Offering low in-band phase noise and fast settling time, a ΔΣ fractional-N synthesizer plays a key role in 3G multi-mode radio transceivers. However, the wideband ΔΣ fractional-N synthesizer often suffers from out-of-band phase noise since quantization noise from the ΔΣ modulator can be dominant over VCO phase noise at high frequencies. In most RF applications, frequency synthesizers need to meet both in-band and out-of-band phase noise requirements which are usually specified by a phase noise mask as illustrated in Fig. 1. Since the brick-wall phase noise mask gives a tight margin only at certain corner frequencies, the phase noise performances at those frequencies determine the overall noise performance of the frequency synthesizer. In Fig. 1, the phase noise at 3.5MHz offset frequency should be as low as –124dBc/Hz while –89dBc/Hz at 1MHz offset frequency is acceptable. Hence, meeting the phase noise requirement at 3.5MHz offset frequency is more important for the out-of-band phase noise performance.

In the previous work, a hybrid finite impulse response (FIR) noise filtering technique is proposed to suppress the out-of-band quantization noise [2]. Since the transfer function of the hybrid FIR filter can be designed regardless of the PLL loop dynamics, a customized quantization noise shaping is

possible. In other words, phase noise at certain frequency can be suppressed more by introducing the zero of the FIR filter at that frequency.

In this paper, we introduce a low power hybrid FIR filtering method and present a ΔΣ fractional-N synthesizer design with customized noise shaping for WCDMA/HSDPA applications.

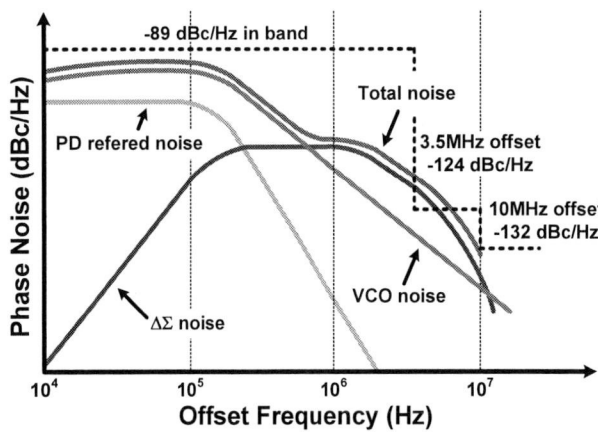

Fig. 1. Quantization noise effect in wideband ΔΣ synthesizer.

II. NOISE SHAPING WITH HYBRID FIR FILTER

Fig. 2 shows a conceptual diagram of the hybrid FIR filter. The multibit output of the ΔΣ modulator is loaded into the first register and then shifted by one clock cycle or several clock cycles per stage. Each multi-modulus divider (MMD) is sequentially controlled by the shifted output of each register. The multiple MMDs in parallel with sequential control bits perform an FIR filtering with respect to the input control bits. Since the input control bits are modulated by the ΔΣ modulator, the FIR filtering reduces the modulator quantization noise without affecting the loop dynamics of the PLL. Different from the all-digital FIR filter, the hybrid FIR filter does not increase quantization noise by having unit dc gain. Also, it is shown to be more immune to analog mismatch and linearity than DAC cancellation methods [2].

However, the hybrid FIR noise filtering method requires multiple MMDs, which is not suitable for low power RF applications. In this work, the multiple MMDs based on a phase shifting method are proposed so that the hybrid FIR

978-1-4244-2018-6/08 $25.00 © 2008 IEEE   801

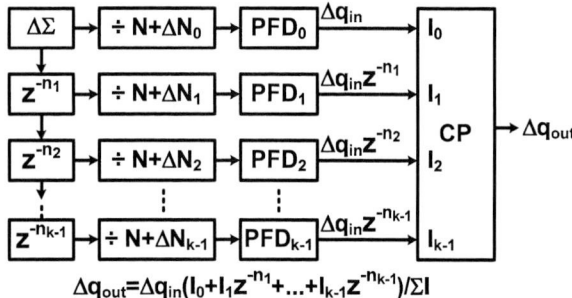

Fig. 2. Conceptual diagram of hybrid FIR filter.

$$\Delta q_{out}=\Delta q_{in}(I_0+I_1z^{-n_1}+\dots+I_{k-1}z^{-n_{k-1}})/\Sigma I$$

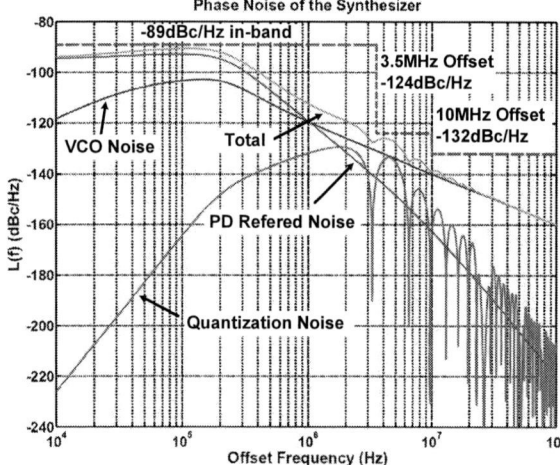

Fig. 3. Quantization noise with FIR filtering.

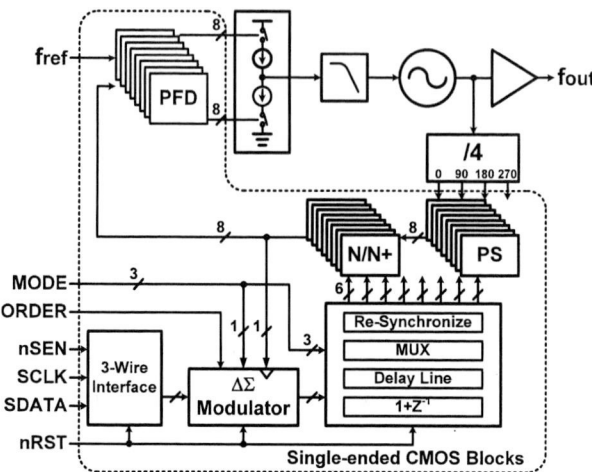

Fig. 4. Frequency synthesizer block diagram.

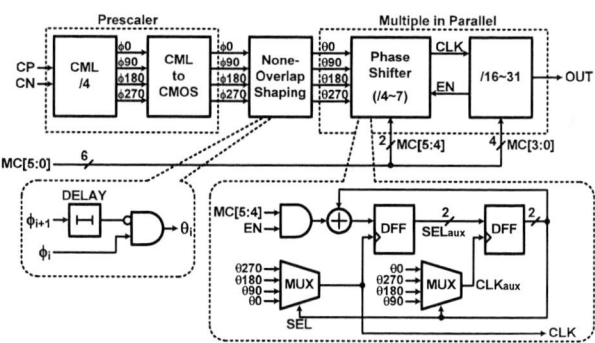

Fig. 5. PS-based multi-modulus divider.

filtering can be done at low frequency with standard CMOS logic.

Fig. 3 shows phase noise contribution from each source including quantization noise from a 4th-order single-loop $\Delta\Sigma$ modulator. The embedded FIR filter suppresses quantization noise without affecting the loop dynamics. Since the reference frequency is 26MHz, an 8-tap FIR filter is designed to have the zero frequency at 3.25MHz so as to alleviate the phase mask requirement specified by WCDMA/HSDPA standards.

III. IMPLEMENTATION

A. PLL Design

Fig. 4 shows a block diagram of the proposed synthesizer. The synthesizer employs a single high frequency CML divider and multiple single-ended phase shifters for hybrid FIR noise filtering. The CML divider has a fixed division ratio of four with quadrature output phases [3]. All of them are fed into a phase shifter (PS) followed by 5-stage single-ended 2/3 prescalers. The synthesizer employs eight PS-based MMDs and phase detectors in parallel, and all the phase errors are summed up at the multiinput charge pump. A singled-ended charge pump is used with an external 3rd-order loop filter resulting in type-2 4th-order PLL design. The LC VCO employs conventional cross-coupled CMOS topology. A 6-bit

binary-weighted capacitor array is added in the LC tank to provide a wide tuning range.

B. PS-based Divider

Fig. 5 illustrates the topology of the phase shifter and how it is connected with the preceding CML divider and succeeding single-ended divider. By cascading two stage of the CML-latch-based /2 divider, the quadrature phases are generated at the output followed by a CML-to-CMOS converter. The quadrature clocks are shaped to non-overlapping clocks before fed into the phase shifter to ease the timing of the logic circuits.

Fig. 6 shows the timing diagram of the phase shifter. The principle of the phase shifting logic is as follows:

- For a certain control word of the multiplexer, two clocks are selected, namely the main clock *CLK* and the auxiliary clock *CLK_aux*. The auxiliary clock leads the main clock by 90 degree in phase.
- At each rising edge of the main clock, the new control word for the multiplexer is loaded into the first register.
- A second register is used for buffering the control word. It is synchronized to the rising edge of the auxiliary clock. The output of this register controls the two multiplexer. In this way, an incorrect phase step between two adjacent clock edges can be avoided.

- The signal *EN* is used for shielding the modulus control word *MC*. It is generated by the succeeding divider to guarantee that the phase steps only once during a whole division period.

It can be noticed that, despite the multiphase operation, the CMOS phase shifter works at 1/4 of the VCO frequency. The state of the registers and other logic circuits in the phase shifter changes once only during the division period. As a result, the power consumption is minimized and it can be further scaled down with technology.

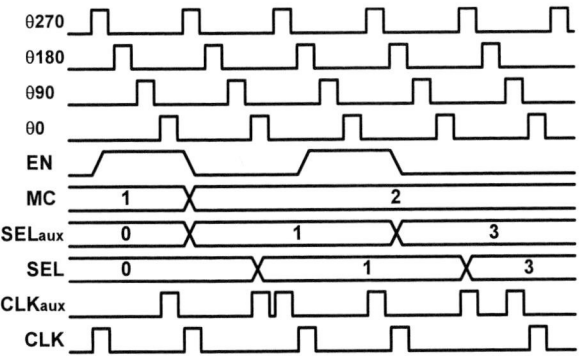

Fig. 6. Timing diagram of the phase shifter.

C. ΔΣ modulator

In the design of the multiphase multiinput charge pump, it is good to reduce maximum peak phase deviation for each input stage. The single-loop ΔΣ modulator has fewer output levels, resulting in smaller instantaneous phase error when the noise transfer function (NTF) is designed with Butterworth poles [4]. In this work, single-loop 4th and 5th order ΔΣ modulators are used with 20-bit input. Note that the Butterworth-based NTF combined with the hybrid FIR filtering makes those high-order modulators possible in the 4th-order PLL. The block diagram of each modulator is shown in Fig. 7. The integrators and adders for each modulator are shared and designed programmable for hardware saving. An optional one-tap digital FIR filter ($1+z^{-1}$) is implemented at the output of the ΔΣ modulator and can be selected for performance comparison.

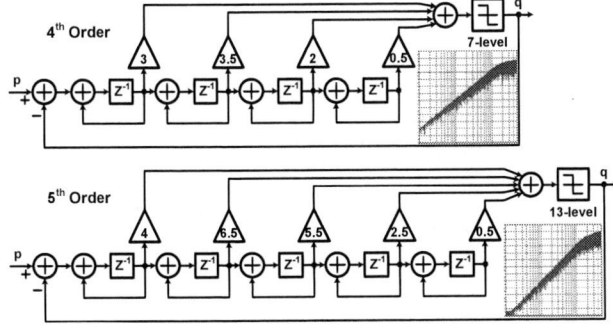

Fig. 7. Block diagram of 4th- and 5th-order ΔΣ modulators.

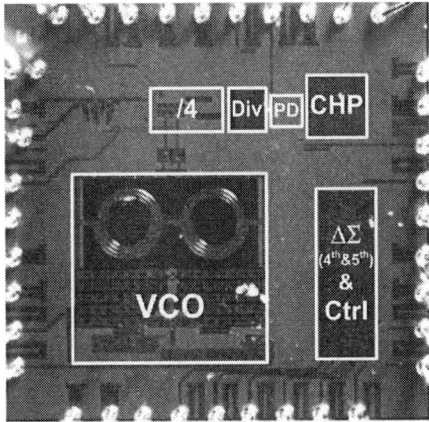

Fig. 8. Die micrograph.

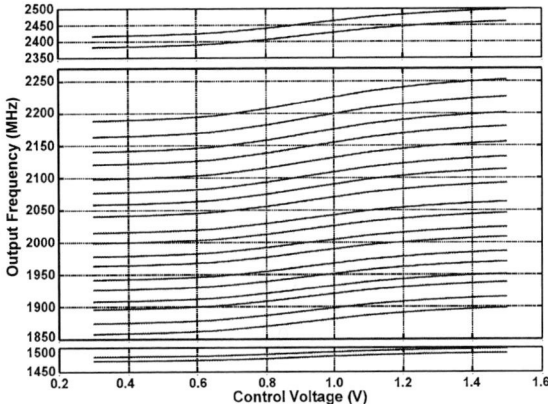

Fig. 9. Measured VCO tuning curve.

IV. TESTING RESULTS

Fig. 8 shows a micrograph of the test chip fabricated in UMC 0.18μm RF CMOS process. The die area is 1.7 x $1.8mm^2$ with the active core occupying $1.5mm^2$. The area of the multiple PS-based dividers is only $0.03mm^2$.

The measured VCO tuning curve is plotted in Fig. 9. The 64 bands cover frequency from 1.47GHz to 2.51GHz. The tuning range of each band is about 40MHz.

Fig. 10 shows the measured phase noise at 1965MHz output frequency. The measured in-band phase noise is less than –92dBc/Hz with the loop bandwidth of about 200kHz. The out-of-band phase noise is –128dBc/Hz at 3.5MHz offset frequency. The phase noise performance at 10MHz offset frequency is limited by the noise floor of the testing equipment.

By moving the 4th pole of the loop to about 9MHz from 1MHz, the quantization noise from the 4th-order modulator becomes more dominant at high frequency so the effect of the FIR filter can be demonstrated more clearly. The results are shown in Fig. 11. The optional $1+z^{-1}$ digital FIR performance is also shown for comparison. Even though the digital FIR filter reduces the noise at 1/2 f_{ref}, it increases quantization noise by 6dB when the offset frequency is less than 1/10 f_{ref}.

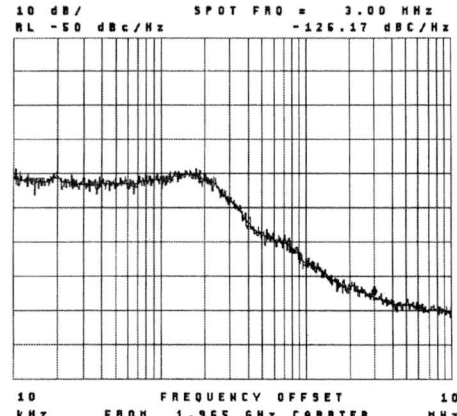

Fig. 10. Measured phase noise performance.

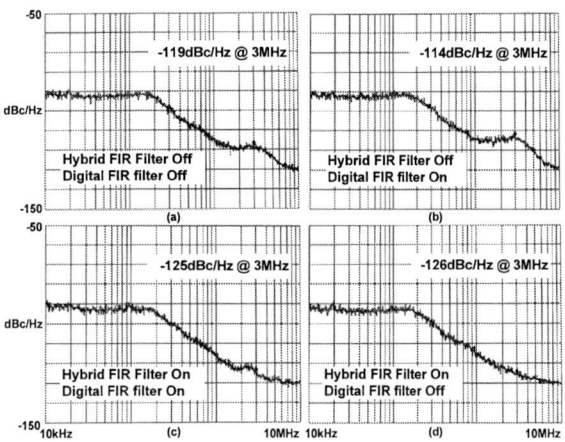

Fig. 11. FIR filtering demonstration with higher 4th-pole frequency

On the contrary, the hybrid FIR filter reduces the quantization noise without affecting noise at low frequencies. From the bottom two plots in Fig. 11, we can clearly see that the hybrid FIR filter provides large suppression to the noise at offset frequency of $k/8 f_{ref}$ (k = 1, 2, ...). It also implies that the PLL bandwidth can be further increased if necessary. The results verify that customized noise shaping can be made by employing the proposed method.

The reference spur at 26MHz offset frequency is less than –62dBc as shown in Fig. 12. The fractional spur performance at 1MHz offset frequency is compared between two modulators; –66dBc and –71dBc with the 4th-order modulator and the 5th-order modulator, respectively.

The prototype synthesizer consumes 17.2mW at 1965MHz output while the multiple PS-based dividers draw 3mW from 1.8V supply. The power and area of the multiple PS-based dividers are technology scalable since they are designed in standard CMOS logic. A summary of the measured results is given in Table 1.

V. CONCLUSION

The 2GHz $\Delta\Sigma$ fractional-N synthesizer employing the hybrid FIR filtering method is implemented in 180nm CMOS.

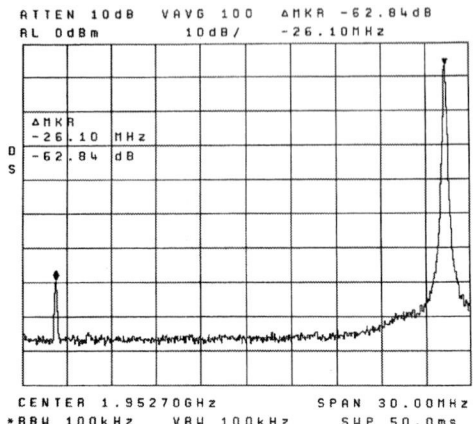

Fig. 12 Measured reference spur performance.

The synthesizer is designed with the single high frequency CML divider and the single-ended MMDs for low power, exhibiting the in-band phase noise of less than –92dBc/Hz with the loop bandwidth of 200kHz and –128dBc/Hz at 3.5MHz offset frequency. The experimental results show that the proposed method is useful to meet the phase noise mask requirements for various RF applications including WCDMA/HSDPA standards.

REFERENCES

[1] D. Kaczman et al., "A single-chip tri-band (2100, 1900, 850/8000 MHz) WCDMA/HSDPA cellular transceiver," *IEEE J. of Solid-State Circuits*, vol. 41, pp. 1122–1132, May 2006.

[2] X. Yu, Y. Sun, L. Zhang, W. Rhee, and Z. Wang, "A 1GHz fractional-N PLL clock generator with low-OSR $\Delta\Sigma$ modulation and FIR-embedded noise filtering," *ISSCC*, Feb. 2008, pp. 346–347.

[3] J. Craninckx and M. Steyaert, "A 1.75-GHz/3-V dual-modulus divide-by-128/129 prescaler in 0.7-μm CMOS," *IEEE J. of Solid-State Circuits*, vol. 31, pp. 890–897, July 1996.

[4] W. Rhee, B. Song, and A. Ali, "A 1.1-GHz CMOS fractional-N frequency synthesizer with a 3-b third-order delta-sigma modulator," *IEEE J. of Solid-State Circuits*, vol. 35, pp. 1453–1460, Oct. 2000.

Table 1. Measured performance summary.

Process	0.18μm RF CMOS
Supply Voltage	1.8V ($\Delta\Sigma$ Modulator with 1.2V)
Power Dissipation	Total: 17.2 mW (8 x PS-Based Divider : ~ 3 mW)
Occupied Area	Active Core : ~ 1.5 mm^2 (8 x PS-Based Divider : ~ 0.03 mm^2)
Tuning Range	1470 ~ 2500 MHz
Frequency Resolution	< 100 Hz (< 0.05 ppm)
Reference Clock	26 MHz
Bandwidth	~ 200 kHz
In-band Phase Noise	< -92 dBc/Hz
Out-band Phase Noise	-128 dBc/Hz @ 3.5 MHz offset
Reference Spur	< -62 dBc
RMS Phase Error	< 1.1 $^\circ_{RMS}$
Integer-boundary Spur @ 1MHz	4th order $\Delta\Sigma$ Modulator: -66 dBc (5th order $\Delta\Sigma$ Modulator: -71 dBc)

AUTHOR INDEX

Agarwal, Kanak 367
Agrawal, Ankur 491
Ahmed, Haseeb...............................433
Ahn, Hyung Ki 801
Ahn, Sunghoon 487
Akopyan, Filipp...........................701
Al-Hashimi, Bashir 571
Allee, David 199
Allstot, David 611
Amrutur, Bharadwaj 145
Annema, Anne Johan........................239
Aparin, Vladimir...........................777
Arakali, Abhijith 475
Arimoto, Kazutami 717
Aroca, Ricardo A. 505
Arslan, Umut 445
Arvind, N. V. 145
Asbeck, Peter 789
Auth, Chris 407
Avanzo, Flavio 615
Ayers, James 629
Bakir, Muhannad 705
Bakkaloglu, Bertan 385
Balakrishnan, Varsha 13
Balamurugan, Ganesh 743
Barale, Francesco 81
Barnaby, Hugh 293, 385
Bashirullah, Rizwan 23
Basu, Arindam 231
Bertholet, Nicole...........................509
Bhargava, Mudit 445
Bhatia, Deepak 23
Bichan, Mike 651
Blaauw, David 453, 543, 579
Borremans, Jonathan 53
Bottarel, Valeria 591

Brama, Riccardo...........................595
Brink, Stephen 231
Brodersen, Robert 323
Burnell, David 771
Butzen, Paulo 133
Cai, Liuchun 727
Cao, Yu 13
Cao, Zhiheng 331
Carlson, Andrew 441
Carnes, Josh 739
Carusone, Anthony Chan 479, 651, 675
Casper, Bryan 743
Chabert, Laurent...........................509
Chandrashekar, Kailash 385
Chang, Hsiang-Hui 797
Chang, Ken 467
Chantre, Alain 415
Chen, Bei 267, 271
Chen, Charles 323
Chen, Fangxiong....................267, 271
Chen, Hong 389
Chen, Jia 389
Chen, Ke- Horng 27, 35
Chen, Poki 227
Chen, Po-Yu 227
Chen, Qiang 285
Chen, X. J. 293
Chen, Yanfei 531
Chen, Yi..................................389
Chen, Young-Kai 335
Cheng, Horace 675
Cheng, Wei 239
Chevalier, Pascal 415
Chiang, Ed 761
Chin, Sing................................319
Chin, Tj 467

AUTHOR INDEX

Chiu, Yun 323
Cho, Choongyeun 223, 667
Cho, Min-Hyung 539
Cho, Minki 305
Cho, Sangyeun 223
Choi, Minseok 721
Choi, Moo-Yeol 99
Chu, Hyunho 587
Chu, Kun-Da 513
Chun, Jung-Hoon 467
Chung, Hayun 599
Chung, Hoon Hee 385
Chung, Kyusik 393
Cicek, Paul-Vahé 207
Ciftcioglu, Berkehan 603
Clark, Lawrence 199
Cole Jr., Edward 57
Collados, Manel 781
Cong, Peng 559
Copani, Tino 385
Cressler, John 63
Croon, Jeroen 239
Dai, Fa Foster 267, 271, 525
Dallago, Enrico 591
Dang, Bing 705
Daniel, Erik 663
Das, Bishnu Prasad 145
Davis, W. Rhett 697
Dawn, Debasis 81
De Nicola, Paolo 713
De Paola, Francesco M. 607, 615
Debnath, Chandrajit 327
Dehos, Cedric 509
Deligoz, Ilker 385
Depaoli, Emanuele 595
Dewey, Bill 301

Dhanasekaran, Vijay 749
Diorio, Chris 357
Doi, Yoshiyasu 505
Dong, Qiaohua 457
Dong, Yunzhi 625
Dusatko, Tomas A. 203, 361
El-Gamal, Mourad 207, 619
El-Gamal, Mourad N. 203, 361
Elkhatib, Tamer 551
English, George 463
Erba, Simone 595
Esqueda, Ivan 293
Eum, Nakwoong 721
Evans, William 771
Fang, David 701
Feng, Guoyou 457
Fiez, Terri 629, 633
Flynn, Michael 373, 757
Foo, Zhiyoong 579
Frattini, Giovanni 591
Friedman, Daniel 463
Friedman, Eby 693
Fritz, Karl 663
Fujita, Hiroyuki 93
Fukuzawa, Fumitaka 717
Funaki, Hideyuki 93
Gambhir, Manisha 749
Garcia, Patrice 415, 509
Garnier, Christophe 415
Genesi, Raffaella 607
Gerosa, Gian 433
Gerrits, John F. M. 625
Gierkink, Sander 471
Gilbert, Barry 663
Goethals, Mieke 399
Gondi, Srikanth 475

AUTHOR INDEX

Goo, Jung-Suk 285
Gord, John 341
Goswami, Sushmit 385
Greyling, Braam 713
Groeseneken, Guido 53
Gronheid, Roel 399
Gu, Ben 247
Gu, Ming 267, 271
Gubbins, David 187
Gullapalli, Kiran 247
Guo, Lu 457
Guo, Song 547
Guo, Zheng 441
Hamada, Mototsugu 31
Hamada, Takayuki 531
Hamoui, Anas 555
Hanson, Scott 453
Hanumolu, Pavan 739
Hanumolu, Pavan Kumar 187, 475, 491, 753
Hara, Hiroyuki 31
Harjani, Ramesh 243, 727
Hasan, Mohammad 433
Hashemi, Hossein 75
Hashimoto, Eiri 689
Hasler, Paul 231
Hendrickx, Eric 399
Hess, Phil 341
Heydari, Payam 521
Holleman, Jeremy 357
Honda, Kazutaka 127
Horowitz, Mark 219
Hossain, Masum 479
Hoyos, Sebastian 323
Hsiao, Yuan-Wen 251
Hsieh, Bing-Yu 797
Hsieh, Ping-Hsuan 567

Hsieh, Yao-Jen 767
Hsu, Klaus Yune-Jane 263
Hu, Xueqing 267, 271
Huang, Gang 705
Huang, Hong-Wei 27, 35
Huang, Ji-Chen 263
Huang, Juin-Wei 513
Huang, Tian-Wei 497
Humble, James 663
Hung, B. 761
Hwang, Myeong-Eun 449
Inaoka, Kazuhiro 717
Irwin, J. David 525
Ishida, Koichi 149
Ishimi, Koichi 717
Itaya, Kazuhiko 93
Ito, Minoru 41
Itoh, Shinya 127
Jackson, Sandra 701
Jaeger, Richard 267, 271
Jaeger, Richard C. 525
Jain, Anuj 385
Jain, Vipul 521
Jamadagni, H. S. 145
Jansen, Philippe 53, 399
Jaussi, James 743
Javid, Babak 521
Jee, Dong-Woo 535
Jenkal, Ravi 697
Jou, Shyh-Jye 377
Jun, Young-Hyun 679
Jung, Inhwa 487
Kajita, Mikihiro 731
Kang, Joshua 373, 757
Karaki, Habib 385
Kasai, Naoki 381

AUTHOR INDEX

Kaviani, Kambiz 467

Kawahito, Shoji 127

Keane, John 133

Kennedy, Joseph 743

Kent, John 423

Ker, Ming-Dou 251, 767

Kertis, Robert 663

Keskin, Gökçe 9

Khaligh, Alireza 195

Kiaei, Sayfe 385

Kibune, Masaya 531

Kim, Bongjin 255

Kim, Chiwon 255

Kim, Chris 133

Kim, Chris H. 437

Kim, Chulwoo..............................487, 587

Kim, Daeik 223, 667

Kim, Dong Woon 579

Kim, Donghyun 393, 575

Kim, Duckhwan 721

Kim, Hyungoo 255

Kim, Jaeha 735

Kim, Jaewhui....................99, 123, 183, 255

Kim, Jinkyu 721

Kim, Jiyoung 255

Kim, Jongdae 539

Kim, Jonghae.............................223, 667

Kim, Jongsik 641

Kim, Ju-Hwa 123

Kim, Jungmoon 587

Kim, Kihong 641

Kim, Kinam 679

Kim, Kwi-Dong 539

Kim, Lee-Sup..........................393, 575, 679

Kim, Namsoo 777

Kim, Seon Wook 587

Kim, Seungsoo 641

Kim, Soo-Won 353

Kim, Sungdo 721

Kim, Tae-Hyoung 437

Kim, Yi-Gyeong 539

Kim, Yontae 487

Kim, Youngcho 641

Kimball, Donald 789

King, Calvin 705

Kinget, Peter 761

Kitahara, Takeshi 31

Kitao, Masaya 717

Kitchen, Jennifer 385

Klaassen, Dick 239

Ko, Wen 559

Kondo, Hiroyuki 717

Koo, Bontae.....................................721

Koziol, Scott 231

Krishnan, Srikanth 13

Kristan, Ernst 713

Kulkarni, Jaydeep P. 137

Kumar, Ashish 327

Kuo, Chung-Hung 767

Kuo, Tzu-Chieh 563

Kurahashi, Peter 753

Kuroda, Tadahiro...............................531

Kwon, Jong-Kee 539

Kwon, Sunwoo 111

Kwon, Youngsu 721

Kyung, Moon- Jung 175

Lai, Juan-Shan 227

Lai, Yu-Sheng 263

Larson, Larry 789

Larson, Lawrence E. 777

Laskar, Joy 81

Le Tual, Stephane...............................509

AUTHOR INDEX

Lee, Bumha.............................187
Lee, Edward 341
Lee, Haechang 467
Lee, Hae-Seung 99, 123
Lee, Hoi..............................547
Lee, Ho-Young 123
Lee, Hyuckjae 721
Lee, Jaesik 335
Lee, Joohyun 721
Lee, Kang-Jin 183
Lee, Kyehyung.....................103, 107
Lee, Sungjae 285
Lee, Sung-No 99
Lee, Wei-Ming 175
Lei, Yu...............................457
Leibowitz, Brian 735
Lemoine, Dominique 207
Levacq, David.........................583
Leven, Andreas 335
Li, Chun-Huai 767
Li, Jing 211
Li, Li................................757
Li, Pengfei 23
Li, T. L. 761
Li, Taihu 341
Li, Xin 9, 445
Lim, Daihyun 223
Lin, Chia-Hsiang 27, 35
Lin, David 757
Lin, Kuan-Yu 619
Lin, Tsung-Hsien 637
Lin, Xue 23
Lin, Yu- Shiang 543
Linten, Dimitri 53
Liu, Andrew 599
Liu, Chien-Nan 259

Liu, Chun-Ting 767
Liu, Haixin 211
Liu, Jason 437
Liu, Ming............................389
Liu, Tsu-Jae King 441
Liu, Wei-Chang 377
Liu, Yao- Hong 637
Liu, Zheng 127
Loizou, Philipos 547
Long, John R. 625
Long, Syu-Siang 377
Lou, Shuzuo 71
Lu, Hung Wen 259
Lu, Minhua 89
Lu, Ning.............................301
Lu, Ting-Chou 767
Luo, Xian 215
Luong, Howard 71
Ma, Heping 267, 271
Ma, Jirong 215
Ma, Kwang-Sung 183
Madden, Christopher 735
Maebashi, Takanore....................689
Maenhoudt, Mireille 399
Maghari, Nima.........................111
Mai, Ken 9, 445
Maitra, Kingsuk 305
Malik, Rakesh 327
Manohar, Rajit 701
Manstretta, Danilo 607, 615
Mansuri, Mozhgan 743
Marchetti, Maxime 509
Masui, Norio 717
Matei, Eusebiu 341
Matsumoto, Shuuji 31
Matsushita, Hiroaki 41

AUTHOR INDEX

Maxey, Jay...567
Mayaram, Kartikeya 629, 633, 739
McCartney, Mark P...445
McLain, Michael 293
McNally, Iain...571
Meindl, James..705
Melamed, Samson 697
Mellon, Carnegie..9
Meterelliyoz, Mesut..137
Miatton, Daniele...591
Miller, Matt 611
Miller, Matthew R...103
Min, Junki 641
Min, Young- Jae...353
Minakawa, Takuya 583
Mineo, Christopher 697
Mishra, Apurva 357
Mita, Makoto...93
Miyakawa, Nobuaki 689
Moon, Kyoung-Ho 123, 183
Moon, Kyoung-Jun 183
Moon, Un-Ku.....................111, 187, 739, 753
Mooney, Randy 743
Moquillon, Laurence..509
Morche, Domique..509
Mukhopadhyay, Saibal......................................305
Murmann, Boris.......................................115, 179
Nabki, Frederic 203, 207, 361
Naeemi, Azad..705
Nairn, David 311
Nakajima, Masami 717
Nakamoto, Satoshi 731
Nakamura, Natsuo 689
Nakamura, Yasumi 583
Nakayama, Naoya 731
Nakayama, Shigeto 689

Nassif, Sani 1, 367
Nauta, Bram..239
Nebshi, Ryusuke 381
Neihart, Nathan 611
Nercessian, Patrick..341
Nho, H. Henry...219
Nicolson, Sean 415
Nicolson, Sean T. 505
Niiyama, Taro 149, 583
Nikaeen, Parastoo 179
Nikolic, Borivoje 141, 323, 441
Nomura, Shuou 31
Nowak, Edward 285
Nowka, Kevin..367
O, Kenneth 163
Oh, Jaegeun 587
Oh, Kwang-Il 679
Oh, Kyujin 163
Oh, Tae-Hwan 123
Ohta, Jun...349
Ok, Sunghwa 587
Okpisz, Alex 433
Okumura, Naoto 717
Okushima, Mototsugu 53
O'Mahony, Frank 743
Omole, Umaikhe 243
Otani, Sugako 717
Otero, Carlos Tadeo Ortega 701
Otis, Brian 357
Pain, Bedabrata..175
Palaskas, Yorgos...155
Pan, Liyang 215
Pang, Liang-Teck 141, 441
Panitantum, Napong 633
Pareschi, Fabio...483
Park, Byeong-Ha 801

AUTHOR INDEX

Park, Ho-Jin 99, 123

Park, Hong-June 535

Park, Jaehyun 255

Park, Junyoung 373

Park, Kihyuk 721

Park, Kwang-Il 679

Park, Min 579

Park, Piljae 235

Park, Seung- Jin 535

Park, Sunghyun 373

Pavlidis, Vasilis 693

Pavlovic, Nenad 781

Pellerano, Stefano 155

Perego, Rich 467

Perfecto, Eric 89

Perumana, Bevin 81

Petre, Csaba 231

Philpott, Rick 663

Piazzese, Nicolo Ivan 509

Piccolella, Salvatore 713

Pileggi, L. 9, 171, 445

Pinel, Stephane 81

Pisati, Matteo 595

Plouchart, Jean-Olivier 667

Pozzoni, Massimo 595

Prasad, Jagdish 423

Prete, Edoardo 45

Proesel, Jon 9

Proesel, Jonathan 171

Pruvost, Sebastien 415, 509

Puma, Giuseppe Li 713

Rakers, Pat 611

Ramakrishnan, Shubha 231

Ravi, Ashoke 155

Razavi, Behzad 655

Reddy, Vijay 13

Redouté, Jean-Michel 683

Ren, Jihong 735

Rhee, Woogeun 801

Ricotti, Giulio 591

Riley, John 433

Ronse, Kurt 399

Roux, Pascal 335

Rovatti, Riccardo 483

Roy, David 415

Roy, Kaushik 137, 211, 449

Rylyakov, Alexander 463

Ryu, Jongjae 255

Sacho, Yutaka 689

Sadhu, Bodhisatwa 243

Safarian, Zahra 75

Saito, Yoshihiro 717

Sakimura, Noboru 381

Sakugawa, Mamoru 717

Sakurai, Takayasu 149, 583

Salahuddin, Sayeef 211

Salama, Khaled 551

Sanchez-Sinencio, Edgar 749

Sanders, Seth R. 785

Sanduleanu, Mihai A. T. 625

Sankaran, Swaminathan 163

Sanzogni, Davide 595

Sato, Hironori 31

Sato, Masayuki 717

Savidis, Ioannis 693

Sawamura, Shigeki 349

Scaccianoce, Salvatore 509

Schipani, Monica 591

Schlottman, Craig 231

Scholz, Mirko 53

Scuderi, Angelo 509

Sekar, Deepak 705

AUTHOR INDEX

Seki, Keiko.....31
Sen, Padmanava.....81
Seo, Jae-Sun.....453
Seok, Mingoo.....453
Serratore, Alberto.....509
Setti, Gianluca.....483
Shekhar, Sudip.....743
Shen, Jie.....467
Shi, Xudong.....467
Shi, Yin.....267, 271
Shimizu, Toru.....717
Shin, Hyunchol.....641
Shin, Jaewook.....641
Shin, Jongshin.....255
Shiue, Muh-Tian.....377
Shringarpure, Rahul.....199
Shu, Yun-Shiang.....175
Silva-Martinez, Jose.....749
Sim, Jae-Yoon.....535
Singh, Pratap Narayan.....327
Sode, Mikiki.....731
Song, Bang-Sup.....175
Song, Jooyoung.....277
Song, Minyoung.....487
Song, Peilin.....137
Sonkusale, Sameer.....191
Sperling, Michael.....463
Srivastava, Kamalesh.....89
Stauth, Jason T.....785
Stellari, Franco.....137
Steyaert, Michiel.....19, 683
Stover, Howard.....341
Su, Chau Chin.....259
Su, David.....645
Sugibayashi, Tadahiko.....381
Sun, Yuanfeng.....801

Sundareswaran, Savithri.....247
Sundlof, Brian.....89
Suzuki, Kazuhiro.....93
Svelto, Francesco.....595
Swabey, Matthew.....571
Sylvester, Dennis.....453, 543, 579
Tachibana, Fumihiko.....31
Takahashi, Kazuhiro.....93
Takamiya, Makoto.....149, 583
Takeuchi, Kan.....41
Talbot, Gerry.....45
Tamura, Hirotaka.....531
Tanaka, Genichi.....41
Tano, Yasuo.....349
Taufique, Mohammed.....433
Taur, Yuan.....277
Tecpoyotl-Torres, Margarita.....555
Temes, Gabor.....107
Temes, Gabor C.....103
Terasawa, Yasuo.....349
Thacker, Hiren.....705
Thijs, Steven.....53
Tierno, Jose.....463
Tokuda, Takashi.....349
Tokunaga, Carlos.....579
Tomita, Yasumoto.....531
Tomkins, Alexander.....505
Toshiyoshi, Hiroshi.....93
Toyoda, Shinjiro.....689
Trakimas, Michael.....191
Tsang, Cheongyuen.....323
Tseng, Chi-Yao.....377
Tsuboi, Yoshiro.....31
Tsukamoto, Sanroku.....531
Twigg, Christopher M.....231
Uppili, Shrinivas.....199

AUTHOR INDEX

Van Veenendaal, Gerrit.............625
Van Zeijl, Paul T. M. 781
Vandenberghe, Geert.............399
Vanderhaegen, Johan 323
Vannier, Cyril.............713
Varona, Jorge.............555
Venchi, Giuseppe 591
Venkatraman, Shrinivas.............133
Venugopal, Sameer 199
Vermeire, Bert.............385
Vezyrtzis, Christos 761
Viola, Paolo.............595
Visvanathan, V..............145
Voinigescu, Sorin 415
Voinigescu, Sorin P. 505
Vytyaz, Igor 739
Wambacq, Piet 53
Wang, Chao-Shiun 513, 517
Wang, Chorng-Kuang 513, 517
Wang, Huei.............497
Wang, Nan 457
Wang, Ping-Ying.............797
Wang, Wenping.............13
Wang, Yan 107
Wang, Zhihua.............801
Wang, Zi 457
Wang, Zihua.............389
Watanabe, Yoshinori 31
Wei, Gu-Yeon 491, 599
Wei, Ting-Chen 377
Weiner, Joseph 335
Wen, Shon-Hang 517
Wens, Mike.............19
Wilcock, Reuben 571
Williams, Richard 285
Willson, Alan 563

Wilson, Peter.............571
Wiser, Robert.............645
Wolfe, James 341
Wong, Bill 319
Wong, S. Simon 219
Wooley, Bruce.............645
Workman, Glenn 285
Wu, Dong 215
Wu, Hsinta 163
Wu, Hui 603
Wu, Ting 467, 739
Wu, Wanghua 625
Wu, Xiaochun 195
Xiao, Limin 583
Xu, Hua 267, 271
Xu, Jun 215
Xu, Liang.............457
Xu, Qiming.............267, 271
Xu, Yang 195
Yamamoto, Osamu 717
Yamamoto, Takuji 505
Yamane, Fumiyuki 31
Yamashita, Takahiro 31
Yan, Jun 267, 271
Yan, Shouli.............331
Yan, Yaru.............215
Yang, Bo 13
Yang, Chih-Kong Ken 567
Yang, Dayu 525
Yang, Seung- Hee 255
Yao, Xiang.............457
Yasutomi, Keita 127
Yeh, David.............81
Yeum, Wang-Seup 99
Yoon, Gilwon 587
Yoon, Jae-Sung.............575

AUTHOR INDEX

You, Seung-Bin.....................................99

Young, Darrin 559

Young, Michael Wieckowski........................579

Yu, Bo..277

Yu, Chang-Hyo...............................393, 575

Yu, Hao...319

Yu, Peng.....................................267, 271

Yu, Xuefeng 525

Yu, Xueyi 801

Yuan, Fang 267, 271

Yue, C. Patrick 235

Zan, Hsiao-Wen 767

Zargari, Masoud 645

Zhan, Jing-Hong 797

Zhang, Chun 389

Zhang, Lin 603

Zhang, Xuelian 267, 271

Zhang, Yun 247

Zhao, Yi 625

Zhu, Xiaolei 531